BALANIS' ADVANCED ENGINEERING ELECTROMAGNETICS

BALANIS' ADVANCED ENGINEERING ELECTROMAGNETICS

THIRD EDITION

Constantine A. Balanis
Arizona State University

Published by John Wiley & Sons, Inc., Hoboken, New Jersey.
Published simultaneously in Canada.

Edition History
John Wiley & Sons, Inc. (1e, 1989 and 2e, 2012)

For general information on our other products and services or for technical support, please contact our Customer Care Department within the United States at (800) 762-2974, outside the United States at (317) 572-3993 or fax (317) 572-4002.

Wiley also publishes its books in a variety of electronic formats. Some content that appears in print may not be available in electronic formats. For more information about Wiley products, visit our website at www.wiley.com.

Library of Congress Cataloging-in-Publication Data
Names: Balanis, Constantine A., 1938- author.
Title: Balanis' advanced engineering electromagnetics / Constantine A. Balanis.
Description: Third edition. | Hoboken, New Jersey : John Wiley & Sons, [2024] | Includes bibliographical references and index.
Identifiers: LCCN 2023023649 (print) | LCCN 2023023650 (ebook) | ISBN 9781394180011 (hardback) | ISBN 9781394180028 (pdf) | ISBN 9781394180035 (epub)
Subjects: LCSH: Electromagnetism.
Classification: LCC QC760 .B25 2024 (print) | LCC QC760 (ebook) | DDC 537--dc23/eng20230926
LC record available at https://lccn.loc.gov/2023023649
LC ebook record available at https://lccn.loc.gov/2023023650

Cover Image: Courtesy of Dr. Wengang Chen
Cover Design: Wiley

Set in 10/12pt TimesNewRomanMTStd by Integra Software Services Pvt. Ltd, Pondicherry, India

SKY10061632_120823

To my family:

Helen, Renie, Stephanie, Bill, Pete, and Ellie

Στην οικογένεια μου:

Ελένη, Ειρήνη, Στεφανία, Βασίλη, Παναγιώτη, και Ελένη

Contents

Preface   xix

About the Companion Website   xxiii

1   Time-Varying and Time-Harmonic Electromagnetic Fields   1

   1.1   Introduction   1
   1.2   Maxwell's Equations   2
            1.2.1   *Differential Form of Maxwell's Equations*   2
            1.2.2   *Integral Form of Maxwell's Equations*   3
   1.3   Constitutive Parameters and Relations   5
   1.4   Circuit-Field Relations   7
            1.4.1   *Kirchhoff's Voltage Law*   7
            1.4.2   *Kirchhoff's Current Law*   8
            1.4.3   *Element Laws*   10
   1.5   Boundary Conditions   12
            1.5.1   *Finite Conductivity Media*   12
            1.5.2   *Infinite Conductivity Media*   15
            1.5.3   *Sources Along Boundaries*   17
   1.6   Power and Energy   18
   1.7   Time-Harmonic Electromagnetic Fields   21
            1.7.1   *Maxwell's Equations in Differential and Integral Forms*   22
            1.7.2   *Boundary Conditions*   22
            1.7.3   *Power and Energy*   25
   1.8   Multimedia   29
            References   29
            Problems   30

2   Electrical Properties of Matter   41

   2.1   Introduction   41
   2.2   Dielectrics, Polarization, and Permittivity   43
   2.3   Magnetics, Magnetization, and Permeability   50

2.4 Current, Conductors, and Conductivity   57
    2.4.1 *Current*   57
    2.4.2 *Conductors*   58
    2.4.3 *Conductivity*   59
2.5 Semiconductors   61
2.6 Superconductors   66
2.7 Metamaterials   68
2.8 Linear, Homogeneous, Isotropic, and Nondispersive Media   69
2.9 A.C. Variations in Materials   70
    2.9.1 *Complex Permittivity*   70
    2.9.2 *Complex Permeability*   82
    2.9.3 *Ferrites*   83
2.10 Multimedia   92
    References   92
    Problems   93

3 Wave Equation and Its Solutions   **103**

3.1 Introduction   103
3.2 Time-Varying Electromagnetic Fields   103
3.3 Time-Harmonic Electromagnetic Fields   105
3.4 Solution to the Wave Equation   106
    3.4.1 *Rectangular Coordinate System*   107
      A. Source-Free and Lossless Media   107
      B. Source-Free and Lossy Media   112
    3.4.2 *Cylindrical Coordinate System*   114
    3.4.3 *Spherical Coordinate System*   120
3.5 Multimedia   125
    References   125
    Problems   125

4 Wave Propagation and Polarization   **127**

4.1 Introduction   127
4.2 Transverse Electromagnetic Modes   127
    4.2.1 *Uniform Plane Waves in an Unbounded Lossless Medium— Principal Axis*   128
      A. Electric and Magnetic Fields   128
      B. Wave Impedance   131
      C. Phase and Energy (Group) Velocities, Power, and Energy Densities   132
      D. Standing Waves   134
    4.2.2 *Uniform Plane Waves in an Unbounded Lossless Medium— Oblique Angle*   136
      A. Electric and Magnetic Fields   136
      B. Wave Impedance   139
      C. Phase and Energy (Group) Velocities   140
      D. Power and Energy Densities   141
4.3 Transverse Electromagnetic Modes in Lossy Media   142
    4.3.1 *Uniform Plane Waves in an Unbounded Lossy Medium—Principal Axis*   143
      A. Good Dielectrics $[(\sigma/\omega\varepsilon)^2 \ll 1]$ 147
      B. Good Conductors $[(\sigma/\omega\varepsilon)^2 \gg 1]$ 147
    4.3.2 *Uniform Plane Waves in an Unbounded Lossy Medium—Oblique Angle*   148

4.4	Polarization	151
	4.4.1 *Linear Polarization*	152
	4.4.2 *Circular Polarization*	155
	A. Right-Hand (Clockwise) Circular Polarization	155
	B. Left-Hand (Counterclockwise) Circular Polarization	158
	4.4.3 *Elliptical Polarization*	160
	4.4.4 *Poincaré Sphere*	165
4.5	Multimedia	171
	References	171
	Problems	172

5 Reflection and Transmission   **179**

5.1	Introduction	179
5.2	Normal Incidence—Lossless Media	179
5.3	Oblique Incidence—Lossless Media	183
	5.3.1 *Perpendicular (Horizontal or E) Polarization*	184
	5.3.2 *Parallel (Vertical or H) Polarization*	188
	5.3.3 *Total Transmission–Brewster Angle*	192
	A. Perpendicular (Horizontal) Polarization	192
	B. Parallel (Vertical) Polarization	193
	5.3.4 *Total Reflection–Critical Angle*	194
	A. Perpendicular (Horizontal) Polarization	195
	B. Parallel (Vertical) Polarization	204
5.4	Lossy Media	204
	5.4.1 *Normal Incidence: Conductor–Conductor Interface*	205
	5.4.2 *Oblique Incidence: Dielectric–Conductor Interface*	208
	5.4.3 *Oblique Incidence: Conductor–Conductor Interface*	212
5.5	Reflection and Transmission of Multiple Interfaces	212
	5.5.1 *Reflection Coefficient of a Single Slab Layer*	212
	5.5.2 *Reflection Coefficient of Multiple Layers*	220
	A. Quarter-Wavelength Transformer	221
	B. Binomial (Maximally Flat) Design	222
	C. Tschebyscheff (Equal-Ripple) Design	224
	D. Oblique-Wave Incidence	226
5.6	Polarization Characteristics on Reflection	228
5.7	Metamaterials	235
	5.7.1 *Classification of Materials*	236
	5.7.2 *Double Negative (DNG) Materials*	237
	5.7.3 *Historical Perspective*	238
	5.7.4 *Propagation Characteristics of DNG Materials*	239
	5.7.5 *Refraction and Propagation Through DNG Interfaces and Materials*	241
	5.7.6 *Negative-Refractive-Index (NRI) Transmission Lines*	249
5.8	Multimedia	253
	References	254
	Problems	256

6 Auxiliary Vector Potentials, Construction of Solutions, and Radiation and Scattering Equations   **271**

6.1	Introduction	271
6.2	The Vector Potential A	272
6.3	The Vector Potential F	274

6.4	The Vector Potentials A and F		275
6.5	Construction of Solutions		277
	6.5.1	*Transverse Electromagnetic Modes: Source-Free Region*	277
		A. Rectangular Coordinate System	277
		B. Cylindrical Coordinate System	281
	6.5.2	*Transverse Magnetic Modes: Source-Free Region*	285
		A. Rectangular Coordinate System	285
		B. Cylindrical Coordinate System	287
	6.5.3	*Transverse Electric Modes: Source-Free Region*	288
		A. Rectangular Coordinate System	288
		B. Cylindrical Coordinate System	290
6.6	Solution of the Inhomogeneous Vector Potential Wave Equation		291
6.7	Far-Field Radiation		295
6.8	Radiation and Scattering Equations		296
	6.8.1	*Near Field*	296
	6.8.2	*Far Field*	298
		A. Rectangular Coordinate System	302
		B. Cylindrical Coordinate System	311
6.9	Multimedia		317
	References		317
	Problems		318
7	**Electromagnetic Theorems and Principles**		**323**
7.1	Introduction		323
7.2	Duality Theorem		323
7.3	Uniqueness Theorem		325
7.4	Image Theory		327
	7.4.1	*Vertical Electric Dipole*	329
	7.4.2	*Horizontal Electric Dipole*	333
7.5	Reciprocity Theorem		335
7.6	Reaction Theorem		337
7.7	Volume Equivalence Theorem		338
7.8	Surface Equivalence Theorem: Huygens' Principle		340
7.9	Induction Theorem (Induction Equivalent)		345
7.10	Physical Equivalent and Physical Optics Equivalent		349
7.11	Induction and Physical Equivalent Approximations		351
7.12	Multimedia		356
	References		356
	Problems		357
8	**Rectangular Cross-Section Waveguides and Cavities**		**365**
8.1	Introduction		365
8.2	Rectangular Waveguide		366
	8.2.1	*Transverse Electric (TE^z)*	367
	8.2.2	*Transverse Magnetic (TM^z)*	375
	8.2.3	*Dominant TE_{10} Mode*	378
	8.2.4	*Power Density and Power*	386
	8.2.5	*Attenuation*	388
		A. Conduction (Ohmic) Losses	388
		B. Dielectric Losses	392
		C. Coupling	395

8.3	Rectangular Resonant Cavities	396
	8.3.1 *Transverse Electric (TE^z) Modes*	399
	8.3.2 *Transverse Magnetic (TM^z) Modes*	403
8.4	Hybrid (LSE and LSM) Modes	404
	8.4.1 *Longitudinal Section Electric (LSE^y) or Transverse Electric (TE^y) or H^y Modes*	405
	8.4.2 *Longitudinal Section Magnetic (LSM^y) or Transverse Magnetic (TM^y) or E^y Modes*	407
8.5	Partially Filled Waveguide	407
	8.5.1 *Longitudinal Section Electric (LSE^y) or Transverse Electric (TE^y)*	407
	8.5.2 *Longitudinal Section Magnetic (LSM^y) or Transverse Magnetic (TE^y)*	414
8.6	Transverse Resonance Method	419
	8.6.1 *Transverse Electric (TE^y) or Longitudinal Section Electric (LSE^y) or H^y*	421
	8.6.2 *Transverse Magnetic (TE^y) or Longitudinal Section Magnetic (LSM^y) or E^y*	422
8.7	Dielectric Waveguide	422
	8.7.1 *Dielectric Slab Waveguide*	422
	8.7.2 *Transverse Magnetic (TM^z) Modes*	424
	A. TM^z (Even)	425
	B. TM^z (Odd)	428
	C. Summary of TM^z (Even) and TM^z (Odd) Modes	428
	D. Graphical Solution for TM^z_m (Even) and TM^z_m (Odd) Modes	430
	8.7.3 *Transverse Electric (TE^z) Modes*	433
	8.7.4 *Ray-Tracing Method*	437
	A. Transverse Magnetic (TM^z) Modes (Parallel Polarization)	442
	B. Transverse Electric (TE^z) Modes (Perpendicular Polarization)	445
	8.7.5 *Dielectric-Covered Ground Plane*	447
8.8	Stripline and Microstrip Lines	450
	8.8.1 *Stripline*	451
	8.8.2 *Microstrip*	455
	8.8.3 *Microstrip: Boundary-Value Problem*	461
8.9	Ridged Waveguide	461
8.10	Multimedia	464
	References	467
	Problems	468
9	**Circular Cross-Section Waveguides and Cavities**	**479**
9.1	Introduction	479
9.2	Circular Waveguide	479
	9.2.1 *Transverse Electric (TE^z) Modes*	479
	9.2.2 *Transverse Magnetic (TM^z) Modes*	485
	9.2.3 *Attenuation*	491
9.3	Circular Cavity	496
	9.3.1 *Transverse Electric (TE^z) Modes*	499
	9.3.2 *Transverse Magnetic (TM^z) Modes*	500
	9.3.3 *Quality Factor Q*	501
9.4	Radial Waveguides	505
	9.4.1 *Parallel Plates*	505
	A. Transverse Electric (TE^z) Modes	505
	B. Transverse Magnetic (TM^z) Modes	508
	9.4.2 *Wedged Plates*	509

	A.	Transverse Electric (TE^z) Modes	510
	B.	Transverse Magnetic (TM^z) Modes	511
9.5	Dielectric Waveguides and Resonators		512
	9.5.1	*Circular Dielectric Waveguide*	512
	9.5.2	*Circular Dielectric Resonator*	522
	A.	TE^z Modes	524
	B.	TM^z Modes	525
	C.	TE$_{01\delta}$ Mode	526
	9.5.3	*Optical Fiber Cable*	528
	9.5.4	*Dielectric-Covered Conducting Rod*	530
	A.	TM^z Modes	530
	B.	TE^z Modes	536
9.6	Multimedia		537
	References		537
	Problems		539

10   Spherical Transmission Lines and Cavities   **547**

10.1	Introduction		547
10.2	Construction of Solutions		547
	10.2.1	*The Vector Potential F (J = 0, M ≠ 0)*	548
	10.2.2	*The Vector Potential A (J ≠ 0, M = 0)*	550
	10.2.3	*The Vector Potentials F and A*	550
	10.2.4	*Transverse Electric (TE) Modes: Source-Free Region*	551
	10.2.5	*Transverse Magnetic (TM) Modes: Source-Free Region*	553
	10.2.6	*Solution of the Scalar Helmholtz Wave Equation*	554
10.3	Biconical Transmission Line		555
	10.3.1	*Transverse Electric (TE^r) Modes*	555
	10.3.2	*Transverse Magnetic (TM^r) Modes*	557
	10.3.3	*Transverse Electromagnetic (TEM^r) Modes*	557
10.4	The Spherical Cavity		559
	10.4.1	*Transverse Electric (TE^r) Modes*	560
	10.4.2	*Transverse Magnetic (TM^r) Modes*	562
	10.4.3	*Quality Factor Q*	564
10.5	Multimedia		567
	References		567
	Problems		567

11   Scattering   **573**

11.1	Introduction		573
11.2	Infinite Line-Source Cylindrical Wave Radiation		574
	11.2.1	*Electric Line Source*	574
	11.2.2	*Magnetic Line Source*	578
	11.2.3	*Electric Line Source Above an Infinite Plane Electric Conductor*	578
11.3	Plane Wave Scattering by Planar Surfaces		581
	11.3.1	*TM^z Plane Wave Scattering from a Strip*	582
	11.3.2	*TE^x Plane Wave Scattering from a Flat Rectangular Plate*	589
11.4	Cylindrical Wave Transformations and Theorems		597
	11.4.1	*Plane Waves in Terms of Cylindrical Wave Functions*	597
	11.4.2	*Addition Theorem of Hankel Functions*	599
	11.4.3	*Addition Theorem for Bessel Functions*	602
	11.4.4	*Summary of Cylindrical Wave Transformations and Theorems*	604

11.5 Scattering by Circular Cylinders 605
    11.5.1 *Normal Incidence Plane Wave Scattering by a Conducting Circular Cylinder:* TM^z *Polarization* 605
        A. Small Radius Approximation 608
        B. Far-Zone Scattered Field 608
    11.5.2 *Normal Incidence Plane Wave Scattering by a Conducting Circular Cylinder:* TE^z *Polarization* 610
        A. Small Radius Approximation 612
        B. Far-Zone Scattered Field 613
    11.5.3 *Oblique Incidence Plane Wave Scattering by a Conducting Circular Cylinder:* TM^z *Polarization* 615
        A. Far-Zone Scattered Field 619
    11.5.4 *Oblique Incidence Plane Wave Scattering by a Conducting Circular Cylinder:* TE^z *Polarization* 621
        A. Far-Zone Scattered Field 625
    11.5.5 *Line-Source Scattering by a Conducting Circular Cylinder* 626
        A. Electric Line Source (TM^z Polarization) 626
        B. Magnetic Line Source (TE^z Polarization) 630
11.6 Scattering By a Conducting Wedge 637
    11.6.1 *Electric Line-Source Scattering by a Conducting Wedge:* TM^z *Polarization* 637
        A. Far-Zone Field 641
        B. Plane Wave Scattering 642
    11.6.2 *Magnetic Line-Source Scattering by a Conducting Wedge:* TE^z *Polarization* 642
    11.6.3 *Electric and Magnetic Line-Source Scattering by a Conducting Wedge* 646
11.7 Spherical Wave Orthogonalities, Transformations, and Theorems 648
    11.7.1 *Vertical Dipole Spherical Wave Radiation* 648
    11.7.2 *Orthogonality Relationships* 650
    11.7.3 *Wave Transformations and Theorems* 651
11.8 Scattering by a Sphere 653
    11.8.1 *Perfect Electric Conducting (PEC) Sphere* 653
    11.8.2 *Lossy Dielectric Sphere* 661
11.9 Multimedia 663
    References 664
    Problems 666

12 **Integral Equations and the Moment Method** **677**

  12.1 Introduction 677
  12.2 Integral Equation Method 678
    12.2.1 *Electrostatic Charge Distribution* 678
        A. Finite Straight Wire 678
        B. Bent Wire 682
    12.2.2 *Integral Equation* 684
    12.2.3 *Radiation Pattern* 686
    12.2.4 *Point-Matching (Collocation) Method* 687
    12.2.5 *Basis Functions* 689
        A. Subdomain Functions 689
        B. Entire-Domain Functions 691
    12.2.6 *Application of Point Matching* 693
    12.2.7 *Weighting (Testing) Functions* 695
    12.2.8 *Moment Method* 695

12.3 Electric and Magnetic Field Integral Equations 701
  12.3.1 *Electric Field Integral Equation* 702
    A. Two-Dimensional EFIE: TM^z Polarization 703
    B. Two-Dimensional EFIE: TE^z Polarization 707
  12.3.2 *Magnetic Field Integral Equation* 711
    A. Two-Dimensional MFIE: TM^z Polarization 713
    B. Two-Dimensional MFIE: TE^z Polarization 715
    C. Solution of the Two-Dimensional MFIE TE^z Polarization 717
12.4 Finite-Diameter Wires 721
  12.4.1 *Pocklington's Integral Equation* 722
  12.4.2 *Hallén's Integral Equation* 725
  12.4.3 *Source Modeling* 727
    A. Delta Gap 727
    B. Magnetic Frill Generator 727
12.5 Computer Codes 730
  12.5.1 *Two-Dimensional Radiation and Scattering* 730
    A. Strip 731
    B. Circular, Elliptical, or Rectangular Cylinder 731
  12.5.2 *Pocklington's Wire Radiation and Scattering* 732
    A. Radiation 732
    B. Scattering 732
  12.5.3 *Numerical Electromagnetics Code* 732
12.6 Multimedia 733
  References 733
  Problems 735

13 Geometrical Theory of Diffraction **739**

13.1 Introduction 739
13.2 Geometrical Optics 740
  13.2.1 *Amplitude Relation* 743
  13.2.2 *Phase and Polarization Relations* 747
  13.2.3 *Reflection from Surfaces* 749
13.3 Geometrical Theory of Diffraction: Edge Diffraction 759
  13.3.1 *Amplitude, Phase, and Polarization Relations* 759
  13.3.2 *Straight Edge Diffraction: Normal Incidence* 763
    A. Modal Solution 765
    B. High-Frequency Asymptotic Solution 766
    C. Method of Steepest Descent 770
    D. Geometrical Optics and Diffracted Fields 775
    E. Diffraction Coefficients 778
  13.3.3 *Straight Edge Diffraction: Oblique Incidence* 798
  13.3.4 *Curved Edge Diffraction: Oblique Incidence* 806
  13.3.5 *Equivalent Currents in Diffraction* 813
  13.3.6 *Slope Diffraction* 817
  13.3.7 *Multiple Diffractions* 819
    A. Higher-Order Diffractions 820
    B. Self-Consistent Method 822
    C. Overlap Transition Diffraction Region 825
13.4 Computer Codes 827
  13.4.1 *Wedge Diffraction Coefficients* 828

13.4.2 *Fresnel Transition Function* 829
13.4.3 *Slope Wedge Diffraction Coefficients* 829
13.5 Multimedia 829
References 830
Problems 833

14 **Diffraction by a Wedge with Impedance Surfaces** **847**

14.1 Introduction 847
14.2 Impedance Surface Boundary Conditions 849
14.3 Impedance Surface Reflection Coefficients 850
14.4 The Maliuzhinets Impedance Wedge Solution 852
14.5 Geometrical Optics 854
14.6 Surface Wave Terms 863
14.7 Diffracted Fields 865
14.7.1 *Diffraction Terms* 865
14.7.2 *Asymptotic Expansions* 866
14.7.3 *Diffracted Field* 868
14.8 Surface Wave Transition Field 873
14.9 Computations 875
14.10 Multimedia 877
References 878
Problems 881

15 **Green's Functions** **883**

15.1 Introduction 883
15.2 Green's Functions in Engineering 884
15.2.1 *Circuit Theory* 884
15.2.2 *Mechanics* 887
15.3 Sturm-Liouville Problems 889
15.3.1 *Green's Function in Closed Form* 891
15.3.2 *Green's Function in Series* 896
A. Vibrating String 896
B. Sturm-Liouville Operator 897
15.3.3 *Green's Function in Integral Form* 902
15.4 Two-Dimensional Green's Function in Rectangular Coordinates 906
15.4.1 *Static Fields* 906
A. Closed Form 906
B. Series Form 912
15.4.2 *Time-Harmonic Fields* 915
15.5 Green's Identities and Methods 917
15.5.1 *Green's First and Second Identities* 918
15.5.2 *Generalized Green's Function Method* 920
A. Nonhomogeneous Partial Differential Equation with
Homogeneous Dirichlet Boundary Conditions 920
B. Nonhomogeneous Partial Differential Equation with
Nonhomogeneous Dirichlet Boundary Conditions 921
C. Nonhomogeneous Partial Differential Equation with
Homogeneous Neumann Boundary Conditions 921
D. Nonhomogeneous Partial Differential Equation
with Mixed Boundary Conditions 922

15.6	Green's Functions of the Scalar Helmholtz Equation	923
	15.6.1 *Rectangular Coordinates*	923
	15.6.2 *Cylindrical Coordinates*	926
	15.6.3 *Spherical Coordinates*	931
15.7	Dyadic Green's Functions	935
	15.7.1 *Dyadics*	935
	15.7.2 *Green's Functions*	936
15.8	Multimedia	938
	References	938
	Problems	939
16	**Artificial Impedance Surfaces**	**943**
16.1	Introduction	943
16.2	Corrugations	945
16.3	Artificial Magnetic Conductors, Electromagnetic Bandgap, and Photonic Bandgap Surfaces	947
16.4	Design of Mushroom AMC	950
16.5	Surface-Wave Dispersion Characteristics	955
16.6	Limitations of The Design	959
16.7	Applications of AMCs	959
16.8	RCS Reduction Using Checkerboard Metasurfaces	960
	16.8.1 *Introduction*	960
	16.8.2 *Plane Wave Scattering by PEC-PMC Hybrid Surfaces*	961
	16.8.3 *Fundamentals of Conventional Checkerboard Metasurfaces*	964
	16.8.4 *Broadband RCS Reduction Metasurfaces*	967
	A. Conventional Broadband Checkerboard Metasurfaces	967
	B. Judicious Selection of AMC Unit Cells	968
	C. Modified Checkerboard Surfaces	973
	D. Generalized Approach to Synthesize Ultrabroadband RCS Reduction Checkerboard Surfaces	975
	16.8.5 *Broadband RCS Reduction of Complex Targets*	977
	A. Antenna and Antenna Array	977
	B. Cylindrical Structures	978
	C. Corner Reflectors	979
16.9	Antenna Fundamental Parameters and Figures-of-Merit	980
16.10	Antenna Applications	982
	16.10.1 *Monopole*	982
	16.10.2 *Horizontal Dipole*	983
	16.10.3 *Circular Loop*	985
	16.10.4 *Aperture Antenna*	990
	16.10.5 *Microstrip Array*	994
	16.10.6 *Surface-Wave Antennas*	996
16.11	High-Gain Printed Leaky-Wave Antennas Using Metasurfaces	997
	16.11.1 *Floquet-Bloch Modes*	998
16.12	Metasurface Leaky-Wave Antennas	999
	16.12.1 *Holographic Principle on Antennas*	999
	16.12.2 *One-Dimensional Periodic Metasurface LWAs*	1000
	16.12.3 *Two-Dimensional Holographic Metasurface LWAs*	1007
	16.12.4 *Radiation Mechanism of 2-D Holographic Metasurfaces*	1008
	16.12.5 *Polarization-Diverse Holographic Metasurfaces*	1009

	A.	Vertical Polarization	1009
	B.	Horizontal Polarization	1010
	C.	Circular Polarization	1011
	D.	Multiple Polarizations	1011
16.12.6	*Tensor Impedance Surfaces*		1013
16.13	Multimedia		1013
	References		1014
	Problems		1019

Appendix I   **Identities**   **1023**

Appendix II   **Vector Analysis**   **1027**

Appendix III   **Fresnel Integrals**   **1037**

Appendix IV   **Bessel Functions**   **1043**

Appendix V   **Legendre Polynomials and Functions**   **1057**

Appendix VI   **The Method of Steepest Descent (Saddle-Point Method)**   **1073**

Glossary   **1079**

Index   **1085**

Preface

Because of the immense interest in and success of the first two editions, the third edition renamed *Balanis' Advanced Engineering Electromagnetics* has maintained all the attractive features of the first two editions. A new chapter, Chapter 16 on *Artificial Impedance Surfaces* (AIS), contains material on current and advanced EM technologies including *metasurfaces* for:

- Control and broadband RCS reduction using checkerboard designs.
- Optimization of antenna fundamental parameters, such as: input impedance, directivity, realized gain, amplitude radiation pattern.
- Leaky-wave antennas using 1-D and 2-D polarization diverse-holographic high impedance metasurfaces for antenna radiation control and optimization.
- MATLAB programs for the design of checkerboard metasurfaces for RCS reduction, and metasurface printed antennas and holographic LWA for radiation control and optimization.

In addition, smaller inserts have been added throughout the book, including:

- New figures, photos, and tables.
- Additional examples and numerous end-of-chapter problems.
- PPT notes to supplement lectures on design of checkerboard metasurfaces for RCS reduction, antenna fundamental parameters optimization, and Leaky Wave Antennas (LWA) for amplitude radiation scanning.

The book also provides multimedia material (refer to About the Companion Website section for full details), including:

- Over 4,500 multicolor PowerPoint slides for the 16 chapters.
- Fifty-three MATLAB® computer programs (most of them new; the four Fortran programs from the first edition were translated to MATLAB®)

Given the space limitations, the added material supplements, expands, and reinforces the analytical methods that were, and continue to be, the main focus of this book. The analytical methods are the foundation of electromagnetics and provide understanding and physical interpretation of electromagnetic phenomena and interactions. Although numerical and computational methods have, especially in the last four decades, played a key role in the solution of complex electromagnetic problems, they are highly dependent on fundamental principles. Not understanding the basic fundamentals of electromagnetics, represented by analytical methods, may lead to the lack of physical realization, interpretation, and verification of simulated results. In fact, there are a plethora of personal and commercial codes that are now available, and they are expanding very rapidly. Users are now highly dependent on these codes, and we seem to lose focus on the interpretation and physical realization of the simulated results because, possibly, of the lack of understanding of fundamental principles. There are numerous books that address numerical and computational methods, and this author did not want to repeat what is already available in the literature, especially with space limitations. Only the moment method (MM), in support of Integral Equations (IEs), and Diffraction Theory (GTD/UTD) are included in this book. However, to aid in the computation, simulation, and animation of results based on analytical formulations included in this book, even provide some of the data in graphical form, forty-eight basic MATLAB® computer programs have been developed and are included on the website that is part of this book.

The first edition was based on material taught on a yearly basis and notes developed over nearly 20 years. This third edition, based on an additional 20 years of teaching and development of notes and multimedia (for a total of over 40 years of teaching), refined any shortcomings of the first edition and added: a new chapter, two new complete sections, numerous smaller inserts, examples, end-of-chapter problems, and Multimedia (including PPT notes, MATLAB® computer programs for computations, simulations, visualization, and animation). The four Fortran programs from the first edition were translated in MATLAB®, and numerous additional ones were developed only in MATLAB®. These are spread throughout Chapters 4 through 14. The revision of the book also took into account suggestions of nearly 20 reviewers selected by the publisher, some of whom are identified and acknowledged based on their approval. The multicolor PowerPoint (PPT) notes, over 4,500 viewgraphs, can be used as ready-made lectures so that instructors will not have to labor at developing their own notes. Instructors also have the option to add PPT viewgraphs of their own or delete any that do not fit their class objectives. This third edition is based on additional 10 years of teaching, for a total of 40. The additional material included in the third edition is outlined at the beginning of the Preface.

The book can be used for at least a two-semester sequence in electromagnetics, beyond an introduction to basic undergraduate EM. Although the first part of the book is intended for senior undergraduates and beginning graduates in electrical engineering and physics, the later chapters are targeted for advanced graduate students and practicing engineers and scientists. The majority of Chapters 1 through 10 can be covered in the first semester, and most of Chapters 11 through 16 can be covered in the second semester. To cover all of the material in the proposed time frame would be, in many instances, an ambitious task. However, sufficient topics have been included to make the text complete and to allow instructors the flexibility to emphasize, de-emphasize, or omit sections and/or chapters. Some chapters can be omitted without loss of continuity.

The discussion presumes that the student has general knowledge of vector analysis, differential and integral calculus, and electromagnetics from at least either an introductory undergraduate electrical engineering or a physics course. Mathematical techniques required for understanding some advanced topics, mostly in the later chapters, are incorporated in the individual chapters or are included as appendixes.

Like the first and second editions, this third edition is a thorough and detailed student-oriented book. The analytical detail, rigor, and thoroughness allow many of the topics to be traced to their origin, and they are presented in sufficient detail so that the students, and the instructors as well, will follow the analytical developments. In addition to the coverage of traditional classical topics, the book includes state of the art advanced topics on Metamaterials, Artificial Impedance Surfaces

(AIS, EBG, PBG, HIS, AMC, PMC), Integral Equations (IE), Moment Method (MM), Geometrical and Uniform Theory of Diffraction (GTD/UTD) for PEC and impedance surfaces, and Green's functions. Electromagnetic theorems, as applied to the solution of boundary-value problems, are also included and discussed.

The material is presented in a methodical, sequential, and unified manner, and each chapter is subdivided into sections or subsections whose individual headings clearly identify the topics discussed, examined, or illustrated. The examples and end-of-chapter problems have been designed to illustrate basic principles and to challenge the knowledge of the student. An exhaustive list of references is included at the end of each chapter to allow the interested reader to trace each topic. A number of appendixes of mathematical identities and special functions, some represented also in tabular and graphical forms, are included to aid the student in the solution of the examples and assigned end-of-chapter problems. A solutions manual for all end-of-chapter problems is available exclusively to instructors.

In Chapter 1, the book covers classical topics on Maxwell's equations, constitutive parameters and relations, circuit relations, boundary conditions, and power and energy relations. The electrical properties of matter for both direct current and alternating current, including an update on super-conductivity, are covered in Chapter 2. The wave equation and its solution in rectangular, cylindrical, and spherical coordinates are discussed in Chapter 3. Electromagnetic wave propagation and polarization are introduced in Chapter 4. Reflection and transmission at normal and oblique incidences are considered in Chapter 5, along with depolarization of the wave due to reflection and transmission and an introduction to metamaterials (especially those with negative index of refraction, referred to as double negative, DNG). Chapter 6 covers the auxiliary vector potentials and their use toward the construction of solutions for radiation and scattering problems. The theorems of duality, uniqueness, image, reciprocity, reaction, volume and surface equivalences, induction, and physical and physical optics equivalents are introduced and applied in Chapter 7. Rectangular cross-section waveguides and cavities, including dielectric slabs, striplines and microstrips, and ridged waveguides are discussed in Chapter 8. Waveguides and cavities with circular cross section, including the fiber optics cable, are examined in Chapter 9, and those of spherical geometry are introduced in Chapter 10. Scattering by strips, plates, circular cylinders, wedges, and spheres is analyzed in Chapter 11. Chapter 12 covers the basics and applications of Integral Equations (IE) and the Moment Method (MM). The techniques and applications of the Geometrical and Uniform Theory of Diffraction (GTD/UTD) are introduced and discussed in Chapter 13. The PEC GTD/UTD techniques of Chapter 13 are extended in Chapter 14 to wedges with impedance surfaces, utilizing Maliuzhinets functions. The classic topic of Green's functions is introduced and applied in Chapter 15.

The new Chapter 16 addresses Artificial Impedance Surfaces (AIS), also referred to as Metasurfaces; Electromagnetics Band-Gap (EBG) Structures; Photonic Band Gap (PBG) Structures; High Impedance Surfaces (HIS); Artificial Magnetic Conductors (AMC); Perfect Magnetic Conductors (PMC), and others. The aim of this new chapter is to introduce current and fascinating metasurface EM technology for broadband RCS control and reduction, antenna fundamental parameters optimization, and Leaky Wave Antennas (LWA) for amplitude radiation scanning.

Throughout the book an $e^{j\omega t}$ time convention is assumed, and it is suppressed in almost all the chapters. The International System of Units, which is an expanded form of the rationalized MKS system, is used throughout the text. In some instances, the units of length are given in meters (or centimeters) and feet (or inches). Numbers in parentheses ( ) refer to equations, whereas those in brackets [ ] refer to references. For emphasis, the most important equations, once they are derived, are boxed.

I would like to acknowledge the invaluable suggestions and contributions of all who contributed to the first two editions of the book; they are too numerous to mention here. Their names and contributions are explicitly acknowledged in the Prefaces of the respective two editions. It is my pleasure to acknowledge here the contributors to this third edition, especially those who contributed to the new Chapter 16. I am indebted to my most recent graduate students whose PhD dissertations formed the cornerstone of Chapter 16. In particular I will like to thank: Dr. Meshaal A. Alyahya, Dr. Anuj Modi, Dr. Subramanian Ramalingam, Dr. Mohammed Albarbi, Dr. Wengang Chen,

Dr. Mikal Askarian Amiri, and Dr. Sivaseetharaman Pandi. Special recognition to Dr. Meshaal A. Alyahya, who combined the updates from all the other contributors to form the LaTex files of new Chapter 16. My appreciation to Dr. Wengang Chen for the design of the front cover based on research of checkerboard metasurfaces of Chapter 16, which was initiated while he was a PhD graduate student in my research group. Also many thanks to Craig R. Birtcher, colleague and manager of the ASU Anechoic Chamber (EMAC), for the many stimulating discussions, partnership, and collaboration for 35 years.

During my 56-year professional and academic career, I have been influenced and inspired, directly and indirectly, by outstanding book authors and renowned researchers for whom I developed respect and admiration. Many of them I consider mentors, role models, and colleagues. Over the same time I developed professional and social friends and colleagues who supported and encouraged me in advancing and reaching many of my professional objectives, interests, and goals. They are too numerous, and I will not attempt to list them as I may, inadvertently, omit someone. However, I want to sincerely acknowledge their continued interest, support, friendship, collegiality, and comradery. To all my graduate students, dating back to 1970, thank you for your contributions, partnership, and fellowship.

The journey for this book got started with the first edition in the middle 1980s, and became a reality in 1989; we had the second edition in 2012. The impetus of the first, and subsequent editions, may have been the TV/Online classes when I joined Arizona State University in 1983; initially they were referred to as TV classes broadcast locally primarily to the high-tech companies in the Phoenix metro, using a microwave link; later, with the introduction of the internet, they were renamed Online classes, and they were expanded to the national and international audience. I took an active participation in the preparation and presentation of TV/Online classes, both undergraduate and graduate. Eventually I had prepared notes, delivered and taped five (5) TV/Online classes; two antenna classes (one UG and one G), and three graduate EM classes. As a result of the preparation of notes for the graduate EM classes, especially those delivered as TV/Online classes, I developed notes that eventually were converted to a manuscript. It was published in 1989 as a book entitled *Advanced Engineering Electromagnetics.*

I have chosen to remain the sole author for the three editions of *Advanced Engineering Electromagnetics,* renamed *Balanis' Advanced Engineering Electromagnetics* in the third edition by the initiative of the publisher Wiley, and the four editions, of the *Antenna Theory: Analysis and Design,* as long as I was able to complete the tasks, so the books manifest my own fingerprint and stewardship, and reflect my personal philosophy, methodology, and pedagogy. Also I wanted the manuscript to display continuity and consistency, and to control my own destiny, in terms of material to be included and excluded, revisions, deadlines, and timelines. Finally, I wanted to be responsible for the contents of the book. In the words of Frank Sinatra, 'I did it my way.' Each edition presented its own challenges, but each time I cherished and looked forward to the mission and venture.

The interest and support shown my books by the international readership, especially students, instructors, engineers, and scientists, has been a lifelong, rewarding, and fulfilling professional accomplishment. I am most appreciative and grateful for the interest, support, and acknowledgement of those who were influenced and inspired, and hopefully benefitted, in advancing their educational and professional knowledge, objectives, and careers.

I am also grateful to the Staff of John Wiley & Sons, Inc., especially Brett Kurzman, Editor, and Aileen Storry, Publisher, and Becky Cowan, Senior Editorial Assistant.

Finally, I pay tribute and homage to my family (Helen, Renie, Stephanie, Bill, Pete, and Ellie) for their unconditional support, patience, sacrifice, and understanding for the many years of neglect during the completion of all the three editions of this book and four editions of the *Antenna Theory: Analysis and Design.* Each edition has been a pleasant experience although a daunting task.

Constantine A. Balanis
Arizona State University
Tempe, AZ

About the Companion Website

This book is accompanied by a companion website, which includes a number of resources created by the author for students and instructors that you will find helpful.

www.wiley.com/go/balanis/advancedengineeringelectromagnetics3e

The Instructor website includes the following resources for each chapter:

- Solutions Manual
- PowerPoint Notes
- Information about Multimedia
- MATLAB Computer Programs
- Image Gallery
- Art PowerPoints
- FORTRAN Computer Programs

The Student website includes the following resources for each chapter:

- PowerPoint Notes
- Information about Multimedia
- MATLAB Computer Programs

Please note that the resources on the Instructor website are password protected and can only be accessed by instructors who register with the site.

CHAPTER 1

Time-Varying and Time-Harmonic Electromagnetic Fields

1.1 INTRODUCTION

Electromagnetic field theory is a discipline concerned with the study of charges, at rest and in motion, that produce currents and electric-magnetic fields. It is, therefore, fundamental to the study of electrical engineering and physics and indispensable to the understanding, design, and operation of many practical systems using antennas, scattering, microwave circuits and devices, radio-frequency and optical communications, wireless communications, broadcasting, geosciences and remote sensing, radar, radio astronomy, quantum electronics, solid-state circuits and devices, electromechanical energy conversion, and even computers. Circuit theory, a required area in the study of electrical engineering, is a special case of electromagnetic theory, and it is valid when the physical dimensions of the circuit are small compared to the wavelength. Circuit concepts, which deal primarily with lumped elements, must be modified to include distributed elements and coupling phenomena in studies of advanced systems. For example, signal propagation, distortion, and coupling in microstrip lines used in the design of sophisticated systems (such as computers and electronic packages of integrated circuits) can be properly accounted for only by understanding the electromagnetic field interactions associated with them.

The study of electromagnetics includes both theoretical and applied concepts. The theoretical concepts are described by a set of basic laws formulated primarily through experiments conducted during the nineteenth century by many scientists—Faraday, Ampere, Gauss, Lenz, Coulomb, Volta, and others. Although Maxwell had come up with 20 equations with 20 variables, it was Heaviside and Hertz that both independently put them into a consistent and compact vectorial form. Both Heaviside and Hertz named them in honor of Maxwell, and today they are the widely acclaimed *Maxwell's equations*. The applied concepts of electromagnetics are formulated by applying the theoretical concepts to the design and operation of practical systems.

In this chapter, we will review Maxwell's equations (both in differential and integral forms), describe the relations between electromagnetic field and circuit theories, derive the boundary conditions associated with electric and magnetic field behavior across interfaces, relate power and energy concepts for electromagnetic field and circuit theories, and specialize all these equations, relations, conditions, concepts, and theories to the study of time-harmonic fields.

Balanis' Advanced Engineering Electromagnetics, Third Edition. Constantine A. Balanis.
© 2024 John Wiley & Sons, Inc. Published 2024 by John Wiley & Sons, Inc.
Companion Website: www.wiley.com/go/balanis/advancedengineeringelectromagnetics3e

1.2 MAXWELL'S EQUATIONS

In general, electric and magnetic fields are vector quantities that have both magnitude and direction. The relations and variations of the electric and magnetic fields, charges, and currents associated with electromagnetic waves are governed by physical laws, which are known as Maxwell's equations. These equations, as we have indicated, were arrived at mostly through various experiments carried out by different investigators, but they were put in their final form by James Clerk Maxwell, a Scottish physicist and mathematician. These equations can be written either in differential or in integral form.

1.2.1 Differential Form of Maxwell's Equations

The differential form of Maxwell's equations is the most widely used representation to solve boundary-value electromagnetic problems. It is used to describe and relate the field vectors, current densities, and charge densities *at any point in space at any time*. For these expressions to be valid, it is assumed that the field vectors are *single-valued, bounded, continuous* functions of position and time and exhibit *continuous derivatives*. Field vectors associated with electromagnetic waves possess these characteristics except where there exist abrupt changes in charge and current densities. Discontinuous distributions of charges and currents usually occur at interfaces between media where there are discrete changes in the electrical parameters across the interface. The variations of the field vectors across such boundaries (interfaces) are related to the discontinuous distributions of charges and currents by what are usually referred to as the *boundary conditions*. Thus a complete description of the field vectors at any point (including discontinuities) at any time requires not only Maxwell's equations in differential form but also the associated *boundary conditions*.

In differential form, Maxwell's equations can be written as

$$\nabla \times \mathcal{E} = -\mathcal{M}_i - \frac{\partial \mathcal{B}}{\partial t} = -\mathcal{M}_i - \mathcal{M}_d = -\mathcal{M}_t \tag{1-1}$$

$$\nabla \times \mathcal{H} = \mathcal{J}_i + \mathcal{J}_c + \frac{\partial \mathcal{D}}{\partial t} = \mathcal{J}_{ic} + \frac{\partial \mathcal{D}}{\partial t} = \mathcal{J}_{ic} + \mathcal{J}_d = \mathcal{J}_t \tag{1-2}$$

$$\nabla \cdot \mathcal{D} = q_{ev} \tag{1-3}$$

$$\nabla \cdot \mathcal{B} = q_{mv} \tag{1-4}$$

where

$$\mathcal{J}_{ic} = \mathcal{J}_i + \mathcal{J}_c \tag{1-5a}$$

$$\mathcal{J}_d = \frac{\partial \mathcal{D}}{\partial t} \tag{1-5b}$$

$$\mathcal{M}_d = \frac{\partial \mathcal{B}}{\partial t} \tag{1-5c}$$

All these field quantities—$\mathcal{E}$, $\mathcal{H}$, $\mathcal{D}$, $\mathcal{B}$, $\mathcal{J}$, $\mathcal{M}$, and q_v—are assumed to be time-varying, and each is a function of the space coordinates and time, that is, $\mathcal{E} = \mathcal{E}(x, y, z; t)$. The definitions and units of the quantities are

$\mathcal{E}$ = electric field intensity (volts/meter)
$\mathcal{H}$ = magnetic field intensity (amperes/meter)
$\mathcal{D}$ = electric flux density (coulombs/square meter)
$\mathcal{B}$ = magnetic flux density (webers/square meter)

$\mathscr{J}_i = $ impressed (source) electric current density (amperes/square meter)

$\mathscr{J}_c = $ conduction electric current density (amperes/square meter)

$\mathscr{J}_d = $ displacement electric current density (amperes/square meter)

$\mathcal{M}_i = $ impressed (source) magnetic current density (volts/square meter)

$\mathcal{M}_d = $ displacement magnetic current density (volts/square meter)

$q_{ev} = $ electric charge density (coulombs/cubic meter)

$q_{mv} = $ magnetic charge density (webers/cubic meter)

The electric displacement current density $\mathscr{J}_d = \partial \mathscr{D}/\partial t$ was introduced by Maxwell to complete Ampere's law for statics, $\nabla \times \mathcal{H} = \mathscr{J}$. For free space, $\mathscr{J}_d$ was viewed as a motion of bound charges moving in "ether," an ideal weightless fluid pervading all space. Since ether proved to be undetectable and its concept was not totally reasonable with the special theory of relativity, it has since been disregarded. Instead, for dielectrics, part of the displacement current density has been viewed as a motion of bound charges creating a true current. Because of this, it is convenient to consider, even in free space, the entire $\partial \mathscr{D}/\partial t$ term as a displacement current density.

Because of the symmetry of Maxwell's equations, the $\partial \mathscr{B}/\partial t$ term in (1-1) has been designated as a magnetic displacement current density. In addition, impressed (source) magnetic current density $\mathcal{M}_i$ and magnetic charge density q_{mv} have been introduced, respectively, in (1-1) and (1-4) through the "generalized" current concept. Although we have been accustomed to viewing magnetic charges and impressed magnetic current densities as not being physically realizable, they have been introduced to balance Maxwell's equations. Equivalent magnetic charges and currents will be introduced in later chapters to represent physical problems. In addition, impressed magnetic current densities, like impressed electric current densities, can be considered energy sources that generate fields whose expressions can be written in terms of these current densities. For some electromagnetic problems, their solution can often be aided by the introduction of "equivalent" impressed electric and magnetic current densities. The importance of both will become more obvious to the reader as solutions to specific electromagnetic boundary-value problems are considered in later chapters. However, to give the reader an early glimpse of the importance and interpretation of the electric and magnetic current densities, let us consider two familiar circuit examples.

In Figure 1-1a, an electric current source is connected in series to a resistor and a parallel-plate capacitor. The electric current density $\mathscr{J}_i$ can be viewed as the current source that generates the conduction current density $\mathscr{J}_c$ through the resistor and the displacement current density $\mathscr{J}_d$ through the dielectric material of the capacitor. In Figure 1-1b, a voltage source is connected to a wire that, in turn, is wrapped around a high-permeability magnetic core. The voltage source can be viewed as the impressed magnetic current density that generates the displacement magnetic current density through the magnetic material of the core.

In addition to the four Maxwell's equations, there is another equation that relates the variations of the current density $\mathscr{J}_{ic}$ and the charge density q_{ev}. Although not an independent relation, this equation is referred to as the *continuity equation* because it relates the net flow of current out of a small volume (in the limit, a point) to the rate of decrease of charge. It takes the form

$$\nabla \cdot \mathscr{J}_{ic} = -\frac{\partial q_{ev}}{\partial t} \qquad (1\text{-}6)$$

The continuity equation 1-6 can be derived from Maxwell's equations as given by (1-1) through (1-5c).

1.2.2 Integral Form of Maxwell's Equations

The integral form of Maxwell's equations describes the relations of the field vectors, charge densities, and current densities over an *extended region of space*. They have limited applications, and they are usually utilized only to solve electromagnetic boundary-value problems that possess

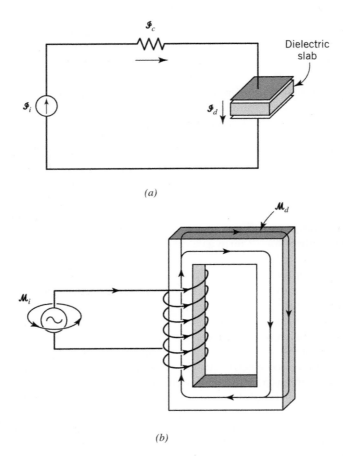

Figure 1-1 Circuits with electric and magnetic current densities. (*a*) Electric current density. (*b*) Magnetic current density.

complete symmetry (such as rectangular, cylindrical, spherical, etc., symmetries). However, *the fields and their derivatives in question do not need to possess continuous distributions.*

The integral form of Maxwell's equations can be derived from its differential form by utilizing the *Stokes'* and *divergence theorems.* For any arbitrary vector **A**, Stokes' theorem states that *the line integral of the vector* **A** *along a closed path C is equal to the integral of the dot product of the curl of the vector* **A** *with the normal to the surface S that has the contour C as its boundary.*

In equation form, Stokes' theorem can be written as

$$\oint_C \mathbf{A} \cdot d\boldsymbol{\ell} = \iint_S (\boldsymbol{\nabla} \times \mathbf{A}) \cdot d\mathbf{s} \tag{1-7}$$

The divergence theorem states that, for any arbitrary vector **A**, *the closed surface integral of the normal component of vector* **A** *over a surface S is equal to the volume integral of the divergence of* **A** *over the volume V enclosed by S.* In mathematical form, the divergence theorem is stated as

$$\oiint_S \mathbf{A} \cdot d\mathbf{s} = \iiint_V \boldsymbol{\nabla} \cdot \mathbf{A} \, dv \tag{1-8}$$

Taking the surface integral of both sides of (1-1), we can write

$$\iint_S (\boldsymbol{\nabla} \times \boldsymbol{\mathscr{E}}) \cdot d\mathbf{s} = -\iint_S \boldsymbol{\mathscr{M}}_i \cdot d\mathbf{s} - \iint_S \frac{\partial \boldsymbol{\mathscr{B}}}{\partial t} \cdot d\mathbf{s} = -\iint_S \boldsymbol{\mathscr{M}}_i \cdot d\mathbf{s} - \frac{\partial}{\partial t} \iint_S \boldsymbol{\mathscr{B}}_i \cdot d\mathbf{s} \tag{1-9}$$

Applying Stokes' theorem, as given by (1-7), on the left side of (1-9) reduces it to

$$\oint_C \mathscr{E} \cdot d\ell = -\iint_S \mathcal{M}_i \cdot d\mathbf{s} - \frac{\partial}{\partial t} \iint_S \mathscr{B} \cdot d\mathbf{s} \qquad (1\text{-}9a)$$

which is referred to as *Maxwell's equation in integral form as derived from Faraday's law*. In the absence of an impressed magnetic current density, *Faraday's law* states that the electromotive force (emf) appearing at the open-circuited terminals of a loop is equal to the time rate of decrease of magnetic flux linking the loop.

Using a similar procedure, we can show that the corresponding integral form of (1-2) can be written as

$$\oint_C \mathscr{H} \cdot d\ell = \iint_S \mathscr{I}_{ic} \cdot d\mathbf{s} + \frac{\partial}{\partial t} \iint_S \mathscr{D} \cdot d\mathbf{s} = \iint_S \mathscr{I}_{ic} \cdot d\mathbf{s} + \iint_S \mathscr{I}_d \cdot d\mathbf{s} \qquad (1\text{-}10)$$

which is usually referred to as *Maxwell's equation in integral form as derived from Ampere's law*. *Ampere's law* states that the line integral of the magnetic field over a closed path is equal to the current enclosed.

The other two Maxwell equations in integral form can be obtained from the corresponding differential forms, using the following procedure. First take the volume integral of both sides of (1-3); that is,

$$\iiint_V \nabla \cdot \mathscr{D} \, dv = \iiint_V q_{ev} \, dv = \mathcal{Q}_e \qquad (1\text{-}11)$$

where $\mathcal{Q}_e$ is the total electric charge. Applying the divergence theorem, as given by (1-8), on the left side of (1-11) reduces it to

$$\oiint_S \mathscr{D} \cdot d\mathbf{s} = \iiint_V q_{ev} \, dv = \mathcal{Q}_e \qquad (1\text{-}11a)$$

which is usually referred to as *Maxwell's electric field equation in integral form as derived from Gauss's law*. *Gauss's law* for the electric field states that the total electric flux through a closed surface is equal to the total charge enclosed.

In a similar manner, the integral form of (1-4) is given in terms of the total magnetic charge $\mathcal{Q}_m$ by

$$\oiint_S \mathscr{B} \cdot d\mathbf{s} = \mathcal{Q}_m \qquad (1\text{-}12)$$

which is usually referred to as *Maxwell's magnetic field equation in integral form as derived from Gauss's law*. Even though magnetic charge does not exist in nature, it is used as an equivalent to represent physical problems. The corresponding integral form of the continuity equation, as given by (1-6) in differential form, can be written as

$$\oiint_S \mathscr{I}_{ic} \cdot d\mathbf{s} = -\frac{\partial}{\partial t} \iiint_V q_{ev} \, dv = -\frac{\partial \mathcal{Q}_e}{\partial t} \qquad (1\text{-}13)$$

Maxwell's equations in differential and integral form are summarized and listed in Table 1-1.

1.3 CONSTITUTIVE PARAMETERS AND RELATIONS

Materials contain charged particles, and when these materials are subjected to electromagnetic fields, their charged particles interact with the electromagnetic field vectors, producing currents and modifying the electromagnetic wave propagation in these media compared to that in free space. A more complete discussion of this is in Chapter 2. To account on a macroscopic scale for the

TABLE 1-1 Maxwell's equations and the continuity equation in differential and integral forms for time-varying fields

Differential form	Integral form
$\nabla \times \mathscr{E} = -\mathscr{M}_i - \dfrac{\partial \mathscr{B}}{\partial t}$	$\oint_C \mathscr{E} \cdot d\ell = -\iint_S \mathscr{M}_i \cdot d\mathbf{s} - \dfrac{\partial}{\partial t} \iint_S \mathscr{B} \cdot d\mathbf{s}$
$\nabla \times \mathscr{H} = \mathscr{I}_i + \mathscr{I}_c + \dfrac{\partial \mathscr{D}}{\partial t}$	$\oint_C \mathscr{H} \cdot d\ell = \iint_S \mathscr{I}_i \cdot d\mathbf{s} + \iint_S \mathscr{I}_c \cdot d\mathbf{s} + \dfrac{\partial}{\partial t} \iint_S \mathscr{D} \cdot d\mathbf{s}$
$\nabla \cdot \mathscr{D} = q_{ev}$	$\oiint_S \mathscr{D} \cdot d\mathbf{s} = \mathcal{Q}_e$
$\nabla \cdot \mathscr{B} = q_{mv}$	$\oiint_S \mathscr{B} \cdot d\mathbf{s} = \mathcal{Q}_m$
$\nabla \cdot \mathscr{I}_{ic} = -\dfrac{\partial q_{ev}}{\partial t}$	$\oiint_S \mathscr{I}_{ic} \cdot d\mathbf{s} = -\dfrac{\partial}{\partial t} \iiint_V q_{ev} dv = -\dfrac{\partial \mathcal{Q}_e}{\partial t}$

presence and behavior of these charged particles, without introducing them in a microscopic lattice structure, we give a set of three expressions relating the electromagnetic field vectors. These expressions are referred to as the *constitutive relations*, and they will be developed in more detail in Chapter 2.

One of the constitutive relations relates *in the time domain* the electric flux density $\mathscr{D}$ to the electric field intensity $\mathscr{E}$ by

$$\mathscr{D} = \hat{\varepsilon} * \mathscr{E} \tag{1-14}$$

where $\hat{\varepsilon}$ is the time-varying permittivity of the medium (farads/meter) and $*$ indicates convolution. For free space

$$\hat{\varepsilon} = \varepsilon_0 = 8.854 \times 10^{-12} \simeq \frac{10^{-9}}{36\pi} \text{ (farads/meter)} \tag{1-14a}$$

and (1-14) reduces to a product.

Another relation equates *in the time domain* the magnetic flux density $\mathscr{B}$ to the magnetic field intensity $\mathscr{H}$ by

$$\mathscr{B} = \hat{\mu} * \mathscr{H} \tag{1-15}$$

where $\hat{\mu}$ is the time-varying permeability of the medium (henries/meter). For free space

$$\hat{\mu} = \mu_0 = 4\pi \times 10^{-7} \text{ (henries/meter)} \tag{1-15a}$$

and (1-15) reduces to a product.

Finally, the conduction current density $\mathscr{I}_c$ is related *in the time domain* to the electric field intensity $\mathscr{E}$ by

$$\mathscr{I}_c = \hat{\sigma} * \mathscr{E} \tag{1-16}$$

where $\hat{\sigma}$ is the time-varying conductivity of the medium (siemens/meter). For free space

$$\hat{\sigma} = 0 \tag{1-16a}$$

In the frequency domain or for frequency nonvarying constitutive parameters, the relations (1-14), (1-15), and (1-16) reduce to products. For simplicity of notation, they will be indicated everywhere from now on as products, and the caret (^) in the time-varying constitutive parameters will be omitted.

Whereas (1-14), (1-15), and (1-16) are referred to as the *constitutive relations*, $\hat{\varepsilon}$, $\hat{\mu}$, and $\hat{\sigma}$ are referred to as the *constitutive parameters*, which are, in general, functions of the applied field strength, the position within the medium, the direction of the applied field, and the frequency of operation.

The constitutive parameters are used to characterize the electrical properties of a material. In general, materials are characterized as *dielectrics (insulators)*, *magnetics*, and *conductors*, depending on whether *polarization* (electric displacement current density), *magnetization* (magnetic displacement current density), or *conduction* (conduction current density) is the predominant phenomenon. Another class of material is made up of *semiconductors*, which bridge the gap between dielectrics and conductors where neither displacement nor conduction currents are, in general, predominant. In addition, materials are classified as *linear* versus *nonlinear*, *homogeneous* versus *nonhomogeneous (inhomogeneous)*, *isotropic* versus *nonisotropic (anisotropic)*, and *dispersive* versus *nondispersive*, according to their lattice structure and behavior. All these types of materials will be discussed in detail in Chapter 2.

If all the constitutive parameters of a given medium are not functions of the applied field strength, the material is known as *linear*; otherwise it is *nonlinear*. Media whose constitutive parameters are not functions of position are known as *homogeneous*; otherwise they are referred to as *nonhomogeneous (inhomogeneous)*. *Isotropic* materials are those whose constitutive parameters are not functions of direction of the applied field; otherwise they are designated as *nonisotropic (anisotropic)*. Crystals are one form of anisotropic material. Material whose constitutive parameters are functions of frequency are referred to as *dispersive*; otherwise they are known as *nondispersive*. All materials used in our everyday life exhibit some degree of dispersion, although the variations for some may be negligible and for others significant. More details concerning the development of the constitutive parameters can be found in Chapter 2.

1.4 CIRCUIT-FIELD RELATIONS

The differential and integral forms of Maxwell's equations were presented, respectively, in Sections 1.2.1 and 1.2.2. These relations are usually referred to as *field equations*, since the quantities appearing in them are all *field quantities*. Maxwell's equations can also be written in terms of what are usually referred to as *circuit quantities*; the corresponding forms are denoted *circuit equations*. The circuit equations are introduced in circuit theory texts, and they are special cases of the more general field equations.

1.4.1 Kirchhoff's Voltage Law

According to Maxwell's equation 1-9a, the left side represents the sum voltage drops (use the convention where positive voltage begins at the start of the path) along a closed path C, which can be written as

$$\sum v = \oint_C \mathscr{E} \cdot d\boldsymbol{\ell} \text{ (volts)} \tag{1-17}$$

The right side of (1-9a) must also have the same units (volts) as its left side. Thus, in the absence of impressed magnetic current densities ($\mathcal{M}_i = 0$), the right side of (1-9a) can be written as

$$-\frac{\partial}{\partial t}\iint_S \mathscr{B} \cdot d\mathbf{s} = -\frac{\partial \psi_m}{\partial t} = -\frac{\partial}{\partial t}(L_s i) = -L_s \frac{\partial i}{\partial t} \text{ (webers/second = volts)} \tag{1-17a}$$

because by definition $\psi_m = L_s i$ where L_s is an inductance (assumed to be constant) and i is the associated current. Using (1-17) and (1-17a), we can write (1-9a) with $\mathcal{M}_i = 0$ as

$$\sum v = -\frac{\partial \psi_m}{\partial t} = -\frac{\partial}{\partial t}(L_s i) = -L_s \frac{\partial i}{\partial t} \tag{1-17b}$$

Equation 1-17b states that the voltage drops along a closed path of a circuit are equal to the time rate of change of the magnetic flux passing through the surface enclosed by the closed path, or equal to the voltage drop across an inductor L_s that is used to represent the *stray inductance* of the circuit. This is the well-known *Kirchhoff loop voltage law*, which is used widely in circuit theory, and its form represents a circuit relation. Thus we can write the following field and circuit relations:

Field Relation   Circuit Relation

$$\oint_C \mathscr{E} \cdot d\ell = -\frac{\partial}{\partial t} \iint_S \mathscr{B} \cdot d\mathbf{s} = -\frac{\partial \psi_m}{\partial t} \Leftrightarrow \sum v = -\frac{\partial \psi_m}{\partial t} = -L_s \frac{\partial i}{\partial t} \qquad (1\text{-}17c)$$

In lumped-element circuit analysis, where usually the wavelength is very large (or the dimensions of the total circuit are small compared to the wavelength) and the stray inductance of the circuit is very small, the right side of (1-17b) is very small and it is usually set equal to zero. In these cases, (1-17b) states that the voltage drops (or rises) along a closed path are equal to zero, and it represents a widely used relation to electrical engineers and many physicists.

To demonstrate Kirchhoff's loop voltage law, let us consider the circuit of Figure 1-2 where a voltage source and three ideal lumped elements (a resistance R, an inductor L, and a capacitor C) are connected in series to form a closed loop. According to (1-17b)

$$-v_s + v_R + v_L + v_C = -L_s \frac{\partial i}{\partial t} = -v_{sL} \qquad (1\text{-}18)$$

where L_s, shown dashed in Figure 1-2, represents the total stray inductance associated with the current and the magnetic flux generated by the loop that connects the ideal lumped elements (we assume that the wire resistance is negligible). If the stray inductance L_s of the circuit and the time rate of change of the current is small (the case for low-frequency applications), the right side of (1-18) is small and can be set equal to zero.

1.4.2 Kirchhoff's Current Law

The left side of the integral form of the continuity equation, as given by (1-13), can be written in circuit form as

$$\sum i = \oiint_S \mathscr{J}_{ic} \cdot d\mathbf{s} \qquad (1\text{-}19)$$

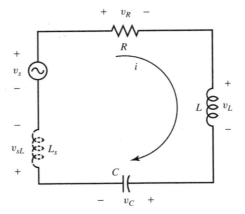

Figure 1-2 RLC series network.

where $\sum i$ represents the sum of the currents passing through closed surface S. Using (1-19) reduces (1-13) to

$$\sum i = -\frac{\partial \mathcal{Q}_e}{\partial t} = -\frac{\partial}{\partial t}(C_s v) = -C_s \frac{\partial v}{\partial t} \tag{1-19a}$$

since by definition $\mathcal{Q}_e = C_s v$ where C_s is a capacitance (assumed to be constant) and v is the associated voltage.

Equation 1-19a states that the sum of the currents crossing a surface that encloses a circuit is equal to the time rate of change of the total electric charge enclosed by the surface, or equal to the current flowing through a capacitor C_s that is used to represent the *stray capacitance* of the circuit. This is the well-known *Kirchhoff node current law,* which is widely used in circuit theory, and its form represents a circuit relation. Thus, we can write the following field and circuit relations:

Field Relation   *Circuit Relation*

$$\oiint_S \mathcal{I}_{ic} \cdot d\mathbf{s} = -\frac{\partial}{\partial t} \iiint_V \mathcal{I}_{ev} \, dv = -\frac{\partial \mathcal{Q}_e}{\partial t} \Leftrightarrow \sum i = -\frac{\partial \mathcal{Q}_e}{\partial t} = -C_s \frac{\partial v}{\partial t} \tag{1-19b}$$

In lumped-element circuit analysis, where the stray capacitance associated with the circuit is very small, the right side of (1-19a) is very small and it is usually set equal to zero. In these cases, (1-19a) states that the currents exiting (or entering) a surface enclosing a circuit are equal to zero. This represents a widely used relation to electrical engineers and many physicists.

To demonstrate Kirchhoff's node current law, let us consider the circuit of Figure 1-3 where a current source and three ideal lumped elements (a resistance R, an inductor L, and a capacitor C) are connected in parallel to form a node. According to (1-19a)

$$-i_s + i_R + i_L + i_C = -C_s \frac{\partial v}{\partial t} = -i_{sC} \tag{1-20}$$

where C_s, shown dashed in Figure 1-3, represents the total stray capacitance associated with the circuit of Figure 1-3. If the stray capacitance C_s of the circuit and the time rate of change of the total charge $\mathcal{Q}_e$ are small (the case for low-frequency applications), the right side of (1-20) is small and can be set equal to zero. *The current i_{sC} associated with the stray capacitance C_s also includes the displacement (leakage) current crossing the closed surface S of Figure 1-3 outside of the wires.*

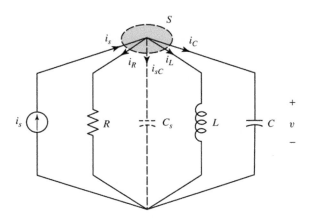

Figure 1-3   RLC parallel network.

1.4.3 Element Laws

In addition to Kirchhoff's loop voltage and node current laws as given, respectively, by (1-17b) and (1-19a), there are a number of current element laws that are widely used in circuit theory. One of the most popular is *Ohm's law* for a resistor (or a conductance G), which states that the *voltage drop v_R across a resistor R is equal to the product of the resistor R and the current i_R flowing through it* ($v_R = Ri_R$ or $i_R = v_R/R = Gv_R$). Ohm's law of circuit theory is a special case of the constitutive relation given by (1-16). Thus

Field Relation   *Circuit Relation*

$$\mathcal{J}_c = \sigma\mathcal{E} \quad \Leftrightarrow \quad i_R = \frac{1}{R}v_R = Gv_R \tag{1-21}$$

Another element law is *associated with an inductor L* and states that *the voltage drop across an inductor is equal to the product of L and the time rate of change of the current through the inductor* ($v_L = L di_L/dt$). Before proceeding to relate the inductor's voltage drop to the corresponding field relation, let us first define inductance. To do this we state that the magnetic flux ψ_m is equal to the product of the inductance L and the corresponding current i. That is $\psi_m = Li$. The corresponding field equation of this relation is (1-15). Thus

Field Relation   *Circuit Relation*

$$\mathcal{B} = \mu\mathcal{H} \quad \Leftrightarrow \quad \psi_m = Li_L \tag{1-22}$$

Using (1-5c) and (1-15), we can write for a homogeneous and non-time-varying medium that

$$\mathcal{M}_d = \frac{\partial\mathcal{B}}{\partial t} = \frac{\partial}{\partial t}(\mu\mathcal{H}) = \mu\frac{\partial\mathcal{H}}{\partial t} \tag{1-22a}$$

where $\mathcal{M}_d$ is defined as the magnetic displacement current density [analogous to the electric displacement current density $\mathcal{J}_d = \partial\mathcal{D}/\partial t = \partial(\varepsilon\mathcal{E})/\partial t = \varepsilon\partial\mathcal{E}/\partial t$]. With the aid of the right side of (1-9a) and the circuit relation of (1-22), we can write

$$\frac{\partial}{\partial t}\iint_S \mathcal{B}\cdot d\mathbf{s} = \frac{\partial\psi_m}{\partial t} = \frac{\partial}{\partial t}(Li_L) = L\frac{\partial i_L}{\partial t} = v_L \tag{1-22b}$$

Using (1-22a) and (1-22b), we can write the following relations:

Field Relation   *Circuit Relation*

$$\mathcal{M}_d = \mu\frac{\partial\mathcal{H}}{\partial t} \quad \Leftrightarrow \quad v_L = L\frac{\partial i_L}{\partial t} \tag{1-22c}$$

Using a similar procedure for a capacitor C, we can write the field and circuit relations analogous to (1-22) and (1-22c):

Field Relation   *Circuit Relation*

$$\mathcal{D} = \varepsilon\mathcal{E} \quad \Leftrightarrow \quad \mathcal{Q}_e = Cv_e \tag{1-23}$$

$$\mathcal{J}_d = \varepsilon\frac{\partial\mathcal{E}}{\partial t} \quad \Leftrightarrow \quad i_C = C\frac{\partial v_C}{\partial t} \tag{1-24}$$

A summary of the field theory relations and their corresponding circuit concepts are listed in Table 1-2.

TABLE 1-2 Relations between electromagnetic field and circuit theories

Field theory	Circuit theory
1. $\mathscr{E}$ (electric field intensity)	1. v (voltage)
2. $\mathscr{H}$ (magnetic field intensity)	2. i (current)
3. $\mathscr{D}$ (electric flux density)	3. q_{ev} (electric charge density)
4. $\mathscr{B}$ (magnetic flux density)	4. q_{mv} (magnetic charge density)
5. $\mathscr{J}$ (electric current density)	5. i_e (electric current)
6. $\mathscr{M}$ (magnetic current density)	6. i_m (magnetic current)
7. $\mathscr{J}_d = \varepsilon \dfrac{\partial \mathscr{E}}{\partial t}$ (electric displacement current density)	7. $i = C \dfrac{dv}{dt}$ (current through a capacitor)
8. $\mathscr{M}_d = \mu \dfrac{\partial \mathscr{H}}{\partial t}$ (magnetic displacement current density)	8. $v = L \dfrac{di}{dt}$ (voltage across an inductor)
9. *Constitutive relations*	9. *Element laws*
(a) $\mathscr{J}_c = \sigma \mathscr{E}$ (electric conduction current density)	(a) $i = Gv = \dfrac{1}{R} v$ (Ohm's law)
(b) $\mathscr{D} = \varepsilon \mathscr{E}$ (dielectric material)	(b) $\mathscr{Q}_e = Cv$ (charge in a capacitor)
(c) $\mathscr{B} = \mu \mathscr{H}$ (magnetic material)	(c) $\psi = Li$ (flux of an inductor)
10. $\oint_C \mathscr{E} \cdot d\boldsymbol{\ell} = -\dfrac{\partial}{\partial t} \iint_S \mathscr{B} \cdot d\mathbf{s}$ (Maxwell–Faraday equation)	10. $\sum v = -L_s \dfrac{\partial i}{\partial t} \simeq 0$ (Kirchhoff's voltage law)
11. $\oiint_S \mathscr{J}_{ic} \cdot d\mathbf{s} = -\dfrac{\partial}{\partial t} \iiint_V q_{ev}\, dv = -\dfrac{\partial \mathscr{Q}_e}{\partial t}$ (continuity equation)	11. $\sum i = -\dfrac{\partial \mathscr{Q}_e}{\partial t} = -C_s \dfrac{\partial v}{\partial t} \simeq 0$ (Kirchhoff's current law)
12. *Power and energy densities*	12. *Power and energy*
(a) $\oiint_S (\mathscr{E} \times \mathscr{H}) \cdot d\mathbf{s}$ (instantaneous power)	(a) $\mathscr{P} = vi$ (power–voltage–current relation)
(b) $\iiint_V \sigma \mathscr{E}^2\, dv$ (dissipated power)	(b) $\mathscr{P}_d = Gv^2 = \dfrac{1}{R} v^2$ (power dissipated in a resistor)
(c) $\dfrac{1}{2} \iiint_V \varepsilon \mathscr{E}^2\, dv$ (electric stored energy)	(c) $\dfrac{1}{2} Cv^2$ (energy stored in a capacitor)
(d) $\dfrac{1}{2} \iiint_V \mu \mathscr{H}^2\, dv$ (magnetic stored energy)	(d) $\dfrac{1}{2} Li^2$ (energy stored in an inductor)

1.5 BOUNDARY CONDITIONS

As previously stated, the differential form of Maxwell's equations is used to solve for the field vectors provided the field quantities are single-valued, bounded, and possess (along with their derivatives) continuous distributions. Along boundaries where the media involved exhibit discontinuities in electrical properties (or there exist sources along these boundaries), the field vectors are also discontinuous and their behavior across the boundaries is governed by the *boundary conditions*.

Maxwell's equations in differential form represent derivatives, with respect to the space coordinates, of the field vectors. At points of discontinuity in the field vectors, the derivatives of the field vectors have no meaning and cannot be properly used to define the behavior of the field vectors across these boundaries. Instead, the behavior of the field vectors across discontinuous boundaries must be handled by examining the field vectors themselves and not their derivatives. The dependence of the field vectors on the electrical properties of the media along boundaries of discontinuity is manifested in our everyday life. It has been observed that cell phone, radio, or television reception deteriorates or even ceases as we move from outside to inside an enclosure (such as a tunnel or a well-shielded building). The reduction or loss of the signal is governed not only by the attenuation as the signal/wave travels through the medium, but also by its behavior across the discontinuous interfaces. Maxwell's equations in integral form provide the most convenient formulation for derivation of the boundary conditions.

1.5.1 Finite Conductivity Media

Initially, let us consider an interface between two media, as shown in Figure 1-4*a*, along which there are no charges or sources. These conditions are satisfied provided that neither of the two media is a perfect conductor or that actual sources are not placed there. Media 1 and 2 are characterized, respectively, by the constitutive parameters ε_1, μ_1, σ_1 and ε_2, μ_2, σ_2.

At a given point along the interface, let us choose a rectangular box whose boundary is denoted by C_0 and its area by S_0. The x, y, z coordinate system is chosen to represent the local geometry of the rectangle. Applying Maxwell's equation 1-9a, with $\mathcal{M}_i = 0$, on the rectangle along C_0 and on S_0, we have

$$\oint_{C_0} \mathcal{E} \cdot d\ell = -\frac{\partial}{\partial t} \iint_{S_0} \mathcal{B} \cdot d\mathbf{s} \tag{1-25}$$

As the height Δy of the rectangle becomes progressively shorter, the area S_0 also becomes vanishingly smaller so that the contributions of the surface integral in (1-25) are negligible. In addition, the contributions of the line integral in (1-25) along Δy are also minimal, so that in the limit $(\Delta y \to 0)$ (1-25) reduces to

$$\mathcal{E}_1 \cdot \hat{\mathbf{a}}_x \, \Delta x - \mathcal{E}_2 \cdot \hat{\mathbf{a}}_x \Delta x = 0$$
$$\mathcal{E}_{1t} - \mathcal{E}_{2t} = 0 \Rightarrow \mathcal{E}_{1t} = \mathcal{E}_{2t} \tag{1-26}$$

or

$$\boxed{\hat{\mathbf{n}} \times (\mathcal{E}_2 - \mathcal{E}_1) = 0} \qquad \sigma_1, \sigma_2 \text{ are finite} \tag{1-26a}$$

In (1-26), $\mathcal{E}_{1t}$ and $\mathcal{E}_{2t}$ represent, respectively, the tangential components of the electric field in media 1 and 2 along the interface. Both (1-26) and (1-26a) state that *the tangential components of the electric field across an interface between two media, with no impressed magnetic current densities along the boundary of the interface, are continuous.*

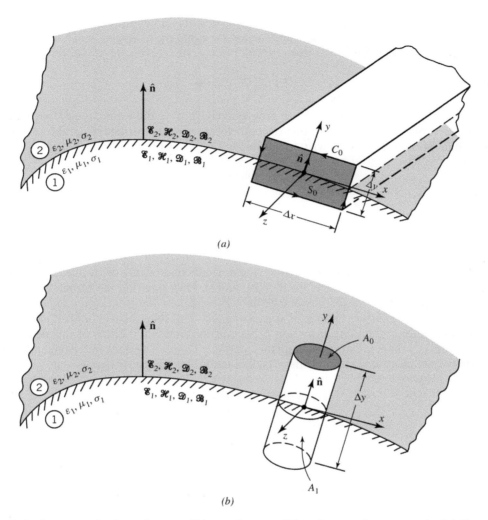

Figure 1-4 Geometry for boundary conditions of tangential and normal components. (*a*) Tangential. (*b*) Normal.

Using a similar procedure on the same rectangle but for (1-10), assuming $\mathcal{J}_i = 0$, we can write that

$$\mathcal{H}_{1t} - \mathcal{H}_{2t} = 0 \Rightarrow \mathcal{H}_{1t} = \mathcal{H}_{2t} \tag{1-27}$$

or

$$\boxed{\hat{\mathbf{n}} \times (\mathcal{H}_2 - \mathcal{H}_1) = 0} \qquad \sigma_1, \sigma_2 \text{ are finite} \tag{1-27a}$$

which state that *the tangential components of the magnetic field across an interface between two media, neither of which is a perfect conductor, are continuous.* This relation also holds if either or both media possess finite conductivity. Equations 1-26a and 1-27a must be modified if either of the two media is a perfect conductor or if there are impressed (source) current densities along the interface. This will be done in the pages that follow.

In addition to the boundary conditions on the tangential components of the electric and magnetic fields across an interface, their normal components are also related. To derive these relations, let us consider the geometry of Figure 1-4*b* where a cylindrical pillbox is chosen at a given point along the interface. If there are no charges along the interface, which is the case when there are no sources or either of the two media is not a perfect conductor, (1-11a) reduces to

$$\oiint_{A_0, A_1} \mathscr{D} \cdot ds = 0 \tag{1-28}$$

As the height Δy of the pillbox becomes progressively shorter, the total circumferential area A_1 also becomes vanishingly smaller, so that the contributions to the surface integral of (1-28) by A_1 are negligible. Thus (1-28) can be written, in the limit $(\Delta y \to 0)$, as

$$\mathscr{D}_2 \cdot \hat{\mathbf{a}}_y A_0 - \mathscr{D}_1 \cdot \hat{\mathbf{a}}_y A_0 = 0$$
$$\mathscr{D}_{2n} - \mathscr{D}_{in} = 0 \Rightarrow \mathscr{D}_{2n} = \mathscr{D}_{1n} \tag{1-29}$$

or

$$\boxed{\hat{\mathbf{n}} \cdot (\mathscr{D}_2 - \mathscr{D}_1) = 0} \qquad \sigma_1, \sigma_2 \text{ are finite} \tag{1-29a}$$

In (1-29), $\mathscr{D}_{1n}$ and $\mathscr{D}_{2n}$ represent, respectively, the normal components of the electric flux density in media 1 and 2 along the interface. Both (1-29) and (1-29a) state that *the normal components of the electric flux density across an interface between two media, both of which are imperfect electric conductors and where there are no sources, are continuous.* This relation also holds if either or both media possess finite conductivity. Equation 1-29a must be modified if either of the media is a perfect conductor or if there are sources along the interface. This will be done in the pages that follow.

In terms of the electric field intensities, (1-29) and (1-29a) can be written as

$$\varepsilon_2 \mathscr{E}_{2n} = \varepsilon_1 \mathscr{E}_{1n} \Rightarrow \mathscr{E}_{2n} = \frac{\varepsilon_1}{\varepsilon_2} \mathscr{E}_{1n} \Rightarrow \mathscr{E}_{1n} = \frac{\varepsilon_2}{\varepsilon_1} \mathscr{E}_{2n} \tag{1-30}$$

$$\boxed{\hat{\mathbf{n}} \cdot (\varepsilon_2 \mathscr{E}_2 - \varepsilon_1 \mathscr{E}_1) = 0} \qquad \sigma_1, \sigma_2 \text{ are finite} \tag{1-30a}$$

which state that *the normal components of the electric field intensity across an interface are discontinuous.*

Using a similar procedure on the same pillbox, but for (1-12) with no charges along the interface, we can write that

$$\mathscr{B}_{2n} - \mathscr{B}_{1n} = 0 \Rightarrow \mathscr{B}_{2n} = \mathscr{B}_{1n} \tag{1-31}$$

$$\boxed{\hat{\mathbf{n}} \cdot (\mathscr{B}_2 - \mathscr{B}_1) = 0} \tag{1-31a}$$

which state that *the normal components of the magnetic flux density, across an interface between two media where there are no sources, are continuous.* In terms of the magnetic field intensities, (1-31) and (1-31a) can be written as

$$\mu_2 \mathscr{H}_{2n} = \mu_1 \mathscr{H}_{1n} \Rightarrow \mathscr{H}_{2n} = \frac{\mu_1}{\mu_2} \mathscr{H}_{1n} \Rightarrow \mathscr{H}_{1n} = \frac{\mu_2}{\mu_1} \mathscr{H}_{2n} \tag{1-32}$$

$$\boxed{\hat{\mathbf{n}} \cdot (\mu_2 \mathscr{H}_2 - \mu_1 \mathscr{H}_1) = 0} \tag{1-32a}$$

which state that *the normal components of the magnetic field intensity across an interface are discontinuous.*

1.5.2 Infinite Conductivity Media

If actual electric sources and charges exist along the interface between the two media, or if either of the two media forming the interface displayed in Figure 1-4 is a perfect electric conductor (PEC), the boundary conditions on the tangential components of the magnetic field [stated by (1-27a)] and on the normal components of the electric flux density or normal components of the electric field intensity [stated by (1-29a) or (1-30a)] must be modified to include the sources and charges or the induced linear electric current density ($\mathcal{J}_s$) and surface electric charge density (q_{es}). Similar modifications must be made to (1-26a), (1-31a), and (1-32a) if magnetic sources and charges exist along the interface between the two media, or if either of the two media is a perfect magnetic conductor (PMC).

To derive the appropriate boundary conditions for such cases, let us refer first to Figure 1-4a and assume that on a very thin layer along the interface there exists an electric surface charge density q_{es} (C/m^2) and linear electric current density $\mathcal{J}_s$ (A/m). Applying (1-10) along the rectangle of Figure 1-4a, we can write that

$$\oint_{C_0} \mathcal{H} \cdot d\ell = \iint_{S_0} \mathcal{J}_{ic} \cdot ds + \frac{\partial}{\partial t} \iint_{S_0} \mathcal{D} \cdot ds \tag{1-33}$$

In the limit as the height of the rectangle is shrinking, the left side of (1-33) reduces to

$$\lim_{\Delta y \to 0} \oint_{C_0} \mathcal{H} \cdot d\ell = (\mathcal{H}_1 - \mathcal{H}_2) \cdot \hat{\mathbf{a}}_x \Delta x \tag{1-33a}$$

Since the electric current density $\mathcal{J}_{ic}$ is confined on a very thin layer along the interface, the first term on the right side of (1-33) can be written as

$$\lim_{\Delta y \to 0} \iint_{S_0} \mathcal{J}_{ic} \cdot ds$$
$$= \lim_{\Delta y \to 0} \left[\mathcal{J}_{ic} \cdot \hat{\mathbf{a}}_z \Delta x \Delta y \right] = \lim_{\Delta y \to 0} \left[(\mathcal{J}_{ic} \Delta y) \cdot \hat{\mathbf{a}}_z \Delta x \right] = \mathcal{J}_s \cdot \hat{\mathbf{a}}_z \Delta x \tag{1-33b}$$

Since S_0 becomes vanishingly smaller as $\Delta y \to 0$, the last term on the right side of (1-33) reduces to

$$\lim_{\Delta y \to 0} \frac{\partial}{\partial t} \iint_{S_0} \mathcal{D} \cdot ds = \lim_{\Delta y \to 0} \frac{\partial}{\partial t} \iint_{S_0} \mathcal{D} \cdot \hat{\mathbf{a}}_z ds = 0 \tag{1-33c}$$

Substituting (1-33a) through (1-33c) into (1-33), we can write it as

$$(\mathcal{H}_1 - \mathcal{H}_2) \cdot \hat{\mathbf{a}}_x \Delta x = \mathcal{J}_s \cdot \hat{\mathbf{a}}_z \Delta x$$

or

$$(\mathcal{H}_1 - \mathcal{H}_2) \cdot \hat{\mathbf{a}}_x - \mathcal{J}_s \cdot \hat{\mathbf{a}}_z = 0 \tag{1-33d}$$

Since

$$\hat{\mathbf{a}}_x = \hat{\mathbf{a}}_y \times \hat{\mathbf{a}}_z \tag{1-34}$$

(1-33d) can be written as

$$(\mathcal{H}_1 - \mathcal{H}_2) \cdot (\hat{\mathbf{a}}_y \times \hat{\mathbf{a}}_z) - \mathcal{J}_s \cdot \hat{\mathbf{a}}_z = 0 \tag{1-35}$$

Using the vector identity

$$\mathbf{A} \cdot \mathbf{B} \times \mathbf{C} = \mathbf{C} \cdot \mathbf{A} \times \mathbf{B} \tag{1-36}$$

on the first term in (1-35), we can then write it as

$$\hat{\mathbf{a}}_z \cdot [(\mathcal{H}_1 - \mathcal{H}_2) \times \hat{\mathbf{a}}_y] - \mathcal{J}_s \cdot \hat{\mathbf{a}}_z = 0 \tag{1-37}$$

or

$$\{[\hat{\mathbf{a}}_y \times (\mathcal{H}_2 - \mathcal{H}_1)] - \mathcal{J}_s\} \cdot \hat{\mathbf{a}}_z = 0 \tag{1-37a}$$

Equation 1-37a is satisfied provided

$$\hat{\mathbf{a}}_y \times (\mathcal{H}_2 - \mathcal{H}_1) - \mathcal{J}_s = 0 \tag{1-38}$$

or

$$\hat{\mathbf{a}}_y \times (\mathcal{H}_2 - \mathcal{H}_1) = \mathcal{J}_s \tag{1-38a}$$

Similar results are obtained if the rectangles chosen are positioned in other planes. Therefore, we can write an expression on the boundary conditions of the tangential components of the magnetic field, using the geometry of Figure 1-4a, as

$$\boxed{\hat{\mathbf{n}} \times (\mathcal{H}_2 - \mathcal{H}_1) = \mathcal{J}_s} \tag{1-39}$$

Equation 1-39 states that *the tangential components of the magnetic field across an interface, along which there exists a surface electric current density $\mathcal{J}_s$ (A/m), are discontinuous by an amount equal to the electric current density.*

If either of the two media is a perfect electric conductor (PEC), (1-39) must be reduced to account for the presence of the conductor. Let us assume that medium 1 in Figure 1-4a possesses an infinite conductivity ($\sigma_1 = \infty$). With such conductivity $\mathcal{E}_1 = 0$, and (1-26a) reduces to

$$\boxed{\hat{\mathbf{n}} \times \mathcal{E}_2 = 0 \Rightarrow \mathcal{E}_{2t} = 0} \tag{1-40}$$

Then (1-1) can be written as

$$\nabla \times \mathcal{E}_1 = 0 = -\frac{\partial \mathcal{B}_1}{\partial t} \Rightarrow \mathcal{B}_1 = 0 \Rightarrow \mathcal{H}_1 = 0 \tag{1-41}$$

provided μ_1 is finite.

In a perfect electric conductor, its free electric charges are confined to a very thin layer on the surface of the conductor, forming a surface charge density q_{es} (with units of coulombs/square meter). This charge density does not include *bound* (polarization) charges (which contribute to the polarization surface charge density) that are usually found inside and on the surface of dielectric media and form atomic dipoles having equal and opposite charges separated by an assumed infinitesimal distance. Here, instead, the surface charge density q_{es} represents actual electric charges separated by finite dimensions from equal quantities of opposite charge.

When the conducting surface is subjected to an applied electromagnetic field, the electric surface charges are subjected to electric field Lorentz forces. These charges are set in motion and thus create a surface electric current density $\mathcal{J}_s$ with units of amperes per meter. The surface current density $\mathcal{J}_s$ also resides in a vanishingly thin layer on the surface of the conductor so that in the limit, as $\Delta y \to 0$ in Figure 1-4a, the volume electric current density $\mathcal{J}$(A/m^2) reduces to

$$\lim_{\Delta y \to 0} (\mathcal{J} \Delta y) = \mathcal{J}_s \tag{1-42}$$

Then the boundary condition of (1-39) reduces, using (1-41) and (1-42), to

$$\boxed{\hat{\mathbf{n}} \times \mathcal{H}_2 = \mathcal{J}_s \Rightarrow \mathcal{H}_{2t} = \mathcal{J}_s} \tag{1-43}$$

which states that *the tangential components of the magnetic field intensity are discontinuous next to a perfect electric conductor by an amount equal to the induced linear electric current density.*

The boundary conditions on the normal components of the electric field intensity, and the electric flux density on an interface along which a surface charge density q_{es} resides on a very thin layer, can be derived by applying the integrals of (1-11a) on a cylindrical pillbox as shown in Figure 1-4b. Then we can write (1-11a) as

$$\lim_{\Delta y \to 0} \oiint_{A_0, A_1} \mathcal{D} \cdot d\mathbf{s} = \lim_{\Delta y \to 0} \iiint_V q_{ev} \, dv \tag{1-44}$$

Since the cylindrical surface A_1 of the pillbox diminishes as $\Delta y \to 0$, its contributions to the surface integral vanish. Thus we can write (1-44) as

$$(\mathcal{D}_2 - \mathcal{D}_1) \cdot \hat{\mathbf{n}} \quad A_0 = \lim_{\Delta y \to 0} \left[(q_{ev} \Delta y) A_0 \right] = q_{es} A_0 \tag{1-45}$$

which reduces to

$$\boxed{\hat{\mathbf{n}} \cdot (\mathcal{D}_2 - \mathcal{D}_1) = q_{es} \Rightarrow \mathcal{D}_{2n} - \mathcal{D}_{1n} = q_{es}} \tag{1-45a}$$

Equation 1-45a states that *the normal components of the electric flux density on an interface, along which a surface charge density resides, are discontinuous by an amount equal to the surface charge density.*

In terms of the normal components of the electric field intensity (1-45a) can be written as

$$\boxed{\hat{\mathbf{n}} \cdot (\varepsilon_2 \mathcal{E}_2 - \varepsilon_1 \mathcal{E}_1) = q_{es}} \tag{1-46}$$

which also indicates that *the normal components of the electric field are discontinuous across a boundary along which a surface charge density resides.*

If either of the media is a perfect electric conductor (PEC) (assuming that medium 1 possesses infinite conductivity $\sigma_1 = \infty$), (1-45a) and (1-46) reduce, respectively, to

$$\boxed{\hat{\mathbf{n}} \cdot \mathcal{D}_2 = q_{es} \Rightarrow \mathcal{D}_{2n} = q_{es}} \tag{1-47a}$$

$$\boxed{\hat{\mathbf{n}} \cdot \mathcal{E}_2 = q_{es} / \varepsilon_2 \Rightarrow \mathcal{E}_{2n} = q_{es} / \varepsilon_2} \tag{1-47b}$$

Both (1-47a) and (1-47b) state that *the normal components of the electric flux density, and corresponding electric field intensity, are discontinuous next to a perfect electric conductor.*

1.5.3 Sources Along Boundaries

If electric and magnetic sources (charges and current densities) are present along the interface between the two media with neither one being a perfect conductor, the boundary conditions on the tangential and normal components of the fields can be written, in general form, as

$$-\hat{\mathbf{n}} \times (\mathcal{E}_2 - \mathcal{E}_1) = \mathcal{M}_s \tag{1-48a}$$

$$\hat{\mathbf{n}} \times (\mathcal{H}_2 - \mathcal{H}_1) = \mathcal{J}_s \tag{1-48b}$$

$$\hat{\mathbf{n}} \cdot (\mathcal{D}_2 - \mathcal{D}_1) = q_{es} \tag{1-48c}$$

$$\hat{\mathbf{n}} \cdot (\mathcal{B}_2 - \mathcal{B}_1) = q_{ms} \tag{1-48d}$$

TABLE 1-3   Boundary conditions on instantaneous electromagnetic fields

	General	Finite conductivity media, no sources or charges $\sigma_1, \sigma_2 \neq \infty$ $\mathcal{J}_s = 0; \; q_{es} = 0$ $\mathcal{M}_s = 0; \; q_{ms} = 0$	Medium 1 of infinite electric conductivity $(\mathcal{E}_1 = \mathcal{H}_1 = 0)$ $\sigma_1 = \infty; \; \sigma_2 \neq \infty$ $\mathcal{M}_s = 0; \; q_{ms} = 0$	Medium 1 of infinite magnetic conductivity $(\mathcal{E}_1 = \mathcal{H}_1 = 0)$ $\mathcal{J}_s = 0; \; q_{es} = 0$
Tangential electric field intensity	$-\hat{\mathbf{n}} \times (\mathcal{E}_2 - \mathcal{E}_1) = \mathcal{M}_s$	$\hat{\mathbf{n}} \times (\mathcal{E}_2 - \mathcal{E}_1) = 0$	$\hat{\mathbf{n}} \times \mathcal{E}_2 = 0$	$-\hat{\mathbf{n}} \times \mathcal{E}_2 = \mathcal{M}_s$
Tangential magnetic field intensity	$\hat{\mathbf{n}} \times (\mathcal{H}_2 - \mathcal{H}_1) = \mathcal{J}_s$	$\hat{\mathbf{n}} \times (\mathcal{H}_2 - \mathcal{H}_1) = 0$	$\hat{\mathbf{n}} \times \mathcal{H}_2 = \mathcal{J}_s$	$\hat{\mathbf{n}} \times \mathcal{H}_2 = 0$
Normal electric flux density	$\hat{\mathbf{n}} \cdot (\mathcal{D}_2 - \mathcal{D}_1) = q_{es}$	$\hat{\mathbf{n}} \cdot (\mathcal{D}_2 - \mathcal{D}_1) = 0$	$\hat{\mathbf{n}} \cdot \mathcal{D}_2 = q_{es}$	$\hat{\mathbf{n}} \cdot \mathcal{D}_2 = 0$
Normal magnetic flux density	$\hat{\mathbf{n}} \cdot (\mathcal{B}_2 - \mathcal{B}_1) = q_{ms}$	$\hat{\mathbf{n}} \cdot (\mathcal{B}_2 - \mathcal{B}_1) = 0$	$\hat{\mathbf{n}} \cdot \mathcal{B}_2 = 0$	$\hat{\mathbf{n}} \cdot \mathcal{B}_2 = q_{ms}$

where $(\mathcal{M}_s, \mathcal{J}_s)$ and (q_{ms}, q_{es}) are the magnetic and electric linear (per meter) current and surface (per square meter) charge densities, respectively. The derivation of (1-48a) and (1-48d) proceeds along the same lines, respectively, as the derivation of (1-48b) and (1-48c) in Section 1.5.2, but begins with (1-9a) and (1-12).

A summary of the boundary conditions on all the field components is found in Table 1-3, which also includes the boundary conditions assuming that medium 1 is a perfect magnetic conductor (PMC). *In general, a magnetic conductor is defined as a material inside of which both time-varying electric and magnetic fields vanish when it is subjected to an electromagnetic field. The tangential components of the magnetic field also vanish next to its surface. In addition, the magnetic charge moves to the surface of the material and creates a magnetic current density that resides on a very thin layer at the surface. Although such materials do not physically exist, they are often used in electromagnetics to develop electrical equivalents that yield the same answers as the actual physical problems. PMCs can be synthesized approximately over a limited frequency range (band-gap); see Section 8.8.*

1.6   POWER AND ENERGY

In a wireless communication system, electromagnetic fields are used to transport information over long distances. To accomplish this, energy must be associated with electromagnetic fields. This transport of energy is accomplished even in the absence of any intervening medium.

To derive the equations that indicate that energy (and forms of it) is associated with electromagnetic waves, let us consider a region V characterized by ε, μ, σ and enclosed by the surface S, as shown in Figure 1-5. Within that region there exist electric and magnetic sources represented, respectively, by the electric and magnetic current densities $\mathcal{J}_i$ and $\mathcal{M}_i$. The fields generated by $\mathcal{J}_i$ and $\mathcal{M}_i$ that exist within S are represented by $\mathcal{E}, \mathcal{H}$. These fields obey Maxwell's equations, and we can write using (1-1) and (1-2) that

$$\nabla \times \mathcal{E} = -\mathcal{M}_i - \frac{\partial \mathcal{B}}{\partial t} = -\mathcal{M}_i - \mu \frac{\partial \mathcal{H}}{\partial t} = -\mathcal{M}_i - \mathcal{M}_d \qquad (1\text{-}49a)$$

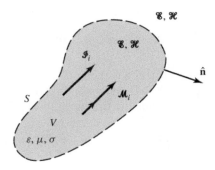

Figure 1-5 Electric and magnetic fields within S generated by $\mathcal{J}_i$ and $\mathcal{M}_i$.

$$\nabla\times\mathcal{H} =\mathcal{J}_i+\mathcal{J}_c +\frac{\partial\mathcal{D}}{\partial t}=\mathcal{J}_i +\sigma\mathcal{E}+\varepsilon\frac{\partial\mathcal{E}}{\partial t}= \mathcal{J}_i+\mathcal{J}_c+\mathcal{J}_d \qquad (1\text{-}49b)$$

Scalar multiplying (1-49a) by $\mathcal{H}$ and (1-49b) by $\mathcal{E}$, we can write that

$$\mathcal{H}\cdot(\nabla\times\mathcal{E})= -\mathcal{H}\cdot(\mathcal{M}_i+\mathcal{M}_d) \qquad (1\text{-}50a)$$

$$\mathcal{E}\cdot(\nabla\times\mathcal{H})= \mathcal{E}\cdot(\mathcal{J}_i+\mathcal{J}_c+\mathcal{J}_d) \qquad (1\text{-}50b)$$

Subtracting (1-50b) from (1-50a) reduces to

$$\mathcal{H}\cdot(\nabla\times\mathcal{E})-\mathcal{E}\cdot(\nabla\times\mathcal{H})= -\mathcal{H}\cdot(\mathcal{M}_i+\mathcal{M}_d)-\mathcal{E}\cdot(\mathcal{J}_i+\mathcal{J}_c+\mathcal{J}_d) \qquad (1\text{-}51)$$

Using the vector identity

$$\nabla\cdot(\mathbf{A}\times\mathbf{B})= \mathbf{B}\cdot(\nabla\times\mathbf{A})-\mathbf{A}\cdot(\nabla\times\mathbf{B}) \qquad (1\text{-}52)$$

on the left side of (1-51), we can write that

$$\nabla\cdot(\mathcal{E}\times\mathcal{H})= -\mathcal{H}\cdot(\mathcal{M}_i+\mathcal{M}_d)-\mathcal{E}\cdot(\mathcal{J}_i+\mathcal{J}_c+\mathcal{J}_d) \qquad (1\text{-}53)$$

or

$$\boxed{\nabla\cdot(\mathcal{E}\times\mathcal{H})+\mathcal{H}\cdot(\mathcal{M}_i+\mathcal{M}_d)+\mathcal{E}\cdot(\mathcal{J}_i+\mathcal{J}_c+\mathcal{J}_d)= 0} \qquad (1\text{-}53a)$$

Integrating (1-53) over the volume V leads to

$$\iiint_V \nabla\cdot(\mathcal{E}\times\mathcal{H})\,dv=-\iiint_V [\mathcal{H}\cdot(\mathcal{M}_i+\mathcal{M}_d)+\mathcal{E}\cdot(\mathcal{J}_i+\mathcal{J}_c+\mathcal{J}_d)]\,dv \qquad (1\text{-}54)$$

Applying the divergence theorem (1-8) on the left side of (1-54) reduces it to

$$\oiint_S (\mathcal{E}\times\mathcal{H})\cdot d\mathbf{s}=-\iiint_V [\mathcal{H}\cdot(\mathcal{M}_i+\mathcal{M}_d)+\mathcal{E}\cdot(\mathcal{J}_i+\mathcal{J}_c+\mathcal{J}_d)]\,dv \qquad (1\text{-}55)$$

or

$$\boxed{\oiint_S (\mathcal{E}\times\mathcal{H})\cdot d\mathbf{s}+\iiint_V [\mathcal{H}\cdot(\mathcal{M}_i+\mathcal{M}_d)+\mathcal{E}\cdot(\mathcal{J}_i+\mathcal{J}_c+\mathcal{J}_d)]\,dv= 0} \qquad (1\text{-}55a)$$

Equations 1-53a and 1-55a can be interpreted, respectively, as the differential and integral forms of the *conservation of energy*. To accomplish this, let us consider each of the terms included in (1-55a).

The integrand, in the first term of (1-55a), has the form

$$\mathcal{S} = \mathcal{E} \times \mathcal{H} \tag{1-56}$$

where $\mathcal{S}$ is known as the Poynting vector. It has the units of power density (watts/square meter), since $\mathcal{E}$ has units of volts/meter and $\mathcal{H}$ has units of ampere/meter, so that the units of $\mathcal{S}$ are volts·ampere/meter^2 = watts/meter^2. Thus the first term of (1-55a), written as

$$\mathcal{P}_e = \oiint_S (\mathcal{E} \times \mathcal{H}) \cdot d\mathbf{s} = \oiint_S \mathcal{S} \cdot d\mathbf{s} \tag{1-57}$$

represents the total power $\mathcal{P}_e$ exiting the volume V bounded by the surface S.

The other terms in (1-55a), which represent the integrand of the volume integral, can be written as

$$p_s = -(\mathcal{H} \cdot \mathcal{M}_i + \mathcal{E} \cdot \mathcal{J}_i) \tag{1-58a}$$

$$\mathcal{H} \cdot \mathcal{M}_d = \mathcal{H} \cdot \frac{\partial \mathcal{B}}{\partial t} = \mu \mathcal{H} \cdot \frac{\partial \mathcal{H}}{\partial t} = \frac{1}{2} \mu \frac{\partial \mathcal{H}^2}{\partial t} = \frac{\partial}{\partial t} \left(\frac{1}{2} \mu \mathcal{H}^2 \right) = \frac{\partial}{\partial t} w_m \tag{1-58b}$$

$$p_d = \mathcal{E} \cdot \mathcal{J}_c = \mathcal{E} \cdot (\sigma \mathcal{E}) = \sigma \mathcal{E}^2 \tag{1-58c}$$

$$\mathcal{E} \cdot \mathcal{J}_d = \mathcal{E} \cdot \frac{\partial \mathcal{D}}{\partial t} = \varepsilon \mathcal{E} \cdot \frac{\partial \mathcal{E}}{\partial t} = \frac{1}{2} \varepsilon \frac{\partial \mathcal{E}^2}{\partial t} = \frac{\partial}{\partial t} \left(\frac{1}{2} \varepsilon \mathcal{E}^2 \right) = \frac{\partial}{\partial t} w_e \tag{1-58d}$$

where

$$w_m = \frac{1}{2} \mu \mathcal{H}^2 = \text{magnetic energy density} \, (\text{J/m}^3) \tag{1-58e}$$

$$w_e = \frac{1}{2} \varepsilon \mathcal{E}^2 = \text{electric energy density} \, (\text{J/m}^3) \tag{1-58f}$$

$$p_s = -(\mathcal{H} \cdot \mathcal{M}_i + \mathcal{E} \cdot \mathcal{J}_i) = \text{supplied power density} \, (\text{W/m}^3) \tag{1-58g}$$

$$p_d = \sigma \mathcal{E}^2 = \text{dissipated power density} \, (\text{W/m}^3) \tag{1-58h}$$

Integrating each of the terms in (1-58a) through (1-58d), we can write the corresponding forms as

$$\mathcal{P}_s = -\iiint_V (\mathcal{H} \cdot \mathcal{M}_i + \mathcal{E} \cdot \mathcal{J}_i) \, dv = \iiint_V p_s \, dv \tag{1-59a}$$

$$\iiint_V (\mathcal{H} \cdot \mathcal{M}_d) \, dv = \frac{\partial}{\partial t} \iiint_V \frac{1}{2} \mu \mathcal{H}^2 \, dv = \frac{\partial}{\partial t} \iiint_V w_m \, dv = \frac{\partial}{\partial t} W_m \tag{1-59b}$$

$$\mathcal{P}_d = \iiint_V (\mathcal{E} \cdot \mathcal{J}_c) \, dv = \iiint_V \sigma \mathcal{E}^2 \, dv = \iiint_V p_d \, dv \tag{1-59c}$$

$$\iiint_V (\mathcal{E} \cdot \mathcal{J}_d) \, dv = \frac{\partial}{\partial t} \iiint_V \frac{1}{2} \varepsilon \mathcal{E}^2 \, dv = \frac{\partial}{\partial t} \iiint_V w_e \, dv = \frac{\partial}{\partial t} W_e \tag{1-59d}$$

where $\mathcal{W}_m$ = magnetic energy (J)
   $\mathcal{W}_e$ = electric energy (J)
   $\mathcal{P}_s$ = supplied power (W)
   $\mathcal{P}_e$ = exiting power (W)
   $\mathcal{P}_d$ = dissipated power (W)

Using (1-57) and (1-59a) through (1-59d), we can write (1-55a) as

$$\mathcal{P}_e - \mathcal{P}_s + \mathcal{P}_d + \frac{\partial}{\partial t}(\mathcal{W}_e + \mathcal{W}_m) = 0 \tag{1-60}$$

or

$$\mathcal{P}_s = \mathcal{P}_e + \mathcal{P}_d + \frac{\partial}{\partial t}(\mathcal{W}_e + \mathcal{W}_m) \tag{1-60a}$$

which is the *conservation of power law*. This law states that within a volume V, bounded by S, the supplied power $\mathcal{P}_s$ is equal to the power $\mathcal{P}_e$ exiting S plus the power $\mathcal{P}_d$ dissipated within that volume plus the rate of change (increase if positive) of the electric ($\mathcal{W}_e$) and magnetic ($\mathcal{W}_m$) energies stored within that same volume.

A summary of the field theory relations and their corresponding circuit concepts is found listed in Table 1-2.

1.7 TIME-HARMONIC ELECTROMAGNETIC FIELDS

Maxwell's equations in differential and integral forms, for general time-varying electromagnetic fields, were presented in Sections 1.2.1 and 1.2.2. In addition, various expressions involving and relating the electromagnetic fields (such as the constitutive parameters and relations, circuit relations, boundary conditions, and power and energy) were also introduced in the preceding sections. However, in many practical systems involving electromagnetic waves, the time variations are of cosinusoidal form and are referred to as *time-harmonic*. In general, such time variations can be represented by[1] $e^{j\omega t}$, and the instantaneous electromagnetic field vectors can be related to their complex forms in a very simple manner. Thus for time-harmonic fields, we can relate the instantaneous fields, current density and charge (represented by script letters) to their complex forms (represented by roman letters) by

$$\mathcal{E}(x,y,z;t) = \mathrm{Re}\left[\mathbf{E}(x,y,z)e^{j\omega t}\right] \tag{1-61a}$$

$$\mathcal{H}(x,y,z;t) = \mathrm{Re}\left[\mathbf{H}(x,y,z)e^{j\omega t}\right] \tag{1-61b}$$

$$\mathcal{D}(x,y,z;t) = \mathrm{Re}\left[\mathbf{D}(x,y,z)e^{j\omega t}\right] \tag{1-61c}$$

$$\mathcal{B}(x,y,z;t) = \mathrm{Re}\left[\mathbf{B}(x,y,z)e^{j\omega t}\right] \tag{1-61d}$$

$$\mathcal{J}(x,y,z;t) = \mathrm{Re}\left[\mathbf{J}(x,y,z)e^{j\omega t}\right] \tag{1-61e}$$

$$q(x,y,z;t) = \mathrm{Re}\left[q(x,y,z)e^{j\omega t}\right] \tag{1-61f}$$

where $\mathcal{E}, \mathcal{H}, \mathcal{D}, \mathcal{B}, \mathcal{J}$, and q represent the instantaneous field vectors, current density and charge, while $\mathbf{E}, \mathbf{H}, \mathbf{D}, \mathbf{B}, \mathbf{J}$, and q represent the corresponding complex spatial forms which are only a function of position. In this book we have chosen to represent the instantaneous quantities by the real part of the product of the corresponding complex spatial quantities with $e^{j\omega t}$. Another option

---

[1] Another representation form of time-harmonic variations is $e^{-j\omega t}$ (most scientists prefer $e^{i\omega t}$ or $e^{-i\omega t}$ where $i = \sqrt{-1}$). Throughout this book, we will use the $e^{j\omega t}$ form, which when it is not stated will be assumed. The $e^{-j\omega t}$ fields are related to those of the $e^{j\omega t}$ form by the complex conjugate.

would be to represent the instantaneous quantities by the imaginary part of the products. It should be stated that throughout this book the magnitudes of the instantaneous fields represent *peak* values that are related to their corresponding root-mean-square (rms) values by the square root of 2 (peak = $\sqrt{2}$ rms). If the complex spatial quantities can be found, it is then a very simple procedure to find their corresponding instantaneous forms by using (1-61a) through (1-61f). In what follows, it will be shown that Maxwell's equations in differential and integral forms for time-harmonic electromagnetic fields can be written in much simpler forms using the complex field vectors.

1.7.1 Maxwell's Equations in Differential and Integral Forms

It is a very simple exercise to show that, by substituting (1-61a) through (1-61f) into (1-1) through (1-4) and (1-6), Maxwell's equations and the continuity equation in differential form for time-harmonic fields can be written in terms of the complex field vectors as shown in Table 1-4. Using a similar procedure, we can write the corresponding integral forms of Maxwell's equations and the continuity equation listed in Table 1-1 in terms of the complex spatial field vectors as shown in Table 1-4. Both of these derivations have been assigned as exercises to the reader at the end of the chapter.

By examining the two forms in Table 1-4, we see that one form can be obtained from the other by doing the following:

1. Replace the instantaneous field vectors by the corresponding complex spatial forms, or vice versa.
2. Replace $\partial/\partial t$ by $j\omega(\partial/\partial t = j\omega)$, or vice versa.

The second step is very similar to that followed in circuit analysis when Laplace transforms are used to analyze RLC a.c. circuits. In these analyses $\partial/\partial t$ is replaced by s ($\partial/\partial t \equiv s$). For steady-state conditions $\partial/\partial t$ is replaced by $j\omega(\partial/\partial t \equiv s \equiv j\omega)$. The reason for using Laplace transforms is to transform differential equations to algebraic equations, which are simpler to solve. The same intent is used here to write Maxwell's equations in forms that are easier to solve. Thus, if it is desired to solve for the instantaneous field vectors of time-harmonic fields, it is easier to use the following two-step procedure, instead of attempting to do it in one step using the general instantaneous forms of Maxwell's equations:

1. Solve for the complex spatial field vectors, current densities and charges (**E, H, D, B, J, M,** q), using Maxwell's equations from Table 1-4 that are written in terms of the complex spatial field vectors, current densities and charges.
2. Determine the corresponding instantaneous field vectors, current densities and charges using (1-61a) through (1-61f).

Step 1 is obviously the most difficult, and it is often the only step needed. Step 2 is straightforward, and it is often omitted. In practice, the time variations of $e^{j\omega t}$ are stated at the outset, but then are suppressed.

1.7.2 Boundary Conditions

The boundary conditions for time-harmonic fields are identical to those of general time-varying fields, as derived in Section 1.5, and they can be expressed simply by replacing the instantaneous field vectors, current densities and charges in Table 1-3 with their corresponding complex spatial field vectors, current densities and charges. A summary of all the boundary conditions for time-harmonic fields, referring to Figure 1-4, is found in Table 1-5.

In addition to the boundary conditions found in Table 1-5, an additional boundary condition on the tangential components of the electric field is often used along an interface when one of the two media is a very good conductor (material that possesses large but finite conductivity).

TABLE 1-4 Instantaneous and time-harmonic forms of Maxwell's equations and continuity equation in differential and integral forms

Instantaneous	Time harmonic
Differential form	
$\nabla \times \boldsymbol{\mathscr{E}} = -\boldsymbol{\mathscr{M}}_i - \dfrac{\partial \boldsymbol{\mathscr{B}}}{\partial t}$	$\nabla \times \mathbf{E} = -\mathbf{M}_i - j\omega \mathbf{B}$
$\nabla \times \boldsymbol{\mathscr{H}} = \boldsymbol{\mathscr{I}}_i + \boldsymbol{\mathscr{I}}_c + \dfrac{\partial \boldsymbol{\mathscr{D}}}{\partial t}$	$\nabla \times \mathbf{H} = \mathbf{J}_i + \mathbf{J}_c + j\omega \mathbf{D}$
$\nabla \cdot \boldsymbol{\mathscr{D}} = q_{ev}$	$\nabla \cdot \mathbf{D} = q_{ev}$
$\nabla \cdot \boldsymbol{\mathscr{B}} = q_{mv}$	$\nabla \cdot \mathbf{B} = q_{mv}$
$\nabla \cdot \boldsymbol{\mathscr{I}}_{ic} = -\dfrac{\partial q_{ev}}{\partial t}$	$\nabla \cdot \mathbf{J}_{ic} = -j\omega q_{ev}$
Integral form	
$\oint_C \boldsymbol{\mathscr{E}} \cdot d\boldsymbol{\ell} = -\iint_S \boldsymbol{\mathscr{M}}_i \cdot d\mathbf{s} - \dfrac{\partial}{\partial t}\iint_S \boldsymbol{\mathscr{B}} \cdot d\mathbf{s}$	$\oint_C \mathbf{E} \cdot d\boldsymbol{\ell} = -\iint_S \mathbf{M}_i \cdot d\mathbf{s} - j\omega\iint_S \mathbf{B} \cdot d\mathbf{s}$
$\oint_C \boldsymbol{\mathscr{H}} \cdot d\boldsymbol{\ell} = \iint_S \boldsymbol{\mathscr{I}}_i \cdot d\mathbf{s} + \iint_S \boldsymbol{\mathscr{I}}_c \cdot d\mathbf{s} + \dfrac{\partial}{\partial t}\iint_S \boldsymbol{\mathscr{D}} \cdot d\mathbf{s}$	$\oint_C \mathbf{H} \cdot d\boldsymbol{\ell} = \iint_S \mathbf{J}_i \cdot d\mathbf{s} + \iint_S \mathbf{J}_c \cdot d\mathbf{s} + j\omega\iint_S \mathbf{D} \cdot d\mathbf{s}$
$\oiint_S \boldsymbol{\mathscr{D}} \cdot d\mathbf{s} = \mathscr{Q}_e$	$\oiint_S \mathbf{D} \cdot d\mathbf{s} = Q_e$
$\oiint_S \boldsymbol{\mathscr{B}} \cdot d\mathbf{s} = \mathscr{Q}_m$	$\oiint_S \mathbf{B} \cdot d\mathbf{s} = Q_m$
$\oiint_S \boldsymbol{\mathscr{I}}_{ic} \cdot d\mathbf{s} = -\dfrac{\partial \mathscr{Q}_e}{\partial t}$	$\oiint_S \mathbf{J}_{ic} \cdot d\mathbf{s} = -j\omega Q_e$

TABLE 1-5   Boundary conditions on time-harmonic electromagnetic fields

	General	Finite conductivity media, no sources or charges $\sigma_1, \sigma_2 \neq \infty$ $J_s = M_s = 0$ $q_{es} = q_{ms} = 0$	Medium 1 of infinite electric conductivity $(E_1 = H_1 = 0)$ $\sigma_1 = \infty; \sigma_2 \neq \infty$ $M_s = 0; q_{ms} = 0$	Medium 1 of infinite magnetic conductivity $(E_1 = H_1 = 0)$ $J_s = 0; q_{es} = 0$
Tangential electric field intensity	$-\hat{n} \times (E_2 - E_1) = M_s$	$\hat{n} \times (E_2 - E_1) = 0$	$\hat{n} \times E_2 = 0$	$-\hat{n} \times E_2 = M_s$
Tangential magnetic field intensity	$\hat{n} \times (H_2 - H_1) = J_s$	$\hat{n} \times (H_2 - H_1) = 0$	$\hat{n} \times H_2 = J_s$	$\hat{n} \times H_2 = 0$
Normal electric flux density	$\hat{n} \cdot (D_2 - D_1) = q_{es}$	$\hat{n} \cdot (D_2 - D_1) = 0$	$\hat{n} \cdot D_2 = q_{es}$	$\hat{n} \cdot D_2 = 0$
Normal magnetic flux density	$\hat{n} \cdot (B_2 - B_1) = q_{ms}$	$\hat{n} \cdot (B_2 - B_1) = 0$	$\hat{n} \cdot B_2 = 0$	$\hat{n} \cdot B_2 = q_{ms}$

This is illustrated in Figure 1-6 where it is assumed that medium 1 is a very good conductor whose surface, as will be shown in Section 4.3.1, exhibits a surface impedance Z_s (ohms) given, approximately, by (4-42) or

$$Z_s = R_s + jX_s = (1+j)\sqrt{\frac{\omega \mu_1}{2\sigma_1}} \qquad (1\text{-}62)$$

with equal real and imaginary (inductive) parts (σ_1 is the conductivity of the conductor). At the surface there exists a linear current density J_s (A/m) related to the tangential magnetic field in medium 2 by

$$J_s \simeq \hat{n} \times H_2 \qquad (1\text{-}63)$$

Since the conductivity is finite (although large), the most intense current density resides at the surface, and it diminishes (in an exponential form) as the observations are made deeper into the conductor. This is demonstrated in Example 5.7 of Section 5.4.1. In addition, the electric field intensity along the interface cannot be zero (although it may be small). Thus, we can write that the

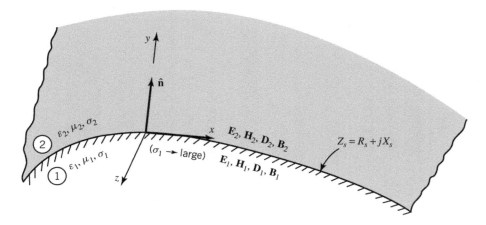

Figure 1-6   Surface impedance along the surface of a very good conductor.

tangential component of the electric field in medium 2, along the interface, is related to the electric current density $\mathbf{J}_s$ and tangential component of the magnetic field by

$$\mathbf{E}_{t2} = Z_s \mathbf{J}_s = Z_s \hat{\mathbf{n}} \times \mathbf{H}_2 = \hat{\mathbf{n}} \times \mathbf{H}_2 \sqrt{\frac{\omega \mu_1}{2\sigma_1}}(1+j) \qquad (1\text{-}64)$$

For time-harmonic fields, the boundary conditions on the normal components are not independent of those on the tangential components, and vice-versa, since they are related through Maxwell's equations. In fact, if the tangential components of the electric and magnetic fields satisfy the boundary conditions, then the normal components of the same fields necessarily satisfy the appropriate boundary conditions. For example, if the tangential components of the electric field are continuous across a boundary, their derivatives (with respect to the coordinates on the boundary surface) are also continuous. This, in turn, ensures continuity of the normal component of the magnetic field.

To demonstrate that, let us refer to the geometry of Figure 1-6 where the local surface along the interface is described by the x, z coordinates with y being normal to the surface. Let us assume that E_x and E_z are continuous, which ensures that their derivatives with respect to x and z $(\partial E_x / \partial x, \partial E_x / \partial z, \partial E_z / \partial x, \partial E_z / \partial z)$ are also continuous. Therefore, according to Maxwell's curl equation of the electric field

$$\nabla \times \mathbf{E} = \nabla \times (\hat{\mathbf{a}}_x E_x + \hat{\mathbf{a}}_z E_z) = \begin{vmatrix} \hat{\mathbf{a}}_x & \hat{\mathbf{a}}_y & \hat{\mathbf{a}}_z \\ \dfrac{\partial}{\partial x} & 0 & \dfrac{\partial}{\partial z} \\ E_x & 0 & E_z \end{vmatrix}$$

$$= \hat{\mathbf{a}}_x(0) + \hat{\mathbf{a}}_y \left(\frac{\partial E_x}{\partial z} - \frac{\partial E_z}{\partial x} \right) + \hat{\mathbf{a}}_z(0)$$

$$\nabla \times \mathbf{E} = \hat{\mathbf{a}}_y \left(\frac{\partial E_x}{\partial z} - \frac{\partial E_z}{\partial x} \right) = -j\omega\mu\mathbf{H} \qquad (1\text{-}65)$$

or

$$B_y = \mu H_y = -\frac{1}{j\omega} \left(\frac{\partial E_x}{\partial z} - \frac{\partial E_z}{\partial x} \right) \qquad (1\text{-}65a)$$

According to (1-65a), B_y, the normal component of the magnetic flux density along the interface, is continuous across the boundary if $\partial E_x / \partial z$ and $\partial E_z / \partial x$ are also continuous across the boundary.

In a similar manner, it can be shown that continuity of the tangential components of the magnetic field ensures continuity of the normal component of the electric flux density (**D**).

1.7.3 Power and Energy

In Section 1.6, it was shown that power and energy are associated with time-varying electromagnetic fields. The conservation-of-energy equation, in differential and integral forms, was stated respectively by (1-53a) and (1-55a). Similar equations can be derived for time-harmonic electromagnetic fields using the complex spatial forms of the field vectors. Before we attempt this, let us first rewrite the instantaneous Poynting vector $\mathscr{P}$ in terms of the complex field vectors.

The instantaneous Poynting vector was defined by (1-56) and is repeated here as

$$\mathscr{P} = \mathscr{E} \times \mathscr{H} \qquad (1\text{-}66)$$

The electric and magnetic fields of (1-61a) and (1-61b) can also be written as

$$\mathscr{E}(x, y, z; t) = \text{Re}\left[\mathbf{E}(x, y, z)e^{j\omega t} \right] = \tfrac{1}{2}\left[\mathbf{E}e^{j\omega t} + (\mathbf{E}e^{j\omega t})^* \right] \qquad (1\text{-}67a)$$

$$\mathscr{H}(x, y, z; t) = \text{Re}\left[\mathbf{H}(x, y, z)e^{j\omega t} \right] = \tfrac{1}{2}\left[\mathbf{H}e^{j\omega t} + (\mathbf{H}e^{j\omega t})^* \right] \qquad (1\text{-}67b)$$

where the asterisk (*) indicates complex conjugate. Substituting (1-67a) and (1-67b) into (1-66), we have that

$$
\begin{aligned}
\mathcal{S} = \mathcal{E} \times \mathcal{H} &= \tfrac{1}{2}(\mathbf{E}e^{j\omega t} + \mathbf{E}^{*}e^{-j\omega t}) \times \tfrac{1}{2}(\mathbf{H}e^{j\omega t} + \mathbf{H}^{*}e^{-j\omega t}) \\
&= \tfrac{1}{2}\left\{ \tfrac{1}{2}\left[\mathbf{E}\times\mathbf{H}^{*} + \mathbf{E}^{*}\times\mathbf{H}\right] + \tfrac{1}{2}\left[\mathbf{E}\times\mathbf{H}e^{j2\omega t} + \mathbf{E}^{*}\times\mathbf{H}^{*}e^{-j2\omega t}\right]\right\} \\
\mathcal{S} &= \tfrac{1}{2}\left\{ \tfrac{1}{2}\left[\mathbf{E}\times\mathbf{H}^{*} + (\mathbf{E}\times\mathbf{H}^{*})^{*}\right] + \tfrac{1}{2}\left[\mathbf{E}\times\mathbf{H}e^{j2\omega t} + (\mathbf{E}\times\mathbf{H}e^{j2\omega t})^{*}\right]\right\}
\end{aligned}
$$

(1-68)

Using the equalities (1-67a) or (1-67b) in reverse order, we can write (1-68) as

$$
\mathcal{S} = \tfrac{1}{2}\left[\mathrm{Re}(\mathbf{E}\times\mathbf{H}^{*}) + \mathrm{Re}(\mathbf{E}\times\mathbf{H}e^{j2\omega t})\right]
$$

(1-69)

Since both $\mathbf{E}$ and $\mathbf{H}$ are not functions of time and the time variations of the second term are twice the frequency of the field vectors, the time-average Poynting vector (average power density) over one period is equal to

$$
\mathcal{S}_{av} = \mathbf{S} = \tfrac{1}{2}\mathrm{Re}\left[\mathbf{E}\times\mathbf{H}^{*}\right]
$$

(1-70)

Since $\mathbf{E}\times\mathbf{H}^{*}$ is, in general, complex and the real part of $\mathbf{E}\times\mathbf{H}^{*}$ represents the real part of the power density, what does the imaginary part represent? As will be seen in what follows, the imaginary part represents the reactive power. With (1-69) and (1-70) in mind, let us now derive the conservation-of-energy equation in differential and integral forms using the complex forms of the field vector.

From Table 1-4, the first two of Maxwell's equations can be written as

$$
\nabla \times \mathbf{E} = -\mathbf{M}_i - j\omega\mu\mathbf{H}
$$

(1-71a)

$$
\nabla \times \mathbf{H} = \mathbf{J}_i + \mathbf{J}_c + j\omega\varepsilon\mathbf{E} = \mathbf{J}_i + \sigma\mathbf{E} + j\omega\varepsilon\mathbf{E}
$$

(1-71b)

Dot multiplying (1-71a) by $\mathbf{H}^{*}$ and the conjugate of (1-71b) by $\mathbf{E}$, we have that

$$
\mathbf{H}^{*}\cdot(\nabla\times\mathbf{E}) = -\mathbf{H}^{*}\cdot\mathbf{M}_i - j\omega\mu\mathbf{H}\cdot\mathbf{H}^{*}
$$

(1-72a)

$$
\mathbf{E}\cdot(\nabla\times\mathbf{H}^{*}) = \mathbf{E}\cdot\mathbf{J}_i^{*} + \sigma\mathbf{E}\cdot\mathbf{E}^{*} - j\omega\varepsilon\mathbf{E}\cdot\mathbf{E}^{*}
$$

(1-72b)

Subtracting (1-72a) from (1-72b), we can write that

$$
\begin{aligned}
\mathbf{E}\cdot(\nabla\times\mathbf{H}^{*}) &- \mathbf{H}^{*}\cdot(\nabla\times\mathbf{E}) \\
&= \mathbf{H}^{*}\cdot\mathbf{M}_i + \mathbf{E}\cdot\mathbf{J}_i^{*} + \sigma\mathbf{E}\cdot\mathbf{E}^{*} - j\omega\varepsilon\mathbf{E}\cdot\mathbf{E}^{*} + j\omega\mu\mathbf{H}\cdot\mathbf{H}^{*}
\end{aligned}
$$

(1-73)

Using the vector identity (1-52) reduces (1-73) to

$$
\nabla\cdot(\mathbf{H}^{*}\times\mathbf{E}) = \mathbf{H}^{*}\cdot\mathbf{M}_i + \mathbf{E}\cdot\mathbf{J}_i^{*} + \sigma\,|\mathbf{E}|^2 + j\omega\mu\,|\mathbf{H}|^2 - j\omega\varepsilon\,|\mathbf{E}|^2
$$

(1-74)

or

$$
-\nabla\cdot(\mathbf{E}\times\mathbf{H}^{*}) = \mathbf{H}^{*}\cdot\mathbf{M}_i + \mathbf{E}\cdot\mathbf{J}_i^{*} + \sigma\,|\mathbf{E}|^2 + j\omega(\mu\,|\mathbf{H}|^2 - \varepsilon\,|\mathbf{E}|^2)
$$

(1-74a)

Dividing both sides by 2, we can write that

$$
\boxed{-\nabla\cdot(\tfrac{1}{2}\mathbf{E}\times\mathbf{H}^{*}) = \tfrac{1}{2}\mathbf{H}^{*}\cdot\mathbf{M}_i + \tfrac{1}{2}\mathbf{E}\cdot\mathbf{J}_i^{*} + \tfrac{1}{2}\sigma\,|\mathbf{E}|^2 + j2\omega(\tfrac{1}{4}\mu\,|\mathbf{H}|^2 - \tfrac{1}{4}\varepsilon\,|\mathbf{E}|^2)}
$$

(1-75)

For time-harmonic fields, (1-75) represents the *conservation-of-energy equation* in differential form.

To verify that (1-75) represents the conservation-of-energy equation in differential form, it is easier to examine its integral form. To accomplish this, let us first take the volume integral of both sides of (1-75) and then apply the divergence theorem (1-8) to the left side. Doing both of these steps reduces (1-75) to

$$-\iiint_V \nabla \cdot (\tfrac{1}{2}\mathbf{E} \times \mathbf{H}^*)\, dv = -\oiint_S (\tfrac{1}{2}\mathbf{E} \times \mathbf{H}^*) \cdot d\mathbf{s}$$
$$= \tfrac{1}{2}\iiint_V (\mathbf{H}^* \cdot \mathbf{M}_i + \mathbf{E} \cdot \mathbf{J}_i^*)\, dv$$
$$+ \tfrac{1}{2}\iiint_V \sigma |\mathbf{E}|^2\, dv + j2\omega \iiint_V (\tfrac{1}{4}\mu |\mathbf{H}|^2 - \tfrac{1}{4}\varepsilon |\mathbf{E}|^2)\, dv$$

or

$$\boxed{\begin{aligned} -\tfrac{1}{2}\iiint_V (\mathbf{H}^* \cdot \mathbf{M}_i + \mathbf{E} \cdot \mathbf{J}_i^*)\, dv &= \oiint_S (\tfrac{1}{2}\mathbf{E} \times \mathbf{H}^*) \cdot d\mathbf{s} + \tfrac{1}{2}\iiint_V \sigma |\mathbf{E}|^2\, dv \\ &+ j2\omega \iiint_V (\tfrac{1}{4}\mu |\mathbf{H}|^2 - \tfrac{1}{4}\varepsilon |\mathbf{E}|^2)\, dv \end{aligned}}$$

(1-76)

which can be written as

$$P_s = P_e + P_d + j2\omega(\overline{W}_m - \overline{W}_e) \tag{1-76a}$$

where

$$P_s = -\frac{1}{2}\iiint_V (\mathbf{H}^* \cdot \mathbf{M}_i + \mathbf{E} \cdot \mathbf{J}_i^*)\, dv = \text{supplied complex power (W)} \tag{1-76b}$$

$$P_e = \oiint_S \left(\frac{1}{2}\mathbf{E} \times \mathbf{H}^*\right) \cdot d\mathbf{s} = \text{exiting complex power (W)} \tag{1-76c}$$

$$P_d = \frac{1}{2}\iiint_V \sigma |\mathbf{E}|^2\, dv = \text{dissipated real power (W)} \tag{1-76d}$$

$$\overline{W}_m = \iiint_V \frac{1}{4}\mu |\mathbf{H}|^2\, dv = \text{time-average magnetic energy (J)} \tag{1-76e}$$

$$\overline{W}_e = \iiint_V \frac{1}{4}\varepsilon |\mathbf{E}|^2\, dv = \text{time-average electric energy (J)} \tag{1-76f}$$

For an electromagnetic source (represented in Figure 1-5 by electric and magnetic current densities $\mathbf{J}_i$ and $\mathbf{M}_i$, respectively) supplying power in a region within S, (1-76) and (1-76a) represent the conservation-of-energy equation in integral form. Now, it is also much easier to accept that (1-75), from which (1-76) was derived, represents the conservation-of-energy equation in differential form. In (1-76a), P_s and P_e are in general complex and P_d is always real, but the last two terms are always imaginary and represent the reactive power associated, respectively, with magnetic and electric fields. It should be stated that for complex permeabilities and permittivities the contributions from their imaginary parts to the integrals of (1-76e) and (1-76f) should both be combined with (1-76d), since they both represent losses associated with the imaginary parts of the permeabilities and permittivities.

It should be stated that the imaginary term of the right side of (1-76), including its signs, which represents the complex stored power (inductive and capacitive), does conform to the notation of conventional circuit theory. For example, defining the complex power P, assuming V and I are peak values, as

$$P = \frac{1}{2}(VI^*) \tag{1-77}$$

the complex power of a series circuit consisting of a resistor R in series with an inductor L, with a current I through both R and L and total voltage V across both the resistor and inductor, can be written, based on (1-77), as

$$P = \frac{1}{2}(VI^*) = \frac{1}{2}(ZI)I^* = \frac{1}{2}Z\,|\,I\,|^2 = \frac{1}{2}(R + j\omega L)\,|\,I\,|^2 \qquad (1\text{-}77\text{a})$$

The imaginary part of (1-77a) is positive. Similarly, for a parallel circuit consisting of a conductor G in parallel with a capacitor C, with a voltage V across G and C and a total current $I\,(I = I_G + I_C$, where I_G is the current through the conductor and I_C is the current through the capacitor), its complex power P, based on (1-77), can be expressed as

$$P = \frac{1}{2}(VI^*) = \frac{1}{2}V(YV)^* = \frac{1}{2}Y^*\,|\,V\,|^2 = \frac{1}{2}(G + j\omega C)^*\,|\,V\,|^2 = \frac{1}{2}(G - j\omega C)\,|\,V\,|^2 \qquad (1\text{-}77\text{b})$$

TABLE 1-6   Relations between time-harmonic electromagnetic field and steady-state a.c. circuit theories

Field theory	Circuit theory
1. **E** (electric field intensity)	1. v (voltage)
2. **H** (magnetic field intensity)	2. i (current)
3. **D** (electric flux density)	3. q_{ev} (electric charge density)
4. **B** (magnetic flux density)	4. q_{mv} (magnetic charge density)
5. **J** (electric current density)	5. i_e (electric current)
6. **M** (magnetic current density)	6. i_m (magnetic current)
7. $\mathbf{J}_d = j\omega\varepsilon\mathbf{E}$ (electric displacement current density)	7. $i = j\omega Cv$ (current through a capacitor)
8. $\mathbf{M}_d = j\omega\mu\mathbf{H}$ (magnetic displacement current density)	8. $v = j\omega Li$ (voltage across an inductor)
9. *Constitutive relations*	9. *Element laws*
(a) $\mathbf{J}_c = \sigma\mathbf{E}$ (electric conduction current density)	(a) $i = Gv = \dfrac{1}{R}v$ (Ohm's law)
(b) $\mathbf{D} = \varepsilon\mathbf{E}$ (dielectric material)	(b) $Q_e = Cv$ (charge in a capacitor)
(c) $\mathbf{B} = \mu\mathbf{H}$ (magnetic material)	(c) $\psi = Li$ (flux of an inductor)
10. $\oint_C \mathbf{E}\cdot d\boldsymbol{\ell} = -j\omega\iint_S \mathbf{B}\cdot d\mathbf{s}$ (Maxwell–Faraday equation)	10. $\sum v = -j\omega L_s i \simeq 0$ (Kirchhoff's voltage law)
11. $\oiint_S \mathbf{J}_{ic}\cdot d\mathbf{s} = -j\omega\iiint_V q_{ev}dv = -\dfrac{\partial Q_e}{\partial t}$ (continuity equation)	11. $\sum i = -j\omega Q_e = -j\omega C_s v \simeq 0$ (Kirchhoff's current law)
12. *Power and energy densities*	12. *Power and energy* (v and i represent peak values)
(a) $\dfrac{1}{2}\oiint_S (\mathbf{E}\times\mathbf{H}^*)\cdot d\mathbf{s}$ (complex power)	(a) $P = \dfrac{1}{2}vi$ (power-voltage-current relation)
(b) $\dfrac{1}{2}\iiint_V \sigma\,|\,\mathbf{E}\,|^2\,dv$ (dissipated real power)	(b) $P_d = \dfrac{1}{2}Gv^2 = \dfrac{1}{2}\dfrac{v^2}{R}$ (power dissipated in a resistor)
(c) $\dfrac{1}{4}\iiint_V \varepsilon\,|\,\mathbf{E}\,|^2\,dv$ (time-average electric stored energy)	(c) $\dfrac{1}{4}Cv^2$ (energy stored in a capacitor)
(d) $\dfrac{1}{4}\iiint_V \mu\,|\,\mathbf{H}\,|^2\,dv$ (time-average magnetic stored energy)	(d) $\dfrac{1}{4}Li^2$ (energy stored in an inductor)

The imaginary part of (1-77b) is negative. Therefore the imaginary parts of (1-77a) and (1-77b) conform, respectively, to the notation (positive and negative) of the imaginary parts of the complex power in (1-76) due to the H and E fields.

The field and circuit theory relations for time-harmonic electromagnetic fields are similar to those found in Table 1-2 for the general time-varying electromagnetic fields, but with the instantaneous field quantities (represented by script letters) replaced by their corresponding complex field quantities (represented by roman letters) and with $\partial/\partial t$ replaced by $j\omega\,(\partial/\partial t \equiv j\omega)$. These are shown listed in Table 1-6.

Over the years many excellent introductory books on electromagnetics [1] through [28], and advanced books [29] through [40], have been published. Some of them can serve both purposes, and a few may not now be in print. Each is contributing to the general knowledge of electromagnetic theory and its applications. The reader is encouraged to consult them for an even better understanding of the subject.

1.8 MULTIMEDIA

On the website that accompanies this book, the following multimedia resources are included for the review, understanding, and presentation of the material of this chapter.

- **PowerPoint (PPT)** viewgraphs, in multicolor.

REFERENCES

1. F. T. Ulaby, E. Michielssen, and U. Ravaioli, *Fundamentals of Applied Electromagnetics*, Sixth Edition, Pearson Education, Inc., Upper Saddle River, NJ, 2010.
2. M. N. Sadiku, *Elements of Electromagnetics*, Fifth Edition, Oxford University Press, Inc., New York, 2010.
3. S. M. Wentworth, *Fundamentals of Electromagnetics with Engineering Applications*, John Wiley & Sons, 2005.
4. C. R. Paul, *Electromagnetics for Engineers: With Applications to Digital Systems and Electromagnetic Interference*, John Wiley & Sons, Inc., 2004.
5. N. Ida, *Engineering Electromagnetics*, Second Edition, Springer, NY, 2004.
6. U. S. Inan and A. S. Inan, *Electromagnetic Waves*, Prentice-Hall, Inc., Upper Saddle River, NJ, 2000.
7. M. F. Iskander, *Electromagnetic Fields & Waves*, Waveland Press, Inc., Long Grove, IL, 2000.
8. K. R. Demarest, *Engineering Electromagnetics*, Prentice-Hall, Inc., Upper Saddle River, NJ, 1998.
9. G. F. Miner, *Lines and Electromagnetic Fields for Engineers*, Oxford University Press, Inc., New York, 1996.
10. D. H. Staelin, A. W. Morgenthaler, and J. A. Kong, *Electromagnetic Waves*, Prentice-Hall, Inc., Englewood Cliffs, NJ, 1994.
11. J. D. Kraus, *Electromagnetics*, Third Edition, McGraw-Hill, New York, 1992.
12. S. Ramo, J. R. Whinnery, and T. Van Duzer, *Fields and Waves in Communication Electronics*, Second Edition, John Wiley & Sons, New York, 1984.
13. W. H. Hayt, Jr. and John A. Buck, *Engineering Electromagnetics*, Sixth Edition, McGraw-Hill, New York, 2001.
14. D. T. Paris and F. K. Hurd, *Basic Electromagnetic Theory*, McGraw-Hill, New York, 1969.
15. C. T. A. Johnk, *Engineering Electromagnetic Fields and Waves*, Second Edition, John Wiley & Sons, New York, 1988.
16. H. P. Neff, Jr., *Basic Electromagnetic Fields*, Second Edition, John Wiley & Sons, New York, 1987.

17. S. V. Marshall, R. E. DuBroff, and G. G. Skitek, *Electromagnetic Concepts and Applications*, Fourth Edition, Prentice-Hall, Englewood Cliffs, NJ, 1996.

18. D. K. Cheng, *Field and Wave Electromagnetics*, Addison-Wesley, Reading, MA, 1983.

19. C. R. Paul and S. A. Nasar, *Introduction to Electromagnetic Fields*, Third Edition, McGraw-Hill, New York, 1998.

20. L. C. Shen and J. A. Kong, *Applied Electromagnetism*, Third Edition, PWS Publishing Co., Boston, MA, 1987.

21. N. N. Rao, *Elements of Engineering Electromagnetics*, Fourth Edition, Prentice-Hall, Englewood Cliffs, NJ, 1994.

22. M. A. Plonus, *Applied Electromagnetics*, McGraw-Hill, New York, 1978.

23. A. T. Adams, *Electromagnetics for Engineers*, Ronald Press, New York, 1971.

24. M. Zahn, *Electromagnetic Field Theory*, John Wiley & Sons, New York, 1979.

25. L. M. Magid, *Electromagnetic Fields, Energy, and Waves*, John Wiley & Sons, New York, 1972.

26. S. Seely and A. D. Poularikas, *Electromagnetics: Classical and Modern Theory and Applications*, Dekker, New York, 1979.

27. D. M. Cook, *The Theory of the Electromagnetic Field*, Prentice-Hall, Englewood Cliffs, NJ, 1975.

28. R. P. Feynman, R. B. Leighton, and M. Sands, *The Feynman Lectures on Physics: Mainly Electromagnetism and Matter*, Volume II, Addison-Wesley, Reading, MA, 1964.

29. W. R. Smythe, *Static and Dynamic Electricity*, McGraw-Hill, New York, 1939.

30. J. A. Stratton, *Electromagnetic Theory*, Wiley-Interscience, New York, 2007.

31. R. E. Collin, *Field Theory of Guided Waves*, IEEE Press, New York, 1991.

32. E. C. Jordan and K. G. Balmain, *Electromagnetic Waves and Radiating Systems*, Second Edition, Prentice-Hall, Englewood Cliffs, NJ, 1968.

33. R. F. Harrington, *Time-Harmonic Electromagnetic Fields*, McGraw-Hill, New York, 1961.

34. J. R. Wait, *Electromagnetic Wave Theory*, Harper & Row, New York, 1985.

35. J. A. Kong, *Theory of Electromagnetic Waves*, John Wiley & Sons, New York, 1975.

36. C. C. Johnson, *Field and Wave Electrodynamics*, McGraw-Hill, New York, 1965.

37. J. D. Jackson, *Classical Electrodynamics*, Third Edition, John Wiley & Sons, New York, 1999.

38. D. S. Jones, *Methods in Electromagnetic Wave Propagation*, Oxford Univ. Press (Clarendon), London/New York, 1979.

39. M. Kline (Ed.), *The Theory of Electromagnetic Waves*, Interscience, New York, 1951.

40. J. Van Bladel, *Electromagnetic Fields*, Second Edition, Wiley-Interscience, New York, 2007.

PROBLEMS

1.1. Derive the differential form of the continuity equation, as given by (1-6), from Maxwell's equations 1-1 through 1-4.

1.2. Derive the integral forms of Maxwell's equations and the continuity equation, as listed in Table 1-1, from the corresponding ones in differential form.

1.3. The electric flux density inside a cube is given by:

(a) $\mathbf{D} = \hat{\mathbf{a}}_x(3 + x)$

(b) $\mathbf{D} = \hat{\mathbf{a}}_y(4 + y^2)$

Find the total electric charge enclosed inside the cubical volume when the cube is in the first octant with three edges coincident with the x, y, z axes and one corner at the origin. Each side of the cube is 1 m long.

1.4. An infinite planar interface between media, as shown in the figure, is formed by having air (medium #1) on the left of the interface and lossless polystyrene (medium #2) (with a dielectric constant of 2.56) to the right of the interface. An electric surface charge density $q_{es} = 0.2$ C/m^2 exists along the entire interface.

The static electric flux density inside the polystyrene is given by

$$\mathbf{D}_2 = 6\hat{\mathbf{a}}_x + 3\hat{\mathbf{a}}_z \text{ C/m}^2$$

Determine the corresponding vector:
(a) Electric field intensity inside the polystyrene.
(b) Electric polarization vector inside the polystyrene.
(c) Electric flux density inside the air medium.
(d) Electric field intensity inside the air medium.
(e) Electric polarization vector inside the air medium.

Leave your answers in terms of ε_0, μ_0.

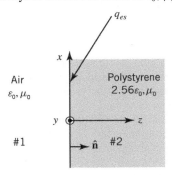

Figure P1-4

1.5. An infinite planar interface between media, as shown in the figure, is formed by having air (medium #1) on the left of the interface and lossless magnetic material (medium #2) (with a relative permeability of 4 and relative permittivity of 2.56) to the right of the interface.

The static magnetic field intensity inside the air is given by

$$\mathbf{H}_1 = 3\hat{\mathbf{a}}_x + 9\hat{\mathbf{a}}_z \ \text{A/m}$$

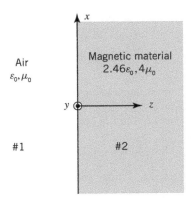

Figure P1-5

Determine the corresponding vector:
(a) Magnetic flux density in the air medium.
(b) Magnetic polarization in the air medium.
(c) Magnetic field intensity in the magnetic material.
(d) Magnetic flux density in the magnetic material.
(e) Magnetic polarization in the magnetic material.

Leave your answers in terms of ε_0, μ_0.

1.6. A static electric field of intensity/strength $\mathbf{E}_0$ is established inside a free-space medium as shown below. The static electric field intensity is oriented at an angle of $30°$ relative to the principal z axis. A semi-infinite dielectric slab of relative permittivity of 4 and relative permeability of unity is immersed into the initially established static electric field, as shown in Figure P1-6.
Determine the:
(a) Total electric field intensity $\mathbf{E}_1$ and total electric flux density $\mathbf{D}_1$ within the dielectric slab. Leave your answers in terms of E_0, ε_0, μ_0, and any constants.
(b) Angle θ (in degrees).

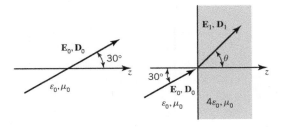

Figure P1-6

1.7. A static magnetic field of field intensity/strength $\mathbf{H}_0$ is established inside a free-space medium as shown in Figure P1-7. The static magnetic field intensity is oriented at an angle of $30°$ relative to the principal z axis. A semi-infinite magnetic slab of relative permeability of 4 and relative permittivity of 9 is immersed into the initially established static magnetic field, as shown in the figure.
Determine the:
(a) Total static magnetic field intensity $\mathbf{H}_1$ and total static magnetic flux density $\mathbf{B}_1$ within the magnetic slab. Leave your answers in terms of H_0, ε_0, μ_0, and any constants.
(b) Angle θ (in degrees).

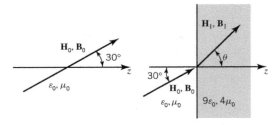

Figure P1-7

1.8. A dielectric slab, with a thickness of 6 cm and dielectric constant of 4, is sandwiched between two different media; free space to the left and another dielectric, with a dielectric constant of 9, to the right. If the electric field in the free-space medium is at an angle of $30°$ at a height of 3 cm at the leading interface, as shown in the figure below, determine the:
  (a) Angle α (in degrees, as measured from the normal to the interface) the electric field will make in the dielectric medium to the right of the center slab.
  (b) Height h (in cm) the electric field will have at the trailing interface.

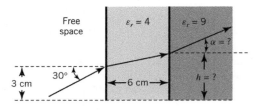

Figure P1-8

1.9. The electric field inside a circular cylinder of radius a and height h is given by

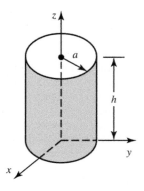

Figure P1-9

$$\mathbf{E} = \hat{\mathbf{a}}_z\left[-\frac{c}{h} + \frac{b}{6\varepsilon_0}(3z^2 - h^2)\right]$$

where c and b are constants. Assuming the medium within the cylinder is free space, find the total charge enclosed within the cylinder.

1.10. The static magnetic field on the inside part of the surface of an infinite length dielectric cylinder of circular cross section of radius $a = 4$ cm and of magnetic material with a relative permittivity and permeability of $\varepsilon_r = 4$, $\mu_r = 9$ is given by

$$\mathbf{H} = \hat{\mathbf{a}}_\rho 3 + \hat{\mathbf{a}}_\phi 6 + \hat{\mathbf{a}}_z\, 8 \text{ A/m at } \rho = 4^- \text{ cm}$$

The cylinder is surrounded on the outside with air. Refer to Figure 3.4 for the cylindrical coordinate system and its unit vectors. Determine the:
  (a) Magnetic flux density on the inside part of the surface of the cylinder ($\rho = 4^-$ cm; magnetic material).
  (b) Magnetic field on the outside part of the cylinder surface ($\rho = 4^+$ cm; air).
  (c) Magnetic flux density on the outside part of the cylinder surface ($\rho = 4^+$ cm; air).

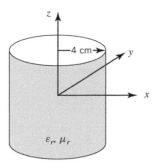

Figure P1-10

1.11. The instantaneous electric field inside a source-free, homogeneous, isotropic, and linear medium is given by

$$\mathscr{E} = \left[\hat{\mathbf{a}}_x A(x + y) + \hat{\mathbf{a}}_y B(x - y)\right]\cos(\omega t)$$

Determine the relations between A and B.

1.12. The magnetic flux density produced on its plane by a current-carrying circular loop of radius $a = 0.1$ m, placed on the xy plane at $z = 0$, is given by

$$\mathscr{B} = \hat{a}_z \frac{10^{-12}}{1+25\rho}\cos(1500\pi t) \text{ Wb/m}^2$$

where ρ is the radial distance in cylindrical coordinates. Find the:
(a) Total flux in the z direction passing through the loop.
(b) Electric field at any point ρ within the loop. Check your answer by using Maxwell's equation 1-1.

1.13. The instantaneous magnetic flux density in free space is given by

$$\mathscr{B} = \hat{a}_x B_x \cos(2y)\sin(\omega t - \pi z)$$
$$+ \hat{a}_y B_y \cos(2x)\cos(\omega t - \pi z)$$

where B_x and B_y are constants. Assuming there are no sources at the observation points x,y, determine the electric displacement current density.

1.14. The displacement current density within a source-free ($\mathscr{I}_i = 0$) cube centered about the origin is given by

$$\mathscr{I}_d = \hat{a}_x yz + \hat{a}_y y^2 + \hat{a}_z xyz$$

Each side of the cube is 1 m long and the medium within it is free space. Find the displacement current leaving, in the outward direction, through the surface of the cube.

1.15. The electric flux density in free space produced by an oscillating electric charge placed at the origin is given by

$$\mathscr{D} = \hat{a}_r \frac{10^{-9}}{4\pi}\frac{1}{r^2}\cos(\omega t - \beta_0 r)$$

where $\beta_0 = \omega\sqrt{\mu_0\varepsilon_0}$. Find the time-average charge that produces this electric flux density.

1.16. The electric field radiated at large distances in free space by a current-carrying small circular loop of radius a, placed on the xy plane at $z = 0$, is given by

$$\mathscr{E} = \hat{a}_\phi E_0 \sin\theta \frac{\cos(\omega t - \beta_0 r)}{r}, \quad r \gg a$$

where E_0 is a constant, $\beta_0 = \omega\sqrt{\mu_0\varepsilon_0}$, r is the radial distance in spherical coordinates, and θ is the spherical angle measured from the z axis that is perpendicular to the plane of the

loop. Determine the corresponding radiated magnetic field at large distances from the loop ($r \gg a$).

1.17. A time-varying voltage source of $v(t) = 10\cos(\omega t)$ is connected across a parallel plate capacitor with polystyrene ($\varepsilon = 2.56\varepsilon_0$, $\sigma = 3.7\times10^{-4}$ S/m) between the plates. Assuming a small plate separation of 2 cm and no field fringing, determine at:
(a) $f = 1$ MHz
(b) $f = 100$ MHz
the maximum values of the conduction and displacement current densities within the polystyrene and compare them.

1.18. A dielectric slab of polystyrene ($\varepsilon = 2.56\varepsilon_0$, $\mu = \mu_0$) of height $2h$ is bounded above and below by free space, as shown in Figure P1-18. Assuming the electric field within the slab is given by

$$\mathscr{E} = (\hat{a}_y 5 + \hat{a}_z 10)\cos(\omega t - \beta x)$$

where $\beta = \omega\sqrt{\mu_0\varepsilon}$, determine the:
(a) Corresponding magnetic field within the slab.
(b) Electric and magnetic fields in free space right above and below the slab.

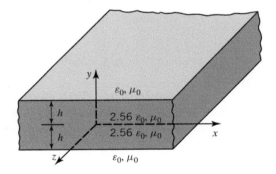

Figure P1-18

1.19. A finite conductivity rectangular strip, shown in Figure P1-19, is used to carry electric current. Because of the strip's lossy nature, the current is nonuniformly distributed over the cross section of the strip. The current density on the *upper and lower* sides is given by

$$\mathscr{I} = \hat{a}_z 10^4 \cos(2\pi\times10^9 t) \text{ A/m}^2$$

and it rapidly decays in an exponential fashion from the lower side toward the center by the factor $e^{-10^6 y}$, or

$$\mathscr{I} = \hat{a}_z 10^4 e^{-10^6 y}\cos(2\pi\times10^9 t) \text{ A/m}^2$$

A similar decay is experienced by the current density from the upper side toward the center. Assuming no variations of the current density with respect to x, determine the total current flowing through the wire.

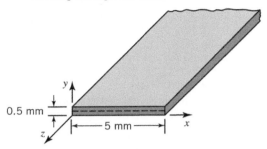

0.5 mm

5 mm

Figure P1-19

1.20. The instantaneous electric field inside a conducting rectangular pipe (waveguide) of width a is given by

$$\mathscr{E} = \hat{\mathbf{a}}_y E_0 \sin\left(\frac{\pi}{a}x\right)\cos(\omega t - \beta_z z)$$

where β_z is the waveguide's phase constant. Assuming there are no sources within the free-space-filled pipe determine the:
(a) Corresponding instantaneous magnetic field components inside the conducting pipe.
(b) Phase constant β_z.
The height of the waveguide is b.

1.21. The instantaneous electric field intensity inside a source-free coaxial line with inner and outer radii of a and b, respectively, that is filled with a homogeneous dielectric of $\varepsilon = 2.25\varepsilon_0$, $\mu = \mu_0$, and $\sigma = 0$, is given by

$$\mathscr{E} = \hat{\mathbf{a}}_\rho\left(\frac{100}{\rho}\right)\cos(10^8 t - \beta z)$$

where β is the phase constant and ρ is the cylindrical radial distance from the center of the coaxial line. Determine the:
(a) Corresponding instantaneous magnetic field $\mathscr{H}$.
(b) Phase constant β.
(c) Displacement current density $\mathscr{J}_d$.

1.22. A coaxial line resonator with inner and outer conductors at $a = 5$ mm and $b = 20$ mm, and with conducting plates at $z = 0$ and $z = \ell$, is filled with a dielectric with $\varepsilon_r = 2.56$, $\mu_r = 1$, and $\sigma = 0$. The instantaneous magnetic field

intensity inside the source-free dielectric medium is given by

$$\mathscr{H} = \hat{\mathbf{a}}_\phi\left(\frac{2}{\rho}\right)\cos\left(\frac{\pi}{\ell}z\right)\cos(4\pi \times 10^8 t)$$

Find the following:
(a) Electric field intensity within the dielectric.
(b) Surface current density $\mathscr{J}_s$ at the conductor surfaces at $\rho = a$ and $\rho = b$.
(c) Displacement current density $\mathscr{J}_d$ at any point within the dielectric.
(d) Total displacement current flowing through the circumferential surface of the resonator.

1.23. Using the instantaneous forms of Maxwell's equation and the continuity equations listed in Tables 1-1 and 1-4, derive the corresponding time-harmonic forms (in differential and integral forms) listed in Table 1-4. Use definitions (1-61a) through (1-61f).

1.24. Show that the electric and magnetic fields (1-61a) and (1-61b) can be written, respectively, as in (1-67a) and (1-67b).

1.25. The time-harmonic instantaneous electric field traveling along the z axis, in a free-space medium, is given by

$$\mathscr{E}(z,t) = \hat{\mathbf{a}}_x E_0 \sin\left[(\omega t - \beta_0 z) + \left(\frac{\pi}{2}\right)\right]$$

where E_0 is a real constant and $\beta_0 = \omega\sqrt{\mu_0\varepsilon_0}$.
(a) Write an expression for the complex spatial electric field intensity $\mathbf{E}(z)$.
(d) Find the corresponding complex spatial magnetic field intensity $\mathbf{H}(z)$.
(c) Determine the time-average Poynting vector (average power density) $\mathbf{S}_{ave}$.

1.26. An electric line source of infinite length and constant current, placed along the z axis, radiates in free space at large distances from the source ($\rho \gg 0$) a time-harmonic complex magnetic field given by

$$\mathbf{H} = \hat{\mathbf{a}}_\phi H_0 \frac{e^{-j\beta_0 \rho}}{\sqrt{\rho}}, \quad \rho \gg 0$$

where H_0 is a constant, $\beta_0 = \omega\sqrt{\mu_0\varepsilon_0}$, and ρ is the radial cylindrical distance. Determine the corresponding electric field for $\rho \gg 0$.

1.27. The time-harmonic complex electric field radiated in free space by a linear radiating element is given by

$$\mathbf{E} = \hat{\mathbf{a}}_r E_r + \hat{\mathbf{a}}_\theta E_\theta$$

$$E_r = E_0 \frac{\cos\theta}{r^2} \left[1 + \frac{1}{j\beta_0 r} \right] e^{-j\beta_0 r}$$

$$E_\theta = jE_0 \frac{\beta_0 \sin\theta}{2r} \left[1 + \frac{1}{j\beta_0 r} - \frac{1}{(\beta_0 r)^2} \right] e^{-j\beta_0 r}$$

where $\hat{\mathbf{a}}_r$ and $\hat{\mathbf{a}}_\theta$ are unit vectors in the spherical directions r and θ, E_0 is a constant, and $\beta_0 = \omega\sqrt{\mu_0\varepsilon_0}$. Determine the corresponding spherical magnetic field components.

1.28. The time-harmonic complex electric field radiated by a current-carrying small circular loop in free space is given by

$$\mathbf{E} = \hat{\mathbf{a}}_\phi E_0 \frac{\sin\theta}{r} \left[1 + \frac{1}{j\beta_0 r} \right] e^{-j\beta_0 r}$$

where $\hat{\mathbf{a}}_\phi$ is the spherical unit vector in the ϕ direction, E_0 is a constant, and $\beta_0 = \omega\sqrt{\mu_0\varepsilon_0}$. Determine the corresponding spherical magnetic field components.

1.29. The complex electric field inside an infinitely long rectangular pipe, with all four vertical walls perfectly electric conducting, as shown in Figure P1-29, is given by

$$\mathbf{E} = \hat{\mathbf{a}}_z (1+j) \sin\left(\frac{\pi}{a}x\right) \sin\left(\frac{\pi}{b}y\right)$$

Assuming that there are no sources within the box and $a = \lambda_0$, $b = 0.5\lambda_0$, and $\mu = \mu_0$, where $\lambda_0 =$ free space, infinite medium wavelength, find the:

(a) Conductivity
(b) Dielectric constant
of the medium within the box.

1.30. A time-harmonic electromagnetic field in free space is perpendicularly incident upon a perfectly conducting semi-infinite planar surface, as shown in Figure P1-30. Assuming the incident $\mathbf{E}^i$ and reflected $\mathbf{E}^r$ complex electric fields on the free-space side of the interface are given by

$$\mathbf{E}^i = \hat{\mathbf{a}}_x e^{-j\beta_0 z}$$

$$\mathbf{E}^r = -\hat{\mathbf{a}}_x e^{+j\beta_0 z}$$

where

$$\beta_0 = \omega\sqrt{\mu_0\varepsilon_0}$$

determine the current density $\mathbf{J}_s$ induced on the surface of the conducting surface. Evaluate all the constants.

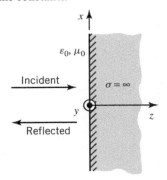

Figure P1-30

1.31. The free-space incident $\mathbf{E}^i$ and reflected $\mathbf{E}^r$ fields of a time-harmonic electromagnetic field obliquely incident upon a perfectly conducting semi-infinite planar surface of Figure P1-31 are given by

$$\mathbf{E}^i = \hat{\mathbf{a}}_y E_0 e^{-j\beta_0(x\sin\theta_i + z\cos\theta_i)}$$

$$\mathbf{E}^r = \hat{\mathbf{a}}_y E_0 \Gamma_h e^{-j\beta_0(x\sin\theta_i - z\cos\theta_i)}$$

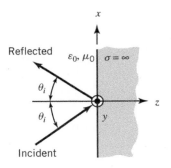

Figure P1-31

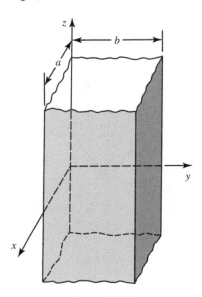

Figure P1-29

where E_0 is a constant and $\beta_0 = \omega\sqrt{\mu_0\varepsilon_0}$. Determine the coefficient Γ_h.

1.32. For Problem 1.31, determine the:
(a) Corresponding incident and reflected magnetic fields.
(b) Electric current density along the interface between the two media.

1.33. Repeat Problem 1.31 when the incident and reflected electric fields are given by

$$\mathbf{E}^i = (\hat{\mathbf{a}}_x \cos\theta_i - \hat{\mathbf{a}}_z \sin\theta_i)$$
$$\times E_0 e^{-j\beta_0(x\sin\theta_i + z\cos\theta_i)}$$

$$\mathbf{E}^r = (\hat{\mathbf{a}}_x \cos\theta_i + \hat{\mathbf{a}}_z \sin\theta_i)$$
$$\times \Gamma_e E_0 e^{-j\beta_0(x\sin\theta_i - z\cos\theta_i)}$$

where E_0 is a constant and $\beta_0 = \omega\sqrt{\mu_0\varepsilon_0}$. Determine the coefficient Γ_e by applying the boundary conditions on the tangential components.

1.34. Repeat Problem 1.33 except that Γ_e should be determined using the boundary conditions on the normal components. Compare the answer with that obtained in Problem 1.33. Explain.

1.35. For Problem 1.33 determine the:
(a) Corresponding incident and reflected magnetic fields.
(b) Electric current density along the interface between the two media.

1.36. A time-harmonic electromagnetic field traveling in free space and perpendicularly incident upon a flat surface of distilled water ($\varepsilon = 81\varepsilon_0$, $\mu = \mu_0$), as shown in Figure P1-36, creates a reflected field on the free-space side of the interface and a transmitted field on the water side of the interface. Assuming the incident ($\mathbf{E}^i$), reflected ($\mathbf{E}^r$),

and transmitted ($\mathbf{E}^t$) electric fields are given, respectively, by

$$\mathbf{E}^i = \hat{\mathbf{a}}_x E_0 e^{-j\beta_0 z}$$
$$\mathbf{E}^r = \hat{\mathbf{a}}_x \Gamma_0 E_0 e^{+j\beta_0 z}$$
$$\mathbf{E}^t = \hat{\mathbf{a}}_x T_0 E_0 e^{-j\beta z}$$

determine the coefficients Γ_0 and T_0. E_0 is a constant, $\beta_0 = \omega\sqrt{\mu_0\varepsilon_0}$, $\beta = \omega\sqrt{\mu_0\varepsilon}$.

1.37. When a time-harmonic electromagnetic field is traveling in free space and is obliquely incident upon a flat surface of distilled water ($\varepsilon = 81\varepsilon_0$, $\mu = \mu_0$), as shown in Figure P1-37, it creates a reflected field on the free-space side of the interface and a transmitted field on the water side of the interface. Assume the incident, reflected, and transmitted electric and magnetic fields are given by

$$\mathbf{E}^i = \hat{\mathbf{a}}_y E_0 e^{-j\beta_0(x\sin\theta_i + z\cos\theta_i)}$$
$$\mathbf{H}^i = (-\hat{\mathbf{a}}_x \cos\theta_i + \hat{\mathbf{a}}_z \sin\theta_i)$$
$$\times \sqrt{\frac{\varepsilon_0}{\mu_0}} E_0 e^{-j\beta_0(x\sin\theta_i + z\cos\theta_i)}$$

$$\mathbf{E}^r = \hat{\mathbf{a}}_y \Gamma_h E_0 e^{-j\beta_0(x\sin\theta_i - z\cos\theta_i)}$$
$$\mathbf{H}^r = (\hat{\mathbf{a}}_x \cos\theta_i + \hat{\mathbf{a}}_z \sin\theta_i)$$
$$\times \sqrt{\frac{\varepsilon_0}{\mu_0}} \Gamma_h E_0 e^{-j\beta_0(x\sin\theta_i + z\cos\theta_i)}$$

$$\mathbf{E}^t = \hat{\mathbf{a}}_y T_h E_0 e^{-j\beta_0\left(x\sin\theta_i + z\sqrt{\frac{\varepsilon}{\varepsilon_0} - \sin^2\theta_i}\right)}$$
$$\mathbf{H}^t = \left(-\hat{\mathbf{a}}_x \sqrt{1 - \frac{\varepsilon_0}{\varepsilon}\sin^2\theta_i} + \hat{\mathbf{a}}_z \sqrt{\frac{\varepsilon_0}{\varepsilon}}\sin\theta_i\right)$$
$$\times \sqrt{\frac{\varepsilon}{\mu_0}} T_h E_0 e^{-j\beta_0\left(x\sin\theta_i + z\sqrt{\frac{\varepsilon}{\varepsilon_0} - \sin^2\theta_i}\right)}$$

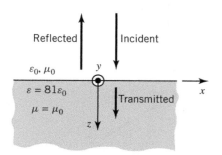

Figure P1-36

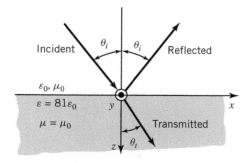

Figure P1-37

where E_0 is a constant and $\beta_0 = \omega\sqrt{\mu_0\varepsilon_0}$. Determine the coefficients Γ_h and T_h by applying the boundary conditions on the tangential components. Evaluate all the constants.

1.38. Repeat Problem 1.37 except that Γ_h and T_h should be determined using the boundary conditions on the normal components. Compare the answers to those obtained in Problem 1.37. Explain.

1.39. Repeat Problem 1.37 when the incident, reflected, and transmitted electric and magnetic fields are given by

$$E^i = (\hat{a}_x \cos\theta_i - \hat{a}_z \sin\theta_i)$$
$$\times E_0 e^{-j\beta_0(x\sin\theta_i + z\cos\theta_i)}$$

$$H^i = \hat{a}_y \sqrt{\frac{\varepsilon_0}{\mu_0}} E_0 e^{-j\beta_0(x\sin\theta_i + z\cos\theta_i)}$$

$$E^r = (\hat{a}_x \cos\theta_i + \hat{a}_z \sin\theta_i)$$
$$\times \Gamma_e E_0 e^{-j\beta_0(x\sin\theta_i - z\cos\theta_i)}$$

$$H^r = -\hat{a}_y \sqrt{\frac{\varepsilon_0}{\mu_0}} \Gamma_e E_0 e^{-j\beta_0(x\sin\theta_i - z\cos\theta_i)}$$

$$E^t = \left[\hat{a}_x \sqrt{1 - \frac{\varepsilon_0}{\varepsilon}\sin^2\theta_i} - \hat{a}_z \sqrt{\frac{\varepsilon_0}{\varepsilon}}\sin\theta_i\right]$$
$$\times T_e E_0 e^{-j\beta\left[x\sin\theta_i + z\sqrt{\frac{\varepsilon}{\varepsilon_0} - \sin^2\theta_i}\right]}$$

$$H^t = \hat{a}_y \sqrt{\frac{\varepsilon}{\mu_0}} T_e E_0 e^{-j\beta\left[x\sin\theta_i + z\sqrt{\frac{\varepsilon}{\varepsilon_0} - \sin^2\theta_i}\right]}$$

Γ_e and T_e should be determined using the boundary conditions on the tangential components.

1.40. Repeat Problem 1.39 except that Γ_e and T_e should be determined using the boundary conditions on the normal components. Compare the answers to those obtained in Problem 1.39. Explain.

1.41. For Problem 1.16 find the:
(a) Average power density at large distances.
(b) Total power exiting through the surface of a large sphere of radius r $(r \gg a)$.

1.42. A uniform plane wave traveling in a free space medium is incident at an oblique angle θ_i upon an infinite and flat perfect electric conductor (PEC, $\sigma = \infty$), as shown in Figure P1-42. The normalized incident and reflected magnetic fields at the surface of the PEC ($y = 0$, on the free space part of the PEC), are given by

$H^{incident}$ (on surface of PEC)
$$= \frac{1}{377}(-\hat{a}_x \cos\theta_i + \hat{a}_z \sin\theta_i)$$

$H^{reflected}$ (on surface of PEC)
$$= \frac{1}{377}(-\hat{a}_x \cos\theta_i - \hat{a}_z \sin\theta_i)$$

Find the total electric current density J_s induced on the surface of the PEC.

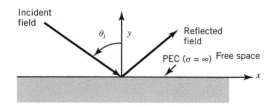

Figure P1-42

1.43. A uniform plane wave, as shown in Figure P1-43, is incident upon a planar perfectly magnetic conducting (PMC) interface. Using a right-hand rectangular coordinate system, the normalized incident and reflected electric and magnetic fields are given by

$$E^i = \left(\hat{a}_y \cos\theta_i + \hat{a}_z \sin\theta_i\right)e^{-j\beta_0(y\sin\theta_i - z\cos\theta_i)}$$

$$H^i = \hat{a}_x \frac{1}{\eta_0} e^{-j\beta_0(y\sin\theta_i - z\cos\theta_i)}$$

$$E^r = \Gamma\left(\hat{a}_y \cos\theta_i - \hat{a}_z \sin\theta_i\right)e^{-j\beta_0(y\sin\theta_i + z\cos\theta_i)}$$

$$H^r = -\Gamma\hat{a}_x \frac{1}{\eta_0} e^{-j\beta_0(y\sin\theta_i - z\cos\theta_i)}$$

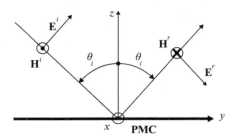

Figure P1-43

Derive:
(a) The value of the reflection coefficient. Show all steps.
(b) Vector expression for the magnetic current density induced on the perfectly magnetic conducting (PMC) surface.

(c) Vector expression for the electric current density induced on the perfectly magnetic conducting (PMC) surface.

Show the appropriate steps that lead to the answer. It is preferable that you also state, in words, what you are doing.

1.44. Repeat Problem 1.43 if the planar reflecting surface is PEC.

1.45. A time-harmonic uniform plane wave traveling in a free space medium is incident, at an oblique angle θ_i, upon an infinite in extent and flat perfect magnetic conductor (PMC). A PMC surface is one where the tangential components of the magnetic field vanish, and it is the dual of a perfectly electric conducting (PEC) surface.

   The normalized incident and reflected magnetic fields, on the free space part of the PMC, are given by

$$\mathbf{E}^{incident} = \hat{\mathbf{a}}_y E_0 e^{-j\beta_0(x\sin\theta_i + z\cos\theta_i)};$$

$$\mathbf{H}^{incident} = \frac{E_0}{\sqrt{\mu/\varepsilon}}(-\hat{\mathbf{a}}_x \cos\theta_i + \hat{\mathbf{a}}_z \sin\theta_i)$$
$$\times\, e^{-j\beta_0(x\sin\theta_i + z\cos\theta_i)}$$

$$\mathbf{E}^{reflected} = \hat{\mathbf{a}}_y \Gamma E_0 e^{-j\beta_0(x\sin\theta_i - z\cos\theta_i)};$$

$$\mathbf{H}^{reflected} = \frac{\Gamma E_0}{\sqrt{\mu/\varepsilon}}(\hat{\mathbf{a}}_x \cos\theta_i + \hat{\mathbf{a}}_z \sin\theta_i)$$
$$\times\, e^{-j\beta_0(x\sin\theta_i - z\cos\theta_i)}$$

where E_0 and Γ are constants.

(a) Determine the constant Γ.

(b) Determine/derive, in vector form, specific expressions applicable for this problem, for the electric $\mathbf{J}_s$ and magnetic $\mathbf{M}_s$ vector current densities induced on the PMC surface.

Express the induced electric $\mathbf{J}_s$ and magnetic $\mathbf{M}_s$ current densities in vector form.

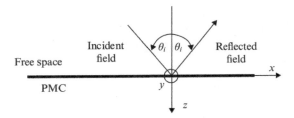

Figure P1-45

1.46. The time-harmonic complex field inside a source-free conducting pipe of rectangular cross section (waveguide), shown in Figure P1-46 filled with free space, is given by

$$\mathbf{E} = \hat{\mathbf{a}}_y E_0 \sin\left(\frac{\pi}{a}x\right)e^{-j\beta_z z},\ 0 \le x \le a,\ 0 \le y \le b$$

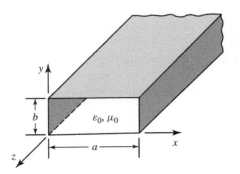

Figure P1-46

where

$$\beta_z = \beta_0\sqrt{1 - \left(\frac{\lambda_0}{2a}\right)^2}$$

E_0 is a constant, and $\beta_0 = 2\pi/\lambda_0 = \omega\sqrt{\mu_0\varepsilon_0}$. For a section of waveguide of length l along the z axis, determine the:

(a) Corresponding complex magnetic field.
(b) Supplied complex power.
(c) Exiting complex power.
(d) Dissipated real power.
(e) Time-average magnetic energy.
(f) Time-average electric energy.
Ultimately verify that the conservation-of-energy equation in integral form is satisfied for this set of fields inside this section of the waveguide.

1.47. For the waveguide and its set of fields of Problem P1.46, verify the conservation-of-energy equation in differential form for any observation point within the waveguide.

1.48. The normalized time-harmonic electric field inside an air-filled, source-free rectangular pipe/waveguide of infinite length and with cross-sectional dimensions of a and b, whose four walls (left-right, top-bottom) are perfect electric conductors (PEC, $\sigma = \infty$), is given by

$$E_x = \cos(\beta_x x)\sin(\beta_y y)$$
$$E_y = \sin(\beta_x x)\cos(\beta_y y)$$

where β_x and β_y are real constants. For non-trivial (nonzero) fields, determine all possible values of β_x in terms of a, and β_y in terms of b.

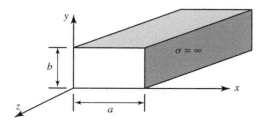

Figure P1-48

1.49. The normalized time-harmonic electric and magnetic field rectangular components, inside an free space-filled, source-free rectangular pipe (waveguide) with four walls *and* with dimensions of a and b, are given by:

Field inside the PEC rectangular waveguide

$$E_x = E_0 \sin(\beta_x x)\sin(\beta_y y)e^{-j\beta_z z}$$

$$E_y = -\left(\frac{\beta_y}{\beta_x}\right)E_0 \cos(\beta_x x)\cos(\beta_y y)e^{-j\beta_z z}$$

$$H_x = \left(\frac{\omega\varepsilon_0\beta_y}{\beta_x\beta_z}\right)E_0 \cos(\beta_x x)\cos(\beta_y y)e^{-j\beta_z z}$$

$$H_y = -\left(\frac{\omega\varepsilon_0}{\beta_z}\right)E_0 \sin(\beta_x x)\sin(\beta_y y)e^{-j\beta_z z}$$

where E_0 β_x, β_y, and β_z are *real* constants. The waveguide, shown in Figure P1-49, is infinite in length.

The top and bottom walls are perfectly electric conducting (PEC) (i.e., tangential components of electric field vanish).

The left and right walls are perfectly magnetic conducting (PMC) (i.e., tangential components of the magnetic field vanish). The PMC walls are the duals of the PEC walls.

Determine all possible expressions/values of the constants. Be specific; indicate all the constants that lead to nontrivial solutions; i.e., electric and magnetic fields not being zero.

(a) β_x in terms of a and any other constants.
(b) β_y in terms of b and any other constants.

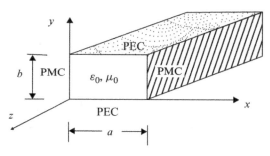

Figure P1-49

1.50. At microwave frequencies, high Q resonant cavities are usually constructed of enclosed conducting pipes (waveguides) of different cross sections. One such cavity is that of rectangular cross section that is enclosed on all six sides, as shown in Figure P1-50. One set of complex fields that can exist inside such a source-free cavity filled with free space is given by

$$\mathbf{E} = \hat{\mathbf{a}}_y E_0 \sin\left(\frac{\pi}{a}x\right)\sin\left(\frac{\pi}{c}z\right)$$

such that

$$\omega = \omega_r = \frac{1}{\sqrt{\mu_0\varepsilon_0}}\sqrt{\left(\frac{\pi}{a}\right)^2 + \left(\frac{\pi}{c}\right)^2}$$

where E_0 is a constant and ω_r is referred to as the resonant radian frequency. Within the cavity, determine the:
(a) Corresponding magnetic field.
(b) Supplied complex power.
(c) Dissipated real power.
(d) Time-average magnetic energy.
(e) Time-average electric energy.
Ultimately verify that the conservation of energy equation in integral form is satisfied for this set of fields inside this resonant cavity.

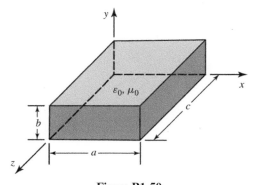

Figure P1-50

CHAPTER 2

Electrical Properties of Matter

2.1 INTRODUCTION

Since the late 1990s a renewed interest has been spurred in the application, integration, modeling, and optimization of materials in a plethora of electromagnetic radiation, guiding, and scattering structures. In particular, material structures whose constitutive parameters (permittivity and permeability) are both negative, often referred to as a Double Negative (DNG), have received considerable interest and attention. Artificial magnetic conductors can also be included in the DNG class of materials. A more inclusive name for all these materials is *metamaterials*. It is the class of metamaterials that has captivated the interest and imagination of many leading researchers and practitioners, scientists, and engineers from academia, industry, and government. When electromagnetic waves interact with such surfaces, they result in some very unique and intriguing characteristics and phenomena that can be used, for example, to optimize the performance of antennas, microwave devices, and other electromagnetic wave guiding structures. While the revitalization of metamaterials introduced a welcomed renewed interest in materials for electromagnetics, especially for applications related to antennas, microwaves, transmission lines, scattering, optics, etc., it also brought along some spirited dialogue that will be discussed in more detail in Section 5.7. The uniqueness of these materials is characterized and demonstrated by their basic constitutive parameters, such as permittivity, permeability, and conductivity. In order to appreciate the behavior of materials, it is very important that engineers and scientists understand the very basics of these constitutive parameters from d.c. to rf frequencies. An in-depth development of models for these constitutive parameters, from their basic atomic structure to their interaction with electromagnetic fields, is undertaken in this chapter.

An *atom* of an element consists of a very small but massive nucleus that is surrounded by a number of negatively charged electrons revolving about the nucleus. The nucleus contains *neutrons*, which are neutral particles, and *protons*, which are positively charged particles. All matter is made up of one or more of the 118 different elements that are now known to exist. Elements 112 to 118 have been discovered but not confirmed. Of this number, only 92 occur naturally. If the substance in question is a compound, it is composed of two or more different elements. The smallest constituent of a compound is a *molecule*, which is composed of one or more atoms held together by the short-range forces of their electrical charges.

Balanis' Advanced Engineering Electromagnetics, Third Edition. Constantine A. Balanis.
© 2024 John Wiley & Sons, Inc. Published 2024 by John Wiley & Sons, Inc.
Companion Website: www.wiley.com/go/balanis/advancedengineeringelectromagnetics3e

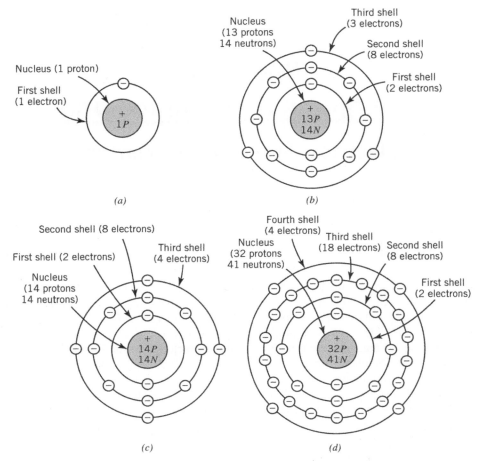

Figure 2-1   Atoms of representative elements of most interest in electronics. (*a*) Hydrogen atom. (*b*) Aluminum atom. (*c*) Silicon atom. (*d*) Germanium atom. (Source: R. R. Wright and H. R. Skutt, *Electronics: Circuits and Devices*, 1965; reprinted by permission of John Wiley and Sons, Inc.)

For a given element, each of its atoms contains the same number of protons in its nucleus. Depending on the element, that number ranges from 1 to 118 and represents the *atomic number* of the element. For an atom in its normal state, the number of electrons is also equal to the atomic number. The revolving electrons that surround the nucleus exist in various shells, and they exert forces of repulsion on each other and forces of attraction on the positive charges of the nucleus. The outer shell of an atom is referred to as the *valence shell* (band) and the electrons occupying that shell are known as *valence electrons*. They are of most interest here. The portrayal of an atom by such a model is referred to as the *Bohr* model [1]. Atoms and their charges for some typical elements of interest in electronics (such as hydrogen, aluminum, silicon, and germanium) are shown in Figure 2-1.

For an atom, all the electrons in a given shell (orbit) exist in the same energy level (fixed state). Since there are several shells (orbits) around the nucleus of an atom, there exist several discrete energy levels (fixed states) each representing a given shell (orbit). In general, there are more energy levels than electrons. Therefore some of the energy levels (orbits, shells, bands) are not occupied by electrons. The Bohr model of an atom states that:

1. Electrons of any atom exist only in *discrete* states and possess only *discrete* amounts of energy corresponding to the *discrete* radii of their corresponding orbital shells.

2. If an electron moves from a lower- to a higher-energy level (orbit), it *absorbs* a *discrete* quantity of energy (referred to as *quanta*).
3. If an electron moves from a higher- to a lower-energy level (orbit), it *radiates* a discrete quantity of energy (referred to as *quanta*).
4. If an electron maintains its energy level (orbit), it neither absorbs nor radiates energy.

When a molecule is formed with two or more atoms, forces between the atoms result in new arrangements of the charges. For an electron to be freed from an atom, it must acquire sufficient energy to allow it to escape its atomic forces and become a free body. This is analogous to the energy required by a projectile to escape the earth's gravity and become a free body.

2.2 DIELECTRICS, POLARIZATION, AND PERMITTIVITY

Dielectrics (insulators) are materials whose dominant charges in atoms and molecules are *bound* negative and positive charges that are held in place by atomic and molecular forces, and they are not free to travel. Thus ideal dielectrics do not contain any free charges (such as in conductors), and their atoms and molecules are macroscopically neutral as shown in Figure 2-2*a*. Furthermore, when external fields are applied, these bound negative and positive charges do not move to the surface of the material, as would be the case for conductors, but their respective centroids can shift slightly in positions (assumed to be an infinitesimal distance) relative to each other, thus creating numerous electric dipoles. This is illustrated in Figure 2-2*b*. In conductors, positive and negative charges are separated by macroscopic distances, and they can be separated by a surface of integration. This is not permissible for bound charges and illustrates a fundamental difference between bound charges in dielectrics and true charges in conductors.

For dielectrics, the formation of the electric dipoles is usually referred to as *orientational polarization*. The effect of each electric dipole can be represented by a dipole, as shown in Figure 2-3, with a dipole moment $d\mathbf{p}_i$ given by

$$d\mathbf{p}_i = Q\boldsymbol{\ell}_i \tag{2-1}$$

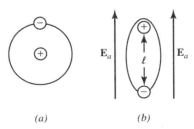

(a)   *(b)*

Figure 2-2   A typical atom. (*a*) Absence of applied field. (*b*) Under applied field.

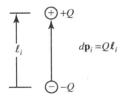

Figure 2-3   Formation of a dipole between two opposite charges of equal magnitude Q.

where Q is the magnitude (in coulombs) of each of the negative and positive charges whose centroids are displaced vectorially by distance ℓ_i.

When a material is subjected to an electric field, the polarization dipoles of the material interact with the applied electromagnetic field. For dielectric (insulating) material, whether they are solids, liquids, or gases, this interaction provides the material the ability to store electric energy, which is accomplished by the shift against restraining forces of their bound charges when they are subjected to external applied forces. This is analogous to stretching a spring or lifting a weight, and it represents potential energy.

The presence of these dipoles can be accounted for by developing a microscopic model in which each individual charge and dipole as represented by (2-1) is considered. Such a procedure, although accurate if performed properly, is very impractical if applied to a dielectric slab because the spatial position of each atom and molecule of the material must be known. Instead, in practice, the behavior of these dipoles and bound charges is accounted for in a qualitative way by introducing an *electric polarization vector* **P** using a macroscopic scale model involving thousands of atoms and molecules.

The total dipole moment $\mathbf{p}_t$ of a material is obtained by summing the dipole moments of all the orientational polarization dipoles, each of which is represented by (2-1). For a volume Δv where there are N_e electric dipoles per unit volume, or a total of $N_e \Delta v$ electric dipoles, we can write that

$$\mathbf{p}_t = \sum_{i=1}^{N_e \Delta v} d\mathbf{p}_i \tag{2-2}$$

The *electric polarization vector* **P** can then be defined as the *dipole moment per unit volume*, or

$$\mathbf{P} = \lim_{\Delta v \to 0} \left[\frac{1}{\Delta v} \mathbf{p}_t \right] = \lim_{\Delta v \to 0} \left[\frac{1}{\Delta v} \sum_{i=1}^{N_e \Delta v} d\mathbf{p}_i \right] (\text{C/m}^2) \tag{2-3}$$

The units of **P** are coulomb-meters per cubic meter or coulombs per square meter, which is representative of a surface charge density. It should be noted that this is a *bound* surface charge density (q_{sp}), and it is not permissible to separate the positive and negative charges by an integration surface. Therefore, within a volume, an integral (whole) number of positive and negative pairs (dipoles) with an overall zero net charge must exist. Hence the bound surface charge should *not* be included in (1-45a) or (1-46) to determine the boundary conditions on the normal components of the electric flux density (or normal components of the electric field intensity).

Assuming an average dipole moment of

$$d\mathbf{p}_i = d\mathbf{p}_{av} = Q\ell_{av} \tag{2-4}$$

per molecule, the electric polarization vector of (2-3) can be written, when all dipoles are aligned in the same direction, as

$$\mathbf{P} = \lim_{\Delta v \to 0} \left[\frac{1}{\Delta v} \sum_{i=1}^{N_e \Delta v} d\mathbf{p}_i \right] = N_e d\mathbf{p}_{av} = N_e Q\ell_{av} \tag{2-5}$$

Electric polarization for dielectrics can be produced by any of the following three mechanisms, as demonstrated in Figure 2-4 [2]. Few materials involve all three of the following mechanisms:

1. *Dipole or Orientational Polarization:* This polarization is evident in materials that, in the absence of an applied field and owing to their structure, possess permanent dipole moments that are randomly oriented. However when an electric field is applied, the dipoles tend to align with the applied field. As will be discussed later, such materials are known as *polar* materials; water is a good example.

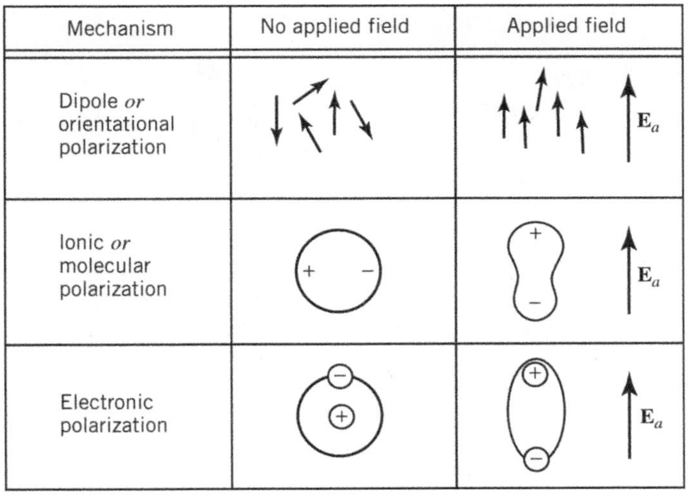

Mechanism	No applied field	Applied field
Dipole *or* orientational polarization		
Ionic *or* molecular polarization		
Electronic polarization		

Figure 2-4   Mechanisms producing electric polarization in dielectrics.

2. *Ionic or Molecular Polarization:* This polarization is evident in materials, such as sodium chloride (NaCl), that possess positive and negative ions and that tend to displace themselves when an electric field is applied.
3. *Electronic Polarization:* This polarization is evident in most materials, and it exists when an applied electric field displaces the electric cloud center of an atom relative to the center of the nucleus.

If the charges in a material, in the absence of an applied electric field $\mathbf{E}_a$, are averaged in such a way that positive and negative charges cancel each other throughout the entire material, then there are no individual dipoles formed and the total dipole moment of (2-2) and the electric polarization vector $\mathbf{P}$ of (2-3) are zero. However, when an electric field is applied, it exhibits a net nonzero polarization. Such a material is referred to as *nonpolar*, and it is illustrated in Figure 2-5a. Polar materials are those whose charges in the absence of an applied electric field $\mathbf{E}_a$ are distributed so that there are individual dipoles formed, each with a dipole moment $\mathbf{p}_i$ as given by (2-1) but with a net total dipole moment $\mathbf{p}_t = 0$ and electric polarization vector $\mathbf{P} = 0$. This is usually a result of the random orientation of the dipoles as illustrated in Figure 2-5b. Typical dipole moments of polar materials are of the order of 10^{-30} C-m. Materials that, in the absence of an applied electric field $\mathbf{E}_a$, possess nonzero net dipole moment and electric polarization vector $\mathbf{P}$ are referred to as *electrets*.

There is also a class of dielectric materials that are usually referred to as *ferroelectrics* [3]. They exhibit a hysteresis loop of polarization (P) versus electric field (E) that is similar to the hysteresis loop of B versus H for ferromagnetic material, and it possesses a *remnant polarization* P_r and *coercive electric field* E_c. At some critical temperature, referred to as *ferroelectric Curie temperature*, the spontaneous polarization in ferroelectrics disappears. Above the Curie temperature the relative permittivity varies according to the Curie–Weiss law; below it the electric flux density D and the polarization P are not linear functions of the electric field E [3]. Barium titanate $(BaTiO_3)$ is one such material.

When an electric field is applied to a nonpolar or polar dielectric material, as shown in Figures 2-5a and 2-5b, the charges in each medium are aligned in such a way that individual dipoles with nonzero dipole moments are formed within the material. However, when we

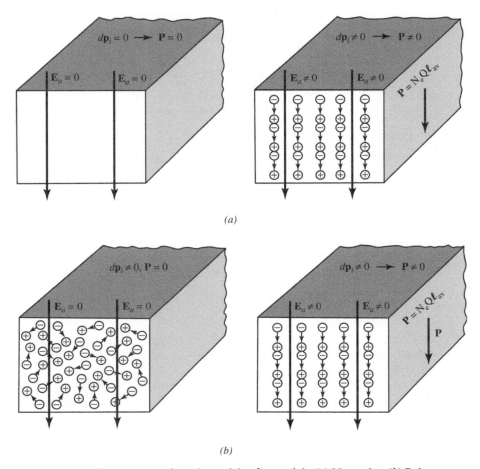

Figure 2-5   Macroscopic scale models of materials. (*a*) Nonpolar. (*b*) Polar.

examine the material on a microscopic scale, the following items become evident from Figures 2-5*a* and 2-5*b*:

1. On the lower surface there exists a net positive surface charge density q_s^+ (representing bound charges).
2. On the upper surface there exists a net negative surface charge density q_s^- (representing bound charges).
3. The volume charge density q_v inside the material is zero because the positive and negative charges of adjacent dipoles cancel each other.

The preceding items can also be illustrated by macroscopically examining Figure 2-6*a* where a d.c. voltage source is connected and remains across two parallel plates separated by distance *s*. Half of the space between the two plates is occupied by a dielectric material, whereas the other half is free space. For a better illustration of this point, let us assume that there are five free charges on each part of the plates separated by free space. The same number appears on the part of the plates separated by the dielectric material. Because of the realignment of the bound charges in the dielectric material and the formation of the electric dipoles and cancellation of adjacent opposite charges shown circled in Figure 2-6*b*, a polarization electric vector **P** is formed within the dielectric material. Thus the polarization vector **P** is a result of the *bound* surface charge

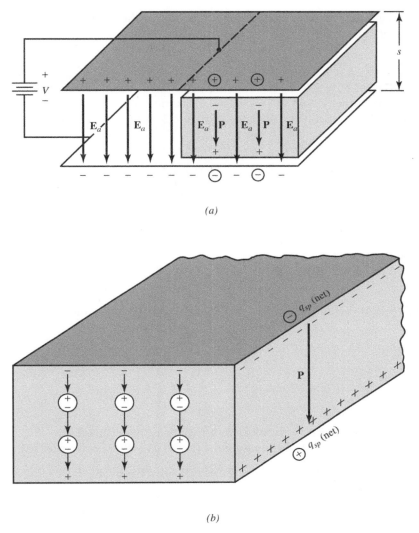

(a)

(b)

Figure 2-6   Dielectric slab subjected to an applied electric field $\mathbf{E}_a$. (a) Total charge. (b) Net charge.

density $-q_{sp}$ found on the upper and $+q_{sp}$ found on the lower surface of the dielectric slab. Let us assume that there are two pairs of bound charges that form the bound surface charge density q_{sp} on the surface of the dielectric slab of Figure 2-6a (negative on top and positive on the bottom). Because the surfaces of the slab are assumed to be in contact with the plates of the capacitor, the two negative bound charges on the top surface will tend to cancel two of the positive free charges on the upper capacitor plate; a similar phenomenon occurs at the bottom. If this were to happen, the net number of charges on the top and bottom plates of the capacitor would diminish to three and the electric field intensity in the dielectric material between the plates would be reduced. Since the d.c. voltage supply is maintained across the plates, the net charge on the upper and lower parts of the capacitor and the electric field intensity in the dielectric material between the plates are also maintained by the introduction of two additional free charges on each of the capacitor plates (positive on top and negative on bottom). For identification purposes, these two induced free charges have been circled in each of the two plates in Figure 2-6.

    In each of the situations discussed previously, the net effect is that between the lower and upper surfaces of the dielectric there is a net electric polarization vector **P** directed from the upper toward the lower surfaces, in the same direction as the applied electric field $\mathbf{E}_a$, whose amplitude is given by

$$P = q_{sp} \tag{2-6}$$

Whereas the applied electric field $\mathbf{E}_a$ maintains its value, the electric flux density inside the dielectric material differs from what would exist were the dielectric material replaced by free space. In the free-space part of the parallel plate capacitor of Figure 2-6, the electric flux density $\mathbf{D}_0$ is given by

$$\mathbf{D}_0 = \varepsilon_0 \mathbf{E}_a \tag{2-7}$$

In the dielectric portion, the electric flux density $\mathbf{D}$ is related to that in free space $\mathbf{D}_0$ by

$$\mathbf{D} = \varepsilon_0 \mathbf{E}_a + \mathbf{P} \tag{2-8}$$

where the magnitude of $\mathbf{P}$ is given by (2-6). The electric flux density $\mathbf{D}$ of (2-8) can also be related to the applied electric field intensity $\mathbf{E}_a$ by a parameter that we designate here as ε_s (farads/meter). Thus we can write that

$$\mathbf{D} = \varepsilon_s \mathbf{E}_a \tag{2-9}$$

Comparing (2-8) and (2-9), it is apparent that $\mathbf{P}$ is also related to $\mathbf{E}_a$ and can be expressed as

$$\mathbf{P} = \varepsilon_0 \chi_e \mathbf{E}_a \tag{2-10}$$

or

$$\chi_e = \frac{1}{\varepsilon_0} \frac{P}{E_a} \tag{2-10a}$$

where χ_e is referred to as the *electric susceptibility* (dimensionless quantity).

Substituting (2-10) into (2-8) and equating the result to (2-9), we can write that

$$\mathbf{D} = \varepsilon_0 \mathbf{E}_a + \varepsilon_0 \chi_e \mathbf{E}_a = \varepsilon_0 \left(1 + \chi_e\right) \mathbf{E}_a = \varepsilon_s \mathbf{E}_a \tag{2-11}$$

or that

$$\varepsilon_s = \varepsilon_0 \left(1 + \chi_e\right) \tag{2-11a}$$

In (2-11a) ε_s is the *static permittivity* of the medium whose relative value ε_{sr} (compared to that of free space ε_0) is given by

$$\varepsilon_{sr} = \frac{\varepsilon_s}{\varepsilon_0} = 1 + \chi_e \tag{2-12}$$

which is usually referred to as the *relative permittivity*, better known in practice as the *dielectric constant*. Scientists and engineers usually designate the square root of the relative permittivity as the *index of refraction*. Typical values of dielectric constants at static frequencies of some prominent dielectric materials are listed in Table 2-1.

Thus the dielectric constant of a dielectric material is a parameter that indicates the relative (compared to free space) charge (energy) storage capabilities of a dielectric material; the larger its value, the greater its ability to store charge (energy). Parallel plate capacitors utilize dielectric material between their plates to increase their charge (energy) storage capacity by forcing extra free charges to be induced on the plates. These free charges neutralize the bound charges on the surface of the dielectric so that the voltage and electric field intensity are maintained constant between the plates.

TABLE 2-1 Approximate static dielectric constants (relative permittivities) of dielectric materials

Material	Static dielectric constant (ε_r)
Air	1.0006
Styrofoam	1.03
Paraffin	2.1
Teflon	2.1
Plywood	2.1
RT/duroid 5880	2.20
Polyethylene	2.26
RT/duroid 5870	2.35
Glass-reinforced teflon (microfiber)	2.32–2.40
Teflon quartz (woven)	2.47
Glass-reinforced teflon (woven)	2.4–2.62
Cross-linked polystyrene (unreinforced)	2.56
Polyphenelene oxide (PPO)	2.55
Glass-reinforced polystyrene	2.62
Amber	3
Soil (dry)	3
Rubber	3
Plexiglas	3.4
Lucite	3.6
Fused silica	3.78
Nylon (solid)	3.8
Quartz	3.8
Sulfur	4
Bakelite	4.8
Formica	5
Lead glass	6
Mica	6
Beryllium oxide (BeO)	6.8–7.0
Marble	8
Sapphire	$\varepsilon_x = \varepsilon_y = 9.4$ $\varepsilon_z = 11.6$
Flint glass	10
Ferrite (Fe_2O_3)	12–16
Silicon (Si)	12
Gallium arsenide (GaAs)	13
Ammonia (liquid)	22
Glycerin	50
Water	81
Rutile (TiO_2)	$\varepsilon_x = \varepsilon_y = 89$ $\varepsilon_z = 173$

Example 2-1

The static dielectric constant of water is 81. Assuming the electric field intensity applied to water is 1 V/m, determine the magnitudes of the electric flux density and electric polarization vector within the water.

Solution: Using (2-9), we have

$$D = \varepsilon_s E_a = 81(8.854 \times 10^{-12})(1) = 7.17 \times 10^{-10} \ C/m^2$$

Using (2-12), we have

$$\chi_e = \varepsilon_{sr} - 1 = 81 - 1 = 80$$

Thus the magnitude of the electric polarization vector is given, using (2-10), by

$$P = \varepsilon_0 \chi_e E_a = 8.854 \times 10^{-12}(80)(1) = 7.08 \times 10^{-10} \ C/m^2$$

The permittivity of (2-11a), or its relative form of (2-12), represents values at static or quasistatic frequencies. These values vary as a function of the alternating field frequency. The variations of the permittivity as a function of the frequency of the applied fields are examined in Section 2.9.1.

2.3   MAGNETICS, MAGNETIZATION, AND PERMEABILITY

Magnetic materials are those that exhibit magnetic polarization when they are subjected to an applied magnetic field. The magnetization phenomenon is represented by the alignment of the magnetic dipoles of the material with the applied magnetic field, similar to the alignment of the electric dipoles of the dielectric material with the applied electric field.

Accurate results concerning the behavior of magnetic material when they are subjected to applied magnetic fields can only be predicted by the use of quantum theory. This is usually quite complex and unnecessary for most engineering applications. Quite satisfactory quantitative results can be obtained, however, by using simple atomic models to represent the atomic lattice structure of the material. The atomic models used here represent the electrons as negative charges orbiting around the positively-charged nucleus, as shown in Figure 2-7a. Each orbiting electron can be modeled by an equivalent small electric current loop of area ds whose current flows in the direction opposite to the electron orbit, as shown in Figure 2-7b. As long as the loop is small, its shape can be circular, square, or any other configuration, as shown in Figures 2-7b and 2-7c. The fields produced by a small loop of electric current at large distances are the same as those produced by a linear bar magnet (magnetic dipole) of length d.

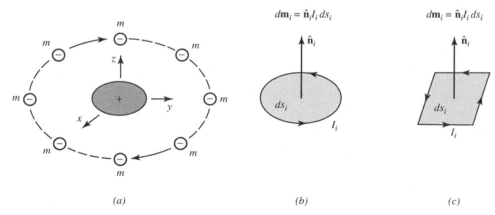

(a)   (b)   (c)

Figure 2-7   Atomic models and their equivalents, representing the atomic lattice structure of magnetic material. (*a*) Orbiting electrons. (*b*) Equivalent circular electric loop. (*c*) Equivalent square electric loop.

By referring to the equivalent loop models of Figure 2-7, the angular momentum associated with an orbiting electron can be represented by a magnetic dipole moment $d\mathbf{m}_i$ of

$$d\mathbf{m}_i = I_i d\mathbf{s}_i = I_i \hat{\mathbf{n}}_i ds_i = \hat{\mathbf{n}}_i I_i ds_i \ (\text{A-m}^3) \tag{2-13}$$

For atoms that possess many orbiting electrons, the total magnetic dipole moment $\mathbf{m}_t$ is equal to the vector sum of all the individual magnetic dipole moments, each represented by (2-13). Thus we can write that

$$\mathbf{m}_t = \sum_{i=1}^{N_m \Delta v} d\mathbf{m}_i = \sum_{i=1}^{N_m \Delta v} \hat{\mathbf{n}}_i I_i \, ds_i \tag{2-14}$$

where N_m is equal to the number of orbiting electrons (equivalent loops) per unit volume. A magnetic polarization (magnetization) vector $\mathbf{M}$ is then defined as

$$\mathbf{M} = \lim_{\Delta v \to 0}\left[\frac{1}{\Delta v}\mathbf{m}_t\right] = \lim_{\Delta v \to 0}\left[\frac{1}{\Delta v}\sum_{i=1}^{N_m \Delta v} d\mathbf{m}_i\right] = \lim_{\Delta v \to 0}\left[\frac{1}{\Delta v}\sum_{i=1}^{N_m \Delta v}\hat{\mathbf{n}}_i I_i \, ds_i\right] (\text{A/m}) \tag{2-15}$$

Assuming for each of the loops an average magnetic moment of

$$d\mathbf{m}_i = d\mathbf{m}_{\text{av}} = \hat{\mathbf{n}}(Ids)_{\text{av}} \tag{2-16}$$

the magnetic polarization vector $\mathbf{M}$ of (2-15) can be written (assuming all the loops are aligned in the parallel planes) as

$$\mathbf{M} = \lim_{\Delta v \to 0}\left[\frac{1}{\Delta v}\sum_{i=1}^{N_m \Delta v} d\mathbf{m}_i\right] = N_m \, d\mathbf{m}_{\text{av}} = \hat{\mathbf{n}} N_m (Ids)_{\text{av}} \tag{2-17}$$

A magnetic material is represented by a number of magnetic dipoles and thus by many magnetic moments. In the absence of an applied magnetic field the magnetic dipoles and their corresponding electric loops are oriented in a random fashion so that on a macroscopic scale the vector sum of the magnetic moments of (2-14) and the magnetic polarization of (2-15) are equal to zero. The random orientation of the magnetic dipoles and loops is illustrated in Figure 2-8a. When the

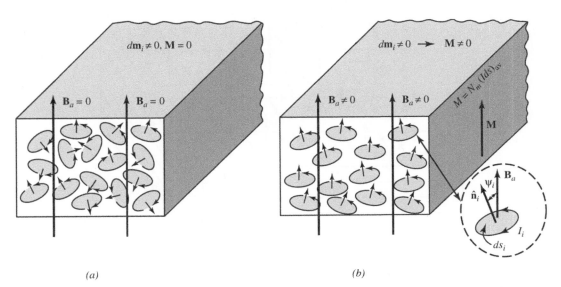

(a)   (b)

Figure 2-8 Orientation and alignment of magnetic dipoles. (a) Random in absence of an applied field. (b) Aligned under an applied field.

magnetic material is subjected to an applied magnetic field, represented by the magnetic flux density $\mathbf{B}_a$ in Figure 2-8b, the magnetic dipoles of most material will tend to align in the direction of the $\mathbf{B}_a$ since a torque given by

$$\left|\Delta\mathbf{T}\right|=\left|d\mathbf{m}_i\times\mathbf{B}_a\right|=\left|d\mathbf{m}_i\right|\left|\mathbf{B}_a\right|\sin(\psi_i)=\left|(\hat{\mathbf{n}}_i\,I_i\,ds_i)\times\mathbf{B}_a\right|=\left|I_i\,ds_i\,B_a\sin(\psi_i)\right| \qquad (2\text{-}18)$$

will be exerted in each of the magnetic dipole moments. This is shown in the inset to Figure 2-8b. Ideally, if there were no other magnetic moments to consider, torque would be exerted. The torque would exist until each of the orbiting electrons shifted in such a way that the magnetic field produced by each of its equivalent electric loops (or magnetic moments) was aligned with the applied field, and its value, represented by (2-18), vanished. Thus the resultant magnetic field at every point in the material would be greater than its corresponding value at the same point when the material is absent.

The magnetization vector $\mathbf{M}$ resulting from the realignment of the magnetic dipoles is better illustrated by considering a slab of magnetic material across which a magnetic field $\mathbf{B}_a$ is applied, as shown in Figure 2-9. Ideally, on a microscopic scale, for most magnetic material all the magnetic dipoles will align themselves so that their individual magnetic moments are pointed in the direction of the applied field, as shown in Figure 2-9. In the limit, as the number of magnetic dipoles and their corresponding equivalent electric loops become very large, the currents of the loops found in the interior parts of the slab are canceled by those of the neighboring loops. On a macroscopic scale a net nonzero equivalent magnetic current, resulting in an equivalent magnetic current surface density (A/m), is found on the exterior surface of the slab. This equivalent magnetic current density $\mathbf{J}_{ms}$ is responsible for the introduction of the magnetization vector $\mathbf{M}$ in the direction of $\mathbf{B}_a$.

The magnetic flux density across the slab is increased by the presence of $\mathbf{M}$ so that the net magnetic flux density at any interior point of the slab is given by

$$\mathbf{B}=\mu_0(\mathbf{H}_a+\mathbf{M}) \qquad (2\text{-}19)$$

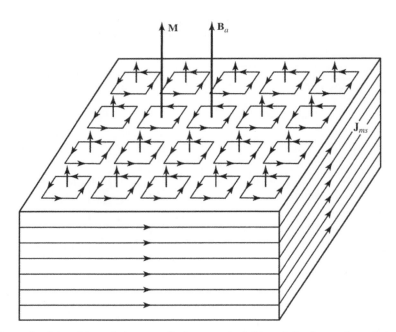

Figure 2-9 Magnetic slab subjected to an applied magnetic field and the formation of the magnetization current density $\mathbf{J}_{ms}$.

It should be pointed out that $\mathbf{M}$, as given by (2-15), has the units of amperes per meter and corresponds to those of the magnetic field intensity. In general, we can relate the magnetic flux density to the magnetic field intensity by a parameter that is designated as μ_s (henries/meter). Thus we can write that

$$\mathbf{B} = \mu_s \mathbf{H}_a \qquad (2\text{-}20)$$

Comparing (2-19) and (2-20) indicates that $\mathbf{M}$ is also related to $\mathbf{H}_a$ by

$$\mathbf{M} = \chi_m \mathbf{H}_a \qquad (2\text{-}21)$$

where χ_m is called the *magnetic susceptibility* (dimensionless quantity).

Substituting (2-21) into (2-19) and equating the result to (2-20) leads to

$$\mathbf{B} = \mu_0(\mathbf{H}_a + \chi_m \mathbf{H}_a) = \mu_0(1 + \chi_m)\mathbf{H}_a = \mu_s \mathbf{H}_a \qquad (2\text{-}22)$$

Therefore we can define

$$\mu_s = \mu_0(1 + \chi_m) \qquad (2\text{-}22a)$$

In (2-22a) μ_s is the *static permeability* of the medium whose relative value μ_{sr} (compared to that of free space μ_0) is given by

$$\mu_{sr} = \frac{\mu_s}{\mu_0} = 1 + \chi_m \qquad (2\text{-}23)$$

Static values of μ_{sr} for some representative materials are listed in Table 2-2.

Within the material, a *bound* magnetic current density $\mathbf{J}_m$ is induced that is related to the magnetic polarization vector $\mathbf{M}$ by

$$\mathbf{J}_m = \nabla \times \mathbf{M} \left(A/m^2 \right) \qquad (2\text{-}24)$$

TABLE 2-2   Approximate static relative permeabilities of magnetic materials

Material	Class	Relative permeability (μ_{sr})
Bismuth	Diamagnetic	0.999834
Silver	Diamagnetic	0.99998
Lead	Diamagnetic	0.999983
Copper	Diamagnetic	0.999991
Water	Diamagnetic	0.999991
Vacuum	Nonmagnetic	1.0
Air	Paramagnetic	1.0000004
Aluminum	Paramagnetic	1.00002
Nickel chloride	Paramagnetic	1.00004
Palladium	Paramagnetic	1.0008
Cobalt	Ferromagnetic	250
Nickel	Ferromagnetic	600
Mild steel	Ferromagnetic	2,000
Iron	Ferromagnetic	5,000
Silicon iron	Ferromagnetic	7,000
Mumetal	Ferromagnetic	100,000
Purified iron	Ferromagnetic	200,000
Supermalloy	Ferromagnetic	1,000,000

To account for this current density, we modify the Maxwell–Ampere equation 1-71b and write it as

$$\mathbf{\nabla}\times\mathbf{H} = \mathbf{J}_i + \mathbf{J}_c + \mathbf{J}_m + \mathbf{J}_d = \mathbf{J}_i + \sigma\mathbf{E} + \mathbf{\nabla}\times\mathbf{M} + j\omega\varepsilon\mathbf{E} \qquad (2\text{-}24a)$$

On the surface of the material, the *bound* magnetization surface current density $\mathbf{J}_{ms}$ is related to the magnetic polarization vector $\mathbf{M}$ at the surface by

$$\mathbf{J}_{ms} = \mathbf{M}\times\hat{\mathbf{n}}\,|_{\text{surface}}\ (\text{A/m}) \qquad (2\text{-}25)$$

where $\hat{\mathbf{n}}$ is a unit vector normal to the surface of the material. The *bound* magnetization current I_m flowing through a cross section S_0 of the material can be obtained by using

$$I_m = \iint_{S_0}\mathbf{J}_m\cdot d\mathbf{s} = \iint_{S_0}(\mathbf{\nabla}\times\mathbf{M})\cdot d\mathbf{s}\ (\text{A}) \qquad (2\text{-}26)$$

In addition to orbiting, the electrons surrounding the nucleus of an atom also spin about their own axis. Therefore magnetic moments of the order of $\pm 9\times10^{-24}$ A-m^2 are also associated with the spinning of the electrons that aid or oppose the applied magnetic field (the $+$ sign is used for addition and the $-$ for subtraction). For atoms that have many electrons in their shells, only the spins associated with the electrons found in shells that are not completely filled will contribute to the magnetic moment of the atoms. A third contributor to the total magnetic moment of an atom is that associated with the spinning of the nucleus, which is referred to as *nuclear spin*. However, this nuclear spin magnetic moment is usually much smaller (typically by a factor of about 10^{-3}) than those attributed to the orbiting and the spinning electrons.

Example 2-2

A bar of magnetic material of finite length, which is placed along the z axis, as shown in Figure 2-10, has a cross section of 0.3 m in the x direction ($0 \le x \le 0.3$) and 0.2 m in the y direction ($0 \le y \le 0.2$). The bar is subjected to a magnetic field so that the magnetization vector inside the bar is given by

$$\mathbf{M} = \hat{\mathbf{a}}_z(4y)$$

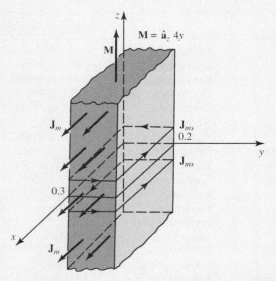

Figure 2-10   Magnetic bar of rectangular cross section subjected to a magnetic field.

Determine the volumetric current density $\mathbf{J}_m$ at any point inside the bar, the surface current density $\mathbf{J}_{ms}$ on the surface of each of the four faces, and the total current I_m per unit length flowing through the bar face that is parallel to the y axis at $x = 0.3$ m.

Solution: Using (2-24), we have

$$\mathbf{J}_m = \nabla \times \mathbf{M} = \hat{\mathbf{a}}_x \frac{\partial M_z}{\partial y} = \hat{\mathbf{a}}_x 4$$

Using (2-25), we have

$$\mathbf{J}_{ms} = \mathbf{M} \times \hat{\mathbf{n}}|_{\text{surface}}$$

Therefore at

$x = 0$:
$$\mathbf{J}_{ms} = (\hat{\mathbf{a}}_z 4y) \times (-\hat{\mathbf{a}}_x)|_{x=0} = -\hat{\mathbf{a}}_y(4y) \qquad \text{for } 0 \le y \le 0.2$$
$y = 0$:
$$\mathbf{J}_{ms} = (\hat{\mathbf{a}}_z 4y) \times (-\hat{\mathbf{a}}_y)|_{y=0} = \hat{\mathbf{a}}_x(4y) = 0 \qquad \text{for } 0 \le x \le 0.3$$
$x = 0.3$:
$$\mathbf{J}_{ms} = (\hat{\mathbf{a}}_z 4y) \times (\hat{\mathbf{a}}_x)|_{x=0.3} = \hat{\mathbf{a}}_y(4y) \qquad \text{for } 0 \le y \le 0.2$$
$y = 0.2$:
$$\mathbf{J}_{ms} = (\hat{\mathbf{a}}_z 4y) \times (\hat{\mathbf{a}}_y)|_{y=0.2} = -\hat{\mathbf{a}}_x(4y) = -\hat{\mathbf{a}}_x 0.8 \quad \text{for } 0 \le x \le 0.3$$

According to (2-26), the current (per unit length) flowing through the bar face at $x = 0.3$ is given by

$x = 0.3$:
$$I_m = \iint_S \mathbf{J}_m \cdot d\mathbf{s} = \int_0^1 \int_0^{0.2} (\hat{\mathbf{a}}_x 4) \cdot (\hat{\mathbf{a}}_x\, dy\, dz) = 4(1)(0.2) = 0.8$$

Consistent with the relative permittivity (dielectric constant), the values of μ, and thus μ_r, vary as a function of frequency. These variations will be discussed in Section 2.9.2. The values of μ_r listed in Table 2-2 are representative of frequencies related to static or quasistatic fields. Excluding ferromagnetic material, it is apparent that most relative permeabilities are very near unity, so that for engineering problems a value of unity is almost always used.

According to the direction in which the net magnetization vector $\mathbf{M}$ is pointing (either aiding or opposing the applied magnetic field), material are classified into two groups, Group A and Group B as shown:

Group A	Group B
Diamagnetic	Paramagnetic
	Ferromagnetic
	Antiferromagnetic
	Ferrimagnetic

In general, for material in Group A the net magnetization vector (although small in magnitude) opposes the applied magnetic field, resulting in a relative permeability slightly smaller than unity. *Diamagnetic* materials fall into that group. For material in Group B the net magnetization vector is aiding the applied magnetic field, resulting in relative permeabilities greater than unity. Some of them (*paramagnetic* and *antiferromagnetic*) have only slightly greater than unity relative permeabilities whereas others (*ferromagnetic* and *ferrimagnetic*) have relative permeabilities much greater than unity.

In the absence of an applied magnetic field, the moments of the electron spins of *diamagnetic* material are opposite to each other as well as to the moments associated with the orbiting electrons so that a zero net magnetic moment m_t is produced on a macroscopic scale. In the presence of an external applied magnetic field, each atom has a net nonzero magnetic moment, and on a macroscopic scale there is a net total magnetic moment for all the atoms that results in a magnetization vector **M**. For diamagnetic material, this vector **M** is very small, opposes the applied magnetic field, leads to a negative magnetic susceptibility χ_m, and results in values of relative permeability that are slightly less than unity. For example, copper is a diamagnetic material with a magnetic susceptibility $\chi_m = -9 \times 10^{-6}$ and a relative permeability $\mu_r = 0.999991$.

In *paramagnetic* material, the magnetic moments associated with the orbiting and spinning electrons of an atom do not quite cancel each other in the absence of an applied magnetic field. Therefore each atom possesses a small magnetic moment. However, because the orientation of the magnetic moment of each atom is random, the net magnetic moment of a large sample (macroscopic scale) of dipoles, and the magnetization vector **M**, are zero when there is no applied field. When the paramagnetic material is subjected to an applied magnetic field, the magnetic dipoles align slightly with the applied field to produce a small nonzero **M** in its direction and a small increase in the magnetic flux density within the material. Thus the magnetic susceptibilities have small positive values and the relative permeabilities are slightly greater than unity. For example, aluminum possesses a susceptibility of $\chi_m = 2 \times 10^{-5}$ and a relative permeability of $\mu_r = 1.00002$.

The individual atoms of *ferromagnetic* material possess, in the absence of an applied magnetic field, very strong magnetic moments caused primarily by uncompensated electron spin moments. The magnetic moments of many atoms (usually as many as five to six) reinforce one another and form regions called *domains*, which have various sizes and shapes. The dimensions of the domains depend on the material's past magnetic state and history, and range from 1μm to a few millimeters. On a macroscopic scale, however, the net magnetization vector **M** in the absence of an applied field is zero because the domains are randomly oriented and the magnetic moments of the various atoms cancel one another. When a ferromagnetic material is subjected to an applied field, there are not only large magnetic moments associated with the individual atoms, but the vector sum of all the magnetic moments and the associated vector magnetization **M** are very large, leading to extreme values of magnetic susceptibility χ_m and relative permeability. Typical values of μ_r for some representative ferromagnetic materials are found in Table 2-2. When the applied field is removed, the magnetic moments of the various atoms do not attain a random orientation and a net nonzero residual magnetic moment remains. Since the magnetic moment of a ferromagnetic material on a macroscopic scale is different after the applied field is removed, its magnetic state depends on the material's past history. Therefore a plot of the magnetic flux density $\mathcal{B}$ versus $\mathcal{H}$ leads to a double-valued curve known as the *hysteresis loop*. Material with such properties are very desirable in the design of transformers, induction cores, and coatings for magnetic recording tapes.

Materials that possess strong magnetic moments, but whose adjacent atoms are about equal in magnitude and opposite in direction, with zero net total magnetic moment in the absence of an applied magnetic field, are called *antiferromagnetic*. The presence of an applied magnetic field has a minor effect on the material and leads to relative permeabilities slightly greater than unity.

If the adjacent opposing magnetic moments of a material are very large in magnitude but greatly unequal in the absence of an applied magnetic field, the material is known as *ferrimagnetic*. The presence of an applied magnetic field has a large effect on the material and leads to large permeabilities (but not as large as those of ferromagnetic material). *Ferrites* make up a group of ferrimagnetic materials that have low conductivities (several orders smaller than those of semiconductors). Because of their large resistances, smaller currents are induced in them that result in lower ohmic losses when they are subjected to alternating fields. They find wide applications in the design of nonreciprocal microwave components (isolators, hybrids, gyrators, phase shifters, etc.) and they will be discussed briefly in Section 2.9.3.

2.4 CURRENT, CONDUCTORS, AND CONDUCTIVITY

The prominent characteristic of dielectric materials is the electric polarization introduced through the formation of electric dipoles between opposite charges of atoms. Magnetic dipoles, modeled by equivalent small electric loops, were introduced to account for the orbiting of electrons in atoms of magnetic material. This phenomenon was designated as magnetic polarization. Conductors are materials whose prominent characteristic is the motion of electric charges and the creation of a current flow.

2.4.1 Current

Let us assume that an electric volume charge density, represented here by q_v, is distributed uniformly in an infinitesimal circular cylinder of cross-sectional area Δs and volume ΔV, as shown in Figure 2-11. The total electric charge ΔQ_e within the volume ΔV is moving in the z direction with a uniform velocity v_z. Thus we can write that

$$\frac{\Delta Q_e}{\Delta t} = q_v \frac{\Delta V}{\Delta t} = q_v \frac{\Delta s \Delta z}{\Delta t} = q_v \Delta s \frac{\Delta z}{\Delta t} \tag{2-27}$$

In the limit as $\Delta t \to 0$, (2-27) is used to define the current ΔI (with units of amperes) that flows through Δs. Thus

$$\Delta I = \lim_{\Delta t \to 0} \left[\frac{\Delta Q_e}{\Delta t} \right] = \lim_{\Delta t \to 0} \left[q_v \Delta s \frac{\Delta z}{\Delta t} \right] = q_v v_z \Delta s \tag{2-28}$$

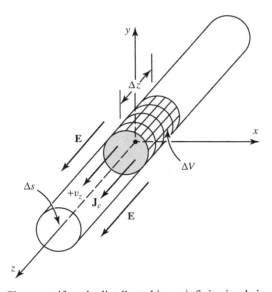

Figure 2-11 Charge uniformly distributed in an infinitesimal circular cylinder.

Dividing both sides of (2-28) by Δs and taking the limit as $\Delta s \to 0$, we can define the current density J_z (with units of amperes per square meter) as

$$J_z = \lim_{\Delta s \to 0} \left[\frac{\Delta I}{\Delta s} \right] = q_v v_z \qquad (2\text{-}29)$$

Using a similar procedure for the x- and y-directed currents, we can write in general that

$$\mathbf{J} = q_v \mathbf{v} \ (\text{A/m}^2) \qquad (2\text{-}30)$$

In (2-30), $\mathbf{J}$ is defined as the *convection current density*. The current density between the cathode and anode of a vacuum tube is a convection current density. It should be noted that for an electric field intensity of $\mathbf{E} = \hat{\mathbf{a}}_z E_z$, a positive charge density $+q_v$ will experience a force that will move it in the $+z$ direction. Thus the current density $\mathbf{J}$ will be directed in the $+z$ direction or

$$\mathbf{J} = +q_v(+\hat{\mathbf{a}}_z v_z) = \hat{\mathbf{a}}_z q_v v_z \qquad (2\text{-}31)$$

If the same electric field $\mathbf{E} = \hat{\mathbf{a}}_z E_z$ is subjected to a negative charge density $-q_v$, the field will force the negative charge to move in the negative z direction ($\mathbf{v} = -\hat{\mathbf{a}}_z v_z$). However, the electric current density $\mathbf{J}$ is still directed along the $+z$ direction,

$$\mathbf{J} = -q_v(-\hat{\mathbf{a}}_z v_z) = \hat{\mathbf{a}}_z q_v v_z \qquad (2\text{-}32)$$

since both the charge density and the velocity are negative. If positive (q_v^+) and negative (q_v^-) charges are present, (2-30) can be written as

$$\mathbf{J} = q_v^+ \mathbf{v}^+ + q_v^- \mathbf{v}^- \qquad (2\text{-}33)$$

2.4.2 Conductors

Conductors are material whose atomic outer shell (valence) electrons are not held very tightly and can migrate from one atom to another. These are known as *free electrons*, and for metal conductors they are very large in number. With no applied external field, these free electrons move with different velocities in random directions producing zero net current through the surface of the conductor.

When free charge q_{v0} is placed inside a conductor that is subjected to a static field, the charge density at that point decays exponentially as

$$q_v(t) = q_{v0} e^{-t/t_r} = q_{v0} e^{-(\sigma/\varepsilon)t} \qquad (2\text{-}34)$$

because the charge migrates toward the surface of the conductor. The time it takes for this to occur depends on the conductivity of the material; for metals it is equal to a few time constants. During this time, charges move, currents flow, and nonstatic conditions exist. The time t_r that it takes for the free charge density placed inside a conductor to decay to $e^{-1} = 0.368$, or 36.8 percent of its initial value, is known as the *relaxation time constant*. Mathematically it is represented by

$$t_r = \frac{\varepsilon}{\sigma} \qquad (2\text{-}35)$$

where
   $\varepsilon =$ permittivity of conductor (F/m)
   $\sigma =$ conductivity of conductor (S/m) (see equation (2-39))

Example 2-3

Find the relaxation time constant for a metal such as copper ($\sigma = 5.76 \times 10^7$ S/m, $\varepsilon = \varepsilon_0$) and a good dielectric such as glass ($\sigma \simeq 10^{-12}$ S/m, $\varepsilon = 6\varepsilon_0$).

Solution: For copper

$$t_r = \frac{\varepsilon}{\sigma} = \frac{8.854 \times 10^{-12}}{5.76 \times 10^7} = 1.54 \times 10^{-19}\,\text{s}$$

which is very short. For glass

$$t_r = \frac{\varepsilon}{\sigma} = 6\left(\frac{8.854 \times 10^{-12}}{10^{-12}}\right) = 53.1\,\text{s} \simeq 1\,\text{min}$$

which is comparatively quite long.

The free charges of a very good conductor ($\sigma \to \infty$), which is subjected to an electric field, migrate very rapidly and distribute themselves as surface charge density q_s to the surface of the conductor within an extremely short period of time (several very short relaxation time constants). The surface charge density q_s will induce on the conductor an electric field intensity $\mathbf{E}_i$, so that the total electric field $\mathbf{E}_t$, within the conductor ($\mathbf{E}_i + \mathbf{E}_a = \mathbf{E}_t$, where $\mathbf{E}_a$ is the applied field) is essentially zero. This is illustrated in Figure 2-12. For perfect conductors ($\sigma = \infty$) the electric field within the conductors is exactly zero.

2.4.3 Conductivity

When a conductor is subjected to an electric field, the electrons still move in random directions but drift slowly (with a drift velocity $\mathbf{v}_e$) in the negative direction of the applied electric field, thus

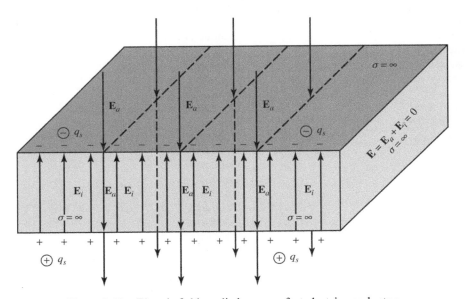

Figure 2-12   Electric field applied on a perfect electric conductor.

TABLE 2-3   Typical conductivities of insulators, semiconductors, and conductors

Material	Class	Conductivity σ (S/m)
Fused quartz	Insulator	$\sim 10^{-17}$
Ceresin wax	Insulator	$\sim 10^{-17}$
Sulfur	Insulator	$\sim 10^{-15}$
Mica	Insulator	$\sim 10^{-15}$
Paraffin	Insulator	$\sim 10^{-15}$
Hard rubber	Insulator	$\sim 10^{-15}$
Porcelain	Insulator	$\sim 10^{-14}$
Glass	Insulator	$\sim 10^{-12}$
Bakelite	Insulator	$\sim 10^{-9}$
Distilled water	Insulator	$\sim 10^{-4}$
Gallium arsenide (GaAs)*	Semiconductor	$\sim 2.38 \times 10^{-7}$
Fused silica*	Semiconductor	$\sim 2.1 \times 10^{-4}$
Cross-linked polystyrene (unreinforced)*	Semiconductor	$\sim 3.7 \times 10^{-4}$
Beryllium oxide (BeO)*	Semiconductor	$\sim 3.9 \times 10^{-4}$
Intrinsic silicon	Semiconductor	$\sim 4.39 \times 10^{-4}$
Sapphire*	Semiconductor	$\sim 5.5 \times 10^{-4}$
Glass-reinforced Teflon (microfiber)*	Semiconductor	$\sim 7.8 \times 10^{-4}$
Teflon quartz (woven)*	Semiconductor	$\sim 8.2 \times 10^{-4}$
Dry soil	Semiconductor	$\sim 10^{-4} - 10^{-3}$
Ferrite(Fe$_2$O$_3$)*	Semiconductor	$\sim 1.3 \times 10^{-3}$
Glass-reinforced polystyrene*	Semiconductor	$\sim 1.45 \times 10^{-3}$
Polyphenelene oxide (PPO)*	Semiconductor	$\sim 2.27 \times 10^{-3}$
Glass-reinforced Teflon (woven)*	Semiconductor	$\sim 2.43 \times 10^{-3}$
Plexiglas*	Semiconductor	$\sim 5.1 \times 10^{-3}$
Wet soil	Semiconductor	$\sim 10^{-3} - 10^{-2}$
Fresh water	Semiconductor	$\sim 10^{-2}$
Human and animal tissue	Semiconductor	$\sim 0.2 - 0.7$
Intrinsic germanium	Semiconductor	~ 2.227
Seawater	Semiconductor	~ 4
Tellurium	Conductor	$\sim 5 \times 10^{-2}$
Carbon	Conductor	$\sim 3 \times 10^{-4}$
Graphite	Conductor	$\sim 3 \times 10^{4}$
Cast iron	Conductor	$\sim 10^{6}$
Mercury	Conductor	10^{6}
Nichrome	Conductor	10^{6}
Silicon steel	Conductor	2×10^{6}
German silver	Conductor	2×10^{6}
Lead	Conductor	5×10^{6}
Tin	Conductor	9×10^{6}
Iron	Conductor	1.03×10^{7}
Nickel	Conductor	1.45×10^{7}
Zinc	Conductor	1.7×10^{7}
Tungsten	Conductor	1.83×10^{7}
Brass	Conductor	2.56×10^{7}
Aluminum	Conductor	3.96×10^{7}
Gold	Conductor	4.1×10^{7}
Copper	Conductor	5.76×10^{7}
Silver	Conductor	6.1×10^{7}

*For most semiconductors the conductivities are representative for a frequency of about 10 GHz.

creating a conduction current in the conductor. The applied electric field $\mathbf{E}$ and drift velocity $\mathbf{v}_e$ of the electrons are related by

$$\mathbf{v}_e = -\mu_e \mathbf{E} \tag{2-36}$$

where μ_e is defined to be the *electron mobility* [positive quantity with units of $m^2/(V\text{-}s)$]. Substituting (2-36) into (2-30), we can write that

$$\mathbf{J} = q_{ve}\mathbf{v}_e = q_{ve}(-\mu_e\mathbf{E}) = -q_{ve}\mu_e\mathbf{E} \tag{2-37}$$

where q_{ve} is the electron charge density. Comparing (2-37) with (1-16), or

$$\mathbf{J} = \sigma_s\mathbf{E} \tag{2-38}$$

we define the static conductivity of a conductor as

$$\sigma_s = -q_{ve}\mu_e \ (S/m) \tag{2-39}$$

Its reciprocal value is called the *resistivity* (ohm-meters).

The conductivity σ_s of a conductor is a parameter that characterizes the free-electron conductive properties of a conductor. As temperature increases, the increased thermal energy of the conductor lattice structure increases the lattice vibration. Thus the possibility of the moving free electrons colliding increases, which results in a decrease in the conductivity of the conductor. Materials with a very low value of conductivity are classified as *dielectrics* (*insulators*). The conductivity of ideal dielectrics is zero.

The conductivity of (2-39) is referred to as the static or d.c. conductivity; typical values of several materials are listed in Table 2-3. The conductivity varies as a function of frequency. These variations, along with the mechanisms that result in them, will be discussed in Section 2.8.1.

2.5 SEMICONDUCTORS

Materials whose conductivities bridge the gap between dielectrics (insulators) and conductors (typically the conductivity being 10^{-3} to unity) are referred to as *intrinsic* (pure) semiconductors. A graph illustrating the range of conductivities, from insulators to conductors, is displayed in Figure 2-13. Two such materials of significant importance to electrical engineering are intrinsic *germanium* and intrinsic *silicon*. In intrinsic (pure) semiconductors there are two common carriers: the *free electrons* and the *bound electrons* (referred to as positive *holes*) [4].

As the temperature rises, the mobilities of semiconducting material decrease but their charge densities increase more rapidly. The increases in the charge density more than offset the decreases in mobilities, resulting in a general increase in the conductivity of semiconducting material with rises in temperature. This is one of the characteristic differences between intrinsic semiconductors and metallic conductors: for semiconductors the conductivity increases with rising temperature whereas for metallic conductors it decreases. Typically the conductivity of germanium will increase by a factor of 10 as the temperature increases from 300 to about 360 K, and it will decrease by the same factor of 10 as the temperature decreases from 300 to 255 K. The conductivity of semiconductors can also be increased by adding impurities to the intrinsic (pure) materials. This process is known as *doping*. Some impurities (such as phosphorus) are called *donors* because they add more electrons and form *n-type* semiconductors, with the electrons being the major carriers. Impurities (such as boron) are called *acceptors* because they add more holes to form *p-type* semiconductors, with the holes being the predominant carriers. When both *n-* and *p*-type regions exist on a single semiconductor, the junction formed between the two regions is used to build diodes and transistors.

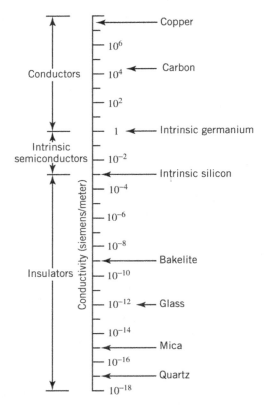

Figure 2-13   Range of conductivities of insulators, intrinsic semiconductors, and conductors.

At temperatures near absolute zero ($0\,\text{K} \simeq -273^\circ\text{C}$), the valence electrons of the outer shell of a semiconducting material are held very tightly and they are not free to travel. Thus the material behaves as an insulator under those conditions. As the temperature rises, thermal vibration of the lattice structure in a semiconductor material increases, and some of the electrons gain sufficient thermal energy to break away from the tight grip of their atom and become free electrons similar to those in a metallic conductor. As was shown in Figure 2-1, the atoms of silicon and germanium have four valence electrons in their outer shell which are held very tightly at temperatures near absolute zero, but some of them may break away as the temperature rises. The valence electrons of any semiconductor must gain sufficient energy to allow them to go from the valence band to the conduction band by jumping over the *forbidden* band, as shown in Figure 2-14. For all semiconductors, the energy gap of the forbidden band is about $E_g = 1.43\,\text{eV} = 2.29 \times 10^{19}\,\text{J}$. The bound electrons must gain at least that much energy, although they sometimes gain more, through increased thermal activity to make the jump.

The electrons that gain sufficient energy to break away from their atoms create vacancies in the shells that they vacate, designated as *holes*, which also move in a random fashion. When the semiconducting material is not subjected to an applied electric field, the net current from the bound electrons (which became free electrons) and the bound holes is zero because the net drift velocity of each type of carrier (electrons and holes) is zero, since they move in a random fashion. When an electric field is applied, the electrons move with a nonzero net drift velocity of $\mathbf{v}_{ed}$ (in the direction opposite to the applied field) while the holes move with a nonzero net drift velocity of

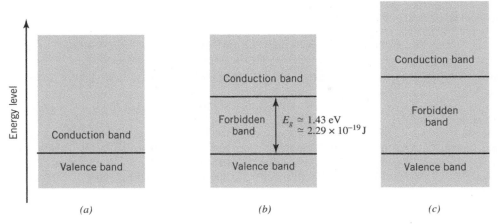

Figure 2-14   Energy levels for: (*a*) Conductors. (*b*) Semiconductors. (*c*) Insulators.

$\mathbf{v}_{hd}$ (in the same direction as the applied field), thus creating a nonzero current. Therefore we write the conduction current density for the two carriers (electrons and holes) as

$$\mathbf{J}_c = q_{ev}\mathbf{v}_{ed} + q_{hv}\mathbf{v}_{hd} = q_{ev}(-\mu_e\mathbf{E}) + q_{hv}(+\mu_h\mathbf{E})$$
$$\mathbf{J}_c = (-q_{ev}\mu_e + q_{hv}\mu_h)\mathbf{E} = (\sigma_{es} + \sigma_{hs})\mathbf{E} = \sigma_s\mathbf{E} \qquad (2\text{-}40)$$

where

μ_e = mobility of electrons $[\text{m}^2/(\text{V-s})]$
μ_h = mobility of holes $[\text{m}^2/(\text{V-s})]$
σ_{es} = static conductivity due to electrons
σ_{hs} = static conductivity due to holes

The static conductivities of the electrons (σ_{es}) and the holes (σ_{hs}) can also be written as

$$\sigma_{es} = -q_{ev}\mu_e = -N_e q_e \mu_e = N_e|q_e|\mu_e \qquad (2\text{-}41\text{a})$$

$$\sigma_{hs} = +q_{hv}\mu_h = +N_h q_h \mu_h = N_h|q_h|\mu_h \qquad (2\text{-}41\text{b})$$

where

N_e = free electron density (electrons per cubic meter)
N_h = bound hole density (holes per cubic meter)
$|q_e| = |q_h|$ = charge of an electron (magnitude) = 1.6×10^{-19} (coulombs)
$q_{ev} = N_e q_e = -N_e|q_e|$
$q_{hv} = N_h q_h = +N_h|q_h| = N_h|q_e|$

For comparison, representative values of charge densities, mobilities, and conductivities for intrinsic silicon, intrinsic germanium, aluminum, copper, silver, and gallium arsenide are given in Table 2-4 [5].

Six different materials were chosen to illustrate the formation of conductivity; their conductivity conditions are shown in Figure 2-15 [6]. These, in order, are representative of a dielectric (insulator), plasma (liquid or gas), conductor (metal), pure semiconductor, *n*-type semiconductor, and *p*-type semiconductor. It is observed that positively charged particles (holes) travel in the direction of the electric field whereas negatively charged particles (electrons) travel opposite to the electric field. However, both add to the total current.

TABLE 2-4   Charge densities, mobilities, and conductivities for silicon, germanium, aluminum, copper, silver, and gallium arsenide at 300 K

	q_{ev} (C/m^3)	q_{hv} (C/m^3)	μ_e [m^2/(V-s)]	μ_h [m^2/(V-s)]	σ (S/m)
Intrinsic silicon	-2.4×10^{-3}	$+2.4\times10^{-3}$	0.135 at 300 K	0.048 at 300 K	0.439×10^{-3}
Intrinsic germanium	-3.84	$+3.84$	0.39 at 300 K	0.19 at 300 K	2.227
Aluminum	-1.8×10^{10}	0	2.2×10^{-3}	0	3.96×10^7
Copper	-1.8×10^{10}	0	3.2×10^{-3}	0	5.76×10^7
Silver	-1.8×10^{10}	0	3.4×10^{-3}	0	6.12×10^7
*Intrinsic gallium arsenide	-2.86×10^{-7}	2.86×10^{-7}	0.8 at 300 K	0.032 at 300 K	2.38×10^{-7}

*O. Madelung (Ed.) *Numerical Data and Fundamental Relationships in Science and Technology*, Springer-Verlag, Berlin, Heidelberg, Germany, 1987.

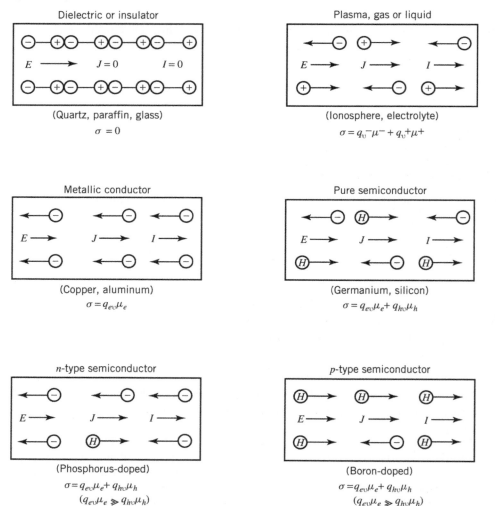

Figure 2-15   Conductivity conditions for six different materials representing dielectrics, plasmas, conductors, and semiconductors. (Source: J. D. Kraus, *Electromagnetics*, 1984, McGraw-Hill Book Co.)

The temperature variations of the mobilities of germanium, silicon, and gallium arsenide are given approximately by

Silicon [5]:

$$\mu_e \simeq (2.1 \pm 0.2) \times 10^5 \, T^{-2.5 \pm 0.1} \qquad 160 \leq T \leq 400 \, \text{K} \tag{2-42a}$$

$$\mu_h \simeq (2.3 \pm 0.1) \times 10^5 \, T^{-2.7 \pm 0.1} \qquad 150 \leq T \leq 400 \, \text{K} \tag{2-42b}$$

Germanium [5]:

$$\mu_e \simeq 4.9 \times 10^3 \, T^{-1.66} \qquad 100 \leq T \leq 300 \, \text{K} \tag{2-43a}$$

$$\mu_h \simeq 1.05 \times 10^5 \, T^{-2.33} \qquad 125 \leq T \leq 300 \, \text{K} \tag{2-43b}$$

Gallium arsenide:

$$\mu_e \simeq 0.8 \left(\frac{300}{T} \right)^{2.3} \tag{2-44a}$$

$$\mu_h \simeq 0.032 \left(\frac{300}{T} \right)^{2.3} \tag{2-44b}$$

Example 2-4

For the semiconducting materials silicon and germanium, determine conductivities at a temperature of 10°F. The electron and hole densities for silicon and germanium are, respectively, equal to about 3.03×10^{16} and 1.47×10^{19} electrons or holes per cubic meter.

Solution: At $T = 10°\text{F}$, the respective temperatures on the Celsius (°C) and Kelvin (K) scales are

$$°\text{C} = \tfrac{5}{9}(°\text{F} - 32) = \tfrac{5}{9}(10 - 32) = -12.2$$
$$\text{K} = °\text{C} + 273.2 = -12.2 + 273.2 = 261$$

The mobilities of silicon and germanium at 10°F (261 K) are approximately equal to

Silicon:

$$\mu_e \simeq 2.1 \times 10^5 T^{-2.5} = 2.1 \times 10^5 (261^{-2.5}) = 0.1908$$
$$\mu_h \simeq 2.3 \times 10^5 T^{-2.7} = 2.3 \times 10^5 (261^{-2.7}) = 0.0687$$

Germanium:

$$\mu_e \simeq 4.9 \times 10^3 T^{-1.66} = 4.9 \times 10^3 (261^{-1.66}) = 0.4771$$
$$\mu_h \simeq 1.05 \times 10^5 T^{-2.33} = 1.05 \times 10^5 (261^{-2.33}) = 0.2457$$

In turn the conductivities are equal to

Silicon:

$$\sigma_e \simeq n_e |q_e| \mu_e = 3.03 \times 10^{16} (1.6 \times 10^{-19})(0.1908) = 0.925 \times 10^{-3} \text{ S/m}$$
$$\sigma_h \simeq n_h |q_h| \mu_h = 3.03 \times 10^{16} (1.6 \times 10^{-19})(0.0687) = 0.333 \times 10^{-3} \text{ S/m}$$
$$\sigma = \sigma_e + \sigma_h \simeq 1.258 \times 10^{-3} \text{ S/m}$$

Germanium:

$$\sigma_e \simeq n_e |q_e| \mu_e = 1.47 \times 10^{19} (1.6 \times 10^{-19})(0.4771) = 1.122 \text{ S/m}$$

$$\sigma_h \simeq n_h |q_h| \mu_h = 1.47 \times 10^{19} (1.6 \times 10^{-19})(0.2457) = 0.578 \text{ S/m}$$

$$\sigma = \sigma_e + \sigma_h \simeq 1.7 \text{ S/m}$$

2.6 SUPERCONDUCTORS

Ideal conductors $(\sigma = \infty)$ are usually understood to be materials within which an electric field **E** cannot exist at any frequency. Through Maxwell's time-varying equations, this absence of an electric field also assures that there is no time-varying magnetic field. For static fields, however, the magnetic field should not be affected by the conductivity (including infinity) of the material. Therefore for static fields $(f = 0)$ a perfect conductor is defined as one that possesses an equipotential on its surface.

In practice no ideal conductors exist. Metallic conductors (such as aluminum, copper, silver, gold, etc.) have very large conductivities (typically $10^7 - 10^8$ S/m), and the rf fields in them decrease very rapidly with depth measured from the surface (being essentially zero at a few skin depths). However, the d.c. resistivity of certain metals essentially vanishes (conductivity becomes extremely large, almost infinity) at temperatures near absolute zero $(T = 0 \text{K}$ or $-273°\text{C})$. Such materials are usually called *superconductors*, and the temperature at which this is achieved is referred to as the *critical temperature* (T_c). Superconductivity was discovered in 1911 by Dutch physicist H. Kamerlingh Onnes, who received the Nobel Prize in 1913. For example, aluminum becomes superconducting at a critical temperature of 1.2 K, niobium (also called columbium) at 9.2 K, and the intermetallic compound niobium-germanium (Nb_3Ge) at 23 K. For temperatures down to 0.05 K, copper and gold do not superconduct. Even for low frequencies, superconductors above 0 K do exhibit a very small level of loss as a result of the presence of two types of carriers, lossless Cooper pairs and normal electrons. The ability of superconductors to expel magnetic fields, now referred to as the *Meissner effect*, was first observed experimentally in 1933 by Meissner and Ochsenfeld [7, 8]. In 1957, Bardeen, Cooper, and Schrieffer developed a theory that was able to accurately simulate the properties of superconductors using only first principles [9].

The electrodynamic response of a superconductor at microwave frequencies above 0 K has a small, but measurable, loss as a result of the presence of a resistive branch from the dissipative normal electrons (R) and an inductive branch because of the lossless Cooper pairs. Although Cooper pairs do not experience dissipation, they exhibit an inductive component from their finite inertia from their momentum (i.e., L_k, a kinetic inductance). Because the superconductor inductive Cooper pairs and normal electrons act in parallel, a.c. losses scale as ω^2, as would be expected from a parallel R-L circuit. Since the superconductor current density is finite, the microwave field will penetrate exponentially with a characteristic length called the *penetration depth* that is frequency independent and much smaller than the skin depth of a normal metal. Because of the smaller interaction volume and the small number of normal electrons, a superconductor will have typically several orders of magnitude smaller surface resistance than a normal metal.

Before 1986, it was accepted that if materials could become superconducting at temperatures of 25 K or greater, there would be a major technological breakthrough. The reason for the breakthrough is that materials can be cooled to these temperatures with relatively inexpensive liquid hydrogen, whose boiling temperature is about 20.4 K. Some of the potential applications of superconductivity would be:

1. supercomputers becoming smaller, faster, and thus more powerful;
2. ultra low-loss microwave communication systems;

3. economical, efficient, pollution-free, and safe-generating power plants using fusion or mag-
   netohydromagnetic technology;
4. virtually loss-free transmission lines and more efficient power transmission;
5. high-field magnets for use in MRI instruments, 300 mph trains levitated on a magnet
   cushion, particle accelerators and laboratory instrumentation; and
6. improved electronic instrumentation.

From 1911 to 1986, a span of 75 years, research into superconductivity yielded more than one thousand superconductive substances, but the increase in critical temperature was moderate and was accomplished at a very slow pace. Prior to January 1986, the record for the highest critical temperature belonged to niobium-germanium (Nb_3Ge), which in 1973 achieved a T_c of 23 K.

In January 1986 a major breakthrough in superconductivity may have provided the spark for which the scientific community had been waiting. Karl Alex Mueller and Johannes Georg Bednorz, IBM Zurich Research Laboratory scientists, observed that a new class of oxide materials exhibited superconductivity at a critical temperature much higher than anyone had observed before [10, 11]. The material was a ceramic copper oxide containing barium and lanthanum, and it had a critical temperature up to about 35 K, which was substantially higher than the 23 K for niobium-germanium.

Before Mueller and Bednorz's discovery, the best superconducting materials were intermetallic compounds, which included niobium-tin, niobium-germanium, and others. However, Mueller and Bednorz were convinced that the critical temperature could not be raised much higher using such compounds. Therefore they turned their attention to oxides with which they were familiar and which they believed to be better candidates for higher-temperature superconductors. For superconductivity to occur in a material, either the number of electrons that are available to transport current (i.e., a high density of states at the Fermi level) must be high or the electron pairs that are responsible for superconductivity must exhibit strong attractive forces [10]. Usually metals are very good candidates for superconductors because they have many available electrons. Oxides, however, have fewer electrons but it was shown that some metallic oxides of nickel and copper exhibited strong attractive electron-pair forces, and others could be found with even stronger pairing forces. Mueller and Bednorz became aware that some copper oxides behave like metals in conducting electricity. This led them to the superconducting copper oxide containing barium and lanthanum with a critical temperature of 35 K.

Since then many other groups have reported even higher superconductivities, up to about 90 K in a number of ternary oxides of rare earth elements [11]. One of the main questions still to be answered is why are they superconducting at such high temperatures. Paul C. W. Chu, from the University of Houston, found that by pressurizing a superconducting copper oxide, lanthanum, and barium he could observe critical temperatures of up to 70 K. He reasoned that the pressure brought the layers of the different elements closer together, leading to the higher superconductivity temperature. He also found that replacing barium with strontium, which is a very similar element but has smaller atoms, brought the layers even closer together and led to even higher temperatures. In February 1987 Dr. Chu also discovered that replacing lanthanum with yttrium resulted in even higher temperatures, up to 92 K. This was considered another major breakthrough because it surpassed the barrier of the boiling point of liquid nitrogen (77 K). Liquid nitrogen is relatively inexpensive (by a factor of 50) compared to liquid helium or hydrogen, which are used with superconducting material at lower temperatures.

On January 22, 1988 researchers at the National Research Institute for Metals, Tsukaba, Japan, reported that a compound of bismuth, calcium, strontium-copper, and oxygen had achieved a critical temperature of 105 K. Three days later Dr. Chu announced an identical compound except that it contained one additional element—aluminum. Dr. Chu has indicated that bismuth contains two superconducting phases (chemical structures). This two-phase superconducting condition causes the resistance to drop drastically between 120 and 110 K, but not to reach zero until about $83\,K$, after a second sudden drop. One of the phases has a transition temperature of about $115\,K$, and the other phase becomes superconducting at $90\,K$. Efforts are underway to isolate the two phases, to keep the lower temperature phase from surrounding the higher one. Although the yttrium–copper oxides are very sensitive to oxygen content and a high temperature anneal is consequently needed after the material is made superconducting, bismuth compounds do not lose oxygen when heated. In addition, the bismuth compounds appear less brittle than the yttrium compounds.

To date, the record T_c is 134 K in $HgBa_2Ca_2Cu_3O_{9-\delta}$ at ambient and at 164 K under pressure. According to the words of Dr. Chu, "The discovery of high temperature superconductivity (HTS) in the non-inter-metallic compounds, $La_{2-x}Ba_xCuO_4$ at 35 K (1986) and $Yba_2Cu_3O_7$ at 93 K (1987), has been ranked as one of the most exciting advancements in modern physics, with profound implications for technologies. In the ensuing 15 years, extensive worldwide research efforts have resulted in great progress in all areas of HTS science and technology. For instance, more than 150 compounds have been discovered with a T_c above 23 K; many anomalous properties have been observed; various models have been proposed to account for the observations; and numerous prototype devices have been made and successfully demonstrated. In spite of the impressive progress, the mechanism responsible for HTS has yet to be identified; a comprehensive theory remains elusive; the highest possible T_c is still to be found, if it exists; and commercialization of HTS devices is not yet realized" [12].

Now the march is on to try to understand better the physics of superconductivity and to see whether the critical temperature can be raised even further. It is even reasonable to expect that superconductivity could be achieved at room temperature. Even though practical superconductivity now seems more of a reality, there are many problems that must be overcome. For example, most superconductive materials are difficult to produce consistently. They seem to be stronger in some directions than in others and in general are too brittle to be used for flexible wires. Moreover, they exhibit certain crystal anisotropies as current flow can vary by a factor of 30, depending on the direction. In addition, properties of materials with high critical temperatures appear to be generally very susceptible to degradation from crystal defects. While critical current densities are high in thin films, bulk superconductor values are orders of magnitude smaller. These critical current densities are believed to be around $10^5\,A/cm^2$, although values of $1.8 \times 10^6\,A/cm^2$ have been reported at Japan's NTT Ibaragi Telecommunication Laboratory [11]. These current densities are about 10 to 100 times greater than reported previously, and they are also about 1,000 times the current density of typical household wiring. These values are reassurance that materials would sustain superconductivity at current density levels required for power transmission and generation, electronic circuits, and electromagnets.

2.7 METAMATERIALS

The decades of the 1990s and 2000s have introduced interest and excitement into the field of electromagnetics, especially as they relate to the integration of special types of artificial dielectric materials, coined *metamaterials*. The word *meta*, in *metamaterials*, is a Greek word that means beyond/after, and the term has been coined to represent materials that are artificially fabricated so that they have electromagnetic properties that go beyond those found readily in nature. In fact, the word has been used to represent materials which usually are constructed to exhibit periodic formations whose period is much smaller than the free-space and/or guided wavelength.

Using a "broad brush," the word metamaterials can encompass *engineered textured surfaces, artificial impedance surfaces, artificial magnetic conductors, Electromagnetic Band-Gap (EBG) structures, double negative (DNG) materials, frequency selective surfaces*, and even *fractals* or *chirals*. Engineered textured surfaces, artificial impedance surfaces, artificial magnetic conductors, and Electromagnetic Band-Gap (EBG) structures are discussed in Section 8.6. Materials whose constitutive parameters (permittivity and permeability) are both negative are often referred to as *Double Negatives* (DNGs). It is the class of DNG materials that has captivated the interest and imagination of many leading researchers and practitioners, scientists, and engineers, from academia, industry, and government; it also introduced a spirited dialogue. The properties and characteristics of DNG materials are discussed in more detail in Section 5.7.

2.8 LINEAR, HOMOGENEOUS, ISOTROPIC, AND NONDISPERSIVE MEDIA

The electrical behavior of materials when they are subjected to electromagnetic fields is characterized by their constitutive parameters (ε, μ, and σ).

Materials whose constitutive parameters are not functions of the applied field are usually known as *linear*; otherwise they are *nonlinear*. In practice, many materials exhibit almost linear characteristics as long as the applied fields are within certain ranges. Beyond those points, the material may exhibit a high degree of nonlinearity. For example, air is nearly linear for applied electric fields up to about 1×10^6 V/m. Beyond that, air breaks down and exhibits a high degree of nonlinearity.

When the constitutive parameters of media are not functions of position, the materials are referred to as *homogeneous*; otherwise they are *inhomogeneous* or *nonhomogeneous*. Almost all materials exhibit some degree of nonhomogeneity; however, for most materials used in practice the nonhomogeneity is so small that the materials are treated as being purely homogeneous.

If the constitutive parameters of a material vary as a function of frequency, they are denoted as being *dispersive*; otherwise they are *nondispersive*. All materials used in practice display some degree of dispersion. The permittivities and the conductivities, especially of dielectric material, and the permeabilities of ferromagnetic material and ferrites exhibit rather pronounced dispersive characteristics. These will be discussed in the text two sections.

Anisotropic or *nonisotropic* materials are those whose constitutive parameters are a function of the direction of the applied field; otherwise they are known as *isotropic*. Many materials, especially crystals, exhibit a rather high degree of anisotropy. For example, dielectric materials in which each component of their electric flux density **D** depends on more than one component of the electric field **E** are called *anisotropic dielectrics*. For such material, the permittivities and susceptibilities cannot be represented by a single value. Instead, for example, [$\overline{\varepsilon}$] takes the form of a 3×3 tensor, which is known as the *permittivity tensor*. The electric flux density **D** and electric field intensity **E** are not parallel to each other, and they are related by the permittivity tensor $\overline{\varepsilon}$ in a form given by

$$\mathbf{D} = \overline{\varepsilon} \cdot \mathbf{E} \tag{2-45}$$

In expanded form (2-45) can be written as

$$\begin{bmatrix} D_x \\ D_y \\ D_z \end{bmatrix} = \begin{bmatrix} \varepsilon_{xx} & \varepsilon_{xy} & \varepsilon_{xz} \\ \varepsilon_{yx} & \varepsilon_{yy} & \varepsilon_{yz} \\ \varepsilon_{zx} & \varepsilon_{zy} & \varepsilon_{zz} \end{bmatrix} \begin{bmatrix} E_x \\ E_y \\ E_z \end{bmatrix} \tag{2-46}$$

which reduces to

$$\begin{aligned} D_x &= \varepsilon_{xx} E_x + \varepsilon_{xy} E_y + \varepsilon_{xz} E_z \\ D_y &= \varepsilon_{yx} E_x + \varepsilon_{yy} E_y + \varepsilon_{yz} E_z \\ D_z &= \varepsilon_{zx} E_x + \varepsilon_{zy} E_y + \varepsilon_{zz} E_z \end{aligned} \tag{2-46a}$$

The permittivity tensor $\bar{\bar{\varepsilon}}$ is written, in general, as a 3×3 matrix of the form

$$[\bar{\bar{\varepsilon}}] = \begin{bmatrix} \varepsilon_{xx} & \varepsilon_{xy} & \varepsilon_{xz} \\ \varepsilon_{yx} & \varepsilon_{yy} & \varepsilon_{yz} \\ \varepsilon_{zx} & \varepsilon_{zy} & \varepsilon_{zz} \end{bmatrix} \tag{2-47}$$

where each entry may be complex. For anisotropic material, not all the entries of the permittivity tensor are necessarily nonzero. For some, only the diagonal terms ($\varepsilon_{xx}, \varepsilon_{yy}, \varepsilon_{zz}$), referred to as the *principal permittivities*, are nonzero. If that is not the case, for some material a set of new axes (x', y', z') can be selected by rotation of coordinates so that the permittivity tensor referenced to this set of axes possesses only diagonal entries (principal permittivities). This process is known as *diagonalization*, and the new set of axes are referred to as the *principal coordinates*. For physically realizable materials, the entries ε_{ij} of the permittivity tensor satisfy the relation

$$\varepsilon_{ij} = \varepsilon_{ji}^* \tag{2-48}$$

Matrices whose entries satisfy (2-48) are referred to as *Hermitian*. If the material is lossless (imaginary parts of ε_{ij} are zero) and the entries of the permittivity tensor satisfy (2-48), then the permittivity tensor is also symmetrical.

2.9   A.C. VARIATIONS IN MATERIALS

It has been shown that when a material is subjected to an applied static electric field, the centroids of the positive and negative charges (representing, respectively, the positive charges found in the nucleus of an atom and the negative electrons found in the shells surrounding the nucleus) are displaced relative to each other forming a linear electric dipole. When a material is examined macroscopically, the presence of all the electric dipoles is accounted for by introducing an electric polarization vector **P** [see (2-3) and (2-10)]. Ultimately, the static permittivity ε_s [see (2-11a)] is introduced to account for the presence of **P**. A similar procedure is used to account for the orbiting and spinning of the electrons of atoms (which are represented electrically by small electric current-carrying loops) when magnetic materials are subjected to applied static magnetic fields. When the material is examined macroscopically, the presence of all the loops is accounted for by introducing the magnetic polarization (magnetization) vector **M** [see (2-15) and (2-21)]. In turn the static permeability μ_s [see (2-22a)] is introduced to account for the presence of **M**.

When the applied fields begin to alternate in polarity, the polarization vectors **P** and **M**, and in turn the permittivities and permeabilities, are affected and they are functions of the frequency of the alternating fields. By this action of the alternating fields, there are simultaneous changes imposed upon the static conductivity σ_s [see (2-39) and (2-40)] of the material. In fact, the incremental changes in the conductivity that are attributable to the reverses in polarity of the applied fields (frequency) are responsible for the heating of materials using microwaves (for example, microwave cooking of food) [13, 18].

In the sections that follow, the variations of ε, σ, and μ as a function of frequency of the applied fields will be examined.

2.9.1   Complex Permittivity

Let us assume that each atom of a material in the absence of an applied electric field (unpolarized atom) is represented by positive (representing the nucleus) and negative (representing the electrons) charges whose respective centroids coincide. The electrical and mechanical equivalents of a typical atom are shown in Figure 2-16a [6]. The large positive sphere of a mass M represents the

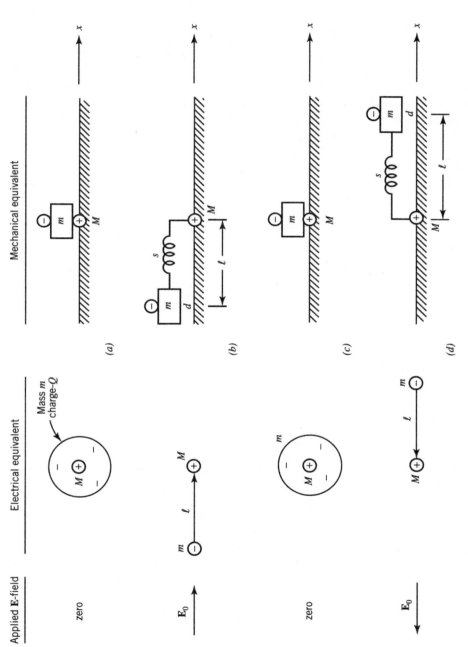

Figure 2-16 Electrical and mechanical equivalents of a typical atom in the absence of and under an applied electric field.

massive nucleus whereas the small negative sphere of mass m and charge $-Q$ represents the electrons. When an electric field is applied, it is assumed that the positive charge remains stationary and the negative charge moves relative to the positive along a platform that exhibits a friction (damping) coefficient d. In addition, the two charges will be connected with a spring whose spring (tension) coefficient is s. The entire mechanical equivalent of a typical atom then consists of the classical mass–spring system moving along a platform with friction.

When an electric field is applied that is directed along the $+x$ direction, the negative charge will be displaced a distance ℓ in the negative x direction, as shown in Figure 2-16b, forming an electric dipole. If the material is not permanently polarized (as are the electrets), the atom will achieve its initial normal position when the applied electric field diminishes to zero, as shown in Figure 2-16c. Now if the applied electric field is polarized in the $-x$ direction, the negative charge will move a distance ℓ in the positive x direction, as shown in Figure 2-16d, forming again an electric dipole in the direction opposite of that in Figure 2-16b.

When a time-harmonic field of angular frequency ω is applied to an atom, the forces of the system that describe the movement of the negative charge of mass m relative to the stationary nucleus and that are opposed by damping (friction) and tension (spring) can be represented by [6, 19]

$$m\frac{d^2\ell}{dt^2} + d\frac{d\ell}{dt} + s\ell = Q\mathscr{E}(t) = QE_0\,e^{j\omega t} \tag{2-49}$$

By dividing both sides of (2-49) by m, we can write it as

$$\frac{d^2\ell}{dt^2} + 2\alpha\frac{d\ell}{dt} + \omega_0^2\ell = \frac{Q}{m}\mathscr{E}(t) = \frac{Q}{m}E_0\,e^{j\omega t} \tag{2-50}$$

where

$$\alpha = \frac{d}{2m} \tag{2-50a}$$

$$\omega_0 = \sqrt{\frac{s}{m}} \tag{2-50b}$$

$$Q = \text{dipole charge} \tag{2-50c}$$

The terms on the left side of (2-49) represent, in order, the forces associated with mass times acceleration, damping times velocity, and spring times displacement. The term on the right side represents the driving force of the time-harmonic applied field (of peak value QE_0). Equations 2-49 and 2-50 are second-order differential equations that are also representative of the natural responses of RLC circuit systems.

For a source-free series RLC network, (2-50) takes the form for the current $i(t)$ of

$$\frac{d^2i}{dt^2} + 2\alpha\frac{di}{dt} + \omega_0^2 i = 0 \tag{2-51}$$

where

$$\alpha = \frac{R}{2L} \tag{2-51a}$$

$$\omega_0 = \frac{1}{\sqrt{LC}} \tag{2-51b}$$

In a similar manner, the voltage $v(t)$ for a parallel source-free *RLC* network can be obtained by writing (2-50) as

$$\frac{d^2v}{dt^2} + 2\alpha\frac{dv}{dt} + \omega_0^2 v = 0 \tag{2-52}$$

where

$$\alpha = \frac{1}{2RC} \tag{2-52a}$$

$$\omega_0 = \frac{1}{\sqrt{LC}} \tag{2-52b}$$

Solutions to (2-51) and (2-52) can be classified as *overdamped, critically damped*, or *underdamped* according to the values of the α/ω_0 ratio. That is, the solution to (2-51) for $i(t)$ or (2-52) for $v(t)$ is considered

Classification of Solution	Criterion	
overdamped	if $\alpha > \omega_0$	(2-53a)
critically damped	if $\alpha = \omega_0$	(2-53b)
underdamped	if $\alpha < \omega_0$	(2-53c)

The solutions to (2-49) can be obtained by first dividing both of its sides by m. Doing this reduces (2-49) to

$$\frac{d^2\ell}{dt^2} + \frac{d}{m}\frac{d\ell}{dt} + \frac{s}{m}\ell = \frac{Q}{m}E_0\,e^{j\omega t} \tag{2-54}$$

The general solution to (2-54) is usually composed of two parts: a complementary solution ℓ_c and a particular solution ℓ_p. The complementary solution represents the transient response of the system and is obtained by setting the driving force equal to zero. Since (2-54) is a quadratic, the general form of the complementary (transient) solution will be in terms of exponentials whose values vanish as $t \to \infty$. The particular solution represents the steady-state response of the system, and it is of interest here. Thus the particular (steady-state) solution of (2-54) can be written as

$$\ell_p(t) = \ell_0\,e^{j\omega t} \tag{2-55}$$

where ℓ_0 is the solution of $\ell_p(t)$ when $t = 0$.

Substituting (2-55) into (2-54) leads to

$$\ell_0 = \frac{\dfrac{Q}{m}E_0}{(\omega_0^2 - \omega^2) + j\omega\left(\dfrac{d}{m}\right)} \tag{2-56}$$

where

$$\omega_0 = \sqrt{\frac{s}{m}} \tag{2-56a}$$

Thus (2-55) can be written as

$$
\ell_p(t) = \ell_0 \, e^{j\omega t} = \frac{\dfrac{Q}{m} E_0 \, e^{j\omega t}}{(\omega_0^2 - \omega^2) + j\omega\left(\dfrac{d}{m}\right)} \tag{2-57}
$$

and it represents the steady-state displacements of the negative charges (electrons) of an atom relative to those of the positive charges (nucleus).

The resonant (natural) angular frequency ω_d of the system is obtained by setting $E_0 = 0$ in (2-54). Doing this and assuming an underdamped system ($\alpha < \omega_0$ or $d < 2\sqrt{sm}$) leads to

$$
\omega_d = \sqrt{\omega_0^2 - \alpha^2} = \sqrt{\frac{s}{m} - \left(\frac{d}{2m}\right)^2} \tag{2-58}
$$

For a frictionless system ($d = 0$) the resonant angular frequency ω_d reduces to

$$
\omega_d \big|_{d=0} = \omega_0 = \sqrt{\frac{s}{m}} \tag{2-58a}
$$

Assuming that the oscillating dipoles, which represent the numerous atoms of a material, are all similar and there is no coupling between the dipoles (atoms), the macroscopic steady-state electric polarization $\mathcal{P}$ of (2-5) can be written using (2-57) as

$$
\mathcal{P} = \mathcal{P}(t) = N_e Q\ell(t) = \frac{N_e\left(\dfrac{Q^2}{m}\right) E_0 \, e^{j\omega t}}{(\omega_0^2 - \omega^2) + j\omega\left(\dfrac{d}{m}\right)} = \frac{N_e\left(\dfrac{Q^2}{m}\right)\mathcal{E}(t)}{(\omega_0^2 - \omega^2) + j\omega\left(\dfrac{d}{m}\right)} \tag{2-59}
$$

where N_e represents the number of dipoles per unit volume. Dividing both sides of (2-59) by $\mathcal{E}(t) = E_0 e^{j\omega t}$ reduces it to

$$
\frac{\mathcal{P}}{\mathcal{E}} = \frac{N_e\left(\dfrac{Q^2}{m}\right)}{(\omega_0^2 - \omega^2) + j\omega\left(\dfrac{d}{m}\right)} \tag{2-60}
$$

In turn the permittivity $\dot{\varepsilon}$ of the medium can be written, using (2-10a) and (2-11a), as

$$
\dot{\varepsilon} = \varepsilon_0 + \frac{\mathcal{P}}{\mathcal{E}} = \varepsilon_0 + \frac{N_e\left(\dfrac{Q^2}{m}\right)}{(\omega_0^2 - \omega^2) + j\omega\left(\dfrac{d}{m}\right)} = \varepsilon' - j\varepsilon'' \tag{2-61}
$$

which is recognized as being complex, as denoted by the dot (with real and imaginary parts, respectively, of ε' and ε''). Equation (2-61) is also referred to as the *dispersion equation* for the complex permittivity.

The relative complex permittivity $\dot{\varepsilon}_r$ of the material is obtained by dividing both sides of (2-61) by ε_0 leading to

$$\dot{\varepsilon}_r = \frac{\dot{\varepsilon}}{\varepsilon_0} = \varepsilon_r' - j\varepsilon_r'' = 1 + \frac{\dfrac{N_e Q^2}{\varepsilon_0 m}}{\left(\omega_0^2 - \omega^2\right) + j\omega \dfrac{d}{m}} \tag{2-62}$$

The real ε_r' and imaginary ε_r'' parts of (2-62) can be written, respectively, as

$$\varepsilon_r' = 1 + \frac{\dfrac{N_e Q^2}{\varepsilon_0 m}(\omega_0^2 - \omega^2)}{(\omega_0^2 - \omega^2)^2 + \left(\omega \dfrac{d}{m}\right)^2} \tag{2-63a}$$

$$\varepsilon_r'' = \frac{N_e Q^2}{\varepsilon_0 m} \left[\frac{\omega \dfrac{d}{m}}{(\omega_0^2 - \omega^2)^2 + \left(\omega \dfrac{d}{m}\right)^2} \right] \tag{2-63b}$$

For nonmagnetic material

$$\dot{\varepsilon}_r = \dot{n}^2 \tag{2-64}$$

where $\dot{n}$ is the complex index of refraction. For materials with no damping $(d/m = 0)$, (2-63a) and (2-63b) reduce to

$$\varepsilon_r' = 1 + \frac{\dfrac{N_e Q^2}{\varepsilon_0 m}}{\omega_0^2 - \omega^2} \tag{2-65a}$$

$$\varepsilon_r'' = 0 \tag{2-65b}$$

Since the permittivity of a medium as given by (2-61) [or its relative value as given by (2-62)] is in general complex, the Maxwell–Ampere equation can be written as

$$\nabla \times \mathbf{H} = \mathbf{J}_i + \mathbf{J}_c + j\omega\dot{\varepsilon}\mathbf{E} = \mathbf{J}_i + \sigma_s \mathbf{E} + j\omega(\varepsilon' - j\varepsilon'')\mathbf{E}$$
$$\nabla \times \mathbf{H} = \mathbf{J}_i + (\sigma_s + \omega\varepsilon'')\mathbf{E} + j\omega\varepsilon'\mathbf{E} = \mathbf{J}_i + \sigma_e \mathbf{E} + j\omega\varepsilon'\mathbf{E} \tag{2-66}$$

where

$$\sigma_e = \text{equivalent conductivity} = \sigma_s + \omega\varepsilon'' = \sigma_s + \sigma_a \tag{2-66a}$$

$$\sigma_a = \text{alternating field conductivity} = \omega\varepsilon'' \tag{2-66b}$$

$$\sigma_s = \text{static field conductivity} \tag{2-66c}$$

$$= \begin{cases} -\mu_e q_{ve} & \text{for conductors} \\ -\mu_e q_{ve} + \mu_h q_{vh} & \text{for semiconductors} \end{cases} \tag{2-66d}$$

In (2-66a) σ_e represents the total (referred to here as the equivalent) conductivity composed of the static portion σ_s and the alternating part σ_a caused by the rotation of the dipoles as they attempt to align with the applied field when its polarity is alternating. The phenomenon (rotation of dipoles) that contributes the alternating conductivity σ_a is referred to as *dielectric hysteresis*.

Many dielectric materials (such as glass and plastic) possess very low values of static σ_s conductivities and behave as good insulators. However, when they are subjected to alternating fields, they exhibit very high values of alternating field σ_a conductivities and they consume considerable energy. The heat generated by this radio frequency process is used for industrial heating processes. The best-known process is that of *microwave cooking* [13–18]. Others include selective heating of human tissue for tumor treatment [20–22] and selective heating of certain compounds in materials that possess conductivities higher than the other constituents. For example, pyrite (a form of sulfur considered to be a pollutant), which exhibits higher conductivities than the other minerals of coal, can be heated selectively. This technique has been used as a process to clean coal by extracting, through microwave heating, its sulfur content.

In (2-66), aside from the impressed (source) electric current density $\mathbf{J}_i$, there are two other components: the effective conduction electric current density $\mathbf{J}_{ce}$ and the effective displacement electric current density $\mathbf{J}_{de}$. Thus we can write the total electric current density $\mathbf{J}_t$ as

$$\mathbf{J}_t = \mathbf{J}_i + \mathbf{J}_{ce} + \mathbf{J}_{de} = \mathbf{J}_i + \sigma_e \mathbf{E} + j\omega\varepsilon'\mathbf{E} \tag{2-67}$$

where

$$\mathbf{J}_t = \text{total electric current density} \tag{2-67a}$$

$$\mathbf{J}_i = \text{impressed (source) electric current density} \tag{2-67b}$$

$$\mathbf{J}_{ce} = \text{effective electric conduction current density}$$
$$= \sigma_e \mathbf{E} = (\sigma_s + \omega\varepsilon'')\mathbf{E} \tag{2-67c}$$

$$\mathbf{J}_{de} = \text{effective displacement electric current density}$$
$$= j\omega\varepsilon'\mathbf{E} \tag{2-67d}$$

The total electric current density of (2-67) can also be written as

$$\mathbf{J}_t = \mathbf{J}_i + \sigma_e \mathbf{E} + j\omega\varepsilon'\mathbf{E} = \mathbf{J}_i + j\omega\varepsilon'\left(1 - j\frac{\sigma_e}{\omega\varepsilon'}\right)\mathbf{E} = \mathbf{J}_i + j\omega\varepsilon'(1 - j\tan\delta_e)\mathbf{E} \tag{2-68}$$

where

$$\tan\delta_e = \text{effective electric loss tangent} = \frac{\sigma_e}{\omega\varepsilon'} = \frac{\sigma_s + \sigma_a}{\omega\varepsilon'} = \frac{\sigma_s}{\omega\varepsilon'} + \frac{\sigma_a}{\omega\varepsilon'}$$

$$\tan\delta_e = \frac{\sigma_s}{\omega\varepsilon'} + \frac{\varepsilon''}{\varepsilon'} = \tan\delta_s + \tan\delta_a = \frac{\varepsilon_e''}{\varepsilon_e'} \tag{2-68a}$$

$$\tan\delta_s = \text{static electric loss tangent} = \frac{\sigma_s}{\omega\varepsilon'} \tag{2-68b}$$

$$\tan\delta_a = \text{alternating electric loss tangent} = \frac{\sigma_a}{\omega\varepsilon'} = \frac{\varepsilon''}{\varepsilon'} \tag{2-68c}$$

The manufacturer of any given material usually specifies either the conductivity (S/m) or the electric loss tangent (tan δ, dimensionless). Although it is usually not stated as such, the specified conductivity σ_e and loss tangent should represent, respectively, the effective conductivity and loss tangent tan δ_e at a given frequency. Typical values of loss tangent for some materials are listed in Table 2-5.

The effective conduction $\mathbf{J}_{ce}$ and displacement $\mathbf{J}_{de}$ current densities of (2-67) can also be written as

$$\mathbf{J}_{cd} = \mathbf{J}_{ce} + \mathbf{J}_{de} = \sigma_e \mathbf{E} + j\omega\varepsilon'\mathbf{E} = j\omega\varepsilon'\left(1 - j\frac{\sigma_e}{\omega\varepsilon'}\right)\mathbf{E} = j\omega\varepsilon'(1 - j\tan\delta_e)\mathbf{E} \tag{2-69}$$

In phasor form, these can be represented as shown in Figure 2-17. It is evident that the conduction and displacement current densities are orthogonal to each other. Material can also be classified as good dielectrics or good conductors according to the values of the $\sigma_e/\omega\varepsilon'$ ratio. That is

1. *Good Dielectrics,* $(\sigma_e/\omega\varepsilon') \ll 1$

$$\mathbf{J}_{cd} = j\omega\varepsilon'\left(1 - j\frac{\sigma_e}{\omega\varepsilon'}\right)\mathbf{E} \overset{\sigma_e/\omega\varepsilon' \ll 1}{\simeq} j\omega\varepsilon'\mathbf{E} \qquad (2\text{-}70\text{a})$$

TABLE 2-5   Dielectric constants and loss tangents of typical dielectric materials

Material	ε'_r	$\tan \delta$
Air	1.0006	
Alcohol (ethyl)	25	0.1
Aluminum oxide	8.8	6×10^{-4}
Bakelite	4.74	22×10^{-3}
Carbon dioxide	1.001	
Germanium	16	
Glass	4–7	1×10^{-3}
Ice	4.2	0.1
Mica	5.4	6×10^{-4}
Nylon	3.5	2×10^{-2}
Paper	3	8×10^{-3}
Plexiglas	3.45	4×10^{-2}
Polystyrene	2.56	5×10^{-5}
Porcelain	6	14×10^{-3}
Pyrex glass	4	6×10^{-4}
Quartz (fused)	3.8	7.5×10^{-4}
Rubber	2.5–3	2×10^{-3}
Silica (fused)	3.8	7.5×10^{-4}
Silicon	11.8	
Snow	3.3	0.5
Sodium chloride	5.9	1×10^{-4}
Soil (dry)	2.8	7×10^{-2}
Styrofoam	1.03	1×10^{-4}
Teflon	2.1	3×10^{-4}
Titanium dioxide	100	15×10^{-4}
Water (distilled)	80	4×10^{-2}
Water (sea)	81	4.64
Wood (dry)	1.5–4	1×10^{-2}

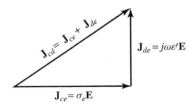

Figure 2-17   Phasor representation of effective conduction and displacement current densities.

For these materials, the displacement current density is much greater than the conduction current density, and the total current density is approximately equal to the displacement current density.

2. *Good Conductors, $(\sigma_e/\omega\varepsilon') \gg 1$*

$$\mathbf{J}_{cd} = j\omega\varepsilon'\left(1 - j\frac{\sigma_e}{\omega\varepsilon'}\right)\mathbf{E} \overset{\sigma_e/\omega\varepsilon' \gg 1}{\simeq} \sigma_e\mathbf{E} \qquad (2\text{-}70\text{b})$$

For these materials, the conduction current density is much greater than the displacement current density, and the total current density is approximately equal to the conduction current density.

As discussed in Section 2.2 and demonstrated in Figure 2-4, the electric polarization for dielectrics, as given by (2-3) or (2-5), can be composed of any combination involving the dipole (orientational), ionic (molecular), and electronic polarizations. As a function of frequency, the electric polarization of (2-10) can be written as

$$\mathbf{P}(\omega) = \varepsilon_0\chi_e(\omega)\mathbf{E}_a(\omega) \qquad (2\text{-}71)$$

where in general

$$\chi_e(\omega) = \chi_e'(\omega) - j\chi_e''(\omega)$$
$$= \left[\chi_{ed}'(\omega) + \chi_{ei}'(\omega) + \chi_{ee}'(\omega)\right] - j\left[\chi_{ed}''(\omega) + \chi_{ei}''(\omega) + \chi_{ee}''((\omega)\right] \qquad (2\text{-}71\text{a})$$

$$\chi_{ed}'(\omega) = \text{dipole real electric susceptibility} \qquad (2\text{-}71\text{b})$$

$$\chi_{ei}'(\omega) = \text{ionic real electric susceptibility} \qquad (2\text{-}71\text{c})$$

$$\chi_{ee}'(\omega) = \text{electronic real electric susceptibility} \qquad (2\text{-}71\text{d})$$

$$\chi_{ed}''(\omega) = \text{dipole loss electric susceptibility} \qquad (2\text{-}71\text{e})$$

$$\chi_{ei}''(\omega) = \text{ionic loss electric susceptibility} \qquad (2\text{-}71\text{f})$$

$$\chi_{ee}''(\omega) = \text{electronic loss electric susceptibility} \qquad (2\text{-}71\text{g})$$

It should be noted that, in general,

$$\chi_e'(-\omega) = \chi_e'(\omega) \qquad (2\text{-}72\text{a})$$

$$\chi_e''(-\omega) = -\chi_e''(\omega) \qquad (2\text{-}72\text{b})$$

A general sketch of the variations of the susceptibilities as a function of frequency is given in Figure 2-18 [25, 26]. It should be stated, however, that this does not represent any one particular material, and very few materials exhibit all three mechanisms. Measurements have been made on many materials, with some up to 90 GHz, using microwave and millimeter wave techniques [25].

Since the relative permittivity (dielectric constant) is related to the electric susceptibility by (2-12), we should expect similar variations of the dielectric constant as a function of frequency. To demonstrate that, we have plotted in Figure 2-19 as a function of frequency ($0 \leq \omega \leq 10$) the relative complex permittivity (real and imaginary parts and magnitude) of (2-62) or (2-63a) and (2-63b) (assuming $N_eQ^2/\varepsilon_0 m = 1$ and $d/m = 1$) for

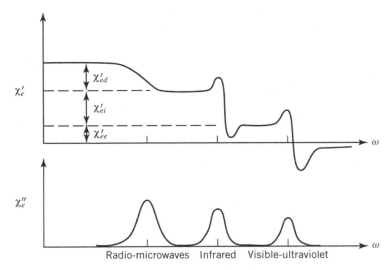

Figure 2-18   Electric susceptibility (real and imaginary) variations as a function of frequency for a typical dielectric.

$$\frac{\alpha}{\omega_0} = \frac{d}{2m}\sqrt{\frac{m}{s}} = \frac{1}{5}\left(\text{underdamped with } \omega_0 = 2.5 \text{ and } \omega_d = \sqrt{6} = 2.449\right)$$

$$\frac{\alpha}{\omega_0} = \frac{d}{2m}\sqrt{\frac{m}{s}} = \frac{1}{10}\left(\text{underdamped with } \omega_0 = 5 \text{ and } \omega_d = \sqrt{99}/2 = 4.975\right)$$

It is observed that the values of ε_r'' peak at the resonant frequencies, which indicates that the medium attains its most lossy state at the resonant frequency. Multiple variations of this type would also be observed in a given curve at other frequencies if the medium possesses multiple resonant frequencies. For frequencies not near one of the resonant frequencies, the curve representing the variations of $|\dot{\varepsilon}_r|$ exhibits a positive slope and is referred to as *normal* dispersion (because it occurs most commonly). Very near the resonant frequencies there is a small range of frequencies for which the variations of $|\dot{\varepsilon}_r|$ exhibit a negative slope that is referred to as *anomalous* (abnormal) dispersion. Although there is nothing abnormal about this type of dispersion, the name was given because it seemed unusual when it was first observed.

When (2-57) and (2-59) to (2-63b) were derived, it was assumed that the medium possessed only one resonant (natural) frequency presented by one type of harmonic oscillator. In general, however, there are several natural frequencies associated with a particular atom. These can be accounted for in our dispersion equations for ε_r' and ε_r'' by introducing several different kinds of oscillators with no coupling between them. This type of modeling allows the contributions from each oscillator to be accounted for by a simple addition. Thus for a medium with p natural frequencies (represented by p independent oscillators), we can write (2-60) to (2-63b) as

$$\frac{\mathscr{P}}{\mathscr{E}} = \sum_{s=1}^{p} \frac{N_e \dfrac{Q^2}{m}}{(\omega_s^2 - \omega^2) + j\dfrac{\omega d}{m}} \tag{2-73a}$$

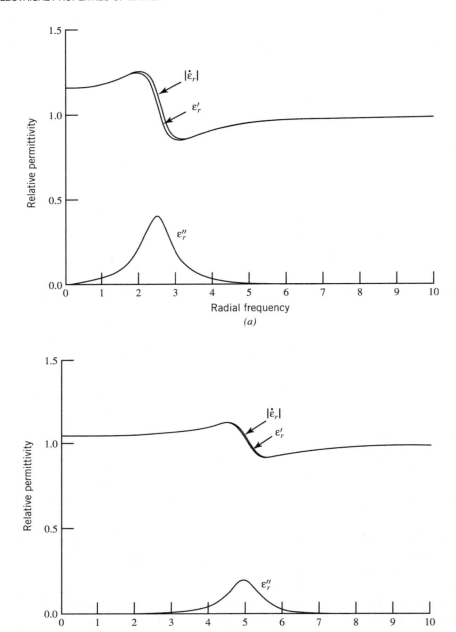

Figure 2-19 Typical frequency variations of real and imaginary parts of relative permittivity of dielectrics. (a) $N_e Q^2 / \varepsilon_0 m = 1$, $d/m = 1$, $\alpha/\omega_0 = 1/5$, $\omega_0 = 2.5$. (b) $N_e Q^2 / \varepsilon_0 m = 1$, $d/m = 1$, $\alpha/\omega_0 = 1/10$, $\omega_0 = 5$.

$$\dot{\varepsilon} = \varepsilon' - j\varepsilon'' = \varepsilon_0 + \sum_{s=1}^{p} \frac{N_e \dfrac{Q^2}{m}}{(\omega_s^2 - \omega^2) + j\dfrac{\omega d}{m}} \tag{2-73b}$$

$$\dot{\varepsilon}_r = \varepsilon_r' - j\varepsilon_r'' = 1 + \sum_{s=1}^{p} \frac{\dfrac{N_e Q^2}{\varepsilon_0 m}}{(\omega_s^2 - \omega^2) + j\dfrac{\omega d}{m}} \tag{2-73c}$$

$$\varepsilon_r' = 1 + \sum_{s=1}^{p} \frac{\dfrac{N_e Q^2}{\varepsilon_0 m}(\omega_s^2 - \omega^2)}{(\omega_s^2 - \omega^2)^2 + \left(\dfrac{\omega d}{m}\right)^2} \tag{2-73d}$$

$$\varepsilon_r'' = \sum_{s=1}^{p} \frac{N_e Q^2}{\varepsilon_0 m} \frac{\omega\dfrac{d}{m}}{(\omega_s^2 - \omega^2)^2 + \left(\dfrac{\omega d}{m}\right)^2} \tag{2-73e}$$

Often the question is asked whether there are any relations between the real and imaginary parts of the complex permittivity. The answer to that is yes. Known as the Kramers–Kronig [26–28] relations, they are given by

$$\varepsilon_r'(\omega) = 1 + \frac{2}{\pi} \int_0^\infty \frac{\omega' \varepsilon_r''(\omega')}{(\omega')^2 - \omega^2} d\omega' \tag{2-74a}$$

$$\varepsilon_r''(\omega) = \frac{2\omega}{\pi} \int_0^\infty \frac{1 - \varepsilon_r'(\omega')}{(\omega')^2 - \omega^2} d\omega' \tag{2-74b}$$

and they are very similar to the frequency relations between resistance and reactance in circuit theory [28].

In addition to the Kramers–Kronig relations of (2-74a) and (2-74b), there are simple relations that allow the calculation of the real and imaginary parts of the complex relative permittivity for many materials as a function of frequency provided that the real part of the complex permittivity is known at zero frequency (denoted by ε_{rs}') and at very large (ideally infinity) frequency (denoted by $\varepsilon_{r\infty}'$). These relations are obtained from the well-known *Debye equation* [19, 23, 24] for the complex dielectric constant, which states that

$$\dot{\varepsilon}_r(\omega) = \varepsilon_r'(\omega) - j\varepsilon_r''(\omega) = \varepsilon_{r\infty}' + \frac{\varepsilon_{rs}' - \varepsilon_{r\infty}'}{1 + j\omega\tau_e} \tag{2-75}$$

where τ_e is a *new relaxation time constant* related to *original relaxation time constant* τ by

$$\tau_e = \tau \frac{\varepsilon_{rs}' + 2}{\varepsilon_{r\infty}' + 2} \tag{2-75a}$$

The Debye equation of (2-75) is derived using the *Clausius–Mosotti equation* [23, 24, 29]. The real and imaginary parts of (2-75) can be written as

$$\varepsilon_r'(\omega) = \varepsilon_{r\infty}' + \frac{\varepsilon_{rs}' - \varepsilon_{r\infty}'}{1 + (\omega\tau_e)^2} \tag{2-76a}$$

$$\varepsilon_r''(\omega) = \frac{(\varepsilon_{rs}' - \varepsilon_{r\infty}')\omega\tau_e}{1+(\omega\tau_e)^2} \tag{2-76b}$$

which can be found at any frequency provided ε_{rs}', $\varepsilon_{r\infty}'$, and τ are known. The relations of (2-76a) and (2-76b) can be used to estimate the real and imaginary parts of the complex relative permittivity (complex dielectric constant) for many gases, liquids, and solids.

2.9.2 Complex Permeability

As discussed in Section 2.3, the permeability of most dielectric material, including diamagnetic, paramagnetic, and antiferromagnetic material, is nearly the same as that of free space μ_0 ($\mu_0 = 4\pi \times 10^{-7}$ H/m). Ferromagnetic and ferrimagnetic materials exhibit much higher permeability than free space, as is demonstrated by the data of Table 2-2. These classes of materials are also magnetically lossy, and their magnetic losses are accounted for by introducing a complex permeability.

In general then, we can write the Maxwell–Faraday equation as

$$\nabla \times \mathbf{E} = -\mathbf{M}_i - j\omega\dot{\mu}\mathbf{H} = -\mathbf{M}_i - j\omega(\mu' - j\mu'')\mathbf{H}$$
$$= -\mathbf{M}_i - j\omega\mu'\mathbf{H} - \omega\mu''\mathbf{H} = -\mathbf{M}_t \tag{2-77}$$

where

$$\mathbf{M}_t = \mathbf{M}_i + j\omega\mu'\mathbf{H} + \omega\mu''\mathbf{H} \tag{2-77a}$$

$$\mathbf{M}_t = \text{total magnetic current density} \tag{2-77b}$$

$$\mathbf{M}_i = \text{impressed (source) magnetic current density} \tag{2-77c}$$

$$\mathbf{M}_d = \text{displacement magnetic current density} = j\omega\mu'\mathbf{H} \tag{2-77d}$$

$$\mathbf{M}_c = \text{conduction magnetic current density} = \omega\mu''\mathbf{H} \tag{2-77e}$$

Another form of (2-77a) is to write it as

$$\mathbf{M}_t = \mathbf{M}_i + j\omega\mu'\left(1 - j\frac{\mu''}{\mu'}\right)\mathbf{H} = \mathbf{M}_i + j\omega\mu'(1 - j\tan\delta_m)\mathbf{H} \tag{2-78}$$

where

$$\tan\delta_m = \text{alternating magnetic loss tangent} = \frac{\mu''}{\mu'} \tag{2-78a}$$

In addition to being complex, the permeability of ferromagnetic and ferrimagnetic material is often a function of frequency. Thus it should, in general, be written as

$$\dot{\mu} = \mu'(\omega) - j\mu''(\omega) \tag{2-79}$$

or

$$\dot{\mu}_r = \frac{\dot{\mu}}{\mu_0} = \mu_r'(\omega) - j\mu_r''(\omega) \tag{2-79a}$$

Most ferromagnetic materials possess very high relative permeabilities (on the order of several thousand) and good conductivities such that there is a minimum interaction between these materials and the electromagnetic waves propagating through them. As such, they will not be discussed further here. There is, however, a class of ferrimagnetic material, referred to as *ferrites*, that finds wide applications in the design of nonreciprocal microwave components (such as isolators, hybrids, gyrators, phase shifters, etc.). Ferrites become attractive for these applications because at microwave frequencies they exhibit strong magnetic effects that result in anisotropic properties and large resistances (good insulators). These resistances limit the current induced in them and in turn result in lower ohmic losses. Because of the appeal of ferrites to microwave circuit design, their magnetic properties will be discussed further in the section that follows.

2.9.3 Ferrites

Ferrites are a class of solid ceramic materials that have crystal structures formed by sintering at high temperatures (typically $1000-1500\,^{\circ}\mathrm{C}$) stoichiometric mixtures of certain metal oxides (such as oxygen and iron, and cadmium, lithium, magnesium, nickel, or zinc, or some combination of them). These materials are ferrimagnetic, and they are considered to be good insulators with high permeabilities, dielectric constants between 10 to 15 or greater, and specific resistivities as much as 10^{14} greater than those of metals. In addition, they possess properties that allow strong interaction between the magnetic dipole moment associated with the electron spin, as discussed in Section 2.3, and the microwave electromagnetic fields [30–32]. In contrast to ferromagnetic materials, ferrites have their magnetic ions distributed over at least two interpenetrating sublattices. Within each sublattice all magnetic moments are aligned, but the sublattices are oppositely directed.

As a result of these interactions, ferrites exhibit nonreciprocal properties such as different phase constants and phase velocities for right- and left-hand circularly polarized waves, transmission coefficients that are functions of direction of travel, and permeabilities that are represented by tensors (in the form of a matrix) rather than by a single scalar. These characteristics become important in the design of nonreciprocal microwave devices [33–35]. Although all ferrimagnetic materials possess these properties, it is only in ferrites that they are pronounced and significant. The properties of ferrites will be discussed here by examining the propagation of microwave electromagnetic waves in an unbounded ferrite material.

There are two possible models that can be used to understand the technical properties of magnetic material: the *phenomenological* model and the *atomic* model [32]. For the purposes of this book, the phenomenological model is sufficient to examine the properties of magnetic oxides. As discussed in Section 2.3, the magnetic material is replaced by an array of magnetic dipoles that are maintained in a permanent and rigid alignment as shown in Figure 2-8a. When a magnetic field is applied, as shown in Figure 2-8b, the magnetic moments of the dipoles can turn freely in space as long as they turn together. Much of the discussion of this section follows that of [32] and [35].

Under an applied magnetic field, each single magnetic dipole rotates with a precession frequency that is referred to as the *Larmor precession frequency*. The precession frequency is altered when one or more dipoles are introduced. The dipoles in the array interact with each other and attempt to achieve an alignment that will minimize the interaction energy. The change in precession frequency is equivalent to introducing an additional demagnetizing field. When many dipole arrays are subjected to d.c, rf, or demagnetizing fields, magnetic resonance is introduced. This is a phenomenon that is of fundamental interest to the design of microwave nonreciprocal components. The discussion here will be that of the phenomenological model.

The magnetic dipole moment **m** of a single magnetic dipole of Figure 2-7a or 2-7b is given by (2-13). When an external magnetic field is applied, as shown in Figure 2-20 for a single dipole, exerted on the dipole is a torque **T** of

$$\mathbf{T} = \mu_0 \mathbf{m} \times \mathbf{H}_0 = \mathbf{m} \times \mathbf{B}_0 \qquad (2\text{-}80)$$

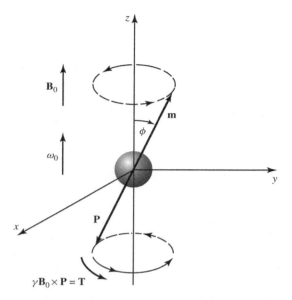

Figure 2-20   Torque on a single magnetic dipole caused by an applied external magnetic field. (Source: R. E. Collin, *Foundations for Microwave Engineering*, 2nd Edition, 1992, McGraw-Hill Book Co.)

where $\mathbf{m} = \hat{\mathbf{n}} \, I \, ds$ = magnetic dipole moment of a single dipole
   $\mathbf{H}_0$ = applied magnetic field
   $\mathbf{B}_0$ = applied magnetic flux density

The torque will cause the dipole to precess about the z axis, which is parallel to $\mathbf{B}_0$, as shown in Figure 2-20.
   The interaction energy W_m between the dipole and the applied field can be expressed as

$$W_m = -\mu_0 m H_0 \cos \phi \tag{2-81}$$

$$T = -\frac{\partial W_m}{\partial \phi} \tag{2-81a}$$

where ϕ is the angle between the applied magnetic field and the magnetic dipole axis. It is observed that the energy is minimum $(W_m = -\mu_0 m H_0)$ when $\phi = 0$ whereas when $\phi = 90°$, T is zero and the dipole is in unstable equilibrium.
   When electrons of a physically realizable dipole are moving, they create a current whose motion is associated with a circulation of mass (angular momentum) as well as charge. Therefore the magnetic dipole moment of a single electron of charge e, which is moving with a velocity v in a circle of radius a, can be also be expressed as

$$m = I \, ds = \frac{ev}{2\pi a}(\pi a^2) = \frac{1}{2}eva \tag{2-82}$$

and the angular momentum P can be written as

$$P = m_e v a \tag{2-83}$$

where m_e is the mass of the electron. The ratio of the magnetic moment [as given by (2-82)] to the angular momentum [as given by (2-83)] is referred to as the *gyromagnetic ratio* γ, and it is equal to

$$\gamma = \frac{m}{P} = \frac{e}{2m_e} \Rightarrow m = \gamma P \tag{2-84}$$

which is negative because of the negative electron charge e. This makes the angular momentum P of the electron antiparallel to the magnetic dipole moment $\mathbf{m}$, as shown in Figure 2-20.

To obtain the equation of motion we set the rate of change (with time) of the angular momentum equal to the torque, that is,

$$\frac{d\mathbf{P}}{dt} = \mathbf{T} = \mu_0 \mathbf{m} \times \mathbf{H}_0 = -\mu_0 |\gamma| \mathbf{P} \times \mathbf{H}_0 = -\mathbf{P} \times \boldsymbol{\omega}_0 = \boldsymbol{\omega}_0 \times \mathbf{P} \tag{2-85}$$

or

$$\mu_0 |\gamma| P H_0 \sin\phi = \omega_0 P \sin\phi = -\mu_0 m H_0 \sin\phi \tag{2-85a}$$

In (2-85) and (2-85a) ω_0 is the vector precession angular velocity which is directed along $\mathbf{H}_0$, as shown in Figure 2-20. For the free precession of a *single* dipole, the angular velocity ω_0 is referred to as the *Larmor precession frequency*, which is given by

$$\omega_0 = |\gamma| \mu_0 H_0 = |\gamma| B_0 \tag{2-86}$$

and it is independent of the angle ϕ.

Let us assume that on the static applied field $\mathbf{B}_0$ a small a.c. magnetic field $\mathbf{B}_1$ is superimposed. This additional applied field will impose a forced precession on the magnetic dipole. To examine the effects of the forced precession, let us assume that the a.c. applied magnetic field $\mathbf{B}_1^{\pm}$ is circularly polarized, either right hand (CW) $\mathbf{B}_1^{+}$ or left hand (CCW) $\mathbf{B}_1^{-}$, and it is directed perpendicular to the z axis. As will be shown in Section 4.4.2 these fields can be written as

$$\mathbf{B}_1^{+} = (\hat{\mathbf{a}}_x - j\hat{\mathbf{a}}_y)B_1^{+}e^{-j\beta z} \quad \text{right-hand (CW)} \tag{2-86a}$$

$$\mathbf{B}_1^{-} = (\hat{\mathbf{a}}_x + j\hat{\mathbf{a}}_y)B_1^{-}e^{-j\beta z} \quad \text{left-hand (CCW)} \tag{2-86b}$$

The corresponding instantaneous fields obtained using (1-61d) rotate, respectively, in the clockwise and counterclockwise directions when viewed from the rear as they travel in the $+z$ direction. This is demonstrated in Figure 2-21. When each of the a.c. signals are superimposed upon the static field $\mathbf{B}_0$ directed along the z axis, the resultant $\mathbf{B}_t^{\pm}$ field will be at angle $\theta^{\pm}$ (measured from the z axis) given by

$$\theta^{\pm} = \tan^{-1}\left(\frac{B_1^{\pm}}{B_0}\right) \tag{2-87}$$

as shown in Figures 2-22a and 2-22b. The resultant magnetic field $\mathbf{B}_t^{\pm}$ will rotate about the z axis at a rate of ω^{+} in the clockwise direction for $\mathbf{B}_t^{+}$ and ω^{-} in the counterclockwise direction for $\mathbf{B}_t^{-}$, as shown in Figure 2-22. The magnetic dipole will be forced to precess at the same rate about the z axis when steady-state conditions prevail.

For the torque to impose a clockwise precession on $\mathbf{B}_t^{+}$ and a counterclockwise precession on $\mathbf{B}_t^{-}$, the precession angle ϕ^{+} must be larger than θ^{+} (as shown in Figure 2-22a) and ϕ^{-} must be smaller than θ^{-} (as shown in Figure 2-22b). Therefore for each case (2-85), the equation of motion, can be written as

$$\frac{d\mathbf{P}^{+}}{dt} = \mathbf{T}^{+} = \mathbf{m}^{+} \times \mathbf{B}_t^{+} = -|\gamma|\mathbf{P}^{+} \times \mathbf{B}_t^{+} = \omega^{+}\hat{\mathbf{a}}_z \times \mathbf{P}^{+} \tag{2-88a}$$

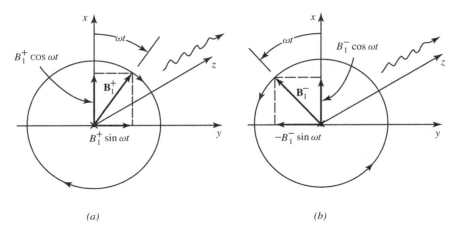

Figure 2-21 Rotation of magnetic field, as a function of time, for CW and CCW polarizations. (*a*) Clockwise. (*b*) Counterclockwise. (Source: R. E. Collin, *Foundations for Microwave Engineering*, Second Edition, 1992, McGraw-Hill Book Co.)

$$\frac{d\mathbf{P}^-}{dt} = \mathbf{T}^- = \mathbf{m}^- \times \mathbf{B}_t^- = -|\gamma|\mathbf{P}^- \times \mathbf{B}_t^- = -\omega^-\hat{\mathbf{a}}_z \times \mathbf{P}^- \qquad (2\text{-}88b)$$

or

$$-|\gamma|P^+B_t^+\sin(\phi^+ - \theta^+) = \omega^+P^+\sin\phi^+ \qquad (2\text{-}89a)$$

$$-|\gamma|P^-B_t^-\sin(\theta^- - \phi^-) = -\omega^-P^-\sin\phi^- \qquad (2\text{-}89b)$$

Expanding (2-89a) and (2-89b) leads to

$$-|\gamma|\left[(B_t^+\sin\phi^+)\cos\theta^+ - (B_t^+\cos\phi^+)\sin\theta^+\right] = \omega^+\sin\phi^+ \qquad (2\text{-}90a)$$

$$-|\gamma|\left[(B_t^-\sin\theta^-)\cos\phi^- - (B_t^-\cos\theta^-)\sin\phi^-\right] = -\omega^-\sin\phi^- \qquad (2\text{-}90b)$$

Since

$$B_t^+\sin\theta^+ = B_1^+ \qquad (2\text{-}91a)$$

$$B_t^+\cos\theta^+ = B_0 \qquad (2\text{-}91b)$$

$$B_t^-\sin\theta^- = B_1^- \qquad (2\text{-}91c)$$

$$B_t^-\cos\theta^- = B_0 \qquad (2\text{-}91d)$$

then (2-90a) and (2-90b) can be reduced, respectively, to

$$\tan\phi^+ = \frac{|\gamma|B_1^+}{|\gamma|B_0 - \omega^+} = \frac{|\gamma|B_1^+}{\omega_0 - \omega^+} \qquad (2\text{-}92a)$$

$$\tan\phi^- = \frac{|\gamma|B_1^-}{|\gamma|B_0 + \omega^-} = \frac{|\gamma|B_1^-}{\omega_0 + \omega^-} \qquad (2\text{-}92b)$$

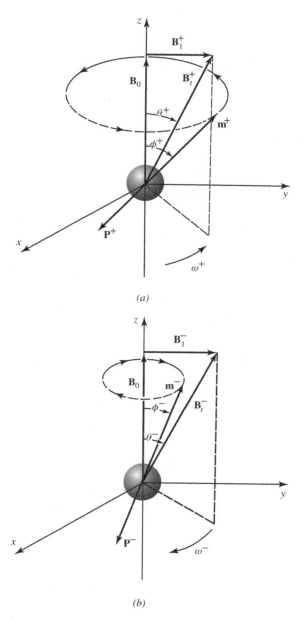

Figure 2-22   Precession of spinning electron caused by applied magnetic field CW and CCW polarizations. (*a*) Clockwise. (*b*) Counterclockwise. (Source: R. E. Collin, *Foundations for Microwave Engineering*, Second Edition, 1992, McGraw-Hill Book Co.)

According to Figure 2-22 the components $m_t^\pm$ of $\mathbf{m}^\pm$ that rotate in synchronism with their respective $\mathbf{B}_1^\pm$, and $m_z^\pm$ that are directed along the z axis, are given, respectively, by

$$m_t^\pm = m^\pm \sin\phi^\pm = m^\pm \cos\phi^\pm \frac{\sin\phi^\pm}{\cos\phi^\pm} = m^\pm \cos\phi^\pm \tan\phi^\pm = m_0^\pm \tan\phi^\pm \qquad (2\text{-}93a)$$

$$m_z^\pm = m^\pm \cos\phi^\pm = m_0^\pm \qquad (2\text{-}93b)$$

where

$$m_0^\pm = m^\pm \cos \phi^\pm \tag{2-93c}$$

Using (2-92a) and (2-92b) we can write the components of $\mathbf{m}^\pm$ that rotate in synchronism with $\mathbf{B}_1^\pm$ as

$$m_t^+ = m_0^+ \tan \phi^+ = \frac{m_0^+ |\gamma| B_1^+}{\omega_0 - \omega^+} \tag{2-94a}$$

$$m_t^- = m_0^- \tan \phi^- = \frac{m_0^- |\gamma| B_1^-}{\omega_0 + \omega^-} \tag{2-94b}$$

In the previous discussion we considered the essential properties of single spinning electrons in a magnetic field that is a superposition of a static magnetic field along the z axis and an a.c. circularly polarized field perpendicular to it. Let us now examine macroscopically the properties of N orbiting electrons per unit volume whose density is uniformly and continuously distributed. Doing this we can represent the total magnetization $\mathbf{M}$ of all N electrons as the product of N times that of a single electron ($\mathbf{M} = N\mathbf{m}$), as given by (2-17). In addition, the magnetic flux density $\mathbf{M}$ will be related to the magnetic field intensity $\mathbf{H}$ and magnetization vector $\mathbf{M}$ by (2-19). Thus we can write (2-19), using (2-94a) and (2-94b) for the magnetization of the N orbiting electrons superimposed with the circularly polarized a.c. signal of $\mathbf{B}_1^\pm$, as

$$\mathbf{B}^+ = \mu_0(\mathbf{H}_1^+ + \mathbf{M}_1^+) = \mu_0(\mathbf{H}_1^+ + N\mathbf{m}_t^+) = \mu_0\left(\mathbf{H}_1^+ + \frac{Nm_0^+ |\gamma| \mathbf{B}_1^+}{\omega_0 - \omega^+}\right)$$

$$= \mu_0\left(1 + \frac{Nm_0^+ |\gamma| \mu_0}{\omega_0 - \omega^+}\right)\mathbf{H}_1^+ = \mu_0\left(1 + \frac{\mu_0 |\gamma| M_0^+}{\omega_0 - \omega^+}\right)\mathbf{H}_1^+ = \mu_e^+ \mathbf{H}_1^+ \tag{2-95a}$$

$$\mathbf{B}^- = \mu_0(\mathbf{H}_1^- + \mathbf{M}_1^-) = \mu_0(\mathbf{H}_1^- + N\mathbf{m}_t^-) = \mu_0\left(\mathbf{H}_1^- + \frac{Nm_0^- |\gamma| \mathbf{B}_1^-}{\omega_0 + \omega^-}\right)$$

$$= \mu_0\left(1 + \frac{Nm_0^- |\gamma| \mu_0}{\omega_0 + \omega^-}\right)\mathbf{H}_1^- = \mu_0\left(1 + \frac{\mu_0 |\gamma| M_0^-}{\omega_0 + \omega^-}\right)\mathbf{H}_1^- = \mu_e^- \mathbf{H}_1^- \tag{2-95b}$$

where

$$M_0^+ = Nm_0^+ \tag{2-95c}$$

$$M_0^- = Nm_0^- \tag{2-95d}$$

$$\boxed{\mu_e^+ = \mu_0\left(1 + \frac{\mu_0 |\gamma| M_0^+}{\omega_0 - \omega^+}\right)} \tag{2-95e}$$

$$\boxed{\mu_e^- = \mu_0\left(1 + \frac{\mu_0 |\gamma| M_0^-}{\omega_0 + \omega^-}\right)} \tag{2-95f}$$

In (2-95e) and (2-95f) μ_e^+ and μ_e^- represent, respectively, the effective permeabilities for clockwise and counterclockwise circularly polarized waves. It is apparent that the two are not equal, which is a fundamental property utilized in the design of nonreciprocal microwave devices.

If the static magnetic field $\mathbf{B}_0$ is much larger than the superimposed a.c. magnetic field $\mathbf{B}_1^\pm \left(B_0 \gg B_1^\pm\right)$ so that the magnetization of the ferrite material is saturated by the static field, then all the spinning dipoles are tightly coupled and the entire material acts as a large single magnetic dipole. In that case the magnetization vector $\mathbf{M}^\pm$ for the positive (CW) and negative (CCW) circularly polarized fields superimposed on the static field can be approximated by

$$\mathbf{M}^\pm = N\mathbf{m}^\pm \simeq \mathbf{M}_s \simeq \mathbf{M}_0 \tag{2-96}$$

where $\mathbf{M}_s$ is the magnetization vector caused by the static field when no time-varying magnetic field is applied. For those cases the effective permeabilities can be approximated by

$$\mu_e^+ \simeq \mu_0 \left(1 + \frac{\mu_0 |\gamma| M_s}{\omega_0 - \omega^+} \right) \tag{2-97a}$$

$$\mu_e^- \simeq \mu_0 \left(1 + \frac{\mu_0 |\gamma| M_s}{\omega_0 + \omega^-} \right) \tag{2-97b}$$

which are not equal. Equations 2-97a and 2-97b are good approximations when the a.c. signals are small compared to the applied static field.

It can be shown (see Chapter 4) that a time-harmonic transverse electromagnetic (TEM) wave can be decomposed into a combination of clockwise and counterclockwise circularly polarized waves. Therefore the implications of (2-97a) and (2-97b) are that when a TEM wave travels through a ferrite material the clockwise circularly polarized portion of the wave will experience the permeability of (2-97a) while the counterclockwise wave will experience that of (2-97b). Since the permeability of a material influences the phase velocity and phase constant (see Chapter 4), the phases associated with (2-97a) and (2-97b) will be different. This is one of the fundamental features of ferrites that is utilized for the design of microwave nonreciprocal devices.

When an unbounded ferrite material is subjected to a static magnetic field $\mathbf{B}_0$ directed along the z axis of

$$\mathbf{B}_0 = \hat{\mathbf{a}}_z B_0 = \hat{\mathbf{a}}_z \mu_0 H_0 \tag{2-98a}$$

and a time-harmonic magnetic field $\mathscr{B}$ of

$$\mathscr{B} = \mu_0 \mathscr{H} \tag{2-98b}$$

each will induce a magnetization per unit volume vector of $\mathbf{M}_s$, and $\mathcal{M}$, respectively. The script is used to indicate time-varying components. Under these conditions, the equation of motion can be written as

$$\frac{d(\mathbf{M}_s + \mathcal{M})}{dt} = \frac{d\mathcal{M}}{dt} = -|\gamma| \big[(\mathbf{M}_s + \mathcal{M}) \times (\mathbf{B}_0 + \mathscr{B}) \big] \tag{2-99}$$

or in expanded form as

$$\frac{d\mathcal{M}}{dt} = -|\gamma| \mu_0 \big[(\mathbf{M}_s + \mathcal{M}) \times (\mathbf{H}_0 + \mathscr{H}) \big]$$

$$\frac{d\mathcal{M}}{dt} = -|\gamma| \mu_0 (\mathbf{M}_s \times \mathbf{H}_0 + \mathbf{M}_s \times \mathscr{H} + \mathcal{M} \times \mathbf{H}_0 + \mathcal{M} \times \mathscr{H}) \tag{2-99a}$$

If the time-harmonic field $\mathscr{B}$ is small such that

$$|\mathcal{M}| \ll |\mathbf{M}_s| \tag{2-100a}$$

$$|\mathscr{H}| \ll |\mathbf{H}_0| \tag{2-100b}$$

and since the applied magnetic field $\mathbf{B}_0$ is in the same direction as the static saturation magnetization vector $\mathbf{M}_s$, or

$$\mathbf{M}_s \times \mathbf{H}_0 = 0 \qquad (2\text{-}101)$$

then (2-99a) can be approximated by

$$\frac{d\mathcal{M}}{dt} \simeq -|\gamma|\mu_0(\mathbf{M}_s \times \mathcal{H} + \mathcal{M} \times \mathbf{H}_0) \qquad (2\text{-}102)$$

If each of the time-harmonic components is written in the form described by (1-61a) through (1-61d), then (2-102) ultimately reduces, using (2-86), to

$$
\begin{aligned}
j\omega\mathbf{M} &\simeq -|\gamma|\mu_0(\mathbf{M}_s \times \mathbf{H} + \mathbf{M} \times \mathbf{H}_0) \\
j\omega\mathbf{M} + |\gamma|\mu_0\mathbf{M} \times \mathbf{H}_0 &\simeq -|\gamma|\mu_0\mathbf{M}_s \times \mathbf{H} \\
j\omega\mathbf{M} + |\gamma|\mathbf{M} \times \mathbf{B}_0 &\simeq -|\gamma|\mu_0\mathbf{M}_s \times \mathbf{H} \\
j\omega\mathbf{M} + \mathbf{M} \times (|\gamma|\mathbf{B}_0) &\simeq -|\gamma|\mu_0\mathbf{M}_s \times \mathbf{H} \\
j\omega\mathbf{M} + \omega_0\mathbf{M} \times \hat{\mathbf{a}}_z &\simeq -|\gamma|\mu_0\mathbf{M}_s \times \mathbf{H}
\end{aligned}
\qquad (2\text{-}103)
$$

Assuming $\mathbf{M}_s$ has only a z component, whereas $\mathbf{H}$ has both x and y components, expanding (2-103) leads to

$$j\omega M_x + \omega_0 M_y \simeq |\gamma|\mu_0 M_s H_y \qquad (2\text{-}104a)$$

$$-\omega_0 M_x + j\omega M_y \simeq -|\gamma|\mu_0 M_s H_x \qquad (2\text{-}104b)$$

$$j\omega M_z \simeq 0 \qquad (2\text{-}104c)$$

Solving (2-104a) through (2-104c) for M_x, M_y, and M_z leads to

$$M_x = \frac{\omega_0|\gamma|\mu_0 M_s H_x + j\omega|\gamma|\mu_0 M_s H_y}{\omega_0^2 - \omega^2} \qquad (2\text{-}105a)$$

$$M_y = \frac{\omega_0|\gamma|\mu_0 M_s H_y - j\omega|\gamma|\mu_0 M_s H_x}{\omega_0^2 - \omega^2} \qquad (2\text{-}105b)$$

$$M_z = 0 \qquad (2\text{-}105c)$$

By introducing the magnetic susceptibility tensor $\bar{\mathcal{X}}$, we can write (2-105a) through (2-105c) using the forms of (2-21) and (2-22) as

$$[M] = [\chi_m][H] \qquad (2\text{-}106)$$

or

$$
\begin{bmatrix} M_x \\ M_y \\ M_z \end{bmatrix} =
\begin{bmatrix} \chi_{xx} & \chi_{xy} & 0 \\ \chi_{yx} & \chi_{yy} & 0 \\ 0 & 0 & 0 \end{bmatrix}
\begin{bmatrix} H_x \\ H_y \\ H_z \end{bmatrix}
\qquad (2\text{-}106a)
$$

$$[B] = \mu_0[[I] + [\chi_m]][H] \qquad (2\text{-}107)$$

or

$$
\begin{bmatrix} B_x \\ B_y \\ B_z \end{bmatrix} = \mu_0
\begin{bmatrix} 1 + \chi_{xx} & \chi_{xy} & 0 \\ \chi_{yx} & 1 + \chi_{yy} & 0 \\ 0 & 0 & 1 \end{bmatrix}
\begin{bmatrix} H_x \\ H_y \\ H_z \end{bmatrix}
\qquad (2\text{-}107a)
$$

where

$$\chi_{xx} = \chi_{yy} = \frac{\omega_0|\gamma|\mu_0 M_s}{\omega_0^2 - \omega^2} \qquad (2\text{-}107b)$$

$$\chi_{xy} = -\chi_{yx} = j\frac{\omega|\gamma|\mu_0 M_s}{\omega_0^2 - \omega^2} \qquad (2\text{-}107c)$$

In (2-106) through (2-107c) χ_{xx}, χ_{yy}, χ_{xy}, and χ_{yx} represent the entries of the susceptibility tensor $\overline{\overline{\chi}}$ for the ferrite material and $[I]$ is the unit matrix. Equation 2-107a can also be written in a more general form as

$$\mathbf{B} = \overline{\overline{\mu}} \cdot \mathbf{H} \qquad (2\text{-}108)$$

where $\overline{\overline{\mu}}$ is the permeability tensor written, in general, as a 3×3 matrix of the form

$$[\overline{\overline{\mu}}] = \mu_0 \begin{bmatrix} 1 + \chi_{xx} & \chi_{xy} & 0 \\ \chi_{yx} & 1 + \chi_{yy} & 0 \\ 0 & 0 & 1 \end{bmatrix} \qquad (2\text{-}108a)$$

which is a more general form of (2-22a).

Practical ferrite materials also contain magnetic losses. Therefore the permeability of the material will have both real and imaginary parts, as given by (2-79) or (2-79a). A phenomenological model used to derive the variations as a function of frequency of the real $\mu'(\omega)$ and imaginary $\mu''(\omega)$ parts of both (2-95e) and (2-95f) when losses are included is somewhat complex and beyond the treatment presented here for ferrites. However, the development of this can be found in [35] and [36]. A typical plot as a function of ω_0/ω is shown in Figure 2-23 where $\omega_m = \mu_0|\gamma|M_s$. A resonance phenomenon is indicated when $\omega_0/\omega = 1$.

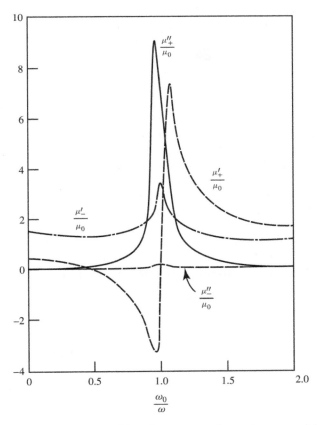

Figure 2-23 Frequency variations of real and imaginary parts of complex permeability for circularly polarized waves in a ferrite ($\omega = 20\pi$ GHz, $\omega_m = 11.2\pi$ GHz, $\alpha = 0.05$). (Source: R. E. Collin, *Foundations for Microwave Engineering*, Second Edition, 1992, McGraw-Hill Book Co.)

2.10 MULTIMEDIA

On the website that accompanies this book, the following multimedia resources are included for the review, understanding, and presentation of the material of this chapter.

- **PowerPoint (PPT)** viewgraphs, in multicolor.

REFERENCES

1. R. R. Wright and H. R. Skutt, *Electronics: Circuits and Devices*, Ronald Press, New York, 1965.
2. S. V. Marshall and G. G. Skitek, *Electromagnetic Concepts and Applications*, Second Edition, Prentice-Hall, Englewood Cliffs, NJ, 1987.
3. R. M. Rose, L. A. Shepard, and J. Wulff, *Electronic Properties*, John Wiley & Sons, New York, p. 262, 1966.
4. M. F. Uman, *Introduction to the Physics of Electronics*, Prentice-Hall, Englewood Cliffs, NJ, 1974.
5. E. M. Conwell, "Properties of silicon and germanium: II," *Proc. IRE*, vol. 46, pp. 1281–1300, June 1958.
6. J. D. Kraus, *Electromagnetics*, Fourth Edition, McGraw-Hill, New York, 1992.
7. F. London, *Superfluids*, vol. 1, Dover, New York, 1961.
8. A. C. Rose-Innes and E. H. Rhoderick, *Introduction to Superconductivity*, Pergamon, Elmsford, NY, 1978.
9. J. Bardeen, L. N. Cooper, and J. R. Schrieffer, "Microscopic theory of superconductivity," *Physical Review*, vol. 106, pp. 162–164, 1957.
10. B. C. Fenton, "Superconductivity breakthroughs," *Radio Electronics*, vol. 59, no. 2, pp. 43–45, Feb. 1988.
11. A. Khurana, "Superconductivity seen above the boiling point of nitrogen," *Physics Today*, vol. 40, no. 4, pp. 17–23, Apr. 1987.
12. P. C. W. Chu, "High Temperature Superconductivity: Past, Present and Future," APS Colloquium, Argonne National Laboratory, Oct. 2, 2002.
13. G. P. de Loor and F. W. Meijboom, "The dielectric constant of foods and other materials with high water content at microwave frequencies," *Journal of Food Technology*, vol. 1, no. 1, pp. 313–322, 1966.
14. W. E. Pace, W. B. Westphal, and S. A. Goldblith, "Dielectric properties of commercial cooking oils," *Journal of Food Science*, vol. 33, p. 30, 1968.
15. D. Van Dyke, D. I. C. Wang, and S. A. Goldblith, "Dielectric loss factor of reconstituted ground beef: The effect of chemical composition," *Journal of Food Technology*, vol. 23, p. 944, 1969.
16. N. E. Bengtsson and P. O. Risman, "Dielectric properties of foods at 3 GHz as determined by a cavity perturbation technique. II. Measurements on food materials," *Journal of Microwave Power*, vol. 6, no. 2, pp. 107–123, 1971.
17. S. S. Stuchly and M. A. K. Hamid, "Physical properties in microwave heating processes," *Journal of Microwave Power*, vol. 7, no. 2, p. 117, 1972.
18. N. E. Bengtsson and T. Ohlsson, "Microwave heating in the food industry," *Proc. IEEE*, vol. 62, no. 1, pp. 44–55, Jan. 1974.
19. P. Debye, *Polar Molecules*, Chem. Catalog Co., New York, 1929.
20. H. F. Cook, "The dielectric behaviour of some types of human tissue at microwave frequencies," *British Journal of Applied Physics*, vol. 2, p. 295, Oct. 1951.
21. A. W. Guy, J. F. Lehmann, and J. B. Stonebridge, "Therapeutic applications of electromagnetic power," *Proc. IEEE*, vol. 62, no. 1, pp. 55–75, Jan. 1974.
22. "Biological Effects of EM Waves," Special Issue, *Radio Science*, vol. 12, no. 6(S), Nov.–Dec. 1977.
23. A. R. von Hippel, *Dielectrics and Waves*, MIT Press, Cambridge, MA, 1954.

24. A. R. von Hippel, *Dielectric Materials and Applications*, John Wiley & Sons, New York, pp. 93–252, 1954.

25. C. A. Balanis, "Dielectric constant and loss tangent measurements at 60 and 90 GHz using the Fabry–Perot interferometer," *Microwave Journal*, vol. 14, pp. 39–44, Mar. 1971.

26. L. D. Landau and E. M. Lifshitz, *Electrodynamics of Continuous Media* (translated by J. B. Sykes and J. S. Bell), Pergamon, Elmsford, NY, Chapter IX, pp. 239–268, 1960.

27. J. D. Jackson, *Classical Electrodynamics*, Second Edition, John Wiley & Sons, New York, p. 311, 1975.

28. S. Ramo, J. R. Whinnery, and T. Van Duzer, *Fields and Waves in Communication*, Second Edition, John Wiley & Sons, New York, pp. 556–558, 671, 1984.

29. M. C. Lovell, A. J. Avery, and M. W. Vernon, *Physical Properties of Materials*, Van Nostrand–Reinhold, Princeton, NJ, p. 161, 1976.

30. J. L. Snoek, "Non-metallic magnetic material for high frequencies," *Philips Technical Review*, vol. 8, no. 12, pp. 353–384, Dec. 1946.

31. D. Polder, "On the theory of ferromagnetic resonance," *Philosophical Magazine*, vol. 40, pp. 99–115, Jan. 1949.

32. W. H. von Aulock, *"Ferrimagnetic materials—phenomenological and atomic models,"* in *Handbook of Microwave Ferrite Materials*, W. H. von Aulock (Ed.), Section I, Chapter 1, Academic, New York, 1965.

33. W. von Aulock and J. H. Rowen, "Measurement of dielectric and magnetic properties of ferromagnetic materials at microwave frequencies," *The Bell System Technical Journal*, vol. 36, pp. 427–448, Mar. 1957.

34. W. H. von Aulock, "Selection of ferrite materials for microwave device applications," *IEEE Trans. on Magnetics*, vol. MAG-2, no. 3, pp. 251–255, Sept. 1966.

35. R. E. Collin, *Foundations for Microwave Engineering*, Second Edition, McGraw-Hill, New York, pp. 286–302, 1992.

36. R. F. Soohoo, *Theory and Application of Ferrites*, Chapter 5, Prentice-Hall, Englewood Cliffs, NJ, 1960.

PROBLEMS

2.1. A dielectric slab, shown in Figure P2-1, exhibits an electric polarization vector of

$$\mathbf{P} = \hat{\mathbf{a}}_y 2.762 \times 10^{-11} \text{ C/m}^2$$

when it is subjected to an electric field of

$$\mathbf{E} = \hat{\mathbf{a}}_y 2 \text{ V/m}$$

Determine:
(a) The bound surface charge density q_{sp} in each of its six faces.
(b) The net bound charge Q_p associated with the slab.
(c) The volume bound charge density q_{vp} *within* the dielectric slab.
(d) The dielectric constant of the material.

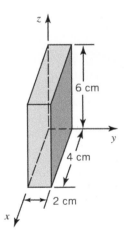

Figure P2-1

2.2. A cylindrical dielectric shell of Figure P2-2 with inner and outer radii, respectively, of $a = 2$ cm and $b = 6$ cm, and of length $\ell = 10$ cm exhibits an electric polarization vector of

$$\mathbf{P} = \hat{\mathbf{a}}_\rho \frac{2}{\rho} \times 10^{-10}\,\text{C/m}^2, \qquad a \le \rho \le b$$

when it is subjected to an electric field of

$$\mathbf{E} = \hat{\mathbf{a}}_\rho \frac{7.53}{\rho}\,\text{V/m}, \quad a \le \rho \le b$$

Neglecting fringing, find:
(a) The bound surface charge density q_{sp} in each of the surfaces.
(b) The net bound charge Q_p at the inner, outer, upper, and lower surfaces.
(c) The volume bound charge density q_{vp} within the dielectric.
(d) The dielectric constant of the material.

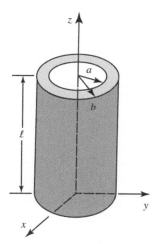

Figure P2-2

2.3. A spherical dielectric shell of Figure P2-3 with inner and outer radii $a = 2$ cm and $b = 4$ cm, respectively, exhibits an electric polarization vector of

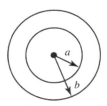

Figure P2-3

$$\mathbf{P} = \hat{\mathbf{a}}_r \frac{31.87}{r^2} \times 10^{-12}\,\text{C/m}^2, \quad a \le r \le b$$

when it is subjected to an electric field of

$$\mathbf{E} = \hat{\mathbf{a}}_r \frac{0.45}{r^2}\,\text{V/m}, \quad a \le r \le b$$

Determine the:
(a) Bound surface charge density q_{sp} in each of the surfaces.
(b) Net bound charge Q_p at the inner and outer surfaces.
(c) Volume bound charge density q_{vp} within the dielectric.
(d) Dielectric constant of the material.

2.4. A parallel-plate capacitor, whose parallel plates are separated by a distance d, is filled with a dielectric medium (polystyrene) whose relative permittivity (dielectric constant) is 2.56. The electric field inside the dielectric medium between the plates is 15×10^3 V/m. Assuming the plates are separated by a distance of 2 cm, determine the:
(a) Electric flux density D (in C/m^2).
(b) Electric susceptibility χ_e.
(c) Electric polarization P (in C/m^2).
(d) Electric surface charge density in each of the plates (in C/m^2).
(e) Voltage between the plates (in V). Assume the plates are separated by 2 cm.

2.5. Two parallel conducting plates, each having a surface area of 2×10^{-2} m^2 on its sides, form a parallel-plate capacitor. Their separation is 1.25 mm and the medium between them is free space. A 100-V d.c. battery is connected across them, and it is maintained there at all times. Then a dielectric sheet, 1 mm thick and with the same shape and area as the plates, is slipped carefully between the plates so that one of its sides touches one of the conducting plates. After the insertion of the slab and neglecting fringing, if the dielectric constant of the dielectric sheet is $\varepsilon_r = 5$, determine the:
(a) Electric field intensity between the plates (inside and outside the slab).
(b) Electric flux density between the plates (inside and outside the slab).
(c) Surface charge density in each of the plates.
(d) Total charge in each of the plates.
(e) Capacitance across the slab, the free space, and both of them.
(f) Energy stored in the slab, the free space, and both of them.

2.6. For Problem 2.5, assume that after the 100-V voltage source charges the conducting plates, it is then removed. Then the dielectric sheet is inserted between the plates as indicated in Problem 2.5. After the insertion of the dielectric sheet, find the:

(a) Total charge Q on the upper and lower plates.

(b) Surface charge density on the upper and lower plates.

(c) Electric flux density in the dielectric slab and free space.

(d) Electric field intensity in the dielectric slab and free space.

(e) Voltage across the slab, the free space, and both of them.

(f) Capacitance across the slab, the free space, and both of them.

(g) Energy stored in the slab, the free space, and both of them.

2.7. A parallel-plate capacitor consists of two identical square PEC plates, each of area 100 cm^2, 2 cm separation, and with a perfect lossless dielectric slab/material of relative permittivity (dielectric constant) of $\varepsilon_r = 2.56$, area 100 cm^2, and thickness 2 cm between them (assume the permeability of the dielectric slab is the same as that of free space). A 10-V d.c source is connected and charges the plates; the d.c. voltage source is maintained connected to the plates at all times. Determine the:

(a) Electric field intensity between the plates (in V/m).

(b) Electric flux density between the plates (in C/m^2).

(c) Electric polarization in the dielectric slab (in C/m^2).

(d) Electric susceptibility of the dielectric slab (dimensionless).

(e) Magnitude of the electric surface charge density on each plate (in C/m^2).

(f) Magnitude of the total electric charge in each plate (in C).

(g) Capacitance of the two-plate capacitor (in farads) using one formula.

(h) Capacitance of the two-plate capacitor (in farads) using an alternate formula.

(i) Are the two answers for the capacitance, in parts (g) and (h), equal or different? Why or why not?

2.8. A parallel-plate capacitor consists of two identical square PEC plates each plate of area 100 cm^2 and the two plates separated by 2 cm.

Initially the medium between the plates is free space and the plates are connected and charged by a 10-V d.c. source. Determine the:

(a) Electric field intensity (in V/m) between the plates when there is free space between them.

(b) Electric flux density (in C/m^2) between the plates when there is free space between them.

(c) Magnitude of electric surface charge density on each plate (in C/m^2) when there is free space between them.

Once the plates, with free space between them, are charged by the 10-V d.c. source, the d.c. source is removed/disconnected and then a perfect lossless dielectric slab/material of relative permittivity (dielectric constant) $\varepsilon_r = 2.56$, area 100 cm^2 and 2 cm thickness is inserted between the plates.

After the removal/disconnection of the d.c. voltage source and the insertion of the dielectric slab, determine in the presence of the dielectric slab the:

(d) Magnitude of the electric surface charge density on each plate (in C/m^2).

(e) Magnitude of the total electric charge in each plate (in C).

(f) Electric flux density between the plates (in C/m^2).

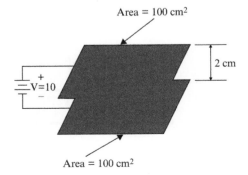

Area = 100 cm²

2 cm

+
V=10
−

Area = 100 cm²

Figure P2-7

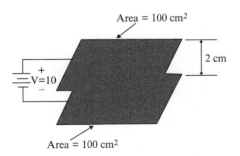

Area = 100 cm²

2 cm

+
V=10
−

Area = 100 cm²

Figure P2-8

(g) Electric field intensity between the plates (in V/m).
(h) Voltage between the plates (in V).
(i) Capacitance of the two-plate capacitor (in farads) using one formula.
(j) Capacitance of the two-plate capacitor (in farads) using an alternate formula.
(k) Are the two answers for the capacitance, in parts (i) and (j), equal or different? Why or why not?

2.9. A parallel-plate capacitor of Figure P2-9, with plates each of area 64 cm^2, separation of 4 cm, and free space between them, is charged by a 8-V d.c. source that is kept across the plates at all times. After the charging of the plates a 4-cm dielectric slab of polystyrene ($\varepsilon_r = 2.56$, $\mu_r = 1$) 4 cm in thickness is inserted between the plates and occupies half of the space between them.
Before insertion of the slab, determine the:
(a) Total charge on the upper and lower plates.
(b) Electric field between the plates.
(c) Electric flux density between the plates.
(d) Capacitance of the capacitor.
(e) Total stored energy in the capacitor.
After insertion of the slab, determine the:
(f) Total charge on the upper and lower plates in the free space and dielectric parts.
(g) Electric field in the free space and dielectric parts.
(h) Electric flux density in the free space and dielectric parts.
(i) Capacitance of each of the free space and dielectric parts.
(j) Total capacitance (combined free space and dielectric parts).
(k) Stored energy in each of the free space and dielectric parts.

(l) Total stored energy (combined free space and dielectric parts). Compare with that of part (e) and if there is a difference, explain why.

2.10. Repeat Problem 2.9 except assume that the voltage source is removed after the charging of the plates and before the insertion of the slab.

2.11. Two parallel PEC plates, each having a total surface area of 2 cm^2, form a parallel plate capacitor. The separation between the PEC plates is 1.25 mm and the medium between the plates is initially free space. A 100-V battery is attached to the plates, charges them, and is then removed. After removal of the battery, a 1 mm thick dielectric slab, with a dielectric constant (relative permittivity) of 5 and an area of 2 cm^2 on each of its sides is inserted between the PEC plates and occupies the lower part of the space between the PEC plates (basically touching the lower PEC plate), as shown in Figure P2-11.
After insertion of the dielectric slab, find the:
(a) Total charge Q on the lower and upper PEC plates (in C).
(b) Surface charge density on the upper and lower PEC plates (in C/m^2).
(c) Electric flux density in the:
   1. Dielectric (in C/m^2)
   2. Free space medium (in C/m^2)
(d) Electric field intensity in the:
   1. Dielectric (in V/m)
   2. Free space medium (in V/m)
(e) Total voltage in the:
   1. Dielectric slab (in V)
   2. Free-space medium (in V)
   3. Between the PEC plates (dielectric slab + free space medium) (in V)
(f) The capacitance across the:
   1. Dielectric slab (in farads)
   2. Free-space medium (in farads)
   3. Between the PEC plates (dielectric slab + free space medium) (in farads)

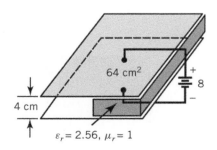

Figure P2-9

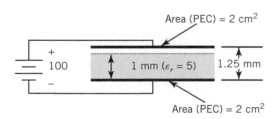

Figure P2-11

2.12. Two different dielectric slabs, with dielectric constants of 2 and 6, respectively, are placed one on top of the other, between two square perfectly electric conducting (PEC) plates, each plate with an area of $1 \, cm^2$, as shown in Figure P2-12. The thickness of each dielectric slab is 1 cm. A 10-V d.c. power supply is placed between the two plates forming a parallel-plate capacitor, and is maintained connected at all times. Find (in terms of ε_0 when applicable) the following:

(a) Electric field in each slab (in V/meter).
(b) Electric flux density in each slab (in C/m^2).
(c) Total charge in each of the two PEC plates (in C).
(d) Total capacitance of the parallel-plate capacitor (in farads) using its definition based on the charge and voltage.
(e) Capacitance of each slab (in farads) based on the definition of capacitance (using plate area, separation and permittivity of the medium).
(f) Total capacitance (in farads) of the parallel-plate capacitor, using the capacitances of part (e). How does this capacitance compares with that of part (d)? Are they the same or different? Explain.

(c) Total charge density in each of the two PEC plates (q_1 and q_2) (in C/m^2).
(d) Total charge in each of the two PEC plates (Q_1 and Q_2) (in C).
(e) Capacitance in each of the parallel-plate capacitors (C_1 and C_2) (in farads) using the definition based on the charge and voltage.
(f) Total capacitance C_T (in farads) of the parallel-plate capacitor using the capacitances from part (e).
(g) Capacitance of each capacitor (in farads) based on the definition of capacitance for each (using plate area, separation and permittivity of the medium).
(h) Total capacitance C_T (in farads) of the parallel plate capacitors, using the capacitances of part (g).
(i) How do the capacitances of parts (f) and (h) compare? Are they the same or different? Should they be the same or different? Explain.

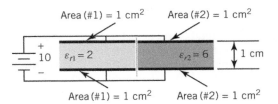

Figure P2-13

2.14. A 10-V d.c. power supply, placed between the four plates as shown in Figure P2-13, with air between the plates which form two separate parallel-plate capacitors, charges the plates of the two capacitors. The power supply is then disconnected; the two capacitors are not connected to each other.
After the power supply is disconnected and the two capacitors are not connected to each other, two different dielectric slabs with dielectric constants of 2 and 6, respectively, are inserted side-by-side between the two rectangular PEC plates. Each dielectric slab is square (1 cm by 1 cm; area $= 1 \, cm^2$) and each with a thickness of 1 cm. Each of the top and bottom PEC plates has dimensions of $1 \, cm \times 1 \, cm$, as shown in Figure P2-14.
For each capacitor, in the presence of the dielectric slabs but after the removal of the power supply, find the following:
(a) Total electric charge density in each of the four PEC plates (q_1 and q_2) (in C/m^2).

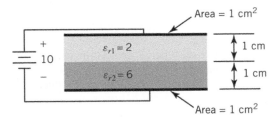

Figure P2-12

2.13. Two different dielectric slabs, with dielectric constants of 2 and 6, respectively, are placed side-by-side between four rectangular PEC plates. Each dielectric slab is square (1 cm by 1 cm; area $= 1 \, cm^2$) and each with a thickness of 1 cm. Each of the top and bottom PEC plates has dimensions of 1 cm by 1 cm, as shown in Figure P2-13. A 10-V d.c. power supply is placed between the two plates forming two separate parallel-plate capacitors.
For each capacitor, find the following:
(a) Electric field in each slab (E_1 and E_2) (in V/meter).
(b) Electric flux density in each slab (D_1 and D_2) (in C/m^2).

(b) Electric flux density in each slab (D_1 and D_2) (in C/m^2).

(c) Total charge in each of the four PEC plates (Q_1 and Q_2) (in C).

(d) Electric field in each slab (E_1 and E_2) (in V/meter).

(e) Voltage across each of the parallel plate capacitors (V_1 and V_2) (in V).

(f) Capacitance in each of the parallel-plate capacitors (C_1 and C_2) (in farads) based on the geometry of each capacitor (area of plates, separation of plates, permittivity of medium).

(g) Capacitance in each of the parallel-plate capacitors (C_1 and C_2) (in farads) based on the results of parts (c) and (e) (charge and voltage).

(h) Are the corresponding results/answers in parts (f) and (g) the same or different? Explain.

2.15. A 10-V d.c. voltage source, placed across the inner and outer conductors of a coaxial cylinder as shown in Figure P2-15, is used to charge the conductors and is then removed. The total charge in each conductor is $\pm Q$. The inner conductor has a radius of $a = 2$ cm, the radius of the outer conductor is 4 cm, and the length of the cylinder is $\ell = 6$ cm. Assuming no field fringing and free space between the conductors, find the:

(a) Electric field intensity between the conductors in terms of Q.

(b) Total charge Q on the inner and outer conductors.

(c) Surface charge density on the inner and outer conductors.

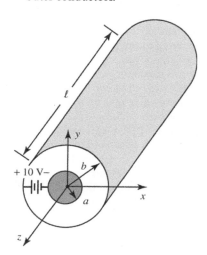

(d) Electric flux density between the conductors.

(e) Capacitance between the conductors.

(f) Energy stored between the conductors.

2.16. For Problem 2.15, assume that after the 10-V source charges the conductors and is removed, a cylindrical dielectric jacket of polystyrene ($\varepsilon_r = 2.56$) of inner radius $a = 2$ cm and outer radius $b = 3$ cm is inserted over the inner conductor of the coaxial cylinder. After the insertion of the jacket and neglecting fringing, find the:

(a) Total charge Q on the inner and outer conductors.

(b) Surface charge density on the inner and outer conductors.

(c) Electric flux density between the conductors in the dielectric and free space.

(d) Electric field intensity between the conductors in the dielectric and free space.

(e) Voltage between the conductors.

(f) Total capacitance between the conductors.

(g) Total energy stored between the conductors.

2.17. For Problem 2.16 assume that the 10-V source that charges the conductors remains connected at all times. Neglecting fringing, determine the:

(a) Electric field intensity between the conductors inside and outside the dielectric jacket.

(b) Electric flux density between the conductors inside and outside the dielectric jacket.

(c) Surface charge density in each of the plates.

(d) Total charge in each of the conductors.

(e) Total capacitance between the conductors.

(f) Total energy stored between the conductors.

2.18. A 10-V d.c. battery is connected between two concentric PEC (perfectly electric conducting) spheres. The inner PEC sphere has a radius a of $a = 1$ cm while the outer one has a radius b of $b = 2$ cm. The medium between the two spheres is free space. Assuming the total charge Q on the surface of one each of the two PEC spheres is uniformly distributed, determine the:

(a) Total charge Q (in C) on the:

• Outer surface of the inner sphere ($a = 1$ cm).

• Inner surface of the outer sphere ($b = 2$ cm).

(c) The surface charge density (in C/m^2):

• Outer surface of the inner sphere ($a = 1$ cm).

Figure P2-15

• Inner surface of the outer sphere ($b = 2$ cm).

P. S. r is the spherical radial distance from the center of the coordinate system toward the radial direction.

2.19. The electric flux density $\mathbf{D}_1$ within the dielectric circular cylinder of Figure P2-19, in cylindrical coordinates, is

$$\mathbf{D}_1 = 2\hat{\mathbf{a}}_\rho + 4\hat{\mathbf{a}}_\phi + 3\hat{\mathbf{a}}_z; \qquad \rho \leq a = 2 \text{ cm}$$

The dielectric circular cylinder has a radius $a = 2$ cm and a dielectric constant (relative permittivity) of $\varepsilon_r = 2.56$. Assume the permeability of the cylinder dielectric medium is the same as that of free space. The cylinder is surrounded by free space. Determine the:
(a) Vector electric flux density $\mathbf{D}_2$ outside the cylinder ($\rho > a$).
(b) Vector electric polarization $\mathbf{P}_1$ inside the dielectric cylinder ($\rho \leq a$).
(c) Vector electric polarization $\mathbf{P}_2$ outside the dielectric cylinder ($\rho > a$).
Write each expression for (a)–(c) in vector form.

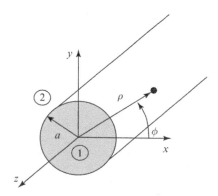

Figure P2-19

2.20. A 100-V d.c. voltage source is placed across two parallel-plate sets that are connected in parallel. Each conductor in each parallel-plate set has a surface area of 2×10^{-2} m^2 on each of its sides which are separated by 4 cm. For one parallel-plate set the medium between them is free space, whereas for the other it is lossless polystyrene ($\varepsilon_r = 2.56$, $\mu_r = 1$). For each parallel-plate set, by neglecting fringing, determine the:
(a) Electric field intensity between the plates.
(b) Electric flux density between the plates.
(c) Total charge on the upper and lower plates.

(d) Total energy stored between the plates.
For the two-set parallel-plate combination, determine the total:
(e) Charge on the two upper and two lower plates.
(f) Capacitance between the upper and lower plates.
(g) Energy stored between the plates.

2.21. For the coaxial cylinder of Problem 2.15 assume that once the 10-V voltage source charges the conductors and is removed, a curved dielectric slab of polystyrene ($\varepsilon_r = 2.56$, $\mu_r = 1$) of thickness equal to the spacing between the conductors is inserted between the conductors and occupies half of the space ($\pi \leq \phi \leq 2\pi$); the other half, $0 \leq \phi \leq \pi$, is still occupied by free space. By neglecting fringing, determine the:
(a) Total charge on the inner and outer conductors in free space and in polystyrene.
(b) Surface charge density on inner and outer conductors in free space and in polystyrene.
(c) Electric flux density between the conductors in free space and in polystyrene.
(d) Electric field intensity between the conductors in free space and in polystyrene.
(e) Voltage between the conductors in free space and in polystyrene.
(f) Capacitance between the conductors in free space, in polystyrene, and total.
(g) Energy stored between the conductors in free space, in polystyrene, and total.

2.22. For Problem 2.21 assume that the 10-V charging source is maintained across the conductors at all times. By neglecting fringing, determine the:
(a) Electric field intensity between the conductors in free space and in polystyrene.
(b) Electric flux density between the conductors in free space and in polystyrene.
(c) Charge density in each of the conductors in free space and in polystyrene.
(d) Total charge in each of the conductors in free space and in polystyrene.
(e) Capacitance between the conductors in free space, in polystyrene, and total.
(f) Energy stored between the conductors in free space, in polystyrene, and total.

2.23. The time-varying electric field inside a lossless dielectric material of polystyrene, of infinite dimensions and with a relative permittivity (dielectric constant) of 2.56, is

$$\mathscr{E} = \hat{\mathbf{a}}_z 10^{-3} \sin(2\pi \times 10^7 t) \text{ V/m}$$

Determine the corresponding:
(a) Electric susceptibility of the dielectric material.
(b) Time-harmonic electric flux density vector.
(c) Time-harmonic electric polarization vector.
(d) Time-harmonic displacement current density vector.
(e) Time-harmonic polarization current density vector defined as the partial derivative of the corresponding electric polarization vector.

Leave your answers in terms of ε_0, μ_0.

2.24. A rectangular slab of ferrimagnetic material as shown in Figure P2-24 exhibits a magnetization vector of

$$\mathbf{M} = \hat{\mathbf{a}}_z 1.245 \times 10^6 \text{ A/m}$$

when it is subjected to a magnetic field intensity of

$$\mathbf{H} = \hat{\mathbf{a}}_z 5 \times 10^3 \text{ A/m}$$

Find the:
(a) Bound magnetization surface current density in all its six faces.
(b) Bound magnetization volume current density within the slab.
(c) Net bound magnetization current associated with the slab.
(d) Relative permeability of the slab.

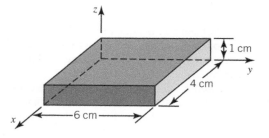

Figure P2-24

2.25. A coaxial line of length ℓ with inner and outer conductor radii of 1 and 3 cm, respectively, is filled with a ferromagnetic material, as shown in Figure P2-25. When the material is subjected to a magnetic field intensity of

$$\mathbf{H} = \hat{\mathbf{a}}_\phi \frac{0.3183}{\rho} \text{ A/m}$$

it induces a magnetization vector potential of

$$\mathbf{M} = \hat{\mathbf{a}}_\phi \frac{190.67}{\rho} \text{ A/m}$$

Determine the:
(a) Bound magnetization surface current density in all surfaces.
(b) Bound magnetization volume current density within the material.
(c) Net bound magnetization current associated with the coaxial line.
(d) Relative permeability of the material.

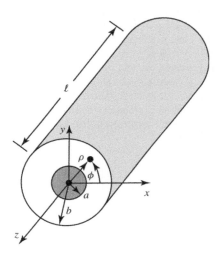

Figure P2-25

2.26. The magnetization vector inside a cylindrical magnetic bar of infinite length and circular cross section of radius $a = 1$ m, as shown in Figure P2-26, is given by

$$\mathbf{M} = \hat{\mathbf{a}}_\phi 10 \text{ A/m}$$

Find the:
(a) Magnetic surface current density at the outside circumferential surface of the bar.
(b) Magnetic volume current density at any point inside the bar.
(c) Total current that flows through the cross section of the bar.

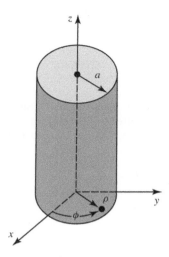

Figure P2-26

2.27. The current density through a cylindrical wire of square cross section as shown in Figure P2-27 is given by

$$\mathbf{J} \simeq \hat{\mathbf{a}}_z J_0 \, e^{-10^2[(a-|x|)+(a-|y|)]}$$

where J_0 is a constant. Assuming that each side of the wire is 2×10^{-2} m, find the current flow through the cross section of the wire.

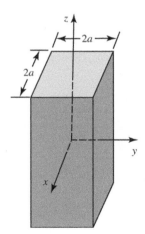

Figure P2-27

2.28. A 10-A current is pushed through a circular cross section of wire of infinite length as shown in Figure P2-28. Assuming that the current density over the cross section of the wire decays from its surface toward its center as

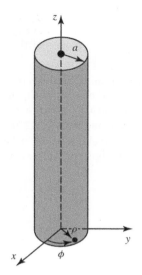

Figure P2-28

$$\mathbf{J} = \hat{\mathbf{a}}_z J_0 \, e^{-10^4(a-\rho)} \, \text{A/m}^2$$

where J_0 is the current density at the surface and the wire radius is $a = 10^{-2}$ m, determine the:

(a) Current density at the surface of the wire.
(b) Depth from the surface of the wire through which the current density has decayed to 36.8 percent of its value at the surface.

2.29. Show that the relaxation time constant for copper ($\sigma = 5.76 \times 10^7$ S/m) is much smaller than the period of waves in the microwave (1–10 GHz) region and is comparable to the period of x-rays $\left[\lambda \simeq 1 - 10\,\text{Å} = (1-10) \times 10^{-8}\,\text{cm}\right]$. Consequently, conductors cannot maintain a charge configuration long enough to permit propagation of the wave more than a short distance into the conductor at microwave frequencies. However x-ray propagation is possible because the relaxation time constant is comparable to the period of the wave.

2.30. Aluminum has a static conductivity of about $\sigma = 3.96 \times 10^7$ S/m and an electron mobility of $\mu_e = 2.2 \times 10^{-3}\,\text{m}^2/(\text{V-s})$. Assuming that an electric field of $\mathbf{E} = \hat{\mathbf{a}}_x 2$ V/m is applied perpendicularly to the square area of an aluminum wafer with cross-sectional area of about 10 cm^2, find the:

(a) Electron charge density q_{ve}.
(b) Electron drift velocity $\mathbf{v}_e$.

(c) Electric current density $\mathbf{J}$.

(d) Electric current flowing through the square cross section of the wafer.

(e) Electron density N_e.

2.31. Gallium arsenide is a very popular material used in electronics, especially in the solid state area. Gallium arsenide has a volume charge q_v for electrons and holes of $\pm 2.86 \times 10^{-7} \, C/m^3$, a mobility for electrons of $0.8 \, m^2/V\text{-}s$, and a mobility for holes of $0.032 \, m^2/V\text{-}s$.

Assume that the relative real part of its permittivity ($\varepsilon_r' = \varepsilon'/\varepsilon_0$) is 12, and its alternating loss tangent is 10^{-4}.

Determine, at 10 GHz, the:

(a) Static conductivity (S/m).

(b) Static loss tangent.

(c) Alternating conductivity (S/m).

(d) Total effective loss tangent.

(e) Total effective conductivity (S/m).

CHAPTER 3

Wave Equation and Its Solutions

3.1 INTRODUCTION

The electromagnetic fields of boundary-value problems are obtained as solutions to Maxwell's equations, which are first-order partial differential equations. However, Maxwell's equations are coupled partial differential equations, which means that each equation has more than one unknown field. These equations can be uncoupled only at the expense of raising their order. For each of the fields, following such a procedure leads to an uncoupled second-order partial differential equation that is usually referred to as the *wave equation*. Therefore electric and magnetic fields for a given boundary-value problem can be obtained either as solutions to Maxwell's or the wave equations. The choice of equations is related to individual problems by convenience and ease of use. In this chapter we will develop the vector wave equations for each of the fields, and then we will demonstrate their solutions in the rectangular, cylindrical, and spherical coordinate systems.

3.2 TIME-VARYING ELECTROMAGNETIC FIELDS

The first two of Maxwell's equations in differential form, as given by (1-1) and (1-2), are first-order, coupled differential equations; that is, both the unknown fields ($\mathscr{E}$ and $\mathscr{H}$) appear in each equation. Usually it is very desirable, for convenience in solving for $\mathscr{E}$ and $\mathscr{H}$, to uncouple these equations. This can be accomplished at the expense of increasing the order of the differential equations to second order. To do this, we repeat (1-1) and (1-2), that is,

$$\nabla \times \mathscr{E} = -\mathscr{M}_i - \mu \frac{\partial \mathscr{H}}{\partial t} \tag{3-1}$$

$$\nabla \times \mathscr{H} = \mathscr{J}_i + \sigma \mathscr{E} + \varepsilon \frac{\partial \mathscr{E}}{\partial t} \tag{3-2}$$

Balanis' Advanced Engineering Electromagnetics, Third Edition. Constantine A. Balanis.
© 2024 John Wiley & Sons, Inc. Published 2024 by John Wiley & Sons, Inc.
Companion Website: www.wiley.com/go/balanis/advancedengineeringelectromagnetics3e

where it is understood, in the remaining part of the book, that σ represents the effective conductivity σ_ε and ε represents ε'. Taking the curl of both sides of each of equations 3-1 and 3-2 and assuming a homogeneous medium, we can write that

$$\nabla \times \nabla \times \mathscr{E} = -\nabla \times \mathscr{M}_i - \mu \nabla \times \left(\frac{\partial \mathscr{H}}{\partial t} \right) = -\nabla \times \mathscr{M}_i - \mu \frac{\partial}{\partial t} (\nabla \times \mathscr{H}) \tag{3-3}$$

$$\nabla \times \nabla \times \mathscr{H} = \nabla \times \mathscr{J}_i + \sigma \nabla \times \mathscr{E} + \varepsilon \nabla \times \left(\frac{\partial \mathscr{E}}{\partial t} \right)$$

$$= \nabla \times \mathscr{J}_i + \sigma \nabla \times \mathscr{E} + \varepsilon \frac{\partial}{\partial t} (\nabla \times \mathscr{E}) \tag{3-4}$$

Substituting (3-2) into the right side of (3-3) and using the vector identity

$$\nabla \times \nabla \times \mathbf{F} = \nabla(\nabla \cdot \mathbf{F}) - \nabla^2 \mathbf{F} \tag{3-5}$$

into the left side, we can rewrite (3-3) as

$$\nabla(\nabla \cdot \mathscr{E}) - \nabla^2 \mathscr{E} = -\nabla \times \mathscr{M}_i - \mu \frac{\partial}{\partial t} \left[\mathscr{J}_i + \sigma \mathscr{E} + \varepsilon \frac{\partial \mathscr{E}}{\partial t} \right]$$

$$\nabla(\nabla \cdot \mathscr{E}) - \nabla^2 \mathscr{E} = -\nabla \times \mathscr{M}_i - \mu \frac{\partial \mathscr{J}_i}{\partial t} - \mu \sigma \frac{\partial \mathscr{E}}{\partial t} - \mu \varepsilon \frac{\partial^2 \mathscr{E}}{\partial t^2} \tag{3-6}$$

Substituting Maxwell's equation 1-3, or

$$\nabla \cdot \mathscr{D} = \varepsilon \nabla \cdot \mathscr{E} = \mathscr{q}_{ev} \Rightarrow \nabla \cdot \mathscr{E} = \frac{\mathscr{q}_{ev}}{\varepsilon} \tag{3-7}$$

into (3-6) and rearranging its terms, we have that

$$\boxed{ \nabla^2 \mathscr{E} = \nabla \times \mathscr{M}_i + \mu \frac{\partial \mathscr{J}_i}{\partial t} + \frac{1}{\varepsilon} \nabla \mathscr{q}_{ev} + \mu \sigma \frac{\partial \mathscr{E}}{\partial t} + \mu \varepsilon \frac{\partial^2 \mathscr{E}}{\partial t^2} } \tag{3-8}$$

which is recognized as an uncoupled second-order differential equation for $\mathscr{E}$.

In a similar manner, by substituting (3-1) into the right side of (3-4) and using the vector identity of (3-5) in the left side of (3-4), we can rewrite it as

$$\nabla(\nabla \cdot \mathscr{H}) - \nabla^2 \mathscr{H} = \nabla \times \mathscr{J}_i + \sigma \left(-\mathscr{M}_i - \mu \frac{\partial \mathscr{H}}{\partial t} \right) + \varepsilon \frac{\partial}{\partial t} \left(-\mathscr{M}_i - \mu \frac{\partial \mathscr{H}}{\partial t} \right)$$

$$\nabla(\nabla \cdot \mathscr{H}) - \nabla^2 \mathscr{H} = \nabla \times \mathscr{J}_i - \sigma \mathscr{M}_i - \mu \sigma \frac{\partial \mathscr{H}}{\partial t} - \varepsilon \frac{\partial \mathscr{M}_i}{\partial t} - \mu \varepsilon \frac{\partial^2 \mathscr{H}}{\partial t^2} \tag{3-9}$$

Substituting Maxwell's equation

$$\nabla \cdot \mathscr{B} = \mu \nabla \cdot \mathscr{H} = \mathscr{q}_{mv} \Rightarrow \nabla \cdot \mathscr{H} = \left(\frac{\mathscr{q}_{mv}}{\mu} \right) \tag{3-10}$$

into (3-9), we have that

$$\nabla^2 \mathcal{H} = -\nabla \times \mathcal{J}_i + \sigma \mathcal{M}_i + \frac{1}{\mu} \nabla(\mathcal{G}_{mv}) + \varepsilon \frac{\partial \mathcal{M}_i}{\partial t} + \mu\sigma \frac{\partial \mathcal{H}}{\partial t} + \mu\varepsilon \frac{\partial^2 \mathcal{H}}{\partial t^2} \qquad (3\text{-}11)$$

which is recognized as an uncoupled second-order differential equation for $\mathcal{H}$. Thus (3-8) and (3-11) form a pair of uncoupled second-order differential equations that are a by-product of Maxwell's equations as given by (1-1) through (1-4).

Equations 3-8 and 3-11 are referred to as the *vector wave equations* for $\mathcal{E}$ and $\mathcal{H}$. For solving an electromagnetic boundary-value problem, the equations that must be satisfied are Maxwell's equations as given by (1-1) through (1-4) or the wave equations as given by (3-8) and (3-11). Often, the forms of the wave equations are preferred over those of Maxwell's equations.

For source-free regions ($\mathcal{J}_i = \mathcal{G}_{ev} = 0$ and $\mathcal{M}_i = \mathcal{G}_{mv} = 0$), the wave equations 3-8 and 3-11 reduce, respectively, to

$$\nabla^2 \mathcal{E} = \mu\sigma \frac{\partial \mathcal{E}}{\partial t} + \mu\varepsilon \frac{\partial^2 \mathcal{E}}{\partial t^2} \qquad (3\text{-}12)$$

$$\nabla^2 \mathcal{H} = \mu\sigma \frac{\partial \mathcal{H}}{\partial t} + \mu\varepsilon \frac{\partial^2 \mathcal{H}}{\partial t^2} \qquad (3\text{-}13)$$

For source-free ($\mathcal{J}_i = \mathcal{G}_{ev} = 0$ and $\mathcal{M}_i = \mathcal{G}_{mv} = 0$) and lossless media ($\sigma = 0$), the wave equations 3-8 and 3-11 or 3-12 and 3-13 simplify to

$$\nabla^2 \mathcal{E} = \mu\varepsilon \frac{\partial^2 \mathcal{E}}{\partial t^2} \qquad (3\text{-}14)$$

$$\nabla^2 \mathcal{H} = \mu\varepsilon \frac{\partial^2 \mathcal{H}}{\partial t^2} \qquad (3\text{-}15)$$

Equations 3-14 and 3-15 represent the simplest forms of the vector wave equations.

3.3 TIME-HARMONIC ELECTROMAGNETIC FIELDS

For time-harmonic fields (time variations of the form $e^{j\omega t}$), the wave equations can be derived using a similar procedure as in Section 3.2 for the general time-varying fields, starting with Maxwell's equations as given in Table 1-4. However, instead of going through this process, we find, by comparing Maxwell's equations for the general time-varying fields with those for the time-harmonic fields (both are displayed in Table 1-4), that one set can be obtained from the other by replacing $\partial/\partial t \equiv j\omega$, $\partial^2/\partial t^2 \equiv (j\omega)^2 = -\omega^2$, and the instantaneous fields $(\mathcal{E}, \mathcal{H}, \mathcal{D}, \mathcal{B})$, respectively, with the complex fields $(\mathbf{E}, \mathbf{H}, \mathbf{D}, \mathbf{B})$ and vice versa. Doing this for the wave equations 3-8, 3-11, 3-12, and 3-13, we can write each, respectively, as

$$\nabla^2 \mathbf{E} = \nabla \times \mathbf{M}_i + j\omega\mu \mathbf{J}_i + \frac{1}{\varepsilon}\nabla q_{ev} + j\omega\mu\sigma \mathbf{E} - \omega^2 \mu\varepsilon \mathbf{E} \qquad (3\text{-}16a)$$

$$\nabla^2 \mathbf{H} = -\nabla \times \mathbf{J}_i + \sigma \mathbf{M}_i + j\omega\varepsilon \mathbf{M}_i + \frac{1}{\mu}\nabla q_{mv} + j\omega\mu\sigma \mathbf{H} - \omega^2 \mu\varepsilon \mathbf{H} \qquad (3\text{-}16b)$$

$$\nabla^2 \mathbf{E} = j\omega\mu\sigma \mathbf{E} - \omega^2 \mu\varepsilon \mathbf{E} = \gamma^2 \mathbf{E} \qquad (3\text{-}17a)$$

$$\nabla^2 \mathbf{H} = j\omega\mu\sigma \mathbf{H} - \omega^2 \mu\varepsilon \mathbf{H} = \gamma^2 \mathbf{H} \qquad (3\text{-}17b)$$

where

$$\gamma^2 = j\omega\mu\sigma - \omega^2 \mu\varepsilon = j\omega\mu(\sigma + j\omega\varepsilon) \qquad (3\text{-}17c)$$

$$\gamma = \alpha + j\beta = \text{propagation constant} \qquad (3\text{-}17d)$$

$$\alpha = \text{attenuation constant (Np/m)} \qquad (3\text{-}17e)$$

$$\beta = \text{phase constant (rad/m)} \qquad (3\text{-}17f)$$

The constants α, β, and γ will be discussed in more detail in Section 4.3 where α and β are expressed by (4-28c) and (4-28d) in terms of ω, ε, μ, and σ.

Similarly (3-14) and (3-15) can be written, respectively, as

$$\nabla^2 \mathbf{E} = -\omega^2 \mu\varepsilon \mathbf{E} = -\beta^2 \mathbf{E} \qquad (3\text{-}18a)$$

$$\nabla^2 \mathbf{H} = -\omega^2 \mu\varepsilon \mathbf{H} = -\beta^2 \mathbf{H} \qquad (3\text{-}18b)$$

where

$$\beta^2 = \omega^2 \mu\varepsilon \qquad (3\text{-}18c)$$

In the literature the phase constant β is also represented by k.

3.4 SOLUTION TO THE WAVE EQUATION

The time variations of most practical problems are of the time-harmonic form. Fourier series can be used to express time variations of other forms in terms of a number of time-harmonic terms. Electromagnetic fields associated with a given boundary-value problem must satisfy Maxwell's equations or the vector wave equations. For many cases, the vector wave equations reduce to a number of scalar Helmholtz (wave) equations, and the general solutions can be constructed once solutions to each of the scalar Helmholtz equations are found.

In this section we want to demonstrate at least one method that can be used to solve the scalar Helmholtz equation in rectangular, cylindrical, and spherical coordinates. The method is known as the *separation of variables* [1, 2], and the general solution to the scalar Helmholtz equation using this method can be constructed in 11 three-dimensional orthogonal coordinate systems (including the rectangular, cylindrical, and spherical systems) [3].

The solutions for the instantaneous time-harmonic electric and magnetic field intensities can be obtained by considering the forms of the vector wave equations given in either Section 3.2 or Section 3.3. The approach chosen here will be to use those of Section 3.3 to solve for the complex field intensities **E** and **H** first. The corresponding instantaneous quantities can then be formed using the relations (1-61a) through (1-61f) between the instantaneous time-harmonic fields and their complex counterparts.

3.4.1 Rectangular Coordinate System

In a rectangular coordinate system, the vector wave equations 3-16a through 3-18c can be reduced to three scalar wave (Helmholtz) equations. First, we will consider the solutions for source-free and lossless media. This will be followed by solutions for source-free but lossy media.

A. Source-Free and Lossless Media   For source-free ($\mathbf{J}_i = \mathbf{M}_i = q_{ve} = q_{vm} = 0$) and lossless ($\sigma = 0$) media, the vector wave equations for the complex electric and magnetic field intensities are those given by (3-18a) through (3-18c). Since (3-18a) and (3-18b) are of the same form, let us examine the solution to one of them. The solution to the other can then be written by an interchange of **E** with **H** or **H** with **E**. We will begin by examining the solution for **E**.

In rectangular coordinates, a general solution for **E** can be written as

$$\mathbf{E}(x,y,z) = \hat{\mathbf{a}}_x E_x(x,y,z) + \hat{\mathbf{a}}_y E_y(x,y,z) + \hat{\mathbf{a}}_z E_z(x,y,z) \tag{3-19}$$

where x, y, z are the rectangular coordinates, as illustrated in Figure 3-1. Substituting (3-19) into (3-18a) we can write that

$$\nabla^2 \mathbf{E} + \beta^2 \mathbf{E} = \nabla^2 (\hat{\mathbf{a}}_x E_x + \hat{\mathbf{a}}_y E_y + \hat{\mathbf{a}}_z E_z) + \beta^2 (\hat{\mathbf{a}}_x E_x + \hat{\mathbf{a}}_y E_y + \hat{\mathbf{a}}_z E_z) = 0 \tag{3-20}$$

which reduces to three scalar wave equations of

$$\nabla^2 E_x(x,y,z) + \beta^2 E_x(x,y,z) = 0 \tag{3-20a}$$

$$\nabla^2 E_y(x,y,z) + \beta^2 E_y(x,y,z) = 0 \tag{3-20b}$$

$$\nabla^2 E_z(x,y,z) + \beta^2 E_z(x,y,z) = 0 \tag{3-20c}$$

because

$$\nabla^2 (\hat{\mathbf{a}}_x E_x + \hat{\mathbf{a}}_y E_y + \hat{\mathbf{a}}_z E_z) = \hat{\mathbf{a}}_x \nabla^2 E_x + \hat{\mathbf{a}}_y \nabla^2 E_y + \hat{\mathbf{a}}_z \nabla^2 E_z \tag{3-21}$$

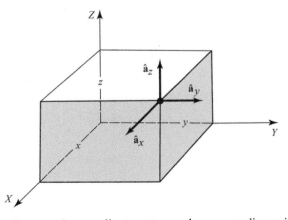

Figure 3-1   Rectangular coordinate system and corresponding unit vectors.

Equations 3-20a through 3-20c are all of the same form; once a solution of any one of them is obtained, the solutions to the others can be written by inspection. We choose to work first with that for E_x as given by (3-20a).

In expanded form (3-20a) can be written as

$$\nabla^2 E_x + \beta^2 E_x = \frac{\partial^2 E_x}{\partial x^2} + \frac{\partial^2 E_x}{\partial y^2} + \frac{\partial^2 E_x}{\partial z^2} + \beta^2 E_x = 0 \tag{3-22}$$

Using the *separation-of-variables method*, we assume that a solution for $E_x(x,y,z)$ can be written as

$$E_x(x,y,z) = f(x)g(y)h(z) \tag{3-23}$$

where the x, y, z variations of E_x are separable (hence the name). If any inconsistencies are encountered with assuming such a form of solution, another form must be attempted. This is the procedure usually followed in solving differential equations. Substituting (3-23) into (3-22), we can write that

$$gh\frac{\partial^2 f}{\partial x^2} + fh\frac{\partial^2 g}{\partial y^2} + fg\frac{\partial^2 h}{\partial z^2} + \beta^2 fgh = 0 \tag{3-24}$$

Since $f(x)$, $g(y)$, and $h(z)$ are each a function of only one variable, we can replace the partials in (3-24) by ordinary derivatives. Doing this and dividing each term by fgh, we can write that

$$\frac{1}{f}\frac{d^2 f}{dx^2} + \frac{1}{g}\frac{d^2 g}{dy^2} + \frac{1}{h}\frac{d^2 h}{dz^2} + \beta^2 = 0 \tag{3-25}$$

or

$$\frac{1}{f}\frac{d^2 f}{dx^2} + \frac{1}{g}\frac{d^2 g}{dy^2} + \frac{1}{h}\frac{d^2 h}{dz^2} = -\beta^2 \tag{3-25a}$$

Each of the first three terms in (3-25a) is a function of only a single independent variable; hence the sum of these terms can equal $-\beta^2$ only if each term is a constant. Thus (3-25a) separates into three equations of the form

$$\frac{1}{f}\frac{d^2 f}{dx^2} = -\beta_x^2 \Rightarrow \frac{d^2 f}{dx^2} = -\beta_x^2 f \tag{3-26a}$$

$$\frac{1}{g}\frac{d^2 g}{dy^2} = -\beta_y^2 \Rightarrow \frac{d^2 g}{dy^2} = -\beta_y^2 g \tag{3-26b}$$

$$\frac{1}{h}\frac{d^2 h}{dz^2} = -\beta_z^2 \Rightarrow \frac{d^2 h}{dz^2} = -\beta_z^2 h \tag{3-26c}$$

where, in addition,

$$\beta_x^2 + \beta_y^2 + \beta_z^2 = \beta^2 \tag{3-27}$$

Equation 3-27 is referred to as the *constraint (dispersion)* equation. In addition β_x, β_y, β_z are known as the wave constants (numbers) in the x, y, z directions, respectively, that will be determined using boundary conditions.

The solution to each of (3-26a), (3-26b), or (3-26c) can take different forms. Some typical valid solutions for $f(x)$ of (3-26a) would be

$$f_1(x) = A_1 e^{-j\beta_x x} + B_1 e^{+j\beta_x x} \tag{3-28a}$$

or

$$f_2(x) = C_1 \cos(\beta_x x) + D_1 \sin(\beta_x x) \tag{3-28b}$$

Similarly the solutions to (3-26b) and (3-26c) for $g(y)$ and $h(z)$ can be written, respectively, as

$$g_1(y) = A_2 e^{-j\beta_y y} + B_2 e^{+j\beta_y y} \tag{3-29a}$$

or

$$g_2(y) = C_2 \cos(\beta_y y) + D_2 \sin(\beta_y y) \tag{3-29b}$$

and

$$h_1(z) = A_3 e^{-j\beta_z z} + B_3 e^{+j\beta_z z} \tag{3-30a}$$

or

$$h_2(z) = C_3 \cos(\beta_z z) + D_3 \sin(\beta_z z) \tag{3-30b}$$

Although all the aforementioned solutions are valid for $f(x)$, $g(y)$, and $h(z)$, the most appropriate form should be chosen to simplify the complexity of the problem at hand. In general, the solutions of (3-28a), (3-29a), and (3-30a) in terms of complex exponentials represent *traveling waves* and the solutions of (3-28b), (3-29b), and (3-30b) represent *standing waves*. Wave functions representing various wave types in rectangular coordinates are found listed in Table 3-1. In Chapter 8 we will consider specific examples and the appropriate solution forms for $f(x)$, $g(y)$, and $h(z)$.

Once the appropriate forms for $f(x)$, $g(y)$, and $h(z)$ have been decided, the solution for the scalar function $E_x(x, y, z)$ of (3-22) can be written as the product of fgh as stated by (3-23). To demonstrate that, let us consider a specific example in which it will be assumed that the appropriate solutions for f, g, and h are given, respectively, by (3-28b), (3-29b), and (3-30a). Thus we can write that

$$E_x(x, y, z) = \left[C_1 \cos(\beta_x x) + D_1 \sin(\beta_x x)\right]$$
$$\left[C_2 \cos(\beta_y y) + D_2 \sin(\beta_y y)\right] \times \left[A_3 e^{-j\beta_z z} + B_3 e^{+j\beta_z z}\right] \tag{3-31}$$

This is an appropriate solution for any of the electric or magnetic field components inside a rectangular pipe (waveguide), shown in Figure 3-2, that is bounded in the x and y directions and has its length along the z axis. Because the waveguide is bounded in the x and y directions, standing waves, represented by cosine and sine functions, have been chosen as solutions for $f(x)$ and $g(y)$ functions. However, because the waveguide is not bounded in the z direction, traveling waves, represented by complex exponential functions, have been chosen as solutions for $h(z)$. A complete discussion of the fields inside a rectangular waveguide can be found in Chapter 8.

For $e^{j\omega t}$ time variations, which are assumed throughout this book, the first complex exponential term in (3-31) represents a wave that travels in the $+z$ direction; the second exponential represents a wave that travels in the $-z$ direction. To demonstrate this, let us examine the instantaneous form $\mathscr{E}_x(x, y, z; t)$ of the scalar complex function $E_x(x, y, z)$. Since the solution of (3-31) represents the complex form of E_x, its instantaneous form can be written as

TABLE 3-1 Wave functions, zeroes, and infinities of plane wave functions in rectangular coordinates

Wave type	Wave functions	Zeroes of wave functions	Infinities of wave functions
Traveling waves	$e^{-j\beta x}$ for $+x$ travel $e^{+j\beta x}$ for $-x$ travel	$\beta x \rightarrow -j\infty$ $\beta x \rightarrow +j\infty$	$\beta x \rightarrow +j\infty$ $\beta x \rightarrow -j\infty$
Standing waves	$\cos(\beta x)$ for $\pm x$ $\sin(\beta x)$ for $\pm x$	$\beta x = \pm\left(n+\frac{1}{2}\right)\pi$ $\beta x = \pm n\pi$ $n = 0, 1, 2,...$	$\beta x \rightarrow \pm j\infty$ $\beta x \rightarrow \pm j\infty$
Evanescent waves	$e^{-\alpha x}$ for $+x$ $e^{+\alpha x}$ for $-x$ $\cosh(\alpha x)$ for $\pm x$ $\sinh(\alpha x)$ for $\pm x$	$\alpha x \rightarrow +\infty$ $\alpha x \rightarrow -\infty$ $\alpha x = \pm j\left(n+\frac{1}{2}\right)\pi$ $\alpha x = \pm jn\pi$ $n = 0, 1, 2,...$	$\alpha x \rightarrow -\infty$ $\alpha x \rightarrow +\infty$ $\alpha x \rightarrow \pm\infty$ $\alpha x \rightarrow \pm\infty$
Attenuating traveling waves	$e^{-\gamma x} = e^{-\alpha x}e^{-j\beta x}$ for $+x$ travel $e^{+\gamma x} = e^{+\alpha x}e^{+j\beta x}$ for $-x$ travel	$\gamma x \rightarrow +\infty$ $\gamma x \rightarrow -\infty$	$\gamma x \rightarrow -\infty$ $\gamma x \rightarrow +\infty$
Attenuating standing waves	$\cosh(\gamma x) = \cosh(\alpha x)\cos(\beta x)$ $\quad + j\sinh(\alpha x)\sin(\beta x)$ $\quad\quad$ for $\pm x$ $\sinh(\gamma x) = \sinh(\alpha x)\cos(\beta x)$ $\quad + j\cosh(\alpha x)\sin(\beta x)$ $\quad\quad$ for $\pm x$	$\gamma x = \pm j\left(n+\frac{1}{2}\right)\pi$ $\gamma x = \pm jn\pi$ $n = 0, 1, 2,...$	$\gamma x \rightarrow \pm j\infty$ $\gamma x \rightarrow \pm j\infty$

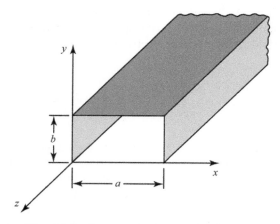

Figure 3-2 Rectangular waveguide geometry.

$$\mathcal{E}_x(x,y,z;t) = \mathrm{Re}\left[E_x(x,y,z)e^{j\omega t}\right] \qquad (3\text{-}32)$$

Considering only the first exponential term of (3-31) and assuming all constants are real, we can write the instantaneous form of the $\mathcal{E}_x$ function for that term as

$$\mathcal{E}_x^+(x, y, z; t) = \mathrm{Re}\left[E_x^+(x, y, z)e^{j\omega t}\right]$$

$$= \mathrm{Re}\left\{\left[C_1 \cos(\beta_x x) + D_1 \sin(\beta_x x)\right]\right.$$

$$\left.\times\left[C_2 \cos(\beta_y y) + D_2 \sin(\beta_y y)\right]A_3 e^{j(\omega t - \beta_z z)}\right\} \quad (3\text{-}33)$$

or, if the constants C_1, D_1, C_2, D_2, and A_3 are real, as

$$\mathcal{E}_x^+(x, y, z; t) = \left[C_1 \cos(\beta_x x) + D_1 \sin(\beta_x x)\right]$$

$$\times\left[C_2 \cos(\beta_y y) + D_2 \sin(\beta_y y)\right]A_3 \cos(\omega t - \beta_z z) \quad (3\text{-}33a)$$

where the superscript plus is used to denote a positive traveling wave.

A plot of the normalized $\mathcal{E}_x^+(x, y, z; t)$ as a function of z for different times $(t = t_0, t_1, \ldots, t_n, t_{n+1})$ is shown in Figure 3-3. It is evident that as time increases $(t_{n+1} > t_n)$, the waveform of $\mathcal{E}_x^+$ is essentially the same, with the exception of an apparent shift in the $+z$ direction indicating a wave traveling in the $+z$ direction. This shift in the $+z$ direction can also be demonstrated by examining what happens to a given point z_p in the waveform of $\mathcal{E}_x^+$ for $t = t_0, t_1, \ldots, t_n, t_{n+1}$. To follow the point z_p for different values of t, we must maintain constant the amplitude of the last cosine term in (3-33a). This is accomplished by keeping its argument $\omega t - \beta_z z_p$ constant, that is,

$$\omega t - \beta_z z_p = C_0 = \text{constant} \quad (3\text{-}34)$$

which when differentiated with respect to time reduces to

$$\omega(1) - \beta_z \frac{dz_p}{dt} = 0 \Rightarrow \frac{dz_p}{dt} = v_p = +\frac{\omega}{\beta_z} \quad (3\text{-}35)$$

The point z_p is referred to as an *equiphase* point and its velocity is denoted as the *phase velocity*. A similar procedure can be used to demonstrate that the second complex exponential term in (3-31) represents a wave that travels in the $-z$ direction.

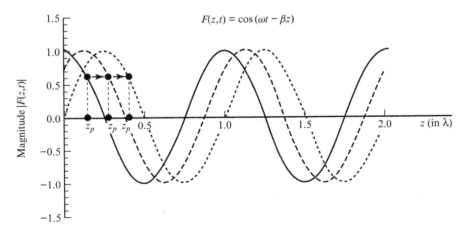

Figure 3-3 Variations as a function of distance for different times of positive traveling wave. —— time $t_0 = 0$; ---- time $t_1 = T/8$; ---- time $t_2 = T/4$.

B. Source-Free and Lossy Media   When the media in which the waves are traveling are lossy ($\sigma \neq 0$) but source-free ($\mathbf{J}_i = \mathbf{M}_i = q_{ve} = q_{vm} = 0$), the vector wave equations that the complex electric $\mathbf{E}$ and magnetic $\mathbf{H}$ field intensities must satisfy are (3-17a) and (3-17b). As for the lossless case, let us examine the solution to one of them; the solution to the other can then be written by inspection once the solution to the first has been obtained. We choose to consider the solution for the electric field intensity $\mathbf{E}$, which must satisfy (3-17a). An extended presentation of electromagnetic wave propagation in lossy media can be found in [4].

In a rectangular coordinate system, the general solution for $\mathbf{E}(x,y,z)$ can be written as

$$\mathbf{E}(x,y,z) = \hat{\mathbf{a}}_x E_x(x,y,z) + \hat{\mathbf{a}}_y E_y(x,y,z) + \hat{\mathbf{a}}_z E_z(x,y,z) \tag{3-36}$$

When (3-36) is substituted into (3-17a), we can write that

$$\nabla^2 \mathbf{E} - \gamma^2 \mathbf{E} = \nabla^2(\hat{\mathbf{a}}_x E_x + \hat{\mathbf{a}}_y E_y + \hat{\mathbf{a}}_z E_z) - \gamma^2(\hat{\mathbf{a}}_x E_x + \hat{\mathbf{a}}_y E_y + \hat{\mathbf{a}}_z E_z) = 0 \tag{3-37}$$

which reduces to three scalar wave equations of

$$\nabla^2 E_x(x,y,z) - \gamma^2 E_x(x,y,z) = 0 \tag{3-37a}$$

$$\nabla^2 E_y(x,y,z) - \gamma^2 E_y(x,y,z) = 0 \tag{3-37b}$$

$$\nabla^2 E_z(x,y,z) - \gamma^2 E_z(x,y,z) = 0 \tag{3-37c}$$

where

$$\gamma^2 = j\omega\mu(\sigma + j\omega\varepsilon) \tag{3-37d}$$

If we were to allow positive and negative values of σ

$$\gamma = \pm\sqrt{j\omega\mu(\sigma + j\omega\varepsilon)} = \begin{cases} \pm(\alpha + j\beta) & \text{for } +\sigma \\ \pm(\alpha - j\beta) & \text{for } -\sigma \end{cases} \tag{3-37e}$$

In (3-37e),

$$\gamma = \text{propagation constant}$$
$$\alpha = \text{attenuation constant (Np/m)}$$
$$\beta = \text{phase constant (rad/m)}$$

where α and β are assumed to be real and positive. Although some authors choose to represent the phase constant by k, the symbol β will be used throughout this book.

Examining (3-37e) reveals that there are four possible combinations for the form of γ. That is,

$$\gamma = \begin{cases} +(\alpha + j\beta) & \text{(3-38a)} \\ -(\alpha + j\beta) & \text{(3-38b)} \\ +(\alpha - j\beta) & \text{(3-38c)} \\ -(\alpha - j\beta) & \text{(3-38d)} \end{cases}$$

Of the four combinations, only one will be appropriate for our solution. That form will be selected once the solutions to any of (3-37a) through (3-37c) have been decided.

Since all three equations represented by (3-37a) through (3-37c) are of the same form, let us examine only one of them. We choose to work first with (3-37a) whose solution can be derived using the method of *separation of variables*. Using a similar procedure as for the lossless case, we can write that

$$E_x(x,y,z) = f(x)g(y)h(z) \tag{3-39}$$

where it can be shown that $f(x)$ has solutions of the form

$$f_1(x) = A_1 e^{-\gamma_x x} + B_1 e^{+\gamma_x x} \tag{3-40a}$$

or

$$f_2(x) = C_1 \cosh(\gamma_x x) + D_1 \sinh(\gamma_x x) \tag{3-40b}$$

and $g(y)$ can be expressed as

$$g_1(y) = A_2 e^{-\gamma_y y} + B_2 e^{+\gamma_y y} \tag{3-41a}$$

or

$$g_2(y) = C_2 \cosh(\gamma_y y) + D_2 \sinh(\gamma_y y) \tag{3-41b}$$

and $h(z)$ as

$$h_1(z) = A_3 e^{-\gamma_z z} + B_3 e^{+\gamma_z z} \tag{3-42a}$$

or

$$h_2(z) = C_3 \cosh(\gamma_z z) + D_3 \sinh(\gamma_z z) \tag{3-42b}$$

Whereas (3-40a) through (3-42b) are appropriate solutions for f, g, and h of (3-39), which satisfy (3-37a), the *constraint (dispersion) equation* takes the form of

$$\boxed{\gamma_x^2 + \gamma_y^2 + \gamma_z^2 = \gamma^2} \tag{3-43}$$

The appropriate forms of f, g, and h chosen to represent the solution of $E_x(x,y,z)$, as given by (3-39), must be made by examining the geometry of the problem in question. As for the lossless case, the exponentials represent attenuating traveling waves and the hyperbolic cosines and sines represent attenuating standing waves. These and other waves types are listed in Table 3-1.

To decide on the appropriate form for any of the γ's (whether it be γ_x, γ_y, γ_z, or γ), let us choose the form of γ_z by examining one of the exponentials in (3-42a). We choose to work with the first one. The four possible combinations for γ_z, according to (3-38a) through (3-38d) will be

$$\gamma_z = \begin{cases} +(\alpha_z + j\beta_z) & \text{(3-44a)} \\ -(\alpha_z + j\beta_z) & \text{(3-44b)} \\ +(\alpha_z - j\beta_z) & \text{(3-44c)} \\ -(\alpha_z - j\beta_z) & \text{(3-44d)} \end{cases}$$

If we want the first exponential in (3-42a) to represent a decaying wave which travels in the $+z$ direction, then by substituting (3-44a) through (3-44d) into it we can write that

$$h_1^+(z) = \begin{cases} A_3 e^{-\gamma_z z} = A_3 e^{-\alpha_z z} \, e^{-j\beta_z z} & \text{(3-45a)} \\ A_3 e^{-\gamma_z z} = A_3 e^{+\alpha_z z} \, e^{+j\beta_z z} & \text{(3-45b)} \\ A_3 e^{-\gamma_z z} = A_3 e^{-\alpha_z z} \, e^{+j\beta_z z} & \text{(3-45c)} \\ A_3 e^{-\gamma_z z} = A_3 e^{+\alpha_z z} \, e^{-j\beta_z z} & \text{(3-45d)} \end{cases}$$

By examining (3-45a) through (3-45d) and assuming $e^{j\omega t}$ time variations, the following statements can be made:

1. Equation 3-45a represents a wave that travels in the $+z$ direction, as determined by $e^{-j\beta_z z}$, and it decays in that direction, as determined by $e^{-\alpha_z z}$.
2. Equation 3-45b represents a wave that travels in the $-z$ direction, as determined by $e^{+j\beta_z z}$, and it decays in that direction, as determined by $e^{+\alpha_z z}$.
3. Equation 3-45c represents a wave that travels in the $-z$ direction, as determined by $e^{+j\beta_z z}$, and it is increasing in that direction, as determined by $e^{-\alpha_z z}$.
4. Equation 3-45d represents a wave that travels in the $+z$ direction, as determined by $e^{-j\beta_z z}$, and it is increasing in that direction, as determined by $e^{+\alpha_z z}$.

From the preceding statements it is apparent that for $e^{-\gamma_z z}$ to represent a wave that travels in the $+z$ direction and that concurrently also decays (to represent propagation in passive lossy media), and to satisfy the conservation of energy laws, the only correct form of γ_z is that of (3-44a). The same conclusion will result if the second exponential of (3-42a) represents a wave that travels in the $-z$ direction and that concurrently also decays. Thus the general form of any γ_i (whether it be γ_x, γ_y, γ_z, or γ), as given by (3-38a) through (3-38d), is

$$\gamma_i = \alpha_i + j\beta_i \tag{3-46}$$

Whereas the forms of f, g, and h [as given by (3-40a) through (3-42b)] are used to arrive at the solution for the complex form of E_x as given by (3-39), the instantaneous form of $\mathscr{E}_x$ can be obtained by using the relation of (3-32). A similar procedure can be used to derive the solutions of the other components of $\mathbf{E}$ (E_y and E_z), all those of $\mathbf{H}$ (H_x, H_y, and H_z), and of their instantaneous counterparts.

3.4.2 Cylindrical Coordinate System

If the geometry of the system is of a cylindrical configuration, it would be very advisable to solve the boundary-value problem for the $\mathbf{E}$ and $\mathbf{H}$ fields using cylindrical coordinates. Maxwell's equations and the vector wave equations, which the $\mathbf{E}$ and $\mathbf{H}$ fields must satisfy, should be solved using cylindrical coordinates. Let us first consider the solution for $\mathbf{E}$ for a source-free and lossless medium. A similar procedure can be used for $\mathbf{H}$. To maintain some simplicity in the mathematics, we will examine only lossless media.

In cylindrical coordinates a general solution to the vector wave equation for source-free and lossless media, as given by (3-18a), can be written as

$$\mathbf{E}(\rho,\phi,z) = \hat{\mathbf{a}}_\rho E_\rho(\rho,\phi,z) + \hat{\mathbf{a}}_\phi E_\phi(\rho,\phi,z) + \hat{\mathbf{a}}_z E_z(\rho,\phi,z) \tag{3-47}$$

where ρ, ϕ, and z are the cylindrical coordinates as illustrated in Figure 3-4. Substituting (3-47) into (3-18a), we can write that

$$\nabla^2(\hat{\mathbf{a}}_\rho E_\rho + \hat{\mathbf{a}}_\phi E_\phi + \hat{\mathbf{a}}_z E_z) = -\beta^2(\hat{\mathbf{a}}_\rho E_\rho + \hat{\mathbf{a}}_\phi E_\phi + \hat{\mathbf{a}}_z E_z) \tag{3-48}$$

which does not reduce to three simple scalar wave equations, similar to those of (3-20a) through (3-20c) for (3-20), because

$$\nabla^2(\hat{\mathbf{a}}_\rho E_\rho) \neq \hat{\mathbf{a}}_\rho \nabla^2 E_\rho \tag{3-49a}$$

$$\nabla^2(\hat{\mathbf{a}}_\phi E_\phi) \neq \hat{\mathbf{a}}_\phi \nabla^2 E_\phi \tag{3-49b}$$

However, because

$$\nabla^2(\hat{\mathbf{a}}_z E_z) = \hat{\mathbf{a}}_z \nabla^2 E_z \tag{3-49c}$$

one of the three scalar equations to which (3-48) reduces is

$$\nabla^2 E_z + \beta^2 E_z = 0 \tag{3-50}$$

The other two are of more complex form and they will be addressed in what follows.

Before we derive the other two scalar equations [in addition to (3-50)] to which (3-48) reduces, let us attempt to give a physical explanation of (3-49a), (3-49b), and (3-49c). By examining two different points (ρ_1, ϕ_1, z_1) and (ρ_2, ϕ_2, z_2) and their corresponding unit vectors on a cylindrical surface (as shown in Figure 3-4), we see that the directions of $\hat{\mathbf{a}}_\rho$ and $\hat{\mathbf{a}}_\phi$ have changed from one point to another (they are not parallel) and therefore cannot be treated as constants but rather are functions of ρ, ϕ, and z. In contrast, the unit vector $\hat{\mathbf{a}}_z$ at the two points is pointed in the same direction (is parallel). The same is true for the unit vectors $\hat{\mathbf{a}}_x$ and $\hat{\mathbf{a}}_y$ in Figure 3-1.

Let us now return to the solution of (3-48). Since (3-48) does not reduce to (3-49a) and (3-49b), although it does satisfy (3-49c), how do we solve (3-48)? The procedure that follows can be used to reduce (3-48) to three scalar partial differential equations.

The form of (3-48) written in general as

$$\nabla^2 \mathbf{E} = -\beta^2 \mathbf{E} \tag{3-51}$$

was placed in this form by utilizing the vector identity of (3-5) during its derivation. Generally we are under the impression that we do not know how to perform the Laplacian of a vector ($\nabla^2 \mathbf{E}$) as given by the left side of (3-51). However, by utilizing (3-5) we can rewrite the left side of (3-51) as

$$\nabla^2 \mathbf{E} = \nabla(\nabla \cdot \mathbf{E}) - \nabla \times \nabla \times \mathbf{E} \tag{3-52}$$

whose terms can be expanded in any coordinate system. Using (3-52) we can write (3-51) as

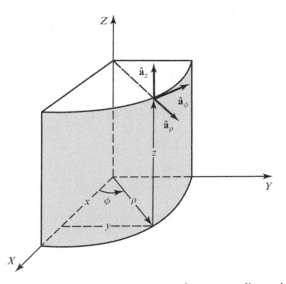

Figure 3-4 Cylindrical coordinate system and corresponding unit vectors.

$$\nabla(\nabla\cdot\mathbf{E})-\nabla\times\nabla\times\mathbf{E}=-\beta^2\mathbf{E} \tag{3-53}$$

which is an alternate form, but not as commonly recognizable, of the vector wave equation for the electric field in source-free and lossless media.

Assuming a solution for the electric field of the form given by (3-47), we can expand (3-53) and reduce it to three scalar partial differential equations of the form

$$\nabla^2 E_\rho + \left(-\frac{E_\rho}{\rho^2}-\frac{2}{\rho^2}\frac{\partial E_\phi}{\partial\phi}\right)=-\beta^2 E_\rho \tag{3-54a}$$

$$\nabla^2 E_\phi + \left(-\frac{E_\phi}{\rho^2}+\frac{2}{\rho^2}\frac{\partial E_\rho}{\partial\phi}\right)=-\beta^2 E_\phi \tag{3-54b}$$

$$\nabla^2 E_z = -\beta^2 E_z \tag{3-54c}$$

In each of (3-54a) through (3-54c) $\nabla^2\psi(\rho,\phi,z)$ is the Laplacian of a scalar that in cylindrical coordinates takes the form of

$$\nabla^2\psi(\rho,\phi,z)=\frac{1}{\rho}\frac{\partial}{\partial\rho}\left(\rho\frac{\partial\psi}{\partial\rho}\right)+\frac{1}{\rho^2}\frac{\partial^2\psi}{\partial\phi^2}+\frac{\partial^2\psi}{\partial z^2}$$

$$=\frac{\partial^2\psi}{\partial\rho^2}+\frac{1}{\rho}\frac{\partial\psi}{\partial\rho}+\frac{1}{\rho^2}\frac{\partial^2\psi}{\partial\phi^2}+\frac{\partial^2\psi}{\partial z^2} \tag{3-55}$$

Equations 3-54a and 3-54b are *coupled* (each contains more than one electric field component) second-order partial differential equations, which are the most difficult to solve. However, (3-54c) is an *uncoupled* second-order partial differential equation whose solution will be most useful in the construction of TE^z and TM^z mode solutions of boundary-value problems, as discussed in Chapters 6 and 9.

In expanded form (3-54c) can then be written as

$$\frac{\partial^2\psi}{\partial\rho^2}+\frac{1}{\rho}\frac{\partial\psi}{\partial\rho}+\frac{1}{\rho^2}\frac{\partial^2\psi}{\partial\phi^2}+\frac{\partial^2\psi}{\partial z^2}=-\beta^2\psi \tag{3-56}$$

where $\psi(\rho,\phi,z)$ is a scalar function that can represent a field or a vector potential component. Assuming a separable solution for $\psi(\rho,\phi,z)$ of the form

$$\psi(\rho,\phi,z)=f(\rho)g(\phi)h(z) \tag{3-57}$$

and substituting it into (3-56), we can write that

$$gh\frac{\partial^2 f}{\partial\rho^2}+gh\frac{1}{\rho}\frac{\partial f}{\partial\rho}+fh\frac{1}{\rho^2}\frac{\partial^2 g}{\partial\phi^2}+fg\frac{\partial^2 h}{\partial z^2}=-\beta^2 fgh \tag{3-58}$$

Dividing both sides of (3-58) by *fgh* and replacing the partials by ordinary derivatives reduces (3-58) to

$$\frac{1}{f}\frac{d^2 f}{d\rho^2}+\frac{1}{f}\frac{1}{\rho}\frac{df}{d\rho}+\frac{1}{g}\frac{1}{\rho^2}\frac{d^2 g}{d\phi^2}+\frac{1}{h}\frac{d^2 h}{dz^2}=-\beta^2 \tag{3-59}$$

The last term on the left side of (3-59) is only a function of z. Therefore, using the discussion of Section 3.4.1, we can write that

$$\frac{1}{h}\frac{d^2h}{dz^2} = -\beta_z^2 \Rightarrow \frac{d^2h}{dz^2} = -\beta_z^2 h \qquad (3\text{-}60)$$

where β_z is a constant. Substituting (3-60) into (3-59) and multiplying both sides by ρ^2 reduces it to

$$\frac{\rho^2}{f}\frac{d^2f}{d\rho^2} + \frac{\rho}{f}\frac{df}{d\rho} + \frac{1}{g}\frac{d^2g}{d\phi^2} + (\beta^2 - \beta_z^2)\rho^2 = 0 \qquad (3\text{-}61)$$

Since the third term on the left side of (3-61) is only a function of ϕ, it can be set equal to a constant $-m^2$. Thus we can write that

$$\frac{1}{g}\frac{d^2g}{d\phi^2} = -m^2 \Rightarrow \frac{d^2g}{d\phi^2} = -m^2 g \qquad (3\text{-}62)$$

Letting

$$\beta^2 - \beta_z^2 = \beta_\rho^2 \Rightarrow \beta_\rho^2 + \beta_z^2 = \beta^2 \qquad (3\text{-}63)$$

then using (3-62), and multiplying both sides of (3-61) by f, we can reduce (3-61) to

$$\rho^2\frac{d^2f}{d\rho^2} + \rho\frac{df}{d\rho} + \left[(\beta_\rho\rho)^2 - m^2\right]f = 0 \qquad (3\text{-}64)$$

Equation 3-63 is referred to as the *constraint (dispersion)* equation for the solution to the wave equation in cylindrical coordinates, and (3-64) is recognized as the classic *Bessel differential equation* [1–3, 5–10].

In summary then, the partial differential equation 3-56 whose solution was assumed to be separable of the form given by (3-57) reduces to the three differential equations 3-60, 3-62, 3-64 and the constraint equation 3-63. Thus

$$\nabla^2\psi(\rho, \phi, z) = \frac{\partial^2\psi}{\partial\rho^2} + \frac{1}{\rho}\frac{\partial\psi}{\partial\rho} + \frac{1}{\rho^2}\frac{\partial^2\psi}{\partial\phi^2} + \frac{\partial^2\psi}{\partial z^2} = -\beta^2\psi \qquad (3\text{-}65)$$

where

$$\psi(\rho, \phi, z) = f(\rho)g(\phi)h(z) \qquad (3\text{-}65a)$$

reduces to

$$\boxed{\rho^2\frac{d^2f}{d\rho^2} + \rho\frac{df}{d\rho} + \left[(\beta_\rho\rho)^2 - m^2\right]f = 0} \qquad (3\text{-}66a)$$

$$\boxed{\frac{d^2g}{d\phi^2} = -m^2 g} \qquad (3\text{-}66b)$$

$$\boxed{\frac{d^2h}{dz^2} = -\beta_z^2 h} \qquad (3\text{-}66c)$$

with

$$\boxed{\beta_\rho^2 + \beta_z^2 = \beta^2}$$
(3-66d)

Solutions to (3-66a), (3-66b), and (3-66c) take the form, respectively, of

$$f_1(\rho) = A_1 J_m(\beta_\rho \rho) + B_1 Y_m(\beta_\rho \rho)$$
(3-67a)

or

$$f_2(\rho) = C_1 H_m^{(1)}(\beta_\rho \rho) + D_1 H_m^{(2)}(\beta_\rho \rho)$$
(3-67b)

and

$$g_1(\phi) = A_2 e^{-jm\phi} + B_2 e^{+jm\phi}$$
(3-68a)

or

$$g_2(\phi) = C_2 \cos(m\phi) + D_2 \sin(m\phi)$$
(3-68b)

and

$$h_1(z) = A_3 e^{-j\beta_z z} + B_3 e^{+j\beta_z z}$$
(3-69a)

or

$$h_2(z) = C_3 \cos(\beta_z z) + D_3 \sin(\beta_z z)$$
(3-69b)

In (3-67a) $J_m(\beta_\rho \rho)$ and $Y_m(\beta_\rho \rho)$ represent, respectively, the Bessel functions of the first and second kind; $H_m^{(1)}(\beta_\rho \rho)$ and $H_m^{(2)}(\beta_\rho \rho)$ in (3-67b) represent, respectively, the Hankel functions of the first and second kind. A more detailed discussion of Bessel and Hankel functions is found in Appendix IV.

Although (3-67a) through (3-69b) are valid solutions for $f(\rho)$, $g(\phi)$, and $h(z)$, the most appropriate form will depend on the problem in question. For example, for the cylindrical waveguide of Figure 3-5 the most convenient solutions for $f(\rho)$, $g(\phi)$, and $h(z)$ are those given, respectively, by (3-67a), (3-68b), and (3-69a). Thus we can write

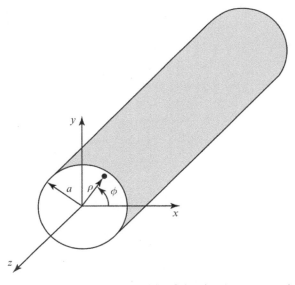

Figure 3-5   Cylindrical waveguide of the circular cross section.

$$\psi_1(\rho,\phi,z) = f(\rho)g(\phi)h(z)$$
$$= \left[A_1 J_m(\beta_\rho\rho) + B_1 Y_m(\beta_\rho\rho)\right]$$
$$\times \left[C_2 \cos(m\phi) + D_2 \sin(m\phi)\right]\left[A_3 e^{-j\beta_z z} + B_3 e^{+j\beta_z z}\right] \qquad (3\text{-}70)$$

These forms for $f(\rho)$, $g(\phi)$, and $h(z)$ were chosen in cylindrical coordinates for the following reasons.

1. Bessel functions of (3-67a) are used to represent standing waves, whereas Hankel functions of (3-67b) represent traveling waves.
2. Exponentials of (3-68a) represent traveling waves, whereas the cosines and sines of (3-68b) represent periodic waves.
3. Exponentials of (3-69a) represent traveling waves, whereas the cosines and sines of (3-69b) represent standing waves.

Wave functions representing various radial waves in cylindrical coordinates are found listed in Table 3-2.

Within the circular waveguide of Figure 3-5 standing waves are created in the radial (ρ) direction, periodic waves in the phi (ϕ) direction, and traveling waves in the z direction. For the fields to be finite at $\rho = 0$, where $Y_m(\beta_\rho\rho)$ possesses a singularity, (3-70) reduces to

$$\psi_1(\rho,\phi,z) = A_1 J_m(\beta_\rho\rho)\left[C_2 \cos(m\phi) + D_2 \sin(m\phi)\right]\left[A_3 e^{-j\beta_z z} + B_3 e^{+j\beta_z z}\right] \qquad (3\text{-}70a)$$

TABLE 3-2   **Wave functions, zeroes, and infinities for radial wave functions in cylindrical coordinates**

Wave type	Wave functions	Zeroes of wave functions	Infinities of wave functions
Traveling waves	$H_m^{(1)}(\beta\rho) = J_m(\beta\rho) + jY_m(\beta\rho)$ for $-\rho$ travel	$\beta\rho \rightarrow +j\infty$	$\beta\rho = 0$ $\beta\rho \rightarrow -j\infty$
	$H_m^{(2)}(\beta\rho) = J_m(\beta\rho) - jY_m(\beta\rho)$ for $+\rho$ travel	$\beta\rho \rightarrow -j\infty$	$\beta\rho = 0$ $\beta\rho \rightarrow +j\infty$
Standing waves	$J_m(\beta\rho)$ for $\pm\rho$	Infinite number (see Table 9-2)	$\beta\rho \rightarrow \pm j\infty$
	$Y_m(\beta\rho)$ for $\pm\rho$	Infinite number	$\beta\rho = 0$ $\beta\rho \rightarrow \pm j\infty$
Evanescent waves	$K_m(\alpha\rho) = \dfrac{\pi}{2}(-j)^{m+1} H_m^{(2)}(-j\alpha\rho)$ for $+\rho$	$\alpha\rho \rightarrow +\infty$	$\alpha\rho \rightarrow 0$
	$I_m(\alpha\rho) = j^m J_m(-j\alpha\rho)$ for $-\rho$		$\alpha\rho \rightarrow +\infty$ for integer orders
Attenuating traveling waves	$H_m^{(1)}(-j\gamma\rho) = H_m^{(1)}(-j\alpha\rho + \beta\rho)$ for $-\rho$ travel	$\gamma\rho \rightarrow -\infty$	$\gamma\rho \rightarrow +\infty$
	$H_m^{(2)}(-j\gamma\rho) = H_m^{(2)}(-j\alpha\rho + \beta\rho)$ for $+\rho$ travel	$\gamma\rho \rightarrow +\infty$	$\gamma\rho \rightarrow -\infty$
Attenuating standing waves	$J_m(-j\gamma\rho) = J_m(-j\alpha\rho + \beta\rho)$ for $\pm\rho$	Infinite number	$\gamma\rho \rightarrow \pm j\infty$
	$Y_m(-j\gamma\rho) = Y_m(-j\alpha\rho + \beta\rho)$ for $\pm\rho$	Infinite number	$\gamma\rho \rightarrow \pm j\infty$

To represent the fields in the region outside the cylinder, like scattering by the cylinder, a typical solution for $\psi(\rho, \phi, z)$ would take the form of

$$\psi_2(\rho, \phi, z) = B_1 H_m^{(2)}(\beta_\rho \rho)\left[C_2 \cos(m\phi) + D_2 \sin(m\phi)\right]\left[A_3 e^{-j\beta_z z} + B_3 e^{+j\beta_z z}\right] \qquad (3\text{-}70\text{b})$$

whereby the Hankel function of the second kind $H_m^{(2)}(\beta_\rho \rho)$ has replaced the Bessel function of the first kind $J_m(\beta_\rho \rho)$ because outward traveling waves are formed outside the cylinder, in contrast to the standing waves inside the cylinder.

More details concerning the application and properties of Bessel and Hankel functions can be found in Chapters 9 and 11.

3.4.3 Spherical Coordinate System

Spherical coordinates should be utilized in solving problems that exhibit spherical geometries. As for the rectangular and cylindrical geometries, the electric and magnetic fields of a spherical geometry boundary-value problem must satisfy the corresponding vector wave equation, which is most conveniently solved in spherical coordinates as illustrated in Figure 3-6.

To simplify the problem, let us assume that the space in which the electric and magnetic fields must be solved is source-free and lossless. A general solution for the electric field can then be written as

$$\mathbf{E}(r, \theta, \phi) = \hat{\mathbf{a}}_r E_r(r, \theta, \phi) + \hat{\mathbf{a}}_\theta E_\theta(r, \theta, \phi) + \hat{\mathbf{a}}_\phi E_\phi(r, \theta, \phi) \qquad (3\text{-}71)$$

Substituting (3-71) into the vector wave equation of (3-18a), we can write that

$$\nabla^2(\hat{\mathbf{a}}_r E_r + \hat{\mathbf{a}}_\theta E_\theta + \hat{\mathbf{a}}_\phi E_\phi) = -\beta^2(\hat{\mathbf{a}}_r E_r + \hat{\mathbf{a}}_\theta E_\theta + \hat{\mathbf{a}}_\phi E_\phi) \qquad (3\text{-}72)$$

Since

$$\nabla^2(\hat{\mathbf{a}}_r E_r) \neq \hat{\mathbf{a}}_r \nabla^2 E_r \qquad (3\text{-}73\text{a})$$

$$\nabla^2(\hat{\mathbf{a}}_\theta E_\theta) \neq \hat{\mathbf{a}}_\theta \nabla^2 E_\theta \qquad (3\text{-}73\text{b})$$

$$\nabla^2(\hat{\mathbf{a}}_\phi E_\phi) \neq \hat{\mathbf{a}}_\phi \nabla^2 E_\phi \qquad (3\text{-}73\text{c})$$

(3-72) does not reduce to three simple scalar wave equations, similar to those of (3-20a) through (3-20c) for (3-20). Therefore the reduction of (3-72) to three scalar partial differential equations must proceed in a different manner. In fact, the method used here will be similar to that utilized in cylindrical coordinates to reduce the vector wave equation to three scalar partial differential equations.

To accomplish this, we first rewrite the vector wave equation of (3-51) in a form given by (3-53) where now all the operators on the left side can be performed in any coordinate system. Substituting (3-71) into (3-53) shows that, after some lengthy mathematical manipulations, (3-53) reduces to three scalar partial differential equations of the form

$$\nabla^2 E_r - \frac{2}{r^2}\left(E_r + E_\theta \cot\theta + \csc\theta \frac{\partial E_\phi}{\partial \phi} + \frac{\partial E_\theta}{\partial \theta}\right) = -\beta^2 E_r \qquad (3\text{-}74\text{a})$$

$$\nabla^2 E_\theta - \frac{1}{r^2}\left(E_\theta \csc^2\theta - 2\frac{\partial E_r}{\partial \theta} + 2\cot\theta \csc\theta \frac{\partial E_\phi}{\partial \phi}\right) = -\beta^2 E_\theta \qquad (3\text{-}74\text{b})$$

$$\nabla^2 E_\phi - \frac{1}{r^2}\left(E_\phi \csc^2\theta - 2\csc\theta \frac{\partial E_r}{\partial \phi} - 2\cot\theta \csc\theta \frac{\partial E_\theta}{\partial \phi}\right) = -\beta^2 E_\phi \qquad (3\text{-}74\text{c})$$

Unfortunately, all three of the preceding partial differential equations are coupled. This means each contains more than one component of the electric field and would be most difficult to solve in its present form. However, as will be shown in Chapter 10, TE^r and TM^r wave mode solutions can be formed that in spherical coordinates must satisfy the scalar wave equation of

$$\nabla^2 \psi(r, \theta, \phi) = -\beta^2 \psi(r, \theta, \phi) \tag{3-75}$$

where $\psi(r, \theta, \phi)$ is a scalar function that can represent a field or a vector potential component. Therefore, it would be advisable here to demonstrate the solution to (3-75) in spherical coordinates.

Assuming a separable solution for $\psi(r, \theta, \phi)$ of the form

$$\psi(r, \theta, \phi) = f(r) g(\theta) h(\phi) \tag{3-76}$$

we can write the expanded form of (3-75)

$$\frac{1}{r^2} \frac{\partial}{\partial r} \left\{ r^2 \frac{\partial \psi}{\partial r} \right\} + \frac{1}{r^2 \sin\theta} \frac{\partial}{\partial \theta} \left\{ \sin\theta \frac{\partial \psi}{\partial \theta} \right\} + \frac{1}{r^2 \sin^2\theta} \frac{\partial^2 \psi}{\partial \phi^2} = -\beta^2 \psi \tag{3-77}$$

as

$$gh \frac{1}{r^2} \frac{\partial}{\partial r} \left\{ r^2 \frac{\partial f}{\partial r} \right\} + fh \frac{1}{r^2 \sin\theta} \frac{\partial}{\partial \theta} \left\{ \sin\theta \frac{\partial g}{\partial \theta} \right\} + fg \frac{1}{r^2 \sin^2\theta} \frac{\partial^2 h}{\partial \phi^2} = -\beta^2 fgh \tag{3-78}$$

Dividing both sides by fgh, multiplying by $r^2 \sin^2\theta$, and replacing the partials by ordinary derivatives reduces (3-78) to

$$\frac{\sin^2\theta}{f} \frac{d}{dr} \left\{ r^2 \frac{df}{dr} \right\} + \frac{\sin\theta}{g} \frac{d}{d\theta} \left\{ \sin\theta \frac{dg}{d\theta} \right\} + \frac{1}{h} \frac{d^2 h}{d\phi^2} = -(\beta r \sin\theta)^2 \tag{3-79}$$

Since the last term on the left side of (3-79) is only a function of ϕ, it can be set equal to

$$\frac{1}{h} \frac{d^2 h}{d\phi^2} = -m^2 \Rightarrow \frac{d^2 h}{d\phi^2} = -m^2 h \tag{3-80}$$

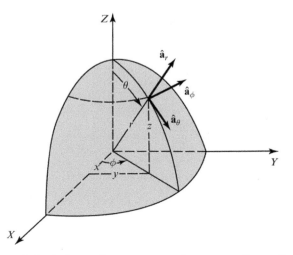

Figure 3-6 Spherical coordinate system and corresponding unit vectors.

where m is a constant.

Substituting (3-80) into (3-79), dividing both sides by $\sin^2\theta$, and transposing the term from the right to the left side reduces (3-79) to

$$\frac{1}{f}\frac{d}{dr}\left\{r^2\frac{df}{dr}\right\}+(\beta r)^2+\frac{1}{g\sin\theta}\frac{d}{d\theta}\left\{\sin\theta\frac{dg}{d\theta}\right\}-\left\{\frac{m}{\sin\theta}\right\}^2=0 \qquad (3\text{-}81)$$

Since the last two terms on the left side of (3-81) are only a function of θ, we can set them equal to

$$\frac{1}{g\sin\theta}\frac{d}{d\theta}\left\{\sin\theta\frac{dg}{d\theta}\right\}-\left\{\frac{m}{\sin\theta}\right\}^2=-n(n+1) \qquad (3\text{-}82)$$

where n is usually an integer. Equation 3-82 is closely related to the well-known *Legendre differential equation* (see Appendix V) [1–3, 6–10].

Substituting (3-82) into (3-81) reduces it to

$$\frac{1}{f}\frac{d}{dr}\left\{r^2\frac{df}{dr}\right\}+(\beta r)^2-n(n+1)=0 \qquad (3\text{-}83)$$

which is closely related to the Bessel differential equation (see Appendix IV).

In summary then, the scalar wave equation 3-75 whose expanded form in spherical coordinates can be written as

$$\frac{1}{r^2}\frac{\partial}{\partial r}\left\{r^2\frac{\partial\psi}{\partial r}\right\}+\frac{1}{r^2\sin\theta}\frac{\partial}{\partial\theta}\left\{\sin\theta\frac{\partial\psi}{\partial\theta}\right\}+\frac{1}{r^2\sin^2\theta}\frac{\partial^2\psi}{\partial\phi^2}=-\beta^2\psi \qquad (3\text{-}84)$$

and whose separable solution takes the form of

$$\boxed{\psi(r,\theta,\phi)=f(r)g(\theta)h(\phi)} \qquad (3\text{-}85)$$

reduces to the three scalar differential equations

$$\boxed{\frac{d}{dr}\left\{r^2\frac{df}{dr}\right\}+\left[(\beta r)^2-n(n+1)\right]f=0} \qquad (3\text{-}86a)$$

$$\boxed{\frac{1}{\sin\theta}\frac{d}{d\theta}\left\{\sin\theta\frac{dg}{d\theta}\right\}+\left[n(n+1)-\left\{\frac{m}{\sin\theta}\right\}^2\right]g=0} \qquad (3\text{-}86b)$$

$$\boxed{\frac{d^2h}{d\phi^2}=-m^2h} \qquad (3\text{-}86c)$$

where m and n are constants (usually integers).

Solutions to (3-86a) through (3-86c) take the forms, respectively, of

$$f_1(r) = A_1 j_n(\beta r) + B_1 y_n(\beta r) \tag{3-87a}$$

or

$$f_2(r) = C_1 h_n^{(1)}(\beta r) + D_1 h_n^{(2)}(\beta r) \tag{3-87b}$$

and

$$g_1(\theta) = A_2 P_n^m(\cos\theta) + B_2 P_n^m(-\cos\theta) \qquad n \neq \text{integer} \tag{3-88a}$$

or

$$g_2(\theta) = C_2 P_n^m(\cos\theta) + D_2 Q_n^m(\cos\theta) \qquad n = \text{integer} \tag{3-88b}$$

and

$$h_1(\phi) = A_3 e^{-jm\phi} + B_3 e^{+jm\phi} \tag{3-89a}$$

or

$$h_2(\phi) = C_3 \cos(m\phi) + D_3 \sin(m\phi) \tag{3-89b}$$

In (3-87a) $j_n(\beta r)$ and $y_n(\beta r)$ are referred to, respectively, as the *spherical Bessel functions* of the first and second kind. They are used to represent radial standing waves, and they are related, respectively, to the corresponding regular Bessel functions $J_{n+1/2}(\beta r)$ and $Y_{n+1/2}(\beta r)$ by

$$j_n(\beta r) = \sqrt{\frac{\pi}{2\beta r}} J_{n+1/2}(\beta r) \tag{3-90a}$$

$$y_n(\beta r) = \sqrt{\frac{\pi}{2\beta r}} Y_{n+1/2}(\beta r) \tag{3-90b}$$

In (3-87b) $h_n^{(1)}(\beta r)$ and $h_n^{(2)}(\beta r)$ are referred to, respectively, as the *spherical Hankel functions* of the first and second kind. They are used to represent radial traveling waves, and they are related, respectively, to the regular Hankel functions $H_{n+1/2}^{(1)}(\beta r)$ and $H_{n+1/2}^{(2)}(\beta r)$ by

TABLE 3-3 Wave functions, zeroes, and infinities for radial waves in spherical coordinates

Wave type	Wave functions	Zeroes of wave functions	Infinities of wave functions
Traveling waves	$h_n^{(1)}(\beta r) = j_n(\beta r) + j y_n(\beta r)$ for $-r$ travel	$\beta r \to +j\infty$	$\beta r = 0$ $\beta r \to -j\infty$
	$h_n^{(2)}(\beta r) = j_n(\beta r) - j y_n(\beta r)$ for $+r$ travel	$\beta r \to -j\infty$	$\beta r = 0$ $\beta r \to +j\infty$
Standing waves	$j_n(\beta r)$ for $\pm r$ $y_n(\beta r)$ for $\pm r$	Infinite number Infinite number	$\beta r \to \pm j\infty$ $\beta r = 0$ $\beta r \to \pm j\infty$

$$h_n^{(1)}(\beta r) = \sqrt{\frac{\pi}{2\beta r}} H_{n+1/2}^{(1)}(\beta r) \tag{3-91a}$$

$$h_n^{(2)}(\beta r) = \sqrt{\frac{\pi}{2\beta r}} H_{n+1/2}^{(2)}(\beta r) \tag{3-91b}$$

Wave functions used to represent radial traveling and standing waves in spherical coordinates are listed in Table 3-3. More details on the spherical Bessel and Hankel functions can be found in Chapters 10 and 11 and Appendix IV.

In (3-88a) and (3-88b) $P_n^m(\cos\theta)$ and $Q_n^m(\cos\theta)$ are referred to, respectively, as the *associated Legendre functions* of the first and second kind (more details can be found in Chapter 10 and Appendix V).

The appropriate solution forms of f, g, and h will depend on the problem in question. For example, a typical solution for $\psi(r,\theta,\phi)$ of (3-85) to represent the fields within a sphere as shown in Figure 3-7 may take the form

$$\psi_1(r,\theta,\phi) = [A_1 j_n(\beta r) + B_1 y_n(\beta r)]$$
$$\times [C_2 P_n^m(\cos\theta) + D_2 Q_n^m(\cos\theta)][C_3 \cos(m\phi) + D_3 \sin(m\phi)] \tag{3-92}$$

For the fields to be finite at $r = 0$, where $y_n(\beta r)$ possesses a singularity, and for any value of θ, including $\theta = 0, \pi$ where $Q_n^m(\cos\theta)$ possesses singularities, (3-92) reduces to

$$\psi_1(r,\theta,\phi) = A_{mn} j_n(\beta r) P_n^m(\cos\theta)[C_3 \cos(m\phi) + D_3 \sin(m\phi)] \tag{3-92a}$$

To represent the fields outside a sphere, like for scattering, a typical solution for $\psi(r,\theta,\phi)$ would take the form of

$$\psi_2(r,\theta,\phi) = B_{mn} h_n^{(2)}(\beta r) P_n^m(\cos\theta)[C_3 \cos(m\phi) + D_3 \sin(m\phi)] \tag{3-92b}$$

whereby the spherical Hankel function of the second kind $h_n^{(2)}(\beta r)$ has replaced the spherical Bessel function of the first kind $j_n(\beta r)$ because outward traveling waves are formed outside the sphere, in contrast to the standing waves inside the sphere.

Other spherical Bessel and Hankel functions that are most often encountered in boundary-value electromagnetic problems are those utilized by Schelkunoff [3, 11]. These spherical Bessel and Hankel functions, denoted in general by $\hat{B}_n(\beta r)$ to represent any of them, must satisfy the differential equation

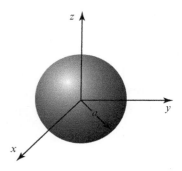

Figure 3-7   Geometry of a sphere of radius a.

$$\frac{d^2\hat{B}_n}{dr^2} + \left[\beta^2 - \frac{n(n+1)}{r^2}\right]\hat{B}_n = 0 \tag{3-93}$$

The spherical Bessel and Hankel functions that are solutions to this equation are related to other spherical Bessel and Hankel functions of (3-90a) through (3-91b), denoted here by $b_n(\beta r)$, and to the regular Bessel and Hankel functions, denoted here by $B_{n+1/2}(\beta r)$, by

$$\hat{B}_n(\beta r) = \beta r\, b_n(\beta r) = \beta r \sqrt{\frac{\pi}{2\beta r}} B_{n+1/2}(\beta r) = \sqrt{\frac{\pi\beta r}{2}} B_{n+1/2}(\beta r) \tag{3-94}$$

More details concerning the application and properties of the spherical Bessel and Hankel functions can be found in Chapter 10.

3.5 MULTIMEDIA

On the website that accompanies this book, the following multimedia resources are included for the review, understanding, and presentation of the material of this chapter.

- **PowerPoint (PPT)** viewgraphs, in multicolor.

REFERENCES

1. F. B. Hildebrand, *Advanced Calculus for Applications*, Prentice-Hall, Englewood Cliffs, NJ, 1962.
2. C. R. Wylie, Jr., *Advanced Engineering Mathematics*, McGraw-Hill, New York, 1960.
3. R. F. Harrington, *Time-Harmonic Electromagnetic Fields*, McGraw-Hill, New York, 1961.
4. R. B. Adler, L. J. Chu, and R. M. Fano, *Electromagnetic Energy Transmission and Radiation*, Chapter 8, John Wiley & Sons, New York, 1960.
5. G. N. Watson, *A Treatise on the Theory of Bessel Functions*, Cambridge Univ. Press, London, 1948.
6. W. R. Smythe, *Static and Dynamic Electricity*, McGraw-Hill, New York, 1941.
7. J. A. Stratton, *Electromagnetic Theory*, McGraw-Hill, New York, 1960.
8. P. M. Morse and H. Feshbach, *Methods of Theoretical Physics*, Parts I and II, McGraw-Hill, New York, 1953.
9. M. Abramowitz and I. A. Stegun (eds.), *Handbook of Mathematical Functions with Formulas, Graphs, and Mathematical Tables*, National Bureau of Standards Applied Mathematics Series-55, U.S. Gov. Printing Office, Washington, DC, 1966.
10. M. R. Spiegel, *Mathematical Handbook of Formulas and Tables*, Schaum's Outline Series, McGraw-Hill, New York, 1968.
11. S. A. Schelkunoff, *Electromagnetic Waves*, Van Nostrand, Princeton, NJ, 1943.

PROBLEMS

3.1. Derive the vector wave equations 3-16a and 3-16b for time-harmonic fields using the Maxwell equations of Table 1-4 for time-harmonic fields.

3.2. Verify that (3-28a) and (3-28b) are solutions to (3-26a).

3.3. Show that the second complex exponential in (3-31) represents a wave traveling in the $-z$ direction. Determine its phase velocity.

3.4. Using the method of separation of variables show that a solution to (3-37a) of the form (3-39) can be represented by (3-40a) through (3-43).

3.5. Show that the vector wave equation of (3-53) reduces, when **E** has a solution of the form (3-47), to the three scalar wave equations (3-54a) through 3-54c.

3.6. Reduce (3-51) to (3-54a) through (3-54c) by expanding $\nabla^2 \mathbf{E}$. Do not use (3-52); rather use the scalar Laplacian in cylindrical coordinates and treat **E** as a vector given by (3-47). Use that

$$\frac{\partial \hat{\mathbf{a}}_\rho}{\partial \rho} = \frac{\partial \hat{\mathbf{a}}_\phi}{\partial \rho} = \frac{\partial \hat{\mathbf{a}}_z}{\partial \rho} = 0 = \frac{\partial \hat{\mathbf{a}}_z}{\partial \phi} = \frac{\partial \hat{\mathbf{a}}_\rho}{\partial z}$$

$$= \frac{\partial \hat{\mathbf{a}}_\phi}{\partial z} = \frac{\partial \hat{\mathbf{a}}_z}{\partial z}$$

$$\frac{\partial \hat{\mathbf{a}}_\rho}{\partial \phi} = \hat{\mathbf{a}}_\phi \qquad \frac{\partial \hat{\mathbf{a}}_\phi}{\partial \phi} = -\hat{\mathbf{a}}_\rho$$

3.7. Using large argument asymptotic forms, show that Bessel and Hankel functions represent, respectively, standing and traveling waves in the radial direction.

3.8. Using large argument asymptotic forms and assuming $e^{j\omega t}$ time convention, show that Hankel functions of the first kind represent traveling waves in the $-\rho$ direction whereas Hankel functions of the second kind represent traveling waves in the $+\rho$ direction. The opposite would be true were the time variations of the $e^{-j\omega t}$ form.

3.9. Using large argument asymptotic forms, show that Bessel functions of complex argument represent attenuating standing waves.

3.10. Assuming time variations of $e^{j\omega t}$ and using large argument asymptotic forms, show that Hankel functions of the first and second kind with complex arguments represent, respectively, attenuating traveling waves in the $-\rho$ and $+\rho$ directions.

3.11. Show that when **E** can be expressed as (3-71), the vector wave equation 3-53 reduces to the three scalar wave equations 3-74a through 3-74c.

3.12. Reduce (3-51) to (3-74a) through (3-74c) by expanding $\nabla^2 \mathbf{E}$. Do not use (3-52); rather use the scalar Laplacian in spherical coordinates and treat **E** as a vector given by (3-71). Use that

$$\frac{\partial \hat{\mathbf{a}}_r}{\partial r} = \frac{\partial \hat{\mathbf{a}}_\theta}{\partial r} = \frac{\partial \hat{\mathbf{a}}_\phi}{\partial r} = 0$$

$$\frac{\partial \hat{\mathbf{a}}_r}{\partial \theta} = \hat{\mathbf{a}}_\theta \qquad \frac{\partial \hat{\mathbf{a}}_\theta}{\partial \theta} = -\hat{\mathbf{a}}_r \qquad \frac{\partial \hat{\mathbf{a}}_\phi}{\partial \theta} = 0$$

$$\frac{\partial \hat{\mathbf{a}}_r}{\partial \phi} = \sin\theta \hat{\mathbf{a}}_\phi \qquad \frac{\partial \hat{\mathbf{a}}_\theta}{\partial \phi} = \cos\theta \hat{\mathbf{a}}_\phi$$

$$\frac{\partial \hat{\mathbf{a}}_\phi}{\partial \phi} = -\sin\theta \hat{\mathbf{a}}_r - \cos\theta \hat{\mathbf{a}}_\theta$$

3.13. Using large argument asymptotic forms, show that spherical Bessel functions represent standing waves in the radial direction.

3.14. Show that spherical Hankel functions of the first and second kind represent, respectively, radial traveling waves in the $-r$ and $+r$ directions. Assume time variations of $e^{j\omega t}$ and large argument asymptotic expansions for the spherical Hankel functions.

3.15. Justify that associated Legendre functions represent standing waves in the θ direction of the spherical coordinate system.

3.16. Verify the relation (3-94) between the various forms of the spherical Bessel and Hankel functions and the regular Bessel and Hankel functions.

CHAPTER 4

Wave Propagation and Polarization

4.1 INTRODUCTION

In Chapter 3 we developed the vector wave equations for the electric and magnetic fields in lossless and lossy media. Solutions to the wave equations were also demonstrated in rectangular, cylindrical, and spherical coordinates using the method of *separation of variables*. In this chapter we want to consider solutions for the electric and magnetic fields of time-harmonic waves that travel in infinite lossless and lossy media. In particular, we want to develop expressions for *transverse electromagnetic* (TEM) waves (or modes) traveling along principal axes and oblique angles. The parameters of wave impedance, phase and group velocities, and power and energy densities will be discussed for each.

The concept of wave polarization will be introduced, and the necessary and sufficient conditions to achieve linear, circular, and elliptical polarizations will be discussed and illustrated. The sense of rotation, clockwise (right-hand) or counterclockwise (left-hand), will also be introduced.

4.2 TRANSVERSE ELECTROMAGNETIC MODES

A *mode* is a particular field configuration. For a given electromagnetic boundary-value problem, many field configurations that satisfy the wave equations, Maxwell's equations, and the boundary conditions usually exist. All these different field configurations (solutions) are usually referred to as *modes*.

A TEM mode is one whose field intensities, both **E** (electric) and **H** (magnetic), at every point in space are contained on a local plane, referred to as *equiphase plane*, that is independent of time. In general, the orientations of the local planes associated with the TEM wave are different at different points in space. In other words, at point (x_1, y_1, z_1) all the field components are contained on a plane. At another point (x_2, y_2, z_2) all field components are again contained on a plane; however, the two planes need not be parallel. This is illustrated in Figure 4-1a.

If the space orientation of the planes for a TEM mode is the same (equiphase planes are parallel), as shown in Figure 4-1b, then the fields form *plane waves*. In other words, the equiphase

Balanis' Advanced Engineering Electromagnetics, Third Edition. Constantine A. Balanis.
© 2024 John Wiley & Sons, Inc. Published 2024 by John Wiley & Sons, Inc.
Companion Website: www.wiley.com/go/balanis/advancedengineeringelectromagnetics3e

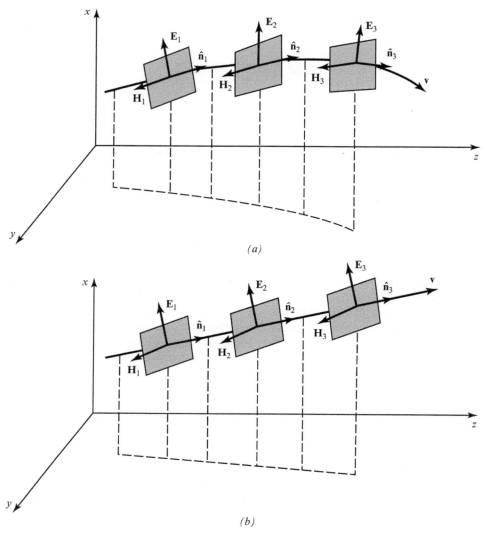

Figure 4-1 Phase fronts of waves. (*a*) TEM. (*b*) Plane.

surfaces are parallel planar surfaces. If in addition to having planar equiphases the field has equi-amplitude planar surfaces (the amplitude is the same over each plane), then it is called a *uniform plane wave*; that is, the field is not a function of the coordinates that form the equiphase and equi-amplitude planes.

4.2.1 Uniform Plane Waves in an Unbounded Lossless Medium—Principal Axis

In this section we will write expressions for the electric and magnetic fields of a uniform plane wave traveling in an unbounded medium. In addition the wave impedance, phase and energy (group) velocities, and power and energy densities of the wave will be discussed.

A. Electric and Magnetic Fields Let us assume that a time-harmonic uniform plane wave is traveling in an unbounded lossless medium (ε, μ) in the z direction (either positive or negative), as

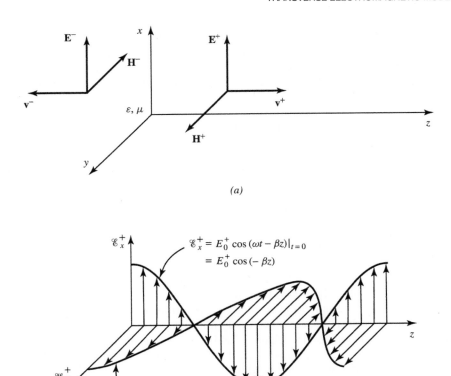

(a)

(b)

Figure 4-2   Uniform plane wave fields. (*a*) Complex. (*b*) Instantaneous.

shown in Figure 4-2*a*. In addition, for simplicity, let us assume the electric field of the wave has only an *x* component. We want to write expressions for the electric and magnetic fields associated with this wave.

For the electric and magnetic field components to be valid solutions of a time-harmonic electromagnetic wave, they must satisfy Maxwell's equations as given in Table 1-4 or the corresponding wave equations as given, respectively, by (3-18a) and (3-18b). Here the approach will be to initiate the solution by solving the wave equation for either the electric or magnetic field and then finding the other field using Maxwell's equations. An alternate procedure, which has been assigned as an end-of-chapter problem, would be to follow the entire solution using only Maxwell's equations.

Since the electric field has only an *x* component, it must satisfy the scalar wave equation of (3-20a) or (3-22), whose general solution is given by (3-23). Because the wave is a uniform plane wave that travels in the *z* direction, its solution is not a function of *x* and *y*. Therefore (3-23) reduces to

$$E_x(z) = h(z) \qquad (4\text{-}1)$$

The solutions of $h(z)$ are given by (3-30a) or (3-30b). Since the wave in question is a traveling wave, instead of a standing wave, its most appropriate solution is that given by (3-30a). The first term in (3-30a) represents a wave that travels in the $+z$ direction and the second term represents a wave that travels in the $-z$ direction. Therefore the solution of (4-1), using (3-30a), can be written as

$$E_x(z) = A_3 e^{-j\beta z} + B_3 e^{+j\beta z} = E_x^+ + E_x^- \tag{4-2}$$

or

$$E_x(z) = E_0^+ e^{-j\beta z} + E_0^- e^{+j\beta z} = E_x^+ + E_x^- \tag{4-2a}$$

$$E_x^+(z) = E_0^+ e^{-j\beta z} \tag{4-2b}$$

$$E_x^-(z) = E_0^- e^{+j\beta z} \tag{4-2c}$$

since $\beta_z = \beta$ because $\beta_x = \beta_y = 0$. E_0^+ and E_0^- represent, respectively, the amplitudes of the positive and negative (in the z direction) traveling waves.

The corresponding magnetic field must also be a solution of its wave equation 3-18b, and its form will be similar to (4-2). However, since we do not know which components of magnetic field coexist with the x component of the electric field, they are most appropriately determined by using one of Maxwell's equations as given in Table 1-4. Since the electric field is known, as given by (4-2), the magnetic field can best be found using

$$\nabla \times \mathbf{E} = -j\omega\mu\mathbf{H} \tag{4-3}$$

or

$$\mathbf{H} = -\frac{1}{j\omega\mu}\nabla \times \mathbf{E} = -\frac{1}{j\omega\mu}\begin{vmatrix} \hat{\mathbf{a}}_x & \hat{\mathbf{a}}_y & \hat{\mathbf{a}}_z \\ \dfrac{\partial}{\partial x} & \dfrac{\partial}{\partial y} & \dfrac{\partial}{\partial z} \\ E_x & 0 & 0 \end{vmatrix} \tag{4-3a}$$

which, using (4-2a), reduces to

$$\mathbf{H} = -\hat{\mathbf{a}}_y \frac{1}{j\omega\mu}\left\{\frac{\partial E_x}{\partial z}\right\} = \hat{\mathbf{a}}_y \frac{\beta}{\omega\mu}\left\{E_0^+ e^{-j\beta z} - E_0^- e^{+j\beta z}\right\}$$

$$\mathbf{H} = \hat{\mathbf{a}}_y \frac{1}{\sqrt{\mu/\varepsilon}}\left\{E_0^+ e^{-j\beta z} - E_0^- e^{+j\beta z}\right\} = \hat{\mathbf{a}}_y \frac{1}{\sqrt{\mu/\varepsilon}}\left\{E_x^+ - E_x^-\right\} = \hat{\mathbf{a}}_y\left\{H_y^+ + H_y^-\right\} \tag{4-3b}$$

where

$$H_y^+ = \frac{1}{\sqrt{\mu/\varepsilon}} E_x^+ \tag{4-3c}$$

$$H_y^- = -\frac{1}{\sqrt{\mu/\varepsilon}} E_x^- \tag{4-3d}$$

Plots of the instantaneous *positive* traveling electric and magnetic fields at $t = 0$ as a function of z are shown in Figure 4-2b. Similar plots can be drawn for the negative traveling fields.

B. Wave Impedance   Since each term for the magnetic field (A / m) in (4-3c) and (4-3d) is individually identical to the corresponding term for the electric field (V / m) in (4-2a), the factor $\sqrt{\mu / \varepsilon}$ in the denominator in (4-3c) and (4-3d) must have units of ohms (V / A). Therefore the factor $\sqrt{\mu / \varepsilon}$ is known as the *wave impedance*, Z_w, denoted by the ratio of the electric to magnetic field, and it is usually represented by η

$$Z_w = \frac{E_x^+}{H_y^+} = -\frac{E_x^-}{H_y^-} = \eta = \sqrt{\frac{\mu}{\varepsilon}} \qquad (4\text{-}4)$$

The wave impedance of (4-4) is identical to a quantity that is referred to as the *intrinsic impedance* $\eta = \sqrt{\mu / \varepsilon}$ of the medium. In general, this is true not only for uniform plane waves but also for plane and TEM waves; however, it is not true for TE or TM modes.

In (4-3d) it is also observed that a negative sign is found in front of the magnetic field component that travels in the $-z$ direction; a positive sign is noted in front of the positive traveling wave. The general procedure that can be followed to find the magnetic field components, given the electric field components, or to find the electric field components, given the magnetic field components, is the following:

1. Place the fingers of your right hand in the direction of the electric field component.
2. Direct your thumb toward the direction of wave travel (power flow).
3. Rotate your fingers 90° in a direction so that a right-hand screw is formed.
4. The new direction of your fingers is the direction of the magnetic field component.
5. Divide the electric field component by the wave impedance to obtain the corresponding magnetic field component.

The foregoing procedure must be followed for each term of each component of an electric or magnetic field. The results are identical to those that would be obtained by using Maxwell's equations. If the wave impedance is known in advance, as it is for TEM waves, this procedure is simpler and much more rapid than using Maxwell's equations. By following this procedure, the answers (including the signs) in (4-3c) and (4-3d) given (4-2b) and (4-2c) are obvious.

To illustrate the procedure, let us consider another example.

Example 4-1

The electric field of a uniform plane wave traveling in free space is given by

$$\mathbf{E} = \hat{\mathbf{a}}_y \left(E_0^+ e^{-j\beta z} + E_0^- e^{+j\beta z} \right) = \hat{\mathbf{a}}_y \left(E_y^+ + E_y^- \right)$$

where E_0^+ and E_0^- are constants. Find the corresponding magnetic field using the outlined procedure.

Solution: For the electric field component that is traveling in the $+z$ direction, the corresponding magnetic field component is given by

$$\mathbf{H}^+ = -\hat{\mathbf{a}}_x \frac{E_0^+}{\eta_0} e^{-j\beta z} \simeq -\hat{\mathbf{a}}_x \frac{E_0^+}{377} e^{-j\beta z}$$

where

$$\eta_0 = Z_w = \sqrt{\frac{\mu_0}{\varepsilon_0}} \simeq 377 \text{ ohms}$$

Similarly, for the wave that is traveling in the $-z$ direction we can write that

$$\mathbf{H}^- = \hat{\mathbf{a}}_x \frac{E_0^-}{\eta_0} e^{+j\beta z} \simeq \hat{\mathbf{a}}_x \frac{E_0^-}{377} e^{+j\beta z}$$

Therefore the total magnetic field is equal to

$$\mathbf{H} = \mathbf{H}^+ + \mathbf{H}^- = \hat{\mathbf{a}}_x \frac{1}{\eta_0}\left(-E_0^+ e^{-j\beta z} + E_0^- e^{+j\beta z} \right)$$

The same answer would be obtained if Maxwell's equations were used, and it is assigned as an end-of-chapter problem.

The term in the expression for the electric field in (4-2a) that identifies the direction of wave travel can also be written in vector notation. This is usually more convenient to use when dealing with waves traveling at oblique angles. Equation 4-2a can therefore take the more general form of

$$E_x(z) = E_0^+ e^{-j\boldsymbol{\beta}^+ \cdot \mathbf{r}} + E_0^- e^{-j\boldsymbol{\beta}^- \cdot \mathbf{r}} \tag{4-5}$$

where

$$\boldsymbol{\beta}^+ = \hat{\boldsymbol{\beta}}^+ \beta = \hat{\mathbf{a}}_x \beta_x^+ + \hat{\mathbf{a}}_y \beta_y^+ + \hat{\mathbf{a}}_z \beta_z^+ \mid_{\substack{\beta_x^+ = \beta_y^+ = 0 \\ \beta_z^+ = \beta}} = \hat{\mathbf{a}}_z \beta \tag{4-5a}$$

$$\boldsymbol{\beta}^- = \hat{\boldsymbol{\beta}}^- \beta = \hat{\mathbf{a}}_x \beta_x^- + \hat{\mathbf{a}}_y \beta_y^- - \hat{\mathbf{a}}_z \beta_z^- \mid_{\substack{\beta_x^- = \beta_y^- = 0 \\ \beta_z^- = \beta}} = -\hat{\mathbf{a}}_z \beta \tag{4-5b}$$

$$\mathbf{r} = \text{position vector} = \hat{\mathbf{a}}_x x + \hat{\mathbf{a}}_y y + \hat{\mathbf{a}}_z z \tag{4-5c}$$

In (4-5a) through (4-5c), β_x, β_y, β_z represent, respectively, the phase constants of the wave in the x, y, z directions, $\mathbf{r}$ represents the position vector in rectangular coordinates, and $\hat{\boldsymbol{\beta}}^+$ and $\hat{\boldsymbol{\beta}}^-$ represent unit vectors in the directions of $\boldsymbol{\beta}^+$ and $\boldsymbol{\beta}^-$. The notation used in (4-5) through (4-5c) to represent the wave travel will be most convenient to express wave travel at oblique angles, as will be the case in Section 4.2.2.

C. Phase and Energy (Group) Velocities, Power, and Energy Densities   The expressions for the electric and magnetic fields, as given by (4-2a) and (4-3b), represent the spatial variations of the field intensities. The corresponding instantaneous forms of each can be written, using (1-61a) and (1-61b) and assuming E_0^+ and E_0^- are real constants, as

$$\mathscr{E}_x(z;t) = \mathscr{E}_x^+(z;t) + \mathscr{E}_x^-(z;t) = \text{Re}\left[E_0^+ e^{-j\beta z} e^{j\omega t} \right] + \text{Re}\left[E_0^- e^{+j\beta z} e^{j\omega t} \right]$$

$$= E_0^+ \cos(\omega t - \beta z) + E_0^- \cos(\omega t + \beta z) \tag{4-6a}$$

$$\mathscr{H}_y(z;t) = \mathscr{H}_y^+(z;t) + \mathscr{H}_y^-(z;t)$$

$$= \frac{1}{\sqrt{\mu/\varepsilon}}\left[E_0^+ \cos(\omega t - \beta z) - E_0^- \cos(\omega t + \beta z) \right] \tag{4-6b}$$

In each of the fields, as given by (4-6a) and (4-6b), the first term represents, according to (3-34) through (3-35) and Figure 3-3, a wave that travels in the $+z$ direction; the second term represents a wave that travels in the $-z$ direction. To maintain a constant phase in the first term of (4-6a), the velocity must be equal, according to (3-35), to

$$v_p^+ = +\frac{dz}{dt} = \frac{\omega}{\beta} = \frac{\omega}{\omega\sqrt{\mu\varepsilon}} = \frac{1}{\sqrt{\mu\varepsilon}} \qquad (4\text{-}7)$$

The corresponding velocity of the second term in (4-6a) is identical in magnitude to (4-7) but with a negative sign to reflect the direction of wave travel. The velocity of (4-7) is referred to as the *phase velocity*, and it represents the velocity that must be maintained in order to keep in step with a constant phase front of the wave. As will be shown for oblique traveling waves, the phase velocity of such waves can exceed the velocity of light. This is only a hypothetical speed, as will be explained in Section 4.2.2C. Aside of nonuniform plane waves, also referred to as slow surface waves (see Section 5.3.4A), in general the phase velocity can be equal to or even greater than the speed of light. Variations of the instantaneous positive traveling electric $\mathcal{E}_x^+(z;t)$ and magnetic $\mathcal{H}_y^+(z;t)$ fields as a function of z for $t = 0$ are shown in Figure 4-2b. As time increases, both curves will shift in the positive z direction. A similar set of curves can be drawn for the negative traveling electric $\mathcal{E}_x^-(z;t)$ and magnetic $\mathcal{H}_y^-(z;t)$ fields.

The electric and magnetic energies (W-s/m^3) and power densities (W/m^2) associated with the positive traveling waves of (4-6a) and (4-6b) can be written, according to (1-58f) and (1-58e), as

$$w_e^+ = \frac{1}{2}\varepsilon\mathcal{E}_x^{+2} = \frac{1}{2}\varepsilon E_0^{+2}\cos^2(\omega t - \beta z) \qquad (4\text{-}8a)$$

$$w_m^+ = \frac{1}{2}\mu\mathcal{H}_y^{+2} = \frac{1}{2}\mu[(\varepsilon/\mu)E_0^{+2}\cos^2(\omega t - \beta z)] = \frac{1}{2}\varepsilon E_0^{+2}\cos^2(\omega t - \beta z) \qquad (4\text{-}8b)$$

$$\mathcal{S}^+ = \mathcal{E}^+ \times \mathcal{H}^+ = \hat{\mathbf{a}}_x E_0^+ \cos(\omega t - \beta z) \times \left[\hat{\mathbf{a}}_y \left(1/\sqrt{\mu/\varepsilon}\right) E_0^+ \cos(\omega t - \beta z)\right]$$

$$= \hat{\mathbf{a}}_z \mathcal{S}^+ = \hat{\mathbf{a}}_z \left(1/\sqrt{\mu/\varepsilon}\right) E_0^{+2} \cos^2(\omega t - \beta z) \qquad (4\text{-}8c)$$

The ratio formed by dividing the power density $\mathcal{S}$(W/m^2) by the total energy density $w = w_e + w_m$ (J/m^3 = W-s/m^3) is referred to as the *energy (group) velocity* v_e, and it is given by

$$v_e^+ = \frac{\mathcal{S}^+}{w^+} = \frac{\mathcal{S}^+}{w_e^+ + w_m^+} = \frac{(1/\sqrt{\mu/\varepsilon})E_0^{+2}\cos^2(\omega t - \beta z)}{\varepsilon E_0^{+2}\cos^2(\omega t - \beta z)} = \frac{1}{\sqrt{\mu\varepsilon}} \qquad (4\text{-}9)$$

The energy velocity represents the velocity with which the wave energy is transported. It is apparent that (4-9) is identical to (4-7). In general that is not the case. In fact, the energy velocity v_e^+ can be equal to, but not exceed, the speed of light, and the product of the phase velocity v_p and energy velocity v_e must always be equal to

$$\boxed{v_p^+ v_e^+ = (v^+)^2 = \frac{1}{\mu\varepsilon}} \qquad (4\text{-}10)$$

where $v^+ = 1/\sqrt{\mu\varepsilon}$ is the speed of light. The same holds for the negative traveling waves.

The time-average power density (Poynting vector) associated with the positive traveling wave can be written, using (1-70) and the first terms of (4-2a) and (4-3b), as

$$\mathcal{S}_{av}^+ = \frac{1}{2}\,\mathrm{Re}\left(\mathbf{E}^+ \times \mathbf{H}^{+*}\right) = \hat{\mathbf{a}}_z \frac{1}{2\sqrt{\mu/\varepsilon}}|E_x^+|^2 = \hat{\mathbf{a}}_z \frac{|E_0^+|^2}{2\sqrt{\mu/\varepsilon}} = \hat{\mathbf{a}}_z \frac{|E_0^+|^2}{2\eta} \qquad (4\text{-}11)$$

A similar expression is derived for the negative traveling wave.

D. Standing Waves   Each of the terms in (4-2a) and (4-3b) represents individually *traveling* waves, the first traveling in the positive z direction and the second in the negative z direction. The two together form a so-called *standing wave*, which is comprised of two oppositely traveling waves.

To examine the characteristics of a standing wave, let us rewrite (4-2a) as

$$E_x(z) = E_0^+ e^{-j\beta z} + E_0^- e^{+j\beta z}$$
$$= E_0^+ \left[\cos(\beta z) - j \sin(\beta z) \right] + E_0^- \left[\cos(\beta z) + j \sin(\beta z) \right]$$
$$= \left(E_0^+ + E_0^- \right) \cos(\beta z) - j \left(E_0^+ - E_0^- \right) \sin(\beta z)$$

$$E_x(z) = \sqrt{\left(E_0^+ + E_0^- \right)^2 \cos^2(\beta z) + \left(E_0^+ - E_0^- \right)^2 \sin^2(\beta z)}$$
$$\times \exp\left\{ -j \tan^{-1} \left[\frac{\left(E_0^+ - E_0^- \right) \sin(\beta z)}{\left(E_0^+ + E_0^- \right) \cos(\beta z)} \right] \right\}$$

$$E_x(z) = \sqrt{\left(E_0^+ \right)^2 + \left(E_0^- \right)^2 + 2 E_0^+ E_0^- \cos(2\beta z)}$$
$$\times \exp\left\{ -j \tan^{-1} \left[\frac{\left(E_0^+ - E_0^- \right)}{\left(E_0^+ + E_0^- \right)} \tan(\beta z) \right] \right\} \qquad (4\text{-}12)$$

The amplitude of the waveform given by (4-12) is equal to

$$|E_x(z)| = \sqrt{(E_0^+)^2 + (E_0^-)^2 + 2 E_0^+ E_0^- \cos(2\beta z)} \qquad (4\text{-}12a)$$

By examining (4-12a), it is evident that its maximum and minimum values are given, respectively, by

$$|E_x(z)|_{\max} = |E_0^+| + |E_0^-| \quad \text{when } \beta z = m\pi, \ m = 0, 1, 2,\dots \qquad (4\text{-}13a)$$

and for $|E_0^+| > |E_0^-|$,

$$|E_x(z)|_{\min} = |E_0^+| - |E_0^-| \quad \text{when } \beta z = \frac{(2m+1)\pi}{2}, \ m = 0, 1, 2,\dots \qquad (4\text{-}13b)$$

Neighboring maximum and minimum values are separated by a distance of $\lambda/4$ or successive maxima or minima are separated by $\lambda/2$.

The instantaneous field of (4-12) can also be written as

$$\mathcal{E}_x(z;t) = \text{Re}\left[E_x(z) e^{j\omega t} \right]$$
$$= \sqrt{(E_0^+)^2 + (E_0^-)^2 + 2 E_0^+ E_0^- \cos(2\beta z)}$$
$$\times \cos\left[\omega t - \tan^{-1}\left\{ \frac{E_0^+ - E_0^-}{E_0^+ + E_0^-} \tan(\beta z) \right\} \right] \qquad (4\text{-}14)$$

It is apparent that (4-12a) represents the envelope of the maximum values the instantaneous field of (4-14) will achieve as a function of time at a given position. Since this envelope of maximum values does not move (change) in position as a function of time, it is referred to as the *standing wave pattern* and the associated wave of (4-12) or (4-14) is referred to as the *standing wave*.

The ratio of the maximum/minimum values of the standing wave pattern of (4-12a), as given by (4-13a) and (4-13b), is referred to as the standing wave ratio (SWR), and it is given by

$$\text{SWR} = \frac{|E_x(z)|_{\max}}{|E_x(z)|_{\min}} = \frac{|E_0^+| + |E_0^-|}{|E_0^+| - |E_0^-|} = \frac{1 + \frac{|E_0^-|}{|E_0^+|}}{1 - \frac{|E_0^-|}{|E_0^+|}} = \frac{1 + |\Gamma|}{1 - |\Gamma|} \qquad (4\text{-}15)$$

where Γ is the reflection coefficient. Since in transmission lines we usually deal with voltages and currents (instead of electric and magnetic fields), the SWR is usually referred to as the VSWR (voltage standing wave ratio). Plots of the standing wave pattern in terms of E_0^+ as a function of $z(-\lambda \leq z \leq \lambda)$ for $|\Gamma| = 0$, 0.2, 0.4, 0.6, 0.8, and 1 are shown in Figure 4-3.

The SWR is a quantity that can be measured with instrumentation [1, 2]. SWR has values in the range of $1 \leq \text{SWR} \leq \infty$. The value of the SWR indicates the amount of interference between the two opposite traveling waves; the smaller the SWR value, the lesser the interference. The minimum SWR value of unity occurs when $|\Gamma| = E_0^- / E_0^+ = 0$, and it indicates that no

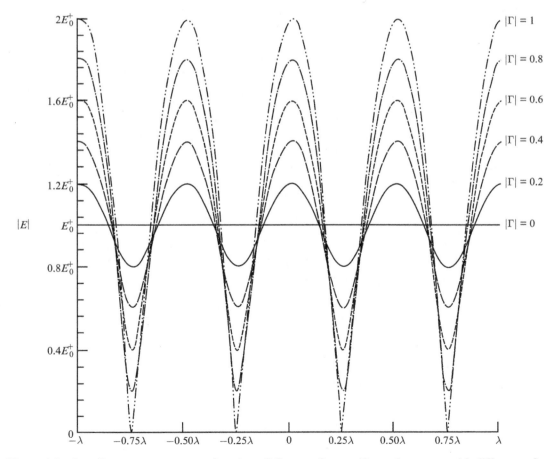

Figure 4-3 Standing wave pattern as a function of distance for a uniform plane wave with different reflection coefficients.

interference is formed. Thus the standing wave reduces to a pure traveling wave. The maximum SWR of infinity occurs when $|\Gamma| = E_0^- / E_0^+ = 1$, and it indicates that the negative traveling wave is of the same intensity as the positive traveling wave. This provides the maximum interference, and the wave forms a pure standing wave pattern given by

$$\left| E_x(z) \right|_{E_0^+ = E_0^-} = 2E_0^+ \left| \cos(\beta z) \right| = 2E_0^- \left| \cos(\beta z) \right| \qquad (4\text{-}16)$$

The pattern of this is a rectified cosine function, and it is represented in Figure 4-3 by the $|\Gamma| = 1$ curve. The pattern exhibits pure nulls and peak values of twice the amplitude of the incident wave.

4.2.2   Uniform Plane Waves in an Unbounded Lossless Medium—Oblique Angle

In this section, expressions for the electric and magnetic fields, wave impedance, phase and group velocities, and power and energy densities will be written for uniform plane waves traveling at oblique angles in an unbounded medium. All of these will be done for waves that are uniform plane waves to the direction of travel.

A. Electric and Magnetic Fields   Let us assume that a uniform plane wave is traveling in an unbounded medium in a direction shown in Figure 4-4a. The amplitudes of the positive and negative traveling electric fields are E_0^+ and E_0^-, respectively, and the assumed directions of each are also illustrated in Figure 4-4a. It is desirable to write expressions for the positive and negative traveling electric and magnetic field components.

Since the electric field of the wave of Figure 4-4a does not have a y component, the field configuration is referred to as *transverse electric to y* (TE^y). More detailed discussion on the construction of *transverse electric* (TE) and *transverse magnetic* (TM) field configurations, as well as *transverse electromagnetic* (TEM), can be found in Chapter 6.

Because for the TE^y wave of Figure 4-4a the electric field is pointing along a direction that does not coincide with any of the principal axes, it can be decomposed into components coincident with the principal axes. According to the geometry of Figure 4-4a, it is evident that the electric field can be written as

$$\mathbf{E} = \mathbf{E}^+ + \mathbf{E}^- = E_0^+ (\hat{\mathbf{a}}_x \cos\theta_i - \hat{\mathbf{a}}_z \sin\theta_i) e^{-j\boldsymbol{\beta}^+ \cdot \mathbf{r}}$$
$$+ E_0^- (\hat{\mathbf{a}}_x \cos\theta_i - \hat{\mathbf{a}}_z \sin\theta_i) e^{-j\boldsymbol{\beta}^+ \cdot \mathbf{r}} \qquad (4\text{-}17)$$

where $\mathbf{r}$ is the position vector of (4-5c), and it is displayed graphically in Figure 4-5. Since the phase constants $\boldsymbol{\beta}^+$ and $\boldsymbol{\beta}^-$ can be written, respectively, as

$$\boldsymbol{\beta}^+ = \hat{\boldsymbol{\beta}}^+ \beta = \hat{\mathbf{a}}_x \beta_x^+ + \hat{\mathbf{a}}_z \beta_z^+ = \beta(\hat{\mathbf{a}}_x \sin\theta_i + \hat{\mathbf{a}}_z \cos\theta_i) \qquad (4\text{-}17a)$$

$$\boldsymbol{\beta}^- = \hat{\boldsymbol{\beta}}^- \beta = \hat{\mathbf{a}}_x \beta_x^- + \hat{\mathbf{a}}_z \beta_z^- = -\beta(\hat{\mathbf{a}}_x \sin\theta_i + \hat{\mathbf{a}}_z \cos\theta_i) \qquad (4\text{-}17b)$$

(4-17) can be expressed as

$$\mathbf{E} = E_0^+ (\hat{\mathbf{a}}_x \cos\theta_i - \hat{\mathbf{a}}_z \sin\theta_i) \, e^{-j\beta(x \sin\theta_i + z \cos\theta_i)}$$
$$+ E_0^- (\hat{\mathbf{a}}_x \cos\theta_i - \hat{\mathbf{a}}_z \sin\theta_i) \, e^{+j\beta(x \sin\theta_i + z \cos\theta_i)} \qquad (4\text{-}18a)$$

Since the wave is a uniform plane wave, the amplitude of its magnetic field is related to the amplitude of its electric field by the wave impedance (in this case also by the intrinsic impedance)

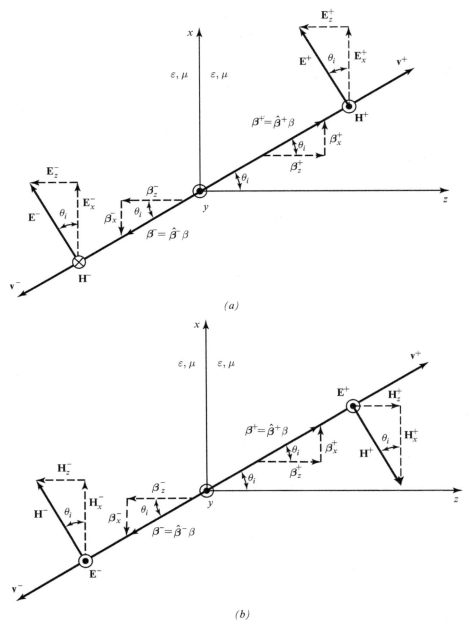

(a)

(b)

Figure 4-4 Transverse electric and magnetic uniform plane waves in an unbounded medium at an oblique angle. (a) TE^y mode. (b) TM^y mode.

as given by (4-4). Since the magnetic field is traveling in the same direction as the electric field, the exponentials used to indicate its directions of travel are the same as those of the electric field as given in (4-18a). The directions of the magnetic field can be found using the right-hand procedure outlined in Section 4.2.1 and illustrated graphically in Figure 4-2b for the positive traveling wave. Using all of the preceding information, it is evident that the magnetic field corresponding to the electric field of (4-18a) can be written as

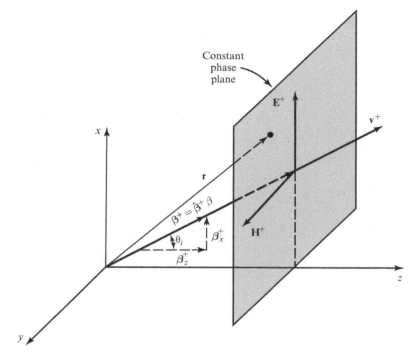

Figure 4-5   Phase front of a TEM wave traveling in a general direction.

$$\mathbf{H} = \mathbf{H}^+ + \mathbf{H}^- = \hat{\mathbf{a}}_y \left[\frac{E_0^+}{\eta} e^{-j\beta(x\sin\theta_i + z\cos\theta_i)} - \frac{E_0^-}{\eta} e^{+j\beta(x\sin\theta_i + z\cos\theta_i)} \right] \tag{4-18b}$$

In vector form, (4-18b) can also be written as

$$\mathbf{H} = \frac{1}{\eta} \left[\hat{\boldsymbol{\beta}}^+ \times \mathbf{E}^+ + \hat{\boldsymbol{\beta}}^- \times \mathbf{E}^- \right] \tag{4-18c}$$

The same form can be used to relate the $\mathbf{E}$ and $\mathbf{H}$ for any TEM wave traveling in any direction. It is apparent that when $\theta_i = 0$ (4-18a) and (4-18b) reduce to (4-2a) and (4-3b), respectively. The same answer for the magnetic field of (4-18b) can be obtained by applying Maxwell's equation 4-3 to the electric field of (4-18a). This is left for the reader as an end-of-the-chapter exercise.

The planes of constant phase at any time t are obtained by setting the phases of (4-18a) or (4-18b) equal to a constant, that is

$$\boldsymbol{\beta}^+ \cdot \mathbf{r} = \beta_x^+ x + \beta_y^+ y + \beta_z^+ z \big|_{y=0} = \beta(x\sin\theta_i + z\cos\theta_i) = C^+ \tag{4-19a}$$

$$\boldsymbol{\beta}^- \cdot \mathbf{r} = \beta_x^- x + \beta_y^- y + \beta_z^- z \big|_{y=0} = -\beta(x\sin\theta_i + z\cos\theta_i) = C^- \tag{4-19b}$$

Each of (4-19a) and (4-19b) are equations of a plane in either the spherical or rectangular coordinates with unit vectors $\hat{\boldsymbol{\beta}}^+$ and $\hat{\boldsymbol{\beta}}^-$ normal to each of the respective surfaces. The respective phase velocities in any direction (r, x, or z) are obtained by letting

$$\boldsymbol{\beta}^+ \cdot \mathbf{r} - \omega t = \beta(x\sin\theta_i + z\cos\theta_i) - \omega t = C_0^+ \tag{4-19c}$$

$$\boldsymbol{\beta}^- \cdot \mathbf{r} - \omega t = -\beta\,(x\sin\theta_i + z\cos\theta_i) - \omega t = C_0^- \tag{4-19d}$$

and taking a derivative with respect to time.

Example 4-2

Another exercise of interest is that in which the electric field is directed along the $+y$ direction and the wave is traveling along an oblique angle θ_i, as shown in Figure 4-4b. This is referred to as a TM^y wave. The objective here is again to write expressions for the positive and negative electric and magnetic field components, assuming the amplitudes of the positive and negative electric field components are E_0^+ and E_0^-, respectively.

Solution: Since this wave only has a y electric field component, and it is traveling in the same direction as that of Figure 4-4a, we can write the electric field as

$$\mathbf{E} = \mathbf{E}^+ + \mathbf{E}^- = \hat{\mathbf{a}}_y \left[E_0^+ e^{-j\beta(x\sin\theta_i + z\cos\theta_i)} + E_0^- e^{+j\beta(x\sin\theta_i + z\cos\theta_i)} \right]$$

Using the right-hand procedure outlined in Section 4.2.1, the corresponding magnetic field components are pointed along directions indicated in Figure 4-4b. Since the magnetic field is not directed along any of the principal axes, it can be decomposed into components that coincide with the principal axes, as shown in Figure 4-4b. Doing this and relating the amplitude of the electric and magnetic fields by the intrinsic impedance, we can write the magnetic field as

$$\mathbf{H} = \mathbf{H}^+ + \mathbf{H}^- = \frac{E_0^+}{\eta}(-\hat{\mathbf{a}}_x \cos\theta_i + \hat{\mathbf{a}}_z \sin\theta_i)\,e^{-j\beta(x\sin\theta_i + z\cos\theta_i)}$$

$$+ \frac{E_0^-}{\eta}(\hat{\mathbf{a}}_x \cos\theta_i - \hat{\mathbf{a}}_z \sin\theta_i)\,e^{+j\beta(x\sin\theta_i + z\cos\theta_i)}$$

The same answers could have been obtained if Maxwell's equation 4-3 were used. Since the magnetic field does not have any y components, this field configuration is referred to as *transverse magnetic to y* (TM^y), which will be discussed in more detail in Chapter 6.

B. Wave Impedance Since the TE^y and TM^y fields of Section 4.2.2A were TEM to the direction of travel, the wave impedance of each in the direction $\boldsymbol{\beta}$ of wave travel is the same as the intrinsic impedance of the medium. However, there are other directional impedances toward the x and z directions. These impedances are obtained by dividing the electric field component by the corresponding orthogonal magnetic field component. These two components are chosen so that the cross product of the electric field and the magnetic field, which corresponds to the direction of power flow, is in the direction of the wave travel.

Following the aforementioned procedure, the directional impedances for the TE^y fields of (4-18a) and (4-18b) can be written as

$$TE^y$$

$$Z_x^+ = -\frac{E_z^+}{H_y^+} = \eta \sin\theta_i = Z_x^- = \frac{E_z^-}{H_y^-} \tag{4-20a}$$

$$Z_z^+ = \frac{E_x^+}{H_y^+} = \eta \cos\theta_i = Z_z^- = -\frac{E_x^-}{H_y^-} \tag{4-20b}$$

In the same manner, the directional impedances of the TM^y fields of Example 4-2 can be written as

$$\text{TM}^y$$

$$Z_x^+ = \frac{E_y^+}{H_z^+} = \frac{\eta}{\sin\theta_i} = Z_x^- = -\frac{E_y^-}{H_z^-} \tag{4-21a}$$

$$Z_z^+ = -\frac{E_y^+}{H_x^+} = \frac{\eta}{\cos\theta_i} = Z_z^- = \frac{E_y^-}{H_x^-} \tag{4-21b}$$

It is apparent from the preceding results that the directional impedances of the TE^y oblique incidence traveling waves are equal to or smaller than the intrinsic impedance and those of the TM^y are equal to or larger than the intrinsic impedance. In addition, the positive and negative directional impedances of the same orientation are the same. This is the main principle of the *transverse resonance method* (see Section 8.6), which is used to analyze microwave circuits and antenna systems [3, 4].

C. Phase and Energy (Group) Velocities   The wave velocity v_r of the fields given by (4-18a) and (4-18b) in the direction β of travel is equal to the speed of light v. Since the wave is a plane wave to the direction β of travel, the planes over which the phase is constant (constant phase planes) are perpendicular to the direction β of wave travel. This is illustrated graphically in Figure 4-6. To maintain a constant phase (or to keep in step with a constant phase plane), a velocity equal to the speed of light must be maintained in the direction β of travel. This is referred to as the phase velocity v_{pr} along the direction β of travel. Since the energy also is being transported with the same speed, the energy velocity v_{er} in the direction β of travel is also equal to the speed of light. Thus

$$v_r = v_{pr} = v_{er} = v = \frac{1}{\sqrt{\mu\varepsilon}} \tag{4-22}$$

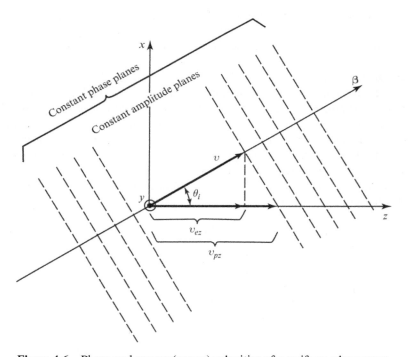

Figure 4-6   Phase and energy (group) velocities of a uniform plane wave.

where

v_r = wave velocity in the direction of wave travel
v_{pr} = phase velocity in the direction of wave travel
v_{er} = energy (group) velocity in the direction of wave travel
v = speed of light

To keep in step with a constant phase plane of the wave of Figure 4-6, a velocity in the z direction equal to

$$v_{pz} = \frac{v}{\cos\theta_i} = \frac{1}{\sqrt{\mu\varepsilon}\,\cos\theta_i} \geq v \tag{4-23}$$

must be maintained. This is referred to as the phase velocity v_{pz} in the z direction, and it is greater than the speed of light. Since nothing travels with speeds greater than the speed of light, it must be remembered that this is a hypothetical velocity that must be maintained in order to keep in step with a constant phase plane of the wave that itself travels with the speed of light in the direction β of travel. The phase velocities of (4-22) and (4-23) can be obtained, respectively, by using (4-19c) and (4-19d). These are left as end-of-chapter exercises for the reader.

Whereas a velocity greater than the speed of light must be maintained in the z direction to keep in step with a constant phase plane of Figure 4-6, the energy is transported in the z direction with a velocity that is equal to

$$v_{ez} = v\cos\theta_i = \frac{\cos\theta_i}{\sqrt{\mu\varepsilon}} \leq v \tag{4-24}$$

This is referred to as the energy (group) velocity v_{ez} in the z direction, and it is equal to or smaller than the speed of light. Graphically this is illustrated in Figure 4-6.

For any wave, the product of the phase and energy velocities in any direction must be equal to the speed of light squared or

$$v_{pr}v_{er} = v_{pz}v_{ez} = v^2 = \frac{1}{\mu\varepsilon} \tag{4-25}$$

This obviously is satisfied by the previously derived results.

The energy velocity of (4-24) can be derived using (4-18a) and (4-18b) along with the definition (4-9). This is left for the reader as an end-of-chapter exercise.

Since the fields of (4-18a) and (4-18b) form a uniform plane wave, the planes over which the amplitude is maintained constant are also constant planes that are perpendicular to the direction β of travel. These are illustrated in Figure 4-6 and coincide with the constant phase planes. For other types of waves, the constant phase and amplitude planes do not in general coincide.

D. Power and Energy Densities The average power density associated with the fields of (4-18a) and (4-18b) that travel in the β^+ direction is given by

$$\left(\mathbf{S}_{av}^+\right)_r = \frac{1}{2}\text{Re}\left[\left(\mathbf{E}^+\right)\times\left(\mathbf{H}^+\right)^*\right]$$

$$= \frac{1}{2}\text{Re}\left[E_0^+\left(\hat{\mathbf{a}}_x\cos\theta_i - \hat{\mathbf{a}}_z\sin\theta_i\right)e^{-j\beta(x\sin\theta_i + z\cos\theta_i)}\right.$$

$$\left. \times\,\hat{\mathbf{a}}_y\,\frac{E_0^{+*}}{\eta}e^{+j\beta(x\sin\theta_i + z\cos\theta_i)}\right]$$

$$\left(\mathbf{S}_{av}^+\right)_r = \left(\hat{\mathbf{a}}_x\sin\theta_i + \hat{\mathbf{a}}_z\cos\theta_i\right)\frac{|E_0^+|^2}{2\eta} = \hat{\mathbf{a}}_r\frac{|E_0^+|^2}{2\eta} = \hat{\mathbf{a}}_x\left(S_{av}^+\right)_x + \hat{\mathbf{a}}_z\left(S_{av}^+\right)_z \tag{4-26}$$

where

$$(S_{av}^+)_x = \sin\theta_i \frac{|E_0^+|^2}{2\eta} = \sin\theta_i (S_{av}^+)_r \qquad (4\text{-}26a)$$

$$(S_{av}^+)_z = \cos\theta_i \frac{|E_0^+|^2}{2\eta} = \cos\theta_i (S_{av}^+)_r \qquad (4\text{-}26b)$$

$(S_{av}^+)_r$ represents the average power density along the principal β^+ direction of travel and $(S_{av}^+)_x$ and $(S_{av}^+)_z$ represent the directional power densities of the wave in the $+x$ and $+z$ directions, respectively. Similar expressions can be derived for the wave that travels along the β^- direction.

Example 4-3

For the TM^y fields of Example 4-2, derive expressions for the average power density along the principal β^+ direction of travel and for the directional power densities along the $+x$ and $+z$ directions.

Solution: Using the electric and magnetic fields of the solution of Example 4-2 and following the procedure used to derive (4-26) through (4-26b), it can be shown that

$$(\mathbf{S}_{av}^+)_r = \frac{1}{2}\mathrm{Re}\left[\left(\mathbf{E}^+\right)\times\left(\mathbf{H}^+\right)^*\right]$$

$$= \frac{1}{2}\mathrm{Re}\Big[\hat{\mathbf{a}}_y E_0^+ e^{-j\beta(x\cos\theta_i + y\sin\theta_i)}$$

$$\times \frac{\left(E_0^+\right)^*}{\eta}(-\hat{\mathbf{a}}_x \cos\theta_i + \hat{\mathbf{a}}_z \sin\theta_i)e^{+j\beta(x\cos\theta_i + y\sin\theta_i)}\Big]$$

$$(\mathbf{S}_{av}^+)_r = (\hat{\mathbf{a}}_x \sin\theta_i + \hat{\mathbf{a}}_z \cos\theta_i)\frac{|E_0^+|^2}{2\eta} = \hat{\mathbf{a}}_r \frac{|E_0^+|^2}{2\eta}$$

$$= \hat{\mathbf{a}}_x (S_{av}^+)_x + \hat{\mathbf{a}}_z (S_{av}^+)_z$$

where

$$(S_{av}^+)_x = \sin\theta_i \frac{|E_0^+|^2}{2\eta} = \sin\theta_i (S_{av}^+)_r$$

$$(S_{av}^+)_z = \cos\theta_i \frac{|E_0^+|^2}{2\eta} = \cos\theta_i (S_{av}^+)_r$$

(S_{av}^+), $(S_{av}^+)_x$, and $(S_{av}^+)_z$ of this TM^y wave are identical to the corresponding ones of the TE^y wave, given by (4-26) through (4-26b).

4.3 TRANSVERSE ELECTROMAGNETIC MODES IN LOSSY MEDIA

In addition to the accumulation of phase, electromagnetic waves that travel in lossy media undergo attenuation. To account for the attenuation, an attenuation constant is introduced as discussed in Chapter 3, Section 3.4.1B. In this section we want to discuss the solution for the electric and magnetic fields of uniform plane waves as they travel in lossy media [5].

4.3.1   Uniform Plane Waves in an Unbounded Lossy Medium—Principal Axis

As for the electromagnetic wave of Section 4.2.1, let us assume that a uniform plane wave is traveling in a lossy medium. Using the coordinate system of Figure 4-1, the electric field is assumed to have an x component and the wave is traveling in the $\pm z$ direction. Since the electric field must satisfy the wave equation for lossy media, its expression takes, according to (3-42a), the form

$$\mathbf{E}(z) = \hat{\mathbf{a}}_x E_x(z) = \hat{\mathbf{a}}_x (E_0^+ e^{-\gamma z} + E_0^- e^{+\gamma z}) = \hat{\mathbf{a}}_x (E_0^+ e^{-\alpha z} e^{-j\beta z} + E_0^- e^{+\alpha z} e^{+j\beta z}) \qquad (4\text{-}27)$$

where $\gamma_x = \gamma_y = 0$ and $\gamma_z = \gamma$. The first term represents the positive traveling wave and the second term represents the negative traveling wave. In (4-27) γ is the propagation constant whose real α and imaginary β parts are defined, respectively, as the attenuation and phase constants. According to (3-27e) and (3-46), γ takes the form

$$\gamma = \alpha + j\beta = \sqrt{j\omega\mu(\sigma + j\omega\varepsilon)} = \sqrt{-\omega^2\mu\varepsilon + j\omega\mu\sigma} \qquad (4\text{-}28)$$

Squaring (4-28) and equating real and imaginary from both sides reduces it to

$$\alpha^2 - \beta^2 = -\omega^2\mu\varepsilon \qquad (4\text{-}28a)$$

$$2\alpha\beta = \omega\mu\sigma \qquad (4\text{-}28b)$$

Solving (4-28a) and (4-28b) simultaneously, we can write α and β as

$$\alpha = \omega\sqrt{\mu\varepsilon}\left\{\frac{1}{2}\left[\sqrt{1 + \left(\frac{\sigma}{\omega\varepsilon}\right)^2} - 1\right]\right\}^{1/2} \quad \text{Np/m} \qquad (4\text{-}28c)$$

$$\beta = \omega\sqrt{\mu\varepsilon}\left\{\frac{1}{2}\left[\sqrt{1 + \left(\frac{\sigma}{\omega\varepsilon}\right)^2} + 1\right]\right\}^{1/2} \quad \text{rad/m} \qquad (4\text{-}28d)$$

In the literature, the phase constant β is also represented by k.

The attenuation constant α is often expressed in decibels per meter (dB/m). The conversion between nepers per meter and decibels per meter is obtained by examining the real exponential in (4-27) that represents the attenuation factor of the wave in a lossy medium. Since that factor represents the relative attenuation of the electric or magnetic field, its conversion to decibels (dB) is obtained by

$$dB = 20\log_{10}(e^{-\alpha z}) = 20\,(-\alpha z)\,\log_{10}(e)$$
$$= 20\,(-\alpha z)\,(0.434) = -8.68(\alpha z) \qquad (4\text{-}28e)$$

or

$$|\alpha\,(\text{Np/m})| = \frac{1}{8.68}\,|\alpha\,(\text{dB/m})| \qquad (4\text{-}28f)$$

The magnetic field associated with the electric field of (4-27) can be obtained using Maxwell's equation (4-3) or (4-3a), that is,

$$\mathbf{H} = -\frac{1}{j\omega\mu}\nabla\times\mathbf{E} = -\hat{\mathbf{a}}_y\frac{1}{j\omega\mu}\frac{\partial E_x}{\partial z} \tag{4-29}$$

Using (4-27) reduces (4-29) to

$$\mathbf{H} = +\hat{\mathbf{a}}_y\frac{\gamma}{j\omega\mu}(E_0^+e^{-\gamma z} - E_0^-e^{+\gamma z})$$

$$= \hat{\mathbf{a}}_y\frac{\sqrt{j\omega\mu(\sigma + j\omega\varepsilon)}}{j\omega\mu}(E_0^+e^{-\gamma z} - E_0^-e^{+\gamma z})$$

$$= \hat{\mathbf{a}}_y\sqrt{\frac{\sigma + j\omega\varepsilon}{j\omega\mu}}(E_0^+e^{-\gamma z} - E_0^-e^{+\gamma z})$$

$$\mathbf{H} = \hat{\mathbf{a}}_y\frac{1}{Z_w}(E_0^+e^{-\gamma z} - E_0^-e^{+yz}) \tag{4-29a}$$

In (4-29a), Z_w is the wave impedance of the wave, and it takes the form

$$Z_w = \sqrt{\frac{j\omega\mu}{\sigma + j\omega\varepsilon}} = \eta_c \tag{4-30}$$

which is also equal to the intrinsic impedance η_c of the lossy medium. The equality between the wave and intrinsic impedances for TEM waves in lossy media is identical to that for lossless media of Section 4.2.1B.

The average power density associated with the positive traveling fields of (4-27) and (4-29a) can be written as

$$\mathbf{S}^+ = \frac{1}{2}\text{Re}(\mathbf{E}^+ \times \mathbf{H}^{+*}) = \frac{1}{2}\text{Re}\left[\hat{\mathbf{a}}_x E_0^+ e^{-\alpha z}e^{-j\beta z} \times \hat{\mathbf{a}}_y\frac{E_0^{+*}}{\eta_c^*}e^{-\alpha z}e^{+j\beta z}\right]$$

$$\mathbf{S}^+ = \hat{\mathbf{a}}_z\frac{|E_0^+|^2}{2}e^{-2\alpha z}\text{Re}\left[\frac{1}{\eta_c^*}\right] \tag{4-31}$$

Individually each term of (4-27) or (4-29a) represents a traveling wave in its respective direction. The magnitude of each term in (4-27) takes the form

$$\left|E_x^+(z)\right| = \left|E_0^+\right|e^{-\alpha z} \tag{4-32a}$$

$$\left|E_x^-(z)\right| = \left|E_0^-\right|e^{+\alpha z} \tag{4-32b}$$

which, when plotted for $-\lambda \leq z \leq +\lambda$ and $|\Gamma| = 0.2$ through 1 (in increments of 0.2), take the form shown in Figure 4-7a.

Collectively, both terms in each of the fields in (4-27) or (4-29a) represent a standing wave. Using the procedure outlined in Section 4.2.1D, (4-27) can also be written as

$$E_x(z) = \sqrt{(E_0^+)^2 e^{-2\alpha z} + (E_0^-)^2 e^{+2\alpha z} + 2E_0^+E_0^-\cos(2\beta z)}$$

$$\times \exp\left\{-j\tan^{-1}\left[\frac{E_0^+e^{-\alpha z} - E_0^-e^{+\alpha z}}{E_0^+e^{-\alpha z} + E_0^-e^{+\alpha z}}\tan(\beta z)\right]\right\} \tag{4-33}$$

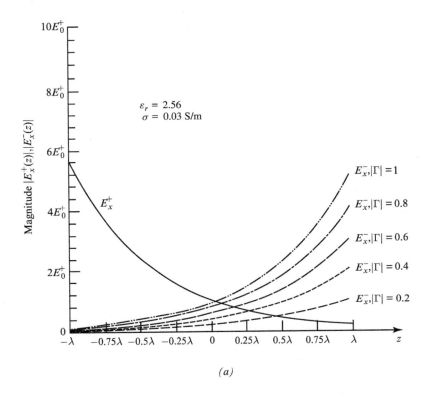

(a)

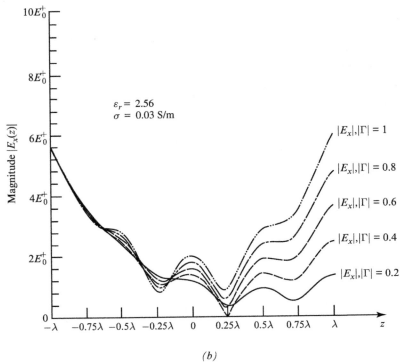

(b)

Figure 4-7   Wave patterns of uniform plane waves in a lossy medium. (a) Traveling. (b) Standing.

The standing wave pattern is given by the amplitude term of

$$|E_x(z)| = \sqrt{(E_0^+)^2 e^{-2\alpha z} + (E_0^-)^2 e^{+2\alpha z} + 2E_0^+ E_0^- \cos(2\beta z)} \tag{4-33a}$$

which for $|\Gamma| = E_0^- / E_0^+ = 0.2$ through 1, in increments of 0.2, is shown plotted in Figure 4-7b in the range $-\lambda \le z \le \lambda$ when $f = 100$ MHz, $\varepsilon_r = 2.56$, $\mu_r = 1$, and $\sigma = 0.03$ S/m.

The distance the wave must travel in a lossy medium to reduce its value to e^{-1} $= 0.368 = 36.8\%$ *is defined as the skin depth δ.* For each of the terms of (4-27) or (4-29a), this distance is

$$\delta = \text{skin depth} = \frac{1}{\alpha} = \frac{1}{\omega\sqrt{\mu\varepsilon}\left\{\frac{1}{2}\left[\sqrt{1+(\sigma/\omega\varepsilon)^2} - 1\right]\right\}^{1/2}} \text{ m} \tag{4-34}$$

In summary, the attenuation constant α, phase constant β, wave Z_w and intrinsic η_c impedances, wavelength λ, velocity v, and skin depth δ for a uniform plane wave traveling in a lossy medium are listed in the second column of Table 4-1. The same expressions are valid for plane and TEM waves. Simpler expressions for each can be derived depending upon the value of the $(\sigma/\omega\varepsilon)^2$ ratio. Media whose $(\sigma/\omega\varepsilon)^2$ is much less than unity [$(\sigma/\omega\varepsilon)^2 \ll 1$] are referred to as *good dielectrics* and those whose $(\sigma/\omega\varepsilon)^2$ is much greater than unity [$(\sigma/\omega\varepsilon)^2 \gg 1$] are referred to as *good conductors* [6]; each will now be discussed.

TABLE 4-1 **Propagation constant, wave impedance, wavelength, velocity, and skin depth of TEM wave in lossy media**

	Exact	Good dielectric $\left(\frac{\sigma}{\omega\varepsilon}\right)^2 \ll 1$	Good conductor $\left(\frac{\sigma}{\omega\varepsilon}\right)^2 \gg 1$
Attenuation constant α	$= \omega\sqrt{\mu\varepsilon}\left\{\frac{1}{2}\left[\sqrt{1+\left(\frac{\sigma}{\omega\varepsilon}\right)^2} - 1\right]\right\}^{1/2}$	$\simeq \frac{\sigma}{2}\sqrt{\frac{\mu}{\varepsilon}}$	$\simeq \sqrt{\frac{\omega\mu\sigma}{2}}$
Phase constant β	$= \omega\sqrt{\mu\varepsilon}\left\{\frac{1}{2}\left[\sqrt{1+\left(\frac{\sigma}{\omega\varepsilon}\right)^2} + 1\right]\right\}^{1/2}$	$\simeq \omega\sqrt{\mu\varepsilon}$	$\simeq \sqrt{\frac{\omega\mu\sigma}{2}}$
Wave Z_w intrinsic η_c impedances $Z_w = \eta_c$	$= \sqrt{\frac{j\omega\mu}{\sigma + j\omega\varepsilon}}$	$\simeq \sqrt{\frac{\mu}{\varepsilon}}$	$\simeq \sqrt{\frac{\omega\mu}{2\sigma}}(1+j)$
Wavelength λ	$= \frac{2\pi}{\beta}$	$\simeq \frac{2\pi}{\omega\sqrt{\mu\varepsilon}}$	$\simeq 2\pi\sqrt{\frac{2}{\omega\mu\sigma}}$
Velocity v	$= \frac{\omega}{\beta}$	$\simeq \frac{1}{\sqrt{\mu\varepsilon}}$	$\simeq \sqrt{\frac{2\omega}{\mu\sigma}}$
Skin depth δ	$= \frac{1}{\alpha}$	$\simeq \frac{2}{\sigma}\sqrt{\frac{\varepsilon}{\mu}}$	$\simeq \sqrt{\frac{2}{\omega\mu\sigma}}$

A. Good Dielectrics $\left[(\sigma/\omega\varepsilon)^2 \ll 1\right]$ For source-free lossy media, Maxwell's equation in differential form as derived from Ampère's law takes the form, by referring to Table 1-4, of

$$\nabla \times \mathbf{H} = \mathbf{J}_c + \mathbf{J}_d = \sigma\mathbf{E} + j\omega\varepsilon\mathbf{E} = (\sigma + j\omega\varepsilon)\mathbf{E} \qquad (4\text{-}35)$$

where $\mathbf{J}_c$ and $\mathbf{J}_d$ represent, respectively, the conduction and displacement current densities. When $\sigma/\omega\varepsilon \ll 1$, the displacement current density is much greater than the conduction current density; when $\sigma/\omega\varepsilon \gg 1$ the conduction current density is much greater than the displacement current density. For each of these two cases, the exact forms of the field parameters of Table 4-1 can be approximated by simpler forms. This will be demonstrated next.

For a good dielectric [when $(\sigma/\omega\varepsilon)^2 \ll 1$], the exact expression for the attenuation constant of (4-28c) can be written using the binomial expansion and it takes the form

$$\alpha = \omega\sqrt{\mu\varepsilon}\left\{\frac{1}{2}\left[\sqrt{1 + \left(\frac{\sigma}{\omega\varepsilon}\right)^2} - 1\right]\right\}^{1/2}$$

$$\alpha = \omega\sqrt{\mu\varepsilon}\left\{\frac{1}{2}\left[\left[1 + \frac{1}{2}\left(\frac{\sigma}{\omega\varepsilon}\right)^2 - \frac{1}{8}\left(\frac{\sigma}{\omega\varepsilon}\right)^4 \cdots\right] - 1\right]\right\}^{1/2} \qquad (4\text{-}36)$$

Retaining only the first two terms of the infinite series, (4-36) can be approximated by

$$\alpha \simeq \omega\sqrt{\mu\varepsilon}\left[\frac{1}{4}\left(\frac{\sigma}{\omega\varepsilon}\right)^2\right]^{1/2} = \frac{\sigma}{2}\sqrt{\frac{\mu}{\varepsilon}} \qquad (4\text{-}36a)$$

In a similar manner it can be shown that by following the same procedure but only retaining the first term of the infinite series, the exact expression for β of (4-28d) can be approximated by

$$\beta \simeq \omega\sqrt{\mu\varepsilon} \qquad (4\text{-}37)$$

For good dielectrics, the wave and intrinsic impedances of (4-30) can be approximated by

$$Z_w = \eta_c = \sqrt{\frac{j\omega\mu}{\sigma + j\omega\varepsilon}} = \sqrt{\frac{j\omega\mu/j\omega\varepsilon}{\sigma/j\omega\varepsilon + 1}} \simeq \sqrt{\frac{\mu}{\varepsilon}} \qquad (4\text{-}38)$$

while the skin depth can be represented by

$$\delta = \frac{1}{\alpha} \simeq \frac{2}{\sigma}\sqrt{\frac{\varepsilon}{\mu}} \qquad (4\text{-}39)$$

These and other approximate forms for the parameters of good dielectrics are summarized on the third column of Table 4-1.

B. Good Conductors $\left[(\sigma/\omega\varepsilon)^2 \gg 1\right]$ For good conductors, the exact expression for the attenuation constant of (4-28c) can be written using the binomial expansion and takes the form

$$\alpha = \omega\sqrt{\mu\varepsilon}\left\{\frac{1}{2}\left[\sqrt{\left(\frac{\sigma}{\omega\varepsilon}\right)^2 + 1} - 1\right]\right\}^{1/2} = \omega\sqrt{\mu\varepsilon}\left\{\frac{1}{2}\left[\frac{\sigma}{\omega\varepsilon}\left(1 + \frac{1}{(\sigma/\omega\varepsilon)^2}\right)^{1/2} - 1\right]\right\}^{1/2}$$

$$\alpha = \omega\sqrt{\mu\varepsilon}\left\{\frac{1}{2}\left[\frac{\sigma}{\omega\varepsilon} + \frac{1}{2}\frac{1}{\sigma/\omega\varepsilon} - \frac{1}{8}\frac{1}{(\sigma/\omega\varepsilon)^3} + \cdots - 1\right]\right\}^{1/2} \qquad (4\text{-}40)$$

Retaining only the first term of the infinite series expansion, (4-40) can be approximated by

$$\alpha \simeq \omega\sqrt{\mu\varepsilon}\left(\frac{1}{2}\frac{\sigma}{\omega\varepsilon}\right)^{1/2} = \sqrt{\frac{\omega\mu\sigma}{2}} \qquad (4\text{-}40a)$$

Following a similar procedure, the phase constant of (4-28d) can be approximated by

$$\beta \simeq \sqrt{\frac{\omega\mu\sigma}{2}} \qquad (4\text{-}41)$$

which is identical to the approximate expression for the attenuation constant of (4-40a).

For good conductors, the wave and intrinsic impedances of (4-30) can be approximated by

$$Z_w = \eta_c = \sqrt{\frac{j\omega\mu}{\sigma + j\omega\varepsilon}} = \sqrt{\frac{j\omega\mu/\omega\varepsilon}{\sigma/\omega\varepsilon + j}} \simeq \sqrt{j\frac{\omega\mu}{\sigma}} = \sqrt{\frac{\omega\mu}{2\sigma}}(1+j) \qquad (4\text{-}42)$$

whose real and imaginary parts are identical. For the same conditions, the skin depth can be approximated by

$$\delta = \frac{1}{\alpha} \simeq \sqrt{\frac{2}{\omega\mu\sigma}} \qquad (4\text{-}43)$$

This is the most widely recognized form for the skin depth.

4.3.2   Uniform Plane Waves in an Unbounded Lossy Medium—Oblique Angle

For lossy media the difference between principal axes propagation and propagation at oblique angles is that the propagation constant γ_r along the direction β of propagation must be decomposed into its directional components along the principal axes of the coordinate system. In addition, since the propagation constant γ has real (α) and imaginary (β) parts, constant amplitude and constant phase planes are associated with the wave. As discussed in Section 4.2.2C and illustrated in Figure 4-6, the constant phase planes for a uniform plane wave are planes that are parallel to each other, perpendicular to the direction of propagation, and coincide with the constant amplitude planes. The constant amplitude planes are planes over which the amplitude remains constant. For a uniform plane wave traveling in a lossy medium, the constant amplitude planes are also parallel to each other, are perpendicular to the direction of travel, and coincide with the constant phase planes. This is illustrated in Figure 4-6 for a uniform plane wave traveling at an oblique angle in a lossless medium.

Let us assume that a uniform plane wave that is also TE^y is traveling in a lossy medium at an angle θ_i, as shown in Figure 4-4a. Following a procedure similar to the lossless case and referring to (4-17a) and (4-17b), the propagation constant of (4-28) can now be written for the positive and negative traveling waves as

$$\gamma^+ = \gamma\,(\hat{\mathbf{a}}_x \sin\theta_i + \hat{\mathbf{a}}_z \cos\theta_i) = (\alpha + j\beta)(\hat{\mathbf{a}}_x \sin\theta_i + \hat{\mathbf{a}}_z \cos\theta_i) \qquad (4\text{-}44a)$$

$$\gamma^- = -\gamma\,(\hat{\mathbf{a}}_x \sin\theta_i + \hat{\mathbf{a}}_z \cos\theta_i) = -(\alpha + j\beta)(\hat{\mathbf{a}}_x \sin\theta_i + \hat{\mathbf{a}}_z \cos\theta_i) \qquad (4\text{-}44b)$$

where the real (α) and imaginary (β) parts of γ are given by (4-28c) and (4-28d), respectively. Using (4-44a) and (4-44b), the electric and magnetic fields can be written, by referring to (4-17) through (4-18c), as

$$\mathbf{E} = E_0^+ \left(\hat{\mathbf{a}}_x \cos\theta_i - \hat{\mathbf{a}}_z \sin\theta_i \right) e^{-\gamma^+ \cdot \mathbf{r}} + E_0^- \left(\hat{\mathbf{a}}_x \cos\theta_i - \hat{\mathbf{a}}_z \sin\theta_i \right) e^{-\gamma^- \cdot \mathbf{r}}$$

$$\mathbf{E} = E_0^+ \left(\hat{\mathbf{a}}_x \cos\theta_i - \hat{\mathbf{a}}_z \sin\theta_i \right) e^{-(\alpha + j\beta)(x \sin\theta_i + z \cos\theta_i)}$$

$$+ E_0^- \left(\hat{\mathbf{a}}_x \cos\theta_i - \hat{\mathbf{a}}_z \sin\theta_i \right) e^{+(\alpha + j\beta)(x \sin\theta_i + z \cos\theta_i)} \tag{4-45a}$$

$$\mathbf{H} = \hat{\mathbf{a}}_y \left[\frac{E_0^+}{\eta_c} e^{-(\alpha + j\beta)(x \sin\theta_i + z \cos\theta_i)} - \frac{E_0^-}{\eta_c} e^{+(\alpha + j\beta)(x \sin\theta_i + z \cos\theta_i)} \right] \tag{4-45b}$$

Because the wave is a uniform plane wave in the β direction of propagation, the wave impedance Z_{wr} in the direction of propagation is equal to the intrinsic impedance η_c of the lossy medium given by (4-30) or

$$Z_{wr} = \eta_c = \sqrt{\frac{j\omega\mu}{\sigma + j\omega\varepsilon}} \tag{4-46}$$

However, the directional impedances in the x and z directions are given, by referring to (4-20a) and (4-20b), by

$$Z_x^+ = -\frac{E_z^+}{H_y^+} = \eta_c \sin\theta_i = Z_x^- = \frac{E_z^-}{H_y^-} \tag{4-47a}$$

$$Z_z^+ = \frac{E_x^+}{H_y^+} = \eta_c \cos\theta_i = Z_z^- = -\frac{E_x^-}{H_y^-} \tag{4-47b}$$

According to (4-22) through (4-24) the phase and energy velocities in the principal β direction of travel and in the z direction are given, respectively, by

$$v_r = v_{pr} = v_{er} = v = \frac{\omega}{\beta} \tag{4-48a}$$

$$v_{pz} = \frac{v}{\cos\theta_i} = \frac{\omega}{\beta \cos\theta_i} \geq v = \frac{\omega}{\beta} \tag{4-48b}$$

$$v_{ez} = v \cos\theta_i = \frac{\omega}{\beta} \cos\theta_i \leq v = \frac{\omega}{\beta} \tag{4-48c}$$

where β for a lossy medium is given by (4-28d) or

$$\beta = \omega \sqrt{\mu\varepsilon} \left\{ \frac{1}{2} \left[\sqrt{1 + \left(\frac{\sigma}{\omega\varepsilon} \right)^2} + 1 \right] \right\}^{1/2} \tag{4-48d}$$

As for the lossless medium, the product of the phase and energy velocities is equal to the square of the velocity of light v in the lossy medium, or

$$v_{pr} v_{er} = v_{pz} v_{ez} = v^2 \tag{4-48e}$$

Using the procedure followed to derive (4-26) through (4-26b) and (4-31), the average power density along the principal direction β of travel and the directional power densities along the x and z directions can be written for the fields of (4-45a) and (4-45b) as

$$(\mathbf{S}_{av}^+)_r = (\hat{\mathbf{a}}_x \sin\theta_i + \hat{\mathbf{a}}_z \cos\theta_i) \frac{|E_0^+|^2}{2} e^{-2\alpha(x\sin\theta_i + z\cos\theta_i)} \operatorname{Re}\left[\frac{1}{\eta_c^*}\right]$$

$$= \hat{\mathbf{a}}_r \frac{|E_0^+|^2}{2} e^{-2\alpha r} \operatorname{Re}\left[\frac{1}{\eta_c^*}\right] \tag{4-49a}$$

$$(\mathbf{S}_{av}^+)_x = \sin\theta_i \frac{|E_0^+|^2}{2} e^{-2\alpha(x\sin\theta_i + z\cos\theta_i)} \operatorname{Re}\left[\frac{1}{\eta_c^*}\right]$$

$$= \sin\theta_i \frac{|E_0^+|^2}{2} e^{-2\alpha r} \operatorname{Re}\left[\frac{1}{\eta_c^*}\right] \tag{4-49b}$$

$$(\mathbf{S}_{av}^+)_z = \cos\theta_i \frac{|E_0^+|^2}{2} e^{-2\alpha(x\sin\theta_i + z\cos\theta_i)} \operatorname{Re}\left[\frac{1}{\eta_c^*}\right]$$

$$= \cos\theta_i \frac{|E_0^+|^2}{2} e^{-2\alpha r} \operatorname{Re}\left[\frac{1}{\eta_c^*}\right] \tag{4-49c}$$

Example 4-4

For a TM^y wave traveling in a lossy medium at an oblique angle θ_i, derive expressions for the fields, wave impedances, phase and energy velocities, and average power densities.

Solution: The solution to this problem can be accomplished by following the procedure used to derive the expressions of the fields and other wave characteristics of a TE^y wave traveling at an oblique angle in a lossy medium, as outlined in this section, and referring to the solution of Examples 4-2 and 4-3. Doing this we can write the fields of a TM^y traveling in a lossy medium at an oblique angle θ_i, the coordinate system of which is illustrated in Figure 4-4b, as

$$\mathbf{E} = \mathbf{E}^+ + \mathbf{E}^- = \hat{\mathbf{a}}_y \left[E_0^+ e^{-(\alpha+j\beta)(x\sin\theta_i + z\cos\theta_i)} + E_0^- e^{+(\alpha+j\beta)(x\sin\theta_i + z\cos\theta_i)} \right]$$

$$\mathbf{H} = \mathbf{H}^+ + \mathbf{H}^- = \frac{E_0^+}{\eta_c}(-\hat{\mathbf{a}}_x \cos\theta_i + \hat{\mathbf{a}}_z \sin\theta_i) e^{-(\alpha+j\beta)(x\sin\theta_i + z\cos\theta_i)}$$

$$+ \frac{E_0^-}{\eta_c}(\hat{\mathbf{a}}_x \cos\theta_i - \hat{\mathbf{a}}_z \sin\theta_i) e^{+(\alpha+j\beta)(x\sin\theta_i + z\cos\theta_i)}$$

In addition, the wave impedances are given, by referring to (4-21a) and (4-21b), by

$$Z_x^+ = \frac{E_y^+}{H_z^+} = \frac{\eta_c}{\sin\theta_i} = Z_x^- = -\frac{E_y^-}{H_z^-}$$

$$Z_z^+ = -\frac{E_y^+}{H_x^+} = \frac{\eta_c}{\cos\theta_i} = Z_z^- = \frac{E_y^-}{H_x^-}$$

The phase and energy velocities, and their relationships, are the same as those for the TE^y wave, as given by (4-48a) through (4-48e). Similarly, the average power densities are those given by (4-49a) through (4-49c).

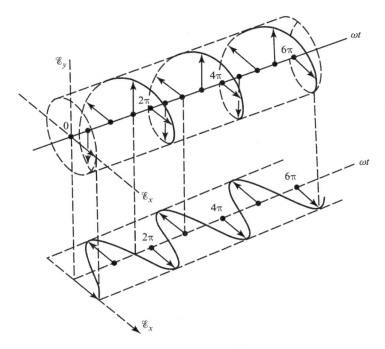

Figure 4-8 Rotation of a plane electromagnetic wave at $z = 0$ as a function of time. (Source: C. A. Balanis, *Antenna Theory: Analysis and Design*, Third Edition. Copyright © 2005, John Wiley & Sons, Inc. Reprinted by permission of John Wiley & Sons, Inc.)

4.4 POLARIZATION

According to the *IEEE Standard Definitions for Antennas* [7, 8], the *polarization of a radiated wave* is defined as "that property of a radiated electromagnetic wave describing the time-varying direction and relative magnitude of the electric field vector; specifically, the figure traced as a function of time by the extremity of the vector at a fixed location in space, and the sense in which it is traced, as observed along the direction of propagation." In other words, polarization is the curve traced out, at a given observation point as a function of time, by the end point of the arrow representing the instantaneous electric field. The field must be observed along the direction of propagation. A typical trace as a function of time is shown in Figure 4-8 [8].

Polarization may be classified into three categories: *linear, circular, and elliptical* [8]. If the vector that describes the electric field at a point in space as a function of time is always directed along a line, which is normal to the direction of propagation, the field is said to be *linearly* polarized. In general, however, the figure that the electric field traces is an ellipse, and the field is said to be *elliptically* polarized. Linear and circular polarizations are special cases of elliptical, and they can be obtained when the ellipse becomes a straight line or a circle, respectively. The figure of the electric field is traced in a *clockwise* (CW) or *counterclockwise* (CCW) sense. Clockwise rotation of the electric field vector is also designated as *right-hand polarization* and counterclockwise as *left-hand polarization*. In Figure 4-9 we show the figure traced by the extremity of the time-varying field vector for linear, circular, and elliptical polarizations.

The mathematical details for defining linear, circular, and elliptical polarizations follow.

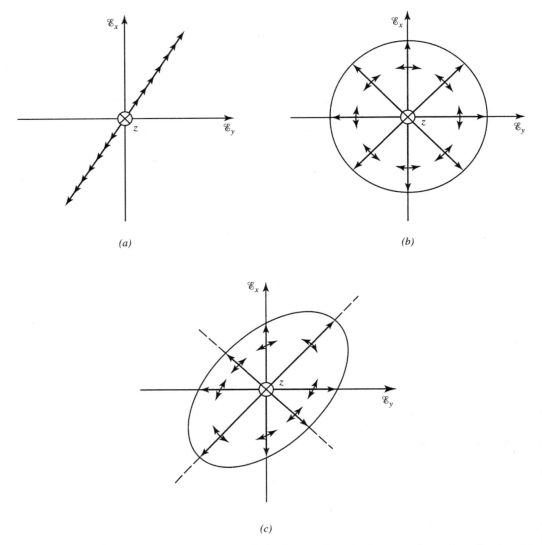

Figure 4-9 Polarization figure traces of an electric field extremity *as a function of time for a fixed position.* (*a*) Linear. (*b*) Circular. (*c*) Elliptical.

4.4.1   Linear Polarization

Let us consider a harmonic plane wave, with x and y electric field components, traveling in the positive z direction (into the page), as shown in Figure 4-10 [8]. The instantaneous electric and magnetic fields are given by

$$\mathscr{E} = \hat{\mathbf{a}}_x \mathscr{E}_x + \hat{\mathbf{a}}_y \mathscr{E}_y = \text{Re}\left[\hat{\mathbf{a}}_x E_x^+ e^{j(\omega t - \beta z)} + \hat{\mathbf{a}}_y E_y^+ e^{j(\omega t - \beta z)}\right]$$

$$= \hat{\mathbf{a}}_x E_{x_0}^+ \cos(\omega t - \beta z + \phi_x) + \hat{\mathbf{a}}_y E_{y_0}^+ \cos(\omega t - \beta z + \phi_y) \tag{4-50a}$$

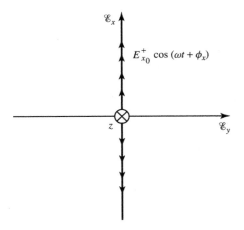

Figure 4-10   Linearly polarized field in the x direction.

$$\mathcal{H} = \hat{\mathbf{a}}_y \mathcal{H}_y + \hat{\mathbf{a}}_x \mathcal{H}_x = \mathrm{Re}\left[\hat{\mathbf{a}}_y \frac{E_x^+}{\eta} e^{j(\omega t - \beta z)} - \hat{\mathbf{a}}_x \frac{E_y^+}{\eta} e^{j(\omega t - \beta z)}\right]$$

$$= \hat{\mathbf{a}}_y \frac{E_{x_0}^+}{\eta} \cos(\omega t - \beta z + \phi_x) - \hat{\mathbf{a}}_x \frac{E_{y_0}^+}{\eta} \cos(\omega t - \beta z + \phi_y) \qquad (4\text{-}50\mathrm{b})$$

where E_x^+, E_y^+ are complex and $E_{x_0}^+$, $E_{y_0}^+$ are real.

Let us now examine the variation of the instantaneous electric field vector $\mathcal{E}$ as given by (4-50a) at the $z = 0$ plane. Other planes may be considered, but the $z = 0$ plane is chosen for convenience and simplicity. For the first example, let

$$E_{y_0}^+ = 0 \qquad (4\text{-}51)$$

in (4-50a). Then

$$\mathcal{E}_x = E_{x_0}^+ \cos(\omega t + \phi_x)$$
$$\mathcal{E}_y = 0 \qquad (4\text{-}51\mathrm{a})$$

The locus of the instantaneous electric field vector is given by

$$\mathcal{E} = \hat{\mathbf{a}}_x E_{x_0}^+ \cos(\omega t + \phi_x) \qquad (4\text{-}51\mathrm{b})$$

which is a straight line, and it will always be directed along the x axis at all times, as shown in Figure 4-10. The field is said to be *linearly polarized in the x direction*.

Example 4-5

Determine the polarization of the wave given by (4-50a) when $E_{x_0}^+ = 0$.

Solution: Since

$$E_{x_0}^+ = 0$$

then

$$\mathscr{E}_x = 0$$
$$\mathscr{E}_y = E_{y_0}^+ \cos(\omega t + \phi_y)$$

The locus of the instantaneous electric field vector is given by

$$\mathscr{E} = \hat{\mathbf{a}}_y E_{y_0}^+ \cos(\omega t + \phi_y)$$

which again is a straight line but directed along the y axis at all times, as shown in Figure 4-11. The field is said to be *linearly polarized in the y direction*.

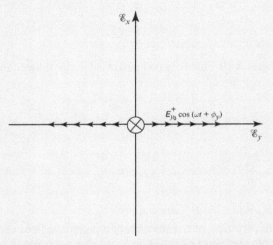

Figure 4-11 Linearly polarized field in the y direction.

Example 4-6

Determine the polarization and direction of polarization of the wave given by (4-50a) when $\phi_x = \phi_y = \phi$.

Solution: Since

$$\phi_x = \phi_y = \phi$$

then

$$\mathscr{E}_x = E_{x_0}^+ \cos(\omega t + \phi)$$
$$\mathscr{E}_y = E_{y_0}^+ \cos(\omega t + \phi)$$

The amplitude of the electric field vector is given by

$$\mathscr{E} = \sqrt{\mathscr{E}_x^2 + \mathscr{E}_y^2} = \sqrt{(E_{x_0}^+)^2 + (E_{y_0}^+)^2} \cos(\omega t + \phi)$$

which is a straight line directed at all times along a line that makes an angle ψ with the x axis as shown in Figure 4-12. The angle ψ is given by

$$\psi = \tan^{-1}\left[\frac{\mathcal{E}_y}{\mathcal{E}_x}\right] = \tan^{-1}\left[\frac{E_{y_0}^+}{E_{x_0}^+}\right]$$

The field is said to be *linearly polarized in the ψ direction.*

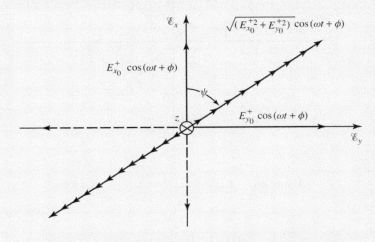

Figure 4-12 Linearly polarized field in the ψ direction.

It is evident from the preceding examples that *a time-harmonic field is linearly polarized at a given point in space if the electric field (or magnetic field) vector at that point is oriented along the same straight line at every instant of time.* This is accomplished if the field vector (electric or magnetic) possesses (a) only one component or (b) two orthogonal linearly polarized components that are in time phase or integer multiples of 180° out of phase.

4.4.2 Circular Polarization

A wave is said to be *circularly polarized if the tip of the electric field vector traces out a circular locus in space.* At various instants of time, the electric field intensity of such a wave always has the same amplitude and the orientation in space of the electric field vector changes continuously with time in such a manner as to describe a circular locus [8, 9].

A. Right-Hand (Clockwise) Circular Polarization A wave has *right-hand circular polarization* if its electric field vector has a clockwise sense of rotation *when it is viewed along the axis of propagation.* In addition, the electric field vector must trace a circular locus if the wave is to have also a circular polarization.

Let us examine the locus of the instantaneous electric field vector ($\mathcal{E}$) at the $z = 0$ plane at all times. For this particular example, let in (4-50a)

$$\phi_x = 0$$
$$\phi_y = -\pi/2$$
$$E_{x_0}^+ = E_{y_0}^+ = E_R \qquad (4\text{-}52)$$

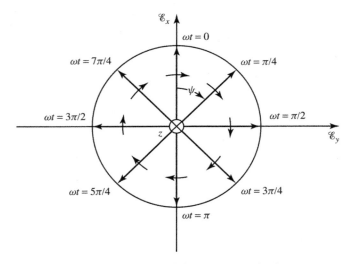

Figure 4-13 Right-hand circularly polarized wave.

Then

$$\mathcal{E}_x = E_{\mathrm{R}}\cos(\omega t)$$

$$\mathcal{E}_y = E_{\mathrm{R}}\cos\left(\omega t - \frac{\pi}{2}\right) = E_{\mathrm{R}}\sin(\omega t) \qquad (4\text{-}52\mathrm{a})$$

The locus of the amplitude of the electric field vector is given by

$$\mathcal{E} = \sqrt{\mathcal{E}_x^2 + \mathcal{E}_y^2} = \sqrt{E_{\mathrm{R}}^2(\cos^2\omega t + \sin^2\omega t)} = E_{\mathrm{R}} \qquad (4\text{-}52\mathrm{b})$$

and it is directed along a line making an angle ψ with the x axis, which is given by

$$\psi = \tan^{-1}\left[\frac{\mathcal{E}_y}{\mathcal{E}_x}\right] = \tan^{-1}\left[\frac{E_{\mathrm{R}}\sin(\omega t)}{E_{\mathrm{R}}\cos(\omega t)}\right] = \tan^{-1}\left[\tan(\omega t)\right] = \omega t \qquad (4\text{-}52\mathrm{c})$$

If we plot the locus of the electric field vector for various times at the $z = 0$ plane, we see that it forms a circle of radius E_{R} and it rotates clockwise with an angular frequency ω, as shown in Figure 4-13. Thus the wave is said to have a *right-hand circular polarization*. Remember that the rotation is viewed from the "rear" of the wave in the direction of propagation. In this example, the wave is traveling in the positive z direction (into the page) so that the rotation is examined from an observation point looking into the page and perpendicular to it.

We can write the instantaneous electric field vector as

$$\mathcal{E} = \mathrm{Re}\left[\hat{\mathbf{a}}_x E_{\mathrm{R}} e^{j(\omega t - \beta z)} + \hat{\mathbf{a}}_y E_{\mathrm{R}} e^{j(\omega t - \beta z - \pi/2)}\right]$$

$$= E_{\mathrm{R}}\,\mathrm{Re}\left\{\left[\hat{\mathbf{a}}_x - j\hat{\mathbf{a}}_y\right]e^{j(\omega t - \beta z)}\right\} \qquad (4\text{-}52\mathrm{d})$$

We note that there is a 90° phase difference between the two orthogonal components of the electric field vector.

Example 4-7

If $\phi_x = +\pi/2$, $\phi_y = 0$, and $E_{x_0}^+ = E_{y_0}^+ = E_R$, determine the polarization and sense of rotation of the wave of (4-50a).

Solution: Since

$$\phi_x = +\frac{\pi}{2}$$

$$\phi_y = 0$$

$$E_{x_0}^+ = E_{y_0}^+ = E_R$$

then

$$\mathscr{E}_x = E_R \cos\left(\omega t + \frac{\pi}{2}\right) = -E_R \sin \omega t$$

$$\mathscr{E}_y = E_R \cos(\omega t)$$

and the locus of the amplitude of the electric field vector is given by

$$\mathscr{E} = \sqrt{\mathscr{E}_x^2 + \mathscr{E}_y^2} = \sqrt{E_R^2(\cos^2 \omega t + \sin^2 \omega t)} = E_R$$

The angle ψ along which the field is directed is given by

$$\psi = \tan^{-1}\left[\frac{\mathscr{E}_y}{\mathscr{E}_x}\right] = \tan^{-1}\left[-\frac{E_R \cos(\omega t)}{E_R \sin(\omega t)}\right] = \tan^{-1}[-\cot(\omega t)] = \omega t + \frac{\pi}{2}$$

The locus of the field vector is a circle of radius E_R, and it rotates clockwise with an angular frequency ω as shown in Figure 4-14; *hence, it is a right-hand circular polarization*.

The expression for the instantaneous electric field vector is

$$\mathscr{E} = \text{Re}\left[\hat{\mathbf{a}}_x E_R\, e^{j(\omega t - \beta z + \pi/2)} + \hat{\mathbf{a}}_y E_R\, e^{j(\omega t - \beta z)}\right]$$

$$= E_R\, \text{Re}\left\{\left[j\hat{\mathbf{a}}_x + \hat{\mathbf{a}}_y\right] e^{j(\omega t - \beta z)}\right\}$$

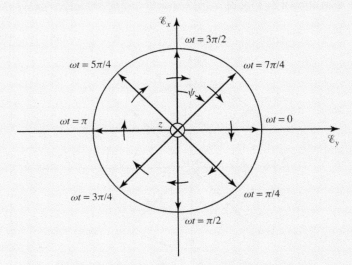

Figure 4-14 Right-hand circularly polarized wave.

Again we note a 90° phase difference between the orthogonal components.

From the previous discussion we see that a right-hand *circular polarization* can be achieved *if and only if its two orthogonal linearly polarized components have equal amplitudes and a 90° phase difference of one relative to the other. The sense of rotation (clockwise here) is determined by rotating the phase-leading component (in this instance $\mathscr{E}_x$) toward the phase-lagging component (in this instance $\mathscr{E}_y$). The field rotation must be viewed as the wave travels away from the observer.*

B. Left-Hand (Counterclockwise) Circular Polarization

If the electric field vector has a counterclockwise sense of rotation, the polarization is designated as *left-hand polarization*. To demonstrate this, let in (4-50a)

$$\phi_x = 0$$

$$\phi_y = \frac{\pi}{2}$$

$$E_{x_0}^+ = E_{y_0}^+ = E_L \tag{4-53}$$

then

$$\mathscr{E}_x = E_L \cos(\omega t)$$

$$\mathscr{E}_y = E_L \cos\left(\omega t + \frac{\pi}{2}\right) = -E_L \sin(\omega t) \tag{4-53a}$$

and the locus of the amplitude is

$$\mathscr{E} = \sqrt{\mathscr{E}_x^2 + \mathscr{E}_y^2} = \sqrt{E_L^2 (\cos^2 \omega t + \sin^2 \omega t)} = E_L \tag{4-53b}$$

The angle ψ is given by

$$\psi = \tan^{-1}\left[\frac{\mathscr{E}_y}{\mathscr{E}_x}\right] = \tan^{-1}\left[\frac{-E_L \sin(\omega t)}{E_L \cos(\omega t)}\right] = -\omega t \tag{4-53c}$$

The locus of the field vector is a circle of radius E_L, and it rotates counterclockwise with an angular frequency ω as shown in Figure 4-15; *hence, it is a left-hand circular polarization.*

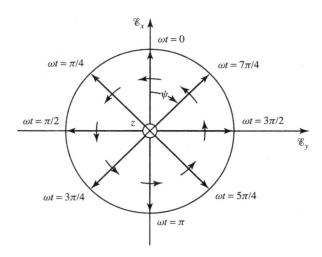

Figure 4-15   Left-hand circularly polarized wave.

The instantaneous electric field vector can be written as

$$\mathscr{E} = \text{Re}\left[\hat{\mathbf{a}}_x E_{\text{L}} e^{j(\omega t - \beta z)} + \hat{\mathbf{a}}_y E_{\text{L}} e^{j(\omega t - \beta z + \pi/2)}\right]$$

$$= E_{\text{L}}\, \text{Re}\left\{\left[\hat{\mathbf{a}}_x + j\hat{\mathbf{a}}_y\right] e^{j(\omega t - \beta z)}\right\} \qquad (4\text{-}53\text{d})$$

In (4-53d) we note a 90° phase advance of the $\mathscr{E}_y$ component relative to the $\mathscr{E}_x$ component.

Example 4-8

Determine the polarization and sense of rotation of the wave given by (4-50a) if $\phi_x = -\pi/2$, $\phi_y = 0$, and $E_{x_0}^+ = E_{y_0}^+ = E_{\text{L}}$.

Solution: Since

$$\phi_x = -\frac{\pi}{2}$$
$$\phi_y = 0$$
$$E_{x_0}^+ = E_{y_0}^+ = E_{\text{L}}$$

then

$$\mathscr{E}_x = E_{\text{L}} \cos\left(\omega t - \frac{\pi}{2}\right) = E_{\text{L}} \sin(\omega t)$$
$$\mathscr{E}_y = E_{\text{L}} \cos(\omega t)$$

and the locus of the amplitude is

$$\mathscr{E} = \sqrt{\mathscr{E}_x^2 + \mathscr{E}_y^2} = \sqrt{E_{\text{L}}^2(\sin^2 \omega t + \cos^2 \omega t)} = E_{\text{L}}$$

The angle ψ is given by

$$\psi = \tan^{-1}\left[\frac{\mathscr{E}_y}{\mathscr{E}_x}\right] = \tan^{-1}\left[\frac{E_{\text{L}} \cos(\omega t)}{E_{\text{L}} \sin(\omega t)}\right] = \tan^{-1}[\cot(\omega t)] = \frac{\pi}{2} - \omega t$$

The locus of the electric field vector is a circle of radius E_{L}, and it rotates counterclockwise with an angular frequency ω as shown in Figure 4-16; *hence, it is a left-hand circular polarization.* For this case we can write the electric field as

$$\mathscr{E} = \text{Re}\left[\hat{\mathbf{a}}_x E_{\text{L}} e^{j(\omega t - \beta z - \pi/2)} + \hat{\mathbf{a}}_y E_{\text{L}} e^{j(\omega t - \beta z)}\right]$$

$$= E_{\text{L}}\, \text{Re}\left\{\left[-j\hat{\mathbf{a}}_x + \hat{\mathbf{a}}_y\right] e^{j(\omega t - \beta z)}\right\}$$

and we note a 90° phase delay of the $\mathscr{E}_x$ component relative to $\mathscr{E}_y$.

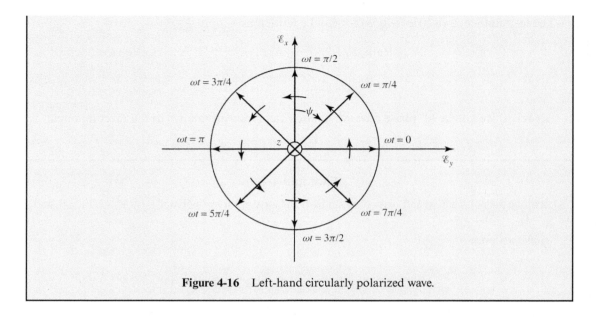

Figure 4-16   Left-hand circularly polarized wave.

From the previous discussion we see that *left-hand circular polarization* can be achieved *if and only if its two orthogonal components have equal amplitudes and odd multiples of 90° phase difference of one component relative to the other. The sense of rotation (counterclockwise here) is determined by rotating the phase-leading component (in this instance $\mathcal{E}_y$) toward the phase-lagging component (in this instance $\mathcal{E}_x$). The field rotation must be viewed as the wave travels away from the observer.*

The *necessary and sufficient* conditions for circular polarization are the following:

1. The field must have two orthogonal linearly polarized components.
2. The two components must have the same magnitude.
3. The two components must have a time-phase difference of odd multiples of 90°.

The sense of rotation is always determined by rotating the phase-leading component toward the phase-lagging component and observing the field rotation as the wave is traveling away from the observer. The rotation of the phase-leading component toward the phase-lagging component should be done along the angular separation between the two components that is less than 180°. Phases equal to or greater than 0° and less than 180° should be considered leading whereas those equal to or greater than 180° and less than 360° should be considered lagging.

4.4.3 Elliptical Polarization

A wave is said to be elliptically polarized if the tip of the electric field vector traces, as a function of time, an elliptical locus in space. At various instants of time the electric field vector changes continuously with time in such a manner as to describe an elliptical locus. It is right-hand elliptically polarized if the electric field vector of the ellipse rotates clockwise, and it is left-hand elliptically polarized if the electric field vector of the ellipse rotates counterclockwise [8, 10–14].

Let us examine the locus of the instantaneous electric field vector ($\mathscr{E}$) at the $z = 0$ plane at all times. For this particular example, let in (4-50a)

$$\phi_x = \frac{\pi}{2}$$
$$\phi_y = 0$$
$$E_{x_0}^+ = (E_R + E_L)$$
$$E_{y_0}^+ = (E_R - E_L) \tag{4-54}$$

Then,

$$\mathscr{E}_x = (E_R + E_L)\, \cos\left(\omega t + \frac{\pi}{2}\right) = -(E_R + E_L)\sin \omega t$$

$$\mathscr{E}_y = (E_R - E_L)\, \cos(\omega t) \tag{4-54a}$$

We can write the locus for the amplitude of the electric field vector as

$$\begin{aligned}
\mathscr{E}^2 = \mathscr{E}_x^2 + \mathscr{E}_y^2 &= (E_R + E_L)^2 \sin^2 \omega t + (E_R - E_L)^2 \cos^2 \omega t \\
&= E_R^2 \sin^2 \omega t + E_L^2 \sin^2 \omega t + 2 E_R E_L \sin^2 \omega t \\
&\quad + E_R^2 \cos^2 \omega t + E_L^2 \cos^2 \omega t - 2 E_R E_L \cos^2 \omega t \\
\mathscr{E}_x^2 + \mathscr{E}_y^2 &= E_R^2 + E_L^2 + 2 E_R E_L \left[\sin^2 \omega t - \cos^2 \omega t\right]
\end{aligned} \tag{4-54b}$$

However,

$$\sin \omega t = -\mathscr{E}_x / (E_R + E_L)$$
$$\cos \omega t = \mathscr{E}_y / (E_R - E_L) \tag{4-54c}$$

Substituting (4-54c) into (4-54b) reduces to

$$\left\{\frac{\mathscr{E}_x}{E_R + E_L}\right\}^2 + \left\{\frac{\mathscr{E}_y}{E_R - E_L}\right\}^2 = 1 \tag{4-54d}$$

which is the equation for an ellipse with the major axis $|\mathscr{E}|_{max} = |E_R + E_L|$ and the minor axis $|\mathscr{E}|_{min} = |E_R - E_L|$. As time elapses, the electric vector rotates and its length varies with its tip tracing an ellipse, as shown in Figure 4-17. The maximum and minimum lengths of the electric vector are the major and minor axes, given by

$$|\mathscr{E}|_{max} = |E_R + E_L|, \quad \text{when } \omega t = (2n+1)\frac{\pi}{2}, \ n = 0,\ 1,\ 2,\ldots \tag{4-54e}$$

$$|\mathscr{E}|_{min} = |E_R - E_L|, \quad \text{when } \omega t = n\pi, \ n = 0,\ 1,\ 2,\ldots \tag{4-54f}$$

The axial ratio (AR) is defined to be the ratio of the major axis (including its sign) of the polarization ellipse to the minor axis, or

$$AR = -\frac{\mathscr{E}_{max}}{\mathscr{E}_{min}} = -\frac{2(E_R + E_L)}{2(E_R - E_L)} = -\frac{(E_R + E_L)}{(E_R - E_L)} \tag{4-54g}$$

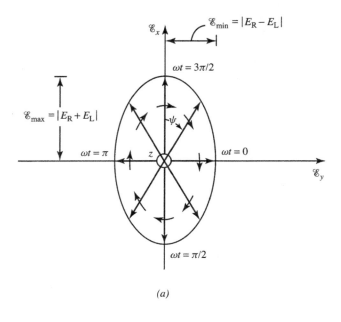

(a)

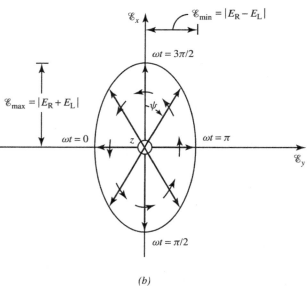

(b)

Figure 4-17 Right- and left-hand elliptical polarizations with major axis along the x axis. (a) Right-hand (clockwise) when $E_R > E_L$. (b) Left-hand (counterclockwise) when $E_R < E_L$.

where E_R and E_L are positive real quantities. As defined in (4-54g), the axial ratio AR can take positive (for left-hand polarization) or negative (for right-hand polarization) values in the range $1 \leq |AR| \leq \infty$. The instantaneous electric field vector can be written as

$$\mathscr{E} = \mathrm{Re}\left\{\hat{\mathbf{a}}_x [E_R + E_L] e^{j(\omega t - \beta z + \pi/2)} + \hat{\mathbf{a}}_y [E_R - E_L] e^{j(\omega t - \beta z)}\right\}$$

$$= \mathrm{Re}\left\{[\hat{\mathbf{a}}_x j(E_R + E_L) + \hat{\mathbf{a}}_y (E_R - E_L)] e^{j(\omega t - \beta z)}\right\}$$

$$\mathscr{E} = \mathrm{Re}\left\{[E_R(j\hat{\mathbf{a}}_x + \hat{\mathbf{a}}_y) + E_L(j\hat{\mathbf{a}}_x - \hat{\mathbf{a}}_y)] e^{j(\omega t - \beta z)}\right\} \tag{4-54h}$$

From (4-54h) we see that we can represent an elliptical wave as the sum of right-hand [first term of (4-54h)] and left-hand [second term of (4-54h)] circularly polarized waves with amplitudes E_R and E_L, respectively. If $E_R > E_L$, the axial ratio will be negative and the right-hand circular component will be stronger than the left-hand circular component. Thus, the electric vector rotates in the same direction as that of the right-hand circularly polarized wave, producing a *right-hand elliptically polarized wave*, as shown in Figure 4-17a. If $E_L > E_R$, the axial ratio will be positive and the left-hand circularly polarized component will be stronger than the right-hand circularly polarized component. The electric field vector will rotate in the same direction as that of the left-hand circularly polarized component, producing a *left-hand elliptically polarized wave*, as shown in Figure 4-17b. The sign of the axial ratio carries information on the direction of rotation of the electric field vector.

An analogous situation exists when

$$\phi_x = \frac{\pi}{2}$$
$$\phi_y = 0$$
$$E_{x_0}^+ = (E_R - E_L) \tag{4-55}$$
$$E_{y_0}^+ = (E_R + E_L)$$

The polarization loci are shown in Figure 4-18a and 4-18b when $E_R > E_L$ and $E_R < E_L$, respectively.

From (4-54e) and (4-54f), it can be seen that the component of $\mathscr{E}$ measured along the major axis of the polarization ellipse is 90° out of phase with the component of $\mathscr{E}$ measured along the minor axis. Also with the aid of (4-54b), it can be shown that the electric vector rotates through 90° in space between the instants of time given by (4-54e) and (4-54f) when the vector has maximum and minimum lengths, respectively. Thus the major and minor axes of the polarization ellipse are orthogonal in space, just as we might anticipate.

Since linear polarization is a special kind of elliptical polarization, we can represent a linear polarization as the sum of a right- and a left-hand circularly polarized component of *equal amplitudes*. We see that for this case ($E_R = E_L$), (4-54h) will degenerate into a linear polarization.

A more general orientation of an elliptically polarized locus is the tilted ellipse of Figure 4-19. This is representative of the fields of (4-50a) when

$$\Delta\phi = \phi_x - \phi_y \neq \frac{n\pi}{2}, \quad n = 0, 2, 4 \ldots$$

$$\geq 0 \begin{cases} \text{for CW if } E_R > E_L \\ \text{for CCW if } E_R < E_L \end{cases} \tag{4-56a}$$

$$\leq 0 \begin{cases} \text{for CW if } E_R < E_L \\ \text{for CCW if } E_R > E_L \end{cases} \tag{4-56b}$$

$$E_{x_0}^+ = E_R + E_L$$
$$E_{y_0}^+ = E_R - E_L \tag{4-56c}$$

Thus the major and minor axes of the ellipse do not, in general, coincide with the principal axes of the coordinate system unless the magnitudes are not equal and the phase difference between the two orthogonal components is equal to odd multiples of ±90°.

The ratio of the major to the minor axes, which is defined as the axial ratio (AR), is equal to [8]

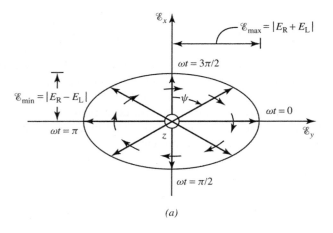

(a)

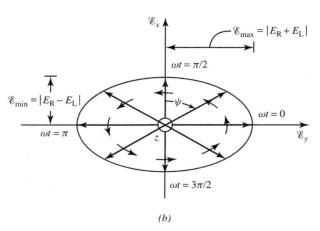

(b)

Figure 4-18 Right- and left-hand elliptical polarizations with major axis along the y axis. (*a*) Right-hand (clockwise) when $E_R > E_L$. (*b*) Left-hand (counterclockwise) when $E_R < E_L$.

$$AR = \pm\frac{\text{major axis}}{\text{minor axis}} = \pm\frac{OA}{OB}, \qquad 1 \le |AR| \le \infty \qquad (4\text{-}57)$$

where

$$OA = \left[\frac{1}{2}\left\{(E_{x_0}^+)^2 + (E_{y_0}^+)^2 + \left[(E_{x_0}^+)^4 + (E_{y_0}^+)^4 + 2(E_{x_0}^+)^2(E_{y_0}^+)^2\cos(2\Delta\phi)\right]^{1/2}\right\}\right]^{1/2} \qquad (4\text{-}57a)$$

$$OB = \left[\frac{1}{2}\left\{(E_{x_0}^+)^2 + (E_{y_0}^+)^2 - \left[(E_{x_0}^+)^4 + (E_{y_0}^+)^4 + 2(E_{x_0}^+)^2(E_{y_0}^+)^2\cos(2\Delta\phi)\right]^{1/2}\right\}\right]^{1/2} \qquad (4\text{-}57b)$$

$E_{x_0}^+$ and $E_{y_0}^+$ are given by (4-56c). The plus ($+$) sign in (4-57) is for left-hand and the minus ($-$) sign is for right-hand polarization.

The tilt of the ellipse, *relative to the x axis*, is represented by the angle τ given by

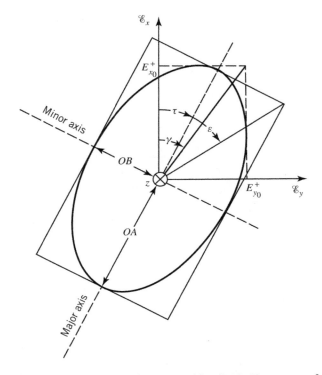

Figure 4-19 Rotation of a plane electromagnetic wave and its tilted ellipse at $z = 0$ as a function of time.

$$\tau = \frac{\pi}{2} - \frac{1}{2}\tan^{-1}\left[\frac{2E_{x_0}^+ E_{y_0}^+}{(E_{x_0}^+)^2 - (E_{y_0}^+)^2}\cos(\Delta\phi)\right] \qquad (4\text{-}57\text{c})$$

4.4.4 Poincaré Sphere

The polarization state, defined here as P, of any wave can be uniquely represented by a point on the surface of a sphere [15–19]. This is accomplished by either of the two pairs of angles (γ, δ) or (ε, τ). By referring to (4-50a) and Figure 4-20a, we can define the two pairs of angles:

$$(\gamma, \delta) \ \text{set}$$

$$\gamma = \tan^{-1}\left[\frac{E_{y_0}^+}{E_{x_0}^+}\right] \quad \text{or} \quad \gamma = \tan^{-1}\left[\frac{E_{x_0}^+}{E_{y_0}^+}\right], \qquad 0° \le \gamma \le 90° \qquad (4\text{-}58\text{a})$$

$$\delta = \phi_y - \phi_x = \text{phase difference between } \mathscr{E}_y \text{ and } \mathscr{E}_x, \quad -180° \le \delta \le 180° \qquad (4\text{-}58\text{b})$$

where 2γ is the great-circle angle drawn from a reference point on the equator and δ is the equator to great-circle angle;

$$(\varepsilon, \tau) \ \text{set}$$

$$\varepsilon = \cot^{-1}(\text{AR}) \Rightarrow \text{AR} = \cot(\varepsilon), \qquad -45° \le \varepsilon \le +45° \qquad (4\text{-}59\text{a})$$

$$\tau = \text{tilt angle}, \qquad 0° \le \tau \le 180° \qquad (4\text{-}59\text{b})$$

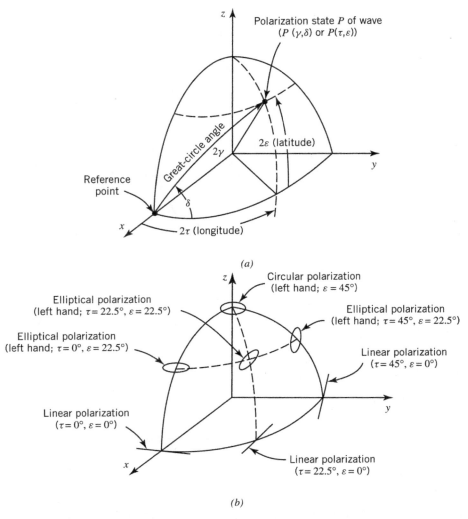

Figure 4-20 Poincaré sphere for the polarization state of an electromagnetic wave. (*a*) Poincaré sphere. (*b*) Polarization state. (Source: J. D. Kraus, *Electromagnetics*, McGraw-Hill, New York, 1984.)

where

$$2\varepsilon = \text{latitude}$$
$$2\tau = \text{longitude}$$

In (4-58a) the appropriate ratio is the one that satisfies the angular limits of all the Poincaré sphere angles (especially those of ε). The axial ratio AR is positive for left-hand polarization and negative for right-hand polarization. Some polarization states are displayed on the first octant of the Poincaré sphere in Figure 4-20b. The polarization states on a planar surface representation (projection) of the Poincaré sphere ($-45° \leq \varepsilon \leq +45°$, $0° \leq \tau \leq 180°$) are shown in Figure 4-21.

For the polarization ellipse of Figure 4-19, the two sets of angles are related geometrically as shown in Figure 4-20. Analytically, it can be shown through spherical trigonometry [20] that the two pairs of angles (γ, δ) and (ε, τ) are related by

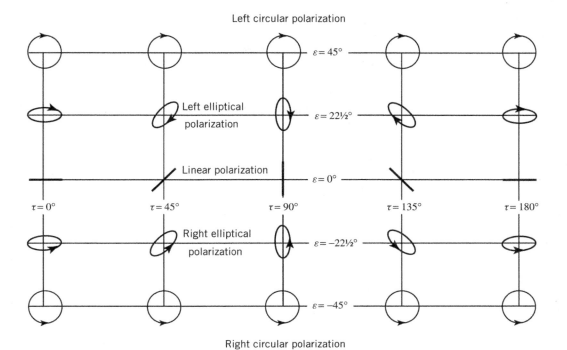

Left circular polarization

Right circular polarization

Figure 4-21 Polarization states of electromagnetic waves on a planar surface projection of a Poincaré sphere. (Source: J. D. Kraus, *Electromagnetics*, 1984, McGraw-Hill Book Co.)

$$\cos(2\gamma) = \cos(2\varepsilon)\cos(2\tau) \qquad \text{(4-60a)}$$

$$\tan(\delta) = \frac{\tan(2\varepsilon)}{\sin(2\tau)} \qquad \text{(4-60b)}$$

or

$$\sin(2\varepsilon) = \sin(2\gamma)\sin(\delta) \qquad \text{(4-61a)}$$

$$\tan(2\tau) = \tan(2\gamma)\cos(\delta) \qquad \text{(4-61b)}$$

Thus one set can be obtained by knowing the other.

It is apparent from Figure 4-20 that the linear polarization is always found along the equator; the right-hand circular resides along the south pole and the left-hand circular along the north pole. The remaining surface of the sphere is used to represent elliptical polarization with left-hand elliptical in the upper hemisphere and right-hand elliptical on the lower hemisphere.

Because the Poincaré sphere parameter pairs (γ, δ) and (ε, τ) are related by transcendental functions, of (4-60a) and 4-60b), there may be some ambiguity at which quadrant should the angles be chosen. The angles should be selected to each satisfy respectively the range of values given by (4-58a) and (4-58b), and (4-57c), and each set should represent the same point on the Poincaré sphere. Also the range of values of the axial ratio (AR) should be $1 \leq |AR| \leq \infty$, with positive values to represent CCW (left-hand) polarization and negative values to represent CW (right-hand) polarization. A MATLAB computer program, **Polarization_Propag**, has been written and it is part of the website that accompanies this book.

Example 4-9

Determine the point on the Poincaré sphere of Figure 4-20 when the wave represented by (4-50a) is such that

$$\mathscr{E}_x = E_{x_0}^+ \cos(\omega t - \beta z + \phi_x)$$
$$\mathscr{E}_y = 0$$

Solution: Using (4-58a) and (4-59b)

$$\gamma = \tan^{-1}\left[\frac{E_{y_0}^+}{E_{x_0}^+}\right] = \tan^{-1}\left[\frac{0}{E_{x_0}^+}\right] = 0°$$

and δ could be of any value, i.e., $-180° \le \delta \le 180°$. The values of ε and τ can now be obtained from (4-61a) and (4-61b), and they are equal to

$$2\varepsilon = \sin^{-1}[\sin(2\gamma)\sin(\delta)] = \sin^{-1}(0) = 0°$$
$$2\tau = \tan^{-1}[\tan(2\gamma)\cos(\delta)] = \tan^{-1}(0) = 0°$$

It is apparent that for this wave, which is obviously linearly polarized, the polarization state (point) is at the reference point of Figure 4-20. The axial ratio is obtained from (4-59a), and it is equal to

$$\mathrm{AR} = \cot(\varepsilon) = \cot(0) = \infty$$

An axial ratio of infinity always represents linear polarization.

Example 4-10

Repeat Example 4-9 when the wave of (4-50a) is such that

$$\mathscr{E}_x = 0$$
$$\mathscr{E}_y = E_{y_0}^+ \cos(\omega t - \beta z + \phi_y)$$

Solution: Using (4-58a) and (4-58b),

$$\gamma = \tan^{-1}\left[\frac{E_{y_0}^+}{E_{x_0}^+}\right] = \tan^{-1}(\infty) = 90°$$

and δ could be of any value, i.e., $-180° \le \delta \le 180°$. The values of ε and τ can now be obtained from (4-61a) and (4-61b), and they are equal to

$$2\varepsilon = \sin^{-1}[\sin(2\gamma)\sin(\delta)] = \sin^{-1}(0) = 0°$$
$$2\tau = \tan^{-1}[\tan(2\gamma)\cos(\delta)] = \tan^{-1}(0) = 180°$$

The polarization state (point) of this linearly polarized wave is diametrically opposed to that in Example 4-9. The axial ratio is also infinity.

Example 4-11

Determine the polarization state (point) on the Poincaré sphere of Figure 4-20 when the wave of (4-50a) is such that

$$\mathscr{E}_x = E_{x_0}^+ \cos(\omega t - \beta z + \phi_x) = 2E_0 \cos\left(\omega t - \beta z + \frac{\pi}{2}\right)$$

$$\mathscr{E}_y = E_{y_0}^+ \cos(\omega t - \beta z + \phi_y) = E_0 \cos(\omega t - \beta z)$$

Solution: Using (4-58a) and (4-58b),

$$\gamma = \tan^{-1}\left[\frac{E_{y_0}^+}{E_{x_0}^+}\right] = \tan^{-1}\left[\frac{E_0}{2E_0}\right] = 26.56°$$

$$\delta = \phi_y - \phi_x = -90°$$

The values of ε and τ can now be obtained from (4-61a) and (4-61b), and they are equal to

$$2\varepsilon = \sin^{-1}[\sin(2\gamma)\sin(\delta)] = \sin^{-1}[-\sin(2\gamma)] = -2\gamma = -53.12°$$

$$2\tau = \tan^{-1}[\tan(2\gamma)\cos(\delta)] = \tan^{-1}(0) = 0°$$

Therefore, this point is situated on the principal xz plane at an angle of $2\gamma = -2\varepsilon = 53.12°$ from the reference point of the x axis of Figure 4-20. The axial ratio is obtained using (4-59a), and it is equal to

$$AR = \cot(\varepsilon) = \cot(-26.56°) = -2$$

The negative sign indicates that the wave has a right-hand (clockwise) polarization. Therefore the wave is right-hand elliptically polarized with $AR = -2$.

In general, points on the principal xz elevation plane, aside from the two intersecting points on the equator and the north and south poles, are used to represent elliptical polarization when the major and minor axes of the polarization ellipse of Figure 4-19 coincide with the principal axes.

If the polarization state of a wave is defined as P_w and that of an antenna as P_a, then the voltage response of the antenna due to the wave is obtained by [10, 19]

$$V = C\cos\left[\frac{P_w P_a}{2}\right] \tag{4-62}$$

where

C = constant that is a function of the antenna size and field strength of the wave
P_w = polarization state of the wave
P_a = polarization state of the antenna
$P_w P_a$ = angle subtended by a great-circle arc from polarization P_w to P_a

Remember that the polarization of a wave, by IEEE standards [7, 8], is determined as the wave is observed from the rear (is receding). Therefore the polarization of the antenna is determined by its radiated field in the transmitting mode.

Example 4-12

If the polarization states of the wave and antenna are given, respectively, by those of Examples 4-9 and 4-10, determine the voltage response of the antenna due to that wave.

Solution: Since the polarization state P_w of the wave is at the $+x$ axis and that of the antenna P_a is at the $-x$ axis of Figure 4-20, then the angle $P_w P_a$ subtended by a great-circle arc from P_w to P_a is equal to

$$P_w P_a = 180°$$

Therefore the voltage response of the antenna is, according to (4-62), equal to

$$V = C\cos\left[\frac{P_w P_a}{2}\right] = C\cos(90°) = 0$$

This is expected since the fields of the wave and those of the antenna are orthogonal (cross-polarized) to each other.

Example 4-13

The polarization of a wave that impinges upon a left-hand (counterclockwise) circularly polarized antenna is circularly polarized. Determine the response of the antenna when the sense of rotation of the incident wave is

1. Left-hand (counterclockwise).
2. Right-hand (clockwise).

Solution:

1. Since the antenna is left-hand circularly polarized, its polarization state (point) on the Poincaré sphere is on the north pole ($2\gamma = \delta = 90°$). When the wave is also left-hand circularly polarized, its polarization state (point) is also on the north pole ($2\gamma = \delta = 90°$). Therefore, the subtended angle $P_w P_a$ between the two polarization states is equal to

$$P_w P_a = 0°$$

and the voltage response of the antenna, according to (4-62), is equal to

$$V = C\cos\left[\frac{P_w P_a}{2}\right] = C\cos(0) = C$$

This represents the maximum response of the antenna, and it occurs when the polarization (including sense of rotation) of the wave is the same as that of the antenna.

2. When the sense of rotation of the wave is right-hand circularly polarized, its polarization state (point) is on the south pole ($2\gamma = 90°$, $\delta = -90°$). Therefore, the subtended angle $P_w P_a$ between the two polarization states is equal to

$$P_w P_a = 180°$$

and the response of the antenna, according to (4-62), is equal to

$$V = C\cos\left[\frac{P_w P_a}{2}\right] = C\cos\left[\frac{180°}{2}\right] = C\cos(90°) = 0$$

This represents a null response of the antenna, and it occurs when the sense of rotation of the circularly polarized wave is opposite to that of the circularly polarized antenna. This is one technique, in addition to those shown in Example 4-12, that can be used to null the response of an antenna system.

4.5 MULTIMEDIA

On the website that accompanies this book, the following multimedia resources are included for the review, understanding, and presentation of the material of this chapter.

- **MATLAB** computer programs:
  a. **Polarization_Diagram_Ellipse_Animation:** Animates the 3-D polarization diagram of a rotating electric field vector (Figure 4-8). It also animates the 2-D polarization ellipse (Figure 4-19) for linear, circular, and elliptical polarized waves, and sense of rotation. It also computes the axial ratio (AR).
  b. **Polarization_Propag:** Computes the Poincaré sphere angles, and thus the polarization wave traveling in an infinite homogeneous medium.
- **PowerPoint (PPT)** viewgraphs, in multicolor.

REFERENCES

1. S. F. Adam, *Microwave Theory and Applications*, Prentice-Hall, Englewood Cliffs, N.J., 1969.
2. A. L. Lance, *Introduction to Microwave Theory and Measurements*, McGraw-Hill, New York, 1964.
3. N. Marcuvitz (Ed.), *Waveguide Handbook*, McGraw-Hill, New York, 1951, Chapter 8, pp. 387–413.
4. C. H. Walter, *Traveling Wave Antennas*, McGraw-Hill, New York, 1965, pp. 172–187.
5. R. B. Adler, L. J. Chu, and R. M. Fano, *Electromagnetic Energy Transmission and Radiation*, John Wiley & Sons, New York, 1960, Chapter 8.
6. D. T. Paris and F. K. Hurd, *Basic Electromagnetic Theory*, McGraw-Hill, New York, 1969.
7. "IEEE Standard 145-1983, IEEE Standard Definitions of Terms for Antennas," reprinted in *IEEE Trans. Antennas Propagat.*, vol. AP-31, no. 6, part II, pp. 1–29, Nov. 1983.
8. C. A. Balanis, *Antenna Theory: Analysis and Design*, Third Edition. John Wiley & Sons, New York, 2005.
9. W. Sichak and S. Milazzo, "Antennas for circular polarization," *Proc. IEEE*, vol. 36, pp. 997–1002, Aug. 1948.
10. G. Sinclair, "The transmission and reception of elliptically polarized waves," *Proc. IRE*, vol. 38, pp. 148–151, Feb. 1950.
11. V. H. Rumsey, G. A. Deschamps, M. L. Kales, and J. I. Bohnert, "Techniques for handling elliptically polarized waves with special reference to antennas," *Proc. IRE*, vol. 39, pp. 533–534, May 1951.
12. V. H. Rumsey, "Part I—Transmission between elliptically polarized antennas," *Proc. IRE*, vol. 39, pp. 535–540, May 1951.
13. M. L. Kales, "Part III—Elliptically polarized waves and antennas," *Proc. IRE*, vol. 39, pp. 544–549, May 1951.
14. J. I. Bohnert, "Part IV—Measurements on elliptically polarized antennas," *Proc. IRE*, vol. 39, pp. 549–552, May 1951.
15. H. Poincaré, *Théorie Mathématique de la Limiére*, Georges Carré, Paris, France, 1892.
16. G. A. Deschamps, "Part II—Geometrical representation of the polarization of a plane electromagnetic wave," *Proc. IRE*, vol. 39, pp. 540–544, May 1951.
17. E. F. Bolinder, "Geometric analysis of partially polarized electromagnetic waves," *IEEE Trans. Antennas Propagat.*, vol. AP-15, no. 1, pp. 37–40, Jan. 1967.
18. G. A. Deschamps and P. E. Mast, "Poincaré sphere representation of partially polarized fields," *IEEE Trans. Antennas Propagat.*, vol. AP-21, no. 4, pp. 474–478, July 1973.
19. J. D. Kraus, *Electromagnetics*, Third Edition, McGraw-Hill, New York, 1984.
20. M. Born and E. Wolf, *Principles of Optics*, Macmillan Co., New York, pp. 24–27, 1964.

PROBLEMS

4.1. A uniform plane wave having only an x component of the electric field is traveling in the $+z$ direction in an unbounded lossless, source-free region. Using Maxwell's equations write expressions for the electric and corresponding magnetic field intensities. Compare your answers to those of (4-2b) and (4-3c).

4.2. Using Maxwell's equations, find the magnetic field components for the wave whose electric field is given in Example 4-1. Compare your answer with that obtained in the solution of Example 4-1.

4.3. The complex **H** field of a uniform plane wave, traveling in an unbounded source-free medium of free space, is given by

$$\mathbf{H} = \frac{1}{120\pi}(\hat{\mathbf{a}}_x - 2\hat{\mathbf{a}}_y)e^{-j\beta_0 z}$$

Find the:
(a) Corresponding electric field.
(b) Instantaneous power density vector.
(c) Time-average power density.

4.4. The complex **E** field of a uniform plane wave is given by

$$\mathbf{E} = (\hat{\mathbf{a}}_x + j\hat{\mathbf{a}}_z)e^{-j\beta_0 y} + (2\hat{\mathbf{a}}_x - j\hat{\mathbf{a}}_z)e^{+j\beta_0 y}$$

Assuming an unbounded source-free, free-space medium, find the:
(a) Corresponding magnetic field.
(b) Time-average power density flowing in the $+y$ direction.
(c) Time-average power density flowing in the $-y$ direction.

4.5. The magnetic field of a uniform plane wave in a source-free region is given by

$$\mathbf{H} = 10^{-6}\left[-\hat{\mathbf{a}}_x(2+j) + \hat{\mathbf{a}}_z(1+j3)\right]e^{+j\beta y}$$

Assuming that the medium is free space, determine the:
(a) Corresponding electric field.
(b) Time-average power density.

4.6. The electric field of a uniform plane wave traveling in a source-free region of free space is given by

$$\mathbf{E} = 10^{-3}(\hat{\mathbf{a}}_x + j\hat{\mathbf{a}}_y)\sin(\beta_0 z)$$

(a) Is this a traveling or a standing wave?
(b) Identify the traveling wave(s) of the electric field and the direction(s) of travel.
(c) Find the corresponding magnetic field.
(d) Determine the time-average power density of the wave.

4.7. The magnetic field of a uniform plane wave traveling in a source-free, free-space region is given by

$$\mathbf{H} = 10^{-6}(\hat{\mathbf{a}}_y + j\hat{\mathbf{a}}_z)\cos(\beta_0 x)$$

(a) Is this a traveling or a standing wave?
(b) Identify the traveling wave(s) of the magnetic field and the direction(s) of travel.
(c) Find the corresponding electric field.
(d) Determine the time-average power density of the wave.

4.8. A uniform plane wave is traveling in the $-z$ direction inside an unbounded source-free, free-space region. Assuming that the electric field has only an E_x component, its value at $z = 0$ is 4×10^{-3} V/m, and its frequency of operation is 300 MHz, write expressions for the:
(a) Complex electric and magnetic fields.
(b) Instantaneous electric and magnetic fields.
(c) Time-average and instantaneous power densities.
(d) Time-average and instantaneous electric and magnetic energy densities.

4.9. A uniform plane wave traveling inside an unbounded free-space medium has peak electric and magnetic fields given by

$$\mathbf{E} = \hat{\mathbf{a}}_x E_0\, e^{-j\beta_0 z}$$
$$\mathbf{H} = \hat{\mathbf{a}}_y H_0\, e^{-j\beta_0 z}$$

where $E_0 = 1$ mV/m.
(a) Evaluate H_0.
(b) Find the corresponding average power density. Evaluate all the constants.
(c) Determine the volume electric and magnetic energy densities. Evaluate all the constants.

4.10. The complex electric field of a uniform plane wave traveling in an unbounded nonferromagnetic dielectric medium is given by

$$\mathbf{E} = \hat{\mathbf{a}}_y 10^{-3} e^{-j2\pi z}$$

where z is measured in meters. Assuming that the frequency of operation is 100 MHz, find the:

(a) Phase velocity of the wave (give units).
(b) Dielectric constant of the medium.
(c) Wavelength (in meters).
(d) Time-average power density.
(e) Time-average total energy density.

4.11. The complex electric field of a time-harmonic field in free space is given by

$$\mathbf{E} = \hat{\mathbf{a}}_z 10^{-3}(1+j)e^{-j(2/3)\pi x}$$

Assuming the distance x is measured in meters, find the:

(a) Wavelength (in meters).
(b) Frequency.
(c) Associated magnetic field.

4.12. A uniform plane wave is traveling inside the earth, which is assumed to be a perfect dielectric infinite in extent. If the relative permittivity of the earth is 9, find, at a frequency of 1 MHz, the:

(a) Phase velocity.
(b) Wave impedance.
(c) Intrinsic impedance.
(d) Wavelength of the wave inside the earth.

4.13. An 11 GHz transmitter radiates its power isotropically in a free-space medium. Assuming its total radiated power is 50 mW, at a distance of 3 km, find the:

(a) Time-average power density.
(b) RMS electric and magnetic fields.
(c) Total time-average volume energy densities.
In all cases, specify the units.

4.14. The electric field of a time-harmonic wave traveling in free space is given by

$$\mathbf{E} = \hat{\mathbf{a}}_x 10^{-4}(1+j)e^{-j\beta_0 z}$$

Find the amount of real power crossing a rectangular aperture whose cross section is perpendicular to the z axis. The area of the aperture is 20 cm^2.

4.15. The following complex electric field of a time-harmonic wave traveling in a source-free, free-space region is given by

$$\mathbf{E} = 5 \times 10^{-3}(4\hat{\mathbf{a}}_y + 3\hat{\mathbf{a}}_z)e^{j(6y-8z)}$$

Assuming y and z represent their respective distances in meters, determine the:

(a) Angle of wave travel (relative to the z axis).
(b) Three phase constants of the wave along its oblique direction of travel, the y axis, and the z axis (in radians per meter).
(c) Three wavelengths of the wave along its oblique direction of travel, the y axis, and the z axis (in meters).
(d) Three phase velocities of the wave along the oblique direction of travel, the y axis, and the z axis (in meters per second).
(e) Three energy velocities of the wave along the oblique direction of travel, the y axis, and the z axis (in meters per second).
(f) Frequency of the wave.
(g) Associated magnetic field.

4.16. Using Maxwell's equations, determine the magnetic field of (4-18b) given the electric field of (4-18a).

4.17. Given the electric field of Example 4-2 and using Maxwell's equations, determine the magnetic field. Compare it with that found in the solution of Example 4-2.

4.18. Given (4-19a) and (4-19c), determine the phase velocities of (4-22) and (4-23).

4.19. Derive the energy velocity of (4-24) using the definition of (4-9), (4-18a), and (4-18b).

4.20. A uniform plane wave of 3 GHz is incident upon an unbounded conducting medium of copper that has a conductivity of 5.76×10^7 S/m, $\varepsilon = \varepsilon_0$, and $\mu = \mu_0$. Find the approximate:

(a) Intrinsic impedance of copper.
(b) Skin depth (in meters).

4.21. The magnetic field intensity of a plane wave traveling in a lossy earth is given by

$$\mathbf{H} = (\hat{\mathbf{a}}_y + j2\hat{\mathbf{a}}_z)H_0 e^{-\alpha x}e^{-j\beta x}$$

where $H_0 = 1$ μA/m. Assuming the lossy earth has a conductivity of 10^{-4} S/m, a dielectric constant of 9, and the frequency of operation is 1 GHz, find inside the earth the:

(a) Corresponding electric field vector.
(b) Average power density vector.
(c) Phase constant (radians per meter).
(d) Phase velocity (meters per second).
(e) Wavelength (meters).
(f) Attenuation constant (nepers per meter).
(g) Skin depth (meters).

4.22. Sea water is an important medium in communication between submerged submarines or between submerged submarines and receiving and transmitting stations located above the surface of the sea. Assuming the constitutive electrical parameters of the sea are $\sigma = 4$ S/m, $\varepsilon_r = 81$, $\mu_r = 1$, and $f = 10^4$ Hz, find the:
(a) Complex propagation constant (per meter).
(b) Phase velocity (meters per second).
(c) Wavelength (meters).
(d) Attenuation constant (nepers per meter).
(e) Skin depth (meters).

4.23. The electrical constitutive parameters of moist earth at a frequency of 1 MHz are $\sigma = 10^{-1}$ S/m, $\varepsilon_r = 4$, and $\mu_r = 1$. Assuming that the electric field of a uniform plane wave at the interface (on the side of the earth) is 3×10^{-2} V/m, find the:
(a) Distance through which the wave must travel before the magnitude of the electric field reduces to 1.104×10^{-2} V/m.
(b) Attenuation the electric field undergoes in part (a) (in decibels).
(c) Wavelength inside the earth (in meters).
(d) Phase velocity inside the earth (in meters per second).
(e) Intrinsic impedance of the earth.

4.24. The complex electric field of a uniform plane wave is given by

$$\mathbf{E} = 10^{-2}\left[\hat{\mathbf{a}}_x\sqrt{2} + \hat{\mathbf{a}}_z(1+j)e^{j\pi/4}\right]e^{-j\beta y}$$

(a) Find the polarization of the wave (linear, circular, or elliptical).
(b) Determine the sense of rotation (clockwise or counterclockwise).
(c) Sketch the figure the electric field traces as a function of ωt.

4.25. A uniform plane wave is traveling along the $+z$ axis in a lossy dielectric, with complex permittivity $\dot{\varepsilon} = \varepsilon' - j\varepsilon''$. It is desired to select the dielectric properties (conductivity σ and relative permittivity of the real part (dielectric constant) $\varepsilon_r' = \varepsilon'/\varepsilon_0$ of the lossy dielectric such that at 10 GHz the wave undergoes phase variations of 240 degrees/centimeter and an attenuation of 2 dB/meter. Assuming the relative permeability of the dielectric is unity, determine the:
(a) Relative permittivity of the real part (dielectric constant) $\varepsilon_r' = \varepsilon'/\varepsilon_0$ of the material.
(b) Conductivity σ (in siemens/meter) of the material.

(c) Distance d (in meters) over which the wave has to travel in order for its magnitude to be reduced to 36.788% of its value at the reference point $d = 0$.
(d) Intrinsic impedance of the lossy medium (in ohms).

4.26. The normalized vector magnetic field radiated in free space by an antenna is given by

$$\mathbf{H} = (-\hat{\mathbf{a}}_y + j2\hat{\mathbf{a}}_z)e^{-j\beta_0 x}$$

(a) Write an expression for the corresponding vector electric field.
(b) Determine the polarization of the wave:
• linear, circular, elliptical
• Axial ratio (AR, including sign)
• Sense of rotation (CW or CCW), if applicable
based on the rules established in the book (be able to determine the polarization based on the field's number of components, their relative magnitude, phase difference, etc.). State all the necessary and sufficient conditions for that polarization to justify your answer.

4.27. The complex magnetic field of a uniform plane wave is given by

$$\mathbf{H} = \frac{10^{-3}}{120\pi}(\hat{\mathbf{a}}_x - j\hat{\mathbf{a}}_z)e^{+j\beta y}$$

(a) Find the polarization of the wave (linear, circular, or elliptical).
(b) State the direction of rotation (clockwise or counterclockwise). Justify your answer.
(c) Sketch the polarization curve denoting the $\mathcal{H}$-field amplitude, and direction of rotation. Indicate on the curve the various times for the rotation of the vector.

4.28. In a source-free, free-space region, the complex magnetic field of a time-harmonic field is represented by

$$\mathbf{H} = \left[\hat{\mathbf{a}}_x(1+j) + \hat{\mathbf{a}}_z\sqrt{2}e^{j\pi/4}\right]\frac{E_0}{\eta_0}e^{-j\beta_0 y}$$

where E_0 is a constant and η_0 is the intrinsic impedance of free space. Determine the:
(a) Polarization of the wave (linear, circular, or elliptical). Justify your answer.
(b) Sense of rotation, if any.
(c) Corresponding electric field.

4.29. Show that any linearly polarized wave can be decomposed into two circularly polarized waves (one CW and the other CCW) but both traveling in the same direction as the linearly polarized wave.

4.30. The electric field of a $f = 10$ GHz time-harmonic uniform plane wave traveling in a perfect dielectric medium is given by

$$\mathbf{E} = (\hat{\mathbf{a}}_x + j2\hat{\mathbf{a}}_y)e^{-j600\pi z}$$

where z is in meters. Determine, assuming the permeability of the medium is the same as that of free space, the:
(a) Wavelength of the wave (in meters).
(b) Velocity of the wave (in meters/sec).
(c) Dielectric constant (relative permittivity) of the medium (dimensionless).
(d) Intrinsic impedance of the medium (in ohms).
(e) Wave impedance of the medium (in ohms).
(f) Vector magnetic field of the wave.
(g) Polarization of the wave (linear, circular, elliptical; AR; and sense of rotation).

4.31. The spatial variations of the electric field of a time-harmonic wave traveling in free space are given by

$$\mathbf{E}(x) = \hat{\mathbf{a}}_y e^{-j(\beta_0 x - \frac{\pi}{4})} + \hat{\mathbf{a}}_z e^{-j(\beta_0 x - \frac{\pi}{2})}$$

Determine, using the necessary and sufficient conditions of the wave, the:
(a) Direction of wave travel $(+x, -x, +y, -y, +z,$ or $-z)$ based on $e^{+j\omega t}$ time.
(b) Polarization of the wave (linear, circular, or elliptical). Justify your answer.
(c) Sense of rotation (CW or CCW), if any, of the wave. Justify your answer.

4.32. The spatial variations of the electric field of a time-harmonic wave traveling in free space are given by

$$\mathbf{E}(z) = \hat{\mathbf{a}}_x 2e^{-j(\beta_0 z - \frac{\pi}{4})} + \hat{\mathbf{a}}_y e^{-j(\beta_0 z - \frac{3\pi}{4})}$$

Determine the:
(a) Direction of wave travel $(+x, -x, +y, -y, +z,$ or $-z)$ based on $e^{+j\omega t}$ time.
(b) Two pairs of Poincaré sphere polarization parameters (γ, δ) and (ε, τ).
(c) Based on either one of the two pairs of parameters from part (b), state the:
  • Polarization of the wave (linear, circular, or elliptical). Justify your answer.
  • Sense of rotation (CW or CCW) of the wave. Justify your answer.
  • Axial ratio. Justify your answer.

4.33. The time-harmonic electric field traveling inside an infinite lossless dielectric medium is given by

$$\mathbf{E}^i(z) = (j2\hat{\mathbf{a}}_x + 5\hat{\mathbf{a}}_y)E_0 e^{-j\beta z}$$

where β and E_0 are real constants. Assuming a $e^{+j\omega t}$ time convention, determine the:
(a) Polarization of the wave (linear, circular, or elliptical). You must justify your answer. Be specific.
(b) Sense of rotation (CW or CCW). You must justify your answer. Be specific.
(c) Axial ratio (AR) based on the expression of the electric field. You must justify your answer. Be specific.
(d) Poincaré sphere angles (in degrees):
  • γ and δ
  • ε and τ
  Make sure that the polarization point on the Poincaré sphere based on the pair of angles (γ, δ) is the same as that based on the set of angles (ε, τ).
(e) Axial ratio (AR) based on the Poincaré sphere angles. Compare with that in part (c).

4.34. The magnetic field of a uniform plane wave traveling in free space is given by:

$$\mathbf{H} = \frac{E_0}{377}(-j\hat{\mathbf{a}}_x + 2\hat{\mathbf{a}}_y)e^{-j\beta_0 z}$$

where E_0 is a constant.
Determine the:
(a) Corresponding electric field.
(b) Polarization of the wave (linear, circular or elliptical). Justify as to why.
(c) Sense of rotation of the wave (CW or CCW). Justify as to why.
(d) Poincaré sphere polarization parameters $(\gamma, \delta,$ and ε; in degrees) of the wave.
(e) Axial ratio (AR) of the wave (including its sign to represent its sense of rotation). Justify.

4.35. The normalized vector electric field radiated in free space by an antenna is given by

$$\mathbf{E} = (j2\hat{\mathbf{a}}_y + \hat{\mathbf{a}}_z)e^{-j\beta_0 x}$$

(a) Determine, analytically, the polarization of the field based on the Poincaré sphere

parameters (must compute both sets of Poincaré sphere parameters; γ, δ and ε, τ).
(b) Do the two sets of Poincaré sphere parameters (γ, δ and ε, τ) lead to the same polarization point on the Poincaré sphere? If not, why not?
(c) Compare the polarization of this wave with that of Problem 4.34. Are the polarizations the same or different? Why yes or why not? Be specific.

4.36. The normalized electric field radiated by a source is represented by

$$\mathbf{E} = E_0(4\hat{\mathbf{a}}_x + j2\hat{\mathbf{a}}_y)e^{-j\beta z}$$

where E_0 is a constant.
(a) Write an expression for the corresponding vector magnetic field.
(b) By inspection, determine the polarization of the wave:
- Linear, circular or elliptical, and why?
- Sense of rotation, and why?
- Axial ratio (AR) including sign, and why?
- Tilt angle, and why?
(c) Repeating part (b), determine the polarization, based on the Poincaré sphere parameters.
- Linear, circular or elliptical, and why?
- Sense of rotation, and why?
- Axial ratio (AR) including sign, and why?
- Tilt angle, and why?
For parts (b) and (c), state which Poincaré sphere parameters you used, and their corresponding values, to determine the answers.
(d) Are the answers to part (c) the same as those to part (b)? If not, why not?

4.37. In a source-free, free-space region the complex magnetic field is given by

$$\mathbf{H} = j(\hat{\mathbf{a}}_y - j\hat{\mathbf{a}}_z)\frac{E_0}{\eta_0}e^{+j\beta_0 x}$$

where E_0 is a constant and η_0 is the intrinsic impedance of free space. Find the:
(a) Polarization of the wave (linear, circular, or elliptical). Justify your answer.
(b) Sense of rotation, if any (CW or CCW). Justify your answer.
(c) Time-average power density.
(d) Polarization of the wave on the Poincaré sphere.

4.38. The electric field of a time-harmonic wave is given by

$$\mathbf{E} = 2\times10^{-3}(\hat{\mathbf{a}}_x + \hat{\mathbf{a}}_y)e^{-j2z}$$

(a) State the polarization of the wave (linear, circular, or elliptical).
(b) Find the polarization on the Poincaré sphere by identifying the angles δ, γ, τ, and ε (in degrees).
(c) Locate the polarization point on the Poincaré sphere.

4.39. For a uniform plane wave represented by the electric field

$$\mathbf{E} = E_0(\hat{\mathbf{a}}_x - j2\hat{\mathbf{a}}_y)e^{-j\beta z}$$

where E_0 is constant, do the following.
(a) Determine the longitude angle 2τ, latitude angle 2ε, great-circle angle 2γ, and equator to great-circle angle δ (all in degrees) that are used to identify and locate the polarization of the wave on the Poincaré sphere.
(b) Using the answers from part (a), state the polarization of the wave (linear, circular, or elliptical), its sense of rotation (CW or CCW), and its Axial Ratio.
(c) Find the signal loss (in decibels) when the wave is received by a right-hand circularly polarized antenna.

4.40. The electric field of (4-50a) has an axial ratio of infinity and a great-circle angle of $2\gamma = 109.47°$.
(a) Find the relative magnitude (ratio) of $E_{y_0}^+$ to $E_{x_0}^+$. Which component is more dominant, E_x or E_y? Use the first definition of γ in (4-58a).
(b) Identify the polarization point on the Poincaré sphere (i.e., find δ, τ, and ε; in degrees).
(c) State the polarization of the wave (linear, circular, or elliptical).

4.41. A uniform plane wave is traveling along the $+z$ axis and its electric field is given by

$$\mathbf{E}_w = (\hat{\mathbf{a}}_x + j\hat{\mathbf{a}}_y)e^{-j\beta z}E_0$$

This incident plane wave impinges upon an antenna whose field radiated along the z axis is given by
(a) $\mathbf{E}_{aa} = (\hat{\mathbf{a}}_x + j\hat{\mathbf{a}}_y)e^{+j\beta z}E_0$
(b) $\mathbf{E}_{ab} = (\hat{\mathbf{a}}_x - j\hat{\mathbf{a}}_y)e^{+j\beta z}E_0$

Determine the:

1. Polarization of the incident wave (linear, circular, elliptical; sense of rotation; and AR).
2. Polarization of antenna of part (a) (linear, circular, elliptical; sense of rotation; and AR).
3. Polarization of antenna of part (b) (linear, circular, elliptical; sense of rotation; and AR).
4. Normalized output voltage when the incident wave impinges upon the antenna whose electric field is that of part (a).
5. Normalized output voltage when the incident wave impinges upon the antenna whose electric field is that of part (b).

4.42. The field radiated by an antenna has electric field components represented by (4-50a) such that $E_{x_0}^+ = E_{y_0}^+$ and its axial ratio is infinity.

(a) Identify the polarization point on the Poincaré sphere (i.e., find γ, δ, τ, and ε; in degrees).

(b) If this antenna is used to receive the wave of Problem 4.40, find the polarization loss (in decibels). To do this part, use the Poincaré sphere parameters.

CHAPTER 5

Reflection and Transmission

5.1 INTRODUCTION

In the previous chapter we discussed solutions to TEM waves in unbounded media. In real-world problems, however, the fields encounter boundaries, scatterers, and other objects. Therefore the fields must be found by taking into account these discontinuities.

In this chapter we want to discuss TEM field solutions in two semi-infinite lossless and lossy media bounded by a planar boundary of infinite extent. Reflection and transmission coefficients will be derived to account for the reflection and transmission of the fields by the boundary. These coefficients will be functions of the constitutive parameters of the two media, the direction of wave travel (angle of incidence), and the direction of the electric and magnetic fields (wave polarization).

In general, the reflection and transmission coefficients are complex quantities. It will be demonstrated that their amplitudes and phases can be varied by controlling the direction of wave travel (angle of incidence). In fact, for one wave polarization (parallel polarization) the reflection coefficient can be made equal to zero. When this occurs, the angle of incidence is known as the *Brewster angle*. This principle is used in the design of many instruments (such as binoculars).

The magnitude of the reflection coefficient can also be made equal to unity by properly selecting the wave incidence angle. This angle is known as the *critical angle*, and it is independent of wave polarization; however, in order for this angle to occur, the incident wave must exist in the denser medium. The critical angle concept plays a crucial role in the design of transmission lines (such as optical fiber, slab wave-guides, and coated conductors; the microstrip is one example).

5.2 NORMAL INCIDENCE—LOSSLESS MEDIA

We begin the discussion of reflection and transmission from planar boundaries of lossless media by assuming the wave travels perpendicular (*normal incidence*) to the planar interface formed by two semi-infinite lossless media, as shown in Figure 5-1, each characterized by the constitutive

Balanis' Advanced Engineering Electromagnetics, Third Edition. Constantine A. Balanis.
© 2024 John Wiley & Sons, Inc. Published 2024 by John Wiley & Sons, Inc.
Companion Website: www.wiley.com/go/balanis/advancedengineeringelectromagnetics3e

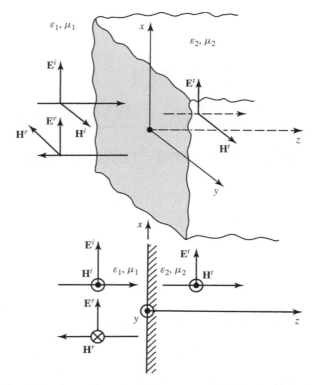

Figure 5-1   Wave reflection and transmission at normal incidence by a planar interface.

parameters of ε_1, μ_1 and ε_2, μ_2. When the incident wave encounters the interface, a fraction of the wave intensity will be reflected into medium 1 and part will be transmitted into medium 2.

Assuming the incident electric field of amplitude E_0 is polarized in the x direction, we can write expressions for its incident, reflected, and transmitted electric field components, respectively, as

$$\mathbf{E}^i = \hat{\mathbf{a}}_x E_0 e^{-j\beta_1 z} \tag{5-1a}$$

$$\mathbf{E}^r = \hat{\mathbf{a}}_x \Gamma^b E_0 e^{+j\beta_1 z} \tag{5-1b}$$

$$\mathbf{E}^t = \hat{\mathbf{a}}_x T^b E_0 e^{-j\beta_2 z} \tag{5-1c}$$

where Γ^b and T^b are used here to represent, respectively, the reflection and transmission coefficients *at the interface*. Presently these coefficients are unknowns and will be determined by applying boundary conditions on the fields along the interface. Since the incident fields are linearly polarized and the reflecting surface is planar, the reflected and transmitted fields will also be linearly polarized. Because we do not know the direction of polarization (positive or negative) of the reflected and transmitted electric fields, they are assumed here to be in the same direction (positive) as the incident electric fields. If that is not the case, it will be corrected by the appropriate signs on the reflection and transmission coefficients.

Using the right-hand procedure outlined in Section 4.2.1 or Maxwell's equations 4-3 or 4-3a, the magnetic field components corresponding to (5-1a) through (5-1c) can be written as

$$\mathbf{H}^i = \hat{\mathbf{a}}_y \frac{E_0}{\eta_1} e^{-j\beta_1 z} \tag{5-2a}$$

$$\mathbf{H}^r = -\hat{\mathbf{a}}_y \frac{\Gamma^b E_0}{\eta_1} e^{+j\beta_1 z} \tag{5-2b}$$

$$\mathbf{H}^t = \hat{\mathbf{a}}_y \frac{T^b E_0}{\eta_2} e^{-j\beta_2 z} \tag{5-2c}$$

The reflection and transmission coefficients will now be determined by enforcing continuity of the tangential components of the electric and magnetic fields across the interface. Using (5-1a) through (5-2c), continuity of the tangential components of the electric and magnetic fields at the interface ($z = 0$) leads, respectively, to

$$1 + \Gamma^b = T^b \tag{5-3a}$$

$$\frac{1}{\eta_1}(1 - \Gamma^b) = \frac{1}{\eta_2} T^b \tag{5-3b}$$

Solving these two equations for Γ^b and T^b, we can write that

$$\boxed{\Gamma^b = \frac{\eta_2 - \eta_1}{\eta_2 + \eta_1} = \frac{E^r}{E^i} = -\frac{H^r}{H^i}} \tag{5-4a}$$

$$\boxed{T^b = \frac{2\eta_2}{\eta_1 + \eta_2} = 1 + \Gamma^b = \frac{E^t}{E^i} = \frac{\eta_2}{\eta_1}\frac{H^t}{H^i}} \tag{5-4b}$$

Therefore the plane wave reflection and transmission coefficients of a planar interface for normal incidence are functions of the constitutive properties, and they are given by (5-4a) and (5-4b). Since the angle of incidence is fixed at normal, the reflection coefficient cannot be equal to zero unless $\eta_2 = \eta_1$. For most dielectric material, aside from ferromagnetics, this implies that $\varepsilon_2 = \varepsilon_1$ since for them $\mu_1 \simeq \mu_2$.

Away from the interface the reflection Γ and transmission T coefficients are related to those at the boundary (Γ^b, T^b) and can be written, respectively, as

$$\boxed{\Gamma(z = -\ell_1) = \frac{E^r(z)}{E^i(z)}\bigg|_{z=-\ell_1} = \frac{\Gamma^b E_0 e^{+j\beta_1 z}}{E_0 e^{-j\beta_1 z}}\bigg|_{z=-\ell_1} = \Gamma^b e^{-j2\beta_1\ell_1}} \tag{5-5a}$$

$$\boxed{T\binom{z_2 = \ell_2,}{z_1 = -\ell_1} = \frac{E^t(z_2)|_{z_2=\ell_2}}{E^i(z_1)|_{z_1=-\ell_1}} = \frac{T^b E_0 e^{-j\beta_2\ell_2}}{E_0 e^{+j\beta_1\ell_1}} = T^b e^{-j(\beta_2\ell_2 + \beta_1\ell_1)}} \tag{5-5b}$$

where ℓ_1 and ℓ_2 are positive distances measured from the interface to media 1 and 2, respectively.

Associated with the electric and magnetic fields (5-1a) through (5-2c) are corresponding average power densities that can be written as

$$\mathbf{S}_{av}^i = \frac{1}{2}\text{Re}(\mathbf{E}^i \times \mathbf{H}^{i*}) = \hat{\mathbf{a}}_z \frac{|E_0|^2}{2\eta_1} \tag{5-6a}$$

$$\mathbf{S}_{av}^{r} = \frac{1}{2}\text{Re}(\mathbf{E}^{r} \times \mathbf{H}^{r*}) = -\hat{\mathbf{a}}_{z}\,|\Gamma^{b}|^{2}\,\frac{|E_{0}|^{2}}{2\eta_{1}} = -\hat{\mathbf{a}}_{z}\,|\Gamma^{b}|^{2}\,S_{av}^{i} \qquad (5\text{-}6\text{b})$$

$$\mathbf{S}_{av}^{t} = \frac{1}{2}\text{Re}(\mathbf{E}^{t} \times \mathbf{H}^{t*}) = \hat{\mathbf{a}}_{z}\,|T^{b}|^{2}\,\frac{|E_{0}|^{2}}{2\eta_{2}} = \hat{\mathbf{a}}_{z}\,|T^{b}|^{2}\,\frac{\eta_{1}}{\eta_{2}}\frac{|E_{0}|^{2}}{2\eta_{1}}$$

$$= \hat{\mathbf{a}}_{z}\,|T^{b}|^{2}\,\frac{\eta_{1}}{\eta_{2}}S_{av}^{i} = \hat{\mathbf{a}}_{z}\left(1 - |\Gamma^{b}|^{2}\right)S_{av}^{i} \qquad (5\text{-}6\text{c})$$

It is apparent that the ratio of the reflected to the incident power densities is equal to the square of the magnitude of the reflection coefficient. However, the ratio of the transmitted to the incident power density is not equal to the square of the magnitude of the transmission coefficient; this is one of the most common errors. Instead the ratio is proportional to the magnitude of the transmission coefficient squared and weighted by the intrinsic impedances of the two media, as given by (5-6c). Remember that the reflection and transmission coefficients relate the reflected and transmitted field intensities to the incident field intensity. Since the total tangential components of these field intensities on either side must be continuous across the boundary, the transmitted field could be greater than the incident field, which would require a transmission coefficient greater than unity. However, by the conservation of power, it is well known that the transmitted power density cannot exceed the incident power density.

Example 5-1

A uniform plane wave traveling in free space is incident normally upon a flat semi-infinite lossless medium with a dielectric constant of 2.56 (being representative of polystyrene). Determine the reflection and transmission coefficients as well as the incident, reflected, and transmitted power densities. Assume that the amplitude of the incident electric field at the interface is 1 mV/m.

Solution: Since $\varepsilon_{1} = \varepsilon_{0}$ and $\varepsilon_{2} = 2.56\varepsilon_{0}$,

$$\mu_{1} = \mu_{2} = \mu_{0}$$

then

$$\eta_{1} = \sqrt{\frac{\mu_{1}}{\varepsilon_{1}}} = \sqrt{\frac{\mu_{0}}{\varepsilon_{0}}}$$

$$\eta_{2} = \sqrt{\frac{\mu_{2}}{\varepsilon_{2}}} = \sqrt{\frac{\mu_{0}}{2.56\varepsilon_{0}}} = \frac{1}{1.6}\sqrt{\frac{\mu_{0}}{\varepsilon_{0}}} = \frac{\eta_{1}}{1.6}$$

Thus according to (5-4a) and (5-4b)

$$\Gamma^{b} = \frac{\eta_{2} - \eta_{1}}{\eta_{2} + \eta_{1}} = \frac{\dfrac{1}{1.6} - 1}{\dfrac{1}{1.6} + 1} = \frac{1 - 1.6}{1 + 1.6} = -0.231$$

$$T^{b} = \frac{2\eta_{2}}{\eta_{1} + \eta_{2}} = \frac{2\left(\dfrac{1}{1.6}\right)}{1 + \dfrac{1}{1.6}} = \frac{2}{2.6} = 0.769$$

In addition, the incident, reflected, and transmitted power densities are obtained using, respectively, (5-6a), (5-6b), and (5-6c). Thus

$$S_{av}^i = \frac{|E_0|^2}{2\eta_1} = \frac{(10^{-3})^2}{2(376.73)} = 1.327 \times 10^{-9} \text{ W/m}^2 = 1.327 \text{ nW/m}^2$$

$$S_{av}^r = |\Gamma^b|^2 \, S_{av}^i = |-0.231|^2 (1.327) \times 10^{-9} = 0.071 \text{ nW/m}^2$$

$$S_{av}^t = |T^b|^2 \frac{\eta_1}{\eta_2} S_{av}^i = |0.769|^2 \frac{1}{1/1.6}(1.327) \times 10^{-9} = 1.256 \text{ nW/m}^2$$

or

$$S_{av}^t = \left(1 - |\Gamma^b|^2\right) S_{av}^i = \left(1 - |0.231|^2\right)(1.327) \times 10^{-9} = 1.256 \text{ nW/m}^2$$

In medium 1, the total field is equal to the sum of the incident and reflected fields. Thus, for the total electric and magnetic fields in medium 1, we can write that

$$\mathbf{E}^1 = \mathbf{E}^i + \mathbf{E}^r = \hat{\mathbf{a}}_x \underbrace{E_0 e^{-j\beta_1 z}}_{\substack{\text{traveling} \\ \text{wave}}} \underbrace{(1 + \Gamma^b e^{+j2\beta_1 z})}_{\substack{\text{standing} \\ \text{wave}}} = \hat{\mathbf{a}}_x E_0 e^{-j\beta_1 z} \left[1 + \Gamma(z)\right] \qquad (5\text{-}7a)$$

$$\mathbf{H}^1 = \mathbf{H}^i + \mathbf{H}^r = \hat{\mathbf{a}}_y \underbrace{(E_0/\eta_1) e^{-j\beta_1 z}}_{\substack{\text{traveling} \\ \text{wave}}} \underbrace{(1 - \Gamma^b e^{+j2\beta_1 z})}_{\substack{\text{standing} \\ \text{wave}}} = \hat{\mathbf{a}}_y \frac{E_0}{\eta_1} e^{-j\beta_1 z} \left[1 - \Gamma(z)\right] \qquad (5\text{-}7b)$$

In each expression the factors outside the parentheses represent the *traveling wave part* of the wave and those within the parentheses represent the *standing wave part*. Therefore the total field of two waves is the product of one of the waves times a factor that in this case is the standing wave pattern. This is analogous to the *array multiplication rule* in antennas where the total field of an array of identical elements is equal to the product of the field of a single element times a factor that is referred to as the array factor [1].

As discussed in Section 4.2.1D, the ratio of the maximum value of the electric field magnitude to that of the minimum is defined as the standing wave ratio (SWR), and it is given here by

$$\text{SWR} = \frac{\left|\mathbf{E}^1\right|_{max}}{\left|\mathbf{E}^1\right|_{min}} = \frac{1 + |\Gamma^b|}{1 - |\Gamma^b|} = \frac{1 + \left|\dfrac{\eta_2 - \eta_1}{\eta_2 + \eta_1}\right|}{1 - \left|\dfrac{\eta_2 - \eta_1}{\eta_2 + \eta_1}\right|} \qquad (5\text{-}8)$$

For two media with identical permeabilities $(\mu_1 = \mu_2)$, the SWR can be written as

$$\text{SWR} = \frac{\left|\sqrt{\varepsilon_1} + \sqrt{\varepsilon_2}\right| + \left|\sqrt{\varepsilon_1} - \sqrt{\varepsilon_2}\right|}{\left|\sqrt{\varepsilon_1} + \sqrt{\varepsilon_2}\right| - \left|\sqrt{\varepsilon_1} - \sqrt{\varepsilon_2}\right|} = \begin{cases} \sqrt{\dfrac{\varepsilon_1}{\varepsilon_2}}, & \varepsilon_1 > \varepsilon_2 & (5\text{-}9a) \\[3mm] \sqrt{\dfrac{\varepsilon_2}{\varepsilon_1}}, & \varepsilon_2 > \varepsilon_1 & (5\text{-}9b) \end{cases}$$

5.3 OBLIQUE INCIDENCE—LOSSLESS MEDIA

To analyze reflections and transmissions at oblique wave incidence, we need to introduce the *plane of incidence*, which is defined as *the plane formed by a unit vector normal to the reflecting interface and the vector in the direction of incidence*. For a wave whose wave vector is on the xz plane and is incident upon an interface that is parallel to the xy plane, as shown in Figure 5-2, the plane of incidence is the xz plane.

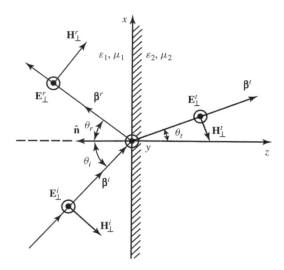

Figure 5-2 Perpendicular (horizontal) polarized uniform plane wave incident at an oblique angle on an interface.

To examine reflections and transmissions at oblique angles of incidence for a general wave polarization, it is most convenient to decompose the electric field into its *perpendicular* and *parallel* components (relative to the plane of incidence) and analyze each one of them individually. The total reflected and transmitted field will be the vector sum of these two polarizations.

When the electric field is perpendicular to the plane of incidence, the polarization of the wave is referred to as *perpendicular polarization*. Since the electric field is parallel to the interface, it is also known as *horizontal* or *E polarization*. When the electric field is parallel to the plane of incidence, the polarization is referred to as *parallel polarization*. Because a component of the electric field is also perpendicular to the interface when the magnetic field is parallel to the interface, it is also known as *vertical* or *H polarization*. Each type of polarization will be further examined.

5.3.1   Perpendicular (Horizontal or E) Polarization

Let us now assume that the electric field of the uniform plane wave incident on a planar interface at an oblique angle, as shown in Figure 5-2, is oriented perpendicularly to the plane of incidence. As previously stated, this is referred to as the perpendicular polarization.

Using the techniques outlined in Section 4.2.2, the incident electric and magnetic fields can be written as

$$\mathbf{E}_{\perp}^{i} = \hat{\mathbf{a}}_y E_{\perp}^{i} e^{-j\boldsymbol{\beta}^i \cdot \mathbf{r}} = \hat{\mathbf{a}}_y E_0 e^{-j\beta_1(x\sin\theta_i + z\cos\theta_i)} \tag{5-10a}$$

$$\mathbf{H}_{\perp}^{i} = \left(-\hat{\mathbf{a}}_x \cos\theta_i + \hat{\mathbf{a}}_z \sin\theta_i\right) H_{\perp}^{i} e^{-j\boldsymbol{\beta}^i \cdot \mathbf{r}}$$
$$= \left(-\hat{\mathbf{a}}_x \cos\theta_i + \hat{\mathbf{a}}_z \sin\theta_i\right) \frac{E_0}{\eta_1} e^{-j\beta_1(x\sin\theta_i + z\cos\theta_i)} \tag{5-10b}$$

where

$$E_{\perp}^{i} = E_0 \tag{5-10c}$$

$$H_{\perp}^{i} = \frac{E_{\perp}^{i}}{\eta_1} = \frac{E_0}{\eta_1} \tag{5-10d}$$

Similarly, the reflected fields can be expressed as

$$\mathbf{E}_\perp^r = \hat{\mathbf{a}}_y E_\perp^r e^{-j\boldsymbol{\beta}^r \cdot \mathbf{r}} = \hat{\mathbf{a}}_y \Gamma_\perp^b E_0 e^{-j\beta_1(x\sin\theta_r - z\cos\theta_r)} \tag{5-11a}$$

$$\mathbf{H}_\perp^r = (\hat{\mathbf{a}}_x \cos\theta_r + \hat{\mathbf{a}}_z \sin\theta_r)H_\perp^r e^{-j\boldsymbol{\beta}^r \cdot \mathbf{r}}$$

$$= (\hat{\mathbf{a}}_x \cos\theta_r + \hat{\mathbf{a}}_z \sin\theta_r)\frac{\Gamma_\perp^b E_0}{\eta_1} e^{-j\beta_1(x\sin\theta_r - z\cos\theta_r)} \tag{5-11b}$$

where

$$E_\perp^r = \Gamma_\perp^b E^i = \Gamma_\perp^b E_0 \tag{5-11c}$$

$$H_\perp^r = \frac{E_\perp^r}{\eta_1} = \frac{\Gamma_\perp^b E_0}{\eta_1} \tag{5-11d}$$

Also the transmitted fields can be written as

$$\mathbf{E}_\perp^t = \hat{\mathbf{a}}_y E_\perp^t e^{-j\boldsymbol{\beta}^t \cdot \mathbf{r}} = \hat{\mathbf{a}}_y T_\perp^b E_0 e^{-j\beta_2(x\sin\theta_t + z\cos\theta_t)} \tag{5-12a}$$

$$\mathbf{H}_\perp^t = (-\hat{\mathbf{a}}_x \cos\theta_t + \hat{\mathbf{a}}_z \sin\theta_t)H_\perp^t e^{-j\boldsymbol{\beta}^t \cdot \mathbf{r}}$$

$$= (-\hat{\mathbf{a}}_x \cos\theta_t + \hat{\mathbf{a}}_z \sin\theta_t)\frac{T_\perp^b E_0}{\eta_2} e^{-j\beta_2(x\sin\theta_t + z\cos\theta_t)} \tag{5-12b}$$

where

$$E_\perp^t = T_\perp^b E_\perp^i = T_\perp^b E_0 \tag{5-12c}$$

$$H_\perp^t = \frac{E_\perp^t}{\eta_2} = \frac{T_\perp^b E_0}{\eta_2} \tag{5-12d}$$

The reflection $\Gamma_\perp^b$ and transmission $T_\perp^b$ coefficients, and the relation between the incident θ_i, reflected θ_r, and transmission (refracted) θ_t angles can be obtained by applying the boundary conditions on the continuity of the tangential components of the electric and magnetic fields. That is

$$(\mathbf{E}_\perp^i + \mathbf{E}_\perp^r)\big|_{\substack{\text{tan}\\z=0}} = (\mathbf{E}_\perp^t)\big|_{\substack{\text{tan}\\z=0}} \tag{5-13a}$$

$$(\mathbf{H}_\perp^i + \mathbf{H}_\perp^r)\big|_{\substack{\text{tan}\\z=0}} = (\mathbf{H}_\perp^t)\big|_{\substack{\text{tan}\\z=0}} \tag{5-13b}$$

Using the appropriate terms of (5-10a) through (5-12d), (5-13a) and (5-13b) can be written, respectively, as

$$e^{-j\beta_1 x\sin\theta_i} + \Gamma_\perp^b e^{-j\beta_1 x\sin\theta_r} = T_\perp^b e^{-j\beta_2 x\sin\theta_t} \tag{5-14a}$$

$$\frac{1}{\eta_1}\left(-\cos\theta_i e^{-j\beta_1 x\sin\theta_i} + \Gamma_\perp^b \cos\theta_r e^{-j\beta_1 x\sin\theta_r}\right) = -\frac{T_\perp^b}{\eta_2}\cos\theta_t e^{-j\beta_2 x\sin\theta_t} \tag{5-14b}$$

Whereas (5-14a) and (5-14b) represent two equations with four unknowns ($\Gamma_\perp^b, T_\perp^b, \theta_r, \theta_t$), it should be noted that each equation is complex. By equating the corresponding real and imaginary parts of each side, each can be reduced to two equations (a total of four). If this procedure is utilized, it will be concluded that (5-14a) and (5-14b) lead to the following two relations:

$$\theta_r = \theta_i \qquad \text{(Snell's law of reflection)} \tag{5-15a}$$

$$\beta_1 \sin \theta_i = \beta_2 \sin \theta_t \quad \text{(Snell's law of refraction)} \tag{5-15b}$$

Using (5-15a) and (5-15b) reduces (5-14a) and (5-14b) to

$$1 + \Gamma_\perp^b = T_\perp^b \tag{5-16a}$$

$$\frac{\cos \theta_i}{\eta_1} \left(-1 + \Gamma_\perp^b \right) = -\frac{\cos \theta_t}{\eta_2} T_\perp^b \tag{5-16b}$$

Solving (5-16a) and (5-16b) simultaneously for $\Gamma_\perp^b$ and $T_\perp^b$ leads to

$$\Gamma_\perp^b = \frac{E_\perp^r}{E_\perp^i} = \frac{\eta_2 \cos \theta_i - \eta_1 \cos \theta_t}{\eta_2 \cos \theta_i + \eta_1 \cos \theta_t} = \frac{\sqrt{\dfrac{\mu_2}{\varepsilon_2}} \cos \theta_i - \sqrt{\dfrac{\mu_1}{\varepsilon_1}} \cos \theta_t}{\sqrt{\dfrac{\mu_2}{\varepsilon_2}} \cos \theta_i + \sqrt{\dfrac{\mu_1}{\varepsilon_1}} \cos \theta_t} \tag{5-17a}$$

$$T_\perp^b = \frac{E_\perp^t}{E_\perp^i} = \frac{2\eta_2 \cos \theta_i}{\eta_2 \cos \theta_i + \eta_1 \cos \theta_t} = \frac{2\sqrt{\dfrac{\mu_2}{\varepsilon_2}} \cos \theta_i}{\sqrt{\dfrac{\mu_2}{\varepsilon_2}} \cos \theta_i + \sqrt{\dfrac{\mu_1}{\varepsilon_1}} \cos \theta_t} \tag{5-17b}$$

$\Gamma_\perp^b$ and $T_\perp^b$ of (5-17a) and (5-17b) are usually referred to, respectively, as the plane wave *Fresnel reflection and transmission coefficients* for perpendicular polarization.

Since for most dielectric media (excluding ferromagnetic material) $\mu_1 \simeq \mu_2 \simeq \mu_0$, (5-17a) and (5-17b) reduce, by also utilizing (5-15b), to

$$\Gamma_\perp^b \big|_{\mu_1 = \mu_2} = \frac{\cos \theta_i - \sqrt{\dfrac{\varepsilon_2}{\varepsilon_1}} \sqrt{1 - \left(\dfrac{\varepsilon_1}{\varepsilon_2} \right) \sin^2 \theta_i}}{\cos \theta_i + \sqrt{\dfrac{\varepsilon_2}{\varepsilon_1}} \sqrt{1 - \left(\dfrac{\varepsilon_1}{\varepsilon_2} \right) \sin^2 \theta_i}} \tag{5-18a}$$

$$T_\perp^b \big|_{\mu_1 = \mu_2} = \frac{2 \cos \theta_i}{\cos \theta_i + \sqrt{\dfrac{\varepsilon_2}{\varepsilon_1}} \sqrt{1 - \left(\dfrac{\varepsilon_1}{\varepsilon_2} \right) \sin^2 \theta_i}} \tag{5-18b}$$

Plots of $\left| \Gamma_\perp^b \right|$ and $\left| T_\perp^b \right|$ of (5-18a) and (5-18b) for $\varepsilon_2 / \varepsilon_1 = 2.56, 4, 9, 16, 25,$ and 81 as a function of θ_i are shown in Figure 5-3. It is apparent that as the relative ratio of $\varepsilon_2 / \varepsilon_1$ increases, the magnitude of the reflection coefficient increases, whereas that of the transmission coefficient decreases. This is expected since large ratios of $\varepsilon_2 / \varepsilon_1$ project larger discontinuities in the dielectric properties of the media along the interface. Also it is observed that for $\varepsilon_2 > \varepsilon_1$ the magnitude of the reflection coefficient never vanishes regardless of the $\varepsilon_2 / \varepsilon_1$ ratio or the angle of incidence.

For $\varepsilon_2 / \varepsilon_1 > 1$, both $\Gamma_\perp^b$ and $T_\perp^b$ are real with $\Gamma_\perp^b$ being negative and $T_\perp^b$ being positive for all angles of incidence. Therefore, as a function of θ_i, the phase of $\Gamma_\perp^b$ is equal to 180° and that of the transmission coefficient $T_\perp^b$ is zero. When $\varepsilon_2 / \varepsilon_1 = 1$ the reflection coefficient vanishes and the transmission coefficient reduces to unity. When $\varepsilon_2 / \varepsilon_1 < 1$, both $\Gamma_\perp^b$ and $T_\perp^b$ are real when the incidence angle $\theta_i \leq \theta_c$; for $\theta_i > \theta_c$, they become complex. The angle θ_i for which $\left| \Gamma_\perp^b \right|_{\varepsilon_2/\varepsilon_1 < 1} (\theta_i = \theta_c) = 1$ is referred to as the *critical* angle, and it represents conditions of total internal reflection. More

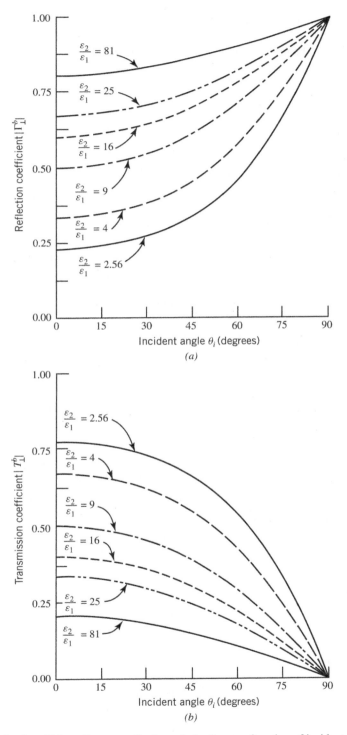

Figure 5-3   Magnitude of coefficients for perpendicular polarization as a function of incident angle. (*a*) Reflection. (*b*) Transmission.

discussion on the critical angle $(\theta_i = \theta_c)$ and the wave propagation for $\theta_i > \theta_c$ can be found in Section 5.3.4.

In medium 1 the total electric field can be written as

$$\mathbf{E}_\perp^1 = \mathbf{E}_\perp^i + \mathbf{E}_\perp^r = \hat{\mathbf{a}}_y \underbrace{E_0 e^{-j\beta_1(x\sin\theta_i + z\cos\theta_i)}}_{\text{traveling wave}} \underbrace{\left[1 + \Gamma_\perp^b e^{+j2\beta_1 z\cos\theta_i}\right]}_{\text{standing wave}}$$

$$= \hat{\mathbf{a}}_y E_0 e^{-j\beta_1(x\sin\theta_i + z\cos\theta_i)}\left[1 + \Gamma_\perp(z)\right] \qquad (5\text{-}19)$$

where

$$\Gamma_\perp(z) = \Gamma_\perp^b e^{+j2\beta_1 z\cos\theta_i} \qquad (5\text{-}19a)$$

5.3.2   Parallel (Vertical or H) Polarization

For this polarization the electric field is parallel to the plane of incidence and it impinges upon a planar interface as shown in Figure 5-4. The directions of the incident, reflected, and transmitted electric and magnetic fields in Figure 5-4 are chosen so that for the special case of $\theta_i = 0$ they reduce to those of Figure 5-1.

Using the techniques outlined in Section 4.2.2, we can write that

$$\mathbf{E}_\parallel^i = \left(\hat{\mathbf{a}}_x \cos\theta_i - \hat{\mathbf{a}}_z \sin\theta_i\right) E_0 e^{-j\boldsymbol{\beta}^i \cdot \mathbf{r}}$$

$$= \left(\hat{\mathbf{a}}_x \cos\theta_i - \hat{\mathbf{a}}_z \sin\theta_i\right) E_0 e^{-j\beta_1(x\sin\theta_i + z\cos\theta_i)} \qquad (5\text{-}20a)$$

$$\mathbf{H}_\parallel^i = \hat{\mathbf{a}}_y H_\parallel^i e^{-j\boldsymbol{\beta}^i \cdot \mathbf{r}} = \hat{\mathbf{a}}_y \frac{E_0}{\eta_1} e^{-j\beta_1(x\sin\theta_i + z\cos\theta_i)} \qquad (5\text{-}20b)$$

where

$$E_\parallel^i = E_0 \qquad (5\text{-}20c)$$

$$H_\parallel^i = \frac{E_\parallel^i}{\eta_1} = \frac{E_0}{\eta_1} \qquad (5\text{-}20d)$$

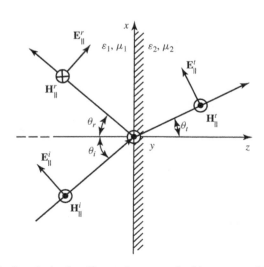

Figure 5-4   Parallel (vertical) polarized uniform plane wave incident at an oblique angle on an interface.

Similarly,

$$\mathbf{E}_\parallel^r = \left(\hat{\mathbf{a}}_x \cos\theta_r + \hat{\mathbf{a}}_z \sin\theta_r\right) E^r e^{-j\boldsymbol{\beta}^r \cdot \mathbf{r}}$$
$$= \left(\hat{\mathbf{a}}_x \cos\theta_r + \hat{\mathbf{a}}_z \sin\theta_r\right) \Gamma_\parallel^b E_0 e^{-j\beta_1(x\sin\theta_r - z\cos\theta_r)} \tag{5-21a}$$

$$\mathbf{H}_\parallel^r = -\hat{\mathbf{a}}_y H_\parallel^r e^{-j\boldsymbol{\beta}^r \cdot \mathbf{r}} = -\hat{\mathbf{a}}_y \frac{\Gamma_\parallel^b E_0}{\eta_1} e^{-j\beta_1(x\sin\theta_r - z\cos\theta_r)} \tag{5-21b}$$

where

$$E_\parallel^r = \Gamma_\parallel^b E^i = \Gamma_\parallel^b E_0 \tag{5-21c}$$

$$H_\parallel^r = \frac{E_\parallel^r}{\eta_1} = \frac{\Gamma_\parallel^b E_0}{\eta_1} \tag{5-21d}$$

Also,

$$\mathbf{E}_\parallel^t = \left(\hat{\mathbf{a}}_x \cos\theta_t - \hat{\mathbf{a}}_z \sin\theta_t\right) E_\parallel^t e^{-j\boldsymbol{\beta}^t \cdot \mathbf{r}}$$
$$= \left(\hat{\mathbf{a}}_x \cos\theta_t - \hat{\mathbf{a}}_z \sin\theta_t\right) T_\parallel^b E_0 e^{-j\beta_2(x\sin\theta_t + z\cos\theta_t)} \tag{5-22a}$$

$$\mathbf{H}_\parallel^t = \hat{\mathbf{a}}_y H_\parallel^t e^{-j\boldsymbol{\beta}^t \cdot \mathbf{r}} = \hat{\mathbf{a}}_y \frac{T_\parallel^b E_0}{\eta_2} e^{-j\beta_2(x\sin\theta_t + z\cos\theta_t)} \tag{5-22b}$$

where

$$E_\parallel^t = T_\parallel^b E^i = T_\parallel^b E_0 \tag{5-22c}$$

$$H_\parallel^t = \frac{E_\parallel^t}{\eta_2} = \frac{T_\parallel^b E_0}{\eta_2} \tag{5-22d}$$

As before, the reflection $\Gamma_\parallel^b$ and transmission $T_\parallel^b$ coefficients, and the reflection θ_r and transmission (refraction) θ_t angles are the four unknowns. These can be determined and expressed in terms of the incident angle θ_i and the constitutive parameters of the two media by applying the boundary conditions on the continuity across the interface $(z = 0)$ of the tangential components of the electric and magnetic fields as given by (5-13a) and (5-13b) and applied to parallel polarization. Using the appropriate terms of (5-20a) through (5-22d), we can write (5-13a) and (5-13b) as applied to parallel polarization, respectively, as

$$\cos\theta_i \, e^{-j\beta_1 x\sin\theta_i} + \Gamma_\parallel^b \cos\theta_r \, e^{-j\beta_1 x\sin\theta_r} = T_\parallel^b \cos\theta_t \, e^{-j\beta_2 x\sin\theta_t} \tag{5-23a}$$

$$\frac{1}{\eta_1}\left(e^{-j\beta_1 x\sin\theta_i} - \Gamma_\parallel^b e^{-j\beta_1 x\sin\theta_r}\right) = \frac{1}{\eta_2} T_\parallel^b e^{-j\beta_2 x\sin\theta_t} \tag{5-23b}$$

Following the procedure outlined in Section 5.3.1 for the solution of (5-14a) and (5-14b), it can be shown that (5-23a) and (5-23b) reduce to

$$\theta_r = \theta_i \qquad \textit{(Snell's law of reflection)} \tag{5-24a}$$

$$\beta_1 \sin\theta_i = \beta_2 \sin\theta_t \quad \textit{(Snell's law of refraction)} \tag{5-24b}$$

$$\boxed{\Gamma_\parallel^b = \frac{-\eta_1\cos\theta_i + \eta_2\cos\theta_t}{\eta_1\cos\theta_i + \eta_2\cos\theta_t} = \frac{-\sqrt{\dfrac{\mu_1}{\varepsilon_1}}\cos\theta_i + \sqrt{\dfrac{\mu_2}{\varepsilon_2}}\cos\theta_t}{\sqrt{\dfrac{\mu_1}{\varepsilon_1}}\cos\theta_i + \sqrt{\dfrac{\mu_2}{\varepsilon_2}}\cos\theta_t}} \tag{5-24c}$$

$$T_{\parallel}^{b} = \frac{2\eta_2 \cos\theta_i}{\eta_1\cos\theta_i + \eta_2\cos\theta_t} = \frac{2\sqrt{\dfrac{\mu_2}{\varepsilon_2}}\cos\theta_i}{\sqrt{\dfrac{\mu_1}{\varepsilon_1}}\cos\theta_i + \sqrt{\dfrac{\mu_2}{\varepsilon_2}}\cos\theta_t} \qquad (5\text{-}24\text{d})$$

$\Gamma_{\parallel}^{b}$ and $T_{\parallel}^{b}$ of (5-24c) and (5-24d) are usually referred to, respectively, as the plane wave *Fresnel reflection and transmission coefficients* for parallel polarization.

Excluding ferromagnetic material, (5-24c) and (5-24d) reduce, using also (5-24b), to

$$\Gamma_{\parallel}^{b}\big|_{\mu_1=\mu_2} = \frac{-\cos\theta_i + \sqrt{\dfrac{\varepsilon_1}{\varepsilon_2}}\sqrt{1 - \left(\dfrac{\varepsilon_1}{\varepsilon_2}\right)\sin^2\theta_i}}{\cos\theta_i + \sqrt{\dfrac{\varepsilon_1}{\varepsilon_2}}\sqrt{1 - \left(\dfrac{\varepsilon_1}{\varepsilon_2}\right)\sin^2\theta_i}} \qquad (5\text{-}25\text{a})$$

$$T_{\parallel}^{b}\big|_{\mu_1=\mu_2} = \frac{2\sqrt{\dfrac{\varepsilon_1}{\varepsilon_2}}\cos\theta_i}{\cos\theta_i + \sqrt{\dfrac{\varepsilon_1}{\varepsilon_2}}\sqrt{1 - \left(\dfrac{\varepsilon_1}{\varepsilon_2}\right)\sin^2\theta_i}} \qquad (5\text{-}25\text{b})$$

Plots of $\left|\Gamma_{\parallel}^{b}\right|$ and $\left|T_{\parallel}^{b}\right|$ of (5-25a) and (5-25b) for $\varepsilon_2/\varepsilon_1 = 2.56, 4, 9, 16, 25$, and 81 as a function of θ_i are shown in Figure 5-5. It is observed in Figure 5-5a that for this polarization there is an angle where the reflection coefficient does vanish. The angle where the reflection coefficient vanishes is referred to as the *Brewster angle*, θ_B, and it increases toward 90° as the ratio $\varepsilon_2/\varepsilon_1$ becomes larger. More discussion on the Brewster angle can be found in the next section (Section 5.3.3).

For $\varepsilon_2/\varepsilon_1 > 1$, $\Gamma_{\parallel}^{b}$ and $T_{\parallel}^{b}$ are both real. For angles of incidence less than the Brewster angle ($\theta_i < \theta_B$), $\Gamma_{\parallel}^{b}$ is negative, indicating a 180° phase as a function of the incident angle; for $\theta_i > \theta_B$, $\Gamma_{\parallel}^{b}$ is positive, representing a 0° phase. The transmission coefficient $T_{\parallel}^{b}$ is positive for all values of θ_i, indicating a 0° phase. When $\varepsilon_2/\varepsilon_1 = 1$, the reflection coefficient vanishes and the transmission coefficient reduces to unity. As for the perpendicular polarization, when $\varepsilon_2/\varepsilon_1 < 1$ both $\Gamma_{\parallel}^{b}$ and $T_{\parallel}^{b}$ are real when the incident angle $\theta_i \leq \theta_c$; after that, they become complex. The angle for which $\left|\Gamma_{\parallel}^{b}\right|_{\varepsilon_2/\varepsilon_1<1} (\theta_i = \theta_c) = 1$ is again referred to as *critical* angle, and it represents conditions of total internal reflection. Further discussion of the critical angle ($\theta_i = \theta_c$) and the wave propagation for $\theta_i > \theta_c$ can be found in Section 5.3.4. It is evident that the critical angle is not a function of polarization; it occurs only when the wave propagates from the more dense to the less dense medium.

The total electric field in medium 1 can be written as

$$\mathbf{E}_{\parallel}^{1} = \mathbf{E}_{\parallel}^{i} + \mathbf{E}_{\parallel}^{r} = \hat{\mathbf{a}}_x \cos\theta_i \underbrace{E_0 e^{-j\beta_1(x\sin\theta_i + z\cos\theta_i)}}_{\text{traveling wave}} \underbrace{\left[1 + \Gamma_{\parallel}^{b} e^{+j2\beta_1 z\cos\theta_i}\right]}_{\text{standing wave}}$$

$$-\hat{\mathbf{a}}_z \sin\theta_i \underbrace{E_0 e^{-j\beta_1(x\sin\theta_i + z\cos\theta_i)}}_{\text{traveling wave}} \underbrace{\left[1 - \Gamma_{\parallel}^{b} e^{+j2\beta_1 z\cos\theta_i}\right]}_{\text{standing wave}}$$

$$\mathbf{E}_{\parallel}^{1} = \mathbf{E}_x^{1} + \mathbf{E}_z^{1} = \hat{\mathbf{a}}_x \cos\theta_i E_0 e^{-j\beta_1(x\sin\theta_i + z\cos\theta_i)}\left[1 + \Gamma_{\parallel}(z)\right]$$

$$-\hat{\mathbf{a}}_z \sin\theta_i E_0 e^{-j\beta_1(x\sin\theta_i + x\cos\theta_i)}\left[1 - \Gamma_{\parallel}(z)\right] \qquad (5\text{-}26)$$

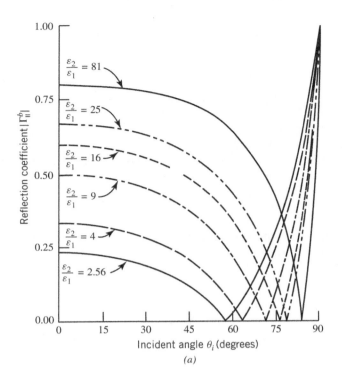

(a)

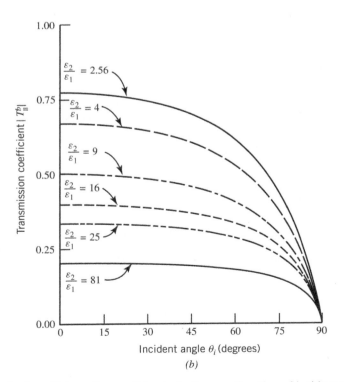

(b)

Figure 5-5 Magnitude of coefficients for parallel polarization as a function of incident angle. (*a*) Reflection. (*b*) Transmission.

where

$$\Gamma_\|(z) = \Gamma_\|^b e^{+j2\beta_1 z \cos\theta_i} \tag{5-26a}$$

5.3.3 Total Transmission–Brewster Angle

The reflection and transmission coefficients for both perpendicular and parallel polarizations are functions of the constitutive parameters of the two media forming the interface, the angle of incidence, and the angle of refraction that is related to the angle of incidence through Snell's law of refraction. One may ask: "For a given set of constitutive parameters of two media forming an interface, is there an incidence angle that allows no reflection, i.e., $\Gamma = 0$?" To answer this we need to refer back to the expressions for the reflection coefficients as given by (5-17a) and (5-24c).

A. Perpendicular (Horizontal) Polarization To see the conditions under which the reflection coefficient of (5-17a) will vanish, we set it equal to zero, which leads to

$$\Gamma_\perp^b = \frac{\sqrt{\dfrac{\mu_2}{\varepsilon_2}}\cos\theta_i - \sqrt{\dfrac{\mu_1}{\varepsilon_1}}\cos\theta_t}{\sqrt{\dfrac{\mu_2}{\varepsilon_2}}\cos\theta_i + \sqrt{\dfrac{\mu_1}{\varepsilon_1}}\cos\theta_t} = 0 \tag{5-27}$$

or

$$\cos\theta_i = \sqrt{\frac{\mu_1}{\mu_2}\left(\frac{\varepsilon_2}{\varepsilon_1}\right)}\cos\theta_t \tag{5-27a}$$

Using Snell's law of refraction, as given by (5-15b), (5-27a) can be written as

$$(1 - \sin^2\theta_i) = \frac{\mu_1}{\mu_2}\left(\frac{\varepsilon_2}{\varepsilon_1}\right)(1 - \sin^2\theta_t)$$

$$(1 - \sin^2\theta_i) = \frac{\mu_1}{\mu_2}\left(\frac{\varepsilon_2}{\varepsilon_1}\right)\left[1 - \frac{\mu_1}{\mu_2}\left(\frac{\varepsilon_1}{\varepsilon_2}\right)\sin^2\theta_i\right] \tag{5-28}$$

or

$$\boxed{\sin\theta_i = \sqrt{\frac{\dfrac{\varepsilon_2}{\varepsilon_1} - \dfrac{\mu_2}{\mu_1}}{\dfrac{\mu_1}{\mu_2} - \dfrac{\mu_2}{\mu_1}}}} \tag{5-28a}$$

Since the sine function cannot exceed unity, (5-28a) exists only if

$$\left|\frac{\varepsilon_2}{\varepsilon_1} - \frac{\mu_2}{\mu_1}\right| \le \left|\frac{\mu_1}{\mu_2} - \frac{\mu_2}{\mu_1}\right| \tag{5-29}$$

which reduces to

$$\frac{\varepsilon_2}{\varepsilon_1} \le \frac{\mu_1}{\mu_2} \text{ for } \mu_1 \ge \mu_2 \text{ or } \frac{\varepsilon_2}{\varepsilon_1} \ge \frac{\mu_1}{\mu_2} \text{ for } \mu_2 \ge \mu_1 \tag{5-29a}$$

If $\mu_1 = \mu_2$, (5-28a) indicates that

$$\sin\theta_i\,|_{\mu_1=\mu_2} = \infty \tag{5-29b}$$

Therefore there exists no real angle θ_i under this condition that will reduce the reflection coefficient to zero. Since the permeability for most dielectric material (aside from ferromagnetics) is almost the same and equal to that of free space ($\mu_1 \simeq \mu_2 \simeq \mu_0$), for *these materials there exists no real incidence angle that will reduce the reflection coefficient for perpendicular polarization to zero.*

B. Parallel (Vertical) Polarization   To examine the conditions under which the reflection coefficient for parallel polarization will vanish, we set (5-24c) equal to zero; that is

$$\Gamma_\parallel^b = \frac{-\sqrt{\dfrac{\mu_1}{\varepsilon_1}}\cos\theta_i + \sqrt{\dfrac{\mu_2}{\varepsilon_2}}\cos\theta_t}{\sqrt{\dfrac{\mu_1}{\varepsilon_1}}\cos\theta_i + \sqrt{\dfrac{\mu_2}{\varepsilon_2}}\cos\theta_t} = 0 \tag{5-30}$$

or

$$\cos\theta_i = \sqrt{\frac{\mu_2}{\mu_1}\left(\frac{\varepsilon_1}{\varepsilon_2}\right)}\cos\theta_t \tag{5-30a}$$

Using Snell's law of refraction, as given by (5-24b), (5-30a) can be written as

$$\left(1 - \sin^2\theta_i\right) = \frac{\mu_2}{\mu_1}\left(\frac{\varepsilon_1}{\varepsilon_2}\right)(1 - \sin^2\theta_t)$$

$$(1 - \sin^2\theta_i) = \frac{\mu_2}{\mu_1}\left(\frac{\varepsilon_1}{\varepsilon_2}\right)\left[1 - \frac{\mu_1}{\mu_2}\left(\frac{\varepsilon_1}{\varepsilon_2}\right)\sin^2\theta_i\right] \tag{5-31}$$

or

$$\sin\theta_i = \sqrt{\frac{\dfrac{\varepsilon_2}{\varepsilon_1} - \dfrac{\mu_2}{\mu_1}}{\dfrac{\varepsilon_2}{\varepsilon_1} - \dfrac{\varepsilon_1}{\varepsilon_2}}} \tag{5-31a}$$

Since the sine function cannot exceed unity, (5-31a) exists only if

$$\left|\frac{\varepsilon_2}{\varepsilon_1} - \frac{\mu_2}{\mu_1}\right| \leq \left|\frac{\varepsilon_2}{\varepsilon_1} - \frac{\varepsilon_1}{\varepsilon_2}\right| \tag{5-32}$$

which reduces to

$$\frac{\mu_2}{\mu_1} \geq \frac{\varepsilon_1}{\varepsilon_2} \text{ for } \varepsilon_2 \geq \varepsilon_1 \quad \text{or} \quad \frac{\mu_2}{\mu_1} \leq \frac{\varepsilon_1}{\varepsilon_2} \text{ for } \varepsilon_1 \geq \varepsilon_2 \tag{5-32a}$$

If, $\mu_1 = \mu_2$, (5-31a) leads to

$$\theta_i = \theta_B = \sin^{-1}\left(\sqrt{\frac{\varepsilon_2}{\varepsilon_1 + \varepsilon_2}}\right) \tag{5-33}$$

The incident angle θ_i, as given by (5-31a) or (5-33), *which reduces the reflection coefficient for parallel polarization to zero, is referred to as the Brewster angle, θ_B.* It should be noted that *when $\mu_1 = \mu_2$, the incidence Brewster angle $\theta_i = \theta_B$ of (5-33) exists only if the polarization of the wave is parallel (vertical).*

Other forms of the Brewster angle, besides that given by (5-33), are

$$\boxed{\theta_i = \theta_{\mathrm{B}} = \cos^{-1}\left(\sqrt{\frac{\varepsilon_1}{\varepsilon_1 + \varepsilon_2}}\right)} \tag{5-33a}$$

$$\boxed{\theta_i = \theta_{\mathrm{B}} = \tan^{-1}\left(\sqrt{\frac{\varepsilon_2}{\varepsilon_1}}\right)} \tag{5-33b}$$

Example 5-2

A parallel polarized electromagnetic wave radiated from a submerged submarine impinges upon a water–air planar interface. Assuming the water is lossless, its dielectric constant is 81, and the wave approximates a plane wave at the interface, determine the angle of incidence to allow complete transmission of the energy.

Solution: The angle of incidence that allows complete transmission of the energy is the Brewster angle. Using (5-33b), the Brewster angle of the water–air interface is

$$\theta_{i\mathrm{wa}} = \theta_{\mathrm{Bwa}} = \tan^{-1}\left(\sqrt{\frac{\varepsilon_0}{81\varepsilon_0}}\right) = \tan^{-1}\left(\frac{1}{9}\right) = 6.34°$$

This indicates that the Brewster angle is close to the normal to the interface.

Example 5-3

Repeat the problem of Example 5-2 assuming that the same wave is radiated from a spacecraft in air, and it impinges upon the air–water interface.

Solution: The Brewster angle for an air–water interface is

$$\theta_{i\mathrm{aw}} = \theta_{\mathrm{Baw}} = \tan^{-1}\left(\sqrt{\frac{81\varepsilon_0}{\varepsilon_0}}\right) = \tan^{-1}(9) = 83.66°$$

It is apparent that the sum of the Brewster angle of Example 5-2 (water–air interface) plus that of Example 5-3 (air–water interface) is equal to 90°. That is

$$\theta_{\mathrm{Bwa}} + \theta_{\mathrm{Baw}} = 6.34° + 83.66° = 90°$$

From trigonometry, it is obvious that the preceding relation is always going to hold, no matter what two media form the interface.

5.3.4 Total Reflection–Critical Angle

In Section 5.3.3 we found the angles that allow total transmission for perpendicular, (5-28a), and parallel, (5-31a), polarizations. When the permeabilities of the two media forming the interface are the same ($\mu_1 = \mu_2$), only parallel polarized fields possess an incidence angle that allows total transmission. As before, that angle is known as the Brewster angle, and it is given by either (5-33), (5-33a), or (5-33b).

The next question we will consider is: "Is there an incident angle that allows total reflection of energy at a planar interface?" If this is possible, then $|\Gamma| = 1$. To determine the conditions under which this can be accomplished, we proceed in a similar manner as for the total transmission case of Section 5.3.3.

A. Perpendicular (Horizontal) Polarization
To see the conditions under which the magnitude of the reflection coefficient is equal to unity, we set the magnitude of (5-17a) equal to

$$\frac{\left|\sqrt{\frac{\mu_2}{\varepsilon_2}}\cos\theta_i - \sqrt{\frac{\mu_1}{\varepsilon_1}}\cos\theta_t\right|}{\left|\sqrt{\frac{\mu_2}{\varepsilon_2}}\cos\theta_i + \sqrt{\frac{\mu_1}{\varepsilon_1}}\cos\theta_t\right|} = 1 \tag{5-34}$$

This is satisfied provided the second term in the numerator and denominator is imaginary. Using Snell's law of refraction, as given by (5-15b), the second term in the numerator and denominator can be imaginary if

$$\cos\theta_t = \sqrt{1 - \sin^2\theta_t} = \sqrt{1 - \frac{\mu_1\varepsilon_1}{\mu_2\varepsilon_2}\sin^2\theta_i} = -j\sqrt{\frac{\mu_1\varepsilon_1}{\mu_2\varepsilon_2}\sin^2\theta_i - 1} \tag{5-35}$$

In order for (5-35) to hold

$$\frac{\mu_1\varepsilon_1}{\mu_2\varepsilon_2}\sin^2\theta_i \geq 1 \tag{5-35a}$$

or

$$\boxed{\theta_i \geq \theta_c = \sin^{-1}\left(\sqrt{\frac{\mu_2\varepsilon_2}{\mu_1\varepsilon_1}}\right)} \tag{5-35b}$$

The incident angle θ_i of (5-35b) that allows total reflection is known as the *critical angle*. Since the argument of the inverse sine function cannot exceed unity, then

$$\mu_2\varepsilon_2 \leq \mu_1\varepsilon_1 \tag{5-35c}$$

in order for the critical angle (5-35b) to be physically realizable.

If the permeabilities of the two media are the same ($\mu_1 = \mu_2$), then (5-35b) reduces to

$$\boxed{\theta_i \geq \theta_c = \sin^{-1}\left(\sqrt{\frac{\varepsilon_2}{\varepsilon_1}}\right)} \tag{5-36}$$

which leads to a physically realizable angle provided

$$\varepsilon_2 \leq \varepsilon_1 \tag{5-36a}$$

Therefore for two media with identical permeabilities (which is the case for most dielectrics, aside from ferromagnetic material), *the critical angle exists only if the wave propagates from a more dense to a less dense medium, as stated by* (5-36a).

Example 5-4

A perpendicularly polarized wave radiated from a submerged submarine impinges upon a water–air interface. Assuming the water is lossless, its dielectric constant is 81, and the wave approximates a plane wave at the interface, determine the angle of incidence that will allow complete reflection of the energy at the interface.

Solution: The angle of incidence that allows complete reflection of energy is the critical angle. Since for water $\mu_2 = \mu_0$, the critical angle is obtained using (5-36), which leads to

$$\theta_i \geq \theta_c = \sin^{-1}\left(\sqrt{\frac{\varepsilon_0}{81\varepsilon_0}}\right) = 6.38°$$

Since there is a large difference between the permittivities of the two media forming the interface, the critical angle of this example is very nearly the same as the Brewster angle of Example 5-2.

The next question we will answer is: "What happens to the angle of refraction and to the propagation of the wave when the angle of incidence is equal to or greater than the critical angle?"

When the angle of incidence is equal to the critical angle, the angle of refraction reduces, through Snell's law of refraction (5-15b) and (5-35b), to

$$\theta_t = \sin^{-1}\left(\sqrt{\frac{\mu_1\varepsilon_1}{\mu_2\varepsilon_2}}\sin\theta_i\right)\Bigg|_{\theta_i=\theta_c} = \sin^{-1}\left(\sqrt{\frac{\mu_1\varepsilon_1}{\mu_2\varepsilon_2}}\sqrt{\frac{\mu_2\varepsilon_2}{\mu_1\varepsilon_1}}\right) = \sin^{-1}(1) = 90° \tag{5-37}$$

In turn the reflection and transmission coefficients reduce to

$$\Gamma_\perp^b\big|_{\theta_i=\theta_c} = 1 \tag{5-38a}$$

$$T_\perp^b\big|_{\theta_i=\theta_c} = 2 \tag{5-38b}$$

Also the transmitted fields of (5-12a) and (5-12b) can be written as

$$\mathbf{E}_\perp^t = \hat{\mathbf{a}}_y 2E_0 e^{-j\beta_2 x} \tag{5-39a}$$

$$\mathbf{H}_\perp^t = \hat{\mathbf{a}}_z \frac{2E_0}{\eta_2} e^{-j\beta_2 x} \tag{5-39b}$$

which represent a plane wave that travels parallel to the interface in the $+x$ direction as shown in Figure 5-6a. The constant phase planes of the wave are parallel to the z axis. This wave is referred to as a *surface wave* [2].

The average power density associated with the transmitted fields is given by

$$\mathbf{S}_{av}^t\big|_{\theta_i=\theta_c} = \frac{1}{2}\mathrm{Re}\left(\mathbf{E}_\perp^t \times \mathbf{H}_\perp^{t*}\right)\Bigg|_{\theta_i=\theta_c} = \hat{\mathbf{a}}_x \frac{2|E_0|^2}{\eta_2} \tag{5-40}$$

and it does not contain any component normal to the interface. Therefore, there is no transfer of real power across the interface in a direction normal to the boundary; thus, all power must be reflected. This is also evident by examining the magnitude of the incident and reflected average power densities associated with the fields (5-10a) through (5-11d) under critical angle incidence. These are obviously identical and are given by

$$|\mathbf{S}_{av}^i|_{\theta_i=\theta_c} = \left|\frac{1}{2}\mathrm{Re}\left(\mathbf{E}_\perp^i \times \mathbf{H}_\perp^{i*}\right)\right|_{\theta_i=\theta_c} = \frac{|E_0|^2}{2\eta_i}\left|\hat{\mathbf{a}}_x \sin\theta_i + \hat{\mathbf{a}}_z \cos\theta_i\right| = \frac{|E_0|^2}{2\eta_1} \tag{5-41a}$$

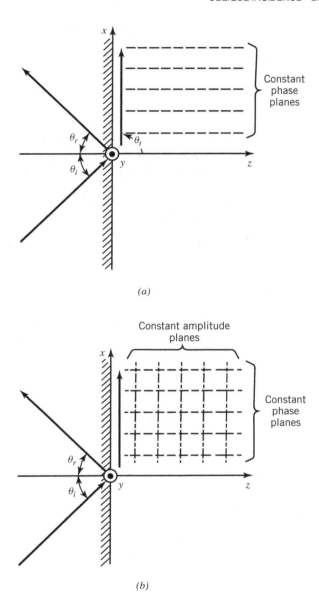

(a)

(b)

Figure 5-6   Constant phase and amplitude planes for incident angles. (a) Critical ($\theta_i = \theta_c$). (b) Above critical ($\theta_i > \theta_c$).

$$|\mathbf{S}_{av}^r|_{\theta_i=\theta_c} = \left|\frac{1}{2}\text{Re}\left(\mathbf{E}_{\perp}^r \times \mathbf{H}_{\perp}^{r*}\right)\right|_{\theta_i=\theta_c} = \frac{|E_0|^2}{2\eta_1}\left|\hat{\mathbf{a}}_x \sin\theta_i - \hat{\mathbf{a}}_z \cos\theta_i\right| = \frac{|E_0|^2}{2\eta_1} \qquad (5\text{-}41\text{b})$$

When the angle of incidence θ_i is greater than the critical angle θ_c ($\theta_i > \theta_c$), Snell's law of refraction can be written as [3]

$$\sin\theta_t\big|_{\theta_i>\theta_c} = \frac{\beta_1}{\beta_2}\sin\theta_i\bigg|_{\theta_i>\theta_c} = \sqrt{\frac{\mu_1\varepsilon_1}{\mu_2\varepsilon_2}}\sin\theta_i\bigg|_{\theta_i>\theta_c} > 1 \qquad (5\text{-}42\text{a})$$

which can only be satisfied provided θ_t is complex, that is, $\theta_t = \theta_R + j\theta_X$, where $\theta_X \neq 0$. Also

$$\cos\theta_t\big|_{\theta_i>\theta_c} = \sqrt{1-\sin^2\theta_t}\,\bigg|_{\theta_i>\theta_c} = \sqrt{1-\frac{\mu_1\varepsilon_1}{\mu_2\varepsilon_2}\sin^2\theta_i}\,\bigg|_{\theta_i>\theta_c}$$

$$= \pm j\sqrt{\frac{\mu_1\varepsilon_1}{\mu_2\varepsilon_2}\sin^2\theta_i-1}\,\bigg|_{\theta_i>\theta_c} \tag{5-42b}$$

which again indicates that θ_t is complex.

Therefore when $\theta_i>\theta_c$, there is no physically realizable angle θ_t. If not, what really does happen to the wave propagation? Since under this condition θ_t is complex and not physically realizable, this may be a clue that the wave in medium 2 is again a surface wave. To see this, let us examine the field in medium 2, the reflection and transmission coefficients, and the average power densities.

When the angle of incidence exceeds the critical angle ($\theta_i>\theta_c$), the transmitted **E** field of (5-12a) can be written, using (5-15b) and (5-35b), as

$$\mathbf{E}_\perp^t\big|_{\theta_i>\theta_c} = \hat{\mathbf{a}}_y T_\perp^b E_0 \exp(-j\beta_2 x\sin\theta_t)\exp(-j\beta_2 z\cos\theta_t)\big|_{\theta_i>\theta_c}$$

$$= \hat{\mathbf{a}}_y T_\perp^b E_0 \exp\left[-j\beta_2 x\left(\sqrt{\frac{\mu_1\varepsilon_1}{\mu_2\varepsilon_2}}\sin\theta_i\right)\right]\exp\left(-j\beta_2 z\sqrt{1-\sin^2\theta_t}\right)\bigg|_{\theta_i>\theta_c}$$

$$\mathbf{E}_\perp^t\big|_{\theta_i>\theta_c} = \hat{\mathbf{a}}_y T_\perp^b E_0 \exp\left[-j\beta_2 x\left(\sqrt{\frac{\mu_1\varepsilon_1}{\mu_2\varepsilon_2}}\sin\theta_i\right)\right]\exp\left(-j\beta_2 z\sqrt{1-\frac{\mu_1\varepsilon_1}{\mu_2\varepsilon_2}\sin^2\theta_i}\right)\bigg|_{\theta_i>\theta_c}$$

$$= \hat{\mathbf{a}}_y T_\perp^b E_0 \exp\left[-j\beta_2 x\left(\sqrt{\frac{\mu_1\varepsilon_1}{\mu_2\varepsilon_2}}\sin\theta_i\right)\right]\exp\left(-\beta_2 z\sqrt{\frac{\mu_1\varepsilon_1}{\mu_2\varepsilon_2}\sin^2\theta_i-1}\right)\bigg|_{\theta_i>\theta_c}$$

$$= \hat{\mathbf{a}}_y T_\perp^b E_0 \exp\left[-\beta_2 z\left(\sqrt{\frac{\mu_1\varepsilon_1}{\mu_2\varepsilon_2}\sin^2\theta_i-1}\right)\right]\exp\left[-j\beta_2 x\left(\sqrt{\frac{\mu_1\varepsilon_1}{\mu_2\varepsilon_2}}\sin\theta_i\right)\right]\bigg|_{\theta_i>\theta_c}$$

$$\mathbf{E}_\perp^t\big|_{\theta_i>\theta_c} = \hat{\mathbf{a}}_y T_\perp^b E_0 e^{-\alpha_e z}e^{-j\beta_e x} \tag{5-43}$$

where

$$\alpha_e = \beta_2\sqrt{\frac{\mu_1\varepsilon_1}{\mu_2\varepsilon_2}\sin^2\theta_i-1}\,\bigg|_{\theta_i>\theta_c} = \omega\sqrt{\mu_1\varepsilon_1\sin^2\theta_i-\mu_2\varepsilon_2}\,\big|_{\theta_i>\theta_c} \tag{5-43a}$$

$$\beta_e = \beta_2\sqrt{\frac{\mu_1\varepsilon_1}{\mu_2\varepsilon_2}}\sin\theta_i\,\bigg|_{\theta_i>\theta_c} = \omega\sqrt{\mu_1\varepsilon_1}\sin\theta_i\,\big|_{\theta_i>\theta_c} \tag{5-43b}$$

$$v_{pe} = \frac{\omega}{\beta_e} = \frac{\omega}{\beta_2\sqrt{\frac{\mu_1\varepsilon_1}{\mu_2\varepsilon_2}}\sin\theta_i}\,\Bigg|_{\theta_i>\theta_c} = \frac{v_{p2}}{\sqrt{\frac{\mu_1\varepsilon_1}{\mu_2\varepsilon_2}}\sin\theta_i}\,\Bigg|_{\theta_i>\theta_c} = \frac{1}{\sqrt{\mu_1\varepsilon_1}\sin\theta_i} < v_{p2} \tag{5-43c}$$

The wave associated with (5-43) also propagates parallel to the interface with constant phase planes that are parallel to the z axis, as shown in Figure 5-6b. The effective phase velocity v_{pe} of the wave is given by (5-43c), and it is less than v_{p2} of an ordinary wave in medium 2. The wave also possesses constant amplitude planes that are parallel to the x axis, as shown in Figure 5-6b. The effective attenuation constant α_e of the wave in the z direction is that given by (5-43a). Its values are such that the wave decays very rapidly, and in a few wavelengths it essentially vanishes. This wave is also a *surface wave*. Since its phase velocity is less than the speed of light, it is a *slow* surface wave. Also since it decays very rapidly in a direction normal to the interface, it is *tightly bound* to the surface—i.e., it is a *tightly bound slow surface wave*.

Phase velocities *greater* than the intrinsic phase velocity of an ordinary plane wave in a given medium can be achieved by uniform plane waves at *real oblique angles* of propagation, as illustrated in Section 4.2.2C; phase velocities *smaller* than the intrinsic velocity can only be achieved by uniform plane waves at *complex angles* of propagation. Waves traveling at complex angles are *nonuniform* plane waves oriented so as to provide small phase velocities or large rates of change of phase in a given direction. The price for such large rates of change of phase or small velocities in one direction is associated with large attenuation at perpendicular directions.

Example 5-5

Assume that $\theta_i > \theta_c$ (so the angle of refraction $\theta_t = \theta_R + j\theta_X$ is complex, i.e. $\theta_X \neq 0$). Determine the real (θ_R) and imaginary (θ_X) parts of θ_t in terms of the constitutive parameters of the two media and the angle of incidence.

Solution: Using (5-42a)

$$\sin\theta_t = \sin(\theta_R + j\theta_X) = \sqrt{\frac{\mu_1\varepsilon_1}{\mu_2\varepsilon_2}}\sin\theta_i$$

or

$$\sin(\theta_R)\cosh(\theta_X) + j\cos(\theta_R)\sinh(\theta_X) = \sqrt{\frac{\mu_1\varepsilon_1}{\mu_2\varepsilon_2}}\sin\theta_i$$

Since the right side is real, then the only solution that exists is for the imaginary part of the left side to vanish and the real part to be equal to the real part of the right side. Thus

$$\cos(\theta_R)\sinh(\theta_X) = 0 \Rightarrow \theta_R = \frac{\pi}{2}$$

$$\sin(\theta_R)\cosh(\theta_X) = \sqrt{\frac{\mu_1\varepsilon_1}{\mu_2\varepsilon_2}}\sin\theta_i \Rightarrow \theta_X = \cosh^{-1}\left(\sqrt{\frac{\mu_1\varepsilon_1}{\mu_2\varepsilon_2}}\sin\theta_i\right)$$

In turn $\cos\theta_t$ is defined as

$$\cos\theta_t = \cos(\theta_R + j\theta_X) = \cos(\theta_R)\cosh(\theta_X) - j\sin(\theta_R)\sinh(\theta_X)$$

or

$$\cos\theta_t = -j\sinh(\theta_X)$$

which again is shown to be complex as was in (5-42b). When these expressions for $\sin\theta_t$ and $\cos\theta_t$ are used to represent the fields in medium 2, it will be shown that the fields are nonuniform plane waves as illustrated by (5-43).

Under the conditions where the angle of incidence is equal to or greater than the critical angle, the reflection $\Gamma_\perp^b$ and transmission $T_\perp^b$ coefficients of (5-17a) and (5-17b) reduce, respectively, to [3]

$$\Gamma_\perp^b\big|_{\theta_i \geq \theta_c} = \left.\frac{\sqrt{\frac{\mu_2}{\varepsilon_2}}\cos\theta_i - \sqrt{\frac{\mu_1}{\varepsilon_1}}\cos\theta_t}{\sqrt{\frac{\mu_2}{\varepsilon_2}}\cos\theta_i + \sqrt{\frac{\mu_1}{\varepsilon_1}}\cos\theta_t}\right|_{\theta_i \geq \theta_c}$$

$$= \left.\frac{\sqrt{\frac{\mu_2}{\varepsilon_2}}\cos\theta_i - \sqrt{\frac{\mu_1}{\varepsilon_1}}\sqrt{1 - \sin^2\theta_t}}{\sqrt{\frac{\mu_2}{\varepsilon_2}}\cos\theta_i + \sqrt{\frac{\mu_1}{\varepsilon_1}}\sqrt{1 - \sin^2\theta_t}}\right|_{\theta_i \geq \theta_c}$$

$$\Gamma_\perp^b \big|_{\theta_i \geq \theta_c} = \frac{\sqrt{\dfrac{\mu_2}{\varepsilon_2}}\cos\theta_i - \sqrt{\dfrac{\mu_1}{\varepsilon_1}}\sqrt{1 - \dfrac{\mu_1 \varepsilon_1}{\mu_2 \varepsilon_2}\sin^2\theta_i}}{\sqrt{\dfrac{\mu_2}{\varepsilon_2}}\cos\theta_i + \sqrt{\dfrac{\mu_1}{\varepsilon_1}}\sqrt{1 - \dfrac{\mu_1 \varepsilon_1}{\mu_2 \varepsilon_2}\sin^2\theta_i}}\Bigg|_{\theta_i \geq \theta_c}$$

$$= \frac{\sqrt{\dfrac{\mu_2}{\varepsilon_2}}\cos\theta_i + j\sqrt{\dfrac{\mu_1}{\varepsilon_1}}\sqrt{\dfrac{\mu_1 \varepsilon_1}{\mu_2 \varepsilon_2}\sin^2\theta_i - 1}}{\sqrt{\dfrac{\mu_2}{\varepsilon_2}}\cos\theta_i - j\sqrt{\dfrac{\mu_1}{\varepsilon_1}}\sqrt{\dfrac{\mu_1 \varepsilon_1}{\mu_2 \varepsilon_2}\sin^2\theta_i - 1}}\Bigg|_{\theta_i \geq \theta_c}$$

$$\Gamma_\perp^b \big|_{\theta_i \geq \theta_c} = \left|\Gamma_\perp^b\right| e^{j2\psi_\perp} = e^{j2\psi_\perp} \qquad (5\text{-}44)$$

where

$$\left|\Gamma_\perp^b\right| = 1 \qquad (5\text{-}44\text{a})$$

$$\psi_\perp = \tan^{-1}\left[\frac{X_\perp}{R_\perp}\right] \qquad (5\text{-}44\text{b})$$

$$X_\perp = \sqrt{\frac{\mu_1}{\varepsilon_1}}\sqrt{\frac{\mu_1 \varepsilon_1}{\mu_2 \varepsilon_2}\sin^2\theta_i - 1} \qquad (5\text{-}44\text{c})$$

$$R_\perp = \sqrt{\frac{\mu_2}{\varepsilon_2}}\cos\theta_i \qquad (5\text{-}44\text{d})$$

$$T_\perp^b \big|_{\theta_i \geq \theta_c} = \frac{2\sqrt{\dfrac{\mu_2}{\varepsilon_2}}\cos\theta_i}{\sqrt{\dfrac{\mu_2}{\varepsilon_2}}\cos\theta_i + \sqrt{\dfrac{\mu_1}{\varepsilon_1}}\cos\theta_t}\Bigg|_{\theta_i \geq \theta_c}$$

$$= \frac{2\sqrt{\dfrac{\mu_2}{\varepsilon_2}}\cos\theta_i}{\sqrt{\dfrac{\mu_2}{\varepsilon_2}}\cos\theta_i + \sqrt{\dfrac{\mu_1}{\varepsilon_1}}\sqrt{1 - \sin^2\theta_t}}\Bigg|_{\theta_i \geq \theta_c}$$

$$= \frac{2\sqrt{\dfrac{\mu_2}{\varepsilon_2}}\cos\theta_i}{\sqrt{\dfrac{\mu_2}{\varepsilon_2}}\cos\theta_i + \sqrt{\dfrac{\mu_1}{\varepsilon_1}}\sqrt{1 - \dfrac{\mu_1 \varepsilon_1}{\mu_2 \varepsilon_2}\sin^2\theta_i}}\Bigg|_{\theta_i \geq \theta_c}$$

$$= \frac{2\sqrt{\dfrac{\mu_2}{\varepsilon_2}}\cos\theta_i}{\sqrt{\dfrac{\mu_2}{\varepsilon_2}}\cos\theta_i - j\sqrt{\dfrac{\mu_1}{\varepsilon_1}}\sqrt{\dfrac{\mu_1 \varepsilon_1}{\mu_2 \varepsilon_2}\sin^2\theta_i - 1}}\Bigg|_{\theta_i \geq \theta_c}$$

$$T_\perp^b \big|_{\theta_i \geq \theta_c} = \left|T_\perp^b\right| e^{j\psi_\perp} \qquad (5\text{-}45)$$

where

$$|T_\perp^b| = \frac{2R_\perp}{\sqrt{R_\perp^2 + X_\perp^2}} \tag{5-45a}$$

In addition, the transmitted average power density can now be written, using (5-12a) through (5-12b) and the modified forms (5-43) through (5-43b) for the fields when the incidence angle is equal to or greater than the critical angle, as

$$\mathbf{S}_{av}^t |_{\theta_i \geq \theta_c} = \frac{1}{2} \mathrm{Re}(\mathbf{E}^t \times \mathbf{H}^{t*})_{\theta_i \geq \theta_c}$$

$$= \frac{1}{2} \mathrm{Re} \left[\left(\hat{\mathbf{a}}_y T_\perp^b E_0 e^{-\alpha_e z} e^{-j\beta_e x} \right) \times (-\hat{\mathbf{a}}_x \cos\theta_t + \hat{\mathbf{a}}_z \sin\theta_t)^* \frac{(T_\perp^b)^* E_0^*}{\eta_2} e^{-\alpha_e z} e^{+j\beta_e x} \right]_{\theta_i \geq \theta_c}$$

$$= \frac{1}{2} \mathrm{Re} \left\{ \left[\hat{\mathbf{a}}_z (\cos\theta_t)^* + \hat{\mathbf{a}}_x (\sin\theta_t)^* \right] \frac{|T_\perp^b|^2 |E_0|^2}{\eta_2} e^{-2\alpha_e z} \right\}_{\theta_i \geq \theta_c}$$

$$\mathbf{S}_{av}^t |_{\theta_i \geq \theta_c} = \frac{1}{2} \mathrm{Re} \left\{ \left[\hat{\mathbf{a}}_z \left(\sqrt{1 - \sin^2\theta_t} \right)^* + \hat{\mathbf{a}}_x (\sin\theta_t)^* \right] \frac{|T_\perp^b|^2 |E_0|^2}{\eta_2} e^{-2\alpha_e z} \right\}_{\theta_i \geq \theta_c}$$

$$\mathbf{S}_{av}^t |_{\theta_i \geq \theta_c} = \frac{1}{2} \mathrm{Re} \left\{ \left[\hat{\mathbf{a}}_z \left(\sqrt{1 - \frac{\mu_1 \varepsilon_1}{\mu_2 \varepsilon_2} \sin^2\theta_i} \right)^* \right. \right.$$

$$\left. \left. + \hat{\mathbf{a}}_x \left(\sqrt{\frac{\mu_1 \varepsilon_1}{\mu_2 \varepsilon_2}} \sin\theta_i \right)^* \right] \frac{|T_\perp^b|^2 |E_0|^2}{\eta_2} e^{-2\alpha_e z} \right\}_{\theta_i \geq \theta_c}$$

$$= \frac{1}{2} \mathrm{Re} \left\{ \left[\hat{\mathbf{a}}_z \left(-j\sqrt{\frac{\mu_1 \varepsilon_1}{\mu_2 \varepsilon_2} \sin^2\theta_i - 1} \right) \right. \right.$$

$$\left. \left. + \hat{\mathbf{a}}_x \left(\sqrt{\frac{\mu_1 \varepsilon_1}{\mu_2 \varepsilon_2}} \sin\theta_i \right) \right] \frac{|T_\perp^b|^2 |E_0|^2}{\eta_2} e^{-2\alpha_e z} \right\}_{\theta_i \geq \theta_c}$$

$$\mathbf{S}_{av}^t |_{\theta_i \geq \theta_c} = \hat{\mathbf{a}}_x \sqrt{\frac{\mu_1 \varepsilon_1}{\mu_2 \varepsilon_2}} \sin\theta_i \frac{|T_\perp^b|^2 |E_0|^2}{2\eta_2} e^{-2\alpha_e z} \Bigg|_{\theta_i \geq \theta_c} \tag{5-46}$$

Again, from (5-46), it is apparent that there is no real power transfer across the interface in a direction normal to the boundary. Therefore all the power must be reflected into medium 1. This can also be verified by formulating and examining the incident and reflected average power densities. Doing this, using the fields (5-10a) through (5-11b) where the reflection coefficient is that of (5-44), shows that the magnitudes of the incident and reflected average power densities are those of (5-41a) and (5-41b), which are identical.

The propagation of a wave from a medium with higher density to one with lower density ($\varepsilon_2 < \varepsilon_1$ when $\mu_1 = \mu_2$) under oblique incidence can be summarized as follows.

1. When the angle of incidence is smaller than the critical angle ($\theta_i < \theta_c = \sin^{-1}(\sqrt{\varepsilon_2 / \varepsilon_1})$), a wave is transmitted into medium 2 at an angle θ_t, which is greater than the incident angle θ_i. Real power is transferred into medium 2, and it is directed along angle θ_t as shown in Figure 5-7a.

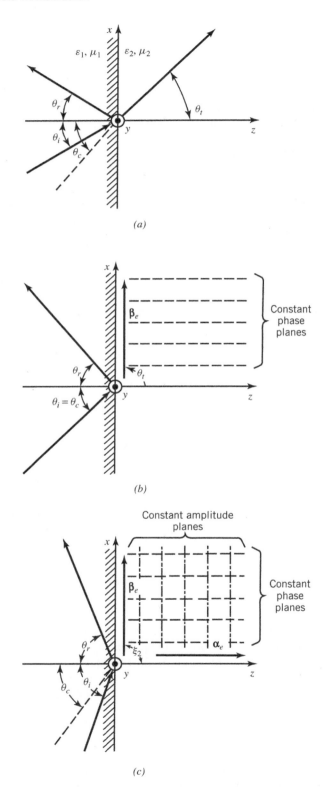

(a)

(b)

(c)

Figure 5-7   Critical angle wave propagation along an interface.

2. As the angle of incidence increases and reaches the critical angle $\theta_i = \theta_c = \sin^{-1}(\sqrt{\varepsilon_2/\varepsilon_1})$, the refracted angle θ_t, which varies more rapidly than the incident angle θ_i, approaches $90°$. Although a wave into medium 2 exists under this condition (which is necessary to satisfy the boundary conditions), the fields form a surface wave that is directed along the x axis (which is parallel to the interface). There is no real power transfer normal to the boundary into medium 2, and all the power is reflected in medium 1 along reflected angle θ_r as shown in Figure 5-7b. The constant phase planes are parallel to the z axis.

3. When the incident angle θ_i exceeds the critical angle $\theta_c \left[\theta_i > \theta_c = \sin^{-1}\left(\sqrt{\varepsilon_2/\varepsilon_1}\right)\right]$, a wave into medium 2 still exists, which travels along the x axis (which is parallel to the interface) and is heavily attenuated in the z direction (which is normal to the interface). There is no real power transfer normal to the boundary into medium 2, and all power is reflected into medium 1 along reflection angle θ_r, as shown in Figure 5-7c. Although there is no power transferred into medium 2, a wave exists there that is necessary to satisfy the boundary conditions on the continuity of the tangential components of the electric and magnetic fields. The wave in medium 2 travels parallel to the interface with a phase velocity that is less than that of an ordinary wave in the same medium [as given by (5-43c)], and it is rapidly attenuated in a direction normal to the interface with an effective attenuation constant given by (5-43a). This wave is *tightly bound* to the surface, and it is referred to as a *tightly bound slow surface wave*.

The critical angle is used to design many practical instruments and transmission lines, such as binoculars, dielectric covered ground plane (surface wave) transmission lines, fiber optic cables, etc. To see how the critical angle may be utilized, let us consider an example.

Example 5-6

Determine the range of values of the dielectric constant of a dielectric slab of thickness t so that, when a wave is incident on it from one of its ends at an oblique angle $0° \leq \theta_i \leq 90°$, the energy of the wave in the dielectric is contained within the slab. The geometry of the problem is shown in the Figure 5-8.

Solution: We assume that the slab width is infinite (two-dimensional geometry). To contain the energy of the wave within the slab, the reflection angle θ_r of the wave bouncing within the slab must be equal to or greater than the critical angle θ_c. By referring to Figure 5-8, the critical angle can be related to the refraction angle θ_t by

$$\sin\theta_r = \sin\left(\frac{\pi}{2} - \theta_t\right) = \cos\theta_t \geq \sin\theta_c = \sqrt{\frac{\varepsilon_0}{\varepsilon_r\varepsilon_0}} = \frac{1}{\sqrt{\varepsilon_r}}$$

or

$$\cos\theta_t \geq \frac{1}{\sqrt{\varepsilon_r}}$$

At the interface formed at the leading edge, Snell's law of refraction must be satisfied. That is,

$$\beta_0 \sin\theta_i = \beta_1 \sin\theta_t \Rightarrow \sin\theta_t = \frac{\beta_0}{\beta_1}\sin\theta_i = \frac{1}{\sqrt{\varepsilon_r}}\sin\theta_i$$

Using this, we can write the aforementioned $\cos\theta_t$ as

$$\cos\theta_t = \sqrt{1 - \sin^2\theta_t} = \sqrt{1 - \frac{1}{\varepsilon_r}\sin^2\theta_i} \geq \frac{1}{\sqrt{\varepsilon_r}}$$

or

$$\sqrt{1 - \frac{1}{\varepsilon_r} \sin^2 \theta_i} \geq \frac{1}{\sqrt{\varepsilon_r}}$$

Solving this leads to

$$\varepsilon_r - \sin^2 \theta_i \geq 1$$

or

$$\varepsilon_r \geq 1 + \sin^2 \theta_i$$

To accommodate all possible angles, the dielectric constant must be

$$\varepsilon_r \geq 2$$

since the smallest and largest values of θ_i, are, respectively, $0°$ and $90°$. This is achievable by many practical dielectric materials such as Teflon ($\varepsilon_r \simeq 2.1$), polystyrene ($\varepsilon_r \simeq 2.56$), and many others.

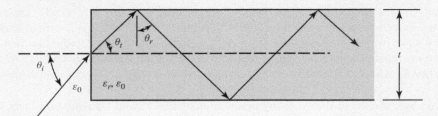

Figure 5-8   Dielectric slab of thickness t and wave containment within.

B. Parallel (Vertical) Polarization   The procedure used to derive the critical angle and to examine the properties for perpendicular (horizontal) polarization can be repeated for parallel (vertical) polarization. However, it can be shown that the critical angle is not a function of polarization, and that it exists for both parallel and perpendicular polarizations. The only limitation of the critical angle is that the wave propagation be to a less dense medium ($\mu_2 \varepsilon_2 < \mu_1 \varepsilon_1$ or $\varepsilon_2 < \varepsilon_1$ when $\mu_1 = \mu_2$).

The expression for the critical angle for parallel polarization is the same as that for perpendicular polarization as given by (5-35b) or (5-36). In addition, the wave propagation phenomena that occur for perpendicular polarization when the incidence angle is less than, equal to, or greater than the critical angle are also identical to those for parallel polarization. Although the formulas for the reflection and transmission coefficients, $\Gamma_\parallel^b$ and $T_\parallel^b$ respectively, and transmitted average power density $S_\parallel^t$ for parallel polarization are not identical to those of perpendicular polarization as given by (5-44) through (5-46), the principles stated previously are identical here. The derivation of the specific formulas for the parallel polarization for critical angle propagation are left as an end-of-chapter exercise for the reader.

5.4   LOSSY MEDIA

In the previous sections we examined wave reflection and transmission under normal and oblique wave incidence when both media forming the interface are lossless. Let us now examine the reflection and transmission of waves under normal and oblique incidence when either one or both media are lossy [4]. Although in some cases the formulas will be the same as for the lossless cases, there are differences, especially under oblique wave incidence.

5.4.1 Normal Incidence: Conductor–Conductor Interface

When a uniform plane wave is normally incident upon a planar interface formed by two lossy media (as shown in Figure 5-1 but allowing for losses in both media through the conductivity σ), the incident, reflected, and transmitted fields, reflection and transmission coefficients, and average power densities are identical to (5-1a) through (5-6c) except that (a) an attenuation constant must be included in each field and (b) the intrinsic impedances, and attenuation and phases constants must be modified to include the conductivities of the media. Thus we can summarize the results here as

$$\mathbf{E}^i = \hat{\mathbf{a}}_x E_0 e^{-\alpha_1 z} e^{-j\beta_1 z} \tag{5-47a}$$

$$\mathbf{H}^i = \hat{\mathbf{a}}_y \frac{E_0}{\eta_1} e^{-\alpha_1 z} e^{-j\beta_1 z} \tag{5-47b}$$

$$\mathbf{E}^r = \hat{\mathbf{a}}_x \Gamma^b E_0 e^{+\alpha_1 z} e^{+j\beta_1 z} \tag{5-48a}$$

$$\mathbf{H}^r = -\hat{\mathbf{a}}_y \frac{\Gamma^b E_0}{\eta_1} e^{+\alpha_1 z} e^{+j\beta_1 z} \tag{5-48b}$$

$$\mathbf{E}^t = \hat{\mathbf{a}}_x T^b E_0 e^{-\alpha_2 z} e^{-j\beta_2 z} \tag{5-49a}$$

$$\mathbf{H}^t = \hat{\mathbf{a}}_y \frac{T^b E_0}{\eta_2} e^{-\alpha_2 z} e^{-j\beta_2 z} \tag{5-49b}$$

$$\Gamma^b = \frac{\eta_2 - \eta_1}{\eta_2 + \eta_1} \tag{5-50a}$$

$$T^b = \frac{2\eta_2}{\eta_2 + \eta_1} \tag{5-50b}$$

$$\mathbf{S}^i_{av} = \hat{\mathbf{a}}_z \frac{|E_0|^2}{2} e^{-2\alpha_1 z} \operatorname{Re}\left(\frac{1}{\eta_1^*}\right) \tag{5-51a}$$

$$\mathbf{S}^r_{av} = -\hat{\mathbf{a}}_z |\Gamma^b|^2 \frac{|E_0|^2}{2} e^{+2\alpha_1 z} \operatorname{Re}\left(\frac{1}{\eta_1^*}\right) \tag{5-51b}$$

$$\mathbf{S}^t_{av} = \hat{\mathbf{a}}_z |T^b|^2 \frac{|E_0|^2}{2} e^{-2\alpha_2 z} \operatorname{Re}\left(\frac{1}{\eta_2^*}\right) \tag{5-51c}$$

For each lossy medium the attenuation constants α_i, phase constants β_i, and intrinsic impedances η_i are related to the corresponding constitutive parameters ε_i, μ_i, and σ_i, by the expressions in Table 4-1.

The total electric and magnetic fields in medium 1 can be written as

$$\mathbf{E}^1 = \mathbf{E}^i + \mathbf{E}^r = \hat{\mathbf{a}}_x \underbrace{E_0 e^{-\alpha_1 z} e^{-j\beta_1 z}}_{\text{traveling wave}} \underbrace{(1 + \Gamma^b e^{+2\alpha_1 z} e^{+j2\beta_1 z})}_{\text{standing wave}} \tag{5-52a}$$

$$\mathbf{H}^1 = \mathbf{H}^i + \mathbf{H}^r = \hat{\mathbf{a}}_y \underbrace{(E_0/\eta_1) e^{-\alpha_1 z} e^{-j\beta_1 z}}_{\text{traveling wave}} \underbrace{(1 - \Gamma^b e^{+2\alpha_1 z} e^{+j2\beta_1 z})}_{\text{standing wave}} \tag{5-52b}$$

In each field the factors outside the parentheses form the *traveling wave part* of the total wave; those within the parentheses form the *standing wave part*.

Example 5-7

A uniform plane wave, whose incident electric field has an x component with an amplitude at the interface of 10^{-3} V/m, is traveling in a free-space medium and is normally incident upon a lossy flat earth as shown in Figure 5-9. Assuming that the constitutive parameters of the earth are $\varepsilon_2 = 9\varepsilon_0$, $\mu_2 = \mu_0$, and $\sigma_2 = 10^{-1}$ S/m, determine the variation of the conduction current density in the earth at a frequency of 1 MHz.

Solution: At $f = 10^6$ Hz

$$\frac{\sigma_2}{\omega\varepsilon_2} = \frac{10^{-1}}{2\pi \times 10^6 (9 \times 10^{-9}/36\pi)} = 2 \times 10^2 \gg 1$$

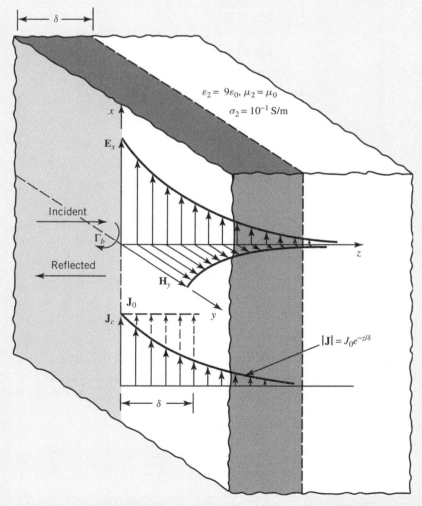

Figure 5-9 Electric and magnetic field intensities, and electric current density distributions in a lossy earth.

which classifies the material as a very good conductor.

On either side of the interface, the total electric field is equal to

$$\mathbf{E}^{\text{total}}\,|_{z=0} = \hat{\mathbf{a}}_x \times 10^{-3}\left|1 + \Gamma^b\right|$$

where

$$\Gamma^b = \frac{\eta_2 - \eta_1}{\eta_2 + \eta_1} = \frac{\eta_2 - \eta_0}{\eta_2 + \eta_0}$$

$$\eta_2 \simeq \sqrt{\frac{\omega\mu}{2\sigma}}(1+j) = \sqrt{\frac{2\pi \times 10^6 (4\pi \times 10^{-7})}{2 \times 10^{-1}}}(1+j) = 2\pi(1+j)$$

Thus

$$\Gamma^b = \frac{2\pi(1+j) - 377}{2\pi(1+j) + 377} = \frac{-370.72 + j2\pi}{383.28 + j2\pi}$$

$$= \frac{370.77\,\underline{/179.04°}}{383.33\,\underline{/0.94°}} = 0.967\,\underline{/178.1°}$$

and

$$\mathbf{E}^{\text{total}}\,|_{z=0} = \hat{\mathbf{a}}_x \times 10^{-3}\,|1 + 0.967\,\underline{/178.1°}\,|$$

$$= \hat{\mathbf{a}}_x \times 10^{-3}|0.0335 + j0.0321| = \hat{\mathbf{a}}_x(4.64 \times 10^{-5})$$

The conduction current density at the surface of the earth is equal to

$$\mathbf{J}_c|_{z=0} = \hat{\mathbf{a}}_x J_0 = \hat{\mathbf{a}}_x \sigma E^{\text{total}}\,|_{z=0} = \hat{\mathbf{a}}_x \times 10^{-1}(4.64 \times 10^{-5})$$

$$= \hat{\mathbf{a}}_x(4.64 \times 10^{-6})$$

or

$$J_0 = 4.64\ \mu\text{A/m}^2$$

The magnitude of the current density varies inside the earth as

$$|J_c| = J_0 \left|e^{-\alpha_2 z}e^{-j\beta_2 z}\right| = J_0 e^{-\alpha_2 z} = J_0 e^{-z/\delta_2}$$

where

$$\delta_2 = \text{skin depth} = \sqrt{\frac{2}{\omega\mu_2\sigma_2}} = \sqrt{\frac{2}{2\pi \times 10^6 (4\pi \times 10^{-7}) \times 10^{-1}}}$$

$$= \frac{10}{2\pi} = 1.5915\ \text{m}$$

The magnitude variations of the current density inside the earth are shown in Figure 5-9 and they exhibit an exponential decay. At one skin depth ($z = \delta_2 = 1.5915$ m), the current density has been reduced to

$$|\mathbf{J}_c|_{z=\delta_2} = J_0 e^{-1} = 0.3679 J_0 = 0.3679(4.64 \times 10^{-6}) = 1.707\ \mu\text{A/m}^2$$

Therefore, at one skin depth the current is reduced to 36.79% of its value at the surface.

If the area under the current density curve is found, it is shown to be equal to

$$J_s = \int_0^\infty |J_c|\,dz = \int_0^\infty J_0 e^{-z/\delta_2}\,dz = -\delta_2 J_0 e^{-z/\delta_2}\Big|_0^\infty = \delta_2 J_0$$

The same answer can be obtained by assuming that the current density maintains a constant surface value J_0 to a depth equal to the skin depth and equal to zero thereafter, as shown by the dashed curve in Figure 5-9.

The area under the curve can then be interpreted as the total current density J_s (A/m) per unit width in the y direction. It can be obtained by finding the area formed by maintaining constant surface current density J_0 (A/m^2) through a depth equal to the skin depth.

5.4.2   Oblique Incidence: Dielectric–Conductor Interface

Let us assume that a uniform plane wave is obliquely incident upon a planar interface where medium 1 is a perfect dielectric and medium 2 is lossy, as shown in Figure 5-10 [3]. For either the perpendicular or parallel polarization, the transmitted electric field into medium 2 can be written, using modified forms of either (5-12a) or (5-22a), as

$$\mathbf{E}^t = \mathbf{E}_2 \exp\left[-\gamma_2(x \sin\theta_t + z \cos\theta_t)\right] = \mathbf{E}_2 \exp\left[-(\alpha_2 + j\beta_2)(x \sin\theta_t + z \cos\theta_t)\right] \qquad (5\text{-}53)$$

It can be shown that for lossy media, Snell's law of refraction can be written as

$$\gamma_1 \sin\theta_i = \gamma_2 \sin\theta_t \qquad (5\text{-}54)$$

Therefore, for the geometry of Figure 5-10,

$$\sin\theta_t = \frac{\gamma_1}{\gamma_2}\sin\theta_i = \frac{j\beta_1}{\alpha_2 + j\beta_2}\sin\theta_i \qquad (5\text{-}55a)$$

and

$$\cos\theta_t = \sqrt{1 - \sin^2\theta_t} = \sqrt{1 - \left(\frac{j\beta_1}{\alpha_2 + j\beta_2}\right)^2 \sin^2\theta_i} = s\,e^{j\zeta} = s(\cos\zeta + j\sin\zeta) \qquad (5\text{-}55b)$$

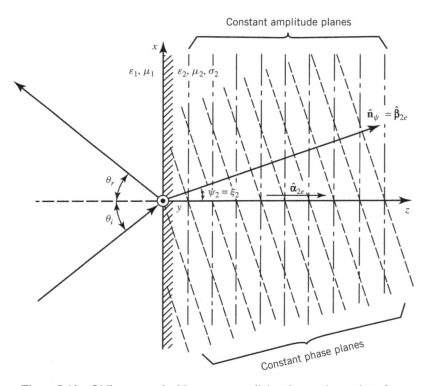

Figure 5-10   Oblique wave incidence upon a dielectric–conductor interface.

Using (5-55a) and (5-55b) we can write (5-53) as

$$\mathbf{E}^t = \mathbf{E}_2 \exp\left\{-(\alpha_2 + j\beta_2)\left[x\frac{j\beta_1}{\alpha_2 + j\beta_2}\sin\theta_i + zs(\cos\zeta + j\sin\zeta)\right]\right\} \tag{5-56}$$

which reduces to

$$\mathbf{E}^t = \mathbf{E}_2 \exp\left[-zs(\alpha_2\cos\zeta - \beta_2\sin\zeta)\right]$$
$$\times \exp\left\{-j\left[\beta_1 x\sin\theta_1 + zs(\alpha_2\sin\zeta + \beta_2\cos\zeta)\right]\right\}$$
$$\mathbf{E}^t = \mathbf{E}_2 e^{-zp}\exp\left[-j(\beta_1 x\sin\theta_i + zq)\right] \tag{5-57}$$

where

$$p = s(\alpha_2\cos\zeta - \beta_2\sin\zeta) = \alpha_{2e} \tag{5-57a}$$

$$q = s(\alpha_2\sin\zeta + \beta_2\cos\zeta) \tag{5-57b}$$

It is apparent that (5-57) represents a nonuniform wave.
    The instantaneous field of (5-57) can be written, assuming $\mathbf{E}_2$ is real, as

$$\mathscr{E}^t = \operatorname{Re}(\mathbf{E}^t e^{j\omega t}) = \mathbf{E}_2 e^{-zp}\operatorname{Re}\left(\exp\left\{j\left[\omega t - (\beta_1 x\sin\theta_i + zq)\right]\right\}\right)$$
$$\mathscr{E}^t = \mathbf{E}_2 e^{-zp}\cos\left[\omega t - (\beta_1 x\sin\theta_i + zq)\right] \tag{5-58}$$

The constant amplitude planes ($z = $ constant) of (5-58) are parallel to the interface, and they are shown dashed-dotted in Figure 5-10. The constant phase planes $\left[\omega t - (\beta_1 x\sin\theta_i + zq) = \text{constant}\right]$ are inclined at an angle ψ_2 that is no longer θ_t, and they are indicated by the dashed lines in Figure 5-10.
    To determine the constant phase we write the argument of the exponential or of the cosine function in (5-58) as

$$\omega t - (\beta_1 x\sin\theta_i + zq) = \omega t - \sqrt{(\beta_1\sin\theta_i)^2 + q^2}$$
$$\times\left[\frac{(\beta_1\sin\theta_i)x}{\sqrt{(\beta_1\sin\theta_i)^2 + q^2}} + \frac{qz}{\sqrt{(\beta_1\sin\theta_i)^2 + q^2}}\right] \tag{5-59}$$

If we define an angle ψ_2 such that

$$u = \beta_1\sin\theta_i \tag{5-60a}$$

$$\sin\psi_2 = \frac{\beta_1\sin\theta_i}{\sqrt{(\beta_1\sin\theta_i)^2 + q^2}} = \frac{u}{\sqrt{u^2 + q^2}} \tag{5-60b}$$

$$\cos\psi_2 = \frac{q}{\sqrt{(\beta_1\sin\theta_i)^2 + q^2}} = \frac{q}{\sqrt{u^2 + q^2}} \tag{5-60c}$$

or

$$\psi_2 = \tan^{-1}\left(\frac{\beta_1\sin\theta_i}{q}\right) = \tan^{-1}\left(\frac{u}{q}\right) \tag{5-60d}$$

we can write (5-59), and in turn (5-58), as

$$\mathscr{E}^t = \mathbf{E}_2 e^{-zp} \operatorname{Re}\left(\exp\left\{j\left[\omega t - \sqrt{u^2+q^2}\left(\frac{ux}{\sqrt{u^2+q^2}} + \frac{qz}{\sqrt{u^2+q^2}}\right)\right]\right\}\right)$$

$$= \mathbf{E}_2 e^{-zp} \operatorname{Re}\left(\exp\left\{j\left[\omega t - \beta_{2e}(x\sin\psi_2 + z\cos\psi_2)\right]\right\}\right)$$

$$\mathscr{E}^t = \mathbf{E}_2 e^{-zp} \operatorname{Re}\left(\exp\left\{j\left[\omega t - \beta_{2e}(\hat{\mathbf{n}}_\psi \cdot \mathbf{r})\right]\right\}\right) \tag{5-61}$$

where

$$\hat{\mathbf{n}}_\psi = \hat{\mathbf{a}}_x \sin\psi_2 + \hat{\mathbf{a}}_z \cos\psi_2 \tag{5-61a}$$

$$\beta_{2e} = \sqrt{u^2+q^2} \tag{5-61b}$$

It is apparent from (5-60a) through (5-61a) that

1. The true angle of refraction is ψ_2 and not θ_t (θ_t is complex).
2. The wave travels along a direction defined by unit vector $\hat{\mathbf{n}}_\psi$.
3. The constant phase planes are perpendicular to unit vector $\hat{\mathbf{n}}_\psi$, and they are shown as dashed lines in Figure 5-10.

The phase velocity of the wave in medium 2 is obtained by setting the exponent of (5-61) to a constant and differentiating it with respect to time. Doing this, we can write the phase velocity v_p of the wave as

$$\omega(1) - \sqrt{u^2+q^2}\left(\hat{\mathbf{n}}_\psi \cdot \frac{d\mathbf{r}}{dt}\right) = 0$$

$$\omega(1) - \sqrt{u^2+q^2}\left(\hat{\mathbf{n}}_\psi \cdot \frac{d\mathbf{r}}{dt}\right) = \omega - \beta_{2e}(\hat{\mathbf{n}}_\psi \cdot \mathbf{v}_p) = 0 \tag{5-62}$$

or

$$v_{pr} = \frac{\omega}{\beta_{2e}} = \frac{\omega}{\sqrt{u^2+q^2}} = \frac{\omega}{\sqrt{(\beta_1 \sin\theta_i)^2 + q^2}} \tag{5-62a}$$

It is evident that the phase velocity is a function of the incidence angle θ_i and the constitutive parameters of the two media.

Example 5-8

A plane wave of either perpendicular or parallel polarization traveling in air is obliquely incident upon a planar interface of copper $\left(\sigma = 5.76\times10^7 \text{ S/m}\right)$. At a frequency of 10 GHz, determine the angle of refraction and reflection coefficients for each of the two polarizations.

Solution: For copper

$$\frac{\sigma_2}{\omega\varepsilon_2} = \frac{5.8\times10^7(36\pi)}{(2\pi\times10^{10})\times10^{-9}} = 1.037\times10^8 \gg 1$$

Therefore according to Table 4-1

$$\alpha_2 \simeq \beta_2 \simeq \sqrt{\frac{\omega \mu_2 \sigma_2}{2}}$$

Using (5-55a)

$$\sin \theta_t = \frac{j\beta_1}{\alpha_2 + j\beta_2} \sin \theta_i \simeq \frac{j\beta_1}{\sqrt{\frac{\omega \mu_2 \sigma_2}{2}}(1+j)} \sin \theta_i \overset{\sigma_2 \gg 1}{\simeq} 0 \Rightarrow \theta_t \simeq 0$$

Therefore (5-55b), (5-57a), and (5-57b) reduce to

$$\cos \theta_t = 1 = s e^{j\zeta} \Rightarrow s = 1 \quad \zeta = 0$$

$$p = s\,(\alpha_2 \cos \zeta - \beta_2 \sin \zeta) \simeq \alpha_2 = \sqrt{\frac{\omega \mu_2 \sigma_2}{2}}$$

$$q = s\,(\alpha_2 \sin \zeta + \beta_2 \cos \zeta) \simeq \beta_2 = \sqrt{\frac{\omega \mu_2 \sigma_2}{2}}$$

Using (5-60d), the true angle of refraction is

$$\psi_2 = \tan^{-1}\left(\frac{u}{q}\right) \simeq \tan^{-1}\left(\frac{\beta_1 \sin \theta_i}{\beta_2}\right) = \tan^{-1}\left(\frac{\omega \sqrt{\mu_0 \varepsilon_0}}{\sqrt{\frac{\omega \mu_0 \sigma_2}{2}}} \sin \theta_i\right)$$

$$= \tan^{-1}\left(\sqrt{\frac{2\omega \varepsilon_0}{\sigma_2}} \sin \theta_i\right) \leq \tan^{-1}\left(\sqrt{\frac{2\omega \varepsilon_0}{\sigma_2}}\right) = \tan^{-1}(0.139 \times 10^{-3})$$

$$\psi_2 = \tan^{-1}(0.139 \times 10^{-3} \sin \theta_i) \leq 0.139 \times 10^{-3}\,\text{rad} = (7.96 \times 10^{-3})^\circ$$

Using (5-17a) and (5-24c), the reflection coefficients for perpendicular and parallel polarizations reduce to

$$\Gamma_\perp^b = \frac{\eta_2 \cos \theta_i - \eta_1 \cos \theta_t}{\eta_2 \cos \theta_i + \eta_1 \cos \theta_t} \simeq \frac{\eta_2 \cos \theta_i - \eta_1}{\eta_2 \cos \theta_i + \eta_1} = \frac{\cos \theta_i - \eta_1/\eta_2}{\cos \theta_i + \eta_1/\eta_2}$$

$$\Gamma_\parallel^b = \frac{-\eta_1 \cos \theta_i + \eta_2 \cos \theta_t}{\eta_1 \cos \theta_i + \eta_2 \cos \theta_t} \simeq \frac{-\eta_1 \cos \theta_i + \eta_2}{\eta_1 \cos \theta_i + \eta_2} = \frac{-\cos \theta_i + \eta_2/\eta_1}{\cos \theta_i + \eta_2/\eta_1}$$

Since

$$\frac{\eta_1}{\eta_2} = \frac{\sqrt{\frac{\mu_1}{\varepsilon_1}}}{\sqrt{\frac{j\omega \mu_2}{\sigma_2 + j\omega \varepsilon_2}}} \simeq \frac{\sqrt{\frac{\mu_0}{\varepsilon_0}}}{\sqrt{\frac{j\omega \mu_0}{\sigma_2}}} = \sqrt{\frac{\sigma_2}{j\omega \varepsilon_0}}$$

$$\frac{\eta_1}{\eta_2} \simeq 1.02 \times 10^4 e^{-j\pi/4} \gg 1 \geq \cos \theta_i$$

Then

$$\Gamma_\perp^b \simeq \frac{\cos \theta_i - \eta_1/\eta_2}{\cos \theta_i + \eta_1/\eta_2} \simeq -1$$

$$\Gamma_\parallel^b \simeq \frac{-\cos \theta_i + \eta_2/\eta_1}{\cos \theta_i + \eta_2/\eta_1} \simeq -1$$

Thus for a very good conductor, such as copper, the angle of refraction approaches zero and the magnitude of the reflection coefficients for perpendicular and parallel polarizations approach unity, and they are all essentially independent of the angle of incidence. The same will be true for all other good conductors.

5.4.3   Oblique Incidence: Conductor–Conductor Interface

In Section 5.3.4 it was shown that when a uniform plane wave is incident upon a dielectric–dielectric planar interface at an incidence angle θ_i equal to or greater than the critical angle θ_c, the transmitted wave produced into medium 2 is a nonuniform plane wave. For this plane wave, the constant amplitude planes (which are perpendicular to the α_{2e} vector) of Figure 5-7 are perpendicular to the constant phase planes (which are perpendicular to the β_{2e} vector), or the angle ξ_2 between the α_{2e} and β_{2e} vectors is 90°.

In Section 5.4.2 it was demonstrated that a uniform plane wave traveling in a lossless medium and obliquely incident upon a lossy medium also produces a nonuniform plane wave where the angle ξ_2 between the α_{2e} and β_{2e} vectors in Figure 5-10 is greater than 0° but less than 90°. In fact, for a very good conductor the angle ξ_2 between α_{2e} and β_{2e} is almost zero [for copper with $\sigma = 5.76 \times 10^7$ S/m, $\xi_2 \leq (8 \times 10^{-3})°$]. As the conducting medium becomes less lossy, the angle ξ_2 increases and in the limit it approaches 90° for a lossless medium. In fact *for all lossless media, the angle between the effective attenuation constant α_{2e} and phase constant β_{2e} should always be 90°, with reactive power flowing along α_{2e} and positive real power along β_{2e}* [4]. This is necessary since there are no real losses associated with the wave propagation along β_{2e}. This was well illustrated in Section 5.3.4 for the nonuniform wave produced in a lossless medium when the incidence angle was equal to or greater than the critical angle.

It is very interesting to investigate the field characteristics of uniform or nonuniform plane waves that are obliquely incident upon interfaces comprised of lossy–lossy interfaces. These types of waves have been examined [5–6], but, because of the general complexity of the formulations, they will not be repeated here. The reader is referred to the literature. An excellent discussion of uniform and nonuniform plane waves propagating in lossless and lossy media and associated interfaces is found in Chapters 7 and 8 of [4].

5.5   REFLECTION AND TRANSMISSION OF MULTIPLE INTERFACES

Many applications require dielectric interfaces that exhibit specific characteristics as a function of frequency. Accomplishing this often requires multiple interfaces. The objective of this section is to analyze the characteristics of multiple layer interfaces. To reduce the complexity of the problem, we will consider only normal incidence and restrict most of our attention to lossless media. A general formulation for lossy media will also be stated.

5.5.1   Reflection Coefficient of a Single Slab Layer

Section 5.2 showed that for normal incidence the reflection coefficient Γ^b at the boundary of a single planar interface is given by (5-4a) or

$$\Gamma^b = \frac{\eta_2 - \eta_1}{\eta_2 + \eta_1} \tag{5-63}$$

and at a distance $z = -\ell$ from the boundary it is given by (5-5a) or

$$\Gamma_{\text{in}}(z = -\ell) = \Gamma^b e^{-j2\beta_1 \ell} \tag{5-64}$$

Just to the right of the boundary the input impedance in the $+z$ direction is equal to the intrinsic impedance η_2 of medium 2, that is,

$$Z_{\text{in}}(z = 0^+) = \eta_2 = \sqrt{\frac{\mu_2}{\varepsilon_2}} \tag{5-65}$$

The input impedance at $z = -\ell$ can be found by using the field expressions (5-1a) through (5-2c). By definition $Z_{in}(z = -\ell)$ is equal to

$$Z_{in}\big|_{z=-\ell} = \frac{E^{total}\big|_{z=-\ell}}{H^{total}\big|_{z=-\ell}} \tag{5-66}$$

where

$$E^{total}\big|_{z=-\ell} = (E^i + E^r)\big|_{z=-\ell} = E_0 e^{+j\beta_1\ell}(1 + \Gamma^b e^{-j2\beta_1\ell}) = E_0 e^{+j\beta_1\ell}[1 + \Gamma_{in}(\ell)] \tag{5-66a}$$

$$H^{total}\big|_{z=-\ell} = (H^i - H^r)\big|_{z=-\ell} = \frac{E_0}{\eta_1} e^{+j\beta_1\ell}(1 - \Gamma^b e^{-j2\beta_1\ell}) = \frac{E_0}{\eta_1} e^{+j\beta_1\ell}[1 - \Gamma_{in}(\ell)] \tag{5-66b}$$

Therefore

$$Z_{in}\big|_{z=-\ell} = \eta_1\left(\frac{1 + \Gamma^b e^{-j2\beta_1\ell}}{1 - \Gamma^b e^{-j2\beta_1\ell}}\right) = \eta_1\left(\frac{1 + \Gamma_{in}(\ell)}{1 - \Gamma_{in}(\ell)}\right) \tag{5-66c}$$

which by using (5-63) can also be written as

$$\boxed{Z_{in}\big|_{z=-\ell} = \eta_1\left(\frac{1 + \Gamma^b e^{-j2\beta_1\ell}}{1 - \Gamma^b e^{-j2\beta_1\ell}}\right) = \eta_1\left(\frac{1 + \Gamma_{in}(\ell)}{1 - \Gamma_{in}(\ell)}\right) = \eta_1\left(\frac{\eta_2 + j\eta_1\tan(\beta_1\ell)}{\eta_1 + j\eta_2\tan(\beta_1\ell)}\right)} \tag{5-66d}$$

Equation (5-66d) is analogous to the well-known impedance transfer equation that is widely used in transmission line theory [7].

Using the foregoing procedure for normal wave incidence, we can derive expressions for multiple layer interfaces [8]. Referring to Figure 5-11a the input impedance at $z = 0^+$ is equal to the intrinsic impedance η_3 of medium 3, that is

$$Z_{in}(z = 0^+) = \eta_3 \tag{5-67}$$

In turn, the input reflection coefficient at the same interface can be written as

$$\Gamma_{in}(z = 0^-) = \frac{Z_{in}(0^+) - \eta_2}{Z_{in}(0^+) + \eta_2} = \frac{\eta_3 - \eta_2}{\eta_3 + \eta_2} \tag{5-67a}$$

At $z = -d^+$ the input impedance can be written using (5-66d) as

$$Z_{in}(z = -d^+) = \eta_2\left(\frac{1 + \Gamma_{in}(z = 0^-)e^{-j2\beta_2 d}}{1 - \Gamma_{in}(z = 0^-)e^{-j2\beta_2 d}}\right) = \eta_2\left(\frac{(\eta_3 + \eta_2) + (\eta_3 - \eta_2)e^{-j2\beta_2 d}}{(\eta_3 + \eta_2) - (\eta_3 - \eta_2)e^{-j2\beta_2 d}}\right) \tag{5-67b}$$

and the input reflection coefficient at $z = -d^-$ can be expressed as

$$\Gamma_{in}(z = -d^-) = \frac{Z_{in}(z = -d^+) - \eta_1}{Z_{in}(z = -d^+) + \eta_1}$$

$$= \frac{\eta_2\left[(\eta_3 + \eta_2) + (\eta_3 - \eta_2)e^{-j2\beta_2 d}\right] - \eta_1\left[(\eta_3 + \eta_2) - (\eta_3 - \eta_2)e^{-j2\beta_2 d}\right]}{\eta_2\left[(\eta_3 + \eta_2) + (\eta_3 - \eta_2)e^{-j2\beta_2 d}\right] + \eta_1\left[(\eta_3 + \eta_2) - (\eta_3 - \eta_2)e^{-j2\beta_2 d}\right]} \tag{5-67c}$$

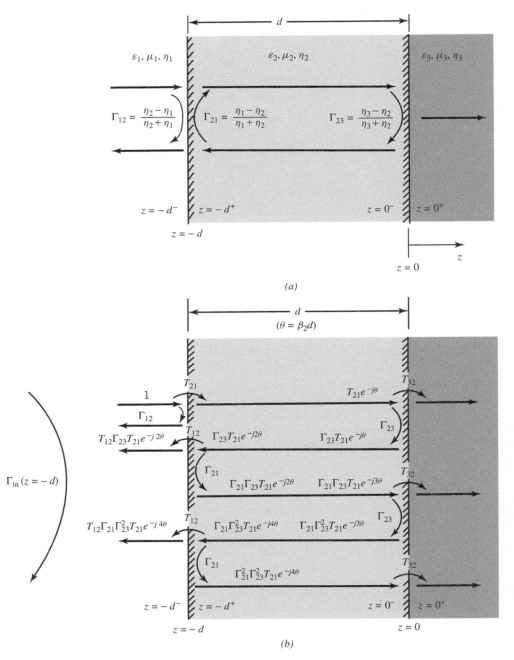

Figure 5-11 Impedances and reflection and transmission coefficients for wave propagation in dielectric slab. (*a*) Dielectric slab. (*b*) Reflection and transmission coefficients.

In Figure 5-11*a* we have defined individual reflection coefficients at each of the boundaries. Here these coefficients are referred to as *intrinsic* reflection coefficients, and they would exist at each boundary if two semi-infinite media form each of the boundaries (neglecting the presence of the other boundaries). Using the intrinsic reflection coefficients defined in Figure 5-11*a*, the input reflection coefficient of (5-67c) can also be written as

$$\boxed{\Gamma_{in}(z=-d^-) = \frac{\Gamma_{12}+\Gamma_{23}e^{-j2\beta_2 d}}{1+\Gamma_{12}\Gamma_{23}e^{-j2\beta_2 d}}} \tag{5-67d}$$

Equation (5-67d) can also be derived using the ray-tracing model of Figure 5-11b. At the leading interface of Figure 5-11b, Γ_{12} represents the intrinsic reflection coefficient of the initial reflection and $T_{12}\Gamma_{23}T_{21}e^{-j2\theta}$, etc., are the contributions to the input reflection due to the multiple bounces within the medium 2 slab. The total input reflection coefficient can be written as a geometric series that takes the form

$$\Gamma_{in}(z=-d^-) = \Gamma_{12}+T_{12}\Gamma_{23}T_{21}e^{-j2\theta}+T_{12}\Gamma_{21}\Gamma_{23}^2 T_{21}e^{-j4\theta}+\cdots$$

$$\Gamma_{in}(z=-d^-) = \Gamma_{12}+T_{12}\Gamma_{23}T_{21}e^{-j2\theta}\left[1+\Gamma_{21}\Gamma_{23}e^{-j2\theta}+\left(\Gamma_{21}\Gamma_{23}e^{-j2\theta}\right)^2+\cdots\right]$$

$$\Gamma_{in}(z=-d^-) = \Gamma_{12}+\frac{T_{12}T_{21}\Gamma_{23}e^{-j2\theta}}{1-\Gamma_{21}\Gamma_{23}e^{-j2\theta}} \tag{5-68}$$

where

$$\theta = \beta_2 d \tag{5-68a}$$

Since according to (5-4a) and (5-4b)

$$\Gamma_{21} = -\Gamma_{12} \tag{5-69a}$$

$$T_{12} = 1+\Gamma_{21} = 1-\Gamma_{12} \tag{5-69b}$$

$$T_{21} = 1+\Gamma_{12} \tag{5-69c}$$

(5-68) can be rewritten and reduced to the form of (5-67d).

If the magnitudes of the intrinsic reflection coefficients $|\Gamma_{12}|$ and $|\Gamma_{23}|$ are low compared to unity, (5-67d) can be approximated by the numerator

$$\Gamma_{in}(z=-d^-) = \frac{\Gamma_{12}+\Gamma_{23}e^{-j2\beta_2 d}}{1+\Gamma_{12}\Gamma_{23}e^{-j2\beta_2 d}} \underset{\substack{|\Gamma_{12}|\ll 1 \\ |\Gamma_{23}|\ll 1}}{\simeq} \Gamma_{12}+\Gamma_{23}e^{-j2\beta_2 d} \tag{5-70}$$

The approximate form of (5-70) yields good results if the individual intrinsic reflection coefficients are low. Typically when $|\Gamma_{12}|=|\Gamma_{23}|\leq 0.2$, the error of the approximate form of (5-70) is equal to or less than about 4 percent. The approximate form of (5-70) will be very convenient for representing the input reflection coefficient of multiple interfaces (>2) when the individual intrinsic reflection coefficients at each interface are low compared to unity.

Example 5-9

A uniform plane wave at a frequency of 10 GHz is incident normally on a dielectric slab of thickness d and bounded on both sides by air. Assume that the dielectric constant of the slab is 2.56.

1. Determine the thickness of the slab so that the input reflection coefficient at 10 GHz is zero.
2. Plot the magnitude of the reflection coefficient as a function of frequency between $5\,\text{GHz}\leq f \leq 15\,\text{GHz}$ when the dielectric slab has a thickness of 0.9375 cm.

Solution:

1. For the input reflection coefficient to be equal to zero, the reflection coefficient of (5-70) must be set equal to zero. This can be accomplished if

$$\left| \Gamma_{12} + \Gamma_{23} e^{-j2\beta_2 d} \right| = 0$$

Since

$$\Gamma_{23} = -\Gamma_{12} = \frac{\eta_1 - \eta_2}{\eta_1 + \eta_2}$$

then

$$|\Gamma_{12}|\,|1 - e^{-j2\beta_2 d}| = 0 \Rightarrow 2\beta_2 d = 2n\pi, \quad n = 0, 1, 2, \dots$$

For nontrivial solutions, the thickness must be

$$d = \frac{n\pi}{\beta_2} = \frac{n}{2}\lambda_2, \quad n = 1, 2, 3, \dots$$

where λ_2 is the wavelength inside the dielectric slab. Thus the thickness of the slab must be an integral number of half wavelengths inside the dielectric. At a frequency of 10 GHz and a dielectric constant of 2.56, the wavelength inside the dielectric is

$$\lambda_2 = \frac{30 \times 10^9}{10 \times 10^9 \sqrt{2.56}} = 1.875 \text{ cm}$$

2. At a frequency of 5 GHz, the dielectric slab of thickness 0.9375 cm is equal to

$$d = \frac{0.9375\sqrt{2.56}\lambda_2}{30 \times 10^9 / 5 \times 10^9} = 0.25\lambda_2 \Rightarrow 2\beta_2 d = \frac{4\pi}{\lambda_2}\left(\frac{\lambda_2}{4}\right) = \pi$$

and at 15 GHz it is equal to

$$d = \frac{0.9375\sqrt{2.56}\lambda_2}{30 \times 10^9 / 15 \times 10^9} = 0.75\lambda_2 \Rightarrow 2\beta_2 d = \frac{4\pi}{\lambda_2}\left(\frac{3\lambda_2}{4}\right) = 3\pi$$

Since

$$\Gamma_{12} = -\Gamma_{23} = \frac{\eta_2 - \eta_1}{\eta_2 + \eta_1} = \frac{\eta_2/\eta_1 - 1}{\eta_2/\eta_1 + 1} = \frac{1 - \sqrt{\varepsilon_r}}{1 + \sqrt{\varepsilon_r}} = -\frac{0.6}{2.6} = -0.231$$

the input reflection coefficient of (5-70), at $f = 5$ and 15 GHz, achieves the maximum magnitude of

$$\left| \Gamma_{in}(z = -d^-) \right| = \left| \frac{-0.231 - 0.231}{1 - (-0.231)(0.231)} \right| = \frac{2(0.231)}{1 + (0.231)^2} = 0.438$$

A complete plot of $\left| \Gamma_{in}(z = -d^-) \right|$ for 5 GHz $\leq f \leq$ 15 GHz is shown in Figure 5-12.

Using the approximate form of (5-70), the magnitude of the input reflection coefficient is equal to

$$\left| \Gamma_{in}(z = -d^-) \right| \big|_{\substack{f=5, \\ 15 \text{ GHz}}} \simeq \left| -0.231 - (0.231) \right| = 0.462$$

The percent error of this is

$$\text{percent error} = \left(\frac{-0.438 + 0.462}{0.438} \right) \times 100 = 5.48$$

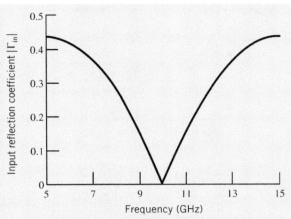

Figure 5-12 Input reflection coefficient, as a function of frequency, for wave propagation through a dielectric slab.

Example 5-10

A uniform plane wave is incident normally upon a dielectric slab whose thickness at $f_0 = 10$ GHz is $\lambda_{20}/4$ where λ_{20} is the wavelength in the dielectric slab. The slab is bounded on the left side by air and on the right side by a semi-infinite medium of dielectric constant $\varepsilon_{r3} = 4$.

1. Determine the intrinsic impedance η_2 and dielectric constant ε_{r2} of the sandwiched slab so that the input reflection coefficient at $f_0 = 10$ GHz is zero.
2. Plot the magnitude response of the input reflection coefficient for $0 \leq f \leq 20$ GHz when the intrinsic impedance and physical thickness of the slab are those found in part 1.
3. Using the ray-tracing model of Figure 5-11b, at $f_0 = 10$ GHz determine the first and next two higher-order terms that contribute to the overall input reflection coefficient. What is the input reflection coefficient using these three terms?

Solution:

1. In order for the input reflection coefficient to vanish, the magnitude of (5-70) must be equal to zero, that is

$$\left|\Gamma_{12} + \Gamma_{23}e^{-j2\beta_2 d}\right| = 0$$

Since at $f_0 = 10$ GHz, $d = \lambda_{20}/4$, then

$$2\beta_2 d\big|_{f=10\ \text{GHz}} = 2\left(\frac{2\pi}{\lambda_{20}}\right)\left(\frac{\lambda_{20}}{4}\right) = \pi$$

Also

$$\Gamma_{12} = \frac{\eta_2 - \eta_1}{\eta_2 + \eta_1}$$

and

$$\Gamma_{23} = \frac{\eta_3 - \eta_2}{\eta_3 + \eta_2}$$

Thus

$$\left|\Gamma_{12} + \Gamma_{23}e^{-j2\beta_2 d}\right|_{\substack{d=\lambda_{20}/4 \\ f=10 \text{ GHz}}} = \left|\frac{\eta_2 - \eta_1}{\eta_2 + \eta_1} - \frac{\eta_3 - \eta_2}{\eta_3 + \eta_2}\right|$$

$$= \left|\frac{(\eta_2 - \eta_1)(\eta_3 + \eta_2) - (\eta_3 - \eta_2)(\eta_2 + \eta_1)}{(\eta_2 + \eta_1)(\eta_3 + \eta_2)}\right| = 0$$

or

$$2\left|\eta_2^2 - \eta_1\eta_3\right| = 0 \Rightarrow \eta_2 = \sqrt{\eta_1\eta_3}$$

Since $\eta_1 = \sqrt{\dfrac{\mu_0}{\varepsilon_0}} = 377$ ohms and $\eta_3 = \sqrt{\dfrac{\mu_0}{4\varepsilon_0}} = \dfrac{1}{2}\eta_1 = 188.5$ ohms, then

$$\eta_2 = \sqrt{\eta_1\eta_3} = \frac{\eta_1}{\sqrt{2}} = 0.707\eta_1 = 0.707(377) = 266.5 \text{ ohms}$$

The dielectric constant of the slab must be equal to

$$\varepsilon_{r2} = 2$$

whereas the physical thickness of the dielectric is

$$d = \frac{\lambda_{20}}{4} = \frac{30 \times 10^9}{4(10 \times 10^9)\sqrt{2}} = 0.53 \text{ cm}$$

It is apparent then that whenever the dielectric is bounded by two semi-infinite media and its thickness is a quarter of a wavelength in the dielectric, its intrinsic impedance must always be equal to the square root of the product of the intrinsic impedances of the two media on each of its sides in order for the input reflection coefficient to vanish. This is referred to as the quarter-wavelength transformer that is so popular in transmission line design.

2. Since at $f_0 = 10$ GHz, $d = \lambda_{20}/4 = 0.53$ cm, then in the frequency range $0 \leq f \leq 20$ GHz

$$2\beta_2 d = 2\left(\frac{2\pi}{\lambda_2}\right)\left(\frac{\lambda_{20}}{4}\right) = \pi\left(\frac{f}{f_0}\right)$$

also

$$\Gamma_{12} = \frac{\eta_2 - \eta_1}{\eta_2 + \eta_1} = \frac{\eta_2/\eta_1 - 1}{\eta_2/\eta_1 + 1} = \frac{1 - \sqrt{2}}{1 + \sqrt{2}}$$

$$\Gamma_{23} = \frac{\eta_3 - \eta_2}{\eta_3 + \eta_2} = \frac{\eta_3/\eta_2 - 1}{\eta_3/\eta_2 + 1} = \frac{1 - \sqrt{2}}{1 + \sqrt{2}} = \Gamma_{12}$$

Therefore, the magnitude of the input reflection coefficient of (5-70) can be written now as

$$\left|\Gamma_{in}(z = -d^-)\right| = \left|\frac{\Gamma_{12}(1 + e^{-j\pi f/f_0})}{1 + (\Gamma_{12})^2 e^{-j\pi f/f_0}}\right|$$

whose maximum value, which occurs when $f = 0$ and $2f_0 = 20$ GHz, is approximately equal to

$$\left|\Gamma_{in}(z = -d^-)\right|_{max} = \frac{2|\Gamma_{12}|}{(1 + |\Gamma_{12}|^2)} = |\Gamma_{13}| = \left|\frac{\eta_3 - \eta_1}{\eta_3 + \eta_1}\right|$$

$$= 0.333 \simeq 2|\Gamma_{12}| = 0.3431$$

A complete plot of $\left|\Gamma_{in}(z = -d^-)\right|_{d=\lambda_{20}/4}$ when $0 \leq f \leq 20$ GHz is shown in the Figure 5-13.

It is interesting to note that the magnitude of the input reflection coefficient monotonically decreases from $f = 0$ to f_0, and it monotonically increases from f_0 to $2f_0$. It can also be noted that the bandwidth of the response curve near f_0 is very small, and any deviations of the frequency from f_0 will cause the reflection coefficient to rise sharply.

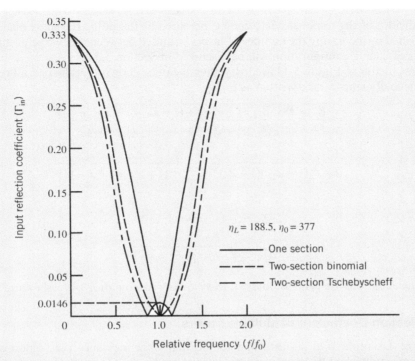

Figure 5-13 Responses of single-section, two-section binomial, and two-section Tschebyscheff quarter-wavelength transformers. (Source: C. A. Balanis, *Antenna Theory: Analysis and Design*, Third Edition. Copyright © 2005, John Wiley & Sons, Inc. Reprinted by permission of John Wiley & Sons, Inc.)

3. According to Figure 5-11b, the first-order term of the input reflection coefficient is

$$\Gamma_{12} = \frac{\eta_2 - \eta_1}{\eta_2 + \eta_1} = \frac{266.5 - 377}{266.5 + 377} = -0.1717$$

The next two higher terms are equal to

$$T_{12}\Gamma_{23}T_{21}e^{-j2\beta_2 d} = \frac{2\eta_1}{\eta_1 + \eta_2}\left(\frac{\eta_3 - \eta_2}{\eta_3 + \eta_2}\right)\left(\frac{2\eta_2}{\eta_1 + \eta_2}\right)e^{-j\pi}$$

$$= -\frac{2(377)}{377 + 266.5}\left(\frac{188.5 - 266.5}{188.5 + 266.5}\right)\frac{2(266.5)}{377 + 266.5} = +0.1664$$

$$T_{12}\Gamma_{21}\Gamma_{23}^2 T_{21}e^{-j4\beta_2 d} = \frac{2\eta_1}{\eta_1 + \eta_2}\left(\frac{\eta_1 - \eta_2}{\eta_1 + \eta_2}\right)\left(\frac{\eta_3 - \eta_2}{\eta_3 + \eta_2}\right)^2\left(\frac{2\eta_2}{\eta_1 + \eta_2}\right)e^{-j2\pi}$$

$$T_{12}\Gamma_{21}\Gamma_{23}^2 T_{21}e^{-j4\beta_2 d} = \frac{2(337)}{377 + 266.5}\left(\frac{377 - 266.5}{377 + 266.5}\right)\left(\frac{188.5 - 266.5}{188.5 + 266.5}\right)^2\frac{2(266.5)}{377 + 266.5}$$

$$= 0.0049$$

$$\Gamma_{in} \simeq \Gamma_{12} + T_{12}\Gamma_{23}T_{21}e^{-j2\beta_2 d} + T_{12}\Gamma_{21}\Gamma_{23}^2 T_{21}\, e^{-j4\beta_2 d}$$

$$= -0.1717 + 0.1664 + 0.0049$$

$$\Gamma_{in} \simeq -4 \times 10^{-4} \simeq 0$$

Thus, the first three terms, or even the first two terms, provide an excellent approximation to the exact value of zero.

The bandwidth of the response curve can be increased by flattening the curve near f_0. This can be accomplished by increasing the number of layers bounded between the two semi-infinite media. The analysis of such a configuration will be discussed in Section 5.5.2.

If the three media of Figure 5-11 are lossy, then it can be shown that the overall reflection and transmission coefficients can be written as [3]

$$\Gamma_{\text{in}} = \frac{E^r}{E^i} = \frac{(1-Z_{12})(1+Z_{23})+(1+Z_{12})(1-Z_{23})e^{-2\gamma_2 d}}{(1+Z_{12})(1+Z_{23})+(1-Z_{12})(1-Z_{23})e^{-2\gamma_2 d}} \qquad (5\text{-}71\text{a})$$

$$T = \frac{E^t}{E_i} = \frac{4}{(1-Z_{12})(1-Z_{23})e^{-\gamma_2 d}+(1+Z_{12})(1+Z_{23})e^{\gamma_2 d}} \qquad (5\text{-}71\text{b})$$

where

$$Z_{ij} = \frac{\mu_i \gamma_j}{\mu_j \gamma_i} \quad i,j = 1, 2, 3 \qquad (5\text{-}71\text{c})$$

$$\gamma_k = \pm \sqrt{j\omega \mu_k (\sigma_k + j\omega \varepsilon_k)} \qquad (5\text{-}71\text{d})$$

The preceding equations are valid for lossless, lossy, or any combination of lossless and lossy media.

5.5.2   Reflection Coefficient of Multiple Layers

The results of Example 5-10 indicate that for normal wave incidence the response of a single dielectric layer sandwiched between two semi-infinite media did not exhibit very broad characteristics around the center frequency f_0, and its overall response was very sensitive to frequency changes. The characteristics of such a response are very similar to the bandstop characteristics of a single section filter or single section quarter-wavelength impedance transformer. To increase the bandwidth of the system under normal wave incidence, multiple layers of dielectric slabs, each with different dielectric constant, must be inserted between the two semi-infinite media. Multiple section dielectric layers can be used to design dielectric filters [9]. Coating radar targets with multilayer slabs can also be used to reduce or enhance their scattering characteristics.

When N layers, each with its own thickness and constitutive parameters, are sandwiched between two semi-infinite media as shown in Figure 5-14, the analysis for the overall reflection and transmission coefficients is quite cumbersome, although it is straightforward. However, an approximate form of the input reflection coefficient for the entire system under normal wave incidence can be obtained by utilizing the approximation first introduced to represent (5-70). With this in mind, the input reflection coefficient under normal wave incidence for the system of Figure 5-14, referenced at the boundary of the leading interface, can be written approximately as [1, 8]

$$\Gamma_{\text{in}} \simeq \Gamma_0 + \Gamma_1 e^{-j2\beta_1 d_1} + \Gamma_2 e^{-j2(\beta_1 d_1 + \beta_2 d_2)} + \cdots + \Gamma_N e^{-j2(\beta_1 d_1 + \beta_2 d_2 + \cdots + \beta_N d_N)} \qquad (5\text{-}72)$$

where

$$\Gamma_0 = \frac{\eta_1 - \eta_0}{\eta_1 + \eta_0} \qquad (5\text{-}72\text{a})$$

$$\Gamma_1 = \frac{\eta_2 - \eta_1}{\eta_2 + \eta_1} \qquad (5\text{-}72\text{b})$$

$$\Gamma_2 = \frac{\eta_3 - \eta_2}{\eta_3 + \eta_2} \qquad (5\text{-}72\text{c})$$

$$\vdots$$

$$\Gamma_N = \frac{\eta_L - \eta_N}{\eta_L + \eta_N} \qquad (5\text{-}72\text{d})$$

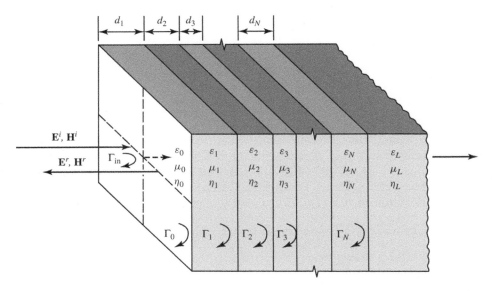

Figure 5-14   Normal wave propagation through N layers sandwiched between two media.

Expression (5-72) is accurate provided that at each boundary the intrinsic reflection coefficients of (5-72a) through (5-72d) are small in comparison to unity.

A. Quarter-Wavelength Transformer   Example 5-10 demonstrated that when a lossless dielectric slab of thickness $\lambda_{20}/4$ at a frequency f_0 is sandwiched between two lossless semi-infinite dielectric media, the input reflection coefficient at f_0 is zero provided its intrinsic impedance η_1 is equal to

$$\eta_1 = \sqrt{\eta_0 \eta_L} \tag{5-73}$$

where

$\eta_1 =$ intrinsic impedance of dielectric slab
$\eta_0 =$ intrinsic impedance of the input semi-infinite medium
$\eta_L =$ intrinsic impedance of the load semi-infinite medium

However, as was illustrated in Figure 5-13, the response of the input reflection coefficient as a function of frequency was not very broad near the center frequency f_0.

Matchings that are less sensitive to frequency variations and that provide broader bandwidths require multiple $\lambda/4$ sections. In fact the number of sections and the intrinsic impedance of each section can be designed so that the reflection coefficient follows, within the desired frequency bandwidth, prescribed variations that are symmetrical about the center frequency. This design assumes that the semi-infinite media and the dielectric slabs are all lossless so that their intrinsic impedances are all real. The discussion that follows parallels that of [1] and [8].

Referring to Figure 5-14, the total input reflection coefficient Γ_{in} for an N-section quarter-wavelength transformer with $\eta_L > \eta_0$ can be written, using an extension of the approximation used to represent (5-70), as [1, 8]

$$\Gamma_{in}(f) \simeq \Gamma_0 + \Gamma_1 e^{-j2\theta} + \Gamma_2 e^{-j4\theta} + \ldots + \Gamma_N e^{-j2N\theta} = \sum_{n=0}^{N} \Gamma_n e^{-j2n\theta} \tag{5-74}$$

where Γ_n and θ are represented, respectively, by

$$\Gamma_n = \frac{\eta_{n+1} - \eta_n}{\eta_{n+1} + \eta_n} \tag{5-74a}$$

$$\theta = \beta_n d_n = \frac{2\pi}{\lambda_n}\left(\frac{\lambda_{n_0}}{4}\right) = \frac{\pi}{2}\left(\frac{f}{f_0}\right) \tag{5-74b}$$

In (5-74) Γ_n represents the reflection coefficient at the junction of two infinite lines that have intrinsic impedances η_n and η_{n+1}, f_0 represents the designed center frequency, and f represents the operating frequency. Equation 5-74 is valid provided the Γ_n's at each junction are small (the requirements will be met if $\eta_L \simeq \eta_0$). For lossless dielectrics, the η_n's and Γ_n's will all be real.

For a symmetrical transformer ($\Gamma_0 = \Gamma_N$, $\Gamma_1 = \Gamma_{N-1}$, etc.), (5-74) reduces to

$$\Gamma_{\text{in}}(f) \simeq 2e^{-jN\theta}\left[\Gamma_0 \cos N\theta + \Gamma_1 \cos(N-2)\theta + \Gamma_2 \cos(N-4)\theta + ...\right] \tag{5-75}$$

The last term in (5-75) should be

$$\Gamma_{[(N-1)/2]} \cos\theta \qquad \text{for } N = \text{odd integer} \tag{5-75a}$$

$$\tfrac{1}{2}\Gamma_{(N/2)} \qquad\qquad \text{for } N = \text{even integer} \tag{5-75b}$$

B. Binomial (Maximally Flat) Design   One technique, used to design an N-section $\lambda/4$ transformer, requires that the input reflection coefficient (5-74) have maximally flat passband characteristics. For this method, the junction reflection coefficients (Γ_n's) are derived using the binomial expansion and we can equate (5-74) to [1, 8]

$$\Gamma_{\text{in}}(f) \simeq \sum_{n=0}^{N} \Gamma_n e^{-j2n\theta} = e^{-jN\theta}\frac{\eta_L - \eta_0}{\eta_L + \eta_0}\cos^N(\theta)$$
$$\simeq 2^{-N}\frac{\eta_L - \eta_0}{\eta_L + \eta_0}\sum_{n=0}^{N}C_n^N e^{-j2n\theta} \tag{5-76}$$

where

$$C_n^N = \frac{N!}{(N-n)!n!} \qquad n = 0, 1, 2, ..., N \tag{5-76a}$$

From (5-76)

$$\Gamma_n = 2^{-N}\frac{\eta_L - \eta_0}{\eta_L + \eta_0}C_n^N \tag{5-77}$$

For this type of design, the fractional bandwidth $\Delta f/f_0$ is given by

$$\frac{\Delta f}{f_0} = 2\frac{f_0 - f_m}{f_0} = 2\left(1 - \frac{f_m}{f_0}\right) = 2\left(1 - \frac{2}{\pi}\theta_m\right) \tag{5-78}$$

Since

$$\theta_m = \frac{2\pi}{\lambda_m}\left(\frac{\lambda_0}{4}\right) = \frac{\pi}{2}\left(\frac{f_m}{f_0}\right) \tag{5-79}$$

(5-78) reduces, using (5-76), to

$$\frac{\Delta f}{f_0} = 2 - \frac{4}{\pi} \cos^{-1} \left| \frac{\Gamma_m}{(\eta_L - \eta_0)/(\eta_L + \eta_0)} \right|^{1/N} \tag{5-80}$$

where Γ_m is the magnitude of the maximum value of reflection coefficient that can be tolerated within the bandwidth.

The usual design procedure is to specify

1. the load intrinsic impedance η_L
2. the input intrinsic impedance η_0
3. the number of sections N
4. the maximum tolerable reflection coefficient Γ_m (or fractional bandwidth $\Delta f/f_0$)

and to find

1. the intrinsic impedance of each section
2. the fractional bandwidth $\Delta f/f_0$ (or maximum tolerable reflection coefficient Γ_m)

To illustrate the principle, let us consider an example.

Example 5-11

Two lossless dielectric slabs each of thickness $\lambda_0/4$ at a center frequency $f_0 = 10$ GHz are sandwiched between air to the left and a lossless semi-infinite medium of dielectric constant $\varepsilon_L = 4$ to the right. Assuming a fractional bandwidth of 0.375 and a binomial design:

1. Determine the intrinsic impedances, dielectric constants, and thicknesses of the sandwiched slabs so that the input reflection coefficient at $f_0 = 10$ GHz is zero.
2. Determine the maximum reflection coefficient and SWR within the fractional bandwidth.
3. Plot the response of the input reflection coefficient for $0 \leq f \leq 20$ GHz when the intrinsic impedances and physical thicknesses of the slabs are those found in part 1. Compare the response of the two-section binomial design with that of the single section of Example 5-10.

Solution:

1. Using (5-76a) and (5-77)

$$\Gamma_n = 2^{-N} \frac{\eta_L - \eta_0}{\eta_L + \eta_0} C_n^N = 2^{-N} \frac{\eta_L - \eta_0}{\eta_L + \eta_0} \frac{N!}{(N-n)!n!}$$

Since the input dielectric is air and the load dielectric has a dielectric constant $\varepsilon_L = 4$, then

$$\eta_0 = 377$$

$$\eta_L = \sqrt{\frac{\mu_0}{\varepsilon_L \varepsilon_0}} = \frac{377}{2} \doteq 188.5$$

Therefore,

$$n = 0 : \Gamma_0 = \frac{\eta_1 - \eta_0}{\eta_1 + \eta_0} = 2^{-2} \left(\frac{188.5 - 377}{188.5 + 377} \right) \frac{2!}{2!0!} = -\frac{1}{12}$$

$$\Rightarrow \eta_1 = \eta_0 \left(\frac{1 - 1/12}{1 + 1/12} \right) = 0.846 \eta_0 = 318.94 \text{ ohms}$$

$$\Rightarrow \varepsilon_{r1} = 1.40 \quad d_1 = \frac{\lambda_{10}}{4} = 0.634 \text{ cm}$$

$$n = 1 : \Gamma_1 = \frac{\eta_2 - \eta_1}{\eta_2 + \eta_1} = 2^{-2} \left(\frac{188.5 - 377}{188.5 + 377} \right) \frac{2!}{1!1!} = -\frac{1}{6}$$

$$\Rightarrow \eta_2 = \eta_1 \left(\frac{1 - 1/6}{1 + 1/6} \right) = 0.714 \eta_1 = 227.72 \text{ ohms}$$

$$\Rightarrow \varepsilon_{r2} = 2.74 \quad d_2 = \lambda_{20}/4 = 0.453 \text{ cm}$$

2. For a fractional bandwidth of 0.375, the magnitude of the maximum reflection coefficient Γ_m is obtained using (5-80) or

$$\frac{\Delta f}{f_0} = 0.375 = 2 - \frac{4}{\pi} \cos^{-1} \left| \frac{\Gamma_m}{(\eta_L - \eta_0)/(\eta_L + \eta_0)} \right|^{1/2}$$

which for $\eta_L = 188.5$ and $\eta_0 = 377$ leads to

$$\Gamma_m = 0.028$$

The maximum standing wave ratio is

$$\text{SWR}_m = \frac{1 + \Gamma_m}{1 - \Gamma_m} = \frac{1 + 0.028}{1 - 0.028} = 1.058$$

3. The magnitude of the input reflection coefficient is given by (5-76) as

$$|\Gamma_{\text{in}}| = \left| \frac{\eta_L - \eta_0}{\eta_L + \eta_0} \right| \cos^2 \theta = \frac{1}{3} \cos^2 \theta = \frac{1}{3} \cos^2 \left[\frac{\pi}{2} \left(\frac{f}{f_0} \right) \right]$$

which is shown plotted in Figure 5-13 where it is also compared with that of the one- and two-section Tschebyscheff design to be discussed next.

C. Tschebyscheff (Equal-Ripple) Design The reflection coefficient can be made to vary within the bandwidth in an oscillatory manner and have equal-ripple characteristics [10–12]. This can be accomplished by making Γ_{in} vary similarly as a Tschebyscheff (Chebyshev) polynomial. For the Tschebyscheff design, the equation that corresponds to (5-76) is [1, 8]

$$\boxed{\Gamma_{\text{in}}(f) = e^{-jN\theta} \frac{\eta_L - \eta_0}{\eta_L + \eta_0} \frac{T_N(\sec\theta_m \cos\theta)}{T_N(\sec\theta_m)}} \tag{5-81}$$

where $T_N(z)$ is the Tschebyscheff polynomial of order N.

The maximum allowable reflection coefficient occurs at the edges of the passband where $\theta = \theta_m$ and $|T_N(\sec\theta_m \cos\theta)|_{\theta=\theta_m} = 1$. Thus,

$$\rho_m = \left| \frac{\eta_L - \eta_0}{\eta_L + \eta_0} \frac{1}{T_N(\sec\theta_m)} \right| \tag{5-82}$$

or

$$|T_N(\sec\theta_m)| = \left| \frac{1}{\rho_m} \frac{\eta_L - \eta_0}{\eta_L + \eta_0} \right| \tag{5-82a}$$

Using (5-82), we can write (5-81) as

$$\Gamma_{\text{in}}(f) = e^{-jN\theta} \rho_m T_N (\sec\theta_m \cos\theta) \tag{5-83}$$

and its magnitude as

$$\left|\Gamma_{\text{in}}(f)\right| = \rho_{in}(f) = \left|\rho_m T_N (\sec\theta_m \cos\theta)\right| \tag{5-83a}$$

For this type of a design, the fractional bandwidth $\Delta f / f_0$ is also given by (5-78).

To be physical, ρ_m must be smaller than the reflection coefficient when there are no matching layers. Therefore, from (5-82),

$$\rho_m = \left|\frac{\eta_L - \eta_0}{\eta_L + \eta_0} \frac{1}{T_N(\sec\theta_m)}\right| < \left|\frac{\eta_L - \eta_0}{\eta_L + \eta_0}\right| \tag{5-84}$$

or

$$\left|T_N(\sec\theta_m)\right| > 1 \tag{5-84a}$$

The Tschebyscheff polynomial can be expressed by either (6-71a) or (6-71b) of [1], or

$$T_m(z) = \cos[m\cos^{-1}(z)] \qquad -1 \le z \le +1 \tag{5-85a}$$

$$T_m(z) = \cosh[m\cosh^{-1}(z)] \qquad z < -1, z > +1 \tag{5-85b}$$

Since $\left|T_N(\sec\theta_m)\right| > 1$, using (5-85b) we can express $T_N(\sec\theta_m)$ as

$$T_N(\sec\theta_m) = \cosh\left[N\cosh^{-1}(\sec\theta_m)\right] \tag{5-86}$$

or by using (5-82a), as

$$\left|T_N(\sec\theta_m)\right| = \left|\cosh\left[N\cosh^{-1}(\sec\theta_m)\right]\right| = \left|\frac{1}{\rho_m}\frac{\eta_L - \eta_0}{\eta_L + \eta_0}\right| \tag{5-86a}$$

Thus,

$$\sec\theta_m = \cosh\left[\frac{1}{N}\cosh^{-1}\left(\left|\frac{1}{\rho_m}\frac{\eta_L - \eta_0}{\eta_L + \eta_0}\right|\right)\right] \tag{5-87}$$

or

$$\theta_m = \sec^{-1}\left\{\cosh\left[\frac{1}{N}\cosh^{-1}\left(\left|\frac{1}{\rho_m}\frac{\eta_L - \eta_0}{\eta_L + \eta_0}\right|\right)\right]\right\} \tag{5-87a}$$

Using (5-83) we can write the reflection coefficient of (5-75) as

$$\Gamma_{\text{in}}(\theta) = 2e^{-jN\theta}\left\{\rho_0\cos(N\theta) + \rho_1\cos[(N-2)\theta] + ...\right\}$$
$$= e^{-jN\theta}\rho_m T_N(\sec\theta_m \cos\theta) \tag{5-88}$$

For a given N, replace $T_N(\sec\theta_m \cos\theta)$ in (5-88) by its polynomial series of (6-69) of [1] and then match terms. This will allow you to determine the intrinsic reflection coefficients $\rho_n's$ and subsequently the $\eta_n's$. The design procedure for the Tschebyscheff design is the same as that of the binomial design, as outlined previously.

The first few Tschebyscheff polynomials can be found in [1, 8]. For $z = \sec\theta_m \cos\theta$, the first three polynomials reduce to

$$T_1(\sec\theta_m \cos\theta) = \sec\theta_m \cos\theta$$

$$T_2(\sec\theta_m \cos\theta) = 2(\sec\theta_m \cos\theta)^2 - 1 = \sec^2\theta_m \cos 2\theta + (\sec^2\theta_m - 1)$$

$$T_3(\sec\theta_m \cos\theta) = 4(\sec\theta_m \cos\theta)^3 - 3(\sec\theta_m \cos\theta)$$

$$= \sec^3\theta_m \cos 3\theta + 3(\sec^3\theta_m - \sec\theta_m)\cos\theta \tag{5-89}$$

The remaining details of the analysis are found in [1, 8].

The design of Example 5-11 using a Tschebyscheff transformer is assigned as an exercise to the reader. However, its response is plotted in Figure 5-13 for comparison.

In general, multiple sections (either binomial or Tschebyscheff) provide greater bandwidths than a single section. As the number of sections increases, the bandwidth also increases. The advantage of the binomial design is that the reflection coefficient values within the bandwidth monotonically decrease from both ends toward the center. Thus the values are always smaller than an acceptable and designed value that occurs at the "skirts" of the bandwidth. For the Tschebyscheff design, the reflection coefficient values within the designed bandwidth are equal to or smaller than an acceptable and designed value. The number of times the reflection coefficient reaches the maximum value within the bandwidth is determined by the number of sections. In fact, for an even number of sections the reflection coefficient at the designed center frequency is equal to the maximum allowable value, whereas for an odd number of sections it is zero. For a maximum tolerable reflection coefficient, the N-section Tschebyscheff transformer provides a larger bandwidth than a corresponding N-section binomial design, or for a given bandwidth the maximum tolerable reflection coefficient is smaller for a Tschebyscheff design.

D. Oblique-Wave Incidence   A more general formulation of the reflection and transmission coefficients can be developed by considering the geometry of Figure 5-15 where a uniform plane wave is incident at an oblique angle upon N layers of planar slabs that are bordered on either side by free space. This type of a geometry can be used to approximate the configuration of a radome whose radius of curvature is large in comparison to the wavelength. It can be shown that the

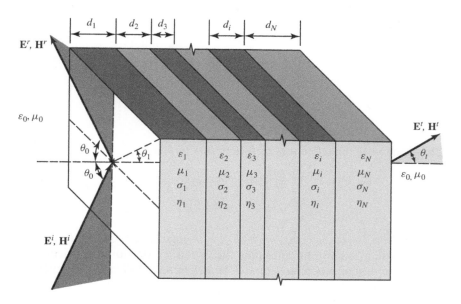

Figure 5-15   Oblique wave propagation through N layers of dielectric slabs.

overall reflection and transmission coefficients for perpendicular (horizontal) and parallel (vertical) polarizations can be written as [3]

Perpendicular (Horizontal)

$$\Gamma_\perp = \frac{E_\perp^r}{E_\perp^i} = \frac{B_0}{A_0} \tag{5-90a}$$

$$T_\perp = \frac{E_\perp^t}{E_\perp^i} = \frac{1}{A_0} \tag{5-90b}$$

Parallel (Vertical)

$$\Gamma_\parallel = \frac{E_\parallel^r}{E_\parallel^i} = \frac{D_0}{C_0} \tag{5-91a}$$

$$T_\parallel = \frac{E_\parallel^t}{E_\parallel^i} = \frac{1}{C_0} \tag{5-91b}$$

The functions A_0, B_0, C_0, and D_0 are found using the recursive formulas

$$A_j = \frac{e^{\psi_j}}{2} \left[A_{j+1}(1 + Y_{j+1}) + B_{j+1}(1 - Y_{j+1}) \right] \tag{5-92a}$$

$$B_j = \frac{e^{-\psi_j}}{2} \left[A_{j+1}(1 - Y_{j+1}) + B_{j+1}(1 + Y_{j+1}) \right] \tag{5-92b}$$

$$C_j = \frac{e^{\psi_j}}{2} \left[C_{j+1}(1 + Z_{j+1}) + D_{j+1}(1 - Z_{j+1}) \right] \tag{5-92c}$$

$$D_j = \frac{e^{-\psi_j}}{2} \left[C_{j+1}(1 - Z_{j+1}) + D_{j+1}(1 + Z_{j+1}) \right] \tag{5-92d}$$

where

$$A_{N+1} = C_{N+1} = 1 \tag{5-92e}$$

$$B_{N+1} = D_{N+1} = 0 \tag{5-92f}$$

$$Y_{j+1} = \frac{\cos\theta_{j+1}}{\cos\theta_j} \sqrt{\frac{\varepsilon_{j+1}(1 - j\tan\delta_{j+1})\mu_j}{\varepsilon_j(1 - j\tan\delta_j)\mu_{j+1}}} \tag{5-92g}$$

$$Z_{j+1} = \frac{\cos\theta_{j+1}}{\cos\theta_j} \sqrt{\frac{\varepsilon_j(1 - j\tan\delta_j)\mu_{j+1}}{\varepsilon_{j+1}(1 - j\tan\delta_{j+1})\mu_j}} \tag{5-92h}$$

$$\psi_j = d_j \gamma_j \cos\theta_j \tag{5-92i}$$

$$\gamma_j = \pm\sqrt{j\omega\mu_j(\sigma_j + j\omega\varepsilon_j)} \tag{5-92j}$$

$$\theta_j = \text{complex angle of refraction in the } j^{\text{th}} \text{ layer} \tag{5-92k}$$

where d_0 is the distance from the leading interface, which serves as the reference for the reflection and transmission coefficients [see (5-5a) and (5-5b)].

5.6   POLARIZATION CHARACTERISTICS ON REFLECTION

When linearly polarized fields are reflected from smooth flat surfaces, the reflected fields maintain their linear polarization characteristics. However, when the reflected surfaces are curved or rough, a linearly polarized component orthogonal to that of the incident field is introduced during reflection. Therefore, the total field exhibits two components: one with the same polarization as the incident field (main polarization) and one orthogonal to it (cross polarization). During this process, the field is depolarized due to reflection.

Circularly polarized fields in free space incident upon flat surfaces:

1. Maintain their circular polarization but reverse their sense of rotation when the reflecting surface is perfectly conducting.
2. Are transformed to elliptically polarized fields of opposite sense of rotation when the flat surface is a lossless dielectric and the angle of incidence is smaller than the Brewster angle.

Similarly, elliptically polarized fields in free space upon reflection from flat surfaces

1. Maintain their elliptical polarization and magnitude of axial ratio but reverse their sense of rotation when reflected from a perfectly conducting surface.
2. Maintain their elliptical polarization but change their axial ratio and sense of rotation when the reflecting surface is a dielectric and the angle of incidence is smaller than the Brewster angle.

To analyze the polarization properties of a wave when it is reflected by a surface, let us assume that an elliptically polarized wave is obliquely incident upon a flat surface of infinite extent as shown in Figure 5-16 [7]. Using the localized coordinate system (x', y, z') of Figure 5-16, the incident electric field components can be written as

$$\mathbf{E}_{\parallel}^{i} = \hat{\mathbf{a}}_{x'} E_{\parallel}^{i} e^{-j\boldsymbol{\beta}^{i}\cdot\mathbf{r}} = \hat{\mathbf{a}}_{x'} E_{\parallel}^{0} e^{-j\boldsymbol{\beta}^{i}\cdot\mathbf{r}} \tag{5-93a}$$

$$\mathbf{E}_{\perp}^{i} = \hat{\mathbf{a}}_{y} E_{\perp}^{i} e^{-j\boldsymbol{\beta}^{i}\cdot\mathbf{r}} = \hat{\mathbf{a}}_{y} E_{\perp}^{0} e^{-j(\boldsymbol{\beta}^{i}\cdot\mathbf{r}-\phi_{\perp}^{i})} \tag{5-93b}$$

where $E_{\parallel}^{0}$ and $E_{\perp}^{0}$ are assumed to be real.

For this set of field components, the Poincaré sphere angles (4-58a) through (4-59b) can be written [assuming that the ratio in (4-58a), selected here to demonstrate the procedure, satisfies the angular limits of all the Poincaré sphere angles] as

$$\gamma^{i} = \tan^{-1}\left(\frac{|E_{\perp}^{0}|}{|E_{\parallel}^{0}|}\right) \tag{5-94a}$$

$$\delta^{i} = \phi_{\perp}^{i} - \phi_{\parallel}^{i} = \phi_{\perp}^{i} \tag{5-94b}$$

$$\varepsilon^{i} = \cot^{-1}(\mathrm{AR}^{i}) \tag{5-94c}$$

$$\tau^{i} = \text{tilt angle of incident wave} \tag{5-94d}$$

where δ^i is the phase angle by which the perpendicular component of the incident field leads the parallel component. It is assumed that (AR^i) is positive for left-hand and negative for right-hand polarized fields. These two sets of angles are related to each other by (4-60a) through (4-61b), or

$$\cos(2\gamma^i) = \cos(2\varepsilon^i)\cos(2\tau^i) \tag{5-95a}$$

$$\tan(\delta^i) = \frac{\tan(2\varepsilon^i)}{\sin(2\tau^i)} \tag{5-95b}$$

or

$$\sin(2\varepsilon^i) = \sin(2\gamma^i)\sin(\delta^i) \tag{5-95c}$$

$$\tan(2\tau^i) = \tan(2\gamma^i)\cos(\delta^i) \tag{5-95d}$$

In a similar manner, the reflected fields of the elliptically polarized wave can be written according to the localized coordinate system (x'', y, z'') of Figure 5-16 as

$$\begin{aligned}
\mathbf{E}_{\parallel}^{r} &= \hat{\mathbf{a}}_{x''} E_{\parallel}^{r} e^{-j\boldsymbol{\beta}^r \cdot \mathbf{r}} = -\hat{\mathbf{a}}_{x''} \Gamma_{\parallel}^{b} E_{\parallel}^{0} e^{-j\boldsymbol{\beta}^r \cdot \mathbf{r}} = \hat{\mathbf{a}}_{x''} \left|\Gamma_{\parallel}^{b}\right| E_{\parallel}^{0} e^{-j(\boldsymbol{\beta}^r \cdot \mathbf{r} - \pi - \zeta_{\parallel}^r)} \\
&= \hat{\mathbf{a}}_{x''} \left|\Gamma_{\parallel}^{b}\right| E_{\parallel}^{0} e^{-j(\boldsymbol{\beta}^r \cdot \mathbf{r} - \phi_{\parallel}^r)}
\end{aligned} \tag{5-96a}$$

$$\begin{aligned}
\mathbf{E}_{\perp}^{r} &= \hat{\mathbf{a}}_{y} E_{\perp}^{r} e^{-j\boldsymbol{\beta}^r \cdot \mathbf{r}} = \hat{\mathbf{a}}_{y} \Gamma_{\perp}^{b} E_{\perp}^{0} e^{-j(\boldsymbol{\beta}^r \cdot \mathbf{r} - \phi_{\perp}^i)} = \hat{\mathbf{a}}_{y} \left|\Gamma_{\perp}^{b}\right| E_{\perp}^{0} e^{-j(\boldsymbol{\beta}^r \cdot \mathbf{r} - \delta^i - \zeta_{\perp}^r)} \\
&= \hat{\mathbf{a}}_{y} \left|\Gamma_{\perp}^{b}\right| E_{\perp}^{0} e^{-j(\boldsymbol{\beta}^r \cdot \mathbf{r} - \phi_{\perp}^r)}
\end{aligned} \tag{5-96b}$$

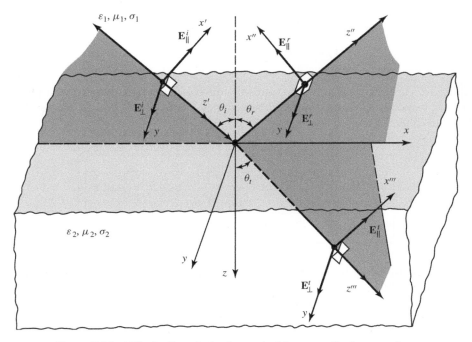

Figure 5-16 Elliptically polarized wave incident on a flat lossy surface.

where $\zeta_\parallel^r$ and $\zeta_\perp^r$ are the phases of the reflection coefficients for parallel and perpendicular polarizations, respectively. The Poincaré sphere angles γ^r and δ^r of the reflected field can now be written by referring to (5-96a) and (5-96b) as

$$\gamma^r = \tan^{-1}\left(\frac{|\mathbf{E}_\perp^r|}{|\mathbf{E}_\parallel^r|}\right) = \tan^{-1}\left(\frac{|\Gamma_\perp^b|E_\perp^0|}{|\Gamma_\parallel^b|E_\parallel^0|}\right) = \tan^{-1}\left(\frac{|\Gamma_\perp^b|}{|\Gamma_\parallel^b|}\tan\gamma^i\right) \tag{5-97a}$$

$$\delta^r = \phi_\perp^r - \phi_\parallel^r = (\delta^i + \zeta_\perp^r) - (\pi + \zeta_\parallel^r) = (\delta^i - \pi) + (\zeta_\perp^r - \zeta_\parallel^r) \tag{5-97b}$$

where δ^r is the phase angle by which the perpendicular (y) component leads the parallel (x'') component of the reflected field. Using the angles γ^r and δ^r of (5-97a) and (5-97b), the corresponding Poincaré sphere angles ε^r, τ^r (tilt angle of ellipse) and axial ratio $(AR)^r$ of the reflected field can be found using the relations

$$\sin(2\varepsilon^r) = \sin(2\gamma^r)\sin(\delta^r) \tag{5-98a}$$

$$\tan(2\tau^r) = \tan(2\gamma^r)\cos(\delta^r) \tag{5-98b}$$

$$(AR)^r = \cot(\varepsilon^r) \tag{5-98c}$$

Following a similar procedure, the transmitted fields can be expressed as

$$\mathbf{E}_\parallel^t = \hat{\mathbf{a}}_{x''}E_\parallel^t e^{-j\boldsymbol{\beta}^t\cdot\mathbf{r}} = \hat{\mathbf{a}}_{x''}T_\parallel^b E_\parallel^0 e^{-j\boldsymbol{\beta}^t\cdot\mathbf{r}} = \hat{\mathbf{a}}_{x''}|T_\parallel^b|E_\parallel^0 e^{-j(\boldsymbol{\beta}^t\cdot\mathbf{r}-\xi_\parallel^t)}$$
$$= \hat{\mathbf{a}}_{x''}|T_\parallel^b|E_\parallel^0 e^{-j(\boldsymbol{\beta}^t\cdot\mathbf{r}-\phi_\parallel^t)} \tag{5-99a}$$

$$\mathbf{E}_\perp^t = \hat{\mathbf{a}}_y E_\perp^t e^{-j\boldsymbol{\beta}^t\cdot\mathbf{r}} = \hat{\mathbf{a}}_y T_\perp^b E_\perp^0 e^{-j(\boldsymbol{\beta}^t\cdot\mathbf{r}-\phi_\perp^t)} = \hat{\mathbf{a}}_y|T_\perp^b|E_\perp^0 e^{-j(\boldsymbol{\beta}^t\cdot\mathbf{r}-\delta^i-\xi_\perp^t)}$$
$$= \hat{\mathbf{a}}_y|T_\perp^b|E_\perp^0 e^{-j(\boldsymbol{\beta}^t\cdot\mathbf{r}-\phi_\perp^t)} \tag{5-99b}$$

where $\xi_\parallel^t$, and $\xi_\perp^t$ are the phases of the transmission coefficients for parallel and perpendicular polarizations, respectively. The Poincaré sphere angles δ^t and γ^t can now be written by referring to (5-99a) and (5-99b) as

$$\gamma^t = \tan^{-1}\left(\frac{|\mathbf{E}_\perp^t|}{|\mathbf{E}_\parallel^t|}\right) = \tan^{-1}\left(\frac{|T_\perp^b|E_\perp^0|}{|T_\parallel^b|E_\parallel^0|}\right) = \tan^{-1}\left(\frac{|T_\perp^b|}{|T_\parallel^b|}\tan\gamma^i\right) \tag{5-100a}$$

$$\delta^t = \phi_\perp^t - \phi_\parallel^t = (\delta^i + \xi_\perp^t) - \xi_\parallel^t = \delta^i + (\xi_\perp^t - \xi_\parallel^t) \tag{5-100b}$$

where δ^t is the phase angle by which the perpendicular (y) component of the transmitted field leads the parallel (x''') component of the transmitted field. Using the angles γ^t and δ^t of (5-100a) and (5-100b), the corresponding Poincaré sphere angles ε^t, τ^t (tilt angle of ellipse) and axial ratio $(AR)^t$ of the transmitted field can be found using the relations

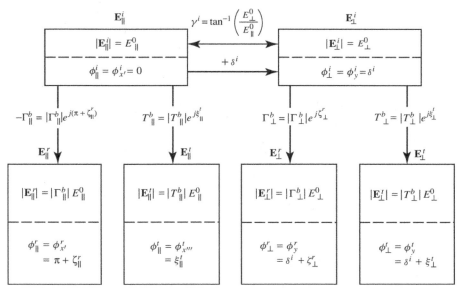

Figure 5-17 Block diagram for polarization analysis of reflected and transmitted waves.

$$\sin(2\varepsilon^t) = \sin(2\gamma^t)\sin(\delta^t)$$ (5-101a)

$$\tan(2\tau^t) = \tan(2\gamma^t)\cos(\delta^t)$$ (5-101b)

$$(AR)^t = \cot(\varepsilon^t)$$ (5-101c)

The set of (5-96a) through (5-98c) and (5-99a) through (5-101c) can be used to find, respectively, the polarization of the reflected and transmitted fields once the polarization of the incident fields of (5-93a) through (5-94d) has been stated. A block diagram of the relations between the incident, reflected, and transmitted fields is shown in Figure 5-17. The parallel component of the incident field is taken as the reference for the phase of all of the other components.

Example 5-12

A left-hand (CCW) circularly polarized field traveling in free space at an angle of $\theta_i = 30°$ is incident on a flat perfect electric conductor of infinite extent. Find the polarization of the reflected wave.

Solution: A circularly polarized wave is made of two orthogonal linearly polarized components with a 90° phase difference between them. Therefore we can assume that these two orthogonal linearly polarized components represent the perpendicular and parallel polarizations. Since the reflecting surface is perfectly conducting ($\eta_2 = 0$), the reflection coefficients of (5-17a) and (5-24c) reduce to

$$\Gamma_\perp^b = -1 = 1\underline{/\pi} \Rightarrow \left|\Gamma_\perp^b\right| = 1, \quad \zeta_\perp^r = \pi$$

$$\Gamma_\parallel^b = -1 = 1\underline{/\pi} \Rightarrow \left|\Gamma_\parallel^b\right| = 1, \quad \zeta_\parallel^r = \pi$$

Since the incident field is left-hand circularly polarized, then according to (5-93a) through (5-94b)

$$E_\parallel^0 = E_\perp^0$$

$$\delta^i = \phi_\perp^i = \frac{\pi}{2}$$

$$\gamma^i = \tan^{-1}\left(\frac{E_\perp^0}{E_\parallel^0}\right) = \frac{\pi}{4} \Rightarrow \tan\gamma^i = 1$$

Thus according to (5-97a) and (5-97b)

$$\gamma^r = \tan^{-1}\left(\frac{\left|\Gamma_\perp^b\right|}{\left|\Gamma_\parallel^b\right|}\tan\gamma^i\right) = \frac{\pi}{4}$$

$$\delta^r = \delta^i - \pi + (\zeta_\perp^r - \zeta_\parallel^r) = \frac{\pi}{2} - \pi + (\pi - \pi) = -\frac{\pi}{2}$$

On the Poincaré sphere of Figure 4-20 the angles $\gamma^r = \pi/4$ and $\delta^r = -\pi/2$ define the south pole, which represents right-hand (CW) circular polarization. Therefore, the reflected field is right-hand (CW) circularly polarized, and it is opposite in rotation to that of the incident field as shown in Figure 5-18a.

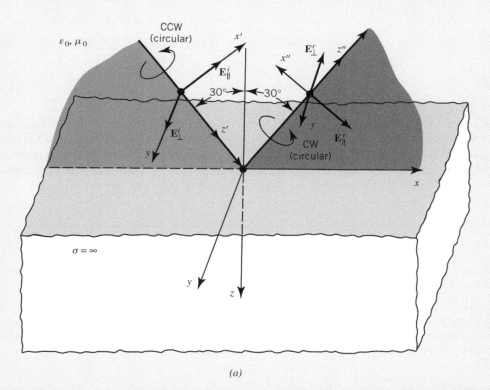

(a)

Figure 5-18 Circularly polarized wave incident upon flat surfaces with infinite and zero conductivities. (a) Infinite conductivity. (b) Lossless ocean.

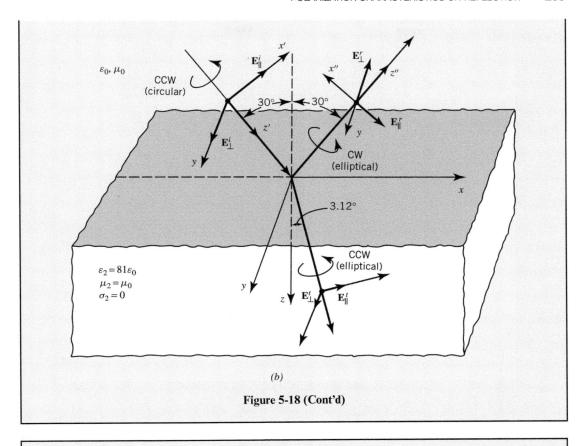

(b)

Figure 5-18 (Cont'd)

Example 5-13

A left-hand (CCW) circularly polarized field traveling in free space at an angle of $\theta_i = 30°$ is incident on a flat lossless ($\sigma_2 = 0$) ocean ($\varepsilon_2 = 81\varepsilon_0$, $\mu_2 = \mu_0$) of infinite extent. Find the polarization of the reflected and transmitted fields.

Solution: Since the incident field is left-hand circularly polarized, then according to (5-93a) through (5-94b)

$$E_\parallel^0 = E_\perp^0$$

$$\delta^i = \phi_\perp^i = \frac{\pi}{2}$$

$$\gamma^i = \tan^{-1}\left(\frac{E_\perp^0}{E_\parallel^0}\right) = \frac{\pi}{4} \Rightarrow \tan\gamma^i = 1$$

To find the polarization of the reflected field, we proceed as follows. Using (5-18a)

$$\Gamma_\perp^b = \frac{\cos(30°) - \sqrt{81}\sqrt{1 - \left(\frac{1}{81}\right)\sin^2(30°)}}{\cos(30°) + \sqrt{81}\sqrt{1 - \left(\frac{1}{81}\right)\sin^2(30°)}} = \frac{0.866 - 9\sqrt{1 - \frac{1}{81}\left(\frac{1}{4}\right)}}{0.866 + 9\sqrt{1 - \frac{1}{81}\left(\frac{1}{4}\right)}}$$

$$= \frac{0.866 - 8.986}{0.866 + 8.986}$$

$$\Gamma_\perp^b = -0.824 \Rightarrow \left|\Gamma_\perp^b\right| = 0.824, \ \zeta_\perp^r = \pi$$

Using (5-25a)

$$\Gamma_\parallel^b = \frac{-\cos(30°) + \sqrt{\dfrac{1}{81}}\sqrt{1 - \left(\dfrac{1}{81}\right)\sin^2(30°)}}{\cos(30°) + \sqrt{\dfrac{1}{81}}\sqrt{1 - \left(\dfrac{1}{81}\right)\sin^2(30°)}} = \frac{-0.866 + \dfrac{1}{9}\sqrt{1 - \dfrac{1}{81}\left(\dfrac{1}{4}\right)}}{0.866 + \dfrac{1}{9}\sqrt{1 - \dfrac{1}{81}\left(\dfrac{1}{4}\right)}}$$

$$= \frac{-0.866 + 0.111}{0.866 + 0.111}$$

$$\Gamma_\parallel^b = -0.773 \Rightarrow \left|\Gamma_\parallel^b\right| = 0.773, \ \zeta_\parallel^r = \pi$$

According to (5-97a) and (5-97b)

$$\gamma^r = \tan^{-1}\left(\frac{\left|\Gamma_\perp^b\right|}{\left|\Gamma_\parallel^b\right|}\tan\gamma^i\right) = \tan^{-1}\left(\frac{0.824}{0.773}\right) = 46.83° = 0.817 \ \text{rad}$$

$$\delta^r = \delta^i - \pi + (\zeta_\perp^r - \zeta_\parallel^r) = \frac{\pi}{2} - \pi + (\pi - \pi) = -\frac{\pi}{2}$$

Using (5-98a) through (5-98c)

$$2\varepsilon^r = \sin^{-1}[\sin(2\gamma^r)\sin(\delta^r)]$$

$$= \sin^{-1}\left[\sin(93.66°)\sin\left(-\frac{\pi}{2}\right)\right] = -86.34°$$

$$\Rightarrow \varepsilon^r = -43.17°$$

$$2\tau^r = \tan^{-1}[\tan(2\gamma^r)\cos(\delta^r)]$$

$$= \tan^{-1}\left[\tan(93.66°)\cos\left(-\frac{\pi}{2}\right)\right] = 180°$$

$$\Rightarrow \tau^r = 90°$$

$$(AR)^r = \cot(\varepsilon^r) = \cot(-43.17°) = -1.066$$

On the Poincaré sphere of Figure 4-20 the angles $\gamma^r = 0.817$ and $\delta^r = -\pi/2$ locate a point on the lower hemisphere on the principal xz plane. Therefore the reflected field is right-hand (CW) elliptically polarized, and it has an opposite sense of rotation compared to the left-hand (CCW) circularly polarized incident field as shown in Figure 5-18b. Its axial ratio is -1.066.

To find the polarization of the transmitted field we proceed as follows. Using (5-18b)

$$T_\perp^b = \frac{2\cos(30°)}{\cos(30°) + \sqrt{81}\sqrt{1 - \left(\dfrac{1}{81}\right)\sin^2(30°)}} = \frac{2(0.866)}{0.866 + 8.986}$$

$$= 0.1758 \Rightarrow \left|T_\perp^b\right| = 0.1758, \ \xi_\perp^t = 0$$

Using (5-25b)

$$T_{\parallel}^b = \frac{2\sqrt{\dfrac{1}{81}}\cos(30°)}{\cos(30°)+\sqrt{\dfrac{1}{81}}\sqrt{1-\left(\dfrac{1}{81}\right)\sin^2(30°)}} = \frac{2\left(\dfrac{1}{9}\right)0.866}{0.866+0.111}$$

$$= 0.197 \Rightarrow \left|T_{\parallel}^b\right| = 0.197, \quad \xi_{\parallel}^t = 0$$

According to (5-100a) and (5-100b)

$$\gamma^t = \tan^{-1}\left(\frac{\left|T_{\perp}^b\right|}{\left|T_{\parallel}^b\right|}\tan\gamma^i\right) = \tan^{-1}\left(\frac{0.1758}{0.197}\right) = 41.75° = 0.729 \text{ rad}$$

$$\delta^t = \delta^i + (\xi_{\perp}^t - \xi_{\parallel}^t) = \frac{\pi}{2} + (0-0) = \frac{\pi}{2}$$

Using (5-101a) through (5-101c)

$$2\varepsilon^t = \sin^{-1}[\sin(2\gamma^t)\sin(\delta^t)] = \sin^{-1}[\sin(83.5°)\sin(90°)] = 83.5°$$

$$\Rightarrow \varepsilon^t = 41.75°$$

$$2\tau^t = \tan^{-1}[\tan(2\gamma^t)\cos(\delta^t)] = \tan^{-1}[\tan(83.5°)\cos(90°)] = 0$$

$$\Rightarrow \tau^t = 0°$$

$$(AR)^t = \cot(\varepsilon^t) = \cot(41.75°) = 1.12$$

On the Poincaré sphere of Figure 4-20 the angles $\gamma^t = 0.729$ and $\delta^t = \pi/2$ locate a point on the upper hemisphere on the principal xz plane. Therefore the transmitted field is left-hand (CCW) elliptically polarized, and it is of the same sense of rotation as the left-hand (CCW) circularly polarized incident field as shown in Figure 5-18b. Its axial ratio is 1.12.

5.7 METAMATERIALS

The decades of the 1990s and 2000s saw renewed interest and excitement in the field of electro-magnetics, especially as they relate to the integration of a special type of artificial dielectric materials, coined *metamaterials* [13–18]. Using a 'broad brush,' the word metamaterials can encompass *engineered textured surfaces, artificial impedance surfaces, artificial magnetic conductors, double negative materials, frequency selective surfaces, Photonic Band-Gap (PBG) surfaces, Electromagnetic Band-Gap (EBG) surfaces/structures,* and even *fractals* or *chirals.* Artificial impedance surfaces are discussed in Section 8.8. In this section we want to focus more on material structures whose constitutive parameters (permittivity and permeability) are both negative, often referred to as Double Negative (DNG). Artificial magnetic conductors can also be included in the DNG class of materials. It is the class of DNG materials that has captivated the interest and imagination of many leading researchers and practitioners, scientists and engineers, from academia, industry, and government. When electromagnetic waves interact with such materials, they exhibit some very unique and intriguing characteristics and phenomena that can be used, for example, to optimize the performance of antennas, microwave components and circuits, transmission lines, scatterers, and optical devices such as lenses. While the revitalization of metamaterials introduced welcomed renewed interest in materials for electromagnetics, it also brought along some spirited dialogue, which will be referred to in the pages that follow.

The word *meta*, in *metamaterials*, is a Greek word that means beyond/after. The term *metamaterials* was coined in 1999 by Dr. Rodger Walser, of the University of Texas-Austin and Metamaterial, Inc., to present materials that are artificially fabricated so that they have electromagnetic properties that go beyond those found readily in nature. In fact, the word has been used to represent materials that microscopically are intrinsically inhomogeneous and constructed from metallic arrangements that exhibit periodic formations whose period is much smaller than the free-space and/or guided wavelength. Using Dr. Walser's own words, he defined metamaterials as 'Macroscopic composites having man-made, three-dimensional, periodic cellular architecture designed to produce an optimized combination, not available in nature, of two or more responses to specific excitation' [19]. Because of the very small period, such structures can be treated as homogeneous materials, similarly to materials found in nature, and they can then be represented using bulk constitutive parameters, such as permittivity and permeability. When the period is not small compared to the free-space or guided wavelength, then such materials can be examined using periodic analysis (i.e., the *Floquet Theorem*). Typically the construction of metamaterials is usually performed by embedding inclusions or inhomogeneities in the host medium, as shown in Figure 5-19 [13].

5.7.1 Classification of Materials

In general, materials, using their constitutive parameters ε (permittivity) and μ (permeability) as a reference, can be classified into four categories. They are those that exhibit:

- Negative ε and positive μ; they are usually coined as **ENG** (epsilon negative) material.
- Positive ε and positive μ; they are usually coined as **DPS** (double positive) material.
- Negative ε and negative μ; they are usually coined as **DNG** (double negative) material.
- Positive ε and negative μ; they are usually coined as **MNG** (mu negative) material.

These are shown schematically in Figure 5-20.

Of the materials shown in Figure 5-20, the ones that usually are encountered in nature are those of **DPS** (double positive; first quadrant, like dielectrics such as water, glass, plastics, etc.), **ENG** (epsilon negative; second quadrant, like plasmas) and **MNG** (mu negative; fourth quadrant, like magnetic materials). Obviously the one set that is most widely familiar and used in applications is that of **DPS**, although the other two, **ENG** and **MNG**, are used in a wide range of applications.

Figure 5-19 Metamaterial representation using embedded periodic inclusions (after [13]).

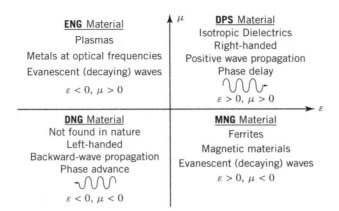

Figure 5-20 Characterization of materials according to the values of their permittivity and permeability (after [13], [17]).

5.7.2   Double Negative (DNG) Materials

The materials that have recently captured the attention and imagination of electromagnetic engineers and scientists are the **DNG**, which, as indicated, are not found in nature but may be artificially realizable. The **DNG** materials are also referred to as **NRI** (negative refractive index), **NIM** (negative index material), **BW** (backward) media, and left-handed (**LH**) media, to name a few. For clarity and simplicity, we will stay with the **DNG** designation. The **DNG** class has created an intense activity as many have attempted to incorporate material with such characteristics to design, enhance, or increase the performance of lenses, microwave circuits, transmission lines, antennas, phase shifters, broadband power dividers, backward and forward leaky-wave antennas, electrically small ring antennas, cloaking, plasmonic nanowires, photonic crystals, and miniaturization [13–21]. More specifically, using antennas as an example, it has been reported that the integration of materials with radiating elements can increase the radiated power, enhance the gain, and tune the frequency of operation.

While there has been a lot of activity since the recent revival of metamaterials, their introduction has also created some spirited dialogue about the negative index-of-refraction, negative refraction angle, and phase advancement [19–21]. What may have elevated this dialogue to a greater level is that some of the reported results using DNG metamaterials may have been overstated, and lacked verification, interpretation, and practical physical realization; see [22] and Appendix C of [23]. However, within the broader definition of metamaterials, there have been metamaterial structures whose performance, when combined with devices and circuits, has been validated not only by simulations but also by careful experimentation. For such structures not only good agreement between simulations and measurements has been found, but also the results have been within limits of physical reality and interpretation. Some of these have been acknowledged for their validity, and they have also often been referred to as engineered textured surfaces, artificial impedance surfaces (AIS), artificial magnetic conductors (AMC), photonic band-gap structures (PBG), and electromagnetic band-gap structures (EBG). This class of metamaterials is discussed in Section 8.8, and the reader is referred to that section for details and references.

Because of the interest in the electromagnetic community, it is important that the topic of metamaterials be introduced to graduate students, and maybe even to undergraduates, but presented in the proper context. Because of space limitations, only an introductory overview of the subject is included in this book. A succinct chronological sequence of the basic events that led to this immense interest in metamaterials is also presented. The reader is referred to the literature for an in-depth presentation of the topic and its applications.

5.7.3 Historical Perspective

The origins of metamaterials can be traced back to the end of the 19th century, and they are outlined in many publications. Since metamaterials is a rather new designation, it is a branch of artificial dielectrics. In fact, it was indicated in 1898 that Jagadish Chandra Bose may have emulated chiral media by using man-made twisted fibers to rotate the polarization of electromagnetic waves [24]. In 1914, Karl F. Lindman examined artificial chiral media when he attempted to embed into the material an ensemble of randomly oriented small wire helices [25]. In 1948, Winston E. Kock of Bell Laboratories introduced the basic principles of artificial dielectrics to design lightweight lenses in the microwave frequency range (around 3–5 GHz) [26]. His attempt was to replace at these frequencies, where the wavelength is 10-6 centimeters, heavy and bulky lenses made of natural dielectric materials. He realized his concept of artificial dielectrics by controlling the effective index-of-refraction of the materials by embedding into them, and arranging periodically, metallic disks and spheres in a concave lens shape.

The paper that revived the interest in the special class of artificial materials, now coined *metamaterials* and not found in nature, was that of Victor Veselago in 1968 who analyzed the propagation of uniform waves in materials that exhibited, simultaneously, both negative permittivity and permeability (DNG; double negative) [27]. Although Veselago may not have been interested in dielectric materials, he examined analytically the wave propagation through materials that exhibited, simultaneously, negative ε and negative μ. One of the materials that can be created in nature is plasma, which can exhibit negative permittivity. Plasma is an ionized gas of which a significant number of its charged particles interact strongly with electromagnetic fields and make it electrically conductive. For those that lived through the birth of the U.S. space program in the mid-1960s, led by NASA, there was a lot of interest and research in plasmas, formed beneath and around the nose of the spacecraft during re-entry that caused loss of communication with the astronauts during the final 10–15 minutes of landing. To attempt to alleviate this loss of communication (referred to then as *blackout*), due to the formed plasma sheath near the nose and belly of the spacecraft, NASA initiated and carried out an intense research program on plasma. The plasma was modeled with a negative dielectric constant (negative permittivity), and it was verified through many experiments.

Although Veselago may have known that negative ε can be obtained by plasma-type materials, he did not speculate, at least in [27], how and what kind of materials may exhibit DNG properties. However, he was able to show and conclude, through analytical formulation, that for wave propagation through DNG type of materials, the direction of the power density flow (Poynting vector) is opposite to the wave propagation (phase vector). He referred to such materials as *left-handed*. Based on his conclusions, the directions of power density flow and phase velocity for DPS materials (double positive, which are conventional dielectrics) and DNG materials (double negative, not found in nature) are illustrated graphically in Figure 5-21, where a uniform plane wave propagates in DPS (Figure 5-21a) and DNG (Figure 5-21b) materials. The DPS materials are also dubbed Right-Handed Materials (RHM) while the DNG materials are dubbed Left-Handed Materials (LHM). The solid arrows represent the directions of wave vectors (phase velocities) while the dashed arrows represent power flow (Poynting vectors). While the arrows in Figure 5-21a illustrate the directions that we expect from conventional dielectrics, the arrows in Figure 5-21b point in the opposite direction, which will indicate that there is a phase advance (phase wave fronts move toward the source) for the wave in Figure 5-21b and a phase delay for the wave in Figure 5-21a, which is what we are accustomed to from conventional dielectrics. To get the phase advance of Figure 5-21b requires that the phase constant (wave number) is negative. This is accomplished by defining both the permittivity and permeability negative; thus the name of DNG material. These concepts will be presented here analytically, but first an outline will be created to lay the groundwork of metamaterials, at least as of this writing.

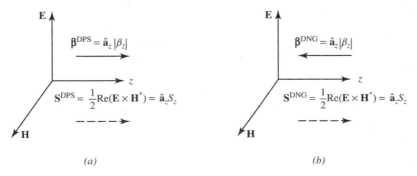

Figure 5-21 Direction of phase vector (β) and Poynting vector (**S**) for uniform wave propagation in double positive (DPS) and double negative (DNG) materials. (*a*) RHM: double positive material (DPS). (*b*) LHM: double negative material (DNG).

5.7.4 Propagation Characteristics of DNG Materials

Veselago in his seminal paper showed, using a slab of DNG material embedded into a host DPS medium (the same DPS to the left and to the right of the DNG slab), that an impinging wave emanating from a source to the left of the DNG slab will focus, creating caustics at two different points (one within the DNG slab and the other one to the right of the DNG slab), as long as the slab is sufficiently thick. This is accomplished by using, for the DNG slab, permittivity and permeability that are of the same magnitudes but opposite signs as those of the host DPS medium ($\varepsilon_2 = -\varepsilon_1$, $\mu_2 = -\mu_1$; index-of-refraction $n_2 = -n_1$). This is shown graphically in Figure 5-22, and it is often referred to as the *Veselago planar lens*. This, of course, seemed very attractive and was probably one of the reasons for the genesis of the renewed interest of modern metamaterials. However, the Veselago planar lens was also analyzed using a classical method based on Fourier transforms in the frequency domain, and the sinusoidal field exciting the lens expressed in terms of even and odd resonant surface wave modes whose amplitudes were evaluated by residues at the poles [28], Appendix D of [23]. Based on this analytical approach, the following observations were made in [28] and Appendix D of [23]: A CW sinusoidal source solution to "a lossless Veselago flat lens with super resolution is not physically possible" because of the presence of surface waves that

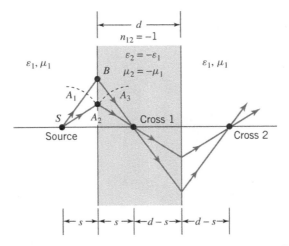

Figure 5-22 Veselago's planar/flat lens: focusing by a DNG slab between two DPS materials [19]. (Reprinted with permission from John Wiley & Sons, Inc.)

produce divergent fields over a region within and near the Veselago lens. If losses are included, the excited interfering surface wave modes will decay in a short time interval; however, the lens resolution will depend on the losses, and it will be substantially reduced if they are moderate to large [28], Appendix D of [23]. The analysis assumes that the incident field has a finite continuous frequency spectrum, and the negative epsilon and mu are frequency dispersive, which Veselago indicates are necessary for the field energy to be positive.

The time-domain solution to a frequency dispersive Veselago lens illuminated by a sinusoidal source that begins at $t = 0$ has also been determined [29]. The time-domain fields remain finite everywhere for finite time t and approach the fields of a CW source only as $t \to \infty$. In particular, the divergent fields encountered in the CW solution to the lossless Veselago lens are caused by the infinite CW energy imparted (during the infinite amount of time between $t \to -\infty$ and the present time t) to the evanescent fields in the vicinity of the slab; analogous to the divergent fields produced by a CW source inside a lossless cavity at a resonant frequency.

The work of Veselago remained dormant for about 30 years, and it was not until the late 1990s when J. B. Pendry and his colleagues suggested that DNG materials could be created artificially by using periodic structures [30–33]. Not long after Pendry, D. R. Smith and his collaborators [34–38] built materials that exhibited DNG characteristics. This was accomplished by the use of a structure consisting of split-ring resonators and wires, a unit cell of which is shown in Figure 5-23. It was suggested that the split-ring element, of the type shown in Figure 5-23a, will contribute a negative permeability while the infinite length wire of Figure 5-23b will contribute a negative permittivity; the combination of the two will, in a periodic structure, contribute a negative index-of-refraction. An experimental array of split-ring resonators and wires is shown in Figure 5-24. In fact, Smith and his team claimed to have observed experimentally negative refraction. In [19] this phenomenon was claimed to be radiation from either a surface wave characteristic of finite periodic structures or possibly a sidelobe from the main beam [39].

Because of the immense interest in DNG materials, with negative permittivity and permeability, there were a number of subsequent experiments, in addition to that in [38], to attempt to verify the negative permittivity and permeability, and thus negative index-of-refraction. Some of these experiments, along with the corresponding references, are summarized in [40]. For the simulations, a frequency-dispersive Drude model [13] was used to represent the negative permittivity of the infinite wires while a frequency-dispersive Lorentz model [32] was utilized for the representation of the negative permeability of the split-rings of Figure 5-23. The experiments consisted of parallel plate waveguide techniques utilizing both metamaterial slabs and prisms [40], and most of

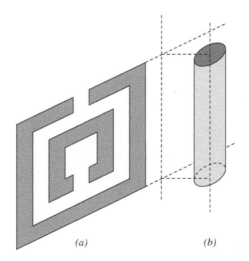

(a)   *(b)*

Figure 5-23 Simulation of DNG material (negative refraction) using split-ring resonators and wires. (*a*) Split ring. (*b*) Wire.

Figure 5-24 Simulation of DNG material (negative refraction) using split-ring resonators and wires [38]. From R. A. Shelby, D. R. Smith, S. Schultz, 'Experimental verification of a negative index-of-refraction,' *Science*, vol. 292, pp. 77–79, April 2001. (Reprinted with permission from AAAS.)

the measurements were carried out in the 4–20 GHz region. The refraction could be observed by having the slab samples rotated or by having the plane wave incident at an oblique angle. While the nearly plane wave incidence was easier to implement experimentally, the rotation of the samples yield good experimental results. The use of prisms was also an alternative and popular experiment. The metamaterial slabs and prisms were fabricated by embedding various geometrical shapes to represent the characteristics of both wire and different shape split-ring inclusions. In some of the experiments, the metamaterials included only split-ring type of inclusions to verify the negative permeability. The use of an S-shaped unit cell in the metamaterial structure provided an alternative geometry that simulated both a negative permittivity and permeability, and thus did not require the straight wire to represent the negative permittivity; alternate S-ring designs could also be used to possibly achieve dual frequency bands [40]. Gaussian beams and nearly simulated plane waves were used to perform transmission and focusing experiments to validate the negative index-of-refraction, using both dielectric and solid state structures. The solid state metamaterial structures were introduced to minimize the mismatch losses (which were greater for dielectric structures and led to low power levels), improve the mechanical fragility, and make metamaterials more attractive for industrial applications [40]. It was reported that both the transmission and focusing experiments produced results that indicated negative permittivity and permeability, and thus, the creation of a negative effective index-of-refraction [40].

The attractive performance of devices and systems that incorporated metamaterials led to the genesis of the enormous interest on the subject by many teams around the world, and the avalanche of papers published in transactions and journals, presented in symposia and conferences, and applied to numerous problems with exotic characteristics and performances. The word metamaterials became a 'household' word in the electromagnetic community in the 2000–2010 time period. This type of materials exhibit narrow bandwidths, which may have limited its applications.

5.7.5 Refraction and Propagation Through DNG Interfaces and Materials

Now that a brief historical and chronological background of the evolution of metamaterials has been outlined, we will present a special case of what initially were referred to as artificial dielectrics, the basics from the analytical point of view as well as from a sample of simulations, and experiments. It should be pointed out, however, that what ensued after the work by Pendry and Smith was a plethora of publications which are too numerous to include here. Up to this point an attempt was made to reference some of the most basic books and papers. The reader is referred to the technical transactions, journals, and letters where most of these ensuing papers were published or presented at leading international conferences and symposia. Most of these can be found in references [41–46].

The greatest potential of the DNG materials is the creation of a structure with a negative index-of-refraction n defined as

$$n^2 = \varepsilon_r \mu_r \Rightarrow n = \pm\sqrt{\varepsilon_r \mu_r} = \pm\sqrt{-|\varepsilon_r|(-|\mu_r|)} = \pm\,(j\sqrt{|\varepsilon_r|})(j\sqrt{|\mu_r|}) = \pm j^2\sqrt{|\varepsilon_r \mu_r|}$$

$$n = \mp\sqrt{|\varepsilon_r \mu_r|} \qquad (5\text{-}102)$$

Which sign of n should be chosen for DNG materials (with both ε_r and μ_r negative)? It seems from (5-102) that there are two basic choices; either negative or positive n. If a positive n is selected, that resorts back to the DPS representation. If the negative value of n in (5-102) is selected, then that is the basis of DNG materials.

Materials with negative index-of-refraction have some interesting properties, some of which have been mentioned and illustrated in Figure 5-21. Now let us examine two interface options using Snell's law of refraction which is the manifest of phase match across the interface. Of particular interest are materials with negative index-of-refraction.

- Snell's law of refraction, represented by (5-15b) and (5-24b), or

$$\beta_1 \sin\theta_i = \omega\sqrt{\mu_1 \varepsilon_1}\,\sin\theta_i \equiv \beta_2 \sin\theta_t = \omega\sqrt{\mu_2 \varepsilon_2}\,\sin\theta_t \qquad (5\text{-}103)$$

  can also be written as

$$n_1 \sin\theta_i = n_2 \sin\theta_t \qquad (5\text{-}104)$$

  When the index-of-refraction of both materials forming the interface is positive, then the refracted ray (transmitted wave) will be, as expected for conventional materials, on the same side (relative to the normal to the interface) as the reflected ray, as illustrated in Figure 5-25a. However, when the index-of-refraction of one material is positive while that of the other is negative, the refracted ray (transmitted wave) will be in the opposite direction of the reflected ray, as illustrated in Figure 5-25b.

- For DNG materials with a negative index-of-refraction the phase constant (wave number) of the wave traveling in the DNG material is negative, or based on the definition of (5-103)

$$\beta_2 = \omega\sqrt{\mu_2 \varepsilon_2} = -\omega\sqrt{|\mu_2||\varepsilon_2|} \qquad (5\text{-}105)$$

  This implies that, for positive time, there will be a phase advance (phase wavefronts move toward the source), instead of a phase delay that we have been accustomed to. This is an interesting phenomenon, which has been part of the spirited dialogue.

So, based on the above, a negative index-of-refraction leads to:

- A refracted angle that is on the same side, relative to the normal to the interface, as the incident angle, and the power flow (Poynting vector) is outward (as expected); however, the phase vector in inward (opposite to the Poynting vector).
- Phase advance, instead of phase delay that is typical of DPS materials.

Based on the above, let us examine through an example a more general case of the planar lens that was illustrated in Figure 5-22.

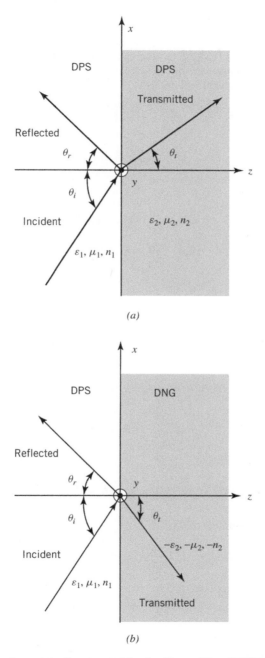

Figure 5-25   Refraction by planar interface created by double positive (DPS) and double negative (DNG) materials. (*a*) DPS-DPS. (*b*) DPS-DNG.

Example 5-14

Figure 5-22 displays Veselago's planar/flat lens. A more general one is the one of Figure 5-26 where a DNG slab is sandwiched within free space. Given the dimensions of the DNG slab of thickness d and the source position s, as shown in the Figure 5-26, determine the location of the foci (caustics) f_0 and f_1 (one within the DNG slab and one outside it) in terms of the incidence angle θ_i, position of the source s, and thickness d and index-of-refraction n_1 of the DNG slab. Assume the DNG slab possesses negative permittivity $-\varepsilon_1$, negative permeability $-\mu_1$, and negative index-of-refraction $-n_1$. Furthermore, let us assume that we are looking for a solution based on geometrical optics.

Solution: Using (5-103) through (5-105), we can write for the leading interface between free space and the DNG slab that

$$\theta_1 = \sin^{-1}\left(\frac{1}{|n_1|}\sin\theta_i\right)$$

Also from Figure 5-26

$$\tan\theta_i = \frac{h_1}{s} \Rightarrow h_1 = s\,\tan\theta_i$$

$$\tan\theta_1 = \frac{h_1}{f_0} \Rightarrow h_1 = f_0\,\tan\theta_1$$

Equating the two previous equations leads to

$$s\,\tan\theta_i = f_0\,\tan\theta_1 \Rightarrow f_0 = s\frac{\tan\theta_i}{\tan\theta_1} \Rightarrow \tan\theta_1 = \frac{s}{f_0}\tan\theta_i$$

From Figure 5-26

$$\tan\theta_0 = \frac{h_2}{f_1} \Rightarrow h_2 = f_1\,\tan\theta_0$$

$$\tan\theta_1 = \frac{h_2}{d - f_0} \Rightarrow h_2 = (d - f_0)\tan\theta_1$$

Equating the last two equations leads to

$$f_1\,\tan\theta_0 = (d - f_0)\tan\theta_1 \Rightarrow f_1 = (d - f_0)\frac{\tan\theta_1}{\tan\theta_0}$$

which can also be expressed, assuming $d > f_0$, as

$$f_1 = (d - f_0)\frac{\tan\theta_1}{\tan\theta_0} = (d - f_0)\frac{s}{f_0}\frac{\tan\theta_i}{\tan\theta_0}$$

Since $\theta_0 = \theta_i$, the above equation reduces to

$$\boxed{f_1 = (d - f_0)\frac{s}{f_0}}$$

As the magnitude of $-\varepsilon_1$ approaches that of free space (that is $|-\varepsilon_1| \to |\varepsilon_0| \Rightarrow |-n_1| \to |n_0| = 1$), the focal distance $f_0 f$ approaches $s(f_0 \to s)$ and f_1 approaches $d - s(f_1 \to d - s)$. Then Figure 5-26 reduces, in this limiting case, to Figure 5-22. When s becomes very large (approaching infinity), the incident wave reduces to near normal incidence. In this case the focusing moves toward infinity (ideally no focusing).

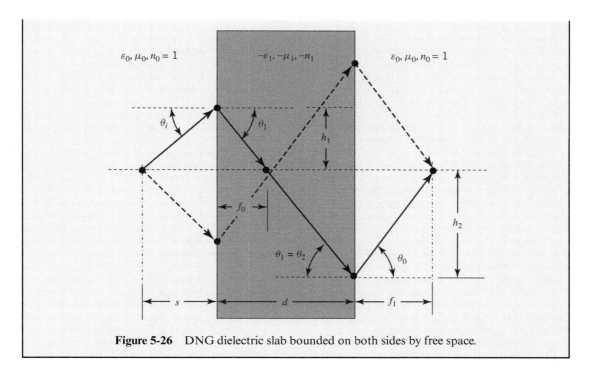

Figure 5-26 DNG dielectric slab bounded on both sides by free space.

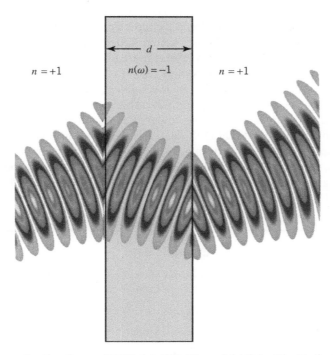

Figure 5-27 Negative refraction from a DNG slab [48]. (Copyright © by The Optical Society of America. Permission and courtesy of R. W. Ziolkowski.)

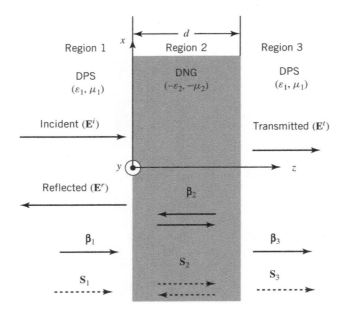

Figure 5-28 Reflection and transmission through a DNG slab.

To illustrate the DNG refraction, a simulation has been performed, using the Finite-Difference Time-Domain method, of a 30 GHz perpendicularly polarized CW Gaussian beam incident at 20° on a DNG slab bordered from the left and right by free space, as shown in Figure 5-27 [48]. Because the incident wave is a plane wave, there is no focusing. The index-of-refraction of the DNG slab is $n = -1$, and it was chosen to minimize reflections. Identical electric and magnetic Drude models were selected with parameters chosen so that only small losses were considered [13, 48]. Assuming the stated parameters of the media, the negative refraction is visible at the leading and trailing interfaces.

Another interesting observation will be to illustrate, through an example, the propagation of a plane wave through a slab of metamaterial, of thickness d, when it is embedded into a conventional dielectric material, as shown in Figure 5-28. This is similar to the problem for ordinary dielectrics, illustrated in Figure 5-11. For convenience, it is assumed that in Figure 5-28 the media to the left and right of the metamaterial DNG slab are both conventional dielectrics and *identical*. Also, at first we examine wave propagation at normal incidence, which is similar to that of conventional dielectrics, shown in Figure 5-11. The phase vectors β ($\longleftrightarrow$) and Poynting vectors **S** ($\centerdot\centerdot\centerdot\centerdot\!\!\blacktriangleright$) in each region are also indicated by their respective arrows. The analytical formulation of the reflection and transmission coefficients follows.

Example 5-15

For the DNG geometry of Figure 5-28, derive a simplified expression for the total input reflection at the initial interface and the total transmission coefficient through the entire DNG slab.

Solution: Using (5-67d), the total input reflection coefficient at the leading edge of the slab can be written as

$$\Gamma = \frac{E^r}{E^t} = \frac{\Gamma_{12} + \Gamma_{23}e^{-j2\beta_2 d}}{1 + \Gamma_{12}\Gamma_{23}e^{-j2\beta_2 d}} \xrightarrow[\eta_3 = \eta_1]{\Gamma_{23} = -\Gamma_{12}} \frac{\Gamma_{12}\left(1 - e^{-j2\beta_2 d}\right)}{1 - (\Gamma_{12})^2 e^{-j2\beta_2 d}}$$

which for a DNG slab, based on (5-105), reduces to

$$\Gamma = \frac{E^r}{E^t} = \frac{\Gamma_{12} + \Gamma_{23}e^{+j2|\beta_2|d}}{1 + \Gamma_{12}\Gamma_{23}e^{+j2|\beta_2|d}} \xrightarrow[\eta_3 = \eta_1]{\Gamma_{23} = -\Gamma_{12}} \frac{\Gamma_{12}\left(1 - e^{+j2|\beta_2|d}\right)}{1 - (\Gamma_{12})^2 e^{+j2|\beta_2|d}}$$

since

$$\Gamma_{12} = \left[\frac{\eta_2 - \eta_1}{\eta_2 + \eta_1}\right] = -\Gamma_{23}$$

Similarly, it can be shown that the transmission coefficient can be written as [13]

$$T = \frac{E^t}{E^i} = \frac{4\eta_2\eta_3 e^{-j\beta_2 d}}{(\eta_1 + \eta_2)(\eta_2 + \eta_3)}\frac{1}{\left(1 + \Gamma_{12}\Gamma_{23}e^{-j\beta_2 d}\right)}$$

$$T = \frac{E^t}{E^i} \xrightarrow[\eta_3 = \eta_1]{\Gamma_{23} = -\Gamma_{12}} \frac{4\eta_2\eta_1 e^{-j\beta_2 d}}{(\eta_1 + \eta_2)^2}\frac{1}{\left[1 - (\Gamma_{12})^2 e^{-j\beta_2 d}\right]}$$

which for the DNG slab reduces to

$$T = \frac{E^t}{E^i} \xrightarrow[\eta_3 = \eta_1]{\Gamma_{23} = -\Gamma_{12}} \frac{4\eta_2\eta_1 e^{+j|\beta_2|d}}{(\eta_1 + \eta_2)^2}\frac{1}{\left[1 - (\Gamma_{12})^2 e^{+j2|\beta_2|d}\right]}$$

An interesting observation is made if the DNG dielectric slab of Example 5-15 is matched to the medium it is embedded; that is, if $\eta_2 = \eta_1$. For this case, $\Gamma_{12} = 0$, and the total input reflection and the transmission coefficients of Example 5-15 reduce, respectively, to

$$\Gamma = 0 \qquad\qquad (5\text{-}106a)$$

$$T = e^{+j|\beta_2|d} \qquad\qquad (5\text{-}106b)$$

The transmission coefficient of (5-106b) indicates a phase advance (phase wavefront moving toward the source), instead of a phase delay as we are accustomed for wave propagation through conventional materials. This wave propagation through DNG materials is a unique feature that can be taken advantage of in various applications. As an example, the usual phase delay in conventional dielectric slabs and/or transmission lines can be compensated by phase advance in DNG type of slabs and/or transmission lines [13, 15, 16, 47] and others.

Now consider a uniform plane wave propagating at oblique incidence through a planar interface consisting of two materials. The case where both media are DPS has been treated in Section 5.3.1 for perpendicular polarization (Figure 5-2) and in Section 5.3.2 for parallel polarization (Figure 5-4). Now we will examine the wave propagation through a DNG medium; in this case medium 2 is DNG, when the first medium is DPS. However, before this is done, the interface formed by two DPS materials will be examined first. The planar interface formed by one DPS and one DNG material is examined afterwards. Only the perpendicular polarization of Figure 5-2 is considered. The same procedure can be applied to Figure 5-4 for the parallel polarization.

Based on the geometry of Figure 5-2, the vector wavenumbers for the incident, reflected, and transmitted fields can be written as

$$\boldsymbol{\beta}_i = \beta_1\left(\hat{\mathbf{a}}_x \sin\theta_i + \hat{\mathbf{a}}_z \cos\theta_i\right) = n_1 \frac{\omega}{v_0}\left(\hat{\mathbf{a}}_x \sin\theta_i + \hat{\mathbf{a}}_z \cos\theta_i\right) \tag{5-107a}$$

$$\boldsymbol{\beta}_r = \beta_1\left(\hat{\mathbf{a}}_x \sin\theta_i - \hat{\mathbf{a}}_z \cos\theta_i\right) = n_1 \frac{\omega}{v_0}\left(\hat{\mathbf{a}}_x \sin\theta_i - \hat{\mathbf{a}}_z \cos\theta_i\right) \tag{5-107b}$$

$$\boldsymbol{\beta}_t = \beta_2\left(\hat{\mathbf{a}}_x \sin\theta_t + \hat{\mathbf{a}}_z \cos\theta_t\right) = n_2 \frac{\omega}{v_0}\left(\hat{\mathbf{a}}_x \sin\theta_t + \hat{\mathbf{a}}_z \cos\theta_t\right) \tag{5-107c}$$

Using the expressions for the electric and magnetic fields of (5-10a)–(5-12b), the Poynting vectors for the respective three fields (incident, reflected, and refracted) can be written as

$$\mathbf{S}^i = \frac{1}{2}\frac{|E_0|^2}{\eta_1}(\hat{\mathbf{a}}_x \sin\theta_i + \hat{\mathbf{a}}_z \cos\theta_i) \tag{5-108a}$$

$$\mathbf{S}^r = \frac{1}{2}\frac{|\Gamma E_0|^2}{\eta_1}(\hat{\mathbf{a}}_x \sin\theta_i - \hat{\mathbf{a}}_z \cos\theta_i) \tag{5-108b}$$

$$\mathbf{S}^t = \frac{1}{2}\frac{|T E_0|^2}{\eta_2}(\hat{\mathbf{a}}_x \sin\theta_t + \hat{\mathbf{a}}_z \cos\theta_t) \tag{5-108c}$$

This is left as end-of-the-chapter exercises for the reader. It is apparent, from the vectors within the parentheses in (5-107a) through (5-108c), that for a DPS-DPS interface the phase vectors and the Poynting vectors for all three fields (incident, reflected, and refracted) are all parallel to each other and in the same directions.

Now let us consider the same oblique incidence upon a DPS-DNG interface, as shown in Figure 5-29. Snell's law of refraction, which is given by (5-103) and (5-104), can be expressed as

$$\sin\theta_t = \frac{\omega\sqrt{\mu_1\varepsilon_1}}{\omega\sqrt{\mu_2\varepsilon_2}}\sin\theta_i = \frac{n_1}{n_2}\sin\theta_i \Rightarrow \theta_t = \sin^{-1}\left(\frac{n_1}{n_2}\sin\theta_i\right) \tag{5-109}$$

For positive n_1 and n_2, the angle θ_t is positive, and everything follows what we already have experienced with DPS materials. However, when n_1 and n_2 have opposite signs, the angle θ_t is negative, as indicated in Figures 5-25, 5-26, and 5-29, and simulated in Figure 5-27. Based on these figures, whose interface is formed by a DPS and a DNG material (which leads to a negative angle of refraction), we will examine the directions of the phase vectors of (5-107) and Poynting vectors of (5-108) for the perpendicular polarization. The same can be done for the parallel polarization. This is left as an end-of-the-chapter exercise for the reader.

Since for the interface of Figure 5-29 the index-of-refraction of medium 2 is negative and the wavenumber is also negative, as expressed by (5-105), the wave vectors of (5-107a) and the Poynting vectors of (5-108a) can now be written, respectively, as

$$\boldsymbol{\beta}_i = \beta_1(\hat{\mathbf{a}}_x \sin\theta_i + \hat{\mathbf{a}}_z \cos\theta_i) = n_1 \frac{\omega}{v_0}(\hat{\mathbf{a}}_x \sin\theta_i + \hat{\mathbf{a}}_z \cos\theta_i) \tag{5-110a}$$

$$\boldsymbol{\beta}_r = \beta_1(\hat{\mathbf{a}}_x \sin\theta_i - \hat{\mathbf{a}}_z \cos\theta_i) = n_1 \frac{\omega}{v_0}(\hat{\mathbf{a}}_x \sin\theta_i - \hat{\mathbf{a}}_z \cos\theta_i) \tag{5-110b}$$

$$\boldsymbol{\beta}_t = |\beta_2|\left(\hat{\mathbf{a}}_x \sin|\theta_t| - \hat{\mathbf{a}}_z \cos|\theta_t|\right) = |n_2|\frac{\omega}{v_0}\left(\hat{\mathbf{a}}_x \sin|\theta_t| - \hat{\mathbf{a}}_z \cos|\theta_t|\right) \tag{5-110c}$$

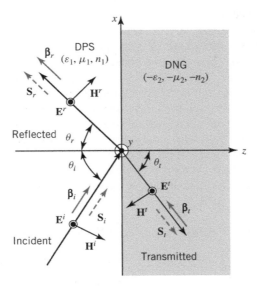

Figure 5-29 Uniform plane wave reflection and refraction of perpendicular polarization by a planar interface formed by DPS and DNG materials.

$$\mathbf{S}^i = \frac{1}{2}\frac{|E_0|^2}{\eta_1}(\hat{\mathbf{a}}_x \sin\theta_i + \hat{\mathbf{a}}_z \cos\theta_i) \tag{5-111a}$$

$$\mathbf{S}^r = \frac{1}{2}\frac{|\Gamma E_0|^2}{\eta_1}(\hat{\mathbf{a}}_x \sin\theta_i - \hat{\mathbf{a}}_z \cos\theta_i) \tag{5-111b}$$

$$\mathbf{S}^t = \frac{1}{2}\frac{|TE_0|^2}{\eta_2}(-\hat{\mathbf{a}}_x \sin|\theta_t| + \hat{\mathbf{a}}_z \cos|\theta_t|) \tag{5-111c}$$

While the wave and Poynting vectors of the incident and reflected fields are unaffected by the presence of the DNG material forming the interface in Figure 5-29 [they are the same as in (5-107) and (5-108)], those of the transmitted fields, as represented by (5-110c) and (5-111c), are different from the corresponding ones of (5-107c) and (5-108c) in two ways.

The first difference is that the wave vector of (5-110c) is antiparallel to the Poynting vector of (5-111c), whereas they were parallel for (5-107c) and (5-108c). Also, for positive time, the wavenumber of (5-107c) leads to a phase delay, but the wavenumber of (5-110c) leads to a phase advance. In addition, while the phase vector of (5-107c) and the Poynting vector of (5-108c) are both directed away from the source (point of refraction in the first quadrant), the Poynting vector of (5-111c) is also directed away from the source, but in the fourth quadrant. These are also illustrated graphically in Figures 5-25a, 5-25b, and 5-29. These are some of the similarities and differences in the transmitted fields for DPS-DPS and DPS-DNG interfaces.

5.7.6 Negative-Refractive-Index (NRI) Transmission Lines

Another application of the DNG material is the design of Negative-Refractive-Index Transmission Lines (NRI-TL) [15, 16, 47]. This concept can be used to design:

- nonradiating phase-shifting lines that can produce either positive or negative phase shift
- broadband series power dividers
- forward leaky-wave antennas

and other applications [16]. When a wave propagates through a DPS medium, like in a conventional dielectric slab of thickness d_1, it will accumulate phase lag $|\phi_1|$ of $\beta_1 d_1 (\phi_1 = -\beta_1 d_1)$, also referred to as negative phase shift, where β_1 is the phase constant (wavenumber). This negative phase shift can be compensated by a positive phase shift $\phi_2 (\phi_2 = +|\beta_2|d_2)$ through a DNG slab that follows the DPS slab. In fact, ideally, the negative phase shift accumulated through propagation in the DPS slab $(\phi_1 = -\beta_1 d_1)$ can be totally eliminated if the positive phase $\phi_2 (\phi_2 = +|\beta_2|d_2)$ can be created by propagation through the DNG slab such that $|\phi_1| = |\phi_2|$ so that the total phase ϕ by wave propagation through both slabs is equal to zero $(\phi = \phi_1 + \phi_2 = 0)$. Such an arrangement is shown graphically in Figure 5-30 where the arrows are used to designate the directions of the phase vectors β and the Poynting vectors **S**. This phase compensation can also be used to create any other desired total phase shift by appropriately choosing the phase constants and thicknesses of the DPS and DNG slabs. The special case of zero phase shift of wave propagation through both slabs is accomplished provided

$$|\phi_1| = \omega\sqrt{\mu_1 \varepsilon_1}\, d_1 = |\phi_2| = \omega\sqrt{|\mu_2||\varepsilon_2|}\, d_2 \Rightarrow n_1 d_1 = n_2 d_2 \Rightarrow \frac{d_1}{d_2} = \frac{n_1}{n_2} \qquad (5\text{-}112)$$

A graphical illustration of such phase compensation of the electric field intensity of a perpendicularly polarized field, simulated using the FDTD method, is exhibited in Figure 5-31 [13]. The incident field is a Gaussian beam traveling in a free-space medium and normally incident upon the DPS slab followed by a DNG slab. The indices of refraction were chosen to be $n_{\text{real}}(\omega) = +3$ for the DPS slab and $n_{\text{real}}(\omega) = -3$ for the DNG slab. Observing the phase fronts of the beam inside the two slabs, it is evident that the beam expands (diverges) in the DPS slab while it refocuses (converges) in the DNG slab. Ultimately, the phase fronts of the exiting beam in the free-space medium to the right of the DNG slab begin to expand and match those of the incident field to the left of the DPS slab. According to [13], there was only 0.323 dB attenuation of wave propagation through the two slabs that span a total distance of $4\lambda_0$. However, the total phase accumulation from the leading edge of the DPS slab to the trailing edge of the DNG slab is zero. Thus, the output field exits the trailing edge, along the symmetry line of the source/beam which is perpendicular to the

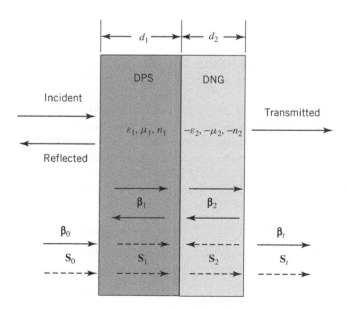

Figure 5-30   Wave propagation through two successive dielectric slabs, one made of DPS material and the other of DNG material, for phase wave compensation.

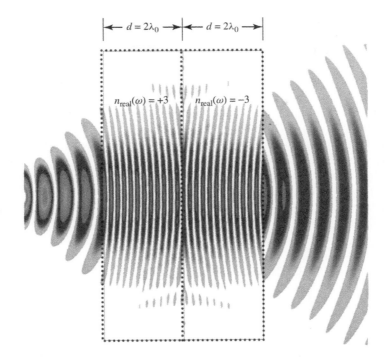

$\leftarrow d = 2\lambda_0 \rightarrow|\leftarrow d = 2\lambda_0 \rightarrow|$

$n_{\text{real}}(\omega) = +3$ $n_{\text{real}}(\omega) = -3$

Figure 5-31 Phase compensation by successive conventional DPS and DNG slabs [13]. (Reprinted with permission from John Wiley & Sons, Inc. Original courtesy of R. W. Ziolkowski.)

interface, with the same phase as the input field and with only a slight attenuation in the peak value of about of 0.323 dB, which is due to a small loss in the medium and to the Gaussian beam diverging from the source. While the negative (second) layer refocuses the beam, the small loss by the first layer is not totally compensated by the second layer and leads to the slight attenuation at the output face of the system. Such an arrangement of slabs is usually referred to, for obvious reasons, as a *beam translator* [13].

This phase compensation concept can also be applied to compensate for negative phase shift by wave propagation through a conventional DPS transmission line followed by a NRI line with DNG material, often referred to as a BW (backward-wave) line, as shown graphically in Figure 5-32 [16].

In Figure 5-32*b* the equivalent circuit of BW line indicates that the phase advance through the unit cell of a BW line is given by

$$\phi_{\text{BW}} = \frac{1}{\omega\sqrt{L_0 C_0}} \tag{5-113}$$

which is representative of the phase through a high-pass LC filter of the type shown in the unit cell of the BW line in Figure 5-32*b*. Such a backward type of a wave, for the equivalent circuit of the backward section of the line, has also been addressed in [49], which states that "a wave in which the phase velocity and group velocity have opposite signs is known as a backward wave. Conditions for these may seem unexpected or rare, but they are not." In fact, it is also stated in [49] that many filter type of lines have backward waves and that periodic circuits exhibit an equal number of forward and backward "space harmonics."

The low-pass filter (regular transmission line) and high-pass filter (backward-wave line) characteristics can be verified using the Brillouin dispersion diagram [49, 50], which is a plot of ω vs. β with the phase velocity defined as

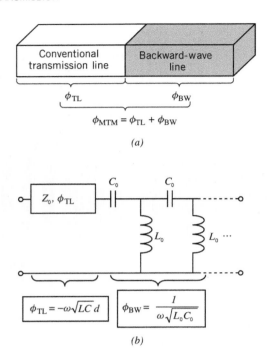

$$\phi_{\mathrm{MTM}} = \phi_{\mathrm{TL}} + \phi_{\mathrm{BW}}$$

(a)

$$\phi_{\mathrm{TL}} = -\omega\sqrt{LC}\,d \qquad \phi_{\mathrm{BW}} = \dfrac{1}{\omega\sqrt{L_0 C_0}}$$

(b)

Figure 5-32   Phase compensation by successive conventional and backward-wave transmission lines [16]. (Reprinted with permission from John Wiley & Sons, Inc. Originals courtesy of G. V. Eleftheriades and M. Antoniades.) (a) Conventional transmission line followed by a backward-wave line. (b) Equivalent circuit of conventional transmission line followed by a backward-wave line.

$$v_p = \frac{\omega}{\beta} \tag{5-114}$$

while the group velocity is defined as

$$v_g = \frac{\partial \omega}{\partial \beta} \tag{5-115}$$

For the regular transmission type line v_p and v_g have the same sign while for the backward-wave type of line, v_p and v_g have opposite signs.

Therefore, it seems that in Figure 5-32 there is a low-pass filter (conventional) line followed by a high-pass filter (BW line) with a total phase shift for the two of

$$\phi_{\mathrm{MTM}} = \phi_{\mathrm{TL}} + \phi_{\mathrm{BW}} = -\omega\sqrt{LC}d + \frac{1}{\omega\sqrt{L_0 C_0}} \tag{5-116}$$

The transmission line is of the delay type while the backward-wave line is of the phase advance type.

Various one-dimensional phase-shifting lines were constructed at 0.9 GHz using coplanar waveguide (CPW) technology [16]. Two such units, one a two-stage and the other a four-stage phase shifter, are shown in Figure 5-33a. The corresponding simulated and measured phase responses of both units are shown in Figure 5-33b where they are compared with the phase responses of a conventional $-360°$ TL line and a $-360°$ low-pass loaded line. The corresponding magnitudes of both units of $0°$ phase shift are also indicated in Figure 5-33b. A good comparison is observed between the simulated and measured results and confirms the broadband nature of the phase shifting lines which also exhibit rather small losses [16].

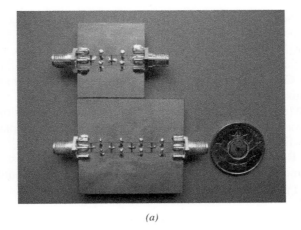

(a)

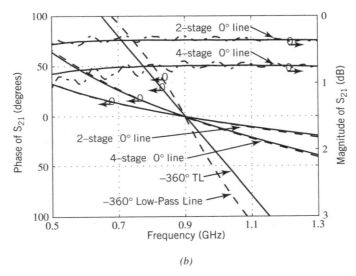

(b)

Figure 5-33 Experimental units, and simulated and measured responses of two- and four-stage phase shifting lines. (Source: B. MacA. Thomas, "Design of corrugated conical horns," *IEEE Trans. Antennas Propagat.*, vol. 26, no. 2, pp. 367–372, Mar. 1978. Reproduced with permission from John Wiley & Sons. Originals courtesy of G. V. Eleftheriades and M. Antoniades.) (*a*) Two-stage phase shifting line (16 mm) (top) and a four-stage phase-shifting line (32 mm) both at 0.9 GHz [16]. (*b*) Phase and magnitude responses of a two-stage and four-stage phase-shifting lines compared to conventional −360°TL and a −360° low-pass loaded line at 0.9 GHz [16]. Phase: − − − − − Measured ——— Simulated (Agilent ADS) Magnitude: −·—·—· Measured ——— Simulated (Agilent ADS)

5.8 MULTIMEDIA

On the website that accompanies this book, the following multimedia resources are included for the review, understanding, and presentation of the material of this chapter.

- **MATLAB** computer programs:
  - a. **SWR_Animation_Γ_SWR_Impedance:** Animates the standing wave pattern of a plane wave traveling in a semi-infinite lossless medium and impinging, at normal incidence, upon a planar interface formed by two semi-infinite planar media; the second medium

can be lossy (see Figure 5-1). It also computes the input reflection coefficient Γ, SWR, and input impedance.

  b. **QuarterWave_Match:** Designs a quarter-wavelength impedance transformer of N slabs to match a given semi-infinite medium (input) to another semi-infinite medium (load).

  c. **Single_Slab:** Characterizes the reflection and transmission characteristics of a single layer slab bounded on both sides by two semi-infinite media.

  d. **Refl_Trans_Multilayer:** Computes the reflection and transmission coefficients of a uniform plane wave incident at oblique angle upon N layers of planar slabs bordered on either side by free space.

  e. **Polarization_Refl_Trans:** Computes the Poincaré sphere angles, and thus, the polarization, of a plane wave incident at oblique angles upon a planar interface.

- **PowerPoint (PPT)** viewgraphs, in multicolor.

REFERENCES

1. C. A. Balanis, *Antenna Theory: Analysis and Design*, Third Edition, John Wiley & Sons, New York, 2005.

2. M. A. Plonus, *Applied Electromagnetics*, McGraw-Hill, New York, 1978.

3. D. T. Paris and F. K. Hurd, *Basic Electromagnetic Theory*, McGraw-Hill, New York, 1969.

4. R. B. Adler, L. J. Chu, and R. M. Fano, *Electromagnetic Energy Transmission and Radiation*, Chapters 7 and 8, John Wiley & Sons, New York, 1960.

5. J. J. Holmes and C. A. Balanis, "Refraction of a uniform plane wave incident on a plane boundary between two lossy media," *IEEE Trans. Antennas Propagat.*, vol. AP-26, no. 5, pp. 738–741, Sept. 1978.

6. R. D. Radcliff and C. A. Balanis, "Modified propagation constants for nonuniform plane wave transmission through conducting media," *IEEE Trans. Geoscience Remote Sensing*, vol. GE-20, no. 3, pp. 408–411, July 1982.

7. J. D. Kraus, *Electromagnetics*, Fourth Edition, McGraw-Hill, New York, 1992.

8. R. E. Collin, *Foundations for Microwave Engineering*, Second Edition, McGraw-Hill, New York, 1992.

9. R. E. Collin and J. Brown, "The design of quarter-wave matching layers for dielectric surfaces," *Proc. IEE*, vol. 103, Part C, pp. 153–158, Mar. 1956.

10. S. B. Cohn, "Optimum design of stepped transmission line transformers," *IRE Trans. Microwave Theory Tech.*, vol. MTT-3, pp. 16–21, Apr. 1955.

11. L. Young, "Optimum quarter-wave transformers," *IRE Trans. Microwave Theory Tech.*, vol. MTT-8, pp. 478–482, Sept. 1960.

12. G. L. Matthaei, L. Young, and E. M. T. Jones, *Microwave Filters, Impedance-Matching Networks and Coupling Structures*, McGraw-Hill, New York, 1964.

13. N. Engheta and R. W. Ziolkowski (Eds.), *Metamaterials: Physics and Engineering Explorations*, N. Engheta, R. W. Ziolkowski (Eds.), IEEE Press, Wiley Inter-Science, New York, 2006.

14. *IEEE Trans. Antennas Propagat.*, Special Issue on Metamaterials, vol. 51, no. 10, Oct. 2003.

15. G. V. Eleftheriades and K. G. Balmain (editors), *Negative-Refraction Metamaterials: Fundamental Principles and Applications*, John Wiley & Sons, New York, 2005.

16. G. V. Eleftheriades and M. A. Antoniades, "*Antenna Application of Negative Refractive Index Transmission Line (NRI-TL) Metamaterials,*" Chapter 14, in *Modern Antenna Handbook*, C. A. Balanis (Ed.), John Wiley & Sons, pp. 677–736, 2008.

17. C. Caloz and T. Itoh, *Electromagnetic Metamaterials: Transmission Line Theory and Microwave Applications*, John Wiley & Sons, New York, 2006.

18. R. Marques, F. Martin, and M. Sorolla, *Metamaterials with Negative Parameters: Theory, Design and Microwave Applications*, John Wiley & Sons, New York, 2008.

19. B. A. Munk, *Metamaterials: Critique and Alternatives*, John Wiley & Sons, New York, 2009.

20. B. E. Spielman, S. Amari, C. Caloz, G. V. Eleftheriades, T. Itoh, D. R. Jackson, R. Levy, J. D. Rhodes, and R. V. Snyder, "Metamaterials face-off. Metamaterials: A rich opportunity for discovery or an overhyped gravy train!," *IEEE Microwave Mag.*, pp. 8–10, 12, 14, 16–17, 22, 26, 28, 30, 32, 34, 36, 38, 42, May 2009.

21. P. M. Valanju, R. M. Walser, and A. P. Valanju, "Wave refraction in negative-index media: always positive and very inhomogeneous," *Phys. Rev. Lett.*, vol. 88, no. 18, 187401:1–4, May 2002.

22. G. K. Karawas and R. E. Collin, "Spherical shell of ENG metamaterial surrounding a dipole," *Military Communications Conference 2008 (MILCOM 2008)*, pp. 1–7, San Diego, CA, Nov. 2008.

23. R. C. Hansen and R. E. Collin, *Small Antenna Handbook*, John Wiley & Sons, Hoboken, NJ, 2011.

24. J. C. Bose, "On the rotation of plane of polarization of electric wave by a twisted structure," *Proc. Roy. Soc.*, vol. 63, pp. 146–152, 1898.

25. I. V. Lindell, A. H. Sihvola, and J. Kurkijarvi, "Karl F. Lindman: The last Hertzian and a Harbinger of electromagnetic chirality," *IEEE Antennas Propagat. Mag.*, vol. 34, no. 3, pp. 24–30, 1992.

26. W. E Kock, "Metallic delay lines," *Bell Sys. Tech. J.*, Vol. 27, pp. 58–82, 1948.

27. V. G. Veselago, "The electrodynamics of substances with simultaneous negative values of ε and μ," *Sov. Phys.-Usp.*, vol. 47, pp. 509–514, Jan.–Feb. 1968.

28. R. E. Collin, "Frequency dispersion limits resolution in Veselago lens," *Prog. Electromagn. Res.*, vol. 19, pp. 233–261, 2010.

29. A. D. Yaghjian and T. B. Hansen, "Plane-wave solutions to frequency-domain and time-domain scattering from magnetodielectric slabs," *Phys. Rev. E*, 73, 046608, Apr. 2006; erratum, 76, 049903, Oct. 2007.

30. J. B. Pendry, A. J. Holden, W. J. Stewart, and I. Youngs, "Extremely, low-frequency plasmons in metallic mesostructure," *Phys. Rev. Letters.*, vol. 76, pp. 4773–4776, June 1996.

31. J. B. Pendry, A. J. Holden, D. J. Robbins, and W. J. Stewart, "Low-frequency plasmons in thin wire structures," *J. Phys., Condens. Matter*, vol. 10, pp. 4785–4809, 1998.

32. J. B. Pendry, A. J. Holden, D. J. Robbins, and W. J. Stewart, "Magnetism from conductors and enhanced nonlinear phenomena," *IEEE Trans. Microw. Theory Tech.*, vol. 47, no. 11, pp. 2075–2081, Nov. 1999.

33. J. B. Pendry, "Negative refraction makes a perfect lens," *Phys. Rev. Lett.*, vol. 85, pp. 3966–3969, Oct. 2000.

34. D. R. Smith, W. J. Padilla, D. C. Vier, S. C. Nemat-Nasser, and S. Schultz, "Composite medium with simultaneously negative permeability and permittivity," *Phys. Rev. Lett.*, vol. 84, pp. 4184–4187, May 2000.

35. D. R. Smith, D. C. Vier, N. Kroll, and S. Schultz, "Direct calculation of the permeability and permittivity for left-handed metamaterials," *Appl. Phys. Lett.*, vol. 77, pp. 2246–2248, Oct. 2000.

36. D. R. Smith and N. Kroll, "Negative refractive index in left-handed materials," *Phys. Rev. Lett.*, vol. 85, pp. 2933–2936, Oct. 2000.

37. R. A. Shelby, D. R. Smith, S. C. Nemat-Nasser, and S. Schultz, "Microwave transmission through a two-dimensional, isotropic, left-handed metamaterial," *Appl. Phys. Lett.*, 78, pp. 489–491, Jan. 2001.

38. A. Shelby, D. R. Smith, and S. Schultz, "Experimental verification of a negative index-of-refraction," *Science*, vol. 292, pp. 77–79, Apr. 2001.

39. B. A. Munk, *Finite Antenna Arrays and FSS*, John Wiley & Sons, Hoboken, NJ, 2003.

40. T. M. Grzegorczyk, J. A. Kong, and R. Lixin, "Refraction experiments in waveguide environments," Chapter 4 in *Metamaterials: Physics and Engineering Explorations*, N. Engeta and R. W. Ziolkowski (Eds.), Wiley-Interscience, 2006.

41. *IEEE Transactions on Antennas and Propagation*.

42. *IEEE Transactions on Microwave Theory and Techniques*.

43. *IEEE Antennas and Wireless Propagation Letters*.

44. *IEEE Microwave and Wireless Components Letters*.

45. *Physical Review Letters*.

46. *Journal of Applied Physics*.

47. G. V. Eleftheriades, "EM transmission-line metamaterials," *Materials Today*, vol. 12, no. 3, pp. 30–41, Mar. 2009.

48. R. W. Ziolkowski, "Pulsed and CW Gaussian beam interactions with double negative material slabs," *Opt. Express*, Vol. 11, pp. 662–681, Apr. 2003.

49. S. Ramo, J. R. Whinnery, and T. Van Duzer, *Fields and Waves in Communication Electronics*, Third Edition, John Wiley & Sons, New York, 1994.

50. D. M. Pozar, *Microwave Engineering*, Second Edition, John Wiley & Sons, New York, 2004.

PROBLEMS

5.1. A uniform plane wave traveling in a dielectric medium with $\varepsilon_r = 4$ and $\mu_r = 1$ is incident normally upon a free-space medium. If the incident electric field is given by

$$\mathbf{E}^i = \hat{\mathbf{a}}_y 2 \times 10^{-3} e^{-j\beta z} \text{ V/m}$$

write the:

(a) Corresponding incident magnetic field.
(b) Reflection and transmission coefficients.
(c) Reflected and transmitted electric and magnetic fields.
(d) Incident, reflected, and transmitted power densities.

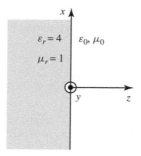

Figure P5-1

5.2. The dielectric constant of water is 81. Calculate the percentage of power density reflected and transmitted when a uniform plane wave traveling in air is incident normally upon a calm lake. Assume that the water in the lake is lossless.

5.3. A uniform plane wave propagating in a medium with relative permittivity of 4 is incident normally upon a dielectric medium with dielectric constant of 9. Assuming both media are nonferromagnetic and lossless, determine the:

(a) Reflection and transmission coefficients.
(b) Percentage of incident power density that is reflected and transmitted.

5.4. A vertical interface is formed by having free space to its left and a lossless dielectric

medium to its right with $\varepsilon = 4\varepsilon_0$ and $\mu = \mu_0$, as shown in Figure P5-4. The incident electric field of a uniform plane wave traveling in the free-space medium and incident normally upon the interface has a value of 2×10^{-3} V/m right before it strikes the boundary. At a frequency of 3 GHz, find the:

(a) Reflection coefficient.
(b) SWR in the free-space medium.
(c) Positions (in meters) in the free-space medium where the electric field maxima and minima occur.
(d) Maximum and minimum values of the electric field in the free-space medium.

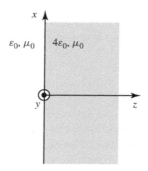

Figure P5-4

5.5. A uniform plane wave traveling in air is incident upon a flat, lossless, and infinite in extent dielectric interface with a dielectric constant of 4. In the air medium, a standing wave is formed. If the normalized magnitude of the incident E-field is $E_0 = 1$, determine the:

(a) Maximum value of the E-field standing wave pattern in air.
(b) Shortest distance l (in λ_0) from the interface where the first maximum in the E-field standing wave pattern will occur (normalized to the incident field).
(c) Minimum value of the E-field standing wave pattern in air (normalized to the incident field).
(d) Shortest distance l (in λ_0) in air from the interface where the first minimum in the

E-field standing wave pattern will occur (normalized to the incident field).

(e) Standing Wave Ratio (SWR) measured in the air medium.

(f) Input wave impedance inside the air medium where the:

1. First maximum in the E-field standing wave pattern occurs.
2. First minimum in the E-field standing wave pattern occurs.

5.6. A uniform plane wave, traveling in free space along the $+x$ direction, is normally incident upon a planar interface formed by the free space and a perfect magnetic conductor (PMC). The interface formed by free space and the PMC is on the yz plane at $x = 0$. The normalized incident magnetic field of the wave, that is traveling in the free-space, is given by

$$\mathbf{H}^i = (-2\hat{\mathbf{a}}_y + j\hat{\mathbf{a}}_z)e^{-j\beta_0 x}$$

Without using the Poincare sphere parameters (do it by inspecting the field expressions):

(a) Write an expression for the corresponding incident vector electric field.

(b) Determine the polarization (linear, circular, or elliptical; axial ratio AR; and the sense of rotation, CW or CCW, if pertinent) of the incident wave. Justify your answers concerning the polarization of the incident wave.

(c) Write an expression for the corresponding reflected vector electric field.

(d) Determine the polarization (linear, circular, or elliptical; axial ratio AR; and the sense of rotation, CW or CCW, if pertinent) of the reflected wave. Justify your answers concerning the polarization of the reflected wave.

5.7. A uniform plane wave traveling in free space is incident upon a planar interface formed by free space and a lossless dielectric with dielectric constant of 9. Determine the:

(a) Total input reflection coefficient, magnitude and phase, at the boundary (just to the left of the interface).

(b) SWR at the boundary (just to the left of the interface).

(c) Input impedance at the boundary (just to the left of the interface).

(d) Input reflection coefficient, magnitude and phase, at a distance of $0.75\lambda_0$ to the left of the interface (λ_0 with the free-space wavelength).

(e) SWR at a distance of $0.75\lambda_0$ to the left of the interface.

(f) Total input impedance at a distance of $0.75\lambda_0$ to the left of the interface.

5.8. A uniform plane wave is incident upon a planar interface formed by free space (upper half) and a magneto-dielectric material (lower half) whose relative permittivity is unity and relative permeability is very large; ideally approaching infinity ($\mu_{r1} \to \infty$). See Figure P5-8. The normalized incident magnetic field is given by

$$\mathbf{H}^i = \hat{\mathbf{a}}_y e^{+j\beta_0 z}$$

(a) Determine the reflection coefficient (including its sign), as defined in the book.

(b) Does the interface, looking from the free-space side, simulate a PEC or PMC?

(c) Determine the SWR in the free-space medium.

(d) Write an expression for the corresponding reflected vector magnetic field.

(e) Write an expression for the total vector magnetic field in the free-space medium.

(f) Write an expression for the magnitude of the magnetic field standing wave (SW) pattern in the free-space medium.

(g) Plot/sketch the magnitude of the magnetic field standing wave (SW) pattern as a function of z (in free-space wavelengths λ_0) for $0 \le z \le \lambda_0$.

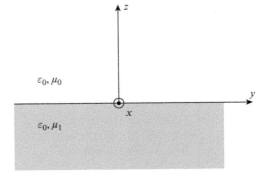

Figure P5-8

5.9. A time-harmonic electromagnetic field in the free-space region is perpendicularly incident upon a perfect electric conducting (PEC) planar interface, as shown in Figure P5-9.

Assuming the incident $\mathbf{E}^i$ and reflected $\mathbf{E}^r$ complex electric fields on the free-space side of the interface are given by

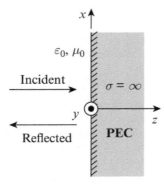

Figure P5-9

$$\mathbf{E}^i = \hat{\mathbf{a}}_x e^{-j\beta_0 z}$$

$$\mathbf{E}^r = -\hat{\mathbf{a}}_x e^{+j\beta_0 z}$$

where β_0 is the free-space phase constant ($\beta_0 = \omega\sqrt{\mu_0 \varepsilon_0}$).

Determine the:
(a) Vector incident and reflected magnetic fields.
(b) Vector complex power density $\mathbf{S}$ associated with each wave, incident and reflected.
(c) Total vector electric $\mathbf{J}_s$ current density induced on the PEC surface.
(d) Total vector electric $\mathbf{M}_s$ current density induced on the PEC surface.

5.10. An electromagnetic wave, traveling in free space along the $+z$ axis, is normally incident upon a planar and infinite in extent interface, as shown in Figure P5-9. Assuming the incident electric field is given by:

$$\mathbf{E} = E_0(\hat{\mathbf{a}}_x + j2\hat{\mathbf{a}}_y) e^{-j\beta_0 z}$$

where E_0 is a constant,
(a) Determine the:
  • Polarization, including sense of rotation (if any), of the incident electric field. Justify the answer.
  • Axial ratio (AR) of the incident electric field (including the $+$ or $-$ sign).
(b) Assuming the planar reflecting surface is a PEC:
  • Write an expression for the reflected electric field.
  • Determine the polarization, including sense of rotation (if any), of the reflected wave.

• What is the axial ratio (AR) of the reflected electromagnetic wave (including the $+$ or $-$ sign)?

5.11. For the electric field of Problem 5.10 and assuming the planar reflecting surface is PMC,
  • Write an expression for the incident magnetic field.
  • Write an expression for the reflected magnetic field.
  • Determine the polarization, including sense of rotation (if any), of the reflected wave.
  • What is the axial ratio (AR) of the reflected electromagnetic wave (including the $+$ or $-$ sign)?

5.12. A uniform plane wave, traveling in free space along the $+z$ direction, is normally incident upon a planar interface, as shown in Figure P5-9, formed by the free space and a dielectric medium whose dielectric constant $\varepsilon_{r2} \gg 1$ and whose permeability is the same as that of free space.
Assuming the normalized incident electric field is represented by:

$$\mathbf{E}^i = (\hat{\mathbf{a}}_x + j\hat{\mathbf{a}}_y) e^{-j\beta_0 z}$$

(a) What is the polarization (linear, circular, or elliptical), sense of rotation (CW or CCW), and axial ratio (AR) of the incident wave?
(b) Write an expression for the reflection coefficient in terms of ε_{r2}; then evaluate the expression when $\varepsilon_{r2} = \infty$.
(c) Write an expression for the reflected electric field based on the reflection coefficient of part (b) when $\varepsilon_{r2} = \infty$.
(d) What are the polarization (linear, circular, or elliptical), sense of rotation (CW or CCW) and axial ratio (AR) of the reflected wave?
(e) If the incident electric represents the polarization of the field radiated by a transmitting antenna and the reflected electric field represents the wave which impinges upon the transmitting antenna, what is the normalized voltage response (output) of the antenna due to the impinging reflected wave?

5.13. A uniform plane wave, traveling in free space along the $+z$ direction, is normally incident upon a planar interface, as shown in Figure P5-9, formed by the free space and a magnetic medium whose relative permeability $\mu_{r2} \gg 1$ and whose permittivity is the same as that of free space.

Assuming the normalized incident electric field is represented by:

$$\mathbf{E}^i = (\hat{\mathbf{a}}_x + j\hat{\mathbf{a}}_y)e^{-j\beta_0 z}$$

(a) What are the polarization (linear, circular or elliptical), sense of rotation (CW or CCW), and axial ratio (AR) of the incident wave?

(b) Write an expression for the reflection coefficient in terms of μ_{r2}; then evaluate the expression when $\mu_{r2} = \infty$.

(c) Write an expression for the reflected electric field based on the reflection coefficient of part (b) when $\mu_{r2} = \infty$.

(d) What is the polarization (linear, circular, or elliptical), sense of rotation (CW or CCW), and axial ratio (AR) of the reflected wave?

(e) If the above stated incident electric field represents the polarization of the field radiated by a transmitting antenna and the reflected electric field of part (c) represents the wave which impinges upon the transmitting antenna, what is the normalized voltage response (output) of the antenna due to the impinging reflected wave?

5.14. For Problem 5.13 using the procedure outlined in Section 5.6:

(a) Determine the Poincaré sphere parameters (γ, δ, and AR) for the:
  • Incident wave
  • Reflected wave

(b) Based upon these Poincaré sphere parameters for each of the waves, incident and reflected, state the polarization of each of the waves:
  • Linear, circular, elliptical
  • Axial Ratio (AR, including sign)
  • Sense of rotation (CW or CCW), if applicable

(c) Using the procedure of Section 5.6, are the answers (polarization, AR, and sense of rotation) of the incident and reflected waves the same as or different from those in Problem 5.13? Compare and indicate

whether they are the same or not. If they are not the same, justify why.

5.15. The normalized vector electric field radiated in free space by an antenna is given by

$$\mathbf{E} = (\hat{\mathbf{a}}_x + j\hat{\mathbf{a}}_y)e^{-j\beta_0 z}$$

and it is incident upon a flat and infinite in extent perfect electric conducting (PEC) surface, as shown in Figure P5-15.

(a) Write an expression for the corresponding incident vector magnetic field.

(b) Determine the polarization of the incident wave:
  • Linear, circular, elliptical
  • Axial ratio (AR, including sign)
  • Sense of rotation (CW or CCW), if applicable

(c) Write an expression for the reflected electric field vector.

(d) Write an expression for the reflected magnetic field vector.

(e) Determine the polarization of the reflected wave:
  • Linear, circular, elliptical
  • Axial ratio (AR, including sign)
  • Sense of rotation (CW or CCW), if applicable

(f) Assuming the polarization of the radar antenna receiving the reflected wave is the same as its polarization in the transmission mode, which produced the incident wave, what will be the normalized output loss of the radar antenna due to polarization match or mismatch, in dB (minimum loss 0 dB), of the antenna based on receiving the reflected wave?

For parts (a)–(e), the polarization of the incident and reflected waves must be determined based on the rules that were established in the book (be able to determine the polarization based on the field's number of components, their relative magnitude, phase difference, etc.). State all the necessary and sufficient conditions for that polarization to have been met to justify the stated answer.

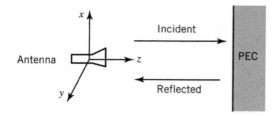

Figure P5-15

5.16. Repeat all the parts of Problem 5.15 if the reflecting surface is perfect magnetic conducting (PMC).

5.17. A CW circularly-polarized wave of $f = 100$ MHz of the form

$$\mathbf{E}^i(z) = \left(\hat{\mathbf{a}}_x - j\hat{\mathbf{a}}_y\right)e^{-j6\pi z}$$

where z is in meters, is traveling inside a lossless dielectric medium and is normally incident upon a flat planar interface formed by the dielectric medium and air. The interface is on the xy-plane. Assuming the permeability of the dielectric medium is the same as free space, determine the:

(a) Dielectric constant (relative permittivity) of the dielectric medium.
(b) Reflection coefficients for the $\hat{\mathbf{a}}_x$ and $\hat{\mathbf{a}}_y$ components.
(c) Transmission coefficients for the $\hat{\mathbf{a}}_x$ and $\hat{\mathbf{a}}_y$ components.
(d) Polarization (linear, circular or elliptical) of the reflected field.
(e) Sense of polarization rotation, if any, of the reflected field.
(f) Polarization (linear, circular, or elliptical) of the transmitted field.
(g) Sense of polarization rotation, if any, of the transmitted field.

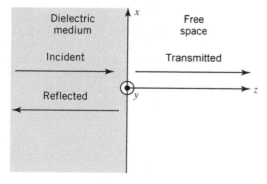

Figure P5-17

5.18. The field radiated by an antenna along the $+z$ axis is a uniform plane wave whose polarization is right-hand circularly-polarized (RHC). The field radiated by the antenna impinges, at normal incidence, upon a perfectly electric conducting (PEC) flat and infinite in extent ground plane. Determine the:
(a) Polarization of the field reflected by the ground plane toward the antenna, including the sense of rotation (if any). Justify your answer.
(b) Normalized output voltage (dimensionless and in dB) at the transmitting antenna, which is now acting as a receiving antenna, based on its reception of the reflected field. Justify your answer. Is it what you are expecting or is it a surprise?

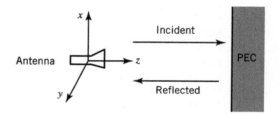

Figure P5-18

5.19. A time-harmonic electromagnetic wave traveling in free space is incident normally upon a perfect conducting planar surface, as shown in Figure P5-19. Assuming the incident electric field is given by

$$\mathbf{E}^i = \hat{\mathbf{a}}_x E_0 e^{-j\beta_0 z}$$

find the (a) reflected electric field, (b) incident and reflected magnetic fields, and (c) current density $\mathbf{J}_s$ induced on the conducting surface.

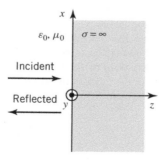

Figure P5-19

5.20. A uniform plane wave traveling in air is incident normally on a half space occupied by a lossless dielectric medium of relative permittivity of 4. The reflections can be eliminated by placing another dielectric slab, $\lambda_1/4$ thick, between the air and the original dielectric medium, as shown in Figure P5-20. To accomplish this, the intrinsic impedance η_1 of the slab must be equal to $\sqrt{\eta_0\eta_2}$ where η_0 and η_2 are, respectively, the intrinsic impedances of air and the original dielectric medium. Assuming that the relative permeabilities of all the media are unity, what should the relative permittivity of the dielectric slab be to accomplish this?

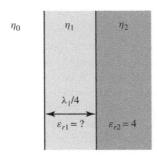

Figure P5-20

5.21. A uniform plane wave traveling in free space is incident normally upon a lossless dielectric slab of thickness t, as shown in Figure P5-21. Free space is found on the other side of the slab. Derive expressions for the total reflection and transmission coefficients in terms of the media constitutive electrical parameters and thickness of the slab.

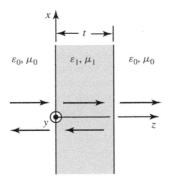

Figure P5-21

5.22. The vertical height from the ground to a person's eyes is h, and from his eyes to the top of his head is Δh. A flat mirror of height y is hung vertically at a distance x from the person. The top of the mirror is at a height of $h+(\Delta h/2)$ from the ground, as shown in Figure P5-22. What is the minimum length of the mirror in the vertical direction so that the person *only* sees his entire image in the mirror?

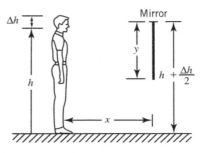

Figure P5-22

5.23. A linearly polarized wave is incident on an isosceles right triangle (prism) of glass, and it exits as shown in Figure P5-23. Assuming that the dielectric constant of the prism is 2.25, find the ratio of the exited average power density S_e to that of the incident S_i.

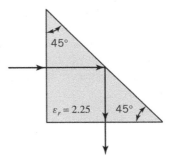

Figure P5-23

5.24. A uniform plane wave is obliquely incident at an angle of $30°$ on a dielectric slab of thickness d with $\varepsilon = 4\varepsilon_0$ and $\mu = \mu_0$ that is embedded in free space, as shown in Figure P5-24. Find the angles θ_2 and θ_3 (in degrees).

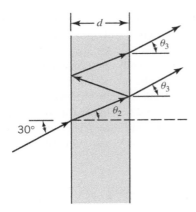

Figure P5-24

5.25. A perpendicularly polarized uniform plane wave traveling in free space is obliquely incident on a dielectric with a relative permittivity of 4, as shown in Figure 5-2. What should the incident angle be so that the reflected power density is 25% of the incident power density?

5.26. Repeat Problem 5-25 for a parallel polarized uniform plane wave.

5.27. Find the Brewster angles for the interfaces whose reflection coefficients are plotted in Figure 5-5.

5.28. A parallel-polarized uniform plane wave is incident obliquely on a lossless dielectric slab that is embedded in a free-space medium, as shown in Figure P5-28. Derive expressions for the total reflection and transmission coefficients in terms of the electrical constitutive parameters, thickness of the slab, and angle of incidence.

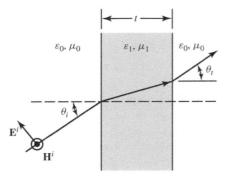

Figure P5-28

5.29. Repeat Problem 5-28 for a perpendicularly polarized plane wave, as shown in Figure P5-29.

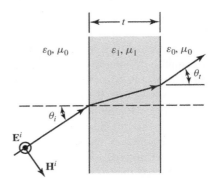

Figure P5-29

5.30. A perpendicularly polarized plane wave traveling in a dielectric medium with relative permittivity of 9 is obliquely incident on another dielectric with relative permittivity of 4. Assuming that the permeabilities of both media are the same, find the incident angle (measured from the normal to the interface) that results in total reflection.

5.31. Calculate the Brewster and critical angles for a parallel-polarized wave when the plane interface is:
(a) Water to air (ε_r of water is 81).
(b) Air to water.
(c) High-density glass to air (ε_r of glass is 9).

5.32. A uniform plane wave traveling in a lossless dielectric is incident normally on a flat interface formed by the presence of air. For ε_r's of 2.56, 4, 9, 16, 25, and 81:
(a) Determine the critical angles.
(b) Find the Brewster angles if the wave is of parallel polarization.
(c) Compare the critical and Brewster angles found in parts (a) and (b).
(d) Plot the magnitudes of the reflection coefficients for both perpendicular, $|\Gamma_\perp|$, and parallel, $|\Gamma_\parallel|$, polarizations versus incidence angle.
(e) Plot the phase (in degrees) of the reflection coefficients for both perpendicular and parallel polarizations versus incidence angle.

5.33. The transmitting antenna of a ground-to-air communication system is placed at a height of 10 m above the water, as shown in Figure P5-33. For a ground separation of 10 km between the transmitter and the receiver, which is placed on an airborne platform, find the height h_2 above water of the receiving system so that the wave reflected by the water does not possess a parallel polarized component. Assume that the water surface is flat and lossless.

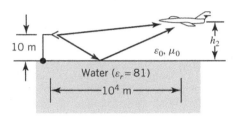

Figure P5-33

5.34. For the geometry of Problem 5.33, the transmitter is radiating a right-hand circularly polarized wave. Assuming the aircraft is at a height of 1,101.11 m, give the polarization (linear, circular, or elliptical) and sense of rotation (right or left hand) of the following.
(a) A wave reflected by the sea and intercepted by the receiving antenna.
(b) A wave transmitted, at the same reflection point as in part (a), into the sea.

5.35. The heights above the earth of a transmitter and receiver are, respectively, 100 and 10 m, as shown in Figure P5-35. Assuming that the transmitter radiates both perpendicular and parallel polarizations, how far apart (in meters) should the transmitter and receiver be placed so that the reflected wave has no parallel polarization? Assume that the reflecting medium is a lossless flat earth with a dielectric constant of 16.

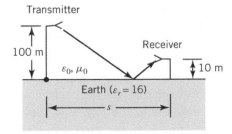

Figure P5-35

5.36. A light source that shines isotropically is submerged at a depth d below the surface of water, as shown in Figure P5-36. How far in the x direction (both positive and negative) can an observer (just above the water interface) go and still see the light? Assume that the water is flat and lossless with a dielectric constant of 81.

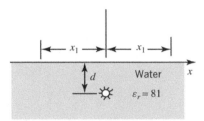

Figure P5-36

5.37. The 30° to 60° dielectric prism shown in Figure P5-37 is surrounded by free space.
(a) What is the minimum value of the prism's dielectric constant so that there is no time-average power density transmitted across the hypotenuse when a plane wave is incident on the prism, as shown in the figure?

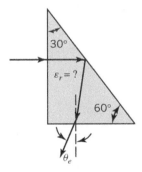

Figure P5-37

(b) What is the exiting angle θ_e if the dielectric constant of the prism is that found in part (a)?

5.38. A uniform plane wave of parallel polarization, traveling in a lossless dielectric medium with relative permittivity of 4, is obliquely incident on a free-space medium. What is the angle of incidence so that the wave results in a complete (a) transmission into the free-space medium and (b) reflection from the free-space medium?

5.39. A fish is swimming in water beneath a circular boat of diameter D, as shown in Figure P5-39.

(a) Find the largest included angle $2\theta_c$ of an imaginary cone within which the fish can swim and not be seen by an observer at the surface of the water.

(b) Find the smallest height of the cone. Assume that light strikes the boat at grazing incidence $\theta_i = \pi/2$ and refracts into the water.

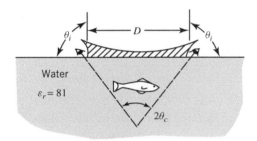

Figure P5-39

5.40. Any object above absolute zero temperature (0 K or $-273°$ C) emits electromagnetic radiation. According to the reciprocity theorem, the amount of electromagnetic energy emitted by the object toward an angle θ_i is equal to the energy received by the object when an electromagnetic wave is incident at an angle θ_i, as shown in Figure P5-40. The electromagnetic power emitted by the object is sensed by a microwave remote detection system as a brightness temperature T_B given by

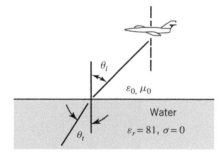

Figure P5-40

$$T_B = eT_m = (1 - |\Gamma|^2)T_m$$

where

e = emissivity of the object (dimensionless)

Γ = reflection coefficient for the interface

T_m = thermal (molecular) temperature of object (water)

It is desired to make the brightness temperature T_B equal to the thermal (molecular) temperature T_m.

(a) State the polarization (perpendicular, parallel, or both) that will accomplish this.

(b) At what angle θ_i (in degrees) will this occur when the object is a flat water surface?

5.41. A uniform plane wave at a frequency of 10^4 Hz is traveling in air, and it is incident normally on a large body of salt water with constants of $\sigma = 3$ S/m and $\varepsilon_r = 81$. If the magnitude of the electric field on the salt water side of the interface is 10^{-3} V/m, find the depth (in meters) inside the salt water at which the magnitude of the electric field has been reduced to 0.368×10^{-3} V/m.

5.42. At large observation distances, the field radiated by a satellite antenna that is attempting to communicate with a submerged submarine is locally TEM (also assume uniform plane wave), as shown in Figure P5-42. Assuming the incident electric field before it impinges on the water is 1 mV/m and the submarine is directly below the satellite, find at 1 MHz, the:

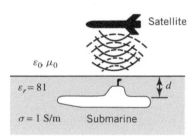

Figure P5-42

(a) Intensity of the reflected E field.

(b) SWR created in air.

(c) Incident and reflected power densities.

(d) Intensity of the transmitted E field.

(e) Intensity of the transmitted power density.

(f) Depth d (in meters) of the submarine where the intensity of the transmitted electric field is 0.368 of its value immediately after it enters the water.

(g) Depth (in meters) of the submarine so that the distance from the surface of the ocean to the submarine is 20λ (λ in water).

(h) Time (in seconds) it takes the wave to travel from the surface of the ocean to the submarine at a depth of 100 m.

(i) Ratio of velocity of the wave in water to that in air (v/v_0).

5.43. A uniform plane wave traveling inside a good conductor with conductivity σ_1 is incident normally on another good conductor with conductivity σ_2, where $\sigma_1 > \sigma_2$. Determine the ratio of σ_1/σ_2 so that the SWR inside medium 1 near the interface is 1.5.

5.44. A right-hand circularly polarized uniform plane wave traveling in air is incident normally on a flat and smooth water surface with $\varepsilon_r = 81$ and $\sigma = 0.1\,\text{S/m}$, as shown in Figure P5-44. Assuming a frequency of 1 GHz and an incident electric field of

$$\mathbf{E}^i = (\hat{\mathbf{a}}_y + \hat{\mathbf{a}}_z e^{j\psi})E_0 e^{j\beta_0 x}$$

do the following.
(a) Determine the value of ψ.
(b) Write an expression for the corresponding incident magnetic field.
(c) Write expressions for the reflected electric and magnetic fields.
(d) Determine the polarization (including sense of rotation) of the reflected wave.
(e) Write expressions for the transmitted electric and magnetic fields.
(f) Determine the polarization (including sense of rotation) of the transmitted wave.
(g) Determine the percentage (compared to the incident) of the reflected and transmitted power densities.

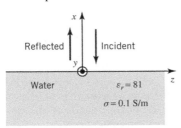

Figure P5-44

5.45. A right-hand circularly polarized wave is incident normally on a perfect conducting flat surface $(\sigma = \infty)$.
(a) What are the polarization and sense of rotation of the reflected field?
(b) What is the normalized (maximum unity) output voltage if the reflected wave is received by a right-hand circularly polarized antenna?
(c) Repeat part (b) if the receiving antenna is left-hand circularly polarized.

5.46. Repeat Problem 5.45 if the reflecting surface is water $(f = 10\,\text{MHz}, \varepsilon_r = 81, \text{and } \sigma = 4\,\text{S/m})$.

5.47. A parallel polarized plane wave traveling in a dielectric medium with ε_1, μ_1 is incident obliquely on a planar interface formed by the dielectric medium with ε_2, μ_2 such that $\varepsilon_2\mu_2 < \varepsilon_1\mu_1$. Assuming that the incident angle θ_i is equal to

or greater than the critical angle θ_c of (5-35b), derive expressions for the reflection coefficient $\Gamma_\parallel^b$ and transmission coefficient $T_\parallel^b$, and the incident $\mathbf{S}_\parallel^i$, reflected $\mathbf{S}_\parallel^r$, and transmitted $\mathbf{S}_\parallel^t$ average power densities, respectively.

5.48. A perpendicularly polarized uniform plane wave traveling inside a free-space medium is obliquely incident, at an incident angle $\theta_i = 60°$, upon a planar dielectric medium with constitutive parameters of $\varepsilon_2 = 4\varepsilon_0, \mu_2 = \mu_0$. Using Figure 5-2 as a reference geometry, determine the:
(a) Wave impedance of the:
  • Incident wave
  • Reflected wave
  • Transmitted wave
(b) Directional impedance in the $+z$ and $+x$ directions, respectively, of the:
  • Incident wave Z_{t0}^{+z}, Z_{t0}^{+x}
  • Transmitted wave Z_{t2}^{+z}, Z_{t2}^{+x}
(c) Reflection coefficient Γ_{in}^{+z} in the $+z$ direction (*magnitude* and *phase*) inside the free-space medium based on:
  • The directional impedances
  • An alternate equation
  • Compare the two answers. Are the answers the same or different in both magnitude and phase? Should they be the same or different in magnitude and phase?
(d) SWR inside the free-space medium.

5.49. A uniform plane wave of either parallel or perpendicular polarization, as shown respectively in Figures 5-2 and 5-4, traveling in free space is incident upon a dielectric/magnetic material such that the product of the relative permittivity and permeability of the dielectric/magnetic material is much greater than unity; that is

$$\varepsilon_r\mu_r \gg 1$$

The intrinsic impedances of the two media are, respectively, η_0 (free space) and η (dielectric/magnetic material).
(a) Determine an approximate value of the refraction angle θ_t (in degrees) for:
  1. Perpendicular polarization.
  2. Parallel polarization.
(b) Obtain simplified expressions, in terms η_0 and η, of the Brewster angle $\theta_i = \theta_B$ for:
  1. Perpendicular polarization.
  2. Parallel polarization.

5.50. A dielectric slab of polystyrene $(\varepsilon_r = 2.56)$, of any thickness, is bounded on both of its sides

by air. In order to eliminate reflections on each of its interfaces, the slab is covered on each of its faces with a dielectric material.
At a frequency of 10 GHz, determine, for each dielectric material that must cover each of the faces of the slab, the:
(a) Thickness (in λ_i; wavelength in the corresponding dielectric).
(b) Thickness (in cm).
(c) Dielectric constant.
(d) Intrinsic impedance of its medium.
(e) SWR created in air when a plane wave impinges at normal incidence from one of its sides when the slab is covered with the selected cover material.

5.51. For Example 5-10, determine the bandwidth, and the lower and upper frequencies of the bandwidth, over which the system can operate so that the magnitude of the reflection coefficient is equal to:
(a) 0.05
(b) 0.10
Assume a center frequency of 10 GHz within the bandwidth.

5.52. For the one-slab reflection problem of Figure 5-11a, write the expressions for the:
(a) Exact transmission-line model.
(b) Exact ray-tracing model.
(c) Approximate ray-tracing model.
For Example 5-9, when $d = 0.9375$ cm, plot the magnitude of the input reflection coefficient for 5 GHz $\leq f \leq$ 15 GHz using the:
(d) Exact transmission line-model.
(e) Exact ray-tracing model.
(f) Approximate ray-tracing model.
For Example 5-10, when $d = \lambda_{20}/4$ at the center frequency $f_0 = 10$ GHz, plot the magnitude of the input reflection coefficient for 5 GHz $\leq f \leq$ 15 GHz using the:
(g) Exact transmission line-model.
(h) Exact ray-tracing model.
(i) Approximate ray-tracing model.

5.53. A dielectric slab of thickness d, as shown in Figure 5-11a, is surrounded with air on its left and with a dielectric material, whose dielectric constant (relative permittivity) is 16, on its right. You are asked as an electromagnetic engineer/scientist to design a dielectric slab with the smallest nonzero thickness that will reduce the input reflection coefficient, at normal incidence, to zero at a frequency of 1 GHz.
What should one set of parameters of the dielectric slab be that will reduce the reflection coefficient to zero? State the:

(a) Smallest thickness of the slab in terms of the wavelength in the dielectric slab.
(b) Smallest thickness of the slab, in cm, at 1 GHz.
(c) Dielectric constant of the dielectric material of the slab.
Justify your answers. Assume that the permeability of all three media is the same as that of free space.

5.54. A symmetrical three-layer dielectric slab is bounded at both sides by air, and it is designed to filter the signal that can pass through it. The dielectric constant of all the five media, including the medium to the left (air), the three slabs, and the medium to the right (air) are, respectively, $\varepsilon_{r0} = 1, \varepsilon_{r1} = 4, \varepsilon_{r2} = 9, \varepsilon_{r3} = 4, \varepsilon_{r4} = 1$.
Assuming that at the operating frequency the width d_n, $n = 1, 2, 3$, of each layer is one quarter of a wavelength in its respective medium, determine the:
(a) Corresponding intrinsic reflection coefficients at each interface ($\Gamma_{01}, \Gamma_{12}, \Gamma_{23}, \Gamma_{30}$).
(b) Approximate total input reflection coefficient at the leading interface between air and the first layer (Γ_{in}) at the center operating frequency.

5.55. A uniform plane wave traveling in air, whose amplitude of the electric field is E_0, is incident normally upon a perfect electric conductor that is coated with a lossless dielectric material with $\varepsilon = 4\varepsilon_0$, $\mu = \mu_0$, $\sigma = 0$, and thickness of $\lambda/8$ (λ is the wavelength in the dielectric). Just to the left of the air side of the air–dielectric interface, determine the:
(a) Exact reflection coefficient looking normally just to the left of the air/dielectric interface ($z = -d^-$, i.e., toward the conductor).
(b) SWR looking normally just to the left of the air/dielectric interface ($z = -d^-$, i.e., toward the conductor).

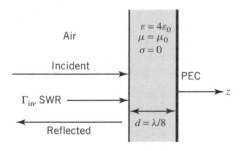

Figure P5-55

5.56. A TE^y uniform plane wave is traveling in a free-space medium on the x-z plane along an oblique angle θ_i, as shown in Figure 4-4(a).

(a) Write expressions for the two directional impedances, one along +x direction and one along +z direction. The directional impedances should be in terms of the incidence angle θ_i and the intrinsic impedance η_0 of the free-space medium.

(b) Now we introduce a planar interface, as shown in Figure 5-4, formed by the free-space medium (with intrinsic impedance η_0) and another medium (with an intrinsic impedance η). Defining the reflection coefficient along the +z direction as

$$\Gamma_z^+ = \frac{\eta - (\text{choose the appropriate directional impedance from part a})}{\eta + (\text{choose the appropriate directional impedance from part a})}$$

write an expression for the reflection coefficient using the appropriate directional impedance from part (a).

(c) Based on the expression of Γ_z^+ from part (b), derive an expression for the incidence angle θ_i for which the reflection coefficient is equal to zero. Assume that the permeabilities of the two media are the same. The equation should only be a function of ε_r ($\varepsilon_r = \varepsilon/\varepsilon_0$, where ε and ε_0 are, respectively, the permittivities of the scattering/reflecting medium and free space).

(d) Based on the expression of part (c), compute the incidence angle θ_i (in degrees) if the dielectric constant of the medium that creates the scattering/reflection is 4.

(e) Repeat part (d). Find the incidence angle (in degrees) but based on the equation for the Brewster Angle. State the polarization (parallel or perpendicular). Is the answer for the incidence angle, using the equation for the Brewster Angle, the same as or different from that of part (d)? If different, why?

5.57. A TM^y (perpendicular polarization) uniform plane wave is traveling in a free-space medium on the x-z plane along an oblique angle θ_i, as shown in Figure 4-4(b).

(a) Write expressions for the two directional impedances one along +x direction and one along +z direction. The directional impedances should be in terms of the incidence angle θ_i and the intrinsic impedance η_0 of the free-space medium.

(b) Now we introduce a planar interface, as shown in Figure 5-2, formed by the free-space medium (with intrinsic impedance η_0) and another medium (with an intrinsic impedance η). Defining the reflection coefficient along the +z direction as

$$\Gamma_z^+ = \frac{\eta - (\text{choose the appropriate directional impedance from part a})}{\eta + (\text{choose the appropriate directional impedance from part a})}$$

write an expression for the reflection coefficient using the appropriate directional impedance from part (a).

(c) Based on the expression of Γ_z^+ from part (b), derive an expression for the incidence angle θ_i for which the reflection coefficient is equal to zero. Assume that the permeabilities of the two media are the same. The equation should only be a function of ε_r ($\varepsilon_r = \varepsilon/\varepsilon_0$, where ε and ε_0 are, respectively, the permittivities of the scattering/reflecting medium and free space).

(d) Based on the expression of part (c), compute the incidence angle θ_i (in degrees) if the dielectric constant of the medium that creates the scattering/reflection is 4.

(e) Repeat part (d). Find the incidence angle (in degrees) but based on the equation for the Brewster Angle. State the polarization (parallel or perpendicular). Is the answer for the incidence angle, using the equation for the Brewster Angle, the same or different than that of part d? If different, why?

5.58. A uniform plane wave traveling, from left to right, is incident at normal incidence upon a 'sandwich' of three lossless dielectric slabs which are bounded to the left by free space and to their right by a semi-infinite medium of dielectric constant of $\varepsilon_{r4} = 25$, as shown in Figure P5-58. The dielectric constants of the three slabs, from left to right, are: first, ε_{r1} = unknown (to be determined); second, $\varepsilon_{r2} = 9$; and third, $\varepsilon_{r3} = 16$. Their corresponding thicknesses/widths, from left to right, are $d_1 = \lambda_1/4$ (λ_1 is the wavelength within that dielectric, $d_2 = 0.5$ cm, and $d_3 = 0.375$ cm. Assuming a frequency of f = 10 GHz determine for the first slab its:

(a) Intrinsic impedance (η_1 in ohms).
(b) Dielectric constant (ε_{r1}).
(c) Width (d_1 in cm).

so that the total input reflection coefficient the incident wave sees at the leading edge of the first slab is zero.

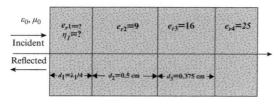

Figure P5-58

5.59. Two vertical lossless dielectric slabs, each of thickness equal to $\lambda_0/4$ at a center frequency of $f_0 = 2$ GHz, are sandwiched between a lossless semi-infinite medium of dielectric constant $\varepsilon_r = 2.25$ to the left and air to the right. Assume a fractional bandwidth of 0.5 and a binomial design.
  (a) Find the magnitude of the maximum reflection coefficient within the allowable bandwidth.
  (b) Determine the magnitude of the reflection coefficients at each interface (junction).
  (c) Compute the intrinsic impedances, dielectric constants, and thickness (in centimeters) of each dielectric slab.
  (d) Determine the lower and upper frequencies of the bandwidth.
  (e) Plot the magnitude of the reflection coefficient inside the dielectric medium with $\varepsilon_r = 2.25$ as a function of frequency (within $0 \le f / f_0 \le 2$).

5.60. It is desired to design a three-layer (each layer of $\lambda_0/4$ thickness) impedance transformer to match a semi-infinite dielectric medium of $\varepsilon_r = 9$ on one of its sides and one with $\varepsilon_r = 2.25$ on the other side. The maximum SWR that can be tolerated inside the dielectric medium with $\varepsilon_r = 9$ is 1.1. Assume a center frequency of $f_0 = 3$ GHz and a binomial design.
  (a) Determine the allowable fractional bandwidth and the lower and upper frequencies of the bandwidth.
  (b) Find the magnitude of reflection coefficients at each junction.
  (c) Compute the magnitude of the maximum reflection coefficient within the bandwidth.
  (d) Determine the intrinsic impedances, dielectric constants, and thicknesses (in centimeters) of each dielectric slab.
  (e) Plot the magnitude of the reflection coefficient inside the dielectric medium with

$\varepsilon_r = 9$ as a function of frequency (within $0 \le f / f_0 \le 2$).

5.61. Design a three-section binomial impedance transformer to match free space to a semi-infinite dielectric medium with a relative permittivity (dielectric constant) of 4. Assume the permeabilities of all the media are the same as those of free space. The thickness of each of the slabs/dielectrics placed between them should be $\lambda_0/4$ at their respective dielectric constants and at a center frequency of $f_0 = 10$ GHz. The magnitude of the maximum tolerable reflection coefficient within the bandwidth is $\rho_m = 0.2$.
  • What is the maximum SWR associated with the maximum tolerable maximum reflection coefficient of $\rho_m = 0.2$?
  • What is the fractional bandwidth within which the maximum tolerable reflection coefficient does not exceed 0.2?
  • Determine the intrinsic impedance *and* associated relative permittivity (dielectric constant) for each of the three slabs that must be inserted between free space and the semi-infinite dielectric medium with a dielectric constant of 4.

5.62. Repeat Example 5-11 using a Tschebyscheff design.

5.63. Repeat Problem 5.59 using a Tschebyscheff design.

5.64. Repeat Problem 5.60 using a Tschebyscheff design.

5.65. A right-hand (CW) elliptically polarized wave traveling in free space is obliquely incident at an angle $\theta_i = 30°$, measured from the normal, on a flat perfect electric conductor of infinite extent. If the incident field has an axial ratio of -2, determine the polarization of the reflected field. This is to include the axial ratio as well as its sense of rotation. Assume that the time-phase difference between the components of the incident field is $90°$.

5.66. Repeat Problem 5.65 if the reflecting surface is a flat lossless ($\sigma_2 = 0$) ocean ($\varepsilon_2 = 81\varepsilon_0$ and $\mu_2 = \mu_0$) of infinite extent. Also find the polarization of the wave transmitted into the water.

5.67. A uniform plane wave is normally incident upon a Perfect Electric Conducting (PEC) surface. The incident electric field is given by

$$\mathbf{E}^i(x) = \left(\hat{\mathbf{a}}_z + j2\hat{\mathbf{a}}_y\right) E_0 e^{-j\beta_0 x}$$

where β_0 and E_0 are real constants. Assuming a $e^{+j\omega t}$ time convention:

(a) Write an expression for the reflected electric field.
(b) For the incident wave, determine the:
 - Polarization (linear, circular, or elliptical). Justify your answer.
 - Sense of rotation of the incident wave (CW or CCW). Justify your answer.
 - Axial ratio (AR). Justify your answer.
(c) For the reflected wave, determine the:
 - Polarization (linear, circular or elliptical). Justify your answer.
 - Sense of rotation of the incident wave (CW or CCW). Justify your answer.
 - Axial ratio (AR). Justify your answer.

For all of the above, be sure to justify your answers. Also verify with the MATLAB computer program **Polarization_Refl_Trans**.

5.68. A uniform plane wave is normally incident upon a Perfect Magnetic Conducting (PMC) surface. The incident electric field is given by

$$\mathbf{E}^i(y) = (2\hat{\mathbf{a}}_x - j\hat{\mathbf{a}}_z) E_0 e^{-j\beta_0 y}$$

where β_0 and E_0 are real constants. Assuming a $e^{+j\omega t}$ time convention:

(a) Write an expression for the reflected electric field.
(b) For the incident wave, determine the:
 - Polarization (linear, circular, or elliptical). Justify your answer.
 - Sense of rotation of the incident wave (CW or CCW). Justify your answer.
 - Axial ratio (AR). Justify your answer.
(c) For the reflected wave, determine the:
 - Polarization (linear, circular, or elliptical). Justify your answer.
 - Sense of rotation of the incident wave (CW or CCW). Justify your answer.
 - Axial ratio (AR). Justify your answer.

For all of the above, justify your answers. Also verify with the MATLAB computer program **Polarization_Refl_Trans**.

5.69. A left-hand (CCW) circularly polarized wave traveling inside a lossless earth, with a dielectric constant of 9, is incident upon a planar interface formed by the earth and air. The angle of incidence is 18.43495°. Determine the:

(a) Polarization of the reflected wave (linear, circular, elliptical).
(b) Sense of rotation of the reflected wave; (CW or CCW), if appropriate.
(c) Polarization of the transmitted wave (linear, circular, elliptical).
(d) Sense of rotation of the transmitted wave; (CW, CCW), if appropriate.

As an option, you do not have to use too many analytical equations as long as you can justify the correct answers using words/text (you can keep the formulations to a minimum).

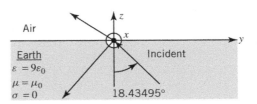

Figure P5-69

5.70. Repeat Problem 5.69 when the incident wave is right-hand (CW) circularly polarized.

5.71. Derive the transmission coefficient for the dielectric slab of Example 5-15.

5.72. For a planar interface formed by DPS-DNG materials and assuming parallel polarization wave incidence, write expressions for the wavenumbers and Poynting vectors, similar in form to the ones of Figure 5-29, (5-110a) – (5-110c), and (5-111a) – (5-111c). Examine the directions of the wavenumbers and Poynting vectors of the transmitted wave and compare with those for a DPS-DPS interface.

CHAPTER **6**

Auxiliary Vector Potentials, Construction of Solutions, and Radiation and Scattering Equations

6.1 INTRODUCTION

It is common practice in the analysis of electromagnetic boundary-value problems to use auxiliary vector potentials as aids in obtaining solutions for the electric ($\mathbf{E}$) and magnetic ($\mathbf{H}$) fields. The most common vector potential functions are the $\mathbf{A}$, magnetic vector potential, and $\mathbf{F}$, electric vector potential. They are the same pair that were introduced and used extensively in the solution of antenna radiation problems [1]. *Although the electric and magnetic field intensities ($\mathbf{E}$ and $\mathbf{H}$) represent physically measurable quantities, for most engineers the vector potentials are strictly mathematical tools.* The introduction of the potentials often simplifies the solution, even though it may require determination of additional functions. Much of the discussion in this chapter is borrowed from [1].

The Hertz vector potentials $\prod_e$ and $\prod_h$ make up another pair. The Hertz vector potential $\prod_e$ is analogous to $\mathbf{A}$ and $\prod_h$ is analogous to $\mathbf{F}$. The functional relation between them is a proportionality constant that is a function of the frequency and the constitutive parameters of the medium. In the solution of a problem, only one set, $\mathbf{A}$ and $\mathbf{F}$ or $\prod_e$ and $\prod_h$, is required. The author prefers $\mathbf{A}$ and $\mathbf{F}$, and they will be used throughout this book.

The main objective of this book is to obtain electromagnetic field configurations (modes) of boundary-value propagation, radiation, and scattering problems. These field configurations must satisfy Maxwell's equations or the wave equation, as well as the appropriate boundary conditions. The procedure is to specify the electromagnetic boundary-value problem, which may or may not contain sources, and to obtain the field configurations that can exist within the region of the boundary-value problem. This can be accomplished in either of two ways, as shown in Figure 6-1.

One procedure for obtaining the electric and magnetic fields of a desired boundary-value problem is to use Maxwell's or the wave equations. This is accomplished essentially in one step, and it is represented in Figure 6-1 by path 1. The formulation using such a procedure is assigned to the reader as an end-of-chapter problem.

The other procedure requires two steps. In the first step, the vector potentials $\mathbf{A}$ and $\mathbf{F}$ (or $\prod_e$ and $\prod_h$) are found, once the boundary-value problem is specified. In the second step, the electric and magnetic fields are found, after the vector potentials are determined. The electric and magnetic fields are functions of the vector potentials. This procedure is represented by path 2 of

Balanis' Advanced Engineering Electromagnetics, Third Edition. Constantine A. Balanis.
© 2024 John Wiley & Sons, Inc. Published 2024 by John Wiley & Sons, Inc.
Companion Website: www.wiley.com/go/balanis/advancedengineeringelectromagnetics3e

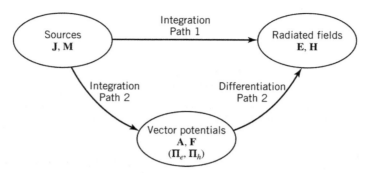

Figure 6-1   Block diagram for computing radiated fields from electric and magnetic sources. (Source: C. A. Balanis, *Antenna Theory: Analysis and Design*, Third Edition. Copyright © 2005, John Wiley & Sons, Inc. Reprinted by permission of John Wiley & Sons, Inc.)

Figure 6-1, and, although it requires two steps, it is often simpler and more straightforward; hence it is often preferred. The mathematical equations of this procedure will be developed next, and they will be utilized in this book to illustrate solutions of boundary-value electromagnetic problems.

In a homogeneous medium, any solution for the time-harmonic electric and magnetic fields must satisfy Maxwell's equations

$$\nabla \times \mathbf{E} = -\mathbf{M} - j\omega\mu\mathbf{H} \tag{6-1a}$$

$$\nabla \times \mathbf{H} = \mathbf{J} + j\omega\varepsilon\mathbf{E} \tag{6-1b}$$

$$\nabla \cdot \mathbf{E} = \frac{q_{ev}}{\varepsilon} \tag{6-1c}$$

$$\nabla \cdot \mathbf{H} = \frac{q_{mv}}{\mu} \tag{6-1d}$$

or the vector wave equations

$$\nabla^2\mathbf{E} + \beta^2\mathbf{E} = \nabla \times \mathbf{M} + j\omega\mu\mathbf{J} + \frac{1}{\varepsilon}\nabla q_{ev} \tag{6-2a}$$

$$\nabla^2\mathbf{H} + \beta^2\mathbf{H} = -\nabla \times \mathbf{J} + j\omega\varepsilon\mathbf{M} + \frac{1}{\mu}\nabla q_{mv} \tag{6-2b}$$

where

$$\beta^2 = \omega^2\mu\varepsilon \tag{6-2c}$$

In regions where there are no sources, $\mathbf{J} = \mathbf{M} = q_{ev} = q_{mv} = 0$. In these regions, the preceding equations are of simpler form. Whereas the electric current density $\mathbf{J}$ may represent either actual or equivalent sources, the magnetic current density $\mathbf{M}$ can only represent equivalent sources. Although all of these equations will still be satisfied, an alternate procedure is developed next for the solution of the electric and magnetic fields in terms of the auxiliary vector potentials, $\mathbf{A}$ and $\mathbf{F}$.

6.2   THE VECTOR POTENTIAL A

In a source-free region, the magnetic flux density $\mathbf{B}$ is always solenoidal, that is, $\nabla \cdot \mathbf{B} = 0$. Therefore, it can be represented as the curl of another vector because it obeys the vector identity

$$\nabla \cdot (\nabla \times \mathbf{A}) = 0 \tag{6-3}$$

where **A** is an arbitrary vector. Thus, we define

$$\mathbf{B}_A = \mu \mathbf{H}_A = \nabla \times \mathbf{A} \tag{6-4}$$

or

$$\boxed{\mathbf{H}_A = \frac{1}{\mu} \nabla \times \mathbf{A}} \tag{6-4a}$$

where subscript A indicates the fields due to the **A** potential. Substituting (6-4a) into Maxwell's curl equation

$$\nabla \times \mathbf{E}_A = -j\omega\mu \mathbf{H}_A \tag{6-5}$$

reduces it to

$$\nabla \times \mathbf{E}_A = -j\omega\mu \mathbf{H}_A = -j\omega \nabla \times \mathbf{A} \tag{6-6}$$

which can also be written as

$$\nabla \times [\mathbf{E}_A + j\omega \mathbf{A}] = 0 \tag{6-7}$$

From the vector identity

$$\nabla \times (-\nabla \phi_e) = 0 \tag{6-8}$$

and (6-7), it follows that

$$\mathbf{E}_A + j\omega \mathbf{A} = -\nabla \phi_e \tag{6-9}$$

or

$$\boxed{\mathbf{E}_A = -\nabla \phi_e - j\omega \mathbf{A}} \tag{6-9a}$$

ϕ_e represents an arbitrary electric scalar potential that is a function of position. Taking the curl of both sides of (6-4) and using the vector identity

$$\nabla \times \nabla \times \mathbf{A} = \nabla(\nabla \cdot \mathbf{A}) - \nabla^2 \mathbf{A} \tag{6-10}$$

leads to

$$\nabla \times (\mu \mathbf{H}_A) = \nabla(\nabla \cdot \mathbf{A}) - \nabla^2 \mathbf{A} \tag{6-10a}$$

For a homogeneous medium, (6-10a) reduces to

$$\mu \nabla \times \mathbf{H}_A = \nabla(\nabla \cdot \mathbf{A}) - \nabla^2 \mathbf{A} \tag{6-11}$$

Equating Maxwell's equation

$$\boxed{\nabla \times \mathbf{H}_A = \mathbf{J} + j\omega\varepsilon \mathbf{E}_A} \tag{6-12}$$

to (6-11) leads to

$$\mu \mathbf{J} + j\omega\mu\varepsilon \mathbf{E}_A = \nabla(\nabla \cdot \mathbf{A}) - \nabla^2 \mathbf{A} \tag{6-13}$$

Substituting (6-9a) into (6-13) reduces it to

$$\nabla^2 \mathbf{A} + \beta^2 \mathbf{A} = -\mu \mathbf{J} + \nabla(\nabla \cdot \mathbf{A}) + \nabla(j\omega\mu\varepsilon\phi_e) = -\mu \mathbf{J} + \nabla(\nabla \cdot \mathbf{A} + j\omega\mu\varepsilon\phi_e) \tag{6-14}$$

where $\beta^2 = \omega^2 \mu\varepsilon$.

In (6-4), the curl of **A** was defined. Now we are at liberty to define the divergence of **A**, which is independent of its curl. Both are required to uniquely define **A**. In order to simplify (6-14), let

$$\boxed{\boldsymbol{\nabla}\cdot\mathbf{A} = -j\omega\varepsilon\mu\phi_e \Rightarrow \phi_e = -\frac{1}{j\omega\mu\varepsilon}\boldsymbol{\nabla}\cdot\mathbf{A}}$$

(6-15)

which is known as the *Lorenz condition* (or *gauge*). Other gauges may be defined. Substituting (6-15) into (6-14) leads to

$$\boxed{\nabla^2\mathbf{A} + \beta^2\mathbf{A} = -\mu\mathbf{J}}$$

(6-16)

In addition, (6-9a) reduces to

$$\boxed{\mathbf{E}_A = -\boldsymbol{\nabla}\phi_e - j\omega\mathbf{A} = -j\omega\mathbf{A} - j\frac{1}{\omega\mu\varepsilon}\boldsymbol{\nabla}(\boldsymbol{\nabla}\cdot\mathbf{A})}$$

(6-17)

Once **A** is known, $\mathbf{H}_A$ can be found from (6-4a) and $\mathbf{E}_A$ from (6-17). $\mathbf{E}_A$ can just as easily be found from Maxwell's equation 6-12 by setting $\mathbf{J} = 0$. Since (6-16) is a vector wave equation, solutions for **A** in rectangular, cylindrical, and spherical coordinate systems are similar to those for **E** in Sections 3.4.1A, 3.4.2, and 3.4.3, respectively.

6.3   THE VECTOR POTENTIAL F

In a source-free region, the electric flux density **D** is always solenoidal, that is, $\boldsymbol{\nabla}\cdot\mathbf{D} = 0$. Therefore, it can be represented as the curl of another vector because it obeys the vector identity

$$\boldsymbol{\nabla}\cdot(-\boldsymbol{\nabla}\times\mathbf{F}) = 0$$

(6-18)

where **F** is an arbitrary vector. Thus we can define $\mathbf{D}_F$ by

$$\mathbf{D}_F = -\boldsymbol{\nabla}\times\mathbf{F}$$

(6-19)

or

$$\boxed{\mathbf{E}_F = -\frac{1}{\varepsilon}\boldsymbol{\nabla}\times\mathbf{F}}$$

(6-19a)

where the subscript F indicates the fields due to the **F** potential. Substituting (6-19a) into Maxwell's curl equation

$$\boldsymbol{\nabla}\times\mathbf{H}_F = j\omega\varepsilon\mathbf{E}_F$$

(6-20)

reduces it to

$$\boldsymbol{\nabla}\times(\mathbf{H}_F + j\omega\mathbf{F}) = 0$$

(6-21)

From the vector identity (6-8), it follows that

$$\boxed{\mathbf{H}_F = -\boldsymbol{\nabla}\phi_m - j\omega\mathbf{F}}$$

(6-22)

where ϕ_m represents an arbitrary magnetic scalar potential that is a function of position. Taking the curl of (6-19a)

$$\boldsymbol{\nabla}\times\mathbf{E}_F = -\frac{1}{\varepsilon}\boldsymbol{\nabla}\times\boldsymbol{\nabla}\times\mathbf{F} = -\frac{1}{\varepsilon}[\boldsymbol{\nabla}(\boldsymbol{\nabla}\cdot\mathbf{F}) - \nabla^2\mathbf{F}]$$

(6-23)

and equating it to Maxwell's equation

$$\boxed{\nabla \times \mathbf{E}_F = -\mathbf{M} - j\omega\mu\mathbf{H}_F}$$ (6-24)

lead to

$$\nabla^2 \mathbf{F} + j\omega\mu\varepsilon\mathbf{H}_F = \nabla(\nabla \cdot \mathbf{F}) - \varepsilon\mathbf{M}$$ (6-25)

Substituting (6-22) into (6-25) reduces it to

$$\nabla^2 \mathbf{F} + \beta^2 \mathbf{F} = -\varepsilon\mathbf{M} + \nabla(\nabla \cdot \mathbf{F} + j\omega\mu\varepsilon\phi_m)$$ (6-26)

Letting

$$\boxed{\nabla \cdot \mathbf{F} = -j\omega\mu\varepsilon\phi_m \Rightarrow \phi_m = -\frac{1}{j\omega\mu\varepsilon}\nabla \cdot \mathbf{F}}$$ (6-27)

reduces (6-26) to

$$\boxed{\nabla^2 \mathbf{F} + \beta^2 \mathbf{F} = -\varepsilon\mathbf{M}}$$ (6-28)

and (6-22) to

$$\boxed{\mathbf{H}_F = -j\omega\mathbf{F} - \frac{j}{\omega\mu\varepsilon}\nabla(\nabla \cdot \mathbf{F})}$$ (6-29)

Once $\mathbf{F}$ is known, $\mathbf{E}_F$ can be found from (6-19a) and $\mathbf{H}_F$ from (6-29) or (6-24) by setting $\mathbf{M} = 0$. Since (6-28) is a vector wave equation, solutions for $\mathbf{F}$ in rectangular, cylindrical, and spherical coordinate systems are similar to those for $\mathbf{E}$ in Sections 3.4.1A, 3.4.2, and 3.4.3, respectively.

6.4 THE VECTOR POTENTIALS A AND F

In the previous two sections, we derived expressions for the $\mathbf{E}$ and $\mathbf{H}$ fields in terms of the vector potentials $\mathbf{A}$ ($\mathbf{E}_A, \mathbf{H}_A$) and $\mathbf{F}$ ($\mathbf{E}_F, \mathbf{H}_F$). In addition, expressions that $\mathbf{A}$ and $\mathbf{F}$ must satisfy were also derived. The total $\mathbf{E}$ and $\mathbf{H}$ fields are obtained by the superposition of the individual fields due to $\mathbf{A}$ and $\mathbf{F}$.

The procedure that can be used to find the fields of path 2 of Figure 6-1 is as follows.

SUMMARY

1. Specify the electromagnetic boundary-value problem, which may or may not contain any sources within its boundaries, and its desired field configurations (modes).
2. a. Solve for $\mathbf{A}$ using (6-16),

$$\boxed{\nabla^2 \mathbf{A} + \beta^2 \mathbf{A} = -\mu\mathbf{J}} \quad \text{where } \beta^2 = \omega^2\mu\varepsilon$$ (6-30)

Depending on the problem, solutions for $\mathbf{A}$ in rectangular, cylindrical, and spherical coordinate systems take the forms found in Sections 3.4.1A, 3.4.2, and 3.4.3, respectively.

b. Solve for **F** using (6-28),

$$\boxed{\nabla^2\mathbf{F} + \beta^2\mathbf{F} = -\varepsilon\mathbf{M}} \quad \text{where } \beta^2 = \omega^2\mu\varepsilon \tag{6-31}$$

Depending on the problem, solutions for **F** in rectangular, cylindrical, and spherical coordinate systems take the forms found in Sections 3.4.1A, 3.4.2, and 3.4.3, respectively.

3. a. Find $\mathbf{H}_A$ using (6-4a) and $\mathbf{E}_A$ using (6-17). $\mathbf{E}_A$ can also be found using (6-12) by letting $\mathbf{J} = 0$.

$$\boxed{\mathbf{H}_A = \frac{1}{\mu}\nabla\times\mathbf{A}} \tag{6-32a}$$

$$\boxed{\mathbf{E}_A = -j\omega\mathbf{A} - j\frac{1}{\omega\mu\varepsilon}\nabla(\nabla\cdot\mathbf{A})} \tag{6-32b}$$

or

$$\boxed{\mathbf{E}_A = \frac{1}{j\omega\varepsilon}\nabla\times\mathbf{H}_A} \tag{6-32c}$$

b. Find $\mathbf{E}_F$ using (6-19a) and $\mathbf{H}_F$ using (6-29). $\mathbf{H}_F$ can also be found using (6-24) by letting $\mathbf{M} = 0$.

$$\boxed{\mathbf{E}_F = -\frac{1}{\varepsilon}\nabla\times\mathbf{F}} \tag{6-33a}$$

$$\boxed{\mathbf{H}_F = -j\omega\mathbf{F} - j\frac{1}{\omega\mu\varepsilon}\nabla(\nabla\cdot\mathbf{F})} \tag{6-33b}$$

or

$$\boxed{\mathbf{H}_F = -\frac{1}{j\omega\mu}\nabla\times\mathbf{E}_F} \tag{6-33c}$$

4. The total fields are then found by the superposition of those given in step 3, that is,

$$\boxed{\mathbf{E} = \mathbf{E}_A + \mathbf{E}_F = -j\omega\mathbf{A} - j\frac{1}{\omega\mu\varepsilon}\nabla(\nabla\cdot\mathbf{A}) - \frac{1}{\varepsilon}\nabla\times\mathbf{F}} \tag{6-34}$$

or

$$\boxed{\mathbf{E} = \mathbf{E}_A + \mathbf{E}_F = \frac{1}{j\omega\varepsilon}\nabla\times\mathbf{H}_A - \frac{1}{\varepsilon}\nabla\times\mathbf{F}} \tag{6-34a}$$

and

$$\boxed{\mathbf{H} = \mathbf{H}_A + \mathbf{H}_F = \frac{1}{\mu}\nabla\times\mathbf{A} - j\omega\mathbf{F} - j\frac{1}{\omega\mu\varepsilon}\nabla(\nabla\cdot\mathbf{F})} \tag{6-35}$$

or

$$\boxed{\mathbf{H} = \mathbf{H}_A + \mathbf{H}_F = \frac{1}{\mu}\nabla\times\mathbf{A} - \frac{1}{j\omega\mu}\nabla\times\mathbf{E}_F} \tag{6-35a}$$

Whether (6-32b) or (6-32c) is used to find $\mathbf{E}_A$ and (6-33b) or (6-33c) to find $\mathbf{H}_F$ depends largely on the nature of the problem. In many instances, one may be more complex than the other. For computing radiation fields in the far zone, it will be easier to use (6-32b) for $\mathbf{E}_A$ and (6-33b) for $\mathbf{H}_F$ because, as it will be shown, the second term in each expression becomes negligible in this region. The same solution should be obtained using either of the two choices in each case.

6.5 CONSTRUCTION OF SOLUTIONS

For many electromagnetic boundary-value problems, there are usually many field configurations (modes) that are solutions that satisfy Maxwell's equations and the boundary conditions. The most widely known modes are those that are referred to as Transverse Electromagnetic (TEM), Transverse Electric (TE), and Transverse Magnetic (TM).

TEM modes are field configurations whose *electric and magnetic* field components are transverse to a given direction. Often, but not necessarily, that direction is the path that the wave is traveling. TE modes are field configurations whose *electric* field components are transverse to a given direction; for TM modes the *magnetic* field components are transverse to a given direction. Here, we will illustrate methods that utilize the vector potentials to construct TEM, TE, and TM modes.

6.5.1 Transverse Electromagnetic Modes: Source-Free Region

TEM modes are usually the simplest forms of field configurations, and they are usually referred to as the *lowest-order modes*. For these field configurations, both the electric *and* magnetic field components are transverse to a given direction. To see how these modes can be constructed using the vector potentials, let us illustrate the procedure using the rectangular and cylindrical coordinate systems.

A. Rectangular Coordinate System According to (6-34), the electric field in terms of the vector potentials $\mathbf{A}$ and $\mathbf{F}$ is given by

$$\mathbf{E} = \mathbf{E}_A + \mathbf{E}_F = -j\omega\mathbf{A} - j\frac{1}{\omega\mu\varepsilon}\nabla(\nabla\cdot\mathbf{A}) - \frac{1}{\varepsilon}\nabla\times\mathbf{F} \tag{6-36}$$

Assuming the vector potentials $\mathbf{A}$ and $\mathbf{F}$ have solutions of the form

$$\mathbf{A}(x,y,z) = \hat{\mathbf{a}}_x A_x(x,y,z) + \hat{\mathbf{a}}_y A_y(x,y,z) + \hat{\mathbf{a}}_z A_z(x,y,z) \tag{6-37}$$

which satisfies (6-30) with $\mathbf{J} = 0$

$$\nabla^2\mathbf{A} + \beta^2\mathbf{A} = 0 \tag{6-38}$$

or

$$\nabla^2 A_x + \beta^2 A_x = 0 \tag{6-38a}$$

$$\nabla^2 A_y + \beta^2 A_y = 0 \tag{6-38b}$$

$$\nabla^2 A_z + \beta^2 A_z = 0 \tag{6-38c}$$

and

$$\mathbf{F}(x,y,z)=\hat{\mathbf{a}}_x F_x(x,y,z)+\hat{\mathbf{a}}_y F_y(x,y,z)+\hat{\mathbf{a}}_z F_z(x,y,z) \tag{6-39}$$

which satisfies (6-31) with $\mathbf{M}=0$

$$\nabla^2 \mathbf{F}+\beta^2 \mathbf{F}=0 \tag{6-40}$$

or

$$\nabla^2 F_x+\beta^2 F_x=0 \tag{6-40a}$$

$$\nabla^2 F_y+\beta^2 F_y=0 \tag{6-40b}$$

$$\nabla^2 F_z+\beta^2 F_z=0 \tag{6-40c}$$

(6-36), when expanded, can be written as

$$
\begin{aligned}
\mathbf{E}=\;&\hat{\mathbf{a}}_x\left[-j\omega A_x-j\frac{1}{\omega\mu\varepsilon}\left(\frac{\partial^2 A_x}{\partial x^2}+\frac{\partial^2 A_y}{\partial x\,\partial y}+\frac{\partial^2 A_z}{\partial x\,\partial z}\right)-\frac{1}{\varepsilon}\left(\frac{\partial F_z}{\partial y}-\frac{\partial F_y}{\partial z}\right)\right]\\
+&\hat{\mathbf{a}}_y\left[-j\omega A_y-j\frac{1}{\omega\mu\varepsilon}\left(\frac{\partial^2 A_x}{\partial x\,\partial y}+\frac{\partial^2 A_y}{\partial y^2}+\frac{\partial^2 A_z}{\partial y\,\partial z}\right)-\frac{1}{\varepsilon}\left(\frac{\partial F_x}{\partial z}-\frac{\partial F_z}{\partial x}\right)\right]\\
+&\hat{\mathbf{a}}_z\left[-j\omega A_z-j\frac{1}{\omega\mu\varepsilon}\left(\frac{\partial^2 A_x}{\partial x\,\partial z}+\frac{\partial^2 A_y}{\partial y\,\partial z}+\frac{\partial^2 A_z}{\partial z^2}\right)-\frac{1}{\varepsilon}\left(\frac{\partial F_y}{\partial x}-\frac{\partial F_x}{\partial y}\right)\right]
\end{aligned} \tag{6-41}
$$

Similarly, (6-35)

$$\mathbf{H}=\mathbf{H}_A+\mathbf{H}_F=\frac{1}{\mu}\nabla\times\mathbf{A}-j\omega\mathbf{F}-j\frac{1}{\omega\mu\varepsilon}\nabla(\nabla\cdot\mathbf{F}) \tag{6-42}$$

when expanded using (6-37) and (6-39) can be written as

$$
\begin{aligned}
\mathbf{H}=\;&\hat{\mathbf{a}}_x\left[-j\omega F_x-j\frac{1}{\omega\mu\varepsilon}\left(\frac{\partial^2 F_x}{\partial x^2}+\frac{\partial^2 F_y}{\partial x\,\partial y}+\frac{\partial^2 F_z}{\partial x\,\partial z}\right)+\frac{1}{\mu}\left(\frac{\partial A_z}{\partial y}-\frac{\partial A_y}{\partial z}\right)\right]\\
+&\hat{\mathbf{a}}_y\left[-j\omega F_y-j\frac{1}{\omega\mu\varepsilon}\left(\frac{\partial^2 F_x}{\partial x\,\partial y}+\frac{\partial^2 F_y}{\partial y^2}+\frac{\partial^2 F_z}{\partial y\,\partial z}\right)+\frac{1}{\mu}\left(\frac{\partial A_x}{\partial z}-\frac{\partial A_z}{\partial x}\right)\right]\\
+&\hat{\mathbf{a}}_z\left[-j\omega F_z-j\frac{1}{\omega\mu\varepsilon}\left(\frac{\partial^2 F_x}{\partial x\,\partial z}+\frac{\partial^2 F_y}{\partial y\,\partial z}+\frac{\partial^2 F_z}{\partial z^2}\right)+\frac{1}{\mu}\left(\frac{\partial A_y}{\partial x}-\frac{\partial A_x}{\partial y}\right)\right]
\end{aligned} \tag{6-43}
$$

Example 6-1

Using (6-41) and (6-43) derive expressions for the **E** and **H** fields, in terms of the components of the **A** and **F** potentials, that are TEM to the z direction (TEM^z).

Solution: It is apparent by examining (6-41) and (6-43) that TEM^z ($E_z=H_z=0$) modes can be obtained by any of the following three combinations.

1. Letting

$$A_x=A_y=F_x=F_y=0 \qquad A_z\neq0 \qquad F_z\neq0 \qquad \partial/\partial x\neq0 \qquad \partial/\partial y\neq0$$

For this combination, according to (6-41)

$$E_z = -j\omega A_z - j\frac{1}{\omega\mu\varepsilon}\frac{\partial^2 A_z}{\partial z^2} = -j\frac{1}{\omega\mu\varepsilon}\left(\frac{\partial^2}{\partial z^2} + \omega^2\mu\varepsilon\right)A_z = 0$$

provided

$$A_z(x,y,z) = A_z^+(x,y)e^{-j\beta z} + A_z^-(x,y)e^{+j\beta z}$$

Similarly, according to (6-43)

$$H_z = -j\omega F_z - j\frac{1}{\omega\mu\varepsilon}\frac{\partial^2 F_z}{\partial z^2} = -j\frac{1}{\omega\mu\varepsilon}\left(\frac{\partial^2}{\partial z^2} + \omega^2\mu\varepsilon\right)F_z = 0$$

provided

$$F_z(x,y,z) = F_z^+(x,y)e^{-j\beta z} + F_z^-(x,y)e^{+j\beta z}$$

Also according to (6-41) and (6-43)

$$E_x = \left(-\frac{1}{\sqrt{\mu\varepsilon}}\frac{\partial A_z^+}{\partial x} - \frac{1}{\varepsilon}\frac{\partial F_z^+}{\partial y}\right)e^{-j\beta z} + \left(\frac{1}{\sqrt{\mu\varepsilon}}\frac{\partial A_z^-}{\partial x} - \frac{1}{\varepsilon}\frac{\partial F_z^-}{\partial y}\right)e^{+j\beta z} = E_x^+ + E_x^-$$

$$E_y = \left(-\frac{1}{\sqrt{\mu\varepsilon}}\frac{\partial A_z^+}{\partial y} + \frac{1}{\varepsilon}\frac{\partial F_z^+}{\partial x}\right)e^{-j\beta z} + \left(\frac{1}{\sqrt{\mu\varepsilon}}\frac{\partial A_z^-}{\partial y} + \frac{1}{\varepsilon}\frac{\partial F_z^-}{\partial x}\right)e^{+j\beta z} = E_y^+ + E_y^-$$

$$H_x = -\sqrt{\frac{\varepsilon}{\mu}}\left(-\frac{1}{\sqrt{\mu\varepsilon}}\frac{\partial A_z^+}{\partial y} + \frac{1}{\varepsilon}\frac{\partial F_z^+}{\partial x}\right)e^{-j\beta z}$$

$$+ \sqrt{\frac{\varepsilon}{\mu}}\left(\frac{1}{\sqrt{\mu\varepsilon}}\frac{\partial A_z^-}{\partial y} + \frac{1}{\varepsilon}\frac{\partial F_z^-}{\partial x}\right)e^{+j\beta z} = H_x^+ + H_x^-$$

$$H_x = -\sqrt{\frac{\varepsilon}{\mu}}(E_y^+) + \sqrt{\frac{\varepsilon}{\mu}}(E_y^-)$$

$$H_y = \sqrt{\frac{\varepsilon}{\mu}}\left(-\frac{1}{\sqrt{\mu\varepsilon}}\frac{\partial A_z^+}{\partial x} - \frac{1}{\varepsilon}\frac{\partial F_z^+}{\partial y}\right)e^{-j\beta z}$$

$$- \sqrt{\frac{\varepsilon}{\mu}}\left(\frac{1}{\sqrt{\mu\varepsilon}}\frac{\partial A_z^-}{\partial x} - \frac{1}{\varepsilon}\frac{\partial F_z^-}{\partial y}\right)e^{+j\beta z} = H_y^+ + H_y^-$$

$$H_y = \sqrt{\frac{\varepsilon}{\mu}}(E_x^+) - \sqrt{\frac{\varepsilon}{\mu}}(E_x^-)$$

Also

$$Z_w^+ = \frac{E_x^+}{H_y^+} = -\frac{E_y^+}{H_x^+} = \sqrt{\frac{\mu}{\varepsilon}}$$

$$Z_w^- = -\frac{E_x^-}{H_y^-} = \frac{E_y^-}{H_x^-} = \sqrt{\frac{\mu}{\varepsilon}}$$

2. Letting

$$A_x = A_y = A_z = F_x = F_y = 0 \qquad F_z \neq 0 \qquad \partial/\partial x \neq 0 \qquad \partial/\partial y \neq 0$$

For this combination, according to (6-41) and (6-43)

$$E_z = 0$$

$$H_z = -j\omega F_z - j\frac{1}{\omega\mu\varepsilon}\frac{\partial^2 F_z}{\partial z^2} = -j\frac{1}{\omega\mu\varepsilon}\left(\frac{\partial^2}{\partial z^2} + \omega^2\mu\varepsilon\right)F_z = 0$$

provided

$$F_z(x,y,z) = F_z^+(x,y)e^{-j\beta z} + F_z^-(x,y)e^{+j\beta z}$$

Also according to (6-41) and (6-43)

$$E_x = -\frac{1}{\varepsilon}\frac{\partial F_z^+}{\partial y}e^{-j\beta z} - \frac{1}{\varepsilon}\frac{\partial F_z^-}{\partial y}e^{+j\beta z} = E_x^+ + E_x^-$$

$$E_y = +\frac{1}{\varepsilon}\frac{\partial F_z^+}{\partial x}e^{-j\beta z} + \frac{1}{\varepsilon}\frac{\partial F_z^-}{\partial x}e^{+j\beta z} = E_y^+ + E_y^-$$

$$H_x = -\sqrt{\frac{\varepsilon}{\mu}}\left(\frac{1}{\varepsilon}\frac{\partial F_z^+}{\partial x}\right)e^{-j\beta z} + \sqrt{\frac{\varepsilon}{\mu}}\left(\frac{1}{\varepsilon}\frac{\partial F_z^-}{\partial x}\right)e^{+j\beta z} = H_x^+ + H_x^-$$

$$= -\sqrt{\frac{\varepsilon}{\mu}}(E_y^+) + \sqrt{\frac{\varepsilon}{\mu}}(E_y^-)$$

$$H_y = \sqrt{\frac{\varepsilon}{\mu}}\left(-\frac{1}{\varepsilon}\frac{\partial F_z^+}{\partial y}\right)e^{-j\beta z} - \sqrt{\frac{\varepsilon}{\mu}}\left(-\frac{1}{\varepsilon}\frac{\partial F_z^-}{\partial y}\right)e^{+j\beta z} = H_y^+ + H_y^-$$

$$= +\sqrt{\frac{\varepsilon}{\mu}}(E_x^+) - \sqrt{\frac{\varepsilon}{\mu}}(E_x^-)$$

Also

$$Z_w^+ = \frac{E_x^+}{H_y^+} = -\frac{E_y^+}{H_x^+} = \sqrt{\frac{\mu}{\varepsilon}}$$

$$Z_w^- = -\frac{E_x^-}{H_y^-} = \frac{E_y^-}{H_x^-} = \sqrt{\frac{\mu}{\varepsilon}}$$

3. Letting

$$A_x = A_y = F_x = F_y = F_z = 0 \quad A_z \neq 0 \quad \partial/\partial x \neq 0 \quad \partial/\partial y \neq 0$$

For this combination, according to (6-41) and (6-43)

$$H_z = 0$$

$$E_z = -j\omega A_z - j\frac{1}{\omega\mu\varepsilon}\frac{\partial^2 A_z}{\partial z^2} = -j\frac{1}{\omega\mu\varepsilon}\left(\frac{\partial^2}{\partial z^2} + \omega^2\mu\varepsilon\right)A_z = 0$$

provided

$$A_z(x,y,z) = A_z^+(x,y)e^{-j\beta z} + A_z^-(x,y)e^{+j\beta z}$$

Also according to (6-41) and (6-43)

$$E_x = -\frac{1}{\sqrt{\mu\varepsilon}}\frac{\partial A_z^+}{\partial x}e^{-j\beta z} + \frac{1}{\sqrt{\mu\varepsilon}}\frac{\partial A_z^-}{\partial x}e^{+j\beta z} = E_x^+ + E_x^-$$

$$E_y = -\frac{1}{\sqrt{\mu\varepsilon}}\frac{\partial A_z^+}{\partial y}e^{-j\beta z} + \frac{1}{\sqrt{\mu\varepsilon}}\frac{\partial A_z^-}{\partial y}e^{+j\beta z} = E_y^+ + E_y^-$$

$$H_x = -\sqrt{\frac{\varepsilon}{\mu}}\left(-\frac{1}{\sqrt{\mu\varepsilon}}\frac{\partial A_z^+}{\partial y}\right)e^{-j\beta z} + \sqrt{\frac{\varepsilon}{\mu}}\left(\frac{1}{\sqrt{\mu\varepsilon}}\frac{\partial A_z^-}{\partial y}\right)e^{+j\beta z} = H_x^+ + H_x^-$$

$$= -\sqrt{\frac{\varepsilon}{\mu}}(E_y^+) + \sqrt{\frac{\varepsilon}{\mu}}(E_y^-)$$

$$H_y = \sqrt{\frac{\varepsilon}{\mu}} \left(-\frac{1}{\sqrt{\mu\varepsilon}} \frac{\partial A_z^+}{\partial x} \right) e^{-j\beta z} - \sqrt{\frac{\varepsilon}{\mu}} \left(\frac{1}{\sqrt{\mu\varepsilon}} \frac{\partial A_z^-}{\partial x} \right) e^{+j\beta z} = H_y^+ + H_y^-$$

$$= \sqrt{\frac{\varepsilon}{\mu}}(E_x^+) - \sqrt{\frac{\varepsilon}{\mu}}(E_x^-)$$

Also

$$Z_w^+ = \frac{E_x^+}{H_y^+} = -\frac{E_y^+}{H_x^+} = \sqrt{\frac{\mu}{\varepsilon}}$$

$$Z_w^- = -\frac{E_x^-}{H_y^-} = \frac{E_y^-}{H_x^-} = \sqrt{\frac{\mu}{\varepsilon}}$$

SUMMARY From the results of Example 6.1, it is evident that TEM^z modes can be obtained by any of the following three combinations:

$$\underline{\text{TEM}^z}$$

$$\boxed{A_x = A_y = F_x = F_y = 0 \quad \partial/\partial x \neq 0 \quad \partial/\partial y \neq 0} \tag{6-44}$$

$$A_z = A_z^+(x,y)e^{-j\beta z} + A_z^-(x,y)e^{+j\beta z} \tag{6-44a}$$

$$F_z = F_z^+(x,y)e^{-j\beta z} + F_z^-(x,y)e^{+j\beta z} \tag{6-44b}$$

$$\boxed{A_x = A_y = A_z = F_x = F_y = 0 \quad \partial/\partial x \neq 0 \quad \partial/\partial y \neq 0} \tag{6-45}$$

$$F_z = F_z^+(x,y)e^{-j\beta z} + F_z^-(x,y)e^{+j\beta z} \tag{6-45a}$$

$$\boxed{A_x = A_y = F_x = F_y = F_z = 0 \quad \partial/\partial x \neq 0 \quad \partial/\partial y \neq 0} \tag{6-46}$$

$$A_z = A_z^+(x,y)e^{-j\beta z} + A_z^-(x,y)e^{+j\beta z} \tag{6-46a}$$

A similar procedure can be used to derive TEM modes in other directions such as TEM^x and TEM^y.

B. Cylindrical Coordinate System To derive expressions for TEM modes in a cylindrical coordinate system, a procedure similar to that in the rectangular coordinate system can be used. When (6-34)

$$\mathbf{E} = \mathbf{E}_A + \mathbf{E}_F = -j\omega\mathbf{A} - j\frac{1}{\omega\mu\varepsilon}\nabla(\nabla\cdot\mathbf{A}) - \frac{1}{\varepsilon}\nabla\times\mathbf{F} \tag{6-47}$$

and (6-35)

$$\mathbf{H} = \mathbf{H}_A + \mathbf{H}_F = \frac{1}{\mu}\nabla\times\mathbf{A} - j\omega\mathbf{F} - j\frac{1}{\omega\mu\varepsilon}\nabla(\nabla\cdot\mathbf{F}) \tag{6-48}$$

are expanded using

$$\mathbf{A}(\rho,\phi,z) = \hat{\mathbf{a}}_\rho A_\rho(\rho,\phi,z) + \hat{\mathbf{a}}_\phi A_\phi(\rho,\phi,z) + \hat{\mathbf{a}}_z A_z(\rho,\phi,z) \tag{6-49a}$$

$$\mathbf{F}(\rho,\phi,z) = \hat{\mathbf{a}}_\rho F_\rho(\rho,\phi,z) + \hat{\mathbf{a}}_\phi F_\phi(\rho,\phi,z) + \hat{\mathbf{a}}_z F_z(\rho,\phi,z) \tag{6-49b}$$

as solutions, they can be written as

$$
\mathbf{E} = \hat{\mathbf{a}}_\rho \left\{ -j\omega A_\rho - j\frac{1}{\omega\mu\varepsilon}\frac{\partial}{\partial\rho}\left[\frac{1}{\rho}\frac{\partial}{\partial\rho}(\rho A_\rho) + \frac{1}{\rho}\frac{\partial A_\phi}{\partial\phi} + \frac{\partial A_z}{\partial z}\right] - \frac{1}{\varepsilon}\left(\frac{1}{\rho}\frac{\partial F_z}{\partial\phi} - \frac{\partial F_\phi}{\partial z}\right)\right\}
$$

$$
+ \hat{\mathbf{a}}_\phi \left\{ -j\omega A_\phi - j\frac{1}{\omega\mu\varepsilon}\frac{1}{\rho}\frac{\partial}{\partial\phi}\left[\frac{1}{\rho}\frac{\partial}{\partial\rho}(\rho A_\rho) + \frac{1}{\rho}\frac{\partial A_\phi}{\partial\phi} + \frac{\partial A_z}{\partial z}\right] - \frac{1}{\varepsilon}\left(\frac{\partial F_\rho}{\partial z} - \frac{\partial F_z}{\partial\rho}\right)\right\}
$$

$$
+ \hat{\mathbf{a}}_z \left\{ -j\omega A_z - j\frac{1}{\omega\mu\varepsilon}\frac{\partial}{\partial z}\left[\frac{1}{\rho}\frac{\partial}{\partial\rho}(\rho A_\rho) + \frac{1}{\rho}\frac{\partial A_\phi}{\partial\phi} + \frac{\partial A_z}{\partial z}\right] - \frac{1}{\varepsilon\rho}\left[\frac{\partial}{\partial\rho}(\rho F_\phi) - \frac{\partial F_\rho}{\partial\phi}\right]\right\} \quad (6\text{-}50)
$$

$$
\mathbf{H} = \hat{\mathbf{a}}_\rho \left\{ -j\omega F_\rho - j\frac{1}{\omega\mu\varepsilon}\frac{\partial}{\partial\rho}\left[\frac{1}{\rho}\frac{\partial}{\partial\rho}(\rho F_\rho) + \frac{1}{\rho}\frac{\partial F_\phi}{\partial\phi} + \frac{\partial F_z}{\partial z}\right] + \frac{1}{\mu}\left(\frac{1}{\rho}\frac{\partial A_z}{\partial\phi} - \frac{\partial A_\phi}{\partial z}\right)\right\}
$$

$$
+ \hat{\mathbf{a}}_\phi \left\{ -j\omega F_\phi - j\frac{1}{\omega\mu\varepsilon}\frac{1}{\rho}\frac{\partial}{\partial\phi}\left[\frac{1}{\rho}\frac{\partial}{\partial\rho}(\rho F_\rho) + \frac{1}{\rho}\frac{\partial F_\phi}{\partial\phi} + \frac{\partial F_z}{\partial z}\right] + \frac{1}{\mu}\left(\frac{\partial A_\rho}{\partial z} - \frac{\partial A_z}{\partial\rho}\right)\right\}
$$

$$
+ \hat{\mathbf{a}}_z \left\{ -j\omega F_z - j\frac{1}{\omega\mu\varepsilon}\frac{\partial}{\partial z}\left[\frac{1}{\rho}\frac{\partial}{\partial\rho}(\rho F_\rho) + \frac{1}{\rho}\frac{\partial F_\phi}{\partial\phi} + \frac{\partial F_z}{\partial z}\right] + \frac{1}{\mu\rho}\left[\frac{\partial}{\partial\rho}(\rho A_\phi) - \frac{\partial A_\rho}{\partial\phi}\right]\right\} \quad (6\text{-}51)
$$

Example 6-2

Using (6-50) and (6-51), derive expressions for the **E** and **H** fields, in terms of the components of the **A** and **F** potentials, that are TEM to the ρ direction (TEM^ρ).

Solution: It is apparent by examining (6-50) and (6-51) that TEM^ρ $(E_\rho = H_\rho = 0)$ modes can be obtained by any of the following three combinations:

1. Letting

$$
A_\phi = A_z = F_\phi = F_z = 0 \quad A_\rho \neq 0 \quad F_\rho \neq 0 \quad \partial/\partial\phi \neq 0 \quad \partial/\partial z \neq 0
$$

For this combination, according to (6-50) and (6-51)

$$
E_\rho = -j\omega A_\rho - j\frac{1}{\omega\mu\varepsilon}\frac{\partial}{\partial\rho}\left[\frac{1}{\rho}\frac{\partial}{\partial\rho}(\rho A_\rho)\right] = -j\frac{1}{\omega\mu\varepsilon}\left\{\frac{\partial}{\partial\rho}\left[\frac{1}{\rho}\frac{\partial}{\partial\rho}(\rho A_\rho)\right] + \omega^2\mu\varepsilon A_\rho\right\}
$$

$$
= -j\frac{1}{\omega\mu\varepsilon}\left[\frac{\partial}{\partial\rho}\left(\frac{\partial A_\rho}{\partial\rho} + \frac{A_\rho}{\rho}\right) + \omega^2\mu\varepsilon A_\rho\right]
$$

$$
= -j\frac{1}{\omega\mu\varepsilon}\left(\frac{\partial^2 A_\rho}{\partial\rho^2} + \frac{1}{\rho}\frac{\partial A_\rho}{\partial\rho} - \frac{A_\rho}{\rho^2} + \omega^2\mu\varepsilon A_\rho\right)
$$

$$
E_\rho = -j\frac{1}{\omega\mu\varepsilon}\left(\frac{\partial^2}{\partial\rho^2} + \frac{1}{\rho}\frac{\partial}{\partial\rho} - \frac{1}{\rho^2} + \beta^2\right)A_\rho = 0
$$

provided

$$
A_\rho(\rho,\phi,z) = A_\rho^+(\phi,z)H_1^{(2)}(\beta\rho) + A_\rho^-(\phi,z)H_1^{(1)}(\beta\rho)
$$

Also

$$H_\rho = -j\omega F_\rho - j\frac{1}{\omega\mu\varepsilon}\frac{\partial}{\partial\rho}\left[\frac{1}{\rho}\frac{\partial}{\partial\rho}(\rho F_\rho)\right]$$

$$= -j\frac{1}{\omega\mu\varepsilon}\left(\frac{\partial^2}{\partial\rho^2}+\frac{1}{\rho}\frac{\partial}{\partial\rho}+\frac{1}{\rho^2}+\beta^2\right)F_\rho = 0$$

provided

$$F_\rho(\rho,\phi,z) = F_\rho^+(\phi,z)H_1^{(2)}(\beta\rho) + F_\rho^-(\phi,z)H_1^{(1)}(\beta\rho)$$

In addition,

$$E_\phi = -j\frac{1}{\omega\mu\varepsilon}\frac{1}{\rho}\frac{\partial}{\partial\phi}\left[\frac{1}{\rho}\frac{\partial}{\partial\rho}(\rho A_\rho)\right] - \frac{1}{\varepsilon}\frac{\partial F_\rho}{\partial z}$$

$$E_z = -j\frac{1}{\omega\mu\varepsilon}\frac{\partial}{\partial z}\left[\frac{1}{\rho}\frac{\partial}{\partial\rho}(\rho A_\rho)\right] - \frac{1}{\varepsilon}\left(-\frac{1}{\rho}\frac{\partial F_\rho}{\partial\phi}\right)$$

$$H_\phi = -j\frac{1}{\omega\mu\varepsilon}\frac{1}{\rho}\frac{\partial}{\partial\phi}\left[\frac{1}{\rho}\frac{\partial}{\partial\rho}(\rho F_\rho)\right] + \frac{1}{\mu}\frac{\partial A_\rho}{\partial z}$$

$$H_z = -j\frac{1}{\omega\mu\varepsilon}\frac{\partial}{\partial z}\left[\frac{1}{\rho}\frac{\partial}{\partial\rho}(\rho F_\rho)\right] + \frac{1}{\mu}\left(-\frac{1}{\rho}\frac{\partial A_\rho}{\partial\phi}\right)$$

2. Letting

$$A_\rho = A_\phi = A_z = F_\phi = F_z = 0 \qquad F_\rho \neq 0 \qquad \partial/\partial\phi \neq 0 \qquad \partial/\partial z \neq 0$$

For this combination, according to (6-50) and (6-51)

$$E_\rho = 0$$

$$H_\rho = -j\omega F_\rho - j\frac{1}{\omega\mu\varepsilon}\frac{\partial}{\partial\rho}\left[\frac{1}{\rho}\frac{\partial}{\partial\rho}(\rho F_\rho)\right]$$

$$= -j\frac{1}{\omega\mu\varepsilon}\left(\frac{\partial^2}{\partial\rho^2}+\frac{1}{\rho}\frac{\partial}{\partial\rho}-\frac{1}{\rho^2}+\beta^2\right)F_\rho = 0$$

provided

$$F_\rho(\rho,\phi,z) = F_\rho^+(\phi,z)H_1^{(2)}(\beta\rho) + F_\rho^-(\phi,z)H_1^{(1)}(\beta\rho)$$

In addition,

$$E_\phi = -\frac{1}{\varepsilon}\frac{\partial F_\rho}{\partial z}$$

$$E_z = -\frac{1}{\varepsilon}\left(-\frac{1}{\rho}\frac{\partial F_\rho}{\partial\phi}\right)$$

$$H_\phi = -j\frac{1}{\omega\mu\varepsilon}\frac{1}{\rho}\frac{\partial}{\partial\phi}\left[\frac{1}{\rho}\frac{\partial}{\partial\rho}(\rho F_\rho)\right]$$

$$H_z = -j\frac{1}{\omega\mu\varepsilon}\frac{\partial}{\partial z}\left[\frac{1}{\rho}\frac{\partial}{\partial\rho}(\rho F_\rho)\right]$$

3. Letting

$$A_\phi = A_z = F_\rho = F_\phi = F_z = 0 \qquad A_\rho \neq 0 \qquad \partial/\partial\phi \neq 0 \qquad \partial/\partial z \neq 0$$

For this combination, according to (6-50) and (6-51)

$$H_\rho = 0$$

$$E_\rho = -j\omega A_\rho - j\frac{1}{\omega\mu\varepsilon}\frac{\partial}{\partial\rho}\left[\frac{1}{\rho}\frac{\partial}{\partial\rho}(\rho A_\rho)\right]$$

$$= -j\frac{1}{\omega\mu\varepsilon}\left(\frac{\partial^2}{\partial\rho^2} + \frac{1}{\rho}\frac{\partial}{\partial\rho} - \frac{1}{\rho^2} + \beta^2\right)A_\rho = 0$$

provided

$$A_\rho(\rho,\phi,z) = A_\rho^+(\phi,z)H_1^{(2)}(\beta\rho) + A_\rho^-(\phi,z)H_1^{(1)}(\beta\rho)$$

In addition,

$$E_\phi = -j\frac{1}{\omega\mu\varepsilon}\frac{1}{\rho}\frac{\partial}{\partial\phi}\left[\frac{1}{\rho}\frac{\partial}{\partial\rho}(\rho A_\rho)\right]$$

$$E_z = -j\frac{1}{\omega\mu\varepsilon}\frac{\partial}{\partial z}\left[\frac{1}{\rho}\frac{\partial}{\partial\rho}(\rho A_\rho)\right]$$

$$H_\phi = \frac{1}{\mu}\left(\frac{\partial A_\rho}{\partial z}\right)$$

$$H_z = \frac{1}{\mu}\left(-\frac{1}{\rho}\frac{\partial A_\rho}{\partial\phi}\right)$$

SUMMARY   From the results of Example 6-2, it is evident that TEM$^\rho$ modes can be obtained by any of the following three combinations:

$$A_\phi = A_z = F_\phi = F_z = 0 \quad \partial/\partial\phi \neq 0 \quad \partial/\partial z \neq 0 \tag{6-52}$$

$$A_\rho(\rho,\phi,z) = A_\rho^+(\phi,z)H_1^{(2)}(\beta\rho) + A_\rho^-(\phi,z)H_1^{(1)}(\beta\rho) \tag{6-52a}$$

$$F_\rho(\rho,\phi,z) = F_\rho^+(\phi,z)H_1^{(2)}(\beta\rho) + F_\rho^-(\phi,z)H_1^{(1)}(\beta\rho) \tag{6-52b}$$

$$A_\rho = A_\phi = A_z = F_\phi = F_z = 0 \quad \partial/\partial\phi \neq 0 \quad \partial/\partial z \neq 0 \tag{6-53}$$

$$F_\rho(\rho,\phi,z) = F_\rho^+(\phi,z)H_1^{(2)}(\beta\rho) + F_\rho^-(\phi,z)H_1^{(1)}(\beta\rho) \tag{6-53a}$$

$$A_\phi = A_z = F_\rho = F_\phi = F_z = 0 \quad \partial/\partial\phi \neq 0 \quad \partial/\partial z \neq 0 \tag{6-54}$$

$$A_\rho(\rho,\phi,z) = A_\rho^+(\phi,z)H_1^{(2)}(\beta\rho) + A_\rho^-(\phi,z)H_1^{(1)}(\beta\rho) \tag{6-54a}$$

A similar procedure can be used to derive TEM modes in other directions such as TEM$^\phi$ and TEM^z.

6.5.2 Transverse Magnetic Modes: Source-Free Region

Often we seek solutions of higher-order modes, other than transverse electromagnetic (TEM). Some of the higher-order modes, often required to satisfy boundary conditions, are designated as transverse magnetic (TM) and transverse electric (TE). Classical examples of the need for TM and TE modes are modes of propagation in waveguides [2].

Transverse magnetic modes (often also known as transverse magnetic fields) are field configurations whose magnetic field components lie in a plane that is transverse to a given direction. That direction is often chosen to be the path of wave propagation. For example, if the desired fields are TM to z (TM^z), this implies that $H_z = 0$. Each of the other two magnetic field components (H_x and H_y) and three electric field components (E_x, E_y, and E_z) may or may not exist.

By examining (6-43) and (6-51) it is evident that *to derive the field expressions that are TM to a given direction, independent of the coordinate system, it is sufficient to let the vector potential* **A** *have only a component in the direction in which the fields are desired to be TM. The remaining components of* **A** *as well as all of* **F** *are set equal to zero.*

A. Rectangular Coordinate System

$$\text{TM}^z$$

To demonstrate the aforementioned procedure, let us assume that we wish to derive field expressions that are TM to z (TM^z). To accomplish this, we let

$$\boxed{\begin{aligned} \mathbf{A} &= \hat{\mathbf{a}}_z A_z(x, y, z) \\ \mathbf{F} &= 0 \end{aligned}}$$

(6-55a)
(6-55b)

The vector potential **A** must satisfy (6-30), which reduces from a vector wave equation to a scalar wave equation

$$\nabla^2 A_z(x, y, z) + \beta^2 A_z(x, y, z) = 0 \tag{6-56}$$

Since (6-56) is of the same form as (3-20a), its solution using the *separation-of-variables method* can be written, according to (3-23), as

$$A_z(x, y, z) = f(x) g(y) h(z) \tag{6-57}$$

The solutions of $f(x)$, $g(y)$, and $h(z)$ take the forms given by (3-28a) through (3-30b). The most appropriate forms for $f(x)$, $g(y)$, and $h(z)$ must be chosen judiciously to reduce the complexity of the problem, and they will depend on the configuration of the problem. For the rectangular waveguide of Figure 3-2, for example, the most appropriate forms for $f(x)$, $g(y)$, and $h(z)$ are those given, respectively, by (3-28b), (3-29b), and (3-30a). Thus, for the rectangular waveguide, (6-57) can be written as

$$A_z(x, y, z) = [C_1 \cos(\beta_x x) + D_1 \sin(\beta_x x)][C_2 \cos(\beta_y y) + D_2 \sin(\beta_y y)]$$
$$\times (A_3 e^{-j\beta_z z} + B_3 e^{+j\beta_z z}) \tag{6.58}$$

where

$$\beta_x^2 + \beta_y^2 + \beta_z^2 = \beta^2 = \omega^2 \mu \varepsilon \tag{6-58a}$$

Once A_z is found, the next step is to use (6-41) and (6-43) to find the **E** and **H** field components. Doing this, it can be shown that by using (6-55a) and (6-55b) we can reduce (6-41) and (6-43) to

---

TM^z Rectangular Coordinate System

$$E_x = -j\frac{1}{\omega\mu\varepsilon}\frac{\partial^2 A_z}{\partial x\,\partial z} \qquad\qquad H_x = \frac{1}{\mu}\frac{\partial A_z}{\partial y}$$

$$E_y = -j\frac{1}{\omega\mu\varepsilon}\frac{\partial^2 A_z}{\partial y\,\partial z} \qquad\qquad H_y = -\frac{1}{\mu}\frac{\partial A_z}{\partial x}$$

$$E_z = -j\frac{1}{\omega\mu\varepsilon}\left(\frac{\partial^2}{\partial z^2}+\beta^2\right)A_z \qquad H_z = 0$$

(6-59)

---

which satisfy the definition of TM^z (i.e., $H_z = 0$).

For the specific example for which the solution of A_z as given by (6-58) is applicable, the unknown constants C_1, D_1, C_2, D_2, A_3, B_3, β_x, β_y, and β_z can be evaluated by substituting A_z of (6-58) into the expressions for **E** and **H** in (6-59) and enforcing the appropriate boundary conditions on the **E** and **H** field components. This will be demonstrated in Chapter 8, and elsewhere, where specific problem configurations are attempted. Following these or similar procedures should lead to the solution of the problem in question.

Expressions for the **E** and **H** field components that are TM^x and TM^y are given, respectively, by

TM^x Rectangular Coordinate System

Let

$$\mathbf{A} = \hat{\mathbf{a}}_x A_x(x,y,z)$$ (6-60a)

$$\mathbf{F} = 0$$ (6-60b)

Then

$$E_x = -j\frac{1}{\omega\mu\varepsilon}\left(\frac{\partial^2}{\partial x^2}+\beta^2\right)A_x \qquad H_x = 0$$

$$E_y = -j\frac{1}{\omega\mu\varepsilon}\frac{\partial^2 A_x}{\partial x\,\partial y} \qquad\qquad H_y = \frac{1}{\mu}\frac{\partial A_x}{\partial z}$$

$$E_z = -j\frac{1}{\omega\mu\varepsilon}\frac{\partial^2 A_x}{\partial x\,\partial z} \qquad\qquad H_z = -\frac{1}{\mu}\frac{\partial A_x}{\partial y}$$

(6-61)

---

where A_x must satisfy the scalar wave equation

$$\nabla^2 A_x(x,y,z)+\beta^2 A_x(x,y,z)=0$$ (6-62)

TM^y Rectangular Coordinate System

Let

$$\mathbf{A} = \hat{\mathbf{a}}_y A_y(x,y,z)$$ (6-63a)

$$\mathbf{F} = 0$$ (6-63b)

Then

$$
\begin{array}{ll}
E_x = -j\dfrac{1}{\omega\mu\varepsilon}\dfrac{\partial^2 A_y}{\partial x\,\partial y} & H_x = -\dfrac{1}{\mu}\dfrac{\partial A_y}{\partial z} \\[3mm]
E_y = -j\dfrac{1}{\omega\mu\varepsilon}\left(\dfrac{\partial^2}{\partial y^2}+\beta^2\right)A_y & H_y = 0 \\[3mm]
E_z = -j\dfrac{1}{\omega\mu\varepsilon}\dfrac{\partial^2 A_y}{\partial y\,\partial z} & H_z = \dfrac{1}{\mu}\dfrac{\partial A_y}{\partial x}
\end{array}
\tag{6-64}
$$

where A_y must satisfy the scalar wave equation of

$$
\nabla^2 A_y(x,y,z)+\beta^2 A_y(x,y,z)=0 \tag{6-65}
$$

The derivations of (6-61) and (6-64) are left to the reader as end-of-chapter assignments.

The expressions of (6-59), (6-61), and (6-64) are valid forms for the **E** and **H** field components of any problem in a rectangular coordinate system, which are, respectively, TM^z, TM^x, and TM^y. A similar procedure can be used to find expressions for the **E** and **H** field components that are TM to any direction in any coordinate system.

B. Cylindrical Coordinate System   In terms of complexity, the next higher-order coordinate system is that of the cylindrical coordinate system. We will derive expressions that will be valid for TM^z. TM$^\rho$ and TM$^\phi$ are more difficult and are not usually utilized. Therefore, they will not be attempted here. The procedure for TM^z in a cylindrical coordinate system is the same as that used for the rectangular coordinate system, as outlined previously in this section.

To accomplish this, let

$$
\mathbf{A} = \hat{\mathbf{a}}_z A_z(\rho,\phi,z) \tag{6-66a}
$$
$$
\mathbf{F} = 0 \tag{6-66b}
$$

The vector potential **A** must satisfy (6-30) with $\mathbf{J}=0$, which reduces from its vector form to the scalar wave equation

$$
\nabla^2 A_z(\rho,\phi,z)+\beta^2 A_z(\rho,\phi,z)=0 \tag{6-67}
$$

Since (6-67) is of the same form as (3-54c), its solution using the *separation-of-variables method* can be written, according to (3-57), as

$$
A_z(\rho,\phi,z)=f(\rho)g(\phi)h(z) \tag{6-68}
$$

The solutions of $f(\rho)$, $g(\phi)$, and $h(z)$ take the forms given by (3-67a) through (3-69b). The most appropriate forms for $f(\rho)$, $g(\phi)$, and $h(z)$ must be chosen judiciously to reduce the complexity of the problem, and they will depend upon the configuration of the problem. For the cylindrical waveguide of Figure 3-5, for example, the most appropriate forms for $f(\rho)$, $g(\phi)$, and $h(z)$ are those given, respectively, by (3-67a), (3-68b), and (3-69a). Thus, for the cylindrical waveguide, (6-68) can be written as

$$
A_z(\rho,\phi,z)=[\,A_1 J_m(\beta_\rho\rho)+B_1 Y_m(\beta_\rho\rho)][\,C_2\cos(m\phi)+D_2\sin(m\phi)\,]
$$
$$
\times(A_3 e^{-j\beta_z z}+B_3 e^{+j\beta_z z}) \tag{6-69}
$$

where

$$\beta_\rho^2 + \beta_z^2 = \beta^2 \tag{6-69a}$$

Once A_z is found, the next step is to use (6-50) and (6-51) to find the **E** and **H** field components. Then we can show that by using (6-66a) and (6-66b), (6-50) and (6-51) can be reduced to

$$
\begin{array}{|ll|}
\hline
\multicolumn{2}{|c|}{\textbf{TM}^z \textbf{ Cylindrical Coordinate System}} \\
\\
E_\rho = -j\dfrac{1}{\omega\mu\varepsilon}\dfrac{\partial^2 A_z}{\partial\rho\,\partial z} & \qquad H_\rho = \dfrac{1}{\mu}\dfrac{1}{\rho}\dfrac{\partial A_z}{\partial\phi} \\
\\
E_\phi = -j\dfrac{1}{\omega\mu\varepsilon}\dfrac{1}{\rho}\dfrac{\partial^2 A_z}{\partial\phi\,\partial z} & \qquad H_\phi = -\dfrac{1}{\mu}\dfrac{\partial A_z}{\partial\rho} \\
\\
E_z = -j\dfrac{1}{\omega\mu\varepsilon}\left(\dfrac{\partial^2}{\partial z^2}+\beta^2\right)A_z & \qquad H_z = 0 \\
\hline
\end{array}
\tag{6-70}
$$

which also satisfies the TM^z definition (i.e., $H_z = 0$).

For the specific example for which the solution of A_z as given by (6-69) is applicable, the unknown constants A_1, B_1, C_2, D_2, A_3, B_3, β_ρ, and β_z can be evaluated by substituting A_z of (6-69) into the expressions for **E** and **H** in (6-70) and enforcing the appropriate boundary conditions on the **E** and **H** field components. This will be demonstrated in Chapter 9, and elsewhere, where specific problem configurations are attempted. Following these or similar procedures should lead to the solution of the problem in question.

It should be stated that the same TM mode field constructions can be obtained by initiating the procedure with a solution to the scalar wave equation for the electric field component in the direction in which TM mode fields are desired. For example, if TM^z modes are desired, assume a solution for E_z of the same form as the vector potential component A_z. It can then be shown through Maxwell's equations that all the remaining electric and magnetic field components (with $H_z = 0$) can be expressed in terms of E_z. The same can be done for other TM^i modes by beginning with a solution for E_i having the same form as the vector potential component A_i. The only difference between the two formulations, one of which uses the vector potentials adopted in this book and the other that uses the fields themselves, is a normalization constant. For TM^z modes, for example, this normalization constant according to (6-59) is equal to $-j(\partial^2/\partial z^2 + \beta^2)/\omega\mu\varepsilon = -j(\beta^2 - \beta_z^2)/\omega\mu\varepsilon$. The preceding procedure is a very popular method used by many authors, and it is assigned to the reader as end-of-chapter exercises.

6.5.3 Transverse Electric Modes: Source-Free Region

Transverse electric (TE) modes can be derived in a fashion similar to the TM fields of Section 6.5.2. This time, however, we let the **F** vector potential have a nonvanishing component in the direction in which the TE fields are desired, and all the remaining components of **F** and **A** are set equal to zero. Without going through any of the details, we will list the expressions for the **E** and **H** field components for TE^z, TE^x, and TE^y in rectangular coordinates and TE^z in cylindrical coordinates. The details are left as exercises for the reader.

A. Rectangular Coordinate System   Modes that are TE^z, TE^x, and TE^y are obtained as follows.

TE^z Rectangular Coordinate System

Let

$$\mathbf{A} = 0 \tag{6-71a}$$
$$\mathbf{F} = \hat{\mathbf{a}}_z F_z(x,y,z) \tag{6-71b}$$

Then

$$
\begin{array}{ll}
E_x = -\dfrac{1}{\varepsilon}\dfrac{\partial F_z}{\partial y} & H_x = -j\dfrac{1}{\omega\mu\varepsilon}\dfrac{\partial^2 F_z}{\partial x\,\partial z} \\[2ex]
E_y = \dfrac{1}{\varepsilon}\dfrac{\partial F_z}{\partial x} & H_y = -j\dfrac{1}{\omega\mu\varepsilon}\dfrac{\partial^2 F_z}{\partial y\,\partial z} \\[2ex]
E_z = 0 & H_z = -j\dfrac{1}{\omega\mu\varepsilon}\left(\dfrac{\partial^2}{\partial z^2}+\beta^2\right)F_z
\end{array}
\tag{6-72}
$$

where F_z must satisfy the scalar wave equation

$$\nabla^2 F_z(x,y,z) + \beta^2 F_z(x,y,z) = 0 \tag{6-73}$$

TE^x Rectangular Coordinate System

Let

$$\mathbf{A} = 0 \tag{6-73a}$$
$$\mathbf{F} = \hat{\mathbf{a}}_x F_x(x,y,z) \tag{6-73b}$$

Then

$$
\begin{array}{ll}
E_x = 0 & H_x = -j\dfrac{1}{\omega\mu\varepsilon}\left(\dfrac{\partial^2}{\partial x^2}+\beta^2\right)F_x \\[2ex]
E_y = -\dfrac{1}{\varepsilon}\dfrac{\partial F_x}{\partial z} & H_y = -j\dfrac{1}{\omega\mu\varepsilon}\dfrac{\partial^2 F_x}{\partial x\,\partial y} \\[2ex]
E_z = \dfrac{1}{\varepsilon}\dfrac{\partial F_x}{\partial y} & H_z = -j\dfrac{1}{\omega\mu\varepsilon}\dfrac{\partial^2 F_x}{\partial x\,\partial z}
\end{array}
\tag{6-74}
$$

where F_x must satisfy the scalar wave equation

$$\nabla^2 F_x(x,y,z) + \beta^2 F_x(x,y,z) = 0 \tag{6-75}$$

TE^y Rectangular Coordinate System

Let

$$\mathbf{A} = 0 \tag{6-76a}$$
$$\mathbf{F} = \hat{\mathbf{a}}_y F_y(x,y,z) \tag{6-76b}$$

Then

$$
\begin{array}{ll}
E_x = \dfrac{1}{\varepsilon}\dfrac{\partial F_y}{\partial z} & H_x = -j\dfrac{1}{\omega\mu\varepsilon}\dfrac{\partial^2 F_y}{\partial x\,\partial y} \\[3mm]
E_y = 0 & H_y = -j\dfrac{1}{\omega\mu\varepsilon}\left(\dfrac{\partial^2}{\partial y^2}+\beta^2\right)F_y \\[3mm]
E_z = -\dfrac{1}{\varepsilon}\dfrac{\partial F_y}{\partial x} & H_z = -j\dfrac{1}{\omega\mu\varepsilon}\dfrac{\partial^2 F_y}{\partial y\,\partial z}
\end{array}
\tag{6-77}
$$

where F_y must satisfy the scalar wave equation

$$
\nabla^2 F_y(x,y,z)+\beta^2 F_y(x,y,z)=0 \tag{6-78}
$$

B. Cylindrical Coordinate System   Modes that are TE^z are obtained as follows.

TE^z Cylindrical Coordinate System

Let

$$
\mathbf{A}=0 \tag{6-79a}
$$
$$
\mathbf{F}=\hat{\mathbf{a}}_z F_z(\rho,\phi,z) \tag{6-79b}
$$

Then

$$
\begin{array}{ll}
E_\rho = -\dfrac{1}{\varepsilon\rho}\dfrac{\partial F_z}{\partial \phi} & H_\rho = -j\dfrac{1}{\omega\mu\varepsilon}\dfrac{\partial^2 F_z}{\partial \rho\,\partial z} \\[3mm]
E_\phi = \dfrac{1}{\varepsilon}\dfrac{\partial F_z}{\partial \rho} & H_\phi = -j\dfrac{1}{\omega\mu\varepsilon\rho}\dfrac{\partial^2 F_z}{\partial \phi\,\partial z} \\[3mm]
E_z = 0 & H_z = -j\dfrac{1}{\omega\mu\varepsilon}\left(\dfrac{\partial^2}{\partial z^2}+\beta^2\right)F_z
\end{array}
\tag{6-80}
$$

where F_z must satisfy the scalar wave equation

$$
\nabla^2 F_z(\rho,\phi,z)+\beta^2 F_z(\rho,\phi,z)=0 \tag{6-81}
$$

As was suggested earlier for the TM modes, an alternate procedure for construction of TE^i field configurations will be to initiate the procedure with a solution for the H_i component with the same form as the vector potential component F_i. For example, if TE^z modes are desired, assume a solution for H_z of the same form as the vector potential component F_z. It can then be shown through Maxwell's equations that all the remaining electric and magnetic fields (with $E_z=0$) can be expressed in terms of H_z. The only difference between the two formulations, one that uses the vector potentials adopted in this book and the other that uses the fields themselves, is a normalization. For TE^z modes, for example, this normalization constant according to (6-72) is equal to $-j(\partial^2/\partial z^2+\beta^2)/\omega\mu\varepsilon=-j(\beta^2-\beta_z^2)/\omega\mu\varepsilon$. The preceding procedure is also a very popular method used by many authors, and it is assigned to the reader as end-of-chapter exercises.

6.6 SOLUTION OF THE INHOMOGENEOUS VECTOR POTENTIAL WAVE EQUATION

In Sections 6.2 and 6.3, we derived the inhomogeneous vector wave equations 6-16 and 6-28. In this section, we want to derive the solutions to each equation.

Let us assume that a source with current density J_z, which in the limit is an infinitesimal point source, is placed at the origin of an x, y, z coordinate system, as shown in Figure 6-2a. Since the current density J_z is directed along the z axis, only an A_z component will exist. Thus, we can write (6-16) as

$$\nabla^2 A_z + \beta^2 A_z = -\mu J_z \qquad (6\text{-}82)$$

At points removed from the source ($J_z = 0$), the wave equation reduces to

$$\nabla^2 A_z + \beta^2 A_z = 0 \qquad (6\text{-}83)$$

Since in the limit the source is a point, it requires that A_z is not a function of direction (θ and ϕ); in a spherical coordinate system, $A_z = A_z(r)$ where r is the radial distance. Thus, (6-83) can be written as

$$\nabla^2 A_z(r) + \beta^2 A_z(r) = \frac{1}{r^2} \frac{\partial}{\partial r} \left[r^2 \frac{\partial A_z(r)}{\partial r} \right] + \beta^2 A_z(r) = 0 \qquad (6\text{-}84)$$

which when expanded reduces to

$$\frac{d^2 A_z(r)}{dr^2} + \frac{2}{r} \frac{dA_z(r)}{dr} + \beta^2 A_z(r) = 0 \qquad (6\text{-}84a)$$

The partial derivatives have been replaced by the ordinary derivative since A_z is only a function of the radial coordinate.

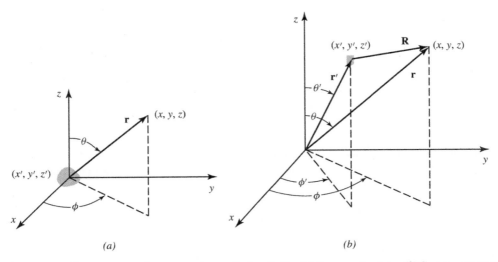

Figure 6-2 Coordinate systems for computing radiation fields. (*a*) Source at origin. (*b*) Source not at origin. (Source: C. A. Balanis, *Antenna Theory: Analysis and Design*, Third Edition. Copyright © 2005, John Wiley & Sons, Inc. Reprinted by permission of John Wiley & Sons, Inc.)

The differential equation 6-84a has two independent solutions

$$A_{z1} = C_1 \frac{e^{-j\beta r}}{r} \qquad (6\text{-}85a)$$

$$A_{z2} = C_2 \frac{e^{+j\beta r}}{r} \qquad (6\text{-}85b)$$

Equation 6-85a represents an outwardly (in the radial direction) traveling wave and (6-85b) describes an inwardly traveling wave, assuming $e^{j\omega t}$ time variations. For this problem, the source is placed at the origin with the radiated fields traveling in the outward radial direction. Therefore, we choose the solution of (6-85a), or

$$A_z = A_{z1} = C_1 \frac{e^{-j\beta r}}{r} \qquad (6\text{-}86)$$

In the static case, $\omega = 0$, $\beta = 0$, so (6-86) simplifies to

$$A_z = \frac{C_1}{r} \qquad (6\text{-}86a)$$

which is a solution to the wave equation 6-83, 6-84, or 6-84a when $\beta = 0$. Thus, at points removed from the source, the time-varying and the static solutions of (6-86) and (6-86a) differ only by the $e^{-j\beta r}$ factor, or the time-varying solution of (6-86) can be obtained by multiplying the static solution of (6-86a) by $e^{-j\beta r}$.

In the presence of the source ($J_z \neq 0$) and with $\beta = 0$, the wave equation 6-82 reduces to

$$\nabla^2 A_z = -\mu J_z \qquad (6\text{-}87)$$

This equation is recognized as Poisson's equation whose solution is widely documented. The most familiar equation with Poisson form is that relating the scalar electric potential ϕ to the electric charge density q. This is given by

$$\nabla^2 \phi = -\frac{q}{\varepsilon} \qquad (6\text{-}88)$$

whose solution is

$$\phi = \frac{1}{4\pi\varepsilon} \iiint_V \frac{q}{r} dv' \qquad (6\text{-}89)$$

where r is the distance from any point on the charge density to the observation point. Since (6-87) is similar in form to (6-88), its solution is similar to (6-89), or

$$A_z = \frac{\mu}{4\pi} \iiint_V \frac{J_z}{r} dv' \qquad (6\text{-}90)$$

Equation 6-90 represents the solution to (6-82) when $\beta = 0$, the static case. Using the comparative analogy between (6-86) and (6-86a), the time-varying solution of (6-82) can be obtained by multiplying the static solution of (6-90) by $e^{-j\beta r}$. Thus,

$$\boxed{A_z = \frac{\mu}{4\pi} \iiint_V J_z \frac{e^{-j\beta r}}{r} dv'} \qquad (6\text{-}91)$$

which is a solution to (6-82).

If the current densities were in the x and y directions (J_x and J_y), the wave equation for each would reduce to

$$\nabla^2 A_x + \beta^2 A_x = -\mu J_x \qquad (6\text{-}92a)$$

$$\nabla^2 A_y + \beta^2 A_y = -\mu J_y \tag{6-92b}$$

with corresponding solutions similar in form to (6-91), or

$$A_x = \frac{\mu}{4\pi} \iiint_V J_x \frac{e^{-j\beta r}}{r} dv' \tag{6-93a}$$

$$A_y = \frac{\mu}{4\pi} \iiint_V J_y \frac{e^{-j\beta r}}{r} dv' \tag{6-93b}$$

The solutions of (6-91), (6-93a), and (6-93b) allow us to write the solution to the vector wave equation 6-16 as

$$\mathbf{A} = \frac{\mu}{4\pi} \iiint_V \mathbf{J} \frac{e^{-j\beta r}}{r} dv' \tag{6-94}$$

If the source is removed from the origin and placed at a position represented by the primed coordinates (x', y', z'), as shown in Figure 6-2b, (6-94) can be written as

$$\mathbf{A}(x,y,z) = \frac{\mu}{4\pi} \iiint_V \mathbf{J}(x',y',z') \frac{e^{-j\beta R}}{R} dv' \tag{6-95a}$$

where the primed coordinates represent the source, the unprimed coordinates represent the observation point, and R represents the distance from any point in the source to the observation point. In a similar fashion, we can show that the solution of (6-28) is given by

$$\mathbf{F}(x,y,z) = \frac{\varepsilon}{4\pi} \iiint_V \mathbf{M}(x',y',z') \frac{e^{-j\beta R}}{R} dv' \tag{6-95b}$$

If $\mathbf{J}$ and $\mathbf{M}$ represent linear densities (m^{-1}), (6-95a) and (6-95b) reduce, respectively, to the following surface integrals.

$$\mathbf{A} = \frac{\mu}{4\pi} \iint_S \mathbf{J}_s(x',y',z') \frac{e^{-j\beta R}}{R} ds' \tag{6-96a}$$

$$\mathbf{F} = \frac{\varepsilon}{4\pi} \iint_S \mathbf{M}_s(x',y',z') \frac{e^{-j\beta R}}{R} ds' \tag{6-96b}$$

For electric and magnetic currents $\mathbf{I}_e$ and $\mathbf{I}_m$, they in turn reduce to line integrals of the form

$$\mathbf{A} = \frac{\mu}{4\pi} \int_C \mathbf{I}_e(x',y',z') \frac{e^{-j\beta R}}{R} dl' \tag{6-97a}$$

$$\mathbf{F} = \frac{\varepsilon}{4\pi} \int_C \mathbf{I}_m(x',y',z') \frac{e^{-j\beta R}}{R} dl' \tag{6-97b}$$

Example 6-3

A very thin linear electric current element of very short length ($\ell \ll \lambda$) and with a constant current

$$\mathbf{I}_e(z') = \hat{\mathbf{a}}_z I_e$$

such that $I_e\ell = \text{constant}$, is positioned symmetrically at the origin and oriented along the z axis, as shown in Figure 6-2a. Such an element is usually referred to as an *infinitesimal dipole* [1]. Determine the electric and magnetic fields radiated by the dipole.

Solution: The solution will be obtained using the procedure summarized in Section 6.4. Since the element (source) carries only an electric current $\mathbf{I}_e$, the magnetic current $\mathbf{I}_m$ and the vector potential $\mathbf{F}$ of (6-97b) are both zero. The vector potential $\mathbf{A}$ of (6-97a) is then written as

$$\mathbf{A}(x,y,z) = \frac{\mu}{4\pi} \int_{-\ell/2}^{+\ell/2} \hat{\mathbf{a}}_z I_e \frac{e^{-j\beta R}}{R} dz'$$

where R is the distance from any point on the element, $-\ell/2 \leq z' \leq \ell/2$, to the observation point. Since in the limit as $\ell \to 0$ ($\ell \ll \lambda$),

$$R = r$$

then

$$\mathbf{A}(x,y,z) = \hat{\mathbf{a}}_z \frac{\mu I_e e^{-j\beta r}}{4\pi r} \int_{-\ell/2}^{+\ell/2} dz' = \hat{\mathbf{a}}_z \frac{\mu I_e \ell}{4\pi r} e^{-j\beta r}$$

Transforming the vector potential $\mathbf{A}$ from rectangular to spherical components using the inverse (in this case also transpose) transformation of (II-12) from Appendix II, we can write

$$A_r = A_z \cos\theta = \frac{\mu I_e \ell e^{-j\beta r}}{4\pi r} \cos\theta$$

$$A_\theta = -A_z \sin\theta = -\frac{\mu I_e \ell e^{-j\beta r}}{4\pi r} \sin\theta$$

$$A_\phi = 0$$

Using the symmetry of the problem, that is, no variations in ϕ, (6-32a) can be expanded in spherical coordinates and written in simplified form as

$$\mathbf{H} = \hat{\mathbf{a}}_\phi \frac{1}{\mu r} \left[\frac{\partial}{\partial r}(r A_\theta) - \frac{\partial A_r}{\partial \theta} \right]$$

which reduces to

$$H_r = H_\theta = 0$$

$$H_\phi = j \frac{\beta I_e \ell \sin\theta}{4\pi r} \left(1 + \frac{1}{j\beta r}\right) e^{-j\beta r}$$

The electric field $\mathbf{E}$ can be found using either (6-32b) or (6-32c), that is,

$$\mathbf{E} = -j\omega \mathbf{A} - j\frac{1}{\omega\mu\varepsilon}\nabla(\nabla\cdot\mathbf{A}) = \frac{1}{j\omega\varepsilon}\nabla\times\mathbf{H}$$

and either leads to

$$E_r = \eta \frac{I_e \ell \cos\theta}{2\pi r^2}\left(1 + \frac{1}{j\beta r}\right) e^{-j\beta r}$$

$$E_\theta = j\eta \frac{\beta I_e \ell \sin\theta}{4\pi r}\left[1 + \frac{1}{j\beta r} - \frac{1}{(\beta r)^2}\right] e^{-j\beta r}$$

$$E_\phi = 0$$

The $\mathbf{E}$- and $\mathbf{H}$-field components are valid everywhere except on the source itself.

6.7 FAR-FIELD RADIATION

The fields radiated by antennas of finite dimensions are spherical waves. For these radiators, a general solution to the vector wave equation 6-16 in spherical components, each as a function of r, θ, and ϕ, takes the general form

$$\mathbf{A} = \hat{\mathbf{a}}_r A_r(r,\theta,\phi) + \hat{\mathbf{a}}_\theta A_\theta(r,\theta,\phi) + \hat{\mathbf{a}}_\phi A_\phi(r,\theta,\phi) \qquad (6\text{-}98)$$

The amplitude variations of r in each component of (6-98) are of the form $1/r^n$, $n = 1, 2,...$ [1]. Neglecting higher-order terms of $1/r^n$ ($1/r^n = 0$, $n = 2, 3,...$) reduces (6-98) to

$$\mathbf{A} \simeq [\hat{\mathbf{a}}_r A_r'(\theta,\phi) + \hat{\mathbf{a}}_\theta A_\theta'(\theta,\phi) + \hat{\mathbf{a}}_\phi A_\phi'(\theta,\phi)] \frac{e^{-j\beta r}}{r} \qquad r \to \infty \qquad (6\text{-}99)$$

The r variations are separable from those of θ and ϕ. This will be demonstrated by many examples in the chapters that follow.

Substituting (6-99) into (6-17) reduces it to

$$\mathbf{E} = \frac{1}{r}\left\{-j\omega e^{-j\beta r}[\hat{\mathbf{a}}_r(0) + \hat{\mathbf{a}}_\theta A_\theta'(\theta,\phi) + \hat{\mathbf{a}}_\phi A_\phi'(\theta,\phi)]\right\} + \frac{1}{r^2}\{\cdots\} + \cdots \qquad (6\text{-}100\text{a})$$

The radial E-field component has no $1/r$ terms because its contributions from the first and second terms of (6-17) cancel each other.

Similarly, by using (6-99), we can write (6-4a) as

$$\mathbf{H} = \frac{1}{r}\left\{-j\frac{\omega}{\eta} e^{-j\beta r}[\hat{\mathbf{a}}_r(0) + \hat{\mathbf{a}}_\theta A_\theta'(\theta,\phi) - \hat{\mathbf{a}}_\phi A_\phi'(\theta,\phi)]\right\} + \frac{1}{r^2}\{\cdots\} + \cdots \qquad (6\text{-}100\text{b})$$

where $\eta = \sqrt{\mu/\varepsilon}$ is the intrinsic impedance of the medium.

Neglecting higher-order terms of $1/r^n$, the radiated E and H fields have only θ and ϕ components. They can be expressed as

Far-Field Region

$$\left.\begin{array}{l} E_r \simeq 0 \\ E_\theta \simeq -j\omega A_\theta \\ E_\phi \simeq -j\omega A_\phi \end{array}\right\} \Rightarrow \boxed{\mathbf{E}_A \simeq -j\omega\mathbf{A}} \quad \begin{array}{l} \text{(for the } \theta \text{ and } \phi \text{ components} \\ \text{only since } E_r \simeq 0) \end{array} \qquad (6\text{-}101\text{a})$$

$$\left.\begin{array}{l} H_r \simeq 0 \\ H_\theta \simeq +j\dfrac{\omega}{\eta} A_\phi = -\dfrac{E_\phi}{\eta} \\ H_\phi \simeq -j\dfrac{\omega}{\eta} A_\theta = +\dfrac{E_\theta}{\eta} \end{array}\right\} \Rightarrow \boxed{\mathbf{H}_A \simeq \dfrac{\hat{\mathbf{a}}_r}{\eta} \times \mathbf{E}_A = -j\dfrac{\omega}{\eta}\hat{\mathbf{a}}_r \times \mathbf{A}} \quad \begin{array}{l} \text{(for the } \theta \text{ and } \phi \text{ components} \\ \text{only since } H_r \simeq 0) \end{array} \qquad (6\text{-}101\text{b})$$

Radial field components exist only for higher-order terms of $1/r^n$.

In a similar manner, the far-zone fields that are due to a magnetic source **M** (potential **F**) can be written as

Far-Field Region

$$\left.\begin{array}{l} H_r \simeq 0 \\ H_\theta \simeq -j\omega F_\theta \\ H_\phi \simeq -j\omega F_\phi \end{array}\right\} \Rightarrow \boxed{\mathbf{H_F} \simeq -j\omega\mathbf{F}} \text{ (for the } \theta \text{ and } \phi \text{ components} \quad \text{(6-102a)}$$

$$\text{only since } H_r \simeq 0)$$

$$\left.\begin{array}{l} E_r \simeq 0 \\ E_\theta \simeq -j\omega\eta F_\phi = +\eta H_\phi \\ E_\phi \simeq +j\omega\eta F_\theta = -\eta H_\theta \end{array}\right\} \Rightarrow \boxed{\mathbf{E_F} = -\eta\hat{\mathbf{a}}_r \times \mathbf{H_F} = j\omega\eta\hat{\mathbf{a}}_r \times \mathbf{F}} \text{ (for the } \theta \text{ and } \phi \text{ components} \quad \text{(6-102b)}$$

$$\text{only since } E_r \simeq 0)$$

Simply stated, the corresponding far-zone **E**- and **H**-field components are orthogonal to each other and form TEM (to r) mode fields. This is a very useful relation, and it will be adopted in the following chapters for the solution of the far-zone radiated fields. The far-zone (far-field) region for a radiator is defined as the region whose smallest radial distance is $2D^2/\lambda$ where D is the largest dimension of the radiator (provided D is large compared to the wavelength) [1].

6.8 RADIATION AND SCATTERING EQUATIONS

In Sections 6.4 and 6.6, it was stated that the fields radiated by sources represented by **J** and **M** in an unbounded medium can be computed using (6-32a) through (6-35a), where **A** and **F** are found using (6-95a) and (6-95b). For (6-95a) and (6-95b), the integration is performed over the entire space occupied by **J** and **M** of Figure 6-2b [or $\mathbf{J}_s$ and $\mathbf{M}_s$ of (6-96a) and (6-96b), or $\mathbf{I}_e$ and $\mathbf{I}_m$ of (6-97a) and (6-97b)]. These equations yield valid solutions for all observation points. For most problems, the main difficulty is the inability to perform the integrations in (6-95a) and (6-65b), (6-96a) and (6-96b), or (6-97a) and (6-97b). However, for far-field observations the complexity of the formulation can be reduced.

6.8.1 Near Field

According to Figure 6-2b and equation 6-95a, the vector potential **A** that is due to current density **J** is given by

$$\mathbf{A}(x,y,z) = \frac{\mu}{4\pi}\iiint_V \mathbf{J}(x',y',z')\frac{e^{-j\beta R}}{R}dv' \quad \text{(6-103)}$$

where the primed coordinates (x', y', z') represent the source and the unprimed coordinates (x, y, z) represent the observation point. Here, we intend to write expressions for the **E** and **H** fields that are due to the potential of (6-103), which would be valid everywhere [3, 4], The equations will not be in closed form, but will be convenient for computational purposes. The development will be restricted to the rectangular coordinate system.

The magnetic field due to the potential of (6-103) is given by (6-32a) as

$$\mathbf{H}_A = \frac{1}{\mu}\nabla\times\mathbf{A} = \frac{1}{4\pi}\nabla \times \iiint_V \mathbf{J}(x',y',z')\frac{e^{-j\beta R}}{R}dv' \quad \text{(6-104)}$$

Interchanging integration and differentiation, we can write (6-104) as

$$\mathbf{H}_A = \frac{1}{4\pi}\iiint_V \nabla \times \left[\mathbf{J}(x',y',z')\frac{e^{-j\beta R}}{R}\right]dv' \quad \text{(6-104a)}$$

Using the vector identity

$$\nabla \times (g\mathbf{F}) = (\nabla g) \times \mathbf{F} + g(\nabla \times \mathbf{F}) \tag{6-105}$$

we can write

$$\nabla \times \left[\frac{e^{-j\beta R}}{R} \mathbf{J}(x', y', z') \right] = \nabla \left(\frac{e^{-j\beta R}}{R} \right) \times \mathbf{J}(x', y', z') + \frac{e^{-j\beta R}}{R} \nabla \times \mathbf{J}(x', y', z') \tag{6-106}$$

Since $\mathbf{J}$ is only a function of the primed coordinates and ∇ is a function of the unprimed coordinates,

$$\nabla \times \mathbf{J}(x', y', z') = 0 \tag{6-106a}$$

Also

$$\nabla \left(\frac{e^{-j\beta R}}{R} \right) = -\hat{\mathbf{R}} \left(\frac{1 + j\beta R}{R^2} \right) e^{-j\beta R} \tag{6-106b}$$

where $\hat{\mathbf{R}}$ is a unit vector directed along the line joining any point of the source and the observation point. Using (6-106) through (6-106b), we can write (6-104a) as

$$\mathbf{H}_A(x, y, z) = -\frac{1}{4\pi} \iiint_V (\hat{\mathbf{R}} \times \mathbf{J}) \frac{1 + j\beta R}{R^2} e^{-j\beta R} dx' dy' dz' \tag{6-107}$$

which can be expanded in its three rectangular components [3, 4]

$$H_{Ax} = \frac{1}{4\pi} \iiint_V \left[(z - z')J_y - (y - y')J_z \right] \frac{1 + j\beta R}{R^3} e^{-j\beta R} dx' \, dy' \, dz' \tag{6-107a}$$

$$H_{Ay} = \frac{1}{4\pi} \iiint_V \left[(x - x')J_z - (z - z')J_x \right] \frac{1 + j\beta R}{R^3} e^{-j\beta R} dx' \, dy' \, dz' \tag{6-107b}$$

$$H_{Az} = \frac{1}{4\pi} \iiint_V \left[(y - y')J_x - (x - x')J_y \right] \frac{1 + j\beta R}{R^3} e^{-j\beta R} dx' \, dy' \, dz' \tag{6-107c}$$

Using (6-32b) or Maxwell's equation 6-32c, we can write the corresponding electric field components as

$$\mathbf{E}_A = \hat{\mathbf{a}}_x E_{Ax} + \hat{\mathbf{a}}_y E_{Ay} + \hat{\mathbf{a}}_z E_{Az} = -j\omega\mathbf{A} - j\frac{1}{\omega\mu\varepsilon} \nabla(\nabla \cdot \mathbf{A}) = \frac{1}{j\omega\varepsilon} \nabla \times \mathbf{H}_A \tag{6-108}$$

which with the aid of (6-107a) through (6-107c) reduce to

$$E_{Ax} = -\frac{j\eta}{4\pi\beta} \iiint_V \{G_1 J_x + (x - x')G_2$$
$$\times [(x - x')J_x + (y - y')J_y + (z - z')J_z]\} e^{-j\beta R} dx' \, dy' \, dz' \tag{6-108a}$$

$$E_{Ay} = -\frac{j\eta}{4\pi\beta} \iiint_V \{G_1 J_y + (y - y')G_2$$
$$\times [(x - x')J_x + (y - y')J_y + (z - z')J_z]\} e^{-j\beta R} dx' \, dy' \, dz' \tag{6-108b}$$

$$E_{Az} = -\frac{j\eta}{4\pi\beta} \iiint_V \{G_1 J_z + (z - z')G_2$$
$$\times [(x - x')J_x + (y - y')J_y + (z - z')J_z]\} e^{-j\beta R} dx' \, dy' \, dz' \tag{6-108c}$$

where

$$G_1 = \frac{-1 - j\beta R + \beta^2 R^2}{R^3} \tag{6-108d}$$

$$G_2 = \frac{3 + j3\beta R - \beta^2 R^2}{R^5} \tag{6-108e}$$

In the same manner, we can write for the vector potential of (6-95b)

$$\mathbf{F}(x, y, z) = \frac{\varepsilon}{4\pi} \iiint_V \mathbf{M}(x', y', z') \frac{e^{-j\beta R}}{R} dv' \tag{6-109}$$

the electric field components using (6-33a),

$$\mathbf{E}_F = -\frac{1}{\varepsilon} \nabla \times \mathbf{F} \tag{6-110}$$

as

$$E_{Fx} = -\frac{1}{4\pi} \iiint_V \left[(z - z')M_y - (y - y')M_z \right] \frac{1 + j\beta R}{R^3} e^{-j\beta R} dx' dy' dz' \tag{6-110a}$$

$$E_{Fy} = -\frac{1}{4\pi} \iiint_V \left[(x - x')M_z - (z - z')M_x \right] \frac{1 + j\beta R}{R^3} e^{-j\beta R} dx' dy' dz' \tag{6-110b}$$

$$E_{Fz} = -\frac{1}{4\pi} \iiint_V \left[(y - y')M_x - (x - x')M_y \right] \frac{1 + j\beta R}{R^3} e^{-j\beta R} dx' dy' dz' \tag{6-110c}$$

Similarly, the corresponding magnetic field components can be written using (6-33b) or (6-33c)

$$\mathbf{H}_F = -j\omega\mathbf{F} - j\frac{1}{\omega\mu\varepsilon} \nabla(\nabla \cdot \mathbf{F}) = -\frac{1}{j\omega\mu} \nabla \times \mathbf{E}_F \tag{6-111}$$

as

$$H_{Fx} = -\frac{j}{4\pi\beta\eta} \iiint_V \{ G_1 M_x + (x - x')G_2 \\
\times [(x - x')M_x + (y - y')M_y + (z - z')M_z] \} e^{-j\beta R} dx' dy' dz' \tag{6-111a}$$

$$H_{Fy} = -\frac{j}{4\pi\beta\eta} \iiint_V \{ G_1 M_y + (y - y')G_2 \\
\times [(x - x')M_x + (y - y')M_y + (z - z')M_z] \} e^{-j\beta R} dx' dy' dz' \tag{6-111b}$$

$$H_{Fz} = -\frac{j}{4\pi\beta\eta} \iiint_V \{ G_1 M_z + (z - z')G_2 \\
\times [(x - x')M_x + (y - y')M_y + (z - z')M_z] \} e^{-j\beta R} dx' dy' dz' \tag{6-111c}$$

where G_1 and G_2 are given by (6-108d) and (6-108e).

6.8.2 Far Field

It was shown in Section 6.7 that the field equations for far-field ($\beta r \gg 1$) observations simplify considerably. Also in the far zone the **E**- and **H**-field components are orthogonal to each other and form TEM (to r) mode fields. Although the field equations in the far zone simplify, integrations still need to be performed to find the vector potentials of **A** and **F** given, respectively, by

(6-95a) and (6-95b), or (6-96a) and (6-96b), or (6-97a) and (6-97b). However, the integrations, as will be shown next, can be simplified if the observations are made in the far field.

If the observations are made in the far field ($\beta r \gg 1$), it can be shown [1] that the radial distance **R** of Figure 6-3a from any point on the source or scatterer to the observation point can be assumed to be parallel to the radial distance **r** from the origin to the observation point, as shown in Figure 6-3b. In such cases, the relation between the magnitudes of **R** and **r** of Figure 6-3a, given by

$$R = \left[r^2 + (r')^2 - 2rr' \cos \psi \right]^{1/2} \tag{6-112}$$

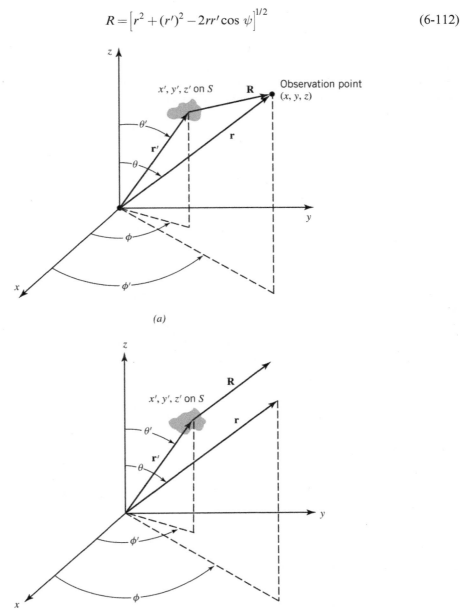

(a)

(b)

Figure 6-3 Coordinate system for antenna analysis. (a) Near and far fields. (b) Far field. (Source: C. A. Balanis, *Antenna Theory: Analysis and Design*, Third Edition. Copyright © 2005, John Wiley & Sons, Inc. Reprinted by permission of John Wiley & Sons, Inc.)

can be approximated, according to Figure 6-3b, most commonly by [1]

$$R = \begin{cases} r - r' \cos \psi & \text{for phase variations} \qquad (6\text{-}112a) \\ r & \text{for amplitude variations} \qquad (6\text{-}112b) \end{cases}$$

where ψ is the angle between $\mathbf{r}$ and $\mathbf{r}'$. These approximations yield a maximum phase error of $\pi/8$ (22.5°) provided the observations are made at distances

$$r \geq \frac{2D^2}{\lambda} \qquad (6\text{-}113)$$

where D is the largest dimension of the radiator or scatterer. The distance (6-113) represents the *minimum* distance to the far-field region. The derivation of (6-113), as well as distances for other zones, can be found in [1]. Using (6-112a) and (6-112b), we can write (6-96a) and (6-96b), assuming the current densities reside on the surface of the source, as

$$\mathbf{A} = \frac{\mu}{4\pi} \iint_S \mathbf{J}_s \frac{e^{-j\beta R}}{R} ds' \simeq \frac{\mu e^{-j\beta r}}{4\pi r} \mathbf{N} \qquad (6\text{-}114a)$$

$$\mathbf{F} = \frac{\varepsilon}{4\pi} \iint_S \mathbf{M}_s \frac{e^{-j\beta R}}{R} ds' \simeq \frac{\varepsilon e^{-j\beta r}}{4\pi r} \mathbf{L} \qquad (6\text{-}114b)$$

where

$$\mathbf{N} = \iint_S \mathbf{J}_s e^{j\beta r' \cos\psi} ds' \qquad (6\text{-}114c)$$

$$\mathbf{L} = \iint_S \mathbf{M}_s e^{j\beta r' \cos\psi} ds' \qquad (6\text{-}114d)$$

It was shown in Section 6.7 that in the far field only the θ and ϕ components of the $\mathbf{E}$ and $\mathbf{H}$ fields are dominant. Although the radial components are not necessarily zero, they are negligible compared to the θ and ϕ components. Also it was shown that for (6-32b) and (6-33b), or

$$\mathbf{E}_A = -j\omega \left[\mathbf{A} + \frac{1}{\beta^2} \nabla (\nabla \cdot \mathbf{A}) \right] \qquad (6\text{-}115a)$$

$$\mathbf{H}_F = -j\omega \left[\mathbf{F} + \frac{1}{\beta^2} \nabla (\nabla \cdot \mathbf{F}) \right] \qquad (6\text{-}115b)$$

where $\mathbf{A}$ and $\mathbf{F}$ are given by (6-114a) and (6-114b), the second terms within the brackets only contribute variations of the order $1/r^2$, $1/r^3$, $1/r^4$, etc. Since observations are made in the far field, the dominant variation is of the order $1/r$ and it is contained in the first term of (6-115a) and (6-115b). Thus, for far-field observations, (6-115a) and (6-115b) reduce to

$$\mathbf{E}_A \simeq -j\omega \mathbf{A} \quad (\theta \text{ and } \phi \text{ components only}) \qquad (6\text{-}116a)$$

$$\mathbf{H}_F \simeq -j\omega \mathbf{F} \quad (\theta \text{ and } \phi \text{ components only}) \qquad (6\text{-}116b)$$

which can be expanded and written as

$$(E_A)_\theta \simeq -j\omega A_\theta \qquad (6\text{-}117a)$$

$$(E_A)_\phi \simeq -j\omega A_\phi \qquad (6\text{-}117b)$$

$$(H_F)_\theta \simeq -j\omega F_\theta \qquad (6\text{-}117c)$$

$$(H_F)_\phi \simeq -j\omega F_\phi \qquad (6\text{-}117d)$$

The radial components are neglected because they are very small compared to the θ and ϕ components.

To find the remaining **E** and **H** fields contributed by the **F** and **A** potentials, that is, $\mathbf{E}_F$ and $\mathbf{H}_A$, we can use (6-33a) and (6-32a), or

$$\mathbf{E}_F = -\frac{1}{\varepsilon} \nabla \times \mathbf{F} \tag{6-118a}$$

$$\mathbf{H}_A = \frac{1}{\mu} \nabla \times \mathbf{A} \tag{6-118b}$$

However, we resort instead to (6-117a) through (6-117d). Since the observations are made in the far field and we know that the **E**- and **H**-field components are orthogonal to each other and to the radial direction (plane waves) and are related by the intrinsic impedance of the medium, we can write, using (6-117a) through (6-117d),

$$(E_F)_\theta \simeq +\eta (H_F)_\phi = -j\omega\eta F_\phi \tag{6-119a}$$

$$(E_F)_\phi \simeq -\eta (H_F)_\theta = +j\omega\eta F_\theta \tag{6-119b}$$

$$(H_A)_\theta \simeq -\frac{(E_A)_\phi}{\eta} = +j\omega\frac{A_\phi}{\eta} \tag{6-119c}$$

$$(H_A)_\phi \simeq +\frac{(E_A)_\theta}{\eta} = -j\omega\frac{A_\theta}{\eta} \tag{6-119d}$$

Combining (6-117a) through (6-117d) with (6-119a) through (6-119d) and remembering that the radial components are negligible, we can write the **E**- and **H**-field components in the far field as

$$E_r \simeq 0 \tag{6-120a}$$

$$E_\theta \simeq (E_A)_\theta + (E_F)_\theta = -j\omega\left[A_\theta + \eta F_\phi\right] \tag{6-120b}$$

$$E_\phi \simeq (E_A)_\phi + (E_F)_\phi = -j\omega\left[A_\phi - \eta F_\theta\right] \tag{6-120c}$$

$$H_r \simeq 0 \tag{6-120d}$$

$$H_\theta \simeq (H_A)_\theta + (H_F)_\theta = +\frac{j\omega}{\eta}\left[A_\phi - \eta F_\theta\right] \tag{6-120e}$$

$$H_\phi \simeq (H_A)_\phi + (H_F)_\phi = -\frac{j\omega}{\eta}\left[A_\theta + \eta F_\phi\right] \tag{6-120f}$$

Using A_θ, A_ϕ, F_θ, and F_ϕ from (6-114a) through (6-114d), that is,

$$A_\theta = \frac{\mu e^{-j\beta r}}{4\pi r} N_\theta \tag{6-121a}$$

$$A_\phi = \frac{\mu e^{-j\beta r}}{4\pi r} N_\phi \tag{6-121b}$$

$$F_\theta = \frac{\varepsilon e^{-j\beta r}}{4\pi r} L_\theta \tag{6-121c}$$

$$F_\phi = \frac{\varepsilon e^{-j\beta r}}{4\pi r} L_\phi \tag{6-121d}$$

we can reduce (6-120a) through (6-120f) to

$$E_r \simeq 0 \tag{6-122a}$$

$$E_\theta \simeq -\frac{j\beta e^{-j\beta r}}{4\pi r}(L_\phi + \eta N_\theta) \tag{6-122b}$$

$$E_\phi \simeq +\frac{j\beta e^{-j\beta r}}{4\pi r}(L_\theta - \eta N_\phi) \tag{6-122c}$$

$$H_r \simeq 0 \tag{6-122d}$$

$$H_\theta \simeq +\frac{j\beta e^{-j\beta r}}{4\pi r}\left(N_\phi - \frac{L_\theta}{\eta}\right) \tag{6-122e}$$

$$H_\phi \simeq -\frac{j\beta e^{-j\beta r}}{4\pi r}\left(N_\theta + \frac{L_\phi}{\eta}\right) \tag{6-122f}$$

A. Rectangular Coordinate System   To find the fields of (6-122a) through (6-122f), the functions N_θ, N_ϕ, L_θ, and L_ϕ must be evaluated from (6-114c) and (6-114d). The evaluation of (6-114c) and (6-114d) can best be accomplished if the most convenient coordinate system is chosen.

For radiators or scatterers whose geometries are most conveniently represented by rectangular coordinates, (6-114c) and (6-114d) can best be expressed as

$$\mathbf{N} = \iint_S \mathbf{J}_s e^{+j\beta r' \cos\psi}\,ds' = \iint_S (\hat{\mathbf{a}}_x J_x + \hat{\mathbf{a}}_y J_y + \hat{\mathbf{a}}_z J_z)e^{+j\beta r' \cos\psi}\,ds' \tag{6-123a}$$

$$\mathbf{L} = \iint_S \mathbf{M}_s e^{+j\beta r' \cos\psi}\,ds' = \iint_S (\hat{\mathbf{a}}_x M_x + \hat{\mathbf{a}}_y M_y + \hat{\mathbf{a}}_z M_z)e^{+j\beta r' \cos\psi}\,ds' \tag{6-123b}$$

Using the rectangular-to-spherical component transformation of (II-13a)

$$\begin{bmatrix} \hat{\mathbf{a}}_x \\ \hat{\mathbf{a}}_y \\ \hat{\mathbf{a}}_z \end{bmatrix} = \begin{bmatrix} \sin\theta\cos\phi & \cos\theta\cos\phi & -\sin\phi \\ \sin\theta\sin\phi & \cos\theta\sin\phi & \cos\phi \\ \cos\theta & -\sin\theta & 0 \end{bmatrix} \begin{bmatrix} \hat{\mathbf{a}}_r \\ \hat{\mathbf{a}}_\theta \\ \hat{\mathbf{a}}_\phi \end{bmatrix} \tag{6-124}$$

we can reduce (6-123a) and (6-123b) for the θ and ϕ components to

$$N_\theta = \iint_S (J_x \cos\theta\cos\phi + J_y \cos\theta\sin\phi - J_z \sin\theta)e^{+j\beta r' \cos\psi}\,ds' \tag{6-125a}$$

$$N_\phi = \iint_S (-J_x \sin\phi + J_y \cos\phi)e^{+j\beta r' \cos\psi}\,ds' \tag{6-125b}$$

$$L_\theta = \iint_S (M_x \cos\theta\cos\phi + M_y \cos\theta\sin\phi - M_z \sin\theta)e^{+j\beta r' \cos\psi}\,ds' \tag{6-125c}$$

$$L_\phi = \iint_S (-M_x \sin\phi + M_y \cos\phi)e^{+j\beta r' \cos\psi}\,ds' \tag{6-125d}$$

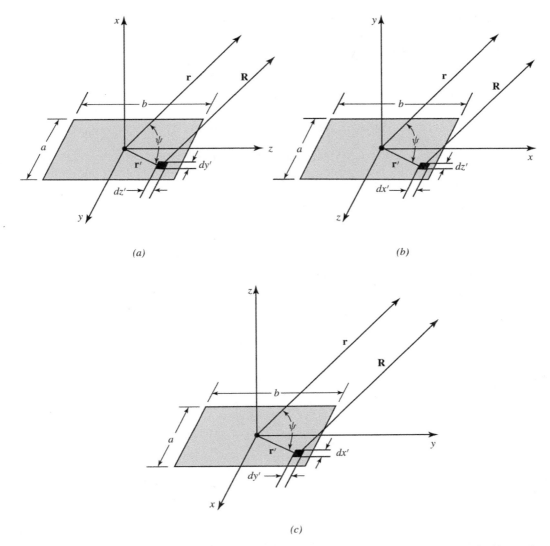

Figure 6-4 Rectangular aperture and plate positions for antenna and scattering system analysis. (*a*) *yz* plane. (*b*) *xz* plane. (*c*) *xy* plane. (Source: C. A. Balanis, *Antenna Theory: Analysis and Design*, Third Edition. Copyright © 2005, John Wiley & Sons, Inc. Reprinted by permission of John Wiley & Sons, Inc.)

Some of the most common and practical radiators and scatterers are represented by rectangular geometries. Because of their configuration, the most convenient coordinate system for expressing the fields or current densities on the structure, and performing the integration over it, would be the rectangular. The three most common and convenient coordinate positions used for the solution of the problem are shown in Figure 6-4. Figures 6-4*a*, 6-4*b*, and 6-4*c* show, respectively, the structure in the *yz* plane, in the *xz* plane, and in the *xy* plane. For a given field or current density distribution, the analytical forms for the radiated or scattered fields for each of the arrangements would not be the same. However, the computed values will be the same because the problem is physically identical.

For each of the geometries shown in Figure 6-4, the only difference in the analysis will be in the following formulations.

1. The components of the equivalent currents, J_x, J_y, J_z, M_x, M_y, and M_z.
2. The difference in paths from the source to the observation point, $r' \cos \psi$.
3. The differential area ds'.

In general, the nonzero components of $\mathbf{J}_s$ and $\mathbf{M}_s$ will be

$$J_y, J_z, M_y, \text{ and } M_z \quad \text{(Fig. 6-4a)} \tag{6-126a}$$

$$J_x, J_z, M_x, \text{ and } M_z \quad \text{(Fig. 6-4b)} \tag{6-126b}$$

$$J_x, J_y, M_x, \text{ and } M_y \quad \text{(Fig. 6-4c)} \tag{6-126c}$$

The differential paths will be of the form

$$r' \cos \psi = \mathbf{r}' \cdot \hat{\mathbf{a}}_r = (\hat{\mathbf{a}}_y y' + \hat{\mathbf{a}}_z z') \cdot (\hat{\mathbf{a}}_x \sin\theta \, \cos\phi + \hat{\mathbf{a}}_y \sin\theta \, \sin\phi + \hat{\mathbf{a}}_z \cos\theta$$
$$= y' \sin\theta \, \sin\phi + z' \cos\theta \quad \text{(Fig. 6-4a)} \tag{6-127a}$$

$$r' \cos \psi = \mathbf{r}' \cdot \hat{\mathbf{a}}_r = (\hat{\mathbf{a}}_x x' + \hat{\mathbf{a}}_z z') \cdot (\hat{\mathbf{a}}_x \sin\theta \, \cos\phi + \hat{\mathbf{a}}_y \sin\theta \, \sin\phi + \hat{\mathbf{a}}_z \cos\theta)$$
$$= x' \sin\theta \, \cos\phi + z' \cos\theta \quad \text{(Fig. 6-4b)} \tag{6-127b}$$

$$r' \cos \psi = \mathbf{r}' \cdot \hat{\mathbf{a}}_r = (\hat{\mathbf{a}}_x x' + \hat{\mathbf{a}}_y y') \cdot (\hat{\mathbf{a}}_x \sin\theta \, \cos\phi + \hat{\mathbf{a}}_y \sin\theta \, \sin\phi + \hat{\mathbf{a}}_z \cos\theta)$$
$$= x' \sin\theta \, \cos\phi + y' \sin\theta \, \sin\phi \quad \text{(Fig. 6-4c)} \tag{6-127c}$$

and the differential areas of

$$ds' = dy' dz' \quad \text{(Fig. 6-4a)} \tag{6-128a}$$

$$ds' = dx' dz' \quad \text{(Fig. 6-4b)} \tag{6-128b}$$

$$ds' = dx' dy' \quad \text{(Fig. 6-4c)} \tag{6-128c}$$

SUMMARY   To summarize the results, we will outline the procedure that must be followed to solve a problem using the radiation or scattering integrals. Figure 6-3 is used to indicate the geometry.

1. Select a closed surface over which the actual current density $\mathbf{J}_s$ or the equivalent current densities $\mathbf{J}_s$ and $\mathbf{M}_s$ exist.
2. Specify the actual current density $\mathbf{J}_s$ or form the equivalent currents $\mathbf{J}_s$ and $\mathbf{M}_s$ over S using [1, 3, 5]

$$\mathbf{J}_s = \hat{\mathbf{n}} \times \mathbf{H}_a \tag{6-129a}$$

$$\mathbf{M}_s = -\hat{\mathbf{n}} \times \mathbf{E}_a \tag{6-129b}$$

where $\hat{\mathbf{n}}$ = unit vector normal to the surface S
$\quad \mathbf{E}_a$ = total electric field over the surface S
$\quad \mathbf{H}_a$ = total magnetic field over the surface S
3. (*Optional*) Determine the potentials $\mathbf{A}$ and $\mathbf{F}$ using, respectively, (6-103) and (6-109) where the integration is over the surface S of the sources.
4. Determine the corresponding $\mathbf{E}$- and $\mathbf{H}$-field components that are due to $\mathbf{J}_s$ and $\mathbf{M}_s$ using (6-107a) through (6-107c), (6-108a) through (6-108e), (6-110a) through (6-110c), and (6-111a) through (6-111c). Combine the $\mathbf{E}$- and $\mathbf{H}$-field components that are due to both $\mathbf{J}_s$ and $\mathbf{M}_s$ to find the total $\mathbf{E}$ and $\mathbf{H}$ fields.

These steps are valid for all regions (near field and far field) outside the surface S. If, however, the observation point is in the far field, steps 3 and 4 can be replaced by 3' and 4'.

3'. Determine N_θ, N_ϕ, L_θ, and L_ϕ using (6-125a) through (6-125d).
4'. Determine the radiated **E** and **H** fields using (6-122a) through (6-122f).

This procedure can be used to analyze radiation and scattering problems. The radiation problems most conducive to this procedure are aperture antennas, such as waveguides, horns, reflectors, and others. These aperture antennas are usually best represented by specifying their fields over their apertures.

Example 6-4

The tangential **E** and **H** fields over a rectangular aperture of dimensions a and b, shown in Figure 6-5, are given by

$$\left. \begin{array}{l} \mathbf{E}_a = \hat{\mathbf{a}}_y E_0 \\[2mm] \mathbf{H}_a = -\hat{\mathbf{a}}_x \dfrac{E_0}{\eta} \end{array} \right\} \quad \begin{array}{l} -\dfrac{a}{2} \le x' \le \dfrac{a}{2} \\[2mm] -\dfrac{b}{2} \le y' \le \dfrac{b}{2} \end{array}$$

$$\mathbf{E}_a \simeq \mathbf{H}_a \simeq 0 \quad \text{elsewhere}$$

Find the far-zone fields radiated by the aperture, and plot the three-dimensional pattern when $a = 3\lambda$ and $b = 2\lambda$. The fields over the aperture and elsewhere have been simplified in order to reduce the complexity of the problem and to avoid having the analytical formulations obscure the analysis procedure.

Solution:

1. The surface of the radiator is defined by $-a/2 \le x \le a/2$ and $-b/2 \le y \le b/2$.
2. Since the electric and magnetic fields exist only over the bounds of the aperture, the equivalent current densities $\mathbf{J}_s$ and $\mathbf{M}_s$ representing the aperture exist only over the bounds of the aperture as well. This is a good approximation, and it is derived by the equivalence principle in Chapter 7 [1, 3, 5]. Using (6-129a) and (6-129b) the current densities $\mathbf{J}_s$ and $\mathbf{M}_s$ can be written, by referring to Figure 6-5, as

$$\mathbf{J}_s = \hat{\mathbf{n}} \times \mathbf{H}_a = \hat{\mathbf{a}}_z \times \left(-\hat{\mathbf{a}}_x \frac{E_0}{\eta}\right) = -\hat{\mathbf{a}}_y \frac{E_0}{\eta} \Rightarrow J_x = J_z = 0 \qquad J_y = -\frac{E_0}{\eta}$$

$$\mathbf{M}_s = -\hat{\mathbf{n}} \times \mathbf{E}_a = -\hat{\mathbf{a}}_z \times \hat{\mathbf{a}}_y E_0 = \hat{\mathbf{a}}_x E_0 \Rightarrow M_x = E_0 \qquad M_y = M_z = 0$$

3. Using (6-125a), (6-127c), and (6-128c), we can reduce N_θ to

$$N_\theta = \iint_S \left[J_x \cos\theta \, \cos\phi + J_y \cos\theta \, \sin\phi - J_z \sin\theta \right] e^{j\beta r' \cos\psi} ds'$$

$$= -\frac{E_0}{\eta} \cos\theta \, \sin\phi \int_{-b/2}^{b/2} \int_{-a/2}^{a/2} e^{j\beta(x' \sin\theta \, \cos\phi + y' \sin\theta \, \sin\phi)} dx' dy'$$

Using the integral

$$\int_{-c/2}^{c/2} e^{j\alpha z} dz = c \left| \frac{\sin\left(\dfrac{\alpha}{2} c\right)}{\dfrac{\alpha}{2} c} \right|$$

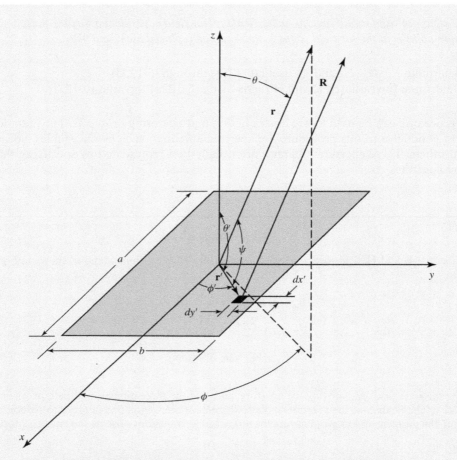

Figure 6-5   Rectangular aperture geometry for radiation problem.

reduces N_θ to

$$N_\theta = -\frac{abE_0}{\eta}\left\{\cos\theta\,\sin\phi\left[\frac{\sin(X)}{X}\right]\left[\frac{\sin(Y)}{Y}\right]\right\}$$

where $X = \dfrac{\beta a}{2}\sin\theta\,\cos\phi$

$\qquad Y = \dfrac{\beta b}{2}\sin\theta\,\sin\phi$

In a similar manner, N_ϕ, L_θ, and L_ϕ of (6-125b), (6-125c), and (6-125d) can be written as

$$N_\phi = \iint_S \left(-J_x\sin\phi + J_y\cos\phi\right)e^{j\beta r'\cos\psi}ds'$$

$$\quad = -\frac{abE_0}{\eta}\left\{\cos\phi\left[\frac{\sin(X)}{X}\right]\left[\frac{\sin(Y)}{Y}\right]\right\}$$

$$L_\theta = \iint_S \left(M_x\cos\theta\cos\phi + M_y\cos\theta\sin\phi - M_z\sin\theta\right)e^{j\beta r'\cos\psi}ds'$$

$$\quad = abE_0\left\{\cos\theta\cos\phi\left[\frac{\sin(X)}{X}\right]\left[\frac{\sin(Y)}{Y}\right]\right\}$$

$$L_\phi = \iint_S \left[-M_x\sin\phi + M_y\cos\phi\right]e^{j\beta r'\cos\psi}ds'$$

$$\quad = -abE_0\left\{\sin\phi\left[\frac{\sin(X)}{X}\right]\left[\frac{\sin(Y)}{Y}\right]\right\}$$

The corresponding far-zone **E**- and **H**-field components radiated by the aperture are obtained using (6-122a) through (6-122f), and they can be written as

$$E_r \simeq H_r \simeq 0$$

$$E_\theta \simeq \frac{C}{2} \sin\phi \, (1+\cos\theta) \left[\frac{\sin(X)}{X}\right]\left[\frac{\sin(Y)}{Y}\right]$$

$$E_\phi \simeq \frac{C}{2} \cos\phi \, (1+\cos\theta) \left[\frac{\sin(X)}{X}\right]\left[\frac{\sin(Y)}{Y}\right]$$

$$H_\theta \simeq -\frac{E_\phi}{\eta}$$

$$H_\phi \simeq +\frac{E_\theta}{\eta}$$

$$C = j\frac{ab\beta E_0 e^{-j\beta r}}{2\pi r}$$

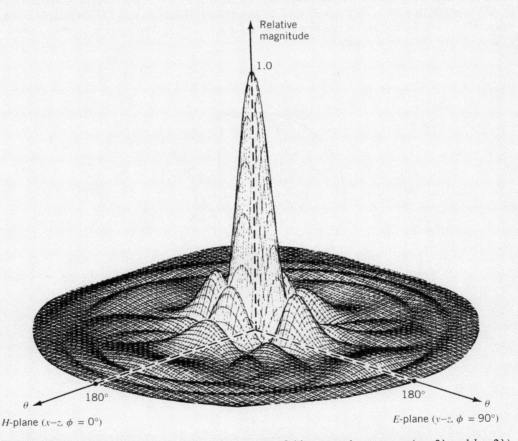

Figure 6-6 Three-dimensional field pattern of a constant field rectangular aperture ($a = 3\lambda$ and $b = 2\lambda$). (Source: C. A. Balanis, *Antenna Theory: Analysis and Design*, Third Edition. Copyright © 2005, John Wiley & Sons, Inc. Reprinted by permission of John Wiley & Sons, Inc.)

In the principal E and H planes, the electric field components reduce to

$$\text{\underline{E \textbf{Plane}$(\phi=\pi/2)$}}$$

$$E_r \simeq E_\phi = 0$$

$$E_\theta = \frac{C}{2}(1+\cos\theta)\frac{\sin\left(\dfrac{\beta b}{2}\sin\theta\right)}{\dfrac{\beta b}{2}\sin\theta}$$

$$\text{\underline{H \textbf{Plane}$(\phi=0)$}}$$

$$E_r \simeq E_\theta = 0$$

$$E_\phi = \frac{C}{2}(1+\cos\theta)\frac{\sin\left(\dfrac{\beta a}{2}\sin\theta\right)}{\dfrac{\beta a}{2}\sin\theta}$$

A three-dimensional plot of the normalized magnitude of the total electric field intensity $E\left(E \simeq \sqrt{E_\theta^2 + E_\phi^2}\right)$ for an aperture with $a=3\lambda$, $b=2\lambda$ as a function of θ, and ϕ ($0° \leq \theta \leq 180°, 0° \leq \phi \leq 360°$) is shown plotted in Figure 6-6. Because the aperture is larger in the x direction ($a=3\lambda$), its pattern in the xz plane exhibits a larger number of lobes compared to the yz plane, as shown also in the two-dimensional E- and H-plane patterns in Figure 6-7.

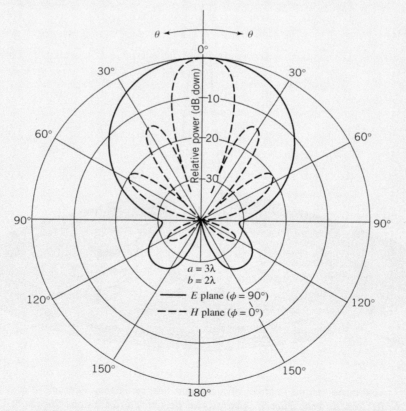

Figure 6-7 E-($\phi=90°$) and H-plane ($\phi=0°$) power patterns of a rectangular aperture with a uniform field distribution.

To demonstrate the application of the techniques to scattering, let us consider a scattering problem.

Example 6-5

A parallel polarized uniform plane wave traveling on the yz plane at an angle θ_i from the z axis is incident upon a rectangular electric perfectly conducting flat plate of dimensions a and b, as shown in Figure 6-8. Assuming that the induced current density on the plate is the same as that on an infinite conducting flat plate, find the far-zone spherical scattered electric and magnetic field components in directions specified by θ_s, ϕ_s. Plot the three-dimensional scattering pattern when $a = 3\lambda$ and $b = 2\lambda$.

Solution: Since the incident wave is a parallel polarized uniform plane wave, the incident electric and magnetic fields can be written as

$$\mathbf{E}^i = E_0(\hat{\mathbf{a}}_y \cos\theta_i + \hat{\mathbf{a}}_z \sin\theta_i)\, e^{-j\beta(y' \sin\theta_i - z' \cos\theta_i)}$$

$$\mathbf{H}^i = \frac{E_0}{\eta}\hat{\mathbf{a}}_x e^{-j\beta(y' \sin\theta_i - z' \cos\theta_i)}$$

The electric current density induced on the surface of the plate is given by

$$\mathbf{J}_s = \hat{\mathbf{n}} \times \mathbf{H}^{\text{total}}|_{z=0} = \hat{\mathbf{a}}_z \times (\mathbf{H}^i + \mathbf{H}^r)|_{z=0}$$

According to Figure 5-4 and (5-24c), the reflected magnetic field of (5-21b) can be written as

$$\mathbf{H}^r = -\Gamma_\parallel \mathbf{H}^i = -(-\mathbf{H}^i) = \mathbf{H}^i$$

Then

$$\mathbf{J}_s = \hat{\mathbf{a}}_z \times (\mathbf{H}^i + \mathbf{H}^i)|_{z=0} = 2\hat{\mathbf{a}}_y H^i|_{z=0} = \hat{\mathbf{a}}_y 2\frac{E_0}{\eta} e^{-j\beta y' \sin\theta_i}$$

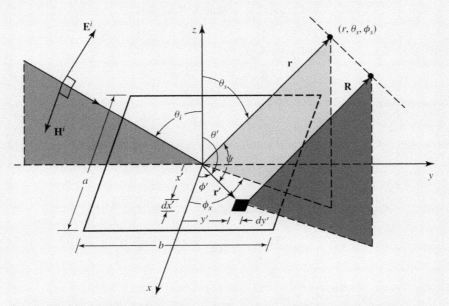

Figure 6-8 Uniform plane wave incident on a rectangular conducting plate.

or

$$J_x = J_z = 0 \qquad \text{everywhere}$$

$$J_y = 2\frac{E_0}{\eta}e^{-j\beta y' \sin\theta_i} \quad \text{for} \begin{cases} -a/2 \le x' \le a/2 \\ -b/2 \le y' \le b/2 \end{cases} \quad \text{and zero elsewhere}$$

Because the geometry of the plate corresponds to the coordinate system of Figure 6-4c, equation 6-125a can be written using (6-127c) and (6-128c) as

$$N_\theta = \iint_S \left[J_x \cos\theta_s \cos\phi_s + J_y \cos\theta_s \sin\phi_s - J_z \sin\theta_s \right] e^{j\beta r' \cos\psi} ds'$$

$$= 2\frac{E_0}{\eta}\cos\theta_s \sin\phi_s \int_{-b/2}^{b/2} \int_{-a/2}^{a/2} e^{j\beta x' \sin\theta_s \cos\phi_s} e^{j\beta y'(\sin\theta_s \sin\phi_s - \sin\theta_i)} dx'dy'$$

Using the integral

$$\int_{-c/2}^{c/2} e^{j\alpha z} dz = c \left| \frac{\sin\left(\frac{\alpha}{2}c\right)}{\frac{\alpha}{2}c} \right|$$

reduces N_θ to

$$N_\theta = 2ab\frac{E_0}{\eta}\left\{ \cos\theta_s \sin\phi_s \left[\frac{\sin(X)}{X} \right]\left[\frac{\sin(Y)}{Y} \right] \right\}$$

where $X = \dfrac{\beta a}{2}\sin\theta_s \cos\phi_s$

$$Y = \frac{\beta b}{2}(\sin\theta_s \sin\phi_s - \sin\theta_i)$$

Similarly, according to (6-125b), N_ϕ can be written as

$$N_\phi = \iint_S \left(-J_x \sin\phi_s + J_y \cos\phi_s \right) e^{j\beta r' \cos\psi} ds'$$

$$= 2ab\frac{E_0}{\eta}\left\{ \cos\phi_s \left[\frac{\sin(X)}{X} \right]\left[\frac{\sin(Y)}{Y} \right] \right\}$$

Because the plate is a perfect electric conductor,

$$M_x = M_y = M_z = 0 \quad \text{everywhere}$$

Therefore, according to (6-125c) and (6-125d),

$$L_\theta = L_\phi = 0$$

Thus the scattered electric and magnetic field components can be reduced according to (6-122a) through (6-122f) to

$$E_r^s \simeq H_r^s \simeq 0$$

$$E_\theta^s \simeq -jab\frac{\beta E_0 e^{-j\beta r}}{2\pi r}\left\{ \cos\theta_s \sin\phi_s \left[\frac{\sin(X)}{X} \right]\left[\frac{\sin(Y)}{Y} \right] \right\}$$

$$E_\phi^s \simeq -jab\frac{\beta E_0 e^{-j\beta r}}{2\pi r}\left\{ \cos\phi_s \left[\frac{\sin(X)}{X} \right]\left[\frac{\sin(Y)}{Y} \right] \right\}$$

$$H_\theta^s \simeq -\frac{E_\phi^s}{\eta}$$

$$H_\phi^s \simeq +\frac{E_\theta^s}{\eta}$$

In the principal E and H planes, the electric field components reduce, respectively, as follows.

$$\textbf{\textit{E} Plane}(\phi_s = \pi/2, 3\pi/2)$$

$$E_r^s \simeq E_\phi^s \simeq 0$$

$$E_\theta^s \simeq -jab\frac{\beta E_0 e^{-j\beta r}}{2\pi r}\cos\theta_s\frac{\sin\left[\frac{\beta b}{2}(\pm\sin\theta_s - \sin\theta_i)\right]}{\frac{\beta b}{2}(\pm\sin\theta_s - \sin\theta_i)}\qquad\begin{array}{l}+\textit{ for }\phi_s = \pi/2\\-\textit{ for }\phi_s = 3\pi/2\end{array}$$

$$\textbf{\textit{H} Plane}\quad(\phi_s = 0, \pi)$$

$$E_r^s \simeq E_\theta^s \simeq 0$$

$$E_\phi^s \simeq -jab\frac{\beta E_0 e^{-j\beta r}}{2\pi r}\frac{\sin\left(\frac{\beta a}{2}\sin\theta_s\right)}{\frac{\beta a}{2}\sin\theta_s}\frac{\sin\left(\frac{\beta b}{2}\sin\theta_i\right)}{\frac{\beta b}{2}\sin\theta_i}$$

A three-dimensional plot of the normalized magnitude of the total electric field $E^s \left(E^s = \sqrt{(E_\theta^s)^2 + (E_\phi^s)^2}\right)$ for a plate of dimensions $a = 3\lambda$ and $b = 2\lambda$ when the incidence angle $\theta_i = 30°$ is shown in Figure 6-9. Its corresponding two-dimensional pattern in the yz plane ($\phi_s = 90°, 270°$) is exhibited in Figure 6-10. It can be observed that the maximum scattered field is directed near $\theta_s = 30°$, which is near the direction of specular reflection (defined as the direction along which the angle of reflection is equal to the angle of incidence). For more details see Section 11.3.

By using such a procedure for plates of finite size, the scattering fields are accurate at and near the specular direction. The angular extent over which the accuracy is acceptable increases as the size of the scatterer increases. Other techniques, such as those discussed in Chapters 12 and 13, can be used to improve the accuracy everywhere.

B. Cylindrical Coordinate System   When the radiating or scattering structure is of circular geometry, the radiation or scattering fields can still be found using (6-122a) through (6-122f). The N_θ, N_ϕ, L_θ, and L_ϕ functions must still be obtained from (6-114c) and (6-114d) but must be expressed in a form that is convenient for cylindrical geometries. Although the general procedure of analysis for circular geometry is identical to that of the rectangular, as outlined in the previous section, the primary differences lie in the following.

1. The formulation of the equivalent currents, J_x, J_y, J_z, M_x, M_y, and M_z.
2. The differential paths from the source to the observation point, $r'\cos\psi$.
3. The differential area ds'.

Before we consider an example, we will reformulate these differences for the circular aperture.

Because of the circular profile of the aperture, it is often convenient and desirable to adopt cylindrical coordinates for the solution of the fields. In most cases, therefore, the radiated or scattered electric and magnetic field components over the circular geometry will be known in cylindrical form, that is, E_ρ, E_ϕ, E_z, H_ρ, H_ϕ, and H_z. Thus, the components of the equivalent currents $\mathbf{M}_s$ and $\mathbf{J}_s$ would also be conveniently expressed in cylindrical form, M_ρ, M_ϕ, M_z, J_ρ,

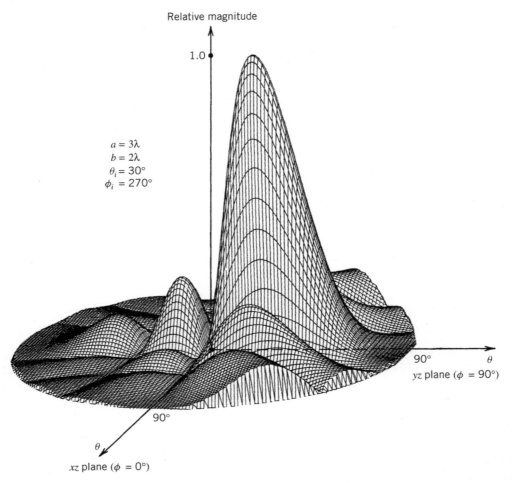

Figure 6-9   Three-dimensional normalized scattering field pattern of a plane wave incident on a rectangular ground plane.

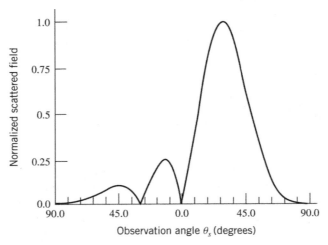

Figure 6-10   Two-dimensional normalized electric field scattering pattern for a plane wave incident ($\theta_i = 30°$ and $\phi_s = 90°, 270°$) on a flat conducting plate with $a = 3\lambda$ and $b = 2\lambda$.

J_ϕ, and J_z. In addition, the required integration over the aperture to find N_θ, N_ϕ, L_θ, and L_ϕ of (6-125a) through (6-125d) should also be done in cylindrical coordinates. It is then desirable to reformulate $r' \cos \psi$ and ds', as given by (6-127a) through (6-128c).

The most convenient position for placing the structure is that shown in Figure 6-11 (structure on xy plane). The transformation between the rectangular and cylindrical components of $\mathbf{J}_s$ is given in Appendix II, equation II-7a, or

$$\begin{bmatrix} J_x \\ J_y \\ J_z \end{bmatrix} = \begin{bmatrix} \cos \phi' & -\sin \phi' & 0 \\ \sin \phi' & \cos \phi' & 0 \\ 0 & 0 & 1 \end{bmatrix} \begin{bmatrix} J_\rho \\ J_\phi \\ J_z \end{bmatrix} \tag{6-130a}$$

A similar transformation exists for the components of $\mathbf{M}_s$. The rectangular and cylindrical coordinates are related by (see Appendix II)

$$\begin{aligned} x' &= \rho' \cos \phi' \\ y' &= \rho' \sin \phi' \\ z' &= z' \end{aligned} \tag{6-130b}$$

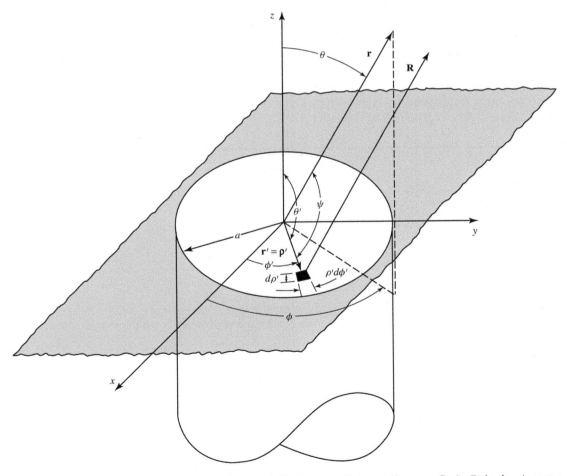

Figure 6-11 Circular aperture mounted on an infinite ground plane. (Source: C. A. Balanis, *Antenna Theory: Analysis and Design*, Third Edition. Copyright © 2005, John Wiley & Sons, Inc. Reprinted by permission of John Wiley & Sons, Inc.)

Using (6-130a), equations 6-125a through 6-125d can be written as

$$N_\theta = \iint_S \left[J_\rho \cos\theta \cos(\phi - \phi') + J_\phi \cos\theta \sin(\phi - \phi') - J_z \sin\theta \right] e^{j\beta r' \cos\psi} \, ds' \tag{6-31a}$$

$$N_\phi = \iint_S \left[-J_\rho \sin(\phi - \phi') + J_\phi \cos(\phi - \phi') \right] e^{j\beta r' \cos\psi} \, ds' \tag{6-31b}$$

$$L_\theta = \iint_S \left[M_\rho \cos\theta \cos(\phi - \phi') + M_\phi \cos\theta \sin(\phi - \phi') - M_z \sin\theta \right] e^{j\beta r' \cos\psi} \, ds' \tag{6-31c}$$

$$L_\phi = \iint_S \left[-M_\rho \sin(\phi - \phi') + M_\phi \cos(\phi - \phi') \right] e^{j\beta r' \cos\psi} \, ds' \tag{6-31d}$$

where $r' \cos\psi$ and ds' can be written, using (6-127c), (6-128c), and (6-130b), as

$$r' \cos\psi = x' \sin\theta \cos\phi + y' \sin\theta \sin\phi = \rho' \sin\theta \cos(\phi - \phi') \tag{6-132a}$$
$$ds' = dx' dy' = \rho' d\rho' d\phi' \tag{6-132b}$$

In summary, for a circular aperture antenna the fields radiated can be obtained by either of the following methods.

1. If the fields over the aperture are known in rectangular components, use the same procedure as for the rectangular aperture except that (6-132a) and (6-132b) should be substituted in (6-125a) through (6-125d).
2. If the fields over the aperture are known in cylindrical components, use the same procedure as for the rectangular aperture with (6-131a) through (6-131d), along (6-132a) and (6-132b), taking the place of (6-125a) through (6-125d).

Example 6-6

To demonstrate the methods, the field radiated by a circular aperture mounted on an infinite ground plane will be formulated. To simplify the mathematical details, the field over the aperture of Figure 6-11 will be assumed to be

$$\left. \begin{array}{l} \mathbf{E}_a = \hat{\mathbf{a}}_y E_0 \\[2mm] \mathbf{H}_a = -\hat{\mathbf{a}}_x \dfrac{E_0}{\eta} \end{array} \right\} \quad \rho' \leq a$$

The objective is to find the far-zone fields radiated by the aperture. The fields over the aperture have been simplified in order to reduce the complexity of the problem and to avoid having the analytical formulations obscure the analysis procedure.

Solution:

1. The surface of the radiating aperture is that defined by $\rho' \leq a$.
2. Since the aperture is mounted on an infinite ground plane, it is shown, by the equivalence principle in Chapter 7 and elsewhere [1, 3, 5], that the equivalent current densities that lead to the appropriate radiated fields are given by

$$\mathbf{M}_s = \begin{cases} -2\hat{\mathbf{n}} \times \mathbf{E}_a = \hat{\mathbf{a}}_x 2E_0 & \rho' \leq a \\ 0 & \text{elsewhere} \end{cases}$$

$$\mathbf{J}_s = 0 \quad \text{elsewhere}$$

This equivalent model for the current densities is valid for any aperture mounted on an infinite perfectly conducting electric ground plane. Thus, according to (6-125a) and (6-125b),

$$N_\theta = N_\phi = 0$$

Using (6-125c), (6-132a), and (6-132b)

$$L_\theta = 2E_0 \cos\theta \cos\phi \int_0^a \rho' \left[\int_0^{2\pi} e^{+j\beta\rho'\sin\theta\cos(\phi-\phi')} d\phi' \right] d\rho'$$

Because

$$\int_0^{2\pi} e^{+j\beta\rho'\sin\theta\cos(\phi-\phi')} d\phi' = 2\pi J_0(\beta\rho' \sin\theta)$$

we can write L_θ as

$$L_\theta = 4\pi E_0 \cos\theta \cos\phi \int_0^a J_0(\beta\rho' \sin\theta)\rho' d\rho'$$

where $J_0(t)$ is the Bessel function of the first kind of order zero. Making the substitution

$$t = \beta\rho' \sin\theta$$
$$dt = \beta \sin\theta d\rho'$$

reduces L_θ to

$$L_\theta = \frac{4\pi E_0 \cos\theta \cos\phi}{(\beta \sin\theta)^2} \int_0^{\beta a \sin\theta} tJ_0(t)dt$$

Since

$$\int_0^\delta zJ_0(z)dz = zJ_1(z) \big|_0^\delta = \delta J_1(\delta)$$

where $J_1(\delta)$ is the Bessel function of order 1, L_θ takes the form

$$L_\theta = 4\pi a^2 E_0 \left\{ \cos\theta \cos\phi \left[\frac{J_1(\beta a \sin\theta)}{\beta a \sin\theta} \right] \right\}$$

Similarly, L_ϕ of (6-125d) reduces to

$$L_\phi = -4\pi a^2 E_0 \left\{ \sin\phi \left[\frac{J_1(\beta a \sin\theta)}{\beta a \sin\theta} \right] \right\}$$

Using N_θ, N_ϕ, L_θ, and L_ϕ previously derived, the electric field components of (6-122a) through (6-122c) can be written as

$$E_r = 0$$
$$E_\theta = j\frac{\beta a^2 E_0 e^{-j\beta r}}{r} \left\{ \sin\phi \left[\frac{J_1(\beta a \sin\theta)}{\beta a \sin\theta} \right] \right\}$$
$$E_\phi = j\frac{\beta a^2 E_0 e^{-j\beta r}}{r} \left\{ \cos\theta \cos\phi \left[\frac{J_1(\beta a \sin\theta)}{\beta a \sin\theta} \right] \right\}$$

In the principal E and H planes, the electric field components simplify to

E Plane $(\phi = \pi/2)$

$$E_r = E_\phi = 0$$
$$E_\theta = j\frac{\beta a^2 E_0 e^{-j\beta r}}{r} \left[\frac{J_1(\beta a \sin\theta)}{\beta a \sin\theta} \right]$$

_H_ Plane ($\phi = 0$)

$$E_r = E_\theta = 0$$

$$E_\phi = j \frac{\beta a^2 E_0 e^{-j\beta r}}{r} \left\{ \cos\theta \left[\frac{J_1(\beta a \sin\theta)}{\beta a \sin\theta} \right] \right\}$$

A three-dimensional plot of the normalized magnitude of the total electric field intensity E $\left(E \simeq \sqrt{E_\theta^2 + E_\phi^2} \right)$ for an aperture of $a = 1.5\lambda$ as a function of θ and ϕ ($0° \le \theta \le 90°$ and $0° \le \phi \le 360°$) is shown plotted in Figure 6-12, and it seems to be symmetrical. However, closer observation, especially through the two-dimensional _E_- and _H_-plane patterns of Figure 6-13, reveals that the pattern is not symmetrical. It does, however, possess characteristics that are almost identical.

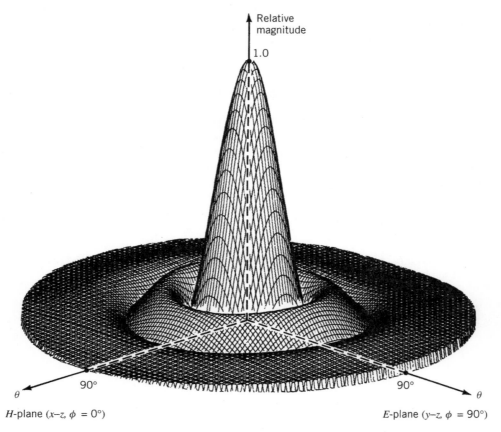

Figure 6-12   Three-dimensional field pattern of a constant field circular aperture mounted on an infinite ground plane ($a = 1.5\lambda$). (Source: C. A. Balanis, _Antenna Theory: Analysis and Design_, Third Edition. Copyright © 2005, John Wiley & Sons, Inc. Reprinted by permission of John Wiley & Sons, Inc.)

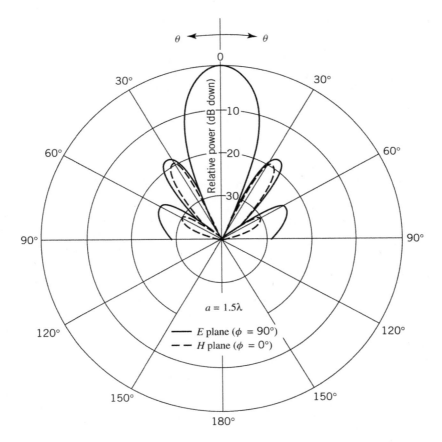

Figure 6-13 E-($\phi = 90°$) and H-plane ($\phi = 0°$) power patterns of a circular aperture with a uniform field distribution.

6.9 MULTIMEDIA

On the website that accompanies this book, the following multimedia resources are included for the review, understanding, and presentation of the material of this chapter.

- **PowerPoint (PPT)** viewgraphs, in multicolor.

REFERENCES

1. C. A. Balanis, *Antenna Theory: Analysis and Design*, Third Edition, John Wiley & Sons, New York, 2005.
2. R. F. Harrington, *Time-Harmonic Electromagnetic Fields*, McGraw-Hill, New York, 1961.
3. J. H. Richmond, The Basic Theory of Harmonic Fields, Antennas and Scattering, Unpublished Notes.
4. R. Mittra (ed.), *Computer Techniques for Electromagnetics*, Chapter 2, Pergamon Press, 1973, pp. 7–95.
5. S. A. Schelkunoff, "Some equivalence theorems of electromagnetics and their application to radiation problems," *Bell System Tech. J.*, vol. 15, pp. 92–112, 1936.

PROBLEMS

6.1. If $\mathbf{H}_e = j\omega\varepsilon\nabla\times\Pi_e$, where Π_e is the electric Hertzian potential, show that
(a) $\nabla^2\Pi_e + \beta^2\Pi_e = j(1/\omega\varepsilon)\mathbf{J}$.
(b) $\mathbf{E}_e = \beta^2\Pi_e + \nabla(\nabla\cdot\Pi_e)$.
(c) $\Pi_e = -j(1/\omega\mu\varepsilon)\mathbf{A}$.

6.2. If $\mathbf{E}_h = -j\omega\mu\nabla\times\Pi_h$, where Π_h is the magnetic Hertzian potential, show that
(a) $\nabla^2\Pi_h + \beta^2\Pi_h = j(1/\omega\mu)\mathbf{M}$.
(b) $\mathbf{H}_h = \beta^2\Pi_h + \nabla(\nabla\cdot\Pi_h)$.
(c) $\Pi_h = -j(1/\omega\mu\varepsilon)\mathbf{F}$.

6.3. Develop expressions for $\mathbf{E}_A$ and $\mathbf{H}_A$ in terms of $\mathbf{J}$ using the path-1 procedure of Figure 6-1. These expressions should be valid everywhere.

6.4. Develop expressions for $\mathbf{E}_F$ and $\mathbf{H}_F$ in terms of $\mathbf{M}$ using the path-1 procedure of Figure 6-1. These expressions should be valid everywhere.

6.5. In rectangular coordinates derive expressions for $\mathbf{E}$ and $\mathbf{H}$, in terms of the components of the $\mathbf{A}$ and $\mathbf{F}$ potentials, that are TEM^x and TEM^y. The procedure should be similar to that of Example 6.1, and it should state all the combinations that lead to the desired modes.

6.6. In cylindrical coordinates derive expressions for $\mathbf{E}$ and $\mathbf{H}$, in terms of the components of the $\mathbf{A}$ and $\mathbf{F}$ potentials, that are TEM^ϕ and TEM^z. The procedure should be similar to that of Example 6-2, and it should state all the combinations that lead to the desired modes.

6.7. Derive the expressions for the components of $\mathbf{E}$ and $\mathbf{H}$ of (6-61) and (6-64), in terms of the components of $\mathbf{A}$ and $\mathbf{F}$, so that the fields are TM^x and TM^y.

6.8. Select one component of $\mathbf{E}$ and write the other components of $\mathbf{E}$ and all of $\mathbf{H}$ in terms of the initial component of $\mathbf{E}$ so that the fields are TM^x, TM^y, and TM^z. Do this in rectangular coordinates.

6.9. For the TE^z modes ($E_z = 0$) in rectangular coordinates, with z variations of the form $e^{-j\beta_z z}$, derive expressions for the $\mathbf{E}$- and $\mathbf{H}$-field rectangular components in terms of $H_z = f(x)g(y)e^{-j\beta_z z}$, where $\nabla^2 H_z(x,y,z) + \beta^2 H_z(x,y,z) = 0$. In other words, instead of expressing the electric and magnetic field components for TE^z modes ($E_z = 0$) in terms

of F_z [as is done in (6-72)], this time you start with H_z not being equal to zero and express all the electric and magnetic field components, except H_z, in terms of H_z. This is an alternate way of finding the TE^z modes. Simplify the expressions. They should be in a form similar to those of (6-72).
Hint: You should use Maxwell's equations back and forth.

6.10. For the TM^z modes ($H_z = 0$) in rectangular coordinates, with z variations of the form $e^{-j\beta_z z}$, derive expressions for the $\mathbf{E}$- and $\mathbf{H}$-field rectangular components in terms of $E_z = f(x)g(y)e^{-j\beta_z z}$, where $\nabla^2 E_z(x,y,z) + \beta^2 E_z(x,y,z) = 0$. In other words, instead of expressing the electric and magnetic field components for TM^z modes ($H_z = 0$) in terms of A_z [as is done in (6-59)], this time you start with E_z not being equal to zero and express all the electric and magnetic field components, except E_z, in terms of E_z. This is an alternate way of finding the TM^z modes. Simplify the expressions. They should be in a form similar to those of (6-59).
Hint: You should use Maxwell's equations back and forth.

6.11. In cylindrical coordinates, select one component of $\mathbf{E}$ and write the other components of $\mathbf{E}$ and all of $\mathbf{H}$ in terms of the initial component of $\mathbf{E}$ so that the fields are TM^z.

6.12. In rectangular coordinates, derive the expressions for the components of $\mathbf{E}$ and $\mathbf{H}$ as given by (6-72), (6-74), and (6-77) that are TE^z, TE^x, and TE^y.

6.13. In cylindrical coordinates, derive the expressions for $\mathbf{E}$ and $\mathbf{H}$ as given by (6-80), which are TE^z.

6.14. Select one component of $\mathbf{H}$ and write the other components of $\mathbf{H}$ and all of $\mathbf{E}$ in terms of the initial component of $\mathbf{H}$ so that the fields are TE^x, TE^y, and TE^z. Do this in rectangular coordinates.

6.15. In cylindrical coordinates, select one component of $\mathbf{H}$ and write the other components of $\mathbf{H}$ and all of $\mathbf{E}$ in terms of the initial component of $\mathbf{H}$ so that the fields are TE^z.

6.16. Verify that (6-85a) and (6-85b) are solutions to (6-84a).

6.17. Show that (6-90) is a solution to (6-87) and that (6-91) is a solution to (6-82).

6.18. For Example 6-3 derive the components of **E**, given the components of **H**.

6.19. Show that for observations made at very large distance $(\beta r \gg 1)$ the electric and magnetic fields of Example 6-3 reduce to

$$E_\theta = j\eta \frac{\beta I_e \ell e^{-j\beta r}}{4\pi r} \sin\theta$$

$$H_\phi \simeq \frac{E_\theta}{\eta}$$

$$E_r \simeq 0$$

$$E_\phi = H_r = H_\theta = 0$$

6.20. For Problem 6.19, show that the:
- Time-average power density is

$$\mathbf{S}_{av} = \frac{1}{2}\text{Re}\left[\mathbf{E}\times\mathbf{H}^*\right] = \hat{\mathbf{a}}_r W_{av} = \hat{\mathbf{a}}_r W_r$$

$$= \hat{\mathbf{a}}_r \frac{\eta}{8}\left|\frac{I_0\ell}{\lambda}\right|^2 \frac{\sin^2\theta}{r^2}$$

- Radiation intensity is

$$U = r^2 S_{av} = \frac{\eta}{8}\left|\frac{I_0\ell}{\lambda}\right|^2 \sin^2\theta$$

- Radiated power is

$$P_{rad} = \int_0^{2\pi}\int_0^\pi U(\theta,\phi)$$

$$\times \sin\theta \, d\theta \, d\phi = \eta\left(\frac{\pi}{3}\right)\left|\frac{I_0\ell}{\lambda}\right|^2$$

- Directivity is $D_0 = \dfrac{4\pi U_{max}(\theta,\phi)}{P_{rad}}$

$$= \frac{3}{2}\text{ (dimensionless)} = 1.761\,\text{dB}$$

- Radiation resistance is

$$R_r = \frac{2P_{rad}}{|I_0|^2} = 80\pi^2\left(\frac{\ell}{\lambda}\right)^2$$

6.21. An infinitesimal electric dipole of length ℓ and constant current I_0 is placed symmetrically about the origin and it is directed along the x axis. Using the procedure outlined in Section 6.7, derive the following expressions for the far zone:
- Magnetic vector potential components (A_r, A_θ, A_ϕ).
- Electric field components (E_r, E_θ, E_ϕ).
- Magnetic field components (H_r, H_θ, H_ϕ).
- Time-average power density as defined in Problem 6.20.
- Radiation intensity as defined in Problem 6.20.
- Directivity as defined in Problem 6.20.
- Radiation resistance as defined in Problem 6.20.

6.22. Repeat the procedure of Problem 6.21 when the electric dipole is directed along the y axis.

6.23. Verify (6-100a) and (6-100b).

6.24. Show that (6-112) reduces to (6-112a) provided $r \geq 2D^2/\lambda$, where D is the largest dimension of the radiator or scatterer. Such an approximation leads to a phase error that is equal to or smaller than 22.5°.

6.25. The current distribution on a very thin wire dipole antenna of overall length ℓ is given by

$$\mathbf{I}_e = \begin{cases} \hat{\mathbf{a}}_z I_0 \sin\left[\beta\left(\frac{\ell}{2}-z'\right)\right] & 0\leq z'\leq\frac{\ell}{2} \\ \hat{\mathbf{a}}_z I_0 \sin\left[\beta\left(\frac{\ell}{2}+z'\right)\right] & -\frac{\ell}{2}\leq z'\leq 0 \end{cases}$$

where I_0 is a constant. Representing the distance R of (6-112) by the far-field approximations of (6-112a) through (6-112b), derive the far-zone electric and magnetic fields radiated by the dipole using (6-97a) and the far-field formulations of Section 6.7.

6.26. Show that the radiated far-zone electric and magnetic fields derived in Problem 6.25 reduce for a half-wavelength dipole $(\ell = \lambda/2)$ to

$$E_\theta \simeq j\eta\frac{I_0 e^{-j\beta r}}{2\pi r}\left[\frac{\cos\left(\frac{\pi}{2}\cos\theta\right)}{\sin\theta}\right]$$

$$H_\phi \simeq \frac{E_\theta}{\eta}$$

$$E_r \simeq E_\phi \simeq H_r \simeq H_\theta \simeq 0$$

6.27. Using the appropriate EM theorems and principles and the solution of Example 6.3:

(a) Write, by inspection, the electric and magnetic fields of a very short $(l \ll \lambda)$ magnetic dipole with a constant magnetic field I_m. State the basic theorem or principle you are using.

(b) Write, by inspection, the electric and magnetic fields of a small magnetic circular loop of radius $a(a \ll \lambda)$ and uniform magnetic current I_m. The fields of a small electric electric circular loop of radius $a(a \ll \lambda)$ and uniform current I_e are given by

$$E_r = E_\theta = H_\phi = 0$$

$$E_\phi = \eta \frac{(\beta a)^2 I_e \sin\theta}{4r}\left[1 + \frac{1}{j\beta r}\right]e^{-j\beta r}$$

$$H_r = j \frac{\beta a^2 I_e \cos\theta}{2r^2}\left[1 + \frac{1}{j\beta r}\right]e^{-j\beta r}$$

$$H_\theta = -\frac{(\beta a)^2 I_e \sin\theta}{4r}\left[1 + \frac{1}{j\beta r} - \frac{1}{(\beta r)^2}\right]e^{-j\beta r}$$

(State the basic theorem or principle you are using.)

6.28. Simplify the expressions of Problem 6.3, if the observations are made in the far field.

6.29. Simplify the expressions of Problem 6.4, if the observations are made in the far field.

6.30. The rectangular aperture of Figure 6-4a is mounted on an infinite ground plane that coincides with the yz plane. Assuming that the tangential field over the aperture is given by

$$E_a = \hat{a}_z E_0, \text{ for } -a/2 \le y' \le a/2$$
$$-b/2 \le z' \le b/2$$

and the equivalent currents are

$$M_s = \begin{cases} -2\hat{n} \times E_a, \text{ for } & -a/2 \le y' \le a/2 \\ & -b/2 \le z' \le b/2 \\ 0 & \text{elsewhere} \end{cases}$$

$$J_s = 0 \quad \text{everywhere}$$

find the far-zone spherical electric and magnetic field components radiated by the aperture.

6.31. Repeat Problem 6.30 when the same aperture is not mounted on an infinite PEC ground plane. For this problem, use both electric and magnetic current densities over the aperture, as was done for Example 6-4. The E- and H-fields at the aperture are related by the

intrinsic impedance, as in Example 6-4, such that $E \times H$ is in the $+x$ direction.

6.32. Repeat Problem 6.30 when the same aperture is analyzed using the coordinate system of Figure 6-4b. The tangential aperture field distribution is given by

$$E_a = \hat{a}_x E_0 \text{ for } -b/2 \le x' \le b/2$$
$$-a/2 \le z' \le a/2$$

and the equivalent currents are

$$M_s = \begin{cases} -2\hat{n} \times E_a, \text{ for } & -b/2 \le x' \le b/2 \\ & -a/2 \le z' \le a/2 \\ 0 & \text{elsewhere} \end{cases}$$

$$J_s = 0 \quad \text{everywhere}$$

6.33. Repeat Problem 6.30 when the same aperture is analyzed using the coordinate system of Figure 6-4c. The tangential aperture field distribution is given by

$$E_a = \hat{a}_x E_0 \text{ for } -a/2 \le x' \le a/2$$
$$-b/2 \le y' \le b/2$$

and the equivalent currents are

$$M_s = \begin{cases} -2\hat{n} \times E_a, & -a/2 \le x' \le a/2 \\ & -b/2 \le y' \le b/2 \\ 0 & \text{elsewhere} \end{cases}$$

$$J_s = 0 \quad \text{everywhere}$$

6.34. The electric and magnetic field components over a circular waveguide aperture of radius a, as shown in Figure 6-11, are given by (E_0 is a constant):

$$\left.\begin{array}{l} E_a = \hat{a}_y E_0 \\ H_a = -\hat{a}_x \dfrac{E_0}{\eta_0} \end{array}\right\} \quad \rho \le a$$

$$E_a = H_a = 0 \qquad \rho > a$$

Assuming the aperture is not mounted on a ground plane, determine the:

a. Approximate vector equivalent electric J_s and magnetic M_s current densities that can be used to represent the waveguide and to find the fields radiated by the aperture.

b. Far-zone electric and magnetic field components radiated by the aperture.

6.35. Repeat Problem 6.30 when the aperture field distribution is given by

$$E_a = \hat{a}_z E_0 \cos\left(\frac{\pi}{a}y'\right), \quad -a/2 \le y' \le a/2$$
$$-b/2 \le z' \le b/2$$

6.36. Repeat Problem 6.32 when the aperture field distribution is given by

$$\mathbf{E}_a = \hat{\mathbf{a}}_x E_0 \cos\left(\frac{\pi}{a} z'\right), \quad -b/2 \le x' \le b/2$$
$$-a/2 \le z' \le a/2$$

6.37. Repeat Problem 6.33 when the aperture field distribution is given by

$$\mathbf{E}_a = \hat{\mathbf{a}}_y E_0 \cos\left(\frac{\pi}{a} x'\right), \quad -a/2 \le x' \le a/2$$
$$-b/2 \le y' \le b/2$$

6.38. For the aperture of Example 6-4, find the angular separation (in degrees) between two points whose radiated electric field value is 0.707 of the maximum (half-power beamwidth). Do this for the radiated fields in the (a) E plane ($\phi = \pi/2$) and (b) H plane ($\phi = 0$). Assume the aperture has dimensions $a = 4\lambda$ and $b = 3\lambda$.

6.39. For the circular aperture of Figure 6-11, derive expressions for the far-zone radiated spherical fields when the aperture field distribution is given by
(a) $\mathbf{E}_a = \hat{\mathbf{a}}_y E_0 [1 - (\rho'/a)^2], \quad \rho' \le a.$
(b) $\mathbf{E}_a = \hat{\mathbf{a}}_y E_0 [1 - (\rho'/a)^2]^2, \quad \rho' \le a.$
For both cases use equivalent currents $\mathbf{M}_s$ and $\mathbf{J}_s$ such that

$$\mathbf{M}_s = \begin{cases} -2\hat{\mathbf{n}} \times \mathbf{E}_a, & \rho' \le a, \\ 0 & \text{elsewhere} \end{cases}$$
$$\mathbf{J}_s = 0 \quad \text{everywhere}$$

6.40. A coaxial line of inner and outer radii a and b, respectively, is mounted on an infinite conducting ground plane. Assuming that the electric field over the aperture of the coax is

$$\mathbf{E}_a = -\hat{\mathbf{a}}_\rho \frac{V}{\varepsilon \ln (b/a)} \frac{1}{\rho'}, \quad a \le \rho' \le b$$

where V is the applied voltage and ε is the permittivity of medium in the coax, find the far-zone spherical electric and magnetic field components radiated by the aperture.

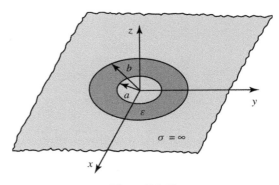

Figure P6-40

Use equivalent currents $\mathbf{M}_s$ and $\mathbf{J}_s$ such that

$$\mathbf{M}_s = \begin{cases} -2\hat{\mathbf{n}} \times \mathbf{E}_a, & a \le \rho' \le b \\ 0 & \text{elsewhere} \end{cases}$$
$$\mathbf{J}_s = 0 \quad \text{everywhere}$$

6.41. Repeat Example 6-5 when the linearly parallel polarized uniform plane wave traveling in free space on the yz plane at an angle θ_i from the z axis is incident upon a rectangular perfect magnetic conductor (PMC) flat plate of dimensions a and b, as shown in the geometry of Figure 6-8. Assuming the induced equivalent current densities, electric $\mathbf{J}_s$ and magnetic $\mathbf{M}_s$, on the plate are the same as those of an infinite PMC flat plate:
- Write simplified expressions for the equivalent current densities $\mathbf{J}_s$ and $\mathbf{M}_s$ on the PMC plate.
- Using the equivalent current densities $\mathbf{J}_s$ and $\mathbf{M}_s$ over the PMC flat plate found in the previous part, determine the far-zone spherical electric and magnetic field components in directions θ_s, ϕ_s scattered by the flat PMC plate.

6.42. For the aperture of Example 6-6, find the angular separation (in degrees) between two points whose radiated electric field value is 0.707 of the maximum (half-power beamwidth). Do this for the radiated fields in the (a) E plane ($\phi = \pi/2$) and (b) H plane ($\phi = 0$). Assume that the radius of the aperture is 3λ.

CHAPTER 7

Electromagnetic Theorems and Principles

7.1 INTRODUCTION

In electromagnetics there are a number of theorems and principles that are fundamental to the understanding of electromagnetic generation, radiation, propagation, scattering, and reception. Many of these are often used to facilitate the solution of interrelated problems. Those that will be discussed here are the theorems of *duality, uniqueness, image, reciprocity, reaction, volume equivalence, surface equivalence, induction,* and *physical equivalent* (*physical optics*). When appropriate, examples will be given to illustrate the principles.

7.2 DUALITY THEOREM

When two equations that describe the behavior of two different variables are of the same mathematical form, their solutions will also be identical. The variables in the two equations that occupy identical positions are known as *dual* quantities, and a solution for one can be formed by a systematic interchange of symbols with the other. This concept is known as the *duality theorem*.

Comparison of (6-30), (6-32a), (6-32b), (6-32c), and (6-95a), respectively, to (6-31), (6-33a), (6-33b), (6-33c), and (6-95b), shows that they are dual equations and their variables are dual quantities. Thus, if we know the solutions to one set ($\mathbf{J} \neq 0$, $\mathbf{M} = 0$), the solutions to the other set ($\mathbf{J} = 0$, $\mathbf{M} \neq 0$) can be formed by a proper interchange of quantities. The dual equations and their dual quantities are listed in Tables 7-1 and 7-2 for electric and magnetic sources, respectively. Duality only serves as a guide to forming mathematical solutions. It can be used in an abstract manner to explain the motion of magnetic charges giving rise to magnetic currents, when compared to their dual quantities of moving electric charges creating electric currents [1]. It must, however, be emphasized that this is purely mathematical in nature since at present there are no known magnetic charges or currents in nature.

Balanis' Advanced Engineering Electromagnetics, Third Edition. Constantine A. Balanis.
© 2024 John Wiley & Sons, Inc. Published 2024 by John Wiley & Sons, Inc.
Companion Website: www.wiley.com/go/balanis/advancedengineeringelectromagnetics3e

---

Example 7-1

A very thin linear magnetic current element of very small length $(\ell \ll \lambda)$, although nonphysically real-izable, is often used to represent the fields of a very small electric loop radiator. It can be shown that the fields radiated by a small linear magnetic current element are identical to those radiated by a small loop whose area is perpendicular to the length of the dipole [2]. Assume that the magnetic dipole is placed at the origin and is symmetric along the z axis with a constant magnetic current of

$$\mathbf{I}_m = \hat{\mathbf{a}}_z I_m$$

Find the fields radiated by the dipole using duality.

Solution: Since the linear magnetic dipole is the dual of the linear electric dipole of Example 6-3, the fields radiated by the magnetic dipole can be written, using the dual quantities of Table 7-2 and the solution of Example 6-3, as

$$E_r = E_\theta = 0$$

$$E_\phi = -j\frac{\beta I_m \ell \sin\theta}{4\pi r}\left(1 + \frac{1}{j\beta R}\right)e^{-j\beta r}$$

$$H_r = \frac{1}{\eta}\frac{I_m \ell \cos\theta}{2\pi r^2}\left(1 + \frac{1}{j\beta r}\right)e^{-j\beta r}$$

$$H_\theta = j\frac{1}{\eta}\frac{\beta I_m \ell \sin\theta}{4\pi r}\left[1 + \frac{1}{j\beta r} - \frac{1}{(\beta r)^2}\right]e^{-j\beta r}$$

$$H_\phi = 0$$

TABLE 7-1   Dual equations for electric ($\mathbf{J}$) and magnetic ($\mathbf{M}$) current sources

Electric sources ($\mathbf{J} \neq 0$, $\mathbf{M} = 0$)	Magnetic sources ($\mathbf{J} = 0$, $\mathbf{M} \neq 0$)
$\nabla \times \mathbf{E}_A = -j\omega\mu\mathbf{H}_A$	$\nabla \times \mathbf{H}_F = j\omega\varepsilon\mathbf{E}_F$
$\nabla \times \mathbf{H}_A = \mathbf{J} + j\omega\varepsilon\mathbf{E}_A$	$-\nabla \times \mathbf{E}_F = \mathbf{M} + j\omega\mu\mathbf{H}_F$
$\nabla^2\mathbf{A} + \beta^2\mathbf{A} = -\mu\mathbf{J}$	$\nabla^2\mathbf{F} + \beta^2\mathbf{F} = -\varepsilon\mathbf{M}$
$\mathbf{A} = \dfrac{\mu}{4\pi}\iiint_V \mathbf{J}\dfrac{e^{-j\beta R}}{R}dv'$	$\mathbf{F} = \dfrac{\varepsilon}{4\pi}\iiint_V \mathbf{M}\dfrac{e^{-j\beta R}}{R}dv'$
$\mathbf{H}_A = \dfrac{1}{\mu}\nabla \times \mathbf{A}$	$\mathbf{E}_F = -\dfrac{1}{\varepsilon}\nabla \times \mathbf{F}$
$\mathbf{H}_A = -j\omega\mathbf{A} - j\dfrac{1}{\omega\mu\varepsilon}\nabla(\nabla \cdot \mathbf{A})$	$\mathbf{H}_F = -j\omega\mathbf{F} - j\dfrac{1}{\omega\mu\varepsilon}\nabla(\nabla \cdot \mathbf{F})$

TABLE 7-2   Dual quantities for electric ($\mathbf{J}$) and magnetic ($\mathbf{M}$) current sources

Electric sources ($\mathbf{J} \neq 0$, $\mathbf{M} = 0$)	Magnetic sources ($\mathbf{J} = 0$, $\mathbf{M} \neq 0$)
$\mathbf{E}_A$	$\mathbf{H}_F$
$\mathbf{H}_A$	$-\mathbf{E}_F$
$\mathbf{J}$	$\mathbf{M}$
$\mathbf{A}$	$\mathbf{F}$
ε	μ
μ	ε
β	β
η	$1/\eta$
$1/\eta$	η

7.3 UNIQUENESS THEOREM

Whenever a problem is solved, it is always gratifying to know that the obtained solution is unique, that is, it is the only solution. If so, we would like to know what conditions or what information is needed to obtain such solutions.

Given the electric and magnetic sources $\mathbf{J}_i$ and $\mathbf{M}_i$, let us assume that the fields generated in a lossy medium of complex constitutive parameters $\dot{\varepsilon}$ and $\dot{\mu}$ within a surface S are $(\mathbf{E}^a, \mathbf{H}^a)$ and $(\mathbf{E}^b, \mathbf{H}^b)$. Each set must satisfy Maxwell's equations

$$-\nabla\times\mathbf{E} = \mathbf{M}_i + j\omega\mu\mathbf{H} \qquad \nabla\times\mathbf{H} = \mathbf{J}_i + \mathbf{J}_c + j\omega\varepsilon\mathbf{E} \tag{7-1}$$

or

$$-\nabla\times\mathbf{E}^a = \mathbf{M}_i + j\omega\dot{\mu}\mathbf{H}^a \qquad \nabla\times\mathbf{H}^a = \mathbf{J}_i + \mathbf{J}_c^a + j\omega\dot{\varepsilon}\mathbf{E}^a \tag{7-1a}$$

$$-\nabla\times\mathbf{E}^b = \mathbf{M}_i + j\omega\dot{\mu}\mathbf{H}^b \qquad \nabla\times\mathbf{H}^b = \mathbf{J}_i + \mathbf{J}_c^b + j\omega\dot{\varepsilon}\mathbf{E}^b \tag{7-1b}$$

Subtracting (7-1b) from (7-1a), we have that

$$-\nabla\times(\mathbf{E}^a - \mathbf{E}^b) = j\omega\dot{\mu}(\mathbf{H}^a - \mathbf{H}^b) \qquad \nabla\times(\mathbf{H}^a - \mathbf{H}^b) = (\sigma + j\omega\dot{\varepsilon})(\mathbf{E}^a - \mathbf{E}^b) \tag{7-2}$$

or

$$\left.\begin{array}{l} -\nabla\times\delta\mathbf{E} = j\omega\dot{\mu}\delta\mathbf{H} = \delta\mathbf{M}_t \\ +\nabla\times\delta\mathbf{H} = (\sigma + j\omega\dot{\varepsilon})\delta\mathbf{E} = \delta\mathbf{J}_t \end{array}\right\} \text{ within } S \tag{7-2a}$$

Thus, the difference fields satisfy the source-free field equations within S. The conditions for uniqueness are those for which $\delta\mathbf{E} = \delta\mathbf{H} = 0$ or $\mathbf{E}^a = \mathbf{E}^b$ and $\mathbf{H}^a = \mathbf{H}^b$.

Let us now apply the conservation-of-energy equation 1-55a using S as the boundary and $\delta\mathbf{E}$, $\delta\mathbf{H}$, $\delta\mathbf{J}_t$, and $\delta\mathbf{M}_t$ as the sources [1]. For a time-harmonic field, (1-55a) can be written as

$$\oiint_S \mathbf{E}\times\mathbf{H}^* \cdot d\mathbf{s} + \iiint_V (\mathbf{E}\cdot\mathbf{J}_t^* + \mathbf{H}^*\cdot\mathbf{M}_t)dv' = 0 \tag{7-3}$$

which for our case must be

$$\oiint_S (\delta\mathbf{E}\times\delta\mathbf{H}^*)\cdot d\mathbf{s} + \iiint_V [\delta\mathbf{E}\cdot(\sigma + j\omega\dot{\varepsilon})^*\delta\mathbf{E}^* + \delta\mathbf{H}^*\cdot(j\omega\dot{\mu})\delta\mathbf{H}]dv' = 0 \tag{7-4}$$

or

$$\oiint_S (\delta\mathbf{E}\times\delta\mathbf{H}^*)\cdot d\mathbf{s} + \iiint_V \left[(\sigma + j\omega\dot{\varepsilon})^*|\delta\mathbf{E}|^2 + (j\omega\dot{\mu})|\delta\mathbf{H}|^2\right]dv' = 0 \tag{7-4a}$$

where

$$(\sigma + j\omega\dot{\varepsilon})^* = [\sigma + j\omega(\varepsilon' - j\varepsilon'')]^* = [(\sigma + \omega\varepsilon'') + j\omega\varepsilon']^* = (\sigma + \omega\varepsilon'') - j\omega\varepsilon' \tag{7-4b}$$

$$j\omega\dot{\mu} = j\omega(\mu' - j\mu'') = \omega\mu'' + j\omega\mu' \tag{7-4c}$$

If we can show that

$$\oiint_S (\delta\mathbf{E}\times\delta\mathbf{H}^*)\cdot d\mathbf{s} = 0 \tag{7-5}$$

then the volume integral must also be zero, or

$$\iiint_V \left[(\sigma + j\omega\dot\varepsilon)^* |\delta\mathbf{E}|^2 + (j\omega\dot\mu)|\delta\mathbf{H}|^2\right] dv'$$

$$= \mathrm{Re} \iiint_V \left[(\sigma + j\omega\dot\varepsilon)^* |\delta\mathbf{E}|^2 + (j\omega\dot\mu)|\delta\mathbf{H}|^2\right] dv'$$

$$+ \mathrm{Im} \iiint_V \left[(\sigma + j\omega\dot\varepsilon)^* |\delta\mathbf{E}|^2 + (j\omega\dot\mu)|\delta\mathbf{H}|^2\right] dv' = 0 \qquad (7\text{-}6)$$

Using (7-4b) and (7-4c), reduce (7-6) to

$$\iiint_V [(\sigma + \omega\varepsilon'')|\delta\mathbf{E}|^2 + \omega\mu''|\delta\mathbf{H}|^2] \, dv' = 0 \qquad (7\text{-}6\text{a})$$

$$\iiint_V [-\omega\varepsilon'|\delta\mathbf{E}|^2 + \omega\mu'|\delta\mathbf{H}|^2] \, dv' = 0 \qquad (7\text{-}6\text{b})$$

Since $\sigma + \omega\varepsilon''$ and $\omega\mu''$ are positive for dissipative media, the only way for (7-6a) to be zero would be for $|\delta\mathbf{E}|^2 = |\delta\mathbf{H}|^2 = 0$ or $\delta\mathbf{E} = \delta\mathbf{H} = 0$. Therefore, we have proved uniqueness. However, all these were based upon the premise that (7-5) applies [1]. Using the vector identity

$$\mathbf{A} \cdot \mathbf{B} \times \mathbf{C} = \mathbf{B} \cdot \mathbf{C} \times \mathbf{A} = \mathbf{C} \cdot \mathbf{A} \times \mathbf{B} \qquad (7\text{-}7)$$

we can write (7-5) as

$$\oiint_S (\delta\mathbf{E} \times \delta\mathbf{H}^*) \cdot \hat{\mathbf{n}} \, da = \oiint_S (\hat{\mathbf{n}} \times \delta\mathbf{E}) \cdot \delta\mathbf{H}^* da = \oiint_S (\delta\mathbf{H}^* \times \hat{\mathbf{n}}) \cdot \delta\mathbf{E} \, da = 0 \qquad (7\text{-}8)$$

If we can state the conditions under which (7-8) is satisfied, then will we have proved uniqueness. This, however, will only be applicable for dissipative media. *We can treat lossless media as special cases of dissipative media as the losses diminish.*

Let us examine some of the important cases where (7-8) is satisfied and uniqueness is obtained in lossy media.

1. A field $(\mathbf{E}, \mathbf{H})$ is unique when $\hat{\mathbf{n}} \times \mathbf{E}$ is specified on S; then $\hat{\mathbf{n}} \times \delta\mathbf{E} = 0$ over S. This results from exact specification of the tangential components of $\mathbf{E}$ and satisfaction of (7-8). No specification of the normal components is necessary.
2. A field $(\mathbf{E}, \mathbf{H})$ is unique when $\hat{\mathbf{n}} \times \mathbf{H}$ is specified on S; then $\hat{\mathbf{n}} \times \delta\mathbf{H} = 0$ over S. This results from exact specification of the tangential components of $\mathbf{H}$ and satisfaction of (7-8). No specification of the normal components is necessary.
3. A field $(\mathbf{E}, \mathbf{H})$ is unique when $\hat{\mathbf{n}} \times \mathbf{E}$ is specified over part of S and $\hat{\mathbf{n}} \times \mathbf{H}$ is specified over the rest of S. No specification on the normal components is necessary.

SUMMARY   *A field in a lossy region, created by sources $\mathbf{J}_i$ and $\mathbf{M}_i$, is unique within the region when one of the following alternatives is specified.*

1. *The tangential components of $\mathbf{E}$ over the boundary.*
2. *The tangential components of $\mathbf{H}$ over the boundary.*
3. *The former over part of the boundary and the latter over the rest of the boundary.*

Note: In general, the uniqueness theorem breaks down for lossless media. To justify uniqueness in this case, the fields in a lossless medium, as the dissipation approaches zero, can be considered to be the limit of the corresponding fields in a lossy medium. In some cases, however, unique solutions for lossless problems can be obtained on their own merits without treating them as special cases of lossy solutions.

7.4 IMAGE THEORY

The presence of an obstacle, especially when it is near the radiating element, can significantly alter the overall radiation properties of the radiating system, as illustrated in Chapter 5. In practice, the most common obstacle that is always present, even in the absence of anything else, is the ground. Any energy from the radiating element directed toward the ground undergoes reflection. The amount of reflected energy and its direction are controlled by the geometry and constitutive parameters of the ground.

In general, the ground is a lossy medium $(\sigma \neq 0)$ whose effective conductivity increases with frequency. Therefore, it should be expected to act as a very good conductor above a certain frequency, depending primarily upon its moisture content. To simplify the analysis, we will assume that the ground is a perfect electric conductor, flat, and infinite in extent. The same procedure can also be used to investigate the characteristics of any radiating element near any other infinite, flat, perfect electric conductor. In practice, it is impossible to have infinite dimensions but we can simulate (electrically) very large obstacles. The effects that finite dimensions have on the radiation properties of a radiating element will be discussed in Chapters 12 through 14.

To analyze the performance of a radiating element near an infinite plane conductor, we will introduce virtual sources (images) that account for the reflections. The discussion here follows that of [2]. As the name implies, these are not real sources but imaginary ones that, in combination with the real sources, form an equivalent system that replaces the actual system for analysis purposes only and gives the same radiated field above the conductor as the actual system itself. Below the conductor the equivalent system does not give the correct field; however, the field there is zero and the equivalent model is not necessary.

To begin our discussion, let us assume that a vertical electric dipole is placed a distance h above an infinite, flat, perfect electric conductor, as shown in Figure 7-1a. Assuming that there is no mutual coupling, energy from the actual source is radiated in all directions in a manner determined by its unbounded medium directional properties. For an observation point P_1, there is a direct wave. In addition, a wave from the actual source radiated toward point Q_{R1} of the interface will undergo reflection with a direction determined by the law of reflection, $\theta_1^r = \theta_1^i$. This follows from the fact that energy in homogeneous media travels in straight lines along the shortest paths. The wave will pass through the observation point P_1 and, by extending its actual path below the interface, it will seem to originate from a virtual source positioned a distance h below the boundary. For another observation point P_2, the point of reflection is Q_{R2} but the virtual source is the same as before. The same conclusions can be drawn for all other points above the interface.

The amount of reflection is generally determined by the constitutive parameters of the medium below the interface relative to those above. For a perfect electric conductor below the interface, the incident wave is completely reflected with zero fields below the boundary. According to the boundary conditions, the tangential components of the electric field must vanish at all points along the interface. This condition is used to determine the polarization of the reflected field, compared to the direct wave, as shown in Figure 7-1b. To excite the polarization of the reflected waves, the virtual source must also be vertical and have polarity in the same direction as the actual source. Thus, a reflection coefficient of $+1$ is required. Since the boundary conditions

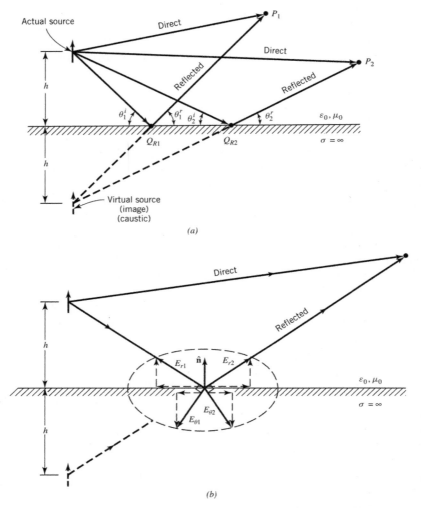

Figure 7-1   Vertical and horizontal dipoles, and their image, for reflection from a flat conducting surface of infinite extent. (*a*) Vertical: Actual source and its image. (*b*) Vertical: Field components at point of reflection. (*c*) Horizontal: Direct and reflected components.

on the tangential electric field components are satisfied over a closed surface, in this case along the interface from $-\infty$ to $+\infty$, then the solution is unique according to the uniqueness theorem of Section 7.3.

Another source orientation is to have the radiating element in a horizontal position, as shown in Figure 7-1*c*. If we follow a procedure similar to that of the vertical dipole, we see that the virtual source (image) is also placed a distance h below the interface but with a 180° polarity difference relative to the actual source, thus requiring a reflection coefficient of -1. Again according to the uniqueness theorem of Section 7.3, the solution is unique because the boundary conditions are satisfied along the closed surface, this time again along the interface extending from $-\infty$ to $+\infty$.

In addition to electric sources, we have equivalent "magnetic" sources and magnetic conductors, such that tangential components of the magnetic field vanish next to their surface. In Figure 7-2*a*,

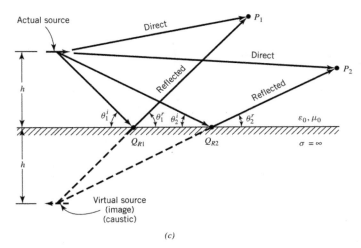

Figure 7-1   (Cont'd)

we have sketched the sources and their images for an electric plane conductor [2]. The single arrow indicates an electric element and the double arrow signifies a magnetic element. The direction of the arrow identifies the polarity. Since many problems can be solved using duality, in Figure 7-2b, we have sketched the sources and their images when the obstacle is an infinite, flat, perfect "magnetic" conductor.

7.4.1   Vertical Electric Dipole

In the previous section we graphically illustrated the analysis procedure, using image theory, for vertical and horizontal electric and magnetic elements near infinite electric and magnetic plane conductors. In this section, we want to derive the mathematical expressions for the fields of a vertical linear element near a perfect electric conductor, and the derivation will be based on the image solution of Figure 7-1a. For simplicity, only far-field [2] observations will be considered.

Let us refer now to the geometry of Figure 7-3a. The far-zone direct component of the electric field, of the infinitesimal dipole of length ℓ, constant current I_0, and observation point P_1, is given according to the dominant terms $(\beta r \gg 1)$ of the fields in Example 6-3 by

$$E_\theta^{d} \overset{r \gg \lambda}{\simeq} j\eta \frac{\beta I_0 \ell e^{-j\beta r_1}}{4\pi r_1} \sin\theta_1 \tag{7-9}$$

The reflected component can be accounted for by the introduction of the virtual source (image), as shown in Figure 7-3a, and we can write it as

$$E_\theta^{r} \overset{r \gg \lambda}{\simeq} jR_v\eta \frac{\beta I_0 \ell e^{-j\beta r_2}}{4\pi r_2} \sin\theta_2$$

$$E_\theta^{r} \overset{r \gg \lambda}{\simeq} j\eta \frac{\beta I_0 \ell e^{-j\beta r_2}}{4\pi r_2} \sin\theta_2 \tag{7-10}$$

since the reflection coefficient R_v is equal to unity.

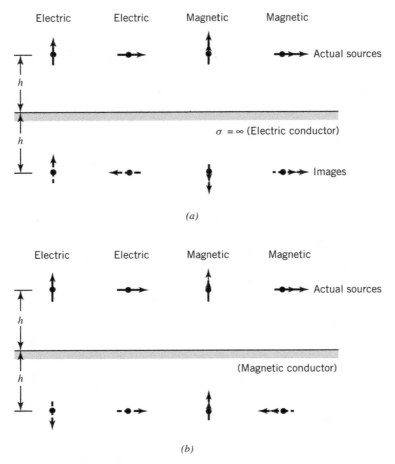

Figure 7-2   Electric and magnetic sources and their images near (a) electric and (b) magnetic conductors. (Source: C. A. Balanis, *Antenna Theory: Analysis and Design*. Third Edition. Copyright © 2005, John Wiley & Sons, Inc. Reprinted by permission of John Wiley & Sons, Inc.)

The total field above the interface ($z \geq 0$) is equal to the sum of the incident and reflected components, as given by (7-9) and (7-10). Since an electric field cannot exist inside a perfect electric conductor, it is equal to zero below the interface. To simplify the expression for the total electric field, we would like to refer it to the origin of the coordinate system ($z = 0$) and express it in terms of r and θ. In general, we can write that

$$r_1 = (r^2 + h^2 - 2rh \cos\theta)^{1/2} \tag{7-11a}$$

$$r_2 = [r^2 + h^2 - 2rh \cos(\pi - \theta)]^{1/2} \tag{7-11b}$$

However, for $r \gg h$ we can simplify and, using the binomial expansion, write [2]

$$\left. \begin{array}{l} r_1 \simeq r - h \cos\theta \\ r_2 \simeq r + h \cos\theta \end{array} \right\} \text{ for phase variations} \tag{7-12a}$$
$$\tag{7-12b}$$

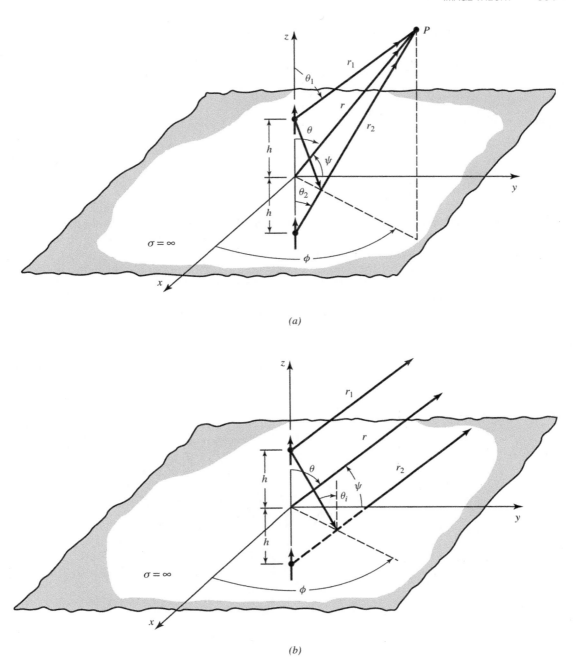

(a)

(b)

Figure 7-3 Vertical electric dipole above an infinite electric conductor. (*a*) Vertical electric dipole. (*b*) Far-field observations. (Source: C. A. Balanis, *Antenna Theory: Analysis and Design*. Third Edition. Copyright © 2005, John Wiley & Sons, Inc. Reprinted by permission of John Wiley & Sons, Inc.)

$$\theta_1 \simeq \theta_2 \simeq \theta \tag{7-12c}$$

As shown in Figure 7-3*b*, (7-12a) and (7-12b) geometrically represent parallel lines. Since the amplitude variations are not as critical,

$$r_1 \simeq r_2 \simeq r \quad \text{for amplitude variations} \tag{7-12d}$$

Use of (7-12a) through (7-12d) allows us to write the sum of (7-9) and (7-10) as

$$E_\theta = E_\theta^d + E_\theta^r \simeq j\eta \frac{\beta I_0 \ell e^{-j\beta r}}{4\pi r} \sin\theta \left(e^{+j\beta h \cos\theta} + e^{-j\beta h \cos\theta} \right) \quad z \geq 0 \tag{7-13}$$

$$E_\theta = 0 \qquad\qquad z < 0$$

which can be reduced to

$$E_\theta = j\eta \frac{\beta I_0 \ell e^{-j\beta r}}{4\pi r} \sin\theta \left[2\cos(\beta h \cos\theta) \right] \quad z \geq 0 \tag{7-13a}$$

$$E_\theta = 0 \qquad\qquad z < 0$$

It is evident that the total electric field is equal to the product of the field of a single source and a factor [within the brackets in (7-13a)] that is a function of the element height, h, and the observation point θ. This product is referred to as the *pattern multiplication* rule, and the factor is known as the *array factor*. More details can be found in Chapter 6 of [2].

The shape and amplitude of the field is not only controlled by the single element but also by the positioning of the element relative to the ground. To examine the field variations as a function of the height h, we have plotted the power patterns for $h = 2\lambda$ and 5λ in Figure 7-4. Because of symmetry, only half of each pattern is shown. It is apparent that the total field pattern is altered

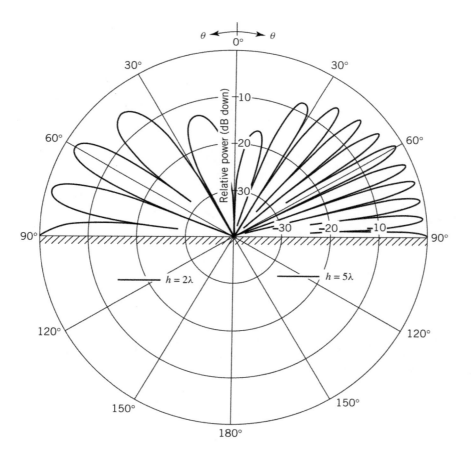

Figure 7-4   Elevation plane amplitude patterns of a vertical infinitesimal electric dipole for heights of 2λ and 5λ above an infinite plane electric conductor.

appreciably by the presence of the ground plane. The height of the element above the interface plays a major role [3, 4]. More details concerning this system configuration can be found in [2].

7.4.2 Horizontal Electric Dipole

Another system configuration is to have the linear antenna placed horizontally relative to the infinite electric ground plane, as shown in Figure 7-5a. The analysis procedure is identical to that of

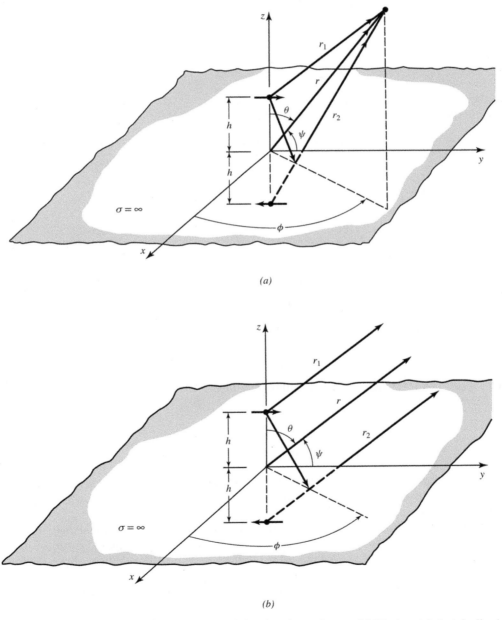

(a)

(b)

Figure 7-5 Horizontal electric dipole above an infinite electric conductor. (*a*) Horizontal electric dipole. (*b*) Far-field observations. (Source: C. A. Balanis, *Antenna Theory: Analysis and Design*. Third Edition. Copyright © 2005, John Wiley & Sons, Inc. Reprinted by permission of John Wiley & Sons, Inc.)

the vertical dipole. By introducing an image and assuming far-field observations, as shown in Figure 7-5*b*, we can write that the dominant terms of the direct component are given by [2]

$$E_\psi^d \overset{r\gg\lambda}{\simeq} j\eta \frac{\beta I_0 \ell e^{-j\beta r_1}}{4\pi r_1} \sin\psi \tag{7-14}$$

and the reflected terms by

$$E_\psi^r \overset{r\gg\lambda}{\simeq} jR_h\eta \frac{\beta I_0 \ell e^{-j\beta r_2}}{4\pi r_2} \sin\psi$$

$$E_\psi^r \overset{r\gg\lambda}{\simeq} -j\eta \frac{\beta I_0 \ell e^{-j\beta r_2}}{4\pi r_2} \sin\psi \tag{7-15}$$

since the reflection coefficient is equal to $R_h = -1$.

To find the angle ψ, which is measured from the y axis toward the observation point, we first form

$$\cos\psi = \hat{\mathbf{a}}_y \cdot \hat{\mathbf{a}}_r = \hat{\mathbf{a}}_y \cdot (\hat{\mathbf{a}}_x \sin\theta\cos\phi + \hat{\mathbf{a}}_y \sin\theta\sin\phi + \hat{\mathbf{a}}_z \cos\theta)$$

$$\cos\psi = \sin\theta\sin\phi \tag{7-16}$$

from which we find

$$\sin\psi = \sqrt{1 - \cos^2\psi} = \sqrt{1 - \sin^2\theta\sin^2\phi} \tag{7-16a}$$

Since for far-field observations

$$\left. \begin{array}{l} r_1 \simeq r - h\cos\theta \\ r_2 \simeq r + h\cos\theta \end{array} \right\} \text{ for phase variations} \tag{7-16b}$$

$$\theta_1 \simeq \theta_2 \simeq \theta \tag{7-16c}$$

$$r_1 \simeq r_2 \simeq r \quad \text{for amplitude variations} \tag{7-16d}$$

we can write the total field, which is valid only above the ground plane ($z \ge 0$, $0 \le \theta \le \pi/2$, $0 \le \phi \le 2\pi$), as

$$E_\psi = E_\psi^d + E_\psi^r = j\eta \frac{\beta I_0 \ell e^{-j\beta r}}{4\pi r} \sqrt{1 - \sin^2\theta \sin^2\phi} [2j\sin(\beta h \cos\theta)] \tag{7-17}$$

Equation 7-17 again is recognized to consist of the product of the field of a single isolated element placed at the origin and a factor (within the brackets) known as the array factor. This, again, is the pattern multiplication rule.

To examine the variations of the total field as a function of the element height above the ground plane, in Figure 7-6 we have plotted two-dimensional elevation plane patterns for $\phi = 90°$ (*yz* plane) when $h = 2\lambda$ and 5λ. Again, we see that the height of the element above the interface plays a significant role in the radiation pattern formation of the radiating system.

Problems that require multiple images, such as corner reflectors, are assigned as exercises at the end of the chapter.

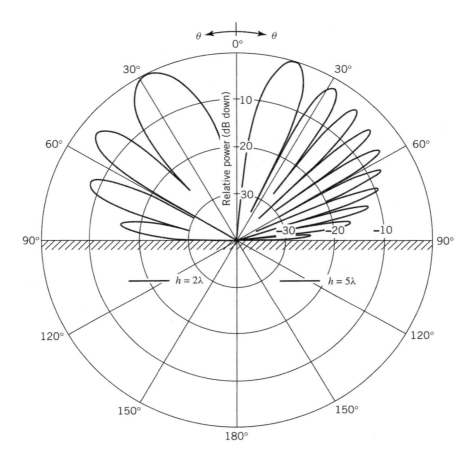

Figure 7-6 Elevation plane ($\phi = 90°$) amplitude patterns of a horizontal infinitesimal dipole for heights of 2λ and 5λ above an infinite plane electric conductor.

7.5 RECIPROCITY THEOREM

We are all well familiar with the reciprocity theorem, as applied to circuits, which states that *in any physical linear network, the positions of an ideal voltage source (zero internal impedance) and an ideal ammeter (infinite internal impedance) can be interchanged without affecting their readings* [5]. Now, we want to discuss the reciprocity theorem as it applies to electromagnetic theory [6]. This is done best by using Maxwell's equations. The reciprocity theorem has many applications; one of the most common relates to the transmitting and receiving properties of radiating systems [2].

Let us assume that within a linear, isotropic medium, which is not necessarily homogeneous, there exist two sets of sources $(\mathbf{J}_1, \mathbf{M}_1)$ and $(\mathbf{J}_2, \mathbf{M}_2)$ that are allowed to radiate simultaneously or individually inside the same medium at the same frequency and produce fields $(\mathbf{E}_1, \mathbf{H}_1)$ and $(\mathbf{E}_2, \mathbf{H}_2)$, respectively. For the fields to be valid, they must satisfy Maxwell's equations

$$\left.\begin{aligned}\nabla \times \mathbf{E}_1 &= -\mathbf{M}_1 - j\omega\mu\mathbf{H}_1\\ \nabla \times \mathbf{H}_1 &= \mathbf{J}_1 + j\omega\varepsilon\mathbf{E}_1\end{aligned}\right\} \text{ for sources } \mathbf{J}_1, \mathbf{M}_1 \qquad \begin{aligned}&(7\text{-}18a)\\ &(7\text{-}18b)\end{aligned}$$

$$\left.\begin{aligned}\nabla \times \mathbf{E}_2 &= -\mathbf{M}_2 - j\omega\mu\mathbf{H}_2\\ \nabla \times \mathbf{H}_2 &= \mathbf{J}_2 + j\omega\varepsilon\mathbf{E}_2\end{aligned}\right\} \text{ for sources } \mathbf{J}_2, \mathbf{M}_2 \qquad \begin{aligned}&(7\text{-}19a)\\ &(7\text{-}19b)\end{aligned}$$

If we dot multiply (7-18a) by $\mathbf{H}_2$ and (7-19b) by $\mathbf{E}_1$, we can write that

$$\mathbf{H}_2 \cdot \nabla \times \mathbf{E}_1 = -\mathbf{H}_2 \cdot \mathbf{M}_1 - j\omega\mu\mathbf{H}_2 \cdot \mathbf{H}_1 \tag{7-20a}$$

$$\mathbf{E}_1 \cdot \nabla \times \mathbf{H}_2 = \mathbf{E}_1 \cdot \mathbf{J}_2 + j\omega\varepsilon\mathbf{E}_1 \cdot \mathbf{E}_2 \tag{7-20b}$$

Subtracting (7-20a) from (7-20b) yields

$$\mathbf{E}_1 \cdot \nabla \times \mathbf{H}_2 - \mathbf{H}_2 \cdot \nabla \times \mathbf{E}_1 = \mathbf{E}_1 \cdot \mathbf{J}_2 + \mathbf{H}_2 \cdot \mathbf{M}_1 + j\omega\varepsilon\mathbf{E}_1 \cdot \mathbf{E}_2 + j\omega\mu\mathbf{H}_2 \cdot \mathbf{H}_1 \tag{7-21}$$

which by use of the vector identity

$$\nabla \cdot (\mathbf{A} \times \mathbf{B}) = \mathbf{B} \cdot (\nabla \times \mathbf{A}) - \mathbf{A} \cdot (\nabla \times \mathbf{B}) \tag{7-22}$$

can be written as

$$\nabla \cdot (\mathbf{H}_2 \times \mathbf{E}_1) = -\nabla \cdot (\mathbf{E}_1 \times \mathbf{H}_2)$$
$$= \mathbf{E}_1 \cdot \mathbf{J}_2 + \mathbf{H}_2 \cdot \mathbf{M}_1 + j\omega\varepsilon\mathbf{E}_1 \cdot \mathbf{E}_2 + j\omega\mu\mathbf{H}_2 \cdot \mathbf{H}_1 \tag{7-23}$$

In a similar manner, if we dot multiply (7-18b) by $\mathbf{E}_2$ and (7-19a) by $\mathbf{H}_1$, we can write

$$\mathbf{E}_2 \cdot \nabla \times \mathbf{H}_1 = \mathbf{E}_2 \cdot \mathbf{J}_1 + j\omega\varepsilon\mathbf{E}_2 \cdot \mathbf{E}_1 \tag{7-24a}$$

$$\mathbf{H}_1 \cdot \nabla \times \mathbf{E}_2 = -\mathbf{H}_1 \cdot \mathbf{M}_2 - j\omega\mu\mathbf{H}_1 \cdot \mathbf{H}_2 \tag{7-24b}$$

Subtraction of (7-24b) from (7-24a) leads to

$$\mathbf{E}_2 \cdot \nabla \times \mathbf{H}_1 - \mathbf{H}_1 \cdot \nabla \times \mathbf{E}_2 = \mathbf{E}_2 \cdot \mathbf{J}_1 + \mathbf{H}_1 \cdot \mathbf{M}_2 + j\omega\varepsilon\mathbf{E}_2 \cdot \mathbf{E}_1 + j\omega\mu\mathbf{H}_1 \cdot \mathbf{H}_2 \tag{7-25}$$

which by use of (7-22) can be written as

$$\nabla \cdot (\mathbf{H}_1 \times \mathbf{E}_2) = -\nabla \cdot (\mathbf{E}_2 \times \mathbf{H}_1)$$
$$= \mathbf{E}_2 \cdot \mathbf{J}_1 + \mathbf{H}_1 \cdot \mathbf{M}_2 + j\omega\mathbf{E}_2 \cdot \mathbf{E}_1 + j\omega\mu\mathbf{H}_1 \cdot \mathbf{H}_2 \tag{7-26}$$

Subtraction of (7-26) from (7-23) leads to

$$\boxed{-\nabla \cdot (\mathbf{E}_1 \times \mathbf{H}_2 - \mathbf{E}_2 \times \mathbf{H}_1) = \mathbf{E}_1 \cdot \mathbf{J}_2 + \mathbf{H}_2 \cdot \mathbf{M}_1 - \mathbf{E}_2 \cdot \mathbf{J}_1 - \mathbf{H}_1 \cdot \mathbf{M}_2} \tag{7-27}$$

which is called the *Lorentz reciprocity theorem* in differential form [7].

By taking a volume integral of both sides of (7-27) and using the divergence theorem on the left side, we can write (7-27) as

$$\boxed{\begin{aligned} -&\oiint_S (\mathbf{E}_1 \times \mathbf{H}_2 - \mathbf{E}_2 \times \mathbf{H}_1) \cdot d\mathbf{s}' \\ &= \iiint_V (\mathbf{E}_1 \cdot \mathbf{J}_2 + \mathbf{H}_2 \cdot \mathbf{M}_1 - \mathbf{E}_2 \cdot \mathbf{J}_1 - \mathbf{H}_1 \cdot \mathbf{M}_2) \, dv' \end{aligned}} \tag{7-28}$$

which is known as the Lorentz reciprocity theorem in integral form.

For a source-free ($\mathbf{J}_1 = \mathbf{J}_2 = \mathbf{M}_1 = \mathbf{M}_2 = 0$) region (7-27) and (7-28) reduce, respectively, to

$$\boxed{\nabla \cdot (\mathbf{E}_1 \times \mathbf{H}_2 - \mathbf{E}_2 \times \mathbf{H}_1) = 0} \tag{7-29}$$

$$\boxed{\oiint_S (\mathbf{E}_1 \times \mathbf{H}_2 - \mathbf{E}_2 \times \mathbf{H}_1) \cdot d\mathbf{s}' = 0} \tag{7-30}$$

Equations 7-29 and 7-30 are special cases of the Lorentz reciprocity theorem and must be satisfied in source-free regions.

As an example of where (7-29) and (7-30) may be applied and what they would represent, consider a section of a waveguide where two different modes exist with fields $(\mathbf{E}_1, \mathbf{H}_1)$ and $(\mathbf{E}_2, \mathbf{H}_2)$. For the expressions of the fields for the two modes to be valid, they must satisfy (7-29) and/or (7-30).

Another useful form of (7-28) is to consider that the fields $(\mathbf{E}_1, \mathbf{H}_1, \mathbf{E}_2, \mathbf{H}_2)$ and the sources $(\mathbf{J}_1, \mathbf{M}_1, \mathbf{J}_2, \mathbf{M}_2)$ are within a medium that is enclosed by a sphere of infinite radius. Assume that the sources are positioned within a finite region and that the fields are observed in the far field (ideally at infinity). Then the left side of (7-28) is equal to zero, or

$$\oiint_S (\mathbf{E}_1 \times \mathbf{H}_2 - \mathbf{E}_2 \times \mathbf{H}_1) \cdot d\mathbf{s}' = 0 \tag{7-31}$$

which reduces (7-28) to

$$\iiint_V (\mathbf{E}_1 \cdot \mathbf{J}_2 + \mathbf{H}_2 \cdot \mathbf{M}_1 - \mathbf{E}_2 \cdot \mathbf{J}_1 - \mathbf{H}_1 \cdot \mathbf{M}_2) \, dv' = 0 \tag{7-32}$$

Equation 7-32 can also be written as

$$\boxed{\iiint_V (\mathbf{E}_1 \cdot \mathbf{J}_2 - \mathbf{H}_1 \cdot \mathbf{M}_2) \, dv' = \iiint_V (\mathbf{E}_2 \cdot \mathbf{J}_1 - \mathbf{H}_2 \cdot \mathbf{M}_1) \, dv'} \tag{7-32a}$$

The reciprocity theorem, as expressed by (7-32a), is the most useful form.

7.6 REACTION THEOREM

Close observation of (7-28) reveals that it does not, in general, represent relations of power because no conjugates appear. The same is true for (7-30) and (7-32a). Each of the integrals in (7-32a) can be interpreted as a coupling between a set of fields and a set of sources, which produce another set of fields. This coupling has been defined as *reaction* [8, 9] and each of the integrals in (7-32a) has been denoted by

$$\langle 1,2 \rangle = \iiint_V (\mathbf{E}_1 \cdot \mathbf{J}_2 - \mathbf{H}_1 \cdot \mathbf{M}_2) \, dv' \tag{7-33a}$$

$$\langle 2,1 \rangle = \iiint_V (\mathbf{E}_2 \cdot \mathbf{J}_1 - \mathbf{H}_2 \cdot \mathbf{M}_1) \, dv' \tag{7-33b}$$

The relation $\langle 1, 2 \rangle$ relates the reaction (coupling) of the fields $(\mathbf{E}_1, \mathbf{H}_1)$, which are produced by sources $(\mathbf{J}_1, \mathbf{M}_1)$, to the sources $(\mathbf{J}_2, \mathbf{M}_2)$, which produce fields $(\mathbf{E}_2, \mathbf{H}_2)$; $\langle 2, 1 \rangle$ relates the reaction (coupling) of the fields $(\mathbf{E}_2, \mathbf{H}_2)$ to the sources $(\mathbf{J}_1, \mathbf{M}_1)$. A requirement for reciprocity to hold is that the reactions (couplings) of the sources with their corresponding fields must be equal. In equation form

$$\langle 1, 2 \rangle = \langle 2, 1 \rangle \tag{7-34}$$

The reaction theorem can also be expressed in terms of the voltages and currents induced in one antenna by another [9]. In a general form, it can be written as

$$\langle i, j \rangle = V_j I_{ji} = \langle j, i \rangle = V_i I_{ij} \tag{7-34a}$$

where V_i = voltage of source $i(j)$
I_{ij} = current through source j due to source at i

The reactions forms of (7-34) and (7-34a) are most convenient to calculate the mutual impedance and admittance between aperture antennas.

Example 7-2

Derive an expression for the mutual admittance between two aperture antennas. The expression should be in terms of the electric and magnetic fields on the apertures and radiated by the apertures.

Solution: In a multiport network, the Y-parameter matrix can be written as

$$[I_i] = [Y_{ij}][V_j]$$

Assuming that the voltages at all ports other than port j are zero, we can write that the current I_{ij} at port i due to the voltage at port j can be written as

$$I_{ij} = Y_{ij}V_j \Rightarrow Y_{ij} = \frac{I_{ij}}{V_j}$$

Using (7-34a), we can write that

$$I_{ij} = \frac{\langle i, j \rangle}{V_i}$$

This allows us to write the mutual admittance as

$$Y_{ij} = \frac{I_{ij}}{V_j} = \frac{\langle i, j \rangle}{V_i V_j}$$

which by using (7-33a) or (7-33b) can be expressed as

$$Y_{ij} = \frac{\langle i, j \rangle}{V_i V_j} = \frac{1}{V_i V_j} \iiint_V (\mathbf{E}_i \cdot \mathbf{J}_j - \mathbf{H}_i \cdot \mathbf{M}_j)\, dv'$$

Since aperture antennas can be represented by magnetic equivalent currents, then

$$\mathbf{J}_j = 0$$
$$\mathbf{M}_j = -\hat{\mathbf{n}} \times \mathbf{E}_j$$

Using these and reducing the volume integral to a surface integral over the aperture of the antenna, we can write the mutual admittance as

$$Y_{ij} = -\frac{1}{V_i V_j} \iint_{S_a} (\mathbf{H}_i \cdot \mathbf{M}_j)\, ds' = -\frac{1}{V_i V_j} \iint_{S_a} [\mathbf{H}_i \cdot (-\hat{\mathbf{n}} \times \mathbf{E}_j)]\, ds'$$

$$Y_{ij} = \frac{1}{V_i V_j} \iint_{S_a} (\mathbf{E}_j \times \mathbf{H}_i) \cdot \hat{\mathbf{n}}\, ds'$$

where $\mathbf{E}_j$ = electric field in aperture j with aperture i shorted
$\quad\ \mathbf{H}_i$ = magnetic field at shorted aperture i due to excitation of aperture j
$\quad\ V_{i(j)}$ = voltage amplitudes at each aperture in the absence of the other

---

7.7 VOLUME EQUIVALENCE THEOREM

Through use of the equivalent electric and magnetic current sources, the volume equivalence theorem can be used to determine the scattered fields when a material obstacle is introduced in a free-space environment where fields $(\mathbf{E}_0, \mathbf{H}_0)$ were previously generated by sources $(\mathbf{J}_i, \mathbf{M}_i)$ [7, 10].

To derive the volume equivalence theorem, let us assume that in the free-space environment sources $(\mathbf{J}_i, \mathbf{M}_i)$ generate fields $(\mathbf{E}_0, \mathbf{H}_0)$. These sources and fields must satisfy Maxwell's equations

$$\nabla \times \mathbf{E}_0 = -\mathbf{M}_i - j\omega\mu_0\mathbf{H}_0 \tag{7-35a}$$

$$\nabla \times \mathbf{H}_0 = \mathbf{J}_i + j\omega\varepsilon_0\mathbf{E}_0 \tag{7-35b}$$

When the same sources $(\mathbf{J}_i, \mathbf{M}_i)$ radiate in a medium represented by (ε, μ), they generate fields $(\mathbf{E}, \mathbf{H})$ that satisfy Maxwell's equations

$$\nabla \times \mathbf{E} = -\mathbf{M}_i - j\omega\mu\mathbf{H} \tag{7-36a}$$

$$\nabla \times \mathbf{H} = \mathbf{J}_i + j\omega\varepsilon\mathbf{E} \tag{7-36b}$$

Subtraction of (7-35a) from (7-36a) and (7-35b) from (7-36b) allows us to write that

$$\nabla \times (\mathbf{E} - \mathbf{E}_0) = -j\omega(\mu\mathbf{H} - \mu_0\mathbf{H}_0) \tag{7-37a}$$

$$\nabla \times (\mathbf{H} - \mathbf{H}_0) = j\omega(\varepsilon\mathbf{E} - \varepsilon_0\mathbf{E}_0) \tag{7-37b}$$

Let us define the difference between the fields $\mathbf{E}$ and $\mathbf{E}_0$, and $\mathbf{H}$ and $\mathbf{H}_0$ as the *scattered* (disturbance) fields $\mathbf{E}^s$ and $\mathbf{H}^s$, that is,

$$\mathbf{E}^s = \mathbf{E} - \mathbf{E}_0 \Rightarrow \mathbf{E}_0 = \mathbf{E} - \mathbf{E}^s \tag{7-38a}$$

$$\mathbf{H}^s = \mathbf{H} - \mathbf{H}_0 \Rightarrow \mathbf{H}_0 = \mathbf{H} - \mathbf{H}^s \tag{7-38b}$$

By using the definitions for the scattered fields of (7-38a) and (7-38b), we can write (7-37a) and (7-37b) as

$$\nabla \times \mathbf{E}^s = -j\omega[\mu\mathbf{H} - \mu_0(\mathbf{H} - \mathbf{H}^s)] = -j\omega(\mu - \mu_0)\mathbf{H} - j\omega\mu_0\mathbf{H}^s \tag{7-39a}$$

$$\nabla \times \mathbf{H}^s = j\omega[\varepsilon\mathbf{E} - \varepsilon_0(\mathbf{E} - \mathbf{E}^s)] = j\omega(\varepsilon - \varepsilon_0)\mathbf{E} + j\omega\varepsilon_0\mathbf{E}^s \tag{7-39b}$$

By defining volume equivalent electric $\mathbf{J}_{eq}$ and magnetic $\mathbf{M}_{eq}$ current densities

$$\boxed{\mathbf{J}_{eq} = j\omega(\varepsilon - \varepsilon_0)\mathbf{E}} \tag{7-40a}$$

$$\boxed{\mathbf{M}_{eq} = j\omega(\mu - \mu_0)\mathbf{H}} \tag{7-40b}$$

which exist only in the region where $\varepsilon \neq \varepsilon_0$ and $\mu \neq \mu_0$ (only in the material itself), we can express (7-39a) and (7-39ab) as

$$\boxed{\nabla \times \mathbf{E}^s = -\mathbf{M}_{eq} - j\omega\mu_0\mathbf{H}^s} \tag{7-41a}$$

$$\boxed{\nabla \times \mathbf{H}^s = \mathbf{J}_{eq} + j\omega\varepsilon_0\mathbf{E}^s} \tag{7-41b}$$

Equations 7-41a and 7-41b state that the electric $\mathbf{E}^s$ and magnetic $\mathbf{H}^s$ fields scattered by a material obstacle can be generated by using equivalent electric $\mathbf{J}_{eq}(\mathrm{A/m}^2)$ and magnetic $\mathbf{M}_{eq}(\mathrm{V/m}^2)$ volume current densities, which are given by (7-40a) and (7-40b), that exist only within the material and radiate in a free-space environment. Although, in principle, the formulation of the problem seems to have been simplified, it is still very difficult to solve because the equivalent current densities are in terms of $\mathbf{E}$ and $\mathbf{H}$, which are unknown. However, the formulation does provide some physical interpretation of scattering and lends itself to development of integral equations for the solution of $\mathbf{E}^s$ and $\mathbf{H}^s$, which are discussed in Chapter 12. The volume equivalent current densities are most useful for finding the fields scattered by dielectric obstacles. The fields scattered by perfectly conducting surfaces can best be determined using surface equivalent densities, especially those discussed in Sections 7.9 and 7.10. The surface equivalence theorem that follows is, usually, best utilized for analysis of antenna aperture radiation.

7.8   SURFACE EQUIVALENCE THEOREM: HUYGENS' PRINCIPLE

The *surface equivalence* theorem is a principle by which actual sources, such as an antenna and transmitter, are replaced by equivalent sources. The fictitious sources are said to be equivalent within a region because they produce within that region the same fields as the actual sources. The formulations of scattering and diffraction problems by the surface equivalence theorem are more suggestive of approximations.

The surface equivalence was introduced in 1936 by Schelkunoff [11], and it is a more rigorous formulation of Huygens' principle [12], which states [13] that "each point on a primary wavefront can be considered to be a new source of a secondary spherical wave and that a secondary wavefront can be constructed as the envelope of these secondary spherical waves." The surface equivalence theorem is based on the uniqueness theorem of Section 7.3, which states [1] that "a field in a lossy region is uniquely specified by the sources within the region plus the tangential components of the electric field over the boundary, or the tangential components of the magnetic field over the boundary, or the former over part of the boundary and the latter over the rest of the boundary." The fields in a lossless medium are considered to be the limit, as the losses go to zero, of the corresponding fields in a lossy medium. Thus, if the tangential electric and magnetic fields are completely known over a closed surface, the fields in the source-free region can be determined.

By the surface equivalence theorem, the fields outside an imaginary closed surface are obtained by placing, over the closed surface, suitable electric and magnetic current densities that satisfy the boundary conditions. The current densities are selected so that the fields inside the closed surface are zero and outside are equal to the radiation produced by the actual sources. Thus, the technique can be used to obtain the fields radiated outside a closed surface by sources enclosed within it. The formulation is exact but requires integration over the closed surface. The degree of accuracy depends on the knowledge of the tangential components of the fields over the closed surface.

In the majority of applications, the closed surface is selected so that most of it coincides with the conducting parts of the physical structure. This is preferred because the tangential electric field components vanish over the conducting parts of the surface, which results in reduction of the physical limits of integration.

The surface equivalence theorem is developed by considering an actual radiating source, which is represented electrically by current densities $\mathbf{J}_1$ and $\mathbf{M}_1$, as shown in Figure 7-7a. The source radiates fields $\mathbf{E}_1$ and $\mathbf{H}_1$ everywhere. However, we wish to develop a method that will yield the fields outside a closed surface. To accomplish this, a closed surface S is chosen, shown dashed in Figure 7-7a, which encloses the current densities $\mathbf{J}_1$ and $\mathbf{M}_1$. The volume within S is denoted by V_1 and outside S by V_2. The primary task is to replace the original problem, shown in Figure 7-7a, with an equivalent that will yield the same fields $\mathbf{E}_1$ and $\mathbf{H}_1$ outside S (within V_2). The formulation of the problem can be aided immensely if the closed surface is chosen judiciously so that fields over most, if not the entire surface, are known *a priori*.

An equivalent problem to Figure 7-7a is shown in Figure 7-7b. The original sources $\mathbf{J}_1$ and $\mathbf{M}_1$ are removed, and we assume that there exist fields $(\mathbf{E}, \mathbf{H})$ inside S and fields $(\mathbf{E}_1, \mathbf{H}_1)$ outside S. For these fields to exist within and outside S, they must satisfy the boundary conditions on the

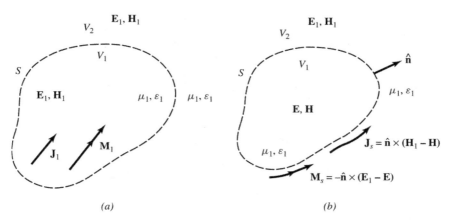

Figure 7-7 (a) Actual and (b) equivalent problem models. (Source: C. A. Balanis, *Antenna Theory: Analysis and Design*. Third Edition. Copyright © 2005, John Wiley & Sons, Inc. Reprinted with permission of John Wiley & Sons, Inc.)

tangential electric and magnetic field components of Table 1-5. Thus, on the imaginary surface S there must exist the equivalent sources

$$\mathbf{J}_s = \hat{\mathbf{n}} \times (\mathbf{H}_1 - \mathbf{H}) \tag{7-42a}$$

$$\mathbf{M}_s = -\hat{\mathbf{n}} \times (\mathbf{E}_1 - \mathbf{E}) \tag{7-42b}$$

which radiate into an unbounded space (same medium everywhere). The current densities of (7-42a) and (7-42b) are said to be equivalent only within V_2, because they will produce the original field $(\mathbf{E}_1, \mathbf{H}_1)$ only outside S. A field $(\mathbf{E}, \mathbf{H})$, different from the original $(\mathbf{E}_1, \mathbf{H}_1)$, will result within V_1. Since the currents of (7-42a) and (7-42b) radiate in an unbounded space, the fields can be determined using (6-30) through (6-35a) and the geometry of Figure 6-3a. In Figure 6-3a, R is the distance from any point on the surface S, where $\mathbf{J}_s$ and $\mathbf{M}_s$ exist, to the observation point.

So far, the tangential components of both $\mathbf{E}$ and $\mathbf{H}$ have been used to set up the equivalent problem. From electromagnetic uniqueness concepts, we know that the tangential components of only $\mathbf{E}$ or $\mathbf{H}$ are needed to determine the field. It will be demonstrated that equivalent problems that require only magnetic currents (tangential $\mathbf{E}$) or only electric currents (tangential $\mathbf{H}$) can be found. This will require modifications to the equivalent problem of Figure 7-7b.

Since the fields $(\mathbf{E}, \mathbf{H})$ within S, which is not the region of interest, can be anything, it can be assumed that they are zero. Then the equivalent problem of Figure 7-7b reduces to that of Figure 7-8a with equivalent current densities equal to

$$\mathbf{J}_s = \hat{\mathbf{n}} \times (\mathbf{H}_1 - \mathbf{H})|_{\mathbf{H}=0} = \hat{\mathbf{n}} \times \mathbf{H}_1 \tag{7-43a}$$

$$\mathbf{M}_s = -\hat{\mathbf{n}} \times (\mathbf{E}_1 - \mathbf{E})|_{\mathbf{E}=0} = -\hat{\mathbf{n}} \times \mathbf{E}_1 \tag{7-43b}$$

This form of the field equivalence principle is known as *Love's equivalence principle* [7, 14]. Since the current densities of (7-43a) and (7-43b) radiate in an unbounded medium, that is, have the same μ_1, ε_1 everywhere, they can be used in conjunction with (6-30) through (6-35a) to find the fields everywhere.

Love's equivalence principle in Figure 7-8a produces a null field within the imaginary surface S. Since the value of the $\mathbf{E} = \mathbf{H} = 0$ within S cannot be disturbed if the properties of the medium within it are changed, let us assume that it is replaced by a perfect electric conductor (PEC, $\sigma = \infty$). The introduction of the perfect conductor will have an effect on the equivalent source $\mathbf{J}_s$, and it will

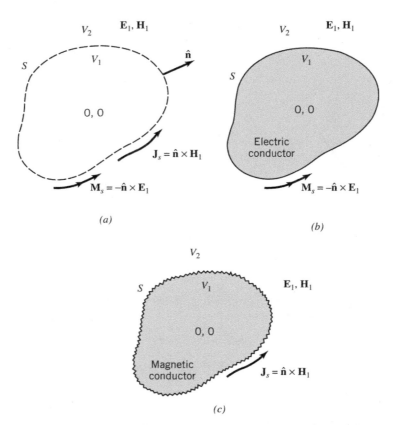

Figure 7-8 Equivalence principle models. (*a*) Love's equivalent. (*b*) Electric conductor equivalent. (*c*) Magnetic conductor equivalent. (Source: C. A. Balanis, *Antenna Theory: Analysis and Design*. Third Edition. Copyright © 2005, John Wiley & Sons, Inc. Reprinted with permission of John Wiley & Sons, Inc.)

prohibit the use of (6-30) through (6-35a) because the current densities no longer radiate into an unbounded medium. Imagine that the geometrical configuration of the electric conductor is identical to the profile of the imaginary surface S, over which $\mathbf{J}_s$ and $\mathbf{M}_s$ exist. As the electric conductor takes its place, as shown in Figure 7-8*b*, *according to the Uniqueness Theorem of Section 7.3*, the equivalent problem of Figure 7-8*a* reduces to that of Figure 7-8*b*. Only a magnetic current density $\mathbf{M}_s$ (tangential component of electric field) is necessary over the entire S, and it radiates in the presence of the electric conductor producing the original fields $\mathbf{E}_1$, $\mathbf{H}_1$ outside S. Within S the fields are zero but, as before, this is not a region of interest. The difficulty in trying to use the equivalent problem of Figure 7-8*b* is that (6-30) through (6-35a) cannot be used, because the current densities do not radiate into an unbounded medium. The problem of a magnetic current radiating in the presence of an electric conducting surface must be solved. Therefore, it seems that the equivalent problem is just as difficult as the original problem.

Before some special simple geometries are considered and some suggestions are made for approximating complex geometries, let us introduce another equivalent problem. Refer to Figure 7-8*a* and assume that instead of placing a perfect electric conductor within S, we introduce a perfect magnetic conductor (PMC). Again, *according to the Uniqueness Theorem of Section 7.3*, the equivalent problem of Figure 7-8*a* reduces to that shown in Figure 7-8*c* (requires only a $\mathbf{J}_s$ over the entire surface S, i.e., tangential components of the magnetic field). Coincident with the equivalent problem of Figure 7-8*b*, (6-30) through (6-35a) cannot be used with Figure 7-8*c*, and the problem is just as difficult as that of Figure 7-8*b* or the original Figure 7-7*a*.

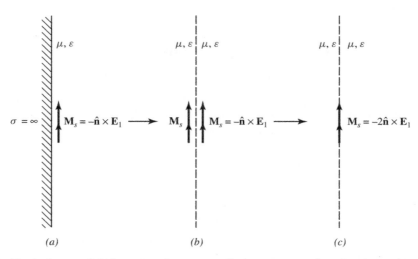

Figure 7-9 Equivalent models for magnetic source radiation near a perfect electric conductor. (Source: C. A. Balanis, *Antenna Theory: Analysis and Design*. Third Edition. Copyright © 2005, John Wiley & Sons, Inc. Reprinted with permission of John Wiley & Sons, Inc.)

To initiate awareness of the utility of the field equivalence principle, especially that of Figure 7-8b, let us assume that the surface of the electric conductor is flat and extends to infinity, as shown in Figure 7-9a. For this geometry, the problem is to determine how a magnetic source radiates in the presence of a flat electric conductor. From image theory, this problem reduces to that of Figure 7-9b, where an imaginary source is introduced on the side of the conductor and takes its place (removes the conductor). Since the imaginary source is in the same direction as the equivalent source, the equivalent problem of Figure 7-9b reduces to that of Figure 7-9c. The magnetic current is doubled, it radiates in an unbounded medium, and (6-30) through (6-35a) can be used. The equivalent problem of Figure 7-9c will yield the correct $(\mathbf{E}, \mathbf{H})$ fields to the right side of the interface. If the surface of the obstacle is not flat and infinite, but its curvature is large compared to the wavelength, a good approximation will be the equivalent problem of Figure 7-9c.

SUMMARY   In the analysis of electromagnetic problems, many times it is easier to form equivalent problems that will yield the same solution within a region of interest. This is true for scattering, diffraction, and aperture antenna radiation problems. In this section, the main emphasis is on aperture antennas, and concepts will be demonstrated by examples.

The following steps must be used to form an equivalent problem to solve an aperture problem.

1. Select an imaginary surface that encloses the actual sources (the aperture). The surface must be chosen judiciously so that the tangential components of the electric and/or the magnetic field are known, ideally exactly (or approximately), over its entire span. In many cases this surface is a flat plane extending to infinity.
2. Over the imaginary surface, form equivalent current densities $\mathbf{J}_s$, $\mathbf{M}_s$, that take one of the following forms.
   a. $\mathbf{J}_s$ and $\mathbf{M}_s$ over S assuming that the $\mathbf{E}$ and $\mathbf{H}$ fields within S are not zero.
   b. $\mathbf{J}_s$ and $\mathbf{M}_s$ over S assuming that the $\mathbf{E}$ and $\mathbf{H}$ fields within S are zero (Love's theorem).
   c. $\mathbf{M}_s$ over S ($\mathbf{J}_s = 0$) assuming that within S the medium is a perfect electric conductor.
   d. $\mathbf{J}_s$ over S ($\mathbf{M}_s = 0$) assuming that within S the medium is a perfect magnetic conductor.

3. Solve the equivalent problem. For equivalents (a) and (b), equations 6-30 through 6-35a can be used. For form (c), the problem of a magnetic current source next to a perfect electric conductor must be solved [(6-30) through (6-35a) cannot be used directly, because the current density does not radiate into an unbounded medium]. If the electric conductor is an infinite flat plane, the problem can be solved exactly by image theory. For form (d), the problem of an electric current source next to a perfect magnetic conductor must be solved. Again (6-30) through (6-35a) cannot be used directly. If the magnetic conductor is an infinite flat plane, the problem can be solved exactly by image theory.

To demonstrate the usefulness and application of the field equivalence theorem to aperture antenna radiation, we consider the following example.

Example 7-3

A waveguide aperture is mounted on an infinite ground plane, as shown in Figure 7-10a. Assume that the tangential components of the electric field over the aperture are known and are given by $\mathbf{E}_a$. Then find an equivalent problem that will yield the same fields $\mathbf{E}$, $\mathbf{H}$ radiated by the aperture to the right side of the interface.

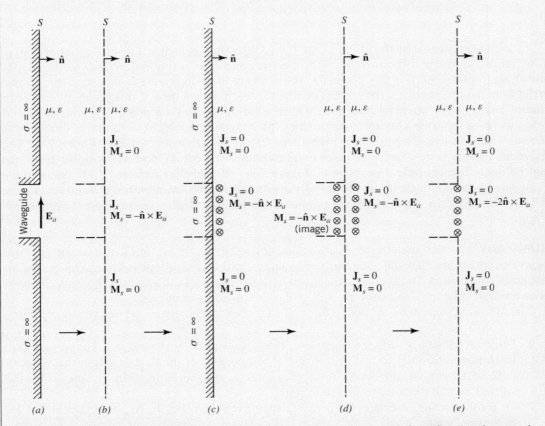

Figure 7-10 Equivalent models for a waveguide aperture mounted on an infinite flat electric ground plane. (Source: C. A. Balanis, *Antenna Theory: Analysis and Design.* Third Edition. Copyright © 2005, John Wiley & Sons, Inc. Reprinted with permission of John Wiley & Sons, Inc.)

Solution: First an imaginary closed surface is chosen. For this problem it is appropriate to select a flat plane extending from $-\infty$ to $+\infty$ as shown in Figure 7-10*b*. Over the infinite plane, the equivalent current densities $\mathbf{J}_s$ and $\mathbf{M}_s$ are formed. Since the tangential components of $\mathbf{E}$ do not exist outside the aperture, because of vanishing boundary conditions, the magnetic current density $\mathbf{M}_s$ is only nonzero over the aperture. The electric current density $\mathbf{J}_s$ is nonzero everywhere and is yet unknown. Now let us assume that an imaginary flat electric conductor is placed next to the surface S and it shorts out the current density $\mathbf{J}_s$ everywhere. $\mathbf{M}_s$ exists only over the space occupied originally by the aperture, and it radiates in the presence of the conductor (see Figure 7-10*c*). By image theory, the conductor can be removed and replaced by an imaginary (equivalent) source $\mathbf{M}_s$, as shown in Figure 7-10*d*, which is analogous to Figure 7-9*b*. Finally, the equivalent problem of Figure 7-10*d* reduces to that of Figure 7-10*e*, which is analogous to that of Figure 7-9*c*. The original problem has been reduced to a very simple equivalent, and (6-30) through (6-35a) can be utilized for its solution. For far-field observations, the radiation integrals of Section 6.8.2 can be used instead.

7.9 INDUCTION THEOREM (INDUCTION EQUIVALENT)

Let us now consider a theorem that is closely related to the surface equivalence theorem. It is, however, used more for scattering than for aperture radiation. Equivalent electric and magnetic current densities are introduced to replace physical obstacles. Figure 7-11*a* shows sources $(\mathbf{J}_1, \mathbf{M}_1)$ in an unbounded medium with constitutive parameters μ_1 and ε_1 and radiating fields $(\mathbf{E}_1, \mathbf{H}_1)$ everywhere, including the region V_1 enclosed by the imaginary surface S_1.

Now let us assume that the space within the imaginary surface S_1 is being replaced by another medium with constitutive parameters μ_2, ε_2, which are different from those of the medium outside S_1, as shown in Figure 7-11*b*. The same sources $(\mathbf{J}_1, \mathbf{M}_1)$, embedded in the original medium (μ_1, ε_1) outside S_1 are now allowed to radiate in the presence of the obstacle that is occupying region V_1. The total field outside region V_1, produced by the sources $(\mathbf{J}_1, \mathbf{M}_1)$, is $(\mathbf{E}, \mathbf{H})$ and inside V_1 is $(\mathbf{E}^t, \mathbf{H}^t)$.

The total field outside V_1 is equal to the original field in the absence of the obstacle $(\mathbf{E}_1, \mathbf{H}_1)$ plus a perturbation field $(\mathbf{E}^s, \mathbf{H}^s)$, usually referred to as *scattered field*, introduced by the obstacle.

In equation form, we can write

$$\mathbf{E} = \mathbf{E}_1 + \mathbf{E}^s \tag{7-44a}$$

$$\mathbf{H} = \mathbf{H}_1 + \mathbf{H}^s \tag{7-44b}$$

where $\mathbf{E}, \mathbf{H}$ = total electric and magnetic fields in the presence of the obstacle

$\quad\quad \mathbf{E}_1, \mathbf{H}_1$ = total electric and magnetic fields in the absence of the obstacle

$\quad\quad \mathbf{E}^s, \mathbf{H}^s$ = scattered (perturbed) electric and magnetic fields due to the obstacle

It is assumed here that the original fields $(\mathbf{E}_1, \mathbf{H}_1)$, in the absence of the obstacle, can be found everywhere. Here we intend to compute $(\mathbf{E}, \mathbf{H})$ outside V_1 and $(\mathbf{E}^t, \mathbf{H}^t)$ inside V_1. It should be pointed out, however, that total $(\mathbf{E}, \mathbf{H})$ can be found if we can determine $(\mathbf{E}^s, \mathbf{H}^s)$, which when added to $(\mathbf{E}_1, \mathbf{H}_1)$ will give $(\mathbf{E}, \mathbf{H})$ [through (7-44a) and (7-44b)].

Let us now formulate an *equivalent* problem that will allow us to determine $(\mathbf{E}^s, \mathbf{H}^s)$ outside V_1 and $(\mathbf{E}^t, \mathbf{H}^t)$ inside V_1. Figure 7-11*c* shows the obstacle occupying region V_1 with fields $(\mathbf{E}^s, \mathbf{H}^s)$ and $(\mathbf{E}^t, \mathbf{H}^t)$ outside and inside V_1, respectively. To support such fields and satisfy the boundary conditions of Table 1-5, we must introduce equivalent current densities $(\mathbf{J}_i, \mathbf{M}_i)$ *on the boundary* such that

$$\mathbf{J}_i = \hat{\mathbf{n}} \times (\mathbf{H}^s - \mathbf{H}^t) \tag{7-45a}$$

$$\mathbf{M}_i = -\hat{\mathbf{n}} \times (\mathbf{E}^s - \mathbf{E}^t) \tag{7-45b}$$

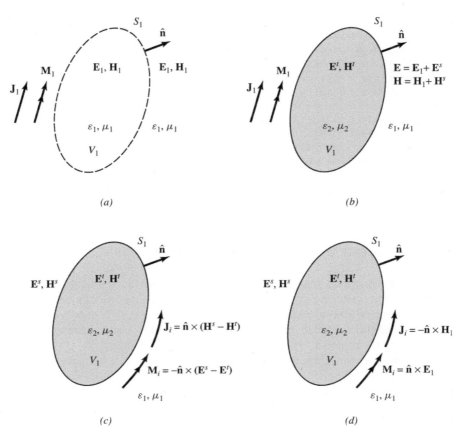

Figure 7-11 Field geometry for the induction theorem (equivalent).

Remember that $(\mathbf{E}^s, \mathbf{H}^s)$ are solutions to Maxwell's equations outside V_1 and $(\mathbf{E}^t, \mathbf{H}^t)$ are solutions within V_1. Therefore, we retain the corresponding media outside and inside V_1.

From Figure 7-11b, we also know that the tangential components of $\mathbf{E}$ and $\mathbf{H}$ must be continuous across the boundary, that is,

$$\mathbf{E}_1\,|_{\tan} + \mathbf{E}^s\,|_{\tan} = \mathbf{E}^t\,|_{\tan} \Rightarrow \hat{\mathbf{n}} \times (\mathbf{E}_1 + \mathbf{E}^s) = \hat{\mathbf{n}} \times \mathbf{E}^t \tag{7-46a}$$

$$\mathbf{H}_1\,|_{\tan} + \mathbf{H}^s\,|_{\tan} = \mathbf{H}^t\,|_{\tan} \Rightarrow \hat{\mathbf{n}} \times (\mathbf{H}_1 + \mathbf{H}^s) = \hat{\mathbf{n}} \times \mathbf{H}^t \tag{7-46b}$$

which can also be written as

$$\mathbf{E}^s\,|_{\tan} - \mathbf{E}^t\,|_{\tan} = -\mathbf{E}_1\,|_{\tan} \Rightarrow \hat{\mathbf{n}} \times (\mathbf{E}^s - \mathbf{E}^t) = -\hat{\mathbf{n}} \times \mathbf{E}_1 \tag{7-47a}$$

$$\mathbf{H}^s\,|_{\tan} - \mathbf{H}^t\,|_{\tan} = -\mathbf{H}_1\,|_{\tan} \Rightarrow \hat{\mathbf{n}} \times (\mathbf{H}^s - \mathbf{H}^t) = -\hat{\mathbf{n}} \times \mathbf{H}_1 \tag{7-47b}$$

Substitution of (7-47a) into (7-45b) and (7-47b) into (7-45a), allows us to write the equivalent currents as

$$\mathbf{J}_i = -\hat{\mathbf{n}} \times \mathbf{H}_1 \tag{7-48a}$$

$$\mathbf{M}_i = \hat{\mathbf{n}} \times \mathbf{E}_1 \tag{7-48b}$$

Now it is quite clear that the equivalent sources of Figure 7-11c have been written, as shown in (7-48a) and (7-48b), in terms of the tangential components $(-\hat{\mathbf{n}} \times \mathbf{H}_1, \hat{\mathbf{n}} \times \mathbf{E}_1)$ of the known fields $\mathbf{E}_1$ and $\mathbf{H}_1$ over the surface occupied by the obstacle.

The equivalent problem of Figure 7-11c is then further reduced to the equivalent problem shown in Figure 7-11d. In words, the equivalent problem of Figure 7-11d states that the scattered fields $(\mathbf{E}^s, \mathbf{H}^s)$ outside V_1 and the transmitted fields $(\mathbf{E}^t, \mathbf{H}^t)$ inside V_1 can be computed by placing, along the boundary of the obstacle, equivalent current densities given by (7-48a) and (7-48b) that *radiate in the presence of the obstacle that is occupying region V_1 and that outside V_1 have the original medium (μ_1, ε_1)*. The equivalent problem of Figure 7-11d is now no simpler to solve than the original problem because we cannot use (6-30) through (6-35a), which assume that we have the same medium everywhere. The equivalent of Figure 7-11d has two media: ε_2, μ_2 inside and ε_1, μ_1 outside. However, even though the equivalent problem of Figure 7-11d is just as difficult to solve exactly as that of Figure 7-11b, it does suggest approximate solutions as will be shown later. We call the problem of Figure 7-11d an *induction equivalent* [1].

Let us now assume that the obstacle occupying region V_1 is a perfect electric conductor (PEC) with $\sigma = \infty$. Again we have the medium with parameters (μ_1, ε_1) outside V_1 and the sources $(\mathbf{J}_1, \mathbf{M}_1)$ radiating in the presence of the conductor (obstacle), as shown in Figure 7-12a. We now need to determine the scattered fields $(\mathbf{E}^s, \mathbf{H}^s)$, which, when added to $(\mathbf{E}_1, \mathbf{H}_1)$, will allow us to determine the total fields $(\mathbf{E}, \mathbf{H})$.

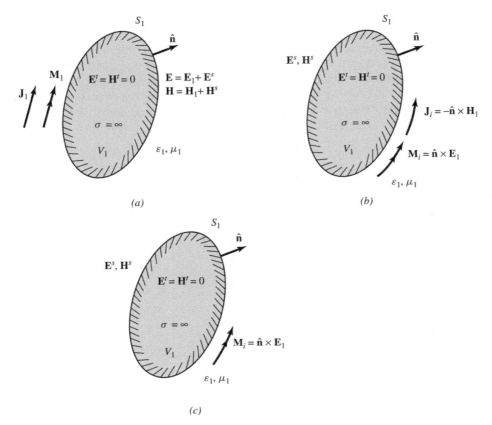

Figure 7-12   Induction equivalents for perfect electric conductor (PEC) scattering.

To compute $(\mathbf{E}^s, \mathbf{H}^s)$ outside V_1 and $\mathbf{E}^t = \mathbf{H}^t = 0$ inside V_1, we form the equivalent problem of Figure 7-12b, analogous to that of Figure 7-11d, with equivalent sources $\mathbf{J}_i = -\hat{\mathbf{n}} \times \mathbf{H}_1$ and $\mathbf{M}_i = \hat{\mathbf{n}} \times \mathbf{E}_1$ over the boundary. The equivalent problem of Figure 7-12b states that the perturbed fields $\mathbf{E}^s$ and $\mathbf{H}^s$ scattered by the perfect conductor of Figure 7-12a can be computed by placing equivalent current densities $\mathbf{J}_i$ and $\mathbf{M}_i$ given by

$$\mathbf{J}_i = -\hat{\mathbf{n}} \times \mathbf{H}_1 \qquad (7\text{-}49a)$$

$$\mathbf{M}_i = \hat{\mathbf{n}} \times \mathbf{E}_1 \qquad (7\text{-}49b)$$

along the boundary of the conductor and radiating in its presence. However, due to the uniqueness theorem, we do not need to specify both the electric current density $\mathbf{J}_i$ (tangential magnetic field) and the magnetic current density $\mathbf{M}_i$ (tangential electric field). Therefore keeping only the magnetic current density $\mathbf{M}_i$, the equivalent of Figure 7-12b reduces to that of Figure 7-12c. The problem of 7-12c is an *induction equivalent* for a perfect electric conductor scatterer.

When the surface S_1 is of complex geometry, the exact solution to the equivalent problem of Figure 7-12c is no easier to compute than the original one shown in Figure 7-12a. However, if the obstacle is an infinite, flat, perfect electric conductor (infinite ground plane), then the equivalent problem for computing the scattered fields is that shown in Figure 7-13a. The exact solution to the equivalent problem of Figure 7-13a of infinite dimension can be obtained by image theory, which allows us to reduce the equivalent problem of Figure 7-13a to that of Figure 7-13b. The equivalent problem of Figure 7-13b permits the solution for the scattered field $\mathbf{E}^s$, $\mathbf{H}^s$ reflected by the perfect electric conductor. The fields radiated by the equivalent source of Figure 7-13b can be obtained by using (6-30) through (6-35a) since we have one medium (μ_1, ε_1 everywhere). The fields obtained using the equivalent problem of Figure 7-13b give nearly the correct answers for the scattered field, for a finite but electrically large plate, only for the region to the left of the boundary S_1 since the flat plate has a strong backscattered field toward that direction. To find the field everywhere and more accurately using the *Induction Equivalent*, for the PEC plate of Figure 7-13a but with finite dimensions, a current density must be placed on each side of the PEC plate, as shown in Figure 7-13c; the current densities must have opposite directions.

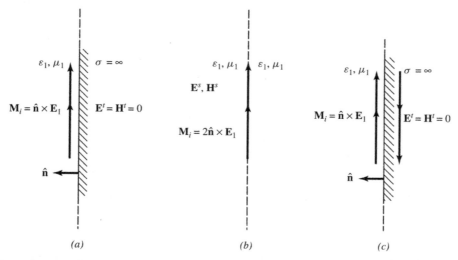

Figure 7-13   Induction equivalent for scattering by flat conducting surface of infinite extent.

7.10 PHYSICAL EQUIVALENT AND PHYSICAL OPTICS EQUIVALENT

The problem of Figure 7-12a, scattering of $(\mathbf{E}, \mathbf{H})$ by a perfect electric conducting obstacle (PEC), is of much practical concern and will also be formulated by an alternate method known as *physical equivalent* [1]. The solutions of the physical equivalent will be compared with those of the induction theorem (induction equivalent) that was discussed in the previous section.

Let us again postulate the problem of Figure 7-12a. In the absence of the obstacle, the fields produced by $(\mathbf{J}_1, \mathbf{M}_1)$ are $(\mathbf{E}_1, \mathbf{H}_1)$, which we assume can be calculated. In the presence of the obstacle (perfect conductor in this case), the fields outside the obstacle are $(\mathbf{E}, \mathbf{H})$ and inside the obstacle are equal to zero. The fields $(\mathbf{E}, \mathbf{H})$ are related to $(\mathbf{E}_1, \mathbf{H}_1)$ by

$$\mathbf{E} = \mathbf{E}_1 + \mathbf{E}^s \tag{7-50a}$$

$$\mathbf{H} = \mathbf{H}_1 + \mathbf{H}^s \tag{7-50b}$$

The original problem is again shown in Figure 7-14a. Again, due to the uniqueness theorem, we do not need to specify both the electric current density $\mathbf{J}_1$ (tangential magnetic field) and the magnetic current density $\mathbf{M}_1$ (tangential electric field). Therefore, keeping only the electric current density $\mathbf{J}_1$, the equivalent of Figure 7-14a reduces to that of Figure 7-14b ($\mathbf{J}_1 = \mathbf{J}_p$). The magnetic current density $\mathbf{M}_1$ of 7-14a is set to zero in 7-14b ($\mathbf{M}_1 = \mathbf{M}_p = 0$). Therefore, the total tangential components of the $\mathbf{H}$ field are equal to the induced current density $\mathbf{J}_p$. In equation form, we have over S_1,

$$\mathbf{M}_p = -\hat{\mathbf{n}} \times (\mathbf{E} - \mathbf{E}^t) = -\hat{\mathbf{n}} \times \mathbf{E} = -\hat{\mathbf{n}} \times (\mathbf{E}_1 + \mathbf{E}^s) = 0 \tag{7-51a}$$

or

$$-\hat{\mathbf{n}} \times \mathbf{E}_1 = \hat{\mathbf{n}} \times \mathbf{E}^s \tag{7-51b}$$

and

$$\mathbf{J}_p = \hat{\mathbf{n}} \times (\mathbf{H} - \mathbf{H}^t) = \hat{\mathbf{n}} \times \mathbf{H} = \hat{\mathbf{n}} \times (\mathbf{H}_1 + \mathbf{H}^s) \tag{7-52a}$$

or

$$\mathbf{J}_p = \hat{\mathbf{n}} \times \mathbf{H}_1 + \hat{\mathbf{n}} \times \mathbf{H}^s \tag{7-52b}$$

Therefore, the equivalent to the problem of Figure 7-14a, computation of $(\mathbf{E}^s, \mathbf{H}^s)$ outside of S_1, is that of Figure 7-14b. *Remember that $(\mathbf{E}_1, \mathbf{H}_1)$ and $(\mathbf{E}^s, \mathbf{H}^s)$ are solutions to Maxwell's equations outside V_1, so in the equivalent problems we retain the same medium μ_1, ε_1 inside and*

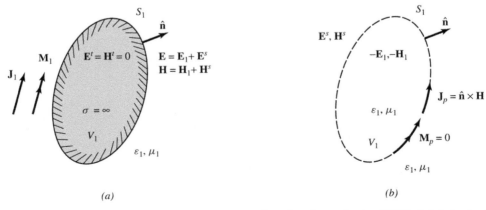

(a)   (b)

Figure 7-14 Physical equivalent for scattering by a perfect electric conductor (PEC). (a) Actual problem. (b) Physical equivalent.

outside V_1. The equivalent of Figure 7-14b will give $(\mathbf{E}^s, \mathbf{H}^s)$ outside of S_1 and $(-\mathbf{E}_1, -\mathbf{H}_1)$ inside of S_1 because

$$\mathbf{J}_p = \hat{\mathbf{n}} \times \mathbf{H} = \hat{\mathbf{n}} \times (\mathbf{H}^s + \mathbf{H}_1) = \hat{\mathbf{n}} \times \left[\mathbf{H}^s - (-\mathbf{H}_1) \right] \qquad (7\text{-}53a)$$

$$\mathbf{M}_p = -\hat{\mathbf{n}} \times \mathbf{E} = -\hat{\mathbf{n}} \times (\mathbf{E}^s + \mathbf{E}_1) = -\hat{\mathbf{n}} \times \left[\mathbf{E}^s - (-\mathbf{E}_1) \right] = 0 \qquad (7\text{-}53b)$$

We call the problem of Figure 7-14b the *physical equivalent*. It can be solved by using (6-30) through (6-35a) since we assume that $\mathbf{J}_p$ radiates in one medium (μ_1, ε_1 everywhere). To form $\mathbf{J}_p$ on S_1 we must know the tangential components of $\mathbf{H}$ on S_1, which are unknown. So the equivalent problem of Figure 7-14b has not aided us in solving the problem of Figure 7-14a. The exact solution of the problem of Figure 7-14b is just as difficult as that of Figure 7-14a. However, as will be discussed later, the formulation of Figure 7-14b is more suggestive when it comes time to make approximations.

The physical equivalent of Figure 7-14b is used in Sections 12.3.1 and 12.3.2 to develop electric and magnetic field integral equations designated, respectively, as EFIE and MFIE. These integral equations are then solved for the unknown current density $\mathbf{J}_p$ by representing it with a series of finite terms of known functions (referred to as basis functions) but with unknown amplitude coefficients. This then allows the reduction of the integral equation to a number of algebraic equations that are usually solved by use of either matrix or iterative techniques. To date, the most popular numerical technique in applied electromagnetics for solving these integral equations is the moment method [15] which is discussed in Sections 12.2.4 through 12.2.8. In particular, in Section 12.3.1 the scattered electric field $\mathbf{E}^s$ is written in terms of $\mathbf{J}_p$. When the observations are restricted to the surface of the electric conducting target, the tangential components of $\mathbf{E}^s$ are related to the negative of the tangential components of $\mathbf{E}_1$, as represented by (7-51b). This allows the development of the electric field integral equation (EFIE) for the unknown current density $\mathbf{J}_p$ in terms of the known tangential components of the electric field $\mathbf{E}_1$, as represented by (12-54). In Section 12.3.2, the equivalent of Figure 7-14b, and in particular the relation of (7-52a) or (7-53a), is used to write an expression for the scattered magnetic field $\mathbf{H}^s$ in terms of the tangential components of the magnetic field $\mathbf{H}_1$. This allows the development of the magnetic field integral equation (MFIE) for the unknown current density $\mathbf{J}_p$, as represented by (12-59a).

If the conducting obstacle of Figure 7-14a is an infinite, flat, perfect electric conductor (infinite ground plane), then the physical equivalent problem of Figure 7-14b is that of Figure 7-15 where

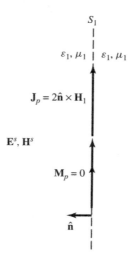

Figure 7-15   Physical equivalent of a flat conducting surface of infinite extent.

the electric current $\mathbf{J}_p$ is equal to

$$\mathbf{J}_p = \hat{\mathbf{n}} \times \mathbf{H} = \hat{\mathbf{n}} \times (\mathbf{H}_1 + \mathbf{H}^s) = 2\hat{\mathbf{n}} \times \mathbf{H}_1 \qquad (7\text{-}54)$$

since the tangential components of the scattered $\mathbf{H}^s$ field ($\mathbf{H}^s|_{\text{tan}}$) are in phase and equal in amplitude to the tangential components of the $\mathbf{H}_1$ field ($\mathbf{H}_1|_{\text{tan}}$). The equivalent of Figure 7-15 is also referred to as the *physical optics* [16].

We have until now discussed two different methods, *induction equivalent* and *physical equivalent*, for the solution of the same problem, that is, the determination of the field scattered by a perfect electric conductor. The induction equivalent is shown in Figure 7-12c and the physical equivalent in Figure 7-14b. The question now is whether both give the same result. The answer to this is yes. However, it must be pointed out that when the geometry of the obstacle is complex, neither of the equivalents is easy to use to obtain convenient results. The next question may then be: Why bother introducing the equivalents if they are not easy to apply? There is a two-part answer to this. The first part of the answer is that when the obstacle is an infinite, flat, perfect electric conductor, the solution to each equivalent is easy to formulate by using "image theory," shown in Figures 7-13a and 7-13b, for the induction equivalent, and in Figure 7-15 for the physical equivalent. The second part of the answer is that the induction and physical equivalent modelings suggest more appropriate approximations or simplifications that can be made when we attempt to solve a problem whose exact solution is not easily obtainable.

The last question then may be stated as follows: "When making approximations or simplifications to solve an otherwise intractable problem, do both of the equivalents lead to identical approximate results or is one superior to the other? The answer is that the induction equivalent and the physical equivalent do *not*, in general, lead to identical results when simplifications or approximations are made to a given problem. For some special approximations, to be discussed later, they give identical results only when the source and the observer are at the same location (backscattering). However, for any general approximation, *they do not yield identical results even for backscattering. One should then use the method that results in the best approximation for the allowable degree of complexity.*

In an attempt to make use of the equivalents of Figure 7-12c and 7-14b to solve a scattering problem, difficulties are encountered. Here we will summarize these difficulties, and in the next section we will discuss appropriate simplifications that allow us to obtain approximate solutions. The induction equivalent of Figure 7-12c is represented by a *known current* ($\mathbf{M}_i = \hat{\mathbf{n}} \times \mathbf{E}_1$) that is placed on the surface of the obstacle and that radiates in its presence. Because the medium within and outside the obstacle is not the same, we *cannot* use (6-30) through (6-35a) to solve for the scattered fields. We must solve a new boundary-value problem, which may be as difficult as the original problem, even though we know the currents on the surface of the obstacle. In other words, we must derive new formulas that will allow us to compute the scattered fields. The physical equivalent of Figure 7-14b is represented by an *unknown current density* ($\mathbf{J}_p = \hat{\mathbf{n}} \times \mathbf{H}$) that is placed on the imaginary surface S_1, which represents the geometry of the obstacle. In this case, however, we *can* use (6-30) through (6-35a) to solve for the scattered fields because we have the same medium within and outside S_1. The difficulty here is that we *do not* know the current density on the surface of the obstacle, which in most cases is just as difficult to find as the solution of the original problem, because it requires knowledge of the total $\mathbf{H}$ field, which is the answer to the original problem.

7.11 INDUCTION AND PHYSICAL EQUIVALENT APPROXIMATIONS

Let us now concentrate on suggesting appropriate simplifications to be made in the induction and physical equivalent formulations so that we obtain approximate solutions when exact solutions

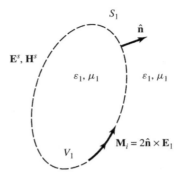

Figure 7-16 Approximate induction equivalent for scattering from a perfect electric conductor (PEC).

are not feasible. In many cases the approximate solutions will lead to results that are well within measuring accuracies of laboratory experiments.

In the induction equivalent form, the difficulty in obtaining a solution arises from the lack of equations that can be used with the known current density. The crudest approximation to the problem is the assumption that the obstacle is large electrically and so we can use image theory to solve the problem. This assumes that locally on the surface of the obstacle each point and its immediate neighbors form a flat surface. The best results with this simplification will be for scatterers whose electrical dimensions are large in comparison to the wavelength. Thus, the induction equivalent of Figure 7-12c can be approximated by that in Figure 7-16. Now (6-30) through (6-35a) can be used to compute the scattered fields because we have the same medium inside and outside S_1.

In many cases, even this approximation may not be amenable to a closed-form solution because of the inability to integrate over the entire closed surface. To simplify this even further, we may restrict our integration over only part of the surface where the current density is more intense and will provide the major contributions to the scattered field. This surface is usually the part that is "visible" by the transmitter (sources $\mathbf{J}_1$ and $\mathbf{M}_1$).

In the physical equivalent, the difficulty in solving the problem arises because we do not know the current density $\mathbf{J}_p$ ($\mathbf{J}_p = \hat{\mathbf{n}} \times \mathbf{H}$) that must be placed along the surface S_1 (see Figure 7-14b). Once we decide on an approximation for the current density, the solution can be carried out because we can use (6-30) through (6-35a). The crudest approximation for this problem is the assumption that the total tangential $\mathbf{H}$ field on the surface of the conductor of Figure 7-14a is equal to twice that of the tangential $\mathbf{H}_1$. Thus, the current density to be placed on the surface of the physical equivalent of Figure 7-14b is

$$\mathbf{J}_p \simeq 2\hat{\mathbf{n}} \times \mathbf{H}_1 \qquad (7\text{-}55)$$

which is a good approximation provided that the scatterer is large electrically (in the limit infinite, flat, perfect conductor). In the shadow region of the scatterer, the physical equivalent current density $\mathbf{J}_p$ is set to zero. We can then approximate the physical equivalent of Figure 7-14b by that of Figure 7-17. This is usually referred to as the *physical optics approximation* [16], because it is similar to the formulation of the infinite, flat, ground plane. Thus, *physical optics approximate the boundary conditions that concern only the fields on the closed surface S_1*. If a closed-form solution still cannot be obtained because of the inability to integrate over the entire surface, then integration over a part of the scattering surface may be sufficient, as was discussed for the induction equivalent.

It should be pointed out that making the aforementioned crude approximations (image theory for the induction equivalent and physical optics for the physical equivalent), the two methods lead to

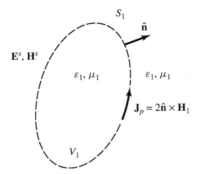

Figure 7-17 Approximate physical equivalent for scattering from a perfect electric conductor (PEC).

identical results only for backscattering. Any further simplifications may lead to solutions that may not be identical even for backscattering. This is discussed in more detail in [17]. The theory can be extended to include imperfect conductors and dielectrics but the formulations become quite complex even when approximations are made.

The best way to illustrate the two different methods, when approximations are made, is to solve the same problem using both methods and compare the results.

Example 7-4

A parallel polarized uniform plane wave on the xy plane, in a free-space medium, is obliquely incident upon a rectangular, flat, perfectly conducting ($\sigma = \infty$) plate, as shown in Figure 7-18a. The dimensions of the plate are a in the y direction and b in the z direction.

Find the electric and magnetic fields scattered by the flat plate, assuming that observations are made in the far zone. Solve the problem by using the *induction equivalent* and *physical equivalent*. Make appropriate simplifications, and compare the results.

Solution: Induction Equivalent: The simplification to be made in the use of induction equivalent modeling is to assume that the dominant part of the magnetic current $\mathbf{M}_i$ resides only in the front face of the plate and that image theory holds for a finite plate. With these approximations, we reduce the equivalent to that of Figure 7-18b, where the magnetic current exists only over the area occupied by the plate. Thus, we can write the **E** and **H** fields as

$$\mathbf{H}^i = \hat{\mathbf{a}}_z H_0 e^{+j\beta(x\cos\phi_i + y\sin\phi_i)}$$

$$\mathbf{E}^i = \eta_0 H_0 \left[\hat{\mathbf{a}}_x \sin\phi_i - \hat{\mathbf{a}}_y \cos\phi_i \right] e^{+j\beta(x\cos\phi_i + y\sin\phi_i)}$$

and the magnetic current density as

$$\mathbf{M}_i = 2\hat{\mathbf{n}} \times \mathbf{E}^i \big|_{x=0} = -\hat{\mathbf{a}}_z 2\eta_0 H_0 \cos\phi_i e^{+j\beta y'\sin\phi_i}$$

or

$$M_x = M_y = 0, \quad M_z = -2\eta_0 H_0 \cos\phi_i e^{+j\beta y'\sin\phi_i}$$

The scattered electric and magnetic fields in the far zone can be written, according to (6-122a) through (6-122f), (6-125a) through (6-125d), (6-127a), and (6-128a), as

$$E_\theta^s = 0$$

$$E_\phi^s = +j\frac{\beta e^{-j\beta r}}{4\pi r} L_\theta$$

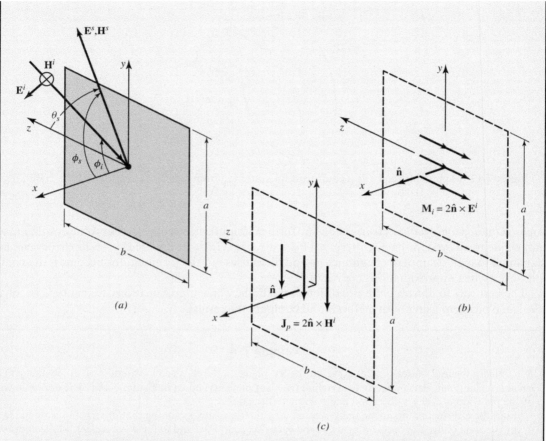

Figure 7-18 Plane wave scattering from a flat rectangular conducting plate. (*a*) Actual problem. (*b*) Induction equivalent. (*c*) Physical equivalent.

where

$$L_\theta = \int_{-a/2}^{+a/2} \int_{-b/2}^{+b/2} -M_z \sin\theta_s e^{+j\beta(y' \sin\theta_s \sin\phi_s + z' \cos\theta_s)} \, dz'dy'$$

$$L_\theta = 2ab\eta_0 H_0 \cos\phi_i \sin\theta_s \left(\frac{\sin Y}{Y}\right)\left(\frac{\sin Z}{Z}\right)$$

$$Y = \frac{\beta a}{2}(\sin\theta_s \sin\phi_s + \sin\phi_i)$$

$$Z = \frac{\beta b}{2}\cos\theta_s$$

In summary,

$$E_\theta^s = 0$$

$$E_\phi^s = j\frac{ab\beta\eta_0 H_0 e^{-j\beta R}}{2\pi r}\left[\cos\phi_i \sin\theta_s \left(\frac{\sin Y}{Y}\right)\left(\frac{\sin Z}{Z}\right)\right]$$

$$H_\theta^s = -\frac{E_\phi^s}{\eta_0}$$

$$H_\phi^s = \frac{E_\theta^s}{\eta_0} = 0$$

For backscattering observations ($\theta_s = \pi/2$, $\phi_s = \phi_i$), the fields reduce to

$$E_\theta^s = 0$$

$$E_\phi^s = j\frac{ab\beta\eta_0 H_0 e^{-j\beta R}}{2\pi r}\left\{\cos\phi_i\left[\frac{\sin(\beta a\,\sin\phi_i)}{\beta a\,\sin\phi_i}\right]\right\}$$

$$H_\theta^s = -\frac{E_\phi^s}{\eta_0}$$

$$H_\phi^s = 0$$

Physical Equivalent: The simplifications for the physical equivalent will be similar to those for the induction equivalent. That is, we will assume that the major contributing current density $\mathbf{J}_p$ resides in the front face of the plate for which the physical equivalent reduces to that of Figure 7-18c. Thus, we can write the current density as

$$\mathbf{J}_p = 2\hat{\mathbf{n}} \times \mathbf{H}^i\,|_{x=0} = -\hat{\mathbf{a}}_y 2H_0 e^{+j\beta y'\,\sin\phi_i}$$

and the scattered **E** and **H** fields, according to (6-122a) through (6-122f), (6-125a) through (6-125d), (6-127a), and (6-128a), as

$$E_\theta^s = -j\frac{\beta\eta_0 e^{-j\beta r}}{4\pi r}N_\theta$$

$$E_\phi^s = -j\frac{\beta\eta_0 e^{-j\beta r}}{4\pi r}N_\phi$$

where

$$N_\theta = \int_{-a/2}^{a/2}\int_{-b/2}^{b/2} J_y\cos\theta_s\sin\phi_s e^{+j\beta(y'\sin\theta_s\,\sin\phi_s + z'\cos\theta_s)}\,dz'dy'$$

$$N_\theta = -2abH_0\left\{\cos\theta_s\sin\phi_s\left[\frac{\sin(Y)}{Y}\right]\left[\frac{\sin(Z)}{Z}\right]\right\}$$

$$N_\phi = \int_{-a/2}^{a/2}\int_{-b/2}^{b/2} J_y\cos\phi_s\,e^{+j\beta(y'\sin\theta_s\,\sin\phi_s + z'\cos\theta_s)}\,dz'dy'$$

$$N_\phi = -2abH_0\left\{\cos\phi_s\left[\frac{\sin(Y)}{Y}\right]\left[\frac{\sin(Z)}{Z}\right]\right\}$$

$$Y = \frac{\beta a}{2}(\sin\theta_s\,\sin\phi_s + \sin\phi_i)$$

$$Z = \frac{\beta b}{2}\cos\theta_s$$

In summary,

$$E_\theta^s = j\frac{ab\beta\eta_0 H_0 e^{-j\beta r}}{2\pi r}\left\{\cos\theta_s\sin\phi_s\left[\frac{\sin(Y)}{Y}\right]\left[\frac{\sin(Z)}{Z}\right]\right\}$$

$$E_\phi^s = j\frac{ab\beta\eta_0 H_0 e^{-j\beta r}}{2\pi r}\left\{\cos\phi_s\left[\frac{\sin(Y)}{Y}\right]\left[\frac{\sin(Z)}{Z}\right]\right\}$$

$$H_\theta^s = -\frac{E_\phi^s}{\eta_0}$$

$$H_\phi^s = \frac{E_\theta^s}{\eta_0}$$

For backscattering observations ($\theta_s = \pi/2$, $\phi_s = \phi_i$), the fields reduce to

$$E_\theta^s = 0$$

$$E_\phi^s = j\frac{ab\beta\eta_0 H_0 e^{-j\beta r}}{2\pi r}\left\{\cos\phi_i\left[\frac{\sin(\beta a\sin\phi_i)}{\beta a\sin\phi_i}\right]\right\}$$

$$H_\theta^s = -\frac{E_\phi^s}{\eta_0}$$

$$H_\phi^s = 0$$

It is quite clear that the solutions of the two different methods do not lead to identical results except for backscatter observations. It seems that the physical equivalent solution gives the best results for general observations because it requires the least simplification in the formulation.

7.12 MULTIMEDIA

On the website that accompanies this book, the following multimedia resources are included for the review, understanding, and presentation of the material of this chapter.

- **PowerPoint (PPT)** viewgraphs, in multicolor.

REFERENCES

1. R. F. Harrington, *Time-Harmonic Electromagnetic Fields*, McGraw-Hill, New York, 1961.
2. C. A. Balanis, *Antenna Theory: Analysis and Design*, Third Edition, John Wiley & Sons, New York, 2005.
3. J. R. Wait, "Characteristics of antennas over lossy earth," in *Antenna Theory Part 2*, Chapter 23, R. E. Collin and F. J. Zucker, (Eds.), McGraw-Hill, New York, 1969.
4. L. E. Vogler and J. L. Noble, *"Curves of input impedance change due to ground for dipole antennas,"* U.S. National Bureau of Standards, Monograph 72, Jan. 31, 1964.
5. W. H. Hayt, Jr., and J. E. Kimmerly, *Engineering Circuit Analysis*, Third Edition, McGraw-Hill, New York, 1978.
6. J. R. Carson, "Reciprocal theorems in radio communication," *Proc. IRE*, vol. 17, pp. 952–956, Jun. 1929.
7. J. H. Richmond, The Basic Theory of Harmonic Fields, Antennas and Scattering, Unpublished Notes.
8. V. H. Rumsey, "Reaction concept in electromagnetic theory," *Physical Rev.*, vol. 94, no. 6, pp. 1483–1491, June 15, 1954.
9. J. H. Richmond, "A reaction theorem and its application to antenna impedance calculations," *IRE Trans. Antennas Propagat.*, vol. AP-9, no. 6, pp. 515–520, Nov. 1961.
10. R. Mittra (ed.), *Computer Techniques for Electromagnetics*, Chapter 2, Pergamon, Elmsford, NY, 1973.
11. S. A. Schelkunoff, "Some equivalence theorems of electromagnetics and their application to radiation problems," *Bell System Tech. J.*, vol. 15, pp. 92–112, 1936.
12. C. Huygens, Traite de la Lumiere, Leyeden, 1690. Translated into English by S. P. Thompson, London, 1912 and reprinted by the University of Chicago Press.
13. J. D. Kraus and K. R. Carver, *Electromagnetics*, Second Edition, McGraw-Hill, New York, 1973, pp. 464–467.
14. A. E. H. Love, "The integration of the equations of propagation of electric waves," *Phil. Trans. Roy. Soc. London, Ser. A*, vol. 197, pp. 1–45, 1901.

15. R. F. Harrington, *Field Computation by Moment Methods*, Macmillan, New York, 1968.

16. P. Beckmann, *The Depolarization of Electromagnetic Waves*, The Golem Press, Boulder, CO, 1968, pp. 76–92.

17. R. F. Harrington, "On scattering by large conducting bodies," *IRE Trans. Antennas Propagat.*, vol. AP-7, no. 2, pp. 150–153, Apr. 1959.

PROBLEMS

7.1. For an infinitesimal vertical electric dipole, a height h above a PEC ground plane, whose far-zone electric field is given by (7-13a):
  (a) Find the corresponding magnetic field.
  (b) Determine the corresponding time-average power density.
  (c) Show that the radiated power, obtained by integrating the power density of part (b) over a sphere of radius r, can be written as

$$P_{\text{rad}} = \pi\eta \left|\frac{I_0\ell}{\lambda}\right|^2$$

$$\times \left[\frac{1}{3} - \frac{\cos(2\beta h)}{(2\beta h)^2} + \frac{\sin(2\beta h)}{(2\beta h)^3}\right]$$

7.2. For Problem 7.1, show the following:
  (a) The radiation intensity U, defined in the far field as $U \approx r^2 S_{\text{av}}$, where S_{av} is the far-field time-average power density, can be written as

$$U = \frac{\eta}{2}\left|\frac{I_0\ell}{\lambda}\right|^2 \sin^2\theta \cos^2(\beta h \cos\theta)$$

  (b) The maximum directivity D_0 of the element, defined as

$$D_0 = \frac{4\pi U_{\text{max}}}{P_{\text{rad}}}$$

  where U_{max} is the maximum radiation intensity, can be written as

$$D_0 = \frac{2}{F(\beta h)}$$

$$F(\beta h) = \left[\frac{1}{3} - \frac{\cos(2\beta h)}{(2\beta h)^2} + \frac{\sin(2\beta h)}{(2\beta h)^3}\right]$$

  (c) The radiation resistance, defined as

$$R_r = \frac{2P_{\text{rad}}}{|I_0|^2}$$

  can be expressed as

$$R_r = 2\pi\eta\left(\frac{\ell}{\lambda}\right)^2 F(\beta h)$$

  where $F(\beta h)$ is that given in part (b).

7.3. An infinitesimal vertical magnetic dipole of length l and constant current I_m is placed symmetrically about the origin and it is directed along the z axis, as shown in Figure 6-2a. Derive expressions valid everywhere, near and far field, for the:
  • Electric vector potential components (F_r, F_θ, F_ϕ).
  • Electric field components (E_r, E_θ, E_ϕ).
  • Magnetic field components (H_r, H_θ, H_ϕ).
  • Time-average power density, defined as

$$S_{\text{av}} = \frac{1}{2}\text{Re}\left[\mathbf{E} \times \mathbf{H}^*\right].$$

  • Radiation intensity, defined in the far field as $U \approx r^2 S_{\text{av}}$.
  • Power radiated, defined as

$$P_{\text{rad}} = \int_0^{2\pi}\int_0^\pi U(\theta,\phi)\sin\theta \, d\theta \, d\phi.$$

  • Maximum directivity, defined as

$$D_0 = \frac{4\pi U_{\text{max}}(\theta,\phi)}{P_{\text{rad}}}$$

  • Radiation resistance, defined as

$$R_r = \frac{2P_{\text{rad}}}{|I_m|^2}$$

  You can minimize the derivations as long as you justify the procedure.

7.4. For the infinitesimal vertical magnetic dipole of Problem 7.3, simplify the expressions for the electric vector potential, and electric and magnetic fields, when the observations are made in the far field.

7.5. An infinitesimal magnetic dipole of length l and constant current I_m is placed symmetrically about the origin and it is directed along the x axis. Derive the following expressions for the far zone:
  • Electric vector potential components (F_r, F_θ, F_ϕ).
  • Electric field components (E_r, E_θ, E_ϕ).
  • Magnetic field components (H_r, H_θ, H_ϕ).
  • Time-average power density as defined in Problem 7.3.

- Radiation intensity as defined in Problem 7.3.
- Directivity as defined in Problem 7.3.
- Radiation resistance as defined in Problem 7.3.

7.6. Repeat Problem 7.4 for an infinitesimal magnetic dipole of length l and constant current I_m but directed along the y axis.

7.7. Repeat Problem 7.1 for the horizontal infinitesimal electric dipole of Section 7.4.2 and Figure 7-5, and show that the:

- Radiation intensity is

$$U(\theta,\phi) = \frac{\eta}{2}|I_0|^2 \left(\frac{l}{\lambda}\right)^2$$
$$\times \left(\cos^2\theta \, \sin^2\phi + \cos^2\phi\right)$$
$$\times \sin^2(\beta h \, \cos\theta), \quad 0 \le \theta \le \frac{\pi}{2}$$

- Power radiated is

$$P_{\text{rad}} = \int_0^{2\pi}\int_0^{\pi/2} U(\theta,\phi)\sin\theta \, d\theta \, d\phi$$
$$= \eta\frac{\pi}{2}|I_0|^2\left(\frac{l}{\lambda}\right)^2\left[\frac{2}{3} - \frac{\sin(2\beta h)}{2\beta h}\right.$$
$$\left. - \frac{\cos(2\beta h)}{(2\beta h)^2} + \frac{\sin(2\beta h)}{(2\beta h)^3}\right]$$

- Maximum directivity is

$$D_0 = \frac{4\pi U_{\max}}{P_{\text{rad}}}$$
$$= \begin{cases} \dfrac{4\sin^2(\beta h)}{R(\beta h)} & \beta h \le \pi/2 \ (h \le \lambda/4) \\[3mm] \dfrac{4}{R(\beta h)} & \beta h > \pi/2 \ (h > \lambda/4) \end{cases}$$

$$R(\beta h) = \left[\frac{2}{3} - \frac{\sin(2\beta h)}{2\beta h} - \frac{\cos(2\beta h)}{(2\beta h)^2}\right.$$
$$\left. + \frac{\sin(2\beta h)}{(2\beta h)^3}\right]$$

- Radiation resistance is

$$R_r = \frac{2P_{\text{rad}}}{|I_0|^2} = \eta\pi\left(\frac{l}{\lambda}\right)^2\left[\frac{2}{3} - \frac{\sin(2\beta h)}{2\beta h}\right.$$
$$\left. - \frac{\cos(2\beta h)}{(2\beta h)^2} + \frac{\sin(2\beta h)}{(2\beta h)^3}\right]$$

7.8. Using the electric field of (7-13a), where r is fixed, plot the normalized radiation pattern (in dB) versus the angle θ when the height h of the element above the ground is $h = 0$, $\lambda/8$, $\lambda/4$, $3\lambda/8$, $\lambda/2$, and λ.

7.9. A quarter-wavelength $(\ell/2 = \lambda/4)$ wire radiator is placed vertically above an infinite electric ground plane and it is fed at its base, as shown in Figure P7-9. This is usually referred to as a $\lambda/4$ monopole. Assume that the current on the wire is represented by

$$\mathbf{I} = \hat{\mathbf{a}}_z I_0 \sin\left[\beta\left(\frac{\ell}{2} - z'\right)\right], \quad 0 \le z' \le \ell/2$$

where z' is any point on the monopole and show, using image theory, (6-97a), (6-112a) and (6-112b), and the formulations of Section 6.7, that the far-zone electric and magnetic fields radiated by the element *above the ground plane* are given by

$$E_r \simeq E_\phi \simeq H_r \simeq H_\theta \simeq 0$$

$$E_\theta \simeq j\eta\frac{I_0 e^{-j\beta R}}{2\pi r}\left[\frac{\cos\left(\frac{\pi}{2}\cos\theta\right)}{\sin\theta}\right]$$

$$H_\phi \simeq \frac{E_\phi}{\eta}$$

These expressions are identical to those of Problem 6.26.

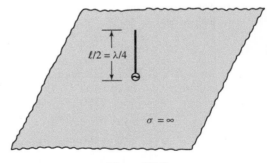

Figure P7-9

7.10. A very small $(\ell \ll \lambda)$ linear radiating current element is placed between two infinite plates forming a 90° corner reflector. Assume that the length of the element is placed parallel to the plates of the corner reflector.

(a) Determine the number of images, their polarizations, and their positions that are necessary to account for all the reflections from the plates of the reflector and to find the radiated fields within the internal space of the reflector.

(b) Show that the total far-zone radiated fields within the internal region of the reflector can be written as

$$E_\theta^t = E_\theta^0 F(\beta s)$$

$$E_\theta^0 = jn\frac{\beta I_0 \ell e^{-j\beta R}}{4\pi r}\sin\theta$$

$$F(\beta s) = 2[\cos(\beta s\,\sin\theta\,\cos\phi)$$
$$-\cos(\beta s\,\sin\theta\,\sin\phi)]$$

$$0° \leq \theta \leq 180°$$
$$315° \leq \phi \leq 360°,\ 0 \leq \phi \leq 45°$$

where θ is measured from the z axis toward the observation point. E_θ^0 is the far-zone field radiated by a very small ($\ell \ll \lambda$) linear element radiating in an unbounded medium (see Example 6-3) and $F(\beta s)$ is referred to as the array factor representing the array of elements that includes the actual radiating element and its associated images.

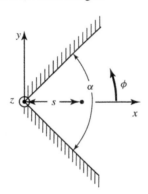

Figure P7-10

7.11. For Problem 7.10 plot the magnitude of $F(\beta s)$ as a function of s ($0 \leq s \leq 10\lambda$) when $\theta = 90°$ and $\phi = 0°$. What is the maximum value of $|F(\beta s)|$? Is the function periodic? If so, what is the period?

7.12. For Problem 7.10 plot the normalized value of the magnitude of $F(\beta s)$ (in dB) as a function of ϕ ($315° \leq \phi \leq 360°, 0 \leq \phi \leq 45°$) when $\theta = 90°$. Do this when $s = 0.1\lambda$, $s = 0.7\lambda$, $s = 0.8\lambda$, $s = 0.9\lambda$, and $s = 1.0\lambda$.

7.13. Repeat Problem 7.10 when the included angle α of the corner reflector is 60°, 45°, and 30°, and show that $F(\beta s)$ takes the following forms.

(a) $\alpha = 60°$

$$F(\beta s) = 4\sin\left(\frac{X}{2}\right)$$
$$\times\left[\cos\left(\frac{X}{2}\right) - \cos\left(\sqrt{3}\frac{Y}{2}\right)\right]$$

(b) $\alpha = 45°$

$$F(\beta s) = \left[2\cos(X) + \cos(Y) - \right.$$
$$\left. 2\cos\left(\frac{X}{\sqrt{2}}\right)\cos\left(\frac{Y}{\sqrt{2}}\right)\right]$$

(c) $\alpha = 30°$

$$F(\beta s) = 2\left[\cos(X) - 2\cos\left(\sqrt{3}\frac{X}{2}\right)\cos\left(\frac{Y}{2}\right)\right.$$
$$\left. -\cos(Y) + 2\cos\left(\frac{X}{2}\right)\cos\left(\sqrt{3}\frac{X}{2}\right)\right]$$

where $X = \beta s\,\sin\theta\,\cos\phi$, $Y = \beta s\,\sin\theta\,\sin\phi$

7.14. For Problem 7.13 and the three values ($\alpha = 60°$, 45°, and 30°), plot the magnitude of $F(\beta s)$ as a function of s ($0 \leq s \leq 10\lambda$) when $\theta = 90°$ and $\phi = 0°$. What is the maximum value $F(\beta s)$ will ever achieve if plotted as a function of s? Is the function periodic? If so, what is the period?

7.15. An infinitesimal electric dipole is placed at an angle of 30° at a height h above a perfectly conducting electric ground plane. Determine

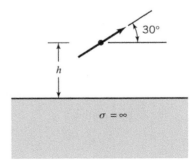

Figure P7-15

the location and orientation of its image. Do this by sketching the image.

7.16. A small circular loop of radius a is placed vertically at a height h above a perfectly conducting electric ground plane. Determine the location and direction of current flow of its image. Do this by sketching the image.

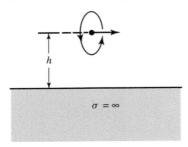

Figure P7-16

7.17. A very small electrical loop antenna ($a \ll \lambda$) of constant electrical current I_0 is placed height h above a flat and infinite in extent ground plane.

(a) The area of the loop is parallel to the interface (x-y plane), as shown in Figure P7-17a. Determine the amplitude (equal or unequal) and phase ($0°$ or $180°$) of the image, compared to the actual/physical element, to account for the reflections. Assume the ground plane is:
  - PEC (amplitude and phase of the image)
  - PMC (amplitude and phase of the image)

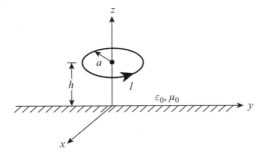

Figure P7-17a

(b) Repeat part (a), if the area of the loop is parallel to the x-z plane, as shown Figure P7-17b that follows.

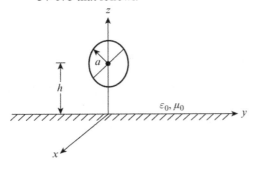

Figure P7-17b

7.18. A small electric circular loop antenna, of radius a and constant electric current I_0, is placed in parallel (parallel to the xy plane) a height h above a perfect magnetic conducting (PMC), flat and infinite in extent, horizontal ground plane. The electric current flowing in the loop antenna is in the counterclockwise (CCW) direction, as viewed from the top (looking from the top downwards, i.e., the $-z$ direction).

(a) To account for the direct field, and the reflected one from the PMC, determine the equivalent problem that will account for the total field (direct and reflected) on and above the PMC ground plane. State, in words, the magnitude (*equal* or *unequal*) and relative phase (in degrees) of the image loop compared to those of the actual loop.

(b) In what direction, as viewed again from the top (looking downward), is the electric current flowing in the image electric loop? CCW or CW?

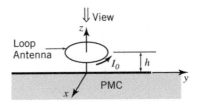

Figure P7-18

7.19. A horizontal magnetic dipole is placed above a planar Perfect Electric Conductor (PEC) of infinite extent, as shown in Figures P7-19 and 7-2.

In order to maintain the maximum total radiation, due to the magnetic dipole itself and its image (to account for reflections), toward the z axis (perpendicular to the interface), what is the smallest nonzero height h that the magnetic dipole should be placed above the PEC:

- In wavelengths?
- In cm, for a frequency of 10 GHz?

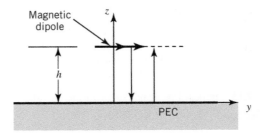

Figure P7-19

7.20. Repeat Problem 7.19 when the element is a magnetic dipole and the ground plane is a PMC.

7.21. Repeat Problem 7.19 when the element is an electric dipole and the ground plane is a PEC.

7.22. Repeat Problem 7.19 when the element is an electric dipole and the ground plane is a PMC.

7.23. A linearly polarized uniform plane wave traveling in free space is incident normally upon a flat dielectric surface. Assume that the incident electric field is given by

$$\mathbf{E} = \hat{\mathbf{a}}_x E_0\, e^{-j\beta_0 z}$$

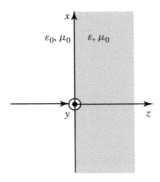

Figure P7-23

Derive expressions for the equivalent volume electric and magnetic current densities, and the regions over which they exist. These current densities can then be used, in principle, to find the fields scattered by the dielectric surface.

7.24. A uniform plane wave traveling in free space is incident, at normal incidence angle, upon an infinite PMC plate, which is parallel to the xy-plane, as shown in the Figure P7-24.
 (a) Write an expression, in vector form, of the incident magnetic field in terms of the incident electric field E_0 and whatever else is needed.
 (b) To determine the scattered field to the left of the PMC infinite plate ($-z$ direction), formulate (do not have to derive) in vector form, the Induction Equivalent electric $\mathbf{J}_i$ or magnetic $\mathbf{M}_i$ current density that must be used to determine the fields scattered to the left ($-z$ direction) of the PMC. The current density, $\mathbf{J}_i$ or $\mathbf{M}_i$ (only

one of them), must be in terms of E_0 and whatever else is needed.

$$\mathbf{E}^i = \hat{\mathbf{a}}_x E_0\, e^{-j\beta_0 z}$$

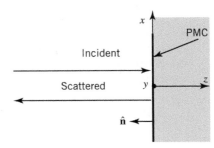

Figure P7-24

7.25. Repeat Example 7-4 when:

$$\mathbf{E}^i = -\hat{\mathbf{a}}_z E_0\, e^{j\beta(x\,\cos\phi_i + y\,\sin\phi_i)}$$

7.26. For the aperture shown in Figure 6-4a and assuming it is mounted on an infinite PEC ground plane:
 (a) Form the most practical, exact or approximate (when necessary to solve the problem), equivalent currents $\mathbf{J}_s$ and $\mathbf{M}_s$.
 (b) Find the far-zone electric and magnetic fields. The electric field distribution at the aperture is given by (E_0 is a constant)

$$\mathbf{E}_a = \hat{\mathbf{a}}_z E_0$$
$$-a/2 \le y' \le a/2;\ -b/2 \le z' \le b/2$$

7.27. Repeat Problem 7.26 when the aperture is mounted on an infinite PMC surface.

7.28. Repeat Problem 7.26 when the aperture is not mounted on a PEC ground plane. Assume the tangential electric and magnetic fields are related by the intrinsic impedance.

7.29. Repeat Problem 7.26 when the aperture is mounted on an infinite PEC but its tangential electric field at the aperture is given by (E_0 is a constant)

$$\mathbf{E}_a = \hat{\mathbf{a}}_z E_0\, \cos\left(\frac{\pi}{a} y'\right)$$
$$-a/2 \le y' \le a/2;\ -b/2 \le z' \le b/2$$

7.30. Repeat Problem 7.29 when the aperture is mounted on an infinite PMC surface.

7.31. Repeat Problem 7.29 when the aperture is not mounted on a PEC ground plane. Assume the tangential electric and magnetic fields are related by the intrinsic impedance.

7.32. For the aperture shown in Figure 6-4*b* and assuming it is mounted on an infinite PEC ground plane:

(a) Form the most practical, exact or approximate (when necessary to solve the problem), equivalent currents $\mathbf{J}_s$ and $\mathbf{M}_s$.

(b) Find the far-zone electric and magnetic fields.

The electric field distribution at the aperture is given by (E_0 is a constant):

$$\mathbf{E}_a = \hat{\mathbf{a}}_x E_0$$
$$-b/2 \le x' \le b/2; \ -a/2 \le z' \le a/2$$

7.33. Repeat Problem 7.32 when the aperture is mounted on an infinite PMC surface.

7.34. Repeat Problem 7.32 when the aperture is not mounted on a PEC ground plane. Assume the tangential electric and magnetic fields are related by the intrinsic impedance.

7.35. Repeat Problem 7.32 when the aperture is mounted on an infinite PEC but its tangential electric field at the aperture is given by

$$\mathbf{E}_a = \hat{\mathbf{a}}_x E_0 \cos\left(\frac{\pi}{a} z'\right)$$
$$-b/2 \le y' \le b/2; \ -a/2 \le z' \le a/2$$

7.36. Repeat Problem 7.35 when the aperture is mounted on a PMC surface.

7.37. Repeat Problem 7.35 when the aperture is not mounted on a PEC ground plane. Assume the tangential electric and magnetic fields are related by the intrinsic impedance.

7.38. For the aperture shown in Figure 6-4*c* and assuming it is mounted on an infinite PEC ground plane:

(a) Form the most practical, exact or approximate (when necessary to solve the problem), equivalent currents $\mathbf{J}_s$ and $\mathbf{M}_s$.

(b) Find the far-zone electric and magnetic fields. The electric field distribution at the aperture is given by (E_0 is a constant)

$$\mathbf{E}_a = \hat{\mathbf{a}}_y E_0$$
$$-a/2 \le x' \le a/2; \ -b/2 \le y' \le b/2$$

7.39. Repeat Problem 7.38 when the aperture is mounted on an infinite PMC surface.

7.40. Repeat Problem 7.38 when the aperture is not mounted on a PEC ground plane. Assume the tangential electric and magnetic fields are related by the intrinsic impedance.

7.41. Repeat Problem 7.38 when the aperture is mounted on an infinite PEC but its tangential electric field at the aperture is given by (E_0 is a constant)

$$\mathbf{E}_a = \hat{\mathbf{a}}_y E_0 \cos\left(\frac{\pi}{a} x'\right)$$
$$-a/2 \le x' \le a/2; \ -b/2 \le y' \le b/2$$

7.42. Repeat Problem 7.41 when the aperture is mounted on an infinite PMC surface.

7.43. Repeat Problem 7.41 when the aperture is not mounted on a PEC ground plane. Assume the tangential electric and magnetic fields are related by the intrinsic impedance.

7.44. The electric and magnetic fields at the aperture of a circular waveguide are given by

$$\left.\begin{array}{l} \mathbf{E}_a = \hat{\mathbf{a}}_\rho E_\rho + \hat{\mathbf{a}}_\phi E_\phi \\[2mm] E_\rho = E_0 J_1\left(\dfrac{\chi'_{11}}{a}\rho'\right)\dfrac{\sin\phi'}{\rho'} \\[4mm] E_\phi = E_0 J_1'\left(\dfrac{\chi'_{11}}{a}\rho'\right)\cos\phi' \end{array}\right\}\ \begin{array}{l} \rho' \le a \\[2mm] \chi'_{11} = 1.841 \\[4mm] ' = \dfrac{\partial}{\partial\rho'} \end{array}$$

$$\mathbf{E}_a = 0 \text{ elsewhere}$$

Develop the surface equivalent that can be used to find the fields radiated by the aperture. State the equivalent by giving expressions for

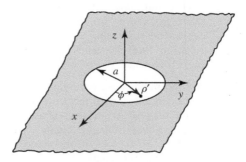

Figure P7-44

the electric $\mathbf{J}_s$ and magnetic $\mathbf{M}_s$ surface current densities and the regions over which they exist.

7.45. A uniform plane wave traveling in an infinite free space with the electric and magnetic fields represented by

$$\mathbf{E}^i = \hat{\mathbf{a}}_x E_0 e^{-j\beta_0 z}$$
$$\mathbf{H}^i = \hat{\mathbf{a}}_y \frac{E_0}{\eta_0} e^{-j\beta_0 z}$$

The above fields are now normally incident upon a planar and infinite in extent perfect magnetic conductor (PMC), as shown in Figure 5-1 whereby medium #2 is the PMC, and they create scattered fields $\mathbf{E}^s$, $\mathbf{H}^s$ to the left of the interface. There are many different ways to solve for the scattered electric $\mathbf{E}^s$ and magnetic $\mathbf{H}^s$ to the left of the interface, in the free-space medium, introduced by the presence of the PMC surface. One way is to use the Induction Equivalent.

(a) Based on the Induction Equivalent, write exact and specific electric $\mathbf{J}_i$ and magnetic $\mathbf{M}_i$ current densities, based on the fields $\mathbf{E}^i$, $\mathbf{H}^i$ above, placed on the PMC surface, and radiating in the presence of the PMC interface, which can be used to find the scattered electric $\mathbf{E}^s$ and magnetic $\mathbf{H}^s$ to the left of the PMC interface; i.e., in the free-space medium.

(b) Part (a) can be reduced to only one equivalent which involves either $\mathbf{J}_i$ or $\mathbf{M}_i$ or $\mathbf{J}_i + \mathbf{M}_i$ radiating in an infinite free-space medium which gives the correct scattered $\mathbf{E}^s$, $\mathbf{H}^s$ to the left of the interface. Choose the correct Induction Equivalent (only one is correct) and write specific expressions, in terms of the incident fields $\mathbf{E}^i$, $\mathbf{H}^i$ from above, the appropriate current density or densities that are radiating in an infinite free-space medium and give the correct $\mathbf{E}^s$, $\mathbf{H}^s$ to the left of the PMC interface.

7.46. (a) A uniform plane of perpendicular polarization traveling is free space, as shown in Figure 5-2, is incident upon a planar interface formed by a perfect electric conductor (PEC). Assume that the electric $\mathbf{J}_s^i$ and magnetic $\mathbf{M}_s^i$ current densities in the presence of the PEC, based on the incident electric and magnetic fields, are given by:

$$\mathbf{J}_s^i = \hat{\mathbf{n}} \times \mathbf{H}^i$$
$$\mathbf{M}_s^i = -\hat{\mathbf{n}} \times \mathbf{E}^i$$

Write:
- Specific expressions for $\mathbf{J}_s^i$ and $\mathbf{M}_s^i$ on the PEC, based on the appropriate incident electric and magnetic fields.
- Specific expressions for the total $\mathbf{J}_s^i$ and $\mathbf{M}_s^i$ which will allow you to remove the PEC and have the total $\mathbf{J}_s^i$ and $\mathbf{M}_s^i$ radiate

into an infinite free-space medium to compute the scattered field.

(b) A uniform plane of perpendicular polarization traveling is free space, as shown in Figure 5-2, is incident upon a planar interface formed by a perfect magnetic conductor (PMC). Assume that the electric $\mathbf{J}_s^i$ and magnetic $\mathbf{M}_s^i$ current densities in the presence of the PMC, based on the incident electric and magnetic fields, are given by:

$$\mathbf{J}_s^i = \hat{\mathbf{n}} \times \mathbf{H}^i$$
$$\mathbf{M}_s^i = -\hat{\mathbf{n}} \times \mathbf{E}^i$$

Write:
- Specific expressions for $\mathbf{J}_s^i$ and $\mathbf{M}_s^i$ on the PMC, based on the appropriate incident electric and magnetic fields.
- Specific expressions for the total $\mathbf{J}_s^i$ and $\mathbf{M}_s^i$ which will allow you to remove the PMC and have the total $\mathbf{J}_s^i$ and $\mathbf{M}_s^i$ radiate into an infinite free-space medium to compute the scattered field.

7.47. A uniform plane wave on the yz plane is incident upon a flat circular PEC plate of radius a. Assume that the incident electric field is given by

$$\mathbf{E}^i = \hat{\mathbf{a}}_x E_0 e^{-j\beta_0(y\sin\theta_i - z\cos\theta_i)}$$

Determine the scattered field using the
(a) Induction equivalent.
(b) Physical equivalent.
Reduce and compare the expressions for backscatter observations.

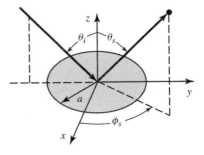

Figure P7-47

7.48. Repeat Problem 7.47 when the incident magnetic field is given by

$$\mathbf{H}^i = \hat{\mathbf{a}}_x H_0 e^{-j\beta_0(y\sin\theta_i - z\cos\theta_i)}$$

CHAPTER 8

Rectangular Cross-Section Waveguides
and Cavities

8.1 INTRODUCTION

Rectangular waveguides became popular during and after World War II because microwave sources and amplifiers, such as klystrons, magnetrons, and traveling-wave tube amplifiers, were developed before, during, and after that period. Because the physical cross-section dimensions are typically around half of a guide wavelength ($\lambda_g/2$), such transmission lines are most commonly used in the microwave region—typically 1 GHz and above (most commonly several GHz's), although they have also been used in the UHF band. The dimensions of the rigid rectangular waveguide have been standardized according to different bands, whose designations and characteristics are listed on Table 8-4. The standard X-band (8.2–12.4 GHz) was one of the most widely used bands, and it was the band of communication, at least in the 1960s, of the NASA space program (actually it started with S-band and transitioned to the X-band). The inner dimensions of an X-band waveguide are 0.9 inches (2.286 cm) by 0.4 inches (1.016 cm), whereas those of the Ku-band (12.4–18 GHz) are 0.622 inches (1.580 cm) by 0.311 inches (0.790 cm). Two such waveguides are shown in Figure 8-1 along with two standard flanges on each end to be connected to other devices, such as isolators, attenuators, phase shifters, circulators, and microwave sources. One very popular X-band microwave source, with an output of about 100 mW, was the Varian X-13 klystron shown in Figures 8-2a and 8-2b. The knob dial is used to vary the frequency by changing the klystron's inner cavity dimensions, which can also be altered slightly by controlling the reflector voltage.

Eventually, in the 1960s, solid-state microwave sources began to appear. One such source was the Gunn diode oscillator, a transferred electron device invented by J. B. Gunn in 1963, with an output power, depending on the frequency, of several milliwatts; maybe as high as 1 watt around 10 GHz. This device is very compact, and it only needs a dc voltage bias of a few volts (typically 10–15 volts) to oscillate and to convert to RF power. A Gunn diode mounted on an X-band wafer is shown in Figure 8-2c. Another solid-state source is the IMPATT diode, which uses a reverse-biased *pn* junction, typically of silicon or gallium arsenide, to generate RF power. Today waveguides are still popular, and they are widely used as transmission lines in communication systems operating at even higher frequencies, such as Ku-band (12.4–18 GHz), K-band (18–26.5 GHz),

Balanis' Advanced Engineering Electromagnetics, Third Edition. Constantine A. Balanis.
© 2024 John Wiley & Sons, Inc. Published 2024 by John Wiley & Sons, Inc.
Companion Website: www.wiley.com/go/balanis/advancedengineeringelectromagnetics3e

Figure 8-1   Two rectangular waveguides (Ku-band and X-band) with flanges.

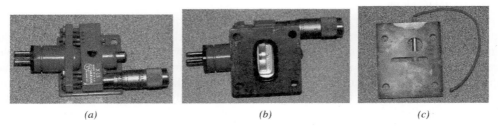

(a)   *(b)*   *(c)*

Figure 8-2   X-band microwave sources: X-13 klystron and Gunn diode wafer. (*a*) Rear view of X-13. (*b*) Front view of X-13. (*c*) Gunn diode wafer.

Ka-band (26.5–40 GHz), etc. It is then very important that we understand the field configurations (modes), and their characteristics, that such transmission lines can support and sustain.

Rectangular transmission lines (such as rectangular waveguides, dielectric slab lines, striplines, and microstrips) and their corresponding cavities represent a significant section of lines used in many practical radio-frequency systems. The objective in this chapter is to introduce and analyze some of them, and to present some data on their propagation characteristics. The parameters of interest include field configurations (modes) that can be supported by such structures and their corresponding cutoff frequencies, guide wavelengths, wave impedances, phase and attenuation constants, and quality factors Q. Because of their general rectilinear geometrical shapes, it is most convenient to use the rectangular coordinate system for the analyses. The field configurations that can be supported by these structures must satisfy Maxwell's equations or the wave equation, and the corresponding boundary conditions.

8.2   RECTANGULAR WAVEGUIDE

Let us consider a rectangular waveguide of lateral dimensions a and b, as shown in Figure 8-3. Initially assume that the waveguide is of infinite length and is empty. It is our purpose to determine the various field configurations (modes) that can exist inside the guide. Although a TEM^z field configuration is of the simplest structure, it cannot satisfy the boundary conditions on the waveguide walls. Therefore, it is not a valid solution. It can be shown that modes TE^x, TM^x, TE^y, TM^y, TE^z, and TM^z satisfy the boundary conditions and are therefore appropriate modes (field configurations) for the rectangular waveguide. We will initially consider TE^z and TM^z; others will follow.

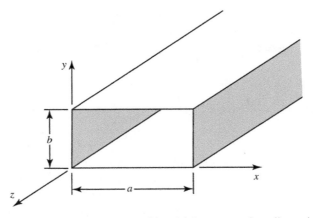

Figure 8-3   Rectangular waveguide with its appropriate dimensions.

8.2.1   Transverse Electric (TE^z)

According to (6-71a) through (6-72), TE^z electric and magnetic fields satisfy the following set of equations:

$$E_x = -\frac{1}{\varepsilon}\frac{\partial F_z}{\partial y} \qquad H_x = -j\frac{1}{\omega\mu\varepsilon}\frac{\partial^2 F_z}{\partial x\partial z}$$

$$E_y = \frac{1}{\varepsilon}\frac{\partial F_z}{\partial x} \qquad H_y = -j\frac{1}{\omega\mu\varepsilon}\frac{\partial^2 F_z}{\partial y\partial z}$$

$$E_z = 0 \qquad H_z = -j\frac{1}{\omega\mu\varepsilon}\left(\frac{\partial^2}{\partial z^2}+\beta^2\right)F_z \qquad (8\text{-}1)$$

where $F_z(x,y,z)$ is a scalar potential function, and it represents the z component of the vector potential function $\mathbf{F}$. The potential $\mathbf{F}$, and in turn F_z, must satisfy (6-73) or

$$\nabla^2 F_z(x,y,z)+\beta^2 F_z(x,y,z)=0 \qquad (8\text{-}2)$$

which can be reduced to

$$\frac{\partial^2 F_z}{\partial x^2}+\frac{\partial^2 F_z}{\partial y^2}+\frac{\partial^2 F_z}{\partial z^2}+\beta^2 F_z=0 \qquad (8\text{-}2a)$$

The solution to (8-2) or (8-2a) is obtained by using the separation-of-variables method outlined in Section 3.4.1. In general, the solution to $F_z(x,y,z)$ can be written initially as

$$F_z(x,y,z)=f(x)g(y)h(z) \qquad (8\text{-}3)$$

The objective here is to choose judiciously the most appropriate forms for $f(x)$, $g(y)$, and $h(z)$ from (3-28a) through (3-30b).

Since the waveguide is bounded in the x and y directions, the forms of $f(x)$ and $g(y)$ must be chosen to represent standing waves. The most appropriate forms are those of (3-28b) and (3-29b). Thus,

$$f(x)=f_2(x)=C_1\cos(\beta_x x)+D_1\sin(\beta_x x) \qquad (8\text{-}4a)$$

$$g(y)=g_2(y)=C_2\cos(\beta_y y)+D_2\sin(\beta_y y) \qquad (8\text{-}4b)$$

Because the waveguide is infinite in length, the variations of the fields in the z direction must represent traveling waves as given by (3-30a). Thus,

$$h(z) = h_1(z) = A_3 e^{-j\beta_z z} + B_3 e^{+j\beta_z z} \tag{8-5}$$

Substituting (8-4a) through (8-5) into (8-3), we can write that

$$F_z(x,y,z) = [C_1 \cos(\beta_x x) + D_1 \sin(\beta_x x)][C_2 \cos(\beta_y y) + D_2 \sin(\beta_y y)]$$
$$\times [A_3 e^{-j\beta_z z} + B_3 e^{+j\beta_z z}] \tag{8-6}$$

The first exponential in (8-6) represents waves traveling in the $+z$ direction (assuming an $e^{j\omega t}$ time variation) and the second term designates waves traveling in the $-z$ direction. To simplify the notation, assume that the source is located such that the waves are traveling only in the $+z$ direction. Then the second term is not present, so $B_3 = 0$. If the waves are traveling in the $-z$ direction, then the second exponential in (8-6) is appropriate and $A_3 = 0$. If the waves are traveling in both directions, superposition can be used to sum the field expressions for the $+z$ and $-z$ traveling waves.

For $+z$ traveling waves, F_z of (8-6) reduces with $B_3 = 0$ to

$$F_z^+(x,y,z) = [C_1 \cos(\beta_x x) + D_1 \sin(\beta_x x)]$$
$$\times [C_2 \cos(\beta_y y) + D_2 \sin(\beta_y y)] A_3 e^{-j\beta_z z} \tag{8-7}$$

where, according to (3-27), the *constraint (dispersion) equation* is

$$\beta_x^2 + \beta_y^2 + \beta_z^2 = \beta^2 = \omega^2 \mu\varepsilon \tag{8-7a}$$

C_1, D_1, C_2, D_2, A_3, β_x, β_y, and β_z are constants that will be evaluated by substituting (8-7) into (8-1) and applying the appropriate boundary conditions on the walls of the waveguide.

For the waveguide structure of Figure 8-3, the necessary and sufficient boundary conditions are those that require the tangential components of the electric field to vanish on the walls of the waveguide. Thus, in general, on the bottom and top walls

$$E_x(0 \le x \le a, y=0, z) = E_x(0 \le x \le a, y=b, z) = 0 \tag{8-8a}$$

$$E_z(0 \le x \le a, y=0, z) = E_z(0 \le x \le a, y=b, z) = 0 \tag{8-8b}$$

and on the left and right walls

$$E_y(x=0, 0 \le y \le b, z) = E_y(x=a, 0 \le y \le b, z) = 0 \tag{8-8c}$$

$$E_z(x=0, 0 \le y \le b, z) = E_z(x=a, 0 \le y \le b, z) = 0 \tag{8-8d}$$

For the TE^z modes, $E_z = 0$, and the boundary conditions of (8-8b) and (8-8d) are automatically satisfied. However, in general, the boundary conditions of (8-8b) and (8-8d) are not independent, but they represent the same conditions as given, respectively, by (8-8a) and (8-8c). Therefore, the necessary and sufficient independent boundary conditions, in general, will be to enforce either (8-8a) or (8-8b) and either (8-8c) or (8-8d).

Substituting (8-7) into (8-1), we can write the x component of the electric field as

$$E_x^+(x,y,z) = -A_3 \frac{\beta_y}{\varepsilon}[C_1 \cos(\beta_x x) + D_1 \sin(\beta_x x)]$$
$$\times [-C_2 \sin(\beta_y y) + D_2 \cos(\beta_y y)] e^{-j\beta_z z} \tag{8-9}$$

Enforcing on (8-9) the boundary conditions (8-8a) on the bottom wall, we have that

$$E_x^+(0 \le x \le a, y=0,z) = -A_3 \frac{\beta_y}{\varepsilon}[C_1 \cos(\beta_x x) + D_1 \sin(\beta_x x)]$$
$$\times[-C_2(0) + D_2(1)]e^{-j\beta_z z} = 0 \tag{8-10}$$

The only way for (8-10) to be satisfied and not lead to a trivial solution will be for $D_2 = 0$. Thus,

$$D_2 = 0 \tag{8-10a}$$

Now by enforcing on (8-9) the boundary conditions (8-8a) on the top wall, and using (8-10a), we can write that

$$E_x^+(0 \le x \le a, y=b,z) = -A_3 \frac{\beta_y}{\varepsilon}[C_1 \cos(\beta_x x) + D_1 \sin(\beta_x x)]$$
$$\times[-C_2 \sin(\beta_y b)]e^{-j\beta_z z} = 0 \tag{8-11}$$

For nontrivial solutions, (8-11) can only be satisfied provided that

$$\sin(\beta_y b) = 0 \tag{8-12}$$

which leads to

$$\beta_y b = \sin^{-1}(0) = n\pi, \qquad n=0, 1, 2, \dots \tag{8-12a}$$

or

$$\beta_y = \frac{n\pi}{b}, \qquad n=0, 1, 2, \dots \tag{8-12b}$$

Equation 8-12 is usually referred to as the *eigenfunction* and (8-12b) as the *eigenvalue*.

In a similar manner, we can enforce the boundary conditions on the left and right walls as given by (8-8c). By doing this, it can be shown that

$$D_1 = 0 \tag{8-13}$$

and

$$\beta_x = \frac{m\pi}{a}, \qquad m=0, 1, 2, \dots \tag{8-13a}$$

Use of (8-10a), (8-12b), (8-13), and (8-13a) reduces (8-7) to

$$F_z^+(x,y,z) = C_1 C_2 A_3 \cos(\beta_x x)\cos(\beta_y y)e^{-j\beta_z z} \tag{8-14}$$

or, by combining $C_1 C_2 A_3 = A_{mn}$, to

$$F_z^+(x,y,z) = A_{mn} \cos(\beta_x x)\cos(\beta_y y)e^{-j\beta_z z} \tag{8-14a}$$

with

$$\begin{array}{|ll|}
\hline
\beta_x = \dfrac{m\pi}{a} = \dfrac{2\pi}{\lambda_x} \Rightarrow \lambda_x = \dfrac{2a}{m} & m=0, 1, 2, \dots \\
 & n=0, 1, 2, \dots \\
\beta_y = \dfrac{n\pi}{b} = \dfrac{2\pi}{\lambda_y} \Rightarrow \lambda_y = \dfrac{2b}{n} & m \text{ and } n \text{ not zero simultaneously} \\
\hline
\end{array} \tag{8-14b}$$

In (8-14b) combination $m=n=0$ is excluded because for that combination, F_z of (8-14a) is a constant and all the components of **E** and **H** as given by (8-1) vanish; thus, it is a trivial solution. Since individually C_1, C_2, and A_3 are constants, their product A_{mn} is also a constant. The subscripts m and n are used to designate the eigenvalues of β_x and β_y and in turn the field configurations (modes). Thus, a given combination of m and n in (8-14b) designates a given TE_{mn}^z mode. Since there are infinite combinations of m and n, there are an infinite number of TE_{mn}^z modes.

In (8-14b), β_x and β_y represent the mode wave numbers (eigenvalues) in the x and y directions, respectively. These are related to the wave number in the z direction (β_z) and to that of the unbounded medium (β) by (8-7a). In (8-14b), λ_x and λ_y represent, respectively, the wavelengths of the wave inside the guide in the x and y directions. These are related to the wavelength in the z direction ($\lambda_z = \lambda_g$) and to that in an unbounded medium (λ), according to (8-7a), by

$$\frac{1}{\lambda_x^2} + \frac{1}{\lambda_y^2} + \frac{1}{\lambda_z^2} = \frac{1}{\lambda^2} \tag{8-14c}$$

In summary then, the appropriate expressions for the TE_{mn}^z modes are, according to (8-1), (8-14a), and (8-14b),

$$TE_{mn}^{+z}$$

$$E_x^+ = A_{mn} \frac{\beta_y}{\varepsilon} \cos(\beta_x x) \sin(\beta_y y) e^{-j\beta_z z} \tag{8-15a}$$

$$E_y^+ = -A_{mn} \frac{\beta_x}{\varepsilon} \sin(\beta_x x) \cos(\beta_y y) e^{-j\beta_z z} \tag{8-15b}$$

$$E_z^+ = 0 \tag{8-15c}$$

$$H_x^+ = A_{mn} \frac{\beta_x \beta_z}{\omega\mu\varepsilon} \sin(\beta_x x) \cos(\beta_y y) e^{-j\beta_z z} \tag{8-15d}$$

$$H_y^+ = A_{mn} \frac{\beta_y \beta_z}{\omega\mu\varepsilon} \cos(\beta_x x) \sin(\beta_y y) e^{-j\beta_z z} \tag{8-15e}$$

$$H_z^+ = -jA_{mn} \frac{\beta_c^2}{\omega\mu\varepsilon} \cos(\beta_x x) \cos(\beta_y y) e^{-j\beta_z z} \tag{8-15f}$$

where

$$\beta_c^2 \equiv \left(\frac{2\pi}{\lambda_c}\right)^2 = \beta^2 - \beta_z^2 = \beta_x^2 + \beta_y^2 = \left(\frac{m\pi}{a}\right)^2 + \left(\frac{n\pi}{b}\right)^2 \tag{8-15g}$$

The constant β_c is the value of β when $\beta_z = 0$, and it will be referred to as the *cutoff wave number*. Thus,

$$\beta_c = \beta|_{\beta_z=0} = \omega\sqrt{\mu\varepsilon}\,|_{\beta_z=0} = \omega_c\sqrt{\mu\varepsilon} = 2\pi f_c\sqrt{\mu\varepsilon} = \sqrt{\left(\frac{m\pi}{a}\right)^2 + \left(\frac{n\pi}{b}\right)^2}$$

or

$$\boxed{(f_c)_{mn} = \frac{1}{2\pi\sqrt{\mu\varepsilon}} \sqrt{\left(\frac{m\pi}{a}\right)^2 + \left(\frac{n\pi}{b}\right)^2}} \quad \begin{array}{l} m = 0, 1, 2, \ldots \\ n = 0, 1, 2, \ldots \\ m \text{ and } n \text{ not zero simultaneously} \end{array} \tag{8-16}$$

where $(f_c)_{mn}$ represents the cutoff frequency of a given mn mode. Modes that have the same cutoff frequency are called *degenerate*.

To determine the significance of the cutoff frequency, let us examine the values of β_z. Using (8-15g), we can write that

$$\beta_z^2 = \beta^2 - \beta_c^2 = \beta^2 - \left[\left(\frac{m\pi}{a}\right)^2 + \left(\frac{n\pi}{b}\right)^2\right] \tag{8-17}$$

or

$$(\beta_z)_{mn} = \begin{cases} \pm\sqrt{\beta^2 - \beta_c^2} = \pm\beta\sqrt{1-\left(\frac{\beta_c}{\beta}\right)^2} \\ \quad = \pm\beta\sqrt{1-\left(\frac{\lambda}{\lambda_c}\right)^2} = \pm\beta\sqrt{1-\left(\frac{f_c}{f}\right)^2} \quad \text{for } \beta > \beta_c, f > f_c \quad (8\text{-}17a) \\[4pt] 0 \qquad\qquad\qquad\qquad\qquad\qquad\qquad\quad \text{for } \beta = \beta_c, f = f_c \quad (8\text{-}17b) \\[4pt] \pm j\sqrt{\beta_c^2 - \beta^2} = \pm j\beta\sqrt{\left(\frac{\beta_c}{\beta}\right)^2 - 1} \\ \quad = \pm j\beta\sqrt{\left(\frac{\lambda}{\lambda_c}\right)^2 - 1} = \pm j\beta\sqrt{\left(\frac{f_c}{f}\right)^2 - 1} \quad \text{for } \beta < \beta_c, f < f_c \quad (8\text{-}17c) \end{cases}$$

In order for the waves to be traveling in the $+z$ direction, the expressions for β_z as given by (8-17a) through (8-17c) reduce to

$$(\beta_z)_{mn} = \begin{cases} \beta\sqrt{1-\left(\frac{\lambda}{\lambda_c}\right)^2} = \beta\sqrt{1-\left(\frac{f_c}{f}\right)^2} \quad & \text{for } f > f_c \quad (8\text{-}18a) \\[4pt] 0 & \text{for } f = f_c \quad (8\text{-}18b) \\[4pt] -j\beta\sqrt{\left(\frac{\lambda}{\lambda_c}\right)^2 - 1} = -j\beta\sqrt{\left(\frac{f_c}{f}\right)^2 - 1} & \text{for } f < f_c \quad (8\text{-}18c) \end{cases}$$

Substituting the expressions for β_z as given by (8-18a) through (8-18c) in the expressions for **E** and **H** as given by (8-15a) through (8-15f), it is evident that (8-18a) leads to propagating waves, (8-18b) to standing waves, and (8-18c) to *evanescent* (reactive) or nonpropagating waves. Evanescent fields are exponentially decaying fields that do not possess real power. Thus, (8-18b) serves as the boundary between propagating and nonpropagating waves, and it is usually referred to as the cutoff, which occurs when $\beta_z = 0$. When the frequency of operation is selected to be higher than the value of $(f_c)_{mn}$ for a given mn mode, as given by (8-16), then the fields propagate unattenuated. If, however, f is selected to be smaller than $(f_c)_{mn}$, then the fields are attenuated. Thus, the waveguide serves as a high-pass filter.

The ratios of E_x/H_y and $-E_y/H_x$ have the units of impedance. Use of (8-15a) through (8-15f) shows that

$$Z_w^{+z}(\text{TE}_{mn}^z) \equiv \frac{E_x}{H_y} = -\frac{E_y}{H_x} = \frac{\omega\mu}{\beta_z} \tag{8-19}$$

which can be written by using (8-18a) through (8-18c) as

$$
Z_w^{+z}(TE_{mn}^z) = \begin{cases}
\dfrac{\omega\mu}{\beta\sqrt{1-\left(\dfrac{f_c}{f}\right)^2}} = \dfrac{\sqrt{\dfrac{\mu}{\varepsilon}}}{\sqrt{1-\left(\dfrac{f_c}{f}\right)^2}} = \dfrac{\eta}{\sqrt{1-\left(\dfrac{f_c}{f}\right)^2}} & \text{for } f > f_c \quad \text{(8-20a)} \\[3ex]
\infty & \text{for } f = f_c \quad \text{(8-20b)} \\[3ex]
+j\dfrac{\omega\mu}{\beta\sqrt{\left(\dfrac{f_c}{f}\right)^2 - 1}} = +j\dfrac{\sqrt{\dfrac{\mu}{\varepsilon}}}{\sqrt{\left(\dfrac{f_c}{f}\right)^2 - 1}} = +j\dfrac{\eta}{\sqrt{\left(\dfrac{f_c}{f}\right)^2 - 1}} & \text{for } f < f_c \quad \text{(8-20c)}
\end{cases}
$$

Z_w^{+z} in (8-20a) through (8-20c) is referred to as the *wave impedance* in the $+z$ direction, which is real and greater than the intrinsic impedance η of the medium inside the guide for values of $f > f_c$, infinity at $f = f_c$, and reactively inductive for $f < f_c$. Thus, the waveguide for TE_{mn}^z modes behaves as an inductive storage element for $f < f_c$. A plot of Z_w^{+z} for any TE_{mn} mode in the range of $0 \le f/f_c \le 3$ is shown in Figure 8-4.

The expressions of (8-18a) through (8-18c) for β_z can also be used to define a wavelength along the axis of the guide. Thus, we can write that

$$
\beta_z \equiv \frac{2\pi}{\lambda_z} \Rightarrow \lambda_z = \lambda_g = \frac{2\pi}{\beta_z} \tag{8-21}
$$

or

$$
(\lambda_z)_{mn} = (\lambda_g)_{mn} = \begin{cases}
\dfrac{2\pi}{\beta\sqrt{1-\left(\dfrac{f_c}{f}\right)^2}} = \dfrac{\lambda}{\sqrt{1-\left(\dfrac{f_c}{f}\right)^2}} = \dfrac{\lambda}{\sqrt{1-\left(\dfrac{\lambda}{\lambda_c}\right)^2}} & \text{for } f > f_c \quad \text{(8-21a)} \\[3ex]
\infty & \text{for } f = f_c \quad \text{(8-21b)} \\[3ex]
+j\dfrac{2\pi}{\beta\sqrt{\left(\dfrac{f_c}{f}\right)^2 - 1}} = +j\dfrac{\lambda}{\sqrt{\left(\dfrac{f_c}{f}\right)^2 - 1}} = +j\dfrac{\lambda}{\sqrt{\left(\dfrac{\lambda}{\lambda_c}\right)^2 - 1}} & \text{(8-21c)} \\[3ex]
\qquad\qquad \text{(nonphysical) for } f < f_c
\end{cases}
$$

In (8-21a) through (8-21c) λ_z represents the wavelength of the wave along the axis of the guide, and it is referred to as the *guide wavelength* λ_g. In the same expressions, λ refers to the wavelength of the wave at the same frequency but traveling in an unbounded medium whose electrical parameters ε and μ are the same as those of the medium inside the waveguide. Cutoff wavelength corresponding to the cutoff frequency f_c is represented by λ_c.

Inspection of (8-21a) through (8-21c) indicates that the guide wavelength λ_g is greater than the unbounded medium wavelength λ for $f > f_c$, it is infinity for $f = f_c$, and has no physical meaning for $f < f_c$ since it is purely imaginary. A plot of λ_g/λ for $1 \le f/f_c \le 3$ is shown in Figure 8-5.

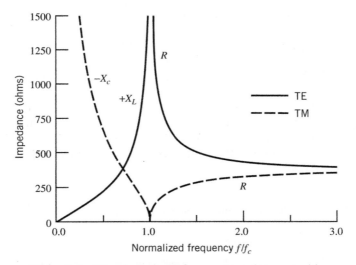

Figure 8-4 Wave impedance for a rectangular waveguide.

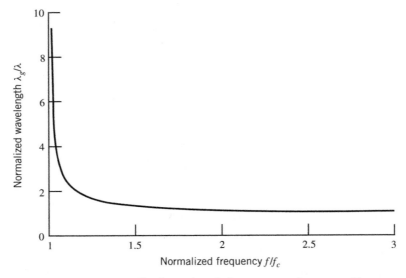

Figure 8-5 Normalized wavelength for a rectangular waveguide.

The values of $(f_c)_{mn}$ for different combinations of m and n but fixed values of ε, μ, a, and b determine the cutoff frequencies and order of existence of each mode above its corresponding cutoff frequency. Assuming $a > b$, the mode with the smallest cutoff frequency is that of TE_{10}. Its cutoff frequency is equal to

$$(f_c)_{10} = \frac{1}{2\pi\sqrt{\mu\varepsilon}}\frac{\pi}{a} = \frac{1}{2a\sqrt{\mu\varepsilon}}$$

(8-22)

In general, the mode with the smallest cutoff frequency is referred to as the *dominant mode*. Thus for a waveguide with $a > b$, the dominant mode is the TE_{10} mode. (If $b > a$, the dominant mode is the TE_{01}.)

The ratio $R_{mn} = (f_c)_{mn}^{TE}/(f_c)_{10}^{TE}$ can be written as

$$R_{mn} = \frac{(f_c)_{mn}^{TE}}{(f_c)_{10}^{TE}} = \sqrt{(m)^2 + \left(\frac{na}{b}\right)^2}$$

$$\begin{array}{l} m = 0, 1, 2, 3, \dots \\ n = 0, 1, 2, 3, \dots \\ m \text{ and } n \text{ not zero simultaneously} \end{array}$$

(8-23)

whose values for $a/b = 10, 5, 2.25, 2$, and 1, for the allowable values of m and n, are listed in Table 8-1. The ratio value R_{mn} of a given m, n combination represents the relative frequency range over which the TE_{10} mode can operate before that m, n mode will begin to appear. For a given a/b ratio, the smallest value of (8-23), above unity, indicates the relative frequency range over which the waveguide can operate in a single TE_{10} mode.

TABLE 8-1 Ratio of cutoff frequency of TE_{mn}^z mode to that of TE_{10}^z

$$R_{mn} = \frac{(f_c)_{mn}^{TE^z}}{(f_c)_{10}^{TE^z}} = \sqrt{m^2 + \left(\frac{na}{b}\right)^2}$$

$$\begin{array}{l} m = 0, 1, 2, \dots \\ n = 0, 1, 2, \dots \\ m \text{ and } n \text{ not zero simultaneously} \end{array}$$

a/b	10	5	2.25	2	1
m,n	1,0	1,0	1,0	1,0	1,0; 0,1
R_{mn}	1	1	1	1	1
m,n	2,0	2,0	2,0	2,0;0,1	1,1
R_{mn}	2	2	2	2	1.414
m,n	3,0	3,0	0,1	1,1	2,0
R_{mn}	3	3	2.25	2.236	2
m,n	4,0	4,0	1,1	2,1	2,1;1,2
R_{mn}	4	4	2.462	2.828	2.236
m,n	5,0	5,0;0,1	3,0	3,0	2,2
R_{mn}	5	5	3	3	2.828
m,n	6,0	1,1	2,1	3,1	3,0;0,3
R_{mn}	6	5.099	3.010	3.606	3
m,n	7,0	2,1	3,1	4,0;0,2	3,1;1,3
R_{mn}	7	5.385	3.75	4	3.162
m,n	8,0	3,1	4,0	1,2	3,2;2,3
R_{mn}	8	5.831	4	4.123	3.606
m,n	9,0	6,0	0,2	4,1;2,2	4,0;0,4
R_{mn}	9	6	4.5	4.472	4
m,n	10,0;0,1	4,1	4,1	5,0;3,2	4,1;1,4
R_{mn}	10	6.403	4.589	5	4.123

Example 8-1

A rectangular waveguide of dimensions a and b ($a > b$), as shown in Figure 8-3, is to be operated in a single mode. Determine the smallest ratio of the a/b dimensions that will allow the largest bandwidth of the single-mode operation. State the dominant mode and its largest bandwidth of single-mode operation.

Solution: According to (8-16), the dominant mode for $a > b$ is the TE_{10} whose cutoff frequency is given by (8-22), i.e.,

$$(f_c)_{10} = \frac{1}{2a\sqrt{\mu\varepsilon}}$$

The mode with the next higher cutoff frequency would be either the TE_{20} or TE_{01} mode whose cutoff frequencies are given, respectively, by

$$(f_c)_{20} = \frac{1}{a\sqrt{\mu\varepsilon}} = 2(f_c)_{10}$$

$$(f_c)_{01} = \frac{1}{2b\sqrt{\mu\varepsilon}}$$

It is apparent that the largest bandwidth of single TE_{10} mode operation would be

$$(f_c)_{10} \le f \le 2(f_c)_{10} = (f_c)_{20} \le (f_c)_{01}$$

and would occur provided

$$2b \le a \Rightarrow 2 \le a/b$$

8.2.2 Transverse Magnetic (TM^z)

A procedure similar to that used for the TE^z modes can be used to derive the TM^z fields and the other appropriate parameters for a rectangular waveguide of the geometry shown in Figure 8-3. According to (6-55a) and (6-55b) these can be obtained by letting $\mathbf{A} = \hat{\mathbf{a}}_z A_z(x, y, z)$ and $\mathbf{F} = 0$. Without repeating the entire procedure, the most important equations 6-59, 6-56, and 6-58 are summarized:

$$
\begin{aligned}
E_x &= -j\frac{1}{\omega\mu\varepsilon}\frac{\partial^2 A_z}{\partial x \partial z} & H_x &= \frac{1}{\mu}\frac{\partial A_z}{\partial y} \\
E_y &= -j\frac{1}{\omega\mu\varepsilon}\frac{\partial^2 A_z}{\partial y \partial z} & H_y &= -\frac{1}{\mu}\frac{\partial A_z}{\partial x} \\
E_z &= -j\frac{1}{\omega\mu\varepsilon}\left(\frac{\partial^2}{\partial z^2} + \beta^2\right)A_z & H_z &= 0
\end{aligned}
\tag{8-24}
$$

$$\nabla^2 A_z + \beta^2 A_z = \frac{\partial^2 A_z}{\partial x^2} + \frac{\partial^2 A_z}{\partial y^2} + \frac{\partial^2 A_z}{\partial z^2} + \beta^2 A_z = 0 \tag{8-25}$$

$$A_z(x, y, z) = [C_1\cos(\beta_x x) + D_1\sin(\beta_x x)][C_2\cos(\beta_y y) + D_2\sin(\beta_y y)]$$
$$\times [A_3 e^{-j\beta_z z} + B_3 e^{+j\beta_z z}] \tag{8-26}$$

For waves that travel in the $+z$ direction and satisfy the boundary conditions of Figure 8-3, as outlined by (8-8a) through (8-8d), (8-26) reduces to

$$A_z^+(x, y, z) = D_1 D_2 A_3 \sin(\beta_x x)\sin(\beta_y y)e^{-j\beta_z z}$$
$$= B_{mn}\sin(\beta_x x)\sin(\beta_y y)e^{-j\beta_z z} \tag{8-26a}$$

where

$$\beta_x \equiv \frac{2\pi}{\lambda_x} = \frac{m\pi}{a} \Rightarrow \lambda_x = \frac{2a}{m}, \quad m = 1, 2, 3, \dots \tag{8-27a}$$

$$\beta_y \equiv \frac{2\pi}{\lambda_y} = \frac{n\pi}{b} \Rightarrow \lambda_y = \frac{2b}{n}, \quad n = 1, 2, 3, \dots \tag{8-27b}$$

$m = 0$ and $n = 0$ are not allowable eigenvalues; they are needed as trivial solutions.

Use of (8-26) allows the fields of (8-24) to be written as

$$E_x^+ = -B_{mn} \frac{\beta_x \beta_z}{\omega\mu\varepsilon} \cos(\beta_x x)\sin(\beta_y y)e^{-j\beta_z z} \tag{8-28a}$$

$$E_y^+ = -B_{mn} \frac{\beta_y \beta_z}{\omega\mu\varepsilon} \sin(\beta_x x)\cos(\beta_y y)e^{-j\beta_z z} \tag{8-28b}$$

$$E_z^+ = -jB_{mn} \frac{\beta_c^2}{\omega\mu\varepsilon} \sin(\beta_x x)\sin(\beta_y y)e^{-j\beta_z z} \tag{8-28c}$$

$$H_x^+ = B_{mn} \frac{\beta_y}{\mu} \sin(\beta_x x)\cos(\beta_y y)e^{-j\beta_z z} \tag{8-28d}$$

$$H_y^+ = -B_{mn} \frac{\beta_x}{\mu} \cos(\beta_x x)\sin(\beta_y y)e^{-j\beta_z z} \tag{8-28e}$$

$$H_z^+ = 0 \tag{8-28f}$$

In turn, the wave impedance, propagation constant, cutoff frequency, and guide wavelength can be expressed as

$$
Z_w^{+z}(\mathrm{TM}_{mn}^z) \equiv \frac{E_x^+}{H_y^+} = -\frac{E_y^+}{H_x^+} = \frac{\beta_z}{\omega\varepsilon} =
\begin{cases}
+\eta\sqrt{1-\left(\dfrac{f_c}{f}\right)^2} & \text{for } f > f_c \tag{8-29a} \\[2ex]
0 & \text{for } f = f_c \tag{8-29b} \\[2ex]
-j\eta\sqrt{\left(\dfrac{f_c}{f}\right)^2 - 1} & \text{for } f < f_c \tag{8-29c}
\end{cases}
$$

$$
(\beta_z)_{mn} \equiv \frac{2\pi}{\lambda_z} =
\begin{cases}
\beta\sqrt{1-\left(\dfrac{f_c}{f}\right)^2} & \text{for } f > f_c \tag{8-30a} \\[2ex]
0 & \text{for } f = f_c \tag{8-30b} \\[2ex]
-j\beta\sqrt{\left(\dfrac{f_c}{f}\right)^2 - 1} & \text{for } f < f_c \tag{8-30c}
\end{cases}
$$

$$\beta_c^2 \equiv \left(\frac{2\pi}{\lambda_c}\right)^2 = \beta^2 - \beta_z^2 = \beta_x^2 + \beta_y^2 = \left(\frac{m\pi}{a}\right)^2 + \left(\frac{n\pi}{b}\right)^2 \tag{8-31}$$

$$(f_c)_{mn} = \frac{1}{2\pi\sqrt{\mu\varepsilon}} \sqrt{\left(\frac{m\pi}{a}\right)^2 + \left(\frac{n\pi}{b}\right)^2} \qquad \begin{matrix} m = 1, 2, 3, \ldots \\ n = 1, 2, 3, \ldots \end{matrix} \tag{8-32}$$

$$(\lambda_z)_{mn} = (\lambda_g)_{mn} = \begin{cases} \dfrac{\lambda}{\sqrt{1 - \left(\dfrac{f_c}{f}\right)^2}} = \dfrac{\lambda}{\sqrt{1 - \left(\dfrac{\lambda}{\lambda_c}\right)^2}} & \text{for } f > f_c \quad \text{(8-33a)} \\[3em] \infty & \text{for } f = f_c \quad \text{(8-33b)} \\[3em] j\dfrac{\lambda}{\sqrt{\left(\dfrac{f_c}{f}\right)^2 - 1}} = j\dfrac{\lambda}{\sqrt{\left(\dfrac{\lambda}{\lambda_c}\right)^2 - 1}} & \\[2em] & \text{(nonphysical) for } f < f_c \quad \text{(8-33c)} \end{cases}$$

It is apparent from (8-29c) that below cutoff ($f < f_c$) the waveguide for TM_{mn}^z modes behaves as a capacitive storage element. A plot of Z_w^{+z} for any TM_{mn}^z mode in the range of $0 \le f/f_c \le 3$ is shown in Figure 8-4.

For TM^z, we can classify the modes according to the order of their cutoff frequency. The TM^z mode with the smallest cutoff frequency, according to (8-32), is the TM_{11}^z whose cutoff frequency is equal to

$$(f_c)_{11} = \frac{1}{2\sqrt{\mu\varepsilon}}\sqrt{\left(\frac{1}{a}\right)^2 + \left(\frac{1}{b}\right)^2} = \frac{1}{2a\sqrt{\mu\varepsilon}}\sqrt{1 + \left(\frac{a}{b}\right)^2} > \frac{1}{2a\sqrt{\mu\varepsilon}} \qquad (8\text{-}34)$$

Since the cutoff frequency of the TM_{11}^z mode, as given by (8-34), is greater than the cutoff frequency of the TE_{10}^z, as given by (8-22), then the TE_{10}^z mode is always the dominant mode if $a > b$. If $a = b$, the dominant modes are the TE_{10}^z and TE_{01}^z modes (degenerate), and if $a < b$ the dominant mode is the TE_{01}^z mode.

The order in which the TM_{mn}^z modes occur, relative to the TE_{10}^z mode, can be determined by forming the ratio T_{mn} of the cutoff frequency of any TM_{mn}^z mode to the cutoff frequency of the TE_{10}^z mode. Then we use (8-32) and (8-22) to write that

$$T_{mn} = \frac{(f_c)_{mn}^{\mathrm{TM}}}{(f_c)_{10}^{\mathrm{TE}}} = \sqrt{m^2 + \left(\frac{na}{b}\right)^2} \qquad \begin{array}{l} m = 1, 2, 3, \dots \\ n = 1, 2, 3, \dots \end{array} \qquad (8\text{-}35)$$

The values of T_{mn} for $a/b = 10$, 5, 2.25, 2, and 1, for the allowable values of m and n, are listed in Table 8-2. Each value of T_{mn} in Table 8-2 represents the relative frequency range over which the TE_{10} mode can operate before that m, n mode will begin to appear.

For a given ratio of a/b, the values of R_{mn} of (8-23) and Table 8-1, and those of T_{mn} of (8-35) and Table 8-2 represent the order, in terms of ascending cutoff frequencies, in which the TE_{mn}^z and TM_{mn}^z modes occur relative to the dominant TE_{10}^z mode.

The xy cross-section field distributions for the first 18 modes [1] of a rectangular waveguide with cross-sectional dimensions $a/b = 2$ are plotted in Figure 8-6. Field configurations of another 18 modes plus the first 30 for a square waveguide ($a/b = 1$) can be found in [1].

Example 8-2

The inner dimensions of an X-band WR90 rectangular waveguide are $a = 0.9$ in. (2.286 cm) and $b = 0.4$ in. (1.016 cm). Assume free space within the guide and determine (in GHz) the cutoff frequencies, in ascending order, of the first 10 TE^z and/or TM^z modes.

Solution: Since $a/b = 0.9/0.4 = 2.25$, then according to Tables 8-1 and 8-2, the cutoff frequencies of the first 10 TE_{mn}^z and/or TM_{mn}^z modes in order of ascending frequency are

1. $TE_{10} = 6.562$ GHz
2. $TE_{20} = 13.124$ GHz
3. $TE_{01} = 14.764$ GHz
4, 5. $TE_{11} = TM_{11} = 16.16$ GHz
6. $TE_{30} = 19.685$ GHz
7, 8. $TE_{21} = TM_{21} = 19.754$ GHz
9, 10. $TE_{31} = TM_{31} = 24.607$ GHz

8.2.3 Dominant TE₁₀ Mode

From the discussion and analysis of the previous two sections, it is evident that there are an infinite number of TE_{mn}^z and TM_{mn}^z modes that satisfy Maxwell's equations and the boundary conditions, and that they can exist inside the rectangular waveguide of Figure 8-3. In addition, other modes, such as TE^x, TM^x, TE^y, and TM^y, can also exist inside that same waveguide. The analysis of the TE^x and TM^x modes has been assigned to the reader as an end-of-chapter problem while the TE^y and TM^y modes are analyzed in Sections 8.5.1 and 8.5.2.

In a given system, the modes that can exist inside a waveguide depend upon the dimensions of the waveguide, the medium inside it which determines the cutoff frequencies of the different modes, and the excitation and coupling of energy from the source (oscillator) to the waveguide.

TABLE 8-2 Ratio of cutoff frequency of TM_{mn}^z mode to that of TE_{10}^z

$$T_{mn} = \frac{(f_c)_{mn}^{TM^z}}{(f_c)_{10}^{TE^z}} = \sqrt{m^2 + \left(\frac{na}{b}\right)^2} \qquad \begin{array}{l} m = 1, 2, 3, \ldots \\ n = 1, 2, 3, \ldots \end{array}$$

$a/b \Rightarrow$	10	5	2.25	2	1
$m,n \Rightarrow$	1,1	1,1	1,1	1,1	1,1
$T_{mn} \Rightarrow$	10.05	5.10	2.46	2.23	1.414
$m,n \Rightarrow$	2,1	2,1	2,1	2,1	2,1;1,2
$T_{mn} \Rightarrow$	10.19	5.38	3.01	2.83	2.236
$m,n \Rightarrow$	3,1	3,1	3,1	3,1	2,2
$T_{mn} \Rightarrow$	10.44	6.00	3.75	3.61	2.828
$m,n \Rightarrow$	4,1	4,1	4,1	1,2	3,1;1,3
$T_{mn} \Rightarrow$	10.77	6.40	4.59	4.12	3.162
$m,n \Rightarrow$	5,1	5,1	1,2	4,1;2,2	3,2;2,3
$T_{mn} \Rightarrow$	11.18	7.07	5.09	4.47	3.606
$m,n \Rightarrow$	6,1	6,1	2,2	3,2	4,1;1,4
$T_{mn} \Rightarrow$	11.66	7.81	5.38	5.00	4.123
$m,n \Rightarrow$	7,1	7,1	3,2	5,1	3,3
$T_{mn} \Rightarrow$	12.21	8.60	5.41	5.39	4.243
$m,n \Rightarrow$	8,1	8,1	5,1	4,2	4,2;2,4
$T_{mn} \Rightarrow$	12.81	9.43	5.48	5.66	4.472
$m,n \Rightarrow$	9,1	1,2	4,2	1,3	4,3;3,4
$T_{mn} \Rightarrow$	13.82	10.04	6.40	6.08	5.00
$m,n \Rightarrow$	10,1	2,2	6,1	2,3	5,1;1,5
$T_{mn} \Rightarrow$	14.14	10.20	6.41	6.32	5.09

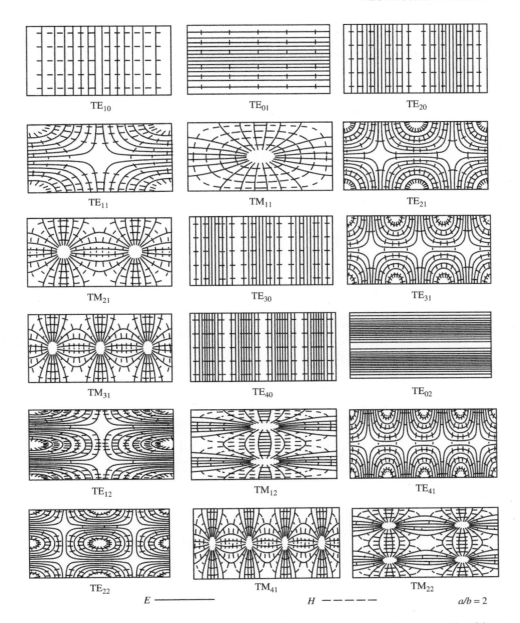

Figure 8-6 Field patterns for the first 18 TE^z and/or TM^z modes in a rectangular waveguide with $a/b = 2$. (Source: C. S. Lee, S. W. Lee, and S. L. Chuang, "Plot of modal field distribution in rectangular and circular waveguides," *IEEE Trans. Microwave Theory Tech.,* © 1985, IEEE).

Since in a multimode waveguide operation the total power is distributed among the existing modes (this will be shown later in this chapter) and the instrumentation (detectors, probes, etc.) required to detect the total power of multimodes is more complex and expensive, it is often most desirable to operate the waveguide in a single mode.

The order in which the different modes enter a waveguide depends upon their cutoff frequency. The modes with cutoff frequencies equal to or smaller than the operational frequency can exist inside the waveguide. Because, in practice, most systems that utilize a rectangular waveguide design require excitation and detection instrumentation for a dominant TE_{10} mode operation, it is prudent at this time to devote some extra effort to examination of the characteristics of this mode.

For the TE_{10}^z mode the pertinent expressions for the field intensities and the various characteristic parameters are obtained from Section 8.2.1 by letting $m=1$ and $n=0$. Then we can write the following summary.

$$\underline{TE_{10}^z \ \text{Mode} \ (m=1, \ n=0)}$$

$$F_z^+(x, z) = A_{10} \cos\left(\frac{\pi}{a}x\right)e^{-j\beta_z z} \tag{8-36}$$

$$\beta_x = \frac{\pi}{a} = \frac{2\pi}{\lambda_x} \ \Rightarrow \ \lambda_x = 2a \tag{8-37a}$$

$$\beta_y = 0 = \frac{2\pi}{\lambda_y} \ \Rightarrow \ \lambda_y = \infty \tag{8-37b}$$

$$\beta_c = \beta_x = \frac{\pi}{a} = \frac{2\pi}{\lambda_c} \ \Rightarrow \ \lambda_c = 2a \tag{8-37c}$$

$$\beta_z = \begin{cases} \sqrt{\beta^2 - \left(\frac{\pi}{a}\right)^2} = \beta\sqrt{1 - \left(\frac{\lambda}{2a}\right)^2} \\ = \beta\sqrt{1 - \left(\frac{\lambda}{\lambda_c}\right)^2} = \beta\sqrt{1 - \left(\frac{f_c}{f}\right)^2} & \text{for } f > f_c \quad (8\text{-}38a) \\ 0 & \text{for } f = f_c \quad (8\text{-}38b) \\ -j\sqrt{\left(\frac{\pi}{a}\right)^2 - \beta^2} = -j\beta\sqrt{\left(\frac{\lambda}{2a}\right)^2 - 1} \\ = -j\beta\sqrt{\left(\frac{\lambda}{\lambda_c}\right)^2 - 1} = -j\beta\sqrt{\left(\frac{f_c}{f}\right)^2 - 1} & \text{for } f < f_c \quad (8\text{-}38c) \end{cases}$$

$$E_x^+ = 0 \tag{8-39a}$$

$$E_y^+ = -\frac{A_{10}}{\varepsilon}\frac{\pi}{a}\sin\left(\frac{\pi}{a}x\right)e^{-j\beta_z z} \tag{8-39b}$$

$$E_z^+ = 0 \tag{8-39c}$$

$$H_x^+ = A_{10}\frac{\beta_z}{\omega\mu\varepsilon}\frac{\pi}{a}\sin\left(\frac{\pi}{a}x\right)e^{-j\beta_z z} \tag{8-39d}$$

$$H_y^+ = 0 \tag{8-39e}$$

$$H_z^+ = -j\frac{A_{10}}{\omega\mu\varepsilon}\left(\frac{\pi}{a}\right)^2\cos\left(\frac{\pi}{a}x\right)e^{-j\beta_z z} \tag{8-39f}$$

$$\mathbf{J}^+ = \hat{\mathbf{n}} \times \mathbf{H}^+|_{\text{wall}} = \begin{cases} \hat{\mathbf{a}}_y \times (\hat{\mathbf{a}}_x H_x^+ + \hat{\mathbf{a}}_z H_z^+)|_{y=0} = (+\hat{\mathbf{a}}_x H_z^+ - \hat{\mathbf{a}}_z H_x^+)|_{y=0} \\ = -\hat{\mathbf{a}}_x j\frac{A_{10}}{\omega\mu\varepsilon}\left(\frac{\pi}{a}\right)^2\cos\left(\frac{\pi}{a}x\right)e^{-j\beta_z z} - \hat{\mathbf{a}}_z A_{10}\frac{\beta_z}{\omega\mu\varepsilon}\frac{\pi}{a}\sin\left(\frac{\pi}{a}x\right)e^{-j\beta_z z} \\ \hline \hspace{3cm} \text{for the bottom wall} \hspace{3cm} (8\text{-}39g) \\ \\ = \hat{\mathbf{a}}_x \times \hat{\mathbf{a}}_z H_z^+|_{x=0} = -\hat{\mathbf{a}}_y H_z^+|_{x=0} = \hat{\mathbf{a}}_y j\frac{A_{10}}{\omega\mu\varepsilon}\left(\frac{\pi}{a}\right)^2 e^{-j\beta_z z} \\ \hline \hspace{3cm} \text{for the left wall} \hspace{3cm} (8\text{-}39h) \end{cases}$$

$$(\lambda_z)_{10} = (\lambda_g)_{10} = \begin{cases} \dfrac{\lambda}{\sqrt{1-\left(\frac{f_c}{f}\right)^2}} = \dfrac{\lambda}{\sqrt{1-\left(\frac{\lambda}{\lambda_c}\right)^2}} \\ \qquad\qquad = \dfrac{\lambda}{\sqrt{1-\left(\frac{\lambda}{2a}\right)^2}} \quad \text{for } f > f & \text{(8-40a)} \\ \infty \qquad\qquad\qquad\qquad \text{for } f = f_c & \text{(8-40b)} \\ j\dfrac{\lambda}{\sqrt{\left(\frac{f_c}{f}\right)^2-1}} = j\dfrac{\lambda}{\sqrt{\left(\frac{\lambda}{\lambda_c}\right)^2-1}} \\ \qquad\qquad = j\dfrac{\lambda}{\sqrt{\left(\frac{\lambda}{2a}\right)^2-1}} \quad \text{(nonphysical) for } f < f_c & \text{(8-40c)} \end{cases}$$

$$(f_c)_{10} = \frac{1}{2a\sqrt{\mu\varepsilon}} = \frac{v}{2a} = \frac{v}{(\lambda_c)_{10}} \tag{8-41}$$

$$Z_w^{+z}(\mathrm{TE}_{10}^z) = \begin{cases} \dfrac{\eta}{\sqrt{1-\left(\frac{f_c}{f}\right)^2}} & \text{for } f > f_c & \text{(8-42a)} \\ \infty & \text{for } f = f_c & \text{(8-42b)} \\ j\dfrac{\eta}{\sqrt{\left(\frac{f_c}{f}\right)^2-1}} & \text{for } f < f_c & \text{(8-42c)} \end{cases}$$

For the TE_{10} mode at cutoff ($\beta_z = 0 \Rightarrow \lambda_z = \infty$) the wavelength of the wave inside the guide in the x direction (λ_x) is, according to (8-14c), equal to the wavelength of the wave in an unbounded medium (λ). That is ($\lambda_x)_{10} = \lambda$ at cutoff.

From the preceding information, it is evident that the electric field intensity inside the guide has only one component, E_y. The E- and H-field variations on the top, front, and side views of the guide are shown graphically in Figure 8-7, and the current density and H-field lines on the top and side views are shown in Figure 8-8 [2]. It is instructive at this time to examine the electric field intensity a little closer and attempt to provide some physical interpretation of the propagation characteristics of the waveguide. The total electric field of (8-39a) through (8-39c) can also be written, by representing the sine function with exponentials, as

$$\mathbf{E}^+(x,z) = \hat{\mathbf{a}}_y E_y^+(x,\,z) = -\hat{\mathbf{a}}_y \frac{A_{10}}{\varepsilon}\frac{\pi}{a}\sin\left(\frac{\pi}{a}x\right)e^{-j\beta_z z}$$

$$= -\hat{\mathbf{a}}_y \frac{A_{10}}{\varepsilon}\frac{\pi}{a}\left[\frac{e^{j[(\pi/a)x-\beta_z z]}-e^{-j[(\pi/a)x+\beta_z z]}}{2j}\right]$$

$$\mathbf{E}^+(x,z) = \hat{\mathbf{a}}_y j\frac{A_{10}}{2\varepsilon}\frac{\pi}{a}\left[e^{j[(\pi/a)x-\beta_z z]}-e^{-j[(\pi/a)x+\beta_z z]}\right] \tag{8-43}$$

Letting

$$\frac{\pi}{a} = \beta_x = \beta \sin \psi \tag{8-44a}$$

$$\beta_z = \beta \cos \psi \tag{8-44b}$$

which satisfy the constraint equation 8-7a, or

$$\beta_x^2 + \beta_y^2 + \beta_z^2 = \beta^2 \sin^2 \psi + \beta^2 \cos^2 \psi = \beta^2 \tag{8-45}$$

We can write (8-43), using (8-44a) and (8-44b), as

$$\mathbf{E}^+(x,z) = \hat{\mathbf{a}}_y j \frac{A_{10}}{2\varepsilon} \frac{\pi}{a} \left[e^{j\beta(x\sin\psi - z\cos\psi)} - e^{-j\beta(x\sin\psi + z\cos\psi)} \right] \tag{8-46}$$

A close inspection of the two exponential terms inside the brackets indicates, by referring to the contents of Section 4.2.2, that each represents a uniform plane wave traveling in a direction determined by the angle ψ. In Figure 8-9a, which represents a top view of the waveguide of Figure 8-3, the two plane waves representing (8-46) or

$$\mathbf{E}^+(x,z) = \mathbf{E}_1^+(x,z) + \mathbf{E}_2^+(x,z) \tag{8-47}$$

where

$$\mathbf{E}_1^+(x,z) = \hat{\mathbf{a}}_y j \frac{A_{10}}{2\varepsilon} \frac{\pi}{a} [e^{j\beta(x\sin\psi - z\cos\psi)}] \tag{8-47a}$$

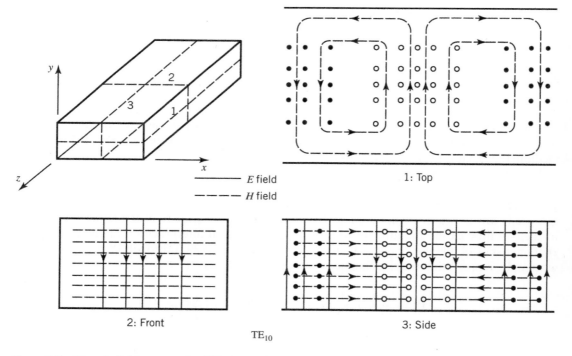

—— E field
----- H field

1: Top

2: Front

3: Side

TE_{10}

Figure 8-7   Electric field patterns for TE_{10} mode in a rectangular waveguide. (Source: S. Ramo, J. R. Whinnery, and T. Van Duzer, *Fields and Waves in Communication Electronics*, 1984. Reprinted with permission of John Wiley & Sons, Inc.)

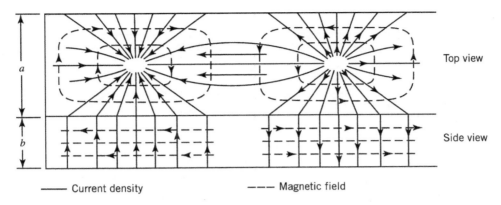

— Current density   ---- Magnetic field

Figure 8-8 Magnetic field and electric current density patterns for the TE_{10} mode in a rectangular wave-guide. (Source: S. Ramo, J. R. Whinnery, and T. Van Duzer, *Fields and Waves in Communication Electronics*, 1984. Reprinted with permission of John Wiley & Sons, Inc.)

$$\mathbf{E}_2^+(x,z) = -\hat{\mathbf{a}}_y j \frac{A_{10}}{2\varepsilon}\frac{\pi}{a}[e^{-j\beta(x\sin\psi + z\cos\psi)}] \tag{8-47b}$$

are indicated as two plane waves that bounce back and forth between the side walls of the wave-guide at an angle ψ. There is a 180° phase reversal between the two, which is also indicated in Figure 8-9a.

According to (8-44b)

$$\beta_z = \beta\cos\psi \Rightarrow \psi = \cos^{-1}\left(\frac{\beta_z}{\beta}\right) \tag{8-48}$$

By using (8-38a), we can write (8-48) as

$$\psi = \begin{cases} 0° & \text{for } f = \infty \tag{8-49a} \\ \cos^{-1}\left[\sqrt{1-\left(\frac{f_c}{f}\right)^2}\right] & \text{for } f_c \le f < \infty \tag{8-49b} \\ 90° & \text{for } f = f_c \tag{8-49c} \end{cases}$$

It is apparent that as $f \to f_c$, the angle ψ approaches 90° and exactly at cutoff ($f = f_c \Rightarrow \psi = 90°$) the plane waves bounce back and forth between the side walls of the waveguide without moving in the z direction. This reduces the fields into standing waves at cutoff.

By using (8-44b), the guide wavelength can be written as

$$\beta_z = \frac{2\pi}{\lambda_z} = \frac{2\pi}{\lambda_g} = \beta\cos\psi \Rightarrow \lambda_g = \frac{2\pi}{\beta\cos\psi} = \frac{\lambda}{\cos\psi} \tag{8-50}$$

which indicates that as cutoff approaches ($\psi \to 90°$), the guide wavelength approaches infinity. In addition, the phase velocity v_p can also be obtained using (8-44b), that is

$$\beta_z \equiv \frac{\omega}{v_p} = \beta\cos\psi = \frac{\omega}{v}\cos\psi \Rightarrow v_p = \frac{v}{\cos\psi} \tag{8-51}$$

where v is the velocity with which the plane wave travels along the direction determined by ψ. Since the phase velocity, as given by (8-51), is greater than the velocity of light, it may be

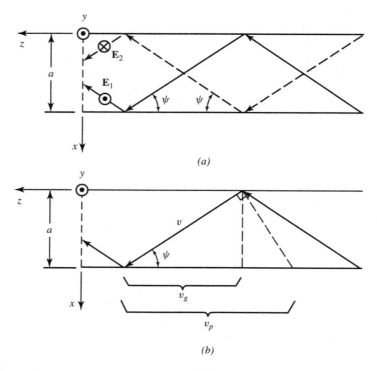

Figure 8-9 Uniform plane wave representation of the TE_{10} mode electric field inside a rectangular wave-guide. (*a*) Two uniform plane waves. (*b*) Phase and group velocities.

appropriate to illustrate graphically its meaning. By referring to Figure 8-9*b*, it is evident that whereas v is the velocity of the uniform plane wave along the direction determined by ψ, v_p ($v_p \geq v$) is the phase velocity, that is, the velocity that must be maintained to keep in step with a constant phase front of the wave, and v_g ($v_g \leq v$) is the group velocity, that is, the velocity with which a uniform plane wave travels along the z direction. According to Figure 8-9*b*

$$v_p = \frac{v}{\cos \psi} \tag{8-52a}$$

$$v_g = v \cos \psi \tag{8-52b}$$

and

$$v_p v_g = v^2 \tag{8-52c}$$

These are the same interpretations given to the oblique plane wave propagation in Section 4.2.2C and Figure 4-6. A plot of v_p and v_g as a function of frequency in the range $0 \leq f/f_c \leq 3$ is shown in Figure 8-10.

Above cutoff, the wave impedance of (8-19) can be written in terms of the angle ψ as

$$Z_w^{+z}(TE_{10}) = \frac{\omega\mu}{\beta_z} = \frac{\omega\mu}{\beta \cos \psi} = \frac{\eta}{\cos \psi} \tag{8-53}$$

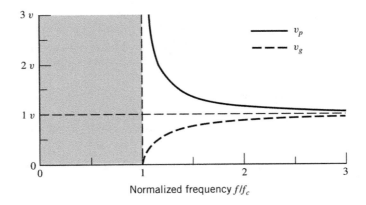

Figure 8-10 Phase and group (energy) velocities for the TE_{10} mode in a rectangular waveguide.

whose values are equal to or greater than the intrinsic impedance η of the medium inside the waveguide.

Example 8-3

Design an air-filled rectangular waveguide with dimensions a and b ($a > b$) that will operate in the dominant TE_{10} mode at $f = 10$ GHz. The dimensions a and b of the waveguide should be chosen so that at $f = 10$ GHz the waveguide not only operates on the single TE_{10} mode but also that $f = 10$ GHz is *simultaneously* 25% above the cutoff frequency of the dominant TE_{10} mode and 25% below the next higher-order TE_{01} mode.

Solution: According to (8-22), the cutoff frequency of the TE_{10} mode with a free-space medium in the guide is

$$(f_c)_{10} = \frac{1}{2a\sqrt{\mu_0\varepsilon_0}} = \frac{30 \times 10^9}{2a}$$

Since $f = 10$ GHz must be greater by 25% above the cutoff frequency of the TE_{10} mode, then

$$10 \times 10^9 \geq 1.25\left(\frac{30 \times 10^9}{2a}\right) \implies a \geq 1.875 \text{ cm} = 0.738 \text{ in.}$$

Since the next higher-order mode is the TE_{01} mode, whose cutoff frequency with a free-space medium in the guide is

$$(f_c)_{01} = \frac{1}{2b\sqrt{\mu_0\varepsilon_0}} = \frac{30 \times 10^9}{2b}$$

then

$$10 \times 10^9 \leq 0.75\left(\frac{30 \times 10^9}{2b}\right) \implies b \leq 1.125 \text{ cm} = 0.443 \text{ in.}$$

Example 8-4

Design a rectangular waveguide with dimensions a and b $(a > b)$ that will operate in a single mode between 9 and 14 GHz. Assuming free space inside the waveguide, determine the waveguide dimensions that will ensure single-mode operation over that band.

Solution: Since $a > b$, the dominant mode is the TE_{10}, whose cutoff frequency must be

$$(f_c)_{10} = \frac{1}{2a\sqrt{\mu_0 \varepsilon_0}} = \frac{30 \times 10^9}{2a} = 9 \times 10^9 \quad \Rightarrow \quad a = 1.667 \text{ cm} = 0.656 \text{ in.}$$

The cutoff frequency of the TE_{20} mode is 18 GHz. Therefore, the next higher-order mode is TE_{01}, whose cutoff frequency must be

$$(f_c)_{01} = \frac{1}{2b\sqrt{\mu_0 \varepsilon_0}} = \frac{30 \times 10^9}{2b} = 14 \times 10^9 \quad \Rightarrow \quad b = 1.071 \text{ cm} = 0.422 \text{ in.}$$

8.2.4   Power Density and Power

The fields that are created and propagating inside the waveguide have power associated with them. To find the power flowing down the guide, it is first necessary to find the average power density directed along the axis of the waveguide. The power flowing along the guide can then be found by integrating the axial directed power density over the cross section of the waveguide.

For the waveguide geometry of Figure 8-3, the z-directed power density can be written as

$$(\mathbf{S}_z)_{mn} = \hat{\mathbf{a}}_z S_z = \frac{1}{2} \text{Re} \left[\left(\hat{\mathbf{a}}_x E_x + \hat{\mathbf{a}}_y E_y \right) \times \left(\hat{\mathbf{a}}_x H_x + \hat{\mathbf{a}}_y H_y \right)^* \right]$$

$$(\mathbf{S}_z)_{mn} = \hat{\mathbf{a}}_z S_z = \hat{\mathbf{a}}_z \frac{1}{2} \text{Re} \left[E_x H_y^* - E_y H_x^* \right] \tag{8-54}$$

$$\underline{TE_{mn}^z \text{ Modes}}$$

Use of the field expressions (8-15a) through (8-15f) allows the z-directed power density of (8-54) for the TE_{mn}^z modes to be written as

$$(\mathbf{S}_z)_{mn} = \hat{\mathbf{a}}_z S_z = \hat{\mathbf{a}}_z \frac{|A_{mn}|^2}{2} \text{Re} \left[\frac{\beta_y^2 \beta_z}{\omega \mu \varepsilon^2} \cos^2 (\beta_x x) \sin^2 (\beta_y y) \right.$$

$$\left. + \frac{\beta_x^2 \beta_z}{\omega \mu \varepsilon^2} \sin^2 (\beta_x x) \cos^2 (\beta_y y) \right]$$

$$(\mathbf{S}_z)_{mn} = \hat{\mathbf{a}}_z S_z = \hat{\mathbf{a}}_z |A_{mn}|^2 \frac{\beta_z}{2\omega \mu \varepsilon^2} \left[\beta_y^2 \cos^2 (\beta_x x) \sin^2 (\beta_y y) \right.$$

$$\left. + \beta_x^2 \sin^2 (\beta_x x) \cos^2 (\beta_y y) \right] \tag{8-55}$$

The associated power is obtained by integrating (8-55) over a cross section A_0 of the guide, or

$$P_{mn} = \iint_{A_0} (\mathbf{S}_z)_{mn} \cdot d\mathbf{s} = \int_0^b \int_0^a (\hat{\mathbf{a}}_z S_z) \cdot (\hat{\mathbf{a}}_z \, dx \, dy) = \int_0^b \int_0^a S_z \, dx \, dy \tag{8-56}$$

Since

$$\int_0^a \cos^2\left(\frac{m\pi}{a}x\right)dx = \begin{cases} a/2 & m \neq 0 \\ a & m = 0 \end{cases} \tag{8-56a}$$

$$\int_0^a \sin^2\left(\frac{m\pi}{a}x\right)dx = \begin{cases} a/2 & m \neq 0 \\ 0 & m = 0 \end{cases} \tag{8-56b}$$

and similar equalities exist for the y variations, (8-56) reduces by using (8-55), (8-56a), and (8-56b) to

$$P_{mn}^{\mathrm{TE}^z} = |A_{mn}|^2 \frac{\beta_z}{2\omega\mu\varepsilon^2}\left[\beta_y^2\left(\frac{a}{\varepsilon_{0m}}\right)\left(\frac{b}{\varepsilon_{0n}}\right) + \beta_x^2\left(\frac{a}{\varepsilon_{0m}}\right)\left(\frac{b}{\varepsilon_{0n}}\right)\right]$$

$$= |A_{mn}|^2 \frac{\beta_z}{2\omega\mu\varepsilon^2}\left(\frac{a}{\varepsilon_{0m}}\right)\left(\frac{b}{\varepsilon_{0n}}\right)(\beta_x^2 + \beta_y^2)$$

$$P_{mn}^{\mathrm{TE}^z} = |A_{mn}|^2 \frac{\beta_z\beta_c^2}{2\omega\mu\varepsilon^2}\left(\frac{a}{\varepsilon_{0m}}\right)\left(\frac{b}{\varepsilon_{0n}}\right) = |A_{mn}|^2 \frac{\beta_c^2}{2\eta\varepsilon^2}\left(\frac{a}{\varepsilon_{0m}}\right)\left(\frac{b}{\varepsilon_{0n}}\right)\sqrt{1 - \left(\frac{f_{c,mn}}{f}\right)^2} \tag{8-57}$$

where

$$\varepsilon_{0q} = \begin{cases} 1 & q = 0 \\ 2 & q \neq 0 \end{cases} \tag{8-57a}$$

TM$_{mn}^z$ Modes

Use of a similar procedure, with (8-28a) through (8-28f), allows us to write that for the TM$_{mn}^z$ modes

$$(\mathbf{S}_z)_{mn} = \hat{\mathbf{a}}_z S_z = \hat{\mathbf{a}}_z |B_{mn}|^2 \frac{\beta_z}{2\omega\varepsilon\mu^2}\left[\beta_x^2\cos^2(\beta_x x)\sin^2(\beta_y y)\right.$$

$$\left. + \beta_y^2\sin^2(\beta_x x)\cos^2(\beta_y y)\right] \tag{8-58}$$

$$P_{mn}^{\mathrm{TM}^z} = |B_{mn}|^2 \frac{\beta_c^2\eta}{2\mu^2}\left(\frac{a}{2}\right)\left(\frac{b}{2}\right)\sqrt{1 - \left(\frac{f_{c,mn}}{f}\right)^2} \tag{8-59}$$

By the use of superposition, the total power associated with a wave is equal to the sum of all the power components associated with each mode that exists inside the waveguide. Thus

$$P_{\text{total}} = \sum_{m,n} P_{mn}^{\mathrm{TE}^z} + \sum_{m,n} P_{mn}^{\mathrm{TM}^z} \tag{8-60}$$

where $P_{mn}^{\mathrm{TE}^z}$ and $P_{mn}^{\mathrm{TM}^z}$ are given, respectively, by (8-57) and (8-59).

Example 8-5

The inside dimensions of an X-band WR90 waveguide are $a = 0.9$ in. (2.286 cm) and $b = 0.4$ in. (1.016 cm). Assume that the waveguide is air-filled and operates in the dominant TE$_{10}$ mode, and that the air will break down when the maximum electric field intensity is 3×10^6 V/m. Find the maximum power that can be transmitted at $f = 9$ GHz in the waveguide before air breakdown occurs.

Solution: Since air will break down when the maximum electric field intensity in the waveguide reaches 3×10^6 V/m, then according to (8-39b)

$$\left. \left| E_y \right| \right|_{max} = \frac{\left| A_{10} \right|}{\varepsilon_0} \frac{\pi}{a} \left| \sin\left(\frac{\pi}{a}x\right) \right|_{max} = \frac{A_{10}}{\varepsilon_0} \frac{\pi}{a} = 3 \times 10^6 \Rightarrow A_{10} = 1.933 \times 10^{-7}$$

By using (8-57)

$$P_{10}^{TE} = \left| A_{10} \right|^2 \frac{(\beta_c)_{10}^2}{2\eta_0 \varepsilon_0^2} \left(\frac{a}{\varepsilon_{01}}\right)\left(\frac{b}{\varepsilon_{00}}\right) \sqrt{1 - \left[\frac{(f_c)_{10}}{f}\right]^2}$$

Since the cutoff frequency of the dominant TE_{10} mode is

$$(f_c)_{10} = \frac{1}{2a\sqrt{\mu_0 \varepsilon_0}} = \frac{30 \times 10^9}{2(2.286)} = 6.562 \text{ GHz}$$

and

$$(\beta_c)_{10} = \frac{\pi}{a}$$

then

$$P_{10}^{TE} = (1.933 \times 10^{-7})^2 \frac{(\pi / 2.286 \times 10^{-2})^2}{2(377)(8.854 \times 10^{-12})^2} \left(\frac{2.286 \times 10^{-2}}{2}\right) \times \left(\frac{1.016 \times 10^{-2}}{1}\right) \sqrt{1 - \left(\frac{6.562}{9}\right)^2}$$

$$P_{10}^{TE} = 948.9 \times 10^3 \text{ W} = 948.9 \text{ kW}$$

8.2.5 Attenuation

Ideally, if the waveguide were made out of a perfect conductor, there would not be any attenuation associated with the guide above cutoff. Below cutoff, the fields reduce to evanescent (nonpropagating) waves that are highly attenuated. In practice, however, no perfect conductors exist, although many (such as metals) are very good conductors, with conductivities on the order of $10^7 - 10^8$ S/m. For waveguides made out of such conductors (metals), there must be some attenuation due to the *conduction (ohmic) losses* in the waveguides themselves. This is accounted for by introducing an attenuation coefficient α_c. Another factor that contributes to the waveguide attenuation is the losses associated with lossy dielectric materials that are inserted inside the guide. These losses are referred to as *dielectric losses* and are accounted for by introducing an attenuation coefficient α_d.

A. Conduction (Ohmic) Losses
To find the losses associated with a waveguide whose walls are not perfectly conducting, a new boundary-value problem must be solved. That problem would be the same one shown in Figure 8-3 but with nonperfectly conducting walls. To solve such a problem exactly is an ambitious and complicated task. Instead an alternate procedure is almost always used whereby the solution is obtained using a perturbational method. With that method, it is assumed that the fields inside the waveguide with lossy walls, but of very high conductivity, are slightly perturbed from those of perfectly conducting walls. The differences between the two sets of fields are so small that the fields are usually assumed to be essentially the same. However, the walls themselves are considered as lossy surfaces represented by a surface impedance Z_s given by (4-42), or

$$Z_s = R_s + jX_s = \sqrt{\frac{j\omega\mu}{\sigma + j\omega\varepsilon}} \overset{\sigma \gg \omega\varepsilon}{\simeq} \sqrt{\frac{j\omega\mu}{\sigma}} = \sqrt{\frac{\omega\mu}{2\sigma}}(1+j) \qquad (8\text{-}61)$$

The power P_c absorbed and dissipated as heat by each surface (wall) A_m of the waveguide is obtained using an expression analogous to $I^2R/2$ used in lumped-circuit theory, that is,

$$P_c = \frac{R_s}{2} \iint_{A_m} \mathbf{J}_s \cdot \mathbf{J}_s^* \, ds \tag{8-62}$$

where

$$\mathbf{J}_s \simeq \hat{\mathbf{n}} \times \mathbf{H}|_{\text{surface}} \tag{8-62a}$$

In (8-62a), $\mathbf{J}_s$ represents the linear current density (in A/m) induced on the surface of a lossy conductor, as discussed in Example 5-7, and illustrated graphically in Figure 5-11.

Once the total conduction power P_c dissipated as heat on the waveguide has been found by applying (8-62) on all four walls of the guide, the next step is to define and derive an expression for the attenuation coefficient α_c. This can be accomplished by referring to Figure 8-11, which represents the lossy waveguide in its axial direction. If P_0 represents the power at some reference point (e.g., $z = 0$), then the power P_{mn} at some other point z is related to P_0 (P_{mn} at $z = 0$) by

$$P_{mn}(z) = P_{mn}|_{z=0} \, e^{-2\alpha_c z} = P_0 e^{-2\alpha_c z} \tag{8-63}$$

The negative rate of change (with respect to z) of P_{mn} represents the dissipated power per unit length. Thus, for a length ℓ of a waveguide, the total dissipated power P_c is found using

$$P_c = -z \frac{dP_{mn}}{dz}\bigg|_{z=\ell} = -z \frac{d}{dz}\left(P_0 e^{-2\alpha_c z}\right)\bigg|_{z=\ell} = 2\alpha_c z P_0 e^{-2\alpha_c z}\big|_{z=\ell} \tag{8-64}$$

$$P_c = 2\alpha_c \ell P_{mn}$$

or

$$\boxed{\alpha_c = \frac{P_c/\ell}{2P_{mn}}} \tag{8-64a}$$

In (8-64a), P_c is obtained using (8-62) and P_{mn} is represented by (8-57) or (8-59).

The derivation of P_c, as given by (8-62), for any m, n TE^z or TM^z mode is a straightforward but tedious process. To reduce the complexity, we will illustrate the procedure for the TE$_{10}$ mode. The derivation for the m, n mode is assigned as an end-of-chapter problem.

Use of the geometry of Figure 8-3 allows the total power P_c dissipated on the walls of the waveguide to be written as

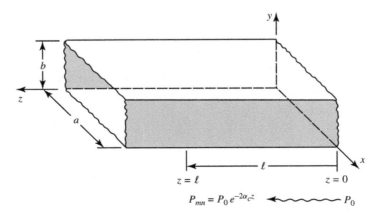

Figure 8-11 Rectangular waveguide geometry for attenuation constant derivation.

$$(P_c)_{10} = 2\left[\frac{R_s}{2}\iint_{\substack{\text{bottom wall} \\ (y=0)}}\mathbf{J}_{sb}\cdot\mathbf{J}_{sb}^*\,ds + \frac{R_s}{2}\iint_{\substack{\text{left wall} \\ (x=0)}}\mathbf{J}_{s\ell}\cdot\mathbf{J}_{s\ell}^*\,ds\right]$$

$$(P_c)_{10} = R_s\left[\iint_{\substack{\text{bottom wall} \\ (y=0)}}\mathbf{J}_{sb}\cdot\mathbf{J}_{sb}^*\,ds + \iint_{\substack{\text{left wall} \\ (x=0)}}\mathbf{J}_{s\ell}\cdot\mathbf{J}_{s\ell}^*\,ds\right] \qquad \text{(8-65)}$$

Since the losses on the top wall are the same as those on the bottom and those on the right wall are the same as those on the left, a factor of 2 was used in (8-65) to multiply the losses of the bottom and left walls.

In (8-65), $\mathbf{J}_{sb}$ and $\mathbf{J}_{s\ell}$ represent the linear current densities on the bottom and left walls of the guide, which are equal to

$$\mathbf{J}_{sb} = \hat{\mathbf{n}}\times\mathbf{H}|_{y=0} = \hat{\mathbf{a}}_y\times(\hat{\mathbf{a}}_x H_x + \hat{\mathbf{a}}_z H_z)|_{y=0} = (\hat{\mathbf{a}}_x H_z - \hat{\mathbf{a}}_z H_x)|_{y=0}$$

$$= -\hat{\mathbf{a}}_x j\frac{A_{10}}{\omega\mu\varepsilon}\left(\frac{\pi}{a}\right)^2\cos\left(\frac{\pi}{a}x\right)e^{-j\beta_z z} - \hat{\mathbf{a}}_z A_{10}\frac{\beta_z}{\omega\mu\varepsilon}\left(\frac{\pi}{a}\right)\sin\left(\frac{\pi}{a}x\right)e^{-j\beta_z z} \qquad \text{(8-66a)}$$

$$\mathbf{J}_{s\ell} = \hat{\mathbf{n}}\times\mathbf{H}|_{x=0} = \hat{\mathbf{a}}_x\times(\hat{\mathbf{a}}_x H_x + \hat{\mathbf{a}}_z H_z)|_{x=0} = -\hat{\mathbf{a}}_y H_z|_{x=0}$$

$$= \hat{\mathbf{a}}_y j\frac{A_{10}}{\omega\mu\varepsilon}\left(\frac{\pi}{a}\right)^2 e^{-j\beta_z z} \qquad \text{(8-66b)}$$

Use of (8-66a) and (8-66b) allows us to write the losses associated with the bottom–top and left–right walls, as given by (8-65), as

$$R_s\iint_{\substack{\text{bottom wall} \\ (y=0)}}\mathbf{J}_{sb}\cdot\mathbf{J}_{sb}^*\,ds = R_s\,|A_{10}|^2\left(\frac{\pi}{a}\right)^2\frac{1}{(\omega\mu\varepsilon)^2}\left\{\left(\frac{\pi}{a}\right)^2\int_0^\ell\int_0^a\cos^2\left(\frac{\pi}{a}x\right)dx\,dz\right.$$
$$\left. + \beta_z^2\int_0^\ell\int_0^a\sin^2\left(\frac{\pi}{a}x\right)dx\,dz\right\}$$

$$= \ell R_s\frac{|A_{10}|^2}{(\omega\mu\varepsilon)^2}\frac{a}{2}\left[\left(\frac{\pi}{a}\right)^2 + \beta_z^2\right]\left(\frac{\pi}{a}\right)^2 \qquad \text{(8-67a)}$$

$$R_s\iint_{\substack{\text{left wall} \\ (x=0)}}\mathbf{J}_{s\ell}\cdot\mathbf{J}_{s\ell}^*\,ds = R_s\,|A_{10}|^2\frac{b\ell}{(\omega\mu\varepsilon)^2}\left(\frac{\pi}{a}\right)^4 \qquad \text{(8-67b)}$$

The total dissipated power per unit length can be obtained by combining (8-67a) and (8-67b). Thus, it can be shown that

$$\frac{(P_c)_{10}}{\ell} = \frac{aR_s}{2\eta^2}\frac{|A_{10}|^2}{\varepsilon^2}\left(\frac{\pi}{a}\right)^2\left[1 + \frac{2b}{a}\left(\frac{f_c}{f}\right)^2\right] \qquad \text{(8-68)}$$

By using (8-57) for $m=1$ and $n=0$, and (8-68), the attenuation coefficient of (8-64a) for the TE_{10} mode can be written as

$$(\alpha_c)_{10} = \left[\frac{(P_c)_{10}/\ell}{2P_{mn}}\right]_{\substack{m=1 \\ n=0}} = \frac{\dfrac{aR_s}{2\eta^2}\dfrac{|A_{10}|^2}{\varepsilon^2}\left(\dfrac{\pi}{a}\right)^2\left[1 + \dfrac{2b}{a}\left(\dfrac{f_c}{f}\right)^2\right]}{|A_{10}|^2\dfrac{\beta_z^2}{\eta\varepsilon^2}\left(\dfrac{a}{2}\right)(b)\sqrt{1 - \left(\dfrac{f_c}{f}\right)^2}} \qquad \text{(8-69)}$$

which reduces to

$$(\alpha_c)_{10} = \frac{R_s}{\eta b} \frac{\left[1 + \frac{2b}{a}\left(\frac{f_c}{f}\right)^2\right]}{\sqrt{1 - \left(\frac{f_c}{f}\right)^2}} \quad (\text{in Np/m}) \tag{8-69a}$$

For an X-band WR waveguide with inner dimensions $a = 0.9$ in. (2.286 cm) and $b = 0.4$ in. (1.016 cm), made of copper ($\sigma = 5.7 \times 10^7$ S/m) and filled with a lossless dielectric, we can plot the attenuation coefficient $(\alpha_c)_{10}$ (in Np/m and dB/m) for $\varepsilon = \varepsilon_0$, $2.56\varepsilon_0$, and $4\varepsilon_0$ as shown in Figure 8-12. The attenuation coefficient for any mode (TE_{mn} or TM_{mn}) is given by

$$\underline{\text{TE}_{mn}^z}$$

$$(\alpha_c)_{mn} = \frac{2R_s}{\varepsilon_m \varepsilon_n b \eta \sqrt{1 - \left(\frac{f_{c,mn}}{f}\right)^2}} \left\{ \left(\varepsilon_m + \varepsilon_n \frac{b}{a}\right)\left(\frac{f_{c,mn}}{f}\right)^2 \right.$$
$$\left. + \frac{b}{a}\left[1 - \left(\frac{f_{c,mn}}{f}\right)^2\right]\frac{m^2 ab + (na)^2}{(mb)^2 + (na)^2} \right\} \tag{8-70a}$$

where

$$\varepsilon_p = \begin{cases} 2 & p = 0 \\ 1 & p \neq 0 \end{cases} \tag{8-70b}$$

$$\underline{\text{TM}_{mn}^z}$$

$$(\alpha_c)_{mn} = \frac{2R_s}{ab\eta\sqrt{1 - \left(\frac{f_{c,mn}}{f}\right)^2}} \frac{m^2 b^3 + n^2 a^3}{(mb)^2 + (na)^2} \tag{8-70c}$$

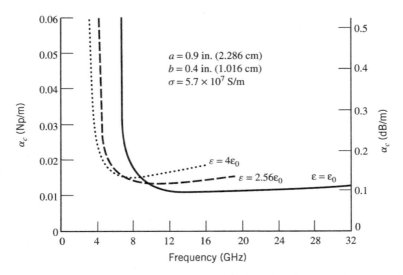

Figure 8-12 TE_{10} mode attenuation constant for the X-band rectangular waveguide.

The most pertinent equations used to describe the characteristics of TE_{mn} and TM_{mn} modes inside a rectangular waveguide are summarized in Table 8-3.

B. Dielectric Losses When waveguides are filled with lossy dielectric material, an additional attenuation constant must be introduced to account for losses in the dielectric material and are usually designated as *dielectric losses*. Thus, the total attenuation constant α_t for the waveguide above cutoff is given by

$$\alpha_t = \alpha_c + \alpha_d \tag{8-71}$$

where

α_t = total attenuation constant
α_c = ohmic losses attenuation constant [(8-70a) and (8-70c)]
α_d = dielectric losses attenuation constant

To derive α_d, let us refer to the constraint equation (8-7a), which, for a lossy medium ($\beta = \dot\beta_e$), can be written for the complex $\dot\beta_z$ as

$$\dot\beta_z^2 = \dot\beta_e^2 - (\beta_x^2 + \beta_y^2) = \dot\beta_e^2 - \beta_c^2 \tag{8-72}$$

or

$$\dot\beta_z = \sqrt{\dot\beta_e^2 - \beta_c^2} = \sqrt{\omega^2 \mu \dot\varepsilon_e - \beta_c^2} = \sqrt{\omega^2 \mu(\varepsilon_e' - j\varepsilon_e'') - \beta_c^2}$$

$$= \sqrt{(\omega^2 \mu \varepsilon_e' - \beta_c^2) - j\omega^2 \mu \varepsilon_e''} = \sqrt{(\beta_c'^2 - \beta_c^2) - j\omega\mu\sigma_e}$$

$$\dot\beta_z = \sqrt{\beta_c'^2 - \beta_c^2}\left[1 - j\frac{\omega\mu\sigma_e}{\beta_c'^2 - \beta_c^2}\right]^{1/2} \tag{8-72a}$$

where

$$\beta' = \omega\sqrt{\mu\varepsilon_e'} \tag{8-72b}$$

By using the binomial expansion, (8-72a) can be approximated by

$$\dot\beta_z \simeq \sqrt{\beta_e'^2 - \beta_c^2}\left[1 - j\frac{\omega\mu\sigma_e}{2(\beta_e'^2 - \beta_c^2)}\right] = \sqrt{\beta_e'^2 - \beta_c^2} - j\frac{\omega\mu\sigma_e}{2\sqrt{\beta_e'^2 - \beta_c^2}}$$

$$\simeq \beta_e'\sqrt{1 - \left(\frac{f_c}{f}\right)^2} - j\frac{\omega\mu\sigma_e}{2\beta_e'\sqrt{1 - \left(\frac{f_c}{f}\right)^2}}$$

$$\dot\beta_z \simeq \omega\sqrt{\mu\varepsilon_e'}\sqrt{1 - \left(\frac{f_c}{f}\right)^2} - j\frac{\omega\mu\sigma_e}{2\omega\sqrt{\mu\varepsilon_e'}\sqrt{1 - \left(\frac{f_c}{f}\right)^2}}$$

$$\dot\beta_z \simeq \omega\sqrt{\mu\varepsilon_e'}\sqrt{1 - \left(\frac{f_c}{f}\right)^2} - j\frac{\eta_e'}{2}\frac{\sigma_e}{\sqrt{1 - \left(\frac{f_c}{f}\right)^2}} \tag{8-73}$$

where

$$\eta_e' = \sqrt{\frac{\mu}{\varepsilon_e'}} \tag{8-73a}$$

TABLE 8-3   Summary of TE^z_{mn} and TM^z_{mn} mode characteristics of rectangular waveguide

	$\text{TE}^z_{mn}\begin{pmatrix} m = 0, 1, 2, \dots \\ n = 0, 1, 2, \dots \\ m \text{ and } n \text{ not both zero simultaneously} \end{pmatrix}$	$\text{TM}^z_{mn}\begin{pmatrix} m = 1, 2, 3, \dots \\ n = 1, 2, 3, \dots \end{pmatrix}$
E_x^+	$A_{mn}\dfrac{n\pi}{b\varepsilon}\cos\left(\dfrac{m\pi}{a}x\right)\sin\left(\dfrac{n\pi}{b}y\right)e^{-j\beta_z z}$	$-B_{mn}\dfrac{m\pi\beta_z}{a\omega\mu\varepsilon}\cos\left(\dfrac{m\pi}{a}x\right)\sin\left(\dfrac{n\pi}{b}y\right)e^{-j\beta_z z}$
E_y^+	$-A_{mn}\dfrac{m\pi}{a\varepsilon}\sin\left(\dfrac{m\pi}{a}x\right)\cos\left(\dfrac{n\pi}{b}y\right)e^{-j\beta_z z}$	$-B_{mn}\dfrac{n\pi\beta_z}{b\omega\mu\varepsilon}\sin\left(\dfrac{m\pi}{a}x\right)\cos\left(\dfrac{n\pi}{b}y\right)e^{-j\beta_z z}$
E_z^+	0	$-jB_{mn}\dfrac{\beta_c^2}{\omega\mu\varepsilon}\sin\left(\dfrac{m\pi}{a}x\right)\sin\left(\dfrac{n\pi}{b}y\right)e^{-j\beta_z z}$
H_x^+	$A_{mn}\dfrac{m\pi\beta_z}{a\omega\mu\varepsilon}\sin\left(\dfrac{m\pi}{a}x\right)\cos\left(\dfrac{n\pi}{b}y\right)e^{-j\beta_z z}$	$B_{mn}\dfrac{n\pi}{b\mu}\sin\left(\dfrac{m\pi}{a}x\right)\cos\left(\dfrac{n\pi}{b}y\right)e^{-j\beta_z z}$
H_y^+	$A_{mn}\dfrac{n\pi\beta_z}{b\omega\mu\varepsilon}\cos\left(\dfrac{m\pi}{a}x\right)\sin\left(\dfrac{n\pi}{b}y\right)e^{-j\beta_z z}$	$-B_{mn}\dfrac{m\pi}{a\mu}\cos\left(\dfrac{m\pi}{a}x\right)\sin\left(\dfrac{n\pi}{b}y\right)e^{-j\beta_z z}$
H_z^+	$-jA_{mn}\dfrac{\beta_c^2}{\omega\mu\varepsilon}\cos\left(\dfrac{m\pi}{a}x\right)\cos\left(\dfrac{n\pi}{b}y\right)e^{-j\beta_z z}$	0
β_c	$\sqrt{\beta_x^2+\beta_y^2}=\sqrt{\left(\dfrac{m\pi}{a}\right)^2+\left(\dfrac{n\pi}{b}\right)^2}$	
f_c	$\dfrac{1}{2\pi\sqrt{\mu\varepsilon}}\sqrt{\left(\dfrac{m\pi}{a}\right)^2+\left(\dfrac{n\pi}{b}\right)^2}$	
λ_c	$\dfrac{2\pi}{\sqrt{\left(\dfrac{m\pi}{a}\right)^2+\left(\dfrac{n\pi}{b}\right)^2}}$	
$\beta_z(f \geq f_c)$	$\beta\sqrt{1-\left(\dfrac{f_c}{f}\right)^2}$	
$\lambda_g(f \geq f_c)$	$\dfrac{\lambda}{\sqrt{1-\left(\dfrac{f_c}{f}\right)^2}}=\dfrac{\lambda}{\sqrt{1-\left(\dfrac{\lambda}{\lambda_c}\right)^2}}$	
$v_p(f \geq f_c)$	$\dfrac{v}{\sqrt{1-\left(\dfrac{f_c}{f}\right)^2}}=\dfrac{v}{\sqrt{1-\left(\dfrac{\lambda}{\lambda_c}\right)^2}}$	
$Z_w(f \geq f_c)$	$\dfrac{\eta}{\sqrt{1-\left(\dfrac{f_c}{f}\right)^2}}=\dfrac{\eta}{\sqrt{1-\left(\dfrac{\lambda}{\lambda_c}\right)^2}}$	$\eta\sqrt{1-\left(\dfrac{f_c}{f}\right)^2}=\eta\sqrt{1-\left(\dfrac{\lambda}{\lambda_c}\right)^2}$

TABLE 8-3 (Continued)

$\text{TE}^z_{mn}\begin{pmatrix} m=0,1,2,\dots \\ n=0,1,2,\dots \\ m \text{ and } n \text{ not both zero simultaneously} \end{pmatrix}$	$\text{TM}^z_{mn}\begin{pmatrix} m=1,2,3,\dots \\ n=1,2,3,\dots \end{pmatrix}$
$Z_w(f \le f_c)$ $j\dfrac{\eta}{\sqrt{\left(\dfrac{f_c}{f}\right)^2-1}}=j\dfrac{\eta}{\sqrt{\left(\dfrac{\lambda}{\lambda_c}\right)^2-1}}$	$-j\eta\sqrt{\left(\dfrac{f_c}{f}\right)^2-1}=-j\eta\sqrt{\left(\dfrac{\lambda}{\lambda_c}\right)^2-1}$

$$(\alpha_c)_{mn} \quad \dfrac{2R_s}{\varepsilon_m\varepsilon_n b\eta\sqrt{1-\left(\dfrac{f_{c,mn}}{f}\right)^2}}\left\{\left(\varepsilon_m+\varepsilon_n\dfrac{b}{a}\right)\left(\dfrac{f_{c,mn}}{f}\right)^2\right.$$

$$\quad\quad \dfrac{2R_s}{abn\sqrt{1-\left(\dfrac{f_{c,mn}}{f}\right)^2}}\dfrac{m^2b^3+n^2a^3}{(mb)^2+(na)^2}$$

$$\left.+\dfrac{b}{a}\left[1-\left(\dfrac{f_{c,mn}}{f}\right)^2\right]\dfrac{m^2ab+(na)^2}{(mb)^2+(na)^2}\right\}$$

$$\text{where } \varepsilon_p=\begin{cases}2 & p=0 \\ 1 & p\ne 0\end{cases}$$

Let us define the complex $\dot{\beta}_z$ as

$$\dot{\beta}_z \equiv \beta_d - j\alpha_d \simeq \beta'_e\sqrt{1-\left(\dfrac{f_c}{f}\right)^2}-j\dfrac{\eta'_e}{2}\dfrac{\sigma_e}{\sqrt{1-\left(\dfrac{f_c}{f}\right)^2}} \tag{8-74}$$

or

$$\alpha_d \equiv \text{attenuation constant} \simeq \dfrac{\eta'_e}{2}\dfrac{\sigma_e}{\sqrt{1-\left(\dfrac{f_c}{f}\right)^2}} \tag{8-74a}$$

$$\beta_d \equiv \text{phase constant} \simeq \beta'_e\sqrt{1-\left(\dfrac{f_c}{f}\right)^2} \tag{8-74b}$$

Another form of (8-74a) would be to write it as

$$\alpha_d \simeq \dfrac{\eta'_e}{2}\dfrac{\sigma_e}{\sqrt{1-\left(\dfrac{f_c}{f}\right)^2}}=\dfrac{1}{2}\sqrt{\dfrac{\mu}{\varepsilon'_e}}\dfrac{\omega\varepsilon''_e}{\sqrt{1-\left(\dfrac{f_c}{f}\right)^2}}$$

$$\simeq \dfrac{1}{2}\dfrac{\varepsilon''_e}{\varepsilon'_e}\dfrac{\omega\sqrt{\mu\varepsilon'_e}}{\sqrt{1-\left(\dfrac{f_c}{f}\right)^2}}=\dfrac{1}{2}\dfrac{\varepsilon''_e}{\varepsilon'_e}\dfrac{\beta'_e}{\sqrt{1-\left(\dfrac{f_c}{f}\right)^2}}$$

$$\simeq \dfrac{1}{2}\dfrac{\varepsilon''_e}{\varepsilon'_e}\dfrac{2\pi}{\lambda\sqrt{1-\left(\dfrac{f_c}{f}\right)^2}}=\dfrac{\pi}{\lambda}\dfrac{\varepsilon''_e}{\varepsilon'_e}\dfrac{1}{\sqrt{1-\left(\dfrac{f_c}{f}\right)^2}}$$

$$\alpha_d \simeq \dfrac{\varepsilon''_e}{\varepsilon'_e}\dfrac{\pi}{\lambda^2}\dfrac{\lambda}{\sqrt{1-\left(\dfrac{f_c}{f}\right)^2}}=\dfrac{\varepsilon''_e}{\varepsilon'_e}\dfrac{\pi}{\lambda}\left(\dfrac{\lambda_g}{\lambda}\right), \quad \text{in Np/m} \quad (\lambda \text{ in meters}) \tag{8-75}$$

where λ is the wavelength inside an unbounded infinite lossy dielectric medium and λ_g is the guide wavelength filled with the lossy dielectric material. In decibels, α_d of (8-75) can be written as

$$\alpha_d \simeq 8.68 \left(\frac{\varepsilon_e''}{\varepsilon_e'}\right) \frac{\pi}{\lambda}\left(\frac{\lambda_g}{\lambda}\right) = \frac{27.27}{\lambda}\left(\frac{\varepsilon''}{\varepsilon_e'}\right)\left(\frac{\lambda_g}{\lambda}\right), \quad \text{in dB/m} \quad (\lambda \text{ in meters}) \qquad (8\text{-}75a)$$

Example 8-6

An X-band (8.2–12.4 GHz) rectangular waveguide is filled with polystyrene whose electrical properties at 6 GHz are $\varepsilon_r' = 2.56$ and $\tan \delta_e = 2.55 \times 10^{-4}$. Determine the dielectric attenuation at a frequency of $f = 6$ GHz when the inside dimensions of the waveguide are $a = 0.9$ in. (2.286 cm) and $b = 0.4$ in. (1.016 cm). Assume TE_{10} mode propagation.

Solution: According to (2-68a)

$$\tan \delta_e = \frac{\varepsilon_e''}{\varepsilon_e'} = 2.55 \times 10^{-4}$$

At $f = 6$ GHz,

$$\lambda_0 = \frac{30 \times 10^9}{6 \times 10^9} = 5 \text{ cm} \qquad \lambda = \frac{\lambda_0}{\sqrt{\varepsilon_r}} = \frac{5}{\sqrt{2.56}} = 3.125 \text{ cm}$$

$$(f_c)_{10} = \frac{1}{2a\sqrt{\mu\varepsilon}} = \frac{30 \times 10^9}{2(2.286)(1.6)} = 4.10 \text{ GHz}$$

$$\lambda_g = \frac{\lambda}{\sqrt{1-\left(\frac{f_c}{f}\right)^2}} = \frac{3.125}{\sqrt{1-\left(\frac{4.10}{6}\right)^2}} = 4.280 \text{ cm}$$

Thus, the dielectric attenuation of (8-75) is equal to

$$\alpha_d = \frac{\varepsilon_e''}{\varepsilon_e'}\frac{\pi}{\lambda}\left(\frac{\lambda_g}{\lambda}\right) = 2.55 \times 10^{-4}\left(\frac{\pi}{3.125}\right)\left(\frac{4.280}{3.125}\right) = 3.511 \times 10^{-4} \text{ Np/cm}$$

$$= 3.511 \times 10^{-2} \text{ Np/m} = 30.476 \times 10^{-2} \text{ dB/m}$$

C. Coupling   Whenever a given mode is to be excited or detected, the excitation or detection scheme must be such that it maximizes the energy exchange or transfer between the source and the guide or the guide and the receiver. Typically, there are a number of techniques that can be used to accomplish this. Some suggested methods that are popular in practice are the following.

1. If the energy exchange is from one waveguide to another, use an iris or hole placed in a location and orientation so that the field distribution of both guides over the extent of the hole or iris are almost identical.
2. If the energy exchange is from a transmission line, such as a coaxial line, to a waveguide, or vice versa, use a linear probe or antenna oriented so that its length is parallel to the electric field lines in the waveguide and placed near the maximum of the electric field mode pattern, as shown in Figure 8-13a. This is usually referred to as *electric field coupling*. Sometimes the position is varied slightly to achieve better impedance matching.
3. If the energy transfer is from a transmission line, such as a coaxial line, to a waveguide, or vice versa, use a loop antenna oriented so that the plane of the loop is perpendicular to the magnetic field lines, as shown in Figure 8-13c. This is usually referred to as *magnetic field coupling*.

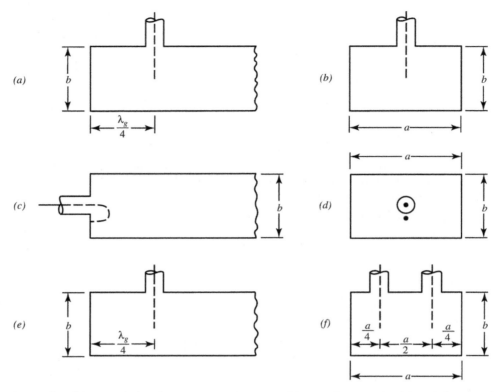

Figure 8-13 Coaxial transmission line to rectangular waveguide coupling. (*a*) The side and (*b*) the end views of the coax to waveguide electric field coupling for the TE_{10} mode. (*c*) The side and (*d*) the end view for the coax to waveguide magnetic field coupling for the TE_{10} mode. (*e*) The side and (*f*) the end view for the coax to waveguide electric field coupling for the TE_{20} mode.

4. If the energy transfer is from a transmission line, such as a two-conductor line, or other sources to a waveguide, or vice versa, use the transmission line or other sources so that they excite currents on the waveguide that match those of the desired modes in the guide.
5. A number of probes, antennas, or transmission lines properly phased can also be used to excite or detect any mode, especially higher-order modes. Shown in Figure 8-13 are some typical arrangements for coupling energy from a transmission line, such as a coax, to a rectangular waveguide to excite or detect the TE_{10} and TE_{20} modes.

The sizes, flanges, frequency bands, and other parameters pertaining to rectangular waveguides have been standardized so that uniformity is maintained throughout the industry. Standardized reference data on rectangular waveguides are displayed in Table 8-4.

8.3 RECTANGULAR RESONANT CAVITIES

Waveguide cavities represent a very important class of microwave components. Their applications are numerous and range from use as frequency meters to cavities for measuring the electrical properties of material. The attractive characteristics of waveguide cavities are their very high quality factors Q, typically on the order of 5,000–10,000, and their simplicity of construction and use. The most common geometries of cavities are rectangular, cylindrical, and spherical. The rectangular geometry will be discussed in this chapter, whereas the cylindrical will be examined in Chapter 9, and the spherical will be analyzed in Chapter 10.

TABLE 8-4   Reference table of rigid rectangular waveguide data and fittings

				Waveguide						
				Dimensions (in.)					Recommended operating range for TE$_{10}$ mode	
EIA designation WR ( )	MDL designation ( ) band	JAN designation RG ( )/U	Material alloy	Inside	Tol.	Outside	Tol.	Wall thickness nominal	Frequency (GHz)	Wavelength (cm)
2300	2300		Alum.	23.000–11.500	±0.020	23.250–11.750	±0.020	0.125	0.32–0.49	93.68–61.18
2100	2100		Alum.	21.000–10.500	±0.020	21.250–10.750	±0.020	0.125	0.35–0.53	85.65–56.56
1800	1800	201	Alum.	18.000–9.000	±0.020	18.250–9.250	±0.020	0.125	0.41–0.625	73.11–47.96
1500	1500	202	Alum.	15.000–7.500	±0.015	15.250–7.750	±0.015	0.125	0.49–0.75	61.18–39.97
1150	1150	203	Alum.	11.500–5.750	±0.015	11.750–6.000	±0.015	0.125	0.64–0.96	46.84–31.23
975	975	204	Alum.	9.750–4.875	±0.010	10.000–5.125	±0.010	0.125	0.75–1.12	39.95–26.76
770	770	205	Alum.	7.700–3.850	±0.005	7.950–4.100	±0.005	0.125	0.96–1.45	31.23–20.67
650	L	69 103	Copper Alum.	6.500–3.250	±0.005	6.660–3.410	±0.005	0.080	1.12–1.70	26.76–17.63
510	510			5.100–2.550	±0.005	5.260–2.710	±0.005	0.080	1.45–2.20	20.67–13.62
430	W	104 105	Copper Alum.	4.300–2.150	±0.005	4.460–2.310	±0.005	0.080	1.70–2.60	17.63–11.53
340	340	112 113	Copper Alum.	3.400–1.700	±0.005	3.560–1.860	±0.005	0.080	2.20–3.30	13.63–9.08
284	S	48 75	Copper Alum.	2.840–1.340	±0.005	3.000–1.500	±0.005	0.080	2.60–3.95	11.53–7.59
229	229			2.290–1.145	±0.005	2.418–1.273	±0.005	0.064	3.30–4.90	9.08–6.12
187	C	49 95	Copper Alum.	1.872–0.872	±0.005	2.000–1.000	±0.005	0.064	3.95–5.85	7.59–5.12
159	159			1.590–0.795	±0.004	1.718–0.923	±0.004	0.064	4.90–7.05	6.12–4.25
137	X_B	50 106	Copper Alum.	1.372–0.622	±0.004	1.500–0.750	±0.004	0.064	5.85–8.20	5.12–3.66
112	X_L	51 68	Copper Alum.	1.122–0.497	±0.004	1.250–0.625	±0.004	0.064	7.05–10.00	4.25–2.99
90	X	52 67	Copper Alum.	0.900–0.400	±0.003	1.000–0.500	±0.003	0.050	8.20–12.40	3.66–2.42
75	75			0.750–0.375	±0.003	0.850–0.475	±0.003	0.050	10.00–15.00	2.99–2.00
62	K_U	91 107	Copper Alum. Silver	0.622–0.311	±0.0025	0.702–0.391	±0.003	0.040	12.4–18.00	2.42–1.66
51	51			0.510–0.255	±0.0025	0.590–0.335	±0.003	0.040	15.00–22.00	2.00–1.36
42	K	53 121 66	Copper Alum. Silver	0.420–0.170	±0.0020	0.500–0.250	±0.003	0.040	18.00–26.50	1.66–1.13
34	34			0.340–0.170	±0.0020	0.420–0.250	±0.003	0.040	22.00–33.00	1.36–0.91
28	K_A	96	Copper Alum. Silver	0.280–0.140	±0.0015	0.360–0.220	±0.002	0.040	26.50–40.00	1.13–0.75
22	Q	97	Copper Silver	0.224–0.112	±0.0010	0.304–0.192	±0.002	0.040	33.00–50.00	0.91–0.60
19	19			0.188–0.094	±0.0010	0.268–0.174	±0.002	0.040	40.00–60.00	0.75–0.50
15	V	98	Copper Silver	0.148–0.074	±0.0010	0.228–0.154	±0.002	0.040	50.00–75.00	0.60–0.40
12	12	99	Copper Silver	0.122–0.061	±0.0005	0.202–0.141	±0.002	0.040	60.00–90.00	0.50–0.33
10	10			0.100–0.050	±0.0005	0.180–0.130	±0.002	0.040	75.00–110.00	0.40–0.27

Source: Microwave Development Laboratories, Inc.
[a]This is an MDL Range Number.

	Waveguide					Fittings		
Cutoff for TE$_{10}$ mode				Theoretical attenuation lowest to highest frequency (dB/100 ft)	Theoretical C/W power rating lowest to highest frequency (MW)	Flange		EJA designation WR ( )
Frequency (GHz)	Wavelength (cm)	Range in $2\lambda/\lambda_c$	Range in λ_g/λ			Choke UG( )/U	Cover UG( )/U	
0.256	116.84	1.60–1.05	1.68–1.17	0.051–0.031	153.0–212.0			2300
0.281	106.68	1.62–1.06	1.68–1.18	0.054–0.034	120.0–173.0		FA168A[a]	2100
0.328	91.44	1.60–1.05	1.67–1.18	0.056–0.038	93.4–131.9			1800
0.393	76.20	1.61–1.05	1.62–1.17	0.069–0.050	67.6–93.3			1500
0.513	58.42	1.60–1.07	1.82–1.18	0.128–0.075	35.0–53.8			1150
0.605	49.53	1.61–1.08	1.70–1.19	0.137–0.095	27.0–38.5			975
0.766	39.12	1.60–1.06	1.66–1.18	0.201–0.136	17.2–24.1			770
0.908	33.02	1.62–1.07	1.70–1.18	0.317–0.212 0.269–0.178	11.9–17.2		417A 418A	650
1.157	25.91	1.60–1.05	1.67–1.18					510
1.372	21.84	1.61–1.06	1.70–1.18	0.588–0.385 0.501–0.330	5.2–7.5		435A 437A	430
1.736	17.27	1.58–1.05	1.78–1.22	0.877–0.572 0.751–0.492	3.1–4.5		553 554	340
2.078	14.43	1.60–1.05	1.67–1.17	1.102–0.752 0940–0.641	2.2–3.2	54A 585	53 584	284
2.577	11.63	1.56–1.05	1.62–1.17					229
3.152	9.510	1.60–1.08	1.67–1.19	2.08–1.44 1.77–1.12	1.4–2.0	148B 406A	149A 407	187
3.711	8.078	1.51–1.05	1.52–1.19					159
4.301	6.970	1.47–1.05	1.48–1.17	2.87–2.30 2.45–1.94	0.56–0.71	343A 440A	344 441	137
5.259	5.700	1.49–1.05	1.51–1.17	4.12–3.21 3.50–2.74	0.35–0.46	52A 137A	51 138	112
6.557	4.572	1.60–1.06	1.68–1.18	6.45–4.48 5.49–3.83	0.20–0.29	40A 136A	39 135	90
7.868	3.810	1.57–1.05	1.64–1.17					75
9.486	3.160	1.53–1.05	1.55–1.18	9.51–8.31 6.14–5.36	0.12–0.16	541 FA190A[a]	419 FA191A[a]	62
11.574	2.590	1.54–1.05	1.58–1.18					51
14.047	2.134	1.56–1.06	1.60–1.18	20.7–14.8 17.6–12.6 13.3–9.5	0.043–0.058	596 598	595 597	42
17.328	1.730	1.57–1.05	1.62–1.18					34
21.081	1.422	1.59–1.05	1.65–1.17	21.9–15.0	0.022–0.031	600 FA1241A[a]	599 FA1242A[a]	28
26.342	1.138	1.60–1.05	1.67–1.17	31.0–20.9	0.014–0.020		383	22
31.357	0.956	1.57–1.05	1.63–1.16				385	19
39.863	0.752	1.60–1.06	1.67–1.17	52.9–39.1	0.0063–0.0090			15
48.350	0.620	1.61–1.06	1.68–1.18	93.3–52.2	0.0042–0.060		387	12
59.010	0.508	1.57–1.06	1.61–1.18					10

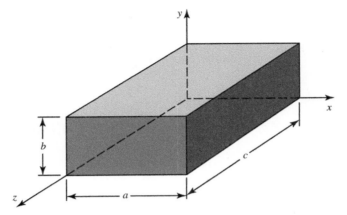

Figure 8-14 Geometry for the rectangular cavity.

A rectangular waveguide cavity is formed by taking a section of a waveguide and enclosing its front and back faces with conducting plates, as shown in Figure 8-14. Coupling into and out of the cavities is done through coupling probes or holes. The coupling probes may be either electric or magnetic, as shown in Figures 8-13a and 8-13c, whereas coupling holes may also be used either on the front, back, top, or bottom walls.

The field configurations inside the rectangular cavity of Figure 8-14 may be either TE^z or TM^z, or any other TE or TM mode, and they are derived in a manner similar to those of the waveguide. The only differences are that we must allow for standing waves, instead of traveling waves, along the length (z axis) of the waveguide, and we must impose additional boundary conditions along the front and back walls. The field forms along the x and y directions and the boundary conditions on the left, right, top, and bottom walls are identical to those of the rectangular waveguide.

8.3.1 Transverse Electric (TE^z) Modes

Since TE^z modes for a rectangular cavity must be derived in a manner similar to those of a rectangular waveguide, they must satisfy (8-1) and (8-2). Therefore, $F_z(x,y,z)$ must take a form similar to (8-6) except that standing wave functions (sines and cosines) must be used to represent the variations in the z direction. Since the boundary conditions on the bottom, top, left, and right walls are, respectively, (8-8a) to (8-8d), the $F_z(x,y,z)$ function for the rectangular cavity can be written as

$$F_z(x,y,z) = A_{mn} \cos(\beta_x x)\cos(\beta_y y)[C_3 \cos(\beta_z z) + D_3 \sin(\beta_z z)] \tag{8-76}$$

$$\beta_x = \frac{m\pi}{a} \quad m = 0, 1, 2, \ldots$$
$$\beta_y = \frac{n\pi}{b} \quad \begin{array}{l} n = 0, 1, 2, \ldots \\ m \text{ and } n \text{ not both zero simultaneously} \end{array} \tag{8-76a}$$

which are similar to (8-14a) and (8-14b) except for the standing wave functions representing the z variations. The additional boundary conditions on the front and back walls of the cavity are

$$E_x(0 \le x \le a, 0 \le y \le b, z=0) = E_x(0 \le x \le a, 0 \le y \le b, z=c) = 0 \tag{8-77a}$$

$$E_y(0 \le x \le a, 0 \le y \le b, z=0) = E_y(0 \le x \le a, 0 \le y \le b, z=c) = 0 \tag{8-77b}$$

The boundary conditions (8-77a) and (8-77b) are not independent and either will be sufficient. By using (8-76), we can write the E_x component, according to (8-1), as

$$E_x(x,y,z) = -\frac{1}{\varepsilon}\frac{\partial F_z}{\partial y} = \frac{\beta_y}{\varepsilon}A_{mn}\cos(\beta_x x)\sin(\beta_y y)[C_3\cos(\beta_z z) + D_3\sin(\beta_z z)] \qquad (8\text{-}78)$$

By applying (8-77a) to (8-78), we can write that

$$E_x(0 \le x \le a,\, 0 \le y \le b,\, z = 0) = \frac{\beta_y}{\varepsilon}A_{mn}\cos(\beta_x x)\sin(\beta_y y)[C_3(1) + D_3(0)] = 0$$

$$\Rightarrow C_3 = 0 \qquad (8\text{-}79a)$$

$$E_x(0 \le x \le a,\, 0 \le y \le b,\, z = c) = \frac{\beta_y}{\varepsilon}A_{mn}\cos(\beta_x x)\sin(\beta_y y)D_3\sin(\beta_z c) = 0$$

$$\Rightarrow \sin(\beta_z c) = 0 \;\Rightarrow\; \beta_z c = \sin^{-1}(0) = p\pi$$

$$\Rightarrow \beta_z = \frac{p\pi}{c}, \qquad p = 1, 2, 3, \ldots \qquad (8\text{-}79b)$$

Thus, (8-76) reduces to

$$F_z(x,y,z) = A_{mn}D_3\cos(\beta_x x)\cos(\beta_y y)\sin(\beta_z z)$$
$$F_z(x,y,z) = A_{mnp}\cos(\beta_x x)\cos(\beta_y y)\sin(\beta_z z) \qquad (8\text{-}80)$$

where

$$\left.\begin{array}{ll} \beta_x = \dfrac{m\pi}{a}, & m = 0, 1, 2, \ldots \\[2mm] \beta_y = \dfrac{n\pi}{b}, & n = 0, 1, 2, \ldots \\[2mm] \beta_z = \dfrac{p\pi}{c}, & p = 1, 2, 3, \ldots \end{array}\right\} m \text{ and } n \text{ not both zero simultaneously} \qquad (8\text{-}80a)$$

Thus, for each mode, the dimensions of the cavity in each direction must be an integral number of half wavelengths of the wave in that direction. In addition, the electric and magnetic field components of (8-1) can be expressed as

$$E_x = \frac{\beta_y}{\varepsilon}A_{mnp}\cos(\beta_x x)\sin(\beta_y y)\sin(\beta_z z) \qquad (8\text{-}81a)$$

$$E_y = -\frac{\beta_x}{\varepsilon}A_{mnp}\sin(\beta_x x)\cos(\beta_y y)\sin(\beta_z z) \qquad (8\text{-}81b)$$

$$E_z = 0 \qquad (8\text{-}81c)$$

$$H_x = j\frac{\beta_x\beta_y}{\omega\mu\varepsilon}A_{mnp}\sin(\beta_x x)\cos(\beta_y y)\cos(\beta_z z) \qquad (8\text{-}81d)$$

$$H_y = j\frac{\beta_y\beta_z}{\omega\mu\varepsilon}A_{mnp}\cos(\beta_x x)\sin(\beta_y y)\cos(\beta_z z) \qquad (8\text{-}81e)$$

$$H_z = -j\frac{A_{mnp}}{\omega\mu\varepsilon}(-\beta_z^2 + \beta^2)\cos(\beta_x x)\cos(\beta_y y)\sin(\beta_z z) \qquad (8\text{-}81f)$$

By using (8-80a) we can write (8-7a) as

$$\beta_x^2 + \beta_y^2 + \beta_z^2 = \left(\frac{m\pi}{a}\right)^2 + \left(\frac{n\pi}{b}\right)^2 + \left(\frac{p\pi}{c}\right)^2 = \beta_r^2 = \omega_r^2\mu\varepsilon = (2\pi f_r)^2\mu\varepsilon \qquad (8\text{-}82)$$

or

$$(f_r)_{mnp}^{\text{TE}} = \frac{1}{2\pi\sqrt{\mu\varepsilon}} \sqrt{\left(\frac{m\pi}{a}\right)^2 + \left(\frac{n\pi}{b}\right)^2 + \left(\frac{p\pi}{c}\right)^2}$$

$m = 0, 1, 2, \ldots$
$n = 0, 1, 2, \ldots$
$p = 1, 2, 3, \ldots$

m and n not both zero simultaneously (8-82a)

In (8-82a), $(f_r)_{mnp}$ represents the resonant frequency for the TE_{mnp}^z mode. If $c > a > b$, the mode with the lowest order is the TE_{101}^z mode, whose resonant frequency is represented by

$$(f_r)_{101}^{\text{TE}} = \frac{1}{2\sqrt{\mu\varepsilon}} \sqrt{\left(\frac{1}{a}\right)^2 + \left(\frac{1}{c}\right)^2} \tag{8-83}$$

In addition to its resonant frequency, one of the most important parameters of a resonant cavity is its quality factor Q defined as

$$Q \equiv \omega \frac{\text{stored energy}}{\text{dissipated power}} = \omega \frac{W_t}{P_d} = \omega \frac{W_e + W_m}{P_d} = \omega \frac{2W_e}{P_d} = \omega \frac{2W_m}{P_d} \tag{8-84}$$

which is proportional to volume and inversely proportional to surface. By using the field expressions of (8-81a) through (8-81f) for the $m = 1$, $n = 0$, and $p = 1$ (101) mode, the total stored energy can be written as

$$W = 2W_e = 2\left[\frac{\varepsilon}{4}\iiint_V |\mathbf{E}|^2 \, dv\right] = \frac{\varepsilon}{2}\left|\frac{|A_{101}|}{\varepsilon}\frac{\pi}{a}\right|^2 \int_0^c \int_0^b \int_0^a \sin^2\left(\frac{\pi}{a}x\right)\sin^2\left(\frac{\pi}{c}z\right) dx \, dy \, dz$$

$$W = \frac{|A_{101}|^2}{\varepsilon}\left(\frac{\pi}{a}\right)^2 \frac{abc}{8} \tag{8-85}$$

The total dissipated power is found by adding the power that is dissipated in each of the six walls of the cylinder. Since the dissipated power on the top wall is the same as that on the bottom, that on the right wall is the same as that on the left, and that on the back is the same as that on the front, we can write the total dissipated power as

$$P_d = \frac{R_s}{2}\left\{2\iint_{\text{bottom}} \mathbf{J}_b \cdot \mathbf{J}_b^* \, ds + 2\iint_{\text{left}} \mathbf{J}_\ell \cdot \mathbf{J}_\ell^* \, ds + 2\iint_{\text{front}} \mathbf{J}_f \cdot \mathbf{J}_f^* \, ds\right\}$$

$$= R_s\left\{\iint_{\text{bottom}} \mathbf{J}_b \cdot \mathbf{J}_b^* \, ds + \iint_{\text{left}} \mathbf{J}_\ell \cdot \mathbf{J}_\ell^* \, ds + \iint_{\text{front}} \mathbf{J}_f \cdot \mathbf{J}_f^* \, ds\right\}$$

$$P_d = P_b + P_\ell + P_f \tag{8-86}$$

where

$$P_b = R_s \iint_{\text{bottom}} \mathbf{J}_b \cdot \mathbf{J}_b^* \, ds = R_s \int_0^a \int_0^a |\mathbf{J}_b|^2 \, dx \, dz \tag{8-86a}$$

$$P_\ell = R_s \iint_{\text{left}} \mathbf{J}_\ell \cdot \mathbf{J}_\ell^* \, ds = R_s \int_0^c \int_0^b |\mathbf{J}_\ell|^2 \, dy \, dz \tag{8-86b}$$

$$P_f = R_s \iint_{\text{front}} \mathbf{J}_f \cdot \mathbf{J}_f^* \, ds = R_s \int_0^b \int_0^a |\mathbf{J}_f|^2 \, dx \, dy \tag{8-86c}$$

$$\mathbf{J}_b = \hat{\mathbf{n}} \times \mathbf{H}|_{y=0} = -\hat{\mathbf{a}}_z j \frac{\pi}{a}\frac{\pi}{c}\frac{A_{101}}{\omega\mu\varepsilon}\sin\left(\frac{\pi}{a}x\right)\cos\left(\frac{\pi}{c}z\right)$$

$$- \hat{\mathbf{a}}_x j\left(\frac{\pi}{a}\right)^2 \frac{A_{101}}{\omega\mu\varepsilon}\cos\left(\frac{\pi}{a}x\right)\sin\left(\frac{\pi}{c}z\right) \tag{8-86d}$$

$$\mathbf{J}_\ell = \hat{\mathbf{n}} \times \mathbf{H}|_{x=0} = \hat{\mathbf{a}}_y j \frac{A_{101}}{\omega\mu\varepsilon} \left(\frac{\pi}{a}\right)^2 \sin\left(\frac{\pi}{c} z\right) \tag{8-86e}$$

$$\mathbf{J}_f = \hat{\mathbf{n}} \times \mathbf{H}|_{z=c} = -\hat{\mathbf{a}}_y j \frac{\pi}{a} \frac{\pi}{c} \frac{A_{101}}{\omega\mu\varepsilon} \sin\left(\frac{\pi}{a} x\right) \tag{8-86f}$$

Application of the fields of (8-81a) through (8-81f) for $m=1$, $n=0$, and $p=1$ in (8-86) through (8-86f) leads to

$$P_b = R_s \left[\frac{\pi^2}{ac} \frac{A_{101}}{\omega\mu\varepsilon}\right]^2 \left(\frac{c}{2}\right)\left(\frac{a}{2}\right) + R_s \left[\left(\frac{\pi}{a}\right)^2 \frac{A_{101}}{\omega\mu\varepsilon}\right]^2 \left(\frac{c}{2}\right)\left(\frac{a}{2}\right) \tag{8-87a}$$

$$P_\ell = R_s \left[\left(\frac{\pi}{a}\right)^2 \frac{A_{101}}{\omega\mu\varepsilon}\right]^2 (c)\left(\frac{b}{2}\right) \tag{8-87b}$$

$$P_f = R_s \left[\frac{\pi^2}{ac} \frac{A_{101}}{\omega\mu\varepsilon}\right]^2 (b)\left(\frac{a}{2}\right) \tag{8-87c}$$

$$P_d = \frac{R_s}{4} \frac{|A_{101}|^2}{(\varepsilon\eta)^2} \left(\frac{\pi}{a}\right)^2 \frac{1}{a^2+c^2} [ac(a^2+c^2) + 2b(a^3+c^3)] \tag{8-87d}$$

Ultimately then, the Q of (8-84) can be expressed, using (8-85) and (8-87d), as

$$(Q)_{101}^{TE} = \frac{\pi\eta}{2R_s}\left[\frac{b(a^2+c^2)^{3/2}}{ac(a^2+c^2) + 2b(a^3+c^3)}\right] \tag{8-88}$$

For a square-based ($a=c$) cavity

$$(Q)_{101}^{TE} = \frac{\pi\eta}{2\sqrt{2}R_s}\left[\frac{1}{1+\dfrac{a}{2b}}\right] = 1.1107\frac{\eta}{R_s}\left[\frac{1}{1+\dfrac{a/2}{b}}\right] \tag{8-88a}$$

Example 8-7

A square-based ($a=c$) cavity of rectangular cross section is constructed of an X-band (8.2–12.4 GHz) copper ($\sigma = 5.7\times10^7$ S/m) waveguide that has inner dimensions of $a=0.9$ in. (2.286 cm) and $b=0.4$ in. (1.016 cm). For the dominant TE_{101} mode, determine the Q of the cavity. Assume a free-space medium inside the cavity.

Solution: According to (8-82a), the resonant frequency of the TE_{101} mode for the square-based ($a=c$) cavity is

$$(f_r)_{101} = \frac{1}{2\pi\sqrt{\mu\varepsilon}}\sqrt{\left(\frac{\pi}{a}\right)^2 + \left(\frac{\pi}{c}\right)^2} = \frac{\sqrt{2}}{2a\sqrt{\mu\varepsilon}}$$

$$= \frac{1}{\sqrt{2}a\sqrt{\mu\varepsilon}} = \frac{30\times10^9}{\sqrt{2}(2.286)} = 9.28\,\text{GHz}$$

Thus, the surface resistance R_s of (8-61) is equal to

$$R_s = \sqrt{\frac{\omega_r\mu}{2\sigma}} = \sqrt{\frac{2\pi(9.28\times10^9)(4\pi\times10^{-7})}{2(5.7\times10^7)}}$$

$$= 2\pi\sqrt{\frac{92.8}{5.7}}\times10^{-3} = 0.0254\,\text{ohms}$$

Therefore, the Q of (8-88a) reduces to

$$(Q)_{101} = 1.1107 \frac{377}{0.0254} \left| \frac{1}{1 + \dfrac{2.286}{2(1.016)}} \right| = 7{,}757.9 \simeq 7{,}758$$

8.3.2 Transverse Magnetic (TM^z) Modes

In addition to TE$_{mnp}^z$ modes inside a rectangular cavity, TM$_{mnp}^z$ modes can also be supported by such a structure. These modes can be derived in a manner similar to the TE$_{mnp}^z$ field configurations.

Using the results of Section 8.2.2, we can write the vector potential component $A_z(x,y,z)$ of (8-26a) for the TM$_{mnp}^z$ modes of Figure 8-14 without applying the boundary conditions on the front and back walls, as

$$A_z(x,y,z) = B_{mn} \sin(\beta_x x) \sin(\beta_y y)[C_3 \cos(\beta_z z) + D_3 \sin(\beta_z z)] \tag{8-89}$$

where

$$\beta_x = \frac{m\pi}{a}, \qquad m = 1,2,3,\dots \tag{8-89a}$$

$$\beta_y = \frac{n\pi}{b}, \qquad n = 1,2,3,\dots \tag{8-89b}$$

The boundary conditions that have not yet been applied on (8-24) are (8-77a) or (8-77b). Using (8-89), we can write the E_x component of (8-24) as

$$E_x(x,y,z) = -j\frac{1}{\omega\mu\varepsilon}\frac{\partial^2 A_z}{\partial x \partial z}$$

$$= -j\frac{\beta_x\beta_z}{\omega\mu\varepsilon} B_{mn} \cos(\beta_x x)\sin(\beta_y y)[-C_3\sin(\beta_z z) + D_3\cos(\beta_z z)] \tag{8-90}$$

Applying the boundary conditions (8-77a), we can write that

$$E_x(0 \le x \le a, 0 \le y \le b, z=0) = -j\frac{\beta_x\beta_z}{\omega\mu\varepsilon} B_{mn} \cos(\beta_x x)\sin(\beta_y y)$$

$$\times [-C_3(0) + D_3(1)] = 0 \;\Rightarrow\; D_3 = 0 \tag{8-91a}$$

$$E_x(0 \le x \le a, 0 \le y \le b, z=c) = -j\frac{\beta_x\beta_z}{\omega\mu\varepsilon} B_{mn} \cos(\beta_x x)\sin(\beta_y y)$$

$$\times [-C_3\sin(\beta_z c)] = 0$$

$$\Rightarrow \sin(\beta_z c) = 0 \;\Rightarrow\; \beta_z c = \sin^{-1}(0) = p\pi$$

$$\beta_z = \frac{p\pi}{c}, \qquad p = 0,1,2,3,\dots \tag{8-91b}$$

Thus, (8-89) reduces to

$$A_z(x,y,z) = B_{mn}C_3 \sin(\beta_x x)\sin(\beta_y y)\cos(\beta_z z)$$

$$A_z(x,y,z) = B_{mnp} \sin(\beta_x x)\sin(\beta_y y)\cos(\beta_z z) \tag{8-92}$$

where

$$\beta_x = \frac{m\pi}{a}, \qquad m = 1,2,3,\dots$$

$$\beta_y = \frac{n\pi}{b}, \qquad n = 1,2,3,\dots \tag{8-92a}$$

$$\beta_z = \frac{p\pi}{c}, \qquad p = 0, 1, 2, \ldots$$

Using (8-7a) and (8-92a), the corresponding resonant frequency can be written as

$$(f_r)_{mnp}^{\mathrm{TM}} = \frac{1}{2\pi\sqrt{\mu\varepsilon}} \sqrt{\left(\frac{m\pi}{a}\right)^2 + \left(\frac{n\pi}{b}\right)^2 + \left(\frac{p\pi}{c}\right)^2} \quad \begin{matrix} m = 1, 2, 3, \ldots \\ n = 1, 2, 3, \ldots \\ p = 0, 1, 2, \ldots \end{matrix} \qquad (8\text{-}93)$$

Since the expression for the resonant frequency of the TM_{mnp} modes is the same as for the TE_{mnp}, the order in which the modes occur can be found by forming the ratio of the resonant frequency of any mnp mode (TE or TM) to that of the TE_{101}, that is,

$$R_{101}^{mnp} = \frac{(f_r)_{mnp}}{(f_r)_{101}^{\mathrm{TE}^z}} = \sqrt{\frac{\left(\frac{m}{a}\right)^2 + \left(\frac{n}{b}\right)^2 + \left(\frac{p}{c}\right)^2}{\left(\frac{1}{a}\right)^2 + \left(\frac{1}{c}\right)^2}} \qquad (8\text{-}94)$$

whose values for $c \geq a \geq b$ and different ratios of a/b and c/b are found listed in Table 8-5.

8.4 HYBRID (LSE AND LSM) MODES

For some waveguide configurations, such as partially filled waveguides with the material interface perpendicular to the x or y axis of Figure 8-3, TE^z or TM^z modes cannot satisfy the boundary conditions of the structure. This will be discussed in the next section. Therefore, some other mode configurations may exist within such a structure. It will be shown that field configurations that are combinations of TE^z and TM^z modes can be solutions and satisfy the boundary conditions of such a partially filled waveguide [3]. The modes are referred to as *hybrid modes*, or *longitudinal section electric* (LSE) or *longitudinal section magnetic* (LSM), or *H* or *E* modes [4].

In the next section it will be shown that for a partially filled waveguide of the form shown in Figure 8-15a, the hybrid modes that are solutions and satisfy the boundary conditions are TE^y (LSE^y or H^y) and/or TM^y (LSM^y or E^y). Here the modes are LSE and/or LSM to a direction that is perpendicular to the interface. Similarly, for the configuration of Figure 8-15b, the appropriate hybrid modes will be TE^x (LSE^x) and/or TM^x (LSM^x). Before proceeding with the analysis of these waveguide configurations, let us examine TE^y (LSE^y) and TM^y (LSM^y) modes for the empty waveguide of Figure 8-3.

TABLE 8-5   Values of R_{101}^{mnp} for a rectangular cavity

$\dfrac{a}{b}$	$\dfrac{c}{b}$	TE_{101}	TE_{011}	TM_{110}	TE_{111} TM_{111}	TE_{102}	TE_{201}	TE_{021}	TE_{012}	TM_{210}	TM_{120}	TE_{112} TM_{112}
1	1	1	1	1	1.22	1.58	1.58	1.58	1.58	1.58	1.58	1.73
1	2	1	1	1.26	1.34	1.26	1.84	1.84	1.26	2.00	2.00	1.55
2	2	1	1.58	1.58	1.73	1.58	1.58	2.91	2.00	2.00	2.91	2.12
2.25	2.25	1	1.74	1.74	1.88	1.58	1.58	3.26	2.13	2.13	3.26	2.24
2	4	1	1.84	2.00	2.05	1.26	1.84	3.60	2.00	2.53	3.68	2.19
2.25	4	1	2.02	2.15	2.20	1.31	1.81	3.95	2.19	2.62	4.02	2.36
4	4	1	2.91	2.91	3.00	1.58	1.58	5.71	3.16	3.16	5.71	3.24
4	8	1	3.62	3.65	3.66	1.26	1.84	7.20	3.65	4.03	7.25	3.82

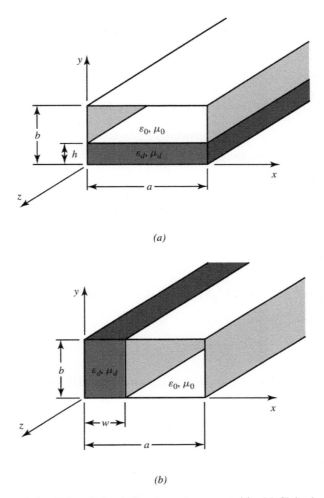

(a)

(b)

Figure 8-15 Geometry of the dielectric loaded rectangular waveguide. (*a*) Slab along broad wall. (*b*) Slab along narrow wall.

8.4.1   Longitudinal Section Electric (LSE^y) or Transverse Electric (TE^y) or H^y Modes

Just as for other transverse electric modes, TE^y modes are derived using 6-77, i.e.,

$$E_x = \frac{1}{\varepsilon}\frac{\partial F_y}{\partial z} \qquad\qquad H_x = -j\frac{1}{\omega\mu\varepsilon}\frac{\partial^2 F_y}{\partial x\,\partial y}$$

$$E_y = 0 \qquad\qquad H_y = -j\frac{1}{\omega\mu\varepsilon}\left(\frac{\partial^2}{\partial y^2}+\beta^2\right)F_y \qquad\qquad (8\text{-}95)$$

$$E_z = -\frac{1}{\varepsilon}\frac{\partial F_y}{\partial x} \qquad\qquad H_z = -j\frac{1}{\omega\mu\varepsilon}\frac{\partial^2 F_y}{\partial y\,\partial z}$$

where for the $+z$ traveling wave

$$F_y^+(x,y,z)=[C_1\,\cos(\beta_x x)+D_1\sin(\beta_x x)]$$
$$\times[C_2\,\cos(\beta_y y)+D_2\sin(\beta_y y)]\,A_3\,e^{-j\beta_z z} \qquad (8\text{-}95a)$$

$$\beta_x^2 + \beta_y^2 + \beta_z^2 = \beta^2 = \omega^2 \mu\varepsilon \tag{8-95b}$$

The boundary conditions are (8-8a) through (8-8d).

By using (8-95a), we can write the E_z of (8-95) as

$$E_z^+(x,y,z) = -\frac{\beta_x}{\varepsilon}[-C_1 \sin(\beta_x x) + D_1 \cos(\beta_x x)]$$
$$\times [C_2 \cos(\beta_y y) + D_2 \sin(\beta_y y)]A_3 e^{-j\beta_z z} \tag{8-96}$$

Application of the boundary condition (8-8b) on (8-96) gives

$$E_z^+(0 \le x \le a, y = 0, z)$$
$$= -\frac{\beta_x}{\varepsilon}[-C_1 \sin(\beta_x x) + D_1 \cos(\beta_x x)][C_2(1) + D_2(0)] A_3 e^{-j\beta_z z} = 0$$
$$\Rightarrow C_2 = 0 \tag{8-97a}$$

$$E_z^+(0 \le x \le a, y = b, z)$$
$$= -\frac{\beta_x}{\varepsilon}[-C_1 \sin(\beta_x x) + D_1 \cos(\beta_x x)]D_2 \sin(\beta_y b) A_3 e^{-j\beta_z z} = 0$$
$$\Rightarrow \sin(\beta_y b) = 0 \Rightarrow \beta_y b = \sin^{-1}(0) = n\pi$$
$$\beta_y = \frac{n\pi}{b}, \quad n = 1, 2, 3, \dots \tag{8-97b}$$

By following the same procedure, the boundary condition (8-8d) leads to

$$E_z^+(x = 0, 0 \le y \le b, z)$$
$$= -\frac{\beta_x}{\varepsilon}[-C_1(0) + D_1(1)]D_2 A_3 \sin(\beta_y y)e^{-j\beta_z z} = 0 \Rightarrow D_1 = 0 \tag{8-98a}$$

$$E_z^+(x = a, 0 \le y \le b, z) = -\frac{\beta_x}{\varepsilon}[-C_1 \sin(\beta_x a)]D_2 A_3 e^{-j\beta_z z} = 0$$
$$\Rightarrow \sin(\beta_x a) = 0 \Rightarrow \beta_x = \frac{m\pi}{a}, \quad m = 0, 1, 2, \dots \tag{8-98b}$$

Therefore, (8-95a) reduces to

$$F_y^+(x,y,z) = C_1 D_2 A_3 \cos(\beta_x x) \sin(\beta_y y)e^{-j\beta_z z} = A_{mn} \cos(\beta_x x) \sin(\beta_y y)e^{-j\beta_z z} \tag{8-99}$$

where

$$\beta_x = \frac{m\pi}{a}, \quad m = 0, 1, 2, \dots \tag{8-99a}$$

$$\beta_y = \frac{n\pi}{b}, \quad n = 1, 2, 3, \dots \tag{8-99b}$$

By using (8-95b),

$$\beta_z = \pm\sqrt{\beta^2 - (\beta_x^2 + \beta_y^2)} = \pm\sqrt{\beta^2 - \beta_c^2} \tag{8-100}$$

where

$$\beta_c^2 = \omega_c^2 \mu\varepsilon = (2\pi f_c)^2 \mu\varepsilon = \beta_x^2 + \beta_y^2 = \left(\frac{m\pi}{a}\right)^2 + \left(\frac{n\pi}{b}\right)^2$$

or

$$\boxed{(f_c)_{mn}^{TE^y} = \frac{1}{2\pi\sqrt{\mu\varepsilon}}\sqrt{\left(\frac{m\pi}{a}\right)^2 + \left(\frac{n\pi}{b}\right)^2}} \quad \begin{array}{l} m = 0, 1, 2, \dots \\ n = 1, 2, 3, \dots \end{array} \tag{8-100a}$$

The dominant mode is the TE_{01}^y whose cutoff frequency is

$$\boxed{(f_c)_{01}^{\text{TE}^y} = \frac{1}{2b\sqrt{\mu\varepsilon}}}$$

(8-100b)

8.4.2 Longitudinal Section Magnetic (LSM^y) or Transverse Magnetic (TM^y) or E^y Modes

By following a procedure similar to that for the TE^y modes of the previous section, it can be shown that for the TM^y modes of Figure 8-3 the field components of (6-64), i.e.,

$$E_x = -j\frac{1}{\omega\mu\varepsilon}\frac{\partial^2 A_y}{\partial x\,\partial y} \qquad\qquad H_x = -\frac{1}{\mu}\frac{\partial A_y}{\partial z}$$

$$E_y = -j\frac{1}{\omega\mu\varepsilon}\left(\frac{\partial^2}{\partial y^2}+\beta^2\right)A_y \qquad H_y = 0$$

(8-101)

$$E_z = -j\frac{1}{\omega\mu\varepsilon}\frac{\partial^2 A_y}{\partial y\,\partial z} \qquad\qquad H_z = \frac{1}{\mu}\frac{\partial A_y}{\partial x}$$

and the boundary conditions (8-8a) through (8-8d) lead to

$$A_y(x,y,z) = B_{mn}\,\sin(\beta_x x)\,\cos(\beta_y y)e^{-j\beta_z z}$$

(8-101a)

$$\beta_x = \frac{m\pi}{a}, \qquad m = 1, 2, 3, \ldots$$

(8-101b)

$$\beta_y = \frac{n\pi}{b}, \qquad n = 0, 1, 2, \ldots$$

(8-101c)

$$\boxed{(f_c)_{mn}^{\text{TM}^y} = \frac{1}{2\pi\sqrt{\mu\varepsilon}}\sqrt{\left(\frac{m\pi}{a}\right)^2+\left(\frac{n\pi}{b}\right)^2}}$$

(8-101d)

The dominant mode is the TM_{10}^y whose cutoff frequency is

$$\boxed{(f_c)_{10}^{\text{TM}^y} = \frac{1}{2a\sqrt{\mu\varepsilon}}}$$

(8-101e)

8.5 PARTIALLY FILLED WAVEGUIDE

Let us now consider in detail the analysis of the field configurations in the partially filled waveguide of Figure 8-15a. The analysis of the configuration of Figure 8-15b is left as an end-of-chapter exercise. It can be shown that for either waveguide configuration, neither TE^z nor TM^z modes individually can satisfy the boundary conditions. In fact, for the configuration of Figure 8-15a, TE^y(LSE^y) or TM^y(LSM^y) are the appropriate modes, whereas TE^x(LSE^x) or TM^x(LSM^x) satisfy the boundary conditions of Figure 8-15b. For either configuration, the appropriate modes are LSE or LSM to a direction that is perpendicular to the material interface.

8.5.1 Longitudinal Section Electric (LSE^y) or Transverse Electric (TE^y)

For the configuration of Figure 8-15a, there are two sets of fields: one for the dielectric region $(0 \le x \le a, 0 \le y \le h, z)$, designated by superscript d, and the other for the free-space region

$(0 \le x \le a, h \le y \le b, z)$, designated by superscript 0. For each region, the TE^y field components are those of (8-95) and the corresponding potential functions are

$$F_y^d(x, 0 \le y \le h, z) = \left[C_1^d \cos(\beta_{xd} x) + D_1^d \sin(\beta_{xd} x) \right]$$
$$\times \left[C_2^d \cos(\beta_{yd} y) + D_2^d \sin(\beta_{yd} y) \right] A_3^d \, e^{-j\beta_z z} \tag{8-102}$$

$$\beta_{xd}^2 + \beta_{yd}^2 + \beta_z^2 = \beta_d^2 = \omega^2 \mu_d \varepsilon_d \tag{8-102a}$$

for the dielectric region, and

$$F_y^0(x, h \le y \le b, z) = \left[C_1^0 \cos(\beta_{x0} x) + D_1^0 \sin(\beta_{x0} x) \right]$$
$$\times \left\{ C_2^0 \cos[\beta_{y0}(b-y)] + D_2^0 \sin[\beta_{y0}(b-y)] \right\} A_3^0 \, e^{-j\beta_z z} \tag{8-103}$$

$$\beta_{x0}^2 + \beta_{y0}^2 + \beta_z^2 = \beta_0^2 = \omega^2 \mu_0 \varepsilon_0 \tag{8-103a}$$

for the free-space region. In both sets of fields, β_z is the same, since for propagation along the interface both sets of fields must be common.

For this waveguide configuration, the appropriate independent boundary conditions are

$$E_z^d(x=0, 0 \le y \le h, z) = E_z^d(x=a, 0 \le y \le h, z) = 0 \tag{8-104a}$$
$$E_z^d(0 \le x \le a, y=0, z) = 0 \tag{8-104b}$$
$$E_z^d(0 \le x \le a, y=h, z) = E_z^0(0 \le x \le a, y=h, z) \tag{8-104c}$$
$$E_z^0(x=0, h \le y \le b, z) = E_z^0(x=a, h \le y \le b, z) = 0 \tag{8-104d}$$
$$E_z^0(0 \le x \le a, y=b, z) = 0 \tag{8-104e}$$
$$H_z^d(0 \le x \le a, y=h, z) = H_z^0(0 \le x \le a, y=h, z) \tag{8-104f}$$

Another set of dependent boundary conditions is

$$E_y^d(x=0, 0 \le y \le h, z) = E_y^d(x=a, 0 \le y \le h, z) = 0 \tag{8-105a}$$
$$E_x^d(0 \le x \le a, y=0, z) = 0 \tag{8-105b}$$
$$E_x^d(0 \le x \le a, y=h, z) = E_x^0(0 \le x \le a, y=h, z) \tag{8-105c}$$
$$E_y^0(x=0, h \le y \le b, z) = E_y^0(x=a, h \le y \le b, z) = 0 \tag{8-105d}$$
$$E_x^0(0 \le x \le a, y=b, z) = 0 \tag{8-105e}$$
$$H_x^d(0 \le x \le a, y=h, z) = H_x^0(0 \le x \le a, y=h, z) \tag{8-105f}$$

By using (8-95) and (8-103), we can write that

$$E_z^0 = -\frac{1}{\varepsilon_0} \frac{\partial F_y^0}{\partial x} = -\frac{\beta_{x0}}{\varepsilon_0} \left[-C_1^0 \sin(\beta_{x0} x) + D_1^0 \cos(\beta_{x0} x) \right]$$
$$\times \left\{ C_2^0 \cos[\beta_{y0}(b-y)] + D_2^0 \sin[\beta_{y0}(b-y)] \right\} A_3^0 \, e^{-j\beta_z z} \tag{8-106}$$

Application of boundary condition (8-104d) leads to

$$E_z^0 (x=0, h \le y \le b, z)$$

$$= -\frac{\beta_{x0}}{\varepsilon_0} \left[-C_1^0 (0) + D_1^0 (1) \right]$$

$$\times \left\{ C_2^0 \cos\left[\beta_{y0}(b-y) \right] + D_2^0 \sin\left[\beta_{y0}(b-y) \right] \right\} A_3^0 \, e^{-j\beta_z z} = 0$$

$$\Rightarrow D_1^0 = 0 \qquad\qquad (8\text{-}106a)$$

$$E_z^0 (x=a, h \le y \le b, z)$$

$$= -\frac{\beta_{x0}}{\varepsilon_0} \left[-C_1^0 \sin(\beta_{x0}a) \right]$$

$$\times \left\{ C_2^0 \cos\left[\beta_{y0}(b-y) \right] + D_2^0 \sin\left[\beta_{y0}(b-y) \right] \right\} A_3^0 \, e^{-j\beta_z z} = 0$$

$$\Rightarrow \sin(\beta_{x0}a) = 0 \quad \Rightarrow \quad \beta_{x0} = \frac{m\pi}{a}, \quad m = 0, 1, 2, \dots \qquad (8\text{-}106b)$$

Application of (8-104e) leads to

$$E_z^0 (0 \le x \le a, y=b, z)$$

$$= -\frac{\beta_{x0}}{\varepsilon_0} \left[-C_1^0 \sin(\beta_{x0}x) \right] \left\{ C_2^0 (1) + D_2^0 (0) \right\} A_3^0 \, e^{-j\beta_z z} = 0$$

$$\Rightarrow C_2^0 = 0 \qquad\qquad (8\text{-}106c)$$

Thus, (8-103) reduces to

$$F_y^0 = A_{mn}^0 \cos(\beta_{x0}x) \sin\left[\beta_{y0}(b-y) \right] e^{-j\beta_z z} \qquad (8\text{-}107)$$

$$\beta_{x0} = \frac{m\pi}{a}, \quad m = 0, 1, 2, \dots \qquad (8\text{-}107a)$$

$$\beta_{x0}^2 + \beta_{y0}^2 + \beta_z^2 = \left(\frac{m\pi}{a} \right)^2 + \beta_{y0}^2 + \beta_z^2 = \beta_0^2 = \omega^2 \mu_0 \varepsilon_0 \qquad (8\text{-}107b)$$

with

$$E_z^0 = -\frac{1}{\varepsilon_0} \frac{\partial F_y^0}{\partial x} = \frac{\beta_{x0}}{\varepsilon_0} A_{mn}^0 \sin(\beta_{x0}x) \sin\left[\beta_{y0}(b-y) \right] e^{-j\beta_z z} \qquad (8\text{-}108)$$

Use of (8-95) and (8-102) gives

$$E_z^d = -\frac{1}{\varepsilon_d} \frac{\partial F_y^d}{\partial x} = -\frac{\beta_{xd}}{\varepsilon_d} \left[-C_1^d \sin(\beta_{xd}x) + D_1^d \cos(\beta_{xd}x) \right]$$

$$\times \left[C_2^d \cos(\beta_{yd}y) + D_2^d \sin(\beta_{yd}y) \right] A_3^d e^{-j\beta_z z} \qquad (8\text{-}109)$$

Application of boundary condition (8-104a) leads to

$$E_z^d (x=0, 0 \le y \le h, z)$$

$$= -\frac{\beta_{xd}}{\varepsilon_d} \left[-C_1^d (0) + D_1^d (1) \right] \left[C_2^d \cos(\beta_{yd}y) + D_2^d \sin(\beta_{yd}y) \right] A_3^d \, e^{-j\beta_z z} = 0$$

$$\Rightarrow D_1^d = 0 \qquad\qquad (8\text{-}109a)$$

$$E_z^d (x=a, 0 \le y \le h, z)$$

$$= -\frac{\beta_{xd}}{\varepsilon_d} \left[-C_1^d \sin(\beta_{xd}a) \right] \left[C_2^d \cos(\beta_{yd}y) + D_2^d \sin(\beta_{yd}y) \right] A_3^d \, e^{-j\beta_z z} = 0$$

$$\Rightarrow \sin(\beta_{xd}a) = 0 \quad \Rightarrow \quad \beta_{xd} = \frac{m\pi}{a}, \quad m = 0, 1, 2, \dots \qquad (8\text{-}109b)$$

Application of (8-104b) leads to

$$E_z^d\left(0 \le x \le a, y = 0, z\right) = -\frac{\beta_{xd}}{\varepsilon_d}\left[-C_1^d \sin\left(\beta_{xd}x\right)\right]\left[C_2^d(1) + D_2^d(0)\right]A_3^d e^{-j\beta_z z} = 0$$

$$\Rightarrow C_2^d = 0 \tag{8-109c}$$

Thus, (8-102) reduces to

$$F_y^d = A_{mn}^d \cos\left(\beta_{xd}x\right)\sin\left(\beta_{yd}y\right)e^{-j\beta_z z} \tag{8-110}$$

$$\beta_{xd} = \frac{m\pi}{a} = \beta_{x0}, \qquad m = 0, 1, 2, \ldots \tag{8-110a}$$

$$\beta_{xd}^2 + \beta_{yd}^2 + \beta_z^2 = \left(\frac{m\pi}{a}\right)^2 + \beta_{yd}^2 + \beta_z^2 = \beta_d^2 = \omega^2 \mu_d \varepsilon_d \tag{8-110b}$$

with

$$E_z^d = -\frac{1}{\varepsilon_d}\frac{\partial F_y^d}{\partial x} = \frac{\beta_{xd}}{\varepsilon_d} A_{mn}^d \sin\left(\beta_{xd}x\right)\sin\left(\beta_{yd}y\right)e^{-j\beta_z z}$$

$$= \frac{\beta_{x0}}{\varepsilon_d} A_{mn}^d \sin\left(\beta_{x0}x\right)\sin\left(\beta_{yd}y\right)e^{-j\beta_z z} \tag{8-111}$$

Application of boundary condition (8-104c) and use of (8-108) and (8-111) leads to

$$\frac{\beta_{x0}}{\varepsilon_0} A_{mn}^0 \sin\left(\beta_{x0}x\right)\sin\left[\beta_{y0}(b-h)\right]e^{-j\beta_z z} = \frac{\beta_{x0}}{\varepsilon_d} A_{mn}^d \sin\left(\beta_{xd}x\right)\sin\left(\beta_{yd}h\right)e^{-j\beta_z z}$$

$$\frac{1}{\varepsilon_0} A_{mn}^0 \sin\left[\beta_{y0}(b-h)\right] = \frac{1}{\varepsilon_d} A_{mn}^d \sin\left(\beta_{yd}h\right) \tag{8-112}$$

By using (8-107) and (8-110), the z component of the H field from (8-95) can be written as

$$H_z^0 = -j\frac{1}{\omega\mu_0\varepsilon_0}\frac{\partial^2 F_y^0}{\partial y \partial z} = \frac{\beta_{y0}\beta_z}{\omega\mu_0\varepsilon_0} A_{mn}^0 \cos\left(\beta_{x0}x\right)\cos\left[\beta_{y0}(b-y)\right]e^{-j\beta_z z} \tag{8-113a}$$

$$H_z^d = -j\frac{1}{\omega\mu_d\varepsilon_d}\frac{\partial^2 F_y^d}{\partial y \partial z} = -\frac{\beta_{yd}\beta_z}{\omega\mu_d\varepsilon_d} A_{mn}^d \cos\left(\beta_{xd}x\right)\cos\left(\beta_{yd}y\right)e^{-j\beta_z z} \tag{8-113b}$$

Application of the boundary condition of (8-104f) reduces, with $\beta_{xd} = \beta_{x0}$, to

$$\frac{\beta_{y0}\beta_z}{\omega\mu_0\varepsilon_0} A_{mn}^0 \cos\left(\beta_{x0}x\right)\cos\left[\beta_{y0}(b-h)\right]e^{-j\beta_z z}$$

$$= -\frac{\beta_{yd}\beta_z}{\omega\mu_d\varepsilon_d} A_{mn}^d \cos\left(\beta_{xd}x\right)\cos\left(\beta_{yd}h\right)e^{-j\beta_z z}$$

$$\frac{\beta_{y0}}{\mu_0\varepsilon_0} A_{mn}^0 \cos\left[\beta_{y0}(b-h)\right] = -\frac{\beta_{yd}}{\mu_d\varepsilon_d} A_{mn}^d \cos\left(\beta_{yd}h\right) \tag{8-114}$$

Division of (8-114) by (8-112) leads to

$$\frac{\beta_{y0}}{\mu_0}\cot\left[\beta_{y0}(b-h)\right]=-\frac{\beta_{yd}}{\mu_d}\cot\left(\beta_{yd}h\right) \tag{8-115}$$

$$\beta_{x0}^2+\beta_{y0}^2+\beta_z^2=\left(\frac{m\pi}{a}\right)^2+\beta_{y0}^2+\beta_z^2=\beta_0^2=\omega^2\mu_0\varepsilon_0 \qquad m=0,1,2,\dots \tag{8-115a}$$

$$\beta_{xd}^2+\beta_{yd}^2+\beta_z^2=\left(\frac{m\pi}{a}\right)^2+\beta_{yd}^2+\beta_z^2=\beta_d^2=\omega^2\mu_d\varepsilon_d \qquad m=0,1,2,\dots \tag{8-115b}$$

Whereas $\beta_{x0}=\beta_{xd}=m\pi/a$, $m=0,1,2,\dots$, have been determined, β_{y0}, β_{yd}, and β_z have not yet been found. They can be determined for each mode using (8-115) through (8-115b), and their values vary as a function of frequency. Thus, for each frequency a new set of values for β_{y0}, β_{yd}, and β_z must be found that satisfy (8-115) through (8-115b). One procedure that can be used to accomplish this will be to solve (8-115a) for β_{y0} (as a function of β_z and β_0) and (8-115b) for β_{yd} (as a function of β_z and β_d), and then substitute these expressions in (8-115) for β_{y0} and β_{yd}. The new form of (8-115) will be a function of β_z, β_0, and β_d. Thus, for a given mode, determined by the value of m, at a given frequency, a particular value of β_z will satisfy the new form of the transcendental equation 8-115; that value of β_z can be found iteratively. The range of β_z will be $\beta_z^0<\beta_z<\beta_z^d$ where β_z^0 represents the values of the same mode of an air-filled waveguide and β_z^d represents the values of the same mode of a waveguide completely filled with the dielectric. Once β_z has been found at a given frequency for a given mode, the corresponding values of β_{y0} and β_{yd} for the same mode at the same frequency can be determined by using, respectively, (8-115a) and (8-115b). It must be remembered that for each value of m there are infinite values of $n(n=1,2,3,\dots)$. Thus, the dominant mode is the one for which $m=0$ and $n=1$, i.e., TE_{01}^y.

For $m=0$, the modes will be denoted as TE_{0n}. For these modes, (8-115a) and (8-115b) reduce to

$$\beta_{y0}^2+\beta_z^2=\omega^2\mu_0\varepsilon_0 \Rightarrow \beta_z=\pm\sqrt{\omega^2\mu_0\varepsilon_0-\beta_{y0}^2} \tag{8-116a}$$

$$\beta_{yd}^2+\beta_z^2=\omega^2\mu_d\varepsilon_d \Rightarrow \beta_z=\pm\sqrt{\omega^2\mu_d\varepsilon_d-\beta_{yd}^2} \tag{8-116b}$$

Cutoff occurs when $\beta_z=0$. Thus, at cutoff (8-116a) and (8-116b) reduce to

$$\beta_z=0=\pm\sqrt{\omega^2\mu_0\varepsilon_0-\beta_{y0}^2}\,|_{\omega=\omega_c} \Rightarrow \omega_c^2\mu_0\varepsilon_0=\beta_{y0}^2 \Rightarrow \beta_{y0}=\omega_c\sqrt{\mu_0\varepsilon_0} \tag{8-117a}$$

$$\beta_z=0=\pm\sqrt{\omega^2\mu_d\varepsilon_d-\beta_{yd}^2}\,|_{\omega=\omega_c} \Rightarrow \omega_c^2\mu_d\varepsilon_d=\beta_{yd}^2 \Rightarrow \beta_{yd}=\omega_c\sqrt{\mu_d\varepsilon_d} \tag{8-117b}$$

which can be used to find β_{y0} and β_{yd} at cutoff (actually slightly above), once the cutoff frequency has been determined. By using (8-117a) and (8-117b), we can write (8-115) as

$$\frac{\omega_c\sqrt{\mu_0\varepsilon_0}}{\mu_0}\cot\left[\omega_c\sqrt{\mu_0\varepsilon_0}(b-h)\right]=-\frac{\omega_c\sqrt{\mu_d\varepsilon_d}}{\mu_d}\cot\left(\omega_c\sqrt{\mu_d\varepsilon_d}h\right)$$

or

$$\sqrt{\frac{\varepsilon_0}{\mu_0}}\cot\left[\omega_c\sqrt{\mu_0\varepsilon_0}(b-h)\right]=-\sqrt{\frac{\varepsilon_d}{\mu_d}}\cot\left(\omega_c\sqrt{\mu_d\varepsilon_d}h\right) \tag{8-118}$$

which can be used to find the cutoff frequencies of the TE_{0n}^y modes in a partially filled waveguide. A similar expression must be written for the other modes.

For a rectangular waveguide filled completely either with free space or with a dielectric material with ε_d, μ_d, the cutoff frequency of the TE_{01}^y mode is given, respectively, according to (8-100b), by

$$(f_c^0)_{01}^{\text{TE}^y} = \frac{1}{2b\sqrt{\mu_0\varepsilon_0}} \tag{8-119a}$$

$$(f_c^d)_{01}^{\text{TE}^y} = \frac{1}{2b\sqrt{\mu_d\varepsilon_d}} \tag{8-119b}$$

Use of perturbational techniques shows that, in general, the cutoff frequency of the partially filled waveguide (part free space and part dielectric) is greater than the cutoff frequency of the same mode in the same waveguide filled with a dielectric material with ε_d, μ_d and is smaller than the cutoff frequency of the same waveguide filled with free space. Thus, the cutoff frequency of the TE_{01}^y mode of a partially filled waveguide (part free space and part dielectric) is greater than (8-119b) and smaller than (8-119a), that is,

$$\boxed{\frac{1}{2b\sqrt{\mu_d\varepsilon_d}} \leq (f_c)_{01}^{\text{TE}^y} \leq \frac{1}{2b\sqrt{\mu_0\varepsilon_0}}} \tag{8-120}$$

or

$$\boxed{\frac{\pi}{b\sqrt{\mu_d\varepsilon_d}} \leq (\omega_c)_{01}^{\text{TE}^y} \leq \frac{\pi}{b\sqrt{\mu_0\varepsilon_0}}} \tag{8-120a}$$

With this permissible range, the exact values can be found using (8-118). The propagation constant β_z must be solved at each frequency on an individual basis using (8-116a) or (8-116b).

Example 8-8

A WR90 X-band (8.2–12.4 GHz) waveguide of Figure 8-15a with inner dimensions of $a = 0.9$ in. (2.286 cm), $b = 0.4$ in. (1.016 cm), and $a/b = 2.25$, is partially filled with free space and polystyrene ($\varepsilon_d = 2.56\varepsilon_0$, $\mu_d = \mu_0$, and $h = b/3$). For $m = 0$ determine the following.

1. The cutoff frequencies of the hybrid TE_{0n}^y (LSE_{0n}^y) modes for $n = 1, 2, 3$.
2. The values of β_{y0} and β_{yd} for each mode at sightly above their corresponding cutoff frequencies.
3. The corresponding values of β_{y0}, β_{yd}, and β_z for the TE_{01}^y mode in the frequency range $(f_c)_{01} \leq f \leq 2(f_c)_{01}$.

How do the cutoff frequencies of the first three TE_{0n}^y modes ($n = 1, 2, 3$) of the partially filled waveguide compare with those of the TE_{0n}^y modes of the empty waveguide?

Solution:

1. The cutoff frequencies of the partially filled waveguide are found using (8-118). According to (8-120) and (8-100a), the cutoff frequencies for each of the desired modes must fall in the ranges

$$9.23\,\text{GHz} \leq (f_c)_{01} \leq 14.76\,\text{GHz}$$
$$18.45\,\text{GHz} \leq (f_c)_{02} \leq 29.53\,\text{GHz}$$
$$27.68\,\text{GHz} \leq (f_c)_{03} \leq 44.29\,\text{GHz}$$

The actual frequencies are listed in Table 8-6.

2. Once the cutoff frequencies are found, the corresponding wave numbers β_z, β_{y0}, and β_{yd} *slightly above cutoff* can be found by using (8-117a) and (8-117b). These are also listed in Table 8-6. Also listed in Table 8-6 are the values of the cutoff frequencies, and at slightly above their corresponding cutoff frequencies, the wave numbers for the air-filled and dielectric-filled waveguides.

3. Finally, the wave numbers β_{y0}, β_{yd}, and β_z for each frequency in the range $(f_c)_{0n} \leq f \leq 2(f_c)_{0n}$ are found by solving (8-115) through (8-115b) as outlined previously. These are shown plotted in Figure 8-16 for the TE_{01}^y mode where they are compared with those of the waveguide filled completely with air (β_z^0) or with the dielectric (β_z^d). The others for the TE_{02}^y, and TE_{03}^y modes are assigned as an end-of-chapter exercise.

TABLE 8-6 Cutoff frequencies and phase constants of partially filled, air-filled, and dielectric-filled rectangular waveguide.*

TE_{0n}^y modes		$n=1$	$n=2$	$n=3$
Partially filled waveguide	$(f_c)_{0n}$ (GHz)	12.61	24.02	37.68
	β_z (rad/m)	11.56	18.11	19.19
	$(\beta_{y0})_{0n}$ at $(f_c)_{0n}$ (rad/m)	264.32	503.38	789.75
	$(\beta_{yd})_{0n}$ at $(f_c)_{0n}$ (rad/m)	422.91	805.41	1263.60
Air-filled waveguide	$(f_c^0)_{0n}$ (GHz)	14.75	29.51	44.26
	β_z (rad/m)	9.02	18.04	18.53
	$(\beta_y^0)_{0n}$ (rad/m)	309.21	618.42	927.64
Dielectric-filled waveguide	$(f_c^d)_{0n}$ (GHz)	9.22	18.44	27.66
	β_z (rad/m)	13.61	18.04	20.52
	$(\beta_y^d)_{0n}$ (rad/m)	309.21	618.42	927.64

*$a=0.9$ in. (2.286 cm), $b=0.4$ in. (1.016 cm), $h=b/3, \mu_d=\mu_d$, and $\varepsilon_d=2.56\varepsilon_0$.

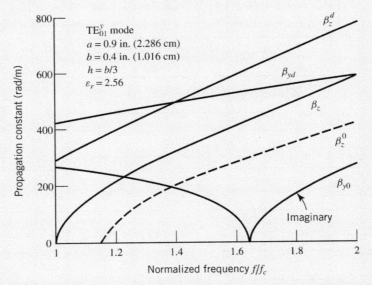

Figure 8-16 Propagation constants of TE_{01}^y modes for a partially filled rectangular waveguide.

8.5.2   Longitudinal Section Magnetic (LSM^y) or Transverse Magnetic (TM^y)

For the waveguide configuration of Figure 8-15a, the TM^y field components are those of (8-101), where the corresponding vector potentials in the dielectric and free-space regions for the waves traveling in the $+z$ direction are given, respectively, by

$$A_y^d = [C_1^d \cos(\beta_{xd}x) + D_1^d \sin(\beta_{xd}x)][C_2^d \cos(\beta_{yd}y) + D_2^d \sin(\beta_{yd}y)]A_3^d \, e^{-j\beta_z z} \tag{8-121}$$

$$\beta_{xd}^2 + \beta_{yd}^2 + \beta_z^2 = \beta_d^2 = \omega^2 \mu_d \varepsilon_d \tag{8-121a}$$

$$A_y^0 = [C_1^0 \cos(\beta_{x0}x) + D_1^0 \sin(\beta_{x0}x)]$$
$$\times \left\{ C_2^0 \cos[\beta_{y0}(b-y)] + D_2^0 \sin[\beta_{y0}(b-y)] \right\} A_3^0 \, e^{-j\beta_z z} \tag{8-122}$$

$$\beta_{x0}^2 + \beta_{y0}^2 + \beta_z^2 = \beta_0^2 = \omega^2 \mu_0 \varepsilon_0 \tag{8-122a}$$

The appropriate boundary conditions are (8-104a) through (8-105f).

Application of the boundary conditions (8-104a) through (8-105f) shows that the following relations follow:

$$A_y^0 = B_{mn}^0 \sin(\beta_{x0}x) \cos[\beta_{y0}(b-y)]e^{-j\beta_z z} \tag{8-123}$$

$$\beta_{x0} = \frac{m\pi}{a}, \quad m = 1, 2, 3, \dots \tag{8-123a}$$

$$\beta_{x0}^2 + \beta_{y0}^2 + \beta_z^2 = \left(\frac{m\pi}{a}\right)^2 + \beta_{y0}^2 + \beta_z^2 = \beta_0^2 = \omega^2 \mu_0 \varepsilon_0 \tag{8-123b}$$

$$A_y^d = B_{mn}^d \sin(\beta_{xd}x) \cos(\beta_{yd}y)e^{-j\beta_z z} \tag{8-124}$$

$$\beta_{xd} = \frac{m\pi}{a}, \quad m = 1, 2, 3, \dots \tag{8-124a}$$

$$\beta_{xd}^2 + \beta_{yd}^2 + \beta_z^2 = \left(\frac{m\pi}{a}\right)^2 + \beta_{yd}^2 + \beta_z^2 = \beta_d^2 = \omega^2 \mu_d \varepsilon_d \tag{8-124b}$$

$$-\frac{\beta_{y0}}{\mu_0 \varepsilon_0} B_{mn}^0 \sin[\beta_{y0}(b-h)] = \frac{\beta_{yd}}{\mu_d \varepsilon_d} B_{mn}^d \sin(\beta_{yd}h) \tag{8-125}$$

$$\frac{1}{\mu_0} B_{mn}^0 \cos[\beta_{y0}(b-h)] = \frac{1}{\mu_d} B_{mn}^d \cos(\beta_{yd}h) \tag{8-126}$$

$$\frac{\beta_{y0}}{\varepsilon_0} \tan[\beta_{y0}(b-h)] = -\frac{\beta_{yd}}{\varepsilon_d} \tan(\beta_{yd}h) \tag{8-127}$$

$$\beta_{x0}^2 + \beta_{y0}^2 + \beta_z^2 = \left(\frac{m\pi}{a}\right)^2 + \beta_{y0}^2 + \beta_z^2 = \beta_0^2 = \omega^2 \mu_0 \varepsilon_0, \quad m = 1, 2, 3, \dots \tag{8-127a}$$

$$\beta_{xd}^2 + \beta_{yd}^2 + \beta_z^2 = \left(\frac{m\pi}{a}\right)^2 + \beta_{yd}^2 + \beta_z^2 = \beta_d^2 = \omega^2 \mu_d \varepsilon_d, \quad m = 1, 2, 3, \dots \tag{8-127b}$$

Whereas $\beta_{x0} = \beta_{xd} = m\pi/a$, $m = 1, 2, 3, \dots$, have been determined, β_{y0}, β_{yd}, and β_z have not yet been found. They can be determined by using (8-127) through (8-127b) and following a procedure similar to that outlined in the previous section for the TE^y modes. For each value of m, there are infinite values of $n(n = 0, 1, 2,...)$. Thus the dominant mode is that for which $m = 1$ and $n = 0$, i.e., the dominant mode is the TM$^y_{10}$.

For $m = 1$, the modes will be denoted as TM$^y_{1n}$. For these modes, (8-127a) and (8-127b) reduce to

$$\left(\frac{\pi}{a}\right)^2 + \beta_{y0}^2 + \beta_z^2 = \omega^2 \mu_0 \varepsilon_0 \;\Rightarrow\; \beta_z = \pm \sqrt{\omega^2 \mu_0 \varepsilon_0 - \left[\beta_{y0}^2 + \left(\frac{\pi}{a}\right)^2\right]} \tag{8-128a}$$

$$\left(\frac{\pi}{a}\right)^2 + \beta_{yd}^2 + \beta_z^2 = \omega^2 \mu_d \varepsilon_d \;\Rightarrow\; \beta_z = \pm \sqrt{\omega^2 \mu_d \varepsilon_d - \left[\beta_{yd}^2 + \left(\frac{\pi}{a}\right)^2\right]} \tag{8-128b}$$

Cutoff occurs when $\beta_z = 0$. Thus, at cutoff, (8-128a) and (8-128b) reduce to

$$\omega_c^2 \mu_0 \varepsilon_0 = \beta_{y0}^2 + \left(\frac{\pi}{a}\right)^2 \;\Rightarrow\; \beta_{y0} = \sqrt{\omega_c^2 \mu_0 \varepsilon_0 - \left(\frac{\pi}{a}\right)^2} \tag{8-129a}$$

$$\omega_c^2 \mu_d \varepsilon_d = \beta_{yd}^2 + \left(\frac{\pi}{a}\right)^2 \;\Rightarrow\; \beta_{yd} = \sqrt{\omega_c^2 \mu_d \varepsilon_d - \left(\frac{\pi}{a}\right)^2} \tag{8-129b}$$

which can be used to find β_{y0} and β_{yd} *slightly above cutoff*, once the cutoff frequency has been determined. By using (8-129a) and (8-129b), we can write (8-127) as

$$\frac{1}{\varepsilon_0} \sqrt{\omega_c^2 \mu_0 \varepsilon_0 - \left(\frac{\pi}{a}\right)^2} \, \tan\left[\sqrt{\omega_c^2 \mu_0 \varepsilon_0 - \left(\frac{\pi}{a}\right)^2} \,(b-h)\right]$$

$$= -\frac{1}{\varepsilon_d} \sqrt{\omega_c^2 \mu_d \varepsilon_d - \left(\frac{\pi}{a}\right)^2} \, \tan\left[h\sqrt{\omega_c^2 \mu_d \varepsilon_d - \left(\frac{\pi}{a}\right)^2}\right]$$

or

$$\boxed{\begin{aligned}\frac{\varepsilon_d}{\varepsilon_0} \sqrt{\omega_c^2 \mu_0 \varepsilon_0 - \left(\frac{\pi}{a}\right)^2} &\, \tan\left[\sqrt{\omega_c^2 \mu_0 \varepsilon_0 - \left(\frac{\pi}{a}\right)^2}\,(b-h)\right] \\ &= -\sqrt{\omega_c^2 \mu_d \varepsilon_d - \left(\frac{\pi}{a}\right)^2} \, \tan\left[h\sqrt{\omega_c^2 \mu_d \varepsilon_d - \left(\frac{\pi}{a}\right)^2}\right]\end{aligned}} \tag{8-130}$$

which can be used to find the cutoff frequencies of the TM$^y_{1n}$ modes in a partially filled waveguide.

For a rectangular waveguide filled completely either with free space (μ_0, ε_0) or with a dielectric material (ε_d, μ_d), the cutoff frequency of the hybrid TM$^y_{10}$ mode is given, respectively, according to (8-101d) by

$$\left(f_c^0\right)_{10}^{TM^y} = \frac{1}{2a\sqrt{\mu_0 \varepsilon_0}} \tag{8-131a}$$

$$\left(f_c^d\right)_{10}^{TM^y} = \frac{1}{2a\sqrt{\mu_d \varepsilon_d}} \tag{8-131b}$$

By using perturbational techniques, it can be shown that, in general, the cutoff frequency of the partially filled waveguide (part free space and part dielectric) is greater than the cutoff frequency of the same mode in the same waveguide filled with a dielectric material with ε_d, μ_d and smaller than the cutoff frequency of the same waveguide filled with free space. Thus, the cutoff frequency of the TM_{10}^y mode of a partially filled waveguide (part free space and part dielectric) is greater than (8-131b) and smaller than (8-131a), that is

$$\frac{1}{2a\sqrt{\mu_d\varepsilon_d}} \le (f_c)_{10}^{TM^y} \le \frac{1}{2a\sqrt{\mu_0\varepsilon_0}} \tag{8-132}$$

or

$$\frac{\pi}{a\sqrt{\mu_d\varepsilon_d}} \le (\omega_c)_{10}^{TM^y} \le \frac{\pi}{a\sqrt{\mu_0\varepsilon_0}} \tag{8-132a}$$

With this permissible range, the exact values can be found by using (8-130). The propagation constant β_z must be solved at each frequency on an individual basis using (8-128a) or (8-128b) once the values of β_{y0} and β_{yd} have been determined at that frequency.

Example 8-9

A WR90 X-band (8.2–12.4 GHz) waveguide of Figure 8-15a with inner dimensions of $a = 0.9$ in. (2.286 cm), $b = 0.4$ in. (1.016 cm), and $a/b = 2.25$, is partially filled with free space and polystyrene ($\varepsilon_d = 2.56$, $\mu_d = \mu_0$, and $h = b/3$). For $m = 1$ determine the following.

1. The cutoff frequencies of the hybrid $TM_{1n}^y (LSM_{1n}^y)$ modes for $n = 0, 1, 2$.
2. The values of β_{y0} and β_{yd} for each model at *slightly above* their corresponding cutoff frequencies.
3. The corresponding values of β_{y0}, β_{yd}, and β_z for the TM_{10}^y mode in the frequency range $(f_c)_{10} \le f \le 2(f_c)_{10}$.

How do the cutoff frequencies of the first three $TM_{1n}^y (n = 0, 1, 2)$ of the partially filled waveguide compare with those of the TM_{1n}^y of the empty waveguide?

Solution: Follow a procedure similar to that for the TE_{0n}^y modes, as was done for the solution of Example 8-8. Then the parameters and their associated values are obtained as listed in Table 8-7 and

TABLE 8-7   Cutoff frequencies and phase constants of partially filled, air-filled, and dielectric-filled rectangular waveguide*

TM_{1n}^y modes		$n = 0$	$n = 1$	$n = 2$
Partially filled waveguide	$(f_c)_{1n}$ (GHz)	5.78	13.67	24.73
	β_z (rad/m)	6.58	13.94	12.07
	$(\beta_{y0})_{1n}$ at $(f_c)_{1n}$ (rad/m)	$\pm j64.76$	251.37	499.84
	$(\beta_{yd})_{1n}$ at $(f_c)_{1n}$ (rad/m)	136.84	437.29	817.96
Air-filled waveguide	$(f_c^0)_{1n}$ (GHz)	6.56	16.15	30.23
	β_z (rad/m)	8.59	14.46	18.41
	$(\beta_y^0)_{1n}$ (rad/m)	0	309.21	618.42
Dielectric-filled waveguide	$(f_c^d)_{1n}$ (GHz)	4.10	10.09	18.89
	β_z (rad/m)	10.41	14.46	18.41
	$(\beta_y^d)_{1n}$ (rad/m)	0	309.21	618.42

*$a = 0.9$ in. (2.286 cm), $b = 0.4$ in. (1.016 cm), $h = b/3$, $\mu_d = \mu_0$, and $\varepsilon_d = 2.56\varepsilon_0$.

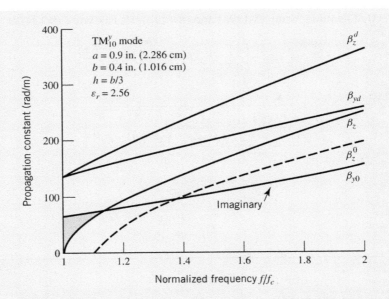

Figure 8-17 Propagation constants of TM_{10}^y modes for a partially filled rectangular waveguide.

shown in Figure 8-17. The parameters versus frequency for the TM_{11}^y and TM_{12}^y modes are assigned as an end-of-chapter exercise. The cutoff frequencies of the partially filled waveguide are found using (8-130). According to (8-132) and (8-101d), the cutoff frequencies for each of the desired modes must fall in the ranges

$$4.101\,\text{GHz} \le (f_c)_{10} \le 6.56\,\text{GHz}$$
$$10.098\,\text{GHz} \le (f_c)_{11} \le 16.16\,\text{GHz}$$
$$18.905\,\text{GHz} \le (f_c)_{12} \le 30.25\,\text{GHz}$$

The actual frequencies, as well as the other desired parameters of this problem, are listed in Table 8-7.

When the dielectric properties of the dielectric material inserted into the waveguide are such that $\varepsilon_d \simeq \varepsilon_0$ and $\mu_d \simeq \mu_0$, the wave constants of the TM_{mn} modes of the partially filled waveguide are approximately equal to the corresponding wave constants of the TM_{mn}^y modes of the totally filled waveguide. Thus, according to (8-101c)

$$(\beta_{y0})_{m0} \simeq (\beta_{yd})_{m0} \simeq \text{small} \simeq 0 \tag{8-133}$$

Therefore, (8-127) can be approximated by

$$\frac{\beta_{y0}}{\varepsilon_0} \tan\left[\beta_{y0}(b-h)\right] \simeq \frac{\beta_{y0}}{\varepsilon_0}\left[\beta_{y0}(b-h)\right] = -\frac{\beta_{yd}}{\varepsilon_d}\tan(\beta_{yd}h) \simeq -\frac{\beta_{yd}}{\varepsilon_d}(\beta_{yd}h)$$

$$\frac{\beta_{y0}^2}{\varepsilon_0}(b-h) \simeq -\frac{\beta_{yd}^2}{\varepsilon_d}h$$

$$\frac{\varepsilon_d}{\varepsilon_0}\beta_{y0}^2(b-h) \simeq -\beta_{yd}^2 h \tag{8-134}$$

At cutoff, $\beta_z = 0$. Therefore, using (8-129a) and (8-129b), we can write (8-134) as

$$\frac{\varepsilon_d}{\varepsilon_0}\left[\omega_c^2 \mu_0 \varepsilon_0 - \left(\frac{\pi}{a}\right)^2\right](b-h) \simeq -\left[\omega_c^2 \mu_d \varepsilon_d - \left(\frac{\pi}{a}\right)^2\right]h \qquad (8\text{-}134a)$$

which reduces to

$$\boxed{(\omega_c)_{10}^{\text{TM}^y} \simeq \frac{\pi}{a\sqrt{\varepsilon_r \mu_0 \varepsilon_0}} \sqrt{\frac{h + \varepsilon_r(b-h)}{(b-h) + \mu_r h}}} \qquad (8\text{-}135)$$

where

$$\varepsilon_r = \frac{\varepsilon_d}{\varepsilon_0} \qquad (8\text{-}135a)$$

$$\mu_r = \frac{\mu_d}{\mu_0} \qquad (8\text{-}135b)$$

Example 8-10

By using the approximate expression of (8-135), determine the cutoff frequency of the dominant TM_{10}^y hybrid mode for the following cases.

1. $\varepsilon_r = 1$ and $\mu_r = 1$.
2. $h = 0$.
3. $h = b$.
4. $\varepsilon_r = 2.56$, $\mu_r = 1$, $a = 0.9$ in. (2.286 cm), $b = 0.4$ in. (1.016 cm), $a/b = 2.25$, and $h = b/3$.

Solution:

1. When $\varepsilon_r = 1$ and $\mu_r = 1$, (8-135) reduces to

$$(\omega_c)_{10}^{\text{TM}^y} \simeq \frac{\pi}{a\sqrt{\mu_0 \varepsilon_0}}$$

which is equal to the exact value as predicted by (8-101d).

2. When $h = 0$, (8-135) reduces to

$$(\omega_c)_{10}^{\text{TM}^y} \simeq \frac{\pi}{a\sqrt{\mu_0 \varepsilon_0}}$$

which again is equal to the exact value predicted by (8-101d).

3. When $h = b$, (8-135) reduces to

$$(\omega_c)_{10}^{\text{TM}^y} \simeq \frac{\pi}{a\sqrt{\mu_0 \mu_r \varepsilon_0 \varepsilon_r}} = \frac{\pi}{a\sqrt{\mu_d \varepsilon_d}}$$

which again is equal to the exact value predicted by (8-101d).

4. When $\varepsilon_r = 2.56$, $\mu_r = 1$, $a = 2.286$ cm, $b = 1.016$ cm, and $h = b/3$ (8-135) reduces to

$$(\omega_c)_{10}^{\text{TM}^y} \simeq \frac{\pi}{a\sqrt{\varepsilon_r \mu_0 \varepsilon_0}} \sqrt{\frac{b/3 + 2.56(b - b/3)}{(b - b/3) + b/3}}$$

$$= \sqrt{\frac{2.04}{2.56}} \frac{\pi}{a\sqrt{\mu_0 \varepsilon_0}} = 0.8927 \frac{\pi}{a\sqrt{\mu_0 \varepsilon_0}}$$

$$(f_c)_{10}^{\text{TM}^y} \simeq 5.8576 \text{ GHz}$$

whose exact value, according to Example 8-9, is equal to 5.786 GHz. It should be noted that the preceding approximate expression for $(\omega_c)_{10}$ of

$$(\omega_c)_{10} \simeq 0.8927 \frac{\pi}{a\sqrt{\mu_0 \varepsilon_0}}$$

falls in the permissible range of

$$\frac{\pi}{a\sqrt{\mu_0 \varepsilon_r \varepsilon_0}} = \frac{\pi}{\sqrt{2.56}\, a\sqrt{\mu_0 \varepsilon_0}} = 0.6250 \frac{\pi}{a\sqrt{\mu_0 \varepsilon_0}} \leq (\omega_c)_{10}^{TM^y} \leq \frac{\pi}{a\sqrt{\mu_0 \varepsilon_0}}$$

as given by (8-132a).

8.6 TRANSVERSE RESONANCE METHOD

The transverse resonance method (TRM) is a technique that can be used to find the propagation constant of many practical composite waveguide structures [5, 6], as well as many traveling wave antenna systems [6–8]. By using this method, the cross section of the waveguide or traveling wave antenna structure is represented as a transmission line system. The fields of such a structure must satisfy the transverse wave equation, and the resonances of this transverse network will yield expressions for the propagation constants of the waveguide or antenna structure. Whereas the formulations of this method are much simpler when applied to finding the propagation constants, they do not contain the details for finding other parameters of interest (such as field distributions, wave impedances, etc.).

The objective here is to analyze the waveguide geometry of Figure 8-15a using the transverse resonance method. Although the method will not yield all the details of the analysis of Sections 8.5.1 and 8.5.2, it will lead to the same characteristic equations 8-115 and 8-127. The problem will be modeled as a two-dimensional structure represented by two transmission lines; one dielectric-filled ($0 \leq y \leq h$) with characteristic impedance Z_{cd} and wave number β_{td} and the other air-filled ($h \leq y \leq b$) with characteristic impedance Z_{c0} and wave number β_{t0}, as shown in Figure 8-18. Each line is considered shorted at its load, that is, $Z_L = 0$ at $y = 0$ and $y = b$.

It was shown in Section 3.4.1A that the solution to the scalar wave equation for any of the electric field components, for example, that for E_x of (3-22) as given by (3-23), takes the general form of

$$\nabla^2 \psi + \beta^2 \psi = \frac{\partial^2 \psi}{\partial x^2} + \frac{\partial^2 \psi}{\partial y^2} + \frac{\partial^2 \psi}{\partial z^2} + \beta^2 \psi = 0 \tag{8-136}$$

where

$$\psi(x,y,z) = f(x)g(y)h(z) \tag{8-136a}$$

The scalar function ψ represents any of the electric or magnetic field components. For waves traveling in the z direction, the variations of $h(z)$ are represented by exponentials of the form $e^{\pm j\beta_z z}$. Therefore, for such waves, (8-136) reduces to

$$\nabla^2 \psi + \beta^2 \psi = \left(\nabla_t^2 + \frac{\partial^2}{\partial z^2}\right)\psi + \beta^2 \psi = (\nabla_t^2 - \beta_z^2)\psi + \beta^2 \psi = 0$$

$$\nabla_t^2 \psi + (\beta^2 - \beta_z^2)\psi = 0 \tag{8-137}$$

where

$$\nabla_t^2 = \frac{\partial^2}{\partial x^2} + \frac{\partial^2}{\partial y^2} \tag{8-137a}$$

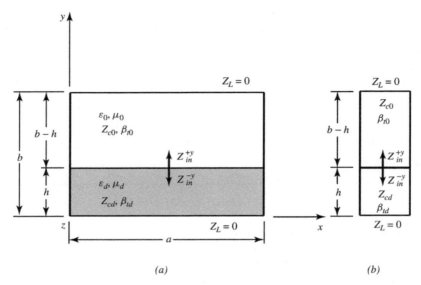

Figure 8-18   (a) Cross section of rectangular waveguide. (b) Transmission line equivalent for transverse resonance method (TRM).

The wave numbers associated with (8-137) are related by

$$\left(\beta_x^2 + \beta_y^2\right) + \beta_z^2 = \beta_t^2 + \beta_z^2 = \beta^2 \tag{8-138}$$

where

$$\beta_t^2 = \beta_x^2 + \beta_y^2 \tag{8-138a}$$

The constant β_t is referred to as the *transverse direction wave number* and (8-137) is referred to as the *transverse wave equation*.

Each of the electric and magnetic field components in the dielectric- and air-filled sections of the two-dimensional structure of Figure 8-18 must satisfy the transverse wave equation (8-137) with corresponding transverse wave numbers of β_{td} and β_{t0}, where

$$\beta_{td}^2 + \beta_z^2 = \beta_d^2 = \omega^2 \mu_d \varepsilon_d \tag{8-139a}$$

$$\beta_{t0}^2 + \beta_z^2 = \beta_0^2 = \omega^2 \mu_0 \varepsilon_0 \tag{8-139b}$$

In Section 4.2.2B it was shown by (4-20a) through (4-21b) that the wave impedances of the waves in the positive and negative directions are equal. However, the ratios of the corresponding electric/magnetic field component magnitudes were equal but opposite in direction. Since the input impedance of a line is defined as the ratio of the electric/magnetic field components (or voltage/current), then at any point along the transverse direction of the waveguide structure, the input impedance of the transmission line network looking in the positive y direction is equal in magnitude but opposite in phase to that looking in the negative y direction. This follows from the boundary conditions that require continuous tangential components of the electric (**E**) and magnetic (**H**) fields at any point on a plane orthogonal to the transverse structure of the waveguide.

For the transmission line model of Figure 8-18b, the input impedance at the interface looking in the $+y$ direction of the air-filled portion toward the shorted load is given, according to the impedance transfer equation (5-66d), as

$$Z_{in}^{+y} = Z_{c0} \left[\frac{Z_L + jZ_{c0} \tan[\beta_{t0}(b-h)]}{Z_{c0} + jZ_L \tan[\beta_{t0}(b-h)]} \right]_{Z_L=0} = jZ_{c0} \tan[\beta_{t0}(b-h)] \qquad (8\text{-}140a)$$

In a similar manner, the input impedance at the interface looking in the $-y$ direction of the dielectric-filled portion toward the shorted load is given by

$$Z_{in}^{-y} = jZ_{cd} \tan(\beta_{td}h) \qquad (8\text{-}140b)$$

Since these two impedances must be equal in magnitude but of opposite signs, then

$$Z_{in}^{+y} = -Z_{in}^{-y} = jZ_{c0} \tan[\beta_{t0}(b-h)] = -jZ_{cd} \tan(\beta_{td}h)$$

$$\boxed{Z_{c0} \tan[\beta_{t0}(b-h)] = -Z_{cd} \tan(\beta_{td}h)} \qquad (8\text{-}141)$$

The preceding equation is applicable for both TE and TM modes. It will be applied in the next two sections to examine the TE^y and TM^y modes of the partially filled waveguide of Figure 8-18a.

8.6.1 Transverse Electric (TE^y) or Longitudinal Section Electric (LSE^y) or H^y

The characteristic equation 8-141 will now be applied to examine the TE^y modes of the partially filled waveguide of Figure 8-18a. It was shown in Section 8.2.1 that the wave impedance of the TE_{mn}^z modes is given by (8-19), i.e.,

$$Z_w^{\text{TE}^z} = \frac{\omega\mu}{\beta_z} \qquad (8\text{-}142)$$

Allow the characteristic impedances for the TE^y modes of the dielectric- and air-filled sections of the waveguide, represented by the two-section transmission line of Figure 8-18b, to be of the same form as (8-142), or

$$Z_{cd} = Z_d^h = \frac{\omega\mu_d}{\beta_{yd}} \qquad (8\text{-}143a)$$

$$Z_{c0} = Z_0^h = \frac{\omega\mu_0}{\beta_{y0}} \qquad (8\text{-}143b)$$

$$\beta_{td} = \beta_{yd} \qquad (8\text{-}143c)$$

$$\beta_{t0} = \beta_{y0} \qquad (8\text{-}143d)$$

Then (8-141) reduces to

$$\frac{\omega\mu_0}{\beta_{y0}} \tan[\beta_{y0}(b-h)] = -\frac{\omega\mu_d}{\beta_{yd}} \tan(\beta_{yd}h)$$

or

$$\boxed{\frac{\beta_{y0}}{\mu_0} \cot[\beta_{y0}(b-h)] = -\frac{\beta_{yd}}{\mu_d} \cot(\beta_{yd}h)} \qquad (8\text{-}144)$$

Equation (8-144) is identical to (8-115), and it can be solved using the same procedures used in Section 8.5.1 to solve (8-115).

8.6.2   Transverse Magnetic (TM^y) or Longitudinal Section Magnetic (LSM^y) or E^y

The same procedure used in Section 8.6.1 for the TE^y modes can also be used to examine the TM^y modes of the partially filled waveguide of Figure 8-18a. According to (8-29a), the wave impedance of TM$^z_{mn}$ modes is given by

$$Z_w^{TM^z} = \frac{\beta_z}{\omega\varepsilon} \tag{8-145}$$

Allow the characteristic impedances for the TM^y modes of the dielectric- and air-filled sections of the waveguide to be of the same form as (8-145), or

$$Z_{cd} = Z_d^e = \frac{\beta_{yd}}{\omega\varepsilon_d} \tag{8-146a}$$

$$Z_{c0} = Z_0^e = \frac{\beta_{y0}}{\omega\varepsilon_0} \tag{8-146b}$$

$$\beta_{td} = \beta_{yd} \tag{8-146c}$$

$$\beta_{t0} = \beta_{y0} \tag{8-146d}$$

Then (8-141) reduces to

$$\frac{\beta_{y0}}{\omega\varepsilon_0}\tan\left[\beta_{y0}(b-h)\right] = -\frac{\beta_{yd}}{\omega\varepsilon_d}\tan(\beta_{yd}h)$$

or

$$\boxed{\frac{\beta_{y0}}{\varepsilon_0}\tan\left[\beta_{y0}(b-h)\right] = -\frac{\beta_{yd}}{\varepsilon_d}\tan(\beta_{yd}h)} \tag{8-147}$$

Equation (8-147) is identical to (8-127), and it can be solved using the same procedures used in Section 8.5.2 to solve (8-127).

The transverse resonance method can be used to solve other transmission line discontinuity problems [5] as well as many traveling wave antenna systems [6].

8.7   DIELECTRIC WAVEGUIDE

Transmission lines are used to contain the energy associated with a wave within a given space and guide it in a given direction. Typically, many people associate these types of transmission lines with either coaxial and twin lead lines or metal pipes (usually referred to as waveguides) with part or all of their structure being metal. However, dielectric slabs and rods, with or without any associated metal, can also be used to guide waves and serve as transmission lines. Usually these are referred to as *dielectric waveguides*, and the field modes that they can support are known as *surface wave modes* [9].

8.7.1   Dielectric Slab Waveguide

One type of dielectric waveguide is a dielectric slab of height $2h$, as shown in Figure 8-19. To simplify the analysis of the structure, we reduce the problem to a two-dimensional one (its width in the x direction is infinite) so that $\partial/\partial x = 0$. Although in practice the dimensions of the structure are finite, the two-dimensional approximation not only simplifies the analysis but also sheds insight into the characteristics of the structure. Typically, the cross section of the slab in

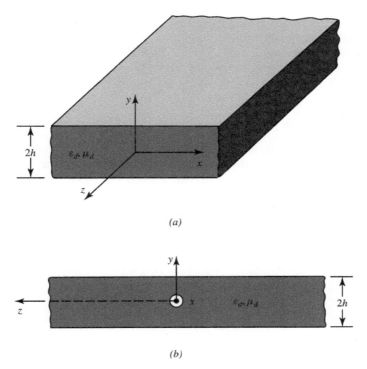

(a)

(b)

Figure 8-19   Geometry for dielectric slab waveguide. (*a*) Perspective. (*b*) Side view.

Figure 8-19*a* would be rectangular with height 2*h* and finite width *a*. We also assume that the waves are traveling in the ±*z* directions, and the structure is infinite in that direction, as illustrated in Figure 8-19*b*.

Another practical configuration for a dielectric transmission line is a dielectric rod of circular cross section. Because of the cylindrical geometry of the structure, the field solutions will be in terms of Bessel functions. Therefore, the discussion of this line will be postponed until Chapter 9.

A very popular dielectric rod waveguide is the fiber optics cable. Typically, this cable is made of two different materials, one that occupies the center core and the other that serves as a cladding to the center core. This configuration is usually referred to as the *step index*, and the index of refraction of the center core is slightly greater than that of the cladding. Another configuration has the index of refraction distribution along the cross section of the line graded so that there is a smooth variation in the radial direction from the larger values at the center toward the smaller values at the periphery. This is referred to as the *graded index*. This line is discussed in more detail in Section 9.5.3.

The objective in a dielectric slab waveguide, or any type of waveguide, is to contain the energy within the structure and direct it toward a given direction. For the dielectric slab waveguide this is accomplished by having the wave bounce back and forth between its upper and lower interfaces at an incidence angle greater than the critical angle. When this is accomplished, the refracted fields outside the dielectric form evanescent (decaying) waves and all the real energy is reflected and contained within the structure. The characteristics of this line can be analyzed by treating the structure as a boundary-value problem whose modal solution is obtained by solving the wave equation and enforcing the boundary conditions. The other approach is to examine the characteristics of the line using ray-tracing (geometrical optics) techniques. This approach is simpler and sheds more physical insight onto the propagation characteristics of the line but does not provide the details of the more cumbersome modal solution. Both methods will be examined here. We will begin with the modal solution approach.

It can be shown that the waveguide structure of Figure 8-19 can support TE^z, TM^z, TE^y, and TM^y modes. We will examine here both the TM^z and TE^z modes. We will treat TM^z in detail and then summarize the TE^z.

8.7.2   Transverse Magnetic (TM^z) Modes

The TM^z mode fields that can exist within and outside the dielectric slab of Figure 8-19 must satisfy (8-24), where A_z is the potential function representing the fields either within or outside the dielectric slab. Inside and outside the dielectric region, the fields can be represented by a combination of even and odd modes, as shown in Figure 8-20 [10].

For the fields within the dielectric slab, the potential function A_z takes the following form:

$$-h \leq y \leq h$$

$$A_z^d = \left[C_2^d \cos(\beta_{yd} y) + D_2^d \sin(\beta_{yd} y) \right] A_3^d e^{-j\beta_z z} = A_{ze}^d + A_{z0}^d \tag{8-148}$$

where

$$A_{ze}^d = C_2^d A_3^d \cos(\beta_{yd} y) e^{-j\beta_z z} = A_{me}^d \cos(\beta_{yd} y) e^{-j\beta_z z} \tag{8-148a}$$

$$A_{z0}^d = D_2^d A_3^d \sin(\beta_{yd} y) e^{-j\beta_z z} = A_{m0}^d \sin(\beta_{yd} y) e^{-j\beta_z z} \tag{8-148b}$$

$$\beta_{yd}^2 + \beta_z^2 = \beta_d^2 = \omega^2 \mu_d \varepsilon_d \tag{8-148c}$$

In (8-148), A_{ze}^d and A_{z0}^d represent, respectively, the even and odd modes. For the slab to function as a waveguide, the fields outside the dielectric slab must be of evanescent form. Therefore, the

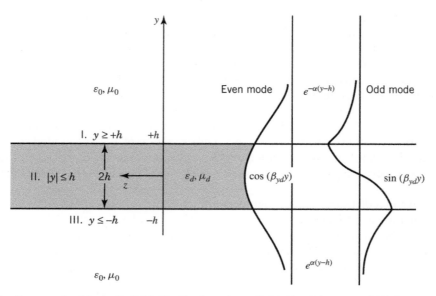

Figure 8-20   Even and odd mode field distributions in a dielectric slab waveguide. (Source: M. Zahn, *Electromagnetic Field Theory*, 1979. Reprinted with permission of John Wiley & Sons, Inc.)

potential function A_z takes the following form:

$$\underline{y \geq h}$$

$$A_z^{0+} = \left(A_{2e}^{0+} e^{-j\beta_{y0}y} + B_{20}^{0+} e^{-j\beta_{y0}y} \right) A_{30}^{0+} e^{-j\beta_z z} = A_{ze}^{0+} + A_{z0}^{0+} \qquad (8\text{-}149)$$

where

$$A_{ze}^{0+} = A_{2e}^{0+} A_{30}^{0+} e^{-j\beta_{y0}y} e^{-j\beta_z z} = A_{me}^{0+} e^{-\alpha_{y0}y} e^{-j\beta_z z} \qquad (8\text{-}149a)$$

$$A_{z0}^{0+} = B_{20}^{0+} A_{30}^{0+} e^{-j\beta_{y0}y} e^{-j\beta_z z} = B_{m0}^{0+} e^{-\alpha_{y0}y} e^{-j\beta_z z} \qquad (8\text{-}149b)$$

$$\beta_{y0}^2 + \beta_z^2 = -\alpha_{y0}^2 + \beta_z^2 = \beta_0^2 = \omega^2 \mu_0 \varepsilon_0 \qquad (8\text{-}149c)$$

$$\underline{y \leq -h}$$

$$A_z^{0-} = \left(A_{2e}^{0-} e^{+j\beta_{y0}y} + B_{20}^{0-} e^{+j\beta_{y0}y} \right) A_{30}^{0-} e^{-j\beta_z z} = A_{ze}^{0-} + A_{z0}^{0-} \qquad (8\text{-}150)$$

where

$$A_{ze}^{0-} = A_{2e}^{0-} A_{30}^{0-} e^{+j\beta_{y0}y} e^{-j\beta_z z} = A_{me}^{0-} e^{+\alpha_{y0}y} e^{-j\beta_z z} \qquad (8\text{-}150a)$$

$$A_{z0}^{0-} = B_{20}^{0-} A_{30}^{0-} e^{+j\beta_{y0}y} e^{-j\beta_z z} = B_{me}^{0-} e^{+\alpha_{y0}y} e^{-j\beta_z z} \qquad (8\text{-}150b)$$

$$\beta_{y0}^2 + \beta_z^2 = -\alpha_{y0}^2 + \beta_z^2 = \beta_0^2 = \omega^2 \mu_0 \varepsilon_0 \qquad (8\text{-}150c)$$

For the fields of (8-149) through (8-150c) to be of evanescent form, α_{y0} must be real and positive.
   Since the fields within and outside the slab have been separated into even and odd modes, we can examine them separately and then apply superposition. We will examine the even modes first and then the odd. For each mode (even or odd), a number of dependent and independent boundary conditions must be satisfied. A sufficient set of independent boundary conditions chosen here are

$$E_z^d(y = h, z) = E_z^{0+}(y = h, z) \qquad (8\text{-}151a)$$

$$E_z^d(y = -h, z) = E_z^{0-}(y = -h, z) \qquad (8\text{-}151b)$$

$$H_x^d(y = h, z) = H_x^{0+}(y = h, z) \qquad (8\text{-}151c)$$

$$H_x^d(y = -h, z) = H_x^{0-}(y = -h, z) \qquad (8\text{-}151d)$$

A. TM^z (Even)   By using (8-24) along with the appropriate potential function of (8-148) through (8-150c), we can write the field components as follows.

$$\underline{-h \leq y \leq h}$$

$$E_{xe}^d = -j \frac{1}{\omega \mu_d \varepsilon_d} \frac{\partial^2 A_{ze}^d}{\partial x \partial z} = 0 \qquad (8\text{-}152a)$$

$$E_{ye}^d = -j \frac{1}{\omega \mu_d \varepsilon_d} \frac{\partial^2 A_{ze}^d}{\partial y \partial z} = \frac{\beta_{yd} \beta_z}{\omega \mu_d \varepsilon_d} A_{me}^d \sin(\beta_{yd} y) e^{-j\beta_z z} \qquad (8\text{-}152b)$$

$$E_{ze}^d = -j \frac{1}{\omega \mu_d \varepsilon_d} \left(\frac{\partial^2}{\partial z^2} + \beta_d^2 \right) A_{ze}^d = -j \frac{\beta_d^2 - \beta_z^2}{\omega \mu_d \varepsilon_d} A_{me}^d \cos(\beta_{yd} y) e^{-j\beta_z z} \qquad (8\text{-}152c)$$

$$H_{xe}^d = \frac{1}{\mu_d} \frac{\partial A_{ze}^d}{\partial y} = -\frac{\beta_{yd}}{\mu_d} A_{me}^d \sin(\beta_{yd} y) e^{-j\beta_z z} \tag{8-152d}$$

$$H_{ye}^d = -\frac{1}{\mu_d} \frac{\partial A_{ze}^d}{\partial x} = 0 \tag{8-152e}$$

$$H_{ze}^d = 0 \tag{8-152f}$$

$y \geq +h$

$$E_{xe}^{0+} = -j \frac{1}{\omega\mu_0\varepsilon_0} \frac{\partial^2 A_{ze}^{0+}}{\partial x \partial z} = 0 \tag{8-153a}$$

$$E_{ye}^{0+} = -j \frac{1}{\omega\mu_0\varepsilon_0} \frac{\partial^2 A_{ze}^{0+}}{\partial y \partial z} = \frac{\alpha_{y0}\beta_z}{\omega\mu_0\varepsilon_0} A_{me}^{0+} e^{-\alpha_{y0} y} e^{-j\beta_z z} \tag{8-153b}$$

$$E_{ze}^{0+} = -j \frac{1}{\omega\mu_0\varepsilon_0} \left(\frac{\partial^2}{\partial z^2} + \beta_0^2 \right) A_{ze}^{0+} = -j \frac{\beta_0^2 - \beta_z^2}{\omega\mu_0\varepsilon_0} A_{me}^{0+} e^{-\alpha_{y0} y} e^{-j\beta_z z} \tag{8-153c}$$

$$H_{xe}^{0+} = \frac{1}{\mu_0} \frac{\partial A_{ze}^{0+}}{\partial y} = -\frac{\alpha_{y0}}{\mu_0} A_{me}^{0+} e^{-\alpha_{y0} y} e^{-j\beta_z z} \tag{8-153d}$$

$$H_{ye}^{0+} = -\frac{1}{\mu_0} \frac{\partial A_{ze}^{0+}}{\partial x} = 0 \tag{8-153e}$$

$$H_{ze}^{0+} = 0 \tag{8-153f}$$

$y \leq -h$

$$E_{xe}^{0-} = -j \frac{1}{\omega\mu_0\varepsilon_0} \frac{\partial^2 A_{ze}^{0-}}{\partial x \partial z} = 0 \tag{8-154a}$$

$$E_{ye}^{0-} = -j \frac{1}{\omega\mu_0\varepsilon_0} \frac{\partial^2 A_{ze}^{0-}}{\partial y \partial z} = -\frac{\alpha_{y0}\beta_z}{\omega\mu_0\varepsilon_0} A_{me}^{0-} e^{+\alpha_{y0} y} e^{-j\beta_z z} \tag{8-154b}$$

$$E_{ze}^{0-} = -j \frac{1}{\omega\mu_0\varepsilon_0} \left(\frac{\partial^2}{\partial z^2} + \beta_0^2 \right) A_{ze}^{0-} = -j \frac{\beta_0^2 - \beta_z^2}{\omega\mu_0\varepsilon_0} A_{me}^{0-} e^{+\alpha_{y0} y} e^{-j\beta_z z} \tag{8-154c}$$

$$H_{xe}^{0-} = \frac{1}{\mu_0} \frac{\partial A_{ze}^{0-}}{\partial y} = \frac{\alpha_{y0}}{\mu_0} A_{me}^{0-} e^{+\alpha_{y0} y} e^{-j\beta_z z} \tag{8-154d}$$

$$H_{ye}^{0-} = -\frac{1}{\mu_0} \frac{\partial A_{ze}^{0-}}{\partial x} = 0 \tag{8-154e}$$

$$H_{ze}^{0-} = 0 \tag{8-154f}$$

Applying the boundary condition (8-151a) and using (8-148c) and (8-149c) yields

$$-j \frac{\beta_d^2 - \beta_z^2}{\omega\mu_d\varepsilon_d} A_{me}^d \cos(\beta_{yd} h) e^{-j\beta_z z} = -j \frac{\beta_0^2 - \beta_z^2}{\omega\mu_0\varepsilon_0} A_{me}^{0+} e^{-\alpha_{y0} h} e^{-j\beta_z z}$$

$$\frac{\beta_d^2 - \beta_z^2}{\mu_d\varepsilon_d} A_{me}^d \cos(\beta_{yd} h) = \frac{\beta_0^2 - \beta_z^2}{\mu_0\varepsilon_0} A_{me}^{0+} es^{-\alpha_{y0} h}$$

$$\frac{\beta_{yd}^2}{\mu_d\varepsilon_d} A_{me}^d \cos(\beta_{yd} h) = -\frac{\alpha_{y0}^2}{\mu_0\varepsilon_0} A_{me}^{0+} e^{-\alpha_{y0} h} \tag{8-155a}$$

In a similar manner, enforcing (8-151b) and using (8-148c) and (8-149c) yields

$$\frac{\beta_{yd}^2}{\mu_d \varepsilon_d} A_{me}^d \cos(\beta_{yd}h) = -\frac{\alpha_{y0}^2}{\mu_0 \varepsilon_0} A_{me}^{0-} e^{-\alpha_{y0}h} \tag{8-155b}$$

Comparison of (8-155a) and (8-155b) makes it apparent that

$$A_{me}^{0+} = A_{me}^{0-} = A_{me}^0 \tag{8-155c}$$

Thus, (8-155a) and (8-155b) are the same and both can be represented by

$$\boxed{\frac{\beta_{yd}^2}{\mu_d \varepsilon_d} A_{me}^d \cos(\beta_{yd}h) = -\frac{\alpha_{y0}^2}{\mu_0 \varepsilon_0} A_{me}^0 e^{-\alpha_{y0}h}} \tag{8-156}$$

Follow a similar procedure by applying (8-151c) and (8-151d) and using (8-155c). Then we arrive at

$$\boxed{\frac{\beta_{yd}}{\mu_d} A_{me}^d \sin(\beta_{yd}h) = \frac{\alpha_{y0}}{\mu_0} A_{me}^0 e^{-\alpha_{y0}h}} \tag{8-157}$$

Division of (8-157) by (8-156) allows us to write that

$$\frac{\varepsilon_d}{\beta_{yd}} \tan(\beta_{yd}h) = -\frac{\varepsilon_0}{\alpha_{y0}}$$

$$\beta_{yd} \cot(\beta_{yd}h) = -\frac{\varepsilon_d}{\varepsilon_0} \alpha_{y0}$$

$$\boxed{-\frac{\varepsilon_0}{\varepsilon_d}(\beta_{yd}h) \cot(\beta_{yd}h) = \alpha_{y0}h}$$

where according to (8-148c) and (8-149c)

$$\boxed{\beta_{yd}^2 + \beta_z^2 = \beta_d^2 = \omega^2 \mu_d \varepsilon_d \ \Rightarrow \ \beta_{yd}^2 = \beta_d^2 - \beta_z^2 = \omega^2 \mu_d \varepsilon_d - \beta_z^2} \tag{8-158a}$$

$$\boxed{-\alpha_{y0}^2 + \beta_z^2 = \beta_0^2 = \omega^2 \mu_0 \varepsilon_0 \ \Rightarrow \ \alpha_{y0}^2 = \beta_z^2 - \beta_0^2 = \beta_z^2 - \omega^2 \mu_0 \varepsilon_0} \tag{8-158b}$$

From the free space looking down the slab we can define an impedance, which, by using (8-153a) through (8-153f) and (8-158b), can be written as

$$\boxed{Z_w^{-y0} = -\frac{E_{ze}^{0+}}{H_{xe}^{0+}} = \frac{E_{ze}^{0-}}{H_{xe}^{0-}} = -j\frac{\beta_0^2 - \beta_z^2}{\omega \varepsilon_0 \alpha_{y0}} = j\frac{\alpha_{y0}}{\omega \varepsilon_0}} \tag{8-158c}$$

which is inductive, and it indicates that *TM mode surface waves are supported by inductive surfaces*. In fact, surfaces with inductive impedance characteristics, such as dielectric slabs, dielectric-covered ground planes, and corrugated surfaces with certain heights and constitutive parameters, are designed to support TM surface waves.

B. TM^z (Odd)   By following a procedure similar to that used for the TM^z (even), utilizing the odd mode TM^z potential functions (8-148) through (8-150c), it can be shown that the expression corresponding to (8-158) is

$$\boxed{\frac{\varepsilon_0}{\varepsilon_d}(\beta_{yd}h)\tan(\beta_{yd}h) = \alpha_{y0}h} \tag{8-159}$$

where (8-158a) and (8-158b) also apply for the TM^z odd modes.

C. Summary of TM^z (Even) and TM^z (Odd) Modes   The most important expressions that are applicable for TM^z even and odd modes for a dielectric slab waveguide are (8-158) through (8-159), which are summarized here.

$$\boxed{-\frac{\varepsilon_0}{\varepsilon_d}(\beta_{yd}h)\cot(\beta_{yd}h) = \alpha_{y0}h} \quad \text{TM}^z \text{ (even)} \tag{8-160a}$$

$$\boxed{\frac{\varepsilon_0}{\varepsilon_d}(\beta_{yd}h)\tan(\beta_{yd}h) = \alpha_{y0}h} \quad \text{TM}^z \text{ (odd)} \tag{8-160b}$$

$$\boxed{\beta_{yd}^2 + \beta_z^2 = \beta_d^2 = \omega^2\mu_d\varepsilon_d \;\Rightarrow\; \beta_{yd}^2 = \beta_d^2 - \beta_z^2 = \omega^2\mu_d\varepsilon_d - \beta_z^2}$$
$$\text{TM}^z \text{ (even and odd)} \tag{8-160c}$$

$$\boxed{-\alpha_{y0}^2 + \beta_z^2 = \beta_0^2 = \omega^2\mu_0\varepsilon_0 \;\Rightarrow\; \alpha_{y0}^2 = \beta_z^2 - \beta_0^2 = \beta_z^2 - \omega^2\mu_0\varepsilon_0}$$
$$\text{TM}^z \text{ (even and odd)} \tag{8-160d}$$

$$\boxed{Z_w^{-y0} = -\frac{E_z^{0+}}{H_x^{0+}} = \frac{E_z^{0-}}{H_x^{0-}} = j\frac{\alpha_{y0}}{\omega\varepsilon_0}} \quad \text{TM}^z \text{ (even and odd)} \tag{8-160e}$$

The objective here is to determine which modes can be supported by the dielectric slab when it is used as a waveguide, and to solve for β_{yd}, α_{y0}, β_z, and the cutoff frequencies for each of these modes by using (8-160a) through (8-160d). We will begin by determining the modes and their corresponding frequencies.

It is apparent from (8-160c) and (8-160d) that if β_z is real, then

1. $\beta_z < \beta_0 < \beta_d$:

$$\beta_{yd} = \pm\sqrt{\beta_d^2 - \beta_z^2} = \text{real} \tag{8-161a}$$

$$\alpha_{y0} = \pm j\sqrt{\beta_0^2 - \beta_z^2} = \text{imaginary} \tag{8-161b}$$

2. $\beta_z > \beta_d > \beta_0$:

$$\beta_{yd} = \pm j\sqrt{\beta_z^2 - \beta_d^2} = \text{imaginary} \tag{8-162a}$$

$$\alpha_{y0} = \pm\sqrt{\beta_z^2 - \beta_0^2} = \text{real} \tag{8-162b}$$

3. $\beta_0 < \beta_z < \beta_d$:

$$\beta_{yd} = \pm\sqrt{\beta_d^2 - \beta_z^2} = \text{real} \tag{8-163a}$$

$$\alpha_{y0} = \pm\sqrt{\beta_z^2 - \beta_0^2} = \text{real} \tag{8-163b}$$

For the dielectric slab to perform as a lossless transmission line, β_{yd}, α_{y0}, and β_z *must all be real*. Therefore, for this to occur,

$$\boxed{\omega\sqrt{\mu_0\varepsilon_0} = \beta_0 < \beta_z < \beta_d = \omega\sqrt{\mu_d\varepsilon_d}} \tag{8-164}$$

The lowest frequency for which unattenuated propagation occurs is called the cutoff frequency. For the dielectric slab this occurs when $\beta_z = \beta_0$. Thus, at cutoff, $\beta_z = \beta_0$, and (8-158a) and (8-158b) reduce to

$$\beta_{yd}|_{\beta_z=\beta_0} = \pm\sqrt{\omega^2\mu_d\varepsilon_d - \beta_z^2}\,|_{\beta_z=\beta_0} = \pm\omega_c\sqrt{\mu_d\varepsilon_d - \mu_0\varepsilon_0} = \pm\omega_c\sqrt{\mu_0\varepsilon_0}\,\sqrt{\mu_r\varepsilon_r - 1} \tag{8-165a}$$

$$\alpha_{y0}|_{\beta_z=\beta_0} = \pm\sqrt{\beta_z^2 - \omega^2\mu_0\varepsilon_0}\,|_{\beta_z=\beta_0} = 0 \tag{8-165b}$$

Through the use of (8-165a) and (8-165b), the nonlinear transcendental equations 8-160a and 8-160b are satisfied, respectively, when the following equations hold.

$$\text{TM}_m^z \text{ (even)}$$

$$\cot(\beta_{yd}h) = 0 \Rightarrow \beta_{yd}h = \omega_c h\sqrt{\mu_d\varepsilon_d - \mu_0\varepsilon_0} = \frac{m\pi}{2}$$

$$\boxed{(f_c)_m = \frac{m}{4h\sqrt{\mu_d\varepsilon_d - \mu_0\varepsilon_0}}, \quad m = 1, 3, 5, \ldots} \tag{8-166a}$$

$$\text{TM}_m^z \text{ (odd)}$$

$$\tan(\beta_{yd}h) = 0 \Rightarrow \beta_{yd}h = \omega_c h\sqrt{\mu_d\varepsilon_d - \mu_0\varepsilon_0} = \frac{m\pi}{2}$$

$$\boxed{(f_c)_m = \frac{m}{4h\sqrt{\mu_d\varepsilon_d - \mu_0\varepsilon_0}}, \quad m = 0, 2, 4, \ldots} \tag{8-166b}$$

It is apparent that the cutoff frequency of a given mode is a function of the electrical constitutive parameters of the dielectric slab and its height. The modes are referred to as odd TM_m^z (when $m = 0, 2, 4, \ldots$), and even TM_m^z (when $m = 1, 3, 5, \ldots$). The dominant mode is the TM_0, which is an odd mode and its cutoff frequency is zero. This means that the TM_0 mode will always propagate unattenuated no matter what the frequency of operation. Other higher-order modes can be cut off by selecting a frequency of operation smaller than their cutoff frequencies.

Now that the TM_m^z (even) and TM_m^z (odd) modes and their corresponding cutoff frequencies have been determined, the next step is to find β_{yd}, a_{y0}, and β_z for any TM^z even or odd mode at any frequency above its corresponding cutoff frequency. This is accomplished by solving the transcendental equations 8-160a and 8-160b.

Assume that $\varepsilon_0/\varepsilon_d$, h, and the frequency of operation f are specified. Then (8-160a) and (8-160b) can be solved numerically through the use of iterative techniques by selecting values of β_{yd} and α_{y0}

that balance them. Since multiple combinations of β_{yd} and α_{y0} are possible solutions of (8-160a) and (8-160b), each combination corresponds to a given mode. Once the combination of β_{yd} and α_{y0} values that correspond to a given mode is found, the corresponding value of the phase constant β_z is found by using either (8-160c) or (8-160d).

The solution of (8-160a) through (8-160d) for the values of β_{y0}, α_{y0}, and β_z of a given TM^z mode, once $\varepsilon_0/\varepsilon_d$, h, and f are specified, can also be accomplished graphically. Although such a procedure is considered approximate (its accuracy will depend upon the size of the graph), it does shed much more physical insight onto the radiation characteristics of the modes for the dielectric slab waveguide. With such a procedure it becomes more apparent what must be done to limit the number of unattenuated modes that can be supported by the structure and how to control their characteristics. Let us now demonstrate the graphical solution of (8-160a) through (8-160d).

D. Graphical Solution for TM$_m^z$ (Even) and TM$_m^z$ (Odd) Modes

Equations 8-160a through 8-160d can be solved graphically for the characteristics of the TM^z even and odd modes. This is accomplished by referring to Figure 8-21 where the abscissa represents $\beta_{yd}h$ and the ordinate, $\alpha_{y0}h$. The procedure can best be illustrated by considering a specific value of $\varepsilon_0/\varepsilon_d$.

Let us assume that $\varepsilon_0/\varepsilon_d = 1/2.56$. With this value of $\varepsilon_0/\varepsilon_d$ (8-160a) and (8-160b) are plotted for $\alpha_{y0}h$ (ordinate) as a function of $\beta_{yd}h$ (abscissa), as shown in Figure 8-21. The next step is to solve graphically (8-160c) and (8-160d). By combining (8-160c) and (8-160d), we can write that

$$\alpha_{y0}^2 + \beta_{yd}^2 = \beta_d^2 - \beta_0^2 = \omega^2(\mu_d\varepsilon_d - \mu_0\varepsilon_0) \tag{8-167}$$

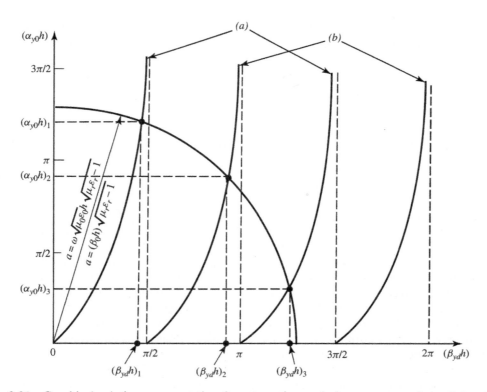

Figure 8-21   Graphical solution representation for attenuation and phase constants for a dielectric slab waveguide. (a) TM$_m^{z0}$ odd, $\varepsilon_0/\varepsilon_d(\beta_{yd}h)\tan(\beta_{yd}h)$, $m = 0, 2, 4, \ldots$ (b) TM$_m^{ze}$ even, $-\varepsilon_0/\varepsilon_d(\beta_{yd}h)\cot(\beta_{yd}h)$, $m = 1, 3, 5, \ldots$

By multiplying both sides by h^2, we can write (8-167) as

$$(\alpha_{y0}h)^2 + (\beta_{yd}h)^2 = (\omega h)^2(\mu_d\varepsilon_d - \mu_0\varepsilon_0) = (\omega h)^2\mu_0\varepsilon_0(\mu_r\varepsilon_r - 1)$$

$$\boxed{(\alpha_{y0}h)^2 + (\beta_{yd}h)^2 = a^2} \quad TM^z \text{ (even and odd)}$$

(8-168)

where

$$\boxed{a = \omega h\sqrt{\mu_0\varepsilon_0}\sqrt{\mu_r\varepsilon_r - 1} = \beta_0 h\sqrt{\mu_r\varepsilon_r - 1}} \quad TM^z \text{(even and odd)}$$

(8-168a)

It is recognized that by using the axes $\alpha_{y0}h$ (ordinate) and $\beta_{yd}h$ (abscissa), (8-168) represents a circle with a radius a determined by (8-168a). The radius is determined by the frequency of operation, the height, and the constitutive electrical parameters of the dielectric slab. The intersections of the circle of (8-168) and (8-168a) with the curves representing (8-160a) and (8-160b), as illustrated in Figure 8-21, determine the modes that propagate unattenuated within the dielectric slab waveguide. For a given intersection representing a given mode, the point of intersection is used to determine the values of $\beta_{yd}h$ and $\alpha_{y0}h$, or β_{yd} and α_{y0} for a specified h, for that mode, as shown in Figure 8-21. Once this is accomplished, the corresponding values of β_z are determined using either (8-160c) or (8-160d). This procedure is followed for each intersection point between the curves representing (8-160a), (8-160b), (8-168), and (8-168a). To illustrate the principles, let us consider a specific example.

Example 8-11

A dielectric slab of polystyrene of half thickness $h = 0.125$ in. (0.3175 cm) and with electrical properties of $\varepsilon_r = 2.56$ and $\mu_r = 1$ is bounded above and below by air. The frequency of operation is 30 GHz.

1. Determine the TM_m^z, modes, and their corresponding cutoff frequencies, that propagate unattenuated.
2. Calculate β_{yd} (rad/cm), α_{y0} (Np/cm), β_z (rad/cm), and $(\beta_z/\beta_0)^2$ for the unattenuated TM_m^z modes.

Solution:

1. By using (8-166a) and (8-166b), the cutoff frequencies of the TM_m^z modes that are lower than 30 GHz are

$$(f_c)_0 = 0 \quad TM_0^z \text{ (odd)}$$

$$(f_c)_1 = \frac{30 \times 10^9}{4(0.3175)\sqrt{2.56 - 1}} = 18.913\,\text{GHz} \quad TM_1^z \text{ (even)}$$

The remaining modes have cutoff frequencies that are higher than the desired operational frequency.

2. The corresponding wave numbers for these two modes [TM_0^z (odd) and TM_1^z (even)] will be found by referring to Figure 8-22. Since (8-160a) and (8-160b) are plotted in Figure 8-22 for $\varepsilon_0/\varepsilon_d = 1/2.56$, the only thing that remains is to plot (8-168) where the radius a is given by (8-168a).

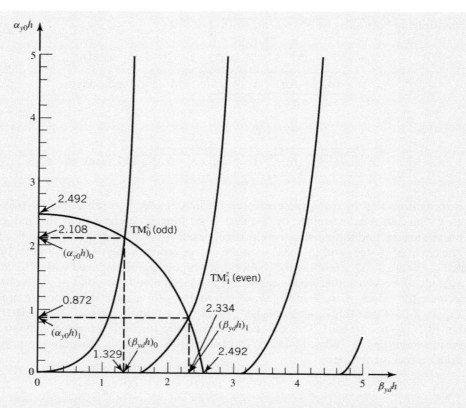

Figure 8-22 Graphical solution for attenuation and phase constants of TM_m^z modes in a dielectric slab waveguide ($\varepsilon_r = 2.56$, $\mu_r = 1$, $h = 0.3175$ cm, $f = 30$ GHz).

For $f = 30$ GHz, the radius of (8-168a) the circle (8-168) is equal to

$$a = \frac{2\pi(30 \times 10^9)(0.3175)}{30 \times 10^9}\sqrt{2.56 - 1} = 2.492$$

This is also plotted in Figure 8-22. The projections from each intersection point to the abscissa ($\beta_{yd}h$ axis) and ordinate ($\alpha_{y0}h$ axis) allow the determination of the corresponding wave numbers. From Figure 8-22

$$\underline{TM_0^z \text{ (odd)}}$$

$$(\beta_{yd}h)_0 = 1.329 \Rightarrow \beta_{yd} = 4.186 \text{ rad/cm}$$

$$(\alpha_{y0}h)_0 = 2.108 \Rightarrow \alpha_{y0} = 6.639 \text{ Np/cm}$$

When these values are substituted in (8-160c) or (8-160d), they lead to

$$\beta_z = 9.140 \text{ rad/cm } (\beta_z/\beta_0)^2 = 2.116$$

$$\underline{TM_1^z \text{ (even)}}$$

$$(\beta_{yd}h)_1 = 2.334 \Rightarrow \beta_{yd} = 7.351 \text{ rad/cm}$$

$$(\alpha_{y0}h)_1 = 0.872 \Rightarrow \alpha_{y0} = 2.747 \text{ Np/cm}$$

When these values are substituted in (8-160c) or (8-160d), they lead to

$$\beta_z = 6.857 \text{ rad/cm}, (\beta_z/\beta_0)^2 = 1.191$$

Equations 8-160a through 8-160d can also be solved simultaneously and analytically for α_{y0}, β_{yd}, and β_z through use of a procedure very similar to that outlined in Section 8.5.1 for the TE^y modes of a partially filled rectangular waveguide. The results are shown, respectively, in Figures 8-23a and 8-23b for the TM_0^z and TM_1^z modes of Example 8-11 in the frequency range $0 \le f \le 2(f_c)_1$, where $(f_c)_1$ is the cutoff frequency of the TM_1^z mode.

Curves similar to those of Figures 8-21 and 8-22 were generated for $\varepsilon_r = 1$, 2.56, 4, 9, 16, and 25 and are shown in Figure 8-24. These can be used for the solution of TM_m^z, problems where the values in the curves will be representing μ_r's instead of ε_r's. This will be seen in the next section.

8.7.3 Transverse Electric (TE^z) Modes

By following a procedure similar to that for the TM^z modes, it can be shown (by leaving out the details) that the critical expressions for the TE^z modes that correspond to those of the TM^z modes of (8-160a) through (8-160d), (8-166a) through (8-166b), (8-168), (8-168a), and (8-160e) are

$$\underline{TE^z \text{ (even) and } TE^z \text{ (odd)}}$$

$$\boxed{-\frac{\mu_0}{\mu_d}(\beta_{yd}h)\cot(\beta_{yd}h) = \alpha_{y0}h} \quad TE^z \text{ (even)} \qquad (8\text{-}169a)$$

$$\boxed{\frac{\mu_0}{\mu_d}(\beta_{yd}h)\tan(\beta_{yd}h) = \alpha_{y0}h} \quad TE^z \text{ (odd)} \qquad (8\text{-}169b)$$

$$\boxed{\beta_{yd}^2 + \beta_z^2 = \beta_d^2 = \omega^2\mu_d\varepsilon_d \;\Rightarrow\; \beta_{yd}^2 = \beta_d^2 - \beta_z^2 = \omega^2\mu_d\varepsilon_d - \beta_z^2}$$
$$TE^z \text{ (even and odd)} \qquad (8\text{-}169c)$$

$$\boxed{-\alpha_{y0}^2 + \beta_z^2 = \beta_0^2 = \omega^2\mu_0\varepsilon_0 \;\Rightarrow\; \alpha_{y0}^2 = \beta_z^2 - \beta_0^2 = \beta_z^2 - \omega^2\mu_0\varepsilon_0}$$
$$TE^z \text{ (even and odd)} \qquad (8\text{-}169d)$$

$$\boxed{(f_c)_m = \frac{m}{4h\sqrt{\mu_d\varepsilon_d - \mu_0\varepsilon_0}}} \quad \begin{array}{l} m = 1,3,5,\dots, \;\; TE^z \text{ (even)} \\ m = 0,2,4,\dots, \;\; TE^z \text{ (odd)} \end{array} \qquad \begin{array}{l}(8\text{-}169e)\\(8\text{-}169f)\end{array}$$

$$\boxed{(\alpha_{y0}h)^2 + (\beta_{yd}h)^2 = a^2} \quad TE^z \text{ (even and odd)} \qquad (8\text{-}169g)$$

$$\boxed{a = \omega h\sqrt{\mu_0\varepsilon_0}\sqrt{\mu_r\varepsilon_r - 1} = \beta_0 h\sqrt{\mu_r\varepsilon_r - 1}} \quad TE^z \text{ (even and odd)} \qquad (8\text{-}169h)$$

$$\boxed{Z_w^{-y0} = \frac{E_x^{0+}}{H_z^{0+}} = -\frac{E_x^{0-}}{H_z^{0-}} = -j\frac{\omega\mu_0}{\alpha_{y0}}} \quad TE^z \text{ (even and odd)} \qquad (8\text{-}169i)$$

Therefore, *TE surface waves are capacitive and are supported by capacitive surfaces*, whether they are dielectric slabs, dielectric covered ground planes, or corrugated surfaces.

The solution of these proceeds in the same manner as before. The curves shown in Figure 8-24 must be used and the appropriate value of $\mu_r = \mu_d/\mu_0$ must be selected.

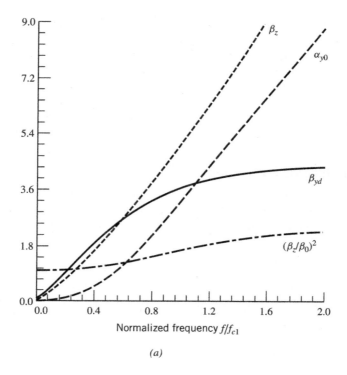

(a)

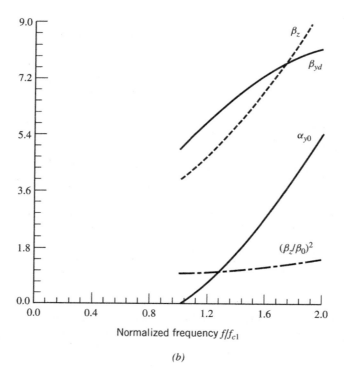

(b)

Figure 8-23   Attenuation and phase constants of TM_m^z modes in a dielectric slab waveguide ($\varepsilon_r = 2.56$, $\mu_r = 1$, $h = 0.3175$ cm, $f = 30$ GHz). (a) TM_0^z mode. (b) TM_1^z mode.

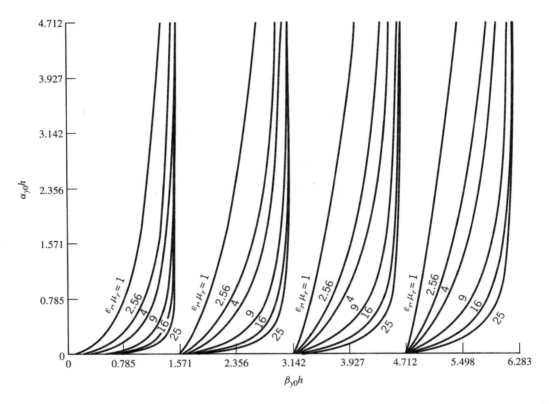

Figure 8-24 Curves to be used for graphical solution of attenuation and phase constants for TM_m^z and TE_m^z modes in a dielectric slab waveguide.

Example 8-12

Repeat the problem of Example 8-11 for the TE_m^z modes.

Solution:

1. By using (8-169e) and (8-169f), the cutoff frequencies of the TE_m^z modes that are smaller than 30 GHz are

$$(f_c)_0 = 0 \quad \mathrm{TE}_0^z \text{ (odd)}$$

$$(f_c)_1 = \frac{30 \times 10^9}{4(0.3175)\sqrt{2.56 - 1}} = 18.913 \text{ GHz} \quad \mathrm{TE}_1^z \text{ (even)}$$

These correspond to the cutoff frequencies of the TM_m^z modes of Example 8-11.

2. The corresponding wave numbers of the two modes TE_0^z (odd) and TE_1^z (even) are obtained using Figure 8-25. For $f = 30$ GHz the radius of (8-169h), which defines the circle (8-169g), is the same as that of the TM_m^z modes of Example 8-11, and it is equal to

$$a = \frac{2\pi \ (30 \times 10^9)(0.3175)}{30 \times 10^9} \sqrt{2.56 - 1} = 2.492$$

This is plotted in Figure 8-25. The projections from each intersection point on the abscissa ($\beta_{yd}h$ axis) and ordinate ($\alpha_{y0}h$ axis) allows the determination of the corresponding wave numbers. From Figure 8-25:

$$\underline{\text{TE}_0^z \text{ (odd)}}$$

$$(\beta_{yd}h)_0 = 1.109 \ \Rightarrow \ \beta_{yd} = 3.494 \text{ rad/cm}$$

$$(\alpha_{y0}h)_0 = 2.231 \ \Rightarrow \ \alpha_{y0} = 7.027 \text{ Np/cm}$$

When these are substituted in (8-169c) or (8-169d),

$$\beta_z = 9.426 \text{ rad/cm}, \quad (\beta_z/\beta_0)^2 = 2.251$$

$$\underline{\text{TE}_1^z \text{ (even)}}$$

$$(\beta_{yd}h)_1 = 2.122 \ \Rightarrow \ \beta_{yd} = 6.684 \text{ rad/cm}$$

$$(\alpha_{y0}h)_1 = 1.306 \ \Rightarrow \ \alpha_{y0} = 4.113 \text{ Np/cm}$$

When these are substituted in (8-169c) or (8-169d),

$$\beta_z = 7.510 \text{ rad/cm}, \quad (\beta_z/\beta_0)^2 = 1.428$$

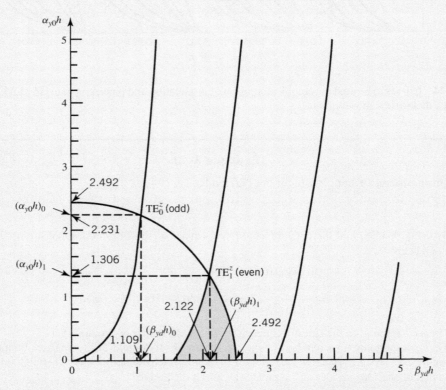

Figure 8-25   Graphical solution for attenuation and phase constants of TE_m^z modes in a dielectric slab waveguide ($\varepsilon_r = 2.56$, $\mu_r = 1$, $h = 0.3175$ cm, $f = 30$ GHz).

For the TE^z modes, (8-169a) through (8-169d) can be solved simultaneously and analytically for α_{y0}, β_{yd}, and β_z using a procedure very similar to that outlined in Section 8.5.1 for the TE^y modes of a partially filled rectangular waveguide. The results are shown in Figures 8-26a and 8-26b for the TE$_0^z$ and TE$_1^z$ modes, respectively, of Example 8-12 in the frequency range $0 \le f \le 2(f_c)_1$ where $(f_c)_1$ is the cutoff frequency of the TE$_1^z$ mode.

8.7.4   Ray-Tracing Method

In Sections 8.7.2, and 8.7.3 we analyzed the dielectric slab waveguide as a boundary-value problem using modal techniques. In this section we want to repeat the analysis of both TE^z and TM^z modes by using a ray-tracing method that sheds more physical insight onto the propagation characteristics of the dielectric slab waveguide but is not as detailed.

A wave beam that is fed into the dielectric slab can propagate into three possible modes [11]. Let us assume that the slab is bounded above by air and below by another dielectric slab, as shown in Figure 8-27, such that $\varepsilon_1 > \varepsilon_2 > \varepsilon_0$. Mathematically the problem involves a solution of Maxwell's equations and the appropriate boundary conditions at the two interfaces, as was done in Sections 8.7.2 and 8.7.3. The wave beam has the following properties.

1. It can radiate from the slab into both air and substrate, referred to as the *air–substrate modes*, as shown in Figure 8-27a.
2. It can radiate from the slab only into the substrate, referred to as the *substrate modes*, as shown in Figure 8-27b.
3. It can be bounded and be guided by the slab, referred to as the *waveguide modes*, as shown in Figure 8-27c.

To demonstrate these properties, let us assume that a wave enters the slab, which is bounded above by air and below by a substrate such that $\varepsilon_1 > \varepsilon_2 > \varepsilon_0$.

1. *Air–Substrate Modes:* Referring to Figure 8-27a, let us increase θ_1 gradually starting at $\theta_1 = 0$. When θ_1 is small, a wave that enters the slab will be refracted and will exit into the air and the substrate provided that $\theta_1 < (\theta_c)_{10} = \sin^{-1}\left(\sqrt{\varepsilon_0/\varepsilon_1}\right) < (\theta_c)_{12} = \sin^{-1}\left(\sqrt{\varepsilon_2/\varepsilon_1}\right)$. The angles $(\theta_c)_{10}$ and $(\theta_c)_{12}$ represent, respectively, the critical angles at the slab–air and slab–substrate interfaces. In this situation, wave energy can propagate freely in all three media (air, slab, and substrate) and can create radiation fields (air–substrate modes).
2. *Substrate Modes:* When θ_1 increases such that it passes the critical angle $(\theta_c)_{10}$ of the slab–air interface but is smaller than the critical angle $(\theta_c)_{12}$ of the slab–substrate interface $\left[(\theta_c)_{12} = \sin^{-1}\left(\sqrt{\varepsilon_2/\varepsilon_1}\right) > \theta_1 > (\theta_c)_{10} = \sin^{-1}\left(\sqrt{\varepsilon_0/\varepsilon_1}\right)\right]$, $\sin\theta_0 > 1$, which indicates that the wave is totally reflected at the slab–air interface. This describes a solution that wave energy in the slab radiates only in the substrate, as shown in Figure 8-27b. These are referred to as *substrate modes*.
3. *Waveguide Modes:* Finally when θ_1 is larger than the critical angle $(\theta_c)_{12}$ of the slab–substrate interface $\left[\theta_1 > (\theta_c)_{12} = \sin^{-1}\left(\sqrt{\varepsilon_2/\varepsilon_1}\right) > (\theta_c)_{10} = \sin^{-1}\left(\sqrt{\varepsilon_0/\varepsilon_1}\right)\right]$, then $\sin\theta_0 > 1$ and $\sin\theta_2 > 1$, which indicate that the wave is totally reflected at both interfaces. These are referred to as *waveguide modes*, as shown in Figure 8-27c. For these modes the energy is trapped inside the slab, and the waves follow the wave motion pattern represented by two wave vectors $\mathbf{A}_1$ and $\mathbf{B}_1$ as shown in Figure 8-28a. These two vectors are decomposed into their horizontal and vertical components $(\mathbf{A}_{1z}, \mathbf{A}_{1y})$ and $(\mathbf{B}_{1z}, \mathbf{B}_{1y})$.

The horizontal wave vector components are equal, which indicates that the waves propagate with a constant velocity in the z direction. However, the vertical components of $\mathbf{A}_1$ and $\mathbf{B}_1$

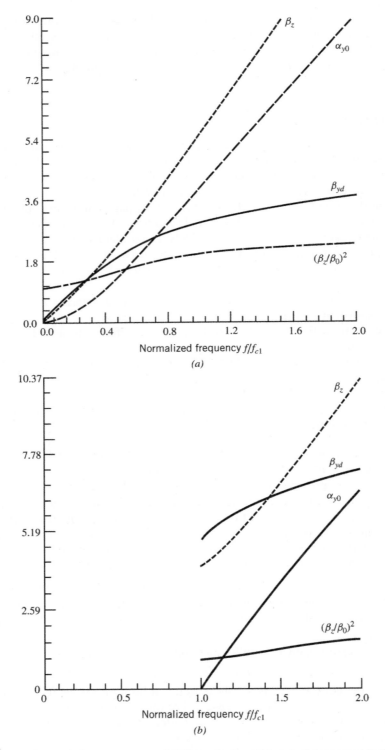

Figure 8-26 Attenuation and phase constants of TE_m^z modes in a dielectric slab waveguide ($\varepsilon_r = 2.56$, $\mu_r = 1$, $h = 0.3175$ cm, $f = 30$ GHz). (a) TE_0^z mode. (b) TE_1^z mode.

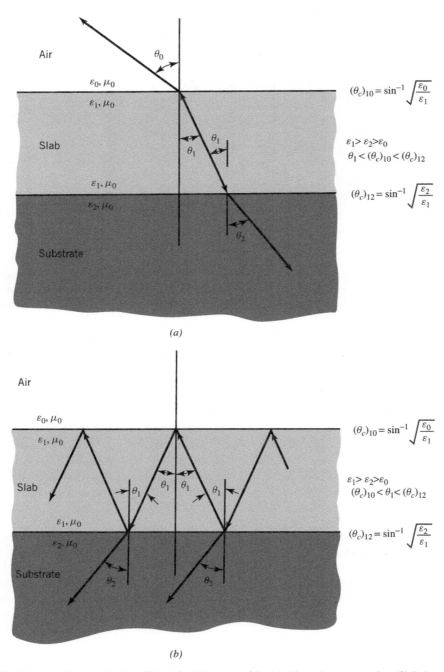

Figure 8-27 Propagation modes in a dielectric slab waveguide. (*a*) Air–substrate modes. (*b*) Substrate modes. (*c*) Waveguide modes.

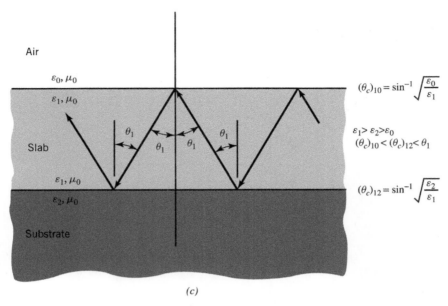

(c)

Figure 8-27 (*Continued*)

represent opposite traveling waves, which when combined form a standing wave. By changing the angle θ_1, we change the direction of $\mathbf{A}_1$ and $\mathbf{B}_1$. This results in changes in the horizontal and vertical components of $\mathbf{A}_1$ and $\mathbf{B}_1$, in the wave velocity in the z direction, and in the standing wave pattern across the slab.

The wave vectors $\mathbf{A}_1$ and $\mathbf{B}_1$ can be thought to represent a plane wave, which bounces back and forth inside the slab. The phase fronts of this plane wave are dashed in Figure 8-28*b*. An observer who moves in a direction parallel to the z axis does not see the horizontal components of the wave vectors. He does, however, observe a plane wave that bounces upward and downward, which folds one directly on top of the other. In order for the standing wave pattern across the slab to remain the same as the observer travels along the z axis, all multiple reflected waves must add in phase. This is accomplished by having the plane wave that makes one round trip, up and down across the slab, experience a phase shift equal to $2m\pi$, where m is an integer [11]. Otherwise if after the first round trip the wave experiences a small differential phase shift of δ away from $2m\pi$, it will experience differential phase shifts 2δ, 3δ,... after the second, third, ... trips. Therefore, these higher-order reflected waves will experience larger differential phase shifts which when added will eventually equal zero and the broadside wave pattern of Figure 8-28*c* will be a function of the axial position.

A one round-trip phase shift must include not only the phase change that is due to the distance traveled by the wave but also the changes in the wave phase that are due to reflections from the upper and lower interfaces. If the phase constant of the plane wave along wave vectors $\mathbf{A}_1$ and $\mathbf{B}_1$ is β_1, then the wave constant along the vertical direction y is $\beta_1 \cos\theta_1$. Therefore, the total phase shift to one round trip (up and down) of wave travel, including the phase changes due to reflection, must be equal to

$$\boxed{4\beta_1 h \cos\theta_1 - \phi_{10} - \phi_{12} = 2m\pi} \quad m = 0, 1, 2, \ldots \tag{8-170}$$

where ϕ_{10} = phase of reflection coefficient at slab–air interface
ϕ_{12} = phase of reflection coefficient at slab–substrate interface

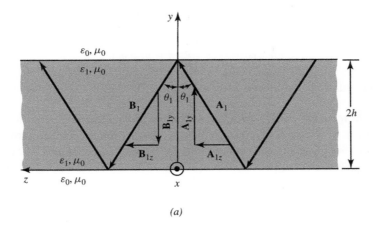

(a)

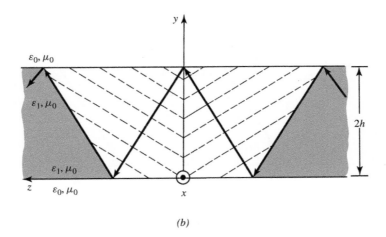

(b)

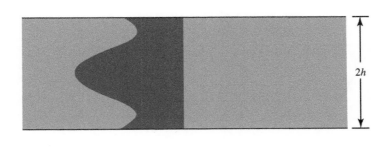

(c)

Figure 8-28   (*a*) Reflecting plane wave representation. (*b*) Phase wavefronts. (*c*) Broadside amplitude pattern of waveguide modes in a dielectric slab waveguide.

The phases of the reflection coefficients are assumed to be leading. Assume that the media above and below the slab are the same. Then $\phi_{10} = \phi_{12} = \phi$. Thus (8-170) reduces to

$$\boxed{4\beta_1 h \cos\theta_1 - 2\phi = 2m\pi} \quad m = 0, 1, 2, \ldots \tag{8-170a}$$

The preceding equations will now be applied to both TM^z and TE^z modes.

A. Transverse Magnetic (TM^z) Modes (Parallel Polarization)

Let us assume that the bouncing plane wave of Figure 8-28a is such that its polarization is TM^z (or parallel polarization), as shown in Figure 8-29a, where the slab is bounded on both sides by air. For the orientation of the fields taken as shown in Figure 8-29a, the reflected electric field $\mathbf{E}_{\parallel}^r$ is related to the incident

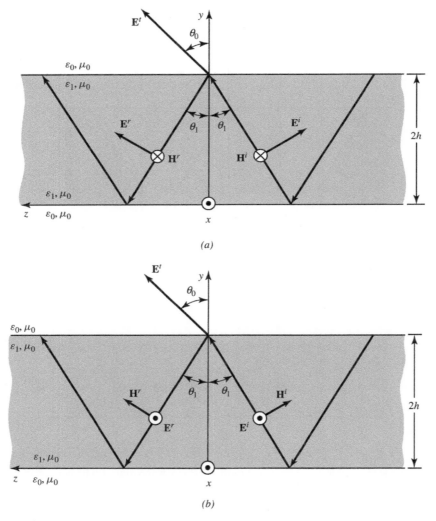

(a)

(b)

Figure 8-29 Modes in a dielectric slab waveguide. (a) TM^z (parallel polarization). (b) TE^z (perpendicular polarization).

electric field $E_\parallel^i$ by

$$\frac{E_\parallel^r}{E_\parallel^i} = -\Gamma_\parallel^b = -\left[\frac{-\eta_1 \cos\theta_1 + \eta_0 \cos\theta_0}{+\eta_1 \cos\theta_1 + \eta_0 \cos\theta_0}\right] = \frac{\eta_1 \cos\theta_1 - \eta_0 \cos\theta_0}{\eta_1 \cos\theta_1 + \eta_0 \cos\theta_0} \qquad (8\text{-}171)$$

where $\Gamma_\parallel^b$ is the reflection coefficient of (5-24c).

For $\mu_1 = \mu_0$ and for an incidence angle θ_1 greater than the critical angle $(\theta_c)_{10}$ $\left[\theta_1 > (\theta_c)_{10} = \sin^{-1}\left(\sqrt{\varepsilon_0/\varepsilon_1}\right)\right]$ (8-171) reduces, using Snell's law of refraction, (5-24b) or (5-35), to

$$-\Gamma_\parallel^b = \frac{\cos\theta_1 + j\sqrt{\dfrac{\varepsilon_1}{\varepsilon_0}}\sqrt{\dfrac{\varepsilon_1}{\varepsilon_0}\sin^2\theta_1 - 1}}{\cos\theta_1 - j\sqrt{\dfrac{\varepsilon_1}{\varepsilon_0}}\sqrt{\dfrac{\varepsilon_1}{\varepsilon_0}\sin^2\theta_1 - 1}} = \left|\Gamma_\parallel^b\right|\angle\phi_\parallel = 1\angle\phi_\parallel \qquad (8\text{-}172)$$

where

$$\phi_\parallel = 2\tan^{-1}\left[\frac{\sqrt{\dfrac{\varepsilon_1}{\varepsilon_0}}\sqrt{\dfrac{\varepsilon_1}{\varepsilon_0}\sin^2\theta_1 - 1}}{\cos\theta_1}\right] \qquad (8\text{-}172\text{a})$$

Therefore, the transcendental equation that governs these modes is derived by using (8-170a) and (8-172a). Thus,

$$4\beta_1 h \cos\theta_1 - 2\phi_\parallel = 4\beta_1 h \cos\theta_1 - 4\tan^{-1}\left[\frac{\sqrt{\dfrac{\varepsilon_1}{\varepsilon_0}}\sqrt{\dfrac{\varepsilon_1}{\varepsilon_0}\sin^2\theta_1 - 1}}{\cos\theta_1}\right] = 2m\pi$$

$$\beta_1 h \cos\theta_1 - \frac{m\pi}{2} = \tan^{-1}\left[\frac{\sqrt{\dfrac{\varepsilon_1}{\varepsilon_0}}\sqrt{\dfrac{\varepsilon_1}{\varepsilon_0}\sin^2\theta_1 - 1}}{\cos\theta_1}\right]$$

$$\tan\left(\beta_1 h \cos\theta_1 - \frac{m\pi}{2}\right) = \left[\frac{\sqrt{\dfrac{\varepsilon_1}{\varepsilon_0}}\sqrt{\dfrac{\varepsilon_1}{\varepsilon_0}\sin^2\theta_1 - 1}}{\cos\theta_1}\right]$$

$$\boxed{\tan\left(\beta_1 h \cos\theta_1 - \frac{m\pi}{2}\right) = \frac{\sqrt{\varepsilon_r}\sqrt{\varepsilon_r \sin^2\theta_1 - 1}}{\cos\theta_1}} \qquad m = 0, 1, 2, \ldots \qquad (8\text{-}173)$$

Example 8-13

The polystyrene dielectric slab of Example 5-11 of half thickness $h = 0.125$ in. (0.3175 cm) and with electrical properties $\varepsilon_r = 2.56$ and $\mu_r = 1$ is bounded above and below by air. For a frequency of operation of 30 GHz determine the TM^z modes and their corresponding angles of incidence within the slab. Plot the incidence angles as a function of frequency in the range $0 \le f/(f_c)_1 \le 2$ where $(f_c)_1$ is the cutoff frequency of the TM_1^z mode.

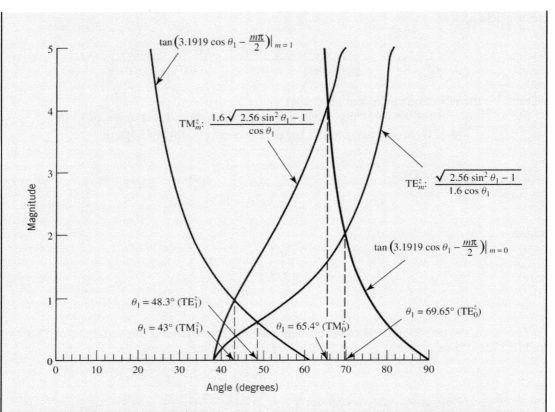

Figure 8-30 Graphical solution for angles of incidence of TM_m^z and TE_m^z modes in a dielectric slab waveguide.

Solution: The solution for this set of modes is governed by (8-173). For $\varepsilon_r = 2.56$ the right side of (8-173) is plotted in Figure 8-30. For $f = 30$ GHz, the height h of the slab is equal to

$$\lambda_1 = \frac{30 \times 10^9}{\sqrt{2.56}\, 30 \times 10^9} = 0.6250 \text{ cm} \quad \Rightarrow \quad h = \frac{0.3175}{0.6250}\lambda_1 = 0.5080\lambda_1$$

Thus (8-173) reduces to

$$\tan\left(\beta_1 h \cos\theta_1 - \frac{m\pi}{2}\right) = \tan\left[\frac{2\pi}{\lambda_1}(0.5080\lambda_1)\cos\theta_1 - \frac{m\pi}{2}\right]$$

$$= \tan\left(3.1919\cos\theta_1 - \frac{m\pi}{2}\right) = \frac{1.6\sqrt{2.56\sin^2\theta_1 - 1}}{\cos\theta_1}$$

The left side is plotted for $m = 0, 1$ in Figure 8-30, and the solutions are intersections of these curves with the curve representing the right side of (8-173). From Figure 8-30, there are two intersections that occur at $\theta_1 = 65.4°$ and $43°$, which represent, respectively, the modes $TM_0^z(\theta_1 = 65.4°)$ and $TM_1^z(\theta_1 = 43°)$. They agree with the modes of Example 8-11. No other modes are present because curves of the left side of (8-173) for higher orders of $m(m = 2, 3, \ldots)$ do not intersect with the curve that represents the right side of (8-173). Remember also that the critical angle for the slab–air interface is equal to

$$(\theta_c)_{10} = \sin^{-1}(\sqrt{\varepsilon_0/\varepsilon_1}) = \sin^{-1}(\sqrt{1/2.56}) = 38.68°$$

and the curve that represents the right side of (8-173) does not exist in Figure 8-30 below $38.68°$.

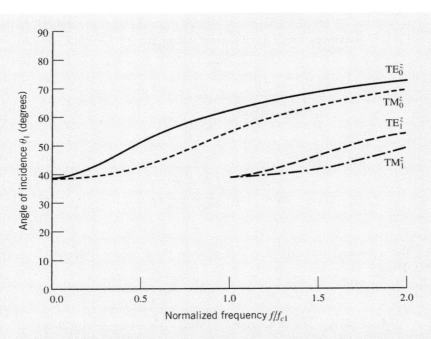

Figure 8-31 Angles of incidence for TM_m^z and TE_m^z modes in a dielectric slab waveguide.

In the slab, the wave number in the direction of incidence is equal to

$$\beta_1 = \frac{2\pi}{\lambda_1} = \frac{2\pi}{0.6250} = 10.053 \text{ rad/cm}$$

Therefore, for each mode, the wave numbers in the y direction are equal to

$$TM_0^z: \qquad \beta_{yd} = \beta_1 \cos\theta_1 = 10.053 \cos\,(65.4°) = 4.185 \text{ rad/cm}$$
$$TM_1^z: \qquad \beta_{yd} = \beta_1 \cos\theta_1 = 10.053 \cos\,(43°) = 7.352 \text{ rad/cm}$$

which closely agree with the corresponding wave numbers obtained graphically in Example 5-11. The angles of incidence as a function of frequency in the range $0 \le f/(f_c)_1 \le 2$, where $(f_c)_1$ is the cutoff frequency of the TM_1^z mode, are shown plotted in Figure 8-31.

B. Transverse Electric (TE^z) Modes (Perpendicular Polarization)   Let us now assume that the bouncing plane wave of Figure 8-28a is such that its polarization is TE^z (or perpendicular polarization) as shown in Figure 8-29b where the slab is bounded on both sides by air. For the orientation of the fields taken as shown in Figure 8-29b, the reflected electric field $\mathbf{E}_\perp^r$ is related to the incident electric field $\mathbf{E}_\perp^i$ by

$$\frac{E_\perp^r}{E_\perp^i} = \Gamma_\perp^b = \frac{\eta_0 \cos\theta_1 - \eta_1 \cos\theta_0}{\eta_0 \cos\theta_1 + \eta_1 \cos\theta_0} \qquad (8\text{-}174)$$

where $\Gamma_\perp^b$ is the reflection coefficient of (5-17a).

For $\mu_1 = \mu_0$ and for an incidence angle θ_1 greater than the critical angle $(\theta_c)_{10} [\theta_1 > (\theta_c)_{10} = \sin^{-1}(\sqrt{\varepsilon_0 / \varepsilon_1})]$, (8-174) reduces, using Snell's law of refraction, (5-15b) or (5-35), to

$$\Gamma_\perp^b = \frac{\cos\theta_1 + j\sqrt{\sin^2\theta_1 - \dfrac{\varepsilon_0}{\varepsilon_1}}}{\cos\theta_1 - j\sqrt{\sin^2\theta_1 - \dfrac{\varepsilon_0}{\varepsilon_1}}} = |\Gamma_\perp^b| \angle \phi_\perp = 1 \angle \phi_\perp \tag{8-175}$$

where

$$\phi_\perp = 2\tan^{-1}\left[\frac{\sqrt{\sin^2\theta_1 - \dfrac{\varepsilon_0}{\varepsilon_1}}}{\cos\theta_1}\right] \tag{8-175a}$$

Therefore, the transcendental equation that governs these modes is derived by using (8-170a) and (8-175a). Thus,

$$4\beta_1 h \cos\theta_1 - 2\phi_\perp = 4\beta_1 h \cos\theta_1 - 4\tan^{-1}\left(\frac{\sqrt{\sin^2\theta_1 - \dfrac{\varepsilon_0}{\varepsilon_1}}}{\cos\theta_1}\right) = 2m\pi$$

$$\beta_1 h \cos\theta_1 - \frac{m\pi}{2} = \tan^{-1}\left(\frac{\sqrt{\sin^2\theta_1 - \dfrac{\varepsilon_0}{\varepsilon_1}}}{\cos\theta_1}\right)$$

$$\tan\left(\beta_1 h \cos\theta_1 - \frac{m\pi}{2}\right) = \frac{\sqrt{\sin^2\theta_1 - \dfrac{\varepsilon_0}{\varepsilon_1}}}{\cos\theta_1}$$

$$\tan\left(\beta_1 h \cos\theta_1 - \frac{m\pi}{2}\right) = \frac{\sqrt{\sin^2\theta_1 - \dfrac{1}{\varepsilon_r}}}{\cos\theta_1}$$

$$\boxed{\tan\left(\beta_1 h \cos\theta_1 - \frac{m\pi}{2}\right) = \frac{\sqrt{\varepsilon_r \sin^2\theta_1 - 1}}{\sqrt{\varepsilon_r} \cos\theta_1}} \qquad m = 0, 1, 2, \ldots \tag{8-176}$$

Example 8-14

For the polystyrene slab of Examples 5-12 and 5-13, determine the TE_m^z modes and their corresponding angles of incidence within the slab. Plot the angles of incidence as a function of frequency in the range $0 \le f/(f_c)_1 \le 2$ where $(f_c)_1$ is the cutoff frequency of the TE_1^z mode.

Solution: The solution for this set of modes is governed by (8-176). From the solution of Example 8-13

$$h = 0.5080\lambda_1$$

and (8-176) reduces to

$$\tan\left(3.1919\cos\theta_1 - \frac{m\pi}{2}\right) = \frac{\sqrt{2.56\sin^2\theta_1 - 1}}{1.6\cos\theta_1}, \quad m = 0, 1, 2, \ldots$$

The right side is plotted in Figure 8-30. The left side for $m = 0$, 1 is also plotted in Figure 8-30. The intersections of these curves represent the solutions that from Figure 8-30 correspond, respectively, to the modes TE_0^z ($\theta_1 = 69.65°$) and TE_1^z ($\theta_1 = 48.3°$). These agree with the modes of Example 8-12. In the slab, the wave number in the direction of incidence is equal to

$$\beta_1 = \frac{2\pi}{\lambda_1} = \frac{2\pi}{0.6250} = 10.053 \text{ rad/cm}$$

Therefore, for each mode, the wave numbers in the y direction are equal to

$$TE_0^z: \quad \beta_{yd} = \beta_1 \cos\theta_1 = 10.053 \cos(69.65°) = 3.496 \text{ rad/cm}$$

$$TE_1^z: \quad \beta_{yd} = \beta_1 \cos\theta_1 = 10.053 \cos(48.3°) = 6.688 \text{ rad/cm}$$

which closely agree with the corresponding wave numbers obtained graphically in Example 8-12. The angles of incidence as a function of frequency in the range $0 \leq f / (f_c)_1$, where $(f_c)_1$ is the cutoff frequency of the TE_1 mode, are shown plotted in Figure 8-31.

8.7.5   Dielectric-Covered Ground Plane

Another type of dielectric waveguide is that of a PEC ground plane covered with a dielectric slab of height h, as shown in Figure 8-32. The field analysis of this is similar to the dielectric slab of Sections 8.7.2 and 8.7.3. However, instead of going through all of the details, the solution can be

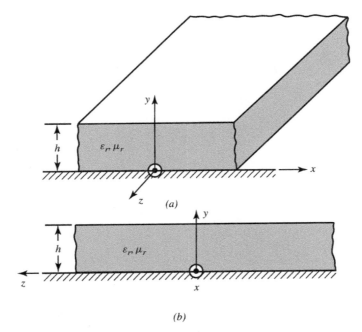

Figure 8-32   Geometry for dielectric-covered ground plane waveguide. (a) Perspective. (b) Side view.

obtained by examining the solutions for the dielectric slab as applied to the dielectric-covered ground plane. For $y \geq h$, the main differences between the two geometries are the additional boundary conditions at $y = 0$ for the dielectric-covered ground plane.

For the dielectric slab of Figure 8-19, the electric field components *within the dielectric slab* for the TM^z and TE^z modes (even and odd) of Figure 8-20 can be written from Sections 8.7.2 and 8.7.3 as

$$\text{TM}^z(\text{even})$$

$$
\left.
\begin{aligned}
E_{xe}^d &= 0 \\
E_{ye}^d &= \frac{\beta_{yd}\beta_z}{\omega\mu_d\varepsilon_d} A_{me}^d \sin(\beta_{yd}y)e^{-j\beta_z z} \\
E_{ze}^d &= -j\frac{\beta_d^2 - \beta_z^2}{\omega\mu_d\varepsilon_d} A_{me}^d \cos(\beta_{yd}y)e^{-j\beta_z z}
\end{aligned}
\right\} \quad |y| \leq h
\qquad (8\text{-}177a)
$$

$$\text{TM}^z(\text{odd})$$

$$
\left.
\begin{aligned}
E_{x0}^d &= 0 \\
E_{y0}^d &= -\frac{\beta_{yd}\beta_z}{\omega\mu_d\varepsilon_d} A_{m0}^d \cos(\beta_{yd}y)e^{-j\beta_z z} \\
E_{z0}^d &= -j\frac{\beta_d^2 - \beta_z^2}{\omega\mu_d\varepsilon_d} A_{m0}^d \sin(\beta_{yd}y)e^{-j\beta_z z}
\end{aligned}
\right\} \quad |y| \leq h
\qquad (8\text{-}177b)
$$

$$\text{TE}^z(\text{even})$$

$$
\left.
\begin{aligned}
E_{xe}^d &= \frac{\beta_{yd}}{\varepsilon_d} B_{me}^d \sin(\beta_{yd}y)e^{-j\beta_z z} \\
E_{ye}^d &= 0 \\
E_{ze}^d &= 0
\end{aligned}
\right\} \quad |y| \leq h
\qquad (8\text{-}177c)
$$

$$\text{TE}^z(\text{odd})$$

$$
\left.
\begin{aligned}
E_{x0}^d &= -\frac{\beta_{yd}}{\varepsilon_d} B_{m0}^d \cos(\beta_{yd}y)e^{-j\beta_z z} \\
E_{y0}^d &= 0 \\
E_{z0}^d &= 0
\end{aligned}
\right\} \quad |y| \leq h
\qquad (8\text{-}177d)
$$

By examining (8-177a) through (8-177d), it is apparent that the tangential electric field components TM^z (odd) of (8-177b) and TE^z (even) of (8-177c) *do* satisfy the boundary conditions of Figure 8-32 at $y = 0$ (vanishing tangential electric components at $y = 0$). However, those TM^z (even) of (8-177a) and TE^z (odd) of (8-177d) *do not* satisfy the boundary conditions of the tangential electric field components at $y = 0$. *Therefore, the geometry of Figure 8-32 supports only modes that are* TM^z *(odd) and* TE^z *(even)*. From Sections 8.7.2 and 8.7.3, the governing equations for the geometry of Figure 8-32 for TM^z (odd) and TE^z (even) modes are

$$\boxed{\frac{\varepsilon_0}{\varepsilon_d}(\beta_{yd}h)\tan(\beta_{yd}h) = (\alpha_{y0}h)} \qquad \text{TM}^z \text{ (odd)} \qquad (8\text{-}178a)$$

$$\boxed{-\frac{\mu_0}{\mu_d}(\beta_{yd}h)\cot(\beta_{yd}h) = (\alpha_{y0}h)} \qquad \text{TE}^z \text{ (even)} \qquad (8\text{-}178b)$$

$$\boxed{\beta_{yd}^2 + \beta_z^2 = \beta_d^2 = \omega^2 \mu_d \varepsilon_d} \qquad \text{TM}^z \text{ (odd), TE}^z \text{ (even)} \qquad (8\text{-}178c)$$

$$\boxed{-\alpha_{y0}^2 + \beta_z^2 = \beta_0^2 = \omega^2 \mu_0 \varepsilon_0} \qquad \text{TM}^z \text{ (odd), TE}^z \text{ (even)} \qquad (8\text{-}178d)$$

$$\boxed{(f_c)_m = \frac{m}{4h\sqrt{\mu_d \varepsilon_d - \mu_0 \varepsilon_0}}} \qquad \begin{array}{l} m = 0, 2, 4, \ldots, \text{TM}^z \text{ (odd)} \\ m = 1, 3, 5, \ldots, \text{TE}^z \text{ (even)} \end{array} \qquad \begin{array}{l}(8\text{-}178e) \\ (8\text{-}178f)\end{array}$$

Thus, the dominant mode is the TM$_0^z$ with a zero cutoff frequency. All the modes in a dielectric-covered ground plane are usually referred to as *surface wave modes*, and their solutions are obtained in the same manner as outlined in Sections 8.7.2, 8.7.3, and 8.7.4 for the dielectric slab waveguide. The only difference is that for the dielectric covered ground plane we only have TM$_m^z$ (odd) and TE$_m^z$ (even) modes. The structure *cannot* support TM$_m^z$ (even) and TE$_m^z$ (odd) modes.

The attenuation rate of the evanescent fields in air above the dielectric cover is determined by the value of α_{y0}, which is found using (8-178d). Above cutoff, it is expressed as

$$\alpha_{y0} = \sqrt{\beta_z^2 - \beta_0^2} = \sqrt{\beta_z^2 - \omega^2 \mu_0 \varepsilon_0} \qquad (8\text{-}179)$$

For very thick dielectrics ($h \to$ large) the phase constant β_z approaches β_d ($\beta_z \to \beta_d$). Thus,

$$\alpha_{y0}|_{h\to\text{large}} = \sqrt{\beta_z^2 - \omega^2 \mu_0 \varepsilon_0} \simeq \omega\sqrt{\mu_d \varepsilon_d - \mu_0 \varepsilon_0} = \omega\sqrt{\mu_0 \varepsilon_0}\sqrt{\frac{\mu_d \varepsilon_d}{\mu_0 \varepsilon_0} - 1} \qquad (8\text{-}180)$$

which is usually very large.

For very thin dielectrics ($h \to$ small), the phase constant β_z approaches β_0 ($\beta_z \to \beta_0$). Thus, from (8-178c) and (8-178d),

$$\beta_{yd}^2|_{h\to\text{small}} = \beta_d^2 - \beta_z^2 \simeq \beta_d^2 - \beta_0^2 = \omega^2(\mu_d \varepsilon_d - \mu_0 \varepsilon_0) = \omega^2 \mu_0 \varepsilon_0 \left(\frac{\mu_d \varepsilon_d}{\mu_0 \varepsilon_0} - 1\right) \qquad (8\text{-}181a)$$

$$\alpha_{y0}^2|_{h\to\text{small}} = \beta_z^2 - \beta_0^2 \simeq \text{small} \qquad (8\text{-}181b)$$

For small values of h, (8-178a) reduces for α_{y0} to

$$\alpha_{y0}|_{h\to\text{small}} = \beta_{yd}\left(\frac{\varepsilon_0}{\varepsilon_d}\right)\tan(\beta_{yd}h) \stackrel{h\to 0}{\simeq} h\left(\frac{\varepsilon_0}{\varepsilon_d}\right)(\beta_{yd})^2 \qquad (8\text{-}182)$$

Substituting (8-181a) into (8-182) reduced it to

$$\alpha_{y0}\,|_{h\to\text{small}} \simeq h\frac{\varepsilon_0}{\varepsilon_d}\left[\omega^2\mu_0\varepsilon_0\left(\frac{\mu_d\varepsilon_d}{\mu_0\varepsilon_0}-1\right)\right] = h\beta_0^2\left(\frac{\mu_d}{\mu_0}-\frac{\varepsilon_0}{\varepsilon_d}\right)$$

$$= 2\pi\beta_0\left(\frac{\mu_d}{\mu_0}-\frac{\varepsilon_0}{\varepsilon_d}\right)\frac{h}{\lambda_0} \qquad\qquad (8\text{-}182a)$$

which is usually very small.

Example 8-15

A ground plane is covered with a dielectric sheet of polystyrene of height h. Determine the distance δ (skin depth) above the dielectric–air interface so that the evanescent fields above the sheet will decay to $e^{-1} = 0.368$ of their value at the interface, when h is very large and h is very small $(= 10^{-3}\lambda_0)$.

Solution: The distance the wave travels and decays to 36.84% of its value is referred to as the skin depth. For h very large, according to (8-180),

$$\delta = \frac{1}{\alpha_{y0}} \simeq \frac{1}{\beta_0\sqrt{\dfrac{\mu_d\varepsilon_d}{\mu_0\varepsilon_0}-1}} = \frac{\lambda_0}{2\pi\sqrt{2.56-1}} = 0.126\lambda_0$$

and the wave is said to be "tightly bound" to the thick dielectric sheet.
For h very small $(h = 10^{-3}\lambda_0)$, according to (8-182a),

$$\delta = \frac{1}{\alpha_{y0}} \simeq \frac{\lambda_0}{2\pi\beta_0\left(\dfrac{\mu_d}{\mu_0}-\dfrac{\varepsilon_0}{\varepsilon_d}\right)h} = \frac{\lambda_0}{(2\pi)^2\left(1-\dfrac{1}{2.56}\right)\times 10^{-3}} = 41.6\lambda_0$$

and this wave is said to be "loosely bound" to the thin dielectric sheet.

8.8 STRIPLINE AND MICROSTRIP LINES

Microwave printed circuit technology has advanced considerably with the introduction of the stripline and microstrip transmission lines [12–27]. These lines are shown, respectively, in Figures 8-33a and 8-33b. The stripline consists of a center conductor embedded in a dielectric material that is sandwiched between two conducting plates. The microstrip consists of a thin conducting strip placed above a dielectric material, usually referred to as the substrate, which is supported on its bottom by a conducting plate. Both of these lines have evolved from the coaxial line in stages illustrated in Figure 8-34. In general the stripline and microstrip are lightweight, miniature, easy to fabricate with integrated circuit techniques, and cost effective. Their principal mode of operation is that of the quasi-TEM mode, although higher-order modes, including surface waves, are evident at higher frequencies. In comparison to other popular transmission lines, such as the coax and the waveguide, the stripline and microstrip possess characteristics that are shown listed in Table 8-8. Each of the lines will be discussed by using the most elementary approach to their basic operation. More advanced techniques of analysis can be found in the literature.

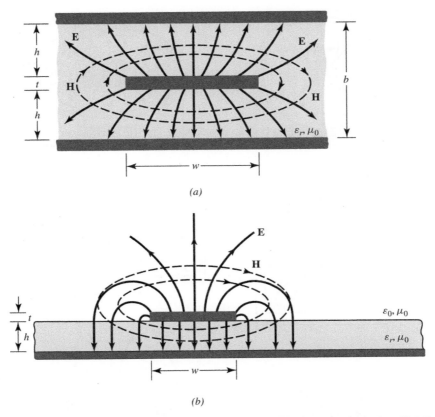

Figure 8-33   Geometries for stripline and microstrip transmission lines. (*a*) Stripline. (*b*) Microstrip.

8.8.1   Stripline

Two of the most important parameters of any transmission line are its characteristic impedance and phase velocity. Since the basic mode of operation is the TEM, its characteristic impedance Z_c and phase velocity v_p can be written, respectively, as

$$Z_c = \sqrt{\frac{L}{C}} \tag{8-183a}$$

$$v_p = \frac{1}{\sqrt{LC}} = \frac{1}{\sqrt{\mu\varepsilon}} \Rightarrow \sqrt{L} = \frac{1}{v_p\sqrt{C}} = \frac{\sqrt{\mu\varepsilon}}{\sqrt{C}} \tag{8-183b}$$

where L = inductance of line per unit length
         C = capacitance of line per unit length

Substituting (8-183b) into (8-183a) reduces it to

$$Z_c = \sqrt{\frac{L}{C}} = \frac{1}{v_p C} = \frac{\sqrt{\mu\varepsilon}}{C} = \frac{\sqrt{\mu_0\varepsilon_0}\sqrt{\mu_r\varepsilon_r}}{C} = \frac{\sqrt{\mu_r\varepsilon_r}}{v_0 C} \tag{8-184}$$

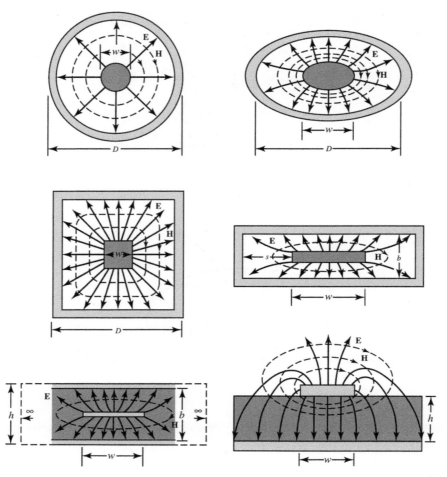

Figure 8-34 Evolution of stripline and microstrip transmission lines.

TABLE 8-8 **Characteristic comparison of popular transmission lines**

Characteristic	Coaxial	Waveguide	Stripline	Microstrip
Line losses	Medium	Low	High	High
Unloaded Q	Medium	High	Low	Low
Power capability	Medium	High	Low	Low
Bandwidth	Large	Small	Large	Large
Miniaturization	Poor	Poor	Very good	Excellent
Volume and weight	Large	Large	Medium	Small
Isolation between neighboring circuits	Very good	Very good	Fair	Poor
Realization of passive circuits	Easy	Easy	Very easy	Very easy
Integration with chip devices	Poor	Poor	Fair	Very good

where v_0 is the speed of light in free space. Therefore, the characteristic impedance can be determined if the capacitance of the line is known.

The total capacitance C_t of a stripline can be modeled as shown in Figure 8-35, and it is given by

$$C_t = 2C_p + 4C_f \tag{8-185}$$

where

$$C_t = \text{total capacitance per unit length} \tag{8-185a}$$
$$C_p = \text{parallel plate capacitance per unit length}$$
$$\text{(in the absence of fringing)}$$

$$C_p = \varepsilon \frac{2w}{b-t} = 2\varepsilon_r\varepsilon_0 \frac{\dfrac{w}{b}}{1-\dfrac{t}{b}} \tag{8-185b}$$

$$C_f = \text{fringing capacitance per unit length} \tag{8-185c}$$

Assume that the dielectric medium between the plates is not ferromagnetic. Then the characteristic impedance of (8-184) can also be written by using (8-185) and (8-185b) as

$$Z_c = \frac{\sqrt{\mu\varepsilon}}{C_t} = \frac{\varepsilon}{C_t}\sqrt{\frac{\mu}{\varepsilon}} = \frac{\varepsilon}{\sqrt{\varepsilon_r}C_t}\sqrt{\frac{\mu_0}{\varepsilon_0}} = \frac{120\pi\varepsilon}{\sqrt{\varepsilon_r}C_t} \tag{8-186}$$

or

$$Z_c\sqrt{\varepsilon_r} = \frac{120\pi}{\dfrac{1}{\varepsilon}C_t} = \frac{120\pi}{\dfrac{1}{\varepsilon}(2C_p+4C_f)} = \frac{30\pi}{\dfrac{w/b}{1-t/b}+\dfrac{C_f}{\varepsilon}} \tag{8-186a}$$

The fringing capacitance of the stripline can be approximated by using

$$\frac{C_f}{\varepsilon} \simeq \frac{1}{\pi}\left\{\frac{2}{1-\dfrac{t}{b}}\ln\left(1+\frac{1}{1-\dfrac{t}{b}}\right) - \left(\frac{1}{1-\dfrac{t}{b}}-1\right)\ln\left[\frac{1}{\left(1-\dfrac{t}{b}\right)^2}-1\right]\right\} \tag{8-187}$$

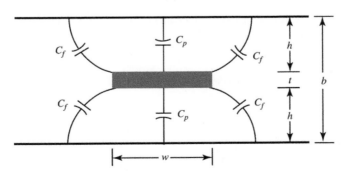

Figure 8-35 Capacitance model for stripline transmission line.

which for zero-thickness center conductor ($t = 0$), reduces to

$$\frac{C_f}{\varepsilon} \simeq \frac{1}{\pi}[2\ln(2)] = 0.4413 \tag{8-187a}$$

For zero-thickness center conductor ($t = 0$), an exact solution based on conformal mapping represents the characteristic impedance of (8-186a) by [51]

$$Z_c\sqrt{\varepsilon_r} = \frac{30\pi}{K(k)/K(k')} = 30\pi\left[\frac{K(k')}{K(k)}\right] \tag{8-188}$$

where $K(k)$ is an elliptic function of the first kind and it is given by

$$K(k) = \int_0^1 \frac{1}{\sqrt{1-q^2}}\frac{1}{\sqrt{1-kq^2}}dq = \int_0^{\pi/2}\frac{1}{\sqrt{1-k\sin^2\psi}}d\psi \tag{8-188a}$$

$$k = \tanh\left(\frac{\pi}{2}\frac{w}{b}\right) \tag{8-188b}$$

$$k' = \sqrt{1-k^2} = \sqrt{1-\tanh^2\left(\frac{\pi}{2}\frac{w}{b}\right)} = \operatorname{sech}\left(\frac{\pi}{2}\frac{w}{b}\right) \tag{8-188c}$$

It can be shown that the ratio of the elliptic functions in (8-188) can be approximated by

$$\frac{K(k)}{K(k')} \simeq \begin{cases} \dfrac{1}{\pi}\ln\left(2\dfrac{1+\sqrt{k}}{1-\sqrt{k}}\right) & \text{when} \quad \dfrac{1}{\sqrt{2}} \le k = \tanh\left(\dfrac{\pi}{2}\dfrac{w}{b}\right) \le 1 \quad (8\text{-}189a) \\[4ex] \dfrac{\pi}{\ln\left(2\dfrac{1+\sqrt{k'}}{1-\sqrt{k'}}\right)} & \text{when} \quad 0 \le k = \tanh\left(\dfrac{\pi}{2}\dfrac{w}{b}\right) \le \dfrac{1}{\sqrt{2}} \quad (8\text{-}189b) \end{cases}$$

Other forms to represent the characteristic impedance of the stripline are available, but the preceding are considered to be sufficiently simple, practical, and accurate.

Example 8-16

Determine the characteristic impedance of a zero-thickness center conductor stripline whose dielectric constant is 2.20 and w/b ratio is $w/b = 1$ and 0.1.

Solution: The solution for the characteristic impedance will be based on the more accurate formulation of (8-188) through (8-189b).

Since $w/b = 1$, then according to (8-188b)

$$k = \tanh\left(\frac{\pi}{2}\right) = 0.91715 < 1$$

Thus, by using (8-189a),

$$\frac{K(k)}{K(k')} = \frac{1}{\pi}\ln\left(2\frac{1+\sqrt{0.91715}}{1-\sqrt{0.91715}}\right) = 1.4411$$

Therefore, the characteristic impedance of (8-188) is equal to

$$Z_c = \frac{30\pi}{1.4411\sqrt{2.2}} = 44.09 \text{ ohms}$$

For $w/b = 0.1$, according to (8-188b),

$$k = \tanh\left[\frac{\pi}{2}(0.1)\right] = 0.1558$$

and from (8-188c),

$$k' = \sqrt{1 - (0.1558)^2} = 0.98779$$

Thus, by using (8-189b),

$$\frac{K(k)}{K(k')} = \frac{\pi}{\ln\left(2\frac{1+\sqrt{0.98779}}{1-\sqrt{0.98779}}\right)} = 0.4849$$

Therefore, the characteristic impedance of (8-188) is equal to

$$Z_c = \frac{30\pi}{0.4849\sqrt{2.2}} = 131.04 \text{ ohms}$$

8.8.2 Microstrip

The early investigations of the microstrip line in the early 1950s did not stimulate its widespread acceptance because of the excitation of radiation and undesired modes caused by lines with discontinuities. However, the rapid rise in miniature microwave circuits, which are usually planar in structure, caused renewed interest in microstrip circuit design. Also the development of high dielectric-constant material began to bind the fringing fields more tightly to the center conductor, thus decreasing radiation losses, and simultaneously shrinking the overall circuit dimensions. These developments, plus the advantages of convenient and economical integrated circuit fabrication techniques, tended to lessen the previous concerns and finally allowed microstrip design methods to achieve widespread application.

Because the upper part of the microstrip is usually exposed, some of the fringing field lines will be in air while others will reside within the substrate. Therefore, overall, the microstrip can be thought of as being a line composed of a homogeneous dielectric whose overall dielectric constant is greater than air but smaller than that of the substrate. The overall dielectric constant is usually referred to as the *effective dielectric constant*. Because most of the field lines reside within the substrate, the effective dielectric constant is usually closer in value to that of the substrate than to that of air; this becomes even more pronounced as the dielectric constant of the substrate increases. Since the microstrip is composed of two different dielectric materials (nonhomogeneous line), it cannot support pure TEM modes. The lowest order modes are quasi-TEM.

There have been numerous investigations of the microstrip ([19–27], and many others). Because of the plethora of information on the microstrip, we will summarize some of the formulations for the characteristic impedance and effective dielectric constant that are simple, accurate, and practical.

At low frequencies, the characteristic parameters of the microstrip can be found by using the following expressions:

$$\frac{w_{\text{eff}}(0)}{h} \leq 1$$

$$Z_c(0) = Z_c(f=0) = \frac{60}{\sqrt{\varepsilon_{r,\text{eff}}(0)}} \ln\left[\frac{8h}{w_{\text{eff}}(0)} + \frac{w_{\text{eff}}(0)}{4h}\right] \tag{8-190a}$$

$$\varepsilon_{r,\text{eff}}(0) = \varepsilon_{r,\text{eff}}(f=0) = \frac{\varepsilon_r + 1}{2} + \frac{\varepsilon_r - 1}{2}$$
$$\times \left\{\left[1 + 12\frac{h}{w_{\text{eff}}(0)}\right]^{-1/2} + 0.04\left[1 - \frac{w_{\text{eff}}(0)}{h}\right]^2\right\} \tag{8-190b}$$

$$\frac{w_{\text{eff}}(0)}{h} > 1$$

$$Z_c(0) = Z_c(f=0) = \frac{\frac{120\pi}{\sqrt{\varepsilon_{r,\text{eff}}(0)}}}{\frac{w_{\text{eff}}(0)}{h} + 1.393 + 0.667\ln\left[\frac{w_{\text{eff}}(0)}{h} + 1.444\right]} \tag{8-191a}$$

$$\varepsilon_{r,\text{eff}}(0) = \varepsilon_{r,\text{eff}}(f=0) = \frac{\varepsilon_r + 1}{2} + \frac{\varepsilon_r - 1}{2}\left[1 + 12\frac{h}{w_{\text{eff}}(0)}\right]^{-1/2} \tag{8-191b}$$

where

$$\frac{w_{\text{eff}}(0)}{h} = \frac{w_{\text{eff}}(f=0)}{h} = \frac{w}{h} + \frac{1.25}{\pi}\frac{t}{h}\left[1 + \ln\left(\frac{2h}{t}\right)\right] \quad \text{for} \quad \frac{w}{h} \geq \frac{1}{2\pi} \tag{8-192a}$$

$$\frac{w_{\text{eff}}(0)}{h} = \frac{w_{\text{eff}}(f=0)}{h} = \frac{w}{h} + \frac{1.25}{\pi}\frac{t}{h}\left[1 + \ln\left(\frac{4\pi w}{t}\right)\right] \quad \text{for} \quad \frac{w}{h} < \frac{1}{2\pi} \tag{8-192b}$$

$\varepsilon_{r,\text{eff}}$ and w_{eff} represent the effective dielectric constant and width of the line, respectively. Plots of the characteristic impedance of (8-190a) or (8-191a) and the effective dielectric constant of (8-190b) or (8-191b) as a function of w/h for three different dielectric constants ($\varepsilon_r = 2.33$, 6.80, and 10.2) are shown, respectively, in Figures 8-36 and 8-37 [28]. These dielectric constants are representative of common substrates such as RT/duroid ($\simeq 2.33$), beryllium oxide ($\simeq 6.8$), and alumina ($\simeq 10.2$) used for microstrips. It is evident that the effective dielectric constant is not very sensitive to the thickness of the center strip.

Example 8-17

For a microstrip line with $w/h = 1$, $\varepsilon_r = 10$, and $t/h = 0$, calculate at $f = 0$ the effective width, effective dielectric constant, and characteristic impedance of the line.

Solution: Since $t/h = 0$, then according to either (8-192a) or (8-192b),

$$\frac{w_{\text{eff}}(0)}{h} = \frac{w}{h} = 1$$

By using (8-190b) the effective dielectric constant is equal to

$$\varepsilon_{r,\text{eff}}(0) = \frac{10+1}{2} + \frac{10-1}{2}[1+12(1)]^{-1/2} = 6.748 < 10$$

The characteristic impedance of (8-190a) is now equal to

$$Z_c(0) = \frac{60}{\sqrt{6.748}} \ln\left[8(1) + \frac{1}{4}(1)\right] = 48.74 \text{ ohms}$$

The microstrip line is considered to be a dispersive transmission line at frequencies about equal to or greater than

$$f_c \geq 0.3\sqrt{\frac{Z_c(0)}{h}\frac{1}{\sqrt{\varepsilon_r - 1}}} \times 10^9 \quad \text{where } h \text{ is in cm} \tag{8-193}$$

For many typical transmission lines this frequency will be in the 3–10 GHz range. This indicates that the effective dielectric constant, phase velocity, and characteristic impedance will be a function of frequency. In addition, pulse wave propagation, whose spectrum spans a wide range of frequencies that depend largely on the width and shape of the pulse, can greatly be affected by the dispersive properties of the line [29–31, 42].

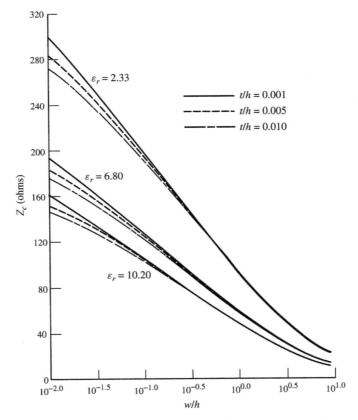

Figure 8-36 Characteristic impedance of microstrip line as a function of w/h and t/h.

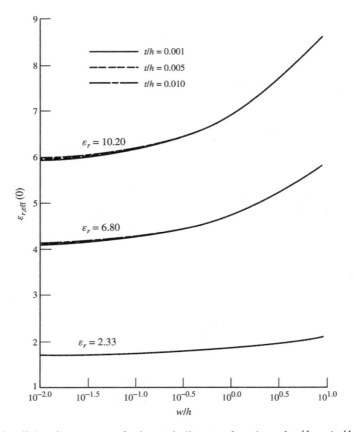

Figure 8-37 Effective dielectric constant of microstrip line as a function of w/h and t/h at zero frequency.

Many models have been developed to predict the dispersive behavior of a microstrip [32–38]. One model, which allows simple, accurate, and practical values, computes the dispersive characteristics using

$$Z_c(f) = Z_c(0)\sqrt{\frac{\varepsilon_{r,\text{eff}}(0)}{\varepsilon_{r,\text{eff}}(f)}} \tag{8-194a}$$

$$v_p(f) = \frac{1}{\sqrt{\mu\varepsilon_{\text{eff}}(f)}} = \frac{1}{\sqrt{\mu_r\mu_0\varepsilon_0\varepsilon_{r,\text{eff}}(f)}} = \frac{v_0}{\sqrt{\mu_r\varepsilon_{r,\text{eff}}(f)}} \tag{8-194b}$$

$$\lambda_g(f) = \frac{v_p(f)}{f} = \frac{v_0}{f\sqrt{\mu_r\varepsilon_{r,\text{eff}}(f)}} = \frac{\lambda_0}{\sqrt{\mu_r\varepsilon_{r,\text{eff}}(f)}} \tag{8-194c}$$

$$\varepsilon_{r,\text{eff}}(f) = \varepsilon_r - \frac{\varepsilon_r - \varepsilon_{r,\text{eff}}(0)}{1 + \dfrac{\varepsilon_{r,\text{eff}}(0)}{\varepsilon_r}\left(\dfrac{f}{f_t}\right)^2} \tag{8-194d}$$

$$f_t = \frac{Z_c(0)}{2\mu_0 h} \tag{8-194e}$$

Typical plots of $\varepsilon_{r,\text{eff}}(f)$ versus frequency for three microstrip lines ($\varepsilon_r = 2.33$, 6.8, and 10.2) are shown in Figures 8-38a and 8-38b for $w/h = 0.2$ and 5 [28]. It is evident that for $w/h \gg 1$ the variations are smaller than those for $w/h \ll 1$.

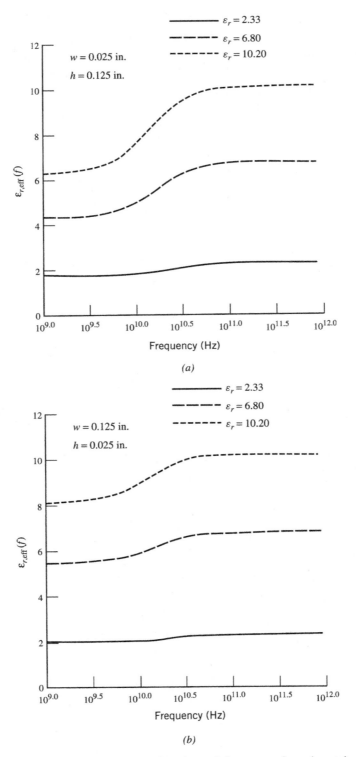

Figure 8-38 Effective dielectric constant as a function of frequency for microstrip transmission line. (a) $w/h = 0.2$. (b) $w/h = 5$.

Example 8-18

For a microstrip line with $w/h = 1$, $h = 0.025$ in. (0.0635 cm), $\varepsilon_r = 10$, and $t/h = 0$, calculate the effective dielectric constant, characteristic impedance, phase velocity, and guide wavelength at $f = 3$ and 10 GHz.

Solution: At zero frequency, from Example 8-17,

$$\varepsilon_{r,\text{eff}}(0) = 6.748$$
$$Z_c(0) = 48.74 \text{ ohms}$$

The critical frequency, where dispersion begins to appear, according to (8-193), is equal to or greater than

$$f_c \geq 0.3 \sqrt{\frac{48.74}{0.0635\sqrt{10-1}}} \times 10^9 = 4.799 \text{ GHz}$$

By using (8-194e),

$$f_t = \frac{48.74}{2(4\pi \times 10^{-7})(6.35 \times 10^{-4})} = 30.54 \times 10^9$$

$f = 3$ GHz: By using (8-194d),

$$\varepsilon_{r,\text{eff}}(f = 3 \text{ GHz}) = 10 - \left[\frac{10 - 6.748}{1 + \left(\dfrac{6.748}{10}\right)\left(\dfrac{3}{30.54}\right)^2} \right] = 6.7691$$

Thus, the characteristic impedance of (8-194a), phase velocity of (8-194b), and guide wavelength of (8-194c) are equal to

$$Z_c(f = 3 \text{ GHz}) = 48.74 \sqrt{\frac{6.748}{6.7691}} = 48.664 \text{ ohms}$$

$$v_p(f = 3 \text{ GHz}) = \frac{3 \times 10^8}{\sqrt{6.7691}} = 1.153 \times 10^8 \text{ m/sec}$$

$$\lambda_g(f = 3 \text{ GHz}) = \frac{3 \times 10^8}{3 \times 10^9 \sqrt{6.7691}} = 0.0384 \text{ m} = 3.84 \text{ cm}$$

$f = 10$ GHz: By repeating the preceding calculations at $f = 10$ GHz, we obtain

$$\varepsilon_{r,\text{eff}}(f = 10 \text{ GHz}) = 10 - \left[\frac{10 - 6.748}{1 + \dfrac{6.748}{10}\left(\dfrac{10}{30.54}\right)^2} \right] = 6.968$$

$$Z_c(f = 10 \text{ GHz}) = 48.74 \sqrt{\frac{6.748}{6.968}} = 47.964 \text{ ohms}$$

$$v_p(f = 10 \text{ GHz}) = \frac{3 \times 10^8}{\sqrt{6.968}} = 1.128 \times 10^8 \text{ m/sec}$$

$$\lambda_g(f = 10 \text{ GHz}) = \frac{3 \times 10^8}{10 \times 10^9 \sqrt{6.968}} = 0.0114 \text{ m} = 1.14 \text{ cm}$$

8.8.3 Microstrip: Boundary-Value Problem

The open microstrip line can be analyzed as a boundary-value problem using modal solutions of the form used for the partially-filled waveguide or dielectric-covered ground plane. In fact, the open microstrip line can be represented as a partially-filled waveguide with the addition of a center conductor placed along the air–dielectric interface, as shown in Figure 8-39. This shielded configuration is considered a good model for the open microstrip provided that the dimensions a and b of the waveguide are equal to or greater than about 10 to 20 times the center conductor width. The fields configurations of this structure that satisfy all the boundary conditions are hybrid modes that are a superposition of TE^z and TM^z modes [38–42].

Initially the vector potential functions used to represent, respectively, the TE^z and TM^z modes are chosen so that individually they satisfy the field boundary conditions along the metallic periphery of the waveguide. Then the total fields, which are due to the superposition of the TE^z and TM^z fields, must be such that they satisfy all the additional boundary conditions along the air–dielectric interface $(y = h)$, including those at the center metallic strip $(y = h, |x| \le w/2)$. The end result of this procedure is an infinite set of coupled homogeneous simultaneous equations that can be solved for the normalized propagation constant along the z direction $(\beta_n = \beta_z / \beta_0)$ through the use of various techniques [39–42]. A complete formulation of this problem is very lengthy, and is assigned to the reader as an end-of-chapter problem.

Another method that can also be used to solve for β_z is to use spectral domain techniques, which transforms the resulting field equations to the spectral domain, and allow for rapid convergence [40–42]. Results obtained with these methods for open and shielded microstrip geometries are shown in Figure 8-40 [42]. The waveguide width and height were chosen to be 10 times greater than the center conductor strip width. As the waveguide width and height are chosen to be even greater, the results of the open and shielded microstrips agree even better [42].

8.9 RIDGED WAVEGUIDE

It was illustrated in Section 8.2.1 that the maximum bandwidth for a dominant single TE_{10} mode operation that can be achieved by a standard rectangular waveguide is 2:1. For some applications, such as coupling, matching, filters, arrays, and so forth, larger bandwidths may be desired. This can be accomplished by using a ridged waveguide.

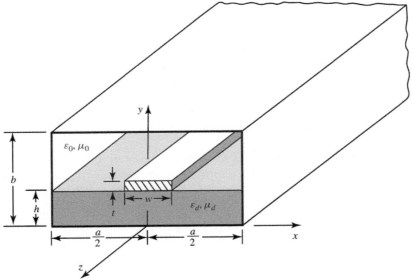

Figure 8-39 Shielded configuration of microstrip transmission line.

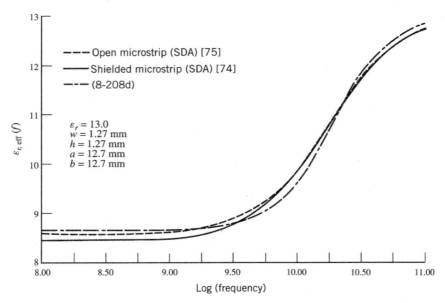

Figure 8-40   Effective dielectric constant as a function of logarithm of the frequency for open and shielded microstrip lines.

A ridged waveguide is formed by placing longitudinal metal strip(s) inside a rectangular waveguide, as shown in Figure 8-41. This has the same effect as placing inward ridges on the walls of the waveguide. The most common configurations of a ridged waveguide are those of single, dual, and quadruple ridges, as illustrated in Figure 8-41. In general, the ridges act as uniform distributed loadings, which tend to lower the phase velocity and reduce (by a factor of 25 or more) the characteristic impedance. The lowering of the phase velocity is accompanied by a reduction (by a factor as large as 5 to 6) of the cutoff frequency of the TE_{10} mode, an increase of the cutoff frequencies of the higher-order modes, an increase in the attenuation due to losses on the boundary walls, and a decrease in the power-handling capability. The increases in the bandwidth and attenuation depend upon the dimensions of the ridge compared to those of the waveguide.

The single, dual, and quadruple ridged waveguides of Figures 8-41 and 8-42 have been investigated by many people [2, 43–46]. Since the ridged waveguide possesses an irregular shape, a very appropriate technique that can be used to analyze it is the transverse resonance method of Section 8.6. At cutoff ($\beta_z = 0$) there are no waves traveling along the length (z direction) of the waveguide, and the waves can be thought of as traveling along the transverse directions (x, y directions) of the guide forming standing waves. For the TE_{10} mode, for example, there are field variations only along the x direction and at cutoff the waveguide has a cutoff frequency that is equal to the resonant frequency of a standing plane wave propagating only in the x direction. The transverse dimension (in the x direction) of the waveguide for the TE_{10} mode at resonance is equal to a half wavelength.

One very approximate equivalent model for representation of the ridged waveguide *at resonance* is that of a parallel LC network [43], shown in Figure 8-42b. The gap between the ridges is represented by the capacitance C, whose value for a waveguide of length ℓ can be found by using

$$C = \varepsilon \left(\frac{A_0}{b_0} \right) = \varepsilon \left(\frac{a_0 \ell}{b_0} \right) \tag{8-195}$$

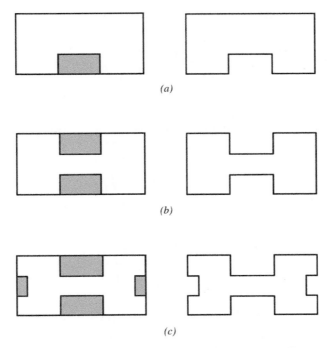

Figure 8-41 Various cross sections of a ridged waveguide. (*a*) Single. (*b*) Dual. (*c*) Quadruple.

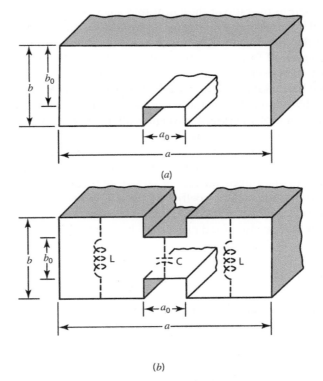

Figure 8-42 Geometry for ridged waveguides. (*a*) Single. (*b*) Dual.

Each side section of the ridged waveguide can be represented by a one-turn solenoidal inductance whose value for a waveguide of length ℓ can be found by using

$$L = \mu\left(\frac{A}{\ell}\right) = \mu\left(\frac{b\dfrac{a-a_0}{2}}{\ell}\right) = \mu\left[\frac{b(a-a_0)}{2\ell}\right] \tag{8-196}$$

Since the total inductance L_t is the parallel combination of the two L's $(L_t = L/2)$, the cutoff frequency is obtained by using

$$\omega_c = 2\pi f_c = \frac{1}{\sqrt{L_t C}} = \sqrt{\frac{2}{LC}} \tag{8-197}$$

Use of (8-195) and (8-196) reduces the cutoff frequency of (8-197) to

$$f_c = \frac{1}{2a\sqrt{\mu\varepsilon}}\left[\frac{2}{\pi}\sqrt{\frac{a}{a_0}\frac{b_0}{b}\frac{1}{1-\dfrac{a_0}{a}}}\right] \tag{8-198}$$

which is more valid for the smaller gaps where the b_0/b ratio is very small. More accurate equivalents can be obtained through use of the transverse resonance method, where the ridge waveguide can be modelled at resonance as a parallel plate waveguide with a capacitance between them that represents the discontinuity of the ridges.

Curves of available bandwidth of a single TE_{10} mode operation for single (Figure 8-42a) and dual (Figure 8-42b) ridged waveguides are shown, respectively, in Figures 8-43a and 8-43b [45]. Bandwidth is defined here as the ratio of the cutoff frequency of the next higher-order mode to that of the TE_{10} mode, and it is not necessarily the useful bandwidth. In many applications the lower and upper frequencies of the useful bandwidth are chosen with about a 15 to 25 percent safety factor from the corresponding cutoff frequencies. It is seen from the data in Figure 8-43 that a single-mode bandwidth of about 6 : 1 is realistic with a ridged rectangular waveguide. However, the penalty in realizing this extended bandwidth is the increase in attenuation. To illustrate this, we have plotted in Figures 8-44a and 8-44b the normalized attenuation α_n for single and dual ridged waveguides, which is defined as the ratio of the ridged waveguide attenuation to that of the rectangular waveguide attenuation, of identical cutoff frequency, evaluated at a frequency of $f = \sqrt{3}f_c$. The curves of Figure 8-44 have been calculated assuming that the ratio b/a of the ridged waveguide, which is 0.45 for the single ridge and 0.5 for the dual ridge, is the same as that of the rectangular waveguide. The actual attenuation of the ridged waveguide at $f = \sqrt{3}f_c$ can be obtained by multiplying the normalized values of the attenuation coefficient from Figure 8-44 by the attenuation of the rectangular waveguide evaluated at $f = \sqrt{3}f_c$. It should be noted that the increase in bandwidth of ridged waveguides is at the expense of reduced power handling capabilities.

8.10   MULTIMEDIA

On the website that accompanies this book, the following multimedia resources are included for the review, understanding, and presentation of the material of this chapter.

- **MATLAB** computer programs:
  a. **Rect_Waveguide:** Computes the propagation characteristics of a rectangular waveguide.

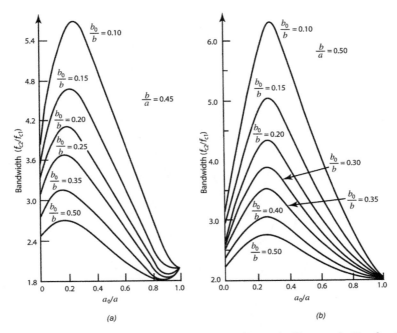

Figure 8-43 Bandwidth for ridged waveguides. (*a*) Single. (*b*) Dual. (Source: S. Hopfer, "The design of ridged waveguides," *IRE Trans. Microwave Theory Tech.*, 1955.)

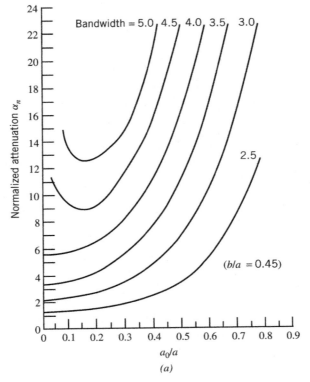

Figure 8-44 Normalized attenuation for ridged waveguides. (*a*) Single. (*b*) Dual. (Source: S. Hopfer, "The design of ridged waveguides," *IRE Trans. Microwave Theory Tech.*, IEEE.)

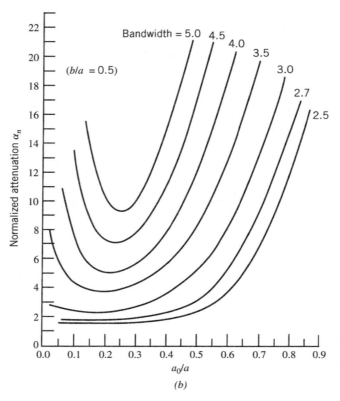

Figure 8-44   (*Continued*)

b. **Rect_Resonator:** Computes the resonant characteristics of a rectangular resonator.
c. **LS_TE_TM_Y:** Computes the TE^y and TM^y modes propagation characteristics of a partially filled rectangular waveguide based on the geometry of Figure 8-15*a*.
d. **LS_TE_TM_X:** Computes the TE^x and TM^x modes propagation characteristics of a partially filled rectangular waveguide based on the geometry of Figure 8-15*b*.
e. **Slab_TE_TM_Graph:** Computes the TE^z and TM^z modes, even and odd, propagation characteristics, using the graphical procedure of Sections 8.7.2 and 8.7.3, of a dielectric slab waveguide based on the geometry of Figure 8-19.
f. **Slab_TE_TM_Ray:** Computes the TE^z and TM^z modes, even and odd, propagation characteristics, using the ray tracing procedure of Section 8.7.4, of a dielectric slab waveguide based on the geometry of Figure 8-19.
g. **Ground_TE_TM_Graph:** Computes the TE^z and TM^z modes, even and odd, propagation characteristics of Section 8.7.5, using the graphical procedure of Sections 8.7.2 and 8.7.3, of a dielectric-covered ground plane based on the geometry of Figure 8-32.
h. **Ground_TE_TM_Ray:** Computes the TE^z and TM^z modes, even and odd, propagation characteristics of Section 8.7.5, using the ray tracing procedure of Section 8.7.4, of a dielectric slab waveguide based on the geometry of Figure 8-32.

- **PowerPoint (PPT)** viewgraphs, in multicolor.

REFERENCES

1. C. S. Lee, S. W. Lee, and S. L. Chuang, "Plot of modal field distribution in rectangular and circular waveguides," *IEEE Trans. Microwave Theory Tech.*, vol. MTT-33, pp. 271–274, Mar. 1985.

2. S. Ramo, J. R. Whinnery, and T. Van Duzer, *Fields and Waves in Communication Electronics*, Second Edition, John Wiley & Sons, New York, 1984.

3. R. F. Harrington, *Time-Harmonic Electromagnetic Fields*, McGraw-Hill, New York, 1961.

4. R. E. Collin, *Field Theory of Guided Waves*, McGraw-Hill, New York, 1960.

5. N. Marcuvitz (Ed.), *Waveguide Handbook*, Chapter 8, McGraw-Hill, New York, 1951, pp. 387–413.

6. C. H. Walter, *Traveling Wave Antennas*, McGraw-Hill, New York, 1965, pp. 172–187.

7. L. O. Goldstone and A. A. Oliner, "Leaky wave antennas I: Rectangular waveguides," *IRE Trans. Antennas Propagat.*, vol. AP-7, pp. 307–309, Oct. 1959.

8. L. O. Goldstone and A. A. Oliner, "Leaky wave antennas II: Circular waveguides," *IRE Trans. Antennas Propagat.*, vol. AP-9, pp. 280–290, May 1961.

9. J. H. Richmond, *Reciprocity Theorems and Plane Surface Waves*, Engineering Experiment Station Bulletin, Ohio State University, vol. XXVIII, no. 4, July 1959.

10. M. Zahn, *Electromagnetic Field Theory*, John Wiley & Sons, New York, 1979.

11. P. K. Tien, "Light waves in thin films and integrated optics," *Applied Optics*, vol. 10, no. 11, pp. 2395–2413, Nov. 1971.

12. R. M. Barrett, "Microwave printed circuits—a historical survey," *IRE Trans. Microwave Theory Tech.*, vol. MTT-3, no. 2, p. 9, Mar. 1955.

13. Special issue on Microwave Strip Circuits, *IRE Trans. Microwave Theory Tech.*, vol. MTT-3, no. 2, Mar. 1955.

14. H. Howe Jr., *Stripline Circuit Design*, Artech House, Dedham, MA, 1974.

15. G. L. Matthaei, L. Young, and E. M. T. Jones, *Microwave Filters Impedance-Matching Networks and Coupling Structures*, McGraw-Hill, New York, 1964.

16. S. B. Cohn, "Characteristic impedance of a shielded-strip transmission line," *IRE Trans. Microwave Theory Tech.*, MTT-2, pp. 52–57, Jul. 1954.

17. S. B. Cohn, "Characteristic impedances of broadside-coupled strip transmission lines," *IRE Trans. Microwave Theory Tech.*, vol. MTT-8, pp. 633–637, Nov. 1960.

18. H. A. Wheeler, "Transmission-line properties of parallel wide strips by a conformal-mapping approximation," *IEEE Trans. Microwave Theory Tech.*, vol. MTT-12, no. 3, pp. 280–289, May 1964.

19. H. A. Wheeler, "Transmission-line properties of parallel strips separated by a dielectric sheet," *IEEE Trans. Microwave Theory Tech.*, vol. MTT-13, no. 2, pp. 172–185, Mar. 1965.

20. T. G. Bryant and J. A. Weiss, "Parameters of microstrip transmission lines and coupled pairs of microstrip lines," *IEEE Trans. Microwave Theory Tech.*, vol. MTT-16, no. 12, pp. 1021–1027, Dec. 1968.

21. M. V. Schneider, "Microstrip lines for microwave integrated circuits," *Bell System Tech. J.*, vol. 48, pp. 1421–1444, May-Jun. 1969.

22. E. O. Hammerstad, "Equations for microstrip circuit design," *Proc. European Microwave Conference*, pp. 268–272, Sept. 1975.

23. H. A. Wheeler, "Transmission-line properties of a strip on a dielectric sheet on a plane," *IEEE Trans. Microwave Theory Tech.*, vol. MTT-25, no. 8, pp. 631–647, Aug. 1977.

24. K. C. Gupta, R. Garg, and I. J. Bahl, *Microstrip Lines and Slotlines*, Artech House, Dedham, MA, 1979.

25. M. V. Schneider, "Dielectric loss in integrated microwave circuits," *Bell System Tech. J.*, vol. 50, pp. 2325–2332, Sept. 1969.

26. R. A. Pucel, D. J. Masse, and C. D. Hartwig, "Losses in microstrip," *IEEE Trans. Microwave Theory Tech.*, vol. MTT-16, pp. 342–350, June 1968; correction vol. MTT-16, p. 1064, Dec. 1968.

27. I. J. Bahl and D. K. Trivedi, "A designer's guide to microstrip line," *Microwaves*, vol. 16, pp. 174–182, May 1977.

28. T. Leung, *"Pulse signal distortions in microstrips,"* MS(EE) Thesis, Department of Electrical and Computer Engineering, Arizona State University, Dec. 1987.

29. R. L. Veghte and C. A. Balanis, "Dispersion of transient signals in microstrip transmission lines," *IEEE Trans. Microwave Theory Tech.*, vol. MTT-34, pp. 1427–1436, Dec. 1986.

30. T. Leung and C. A. Balanis, "Attenuation distortion in microstrips," *IEEE Trans. Microwave Theory Tech.*, vol. MTT-36, no. 4, pp. 765–769, Apr. 1988.

31. E. F. Kuester and D. C. Chang, "An appraisal of methods for computation of the dispersion characteristics of open microstrips," *IEEE Trans. Microwave Theory Tech.*, vol. MTT-27, pp. 691–694, Jul. 1979.

32. M. V. Schneider, "Microstrip dispersion," *IEEE Trans. Microwave Theory Tech.*, vol. MTT-20, pp. 144–146, Jan. 1972.

33. E. J. Denlinger, "A frequency dependent solution for microstrip transmission lines," *IEEE Trans. Microwave Theory Tech.*, vol. MTT-19, no. 1, pp. 30–39, Jan. 1971.

34. W. J. Getsinger, "Microstrip dispersion model," *IEEE Trans. Microwave Theory Tech.*, vol. MTT-22, pp. 34–39, Jan. 1973.

35. H. T. Carlin, "A simplified circuit model for microstrip," *IEEE Trans. Microwave Theory Tech.*, vol. MTT-21, pp. 589–591, Sept. 1973.

36. M. Kobayashi, "Important role of inflection frequency in the dispersive properties of microstrip lines," *IEEE Trans. Microwave Theory Tech.*, vol. MTT-30, pp. 2057–2059, Nov. 1982.

37. P. Pramanick and P. Bhartia, "An accurate description of dispersion in microstrip," *Microwave Journal*, vol. 26, pp. 89–96, Dec. 1983.

38. E. Yamashita, K. Atsuki, and T. Veda, "An approximate dispersion formula of microstrip lines for computer-aided design of microwave integrated circuits," *IEEE Trans. Microwave Theory Tech.*, vol. MTT-27, pp. 1036–1038, Dec. 1979.

39. R. Mittra and T. Itoh, "A new technique for the analysis of the dispersion characteristics of microstrip lines," *IEEE Trans. Microwave Theory Tech.*, vol. MTT-19, no. 1, pp. 47–56, Jan. 1971.

40. T. Itoh and R. Mittra, "Spectral-domain approach for calculating the dispersion characteristics of microstrip lines," *IEEE Trans. Microwave Theory Tech.*, vol. MTT-21, pp. 496–499, Jul. 1973.

41. T. Itoh and R. Mittra, "A technique for computing dispersion characteristics of shielded microstrip lines," *IEEE Trans. Microwave Theory Tech.*, vol. MTT-22, pp. 896–898, Oct. 1974.

42. T. Leung and C. A. Balanis, "Pulse dispersion in open and shielded lines using the spectral-domain method," *IEEE Trans. Microwave Theory Tech.*, vol. MTT-36, no. 7, pp. 1223–1226, Jul. 1988.

43. S. B. Cohn, "Properties of ridge wave guide," *Proc. IRE*, vol. 35, pp. 783–788, Aug. 1947.

44. S. Hopfer, "The design of ridged waveguides," *IRE Trans. Microwave Theory Tech.*, vol. MTT-3, pp. 20–29, Oct. 1955.

45. J. P. Montgomery, "Ridged waveguide phased array elements," *IEEE Trans. Antennas Propagat.*, vol. AP-24, no. 1, pp. 46–53, Jan. 1976.

46. Y. Utsumi, "Variational analysis of ridged waveguide modes," *IEEE Trans. Microwave Theory Tech.*, vol. MTT-33, no. 2, pp. 111–120, Feb. 1985.

PROBLEMS

8.1. An air-filled section of an X-band (8.2–12.4 GHz) rectangular waveguide of length ℓ is used as a delay line. Assume that the inside dimensions of the waveguide are 0.9 in. (2.286 cm) by 0.4 in. (1.016 cm) and that it operates at its dominant mode. Determine its length so that the delay at 10 GHz is 2 μs.

8.2. A standard X-band (8.2–12.4 GHz) rectangular waveguide with inner dimensions of 0.9 in.

(2.286 cm) by 0.4 in. (1.016 cm) is filled with lossless polystyrene ($\varepsilon_r = 2.56$). For the lowest-order mode of the waveguide, determine at 10 GHz the following values:
(a) Cutoff frequency (in GHz).
(b) Guide wavelength (in cm).
(c) Wave impedance.
(d) Phase velocity (in m/s).
(e) Group velocity (in m/s).

8.3. A Ku-band (12.4–18 GHz) lossless rectangular waveguide, operating at the dominant TE_{10} mode, with inner dimensions 0.622 in. by 0.311 in. is used as a customized phase shifter for a particular application. The length of the waveguide is chosen so that the total phase, introduced by the insertion of the section of the waveguide, meets the required specifications of the system design operating at 15 GHz.

    (a) For an air-filled waveguide, what is the length (in cm) of the waveguide sections if the total phase, at 15 GHz, introduced by the insertion of this waveguide section is 300°?

    (b) For the waveguide section, whose length is equal to that found in part a, what is the total phase shift (in degrees), at 15 GHz, if the waveguide section is totally filled with a lossless dielectric material with a dielectric constant of 4?

8.4. Design an X-band rectangular waveguide, with dimensions 2.286 cm and 1.026 cm and filled with a dielectric material with a dielectric constant of 2.25, which is to be used as a *delay line*. What should the length (in meters) of the waveguide be so that the total delay it presents by its insertion at 10 GHz is 2 μs?

8.5. An empty X-band (8.2–12.4 GHz) rectangular waveguide, with dimensions of 2.286 cm by 1.016 cm, is to be connected to an X-band waveguide of the same dimensions but filled with lossless polystyrene ($\varepsilon_r = 2.56$). To avoid reflections, an X-band waveguide (of the same dimensions) quarter-wavelength long section is inserted between the two. Assume dominant-mode propagation and that matching is to be made at 10 GHz. Determine the:

    (a) Wave impedance of the quarter-wavelength section waveguide.

    (b) Dielectric constant of the lossless medium that must be used to fill the quarter-wavelength section waveguide.

    (c) Length (in cm) of the quarter-wavelength section waveguide.

8.6. It is desired to design a rectangular waveguide with dimensions a and b ($b > a$), as shown in Figure 8-3, such that the waveguide is completely filled with a lossless dielectric material. The waveguide is to operate in the dominant mode over a bandwidth of $1.5f_c$, where f_c is the cutoff frequency of the dominant mode. The desired frequency of operation, of the dominant mode, is 5–7.5 GHz.

Assuming $a = 2$ cm, determine, to meet the required specifications, the:

    (a) Dielectric constant of the material to be inserted inside the rectangular waveguide.

    (b) Next higher-order mode, TE_{mn} or TM_{mn}; be very specific and indicate the values of m and n.

    (c) Dimension b (in cm) of the rectangular waveguide.

    (d) Wave impedance (in ohms) of the wave at 6.25 GHz.

    (e) Guide wavelength (in cm) of the wave at 6.25 GHz.

8.7. Design a two-section binomial impedance transformer to match an empty ($\varepsilon_r = 1$) X-band waveguide to a dielectric-filled ($\varepsilon_r = 2.56$) X-band waveguide. Use two intermediate X-band waveguide sections, each quarter-wavelength long. Assume dominant mode excitation, $f_0 = 10$ GHz, and waveguide dimensions of 2.286 cm by 1.016 cm. Determine the:

    (a) Wave impedances of each section.

    (b) Dielectric constants of the lossless media that must be used to fill the intermediate waveguide sections.

    (c) Length (in cm) of each intermediate quarter-wavelength waveguide section.

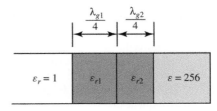

Figure P8-7

8.8. Derive expressions for the attenuation coefficient α_c above cutoff for the rectangular waveguide of Figure 8-3, assuming TE_{mn}^z modes and TM_{mn}^z modes. Compare the answers with those found in Table 8-3.

8.9. A parallel-plate waveguide is formed by placing two infinite planar conductors at $y = 0$ and $y = b$.

    (a) Show that the electric field

$$E_x = E_0 \sin(\beta_y y)e^{-\gamma z}$$

defines a set of TE_n modes where

$$\gamma = \sqrt{\beta_y^2 - \beta_0^2}, \quad \beta_0 = \omega\sqrt{\mu_0\varepsilon_0}$$

(b) For the modes of part (*a*), find the allowable eigenvalues, cutoff frequencies, and power transmitted, per unit width in the *x* direction.

8.10. A rectangular waveguide with dimensions $a = 2.25$ cm and $b = 1.125$ cm, as shown in Figure 8-3, is operating in the dominant mode.

    (a) Assume that the medium inside the guide is free space. Then find the cutoff frequency of the dominant mode.

    (b) Assume that the physical dimensions of the guide stay the same (as stated) and that we want to reduce the cutoff frequency of the dominant mode of the guide by a factor of 3. Then find the dielectric constant of the medium that must be used to fill the guide to accomplish this.

8.11. If the dielectric constant of the material that is used to construct a dielectric rod waveguide is very large (typically 30 or greater), a good approximation to the boundary conditions is to represent the surface as a perfect magnetic conductor (PMC); see Section 9.5.2. For a PMC surface, the tangential components of the magnetic field vanish. Based on such a model for a rectangular cross-section cylindrical dielectric waveguide and TE^z modes, perform the following tasks.

    (a) Write all the boundary conditions on the electric and magnetic fields that must be enforced.

    (b) Derive simplified expressions for the vector potential component, the electric and magnetic fields, and the cutoff frequencies.

    (c) If $a > b$, identify the lowest-order mode.

conducting (PEC) walls at $y = 0$ and $y = b$ and two vertical perfectly magnetic conducting (PMC) walls at $x = 0$ and $x = a$. Derive expressions for the appropriate vector potentials, electric and magnetic fields, eigenvalues, cutoff frequencies, phase constants along the *z* axis, guide wavelengths, and wave impedances for TE^z modes and TM^z modes. Identify the lowest-order mode for each set of modes and the dominant mode for both sets; assume $a > b$.

8.14. Repeat Problem 8.13 for a rectangular waveguide constructed of two horizontal PMC walls at $y = 0$ and $y = b$ and two vertical PEC walls at $x = 0$ and $x = a$.

8.15. A rectangular dielectric waveguide with dimensions a and b $(a > b)$, as shown in Figure P8-15, is used as a transmission line. The dielectric waveguide consists of a dielectric material with very high dielectric constant $(\varepsilon_r \gg 1)$. Also the waveguide has PEC (perfectly electric conducting) plates only on the top and bottom walls. The left and right walls are not covered with anything but can be treated as PMC (perfectly magnetic conducting) walls, where the tangential components of the magnetic field vanish. Assuming TE^z modes only, determine:

    (a) All of the allowable eigenvalues (β_x, β_y) for the TE^z modes for nontrivial solutions.

    (b) The dominant TE^z mode and its cutoff frequency (in GHz) when $a = 0.9$ in. (2.286 cm), $b = 0.4$ in. (1.016 cm), $\varepsilon_r = 81$, and $\mu_r = 1$.

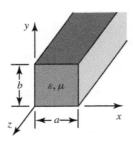

Figure P8-11

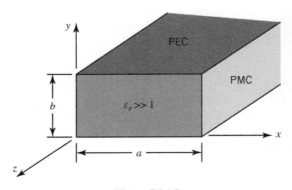

Figure P8-15

8.12. Repeat Problem 8.11 for TM^z modes.

8.13. The rectangular waveguide of Figure 8-3 is constructed of two horizontal perfectly electric

8.16. An X-band waveguide with dimensions of 0.9 in. (2.286 cm) by 0.4 in. (1.016 cm) is made of copper ($\sigma = 5.76 \times 10^7$ S/m) and it is filled with lossy polystyrene ($\varepsilon_r' = 2.56$,

$\tan \delta_e = 4 \times 10^{-4}$). Assume that the frequency of operation is 6.15 GHz. Then determine the attenuation coefficient (in Np/m and dB/m) that accounts for the finite conductivity of the walls and the dielectric losses.

8.17. For the dielectric-filled waveguide of Problem 8.16, assume that the polystyrene is lossless. Determine the following values.
(a) Cutoff frequency of the dominant mode.
(b) Frequency of operation that will allow the plane waves of the dominant mode inside the waveguide to bounce back and forth between its side walls at an angle of 45°.
(c) Guide wavelength (in cm) at the frequency of part (b).
(d) Distance (in cm) the wave must travel along the axis of the waveguide to undergo a 360° phase shift at the frequency of part (b).

8.18. An air-filled X-band waveguide with dimensions of 0.9 in. (2.286 cm) by 0.4 in. (1.016 cm) is operated at 10 GHz and is radiating into free space.
(a) Find the reflection coefficient (magnitude and phase) at the waveguide aperture junction.
(b) Find the standing wave ratio (SWR) inside the waveguide. Assume that the waveguide is made of a perfect electric conductor.
(c) Find the SWR at distances of $z = 0$, $\lambda_g/4$, and $\lambda_g/2$ from the aperture junction when the waveguide walls are made of copper ($\sigma = 5.76 \times 10^7$ S/m).

8.19. A lossless dielectric waveguide (no PEC walls), with $\varepsilon_r \gg 1$, $\mu_r = 1$, of rectangular cross section, as shown in Figure P8-11, is used as an insert line to provide a certain phase shift. Assuming $a = 2.286$ cm, $b = 1.016$ cm, and $\varepsilon_r = 81$, determine the:
(a) Approximate expressions for the cutoff frequency of the TE^z and TM^z modes. Indicate the correct allowable indices of both modes.
(b) Approximate expression for the cutoff frequency of the dominant mode. Identify the mode and its expression. Be very specific.
(c) Cutoff frequency (in GHz) of the dominant mode.
(d) Length of the waveguide (in cm) so that the total phase shift the wave undergoes

is 360° at $f = 2f_c$ as it travels through this length of the waveguide.

8.20. For the rectangular cavity of Figure 8-14, find the length c (in cm) that will resonate the cavity at 10 GHz. Assume dominant mode excitation, $c > a > b$, $a = 2$ cm and $b = 1$ cm, and free space inside the cavity.

8.21. Design a square-based cavity like Figure 8-14, with height one-half the width of the base, to resonate at 1 GHz when the cavity is:
(a) Air-filled.
(b) Filled with polystyrene ($\varepsilon_r = 2.56$).
Assume dominant-mode excitation.

8.22. A rectangular dielectric resonator is composed of dielectric material with $\varepsilon_r \gg 1$. The dimensions of the resonator are: width a in the x direction, height b in the y direction, and length c in the z direction, such that $c > a > b$.
(a) For TE^z modes, write expressions for the allowable eigenvalues β_x, β_y, and β_z (use m for x, n for y, and p for z).
(b) Repeat part (a) for TM^z modes.
(c) Write general expressions for the resonant frequencies for TE^z and TM^z modes.
(d) For $a = 1$ cm, $b = 0.5$ cm, $c = 2$ cm, and $\varepsilon_r = 81$, compute the resonant frequencies of the first two modes with the lowest resonant frequencies (in order of ascending resonant frequency). Identify the modes and their resonant frequencies (in GHz).
You do not have to derive the expressions for any of the parts as long as you justify (in words) your answers.

8.23. The field between the plates is a linearly polarized (in the y direction) uniform plane wave traveling in the z direction.
(a) Assume that the plates are perfect electric conductors. Then find the **E** and **H** field components between the plates. Neglect the edge effects of the finite plates. Referring to Figure P8-23:
(b) Find the separation d between the plates that creates resonance.
(c) Derive an expression for the Q of the cavity assuming a conductivity of σ for the plates. Neglect any radiation losses through the sides of the cavity.
(d) Compute the Q of the cavity for $f = 60$ GHz and $d = 5\lambda$ and 10λ. Assume a plate conductivity of $\sigma = 5.76 \times 10^7$ S/m.

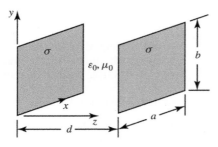

Figure P8-23

8.24. An X-band (8.2–12.4 GHz) rectangular wave-guide of inner dimensions $a = 0.9$ in. (2.286 cm) by $b = 0.4$ in. (1.016 cm) is partially filled with styrofoam ($\varepsilon_r = 1.1 \simeq 1$), as shown in Figure 8-15a. Assume that the height of the styrofoam is $b/4$. Determine the following for the TM_{10}^y mode.
(a) Phase constants (in rad/cm) in the x direction both in the air and in the styro-foam at any frequency above cutoff.
(b) Approximate phase constants (in rad/cm) in the y direction both in the air and in the styrofoam at any frequency above cutoff.
(c) The approximate value of its cutoff frequency.
(d) The phase constant (in rad/cm) in the z direction at a frequency of $f = 1.25\,(f_c)_{10}^{\mathrm{TM}}$.

8.25. For the rectangular waveguide of Figure 8-3, derive expressions for the **E** and **H** fields, eigenvalues, and cutoff frequencies for:
(a) TE^x (LSE^x) modes.
(b) TM^x (LSM^x) modes.

Identify the lowest-order mode for each set of modes and the dominant mode for both sets; assume $a > b$.

8.26. A dielectric waveguide is totally composed of a lossless dielectric medium with a dielectric constant $\varepsilon_r \gg 1$, $\mu_r = 1$. The cross section of the waveguide is rectangular with dimensions a in the x direction, b in the y direction, and infinite in the z direction; similar to Figure 8-3. Assume the wave is traveling in the $+z$ direction. For TM_{mn}^x (m for x, and n for y) modes, determine:
(a) A reduced/simplified expression for the appropriate vector potential.
(b) Expressions for the wavenumbers β_x and β_y. Be specific.

(c) An expression for the cutoff frequencies of the allowable TM_{mn}^x modes.
(d) Cutoff frequency (in GHz) of the dominant mode when $\varepsilon_r = 81$, $a = 0.9$ in. (2.286 cm), and $b = 0.4$ in. (1.016 cm).

8.27. Repeat ALL parts of Problem 8.26 for TE_{mn}^x (m for x, and n for y) modes.

8.28. An X-band waveguide is partially filled with a dielectric material, as shown in Figure 8-15(b). The pertinent equations for TM^x (LSM^x) modes are:

$$\frac{\beta_{x0}}{\varepsilon_0}\tan[\beta_{x0}(a-w)] = -\frac{\beta_{xd}}{\varepsilon_d}\tan(\beta_{xd}w)$$

$$(\beta_{x0})^2 + (\beta_{y0})^2 + (\beta_z)^2 = \beta_0^2 = \omega^2\mu_0\varepsilon_0$$

$$(\beta_{xd})^2 + (\beta_{yd})^2 + (\beta_z)^2 = \beta_d^2 = \omega^2\mu_d\varepsilon_d$$

$$\beta_{y0} = \beta_{yd} = \left(\frac{n\pi}{b}\right),\quad \begin{array}{l} m = 0,1,2,\ldots \\ n = 1,2,3,.. \end{array}$$

Assuming TM^x (LSM^x) modes and $a = 2.286$ cm, $b = 1.1016$ cm, $\varepsilon_r = 1.1$, $\mu_r = 1$, and $w = a/2$, determine for the dominant mode the:
(a) Cutoff frequency (in GHz) when the waveguide is totally filled with free space.
(b) Cutoff frequency (in GHz) when the waveguide is totally filled with the stated dielectric ($\varepsilon_r = 1.1$).
(c) Approximate cutoff frequency (in GHz) for the partially-filled waveguide.
(d) Ascending order (lowest, middle and highest) of the above three cutoff frequencies.

8.29. For the partially-filled waveguide of Figure 8-15b, derive expressions similar to (8-115) through (8-115b) or (8-127) through (8-127b) for:
(a) LSE^x (TE^x) modes.
(b) LSM^x (TM^x) modes.

8.30. For a metallic rectangular waveguide filled with air and with dimensions a and b ($b > a$), as shown in Figure 8-3, and with $a = 0.4$ in. (1.016 cm) and $b = 0.9$ in. (2.286 cm):
(a) Identify the dominant TE^z or TM^z mode and its cutoff frequency (in GHz).
(b) Identify the dominant TE^y or TM^y mode and its cutoff frequency (in GHz).
(c) What is the second mode(s) after the dominant TE^z or TM^z mode? Identify it/them.
(d) To lower the cutoff frequency of the dominant TE^z or TM^z mode to 4 GHz by completely filling the inside of the waveguide with a dielectric material,

what should the dielectric constant of the dielectric material be?

8.31. For the partially-filled waveguide of Figure 8-15a, plot on a single figure $\beta_{y0}, \beta_{yd}, \beta_z, \beta_{z0}$, and β_{zd}, all in rad/m, versus frequency [$(f_c)_{0n}$ $\leq f \leq 2(f_c)_{0n}$, where $(f_c)_{0n}$ is the cutoff frequency of the TE_{0n}^y mode] for the:
(a) TE_{02}^y mode.
(b) TE_{03}^y mode.
Assume $a = 0.9$ in. (2.286 cm), $b = 0.4$ in. (1.016 cm), $h = b/3$, and $\varepsilon_r = 2.56$.

8.32. For the partially-filled waveguide of Figure 8-15a, plot on a single figure $\beta_{y0}, \beta_{yd}, \beta_z, \beta_{z0}$, and β_{zd}, all in rad/m, versus frequency [$(f_c)_{1n}$ $\leq f \leq 2(f_c)_{1n}$, where $(f_c)_{1n}$ is the cutoff frequency of the TM_{1n}^y mode] for the:
(a) TM_{11}^y mode.
(b) TM_{12}^y mode.
Assume $a = 0.9$ in. (2.286 cm), $b = 0.4$ in. (1.016 cm), $h = b/3$, and $\varepsilon_r = 2.56$.

8.33. Use the Transverse Resonance Method (TRM) to derive the basic transcendental eigenvalue and impedance equations of Problem 8.29.

8.34. A metallic rectangular waveguide with dimensions a and b ($a > b$) of Figure 8-15a, with $a = 0.9$ in. (2.286 cm) and $b = 0.4$ in. (1.016 cm), is partially filled with air (ε_0, μ_0) and a ferromagnetic material with $\varepsilon_d = 4\varepsilon_0$ and $\mu_d = 4\mu_0$. The height of the ferromagnetic material is $h = b/3$. For each of the cases below:
(a) Identify the dominant TE_{mn}^y mode (for parts 1, 2, and 3 below).
(b) Write an analytical expression (not graphical or MATLAB solutions) for its cutoff frequency (for parts 1, 2, 3 below).
(c) Compute, based on the analytical expression only, the cutoff frequency (for parts 1, 2, 3) when the waveguide is:
  1. Completely filled with air (in GHz).
  2. Completely filled with ferromagnetic material with $\varepsilon_d = 4\varepsilon_0$ and $\mu_d = 4\mu_0$ (in GHz).
  3. Partially filled with air (ε_0, μ_0) and ferromagnetic material with $\varepsilon_d = 4\varepsilon_0$ and $\mu_d = 4\mu_0$, as shown in the Figure 8-15 (in GHz). Do not give graphical or MATLAB solutions; only solutions based on the analytical expression.

(d) Compare the cutoff frequency of the partially-filled waveguide with the other two cutoff frequencies (completely filled with air and completely filled with the ferromagnetic material); i.e., is it higher, lower, or in between the other two? Is it in the correct frequency range?

8.35. A X-band waveguide is partially filled with a dielectric material, as shown in Figure 8-15b. Assuming TM^x (LSM^x) modes and $a = 2.286$ cm, $b = 1.016$ cm, $\varepsilon_r = 1.1$, $\mu_r = 1$, and $w = a/2$, determine, for the dominant mode, the:
(a) Cutoff frequency (in GHz) when the waveguide is totally filled with free space.
(b) Cutoff frequency (in GHz) when the waveguide is totally filled with the stated dielectric ($\varepsilon_r = 1.1$).
(c) Approximate cutoff frequency (in GHz) for the partially-filled waveguide.
(d) Ascending order (lowest, middle, and highest) of the above three cutoff frequencies.
You do not have to derive the equations as long as you justify them.

8.36. A dielectric slab waveguide, as shown in Figure 8-19, is used to guide electromagnetic energy along its axis. Assume that the slab is 1 cm in height, its dielectric constant is 5, and $\mu = \mu_0$.
(a) Find the modes that can propagate unattenuated at a frequency of 8 GHz. State their cutoff frequencies.
(b) Find the respective attenuation (in Np/m) and phase (in rad/m) constants at 8 GHz for the unattenuated modes.
(c) Find the incidence angles, measured from the normal to the interface, of the bouncing waves within the slab at 8 GHz.

8.37. A guided wave structure consists of an ideal dielectric slab with a dielectric constant of 4 and height $2h$, as shown in Figure 8-19. The dielectric slab is parallel to the xz plane and the wave is traveling in the $+z$ axis. Assuming $2h = 0.50$ cm and $f = 30$ GHz, determine the:
(a) TE_m^z, $m = 0$:
  • Cutoff frequency (in GHz); specify odd, even, both, none; indicate which mode.
  • β_{yd} (in degrees/cm) assuming the incidence angle of the wave within the slab is $\theta_1 = 68.755°$.

(b) TE_m^z, $m=1$:
- Cutoff frequency (in GHz); specify odd, even, both, none; indicate which mode.
- β_{yd} (in degrees/cm) assuming the incidence angle of the wave within the slab is $\theta_1 = 45.578°$.

(c) TM_m^z, $m=0$:
- Cutoff frequency (in GHz); specify odd, even, both, none; indicate which mode.
- β_{yd} (in degrees/cm) assuming the incidence angle of the wave within the slab is $\theta_1 = 63.165°$.

(d) TM_m^z, $m=1$:
- Cutoff frequency (in GHz); specify odd, even, both, none; indicate which mode.
- β_{yd} (in degrees/cm) assuming the incidence angle of the wave within the slab is $\theta_1 = 35.763°$.

8.38. A two-dimensional (infinite in the x and z directions) dielectric slab of height $2h$, whose geometry is shown in Figure P8-38, is used as a guided structure/transmission line of electromagnetic waves traveling in the $+z$ direction. The dielectric constant and relative permeability of the slab material are $\varepsilon_r \gg 1$, $\mu_r = 1$. The medium above and below the dielectric slab is free space. For TE^z modes, write the:
(a) Appropriate simplified vector potential expression.
(b) Expression for the permitted eigenvalues for nontrivial modes; be specific.
(c) Cutoff frequency of dominant mode (in GHz) when $\varepsilon_r = 81$.
(d) Simplified expression for the wave impedance Z_w^{+z} (in the $+z$ direction).

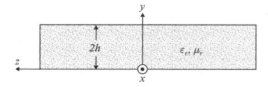

Figure P8-38

8.39. Repeat the TE^z modes of Problem 8.38 when $\mu_r \gg 1$, $\varepsilon_r = 1$. Numerically evaluate part (c) when $\mu_r = 81$.

8.40. A guided waveguide structure consists of two PEC plates placed parallel to each other,

each parallel to the xz plane and separated by a distance b, as shown in Figure P8-40. The plates extend to infinity in the x and z directions, and the wave travels in the $+z$ direction. Assuming the entire medium between the plates is an ideal dielectric of dielectric constant 4 and a total height/separation of the PEC plates of 0.5 cm, determine for the TE^z modes:
(a) The expression for the reduced/simplified vector potential.
(b) The expressions for the reduced/simplified expressions for the electric and magnetic fields.
(c) The expression for all the permissible eigenvalues; be specific.
(d) The expression for the wave impedance Z_w^{+z}.
(e) Dominant mode (be specific) and its cutoff frequency (in GHz).

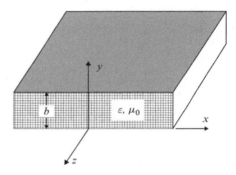

Figure P8-40

8.41. Repeat Problem 8.40 for TM^z modes.

8.42. An electromagnetic wave is traveling, in the z direction, inside a two-dimensional lossless dielectric slab of height d and with dielectric constant ε_{r1} as in Figure P8-42. The slab is bounded above and below by an identical semi-infinite lossless dielectric medium with a dielectric constant of $\varepsilon_{r2} = 4$. What should the range of values of the dielectric constant ε_{r1} be if it is desired to use the dielectric slab as a:
a. Waveguide (structure that guides the wave; keeps the energy within the slab), instead of a radiator/antenna (structure which continuously sheds/losses energy as the wave travels along the slab)?
b. Radiator/antenna (structure that continuously sheds/losses energy as the wave travels along the slab), instead of a waveguide (structure which guides the wave; keeps the energy within the slab)?

For each case, show the mathematical details and/or explain in words (you can make reference, if necessary) why such a range of values of ε_{r1} will accomplish the desired goal in each of the two cases (part a and part b).

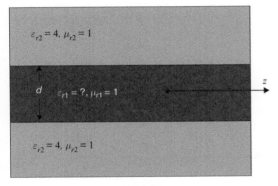

Figure P8-42

8.43. A ground plane is covered with a dielectric material with a dielectric constant of 4. The total height of the dielectric material, of the dielectric cover, is 1.25 cm.

(a) Identify the first two modes with the lowest cutoff frequencies, and determine their corresponding cutoff frequencies.

(b) Determine for both modes of part (a) the phase constant in the dielectric (in rad/cm) in the direction normal to the interface when the incidence angle (measured from the normal to the interface) is twice the critical angle.

8.44. Design a nonferromagnetic lossless dielectric slab of total height 0.5 in. (1.27 cm) bounded above and below by air so that at $f = 10$ GHz the TE_1^z mode operates at 10% above its cutoff frequency. Determine the dielectric constant of the slab and the attenuation α_{y0} (in Np/cm) and β_{yd} (in rad/cm) for the TE_1^z mode at its cutoff frequency.

8.45. A guided wave structure consists of a PEC ground plane covered with a dielectric slab, as shown in Figure 8-32. The dielectric slab is parallel to the xz plane and the wave is traveling in the $+z$ axis. The height of the dielectric slab is $h = 0.25$ cm, the dielectric constant is 4, and the frequency of operation is 30 GHz. Determine the:

(a) TE_m^z, $m = 0$:

- Cutoff frequency (in GHz); specify odd, even, both, none; indicate which mode.
- β_{yd} (in degrees/cm) assuming the incidence angle of the wave within the slab is $\theta_1 = 68.755°$.

(b) TE_m^z, $m = 1$:

- Cutoff frequency (in GHz); specify odd, even, both, none; indicate which mode.
- β_{yd} (in degrees/cm) assuming the incidence angle of the wave within the slab is $\theta_1 = 45.578°$.

(c) TM_m^z, $m = 0$:

- Cutoff frequency (in GHz); specify odd, even, both, none; indicate which mode.
- β_{yd} (in degrees/cm) assuming the incidence angle of the wave within the slab is $\theta_1 = 63.165°$.

(d) TM_m^z, $m = 1$:

- Cutoff frequency (in GHz); specify odd, even, both, none; indicate which mode.
- β_{yd} (in degrees/cm) assuming the incidence angle of the wave within the slab is $\theta_1 = 35.763°$.

8.46. An infinite PMC ground plane is covered with a lossless dielectric slab ($\varepsilon_r, \mu_r = 1$). The slab is of height h in the y direction, infinite in the x and z directions, and the wave is traveling in the $+z$ direction. The geometry,

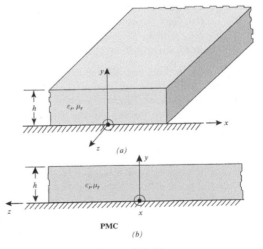

Figure P8-46

shown in Figure P8-46 [(a) 3D view, (b) side view], is the same as that of Figure 8-32 except that the ground plane is PMC instead of PEC. Determine, assuming the dielectric covered PMC ground plane is used as a waveguide (not as an antenna), the:

(a) Allowable TM_m^z modes (even and/or odd). Be specific which ones can and which ones cannot propagate as guided waves within the dielectric-covered PMC ground plane. Show all the steps.

(b) Expression for the cutoff frequencies of the allowable TM^z (even and/or odd) modes. Be specific about all the allowable eigenvalues m for the even and/or odd modes.

(c) Cutoff frequency (in GHz) of the dominant TM^z mode (even and/or odd) when $\varepsilon_r = 4$ and $h = 0.125$ cm. Be specific to identify whether even and/or odd mode(s).

8.47. Repeat ALL parts of Problem 8.46 for TE_m^z modes.

8.48. A planar perfect electric conductor of infinite dimensions is coated with a dielectric medium of thickness h, as shown in Figure 8-32. Assume that the dielectric constant of the coating is 5, its relative permeability is unity, and its thickness is 5.625 cm.

(a) Find the cutoff frequencies of the first four TE^z and/or TM^z modes and specify to which group each one belongs.

(b) For an operating frequency of 1 GHz, find the TE^z modes that can propagate inside the slab unattenuated.

(c) For each of the TE^z modes found in part (b), find the corresponding propagation constant β_z.

The medium above the coating is free space.

8.49. Coupling between distributive microwave and millimeter-wave microstrip circuit elements (such as filters, couplers, antennas, etc.), that are etched on the surface of a grounded dielectric slab—referred to as substrate and considered as a dielectric-covered ground plane—is either through space or surface waves. Space waves are those radiated by the elements and travel through air, while surface waves are those excited and travelling within the substrate. It is desired to design the system so that it will eliminate all the surface-wave modes, other than the dominant mode (static mode) with zero cutoff frequency. The maximum height of the substrate is 0.113 cm.

(a) Determine the dominant surface-wave mode and its cutoff frequency.

(b) Determine the dielectric constant of the substrate so that the designed circuits will operate in a single, dominant surface-wave mode up to 20 GHz.

(c) Identify the next higher-order surface-wave mode and its cutoff frequency.

(d) For the mode of part (c), determine at its cutoff frequency the:
- Attenuation constant α_{y0} (in dB/cm)
- Phase constant β_{yd} (in degrees/cm)

(e) For the mode of part (c), determine at 25 GHz the approximate:
- Attenuation constant α_{y0} (in dB/cm)
- Phase constant β_{yd} (in degrees/cm)

8.50. A transmission line is composed of a dielectric-covered ground plane, as shown in Figure P8-50. The dielectric constant of the dielectric cover is $\varepsilon_{r1} = 2.56$ while its height is 2 cm. It is desired to operate this transmission line in a single, dominant mode over a bandwidth of 3.1 GHz.

(a) Identify the dominant mode that can be supported by the line and its cutoff frequency.

(b) State which mode has the next highest cutoff frequency.

(c) What should the dielectric constant ε_{r2} of the second (unbounded) medium be to meet the desired bandwidth requirements of the dominant-mode operation?

(d) For the second mode, determine at 3.1 GHz the phase constant β_{yd} (in rad/cm) in the dielectric cover with dielectric constant of 2.56, and the attenuation constant α_{y2} (in Np/cm) in the unbounded medium with dielectric constant ε_{r2}

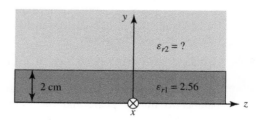

Figure P8-50

8.51. A infinite PMC ground plane is covered with a lossless dielectric slab ($\varepsilon_r, \mu_r = 1$). The slab is of height h in the y direction, infinite in the x and z directions, and the wave is traveling in the $+z$ direction. The geometry is the same as that of Figure 8-32 except that the ground plane is PMC instead of PEC.

Determine, assuming the dielectric-covered PMC ground plane is used as a waveguide (not as an antenna), the:

(a) Allowable TM_m^z modes (even and/or odd).
(b) Expression for the cutoff frequencies of the allowable TM^z (even and/or odd) modes.
(c) Cutoff frequency (in GHz) of the dominant TM^z mode (even and/or odd) when $\varepsilon_r = 4$ and $h = 0.125$ cm.

8.52. Repeat ALL parts of Problem 8.51 for TE_m^z modes.

8.53. For the stripline of Example 8-16, find the characteristic impedances based on the approximate formulas of (8-186a), (8-187), and (8-187a). Compare the answers with the more accurate values obtained in Example 8-16, and comment on the comparisons.

8.54. Design a stripline with a characteristic impedance of 30 ohms whose dielectric constant is 4. Assume the thickness of the center conductor is zero ($t/b = 0$).

8.55. A parallel plate transmission line (waveguide) is formed by two finite width plates placed at $y = 0$ and $y = h$, and it is used to approximate a microstrip. Assume that the electric field between the plates is given by

$$\mathbf{E} \simeq \hat{\mathbf{a}}_y E_0 e^{-j\beta z} \quad \text{provided } w/h \gg 1$$

where E_0 is a constant. Derive, for the conduction losses, an expression for the attenuation constant α_c (in Np/m) in terms of the plate surface resistance R_s, w, h, ε, and μ. The plates are made of metal with conductivity σ, and the medium between the plates is a lossless dielectric.

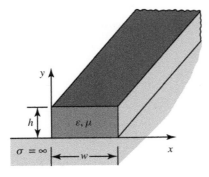

Figure P8-55

8.56. A TEM line is composed of a ground plane and a center conductor of width w and thickness t placed at a height h above the ground plane. Assume that the center conductor thickness t is very small. Then the center conductor can be approximated electrically by a wire whose effective radius is

$$a_e \simeq 0.25w$$

Based upon this approximation, derive an approximate expression for the capacitance and for the characteristic impedance of the line.

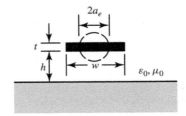

Figure P8-56

8.57. A microstrip line, whose center conductor has zero thickness ($t = 0$), has a dielectric constant of 6 and height of 1 mm. It is desired to design a $\lambda/4$ impedance transformer to match two lines, one input (#1) and the other output (#2), whose center strip widths are, respectively, $w_1 = 1.505$ mm, $w_3 = 0.549$ mm. Determine the:

(a) Effective dielectric constant of both lines.
(b) Characteristic impedance of the input (#1) and output (#2) lines.
(c) Characteristic impedance of the line that is to perform as a $\lambda/4$ impedance transformer.
(d) Length (in cm), at 2 GHz, of the $\lambda/4$ impedance transformer.

8.58. Assume that the fields supported by the microstrip line of Figure 8-33b are a combination of TE^z and TM^z modes. Then derive expressions for the electric and magnetic fields and their associated wave functions and wave numbers by treating the geometry as a boundary-value problem. Do this in the space domain.

8.59. A microstrip transmission line of beryllium oxide ($\varepsilon_r = 6.8$) has a width-to-height ratio of $w/h = 1.5$. Assume that the thickness-to-height ratio is $t/h = 0.01$. Determine the following parameters.

(a) Effective width-to-height ratio at zero frequency.

(b) Effective dielectric constant at zero frequency.

(c) Characteristic impedance at zero frequency.

(d) Approximate frequency where dispersion will begin when $h = 0.05$ cm.

(e) Effective dielectric constant at 15 GHz.

(f) Characteristic impedance at 15 GHz. Compare with the value if dispersion is neglected.

(g) Phase velocity at 15 GHz. Compare with the value if dispersion is neglected.

(h) Guide wavelength at 15 GHz.

8.60. A Dielectric Resonator Antenna (DRA) is made of a dielectric material of different shapes mounted on a PEC ground plane. The cubic DRA is composed of a dielectric cube mounted on a PEC ground plane, as shown in Figure P8-60.

- To simplify the problem, assume the origin ($x = 0$, $y = 0$, $z = 0$) of the rectangular coordinate system is at the bottom left corner of the back side, similar to that of Figure 8-14. However, you must use the coordinate system (directions of x, y and z) and the dimensions a, b, c shown below;

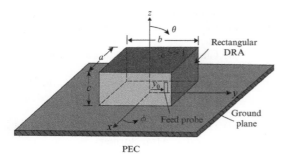

PEC

Figure P8-60

the only difference is to take the origin at the back bottom left corner.

- Neglecting the presence of the feed and treating the five sides of the dielectric cube as PMC ($\varepsilon_r \gg 1$) and the bottom side as PEC, and assuming for TE^z modes a vector potential of:

$$F_z = [C_1 \cos(\beta_x x) + D_1 \sin(\beta_x x)]$$
$$\times [C_2 \cos(\beta_y y) + D_2 \sin(\beta_y y)]$$
$$\times [C_3 \cos(\beta_z z) + D_3 \sin(\beta_z z)]$$

(a) Write a complete set of boundary conditions that will be sufficient to solve the subject problem. Give only one complete set; no duplicates.

(b) Evaluate and state values for the unknown coefficients (C_s and D_s) that can be determined by applying the boundary conditions.

(c) Expression for the reduced/simplified vector potential, after application of boundary conditions.

(d) Expressions for all the eigenvalues β_x, β_y, β_z and m, n, and p (m in x, n in y, and p in z) for nontrivial solution; be specific.

(e) General expression for the resonant frequency $(f_r)_{mnp}$ and associated allowable values of m, n, and p for nontrivial solution.

(f) Dominant mode (be specific in the values of m, n, and p) and the expression of its associated cutoff frequency assuming:

- $a > b > 2c$
- $c > a > b$
- $c > b > a$

8.61. Repeat Problem 8.60 for TM^z modes with vector potential

$$A_z = [C_1 \cos(\beta_x x) + D_1 \sin(\beta_x x)]$$
$$\times [C_2 \cos(\beta_y y) + D_2 \sin(\beta_y y)]$$
$$\times [C_3 \cos(\beta_z z) + D_3 \sin(\beta_z z)]$$

CHAPTER 9

░░░░░░░░░░░░░░░░

Circular Cross-Section Waveguides and Cavities

9.1 INTRODUCTION

Cylindrical transmission lines and cavities are very popular geometrical configurations. Cylindrical structures are those that maintain a uniform cross section along their length. Typical cross sections are rectangular, square, triangular, circular, elliptical, and others. Whereas the rectangular and square cross sections were analyzed in Chapter 8, the circular cross-section geometries will be discussed in this chapter. This will include transmission lines and cavities (resonators) of conducting walls and dielectric material.

9.2 CIRCULAR WAVEGUIDE

A popular waveguide configuration, in addition to the rectangular one discussed in Chapter 8, is the circular waveguide shown in Figure 9-1. This waveguide is very attractive because of its ease in manufacturing and low attenuation of the TE_{0n} modes. An apparent drawback is its fixed bandwidth between modes. Field configurations (modes) that can be supported inside such a structure are TE^z and TM^z.

9.2.1 Transverse Electric (TE^z) Modes

The transverse electric to z (TE^z) modes can be derived by letting the vector potentials $\mathbf{A}$ and $\mathbf{F}$ be equal to

$$\mathbf{A} = 0 \tag{9-1a}$$

$$\mathbf{F} = \hat{\mathbf{a}}_z F_z(\rho, \phi, z) \tag{9-1b}$$

The vector potential $\mathbf{F}$ must satisfy the vector wave equation 3-48, which reduces for the $\mathbf{F}$ of (9-1b) to

Balanis' Advanced Engineering Electromagnetics, Third Edition. Constantine A. Balanis.
© 2024 John Wiley & Sons, Inc. Published 2024 by John Wiley & Sons, Inc.
Companion Website: www.wiley.com/go/balanis/advancedengineeringelectromagnetics3e

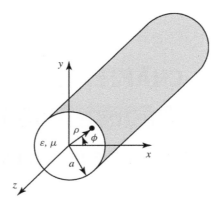

Figure 9-1   Cylindrical waveguide of circular cross section.

$$\nabla^2 F_z(\rho,\phi,z) + \beta^2 F_z(\rho,\phi,z) = 0 \tag{9-2}$$

When expanded in cylindrical coordinates, (9-2) reduces to

$$\frac{\partial^2 F_z}{\partial \rho^2} + \frac{1}{\rho}\frac{\partial F_z}{\partial \rho} + \frac{1}{\rho^2}\frac{\partial^2 F_z}{\partial \phi^2} + \frac{\partial^2 F_z}{\partial z^2} + \beta^2 F_z = 0 \tag{9-3}$$

whose solution for the geometry of Figure 9-1, according to (3-70), is of the form

$$F_z(\rho,\phi,z) = [A_1 J_m(\beta_\rho \rho) + B_1 Y_m(\beta_\rho \rho)]$$
$$\times [C_2 \cos(m\phi) + D_2 \sin(m\phi)]\left[A_3 e^{-j\beta_z z} + B_3 e^{+j\beta_z z}\right] \tag{9-4}$$

where, according to (3-66d), the constraint (dispersion) equation is

$$\beta_\rho^2 + \beta_z^2 = \beta^2 \tag{9-4a}$$

The constants A_1, B_1, C_2, D_2, A_3, B_3, m, β_ρ, and β_z can be found using the boundary conditions

$$E_\phi(\rho = a, \phi, z) = 0 \tag{9-5a}$$

The fields must be finite everywhere $\tag{9-5b}$

The fields must repeat every 2π radians in ϕ $\tag{9-5c}$

According to (9-5b), $B_1 = 0$ since $Y_m(\rho = 0) = \infty$. In addition, according to (9-5c),

$$m = 0, 1, 2, 3, \ldots \tag{9-6}$$

Consider waves that propagate only in the $+z$ direction. Then (9-4) reduces to

$$F_z^+(\rho,\phi,z) = A_{mn} J_m'(\beta_\rho \rho)[C_2 \cos(m\phi) + D_2 \sin(m\phi)]e^{-j\beta_z z} \tag{9-7}$$

Using (6-80) and (9-7), the electric field component of E_ϕ^+ can be written as

$$E_\phi^+ = \frac{1}{\varepsilon}\frac{\partial F_z^+}{\partial \rho} = \beta_\rho \frac{A_{mn}}{\varepsilon} J_m'(\beta_\rho \rho)[C_2 \cos(m\phi) + D_2 \sin(m\phi)]e^{-j\beta_z z} \tag{9-8}$$

TABLE 9-1   Zeroes χ'_{mn} of derivative $J'_m(\chi'_{mn})=0$ $(n=1,2,3,...)$ of the Bessel function $J_m(x)$

	$m=0$	$m=1$	$m=2$	$m=3$	$m=4$	$m=5$	$m=6$	$m=7$	$m=8$	$m=9$	$m=10$	$m=11$
$n=1$	3.8318	1.8412	3.0542	4.2012	5.3175	6.4155	7.5013	8.5777	9.6474	10.7114	11.7708	12.8264
$n=2$	7.0156	5.3315	6.7062	8.0153	9.2824	10.5199	11.7349	12.9324	14.1155	15.2867	16.4479	17.6003
$n=3$	10.1735	8.5363	9.9695	11.3459	12.6819	13.9872	15.2682	16.5294	17.7740	19.0046	20.2230	21.4309
$n=4$	13.3237	11.7060	13.1704	14.5859	15.9641	17.3129	18.6375	19.9419	21.2291	22.5014	23.7607	25.0085
$n=5$	16.4706	14.8636	16.3475	17.7888	19.1960	20.5755	21.9317	23.2681	24.5872	25.8913	27.1820	28.4609

where

$$' \equiv \frac{\partial}{\partial(\beta_\rho \rho)} \tag{9-8a}$$

Apply the boundary condition of (9-5a) in (9-8). Then we have that

$$E_\phi^+(\rho=a,\phi,z) = \beta_\rho \frac{A_{mn}}{\varepsilon} J'_m(\beta_\rho a)[C_2\cos(m\phi)+D_2\sin(m\phi)]e^{-j\beta_z z}=0 \tag{9-9}$$

which is only satisfied provided that

$$J'_m(\beta_\rho a)=0 \;\Rightarrow\; \beta_\rho a = \chi'_{mn} \;\Rightarrow\; \beta_\rho = \frac{\chi'_{mn}}{a} \tag{9-10}$$

In (9-10) χ'_{mn} represents the nth zero $(n=1,2,3,...)$ of the derivative of the Bessel function J_m of the first kind and of order $m(m=0,1,2,3,...)$. An abbreviated list of the zeroes χ'_{mn} of the derivative J'_m of the Bessel function J_m is found in Table 9-1. The smallest value of χ'_{mn} is 1.8412 $(m=1, n=1)$, followed by 3.0542 $(m=2, n=1)$, 3.8318 $(m=0, n=1)$, and so on.

By using (9-4a) and (9-10), β_z of the mn mode can be written as

$$(\beta_z)_{mn} = \begin{cases} \sqrt{\beta^2-\beta_\rho^2}=\sqrt{\beta^2-\left(\frac{\chi'_{mn}}{a}\right)^2} & \text{when } \beta>\beta_\rho=\frac{\chi'_{mn}}{a} \tag{9-11a} \\[2mm] 0 & \text{when } \beta=\beta_c=\beta_\rho=\frac{\chi'_{mn}}{a} \tag{9-11b} \\[2mm] -j\sqrt{\beta_\rho^2-\beta^2}=-j\sqrt{\left(\frac{\chi'_{mn}}{a}\right)^2-\beta^2} & \text{when } \beta<\beta_\rho=\frac{\chi'_{mn}}{a} \tag{9-11c} \end{cases}$$

Cutoff is defined when $(\beta_z)_{mn}=0$. Thus, according to (9-11b),

$$\beta_c = \omega_c\sqrt{\mu\varepsilon}=2\pi f_c\sqrt{\mu\varepsilon}=\beta_\rho=\frac{\chi'_{mn}}{a} \tag{9-12}$$

or

$$\boxed{(f_c)_{mn}=\frac{\chi'_{mn}}{2\pi a\sqrt{\mu\varepsilon}}} \tag{9-12a}$$

By using (9-12) and (9-12a), we can write (9-11a) through (9-11c) as

$$
(\beta_z)_{mn} = \begin{cases} \sqrt{\beta^2 - \beta_\rho^2} = \beta\sqrt{1 - \left(\dfrac{\beta_\rho}{\beta}\right)^2} = \beta\sqrt{1 - \left(\dfrac{\beta_c}{\beta}\right)^2} \\ \qquad = \beta\sqrt{1 - \left(\dfrac{\chi'_{mn}}{\beta a}\right)^2} = \beta\sqrt{1 - \left(\dfrac{f_c}{f}\right)^2} \qquad \text{when } f > f_c = (f_c)_{mn} \quad (9\text{-}13a) \\[4pt] 0 \qquad\qquad\qquad\qquad\qquad\qquad\qquad\qquad \text{when } f = f_c = (f_c)_{mn} \quad (9\text{-}13b) \\[4pt] -j\sqrt{\beta_\rho^2 - \beta^2} = -j\beta\sqrt{\left(\dfrac{\beta_\rho}{\beta}\right)^2 - 1} = -j\beta\sqrt{\left(\dfrac{\beta_c}{\beta}\right)^2 - 1} \\[4pt] \qquad = -j\beta\sqrt{\left(\dfrac{\chi'_{mn}}{\beta a}\right)^2 - 1} = -j\beta\sqrt{\left(\dfrac{f_c}{f}\right)^2 - 1} \quad \text{when } f < f_c = (f_c)_{mn} \quad (9\text{-}13c) \end{cases}
$$

The guide wavelength λ_g is defined as

$$
(\lambda_g)_{mn} = \frac{2\pi}{(\beta_z)_{mn}} \tag{9-14}
$$

which according to (9-13a) and (9-13b) can be written as

$$
(\lambda_g)_{mn} = \begin{cases} \dfrac{2\pi}{\beta\sqrt{1 - \left(\dfrac{f_c}{f}\right)^2}} = \dfrac{\lambda}{\sqrt{1 - \left(\dfrac{f_c}{f}\right)^2}} \quad \text{when } f > f_c = (f_c)_{mn} \quad (9\text{-}14a) \\[6pt] \infty \qquad\qquad\qquad\qquad\qquad\qquad \text{when } f = (f_c)_{mn} \quad (9\text{-}14b) \end{cases}
$$

In (9-14a), λ is the wavelength of the wave in an infinite medium of the kind that exists inside the waveguide. There is no definition of the wavelength below cutoff since the wave is exponentially decaying and there is no repetition of its waveform.

According to (9-12a) and the values of χ'_{mn} in Table 9-1, the order (lower to higher cutoff frequencies) in which the TE^z_{mn} modes occur is TE^z_{11}, TE^z_{21}, TE^z_{01}, etc. It should be noted that for a circular waveguide the order in which the TE^z_{mn} modes occur does not change, and the bandwidth between modes is also fixed. For example, the bandwidth of the first single-mode TE^z_{11} operation is $3.0542/1.8412 = 1.6588{:}1$ which is less than $2:1$. This bandwidth is fixed and cannot be varied, as was the case for the rectangular waveguide where the bandwidth between modes was a function of the a/b ratio. In fact, for a rectangular waveguide the maximum bandwidth of a single dominant mode operation was $2:1$ and it occurred when $a/b \geq 2$; otherwise, for $a/b < 2$, the bandwidth of a single dominant mode operation was less than $2:1$. The reason is that in a rectangular waveguide there are two dimensions a and b (two degrees of freedom) whose relative values can vary; in the circular waveguide there is only one dimension (the radius a) that can vary. A change in the radius only varies, by the same amount, the absolute values of the cutoff frequencies of all the modes, so it does not alter their order or relative bandwidth.

The electric and magnetic field components can be written, using (6-80) and (9-7), as

$$
E_\rho^+ = -\frac{1}{\varepsilon\rho}\frac{\partial F_z^+}{\partial \phi} = -A_{mn}\frac{m}{\varepsilon\rho} J_m(\beta_\rho\rho)[-C_2 \sin(m\phi) + D_2 \cos(m\phi)]e^{-j\beta_z z} \tag{9-15a}
$$

$$E_\phi^+ = \frac{1}{\varepsilon}\frac{\partial F_z^+}{\partial \rho} = A_{mn}\frac{\beta_\rho}{\varepsilon}J_m'(\beta_\rho\rho)[C_2\cos(m\phi)+D_2\sin(m\phi)]e^{-j\beta_z z} \tag{9-15b}$$

$$E_z^+ = 0 \tag{9-15c}$$

$$H_\rho^+ = -j\frac{1}{\omega\mu\varepsilon}\frac{\partial^2 F_z^+}{\partial\rho\,\partial z} = -A_{mn}\frac{\beta_\rho\beta_z}{\omega\mu\varepsilon}J_m'(\beta_\rho\rho)[C_2\cos(m\phi)+D_2\sin(m\phi)]e^{-j\beta_z z} \tag{9-15d}$$

$$H_\phi^+ = -j\frac{1}{\omega\mu\varepsilon}\frac{1}{\rho}\frac{\partial^2 F_z^+}{\partial\phi\,\partial z} = -A_{mn}\frac{m\beta_z}{\omega\mu\varepsilon}\frac{1}{\rho}J_m(\beta_\rho\rho)$$
$$\times[-C_2\sin(m\phi)+D_2\cos(m\phi)]e^{-j\beta_z z} \tag{9-15e}$$

$$H_z^+ = -j\frac{1}{\omega\mu\varepsilon}\left(\frac{\partial^2}{\partial z^2}+\beta^2\right)F_z^+ = -jA_{mn}\frac{\beta_\rho^2}{\omega\mu\varepsilon}J_m(\beta_\rho\rho)$$
$$\times[C_2\cos(m\phi)+D_2\sin(m\phi)]e^{-j\beta_z z} \tag{9-15f}$$

where

$$' \equiv \frac{\partial}{\partial(\beta_\rho\rho)} \tag{9-15g}$$

By using (9-15a) through (9-15f), the wave impedance $(Z_w^{+z})_{mn}^{\text{TE}}$ of the TE$_{mn}^z$ (H_{mn}^z) modes in the $+z$ direction can be written as

$$(Z_w^{+z})_{mn}^{\text{TE}} = \frac{E_\rho^+}{H_\phi^+} = -\frac{E_\phi^+}{H_\rho^+} = \frac{\omega\mu}{(\beta_z)_{mn}} \tag{9-16}$$

With the aid of (9-13a) through (9-13c), the wave impedance of (9-16) reduces to

$$(Z_w^{+z})_{mn}^{\text{TE}} = \begin{cases} \dfrac{\omega\mu}{\beta\sqrt{1-\left(\dfrac{f_c}{f}\right)^2}} = \dfrac{\sqrt{\dfrac{\mu}{\varepsilon}}}{\sqrt{1-\left(\dfrac{f_c}{f}\right)^2}} = \dfrac{\eta}{\sqrt{1-\left(\dfrac{f_c}{f}\right)^2}} & \\ \qquad\qquad\qquad \text{when } f > f_c = (f_c)_{mn} & \text{(9-16a)} \\[4mm] \dfrac{\omega\mu}{0} = \infty \qquad\qquad \text{when } f = f_c = (f_c)_{mn} & \text{(9-16b)} \\[4mm] \dfrac{\omega\mu}{-j\beta\sqrt{\left(\dfrac{f_c}{f}\right)^2-1}} = +j\dfrac{\sqrt{\dfrac{\mu}{\varepsilon}}}{\sqrt{\left(\dfrac{f_c}{f}\right)^2-1}} = +j\dfrac{\eta}{\sqrt{\left(\dfrac{f_c}{f}\right)^2-1}} & \text{(9-16c)} \\ \qquad\qquad\qquad \text{when } f < f_c = (f_c)_{mn} & \end{cases}$$

By examining (9-16a) through (9-16c), we can make the following statements about the impedance.

1. Above cutoff it is real and greater than the intrinsic impedance of the medium inside the waveguide.
2. At cutoff it is infinity.
3. Below cutoff it is imaginary and inductive. This indicates that the waveguide below cutoff behaves as an inductor that is an energy storage element.

The form of Z_w^{+z}, as given by (9-16a) through (9-16c), as a function of f_c/f, and where f_c is the cutoff frequency of that mode, is the same as the Z_w^{+z} for the TE^z modes of a rectangular waveguide, as given by (8-20a) through (8-20c). A plot of (9-16a) through (9-16c) for any one TE$_{mn}^z$ mode as a function of f_c/f is shown in Figure 8-2.

Example 9-1

A circular waveguide of radius $a = 3$ cm that is filled with polystyrene ($\varepsilon_r = 2.56$) is used at a frequency of 2 GHz. For the dominant TE$_{mn}^z$ mode, determine the following:

  a. Cutoff frequency.
  b. Guide wavelength (in cm). Compare it to the infinite medium wavelength λ.
  c. Phase constant β_z (in rad/cm).
  d. Wave impedance.
  e. Bandwidth over single-mode operation (assuming only TE^z modes).

Solution:

  a. The dominant mode is the TE$_{11}$ mode whose cutoff frequency is, according to (9-12a),

$$(f_c)_{11}^{\text{TE}^z} = \frac{1.8412}{2\pi a\sqrt{\mu\varepsilon}} = \frac{1.8412(30\times10^9)}{2\pi(3)\sqrt{2.56}} = 1.8315 \text{ GHz}$$

  b. Since the frequency of operation is 2 GHz, which is greater than the cutoff frequency of 1.8315 GHz, the guide wavelength of (9-14a) for the TE$_{11}$ mode is

$$\lambda_g = \frac{\lambda}{\sqrt{1-\left(\dfrac{f_c}{f}\right)^2}}$$

where

$$\lambda = \frac{\lambda_0}{\sqrt{\varepsilon_r}} = \frac{30\times10^9}{2\times10^9\sqrt{2.56}} = 9.375 \text{ cm}$$

$$\sqrt{1-\left(\frac{f_c}{f}\right)^2} = \sqrt{1-\left(\frac{1.8315}{2}\right)^2} = 0.4017$$

Thus,

$$\lambda_g = \frac{9.375}{0.4017} = 23.34 \text{ cm} \quad \text{where } \lambda = 9.375 \text{ cm}$$

  c. The phase constant β_z of the TE$_{11}$ mode is found using (9-13a), or

$$\beta_z = \beta\sqrt{1-\left(\frac{f_c}{f}\right)^2} = \frac{2\pi}{\lambda}\sqrt{1-\left(\frac{f_c}{f}\right)^2} = \frac{2\pi}{9.375}(0.4017) = 0.2692 \text{ rad/cm}$$

which can also be obtained using

$$\beta_z = \frac{2\pi}{\lambda_g} = \frac{2\pi}{23.34} = 0.2692 \text{ rad/cm}$$

  d. According to (9-16a), the wave impedance of the TE$_{11}$ mode is equal to

$$Z_{11}^{\text{TE}^z} = \frac{\eta}{\sqrt{1-\left(\dfrac{f_c}{f}\right)^2}} = \frac{120\pi/\sqrt{2.56}}{0.4017} = 586.56 \text{ ohms}$$

  e. Since the next higher-order TE$_{mn}$ mode is the TE$_{21}$, the bandwidth of single TE$_{11}$ mode operation is

$$\text{BW} = 3.0542 / 1.8412 : 1 = 1.6588 : 1$$

9.2.2 Transverse Magnetic (TM^z) Modes

The transverse magnetic to z (TM^z) modes can be derived in a similar manner as the TE^z modes of Section 9.2.1 by letting

$$\mathbf{A} = \hat{\mathbf{a}}_z A_z(\rho, \phi, z) \qquad (9\text{-}17a)$$

$$\mathbf{F} = 0 \qquad (9\text{-}17b)$$

The vector potential $\mathbf{A}$ must satisfy the vector wave equation of (3-48), which reduces for the $\mathbf{A}$ of (9-17a) to

$$\nabla^2 A_z(\rho, \phi, z) + \beta^2 A_z(\rho, \phi, z) = 0 \qquad (9\text{-}18)$$

The solution of (9-18) is obtained in a manner similar to that of (9-2), as given by (9-4), and it can be written as

$$A_z(\rho, \phi, z) = [A_1 J_m(\beta_\rho \rho) + B_1 Y_m(\beta_\rho \rho)]$$
$$\times [C_2 \cos(m\phi) + D_2 \sin(m\phi)]\left[A_3 e^{-j\beta_z z} + B_3 e^{+j\beta_z z}\right] \qquad (9\text{-}19)$$

with the constraint (dispersion) equation expressed as

$$\beta_\rho^2 + \beta_z^2 = \beta^2 \qquad (9\text{-}19a)$$

The constants A_1, B_1, C_2, D_2, A_3, B_3, m, β_ρ, and β_z can be found using the following boundary conditions

$$E_\phi(\rho = a, \phi, z) = 0 \qquad (9\text{-}20a)$$

or

$$E_z(\rho = a, \phi, z) = 0 \qquad (9\text{-}20b)$$

$$\text{The fields must be finite everywhere} \qquad (9\text{-}20c)$$

$$\text{The fields must repeat every } 2\pi \text{ radians in } \phi \qquad (9\text{-}20d)$$

According to (9-20c), $B_1 = 0$ since $Y_m(\rho = 0) = \infty$. In addition, according to (9-20d),

$$m = 0, 1, 2, 3,\ldots \qquad (9\text{-}21)$$

Considering waves that propagate only in the $+z$ direction, (9-19) reduces to

$$A_z^+(\rho, \phi, z) = B_{mn} J_m(\beta_\rho \rho)[C_2 \cos(m\phi) + D_2 \sin(m\phi)] e^{-j\beta_z z} \qquad (9\text{-}22)$$

The eigenvalues of β_ρ can be obtained by applying either (9-20a) or (9-20b). Use of (6-70) and (9-22) allows us to write the electric field component E_z^+ as

$$E_z^+ = -j \frac{1}{\omega\mu\varepsilon}\left[\frac{\partial^2}{\partial z^2} + \beta^2\right] A_z^+$$
$$= -j B_{mn} \frac{\beta_\rho^2}{\omega\mu\varepsilon} J_m(\beta_\rho \rho)[C_2 \cos(m\phi) + D_2 \sin(m\phi)] e^{-j\beta_z z} \qquad (9\text{-}23)$$

Application of the boundary condition (9-20b) and use of (9-23) gives

$$E_z^+(\rho = a, \phi, z) = -j B_{mn} \frac{\beta_\rho^2}{\omega\mu\varepsilon} J_m(\beta_\rho a)[C_2 \cos(m\phi) + D_2 \sin(m\phi)] e^{-j\beta_z z} = 0 \qquad (9\text{-}24)$$

which is only satisfied provided that

$$J_m(\beta_\rho a) = 0 \;\Rightarrow\; \beta_\rho a = \chi_{mn} \;\Rightarrow\; \beta_\rho = \frac{\chi_{mn}}{a} \qquad (9\text{-}25)$$

TABLE 9-2   Zeroes χ_{mn} of $J_m(\chi_{mn}) = 0$ $(n = 1, 2, 3, ...)$ of Bessel function $J_m(x)$

	$m=0$	$m=1$	$m=2$	$m=3$	$m=4$	$m=5$	$m=6$	$m=7$	$m=8$	$m=9$	$m=10$	$m=11$
$n=1$	2.4049	3.8318	5.1357	6.3802	7.5884	8.7715	9.9361	11.0864	12.2251	13.3543	14.4755	15.5898
$n=2$	5.5201	7.0156	8.4173	9.7610	11.0647	12.3386	13.5893	14.8213	16.0378	17.2412	18.4335	19.6160
$n=3$	8.6537	10.1735	11.6199	13.0152	14.3726	15.7002	17.0038	18.2876	19.5545	20.8071	22.0470	23.2759
$n=4$	11.7915	13.3237	14.7960	16.2235	17.6160	18.9801	20.3208	21.6415	22.9452	24.2339	25.5095	26.7733
$n=5$	14.9309	16.4706	17.9598	19.4094	20.8269	22.2178	23.5861	24.9349	26.2668	27.5838	28.8874	30.1791

In (9-25), χ_{mn} represents the nth zero $(n = 1, 2, 3, ...)$ of the Bessel function J_m of the first kind and of order m $(m = 0, 1, 2, 3, ...)$. An abbreviated list of the zeroes χ_{mn} of the Bessel function J_m is found in Table 9-2. The smallest value of χ_{mn} is 2.4049 $(m = 0, n = 1)$, followed by 3.8318 $(m = 1, n = 1)$, 5.1357 $(m = 2, n = 1)$, etc.

By using (9-19a) and (9-25), β_z can be written as

$$
(\beta_z)_{mn} = \begin{cases} \sqrt{\beta^2 - \beta_\rho^2} = \sqrt{\beta^2 - \left(\dfrac{\chi_{mn}}{a}\right)^2} & \text{when } \beta > \beta_\rho = \dfrac{\chi_{mn}}{a} \qquad (9\text{-}26a) \\[12pt] 0 & \text{when } \beta = \beta_c = \beta_\rho = \dfrac{\chi_{mn}}{a} \qquad (9\text{-}26b) \\[12pt] -j\sqrt{\beta_\rho^2 - \beta^2} = -j\sqrt{\left(\dfrac{\chi_{mn}}{a}\right)^2 - \beta^2} & \text{when } \beta < \beta_\rho = \dfrac{\chi_{mn}}{a} \qquad (9\text{-}26c) \end{cases}
$$

By following the same procedure as for the TE^z modes, we can write the expressions for the cutoff frequencies $(f_c)_{mn}$, propagation constant $(\beta_z)_{mn}$, and guide wavelength $(\lambda_g)_{mn}$ as

$$
\boxed{(f_c)_{mn} = \frac{\chi_{mn}}{2\pi a \sqrt{\mu\varepsilon}}} \qquad (9\text{-}27)
$$

$$
(\beta_z)_{mn} = \begin{cases} \sqrt{\beta^2 - \beta_\rho^2} = \beta\sqrt{1 - \left(\dfrac{\beta_\rho}{\beta}\right)^2} = \beta\sqrt{1 - \left(\dfrac{\beta_c}{\beta}\right)^2} \\[6pt] \qquad\qquad = \beta\sqrt{1 - \left(\dfrac{\chi_{mn}}{\beta a}\right)^2} = \beta\sqrt{1 - \left(\dfrac{f_c}{f}\right)^2} \qquad\qquad\qquad (9\text{-}28a) \\[6pt] \qquad\qquad\qquad\qquad\qquad \text{when } f > f_c = (f_c)_{mn} \\[6pt] 0 \qquad\qquad\qquad\qquad\qquad \text{when } f = f_c = (f_c)_{mn} \qquad\qquad (9\text{-}28b) \\[6pt] -j\sqrt{\beta_\rho^2 - \beta^2} = -j\beta\sqrt{\left(\dfrac{\beta_\rho}{\beta}\right)^2 - 1} = -j\beta\sqrt{\left(\dfrac{\beta_c}{\beta}\right)^2 - 1} \\[6pt] \qquad\qquad = -j\beta\sqrt{\left(\dfrac{\chi_{mn}}{\beta a}\right)^2 - 1} = -j\beta\sqrt{\left(\dfrac{f_c}{f}\right)^2 - 1} \qquad\qquad (9\text{-}28c) \\[6pt] \qquad\qquad\qquad\qquad\qquad \text{when } f < f_c = (f_c)_{mn} \end{cases}
$$

$$
(\lambda_g)_{mn} = \begin{cases} \dfrac{2\pi}{\beta\sqrt{1 - \left(\dfrac{f_c}{f}\right)^2}} = \dfrac{\lambda}{\sqrt{1 - \left(\dfrac{f_c}{f}\right)^2}} & \text{when } f > f_c = (f_c)_{mn} \qquad (9\text{-}29a) \\[12pt] \infty & \text{when } f = f_c = (f_c)_{mn} \qquad (9\text{-}29b) \end{cases}
$$

According to (9-27) and the values of χ_{mn} of Table 9-2, the order (lower to higher cutoff frequencies) in which the TE^z modes occur is TM$_{01}$, TM$_{11}$, TM$_{21}$, and so forth. The bandwidth of the first single-mode TM$_{01}^z$ operation is $3.8318/2.4049 = 1.5933:1$, which is also less than 2 : 1. Comparing the cutoff frequencies of the TE^z and TM^z modes, as given by (9-12a) and (9-27) along with the data of Tables 9-1 and 9-2, the order of the TE$_{mn}^z$ and TM$_{mn}^z$ modes is that of TE$_{11}$ ($\chi'_{11} = 1.8412$), TM$_{01}$($\chi_{01} = 2.4049$), TE$_{21}$($\chi'_{21} = 3.0542$), TE$_{01}$ ($\chi'_{01} = 3.8318$) = TM$_{11}$($\chi_{11} = 3.8318$), TE$_{31}$ ($\chi'_{31} = 4.2012$), and so forth. The dominant mode is TE$_{11}$ and its bandwidth of single-mode operation is $2.4049/1.8412 = 1.3062:1$, which is much smaller than 2 : 1. Plots of the field configurations over a cross section of the waveguide, both E and H, for the first 30 TE$_{mn}^z$ and/or TM$_{mn}^z$ modes are shown in Figure 9-2 [1].

It is apparent that the cutoff frequencies of the TE$_{0n}$ and TM$_{1n}$ modes are identical; therefore, they are referred to here also as degenerate modes. This is because the zeroes of the derivative of the Bessel function J_0 are identical to the zeroes of the Bessel function J_1. To demonstrate this, let us examine the derivative of $J_0(\beta_\rho\rho)$ evaluated at $\rho = a$. Using (IV-19) we can write that

$$\frac{d}{d(\beta_\rho\rho)} J_0(\beta_\rho\rho)|_{\rho=a} = J_0'(\beta_\rho a) = -J_1(\beta_\rho\rho)|_{\rho=a} = -J_1(\beta_\rho a) \tag{9-30}$$

which vanishes when

$$J_0'(\beta_\rho a) = 0 \Rightarrow \beta_\rho a = \chi'_{0n}, \qquad n = 1, 2, 3, \dots \tag{9-30a}$$

or

$$J_1(\beta_\rho a) = 0 \Rightarrow \beta_\rho a = \chi_{1n}, \qquad n = 1, 2, 3, \dots \tag{9-30b}$$

The electric and magnetic field components can be written, using (6-70) and (9-22), as

$$E_\rho^+ = -j\frac{1}{\omega\mu\varepsilon}\frac{\partial^2 A_z^+}{\partial\rho\,\partial z} = -B_{mn}\frac{\beta_\rho\beta_z}{\omega\mu\varepsilon}J_m'(\beta_\rho\rho)\left[C_2\cos(m\phi)+D_2\sin(m\phi)\right]e^{-j\beta_z z} \tag{9-31a}$$

$$E_\phi^+ = -j\frac{1}{\omega\mu\varepsilon}\frac{1}{\rho}\frac{\partial^2 A_z^+}{\partial\phi\,\partial z} = -B_{mn}\frac{m\beta_z}{\omega\mu\varepsilon\rho}J_m(\beta_\rho\rho)\left[-C_2\sin(m\phi)+D_2\cos(m\phi)\right]e^{-j\beta_z z} \tag{9-31b}$$

$$E_z^+ = -j\frac{1}{\omega\mu\varepsilon}\left(\frac{\partial^2}{\partial z^2}+\beta^2\right)A_z^+$$

$$= -jB_{mn}\frac{\beta_\rho^2}{\omega\mu\varepsilon}J_m(\beta_\rho\rho)\left[C_2\cos(m\phi)+D_2\sin(m\phi)\right]e^{-j\beta_z z} \tag{9-31c}$$

$$H_\rho^+ = \frac{1}{\mu}\frac{1}{\rho}\frac{\partial A_z^+}{\partial\phi} = B_{mn}\frac{m}{\mu}\frac{1}{\rho}J_m(\beta_\rho\rho)\left[-C_2\sin(m\phi)+D_2\cos(m\phi)\right]e^{-j\beta_z z} \tag{9-31d}$$

$$H_\phi^+ = -\frac{1}{\mu}\frac{\partial A_z^+}{\partial\rho} = -B_{mn}\frac{\beta_\rho}{\mu}J_m'(\beta_\rho\rho)\left[C_2\cos(m\phi)+D_2\sin(m\phi)\right]e^{-j\beta_z z} \tag{9-31e}$$

$$H_z^+ = 0 \tag{9-31f}$$

where

$$' \equiv \frac{\partial}{\partial(\beta_\rho\rho)} \tag{9-31g}$$

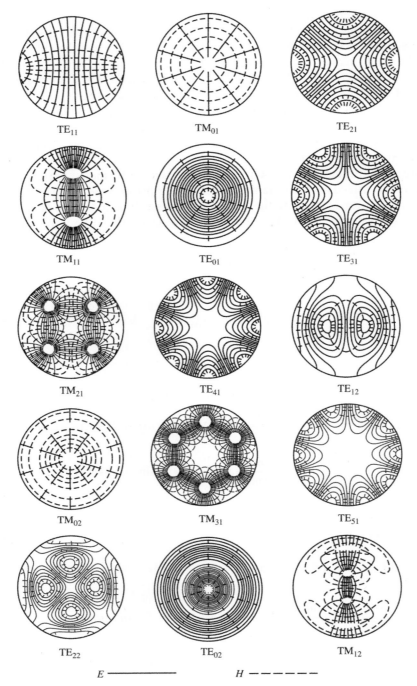

Figure 9-2 Field configurations of the first 30 TE^z and/or TM^z modes in a circular waveguide. (Source: C. S. Lee, S. W. Lee, and S. L. Chuang, "Plot of modal field distribution in rectangular and circular waveguides," *IEEE Trans. Microwave Theory Tech.*, © 1966, IEEE.)

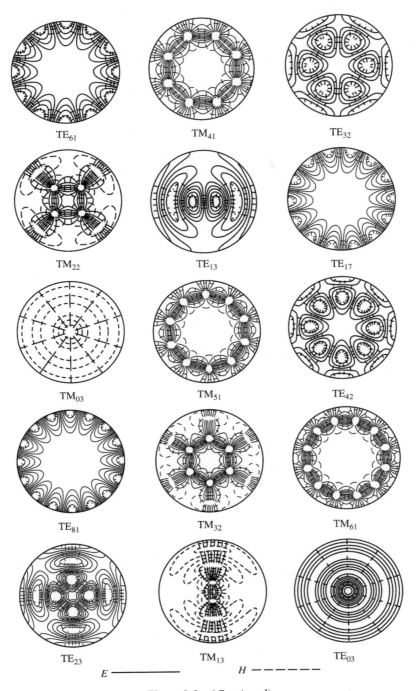

Figure 9-2 (*Continued*)

By using (9-31a) through (9-31f), the wave impedance in the $+z$ direction can be written as

$$\left(Z_w^{+z}\right)_{mn}^{\text{TM}} = \frac{E_\rho^+}{H_\phi^+} = -\frac{E_\phi^+}{H_\rho^+} = \frac{(\beta_z)_{mn}}{\omega\varepsilon} \tag{9-32}$$

With the aid of (9-28a) through (9-28c), the wave impedance of (9-32) reduces to

$$\left(Z_w^{+z}\right)_{mn}^{\text{TM}} = \begin{cases} \dfrac{\beta\sqrt{1-\left(\dfrac{f_c}{f}\right)^2}}{\omega\varepsilon} = \sqrt{\dfrac{\mu}{\varepsilon}}\sqrt{1-\left(\dfrac{f_c}{f}\right)^2} = \eta\sqrt{1-\left(\dfrac{f_c}{f}\right)^2} & \text{(9-32a)} \\[4pt] \qquad\qquad\text{when } f > f_c = (f_c)_{mn} \\[8pt] \dfrac{0}{\omega\varepsilon} = 0 \qquad\qquad \text{when } f = f_c = (f_c)_{mn} & \text{(9-32b)} \\[8pt] \dfrac{-j\beta\sqrt{\left(\dfrac{f_c}{f}\right)^2 - 1}}{\omega\varepsilon} = -j\sqrt{\dfrac{\mu}{\varepsilon}}\sqrt{\left(\dfrac{f_c}{f}\right)^2 - 1} = -j\eta\sqrt{\left(\dfrac{f_c}{f}\right)^2 - 1} & \text{(9-32c)} \\[4pt] \qquad\qquad\text{when } f < f_c = (f_c)_{mn} \end{cases}$$

Examining (9-32a) through (9-32c), we can make the following statements about the wave impedance for the TM^z modes.

1. Above cutoff it is real and smaller than the intrinsic impedance of the medium inside the waveguide.
2. At cutoff it is zero.
3. Below cutoff it is imaginary and capacitive. This indicates that the waveguide below cutoff behaves as a capacitor that is an energy storage element.

The form of $(Z_w^{+z})_{mn}^{\text{TM}}$, as given by (9-32a) through (9-32c), and as a function of f_c/f, where f_c is the cutoff frequency of that mode, is the same as the $(Z_w^{+z})_{mn}^{\text{TM}}$ for the TM_{mn}^z modes of a rectangular waveguide, as given by (8-29a) through (8-29c). A plot of (9-32a) through (9-32c) for any one TM_{mn}^z mode as a function of f_c/f, is shown in Figure 8-2.

Example 9-2

Design a circular waveguide filled with a lossless dielectric medium of dielectric constant 4. The waveguide must operate in a single dominant mode over a bandwidth of 1 GHz.

1. Find its radius (in cm).
2. Determine the lower, center, and upper frequencies of the bandwidth.

Solution:

a. The dominant mode is the TE_{11} mode whose cutoff frequency according to (9-12a) is

$$\left(f_c\right)_{11}^{\text{TE}^z} = \frac{\chi_{11}'}{2\pi a\sqrt{\mu\varepsilon}} = \frac{1.8412(30\times10^9)}{2\pi(a)\sqrt{4}}$$

The next higher-order mode is the TM_{01} mode whose cutoff frequency according to (9-27) is

$$(f_c)_{01}^{TM^z} = \frac{\chi_{01}}{2\pi a\sqrt{\mu\varepsilon}} = \frac{2.4049(30\times10^9)}{2\pi(a)\sqrt{4}}$$

The difference between the two must be 1 GHz. To accomplish this, the radius of the waveguide must be equal to

$$\frac{(2.4049-1.8412)30\times10^9}{2\pi(a)\sqrt{4}} = 1\times10^9 \Rightarrow a = 1.3457 \text{ cm}$$

b. The lower, upper, and center frequencies of the bandwidth are equal to

$$f_\ell = (f_c)_{11}^{TE^z} = \frac{1.8412(30\times10^9)}{2\pi(1.3457)2} = 3.2664\times10^9 = 3.2664 \text{ GHz}$$

$$f_u = (f_c)_{01}^{TM^z} = \frac{2.4049(30\times10^9)}{2\pi(1.3457)2} = 4.2664\times10^9 = 4.2664 \text{ GHz}$$

$$f_0 = f_\ell + 0.5\times10^9 = f_u - 0.5\times10^9 = 3.7664\times10^9 = 3.7664 \text{ GHz}$$

Whenever a given mode is desired, it is necessary to design the proper feed to excite the fields within the waveguide and detect the energy associated with such a mode. Maximization of the energy exchange or transfer is accomplished by designing the feed, which is usually a probe or antenna, so that its field pattern matches that of the field configuration of the desired mode. Usually the probe is placed near the maximum of the field pattern of the desired mode; however, that position may be varied somewhat in order to achieve some desired matching in the excitation and detection systems. Shown in Figure 9-3 are suggested designs to excite and/or detect the TE_{11} and TM_{01} modes in a circular waveguide, to transition between the TE_{10} of a rectangular waveguide and the TE_{11} mode of a circular waveguide, and to couple between the TE_{10} of a rectangular waveguide and TM_{01} mode of a circular waveguide.

9.2.3 Attenuation

The attenuation in a circular waveguide can be obtained by using techniques similar to those for the rectangular waveguide, as outlined and applied in Section 8.2.5. The basic equation is (8-64a), or

$$(\alpha_c)_{mn} = \frac{P_c/\ell}{2P_{mn}} = \frac{P_\ell}{2P_{mn}} \tag{9-33}$$

which is based on the configuration of Figure 8-9.

It has been shown that the attenuation coefficients of the TE_{0n} $(n = 1, 2, ...)$ modes in a circular waveguide monotonically decrease as a function of frequency [2, 3]. This is a very desirable characteristic, and because of this the excitation, propagation, and detection of TE_{0n} modes in a circular waveguide have received considerable attention. It can be shown that the attenuation coefficients for the TE_{mn}^z and TM_{mn}^z modes inside a circular waveguide are given, respectively, by

$$TE_{mn}^z$$

$$(\alpha_c)_{mn}^{TE^z} = \frac{R_s}{a\eta\sqrt{1-\left(\frac{f_c}{f}\right)^2}}\left[\left(\frac{f_c}{f}\right)^2 + \frac{m^2}{(\chi'_{mn})^2 - m^2}\right] \text{ Np/m} \tag{9-34a}$$

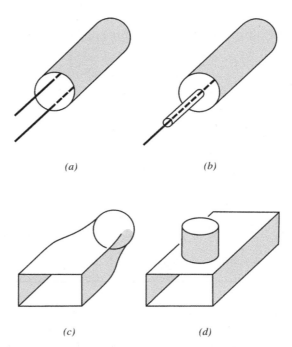

(a)   *(b)*

(c)   *(d)*

Figure 9-3   Excitation of TE$_{mn}$ and TM$_{mn}$ modes in a circular waveguide. (*a*) TE$_{11}$ mode. (*b*) TM$_{01}$ mode. (*c*) TE$_{10}$ (rectangular)-TE$_{11}$ (circular). (*d*) TE$_{10}$ (rectangular)-TM$_{01}$ (circular).

$$\text{TM}_{mn}^{z}$$

$$(\alpha_c)_{mn}^{\text{TM}^z} = \frac{R_s}{a\eta} \frac{1}{\sqrt{1 - \left(\dfrac{f_c}{f}\right)^2}} \quad \text{Np/m} \tag{9-34b}$$

Plots of the attenuation coeficient versus the normalized frequency f/f_c, where f_c is the cutoff frequency of the dominant TE$_{11}$ mode, are shown for six modes in Figures 9-4*a* and 9-4*b* for waveguide radii of 1.5 and 3 cm, respectively. Within the waveguide is free space and its walls are made of copper ($\sigma = 5.7 \times 10^7$ S/m).

Example 9-3

Derive the attenuation coefficient for the TE$_{01}$ mode inside a circular waveguide of radius a.

Solution: According to (9-15a) through (9-15g), the electric and magnetic field components for the TE$_{01}$ ($m = 0$, $n = 1$) mode reduce to

$$E_\rho^+ = E_z^+ = H_\phi^+ = 0$$

$$E_\phi^+ = \beta_\rho \frac{A_{01}}{\varepsilon} J_0'(\beta_\rho\rho) e^{-j\beta_z z}$$

$$H_\rho^+ = -A_{01} \frac{\beta_\rho\beta_z}{\omega\mu\varepsilon} J_0'(\beta_\rho\rho) e^{-j\beta_z z}$$

$$H_z^+ = -jA_{01} \frac{\beta_\rho^2}{\omega\mu\varepsilon} J_0(\beta_\rho\rho) e^{-j\beta_z z}$$

where

$$\beta_\rho = \frac{\chi'_{01}}{a} = \frac{3.8318}{a}$$

Using these equations, the power through a cross section of the waveguide is equal to

$$P_{01} = \frac{1}{2} \iint_{A_0} \text{Re}\left[(\mathbf{E} \times \mathbf{H}^*) \cdot d\mathbf{s}\right]$$

$$= \frac{1}{2} \iint_{A_0} \text{Re}\left[\hat{\mathbf{a}}_\phi E_\phi \times \left(\hat{\mathbf{a}}_\rho H_\rho + \hat{\mathbf{a}}_z H_z\right)^*\right] \cdot \hat{\mathbf{a}}_z \, ds$$

$$P_{01} = -\frac{1}{2} \text{Re} \int_0^{2\pi} \int_0^a \left(E_\phi H_\rho^*\right) \rho \, d\rho \, d\phi = |A_{01}|^2 \frac{\pi \beta_z \beta_\rho^2}{\omega \mu \varepsilon^2} \int_0^a \left[J_0'\left(\frac{\chi'_{01}}{a}\rho\right)\right]^2 \rho \, d\rho$$

Since

$$\frac{dJ_p(cx)}{d(cx)} = -J_{p+1}(cx) + \frac{p}{cx} J_p(cx)$$

then

$$J_0'\left(\frac{\chi'_{01}}{a}\rho\right) = \frac{d}{d(\chi'_{01}\rho/a)} J_0\left(\frac{\chi'_{01}}{a}\rho\right) = -J_1\left(\frac{\chi'_{01}}{a}\rho\right)$$

Thus

$$P_{01} = |A_{01}|^2 \frac{\pi \beta_z}{\omega \mu \varepsilon^2}\left(\frac{\chi'_{01}}{a}\right)^2 \int_0^a J_1^2\left(\frac{\chi'_{01}}{a}\rho\right)\rho \, d\rho$$

Since

$$\int_b^c x J_p^2(cx) dx = \frac{x^2}{2}\left[J_p^2(cx) - J_{p-1}(cx) J_{p+1}(cx)\right]_b^c$$

then

$$\int_0^a \rho J_1^2\left(\frac{\chi'_{01}}{a}\rho\right) d\rho = \frac{a^2}{2}\left[J_1^2\left(\chi'_{01}\right) - J_0\left(\chi'_{01}\right) J_2\left(\chi'_{01}\right)\right]$$

$$= -\frac{a^2}{2} J_0\left(\chi'_{01}\right) J_2\left(\chi'_{01}\right) = \frac{a^2}{2} J_0^2\left(\chi'_{01}\right)$$

because

$$J_1^2\left(\chi'_{01}\right) = J_1^2(3.8318) = 0$$

$$J_2\left(\chi'_{01}\right) = -J_0\left(\chi'_{01}\right)$$

Therefore, the power of the TE$_{01}$ can be written as

$$P_{01} = |A_{01}|^2 \frac{\pi \beta_z}{2\omega \mu \varepsilon^2}\left(\chi'_{01}\right)^2 J_0^2\left(\chi'_{01}\right)$$

The power dissipated on the walls of the waveguide is obtained using

$$P_c = \frac{R_s}{2} \iint_{S_w} (\mathbf{J}_s \cdot \mathbf{J}_s^*)_{\rho=a} ds = \frac{R_s}{2} \int_0^\ell \int_0^{2\pi} |\mathbf{J}_s|_{\rho=a}^2 \, a \, d\phi \, dz$$

where

$$\mathbf{J}_s|_{\rho=a} = \hat{\mathbf{n}} \times \mathbf{H}^+|_{\rho=a} = \hat{\mathbf{a}}_\phi H_z^+|_{\rho=a} = -\hat{\mathbf{a}}_\phi j \frac{\beta_\rho^2}{\omega \mu \varepsilon} A_{01} J_0\left(\beta_\rho a\right) e^{-j\beta_z z}$$

Thus,

$$P_c = |A_{01}|^2 \frac{R_s}{2} \left(\frac{\beta_\rho^2}{\omega\mu\varepsilon} \right)^2 a J_0^2 \left(\chi_{01}' \right) \int_0^\ell \int_0^{2\pi} d\phi \, dz$$

or

$$\frac{P_c}{\ell} = P_\ell = |A_{01}|^2 \frac{\pi R_s}{a^3} \left[\frac{\left(\chi_{01}' \right)^2}{\omega\mu\varepsilon} \right]^2 J_0^2 \left(\chi_{01}' \right)$$

Therefore, the attenuation coefficient of (9-33) for the TE_{01} mode can now be written as

$$\alpha_{01} \left(TE^z \right) = \frac{R_s}{a\eta} \frac{\left(\dfrac{f_c}{f} \right)^2}{\sqrt{1 - \left(\dfrac{f_c}{f} \right)^2}} \quad \text{Np/m}$$

It is evident from the results of the preceding example that as f_c/f becomes smaller, the attenuation coefficient decreases monotonically (as shown in Figure 9-4), which is a desirable characteristic. It should be noted that similar monotonically decreasing variations in the attenuation coefficient are evident in all TE_{0n} modes ($n = 1, 2, 3, \ldots$). According to (9-15a) through (9-15f), the only tangential magnetic field components to the conducting surface of the waveguide for all these TE_{0n} ($m = 0$) modes is the H_z component, while the electric field lines are circular. Therefore, these modes are usually referred to as circular electric modes. For a constant power in the wave, the H_z component decreases as the frequency increases and approaches zero at infinite frequency. Simultaneously, the current density and conductor losses on the waveguide walls also decrease and approach zero. Because of this attractive feature, these modes have received considerable attention for long distance propagation of energy, especially at millimeter wave frequencies. Typically, attenuations as low as 1.25 dB/km (2 dB/mi) have been attained [2]. This is to be compared with attenuations of 120 dB/km for WR90 copper rectangular waveguides, and 3 dB/km at $0.85\,\mu$m, and less than 0.5 dB/km at $1.3\,\mu$m for fiber optics cables.

Although the TE_{0n} modes are very attractive from the attenuation point of view, there are a number of problems associated with their excitation and retention. One of the problems is that the TE_{01} mode, which is the first of the TE_{0n} modes, is not the dominant mode. Therefore, in order for this mode to be above its cutoff frequency and propagate in the waveguide, a number of other modes (such as the TE_{11}, TM_{01}, TE_{21}, and TM_{11}) with lower cutoff frequencies can also exist. Additional modes can also be present if the operating frequency is chosen well above the cutoff frequency of the TE_{01} mode in order to provide a margin of safety from being too close to its cutoff frequency.

To support the TE_{01} mode, the waveguide must be oversized and it can support a number of other modes. One of the problems faced with such a guide is how to excite the desired TE_{01} mode with sufficient purity and suppress the others. Another problem is how to prevent coupling between the TE_{01} mode and undesired modes that can exist since the guide is oversized. The presence of the undesired modes causes not only higher losses but dispersion and attenuation distortion to the signal since each exhibits different phase velocities and attenuation. Irregularities in the inner geometry, surface, and direction (such as bends, nonuniform cross sections, etc.) of the waveguide are the main contributors to the coupling to the undesired modes. However, for the guide to be of any practical use, it must be able to sustain and propagate the desired TE_{01} and other TE_{0n} modes efficiently over bends of reasonable curvature. One technique that has been implemented to

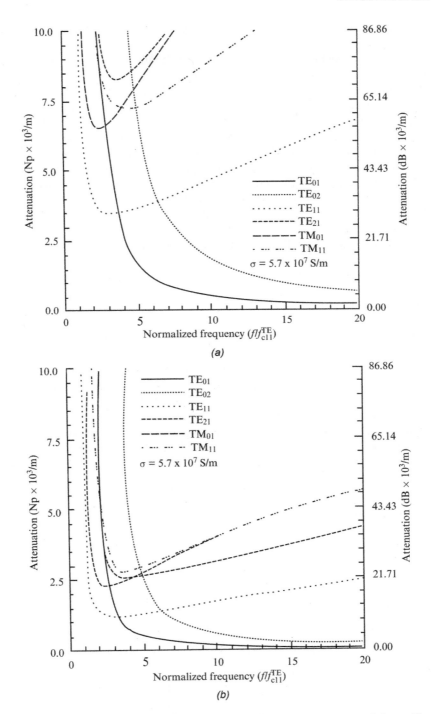

Figure 9-4 Attenuation for TE_{mn}^z and TM_{mn}^z modes in a circular waveguide. (*a*) $a = 1.5$ cm. (*b*) $a = 3$ cm.

achieve this is to use mode conversion before entering the corner and another conversion when exiting to convert back to the desired TE_{0n} mode(s).

Another method that has been used to discriminate against undesired modes and avoid coupling to them is to introduce filters inside the guide that cause negligible attenuation to the desired TE_{0n} mode(s). The basic principle of these filters is to introduce cuts that are perpendicular to the current paths of the undesired modes and parallel to the current direction of the desired mode(s). Since the current path of the undesired modes is along the axis (z direction) of the guide and the path of the desired TE_{0n} modes is along the circumference (ϕ direction), a helical wound wire placed on the inside surface of the guide can serve as a filter that discourages any mode that requires an axial component of current flow but propagates the desired TE_{0n} modes [3, 4].

Another filter that can be used to suppress undesired modes is to introduce within the guide very thin baffles of lossy material that will act as attenuating sheets. The surfaces of the baffles are placed in the radial direction of the guide so that they are parallel to the E_ρ and E_z components of the undesired modes (which will be damped) and normal to the E_ϕ component of the TE_{0n} modes that will remain unaffected. Typically, two baffles are used and are placed in a crossed pattern over the cross section of the guide.

A summary of the pertinent characteristics of the TE_{mn}^z and TM_{mn}^z modes of a circular waveguide are found listed in Table 9-3.

9.3 CIRCULAR CAVITY

As in rectangular waveguides, a circular cavity is formed by closing the two ends of the waveguide with plates, as shown in Figure 9-5. Coupling in and out of the cavity is done using either irises (holes) or probes (antennas), some of which were illustrated in Figure 9-3. Since the boundary conditions along the circumferential surface of the waveguide are the same as those of the cavity, the analysis can begin by modifying only the traveling waves of the z variations of the waveguide in order to obtain the standing waves of the cavity. The radial (ρ) and circumferential (ϕ) variations in both cases will be the same. For the circular cavity, both TE^z and TM^z modes can exist and will be examined here.

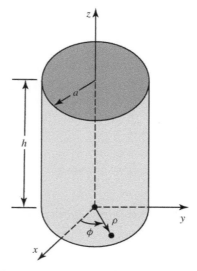

Figure 9-5   Geometry for circular cavity.

TABLE 9-3 Summary of TE_{mn}^z and TM_{mn}^z mode characteristics of circular waveguide

	$\text{TE}_{mn}^z \begin{pmatrix} m=0,1,2,\ldots, \\ n=1,2,3,\ldots \end{pmatrix}$	$\text{TM}_{mn}^z \begin{pmatrix} m=0,1,2,\ldots, \\ n=1,2,3,\ldots \end{pmatrix}$
E_ρ^+	$-A_{mn}\dfrac{m}{\varepsilon\rho}J_m(\beta_\rho\rho)[-C_2\cos(m\phi)+D_2\sin(m\phi)]e^{-j\beta_z z}$	$-B_{mn}\dfrac{\beta_\rho\beta_z}{\omega\mu\varepsilon}J_m'(\beta_\rho\rho)[C_2\cos(m\phi)+D_2\sin(m\phi)]e^{-j\beta_z z}$
E_ϕ^+	$A_{mn}\dfrac{\beta_\rho}{\varepsilon}J_m'(\beta_\rho\rho)[C_2\cos(m\phi)+D_2\sin(m\phi)]e^{-j\beta_z z}$	$-B_{mn}\dfrac{m\beta_z}{\omega\mu\varepsilon}\dfrac{1}{\rho}J_m(\beta_\rho\rho)[-C_2\sin(m\phi)+D_2\cos(m\phi)]e^{-j\beta_z z}$
E_z^+	0	$-jB_{mn}\dfrac{\beta_\rho^2}{\omega\mu\varepsilon}J_m(\beta_\rho\rho)[C_2\cos(m\phi)+D_2\sin(m\phi)]e^{-j\beta_z z}$
H_ρ^+	$-A_{mn}\dfrac{\beta_\rho\beta_z}{\omega\mu\varepsilon}J_m'(\beta_\rho\rho)[C_2\cos(m\phi)+D_2\sin(m\phi)]e^{-j\beta_z z}$	$B_{mn}\dfrac{m}{\mu}\dfrac{1}{\rho}J_m(\beta_\rho\rho)[-C_2\sin(m\phi)+D_2\cos(m\phi)]e^{-j\beta_z z}$
H_ϕ^+	$-A_{mn}\dfrac{m\beta_z}{\omega\mu\varepsilon}\dfrac{1}{\rho}J_m(\beta_\rho\rho)[-C_2\sin(m\phi)+D_2\cos(m\phi)]e^{-j\beta_z z}$	$-B_{mn}\dfrac{\beta_\rho}{\mu}J_m'(\beta_\rho\rho)[C_2\cos(m\phi)+D_2\sin(m\phi)]e^{-j\beta_z z}$
H_z^+	$-jA_{mn}\dfrac{\beta_\rho^2}{\omega\mu\varepsilon}J_m(\beta_\rho\rho)[C_2\cos(m\phi)+D_2\sin(m\phi)]e^{-j\beta_z z}$	0
$'$		$\dfrac{\partial}{\partial(\beta_\rho\rho)}$
$\beta_c=\beta_\rho$	$\dfrac{\chi_{mn}'}{a}$	$\dfrac{\chi_{mn}}{a}$
f_c	$\dfrac{\chi_{mn}'}{2\pi a\sqrt{\mu\varepsilon}}$	$\dfrac{\chi_{mn}}{2\pi a\sqrt{\mu\varepsilon}}$
λ_c	$\dfrac{2\pi a}{\chi_{mn}'}$	$\dfrac{2\pi a}{\chi_{mn}}$

(continued overleaf)

TABLE 9-3 (*Continued*)

	$\mathrm{TE}^z_{mn}\begin{pmatrix} m=0,1,2,\ldots, \\ n=1,2,3,\ldots \end{pmatrix}$	$\mathrm{TM}^z_{mn}\begin{pmatrix} m=0,1,2,\ldots, \\ n=1,2,3,\ldots \end{pmatrix}$
$\beta_z(f \geq f_c)$	$\beta\sqrt{1-\left(\dfrac{f_c}{f}\right)^2}=\beta\sqrt{1-\left(\dfrac{\lambda}{\lambda_c}\right)^2}$	
$\lambda_g(f \geq f_c)$	$\dfrac{\lambda}{\sqrt{1-\left(\dfrac{f_c}{f}\right)^2}}=\dfrac{\lambda}{\sqrt{1-\left(\dfrac{\lambda}{\lambda_c}\right)^2}}$	
$v_p(f \geq f_c)$	$\dfrac{v}{\sqrt{1-\left(\dfrac{f_c}{f}\right)^2}}=\dfrac{v}{\sqrt{1-\left(\dfrac{\lambda}{\lambda_c}\right)^2}}$	
$Z_w(f \geq f_c)$	$\dfrac{\eta}{\sqrt{1-\left(\dfrac{f_c}{f}\right)^2}}=\dfrac{\eta}{\sqrt{1-\left(\dfrac{\lambda}{\lambda_c}\right)^2}}$	$\eta\sqrt{1-\left(\dfrac{f_c}{f}\right)^2}=\eta\sqrt{1-\left(\dfrac{\lambda}{\lambda_c}\right)^2}$
$Z_w(f \leq f_c)$	$j\dfrac{\eta}{\sqrt{\left(\dfrac{f_c}{f}\right)^2-1}}=j\dfrac{\eta}{\sqrt{\left(\dfrac{\lambda}{\lambda_c}\right)^2-1}}$	$-j\eta\sqrt{\left(\dfrac{f_c}{f}\right)^2-1}=-j\eta\sqrt{\left(\dfrac{\lambda}{\lambda_c}\right)^2-1}$
α_c	$\dfrac{R_s}{a\eta\sqrt{1-\left(\dfrac{f_c}{f}\right)^2}}\left[\left(\dfrac{f_c}{f}\right)^2+\dfrac{m^2}{(\chi'_{mn})^2-m^2}\right]$	$\dfrac{R_s}{a\eta}\dfrac{1}{\sqrt{1-\left(\dfrac{f_c}{f}\right)^2}}$

9.3.1 Transverse Electric (TE^z) Modes

We begin the analysis of the TE^z modes by assuming the vector potential F_z is that of (9-7) modified so that the z variations are standing waves instead of traveling waves. Thus, using (9-7), we can write that

$$F_z(\rho,\phi,z) = A_{mn}J_m(\beta_\rho\rho)[C_2\cos(m\phi)+D_2\sin(m\phi)]$$
$$\times[C_3\cos(\beta_z z)+D_3\sin(\beta_z z)] \tag{9-35}$$

where

$$\beta_\rho = \frac{\chi'_{mn}}{a} \tag{9-35a}$$

$$m = 0,1,2,3,\dots \tag{9-35b}$$

To determine the permissible values of β_z, we must apply the additional boundary conditions introduced by the presence of the end plates. These additional boundary conditions are

$$E_\rho(0\le\rho\le a,0\le\phi\le 2\pi,z=0)=E_\rho(0\le\rho\le a,0\le\phi\le 2\pi,z=h)=0 \tag{9-36a}$$

$$E_\phi(0\le\rho\le a,0\le\phi\le 2\pi,z=0)=E_\phi(0\le\rho\le a,0\le\phi\le 2\pi,z=h)=0 \tag{9-36b}$$

Since both boundary conditions are not independent, using either of the two leads to the same results.

Using (6-80) and (9-35), we can write the E_ϕ component as

$$E_\phi = \frac{1}{\varepsilon}\frac{\partial F_z}{\partial\rho}$$

$$= \beta_\rho\frac{A_{mn}}{\varepsilon}J'_m(\beta_\rho\rho)[C_2\cos(m\phi)+D_2\sin(m\phi)][C_3\cos(\beta_z z)+D_3\sin(\beta_z z)] \tag{9-37}$$

where

$$' \equiv \frac{\partial}{\partial(\beta_\rho\rho)} \tag{9-37a}$$

Applying (9-36b) leads to

$$E_\phi(0\le\rho\le a,0\le\phi\le 2\pi,z=0)$$

$$= \beta_\rho\frac{A_{mn}}{\varepsilon}J'_m(\beta_\rho\rho)[C_2\cos(m\phi)+D_2\sin(m\phi)][C_3(1)+D_3(0)]=0$$

$$\Rightarrow C_3=0 \tag{9-38a}$$

$$E_\phi(0\le\rho\le a,0\le\phi\le 2\pi,z=h)$$

$$= \beta_\rho\frac{A_{mn}}{\varepsilon}J'_m(\beta_\rho\rho)[C_2\cos(m\phi)+D_2\sin(m\phi)]D_3\sin(\beta_z h)=0$$

$$\sin(\beta_z h)=0 \;\Rightarrow\; \beta_z h = \sin^{-1}(0)=p\pi$$

$$\beta_z = \frac{p\pi}{h}, \qquad p=1,2,3,\dots \tag{9-38b}$$

Thus, the resonant frequency is obtained using

$$\beta_\rho^2 + \beta_z^2 = \left(\frac{\chi'_{mn}}{a}\right)^2 + \left(\frac{p\pi}{h}\right)^2 = \beta_r^2 = \omega_r^2 \mu\varepsilon \qquad (9\text{-}39)$$

or

$$\boxed{(f_r)_{mnp}^{\text{TE}^z} = \frac{1}{2\pi\sqrt{\mu\varepsilon}}\sqrt{\left(\frac{\chi'_{mn}}{a}\right)^2 + \left(\frac{p\pi}{h}\right)^2} \quad \begin{aligned} & m = 0,1,2,3,\dots \\ & n = 1,2,3,\dots \\ & p = 1,2,3,\dots \end{aligned}} \qquad (9\text{-}39\text{a})$$

The values of χ'_{mn} are found listed in Table 9-1. The final form of F_z of (9-35) is

$$F_z(\rho, \phi, z) = A_{mnp} J_m(\beta_\rho \rho)\, [C_2 \cos(m\phi) + D_2 \sin(m\phi)] \sin(\beta_z z) \qquad (9\text{-}40)$$

9.3.2 Transverse Magnetic (TM^z) Modes

The analysis for the TM^z modes in a circular cavity proceeds in the same manner as for the TE^z modes of the previous section. Using (9-22), we can write that

$$A_z(\rho, \phi, z) = B_{mn} J_m(\beta_\rho \rho)[C_2 \cos(m\phi) + D_2 \sin(m\phi)] \times [C_3 \cos(\beta_z z) + D_3 \sin(\beta_z z)] \quad (9\text{-}41)$$

where

$$\beta_\rho = \frac{\chi_{mn}}{a} \qquad (9\text{-}41\text{a})$$

$$m = 0, 1, 2, \dots \qquad (9\text{-}41\text{b})$$

Using (6-70) and (9-41), we can write the E_ϕ component as

$$\begin{aligned} E_\phi(\rho, \phi, z) &= -j\frac{1}{\omega\mu\varepsilon}\frac{1}{\rho}\frac{\partial^2 A_z}{\partial\phi\,\partial z} \\ &= -jB_{mn}\frac{m\beta_z}{\omega\mu\varepsilon}\frac{1}{\rho} J_m(\beta_\rho\rho)[-C_2 \sin(m\phi) + D_2 \cos(m\phi)] \\ &\quad \times [-C_3 \sin(\beta_z z) + D_3 \cos(\beta_z z)] \end{aligned} \qquad (9\text{-}42)$$

Applying the boundary conditions of (9-36b) leads to

$$E_\phi(0 \le \rho \le a, 0 \le \phi \le 2\pi, z = 0)$$

$$= -jB_{mn}\frac{m\beta_z}{\omega\mu\varepsilon}\frac{1}{\rho} J_m(\beta_\rho\rho)[-C_2 \sin(m\phi) + D_2 \cos(m\phi)]$$

$$\times [-C_3(0) + D_3(1)] = 0 \ \Rightarrow \ D_3 = 0 \qquad (9\text{-}43\text{a})$$

$$E_\phi(0 \le \rho \le a, 0 \le \phi \le 2\pi, z = h)$$

$$= jB_{mn}\frac{m\beta_z}{\omega\mu\varepsilon}\frac{1}{\rho} J_m(\beta_\rho\rho)[-C_2 \sin(m\phi) + D_2 \cos(m\phi)][C_3 \sin(\beta_z h)] = 0$$

$$\sin(\beta_z h) = 0 \ \Rightarrow \ \beta_z h = \sin^{-1}(0) = p\pi$$

$$\beta_z = \frac{p\pi}{h} \qquad p = 0, 1, 2, 3, \dots \qquad (9\text{-}43\text{b})$$

Thus, the resonant frequency is obtained using

$$\beta_\rho^2 + \beta_z^2 = \left(\frac{\chi_{mn}}{a}\right)^2 + \left(\frac{p\pi}{h}\right)^2 = \beta_r^2 = \omega_r^2 \mu\varepsilon \qquad (9\text{-}44)$$

or

$$(f_r)_{mnp}^{TM^z} = \frac{1}{2\pi\sqrt{\mu\varepsilon}}\sqrt{\left(\frac{\chi_{mn}}{a}\right)^2 + \left(\frac{p\pi}{h}\right)^2} \qquad \begin{aligned} m &= 0, 1, 2, 3, \ldots \\ n &= 1, 2, 3, \ldots \\ p &= 0, 1, 2, 3, \ldots \end{aligned} \qquad (9\text{-}45)$$

The values of χ_{mn} are found listed in Table 9-2. The final form of A_z of (9-41) is

$$A_z(\rho, \phi, z) = B_{mnp} J_m(\beta_\rho \rho) [C_2 \cos(m\phi) + D_2 \sin(m\phi)] \cos(\beta_z z) \qquad (9\text{-}46)$$

The resonant frequencies of the TE_{mnp}^z and TM_{mnp}^z modes, as given respectively by (9-39a) and (9-45), are functions of the h/a ratio and they are listed in Table 9-4.

The TE_{mnp}^z mode with the smallest resonant frequency is the TE_{111}, and its cutoff frequency is given by

$$(f_r)_{111}^{TE^z} = \frac{1}{2\pi\sqrt{\mu\varepsilon}}\sqrt{\left(\frac{1.8412}{a}\right)^2 + \left(\frac{\pi}{h}\right)^2} \qquad (9\text{-}47\text{a})$$

Similarly, the TM_{mnp}^z mode with the smallest resonant frequency is the TM_{010}, and its cutoff frequency is given by

$$(f_r)_{010}^{TM^z} = \frac{1}{2\pi\sqrt{\mu\varepsilon}}\sqrt{\left(\frac{2.4049}{a}\right)^2} \qquad (9\text{-}47\text{b})$$

Equating (9-47a) to (9-47b) indicates that the two are identical (degenerate modes) when

$$\frac{h}{a} = 2.03 \simeq 2 \qquad (9\text{-}48)$$

When $h/a < 2.03$ the dominant mode is the TM_{010}, whereas for $h/a > 2.03$ the dominant mode is the TE_{111} mode.

9.3.3 Quality Factor Q

One of the most important parameters of a cavity is its quality factor, better known as the Q, which is defined by (8-84). The Q of the TM_{010} mode, which is the dominant mode when $h/a < 2.03$, is of particular interest and it will be derived here.

For the TM_{010} mode, the potential function of (9-46) reduces to

$$A_z = B_{010} J_0(\beta_\rho \rho) \qquad (9\text{-}49)$$

where

$$\beta_\rho = \frac{\chi_{01}}{a} = \frac{2.4049}{a} \qquad (9\text{-}49\text{a})$$

TABLE 9-4 Resonant frequencies for the TE_{mnp} and TM_{mnp} modes of a circular cavity

$$R_{dom}^{mnp} = \frac{(f_r)_{mnp}}{(f_r)_{dom}}$$

$\dfrac{h}{a}$					R_{dom}^{mnp}					
0	$\dfrac{TM_{010}}{TM_{010}}$	$\dfrac{TM_{110}}{TM_{010}}$	$\dfrac{TM_{210}}{TM_{010}}$	$\dfrac{TM_{020}}{TM_{010}}$	$\dfrac{TM_{310}}{TM_{010}}$	$\dfrac{TM_{120}}{TM_{010}}$	$\dfrac{TM_{410}}{TM_{010}}$	$\dfrac{TM_{220}}{TM_{010}}$	$\dfrac{TM_{030}}{TM_{010}}$	$\dfrac{TM_{510}}{TM_{010}}$
	1.000	1.593	2.136	2.295	2.653	2.917	3.155	3.500	3.598	3.647
0.5	$\dfrac{TM_{010}}{TM_{010}}$	$\dfrac{TM_{110}}{TM_{010}}$	$\dfrac{TM_{210}}{TM_{010}}$	$\dfrac{TM_{020}}{TM_{010}}$	$\dfrac{TM_{310}}{TM_{010}}$	$\dfrac{TE_{111}}{TM_{010}}$	$\dfrac{TM_{011}}{TM_{010}}$	$\dfrac{TE_{211}}{TM_{010}}$	$\dfrac{TM_{120}}{TM_{010}}$	$\dfrac{TE_{011}}{TM_{010}}$
	1.000	1.593	2.136	2.295	2.653	2.722	2.797	2.905	2.917	3.060
1.00	$\dfrac{TM_{010}}{TM_{010}}$	$\dfrac{TE_{111}}{TM_{010}}$	$\dfrac{TM_{110}}{TM_{010}}$	$\dfrac{TM_{011}}{TM_{010}}$	$\dfrac{TE_{211}}{TM_{010}}$	$\dfrac{TM_{111}}{TM_{010}}$	$\dfrac{TE_{011}}{TM_{010}}$	$\dfrac{TM_{210}}{TM_{010}}$	$\dfrac{TE_{311}}{TM_{010}}$	$\dfrac{TM_{020}}{TM_{010}}$
	1.000	1.514	1.593	1.645	1.822	2.060	2.060	2.136	2.181	2.295
2.03	$\dfrac{TM_{010}}{TM_{010},TE_{111}}$	$\dfrac{TE_{111}}{TM_{010},TE_{111}}$	$\dfrac{TM_{011}}{TM_{010},TE_{111}}$	$\dfrac{TE_{211}}{TM_{010},TE_{111}}$	$\dfrac{TE_{212}}{TM_{010},TE_{111}}$	$\dfrac{TM_{110}}{TM_{010},TE_{111}}$	$\dfrac{TM_{012}}{TM_{010},TE_{111}}$	$\dfrac{TE_{011}}{TM_{010},TE_{111}}$	$\dfrac{TM_{111}}{TM_{010},TE_{111}}$	$\dfrac{TE_{212}}{TM_{010},TE_{111}}$
	1.000	1.000	1.189	1.424	1.497	1.593	1.630	1.718	1.718	1.808
3.0	$\dfrac{TE_{111}}{TE_{111}}$	$\dfrac{TM_{010}}{TE_{111}}$	$\dfrac{TM_{011}}{TE_{111}}$	$\dfrac{TE_{112}}{TE_{111}}$	$\dfrac{TM_{012}}{TE_{111}}$	$\dfrac{TE_{211}}{TE_{111}}$	$\dfrac{TE_{113}}{TE_{111}}$	$\dfrac{TE_{212}}{TE_{111}}$	$\dfrac{TM_{110}}{TE_{111}}$	$\dfrac{TM_{013}}{TE_{111}}$
	1.000	1.136	1.238	1.317	1.506	1.524	1.719	1.748	1.809	1.868
4.0	$\dfrac{TE_{111}}{TE_{111}}$	$\dfrac{TM_{010}}{TE_{111}}$	$\dfrac{TE_{112}}{TE_{111}}$	$\dfrac{TM_{011}}{TE_{111}}$	$\dfrac{TM_{012}}{TE_{111}}$	$\dfrac{TE_{113}}{TE_{111}}$	$\dfrac{TE_{211}}{TE_{111}}$	$\dfrac{TM_{013}}{TE_{111}}$	$\dfrac{TE_{212}}{TE_{111}}$	$\dfrac{TE_{114}}{TE_{111}}$
	1.000	1.202	1.209	1.264	1.435	1.494	1.575	1.682	1.717	1.819
5.0	$\dfrac{TE_{111}}{TE_{111}}$	$\dfrac{TE_{112}}{TE_{111}}$	$\dfrac{TM_{010}}{TE_{111}}$	$\dfrac{TM_{011}}{TE_{111}}$	$\dfrac{TE_{113}}{TE_{111}}$	$\dfrac{TM_{012}}{TE_{111}}$	$\dfrac{TM_{013}}{TE_{111}}$	$\dfrac{TE_{114}}{TE_{111}}$	$\dfrac{TE_{211}}{TE_{111}}$	$\dfrac{TE_{212}}{TE_{111}}$
	1.000	1.146	1.236	1.278	1.354	1.395	1.571	1.602	1.603	1.698
10.0	$\dfrac{TE_{111}}{TE_{111}}$	$\dfrac{TE_{112}}{TE_{111}}$	$\dfrac{TE_{113}}{TE_{111}}$	$\dfrac{TE_{114}}{TE_{111}}$	$\dfrac{TM_{010}}{TE_{111}}$	$\dfrac{TE_{115}}{TE_{111}}$	$\dfrac{TM_{011}}{TE_{111}}$	$\dfrac{TM_{012}}{TE_{111}}$	$\dfrac{TM_{013}}{TE_{111}}$	$\dfrac{TE_{116}}{TE_{111}}$
	1.000	1.042	1.107	1.194	1.288	1.296	1.299	1.331	1.383	1.411
∞	$\dfrac{TE_{11p}}{TE_{111}}$	$\dfrac{TM_{01p}}{TE_{111}}$	$\dfrac{TE_{21p}}{TE_{111}}$	$\dfrac{TE_{01p}}{TE_{111}}$	$\dfrac{TM_{11p}}{TE_{111}}$	$\dfrac{TE_{31p}}{TE_{111}}$	$\dfrac{TM_{21p}}{TE_{111}}$	$\dfrac{TE_{41p}}{TE_{111}}$	$\dfrac{TE_{12p}}{TE_{111}}$	$\dfrac{TM_{02p}}{TE_{111}}$
	1.000	1.306	1.659	2.081	2.081	2.282	2.790	2.888	2.896	2.998

These corresponding electric and magnetic fields are obtained using (6-70), or

$$E_\rho = -j \frac{1}{\omega_r \mu \varepsilon} \frac{\partial^2 A_z}{\partial \rho \partial z} = 0$$

$$E_\phi = -j \frac{1}{\omega_r \mu \varepsilon} \frac{1}{\rho} \frac{\partial^2 A_z}{\partial \phi \partial z} = 0$$

$$E_z = -j \frac{1}{\omega_r \mu \varepsilon} \left(\frac{\partial^2}{\partial z^2} + \beta_r^2 \right) A_z = -j \frac{\beta_r^2}{\omega_r \mu \varepsilon} B_{010} J_0\left(\frac{\chi_{01}}{a} \rho \right)$$

$$H_\rho = \frac{1}{\mu} \frac{1}{\rho} \frac{\partial A_z}{\partial \phi} = 0$$ (9-50)

$$H_\phi = -\frac{1}{\mu} \frac{\partial A_z}{\partial \rho} = -\frac{\chi_{01}}{a} \frac{B_{010}}{\mu} J_0'\left(\frac{\chi_{01}}{a} \rho \right)$$

$$H_z = 0$$

The total energy stored in the cavity is given by

$$W = 2W_e = \frac{\varepsilon}{2} \iiint_V |\mathbf{E}|^2 \, dv = |B_{010}|^2 \frac{\varepsilon}{2} \left(\frac{\beta_r^2}{\omega_r \mu \varepsilon} \right)^2 \int_0^h \int_0^{2\pi} \int_0^a J_0^2\left(\frac{\chi_{01}}{a} \rho \right) \rho \, d\rho \, d\phi \, dz$$

$$W = |B_{010}|^2 \, \pi h \varepsilon \left(\frac{\beta_r^2}{\omega_r \mu \varepsilon} \right)^2 \int_0^a J_0^2\left(\frac{\chi_{01}}{a} \rho \right) \rho \, d\rho$$ (9-51)

Since [5],

$$\int_0^a \rho J_0^2\left(\frac{\chi_{01}}{a} \rho \right) d\rho = \frac{a^2}{2} J_1^2(\chi_{01})$$ (9-52)

then (9-51) reduces to

$$W = |B_{010}|^2 \, \frac{\pi h \varepsilon}{2} \left(\frac{a \beta_r^2}{\omega_r \mu \varepsilon} \right)^2 J_1^2(\chi_{01})$$ (9-53)

Because the medium within the cavity is assumed to be lossless, the total power is dissipated on the conducting walls of the cavity. Thus, we can write that

$$P_d = \frac{R_s}{2} \oiint_A |\mathbf{H}|^2 \, ds = \frac{R_s}{2} \left\{ \int_0^{2\pi} \int_0^h |\mathbf{H}|_{\rho=a}^2 a \, d\phi \, dz + 2 \int_0^{2\pi} \int_0^a |\mathbf{H}|_{z=0}^2 \rho \, d\rho \, d\phi \right\}$$

$$= |B_{010}|^2 \frac{R_s}{2\mu^2} \left(\frac{\chi_{01}}{a} \right)^2 \left\{ \int_0^{2\pi} \int_0^h [J_0'(\chi_{01})]^2 a \, dz \, d\phi + 2 \int_0^{2\pi} \int_0^a \left| J_0'\left(\frac{\chi_{01}}{a} \rho \right) \right|^2 \rho \, d\rho \, d\phi \right\}$$

$$P_d = |B_{010}|^2 \frac{\pi R_s}{\mu^2} \left(\frac{\chi_{01}}{a} \right)^2 \left\{ ah[J_0'(\chi_{01})]^2 + 2 \int_0^a \left| J_0'\left(\frac{\chi_{01}}{a} \rho \right) \right|^2 \rho \, d\rho \right\}$$ (9-54)

Because

$$J_0'\left(\frac{\chi_{01}}{a} \rho \right) = \frac{d}{d(\chi_{01}\rho/a)} \left[J_0\left(\frac{\chi_{01}}{a} \rho \right) \right] = -J_1\left(\frac{\chi_{01}}{a} \rho \right)$$ (9-55a)

and at $\rho = a$

$$J_0'(\chi_{01}) = -J_1(\chi_{01})$$ (9-55b)

(9-54) reduces to

$$P_d = |B_{010}|^2 \frac{\pi R_s}{\mu^2} \left(\frac{\chi_{01}}{a}\right)^2 \left\{ ahJ_1^2(\chi_{01}) + 2\int_0^a J_1^2\left(\frac{\chi_{01}}{a}\rho\right)\rho\,d\rho \right\}$$

$$= |B_{010}|^2 \frac{\pi R_s}{\mu^2} \left(\frac{\chi_{01}}{a}\right)^2 \left\{ ahJ_1^2(\chi_{01}) + 2\left[\frac{a^2}{2} J_1^2(\chi_{01})\right] \right\}$$

$$P_d = |B_{010}|^2 \frac{\pi R_s}{\mu^2} \left(\frac{\chi_{01}}{a}\right)^2 a(h+a)J_1^2(\chi_{01}) \tag{9-56}$$

Using the Q definition of (8-84) along with (9-53) and (9-56), we can write that

$$\boxed{Q = \frac{\omega_r W}{P_d} = \frac{\beta_r^4 ha^3}{2\omega_r \varepsilon(h+a)R_s \chi_{01}^2} = \frac{\chi_{01}\sqrt{\frac{\mu}{\varepsilon}}}{2\left(1+\frac{a}{h}\right)R_s} = \frac{1.2025\eta}{R_s\left(1+\frac{a}{h}\right)}} \tag{9-57}$$

since for the TM_{010} mode ($m=0$, $n=1$, and $p=0$)

$$\beta_\rho^2 + \beta_z^2 = \left(\frac{\chi_{01}}{a}\right)^2 = \beta_r^2 \quad \Rightarrow \quad \chi_{01} = \beta_r a \tag{9-57a}$$

Example 9-4

Compare the Q values of a circular cavity operating in the TM_{010} mode to those of a square-based rectangular cavity. The dimensions of each are such that the circular cavity is circumscribed by the square-based rectangular cavity.

Solution: According to (9-57) the Q of a circular cavity of radius a (or diameter d) and height h is given by

$$Q = \frac{1.2025\eta}{R_s\left(1+\frac{a}{h}\right)} = \frac{1.2025\eta}{R_s\left(1+\frac{d/2}{h}\right)}$$

For a square-based rectangular cavity to circumscribe a circular cavity, one of the sides of its base must be equal to the diameter and their heights must be equal. Therefore, with a base of $a=c=d$ on each of its sides and a height $b=h$, according to (8-88a) its Q is equal to

$$Q = \frac{1.1107\eta}{R_s}\left[\frac{1}{\left(1+\frac{a/2}{b}\right)}\right] = \frac{1.1107\eta}{R_s}\left[\frac{1}{\left(1+\frac{d/2}{h}\right)}\right]$$

Compare these two expressions and it is evident that the Q of the circular cavity is greater than that of the square-based cavity by

$$\left(\frac{1.2025-1.1107}{1.1107}\right) \times 100 = 8.26\%$$

This is expected since the circular cavity does not possess as many sharp corners and edges as the square-based cavity whose volume and surface area are not as well utilized by the interior fields. It should be remembered that the Q of a cavity is proportional to volume and inversely proportional to area.

9.4   RADIAL WAVEGUIDES

For a circular waveguide the waves travel in the $\pm z$ directions and their z variations are repre-sented by the factor $e^{\pm j\beta_z z}$. Their constant phase planes (equiphases) are planes that are parallel to each other and perpendicular to the z direction. If the waves were traveling in the $\pm\phi$ direction, their variations in that direction would be represented by $e^{\pm jm\phi}$. Such waves are usually referred to as *circulating waves*, and their equiphase surfaces are constant ϕ planes. For waves that travel in the $\pm\rho$ (radial) direction, their variations in that direction would be represented by either $H_m^{(2)}(\beta_\rho\rho)$ or $H_m^{(1)}(\beta_\rho\rho)$. Such waves are usually referred to as *radial waves*, and their equiphases are constant ρ (radius) planes. The structures that support radial waves are referred to as radial waveguides, and they will be examined here. Examples are parallel plates, wedged plates (representing horn antennas), and others.

9.4.1   Parallel Plates

When two infinite long parallel plates are excited by a line source placed between them at the center, as shown in Figure 9-6, the excited waves travel in the radial direction and form radial waves. We shall examine here both the TE^z and TM^z modes in the region between the plates.

A. Transverse Electric (TE^z) Modes   For the TE^z modes of Figure 9-6, the potential function F_z can be written according to (3-67a) through (3-69b) as

$$F_z(\rho,\phi,z) = [C_1 H_m^{(1)}(\beta_\rho\rho) + D_1 H_m^{(2)}(\beta_\rho\rho)][C_2 \cos(m\phi) + D_2 \sin(m\phi)]$$
$$\times [C_3 \cos(\beta_z z) + D_3 \sin(\beta_z z)] \tag{9-58}$$

where

$$\beta_\rho^2 + \beta_z^2 = \beta^2 \tag{9-58a}$$

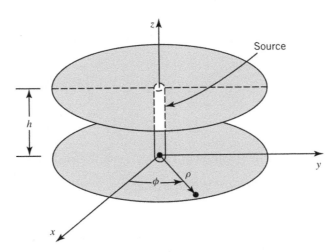

Figure 9-6   Geometry for radial waveguide.

The boundary conditions are

$$E_\rho(0 \leq \rho \leq \infty, 0 \leq \phi \leq 2\pi, z = 0) = E_\rho(0 \leq \rho \leq \infty, 0 \leq \phi \leq 2\pi, z = h) = 0 \qquad (9\text{-}59a)$$

$$E_\phi(0 \leq \rho \leq \infty, 0 \leq \phi \leq 2\pi, z = 0) = E_\phi(0 \leq \rho \leq \infty, 0 \leq \phi \leq 2\pi, z = h) = 0 \qquad (9\text{-}59b)$$

Since both of the preceding boundary conditions are not independent, either of the two leads to the same results.

Using (6-80) and (9-58), the E_ϕ component can be written as

$$
\begin{aligned}
E_\phi(\rho, \phi, z) &= \frac{1}{\varepsilon} \frac{\partial F_z}{\partial \rho} \\
&= \frac{\beta_\rho}{\varepsilon} [C_1 H_m^{(1)'}(\beta_\rho \rho) + D_1 H_m^{(2)'}(\beta_\rho \rho)][C_2 \cos(m\phi) + D_2 \sin(m\phi)] \\
&\quad \times [C_3 \cos(\beta_z z) + D_3 \sin(\beta_z z)]
\end{aligned}
\qquad (9\text{-}60)
$$

Applying (9-59b) leads to

$$
\begin{aligned}
E_\phi(0 \leq \rho \leq \infty, 0 \leq \phi \leq 2\pi, z = 0) &= \frac{\beta_\rho}{\varepsilon} [C_1 H_m^{(1)'}(\beta_\rho \rho) + D_1 H_m^{(2)'}(\beta_\rho \rho)] \\
&\quad \times [C_2 \cos(m\phi) + D_2 \sin(m\phi)][C_3(1) + D_3(0)] = 0 \\
&\Rightarrow C_3 = 0
\end{aligned}
\qquad (9\text{-}61a)
$$

$$
\begin{aligned}
E_\phi(0 \leq \rho \leq a, 0 \leq \phi \leq 2\pi, z = h) &= \frac{\beta_\rho}{\varepsilon} [C_1 H_m^{(1)'}(\beta_\rho \rho) + D_1 H_m^{(2)'}(\beta_\rho \rho)] \\
&\quad \times [C_2 \cos(m\phi) + D_2 \sin(m\phi)] D_3 \sin(\beta_z h) = 0 \\
&\Rightarrow \sin(\beta_z h) = 0 \quad \Rightarrow \quad \beta_z h \sin^{-1}(0) = n\pi \\
&\qquad \beta_z = \frac{n\pi}{h}, \qquad n = 1, 2, 3, \dots
\end{aligned}
\qquad (9\text{-}61b)
$$

Thus, F_z of (9-58) reduces to

$$
\begin{aligned}
F_z(\rho, \phi, z) &= [C_1 H_m^{(1)}(\beta_\rho \rho) + D_1 H_m^{(2)}(\beta_\rho \rho)] \\
&\quad \times [C_2 \cos(m\phi) + D_2 \sin(m\phi)] D_3 \sin(\beta_z z)
\end{aligned}
\qquad (9\text{-}62)
$$

where

$$\beta_\rho^2 + \beta_z^2 = \beta^2 \quad \Rightarrow \quad \beta_\rho = \pm\sqrt{\beta^2 - \beta_z^2} = \pm\sqrt{\beta^2 - \left(\frac{n\pi}{h}\right)^2} \qquad (9\text{-}62a)$$

$$\beta_z = \frac{n\pi}{h}, \quad n = 1, 2, 3, \dots \qquad (9\text{-}62b)$$

$$m = 0, 1, 2, \dots \quad (\text{because of periodicity of the fields in } \phi) \qquad (9\text{-}62c)$$

Cutoff is defined when $\beta_\rho = 0$. Thus, using (9-62a)

$$\beta_\rho = \pm\sqrt{\beta^2 - \left(\frac{n\pi}{h}\right)^2} \Bigg|_{\substack{f=f_c \\ \beta=\beta_c}} = 0 \quad \Rightarrow \quad \beta_c = \frac{n\pi}{h} \qquad (9\text{-}63)$$

or

$$\boxed{(f_c)_n^{TE^z} = \frac{n}{2h\sqrt{\mu\varepsilon}}, \qquad n = 1, 2, 3, \dots} \qquad (9\text{-}63a)$$

Therefore, using (9-63a), β_ρ of (9-62a) takes the following forms above, at, and below cutoff.

$$\beta_\rho = \begin{cases} \sqrt{\beta^2 - \left(\dfrac{n\pi}{h}\right)^2} = \beta\sqrt{1 - \left(\dfrac{f_c}{f}\right)^2} & f > f_c & (9\text{-}64a) \\[3mm] 0 & f = f_c & (9\text{-}64b) \\[3mm] -j\sqrt{\left(\dfrac{n\pi}{h}\right)^2 - \beta^2} = -j\beta\sqrt{\left(\dfrac{f_c}{f}\right)^2 - 1} = -j\alpha & f < f_c & (9\text{-}64c) \end{cases}$$

Let us now examine the outward $(+\rho)$ traveling waves represented by $H_m^{(2)}(\beta_\rho\rho)$ and those represented simultaneously by the $\cos(m\phi)$ variations of (9-62). In that case, (9-62) reduces to

$$F_z^+(\rho,\phi,z) = A_{mn} H_m^{(2)}(\beta_\rho\rho)\cos(m\phi)\sin\left(\frac{n\pi}{h}z\right) \tag{9-65}$$

The corresponding electric and magnetic fields can be written using (6-80) as

$$E_\rho^+ = -\frac{1}{\varepsilon\rho}\frac{\partial F_z^+}{\partial\phi} = A_{mn}\frac{m}{\varepsilon\rho} H_m^{(2)}(\beta_\rho\rho)\sin(m\phi)\sin\left(\frac{n\pi}{h}z\right) \tag{9-65a}$$

$$E_\phi^+ = \frac{1}{\varepsilon}\frac{\partial F_z^+}{\partial\rho} = \beta_\rho\frac{A_{mn}}{\varepsilon} H_m^{(2)'}(\beta_\rho\rho)\cos(m\phi)\sin\left(\frac{n\pi}{h}z\right) \tag{9-65b}$$

$$E_z^+ = 0 \tag{9-65c}$$

$$H_\rho^+ = -j\frac{1}{\omega\mu\varepsilon}\frac{\partial^2 F_z^+}{\partial_\rho\partial z} = -jA_{mn}\beta_\rho\frac{n\pi/h}{\omega\mu\varepsilon} H_m^{(2)'}(\beta_\rho\rho)\cos(m\phi)\cos\left(\frac{n\pi}{h}z\right) \tag{9-65d}$$

$$H_\phi^+ = -j\frac{1}{\omega\mu\varepsilon}\frac{1}{\rho}\frac{\partial^2 F_z^+}{\partial_\phi\partial z} = jA_{mn}\frac{mn\pi/h}{\omega\mu\varepsilon\rho} H_m^{(2)}(\beta_\rho\rho)\sin(m\phi)\cos\left(\frac{n\pi}{h}z\right) \tag{9-65e}$$

$$H_z^+ = -j\frac{1}{\omega\mu\varepsilon}\left(\frac{\partial^2}{\partial z^2} + \beta^2\right)F_z^+ = -jA_{mn}\frac{\beta_\rho^2}{\omega\mu\varepsilon} H_m^{(2)}(\beta_\rho\rho)\cos(m\phi)\sin\left(\frac{n\pi}{h}z\right) \tag{9-65f}$$

$$' \equiv \frac{\partial}{\partial(\beta_\rho\rho)} \tag{9-65g}$$

The impedance of the wave in the $+\rho$ direction is defined and given by

$$Z_w^{+\rho}(\mathrm{TE}_n^z) = \frac{E_\phi}{H_z} = j\frac{\omega\mu}{\beta_\rho}\frac{H_m^{(2)'}(\beta_\rho\rho)}{H_m^{(2)}(\beta_\rho\rho)} \tag{9-66}$$

Since

$$H_m^{(2)'}(\beta_\rho\rho) = \frac{\partial}{\partial(\beta_\rho\rho)}\left[H_m^{(2)}(\beta_\rho\rho)\right] \tag{9-67}$$

(9-66) can be written as

$$Z_w^{+\rho}(\mathrm{TE}_n^z) = j\frac{\omega\mu}{\beta_\rho}\frac{\dfrac{\partial}{\partial(\beta_\rho\rho)}\left[H_m^{(2)}(\beta_\rho\rho)\right]}{H_m^{(2)}(\beta_\rho\rho)} \tag{9-68}$$

Below cutoff $(f < f_c)$ β_ρ is imaginary, and it is given by (9-64c). Therefore, for $f < f_c$

$$H_m^{(2)}(\beta_\rho\rho) = H_m^{(2)}(-j\alpha\rho) \tag{9-69a}$$

$$\frac{d}{d(\beta_\rho\rho)}\left[H_m^{(2)}(\beta_\rho\rho)\right] = \frac{d}{d(-j\alpha\rho)}\left[H_m^{(2)}(-j\alpha\rho)\right] = j\frac{d}{d(\alpha\rho)}\left[H_m^{(2)}(-j\alpha\rho)\right] \tag{9-69b}$$

For complex arguments, the Hankel function $H_m^{(2)}$ of the second kind is related to the modified Bessel function K_m of the second kind by

$$H_m^{(2)}(-j\alpha\rho) = \frac{2}{\pi}j^{m+1}K_m(\alpha\rho) \tag{9-70a}$$

$$\frac{d}{d(\alpha\rho)}\left[H_m^{(2)}(-j\alpha\rho)\right] = \frac{2}{\pi}j^{m+1}\frac{d}{d(\alpha\rho)}\left[K_m(\alpha\rho)\right] \tag{9-70b}$$

Thus, below cutoff $(f < f_c)$, the wave impedance reduces to

$$Z_w^{+\rho}(\mathrm{TE}_n^z)|_{f<f_c} = j\frac{\omega\mu}{-j\alpha}\frac{j\dfrac{d}{d(\alpha\rho)}\left[K_m(\alpha\rho)\right]}{K_m(\alpha\rho)}$$

$$= -j\frac{\omega\mu}{\alpha}\frac{\dfrac{d}{d(\alpha\rho)}\left[K_m(\alpha\rho)\right]}{K_m(\alpha\rho)} \tag{9-71}$$

which is always inductive (for $f < f_c$) since $K_m(\alpha\rho) > 0$ and $d/d(\alpha\rho)[K_m(\alpha\rho)] < 0$. Therefore, below cutoff the modes are nonpropagating (evanescent) since the waveguide is behaving as an inductive storage element.

B. Transverse Magnetic (TM^z) Modes The TM^z modes of the radial waveguide structure of Figure 9-6 with the source at the center are derived in a similar manner. Using such a procedure leads to the following results:

$$A_z(\rho,\phi,z) = \left[C_1'H_m^{(1)}(\beta_\rho\rho) + D_1'H_m^{(2)}(\beta_\rho\rho)\right]$$
$$\times\left[C_2'\cos(m\phi) + D_2'\sin(m\phi)\right]C_3'\cos(\beta_z z) \tag{9-72}$$

$$\beta_\rho^2 + \beta_z^2 = \beta^2 \Rightarrow \beta_\rho = \pm\sqrt{\beta^2 - \beta_z^2} = \pm\sqrt{\beta^2 - \left(\frac{n\pi}{h}\right)^2} \tag{9-72a}$$

$$\beta_z = \frac{n\pi}{h}, \quad n = 0,1,2,\dots \tag{9-72b}$$

$$m = 0,1,2,\dots \quad \text{(because of periodicity of the fields in } \phi) \tag{9-72c}$$

$$\boxed{(f_c)_n^{\mathrm{TM}^z} = \frac{n}{2h\sqrt{\mu\varepsilon}}, \quad n = 0,1,2,\dots} \tag{9-73}$$

$$\beta_\rho = \begin{cases} \sqrt{\beta^2 - \left(\dfrac{n\pi}{h}\right)^2} = \beta\sqrt{1 - \left(\dfrac{f_c}{f}\right)^2} & f > f_c \quad (9\text{-}74a) \\ 0 & f = f_c \quad (9\text{-}74b) \\ -j\sqrt{\left(\dfrac{n\pi}{h}\right)^2 - \beta^2} = -j\beta\sqrt{\left(\dfrac{f_c}{f}\right)^2 - 1} = -j\alpha & f < f_c \quad (9\text{-}74c) \end{cases}$$

For outward $(+\rho)$ traveling waves and only $\cos(m\phi)$ variations

$$A_z^+(\rho,\phi,z) = B_{mn} H_m^{(2)}(\beta_\rho\rho)\, \cos(m\phi)\cos\left(\frac{n\pi}{h}z\right) \tag{9-75}$$

$$E_\rho^+ = -j\frac{1}{\omega\mu\varepsilon}\frac{\partial^2 A_z^+}{\partial\rho\partial z} = jB_{mn}\beta_\rho\frac{n\pi/h}{\omega\mu\varepsilon} H_m^{(2)\prime}(\beta_\rho\rho)\, \cos(m\phi)\sin\left(\frac{n\pi}{h}z\right) \tag{9-75a}$$

$$E_\phi^+ = -j\frac{1}{\omega\mu\varepsilon}\frac{1}{\rho}\frac{\partial^2 A_z^+}{\partial\phi\partial z} = -jB_{mn}\frac{mn\pi/h}{\omega\mu\varepsilon\rho} H_m^{(2)}(\beta_\rho\rho)\, \sin(m\phi)\sin\left(\frac{n\pi}{h}z\right) \tag{9-75b}$$

$$E_z^+ = -j\frac{1}{\omega\mu\varepsilon}\left(\frac{\partial^2}{\partial z^2} + \beta^2\right)A_z^+ = -jB_{mn}\frac{\beta_\rho^2}{\omega\mu\varepsilon} H_m^{(2)}(\beta_\rho\rho)\, \cos(m\phi)\cos\left(\frac{n\pi}{h}z\right) \tag{9-75c}$$

$$H_\rho^+ = \frac{1}{\mu}\frac{1}{\rho}\frac{\partial A_z^+}{\partial\phi} = -B_{mn}\frac{m}{\mu}\frac{1}{\rho} H_m^{(2)}(\beta_\rho\rho)\sin(m\phi)\cos\left(\frac{n\pi}{h}z\right) \tag{9-75d}$$

$$H_\phi^+ = -\frac{1}{\mu}\frac{\partial A_z^+}{\partial\rho} = -\beta_\rho\frac{B_{mn}}{\mu} H_m^{(2)\prime}(\beta_\rho\rho)\cos(m\phi)\cos\left(\frac{n\pi}{h}z\right) \tag{9-75e}$$

$$H_z = 0 \tag{9-75f}$$

$$' \equiv \frac{\partial}{\partial(\beta_\rho\rho)} \tag{9-75g}$$

$$Z_w^{+\rho}(\text{TM}_n^z) = \frac{E_z^+}{-H_\phi^+} = -j\frac{\beta_\rho}{\omega\varepsilon}\frac{H_m^{(2)}(\beta_\rho\rho)}{H_m^{(2)\prime}(\beta_\rho\rho)} \tag{9-76}$$

$$Z_w^{+\rho}(\text{TM}_n^z)\Big|_{f<f_c} = j\frac{\alpha}{\omega\varepsilon}\frac{K_m(\alpha\rho)}{\dfrac{d}{d(\alpha\rho)}[K_m(\alpha\rho)]} \tag{9-76a}$$

which is always capacitive (for $f < f_c$) since $K_m(\alpha\rho) > 0$ and $d/d(\alpha\rho)[K_m(\alpha\rho)] < 0$. Therefore, below cutoff the modes are nonpropagating (evanescent) since the waveguide behaves as a capacitive storage element.

9.4.2 Wedged Plates

Another radial type of waveguide structure is the wedged-plate geometry of Figure 9-7 with plates along $z = 0, h$, and $\phi = 0, \phi_0$. This type of a configuration resembles and can be used to represent the structures of E- and H-plane sectoral horns [6]. In fact, the fields within the horns are found using the procedure outlined here. In general, both TE^z and TM^z modes can exist in the space

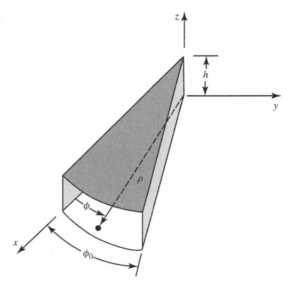

Figure 9-7   Geometry of wedged plate radial waveguide.

between the plates. The independent sets of boundary conditions of this structure that can be used to solve for the TE^z_{pn} and TM^z_{pn} modes are

$$E_\rho(0 \le \rho \le \infty, 0 \le \phi \le \phi_0, z = 0) = E_\rho(0 \le \rho \le \infty, 0 \le \phi \le \phi_0, z = h) = 0 \qquad (9\text{-}77\text{a})$$

$$E_\rho(0 \le \rho \le \infty, \phi = 0, 0 \le z \le h) = E_\rho(0 \le \rho \le \infty, \phi = \phi_0, 0 \le z \le h) = 0 \qquad (9\text{-}77\text{b})$$

or

$$E_\phi(0 \le \rho \le \infty, 0 \le \phi \le \phi_0, z = 0) = E_\phi(0 \le \rho \le \infty, 0 \le \phi \le \phi_0, z = h) = 0 \qquad (9\text{-}78\text{a})$$

$$E_z(0 \le \rho \le \infty, \phi = 0, 0 \le z \le h) = E_z(0 \le \rho \le \infty, \phi = \phi_0, 0 \le z \le h) = 0 \qquad (9\text{-}78\text{b})$$

or appropriate combinations of these. Whichever combination of independent boundary conditions is used, it leads to the same results.

A. Transverse Electric (TE^z) Modes   Since the procedure used to derive this set of TE^z fields is the same as any other TE^z procedure used previously, the results of this set of TE^z_{pn} modes will be summarized here; the details are left as an end-of-chapter exercise for the reader. Only the outward radial $(+\rho)$ parts will be included here.

$$F^+_z(\rho, \phi, z) = A_{pn} H^{(2)}_m(\beta_\rho \rho) \cos(m\phi) \sin(\beta_z z) \qquad (9\text{-}79)$$

$$\boxed{\beta_\rho^2 + \beta_z^2 = \beta^2} \qquad (9\text{-}79\text{a})$$

$$\boxed{\beta_z = \frac{n\pi}{h} \quad n = 1, 2, 3, \dots} \qquad (9\text{-}79\text{b})$$

$$\boxed{m = \frac{p\pi}{\phi_0} \quad p = 0, 1, 2, \dots} \qquad (9\text{-}79\text{c})$$

$$E_\rho^+ = -\frac{1}{\varepsilon}\frac{1}{\rho}\frac{\partial F_z^+}{\partial \phi} = A_{pn}\frac{p\pi/\phi_0}{\varepsilon\rho}H_m^{(2)}(\beta_\rho\rho)\sin\left(\frac{p\pi}{\phi_0}\phi\right)\sin\left(\frac{n\pi}{h}z\right) \qquad (9\text{-}79\text{d})$$

$$E_\phi^+ = \frac{1}{\varepsilon}\frac{\partial F_z^+}{\partial \rho} = \beta_\rho\frac{A_{pn}}{\varepsilon}H_m^{(2)'}(\beta_\rho\rho)\cos\left(\frac{p\pi}{\phi_0}\phi\right)\sin\left(\frac{n\pi}{h}z\right) \qquad (9\text{-}79\text{e})$$

$$E_z^+ = 0 \qquad (9\text{-}79\text{f})$$

$$H_\rho^+ = -j\frac{1}{\omega\mu\varepsilon}\frac{\partial^2 F_z^+}{\partial\rho\,\partial z} = -jA_{pn}\frac{\beta_\rho\beta_z}{\omega\mu\varepsilon}H_m^{(2)'}(\beta_\rho\rho)\cos\left(\frac{p\pi}{\phi_0}\phi\right)\cos\left(\frac{n\pi}{h}z\right) \qquad (9\text{-}79\text{g})$$

$$H_\phi^+ = -j\frac{1}{\omega\mu\varepsilon}\frac{1}{\rho}\frac{\partial^2 F_z^+}{\partial\phi\,\partial z} = jA_{pn}\frac{\beta_z p\pi/\phi_0}{\omega\mu\varepsilon}\frac{1}{\rho}H_m^{(2)}(\beta_\rho\rho)\sin\left(\frac{p\pi}{\phi_0}\phi\right)\cos\left(\frac{n\pi}{h}z\right) \qquad (9\text{-}79\text{h})$$

$$H_z^+ = -j\frac{1}{\omega\mu\varepsilon}\left(\frac{\partial^2}{\partial z^2}+\beta^2\right)F_z^+ = -jA_{pn}\frac{\beta_\rho^2}{\omega\mu\varepsilon}H_m^{(2)}(\beta_\rho\rho)\cos\left(\frac{p\pi}{\phi_0}\phi\right)\sin\left(\frac{n\pi}{h}z\right) \qquad (9\text{-}79\text{i})$$

$$\boxed{Z_w^{+\rho}(\mathrm{TE}_{pn}^z) = \frac{E_\phi^+}{H_z^+} = j\frac{\omega\mu}{\beta_\rho}\frac{H_m^{(2)'}(\beta_\rho\rho)}{H_m^{(2)}(\beta_\rho\rho)}; \qquad {}' \equiv \frac{\partial}{\partial(\beta_\rho\rho)}} \qquad (9\text{-}79\text{j})$$

B. Transverse Magnetic (TM^z) Modes   As for the TE_{pn}^z modes, the procedure for deriving the TM_{pn}^z for the wedged plate radial waveguide is the same as that used for TM^z modes of other waveguide configurations. Therefore, the results will be summarized here, and the details left as an end-of-chapter exercise for the reader. Only the outward radial $(+\rho)$ parts will be included here.

$$A_z^+(\rho,\phi,z) = B_{pn}H_m^{(2)}(\beta_\rho\rho)\sin(m\phi)\cos(\beta_z z) \qquad (9\text{-}80)$$

$$\boxed{\begin{aligned}\beta_\rho^2 + \beta_z^2 &= \beta^2\end{aligned}} \qquad (9\text{-}80\text{a})$$

$$\boxed{\beta_z = \frac{n\pi}{h}, \qquad n = 0,1,2,\ldots} \qquad (9\text{-}80\text{b})$$

$$\boxed{m = \frac{p\pi}{\phi_0}, \qquad p = 1,2,3,\ldots} \qquad (9\text{-}80\text{c})$$

$$E_\rho^+ = -j\frac{1}{\omega\mu\varepsilon}\frac{\partial^2 A_z^+}{\partial\rho\,\partial z} = jB_{pn}\frac{\beta_z\beta_\rho}{\omega\mu\varepsilon}H_m^{(2)'}(\beta_\rho\rho)\sin\left(\frac{p\pi}{\phi_0}\phi\right)\sin\left(\frac{n\pi}{h}z\right) \qquad (9\text{-}80\text{d})$$

$$E_\phi^+ = -j\frac{1}{\omega\mu\varepsilon}\frac{1}{\rho}\frac{\partial^2 A_z^+}{\partial\phi\,\partial z} = jB_{pn}\frac{\beta_z p\pi/\phi_0}{\omega\mu\varepsilon}\frac{1}{\rho}H_m^{(2)}(\beta_\rho\rho)\cos\left(\frac{p\pi}{\phi_0}\phi\right)\sin\left(\frac{n\pi}{h}z\right) \qquad (9\text{-}80\text{e})$$

$$E_z^+ = -j\frac{1}{\omega\mu\varepsilon}\left(\frac{\partial^2}{\partial z^2}+\beta^2\right)A_z^+ = -jB_{pn}\frac{\beta_\rho^2}{\omega\mu\varepsilon}H_m^{(2)}(\beta_\rho\rho)\sin\left(\frac{p\pi}{\phi_0}\phi\right)\cos\left(\frac{n\pi}{h}z\right) \qquad (9\text{-}80\text{f})$$

$$H_\rho^+ = \frac{1}{\mu}\frac{1}{\rho}\frac{\partial A_z^+}{\partial\phi} = B_{pn}\frac{p\pi/\phi_0}{\mu\rho}H_m^{(2)}(\beta_\rho\rho)\cos\left(\frac{p\pi}{\phi_0}\phi\right)\cos\left(\frac{n\pi}{h}z\right) \qquad (9\text{-}80\text{g})$$

$$H_\phi^+ = -\frac{1}{\mu}\frac{\partial A_z^+}{\partial\rho} = -\beta_\rho\frac{B_{pn}}{\mu}H_m^{(2)'}(\beta_\rho\rho)\sin\left(\frac{p\pi}{\phi_0}\phi\right)\cos\left(\frac{n\pi}{h}z\right) \qquad (9\text{-}80\text{h})$$

$$H_z^+ = 0 \tag{9-80i}$$

$$\boxed{Z_w^{+\rho}\left(\mathrm{TM}_{pn}^z\right) = \frac{E_z^+}{-H_\phi^+} = -j\frac{\beta_\rho}{\omega\varepsilon}\frac{H_m^{(2)}(\beta_\rho\rho)}{H_m^{(2)'}(\beta_\rho\rho)}; \qquad ' \equiv \frac{\partial}{\partial(\beta_\rho\rho)}} \tag{9-80j}$$

9.5 DIELECTRIC WAVEGUIDES AND RESONATORS

Guided electromagnetic propagation by dielectric media has been studied since as early as the 1920s by well-known people such as Rayleigh, Sommerfeld, and Debye. Dielectric slabs, strips, and rods have been used as waveguides, resonators, and antennas. Since the 1960s a most well-known dielectric waveguide, the fiber optic cable [7–19], has received attention and has played a key role in the general area of communication. Although the subject is very lengthy and involved, we will consider here simplified theories that give the propagation characteristics of the cylindrical dielectric rod waveguide, fiber optic cable, and the cylindrical dielectric resonator. Extensive material on each of these topics and others can be found in the literature.

9.5.1 Circular Dielectric Waveguide

The cylindrical dielectric waveguide that will be examined here is that of circular cross section, as shown in Figure 9-8. It usually consists of a high permittivity (ε_d) central core dielectric of radius a surrounded by a lower dielectric cladding (which is usually air). For simplicity, we usually assume that both are perfect dielectrics with permeabilities equal to that of free space. Such a structure can support an infinite number of modes. However, for a given set of permittivities and radius a, only a finite number of unattenuated waveguide modes exist with their fields localized in the central dielectric core. Generally, the fields within a dielectric waveguide will be TE and/or TM, as was demonstrated in Section 8.7 for the dielectric slab waveguide. However, for the cylindrical dielectric rod of Figure 9-8 pure TE(H) or TM(E) modes exist only when the field configurations are symmetrical and independent of ϕ. Modes that exhibit angular ϕ variations cannot be pure TE or TM modes. Instead field configurations that are combinations of TE (or H) and TM (or E) modes can be nonsymmetrical and possess angular ϕ variations. Such modes are usually referred to as *hybrid modes*, and are usually designated by IEEE (formerly IRE) Standards [20] as HEM$_{mn}$. In general, mode nomenclature for circular dielectric waveguides and resonators is not

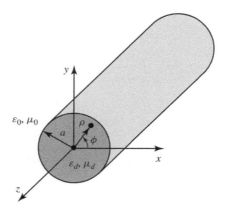

Figure 9-8 Geometry of circular dielectric waveguide.

well defined, and it is quite confusing. Another designation of the hybrid modes is to denote them as HE [when the TE(H) modes predominate] or EH [when the TM(E) modes predominate]. Pure TE or TM, or hybrid HEM (HE or EH) modes exhibit cutoff frequencies, below which unattenuating modes cannot propagate. The cutoff frequency is determined by the minimum electrical radius (a/λ) of the dielectric rod; for small values of a/λ the modes cannot propagate unattenuated within the rod. There is, however, one hybrid mode, namely the HEM$_{11}$ (HE$_{11}$), which does not have a cutoff frequency. Because of its zero cutoff frequency, it is referred to as the *dominant mode* and is most widely used in dielectric rod waveguides and end-fire antennas. This HE$_{11}$ mode is also popularly referred to as the *dipole mode*. An excellent reference on dielectric waveguides and resonators is that of [21].

The TE$_{mn}^z$ and TM$_{mn}^z$ electric and magnetic field components in a cylindrical waveguide are given, respectively, by (9-15a) through (9-15g) and (9-31a) through (9-31g). These expressions include both the cos ($m\phi$) and sin ($m\phi$) angular variations which, in general, both exist. The HEM modes that have only cos ($m\phi$) symmetry are combinations of the TE^z and TM^z modes which also exhibit cos ($m\phi$) symmetry. Use of the expressions (9-15a) through (9-15g) and (9-31a) through (9-31g), and selection of only the terms that possess simultaneously cos ($m\phi$) variations in the E_z component and sin ($m\phi$) variations in the H_z component, allows us to write that

$$\text{HEM Modes}\,(\rho \le a)$$

$$E_\rho^d = -j\frac{1}{\left(\beta_\rho^d\right)^2}\left[m\omega\mu_d A_m \frac{1}{\rho}J_m\left(\beta_\rho^d \rho\right) + \beta_z \beta_\rho^d B_m J_m'\left(\beta_\rho^d \rho\right)\right]\cos\left(m\phi\right)e^{-j\beta_z z} \qquad (9\text{-}81a)$$

$$E_\phi^d = j\frac{1}{\left(\beta_\rho^d\right)^2}\left[\omega\mu_d \beta_\rho^d A_m J_m'\left(\beta_\rho^d \rho\right) + m\beta_z B_m \frac{1}{\rho}J_m\left(\beta_\rho^d \rho\right)\right]\sin\left(m\phi\right)e^{-j\beta_z z} \qquad (9\text{-}81b)$$

$$E_z^d = B_m J_m\left(\beta_\rho^d \rho\right)\cos\left(m\phi\right)e^{-j\beta_z z} \qquad (9\text{-}81c)$$

$$H_\rho^d = -j\frac{1}{\left(\beta_\rho^d\right)^2}\left[\beta_z \beta_\rho^d A_m J_m'\left(\beta_\rho^d \rho\right) + m\omega\varepsilon_d B_m \frac{1}{\rho}J_m\left(\beta_\rho^d \rho\right)\right]\sin\left(m\phi\right)e^{-j\beta_z z} \qquad (9\text{-}81d)$$

$$H_\phi^d = -j\frac{1}{\left(\beta_\rho^d\right)^2}\left[m\beta_z A_m \frac{1}{\rho}J_m\left(\beta_\rho^d \rho\right) + \omega\varepsilon_d \beta_\rho^d B_m J_m'\left(\beta_\rho^d \rho\right)\right]\cos\left(m\phi\right)e^{-j\beta_z z} \qquad (9\text{-}81e)$$

$$H_z^d = A_m J_m\left(\beta_\rho^d \rho\right)\sin\left(m\phi\right)e^{-j\beta_z z} \qquad (9\text{-}81f)$$

where

$$\left(\beta_\rho^d\right)^2 + \beta_z^2 = \beta_d^2 = \omega^2\mu_d\varepsilon_d = \omega^2\mu_0\varepsilon_0\varepsilon_r\mu_r \qquad (9\text{-}81g)$$

$$A_m = -j\frac{\left(\beta_\rho^d\right)^2}{\omega\mu_d\varepsilon_d}A_{mn}D_2 \qquad (9\text{-}81h)$$

$$B_m = -j\frac{\left(\beta_\rho^d\right)^2}{\omega\mu_d\varepsilon_d}B_{mn}C_2 \qquad (9\text{-}81i)$$

$$' \equiv \frac{\partial}{\partial\left(\beta_\rho^d \rho\right)} \qquad (9\text{-}81j)$$

The coefficients A_m and B_m are not independent of each other, and their relationship can be found by applying the appropriate boundary conditions.

For the dielectric rod to act as a waveguide, the fields outside the rod ($\rho \geq a$) must be of the evanescent type that exhibit a decay in the radial direction. The rate of attenuation is a function of the diameter of the rod. As the diameter of the rod decreases, the following changes occur.

1. The attenuation lessens.
2. The distance to which the fields outside the rod can extend is greater.
3. The propagation constant β_z is only slightly greater than β_0.

As the diameter of the rod increases, the following changes occur.

1. The rate of attenuation also increases.
2. The fields are confined closer to the rod.
3. The propagation constant β_z approaches β_d.

Since in all cases β_z is greater than β_0, the phase velocity is smaller than the velocity of light in free space. For small-diameter rods, the surface waves are said to be *loosely bound* to the dielectric surface, whereas for the larger diameters, it is said to be *tightly bound* to the dielectric surface. Therefore, the cylindrical functions that are chosen to represent the radial variations of the fields outside the rod must be cylindrical decaying functions. These functions can be either Hankel functions of order m of the first kind ($H_m^{(1)}$) or second kind ($H_m^{(2)}$) and of imaginary argument, or modified Bessel functions K_m of the second kind of order m. We choose here to use the modified Bessel functions K_m of the second kind, which are related to the Hankel functions of the first and second kind by

$$K_m(\alpha) = \begin{cases} j^{m+1} \dfrac{\pi}{2} H_m^{(1)}(j\alpha) & \text{(9-82a)} \\[2mm] -j^{m+1} \dfrac{\pi}{2} H_m^{(2)}(-j\alpha) & \text{(9-82b)} \end{cases}$$

With (9-81a) through (9-81j), and (9-82a) and (9-82b) as a guide, we can represent the corresponding electric and magnetic field components for the HEM modes outside the dielectric rod ($\rho \geq a$) by

$$\underline{\text{HEM Modes}(\rho \geq a)}$$

$$E_\rho^0 = j \frac{1}{\left(\alpha_\rho^0\right)^2} \left[m\omega\mu_0 C_m \frac{1}{\rho} K_m\left(\alpha_\rho^0 \rho\right) + \beta_z \alpha_\rho^0 D_m K_m'\left(\alpha_\rho^0 \rho\right) \right] \cos(m\phi) e^{-j\beta_z z} \tag{9-83a}$$

$$E_\phi^0 = -j \frac{1}{\left(\alpha_\rho^0\right)^2} \left[\omega\mu_0 \alpha_\rho^0 C_m K_m'\left(\alpha_\rho^0 \rho\right) + m\beta_z D_m \frac{1}{\rho} K_m\left(\alpha_\rho^0 \rho\right) \right] \sin(m\phi) e^{-j\beta_z z} \tag{9-83b}$$

$$E_z^0 = D_m K_m\left(\alpha_\rho^0 \rho\right) \cos(m\phi) e^{-j\beta_z z} \tag{9-83c}$$

$$H_\rho^0 = j \frac{1}{\left(\alpha_\rho^0\right)^2} \left[\beta_z \alpha_\rho^0 C_m K_m'\left(\alpha_\rho^0 \rho\right) + m\omega\varepsilon_0 D_m \frac{1}{\rho} K_m\left(\alpha_\rho^0 \rho\right) \right] \sin(m\phi) e^{-j\beta_z z} \tag{9-83d}$$

$$H_\phi^0 = j \frac{1}{\left(\alpha_\rho^0\right)^2} \left[m\beta_z C_m \frac{1}{\rho} K_m\left(\alpha_\rho^0 \rho\right) + \omega\varepsilon_0 \alpha_\rho^0 D_m K_m'\left(\alpha_\rho^0 \rho\right) \right] \cos(m\phi) e^{-j\beta_z z} \tag{9-83e}$$

$$H_z^0 = C_m K_m\left(\alpha_\rho^0 \rho\right) \sin(m\phi) e^{-j\beta_z z} \tag{9-83f}$$

where

$$\left(j\alpha_\rho^0\right)^2 + \beta_z^2 = -\left(\alpha_\rho^0\right)^2 + \beta_z^2 = \beta_0^2 = \omega^2 \mu_0 \varepsilon_0 \tag{9-83g}$$

$$' \equiv \frac{\partial}{\partial\left(\alpha_\rho^0 \rho\right)} \tag{9-83h}$$

The coefficients C_m and D_m are not independent of each other or from A_m and B_m, and their relations can be found by applying the appropriate boundary conditions.

The relations between the constants A_m, B_m, C_m, and D_m and equation 9-91, which is referred to as the *eigenvalue equation*, can be used to determine the modes that can be supported by the dielectric rod waveguide. These are obtained by applying the following boundary conditions

$$E_\phi^d (\rho = a, 0 \le \phi \le 2\pi, z) = E_\phi^0 (\rho = a, 0 \le \phi \le 2\pi, z) \tag{9-84a}$$

$$E_z^d (\rho = a, 0 \le \phi \le 2\pi, z) = E_z^0 (\rho = a, 0 \le \phi \le 2\pi, z) \tag{9-84b}$$

$$H_\phi^d (\rho = a, 0 \le \phi \le 2\pi, z) = H_\phi^0 (\rho = a, 0 \le \phi \le 2\pi, z) \tag{9-84c}$$

$$H_z^d (\rho = a, 0 \le \phi \le 2\pi, z) = H_z^0 (\rho = a, 0 \le \phi \le 2\pi, z) \tag{9-84d}$$

Doing this leads to

$$\frac{1}{\left(\beta_\rho^d\right)^2}\left[\omega\mu_d\beta_\rho^d A_m J_m'\left(\beta_\rho^d a\right) + m\beta_z B_m \frac{1}{a} J_m\left(\beta_\rho^d a\right)\right]$$

$$= -\frac{1}{\left(\alpha_\rho^0\right)^2}\left[\omega\mu_0\alpha_\rho^0 C_m K_m'\left(\alpha_\rho^0 a\right) + m\beta_z D_m \frac{1}{a} K_m\left(\alpha_\rho^0 a\right)\right] \tag{9-85a}$$

$$B_m J_m\left(\beta_\rho^d a\right) = D_m K_m\left(\alpha_\rho^0 a\right) \tag{9-85b}$$

$$-\frac{1}{\left(\beta_\rho^d\right)^2}\left[m\beta_z A_m \frac{1}{a} J_m\left(\beta_\rho^d a\right) + \omega\varepsilon_d\beta_\rho^d B_m J_m'\left(\beta_\rho^d a\right)\right]$$

$$= \frac{1}{\left(\alpha_\rho^0\right)^2}\left[m\beta_z C_m \frac{1}{a} K_m\left(\alpha_\rho^0 a\right) + \omega\varepsilon_0\alpha_\rho^0 D_m K_m'\left(\alpha_\rho^0 a\right)\right] \tag{9-85c}$$

$$A_m J_m\left(\beta_\rho^d a\right) = C_m K_m\left(\alpha_\rho^0 a\right) \tag{9-85d}$$

where according to (9-81g) and (9-83g)

$$\left(\beta_\rho^d\right)^2 + \beta_z^2 = \beta_d^2 \;\Rightarrow\; \left(\beta_\rho^d a\right)^2 + (\beta_z a)^2 = (\beta_d a)^2 \;\Rightarrow\; \beta_z a = \sqrt{(\beta_d a)^2 - \left(\beta_\rho^d a\right)^2} \tag{9-85e}$$

$$-\left(\alpha_\rho^0\right)^2 + \beta_z^2 = \beta_0^2 \;\Rightarrow\; -\left(\alpha_\rho^0 a\right)^2 + (\beta_z a)^2 = (\beta_0 a)^2 \tag{9-85f}$$

Subtracting (9-85f) from (9-85e) we get that

$$\left(\beta_\rho^d a\right)^2 + \left(\alpha_\rho^0 a\right)^2 = (\beta_d a)^2 - (\beta_0 a)^2 \;\Rightarrow\; \alpha_\rho^0 a = \sqrt{(\beta_d a)^2 - (\beta_0 a)^2 - \left(\beta_\rho^d a\right)^2}$$

$$= \sqrt{(\beta_0 a)^2(\varepsilon_r \mu_r - 1) - \left(\beta_\rho^d a\right)^2} \tag{9-86}$$

Using the abbreviated notation of

$$\chi = \beta_\rho^d a \tag{9-87a}$$

$$\xi = \alpha_\rho^0 a \tag{9-87b}$$

$$\zeta = \beta_z a \tag{9-87c}$$

$$\beta_d = \beta_0 \sqrt{\varepsilon_r \mu_r} \tag{9-87d}$$

we can rewrite (9-85e) and (9-86) as

$$\zeta = \sqrt{(\beta_0 a)^2 \varepsilon_r \mu_r - \chi^2} \tag{9-88a}$$

$$\xi = \sqrt{(\beta_0 a)^2 (\varepsilon_r \mu_r - 1) - \chi^2} \tag{9-88b}$$

With the preceding abbreviated notation and with $\mu_d = \mu_0$ (9-85a) through (9-85d) can be written in matrix form as

$$Fg = 0 \tag{9-89}$$

where F is a 4×4 matrix and g is a column matrix. Each is given by

$$
F =
\begin{bmatrix}
\dfrac{\omega \mu_0 a}{\chi} J_m'(\chi) & \dfrac{m\zeta}{\chi^2} J_m(\chi) & \dfrac{\omega \mu_0 a}{\xi} K_m'(\xi) & \dfrac{m\zeta}{\xi^2} K_m(\xi) \\[2ex]
0 & J_m(\chi) & 0 & -K_m(\xi) \\[2ex]
\dfrac{m\zeta}{\chi^2} J_m(\chi) & \dfrac{\omega \varepsilon_d a}{\chi} J_m'(\chi) & \dfrac{m\zeta}{\xi^2} K_m(\xi) & \dfrac{\omega \varepsilon_0 a}{\xi} K_m'(\xi) \\[2ex]
J_m(\chi) & 0 & -K_m(\xi) & 0
\end{bmatrix}
\tag{9-89a}
$$

$$
g =
\begin{bmatrix}
A_m \\
B_m \\
C_m \\
D_m
\end{bmatrix}
\tag{9-89b}
$$

Equation 9-89 has a nontrivial solution provided that the determinant of F of (9-89a) is equal to zero [i.e., $\det(F) = 0$]. Applying this to (9-89a), it can be shown that it leads to [21]

$$
|F| = \frac{\omega^2 \mu_0 \varepsilon_d a^2}{\chi^2} [J_m'(\chi)]^2 [K_m(\xi)]^2 + \frac{\omega^2 \mu_0 \varepsilon_0 a^2}{\chi \xi} J_m(\chi) J_m'(\chi) K_m(\xi) K_m'(\xi)
$$

$$
- \frac{(m\zeta)^2}{\chi^4} [J_m(\chi)]^2 [K_m(\xi)]^2 - \frac{(m\zeta)^2}{\chi^2 \xi^2} [J_m(\chi)]^2 [K_m(\xi)]^2
$$

$$
+ \frac{\omega^2 \mu_0 \varepsilon_d a^2}{\chi \xi} J_m(\chi) J_m'(\chi) K_m(\xi) K_m'(\xi) + \frac{\omega^2 \mu_0 \varepsilon_0 a^2}{\xi^2} [J_m(\chi)]^2 [K_m'(\xi)]^2
$$

$$
- \frac{(m\zeta)^2}{\chi^2 \xi^2} [J_m(\chi)]^2 [K_m(\xi)]^2 - \frac{(m\zeta)^2}{\xi^4} [J_m(\chi)]^2 [K_m(\xi)]^2 = 0 \tag{9-90}
$$

Dividing all the terms of (9-90) by $\omega^2 \mu_0 \varepsilon_d a^2 [K_m(\xi)]^2$ and regrouping, it can be shown that (9-90) can be placed in the form of [21]

$$G_1(\chi) G_2(\chi) - G_3^2(\chi) = 0 \tag{9-91}$$

where

$$G_1(\chi) = \frac{J_m'(\chi)}{\chi} + \frac{K_m'(\xi)J_m(\chi)}{\varepsilon_r \xi K_m(\xi)} \qquad (9\text{-}91\text{a})$$

$$G_2(\chi) = \frac{J_m'(\chi)}{\chi} + \frac{K_m'(\xi)J_m(\chi)}{\xi K_m(\xi)} \qquad (9\text{-}91\text{b})$$

$$G_3(\chi) = \frac{m\zeta}{\beta_0 a\sqrt{\varepsilon_r}} J_m(\chi) \left(\frac{1}{\chi^2} + \frac{1}{\xi^2}\right) \qquad (9\text{-}91\text{c})$$

$$\zeta = \sqrt{(\beta_0 a)^2 \varepsilon_r - \chi^2} \qquad (9\text{-}91\text{d})$$

$$\xi = \sqrt{(\beta_0 a)^2 (\varepsilon_r - 1) - \chi^2} \qquad (9\text{-}91\text{e})$$

Equation 9-91 is referred to as the *eigenvalue equation* for the dielectric rod waveguide. The values of χ that are solutions to (9-91) are referred to as the *eigenvalues* for the dielectric rod waveguide.

In order for $\xi = \alpha_\rho^0 a$ to remain real and represent decaying fields outside the dielectric rod waveguide, the values of $\chi = \beta_\rho^d a$ should not exceed a certain maximum value. From (9-88b) this maximum value $\chi_{\max}$ is equal to

$$\chi_{\max} = \left(\beta_\rho^d a\right)_{\max} = \beta_0 a\sqrt{\varepsilon_r - 1} = \omega a\sqrt{\mu_0 \varepsilon_0 (\varepsilon_r - 1)} \qquad (9\text{-}92)$$

For values of $\beta_\rho^d a$ greater than $\chi_{\max}$ (see above), the values of $\xi = \alpha_\rho^0 a$ become imaginary and according to (9-82b), the modified Bessel function of the second kind is reduced to a Hankel function of the second kind that represents unattenuated outwardly traveling waves. In this case, the dielectric rod is acting as a cylindrical antenna because of energy loss from its side. Therefore, for a given value of m, there are a finite number n of χ_{mn}'s (eigenvalues) for which the dielectric rod acts as a waveguide. Each combination of allowable values of m, n that determine a given eigenvalue χ_{mn} represent the hybrid mode HEM_{mn}. HEM modes with odd values of the second subscript correspond to HE modes, whereas HEM modes with even values of the second subscript correspond to EH modes. Thus, $\text{HEM}_{m,2n-1}$ $(n = 1, 2, 3, \ldots)$ correspond to HE_{mn} modes and $\text{HEM}_{m,2n}$ $(n = 1, 2, 3, \ldots)$ correspond to EH_{mn} modes. According to (9-88a), if the values of χ exceed $\beta_0 a\sqrt{\varepsilon_r}$ (i.e., $\chi > \beta_0 a\sqrt{\varepsilon_r}$), then $\zeta = \beta_z a$ becomes imaginary and the waves in the dielectric rod become decaying (evanescent) along the axis (z direction) of the rod.

The allowable modes in a dielectric rod waveguide are determined by finding the values of χ, denoted by χ_{mn}, that are solutions to the transcendental equation 9-91. For each value of m there are a finite number of values of n $(n = 1, 2, 3, \ldots)$. Examining (9-91), it is evident that for $m = 0$, the left side of (9-91) vanishes when

$$\text{TM}_{0n}$$

$$G_1(\chi_{0n}) = \frac{J_0'(\chi_{0n})}{\chi_{0n}} + \frac{K_0'(\xi_{0n})J_0(\chi_{0n})}{\varepsilon_r \xi_{0n} K_0(\xi_{0n})} = -\frac{J_1(\chi_{0n})}{\chi_{0n}} - \frac{K_1(\xi_{0n})J_0(\chi_{0n})}{\varepsilon_r \xi_{0n} K_0(\xi_{0n})} = 0 \qquad (9\text{-}93\text{a})$$

$$\text{TE}_{0n}$$

$$G_2(\chi_{0n}) = \frac{J_0'(\chi_{0n})}{\chi_{0n}} + \frac{K_0'(\xi_{0n})J_0(\chi_{0n})}{\xi_{0n} K_0(\xi_{0n})} = -\frac{J_1(\chi_{0n})}{\chi_{0n}} - \frac{K_1(\xi_{0n})J_0(\chi_{0n})}{\xi_{0n} K_0(\xi_{0n})} = 0 \qquad (9\text{-}93\text{b})$$

since $G_3(\chi_{0n}) = 0$, $J_0'(\chi_{0n}) = -J_1(\chi_{0n})$, and $K_0'(\xi_{0n}) = -K_1(\xi_{0n})$. Equation 9-93a is valid for TM_{0n} modes and (9-93b) is applicable for TE_{0n} modes.

The nonlinear equations 9-91 through 9-91e can be solved for the values of χ_{mn} iteratively by assuming values of χ_{mn} examining the sign changes of (9-91). It should be noted that in the allowed range of χ_{mn}'s, G_1, G_2, and G_3 are nonsingular. Computed values of χ_{mn} as a function of $\beta_0 a$ for different HEM_{mn} modes are shown in Figure 9-9 for $\varepsilon_r = 20$ and in Figure 9-10 for $\varepsilon_r = 38$ [21]. It should be noted that for a given mn mode the values of χ_{mn} are nonconstant and vary as a function of the electrical radius of the rod. This is in contrast to the circular waveguide with conducting walls, whose χ_{mn} or χ'_{mn} values (in Tables 9-1 and 9-2) are constant for a given mode.

Once the values of χ_{mn} for a given mode have been found, the corresponding values of $\zeta = \beta_z a$ can be computed using (9-91d). When this is done for a dielectric rod waveguide of polystyrene ($\varepsilon_r = 2.56$), the values of β_z / β_0 for the HE_{11} mode as a function of the radius of the rod are shown in Figure 9-11 [10]. It is apparent that the HE_{11} mode does not possess a cutoff. In the same figure, the values of β_z / β_0 for the axially symmetric TE_{01} and TM_{01} surface wave modes, which possess a finite cutoff, are also displayed. Although in principle the HE_{11} mode has zero cutoff, the rate of attenuation exhibited by the fields outside the slab decreases as the radius of the rod becomes smaller and the wavenumber β_z approaches β_0. Thus, for small radii, the fields outside the rod extend to large distances and are said to be loosely bound to the surface. Practically, then a minimum radius rod is usually utilized, which results in a small but finite cutoff [22]. For the larger dielectric constant material, the fields outside the dielectric waveguide are more tightly bound to the surface since larger values of β_z translate to larger values of the attenuation coefficient α_ρ^0 through (9-83g).

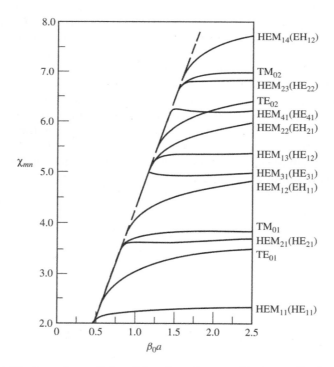

Figure 9-9   The first 13 eigenvalues of the dielectric rod waveguide ($\varepsilon_r = 20$). (Source: D. Kajfez and P. Guillon (Eds.), *Dielectric Resonators*, 1986, Artech House, Inc.)

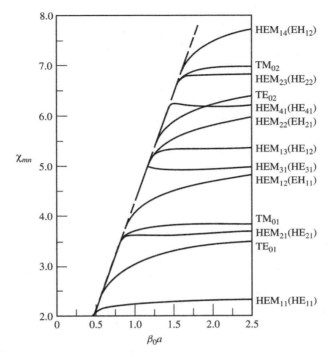

Figure 9-10 The first 13 eigenvalues of the dielectric rod waveguide ($\varepsilon_r = 38$). (Source: D. Kajfez and P. Guillon (Eds.), *Dielectric Resonators*, 1986, Artech House, Inc.)

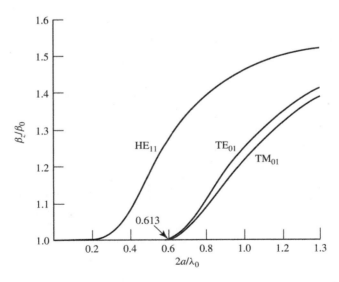

Figure 9-11 Ratio of β_z/β_0, for first three surface-wave modes on a polystyrene rod ($\varepsilon_r = 2.56$). (Source: R. E. Collin, *Field Theory of Guided Waves*, 1960, McGraw-Hill Book Co.)

Typical field patterns in the central core for the HE_{11}, TE_{01}, TM_{01}, HE_{21}, EH_{11}, and HE_{31} modes are shown in Figure 9-12 [23]. Both *E*- and *H*-field lines are displayed over the cross section of the central core and over a cutaway a distance $\lambda_g/2$ along its length. For all the plots, the ratio $\tau = \chi^2/(\chi^2 + \xi^2) = (\beta_\rho^d)^2/[(\beta_\rho^d)^2 + (\alpha_\rho^0)^2] = 0.1$.

For all modes the sum of $\chi^2 + \xi^2$ should be a constant, which will be a function of the radius of the dielectric rod and its dielectric constant. Thus, according to (9-88b),

$$\chi^2 + \xi^2 = (\beta_0 a)^2 (\varepsilon_r - 1) = \text{constant} \tag{9-94}$$

For all modes, excluding the dominant HE_{11}, this constant should always be equal to or greater than $(2.4049)^2 = 5.7835$; that is, excluding the HE_{11} mode,

$$V^2 = \chi^2 + \xi^2 = (\beta_0 a)^2 (\varepsilon_r - 1) \geq (2.4049)^2 \tag{9-94a}$$

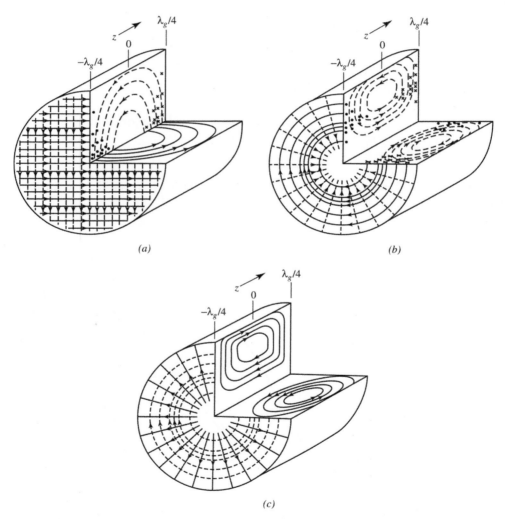

Figure 9-12 Field patterns in the central core of a dielectric rod waveguide (in all cases $\tau = 0.1$; **E**:———, **H**: ----). (a) HEM_{11} (HE_{11}) mode. (b) TE_{01} mode. (c) TM_{01} mode. (Source: T. Okoshi, *Optical Fibers*, 1982, Academic Press.)

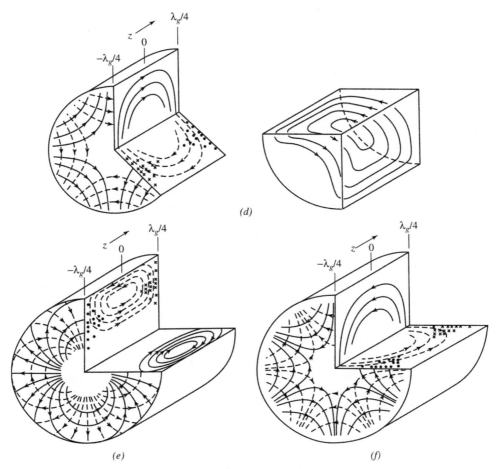

Figure 9-12 (d) HEM$_{21}$ (HE$_{21}$) mode. (e) HEM$_{12}$ (EH$_{11}$) mode. (f) HEM$_{31}$ (HE$_{31}$) mode.

or

$$\frac{2a}{\lambda_0} \geq \frac{1}{\pi} \frac{2.4049}{\sqrt{\varepsilon_r - 1}} \tag{9-94b}$$

The value of 2.4049 is used because one of the next higher-order modes is the TM$_{01}$ mode which, according to Table 9-2, is $\chi_{mn} = \chi_{01} = 2.4049$. For values of $2a/\lambda_0$ smaller than that of (9-94b) only the dominant HE$_{11}$ dipole mode exists as shown in Figure 9-11.

For a dielectric rod waveguide, the first 12 modes (actually first 20 modes if the twofold degeneracy for the HE$_{mn}$ or EH$_{mn}$ is counted), in order of ascending cutoff frequency, along with the vanishing Bessel function and its argument at cutoff, are given:

Mode(s)	$J_m(\chi_{mn}) = 0$	χ_{mn} at cutoff	Total number of propagating modes
HE_{11} (HEM_{11})	$J_1(\chi_{10}) = 0$	$\chi_{10} = 0$	2
TE_{01}, TM_{01}, HE_{21} (HEM_{21})	$J_0(\chi_{01}) = 0$	$\chi_{01} = 2.4049$	6
HE_{12} (HEM_{13}), EH_{11} (HEM_{12}), HE_{31} (HEM_{31})	$J_1(\chi_{11}) = 0$	$\chi_{11} = 3.8318$	12
EH_{21} (HEM_{22}), HE_{41} (HEM_{41})	$J_2(\chi_{21}) = 0$	$\chi_{21} = 5.1357$	16
TE_{02}, TM_{02}, HE_{22} (HEM_{23})	$J_0(\chi_{02}) = 0$	$\chi_{02} = 5.5201$	20

According to (9-94a), a single dominant HE_{11} mode can be maintained within the rod provided the normalized central core radius $V < 2.4049$. This can be accomplished by making the radius a of the central core small and/or choosing, between the central core and the cladding, a small dielectric constant ε_r. However, the smaller the size of the central core, the smaller the rate of attenuation and the less tightly the field outside it is attached to its surface. The normalized diameter over which the $e^{-1} = 0.3679 = 36.79\%$ field point outside the central core extends is shown plotted in Figure 9-13 [16]. Although the fundamental HE_{11} mode does not cutoff as the core diameter shrinks, the fields spread out beyond the physical core and become loosely bound.

9.5.2 Circular Dielectric Resonator

Dielectric resonators are unmetalized dielectric objects (spheres, disks, parallelepipeds, etc.) of high dielectric constant (usually ceramic) and high quality factor Q that can function as energy storage devices. Dielectric resonators were first introduced in 1939 by Richtmyer [24], but for

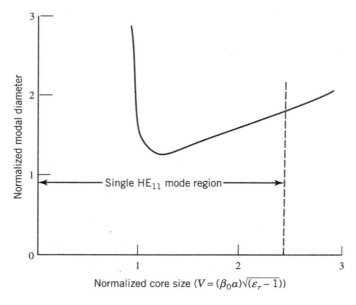

Figure 9-13   Normalized modal diameter as a function of normalized core size for dielectric rod waveguide. (Source: T. G. Giallorenzi, "Optical communications research and technology: Fiber optics," *Proc. IEEE*, © 1978, IEEE.)

almost 25 years his theoretical work failed to generate a continuous and prolonged interest. The introduction in the 1960s of material, such as rutile, of high dielectric constant (around 100) renewed the interest in dielectric resonators [25–30]. However, the poor temperature stability of rutile resulted in large resonant frequency changes and prevented the development of practical microwave components. In the 1970s, low-loss and temperature-stable ceramics, such as barium tetratitanate and $(Zr-Sn)TiO_4$ were introduced and were used for the design of high-performance microwave components such as filters and oscillators. Because dielectric resonators are small, lightweight, temperature stable, high Q, and low cost, they are ideal for design and fabrication of monolithic microwave integrated circuits (MMICs) and general semiconductor devices. Such technology usually requires high Q miniature elements to design and fabricate highly stable frequency oscillators and high-performance narrowband filters. Thus, dielectric resonators have replaced traditional waveguide resonators, especially in MIC applications, and implementations as high as 94 GHz have been reported. The development of higher dielectric constant material (80 or higher) with stable temperature and low-loss characteristics will have a significant impact on MIC design using dielectric resonators.

In order for the dielectric resonator to function as a resonant cavity, the dielectric constant of the material must be large (usually 30 or greater). Under those conditions, the dielectric–air interface acts almost as an open circuit which causes internal reflections and results in the confinement of energy in the dielectric material, thus creating a resonant structure. The plane wave reflection coefficient at the dielectric–air interface is equal to

$$\Gamma = \frac{\eta_0 - \eta}{\eta_0 + \eta} = \frac{\sqrt{\frac{\mu_0}{\varepsilon_0}} - \sqrt{\frac{\mu_0}{\varepsilon}}}{\sqrt{\frac{\mu_0}{\varepsilon_0}} + \sqrt{\frac{\mu_0}{\varepsilon}}} = \frac{\sqrt{\frac{\varepsilon}{\varepsilon_0}} - 1}{\sqrt{\frac{\varepsilon}{\varepsilon_0}} + 1} = \frac{\sqrt{\varepsilon_r} - 1}{\sqrt{\varepsilon_r} + 1} \overset{\varepsilon_r \to \text{large}}{\simeq} +1 \tag{9-95}$$

and it approaches the value of $+1$ as the dielectric constant becomes very large. Under these conditions, the dielectric–air interface can be approximated by a hypothetical perfect magnetic conductor (PMC), which requires that the tangential components of the magnetic field (or normal components of the electric field) vanish (in contrast to the perfect electric conductor, PEC, which requires that the tangential electric field components, or normal components of the magnetic field, vanish). This, of course, is a well known and widely used technique in solving boundary-value electromagnetic problems. It is, however, a first-order approximation, although it usually leads to reasonable results. The magnetic wall model can be used to analyze both the dielectric waveguide and dielectric resonant cavity. Improvements to the magnetic wall approximation have been introduced and resulted in improved data [31, 32].

Although the magnetic wall modeling may not lead to the most accurate data, it will be utilized here because it is simple and instructive not only as a first-order approximation to this problem but also to other problems including antennas (e.g., microstrip antenna). The geometry of the dielectric resonator is that of Figure 9-14a, whose surface is modeled with the PMC walls of Figure 9-14b, which are represented by the independent boundary conditions

$$H_\phi \left(\rho = a, 0 \leq \phi \leq 2\pi, 0 \leq z \leq h \right) = 0 \tag{9-96a}$$

$$H_\phi \left(0 \leq \rho \leq a, 0 \leq \phi \leq 2\pi, z = 0 \right) = H_\phi \left(0 \leq \rho \leq a, 0 \leq \phi \leq 2\pi, z = h \right) = 0 \tag{9-96b}$$

or

$$H_z \left(\rho = a, 0 \leq \phi \leq 2\pi, 0 \leq z \leq h \right) = 0 \tag{9-97a}$$

$$H_\rho \left(0 \leq \rho \leq a, 0 \leq \phi \leq 2\pi, z = 0 \right) = H_\rho \left(0 \leq \rho \leq a, 0 \leq \phi \leq 2\pi, z = h \right) = 0 \tag{9-97b}$$

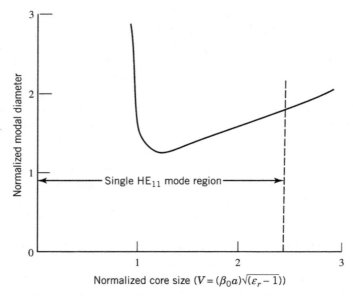

Figure 9-14 Circular dielectric resonator and its PMC modeling. (*a*) Dielectric resonator. (*b*) PMC modeling.

Either of the preceding sets leads to the same results. Since all the boundary conditions of the first set involve only the H_ϕ component, they will be applied here.

A. TE^z Modes   The TE^z modes can be constructed using the vector potential F_z component of (9-35), or

$$F_z(\rho,\phi,z) = A_{mn}J_m\left(\beta_\rho^d\rho\right)\left[C_2\cos(m\phi) + D_2\sin(m\phi)\right]\left[C_3\cos(\beta_z z) + D_3\sin(\beta_z z)\right] \quad (9\text{-}98)$$

$$\text{where} \quad m = 0, 1, 2, \dots \quad (9\text{-}98a)$$

The H_ϕ component is obtained using (6-80), or

$$H_\phi = -j\frac{1}{\omega_r\mu_d\varepsilon_d}\frac{1}{\rho}\frac{\partial^2 F_z}{\partial\phi\,\partial z} = -jA_{mn}\frac{m\beta_z}{\omega_r\mu_d\varepsilon_d}\frac{1}{\rho}J_m\left(\beta_\rho^d\rho\right)\left[-C_2\sin(m\phi) + D_2\cos(m\phi)\right]$$
$$\times\left[-C_3\sin(\beta_z z) + D_3\cos(\beta_z z)\right] \quad (9\text{-}99)$$

Applying (9-96a) leads to

$$H_\phi(\rho = a, 0 \le \phi \le 2\pi, 0 \le z \le h)$$
$$= -jA_{mn}\frac{m\beta_z}{\omega_r\mu_d\varepsilon_d}\frac{1}{a}J_m\left(\beta_\rho^d a\right)\left[-C_2\sin(m\phi) + D_2\cos(m\phi)\right]$$
$$\times\left[-C_3\sin(\beta_z z) + D_3\cos(\beta_z z)\right] = 0 \;\Rightarrow\; J_m(\beta_\rho^d a) = 0 \;\Rightarrow\; \beta_\rho^d a = \chi_{mn}$$
$$\beta_\rho^d = \left(\frac{\chi_{mn}}{a}\right) \quad (9\text{-}100)$$

where χ_{mn} represents the zeroes of the Bessel function of order m, many of which are found in Table 9-2. In a similar manner, the first boundary condition of (9-96b) leads to

$$H_\phi(0 \leq \rho \leq a, 0 \leq \phi \leq 2\pi, z = 0)$$
$$= -jA_{mn} \frac{m\beta_z}{\omega_r \mu_d \varepsilon_d} \frac{1}{\rho} J_m(\beta_\rho^d \rho)[-C_2 \sin(m\phi) + D_2 \cos(m\phi)]$$
$$\times [-C_3(0) + D_3(1)] = 0 \quad \Rightarrow \quad D_3 = 0 \tag{9-100a}$$

while the second boundary condition of (9-96b) leads to

$$H_\phi(0 \leq \rho \leq a, 0 \leq \phi \leq 2\pi, z = h)$$
$$= -jA_{mn} \frac{m\beta_z}{\omega_r \mu_d \varepsilon_d} \frac{1}{\rho} J_m(\beta_\rho^d \rho)$$
$$\times [-C_2 \sin(m\phi) + D_2 \cos(m\phi)][-C_3 \sin(\beta_z h)] = 0$$
$$\Rightarrow \sin(\beta_z h) = 0 \quad \Rightarrow \quad \beta_z h = p\pi$$
$$\beta_z = \frac{p\pi}{h}, \quad p = 0, 1, 2, \dots \tag{9-100b}$$

Using (9-100) and (9-100b), the resonant frequency is obtained by applying (9-4a) at resonance, that is,

$$\beta_r = \omega_r \sqrt{\mu_d \varepsilon_d} = 2\pi f_r \sqrt{\mu_d \varepsilon_d} = \sqrt{\left(\frac{\chi_{mn}}{a}\right)^2 + \left(\frac{p\pi}{h}\right)^2}$$

or

$$\boxed{(f_r)_{mnp}^{TE^z} = \frac{1}{2\pi\sqrt{\mu_d \varepsilon_d}} \sqrt{\left(\frac{\chi_{mn}}{a}\right)^2 + \left(\frac{p\pi}{h}\right)^2}} \qquad \begin{array}{l} m = 0, 1, 2, \dots \\ n = 1, 2, 3, \dots \\ p = 0, 1, 2, \dots \end{array} \tag{9-101}$$

The dominant TE_{mnp}^z mode is the TE_{010}^z whose resonant frequency is equal to

$$(f_r)_{010}^{TE^z} = \frac{\chi_{01}}{2\pi a\sqrt{\mu_d \varepsilon_d}} = \frac{2.4049}{2\pi a\sqrt{\mu_d \varepsilon_d}} \tag{9-101a}$$

B. TM^z Modes   The TM^z modes are obtained using a similar procedure as for the TE^z modes, but starting with the vector potential of (9-41). Doing this leads to the resonant frequency of

$$\boxed{(f_r)_{mnp}^{TM^z} = \frac{1}{2\pi\sqrt{\mu_d \varepsilon_d}} \sqrt{\left(\frac{\chi'_{mn}}{a}\right)^2 + \left(\frac{p\pi}{h}\right)^2}} \qquad \begin{array}{l} m = 0, 1, 2, \dots \\ n = 1, 2, 3, \dots \\ p = 1, 2, 3, \dots \end{array} \tag{9-102}$$

where χ'_{mn} are the zeroes of the derivative of the Bessel function of order m, a partial list of which is found on Table 9-1. The dominant TM^z_{mnp} mode is the TM^z_{111} mode whose resonant frequency is equal to

$$(f_r)^{\text{TM}^z}_{111} = \frac{1}{2\pi\sqrt{\mu_d\varepsilon_d}}\sqrt{\left(\frac{\chi'_{11}}{a}\right)^2 + \left(\frac{\pi}{h}\right)^2} = \frac{1}{2\pi\sqrt{\mu_d\varepsilon_d}}\sqrt{\left(\frac{1.8412}{a}\right)^2 + \left(\frac{\pi}{h}\right)^2} \qquad (9\text{-}102a)$$

A comparison of the resonant frequencies of (9-101) and (9-102) of the circular dielectric resonator modeled by the PMC surface with those of (9-39a) and (9-45) for the circular waveguide resonator with PEC surface shows that the TE^z_{mnp} of one are the TM^z_{mnp} of the other, and vice versa.

C. TE$_{01\delta}$ Mode   Although the dominant mode of the circular dielectric resonator as predicted by the PMC modeling is the TE$_{010}$ mode (provided $h/a < 2.03$), in practice the mode most often used is the TE$_{01\delta}$ where δ is a noninteger value less than 1. This mode can be modeled using Figure 9-15. For resonators with PEC or PMC walls, the third subscript is always an integer (including zero), and it represents the number of half-wavelength variations the field undergoes in the z direction. Since δ is a noninteger less than unity, the dielectric resonator field of a TE$_{01\delta}$ mode undergoes less than one half-wavelength variation along its length h. More accurate modelings of the dielectric resonator [31, 33, 34] indicate that $\beta_z h$ of (9-100b) is

$$\beta_z h \simeq \frac{\psi_1}{2} + \frac{\psi_2}{2} + q\pi, \quad q = 0, 1, 2, \ldots \qquad (9\text{-}103)$$

where

$$\frac{\psi_1}{2} = \tan^{-1}\left[\frac{\alpha_1}{\beta_z}\coth(\alpha_1 h_1)\right] \qquad (9\text{-}103a)$$

$$\frac{\psi_2}{2} = \tan^{-1}\left[\frac{\alpha_2}{\beta_z}\coth(\alpha_2 h_2)\right] \qquad (9\text{-}103b)$$

$$\alpha_1 = \sqrt{\left(\frac{\chi_{01}}{a}\right)^2 - \beta_0^2\varepsilon_{r1}} \qquad (9\text{-}103c)$$

$$\alpha_2 = \sqrt{\left(\frac{\chi_{01}}{a}\right)^2 - \beta_0^2\varepsilon_{r2}} \qquad (9\text{-}103d)$$

When $q = 0$ in (9-103), the mode is the TE$_{01\delta}$ where δ is a noninteger less than unity given by

$$\delta = \frac{1}{\pi}\left(\frac{\psi_1}{2} + \frac{\psi_2}{2}\right) \qquad (9\text{-}104)$$

and it signifies the variations of the field between the ends of the dielectric resonator at $z = 0$ and $z = h$. The preceding equations have been derived by modeling the dielectric resonator as shown in Figure 9-15 where PEC plates have been placed at $z = -h_2$ and $z = h + h_1$ [21]. The medium in regions 1–5 is usually taken to be air. Thus, the dielectric resonator has been sandwiched between two PEC plates, each placed at distances h_1 and h_2 from each of its ends. The distances h_1 and h_2 can be chosen to be $0 \le h_1, h_2 \le \infty$.

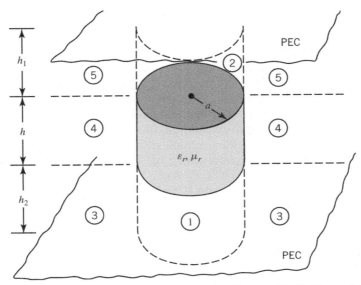

Figure 9-15 Modeling of circular dielectric rod resonator. (Source: D. Kajfez and P. Guillon (Eds.), *Dielectric Resonators*, 1986, Artech House, Inc.)

Example 9-5

Find the resonant $TE_{01\delta}$ mode and its resonant frequency when the distances h_1 and h_2 of Figure 9-15 are both zero ($h_1 = h_2 = 0$); that is, the dielectric resonator has been sandwiched between two PEC plates, each plate touching each of the ends of the dielectric resonator. Assume the resonator has radius of 5.25 mm, height of 4.6 mm, and dielectric constant of $\varepsilon_r = 38$.

Solution: Since $h_1 = h_2 = 0$, then according to (9-103a) and (9-103b)

$$\frac{\psi_1}{2} = \frac{\psi_2}{2} = \tan^{-1}(\infty) = \frac{\pi}{2}$$

Thus, δ of (9-104) is equal to

$$\delta = \frac{1}{\pi}\left(\frac{\pi}{2} + \frac{\pi}{2}\right) = 1$$

and the resonant $TE_{01\delta}$ mode is the TE_{011}. For this mode $\beta_z h$ is, according to (9-103) with $p = 0$, equal to

$$\beta_z h = \pi \Rightarrow \beta_z = \frac{\pi}{h}$$

Its resonant frequency is identical to (9-101) of the PMC modeling, which reduces to

$$(f_r)_{011}^{TE^z} = \frac{1}{2\pi\sqrt{\mu_d \varepsilon_d}}\sqrt{\left(\frac{2.4049}{a}\right)^2 + \left(\frac{\pi}{h}\right)^2}$$

$$= \frac{3 \times 10^{11}}{2\pi\sqrt{38}}\sqrt{\left(\frac{2.4049}{5.25}\right)^2 + \left(\frac{\pi}{4.6}\right)^2} = 6.37 \text{ GHz}$$

When compared to the exact value of 4.82 GHz for a dielectric resonator immersed in free space, the preceding value has an error of +32%. A more accurate result of 4.60 GHz (error of −4.8%) can be obtained using (9-103a)–(9-104) by placing the two PECs at infinity ($h_1 = h_2 = \infty$); then the resonator is isolated in free space.

9.5.3   Optical Fiber Cable

Optical communications, which initially started as a speculative research activity, has evolved into a very practical technique that has brought new dimensions to miniaturization, data handling capabilities, and signal processing methods. This success has been primarily attributed to the development of fiber optics cables, which in 1970 exhibited attenuations of 20 dB/km and in the early 1980s reduced to less than 1 dB/km. The development of suitable solid state diode sources and detectors has certainly eliminated any remaining insurmountable technological barriers and paved the way for widespread implementation of optical communication techniques for commercial and military applications. This technology has spread to the development of integrated optics which provide even greater miniaturization of optical systems, rigid alignment of optical components, and reduced space and weight.

A fiber optics cable is a dielectric waveguide, usually of circular cross section, that guides the electromagnetic wave in discrete modes through internal reflections whose incidence angle at the interface is equal to or greater than the critical angle. The confinement of the energy within the dielectric structure is described analytically by Maxwell's equations and the boundary conditions at dielectric–dielectric boundaries, in contrast to the traditional metallic–dielectric boundaries in metallic waveguides. Such analyses have already been discussed in Section 8.7 for planar structures and Section 9.5.1 for circular geometries.

The most common geometries of fiber cables are those shown in Figure 9-16. They are classified as *step-index multimode, graded-index multimode*, and *single-mode step-index* [14, 16]. Typical dimensions, index-of-refraction distributions, optical ray paths, and pulse spreading by each type of cable are also illustrated in Figure 9-16. Therefore, fiber optics cables can be classified into two cases: *single-mode* and *multimode* fibers.

Single-mode fibers permit wave propagation at a single resolvable angle, whereas multimode cables transmit waves that travel at many resolvable angles all within the central core of the fiber. Both single- and multimode fibers are fabricated with a central core with a high index of refraction (dielectric constant) surrounded by a cladding with a lower index of refraction. The wave is guided in the core by total internal reflection at the core–cladding interface. Single-mode cables usually have a step-index where the diameter of the center core is very small (typically $2-16\ \mu$m) and not much larger than the wavelength of the wave it carries. Structurally, they are the simplest and exhibit abrupt index-of-refraction discontinuities along the core-cladding interface. Usually, the index of refraction of the central core is about 1.471 and that of the cladding is 1.457.

Step-index multimode cables also exhibt a well-defined central core with a constant index of refraction surrounded by a cladding with a lower index of refraction. In contrast to the single-mode step-index, the multimode step-index possesses a central core whose diameter is about $25-150\ \mu$m. When the wavelength of light is of the order of $10\ \mu$m (near the infrared region) and the central core diameter is $100\ \mu$m, such a cable can support as many as 25,000 modes (usually there are about 200 modes). However, as the central core diameter approaches $10\ \mu$m (one wavelength), the number of modes is reduced to one and the cable becomes a single-mode fiber. Because of its multimode field structure, such a cable is very dispersive and provides severe signal distortion to waveforms with broad spectral frequency content. Typically, the index of refraction of the central core is about 1.527 and that of the cladding is about 1.517.

Graded-index multimode cables also possess a relatively large central core (typical diameters of $20-150\ \mu$m) whose index of refraction continuously decreases from the core center toward the core–cladding interface, at which point the core and cladding indexes are identical. For this cable, the waves are still contained within the central core and they are continuously refocused toward the central axis of the core by its continuous lensing action. The number of modes supported by such a cable is usually about 2,000, and such a multimode structure provides waveform distortion especially to transient signals. All cables possess an outside cladding that adds mechanical strength

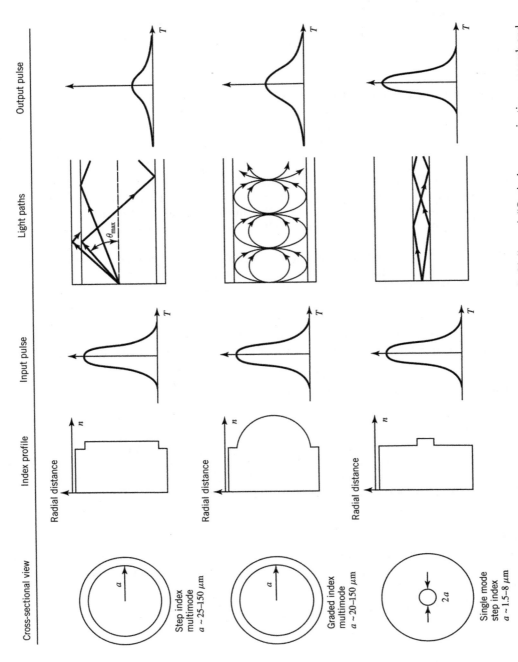

Figure 9-16 Common types and geometries of fiber cables. (Source: T. G. Giallorenzi, "Optical communications research and technology: Fiber optics," *Proc. IEEE,* © 1978, IEEE.)

to the fiber, reduces scattering loss that is due to dielectric discontinuities, and protects the guiding central core from absorbing surface contaminants that the cable may come in contact with.

For the graded-index multimode fiber, the index of refraction $n(\rho) = \sqrt{\varepsilon_r(\rho)}$ of the central core exhibits a nearly parabolic variation in the radial direction from the center of the core toward the cladding. This variation can be represented by [16]

$$n_c(\rho) = n\left[1 + \Delta\left(\frac{a-\rho}{a}\right)^\alpha\right], \quad 0 \le \rho \le a \tag{9-105}$$

where $n_c(\rho) =$ index of refraction of the central core (usually $\simeq 1.562$ on axis)
$\quad\quad n =$ index of refraction of the cladding (usually $\simeq 1.540$)
$\quad\quad a =$ radius of the central core
$\quad\quad \Delta =$ parameter usually much less than unity (usually $0.01 < \Delta < 0.02$)
$\quad\quad \alpha =$ parameter whose value is close to 2 for maximum fiber bandwidth

It has been shown by Example 5-6 that for the dielectric–dielectric interface of the fiber cable to internally reflect the waves of all incidence angles, the ratio of the index of refraction of the central core at the interface to that of the cladding must be equal to or greater than $\sqrt{2}$, that is, $n_c(a)/n \ge \sqrt{2}$. However, this is not necessary if the angles of incidence of the waves are not small.

The modes that can be supported by the step-index cable (either single mode or multimode) can be found using techniques outlined in Section 9.5.1 for the dielectric rod waveguide. These modes are, in general, HEM (HE or EH) hybrid modes. The applicable equations are (9-81a) through (9-94b), where ε_r should represent the square of the ratio of the index of refraction of the central core to that of the cladding [i.e., $\varepsilon_r = (n_c/n)^2$]. The field configurations of the graded-index cable can be analyzed in terms of Hermite-Gaussian functions [35]. Because of the complexity, the analysis will not be presented here. The interested reader is referred to the literature [11, 16].

9.5.4 Dielectric-Covered Conducting Rod

Let us consider the field analysis of a dielectric-covered conducting circular rod, as shown in Figure 9-17 [36–38]. The radius of the conducting rod is a while that of the dielectric cover is b. The thickness of the dielectric sleeving is denoted by t ($t = b - a$). When the radius of the conducting rod is small, the rod will be representative of a dielectric-covered wire. In this section we will consider the TM^z and TE^z modes, which, in general, must co-exist to satisfy the boundary conditions. We will conduct an extended discussion of the modes with azimuthal symmetry—no ϕ variations—will be conducted.

A. TM^z Modes
The fields in the radial direction inside the dielectric sleeving must be represented by standing wave functions while those outside the dielectric cover must be decaying in order for the rod to act as a waveguide. In general, for the TM$_{mn}^z$ modes, the potential function A_z can be written as

$$A_z^d = \left[A_1^d J_m\left(\beta_\rho^d \rho\right) + B_1^d Y_m\left(\beta_\rho^d \rho\right)\right]\left[C_2^d \cos(m\phi) + D_2^d \sin(m\phi)\right]$$
$$\times \left[A_3^d e^{-j\beta_z z} + B_3^d e^{+j\beta_z z}\right], \quad \text{for } a \le \rho \le b \tag{9-106a}$$

$$A_z^0 = A_1^0 K_m\left(\alpha_\rho^0 \rho\right)\left[C_2^0 \cos(m\phi) + D_2^0 \sin(m\phi)\right]$$
$$\times \left[A_3^0 e^{-j\beta_z z} + B_3^0 e^{+j\beta_z z}\right], \quad \text{for } b \le \rho \le \infty \tag{9-106b}$$

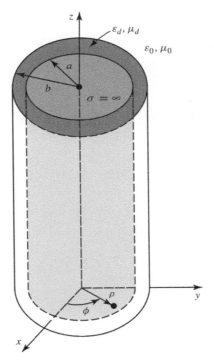

Figure 9-17 Geometry of dielectric-covered conducting rod.

However, the geometry of Figure 9-17 can only support modes with azimuthal symmetry, ($m = 0$). Therefore, for the positive traveling waves of the lowest-order mode ($m = 0$), the vector potentials reduce to

$$A_z^d = \left[A_0^d J_0\left(\beta_\rho^d \rho\right) + B_0^d Y_0\left(\beta_\rho^d \rho\right)\right] e^{-j\beta_z z}, \quad \text{for } a \le \rho \le b \tag{9-107}$$

where

$$\left(\beta_\rho^d\right)^2 + \beta_z^2 = \beta_d^2 = \mu_r \varepsilon_r \beta_0^2 \tag{9-107a}$$

and

$$A_z^0 = A_0^0 K_0\left(\alpha_\rho^0 \rho\right) e^{-j\beta_z z}, \quad \text{for } b \le \rho \le \infty \tag{9-108}$$

where

$$-\left(\alpha_\rho^0\right)^2 + \beta_z^2 = \beta_0^2 \tag{9-108a}$$

The corresponding electric and magnetic fields can be written as

For $a \le \rho \le b$,

$$E_\rho^d = -j\frac{1}{\omega\mu_d\varepsilon_d}\frac{\partial^2 A_z^d}{\partial\rho\,\partial z} = -\frac{\beta_z\beta_\rho^d}{\omega\mu_d\varepsilon_d}\left[A_0^d J_0'\left(\beta_\rho^d \rho\right) + B_0^d Y_0'\left(\beta_\rho^d \rho\right)\right] e^{-j\beta_z z} \tag{9-109a}$$

$$E_\phi^d = -j\frac{1}{\omega\mu_d\varepsilon_d}\frac{\partial^2 A_z^d}{\partial\phi\,\partial z} = 0 \tag{9-109b}$$

$$E_z^d = -j\frac{1}{\omega\mu_d\varepsilon_d}\left(\frac{\partial^2}{\partial z^2}+\beta_d^2\right)A_z^d = -j\frac{\left(\beta_\rho^d\right)^2}{\omega\mu_d\varepsilon_d}\left[A_0^d J_0\left(\beta_\rho^d\rho\right)+B_0^d Y_0\left(\beta_\rho^d\rho\right)\right]e^{-j\beta_z z} \quad \text{(9-109c)}$$

$$H_\rho^d = \frac{1}{\mu_d}\frac{1}{\rho}\frac{\partial A_z^d}{\partial\phi}=0 \quad \text{(9-109d)}$$

$$H_\phi^d = -\frac{1}{\mu_d}\frac{\partial A_z^d}{\partial\rho}=-\frac{\beta_\rho^d}{\mu_d}\left[A_0^d J_0'\left(\beta_\rho^d\rho\right)+B_0^d Y_0'\left(\beta_\rho^d\rho\right)\right]e^{-j\beta_z z} \quad \text{(9-109e)}$$

$$H_z^d = 0 \quad \text{(9-109f)}$$

$$' \equiv \frac{\partial}{\partial\left(\beta_\rho^d\rho\right)} \quad \text{(9-109g)}$$

And for $b \le \rho \le \infty$,

$$E_\rho^0 = -j\frac{1}{\omega\mu_0\varepsilon_0}\frac{\partial^2 A_z^0}{\partial\rho\partial z}=\frac{\alpha_\rho^0\beta_z}{\omega\mu_0\varepsilon_0}A_0^0 K_0'\left(\alpha_\rho^0\rho\right)e^{-j\beta_z z} \quad \text{(9-110a)}$$

$$E_\phi^0 = -j\frac{1}{\omega\mu_0\varepsilon_0}\frac{1}{\rho}\frac{\partial^2 A_z^0}{\partial\phi\partial z}=0 \quad \text{(9-110b)}$$

$$E_z^0 = -j\frac{1}{\omega\mu_0\varepsilon_0}\left(\frac{\partial^2}{\partial z^2}+\beta_0^2\right)A_z^0 = +j\frac{\left(\alpha_\rho^0\right)^2}{\omega\mu_0\varepsilon_0}A_0^0 K_0\left(\alpha_\rho^0\rho\right)e^{-j\beta_z z} \quad \text{(9-110c)}$$

$$H_\rho^0 = \frac{1}{\mu_0}\frac{1}{\rho}\frac{\partial A_z^0}{\partial\phi}=0 \quad \text{(9-110d)}$$

$$H_\phi^0 = -\frac{1}{\mu_0}\frac{\partial A_z^0}{\partial\rho}=-\frac{\alpha_\rho^0}{\mu_0}A_0^0 K_0'\left(\alpha_\rho^0\rho\right)e^{-j\beta_z z} \quad \text{(9-110e)}$$

$$H_z^0 = 0 \quad \text{(9-110f)}$$

$$' \equiv \frac{\partial}{\partial\left(\alpha_\rho^0\rho\right)} \quad \text{(9-110g)}$$

The vanishing of the tangential electric fields at $\rho = a$ and the continuity of the tangential components of the electric and magnetic fields at $\rho = b$ requires that

$$E_z^d(\rho = a, 0 \le \phi \le 2\pi, z) = 0 \quad \text{(9-111a)}$$

$$E_z^d(\rho = b, 0 \le \phi \le 2\pi, z) = E_z^0(\rho = b, 0 \le \phi \le 2\pi, z) \quad \text{(9-111b)}$$

$$H_\phi^d(\rho = b, 0 \le \phi \le 2\pi, z) = H_\phi^0(\rho = b, 0 \le \phi \le 2\pi, z) \quad \text{(9-111c)}$$

Applying (9-111a) leads to

$$A_0^d J_0\left(\beta_\rho^d a\right)+B_0^d Y_0\left(\beta_\rho^d a\right)=0 \quad \text{(9-112)}$$

or

$$B_0^d = -A_0^d\frac{J_0\left(\beta_\rho^d a\right)}{Y_0\left(\beta_\rho^d a\right)} \quad \text{(9-112a)}$$

whereas (9-111b) leads to

$$-\frac{\left(\beta_\rho^d\right)^2}{\omega\mu_d\varepsilon_d}\left[A_0^d J_0\left(\beta_\rho^d b\right)+B_0^d Y_0\left(\beta_\rho^d b\right)\right]=\frac{\left(\alpha_\rho^0\right)^2}{\omega\mu_0\varepsilon_0}A_0^0 K_0\left(\alpha_\rho^0 b\right) \qquad (9\text{-}113)$$

which by using (9-112a) can be written as

$$A_0^0 K_0\left(\alpha_\rho^0 b\right)=-A_0^d\frac{\mu_0}{\mu_d}\frac{\varepsilon_0}{\varepsilon_d}\left(\frac{\beta_\rho^d}{\alpha_\rho^0}\right)^2\left[J_0\left(\beta_\rho^d b\right)-\frac{J_0\left(\beta_\rho^d a\right)Y_0\left(\beta_\rho^d b\right)}{Y_0\left(\beta_\rho^d a\right)}\right] \qquad (9\text{-}113a)$$

The continuity of the tangential magnetic field at $\rho=b$, as stated by (9-111c), leads to

$$-\frac{\beta_\rho^d}{\mu_d}\left[A_0^d J_0'\left(\beta_\rho^d b\right)+B_0^d Y_0'\left(\beta_\rho^d b\right)\right]=-\frac{\alpha_\rho^0}{\mu_0}A_0^0 K_0'\left(\alpha_\rho^0 b\right) \qquad (9\text{-}114)$$

which by using (9-112a) can be written as

$$A_0^0 K_0'\left(\alpha_\rho^0 b\right)=A_0^d\frac{\mu_0}{\mu_d}\frac{\beta_\rho^d}{\alpha_\rho^0}\left[J_0'\left(\beta_\rho^d b\right)-\frac{J_0\left(\beta_\rho^d a\right)Y_0'\left(\beta_\rho^d b\right)}{Y_0\left(\beta_\rho^d a\right)}\right] \qquad (9\text{-}114a)$$

Dividing (9-114a) by (9-113a) leads to

$$\frac{K_0'\left(\alpha_\rho^0 b\right)}{K_0\left(\alpha_\rho^0 b\right)}=-\frac{\varepsilon_d}{\varepsilon_0}\left(\frac{\alpha_\rho^0}{\beta_\rho^d}\right)\frac{\left[J_0'\left(\beta_\rho^d b\right)Y_0\left(\beta_\rho^d a\right)-J_0\left(\beta_\rho^d a\right)Y_0'\left(\beta_\rho^d b\right)\right]}{\left[J_0\left(\beta_\rho^d b\right)Y_0\left(\beta_\rho^d a\right)-J_0\left(\beta_\rho^d a\right)Y_0\left(\beta_\rho^d b\right)\right]} \qquad (9\text{-}115a)$$

Subtracting (9-108a) from (9-107a) leads to

$$\boxed{\left(\beta_\rho^d\right)^2+\left(\alpha_\rho^0\right)^2=\beta_0^2(\mu_r\varepsilon_r-1)} \qquad (9\text{-}115b)$$

which, along with the transcendental Equation 9-115a, can be used to solve for β_ρ^d and α_ρ^0. This can be accomplished using graphical or numerical techniques similar to the ones utilized in Section 8.7 for planar structures.

The technique outlined and implemented in Section 8.7 to solve the wave numbers for planar dielectric waveguides is straightforward but complicated when utilized to solve for the wave numbers β_ρ^d and α_ρ^0 of (9-115a) and (9-115b). An approximate solution can be used to solve (9-115a) and (9-115b) simultaneously when the thickness of the dielectric cladding $t=b-a$ is small. Under these conditions the Bessel functions $J_0(\beta_\rho^d b)$ and $Y_0(\beta_\rho^d b)$ can be expanded in a

Taylor series about the point $\beta_\rho^d a$, that is,

$$J_0\left(\beta_\rho^d b\right) \simeq J_0\left(\beta_\rho^d a\right) + \frac{dJ_0\left(\beta_\rho^d b\right)}{d\left(\beta_\rho^d b\right)}\Bigg|_{\beta_\rho^d b = \beta_\rho^d a} \beta_\rho^d (b-a)$$

$$\simeq J_0\left(\beta_\rho^d a\right) - \beta_\rho^d (b-a) J_1\left(\beta_\rho^d a\right) \tag{9-116a}$$

$$Y_0\left(\beta_\rho^d b\right) \simeq Y_0\left(\beta_\rho^d a\right) + \frac{dY_0\left(\beta_\rho^d b\right)}{d\left(\beta_\rho^d b\right)}\Bigg|_{\beta_\rho^d b = \beta_\rho^d a} \beta_\rho^d (b-a)$$

$$\simeq Y_0\left(\beta_\rho^d a\right) - \beta_\rho^d (b-a) Y_1\left(\beta_\rho^d a\right) \tag{9-116b}$$

$$J_0'\left(\beta_\rho^d b\right) = \frac{dJ_0\left(\beta_\rho^d \rho\right)}{d\left(\beta_\rho^d \rho\right)}\Bigg|_{\rho=b}$$

$$= -J_1\left(\beta_\rho^d b\right) \simeq -\left[J_1\left(\beta_\rho^d a\right) + \frac{dJ_1\left(\beta_\rho^d b\right)}{d\left(\beta_\rho^d b\right)}\Bigg|_{\beta_\rho^d b = \beta_\rho^d a}\beta_\rho^d(b-a)\right]$$

$$\simeq -\left\{J_1\left(\beta_\rho^d a\right) + \beta_\rho^d(b-a)\left[J_0\left(\beta_\rho^d a\right) - \frac{1}{\beta_\rho^d a}J_1\left(\beta_\rho^d a\right)\right]\right\}$$

$$J_0'\left(\beta_\rho^d b\right) \simeq -\left[J_1\left(\beta_\rho^d a\right)\left(1 - \frac{b-a}{a}\right) + \beta_\rho^d(b-a)J_0\left(\beta_\rho^d a\right)\right] \tag{9-116c}$$

$$Y_0'\left(\beta_\rho^d b\right) \simeq -\left[Y_1\left(\beta_\rho^d a\right)\left(1 - \frac{b-a}{a}\right) + \beta_\rho^d(b-a)Y_0\left(\beta_\rho^d a\right)\right] \tag{9-116d}$$

Therefore, the numerator and denominator of (9-115a) can be written, respectively, as

$$J_0'\left(\beta_\rho^d b\right)Y_0\left(\beta_\rho^d a\right) - J_0\left(\beta_\rho^d a\right)Y_0'\left(\beta_\rho^d b\right)$$

$$\simeq -Y_0\left(\beta_\rho^d a\right)\left[J_1\left(\beta_\rho^d a\right)\left(1 - \frac{b-a}{a}\right) + \beta_\rho^d(b-a)J_0\left(\beta_\rho^d a\right)\right]$$

$$+ J_0\left(\beta_\rho^d a\right)\left[Y_1\left(\beta_\rho^d a\right)\left(1 - \frac{b-a}{a}\right) + \beta_\rho^d(b-a)Y_0\left(\beta_\rho^d a\right)\right]$$

$$\simeq \left(1 - \frac{b-a}{a}\right)\left[J_0\left(\beta_\rho^d a\right)Y_1\left(\beta_\rho^d a\right) - Y_0\left(\beta_\rho^d a\right)J_1\left(\beta_\rho^d a\right)\right] \tag{9-117a}$$

$$J_0\left(\beta_\rho^d b\right)Y_0\left(\beta_\rho^d a\right) - J_0\left(\beta_\rho^d a\right)Y_0\left(\beta_\rho^d b\right)$$

$$\simeq Y_0\left(\beta_\rho^d a\right)\left[J_0\left(\beta_\rho^d a\right) - \beta_\rho^d(b-a)J_1\left(\beta_\rho^d a\right)\right]$$

$$- J_0\left(\beta_\rho^d a\right)\left[Y_0\left(\beta_\rho^d a\right) - \beta_\rho^d(b-a)Y_1\left(\beta_\rho^d a\right)\right]$$

$$\simeq \beta_\rho^d(b-a)\left[J_0\left(\beta_\rho^d a\right)Y_1\left(\beta_\rho^d a\right) - Y_0\left(\beta_\rho^d a\right)J_1\left(\beta_\rho^d a\right)\right] \tag{9-117b}$$

Substituting (9-117a) and (9-117b) into (9-115a) leads to

$$
\frac{K_0'\left(\alpha_\rho^0 b\right)}{K_0\left(\alpha_\rho^0 b\right)} \simeq -\frac{\varepsilon_d}{\varepsilon_0}\left(\frac{\alpha_\rho^0}{\beta_\rho^d}\right)\frac{\left(1-\dfrac{b-a}{a}\right)}{\beta_\rho^d\left(b-a\right)} = -\frac{\varepsilon_d}{\varepsilon_0}\frac{\alpha_\rho^0}{\left(\beta_\rho^d\right)^2}\left|\frac{1-\dfrac{b-a}{a}}{b-a}\right|
\tag{9-118}
$$

For small values of a (i.e., $a \ll \lambda_0$) the attenuation constant α_ρ^0 is of the same order of magnitude as β_0 and the wave is loosely bound to the surface so that $\alpha_\rho^0 b$ is small for small values of $b-a$. Under these conditions

$$
K_0'\left(\alpha_\rho^0 b\right) = \left.\frac{dK_0\left(\alpha_\rho^0 \rho\right)}{d\left(\alpha_\rho^0 \rho\right)}\right|_{\rho=b} = -K_1\left(\alpha_\rho^0 b\right) \simeq -\frac{1}{\alpha_\rho^0 b} = -\frac{1}{b\alpha_\rho^0}
\tag{9-119a}
$$

$$
K_0\left(\alpha_\rho^0 b\right) \simeq -\ln\left(0.89\alpha_\rho^0 b\right)
\tag{9-119b}
$$

so that (9-118) reduces for $\varepsilon_d = \varepsilon_r\varepsilon_0$ to

$$
\frac{1}{b\ln\left(0.89\alpha_\rho^0 b\right)} \simeq -\varepsilon_r\left(\frac{\alpha_\rho^0}{\beta_\rho^d}\right)^2\left|\frac{1-\dfrac{b-a}{a}}{b-a}\right|
$$

$$
\left(\beta_\rho^d\right)^2 (b-a) \simeq -\varepsilon_r b\left(1-\frac{b-a}{a}\right)\left(\alpha_\rho^0\right)^2 \ln\left(0.89\alpha_\rho^0 b\right)
\tag{9-120}
$$

Substituting (9-115b) into (9-120) for $(\beta_\rho^d)^2$ leads for $t = b-a \ll a$ to

$$
\left[\beta_0^2(\varepsilon_r - 1) - \left(\alpha_\rho^0\right)^2\right](b-a) \simeq -\varepsilon_r b\left(1-\frac{b-a}{a}\right)\left(\alpha_\rho^0\right)^2 \ln\left(0.89\alpha_\rho^0 b\right)
\tag{9-121}
$$

or

$$
\boxed{\varepsilon_r b\left(\alpha_\rho^0\right)^2 \ln\left(0.89\alpha_\rho^0 b\right) \simeq \left[-\beta_0^2(\varepsilon_r - 1) + \left(\alpha_\rho^0\right)^2\right](b-a)}
\tag{9-121a}
$$

Example 9-6

A perfectly conducting wire of radius $a = 0.09$ cm is covered with a dielectric sleeving of polystyrene ($\varepsilon_r = 2.56$) of radius $b = 0.10$ cm. At a frequency of 9.55 GHz, determine the attenuation constants α_ρ^0, β_z, and β_ρ^d and the relative field strength at $\rho = 2\lambda_0$ compared to that at the outside surface of the dielectric sleeving ($\rho = b$).

Solution: At $f = 9.55$ GHz,

$$
\lambda_0 = \frac{30\times10^9}{9.55\times10^9} = 3.1414 \text{ cm} \quad \Rightarrow \quad \beta_0 = \frac{2\pi}{\lambda_0} = 2 \text{ rad/cm}
$$

$$
\lambda_d = \frac{\lambda_0}{\sqrt{\varepsilon_r}} = \frac{3.1414}{\sqrt{2.56}} = 1.9634 \text{ cm} \quad \Rightarrow \quad \beta_d = \frac{2\pi}{\lambda} = \frac{2\pi}{1.9634} = 3.2 \text{ rad/cm}
$$

Since the thickness t of the dielectric sleeving is much smaller than the wavelength,

$$
t = b-a = 0.10 - 0.09 = 0.01 \text{ cm} < \lambda_d = 1.9634 \text{ cm} < \lambda_0 = 3.1414 \text{ cm}
$$

then the approximate relation of (9-121a) is applicable. Using an iterative procedure, it can be shown that

$$\alpha_\rho^0 = 0.252 \text{ Np/cm}$$

is a solution to (9-121a). Using (9-108a),

$$\beta_z = \sqrt{\beta_0^2 + \left(\alpha_\rho^0\right)^2} = \sqrt{(2)^2 + (0.252)^2} = \sqrt{4.0635} = 2.0158 \text{ rad/cm}$$

and β_ρ^d is found using (9-107a) as

$$\beta_\rho^d = \sqrt{\beta_d^2 - \beta_z^2} = \sqrt{(3.2)^2 - (2.0158)^2} = \sqrt{6.1765} = 2.485 \text{ rad/cm}$$

The field outside the sleeving is of decaying form represented by the modified Bessel function $K_0(\alpha_\rho^0 \rho)$ of (9-108). Thus,

$$\frac{E(\rho = 2\lambda_0)}{E(\rho = b)} = \frac{E(\rho = 2\pi \text{ cm})}{E(\rho = 0.10 \text{ cm})} = \frac{K_0(\rho = 2\pi \text{ cm})}{K_0(\rho = 0.10 \text{ cm})}$$

$$= \frac{\left.\dfrac{C_0 e^{-\alpha_\rho^0 \rho}}{\sqrt{\alpha_\rho^0 \rho}}\right|_{\rho = 2\pi \text{ cm}}}{\left.\dfrac{C_0 e^{-\alpha_\rho^0 \rho}}{\sqrt{\alpha_\rho^0 \rho}}\right|_{\rho = 0.10 \text{ cm}}} = \frac{0.1631}{6.1426} = 0.0266 = 2.66\%$$

Therefore, at a distance of $\rho = 2\lambda_0$ the relative field has been reduced to a very small value; at points further away, the field intensity is even smaller.

For the dielectric-covered wire and the dielectric rod waveguide, we can define an *effective radius* as the radial distance at which point and beyond the relative field intensity is of very low value. If we use the 2.66% value of Example 9-6 as the field value with which we can define the effective radius, then the effective radius for the wire of Example 9-6 is $a_e = 2\lambda_0 = 2\pi$ cm.

The rate of attenuation α_ρ^0 can be increased and the wave can be made more tightly bound to the surface by increasing the dielectric constant and/or thickness of the dielectric sleeving. This, however, results in greater attenuation and larger losses of the wave along the direction of wave travel (axis of the wire) because of the greater field concentration near the conducting boundary.

B. TE^z Modes   Following a procedure similar to that used for the TM$_{mn}^z$ modes, it can be shown that for the dielectric-covered conducting rod of Figure 9-17, the TE$_{mn}^z$ positive traveling waves of the lowest-order mode ($m = 0$) also possess azimuthal symmetry (no ϕ variations, $m = 0$) and can exist individually. Therefore, for the $m = 0$ mode, the vector potentials can be written as

$$F_z^d = \left[A_0^d J_0\left(\beta_\rho^d \rho\right) + B_0^d Y_0\left(\beta_\rho^d \rho\right) \right] e^{-j\beta_z z}, \quad \text{for } a \leq \rho \leq b \tag{9-122}$$

where

$$\left(\beta_\rho^d\right)^2 + \beta_z^2 = \beta_d^2 = \mu_r \varepsilon_r \beta_0^2 \tag{9-122a}$$

$$F_z^0 = A_0^0 K_0\left(\alpha_\rho^0 \rho\right) e^{-j\beta_z z}, \quad \text{for } b \leq \rho \leq \infty \tag{9-123}$$

where

$$-\left(\alpha_\rho^0\right)^2 + \beta_z^2 = \beta_0^2 \tag{9-123a}$$

Leaving the details for the reader as an end-of-chapter exercise, the equations for the TE_{0n}^z modes corresponding to (9-113a), (9-114a), (9-115a), (9-115b), and (9-121a) for the TM_{0n}^z modes, are given by

$$A_0^0 K_0'(\alpha_\rho^0 b) = A_0^d \frac{\varepsilon_0}{\varepsilon_d} \frac{\beta_\rho^d}{\alpha_\rho^0} \left[J_0'(\beta_\rho^d b) - \frac{J_0'(\beta_\rho^d a)Y_0'(\beta_\rho^d b)}{Y_0'(\beta_\rho^d a)} \right] \tag{9-124a}$$

$$A_0 K_0(\alpha_\rho^0 b) = -A_0^d \frac{\mu_0}{\mu_d} \frac{\varepsilon_0}{\varepsilon_d} \left(\frac{\beta_\rho^d}{\alpha_\rho^0}\right)^2 \left[J_0(\beta_\rho^d b) - \frac{J_0'(\beta_\rho^d a)Y_0(\beta_\rho^d b)}{Y_0'(\beta_\rho^d a)} \right] \tag{9-124b}$$

$$\frac{K_0'(\alpha_\rho^0 b)}{K_0(\alpha_\rho^0 b)} = -\frac{\mu_d}{\mu_0} \left(\frac{\alpha_\rho^0}{\beta_\rho^d}\right) \frac{\left[J_0'(\beta_\rho^d b)Y_0'(\beta_\rho^d a) - J_0'(\beta_\rho^d a)Y_0'(\beta_\rho^d b) \right]}{\left[J_0(\beta_\rho^d b)Y_0'(\beta_\rho^d a) - J_0'(\beta_\rho^d a)Y_0(\beta_\rho^d b) \right]} \tag{9-124c}$$

$$(\beta_\rho^d)^2 + (\alpha_\rho^0)^2 = \beta_0^2(\mu_r \varepsilon_r - 1) \tag{9-124d}$$

$$\mu_r b(b-a)(\alpha_\rho^0)^2 \ln(0.89\alpha_\rho^0 b) \simeq 1 \tag{9-124e}$$

9.6 MULTIMEDIA

On the website that accompanies this book, the following multimedia resources are included for the review, understanding, and presentation of the material of this chapter.

- **MATLAB** computer programs:
  a. **Cyl_Waveguide:** Computes the propagation characteristics of a cylindrical waveguide.
  b. **Cyl_Resonator:** Computes the resonant characteristics of a cylindrical resonator.
  c. **CircDielGuide:** Computes the propagation characteristics of a circular dielectric waveguide based on the solution of the eigenvalue equation 9-91.
- **PowerPoint (PPT)** viewgraphs, in multicolor.

REFERENCES

1. C. S. Lee, S. W. Lee, and S. L. Chuang, "Plot of modal field distribution in rectangular and circular waveguides," *IEEE Trans. Microwave Theory Tech.*, vol. MTT-33, no. 3, pp. 271–274, Mar. 1985.
2. S. E. Miller, "Waveguide as a communication medium," *Bell System Tech. J.*, vol. 35, pp. 1347–1384, Nov. 1956.
3. S. P. Morgan and J. A. Young, "Helix waveguide," *Bell System Tech. J.*, vol. 35, pp. 1347–1384, Nov. 1956.
4. S. Ramo, J. R. Whinnery, and T. Van Duzer, *Fields and Waves in Communication Electronics*, John Wiley & Sons, New York, 1965, pp. 429–439.

5. M. R. Spiegel, *Mathematical Handbook, Schaum's Outline Series*, McGraw-Hill, New York, 1968.

6. C. A. Balanis, *Antenna Theory: Analysis and Design* (Third edition), John Wiley & Sons, New York, 2005.

7. C. H. Chandler, "An investigation of dielectric rod as waveguide," *J. Appl. Phys.*, vol. 20, pp. 1188–1192, Dec. 1949.

8. W. M. Elasser, "Attenuation in a dielectric circular rod," *J. Appl. Phys.*, vol. 20, pp. 1193–1196, Dec. 1949.

9. J. W. Duncan and R. H. DuHamel, "A technique for controlling the radiation from dielectric rod waveguides," *IRE Trans. Antennas Propagat.*, vol. AP-5, no. 4, pp. 284–289, Jul. 1957.

10. R. E. Collin, *Field Theory of Guided Waves*, Second Edition, McGraw-Hill, New York, 1991.

11. E. Snitzer, "Cylindrical electric waveguide modes," *J. Optical Soc. America*, vol. 51, no. 5, pp. 491–498, May 1961.

12. A. W. Snyder, "Asymptotic expressions for eigenfunctions and eigenvalues of a dielectric or optical waveguide," *IEEE Trans. Microwave Theory Tech.*, vol. MTT-17, pp. 1130–1138, Dec. 1969.

13. A. W. Snyder, "Excitation and scattering of modes on a dielectric or optical fiber," *IEEE Trans. Microwave Theory Tech.*, vol. MTT-17, pp. 1138–1144, Dec. 1969.

14. S. E. Miller, E. A. J. Marcatili, and T. Li, "Research toward optical-fiber transmission systems. Part I: The transmission medium," *Proc. IEEE*, vol. 61, no. 12, pp. 1703–1726, Dec. 1973.

15. D. Marcuse, *Theory of Dielectric Optical Waveguide*, Academic, New York, 1974.

16. T. G. Giallorenzi, "Optical communications research and technology: Fiber optics," *Proc. IEEE*, vol. 66, no. 7, pp. 744–780, Jul. 1978.

17. J. Kane, "Fiber optic cables compete with mw relays and coax," *Microwave J.*, vol. 26, pp. 16, 61, Jan. 1979.

18. C. Yeh, "Guided-wave modes in cylindrical optical fibers," *IEEE Trans. Education*, vol. E-30, no. 1, Feb. 1987.

19. R. J. Pieper, "A heuristic approach to fiber optics," *IEEE Trans. Education*, vol. E-30, no. 2, pp. 77–82, May 1987.

20. "IRE standards on antennas and waveguides: definitions of terms, 1953," *Proc. IRE*, vol. 41, pp. 1721–1728, Dec. 1953.

21. D. Kajfez and P. Guillon (Eds.), *Dielectric Resonators*, Artech House, Inc., Dedham, MA, 1986.

22. R. F. Harrington, *Time-Harmonic Electromagnetic Fields*, McGraw-Hill, New York, 1961.

23. T. Okoshi, *Optical Fibers*, Academic, New York, 1982.

24. R. D. Richtmyer, "Dielectric resonator," *J. Appl. Phys.*, vol. 10, pp. 391–398, Jun. 1939.

25. A. Okaya, "The rutile microwave resonator," *Proc. IRE*, vol. 48, p. 1921, Nov. 1960.

26. A. Okaya and L. F. Barash, "The dielectric microwave resonator," *Proc. IRE*, vol. 50, pp. 2081–2092, Oct. 1962.

27. H. Y. Yee, "Natural resonant frequencies of microwave dielectric resonators," *IEEE Trans. Microwave Theory Tech.*, vol. MTT-13, p. 256, Mar. 1965.

28. S. J. Fiedziuszko, "Microwave dielectric resonators," *Microwave J.*, pp. 189–200, Sept. 1980.

29. M. W. Pospieszalski, "Cylindrical dielectric resonators and their applications in TEM line microwave circuits," *IEEE Trans. Microwave Theory Tech.*, vol. MTT-27, no. 3, pp. 233–238, Mar. 1979.

30. K. A. Zaki and C. Chen, "Loss mechanisms in dielectric-loaded resonators," *IEEE Trans. Microwave Theory Tech.*, vol. MTT-33, no. 12, pp. 1448–1452, Dec. 1985.

31. S. B. Cohn, "Microwave bandpass filters containing high-Q dielectric resonators," *IEEE Trans. Microwave Theory Tech.*, vol. MTT-16, pp. 218–227, Apr. 1968.

32. T. Itoh and R. S. Rudokas, "New method for computing the resonant frequencies of dielectric resonators," *IEEE Trans. Microwave Theory Tech.*, vol. MTT-25, pp. 52–54, Jan. 1977.

33. A. W. Glisson, D. Kajfez, and J. James, "Evaluation of modes in dielectric resonators using a surface integral equation formulation," *IEEE Trans. Microwave Theory Tech.*, vol. MTT-31, pp. 1023–1029, Dec. 1983.

34. Y. Kobayashi and S. Tanaka, "Resonant modes of a dielectric rod resonator short-circuited at both ends by parallel conducting plates," *IEEE Trans. Microwave Theory Tech.*, vol. MTT-28, no. 10, pp. 1077–1085, Oct. 1978.

35. R. D. Maurer, "Introduction to optical fiber waveguides," in *Introduction to Integrated Optics*, M. Barnoski (Ed.), Plenum, New York, Chapter 8, 1974.

36. G. Goubau, "Surface waves and their application to transmission lines," *J. Appl. Phys.*, vol. 21, pp. 1119–1128, Nov. 1950.

37. G. Goubau, "Single-conductor surface-wave transmission lines," *Proc. IRE*, vol. 39, pp. 619–624, Jun. 1951.

38. R. E. Collin, *Foundations for Microwave Engineering*, McGraw-Hill, New York, 1966.

PROBLEMS

9.1. Design a circular waveguide filled with a lossless dielectric medium whose relative permeability is unity. The waveguide must operate in a single dominant mode over a bandwidth of 1.5 GHz. Assume that the radius of the guide is 1.12 cm. Find the
   (a) Dielectric constant of the medium that must fill the cavity to meet the desired design specifications.
   (b) Lower and upper frequencies of operation.

9.2. A dielectric waveguide, with a dielectric constant of 2.56, is inserted inside a section of a circular waveguide of radius a and length L. This section of the circular waveguide is inserted and designed to be used as a phase shifter. Assuming:

   • Dominant mode propagation
   • Radius $a = 2$ cm
   • Length $L = 5$ cm
   • Frequency = 6 GHz

determine the additional phase shift (in degrees), from what it would have been if the waveguide was filled with air, provided by the presence of the dielectric material in this section of the circular waveguide.

9.3. Design a circular waveguide, with radius of 2 cm and filled with a dielectric material with dielectric constant of 2.25, to be used as a delay line. What should the length (in meters) of the waveguide be so that the total delay it presents by its insertion at 3.5 GHz is 2 microseconds?

9.4. Design a waveguide phase shifter using a circular waveguide with a 2 cm radius, operating at 6 GHz in its dominant mode, and completely filled with a dielectric material with a dielectric constant of 2.25. Determine the length (in cm) of the waveguide section so that the total phase shift introduced by its insertion is 180°.

9.5. An air-filled circular waveguide of radius a has a conducting baffle placed along its length at $\phi = 0$ extending from $\rho = 0$ to $\rho = a$, as shown in Figure P9-5. For TE^z modes, derive simplified expressions for the vector potential component, the electric and magnetic fields, and the cutoff frequencies, eigenvalues, phase constant along the axis of the guide, guide wavelength, and wave impedance.
Also determine the following.
   (a) The cutoff frequencies of the three lowest-order propagating modes in order of ascending cutoff frequency when the radius of the cylinder is 1 cm.
   (b) The wave impedance and guide wavelength (in cm) for the lowest-order mode at $f = 1.5f_c$ where f_c is the cutoff frequency of the lowest-order mode.

   Hint:

   $$J'_{1/2}(x) = 0 \quad \text{for } x = 1.1655, 4.6042$$

   $$J'_{3/2}(x) = 0 \quad \text{for } x = 2.4605, 6.0293$$

   $$J'_{5/2}(x) = 0 \quad \text{for } x = 3.6328$$

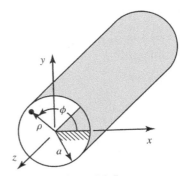

Figure P9-5

9.6. Repeat Problem 9.5 for TM^z modes. *Hint:*

$$J_{1/2}(x)=0 \quad \text{for } x=3.1416, 6.2832$$
$$J_{3/2}(x)=0 \quad \text{for } x=4.4934$$
$$J_{5/2}(x)=0 \quad \text{for } x=5.7635$$

9.7. The cross section of a cylindrical waveguide is a half circle, as shown in Figure P9-7. Derive simplified expressions for the vector potential component, electric and magnetic fields, eigenvalues, and cutoff frequencies for TE^z modes and TM^z modes.

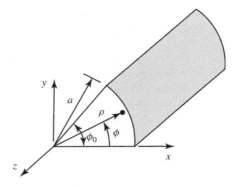

Figure P9-9

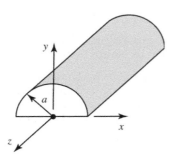

Figure P9-7

9.8. Repeat Problem 9.7 for the waveguide cross section of Figure P9-8.

9.10. The cross section of a cylindrical waveguide is that of a coaxial line with inner radius a and outer radius b, as shown in Figure P9-10. Assume TE^z modes within the waveguide.
(a) Derive simplified expressions for the vector potential component, and the electric and magnetic fields.
(b) Show that the eigenvalues are obtained as solutions to

$$J'_m(\beta_\rho a)Y'_m(\beta_\rho b) - Y'_m(\beta_\rho a)J'_m(\beta_\rho b) = 0$$

where $m = 0, 1, 2, \dots$.

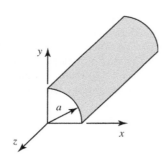

Figure P9-8

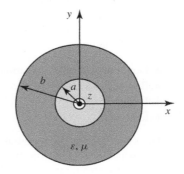

Figure P9-10

9.9. Repeat Problem 9.7 when the waveguide cross section is an angular sector as shown in Figure P9-9. Show that the zeroes of $\beta_\rho a$ are obtained using

(a) TE^z : $J'_m(\beta_\rho a)=0$, $\beta_\rho = \dfrac{\chi'_{mn}}{a}$
$m = p\left(\dfrac{\pi}{\phi_0}\right)$, $p = 0, 1, 2, \dots$

(b) TM^z : $J_m(\beta_\rho a)=0$, $\beta_\rho = \dfrac{\chi_{mn}}{a}$
$m = p\left(\dfrac{\pi}{\phi_0}\right)$, $p = 1, 2, 3, \dots$

9.11. Repeat Problem 9.10 for TM^z and show that the eigenvalues are obtained as solutions to

$$J_m(\beta_\rho a)Y_m(\beta_\rho b) - Y_m(\beta_\rho a)J_m(\beta_\rho b) = 0$$

where $m = 0, 1, 2, \dots$.

9.12. The cross section of a cylindrical waveguide is an annular sector with inner and outer radii of a and b, as shown in Figure 9-23. Assume TE^z modes within the waveguide.
(a) Derive simplified expressions for the vector potential component, and the electric and magnetic fields.

(b) Show that the eigenvalues are determined by solving

$$J_m'(\beta_\rho b)Y_m'(\beta_\rho a) - J_m'(\beta_\rho a)Y_m'(\beta_\rho b) = 0$$

where $m = p(\pi/\phi_0)$, $p = 0, 1, 2, \ldots$.

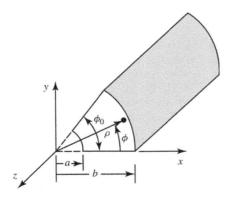

Figure P9-12

9.13. Repeat Problem 9.12 for TM^z modes. The eigenvalues are determined by solving

$$J_m(\beta_\rho b)Y_m(\beta_\rho a) - J_m(\beta_\rho a)Y_m(\beta_\rho b) = 0$$

where $m = p(\pi/\phi_0)$, $p = 1, 2, 3, \ldots$.

9.14. A circular waveguide with radius of 3 cm is made of copper ($\sigma = 5.76 \times 10^7$ S/m). For the dominant TE$_{11}$ and low-loss TE$_{01}$ modes, determine their corresponding cutoff frequencies and attenuation constants (in Np/m and dB/m) at a frequency of 7 GHz. Assume that the waveguide is filled with air.

9.15. Derive the attenuation coefficient α_c for the conduction losses of a circular waveguide of radius a for the TM$_{01}^z$. Show that α_c can be expressed as

$$\alpha_c(\text{TM}_{01}) = \frac{R_s}{\eta a \sqrt{1 - \left(\dfrac{f_c}{f}\right)^2}}$$

where R_s is the surface resistance of the waveguide metal and η is the intrinsic impedance of the medium within the waveguide.

9.16. A circular cavity, as shown in Figure 9-5, has a radius of 6 cm and a height of 10 cm. It is filled with a lossless dielectric with $\varepsilon_r = 4$. Find the:

(a) First four TE^z and/or TM^z modes according to their resonant frequency (in order of ascending values).

(b) Q of the cavity (assuming dominant mode operation). The walls of the cavity are copper ($\sigma = 5.76 \times 10^7$ S/m).

9.17. Design a circular cavity of radius a and height h such that the resonant frequency of the next higher-order mode is 1.5 times greater than the resonant frequency of the dominant mode. Assume that the radius is 4 cm. Find the:

(a) Height of the cavity (in cm).

(b) Resonant frequency of the dominant mode (assume free space within the cavity).

(c) Dielectric constant of the dielectric that must be inserted inside the cavity to reduce the resonant frequency by a factor of 1.5.

9.18. A circular cavity of radius a and height h is completely filled with a dielectric material of dielectric constant ε_r. The height-to-radius ratio is $h/a = 1.9$ where $a = 2$ cm.

(a) Identify the mode with the lowest resonant frequency.

(b) Determine the dielectric constant of the material so that the difference between the resonant frequencies of the lowest to the next lowest order modes is 50 MHz.

9.19. For the radial waveguide of Figure 9-6 determine the maximum spacing h (in m) to insure operation of a single lowest-order mode between the plates up to 300 MHz. Do this individually for TE$_n^z$ modes and TM$_n^z$ modes.

9.20. For the wedged-plate radial waveguide of Figure 9-7, derive for the TE^z modes the expressions for the vector potential component of (9-79), the electric and magnetic fields of (9-79d) through (9-79i), the eigenvalues of (9-79a) through (9-79c), and the wave impedance of (9-79j).

9.21. Given the wedged-plate geometry of Figure 9-7, assume TM^z modes and derive the expressions for the vector potential of (9-80), the electric and magnetic fields of (9-80d) through (9-80i), the eigenvalues of (9-80a) through (9-80c), and the wave impedance of (9-80j).

9.22. If the dielectric constant of the materials that make up the circular dielectric waveguide of Figure 9-8 is very large (usually 30 or greater), the dielectric–air interface along the surface acts almost as an open circuit (see Section 9.5.2). Under these conditions the surface of the dielectric can be approximated by a perfect magnetic conductor (PMC). Assume that the surface of the dielectric rod can be modeled as a PMC and derive, for the TE^z modes, simplified expressions for the vector potential component, electric and magnetic fields, and cutoff frequencies. Assume that the dielectric material is rutile ($\varepsilon_r \simeq 130$). Determine the cutoff frequencies of the lowest two modes when the radius of the rod is 3 cm. Verify with the MATLAB program **CircDielGuide**.

9.23. Repeat Problem 9.22 for the TM^z modes.

9.24. Determine the cutoff frequencies of the first four lowest-order modes (HE, EH, TE, and TM) for a dielectric rod waveguide with radius of 3 cm when the dielectric constant of the material is $\varepsilon_r = 20$, 38, and 130. Verify with the MATLAB program **CircDielGuide**.

9.25. Design a circular dielectric rod waveguide (find its radius in cm) so that the cutoff frequency of the TE_{01}, TM_{01}, and HE_{21} modes is 3 GHz when the dielectric constant of the material is $\varepsilon_r = 2.56$, 4, 9, and 16.

9.26. It is desired to operate a dielectric rod waveguide in the dominant HE_{11} mode over a frequency range of 5 GHz. Design the dielectric rod (find its dielectric constant) to accomplish this when the radius of the rod is $a = 1.315$ and 1.838 cm.

9.27. A dielectric waveguide of circular cross section, radius a, and dielectric constant of 4, has at 3 GHz an effective dielectric constant of $\varepsilon_{reff} = 2.78$. For the dominant mode, determine at 3 GHz the:
   (a) Phase constant along the z and ρ directions inside the waveguide (in rad/cm).
   (b) Attenuation coefficient along the ρ direction outside the waveguide (in Np/cm).

9.28. Design a cylindrical dielectric waveguide of circular cross section so that its dominant single mode operation is 4 GHz. The dielectric constant of the material is 4. Determine, in cm, the radius of the waveguide to accomplish this. Use an exact solution.

9.29. Design a homogeneous dielectric cable to be operated at a dominant, single mode with a center frequency of 50 GHz. The radius of the cable is 1 mm. Determine the:
   (a) Minimum value of the dielectric constant of the cable.
   (b) Frequency range of the dominant, single-mode operation.

9.30. A lossless dielectric waveguide (no PEC walls) of radius a, with $\varepsilon_r \gg 1$, $\mu_r = 1$, has a semi-circular cross section, as shown in Figure P9-30. For the TE^z modes, determine:
   (a) An expression for the cutoff frequency. State the correct eigenvalues and specific allowable indices.
   (b) Cutoff frequency (in GHz) for the lowest order (dominant) mode assuming the radius is $a = 1.5$ cm, $\varepsilon_r = 81$, $\mu_r = 1$.
Show the steps, or explain, as to how you arrive at the specific formulas and corresponding allowable indices you are using.

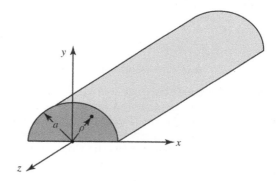

Figure P9-30

9.31. Repeat Problem 9.30 for TM^z modes.

9.32. A cylindrical dielectric waveguide of infinite length is totally composed of a lossless dielectric medium with a dielectric constant $\varepsilon_r \gg 1$, $\mu_r = 1$. The cross section of the waveguide is circular. A very thin PEC baffle is placed along its length at $\phi = 0$ and extends between $0 \le \rho \le a$, as shown in Figure P9-5. Assume the wave is traveling in the $+z$ direction. For TM^z modes, determine:
   (a) A reduced/simplified expression for the appropriate vector potential.
   (b) Expressions for the wavenumbers β_ρ^d and m (representing the ϕ variations) both within the waveguide ($0 \le \rho \le a$). Be specific.

(c) An expression for the cutoff frequencies of the allowable TM^z modes.

(d) Cutoff frequency (in GHz) of the dominant mode when $\varepsilon_r = 81$, $a = 0.45$ in. (1.143 cm).

9.33. Repeat ALL parts of Problem 9.32 for TE^z modes.

9.34. For the dielectric resonator of Figure 9-14a modeled by PMC walls, as shown in Figure 9-14b, derive simplified expressions for the vector potential component, and the electric and magnetic fields when the modes are TE^z and TM^z.

9.35. Assume that the dielectric resonator of Figure 9-14a is modeled by PMC walls as shown in Figure 9-14b.

(a) Determine the lowest TE^z or TM^z mode.

(b) Derive an expression for the Q of the cavity for the lowest-order mode. The only losses associated with the resonator are dielectric losses within the dielectric itself.

(c) Find the resonant frequency.

(d) Compute the Q of the cavity. The resonator material is rutile ($\varepsilon_r \simeq 130$, $\tan\delta_e \simeq 4 \times 10^{-4}$). The radius of the disk is 0.1148 cm and its height is 0.01148 cm.

9.36. A cylindrical dielectric resonator used in microwave integrated circuit (MIC) design is comprised of a PEC ground plane covered with a section of length l of a dielectric rod of semi-circular cross section of radius a (Figure P9-36). Assuming the dielectric constant of the material is much greater than unity ($\varepsilon_r \gg 1$), for TE^z modes, derive the following:

(a) Approximate simplified expression of the appropriate vector potential component.

(b) Determine the allowable nontrivial eigenvalues (be specific).

(c) Compute the resonant frequency of the dominant mode when the radius of the rod section is 2 cm, its length is 4 cm, and the dielectric constant is 81.

9.37. Repeat Problem 9.36 for TM^z modes.

9.38. A lossless dielectric cylindrical resonator antenna (DRA) consists of a dielectric cylinder (no PEC walls), with $\varepsilon_r \gg 1$, $\mu_r = 1$, of circular cross section and height h whose bottom side is resting on a PEC ground plane, as shown in Figure P9-38. Neglecting the presence of the feed and assuming a vector potential $F_z(\rho, \phi, z)$:

$$F_z(\rho,\phi,z) = A_{mn} J_m(\beta_\rho^d \rho)$$
$$\times \left[C_2 \cos(m\phi) + D_2 \sin(m\phi) \right]$$
$$\times \left[C_3 \cos(\beta_z z) + D_3 \sin(\beta_z z) \right]$$

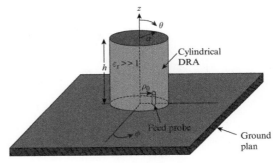

Figure P9-38

(a) Derive an expression for the resonant frequency of the TE$^z_{mnp}$ modes. It is important to indicate the correct allowable indices for m, n, p. You do not have to start the derivation from the very beginning if part of the derivation is already available; indicate the source/expression/equation number from where you start the derivation.

* Assume (start the count) that the lowest value of the indices, m and p, is zero, unless zero is not allowable because it leads to trivial solution.

** *Note*: The correct indices m, n, and p are a major part of this problem.

(b) An expression for the resonant frequency of the dominant mode. Identify the mode and its expression and the indices m, n, and p. Be specific.

* *Note*: The correct indices m, n, and p are a major part of this problem.

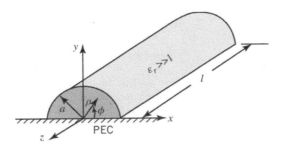

Figure P9-36

(c) The resonant frequency (in GHz) of the dominant mode for $\varepsilon_r = 8.9$, $a = 0.5$ cm, and $h = 0.3$ cm.

9.39. Repeat Problem 9.38 for TM_{mnp}^z.
Use a vector potential $A_z(\rho,\phi,z)$:

$$A_z(\rho,\phi,z) = B_{mn} J_m(\beta_\rho^d \rho)$$
$$\times [C_2 \cos(m\phi) + D_2 \sin(m\phi)]$$
$$\times [C_3 \cos(\beta_z z) + D_3 \sin(\beta_z z)]$$

9.40. A dielectric resonator is made of an angular sector (pie) of radius a, height h, and with a subtended angle of $\phi = \phi_0$. Assuming the dielectric constant of the material is very large ($\varepsilon_r \gg 1$), for TE^z modes:
(a) Derive an approximate simplified expression for the appropriate vector potential.
(b) Determine all the allowable eigenvalues for β_ρ, β_z and m.

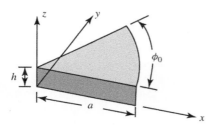

Figure P9-40

9.41. A hybrid dielectric resonator, with a general geometry as shown in Figure 9-14, can be constructed with PEC plates at $z = 0$ and $z = h$ and with open sides. If the dielectric constant of the dielectric is very large, the open dielectric surface can be modeled as a PMC surface. Using such a model for the resonator of Figure 9-14, derive expressions for the resonant frequencies assuming TE^z modes and TM^z modes. This model can also be used as an approximate representation for a circular patch (microstrip) antenna.

9.42. Repeat Problem 9.41 if the top and bottom plates of the resonator are angular sectors each with a subtended angle of ϕ_0, as shown in Figure P9-42. In addition to the resonant frequency, show that the eigenvalues are obtained as solutions to

(a) $\text{TE}^z : J_m(\beta_\rho a) = 0, \quad \beta_\rho = \dfrac{\chi_{mn}}{a}$
$$m = p\left(\frac{\pi}{\phi_0}\right), \quad p = 1, 2, 3, \dots$$

(b) $\text{TM}^z : J_m'(\beta_\rho a) = 0, \quad \beta_\rho = \dfrac{\chi_{mn}'}{a}$
$$m = p\left(\frac{\pi}{\phi_0}\right), \quad p = 0, 1, 2, \dots$$

This model can also be used as an approximate representation for an angular patch (microstrip) antenna.

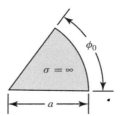

Figure P9-42

9.43. Repeat Problem 9.41 if the top and bottom plates of an annular resonator are annular patches each with inner radius a and outer radius b, as shown in Figure P9-43. In addition to finding the resonant frequency, show that the eigenvalues β_ρ are obtained as solutions to
(a) TE^z :
$$J_m(\beta_\rho a) Y_m(\beta_\rho b) - Y_m(\beta_\rho a) J_m(\beta_\rho b) = 0$$
$$m = 0, 1, 2, \dots$$
(b) TM^z :
$$J_m'(\beta_\rho a) Y_m'(\beta_\rho b) - Y_m'(\beta_\rho a) J_m'(\beta_\rho b) = 0$$
$$m = 0, 1, 2, \dots$$

This model can also be used as an approximate representation for an annular patch (microstrip) antenna.

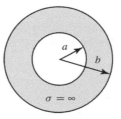

Figure P9-43

9.44. Repeat Problem 9.41 if the top and bottom plates of the resonator are annular sectors each with inner radius a and outer radius b, and subtended angle ϕ_0, as shown in Figure P9-44. In addition to finding the resonant frequency, show that the eigenvalues β_ρ are obtained as solutions to

(a) TE^z :

$$J_m(\beta_\rho a)Y_m(\beta_\rho b) - Y_m(\beta_\rho a)J_m(\beta_\rho b) = 0$$

$$m = p\left(\frac{\pi}{\phi_0}\right), \quad p = 1, 2, 3, \ldots$$

(b) TM^z :

$$J'_m(\beta_\rho a)Y'_m(\beta_\rho b) - Y'_m(\beta_\rho a)J'_m(\beta_\rho b) = 0$$

$$m = p\left(\frac{\pi}{\phi_0}\right), \quad p = 0, 1, 2, \ldots$$

This model can also be used as an approximate representation for an annular patch (microstrip) antenna.

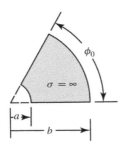

Figure P9-44

9.45. A dielectric resonator of height h has a geometry of an annular sector, with an inner radius a and outer radius b. The material of the resonator has a dielectric constant of $\varepsilon_r \gg 1$. For TE^z modes, derive the characteristic equation that should be used to determine the eigenvalues of the resonator. State any other known eigenvalues.

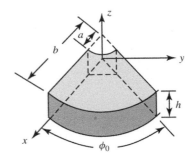

Figure P9-45

9.46. Repeat Problem 9.45 for TM^z modes.

9.47. For the dielectric covered conducting rod of Figure 9-17, assume TE$^z_{mn}$ modes. For within and outside the dielectric sleeving, derive expressions for the vector potential components and the electric and magnetic fields. Also verify (9-124a) through (9-124e).

CHAPTER **10**

Spherical Transmission Lines and Cavities

10.1 INTRODUCTION

Problems involving spherical geometries constitute an important class of electromagnetic boundary-value problems that are used to design transmission lines, cavities, antennas, and scatterers. Some of these may be constructed of metallic walls and others may be of dielectric material. In either case, the field configurations that can be supported by the structure can be obtained by analyzing the structure as a boundary-value problem. We will concern ourselves here with spherical transmission lines and cavities. Scattering by spherical structures will be examined in Chapter 11.

10.2 CONSTRUCTION OF SOLUTIONS

In Chapter 3, Section 3.4.3, we examined the solution of the scalar wave equation in spherical coordinates. It was found that the solution is that of (3-76) where the r, θ, and ϕ variations take the following forms.

1. Radial (r) variations of $f(r)$ can be represented by either:
   a. spherical Bessel functions of the first $[j_n(\beta r)]$ and second $[y_n(\beta r)]$ kind, as given by (3-87a) [these functions are used to represent *standing* waves and are related to the regular Bessel functions by (3-90a) and (3-90b)],
   b. or spherical Hankel functions of the first $[h_n^{(1)}(\beta r)]$ and second $[h_n^{(2)}(\beta r)]$ kind, as given by (3-87b) [these are related to the regular Hankel functions by (3-91a) and (3-91b)].
2. θ variations of $g(\theta)$ can be represented by associated Legendre functions of the first $P_n^m(\cos\theta)$ or second $Q_n^m(\cos\theta)$ kind, as given by either (3-88a) or (3-88b).
3. ϕ variations of $h(\phi)$ can be represented by either complex exponentials or cosinusoids as given, respectively, by (3-89a) and (3-89b).

It must be remembered, however, that the vector wave equation in spherical coordinates, as given by (3-72), does not reduce to three uncoupled scalar Helmholtz wave equations as stated by

Balanis' Advanced Engineering Electromagnetics, Third Edition. Constantine A. Balanis.
© 2024 John Wiley & Sons, Inc. Published 2024 by John Wiley & Sons, Inc.
Companion Website: www.wiley.com/go/balanis/advancedengineeringelectromagnetics3e

(3-73a) through (3-73c). Therefore, using the basic approach outlined in Chapter 6, Section 6.5, we cannot construct field solutions that are TE^r and/or TM^r. Hence, we must look for other approaches for finding field configurations that are supported by spherical structures.

One approach that can be used to find field configurations that are TE^z and/or TM^z will be to represent the potential functions by [1]

$$\text{TE}^z$$

$$\mathbf{A} = 0 \tag{10-1a}$$

$$\mathbf{F} = \hat{\mathbf{a}}_z F_z = (\hat{\mathbf{a}}_r \cos\theta - \hat{\mathbf{a}}_\theta \sin\theta)F_z \tag{10-1b}$$

$$\text{TM}^z$$

$$\mathbf{A} = \hat{\mathbf{a}}_z A_z = (\hat{\mathbf{a}}_r \cos\theta - \hat{\mathbf{a}}_\theta \sin\theta)A_z \tag{10-2a}$$

$$\mathbf{F} = 0 \tag{10-2b}$$

where $F_z(r, \theta, \phi)$ and $A_z(r, \theta, \phi)$ are solutions to the scalar Helmholtz equation in the spherical coordinate system. Other field configurations can be formed as superpositions of TE^z and TM^z modes. Although this is a valid approach to the problem, it will not be pursued here; it is assigned to the reader as an end-of-chapter exercise. Instead, an alternate procedure will be outlined for construction of TE^r and TM^r modes.

The procedure outlined in Chapter 6, Sections 6.5.2 and 6.5.3, for the construction, respectively, of TM and TE field configurations was based on the vector potentials **A** and **F**, as derived in Sections 6.2 and 6.3, respectively, and summarized in Section 6.4. The final forms, which are summarized in Section 6.4, were based on the selection of the *Lorenz conditions* of (6-15) and (6-27), or

$$\psi_e = -\frac{1}{j\omega\mu\varepsilon}\nabla\cdot\mathbf{A} \tag{10-3a}$$

$$\psi_m = -\frac{1}{j\omega\mu\varepsilon}\nabla\cdot\mathbf{F} \tag{10-3b}$$

to represent, respectively, the scalar potential functions ψ_e and ψ_m. If that choice was not made, then the relations of the **E** and **H** fields to the potentials **A** and **F** would take a slightly different form. These forms will be outlined here.

10.2.1   The Vector Potential F (J = 0, M ≠ 0)

According to (6-19), the electric field is related to the potential **F** by

$$\boxed{\mathbf{E}_F = -\frac{1}{\varepsilon}\nabla\times\mathbf{F}} \tag{10-4}$$

Away from the source **M**, the electric and magnetic fields are related by Maxwell's equation

$$\nabla\times\mathbf{E}_F = -j\omega\mu\mathbf{H}_F \tag{10-5}$$

or

$$\boxed{\mathbf{H}_F = -\frac{1}{j\omega\mu}\nabla\times\mathbf{E}_F = \frac{1}{j\omega\mu\varepsilon}\nabla\times\nabla\times\mathbf{F}} \tag{10-5a}$$

Therefore, if the potential **F** can be related to the source (**M**), then $\mathbf{E}_F$ and $\mathbf{H}_F$ can be found using (10-4) and (10-5a). We will attempt to do this next.

Taking the curl of both sides of (10-4) leads to

$$\nabla \times \mathbf{E}_F = -\frac{1}{\varepsilon} \nabla \times \nabla \times \mathbf{F} \qquad (10\text{-}6)$$

Using Maxwell's equation 6-24,

$$\nabla \times \mathbf{E}_F = -\mathbf{M} - j\omega\mu\mathbf{H}_F \qquad (10\text{-}7)$$

and equating it to (10-6) leads to

$$\nabla \times \nabla \times \mathbf{F} = \varepsilon\mathbf{M} + j\omega\mu\varepsilon\mathbf{H}_F \qquad (10\text{-}8)$$

The electric and magnetic fields are also related by Maxwell's equation 6-20,

$$\nabla \times \mathbf{H}_F = j\omega\varepsilon\mathbf{E}_F \qquad (10\text{-}9)$$

Substituting (10-4) into (10-9) and regrouping reduces to

$$\nabla \times \mathbf{H}_F = j\omega\varepsilon\left(-\frac{1}{\varepsilon}\nabla \times \mathbf{F}\right) = -j\omega\nabla \times \mathbf{F} \qquad (10\text{-}10)$$

or

$$\nabla \times (\mathbf{H}_F + j\omega\mathbf{F}) = 0 \qquad (10\text{-}10a)$$

Using the vector identity

$$\nabla \times (-\nabla\psi_m) = 0 \qquad (10\text{-}11)$$

where ψ_m represents an arbitrary scalar potential, and equating it to (10-10a), we can write that

$$\mathbf{H}_F + j\omega\mathbf{F} = -\nabla\psi_m \qquad (10\text{-}12)$$

or

$$\mathbf{H}_F = -j\omega\mathbf{F} - \nabla\psi_m \qquad (10\text{-}12a)$$

Substituting (10-12a) into (10-8) leads to

$$\nabla \times \nabla \times \mathbf{F} = \varepsilon\mathbf{M} + \omega^2\mu\varepsilon\mathbf{F} - j\omega\mu\varepsilon\nabla\psi_m \qquad (10\text{-}13)$$

or

$$\boxed{\nabla \times \nabla \times \mathbf{F} - \omega^2\mu\varepsilon\mathbf{F} = \varepsilon\mathbf{M} - j\omega\mu\varepsilon\nabla\psi_m} \qquad (10\text{-}13a)$$

This is the desired expression, which relates the vector potential $\mathbf{F}$ to the source $\mathbf{M}$ and the associated scalar potential ψ_m. In a source-free ($\mathbf{M} = 0$) region, (10-13a) reduces to

$$\boxed{\nabla \times \nabla \times \mathbf{F} - \omega^2\mu\varepsilon\mathbf{F} = -j\omega\mu\varepsilon\nabla\psi_m} \qquad (10\text{-}14)$$

In a source-free region, the procedure is to solve (10-14) for $\mathbf{F}$, and then use (10-4) and (10-5a) to find, respectively, $\mathbf{E}_F$ and $\mathbf{H}_F$.

10.2.2   The Vector Potential A (J ≠ 0, M = 0)

Following a procedure similar to the one outlined in Section 10.2.1 for the vector potential **F**, it can be shown, by referring also to Section 6.2, that the equations for the vector potential **A** analogous to (10-4), (10-5a), and (10-13a), are

$$\boxed{\mathbf{H}_A = \frac{1}{\mu}\boldsymbol{\nabla}\times\mathbf{A}}$$
(10-15a)

$$\boxed{\mathbf{E}_A = \frac{1}{j\omega\varepsilon}\boldsymbol{\nabla}\times\mathbf{H}_A = \frac{1}{j\omega\mu\varepsilon}\boldsymbol{\nabla}\times\boldsymbol{\nabla}\times\mathbf{A}}$$
(10-15b)

$$\boxed{\boldsymbol{\nabla}\times\boldsymbol{\nabla}\times\mathbf{A} - \omega^2\mu\varepsilon\mathbf{A} = \mu\mathbf{J} - j\omega\mu\varepsilon\boldsymbol{\nabla}\psi_e}$$
(10-15c)

In a source-free region, (10-15c) reduces to

$$\boxed{\boldsymbol{\nabla}\times\boldsymbol{\nabla}\times\mathbf{A} - \omega^2\mu\varepsilon\mathbf{A} = -j\omega\mu\varepsilon\boldsymbol{\nabla}\psi_e}$$
(10-15d)

The details are left as end-of-chapter exercises for the reader.

In a source-free region, the procedure is to solve (10-15d) for **A**, and then use (10-15a) and (10-15b) to find, respectively, $\mathbf{H}_A$ and $\mathbf{E}_A$.

10.2.3   The Vector Potentials F and A

The total fields that are due to both potentials **F** and **A** are found as superpositions of the fields of Sections 10.2.1 and 10.2.2. Doing this, we have that the total fields are obtained using

$$\boxed{\mathbf{E} = \mathbf{E}_F + \mathbf{E}_A = -\frac{1}{\varepsilon}\boldsymbol{\nabla}\times\mathbf{F} + \frac{1}{j\omega\varepsilon}\boldsymbol{\nabla}\times\mathbf{H}_A = -\frac{1}{\varepsilon}\boldsymbol{\nabla}\times\mathbf{F} + \frac{1}{j\omega\mu\varepsilon}\boldsymbol{\nabla}\times\boldsymbol{\nabla}\times\mathbf{A}}$$
(10-16a)

$$\boxed{\mathbf{H} = \mathbf{H}_F + \mathbf{H}_A = -\frac{1}{j\omega\mu}\boldsymbol{\nabla}\times\mathbf{E}_F + \frac{1}{\mu}\boldsymbol{\nabla}\times\mathbf{A} = \frac{1}{j\omega\mu\varepsilon}\boldsymbol{\nabla}\times\boldsymbol{\nabla}\times\mathbf{F} + \frac{1}{\mu}\boldsymbol{\nabla}\times\mathbf{A}}$$
(10-16b)

where **F** and **A** are, respectively, solutions to

$$\boxed{\boldsymbol{\nabla}\times\boldsymbol{\nabla}\times\mathbf{F} - \omega^2\mu\varepsilon\mathbf{F} = \varepsilon\mathbf{M} - j\omega\mu\varepsilon\boldsymbol{\nabla}\psi_m}$$
(10-16c)

$$\boxed{\boldsymbol{\nabla}\times\boldsymbol{\nabla}\times\mathbf{A} - \omega^2\mu\varepsilon\mathbf{A} = \mu\mathbf{J} - j\omega\mu\varepsilon\boldsymbol{\nabla}\psi_e}$$
(10-16d)

which for a source-free region (**M** = **J** = 0) reduce to

$$\boxed{\boldsymbol{\nabla}\times\boldsymbol{\nabla}\times\mathbf{F} - \omega^2\mu\varepsilon\mathbf{F} = -j\omega\mu\varepsilon\boldsymbol{\nabla}\psi_m}$$
(10-16e)

$$\boxed{\boldsymbol{\nabla}\times\boldsymbol{\nabla}\times\mathbf{A} - \omega^2\mu\varepsilon\mathbf{A} = -j\omega\mu\varepsilon\boldsymbol{\nabla}\psi_e}$$
(10-16f)

We will attempt now to form TE^r and TM^r mode field solutions using (10-16a) through (10-16f).

10.2.4 Transverse Electric (TE) Modes: Source-Free Region

It was stated previously in Section 6.5.3 that TE modes to any direction in any coordinate system can be obtained by selecting the vector potential **F** to have only a nonvanishing component in that direction while simultaneously letting **A** = 0. The nonvanishing component of **F** was obtained as a solution to the scalar wave equation 6-31. The same procedure will be used here except that instead of the nonvanishing component of **F** being a solution to (6-31), which does not reduce in spherical coordinates to three scalar noncoupled wave equations, it will be a solution to (10-16e). The nonvanishing component of **F** will be the one that coincides with the direction along which the TE modes are desired. Let us construct solutions that are TE^r in a spherical coordinate system.

TE^r field configurations are constructed by letting the vector potentials **F** and **A** be equal to

$$\boxed{\mathbf{F} = \hat{\mathbf{a}}_r F_r(r, \theta, \phi)} \tag{10-17a}$$

$$\boxed{\mathbf{A} = 0} \tag{10-17b}$$

Since F_r is not a solution to the scalar Helmholtz equation

$$\nabla^2 \mathbf{F} = \nabla^2 (\hat{\mathbf{a}}_r F_r) \neq \hat{\mathbf{a}}_r \nabla^2 F_r \tag{10-18}$$

we will resort, for a source-free region, to (10-16e).

Expanding (10-16e) using (10-17a) leads to

$$\nabla \times \mathbf{F} = \nabla \times (\hat{\mathbf{a}}_r F_r) = \hat{\mathbf{a}}_\theta \frac{1}{r \sin \theta} \frac{\partial F_r}{\partial \phi} - \hat{\mathbf{a}}_\phi \frac{1}{r} \frac{\partial F_r}{\partial \theta} \tag{10-19a}$$

$$\nabla \times \nabla \times \mathbf{F} = \hat{\mathbf{a}}_r \left\{ \frac{1}{r \sin \theta} \left[\frac{\partial}{\partial \theta} \left(-\frac{\sin \theta}{r} \frac{\partial F_r}{\partial \theta} \right) - \frac{\partial}{\partial \phi} \left(\frac{1}{r \sin \theta} \frac{\partial F_r}{\partial \phi} \right) \right] \right\}$$
$$+ \hat{\mathbf{a}}_\theta \left[\frac{1}{r} \left(\frac{\partial^2 F_r}{\partial r \partial \theta} \right) \right] + \hat{\mathbf{a}}_\phi \left(\frac{1}{r \sin \theta} \frac{\partial^2 F_r}{\partial r \partial \phi} \right) \tag{10-19b}$$

$$\nabla \psi_m = \hat{\mathbf{a}}_r \frac{\partial \psi_m}{\partial r} + \hat{\mathbf{a}}_\theta \frac{1}{r} \frac{\partial \psi_m}{\partial \theta} + \hat{\mathbf{a}}_\phi \frac{1}{r \sin \theta} \frac{\partial \psi_m}{\partial \phi} \tag{10-19c}$$

Thus for the r, θ, and ϕ components, (10-16e) reduces to

$$\boxed{\frac{1}{r \sin \theta} \left[-\frac{\partial}{\partial \theta} \left(\frac{\sin \theta}{r} \frac{\partial F_r}{\partial \theta} \right) - \frac{\partial}{\partial \phi} \left(\frac{1}{r \sin \theta} \frac{\partial F_r}{\partial \phi} \right) \right] - \beta^2 F_r = -j \omega \mu \varepsilon \frac{\partial \psi_m}{\partial r}} \tag{10-20a}$$

$$\frac{1}{r} \frac{\partial^2 F_r}{\partial r \partial \theta} = -j \frac{\omega \mu \varepsilon}{r} \frac{\partial \psi_m}{\partial \theta} \Rightarrow \boxed{\frac{\partial^2 F_r}{\partial r \partial \theta} = \frac{\partial}{\partial \theta} \left(\frac{\partial F_r}{\partial r} \right) = \frac{\partial}{\partial \theta} (-j \omega \mu \varepsilon \psi_m)} \tag{10-20b}$$

$$\frac{1}{r \sin \theta} \frac{\partial^2 F_r}{\partial r \partial \phi} = -j \frac{\omega \mu \varepsilon}{r \sin \theta} \frac{\partial \psi_m}{\partial \phi} \Rightarrow \boxed{\frac{\partial^2 F_r}{\partial r \partial \phi} = \frac{\partial}{\partial \phi} \left(\frac{\partial F_r}{\partial r} \right) = \frac{\partial}{\partial \phi} (-j \omega \mu \varepsilon \psi_m)} \tag{10-20c}$$

where $\beta^2 = \omega^2 \mu \varepsilon$. The last two equations, (10-20b) and (10-20c), are satisfied simultaneously if

$$\frac{\partial F_r}{\partial r} = -j \omega \mu \varepsilon \psi_m \Rightarrow \boxed{\psi_m = -\frac{1}{j \omega \mu \varepsilon} \frac{\partial F_r}{\partial r}} \tag{10-21}$$

With the preceding relation for the scalar potential ψ_m, we need to find an *uncoupled* differential equation for F_r. To do this, we substitute (10-21) into (10-20a), which leads to

$$-\frac{1}{r^2 \sin\theta}\frac{\partial}{\partial\theta}\left(\sin\theta\frac{\partial F_r}{\partial\theta}\right) - \frac{1}{r^2 \sin^2\theta}\frac{\partial^2 F_r}{\partial\phi^2} - \beta^2 F_r = \frac{\partial^2 F_r}{\partial r^2} \tag{10-22}$$

or

$$\frac{\partial^2 F_r}{\partial r^2} + \frac{1}{r^2 \sin\theta}\frac{\partial}{\partial\theta}\left(\sin\theta\frac{\partial F_r}{\partial\theta}\right) + \frac{1}{r^2 \sin^2\theta}\frac{\partial^2 F_r}{\partial\phi^2} + \beta^2 F_r = 0 \tag{10-22a}$$

which can also be written in succinct form as

$$\boxed{\left(\nabla^2 + \beta^2\right)\frac{F_r}{r} = 0} \tag{10-22b}$$

Therefore, using this procedure, the ratio F_r/r satisfies the scalar Helmholtz wave equation but not F_r itself. A solution of F_r using (10-22b) allows us to find $\mathbf{E}_F$ and $\mathbf{H}_F$ using, respectively, (10-4) and (10-5a).

The solution of (10-22b) will be pursued in Section 10.2.6. In the meantime, the electric and magnetic field components can be written in terms of F_r by expanding (10-4) and (10-5a):

$$\underline{\text{TE}^r(\mathbf{F} = \hat{\mathbf{a}}_r F_r, \mathbf{A} = 0)}$$

$$\boxed{\mathbf{E}_F = -\frac{1}{\varepsilon}\nabla\times\mathbf{F}} \tag{10-23}$$

or

$$E_r = 0 \tag{10-23a}$$

$$E_\theta = -\frac{1}{\varepsilon}\frac{1}{r\sin\theta}\frac{\partial F_r}{\partial\phi} \tag{10-23b}$$

$$E_\phi = \frac{1}{\varepsilon}\frac{1}{r}\frac{\partial F_r}{\partial\theta} \tag{10-23c}$$

$$\boxed{\mathbf{H}_F = \frac{1}{j\omega\mu\varepsilon}\nabla\times\nabla\times\mathbf{F}} \tag{10-24}$$

or

$$H_r = \frac{1}{j\omega\mu\varepsilon}\left(\frac{\partial^2}{\partial r^2} + \beta^2\right)F_r \tag{10-24a}$$

$$H_\theta = \frac{1}{j\omega\mu\varepsilon}\frac{1}{r}\frac{\partial^2 F_r}{\partial r\,\partial\theta} \tag{10-24b}$$

$$H_\phi = \frac{1}{j\omega\mu\varepsilon}\frac{1}{r\sin\theta}\frac{\partial^2 F_r}{\partial r\,\partial\phi} \tag{10-24c}$$

where F_r/r is a solution to (10-22b).

10.2.5   Transverse Magnetic (TM) Modes: Source-Free Region

Following a procedure similar to the one outlined in the previous section for the TE^r modes, it can be shown that the TM^r fields in spherical coordinates can be constructed by letting the vector potentials **F** and **A** be equal to

$$\boxed{\mathbf{F} = 0}$$
(10-25a)

$$\boxed{\mathbf{A} = \hat{\mathbf{a}}_r A_r(r,\,\theta,\,\phi)}$$
(10-25b)

where the ratio A_r/r, and not A_r, is a solution to the scalar Helmholtz wave equation

$$\boxed{\left(\nabla^2 + \beta^2\right)\frac{A_r}{r} = 0}$$
(10-26)

The solution of (10-26) will be pursued in Section 10.2.6. In the meantime, the electric and magnetic field components can be written in terms of A_r, by expanding (10-15b) and (10-15a), as

$$\text{TM}^r\,(\mathbf{F} = 0, \mathbf{A} = \hat{\mathbf{a}}_r A_r)$$

$$\boxed{\mathbf{E}_A = \frac{1}{j\omega\mu\varepsilon}\nabla\times\nabla\times\mathbf{A}}$$
(10-27)

or

$$E_r = \frac{1}{j\omega\mu\varepsilon}\left(\frac{\partial^2}{\partial r^2} + \beta^2\right)A_r$$
(10-27a)

$$E_\theta = \frac{1}{j\omega\mu\varepsilon}\frac{1}{r}\frac{\partial^2 A_r}{\partial r\,\partial\theta}$$
(10-27b)

$$E_\phi = \frac{1}{j\omega\mu\varepsilon}\frac{1}{r\sin\theta}\frac{\partial^2 A_r}{\partial r\,\partial\phi}$$
(10-27c)

$$\boxed{\mathbf{H}_A = \frac{1}{\mu}\nabla\times\mathbf{A}}$$
(10-28)

or

$$H_r = 0$$
(10-28a)

$$H_\theta = \frac{1}{\mu}\frac{1}{r\sin\theta}\frac{\partial A_r}{\partial\phi}$$
(10-28b)

$$H_\phi = -\frac{1}{\mu}\frac{1}{r}\frac{\partial A_r}{\partial\theta}$$
(10-28c)

where A_r/r is a solution to (10-26).

10.2.6   Solution of the Scalar Helmholtz Wave Equation

To find the TE^r and/or TM^r field of Sections 10.2.4 and 10.2.5 as given, respectively, by (10-23) through (10-24c) and (10-27) through (10-28c), solutions to the scalar Helmholtz wave equations 10-22b and 10-26 must be obtained for F_r/r and A_r/r (and thus, F_r and A_r). Both solutions are of the same form

$$(\nabla^2 + \beta^2)\psi = 0 \tag{10-29}$$

where

$$\psi = \begin{cases} \dfrac{F_r}{r} & \text{for } \text{TE}^r \text{ modes} \tag{10-29a} \\[2ex] \dfrac{A_r}{r} & \text{for } \text{TM}^r \text{ modes} \tag{10-29b} \end{cases}$$

Since the solution of ψ from (10-29) must be multiplied by r to obtain solutions for F_r or A_r, then appropriate solutions for F_r and A_r must be equal to the product of $r\psi$. The solution for F_r or A_r of (10-29) through (10-29b) must take the separable form of

$$\left.\begin{array}{l} F_r(r, \theta, \phi) \\ A_r(r, \theta, \phi) \end{array}\right\} = f(r)g(\theta)h(\phi) \tag{10-30}$$

where $f(r)$, $g(\theta)$, and $h(\phi)$ must be represented by appropriate wave functions that satisfy the wave equation in spherical coordinates. According to (3-88a) or (3-88b), $g(\theta)$ can be represented by associated Legendre functions of the first kind $P_n^m(\cos\theta)$, or second kind $Q_n^m(\cos\theta)$, whereas $h(\phi)$ can be represented by either complex exponentials or cosinusoids as given, respectively, by (3-89a) and (3-89b).

Since the solution of ψ as given by (10-29) must be multiplied by r in order to obtain solutions to F_r and A_r as given by (10-30), it is most convenient to represent $f(r)$ *not* by spherical Bessel $[j_n(\beta r), y_n(\beta r)]$ or Hankel $[h_n^{(1)}(\beta r), h_n^{(2)}(\beta r)]$ functions, *but by another form* of spherical Bessel and Hankel functions denoted by $\hat{B}_n(\beta r)$ [for either $\hat{J}_n(\beta r)$, $\hat{Y}_n(\beta r)$, $\hat{H}_n^{(1)}(\beta r)$, or $\hat{H}_n^{(2)}(\beta r)$]. These are related to the regular spherical Bessel and Hankel functions denoted by $b_n(\beta r)$ [for either $j_n(\beta r)$, $y_n(\beta r)$, $h_n^{(1)}(\beta r)$, or $h_n^{(2)}(\beta r)$] by

$$\hat{B}_n(\beta r) = \beta r\, b_n(\beta r) = \beta r\sqrt{\frac{\pi}{2\beta r}}B_{n+1/2}(\beta r) = \sqrt{\frac{\pi\beta r}{2}}B_{n+1/2}(\beta r) \tag{10-31}$$

where $B_{n+1/2}(\beta r)$ is used to represent the regular cylindrical Bessel or Hankel functions of $J_{n+1/2}(\beta r)$, $Y_{n+1/2}(\beta r)$, $H_{n+1/2}^{(1)}(\beta r)$, and $H_{n+1/2}^{(2)}(\beta r)$. These new spherical Bessel and Hankel functions were introduced by Schelkunoff [2] and satisfy the differential equation

$$\left[\frac{d^2}{dr^2} + \beta^2 - \frac{n(n+1)}{r^2}\right]\hat{B}_n = 0 \tag{10-32}$$

which is obtained by substituting $b_n(\beta r) = \hat{B}_n(\beta r)/\beta r$ in

$$\frac{d}{dr}\left(r^2\frac{db_n}{dr}\right) + [(\beta r)^2 - n(n+1)]b_n = 0 \tag{10-33}$$

Therefore, the solutions for $f(r)$ of (10-30) are of the new form of spherical Bessel or Hankel functions denoted by

$$f_1(r) = A_1\hat{J}_n(\beta r) + B_1\hat{Y}_n(\beta r) \tag{10-34a}$$

or

$$f_2(r) = C_1\hat{H}_n^{(1)}(\beta r) + D_1\hat{H}_n^{(2)}(\beta r) \tag{10-34b}$$

which are related to the regular Bessel and Hankel functions by

$$\hat{J}_n(\beta r) = \sqrt{\frac{\pi \beta r}{2}} J_{n+1/2}(\beta r) \tag{10-35a}$$

$$\hat{Y}_n(\beta r) = \sqrt{\frac{\pi \beta r}{2}} Y_{n+1/2}(\beta r) \tag{10-35b}$$

$$\hat{H}_n^{(1)}(\beta r) = \sqrt{\frac{\pi \beta r}{2}} H_{n+1/2}^{(1)}(\beta r) \tag{10-35c}$$

$$\hat{H}_n^{(2)}(\beta r) = \sqrt{\frac{\pi \beta r}{2}} H_{n+1/2}^{(2)}(\beta r) \tag{10-35d}$$

The total solution for F_r or A_r of (10-30) will be the product of the appropriate spherical wave functions representing $f(r)$, $g(\theta)$, and $h(\phi)$.

10.3 BICONICAL TRANSMISSION LINE

One form of a transmission line whose geometry conforms to the spherical orthogonal coordinate system is the biconical structure of Figure 10-1. Typically, this configuration is also representative of the biconical antenna [3–9] which exhibits very broad band frequency characteristics. Sets of fields that can be supported by such a structure can be either TE^r, TM^r, or TEM^r. Solutions to these will be examined here.

10.3.1 Transverse Electric (TE^r) Modes

According to the procedure established in Section 10.2.4, transverse (to the radial direction) electric modes (TE^r) can be constructed by choosing the potentials $\mathbf{F}$ and $\mathbf{A}$ according to (10-17a)

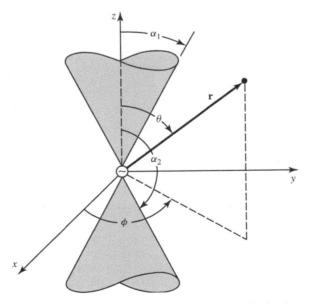

Figure 10-1 Geometry of biconical transmission line. (Source: C. A. Balanis, *Antenna Theory: Analysis and Design*. Third Edition. Copyright © 2005, John Wiley & Sons, Inc. Reprinted by permission of John Wiley & Sons, Inc.)

and (10-17b). The scalar component F_r of the vector potential must satisfy the vector wave equation (10-22b), whose solution takes the form of (3-85), or

$$F_r(r, \theta, \phi) = f(r)g(\theta)h(\phi) \tag{10-36}$$

where

$f(r) =$ a solution to (10-32) as given by either (10-34a) or (10-34b) [the form (10-34b) is chosen here]

$g(\theta) =$ a solution of (3-86b) as given by either (3-88a) or (3-88b) [the form (3-88a) is chosen here]

$h(\phi) =$ is a solution to (3-86c) as given by (3-89a) or (3-89b) [the form (3-89b) is chosen here to represent standing waves]

Therefore, F_r of (10-36) can be written, assuming the source is placed at the apex and is generating outwardly traveling waves [$C_1 = 0$ in (10-34b)], as

$$[F_r(r,\theta,\phi)]_{mn} = D_1 \hat{H}_n^{(2)}(\beta r)[A_2 P_n^m(\cos\theta) + B_2 P_n^m(-\cos\theta)]$$
$$\times [C_3 \cos(m\phi) + D_3 \sin(m\phi)] \tag{10-37}$$

where $m =$ nonnegative integer ($m = 0, 1, 2, \ldots$).

The corresponding electric and magnetic fields can be found using (10-23) through (10-24c) and the eigenvalues of n can be determined by applying the boundary conditions

$$E_\phi(0 \leq r \leq \infty, \theta = \alpha_1, 0 \leq \phi \leq 2\pi) = E_\phi(0 \leq r \leq \infty, \theta = \alpha_2, 0 \leq \phi \leq 2\pi) = 0 \tag{10-38}$$

According to (10-23c),

$$E_\phi = \frac{1}{\varepsilon}\frac{1}{r}\frac{\partial F_r}{\partial \theta} = \frac{D_1}{\varepsilon}\frac{1}{r}\hat{H}_n^{(2)}(\beta r)\left[A_2 \frac{dP_n^m(\cos\theta)}{d\theta} + B_2 \frac{dP_n^m(-\cos\theta)}{d\theta}\right]$$
$$\times [C_3 \cos(m\phi) + D_3 \sin(m\phi)] \tag{10-39}$$

Applying the first boundary condition of (10-38) leads to

$$E_\phi(0 \leq r \leq \infty, \theta = \alpha_1, 0 \leq \phi \leq 2\pi)$$
$$= \frac{D_1}{\varepsilon}\frac{1}{r}\hat{H}_n^{(2)}(\beta r)\left[A_2 \frac{dP_n^m(\cos\theta)}{d\theta} + B_2 \frac{dP_n^m(-\cos\theta)}{d\theta}\right]_{\theta=\alpha_1}$$
$$\times [C_3 \cos(m\phi) + D_3 \sin(m\phi)] = 0$$
$$E_\phi(0 \leq r \leq \infty, \theta = \alpha_1, 0 \leq \phi \leq 2\pi)$$
$$= \frac{D_1}{\varepsilon}\frac{1}{r}\hat{H}_n^{(2)}(\beta r)\left[A_2 \frac{dP_n^m(\cos\alpha_1)}{d\alpha_1} + B_2 \frac{dP_n^m(-\cos\alpha_1)}{d\alpha_1}\right]$$
$$\times [C_3 \cos(m\phi) + D_3 \sin(m\phi)] = 0 \tag{10-40a}$$

and the second boundary condition of (10-38) leads to

$$E_\phi(0 \leq r \leq \infty, \theta = \alpha_2, 0 \leq \phi \leq 2\pi)$$
$$= \frac{D_1}{\varepsilon}\frac{1}{r}\hat{H}_n^{(2)}(\beta r)\left[A_2 \frac{dP_n^m(\cos\alpha_2)}{d\alpha_2} + B_2 \frac{dP_n^m(-\cos\alpha_2)}{d\alpha_2}\right]$$
$$\times [C_3 \cos(m\phi) + D_3 \sin(m\phi)] = 0 \tag{10-40b}$$

Equations 10-40a and 10-40b reduce to

$$A_2 \frac{dP_n^m(\cos\alpha_1)}{d\alpha_1} + B_2 \frac{dP_n^m(-\cos\alpha_1)}{d\alpha_1} = 0 \qquad (10\text{-}41a)$$

$$A_2 \frac{dP_n^m(\cos\alpha_2)}{d\alpha_2} + B_2 \frac{dP_n^m(-\cos\alpha_2)}{d\alpha_2} = 0 \qquad (10\text{-}41b)$$

which are satisfied provided the determinant of (10-41a) and (10-41b) vanishes, that is,

$$\boxed{\frac{dP_n^m(\cos\alpha_1)}{d\alpha_1}\frac{dP_n^m(-\cos\alpha_2)}{d\alpha_2} - \frac{dP_n^m(-\cos\alpha_1)}{d\alpha_1}\frac{dP_n^m(\cos\alpha_2)}{d\alpha_2} = 0} \qquad (10\text{-}42)$$

Therefore, the eigenvalues of n are found as solutions to (10-42), which usually is not necessarily a very easy task.

10.3.2 Transverse Magnetic (TM^r) Modes

Following a procedure similar to that of the previous section and using the formulations of Section 10.2.5, it can be shown that for TM^r modes the potential component A_r of (10-26) reduces to

$$[A_r(r,\theta,\phi)]_{mn} = D_1\hat{H}_n^{(2)}(\beta r)[A_2 P_n^m(\cos\theta) + B_2 P_n^m(-\cos\theta)]$$
$$\times [C_3\cos(m\phi) + D_3\sin(m\phi)] \qquad (10\text{-}43)$$

where $m = $ integer ($m = 0, 1, 2, \ldots$). The values of n are determined by applying the boundary conditions.

The corresponding electric and magnetic fields are obtained using (10-27) through (10-28c). By applying the boundary conditions

$$E_r(0 \le r \le \infty, \theta = \alpha_1, 0 \le \phi \le 2\pi) = E_r(0 \le r \le \infty, \theta = \alpha_2, 0 \le \phi \le 2\pi) = 0 \quad (10\text{-}44a)$$

or

$$E_\phi(0 \le r \le \infty, \theta = \alpha_1, 0 \le \phi \le 2\pi) = E_\phi(0 \le r \le \infty, \theta = \alpha_2, 0 \le \phi \le 2\pi) = 0 \quad (10\text{-}44b)$$

it can be shown that the eigenvalues of n are obtained as solutions to

$$\boxed{P_n^m(\cos\alpha_1)P_n^m(-\cos\alpha_2) - P_n^m(-\cos\alpha_1)P_n^m(\cos\alpha_2) = 0} \qquad (10\text{-}45)$$

This usually is not necessarily a very easy task.

10.3.3 Transverse Electromagnetic (TEM^r) Modes

The lowest-order (dominant) mode of the biconical transmission line is the one for which $m = 0$ and $n = 0$. For this mode both (10-42) and (10-45) are satisfied and the potential components of (10-37) and (10-43) vanish. However, for $m = n = 0$, (10-43) could be redefined as the limit as $n \to 0$. Instead, it is usually more convenient to alternately represent the TEM mode by the TM$_{00}$ which is defined, using (3-88b) to represent $g(\theta)$, by [1]

$$(A_r)_{00} = B_{00} \hat{H}_0^{(2)}(\beta r) Q_0(\cos\theta) \tag{10-46}$$

since $P_0^0(\cos\theta) = P_0(\cos\theta) = 1$. The Legendre polynomial $Q_0(\cos\theta)$ can also be represented by

$$Q_0(\cos\theta) = \ln\left[\cot\left(\frac{\theta}{2}\right)\right] \tag{10-47a}$$

and the spherical Hankel function $\hat{H}_0^{(2)}(\beta r)$ can be replaced by its asymptotic form for large arguments of

$$\hat{H}_0^{(2)}(\beta r) \overset{\beta r \to \text{large}}{\simeq} je^{-j\beta r} \tag{10-47b}$$

Using (10-47a) and (10-47b) reduces (10-46), for large observational distances ($\beta r \to$ large), to

$$(A_r)_{00} \simeq jB_{00} \ln\left[\cot\left(\frac{\theta}{2}\right)\right]e^{-j\beta r} \tag{10-48}$$

The corresponding electric and magnetic field components are given, according to (10-27) through (10-28c), by [3]

$$E_r = \frac{1}{j\omega\mu\varepsilon}\left(\frac{\partial^2}{\partial r^2} + \beta^2\right)A_r \simeq 0 \tag{10-49a}$$

$$E_\theta = \frac{1}{j\omega\mu\varepsilon}\frac{1}{r}\frac{\partial^2 A_r}{\partial r\,\partial\theta} = jB_{00}\frac{\beta}{\omega\mu\varepsilon}\frac{1}{r}\frac{1}{\sin\theta}e^{-j\beta r} \tag{10-49b}$$

$$E_\phi = \frac{1}{j\omega\mu\varepsilon}\frac{1}{r\sin\theta}\frac{\partial^2 A_r}{\partial r\,\partial\phi} = 0 \tag{10-49c}$$

$$H_r = 0 \tag{10-49d}$$

$$H_\theta = \frac{1}{\mu}\frac{1}{r\sin\theta}\frac{\partial A_r}{\partial\phi} = 0 \tag{10-49e}$$

$$H_\phi = -\frac{1}{\mu}\frac{1}{r}\frac{\partial A_r}{\partial\theta} = jB_{00}\frac{1}{\mu r\sin\theta}e^{-j\beta r} \tag{10-49f}$$

Using these equations, we can write the wave impedance in the radial direction as

$$Z_w^{+r} = \frac{E_\theta}{H_\phi} = \frac{\beta}{\omega\varepsilon} = \sqrt{\frac{\mu}{\varepsilon}} = \eta \tag{10-50}$$

which is the same as the intrinsic impedance of the medium.

An impedance of greater interest is the characteristic impedance that is defined in terms of voltages and currents. The voltage between two corresponding points on the cones, a distance r from the origin, is found by

$$V(r) = \int_{\alpha_1}^{\alpha_2} \mathbf{E}\cdot d\ell = \int_{\alpha_1}^{\alpha_2}(\hat{\mathbf{a}}_\theta E_\theta)\cdot(\hat{\mathbf{a}}_\theta r\,d\theta)$$

$$= \int_{\alpha_1}^{\alpha_2} E_\theta r\,d\theta = jB_{00}\frac{\beta e^{-j\beta r}}{\omega\mu\varepsilon}\int_{\alpha_1}^{\alpha_2}\frac{d\theta}{\sin(\theta)}$$

$$V(r) = jB_{00}\frac{\beta e^{-j\beta r}}{\omega\mu\varepsilon}\ln\left[\frac{\cot\left(\dfrac{\alpha_1}{2}\right)}{\cot\left(\dfrac{\alpha_2}{2}\right)}\right] \tag{10-51a}$$

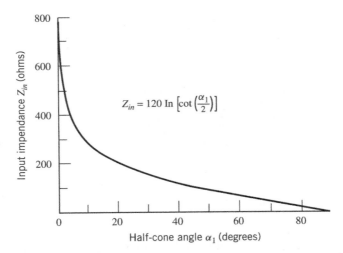

Figure 10-2 Input impedance of biconical transmission line. (Source: C. A. Balanis, *Antenna Theory: Analysis and Design*. Third Edition. Copyright © 2005, John Wiley & Sons, Inc. Reprinted by permission of John Wiley & Sons, Inc.)

The current on the surface of the cones, a distance r from the origin, is found by using (10-49f) as

$$I(r) = \oint_C \mathbf{H} \cdot d\ell = \int_0^{2\pi} (\hat{\mathbf{a}}_\phi H_\phi) \cdot (\hat{\mathbf{a}}_\phi r \sin\theta \, d\phi) = \int_0^{2\pi} H_\phi r \sin\theta \, d\phi = jB_{00} \frac{2\pi e^{-j\beta r}}{\mu} \quad (10\text{-}51\text{b})$$

Taking the ratio of (10-51a) to (10-51b), we can define and write the characteristic impedance as

$$Z_c \equiv \frac{V(r)}{I(r)} = \frac{\beta}{2\pi\omega\varepsilon} \ln \left| \frac{\cot\left(\frac{\alpha_1}{2}\right)}{\cot\left(\frac{\alpha_2}{2}\right)} \right| = \frac{\sqrt{\frac{\mu}{\varepsilon}}}{2\pi} \ln \left| \frac{\cot\left(\frac{\alpha_1}{2}\right)}{\cot\left(\frac{\alpha_2}{2}\right)} \right| \equiv Z_{in} \quad (10\text{-}52)$$

Since the characteristic impedance is not a function of the radial distance r, it also represents the input impedance of the antenna at the feed terminals. For a symmetrical structure ($\alpha_2 = \pi - \alpha_1$), (10-52) reduces to

$$Z_c = \frac{\sqrt{\frac{\mu}{\varepsilon}}}{2\pi} \ln \left[\cot\left(\frac{\alpha_1}{2}\right) \right]^2 = \frac{\sqrt{\frac{\mu}{\varepsilon}}}{\pi} \ln \left[\cot\left(\frac{\alpha_1}{2}\right) \right] = \frac{\eta}{\pi} \ln \left[\cot\left(\frac{\alpha_1}{2}\right) \right] = Z_{in} \quad (10\text{-}52\text{a})$$

It is apparent that the transmission line, or alternately the antenna of Figure 10-1, is a very broad band structure since its characteristic or input impedance is only a function of the included angle of the cone. A plot of (10-52a) as a function of α_1 is shown in Figure 10-2.

There are numerous other transmission lines whose geometry can be represented by the spherical orthogonal coordinate systems. They will not be discussed here but some will be assigned to the reader as end-of-chapter exercises.

10.4 THE SPHERICAL CAVITY

The metallic spherical cavity of Figure 10-3 represents a popular and classic geometry to design resonators. The field configurations that can be supported by such a structure can be TE^r and/or

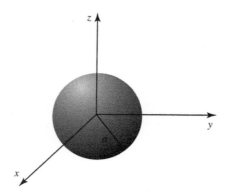

Figure 10-3   Geometry of spherical cavity.

TM^r; both will be examined here. In addition to the field expressions, the resonant frequencies and the quality factors will be the quantities of interest.

10.4.1   Transverse Electric (TE^r) Modes

The TE^r modes in the cavity can be formed by letting the vector potentials **F** and **A** be equal to (10-17a) and (10-17b), respectively. The most appropriate form for the vector potential component F_r is

$$F_r(r, \theta, \phi) = [A_1 \hat{J}_n(\beta r) + B_1 \hat{Y}_n(\beta r)][C_2 P_n^m(\cos\theta) + D_2 Q_n^m(\cos\theta)]$$
$$\times [C_3 \cos(m\phi) + D_3 \sin(m\phi)] \tag{10-53}$$

where m and n are integers. The fields must be finite at $r = 0$; thus, $B_1 = 0$ since $\hat{Y}_n(\beta r)$ possesses a singularity at $r = 0$. Additionally, the fields must also be finite at $\theta = 0, \pi$. Therefore, $D_2 = 0$ since $Q_n^m(\cos\theta)$ possesses a singularity at $\theta = 0, \pi$. Also, because the Legendre polynomial $P_0(w) = 1$, and $P_0^m(w)$ is related to $P_0(w)$ by [10–15]

$$P_0^m(w) = (-1)^m (1 - w^2)^{m/2} \frac{d^m P_0(w)}{dw^m} = 0, \quad m = 1, 2, \ldots \tag{10-54}$$

then, for nontrivial (nonzero) solutions, $n = 1, 2, 3, \ldots$. Therefore, (10-53) reduces to

$$(F_r)_{mnp} = A_{mnp} \hat{J}_n(\beta r) P_n^m(\cos\theta)[C_3 \cos(m\phi) + D_3 \sin(m\phi)] \tag{10-55}$$

It should also be stated that $P_n^m(\cos\theta) = 0$ if $m > n$.

The corresponding electric and magnetic fields are found using (10-23) through (10-24c). Thus, we can write the electric field components of (10-23) through (10-23c), using (10-55), as

$$E_r = 0 \tag{10-56a}$$

$$E_\theta = -\frac{1}{\varepsilon}\frac{1}{r\sin\theta}\frac{\partial F_r}{\partial\phi}$$
$$= -A_{mnp}\frac{m}{\varepsilon}\frac{1}{r\sin\theta}\hat{J}_n(\beta r)P_n^m(\cos\theta)[-C_3\sin(m\phi) + D_3\cos(m\phi)] \tag{10-56b}$$

$$E_\phi = \frac{1}{\varepsilon}\frac{1}{r}\frac{\partial F_r}{\partial\theta} = A_{mnp}\frac{1}{\varepsilon}\frac{1}{r}\hat{J}_n(\beta r)P_n^{m\prime}(\cos\theta)[C_3\cos(m\phi) + D_3\sin(m\phi)] \tag{10-56c}$$

where

$$' \equiv \frac{\partial}{\partial \theta}$$

(10-56d)

The boundary conditions that must be satisfied are

$$E_\theta(r = a, 0 \le \theta \le \pi, 0 \le \phi \le 2\pi) = 0$$ (10-57a)

$$E_\phi(r = a, 0 \le \theta \le \pi, 0 \le \phi \le 2\pi) = 0$$ (10-57b)

Either condition yields the same eigenfunction and corresponding eigenvalues.
Applying (10-57a) to (10-56b) leads to

$$E_\theta(r = a, 0 \le \theta \le \pi, 0 \le \phi \le 2\pi) = -A_{mnp} \frac{m}{\varepsilon} \frac{1}{a \sin\theta} \hat{J}_n(\beta a) P_n^m(\cos\theta)$$

$$\times [-C_3 \sin(m\phi) + D_3 \cos(m\phi)] = 0 \Rightarrow \hat{J}_n(\beta a)|_{\beta=\beta_r} = 0 \Rightarrow \beta_r a = \zeta_{np}$$

$$\beta_r = \frac{\zeta_{np}}{a} \qquad \begin{matrix} n = 1, 2, 3, \dots \\ p = 1, 2, 3, \dots \end{matrix}$$

(10-58)

where ζ_{np} represents the p zeroes of the spherical Bessel function $\hat{J}_n$ of order n. A listing of a
limited, but for most applications sufficient, number of ζ_{np}'s is found in Table 10-1.
    The resonant frequencies are found using (10-58) and can be written as

$$\beta_r = \omega_r \sqrt{\mu\varepsilon} = 2\pi f_r \sqrt{\mu\varepsilon} = \frac{\zeta_{np}}{a}$$

(10-59)

or

$$(f_r)_{mnp}^{TE'} = \frac{\zeta_{np}}{2\pi a \sqrt{\mu\varepsilon}} \qquad \begin{matrix} m = 0, 1, 2, \dots \le n \\ n = 1, 2, 3, \dots \\ p = 1, 2, 3, \dots \end{matrix}$$

(10-59a)

Since the resonant frequencies of (10-59a) obtained using the ζ_{np}'s of Table 10-1 are independent
of the values of m, there are numerous degeneracies (same resonant frequencies) among the
modes; for a given n and p, there are many m's that have the same resonant frequency. To determine how many m modes exist for each set of n and p, remember that $P_n^m = 0$ if $m > n$. Therefore,
for P_n^m to be nonzero, $m \le n$. Thus, the order of degeneracy is equal to $m = n$.

TABLE 10-1   Zeroes ζ_{np} of spherical Bessel function $\hat{J}_n(\zeta_{np}) = 0$

	$n = 1$	$n = 2$	$n = 3$	$n = 4$	$n = 5$	$n = 6$	$n = 7$	$n = 8$
$p = 1$	4.493	5.763	6.988	8.183	9.356	10.513	11.657	12.791
$p = 2$	7.725	9.095	10.417	11.705	12.967	14.207	15.431	16.641
$p = 3$	10.904	12.323	13.698	15.040	16.355	17.648	18.923	20.182
$p = 4$	14.066	15.515	16.924	18.301	19.653	20.983	22.295	
$p = 5$	17.221	18.689	20.122	21.525	22.905			
$p = 6$	20.371	21.854						

According to the values of Table 10-1, the lowest ζ_{np} zeroes in ascending order, along with the number of degenerate and total modes, are

n, p	ζ_{np}	Degenerate modes	Total number of modes
$n=1, \; p=1$	$\zeta_{11} = 4.493$	$m = 0, 1$ (even, odd)	3
$n=2, \; p=1$	$\zeta_{21} = 5.763$	$m = 0, 1, 2$ (even, odd)	8
$n=3, \; p=1$	$\zeta_{31} = 6.988$	$m = 0, 1, 2, 3$ (even, odd)	15
$n=1, \; p=2$	$\zeta_{12} = 7.725$	$m = 0, 1$ (even, odd)	18
$n=4, \; p=1$	$\zeta_{41} = 8.183$	$m = 0, 1, 2, 3, 4$ (even, odd)	27

The even, odd is used to represent either the $\cos(m\phi)$ or $\sin(m\phi)$ variations of (10-55). For example, for $n=1$, $p=1$, (10-55) has a three-fold degeneracy and can be written to represent the following three modes:

$$(F_r)_{011} \quad (\text{even}) = A_{011} C_3 \hat{J}_1(\beta_r r) P_1^0(\cos\theta) = A_{011} C_3 \hat{J}_1 \left(4.493\frac{r}{a}\right)\cos\theta \tag{10-60a}$$

$$(F_r)_{111} \quad (\text{even}) = A_{111} C_3 \hat{J}_1(\beta_r r) P_1^1(\cos\theta)\cos\phi = -A_{111} C_3 \hat{J}_1 \left(4.493\frac{r}{a}\right)\sin\theta\cos\phi \tag{10-60b}$$

$$(F_r)_{111} \quad (\text{odd}) = A_{111} D_3 \hat{J}_1(\beta_r r) P_1^1(\cos\theta)\sin\phi = -A_{111} D_3 \hat{J}_1 \left(4.493\frac{r}{a}\right)\sin\theta\sin\phi \tag{10-60c}$$

since

$$P_1^0(\cos\theta) = P_1(\cos\theta) = \cos\theta \tag{10-60d}$$

$$P_1^1(\cos\theta) = -(1-\cos^2\theta)^{1/2} = -\sin\theta \tag{10-60e}$$

The modes represented by (10-60b) and (10-60c) are the same except that they are rotated 90°, in the ϕ direction, from each other. The same is true between (10-60a) and (10-60b) or (10-60c) except that the rotation is in the θ and ϕ directions.

10.4.2 Transverse Magnetic (TM^r) Modes

Following a procedure and justification similar to that for the TE^r modes, it can be shown that the appropriate vector potential component A_r of (10-25b) takes the form

$$(A_r)_{mnp} = B_{mnp} \hat{J}_n(\beta r) P_n^m(\cos\theta)[C_3 \cos(m\phi) + D_3 \sin(m\phi)] \tag{10-61}$$

The corresponding electric and magnetic fields are found using (10-27) through (10-28c). The boundary conditions are the same as for the TE^r, as given by (10-57a) and (10-57b).

Expanding (10-27b) using (10-61) we can write that

$$E_\theta = \frac{1}{j\omega\mu\varepsilon}\frac{1}{r}\frac{\partial^2 A_r}{\partial r \, \partial\theta} = B_{mnp}\frac{\beta}{j\omega\mu\varepsilon r}\hat{J}_n'(\beta r)P_n^{m'}(\cos\theta)[C_3 \cos(m\phi) + D_3 \sin(m\phi)] \tag{10-62}$$

Applying (10-57a) on (10-62) leads to

$$E_\theta(r=a, 0\le\theta\le\pi, 0\le\phi\le2\pi) = B_{mnp}\frac{\beta}{j\omega\mu\varepsilon a}\hat{J}_n'(\beta_r a)P_n^{m'}(\cos\theta)$$

$$\times[C_3 \cos(m\phi) + D_3 \sin(m\phi)] = 0 \;\Rightarrow\; \hat{J}_n'(\beta a)|_{\beta=\beta_r}=0 \;\Rightarrow\; \beta_r a = \zeta_{np}'$$

$$\beta_r = \frac{\zeta_{np}'}{a} \qquad \begin{matrix} n=1, 2, 3, \dots \\ p=1, 2, 3, \dots \end{matrix} \tag{10-63}$$

TABLE 10-2 Zeroes ζ'_{np} of derivative of spherical Bessel function $\hat{J}'_n(\zeta'_{np})=0$

	$n=1$	$n=2$	$n=3$	$n=4$	$n=5$	$n=6$	$n=7$	$n=8$
$p=1$	2.744	3.870	4.973	6.062	7.140	8.211	9.275	10.335
$p=2$	6.117	7.443	8.722	9.968	11.189	12.391	13.579	14.753
$p=3$	9.317	10.713	12.064	13.380	14.670	15.939	17.190	18.425
$p=4$	12.486	13.921	15.314	16.674	18.009	19.321	20.615	21.894
$p=5$	15.644	17.103	18.524	19.915	21.281	22.626		
$p=6$	18.796	20.272	21.714	23.128				
$p=7$	21.946							

where ζ'_{np} represents the p zeroes of the derivative of the spherical Bessel function $\hat{J}'_n$ of order n. A listing of a limited, but for most applications sufficient, number of ζ'_{np}'s is found in Table 10-2.

The resonant frequencies are found using (10-63) and can be written as

$$\boxed{(f_r)^{\text{TM}'}_{mnp} = \frac{\zeta'_{np}}{2\pi a\sqrt{\mu\varepsilon}}} \qquad \begin{array}{l} m=0,1,2,\dots \le n \\ n=1,2,3,\dots \\ p=1,2,3,\dots \end{array} \qquad (10\text{-}64)$$

As with the TE^r modes, there are numerous degeneracies among the modes since the resonant frequencies determined by (10-64) are independent of m. For a given n, the order of degeneracy is $m=n$.

According to the values of Table 10-2, the lowest ζ'_{np} zeroes in ascending order, along with the number of degenerate and total modes, are

n,p	ζ'_{np}	Degenerate modes	Total number of modes
$n=1,\ p=1$	$\zeta'_{11}=2.744$	$m=0,1$ (even, odd)	3
$n=2,\ p=1$	$\zeta'_{21}=3.870$	$m=0,1,2$ (even, odd)	8
$n=3,\ p=1$	$\zeta'_{31}=4.973$	$m=0,1,2,3$ (even, odd)	15
$n=4,\ p=1$	$\zeta'_{41}=6.062$	$m=0,1,2,3,4$ (even, odd)	24
$n=1,\ p=2$	$\zeta'_{12}=6.117$	$m=0,1$ (even, odd)	27

The lowest-order mode is the one found using $n=1$, $p=1$, and it has a three-fold degeneracy $[m=0$ (even), $m=1$ (even), and $m=1$ (odd)]. For these, (10-61) reduces to

$$(A_r)_{011}\ (\text{even}) = B_{011}C_3\hat{J}_1(\beta_r r)P_1^0(\cos\theta) = -B_{011}C_3\hat{J}_1\left(2.744\frac{r}{a}\right)\cos\theta \qquad (10\text{-}65a)$$

$$(A_r)_{111}\ (\text{even}) = B_{111}C_3\hat{J}_1(\beta_r r)P_1^1(\cos\theta)\cos\phi = -B_{111}C_3\hat{J}_1\left(2.744\frac{r}{a}\right)\sin\theta\cos\phi \qquad (10\text{-}65b)$$

$$(A_r)_{111}\ (\text{odd}) = B_{111}D_3\hat{J}_1(\beta_r r)P_1^1(\cos\theta)\sin\phi = -B_{111}D_3\hat{J}_1\left(2.744\frac{r}{a}\right)\sin\theta\sin\phi \qquad (10\text{-}65c)$$

Example 10-1

For a spherical cavity of a 3-cm radius and filled with air, determine the resonant frequencies (in ascending order) of the first 11 modes (including degenerate modes).

Solution: According to (10-59a) and (10-64), using the values of ζ_{np} and ζ'_{np} from Tables 10-1 and 10-2, and taking into account the degeneracy of the modes in m as well as the even and odd forms in ϕ, we can write the resonant frequencies of the first 11 modes as

1, 2, 3:

$$(f_r)_{011}^{TM} \text{ (even)} = (f_r)_{111}^{TM} \text{ (even)} = (f_r)_{111}^{TM} \text{ (odd)}$$

$$= \frac{2.744(30 \times 10^9)}{2\pi(3)} = 4.367 \times 10^9 \text{ Hz}$$

4, 5, 6, 7, 8:

$$(f_r)_{021}^{TM} \text{ (even)} = (f_r)_{121}^{TM} \text{ (even)} = (f_r)_{121}^{TM} \text{ (odd)} = (f_r)_{221}^{TM} \text{ (even)}$$

$$= (f_r)_{221}^{TM} \text{ (odd)} = \frac{3.870(30 \times 10^9)}{2\pi(3)} = 6.1593 \times 10^9 \text{ Hz}$$

9, 10, 11:

$$(f_r)_{011}^{TE} \text{ (even)} = (f_r)_{111}^{TE} \text{ (even)} = (f_r)_{111}^{TE} \text{ (odd)}$$

$$= \frac{4.493(30 \times 10^9)}{2\pi(3)} = 7.1508 \times 10^9 \text{ Hz}$$

10.4.3 Quality Factor Q

As has already been pointed, the Q of the cavity is probably one of its most important parameters, and it is defined by (8-84). To derive the equation for the Q of any mode of a spherical cavity is a most difficult task. However, it is instructive to consider that of the lowest (dominant) mode, which here is any one of the three-fold degenerate modes TM_{011} (even), TM_{111} (even), or TM_{111} (odd). Let us consider the TM_{011} (even) mode.

For the TM_{011} (even) mode the potential function of (10-61) reduces to that of (10-65a), which can be written as

$$(A_r)_{011} = B'_{011} \hat{J}_1 \left(2.744 \frac{r}{a} \right) \cos\theta \tag{10-66}$$

Since the Q of the cavity is defined by (8-84), it is most convenient to find the stored energy and dissipated power by using the magnetic field, since it has only one nonzero component (the electric field has two).

The magnetic field components of the TM_{011} mode can be written using (10-28a) through (10-28c) and (10-66) as

$$H_r = 0 \tag{10-67a}$$

$$H_\theta = \frac{1}{\mu} \frac{1}{r\sin\theta} \frac{\partial A_r}{\partial\phi} = 0 \tag{10-67b}$$

$$H_\phi = -\frac{1}{\mu} \frac{1}{r} \frac{\partial A_r}{\partial\theta} = B'_{011} \frac{1}{\mu} \frac{1}{r} \hat{J}_1 \left(2.744 \frac{r}{a} \right) \sin\theta \tag{10-67c}$$

Therefore, at resonance, the total stored energy can be found using

$$W = 2W_e = 2W_m = 2 \left[\frac{\mu}{4} \iiint_V |\mathbf{H}|^2 \, dv \right] = \frac{\mu}{2} \int_0^{2\pi} \int_0^\pi \int_0^a |H_\phi|^2 \, r^2 \sin\theta \, dr \, d\theta \, d\phi \tag{10-68}$$

which, by substituting (10-67c), reduces to

$$W = \frac{|B'_{011}|^2}{2\mu} \int_0^{2\pi} \int_0^\pi \int_0^a \hat{J}_1^2\left(2.744\frac{r}{a}\right) \sin^3\theta \, dr \, d\theta \, d\phi$$

$$= \frac{|B'_{011}|^2}{2\mu}(2\pi)\frac{4}{3}\int_0^a \hat{J}_1^2\left(2.744\frac{r}{a}\right) dr \qquad (10\text{-}68\text{a})$$

The integral can be evaluated using the formula

$$\int_0^a \hat{J}_1^2\left(2.744\frac{r}{a}\right) dr = \frac{a}{2}\left[\hat{J}_1^2(2.744) - \hat{J}_0(2.744)\,\hat{J}_2(2.744)\right] \qquad (10\text{-}69)$$

where according to (3-94) or (10-31)

$$\hat{J}_1(2.744) = 2.744\,j_1(2.744) = 2.744(0.3878) = 1.0640 \qquad (10\text{-}69\text{a})$$

$$\hat{J}_0(2.744) = 2.744\,j_0(2.744) = 2.744(0.1428) = 0.3919 \qquad (10\text{-}69\text{b})$$

$$\hat{J}_2(2.744) = 2.744\,j_2(2.744) = 2.744(0.2820) = 0.7738 \qquad (10\text{-}69\text{c})$$

Thus, (10-69) reduces, using that $\beta_r = 2.744/a$, to

$$\int_0^a \hat{J}_1^2\left(2.744\frac{r}{a}\right) dr = \frac{a}{2}[(1.0640)^2 + 0.3919(0.7738)] = \frac{a}{2}(0.8288)$$

$$= \frac{a}{2.744}\frac{(2.744)(0.8288)}{2} = \frac{1.137}{\beta_r}$$

and (10-68a) to

$$W = \frac{|B'_{011}|^2}{2\mu}(2\pi)\left(\frac{4}{3}\right)\frac{1.137}{\beta_r} \qquad (10\text{-}70)$$

The power dissipated on the walls of the cavity can be found using

$$P_d = \frac{R_s}{2} \oiint_S \mathbf{J}_s \cdot \mathbf{J}_s^* ds \qquad (10\text{-}71)$$

where

$$\mathbf{J}_s = \hat{\mathbf{n}} \times \mathbf{H}\,|_{r=a} = -\hat{\mathbf{a}}_r \times \hat{\mathbf{a}}_\phi H_\phi|_{r=a} = \hat{\mathbf{a}}_\theta H_\phi(r=a) = \hat{\mathbf{a}}_\theta B'_{011}\frac{1}{\mu}\frac{1}{a}\hat{J}_1(2.744)\sin\theta \qquad (10\text{-}71\text{a})$$

Thus, (10-71) can be written as

$$P_d = \frac{R_s}{2}\int_0^{2\pi}\int_0^\pi |H_\phi(r=a)|^2\, a^2 \sin\theta \, d\theta \, d\phi$$

$$= \frac{R_s}{2\mu^2}|B'_{011}|^2\, \hat{J}_1^2(2.744)\int_0^{2\pi}\int_0^\pi \sin^3\theta \, d\theta \, d\phi$$

$$= |B'_{011}|^2\frac{R_s}{2\mu^2}(2\pi)\left(\frac{4}{3}\right)\hat{J}_1^2(2.744) = |B'_{011}|^2\frac{R_s}{2\mu^2}(2\pi)\left(\frac{4}{3}\right)(1.0640)^2$$

$$P_d = \frac{|B'_{011}|^2}{2\mu^2}1.132(2\pi)\left(\frac{4}{3}\right)R_s \qquad (10\text{-}72)$$

Using (10-70) and (10-72), the Q of the cavity for the TM_{011}^r mode reduces to

$$Q = \omega_r \frac{W}{P_d} = \omega_r \frac{\frac{|B_{011}'|^2}{2\mu}(2\pi)\left(\frac{4}{3}\right)\frac{1.137}{\beta_r}}{\frac{|B_{011}'|^2}{2\mu^2}(2\pi)\left(\frac{4}{3}\right)(1.132)\,R_s} = \frac{1.137\omega_r\mu}{1.132\beta_r R_s} = 1.004\frac{\omega_r\mu}{\omega_r\sqrt{\mu\varepsilon}R_s}$$

$$Q = 1.004\frac{\sqrt{\frac{\mu}{\varepsilon}}}{R_s} = 1.004\left(\frac{\eta}{R_s}\right) \tag{10-73}$$

Example 10-2

Compare the Q values of a spherical cavity operating in the dominant TM_{011} (even) mode with those of a circular cylinder and cubical cavities. The dimensions of each are such that the cylindrical and spherical cavities are circumscribed by the cubical cavity.

Solution: According to (10-73), the Q of a spherical cavity of radius a operating in the dominant TM_{011} (even) mode is given by

$$Q = 1.004\left(\frac{\eta}{R_s}\right)$$

while that of a circular cavity of diameter d and height h operating in the dominant TM_{010} mode (for $h = d$) is given by (9-57), which reduces to

$$Q = 1.2025\frac{\eta}{R_s}\frac{1}{\left[1+\frac{d/2}{h}\right]} = 0.8017\left(\frac{\eta}{R_s}\right)$$

For a rectangular cavity operating in the dominant TE_{101} mode the Q is given by (8-88), which for a cubical geometry ($a = b = c$) reduces, according to (8-88a), to

$$Q = 1.1107\frac{\eta}{R_s}\frac{1}{\left[1+\frac{a/2}{b}\right]} = \frac{1.1107}{1.5}\left(\frac{\eta}{R_s}\right) = 0.7405\left(\frac{\eta}{R_s}\right)$$

Comparing these three expressions, it is evident that the Q of the spherical cavity is greater than that of the circular cavity with $h = d$ by

$$\frac{1.004 - 0.8017}{0.8017}\times 100\% = 25.23\%$$

and greater than that of the cubical cavity by

$$\frac{1.004 - 0.7405}{0.7405}\times 100\% = 35.58\%$$

This is expected since the spherical cavity does not possess any sharp corners and edges, which are evident in the circular cavity and even more in the cubical cavity. Thus, the volume and surface area of the spherical cavity are better utilized by the interior fields. It should be remembered that the Q of a cavity is proportional to its volume and inversely proportional to its area.

10.5 MULTIMEDIA

On the website that accompanies this book, the following multimedia resources are included for the review, understanding, and presentation of the material of this chapter.

- **MATLAB** computer program **Sphere_Resonator**: Computes the resonant characteristics of a spherical resonator.
- **PowerPoint (PPT)** viewgraphs, in multicolor.

REFERENCES

1. R. F. Harrington, *Time-Harmonic Electromagnetic Fields*, McGraw-Hill, New York, 1961.
2. S. A. Schelkunoff, *Electromagnetic Waves*, Van Nostrand, Princeton, NJ, 1943, pp. 51–52.
3. C. A. Balanis, *Antenna Theory: Analysis and Design*, Third edition, John Wiley & Sons, New York, 2005.
4. S. A. Schelkunoff, "Theory of antennas of arbitrary size and shape," *Proc. IRE*, vol. 29, no. 9, pp. 493–521, Sept. 1941.
5. S. A. Schelkunoff and C. B. Feldman, "On radiation from antennas," *Proc. IRE*, vol. 30, no. 11, pp. 512–516, Nov. 1942.
6. S. A. Schelkunoff and H. T. Friis, *Antennas: Theory and Practice*, John Wiley & Sons, New York, 1952.
7. C. T. Tai, "Application of a variational principle to biconical antennas," *J. Appl. Phys.*, vol. 20, no. 11, pp. 1076–1084, Nov. 1949.
8. P. D. Smith, "The conical dipole of wide angle," *J. Appl. Phys.*, vol. 19, no. 1, pp. 11–23, Jan. 1948.
9. L. Bailin and S. Silver, "Exterior electromagnetic boundary value problems for spheres and cones," *IRE Trans. Antennas Propagat.*, vol. AP-4, no. 1, pp. 5–15, Jan. 1956.
10. W. R. Smythe, *Static and Dynamic Electricity*, McGraw-Hill, New York, 1941.
11. J. A. Stratton, *Electromagnetic Theory*, McGraw-Hill, New York, 1960.
12. P. M. Morse and H. Feshbach, *Methods of Theoretical Physics*, Parts I and II, McGraw-Hill, New York, 1953.
13. M. Abramowitz and I. A. Stegun (Eds.), *Handbook of Mathematical Functions with Formulas, Graphs, and Mathematical Tables*, National Bureau of Standards Applied Mathematical Series-55, U.S. Government Printing Office, Washington, DC 1966.
14. E. Jahnke and F. Emde, *Tables of Functions*, Dover, New York, 1945.
15. M. R. Spiegel, *Mathematical Handbook of Formulas and Tables, Schaum's Outline Series*, McGraw-Hill, New York, 1968.

PROBLEMS

10.1. In Section 11.7.1 it is shown that the magnetic vector potential for an infinitesimal electric dipole of Figure 11-23a and Example 6-3 is given by (11-209)

$$A_z^{(1)} = -\hat{\mathbf{a}}_z j \frac{\mu \beta I_e \Delta \ell}{4\pi} h_0^{(2)}(\beta r)$$

where $h_0^{(2)}(\beta r)$ is the spherical Hankel function of order zero. Assume that two such dipoles of equal amplitude, but 180° out of phase, are displaced along the x axis a distance s apart, as shown in Figure P10-1. The total magnetic potential can be written following a procedure outlined in [1] as

$$A_z^t = A_z^{(1)}\left(x - \frac{s}{2}, y, z\right)$$
$$- A_z^{(1)}\left(x + \frac{s}{2}, y, z\right) \overset{s \to 0}{\simeq} -s \frac{\partial A_z^{(1)}}{\partial x}$$

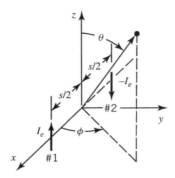

Figure P10-1

Show that the total magnetic potential A_z^t can be reduced to

$$A_z^t \overset{s \to 0}{\simeq} j\frac{\mu\beta^2 s I_e \Delta\ell}{4\pi} h_0^{(2)\prime}(\beta r)\sin\theta\,\cos\phi$$

$$\simeq +j\frac{\mu\beta^2 s I_e \Delta\ell}{4\pi} h_1^{(2)}(\beta r) P_1^1(\cos\theta)\cos\phi$$

where $\quad' \equiv \partial/\partial(\beta r)$.

10.2. Following the procedure of Problem 10.1, show that when the infinitesimal dipoles are displaced along the y axis, as shown in Figure P10-2, the total magnetic potential can be written as

$$A_z^t \overset{s \to 0}{\simeq} j\frac{\mu\beta^2 s I_e \Delta\ell}{4\pi} h_1^{(2)}(\beta r) P_1(\cos\theta)\sin\phi$$

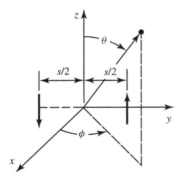

Figure P10-2

10.3. Following the procedure of Problem 10.1, show that when the infinitesimal dipoles are displaced along the z axis, as shown in Figure P10-3, the total magnetic potential can be written as

$$A_z^t \overset{s \to 0}{\simeq} j\frac{\mu\beta^2 s I_e \Delta\ell}{4\pi} h_1^{(2)}(\beta r) P_1(\cos\theta)$$

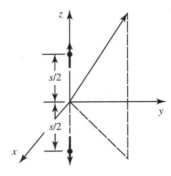

Figure P10-3

10.4. Derive expressions for the electric and magnetic field components, in terms of F_z of (10-1b), that are TE^z. F_z should represent a solution to the scalar Helmholtz equation in spherical coordinates.

10.5. Derive expressions for the electric and magnetic field components, in terms of A_z of (10-2a), that are TM^z. A_z should represent a solution to the scalar Helmholtz equation in spherical coordinates.

10.6. For problems with sources of $\mathbf{J} \neq 0$ and $\mathbf{M} = 0$, show that the electric and magnetic fields and the vector potential $\mathbf{A}$ should satisfy (10-15a) through (10-15d).

10.7. Show that using (10-25a) and (10-25b) for TM^r modes in spherical coordinates reduces (10-16f) to (10-26).

10.8. By applying the boundary conditions of (10-44a) or (10-44b) on the electric field components of (10-27a) or (10-27c), where the vector potential A_r is given by (10-43), show that the eigenvalues of n are obtained as solutions to (10-45).

10.9. Use Maxwell's equations

$$\nabla \times \mathbf{E} = -j\omega\mu\mathbf{H}$$
$$\nabla \times \mathbf{H} = j\omega\varepsilon\mathbf{E}$$

and assume TEM modes for the biconical antenna of Figure 10-1, these two equations reduce to only E_θ and H_ϕ components, each independent of ϕ. Then show that the H_ϕ component must satisfy the partial differential equation

$$\frac{\partial^2}{\partial r^2}(rH_\phi) = -\beta^2(rH_\phi)$$

whose solution must take the form

$$H_\phi = \frac{H_0}{\sin\theta}\frac{e^{-j\beta r}}{r}$$

whereas that of E_θ must then be written as

$$E_\theta = \eta H_\phi = \eta \frac{H_0}{\sin\theta}\frac{e^{-j\beta r}}{r}$$

10.10. Show that by using the electric and magnetic field components of Problem 10.9, the power radiated by the biconical antenna of Figure 10-1 reduces to

$$P_{rad} = \oiint \mathbf{S}_{av} \cdot d\mathbf{s} = 2\pi\eta \, |H_0|^2 \, \ln\left[\cot\left(\frac{\alpha}{2}\right)\right]$$

where $\mathbf{S}_{av}$ represents the average power density and $\alpha = \alpha_1 = \pi - \alpha_2$.

10.11. By using the magnetic field from Problem 10.9, show that the current on the surface of the cone a distance r from the origin is equal to

$$I(r) = 2\pi H_0 e^{-j\beta r}$$

Evaluating the current at the origin $I(r = 0)$, and using the definition for the radiation resistance in terms of the radiated power from Problem 10.10 and the current at the origin, show that the radiation resistance reduces to

$$R_r = \frac{2P_{rad}}{|I(r=0)|^2} = \frac{\eta}{\pi} \ln\left[\cot\left(\frac{\alpha}{2}\right)\right]$$

where $\alpha = \alpha_1 = \pi - \alpha_2$. This is the same as the characteristic impedance of (10-52a) for a symmetrical biconical transmission line.

10.12. Calculate the included angle $\alpha = \alpha_1 = \pi - \alpha_2$ of a symmetrical biconical transmission line so that its characteristic impedance is:
(a) 300 ohms.
(b) 50 ohms.

10.13. For inside ($0 \le \alpha \le \pi/2, 0 \le \theta \le \alpha$) or outside ($\pi/2 \le \alpha \le \pi, 0 \le \theta \le \alpha$) cones shown, respectively, in Figures P10-13a and b for TE^r modes:
(a) Show that the electric vector potential reduces to

$$(F_r)_{mnp} = A_2 P_n^m(\cos\theta)[C_1\hat{H}_n^{(1)}(\beta r) \\ + D_1\hat{H}_n^{(2)}(\beta r)] \times [C_3\cos(m\phi) \\ + D_3\sin(m\phi)], \quad m = 0, 1, 2, \dots$$

(b) Show that the eigenvalues for n are obtained as solutions to

$$\left.\frac{dP_n^m(\cos\theta)}{d\theta}\right|_{\theta=\alpha} = 0$$

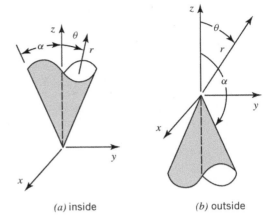

(a) inside   (b) outside

Figure P10-13

10.14. Repeat Problem 10.13 for TM^r modes.
(a) Show that the magnetic vector potential reduces to

$$(A_r)_{mnp} = A_2 P_n^m(\cos\theta)[C_1\hat{H}_n^{(1)}(\beta r) \\ + D_1\hat{H}_n^{(2)}(\beta r)] \times [C_3\cos(m\phi) \\ + D_3\sin(m\phi)], \quad m = 0, 1, 2, \dots$$

(b) Show that the eigenvalues for n are obtained as solutions to

$$P_n^m(\cos\theta)|_{\theta=\alpha} = 0$$

10.15. For the inside ($0 \le \phi \le \alpha$) or outside ($\alpha \le \phi \le 2\pi$) infinite-dimensions wedge of Figure P10-15, derive the reduced vector potential and allowable eigenvalues for TE^r modes.

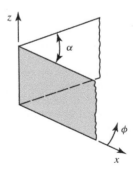

Figure P10-15

10.16. Repeat Problem 10.15 for TM^r modes.

10.17. A lossless dielectric sphere of radius a and dielectric constant ε_r is immersed into a free-space medium with an established static electric field of

$$\mathbf{E}^i = \hat{\mathbf{a}}_z E_0$$

where E_0 is a constant. Using the spherical geometry and coordinates of Figures 10-3 and 11-28, show that the internal (*in*) and external (*ex*) electric field components to the sphere are given, respectively, by

$$E_x^{in} = E_y^{in} = 0; \quad E_z^{in} = E_0\left(\frac{3}{\varepsilon_r + 2}\right)$$

$$E_x^{ex} = E_0 3\left(\frac{\varepsilon_r - 1}{\varepsilon_r + 2}\right)\frac{a^3}{r^5}xz$$

$$E_y^{ex} = E_0 3\left(\frac{\varepsilon_r - 1}{\varepsilon_r + 2}\right)\frac{a^3}{r^5}yz$$

$$E_z^{ex} = E_0 + E_0\frac{a^3}{r^5}\left(2z^2 - x^2 - y^2\right)$$

10.18. Repeat Problem 10.17 when a lossless magnetic sphere of radius a and relative permittivity μ_r is immersed into a free-space medium with an established static magnetic field of

$$\mathbf{H}^i = \hat{\mathbf{a}}_z H_0$$

where H_0 is a constant. Derive the magnetic field components internal and external to the sphere in a form similar to those of the electric field of Problem 10.17.

10.19. Design a spherical cavity (find its radius in cm) so that the resonant frequency of the dominant mode is 1 GHz and that of the next higher-order mode is approximately 1.41 GHz. The lossless medium within the sphere has electric constitutive parameters of $\varepsilon_r = 2.56$ and $\mu_r = 1$.

10.20. Assume a spherical cavity with 2-cm radius and filled with air.
  (a) Determine the resonant frequency of the dominant degenerate modes.
  (b) Find the bandwidth over which the dominant degenerate modes operate before the next higher-order degenerate modes.
  (c) Determine the dielectric constant that must be used to fill the sphere to reduce the resonant frequency of the dominant degenerate modes by a factor of 2.

10.21. Design a spherical cavity totally filled with a lossless dielectric material so that its Q at 10 GHz, while operating in its dominant

mode, is 10,000. The surface of the cavity is made of copper with a conductivity of $\sigma = 5.7 \times 10^7$ S/m. Determine the dielectric constant of the medium that must be used to fill the cavity.

10.22. A spherical cavity, because of its geometrical symmetry, is used to measure the dielectric properties of material samples, which match its geometry. To accomplish this, it is desired that the cavity is operating in the dominant mode and has a very high quality factor. The cavity is constructed of copper, ($\sigma = 5.76 \times 10^7$ S/m).
  (a) Assuming that initially the cavity is filled with air and the desired quality factor is 10,000, determine the:
    • Resonant frequency (in GHz).
    • Radius (in cm) of the cavity.
  (b) While maintaining the same dimensions as in part (a), it is desired to completely fill the cavity with a lossless dielectric material in order to reduce the quality factor of the cavity of part (a) by a factor of 3. Determine the:
    • Dielectric constant of the material that must be used to accomplish this.
    • New resonant frequency (in GHz).

10.23. Derive an expression for the Q of a spherical cavity when the fields within it are those of the dominant TM_{111} (even) mode. Compare the expression with (10-73).

10.24. Derive an expression for the Q of a spherical cavity when the fields within it are those of the dominant TM_{111} (odd) mode. Compare the expression with (10-73).

10.25. Determine the Q of the dominant mode of a spherical cavity of 2-cm radius when the medium within the cavity is:
  (a) Air.
  (b) Polystyrene with a dielectric constant of 2.56.
The cavity is made of copper whose conductivity is 5.76×10^7 S/m. Verify with the MATLAB program **Sphere_Resonator**.

10.26. It is desired to design a spherical cavity whose Q at the resonant frequency of the dominant mode is 10,000. Assume that the cavity is filled with air and it is made of copper ($\sigma = 5.76 \times 10^7$ S/m). Then determine the resonant frequency of the dominant mode and the radius (in cm) of the cavity. Also find the dielectric constant of the medium that must be used to fill the cavity to reduce its Q by a factor of 3. Verify with the MATLAB program **Sphere_Resonator**.

10.27. A spherical cavity of radius a, as shown in Figure 10-3, is filled with air. Its surface is made of a very thin layer of perfect magnetic conductor (PMC). Determine the:
   (a) Dominant TE^r modes. Identify them properly by indicating the appropriate indices (mnp). Must make a statement to justify the answer(s).
   (b) Resonant frequency of the dominant TE^r mode of the cavity.

10.28. Repeat Problem 10.27 for dominant TM^r modes.

10.29. A dielectric spherical cavity, of a geometry shown in Figure 10-3, is made of a dielectric material with $\varepsilon_r \gg 1$.
   (a) Identify the approximate dominant TE^r_{mnp}.
   (b) Compute the resonant frequency for $a = 3$ cm and $\varepsilon_r = 81$.

10.30. Repeat Problem 10.29 for TM^r_{mnp}.

10.31. Assume a hemispherical cavity of Figure P10-31 with radius a.
   (a) Determine the dominant mode and the expression for its resonant frequency.
   (b) Show that the Q of the dominant mode is

$$Q = 0.574 \left(\frac{\eta}{R_s} \right)$$

   (c) Compare the Q of part (b) with that of the spherical cavity, and those of cylindrical and square-based rectangular cavities with the same height-to-diameter ratios.

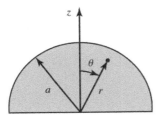

Figure P10-31

10.32. For a hemispherical PEC, air-filled cavity of radius a, as shown in Figure P10-32, determine the:

   (a) Dominant TM^r_{mnp} mode(s); identify them properly.
   (b) Reduced/simplified vector potential. Justify it.
   (c) Lowest resonant frequency (in GHz) when $a = 3$ cm.

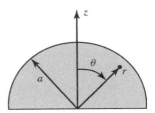

Figure P10-32

10.33. Repeat Problem 10.32 for TE^r_{mnp} modes.

10.34. A hemispherical dielectric resonator, used in microwave integrated circuit (MIC) design, is comprised of a PEC ground plane covered with a dielectric half sphere of radius a, as shown in Figure P10-34. Assuming the dielectric constant of the material is much greater than unity $(\varepsilon_r \gg 1)$:
   (a) Identify all the approximate TE^r_{mnp} mode(s) with the lowest resonant frequency.
   (b) Compute the resonant frequencies (in GHz) when $a = 3$ cm and $\varepsilon_r = 81$.

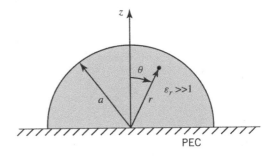

Figure P10-34

10.35. Repeat Problem 10.34 for TM^r_{mnp}.

10.36. A hemispherical (half of a sphere) dielectric resonator (only in upper half sphere) is totally composed of a lossless dielectric medium with a dielectric constant of $\varepsilon_r \gg 1$, $\mu_r = 1$. For TM^r_{mnp} modes (even and/or odd), determine the:
   (a) Expression for the resonant frequencies of the allowable TM^r_{mnp} (even and/or odd)

modes. Be specific about all the allowable eigenvalues m, n, p for the even and/or odd modes. Be specific and justify.

(b) A reduced/simplified expression for the appropriate vector potential. Be specific and justify it.

(c) Resonant frequency (in GHz) of the dominant TM_{mnp}^r modes (even and/or odd, including degenerates; specify values of m, n, and p) when $a = 1.5$ cm, $\varepsilon_r = 81$. Be specific to identify whether even and/or odd mode(s).

10.37. Repeat all parts of Problem 10.36 for TE_{mnp}^r modes.

10.38. A hemispherical PMC cavity, of radius a shown in Figure P10-38, is filled with air. Determine the:

(a) Dominant TE_{mnp}^r mode(s); identify them properly by indicating the appropriate indices (m, n, p).

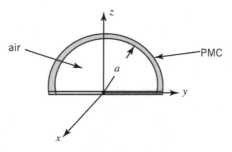

Figure P10-38

(b) Reduced/simplified vector potential expression for the dominant mode(s). Justify it.

(c) Lowest resonant frequency (in GHz) when $a = 3$ cm.

10.39. Repeat Problem 10.38 for TM_{mnp}^r mode(s).

CHAPTER 11

Scattering

11.1 INTRODUCTION

Previously, we have considered wave propagation in unbounded media, semi-infinite media forming planar interfaces, and conducting, dielectric, and surface waveguides. Although wave propagation in unbounded media is somewhat idealistic, it serves as a basic model for examining wave behavior while minimizing mathematical complexities. In general, however, wave propagation must be analyzed when it accounts for the presence of other structures (scatterers), especially when they are in proximity to the wave source and/or receiver.

In this chapter we want to examine wave propagation in the presence of scatterers of various geometries (planar, cylindrical, spherical). This is accomplished by introducing to the total field an additional component, referred to here as the *scattered field*, due to the presence of scatterers. The scattered field ($\mathbf{E}^s, \mathbf{H}^s$) must be such that when it is added, through superposition, to the *incident (direct) field* ($\mathbf{E}^i, \mathbf{H}^i$), the sum represents the total ($\mathbf{E}^t, \mathbf{H}^t$) field, that is,

$$\mathbf{E}^t = \mathbf{E}^i + \mathbf{E}^s \tag{11-1a}$$

$$\mathbf{H}^t = \mathbf{H}^i + \mathbf{H}^s \tag{11-1b}$$

The incident (direct) field $\mathbf{E}^i, \mathbf{H}^i$ will represent the total field produced by the sources *in the absence of any scatterers*.

The direct, scattered, and total fields will be obtained using various techniques. In general, geometrical optics (GO), physical optics (PO), modal techniques (MT), integral equations (IE), and diffraction theory [such as the geometrical theory of diffraction (GTD) and physical theory of diffraction (PTD)] can be used to analyze such problems. Typically, some of the problems are more conveniently analyzed using particular method(s). The fundamentals of physical optics were introduced in Chapter 7 and modal techniques were utilized in Chapters 8, 9, and 10 to analyze waveguide wave propagation. Integral equations are very popular, and they are introduced in Chapter 12. Geometrical optics and diffraction techniques are introduced and applied in Chapter 13.

In this chapter we want to examine scattering primarily by conducting objects. Each scattering problem will be analyzed using image theory, physical optics, or modal techniques. The conveniences and limitations of the applied method to each problem, as well as those of the other methods, will be stated.

Balanis' Advanced Engineering Electromagnetics, Third Edition. Constantine A. Balanis.
© 2024 John Wiley & Sons, Inc. Published 2024 by John Wiley & Sons, Inc.
Companion Website: www.wiley.com/go/balanis/advancedengineeringelectromagnetics3e

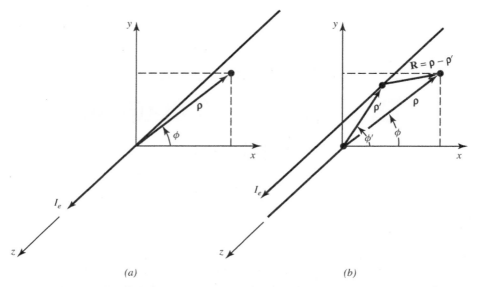

Figure 11-1   Geometry and coordinate system for an infinite electric line source. (*a*) At origin. (*b*) Offset.

11.2   INFINITE LINE-SOURCE CYLINDRICAL WAVE RADIATION

Before we examine the radiation and scattering of sources placed in the presence of scatterers, it is instructive to obtain the fields radiated by an infinite line source (both electric and magnetic) in an unbounded medium. The reason for doing this is that the infinite line source will serve as one type of source for which we will examine radiation properties in the presence of scatterers; its radiation in an unbounded medium will significantly aid in the solution of such problems.

11.2.1   Electric Line Source

The geometry of the line source is that of Figures 11-1*a* and 11-1*b* where it is assumed that its length extends to infinity and the electric current is represented by

$$\mathbf{I}_e(z') = \hat{\mathbf{a}}_z I_e \tag{11-2}$$

where I_e is a constant. Since the current is directed along the z axis, the fields radiated by the line source are TM^z and can be obtained by letting

$$\mathbf{F} = 0$$
$$\mathbf{A} = \hat{\mathbf{a}}_z A_z(\rho, \phi, z)$$
$$= \hat{\mathbf{a}}_z \left[C_1 H_m^{(1)}(\beta_\rho \rho) + D_1 H_m^{(2)}(\beta_\rho \rho) \right]$$
$$\times [C_2 \cos(m\phi) + D_2 \sin(m\phi)](A_3 e^{-j\beta_z z} + B_3 e^{+j\beta_z z}) \tag{11-3}$$

Since the line source is infinite in extent, the fields are two-dimensional (no z variations) so that

$$\beta_z = 0 \Rightarrow \beta_\rho^2 + \beta_z^2 = \beta^2 \Rightarrow \beta_\rho = \beta \tag{11-4}$$

In addition, since the waves radiate only in the outward direction and we choose the lowest-order mode, then

$$C_1 = 0 \tag{11-5a}$$
$$m = 0 \tag{11-5b}$$

Thus, (11-3) reduces to

$$\mathbf{A} = \hat{\mathbf{a}}_z A_z(\rho) = \hat{\mathbf{a}}_z A_0 H_0^{(2)}(\beta\rho) \tag{11-6}$$

whose corresponding electric and magnetic fields can be written using (6-70) as

$$E_\rho = -j\frac{1}{\omega\mu\varepsilon}\frac{\partial^2 A_z}{\partial\rho\,\partial z} = 0 \tag{11-6a}$$

$$E_\phi = -j\frac{1}{\omega\mu\varepsilon}\frac{1}{\rho}\frac{\partial^2 A_z}{\partial\phi\,\partial z} = 0 \tag{11-6b}$$

$$E_z = -j\frac{1}{\omega\mu\varepsilon}\left(\frac{\partial^2}{\partial z^2}+\beta^2\right)A_z = -j\omega A_0 H_0^{(2)}(\beta\rho) \tag{11-6c}$$

$$H_\rho = \frac{1}{\mu}\frac{1}{\rho}\frac{\partial A_z}{\partial\phi} = 0 \tag{11-6d}$$

$$H_\phi = -\frac{1}{\mu}\frac{\partial A_z}{\partial\rho} = -\frac{A_0}{\mu}H_0^{(2)\prime}(\beta\rho) = A_0\frac{\beta}{\mu}H_1^{(2)}(\beta\rho) \tag{11-6e}$$

$$H_z = 0 \tag{11-6f}$$

where $' \equiv \partial/\partial\rho$.

The constant A_0 can be obtained by using

$$I_e = \lim_{\rho\to 0}\oint_C \mathbf{H}\cdot d\mathbf{l} = \lim_{\rho\to 0}\int_0^{2\pi}(\hat{\mathbf{a}}_\phi H_\phi)\cdot(\hat{\mathbf{a}}_\phi\rho\,d\phi) = \lim_{\rho\to 0}\int_0^{2\pi}H_\phi\rho\,d\phi \tag{11-7}$$

Since the integration of (11-7) must be performed in the limit as $\rho \to 0$, it is convenient to represent the Hankel function of (11-6e) by its asymptotic expansion for small arguments. Using (IV-12), we can write that

$$H_1^{(2)}(\beta\rho) = J_1(\beta\rho) - jY_1(\beta\rho) \overset{\beta\rho\to 0}{\simeq} \frac{\beta\rho}{2} + j\frac{2}{\pi}\left(\frac{1}{\beta\rho}\right) \overset{\beta\rho\to 0}{\simeq} j\frac{2}{\pi}\left(\frac{1}{\beta\rho}\right) \tag{11-8}$$

Therefore, (11-7) reduces, using (11-6e) and (11-8), to

$$I_e = \lim_{\rho\to 0}\int_0^{2\pi}\left[A_0\frac{\beta}{\mu}H_1^{(2)}(\beta\rho)\right]\rho\,d\phi \simeq jA_0\frac{2}{\pi\mu}\int_0^{2\pi}\frac{1}{\rho}\rho\,d\phi = jA_0\frac{4}{\mu} \tag{11-9}$$

or

$$A_0 = -j\frac{\mu}{4}I_e \tag{11-9a}$$

Thus, the nonzero electric and magnetic fields of the electric line source reduce to

$$E_z = -I_e\frac{\omega\mu}{4}H_0^{(2)}(\beta\rho) = -I_e\frac{\beta^2}{4\omega\varepsilon}H_0^{(2)}(\beta\rho) \tag{11-10a}$$

$$H_\phi = -jI_e\frac{\beta}{4}H_1^{(2)}(\beta\rho) \tag{11-10b}$$

Each of the field components is proportional to a Hankel function of the second kind whose argument is proportional to the distance from the source to the observation point. If the source is removed from the origin and it is placed as shown in Figure 11-1b, (11-10a) and (11-10b) can be written as

$$E_z = -I_e\frac{\beta^2}{4\omega\varepsilon}H_0^{(2)}(\beta R) = -I_e\frac{\beta^2}{4\omega\varepsilon}H_0^{(2)}(\beta\,|\boldsymbol{\rho}-\boldsymbol{\rho}'|) \tag{11-11a}$$

$$H_\psi = -jI_e\frac{\beta}{4}H_1^{(2)}(\beta R) = -jI_e\frac{\beta}{4}H_1^{(2)}(\beta\,|\boldsymbol{\rho}-\boldsymbol{\rho}'|) \tag{11-11b}$$

where

$$R = |\rho - \rho'| = \sqrt{\rho^2 + (\rho')^2 - 2\rho\rho' \cos(\phi - \phi')} \tag{11-11c}$$

$$\psi = \text{circumferential angle around the source}$$

For observations at far distances such that $\beta\rho \rightarrow$ large, the Hankel functions in (11-10a) and (11-10b) can be approximated by their asymptotic expansions for large argument,

$$H_0^{(2)}(\beta\rho) \overset{\beta r \rightarrow \text{large}}{\simeq} \sqrt{\frac{2j}{\pi\beta\rho}}\, e^{-j\beta\rho} \tag{11-12a}$$

$$H_1^{(2)}(\beta\rho) \overset{\beta r \rightarrow \text{large}}{\simeq} j\sqrt{\frac{2j}{\pi\beta\rho}}\, e^{-j\beta\rho} \tag{11-12b}$$

Thus, (11-10a) and (11-10b) can be simplified for large arguments to

$$E_z = -I_e \frac{\beta^2}{4\omega\varepsilon} H_0^{(2)}(\beta\rho) \overset{\beta\rho \rightarrow \text{large}}{\simeq} -\eta I_e \sqrt{\frac{j\beta}{8\pi}} \frac{e^{-j\beta\rho}}{\sqrt{\rho}} \tag{11-13a}$$

$$H_\phi = -jI_e \frac{\beta}{4} H_1^{(2)}(\beta\rho) \overset{\beta\rho \rightarrow \text{large}}{\simeq} I_e \sqrt{\frac{j\beta}{8\pi}} \frac{e^{-j\beta\rho}}{\sqrt{\rho}} \tag{11-13b}$$

The ratio (11-13a) to (11-13b) is defined as the wave impedance, which reduces to

$$Z_w^{+\rho} = \frac{E_z}{-H_\phi} = \eta \tag{11-14}$$

Since the wave impedance is equal to the intrinsic impedance, the waves radiated by the line source are TEM$^\rho$.

Example 11-1

For a displaced electric line source (at ρ', ϕ'), as shown in Figure 11-1b, of constant current I_e, derive (in terms of the cylindrical coordinates ρ, ρ', ϕ, ϕ'), the:

a. Vector potential A_z.
b. Electric field components (E_ρ, E_ϕ, E_z).
c. Magnetic field components (H_ρ, H_ϕ, H_z).

This is an alternate solution to the expressions of (11-11a) through (11-11c).

Solution: Equations (11-6) through (11-10b) are correct when the electric field line source is located at the origin of the coordinate system, as shown in Figure 11-1a. However, when the electric line source is offset at point (ρ', ϕ', z), as shown in Figure 11-1b, the potential of (11-6) can be written as

$$\mathbf{A} = \hat{\mathbf{a}}_z A_z(\beta|\boldsymbol{\rho} - \boldsymbol{\rho}'|) = \hat{\mathbf{a}}_z A_0 H_0^{(2)}(\beta|\boldsymbol{\rho} - \boldsymbol{\rho}'|) = \hat{\mathbf{a}}_z A_0 H_0^{(2)}\left(\beta\sqrt{\rho^2 + (\rho')^2 - 2\rho\rho'\cos(\phi - \phi')}\right)$$

where $A_0 = -j\dfrac{\mu}{4}I_e$

The electric field of (6-17) or (6-70) can then be written as

$$E_z = -j\frac{1}{\omega\mu\varepsilon}\beta^2 A_z = -j\omega A_z = -j\omega A_0 H_0^{(2)}\left(\beta\sqrt{\rho^2 + (\rho')^2 - 2\rho\rho'\cos(\phi - \phi')}\right)$$

$$E_z = -I_e \frac{\omega\mu}{4} H_0^{(2)}(\beta|\boldsymbol{\rho}-\boldsymbol{\rho}'|)$$

$$\text{where} \quad |\boldsymbol{\rho}-\boldsymbol{\rho}'| = R = \sqrt{\rho^2 + (\rho')^2 - 2\rho\rho'\cos(\phi-\phi')}$$

The magnetic field can now be written based on (6-4a) and (6-70) as

$$\mathbf{H} = \frac{1}{\mu}\nabla\times\mathbf{A} = \hat{\mathbf{a}}_\rho \frac{1}{\mu}\frac{1}{\rho}\frac{\partial A_z}{\partial\phi} + \hat{\mathbf{a}}_\phi\left(-\frac{1}{\mu}\frac{\partial A_z}{\partial\rho}\right)$$

$$H_\rho = \frac{1}{\mu}\frac{1}{\rho}\frac{\partial A_z}{\partial\phi}$$

$$H_\phi = -\frac{1}{\mu}\frac{\partial A_z}{\partial\rho}$$

The individual magnetic field components of H_ρ and H_ϕ can be written using

$$\frac{d}{dx}H_0^{(2)}(\alpha x) = -\alpha H_1^{(2)}(\alpha x)$$

and the following derivatives

$$\frac{\partial A_z}{\partial\phi} = \frac{\partial R}{\partial\phi}\frac{\partial A_z}{\partial R}$$

$$\frac{\partial R}{\partial\phi} = \frac{\partial}{\partial\phi}\left[\sqrt{\rho^2 + (\rho')^2 - 2\rho\rho'\cos(\phi-\phi')}\right] = \frac{\rho\rho'\sin(\phi-\phi')}{\sqrt{\rho^2 + (\rho')^2 - 2\rho\rho'\cos(\phi-\phi')}}$$

$$\frac{\partial A_z}{\partial R} = \frac{\partial}{\partial R}\left[A_0 H_0^{(2)}\left(\beta\sqrt{\rho^2 + (\rho')^2 - 2\rho\rho'\cos(\phi-\phi')}\right)\right]$$

$$\frac{\partial A_z}{\partial R} = -\beta A_0 H_1^{(2)}\left(\beta\sqrt{\rho^2 + (\rho')^2 - 2\rho\rho'\cos(\phi-\phi')}\right)$$

$$\frac{\partial A_z}{\partial\rho} = \frac{\partial R}{\partial\rho}\frac{\partial A_z}{\partial R}$$

$$\frac{\partial R}{\partial\rho} = \frac{\partial}{\partial\rho}\left[\sqrt{\rho^2 + (\rho')^2 - 2\rho\rho'\cos(\phi-\phi')}\right] = \frac{\rho-\rho'\cos(\phi-\phi')}{\sqrt{\rho^2 + (\rho')^2 - 2\rho\rho'\cos(\phi-\phi')}}$$

$$\frac{\partial A_z}{\partial R} = \frac{\partial}{\partial R}\left[A_0 H_0^{(2)}\left(\beta\sqrt{\rho^2 + (\rho')^2 - 2\rho\rho'\cos(\phi-\phi')}\right)\right]$$

$$\frac{\partial A_z}{\partial R} = -\beta A_0 H_1^{(2)}\left(\beta\sqrt{\rho^2 + (\rho')^2 - 2\rho\rho'\cos(\phi-\phi')}\right)$$

as

$$H_\rho = \frac{1}{\mu}\frac{1}{\rho}\frac{\partial A_z}{\partial\phi} = -A_0\frac{\beta\rho'\sin(\phi-\phi')}{\mu}\frac{H_1^{(2)}\left(\beta\sqrt{\rho^2 + (\rho')^2 - 2\rho\rho'\cos(\phi-\phi')}\right)}{\sqrt{\rho^2 + (\rho')^2 - 2\rho\rho'\cos(\phi-\phi')}}$$

$$H_\rho = jI_e \frac{\beta\rho' \sin(\phi-\phi')}{4} \frac{H_1^{(2)}\left(\beta|\mathbf{\rho}-\mathbf{\rho}'|\right)}{|\mathbf{\rho}-\mathbf{\rho}'|}$$

$$H_\phi = -\frac{1}{\mu}\frac{\partial A_z}{\partial \rho} = -A_0 \frac{\beta[\rho-\rho'\cos(\phi-\phi')]}{\mu} \frac{H_1^{(2)}\left(\beta\sqrt{\rho^2 + (\rho')^2 - 2\rho\rho'\cos(\phi-\phi')}\right)}{\sqrt{\rho^2 + (\rho')^2 - 2\rho\rho'\cos(\phi-\phi')}}$$

$$H_\phi = -\frac{1}{\mu}\frac{\partial A_z}{\partial \rho} = jI_e \frac{\beta[\rho-\rho'\cos(\phi-\phi')]}{4} \frac{H_1^{(2)}\left(\beta|\mathbf{\rho}-\mathbf{\rho}'|\right)}{|\mathbf{\rho}-\mathbf{\rho}'|}$$

11.2.2 Magnetic Line Source

Although magnetic sources as presently known are not physically realizable, they are often used to represent virtual sources in equivalent models. This was demonstrated in Chapter 7, Sections 7.7 and 7.8, where the volume and surface fields equivalence theorems were introduced. Magnetic sources can be used to represent radiating apertures.

The fields generated by magnetic sources can be obtained by solutions to Maxwell's equations or the wave equation (subject to the appropriate boundary conditions), or by using the duality theorem of Chapter 7, Section 7.2, once the solution to the same problem but with an electric source excitation is known.

Then, using the duality theorem of Section 7.2, the field generated by an infinite magnetic line source of constant current I_m can be obtained using Tables 7-1 and 7-2, (11-10a) through (11-10b), and (11-13a) through (11-13b). These can then be written as

$$E_\phi = +jI_m \frac{\beta}{4} H_1^{(2)}(\beta\rho) \overset{\beta\rho\to\text{large}}{\simeq} -I_m \sqrt{\frac{j\beta}{8\pi}} \frac{e^{-j\beta\rho}}{\sqrt{\rho}} \tag{11-15a}$$

$$H_z = -I_m \frac{\beta^2}{4\omega\mu} H_0^{(2)}(\beta\rho) \overset{\beta\rho\to\text{large}}{\simeq} -\frac{1}{\eta} I_m \sqrt{\frac{j\beta}{8\pi}} \frac{e^{-j\beta\rho}}{\sqrt{\rho}} \tag{11-15b}$$

which, when the sources are displaced from the origin, can also be expressed according to (11-11a) and (11-11b) as

$$E_\Psi = +jI_m \frac{\beta}{4} H_1^{(2)}(\beta|\mathbf{\rho}-\mathbf{\rho}'|) \tag{11-16a}$$

$$H_z = -I_m \frac{\beta^2}{4\omega\mu} H_0^{(2)}(\beta|\mathbf{\rho}-\mathbf{\rho}'|) \tag{11-16b}$$

11.2.3 Electric Line Source Above an Infinite Plane Electric Conductor

When an infinite electric line source is placed at a height h above an infinite flat electric conductor, as shown in Figure 11-2a, the solution for the field components must include the presence of the conducting plane. This can be accomplished by using Maxwell's equations or the wave equation subject to the radiation conditions at infinity and boundary condition along the air–conductor interface. Instead of doing this, the same solution is obtained by introducing an equivalent model which leads to the same fields in the region of interest. Since the fields below the interface (in the electric conductor, $y < 0$) are known (they are zero), the equivalent model should be valid and

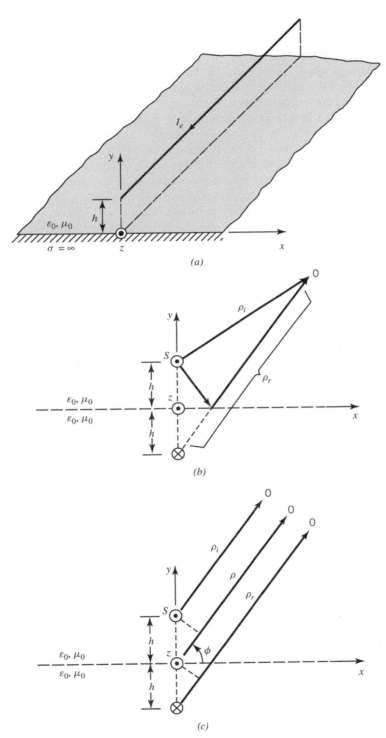

Figure 11-2 Electric line source above a flat and infinite electric ground plane. (*a*) Line source. (*b*) Equivalent (near field). (*c*) Equivalent (far field).

leads to the same fields as the actual physical problem on or above the interface ($y \geq 0$). In this case, as long as the equivalent model satisfies the same boundary conditions as the actual physical problem along a closed surface, according to the uniqueness theorem of Section 7.3, the solution of the equivalent model will be unique and be the same as that of the physical problem. For this problem, the closed surface that will be chosen is that of the air–conductor interface ($y = 0$), which extends on the range $-\infty \leq x \leq +\infty$.

The equivalent problem of Figure 11-2a is that of Figure 11-2b where the ground plane has been replaced by an equivalent source (usually referred to as *image* or *virtual source* or *caustic*). According to the theory of Section 7.4, the image source is introduced to account for the reflections from the surface of the ground plane. The magnitude, phase, polarization, and position of the image source must be such that the boundary conditions of the equivalent problem of Figure (11-2b) along $-\infty \leq x \leq +\infty$ are the same as those of the physical problem of Figure 11-2a. In this situation the image must have the same magnitude as the actual source, its phase must be 180° out of phase from the actual source, it must be placed below the interface at a depth $h(y = -h)$ along a line perpendicular to the interface and passing through the actual source, and its length must also be parallel to the z axis. Such a system configuration, as shown in Figure 11-2b, does lead to zero tangential electric field along $-\infty \leq x \leq +\infty$, which is identical to that of Figure 11-2a along the air–conductor interface.

Therefore, according to Figures 11-2a and 11-2b and (11-13a), the total electric field is equal to

$$\mathbf{E}^t = \mathbf{E}^i + \mathbf{E}^r = \begin{cases} -\hat{\mathbf{a}}_z \dfrac{\beta^2 I_e}{4\omega\varepsilon} \left[H_0^{(2)}(\beta\rho_i) - H_0^{(2)}(\beta\rho_r) \right], & y \geq 0 \quad (11\text{-}17a) \\ 0, & y < 0 \quad (11\text{-}17b) \end{cases}$$

which, for observations at large distances, as shown by Figure 11-2c, reduces using the asymptotic expansion (11-12a) to

$$\mathbf{E}^t = \mathbf{E}^i + \mathbf{E}^r = \begin{cases} -\hat{\mathbf{a}}_z \eta I_e \left(\dfrac{e^{-j\beta\rho_i}}{\sqrt{\rho_i}} - \dfrac{e^{-j\beta\rho_r}}{\sqrt{\rho_r}} \right) \sqrt{\dfrac{j\beta}{8\pi}}, & y \geq 0 \quad (11\text{-}18a) \\ 0, & y < 0 \quad (11\text{-}18b) \end{cases}$$

According to Figure 11-2c, for observations made at large distances ($\rho \gg h$)

$$\left. \begin{aligned} \rho_i &\simeq \rho - h\cos\left(\frac{\pi}{2} - \phi\right) = \rho - h\sin(\phi) \\ \rho_r &\simeq \rho + h\cos\left(\frac{\pi}{2} - \phi\right) = \rho + h\sin(\phi) \end{aligned} \right\} \quad \text{for phase variations} \quad (11\text{-}19a)$$

$$\rho_i \simeq \rho_r \simeq \rho \quad \text{for amplitude variations} \quad (11\text{-}19b)$$

These approximations are usually referred to in antenna and scattering theory as the *far-field approximations* [1]. Using (11-19a) and (11-19b), we can reduce (11-18a) and (11-18b) to

$$\mathbf{E}^t = \mathbf{E}^i + \mathbf{E}^r = \begin{cases} -\hat{\mathbf{a}}_z \eta I_e \sqrt{\dfrac{j\beta}{8\pi}} \left(e^{+j\beta h \sin\phi} - e^{-j\beta h \sin\phi} \right) \dfrac{e^{-j\beta\rho}}{\sqrt{\rho}} \\ = -\hat{\mathbf{a}}_z j\eta I_e \sqrt{\dfrac{j\beta}{2\pi}} \sin(\beta h \sin\phi) \dfrac{e^{-j\beta\rho}}{\sqrt{\rho}}, & y \geq 0 \quad (11\text{-}20a) \\ 0, & y < 0 \quad (11\text{-}20b) \end{cases}$$

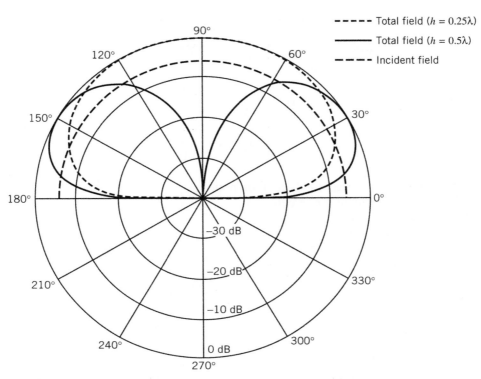

Figure 11-3   Radiation patterns of a line source above an infinite electric ground plane for $h = 0.25\lambda_0$ and $0.5\lambda_0$.

Normalized amplitude patterns (in decibels) for a source placed at a height of $h = 0.25\lambda$ and 0.5λ above the strip are shown in Figure 11-3.

11.3   PLANE WAVE SCATTERING BY PLANAR SURFACES

An important parameter in scattering is the electromagnetic scattering by a target that is usually represented by its *echo area* or *radar cross section* (RCS) (σ). The echo area or RCS is defined as "*the area intercepting the amount of power that, when scattered isotropically, produces at the receiver a density that is equal to the density scattered by the actual target*" [1]. For a two-dimensional target the scattering parameter is referred to as the *scattering width* (SW) or alternatively as the *radar cross section per unit length*. In equation form the scattering width and the radar cross section (σ) of a target take the form of

Scattering Width: Two-Dimensional Target

$$\sigma_{2\text{-}D} = \begin{cases} \lim\limits_{\rho \to \infty} \left[2\pi\rho \dfrac{S^s}{S^i} \right] & (11\text{-}21a) \\[2.5ex] \lim\limits_{\rho \to \infty} \left[2\pi\rho \dfrac{|\mathbf{E}^s|^2}{|\mathbf{E}^i|^2} \right] & (11\text{-}21b) \\[2.5ex] \lim\limits_{\rho \to \infty} \left[2\pi\rho \dfrac{|\mathbf{H}^s|^2}{|\mathbf{H}^i|^2} \right] & (11\text{-}21c) \end{cases}$$

Radar Cross Section: Three-Dimensional Target

$$
\sigma_{3\text{-D}} = \begin{cases} \lim_{r \to \infty} \left[4\pi r^2 \, \dfrac{S^s}{S^i} \right] & (11\text{-}22a) \\[12pt] \lim_{r \to \infty} \left[4\pi r^2 \, \dfrac{|\mathbf{E}^s|^2}{|\mathbf{E}^i|^2} \right] & (11\text{-}22b) \\[12pt] \lim_{r \to \infty} \left[4\pi r^2 \, \dfrac{|\mathbf{H}^s|^2}{|\mathbf{H}^i|^2} \right] & (11\text{-}22c) \end{cases}
$$

where ρ, r = distance from target to observation point
  S^s, S^i = scattered, incident power densities
  $\mathbf{E}^s, \mathbf{E}^i$ = scattered, incident electric fields
  $\mathbf{H}^s, \mathbf{H}^i$ = scattered, incident magnetic fields

For normal incidence, the two- and three-dimensional fields, and scattering width and radar cross sections for a target of length ℓ are related by [2-4]

$$
E_{3\text{-D}} \simeq \left(E_{2\text{-D}} \, \frac{\ell e^{j\pi/4}}{\sqrt{\lambda \rho}} \right)_{\rho = r} \tag{11-22d}
$$

$$
\sigma_{3\text{-D}} \simeq \sigma_{2\text{-D}} \, \frac{2\ell^2}{\lambda} \tag{11-22e}
$$

The unit of the two-dimensional SW is length (meters in the MKS system), whereas that of the three-dimensional RCS is area (meters squared in the MKS system). A most common reference is *one meter* for the two-dimensional SW and *one meter squared* for the three-dimensional RCS. Therefore, a most common designation is dB/m (or dBm) for the two-dimensional SW and dB/(square meter) (or dBsm) for the three-dimensional RCS.

When the transmitter and receiver are at the same location, the RCS is usually referred to as *monostatic* (or *backscattered*), and it is referred to as *bistatic* when the two are at different locations. Observations made toward directions that satisfy Snell's law of reflection are usually referred to as *specular*. Therefore, the RCS of a target is a very important parameter which characterizes its scattering properties. A plot of the RCS as a function of the space coordinates is usually referred to as the *RCS pattern*. The definitions (11-21a) through (11-22c) all indicate that the SW and RCS of targets are defined under plane wave, that in practice can only be approximated when the target is placed in the far field of the source (at least $2D^2 / \lambda$) where D is the largest dimension of the target [1].

In this section, physical optics (PO) techniques will be used to analyze the scattering from conducting strips and plates of finite width, neglecting edge effects. The edge effects will be considered in Chapters 12 to 14, where, respectively, moment method and geometrical theory of diffraction techniques will be utilized. Physical optics techniques are most accurate at specular directions [5].

11.3.1 TM^z Plane Wave Scattering from a Strip

Let us assume that a TM^z uniform plane wave is incident upon an electric conducting strip of width w and infinite length, as shown in Figures 11-4a and 11-4b. The incident electric and magnetic fields can be written as

$$
\mathbf{E}^i = \hat{\mathbf{a}}_z E_0 \, e^{j\beta(x \cos \phi_i + y \sin \phi_i)} \tag{11-23a}
$$

$$
\mathbf{H}^i = \frac{E_0}{\eta} (-\hat{\mathbf{a}}_x \sin \phi_i + \hat{\mathbf{a}}_y \cos \phi_i) e^{j\beta(x \cos \phi_i + y \sin \phi_i)} \tag{11-23b}
$$

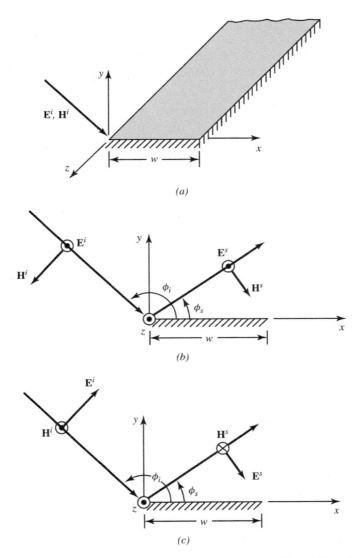

Figure 11-4 Uniform plane wave incident on a finite width strip. (*a*) Finite width strip. (*b*) TM^z polarization. (*c*) TE^z polarization.

where E_0 is a constant, and it represents the magnitude of the incident electric field. In Figures 11-4*a* and 11-4*b* the angle ϕ_i is shown to be greater than 90°. The reflected fields can be expressed as

$$\mathbf{E}^r = \hat{\mathbf{a}}_z \Gamma_\perp E_0 \, e^{-j\beta(x\cos\phi_r + y\sin\phi_r)} \tag{11-24a}$$

$$\mathbf{H}^r = \frac{\Gamma_\perp E_0}{\eta} (\hat{\mathbf{a}}_x \sin\phi_r - \hat{\mathbf{a}}_y \cos\phi_r) e^{-j\beta(x\cos\phi_r + y\sin\phi_r)} \tag{11-24b}$$

where ϕ_r is the reflection angle as determined by enforcing the boundary conditions along the interface, assuming an interface of infinite extent. For the finite-width strip the reflection angle ϕ_r is not the same as the scattering angle $\phi_s (\phi_r \neq \phi_s)$. The two coincide for an infinite-width strip when geometrical optics and physical optics reduce to each other. Thus, we use geometrical optics to determine the reflection coefficient on the surface of the finite-width strip, and then apply physical optics to find the scattered fields. According to (5-17a) and (5-15a), the reflection coefficient for a perfectly conducting surface is equal to $\Gamma_\perp = -1$ and $\phi_r = \pi - \phi_i$. Thus, (11-24a) and (11-24b) reduce to

$$\mathbf{E}^r = -\hat{\mathbf{a}}_z E_0 e^{j\beta(x\cos\phi_i - y\sin\phi_i)} \tag{11-25a}$$

$$\mathbf{H}^r = \frac{E_0}{\eta}(-\hat{\mathbf{a}}_x \sin\phi_i - \hat{\mathbf{a}}_y \cos\phi_i)e^{j\beta(x\cos\phi_i - y\sin\phi_i)} \tag{11-25b}$$

Using physical optics techniques of (7-54), the current density induced on the surface of the strip can be written as

$$\mathbf{J}_s = \hat{\mathbf{n}} \times \mathbf{H}'|_{\substack{y=0 \\ x=x'}} = \hat{\mathbf{n}} \times (\mathbf{H}^i + \mathbf{H}^r)|_{\substack{y=0 \\ x=x'}}$$

$$= 2\hat{\mathbf{n}} \times \mathbf{H}^i|_{\substack{y=0 \\ x=x'}} = \hat{\mathbf{a}}_y \frac{2E_0}{\eta} \times (-\hat{\mathbf{a}}_x \sin\phi_i + \hat{\mathbf{a}}_y \cos\phi_i)e^{j\beta x'\cos\phi_i}$$

$$\mathbf{J}_s = \hat{\mathbf{a}}_z \frac{2E_0}{\eta}\sin\phi_i\, e^{j\beta x'\cos\phi_i} \tag{11-26}$$

and the far-zone scattered field can be found using (6-96a), (6-101a), and (6-101b) or

$$\mathbf{A} = \frac{\mu}{4\pi} \iint_S \mathbf{J}_s(x',y',z')\frac{e^{-j\beta R}}{R}ds' \tag{11-27a}$$

$$\mathbf{E_A} \simeq -j\omega\mathbf{A} \quad \text{(for } \theta \text{ and } \phi \text{ components only)} \tag{11-27b}$$

$$\mathbf{H_A} \simeq \frac{1}{\eta}\hat{\mathbf{a}}_r \times \mathbf{E_A} = -j\frac{\omega}{\eta}\hat{\mathbf{a}}_r \times \mathbf{A} \quad \text{(for } \theta \text{ and } \phi \text{ components only)} \tag{11-27c}$$

where

$$R = \sqrt{(x-x')^2 + (y-y')^2 + (z-z')^2} = \sqrt{(|\boldsymbol{\rho}-\boldsymbol{\rho}'|)^2 + (z-z')^2} \tag{11-27d}$$

Substituting (11-26) into (11-27a) and using (11-27d), we can write that

$$\mathbf{A} = \hat{\mathbf{a}}_z \frac{\mu E_0}{2\pi\eta}\sin\phi_i \int_0^w \left[\int_{-\infty}^{+\infty} \frac{\exp\left[-j\beta\sqrt{(|\boldsymbol{\rho}-\boldsymbol{\rho}'|)^2 + (z-z')^2}\right]}{\sqrt{(|\boldsymbol{\rho}-\boldsymbol{\rho}'|)^2 + (z-z')^2}}dz'\right]e^{j\beta x'\cos\phi_i}dx' \tag{11-28}$$

Since the integral with the infinite limits can be represented by a Hankel function of the second kind of zero order

$$\int_{-\infty}^{+\infty}\frac{e^{-j\alpha\sqrt{x^2+t^2}}}{\sqrt{x^2+t^2}}dt = -j\pi H_0^{(2)}(\alpha x) \tag{11-28a}$$

(11-28) can be reduced to

$$\mathbf{A} = -\hat{\mathbf{a}}_z j\frac{\mu E_0}{2\eta}\sin\phi_i \int_0^w H_0^{(2)}(\beta\,|\boldsymbol{\rho}-\boldsymbol{\rho}'|)e^{j\beta x'\cos\phi_i}dx' \tag{11-28b}$$

For far-zone observations

$$|\boldsymbol{\rho}-\boldsymbol{\rho}'|=\sqrt{\rho^2+(\rho')^2-2\rho\rho'\cos(\phi_s-\phi')}\stackrel{\rho\gg\rho'}{\simeq}\sqrt{\rho^2-2\rho\rho'\cos(\phi_s-\phi')}$$

$$|\boldsymbol{\rho}-\boldsymbol{\rho}'|\stackrel{\rho\gg\rho'}{\simeq}\rho\sqrt{1-2\left(\frac{\rho'}{\rho}\right)\cos(\phi_s-\phi')}\stackrel{\rho\gg\rho'}{\simeq}\rho\left[1-\left(\frac{\rho'}{\rho}\right)\cos(\phi_s-\phi')\right]$$

$$|\boldsymbol{\rho}-\boldsymbol{\rho}'|\stackrel{\rho\gg\rho'}{\simeq}\rho-\rho'\cos(\phi_s-\phi') \tag{11-29}$$

Since the source (here the current density) exists only over the width of the strip that according to Figure 11-4 lies along the x axis, then $\rho'=x'$ and $\phi'=0$. Thus, for far-field observations (11-29) reduces to

$$|\boldsymbol{\rho}-\boldsymbol{\rho}'|\simeq\begin{cases}\rho-\rho'\cos(\phi_s-\phi')=\rho-x'\cos\phi_s & \text{for phase terms} \tag{11-29a}\\ \rho & \text{for amplitude terms} \tag{11-29b}\end{cases}$$

In turn the Hankel function in the integrand of (11-28b) can be expressed, using (11-29a) and (11-29b), as

$$H_0^{(2)}(\beta|\boldsymbol{\rho}-\boldsymbol{\rho}'|)\stackrel{\rho\gg\rho'}{\simeq}\sqrt{\frac{2j}{\pi\beta\rho}}e^{-j\beta(\rho-x'\cos\phi_s)}=\sqrt{\frac{2j}{\pi\beta}}\frac{e^{-j\beta\rho}}{\sqrt{\rho}}e^{j\beta x'\cos\phi_s} \tag{11-30}$$

Substituting (11-30) into (11-28b) reduces it to

$$\mathbf{A}\simeq-\hat{\mathbf{a}}_z j\frac{\mu E_0}{2\eta}\sqrt{\frac{2j}{\pi\beta}}\frac{e^{-j\beta\rho}}{\sqrt{\rho}}\sin\phi_i\int_0^w e^{j\beta x'(\cos\phi_s+\cos\phi_i)}dx'$$

$$\mathbf{A}\simeq-\hat{\mathbf{a}}_z j\frac{\mu w E_0}{\eta}\sqrt{\frac{j}{2\pi\beta}}e^{j(\beta w/2)(\cos\phi_s+\cos\phi_i)}$$

$$\times\left\{\sin\phi_i\left[\frac{\sin\left[\frac{\beta w}{2}(\cos\phi_s+\cos\phi_i)\right]}{\frac{\beta w}{2}(\cos\phi_s+\cos\phi_i)}\right]\frac{e^{-j\beta\rho}}{\sqrt{\rho}}\right\} \tag{11-31}$$

Therefore, the far-zone scattered spherical components of the electric and magnetic fields of (11-27b) and (11-27c) can be written, using (11-31) and (II-12), as

$$E_\theta^s\simeq j\omega A_z\sin\theta_s=-w E_0\sqrt{\frac{j\beta}{2\pi}}e^{j(\beta w/2)(\cos\phi_s+\cos\phi_i)}$$

$$\times\left\{\sin\theta_s\sin\phi_i\left[\frac{\sin\left[\frac{\beta w}{2}(\cos\phi_s+\cos\phi_i)\right]}{\frac{\beta w}{2}(\cos\phi_s+\cos\phi_i)}\right]\frac{e^{-j\beta\rho}}{\sqrt{\rho}}\right\} \tag{11-32a}$$

$$H_\phi^s\simeq\frac{E_\theta^s}{\eta}=-\frac{w E_0}{\eta}\sqrt{\frac{j\beta}{2\pi}}e^{j(\beta w/2)(\cos\phi_s+\cos\phi_i)}$$

$$\times\left\{\sin\theta_s\sin\phi_i\left[\frac{\sin\left[\frac{\beta w}{2}(\cos\phi_s+\cos\phi_i)\right]}{\frac{\beta w}{2}(\cos\phi_s+\cos\phi_i)}\right]\frac{e^{-j\beta\rho}}{\sqrt{\rho}}\right\} \tag{11-32b}$$

The bistatic scattering width is obtained using any of (11-21a) through (11-21c), and it is represented at $\theta_s = 90°$ by

$$\sigma_{2\text{-D}}(\text{bistatic}) = \lim_{\rho \to \infty} \left[2\pi\rho \frac{|\mathbf{E}^s|^2}{|\mathbf{E}^i|^2} \right] = \frac{2\pi w^2}{\lambda} \left\{ \sin\phi_i \left[\frac{\sin\left[\frac{\beta w}{2}(\cos\phi_s + \cos\phi_i)\right]}{\frac{\beta w}{2}(\cos\phi_s + \cos\phi_i)} \right] \right\}^2 \qquad (11\text{-}33)$$

which for the monostatic system configuration ($\phi_s = \phi_i$) reduces to

$$\sigma_{2\text{-D}}(\text{monostatic}) = \frac{2\pi w^2}{\lambda} \left\{ \sin\phi_i \left[\frac{\sin(\beta w \cos\phi_i)}{\beta w \cos\phi_i} \right] \right\}^2 \qquad (11\text{-}33a)$$

Computed patterns of the normalized bistatic SW of (11-33) (in dB), for $0° \le \phi_s \le 180°$ when $\phi_i = 120°$ and $w = 2\lambda$ and 10λ, are shown in Figure 11-5a. It is apparent that the maximum occurs when the $\sin(x)/x$ function reaches its maximum value of unity, that is, when the $x = \beta w (\cos\phi_s + \cos\phi_i)/2 = 0$. For these examples, this occurs when $\phi_s = 180° - \phi_i = 180° - 120° = 60°$, which represents the direction of specular scattering (angle of scattering is equal to the angle of incidence). Away from the direction of maximum radiation, the pattern variations are of $\sin(x)/x$ form. The normalized monostatic SW of (11-33a) (in dB) for $w = 2\lambda$ and 10λ are shown plotted in Figure 11-5b for $0° \le \phi_i \le 180°$, where the maximum occurs when $\phi_i = 90°$, which is the direction of normal incidence (the strip is viewed perpendicularly to its flat surface). Again, away from the maximum radiation, the pattern variations are approximately of $\sin(x)/x$ form.

Example 11-2

Derive the far-zone scattered fields and the associated scattering width when a TE^z uniform plane wave is incident upon a two-dimensional conducting strip of width w, as shown in Figures 11-4a and 11-4c. Use physical optics methods.

Solution: According to Figure 11-4c, the incident electric and magnetic field components for a TE^z uniform plane wave can be written as

$$\mathbf{E}^i = \eta H_0(\hat{\mathbf{a}}_x \sin\phi_i - \hat{\mathbf{a}}_y \cos\phi_i) e^{j\beta(x\cos\phi_i + y\sin\phi_i)}$$

$$\mathbf{H}^i = \hat{\mathbf{a}}_z H_0 e^{j\beta(x\cos\phi_i + y\sin\phi_i)}$$

The current induced on the surface of the finite width strip can be approximated by the physical optics current and is equal to

$$\mathbf{J}_s \simeq 2\hat{\mathbf{n}} \times \mathbf{H}^i \big|_{\substack{y=0 \\ x=x'}} = 2\hat{\mathbf{a}}_y \times \hat{\mathbf{a}}_z H_z^i \big|_{\substack{y=0 \\ x=x'}} = 2\hat{\mathbf{a}}_x H_z^i \big|_{\substack{y=0 \\ x=x'}}$$

$$\mathbf{J}_s \simeq \hat{\mathbf{a}}_x 2 H_0 e^{j\beta x' \cos\phi_i}$$

$$J_y = J_z = 0, \quad J_x = 2 H_0 e^{j\beta x' \cos\phi_i}$$

Using the steps outlined by (11-27a) through (11-31), it can be shown that

$$\mathbf{A} = -\hat{\mathbf{a}}_x j \frac{\mu H_0}{2} \int_0^w H_0^{(2)}(\beta |\rho - \rho'|) e^{j\beta x' \cos\phi_i} \, dx'$$

which for far-zone observations reduces to

$$\mathbf{A} \simeq -\hat{\mathbf{a}}_x j\mu w H_0 \sqrt{\frac{j}{2\pi\beta}} e^{j(\beta w/2)(\cos\phi_s + \cos\phi_i)} \left[\frac{\sin\left[\frac{\beta w}{2}(\cos\phi_s + \cos\phi_i)\right]}{\frac{\beta w}{2}(\cos\phi_s + \cos\phi_i)} \right] \frac{e^{-j\beta\rho}}{\sqrt{\rho}}$$

In spherical components this can be written, according to (II-12), as

$$A_r = A_x \sin\theta_s \cos\phi_s$$
$$A_\theta = A_x \cos\theta_s \cos\phi_s$$
$$A_\phi = -A_x \sin\phi_s$$

which for $\theta_s = 90°$ reduce to

$$A_r = A_x \cos\phi_s$$
$$A_\theta = 0$$
$$A_\phi = -A_x \sin\phi_s$$

Thus, the far-zone electric and magnetic field components in the $\theta_s = 90°$ plane can be written as

$$E_r^s \simeq E_\theta^s \simeq H_r^s \simeq H_\phi^s \simeq 0$$

$$E_\phi^s \simeq -j\omega A_\phi = j\omega A_x \sin\phi_s$$

$$E_\phi^s \simeq -\eta w H_0 \sqrt{\frac{j\beta}{2\pi}}\, e^{j(\beta w/2)(\cos\phi_s + \cos\phi_i)}$$

$$\times \left\{ \sin\phi_s \left[\frac{\sin\left[\dfrac{\beta w}{2}(\cos\phi_s + \cos\phi_i)\right]}{\dfrac{\beta w}{2}(\cos\phi_s + \cos\phi_i)} \right] \frac{e^{-j\beta\rho}}{\sqrt{\rho}} \right\}$$

$$H_\theta^s \simeq -\frac{E_\phi^s}{\eta} = w H_0 \sqrt{\frac{j\beta}{2\pi}}\, e^{j(\beta w/2)(\cos\phi_s + \cos\phi_i)}$$

$$\times \left\{ \sin\phi_s \left[\frac{\sin\left[\dfrac{\beta w}{2}(\cos\phi_s + \cos\phi_i)\right]}{\dfrac{\beta w}{2}(\cos\phi_s + \cos\phi_i)} \right] \frac{e^{-j\beta\rho}}{\sqrt{\rho}} \right\}$$

The bistatic and monostatic (backscattering) scattering widths are given by

$$\sigma_{2\text{-D}}(\text{bistatic}) = \lim_{\rho\to\infty} \left[2\pi\rho \frac{|H^s|^2}{|\mathbf{H}^i|^2} \right]$$

$$= \frac{2\pi w^2}{\lambda} \left\{ \sin\phi_s \left[\frac{\sin\left[\dfrac{\beta w}{2}(\cos\phi_s + \cos\phi_i)\right]}{\dfrac{\beta w}{2}(\cos\phi_s + \cos\phi_i)} \right] \right\}^2$$

$$\sigma_{2\text{-D}}(\text{monostatic}) = \frac{2\pi w^2}{\lambda} \left\{ \sin\phi_i \left[\frac{\sin(\beta w \cos\phi_i)}{\beta w \cos\phi_i} \right] \right\}^2$$

The monostatic SW for the TE^z polarization is identical to that of the TM^z as given by (11-33a), and it is shown plotted in Figure 11-5b for $w = 2\lambda$ and 10λ. The bistatic SW, however, differs from that of (11-33) in that the $\sin^2\phi_i$ term is replaced by $\sin^2\phi_s$. Therefore, the computed normalized bistatic patterns (in dB) for $0° \le \phi_s \le 180°$ when $\phi_i = 120°$ and $w = 2\lambda$, 10λ are shown, respectively, in Figure 11-6. It is evident that the larger the width of the strip, the larger the maximum value of the SW and the larger the number of minor lobes in its pattern.

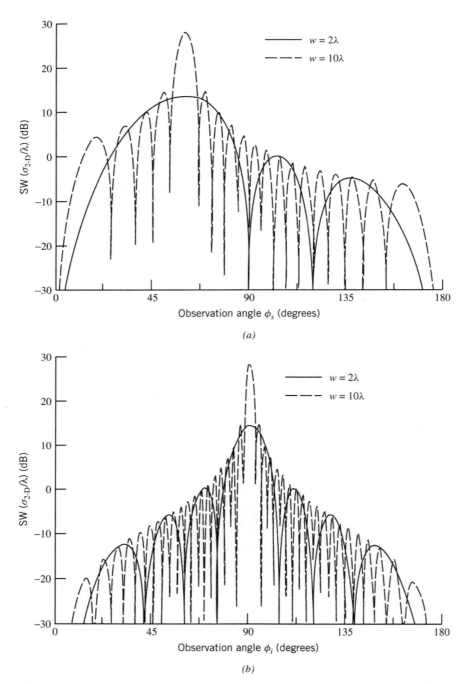

Figure 11-5 Bistatic and monostatic scattering width (SW) for a finite width strip. (*a*) Bistatic ($\phi_i = 120°$): TM^z. (*b*) Monostatic: TM^z and TE^z.

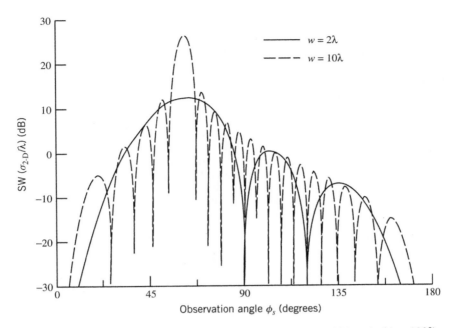

Figure 11-6   Bistatic TE^z scattering width (SW) for a finite width strip ($\phi_i = 120°$).

To see that indeed the bistatic SW patterns of the TM^z and TE^z polarizations are different, we have plotted in Figure 11-7a the two for $w = 2\lambda$ when the incidence angle is $\phi_i = 120°$. It is evident that the two are similar but not identical because the $\sin^2 \phi_i$ term is replaced by $\sin^2 \phi_s$, and vice versa. In addition, whereas the maximum for the TM^z occurs at the specular direction ($\phi_s = 60°$ for this example), that of the TE^z occurs at an angle slightly larger than the specular direction. However, as the size of the target becomes very large, the maximum of the TE^z SW moves closer toward the specular direction and matches that of the TM^z, which always occurs at the specular direction [6]. This is illustrated in Figure 11-7b where the bistatic RCS of the two polarizations has been plotted for $w = 10\lambda$ and $\phi_i = 120°$. When the size of the target is very large electrically, the $[\sin(x)/x]^2$ in the bistatic RCS expression of Example 11-1 varies very rapidly as a function of ϕ_s so that the slowly varying $\sin^2 \phi_s$ is essentially a constant near the maximum of the $[\sin(x)/x]^2$ function. This is not true when the size of the target is small electrically, as was demonstrated by the results of Figure 11-7a.

11.3.2 TE^x Plane Wave Scattering from a Flat Rectangular Plate

Let us now consider scattering from a three-dimensional scatterer, specifically uniform plane-wave scattering from a rectangular plate, as shown in Figure 11-8a. To simplify the details, let us assume that the uniform plane wave is TE^x, and that it lies on the yz plane, as shown in Figure 11-8b. The electric and magnetic fields can now be written as

$$\mathbf{E}^i = \eta H_0(\hat{\mathbf{a}}_y \cos\theta_i + \hat{\mathbf{a}}_z \sin\theta_i) \, e^{-j\beta(y\sin\theta_i - z\cos\theta_i)} \tag{11-34a}$$

$$\mathbf{H}^i = \hat{\mathbf{a}}_x H_0 e^{-j\beta(y\sin\theta_i - z\cos\theta_i)} \tag{11-34b}$$

where H_0 is a constant that represents the magnitude of the incident magnetic field.

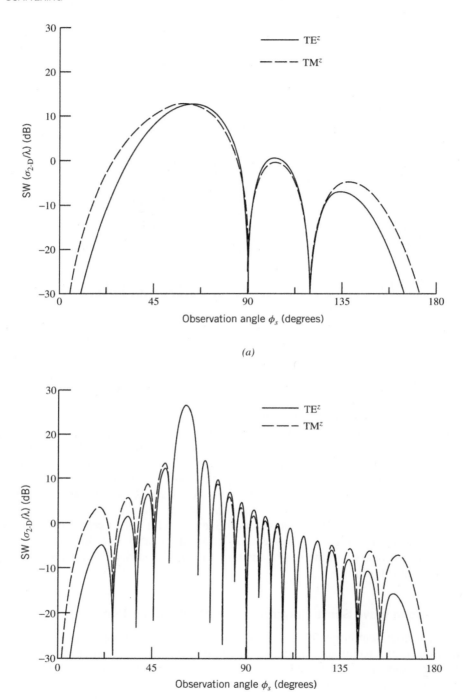

(a)

(b)

Figure 11-7   TE^z and TM^z bistatic scattering widths (SW) for a finite-width strip ($\phi_i = 120°$). (a) $w = 2\lambda$. (b) $w = 10\lambda$.

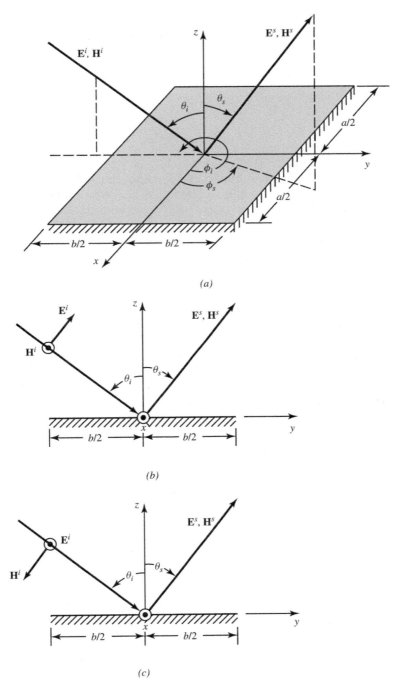

Figure 11-8 Uniform plane wave incident on a rectangular conducting plate. (*a*) Rectangular plate. (*b*) *yz* plane: TE^x polarization. (*c*) *yz* plane: TM^x polarization.

The scattered field can be found, by neglecting edge effects, using physical optics techniques of Section 7.10, where the current density induced at the surface of the plate is represented on the plate by

$$\mathbf{J}_s \simeq 2\hat{\mathbf{n}} \times \mathbf{H}'|_{\substack{z=0 \\ y=y'}} = 2\hat{\mathbf{a}}_z \times \hat{\mathbf{a}}_x H_x |_{\substack{z=0 \\ y=y'}} = \hat{\mathbf{a}}_y 2H_0 e^{-j\beta y' \sin\theta_i} \tag{11-35}$$

Thus,

$$J_x = J_z = 0 \quad \text{and} \quad J_y = 2H_0 e^{-j\beta y' \sin\theta_i} \tag{11-35a}$$

For an infinite plate, the current density of (11-35) or (11-35a) yields exact field solutions. However, for finite-size plates it is approximate, and the corresponding scattered fields obtained using it are more accurate toward the specular direction, where Snell's law of reflection is satisfied. The solutions become less accurate as the observation points are removed further from the specular directions.

The scattered fields are obtained using (6-122a) through (6-122f) and (6-125a) through (6-125b), where the electric current density components are those given by (11-35a). Using (6-125a), (6-127c), (6-128c), and (11-35a), we can write that

$$N_\theta = \iint_S \left[J_x \cos\theta_s \cos\phi_s + J_y \cos\theta_s \sin\phi_s - J_z \sin\theta_s \right]_{J_x=J_z=0}$$
$$\times e^{j\beta(x'\sin\theta_s\cos\phi_s + y'\sin\theta_s\sin\phi_s)} \, dx' \, dy'$$
$$= 2H_0 \cos\theta_s \sin\phi_s \int_{-b/2}^{+b/2} e^{j\beta y'(\sin\theta_s\sin\phi_s - \sin\theta_i)} \, dy' \int_{-a/2}^{+a/2} e^{j\beta x'\sin\theta_s\cos\phi_s} \, dx' \tag{11-36}$$

Since

$$\int_{-c/2}^{+c/2} e^{j\alpha z} dz = c \left[\frac{\sin\left(\frac{\alpha}{2} c\right)}{\frac{\alpha}{2} c} \right] \tag{11-37}$$

(11-36) reduces to

$$N_\theta = 2abH_0 \left\{ \cos\theta_s \sin\phi_s \left[\frac{\sin(X)}{X} \right] \left[\frac{\sin(Y)}{Y} \right] \right\} \tag{11-38}$$

where

$$X = \frac{\beta a}{2} \sin\theta_s \cos\phi_s \tag{11-38a}$$

$$Y = \frac{\beta b}{2} (\sin\theta_s \sin\phi_s - \sin\theta_i) \tag{11-38b}$$

In the same manner, (6-125b) can be written as

$$N_\phi = \iint_S \left[-J_x \sin\phi_s + J_y \cos\phi_s \right]_{J_x=0} e^{j\beta(x'\sin\theta_s\cos\phi_s + y'\sin\theta_s\sin\phi_s)} \, dx' \, dy'$$
$$= 2abH_0 \left\{ \cos\phi_s \left[\frac{\sin(X)}{X} \right] \left[\frac{\sin(Y)}{Y} \right] \right\} \tag{11-39}$$

Therefore, the scattered fields are obtained using (6-122a) through (6-122f) and (11-38) through (11-39), and they can be expressed as

$$E_r^s \simeq 0 \tag{11-40a}$$

$$E_\theta^s \simeq -\frac{j\beta e^{-j\beta r}}{4\pi r}\left(L_\phi + \eta N_\theta\right)_{L_\phi=0} = C\frac{e^{-j\beta r}}{r}\left\{\cos\theta_s\,\sin\phi_s\left[\frac{\sin(X)}{X}\right]\left[\frac{\sin(Y)}{Y}\right]\right\} \tag{11-40b}$$

$$E_\phi^s \simeq +\frac{j\beta e^{-j\beta r}}{4\pi r}\left(L_\theta - \eta N_\phi\right)_{L_\theta=0} = C\frac{e^{-j\beta r}}{r}\left\{\cos\phi_s\left[\frac{\sin(X)}{X}\right]\left[\frac{\sin(Y)}{Y}\right]\right\} \tag{11-40c}$$

$$H_r^s \simeq 0 \tag{11-40d}$$

$$H_\theta^s \simeq -\frac{E_\phi^s}{\eta} \tag{11-40e}$$

$$H_\phi^s \simeq +\frac{E_\theta^s}{\eta} \tag{11-40f}$$

where

$$C = -j\eta\frac{ab\beta H_0}{2\pi} \tag{11-40g}$$

In the principal E plane ($\phi_s = \pi/2$) and H plane ($\theta_s = \theta_i$, $\phi_s = 0$), the electric field components reduce to

$$\underline{E\ \text{Plane}\ (\phi_s = \pi/2)}$$

$$E_r^s \simeq E_\phi^s \simeq 0 \tag{11-41a}$$

$$E_\theta^s \simeq C\frac{e^{-j\beta r}}{r}\left\{\cos\theta_s\left[\frac{\sin\left[\frac{\beta b}{2}(\sin\theta_s - \sin\theta_i)\right]}{\frac{\beta b}{2}(\sin\theta_s - \sin\theta_i)}\right]\right\} \tag{11-41b}$$

$$\underline{H\ \text{Plane}\ (\phi_s = 0)}$$

$$E_r^s \simeq E_\theta^s \simeq 0 \tag{11-42a}$$

$$E_\phi^s \simeq C\frac{e^{-j\beta r}}{r}\left\{\left[\frac{\sin\left[\frac{\beta b}{2}(\sin\theta_i)\right]}{\frac{\beta b}{2}(\sin\theta_i)}\right]\left[\frac{\sin\left[\frac{\beta a}{2}(\sin\theta_s)\right]}{\frac{\beta a}{2}(\sin\theta_s)}\right]\right\} \tag{11-42b}$$

It can be shown that the maximum value of the total scattered field

$$E^s = \sqrt{(E_r^s)^2 + (E_\theta^s)^2 + (E_\phi^s)^2} \simeq \sqrt{(E_\theta^s)^2 + (E_\phi^s)^2} \tag{11-43}$$

for any wave incidence always lies in a scattering plane that is parallel to the incident plane [6]. For the fields of (11-40a) through (11-40f), the scattering plane that contains the maximum scattered field is that defined by $\phi_s = \pi/2,\ 3\pi/2$, and $0 \le \theta_s \le \pi/2$, since the incident plane is that defined by $\phi_i = 3\pi/2$, $0 \le \theta_i \le \pi/2$. The electric field components in the plane that contains the maximum reduce to those of (11-41a) and (11-41b) whose maximum value, when $b \gg \lambda$, occurs approximately when $\theta_s = \theta_i$ (specular reflection). For large values of $b(b \gg \lambda)$ the $\sin(z)/z$ function in (11-41b) varies very rapidly compared to the $\cos\theta_s$ such that the $\cos\theta_s$ function is essentially constant near the maximum of the $\sin(z)/z$ function. For small values of b, the maximum value of (11-41b) can be found iteratively. *Thus, for the TE^x polarization, the maximum of the scattered field from a flat plate does not occur exactly at the specular direction but it approaches that value as the dimensions of the plate become large compared to the wavelength.* This is analogous to the TE^z polarization of the strip of Example 11-2. It will be shown in the example that follows that *for the TM^x polarization, the maximum of the scattered field from a flat plate always occurs at the specular direction no matter what the size of the plate.* This is analogous to the TM^z polarization of the strip of Section 11.3.1.

For the fields of (11-40a) through (11-40g), the radar cross section is obtained using (11-22b) or (11-22c) and can be written as

$$\sigma_{3\text{-D}} = \lim_{r\to\infty}\left[4\pi r^2 \frac{|\mathbf{E}^s|^2}{|\mathbf{E}^i|^2}\right] = \lim_{r\to\infty}\left[4\pi r^2 \frac{|\mathbf{H}^s|^2}{|\mathbf{H}^i|^2}\right]$$

$$= 4\pi\left(\frac{ab}{\lambda}\right)^2 (\cos^2\theta_s \sin^2\phi_s + \cos^2\phi_s)\left[\frac{\sin(X)}{X}\right]^2\left[\frac{\sin(Y)}{Y}\right]^2 \qquad (11\text{-}44)$$

which in the plane that contains the maximum ($\phi_s = \pi/2$) reduces to

$$\underline{\text{Principal Bistatic } (\phi_s = \pi/2)}$$

$$\sigma_{3\text{-D}} = 4\pi\left(\frac{ab}{\lambda}\right)^2 \cos^2\theta_s \frac{\left[\sin\left[\frac{\beta b}{2}(\sin\theta_s \mp \sin\theta_i)\right]\right]^2}{\frac{\beta b}{2}(\sin\theta_s \mp \sin\theta_i)} \quad \begin{array}{l} -\text{ for }\phi_s = \dfrac{\pi}{2},\quad 0\le\theta_s\le\pi/2 \\[6pt] +\text{ for }\phi_s = \dfrac{3\pi}{2},\quad 0\le\theta_s\le\pi/2 \end{array} \quad (11\text{-}44a)$$

while in the backscattering direction ($\phi_s = \phi_i = 3\pi/2$, $\theta_s = \theta_i$) it can be written as

$$\underline{\text{Backscattered}}$$

$$\sigma_{3\text{-D}} = 4\pi\left(\frac{ab}{\lambda}\right)^2 \cos^2\theta_i\left[\frac{\sin(\beta b\sin\theta_i)}{\beta b\sin\theta_i}\right]^2 \qquad (11\text{-}44b)$$

Plots of (11-44a) for $a = b = 5\lambda$ and $\theta_i = 30°$ ($\phi_s = 90°, 270°$ with $0° \le \theta_s \le 90°$) and of (11-44b) for $a = b = 5\lambda$ ($0° \le \theta_i, \le 90°$) are shown, respectively, in Figures 11-9a and 11-9b. It is observed that for Figure 11-9a the maximum occurs when $\phi_s = 90°$ and near $\theta_s \simeq \theta_i = 30°$ while for Figure 11-9b the maximum occurs when $\theta_i = 0°$ (normal incidence).

The monostatic RCS of plates using physical optics techniques is insensitive to polarization, i.e., it is the same for both polarizations [7–10], as has been demonstrated for the strip and the rectangular plate. Measurements, however, have shown that the monostatic RCS is slightly different for the two polarizations. This is one of the drawbacks of physical optics methods. In addition, the predicted RCSs using physical optics are most accurate at and near the specular directions. However, they begin to become less valid away from the specular directions, especially toward grazing incidences. This will be demonstrated in Example 12-3, and Figures 12-13 and 12-14.

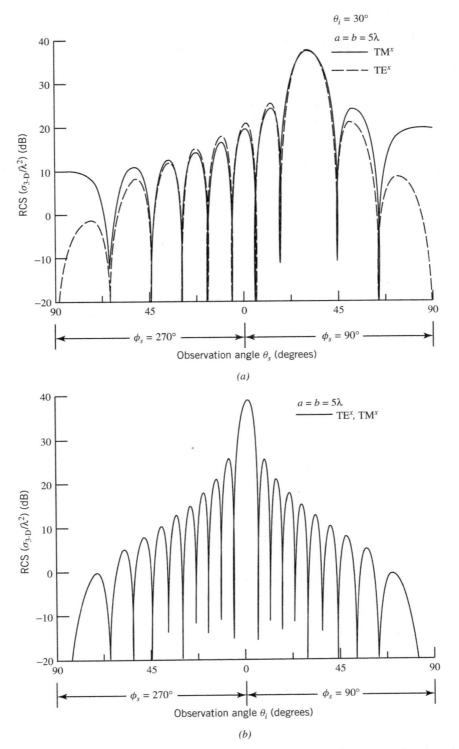

Figure 11-9 Principal plane monostatic and bistatic radar cross sections of a rectangular plate. (*a*) Bistatic ($\theta_i = 30°$, $\phi_i = 270°$, $\phi_s = 90°$, $270°$). (*b*) Monostatic ($\phi_s = 90°$, $270°$).

Example 11-3

Using physical optics techniques, find the scattered fields and radar cross section when a TM^x uniform plane wave is incident upon a flat rectangular plate of dimensions a and b. Assume the incident field lies on the yz plane, as shown in Figure 11-8c.

Solution: According to the geometry of Figure 11-8c, the electric and magnetic field components of the incident uniform plane wave can be written as

$$\mathbf{E}^i = \hat{\mathbf{a}}_x E_0 e^{-j\beta(y\sin\theta_i - z\cos\theta_i)}$$

$$\mathbf{H}^i = -\frac{E_0}{\eta}\left(\hat{\mathbf{a}}_y \cos\theta_i + \hat{\mathbf{a}}_z \sin\theta_i\right)e^{-j\beta(y\sin\theta_i - z\cos\theta_i)}$$

Using physical optics techniques, the current density induced on the plate can be approximated by

$$\mathbf{J}_s \simeq 2\hat{\mathbf{n}} \times \mathbf{H}^i\Big|_{\substack{z=0 \\ y=y'}} = 2\hat{\mathbf{a}}_z \times \left(\hat{\mathbf{a}}_y H_y + \hat{\mathbf{a}}_z H_z\right)\Big|_{\substack{z=0 \\ y=y'}}$$

$$= -2\hat{\mathbf{a}}_x H_y\Big|_{\substack{z=0 \\ y=y'}} = \hat{\mathbf{a}}_x \frac{2E_0}{\eta}\cos\theta_i e^{-j\beta y'\sin\theta_i}$$

or

$$J_y = J_z = 0 \qquad J_x = \frac{2E_0}{\eta}\cos\theta_i e^{-j\beta y'\sin\theta_i}$$

Using (6-125a) and (6-125b), we can write that

$$N_\theta = \iint_S J_x \cos\theta_s \cos\phi_s e^{j\beta(x'\sin\theta_s\cos\phi_s + y'\sin\theta_s\sin\phi_s)}\,dx'\,dy'$$

$$= \frac{2E_0}{\eta}ab\left\{\cos\theta_i \cos\theta_s \cos\phi_s\left[\frac{\sin(X)}{X}\right]\left[\frac{\sin(Y)}{Y}\right]\right\}$$

$$N_\phi = -\iint_S J_x \sin\phi_s e^{j\beta(x'\sin\theta_s\cos\phi_s + y'\sin\theta_s\sin\phi_s)}\,dx'\,dy'$$

$$= -\frac{2E_0}{\eta}ab\left\{\cos\theta_i \sin\theta_s\left[\frac{\sin(X)}{X}\right]\left[\frac{\sin(Y)}{Y}\right]\right\}$$

where

$$X = \frac{\beta a}{2}\sin\theta_s \cos\phi_s$$

$$Y = \frac{\beta b}{2}(\sin\theta_s \sin\phi_s - \sin\theta_i)$$

The scattered fields are obtained using (6-122a) through (6-122f) and can be written as

$$E_r^s \simeq H_r^s \simeq 0$$

$$E_\theta^s \simeq -C_1\frac{e^{-j\beta r}}{r}\left\{\cos\theta_i \cos\theta_s \cos\phi_s\left[\frac{\sin(X)}{X}\right]\left[\frac{\sin(Y)}{Y}\right]\right\}$$

$$E_\phi^s \simeq C_1\frac{e^{-j\beta r}}{r}\left\{\cos\theta_i \sin\phi_s\left[\frac{\sin(X)}{X}\right]\left[\frac{\sin(Y)}{Y}\right]\right\}$$

$$H_\theta^s \simeq -\frac{E_\phi^s}{\eta} \qquad H_\phi^s \simeq +\frac{E_\theta^s}{\eta}$$

$$C_1 = j\frac{ab\beta E_0}{2\pi}$$

In the principal H plane ($\phi_s = \pi/2$), on which the maximum also lies, the electric field components reduce to

$$E_r^s \simeq E_\theta^s \simeq 0$$

$$E_\phi^s \simeq C_1 \frac{e^{-j\beta r}}{r} \left\{ \cos\theta_i \left[\frac{\sin\left[\frac{\beta b}{2}(\sin\theta_s - \sin\theta_i)\right]}{\frac{\beta b}{2}(\sin\theta_s - \sin\theta_i)} \right] \right\}$$

whose maximum value always occurs when $\theta_s = \theta_i$ (no matter what the size of the ground plane). The bistatic and monostatic RCSs can be expressed, respectively, as

$$\sigma_{\text{3-D}}(\text{bistatic}) = 4\pi \left(\frac{ab}{\lambda}\right)^2 [\cos^2\theta_i (\cos^2\theta_s \cos^2\phi_s + \sin^2\phi_s)]$$

$$\times \left[\frac{\sin(X)}{X}\right]^2 \left[\frac{\sin(Y)}{Y}\right]^2$$

$$\sigma_{\text{3-D}}(\text{monostatic}) = 4\pi \left(\frac{ab}{\lambda}\right)^2 \cos^2\theta_i \left[\frac{\sin(\beta b \sin\theta_i)}{\beta b \sin\theta_i}\right]^2$$

They are shown plotted, respectively, in Figures 11-9a and 11-9b for a square plate of $a = b = 5\lambda$.

11.4 CYLINDRICAL WAVE TRANSFORMATIONS AND THEOREMS

In scattering, it is often most convenient to express wave functions of one coordinate system in terms of wave functions of another coordinate system. An example is a uniform plane wave that can be written in a very simple form in terms of rectilinear wave functions. However, when scattering of plane waves by cylindrical structures is considered, it is most desirable to transform the rectilinear form of the uniform plane wave into terms of cylindrical wave functions. This is desirable because the surface of the cylindrical structure is most conveniently defined using cylindrical coordinates. This and other such transformations are referred to as *wave transformations* [11]. Along with these transformations, certain theorems concerning cylindrical wave functions are very desirable in describing the scattering by cylindrical structures.

11.4.1 Plane Waves in Terms of Cylindrical Wave Functions

The scattering of plane waves by cylindrical structures is considered a fundamental problem in scattering theory. To accomplish this, it is first necessary and convenient to express the plane waves by cylindrical wave functions. To demonstrate that, let us assume that a normalized uniform plane wave traveling in the $+x$ direction, as shown in Figure 11-10, can be written as

$$\mathbf{E} = \hat{\mathbf{a}}_z E_z^+ = \hat{\mathbf{a}}_z E_0 e^{-j\beta x} = \hat{\mathbf{a}}_z e^{-j\beta x} \qquad (11\text{-}45)$$

The plane wave can be represented by an infinite sum of cylindrical wave functions of the form

$$E_z^+ = e^{-j\beta x} = e^{-j\beta\rho\cos\phi} = \sum_{n=-\infty}^{+\infty} a_n J_n(\beta\rho) e^{jn\phi} \qquad (11\text{-}45a)$$

since it must be periodic in ϕ and finite at $\rho = 0$. The next step is to determine the amplitude coefficients, a_n.

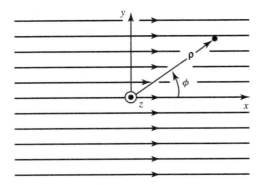

Figure 11-10   Uniform plane wave traveling in the $+x$ direction.

Multiplying both sides of (11-45a) by $e^{-jm\phi}$, where m is an integer, and integrating from 0 to 2π, we have that

$$\int_0^{2\pi} e^{-j(\beta\rho\cos\phi + m\phi)}\, d\phi = \int_0^{2\pi}\left[\sum_{n=-\infty}^{+\infty} a_n J_n(\beta\rho) e^{j(n-m)\phi}\right] d\phi \qquad (11\text{-}46)$$

Interchanging integration and summation, we have that

$$\int_0^{2\pi} e^{-j(\beta\rho\cos\phi + m\phi)}\, d\phi = \sum_{n=-\infty}^{+\infty} a_n J_n(\beta\rho) \int_0^{2\pi} e^{j(n-m)\phi}\, d\phi \qquad (11\text{-}47)$$

Using the orthogonality condition

$$\int_0^{2\pi} e^{j(n-m)\phi}\, d\phi = \begin{cases} 2\pi & n=m \\ 0 & n\neq m \end{cases} \qquad (11\text{-}48)$$

the right side of (11-47) reduces to

$$\sum_{n=-\infty}^{+\infty} a_n J_n(\beta\rho) \int_0^{2\pi} e^{j(n-m)\phi}\, d\phi \overset{n=m}{=} 2\pi a_m J_m(\beta\rho) \qquad (11\text{-}49)$$

Using the integral

$$\int_0^{2\pi} e^{+j(z\cos\phi + n\phi)}\, d\phi = 2\pi j^n J_n(z) \qquad (11\text{-}50)$$

the left side of (11-47) can be written as

$$\int_0^{2\pi} e^{-j(\beta\rho\cos\phi + m\phi)}\, d\phi = 2\pi j^{-m} J_{-m}(-\beta\rho) \qquad (11\text{-}51)$$

Since

$$J_{-m}(x) = (-1)^m J_m(x) \qquad (11\text{-}52a)$$

and

$$J_m(-x) = (-1)^m J_m(x) \qquad (11\text{-}52b)$$

(11-51) can be written as

$$\int_0^{2\pi} e^{-j(\beta\rho\cos\phi+m\phi)}\,d\phi = 2\pi j^{-m}J_{-m}(-\beta\rho) = 2\pi j^{-m}(-1)^m J_m(-\beta\rho)$$

$$= 2\pi j^{-m}(-1)^m(-1)^m J_m(\beta\rho) = 2\pi j^{-m}(-1)^{2m} J_m(\beta\rho)$$

$$\int_0^{2\pi} e^{-j(\beta\rho\cos\phi+m\phi)}\,d\phi = 2\pi j^{-m}J_{-m}(-\beta\rho) = 2\pi j^{-m}J_m(\beta\rho) \qquad (11\text{-}53)$$

Using (11-49) and (11-53), reduce (11-47) to

$$2\pi j^{-m}J_m(\beta\rho) = 2\pi a_m J_m(\beta\rho) \qquad (11\text{-}54)$$

Thus,

$$a_m = j^{-m} \qquad (11\text{-}54a)$$

Therefore, (11-45a) can be written as

$$\boxed{E_z^+ = e^{-j\beta x} = e^{-j\beta\rho\cos\phi} = \sum_{n=-\infty}^{+\infty} a_n J_n(\beta\rho)e^{jn\phi} = \sum_{n=-\infty}^{+\infty} j^{-n}J_n(\beta\rho)e^{jn\phi}} \qquad (11\text{-}55a)$$

In a similar manner it can be shown that

$$\boxed{E_z^- = e^{+j\beta x} = e^{+j\beta\rho\cos\phi} = \sum_{n=+\infty}^{n=-\infty} j^{+n}J_n(\beta\rho)e^{jn\phi}} \qquad (11\text{-}55b)$$

11.4.2 Addition Theorem of Hankel Functions

A transformation that is often convenient and necessary in scattering problems is the *addition theorem of Hankel functions* [11]. Basically, it expresses the fields of a cylindrical line source located away from the origin at a radial distance ρ', which are represented by cylindrical wave functions originating at the source, in terms of cylindrical wave functions originating at the origin ($\rho = 0$) of the coordinate system.

To derive this, let us assume that a line source of electric current I_0 is located at $\rho = \rho'$ and $\phi = \phi'$, as shown in Figure 11-11. According to (11-11a), the fields by the line source are given by

$$E_z(\rho,\phi) = -\frac{\beta^2 I_0}{4\omega\varepsilon} H_0^{(2)}(\beta|\boldsymbol{\rho}-\boldsymbol{\rho}'|) \qquad (11\text{-}56)$$

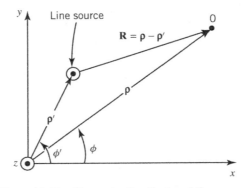

Figure 11-11 Geometry for displaced line source.

where $|\rho - \rho'|$ is the radial distance from the source to the observation point. Using Figure 11-11 and the law of cosines, (11-56) can also be written as

$$E_z(\rho,\phi) = -\frac{\beta^2 I_0}{4\omega\varepsilon} H_0^{(2)}(\beta |\rho - \rho'|)$$

$$= -\frac{\beta^2 I_0}{4\omega\varepsilon} H_0^{(2)}\left[\beta\sqrt{\rho^2 + (\rho')^2 - 2\rho\rho'\cos(\phi - \phi')}\right] \qquad (11\text{-}56a)$$

Equations 11-56 and 11-56a express the electric field in terms of a Hankel function whose radial distance originates at the source ($|\rho - \rho'| = 0$). For scattering problems, it is very convenient to express the field in terms of cylindrical wave functions, such as Bessel and Hankel functions, whose radial distance originates at the origin ($\rho = 0$). Based upon that and because of reciprocity, the field must be symmetric in terms of the primed and unprimed coordinates. Therefore, the permissible wave functions, whose radial distance originates at the origin of the coordinate system, to represent the Hankel function of (11-56a) are of the form

$$\rho \leq \rho'$$

$$f(\beta\rho')J_n(\beta\rho)e^{jn(\phi-\phi')}$$

where n is an integer, because they must be finite at $\rho = 0$ and be periodic with period 2π.

$$\rho \geq \rho'$$

$$g(\beta\rho')H_n^{(2)}(\beta\rho)e^{jn(\phi-\phi')}$$

where n is an integer, because they must represent outward traveling waves and they must be periodic with period 2π.

Thus, the Hankel function of (11-56a) can be written as

$$H_0^{(2)}(\beta |\rho - \rho'|) = \begin{cases} \displaystyle\sum_{n=-\infty}^{+\infty} c_n f(\beta\rho')J_n(\beta\rho)e^{jn(\phi-\phi')} & \text{for } \rho \leq \rho' \qquad (11\text{-}57a) \\[3mm] \displaystyle\sum_{n=-\infty}^{+\infty} d_n g(\beta\rho')H_n^{(2)}(\beta\rho)e^{jn(\phi-\phi')} & \text{for } \rho \geq \rho' \qquad (11\text{-}57b) \end{cases}$$

Since at $\rho = \rho'$ the fields of the two regions must be continuous, then from (11-57a) and (11-57b)

$$c_n f(\beta\rho')J_n(\beta\rho') = d_n g(\beta\rho')H_n^{(2)}(\beta\rho') \qquad (11\text{-}58)$$

which can be satisfied provided

$$c_n = d_n = b_n \qquad (11\text{-}59a)$$

$$f(\beta\rho') = H_n^{(2)}(\beta\rho') \qquad (11\text{-}59b)$$

$$g(\beta\rho') = J_n(\beta\rho') \qquad (11\text{-}59c)$$

Using (11-59a) through (11-59c), we can write (11-57a) and (11-57b) as

$$H_0^{(2)}(\beta |\rho - \rho'|) = \begin{cases} \displaystyle\sum_{n=-\infty}^{+\infty} b_n J_n(\beta\rho)H_n^{(2)}(\beta\rho')e^{jn(\phi-\phi')} & \text{for } \rho \leq \rho' \qquad (11\text{-}60a) \\[3mm] \displaystyle\sum_{n=-\infty}^{+\infty} b_n J_n(\beta\rho')H_n^{(2)}(\beta\rho)e^{jn(\phi-\phi')} & \text{for } \rho \geq \rho' \qquad (11\text{-}60b) \end{cases}$$

The only remaining part is to evaluate b_n. This can be accomplished by returning to (11-56a), according to which

$$H_0^{(2)}(\beta\,|\boldsymbol{\rho}-\boldsymbol{\rho}'|) = H_0^{(2)}\left[\beta\sqrt{\rho^2 + (\rho')^2 - 2\rho\rho'\cos(\phi-\phi')}\right] \qquad (11\text{-}61)$$

Moving the source toward infinity ($\rho' \to \infty$) along $\phi' = 0$, the radial distance in (11-61), as represented by the square root, can be approximated using the binomial expansion

$$\sqrt{\rho^2 + (\rho')^2 - 2\rho\rho'\cos(\phi-\phi')} \underset{\rho'\to\infty}{\overset{\phi'=0}{\simeq}} \sqrt{(\rho')^2 - 2\rho\rho'\cos\phi} = \rho'\sqrt{1 - 2\left(\frac{\rho}{\rho'}\right)\cos\phi}$$

$$\simeq \rho'\left[1 - \frac{\rho}{\rho'}\cos\phi\right] = \rho' - \rho\cos\phi \qquad (11\text{-}62)$$

Using (11-62) allows us to write (11-61) as

$$H_0^{(2)}(\beta\,|\boldsymbol{\rho}-\boldsymbol{\rho}'|) = H_0^{(2)}\left[\beta\sqrt{\rho^2 + (\rho')^2 - 2\rho\rho'\cos(\phi-\phi')}\right]$$

$$\underset{\rho'\to\infty}{\overset{\phi'=0}{\simeq}} H_0^{(2)}\left[\beta(\rho' - \rho\cos\phi)\right] \qquad (11\text{-}63)$$

With the aid of the asymptotic form of the Hankel function for large argument

$$H_0^{(2)}(\alpha x) \overset{\alpha x\to\infty}{\simeq} \sqrt{\frac{2j}{\pi\alpha x}}\,j^n e^{-j\alpha x} \qquad (11\text{-}64)$$

(11-63) reduces to

$$H_0^{(2)}(\beta\,|\boldsymbol{\rho}-\boldsymbol{\rho}'|) \underset{\rho'\to\infty}{\overset{\phi'=0}{\simeq}} H_0^{(2)}\left[\beta(\rho' - \rho\cos\phi)\right] \simeq \sqrt{\frac{2j}{\pi\beta(\rho' - \rho\cos\phi)}}\,e^{-j\beta(\rho' - \rho\cos\phi)}$$

$$\simeq \sqrt{\frac{2j}{\pi\beta\rho'}}\,e^{-j\beta\rho'}e^{+j\beta\rho\cos\phi} \qquad (11\text{-}65)$$

Using (11-55b) allows us to write (11-65) as

$$H_0^{(2)}(\beta\,|\boldsymbol{\rho}-\boldsymbol{\rho}'|) \underset{\rho'\to\infty}{\overset{\phi'=0}{\simeq}} \sqrt{\frac{2j}{\pi\beta\rho'}}\,e^{-j\beta\rho'}\sum_{n=-\infty}^{+\infty} j^n J_n(\beta\rho)e^{jn\phi} \qquad (11\text{-}66)$$

Applying (11-64) to (11-60a) for $\phi' = 0$ and $\rho' \to \infty$ reduces it to

$$H_0^{(2)}(\beta\,|\boldsymbol{\rho}-\boldsymbol{\rho}'|) \overset{\phi'=0}{=} \sum_{n=-\infty}^{+\infty} b_n J_n(\beta\rho)H_n^{(2)}(\beta\rho')e^{jn\phi}$$

$$\underset{\rho'\to\infty}{\overset{\phi'=0}{\simeq}} \sqrt{\frac{2j}{\pi\beta\rho'}}\,e^{-j\beta\rho'}\sum_{n=-\infty}^{+\infty} b_n j^n J_n(\beta\rho)e^{jn\phi} \qquad (11\text{-}67)$$

Comparing (11-66) and (11-67) leads to

$$b_n = 1 \qquad (11\text{-}68)$$

Thus, the final form of (11-60a) and (11-60b) is

$$H_0^{(2)}(\beta|\boldsymbol{\rho}-\boldsymbol{\rho}'|) = \begin{cases} \displaystyle\sum_{n=-\infty}^{+\infty} J_n(\beta\rho)H_n^{(2)}(\beta\rho')e^{jn(\phi-\phi')} & \text{for } \rho \le \rho' \qquad (11\text{-}69a) \\[2em] \displaystyle\sum_{n=-\infty}^{+\infty} J_n(\beta\rho')H_n^{(2)}(\beta\rho)e^{jn(\phi-\phi')} & \text{for } \rho \ge \rho' \qquad (11\text{-}69b) \end{cases}$$

which can be used to write (11-56) as

$$E_z(\rho,\phi) = -\frac{\beta^2 I_0}{4\omega\varepsilon} H_0^{(2)}(\beta|\boldsymbol{\rho}-\boldsymbol{\rho}'|)$$

$$= -\frac{\beta^2 I_0}{4\omega\varepsilon} \begin{cases} \displaystyle\sum_{n=-\infty}^{+\infty} J_n(\beta\rho)H_n^{(2)}(\beta\rho')e^{jn(\phi-\phi')} & \text{for } \rho \le \rho' \qquad (11\text{-}70a) \\[2em] \displaystyle\sum_{n=-\infty}^{+\infty} J_n(\beta\rho')H_n^{(2)}(\beta\rho)e^{jn(\phi-\phi')} & \text{for } \rho \ge \rho' \qquad (11\text{-}70b) \end{cases}$$

The procedure can be repeated to expand $H_0^{(1)}(\beta|\boldsymbol{\rho}-\boldsymbol{\rho}'|)$. However, it is obvious from the results of (11-69a) and (11-69b) that

$$H_0^{(1)}(\beta|\boldsymbol{\rho}-\boldsymbol{\rho}'|) = \begin{cases} \displaystyle\sum_{n=-\infty}^{+\infty} J_n(\beta\rho)H_n^{(1)}(\beta\rho')e^{jn(\phi-\phi')} & \text{for } \rho \le \rho' \qquad (11\text{-}71a) \\[2em] \displaystyle\sum_{n=-\infty}^{+\infty} J_n(\beta\rho')H_n^{(1)}(\beta\rho)e^{jn(\phi-\phi')} & \text{for } \rho \ge \rho' \qquad (11\text{-}71b) \end{cases}$$

11.4.3   Addition Theorem for Bessel Functions

Another theorem that is often useful represents Bessel functions originating at the source, which is located away from the origin, in terms of cylindrical wave functions originating at the origin of the coordinate system. This is usually referred to as the *addition theorem for Bessel functions* [11].

We know that the Hankel functions of the first and second kinds can be written, in terms of the Bessel functions, as

$$H_0^{(1)}(\beta\rho) = J_0(\beta\rho) + jY_0(\beta\rho) \qquad (11\text{-}72a)$$

$$H_0^{(2)}(\beta\rho) = J_0(\beta\rho) - jY_0(\beta\rho) \qquad (11\text{-}72b)$$

Adding the two, we can write that

$$J_0(\beta\rho) = \frac{1}{2}[H_0^{(1)}(\beta\rho) + H_0^{(2)}(\beta\rho)] \qquad (11\text{-}73)$$

Therefore,

$$J_0(\beta|\boldsymbol{\rho}-\boldsymbol{\rho}'|) = \tfrac{1}{2}[H_0^{(1)}(\beta|\boldsymbol{\rho}-\boldsymbol{\rho}'|) + H_0^{(2)}(\beta|\boldsymbol{\rho}-\boldsymbol{\rho}'|)] \qquad (11\text{-}74)$$

With the aid of (11-69a), (11-69b) and (11-71a), (11-71b) we can write (11-74) as

$$
J_0(\beta\,|\boldsymbol{\rho}-\boldsymbol{\rho}'|)=
\begin{cases}
\displaystyle\sum_{n=-\infty}^{+\infty}\tfrac{1}{2}[H_n^{(1)}(\beta\rho')+H_n^{(2)}(\beta\rho')]J_n(\beta\rho)\,e^{jn\,(\phi-\phi')} & \text{for }\rho\leq\rho' \quad (11\text{-}75a)\\[4mm]
\displaystyle\sum_{n=-\infty}^{+\infty}\tfrac{1}{2}[H_n^{(1)}(\beta\rho)+H_n^{(2)}(\beta\rho)]J_n(\beta\rho')\,e^{jn\,(\phi-\phi')} & \text{for }\rho\geq\rho' \quad (11\text{-}75b)
\end{cases}
$$

Using the forms of (11-72a) and (11-72b) for nth order Bessel and Hankel functions, we can write (11-75a) and (11-75b) as

$$
J_0(\beta\,|\boldsymbol{\rho}-\boldsymbol{\rho}'|)=
\begin{cases}
\displaystyle\sum_{n=-\infty}^{+\infty}J_n(\beta\rho')J_n(\beta\rho)e^{jn\,(\phi-\phi')} & \text{for }\rho\leq\rho' \qquad (11\text{-}76a)\\[4mm]
\displaystyle\sum_{n=-\infty}^{+\infty}J_n(\beta\rho)J_n(\beta\rho')e^{jn\,(\phi-\phi')} & \text{for }\rho\geq\rho' \qquad (11\text{-}76b)
\end{cases}
$$

or that

$$
\boxed{\,J_0(\beta\,|\boldsymbol{\rho}-\boldsymbol{\rho}'|)=\sum_{n=-\infty}^{+\infty}J_n(\beta\rho)J_n(\beta\rho')e^{jn\,(\phi-\phi')}\quad\text{for }\rho\lessgtr\rho'\,}
\qquad (11\text{-}77)
$$

Subtracting (11-72b) from (11-72a), we can write that

$$
H_0^{(1)}(\beta\rho)-H_0^{(2)}(\beta\rho)=2jY_0(\beta\rho)
$$

$$
Y_0(\beta\rho)=\frac{1}{2j}\left[H_0^{(1)}(\beta\rho)-H_0^{(2)}(\beta\rho)\right]
\qquad (11\text{-}78a)
$$

or

$$
Y_0(\beta\,|\boldsymbol{\rho}-\boldsymbol{\rho}'|)=\frac{1}{2j}\left[H_0^{(1)}(\beta\,|\boldsymbol{\rho}-\boldsymbol{\rho}'|)-H_0^{(2)}(\beta\,|\boldsymbol{\rho}-\boldsymbol{\rho}'|)\right]
\qquad (11\text{-}78b)
$$

Using (11-69a), (11-69b), (11-71a), and (11-71b), we can write (11-78b) as

$$
Y_0(\beta\,|\boldsymbol{\rho}-\boldsymbol{\rho}'|)=
\begin{cases}
\displaystyle\sum_{n=-\infty}^{\infty}\frac{1}{2}[H_n^{(1)}(\beta\rho')-H_n^{(2)}(\beta\rho')]J_n(\beta\rho)e^{jn\,(\phi-\phi')} & \text{for }\rho\leq\rho' \quad (11\text{-}79a)\\[4mm]
\displaystyle\sum_{n=-\infty}^{+\infty}\frac{1}{2}[H_n^{(1)}(\beta\rho)-H_n^{(2)}(\beta\rho)]J_n(\beta\rho')e^{jn\,(\phi-\phi')} & \text{for }\rho\geq\rho' \quad (11\text{-}79b)
\end{cases}
$$

Using the forms (11-72a) and (11-72b) for nth order Hankel functions, we can write (11-79a) and (11-79b) as

$$
Y_0(\beta|\boldsymbol{\rho}-\boldsymbol{\rho}'|) = \begin{cases}
\displaystyle\sum_{n=-\infty}^{+\infty} \frac{1}{2j}[J_n(\beta\rho')+jY_n(\beta\rho')-J_n(\beta\rho')+jY_n(\beta\rho')]J_n(\beta\rho')e^{jn(\phi-\phi')} \\[2ex]
\displaystyle = \sum_{n=-\infty}^{+\infty} Y_n(\beta\rho')J_n(\beta\rho)e^{jn(\phi-\phi')} \qquad \text{for } \rho \le \rho' \quad \text{(11-80a)} \\[3ex]
\displaystyle\sum_{n=-\infty}^{+\infty} \frac{1}{2j}[J_n(\beta\rho)+jY_n(\beta\rho)-J_n(\beta\rho)+jY_n(\beta\rho)]J_n(\beta\rho')e^{jn(\phi-\phi')} \\[2ex]
\displaystyle = \sum_{n=-\infty}^{+\infty} Y_n(\beta\rho)J_n(\beta\rho')e^{jn(\phi-\phi')} \qquad \text{for } \rho \ge \rho' \quad \text{(11-80b)}
\end{cases}
$$

11.4.4 Summary of Cylindrical Wave Transformations and Theorems

The following are the most prominent cylindrical wave transformations and theorems that are very convenient for scattering from cylindrical scatterers:

$$
e^{-j\beta x} = e^{-j\beta\rho\cos\phi} = \sum_{n=-\infty}^{+\infty} j^{-n}J_n(\beta\rho)e^{jn\phi} \tag{11-81a}
$$

$$
e^{+j\beta x} = e^{+j\beta\rho\cos\phi} = \sum_{n=-\infty}^{+\infty} j^{+n}J_n(\beta\rho)e^{jn\phi} \tag{11-81b}
$$

$$
H_0^{(1,2)}(\beta|\boldsymbol{\rho}-\boldsymbol{\rho}'|) = \begin{cases}
\displaystyle\sum_{n=-\infty}^{+\infty} J_n(\beta\rho)H_n^{(1,2)}(\beta\rho')e^{jn(\phi-\phi')} & \rho \le \rho' \quad \text{(11-82a)} \\[3ex]
\displaystyle\sum_{n=-\infty}^{+\infty} J_n(\beta\rho')H_n^{(1,2)}(\beta\rho)e^{jn(\phi-\phi')} & \rho \ge \rho' \quad \text{(11-82b)}
\end{cases}
$$

$$
J_0(\beta|\boldsymbol{\rho}-\boldsymbol{\rho}'|) = \sum_{n=-\infty}^{+\infty} J_n(\beta\rho')J_n(\beta\rho)e^{jn(\phi-\phi')} \qquad \rho \gtreqless \rho' \tag{11-83}
$$

$$
Y_0(\beta|\boldsymbol{\rho}-\boldsymbol{\rho}'|) = \begin{cases}
\displaystyle\sum_{n=-\infty}^{+\infty} Y_n(\beta\rho')J_n(\beta\rho)e^{jn(\phi-\phi')} & \rho \le \rho' \quad \text{(11-84a)} \\[3ex]
\displaystyle\sum_{n=-\infty}^{+\infty} Y_n(\beta\rho)J_n(\beta\rho')e^{jn(\phi-\phi')} & \rho \ge \rho' \quad \text{(11-84b)}
\end{cases}
$$

11.5 SCATTERING BY CIRCULAR CYLINDERS

Cylinders represent one of the most important classes of geometrical surfaces. The surface of many practical scatterers, such as the fuselage of airplanes, missiles, and so on, can often be represented by cylindrical structures. The circular cylinder, because of its simplicity and the fact that its solution is represented in terms of well known and tabulated functions (such as Bessel and Hankel functions), is probably one of the geometries most widely used to represent practical scatterers [12]. Because of its importance, it will be examined here in some detail. We will consider scattering of both plane and cylindrical waves by circular conducting cylinders of infinite length at normal and oblique incidences. The solutions will be obtained using modal techniques. Scattering from finite length cylinders is obtained by transforming the scattered fields of infinite lengths using approximate relationships. Scattering by dielectric and dielectric covered cylinders are assigned to the reader as end-of-chapter exercises.

11.5.1 Normal Incidence Plane Wave Scattering by a Conducting Circular Cylinder: TM^z Polarization

Let us assume that a TM^z uniform plane wave is normally incident upon a perfectly conducting circular cylinder of radius a, as shown in Figure 11-12a, and the electric field can be written as

$$\mathbf{E}^i = \hat{\mathbf{a}}_z E_z^i = \hat{\mathbf{a}}_z E_0 e^{-j\beta x} = \hat{\mathbf{a}}_z E_0 e^{-j\beta\rho\cos\phi} \tag{11-85}$$

which, according to the transformation (11-55a) or (11-81a), can also be expressed as

$$\mathbf{E}^i = \hat{\mathbf{a}}_z E_z^i = \hat{\mathbf{a}}_z E_0 \sum_{n=-\infty}^{+\infty} j^{-n} J_n(\beta\rho) e^{jn\phi} = \hat{\mathbf{a}}_z E_0 \sum_{n=0}^{\infty} (-j)^n \varepsilon_n J_n(\beta\rho)\cos(n\phi) \tag{11-85a}$$

where

$$\varepsilon_n = \begin{cases} 1 & n = 0 \\ 2 & n \neq 0 \end{cases} \tag{11-85b}$$

The corresponding magnetic field components can be obtained by using Maxwell's Faraday equation, which for this problem reduces to

$$\mathbf{H}^i = -\frac{1}{j\omega\mu}\nabla\times\mathbf{E}^i = -\frac{1}{j\omega\mu}\left[\hat{\mathbf{a}}_\rho \frac{1}{\rho}\frac{\partial E_z^i}{\partial\phi} - \hat{\mathbf{a}}_\phi \frac{\partial E_z^i}{\partial\rho}\right] \tag{11-86}$$

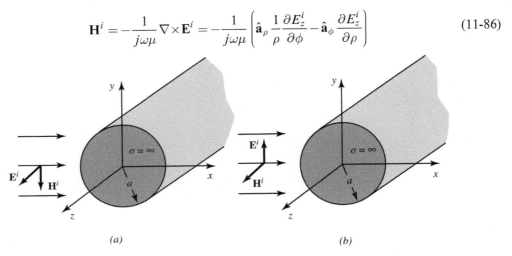

(a)   (b)

Figure 11-12   Uniform plane wave incident on a conducting circular cylinder. (a) TM^z. (b) TE^z.

or

$$H_\rho^i = -\frac{1}{j\omega\mu}\frac{1}{\rho}\frac{\partial E_z^i}{\partial \phi} = -\frac{E_0}{j\omega\mu}\frac{1}{\rho}\sum_{n=-\infty}^{+\infty} nj^{-n+1}J_n(\beta\rho)e^{jn\phi} \tag{11-86a}$$

$$H_\phi^i = \frac{1}{j\omega\mu}\frac{\partial E_z^i}{\partial \rho} = \frac{\beta E_0}{j\omega\mu}\sum_{n=-\infty}^{+\infty} j^{-n}J_n'(\beta\rho)e^{jn\phi} \tag{11-86b}$$

$$' \equiv \frac{\partial}{\partial(\beta\rho)} \tag{11-86c}$$

It should be noted here that throughout this chapter the prime indicates partial derivative with respect to the entire argument of the Bessel or Hankel function.

In the presence of the conducting cylinder, the total field E_z^t according to (11-1a) can be written as

$$\mathbf{E}^t = \mathbf{E}^i + \mathbf{E}^s \tag{11-87}$$

where $\mathbf{E}^s$ is the scattered field. Since the scattered fields travel in the outward direction, they must be represented by cylindrical traveling wave functions. Thus, we choose to represent $\mathbf{E}^s$ by

$$\mathbf{E}^s = \hat{\mathbf{a}}_z E_z^s = \hat{\mathbf{a}}_z E_0 \sum_{n=-\infty}^{+\infty} c_n H_n^{(2)}(\beta\rho) \tag{11-88}$$

where c_n represents the yet unknown amplitude coefficients. Equation 11-88 is chosen to be of similar form to (11-85a) since the two together will be used to represent the total field. This becomes convenient when we attempt to solve for the amplitude coefficients c_n.

The unknown amplitude coefficients c_n can be found by applying the boundary condition

$$\mathbf{E}^t = \hat{\mathbf{a}}_z E_z^t(\rho = a, 0 \le \phi \le 2\pi, z) = 0 \tag{11-89}$$

Using (11-85a), (11-88), and (11-89) we can write that

$$E_z^t(\rho = a, 0 \le \phi \le 2\pi, z) = E_0 \sum_{n=-\infty}^{\infty} \left[j^{-n}J_n(\beta a)e^{jn\phi} + c_n H_n^{(2)}(\beta a) \right] = 0 \tag{11-90}$$

or

$$c_n = -j^{-n}\frac{J_n(\beta a)}{H_n^{(2)}(\beta a)}e^{jn\phi} \tag{11-90a}$$

Thus, the scattered field of (11-88) reduces to

$$\begin{aligned} E_z^s &= -E_0 \sum_{n=-\infty}^{+\infty} j^{-n}\frac{J_n(\beta a)}{H_n^{(2)}(\beta a)} H_n^{(2)}(\beta\rho)e^{jn\phi} \\ &= -E_0 \sum_{n=0}^{+\infty} (-j)^n \varepsilon_n \frac{J_n(\beta a)}{H_n^{(2)}(\beta a)} H_n^{(2)}(\beta\rho)\cos(n\phi) \end{aligned} \tag{11-91}$$

where ε_n is defined by (11-85b).

The corresponding scattered magnetic field components can be obtained by using Maxwell's equation 11-86, which leads to

$$H_\rho^s = -\frac{1}{j\omega\mu}\frac{1}{\rho}\frac{\partial E_z^s}{\partial \phi} = \frac{E_0}{j\omega\mu}\frac{1}{\rho}\sum_{n=-\infty}^{+\infty}nj^{-n+1}\frac{J_n(\beta a)}{H_n^{(2)}(\beta a)}H_n^{(2)}(\beta\rho)e^{jn\phi} \tag{11-92a}$$

$$H_\phi^s = \frac{1}{j\omega\mu}\frac{\partial E_z^s}{\partial \rho} = -\frac{\beta E_0}{j\omega\mu}\sum_{n=-\infty}^{+\infty}j^{-n}\frac{J_n(\beta a)}{H_n^{(2)}(\beta a)}H_n^{(2)'}(\beta\rho)e^{jn\phi} \tag{11-92b}$$

Thus, the total electric and magnetic field components can be written as

$$E_\rho^t = E_\phi^t = H_z^t = 0 \tag{11-93a}$$

$$E_z^t = E_0\sum_{n=-\infty}^{+\infty}j^{-n}\left[J_n(\beta\rho) - \frac{J_n(\beta a)}{H_n^{(2)}(\beta a)}H_n^{(2)}(\beta\rho)\right]e^{jn\phi} \tag{11-93b}$$

$$H_\rho^t = -\frac{E_0}{j\omega\mu}\frac{1}{\rho}\sum_{n=-\infty}^{+\infty}nj^{-n+1}\left[J_n(\beta\rho) - \frac{J_n(\beta a)}{H_n^{(2)}(\beta a)}H_n^{(2)}(\beta\rho)\right]e^{jn\phi} \tag{11-93c}$$

$$H_\phi^t = \frac{\beta E_0}{j\omega\mu}\sum_{n=-\infty}^{+\infty}j^{-n}\left[J_n'(\beta\rho) - \frac{J_n(\beta a)}{H_n^{(2)}(\beta a)}H_n^{(2)'}(\beta\rho)\right]e^{jn\phi} \tag{11-93d}$$

On the surface of the cylinder ($\rho = a$), the total tangential magnetic field can be written as

$$H_\phi^t(\rho = a) = \frac{\beta E_0}{j\omega\mu}\sum_{n=-\infty}^{+\infty}j^{-n}\left[J_n'(\beta a) - \frac{J_n(\beta a)}{H_n^{(2)}(\beta a)}H_n^{(2)'}(\beta a)\right]e^{jn\phi}$$

$$= \frac{\beta E_0}{\omega\mu}\sum_{n=-\infty}^{+\infty}j^{-n}\left[\frac{J_n(\beta a)Y_n'(\beta a) - J_n'(\beta a)Y_n(\beta a)}{H_n^{(2)}(\beta a)}\right]e^{jn\phi} \tag{11-94}$$

Using the Wronskian of Bessel functions

$$J_n(\alpha\rho)Y_n'(\alpha\rho) - Y_n(\alpha\rho)J_n'(\alpha\rho) = \frac{2}{\pi\alpha\rho} \tag{11-95}$$

reduces (11-94) to

$$H_\phi^t(\rho = a) = \frac{2E_0}{\pi a\omega\mu}\sum_{n=-\infty}^{+\infty}j^{-n}\frac{e^{jn\phi}}{H_n^{(2)}(\beta a)} \tag{11-96}$$

Thus, the current induced on the surface of the cylinder can be written as

$$\mathbf{J}_s = \hat{\mathbf{n}}\times\mathbf{H}^t\,|_{\rho=a} = \hat{\mathbf{a}}_\rho\times\left(\hat{\mathbf{a}}_\rho H_\rho^t + \hat{\mathbf{a}}_\phi H_\phi^t\right)|_{\rho=a} = \hat{\mathbf{a}}_z H_\phi^t(\rho = a)$$

$$= \hat{\mathbf{a}}_z\frac{2E_0}{\pi a\omega\mu}\sum_{n=-\infty}^{+\infty}j^{-n}\frac{e^{jn\phi}}{H_n^{(2)}(\beta a)} \tag{11-97}$$

A. Small Radius Approximation   As the radius of the cylinder increases, more terms in the infinite series of (11-97) are needed to obtain convergence. However, for very small cylinders, like a very thin wire ($a \ll \lambda$), the first term ($n = 0$) in (11-97) is dominant and is often sufficient to represent the induced current. Thus, for a very thin wire (11-97) can be approximated by

$$\mathbf{J}_s \overset{a \ll \lambda}{\simeq} \hat{\mathbf{a}}_z \frac{2E_0}{\pi a \omega \mu} \frac{1}{H_0^{(2)}(\beta a)} \tag{11-98}$$

where

$$H_0^{(2)}(\beta a) = J_0(\beta a) - jY_0(\beta a) \overset{a \ll \lambda}{\simeq} 1 - j\frac{2}{\pi}\ln\left(\frac{\gamma \beta a}{2}\right) = 1 - j\frac{2}{\pi}\ln\left(\frac{1.781\beta a}{2}\right)$$

$$\overset{a \ll \lambda}{\simeq} -j\frac{2}{\pi}\ln\left(\frac{1.781\beta a}{2}\right) \tag{11-98a}$$

Thus, for a very thin wire the current density (11-98) can be approximated by

$$\mathbf{J}_s \overset{a \ll \lambda}{\simeq} \hat{\mathbf{a}}_z j \frac{E_0}{a\omega\mu} \frac{1}{\ln\left(\frac{1.781\beta a}{2}\right)} \tag{11-98b}$$

B. Far-Zone Scattered Field

One of the most important parameters in scattering is the scattering width, which is obtained by knowing the scattered field in the far zone. For this problem, it can be accomplished by first reducing the scattered fields for far-zone observations ($\beta\rho \to$ large). Referring to (11-91), the Hankel function can be approximated for observations made in the far field by

$$H_n^{(2)}(\beta\rho) \overset{\beta\rho\to\text{large}}{\simeq} \sqrt{\frac{2j}{\pi\beta\rho}} j^n e^{-j\beta\rho} \tag{11-99}$$

which, when substituted in (11-91), reduces it to

$$E_z^s \overset{\beta\rho\to\infty}{\simeq} -E_0\sqrt{\frac{2j}{\pi\beta}}\frac{e^{-j\beta\rho}}{\sqrt{\rho}}\sum_{n=-\infty}^{+\infty}\frac{J_n(\beta a)}{H_n^{(2)}(\beta a)}e^{jn\phi} \tag{11-100}$$

The ratio of the far-zone scattered electric field to the incident field can then be written as

$$\frac{|E_z^s|}{|E_z^i|} \overset{\beta\rho\to\text{large}}{\simeq} \frac{\left|-E_0\sqrt{\frac{2j}{\pi\beta}}\frac{e^{-j\beta\rho}}{\sqrt{\rho}}\sum_{n=-\infty}^{+\infty}\frac{J_n(\beta a)}{H_n^{(2)}(\beta a)}e^{jn\phi}\right|}{|E_0 e^{-j\beta x}|}$$

$$= \sqrt{\frac{2}{\pi\beta\rho}}\left|\sum_{n=-\infty}^{+\infty}\frac{J_n(\beta a)}{H_n^{(2)}(\beta a)}e^{jn\phi}\right| \tag{11-101}$$

Thus, the scattering width of (11-21b) can be expressed as

$$\boxed{\sigma_{\text{2-D}} = \lim_{\rho\to\infty}\left[2\pi\rho\frac{|E_z^s|^2}{|E_z^i|^2}\right] = \frac{4}{\beta}\left|\sum_{n=+\infty}^{+\infty}\frac{J_n(\beta a)}{H_n^{(2)}(\beta a)}e^{jn\phi}\right|^2}$$

$$\boxed{= \frac{2\lambda}{\pi}\left|\sum_{n=0}^{+\infty}\varepsilon_n\frac{J_n(\beta a)}{H_n^{(2)}(\beta a)}\cos(n\phi)\right|^2} \tag{11-102}$$

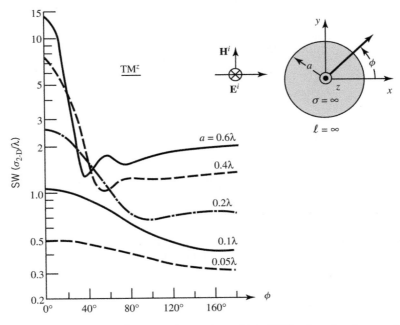

Figure 11-13 Two-dimensional TM^z bistatic scattering width (SW) of a circular conducting cylinder. (Courtesy of J. H. Richmond, Ohio State University.)

where

$$\varepsilon_n = \begin{cases} 1 & n = 0 \\ 2 & n \neq 0 \end{cases} \tag{11-102a}$$

Plots of the bistatic $\sigma_{2\text{-}D} / \lambda$ computed using (11-102) are shown in Figure 11-13 for cylinder radii of $a = 0.05\lambda$, 0.1λ, 0.2λ, 0.4λ, and 0.6λ [13]. The backscattered ($\phi = 180°$) patterns of $\sigma_{2\text{-}D}/\lambda$, as a function of the cylinder radius, are displayed in Figure 11-14 [13].

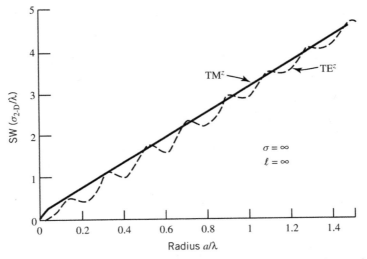

Figure 11-14 Two-dimensional monostatic (backscattered) scattering width for a circular conducting cylinder as a function of its radius. (Courtesy of J. H. Richmond, Ohio State University.)

For small radii $(a \ll \lambda)$, the first term $(n = 0)$ in (11-102) is the dominant term, and it is sufficient to represent the scattered field. Thus, for small radii, the ratio of the Bessel to the Hankel function for $n = 0$ can be approximated using (11-98a) by

$$\frac{J_0(\beta a)}{H_0^{(2)}(\beta a)} \overset{a \ll \lambda}{\simeq} \frac{1}{-j\frac{2}{\pi}\ln(0.89\beta a)} = j\frac{\pi}{2}\frac{1}{\ln(0.89\beta a)} \qquad (11\text{-}103)$$

and (11-102) can then be reduced to

$$\sigma_{2\text{-D}} \overset{a \ll \lambda}{\simeq} \frac{2\lambda}{\pi}\left(\frac{\pi^2}{4}\right)\left|\frac{1}{\ln(0.89\beta a)}\right|^2 = \frac{\pi\lambda}{2}\left|\frac{1}{\ln(0.89\beta a)}\right|^2 \qquad (11\text{-}103a)$$

This is independent of ϕ, which becomes evident in Figure 11-13 by the curves for the smaller values of a.

For a cylinder of finite length ℓ the three-dimensional radar cross section for normal incidence is related to the two-dimensional scattering width by (11-22e). Thus, using (11-102) and (11-103a), we can write the three-dimensional RCS (11-22e) as

$$\sigma_{3\text{-D}} \simeq \frac{4\ell^2}{\pi}\left|\sum_{n=-\infty}^{+\infty}\frac{J_n(\beta a)}{H_n^{(2)}(\beta a)}e^{jn\phi}\right|^2 \qquad (11\text{-}104a)$$

$$\sigma_{3\text{-D}} \overset{a \ll \lambda}{\simeq} \pi\ell^2\left|\frac{1}{\ln(0.89\beta a)}\right|^2 \qquad (11\text{-}104b)$$

11.5.2 Normal Incidence Plane Wave Scattering by a Conducting Circular Cylinder: TE^z Polarization

Now let us assume that a TE^z uniform plane wave traveling in the $+x$ direction is normally incident upon a perfectly conducting circular cylinder of radius a, as shown in Figure 11-12b. The incident magnetic field can be written as

$$\mathbf{H}^i = \hat{\mathbf{a}}_z H_0 e^{-j\beta x} = \hat{\mathbf{a}}_z H_0 e^{-j\beta\rho\cos\phi} = \hat{\mathbf{a}}_z H_0 \sum_{n=-\infty}^{+\infty}j^{-n}J_n(\beta\rho)e^{jn\phi}$$

$$= \hat{\mathbf{a}}_z H_0 \sum_{n=0}^{\infty}(-j)^n\varepsilon_n J_n(\beta\rho)\cos(n\phi) \qquad (11\text{-}105)$$

where ε_n is defined by (11-85b). The corresponding incident electric field can be obtained by using Maxwell's Ampere equation, which for this problem reduces to

$$\mathbf{E}^i = \frac{1}{j\omega\varepsilon}\nabla\times\mathbf{H}^i = +\frac{1}{j\omega\varepsilon}\left[\hat{\mathbf{a}}_\rho\frac{1}{\rho}\frac{\partial H_z^i}{\partial\phi} - \hat{\mathbf{a}}_\phi\frac{\partial H_z^i}{\partial\rho}\right] \qquad (11\text{-}106)$$

and by using (11-105) leads to

$$E_\rho^i = \frac{1}{j\omega\varepsilon}\frac{1}{\rho}\frac{\partial H_z^i}{\partial\phi} = \frac{H_0}{j\omega\varepsilon}\frac{1}{\rho}\sum_{n=-\infty}^{+\infty}nj^{-n+1}J_n(\beta\rho)e^{jn\phi} \qquad (11\text{-}106a)$$

$$E_\phi^i = -\frac{1}{j\omega\varepsilon}\frac{\partial H_z^i}{\partial\rho} = -\frac{\beta H_0}{j\omega\varepsilon}\sum_{n=-\infty}^{+\infty} j^{-n}J_n'(\beta\rho)e^{jn\phi} \qquad (11\text{-}106b)$$

The scattered magnetic field takes a form very similar to that of the scattered electric field of (11-88) for the TM^z polarization, and it can be written as

$$\mathbf{H}^s = \hat{\mathbf{a}}_z H_z^s = \hat{\mathbf{a}}_z H_0 \sum_{n=-\infty}^{+\infty} d_n H_n^{(2)}(\beta\rho) \qquad (11\text{-}107)$$

where d_n represents the yet unknown amplitude coefficients that will be found by applying the appropriate boundary conditions.

Before the boundary conditions on the vanishing of the total tangential electric field on the surface of the cylinder can be applied, it is necessary to first find the corresponding electric fields. This can be accomplished by using Maxwell's equation 11-106 that, for the scattered magnetic field of (11-107), leads to

$$E_\rho^s = \frac{1}{j\omega\varepsilon}\frac{1}{\rho}\frac{\partial H_z^s}{\partial\phi} = \frac{H_0}{j\omega\varepsilon}\frac{1}{\rho}\sum_{n=-\infty}^{+\infty} H_n^{(2)}(\beta\rho)\frac{\partial d_n}{\partial\phi} \qquad (11\text{-}108a)$$

$$E_\phi^s = -\frac{1}{j\omega\varepsilon}\frac{\partial H_z^s}{\partial\rho} = -\frac{\beta H_0}{j\omega\varepsilon}\sum_{n=-\infty}^{+\infty} d_n H_n^{(2)'}(\beta\rho) \qquad (11\text{-}108b)$$

where $'$ indicates a partial derivate with respect to the entire argument of the Hankel function.

Since the cylinder is v electric field must vanish on its surface ($\rho = a$). Thus, using (11-106b) and (11-108b), we can write that

$$E_\phi^t(\rho = a,\ 0 \le \phi \le 2\pi,\ z) = -\frac{\beta H_0}{j\omega\varepsilon}\sum_{n=-\infty}^{+\infty}\left[j^{-n}J_n'(\beta a)e^{jn\phi} + d_n H_n^{(2)'}(\beta a)\right] = 0 \qquad (11\text{-}109)$$

which is satisfied provided

$$d_n = -j^{-n}\frac{J_n'(\beta a)}{H_n^{(2)'}(\beta a)}e^{jn\phi} \qquad (11\text{-}109a)$$

Thus, the scattered electric and magnetic fields can be written, using (11-107) and (11-109a), as

$$E_z^s = H_\rho^s = H_\phi^s = 0 \qquad (11\text{-}110a)$$

$$E_\rho^s = -\frac{H_0}{j\omega\varepsilon}\frac{1}{\rho}\sum_{n=-\infty}^{+\infty} nj^{-n+1}\frac{J_n'(\beta a)}{H_n^{(2)'}(\beta a)}H_n^{(2)}(\beta\rho)e^{jn\phi} \qquad (11\text{-}110b)$$

$$E_\phi^s = \frac{\beta H_0}{j\omega\varepsilon}\sum_{n=-\infty}^{+\infty} j^{-n}\frac{J_n'(\beta a)}{H_n^{(2)'}(\beta a)}H_n^{(2)'}(\beta\rho)e^{jn\phi} \qquad (11\text{-}110c)$$

$$H_z^s = -H_0\sum_{n=-\infty}^{+\infty} j^{-n}\frac{J_n'(\beta a)}{H_n^{(2)'}(\beta a)}H_n^{(2)}(\beta\rho)e^{jn\phi} \qquad (11\text{-}110d)$$

The total electric and magnetic fields can now be expressed [using (11-105), (11-106a), (11-106b), and (11-110a) through (11-110d)] as

$$E_z^t = H_\rho^t = H_\phi^t = 0 \tag{11-111a}$$

$$E_\rho^t = \frac{H_0}{j\omega\varepsilon}\frac{1}{\rho}\sum_{n=-\infty}^{+\infty} nj^{-n+1}\left[J_n(\beta\rho) - \frac{J_n'(\beta a)}{H_n^{(2)'}(\beta a)}H_n^{(2)}(\beta\rho)\right]e^{jn\phi} \tag{11-111b}$$

$$E_\phi^t = -\frac{\beta H_0}{j\omega\varepsilon}\sum_{n=-\infty}^{+\infty} j^{-n}\left[J_n'(\beta\rho) - \frac{J_n'(\beta a)}{H_n^{(2)'}(\beta a)}H_n^{(2)'}(\beta\rho)\right]e^{jn\phi} \tag{11-111c}$$

$$H_z^t = H_0\sum_{n=-\infty}^{+\infty} j^{-n}\left[J_n(\beta\rho) - \frac{J_n'(\beta a)}{H_n^{(2)'}(\beta a)}H_n^{(2)}(\beta\rho)\right]e^{jn\phi} \tag{11-111d}$$

On the surface of the cylinder ($\rho = a$), the total tangential magnetic field can be written as

$$H_z^t(\rho = a) = H_0\sum_{n=-\infty}^{+\infty} j^{-n}\left[J_n(\beta a) - \frac{J_n'(\beta a)}{H_n^{(2)'}(\beta a)}H_n^{(2)}(\beta a)\right]e^{jn\phi}$$

$$= -H_0\sum_{n=-\infty}^{+\infty} j^{-n+1}\left[\frac{J_n(\beta a)Y_n'(\beta a) - J_n'(\beta a)Y_n(\beta a)}{H_n^{(2)'}(\beta\rho)}\right]e^{jn\phi} \tag{11-112}$$

which reduces, using the Wronskian of (11-95), to

$$H_z^t(\rho = a) = -jH_0\frac{2}{\pi\beta a}\sum_{n=-\infty}^{+\infty} j^{-n}\frac{e^{jn\phi}}{H_n^{(2)'}(\beta a)} \tag{11-112a}$$

Thus, the current induced on the surface of the cylinder can be written as

$$\mathbf{J}_s = \hat{\mathbf{n}}\times\mathbf{H}^t|_{\rho=a} = \hat{\mathbf{a}}_\rho\times\hat{\mathbf{a}}_z H_z^t|_{\rho=a} = -\hat{\mathbf{a}}_\phi H_z^t(\rho = a)$$

$$= \hat{\mathbf{a}}_\phi j\frac{2H_0}{\pi\beta a}\sum_{n=-\infty}^{+\infty} j^{-n}\frac{e^{jn\phi}}{H_n^{(2)'}(\beta a)} \tag{11-113}$$

A. Small Radius Approximation As the radius of the cylinder increases, more terms in the infinite series of (11-113) are needed to obtain convergence. However, for very small cylinders, like very thin wires where $a \ll \lambda$, the first three terms ($n = 0$, $n = \pm 1$) in (11-113) are dominant and are sufficient to represent the induced current. Thus, for a very thin wire, (11-113) can be approximated by

$$\mathbf{J}_s \overset{a\ll\lambda}{\simeq} \hat{\mathbf{a}}_\phi j\frac{2H_0}{\pi\beta a}\left[\frac{1}{H_0^{(2)'}(\beta a)} + j^{-1}\frac{e^{j\phi}}{H_1^{(2)'}(\beta a)} + j^{+1}\frac{e^{-j\phi}}{H_{-1}^{(2)'}(\beta a)}\right] \tag{11-114}$$

where

$$H_0^{(2)'}(\beta a) = -H_1^{(2)}(\beta a) = -[J_1(\beta a) - jY_1(\beta a)]$$

$$\overset{a\ll\lambda}{\simeq} -\left[\frac{\beta a}{2} + j\frac{1}{\pi}\left(\frac{2}{\beta a}\right)\right] = -j\frac{2}{\pi\beta a} \tag{11-114a}$$

$$H_1^{(2)'}(\beta a) = -H_2^{(2)}(\beta a) + \frac{1}{\beta a} H_1^{(2)}(\beta a)$$

$$= -[J_2(\beta a) - jY_2(\beta a)] + \frac{1}{\beta a}[J_1(\beta a) - jY_1(\beta a)]$$

$$\overset{a \ll \lambda}{\simeq} -\left[\frac{1}{2}\left(\frac{\beta a}{2}\right)^2 + j\frac{1}{\pi}\left(\frac{2}{\beta a}\right)^2 \right] + \frac{1}{\beta a}\left[\frac{\beta a}{2} + j\frac{1}{\pi}\left(\frac{2}{\beta a}\right) \right]$$

$$\overset{a \ll \lambda}{\simeq} -\left[j\frac{1}{\pi}\left(\frac{2}{\beta a}\right)^2 \right] + \frac{1}{\beta a}\left[j\frac{1}{\pi}\left(\frac{2}{\beta a}\right) \right] = -j\frac{2}{\pi}\frac{1}{(\beta a)^2} \tag{11-114b}$$

$$H_{-1}^{(2)'}(\beta a) = -H_1^{(2)'}(\beta a) \overset{a \ll \lambda}{\simeq} +j\frac{2}{\pi}\frac{1}{(\beta a)^2} \tag{11-114c}$$

Therefore, (11-114) reduces to

$$\mathbf{J}_s \overset{a \ll \lambda}{\simeq} \hat{\mathbf{a}}_\phi j\frac{2H_0}{\pi\beta a}\left[-\frac{\pi\beta a}{j2} + j\frac{\pi}{j2}(\beta a)^2 e^{j\phi} + j\frac{\pi}{j2}(\beta a)^2 e^{-j\phi} \right]$$

$$\overset{a \ll \lambda}{\simeq} \hat{\mathbf{a}}_\phi H_0\left[-1 + j\beta a(e^{j\phi} + e^{-j\phi}) \right] = \hat{\mathbf{a}}_\phi H_0\left[-1 + j2(\beta a)\cos(\phi) \right] \tag{11-114d}$$

B. Far-Zone Scattered Field Since the scattered field, as given by (11-110a), through (11-110d), has two non-vanishing electric field components and only one magnetic field component, it is most convenient to use the magnetic field to find the far-zone scattered field pattern and the radar cross section. However, the same answer can be obtained using the electric field components.

For far-field observations, the scattered magnetic field (11-110d) can be approximated, using the Hankel function approximation (11-99), by

$$H_z^s \overset{\beta\rho\to\infty}{\simeq} -H_0\sqrt{\frac{2j}{\pi\beta}}\frac{e^{-j\beta\rho}}{\sqrt{\rho}}\sum_{n=-\infty}^{+\infty}\frac{J_n'(\beta a)}{H_n^{(2)'}(\beta a)}e^{jn\phi} \tag{11-115}$$

The ratio of the far-zone scattered magnetic field to the incident field can then be written as

$$\frac{|H_z^s|}{|H_z^i|} \overset{\beta\rho\to\infty}{\simeq} \frac{\left| -H_0\sqrt{\dfrac{2j}{\pi\beta}}\dfrac{e^{-j\beta\rho}}{\sqrt{\rho}}\displaystyle\sum_{n=-\infty}^{+\infty}\dfrac{J_n'(\beta a)}{H_n^{(2)'}(\beta a)}e^{jn\phi} \right|}{|H_0 e^{-j\beta x}|}$$

$$= \sqrt{\frac{2}{\pi\beta\rho}}\left| \sum_{n=-\infty}^{+\infty}\frac{J_n'(\beta a)}{H_n^{(2)'}(\beta a)}e^{jn\phi} \right| \tag{11-116}$$

Thus, the scattering width of (11-21c) can be expressed as

$$\sigma_{2\text{-D}} = \lim_{\rho\to\infty}\left[2\pi\rho\frac{|H_z^s|^2}{|H_z^i|^2} \right] = \frac{4}{\beta}\left| \sum_{n=-\infty}^{+\infty}\frac{J_n'(\beta a)}{H_n^{(2)'}(\beta a)}e^{jn\phi} \right|^2$$

$$= \frac{2\lambda}{\pi}\left| \sum_{n=0}^{+\infty}\varepsilon_n\frac{J_n'(\beta a)}{H_n^{(2)'}(\beta a)}\cos(n\phi) \right|^2 \tag{11-117}$$

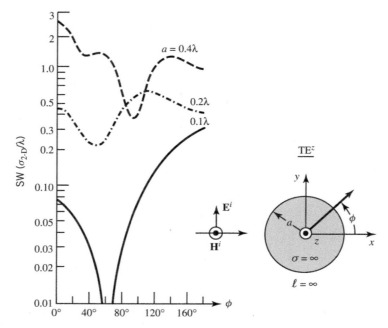

Figure 11-15 Two-dimensional TE^z bistatic scattering width (SW) of a circular conducting cylinder. (Courtesy of J. H. Richmond, Ohio State University.)

where

$$\varepsilon_n = \begin{cases} 1 & n=0 \\ 2 & n\neq 0 \end{cases} \tag{11-117a}$$

Plots of bistatic $\sigma_{2\text{-D}}/\lambda$ computed using (11-117) are shown in Figure 11-15 for cylinder radii of $a=0.1\lambda$, 0.2λ, and 0.4λ while the backscattered patterns as a function of the cylinder radius are displayed in Figure 11-14 [13].

For small radii ($a \ll \lambda$), the first three terms ($n=0$, $n=\pm1$) in (11-117) are the dominant terms, and they are sufficient to represent the scattered field. Thus, for small radii, (11-117) can be approximated by

$$\sigma_{2\text{-D}} \overset{a\ll\lambda}{\simeq} \frac{2\lambda}{\pi}\left| \frac{J_0'(\beta a)}{H_0^{(2)'}(\beta a)} + \frac{J_1'(\beta a)}{H_1^{(2)'}(\beta a)}e^{j\phi} + \frac{J_{-1}'(\beta a)}{H_{-1}^{(2)'}(\beta a)}e^{-j\phi} \right|^2 \tag{11-118}$$

where

$$\frac{J_0'(\beta a)}{H_0^{(2)'}(\beta a)} = \frac{-J_1(\beta a)}{-H_1^{(2)}(\beta a)} \simeq \frac{\dfrac{\beta a}{2}}{\dfrac{\beta a}{2}+j\dfrac{1}{\pi}\left(\dfrac{2}{\beta a}\right)} \simeq \frac{\dfrac{\beta a}{2}}{j\dfrac{1}{\pi}\left(\dfrac{2}{\beta a}\right)} = -j\frac{\pi}{4}(\beta a)^2$$

$$\frac{J_1'(\beta a)}{H_1^{(2)'}(\beta a)} = \frac{-J_2(\beta a)+\dfrac{1}{\beta a}J_1(\beta a)}{-H_2^{(2)}(\beta a)+\dfrac{1}{\beta a}H_1^{(2)}(\beta a)} \tag{11-118a}$$

$$\frac{J_1'(\beta a)}{H_1^{(2)'}(\beta a)} \simeq \frac{-\frac{1}{2}\left(\frac{\beta a}{2}\right)^2 + \frac{1}{\beta a}\left(\frac{\beta a}{2}\right)}{-\left[\frac{1}{2}\left(\frac{\beta a}{2}\right)^2 + j\frac{1}{\pi}\left(\frac{2}{\beta a}\right)^2\right] + \frac{1}{\beta a}\left[\left(\frac{\beta a}{2}\right) + j\frac{1}{\pi}\left(\frac{2}{\beta a}\right)\right]}$$

$$\simeq \frac{-\frac{1}{2}\left(\frac{\beta a}{2}\right)^2 + \frac{1}{2}}{-j\frac{1}{\pi}\left(\frac{2}{\beta a}\right)^2 + \frac{1}{\beta a}\left[j\frac{1}{\pi}\left(\frac{2}{\beta a}\right)\right]} \simeq \frac{\frac{1}{2}}{-j\frac{2}{\pi}\left(\frac{1}{\beta a}\right)^2} = j\frac{\pi}{4}(\beta a)^2 \qquad (11\text{-}118\text{b})$$

$$\frac{J_{-1}'(\beta a)}{H_{-1}^{(2)'}(\beta a)} = \frac{J_1'(\beta a)}{H_1^{(2)'}(\beta a)} \simeq j\frac{\pi}{4}(\beta a)^2 \qquad (11\text{-}118\text{c})$$

Thus, (11-118) reduces to

$$\sigma_{2\text{-D}} \overset{a\ll\lambda}{\simeq} \frac{2\lambda}{\pi}\left| -j\frac{\pi}{4}(\beta a)^2 + j\frac{\pi}{4}(\beta a)^2 e^{j\phi} + j\frac{\pi}{4}(\beta a)^2 e^{-j\phi} \right|^2 \qquad (11\text{-}118\text{d})$$

$$= \frac{\pi\lambda}{8}(\beta a)^4 [1 - 2\cos(\phi)]^2$$

Even for small radii ($a \ll \lambda$), σ is a function of ϕ, as is evident in Figure 11-15 by the curves for small values of a.

For a cylinder of finite length ℓ, the three-dimensional radar cross section for normal incidence is related to the two-dimensional scattering width by (11-22e). Thus, using (11-117) and (11-118d), we can write the three-dimensional RCS (11-22e) as

$$\sigma_{3\text{-D}} \simeq \frac{4\ell^2}{\pi}\left| \sum_{n=-\infty}^{+\infty} \frac{J_n'(\beta a)}{H_n^{(2)'}(\beta a)} e^{jn\phi} \right|^2$$

$$= \frac{4\ell^2}{\pi}\left| \sum_{n=0}^{+\infty} \varepsilon_n \frac{J_n'(\beta a)}{H_n^{(2)'}(\beta a)} \cos(n\phi) \right|^2 \qquad (11\text{-}119\text{a})$$

$$\sigma_{3\text{-D}} \overset{a\ll\lambda}{\simeq} \frac{\pi\ell^2}{4}(\beta a)^4 [1 - 2\cos(\phi)]^2 \qquad (11\text{-}119\text{b})$$

11.5.3 Oblique Incidence Plane Wave Scattering by a Conducting Circular Cylinder: TM^z Polarization

In the previous two sections we analyzed scattering by a conducting cylinder at normal incidence. Scattering at oblique incidence will be considered here. Let us assume that a TM^z plane wave traveling parallel to the xz plane is incident upon a circular cylinder of radius a, as shown in Figure 11-16. The incident electric field can be written as

$$\mathbf{E}^i = E_0(\hat{\mathbf{a}}_x \cos\theta_i + \hat{\mathbf{a}}_z \sin\theta_i)e^{-j\beta x \sin\theta_i}e^{+j\beta z \cos\theta_i} \qquad (11\text{-}120)$$

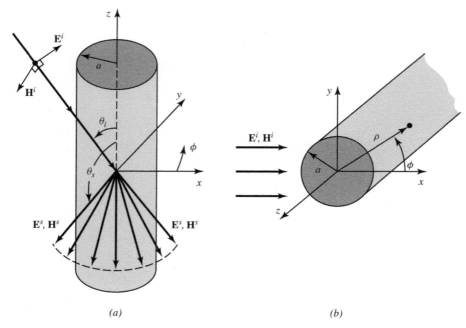

(a)   *(b)*

Figure 11-16   Uniform plane wave obliquely incident on a circular cylinder. (*a*) Side view. (*b*) Top view.

Using the transformation (11-81a), the z component of (11-120) can be expressed as

$$E_z^i = E_0 \sin \theta_i e^{+j\beta z \cos \theta_i} \sum_{n=-\infty}^{+\infty} j^{-n} J_n(\beta \rho \sin \theta_i) e^{jn\phi} \qquad (11\text{-}120a)$$

The tangential component of the total field is composed of two parts: incident and scattered field components. The z component of the scattered field takes a form similar to (11-120a). By referring to (11-88) and Figure 11-16, where θ_s is shown to be greater than 90°, it can be written as

$$E_z^s = E_0 \sin \theta_s e^{-j\beta z \cos \theta_s} \sum_{n=-\infty}^{+\infty} c_n H_n^{(2)}(\beta \rho \sin \theta_s) \qquad (11\text{-}121)$$

The Hankel function was chosen to indicate that the scattered field is a wave traveling in the outward radial direction. It should be stated at this time that smooth perfectly conducting infinite cylinders do not depolarize the oblique incident wave (i.e., do not introduce additional components in the scattered field as compared to the incident field). This, however, is not the case for homogeneous dielectric or dielectric coated cylinders that introduce cross polarization under oblique wave incidences.

When the incident electric field is decomposed into its cylindrical components, the E_x^i component of (11-120) will result in E_ρ^i and E_ϕ^i. Similarly scattered E_ρ^s and E_ϕ^s components will also exist. Therefore, the boundary conditions on the surface of the cylinder are

$$E_z^t(\rho = a, 0 \leq \theta_i, \theta_s \leq \pi, 0 \leq \phi \leq 2\pi) = 0$$

$$= E_z^i(\rho = a, 0 \leq \theta_i \leq \pi, 0 \leq \phi \leq 2\pi) + E_z^s(\rho = a, 0 \leq \theta_s \leq \pi, 0 \leq \phi \leq 2\pi) \qquad (11\text{-}122a)$$

$$E_\phi^t(\rho = a, 0 \leq \theta_i, \theta_s \leq \pi, 0 \leq \phi \leq 2\pi) = 0$$

$$= E_\phi^i(\rho = a, 0 \leq \theta_i \leq \pi, 0 \leq \phi \leq 2\pi) + E_\phi^s(\rho = a, 0 \leq \theta_s \leq \pi, 0 \leq \phi \leq 2\pi) \qquad (11\text{-}122b)$$

Since each is not independent of the other, either one can be used to find the unknown coefficients. Applying (11-122a) leads to

$$E_0 \left[\sin\theta_i e^{+j\beta z \cos\theta_i} \sum_{n=-\infty}^{+\infty} j^{-n} J_n(\beta a \sin\theta_i) e^{jn\phi} \right.$$

$$\left. + \sin\theta_s e^{-j\beta z \cos\theta_s} \sum_{n=-\infty}^{+\infty} c_n H_n^{(2)}(\beta a \sin\theta_s) \right] = 0 \qquad (11\text{-}123a)$$

which is satisfied provided

$$\theta_s = \pi - \theta_i \qquad (11\text{-}123b)$$

$$c_n = -j^{-n} \frac{J_n(\beta a \sin\theta_i)}{H_n^{(2)}(\beta a \sin\theta_i)} e^{jn\phi} = j^{-n} a_n e^{jn\phi} \qquad (11\text{-}123c)$$

$$a_n = -\frac{J_n(\beta a \sin\theta_i)}{H_n^{(2)}(\beta a \sin\theta_i)} \qquad (11\text{-}123d)$$

Thus, the scattered component E_z^s of (11-121) reduces to

$$E_z^s = E_0 \sin\theta_i e^{+j\beta z \cos\theta_i} \sum_{n=-\infty}^{+\infty} j^{-n} a_n H_n^{(2)}(\beta\rho \sin\theta_i) e^{jn\phi} \qquad (11\text{-}124)$$

It is apparent that the scattered field exists for all values of angle ϕ (measured from the x axis) along a cone in the forward direction whose half-angle from the z axis is equal to $\theta_s = \pi - \theta_i$.

To find the remaining $\mathbf{E}^s$ and $\mathbf{H}^s$ scattered field components, we expand Maxwell's curl equations as

$$\nabla \times \mathbf{E}^s = -j\omega\mu\mathbf{H}^s \Rightarrow \mathbf{H}^s = -\frac{1}{j\omega\mu}\nabla \times \mathbf{E}^s \qquad (11\text{-}125)$$

or

$$H_\rho^s = -\frac{1}{j\omega\mu}\left[\frac{1}{\rho}\frac{\partial E_z^s}{\partial\phi} - \frac{\partial E_\phi^s}{\partial z}\right] \qquad (11\text{-}125a)$$

$$H_\phi^s = -\frac{1}{j\omega\mu}\left[\frac{\partial E_\rho^s}{\partial z} - \frac{\partial E_z^s}{\partial\rho}\right] \qquad (11\text{-}125b)$$

$$H_z^s = -\frac{1}{j\omega\mu\rho}\left[\frac{\partial}{\partial\rho}(\rho E_\phi^s) - \frac{\partial E_\rho^s}{\partial\phi}\right] \qquad (11\text{-}125c)$$

and

$$\nabla \times \mathbf{H}^s = j\omega\varepsilon\mathbf{E}^s \Rightarrow \mathbf{E}^s = \frac{1}{j\omega\varepsilon}\nabla \times \mathbf{H}^s \qquad (11\text{-}126)$$

or

$$E_\rho^s = \frac{1}{j\omega\varepsilon}\left[\frac{1}{\rho}\frac{\partial H_z^s}{\partial\phi} - \frac{\partial H_\phi^s}{\partial z}\right] \qquad (11\text{-}126a)$$

$$E_\phi^s = \frac{1}{j\omega\varepsilon}\left[\frac{\partial H_\rho^s}{\partial z} - \frac{\partial H_z^s}{\partial\rho}\right] \qquad (11\text{-}126b)$$

$$E_z^s = \frac{1}{j\omega\varepsilon\rho}\left[\frac{\partial}{\partial\rho}(\rho H_\phi^s) - \frac{\partial H_\rho^s}{\partial\phi}\right] \qquad (11\text{-}126c)$$

Since for the TM^z solution $H_z^s = 0$, (11-126a) reduces to

$$E_\rho^s = -\frac{1}{j\omega\varepsilon}\frac{\partial H_\phi^s}{\partial z} \tag{11-127}$$

When substituted into (11-125b), we can write that

$$
\begin{aligned}
H_\phi^s &= -\frac{1}{j\omega\mu}\left[\frac{\partial}{\partial z}\left(-\frac{1}{j\omega\varepsilon}\frac{\partial H_\phi^s}{\partial z}\right) - \frac{\partial E_z^s}{\partial\rho}\right] \\
&= -\frac{1}{\omega^2\mu\varepsilon}\frac{\partial^2 H_\phi^s}{\partial z^2} + \frac{1}{j\omega\mu}\frac{\partial E_z^s}{\partial\rho}
\end{aligned}
\tag{11-128}
$$

The z variations of all field components are of the same form as in (11-124) (i.e., $e^{+j\beta z\cos\theta_i}$). Thus (11-128) reduces to

$$H_\phi^s = -\frac{(j\beta\cos\theta_i)^2}{\omega^2\mu\varepsilon}H_\phi^s + \frac{1}{j\omega\mu}\frac{\partial E_z^s}{\partial\rho} = \cos^2\theta_i H_\phi^s + \frac{1}{j\omega\mu}\frac{\partial E_z^s}{\partial\rho} \tag{11-129}$$

or

$$H_\phi^s(1-\cos^2\theta_i) = \sin^2\theta_i H_\phi^s = \frac{1}{j\omega\mu}\frac{\partial E_z^s}{\partial\rho} \tag{11-129a}$$

$$
\begin{aligned}
H_\phi^s &= \frac{1}{j\omega\mu}\frac{1}{\sin^2\theta_i}\frac{\partial E_z^s}{\partial\rho} = \frac{\beta}{j\omega\mu}\frac{1}{\sin\theta_i}\frac{\partial E_z^s}{\partial(\beta\rho\sin\theta_i)} \\
&= -jE_0\sqrt{\frac{\varepsilon}{\mu}}e^{+j\beta z\cos\theta_i}\sum_{n=-\infty}^{+\infty}j^{-n}a_n H_n^{(2)'}(\beta\rho\sin\theta_i)e^{jn\phi}
\end{aligned}
\tag{11-129b}
$$

where

$$' \equiv \frac{\partial}{\partial(\beta\rho\sin\theta_i)} \tag{11-129c}$$

In a similar manner we can solve for H_ρ^s by first reducing (11-126b) to

$$E_\phi^s = \frac{1}{j\omega\varepsilon}\left(\frac{\partial H_\rho^s}{\partial z} - \frac{\partial H_z^s}{\partial\rho}\right)_{H_z^s=0} = \frac{1}{j\omega\varepsilon}\frac{\partial H_\rho^s}{\partial z} \tag{11-130}$$

and then substituting it into (11-125a). Thus

$$
\begin{aligned}
H_\rho^s &= -\frac{1}{j\omega\mu}\left[\frac{1}{\rho}\frac{\partial E_z^s}{\partial\phi} - \frac{\partial}{\partial z}\left(\frac{1}{j\omega\varepsilon}\frac{\partial H_\rho^s}{\partial z}\right)\right] = -\frac{1}{j\omega\mu}\frac{1}{\rho}\frac{\partial E_z^s}{\partial\phi} - \frac{1}{\omega^2\mu\varepsilon}\frac{\partial^2 H_\rho^s}{\partial z^2} \\
&= -\frac{1}{j\omega\mu}\frac{1}{\rho}\frac{\partial E_z^s}{\partial\phi} - \frac{(j\beta\cos\theta_i)^2}{\omega^2\mu\varepsilon}H_\rho^s = -\frac{1}{j\omega\mu}\frac{1}{\rho}\frac{\partial E_z^s}{\partial\phi} + \cos^2\theta_i H_\rho^s
\end{aligned}
\tag{11-131}
$$

or

$$(1-\cos^2\theta_i)H_\rho^s = \sin^2\theta_i H_\rho^s = -\frac{1}{j\omega\mu}\frac{1}{\rho}\frac{\partial E_z^s}{\partial\phi} \tag{11-131a}$$

$$H_\rho^s = -\frac{1}{j\omega\mu\rho\sin^2\theta_i}\frac{\partial E_z^s}{\partial\phi} = j\frac{E_0 e^{+j\beta z\cos\theta_i}}{\omega\mu\rho\sin\theta_i}\sum_{n=-\infty}^{+\infty} nj^{-n+1}a_n H_n^{(2)}(\beta\rho\sin\theta_i)e^{jn\phi} \quad (11\text{-}131\text{b})$$

Expressions for E_ρ^s and E_ϕ^s can be written using (11-126a), (11-126b), (11-129b), and (11-131b). Thus

$$E_\rho^s = -\frac{1}{j\omega\varepsilon}\frac{\partial H_\phi^s}{\partial z} = jE_0\cos\theta_i e^{+j\beta z\cos\theta_i}\sum_{n=-\infty}^{+\infty} j^{-n}a_n H_n^{(2)\prime}(\beta\rho\sin\theta_i)e^{jn\phi} \quad (11\text{-}132)$$

$$E_\phi^s = \frac{1}{j\omega\varepsilon}\frac{\partial H_\rho^s}{\partial z} = jE_0\frac{\cot\theta_i}{\beta\rho}e^{+j\beta z\cos\theta_i}\sum_{n=-\infty}^{+\infty} nj^{-n+1}a_n H_n^{(2)}(\beta\rho\sin\theta_i)e^{jn\phi} \quad (11\text{-}133)$$

In summary, the scattered fields can be written as

$$\boxed{\begin{array}{c}
\text{TM}_z \\[4pt]
E_\rho^s = jE_0\cos\theta_i e^{+j\beta z\cos\theta_i}\displaystyle\sum_{n=-\infty}^{+\infty} j^{-n}a_n H_n^{(2)\prime}(\beta\rho\sin\theta_i)e^{jn\phi} \qquad (11\text{-}134\text{a})\\[14pt]
E_\phi^s = jE_0\dfrac{\cot\theta_i}{\beta\rho}e^{+j\beta z\cos\theta_i}\displaystyle\sum_{n=-\infty}^{+\infty} nj^{-n+1}a_n H_n^{(2)}(\beta\rho\sin\theta_i)e^{jn\phi} \qquad (11\text{-}134\text{b})\\[14pt]
E_z^s = E_0\sin\theta_i e^{+j\beta z\cos\theta_i}\displaystyle\sum_{n=-\infty}^{+\infty} j^{-n}a_n H_n^{(2)}(\beta\rho\sin\theta_i)e^{jn\phi} \qquad (11\text{-}134\text{c})
\end{array}}$$

$$\boxed{\begin{array}{c}
H_\rho^s = jE_0\dfrac{e^{+j\beta z\cos\theta_i}}{\omega\mu\rho\sin\theta_i}\displaystyle\sum_{n=-\infty}^{+\infty} nj^{-n+1}a_n H_n^{(2)}(\beta\rho\sin\theta_i)e^{jn\phi} \qquad (11\text{-}134\text{d})\\[14pt]
H_\phi^s = -jE_0\sqrt{\dfrac{\varepsilon}{\mu}}e^{+j\beta z\cos\theta_i}\displaystyle\sum_{n=-\infty}^{+\infty} j^{-n}a_n H_n^{(2)\prime}(\beta\rho\sin\theta_i)e^{jn\phi} \qquad (11\text{-}134\text{e})\\[14pt]
H_z^s = 0 \qquad (11\text{-}134\text{f})\\[10pt]
a_n = -\dfrac{J_n(\beta a\sin\theta_i)}{H_n^{(2)}(\beta a\sin\theta_i)} \qquad (11\text{-}134\text{g})\\[14pt]
{}' = \dfrac{\partial}{\partial(\beta\rho\sin\theta_i)} \qquad (11\text{-}134\text{h})
\end{array}}$$

A. Far-Zone Scattered Field   Often it is desired to know the scattered fields at large distances. This can be accomplished by approximating in (11-134a) through (11-134f) the Hankel function and its derivative by their corresponding asymptotic expressions for large distances, as given by

$$H_n^{(2)}(\alpha x) \overset{\alpha x\to\infty}{\simeq} \sqrt{\frac{2j}{\pi\alpha x}}j^n e^{-j\alpha x} \qquad (11\text{-}135\text{a})$$

$$H_n^{(2)\prime}(\alpha x) = \frac{dH_n^{(2)}(\alpha x)}{d(\alpha x)} \overset{\alpha x\to\infty}{\simeq} -\sqrt{\frac{2j}{\pi\alpha x}}j^{n+1} e^{-j\alpha x} \qquad (11\text{-}135\text{b})$$

Thus, we can reduce the scattered magnetic field expressions of (11-134d) and (11-134e) by

$$H_\rho^s \overset{\rho\to\infty}{\simeq} jE_0 \frac{1}{\omega\mu}\frac{1}{\rho\sin\theta_i}\sqrt{\frac{2j}{\pi\beta\rho\sin\theta_i}}\, e^{+j\beta(z\cos\theta_i-\rho\sin\theta_i)}\sum_{n=-\infty}^{+\infty} na_n e^{jn\phi}$$

$$\overset{\rho\to\infty}{\simeq} jE_0 \frac{1}{\omega\mu}\frac{1}{\rho\sin\theta_i}\sqrt{\frac{2j}{\pi\beta\rho\sin\theta_i}}\, e^{+j\beta(z\cos\theta_i-\rho\sin\theta_i)}\sum_{n=0}^{+\infty} n\varepsilon_n a_n \cos(n\phi) \qquad (11\text{-}136a)$$

$$H_\phi^s \overset{\rho\to\infty}{\simeq} -E_0\sqrt{\frac{\varepsilon}{\mu}}\sqrt{\frac{2j}{\pi\beta\rho\sin\theta_i}}\, e^{+j\beta(z\cos\theta_i-\rho\sin\theta_i)}\sum_{n=-\infty}^{+\infty} a_n e^{jn\phi}$$

$$\overset{\rho\to\infty}{\simeq} -E_0\sqrt{\frac{\varepsilon}{\mu}}\sqrt{\frac{2j}{\pi\beta\rho\sin\theta_i}}\, e^{+j\beta(z\cos\theta_i-\rho\sin\theta_i)}\sum_{n=0}^{+\infty} \varepsilon_n a_n \cos(n\phi) \qquad (11\text{-}136b)$$

where ε_n is defined in (11-102a). A comparison of (11-136a) and (11-136b) indicates that at large distances H_ρ^s is small compared to H_ϕ^s since H_ρ^s varies inversely proportional to $\rho^{3/2}$ while H_ϕ^s is inversely proportional to $\rho^{1/2}$.

The scattering width of (11-21c) can now be expressed as

$$\sigma_{2\text{-D}} = \lim_{\rho\to\infty}\left[2\pi\rho\frac{|H_\phi^s|^2}{|H^i|^2}\right] = \lim_{\rho\to\infty}\left[2\pi\rho\frac{\dfrac{|E_0|^2}{\eta^2}\left(\dfrac{2}{\pi\beta\rho\sin\theta_i}\right)}{\dfrac{|E_0|^2}{\eta^2}}\left|\sum_{n=-\infty}^{+\infty} a_n e^{jn\phi}\right|^2\right]$$

$$\boxed{\sigma_{2\text{-D}} = \frac{4}{\beta}\frac{1}{\sin\theta_i}\left|\sum_{n=-\infty}^{+\infty} a_n e^{jn\phi}\right|^2 = \frac{2\lambda}{\pi}\frac{1}{\sin\theta_i}\left|\sum_{n=0}^{+\infty}\varepsilon_n a_n \cos(n\phi)\right|^2} \qquad (11\text{-}137)$$

where

$$\boxed{a_n = -\frac{J_n(\beta a\sin\theta_i)}{H_n^{(2)}(\beta a\sin\theta_i)}} \qquad (11\text{-}137a)$$

$$\boxed{\varepsilon_n = \begin{cases} 1 & n=0 \\ 2 & n\neq 0\end{cases}} \qquad (11\text{-}137b)$$

which is similar to (11-102) except that β in (11-102) is replaced by $\beta\sin\theta_i$.

From the results of the normal incidence case of Section 11.5.1B we can write, by referring to (11-103a), that for small radii the scattering width of (11-137) reduces to

$$\sigma_{2\text{-D}} \overset{a\ll\lambda}{\simeq} \frac{\pi\lambda}{2\sin\theta_i}\left|\frac{1}{\ln(0.89\beta a\sin\theta_i)}\right|^2 \qquad (11\text{-}138)$$

which is independent of ϕ.

For a cylinder of finite length ℓ, the scattered fields of oblique incidence propagate in all directions, in contrast to the infinitely long cylinder where all the energy is along a conical surface formed in the forward direction whose half-angle is equal to θ_i. However, as the length of the cylinder becomes much larger than its radius ($\ell\gg a$), then the scattered fields along $\theta_s = \pi - \theta_i$ will be much greater than those in other directions. When the length of the cylinder is a multiple of half a wavelength, resonance phenomena are exhibited in the scattered fields [12]. However, as the length increases beyond several wavelengths, the resonance phenomena disappear. For both TM^z and TE^z polarizations, the three-dimensional radar cross section for oblique wave incidence is related approximately to the two-dimensional scattering width, by referring to the geometry of Figure 11-16, by [12, 14]

$$\sigma_{3\text{-}D} = \sigma_{2\text{-}D}\left\{\frac{2\ell^2}{\lambda}\sin^2\theta_{s,i}\left|\frac{\sin\left[\frac{\beta\ell}{2}(\cos\theta_i+\cos\theta_s)\right]}{\frac{\beta\ell}{2}(\cos\theta_i+\cos\theta_s)}\right|^2\right\}, \quad \ell \gg a \qquad (11\text{-}139)$$

where $\sin^2\theta_s$, is used for TM^z and $\sin^2\theta_i$, is used for TE^z. This is analogous to the rectangular plate scattering of Section 11.3.2 and Example 11-3. This indicates that the maximum RCS occurs along the specular direction ($\theta_s = \pi - \theta_i$) and away from it follows the variations exhibited from a flat plate, as given by (11-44a). Equation 11-139 yields reasonable good results even for cylinders with lengths near one wavelength ($\ell \simeq \lambda$). Thus, using (11-139) converts (11-137) and (11-138) for three-dimensional scatterers to

$$\sigma_{3\text{-}D} \simeq \frac{4\ell^2}{\pi}\frac{\sin^2\theta_s}{\sin\theta_i}\left|\sum_{n=0}^{\infty}\varepsilon_n a_n\cos(n\phi)\right|^2\left\{\frac{\sin\left[\frac{\beta\ell}{2}(\cos\theta_i+\cos\theta_s)\right]}{\frac{\beta\ell}{2}(\cos\theta_i+\cos\theta_s)}\right\}^2 \qquad (11\text{-}140a)$$

$$\sigma_{3\text{-}D} \overset{a\ll\lambda}{\simeq} \frac{\pi\ell^2}{\sin\theta_i}\left|\frac{\sin\theta_s}{\ln(0.89\beta a\sin\theta_i)}\right|^2\left\{\frac{\sin\left[\frac{\beta\ell}{2}(\cos\theta_i+\cos\theta_s)\right]}{\frac{\beta\ell}{2}(\cos\theta_i+\cos\theta_s)}\right\}^2 \qquad (11\text{-}140b)$$

11.5.4 Oblique Incidence Plane Wave Scattering by a Conducting Circular Cylinder: TE^z Polarization

TE^z scattering by a cylinder at oblique incidence can be analyzed following a procedure similar to that of TM^z scattering as discussed in the previous section. Using the geometry of Figure 11-16, we can write the incident magnetic field, for a plane wave traveling parallel to the xz plane, as

$$\mathbf{H}^i = H_0(\hat{\mathbf{a}}_x\cos\theta_i + \hat{\mathbf{a}}_z\sin\theta_i)e^{-j\beta x\sin\theta_i}e^{+j\beta z\cos\theta_i} \qquad (11\text{-}141)$$

Using the transformation (11-81a), it can also be expressed as

$$\mathbf{H}^i = H_0(\hat{\mathbf{a}}_x\cos\theta_i + \hat{\mathbf{a}}_z\sin\theta_i)e^{+j\beta z\cos\theta_i}\sum_{n=-\infty}^{+\infty}j^{-n}J_n(\beta\rho\sin\theta_i)e^{jn\phi} \qquad (11\text{-}141a)$$

Using the transformation from rectangular to cylindrical components (II-6), or

$$H_\rho = H_x\cos\phi + H_y\sin\phi = H_x\cos\phi \qquad (11\text{-}142a)$$

$$H_\phi = -H_x\sin\phi + H_y\cos\phi = -H_x\sin\phi \qquad (11\text{-}142b)$$

$$H_z = H_z \qquad (11\text{-}142c)$$

reduces (11-141a) to

$$H_\rho^i = H_0 \cos\theta_i \cos\phi\, e^{+j\beta z \cos\theta_i} \sum_{n=-\infty}^{+\infty} j^{-n} J_n(\beta\rho \sin\theta_i) e^{jn\phi} \tag{11-143a}$$

$$H_\phi^i = -H_0 \cos\theta_i \sin\phi\, e^{+j\beta z \cos\theta_i} \sum_{n=-\infty}^{+\infty} j^{-n} J_n(\beta\rho \sin\theta_i) e^{jn\phi} \tag{11-143b}$$

$$H_z^i = H_0 \sin\theta_i e^{+j\beta z \cos\theta_i} \sum_{n=-\infty}^{+\infty} j^{-n} J_n(\beta\rho \sin\theta_i) e^{jn\phi} \tag{11-143c}$$

In the source-free region, the corresponding electric field components can be obtained using Maxwell's curl equation

$$\nabla \times \mathbf{H}^i = j\omega\varepsilon \mathbf{E}^i \Rightarrow \mathbf{E}^i = \frac{1}{j\omega\varepsilon} \nabla \times \mathbf{H}^i \tag{11-144}$$

or

$$E_\rho^i = \frac{1}{j\omega\varepsilon}\left(\frac{1}{\rho}\frac{\partial H_z^i}{\partial\phi} - \frac{\partial H_\phi^i}{\partial z}\right) \tag{11-144a}$$

$$E_\phi^i = \frac{1}{j\omega\varepsilon}\left(\frac{\partial H_\rho^i}{\partial z} - \frac{\partial H_z^i}{\partial\rho}\right) \tag{11-144b}$$

$$E_z^i = \frac{1}{j\omega\varepsilon}\frac{1}{\rho}\left[\frac{\partial(\rho H_\phi^i)}{\partial\rho} - \frac{\partial H_\rho^i}{\partial\phi}\right] \tag{11-144c}$$

To aid in doing this, we also utilize Maxwell's curl equation

$$\nabla \times \mathbf{E}^i = -j\omega\mu \mathbf{H}^i \Rightarrow \mathbf{H}^i = -\frac{1}{j\omega\mu} \nabla \times \mathbf{E}^i \tag{11-145}$$

which when expanded takes the form, for a TE^z polarization ($E_z^i = 0$), of

$$H_\rho^i = -\frac{1}{j\omega\mu}\left(\frac{1}{\rho}\frac{\partial E_z^i}{\partial\phi} - \frac{\partial E_\phi^i}{\partial z}\right)_{E_z^i=0} = \frac{1}{j\omega\mu}\frac{\partial E_\phi^i}{\partial z} \tag{11-145a}$$

$$H_\phi^i = -\frac{1}{j\omega\mu}\left(\frac{\partial E_\rho^i}{\partial z} - \frac{\partial E_z^i}{\partial\rho}\right)_{E_z^i=0} = -\frac{1}{j\omega\mu}\frac{\partial E_\rho^i}{\partial z} \tag{11-145b}$$

$$H_z^i = -\frac{1}{j\omega\mu}\frac{1}{\rho}\left[\frac{\partial(\rho E_\phi^i)}{\partial\rho} - \frac{\partial E_\rho^i}{\partial\phi}\right] \tag{11-145c}$$

Substituting (11-145a) into (11-144b), we can write that

$$E_\phi^i = \frac{1}{j\omega\varepsilon}\left[\frac{1}{j\omega\mu}\frac{\partial^2 E_\phi^i}{\partial z^2} - \frac{\partial H_z^i}{\partial\rho}\right] = -\frac{1}{\omega^2\mu\varepsilon}\frac{\partial^2 E_\phi^i}{\partial z^2} - \frac{1}{j\omega\varepsilon}\frac{\partial H_z^i}{\partial\rho} \tag{11-146}$$

Since the z variations of all the field components are of the same form (i.e., $e^{+j\beta z \cos\theta_i}$), as given by (11-141), then (11-146) reduces to

$$E_\phi^i = \frac{\beta^2}{\omega^2\mu\varepsilon}\cos^2\theta_i E_\phi^i - \frac{1}{j\omega\varepsilon}\frac{\partial H_z^i}{\partial\rho} = \cos^2\theta_i E_\phi^i - \frac{1}{j\omega\varepsilon}\frac{\partial H_z^i}{\partial\rho} \tag{11-147}$$

or

$$(1 - \cos^2\theta_i)E_\phi^i = \sin^2\theta_i E_\phi^i = -\frac{1}{j\omega\varepsilon}\frac{\partial H_z^i}{\partial\rho}$$

$$E_\phi^i = -\frac{1}{j\omega\varepsilon}\frac{1}{\sin^2\theta_i}\frac{\partial H_z^i}{\partial\rho} = -\frac{\beta}{j\omega\varepsilon}\frac{1}{\sin\theta_i}\frac{\partial H_z^i}{\partial(\beta\rho\sin\theta_i)}$$

$$= j\sqrt{\frac{\mu}{\varepsilon}}H_0 e^{+j\beta z\cos\theta_i}\sum_{n=-\infty}^{+\infty} j^{-n}J_n'(\beta\rho\sin\theta_i)e^{jn\phi} \qquad (11\text{-}147a)$$

In a similar manner, we can solve for E_ρ^i by substituting (11-145b) into (11-144a). Thus,

$$E_\rho^i = \frac{1}{j\omega\varepsilon}\left[\frac{1}{\rho}\frac{\partial H_z^i}{\partial\phi} - \left(-\frac{1}{j\omega\mu}\frac{\partial^2 E_\rho^i}{\partial z^2}\right)\right] = \frac{1}{j\omega\varepsilon}\frac{1}{\rho}\frac{\partial H_z^i}{\partial\phi} - \frac{1}{\omega^2\mu\varepsilon}\frac{\partial^2 E_\rho^i}{\partial z^2}$$

$$= \frac{1}{j\omega\varepsilon}\frac{1}{\rho}\frac{\partial H_z^i}{\partial\phi} - \frac{(j\beta\cos\theta_i)^2}{\omega^2\mu\varepsilon}E_\rho^i = \frac{1}{j\omega\varepsilon}\frac{1}{\rho}\frac{\partial H_z^i}{\partial\phi} + \cos^2\theta_i E_\rho^i \qquad (11\text{-}148)$$

or

$$(1 - \cos^2\theta_i)E_\rho^i = \sin^2\theta_i E_\rho^i = \frac{1}{j\omega\varepsilon}\frac{1}{\rho}\frac{\partial H_z^i}{\partial\phi}$$

$$E_\rho^i = \frac{1}{j\omega\varepsilon\rho}\frac{1}{\sin^2\theta_i}\frac{\partial H_z^i}{\partial\phi} = -j\frac{H_0 e^{+j\beta z\cos\theta_i}}{\omega\varepsilon\rho\sin\theta_i}\sum_{n=-\infty}^{+\infty} nj^{-n+1}J_n(\beta\rho\sin\theta_i)e^{jn\phi} \qquad (11\text{-}148a)$$

Since the z component of the incident $\mathbf{H}$ field is given by (11-143c), its scattered field can be written in a form similar to (11-121) or

$$H_z^s = H_0\sin\theta_s e^{-j\beta z\cos\theta_s}\sum_{n=-\infty}^{+\infty} d_n H_n^{(2)}(\beta\rho\sin\theta_s) \qquad (11\text{-}149)$$

where d_n represents unknown coefficients to be determined by boundary conditions. According to (11-147a), the ϕ component of the scattered field can be written using (11-149) as

$$E_\phi^s = -\frac{1}{j\omega\varepsilon}\frac{1}{\sin^2\theta_s}\frac{\partial H_z^s}{\partial\rho} = -\frac{\beta}{j\omega\varepsilon}\frac{1}{\sin\theta_s}\frac{\partial H_z^s}{\partial(\beta\rho\sin\theta_s)}$$

$$= jH_0\sqrt{\frac{\mu}{\varepsilon}}e^{-j\beta z\cos\theta_s}\sum_{n=-\infty}^{+\infty} d_n H_n^{(2)\prime}(\beta\rho\sin\theta_s) \qquad (11\text{-}150)$$

Applying the boundary condition

$$E_\phi^t(\rho = a, 0\le\theta_i, \theta_s \le\pi, 0\le\phi\le 2\pi) = 0$$
$$= E_\phi^i(\rho = a, 0\le\theta_i \le\pi, 0\le\phi\le 2\pi) + E_\phi^s(\rho = a, 0\le\theta_s \le\pi, 0\le\phi\le 2\pi) \quad (11\text{-}151)$$

leads to

$$jH_0\sqrt{\frac{\mu}{\varepsilon}}\left[e^{+j\beta z\cos\theta_i}\sum_{n=-\infty}^{+\infty} j^{-n}J_n'(\beta a\sin\theta_i)e^{jn\phi}\right.$$

$$\left. + e^{-j\beta z\cos\theta_s}\sum_{n=-\infty}^{+\infty} d_n H_n^{(2)\prime}(\beta a\sin\theta_s)\right] = 0 \qquad (11\text{-}151a)$$

which is satisfied provided

$$\theta_s = \pi - \theta_i \tag{11-151b}$$

$$d_n = -j^{-n} \frac{J_n'(\beta a \sin\theta_i)}{H_n^{(2)'}(\beta a \sin\theta_i)} e^{jn\phi} = j^{-n} b_n e^{jn\phi} \tag{11-151c}$$

$$b_n = -\frac{J_n'(\beta a \sin\theta_i)}{H_n^{(2)'}(\beta a \sin\theta_i)} \tag{11-151d}$$

Thus, the scattered H_z^s component of (11-149) reduces to

$$H_z^s = H_0 \sin\theta_i e^{+j\beta z \cos\theta_i} \sum_{n=-\infty}^{+\infty} j^{-n} b_n H_n^{(2)}(\beta\rho \sin\theta_i) e^{jn\phi} \tag{11-152}$$

Knowing H_z^s, the remaining electric and magnetic field components can be found using (11-148a), (11-147a), and (11-145a) through (11-145c).

In summary, the scattered fields can be written as

$$\underline{\text{TE}^z}$$

$$E_\rho^s = \frac{1}{j\omega\varepsilon\rho} \frac{1}{\sin^2\theta_i} \frac{\partial H_z^s}{\partial\phi} = -j \frac{H_0}{\omega\varepsilon} \frac{e^{+j\beta z \cos\theta_i}}{\sin\theta_i}$$
$$\times \sum_{n=-\infty}^{+\infty} n j^{-n+1} b_n H_n^{(2)}(\beta\rho \sin\theta_i) e^{jn\phi} \tag{11-153a}$$

$$E_\phi^s = -\frac{1}{j\omega\varepsilon} \frac{1}{\sin^2\theta_i} \frac{\partial H_z^s}{\partial\rho} = jH_0 \sqrt{\frac{\mu}{\varepsilon}} e^{+j\beta z \cos\theta_i}$$
$$\times \sum_{n=-\infty}^{+\infty} j^{-n} b_n H_n^{(2)'}(\beta\rho \sin\theta_i) e^{jn\phi} \tag{11-153b}$$

$$E_z^s = 0 \tag{11-153c}$$

$$H_\rho^s = \frac{1}{j\omega\mu} \frac{\partial E_\phi^s}{\partial z} = jH_0 \cos\theta_i e^{+j\beta z \cos\theta_i}$$
$$\times \sum_{n=-\infty}^{+\infty} j^{-n} b_n H_n^{(2)'}(\beta\rho \sin\theta_i) e^{jn\phi} \tag{11-153d}$$

$$H_\phi^s = -\frac{1}{j\omega\mu} \frac{\partial E_\rho^s}{\partial z} = jH_0 \frac{\cot\theta_i}{\beta\rho} e^{+j\beta z \cos\theta_i}$$
$$\times \sum_{n=-\infty}^{+\infty} n j^{-n+1} b_n H_n^{(2)}(\beta\rho \sin\theta_i) e^{jn\phi} \tag{11-153h}$$

$$H_z^s = H_0 \sin\theta_i e^{+j\beta z \cos\theta_i} \sum_{n=-\infty}^{+\infty} j^{-n} b_n H_n^{(2)}(\beta\rho \sin\theta_i) e^{jn\phi} \tag{11-153e}$$

$$b_n = -\frac{J_n'(\beta a \sin\theta_i)}{H_n^{(2)'}(\beta a \sin\theta_i)} \tag{11-153f}$$

$$' \equiv \frac{\partial}{\partial(\beta\rho \sin\theta_i)} \tag{11-153g}$$

A. Far-Zone Scattered Field The scattered electric fields of (11-153a) through (11-153c) can be approximated in the far zone by replacing the Hankel function and its derivative by their asymptotic forms, as given by (11-135a) and (11-135b). Thus,

$$E_\rho^s \simeq H_0 \frac{1}{\omega\varepsilon} \frac{1}{\rho\sin\theta_i} \sqrt{\frac{2j}{\pi\beta\rho\sin\theta_i}} e^{+j\beta(z\cos\theta_i-\rho\sin\theta_i)} \sum_{n=-\infty}^{+\infty} nb_n\, e^{jn\phi}$$

$$\simeq H_0 \frac{1}{\omega\varepsilon} \frac{1}{\rho\sin\theta_i} \sqrt{\frac{2j}{\pi\beta\rho\sin\theta_i}} e^{+j\beta(z\cos\theta_i-\rho\sin\theta_i)} \sum_{n=0}^{+\infty} n\varepsilon_n b_n \cos(n\phi) \qquad (11\text{-}154a)$$

$$E_\phi^s \simeq H_0 \sqrt{\frac{\mu}{\varepsilon}} \sqrt{\frac{2j}{\pi\beta\rho\sin\theta_i}} e^{+j\beta(z\cos\theta_i-\rho\sin\theta_i)} \sum_{n=-\infty}^{+\infty} b_n\, e^{jn\phi}$$

$$\simeq H_0 \sqrt{\frac{\mu}{\varepsilon}} \sqrt{\frac{2j}{\pi\beta\rho\sin\theta_i}} e^{+j\beta(z\cos\theta_i-\rho\sin\theta_i)} \sum_{n=-\infty}^{+\infty} \varepsilon_n b_n \cos(n\phi) \qquad (11\text{-}154b)$$

where ε_n is defined in (11-102a). A comparison of (11-154a) and (11-154b) indicates that at large distances E_ρ^s is small compared to E_ϕ^s since E_ρ^s is inversely proportional to $\rho^{3/2}$, whereas E_ϕ^s is inversely proportional to $\rho^{1/2}$.

The scattering width of (11-21b) can now be expressed using (11-154b) and the incident electric field corresponding to (11-141) as

$$\sigma_{2\text{-}D} = \lim_{\rho\to\infty}\left[2\pi\rho\frac{|E_\phi^s|^2}{|E^i|^2}\right] = \lim_{\rho\to\infty}\left[2\pi\rho\frac{|H_0|^2\dfrac{\mu}{\varepsilon}\left(\dfrac{2}{\pi\beta\rho\sin\theta_i}\right)}{|H_0|^2\dfrac{\mu}{\varepsilon}}\left|\sum_{n=-\infty}^{+\infty} b_n\, e^{jn\phi}\right|^2\right]$$

$$\boxed{\sigma_{2\text{-}D} = \frac{4}{\beta}\frac{1}{\sin\theta_i}\left|\sum_{n=-\infty}^{+\infty} b_n\, e^{jn\phi}\right|^2 = \frac{2\lambda}{\pi}\frac{1}{\sin\theta_i}\left|\sum_{n=0}^{+\infty}\varepsilon_n b_n \cos(n\phi)\right|^2} \qquad (11\text{-}155)$$

where

$$\boxed{b_n = -\frac{J_n'(\beta a\sin\theta_i)}{H_n^{(2)'}(\beta a\sin\theta_i)}} \qquad (11\text{-}155a)$$

$$\boxed{\varepsilon_n = \begin{cases} 1 & n=0 \\ 2 & n\neq 0 \end{cases}} \qquad (11\text{-}155b)$$

which is similar to (11-117) except that β in (11-117) has been replaced by $\beta\sin\theta_i$.

From the results of the normal incidence case of Section 11.5.2B we can write by referring to (11-118d) that for small radii the scattering width of (11-155) reduces to

$$\sigma_{2\text{-}D} \overset{a\ll\lambda}{\simeq} \frac{\pi\lambda}{8}\frac{(\beta a\sin\theta_i)^4}{\sin\theta_i}[1-2\cos(\phi)]^2 \qquad (11\text{-}156)$$

which is dependent on ϕ, even for small-radii cylinders.

Using (11-139), the radar cross section at oblique incidence for a finite length ℓ cylinder can be written using (11-155) and (11-156), and referring to the geometry of Figure 11-16, as

$$\sigma_{3\text{-D}} \simeq \frac{4\ell^2}{\pi} \sin\theta_i \left| \sum_{n=0}^{\infty} \varepsilon_n b_n \cos(n\phi) \right|^2 \left\{ \frac{\sin\left[\frac{\beta\ell}{2}(\cos\theta_i + \cos\theta_s) \right]}{\frac{\beta\ell}{2}(\cos\theta_i + \cos\theta_s)} \right\}^2 \tag{11-157a}$$

$$\sigma_{3\text{-D}} \overset{a\ll\lambda}{\simeq} \frac{\pi\ell^2}{4} \left[(\beta a \sin\theta_i)^4 \sin\theta_i \right] [1 - 2\cos(\phi)]^2 \left\{ \frac{\sin\left[\frac{\beta\ell}{2}(\cos\theta_i + \cos\theta_s) \right]}{\frac{\beta\ell}{2}(\cos\theta_i + \cos\theta_s)} \right\}^2 \tag{11-157b}$$

11.5.5   Line-Source Scattering by a Conducting Circular Cylinder

While in the previous sections we examined plane wave scattering by a conducting circular cylinder, both at normal and oblique wave incidences, a more general problem is that of line-source (both electric and magnetic) scattering. The geometry is that shown in Figure 11-17 where an infinite line of constant current (I_e for electric and I_m for magnetic) is placed in the vicinity of a circular conducting cylinder of infinite length. We will examine here the scattering by the cylinder assuming the source is either electric or magnetic.

A. Electric Line Source (TM^z Polarization)   If the line source of Figure 11-17 is of constant electric current I_e, the field generated everywhere by the source in the absence of the cylinder is

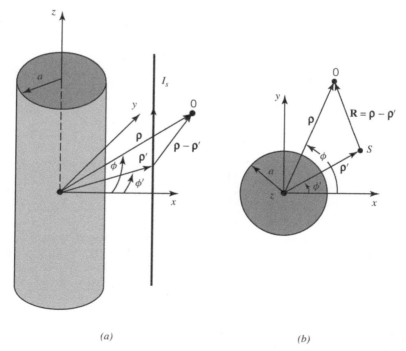

(a)   (b)

Figure 11-17   Electric line source near a circular cylinder. (a) Side view. (b) Top view.

given, according to (11-11a), by

$$E_z^i = -\frac{\beta^2 I_e}{4\omega\varepsilon} H_0^{(2)}(\beta \,|\boldsymbol{\rho} - \boldsymbol{\rho}'|) \tag{11-158}$$

which is referred to here as the incident field. By the addition theorem for Hankel functions (11-69a) and (11-69b), we can write (11-158) as

$$E_z^i = -\frac{\beta^2 I_e}{4\omega\varepsilon} \left\{ \begin{array}{ll} \displaystyle\sum_{n=-\infty}^{\infty} J_n(\beta\rho)H_n^{(2)}(\beta\rho')e^{jn(\phi-\phi')}, & \rho \leq \rho' \qquad (11\text{-}158\text{a}) \\[18pt] \displaystyle\sum_{n=-\infty}^{+\infty} J_n(\beta\rho')H_n^{(2)}(\beta\rho)e^{jn(\phi-\phi')}, & \rho \geq \rho' \qquad (11\text{-}158\text{b}) \end{array} \right.$$

Bessel functions $J_n(\beta\rho)$ were selected to represent the fields for $\rho < \rho'$ because the field must be finite everywhere (including $\rho = 0$) and Hankel functions were chosen for $\rho \geq \rho'$ to represent the traveling nature of the wave.

In the presence of the cylinder, the total field is composed of two parts: incident and scattered fields. The scattered field is produced by the current induced on the surface of the cylinder that acts as a secondary radiator. The scattered field also has only an E_z component (no cross polarized components are produced), and it can be expressed as

$$E_z^s = -\frac{\beta^2 I_e}{4\omega\varepsilon} \sum_{n=-\infty}^{+\infty} c_n H_n^{(2)}(\beta\rho), \qquad a \leq \rho \leq \rho', \;\; \rho \geq \rho' \tag{11-159}$$

The same expression is valid for $\rho \leq \rho'$ and $\rho \geq \rho'$ because the scattered field exists only when the cylinder is present, and it is nonzero only when $\rho \geq a$. Since the scattered field emanates from the surface of the cylinder, the Hankel function of the second kind in (11-159) is chosen to represent the traveling wave nature of the radiation.

The coefficients represented by c_n in (11-159) can be found by applying the boundary condition

$$E_z^t(\rho = a, 0 \leq \phi, \phi' \leq 2\pi, z)$$
$$= E_z^i(\rho = a, 0 \leq \phi, \phi' \leq 2\pi, z) + E_z^s(\rho = a, 0 \leq \phi, \phi' \leq 2\pi, z) = 0 \tag{11-160}$$

which, by using (11-158a) and (11-159), leads to

$$-\frac{\beta^2 I_e}{4\omega\varepsilon} \sum_{n=-\infty}^{+\infty} \left[H_n^{(2)}(\beta\rho')J_n(\beta a)e^{jn(\phi-\phi')} + c_n H_n^{(2)}(\beta a) \right] = 0 \tag{11-161}$$

which is satisfied provided

$$c_n = -H_n^{(2)}(\beta\rho')\frac{J_n(\beta a)}{H_n^{(2)}(\beta a)} e^{jn(\phi-\phi')} \tag{11-161a}$$

Thus, (11-159) can be expressed as

$$E_z^s = +\frac{\beta^2 I_e}{4\omega\varepsilon} \sum_{n=-\infty}^{+\infty} H_n^{(2)}(\beta\rho')\frac{J_n(\beta a)}{H_n^{(2)}(\beta a)} H_n^{(2)}(\beta\rho)e^{jn(\phi-\phi')}, \qquad a \leq \rho \leq \rho', \;\; \rho \geq \rho' \tag{11-162}$$

The total electric field can then be written as

$$E_\rho^t = E_\phi^t = 0 \tag{11-163}$$

$$E_z^t = -\frac{\beta^2 I_e}{4\omega\varepsilon} \left\{ \begin{array}{l} \displaystyle\sum_{n=-\infty}^{+\infty} H_n^{(2)}(\beta\rho') \left[J_n(\beta\rho) - \frac{J_n(\beta a)}{H_n^{(2)}(\beta a)} H_n^{(2)}(\beta\rho) \right] \\[2mm] \quad \times e^{jn(\phi-\phi')} \qquad\qquad a \le \rho \le \rho' \\[4mm] \displaystyle\sum_{n=-\infty}^{+\infty} H_n^{(2)}(\beta\rho) \left[J_n(\beta\rho') - \frac{J_n(\beta a)}{H_n^{(2)}(\beta a)} H_n^{(2)}(\beta\rho') \right] \\[2mm] \quad \times e^{jn(\phi-\phi')} \qquad\qquad \rho \ge \rho' \end{array} \right. \qquad \begin{array}{c} (11\text{-}164a) \\[16mm] (11\text{-}164b) \end{array}$$

where the first terms within the summations and brackets represent the incident fields and the second terms represent the scattered fields. The corresponding magnetic components can be found using Maxwell's equations 11-86 through 11-86c, which can be written as

$$H_\rho^t = -\frac{1}{j\omega\mu}\frac{1}{\rho}\frac{\partial E_z^t}{\partial\phi}$$

$$= -j\frac{I_e}{4\rho} \left\{ \begin{array}{l} \displaystyle\sum_{n=-\infty}^{+\infty} jn H_n^{(2)}(\beta\rho') \left[J_n(\beta\rho) - \frac{J_n(\beta a)}{H_n^{(2)}(\beta a)} H_n^{(2)}(\beta\rho) \right] \\[2mm] \quad \times e^{jn(\phi-\phi')} \qquad\qquad a \le \rho \le \rho' \\[4mm] \displaystyle\sum_{n=-\infty}^{+\infty} jn H_n^{(2)}(\beta\rho) \left[J_n(\beta\rho') - \frac{J_n(\beta a)}{H_n^{(2)}(\beta a)} H_n^{(2)}(\beta\rho') \right] \\[2mm] \quad \times e^{jn(\phi-\phi')} \qquad\qquad \rho \ge \rho' \end{array} \right. \qquad \begin{array}{c} (11\text{-}165a) \\[16mm] (11\text{-}165b) \end{array}$$

$$H_\phi^t = \frac{1}{j\omega\mu}\frac{\partial E_z^t}{\partial\rho}$$

$$= j\frac{\beta I_e}{4} \left\{ \begin{array}{l} \displaystyle\sum_{n=-\infty}^{+\infty} H_n^{(2)}(\beta\rho') \left[J_n'(\beta\rho) - \frac{J_n(\beta a)}{H_n^{(2)}(\beta a)} H_n^{(2)'}(\beta\rho) \right] \\[2mm] \quad \times e^{jn(\phi-\phi')} \qquad\qquad a \le \rho \le \rho' \\[4mm] \displaystyle\sum_{n=-\infty}^{+\infty} H_n^{(2)'}(\beta\rho) \left[J_n(\beta\rho') - \frac{J_n(\beta a)}{H_n^{(2)}(\beta a)} H_n^{(2)}(\beta\rho') \right] \\[2mm] \quad \times e^{jn(\phi-\phi')} \qquad\qquad \rho \ge \rho' \end{array} \right. \qquad \begin{array}{c} (11\text{-}166a) \\[16mm] (11\text{-}166b) \end{array}$$

$$H_z^t = 0$$

On the surface of the cylinder, the current density can be found to be

$$\mathbf{J}_s = \hat{\mathbf{n}} \times \mathbf{H}^t \mid_{\rho=a} = \hat{\mathbf{a}}_\rho \times (\hat{\mathbf{a}}_\rho H_\rho^t + \hat{\mathbf{a}}_\phi H_\phi^t) \mid_{\rho=a} = \mid \hat{\mathbf{a}}_z H_\phi^t \mid_{\rho=a}$$

$$= \hat{\mathbf{a}}_z j\frac{\beta I_e}{4} \sum_{n=-\infty}^{+\infty} H_n^{(2)}(\beta\rho') \left[J_n'(\beta a) - \frac{J_n(\beta a)}{H_n^{(2)}(\beta a)} H_n^{(2)'}(\beta a) \right] e^{jn(\phi-\phi')} \tag{11-167}$$

$$\mathbf{J}_s = -\hat{\mathbf{a}}_z \frac{\beta I_e}{4} \sum_{n=-\infty}^{+\infty} H_n^{(2)}(\beta\rho') \left[\frac{J_n(\beta a)Y_n'(\beta a) - J_n'(\beta a)Y_n(\beta a)}{H_n^{(2)}(\beta a)} \right] e^{jn(\phi-\phi')} \qquad (11\text{-}168)$$

which, by using the Wronskian of (11-95), reduces to

$$\mathbf{J}_s = -\hat{\mathbf{a}}_z \frac{I_e}{2\pi a} \sum_{n=-\infty}^{+\infty} \frac{H_n^{(2)}(\beta\rho')}{H_n^{(2)}(\beta a)} e^{jn(\phi-\phi')} \qquad (11\text{-}168a)$$

For far-field observations ($\beta\rho \gg 1$), the total electric field of (11-164b) can be reduced by replacing the Hankel function $H_n^{(2)}(\beta\rho)$ by its asymptotic expression (11-135a). Doing this reduces (11-164b) to

$$E_z^t \overset{\beta\rho \gg 1}{\simeq} -\frac{\beta^2 I_e}{4\omega\varepsilon} \sqrt{\frac{2j}{\pi\beta}} \frac{e^{-j\beta\rho}}{\sqrt{\rho}} \sum_{n=-\infty}^{+\infty} j^n \left[J_n(\beta\rho') - \frac{J_n(\beta a)}{H_n^{(2)}(\beta a)} H_n^{(2)}(\beta\rho') \right] e^{jn(\phi-\phi')} \qquad (11\text{-}169)$$

which can be used to compute more conveniently far-field patterns of an electric line source near a circular conducting cylinder. Plots of the normalized pattern for $a = 5\lambda$, $\phi' = 0$ with $\rho' = 5.25\lambda$ and 5.5λ are shown, respectively, in Figures 11-18a and 11-18b where they are compared with that of a planar reflector ($a = \infty$) of Figure 11-3. Because of the finite radius of the cylinder, radiation

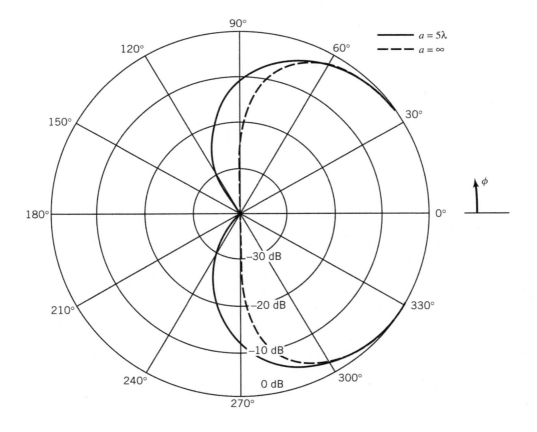

(a)

Figure 11-18 Normalized far-field pattern of an electric line source near a circular conducting cylinder. (a) $\rho' = 5.25\lambda$, $\phi' = 0°$. (b) $\rho' = 5.5\lambda$, $\phi' = 0°$.

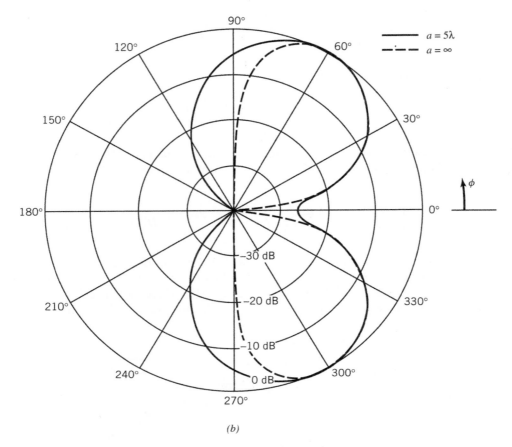

(b)

Figure 11-18   (*Continued*)

is allowed to "leak" around the cylinder in terms of "creeping" waves [15–17]; this is not the case for the planar reflector.

B. Magnetic Line Source (TE^z Polarization)   Magnetic sources, although not physically realizable, are often used as equivalent sources to analyze aperture antennas [1, 18]. If the line source of Figure 11-17 is magnetic and it is allowed to recede to the surface of the cylinder ($\rho' = a$), the total field of the line source in the presence of the cylinder would be representative of a very thin infinite axial slot on the cylinder. Finite-width slots can be represented by a number of line sources with some amplitude and phase distribution across the width. Therefore, knowing the radiation and scattering by a magnetic line source near a cylinder allows us to solve other physical problems by using it as an equivalent.

If the line of Figure 11-17 is magnetic with a current of I_m, the fields that it radiates in the absence of the cylinder can be obtained from those of an electric line source by the use of duality (Section 7.2). Doing this, we can write the incident magnetic field, by referring to (11-158) through (11-158b), as

$$H_z^i = -\frac{\beta^2 I_m}{4\omega\mu} H_0^{(2)}(\beta\,|\boldsymbol{\rho} - \boldsymbol{\rho}'|) \tag{11-170}$$

which can also be expressed as

$$H_z^i = -\frac{\beta^2 I_m}{4\omega\mu} \begin{cases} \displaystyle\sum_{n=-\infty}^{\infty} J_n(\beta\rho) H_n^{(2)}(\beta\rho') e^{jn(\phi-\phi)}, & \rho \le \rho' \qquad (11\text{-}170\text{a}) \\[2em] \displaystyle\sum_{n=-\infty}^{+\infty} J_n(\beta\rho') H_n^{(2)}(\beta\rho) e^{jn(\phi-\phi')}, & \rho \ge \rho' \qquad (11\text{-}170\text{b}) \end{cases}$$

The scattered magnetic field takes a form similar to that of (11-159) and can be written as

$$H_z^s = -\frac{\beta^2 I_m}{4\omega\mu} \sum_{n=-\infty}^{+\infty} d_n H_n^{(2)}(\beta\rho), \qquad a \le \rho \le \rho', \quad \rho \ge \rho' \qquad (11\text{-}171)$$

where d_n is used to represent the coefficients of the scattered field. Thus, the total magnetic field can be expressed, by combining (11-170a) through (11-171), as

$$H_z^t = -\frac{\beta^2 I_m}{4\omega\mu} \begin{cases} \displaystyle\sum_{n=-\infty}^{+\infty} \left[H_n^{(2)}(\beta\rho') J_n(\beta\rho) e^{jn(\phi-\phi')} + d_n H_n^{(2)}(\beta\rho) \right] \\[1em] \hspace{4cm} a \le \rho \le \rho' \qquad (11\text{-}172\text{a}) \\[1em] \displaystyle\sum_{n=-\infty}^{+\infty} \left[J_n(\beta\rho') e^{jn(\phi-\phi')} + d_n \right] H_n^{(2)}(\beta\rho) \\[1em] \hspace{4cm} \rho \ge \rho' \qquad (11\text{-}172\text{b}) \end{cases}$$

The corresponding electric field components can be found using Maxwell's equations 11-106 or 11-106a and 11-106b. Doing this, and utilizing (11-172a) and (11-172b), we can write that

$$E_\rho^t = \frac{1}{j\omega\varepsilon} \frac{1}{\rho} \frac{\partial H_z^t}{\partial \phi}$$

$$= j\frac{I_m}{4\rho} \begin{cases} \displaystyle\sum_{n=-\infty}^{+\infty} \left[jn H_n^{(2)}(\beta\rho') J_n(\beta\rho) e^{jn(\phi-\phi')} + H_n^{(2)}(\beta\rho)\frac{\partial d_n}{\partial\phi} \right] \\[1em] \hspace{4cm} a \le \rho \le \rho' \qquad (11\text{-}173\text{a}) \\[1em] \displaystyle\sum_{n=-\infty}^{+\infty} \left[jn J_n(\beta\rho') e^{jn(\phi-\phi')} + \frac{\partial d_n}{\partial\phi} \right] H_n^{(2)}(\beta\rho) \\[1em] \hspace{4cm} \rho \ge \rho' \qquad (11\text{-}173\text{b}) \end{cases}$$

$$E_\phi^t = -\frac{1}{j\omega\varepsilon} \frac{\partial H_z^t}{\partial \rho}$$

$$= -j\frac{\beta I_m}{4} \begin{cases} \displaystyle\sum_{n=-\infty}^{+\infty} \left[H_n^{(2)}(\beta\rho') J_n'(\beta\rho) e^{jn(\phi-\phi')} + d_n H_n^{(2)'}(\beta\rho) \right] \\[1em] \hspace{4cm} a \le \rho \le \rho' \qquad (11\text{-}174\text{a}) \\[1em] \displaystyle\sum_{n=-\infty}^{+\infty} \left[J_n(\beta\rho') e^{jn(\phi-\phi')} + d_n \right] H_n^{(2)'}(\beta\rho) \\[1em] \hspace{4cm} \rho \ge \rho' \qquad (11\text{-}174\text{b}) \end{cases}$$

Applying the boundary condition

$$E_\phi^t(\rho = a, 0 \leq \phi, \phi' \leq 2\pi, z)$$
$$= E_\phi^i(\rho = a, 0 \leq \phi, \phi' \leq 2\pi, z) + E_\phi^s(\rho = a, 0 \leq \phi, \phi' \leq 2\pi, z) = 0 \qquad (11\text{-}175)$$

on (11-174a) leads to

$$d_n = -H_n^{(2)}(\beta\rho') \frac{J_n'(\beta a)}{H_n^{(2)'}(\beta a)} e^{jn(\phi-\phi')} \qquad (11\text{-}175a)$$

Thus, the total electric and magnetic field components can be written as

$$\underline{\text{TE}^z}$$

$$E_z^t = H_\rho^t = H_\phi^t = 0 \qquad (11\text{-}176a)$$

$$E_\rho^t = -\frac{I_m}{4\rho} \begin{cases} \begin{aligned} &\sum_{n=-\infty}^{+\infty} n H_n^{(2)}(\beta\rho') \left[J_n(\beta\rho) - \frac{J_n'(\beta a)}{H_n^{(2)'}(\beta a)} H_n^{(2)}(\beta\rho) \right] \\ &\times e^{jn(\phi-\phi')} \qquad a \leq \rho \leq \rho' \qquad (11\text{-}176b) \\[1em] &\sum_{n=-\infty}^{+\infty} n H_n^{(2)}(\beta\rho) \left[J_n(\beta\rho') - \frac{J_n'(\beta a)}{H_n^{(2)'}(\beta a)} H_n^{(2)}(\beta\rho') \right] \\ &\times e^{jn(\phi-\phi')} \qquad \rho \geq \rho' \qquad (11\text{-}176c) \end{aligned} \end{cases}$$

$$E_\phi^t = -j\frac{\beta I_m}{4} \begin{cases} \begin{aligned} &\sum_{n=-\infty}^{+\infty} H_n^{(2)}(\beta\rho') \left[J_n'(\beta\rho) - \frac{J_n'(\beta a)}{H_n^{(2)'}(\beta a)} H_n^{(2)'}(\beta\rho) \right] \\ &\times e^{jn(\phi-\phi')} \qquad a \leq \rho \leq \rho' \qquad (11\text{-}176d) \\[1em] &\sum_{n=-\infty}^{+\infty} H_n^{(2)'}(\beta\rho) \left[J_n(\beta\rho') - \frac{J_n'(\beta a)}{H_n^{(2)'}(\beta a)} H_n^{(2)}(\beta\rho') \right] \\ &\times e^{jn(\phi-\phi')} \qquad \rho \geq \rho' \qquad (11\text{-}176e) \end{aligned} \end{cases}$$

$$H_z^t = -\frac{\beta^2 I_m}{4\omega\mu} \begin{cases} \begin{aligned} &\sum_{n=-\infty}^{+\infty} H_n^{(2)}(\beta\rho') \left[J_n(\beta\rho) - \frac{J_n'(\beta a)}{H_n^{(2)'}(\beta a)} H_n^{(2)}(\beta\rho) \right] \\ &\times e^{jn(\phi-\phi')} \qquad a \leq \rho \leq \rho' \qquad (11\text{-}176f) \\[1em] &\sum_{n=-\infty}^{+\infty} H_n^{(2)}(\beta\rho) \left[J_n(\beta\rho') - \frac{J_n'(\beta a)}{H_n^{(2)'}(\beta a)} H_n^{(2)}(\beta\rho') \right] \\ &\times e^{jn(\phi-\phi')} \qquad \rho \geq \rho' \qquad (11\text{-}176g) \end{aligned} \end{cases}$$

where the first terms within the summation and brackets represent the incident fields and the second terms represent the scattered fields.

On the surface of the cylinder, the current density can be found to be

$$\mathbf{J}_s = \hat{\mathbf{n}} \times \mathbf{H}^t\big|_{\rho=a} = \hat{\mathbf{a}}_\rho \times \hat{\mathbf{a}}_z H_z^t\big|_{\rho=a} = -\hat{\mathbf{a}}_\phi H_z^t\big|_{\rho=a}$$

$$= \hat{\mathbf{a}}_\phi \frac{\beta^2 I_m}{4\omega\mu} \sum_{n=-\infty}^{+\infty} H_n^{(2)}(\beta\rho') \left[J_n(\beta a) - \frac{J_n'(\beta a)}{H_n^{(2)'}(\beta a)} H_n^{(2)}(\beta a) \right] e^{jn(\phi-\phi')}$$

$$\mathbf{J}_s = -j\hat{\mathbf{a}}_\phi \frac{\beta^2 I_m}{4\omega\mu} \sum_{n=-\infty}^{+\infty} H_n^{(2)}(\beta\rho') \left[\frac{J_n(\beta a)Y_n'(\beta a) - J_n'(\beta a)Y_n(\beta a)}{H_n^{(2)'}(\beta a)} \right] e^{jn(\phi-\phi')} \qquad (11\text{-}177)$$

which, by using the Wronskian (11-95), reduces to

$$\mathbf{J}_s = -j\hat{\mathbf{a}}_\phi \frac{I_m}{2\eta\pi a} \sum_{n=-\infty}^{+\infty} \frac{H_n^{(2)}(\beta\rho')}{H_n^{(2)'}(\beta a)} e^{jn(\phi-\phi')} \qquad (11\text{-}177a)$$

For far-field observations ($\beta\rho \gg 1$), the total magnetic field (11-176g) can be reduced in form by replacing the Hankel function $H_n^{(2)}(\beta\rho)$ by its asymptotic expression (11-135a). Doing this reduces (11-176g) to

$$H_z^t \overset{\beta\rho\gg1}{\simeq} -\frac{\beta^2 I_m}{4\omega\mu} \sqrt{\frac{2j}{\pi\beta}} \frac{e^{-j\beta\rho}}{\sqrt{\rho}} \sum_{n=-\infty}^{+\infty} j^n \left[J_n(\beta\rho') - \frac{J_n'(\beta a)}{H_n^{(2)'}(\beta a)} H_n^{(2)}(\beta\rho') \right] e^{jn(\phi-\phi')}$$

$$(11\text{-}178)$$

which can be used to compute, more conveniently, far-field patterns of a magnetic line source near a circular electric conducting cylinder. When the line source is moved to the surface of the cylinder ($\rho' = a$), (11-178) reduces, with the aid of the Wronskian (11-95), to

$$H_z^t \overset{\beta\rho\gg1}{\simeq} j\frac{\beta^2 I_m}{4\omega\mu} \sqrt{\frac{2j}{\pi\beta}} \frac{e^{-j\beta\rho}}{\sqrt{\rho}} \sum_{n=-\infty}^{+\infty} j^n \left[\frac{J_n(\beta a)Y_n'(\beta a) - J_n'(\beta a)Y_n(\beta a)}{H_n^{(2)'}(\beta a)} \right] e^{jn(\phi-\phi')}$$

$$H_z^t \overset{\beta\rho\gg1}{\simeq} j\frac{I_m}{\pi}\frac{1}{a} \sqrt{\frac{\varepsilon}{\mu}} \sqrt{\frac{j}{2\pi\beta}} \frac{e^{-j\beta\rho}}{\sqrt{\rho}} \sum_{n=-\infty}^{+\infty} j^n \frac{e^{jn(\phi-\phi')}}{H_n^{(2)'}(\beta a)} \qquad (11\text{-}178a)$$

The pattern of (11-178a) is representative of a very thin (ideally zero width) infinite-length axial slot on a circular conducting cylinder, and its normalized form is shown plotted in Figure 11-19 for $a = 2\lambda$ and 5λ. Because of the larger radius of curvature for the $a = 5\lambda$ radius, which results in larger attenuation, less energy is allowed to "creep" around the cylinder compared to that of $a = 2\lambda$.

Scattering by cylinders of other cross sections, and by dielectric and dielectric-covered cylinders, can be found in the literature [12, 19–31].

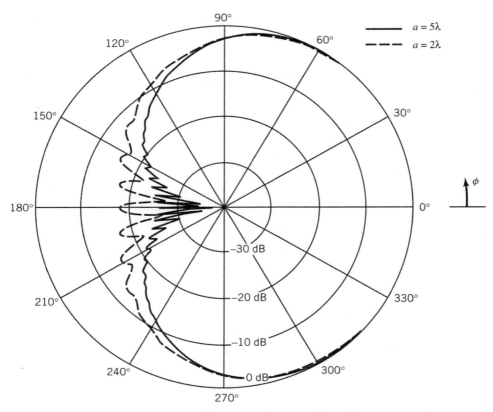

Figure 11-19 Normalized far-field amplitude pattern of a very thin axial slot on a circular conducting cylinder ($\phi' = 0°$).

Example 11-4

A more practical problem is one in which an aperture of finite dimensions is mounted on a circular cylinder. This type of configuration is applicable to apertures mounted on the surface of missiles, fuselages of airplanes, and other similar structures and airframes. Practically, a rectangular aperture can be mounted on the cylinder primarily in two orientations, as shown in Figures 11-20a and 11-20b. The one in Figure 11-20a is usually referred to as a *circumferential* aperture while that in Figure 11-20b is referred to as an *axial* aperture. The choice of orientation is dictated by the desired polarization. One such decision was made in the 1970s in the development of the Microwave Landing System (MLS) [32]. The requirement for the MLS was to select antenna elements that radiate either vertical or horizontal polarization, especially near the forward direction of the aircraft. In using apertures, the choice was either of these two apertures, depending how they were mounted on the fuselage of the aircraft.

a. Assume the electric field expressions on each of the two apertures is given, respectively, by

Circumferential aperture: $E_z = \dfrac{V_0}{h} \cos\left(\dfrac{\pi}{2\phi_0}\phi'\right)$ $\begin{cases} -\phi_0 \leq \phi' \leq +\phi_0 \\ -\dfrac{h}{2} \leq z' \leq +\dfrac{h}{2} \end{cases}$

Axial aperture: $E_\phi = \dfrac{V_0}{2a\phi_0} \cos\left(\dfrac{\pi}{w}z'\right)$ $\begin{cases} -\phi_0 \leq \phi' \leq +\phi_0 \\ -\dfrac{w}{2} \leq z' \leq +\dfrac{w}{2} \end{cases}$

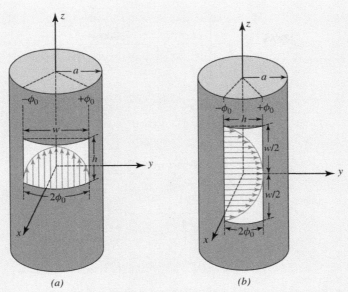

Figure 11-20 Aperture on the surface of a circular cylinder. (*a*) Circumferential. (*b*) Axial.

Write expressions for the electric field spherical components of the far-zone fields radiated by these two apertures.

b. Assuming $f = 10$ GHz, $w = 2.286$ cm, $h = 1.016$ cm (X-band waveguide), compute and plot the:

- H-plane normalized amplitude patterns for the circumferential aperture of Figure 11-20a when it is on a circular cylinder with radii $a = 2\lambda$ and 5λ. Compare them when the same aperture is mounted on an infinite PEC flat ground plane. Compare the results and assess the effect of the cylinder curvature on the H-plane radiation patterns.
- Repeat the previous computations and plots for the E-plane of the axial aperture of Figure 11-20b.

Solution: The fields radiated by the two apertures can be determined by replacing the apertures with equivalent currents densities $\mathbf{J}_s$ and $\mathbf{M}_s$ and then using integration over the aperture and surface of the cylinder. Since $\mathbf{J}_s$ is not known outside the aperture, an approximate equivalent will be to either assume $\mathbf{J}_s$ is small outside the aperture; however, this is not an exact equivalent. Another equivalent is to assume $\mathbf{M}_s$ only over the aperture, as is done for apertures mounted on ground planes [1]. However, for the cylinder, this also is not exact because the surface to which the aperture is mounted is not flat.

Another procedure is to use transform techniques, as it was done in [11, 33-34] where the fields external to the cylinder are expressed as the sum of TE^z and TM^z modes. This is accomplished by writing the corresponding vector potentials F_z and A_z in the transform domain and then the fields are obtained using (6-34) and (6-35). Using such a procedure, it is shown that the far-zone fields radiated by the respective apertures are:

Circumferential Aperture (circular cylinder)

$$E_\theta = -j\frac{V_0}{\pi^2}\frac{e^{-j\beta r}}{r}\left\{\frac{\beta a}{\sin\theta}\sum_{n=0}^{n=+\infty}\frac{\cos\left(\dfrac{n\pi}{2\beta a}\right)}{\left[(\beta a)^2 - n^2\right]}\frac{\varepsilon_n j^n \cos(n\phi)}{H_n^{(2)}(\beta a\sin\theta)}\right\}$$

$$E_\phi = -j\frac{V_0}{\pi^2}\frac{e^{-j\beta r}}{r}\left\{\frac{\cot\theta}{\sin\theta}\sum_{n=1}^{n=+\infty}\frac{\cos\left(\frac{n\pi}{2\beta a}\right)}{\left[(\beta a)^2-n^2\right]}\frac{2nj^n\sin(n\phi)}{H_n^{(2)'}(\beta a\sin\theta)}\right\}$$

$$\varepsilon_n=\begin{cases}1 & n=0\\2 & n>0\end{cases};\quad H_n^{(2)'}(\beta\rho)=\frac{\partial}{\partial\rho}H_n^{(2)}(\beta\rho)$$

$$\text{When}\begin{cases}\theta=\pi/2: & E_\phi=0\\\phi=0: & E_\phi=0\end{cases}\quad\text{as it should be.}$$

Axial Aperture (circular cylinder)

$$E_\phi=\frac{V_0\lambda}{2a\pi^3}\frac{e^{-j\beta r}}{r}\left\{\left[\frac{\cos\left(\frac{\beta w}{2}\cos\theta\right)}{1-\left(\frac{\beta w}{\pi}\cos\theta\right)^2}\right]\sum_{n=0}^{n=+\infty}\frac{\varepsilon_n j^n\cos(n\phi)}{H_n^{(2)'}(\beta a\sin\theta)}\right\}$$

$$H_n^{(2)'}(\beta\rho)=\frac{\partial}{\partial\rho}H_n^{(2)}(\beta\rho)$$

$E_\theta=0$ in all planes, as it should be.

Using the surface equivalence theorem outlined in Section 7.8 (Chapter 7), we can derive the fields when the respective apertures are mounted on infinite ground planes. The same procedure is outlined in Section 12.1 of [1]. Some of these apertures, when mounted on flat ground planes, have been assigned as end-of-the-chapter problems in Chapters 6 and 7. Following such a procedure, it is shown that the fields in the principle H- and E- planes, radiated by these two apertures when mounted on infinite flat PEC ground planes are:

Circumferential Aperture (flat PEC ground plane)
H-Plane ($\theta=90°$)

$$E_\phi=0$$

$$E_\theta=+\frac{\pi}{2}C\left[\cos\phi\frac{\cos\left(\frac{\beta a}{2}\sin\phi\right)}{\left(\frac{\beta a}{2}\sin\phi\right)^2-\left(\frac{\pi}{2}\right)^2}\right]$$

Axial Aperture (flat PEC ground plane)
E-Plane ($\theta=90°$)

$$E_\theta=0$$

$$E_\phi=+\frac{2}{\pi}C\left[\frac{\sin\left(\frac{\beta b}{2}\sin\phi\right)}{\frac{\beta b}{2}\sin\phi}\right]$$

where $C=jab\beta V_0 e^{-j\beta r}/2\pi r$. The expressions for the fields radiated in all space are found in the end-of-chapter exercises of Chapters 6 and 7.

To compute the pattern in the H-plane of the circumferential aperture and E-plane of the axial slot, for both the circular cylinder and flat infinite ground plane, a MATLAB computer program, referred to as **PEC_Cyl_Plate_Rect**, was written and it is included in the multimedia folder associated with this book. The respective patterns are shown in Figure 11-21a for the H-plane of the circumferential aperture and in Figure 11-21b for the E-plane of the axial aperture, where they are compared with those of the flat ground plane. It is apparent, as expected, that the:

- Cylinder allows radiation on the rear region, whereas the PEC flat ground plane does not.
- Larger radius cylinder diminishes more the radiation in the rear region because the creeping waves that travel around the surface of the cylinder attenuate faster.
- Number of lobes in the rear region is greater for the larger cylinder because the two creeping waves that travel in opposite directions around the surface of the cylinder, and radiate tangentially [15–17], have a greater space separation, which allows the formation of a greater number of constructive and destructive interferences, leading to greater number of lobes.

Although the E-plane patterns of the circumferential aperture and the H-plane of the axial aperture are not computed or shown here, those of the cylinder are basically identical to those of the flat ground plane, if both are assumed to be of infinite extent.

11.6   SCATTERING BY A CONDUCTING WEDGE

Scattering of electromagnetic waves by a two-dimensional conducting wedge has received considerable attention since about the middle 1950s. Because the wedge is a canonical problem that can be used to represent locally (near the edge) the scattering of more complex structures, asymptotic forms of its solution have been utilized to solve numerous practical problems. The asymptotic forms of its solution are obtained by taking the infinite series modal solution and first transforming it into an integral by the so-called *Watson transformation* [19, 35, 36]. The integral is then evaluated by the *method of steepest descent (saddle point method)* (see Appendix VI) [37]. The resulting terms of the integral evaluation can be recognized to represent the geometrical optics fields, both incident and reflected geometrical optics fields, and the diffracted fields, both incident and reflected diffracted fields [38, 39]. These forms of the solution have received considerable attention in the geometrical theory of diffraction (GTD), which has become a generic name in the area of antennas and scattering [40–42].

First, we will present the modal solution of the scattering by the wedge. In Chapter 13 we will briefly outline its asymptotic solution, whose form represents the geometrical optics and diffracted fields, and apply it to antenna and scattering problems.

11.6.1   Electric Line-Source Scattering by a Conducting Wedge: TM^z Polarization

Let us assume that an infinite electric line source of electric current I_e is placed near a conducting wedge whose total inner wedge angle is WA $= 2\alpha$, as shown in Figure 11-22a. The incident field produced everywhere by the source, in the absence of the wedge, can be written according to (11-158a) and (11-158b) as

$$E_z^i = -\frac{\beta^2 I_e}{4\omega\varepsilon} = \begin{cases} \displaystyle\sum_{m=-\infty}^{+\infty} J_m(\beta\rho)H_m^{(2)}(\beta\rho')e^{jm(\phi-\phi')}, & \rho \le \rho' \qquad \text{(11-179a)} \\[3mm] \displaystyle\sum_{m=-\infty}^{+\infty} J_m(\beta\rho')H_m^{(2)}(\beta\rho)e^{jm(\phi-\phi')}, & \rho \ge \rho' \qquad \text{(11-179b)} \end{cases}$$

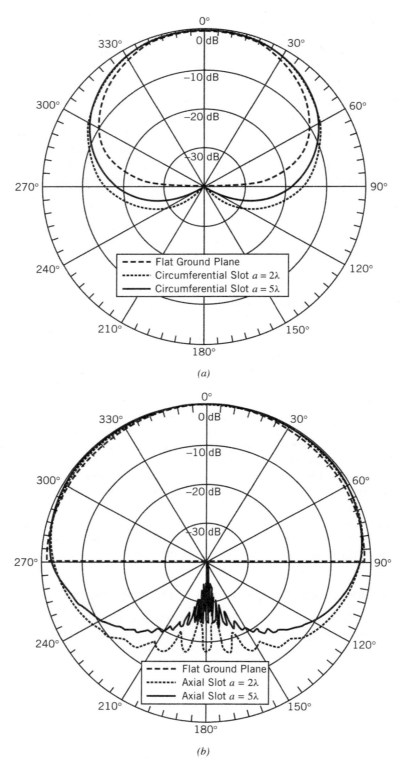

Figure 11-21 Normalized H-plane (for the circumferential aperture) and E-plane (for the axial aperture) amplitude pattern when mounted on a cylinder and flat ground plane. (*a*) H-plane (circumferential). (*b*) E-plane (axial).

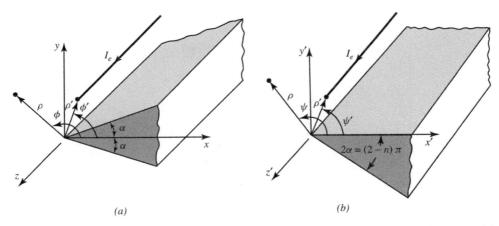

(a)   (b)

Figure 11-22 Electric line source near a two-dimensional conducting wedge. (*a*) Reference at bisector. (*b*) Reference at face.

The corresponding z component of the total scattered field must be chosen so that the sum of the two (incident plus scattered) along the faces of the wedge ($\phi = \alpha$ and $\phi = 2\pi - \alpha$) must vanish and simultaneously satisfy reciprocity (interchanging source and observation points). The ϕ variations must be represented by standing wave functions, since in the ϕ direction the waves bounce back and forth between the plates forming the wedge. It can be shown that expressions for the electric field that satisfy these conditions take the form

$$E_z^t = E_z^i + E_z^s$$

$$= \begin{cases} \displaystyle\sum_v c_v f(\rho') J_v(\beta\rho) \sin[v(\phi'-\alpha)]\sin[v(\phi-\alpha)], & \rho \leq \rho' \quad \text{(11-180a)} \\[2mm] \displaystyle\sum_v d_v g(\rho') H_v^{(2)}(\beta\rho) \sin[v(\phi'-\alpha)]\sin[v(\phi-\alpha)], & \rho \geq \rho' \quad \text{(11-180b)} \end{cases}$$

When $\rho = \rho'$, the two must be identical. Thus,

$$\sum_v c_v f(\rho') J_v(\beta\rho') \sin[v(\phi'-\alpha)]\sin[v(\phi-\alpha)]$$

$$= \sum_v d_v g(\rho') H_v^{(2)}(\beta\rho') \sin[v(\phi'-\alpha)]\sin[v(\phi-\alpha)] \qquad \text{(11-181)}$$

which is satisfied if

$$c_v f(\rho') J_v(\beta\rho') = d_v g(\rho') H_v^{(2)}(\beta\rho') \qquad \text{(11-181a)}$$

or

$$a_v = c_v = d_v \qquad \text{(11-181b)}$$

$$f(\rho') = H_v^{(2)}(\beta\rho') \qquad \text{(11-181c)}$$

$$g(\rho') = J_v(\beta\rho') \qquad \text{(11-181d)}$$

Therefore, the total electric field (11-180a) and (11-180b) can be written as

$$E_z^t = E_z^i + E_z^s = \begin{cases} \sum_v a_v J_v(\beta\rho) H_v^{(2)}(\beta\rho') \sin[v(\phi'-\alpha)] \\ \qquad \times \sin[v(\phi-\alpha)] \qquad\qquad \rho \leq \rho' \\ \sum_v a_v J_v(\beta\rho') H_v^{(2)}(\beta\rho) \sin[v(\phi'-\alpha)] \\ \qquad \times \sin[v(\phi-\alpha)] \qquad\qquad \rho \geq \rho' \end{cases}$$

(11-182a)

(11-182b)

It is evident from (11-182a) and (11-182b) that, when $\phi = \alpha$, the total tangential electric field vanishes. However, when $\phi = 2\pi - \alpha$, the electric field (11-182a) and (11-182b) vanishes when

$$\sin[v(\phi-\alpha)]_{\phi=2\pi-\alpha} = \sin[v(2\pi-2\alpha)] = \sin[2v(\pi-\alpha)] = 0 \qquad (11\text{-}183)$$

or

$$2v(\pi-\alpha) = \sin^{-1}(0) = m\pi$$

$$\boxed{v = \frac{m\pi}{2(\pi-\alpha)}, \quad m = 1, 2, 3\ldots} \qquad (11\text{-}183a)$$

Thus, in (11-182a) and (11-182b), the allowable values of v are those of (11-183a). The values of a_v depend on the type of source.

The magnetic field components can be obtained by using Maxwell's equations 11-86 through 11-86b, so that we can write that

$$H_\rho^t = -\frac{1}{j\omega\mu}\frac{1}{\rho}\frac{\partial E_z^t}{\partial \phi}$$

$$= -\frac{1}{j\omega\mu}\frac{1}{\rho} \begin{cases} \sum_v v a_v J_v(\beta\rho) H_v^{(2)}(\beta\rho') \\ \qquad \times \sin[v(\phi'-\alpha)]\cos[v(\phi-\alpha)] \qquad \rho \leq \rho' \\ \sum_v v a_v J_v(\beta\rho') H_v^{(2)}(\beta\rho) \\ \qquad \times \sin[v(\phi'-\alpha)]\cos[v(\phi-\alpha)] \qquad \rho \geq \rho' \end{cases}$$

(11-184a)

(11-184b)

$$H_\phi^t = \frac{1}{j\omega\mu}\frac{\partial E_z^t}{\partial \rho}$$

$$= \frac{\beta}{j\omega\mu} \begin{cases} \sum_v a_v J_v'(\beta\rho) H_v^{(2)}(\beta\rho') \\ \qquad \times \sin[v(\phi'-\alpha)]\sin[v(\phi-\alpha)] \qquad \rho \leq \rho' \\ \sum_v a_v J_v(\beta\rho') H_v^{(2)'}(\beta\rho) \\ \qquad \times \sin[v(\phi'-\alpha)]\sin[v(\phi-\alpha)] \qquad \rho \geq \rho' \end{cases}$$

(11-185a)

(11-185b)

where

$$' \equiv \frac{\partial}{\partial(\beta\rho)} \qquad (11\text{-}185c)$$

At the source, the current density is obtained using

$$\mathbf{J}_s = \hat{\mathbf{n}} \times \mathbf{H}^t = \hat{\mathbf{a}}_\rho \times (\hat{\mathbf{a}}_\rho H_\rho^t + \hat{\mathbf{a}}_\phi H_\phi^t)_{\rho=\rho_+', \, \rho_-'}$$

$$= \hat{\mathbf{a}}_z H_\phi^t |_{\rho=\rho_+', \rho_-'} = \hat{\mathbf{a}}_z \left[H_\phi^t(\rho_+') - H_\phi^t(\rho_-') \right]$$

$$= \hat{\mathbf{a}}_z \frac{\beta}{j\omega\mu} \sum_v a_v [J_v(\beta\rho') H_v^{(2)'}(\beta\rho') - H_v^{(2)}(\beta\rho') J_v'(\beta\rho')]$$

$$\times \sin[v(\phi' - \alpha)] \sin[v(\phi - \alpha)]$$

$$\mathbf{J}_s = \hat{\mathbf{a}}_z \frac{\beta}{j\omega\mu} \sum_v a_v (-j)[J_v(\beta\rho') Y_v'(\beta\rho') - J_v'(\beta\rho') Y_v(\beta\rho')]$$

$$\times \sin[v(\phi' - \alpha)] \sin[v(\phi - \alpha)] \tag{11-186}$$

which, by using the Wronskian of (11-95), reduces to

$$\mathbf{J}_s = -\hat{\mathbf{a}}_z \frac{2}{\pi\omega\mu\rho'} \sum_v a_v \sin[v(\phi' - \alpha)] \sin[v(\phi - \alpha)] \tag{11-186a}$$

Since the Fourier series for a current impulse of amplitude I_e located at $\rho = \rho'$ and $\phi = \phi'$ is [11]

$$J_z = \frac{I_e}{(\pi - \alpha)\rho'} \sum_v \sin[v(\phi' - \alpha)] \sin[v(\phi - \alpha)] \tag{11-187}$$

then comparing (11-186a) and (11-187) leads to

$$-\frac{2}{\pi\omega\mu} a_v = \frac{I_e}{\pi - \alpha} \Rightarrow \boxed{a_v = -\frac{\pi\omega\mu I_e}{2(\pi - \alpha)}} \tag{11-188}$$

A. Far-Zone Field   When the observations are made in the far zone ($\beta\rho \gg 1$, $\rho > \rho'$) the total electric field (11-182b) can be written, by replacing the Hankel function $H_v^{(2)}(\beta\rho)$ by its asymptotic expression (11-135a), as

$$E_z^t \stackrel{\beta\rho\to\infty}{\simeq} \sqrt{\frac{2j}{\pi\beta\rho}} e^{-j\beta\rho} \sum_v a_v j^v J_v(\beta\rho') \sin[v(\phi' - \alpha)] \sin[v(\phi - \alpha)]$$

$$\stackrel{\beta\rho\to\infty}{\simeq} -I_e \sqrt{\frac{\pi j}{2\beta}} \frac{\omega\mu}{\pi - \alpha} \frac{e^{-j\beta\rho}}{\sqrt{\rho}} \sum_v j^v J_v(\beta\rho') \sin[v(\phi' - \alpha)] \sin[v(\phi - \alpha)]$$

$$\boxed{E_z^t \stackrel{\beta\rho\to\infty}{\simeq} f_e(\rho) \sum_v j^v J_v(\beta\rho') \sin[v(\phi' - \alpha)] \sin[v(\phi - \alpha)]} \tag{11-189}$$

where

$$f_e(\rho) = -I_e \sqrt{\frac{\pi j}{2\beta}} \frac{\omega\mu}{\pi - \alpha} \frac{e^{-j\beta\rho}}{\sqrt{\rho}} \tag{11-189a}$$

Therefore, (11-189) represents the total electric field created in the far-zone region by an electric source of strength I_e located at ρ', ϕ'.

B. Plane Wave Scattering   When the source is placed at far distances ($\beta\rho' \gg 1$ and $\rho' > \rho$) and the observations are made at any point, the total electric field of (11-182a) can be written, by replacing the Hankel function $H_v^{(2)}(\beta\rho')$ by its asymptotic form (11-135a), as

$$E_z^t \overset{\beta\rho'\to\infty}{\simeq} -I_e\sqrt{\frac{\pi j}{2\beta}}\frac{\omega\mu}{\pi-\alpha}\frac{e^{-j\beta\rho'}}{\sqrt{\rho'}}\sum_v j^v J_v(\beta\rho)\sin[v(\phi'-\alpha)]\sin[v(\phi-\alpha)]$$

$$\overset{\beta\rho'\to\infty}{\simeq} g_e(\rho')\sum_v j^v J_v(\beta\rho)\sin[v(\phi'-\alpha)]\sin[v(\phi-\alpha)]$$

$$\boxed{E_z^t \overset{\beta\rho'\to\infty}{\simeq} E_0\sum_v j^v J_v(\beta\rho)\sin[v(\phi'-\alpha)]\sin[v(\phi-\alpha)]} \tag{11-190}$$

where

$$E_0 = g_e(\rho') = -I_e\sqrt{\frac{\pi j}{2\beta}}\frac{\omega\mu}{\pi-\alpha}\frac{e^{-j\beta\rho'}}{\sqrt{\rho'}} \tag{11-190a}$$

It is evident that (11-190) can also be obtained from (11-189) by reciprocity, that is, interchanging source and observation point. This is accomplished by interchanging ρ with ρ', and ϕ with ϕ', or for this problem simply by interchanging ρ and ρ' only.

Equation 11-190 also represents the total electric field of a TM^z uniform plane wave of strength E_0 incident at an angle ϕ' on a conducting wedge of interior angle 2α. When the wedge is a half-plane ($\alpha = 0$) (11-190) reduces to

$$E_z^t \overset{\beta\rho'\to\infty}{\underset{\alpha=0}{\simeq}} E_0\sum_v j^v J_v(\beta\rho)\sin(v\phi')\sin(v\phi) \tag{11-191}$$

which by using (11-183a) can also be expressed as

$$E_z^t \overset{\beta\rho'\to\infty}{\underset{\alpha=0}{\simeq}} E_0\sum_{m=1}^{\infty} j^{m/2} J_{m/2}(\beta\rho)\sin\left(\frac{m}{2}\phi'\right)\sin\left(\frac{m}{2}\phi\right) \tag{11-191a}$$

The normalized scattering patterns at a distance λ ($\rho = \lambda$) from the edge of the wedge formed when a plane wave is incident upon a wedge of $2\alpha = 0°$ ($n = 2$; half-plane) and $90°$ ($n = 1.5$) are shown in Figure 11-23.

11.6.2   Magnetic Line-Source Scattering by a Conducting Wedge: TE^z Polarization

When the line source of Figure 11-22a is magnetic of current I_m, the total magnetic field has only a z component, and it can be written, by referring to (11-182a) and (11-182b), as

$$\boxed{\begin{aligned} H_z^t &= H_z^i + H_z^s \\ &= \begin{cases} \sum_s b_s J_s(\beta\rho)H_s^{(2)}(\beta\rho')\cos[s(\phi'-\alpha)]\cos[s(\phi-\alpha)], & \rho \le \rho' \qquad (11\text{-}192a) \\ \sum_s b_s J_s(\beta\rho')H_s^{(2)}(\beta\rho)\cos[s(\phi'-\alpha)]\cos[s(\phi-\alpha)], & \rho \ge \rho' \qquad (11\text{-}192b) \end{cases} \end{aligned}}$$

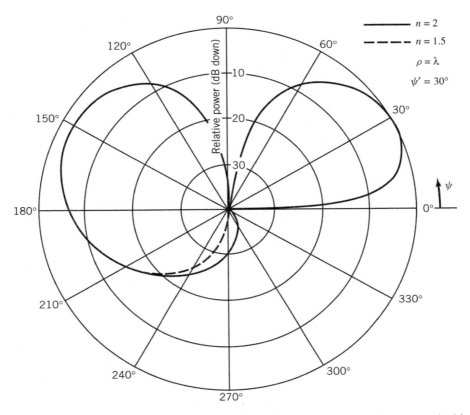

Figure 11-23 Normalized amplitude pattern of a TM^z (soft polarization) plane wave incident on a two-dimensional conducting wedge.

The allowable values of s are obtained by applying the boundary conditions, and the coefficients b_s are determined by the type of source.

To apply the boundary conditions, we first need to find the corresponding electric field components (especially the tangential components). This is accomplished by using Maxwell's equations 11-106 through 11-106b. Thus, the total radial component of the electric field can be written as

$$E_\rho^t = \frac{1}{j\omega\varepsilon} \frac{1}{\rho} \frac{\partial H_z^t}{\partial \phi}$$

$$= -\frac{1}{j\omega\varepsilon} \frac{1}{\rho} \begin{cases} \sum_s s b_s J_s(\beta\rho) H_s^{(2)}(\beta\rho') \\ \quad \times \cos[s(\phi'-\alpha)]\sin[s(\phi-\alpha)], \quad \rho \le \rho' \qquad (11\text{-}193a) \\ \sum_s s b_s J_s(\beta\rho') H_s^{(2)}(\beta\rho) \\ \quad \times \cos[s(\phi'-\alpha)]\sin[s(\phi-\alpha)], \quad \rho \ge \rho' \qquad (11\text{-}193b) \end{cases}$$

The boundary conditions that must be satisfied are

$$E_\rho^t(0 \le \rho, \rho' \le \infty, \phi = \alpha, 0 \le \phi' \le 2\pi)$$
$$= E_\rho^t(0 \le \rho, \rho' \le \infty, \phi = 2\pi - \alpha, 0 \le \phi' \le 2\pi) = 0 \qquad (11\text{-}194)$$

The first boundary condition of (11-194) is always satisfied regardless of the values of s. Applying the second boundary condition leads to

$$E_\rho^t(0 \leq \rho, \, \rho' \leq \infty, \, \phi = 2\pi - \alpha, \, 0 \leq \phi' \leq 2\pi) = 0$$

$$= -\frac{1}{j\omega\varepsilon\rho}\sum_s sb_s J_s(\beta\rho) H_s^{(2)}(\beta\rho') \cos[s(\phi' - \alpha)] \sin[2s(\pi - \alpha)]$$

$$= -\frac{1}{j\omega\varepsilon\rho}\sum_s sb_s J_s(\beta\rho') H_s^{(2)}(\beta\rho) \cos[s(\phi' - \alpha)] \sin[2s(\pi - \alpha)] \qquad (11\text{-}195)$$

which is satisfied provided

$$\sin[2s(\pi - \alpha)] = 0 \Rightarrow 2s(\pi - \alpha) = \sin^{-1}(0) = m\pi$$

$$\boxed{s = \frac{m\pi}{2(\pi - \alpha)}, \qquad m = 0, 1, 2, \ldots} \qquad (11\text{-}195\text{a})$$

Since the source is magnetic, the coefficients b_s take the form of

$$\boxed{b_s = \varepsilon_s \left[\frac{\pi\omega\varepsilon I_m}{4(\pi - \alpha)}\right]} \qquad (11\text{-}196)$$

where

$$\boxed{\varepsilon_s = \begin{cases} 1 & s = 0 \\ 2 & s \neq 0 \end{cases}} \qquad (11\text{-}196\text{a})$$

In the far zone ($\beta\rho \gg 1$), the total field of (11-192b) reduces, by replacing the Hankel function $H_s^{(2)}(\beta\rho)$ with its asymptotic form (11-135a), to

$$H_z^t \overset{\beta\rho\to\infty}{\simeq} I_m \sqrt{\frac{\pi j}{8\beta}} \frac{\omega\varepsilon}{\pi - \alpha} \frac{e^{-j\beta\rho}}{\sqrt{\rho}} \sum_s \varepsilon_s j^s J_s(\beta\rho') \cos[s(\phi' - \alpha)] \cos[s(\phi - \alpha)]$$

$$\boxed{H_z^t \overset{\beta\rho\to\infty}{\simeq} f_h(\rho)\sum_s \varepsilon_s j^s J_s(\beta\rho') \cos[s(\phi' - \alpha)] \cos[s(\phi - \alpha)]} \qquad (11\text{-}197)$$

where

$$f_h(\rho) = I_m \sqrt{\frac{\pi j}{8\beta}} \frac{\omega\varepsilon}{\pi - \alpha} \frac{e^{-j\beta\rho}}{\sqrt{\rho}} \qquad (11\text{-}197\text{a})$$

When the source is removed at far distances ($\beta\rho' \gg 1$ and $\rho' > \rho$), (11-192a) reduces to

$$H_z^t \overset{\beta\rho'\to\infty}{\simeq} g_h(\rho')\sum_s \varepsilon_s j^s J_s(\beta\rho) \cos[s(\phi' - \alpha)] \cos[s(\phi - \alpha)]$$

$$\boxed{H_z^t \overset{\beta\rho'\to\infty}{\simeq} H_0 \sum_s \varepsilon_s j^s J_s(\beta\rho) \cos[s(\phi' - \alpha)] \cos[s(\phi - \alpha)]} \qquad (11\text{-}198)$$

where

$$H_0 = g_h(\rho') = I_m \sqrt{\frac{\pi j}{8\beta}} \frac{\omega\varepsilon}{\pi-\alpha} \frac{e^{-j\beta\rho'}}{\sqrt{\rho'}} \qquad (11\text{-}198\text{a})$$

Equation 11-198) also represents the total magnetic field of a TE^z uniform plane wave of strength H_0 incident at an angle ϕ' on a conducting wedge of interior angle 2α. For a wedge with zero included angle (half-plane $\alpha = 0$), (11-198) reduces to

$$H_z^t \overset{\beta\rho'\to\infty}{\underset{\alpha=0}{\simeq}} H_0 \sum_s \varepsilon_s j^s J_s(\beta\rho) \cos(s\phi') \cos(s\phi) \qquad (11\text{-}199)$$

which, by using (11-195a), can also be expressed as

$$H_z^t \overset{\beta\rho'\to\infty}{\underset{\alpha=0}{\simeq}} H_0 \sum_{m=0}^{\infty} \varepsilon_{m/2} j^{m/2} J_{m/2}(\beta\rho) \cos\left(\frac{m}{2}\phi'\right) \cos\left(\frac{m}{2}\phi\right) \qquad (11\text{-}199\text{a})$$

The normalized scattering patterns at a distance λ ($\rho = \lambda$) from the edge of the wedge formed when a plane wave is incident upon a wedge of $2\alpha = 0°$ ($n = 2$; half-plane) and $90°$ ($n = 1.5$) are shown in Figure 11-24.

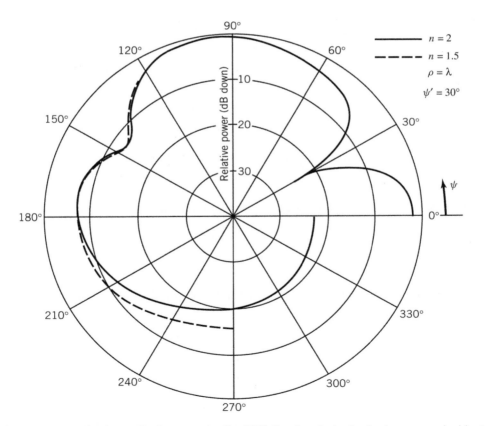

Figure 11-24 Normalized amplitude pattern of a TE^z (hard polarization) plane wave incident on a two-dimensional conducting wedge.

11.6.3 Electric and Magnetic Line-Source Scattering by a Conducting Wedge

The total electric field of an electric line source near a conducting wedge, as given by (11-182a) and (11-182b), and the total magnetic field of a magnetic line source near a conducting wedge, as given by (11-192a) and (11-192b), can both be represented by the same expression by adopting the coordinate system (x', y', z') of Figure 11-22b, instead of the (x, y, z), which is referenced to the side of the wedge that is illuminated by the source. This is usually more convenient because similar forms of the expression can represent either polarization. Also the interior angle of the wedge will be represented by

$$2\alpha = (2 - n)\pi \Rightarrow n = 2 - \frac{2\alpha}{\pi} \tag{11-200}$$

Thus, a given value of n represents a wedge with a specific included angle: values of $n > 1$ represent wedges with included angles less than 180° (referred to as *exterior* wedges) and values of $n < 1$ represent wedges with included angles greater than 180° (referred to as *interior* wedges). Thus the allowable values of v, as given by (11-183a), and those of s, as given by (11-195a), can now be represented by

$$v = \frac{m\pi}{2(\pi - \alpha)}\bigg|_{\alpha = (1 - n/2)\pi} = \frac{m}{n}, \quad m = 1, 2, 3, \dots \tag{11-201a}$$

$$s = \frac{m\pi}{2(\pi - \alpha)}\bigg|_{\alpha = (1 - n/2)\pi} = \frac{m}{n}, \quad m = 0, 1, 2, \dots \tag{11-201b}$$

In addition, the amplitude coefficients a_v, as given by (11-188), and b_s, as given by (11-196), can now be expressed as

$$a_v = -\frac{\pi \omega \mu I_e}{2(\pi - \alpha)}\bigg|_{\alpha = (1 - n/2)\pi} = -\frac{\omega \mu I_e}{2}\left(\frac{2}{n}\right) \tag{11-202a}$$

$$b_s = \varepsilon_s \left[\frac{\pi \omega \varepsilon I_m}{4(\pi - \alpha)}\right]_{\alpha = (1 - n/2)\pi} = \frac{\omega \varepsilon I_m}{2}\left(\frac{\varepsilon_s}{n}\right) \tag{11-202b}$$

Using the new coordinate system (x', y', z') of Figure 11-22b, we can write that

$$\phi' = \psi' + \alpha \tag{11-203a}$$

$$\phi = \psi + \alpha \tag{11-203b}$$

Therefore, the sine functions of (11-182a) and (11-182b) and the cosine functions of (11-192a) and (11-192b) can be written as

$$\sin[v(\phi' - \alpha)] \sin[v(\phi - \alpha)] = \sin\left[\frac{m}{n}(\psi' + \alpha - \alpha)\right] \sin\left[\frac{m}{n}(\psi + \alpha - \alpha)\right]$$

$$= \sin\left(\frac{m}{n}\psi'\right) \sin\left(\frac{m}{n}\psi\right)$$

$$= \frac{1}{2}\left\{\cos\left[\frac{m}{n}(\psi - \psi')\right] - \cos\left[\frac{m}{n}(\psi + \psi')\right]\right\} \tag{11-204a}$$

$$\cos[s(\phi' - \alpha)] \cos[s(\phi - \alpha)] = \cos\left[\frac{m}{n}(\psi' + \alpha - \alpha)\right] \cos\left[\frac{m}{n}(\psi + \alpha - \alpha)\right]$$

$$= \cos\left(\frac{m}{n}\psi'\right) \cos\left(\frac{m}{n}\psi\right)$$

$$= \frac{1}{2}\left\{\cos\left[\frac{m}{n}(\psi - \psi')\right] - \cos\left[\frac{m}{n}(\psi + \psi')\right]\right\} \qquad (11\text{-}204\text{b})$$

Using all these new notations, we can write for the TM^z polarization the total electric field of (11-182a) and (11-182b) and for the TE^z polarization the total magnetic field of (11-192a) and (11-192b) as

<u>TM^z</u>

$$E_z^t = -\frac{\omega\mu I_e}{4}\frac{1}{n}\begin{cases} \displaystyle\sum_{m=0,1,\dots}^{\infty} 2J_{m/n}(\beta\rho)H_{m/n}^{(2)}(\beta\rho') \\[2mm] \quad\times\left\{\cos\left[\frac{m}{n}(\psi - \psi')\right] - \cos\left[\frac{m}{n}(\psi + \psi')\right]\right\}, \quad \rho \le \rho' \qquad (11\text{-}205\text{a}) \\[4mm] \displaystyle\sum_{m=0,1,\dots}^{\infty} 2J_{m/n}(\beta\rho')H_{m/n}^{(2)}(\beta\rho) \\[2mm] \quad\times\left\{\cos\left[\frac{m}{n}(\psi - \psi')\right] - \cos\left[\frac{m}{n}(\psi + \psi')\right]\right\}, \quad \rho \ge \rho' \qquad (11\text{-}205\text{b}) \end{cases}$$

<u>TE^z</u>

$$H_z^t = \frac{\omega\varepsilon I_m}{4}\frac{1}{n}\begin{cases} \displaystyle\sum_{m=0,1,\dots}^{\infty} \varepsilon_m J_{m/n}(\beta\rho)H_{m/n}^{(2)}(\beta\rho') \\[2mm] \quad\times\left\{\cos\left[\frac{m}{n}(\psi - \psi')\right] + \cos\left[\frac{m}{n}(\psi + \psi')\right]\right\}, \quad \rho \le \rho' \qquad (11\text{-}206\text{a}) \\[4mm] \displaystyle\sum_{m=0,1,\dots}^{\infty} \varepsilon_m J_{m/n}(\beta\rho')H_{m/n}^{(2)}(\beta\rho) \\[2mm] \quad\times\left\{\cos\left[\frac{m}{n}(\psi - \psi')\right] + \cos\left[\frac{m}{n}(\psi + \psi')\right]\right\}, \quad \rho \ge \rho' \qquad (11\text{-}206\text{b}) \end{cases}$$

To make the summations in (11-205a) through (11-206b) uniform, the summations of (11-205a) and (11-205b) are noted to begin with $m = 0$, even though the allowable values of m as given by (11-201a) begin with $m = 1$. However, it should be noted that $m = 0$ in (11-205a) and (11-205b) does not contribute, and the expressions are correct as stated.

It is apparent, by comparing (11-205a) and (11-205b) with (11-206a) and (11-206b), that they are of similar forms. Therefore, we can write both as

$$E_z^t = -\frac{\omega\mu I_e}{4}G(\rho, \rho', \psi, \psi', n) \quad \text{for TM}^z \qquad (11\text{-}207\text{a})$$

$$H_z^t = +\frac{\omega\varepsilon I_m}{4}G(\rho, \rho', \psi, \psi', n) \quad \text{for TE}^z \qquad (11\text{-}207\text{b})$$

where

$$G(\rho, \rho', \psi - \psi', n)$$

$$= \frac{1}{n} \begin{cases} \sum_{m=0,1,\dots}^{\infty} \varepsilon_m J_{m/n}(\beta\rho) H_{m/n}^{(2)}(\beta\rho') \\ \qquad \left\{ \cos\left[\frac{m}{n}(\psi - \psi')\right] \pm \cos\left[\frac{m}{n}(\psi + \psi')\right] \right\}, \qquad \rho \leq \rho' \\[2em] \sum_{m=0,1,\dots}^{\infty} \varepsilon_m J_{m/n}(\beta\rho') H_{m/n}^{(2)}(\beta\rho) \\ \qquad \left\{ \cos\left[\frac{m}{n}(\psi - \psi')\right] \pm \cos\left[\frac{m}{n}(\psi + \psi')\right] \right\}, \qquad \rho \geq \rho' \end{cases}$$

(11-208a)

(11-208b)

$$\varepsilon_m = \begin{cases} 1 & m = 0 \\ 2 & m \neq 0 \end{cases}$$

(11-208c)

The plus ($+$) sign between the cosine terms is used for the TE^z polarization and the minus ($-$) sign is used for the TM^z polarization. Again, note that the $m = 0$ terms do not contribute anything for the TM^z polarization.

The forms of (11-207a) through (11-208b) are those usually utilized in the geometrical theory of diffraction (GTD) [40–44] where $G(\rho, \rho', \psi, \psi', n)$ is usually referred to as the *Green's function*. Since the summations in (11-208a) and (11-208b) are poorly convergent when the arguments of the Bessel and/or Hankel functions are large, asymptotic forms of them will be derived in Chapter 13 that are much more computationally efficient. The various terms of the asymptotic forms will be associated with incident and reflected geometrical optics and diffracted fields. It is also convenient in diffraction theory to refer to the TM^z polarization as the *soft* polarization; the TE^z is referred to as the *hard* polarization. This is a convenient designation adopted from acoustics.

11.7   SPHERICAL WAVE ORTHOGONALITIES, TRANSFORMATIONS, AND THEOREMS

When dealing with scattering from structures whose geometry best conforms to spherical coordinates, it is often most convenient to transform wave functions (such as plane waves) from one coordinate system to another. This was done in Section 11.4 where uniform plane wave functions in rectilinear form were transformed and represented by cylindrical wave functions. This allowed convenient examination of the scattering of plane waves by cylindrical structures of circular and wedge cross sections. In addition, certain theorems concerning cylindrical wave functions were introduced, which were helpful in analyzing the scattering by cylindrical structures of circular cross sections of waves emanating from line sources.

In this section we want to introduce some orthogonality relationships, wave transformations, and theorems that are very convenient for examining scattering of plane waves from spherical structures and waves emanating from finite sources placed in the vicinity of spherical scatterers. First of all, let us examine radiation from a finite source radiating in an unbounded medium.

11.7.1   Vertical Dipole Spherical Wave Radiation

There are many sources of spherical wave radiation. In fact, almost all sources used in practice are considered to excite spherical waves. One of the most prominent is that of a finite length wire whose total radiation can be obtained as a superposition of radiation from a very small linear current element of length $\Delta\ell$ and constant electric current $\mathbf{I}_e = \hat{\mathbf{a}}_z I_e$. This is usually referred to as an infinitesimal dipole [1]. The radiation of other sources can be obtained by knowing the radiation from an *infinitesimal dipole*. Therefore, it is important that we briefly examine the radiation from such a source.

It is usually most convenient to place the linear element at the origin of the coordinate system and have its length and current flow along the z axis, as shown in Figure 11-25a. To find the fields radiated by this source, we resort to the techniques of Chapter 6, Sections 6.4 and 6.6, where we first specify the currents I_e and I_m of the source. Then we find the potentials $\mathbf{A}$ and $\mathbf{F}$ [using (6-97a) and (6-97b)], and determine the radiated $\mathbf{E}$ and $\mathbf{H}$ [using (6-34) and (6-35)].

Following such a procedure, the electric and magnetic fields radiated by the infinitesimal dipole of Figure 11-25 were derived in Example 6-3. In terms of spherical wave functions, the vector potential $\mathbf{A}$ can also be written as

$$\mathbf{A} = \hat{\mathbf{a}}_z \frac{\mu I_e \Delta\ell}{4\pi} \frac{e^{-j\beta r}}{r} = -\hat{\mathbf{a}}_z j \frac{\mu \beta I_e \Delta\ell}{4\pi} h_0^{(2)}(\beta r) \tag{11-209}$$

where $h_0^{(2)}(\beta r)$ is the spherical Hankel function of order zero, given by

$$h_0^{(2)}(\beta r) = \frac{e^{-j\beta r}}{-j\beta r} \tag{11-209a}$$

For the dual problem of the linear magnetic current element of current I_m, the vector potential function $\mathbf{F}$ takes the form of

$$\mathbf{F} = \hat{\mathbf{a}}_z \frac{\varepsilon I_m \Delta\ell}{4\pi} \frac{e^{-j\beta r}}{r} = -\hat{\mathbf{a}}_z j \frac{\varepsilon \beta I_m \Delta\ell}{4\pi} h_0^{(2)}(\beta r) \tag{11-210}$$

If the source is removed from the origin, as shown in Figure 11-25b, then the potentials $\mathbf{A}$ and $\mathbf{F}$ of (11-209) and (11-210) take the form of

$$\mathbf{A} = -\hat{\mathbf{a}}_z j \frac{\mu \beta I_e \Delta\ell}{4\pi} h_0^{(2)}(\beta |\mathbf{r} - \mathbf{r}'|) \tag{11-211a}$$

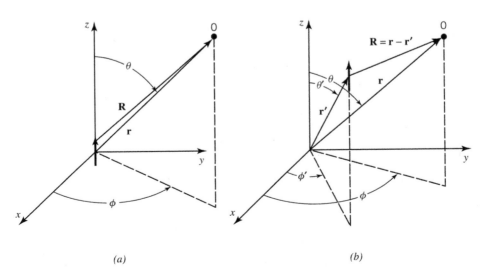

(a)   (b)

Figure 11-25   Geometry and coordinate system for vertical dipole radiation. (*a*) At origin. (*b*) Offset.

$$\mathbf{F} = -\hat{\mathbf{a}}_z j \frac{\varepsilon \beta I_m \Delta \ell}{4\pi} h_0^{(2)}(\beta |\mathbf{r} - \mathbf{r}'|) \tag{11-211b}$$

where

$$h_0^{(2)}(\beta |\mathbf{r} - \mathbf{r}'|) = \frac{e^{-j\beta |\mathbf{r} - \mathbf{r}'|}}{-j\beta |\mathbf{r} - \mathbf{r}'|} \tag{11-211c}$$

11.7.2 Orthogonality Relationships

When solving electromagnetic wave problems dealing with spherical structures (either waveguides, cavities, or scatterers), the θ variations are represented, as illustrated in Chapters 3 (Section 3.4.3) and 10, by Legendre polynomials $P_n(\cos\theta)$ and associated Legendre functions $P_n^m(\cos\theta)$ (see Appendix V).

The Legendre polynomials $P_n(\cos\theta)$ are often called *zonal harmonics* [45, 46], and they form a complete orthogonal set in the interval $0 \leq \theta \leq \pi$. Therefore, in this interval any arbitrary wave function can be represented by a series of Legendre polynomials. This is similar to the representation of any periodic function by a series of sines and cosines (Fourier series), since Legendre polynomials are very similar in form to cosinusoidal functions. In addition, the products of associated Legendre functions $P_n^m(\cos\theta)$ with sines and cosines $[P_n^m(\cos\theta)\cos(m\phi)$ and $P_n^m(\cos\theta)\sin(m\phi)]$ are often referred to as *tesseral harmonics* [45, 46], and they form a complete orthogonal set on the surface of a sphere. Therefore, any wave function that is defined over a sphere can be expressed by a series of tesseral harmonics.

Some of the most important and necessary orthogonality relationships that are necessary to solve wave scattering by spheres will be stated here. The interested reader is referred to [11, 45–48] for more details and derivations.

In the interval $0 \leq \theta \leq \pi$, the integral of the product of Legendre polynomials is equal to

$$\int_0^\pi P_n(\cos\theta)P_m(\cos\theta)\sin\theta \, d\theta = \begin{cases} 0, & n \neq m \tag{11-212a} \\ \dfrac{2}{2n+1}, & n = m \tag{11-212b} \end{cases}$$

Any function $f(\theta)$ defined in the interval of $0 \leq \theta \leq \pi$ can be represented by a series of Legendre polynomials

$$f(\theta) = \sum_{n=0}^\infty a_n P_n(\cos\theta), \qquad 0 \leq \theta \leq \pi \tag{11-213}$$

where

$$a_n = \frac{2n+1}{2} \int_0^\pi f(\theta)P_n(\cos\theta)\sin\theta \, d\theta \tag{11-213a}$$

which is known as the *Fourier–Legendre* series.

Defining the tesseral harmonics by

$$T_{mn}^e(\theta,\phi) = P_n^m(\cos\theta)\cos(m\phi) \tag{11-214a}$$

$$T_{mn}^o(\theta,\phi) = P_n^m(\cos\theta)\sin(m\phi) \tag{11-214b}$$

and because

$$\int_0^{2\pi} \sin(p\phi)\sin(q\phi)d\phi = \int_0^{2\pi} \cos(p\phi)\cos(q\phi)d\phi$$

$$= \begin{cases} 0, & p \neq q \tag{11-215a} \\ \pi, & p = q \neq 0 \tag{11-215b} \end{cases}$$

it can be shown that

$$\int_0^{2\pi} \left[\int_0^{\pi} T_{mn}^e(\theta, \phi) T_{pq}^0(\theta, \phi) \sin\theta \, d\theta \right] d\phi = 0 \tag{11-216a}$$

$$\int_0^{2\pi} \left[\int_0^{\pi} T_{mn}^i(\theta, \phi) T_{pq}^i(\theta, \phi) \sin\theta \, d\theta \right] d\phi = 0 \quad \begin{array}{l} mn \neq pq \\ i = e \text{ or } 0 \end{array} \tag{11-216b}$$

$$\int_0^{2\pi} \left[\int_0^{\pi} \left[T_{mn}^i(\theta, \phi) \right]^2 \sin\theta \, d\theta \right] d\phi$$

$$= \begin{cases} \dfrac{4\pi}{2n+1}, & m = 0, \quad i = e \tag{11-216c} \\[3mm] \dfrac{2\pi}{2n+1} \dfrac{(n+m)!}{(n-m)!}, & m \neq 0, \quad i = e \text{ or } 0 \tag{11-216d} \end{cases}$$

11.7.3 Wave Transformations and Theorems

As was done for cylindrical wave functions in Section 11.4, in scattering it is often most convenient to express wave functions in one coordinate system in terms of wave functions of another coordinate system. The same applies to wave scattering by spherical structures. Therefore, it is convenient for plane wave scattering by spherical geometries to express the plane waves, which are most conveniently written in rectilinear form, in terms of spherical wave functions.

To demonstrate that, let us assume that a uniform plane wave is traveling along the $+z$ direction, as shown in Figure 11-26, and it can be written as

$$\mathbf{E}^+ = \hat{\mathbf{a}}_x E_x^+ = \hat{\mathbf{a}}_x E_0 e^{-j\beta z} = \hat{\mathbf{a}}_x e^{-j\beta z} \tag{11-217}$$

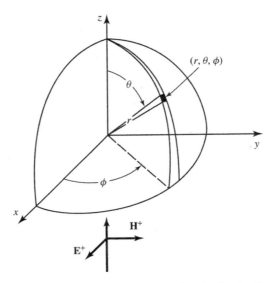

Figure 11-26 Uniform plane wave traveling in the $+z$ direction.

The plane wave can be represented by an infinite sum of spherical wave functions of the form

$$E_x^+ = e^{-j\beta z} = e^{-j\beta r\cos\theta} = \sum_{n=0}^{\infty} a_n j_n(\beta r) P_n(\cos\theta) \tag{11-217a}$$

since it must be independent of ϕ and be finite at the origin ($r=0$). The next step is to determine the amplitude coefficients a_n. This can be accomplished as follows.

Multiplying both sides of (11-217a) by $P_m(\cos\theta)\sin\theta$ and integrating in θ from 0 to π, we have that

$$\int_0^\pi e^{-j\beta r\cos\theta} P_m(\cos\theta)\sin\theta\, d\theta = \int_0^\pi \left[\sum_{n=0}^{\infty} a_n j_n(\beta r) P_n(\cos\theta) P_m(\cos\theta)\sin\theta\right] d\theta \tag{11-218}$$

Interchanging integration and summation, we have that

$$\int_0^\pi e^{-j\beta r\cos\theta} P_m(\cos\theta)\sin\theta\, d\theta = \sum_{n=0}^{\infty} a_n j_n(\beta r)\int_0^\pi P_n(\cos\theta) P_m(\cos\theta)\sin\theta\, d\theta s \tag{11-218a}$$

Using the orthogonality (11-212b) reduces (11-218a) to

$$\int_0^\pi e^{-j\beta r\cos\theta} P_m(\cos\theta)\sin\theta\, d\theta = \frac{2a_m}{2m+1} j_m(\beta r) \tag{11-219}$$

Since the integral of the left side of (11-219) is equal to

$$\int_0^\pi e^{-j\beta r\cos\theta} P_m(\cos\theta)\sin\theta\, d\theta = 2j^{-m} j_m(\beta r) \tag{11-219a}$$

equating (11-219) and (11-219a) leads to

$$\frac{2a_m}{2m+1} j_m(\beta r) = 2j^{-m} j_m(\beta r) \quad\Rightarrow\quad a_m = j^{-m}(2m+1) \tag{11-220}$$

Thus, (11-217a) reduces to

$$\boxed{E_x^+ = e^{-j\beta z} = e^{-j\beta r\cos\theta} = \sum_{n=0}^{\infty} a_n j_n(\beta r) P_n(\cos\theta)} \tag{11-221}$$

where

$$\boxed{a_n = j^{-n}(2n+1)} \tag{11-221a}$$

In a similar manner, it can be shown that

$$\boxed{E_x^- = e^{+j\beta z} = e^{+j\beta r\cos\theta} = \sum_{n=0}^{\infty} b_n j_n(\beta r) P_n(\cos\theta)} \tag{11-222}$$

where

$$\boxed{b_n = j^n(2n+1)} \tag{11-222a}$$

When dealing with spherical wave scattering of waves generated by linear dipole radiators, it is convenient to express their radiation, determined using (11-211a) through (11-211c), in terms of spherical wave functions. This can be accomplished using the *addition theorem* [11] of spherical wave functions, which states that (11-211c) can be expressed by referring to the geometry of Figure 11-25*b* as

$$
h_0^{(2)}(\beta |\mathbf{r} - \mathbf{r}'|) =
\begin{cases}
\displaystyle\sum_{n=0}^{\infty}(2n+1)h_n^{(2)}(\beta r')j_n(\beta r)P_n(\cos\xi), & r < r' \quad \text{(11-223a)} \\[3mm]
\displaystyle\sum_{n=0}^{\infty}(2n+1)h_n^{(2)}(\beta r)j_n(\beta r')P_n(\cos\xi), & r > r' \quad \text{(11-223b)}
\end{cases}
$$

where

$$
\cos\xi = \cos\theta\cos\theta' + \sin\theta\sin\theta'\cos(\phi - \phi') \tag{11-223c}
$$

Similarly,

$$
h_0^{(1)}(\beta |\mathbf{r} - \mathbf{r}'|) =
\begin{cases}
\displaystyle\sum_{n=0}^{\infty}(2n+1)h_n^{(1)}(\beta r')j_n(\beta r)P_n(\cos\xi), & r < r' \quad \text{(11-224a)} \\[3mm]
\displaystyle\sum_{n=0}^{\infty}(2n+1)h_n^{(1)}(\beta r)j_n(\beta r')P_n(\cos\xi), & r > r' \quad \text{(11-224b)}
\end{cases}
$$

11.8   SCATTERING BY A SPHERE

Plane wave scattering by a sphere is a classic problem in scattering and has been addressed by many authors [11, 14, 20, 45, 46, 49–54]. Here we will outline one that parallels that of [11]. Scattering by other sources of excitation such as dipoles, both radial and tangential to the surface of the sphere, has also been addressed. Because of its symmetry, a PEC sphere is often used as a reference scatterer to calibrate and measure the scattering properties (such as RCS) of other radar targets (missiles, airplanes, helicopters, etc.). A set of five aluminum RCS calibration spheres is shown in Figure 11-27. Apart from its symmetry, the RCS of a sphere can be calculated, which allows the user to set on the display, during measurements, a baseline to compare the RCS of other targets to that of the sphere. It is therefore very important that we know and understand the scattering characteristics of a sphere. We will consider both PEC and lossy dielectric spheres.

11.8.1   Perfect Electric Conducting (PEC) Sphere

Let us assume that the electric field of a uniform plane wave is polarized in the x direction, and it is traveling along the z axis, as shown in Figure 11-28. The electric field of the incident wave upon a PEC sphere can then be expressed as

$$
\mathbf{E}^i = \hat{\mathbf{a}}_x E_x^i = \hat{\mathbf{a}}_x E_0 e^{-j\beta z} = \hat{\mathbf{a}}_x E_0 e^{-j\beta r\cos\theta} \tag{11-225}
$$

Using the transformation (II-12), the x component of (11-225) can be transformed in spherical components to

$$
\mathbf{E}^i = \hat{\mathbf{a}}_r E_r^i + \hat{\mathbf{a}}_\theta E_\theta^i + \hat{\mathbf{a}}_\phi E_\phi^i \tag{11-226}
$$

Figure 11-27   A set of five aluminum RCS calibration spheres.

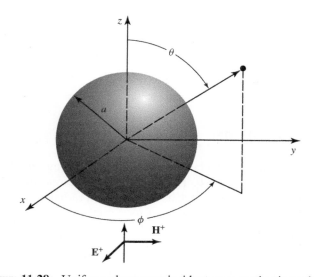

Figure 11-28   Uniform plane wave incident on a conducting sphere.

where

$$E_r^i = E_x^i \sin\theta \cos\phi = E_0 \sin\theta \cos\phi e^{-j\beta r\cos\theta} = E_0 \frac{\cos\phi}{j\beta r} \frac{\partial}{\partial\theta}(e^{-j\beta r\cos\theta}) \tag{11-226a}$$

$$E_\theta^i = E_x^i \cos\theta \cos\phi = E_0 \cos\theta \cos\phi\, e^{-j\beta r\cos\theta} \tag{11-226b}$$

$$E_\phi^i = -E_x^i \sin\phi = -E_0 \sin\phi\, e^{-j\beta r\cos\theta} \tag{11-226c}$$

Each of the spherical components of the preceding incident electric field can be expressed using the transformation (11-221) and (11-221a) as

$$E_r^i = E_0 \frac{\cos\phi}{j\beta r} \sum_{n=0}^{\infty} j^{-n}(2n+1) j_n(\beta r) \frac{\partial}{\partial\theta}[P_n(\cos\theta)] \tag{11-227a}$$

$$E_\theta^i = E_0 \cos\theta \cos\phi \sum_{n=0}^{\infty} j^{-n} (2n+1) j_n(\beta r) P_n(\cos\theta) \qquad (11\text{-}227b)$$

$$E_\phi^i = -E_0 \sin\phi \sum_{n=0}^{\infty} j^{-n} (2n+1) j_n(\beta r) P_n(\cos\theta) \qquad (11\text{-}227c)$$

Since according to (10-31),

$$j_n(\beta r) = \frac{1}{\beta r} \hat{J}_n(\beta r) \qquad (11\text{-}228a)$$

and

$$\frac{\partial P_n}{\partial\theta} = P_n^1(\cos\theta) \qquad (11\text{-}228b)$$

$$P_0^1 = 0 \qquad (11\text{-}228c)$$

we can rewrite (11-227a) through (11-227c) as

$$E_r^i = -jE_0 \frac{\cos\phi}{(\beta r)^2} \sum_{n=1}^{\infty} j^{-n} (2n+1) \hat{J}_n(\beta r) P_n^1(\cos\theta) \qquad (11\text{-}229a)$$

$$E_\theta^i = E_0 \frac{\cos\theta \cos\phi}{\beta r} \sum_{n=0}^{\infty} j^{-n} (2n+1) \hat{J}_n(\beta r) P_n^0(\cos\theta) \qquad (11\text{-}229b)$$

$$E_\phi^i = -E_0 \frac{\sin\phi}{\beta r} \sum_{n=0}^{\infty} j^{-n} (2n+1) \hat{J}_n(\beta r) P_n^0(\cos\theta) \qquad (11\text{-}229c)$$

The incident and scattered fields by the sphere can be expressed as a superposition of TE^r and TM^r as outlined, respectively, in Sections 10.2.4 and 10.2.5. The TE^r fields are constructed by letting the vector potentials $\mathbf{A}$ and $\mathbf{F}$ be equal to $\mathbf{A} = 0$ and $\mathbf{F} = \hat{\mathbf{a}}_r F_r(r,\theta,\phi)$. The TM^r fields are constructed when $\mathbf{A} = \hat{\mathbf{a}}_r A_r(r,\theta,\phi)$ and $\mathbf{F} = 0$. For example, the incident radial electric field component E_r^i can be obtained by expressing it in terms of TM^r modes or A_r^i. Thus, using A_r^i, we can write, according to (10-27a), the incident electric field as

$$E_r^i = \frac{1}{j\omega\mu\varepsilon} \left(\frac{\partial^2}{\partial r^2} + \beta^2 \right) A_r^i \qquad (11\text{-}230)$$

Equating (11-230) to (11-229a), it can be shown that A_r^i takes the form

$$\boxed{A_r^i = E_0 \frac{\cos\phi}{\omega} \sum_{n=1}^{\infty} a_n \hat{J}_n(\beta r) P_n^1(\cos\theta)} \qquad (11\text{-}231)$$

where

$$\boxed{a_n = j^{-n} \frac{(2n+1)}{n(n+1)}} \qquad (11\text{-}231a)$$

This potential component A_r^i will give the correct value of E_r^i, and it will lead to $H_r^i = 0$.

The correct expression for the radial component of the incident magnetic field can be obtained by following a similar procedure but using TE^r modes or F_r^i of Section 10.2.4. This allows us to show that

$$F_r^i = E_0 \frac{\sin\phi}{\omega\eta} \sum_{n=1}^{\infty} a_n \hat{J}_n(\beta r) P_n^1(\cos\theta) \qquad (11\text{-}232)$$

where a_n is given by (11-231a). This expression leads to the correct H_r^i and to $E_r^i = 0$. Therefore, the sum of (11-231) and (11-232) will give the correct E_r^i, H_r^i and the remaining electric and magnetic components.

Since the incident electric and magnetic field components of a uniform plane wave can be represented by TM^r and TE^r modes, that can be constructed using the potentials A_r^i and F_r^i of (11-231) and (11-232), the scattered fields can also be represented by TM^r and TE^r modes and be constructed using potentials A_r^s and F_r^s. The forms of A_r^s and F_r^s are similar to those of A_r^i and F_r^i of (11-231) and (11-232), and we can represent them by

$$A_r^s = E_0 \frac{\cos\phi}{\omega} \sum_{n=1}^{\infty} b_n \hat{H}_n^{(2)}(\beta r) P_n^1(\cos\theta) \qquad (11\text{-}233a)$$

$$F_r^s = E_0 \frac{\sin\phi}{\omega\eta} \sum_{n=1}^{\infty} c_n \hat{H}_n^{(2)}(\beta r) P_n^1(\cos\theta) \qquad (11\text{-}233b)$$

where the coefficients b_n and c_n will be found using the appropriate boundary conditions. In (11-233a) and (11-233b), the spherical Hankel function of the second kind $\hat{H}_n^{(2)}(\beta r)$ has replaced the spherical Bessel function $\hat{J}_n(\beta r)$ in (11-231) and (11-232) in order to represent outward traveling waves. Thus, all the components of the total field, incident plus scattered, can be found using the sum of (10-23) through (10-24c) and (10-27) through (10-28c), or

$$E_r^t = \frac{1}{j\omega\mu\varepsilon}\left(\frac{\partial^2}{\partial r^2} + \beta^2\right) A_r^t \qquad (11\text{-}234a)$$

$$E_\theta^t = \frac{1}{j\omega\mu\varepsilon}\frac{1}{r}\frac{\partial^2 A_r^t}{\partial r \partial\theta} - \frac{1}{\varepsilon}\frac{1}{r\sin\theta}\frac{\partial F_r^t}{\partial\phi} \qquad (11\text{-}234b)$$

$$E_\phi^t = \frac{1}{j\omega\mu\varepsilon}\frac{1}{r\sin\theta}\frac{\partial^2 A_r^t}{\partial r \partial\phi} + \frac{1}{\varepsilon}\frac{1}{r}\frac{\partial F_r^t}{\partial\theta} \qquad (11\text{-}234c)$$

$$H_r^t = \frac{1}{j\omega\mu\varepsilon}\left(\frac{\partial^2}{\partial r^2} + \beta^2\right) F_r^t \qquad (11\text{-}234d)$$

$$H_\theta^t = \frac{1}{\mu}\frac{1}{r\sin\theta}\frac{\partial A_r^t}{\partial\phi} + \frac{1}{j\omega\mu\varepsilon}\frac{1}{r}\frac{\partial^2 F_r^t}{\partial r \partial\theta} \qquad (11\text{-}234e)$$

$$H_\phi^t = -\frac{1}{\mu}\frac{1}{r}\frac{\partial A_r^t}{\partial\theta} + \frac{1}{j\omega\mu\varepsilon}\frac{1}{r\sin\theta}\frac{\partial^2 F_r^t}{\partial r \partial\phi} \qquad (11\text{-}234f)$$

where A_r^t and F_r^t are each equal to the sum of (11-231), (11-232), (11-233a), and (11-233b), or

$$A_r^t = A_r^i + A_r^s = E_0 \frac{\cos\phi}{\omega} \sum_{n=1}^{\infty} \left[a_n \hat{J}_n(\beta r) + b_n \hat{J}_n^{(2)}(\beta r) \right] P_n^1(\cos\theta) \qquad (11\text{-}235a)$$

$$F_r^t = F_r^i + F_r^s = E_0 \frac{\sin\phi}{\omega\eta} \sum_{n-1}^{\infty} \left[a_n \hat{J}_n(\beta r) + c_n \hat{J}_n^{(2)}(\beta r) \right] P_n^1(\cos\theta) \qquad (11\text{-}235b)$$

$$a_n = j^{-n} \frac{2n+1}{n(n+1)} \qquad (11\text{-}235c)$$

To determine the coefficients b_n and c_n, the boundary conditions

$$E_\theta^t (r=a, 0\le\theta\le\pi, 0\le\phi\le 2\pi)=0 \qquad (11\text{-}236a)$$

$$E_\phi^t (r=a, 0\le\theta\le\pi, 0\le\phi\le 2\pi)=0 \qquad (11\text{-}236b)$$

must be applied. Using (11-235a) and (11-235b), we can write (11-234b) as

$$E_\theta^t = +j\frac{E_0}{\omega\mu\varepsilon r \sin\theta} \left\{ \frac{\beta}{\omega} \cos\phi \sum_{n=1}^{\infty} \left[a_n \hat{J}_n'(\beta r) + b_n \hat{H}_n^{(2)'}(\beta r) \right] P_n^{1'}(\cos\theta) \right\}$$

$$- \frac{E_0}{\varepsilon r \sin\theta} \left\{ \frac{1}{\omega\eta} \cos\phi \sum_{n=1}^{\infty} \left[a_n \hat{J}_n(\beta r) + c_n \hat{H}_n^{(2)}(\beta r) \right] P_n^1(\cos\theta) \right\} \qquad (11\text{-}237)$$

where in (11-237)

$$' \equiv \frac{\partial}{\partial(\beta r)} \quad \text{for the spherical Bessel or Hankel function} \qquad (11\text{-}237a)$$

$$' \equiv \frac{\partial}{\partial(\cos\theta)} = -\frac{1}{\sin\theta}\frac{\partial}{\partial\theta} \quad \text{for the associated Legendre functions} \qquad (11\text{-}237b)$$

Using (11-237), the boundary condition of (11-236a) is satisfied provided that

$$a_n \hat{J}_n'(\beta a) + b_n \hat{H}_n^{(2)'}(\beta a)=0 \; \Rightarrow \; b_n = -a_n \frac{\hat{J}_n'(\beta a)}{\hat{H}_n^{(2)'}(\beta a)} \qquad (11\text{-}238a)$$

$$a_n \hat{J}_n(\beta a) + c_n \hat{H}_n^{(2)}(\beta a)=0 \; \Rightarrow \; c_n = -a_n \frac{\hat{J}_n(\beta a)}{\hat{H}_n^{(2)}(\beta a)} \qquad (11\text{-}238b)$$

The scattered electric field components can be written, using (11-233a) and (11-233b), as

$$E_r^s = -jE_0 \cos\phi \sum_{n=1}^{\infty} b_n \left[\hat{H}_n^{(2)''}(\beta r) + \hat{H}_n^{(2)}(\beta r) \right] P_n^1(\cos\theta) \qquad (11\text{-}239a)$$

$$E_\theta^s = \frac{E_0}{\beta r} \cos\phi \sum_{n=1}^{\infty} \left[jb_n \hat{H}_n^{(2)'}(\beta r) \sin\theta P_n^{1'}(\cos\theta) - c_n \hat{H}_n^{(2)}(\beta r) \frac{P_n^1(\cos\theta)}{\sin\theta} \right] \qquad (11\text{-}239b)$$

$$E_\phi^s = \frac{E_0}{\beta r} \sin\phi \sum_{n=1}^{\infty} \left[jb_n \hat{H}_n^{(2)'}(\beta r) \frac{P_n^1(\cos\theta)}{\sin\theta} - c_n \hat{H}_n^{(2)}(\beta r) \sin\theta P_n^{1'}(\cos\theta) \right] \qquad (11\text{-}239c)$$

where in (11-239a) through (11-239c)

$$' \equiv \frac{\partial}{\partial(\beta r)} \quad \text{for the spherical Hankel functions} \tag{11-239d}$$

$$'' \equiv \frac{\partial^2}{\partial(\beta r)^2} \quad \text{for the spherical Hankel functions} \tag{11-239e}$$

$$' \equiv \frac{\partial}{\partial(\cos\theta)} = -\frac{1}{\sin\theta}\frac{\partial}{\partial\theta} \quad \text{for the associated Legendre functions} \tag{11-239f}$$

The spherical Hankel function is related to the regular Hankel function by (10-31) or

$$\hat{H}_n^{(2)}(\beta r) = \sqrt{\frac{\pi\beta r}{2}} H_{n+1/2}^{(2)}(\beta r) \tag{11-240}$$

Since for large values of βr the regular Hankel function can be represented by

$$H_{n+1/2}^{(2)}(\beta r) \overset{\beta r \to \infty}{\simeq} \sqrt{\frac{2j}{\pi\beta r}} j^{n+1/2} e^{-j\beta r} = j\sqrt{\frac{2}{\pi\beta r}} j^n e^{-j\beta r} \tag{11-241}$$

then the spherical Hankel function of (11-240) and its partial derivatives can be approximated by

$$\hat{H}_n^{(2)}(\beta r) \overset{\beta r \to \infty}{\simeq} j^{n+1} e^{-j\beta r} \tag{11-241a}$$

$$\hat{H}_n^{(2)'}(\beta r) = \frac{\partial \hat{H}_n^{(2)}(\beta r)}{\partial(\beta r)} \overset{\beta r \to \infty}{\simeq} -j^2 j^n e^{-j\beta r} = j^n e^{-j\beta r} \tag{11-241b}$$

$$\hat{H}_n^{(2)''}(\beta r) = \frac{\partial^2 H_n^{(2)}(\beta r)}{\partial(\beta r)^2} \overset{\beta r \to \infty}{\simeq} -j^{n+1} e^{-j\beta r} \tag{11-241c}$$

For far-field observations ($\beta r \to$ large), the electric field components of (11-239a) through (11-239c) can be simplified using the approximations (11-241a) through (11-241c). Since the radial component E_r^s of (11-239a) reduces with the approximations of (11-241a) through (11-241c) to zero, then in the far zone (11-239a) through (11-239c) can be approximated by

<u>Far-Field Observations ($\beta r \to$ large)</u>

$$E_r^s \simeq 0 \tag{11-242a}$$

$$E_\theta^s \simeq jE_0 \frac{e^{-j\beta r}}{\beta r} \cos\phi \sum_{n=1}^{\infty} j^n \left[b_n \sin\theta P_n^{1'}(\cos\theta) - c_n \frac{P_n^1(\cos\theta)}{\sin\theta} \right] \tag{11-242b}$$

$$E_\phi^s \simeq jE_0 \frac{e^{-j\beta r}}{\beta r} \sin\phi \sum_{n=1}^{\infty} j^n \left[b_n \frac{P_n^1(\cos\theta)}{\sin\theta} - c_n \sin\theta P_n^{1'}(\cos\theta) \right] \tag{11-242c}$$

where b_n and c_n are given by (11-238a) and (11-238b), respectively.

The bistatic radar cross section is obtained using (11-22b), and it can be written, using (11-225) and (11-242a) through (11-242c), as

$$\sigma(\text{bistatic}) = \lim_{r\to\infty} \left[4\pi r^2 \frac{|\mathbf{E}^s|^2}{|\mathbf{E}^i|^2} \right] = \frac{\lambda^2}{\pi} \left[\cos^2\phi |A_\theta|^2 + \sin^2\phi |A_\phi|^2 \right] \tag{11-243}$$

where

$$|A_\theta|^2 = \left| \sum_{n=1}^{\infty} j^n \left[b_n \sin\theta P_n^{1'}(\cos\theta) - c_n \frac{P_n^1(\cos\theta)}{\sin\theta} \right] \right|^2 \tag{11-243a}$$

$$|A_\phi|^2 = \left| \sum_{n=1}^{\infty} j^n \left[b_n \frac{P_n^1(\cos\theta)}{\sin\theta} - c_n \sin\theta P_n^{'1}(\cos\theta) \right] \right|^2 \tag{11-243b}$$

The monostatic radar cross section can be found by first reducing the field expressions for observations toward $\theta = \pi$. In that direction the scattered electric field of interest is the copolar component, E_x^s, and it can be found using (11-242a) through (11-242c) and the transformation (11-13b), by evaluating either

$$E_x^s = E_\theta^s \cos\theta \cos\phi \Big|_{\substack{\theta=\pi \\ \phi=\pi}} = E_\theta^s \Big|_{\substack{\theta=\pi \\ \phi=\pi}} \tag{11-244a}$$

or

$$E_x^s = -E_\phi^s \sin\phi \Big|_{\substack{\theta=\pi \\ \phi=3\pi/2}} = E_\phi^s \Big|_{\substack{\theta=\pi \\ \phi=3\pi/2}} \tag{11-244b}$$

To accomplish either (11-244a) or (11-244b), we need to first evaluate the associated Legendre function and its derivative when $\theta = \pi$. It can be shown that [11, 45]

$$\frac{P_n^1(\cos\theta)}{\sin\theta} \bigg|_{\theta=\pi} = (-1)^n \frac{n(n+1)}{2} \tag{11-245a}$$

$$\sin\theta P_n^{1'}(\cos\theta)|_{\theta=\pi} = \sin\theta \frac{dP_n^1}{d(\cos\theta)} = -\frac{dP_n^1(\cos\theta)}{d\theta} = (-1)^n \frac{n(n+1)}{2} \tag{11-245b}$$

Thus, (11-242b) can be expressed using (11-235c), (11-238a), (11-238b), (11-245a), and (11-245b), as

$$E_\theta^s \Big|_{\substack{\theta=\pi \\ \phi=\pi}} = jE_0 \frac{e^{-j\beta r}}{\beta r} \sum_{n=1}^{\infty} j^n (-1)^n \frac{n(n+1)}{2} [b_n - c_n]$$

$$= -jE_0 \frac{e^{-j\beta r}}{\beta r} \sum_{n=1}^{\infty} j^n (-1)^n \frac{n(n+1)}{2} a_n \left[\frac{\hat{J}_n'(\beta a)}{\hat{H}_n^{(2)'}(\beta a)} - \frac{\hat{J}_n(\beta a)}{\hat{H}_n^{(2)}(\beta a)} \right]$$

$$E_\theta^s \Big|_{\substack{\theta=\pi \\ \phi=\pi}} = -jE_0 \frac{e^{-j\beta r}}{\beta r} \sum_{n=1}^{\infty} (-1)^n \frac{2(n+1)}{2} \left[\frac{\hat{J}_n'(\beta a)\hat{H}_n^{(2)}(\beta a) - \hat{J}_n(\beta a)\hat{H}_2^{(2)'}(\beta a)}{\hat{H}_n^{(2)'}(\beta a)\hat{H}_n^{(2)}(\beta a)} \right] \tag{11-246}$$

which reduces, using the Wronskian for spherical Bessel functions

$$\hat{J}_n'(\beta a)\hat{H}_n^{(2)}(\beta a) - \hat{J}_n(\beta a)\hat{H}_n^{(2)'}(\beta a) = j[\hat{J}_n(\beta a)\hat{Y}_n'(\beta a) - \hat{J}_n'(\beta a)\hat{Y}_n(\beta a)] = j \tag{11-246a}$$

to

$$E_\theta^s \Big|_{\substack{\theta=\pi \\ \phi=\pi}} = E_0 \frac{e^{-j\beta r}}{2\beta r} \sum_{n=1}^{\infty} \frac{(-1)^n(2n+1)}{\hat{H}_n^{(2)'}(\beta a)\hat{H}_n^{(2)}(\beta a)} \tag{11-246b}$$

Thus, the monostatic radar cross section of (11-22b) can be expressed, using (11-246b), by

$$\sigma_{3\text{-D}}(\text{monostatic}) = \lim_{r\to\infty} \left[4\pi r^2 \frac{|\mathbf{E}^s|^2}{|\mathbf{E}^i|^2} \right] = \frac{\lambda^2}{4\pi} \left| \sum_{n=1}^{\infty} \frac{(-1)^n(2n+1)}{\hat{H}_n^{(2)'}(\beta a)\hat{H}_n^{(2)}(\beta a)} \right|^2 \tag{11-247}$$

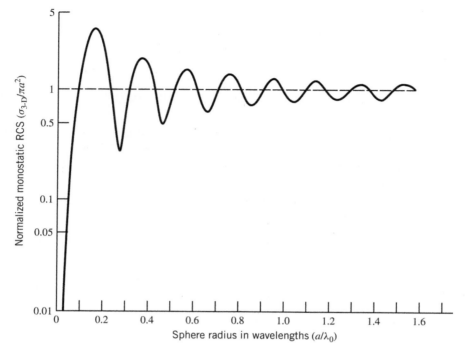

Figure 11-29 Normalized monostatic radar cross section for a conducting sphere as a function of its radius. (Source: G. T. Ruck, D. E. Barrick, W. D. Stuart, and C. K. Krichbaum, *Radar Cross Section Handbook*, Vol. 1, 1970, Plenum Publishing Co.)

A plot of (11-247) as a function of the sphere radius is shown in Figure 11-29 [49]. This is a classic signature that can be found in any literature dealing with electromagnetic scattering. The total curve can be subdivided into three regions; the *Rayleigh*, the *Mie* (or *resonance*), and the *optical* regions. The Rayleigh region represents the part of the curve for small values of the radius ($a < 0.1\lambda$) and the optical region represents the RCS of the sphere for large values of the radius (typically $a > 2\lambda$). The region between those two extremes is the Mie or resonance region. It is apparent that for small values of the radius the RCS is linear, for intermediate values it is oscillatory about πa^2, and for large values it approaches πa^2, which is the physical area of the cross section of the sphere.

For very small values of the radius a, the first term of (11-247) is sufficient to accurately represent the RCS. Doing this, we can approximate (11-247) by

$$\sigma_{3\text{-D}} \text{ (monostatic)} \overset{a \to 0}{\simeq} \frac{\lambda^2}{4\pi} \left| \frac{3}{\hat{H}_1^{(2)'}(\beta a)\hat{H}_1^{(2)}(\beta a)} \right|^2 \tag{11-248}$$

Since

$$\hat{H}_1^{(2)}(\beta a) \overset{a \to 0}{\simeq} -j\hat{Y}_1(\beta a) = -j\sqrt{\frac{\pi\beta a}{2}} Y_{3/2}(\beta a) \simeq -j\sqrt{\frac{\pi\beta a}{2}} \left[-\frac{\frac{1}{2}!}{\pi} \left(\frac{2}{\beta a}\right)^{3/2} \right] = j\frac{1}{\beta a} \tag{11-248a}$$

$$\hat{H}_1^{(2)'}(\beta a) \overset{a \to 0}{\simeq} -j\hat{Y}_1'(\beta a) \simeq -j\frac{1}{(\beta a)^2} \tag{11-248b}$$

$$\frac{1}{2}! = \frac{1}{2}\sqrt{\pi} \tag{11-248c}$$

(11-248) reduces to

$$\boxed{\sigma_{3\text{-D}}(\text{monostatic}) \overset{a \to 0}{\simeq} \frac{9\lambda^2}{4\pi}(\beta a)^6}$$

(11-248d)

which is representative of the Rayleigh region scattering.

For very large values of the radius a, we can approximate the spherical Hankel function and its derivative in (11-247) by their asymptotic forms (11-241a) and (11-241b), or

$$\hat{H}_n^{(2)}(\beta a) \overset{\beta a \to \infty}{\simeq} \frac{e^{-j[\beta a(\sin\alpha - \alpha\cos\alpha) - \pi/4]}}{\sqrt{\sin\alpha}}$$

(11-249a)

$$\hat{H}_n^{(2)'}(\beta a) \overset{\beta a \to \infty}{\simeq} \sqrt{\sin\alpha}\, e^{-j[\beta a(\sin\alpha - \alpha\cos\alpha) + \pi/4]}$$

(11-249b)

$$\cos\alpha = (n + 1/2)/\beta a$$

(11-249c)

Thus, (11-247) reduces for very large values of the radius a to

$$\boxed{\sigma_{3\text{-D}}(\text{monostatic}) = \frac{\lambda^2}{4\pi}\left|\sum_{n=1}^{\infty} \frac{(-1)^n(2n+1)}{\hat{H}_n^{(2)'}(\beta a)\hat{H}_n^{(2)}(\beta a)}\right|^2 \overset{a \to \infty}{\simeq} \pi a^2}$$

(11-250)

which is representative of the optical region scattering, and is also equal to the physical area of the cross section of the sphere.

11.8.2 Lossy Dielectric Sphere

The development of the scattering by a lossy dielectric sphere follows that of a PEC sphere, which was outlined in the previous section. The major difference is that now electric and magnetic fields penetrate the sphere, and we need to write expressions to properly represent them. The expressions for the fields outside the sphere will be of similar forms as those of the PEC sphere. To relate the fields outside and inside the sphere, the appropriate boundary conditions must be applied on the surface of the sphere; continuity of the tangential electric and magnetic fields, in contrast to the vanishing of the tangential electric fields on the surface of the sphere for the PEC case.

To start the development, we will use the geometry of Figure 11-28 and assume that the medium outside the sphere is free space (wave number β_o) and inside is a lossy dielectric (wave number $\dot{\beta}_d$) represented by a relative complex permittivity $\dot{\varepsilon}_r (\dot{\varepsilon}_r = \varepsilon_r' - j\varepsilon_r'')$ and relative complex permeability $\dot{\mu}_r (\dot{\mu}_r = \mu_r' - j\mu_r'')$. The total, incident, and scattered fields outside the sphere can be represented by the vector potentials, (11-235a) through (11-235c) and the corresponding electric and magnetic fields by (11-234a) through (11-234f). Inside the sphere the vector potentials should be similar to those outside the sphere, but chosen to represent standing waves in the radial direction, instead of traveling waves. This is accomplished by choosing the vector potentials for the total fields inside the sphere to be similar to the first terms of (11-235a) and (11-235b), and written as

$$A_r^{t-} = E_0 \frac{\cos\phi}{\omega}\sum_{n=1}^{\infty} d_n \hat{J}_n(\dot{\beta}_d r) P_n^{(1)}(\cos\theta)$$

(11-251a)

$$F_r^{t-} = E_0 \frac{\sin\phi}{\omega\dot{\eta}}\sum_{n=1}^{\infty} e_n \hat{J}_n(\dot{\beta}_d r) P_n^{(1)}(\cos\theta)$$

(11-251b)

The modal coefficients, b_n and c_n (for the fields outside the sphere) and d_n and e_n (for the fields inside the sphere), can be determined by application of the boundary conditions. The superscript minus $(-)$ is used to identify the vector potentials and associated fields on and within the sphere $(r \leq a)$, while the plus $(+)$ is used to identify those on and outside the sphere $(r \geq a)$.

Based on the fields of (11-234a) through (11-234f), there exist two tangential electric (E_θ, E_ϕ) and two tangential magnetic (H_θ, H_ϕ) field components. For the PEC sphere of the previous section, the boundary conditions were (11-236a) and (11-236b). However, the appropriate boundary conditions for the lossy dielectric sphere require the continuity of the tangential electric and magnetic fields on the surface of the sphere, which can be written as

$$E_\theta^{t-}\left(r=a, 0 \leq \theta \leq \pi, 0 \leq \phi \leq 2\pi\right) = E_\theta^{t+}\left(r=a, 0 \leq \theta \leq \pi, 0 \leq \phi \leq 2\pi\right) \quad \text{(11-252a)}$$

$$E_\phi^{t-}\left(r=a, 0 \leq \theta \leq \pi, 0 \leq \phi \leq 2\pi\right) = E_\phi^{t+}\left(r=a, 0 \leq \theta \leq \pi, 0 \leq \phi \leq 2\pi\right) \quad \text{(11-252b)}$$

$$H_\theta^{t-}\left(r=a, 0 \leq \theta \leq \pi, 0 \leq \phi \leq 2\pi\right) = H_\theta^{t+}\left(r=a, 0 \leq \theta \leq \pi, 0 \leq \phi \leq 2\pi\right) \quad \text{(11-252c)}$$

$$H_\phi^{t-}\left(r=a, 0 \leq \theta \leq \pi, 0 \leq \phi \leq 2\pi\right) = H_\phi^{t+}\left(r=a, 0 \leq \theta \leq \pi, 0 \leq \phi \leq 2\pi\right) \quad \text{(11-252d)}$$

The modal coefficients, b_n and c_n (for the fields outside the sphere), and d_n and e_n (for the fields inside the sphere), can be determined by enforcing this set of boundary conditions. To accomplish this, the two tangential electric (E_θ, E_ϕ) and two tangential magnetic (H_θ, H_ϕ) field components, both inside and outside the sphere, must first be written using (11-234b)–(11-234c) and (11-234e)–(11-234f) with the vector potentials of (11-235a)–(11-235b) for outside the sphere and (11-251a)–(11-251b) for inside the sphere.

The enforcement of the boundary conditions (11-252a) through (11-252d) is straightforward but cumbersome. Because of space limitations, and as a practice to the reader, the procedure will not be detailed here but left as an end-of-chapter exercise. Following the procedure outlined here, it can be shown that b_n, c_n, d_n, and e_n can be written and related to a_n by

$$b_n = \frac{-\sqrt{\dot{\varepsilon}_r}\,\hat{J}_n'(\beta_0 a)\,\hat{J}_n(\dot{\beta}_d a) + \sqrt{\dot{\mu}_r}\,\hat{J}_n(\beta_0 a)\,\hat{J}_n'(\dot{\beta}_d a)}{\sqrt{\dot{\varepsilon}_r}\,\hat{H}_n^{(2)'}(\beta_0 a)\,\hat{J}_n(\dot{\beta}_d a) - \sqrt{\dot{\mu}_r}\,\hat{H}_n^{(2)}(\beta_0 a)\,\hat{J}_n'(\dot{\beta}_d a)}\,a_n \quad \text{(11-253a)}$$

$$c_n = \frac{-\sqrt{\dot{\varepsilon}_r}\,\hat{J}_n(\beta_0 a)\,\hat{J}_n'(\dot{\beta}_d a) + \sqrt{\dot{\mu}_r}\,\hat{J}_n'(\beta_0 a)\,\hat{J}_n(\dot{\beta}_d a)}{\sqrt{\dot{\varepsilon}_r}\,\hat{H}_n^{(2)}(\beta_0 a)\,\hat{J}_n'(\dot{\beta}_d a) - \sqrt{\dot{\mu}_r}\,\hat{H}_n^{(2)'}(\beta_0 a)\,\hat{J}_n(\dot{\beta}_d a)}\,a_n \quad \text{(11-253b)}$$

$$d_n = -j\frac{\dot{\mu}_r\sqrt{\dot{\varepsilon}_r}}{\sqrt{\dot{\varepsilon}_r}\,\hat{H}_n^{(2)'}(\beta_0 a)\,\hat{J}_n(\dot{\beta}_d a) - \sqrt{\dot{\mu}_r}\,\hat{H}_n^{(2)}(\beta_0 a)\,\hat{J}_n'(\dot{\beta}_d a)}\,a_n \quad \text{(11-253c)}$$

$$e_n = +j\frac{\dot{\mu}_r\sqrt{\dot{\varepsilon}_r}}{\sqrt{\dot{\varepsilon}_r}\,\hat{H}_n^{(2)}(\beta_0 a)\,\hat{J}_n'(\dot{\beta}_d a) - \sqrt{\dot{\mu}_r}\,\hat{H}_n^{(2)'}(\beta_0 a)\,\hat{J}_n(\dot{\beta}_d a)}\,a_n \quad \text{(11-253d)}$$

where a_n is given by (11-231a). For a dielectric sphere of small radius, the first term $(n=1)$ may be sufficient to represent the fields. It can be shown that for $n=1$ (11-253a) through (11-253d) reduce to

$$b_1 \overset{\beta_0 a \to 0}{\approx} -\left(\beta_0 a\right)^3 \frac{\dot{\varepsilon}_r - 1}{\dot{\varepsilon}_r + 2} \quad \text{(11-254a)}$$

$$c_1 \overset{\beta_0 a \to 0}{\approx} -\left(\beta_0 a\right)^3 \frac{\dot{\mu}_r - 1}{\dot{\mu}_r + 2} \quad \text{(11-254b)}$$

$$d_1 \overset{\beta_0 a \to 0}{\approx} \frac{9}{j2(\dot{\varepsilon}_r + 2)} \quad \text{(11-254c)}$$

$$e_1 \stackrel{\beta_0 a \to 0}{\approx} \frac{9\sqrt{\mu_r}}{j2\sqrt{\dot{\varepsilon}_r}(\mu_r + 2)} \tag{11-254d}$$

Two MATLAB computer programs, **PEC_DIEL_Sphere_Fields** and **Sphere_RCS**, have been written. The first one, **PEC_DIEL_Sphere_Fields**, allows the visualization of the total fields, within and outside the sphere, for both PEC and lossy dielectric spheres. For static fields, a PEC sphere can be represented solely by letting $\dot{\varepsilon}_r \to$ very large (ideally infinity). However, at *rf*, the PEC sphere must be represented as a special case of the lossy dielectric sphere by allowing both $\dot{\varepsilon}_r \to$ very large (ideally infinity) and $\mu_r \to$ very small (ideally zero) so that $\dot{\beta}_d$ remains finite. The second computer program, **Sphere_RCS**, is based on the formulation of Section 11.8.2. It computes and plots the normalized amplitude scattering pattern, and bistatic and monostatic RCSs of a plane wave scattered by PEC and lossy dielectric spheres based on the geometry of Figure 11-28.

11.9 MULTIMEDIA

On the website that accompanies this book, the following multimedia resources are included for the review, understanding, visualization, and presentation of the material of this chapter.

- **MATLAB** computer programs:
  a. **PEC_Strip_SW:** Computes, using PO, the TM^z and TE^z 2-D scattering width (SW), monostatic and bistatic, of a PEC strip of Figure 11-4.
  b. **PEC_Rect_Plate_RCS:** Computes, using PO, the TE^x and TM^x 3-D radar cross section (RCS), monostatic and bistatic, of a PEC rectangular plate of Figure 11-8.
  c. **PEC_Circ_Plate_RCS:** Computes, using PO, the TE^x and TM^x 3-D radar cross section (RCS), monostatic and bistatic, of a PEC circular plate of Figure P11-7.
  d. **PEC_Cyl_Normal_Fields:** Visualizes the TM^z and TE^z scattered fields of a uniform plane wave incident, at normal incidence angles, upon the PEC cylinder of circular cross section of Figure 11-12.
  e. **PEC_Cyl_Normal_SW:** Computes the 2-D scattering width (SW) of a uniform plane TM^z and TE^z wave incident, at normal incidence angles, upon the PEC cylinder of circular cross section of Figure 11-12.
  f. **PEC_Cyl_Oblique_Fields:** Visualizes the TM^z and TE^z scattered fields of a uniform plane wave incident, at oblique incidence angles, upon the PEC cylinder of circular cross section of Figure 11-16.
  g. **PEC_Cyl_Oblique_SW:** Computes the 2-D scattering width (SW) of a uniform plane TM^z and TE^z wave incident, at oblique incidence angles, upon the PEC cylinder of circular cross section of Figure 11-16.
  h. **PEC_Cyl_Oblique_RCS:** Computes and plots the 2-D and 3-D RCS of a uniform plane TM^z and TE^z wave incident, at oblique incidence angles, upon the PEC cylinder of circular cross section and finite length of Figure 11-16.
  i. **Cylinder_RCS:** Computes and plots the normalized amplitude scattering pattern, and bistatic and monostatic RCSs of a TM^z and TE^z plane wave scattered by a:
     - PEC 2-D cylinder, based on geometry of Figures 11-22*a* and 11-22*b* and formulations of Sections 11.5.1 and 11.5.2, respectively.
     - Lossy 2-D dielectric cylinder based on the end-of-chapter Problems 11.58 and 11.60 and associated Figures P11-58 and P11-62, respectively.
  j. **Circum_Axial_Slot:** Computes and plots the radiation patterns of a circumferential and an axial slot/aperture on a PEC cylinder of circular cross section in the H- and E-planes,

respectively. The program also computes the radiation patterns with the ones of a rectangular aperture on a flat PEC ground plane.

    k. **PEC_DIEL_Sphere_Fields:** Visualizes the scattered fields of a uniform plane wave by a lossy dielectric sphere of Figure 11-25.

    l. **Sphere_RCS:** Computes and plots the normalized amplitude scattering pattern, and bistatic and monostatic RCSs of a uniform plane wave scattered by a:

- PEC sphere, based on the geometry of Figure 11-28 and formulations of Section 11.8.1.
- Lossy dielectric sphere based on the geometry of Figure 11.28 and formulations of Section 11.8.2

- **PowerPoint (PPT)** viewgraphs, in multicolor.

REFERENCES

1. C. A. Balanis, *Antenna Theory: Analysis and Design*, Third edition, Wiley, New York, 2005.
2. K. M. Siegel, "Far field scattering from bodies of revolution," *Appl. Sci. Res., Sec. B*, vol. 7, pp. 293–328, 1958.
3. E. F. Knott, V. V. Liepa, and T. B. A. Senior, "Non-specular radar cross section study," Technical Report AFAL-TR-73-70, University of Michigan, Apr. 1973.
4. T. Griesser and C. A. Balanis, "Dihedral corner reflector backscatter using higher-order reflections and diffractions," *IEEE Trans. Antennas Propagat.*, vol. AP-35, no. 11, pp. 1235–1247, Nov. 1987.
5. J. S. Asvestas, "The physical optics method in electromagnetic scattering," *Math. Phys.*, vol. 21, pp. 290–299, 1980.
6. J. S. Asvestas, "Physical optics and the direction of maximization of the far-field average power," *IEEE Trans. Antennas Propagat.*, vol. AP-34, no. 12, pp. 1459–1460, Dec. 1986.
7. E. F. Knott, "RCS reduction of dihedral corners," *IEEE Trans. Antennas Propagat.*, vol. AP-25, no. 3, pp. 406–409, May 1977.
8. R. A. Ross, "Radar cross section of rectangular plates as a function of aspect angle," *IEEE Trans. Antennas Propagat.*, vol. AP-14, no. 3, pp. 329–335, May 1966.
9. T. Griesser and C. A. Balanis, "Backscatter analysis of dihedral corner reflectors using physical optics and the physical theory of diffraction," *IEEE Trans. Antennas Propagat.*, vol. AP-35, no. 10, pp. 1137–1147, Oct. 1987.
10. D. P. Marsland, C. A. Balanis, and S. Brumley, "Higher order diffractions from a circular disk," *IEEE Trans. Antennas Propagat.*, vol. AP-35, no. 12, pp. 1436–1444, Dec. 1987.
11. R. F. Harrington, *Time-Harmonic Electromagnetic Fields*, McGraw-Hill, New York, 1961.
12. D. E. Barrick, "Cylinders," in *Radar Cross Section Handbook*, vol. 1, G. T. Ruck, D. E. Barrick, W. D. Stuart, and C. K. Krichbaum (Eds.), Plenum, New York, 1970, Chapter 4, pp. 205–339.
13. J. H. Richmond, *The Basic Theory of Harmonic Fields, Antennas and Scattering*, Ohio State University, unpublished notes.
14. H. C. Van de Hulst, *Light Scattering by Small Particles*, Wiley, New York, 1957, pp. 304–307.
15. C. A. Balanis and L. Peters, Jr., "Analysis of aperture radiation from an axially slotted circular conducting cylinder using geometrical theory of diffraction," *IEEE Trans. Antennas Propagat.*, vol. AP-17, no. 1, pp. 93–97, Jan. 1969.
16. C. A. Balanis and L. Peters, Jr., "Aperture radiation from an axially slotted elliptical conducting cylinder using geometrical theory of diffraction," *IEEE Trans. Antennas Propagat.*, vol. AP-17, no. 4, pp. 507–513, Jul. 1969.
17. P. H. Pathak and R. G. Kouyoumjian, "An analysis of the radiation from apertures in curved surfaces by the geometrical theory of diffraction," *Proc. IEEE*, vol. 62, no. 11, pp. 1438–1447, Nov. 1974.
18. S. A. Schelkunoff, "Some equivalence theorems of electromagnetics and their application to radiation problems," *Bell Syst. Tech. J.*, vol. 15, pp. 92–112, 1936.

19. J. R. Wait, *Electromagnetic Radiation from Cylindrical Structures*, Pergamon, New York, 1959.

20. J. J. Bowman, T. B. A. Senior, and P. L. E. Uslenghi (Eds.), *Electromagnetic and Acoustic Scattering by Simple Shapes*, North-Holland, Amsterdam, 1969.

21. A. W. Adey, "Scattering of electromagnetic waves by coaxial cylinders," *Can. J. Phys.*, vol. 34, pp. 510–520, May 1956.

22. C. C. H. Tang, "Backscattering from dielectric-coated infinite cylindrical obstacles, *J. Appl. Phys.*, vol. 28, pp. 628–633, May 1957.

23. B. R. Levy, "Diffraction by an elliptic cylinder," *J. Math. Mech.*, vol. 9, pp. 147–165, 1960.

24. R. D. Kodis, "The scattering cross section of a composite cylinder, geometrical optics," *IEEE Trans. Antennas Propagat.*, vol. AP-11, no. 1, pp. 86–93, Jan. 1963.

25. K. Mei and J. Van Bladel, "Low-frequency scattering by a rectangular cylinder," *IEEE Trans. Antennas Propagat.*, vol. AP-11, no. 1, pp. 52–56, Jan. 1963.

26. R. D. Kodis and T. T. Wu, "The optical model of scattering by a composite cylinder," *IEEE Trans. Antennas Propagat.*, vol. AP-11, no. 6, pp. 703–705, Nov. 1963.

27. J. H. Richmond, "Scattering by a dielectric cylinder of arbitrary cross section shape," *IEEE Trans. Antennas Propagat.*, vol. AP-13, no. 3, pp. 334–341, May 1965.

28. W. V. T. Rusch and C. Yeh, "Scattering by an infinite cylinder coated with an inhomogeneous and anisotropic plasma sheath," *IEEE Trans. Antennas Propagat.*, vol. AP-15, no. 3, pp. 452–457, May 1967.

29. J.-C. Sureau, "Reduction of scattering cross section of dielectric cylinder by metallic core loading," *IEEE Trans. Antennas Propagat.*, vol. AP-15, no. 5, pp. 657–662, Sept. 1967.

30. H. E. Bussey and J. H. Richmond, "Scattering by a lossy dielectric circular cylindrical multilayer, numerical values," *IEEE Trans. Antennas Propagat.*, vol. AP-23, no. 5, pp. 723–725, Sept. 1975.

31. R. E. Eaves, "Electromagnetic scattering from a conducting circular cylinder covered with a circumferentially magnetized ferrite," *IEEE Trans. Antennas Propagat.*, vol. AP-24, no. 2, pp. 190–197, Mar. 1976.

32. C. A. Balanis and Y. B. Cheng, "Antenna radiation and modeling for microwave landing system," *IEEE Trans. Antennas Propagat.*, vol. AP-24, pp. 490–497, July 1976.

33. S. Silver and W. K. Saunders, "The external field produced by a slot in an infinite circular cylinder," *J. Appl. Phy.*, vol. 21, pp. 153–158, Feb. 1950.

34. L. L. Bailin, "The radiation field produced by a slot in a large circular cylinder," *IRE Trans. Antennas Propagat.*, vol. AP-3, pp. 128–137, July 1955.

35. G. N. Watson, "The diffraction of electrical waves by the earth," *Proc. Roy. Soc. (London)*, vol. A95, pp. 83–99, 1918.

36. R. G. Kouyoumjian, "Asymptotic high-frequency methods," *Proc. IEEE*, vol. 53, pp. 864–876, Aug. 1965.

37. L. B. Felsen and N. Marcuvitz, *Radiation and Scattering of Waves*, Prentice-Hall, Englewood Cliffs, N.J., 1973.

38. W. Pauli, "On asymptotic series for functions in the theory of diffraction of light," *Phys. Rev.*, vol. 34, pp. 924–931, Dec. 1938.

39. F. Oberhettinger, "On asymptotic series occurring in the theory of diffraction of waves by a wedge," *Math. Phys.*, vol. 34, pp. 245–255, 1956.

40. J. B. Keller, "Geometrical theory of diffraction," *J. Opt. Soc. Amer.*, vol. 52, no. 2, pp. 116–130, Feb. 1962.

41. R. G. Kouyoumjian and P. H. Pathak, "A uniform geometrical theory of diffraction for an edge in a perfectly conducting surface," *Proc. IEEE*, vol. 62, no. 11, pp. 1448–1461, Nov. 1974.

42. G. L. James, *Geometrical Theory of Diffraction for Electromagnetic Waves*, Third Edition Revised, Peregrinus, London, 1986.

43. D. L. Hutchins, "Asymptotic series describing the diffraction of a plane wave by a two-dimensional wedge of arbitrary angle," Ph.D. dissertation, Dept. of Electrical Engineering, Ohio State University, 1967.

44. P. H. Pathak and R. G. Kouyoumjian, "The dyadic diffraction coefficient for a perfectly conducting wedge," Technical Report 2183-4 (AFCRL-69-0546), ElectroScience Laboratory, Ohio State University, Jun. 5, 1970.

45. J. A. Stratton, *Electromagnetic Theory*, McGraw-Hill, New York, 1941.

46. P. M. Morse and H. Feshbach, *Methods of Theoretical Physics*, Part II, McGraw-Hill, New York, 1953.

47. W. R. Smythe, *Static and Dynamic Electricity*, McGraw-Hill, New York, 1939.

48. M. R. Spiegel, *Mathematical Handbook of Formulas and Tables, Schaum's Outline Series*, McGraw-Hill, New York, 1968.

49. A. L. Aden, "Scattering from spheres with sizes comparable to the wavelength," *J. Appl. Phys.*, vol. 12, 1951.

50. S. I. Rubinow and T. T. Wu, "First correction to the geometrical-optics scattering cross section from cylinders and spheres," *J. Appl. Phys.*, vol. 27, pp. 1032–1039, 1956.

51. J. Rheinstein, "Scattering of electromagnetic waves from dielectric coated conducting spheres," *IEEE Trans. Antennas Propagat.*, vol. AP-12, no. 3, pp. 334–340, May 1964.

52. R. G. Kouyoumjian, L. Peters, Jr., and D. T. Thomas, "A modified geometrical optics method for scattering by dielectric bodies," *IEEE Trans. Antennas Propagat.*, vol. AP-11, no. 6, pp. 690–703, Nov. 1963.

53. D. E. Barrick, "Spheres," in *Radar Cross Section Handbook*, vol. 1, G. T. Ruck, D. E. Barrick, W. D. Stuart, and C. K. Krichbaum (Eds.), Plenum, New York, 1970, Chapter 3, pp. 141–204.

54. D. E. Kerr and H. Goldstein, "Radar targets and echoes," in *Propagation of Short Radio Waves*, D. E. Kerr (Ed.), McGraw-Hill, 1951, Chapter 6, pp. 445–469.

PROBLEMS

11.1. Repeat Example 11-1 for a displaced *magnetic* line source of constant current I_m at ρ', ϕ'.

11.2. A magnetic line source of infinite length and constant magnetic current I_m is placed parallel to the z axis at a height h above a PEC ground plane of infinite extent, as shown in Figure 11-2 except that we now have a magnetic line source.
 (a) Determine the total magnetic field at ρ, ϕ for $0 \leq \phi \leq 180°$.
 (b) Simplify the expressions when the observations are made at very large distances (far field).
 (c) Determine the smallest height h (in λ) that will introduce a null in the far field amplitude pattern at:
 • $\phi = 30°$
 • $\phi = 90°$

11.3. Repeat Problem 11.2 for the electric field of an electric line source of infinite length and constant electric current I_e placed parallel to the z axis at a height h above a perfect magnetic conducting (PMC) ground plane of infinite extent, as shown in Figure 11-2, except that we now have a PMC ground plane.

11.4. Repeat Problem 11.3 for the magnetic field of a magnetic line source of constant current I_m above a PMC.

11.5. For the problem of Figure 11-2, where the electric line source is placed at a height $h \ll \lambda$, and observations are made at any point (including near field):
 (a) Show that the corresponding magnetic vector potential of the line source at $y = h$ in the absence of the ground plane is given by

$$A_z^{(1)} = A_z^{(0)}(x, y - h)$$

 where $A_z^{(0)}$ is the magnetic vector potential of an isolated line source at $z = 0$.
 (b) Write a simplified closed-form expression for the total magnetic vector potential of the line source above the ground plane valid for $0 \leq \phi \leq 180°$.
 (c) Determine the angles $\phi (0 \leq \phi \leq 180°)$ where the total far-zone electric field, at a constant observation distance ρ, vanishes.

11.6. Repeat Problem 1.5 for a magnetic line source, and corresponding electric vector potential F_z and magnetic field, above a PMC ground plane.

11.7. Two constant current, infinite length electric line sources, are displaced along the x axis a distance s apart, as shown in Figure P11-7. Use superposition and neglect mutual coupling between the lines.

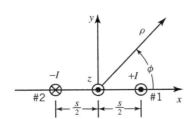

Figure P11-7

(a) Show that the magnetic vector potential for the two lines can be written as

$$A_z^t = A_z^{(1)}\left(x - \frac{s}{2}, y\right) - A_z^{(2)}\left(x + \frac{s}{2}, y\right)$$

(b) Show that for small spacings (in the limit as $s \to 0$), the vector potential of part (a) reduces to

$$A_z^t \overset{s \to 0}{\simeq} \frac{\mu\beta Is}{4j} H_1^{(2)}(\beta\rho)\cos\phi$$

(c) Determine the electric and magnetic field components associated with the two line sources when $s \to 0$.

11.8. Repeat Problem 11.7 for the electric vector potential, and the electric and magnetic fields, when the two sources are magnetic line sources displaced symmetrically along the x axis a distance s apart, as shown in Figure P11-7.

11.9. Two electric line sources, of infinite length and constant current I_e, are displaced along the y axis a distance s apart, as shown in Figure P11-9.
(a) Show that the total magnetic vector potential A_z^t for the two line sources can be written as

$$A_z^t = A_z^{(1)}\left(y - \frac{s}{2}, x\right) - A_z^{(2)}\left(y + \frac{s}{2}, x\right)$$

(b) Show that, for $s \to 0$, the total magnetic vector potential can be written in simplified form as the product of one Hankel function and cosine/sine functions.

(c) Determine the total electric and magnetic field components for the two sources for $s \to 0$.

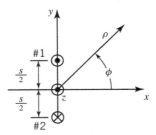

Figure P11-9

11.10. Repeat Problem 11.9 for the electric vector potential F_z^t, and electric and magnetic fields for two magnetic line sources displaced along the y axis a distance s apart.

11.11. Four constant current, infinite length electric line sources, of phase as indicated, are displaced along the x axis, as shown in Figure P11-11. Assume that the spacings s_1 and s_2 are very small.
(a) Find an approximate closed-form expression for the magnetic vector potential for the entire array by using the procedure of Problem 11.7. First consider the pairs on each side as individual arrays and then combine the results to form a new array.

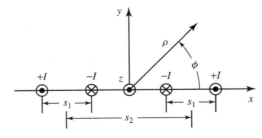

Figure P11-11

(b) Determine in terms of ρ and ϕ the electric and magnetic field components associated with the four line sources when $\rho \gg s_1$ and $\rho \gg s_2$.

11.12. Four constant current, infinite length electric line sources, of phase as indicated, are displaced along the x and y axes, as shown in Figure P11-12. Assume that the spacings s

and h are very small and neglect any mutual coupling between the lines.

(a) Show, by using the procedure of Problem 11.7, that the magnetic vector potential for the entire array can be written as

$$A_z \underset{\substack{s \to 0 \\ h \to 0}}{\simeq} -\frac{\mu\beta shI}{4j} \frac{\partial}{\partial y}[H_1^{(2)}(\beta\rho)\cos\phi]$$

$$= \frac{\mu\beta^2 shI}{8j} H_2^{(2)}(\beta\rho)\sin(2\phi)$$

(b) Determine in terms of ρ and ϕ the electric and magnetic field components associated with the four line sources when $\rho \gg s$ and $\rho \gg h$.

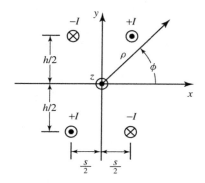

Figure P11-12

11.13. Two constant current, infinite length electric line sources are placed above an infinite electric ground plane, as shown in Figure P11-13. Assume that the spacings s_1 and s_2 are very small and neglect any mutual coupling between the lines.

(a) Show, by using the procedure of Problem 11.7, that the magnetic vector potential for the entire array can be written as

$$A_z \underset{\substack{s \to 0 \\ h \to 0}}{\simeq} -\frac{\mu\beta shI}{4j} \frac{\partial}{\partial y}[H_1^{(2)}(\beta\rho)\cos\phi]$$

$$= \frac{\mu\beta^2 shI}{8j} H_2^{(2)}(\beta\rho)\sin(2\phi)$$

(b) Determine in terms of ρ and ϕ the electric and magnetic field components

associated with the two line sources and ground plane when $\rho \gg s$ and $\rho \gg h$.

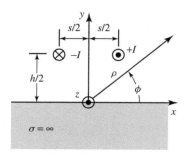

Figure P11-13

11.14. An infinite length and constant current electric line source is placed parallel to the plates of a 90° conducting corner reflector, as shown in Figure P11-14. Assume that the plates of the wedge are infinite in extent and the distance s from the apex to the source is very small ($s \ll \lambda$).

(a) Show that the magnetic vector potential for the line source and corner reflector can be written, using the procedure of Problem 11.7, as

$$A_z = \frac{\mu\beta^2 s^2 I}{4j} H_2^{(2)}(\beta\rho)\sin(2\phi)$$

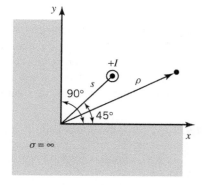

Figure P11-14

(b) Determine the electric and magnetic field components.

(c) Find the angles ϕ of observation (for a constant value of ρ) where the electric field vanishes.

11.15. Two parallel slots, identical, very thin ($w \to 0$), of infinite length and uniform electric field, but directed in opposite directions, are positioned on an infinite PEC ground plane and symmetrically displaced a distance s apart, as shown in Figure P11-15.

(a) Show that the total vector potential for the two line sources can be written as the sum/difference of two magnetic vector potentials. Indicate which vector potential, electric or magnetic, should be used.

(b) Show that for $s \to 0$, the appropriate total vector potential can be written in simplified form as the product of one Hankel function and cosine/sine functions.

(c) Determine the total electric and magnetic field components for the two sources for $s \to 0$.

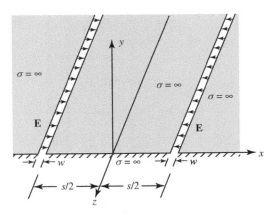

Figure P11-15

11.16. An infinite-length electric line source of constant current I_e is placed a height h above a flat and infinite in extent PEC ground plane, as shown in Figure 11-2. For far-field observations and constant radius $\rho = \rho_0$, it is desired to place a null in the amplitude pattern at an angle $\phi = 60°$. Determine the smallest height $h > 0$ (in λ) that will accomplish this. Assume geometrical optics.

11.17. For Problem 11.16 determine the first two smallest heights (in λ) so that the far-field amplitude pattern at $\phi = 60°$ is -3 dB from the maximum.

11.18. Using the geometry shown in Figure P11-18 of a TE^z of a uniform plane wave incident upon a 2-D PMC strip of infinite width,

(a) Write a complete vector expression for the incident magnetic field in terms of the angle ϕ_i. The magnitude of the incident magnetic field is H_0.

(b) Write a complete vector expression for the reflected magnetic field in terms of the magnitude H_0 of the incident magnetic field, a reflection coefficient Γ, and the angle ϕ_r.

(c) Based on Snell's law, determine the value of the reflection coefficient (magnitude and phase).

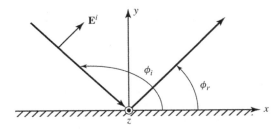

Figure P11-18

11.19. A uniform plane wave is incident upon a flat PEC rectangular plate whose dimensions are very large compared to the wavelength, as shown in Figure 11-8. It is desired to reduce the maximum value of the bistatic RCS of the plate by 10 dB, compared to normal incidence, by illuminating it at an oblique angle. Determine the angle of incidence (in degrees), measured from the normal to the plate, to accomplish this. You can assume that the plate is sufficiently large such that physical optics (PO) is a good approximation. Do this for a polarization:

(a) TE^x (b) TM^x

11.20. For a strip of width $w = 2\lambda$, plot the RCS$/\lambda_0^2$ (in dB) when the length of the strip is $l = 5\lambda$, 10λ, and 20λ (plot all three graphs on the same figure). Use the approximate relation between the 2-D SW and the 3-D RCS. Assume normal incidence.

11.21. Repeat Problem 11.20 by treating the strip of finite length as a rectangular plate. Compare the results of the previous two problems. Are they different? Please comment.

11.22. Design a flat PEC flat plate radar target so that the maximum normalized monostatic RCS (σ/λ^2) at an angle of $30°$ from the normal of the plate is $+20$ dB. Assuming physical optics (PO):
(a) Determine the area (in λ^2) of the plate.
(b) At 10 GHz, what is the:
  - Area of the plate (in cm^2)?
  - RCS in dB/sm (dB/square meter)?

11.23. Show that for normal incidence, the two-dimensional scattering width and three-dimensional RCS are related by (11-22e).

11.24. A uniform plane wave on the yz plane is obliquely incident at an angle θ_i from the vertical z axis upon a perfectly electric conducting circular ground plane of radius a, as shown in Figure P11-24. Assume TE^x polarization for the incident field.
(a) Determine the physical optics current density induced on the plate.
(b) Determine the far-zone bistatic scattered electric and magnetic fields based on the physical optics current density of part (a).
(c) Determine the bistatic and monostatic RCSs of the plate. Plot the normalized monostatic RCS $(\sigma_{3\text{-}D}/\lambda^2)$ in dB for plates with radii of $a=\lambda$ and 5λ.

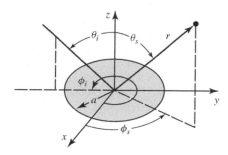

Figure P11-24

11.25. Repeat Problem 11.24 for TM^x plane wave incidence.

11.26. Show that for normal incidence the monostatic RCS of a flat plate of area A and any cross section, based on physical optics, is equal to a $\sigma_{3\text{-}D} = 4\pi(A/\lambda)^2$.

11.27. A uniform plane wave traveling in the $-z$ direction is incident upon a perfectly electric conducting curved surface, as shown in Figure P11-27, with radii of curvature

sufficiently large, usually greater than about one wavelength, so that at each point the surface can be considered locally flat. For such a surface, the induced currents and the fields radiated from each infinitesimal area can be represented if the same area were part of an infinite plane that was tangent to the surface at the same location.

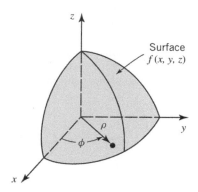

Figure P11-27

(a) Show that the monostatic RCS can be written as

$$\sigma_{3\text{-}D} = \frac{4\pi}{\lambda^2}\left|\iint_A e^{+j2\beta z}\,dA\right|^2$$

where z is any point on the surface of the scatterer and the integration on dA is performed in the xy plane.

(b) If in part (a) the differential area in the xy plane is expressed in terms of the polar coordinates ρ, ϕ, then show that the monostatic RCS can be expressed as

$$\sigma_{3\text{-}D} = \frac{\pi}{\lambda^2}\left|\int_0^{2\pi}\int_0^{z(\phi)}\right.$$
$$\left.\times\,\frac{d\rho^2}{dz}e^{+j2\beta z}\,dz\,d\phi\right|^2$$

11.28. If the scattering conducting curved surface of Problem 11.27, Figure P11-27, is a quadric surface, the integration to find the RCS can be performed in closed form. Assume that the scattering surface is an elliptic paraboloid opening downward along the positive z axis, and that it is represented by

$$\left(\frac{x}{a}\right)^2 + \left(\frac{y}{b}\right)^2 = -\frac{z}{c}$$

where a, b, and c are constants. Show the following.

(a) The radii of curvature a_1 and a_2 in the xz and yz planes are given by

$$a_1 = \frac{a^2}{2c}, \quad a_2 = \frac{b^2}{2c}$$

(b) The equation of the elliptic paraboloid surface transformed to polar coordinates can be expressed as

$$z = -\frac{\rho^2}{2a_1}\left[1-\left(1-\frac{a_1}{a_2}\right)\sin^2\phi\right]$$

(c) The RCS of Problem 11.27 reduces to

$$a_{3\text{-}D} = \pi\,a_1\,a_2\,|e^{-j2\beta h}-1|$$
$$= 4\pi\,a_1\,a_2\,\sin^2(\beta h)$$

(d) The RCS of part (c) reduces to $\sigma_{3\text{-}D} = \pi a_1 a_2$ if the height of the paraboloid is cut very irregular so that the contributions from the last zone(s) would tend to cancel. To account for this, the exponential term in part (c) disappears.

11.29. If the scattering conducting curved surface of Problem 11.27, Figure P11-27, is a closed surface, such as an ellipsoid represented by

$$\left(\frac{x}{a}\right)^2 + \left(\frac{y}{b}\right)^2 + \left(\frac{z}{c}\right)^2 = 1$$

demonstrate the following.
(a) The distance z at any point on the surface can be represented by

$$z^2 = c^2 - \rho^2\frac{c}{a_1}\left[1-\left(1-\frac{a_1}{a_2}\right)\sin^2\phi\right]$$

where $a_1 = a^2/c$ and $a_2 = b^2/c$.
(b) The RCS of Problem 11.27 from the upper part of the ellipsoid from $z=h$ to $z=c$ reduces, neglecting $(2\beta c)^{-1}$ terms, to

$$\sigma_{3\text{-}D} = \pi a_1 a_2 \left\{1+\left(\frac{h}{c}\right)^2\right.$$

$$\left.-2\left(\frac{h}{c}\right)\cos[2\beta(c-h)]\right\}$$

where $h(h<c)$ is the distance along the z axis from the origin to the point of integration. If $h=0$ or if h is very irregular around the periphery of the ellipsoid, the preceding equation reduces to $\sigma_{3\text{-}D} = \pi a_1 a_2$. If $a_1 = a_2 = a$, like for a sphere, then $\sigma_{3\text{-}D} = \pi a^2$.

11.30. Show that $H_{-n}^{(2)}(x)=(-1)^n H_n^{(2)}(x)$. This identity is often used for the computation of fields scattered by circular cylinders based on modal solutions.

11.31. Verify (11-55a).

11.32. Verify (11-71a) and (11-71b).

11.33. Verify that the infinite summation from minus to plus infinity for the incident electric field of (11-85a) can also be written as an infinite summation from $n=0$ to $n=\infty$.

11.34. Write the current density expression of (11-97) as an infinite summation from $n=0$ to $n=\infty$.

11.35. Refer to Figure 11-12a for the TM^z uniform plane wave scattering by a circular conducting cylinder.
(a) Determine the normalized induced current density based on the physical optics approximation of Section 7.10.
(b) Plot and compare for $0 \le \phi \le 180°$ the normalized induced current density based on the physical optics approximation and on the modal solution of (11-97). Do this for cylinders with radii of $a=\lambda$ and 5λ.

11.36. For the TM^z uniform plane wave scattering by a circular conducting cylinder of Figure 11-12a, plot and compare the normalized induced current density based on the exact modal solution of (11-97) and its small argument approximation of (11-98b) for radii of $a = 0.01\lambda$, 0.01λ, and λ.

11.37. Verify that the infinite summation from minus to plus infinity for the scattering width of (11-102) can also be written as an infinite summation from zero to infinity.

11.38. Using the definition (11-21c), instead of (11-21b), show that the TM^z polarization radar cross section reduces to (11-102).

11.39. Write the current density expression of (11-113) as an infinite summation from $n=0$ to $n=\infty$.

11.40. Refer to Figure 11-12b for the TE^z uniform plane wave scattering by a circular conducting cylinder.
(a) Determine the normalized induced current density based on the physical optics approximation of Section 7.10.
(b) Plot and compare for $0° \le \phi \le 180°$ the normalized induced current density

based on the physical optics approximation and on the modal solution of (11-113). Do this for cylinders with radii of $a = \lambda$ and 5λ.

11.41. For the TE^z uniform plane wave scattering by a circular conducting cylinder of Figure 11-12b, plot and compare the normalized induced current density based on the exact modal solution of (11-113) and its small argument approximation of (11-114d) for radii of $a = 0.01\lambda, 0.1\lambda$, and λ.

11.42. Verify that the infinite summation from minus to plus infinity for the scattering width of (11-117) can also be written as an infinite summation from zero to infinity.

11.43. Using the definition (11-21b), instead of (11-21c), show that the TE^z polarization radar cross section reduces to (11-117).

11.44. A right-hand circularly polarized uniform plane wave with an electric field equal to

$$\mathbf{E}^i = \left(\hat{\mathbf{a}}_y + j\hat{\mathbf{a}}_z\right)e^{-j\beta x}$$

is incident, at normal incidence, upon a PEC cylinder of circular cross section and radius a, as shown in Figure 11-12. Assuming far-field observations, determine the:
(a) Cylindrical components of the total scattered electric field.
(b) Polarization of the total scattered electric field toward:
  • $\phi = 0°$.
  • $\phi = 180°$.

11.45. Repeat Problem 11.44 when the circular cylinder is PMC.

11.46. Repeat the plots of Figure 11-13 (both dimensionless and in dB).

11.47. For the same cases of Figure 11-13, plot the magnitude of the induced electric current density (in A/m). Assume $f = 10$ GHz and an incident electric field of 1×10^{-3} V/m.

11.48. Repeat the plots of Fig. 11-15 (both dimensionless and in dB).

11.49. For the same cases of Figure 11-15, plot the magnitude of the induced electric current density (in A/m). Assume $f = 10$ GHz and an incident electric field of 1×10^{-3} V/m.

11.50. A TM^z uniform plane wave is normally incident, upon a very thin PEC wire of radius $a (a \ll \lambda)$, as shown in Figure 11-12a. Determine

the values of ϕ (in degrees) where the bistatic scattering width of the wire is:
(a) Maximum.
(b) Zero.

11.51. Repeat Problem 11.50 when the incident uniform plane wave has TE^z polarization, as shown in Figure 11-12b.

11.52. A PEC wire of circular cross section, and of radius $a \ll \lambda$, is used as a radar target. It is desired to maintain the normalized maximum bistatic SW (σ/λ) at any angle ϕ at a level not greater than -20 dB. When the incident uniform plane wave is TM^z polarized:
(a) What should the maximum radius (in λ) of the wire be to meet the desired specifications?
(b) At what observation angle ϕ would this maximum occur? Identify the angle ϕ (in degrees) graphically.

11.53. Repeat Problem 11.52 for a TE^z polarized incident uniform plane wave.

11.54. A very long (ideally infinite in length) thin (radius $a \ll \lambda$) PEC wire is attached to an airplane and is used as a trailing antenna. In order for the wire not to be very visible to radar, it is desired for the wire to have a normalized scattering width (σ/λ) not to exceed -10 dB. Determine the largest radius of the wire (in λ) when the incident uniform plane wave is TM^z polarized, as shown in Figure 11-12a.

11.55. Repeat Problem 11.54 when the incident uniform plane wave is TE^z polarized, as shown in Figure 11-12b.

11.56. A TM^z uniform plane wave traveling along the $+x$ direction is normally incident upon a PMC cylinder, as shown in Figure 11-12a.
(a) Derive the two-dimensional RCS (SW) expressed as SW / λ.
(b) Derive an expression for electric current density $\mathbf{J}_s$ and numerically evaluate $\mathbf{J}_s$ on the surface of the cylinder $(\rho = a)$. Justify the numerical value; i.e., is it what you were expecting? Should it be that value?

11.57. Repeat Problem 11.56 for TE^z polarization, as shown in Figure 11-12b.

11.58. A TM^z uniform plane wave traveling in the $+x$ direction in free space is incident normally on a lossless dielectric circular cylinder of radius a, as shown in Figure P11-58.

Assume that the incident, scattered, and transmitted (into the cylinder) electric fields can be written as

$$\mathbf{E}^i = \hat{\mathbf{a}}_z E_0 \sum_{n=-\infty}^{+\infty} j^{-n} J_n(\beta_0 \rho) e^{jn\phi}$$

$$\mathbf{E}^s = \hat{\mathbf{a}}_z E_0 \sum_{n=-\infty}^{+\infty} a_n H_n^{(2)}(\beta_0 \rho) e^{jn\phi}$$

$$\mathbf{E}^d = \hat{\mathbf{a}}_z E_0 \sum_{n=-\infty}^{+\infty} [b_n J_n(\beta_1 \rho) + c_n Y_n(\beta_1 \rho)] e^{jn\phi}$$

(a) Derive expressions for the incident, scattered, and transmitted magnetic field components.
(b) Show that the wave amplitude coefficients are equal to

$$c_n = 0$$

$$a_n = j^{-n} \dfrac{\begin{array}{c} J_n'(\beta_0 a) J_n(\beta_1 a) \\ -\sqrt{\varepsilon_r/\mu_r} J_n(\beta_0 a) J_n'(\beta_1 a) \end{array}}{\begin{array}{c} \sqrt{\varepsilon_r/\mu_r} J_n'(\beta_1 a) H_n^{(2)}(\beta_0 a) \\ - J_n(\beta_1 a) H_n^{(2)'}(\beta_0 a) \end{array}}$$

$$b_n = j^{-n} \dfrac{\begin{array}{c} J_n(\beta_0 a) H_n^{(2)'}(\beta_0 a) \\ -J_n'(\beta_0 a) H_n^{(2)}(\beta_0 a) \end{array}}{\begin{array}{c} J_n(\beta_1 a) H_n^{(2)'}(\beta_0 a) - \sqrt{\varepsilon_r/\mu_r} \\ \times J_n'(\beta_1 a) H_n^{(2)'}(\beta_0 a) \end{array}}$$

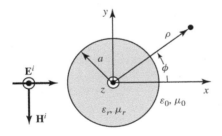

Figure P11-58

11.59. For Problem 11.58,
(a) Derive an expression for the scattering width (SW).
(b) Plot SW$/\lambda_0$ (in dB) for $a = 2\lambda$, dielectric constants of $\varepsilon_r = 4$ and 9 ($0 \le \phi \le 180°$).

11.60. A TM^z uniform plane wave traveling along the $+x$ direction, is normally incident upon a dielectric cylinder with $\varepsilon_r \gg 1$ and $\mu_r = 1$, as

shown in Figure P11-58. Write an expression for the:
(a) Two-dimensional RCS (SW) expressed in SW$/\lambda_0$.
(b) Electric current density $\mathbf{J}_s$ and evaluate $\mathbf{J}_s$ on the surface of the cylinder ($\rho = a$). Is it what you were expecting?
(c) Repeat part (b) for magnetic current density $\mathbf{M}_s$.
For parts (a)–(c), you do not have to derive the equations, but must justify the answers.

11.61. Repeat Problem 11-60 when the plane wave is normally incident upon a dielectric cylinder with $\mu_r \gg 1$ and $\varepsilon_r = 1$, as shown in Figure P11-58.

11.62. A TE^z uniform plane wave traveling in the $+x$ direction in free space is incident normally upon a lossless dielectric circular cylinder of radius a, as shown in Figure P11-62. Assume that the incident, scattered, and transmitted (into the cylinder) magnetic fields can be written as

$$\mathbf{H}^i = \hat{\mathbf{a}}_z H_0 \sum_{n=-\infty}^{+\infty} j^{-n} J_n(\beta_0 \rho) e^{jn\phi}$$

$$\mathbf{H}^s = \hat{\mathbf{a}}_z H_0 \sum_{n=-\infty}^{+\infty} a^n H_n^{(2)}(\beta_0 \rho) e^{jn\phi}$$

$$\mathbf{H}^d = \hat{\mathbf{a}}_z H_0 \sum_{n=-\infty}^{+\infty} [b_n J_n(\beta_1 \rho) + c_n Y_n(\beta_1 \rho)] e^{jn\phi}$$

(a) Derive expressions for the incident, scattered, and transmitted electric field components.
(b) Show that the wave amplitude coefficients are equal to

$$c_n = 0$$

$$a_n = j^{-n} \dfrac{\begin{array}{c} J_n'(\beta_0 a) J_n(\beta_1 a) \\ -\sqrt{\mu_r/\varepsilon_r} J_n(\beta_0 a) J_n'(\beta_1 a) \end{array}}{\begin{array}{c} \sqrt{\mu_r/\varepsilon_r} J_n'(\beta_1 a) H_n^{(2)}(\beta_0 a) \\ - J_n(\beta_1 a) H_n^{(2)'}(\beta_0 a) \end{array}}$$

$$b_n = j^{-n} \dfrac{\begin{array}{c} J_n(\beta_0 a) H_n^{(2)'}(\beta_0 a) \\ -J_n'(\beta_0 a) H_n^{(2)}(\beta_0 a) \end{array}}{\begin{array}{c} J_n(\beta_1 a) H_n^{(2)'}(\beta_0 a) \\ -\sqrt{\mu_r/\varepsilon_r} \times J_n'(\beta_1 a) H_n^{(2)}(\beta_0 a) \end{array}}$$

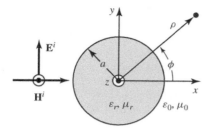

Figure P11-62

11.63. A TE^z uniform plane wave traveling along the $+x$ direction is normally incident upon a dielectric cylinder with $\varepsilon_r \gg 1$ and $\mu_r = 1$, as shown in Figure P11-62. Write an expression for the:
(a) Two-dimensional RCS (SW) expressed in SW $/ \lambda_0$.
(b) Electric current density $\mathbf{J}_s$ and evaluate $\mathbf{J}_s$ on the surface of the cylinder ($\rho = a$). Is it what you were expecting?
(c) Repeat part (b) for magnetic current density $\mathbf{M}_s$.
For parts (a)–(c), you do not have to derive the equations, but must justify the answers.

11.64. Repeat Problem 11.63 when the plane wave is normally incident upon a dielectric cylinder with $\mu \gg 1$ and $\varepsilon_r = 1$, as shown in Figure P11-62.

11.65. A TM^z uniform plane wave traveling in the $+x$ direction in free space is incident normally upon a dielectric-coated conducting circular cylinder of radius a as shown in Figure P11-65. The thickness of the lossless dielectric coating is $b-a$. Assume that the incident, reflected, and transmitted (into the coating) electric fields can be written as shown in Problem 11.58.
(a) Write expressions for the incident, scattered, and transmitted magnetic field components.

(b) Determine the wave amplitude coefficients a_n, b_n, and c_n. Write them in their simplest forms.

11.66. Repeat Problem 11.65 for a TE^z uniform plane wave incidence. Assume that the incident, scattered, and transmitted (into the coating) magnetic fields can be written as shown in Problem 11.62.

11.67. Using the definition of (11-21b), instead of (11-21c), show that the TM^z polarization scattering width reduces to (11-137) through (11-137b).

11.68. Using the definition of (11-21c), instead of (11-21b), show that the TE^z polarization scattering width reduces to (11-155) through (11-155b).

11.69. An electric line source of constant current is placed above a circular PEC cylinder of infinite length, as shown in Figure 11-17 where $\phi' = 90°$. The radius of the cylinder is $a = 50\lambda$. Determine the approximate smallest height (in number of λ) of the line source above the cylinder that will allow the normalized total amplitude pattern to be at $\phi = 90°$:
(a) Maximum.
(b) Minimum.
(c) -3 dB.
Indicate how you arrive at your answers.

11.70. Two infinite length line sources of constant current I and of the same phase are placed near a conducting cylinder along the x axis (one on each side) a distance s from the center of the cylinder, as shown in Figure P11-70.
(a) Neglecting coupling between the sources, write an expression for the total electric field for both sources (assume $\rho > s$).

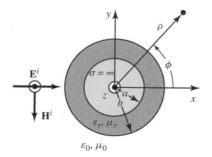

Figure P11-65

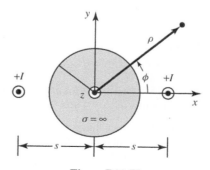

Figure P11-70

(b) Assuming the observations are made at large distances from the cylinder ($\rho \gg s$) and the radius of the cylinder as well as the distance s are very small ($a \ll \lambda$ and $s \ll \lambda$), find the distance s that the sources must be placed so that the electric field at any observation point will vanish. Explain.

11.71. Three infinite length line sources carrying constant magnetic currents of I_m, $2I_m$, and I_m, respectively, are positioned a distance b near a perfect electric conducting cylinder, as shown in Figure P11-71. Neglecting mutual coupling between the sources, find the following.
(a) The total scattered magnetic field when $\rho > b$.
(b) The magnitude of the ratio of the scattered to the incident magnetic field for $\rho > b$.
(c) The normalized total magnetic field pattern when $\beta\rho \to$ large.

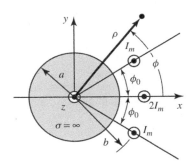

Figure P11-71

11.72. A TM^z uniform plane wave is incident at an angle ϕ' upon a half plane, as shown in Figure P11-72. Show that the current density on the upper side of the half plane is

$$J_z = \frac{E_0}{j2\omega\mu\rho} \sum_{m=1}^{\infty} mj^{m/2} J_{m/2}(\beta\rho)$$

$$\times \sin\left(\frac{m\phi'}{2}\right) \qquad \text{for any } \rho$$

$$J_z \simeq \frac{E_0}{2\eta}\sqrt{\frac{2}{j\pi\beta\rho}} \sin\left(\frac{\phi'}{2}\right) \qquad \text{for } \beta\rho \to 0$$

where E_0 is the amplitude of the incident electric field.

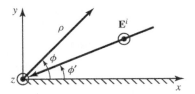

Figure P11-72

11.73. Derive (11-198) from (11-192a).

11.74. Repeat Problem 11.72 for TE^z uniform plane wave incidence. Show that the current density on the upper side of the half plane is

$$J_\rho = H_0 \sum_{m=0}^{\infty} \varepsilon_m j^{m/2} J_{m/2}(\beta\rho)$$

$$\times \cos\left(\frac{m\phi'}{2}\right) \qquad \text{for any } \rho$$

$$J_\rho = H_0 \qquad \text{for } \beta\rho \to 0$$

where H_0 is the amplitude of the incident field and ε_m is defined by (11-196a).

11.75. Verify (11-222) and (11-222a).

11.76. A uniform plane wave is incident upon a conducting sphere of radius a. Assume that the diameter of the sphere is 1.128 m and the frequency is 8.5 GHz.
(a) Determine the monostatic radar cross section of the sphere in decibels per square meter.
(b) Find the area (in square meters) of a flat plate whose normal-incidence monostatic RCS is the same as that of the sphere in part (a).

11.77. A uniform plane wave is normally incident, on a circular PEC ground plane of radius a_p. The same wave is also incident upon a PEC sphere of radius a_s.
(a) Determine the radius of the sphere a_s, in terms of the radius a_p of the circular plate, so that the RCS of the plate and sphere are identical.
(b) If the radius of the plate is 3λ, determine the radius of the sphere so that the plate and sphere will have equal RCS. Assume that the radii of the plate and sphere are sufficiently large, compared to the wavelength, so that both geometrical and physical optics are good approximations.

11.78. A uniform plane wave is incident, at normal incidence, upon a flat PEC plate of area $25\lambda^2$. Determine:
  (a) The 3-D monostatic RCS (in dBsm) of the plate at $f = 10$ GHz based on physical optics.
  (b) The radius (in λ) of a sphere so that it has equal backscattered RCS as the plate. Assume the radius of the sphere is large compared to the wavelength.

11.79. Repeat the calculations of Figure 11-29 by plotting the normalized RCS $[\sigma / (\pi a^2)]$:
  (a) Dimensionless.
  (b) In dB.

11.80. For the scattering of a plane wave by a PEC sphere of radius a, with the incident electric field with only one component (E_x^i), as outlined in Section 11.8:
  (a) Derive an expression, in simplified form, for the cross-polarized component (E_y) of the far-zone scattered electric field in the monostatic direction only.
  (b) Derive an expression, in simplified form, of the 3-D monostatic RCS for the cross-polarized field.
  (c) Plot the normalized RCS $[\sigma / (\pi a^2)]$ of the cross-polarized component E_y for $0 \leq a \leq 2\lambda_0$ (λ_0 is the free-space wavelength), similar to Figure 11-29:
  • Dimensionless.
  • In dB.

If you have any comments to make concerning the monostatic cross-polarized field and associated RCS, please do so.

11.81. Applying the boundary conditions (11-252a) through (11-252d) on the internal and external fields of a dielectric sphere, show that the modal coefficients of the vector potential and fields are those of (11-253a) through (11-253d).

11.82. Show that for $n = 1$, (11-253a) through (11-253d) reduce to (11-254a) through (11-254d).

11.83. A dielectric sphere of radius a, as shown in Figure 11-28, is illuminated by a plane wave with an electric field given by

$$\mathbf{E}^i = \hat{\mathbf{a}}_x E_x^i = \hat{\mathbf{a}}_x E_o e^{-j\beta z} = \hat{\mathbf{a}}_x E_o e^{-j\beta r \cos\theta}$$

The dielectric constant and relative permeability of the material of the sphere are $\varepsilon_r \gg 1$, $\mu_r = 1$. Write an expression for the three-dimensional monostatic RCS of the sphere valid for:
  (a) Any radius.
  (b) Small radius ($a \ll \lambda$; valid only in the so-called Rayleigh region). This expression must be simple.
  (c) Large radius ($a \gg \lambda$; valid only in the so-called optical region). This expression must be simple.

You do not have to derive the expressions, if you think you know what they are. However, justify their use.

CHAPTER **12**

━━━━━━━━━━━━━━━━━━━━━━━━━━━━━━━━

Integral Equations and the Moment Method

12.1 INTRODUCTION

In Chapter 11 we discussed scattering from conducting objects, such as plates, circular cylinders, and spheres, using geometrical optics, physical optics, and modal solutions. For the plates and cylinders, we assumed that their dimensions were of infinite extent. In practice, however, the dimensions of the objects are always finite, although some of them may be very large. Expressions for the radar cross section of finite-size scatterers were introduced in the previous chapter. These, however, represent approximate forms, and more accurate expressions are sometimes desired.

The physical optics method of Chapter 7, Section 7.10, was used in the previous chapter to approximate the current induced on the surface of a finite-size target, such as the strip and rectangular plate. Radiation integrals were then used to find the field scattered by the target. To derive a more accurate representation of the current induced on the surface of the finite-size target, and thus, of the scattered fields, two methods will be examined in this book.

One method, referred to here as the *integral equation* (IE) technique, casts the solution for the induced current in the form of an integral equation (hence its name) where the unknown induced current density is part of the integrand. Numerical techniques, such as the *moment method* (MM) [1–6], can then be used to solve for the current density. Once this is accomplished, the fields scattered by the target can be found using the traditional radiation integrals. The total induced current density will be the sum of the physical optics current density and a *fringe wave* current density [7–13], which can be thought of as a perturbation current density introduced by the edge diffractions of the finite-size structure. This method will be introduced and applied in this chapter.

The other method, referred to here as the *geometrical theory of diffraction* (GTD) [14–17], is an extension of geometrical optics and accounts for the contributions from the edges of the finite structure using diffraction theory. This method will be introduced and applied in Chapters 13 and 14. More extensive discussions of each can be found in the open literature.

Balanis' Advanced Engineering Electromagnetics, Third Edition. Constantine A. Balanis.
© 2024 John Wiley & Sons, Inc. Published 2024 by John Wiley & Sons, Inc.
Companion Website: www.wiley.com/go/balanis/advancedengineeringelectromagnetics3e

12.2 INTEGRAL EQUATION METHOD

The objective of the integral equation (IE) method for scattering is to cast the solution for the unknown current density, which is induced on the surface of the scatterer, in the form of an integral equation where the unknown induced current density is part of the integrand. The integral equation is then solved for the unknown induced current density using numerical techniques such as the *moment method* (MM). To demonstrate the technique, we will initially consider some specific problems. We will start with an electrostatics problem and follow it with time-harmonic problems.

12.2.1 Electrostatic Charge Distribution

In electrostatics, the problem of finding the potential, that is due to a given charge distribution, is often considered. In physical situations, however, it is seldom possible to specify a charge distribution. Whereas we may connect a conducting body to a voltage source, and thus, specify the potential throughout the body, the distribution of charge is obvious only for a few rotationally symmetric canonical geometries. In this section we will consider an integral equation approach to solve for the electric charge distribution, once the electric potential is specified. Some of the material here and in other sections is drawn from [18, 19].

From statics, we know that a linear electric charge distribution $\rho(\mathbf{r}')$ will create an electric potential, $V(\mathbf{r})$, according to [20]

$$V(\mathbf{r}) = \frac{1}{4\pi\varepsilon_0} \int_{\substack{\text{source} \\ \text{(charge)}}} \frac{\rho(\mathbf{r}')}{R} d\ell' \tag{12-1}$$

where $\mathbf{r}'(x', y', z')$ denotes the source coordinates, $\mathbf{r}(x, y, z)$ denotes the observation coordinates, $d\ell'$ is the path of integration, and R is the distance from any point on the source to the observation point, which is generally represented by

$$R(\mathbf{r}, \mathbf{r}') = |\mathbf{r} - \mathbf{r}'| = \sqrt{(x - x')^2 + (y - y')^2 + (z - z')^2} \tag{12-1a}$$

We see that (12-1) may be used to calculate the potentials that are due to any known line charge density. However, the charge distribution on most configurations of practical interest, i.e., complex geometries, is not usually known, even when the potential on the source is given. It is the nontrivial problem of determining the charge distribution, for a specified potential, that is to be solved here using an integral equation approach.

A. Finite Straight Wire Consider a straight wire of length ℓ and radius a, placed along the y axis, as shown in Figure 12-1a. The wire is maintained at a normalized constant electric potential of 1 V.

Note that (12-1) is valid everywhere, including on the wire itself ($V_{\text{wire}} = 1\,\text{V}$). Thus, choosing the observation along the wire axis ($x = z = 0$) and representing the charge density on the surface of the wire by $\rho(y')$, (12-1) can be expressed as

$$1 = \frac{1}{4\pi\varepsilon_0} \int_0^\ell \frac{\rho(y')}{R(y, y')} dy' \qquad 0 \le y \le \ell \tag{12-2}$$

where

$$R(y, y') = R(\mathbf{r}, \mathbf{r}')|_{x=z=0} = \sqrt{(y - y')^2 + [(x')^2 + (z')^2]} = \sqrt{(y - y')^2 + a^2} \tag{12-2a}$$

The observation point is chosen along the wire axis and the charge density is represented along the surface of the wire to avoid $R(y, y') = 0$, which would introduce a singularity in the integrand of (12-2). If the radius of the wire is zero ($a = 0$), then the source (current density) and observation

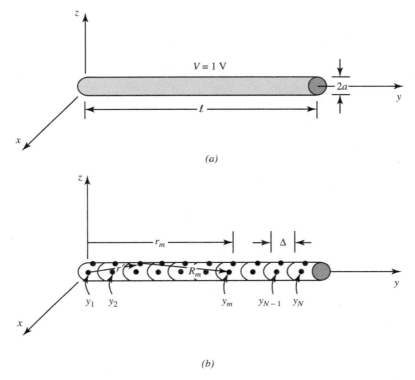

Figure 12-1   (*a*) Straight wire of constant potential. (*b*) Its segmentation.

points will be along the same line (y axis). In that case, to avoid singularities on the distance $R(y, y')$ $[R(y, y') = 0]$, the source (y') and observation point (y) should be chosen not to coincide.

It is necessary to solve (12-2) for the unknown $\rho(y')$ (an inverse problem). Equation 12-2 is an integral equation that can be used to find the charge density $\rho(y')$ based on the 1-V potential. The solution may be reached numerically by reducing (12-2) to a series of linear algebraic equations that may be solved by conventional matrix-equation techniques. To facilitate this, let us approximate the unknown charge distribution $\rho(y')$ by an expansion of N known terms with constant, but unknown, coefficients; that is

$$\rho(y') = \sum_{n=1}^{N} a_n g_n(y') \tag{12-3}$$

Thus, (12-2) may be written, using (12-3), as

$$4\pi\varepsilon_0 = \int_0^\ell \frac{1}{R(y, y')} \left[\sum_{n=1}^{N} a_n g_n(y') \right] dy' \tag{12-4}$$

Because (12-4) is a nonsingular integral, its integration and summation can be interchanged and it can be written as

$$4\pi\varepsilon_0 = \sum_{n=1}^{N} a_n \int_0^\ell \frac{g_n(y')}{\sqrt{(y-y')^2 + a^2}} dy' \tag{12-4a}$$

The wire is now divided into N uniform segments, each of length $\Delta = \ell/N$, as illustrated in Figure 12-1*b*. The $g_n(y')$ functions in the expansion (12-3) are chosen for their ability to accurately model the unknown quantity, while minimizing computation. They are often referred to as *basis* (or expansion) functions, and they will be discussed further in Section 12.2.5. To avoid complexity in this solution, subdomain piecewise constant (or "pulse") functions will be used. These functions, shown in Figure 12-6, are defined to be of a constant value over one segment and zero elsewhere, or

$$g_n(y') = \begin{cases} 0 & y' < (n-1)\Delta \\ 1 & (n-1)\Delta \le y' \le n\Delta \\ 0 & n\Delta < y' \end{cases} \tag{12-5}$$

Many other basis functions are possible, some of which will be introduced later in Section 12.2.5.

Replacing y in (12-4) by a fixed point on the surface of the wire, such as y_m, results in an integrand that is solely a function of y', so the integral may be evaluated. Obviously, (12-4) leads to one equation with N unknowns a_n written as

$$4\pi\varepsilon_0 = a_1 \int_0^\Delta \frac{g_1(y')}{R(y_m, y')} dy' + a_2 \int_\Delta^{2\Delta} \frac{g_2(y')}{R(y_m, y')} dy' + \cdots$$

$$+ a_n \int_{(n-1)\Delta}^{n\Delta} \frac{g_n(y')}{R(y_m, y')} dy' + \cdots + a_N \int_{(N-1)\Delta}^\ell \frac{g_N(y')}{R(y_m, y')} dy' \tag{12-6}$$

In order to obtain a solution for these N amplitude coefficients, N linearly independent equations are necessary. These equations may be produced by choosing an observation point y_m on the surface of the wire and at the center of each Δ length element as shown in Figure 12-1b. This will result in one equation of the form of (12-6) corresponding to each observation point. For N such observation points, we can reduce (12-6) to

$$4\pi\varepsilon_0 = a_1 \int_0^\Delta \frac{g_1(y')}{R(y_1, y')} dy' + \cdots + a_N \int_{(N-1)\Delta}^\ell \frac{g_N(y')}{R(y_1, y')} dy'$$

$$\vdots$$

$$4\pi\varepsilon_0 = a_1 \int_0^\Delta \frac{g_1(y')}{R(y_N, y')} dy' + \cdots + a_N \int_{(N-1)\Delta}^\ell \frac{g_N(y')}{R(y_N, y')} dy' \tag{12-6a}$$

We may write (12-6a) more concisely using matrix notation as

$$[V_m] = [Z_{mn}][I_n] \tag{12-7}$$

where each Z_{mn} term is equal to

$$Z_{mn} = \int_0^\ell \frac{g_n(y')}{\sqrt{(y_m - y')^2 + a^2}} dy' = \int_{(n-1)\Delta}^{n\Delta} \frac{1}{\sqrt{(y_m - y')^2 + a^2}} dy' \tag{12-7a}$$

and

$$[I_n] = [a_n] \tag{12-7b}$$

$$[V_m] = [4\pi\varepsilon_0] \tag{12-7c}$$

The V_m column matrix has all terms equal to $4\pi\varepsilon_0$, and the $I_n = a_n$ values are the unknown charge distribution coefficients. Solving (12-7) for $[I_n]$ gives

$$[I_n] = [a_n] = [Z_{mn}]^{-1}[V_m] \tag{12-8}$$

Either (12-7) or (12-8) may readily be solved on a digital computer by using any of a number of matrix-inversion or equation-solving routines. Whereas the integrals involved here may be evaluated in closed form by making appropriate approximations, this is not usually possible with more complicated problems. Efficient numerical integration computer subroutines are commonly available in easy-to-use forms.

One closed form evaluation of (12-7a) is to reduce the integral and represent it by

$$
Z_{mn} = \begin{cases}
2\ln\left[\dfrac{\dfrac{\Delta}{2} + \sqrt{a^2 + \left(\dfrac{\Delta}{2}\right)^2}}{a}\right] & m = n & \text{(12-9a)} \\[4ex]
\ln\left\{\dfrac{d_{mn}^{+} + [(d_{mn}^{+})^2 + a^2]^{1/2}}{d_{mn}^{-} + [(d_{mn}^{-})^2 + a^2]^{1/2}}\right\} & m \neq n \text{ but } |m-n| \leq 2 & \text{(12-9b)} \\[3ex]
\ln\left(\dfrac{d_{mn}^{+}}{d_{mn}^{-}}\right) & |m-n| > 2 & \text{(12-9c)}
\end{cases}
$$

where

$$ d_{mn}^{+} = \ell_m + \frac{\Delta}{2} \tag{12-9d} $$

$$ d_{mn}^{-} = \ell_m - \frac{\Delta}{2} \tag{12-9e} $$

ℓ_m is the distance between the mth matching point and the center of the nth source point.

In summary, the solution of (12-2) for the charge distribution on a wire has been accomplished by approximating the unknown with some basis functions, dividing the wire into segments, and then sequentially enforcing (12-2) at the center of each segment to form a set of linear equations.

Even for the relatively simple straight wire geometry we have discussed, the exact form of the charge distribution is not intuitively apparent. To illustrate the principles of the numerical solution, an example is now presented.

Example 12-1

A 1-m-long straight wire of radius $a = 0.001$ m is maintained at a constant potential of 1 V. Determine the linear charge distribution on the wire by dividing the length into 5 and 20 uniform segments. Assume subdomain pulse basis functions.

Solution:

1. $N = 5$. When the 1-m-long wire is divided into five uniform segments each of length $\Delta = 0.2$ m, (12-7) reduces to

$$
\begin{bmatrix}
10.60 & 1.10 & 0.51 & 0.34 & 0.25 \\
1.10 & 10.60 & 1.10 & 0.51 & 0.34 \\
0.51 & 1.10 & 10.60 & 1.10 & 0.51 \\
0.34 & 0.51 & 1.10 & 10.60 & 1.10 \\
0.25 & 0.34 & 0.51 & 1.10 & 10.60
\end{bmatrix}
\begin{bmatrix}
a_1 \\ a_2 \\ a_3 \\ a_4 \\ a_5
\end{bmatrix}
=
\begin{bmatrix}
1.11 \times 10^{-10} \\
1.11 \times 10^{-10} \\
\vdots \\
1.11 \times 10^{-10}
\end{bmatrix}
$$

Inverting this matrix leads to the amplitude coefficients and subsequent charge distribution of

$$ a_1 = 8.81 \text{ pC/m} $$
$$ a_2 = 8.09 \text{ pC/m} $$
$$ a_3 = 7.97 \text{ pC/m} $$
$$ a_4 = 8.09 \text{ pC/m} $$
$$ a_5 = 8.81 \text{ pC/m} $$

The charge distribution is shown plotted in Figure 12-2*a*.
2. $N = 20$. Increasing the number of segments to 20 results in a much smoother distribution, as shown plotted in Figure 12-2*b*. As more segments are used, a better approximation of the actual charge distribution is attained, which has smaller discontinuities over the length of the wire.

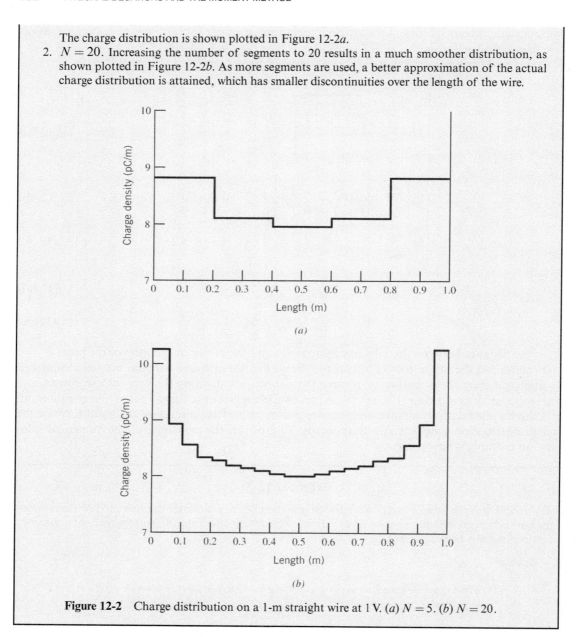

Figure 12-2   Charge distribution on a 1-m straight wire at 1 V. (*a*) $N = 5$. (*b*) $N = 20$.

B. Bent Wire   In order to illustrate the solution for a more complex structure, let us analyze a body composed of two noncollinear straight wires, that is, a bent wire. If a straight wire is bent, the charge distribution will be altered, although the solution to find it will differ only slightly from the straight wire case. We will assume a bend of angle α, which remains in the yz plane, as shown in Figure 12-3.

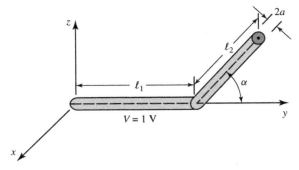

Figure 12-3   Geometry for bent wire.

For the first segment ℓ_1 of the wire, the distance R can be represented by (12-2a). However, for the second segment ℓ_2, we can express the distance as

$$R = \sqrt{(y-y')^2 + (z-z')^2}$$ (12-10)

Also because of the bend, the integral in (12-7a) must be separated into two parts of

$$Z_{mn} = \int_0^{\ell_1} \frac{\rho_n(\ell_1')}{R} d\ell_1' + \int_0^{\ell_2} \frac{\rho_n(\ell_2')}{R} d\ell_2'$$ (12-11)

where ℓ_1 and ℓ_2 are measured along the corresponding straight sections from their left ends.

Example 12-2

Repeat Example 12-1 assuming that the wire has been bent 90° at its midpoint. Subdivide the entire wire into 20 uniform segments.

Solution: The charge distribution for this case, calculated using (12-10) and (12-11), is plotted in Figure 12-4 for $N = 20$ segments. Note that the charge is relatively more concentrated near the ends of this structure than was the case for a straight wire of Figure 12-2b. Further, the overall charge density, and thus capacitance, on the structure has decreased.

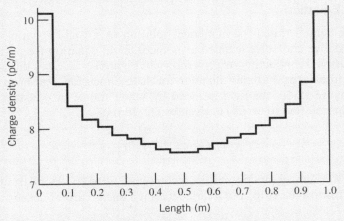

Figure 12-4   Charge distribution on a 1-m bent wire ($\alpha = 90°$, $N = 20$).

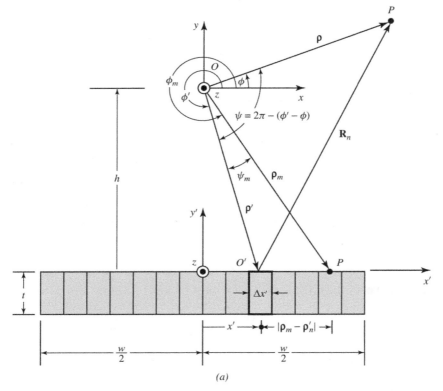

(a)

Figure 12-5   Geometry of a line source above a two-dimensional finite width strip. (*a*) Boundary conditions and integration on the same surface. (*b*) Boundary conditions and integration not on the same surface.

Arbitrary wire configurations, including numerous bends and even curved sections, may be analyzed by the methods already outlined here. As with the simple bent wire, the only alterations generally necessary are those required to describe the geometry analytically.

12.2.2   Integral Equation

Now that we have demonstrated the numerical solution of a well-known electrostatics integral equation, we will derive and solve a time-harmonic integral equation for an infinite line source above a two-dimensional conducting strip, as shown in Figure 12-5a. Once this is accomplished, we will generalize the integral equation formulation for three-dimensional problems in Section 12.3.

Referring to Figure 12-5a, the field radiated by a line source of constant current I_z in the absence of the strip (referred to as E_z^d) is given by (11-10a) or

$$E_z^d(\rho) = -\frac{\beta^2 I_z}{4\omega\varepsilon} H_0^{(2)}(\beta\rho) \tag{12-12}$$

where $H_0^{(2)}(\beta\rho)$ is the Hankel function of the second kind of order zero. Part of the field given by (12-12) is directed toward the strip, and it induces on it a linear current density J_z (in amperes per meter) such that

$$J_z(x')\Delta x' = \Delta I_z(x') \tag{12-13}$$

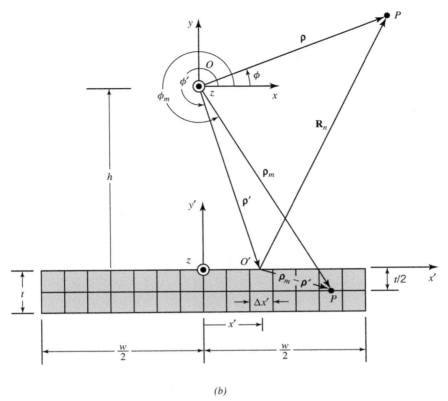

(b)

Figure 12-5   (*Continued*)

which, as $\Delta x' \to 0$, can be written as

$$J_z(x')dx' = dI_z(x') \tag{12-13a}$$

The induced current of (12-13a) reradiates and produces an electric field component that will be referred to as *reflected* (or *scattered*) and designated as $E_z^r(\rho)$ [or $E_z^s(\rho)$]. If the strip is subdivided into N segments, each of width $\Delta x_n'$ as shown in Figure 12-5b, the scattered field can be written, according to (12-12), as

$$E_z^s(\rho) = -\frac{\beta^2}{4\omega\varepsilon}\sum_{n=1}^{N}H_0^{(2)}(\beta R_n)\,\Delta I_z(x_n') = -\frac{\beta^2}{4\omega\varepsilon}\sum_{n=1}^{N}H_0^{(2)}(\beta R_n)J_z(x_n')\,\Delta x_n' \tag{12-14}$$

where x_n' is the position of the nth segment. In the limit, as each segment becomes very small ($\Delta x_n \to 0$), (12-14) can be written as

$$E_z^s(\rho) = -\frac{\beta^2}{4\omega\varepsilon}\int_{\text{strip}}H_0^{(2)}(\beta R)\,dI_z = -\frac{\beta^2}{4\omega\varepsilon}\int_{-w/2}^{w/2}J_z(x')H_0^{(2)}(\beta\,|\rho-\rho'|)dx' \tag{12-15}$$

since

$$R = |\rho - \rho'| = \sqrt{\rho^2 + (\rho')^2 - 2\rho\rho'\cos(\phi - \phi')} \tag{12-15a}$$

The total field at any observation point, including the strip itself, will be the sum of the direct component E_z^d of (12-12) and the scattered component E_z^s of (12-15). However, to determine the scattered component, we need to know the induced current density $J_z(x')$. The objective here then will be to find an equation, which in this case will be in terms of an integral and will be referred to as an *integral equation*, that can be used to determine $J_z(x')$. This can be accomplished by choosing the observation point on the strip itself. Doing this, we have that for any observation point $\rho = \rho_m$ on the strip, the total tangential electric field vanishes and it is given by

$$E_z^t(|\rho = \rho_m|)|_{strip} = [E_z^d(|\rho = \rho_m|) + E_z^s(|\rho = \rho_m|)]_{strip} = 0 \qquad (12\text{-}16)$$

or

$$E_z^d(|\rho = \rho_m|)|_{strip} = -E_z^s(|\rho = \rho_m|)|_{strip} \qquad (12\text{-}16a)$$

Using (12-12) and (12-15), we can write (12-16a) as

$$-\frac{\beta^2 I_z}{4\omega\varepsilon} H_0^{(2)}(\beta |\rho_m|) = +\frac{\beta^2}{4\omega\varepsilon} \int_{-w/2}^{w/2} J_z(x') H_0^{(2)}(\beta |\rho_m - \rho'|) dx' \qquad (12\text{-}17)$$

which for a unit current I_z (i.e., $I_z = 1$) reduces to

$$\boxed{H_0^{(2)}(\beta |\rho_m|) = -\int_{-w/2}^{w/2} J_z(x') H_0^{(2)}(\beta |\rho_m - \rho'|) dx'} \qquad (12\text{-}17a)$$

Equation 12-17a is the *electric field integral equation* (EFIE) for the line source above the strip, and it can be used to find the current density $J_z(x')$ based upon a unit current I_z. If I_z is of any other constant value, then all the values of $J_z(x')$ must be multiplied by that same constant value. Electric field integral equations (EFIE) and magnetic field integral equations (MFIE) are discussed in more general forms in Section 12.3.

12.2.3 Radiation Pattern

Once J_z is found, we can then determine the total field radiated of the entire system for any observation point. The total field is composed of two parts: the field radiated from the line source itself (E_z^d) and that which is scattered (reradiated) from the strip (E_z^s). Thus, using (12-12) and (12-15), we can write the total field as

$$E_z^t(\rho) = E_z^d(\rho) + E_z^s(\rho)$$
$$= -\frac{\beta^2 I_z}{4\omega\varepsilon} H_0^{(2)}(\beta\rho) - \frac{\beta^2}{4\omega\varepsilon} \int_{-w/2}^{w/2} J_z(x') H_0^{(2)}(\beta |\rho - \rho'|) dx' \qquad (12\text{-}18)$$

which for a unit amplitude current I_z ($I_z = 1$) reduces to

$$\boxed{E_z^t(\rho) = -\frac{\beta^2}{4\omega\varepsilon} \left[H_0^{(2)}(\beta\rho) + \int_{-w/2}^{w/2} J_z(x') H_0^{(2)}(\beta |\rho - \rho'|) dx' \right]} \qquad (12\text{-}18a)$$

Equation 12-18a can be used to find the total field at any observation point, near or far field. The current density $J_z(x')$ can be found using (12-17a). However, for far-field observations, (12-18a) can be approximated and written in a more simplified form. In general, the distance

R is given by (12-15a). However, for far-field observations ($\rho \gg \rho'$), (12-15a) reduces, using the binomial expansion, to

$$R \simeq \begin{cases} \rho - \rho'\cos(\phi-\phi') & \text{for phase terms} \qquad\qquad \text{(12-19a)} \\ \rho & \text{for amplitude terms} \qquad \text{(12-19b)} \end{cases}$$

For large arguments, the Hankel functions in (12-18a) can be replaced by their asymptotic form

$$H_n^{(2)}(\beta z) \overset{\beta z \to \infty}{\simeq} \sqrt{\frac{2j}{\pi\beta z}}\, j^n e^{-j\beta z} \tag{12-20}$$

For $n = 0$, (12-20) reduces to

$$H_0^{(2)}(\beta z) \simeq \sqrt{\frac{2j}{\pi\beta z}}\, e^{-j\beta z} \tag{12-20a}$$

Using (12-19a) through (12-20a), we can write the Hankel functions in (12-18a) as

$$H_0^{(2)}(\beta\rho) \simeq \sqrt{\frac{2j}{\pi\beta\rho}}\, e^{-j\beta\rho} \tag{12-21a}$$

$$H_0^{(2)}(\beta|\boldsymbol{\rho}-\boldsymbol{\rho}'|) \simeq \sqrt{\frac{2j}{\pi\beta\rho}}\, e^{-j\beta[\rho-\rho'\cos(\phi-\phi')]}$$

$$\simeq \sqrt{\frac{2j}{\pi\beta\rho}}\, e^{-j\beta\rho+j\beta\rho'\cos(\phi-\phi')} \tag{12-21b}$$

When (12-21a) and (12-21b) are substituted into (12-18a), they reduce it to

$$E_z^t(\rho) \simeq -\frac{\beta^2}{4\omega\varepsilon}\sqrt{\frac{2j}{\pi\beta\rho}}\, e^{-j\beta\rho}\left[1+\int_{-w/2}^{+w/2} J_z(x')\, e^{j\beta\rho'\cos(\phi-\phi')}\, dx'\right] \tag{12-22}$$

which in normalized form can be written as

$$\boxed{E_z^t\,(\text{normalized}) \simeq 1+\int_{-w/2}^{w/2} J_z(x')\, e^{j\beta\rho'\cos(\phi-\phi')} dx'} \tag{12-22a}$$

Equation 12-22a represents the normalized pattern of the line above the strip. It is based on the linear current density $J_z(x')$ that is induced by the source on the strip. The current density can be found using approximate methods or, more accurately, using the electric field integral equation 12-17a.

12.2.4 Point-Matching (Collocation) Method

The next step will be to use a numerical technique to solve the electric field integral equation 12-17a for the unknown current density $J_z(x')$. We first expand $J_z(x')$ into a finite series of the form

$$J_z(x') \simeq \sum_{n=1}^{N} a_n g_n(x') \tag{12-23}$$

where $g_n(x')$ represents *basis (expansion)* functions [1, 2]. When (12-23) is substituted into (12-17a), we can write it as

$$H_0^{(2)}(\beta\,|\rho_m|) = -\int_{-w/2}^{w/2} \sum_{n=1}^{N} a_n g_n(x') H_0^{(2)}(\beta\,|\rho_m - \rho_n'|)\,dx'$$

$$H_0^{(2)}(\beta\,|\rho_m|) = -\sum_{n=1}^{N} a_n \int_{-w/2}^{w/2} g_n(x') H_0^{(2)}(\beta\,|\rho_m - \rho_{ns}'|)\,dx' \qquad (12\text{-}24)$$

which takes the general form

$$h = \sum_{n=1}^{N} a_n F(g_n) \qquad (12\text{-}25)$$

where

$$h = H_0^{(2)}(\beta\,|\rho_m|) \qquad (12\text{-}25a)$$

$$F(g_n) = -\int_{-w/2}^{w/2} g_n(x') H_0^{(2)}(\beta\,|\rho_m - \rho_n'|)\,dx' \qquad (12\text{-}25b)$$

In (12-25), F is referred to as a *linear integral operator*, g_n represents the response function, and h is the known excitation function.

Equation (12-17a) is an electric field integral equation derived by enforcing the boundary conditions of vanishing total tangential electric field on the surface of the conducting strip. A numerical solution of (12-17a) is (12-24) or (12-25) through (12-25b), which, for a given observation point $\rho = \rho_m$, leads to one equation with N unknowns. This can be repeated N times by choosing N observation points. Such a procedure leads to a system of N linear equations each with N unknowns of the form

$$H_0^{(2)}(\beta\,|\rho_m|) = \sum_{n=1}^{N} a_n \left[-\int_{-w/2}^{w/2} g_n(x') H_0^{(2)}(\beta\,|\rho_m - \rho_n'|)\,dx' \right]$$

$$m = 1, 2, \ldots, N \qquad (12\text{-}26)$$

which can also be written as

$$V_m = \sum_{n=1}^{N} I_n Z_{mn} \qquad (12\text{-}27)$$

where

$$V_m = H_0^{(2)}(\beta\,|\rho_m|) \qquad (12\text{-}27a)$$

$$I_n = a_n \qquad (12\text{-}27b)$$

$$Z_{mn} = -\int_{-w/2}^{w/2} g_n(x') H_0^{(2)}(\beta\,|\rho_m - \rho_n'|)\,dx' \qquad (12\text{-}27c)$$

In matrix form, (12-27) can be expressed as

$$\boxed{[V_m] = [Z_{mn}][I_n]}$$

(12-28)

where the unknown is $[I_n]$ and can be found by solving (12-28), or

$$\boxed{[I_n] = [Z_{mn}]^{-1}[V_m]}$$

(12-28a)

Since the system of N linear equations with N unknowns—as given by (12-26), (12-27), or (12-28)—was derived by applying the boundary conditions at N discrete points, the technique is referred to as the *point-matching* (or *collocation*) *method* [1, 2].

Thus, by finding the elements of the $[V]$ and $[Z]$, and then the inverse $[Z]^{-1}$, we can determine the elements a_n of the $[I]$ matrix. This in turn allows us to approximate $J_z(x')$ using (12-23), which can then be used in (12-18a) to find the total field everywhere. However, for far-field observations, the total field can be found more easily using (12-22) or, in normalized form, (12-22a).

12.2.5 Basis Functions

One very important step in any numerical solution is the choice of basis functions. In general, one chooses as basis functions the set that has the ability to accurately represent and resemble the anticipated unknown function, while minimizing the computational effort required to employ it [21–23]. Do not choose basis functions with smoother properties than the unknown being represented.

Theoretically, there are many possible basis sets. However, only a limited number are discussed here. These sets may be divided into two general classes. The first class consists of subdomain functions, which are nonzero only over a part of the domain of the function $g(x')$; its domain is the surface of the structure. The second class contains entire domain functions that exist over the entire domain of the unknown function. The entire domain basis function expansion is analogous to the well-known Fourier series expansion method.

A. Subdomain Functions Of the two types of basis functions, subdomain functions are the most common. Unlike entire domain bases, they may be used without prior knowledge of the nature of the function that they must represent.

The subdomain approach involves subdivision of the structure into N nonoverlapping segments, as illustrated on the axis in Figure 12-6a. For clarity, the segments are shown here to be collinear and of equal length, although neither condition is necessary. The basis functions are defined in conjunction with the limits of one or more of the segments.

Perhaps the most common of these basis functions is the conceptually basic piecewise constant, or "pulse" function, shown in Figure 12-6a. It is defined by

Piecewise Constant

$$g_n(x') = \begin{cases} 1 & x'_{n-1} \leq x' \leq x'_n \\ 0 & \text{elsewhere} \end{cases}$$

(12-29)

Once the associated coefficients are determined, this function will produce a staircase representation of the unknown function, similar to that in Figures 12-6b and 12-6c.

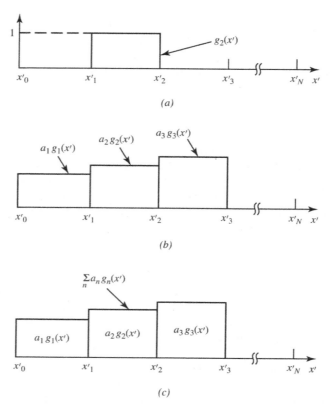

Figure 12-6 Piecewise constant subdomain functions. (*a*) Single. (*b*) Multiple. (*c*) Function representation.

Another common basis set is the piecewise linear, or "triangle," functions seen in Figure 12-7*a*. These are defined by

Piecewise Linear

$$g_n(x') = \begin{cases} \dfrac{x' - x'_{n-1}}{x'_n - x'_{n-1}} & x'_{n-1} \le x' \le x'_n \\ \dfrac{x'_{n+1} - x'}{x'_{n+1} - x'_n} & x'_n \le x' \le x'_{n+1} \\ 0 & \text{elsewhere} \end{cases} \tag{12-30}$$

and are seen to cover two segments, and overlap adjacent functions (Figure 12-7*b*). The resulting representation (Figures 12-7*b* and 12-7*c*) is smoother than that for "pulses," but at the cost of somewhat increased computational complexity.

Increasing the sophistication of subdomain basis functions beyond the level of the "triangle" may not be warranted by the possible improvement in accuracy. However, there are cases where more specialized functions are useful for other reasons. For example, some integral operators may be evaluated without numerical integration when their integrands are multiplied by a $\sin(kx')$ or $\cos(kx')$ function, where x' is the variable of integration. In such examples, considerable advantages in computation time and resistance to errors can be gained by using basis functions

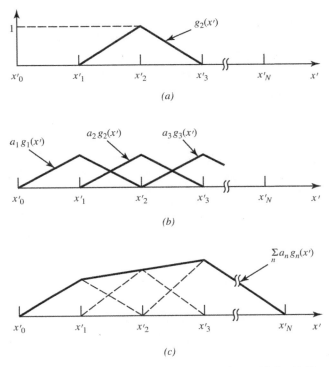

Figure 12-7 Piecewise linear subdomain functions. (*a*) Single. (*b*) Multiple. (*c*) Function representation.

like the piecewise sinusoid of Figure 12-8 or truncated cosine of Figure 12-9. These functions are defined by

Piecewise Sinusoid

$$g_n(x') = \begin{cases} \dfrac{\sin[\beta(x' - x'_{n-1})]}{\sin[\beta(x'_n - x'_{n-1})]} & x'_{n-1} \le x' \le x'_n \\[2mm] \dfrac{\sin[\beta(x'_{n+1} - x')]}{\sin[\beta(x'_{n+1} - x'_n)]} & x'_n \le x' \le x'_{n+1} \\[2mm] 0 & \text{elsewhere} \end{cases} \qquad (12\text{-}31)$$

Truncated Cosine

$$g_n(x') = \begin{cases} \cos\left[\beta\left(x' - \dfrac{x'_n - x'_{n-1}}{2}\right)\right] & x'_{n-1} \le x' \le x'_n \\[2mm] 0 & \text{elsewhere} \end{cases} \qquad (12\text{-}32)$$

B. Entire-Domain Functions   Entire domain basis functions, as their name implies, are defined and are nonzero over the entire length of the structure being considered. Thus, no segmentation is involved in their use.

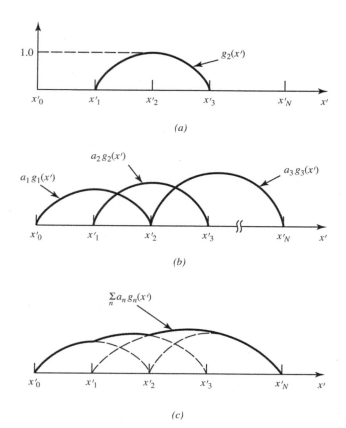

Figure 12-8 Piecewise sinusoids subdomain functions. (*a*) Single. (*b*) Multiple. (*c*) Function representation.

A common entire domain basis set is that of sinusoidal functions, where

Entire Domain

$$g_n(x') = \cos\left[\frac{(2n-1)\pi x'}{\ell}\right] \qquad -\frac{\ell}{2} \le x' \le \frac{\ell}{2} \tag{12-33}$$

Note that this basis set would be particularly useful for modeling the current distribution on a wire dipole, which is known to have primarily sinusoidal distribution. The main advantage of entire domain basis functions lies in problems where the unknown function is known *a priori* to follow a certain pattern. Such entire-domain functions may render an acceptable representation of the unknown while using far fewer terms in the expansion of (12-23) than would be necessary for subdomain bases. Representation of a function by entire domain cosine and/or sine functions is similar to the Fourier series expansion of arbitrary functions.

Because we are constrained to use a finite number of functions (or *modes*, as they are sometimes called), entire domain basis functions usually have difficulty in modeling arbitrary or complicated unknown functions.

Entire domain basis functions, sets like (12-33), can be generated using Tschebyscheff, Maclaurin, Legendre, and Hermite polynomials, or other convenient functions.

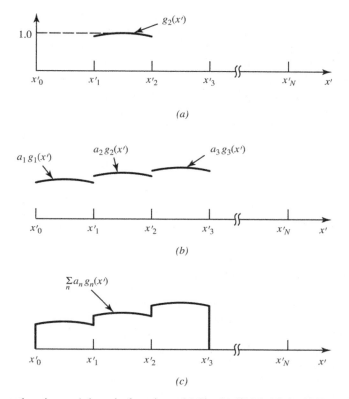

Figure 12-9   Truncated cosines subdomain functions. (*a*) Single. (*b*) Multiple. (*c*) Function representation.

12.2.6   Application of Point Matching

If each of the expansion functions $g_n(x')$ in (12-23) is of the subdomain type, where each exists only over one segment of the structure, then Z_{mn} of (12-27c) reduces to

$$Z_{mn} = -\int_{x_n}^{x_{n+1}} g_n(x')H_0^{(2)}(\beta|\rho_m - \rho_n'|)\,dx' \qquad (12\text{-}34)$$

where x_n and x_{n+1} represent, respectively, the lower and upper limits of the segment over which each of the subdomain expansion functions $g_n(x')$ exists. If, in addition, the g_n are subdomain pulse expansion functions of the form

$$g_n(x') = \begin{cases} 1 & x_n \le x' \le x_{n+1} \\ 0 & \text{elsewhere} \end{cases} \qquad (12\text{-}35)$$

then (12-34) reduces to

$$Z_{mn} = -\int_{x_n}^{x_{n+1}} H_0^{(2)}(\beta|\rho_m - \rho_n'|)\,dx' \qquad (12\text{-}36)$$

The preceding integral cannot be evaluated exactly in closed form. However, there exist various approximations for its evaluation.

In solving (12-17a) using (12-24) or (12-26), there are few problems that must be addressed. Before we do that, let us first state in words what (12-24) and (12-26) represent. Each equation is a solution to (12-17a), which was derived by enforcing the boundary conditions. These conditions required the total tangential electric field to vanish on the surface of the conductor. For each observation point, the total field consists of the sum of the direct (E_z^d) and scattered (E_z^s) components. Thus, to find the total scattered field at *each observation point*, we must add the contributions of the scattered field components *from all* the segments of the strip, which also includes those coming from the segment where the observations are made (referred to as *self-terms*). When the contributions from the segment over which the observation point lies are considered, the distance $R_{mn} = R_{mm} = |\rho_m - \rho'_m|$ used for evaluating the self-term Z_{mm} in (12-27c) will become zero. This introduces a singularity in the integrand of (12-36) because the Hankel function defined as

$$H_0^{(2)}(\beta\rho) = J_0(\beta\rho) - jY_0(\beta\rho) \tag{12-37}$$

is infinite since $Y_0(0) = \infty$.

For finite thickness strips, the easiest way to get around the problem of evaluating the Hankel function for the self-terms will be to choose observation points away from the surface of the strip over which the integration in (12-36) is performed. For example, the observation points can be selected at the center of each segment along a line that divides the thickness of the strip, while the integration is performed along the upper surface of the strip. These points are designated in Figure 12-5b by the distance $|\rho_m - \rho'|$.

Even if the aforementioned procedure is implemented for the evaluation of all the terms of Z_{mn}, including the self-terms, the distance $R_{mn} = |\rho_m - \rho'_n|$ for the self-terms (and some from the neighboring elements) will sometimes be sufficiently small that standard algorithms for computing Bessel functions, and thus, Hankel functions, may not be very accurate. For these cases the Hankel functions can be evaluated using asymptotic expressions for small arguments. That is, for cases where the argument of the Hankel functions in (12-36) is small, which may include the self-terms and some of the neighboring elements, the Hankel function can be computed using [24]

$$H_0^{(2)}(\beta\rho) = J_0(\beta\rho) - jY_0(\beta\rho) \overset{\beta\rho\to 0}{\simeq} 1 - j\frac{2}{\pi}\ln\left(\frac{1.781\beta\rho}{2}\right) \tag{12-38}$$

The integral of (12-36) can be evaluated approximately in closed form, even if the observation and source points are chosen to be along the same line. This can be done not only for diagonal (self, i.e., $m = n$) terms but also for the nondiagonal ($m \neq n$) terms. For the diagonal terms ($m = n$), the Hankel function of (12-36) has an integrable singularity, and the integral can be evaluated analytically in closed form using the small argument approximation of (12-38) for the Hankel function. When (12-38) is used, it can be shown that (12-36) reduces to [2]

<u>Diagonal Terms Approximation</u>

$$Z_{nn} \simeq -\Delta x_n\left[1 - j\frac{2}{\pi}\ln\left(\frac{1.781\beta\Delta x_n}{4e}\right)\right] \tag{12-39}$$

where

$$\Delta x_n = x_{n+1} - x_n \tag{12-39a}$$

$$e \approx 2.718 \tag{12-39b}$$

For evaluation of the nondiagonal terms of (12-36), the crudest approximation would be to consider the Hankel function over each segment to be essentially constant [2]. To minimize the error using such an approximation, it is recommended that the argument of the Hankel function in (12-36) be represented by its average value over each segment. For straight line segments that average value will be representative of the distance from the center of the segment to the observation point. Thus, for the nondiagonal terms, (12-36) can be approximated by

Nondiagonal Terms Approximation

$$Z_{mn} \simeq -\Delta x_n H_0^{(2)}(\beta \, |R_{mn}|_{av}) = -\Delta x_n H_0^{(2)}(\beta \, |\rho_m - \rho_n'|_{av}) \qquad m \neq n \qquad (12\text{-}40)$$

The average-value approximation for the distance R_{mn} in the Hankel function evaluation of (12-36) can also be used for curved surface scattering by approximating each curved segment by a straight line segment. Crude as it may seem, the average-value approximation for the distance yields good results.

12.2.7   Weighting (Testing) Functions

Application of (12-24) for one observation point leads to one equation with N unknowns. It alone is not sufficient to determine the N unknown a_n ($n = 1, 2, ..., N$) constants. To resolve the N constants, it is necessary to have N linearly independent equations. This can be accomplished by evaluating (12-24) (i.e., applying boundary conditions) at N different points, as represented by (12-26). To improve the point-matching solution, an inner product $\langle w, g \rangle$ can be defined, which is a scalar operation satisfying the laws of

$$\langle w, g \rangle = \langle g, w \rangle \qquad (12\text{-}41a)$$
$$\langle bf + cg, w \rangle = b\langle f, w \rangle + c\langle g, w \rangle \qquad (12\text{-}41b)$$
$$\langle g^*, g \rangle > 0 \quad \text{if } g \neq 0 \qquad (12\text{-}41c)$$
$$\langle g^*, g \rangle = 0 \quad \text{if } g = 0 \qquad (12\text{-}41d)$$

where b and c are scalars and the asterisk (*) indicates complex conjugation. Note that the functions w and g can be vectors. A typical, but not unique, inner product is

$$\langle \mathbf{w}, \mathbf{g} \rangle = \iint_S \mathbf{w}^* \cdot \mathbf{g} \, ds \qquad (12\text{-}42)$$

where the w's are the *weighting* (*testing*) functions and S is the surface of the structure being analyzed. This technique is known better as the *moment method* or *method of moments* (MM, MoM) [1, 2].

12.2.8   Moment Method

The collocation (point-matching) method is a numerical technique whose solutions satisfy the electromagnetic boundary conditions (e.g., vanishing tangential electric fields on the surface of an electric conductor) only at discrete points. Between these points the boundary conditions may not be satisfied, and we define the deviation as a *residual* [e.g., residual $= \Delta E|_{tan} = E(\text{scattered})|_{tan} + E(\text{incident})|_{tan} \neq 0$ on the surface of an electric conductor]. For a half-wavelength dipole, a typical residual is shown in Figure 12-10a for pulse-basis functions and point matching and Figure 12-10b exhibits the residual for piecewise sinusoidal Galerkin's method [25]. As expected, the pulse-basis point matching exhibits the most ill-behaved residual and the piecewise sinusoidal Galerkin's method indicates an improved residual. To minimize the residual in such a way that its overall average over the entire structure approaches zero, the method of *weighted residuals* is utilized in conjunction with the inner product of (12-42). This technique, referred to as the

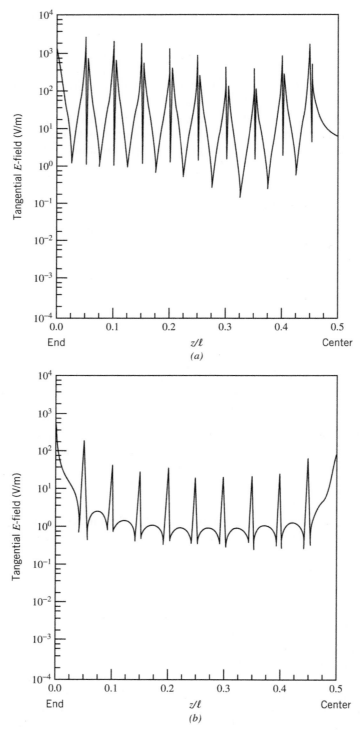

Figure 12-10   Tangential electric field on the conducting surface of the $\lambda/2$ dipole. (a) Pulse basis-point matching. (b) Piecewise sinusoidal Galerkin's method. (Source: E. K. Miller and F. J. Deadrick, "Some computational aspects of thin-wire modeling" in *Numerical and Asymptotic Techniques in Electromagnetics*, 1975, Springer-Verlag.)

Moment Method (MM), does not lead to a vanishing residual at every point on the surface of a conductor, but it forces the boundary conditions to be satisfied in an average sense over the entire surface.

To accomplish this, we define a set of N *weighting* (or testing) functions $\{w_m\}$ $(m=1,2,...,N)$ in the domain of the operator F. Forming the inner product between each of these functions and the excitation function h, (12-25) reduces to

$$\langle w_m, h \rangle = \sum_{n=1}^{N} a_n \langle w_m, F(g_n) \rangle \qquad m=1,2,...,N \tag{12-43}$$

This set of N equations may be written in matrix form as

$$[h_m] = [F_{mn}][a_n] \tag{12-44}$$

where

$$[F_{mn}] = \begin{bmatrix} \langle w_1, F(g_1) \rangle & \langle w_1, F(g_2) \rangle & \cdots \\ \langle w_2, F(g_1) \rangle & \langle w_2, F(g_2) \rangle \\ \vdots & & \vdots \end{bmatrix} \tag{12-44a}$$

$$[a_n] = \begin{bmatrix} a_1 \\ a_2 \\ \vdots \\ a_N \end{bmatrix} \qquad h_m = \begin{bmatrix} \langle w_1, h \rangle \\ \langle w_2, h \rangle \\ \vdots \\ \langle w_N, h \rangle \end{bmatrix} \tag{12-44b}$$

The matrix of (12-44) may be solved for the a_n by inversion, and it can be written as

$$[a_n] = [F_{mn}]^{-1}[h_m] \tag{12-45}$$

The choice of weighting functions is important in that the elements of $\{w_n\}$ must be linearly independent, so that the N equations in (12-43) will be linearly independent [1–3, 22, 23]. Further, it will generally be advantageous to choose weighting functions that minimize the computations required to evaluate the inner products.

The condition of linear independence between elements and the advantage of computational simplicity are also important characteristics of basis functions. Because of this, similar types of functions are often used for both weighting and expansion. A particular choice of functions may be to let the weighting and basis function be the same, that is, $w_n = g_n$. This technique is known as *Galerkin's method* [26].

It should be noted that there are N^2 terms to be evaluated in (12-44a). Each term usually requires two or more integrations; at least one to evaluate each $F(g_n)$, and one to perform the inner product of (12-42). When these integrations are to be done numerically, as is often the case, vast amounts of computation time may be necessary.

There is, however, a unique set of weighting functions that reduce the number of required integrations. This is the set of Dirac delta weighting functions

$$\{w_m\} = \{\delta(p-p_m)\} = \{\delta(p-p_1), \delta(p-p_2), ..., \delta(p-p_N)\} \tag{12-46}$$

where p specifies a position with respect to some reference (origin), and p_m represents a point at which the boundary condition is enforced. Using (12-42) and (12-46) reduces (12-43) to

$$\langle \delta(p-p_m), h \rangle = \sum_n a_n \langle \delta(p-p_m), F(g_n) \rangle, \qquad m = 1, 2, ..., N$$

$$\iint_S \delta(p-p_m) h \, ds = \sum_n a_n \iint_S \delta(p-p_m) F(g_n) \, ds, \qquad m = 1, 2, ..., N$$

$$\boxed{h|_{p=p_m} = \sum_n a_n F(g_n)|_{p=p_m}, \qquad m = 1, 2, ..., N} \qquad (12\text{-}47)$$

Hence, the only remaining integrations are those specified by $F(g_n)$. This simplification may make it possible to obtain some solutions that would be unattainable if other weighting functions were used. Physically, the use of Dirac delta weighting functions is seen as a relaxation of boundary conditions so that they are enforced only at discrete points on the surface of the structure, hence the name *point matching*.

An important consideration when using point matching is the positioning of the N points (p_m). While equally spaced points often yield good results, much depends on the basis functions used. When using subsectional basis functions in conjunction with point matching, one match point should be placed on each segment to maintain linear independence. Placing the points at the center of the segments usually produces the best results. It is important that a match point does not coincide with the "peak" of a triangle or any other point, where the basis function is not differentiable. Ignoring this would cause errors in some situations.

Because it provides acceptable accuracy along with obvious computational advantages, point matching is the most popular testing technique for moment-method solutions to electromagnetics problems. The analysis presented here, along with most problems considered in the literature, proceed via point matching.

For the strip problem, a convenient inner product of the form (12-42) is

$$\langle w_m, g_n \rangle = \int_{-w/2}^{w/2} w_m^*(x) g_n(x) \, dx \qquad (12\text{-}48)$$

Taking the inner product (12-48) with $w_m^*(x)$ on both sides of (12-26), we can write it as

$$V_m' = \sum_{n=1}^{N} I_n Z_{mn}', \qquad m = 1, 2, ..., N \qquad (12\text{-}49)$$

where

$$V_m' = \int_{-w/2}^{w/2} w_m^*(x) H_0^{(2)}(\beta \rho_m) \, dx \qquad (12\text{-}49a)$$

$$Z_{mn}' = -\int_{-w/2}^{w/2} w_m^*(x) \left[\int_{-w/2}^{w/2} g_n(x') H_0^{(2)}(\beta |\rho_m - \rho_n'|) \, dx' \right] dx \qquad (12\text{-}49b)$$

or in matrix form as

$$[V_m'] = [Z_{mn}'][I_n] \qquad (12\text{-}50)$$

If the w_m weighting functions are Dirac delta functions [i.e., $w_m(y) = \delta(y - y_m)$], then (12-49) reduces to (12-27) or

$$V_m' = V_m \qquad (12\text{-}51a)$$

and

$$Z'_{mn} = Z_{mn} \tag{12-51b}$$

The *method of weighted residuals* (moment method) was introduced to minimize the average deviation from the actual values of the boundary conditions over the entire structure. However, it is evident that it has complicated the formulation by requiring an integration in the evaluation of the elements of the V' matrix [as given by (12-49a)] and an additional integration in the evaluation of the elements of the Z'_{mn} matrix [as given by (12-49b)]. Therein lies the penalty that is paid to improve the solution.

If both the expansion g_n and the weighting w_m functions are of the subdomain type, each of which exists only over one of the strip segments, then (12-49b) can be written as

$$Z'_{mn} = -\int_{x_m}^{x_{m+1}} w_m^*(x) \left[\int_{x'_n}^{x'_{n+1}} g_n(x') H_0^{(2)}(\beta \, |\rho_m - \rho'|) \, dx' \right] dx \tag{12-52}$$

where (x_m, x_{m+1}) and (x'_n, x'_{n+1}) represent, respectively, the lower and upper limits of the strip segments over which the weighting w_m and expansion g_n functions exist. To evaluate the *mn*th element of Z'_{mn} from (12-49b) or (12-52), we first choose the weighting function w_m, and the region of the segment over which it exists, and weigh the contributions from the g_n expansion function over the region in which it exists. To find the next element $Z'_{m(n+1)}$, we maintain the same weighting function w_m, and the region over which it exists, and weigh the contributions from the g_{n+1} expansion function. We repeat this until the individual contributions from all the N expansion functions (g_n) are weighted by the w_m weighting function. Then we choose the w_{m+1} weighting function, and the region over which it exists, and we weigh individually the contributions from each of the N expansion functions (g_n). We repeat this until all the N weighting functions (w_m), and the regions of the strip over which they exist, are individually weighted by the N expansion functions (g_n). This procedure allows us to form N linear equations, each with N unknowns, that can be solved using matrix inversion methods.

Example 12-3

For the electric line source of Figure 12-5 with $w = 2\lambda$, $t = 0.001\lambda$, and $h = 0.5\lambda$ perform the following:

1. Compute the equivalent current density induced on the open surface of the strip. This equivalent current density is representative of the vector sum of the current densities that flow on the opposite sides of the strip. Use subdomain pulse expansion functions and point matching. Subdivide the strip into 150 segments.
2. Compare the current density of part 1 with the physical optics current density.
3. Compute the normalized far-field amplitude pattern of (12-22a) using the current densities of parts 1 and 2. Compare these patterns with those obtained using a combination of geometrical optics (GO) and geometrical theory of diffraction (GTD) techniques of Chapter 13 and physical optics (PO) and physical theory of diffraction (PTD) techniques of [13].

Solution:

1. Utilizing (12-27) through (12-27b) and (12-36), the current density of (12-23) is computed using (12-28a). It is plotted in Figure 12-11. It is observed that the current density exhibits singularities toward the edges of the strip.

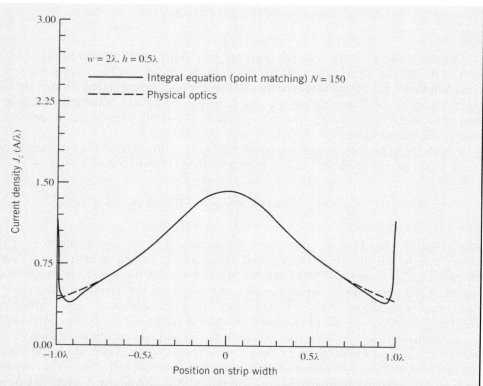

Figure 12-11 Current density on a finite-width strip that is due to the electric line source above the strip.

2. The physical optics current density is found using

$$\mathbf{J}_s^{PO} \simeq 2\hat{\mathbf{n}} \times \mathbf{H}^i$$

which reduces using (11-10b) to

$$\mathbf{J}_s^{PO} \simeq 2\hat{\mathbf{a}}_y \times \hat{\mathbf{a}}_\phi H_\phi^i \big|_{\text{strip}} = 2\hat{\mathbf{a}}_y \times \left(-\hat{\mathbf{a}}_x \sin\phi + \hat{\mathbf{a}}_y \cos\phi\right) H_\phi^i \big|_{\text{strip}}$$

$$= \hat{\mathbf{a}}_z 2\sin\phi H_\phi^i \big|_{\text{strip}} \simeq -j\hat{\mathbf{a}}_z I_z \frac{\beta}{z} \sin\phi_m H_1^{(2)}(\beta\rho_m)$$

$$\mathbf{J}_s^{PO} = -j\hat{\mathbf{a}}_z I_z \frac{\beta}{2}\left(\frac{y_m}{\rho_m}\right) H_1^{(2)}(\beta\rho_m)$$

The normalized value of this has also been plotted in Figure 12-11 so that it can be compared with the more accurate one obtained in part 1 using the integral equation.

3. The far-field amplitude patterns, based on the current densities of parts 1 and 2, are plotted in Figure 12-12. In addition to the normalized radiation patterns obtained using the current densities of parts 1 and 2, the pattern obtained using geometrical optics (GO) plus first-order diffractions by the geometrical theory of diffraction (GTD), to be discussed in Chapter 13, is also displayed in Figure 12-12. There is an excellent agreement between the IE and the GO plus GTD patterns. The pattern obtained using physical optics, supplemented by first-order diffractions of the physical theory of diffraction (PTD) [13], is also displayed in Figure 12-12 for comparison purposes. It also

compares extremely well with the others. As expected, the only one that does not compare well with the others is that of PO. Its largest differences are in the back lobes.

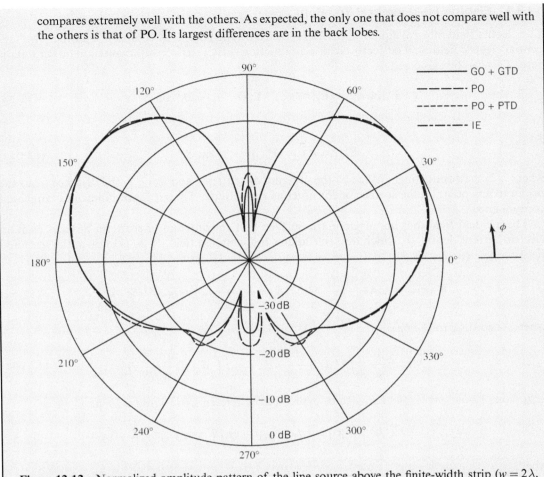

Figure 12-12 Normalized amplitude pattern of the line source above the finite-width strip ($w = 2\lambda$, $h = 0.5\lambda$).

12.3 ELECTRIC AND MAGNETIC FIELD INTEGRAL EQUATIONS

The key to the solution of any antenna or scattering problem is a knowledge of the physical or equivalent current density distributions on the volume or surface of the antenna or scatterer. Once these are known, the radiated or scattered fields can be found using the standard radiation integrals of Chapter 6. A main objective then of any solution method is to be able to predict accurately the current densities over the antenna or scatterer. This can be accomplished by the integral-equation (IE) method. One form of IE, for a two-dimensional structure, was discussed in Section 12.2.2 and represented by the integral equation 12-17a.

In general, there are many forms of integral equations. Two of the most popular forms for time-harmonic electromagnetics are the *electric field integral equation* (EFIE) and the *magnetic field integral equation* (MFIE). The EFIE enforces the boundary condition on the tangential electric field while the MFIE enforces the boundary condition on the tangential components of the magnetic field. Both of these will be discussed here as they apply to perfectly conducting structures.

12.3.1   Electric Field Integral Equation

The electric field integral equation (EFIE) is based on the boundary condition that the total tangential electric field on a perfectly electric conducting (PEC) surface of an antenna or scatterer is zero. This can be expressed as

$$\mathbf{E}_t^t(r = r_s) = \mathbf{E}_t^i(r = r_s) + \mathbf{E}_t^s(r = r_s) = 0 \quad \text{on } S \tag{12-53}$$

or

$$\mathbf{E}_t^s(r = r_s) = -\mathbf{E}_t^i(r = r_s) \quad \text{on } S \tag{12-53a}$$

where S is the conducting surface of the antenna or scatterer and $r = r_s$ is the distance from the origin to any point on the surface of the antenna or scatterer. The subscript t indicates tangential components.

The incident field that impinges on the surface S of the antenna or scatterer induces on it an electric current density $\mathbf{J}_s$ which in turn radiates the scattered field. If $\mathbf{J}_s$ is known, the scattered field everywhere, that is due to $\mathbf{J}_s$, can be found using (6-32b), or

$$\mathbf{E}^s(r) = -j\omega\mathbf{A} - j\frac{1}{\omega\mu\varepsilon}\nabla(\nabla \cdot \mathbf{A}) = -j\frac{1}{\omega\mu\varepsilon}\left[\omega^2\mu\varepsilon\mathbf{A} + \nabla(\nabla \cdot \mathbf{A})\right] \tag{12-54}$$

where, according to (6-96a),

$$\mathbf{A}(r) = \frac{\mu}{4\pi}\iint_S \mathbf{J}_s(r')\frac{e^{-j\beta R}}{R}\,ds' = \mu\iint_S \mathbf{J}_s(r')\frac{e^{-j\beta R}}{4\pi R}\,ds' \tag{12-54a}$$

Equations 12-54 and 12-54a can also be expressed, by referring to Figure 6-2b, as

$$\mathbf{E}^s(r) = -j\frac{\eta}{\beta}\left[\beta^2\iint_S \mathbf{J}_s(r')G(\mathbf{r},\mathbf{r}')\,ds' + \nabla\iint_S \nabla' \cdot \mathbf{J}_s(r')G(\mathbf{r},\mathbf{r}')\,ds'\right] \tag{12-55}$$

where

$$G(\mathbf{r},\mathbf{r}') = \frac{e^{-j\beta R}}{4\pi R} = \frac{e^{-j\beta|\mathbf{r}-\mathbf{r}'|}}{4\pi|\mathbf{r}-\mathbf{r}'|} \tag{12-55a}$$

$$R = |\mathbf{r} - \mathbf{r}'| \tag{12-55b}$$

In (12-55) ∇ and ∇' are, respectively, the gradients with respect to the observation (unprimed) and source (primed) coordinates and $G(\mathbf{r},\mathbf{r}')$ is referred to as the *Green's function* for a three-dimensional radiator or scatterer.

If the observations are restricted on the surface of the antenna or scatterer ($r = r_s$), then (12-55) and (12-55b) can be expressed, using (12-53a), as

$$\left.j\frac{\eta}{\beta}\left[\beta^2\iint_S \mathbf{J}_s(r')G(\mathbf{r}_s,\mathbf{r}')\,ds' + \nabla\iint_S \nabla' \cdot \mathbf{J}_s(r')G(\mathbf{r}_s,\mathbf{r}')\,ds'\right]\right|_t = \mathbf{E}_t^i(r = r_s) \tag{12-56}$$

Because the right side of (12-56) is expressed in terms of the known incident electric field, it is referred to as the *electric field integral equation* (EFIE). It can be used to find the current density $\mathbf{J}_s(r')$ at any point $r = r'$ on the antenna or scatterer. It should be noted that (12-56) is actually an integro-differential equation, but usually it is referred to as an integral equation.

Equation 12-56 can be used for closed or open surfaces. Once $\mathbf{J}_s$ is determined, the scattered field is found using (6-32b) and (6-96a) or (12-54) and (12-54a), which assume that $\mathbf{J}_s$ radiates in one medium. Because of this, $\mathbf{J}_s$ in (12-56) represents the physical equivalent electric current density of (7-53a) in Section 7.10. For open surfaces, $\mathbf{J}_s$ is also the physical equivalent current density that represents the vector sum of the equivalent current densities on the opposite sides of the surface. Whenever this equivalent current density represents open surfaces, a boundary condition supplemental to (12-56) must be enforced to yield a unique solution for the normal component of the current density to vanish on S.

Equation 12-56 is a general surface EFIE for three-dimensional problems, and its form can be simplified for two-dimensional geometries. To demonstrate this, let us derive the two-dimensional EFIEs for both TM^z and TE^z polarizations.

A. Two-Dimensional EFIE: TM^z Polarization

The best way to demonstrate the derivation of the two-dimensional EFIE for TM^z polarization is to consider a specific example. Its form can then be generalized to more complex geometries. The example to be examined here is that of a TM^z uniform plane wave incident on a finite-width strip, as shown in Figure 12-13a.

By referring to Figure 12-13a, the incident electric field can be expressed as

$$\mathbf{E}^i = \hat{\mathbf{a}}_z E_0 e^{-j\boldsymbol{\beta}^i \cdot \mathbf{r}} = \hat{\mathbf{a}}_z E_0 e^{j\beta(x\cos\phi_i + y\sin\phi_i)} \tag{12-57}$$

which at the surface of the strip ($y = 0, 0 \le x \le w$) reduces to

$$\mathbf{E}^i(y = 0, 0 \le x \le w) = \hat{\mathbf{a}}_z E_0 e^{j\beta x \cos\phi_i} \tag{12-57a}$$

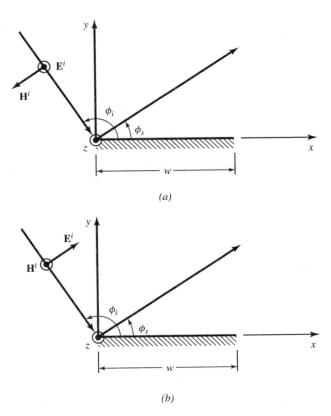

(a)

(b)

Figure 12-13 Uniform plane wave incident on a conducting strip of finite width. (a) TM^z polarization. (b) TE^z polarization.

Since the incident electric field has only a z component, the scattered and total fields each also has only a z component that is independent of z variations (two dimensional). Therefore the scattered field can be found by expanding (12-54), assuming **A** has only a z component that is independent of z variations. Doing this reduces (12-54) to

$$\mathbf{E}^s = -\hat{\mathbf{a}}_z j\omega A_z \tag{12-58}$$

The vector potential component A_z is obtained using (12-54a), which in conjunction with (11-27d), (11-28a) and (12-55b) reduces to

$$A_z = \frac{\mu}{4\pi}\iint_S J_z(x')\frac{e^{-j\beta R}}{R}\,ds' = \frac{\mu}{4\pi}\int_0^w J_z(x')\left[\int_{-\infty}^{+\infty}\frac{e^{-j\beta\sqrt{|\rho-\rho'|^2 + (z-z')^2}}}{\sqrt{|\rho-\rho'|^2 + (z-z')^2}}\,dz'\right]dx'$$

$$= -j\frac{\mu}{4}\int_0^w J_z(x')H_0^{(2)}(\beta|\rho-\rho'|)\,dx' \tag{12-59}$$

where J_z is a linear current density (measured in amperes per meter). Thus, we can write the scattered electric field at any observation point, using the geometry of Figure 12-14, as

$$\mathbf{E}^s = -\hat{\mathbf{a}}_z j\omega A_z = -\hat{\mathbf{a}}_z\frac{\omega\mu}{4}\int_0^w J_z(x')H_0^{(2)}(\beta|\rho-\rho'|)\,dx'$$

$$= -\hat{\mathbf{a}}_z\frac{\beta\eta}{4}\int_0^w J_z(x')H_0^{(2)}(\beta|\rho-x'|)\,dx' \tag{12-60}$$

For far-field observations we can reduce (12-60), using the Hankel function approximation (12-21b) for $\phi' = 0$, to

$$\boxed{\mathbf{E}^s \simeq -\hat{\mathbf{a}}_z\eta\sqrt{\frac{j\beta}{8\pi}}\frac{e^{-j\beta\rho}}{\sqrt{\rho}}\int_0^w J_z(x')e^{j\beta x'\cos\phi}\,dx'} \tag{12-60a}$$

To evaluate the integral in (12-60a) in order to find the scattered field, we must know the induced current density $J_z(x')$ over the extent of the strip ($0 \le x' \le w$). This can be accomplished by observing the field on the surface of the strip ($\rho = x_m$). Under those conditions, the total field over the strip must vanish. Thus,

$$E_z^t(0 \le x_m \le w, y = 0) = E_z^i(0 \le x_m \le w, y = 0) + E_z^s(0 \le x_m \le w, y = 0) = 0 \tag{12-61}$$

or

$$E_z^s(0 \le x_m \le w, y = 0) = -E_z^i(0 \le x_m \le w, y = 0)$$

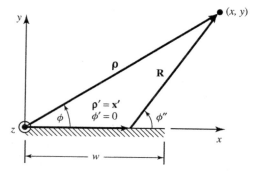

Figure 12-14   Geometry of the finite-width strip for scattering.

Since over the strip $\rho = x_m$, we can write the scattered field over the strip as

$$E_z^s(0 \le x_m \le w, y = 0) = -E_z^i(0 \le x_m \le w, y = 0)$$

$$= -\frac{\beta\eta}{4} \int_0^w J_z(x')H_0^{(2)}(\beta|\mathbf{x}_m - \mathbf{x}'|)\, dx' \qquad (12\text{-}62)$$

or

$$\frac{\beta\eta}{4} \int_0^w J_z(x')H_0^{(2)}(\beta|\mathbf{x}_m - \mathbf{x}'|)\, dx'$$

$$= E_z^i(0 \le x_m \le w, y = 0) = E_0\, e^{j\beta x_m \cos\phi_i} \qquad (12\text{-}62a)$$

For a normalized field of unity amplitude ($E_0 = 1$), (12-62a) reduces to

$$\boxed{\frac{\beta\eta}{4} \int_0^w J_z(x')H_0^{(2)}(\beta|\mathbf{x}_m - \mathbf{x}'|)\, dx' = e^{j\beta x_m \cos\phi_i}} \qquad (12\text{-}63)$$

This is the desired two-dimensional electric field integral equation (EFIE) for the TM^z polarization of the conducting strip, and it is equivalent to (12-56) for the general three-dimensional case. This EFIE can be solved for $J_z(x')$ using techniques similar to those used to solve the EFIE of (12-17a). It must be used to solve for the induced current density $J_z(x')$ over the surface of the strip. Since the surface of the strip is open, the aforementioned $J_z(x')$ represents the equivalent vector current density that flows on the opposite sides of the surface. For a more general geometry, the EFIE of (12-63) can be written as

$$\boxed{\frac{\beta\eta}{4} \int_C J_z(\rho')H_0^{(2)}(\beta|\rho_m - \rho'|)\, dc' = E_z^i(\rho_m)} \qquad (12\text{-}64)$$

where ρ_m = any observation point on the scatterer
$\qquad \rho'$ = any source point on the scatterer
$\qquad C$ = perimeter of the scatterer

The solution of the preceding integral equations for the equivalent linear current density can be accomplished by using either the point-matching (collocation) method of Section 12.2.4 or the weighted residual of the moment method of Section 12.2.8. However, using either method for the solution of the integral equation 12-63 for the strip of Figure 12-13a, we encounter the same problems as for the evaluation of the integral equation 12-17a for the finite strip of Figure 12-5, which are outlined in Section 12.2.6. However, these problems are overcome here using the same techniques that were outlined in Section 12.2.6, namely, choosing the observation points along the bisector of the width of the strip, or using the approximations (12-39) and (12-40).

Example 12-4

For the TM^z plane wave incidence on the conducting strip of Figure 12-13a perform the following:

1. Plot the induced equivalent current density for normal incidence ($\phi_i = 90°$) obtained using the EFIE of (12-63). Assume the strip has a width of $w = 2\lambda$ and 0.001λ thickness. Use subdomain pulse expansion functions and point matching. Subdivide the strip into 250 segments.
2. Compare the current density of part 1 with the physical optics current density.

3. Compute the monostatic scattering width pattern for $0 \leq \phi_i \leq 180°$ using current density obtained using the EFIE of (12-63). Compare this pattern with those obtained with physical optics (PO), geometrical theory of diffraction (GTD) of Chapter 13, and physical optics (PO) plus physical theory of diffraction (PTD) techniques [13].

Solution:

1. Using the EFIE of (12-63) and applying point-matching methods with subdomain pulse expansion functions, the current density of Figure 12-15 for $\phi_i = 90°$ is obtained for a strip of $w = 2\lambda$. It is observed that the current density exhibits singularities toward the edges of the strip.

2. The physical optics current density is represented by

$$\mathbf{J}_s^{\mathrm{PO}} \simeq 2\hat{\mathbf{n}} \times \mathbf{H}^i \big|_{\mathrm{strip}} = 2\hat{\mathbf{a}}_y \times \left(\hat{\mathbf{a}}_x H_x^i + \hat{\mathbf{a}}_y H_y^i \right)\Big|_{\mathrm{strip}}$$

$$= -\hat{\mathbf{a}}_z 2 H_x^i \big|_{\mathrm{strip}} = \hat{\mathbf{a}}_z 2 \frac{E_0}{\eta} \sin\phi_i \, e^{j\beta x \cos\phi_i}$$

which is shown plotted in Figure 12-15 for $\phi_i = 90°$. It is apparent that the PO current density does not compare well with that obtained using the IE, especially toward the edges of the strip. Therefore, it does not provide a good representation of the equivalent current density induced on the strip. In Figure 12-15 we also display the equivalent current density for TE^z polarization which will be discussed in the next section and in Example 12-5.

3. The monostatic scattering width patterns for $0° \leq \phi_i \leq 180°$ obtained using the methods of IE, PO, GTD, and PO plus PTD are all shown in Figure 12-16. As expected, the only one that differs from the others is that due to the PO; the other three are indistinguishable from each other and are represented by the solid curve. The pattern for the TE^z polarization for the IE method is also displayed in Figure 12-16. This will be discussed in the next section and in Example 12-5.

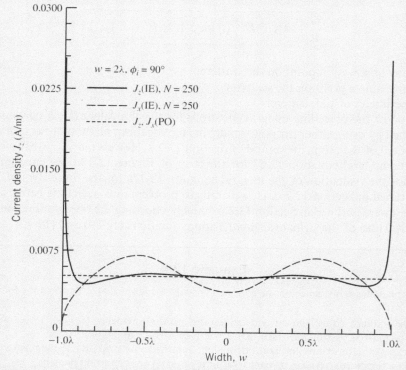

Figure 12-15   Current density induced on a finite-width strip by a plane wave at normal incidence.

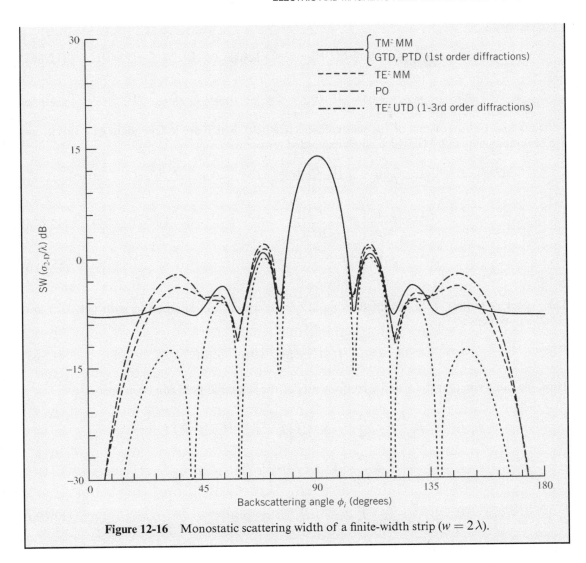

Figure 12-16   Monostatic scattering width of a finite-width strip ($w = 2\lambda$).

B. Two-Dimensional EFIE: TE^z Polarization   As in the previous section, the derivation of the EFIE for TE^z polarization is best demonstrated by considering a uniform plane wave incident on the strip, as shown in Figure 12-13b. Its form can then be generalized to more complex geometries.

By referring to Figure 12-13b, the incident electric field can be expressed as

$$\mathbf{E}^i = E_0(\hat{\mathbf{a}}_x \sin\phi_i - \hat{\mathbf{a}}_y \cos\phi_i)e^{-j\boldsymbol{\beta}_i \cdot \mathbf{r}} = E_0(\hat{\mathbf{a}}_x \sin\phi_i - \hat{\mathbf{a}}_y \cos\phi_i)e^{j\beta(x\cos\phi_i + y\sin\phi_i)} \quad (12\text{-}65)$$

which on the surface of the strip ($y = 0$, $0 \leq x \leq w$) reduces to

$$\mathbf{E}^i = E_0(\hat{\mathbf{a}}_x \sin\phi_i - \hat{\mathbf{a}}_y \cos\phi_i)e^{j\beta x\cos\phi_i} \quad (12\text{-}65a)$$

On the surface of the strip ($0 \leq x \leq w, y = 0$) the tangential components of the total field, incident plus scattered, must vanish. This can be written as

$$\hat{\mathbf{n}} \times \mathbf{E}^t|_{\text{strip}} = \hat{\mathbf{n}} \times (\mathbf{E}^i + \mathbf{E}^s)|_{\text{strip}}$$
$$= \hat{\mathbf{a}}_y \times \left[\left(\hat{\mathbf{a}}_x E_x^i + \hat{\mathbf{a}}_y E_y^i\right) + \left(\hat{\mathbf{a}}_x E_x^s + \hat{\mathbf{a}}_y E_y^s\right)\right]\Big|_{\text{strip}} = 0 \quad (12\text{-}66)$$

which leads to

$$-\hat{\mathbf{a}}_z(E_x^i + E_x^s)|_{\text{strip}} = 0 \Rightarrow E_x^s(0 \le x \le w, y = 0) = -E_x^i(0 \le x \le w, y = 0) \quad (12\text{-}66\text{a})$$

or

$$E_x^s(0 \le x \le w, y = 0) = -E_x^i(0 \le x \le w, y = 0) = -E_0 \sin\phi_i \, e^{j\beta x \cos\phi_i} \quad (12\text{-}66\text{b})$$

The x and y components of the scattered electric field, which are independent of z variations, are obtained using (12-54), which when expanded reduce to

$$E_x^s = -j\omega A_x - j\frac{1}{\omega\mu\varepsilon}\frac{\partial^2 A_x}{\partial x^2} = -j\frac{1}{\omega\mu\varepsilon}\left[\beta^2 A_x + \frac{\partial^2 A_x}{\partial x^2}\right]$$

$$= -j\frac{1}{\omega\mu\varepsilon}\left[\beta^2 + \frac{\partial^2}{\partial x^2}\right]A_x \quad (12\text{-}67\text{a})$$

$$E_y^s = -j\frac{1}{\omega\mu\varepsilon}\frac{\partial^2 A_x}{\partial x \partial y} \quad (12\text{-}67\text{b})$$

The vector potential A_x is obtained using (12-54a), which, in conjunction with (11-27d) and (11-28a), reduces to

$$A_x = -j\frac{\mu}{4}\int_0^w J_x(x')H_0^{(2)}(\beta|\boldsymbol{\rho}-\boldsymbol{\rho}'|)dx' \quad (12\text{-}68)$$

Thus, we can write that the x and y components of the scattered field can be expressed as

$$E_x^s = -j\frac{1}{\omega\mu\varepsilon}\left(-j\frac{\mu}{4}\right)\left[\beta^2 + \frac{\partial^2}{\partial x^2}\right]\int_0^w J_x(x')H_0^{(2)}(\beta|\boldsymbol{\rho}-\boldsymbol{\rho}'|)dx'$$

$$= -\frac{\eta}{4\beta}\left[\beta^2 + \frac{\partial^2}{\partial x^2}\right]\int_0^w J_x(x')H_0^{(2)}(\beta|\boldsymbol{\rho}-\boldsymbol{\rho}'|)dx' \quad (12\text{-}69\text{a})$$

$$E_y^s = -\frac{\eta}{4\beta}\frac{\partial^2}{\partial x \partial y}\int_0^w J_x(x')H_0^{(2)}(\beta|\boldsymbol{\rho}-\boldsymbol{\rho}'|)dx' \quad (12\text{-}69\text{b})$$

Interchanging integration and differentiation and letting $\rho' = x'$, we can rewrite the x and y components as

$$E_x^s = -\frac{\eta}{4\beta}\int_0^w J_x(x')\left[\left(\frac{\partial^2}{\partial x^2} + \beta^2\right)H_0^{(2)}(\beta R)\right]dx' \quad (12\text{-}70\text{a})$$

$$E_y^s = -\frac{\eta}{4\beta}\int_0^w J_x(x')\left[\frac{\partial^2}{\partial x \partial y}H_0^{(2)}(\beta R)\right]dx' \quad (12\text{-}70\text{b})$$

where

$$R = |\boldsymbol{\rho} - x'| \quad (12\text{-}70\text{c})$$

It can be shown, using the geometry of Figure 12-14, that

$$\left[\frac{\partial^2}{\partial x^2} + \beta^2\right]H_0^{(2)}(\beta R) = \frac{\beta^2}{2}\left[H_0^{(2)}(\beta R) + H_2^{(2)}(\beta R)\cos(2\phi'')\right] \quad (12\text{-}71\text{a})$$

$$\frac{\partial^2}{\partial x \partial y}H_0^{(2)}(\beta R) = \frac{\beta^2}{2}H_2^{(2)}(\beta R)\sin(2\phi'') \quad (12\text{-}71\text{b})$$

Thus, the x and y components of the electric field reduce to

$$E_x^s = -\frac{\beta\eta}{8}\int_0^w J_x(x')\left[H_0^{(2)}(\beta R) + H_2^{(2)}(\beta R)\cos(2\phi'')\right]dx' \tag{12-72a}$$

$$E_y^s = -\frac{\beta\eta}{8}\int_0^w J_x(x')H_2^{(2)}(\beta R)\sin(2\phi'')\,dx' \tag{12-72b}$$

The next objective is to solve for the induced current density, that can then be used to find the scattered field. This can be accomplished by applying the boundary conditions on the x component of the electric field. When the observations are restricted to the surface of the strip ($\rho = x_m$), the x component of the scattered field over the strip can be written as

$$E_x^s(0 \le x_m \le w, y = 0)$$
$$= -E_x^i(0 \le x_m \le w, y = 0) = -E_0\sin\phi_i\,e^{j\beta x_m\cos\phi_i}$$
$$= -\frac{\beta\eta}{8}\int_0^w J_x(x')\left[H_0^{(2)}(\beta R_m) + H_2^{(2)}(\beta R_m)\cos(2\phi_m'')\right]dx' \tag{12-73}$$

or

$$\frac{\beta\eta}{8}\int_0^w J_x(x')\left[H_0^{(2)}(\beta R_m) + H_2^{(2)}(\beta R_m)\cos(2\phi_m'')\right]dx' = E_0\sin\phi_i\,e^{j\beta x_m\cos\phi_i} \tag{12-73a}$$

For a normalized field of unity amplitude ($E_0 = 1$), (12-73a) reduces to

$$\boxed{\frac{\beta\eta}{8}\int_0^w J_x(x')\left[H_0^{(2)}(\beta R_m) + H_2^{(2)}(\beta R_m)\cos(2\phi_m'')\right]dx' = \sin\phi_i\,e^{j\beta x_m\cos\phi_i}} \tag{12-74}$$

where

$$R_m = |\boldsymbol{\rho}_m - \mathbf{x}'| \tag{12-74a}$$

This is the desired two-dimensional electric field integral equation (EFIE) for the TE^z polarization of the conducting strip, and it is equivalent to (12-56) for the general three-dimensional case. This EFIE must be used to solve for the induced current density $J_x(x')$ over the surface of the strip using techniques similar to those used to solve the EFIE of (12-17a). For a more general geometry, the EFIE can be written as

$$\boxed{\begin{aligned}\frac{\eta}{4\beta}\Bigg\{\beta^2\int_C J_c(\boldsymbol{\rho}')\left[\hat{\mathbf{c}}_m\cdot\hat{\mathbf{c}}'\,H_0^{(2)}(\beta|\boldsymbol{\rho}_m-\boldsymbol{\rho}'|)\right]dc' \\ +\frac{d}{dc}\left[\nabla\cdot\int_C J_c(\boldsymbol{\rho}')\left[\hat{\mathbf{c}}'H_0^{(2)}(\beta|\boldsymbol{\rho}_m-\boldsymbol{\rho}'|)\right]dc'\right]\Bigg\} = -E_c^i(\boldsymbol{\rho}_m)\end{aligned}} \tag{12-75}$$

where $\boldsymbol{\rho}_m$ = any observation point on the scatterer
$\boldsymbol{\rho}'$ = any source point on the scatterer
C = perimeter of the scatterer
$\hat{\mathbf{c}}_m, \hat{\mathbf{c}}'$ = unit vector tangent to scatterer perimeter at observation, source points

The linear current density J_x is obtained by solving the integral equation 12-74 using either the point-matching (collocation) method of Section 12.2.4 or the weighted residual moment method of Section 12.2.8. Using either method, the solution of the preceding integral equation for J_x is more difficult than that of the TM^z polarization of the previous example. There exist various approaches (either exact or approximate) that can be used to accomplish this.

To demonstrate this, we will discuss one method that can be used to solve the integral equation 12-74. Let us assume that the current density $J_x(x')$ is expanded into a finite series similar to (12-23). Then the integral equation can be written as

$$\sin\phi_i e^{j\beta x_m \cos\phi_i} = \frac{\beta\eta}{8}\sum_{n=1}^{N} a_n \int_0^w g_n(x')\left[H_0^{(2)}(\beta R_m) + H_2^{(2)}(\beta R_m)\cos(2\phi_m'')\right]dx' \qquad (12\text{-}76)$$

If, in addition, the basis functions are subdomain pulse functions, as defined by (12-35), then (12-76) using point matching reduces for each observation point to

$$\sin\phi_i e^{j\beta x_m \cos\phi_i} = \frac{\beta\eta}{8}\sum_{n=1}^{N} a_n \int_{x_n}^{x_{n+1}}\left[H_0^{(2)}(\beta R_{mn}) + H_2^{(2)}(\beta R_{mn})\cos(2\phi_{mn}'')\right]dx' \qquad (12\text{-}77)$$

If N observations are selected, then we can write (12-77) as

$$[\sin\phi_i e^{j\beta x_m \cos\phi_i}] = \sum_{n=1}^{N} a_n\left\{\frac{\beta\eta}{8}\int_{x_n}^{x_{n+1}}\left[H_0^{(2)}(\beta R_{mn}) + H_2^{(2)}(\beta R_{mn})\cos(2\phi_{mn}'')\right]dx'\right\}$$

$$m = 1, 2, \ldots, N \qquad (12\text{-}78)$$

or, in matrix form,

$$[V_m] = [Z_{mn}][I_n] \qquad (12\text{-}78a)$$

where

$$V_m = \sin\phi_i\, e^{j\beta x_m \cos\phi_i} \qquad (12\text{-}78b)$$

$$I_n = a_n \qquad (12\text{-}78c)$$

$$Z_{mn} = \frac{\beta\eta}{8}\int_{x_n}^{x_{n+1}}\left[H_0^{(2)}(\beta R_{mn}) + H_2^{(2)}(\beta R_{mn})\cos(2\phi_{mn}'')\right]dx' \qquad (12\text{-}78d)$$

One of the tasks here will be the evaluation of the integral for Z_{mn}. We will examine one technique that requires Z_{mn} to be evaluated using three different expressions depending upon the position of the segment relative to the observation point. We propose here that Z_{mn} is evaluated using

$$Z_{mn} = \begin{cases} \left|\dfrac{\beta\eta\Delta x_n}{8}\left\{1 - j\dfrac{1}{\pi}\left[3 + 2\ln\left(\dfrac{1.781\beta\Delta x_n}{4e}\right) + \dfrac{16}{(\beta\Delta x_n)^2}\right]\right\}\right| & \\[4pt] \qquad e \approx 2.718 \qquad m = n & (12\text{-}79a) \\[12pt] \dfrac{\beta\eta\Delta x_n}{8}\left\{1 + j\dfrac{4}{\pi\beta^2}\dfrac{1}{|x_m - x_n|^2 - \dfrac{(\Delta x_n)^2}{4}}\right\} & \\[4pt] \qquad |m-n|\le 2, \qquad m\ne n & (12\text{-}79b) \\[12pt] \dfrac{\beta\eta}{4}\int_{-\Delta x_n/2}^{\Delta x_n/2}\dfrac{H_1^{(2)}\left[\beta(|x_m-x_n|+x')\right]}{\beta(|x_m-x_n|+x')}dx' & \\[4pt] \qquad |m-n| > 2 & (12\text{-}79c) \end{cases}$$

where x_m and x_n are measured from the center of their respective segments.

The current density J_x obtained from the preceding integral equation also represents the total current density $\mathbf{J}_s$ induced on the strip. This is evident from the induced current density equation

$$\mathbf{J}_s = \hat{\mathbf{n}} \times \mathbf{H}^t = \hat{\mathbf{a}}_y \times \hat{\mathbf{a}}_z H_z^t = \hat{\mathbf{a}}_x H_z^t = \hat{\mathbf{a}}_x(H_z^i + H_z^s) \qquad (12\text{-}80)$$

Example 12-5

For the TE^z plane wave incident on the conducting strip of Figure 12-13b, perform the following tasks.

1. Plot the induced equivalent current density for normal incidence ($\phi_i = 90°$) obtained using the EFIE of (12-74) or (12-78) through (12-78d). Assume a width of $w = 2\lambda$ and thickness equal to 0.001λ. Use subdomain pulse expansion functions and point matching. Subdivide the strip into 250 segments.
2. Compare the current density of part 1 with the physical optics current density.
3. Compute the monostatic scattering width pattern for $0° \leq \phi_i \leq 180°$ using the current density obtained using the EFIE of (12-78) through (12-78d). Compare this pattern with those obtained with physical optics (PO), geometrical theory of diffraction (GTD) of Chapter 13, and physical optics (PO) plus physical theory of diffraction (PTD) techniques [13].

Solution:

1. Using the EFIE of (12-78) through (12-78d), the current density of Figure 12-15 for $\phi_i = 90°$ is obtained for a strip of $w = 2\lambda$. It is observed that the current density vanishes toward the edges of the strip.
2. The physical optics current density is represented by

$$\mathbf{J}_s^{PO} \simeq 2\hat{\mathbf{n}} \times \mathbf{H}^i \big|_{\text{strip}} = 2\hat{\mathbf{a}}_y \times \hat{\mathbf{a}}_z H_z^i \big|_{\text{strip}} = \hat{\mathbf{a}}_x 2 H_z^i \big|_{\text{strip}} = \hat{\mathbf{a}}_x 2 \frac{E_0}{\eta} e^{j\beta x \cos\phi_i}$$

which for normal incidence ($\phi_i = 90°$) is identical to that for the TM^z polarization, and it is shown plotted in Figure 12-15. As for the TM^z polarization, the PO TE^z polarization current density does not compare well with that obtained using the IE method. Therefore, it does not provide a good representation of the equivalent current density induced on the strip. In Figure 12-15 the TE^z polarization current density is compared with that of the TM^z polarization using the different methods.
3. The monostatic scattering width pattern for $0° \leq \phi_i \leq 180°$ obtained using the IE method is shown plotted in Figure 12-16 where it is compared to those obtained by PO, PO plus PTD (first-order diffractions), and GTD (first-order diffractions) techniques. It is observed that the patterns of PO, PO plus PTD, and GTD (using first-order diffractions only) are insensitive to polarization whereas those of the integral equation with moment method solution vary with polarization. The SW patterns should vary with polarization. Therefore, those obtained using the integral-equation method are more accurate. It can be shown that if higher-order diffractions are included, the patterns of the PO plus PTD, and GTD will also vary with polarization. Higher-order diffractions are greater contributors to the overall scattering pattern for the TE^z polarization than for the TM^z. This is demonstrated by including in Figure 12-16 the monostatic SW for TE^z polarization obtained using higher-order GTD (UTD) diffractions [27]. It is apparent that this pattern agrees quite well with that of the IE method.

12.3.2 Magnetic Field Integral Equation

The magnetic field integral equation (MFIE) is expressed in terms of the known incident magnetic field. It is based on the boundary condition that expresses the total electric current density induced at any point $r = r'$ on a conducting surface S

$$\mathbf{J}_s(r') = \mathbf{J}_s(r = r') = \hat{\mathbf{n}} \times \mathbf{H}^t(r = r') = \hat{\mathbf{n}} \times [\mathbf{H}^i(r = r') + \mathbf{H}^s(r = r')] \qquad (12\text{-}81)$$

Once the current density is known or determined, the scattered magnetic field can be obtained using (6-32a) and (6-96a), or

$$\mathbf{H}^s(r) = \frac{1}{\mu} \nabla \times \mathbf{A} = \nabla \times \iint_S \mathbf{J}_s(r') \frac{e^{-j\beta R}}{4\pi R} ds' = \nabla \times \iint_S \mathbf{J}_s(r') G(\mathbf{r}, \mathbf{r}') ds' \quad (12\text{-}82)$$

where $G(\mathbf{r}, \mathbf{r}')$ is the Green's function of (12-55a). Interchanging differentiation with integration and using the vector identity

$$\nabla \times (\mathbf{J}_s G) = G \nabla \times \mathbf{J}_s - \mathbf{J}_s \times \nabla G \quad (12\text{-}83)$$

where

$$\nabla \times \mathbf{J}_s(r') = 0 \quad (12\text{-}83a)$$
$$\nabla G = -\nabla' G \quad (12\text{-}83b)$$

(12-82) reduces to

$$\mathbf{H}^s(r) = \iint_S \mathbf{J}_s(r') \times [\nabla' G(\mathbf{r}, \mathbf{r}')] ds' \quad (12\text{-}84)$$

On the surface S of the conductor, the tangential magnetic field is discontinuous by the amount of the current density induced on the surface of the conductor. Therefore, the current density is determined by (12-81) but with $\mathbf{H}^s$ found using (12-84). Thus, we can write that

$$\mathbf{J}_s(r') = \hat{\mathbf{n}} \times \mathbf{H}^i(r = r') + \lim_{r \to S}[\hat{\mathbf{n}} \times \mathbf{H}^s(r = r')]$$

$$= \hat{\mathbf{n}} \times \mathbf{H}^i(r = r') + \lim_{r \to S}\left\{\hat{\mathbf{n}} \times \iint_S \mathbf{J}_s(r') \times [\nabla' G(\mathbf{r}, \mathbf{r}')] ds'\right\} \quad (12\text{-}85)$$

or

$$\boxed{\mathbf{J}_s(r') - \lim_{r \to S}\left\{\hat{\mathbf{n}} \times \iint_S \mathbf{J}_s(r') \times [\nabla' G(\mathbf{r}, \mathbf{r}')] ds'\right\} = \hat{\mathbf{n}} \times \mathbf{H}^i(r = r')} \quad (12\text{-}85a)$$

Since in (12-85a)

$$\lim_{r \to S}\left\{\hat{\mathbf{n}} \times \iint_S \mathbf{J}_s(r') \times [\nabla' G(\mathbf{r}, \mathbf{r}')] ds'\right\} = \frac{\mathbf{J}_s(r')}{2} + \hat{\mathbf{n}} \times \iint_S \mathbf{J}_s(r') \times [\nabla' G(\mathbf{r}, \mathbf{r}')] ds' \quad (12\text{-}85b)$$

then (12-85a) can be written, in a more useful form, as

$$\boxed{\mathbf{J}_s(r') = 2\hat{\mathbf{n}} \times \mathbf{H}^i(r = r') + 2\hat{\mathbf{n}} \times \iint_S \mathbf{J}_s(r') \times [\nabla' G(\mathbf{r}, \mathbf{r}')] ds'} \quad (12\text{-}85c)$$

where $r \to S$ indicates that S is approached by r from the outside.

Equation 12-85a is referred to as the *magnetic field integral equation* (MFIE) because its right side is in terms of the incident magnetic field, and it is valid only for *closed* surfaces. Once the current density distribution can be found using (12-85a)–(12-85c), then the scattered fields can be found using standard radiation integrals. It should be noted that the integral of (12-85a) or (12-85b) must be carefully evaluated. The MFIE is the most popular for TE^z polarizations, although it can be used for both TE^z and TM^z cases. Since (12-85a)–(12-85c) are only valid for closed surfaces, the current density obtained using (12-85a)–(12-85b) is the actual current density induced on the surface of the conducting obstacle. Usually, the MFIE is well posed while the EFIE is ill posed.

At this point, before proceeding any further, it may be appropriate to comment as to why the EFIE is valid for *open and closed* surfaces while the MFIE is valid only for *closed* surfaces [28]. This can be accomplished by referring to Figure 12-17, which represents a thin closed surface of

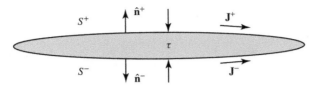

Figure 12-17   Thin closed structure with corresponding current densities, surfaces, and unit vectors [28].

thickness τ with the two sides of the surface and unit vectors represented, respectively, by $S^+, \hat{\mathbf{n}}^+$ and $S^-, \hat{\mathbf{n}}^-$. In the limit, as $\tau \to 0$, $S^+ \to S^-$, and $\hat{\mathbf{n}}^+ \to -\hat{\mathbf{n}}^-$, the surface reduces to an *open* structure, and then the MFIE cannot be used because of what follows.

In applying the EFIE of (12-56) and MFIE of (12-85a) to open thin surfaces, one finds that each equation reduces to the same form on opposite sides of the conducting surface. Hence, neither equation can be used to solve for the two opposite-side surface current densities ($\mathbf{J}^+, \mathbf{J}^-$). In the case of the EFIE, however, the unknown surface conduction current density appears only in the integrand as a sum of the opposite-side surface currents (it does not appear outside the integral), and it is this sum, or total, surface current density that can be solved for. The total equivalent current density is then sufficient to find all fields radiated by the conducting structure. This is different from the MFIE case, in which the sum of the opposite-side surface current densities also appears inside the integral, but their difference appears outside the integral. Thus, the MFIE reduces to an identity relating the surface current densities on opposite sides of the surface, but does not contain sufficient information to completely determine them and, hence, cannot be used to find the surface current densities on open conducting surfaces. The missing information could be provided, for example, by combining the EFIE with the MFIE and solving the two equations as coupled equations for the opposite-side current densities—something that cannot be done using the EFIE alone. There is no problem in applying the MFIE for closed surfaces as long as care is exercised in evaluating the integral in (12-85a) or (12-85b).

An alternate MFIE can be derived using the *null field approach* [28], which utilizes the surface equivalence theorem of Chapter 7, Section 7.8, which requires that the tangential magnetic field vanishes *just inside* the conductor. Therefore, the alternate MFIE, based on the *null field approach*, is derived using $\hat{\mathbf{n}} \times (\mathbf{H}^{inc} + \mathbf{H}^{scat}) = 0$, where the surface S is approached from the *interior* instead of the exterior; when using $\hat{\mathbf{n}} \times (\mathbf{H}^{inc} + \mathbf{H}^{scat}) = \mathbf{J}$, the surface is approached from the *exterior*; hence, the problem for open surfaces.

Whereas (12-85a) is a general MFIE for three-dimensional problems, its form can be simplified for two-dimensional MFIEs for both TM^z and TE^z polarizations.

A. Two-Dimensional MFIE: TM^z Polarization

The best way to demonstrate the derivation of the two-dimensional MFIE for TM^z polarization is to consider a TM^z uniform plane wave incident upon a two-dimensional smooth curved surface, as shown in Figure 12-18.

Since the incident field has only a z component of the electric field, and x and y components of the magnetic field, the electric current density induced on the surface of the scatterer will only have a z component. That is

$$\mathbf{J}_s(\rho) = \hat{\mathbf{a}}_z J_z(\rho)|_C \tag{12-86}$$

On the surface of the scatterer, the current density is related to the incident and scattered magnetic fields by (12-81), which, for the geometry of Figure 12-18, can be written as

$$\mathbf{J}_s(\rho)|_C = \hat{\mathbf{a}}_z J_z(\rho)|_C = \hat{\mathbf{n}} \times (\mathbf{H}^i + \mathbf{H}^s)|_C = \hat{\mathbf{n}} \times \mathbf{H}^i + \lim_{\rho \to C}(\hat{\mathbf{n}} \times \mathbf{H}^s) \tag{12-87}$$

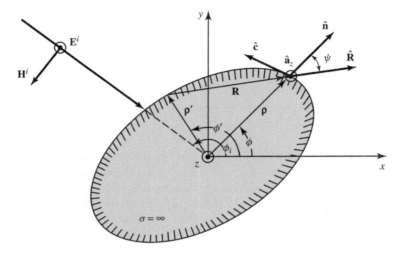

Figure 12-18   Geometry for two-dimensional MFIE TM^z polarization scattering.

where $\rho \to C$ indicates that the boundary C is approached by ρ from the exterior. Since the left side of (12-87) has only a z component, the right side of (12-87) must also have only a z component. Therefore, the only component of $\mathbf{H}^i$ that contributes to (12-87) is that which is tangent to C and coincides with the surface of the scatterer. Thus, we can rewrite (12-87) as

$$J_z(\rho)|_C = H_c^i(\rho)|_C + \lim_{\rho \to C}[\hat{\mathbf{a}}_z \cdot (\hat{\mathbf{n}} \times \mathbf{H}^s)] \tag{12-88}$$

The scattered magnetic field $\mathbf{H}^s$ can be expressed according to (12-82) as

$$\mathbf{H}^s = \frac{1}{\mu} \nabla \times \mathbf{A} = \frac{1}{\mu} \nabla \times \left[\frac{\mu}{4\pi} \int_C \int_{-\infty}^{+\infty} \mathbf{J}_s(\rho') \frac{e^{-j\beta R}}{R} dz' dc' \right]$$

$$= \frac{1}{4\pi} \nabla \times \left\{ \int_C \mathbf{J}_s(\rho') \left[\int_{-\infty}^{+\infty} \frac{e^{-j\beta R}}{R} dz' \right] dc' \right\} = -j\frac{1}{4} \nabla \times \int_C \mathbf{J}_s(\rho') H_0^{(2)}(\beta R) dc'$$

$$\mathbf{H}^s = -j\frac{1}{4} \int_C \nabla \times [\mathbf{J}_s(\rho') H_0^{(2)}(\beta R)] dc' \tag{12-89}$$

Using (12-83) and (12-83a) reduces (12-89) to

$$\mathbf{H}^s = j\frac{1}{4} \int_C \mathbf{J}_s(\rho') \times \nabla H_0^{(2)}(\beta R) dc' \tag{12-90}$$

Since $\mathbf{J}_s(\rho')$ has only a z component, the second term within the brackets on the right side of (12-88) can be written using (12-90) as

$$\hat{\mathbf{a}}_z \cdot (\hat{\mathbf{n}} \times \mathbf{H}^s) = \hat{\mathbf{a}}_z \cdot \left\{ j\frac{1}{4} \hat{\mathbf{n}} \times \int_C [\hat{\mathbf{a}}_z' J_z(\rho')] \times [\nabla H_0^{(2)}(\beta R)] dc' \right\}$$

$$= j\frac{1}{4} \int_C J_z(\rho') \left\{ \hat{\mathbf{a}}_z \cdot \left[\hat{\mathbf{n}} \times \hat{\mathbf{a}}_z \times \nabla H_0^{(2)}(\beta R) \right] \right\} dc' \tag{12-91}$$

since $\hat{\mathbf{a}}_z' = \hat{\mathbf{a}}_z$. Using the vector identity

$$\mathbf{A} \times (\mathbf{B} \times \mathbf{C}) = (\mathbf{A} \cdot \mathbf{C})\mathbf{B} - (\mathbf{A} \cdot \mathbf{B})\mathbf{C} \tag{12-92}$$

we can write that

$$\hat{\mathbf{n}} \times \left[\hat{\mathbf{a}}_z \times \nabla H_0^{(2)}(\beta R) \right] = \hat{\mathbf{a}}_z \left[\hat{\mathbf{n}} \cdot \nabla H_0^{(2)}(\beta R) \right] - (\hat{\mathbf{n}} \cdot \hat{\mathbf{a}}_z) \nabla H_0^{(2)}(\beta R)$$

$$= \hat{\mathbf{a}}_z \left[\hat{\mathbf{n}} \cdot \nabla H_0^{(2)}(\beta R) \right] \tag{12-93}$$

since $\hat{\mathbf{n}} \cdot \hat{\mathbf{a}}_z = 0$. Substituting (12-93) into (12-91) reduces it to

$$\hat{\mathbf{a}}_z \cdot (\hat{\mathbf{n}} \times \mathbf{H}^s) = j \frac{1}{4} \int_C J_z(\rho') \left[\hat{\mathbf{n}} \cdot \nabla H_0^{(2)}(\beta R) \right] dc'$$

$$= j \frac{1}{4} \int_C J_z(\rho') \left[-\beta \cos \psi H_1^{(2)}(\beta R) \right] dc'$$

$$\hat{\mathbf{a}}_z \cdot (\hat{\mathbf{n}} \times \mathbf{H}^s) = -j \frac{\beta}{4} \int_C J_z(\rho') \cos \psi H_1^{(2)}(\beta R) dc' \tag{12-94}$$

where the angle ψ is defined in Figure 12-18. Thus, we can write (12-88), using (12-94), as

$$J_z(\rho)|_C = H_c^i(\rho)|_C + \lim_{\rho \to C} \left[-j \frac{\beta}{4} \int_C J_z(\rho') \cos \psi H_1^{(2)}(\beta R) dc' \right] \tag{12-95}$$

or

$$\boxed{J_z(\rho)|_C + j \frac{\beta}{4} \lim_{\rho \to C} \left[\int_C J_z(\rho') \cos \psi H_1^{(2)}(\beta R) dc' \right] = H_c^i(\rho)|_C} \tag{12-95a}$$

B. Two-Dimensional MFIE: TE^z Polarization To derive the MFIE for the TE^z polarization, let us consider a TE^z uniform plane wave incident upon a two-dimensional curved surface, as shown in Figure 12-19. Since the incident field has only a z component of the magnetic field, the current induced on the surface of the scatterer will have only a component that is tangent to C and it will coincide with the surface of the scatterer. That is

$$\mathbf{J}_s = \hat{\mathbf{c}} J_c(\rho) \tag{12-96}$$

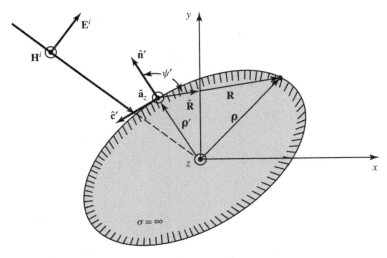

Figure 12-19 Geometry for two-dimensional MFIE TE^z polarization scattering.

On the surface of the scatterer the current density is related to the incident and scattered magnetic fields by (12-81), which for the geometry of Figure 12-19 can be written as

$$\mathbf{J}_s|_C = \hat{\mathbf{c}}\, J_c(\rho)|_C = \hat{\mathbf{n}} \times (\mathbf{H}^i + \mathbf{H}^s)|_C = \hat{\mathbf{n}} \times \mathbf{H}^i + \lim_{\rho \to C}(\hat{\mathbf{n}} \times \mathbf{H}^s)$$

$$= \hat{\mathbf{n}} \times \hat{\mathbf{a}}_z H^i|_C + \lim_{\rho \to C}(\hat{\mathbf{n}} \times \mathbf{H}^s)$$

$$\mathbf{J}_s|_C = \hat{\mathbf{c}}\, J_c(\rho)|_C = -\hat{\mathbf{c}}\, H_z^i|_C + \lim_{\rho \to C}(\hat{\mathbf{n}} \times \mathbf{H}^s) \tag{12-97}$$

where $\rho \to C$ indicates that the boundary C is approached by ρ from the outside. Since the left side and the first term of the right side of (12-97) have only C components, then the second term of the right side of (12-97) must also have only a C component. Thus, we can write (12-97) as

$$J_c(\rho)|_C = -H_z^i(\rho)|_C + \lim_{\rho \to C}[\hat{\mathbf{c}} \cdot (\hat{\mathbf{n}} \times \mathbf{H}^s)] \tag{12-98}$$

Using the scattered magnetic field of (12-90), we can write the second term within the brackets of (12-98) as

$$\hat{\mathbf{c}} \cdot (\hat{\mathbf{n}} \times \mathbf{H}^s) = \hat{\mathbf{c}} \cdot \left\{ j\frac{1}{4}\hat{\mathbf{n}} \times \int_C \left[\hat{\mathbf{c}}' J_c(\rho') \times \nabla H_0^{(2)}(\beta R) \right] dc' \right\}$$

$$= j\frac{1}{4} \int_C J_c(\rho') \left\{ \hat{\mathbf{c}} \cdot \hat{\mathbf{n}} \times \left[\hat{\mathbf{c}}' \times \nabla H_0^{(2)}(\beta R) \right] \right\} dc' \tag{12-99}$$

Since from Figure 12-19

$$\hat{\mathbf{c}}' = -\hat{\mathbf{n}}' \times \hat{\mathbf{a}}_z' = -\hat{\mathbf{n}}' \times \hat{\mathbf{a}}_z \tag{12-100}$$

with the aid of (12-92)

$$\hat{\mathbf{c}}' \times \nabla H_0^{(2)}(\beta R) = (-\hat{\mathbf{n}}' \times \hat{\mathbf{a}}_z) \times \nabla H_0^{(2)}(\beta R) = \nabla H_0^{(2)}(\beta R) \times (\hat{\mathbf{n}}' \times \hat{\mathbf{a}}_z)$$

$$= -\hat{\mathbf{a}}_z \left[\hat{\mathbf{n}}' \cdot \nabla H_0^{(2)}(\beta R) \right] + \hat{\mathbf{n}}' \left[\hat{\mathbf{a}}_z \cdot \nabla H_0^{(2)}(\beta R) \right]$$

$$= -\hat{\mathbf{a}}_z \left[\hat{\mathbf{n}} \cdot \nabla H_0^{(2)}(\beta R) \right] \tag{12-100a}$$

since $\hat{\mathbf{a}}_z \cdot \nabla H_0^{(2)}(\beta R) = 0$. Thus, the terms within the brackets in (12-99) can be written as

$$\hat{\mathbf{c}} \cdot \hat{\mathbf{n}} \times \left[\hat{\mathbf{c}}' \times \nabla H_0^{(2)}(\beta R) \right]$$

$$= -\hat{\mathbf{c}} \cdot (\hat{\mathbf{n}} \times \hat{\mathbf{a}}_z) \left[\hat{\mathbf{n}}' \cdot \nabla H_0^{(2)}(\beta R) \right] = (\hat{\mathbf{c}} \cdot \hat{\mathbf{c}}) \left[\hat{\mathbf{n}}' \cdot \nabla H_0^{(2)}(\beta R) \right]$$

$$= \hat{\mathbf{n}}' \cdot \nabla H_0^{(2)}(\beta R) \tag{12-101}$$

since $-\hat{\mathbf{c}} = \hat{\mathbf{n}} \times \hat{\mathbf{a}}_z$. Substituting (12-101) into (12-99) reduces it to

$$\hat{\mathbf{c}} \cdot \hat{\mathbf{n}} \times \mathbf{H}^s = j\frac{1}{4} \int_C J_c(\rho') \left[\hat{\mathbf{n}}' \cdot \nabla H_0^{(2)}(\beta R) \right] dc' = j\frac{1}{4} \int_C J_c(\rho') \left[-\beta \cos \psi' \, H_1^{(2)}(\beta R) \right] dc'$$

$$\hat{\mathbf{c}} \cdot \hat{\mathbf{n}} \times \mathbf{H}^s = -j\frac{\beta}{4} \int_C J_c(\rho') \cos \psi' \, H_1^{(2)}(\beta R)\, dc' \tag{12-102}$$

where the angle ψ' is defined in Figure 12-19. Thus, we can write (12-98), using (12-102), as

$$J_c(\rho)|_C = -H_z^i(\rho)|_C + \lim_{\rho \to C}\left[-j\frac{\beta}{4} \int_C J_c(\rho') \cos \psi' \, H_1^{(2)}(\beta R)\, dc' \right] \tag{12-103}$$

or

$$\boxed{J_c(\rho)|_C + j\frac{\beta}{4}\lim_{\rho \to C}\left[\int_C J_c(\rho')\cos\psi' H_1^{(2)}(\beta R)dc'\right] = -H_z^i|_C}$$ (12-103a)

C. Solution of the Two-Dimensional MFIE TE^z Polarization

The two-dimensional MFIEs of (12-95a) for TM^z polarization and (12-103a) for TE^z polarization are of identical form and their solutions are then similar. Since TM^z polarizations are very conveniently solved using the EFIE, usually the MFIEs are mostly applied to TE^z polarization problems where the magnetic field has only a z component. Therefore, we will demonstrate here the solution of the TE^z MFIE of (12-103a).

In the evaluation of the scattered magnetic field at $\rho = \rho_m$ from all points on C (including the point $\rho = \rho_m$ where the observation is made), the integral of (12-103a) can be split into two parts; one part coming from ΔC and the other part outside ΔC $(C - \Delta C)$, as shown in Figure 12-20. Thus, we can write the integral of (12-103a) as

$$j\frac{\beta}{4}\lim_{\rho \to C}\int_C J_c(\rho')\cos\psi' H_1^{(2)}(\beta R)dc'$$
$$= j\frac{\beta}{4}\lim_{\rho \to C}\left\{\int_{\Delta C} J_c(\rho')\cos\psi' H_1^{(2)}(\beta R)dc' + \int_{C-\Delta C} J_c(\rho')\cos\psi' H_1^{(2)}(\beta R)dc'\right\}$$ (12-104)

In a solution of (12-103a), where C is subdivided into segments, ΔC would typically represent one segment (the self-term) and $C - \Delta C$ would represent the other segments (the nonself-terms). Let us now examine the evaluation of each of the integrals in (12-104).

At any point, the total magnetic field is equal to the sum of the incident and scattered parts. Within the scattered conducting obstacle, the total magnetic field is zero, whereas above the conducting surface the field is nonzero. The discontinuity of the two along C is used to represent the current density along C. Within the thin rectangular box with dimensions of h and ΔC (as $h \to 0$), the total magnetic field, ($\mathbf{H}_1^t$) above and ($\mathbf{H}_2^t$) below the interface, can be written as

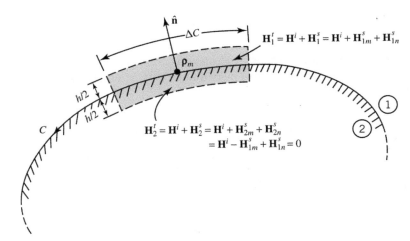

$$\mathbf{H}_1^t = \mathbf{H}^i + \mathbf{H}_1^s = \mathbf{H}^i + \mathbf{H}_{1m}^s + \mathbf{H}_{1n}^s$$

$$\mathbf{H}_2^t = \mathbf{H}^i + \mathbf{H}_2^s = \mathbf{H}^i + \mathbf{H}_{2m}^s + \mathbf{H}_{2n}^s$$
$$= \mathbf{H}^i - \mathbf{H}_{1m}^s + \mathbf{H}_{1n}^s = 0$$

Figure 12-20   Geometry and fields along the scattering surface for a two-dimensional MFIE.

$$\mathbf{H}_1^t = \mathbf{H}^i + \mathbf{H}_1^s = \mathbf{H}^i + (\mathbf{H}_{1m}^s + \mathbf{H}_{1n}^s) \tag{12-105a}$$

$$\mathbf{H}_2^t = \mathbf{H}^i + \mathbf{H}_2^s = \mathbf{H}^i + (\mathbf{H}_{2m}^s + \mathbf{H}_{2n}^s) = \mathbf{H}^i + (-\mathbf{H}_{1m}^s + \mathbf{H}_{1n}^s) = 0 \tag{12-105b}$$

or $\mathbf{H}_{1n}^s = \mathbf{H}_{1m}^s - \mathbf{H}^i$ \hfill (12-105c)

where $\mathbf{H}_{1m}^s$ ($\mathbf{H}_{2m}^s$) = scattered field in region 1 (2) within the box that is due to ΔC (self-term), which is discontinuous across the boundary along ΔC

$\mathbf{H}_{1n}^s$ ($\mathbf{H}_{2n}^s$) = scattered field in region 1 (2) within the box that is due to $C - \Delta C$ (non-self-terms), which is continuous across the boundary along ΔC

It is assumed here that ΔC along C becomes a straight line as the segment becomes small. The current density along ΔC can then be represented using (12-105a) and (12-105b) by

$$\mathbf{J}_c(\rho)|_{\Delta C} = \hat{\mathbf{n}} \times (\mathbf{H}_1^t - \mathbf{H}_2^t)|_{\Delta C} = \hat{\mathbf{n}} \times (\mathbf{H}_{1m}^s + \mathbf{H}_{1m}^s) = 2\hat{\mathbf{n}} \times \mathbf{H}_{1m}^s = -2\hat{\mathbf{c}} H_{1m}^s \tag{12-106}$$

or

$$J_c(\rho_m) = -2H_{1m}^s(\rho_m) \quad \Rightarrow \quad H_{1m}^s(\rho_m) = -\frac{J_c(\rho_m)}{2} \tag{12-106a}$$

Therefore, the integral along ΔC in (12-104), which can be used to represent the scattered magnetic field at $\rho = \rho_m$ that is due to the ΔC, can be replaced by (12-106a). The nonself-terms can be found using the integral along $C - \Delta C$ in (12-104). Thus, using (12-104) and (12-106a), we can reduce (12-103a) for $\rho = \rho_m$ to

$$J_c(\rho_m) - \frac{J_c(\rho_m)}{2} + j\frac{\beta}{4} \int_{C-\Delta C} J_c(\rho') \cos \psi_m' H_1^{(2)}(\beta R_m) \, dc' = -H_z^i(\rho_m) \tag{12-107}$$

or

$$\boxed{\frac{J_c(\rho_m)}{2} + j\frac{\beta}{4} \int_{C-\Delta C} J_c(\rho') \cos \psi_m' H_1^{(2)}(\beta R_m) \, dc' = -H_z^i(\rho_m)} \tag{12-107a}$$

An analogous procedure can be used to reduce (12-95a) to a form similar to that of (12-107a).

Let us now represent the current density $J_c(\rho)$ of (12-107a) by the finite series of (12-23)

$$J_c(\rho) \simeq \sum_{n=1}^N a_n g_n(\rho) \tag{12-108}$$

where $g_n(\rho)$ represents the basis (expansion) functions. Substituting (12-108) into (12-107a) and interchanging integration and summation, we can write that, at any point $\rho = \rho_m$ on C (12-107a) can be written as

$$-H_z^i(\rho_m) = \frac{1}{2} \sum_{n=1}^N a_n g_n(\rho_m) + j\frac{\beta}{4} \sum_{n=1}^N a_n \int_{C-\Delta C} g_n(\rho') \cos \psi_m' H_1^{(2)}(\beta R_m) \, dc' \tag{12-109}$$

If the g_n's are subdomain piecewise constant pulse functions with each basis function existing only over its own segment, then (12-109) reduces to

$$-H_z^i(\rho_m) = \frac{\delta_{mn}}{2} a_n + j\frac{\beta}{4} \sum_{\substack{n=1 \\ n \neq m}}^N a_n \int_{\rho_n}^{\rho_{n+1}} \cos \psi_{mn}' H_1^{(2)}(\beta R_{mn}) \, dc' \tag{12-110}$$

or

$$-H_z^i(\rho_m) = \sum_{n=1}^{N} a_n \left[\frac{\delta_{mn}}{2} + j\frac{\beta}{4} \int_{\rho_n}^{\rho_{n+1}} \cos\psi_{mn}' \, H_1^{(2)}(\beta R_{mn}) dc' \right] \quad (12\text{-}110a)$$

where δ_{mn} is the Kronecker delta function, defined by

$$\delta_{mn} = \begin{cases} 1 & m = n \\ 0 & m \neq n \end{cases} \quad (12\text{-}110b)$$

The Kronecker delta function is used to indicate that for a given observation point m only the segment itself ($n = m$) contributes to the first term on the right side of (12-110a).

If (12-110a) is applied to m points on C, it can be written as

$$\left[-H_z^i(\rho_m) \right] = \sum_{n=1}^{N} a_n \left[\frac{\delta_{mn}}{2} + j\frac{\beta}{4} \int_{\rho_n}^{\rho_{n+1}} \cos\psi_{mn}' \, H_1^{(2)}(\beta R_{mn}) dc' \right]$$
$$m = 1, 2, \ldots, N \quad (12\text{-}111)$$

In general matrix notation, (12-111) can be expressed as

$$[V_m] = [Z_{mn}][I_n] \quad (12\text{-}112)$$

where

$$V_m = -H_z^i(\rho_m) \quad (12\text{-}112a)$$

$$Z_{mn} = \left[\frac{\delta_{mn}}{2} + j\frac{\beta}{4} \int_{\rho_n}^{\rho_{n+1}} \cos\psi_{mn}' \, H_1^{(2)}(\beta R_{mn}) dc' \right] \quad (12\text{-}112b)$$

$$I_n = a_n \quad (12\text{-}112c)$$

To demonstrate the applicability of (12-111), let us consider an example.

Example 12-6

A TE^z uniform plane wave is normally incident upon a circular conducting cylinder of radius a, as shown in Figure 12-21.

1. Using the MFIE of (12-107a), determine and plot the current density induced on the surface of the cylinder when $a = 2\lambda$. Assume the incident magnetic field is of unity amplitude. Use subdomain piecewise constant pulse functions. Subdivide the circumference into 540 segments. Compare the current density obtained using the IE with the exact modal solution of (11-113).
2. Based on the electric current density, derive and then plot the normalized ($\sigma_{2\text{-D}}/\lambda$) bistatic scattering width (in decibels) for $0° \leq \phi \leq 360°$ when $a = 2\lambda$. Compare these values with those obtained using the exact modal solution of (11-117).

Solution:

1. Since for subdomain piecewise constant pulse functions (12-103a) or (12-107a) reduces to (12-111) or (12-112) through (12-112c), then a solution of (12-112) for I_n leads to the current density shown in Figure 12-22. In the same figure we have plotted the current density of (11-113) based on the modal solution, and we can see an excellent agreement between the two. We also have plotted the

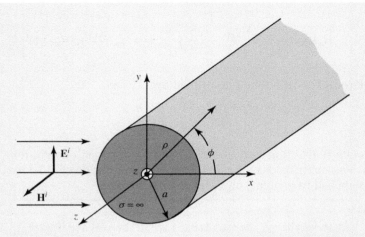

Figure 12-21 TE^z uniform plane wave incident on a circular conducting cylinder.

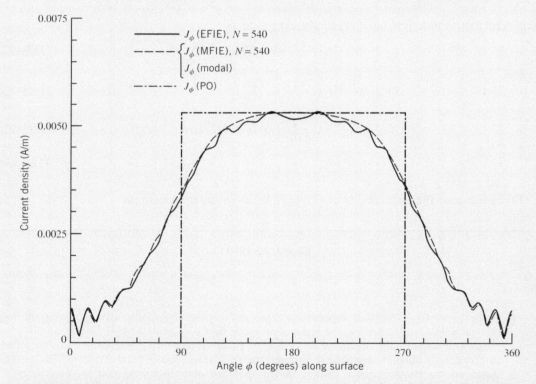

Figure 12-22 Current density induced on the surface of a circular conducting cylinder by TE^z plane wave incidence ($a = 2\lambda$).

current densities based on the EFIE for TE^z polarization of Section 12.3.1B and on the physical optics of (7-54) over the illuminated portion of the cylinder surface. The results of the EFIE do not agree with the modal solution as accurately as those of the MFIE. However, they still are very good. As expected, the physical optics current density is not representative of the true current density.

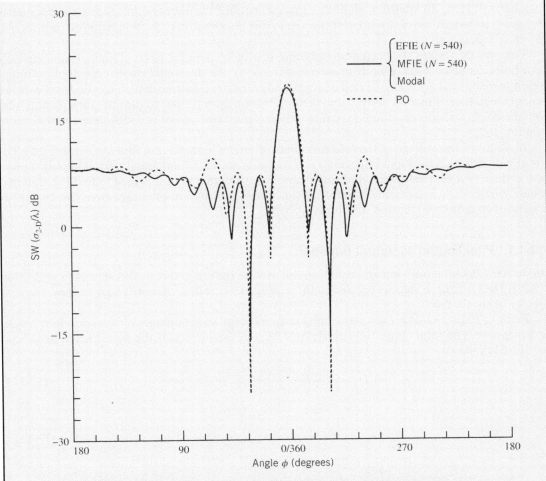

Figure 12-23   TE^z bistatic scattering width of a circular conducting cylinder ($a = 2\lambda$).

2. Based on the current densities obtained in part 1, the far-zone scattered field was derived and the corresponding bistatic scattering width was formulated. The computed SW results are shown in Figure 12-23. Besides the results based on the physical optics approximation, the other three (MFIE, EFIE, and modal solution) give almost indistinguishable data and are indicated in Figure 12-23 by basically one curve.

12.4 FINITE-DIAMETER WIRES

In this section we want to derive and apply two classic three-dimensional integral equations, referred to as *Pocklington's integro-differential equation* and *Hallén's integral equation* [29–37], that can be used most conveniently to find the current distribution on conducting wires. Hallén's equation is usually restricted to the use of a *delta-gap* voltage source model at the feed of a wire antenna. Pocklington's equation, however, is more general and it is adaptable to many types of feed sources (through alteration of its excitation function or excitation matrix), including a magnetic frill [38]. In addition, Hallén's equation requires the inversion of an $N+1$ order matrix (where N is the number of divisions of the wire) while Pocklington's equation requires the inversion of an N order matrix.

For very thin wires, the current distribution is usually assumed to be of sinusoidal form [24]. For finite diameter wires (usually diameters $d > 0.05\lambda$), the sinusoidal current distribution is representative but not accurate. To find a more accurate current distribution on a cylindrical wire, an integral equation is usually derived and solved. Previously, solutions to the integral equation were obtained using iterative methods [31]; presently, it is most convenient to use moment method techniques [1–3].

If we know the voltage at the feed terminals of a wire antenna and find the current distribution, we can obtain the input impedance and radiation pattern. Similarly if a wave impinges upon the surface of a wire scatterer, it induces a current density that in turn is used to find the scattered field. Whereas the linear wire is simple, most of the information presented here can be readily extended to more complicated structures.

12.4.1 Pocklington's Integral Equation

In deriving Pocklington's integral equation, we will use the integral equation approach of Section 12.3.1. However, each step, as applied to the wire scatterer, will be repeated here to show the simplicity of the method.

Refer to Figure 12-24a. Let us assume that an incident wave impinges on the surface of a conducting wire. The total tangential electric field (E_z) at the surface of the wire is given by (12-53) or (12-53a), that is

$$E_z^t(r=r_s) = E_z^i(r=r_s) + E_z^s(r=r_s) = 0 \qquad (12\text{-}113)$$

or

$$E_z^s(r=r_s) = -E_z^i(r=r_s) \qquad (12\text{-}113a)$$

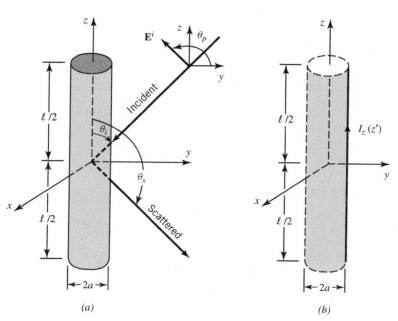

(a)   (b)

Figure 12-24   (a) Uniform plane wave obliquely incident on a conducting wire. (b) Equivalent current.

At any observation point, the field scattered by the induced current density on the surface of the wire is given by (12-54). However, for observations at the wire surface, only the z component of (12-54) is needed, and we can write it as

$$E_z^s(r) = -j\frac{1}{\omega\mu\varepsilon}\left[\beta^2 A_z + \frac{\partial^2 A_z}{\partial z^2}\right] \tag{12-114}$$

According to (12-54a) and neglecting edge effects

$$A_z = \frac{\mu}{4\pi}\iint_S J_z\frac{e^{-j\beta R}}{R}ds' = \frac{\mu}{4\pi}\int_{-\ell/2}^{+\ell/2}\int_0^{2\pi}J_z\frac{e^{-j\beta R}}{R}\,a\,d\phi'\,dz' \tag{12-115}$$

If the wire is very thin, the current density J_z is not a function of the azimuthal angle ϕ, and we can write it as

$$2\pi a J_z = I_z(z') \Rightarrow J_z = \frac{1}{2\pi a}I_z(z') \tag{12-116}$$

where $I_z(z')$ is assumed to be an equivalent filament line-source current located a radial distance $\rho = a$ from the z axis, as shown in Figure 12-24b. Thus, (12-115) reduces to

$$A_z = \frac{\mu}{4\pi}\int_{-\ell/2}^{+\ell/2}\left[\frac{1}{2\pi a}\int_0^{2\pi}I_z(z')\frac{e^{-j\beta R}}{R}\,a\,d\phi'\right]dz' \tag{12-117}$$

$$R = \sqrt{(x-x')^2 + (y-y')^2 + (z-z')^2}$$
$$= \sqrt{\rho^2 + a^2 - 2\rho a\cos(\phi-\phi') + (z-z')^2} \tag{12-117a}$$

where ρ is the radial distance to the observation point and a is the radius.

Because of the symmetry of the scatterer, the observations are not a function of ϕ. For simplicity, let us then choose $\phi = 0$. For observations at the surface ($\rho = a$) of the scatterer (12-117) and (12-117a) reduce to

$$A_z(\rho = a) = \mu\int_{-\ell/2}^{+\ell/2}I_z(z')\left[\frac{1}{2\pi}\int_0^{2\pi}\frac{e^{-j\beta R}}{4\pi R}d\phi'\right]dz' = \mu\int_{-\ell/2}^{+\ell/2}I_z(z')G(z,z')\,dz' \tag{12-118}$$

$$\boxed{G(z,z') = \frac{1}{2\pi}\int_0^{2\pi}\frac{e^{-j\beta R}}{4\pi R}d\phi'} \tag{12-118a}$$

$$\boxed{R(\rho = a) = \sqrt{4a^2\sin^2\left(\frac{\phi'}{2}\right) + (z-z')^2}} \tag{12-118b}$$

Thus, for observations at the surface ($\rho = a$) of the scatterer, the z component of the scattered electric field can be expressed as

$$E_z^s(\rho = a) = -j\frac{1}{\omega\varepsilon}\left(\beta^2 + \frac{d^2}{dz^2}\right)\int_{-\ell/2}^{+\ell/2}I_z(z')G(z,z')dz' \tag{12-119}$$

which by using (12-113a) reduces to

$$-j\frac{1}{\omega\varepsilon}\left(\frac{d^2}{dz^2} + \beta^2\right)\int_{-\ell/2}^{+\ell/2}I_z(z')G(z,z')dz' = -E_z^i(\rho = a) \tag{12-120}$$

or

$$\left(\frac{d^2}{dz^2}+\beta^2\right)\int_{-\ell/2}^{+\ell/2}I_z(z')G(z,z')\,dz'=-j\omega\varepsilon E_z^i(\rho=a) \tag{12-120a}$$

Interchanging integration with differentiation, we can rewrite (12-120a) as

$$\boxed{\int_{-\ell/2}^{+\ell/2}I_z(z')\left[\left(\frac{\partial^2}{\partial z^2}+\beta^2\right)G(z,z')\right]dz'=-j\omega\varepsilon E_z^i(\rho=a)} \tag{12-121}$$

where $G(z,z')$ is given by (12-118a).

Equation 12-121 is referred to as *Pocklington's integro-differential equation* [29], and it can be used to determine the equivalent filamentary line-source current of the wire, and thus current density on the wire, by knowing the incident field on the surface of the wire. It is a simplified form of (12-56) as applied to a wire scatterer, and it could have been derived directly from (12-56).

If we assume that the wire is very thin ($a \ll \lambda$), such that (12-118a) reduces to

$$G(z,z')=G(R)=\frac{e^{-j\beta R}}{4\pi R} \tag{12-122}$$

(12-121) can also be expressed in a more convenient form as [33]

$$\boxed{\int_{-\ell/2}^{+\ell/2}I_z(z')\frac{e^{-j\beta R}}{4\pi R^5}\left[(1+j\beta R)(2R^2-3a^2)+(\beta a R)^2\right]dz'=-j\omega\varepsilon E_z^i(\rho=a)} \tag{12-123}$$

where, for observations along the center of the wire ($\rho=0$),

$$R=\sqrt{a^2+(z-z')^2} \tag{12-123a}$$

In (12-121) or (12-123), $I(z')$ represents the equivalent filamentary line-source current located on the surface of the wire, as shown in Figure 12-24b, and it is obtained by knowing the incident electric field at the surface of the wire. By point-matching techniques, this is solved by matching the boundary conditions at discrete points on the surface of the wire. Often it is easier to choose the matching points to be at the interior of the wire, especially along the axis, as shown in Figure 12-25a, where $I_z(z')$ is located on the surface of the wire. By reciprocity, the configuration of Figure 12-25a is analogous to that of Figure 12-25b, where the equivalent filamentary line-source current is assumed to be located along the center axis of the wire and the matching points are selected on the surface of the wire. Either of the two configurations can be used to determine the equivalent filamentary line-source current $I_z(z')$; the choice is left to the individual.

Pocklington's integral equation of (12-121) is derived methodically based on (12-114). Eventually (12-121), using (12-122), reduces to (12-123). While the derivation of (12-121) and (12-123) is straightforward, their numerical evaluation may be more difficult because it involves double differentiation of the kernel, which leads to a non-integrable singularity, especially of the $1/R^5$ order. However, choosing the observation point along the axis of the wire while the current is on its surface or vice versa, as illustrated in Figure 12-25, mitigates the problem to some extent. Another way to derive such an integral equation, although it may not be as straightforward, will be to represent the electric field as a combination of both the vector **A** and scalar ϕ potentials, as given by (6-9a) ($\mathbf{E}=-j\omega\mathbf{A}-\nabla\phi$), instead of (12-114), and use the continuity equation $\nabla\cdot\mathbf{J}=-j\omega q$ to relate the charge q within ϕ as the divergence of **J** [28]. Using such a procedure, the resulting singularity in the kernel for both the vector **A** and scalar ϕ potentials is no more singular than $1/R$, which can be evaluated numerically more accurately.

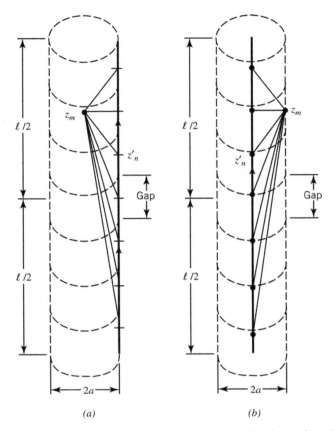

Figure 12-25 Dipole segmentation and its equivalent current. (*a*) On the surface. (*b*) Along its center.

12.4.2 Hallén's Integral Equation

Referring again to Figure 12-24*a*, let us assume that the length of the cylinder is much larger than its radius ($\ell \gg a$) and its radius is much smaller than the wavelength ($a \ll \lambda$), so that the effects of the end faces of the cylinder can be neglected. Therefore, the boundary conditions for a wire with infinite conductivity are those of vanishing total tangential E fields on the surface of the cylinder and vanishing current at the ends of the cylinder [$I_z(z' = \pm \ell/2) = 0$].

Since only an electric current density flows on the cylinder and it is directed along the z axis ($\mathbf{J} = \hat{\mathbf{a}}_z J_z$), according to (6-30) and (6-96a), $\mathbf{A} = \hat{\mathbf{a}}_z A_z(z')$, which for small radii is assumed to be only a function of z'. Thus, (6-34) reduces for $\mathbf{F} = 0$ to

$$E_z^t = -j\omega A_z - j\frac{1}{\omega\mu\varepsilon}\frac{\partial^2 A_z}{\partial z^2} = -j\frac{1}{\omega\mu\varepsilon}\left[\frac{d^2 A_z}{dz^2} + \omega^2\mu\varepsilon A_z\right] \tag{12-124}$$

Since the total tangential electric field E_z^t vanishes on the surface of the cylinder, (12-124) reduces to

$$\frac{d^2 A_z}{dz^2} + \beta^2 A_z = 0 \tag{12-124a}$$

Because the current density on the cylinder is symmetrical [$J_z(z') = J_z(-z')$], the potential A_z is also symmetrical [i.e., $A_z(z') = A_z(-z')$]. Thus, the solution of (12-124a) is given by

$$A_z(z) = -j\sqrt{\mu\varepsilon}\left[B_1\cos(\beta z) + C_1\sin(\beta|z|)\right] \tag{12-125}$$

where B_1 and C_1 are constants. For a current-carrying wire, its potential is also given by (6-97a). Equating (12-125) to (6-97a) leads to

$$\int_{-\ell/2}^{+\ell/2} I_z(z') \frac{e^{-j\beta R}}{4\pi R} dz' = -j\sqrt{\frac{\varepsilon}{\mu}}[B_1\cos(\beta z) + C_1\sin(\beta|z|)] \qquad (12\text{-}126)$$

If a voltage V_i is applied at the input terminals of the wire, it can be shown that the constant $C_1 = V_i/2$. The constant B_1 is determined from the boundary condition that requires the current to vanish at the end points of the wire.

Equation 12-126 is referred to as *Hallén's integral equation* for a perfectly conducting wire. It was derived by solving the differential equation 6-34 or 12-124a with the enforcement of the appropriate boundary conditions.

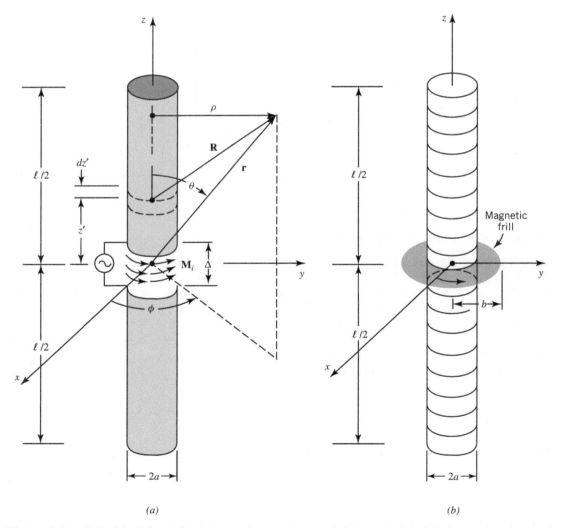

(a)   (b)

Figure 12-26 Cylindrical dipole, its segmentation, and gap modeling. (*a*) Cylindrical dipole. (*b*) Segmented dipole. (Source: C. A. Balanis, *Antenna Theory: Analysis and Design*, Third Edition, copyright © 2005, John Wiley & Sons, Inc. Reprinted by permission of John Wiley & Sons, Inc.)

12.4.3 Source Modeling

Let us assume that the wire of Figure 12-24 is symmetrically fed by a voltage source, as shown in Figure 12-26a, and the element is acting as a dipole antenna. To use, for example, Pocklington's integro-differential equation 12-121 or 12-123, we need to know how to express $E_z^i(\rho = a)$. Traditionally, there have been two methods used to model the excitation to represent $E_z^i(\rho = a, 0 \leq \phi \leq 2\pi, -\ell/2 \leq z \leq +\ell/2)$ at all points on the surface of the dipole: one is referred to as the *delta-gap* excitation and the other as the *equivalent magnetic ring current* (better known as *magnetic frill generator*) [38].

A. Delta Gap The delta-gap source modeling is the simplest and most widely used of the two, but it is also the least accurate, especially for impedances. Usually it is most accurate for smaller width gaps. Using the delta gap, it is assumed that the excitation voltage at the feed terminals is of a constant V_i value, and zero elsewhere. Therefore, the incident electric field $E_z^i(\rho = a, 0 \leq \phi \leq 2\pi, -\ell/2 \leq z \leq +\ell/2)$ is also a constant (V_i/Δ where Δ is the gap width) over the feed gap and zero elsewhere, hence the name delta gap. For the delta-gap model, the feed gap Δ of Figure 12-26a is replaced by a narrow band of strips of equivalent magnetic current density of

$$\mathbf{M}_i = -\hat{\mathbf{n}} \times \mathbf{E}^i = -\hat{\mathbf{a}}_\rho \times \hat{\mathbf{a}}_z \frac{V_i}{\Delta} = \hat{\mathbf{a}}_\phi \frac{V_i}{\Delta}, \qquad -\frac{\Delta}{2} \leq z' \leq \frac{\Delta}{2} \qquad (12\text{-}127)$$

The magnetic current density $\mathbf{M}_i$ is sketched in Figure 12-26a.

B. Magnetic Frill Generator The magnetic frill generator was introduced to calculate the near- as well as the far-zone fields from coaxial apertures [38]. To use this model, the feed gap is replaced with a circumferentially directed magnetic current density that exists over an annular aperture with inner radius a, which is usually chosen to be the radius of the inner wire, and an outer radius b, as shown in Figure 12-26b. Since the dipole is usually fed by transmission lines, the outer radius b of the equivalent annular aperture of the magnetic frill generator is found using the expression for the characteristic impedance of the transmission line.

Over the annular aperture of the magnetic frill generator, the electric field is represented by the TEM mode field distribution of a coaxial transmission line given by

$$\mathbf{E}_f = \hat{\mathbf{a}}_\rho \frac{V_i}{2\rho' \ln(b/a)} \qquad (12\text{-}128)$$

Therefore, the corresponding equivalent magnetic current density $\mathbf{M}_f$ for the magnetic frill generator, used to represent the aperture, is equal to

$$\mathbf{M}_f = -2\hat{\mathbf{n}} \times \mathbf{E}_f = -2\hat{\mathbf{a}}_z \times \hat{\mathbf{a}}_\rho E_\rho = -\hat{\mathbf{a}}_\phi \frac{V_i}{\rho' \ln(b/a)} \qquad (12\text{-}129)$$

The fields generated by the magnetic frill generator of (12-129) on the surface of the wire are found using [38]

$$E_z^i\left(\rho = a, \ 0 \leq \phi \leq 2\pi, \ -\frac{\ell}{2} \leq z \leq \frac{\ell}{2}\right)$$

$$\simeq -V_i \left[\frac{\beta(b^2 - a^2)e^{-j\beta R_0}}{8\ln(b/a)r_0^2} \left\{ 2\left[\frac{1}{\beta R_0} + j\left(1 - \frac{b^2 - a^2}{2R_0^2}\right) \right] \right. \right.$$

$$\left. \left. + \frac{a^2}{R_0}\left[\left(\frac{1}{\beta R_0} + j\left(1 - \frac{(b^2 + a^2)}{2R_0^2}\right) \right)\left(-j\beta - \frac{2}{R_0} \right) + \left(-\frac{1}{\beta R_0^2} + j\frac{b^2 + a^2}{R_0^3} \right) \right] \right\} \right] \qquad (12\text{-}130)$$

where

$$R_0 = \sqrt{z^2 + a^2} \qquad (12\text{-}130\text{a})$$

The fields generated on the surface of the wire computed using (12-130) can be approximated by those found along the axis ($\rho = 0$). Doing this leads to a simpler expression of the form [38]

$$E_z^i \left(\rho = 0, \ -\frac{\ell}{2} \le z \le \frac{\ell}{2} \right) = -\frac{V_i}{2 \ln(b/a)} \left[\frac{e^{-j\beta R_1}}{R_1} - \frac{e^{-j\beta R_2}}{R_2} \right] \qquad (12\text{-}131)$$

where

$$R_1 = \sqrt{z^2 + a^2} \qquad (12\text{-}131\text{a})$$

$$R_2 = \sqrt{z^2 + b^2} \qquad (12\text{-}131\text{b})$$

The following example compares the results obtained using the two source modelings (delta gap and magnetic frill generator).

Example 12-7

Assume a center-fed linear dipole of $\ell = 0.47\lambda$ and $a = 0.005\lambda$.

1. Determine the voltage and normalized current distribution over the length of the dipole using $N = 21$ segments to subdivide the length. Plot the current distribution.
2. Determine the input impedance using segments of $N = 7, 11, 21, 29, 41, 51, 61, 71$, and 79.

Use Pocklington's integro-differential equation 12-123 with piecewise constant subdomain basis functions and point matching to solve the problem, model the gap with one segment, and use both the delta gap and magnetic frill generator to model the excitation. Use (12-131) for the magnetic frill generator. Because the current at the ends of the wire vanishes, the piecewise constant subdomain basis functions are not the most judicious choices. However, because of their simplicity, they are chosen here to illustrate the principles, even though the results are not the most accurate. Assume that the characteristic impedance of the annular aperture is 50 ohms and the excitation voltage V_i is 1 V.

Solution:

1. Since the characteristic impedance of the annular aperture (coaxial line) is 50 ohms, then

$$Z_c = \sqrt{\frac{\mu_0}{\varepsilon_0}} \frac{\ln(b/a)}{2\pi} = 50 \ \Rightarrow \ \frac{b}{a} = 2.3$$

Subdividing the total length ($\ell = 0.47\lambda$) of the dipole into 21 segments makes the gap and each segment equal to

$$\Delta = \frac{0.47\lambda}{21} = 0.0224\lambda$$

   Using (12-131) to compute E_z^i, the corresponding induced voltages, obtained by multiplying the value of $-E_z^t$ at each segment by the length of the segment, are found listed in Table 12-1, where they are compared with those of the delta gap. $N = 1$ represents the outermost segment and $N = 11$ represents the center segment. Because of the symmetry, only values for the center segment and half of the other segments are shown. Although the two distributions are not identical, the magnetic frill distribution voltages decay quite rapidly away from the center segment, and they very quickly reach almost vanishing values.

TABLE 12-1 Unnormalized and normalized dipole induced voltage[a] differences for delta gap and magnetic frill generator ($\ell = 0.47\lambda$, $a = 0.005\lambda$, $N = 21$)

	Delta gap voltage		Magnetic frill generator voltage	
Segment number n	Unnormalized	Normalized	Unnormalized	Normalized
1	0	0	$1.11 \times 10^{-4} \; \underline{/-26.03°}$	$7.30 \times 10^{-5} \; \underline{/-26.03°}$
2	0	0	$1.42 \times 10^{-4} \; \underline{/-20.87°}$	$9.34 \times 10^{-5} \; \underline{/-20.87°}$
3	0	0	$1.89 \times 10^{-4} \; \underline{/-16.13°}$	$1.24 \times 10^{-4} \; \underline{/-16.13°}$
4	0	0	$2.62 \times 10^{-4} \; \underline{/-11.90°}$	$1.72 \times 10^{-4} \; \underline{/-11.90°}$
5	0	0	$3.88 \times 10^{-4} \; \underline{/-8.23°}$	$2.55 \times 10^{-4} \; \underline{/-8.23°}$
6	0	0	$6.23 \times 10^{-4} \; \underline{/-5.22°}$	$4.10 \times 10^{-4} \; \underline{/-5.22°}$
7	0	0	$1.14 \times 10^{-3} \; \underline{/-2.91°}$	$7.5 \times 10^{-4} \; \underline{/-2.91°}$
8	0	0	$2.52 \times 10^{-3} \; \underline{/-1.33°}$	$1.66 \times 10^{-3} \; \underline{/-1.33°}$
9	0	0	$7.89 \times 10^{-3} \; \underline{/-0.43°}$	$5.19 \times 10^{-3} \; \underline{/-0.43°}$
10	0	0	$5.25 \times 10^{-2} \; \underline{/-0.06°}$	$3.46 \times 10^{-2} \; \underline{/-0.06°}$
11	1	1	$1.52 \qquad \underline{/0°}$	$1.0 \qquad \underline{/0°}$

[a]Voltage differences as defined here represent the product of the incident electric field at the center of each segment and the corresponding segment length.

The corresponding unnormalized and normalized currents, obtained using (12-123) with piecewise constant pulse functions and the point-matching technique for both the delta gap and magnetic frill generator, are listed in Table 12-1. The normalized magnitudes of these currents are shown plotted in Figure 12-27. It is apparent that the two distributions are almost identical in shape, and they resemble that of the ideal sinusoidal current distribution which is more valid for very thin wires and very small gaps. The distributions obtained using Pocklington's integral equation do not vanish at the ends because of the use of piecewise constant subdomain basis functions, which make the dipole look longer by either a half or full subdomain length depending whether the reference of each subdomain function is taken at its middle or end point. A more accurate modeling will be to use higher-order subdomain basic functions, such as the piecewise linear function of Figure 12-7 or piecewise sinusoids of Figure 12-8.

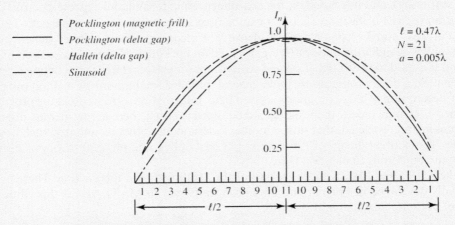

Figure 12-27 Current distribution along a dipole antenna.

2. The input impedances, computed using both the delta gap and the magnetic frill generator, are shown listed in Table 12-2. It is evident that the values begin to stabilize and compare favorably to each other once 61 or more segments are used.

TABLE 12-2   Dipole input impedance for delta gap and magnetic frill generator using Pocklington's integral equation ($\ell = 0.47\lambda$, $a = 0.005\lambda$)

N	Delta gap	Magnetic frill
7	$122.8 + j113.9$	$26.8 + j24.9$
11	$94.2 + j49.0$	$32.0 + j16.7$
21	$77.7 - j0.8$	$47.1 - j0.2$
29	$75.4 - j6.6$	$57.4 - j4.5$
41	$75.9 - j2.4$	$68.0 - j1.0$
51	$77.2 + j2.4$	$73.1 + j4.0$
61	$78.6 + j6.1$	$76.2 + j8.5$
71	$79.9 + j7.9$	$77.9 + j11.2$
79	$80.4 + j8.8$	$78.8 + j12.9$

12.5   COMPUTER CODES

With the advent of the computer there has been a proliferation of computer program development. Many of these programs are based on algorithms that are suitable for efficient computer programming for the analysis and synthesis of electromagnetic boundary-value problems. Some of these computer programs are very sophisticated and can be used to solve complex radiation and scattering problems. Others are much simpler and have limited applications. Many programs are public domain; others are restricted.

Five computer programs based on integral equation formulations and moment method solutions will be described here. The first computes the radiation or scattering by a two-dimensional perfectly electric conducting (PEC) body. It is referred to here as TDRS (two-dimensional radiation and scattering), and it is based on the two-dimensional formulations of the electric field integral equation (EFIE) of Section 12.3.1. It can be used for both electric and magnetic line-source excitation or TM^z and TE^z plane wave incidence. The second program, referred to here as PWRS (Pocklington's wire radiation and scattering) is based on Pocklington's integral equation of Section 12.4.1, and it is used for both radiation and scattering by a perfect electric conducting (PEC) wire.

The remaining three programs are more general, public domain moment method programs. A very brief description of these programs is given here. Information as to where these programs can be obtained is also included. It should be stated, however, that there are numerous other codes, public domain and restricted, that utilize moment method and other techniques, such as geometrical optics, geometrical theory of diffraction, physical optics, and physical theory of diffraction, which are too numerous to mention here.

Both the TDRS and PWRS codes are part of the Multimedia for this chapter. These two codes were initially developed in Fortran and have been translated to MATLAB for this edition. Both versions of each are included in this edition.

12.5.1   Two-Dimensional Radiation and Scattering

The two-dimensional radiation and scattering (TDRS) program is used to analyze four different two-dimensional perfectly electric conducting problems: the strip, and the circular, elliptical, and rectangular cylinders. The algorithm is based on the electric field integral equation of Section 12.3.1, and it is used for both electric and magnetic line-source excitation, or plane wave incidence of arbitrary polarization. For simplicity, piecewise constant pulse expansion functions and point-matching techniques have been adopted.

A. Strip   For the strip problem, the program can analyze either of the following:

1. A line source (electric or magnetic). It computes the electric current density over the width of the strip and the normalized radiation amplitude pattern (in decibels) for $0° \leq \phi \leq 360°$. The user must specify the width of the strip (in wavelengths), the type of line source (either electric or magnetic), and the location x_s, y_s of the source (in wavelengths).
2. Plane wave incidence of arbitrary polarization. The program can analyze either monostatic or bistatic scattering.

For monostatic scattering, the program computes the two-dimensional normalized (with respect to λ) monostatic SW $\sigma_{2\text{-D}}/\lambda$ (in decibels) for all angles of incidence ($0° \leq \phi \leq 360°$). The program starts at $\phi = 0°$ and then completes the entire $360°$ monostatic scattering pattern. The user must specify the width w of the strip (in wavelengths) and the polarization angle θ_p (in degrees) of the incident plane wave. The polarization of the incident wave is specified by the direction θ_p of the incident electric field relative to the z axis ($\theta_p = 0°$ implies TM^z; $\theta_p = 90°$ implies TE^z; other values of θ_p represent an arbitrary polarization). The polarization angle θ_p needs to be specified only when the polarization is neither TM^z nor TE^z.

For bistatic scattering, the program computes for the specified incidence angle the current density over the width of the strip and the two-dimensional normalized (with respect to λ) bistatic SW $\sigma_{2\text{-D}}/\lambda$ (in decibels) for all angles of observation ($0° \leq \phi_s \leq 360°$). The user must specify the width w of the strip (in wavelengths), the angle of incidence ϕ_i (in degrees), and the polarization angle θ_p (in degrees) of the incident plane wave. The polarization angle of the incident wave is specified in the same manner as for the monostatic case.

B. Circular, Elliptical, or Rectangular Cylinder   For the cylinder program, the program can analyze either a line source (electric or magnetic) or plane wave scattering of arbitrary polarization by a two-dimensional circular, elliptical, or rectangular cylinder.

1. For the line-source excitation, the program computes the current distribution over the entire surface of the cylinder and the normalized radiation amplitude pattern (in decibels). The user must specify, for each cylinder, the type of line source (electric or magnetic), the location x_s, y_s of the line source, and the size of the cylinder. For the circular cylinder, the size is specified by its radius a (in wavelengths) and for the elliptical and rectangular cylinders by the principal semiaxes lengths a and b (in wavelengths), with a measured along the x axis and b along the y axis.
2. For the plane wave incidence, the program computes monostatic or bistatic scattering of arbitrary polarization by a circular, elliptical, or rectangular cylinder.

For monostatic scattering, the program computes the two-dimensional normalized (with respect to λ) monostatic SW $\sigma_{2\text{-D}/\lambda}$ (in decibels) for all angles of incidence ($0° \leq \phi \leq 360°$). The program starts at $\phi = 0°$ and then computes the entire $360°$ monostatic scattering pattern. The user must specify the size of the cylinder, as was done for the line-source excitation, and the polarization angle θ_p (in degrees) of the incident plane wave. The polarization of the incident wave is specified by the direction θ_p of the incident electric field relative to the z axis ($\theta_p = 0°$ implies TM^z; $\theta_p = 90°$ implies TE^z; other values of θ_p represent an arbitrary polarization). The polarization angle θ_p needs to be specified only when the polarization is neither TM^z nor TE^z.

For bistatic scattering, the program computes for the specified incidence angle the current density over the entire surface of the cylinder and the two-dimensional normalized (with respect to λ) bistatic SW $\sigma_{2\text{-D}} / \lambda$ (in decibels) for all angles of observation ($0° \leq \phi_s \leq 360°$). The user must specify the size of the cylinder, as was done for the line-source excitation, the incidence angle ϕ_i (in degrees), and the polarization angle θ_p (in degrees) of the incident plane wave. The polarization angle of the incident wave is specified in the same manner as for the monostatic case.

12.5.2 Pocklington's Wire Radiation and Scattering

Pocklington's wire radiation and scattering (PWRS) program computes the radiation characteristics of a center-fed wire antenna and the scattering characteristics of a perfectly electric conducting (PEC) wire, each of radius a and length ℓ. Both are based on Pocklington's integral equation 12-123.

A. Radiation For the wire antenna of Figure 12-26, the excitation is modeled by either a delta gap or a magnetic frill feed modeling, and it computes the current distribution, normalized amplitude radiation pattern, and the input impedance. The user must specify the length of the wire, its radius (both in wavelengths), and the type of feed modeling (delta gap or magnetic frill). A computer program based on Hallén's integral equation can be found in [24].

B. Scattering The geometry for the plane wave scattering by the wire is shown in Figure 12-24(a). The program computes the monostatic or bistatic scattering of arbitrary polarization.

For monostatic scattering the program computes the normalized (with respect to m^2) RCS $\sigma_{3\text{-D}} / \text{m}^2$ (in dBsm) for all angles of incidence ($0° \leq \theta_i \leq 180°$). The program starts at $\theta_i = 0°$ and then computes the entire $180°$ monostatic scattering pattern. The user must specify the length and radius of the wire (both in wavelengths) and the polarization angle θ_p (in degrees) of the incident plane wave. The polarization of the incident wave is specified by the direction θ_p of the incident electric field relative to the plane of incidence, where the plane of incidence is defined as the plane that contains the vector of the incident wave and the wire scatterer ($\theta = 0°$ implies that the electric field is on the plane of incidence; $\theta = 90°$ implies that the electric field is perpendicular to the plane of incidence and to the wire; thus no scattering occurs for this case).

For bistatic scattering, the program computes for the specified incidence angle the current distribution over the length of the wire and the normalized (with respect to m^2) bistatic RCS $\sigma_{3\text{-D}} / \text{m}^2$ (in dBsm) for all angle of observation ($0° \leq \theta_s \leq 180°$). The user must specify the length and radius of the wire (both in wavelengths), the angle of incidence θ_i (in degrees), and the polarization angle θ_p of the incident plane wave. The polarization angle is specified in the same manner as for the monostatic case.

12.5.3 Numerical Electromagnetics Code

Over the years there have been a number of computational electromagnetic codes developed, both personal and commercial, based on the Integral Equation/Method of Moments method. An attempt is made here to indicate the genesis of EM code development. The process started with the development of the Numerical Electromagnetic Code (NEC) [39] and the Mini-Numerical Electromagnetic Code (MININEC) [40, 41]. The NEC code analyzes the interaction of electromagnetic waves with arbitrary structures consisting of conducting wires and surfaces. It uses the EFIE for thin wires and the MFIE for surfaces. The initial MININEC was a user-oriented compact version of the NEC, and it was coded in BASIC. Since the initial introduction of these two codes, there have been various versions of them.

After the NEC code, the Electromagnetic Surface Patch (ESP) code [42] was introduced. The ESP is a method of moments surface patch code based on the piecewise sinusoidal reaction formulation, which is basically equivalent to the EFIE. It can be used for the analysis of the radiation and scattering from 3-D geometries consisting of an interconnection of thin wires, perfectly conducting (or thin dielectric) polygonal plates, thin wires, wire/plate and plate/plate junctions, and polygonal dielectric material volumes. Numerous other codes, including [43], based on IE/MoM have been developed since the NEC and ESP. Even as of the writing of this edition of the book, there are efforts underway for the development of other personal and commercial codes. They are too numerous to mention all of them here. The reader is directed to the internet for the search of such, and other, electromagnetic codes. Some basic student-oriented MATLAB codes have been developed to complement the material of this chapter. They are listed under the 12.6 Multimedia section, and they are available to the reader on the website that accompanies this book.

12.6 MULTIMEDIA

On the website that accompanies this book, the following multimedia resources are included for the review, understanding, and presentation of the material of this chapter.

- **MATLAB** computer programs:
  a. **Wire_Charge:** Computes the charge distribution on a straight or bent PEC wire, of Figures 12-1 and 12-3, based on the Integral Equation (IE) of Section 12.2.
  b. **PEC_Strip_Line_MoM:** Computes the far-zone amplitude radiation pattern and current density of a line source above a PEC strip, of finite width, based on the Integral Equation (IE) of Sections 12.2.2 through 12.2.8 and Physical Optics (PO) of Section 11.2.3, and Figures 12-5 and 11-2.
  c. **PEC_Strip_SW_MoM:** Computes the TM^z and TE^z 2-D scattering width (SW), monostatic and bistatic, and current density of a uniform plane wave incident upon a PEC strip, of finite width, based on the Integral Equations (IE) of Sections 12.3.1 and 12.3.2, and Physical Optics of Section 11.3.1, and Figures 12-13 and 11-4.
  d. **TDRS:** Computes the radiation and scattering of a plane wave incident of a 2-D scatterer (strip, cylinder) based on the Integral Equation (IE) as outlined in Section 12.4.1.
  e. **PWRS:** Computes the radiation characteristics of a symmetrical dipole of Figure 12-25 or scatterer of Figure 12-23, based on Pocklington's Integral Equation (IE) of (12-13) in Section 12.4.1.
- **PowerPoint (PPT)** viewgraphs, in multicolor.

REFERENCES

1. R. F. Harrington, "Matrix methods for field problems," *Proc. IEEE*, vol. 55, no. 2, pp. 136–149, Feb. 1967.
2. R. F. Harrington, *Field Computation by Moment Methods*, Macmillan, New York, 1968.
3. J. H. Richmond, "Digital computer solutions of the rigorous equations for scattering problems," *Proc. IEEE*, vol. 53, pp. 796–804, Aug. 1965.
4. L. L. Tsai, "Moment methods in electromagnetics for undergraduates," *IEEE Trans. on Education*, vol. E-21, no. 1, pp. 14–22, Feb. 1978.
5. R. Mittra (Ed.), *Computer Techniques for Electromagnetics*, Pergamon, New York, 1973.

6. J. Moore and R. Pizer, *Moment Methods in Electromagnetics*, John Wiley & Sons, New York, 1984.

7. P. Y. Ufimtsev, "Method of edge waves in the physical theory of diffraction," translated by U. S. Air Force Foreign Technology Division, Wright-Patterson AFB, Ohio, Sept. 1971.

8. P. Y. Ufimtsev, "Approximate computation of the diffraction of plane electromagnetic waves at certain metal bodies," *Sov. Phys.–Tech. Phys.*, vol. 27, pp. 1708–1718, 1957.

9. P. Y. Ufimtsev, "Secondary diffraction of electromagnetic waves by a strip," *Sov. Phys.–Tech. Phys.*, vol. 3, pp. 535–548, 1958.

10. K. M. Mitzner, "Incremental length diffraction coefficients," Tech. Rep. AFAL-TR-73-296, Northrop Corp., Aircraft Division, Apr. 1974.

11. E. F. Knott and T. B. A. Senior, "Comparison of three high-frequency diffraction techniques," *Proc. IEEE*, vol. 62, no. 11, pp. 1468–1474, Nov. 1974.

12. E. F. Knott, "A progression of high-frequency RCS prediction techniques," *Proc. IEEE*, vol. 73, no. 2, pp. 252–264, Feb. 1985.

13. T. Griesser and C. A. Balanis, "Backscatter analysis of dihedral corner reflectors using physical optics and the physical theory of diffraction," *IEEE Trans. Antennas Propagat.*, vol. AP-35, no. 10, pp. 1137–1147, Oct. 1987.

14. J. B. Keller, "Diffraction by an aperture," *J. Appl. Phys.*, vol. 28, no. 4, pp. 426–444, Apr. 1957.

15. J. B. Keller, "Geometrical theory of diffraction," *J. Opt. Soc. Amer.*, vol. 52, no. 2, pp. 116–130, Feb. 1962.

16. R. G. Kouyoumjian and P. H. Pathak, "A uniform geometrical theory of diffraction for an edge in a prefectly conducting surface," *Proc. IEEE*, vol. 62, no. 11, pp. 1448–1461, Nov. 1974.

17. G. L. James, *Geometrical Theory of Diffraction for Electromagnetic Waves*, Third Edition Revised, Peregrinus, London, 1986.

18. J. D. Lilly, "Application of the moment method to antenna analysis," MSEE Thesis, Department of Electrical Engineering, West Virginia University, 1980.

19. J. D. Lilly and C. A. Balanis, "Current distributions, input impedances, and radiation patterns of wire antennas," North American Radio Science Meeting of URSI, Université Laval, Quebec, Canada, Jun. 2–6, 1980.

20. D. K. Cheng, *Field and Wave Electromagnetics*, Addison-Wesley, Reading, MA, 1983, p. 88.

21. R. Mittra and C. A. Klein, "Stability and convergence of moment method solutions," in *Numerical and Asymptotic Techniques in Electromagnetics*, R. Mittra (Ed.), Springer-Verlag, New York, 1975, Chapter 5, pp. 129–163.

22. T. K. Sarkar, "A note on the choice weighting functions in the method of moments," *IEEE Trans. Antennas Propagat.*, vol. AP-33, no. 4, pp. 436–441, Apr. 1985.

23. T. K. Sarkar, A. R. Djordjević, and E. Arvas, "On the choice of expansion and weighting functions in the numerical solution of operator equations," *IEEE Trans. Antennas Propagat.*, vol. AP-33, no. 9, pp. 988–996, Sept. 1985.

24. C. A. Balanis, *Antenna Theory: Analysis and Design* (Third edition), John Wiley & Sons, New York, 2005.

25. E. K. Miller and F. J. Deadrick, "Some computational aspects of thin-wire modeling," in *Numerical and Asymptotic Techniques in Electromagnetics*, R. Mittra (Ed.), Springer-Verlag, New York, 1975, Chapter 4, pp. 89–127.

26. L. Kantorovich and G. Akilov, *Functional Analysis in Normed Spaces*, Pergamon, Oxford, pp. 586–587, 1964.

27. D. P. Marsland, C. A. Balanis, and S. Brumley, "Higher order diffractions from a circular disk," *IEEE Trans. Antennas Propagat.*, vol. AP-35, no. 12, pp. 1436–1444, Dec. 1987.

28. D. R. Witton, private communication and personal class notes.

29. H. C. Pocklington, "Electrical oscillations in wire," *Cambridge Philos. Soc. Proc.*, vol. 9, pp. 324–332, 1897.

30. E. Hallén, "Theoretical investigations into the transmitting and receiving qualities of antennae," *Nova Acta Regiae Soc. Sci. Upsaliensis*, Ser. IV, no. 4, pp. 1–44, 1938.

31. R. King and C. W. Harrison, Jr., "The distribution of current along a symmetrical center-driven antenna," *Proc. IRE*, vol. 31, pp. 548–567, Oct. 1943.

32. J. H. Richmond, "A wire-grid model for scattering by conducting bodies," *IEEE Trans. Antennas Propagat.*, vol. AP-14, no. 6, pp. 782–786, Nov. 1966.

33. G. A. Thiele, "Wire antennas," in *Computer Techniques for Electromagnetics*, R. Mittra (Ed.), Pergamon, New York, Chapter 2, pp. 7–70, 1973.

34. C. M. Butler and D. R. Wilton, "Evaluation of potential integral at singularity of exact kernel in thin-wire calculations," *IEEE Trans. Antennas Propagat.*, vol. AP-23, no. 2, pp. 293–295, Mar. 1975.

35. L. W. Pearson and C. M. Butler, "Inadequacies of collocation solutions to Pocklington-type models of thin-wire structures," *IEEE Trans. Antennas Propagat.*, vol. AP-23, no. 2, pp. 293–298, Mar. 1975.

36. C. M. Butler and D. R. Wilton, "Analysis of various numerical techniques applied to thin-wire scatterers," *IEEE Trans. Antennas Propagat.*, vol. AP-23, no. 4, pp. 534–540, Jul. 1975.

37. D. R. Wilton and C. M. Butler, "Efficient numerical techniques for solving Pocklington's equation and their relationships to other methods," *IEEE Trans. Antennas Propagat.*, vol. AP-24, no. 1, pp. 83–86, Jan. 1976.

38. L. L. Tsai, "A numerical solution for the near and far fields of an annular ring of magnetic current," *IEEE Trans. Antennas Propagat.*, vol. AP-20, no. 5, pp. 569–576, Sept. 1972.

39. G. J. Burke and A. J. Poggio, "Numerical electromagnetics code (NEC)-method of moments," Technical Document 116, Naval Ocean Systems Center, San Diego, CA, Jan. 1981.

40. A. J. Julian, J. M. Logan, and J. W. Rockway, "MININEC: A mini-numerical electro magnetics code," Technical Document 516, Naval Ocean Systems Center, San Diego, CA, Sept. 6, 1982.

41. J. Rockway, J. Logan, D. Tarn, and S. Li, *The MININEC System: Microcomputer Analysis of Wire Antennas*, Artech House, Inc., 1988.

42. E. H. Newman and D. L. Dilsavor, "A user's manual for the electromagnetic surface patch code: ESP version III," Technical Report No. 716148–19, ElectroScience Laboratory, The Ohio State University, May 1987.

43. http://www.wipl-d.com.

PROBLEMS

12.1. A circular loop of radius $a = 0.2$ m is constructed out of a wire of radius $b = 10^{-3}$ m, as shown in Figure P12-1. The entire loop is maintained at a constant potential of 1 V. Using integral equation techniques, determine and plot for $0° \le \phi \le 360°$ the surface charge density on the wire. Assume that at any given angle the charge is uniformly distributed along the circumference of the wire.

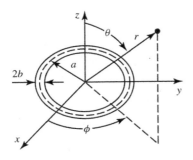

Figure P12-1

12.2. Repeat Problem 12.1 when the loop is split into two parts; one part (from 0 to 180°) is maintained at a constant potential of 1 V and the other part (from 180 to 360°) is maintained at a constant potential of 2 V.

12.3. A linear charge ρ_c is distributed along a very thin wire circular loop of radius a which is discontinuous (open) for $355° \le \phi \le 360°$ and $0° \le \phi \le 5°$. The static potential V_z produced by this charge distribution on a very thin line along the z axis, passing through the origin of the circular loop is given by

$$V_z = \frac{10a}{\sqrt{a^2 + z^2}}, \qquad 0 \le z \le 7 \text{ meters}$$

(a) Subdivide the line for $0 \le z \le 7$ meters in 70 segments and compute the potential V_z at the center of each of the 70 segments when the radius of the loop is $a = 1$ meter.

(b) Plot the potential V_z of part (a) on a linear plot for $0 \le z \le 7$ meters.

(c) Write an integral equation based on Poisson's differential equation to solve this problem.

(d) For a loop of radius $a = 1$ meter, determine the linear charge ρ_c that produces the stated potential along $0 \le z \le 7$ meters that is computed in part (a) and plotted in part (b).

(e) Plot the linear charge distribution of part (d) on a linear plot of $5° \le \phi \le 355°$.

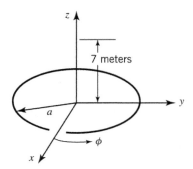

Figure P12-3

Use pulse expansion functions and $N = 70$ segments along the circumference of the loop.

12.4. Repeat Example 12-3 for a strip with $w = 2\lambda$, $h = 0.25\lambda$, and $t = 0.01\lambda$.

12.5. A magnetic line source of constant current I_m is placed a height h above a Perfect Magnetic Conducting (PMC) strip of width w. The geometry is the same as that of Figure 12-5. Write a normalized Integral Equation (IE) that can be used to solve for the linear magnetic current density M_z on the strip. You do not have to derive it as long as you justify it.

12.6. A TE^z uniform plane wave of the form shown in Figure 12-13b, with a normalized z component of the incident magnetic field of magnitude H_0 is incident upon a Perfect Magnetic Conducting (PMC) strip of finite width w. Write a normalized Integral Equation (IE) that can be used to solve for the magnetic current M_z which is induced on the PMC strip. You do not have to derive it as long as you justify it.

12.7. Derive (12-71a) and (12-71b).

12.8. Instead of using the electric field components of (12-72a) and (12-72b) to formulate the two-dimensional SW of a PEC strip for TE^z polarization, derive an integral expression for H_z^s in terms of J_x and then use the definition for SW of (11-21c). This requires only one component of the scattered H-field while using the definition of (11-21b) requires two scattered electric field components as given by (12-71a) and (12-71b).

12.9. An infinite electric line source of constant current I_e is placed next to a circular conducting cylinder of radius a, as shown in Figure P12-9. The line source is positioned a distance $b(b > a)$ from the center of the cylinder. Use the EFIE, piecewise constant subdomain basis functions, and point-matching techniques.

(a) Formulate the current density induced on the surface of the cylinder.

(b) Compute the induced current density when $a = 5\lambda$ and $b = 5.25\lambda$. Assume a unity line-source current. Compare with the modal solution current density of (11-168a).

(c) For part (b), compute the normalized far-zone amplitude pattern (in decibels). Normalize so that the maximum is 0 dB.

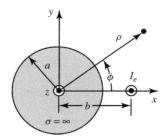

Figure P12-9

12.10. Repeat Problem 12.9 for $a = 5\lambda$ and $b = 5.5\lambda$.

12.11. An infinite electric line source of constant current I_e is placed next to a rectangular cylinder of dimensions a and b, as shown in Figure P12-11. The line source is positioned a distance c $(c > a)$ from the center of the cylinder along the x axis. Use the EFIE and piecewise subdomain basis functions and point-matching techniques, and do the following.

(a) Compute the induced current on the surface of the cylinder when $a = 5\lambda$, $b = 2.5$, and $c = 5.25\lambda$. Assume a unity line-source current.

(b) Compute for part (a) the normalized far-zone amplitude pattern (in decibels). Normalize so that the maximum is 0 dB.

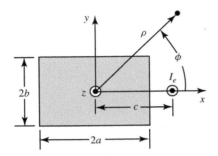

Figure P12-11

12.12. Repeat Problem 12.11 for an electric line source near an elliptic cylinder with $a = 5\lambda$, $b = 2.5\lambda$, and $c = 5.25\lambda$.

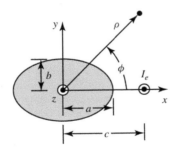

Figure P12-12

12.13. A TM^z uniform plane wave traveling in the $+x$ direction is normally incident upon a conducting circular cylinder of radius a, as shown in Figure P12-13. Using the EFIE, piecewise constant subdomain basis functions, and point-matching techniques, write your own program, and do the following.

(a) Plot the current density induced on the surface of the cylinder when $a = 2\lambda$. Assume the incident field is of unity amplitude. Compare with the modal solution current density of (11-97).

(b) Plot the normalized $\sigma_{2\text{-D}}/\lambda$ bistatic scattering width (in decibels) for $0° \le \phi \le 360°$ when $a = 2\lambda$. Compare with the modal solution of (11-102).

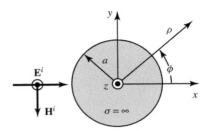

Figure P12-13

12.14. A TM^z uniform plane wave traveling in the $+x$ direction is normally incident upon a conducting rectangular cylinder of dimensions a and b, as shown in Figure P12-14. Using the EFIE, piecewise constant subdomain basis functions, and point-matching techniques, write your own program, and do the following.

(a) Compute the induced current density on the surface of the cylinder when $a = 5\lambda$ and $b = 2.5\lambda$. Assume a unity line-source current.

(b) For part (a), compute and plot the two-dimensional normalized $\sigma_{2\text{-D}}/\lambda$ bistatic scattering width (in dB) for $0° \le \phi \le 360°$.

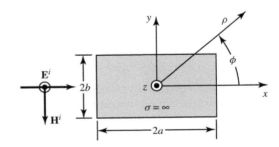

Figure P12-14

12.15. Repeat Problem 12.14 for a TM^z uniform plane wave impinging upon an elliptic conducting cylinder with $a = 5\lambda$ and $b = 2.5\lambda$.

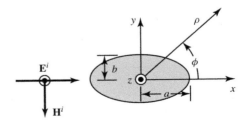

Figure P12-15

12.16. Using the geometry of Figure 12-14, verify (12-71a) and (12-71b), and that (12-70a) reduces to (12-72a) and (12-70b) to (12-72b).

12.17. Show that the integral of (12-78d) can be evaluated using (12-79a) through (12-79c).

12.18. Repeat Problem 12.9 for a magnetic line source of constant current $I_m = 1$ when $a = b = 5\lambda$. This problem is representative of a very thin axial slot on the surface of the cylinder. Compare the current density on the surface of the cylinder from part (b) with that of the modal solution of (11-117a).

12.19. Repeat Problem 12.11 for a magnetic line source of constant current $I_m = 1$ when $a = 5\lambda$, $b = 2.5\lambda$, and $c = 5\lambda$. This problem is representative of a very thin axial slot on the surface of the cylinder.

12.20. Repeat Problem 12.12 for a magnetic line source of constant current $I_m = 1$ when $a = 5\lambda$, $b = 2.5\lambda$, and $c = 5\lambda$. This problem is representative of a very thin axial slot on the surface of the cylinder.

12.21. Repeat Problem 12.13 for a TE^z uniform plane wave of unity amplitude. Compare the current density with the modal solution of (11-113) and the normalized $\sigma_{2\text{-D}}/\lambda$ bistatic scattering width with the modal solution of (11-117).

12.22. Repeat Problem 2.14 for a TE^z uniform plane wave at unity amplitude.

12.23. Repeat Problem 12.15 for a TE^z uniform plane wave of unity amplitude.

12.24. Using the geometry of Figure 12-17, show that

$$\hat{\mathbf{n}} \cdot \nabla H_0^{(2)}(\beta R) = -\beta \cos\psi H_1^{(2)}(\beta R)$$

12.25. Repeat Problem 12.9 using the MFIE.

12.26. Repeat Problem 12.13 using the MFIE. You must write your own computer program to solve this problem.

12.27. Using the geometry of Figure 12-18, show that

$$\hat{\mathbf{n}}' \cdot \nabla H_0^{(2)}(\beta R) = -\beta \cos\psi' H_1^{(2)}(\beta R)$$

12.28. Repeat Problem 12.18 using the MFIE.

12.29. Derive Pocklington's integral equation 12-123 using (12-121) and (12-122).

12.30. Derive the solution of (12-125) to the differential equation (12-124a). Show that Hallén's integral equation can be written as (12-126).

12.31. Show that the incident tangential electric field (E_z^i) generated on the surface of a wire of radius a by a magnetic field generator of (12-129) is given by (12-130).

12.32. Reduce (12-130) to (12-131) valid only along the z axis ($\rho = 0$).

12.33. For the center-fed dipole of Example 12-7, write the $[Z]$ matrix for $N = 21$ using for the gap the delta-gap generator and the magnetic frill generator.

12.34. For an infinitesimal center fed dipole of $\ell = \lambda/50$ and radius $a = 0.005\lambda$, derive the input impedance using Pocklington's integral equation with piecewise constant subdomain basis functions and point matching. Use $N = 21$ and model the gap as a delta-gap generator and as a magnetic-frill generator. Use the MATLAB **PWRS** computer program listed in Multimedia at the end of the chapter.

12.35. A conducting wire of length $\ell = 0.47\lambda$ and radius $a = 0.005\lambda$ is placed symmetrically along the z axis. Assuming a TM^z uniform plane wave is incident on the wire at an angle $\theta_i = 30°$ from the z axis, do the following.
(a) Compute and plot the current induced on the surface of the wire.
(b) Compute and plot the bistatic RCS for $0° \le \theta_s \le 180°$.
(c) Compute and plot the monostatic RCS for $0° \le \theta_i = \theta_s \le 180°$.
The amplitude of the incident electric field is 10^{-3} V/m. Use Pocklington's integral equation and the **PWRS** computer program. Determine the number of segments that leads to a stable solution.

12.36. Repeat Problem 12.35 for a TE^z uniform plane wave incidence.

CHAPTER 13

Geometrical Theory of Diffraction

13.1 INTRODUCTION

The treatment of the radiation and scattering characteristics from radiating and scattering systems using modal solutions is limited to objects whose surfaces can be described by orthogonal curvilinear coordinates. Moreover, most of the solutions are in the form of infinite series, which are poorly convergent when the dimensions of the object are greater than about a wavelength. These limitations, therefore, exclude rigorous analyses of many practical radiating and scattering systems.

A method that describes the solution in the form of an integral equation has received considerable attention. Whereas arbitrary shapes can be handled by this method, it mostly requires the use of a digital computer for numerical computations and therefore, is most convenient for objects that are not too many wavelengths in size because of the capacity limitations of computers. This method is usually referred to as the *integral equation* (IE) method, and its solution is generally accomplished by the *moment method* (MM) [1–4]. These were discussed in Chapter 12.

When the dimensions of the radiating or scattering object are many wavelengths, high-frequency asymptotic techniques can be used to analyze many problems that are otherwise mathematically intractable. Two such techniques, which have received considerable attention in the past few years, are the *geometrical theory of diffraction* (GTD) and the *physical theory of diffraction* (PTD). The GTD, originated by Keller [5, 6] and extended by Kouyoumjian and Pathak [7–10], is an extension of the classical *geometrical optics* (GO) (direct, reflected, and refracted rays), and it overcomes some of the limitations of geometrical optics by introducing a diffraction mechanism [11]. The PTD, introduced by Ufimtsev [12–14], supplements *physical optics* (PO) to provide corrections that are due to diffractions at edges of conducting surfaces. Ufimtsev suggested the existence of nonuniform ("fringe") edge currents in addition to the uniform physical optics surface currents [15–18]. The PTD bears some resemblance to GTD in its method of application.

At high frequencies, diffraction—like reflection and refraction—is a local phenomenon and it depends on two things:

1. The geometry of the object at the point of diffraction (edge, vertex, curved surface).
2. The amplitude, phase, and polarization of the incident field at the point of diffraction.

A field is associated with each diffracted ray, and the total field at a point is the sum of all the rays at that point. Some of the diffracted rays enter the shadow regions and account for the field intensity there. The diffracted field, which is determined by a generalization of *Fermat's principle* [6, 7], is initiated at points on the surface of the object that create a discontinuity in the incident GO field (incident and reflected shadow boundaries).

The phase of the field on a ray is assumed to be equal to the product of the optical length of the ray from some reference point and the wave number of the medium. Appropriate phase jumps must be added as rays pass through caustics (defined in Section 13.2.1). The amplitude is assumed to vary in accordance with the principle of conservation of energy in a narrow tube of rays.

The initial value of the field on a diffracted ray is determined from the incident field with the aid of an appropriate diffraction coefficient that is a dyadic for electromagnetic fields. This is analogous to the manner reflected fields are determined using the reflection coefficient. The rays also follow paths that make the optical distance from the source to the observation point an extremum (usually a minimum). This leads to straight-line propagation within homogeneous media and along geodesics (surface extrema) on smooth surfaces. The field intensity also attenuates exponentially as it travels along surface geodesics.

The diffraction and attenuation coefficients are usually determined from the asymptotic solutions of the simplest boundary-value problems, which have the same local geometry at the points of diffraction as the object at the points of interest. Geometries of this type are referred to as *canonical* problems. One of the simplest geometries that will be discussed in this chapter is a conducting wedge. The primary objective in using the GTD is to resolve each problem to smaller components [19–25], each representing a canonical geometry with a known solution. The ultimate solution is a superposition of the contributions from each canonical problem.

Some of the advantages of GTD are given in the following list.

1. It is simple to apply.
2. It can be used to solve complicated problems that do not have exact solutions.
3. It provides physical insight into the radiation and scattering mechanisms from the various parts of the structure.
4. It yields accurate results that compare quite well with experiments and other methods.
5. It can be combined with other techniques such as the moment method [26–28].

One of the main interests of diffraction by wedges is that engineers and scientists have investigated how the shape and material properties of complex structures affect their backscattering characteristics. The attraction in this area is primarily aimed toward designs of low-profile (stealth) technology by using appropriate shaping along with lossy or coated materials to reduce the radar visibility, as represented by radar cross section (RCS), of complex radar targets, such as aircraft, spacecraft, and missiles. A good example is the F-117 shown in Figure 13-1, whose surface is primarily structured by a number of faceted flat plates and wedges because, as will become evident from the developments, formulations, examples, and problems of this chapter (see also Problem 13.50), the backscatter from exterior wedges is lower than that of convex curved surfaces. In addition, the plates are oriented judiciously so that the maximum scattered field is toward the specular direction and away from the source of detection. While in this chapter we will focus on the diffraction by PEC wedges, the diffraction by wedges with impedance surfaces, to represent lossy and composite wedge surfaces, is the subject of Chapter 14.

13.2 GEOMETRICAL OPTICS

Geometrical optics (GO) is an approximate high-frequency method for determining wave propagation for incident, reflected, and refracted fields. Because it uses ray concepts, it is often referred

Figure 13-1   F-117 Nighthawk. (Printed with permission of Lockheed Martin Corporation © 2010).

to as *ray optics*. Originally, geometrical optics was developed to analyze the propagation of light at sufficiently high frequencies where it was not necessary to consider the wave nature of light. Instead, the transport of energy from one point to another in an isotropic lossless medium is accomplished using the conservation of energy flux in a tube of rays. For reflection problems, geometrical optics approximates the scattered fields only toward specular directions as determined by Snell's law of reflection: the angle of reflection is equal to the angle of incidence. For sufficiently high frequencies, geometrical optics fields may dominate the scattering phenomena and may not require any corrections. This is more evident for backscattering from smooth curved surfaces whose curvature is large compared to the wavelength.

According to classical geometrical optics, the rays between any two points P_1 and P_2 follow a path that makes the optical distance between them an extremum (usually a minimum). In equation form, this is expressed as

$$\delta \int_{P_1}^{P_2} n(s)\, ds = 0 \tag{13-1}$$

where δ represents what is referred to in the calculus of variations as the *variational differential* and $n(s)$ is the index of refraction of the medium, $\beta(s)/\beta_0 = n(s)$. If the medium is homogeneous, $n(s) = n = \text{constant}$, the paths are straight lines. Equation 13-1 is a mathematical representation of *Fermat's principle*. In addition, the light intensity, power per unit solid angle, between any two points is also governed by the conservation of energy flux in a tube of rays.

To demonstrate the principles of geometrical optics, let us consider a *primary wave front* surface ψ_0, as shown in Figure 13-2, formed at $t = t_0$ by the motion of light propagating in an isotropic lossless medium. The objectives here are:

1. To determine the *secondary wave front* surfaces ψ_n formed at $t = t_{n+1} > t_n, n = 0, 1, 2, 3 \ldots$.
2. To relate the power density and field intensity on the secondary wave fronts to those of the primary or previous wave fronts.

The secondary wave fronts can be determined by first selecting a number of discrete points on the primary wave front. If the medium of wave propagation is also assumed to be homogeneous, ray paths from the primary to the secondary wave front are drawn as straight lines that at each point are normal to the surface of the primary wave front.

Since the wave travels in the medium with the speed of light given by $v = c/n$, where c is the speed of light in free space and n is the index of refraction, then at $\Delta t = t_1 - t_0$ $(t_1 > t_0)$ the wave would have traveled a distance $\Delta \ell = v \Delta t$. Along each of the normal rays a distance $\Delta \ell$ is marked,

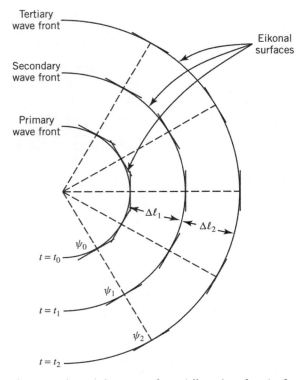

Figure 13-2   Primary and secondary wave front (eikonal surfaces) of a radiated wave.

and a surface perpendicular to each ray is drawn. The surfaces normal to each of the rays are then connected to form the secondary wave front ψ_1, as shown in Figure 13-2. The same procedure can be repeated to determine the subsegment wave front surfaces ψ_2, ψ_3,....

The family of wave front surfaces $\psi_n(x, y, z)$, $n = 0, 1, 2, 3,...$, that are normal to each of the radial rays is referred to as the *eikonal* surfaces, and they can be determined using the *eikonal equation* [7]

$$\|\nabla \psi_n(x, y, z)\|^2 = \left\{\frac{\partial \psi_n}{\partial x}\right\}^2 + \left\{\frac{\partial \psi_n}{\partial y}\right\}^2 + \left\{\frac{\partial \psi_n}{\partial z}\right\}^2 = n^2(s) \tag{13-2}$$

Since the rays normal to the wave fronts and the eikonal surfaces are uniquely related, it is only necessary to know one or the other when dealing with geometrical optics.

Extending this procedure to approximate the wave motion of electromagnetic waves of lower frequencies, it is evident that:

1. The eikonal surfaces for plane waves are planar surfaces perpendicular to the direction of wave travel.
2. The eikonal surfaces for cylindrical waves are cylindrical surfaces perpendicular to the cylindrical radial vectors.
3. The eikonal surfaces for spherical waves are spherical surfaces perpendicular to the spherical radial vectors.

Each of these is demonstrated, respectively, in Figures 13-3a, 13-3b, and 13-3c.

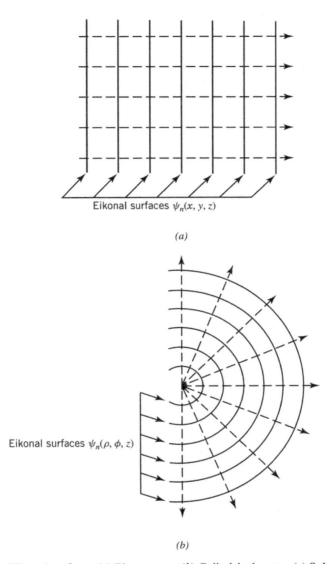

Eikonal surfaces $\psi_n(x, y, z)$

(a)

Eikonal surfaces $\psi_n(\rho, \phi, z)$

(b)

Figure 13-3   Eikonal surfaces. (*a*) Plane waves. (*b*) Cylindrical waves. (*c*) Spherical waves.

13.2.1 Amplitude Relation

In geometrical optics, the light intensity (power per unit solid angle) between two points is also governed by the conservation of energy flux in a tube of rays. To demonstrate that, let us assume that a point source, as shown in Figure 13-4, emanates isotropically spherical waves. Within a tube of rays, the cross-sectional areas at some reference point $s = 0$ and at s are given, respectively, by dA_0 and dA. The radiation density S_0 at $s = 0$ is related to the radiation density S at s by

$$S_0 \, dA_0 = S \, dA \qquad (13\text{-}3)$$

or

$$\frac{S(s)}{S_0(0)} = \frac{dA_0}{dA} \qquad (13\text{-}3a)$$

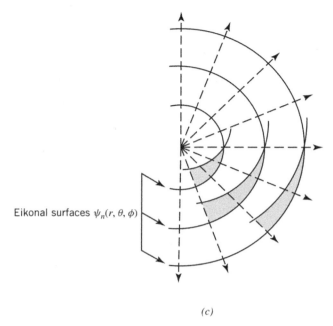

(c)

Figure 13-3 (*Continued*)

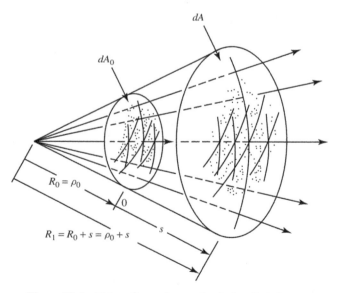

Figure 13-4 Tube of rays for a spherical radiated wave.

It has been assumed that S_0 and S are constant, respectively, throughout the cross-sectional areas dA_0 and dA, and that no power flows across the sides of the conical tube.

For electromagnetic waves, the far-zone electric field $\mathbf{E}(r,\theta,\phi)$ is related to the radiation density $S(r,\theta,\phi)$ by [7]

$$S(r,\theta,\phi) = \frac{1}{2\eta}|\mathbf{E}(r,\theta,\phi)|^2 = \frac{1}{2}\sqrt{\frac{\varepsilon}{\mu}}|\mathbf{E}(r,\theta,\phi)|^2 \qquad (13\text{-}4)$$

Therefore, (13-3a) can also be written, using (13-4), as

$$\frac{|\mathbf{E}|^2}{|\mathbf{E}_0|^2} = \frac{dA_0}{dA} \tag{13-5}$$

or

$$\frac{|\mathbf{E}|}{|\mathbf{E}_0|} = \sqrt{\frac{dA_0}{dA}} \tag{13-5a}$$

Since in the tube of rays in Figure 13-4, the differential surface areas dA_0 and dA are patches of spherical surfaces with radii of $R_0 = \rho_0$ and $R_1 = R_0 + s = \rho_0 + s$, respectively, then (13-5a) can be written in terms of the radii of curvature of the wave fronts at $s = 0$ and s. Thus, (13-5a) reduces to

$$\boxed{\frac{|\mathbf{E}|}{|\mathbf{E}_0|} = \sqrt{\frac{dA_0}{dA}} = \sqrt{\frac{4\pi R_0^2/C_0}{4\pi R_1^2/C_0}} = \frac{R_0}{R_1} = \frac{\rho_0}{\rho_0 + s}} \tag{13-6}$$

and it indicates that the electric field varies, as expected, inversely proportional to the distance of travel; C_0 is a proportionality constant.

If the eikonal surfaces of the radiated fields are cylindrical surfaces, representing the wave fronts of cylindrical waves, then the field relation (13-5a) takes the form

$$\boxed{\frac{|\mathbf{E}|}{|\mathbf{E}_0|} = \sqrt{\frac{dA_0}{dA}} = \sqrt{\frac{2\pi R_0/C_1}{2\pi R_1/C_1}} = \sqrt{\frac{R_0}{R_1}} = \sqrt{\frac{\rho_0}{\rho_0 + s}}} \tag{13-7}$$

where C_1 is a proportionality constant. Relation (13-7) indicates that the electric field for cylindrical waves varies, as expected, inversely to the square root of the distance of travel. For planar eikonal surfaces, representing plane waves, (13-5a) simplifies to

$$\boxed{\frac{|\mathbf{E}|}{|\mathbf{E}_0|} = 1} \tag{13-8}$$

For the previous three cases, the eikonal surfaces were, respectively, spherical, cylindrical, and planar. Let us now consider a more general configuration in which the eikonal surfaces (wave fronts) are not necessarily spherical. This is illustrated in Figure 13-5a where the wave front is represented by a radius of curvature R_1 in the xz and R_2 in the yz planes, which are not equal ($R_1 \neq R_2$). To determine the focusing characteristics of such a surface, let us trace the focusing diagram of rays 1, 2, 3, and 4 from the four corners of the wave front. It is apparent that the rays focus (cross) at different points. For this example, rays 1 and 2 focus at P, rays 3 and 4 focus at P', rays 2 and 3 focus at Q', and rays 1 and 4 focus at Q. This system of a tube of rays is referred to as *astigmatic* (not meeting at a single point) and the lines PP' and QQ' are called *caustics*.[1]

Referring to the geometry of Figure 13-5b, it can be shown that for a wave whose eikonal surface (wave front) forms an astigmatic tube of rays the electric field intensity from one surface relative to that of another, as related by (13-5a), takes the form

$$\boxed{\frac{|\mathbf{E}|}{|\mathbf{E}_0|} = \sqrt{\frac{dA_0}{dA}} = \sqrt{\frac{\rho_1\rho_2}{(\rho_1 + s)(\rho_2 + s)}}} \tag{13-9}$$

---

[1] A *caustic* is a point, a line, or a surface through which all the rays of a wave pass. Examples of it are the focal point of a paraboloid (parabola of revolution) and the focal line of a parabolic cylinder. The field at a caustic is infinite, in principle, because an infinite number of rays pass through it.

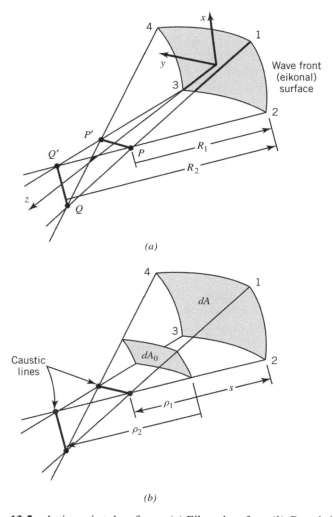

(a)

(b)

Figure 13-5   Astigmatic tube of rays. (*a*) Eikonal surface. (*b*) Caustic lines.

It is apparent that (13-9) reduces to the following equations:

1. (13-6) if the wave front is spherical ($\rho_1 = \rho_2 = \rho_0$).
2. (13-7) if the wave front is cylindrical ($\rho_1 = \infty, \rho_2 = \rho_0$ or $\rho_2 = \infty, \rho_1 = \rho_0$).
3. (13-8) if the wave front is planar ($\rho_1 = \rho_2 = \infty$).

Expressions (13-6) through (13-9) correctly relate the *magnitudes* of the high-frequency electric field at one wave front surface to that of another. These were derived using geometrical optics based on the principle of conservation of energy flux through a tube of rays. Although these may be valid high-frequency approximations for light waves, they are not accurate relations for electromagnetic waves of lower frequencies. Two apparent missing properties in these relations are those of *phase* and *polarization*.

13.2.2 Phase and Polarization Relations

Phase and polarization information can be introduced to the relations (13-6) through (13-9) by examining the approach introduced by Luneberg [29] and Kline [30, 31] to develop high-frequency solutions of electromagnetic problems. The works of Luneberg and Kline, referred to as the *Luneberg–Kline high-frequency expansion*, best bridge the gap between geometrical (ray) optics and wave propagation phenomena.

The Luneberg–Kline series expansion solution begins by assuming that the electric field for large ω can be written as a series

$$\mathbf{E}(\mathbf{R},\omega) = e^{-j\beta_0\psi(\mathbf{R})} \sum_{m=0}^{\infty} \frac{\mathbf{E}_m(\mathbf{R})}{(j\omega)^m} \tag{13-10}$$

where $\mathbf{R}$ = position vector
β_0 = phase constant for free-space

Substituting (13-10) into the wave equation

$$\nabla^2\mathbf{E} + \beta^2\mathbf{E} = 0 \tag{13-11}$$

subject to Maxwell's equation

$$\nabla \cdot \mathbf{E} = 0 \tag{13-12}$$

it can be shown, by equating like powers of ω, that one obtains the following.

1. The eikonal equation 13-2 or

$$\|\nabla\psi\|^2 = n^2 \tag{13-13a}$$

   where ψ = eikonal (wave front) surface
   n = index of refraction

2. The transport equations

$$\frac{\partial\mathbf{E}_0}{\partial s} + \frac{1}{2}\left\{\frac{\nabla^2\psi}{n}\right\}\mathbf{E}_0 = 0 \qquad \text{for first-order terms} \tag{13-13b}$$

$$\frac{\partial\mathbf{E}_m}{\partial s} + \frac{1}{2}\left\{\frac{\nabla^2\psi}{n}\right\}\mathbf{E}_m = \frac{v_p}{2}\nabla^2\mathbf{E}_{m-1} \qquad \text{for higher-order terms} \tag{13-13c}$$

   where $m = 1, 2, 3\ldots$
   v_p = speed of light in medium

3. The conditional equations

$$\hat{\mathbf{s}} \cdot \mathbf{E}_0 = 0 \qquad \text{for first-order terms} \tag{13-13d}$$

$$\hat{\mathbf{s}} \cdot \mathbf{E}_m = v_p \nabla \cdot \mathbf{E}_{m-1} \qquad \text{for higher-order terms}$$
$$m = 1, 2, 3\ldots \tag{13-13e}$$

   where

$$\hat{\mathbf{s}} = \frac{\nabla\psi}{n} = \text{unit vector in the direction path (normal to the wave front } \psi) \tag{13-13f}$$

   s = distance along the ray path

At the present time we are interested mainly in first-order solutions for the electric field of (13-10) that can be approximated and take the form of

$$\mathbf{E}(s) \simeq e^{-j\beta_0 \psi(s)} \, \mathbf{E}_0(s=0) \tag{13-14}$$

Integrating the first-order transport equation 13-13b along s and referring to the geometry of Figure 13-5a, it can be shown that (13-14) can be written as [30, 31]

$$\mathbf{E}(s) \simeq \mathbf{E}_0(0) e^{-j\beta_0 \psi(0)} \sqrt{\frac{\rho_1 \rho_2}{(\rho_1 + s)(\rho_2 + s)}} \, e^{-j\beta s} \tag{13-15}$$

where $s = 0$ is taken as a reference point. Since $\mathbf{E}_0(0)$ is complex, the phase term $e^{-j\beta_0\psi(0)}$ can be combined with $\mathbf{E}_0(0)$ and (13-15) rewritten as

$$
\underbrace{\mathbf{E}(s) = \mathbf{E}_0'(0) e^{j\phi_0(0)}}_{\substack{\text{Field at reference} \\ \text{point } (s=0)}} \quad \underbrace{\sqrt{\frac{\rho_1 \rho_2}{(\rho_1 + s)(\rho_2 + s)}}}_{\substack{\text{Spatial attenuation} \\ \text{(divergence, spreading)} \\ \text{factor}}} \quad \underbrace{e^{-j\beta s}}_{\substack{\text{Phase} \\ \text{factor}}} \tag{13-15a}
$$

where $\mathbf{E}_0'(0)$ = field amplitude at reference point $(s = 0)$
     $\phi_0(0)$ = field phase at reference point $(s = 0)$

Comparing (13-15a) to (13-9), it is evident that the leading term of the Luneberg–Kline series expansion solution for large ω predicts the spatial attenuation relation between the electric fields of two points as obtained by classical geometrical optics, as given by (13-9), which ignores both the polarization and the wave motion (phase) of electromagnetic fields. It also predicts their phase and polarization relations, as given by (13-15a). Obviously, (13-15a) could have been obtained from (13-9) by artificially converting the magnitudes of the fields to vectors (to account for polarization) and by introducing a complex exponential to account for the phase delay of the field from $s = 0$ to s. This was not necessary since (13-15a) was derived here rigorously using the leading term of the Luneberg–Kline expansion series for large ω subject to the wave and Maxwell's equations. It should be pointed out, however, that (13-15a) is only a high-frequency approximation and it becomes more accurate as the frequency approaches infinity. However, for many practical engineering problems, it does predict quite accurate results that compare well with measurements.

In principle, more accurate expressions to the geometrical optics approximation can be obtained by retaining higher-order terms $\mathbf{E}_1(\mathbf{R}_1), \mathbf{E}_2(\mathbf{R}_2),\ldots$ in the Luneberg–Kline series expansion (13-10), and in the transport (13-13c) and conditional (13-13e) equations. However, such a procedure is very difficult. In addition, the resulting terms do not remove the discontinuities introduced by geometrical optics fields along the incident and reflection boundaries, and the method does not lend itself to other improvements in the geometrical optics, such as those of diffraction. Therefore, no such procedure will be pursued here.

It should be noted that when the observation point is chosen so that $s = -\rho_1$ or $s = -\rho_2$, (13-15a) possesses singularities representing the congruence of the rays at the caustic lines PP' and QQ'. Therefore, (13-15a) is not valid along the caustics and not very accurate near them, and it should not be used in those regions. Other methods should be utilized to find the fields at and near caustics [32–36]. In addition, it is observed that when $-\rho_2 < s < -\rho_1$ the sign in the $(\rho_1 + s)$ term of the denominator of (13-15a) changes. Similar changes of sign occur in the $(\rho_1 + s)$ and $(\rho_2 + s)$ terms when $s < -\rho_2 < -\rho_1$. Therefore, (13-15a) correctly predicts $+90°$ phase jumps each time a caustic is crossed in the direction of propagation.

13.2.3 Reflection from Surfaces

Geometrical optics can be used to compute high-frequency approximations to the fields reflected from surfaces, the directions of which are determined by Snell's law of reflection. To demonstrate the procedure, let us assume that a field impinges on a smooth conducting surface S, where it undergoes a reflection at point Q_R. This is illustrated in Figure 13-6a where $\hat{s}^i$ is the unit vector in the direction of incidence, $\hat{s}^r$ is the unit vector in the direction of reflection, $\hat{e}^i_\parallel$, $\hat{e}^r_\parallel$ are unit vectors, for incident and reflected electric fields, parallel to the planes of incidence and reflection, and $\hat{e}^i_\perp$, $\hat{e}^r_\perp$ are unit vectors, for incident and reflected electric fields, perpendicular to the planes of incidence and reflection. The plane of incidence is formed by the unit vector $\hat{n}$ normal to the surface at the point of reflection Q_R and the unit vector $\hat{s}^i$, and the plane of reflection is formed by the unit vectors $\hat{n}$ and $\hat{s}^r$. The angle of incidence θ_i is measured between $\hat{n}$ and $\hat{s}^i$ whereas θ_r is measured between $\hat{n}$ and $\hat{s}^r$, and they are equal $(\theta_i = \theta_r)$.

The polarization unit vectors are chosen so that

$$\hat{e}^i_\perp \times \hat{s}^i = \hat{e}^i_\parallel \tag{13-16a}$$
$$\hat{e}^r_\perp \times \hat{s}^r = \hat{e}^r_\parallel \tag{13-16b}$$

and the incident and reflected electric fields can be expressed as

$$\mathbf{E}^i_0 = \hat{e}^i_\parallel E^i_{0\parallel} + \hat{e}^i_\perp E^i_{0\perp} \tag{13-17a}$$
$$\mathbf{E}^r_0 = \hat{e}^r_\parallel E^r_{0\parallel} + \hat{e}^r_\perp E^r_{0\perp} \tag{13-17b}$$

The incident and reflected fields at the point of reflection can be related by applying the boundary conditions of vanishing tangential components of the electric field at the point of reflection (Q_R). Doing this, we can write that

$$\mathbf{E}^r_0(s=0) = \mathbf{E}^i_0(Q_R) \cdot \bar{\mathbf{R}} = \mathbf{E}^i_0(Q_R) \cdot [\hat{e}^i_\parallel \hat{e}^r_\parallel - \hat{e}^i_\perp \hat{e}^r_\perp] \tag{13-18}$$

where $\mathbf{E}^r_0(s=0)$ = reflected field at the point of reflection (the reference point for the reflected ray is taken on the reflecting surface so that $s = 0$)

$\mathbf{E}^i_0(Q_R)$ = incident field at the point of reflection Q_R

$\bar{\mathbf{R}}$ = dyadic reflection coefficient

In matrix notation, the reflection coefficient can be written as

$$R = \begin{bmatrix} 1 & 0 \\ 0 & -1 \end{bmatrix} \tag{13-19}$$

which is identical to the Fresnel reflection coefficients of electromagnetic plane waves reflected from plane, perfectly conducting surfaces. This is quite acceptable in practice since at high frequencies, reflection, as well as diffraction, is a local phenomenon, and it depends largely on the geometry of the surface in the immediate neighborhood of the reflection point. Therefore, near the reflection point Q_R, the following approximations can be made:

1. The reflecting surface can be approximated by a plane tangent at Q_R.
2. The wave front of the incident field can be assumed to be planar.

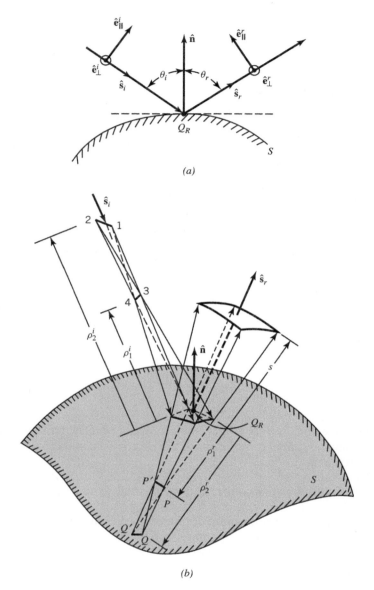

Figure 13-6 Reflection from a curved surface. (*a*) Reflection point. (*b*) Astigmatic tube of rays.

With the aid of (13-15a) and (13-18), it follows that the reflected field $\mathbf{E}^r(s)$ at a distance s from the point of reflection Q_R can be written as

$$\mathbf{E}^r(s) = \underbrace{\mathbf{E}^i(Q_R)}_{\substack{\text{Field at} \\ \text{reference} \\ \text{point}\,(Q_r)}} \cdot \underbrace{\bar{\mathbf{R}}}_{\substack{\text{Reflection} \\ \text{coefficient}}} \underbrace{\sqrt{\frac{\rho_1^r \rho_2^r}{(\rho_1^r + s)(\rho_2^r + s)}}}_{\substack{\text{Spatial attenuation} \\ \text{(divergence, spreading)} \\ \text{factor}}} \underbrace{e^{-j\beta s}}_{\substack{\text{Phase} \\ \text{factor}}} \tag{13-20}$$

where ρ_1^r, ρ_2^r = principal radii of curvature of the reflected wave front at the point of reflection

The astigmatic tube of rays for the reflected fields is shown in Figure 13-6b where the reference surface is taken at the reflecting surface.

The principal radii of curvature of the reflected wave front, ρ_1^r and ρ_2^r, are related to the principal radii of curvature of the incident wave front, ρ_1^i and ρ_2^i, the aspect of wave incidence, and the curvature of the reflecting surface at Q_R. It can be shown that ρ_1^r and ρ_2^r can be expressed as [10]

$$\frac{1}{\rho_1^r} = \frac{1}{2}\left[\frac{1}{\rho_1^i} + \frac{1}{\rho_2^i}\right] + \frac{1}{f_1} \qquad (13\text{-}21a)$$

$$\frac{1}{\rho_2^r} = \frac{1}{2}\left[\frac{1}{\rho_1^i} + \frac{1}{\rho_2^i}\right] + \frac{1}{f_2} \qquad (13\text{-}21b)$$

where ρ_1^i, ρ_2^i = principal radii of curvature of incident wave front ($\rho_1^i = \rho_2^i = s'$ for spherical incident wave front; $\rho_1^i = \rho'$, $\rho_2^i = \infty$ or $\rho_1^i = \infty$, $\rho_2^i = \rho'$ for cylindrical incident wave front, and $\rho_1^i = \rho_2^i = \infty$ for planar incident wave front)

Equations 13-21a and 13-21b are similar in form to the simple lens and mirror formulas of elementary physics. In fact, when the incident ray is spherical ($\rho_1^i = \rho_2^i = s'$), f_1 and f_2 represent focal distances that are independent of the source range that is creating the spherical wave.

When the incident field has a spherical wave front, $\rho_1^i = \rho_2^i = s'$, then f_1 and f_2 simplify to

$$\frac{1}{f_1} = \frac{1}{\cos\theta_i}\left[\frac{\sin^2\theta_2}{R_1} + \frac{\sin^2\theta_1}{R_2}\right] + \sqrt{\frac{1}{\cos^2\theta_i}\left[\frac{\sin^2\theta_2}{R_1} + \frac{\sin^2\theta_1}{R_2}\right]^2 - \frac{4}{R_1 R_2}} \qquad (13\text{-}22a)$$

$$\frac{1}{f_2} = \frac{1}{\cos\theta_i}\left[\frac{\sin^2\theta_2}{R_1} + \frac{\sin^2\theta_1}{R_2}\right] - \sqrt{\frac{1}{\cos^2\theta_i}\left[\frac{\sin^2\theta_2}{R_1} + \frac{\sin^2\theta_1}{R_2}\right]^2 - \frac{4}{R_1 R_2}} \qquad (13\text{-}22b)$$

where R_1, R_2 = radii of curvature of the reflecting surface

θ_1 = angle between the direction of the incident ray $\hat{s}^i$ and $\hat{u}_1$

θ_2 = angle between the direction of the incident ray $\hat{s}^i$ and $\hat{u}_2$

$\hat{u}_1$ = unit vector in the principal direction of S at Q_R with principal radius of curvature R_1

$\hat{u}_2$ = unit vector in the principal direction of S at Q_R with principal radius of curvature R_2

The geometrical arrangement of these is exhibited in Figure 13-7.

If the incident wave form is a plane wave, then $\rho_1^i = \rho_2^i = \infty$ and according to (13-21a) and (13-21b)

$$\frac{1}{\rho_1^r \rho_2^r} = \frac{1}{f_1 f_2} = \frac{4}{R_1 R_2} \qquad (13\text{-}23)$$

or

$$\rho_1^r \rho_2^r = \frac{R_1 R_2}{4} \qquad (13\text{-}23a)$$

The relation of (13-23a) is very useful to calculate the far-zone reflected fields. Now if either R_1 and/or R_2 becomes infinite, as is the case for flat plates or cylindrical scatterers, the geometrical optics field fails to predict the scattered field when the incident field is a plane wave.

To give some physical insight into the principal radii of curvature R_1 and R_2 of a reflecting surface, let us assume that the reflecting surface is well behaved (continuous and smooth). At each point on the surface there exists a unit normal vector. If through that point a plane intersects the reflecting surface, it generates on the reflecting surface a curve as shown in Figure 13-7b. If in addition the intersecting plane contains the unit normal to the surface at that point, the curve generated on the reflecting surface by the intersecting plane is known as the *normal section curve*, as shown in Figure 13-7b. If the intersecting plane is rotated about the surface normal at that point, a number of unique normal sections are generated, one on each orientation of the intersecting plane. Associated with each normal section curve, there is a radius of curvature. It can be shown that for each point on an arbitrary well-behaved curved reflecting surface, there is one intersecting plane that maximizes the radius of curvature of its corresponding normal section curve while there is another intersecting plane at the same point that minimizes the radius of curvature of its corresponding normal section curve. For each point on the reflecting surface there are two normal section radii of curvature, denoted here as R_1 and R_2 and referred to as *principal radii of curvature*, and the two corresponding planes are known as the *principal planes*. For an arbitrary surface, the two principal planes are perpendicular to each other. Also for each principal plane we can define unit vectors $\hat{\mathbf{u}}_1$ and $\hat{\mathbf{u}}_2$ that are tangent to each normal section curve generated

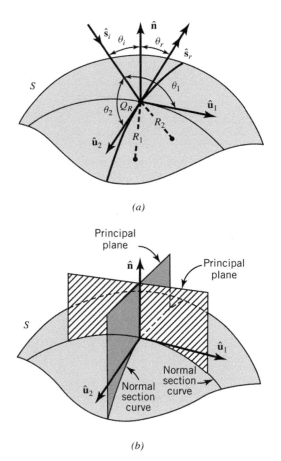

(a)

(b)

Figure 13-7   Geometry for reflection from a three-dimensional curved surface. (*a*) Principal radii of curvature. (*b*) Normal section curves and principal planes.

by each intersecting principal plane. Also the unit vectors $\hat{\mathbf{u}}_1$ and $\hat{\mathbf{u}}_2$, which lie in the principal plane, point along the principal directions whose normal section curves have radii of curvature R_1 and R_2, respectively. Expressions for determining the principal unit vectors $\hat{\mathbf{u}}_1$, $\hat{\mathbf{u}}_2$ and the principal radii of curvature R_1, R_2 of arbitrary surfaces of revolution are given in Problem 13.2 at the end of the chapter.

In general, however, f_1 and f_2 can be obtained using [10]

$$
\frac{1}{f_{1(2)}} = \frac{\cos\theta_i}{|\boldsymbol{\theta}|^2}\left(\frac{\theta_{22}^2 + \theta_{12}^2}{R_1} + \frac{\theta_{21}^2 + \theta_{11}^2}{R_2}\right)
$$

$$
\pm \frac{1}{2}\left\{\left(\frac{1}{\rho_1^i} - \frac{1}{\rho_2^i}\right)^2 + \left(\frac{1}{\rho_1^i} - \frac{1}{\rho_2^i}\right)\frac{4\cos\theta_i}{|\boldsymbol{\theta}|^2}\left(\frac{\theta_{22}^2 - \theta_{12}^2}{R_1} + \frac{\theta_{21}^2 - \theta_{11}^2}{R_2}\right)\right.
$$

$$
\left. + \frac{4\cos^2\theta_i}{|\boldsymbol{\theta}|^4}\left[\left(\frac{\theta_{22}^2 + \theta_{12}^2}{R_1} + \frac{\theta_{21}^2 + \theta_{11}^2}{R_2}\right)^2 - \frac{4|\boldsymbol{\theta}|^2}{R_1 R_2}\right]\right\}^{1/2} \tag{13-24}
$$

where the plus sign is used for f_1 and the minus for f_2. In (13-24), $|\boldsymbol{\theta}|$ is the determinant of

$$
[\boldsymbol{\theta}] = \begin{vmatrix} \hat{\mathbf{X}}_1^i \cdot \hat{\mathbf{u}}_1 & \hat{\mathbf{X}}_1^i \cdot \hat{\mathbf{u}}_2 \\ \hat{\mathbf{X}}_2^i \cdot \hat{\mathbf{u}}_1 & \hat{\mathbf{X}}_2^i \cdot \hat{\mathbf{u}}_2 \end{vmatrix} \tag{13-24a}
$$

or

$$
|\boldsymbol{\theta}| = (\hat{\mathbf{X}}_1^i \cdot \hat{\mathbf{u}}_1)(\hat{\mathbf{X}}_2^i \cdot \hat{\mathbf{u}}_2) - (\hat{\mathbf{X}}_2^i \cdot \hat{\mathbf{u}}_1)(\hat{\mathbf{X}}_1^i \cdot \hat{\mathbf{u}}_2) \tag{13-24b}
$$

and

$$
\theta_{jk} = \hat{\mathbf{X}}_j^i \cdot \hat{\mathbf{u}}_k \tag{13-24c}
$$

The vectors $\hat{\mathbf{X}}_1^i$ and $\hat{\mathbf{X}}_2^i$ represent the principal directions of the incident wave front at the reflection point Q_R with principal radii of curvature ρ_1^i and ρ_2^i.

Equations 13-24 through 13-24c can be used to find single, first-order reflections by a reflecting surface. The process, using basically the same set of equations, must be repeated if second- and higher-order reflections are required. However, to accomplish this the principal plane directions $\hat{\mathbf{X}}_1^r$ and $\hat{\mathbf{X}}_2^r$ of the reflected fields from the previous reflection must be known. For example, second-order reflections can be found provided the principal plane directions $\hat{\mathbf{X}}_1^r$ and $\hat{\mathbf{X}}_2^r$ of the first-order reflected field are found. To do this, we first introduce

$$
\boldsymbol{Q}^r = \begin{vmatrix} Q_{11}^r & Q_{12}^r \\ Q_{12}^r & Q_{22}^r \end{vmatrix} \tag{13-25}
$$

where $\boldsymbol{Q}^r$ is defined as the curvature matrix for the reflected wave front whose entries are

$$
Q_{11}^r = \frac{1}{\rho_1^i} + \frac{2\cos\theta_i}{|\boldsymbol{\theta}|^2}\left(\frac{\theta_{22}^2}{R_1} + \frac{\theta_{21}^2}{R_2}\right) \tag{13-26a}
$$

$$
Q_{12}^r = -\frac{2\cos\theta_i}{|\boldsymbol{\theta}|^2}\left(\frac{\theta_{22}\theta_{12}}{R_1} + \frac{\theta_{11}\theta_{21}}{R_2}\right) \tag{13-26b}
$$

$$
Q_{22}^r = \frac{1}{\rho_2^i} + \frac{2\cos\theta_i}{|\boldsymbol{\theta}|^2}\left(\frac{\theta_{12}^2}{R_1} + \frac{\theta_{11}^2}{R_2}\right) \tag{13-26c}
$$

Then the principal directions $\hat{\mathbf{X}}_1^r$ and $\hat{\mathbf{X}}_2^r$ of the reflected wave front, *with respect to the x_1^r and x_2^r coordinates*, can be written as

$$\hat{\mathbf{X}}_1^r = \frac{\left(Q_{22}^r - \dfrac{1}{\rho_1^r}\right)\hat{\mathbf{x}}_1^r - Q_{12}^r \hat{\mathbf{x}}_2^r}{\sqrt{\left(Q_{22}^r - \dfrac{1}{\rho_1^r}\right)^2 + (Q_{12}^r)^2}} \tag{13-27a}$$

$$\hat{\mathbf{X}}_2^r = -\hat{\mathbf{s}}^r \times \hat{\mathbf{X}}_1^r \tag{13-27b}$$

where $\hat{\mathbf{x}}_1^r$ and $\hat{\mathbf{x}}_2^r$ are unit vectors perpendicular to the reflected ray, and they are determined using

$$\hat{\mathbf{x}}_1^r = \hat{\mathbf{X}}_1^i - 2(\hat{\mathbf{n}} \cdot \hat{\mathbf{X}}_1^i)\hat{\mathbf{n}} \tag{13-28a}$$

$$\hat{\mathbf{x}}_2^r = \hat{\mathbf{X}}_2^i - 2(\hat{\mathbf{n}} \cdot \hat{\mathbf{X}}_2^i)\hat{\mathbf{n}} \tag{13-28b}$$

with $\hat{\mathbf{n}}$ being a unit vector normal to the surface at the reflection point.

To demonstrate the application of these formulations, let us consider a problem that is classified as a classic example in scattering.

Example 13-1

A linearly polarized uniform plane wave of amplitude E_0 is incident on a conducting sphere of radius a, as shown in Figures 11-26 and 11-28. Using geometrical optics methods, determine the:

1. Far-zone ($s \gg \rho_1^r$ and ρ_2^r) fields that are reflected from the surface of the sphere.
2. Backscatter radar cross section.

Solution: For a linearly polarized uniform plane wave incident upon a conducting sphere (13-20) reduces in the far zone to

$$E^r(s) = E_0(-1)\sqrt{\frac{\rho_1^r \rho_2^r}{(\rho_1^r + s)(\rho_2^r + s)}}\, e^{-j\beta s} \overset{s \gg \rho_1^r, \rho_2^r}{\simeq} -E_0 \frac{\sqrt{\rho_1^r \rho_2^r}}{s}\, e^{-j\beta s}$$

According to (13-23a)

$$\rho_1^r \rho_2^r = \frac{a^2}{4}$$

Thus,

$$E^r(s) = -E_0 \frac{a}{2s} e^{-j\beta s} = -\frac{E_0}{2}\left(\frac{a}{s}\right)e^{-j\beta s}$$

In turn, the backscatter radar cross section, according to (11-22b), can be written as

$$\sigma = \lim_{s \to \infty}\left[4\pi s^2 \frac{|E^r(s)|^2}{|E^i(Q_R)|^2}\right] \simeq 4\pi s^2 \frac{\left|-\dfrac{E_0}{2}\left(\dfrac{a}{s}\right)e^{-j\beta s}\right|^2}{|E_0|^2} = \pi a^2$$

It is recognized that the geometrical optics radar cross section of a sphere is equal to its physical cross-sectional area [see Figure 11.29 and (11-250)]. This is a well-known relation, and it is valid when the radius of the sphere is large compared to the wavelength.

The variation of the normalized radar cross section of a sphere as a function of its radius is displayed in Figure 11-29, and it is obtained by solving the wave equation in exact form. The Rayleigh, Mie (resonance), and geometrical optics regions represent three regimes in the figure.

For a cylindrical reflected field with radii of curvature $\rho_1^r = \rho^r$ and $\rho_2^r = \infty$, the reflected field of (13-20) reduces to

$$\mathbf{E}^r(s) = \mathbf{E}^i(Q_R) \cdot \bar{\mathbf{R}} \sqrt{\frac{\rho^r}{\rho^r + s}} \, e^{-j\beta s} \tag{13-29}$$

where ρ^r is the radius of curvature of the reflected field wave front.

An expression for ρ^r can be derived [37] by assuming a cylindrical wave, radiated by a line source at ρ_0, incident upon a two-dimensional curved surface S with positive radius of curvature ρ_a, as shown in Figure 13-8a. The rays that are reflected from the surface S diverge for positive values of ρ_a, as shown in Figure 13-8b, and appear to be emanating from a caustic a distance ρ^r from the reflecting surface S. It can be shown that the wave front curvature of the reflected field can be determined using

$$\frac{1}{\rho^r} = \frac{1}{\rho_0} + \frac{2}{\rho_a \cos\theta_i} \tag{13-30}$$

This is left as an end-of-chapter exercise for the reader.

The expression for ρ^r has been developed for a cylindrical wave incident on a two-dimensional convex scattering surface corresponding to positive values of ρ_a. For this arrangement, the caustic resides within the reflecting surface, and the rays seem to emanate from a virtual source (image) located at the caustic. This is equivalent to the Cassegrain reflector arrangement where the virtual feed focal point for the main reflector is behind the convex subreflector as shown in Figures 15-30 and 15-31 of [38]. If the curved scattering surface is concave, corresponding to negative values of ρ_a, the value of ρ^r is obtained by making ρ_a negative. In this case the caustic resides outside the reflecting surface, and the rays seem to emanate from that location. This is analogous to the Gregorian reflector arrangement where the effective focal point of the main reflector is between the concave subreflector and main reflector, as shown in Figure 15-31 of [38]. Although ρ^r of (13-30) was derived

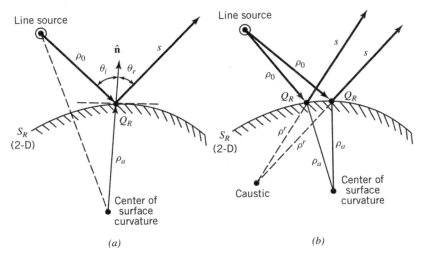

(a)   *(b)*

Figure 13-8 Line source near a two-dimensional curved surface. (*a*) Reflection point. (*b*) Caustic.

for cylindrical wave incidence on a two-dimensional convex curved surface, it can also be used when the plane of incidence coincides with any of the principal planes of curvature of the reflecting surface.

The application of (13-29) and (13-30) can best be demonstrated by an example.

Example 13-2

An electric line source of infinite length and constant current I_0 is placed symmetrically a distance h above an electric conducting strip of width w and infinite length, as shown in Figure 13-9a. The length of the line is placed parallel to the z axis. Assuming a free-space medium and far-field observations ($\rho \gg w$, $\rho \gg h$), derive expressions for the incident and reflected electric field components. Then compute and plot the normalized amplitude distribution (in decibels) of the incident, reflected, and incident plus reflected geometrical optics fields for $h = 0.5\lambda$ when $w = $ infinite and $w = 2\lambda$. Normalize the fields with respect to the maximum of the total geometrical optics field.

Solution: The analysis begins by first determining the incident (direct) field radiated by the source in the absence of the strip, which is given by (11-10a), or

$$E_z^i = -\frac{\beta^2 I_0}{4\omega\varepsilon} H_0^{(2)}(\beta\rho_i)$$

where $H_0^{(2)}(\beta\rho_i)$ is the Hankel function of the second kind and of order zero, and ρ_i is the distance from the source to the observation point.

For far-zone observations ($\beta\rho_i \to$ large) the Hankel function can be replaced by its asymptotic expansion

$$H_0^{(2)}(\beta\rho_i) \overset{\beta\rho_i \to \text{large}}{\simeq} \sqrt{\frac{2j}{\pi\beta\rho_i}} e^{-j\beta\rho_i}$$

This allows us to write the incident field as

$$E_z^i = E_0 \frac{e^{-j\beta\rho_i}}{\sqrt{\rho_i}}$$

where

$$E_0 = \left\{ -\frac{\beta^2 I_0}{4\omega\varepsilon} \sqrt{\frac{2j}{\pi\beta}} \right\}$$

Using (13-29) and referring to Figure 13-9b, the reflected field can be written as

$$E_z^r = E_z^i(\rho_i = s')(-1)\sqrt{\frac{\rho^r}{\rho^r + s}} e^{-j\beta s} = -E_0 \frac{e^{-j\beta s'}}{\sqrt{s'}} \sqrt{\frac{\rho^r}{\rho^r + s}} e^{-j\beta s}$$

The wave front radius of curvature of the reflected field can be found using (13-30), that is,

$$\frac{1}{\rho^r} = \frac{1}{s'} + \frac{2}{\infty \cos\theta_i} = \frac{1}{s'} \Rightarrow \rho^r = s'$$

Therefore, the reflected field can now be written as

$$E_z^r = -E_0 \frac{e^{-j\beta s'}}{\sqrt{s'}} \sqrt{\frac{s'}{s' + s}} e^{-j\beta s} = -E_0 \frac{e^{-j\beta(s + s')}}{\sqrt{s + s'}} = -E_0 \frac{e^{-j\beta\rho_r}}{\sqrt{\rho_r}}$$

It is apparent from Figure 13-9b that the reflected rays seem to emanate from a virtual (image) source which is also a caustic for the reflected fields. The wave front radius of curvature $\rho^r = s'$ of the reflected field also represents the distance of the caustic (image) from the point of reflection. The reflected field

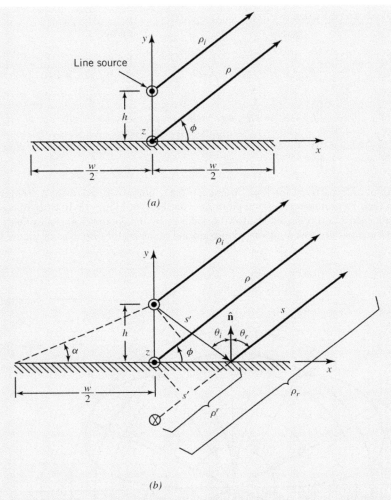

Figure 13-9   Line source above a finite-width strip. (*a*) Coordinate system. (*b*) Reflection geometry.

could also have been obtained very simply by using image theory. However, we chose to use the equations of geometrical optics to demonstrate the principles and applications of geometrical optics.

For far-field observations ($\rho_i \gg w, \rho_i \gg h$),

$$\left.\begin{array}{l} \rho_i = \rho - h\cos\left(\dfrac{\pi}{2}-\phi\right) = \rho - h\,\sin\phi \\[3mm] \rho_r = \rho + h\cos\left(\dfrac{\pi}{2}-\phi\right) = \rho + h\,\sin\phi \end{array}\right\} \quad \text{for phase variations}$$

$$\rho_i \simeq \rho_r \simeq \rho \qquad\qquad \text{for amplitude variations}$$

Therefore, the incident (direct) and reflected fields can be reduced to

$$E_z^i = E_0 e^{+j\beta h\sin\phi}\,\frac{e^{-j\beta\rho}}{\sqrt{\rho}}, \quad 0\le\phi\le\pi+\alpha,\, 2\pi-\alpha\le\phi\le 2\pi$$

$$E_z^r = -E_0 e^{-j\beta h\sin\phi}\,\frac{e^{-j\beta\rho}}{\sqrt{\rho}}, \quad \alpha\le\phi\le\pi-\alpha$$

and the total field can be written as

$$E_z^t = \begin{cases} E_z^i = E_0 e^{+j\beta h \sin\phi} \dfrac{e^{-j\beta\rho}}{\sqrt{\rho}} & \begin{aligned} &0 \le \phi \le \alpha, \\ &\pi - \alpha \le \phi \le \pi + \alpha, \\ &2\pi - \alpha \le \phi \le 2\pi \end{aligned} \\[2em] E_z^i + E_z^r = 2jE_0 \sin(\beta h \sin\phi) \dfrac{e^{-j\beta\rho}}{\sqrt{\rho}} & \alpha \le \phi \le \pi - \alpha \\[1em] 0 & \pi + \alpha \le \phi \le 2\pi - \alpha \end{cases}$$

Normalized amplitude patterns for $w = 2\lambda$ and $h = 0.5\lambda$, computed using the preceding geometrical optics fields in their respective regions, are plotted (in decibels) in Figure 13-10 where discontinuities created along the incident and reflected shadow boundaries by the geometrical optics fields are apparent. The total amplitude pattern assuming an infinite ground plane is also displayed in Figure 13-10.

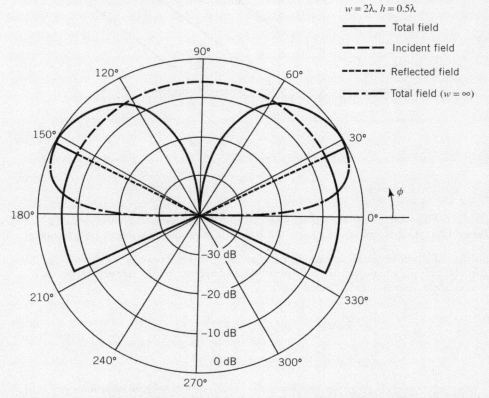

Figure 13-10   Amplitude radiation pattern of an electric line source above a finite-width strip.

In summary, geometrical optics methods approximate the fields by the leading term of the Luneberg–Kline expansion for large ω, but they fail along caustics. Improvements can be incorporated into the solutions by finding higher-order terms $\mathbf{E}_1(\mathbf{R}_1)$, $\mathbf{E}(\mathbf{R}_2)$,... in the Luneberg–Kline series expansion. Higher-order Luneberg–Kline expansions have been derived for fields scattered from spheres, cylinders, and other curved surfaces with simple geometries [30, 31]. These solutions exhibit the following tendencies.

1. They improve the high-frequency field approximations if the observation specular point is not near edges, shadow boundaries, or other surface discontinuities.
2. They become singular as the observation specular point approaches a shadow boundary on the surface.
3. They do not correct for geometrical optics discontinuities along incident and reflection shadow boundaries.
4. They do not describe the diffracted fields in the shadow region.

Because of some of these deficiencies, in addition to being quite complex, higher-order Luneberg–Kline expansion methods cannot be used to treat diffraction. Therefore other approaches, usually somewhat heuristic in nature, must be used to introduce diffraction in order to improve geometrical optics approximations. It should be noted, however, that for sufficiently large ω, geometrical optics fields may dominate the scattering phenomena and may alone provide results that often agree quite well with measurements. This is more evident for backscattering from smooth curved surfaces with large radii of curvature. In those instances, corrections to the fields predicted by geometrical optics methods may not be necessary. However, for other situations where such solutions are inaccurate, corrections are usually provided by including diffraction. Therefore, a combination of geometrical optics and diffraction techniques often leads to solutions of many practical engineering problems whose results agree extremely well with measurements. This has been demonstrated in many applications [19–25, 39]. Because of their extreme importance and ease of application, diffraction techniques will be next introduced, discussed, and applied.

13.3 GEOMETRICAL THEORY OF DIFFRACTION: EDGE DIFFRACTION

Examining high-frequency diffraction problems has revealed that their solutions contain terms of fractional power that are not always included in the geometrical optics expression (13-15a) or in the Luneberg–Kline series solution. For example, geometrical optics fails to account for the energy diffracted into the shadow region when the incident rays are tangent to the surface of a curved object and for the diffracted energy when the surface contains an edge, vertex, or corner. In addition, caustics of the diffracted fields are located at the boundary surface. Therefore, some semi-heuristic approaches must be used to provide correction factors that improve the geometrical optics approximation.

13.3.1 Amplitude, Phase, and Polarization Relations

To introduce diffraction, let us assume that the smooth surface S of Figure 13-6a has a curved edge as shown in Figure 13-11a. When an electromagnetic wave impinges on this curved edge, diffracted rays emanate from the edge whose leading term of the high-frequency solution for the electric field takes the form

$$\mathbf{E}^d(\mathbf{R}) \simeq \frac{e^{-j\beta\psi_d(\mathbf{R})}}{\sqrt{\beta}} \mathbf{A}(\mathbf{R}) \qquad (13\text{-}31)$$

where ψ_d = eikonal surface for the diffracted rays
$\mathbf{A}(\mathbf{R})$ = field factor for the diffracted rays

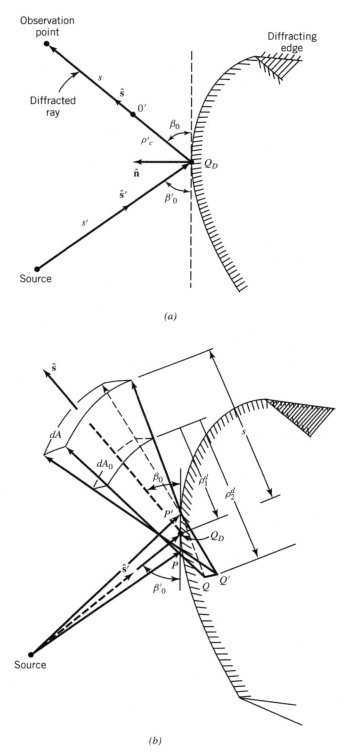

(a)

(b)

Figure 13-11 Geometry for diffraction by a curved edge. (*a*) Diffraction point. (*b*) Astigmatic tube of rays.

Substituting (13-31) into (13-11) and (13-12), and referring to the geometry of Figure 13-11a, it can be shown that the diffracted field can be written as

$$\mathbf{E}^d(s) = \left[\frac{\mathbf{A}(0')}{\sqrt{\beta}} e^{-j\beta\psi_d(0')} \right] \sqrt{\frac{\rho_c'\rho_c}{(\rho_c'+s)(\rho_c+s)}} \, e^{-j\beta s}$$

$$\mathbf{E}^d(s) = [\mathbf{E}^d(0')] \sqrt{\frac{\rho_c'\rho_c}{(\rho_c'+s)(\rho_c+s)}} \, e^{-j\beta s} \qquad (13\text{-}32)$$

where $\mathbf{E}^d(0')$ = diffracted field at the reference point $0'$

s = distance along the diffracted ray from the reference point $0'$

ρ_c' = distance from diffraction point Q_D (first caustic of diffracted field) to reference point $0'$

ρ_c = distance between the second caustic of diffracted field and reference point $0'$

It would have been more convenient to choose the reference point $0'$ to coincide with the diffraction point Q_D, located on the diffracting edge. However, like the geometrical optics rays, the diffracted rays form an astigmatic tube of the form shown in Figure 13-11b where the caustic line PP' coincides with the diffracting edge. Because the diffraction point is a caustic of the diffracted field, it is initially more straightforward to choose the reference point away from the edge diffraction caustic Q_D. However, the diffracted field of (13-32) should be independent of the location of the reference point $0'$, including $\rho_c' = 0$. Therefore, the diffracted field of (13-32) must be such that

$$\lim_{\rho_c'\to 0} \mathbf{E}^d(0')\sqrt{\rho_c'} = \text{finite} \qquad (13\text{-}33)$$

and must be equal to

$$\lim_{\rho_c'\to 0} \mathbf{E}^d(0')\sqrt{\rho_c'} = \mathbf{E}^i(Q_D)\cdot\bar{\mathbf{D}} \qquad (13\text{-}33a)$$

where $\mathbf{E}^i(Q_D)$ = incident field at the point of diffraction

$\bar{\mathbf{D}}$ = dyadic diffraction coefficient (analogous to dyadic reflection coefficient)

Using (13-33a), the diffracted field of (13-32) reduces to

$$\mathbf{E}^d(s) = \lim_{\rho_c'\to 0} \left\{ [\mathbf{E}^d(0')\sqrt{\rho_c'}] \sqrt{\frac{\rho_c}{(\rho_c'+s)(\rho_c+s)}} \, e^{-j\beta s} \right\}$$

$$\mathbf{E}^d(s) = \mathbf{E}^i(Q_D)\cdot\bar{\mathbf{D}} \sqrt{\frac{\rho_c}{s(\rho_c+s)}} \, e^{-j\beta s} \qquad (13\text{-}34)$$

which has the form

$\mathbf{E}^d(s)$	$= \mathbf{E}^i(Q_D)$.	$\bar{\mathbf{D}}$	$A(\rho_c, s)$	$e^{-j\beta s}$
	Field at reference point	Diffraction coefficient (usually a dyadic)	Spatial attenuation (spreading, divergence) factor	Phase factor

$(13\text{-}34a)$

and compares with that of (13-20) for reflection. In (13-34) and (13-34a)

$$A(\rho_c, s) = \sqrt{\frac{\rho_c}{s(\rho_c + s)}}$$

$$= \text{spatial attenuation (spreading, divergence) factor for a curved surface}$$

(13-34b)

$\rho_c = $ distance between the reference point $Q_D(s=0)$ at the edge (also first caustic of the diffracted rays) and the second caustic of the diffracted rays.

In general, ρ_c is a function of the following:

1. Wave front curvature of the incident field.
2. Angles of incidence and diffraction, relative to unit vector normal to edge at the point Q_D of diffraction.
3. Radius of curvature of diffracting edge at point Q_D of diffraction.

An expression for ρ_c is given by (13-100a) in Section 13.3.4.

The diffracted field, which is determined by a generalization of Fermat's principle [7, 9], is initiated at a point on the surface of the object where discontinuities are formed along the incident and reflected shadow boundaries. As represented in (13-34) and (13-34a), the initial value of the field of a diffracted ray is determined from the incident field with the aid of an appropriate diffraction coefficient $\overline{\overline{D}}$, which in general is a dyadic for electromagnetic fields. The amplitude is assumed to vary in accordance with the principle of conservation of energy flux along a tube of rays. Appropriate phase jumps of $+90°$ are added each time a ray passes through a caustic at $s = 0$ and $s = -\rho_c$, as properly accounted for in (13-34) and (13-34a). The phase of the field on a diffracted ray is assumed to be equal to the product of the optical lengths of the ray, from some reference point Q_D, and the phase constant β of the medium.

When the edge is straight, the source is located a distance s' from the point of diffraction, and the observations are made at a distance s from it—as shown in Figure 13-12—the diffracted field can be written as

$$\mathbf{E}^d(s) = \mathbf{E}^i(Q_D) \cdot \overline{\overline{D}} A(s', s) e^{-j\beta s}$$

(13-35)

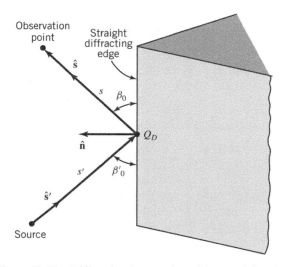

Figure 13-12   Diffraction by a wedge with a straight edge.

where

$$A(s',s) = \begin{cases} \dfrac{1}{\sqrt{s}} & \text{for plane and conical wave incidences} & (13\text{-}35a) \\[2mm] \dfrac{1}{\sqrt{\rho}}, \rho = s\sin\beta_0 & \text{for cylindrical wave incidence} & (13\text{-}35b) \\[2mm] \sqrt{\dfrac{s'}{s(s+s')}} \overset{s\gg s'}{\simeq} \dfrac{\sqrt{s'}}{s} & \text{for spherical wave incidence} & (13\text{-}35c) \end{cases}$$

The $A(s',s)$ formulas of (13-35a) through (13-35c) are obtained by letting ρ_c in (13-34) tend to infinity ($\rho_c = \infty$) for plane, cylindrical, and conical wave incidence and $\rho_c = s'$ for spherical wave incidence.

The diffraction coefficients are usually determined from the asymptotic solutions of canonical problems that have the same local geometry at the points of diffraction as the object(s) of investigation. One of the simplest geometries, which will be discussed in this chapter, is a conducting wedge [40–42]. Another is that of a conducting, smooth, and convex surface [43–46]. The main objectives in the remaining part of this chapter are to introduce and apply the diffraction coefficients for the canonical problem of the conducting wedge. Curved surface diffraction is derived in [43–46].

13.3.2 Straight Edge Diffraction: Normal Incidence

In order to examine the manner in which fields are diffracted by edges, it is necessary to have a diffraction coefficient available. To derive a diffraction coefficient for an edge, we need to consider a canonical problem that has the same local geometry near the edge, like that shown in Figure 13-13a.

Let us begin by assuming that a source is placed near the two-dimensional electric conducting wedge of included angle $\text{WA} = (2-n)\pi$ radians. If observations are made on a circle of constant radius ρ from the edge of the wedge, it is quite clear that, in addition to the direct ray (OP), there are rays that are reflected from the side of the wedge (OQ_RP), which contribute to the intensity at point P. These rays obey Fermat's principle, that is, they minimize the path between points O and P by including points on the side of the wedge, and deduce Snell's law of reflection. It would then seem appropriate to extend the class of such points to include in the trajectory rays that pass through the edge of the wedge (OQ_DP), leading to the generalized Fermat's principle [7]. This class of rays is designated as diffracted rays and they lead to the *law of diffraction*.

By considering rays that obey only geometrical optics radiation mechanisms (direct and reflected), we can separate the space surrounding the wedge into three different field regions. Using the geometrical coordinates of Figure 13-13b, the following geometrical optics fields will contribute to the corresponding regions:

Region I	Region II	Region III
$0 < \phi < \pi - \phi'$	$\pi - \phi' < \phi < \pi < \phi'$	$\pi + \phi' < \phi < n\pi$
Direct	Direct	$\cdots$
Reflected		

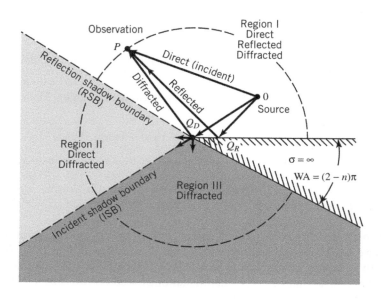

(a)

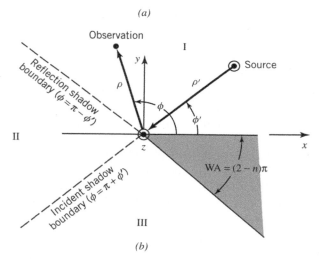

(b)

Figure 13-13   Line source near a two-dimensional conducting wedge. (a) Region separation. (b) Coordinate system. Region I: $0 \leq \phi < \pi - \phi'$. Region II: $\pi - \phi' < \phi < \pi + \phi'$. Region III: $\pi + \phi' < \phi < n\pi$.

With these fields, it is evident that the following will occur:

1. Discontinuities in the field will be formed along the RSB (Reflection Shadow Boundary) separating regions I and II ($\phi = \pi - \phi'$), and along the ISB (Incident Shadow Boundary) separating regions II and III ($\phi = \pi + \phi'$).
2. No field will be present in region III (shadow region).

Since neither of the preceding results should be present in a physically realizable field, modifications and/or additions need to be made.

To *remove* the discontinuities along the boundaries and to *modify* the fields in *all* three regions, diffracted fields must be included. To obtain expressions for the diffracted field distribution, we assume that the source in Figure 13-13 is an infinite line source (either electric or magnetic).

The fields of an electric line source satisfy the homogeneous Dirichlet boundary conditions $E_z = 0$ on both faces of the wedge and the fields of the magnetic line source satisfy the homogeneous Neumann boundary condition $\partial E_z/\partial \phi = 0$ or $H_z = 0$ on both faces of the wedge. The faces of the wedge are formed by two semi-infinite intersecting planes. The infinitely long line source is parallel to the edge of the wedge, and its position is described by the coordinate (ρ', ϕ'). The typical field point is denoted by (ρ, ϕ), as shown in Figure 13-13b. The line source is assumed to have constant current.

Initially, we will consider only normal incidence diffraction by a straight edge, as shown in Figure 13-14. For this situation, the plane of diffraction is perpendicular to the edge of the wedge. Oblique incidence diffraction (Figure 13-31) and curved-edge diffraction (Figure 13-35) will be discussed, respectively, in Sections 13.3.3 and 13.3.4.

The diffraction coefficient for the geometries of Figures 13-13 and 13-14 is obtained by [42]:

1. Finding the Green's function solution in the form of an infinite series using modal techniques and then approximating it for large values of $\beta\rho$ (far-field observations).
2. Converting the infinite series Green's function solution into an integral.
3. Performing, on the integral Green's function, a high-frequency asymptotic expansion (in inverse powers of $\beta\rho$) using standard techniques, such as the *method of steepest descent*.

An abbreviated derivation of this procedure will now be presented.

A. Modal Solution   Using modal techniques, the total radiation electric field for an electric line source of current I_e was found in Chapter 11, Section 11.6.3, to be that of (11-207a), or

$$E_z^e = -\frac{\omega\mu I_e}{4} G \Rightarrow \mathbf{H}^e = -\frac{1}{j\omega\mu} \boldsymbol{\nabla} \times \mathbf{E} \tag{13-36}$$

and the total magnetic field for a magnetic line source of current I_m to be that of (11-207b), or

$$H_z^m = \frac{\omega\varepsilon I_m}{4} G \Rightarrow \mathbf{E}^m = +\frac{1}{j\omega\varepsilon} \boldsymbol{\nabla} \times \mathbf{H} \tag{13-37}$$

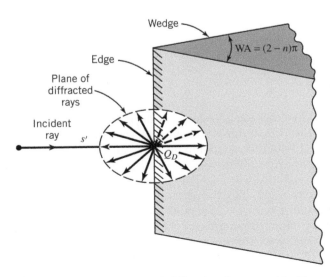

Figure 13-14   Wedge plane of diffraction for normal incidence.

where G is referred to as the Green's function, and it is given by (11-208a) as

$$G = \frac{1}{n} \sum_{m=0}^{\infty} \varepsilon_m J_{m/n}(\beta\rho) H_{m/n}^{(2)}(\beta\rho') \left[\cos \frac{m}{n}(\phi - \phi') \pm \cos \frac{m}{n}(\phi + \phi') \right] \quad \text{for } \rho \leq \rho' \qquad (13\text{-}38)$$

$$\varepsilon_m = \begin{cases} 1 & m = 0 \\ 2 & m \neq 0 \end{cases} \qquad (13\text{-}38a)$$

For the case $\rho \geq \rho'$, ρ and ρ' are interchanged. The plus sign between the two cosine terms is used if the boundary condition is of the homogeneous Neumann type, $\partial G/\partial \phi = 0$, on both faces of the wedge. For the homogeneous Dirichlet boundary condition, $G = 0$, on both faces of the wedge, the minus sign is used. In acoustic terminology, the Neumann boundary condition is referred to as *hard* polarization and the Dirichlet boundary condition is referred to as *soft* polarization. This series is an exact solution to the time-harmonic, inhomogeneous wave equation of a radiating line source and wedge embedded in a linear, isotropic, homogeneous, lossless medium.

B. High-Frequency Asymptotic Solution   Many times it is necessary to determine the total radiation field when the line source is far removed from the vertex of the wedge. In such cases, (13-38) can be simplified by replacing the Hankel function by the first term of its asymptotic expansion, that is, by

$$H_{m/n}^{(2)}(\beta\rho') \overset{\beta\rho' \to \infty}{\simeq} \sqrt{\frac{2}{\pi\beta\rho'}} e^{-j[\beta\rho' - \pi/4 - (m/n)(\pi/2)]} \qquad (13\text{-}39)$$

This substitution reduces G to

$$G = \sqrt{\frac{2}{\pi\beta\rho'}} e^{-j(\beta\rho' - \pi/4)} \frac{1}{n} \sum_{m=0}^{\infty} \varepsilon_m J_{m/n}(\beta\rho) e^{+j(m/n)(\pi/2)}$$

$$\times \left[\cos \frac{m}{n}(\phi - \phi') \pm \cos \frac{m}{n}(\phi + \phi') \right]$$

$$G = \sqrt{\frac{2}{\pi\beta\rho'}} e^{-j(\beta\rho' - \pi/4)} F(\beta\rho) \qquad (13\text{-}40)$$

where

$$F(\beta\rho) = \frac{1}{n} \sum_{m=0}^{\infty} \varepsilon_m J_{m/n}(\beta\rho) e^{+j(m/n)(\pi/2)} \left[\cos \frac{m}{n}(\phi - \phi') \pm \cos \frac{m}{n}(\phi + \phi') \right] \qquad (13\text{-}40a)$$

$F(\beta\rho)$ is used to represent either the normalized total E_z^e, when the source is an electric line (soft polarization), or H_z^m, when the source is a magnetic line (hard polarization).

The infinite series of (13-40a) converges rapidly for small values of $\beta\rho$. For example, if $\beta\rho$ is less than 1, less than 15 terms are required to achieve a five-significant-figure accuracy. However, at least 40 terms should be included when $\beta\rho$ is 10 to achieve the five-significant-figure accuracy. To demonstrate the variations of (13-40a), the patterns of a unit amplitude plane wave of hard polarization incident upon a half-plane ($n = 2$) and 90° wedge ($n = 3/2$), computed using (13-40a), are shown in Figure 13-15. Computed patterns for the soft polarization are shown in Figure 13-16.

Whereas (13-40a) represents the normalized total field, in the space around a wedge of included angle WA $= (2-n)\pi$, of a unity amplitude plane wave incident upon the wedge, the field of a unit amplitude cylindrical wave incident upon the same wedge and observations made

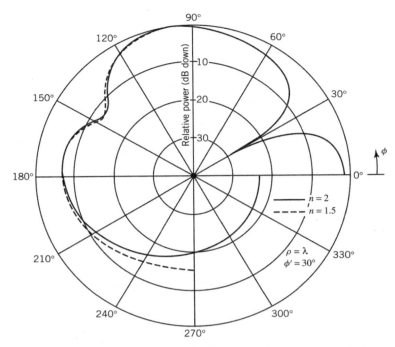

Figure 13-15   Normalized amplitude pattern of a hard polarization plane wave incident normally on a two-dimensional conducting wedge.

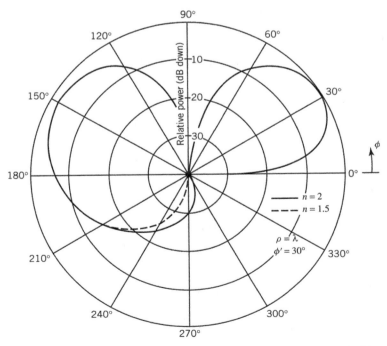

Figure 13-16   Normalized amplitude pattern of a soft polarization plane wave incident normally on a two-dimensional conducting wedge.

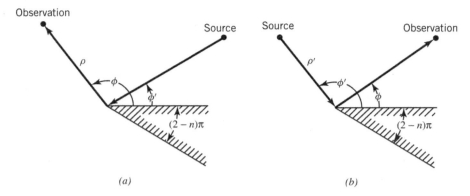

Figure 13-17 Principle of reciprocity in diffraction. (*a*) Plane wave incidence. (*b*) Cylindrical wave incidence.

at very large distances can be obtained by the reciprocity principle. Graphically, this is illustrated in Figures 13-17*a* and 13-17*b*. Analytically, the cylindrical wave incidence fields of Figure 13-17*b* can be obtained from (13-40a) and Figure 13-17*a* by substituting in (13-40a) $\rho = \rho'$, $\phi = \phi'$, and $\phi' = \phi$.

A high-frequency asymptotic expansion for $F(\beta\rho)$ in inverse powers of $\beta\rho$ is very useful for computational purposes, because of the slow convergence of (13-40a) for large values of $\beta\rho$. In order to derive an asymptotic expression for $F(\beta\rho)$ by the *conventional method of steepest descent* for isolated poles and saddle points (see Appendix VI), it must first be transformed into an integral or integrals of the form

$$P(\beta\rho) = \int_C H(z) e^{\beta\rho h(z)} dz \tag{13-41}$$

and then evaluated for large $\beta\rho$ by means of the method of steepest descent [8, 47].

To accomplish this, first the cosine terms are expressed in complex form by

$$\cos\left(\frac{m}{n}\xi^{\mp}\right) = \frac{1}{2}\left[e^{j(m/n)\xi^{\mp}} + e^{-j(m/n)\xi^{\mp}}\right] \tag{13-42}$$

where

$$\xi^{\mp} = \phi \mp \phi' \tag{13-42a}$$

and the Bessel functions are replaced by contour integrals in the complex z plane, of the form

$$J_{m/n}(\beta\rho) = \frac{1}{2\pi}\int_C e^{j[\beta\rho\cos z + m/n(z - \pi/2)]} dz \tag{13-43a}$$

or

$$J_{m/n}(\beta\rho) = \frac{1}{2\pi}\int_{C'} e^{j[\beta\rho\cos z - m/n(z + \pi/2)]} dz \tag{13-43b}$$

where the paths C and C' are shown in Figure 13-18*a*. Doing this and interchanging the order of integration and summation, it can be shown that (13-40a) can be written as

$$F(\beta\rho) = I(\beta\rho, \phi - \phi', n) \pm I(\beta\rho, \phi + \phi', n) \tag{13-44}$$

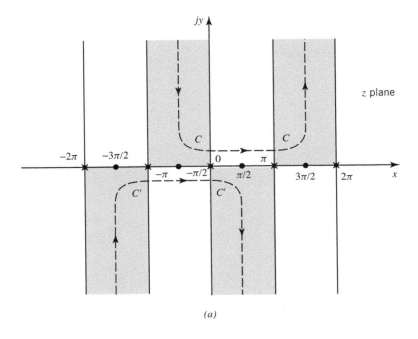

(a)

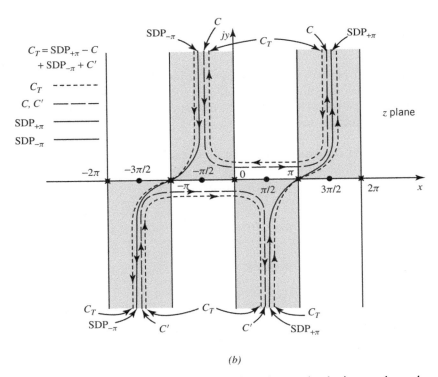

(b)

Figure 13-18   Line contours, steepest descent paths, saddle points, and poles in complex z plane for asymptotic evaluation of wedge diffraction formulas. (*a*) Contours for Bessel function. (*b*) Steepest descent paths, saddle points, and poles.

where

$$I(\beta\rho, \xi^{\mp}, n) = \frac{1}{2\pi n} \int_C e^{j\beta\rho \cos z} \sum_{m=1}^{\infty} e^{+j(m/n)(\xi^{\mp}+z)} dz$$

$$+ \frac{1}{2\pi n} \int_{C'} e^{j\beta\rho \cos z} \sum_{m=0}^{\infty} e^{-j(m/n)(\xi^{\mp}+z)} dz \qquad (13\text{-}44\text{a})$$

with $\xi^{\mp} = \phi \mp \phi'$.

Using the series expansions of

$$-\frac{1}{1-x^{-1}} = x(1+x+x^2+x^3+\cdots) = \sum_{m=1}^{\infty} x^m \qquad (13\text{-}45\text{a})$$

$$\frac{1}{1-x^{-1}} = 1 + x^{-1} + x^{-2} + x^{-3} + \cdots = \sum_{m=0}^{\infty} x^{-m} \qquad (13\text{-}45\text{b})$$

and that

$$\frac{1}{1-e^{-j(\xi^{\mp}+z)/n}} = \frac{e^{j(\xi^{\mp}+z)/2n}}{e^{j(\xi^{\mp}+z)/2n} - e^{-j(\xi^{\mp}+z)/2n}} = \frac{1}{2} + \frac{1}{2j}\cot\left(\frac{\xi^{\mp}+z}{2n}\right) \qquad (13\text{-}46)$$

we can ultimately write (13-44) or (13-40a) as

$$F(\beta\rho) = \frac{1}{4\pi jn} \int_{(C'-C)} \cot\left(\frac{\phi-\phi'+z}{2n}\right) e^{j\beta\rho \cos z} dz$$

$$\pm \frac{1}{4\pi jn} \int_{(C'-C)} \cot\left(\frac{\phi+\phi'+z}{2n}\right) e^{j\beta\rho \cos z} dz \qquad (13\text{-}47)$$

where the negative sign before C indicates that this integration path is to be traversed in the direction opposite to that shown in Figure 13-18a. It is now clear that $F(\beta\rho)$ has been written in an integral of the form of (13-41), which can be evaluated asymptotically by contour integration and by the method of steepest descent.

C. Method of Steepest Descent   Equation 13-47 can also be written as

$$F(\beta\rho) = F_1(\beta\rho) \pm F_2(\beta\rho) \qquad (13\text{-}48)$$

$$F_1(\beta\rho) = \frac{1}{4\pi jn} \int_{(C'-C)} H_1(z) e^{\beta\rho h_1(z)} dz \qquad (13\text{-}48\text{a})$$

$$F_2(\beta\rho) = \frac{1}{4\pi jn} \int_{(C'-C)} H_2(z) e^{\beta\rho h_2(z)} dz \qquad (13\text{-}48\text{b})$$

$$H_1(z) = \cot\left[\frac{(\phi-\phi')+z}{2n}\right] = \cot\left(\frac{\xi^-+z}{2n}\right) \qquad (13\text{-}48\text{c})$$

$$H_2(z) = \cot\left[\frac{(\phi+\phi')+z}{2n}\right] = \cot\left(\frac{\xi^++z}{2n}\right) \qquad (13\text{-}48\text{d})$$

$$h_1(z) = h_2(z) = j\cos(z) \qquad (13\text{-}48\text{e})$$

The evaluation of (13-48) can be accomplished by evaluating separately (13-48a) and (13-48b) and summing the results. Let us examine first the evaluation of (13-48a) in detail. A similar procedure can be used for (13-48b).

Using the complex z-plane closed contour C_T, we can write by referring to Figure 13-18b that

$$\frac{1}{4\pi jn}\oint_{C_T} H_1(z)e^{\beta\rho h_1(z)}dz = \frac{1}{4\pi jn}\int_{(C'-C)} H_1(z)e^{\beta\rho h_1(z)}dz$$

$$+\frac{1}{4\pi jn}\int_{\text{SDP}_{+\pi}} H_1(z)e^{\beta\rho h_1(z)}dz$$

$$+\frac{1}{4\pi jn}\int_{\text{SDP}_{-\pi}} H_1(z)e^{\beta\rho h_1(z)}dz \qquad (13\text{-}49)$$

The closed contour C_T is equal to the sum of

$$C_T = C' + \text{SDP}_{+\pi} - C + \text{SDP}_{-\pi} \qquad (13\text{-}49\text{a})$$

where $\text{SDP}_{\pm\pi}$ is used to represent the steepest descent paths passing through the saddle points $\pm\pi$. We can rewrite (13-49) as

$$F_1(\beta\rho) = \frac{1}{4\pi jn}\int_{(C'-C)} H_1(z)e^{\beta\rho h_1(z)}dz = \frac{1}{4\pi jn}\oint_{C_T} H_1(z)e^{\beta\rho h_1(z)}dz$$

$$-\frac{1}{4\pi jn}\int_{\text{SDP}_{+\pi}} H_1(z)e^{\beta\rho h_1(z)}dz - \frac{1}{4\pi jn}\int_{\text{SDP}_{-\pi}} H_1(z)e^{\beta\rho h_1(z)}dz \qquad (13\text{-}50)$$

which is the same as (13-48a). Therefore, (13-48a) can be integrated by evaluating the three terms on the right side of (13-50). The closed contour of the first term on the right side of (13-50) is evaluated using residue calculus and the other two terms are evaluated using the method of steepest descent. When evaluated, it will be shown that the first term on the right side of (13-50) will represent the incident geometrical optics and the other two terms will represent what will be referred to as the incident diffracted field. A similar interpretation will be given when (13-48b) is evaluated; it represents the reflected geometrical optics and reflected diffracted fields.

Using residue calculus, we can write the first term on the right side of (13-50) as [48]

$$\frac{1}{4\pi jn}\oint_{C_T} H_1(z)e^{\beta\rho h_1(z)}dz = 2\pi j\sum_{p}\text{Res}(z = z_p)$$

$$= 2\pi j\sum_{p}(\text{residues of the poles enclosed by } C_T) \qquad (13\text{-}51)$$

To evaluate (13-51), we first rewrite it as

$$\frac{1}{4\pi jn}\oint_{C_T} H_1(z)e^{\beta\rho h_1(z)}dz$$

$$= \frac{1}{4\pi jn}\oint_{C_T}\cot\left[\frac{(\phi-\phi')+z}{2n}\right]e^{j\beta\rho\,\cos(z)}dz$$

$$= \oint_{C_T}\frac{1}{4\pi jn}\cot\left[\frac{(\phi-\phi')+z}{2n}\right]e^{j\beta\rho\,\cos(z)}dz = \oint_{C_T}\frac{N(z)}{D(z)}dz \qquad (13\text{-}52)$$

where

$$\frac{N(z)}{D(z)} = \frac{\cos\left[\frac{(\phi-\phi')+z}{2n}\right]e^{j\beta\rho\,\cos(z)}}{4\pi\,jn\sin\left[\frac{(\phi-\phi')+z}{2n}\right]} \tag{13-52a}$$

$$N(z) = \cos\left[\frac{(\phi-\phi')+z}{2n}\right]e^{j\beta\rho\,\cos(z)} \tag{13-52b}$$

$$D(z) = 4\pi\,jn\sin\left[\frac{(\phi-\phi')+z}{2n}\right] \tag{13-52c}$$

Equation 13-52 has simple poles that occur when

$$\left[\frac{(\phi-\phi')+z}{2n}\right]_{z=z_p} = \pi N, \quad N = 0, \pm 1, \pm 2, \dots \tag{13-53}$$

or

$$z_p = -(\phi-\phi') + 2\pi nN \tag{13-53a}$$

provided that

$$-\pi \le z_p = -(\phi-\phi') + 2\pi nN \le +\pi \tag{13-53b}$$

Using residue calculus, the residues of (13-51) or (13-52) for simple poles (no branch points, etc.) can be found using [48]

$$\text{Res}(z=z_p) = \left.\frac{N(z)}{\dfrac{dD(z)}{dz}}\right|_{z=z_p} = \left.\frac{N(z)}{D'(z)}\right|_{z=z_p} \tag{13-54}$$

where

$$N(z)|_{z=z_p} = \cos\left[\frac{(\phi-\phi')+z}{2n}\right]e^{j\beta\rho\,\cos(z)}\Bigg|_{z=-(\phi-\phi')+2\pi nN}$$
$$= \cos(\pi N)e^{j\beta\rho\,\cos[-(\phi-\phi')+2\pi nN]} \tag{13-54a}$$

$$D'(z)|_{z=z_p} = 2\pi j\cos\left[\frac{(\phi-\phi')+z}{2n}\right]\Bigg|_{z=-(\phi-\phi')+2\pi nN} = 2\pi j\cos(\pi N) \tag{13-54b}$$

Thus, (13-54) and (13-51) can be written, respectively, as

$$\text{Res}(z=z_p) = \frac{1}{2\pi j}e^{j\beta\rho\,\cos[-(\phi'-\phi')+2\pi nN]} \tag{13-55a}$$

$$\frac{1}{4\pi jn}\oint_{C_T} H_1(z)e^{\beta\rho h_1(z)}dz = e^{j\beta\rho\,\cos[-(\phi-\phi')+2\pi nN]}\,U[\pi-|-(\phi-\phi')+2\pi nN|] \tag{13-55b}$$

The $U(t-t_0)$ function in (13-55b) is a unit step function defined as

$$U(t-t_0) = \begin{cases} 1 & t > t_0 \\ \dfrac{1}{2} & t = t_0 \\ 0 & t < t_0 \end{cases} \tag{13-56}$$

The unit step function is introduced in (13-55b) so that (13-53b) is satisfied. When $z = \pm\pi$, (13-53b) is expressed as

$$\boxed{2\pi nN^+ - (\phi-\phi') = +\pi} \tag{13-57a}$$

for $z_p = +\pi$ and

$$\boxed{2\pi n N^- - (\phi - \phi') = -\pi}$$

(13-57b)

for $z_p = -\pi$. For the principal value of $N(N^\pm = 0)$, (13-55b) reduces to

$$\boxed{F_1(\beta\rho)|_{C_T} = \frac{1}{4\pi jn} \oint_{C_T} H_1(z) e^{\beta\rho h_1(z)} dz = e^{j\beta\rho \cos(\phi-\phi')} U[\pi - |-(\phi-\phi')|]}$$

(13-58)

This is referred to as the *incident geometrical optics* field, and it exists provided $|\phi - \phi'| < \pi$. This completes the evaluation of (13-51). Let us now evaluate the other two terms on the right side of (13-50).

In evaluating the last two terms on the right side of (13-50), the contributions from all the saddle points along the steepest descent paths must be accounted for. In this situation, however, only saddle points at $z = z_s = \pm\pi$ occur. These are found by taking the derivative of (13-48e) and setting it equal to zero. Doing this leads to

$$h_1'(z)|_{z=z_s} = -j\sin(z)|_{z=z_s} = 0 \implies z_s = \pm\pi$$

(13-59)

The form used to evaluate the last two terms on the right side of (13-50) depends on whether the poles of (13-53a), which contribute to the geometrical optics field, are near or far removed from the saddle points at $z = z_s = \pm\pi$. Let us first evaluate each of the last two terms on the right side of (13-50) when the poles of (13-53a) are far removed from the saddle point of (13-59).

When the poles of (13-53a) are far removed from the saddle points of (13-59), then the last two terms on the right side of (13-50) are evaluated using the *conventional steepest descent method* for isolated poles and saddle points. Doing this, we can write the last two terms on the right side of (13-50), using the saddle points of (13-59), as [8, 47]:

$$\frac{1}{4\pi jn} \int_{\text{SDP}_{+\pi}} H_1(z) e^{\beta\rho h_1(z)} dz \overset{\beta\rho \to \text{large}}{\simeq} \frac{1}{4\pi jn} \left| \sqrt{\frac{2\pi}{-\beta\rho h_1''(z_s = +\pi)}} \right| e^{j\pi/4}$$

$$\times H_1(z_s = +\pi) e^{\beta\rho h_1(z_s = \pi)} = \frac{e^{-j\pi/4}}{2n\sqrt{2\pi\beta}} \cot\left[\frac{\pi + (\phi-\phi')}{2n}\right] \frac{e^{-j\beta\rho}}{\sqrt{\rho}}$$

(13-60a)

$$\frac{1}{4\pi jn} \int_{\text{SDP}_{-\pi}} H_1(z) e^{\beta\rho h_1(z)} dz \overset{\beta\rho \to \text{large}}{\simeq} \frac{1}{4\pi jn} \left| \sqrt{\frac{2\pi}{-\beta\rho h_1''(z_s = -\pi)}} \right| e^{-j3\pi/4}$$

$$\times H_1(z_s = -\pi) e^{\beta\rho h_1(z_s = -\pi)} = \frac{e^{-j\pi/4}}{2n\sqrt{2\pi\beta}} \cot\left[\frac{\pi - (\phi-\phi')}{2n}\right] \frac{e^{-j\beta\rho}}{\sqrt{\rho}}$$

(13-60b)

Combining (13-60a) and (13-60b), it can be shown that the sum of the two can be written as

$$\boxed{\begin{aligned} F_1(\beta\rho)|_{\text{SDP}_{\pm\pi}} &= -\frac{1}{4\pi jn} \int_{\text{SDP}_{+\pi}} H_1(z) e^{\beta\rho h_1(z)} dz - \frac{1}{4\pi jn} \int_{\text{SDP}_{-\pi}} H_1(z) e^{\beta\rho h_1(z)} dz \\ &\simeq \frac{e^{-j\pi/4}}{\sqrt{2\pi\beta}} \frac{\frac{1}{n}\sin\left(\frac{\pi}{n}\right)}{\cos\left(\frac{\pi}{n}\right) - \cos\left(\frac{\phi-\phi'}{n}\right)} \frac{e^{-j\beta\rho}}{\sqrt{\rho}} \end{aligned}}$$

(13-61)

This is referred to as the *incident diffracted field*, and its form, as given by (13-61), is valid provided that the poles of (13-53a) are not near the saddle points of (13-59). Another way to say this is that (13-61) is valid provided that the observations are not made at or near the incident shadow boundary of Figure 13-13. When the observations are made at the incident shadow boundary $\phi - \phi' = \pi$, (13-61) becomes infinite. Therefore, another form must be used for such situations.

Following a similar procedure for the evaluation of (13-48b), it can be shown that its contributions along C_T and the saddle points can be written in forms corresponding to (13-58) and (13-61) for the evaluation of (13-48a). Thus, we can write that

$$
\begin{aligned}
F_2(\beta\rho)|_{C_T} &= \frac{1}{4\pi jn} \oint_{C_T} H_2(z)e^{\beta\rho h_2(z)}dz \\
&= e^{j\beta\rho\cos(\phi+\phi')}U[\pi - (\phi+\phi')]
\end{aligned}
\tag{13-62a}
$$

$$
\begin{aligned}
F_2(\beta\rho)|_{\mathrm{SDP}_{\pm\pi}} &= -\frac{1}{4\pi jn}\int_{\mathrm{SDP}_{+\pi}} H_2(z)e^{\beta\rho h_2(z)}dz - \frac{1}{4\pi jn}\int_{\mathrm{SDP}_{-\pi}} H_2(z)e^{\beta\rho h_2(z)}dz \\
&\simeq \frac{e^{-j\pi/4}}{\sqrt{2\pi\beta}}\frac{\dfrac{1}{n}\sin\left(\dfrac{\pi}{n}\right)}{\cos\left(\dfrac{\pi}{n}\right)-\cos\left(\dfrac{\phi+\phi'}{n}\right)}\frac{e^{-j\beta\rho}}{\sqrt{\rho}}
\end{aligned}
\tag{13-62b}
$$

Equation 13-62b is valid provided the observations are not made at or near the reflection shadow boundary of Figure 13-13. When the observations are made at the reflection shadow boundary $\phi + \phi' = \pi$, (13-62b) becomes infinite. Another form must be used for such cases.

If the poles of (13-53a) are near the saddle points of (13-59), then the conventional steepest descent method of (13-60a) and (13-60b) cannot be used for the evaluation of the last two terms of (13-50) for $F_1(\beta\rho)$, and similarly for $F_2(\beta\rho)$ of (13-48b). One method that can be used for such cases is the so-called *Pauli-Clemmow modified method of steepest descent* [8, 47]. The main difference between the two methods is that the solution provided by the Pauli-Clemmow modified method of steepest descent has an additional discontinuous function that compensates for the singularity along the corresponding shadow boundaries introduced by the conventional steepest-descent method for isolated poles and saddle points. This factor is usually referred to as the *transition function*, and it is proportional to a Fresnel integral. Away from the corresponding shadow boundaries these transition functions are nearly unity, and the Pauli-Clemmow modified method of steepest descent reduces to the conventional method of steepest descent.

It can be shown that by using the Pauli-Clemmow modified method of steepest descent, (13-60a) and (13-60b) are evaluated using

$$
\begin{aligned}
\frac{1}{4\pi jn}\int_{\mathrm{SDP}_{+\pi}} H_1(z)e^{\beta\rho h_1(z)}dz &\overset{\beta\rho\to\text{large}}{\simeq} \frac{1}{4\pi jn}\left|\sqrt{\frac{2\pi}{-\beta\rho h_1''(z_s=+\pi)}}\right| \\
&\quad \times e^{j\pi/4}H_1(z_s=+\pi)e^{\beta\rho h_1(z_s=\pi)}F[\beta\rho g^+(\xi^-)] \\
&= \frac{e^{-j\pi/4}}{2n\sqrt{2\pi\beta}}\cot\left[\frac{\pi+(\phi-\phi')}{2n}\right] \\
&\quad \times F[\beta\rho g^+(\phi-\phi')]\frac{e^{-j\beta\rho}}{\sqrt{\rho}}
\end{aligned}
\tag{13-63a}
$$

$$\frac{1}{4\pi jn}\int_{\text{SDP}_{-\pi}} H_1(z)e^{\beta\rho h_1(z)}dz \overset{\beta\rho\rightarrow\text{large}}{\simeq} \frac{1}{4\pi jn}\left|\sqrt{\frac{2\pi}{-\beta\rho h_1''(z_s=-\pi)}}\right|$$

$$\times e^{-j3\pi/4}H_1(z_s=-\pi)e^{\beta\rho h_1(z_s=-\pi)}F[\beta\rho g^-(\xi^-)]$$

$$=\frac{e^{-j\pi/4}}{2n\sqrt{2\pi\beta}}\cot\left[\frac{\pi-(\phi-\phi')}{2n}\right]$$

$$\times F[\beta\rho g^-(\phi-\phi')]\frac{e^{-j\beta\rho}}{\sqrt{\rho}} \qquad (13\text{-}63\text{b})$$

where

$$F[\beta\rho g^\pm(\phi-\phi')]\equiv j[h_1(z_s)-h_1(z_p)]$$

$$\equiv \text{measure of separation between saddle points and poles}$$

$$=2j\left|\sqrt{\beta\rho g^\pm(\phi-\phi')}\right|e^{+j\beta\rho g^\pm}\int_{\sqrt{\beta\rho g^\pm(\phi-\phi')}}^{\infty}e^{-j\tau^2}d\tau \qquad (13\text{-}63\text{c})$$

$$g^\pm(\phi-\phi')=1+\cos[(\phi-\phi')-2\pi nN^\pm] \qquad (13\text{-}63\text{d})$$

$$2\pi nN^+ -(\phi-\phi')=+\pi \qquad (13\text{-}63\text{e})$$

$$2\pi nN^- -(\phi-\phi')=-\pi \qquad (13\text{-}63\text{f})$$

Similar forms are used for the evaluation of $F_2(\beta\rho)$ along the steepest descent path using the Pauli–Clemmow modified method of steepest descent. In (13-63e) and (13-63f), $N^\pm$ represents integer values that most closely satisfy the equalities. Such a procedure accounts for the poles that are nearest to the saddle point at $x=\pm\pi$, either from outside or within $-\pi\leq x\leq+\pi$, of Figure 13-18b. In general, there are two such poles associated with $F_1(\beta\rho)$ and two with $F_2(\beta\rho)$. More details about the transition function will follow.

D. Geometrical Optics and Diffracted Fields

After the contour integration and the method of steepest descent have been applied in the evaluation of (13-47), as discussed in the previous section, it can be shown that for large values of $\beta\rho$, (13-47) is separated into

$$F(\beta\rho)=F_G(\beta\rho)+F_D(\beta\rho) \qquad (13\text{-}64)$$

where $F_G(\beta\rho)$ and $F_D(\beta\rho)$ represent, respectively, the total geometrical optics and total diffracted fields created by the incidence of a unit amplitude plane wave upon a two-dimensional wedge, as shown in Figure 13-13.

In summary, then, the geometrical optics fields (F_G) and the diffracted fields (F_D) are represented, respectively, by

$$
\begin{array}{c}
\text{Geometrical Optics Fields}\\[4pt]
\begin{array}{ccc}
\textit{Incident GO} & \textit{Reflected GO} & \textit{Region}
\end{array}\\[2pt]
F_G(\beta\rho)=\begin{cases}
e^{j\beta\rho\,\cos(\phi-\phi')} \;\pm\; e^{j\beta\rho\,\cos(\phi+\phi')} & 0<\phi<\pi-\phi'\\
e^{j\beta\rho\,\cos(\phi-\phi')} & \pi-\phi'<\phi<\pi+\phi'\\
0 & \pi+\phi'<\phi<n\pi
\end{cases}
\end{array}
\qquad (13\text{-}65)
$$

$$\underline{\text{Diffracted Fields}}$$

$$\textit{Total diffracted field} \qquad = \quad \textit{Incident diffracted} \quad \pm \quad \textit{Reflected diffracted}$$

$$\overbrace{F_D(\beta\rho) = F_D(\rho,\phi,\phi',n) = V_B(\rho,\phi,\phi',n)} \quad = \quad \overbrace{V_B^i(\rho,\phi-\phi',n)} \quad \pm \quad \overbrace{V_B^r(\rho,\phi+\phi',n)}$$

$$(13\text{-}66)$$

where

$$V_B^{i,r}(\rho,\phi\mp\phi',n) = I_{-\pi}(\rho,\phi\mp\phi',n) + I_{+\pi}(\rho,\phi\mp\phi',n) \qquad (13\text{-}66a)$$

$$I_{\pm\pi}(\rho,\phi\mp\phi',n) \simeq \frac{e^{-j(\beta\rho+\pi/4)}}{jn\sqrt{2\pi}}\sqrt{g^{\pm}}\cot\left[\frac{\pi\pm(\phi\pm\phi')}{2n}\right]$$

$$\times e^{+j\beta\rho g^{\pm}}\int_{\sqrt{\beta\rho g^{\pm}}}^{\infty}e^{-j\tau^2}\,d\tau + (\text{higher-order terms}) \qquad (13\text{-}66b)$$

$$g^+ = 1 + \cos[(\phi\mp\phi') - 2n\pi N^+] \qquad (13\text{-}66c)$$

$$g^- = 1 + \cos[(\phi\mp\phi') - 2n\pi N^-] \qquad (13\text{-}66d)$$

with N^+ or N^- being a positive or negative integer or zero that most closely satisfies the equation

$$2n\pi N^+ - (\phi\mp\phi') = +\pi \quad \text{for } g^+ \qquad (13\text{-}66e)$$

$$2n\pi N^- - (\phi\mp\phi') = -\pi \quad \text{for } g^- \qquad (13\text{-}66f)$$

Each of the diffracted fields (incident and reflected) exists in all space surrounding the wedge. Equation 13-66b contains the leading term of the diffracted field plus higher-order terms that are negligible for large values of $\beta\rho$. The integral in (13-66b) is a Fresnel integral (see Appendix III). In (13-65) and (13-66), the plus (+) sign is used for the hard polarization and the minus sign is used for the soft polarization.

If observations are made away from each of the shadow boundaries so that $\beta\rho g^{\pm} \gg 1$, (13-66a) and (13-66b) reduce, according to (13-61) and (13-62b), to

---

$$V_B^{h,s}(\rho,\phi\mp\phi',n) = V_B^i(\rho,\phi-\phi',n) \pm V_B^r(\rho,\phi+\phi',n)$$

$$= \frac{e^{-j\pi/4}}{\sqrt{2\pi\beta}}\frac{1}{n}\sin\left(\frac{\pi}{n}\right)\left[\frac{1}{\cos\left(\dfrac{\pi}{n}\right)-\cos\left(\dfrac{\phi-\phi'}{n}\right)} \pm \frac{1}{\cos\left(\dfrac{\pi}{n}\right)-\cos\left(\dfrac{\phi+\phi'}{n}\right)}\right]\frac{e^{-j\beta\rho}}{\sqrt{\rho}} \qquad (13\text{-}67)$$

---

which is an expression of much simpler form, even for computational purposes. It is quite evident that when the observations are made at the incident shadow boundary (ISB, where $\phi = \pi + \phi'$), $V_B^i(\rho,\phi-\phi',n)$ of (13-67) becomes infinite because $\phi - \phi'$ is equal to π and the two cosine terms in the denominator of (13-67) are identical. Similarly, $V_B^r(\rho,\phi+\phi',n)$ of (13-67) becomes infinite when observations are made at the reflected shadow boundary (RSB, where $\phi = \pi - \phi'$). The incident and reflected diffraction functions of the form in equation 13-67 are referred to as *Keller's diffraction functions* and possess singularities along the incident and reflection shadow boundaries. The diffraction functions of (13-66a) through (13-66f) are representatives of the *uniform theory of diffraction* (UTD). The regions in the neighborhood of the incident and reflection shadow boundaries are referred to as the *transition regions*, and in these regions the fields undergo their most rapid changes.

The functions g^+ and g^- of (13-66c) and (13-66d) are representative of the angular separation between the observation point and the incident or reflection shadow boundary. In fact, when observations are made along the shadow boundaries, the $g^{\pm}$ functions are equal to zero. For *exterior wedges* $(1 \leq n \leq 2)$, the values of N^+ and N^- in (13-66e) and (13-66f) are equal to $N^+ = 0$ or 1 and $N^- = -1, 0,$ or 1. The values of n are plotted, as a function of ξ, where $-2\pi \leq \xi^{\pm} = \phi \pm \phi' \leq 4\pi$; in Figure 13-19a for $N^- = -1, 0, 1$ and in Figure 13-19b for $N^+ = 0, 1$. These integral values of $N^{\pm}$ are particularly important along the shadow boundaries, which are represented by the dotted lines. The variations of $N^{\pm}$ as a function of ϕ near the shadow boundaries are not abrupt, and this is a desirable property. The permissible values of $\xi^{\pm} = \phi \pm \phi'$ for $0 \leq \phi, \phi' \leq n\pi$ when $1 \leq n \leq 2$ are those bounded by the trapezoids formed by the solid straight lines in Figures 13-19a and 13-19b.

In order for (13-67) to be valid, $\beta \rho g^{\pm} \gg 1$. This can be achieved by having one of the following conditions:

1. $\beta \rho$ and $g^{\pm}$ large. This is satisfied if the distance ρ to the observation point is large and the observation angle ϕ is far away from either of the two shadow boundaries.
2. $\beta \rho$ large and $g^{\pm}$ small. This is satisfied if the distance ρ to the observation point is large and the observation angle ϕ is near either one or both of the shadow boundaries.
3. $\beta \rho$ small and $g^{\pm}$ large. This is satisfied if the distance ρ to the observation point is small and the observation angle ϕ is far away from either of the two shadow boundaries.

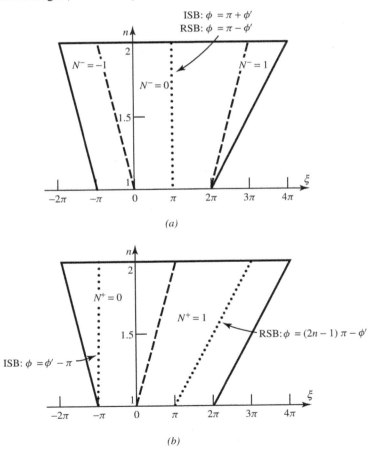

(a)

(b)

Figure 13-19   Graphical representation of (*a*) N^- and (*b*) N^+ as a function of ξ and n. (Source: R. G. Kouyoumjian and P. H. Pathak, "A uniform geometrical theory of diffraction for an edge in a perfectly conducting surface," *Proc. IEEE*, © 1974, IEEE.)

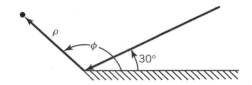

Figure 13-20   Plane wave incident on a half-plane.

To demonstrate, we consider the problem of a plane wave, of incidence angle $\phi' = 30°$, impinging on a half-plane ($n = 2$), shown in Figure 13-20. In Figures 13-21a and 13-21b we have plotted as a function of ϕ, respectively, the magnitude of the incident V_B^i and the reflected V_B^r diffracted fields, as given by (13-66a) through (13-66f) for $\rho = \lambda$, and 100λ when $\phi' = 30°$ and $n = 2$. In the same figures, these results are compared with those obtained using (13-67). It is apparent that as the observation distance ρ increases, the angular sector near the incident and reflected shadow boundaries over which (13-67) becomes invalid decreases; in the limit as $\rho \to \infty$, both give the same results.

The expression of (13-66) represents the diffraction of the unity strength incident plane wave with observations made at $P(\rho, \phi)$, as shown in Figure 13-13. Diffraction solutions of cylindrical waves, with observations at large distances, can be obtained by the use of the reciprocity principle along with the solution of the diffraction of an incident plane wave by a wedge as given by (13-66) through (13-66f). Using the geometry of Figure 13-17b, it is evident that cylindrical wave incidence diffraction can be obtained by substituting in (13-66) through (13-66f) $\rho = \rho'$, $\phi = \phi'$, and $\phi' = \phi$.

E. Diffraction Coefficients   The incident diffraction function V_B^i of (13-66) can also be written, using (13-66a) through (13-66f), as

$$V_B^i(\rho, \phi - \phi', n) = V_B^i(\rho, \xi^-, n) = \frac{e^{-j\beta\rho}}{\sqrt{\rho}} D^i(\rho, \xi^-, n) \qquad (13\text{-}68)$$

where

$$\xi^- = \phi - \phi' \qquad (13\text{-}68a)$$

$$D^i(\rho, \xi^-, n) = -\frac{e^{-j\pi/4}}{2n\sqrt{2\pi\beta}}\{C^+(\xi^-, n)F[\beta\rho g^+(\xi^-)] + C^-(\xi^-, n)F[\beta\rho g^-(\xi^-)]\} \qquad (13\text{-}68b)$$

$$C^+(\xi^-, n) = \cot\left(\frac{\pi + \xi^-}{2n}\right) \qquad (13\text{-}68c)$$

$$C^-(\xi^-, n) = \cot\left(\frac{\pi - \xi^-}{2n}\right) \qquad (13\text{-}68d)$$

$$F[\beta\rho g^+(\xi^-)] = 2j\sqrt{\beta\rho g^+(\xi^-)}\, e^{+j\beta\rho g^+(\xi^-)} \int_{\sqrt{\beta\rho g^+(\xi^-)}}^{\infty} e^{-j\tau^2}\, d\tau \qquad (13\text{-}68e)$$

$$F[\beta\rho g^-(\xi^-)] = 2j\sqrt{\beta\rho g^-(\xi^-)}\, e^{+j\beta\rho g^-(\xi^-)} \int_{\sqrt{\beta\rho g^-(\xi^-)}}^{\infty} e^{-j\tau^2}\, d\tau \qquad (13\text{-}68f)$$

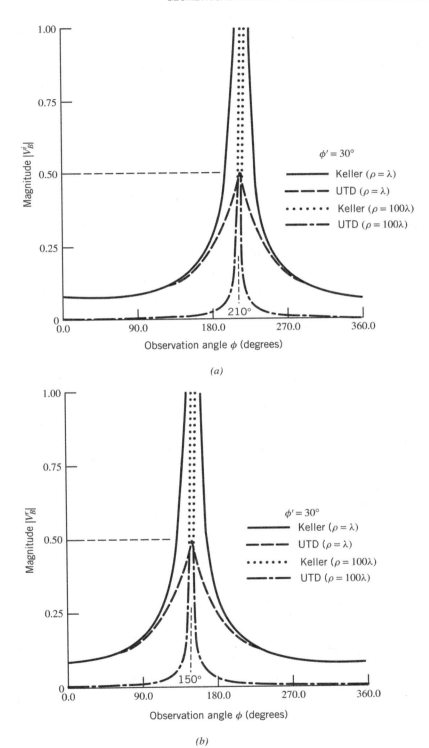

Figure 13-21 Plane wave diffraction by a half-plane. (*a*) Incident diffracted field. (*b*) Reflected diffracted field.

The functions $g^+(\xi^-)$ and $g^-(\xi^-)$ are given by (13-66c) through (13-66f). In (13-68), $D^i(\rho,\xi^-,n)$ is defined as the diffraction coefficient for the incident diffracted field, and it will be referred to as the *incident diffraction coefficient* for a unit amplitude incident plane wave.

In a similar manner, the reflected diffraction function V_B^r of (13-66) can also be written, using (13-66a) through (13-66f), as

$$V_B^r(\rho,\phi+\phi',n)=V_B^r(\rho,\xi^+,n)=\frac{e^{-j\beta\rho}}{\sqrt{\rho}}D^r(\rho,\xi^+,n)$$

(13-69)

where

$$\xi^+=\phi+\phi'$$

(13-69a)

$$D^r(\rho,\,\xi^+,\,n)=-\frac{e^{-j\pi/4}}{2n\sqrt{2\pi\beta}}\left\{C^+(\xi^+,n)F[\beta\rho g^+(\xi^+)]+C^-(\xi^+,n)F[\beta\rho g^-(\xi^+)]\right\}$$

(13-69b)

$$C^+(\xi^+,n)=\cot\left(\frac{\pi+\xi^+}{2n}\right)$$

(13-69c)

$$C^-(\xi^+,n)=\cot\left(\frac{\pi-\xi^+}{2n}\right)$$

(13-69d)

$$F[\beta\rho g^+(\xi^+)]=2j\sqrt{\beta\rho g^+(\xi^+)}\,e^{+j\beta\rho g^+(\xi^+)}\int_{\sqrt{\beta\rho g^+(\xi^+)}}^{\infty}e^{-j\tau^2}\,d\tau$$

(13-69e)

$$F[\beta\rho g^-(\xi^+)]=2j\sqrt{\beta\rho g^-(\xi^+)}\,e^{+j\beta\rho g^-(\xi^+)}\int_{\sqrt{\beta\rho g^-(\xi^+)}}^{\infty}e^{-j\tau^2}\,d\tau$$

(13-69f)

$D^r(\rho,\xi^+,n)$ will be referred to as the *reflection diffraction coefficient* for a unit amplitude incident plane wave.

With the aid of (13-68) and (13-69), the total diffraction function $V_B(\rho,\phi\mp\phi',n)=V_B^i(\rho,\phi-\phi',n)\mp V_B^r(\rho,\phi+\phi',n)$ of (13-66) can now be written as

$$V_B(\rho,\phi,\phi',n)=V_B^i(\rho,\phi-\phi',n)\mp V_B^r(\rho,\phi+\phi',n)$$
$$=\frac{e^{-j\beta\rho}}{\sqrt{\rho}}\left[D^i(\rho,\phi-\phi',n)\mp D^r(\rho,\phi+\phi',n)\right]$$

(13-70)

or

$$V_{Bs}(\rho,\phi,\phi',n)=V_B^i(\rho,\phi-\phi',n)-V_B^r(\rho,\phi+\phi',n)$$
$$=\frac{e^{-j\beta\rho}}{\sqrt{\rho}}\left[D^i(\rho,\phi-\phi',n)-D^r(\rho,\phi+\phi',n)\right]$$
$$=\frac{e^{-j\beta\rho}}{\sqrt{\rho}}D_s(\rho,\phi,\phi',n)$$

(13-70a)

$$
\begin{aligned}
V_{Bh}(\rho,\phi,\phi',n) &= V_B^i(\rho,\phi-\phi',n) + V_B^r(\rho,\phi+\phi',n) \\
&= \frac{e^{-j\beta\rho}}{\sqrt{\rho}}\left[D^i(\rho,\phi-\phi',n) + D^r(\rho,\phi+\phi',n)\right] \\
&= \frac{e^{-j\beta\rho}}{\sqrt{\rho}}D_h(\rho,\phi,\phi',n)
\end{aligned}
\tag{13-70b}
$$

$$
D_s(\rho,\phi,\phi',n) = D^i(\rho,\phi-,\phi',n) - D^r(\rho,\phi+,\phi',n)
\tag{13-70c}
$$

$$
D_h(\rho,\phi,\phi',n) = D^i(\rho,\phi-\phi',n) + D^r(\rho,\phi+\phi',n)
\tag{13-70d}
$$

where $V_{Bs} = V_B^i - V_B^r =$ soft (polarization) diffraction function
$V_{Bh} = V_B^i + V_B^r =$ hard (polarization) diffraction function
$V_B^i =$ incident diffraction function
$V_B^r =$ reflection diffraction function
$D_s = D^i - D^r =$ soft (polarization) diffraction coefficient
$D_h = D^i + D^r =$ hard (polarization) diffraction coefficient
$D^i =$ incident diffraction coefficient
$D^r =$ reflection diffraction coefficient

Using (13-68b) and (13-69b) in expanded form, the soft (D_s) and hard (D_h) diffraction coefficients can ultimately be written, respectively, as

$$
\begin{aligned}
D_s(\rho,\phi,\phi',n) ={}& -\frac{e^{-j\pi/4}}{2n\sqrt{2\pi\beta}} \\
&\times\left(\left\{\cot\left[\frac{\pi+(\phi-\phi')}{2n}\right]F[\beta\rho g^+(\phi-\phi')] + \cot\left[\frac{\pi-(\phi-\phi')}{2n}\right]F[\beta\rho g^-(\phi-\phi')]\right\}\right. \\
&\left. -\left\{\cot\left[\frac{\pi+(\phi+\phi')}{2n}\right]F[\beta\rho g^+(\phi+\phi')] + \cot\left[\frac{\pi-(\phi+\phi')}{2n}\right]F[\beta\rho g^-(\phi+\phi')]\right\}\right)
\end{aligned}
\tag{13-71a}
$$

$$
\begin{aligned}
D_h(\rho,\phi,\phi',n) ={}& -\frac{e^{-j\pi/4}}{2n\sqrt{2\pi\beta}} \\
&\times\left(\left\{\cot\left[\frac{\pi+(\phi-\phi')}{2n}\right]F[\beta\rho g^+(\phi-\phi')] + \cot\left[\frac{\pi-(\phi-\phi')}{2n}\right]F[\beta\rho g^-(\phi-\phi')]\right\}\right. \\
&\left. +\left\{\cot\left[\frac{\pi+(\phi+\phi')}{2n}\right]F[\beta\rho g^+(\phi+\phi')] + \cot\left[\frac{\pi-(\phi+\phi')}{2n}\right]F[\beta\rho g^+(\phi+\phi')]\right\}\right)
\end{aligned}
\tag{13-71b}
$$

where

$$
\phi-\phi' = \xi^-
\tag{13-71c}
$$

$$
\phi+\phi' = \xi^+
\tag{13-71d}
$$

The formulations of (13-68) through (13-71d) are part of the often referred to *uniform theory of diffraction* (UTD) [10], which are extended to include oblique incidence and curved edge diffraction. These will be discussed in Sections 13.3.3 and 13.3.4, respectively.

A Fortran and MATLAB computer program designated WDC, Wedge Diffraction Coefficients, computes the soft and hard polarization diffraction coefficients of (13-71a) and (13-71b) (*actually the diffraction coefficients normalized by* $\sqrt{\lambda}$). The Fortran program was initially developed and reported in [53]. The program also accounts for oblique wave incidence, and it is based on the more general formulation of (13-89a) through (13-90b). The main difference in the two sets of equations is the $\sin \beta_0'$ function found in the denominator of (13-90a) and (13-90b); it has been introduced to account for the oblique wave incidence. In (13-90a) and (13-90b), L is used as the distance parameter and ρ is used in (13-71a) and (13-71b).

Therefore, to use the subroutine WDC to compute (13-71a) and (13-71b) let the oblique incidence angle β_0', referred to as BTD, be 90°. The distance parameter R should represent ρ (in wavelengths). This program uses the complex function FTF (Fresnel transition function) to complete its computations. The FTF program computes (13-68e), (13-68f), (13-69e), and (13-69f) based on the asymptotic expressions of (13-74a), (13-74b), and a linear interpolation for intermediate arguments. Computations of the Fresnel integral can also be made on an algorithm reported in [49] as well as on approximate expressions of [50].

To use the WDC subroutine, the user must specify $R = \rho$ (in wavelengths), PHID $= \phi$ (in degrees), PHIPD $= \phi'$ (in degrees), BTD $= \beta_0'$ (in degrees), and FN $= n$ (dimensionless). The program subroutine computes the normalized (with respect to $\sqrt{\lambda}$) diffraction coefficients CDCS $= D_s$ and CDCH $= D_h$. *The angles represented by ϕ and ϕ' should be referenced from the face of the wedge, as shown in Figure* 13-13*b*. For normal incidence, $\beta_0' = 90°$. This computer subroutine has been used successfully in a multitude of problems; it is very efficient, and the user is encouraged to utilize it effectively.

Following a similar procedure, the incident and reflected diffraction functions and the incident, reflected, soft, and hard diffraction coefficients using Keller's diffraction functions of (13-67) can be written, respectively, as

$$V_B^i(\rho, \phi - \phi', n) = V_B^i(\rho, \xi^-, n) = \frac{e^{-j\beta\rho}}{\sqrt{\rho}} D^i(\rho, \phi - \phi', n) = \frac{e^{-j\beta\rho}}{\sqrt{\rho}} D^i(\rho, \xi^-, n) \qquad (13\text{-}72\text{a})$$

$$V_B^r(\rho, \phi + \phi', n) = V_B^r(\rho, \xi^+, n) = \frac{e^{-j\beta\rho}}{\sqrt{\rho}} D^r(\rho, \phi - \phi', n) = \frac{e^{-j\beta\rho}}{\sqrt{\rho}} D^r(\rho, \xi^+, n) \qquad (13\text{-}72\text{b})$$

$$\boxed{D^i(\rho, \phi - \phi', n) = \frac{e^{-j\pi/4}}{\sqrt{2\pi\beta}} \frac{\dfrac{1}{n}\sin\left(\dfrac{\pi}{n}\right)}{\cos\left(\dfrac{\pi}{n}\right) - \cos\left(\dfrac{\phi - \phi'}{n}\right)}} \qquad (13\text{-}72\text{c})$$

$$\boxed{D^r(\rho, \phi + \phi', n) = \frac{e^{-j\pi/4}}{\sqrt{2\pi\beta}} \frac{\dfrac{1}{n}\sin\left(\dfrac{\pi}{n}\right)}{\cos\left(\dfrac{\pi}{n}\right) - \cos\left(\dfrac{\phi + \phi'}{n}\right)}} \qquad (13\text{-}72\text{d})$$

$$D_s(\rho,\phi,\phi',n) = D^i(\rho,\phi-\phi',n) - D^r(\rho,\phi+\phi',n)$$

$$= \frac{e^{-j\pi/4}\frac{1}{n}\sin\left(\frac{\pi}{n}\right)}{\sqrt{2\pi\beta}}\left[\frac{1}{\cos\left(\frac{\pi}{n}\right)-\cos\left(\frac{\phi-\phi'}{n}\right)} - \frac{1}{\cos\left(\frac{\pi}{n}\right)-\cos\left(\frac{\phi+\phi'}{n}\right)}\right] \qquad (13\text{-}72e)$$

$$D_h(\rho,\phi,\phi',n) = D^i(\rho,\phi-\phi',n) + D^r(\rho,\phi+\phi',n)$$

$$= \frac{e^{-j\pi/4}\frac{1}{n}\sin\left(\frac{\pi}{n}\right)}{\sqrt{2\pi\beta}}\left[\frac{1}{\cos\left(\frac{\pi}{n}\right)-\cos\left(\frac{\phi-\phi'}{n}\right)} + \frac{1}{\cos\left(\frac{\pi}{n}\right)-\cos\left(\frac{\phi+\phi'}{n}\right)}\right] \qquad (13\text{-}72f)$$

The diffraction coefficients of (13-72c) through (13-72f) are referred to as *Keller's diffraction coefficients*, and they possess singularities along the incident and reflection shadow boundaries.

The wedge diffraction coefficients described previously assume that the orientational direction for the incident and diffracted electric and magnetic field components of the soft and hard polarized fields are those shown, respectively, in Figures 13-22a and 13-22b. A negative value in the diffraction coefficients will reverse the directions of the appropriate fields.

The function $F(X)$ of (13-63c), (13-68e), (13-68f), (13-69e), and (13-69f) is known as a *Fresnel transition function* and it involves a Fresnel integral. Its magnitude and phase for $0.001 \leq X \leq 10$ are shown plotted in Figure 13-23. It is evident that

$$\left.\begin{array}{r}|F(X)|\leq 1\\ 0\leq \text{Phase of } F(X)\leq \pi/4\end{array}\right\}, \quad \text{for} \quad 0.001\leq X \leq 10 \qquad (13\text{-}73a)$$

and

$$F(X)\simeq 1, \quad \text{for} \quad X > 10 \qquad (13\text{-}73b)$$

Thus, if the argument X of the transition function exceeds 10, it can be replaced by unity. Then the expressions for the diffraction coefficients of (13-68b), (13-69b), (13-71a), and (13-71b) reduce, respectively, to those of (13-72c) through (13-72f). Asymptotic expressions for the transition function $F(X)$ are [53]:

For small $X(X < 0.3)$

$$F(X)\simeq \left[\sqrt{\pi X} - 2Xe^{j\pi/4} - \frac{2}{3}X^2 e^{-j\pi/4}\right]e^{j(\pi/4 + X)} \qquad (13\text{-}74a)$$

For large $X(X > 5.5)$

$$F(X)\simeq \left[1 + j\frac{1}{2X} - \frac{3}{4}\frac{1}{X^2} - j\frac{15}{8}\frac{1}{X^3} + \frac{75}{16}\frac{1}{X^4}\right] \qquad (13\text{-}74b)$$

To facilitate the reader in the computations, a Fortran and MATLAB computer function program designated as FTF, for Fresnel Transition Function, computes the wedge transition function $F(X)$ of (13-68e), (13-68f), (13-69e), or (13-69f). The algorithm is based on the approximations of

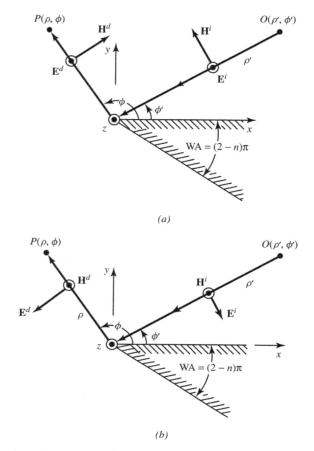

(a)

(b)

Figure 13-22   Polarization of incident and diffracted fields. (*a*) Soft polarization. (*b*) Hard polarization.

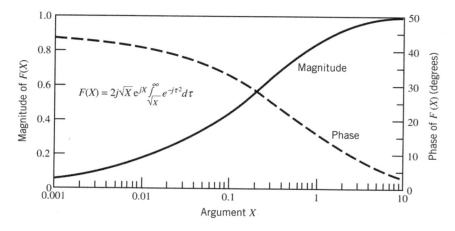

$$F(X) = 2j\sqrt{X}\,e^{jX}\int_{\sqrt{X}}^{\infty}e^{-j\tau^2}d\tau$$

Figure 13-23   Variations of magnitude and phase of transition function $F(X)$ as a function of X. (Source: R. G. Kouyoumjian and P. H. Pathak, "A uniform geometrical theory of diffraction for an edge in a perfectly conducting surface," *Proc. IEEE*, © 1974, IEEE.)

(13-74a) for small arguments $(X < 0.3)$ and on (13-74b) for large arguments $(X > 5.5)$. For intermediate values $(0.3 \leq X \leq 5.5)$, a linear interpolation scheme is used. The program was developed and reported in [53].

To have a better understanding of the UTD diffraction coefficients $D^i(\rho, \phi - \phi', n)$ and $D^r(\rho, \phi + \phi', n)$ of (13-68b) and (13-69b), let us examine their behavior around the incident and reflection shadow boundaries. This will be accomplished by considering only exterior wedges $(n \geq 1)$ and examining separately the geometry where the incident wave illuminates the $\phi = 0$ side of the wedge $[\phi' \leq (n-1)\pi]$, as shown in Figure 13-24a, and the geometry where the incident wave illuminates the $\phi = n\pi$ side of the wedge $[\phi' \geq (n-1)\pi]$, as shown in Figure 13-24b.

Case A $[\phi' \leq (n-1)\pi]$, Figure 13-24a

For this case the incident shadow boundary (ISB) occurs when $\phi = \pi + \phi'$ (or $\phi - \phi' = \pi$) and the reflection shadow boundary (RSB) occurs when $\phi = \pi - \phi'$ (or $\phi + \phi' = \pi$).

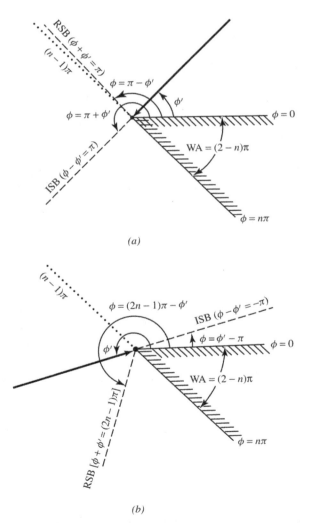

(a)

(b)

Figure 13-24 Incident and reflection shadow boundaries for wedge diffraction. (a) Case A: $\phi' \leq (n-1)\pi$. (b) Case B: $\phi' \geq (n-1)\pi$.

1. Along the ISB ($\phi = \pi + \phi'$ or $\phi - \phi' = \pi$), the second cotangent function of (13-68b) becomes singular and the first cotangent function in (13-68b), and both of them in (13-69b), remain bounded. That is, from (13-68b),

$$C^-(\xi^-,n)|_{\phi-\phi'=\pi} = \cot\left(\frac{\pi-\xi^-}{2n}\right)\bigg|_{\phi-\phi'=\pi} = \cot\left[\frac{\pi-(\phi-\phi')}{2n}\right]\bigg|_{\phi-\phi'=\pi} = \infty \qquad (13\text{-}75a)$$

In addition, the value of N^- from (13-66f) is equal to 0. That is,

$$2n\pi N^- -(\phi-\phi')|_{\phi-\phi'=\pi} = 2n\pi N^- -\pi = -\pi \;\Rightarrow\; N^- = 0 \qquad (13\text{-}75b)$$

2. Along the RSB ($\phi = \pi - \phi'$ or $\phi + \phi' = \pi$), the second cotangent function of (13-69b) becomes singular and the first cotangent function in (13-69b) and both of them in (13-68b) remain bounded. That is, from (13-69b),

$$C^-(\xi^+,n)|_{\phi+\phi'=\pi} = \cot\left(\frac{\pi-\xi^+}{2n}\right)\bigg|_{\phi+\phi'=\pi} = \cot\left[\frac{\pi-(\phi+\phi')}{2n}\right]\bigg|_{\phi+\phi'=\pi} = \infty \qquad (13\text{-}76a)$$

In addition, the value of N^- from (13-66f) is equal to 0. That is,

$$2n\pi N^- -(\phi+\phi')|_{\phi+\phi'=\pi} = 2n\pi N^- -\pi = -\pi \;\Rightarrow\; N^- = 0 \qquad (13\text{-}76b)$$

Case B $[\phi' \geq (n-1)\pi]$, Figure 13-24b

For this case, the incident shadow boundary (ISB) occurs when $\phi = \phi' - \pi$ (or $\phi - \phi' = -\pi$) and the reflection shadow boundary (RSB) occurs when $\phi = (2n-1)\pi - \phi'$ [or $\phi + \phi' = (2n-1)\pi$].

1. Along the ISB ($\phi = \phi' - \pi$ or $\phi - \phi' = -\pi$), the first cotangent function in (13-68b) becomes singular and the second cotangent function of (13-68b), and both of them in (13-69b), remain bounded. That is, from (13-68b),

$$C^+(\xi^-,n)|_{\phi-\phi'=-\pi} = \cot\left(\frac{\pi+\xi^-}{2n}\right)\bigg|_{\phi-\phi'=-\pi}$$
$$= \cot\left[\frac{\pi+(\phi-\phi')}{2n}\right]\bigg|_{\phi-\phi'=-\pi} = \infty \qquad (13\text{-}77a)$$

In addition, the value of N^+ from (13-66e) is equal to 0. That is,

$$2n\pi N^+ -(\phi-\phi')|_{\phi-\phi'=-\pi} = 2n\pi N^+ +\pi = +\pi \;\Rightarrow\; N^+ = 0 \qquad (13\text{-}77b)$$

2. Along the RSB $[\phi = (2n-1)\pi - \phi'$ or $\phi + \phi' = (2n-1)\pi]$, the first cotangent function of (13-69b) becomes singular and the second cotangent function of (13-69b), and both of them in (13-68b), remain bounded. That is, from (13-69b),

$$C^+(\xi^+,n)|_{\phi+\phi'=(2n-1)\pi} = \cot\left(\frac{\pi+\xi^+}{2n}\right)\bigg|_{\phi+\phi'=(2n-1)\pi}$$
$$= \cot\left[\frac{\pi+(\phi+\phi')}{2n}\right]\bigg|_{\phi+\phi'=(2n-1)\pi} = \infty \qquad (13\text{-}78a)$$

In addition, the value of N^+ from (13-66e) is equal to 1. That is,

$$2n\pi N^+ -(\phi+\phi')|_{\phi+\phi'=(2n-1)\pi} = 2n\pi N^+ -(2n-1)\pi = +\pi \;\Rightarrow\; N^+ = 1 \qquad (13\text{-}78b)$$

The results for Cases A and B are summarized in Table 13-1.

TABLE 13-1   Cotangent function behavior and values of $N^\pm$ along the shadow boundaries

	The cotangent function becomes singular when	Value of $N^\pm$ at the shadow boundary
$\cot\left[\dfrac{\pi-(\phi-\phi')}{2n}\right]$	$\phi=\pi+\phi'$ or $\phi-\phi'=\pi$ ISB of Case A Figure 13-24a	$N^-=0$
$\cot\left[\dfrac{\pi-(\phi+\phi')}{2n}\right]$	$\phi=\pi-\phi'$ or $\phi+\phi'=\pi$ RSB of Case A Figure 13-24a	$N^-=0$
$\cot\left[\dfrac{\pi+(\phi-\phi')}{2n}\right]$	$\phi=\phi'-\pi$ or $\phi-\phi'=-\pi$ ISB of Case B Figure 13-24b	$N^+=0$
$\cot\left[\dfrac{\pi+(\phi+\phi')}{2n}\right]$	$\phi=(2n-1)\pi-\phi'$ or $\phi+\phi'=(2n-1)\pi$ RSB of Case B Figure 13-24b	$N^+=1$

Whereas in the diffraction coefficients of UTD one of the cotangent functions becomes singular along the incident or reflection shadow boundary, while the other three cotangent functions are bounded, the product of the cotangent function along with its corresponding Fresnel transition function along that shadow boundary is discontinuous but bounded. It is this finite discontinuity created by the singular cotangent term and its corresponding Fresnel transition function that removes the bounded geometrical optics discontinuity along that boundary.

To demonstrate, let us consider one of the four shadow boundaries created in Figure 13-24. We choose the ISB ($\phi=\pi+\phi'$ or $\phi-\phi'=\pi$) of Figure 13-24a where $\phi'\leq(n-1)\pi$. Similar results are found for the other three choices. At the ISB of Figure 13-24a

$$\xi^-=\phi-\phi'=\pi \tag{13-79}$$

and in the neighborhood of it

$$\xi^-=\phi-\phi'=\pi-\varepsilon \tag{13-79a}$$

where ε is positive on the illuminated side of the incident shadow boundary. Using (13-79a), we can write (13-66f) as

$$2n\pi N^- -(\phi-\phi')=2n\pi N^- -(\pi-\varepsilon)=(2nN^- -1)\pi+\varepsilon=-\pi \tag{13-80}$$

For this situation, the cotangent function that becomes singular is that shown in the first row of Table 13-1 whose N^- value is 0. Therefore, that cotangent function near the ISB of Figure 13-24a can be written using (13-80) with $N^-=0$ (or $\phi-\phi'=\pi-\varepsilon$) as

$$C^-(\phi-\phi',n)=\cot\left[\frac{\pi-(\phi-\phi')}{2n}\right]$$

$$=\cot\left[\frac{\pi-\pi+\varepsilon}{2n}\right]=\cot\left(\frac{\varepsilon}{2n}\right)\simeq\frac{2n}{\varepsilon}=\frac{2n}{|\varepsilon|\,\text{sgn}(\varepsilon)} \tag{13-80a}$$

where sgn is the sign function. According to (13-66d),

$$g^-(\xi^-)=g^-(\phi-\phi')=1+\cos[(\phi-\phi')-2\pi nN^-]=1+\cos(\phi-\phi')$$

$$g^-(\phi-\phi')=1+\cos(\pi-\varepsilon)=1-\cos(\varepsilon)\overset{\varepsilon\to 0}{\simeq}1-\left(1-\frac{\varepsilon^2}{2}\right)=\frac{\varepsilon^2}{2} \tag{13-80b}$$

The transition function of (13-68f) can also be written using (13-80b) as

$$F[\beta\rho g^-(\xi^-)] = F[\beta\rho g^-(\phi-\phi')] = F\left[\beta\rho\left(\frac{\varepsilon^2}{2}\right)\right] \tag{13-80c}$$

which for small values of its argument can be approximated by the first term of its small-argument asymptotic form (13-74a). That is,

$$F\left(\frac{\beta\rho\varepsilon^2}{2}\right) \simeq \sqrt{\frac{\pi\beta\rho\varepsilon^2}{2}}\, e^{j\pi/4} = |\varepsilon|\sqrt{\frac{\pi\beta\rho}{2}}\, e^{j\pi/4} \tag{13-81}$$

Thus, the product of $C^-(\phi-\phi',n)F[\beta\rho g^-(\phi-\phi')]$, as each is given by (13-80a) and (13-81), can be approximated by

$$\cot\left[\frac{\pi-(\phi-\phi')}{2n}\right]F[\beta\rho g^-(\phi-\phi')] = n\sqrt{2\pi\beta\rho}\,\mathrm{sgn}(\varepsilon)e^{j\pi/4} \tag{13-82}$$

It is apparent that (13-82) exhibits a finite discontinuity that is positive along the illuminated side of the incident shadow boundary and negative on the other side.

The corresponding incident diffracted field (13-68) along the incident shadow boundary can be approximated, using (13-82) and only the second term within the brackets in (13-68b), as

$$V_B^i(\rho, \phi-\phi'=\pi-\varepsilon, n) \simeq \frac{e^{-j\beta\rho}}{\sqrt{\rho}}\left[-\frac{e^{-j\pi/4}}{2n\sqrt{2\pi\beta}}n\sqrt{2\pi\beta\rho}\,\mathrm{sgn}(\varepsilon)e^{j\pi/4}\right] = -\frac{e^{-j\beta\rho}}{2}\mathrm{sgn}(\varepsilon) \tag{13-83}$$

Apart from the phase factor, this function is equal to -0.5, on the illuminated side of the incident shadow boundary and $+0.5$ on the other side. Clearly, such a bounded discontinuity possesses the proper magnitude and polarity to compensate for the discontinuity created by the geometrical optics field. A similar procedure can be used to demonstrate the discontinuous nature of the diffracted field along the other shadow boundaries of Figure 13-24.

To illustrate the principles of geometrical optics (GO) and geometrical theory of diffraction (GTD), an example will be considered next.

Example 13-3

A plane wave of unity amplitude is incident upon a half-plane ($n = 2$) at an incidence angle of $\phi' = 30°$, as shown in Figure 13-20. At a distance of one wavelength ($\rho = \lambda$) from the edge of the wedge, compute and plot the amplitude and phase of the following:

1. Total (incident plus reflected) geometrical optics field.
2. Incident diffracted field.
3. Reflected diffracted field.
4. Total field (geometrical optics plus diffracted).

Do these for both *soft* and *hard* polarizations.

Solution: The geometrical optics field components are computed using (13-65); incident and reflected diffracted fields are computed using (13-68) through (13-69f). These are plotted in Figure 13-25a and 13-25b for *soft* and *hard* polarizations, respectively.

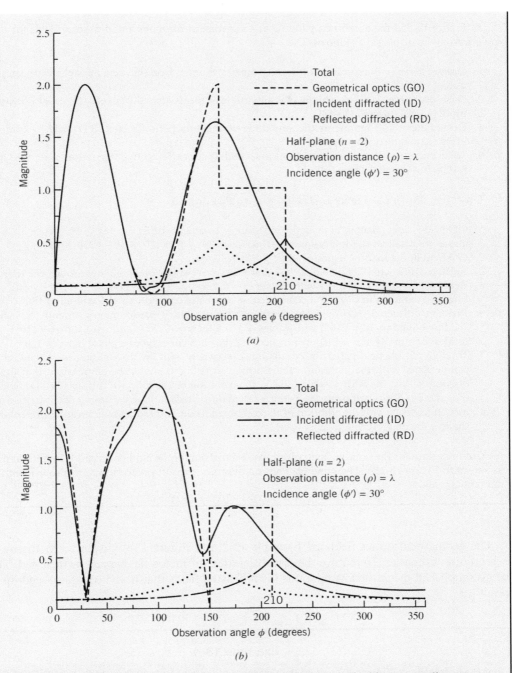

Figure 13-25 Field distribution of various components of a plane wave incident normally on a conducting half-plane. (*a*) Soft polarization. (*b*) Hard polarization. (Source: C. A. Balanis, *Antenna Theory: Analysis and Design*, copyright © 1982, John Wiley & Sons, Inc. Reprinted by permission of John Wiley & Sons, Inc.)

In Figure 13-25a the amplitude patterns of the geometrical optics and diffracted fields for the soft polarization are displayed as follows. The

1. Dashed curve (----) represents the total geometrical optics field (incident and reflected) computed using (13-65).
2. Dash-dot curve (–·–·–) represents the amplitude of the incident diffracted (ID) field V_B^i computed using (13-68) through (13-68f).
3. Dotted curve (····) represents the amplitude of the reflected diffracted (RD) field V_B^r computed using (13-69) through (13-69f).
4. Solid curve (—) represents the total amplitude pattern for soft polarization computed using results from parts 1 through 3.

Observing the data of Figure 13-25a it is evident that the

1. GO field is discontinuous at the reflection shadow boundary (RSB) ($\phi = 180° - \phi' = 180° - 30° = 150°$) and at the incident shadow boundary (ISB) ($\phi = 180° + \phi' = 180° + 30° = 210°$).
2. GO field in the shadow region ($210° < \phi < 360°$) is zero.
3. Reflected diffracted (RD) field, although it exists everywhere, predominates around the reflection shadow boundary ($\phi = 150°$) with values of -0.5 for $\phi = (150°)^-$ and $+0.5$ for $\phi = (150°)^+$. The total discontinuity at $\phi = 150°$ occurs because the phase undergoes a phase jump of $180°$.
4. Incident diffracted (ID) field also exists everywhere but it predominates around the incident shadow boundary ($\phi = 210°$) with values of -0.5 for $\phi = (210°)^-$ and $+0.5$ for $\phi = (210°)^+$. The total discontinuity at $\phi = 210°$ occurs because the phase undergoes a phase jump of $180°$.
5. Total amplitude field pattern is continuous everywhere with the discontinuities of the GO field compensated with the inclusion of the diffracted fields. It should be emphasized that the GO discontinuity at the RSB was removed by the inclusion of the reflected diffracted (RD) field and that at the ISB was compensated by the incident diffracted (ID) field. The GO field was also modified in all space with the addition of the diffracted fields, and radiation intensity is present in the shadow region ($210° < \phi < 360°$).

Computations for the same geometry were also carried out for the hard polarization and the amplitude is shown in Figure 13-25b. The same phenomena observed for soft polarization are also evident for the hard polarization.

The geometrical optics fields of Example 13-2 and Figure 13-9, displayed in Figure 13-10, exhibit discontinuities. To remove the discontinuities, diffracted fields must be included. This can be accomplished using the formulations for diffracted fields that have been developed up to this point.

Example 13-4

For the geometry of Figure 13-9, repeat the formulations of Example 13-2, including the fields diffracted from the edges of the strip.

Solution: According to the solution of Example 13-2, the normalized incident (direct) and reflected fields of the line source above an infinite-width strip are given, respectively, by

$$E_z^i = E_0 \frac{e^{-j\beta\rho_i}}{\sqrt{\rho_i}}$$

and

$$E_z^r = -E_0 \frac{e^{-j\beta\rho_r}}{\sqrt{\rho_r}}$$

where ρ_i and ρ_r are, respectively, the distances from the source and image (caustic) to the observation point, as shown in Figure 13-26a.

To take into account the finite width of the strip, we assume that the far-zone geometrical optics field components (direct and reflected) are the same as for the infinite-width strip and the field intensity at the edges of the strip is the same as for the infinite-width strip.

These assumptions, which become more valid for larger width strips, allow us to determine the diffraction contributions from each of the edges. Because of the geometrical symmetry, we can separate the space surrounding the strip only into four regions, as shown in Figure 13-26a. The angular bounds and the components that contribute to each are as follows:

Region	Angular space	Components
I	$\alpha \leq \phi \leq \pi - \alpha$	Direct, reflected, diffracted (1 and 2)
II	$2\pi - \alpha \leq \phi \leq 2\pi, 0 \leq \phi \leq \alpha$	Direct, diffracted (1 and 2)
III	$\pi + \alpha \leq \phi \leq 2\pi - \alpha$	Diffracted (1 and 2)
IV	$\pi - \alpha \leq \phi \leq \pi + \alpha$	Direct, diffracted (1 and 2)

Because of symmetry, we need only consider half of the total space for computations.

To determine the first-order diffractions from each of the edges, we also assume that each forms a wedge (in this case a half space) that initially is isolated from the other. This allows us to use for each the diffraction properties of the canonical problem (wedge) discussed in the previous section. Thus, the field diffracted from wedge 1 is equal to the product of:

1. The direct (incident) field E_z evaluated at the point of diffraction.
2. The diffraction coefficient as given by (13-70c).
3. The spatial attenuation factor as given by (13-35b).
4. The phase factor as given by (13-34a).

In equation form, it is similar to (13-34a), and it is written as

$$E_{z1}^d(\rho_1,\phi) = E_z^i(\rho_d = s')D_s(s',\psi_1,\alpha,2)A_1(\rho_1)e^{-j\beta\rho_1}$$

where

$$E_z^i(\rho_d,s') = E_0 \frac{e^{-j\beta s'}}{\sqrt{s'}}$$

$$D_s(s',\psi_1,\alpha,2) = D^i(s',\psi_1 - \alpha,2) - D^r(s',\psi_1 + \alpha,2)$$

$$A_1(\rho_1) = \frac{1}{\sqrt{\rho_1}}$$

Using the preceding equations, we can write that

$$E_{z1}^d(\rho_1,\phi) = E_0 \left\{ \frac{e^{-j\beta s'}}{\sqrt{s'}}[D^i(s',\psi_1 - \alpha,2) - D^r(s',\psi_1 + \alpha,2)] \right\} \frac{e^{-j\beta\rho_1}}{\sqrt{\rho_1}}$$

$$E_{z1}^d(\rho_1,\phi) = E_0 [V_B^i(s',\psi_1 - \alpha,2) - V_B^r(s',\psi_1 + \alpha,2)] \frac{e^{-j\beta\rho_1}}{\sqrt{\rho_1}}$$

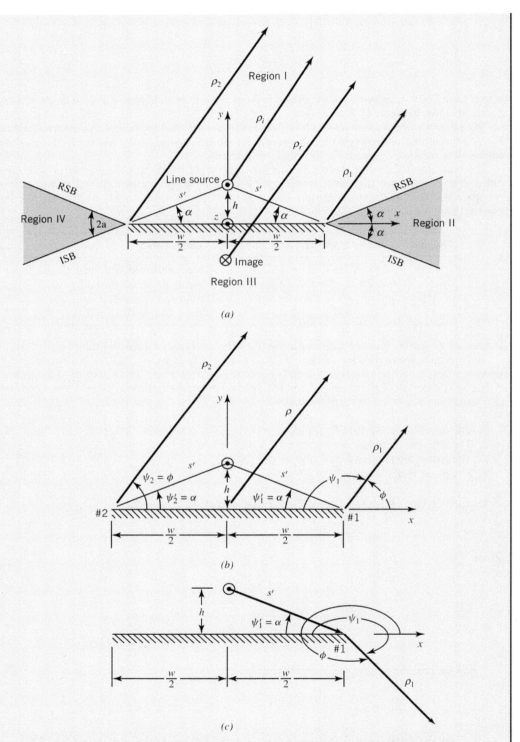

Figure 13-26 Line source above a finite-width strip. (*a*) Region separation. (*b*) Diffraction by edges 1 and 2. (*c*) Diffraction by edge 1 in region III.

where V_B^i and V_B^r are the diffraction functions of (13-68) and (13-69). According to the geometry of Figures 13-26*b* and 13-26*c*

$$\psi_1 = \begin{cases} \pi - \phi & 0 \le \phi \le \pi \text{ (Figure 13-26}b) \\ 3\pi - \phi & \pi \le \phi \le 2\pi \text{ (Figure 13-26}c) \end{cases}$$

In a similar manner, it can be shown that the field diffracted by wedge 2 is given by

$$E_{z2}^d = E_0[V_B^i(s', \psi_2 - \alpha, 2) - V_B^r(s', \psi_2 + \alpha, 2)]\frac{e^{-j\beta\rho_2}}{\sqrt{\rho_2}}$$

$$\psi_2 = \phi \quad \text{for } 0 < \phi < 2\pi \text{ (Figure 13-26}b\text{)}$$

For far-field observations

$$\left.\begin{array}{l} \rho_i \simeq \rho - h\sin\phi \\ \rho_r \simeq \rho + h\sin\phi \\ \rho_1 \simeq \rho - \dfrac{w}{2}\cos\phi \\ \rho_2 \simeq \rho + \dfrac{w}{2}\cos\phi \end{array}\right\} \text{ for phase variations}$$

$$\rho_i \simeq \rho_r \simeq \rho_1 \simeq \rho_2 \simeq \rho \quad \text{for amplitude variations}$$

which allow the fields to be written as

$$E_z^i = E_0 e^{+j\beta h\sin\phi} \quad 0 < \phi < \pi + \alpha, \quad 2\pi - \alpha < \phi < 2\pi \qquad \text{(direct)}$$

$$E_z^r = -E_0 e^{-j\beta h\sin\phi} \quad \alpha < \phi < \pi - \alpha \qquad \text{(reflected)}$$

$$E_{z1}^d = E_0\,[V_B^i(s', \psi_1 - \alpha, 2) - V_B^r(s', \psi_1 + \alpha, 2)]e^{+j(\beta w/2)\cos\phi} \qquad \text{(diffracted from wedge \# 1)}$$

$$\psi_1 = \begin{cases} \pi - \phi & 0 < \phi < \pi \\ 3\pi - \phi & \pi < \phi < 2\pi \end{cases}$$

$$E_{z2}^d = E_0\,[V_B^i(s', \psi_2 - \alpha, 2) - V_B^r(s', \psi_2 + \alpha, 2)]\,e^{-j(\beta w/2)\cos\phi} \qquad \text{(diffracted from wedge \#2)}$$

$$\psi_2 = \phi \qquad 0 < \phi < 2\pi$$

where the $e^{-j\beta\rho}/\sqrt{\rho}$ factor has been suppressed.

It should be stated that the preceding equations represent only first-order diffractions, which usually provide sufficient accuracy for many high-frequency applications. Multiple diffractions between the edges occur and should be included when the strip is electrically small and when more accurate results are required.

For a strip of width $w = 2\lambda$ and with the source at a height of $h = 0.5\lambda$, the normalized pattern computed using these equations is shown in Figure 13-27, where it is compared with GO patterns for infinite- and finite-width strips. It is evident that there is a significant difference between the three, especially in the lower hemisphere, where for the most part the GO pattern exhibits no radiation, and in the regions where there are discontinuities in the GO pattern.

In addition to antenna pattern prediction, diffraction techniques are extremely well suited for scattering problems. To demonstrate the applicability and versatility of diffraction techniques to scattering, let us consider such an example.

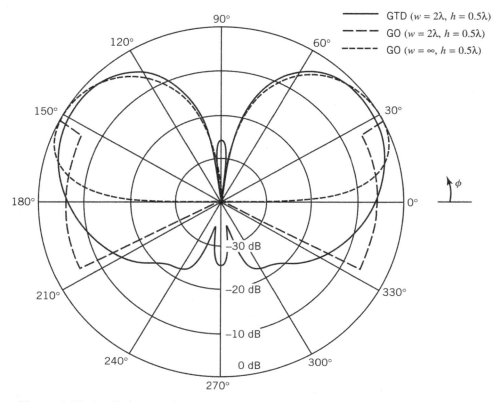

Figure 13-27   Radiation amplitude pattern of electric line source above a finite-width strip.

Example 13-5

A soft polarized uniform plane wave, whose electric field amplitude is E_0, is incident upon a two-dimensional electrically conducting strip of width w, as shown in Figure 13-28a.

1. Determine the backscattered ($\phi = \phi'$) electric field and its backscattered scattering width (SW).
2. Compute and plot the normalized SW ($\sigma_{2\text{-D}} / \lambda$) in dB when $w = 2\lambda$ and the SW ($\sigma_{2\text{-D}}$) in dB/m (dBm) when $w = 2\lambda$ and $f = 10$ GHz.

Solution: For a soft polarized field, the incident electric field can be written, according to the geometry of Figure 13-28a, as

$$\mathbf{E}^i = \hat{\mathbf{a}}_z E_0 e^{-j\boldsymbol{\beta}^i \cdot \mathbf{r}} = \hat{\mathbf{a}}_z E_0 e^{j\beta(x \cos\phi' + y \sin\phi')}$$

The backscattered field diffracted from wedge 1 can be written, by referring to the geometry of Figure 13-28b, as

$$\mathbf{E}_1^d = \mathbf{E}^i(Q_1) \cdot \bar{D}_1^s A_1(\rho_1) e^{-j\beta\rho_1}$$

where

$$E^i(Q_1) = E^i \bigg|_{\substack{x = w/2 \\ y = 0 \\ \phi = \phi'}} = \hat{\mathbf{a}}_z E_0 e^{j(\beta w/2)\cos\phi}$$

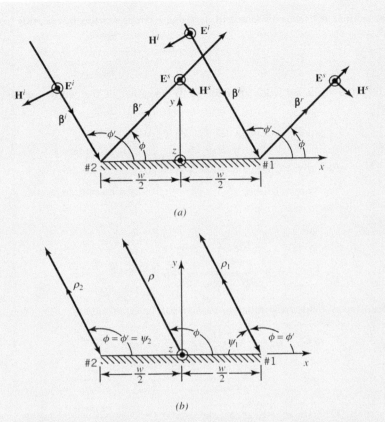

Figure 13-28   Wave diffraction by a finite-width strip. (*a*) Plane wave incidence. (*b*) Plane wave diffraction.

$$\bar{D}_1^s = \hat{a}_z\hat{a}_z \frac{e^{-j\pi/4}\sin\left(\frac{\pi}{n}\right)}{n\sqrt{2\pi\beta}} \left[\frac{1}{\cos\left(\frac{\pi}{n}\right) - \cos\left(\frac{\psi_1 - \psi_1'}{n}\right)} \right.$$

$$\left. - \frac{1}{\cos\left(\frac{\pi}{n}\right) - \cos\left(\frac{\psi_1 + \psi_1'}{n}\right)} \right]_{\substack{n=2 \\ \psi_1 = \psi_1' = \pi - \phi}}$$

$$= -\hat{a}_z\hat{a}_z \frac{e^{-j\pi/4}}{2\sqrt{2\pi\beta}}\left(1 + \frac{1}{\cos\phi}\right)$$

$$A_1(\rho_1) = \frac{1}{\sqrt{\rho_1}}$$

Keller's diffraction form has been used because, at very large distances (ideally infinity), the UTD formulations reduce to those of Keller. Thus, the backscattered field diffracted from wedge 1 reduces to

$$\mathbf{E}_1^d = -\hat{a}_z E_0 \frac{e^{-j\pi/4}e^{j(\beta w/2)\cos\phi}}{2\sqrt{2\pi\beta}}\left(1 + \frac{1}{\cos\phi}\right)\frac{e^{-j\beta\rho_1}}{\sqrt{\rho_1}}$$

In a similar manner, the fields diffracted from wedge 2 can be written, by referring to the geometry of Figure 13-28b, as

$$\mathbf{E}_2^d = -\hat{\mathbf{a}}_z E_0 \frac{e^{-j\pi/4} e^{-j(\beta w/2)\cos\phi}}{2\sqrt{2\pi\beta}} \left(1 - \frac{1}{\cos\phi}\right) \frac{e^{-j\beta\rho_2}}{\sqrt{\rho_2}}$$

When both of the diffracted fields are referred to the center of the coordinate system, they can be written using

$$\left.\begin{array}{l} \rho_1 \simeq \rho + \dfrac{w}{2}\cos(\pi - \phi) = \rho - \dfrac{w}{2}\cos(\phi) \\[2mm] \rho_2 \simeq \rho - \dfrac{w}{2}\cos(\pi - \phi) = \rho + \dfrac{w}{2}\cos(\phi) \end{array}\right\} \quad \text{for phase terms}$$

$$\rho_1 \simeq \rho_2 \simeq \rho \qquad\qquad\qquad \text{for amplitude terms}$$

as

$$\mathbf{E}_1^d = -\hat{\mathbf{a}}_z E_0 \frac{e^{-j\pi/4}}{2\sqrt{2\pi\beta}} \left(1 + \frac{1}{\cos\phi}\right) e^{j\beta w \cos\phi} \frac{e^{-j\beta\rho}}{\sqrt{\rho}}$$

$$\mathbf{E}_2^d = -\hat{\mathbf{a}}_z E_0 \frac{e^{-j\pi/4}}{2\sqrt{2\pi\beta}} \left(1 - \frac{1}{\cos\phi}\right) e^{-j\beta w \cos\phi} \frac{e^{-j\beta\rho}}{\sqrt{\rho}}$$

When the two diffracted fields are combined, the sum can be expressed as

$$\mathbf{E}^d = \mathbf{E}_1^d + \mathbf{E}_2^d = -\hat{\mathbf{a}}_z E_0 \frac{e^{-j\pi/4}}{2\sqrt{2\pi\beta}} [(e^{j\beta w \cos\phi} + e^{-j\beta w \cos\phi})$$

$$+ \frac{1}{\cos\phi}(e^{j\beta w \cos\phi} - e^{-j\beta w \cos\phi})] \frac{e^{-j\beta\rho}}{\sqrt{\rho}}$$

$$\mathbf{E}^d = -\hat{\mathbf{a}}_z E_0 \frac{e^{-j\pi/4}}{\sqrt{2\pi\beta}} \left[\cos(\beta w \cos\phi) + j\beta w \frac{\sin(\beta w \cos\phi)}{(\beta w \cos\phi)}\right] \frac{e^{-j\beta\rho}}{\sqrt{\rho}}$$

Since there are no geometrical optics fields in the backscattered direction (Snell's law is not satisfied) when $\phi = \phi' \neq \pi/2$, the total diffracted field also represents the total field. In the limit as $\phi = \phi' = \pi/2$, each diffracted field exhibits a singularity; however, the total diffracted field is finite because the singularity of one diffracted field compensates for the singularity of the other. This is always evident at normal incidence as long as the edges of the two diffracted wedges are parallel to each other, even though the included angles of the two wedges are not necessarily the same [39]. In addition, the limiting value of the total diffracted field at normal incidence reduces and represents also the geometrical optics scattered (reflected) field.

The two-dimensional backscattered scattering width $\sigma_{2\text{-D}}$ of (11-21b) can now be written as

$$\sigma_{2\text{-D}} = \lim_{\rho \to \infty} \left[2\pi\rho \frac{|\mathbf{E}^s|^2}{|\mathbf{E}^i|^2}\right] = \frac{\lambda}{2\pi} \left|\cos(\beta w \cos\phi) + j\beta w \frac{\sin(\beta w \cos\phi)}{\beta w \cos\phi}\right|^2$$

The limiting value, as $\phi \to \pi/2$, reduces to

$$\sigma_{2\text{-D}}|_{\phi=\pi/2} = \frac{\lambda}{2\pi}|1 + j\beta w|^2 = \frac{\lambda}{2\pi}\left[1 + (\beta w)^2\right] \overset{\beta w \gg 1}{\simeq} \beta w^2$$

which agrees with the physical optics expression. Computed results for $\sigma_{2\text{-D}}/\lambda$ (in decibels) and $\sigma_{2\text{-D}}$ (in decibels per meter or dBm) at $f = 10$ GHz when $w = 2\lambda$ are shown in Figure 13-29.

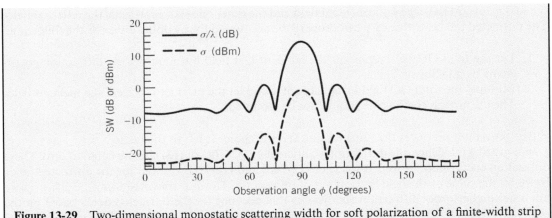

Figure 13-29   Two-dimensional monostatic scattering width for soft polarization of a finite-width strip of $w = 2\lambda$ at $f = 10\,\text{GHz}$.

Before proceeding to discuss other topics in diffraction, such as oblique incidence, curved edge diffraction, equivalent currents, slope diffraction, and multiple diffraction, let us complete our two-dimensional diffraction by addressing some modifications and extensions to the concepts covered in this section.

For a plane wave incidence, ρ in (13-68) through (13-68f) and in (13-69) through (13-69f) represents the distance from the edge of the wedge to the observation point. According to the principle of reciprocity illustrated in Figure 13-17, ρ in (13-68) through (13-69f) must be replaced by ρ' to represent the diffraction of a cylindrical wave whose source is located a distance ρ' from the edge of the wedge and the observations made in the far zone (ideally at infinity). If both the source and observation point are located at finite distances from the edge of the wedge, represented, respectively, by ρ' and ρ, then a better estimate of the distance would be to introduce a so-called *distance parameter* L, which in this case takes the form of

$$L = \frac{\rho\rho'}{\rho+\rho'}\begin{cases} \overset{\rho'\to\infty}{\simeq} \rho \\ \underset{\rho\to\infty}{\simeq} \rho' \end{cases} \tag{13-84}$$

Thus, the incident and reflected diffracted fields and coefficients of (13-68) and (13-69) can be written as

$$V_B^i(L,\phi-\phi',n) = V_B^i(L,\xi^-,n) = \frac{e^{-j\beta\rho}}{\sqrt{\rho}}D^i(L,\xi^-,n) \tag{13-84a}$$

$$V_B^r(L,\phi+\phi',n) = V_B^r(L,\xi^+,n) = \frac{e^{-j\beta\rho}}{\sqrt{\rho}}D^r(L,\xi^+,n) \tag{13-84b}$$

It is observed in (13-68) and (13-69) that for grazing angle incidence, $\phi' = 0$ or $\phi' = n\pi$ of Figure 13-24 (where n represents the wedge angle), then $\xi^- = \xi^+ = \phi - \phi' = \phi - \phi'$ and $V_B^i = V_B^r$, $D^i = D^r$. Therefore, here D_s of (13-70c) or (13-72e) is equal to zero ($D_s = 0$) and D_h of (13-70d) or (13-72f) is equal to twice D^i or twice D^r ($D_h = 2D^i = 2D^r$). Since grazing is a limiting situation, the incident and reflected fields combine to make the total geometrical optics field effectively incident at the observation point. Therefore, one-half of the total field propagating along the face of the wedge toward the edge is the incident field and the other one-half represents the reflected field. The diffracted fields for this case can properly be accounted for by doing *either* of the following:

1. Let the total GO field represent the incident GO field but multiply the diffraction coefficients by a factor of $\frac{1}{2}$.
2. Multiply the total GO field by a factor of $\frac{1}{2}$ and let the product represent the incident field. The diffraction coefficients should not be modified.

Either procedure produces the same results, and the choice is left to the reader.

For grazing incidence, the diffraction coefficient of (13-70c) or (13-72e) are equal to zero. These diffraction coefficients, as well as those of (13-70d) and (13-72f), account for the diffracted fields based on the value of the field at the point of diffraction. This is formulated using (13-34a). There are other higher-order diffraction coefficients that account for the diffracted fields based on the *rate of change (slope)* of the field at the point of diffraction. These diffraction coefficients are referred to as the *slope diffraction coefficients* [52], and they yield nonzero (even though small) fields for soft polarization at grazing incidence. The slope diffraction coefficients exist also for hard polarization, but they are not as dominant as they are for soft polarization.

Up to now, we have restricted our attention to exterior wedge $(1 < n \leq 2)$ diffraction. However, the theory of diffraction can be applied also to interior wedge diffraction $(0 \leq n \leq 1)$. When $n = 1$, the wedge reduces to an infinite flat plate and the diffraction coefficients reduce to zero [as seen better by examining (13-72e) and (13-72f)] since $\sin(\pi/n) = 0$. The incident and reflected fields for $n = 1$ (half-plane), $n = \frac{1}{2}$ (90° interior wedge), and $n = 1/M, M = 3, 4, \ldots$ (acute interior wedges) can be found exactly by image theory. In each of these, the number of finite images is determined by the included angle of the interior wedge [38]. As $n \to 0$ the geometrical optics (incident and reflected) fields become more dominant compared to the nonvanishing diffracted fields.

13.3.3 Straight Edge Diffraction: Oblique Incidence

The normal incidence and diffraction formulations of the previous section are convenient to analyze radiation characteristics of antennas and structures primarily in principal planes. However, a complete analysis of an antenna or scatterer requires examination not only in principal planes but also in nonprincipal planes, as shown in Figure 13-30 for an aperture and a horn antenna each mounted on a finite-size ground plane.

Whereas the diffraction of a normally incident wave discussed in the previous section led to scalar diffraction coefficients, the diffraction of an obliquely incident wave by a two-dimensional wedge can be derived using the geometry of Figure 13-31. To accomplish this, it is most convenient to define ray-fixed coordinate systems (s', β_0', ϕ') for the source and (s, β_0, ϕ) for the observation point [8, 10], in contrast to the edge-fixed coordinate system $(\rho', \phi', z'; \rho, \phi, z)$. By doing this, it can be shown that the diffracted field, in a general form, can be written as

$$\mathbf{E}^d(s) = \mathbf{E}^i(Q_D) \cdot \bar{\mathbf{D}}(L; \phi, \phi'; n; \beta_0') \sqrt{\frac{s'}{s(s'+s)}} \, e^{-j\beta s} \qquad (13\text{-}85)$$

where $\bar{\mathbf{D}}(L; \phi, \phi'; n; \beta_0')$ is the dyadic edge diffraction coefficient for illumination of the wedge by plane, cylindrical, conical, or spherical waves.

Introducing an edge-fixed plane of incidence with the unit vectors $\hat{\beta}_0'$ and $\hat{\phi}'$ parallel and perpendicular to it, and a plane of diffraction with the unit vectors $\hat{\beta}_0'$ and $\hat{\phi}$ parallel

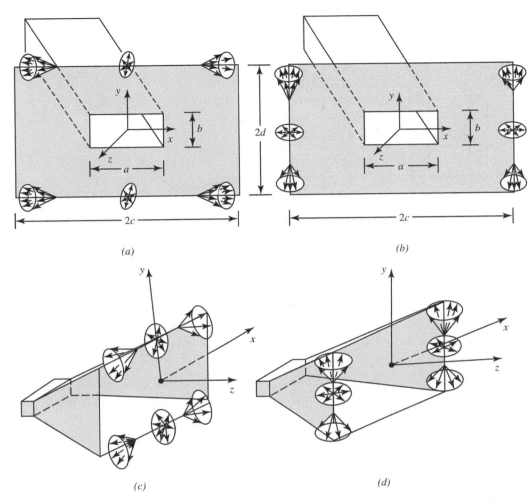

Figure 13-30 *E*- and *H*-plane diffraction by rectangular waveguide and pyramidal horn. Waveguide: (*a*) *E*-plane diffraction. (*b*) *H*-plane diffraction. Horn: (*c*) *E*-plane diffraction. (*d*) *H*-plane diffraction.

and perpendicular to it, we can write the radial unit vectors of incidence and diffraction, respectively, as

$$\hat{\mathbf{s}}' = \hat{\boldsymbol{\phi}}' \times \hat{\boldsymbol{\beta}}_0' \tag{13-86a}$$

$$\hat{\mathbf{s}} = \hat{\boldsymbol{\phi}} \times \hat{\boldsymbol{\beta}}_0 \tag{13-86b}$$

where $\hat{\mathbf{s}}'$ points toward the point of diffraction. With the adoption of the ray-fixed coordinate systems, the dyadic diffraction coefficient can be represented by

$$\bar{\mathbf{D}}(L; \phi, \phi'; n; \beta_0') = -\hat{\boldsymbol{\beta}}_0'\hat{\boldsymbol{\beta}}_0 D_s(L; \phi, \phi'; n; \beta_0') - \hat{\boldsymbol{\phi}}'\hat{\boldsymbol{\phi}} D_h(L; \phi, \phi'; n; \beta_0') \tag{13-87}$$

where D_s and D_h are, respectively, the scalar diffraction coefficients for soft and hard polarizations. If an edge-fixed coordinate system were adopted, the dyadic coefficient would be the sum of seven dyads that in matrix notation would be represented by a 3×3 matrix with seven nonvanishing elements instead of the 2×2 matrix with two nonvanishing elements.

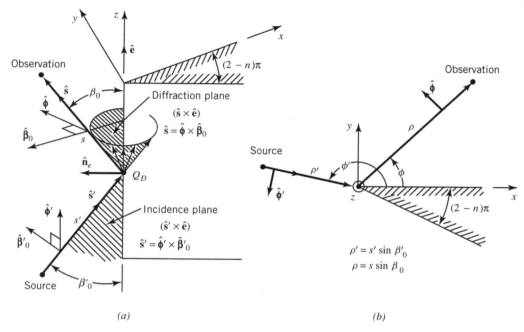

Figure 13-31   Oblique incidence wedge diffraction. (*a*) Oblique incidence. (*b*) Top view.

For the diffraction shown in Figure 13-31, we can write in matrix form the diffracted E-field components that are parallel ($E_{\beta_0}^d$) and perpendicular (E_ϕ^d) to the plane of diffraction as

$$\begin{bmatrix} E_{\beta_0}^d(s) \\ E_\phi^d(s) \end{bmatrix} = -\begin{bmatrix} D_s & 0 \\ 0 & D_h \end{bmatrix}\begin{bmatrix} E_{\beta_0'}^i(Q_D) \\ E_{\phi'}^i(Q_D) \end{bmatrix} A(s',s)e^{-j\beta s} \qquad (13\text{-}88)$$

where

$$\begin{aligned} E_{\beta_0'}^i(Q_D) &= \hat{\beta}_0' \cdot \mathbf{E}^i \\ &= \text{component of the incident } \mathbf{E} \text{ field parallel to the} \\ &\quad \text{plane of incidence at the point of diffraction } Q_D \end{aligned} \qquad (13\text{-}88a)$$

$$\begin{aligned} E_{\phi'}^i(Q_D) &= \hat{\phi}_0' \cdot \mathbf{E}^i \\ &= \text{component of the incident } \mathbf{E} \text{ field perpendicular to the} \\ &\quad \text{plane of incidence at the point of diffraction } Q_D \end{aligned} \qquad (13\text{-}88b)$$

D_s and D_h are the scalar diffraction coefficients that take the form

$$D_s(L;\phi,\phi';n;\beta_0') = D^i(L,\phi-\phi',n,\beta_0') - D^r(L,\phi+\phi',n,\beta_0') \qquad (13\text{-}89a)$$

$$D_h(L;\phi,\phi';n;\beta_0') = D^i(L,\phi-\phi',n,\beta_0') + D^r(L,\phi+\phi',n,\beta_0') \qquad (13\text{-}89b)$$

where

$$
\begin{aligned}
D^i&(L,\phi-\phi',n,\beta_0')\\
&=-\frac{e^{-j\pi/4}}{2n\sqrt{2\pi\beta}\sin\beta_0'}\left\{\cot\left[\frac{\pi+(\phi-\phi')}{2n}\right]F[\beta Lg^+(\phi-\phi')]\right.\\
&\quad\left.+\cot\left[\frac{\pi+(\phi-\phi')}{2n}\right]F[\beta Lg^-(\phi-\phi')]\right\}
\end{aligned}
\tag{13-90a}
$$

$$
\begin{aligned}
D^r&(L,\phi+\phi',n,\beta_0')\\
&=-\frac{e^{-j\pi/4}}{2n\sqrt{2\pi\beta}\sin\beta_0'}\left\{\cot\left[\frac{\pi+(\phi+\phi')}{2n}\right]F[\beta Lg^+(\phi+\phi')]\right.\\
&\quad\left.+\cot\left[\frac{\pi-(\phi+\phi')}{2n}\right]F[\beta Lg^-(\phi+\phi')]\right\}
\end{aligned}
\tag{13-90b}
$$

To facilitate the reader in the computations a Fortran and MATLAB computer subroutine designated **WDC, Wedge Diffraction Coefficient**, computes the normalized (with respect to $\sqrt{\lambda}$) wedge diffraction coefficients based on (13-89a) through (13-90b). The subroutine utilizes the Fresnel transition function program FTF. Both programs were developed and reported in [53].

In general, L is a distance parameter that can be found by satisfying the condition that the total field (the sum of the geometrical optics and the diffracted fields) must be continuous along the incident and reflection shadow boundaries. Doing this, it can be shown that a general form of L is

$$
L=\frac{s(\rho_e^i+s)\rho_1^i\rho_2^i\sin^2\beta_0'}{\rho_e^i(\rho_1^i+s)(\rho_2^i+s)}
\tag{13-91}
$$

where ρ_1^i, ρ_2^i = radii of curvature of the incident wave front at Q_D

ρ_e^i = radius of curvature of the incident wave front in the edge-fixed plane of incidence

For oblique incidence upon a wedge, as shown in Figure 13-31, the distance parameter can be expressed in the ray-fixed coordinate system as

$$
L=\begin{cases}
s\sin^2\beta_0' & \text{plane wave incidence}\\
\dfrac{\rho\rho'}{\rho+\rho'} & \begin{array}{l}\text{cylindrical wave incidence}\\ (\rho=s\sin\beta_0,\rho'=s'\sin\beta_0')\end{array}\\
\dfrac{ss'\sin^2\beta_0'}{s+s'} & \text{conical and spherical wave incidences}
\end{cases}
\tag{13-92}
$$

The spatial attenuation factor $A(s',s)$, which describes how the field intensity varies along the diffracted ray, is given by

$$
A(s',s)=\begin{cases}
\dfrac{1}{\sqrt{s}} & \text{plane and conical wave incidences}\\
\dfrac{1}{\sqrt{\rho}} & \rho=s\sin\beta_0;\ \text{cylindrical wave incidence}\\
\sqrt{\dfrac{s'}{s(s'+s)}} & \text{spherical wave incidence}
\end{cases}
\tag{13-93}
$$

If the observations are made in the far field ($s \gg s'$ or $\rho \gg \rho'$), the distance parameter L and spatial attenuation factor $A(s',s)$ reduce, respectively, to

$$L = \begin{cases} s\,\sin^2\beta_0' & \text{plane wave incidence} \\ \rho' & \text{cylindrical wave incidence} \\ s'\,\sin^2\beta_0' & \text{conical and spherical wave incidences} \end{cases} \tag{13-94}$$

$$A(s',s) = \begin{cases} \dfrac{1}{\sqrt{s}} & \text{plane and conical wave incidences} \\[2mm] \dfrac{1}{\sqrt{\rho}} & \rho = s\sin\beta_0; \text{ cylindrical wave incidence} \\[2mm] \dfrac{\sqrt{s'}}{s} & \text{spherical wave incidence} \end{cases} \tag{13-95}$$

For normal incidence, $\beta_0 = \beta_0' = \pi/2$.

To demonstrate the principles of this section, an example will be considered.

Example 13-6

To determine the far-zone elevation plane pattern, in the principal planes, of a $\lambda/4$ monopole mounted on a finite-size square ground plane of width w on each of its sides, refer to Figure 13-32a. Examine the contributions from all four edges.

Solution: In addition to the direct and reflected field contributions (referred to as geometrical optics, GO), there are diffracted fields from the edges of the ground plane. The radiation mechanisms from the two edges that are perpendicular to the principal plane of observation are illustrated graphically in Figure 13-32b. It is apparent that from these two edges only two points contribute to the radiation in the principal plane. These two points occur at the intersection of the principal plane with the edges.

The incident and reflected fields are obtained by assuming the ground plane is infinite in extent. Using the coordinate system of Figure 13-32a, and the image theory of Section 7.4, the total geometrical optics field of the $\lambda/4$ monopole above the ground plane can be written as [38]

$$E_{\theta G}(r,\theta) = E_0 \left| \frac{\cos\left(\dfrac{\pi}{2}\cos\theta\right)}{\sin\theta} \right| \frac{e^{-j\beta r}}{r}, \quad 0 \le \theta \le \pi/2$$

The field diffracted from wedge 1 can be obtained using the formulation of (13-88) through (13-95). Referring to the geometry of Figure 13-32b, the direct field is incident normally ($\beta_0' = \pi/2$) on the edge of the ground plane along the principal planes, and the diffracted field from wedge 1 can be written as

$$E_{\theta 1}^d(\theta) = +E^i(Q_1) D_h(L, \xi_i^{\pm}, \beta_0' = \pi/2, n=2) A_1(w, r_1) e^{-j\beta r_1}$$

The total field can be assumed to all emanate from the base of the monopole. This is a good approximation whose modeling has agreed well with measurements. Thus,

$$E^i(Q_1) = \frac{1}{2} E_{\theta G}\left(r = \frac{w}{2}, \theta = \frac{\pi}{2}\right) = \frac{E_0}{2} \frac{e^{-j\beta w/2}}{w/2}$$

$$D_h\left(L, \xi_1^{\pm}, \beta_0' = \frac{\pi}{2}, n=2\right) = D^i(L, \xi_1^-, n=2) + D^r(L, \xi_1^+, n=2)$$

Since the incident wave is of spherical waveform and observations are made in the far field, the distance parameter L and spatial attenuation factor $A_1(w, r_1)$ can be expressed, according to (13-94) and (13-95) for $\beta_0' = \pi/2$, as

$$L = s' \sin^2 \beta_0' \Big|_{\substack{s' = w/2 \\ \beta_0' = \pi/2}} = \frac{w}{2}$$

$$A_1(w, r_1) = \frac{\sqrt{s'}}{s} \Big|_{\substack{s' = w/2 \\ s = r_1}} = \frac{\sqrt{w/2}}{r_1}$$

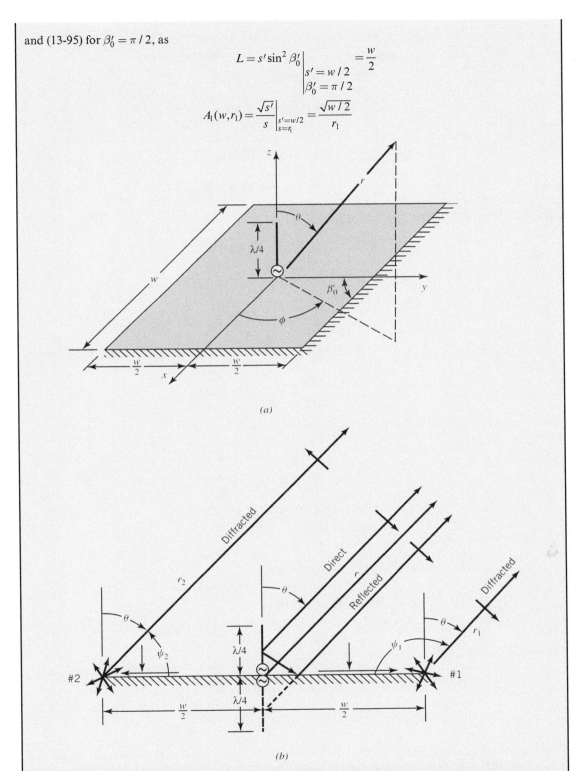

(a)

(b)

Figure 13-32 Vertical monopole on a square ground plane, and reflection and diffraction mechanisms. (a) Monopole on ground plane. (b) Reflection and diffraction mechanisms.

Since the angle of incidence ψ_0 from the main source toward the point of diffraction Q_1 is zero degrees $(\psi_0 = 0)$, then

$$\xi_1^- = \psi_1 - \psi_0 = \psi_1 = \theta + \frac{\pi}{2} = \xi_1$$

$$\xi_1^+ = \psi_1 + \psi_0 = \psi_1 = \theta + \frac{\pi}{2} = \xi_1$$

Therefore,

$$D_h\left(L, \xi_1^\pm, \beta_0' = \frac{\pi}{2}, n = 2\right) = 2D^i\left(\frac{w}{2}, \theta + \frac{\pi}{2}, n = 2\right)$$

$$= 2D^r\left(\frac{w}{2}, \theta + \frac{\pi}{2}, n = 2\right)$$

The total diffracted field can now be written as

$$E_{\theta 1}^d(\theta) = \frac{E_0}{2} \frac{e^{-j\beta w/2}}{w/2} 2D^{i,r}\left(\frac{w}{2}, \theta + \frac{\pi}{2}, n = 2\right)\frac{\sqrt{w/2}}{r_1} e^{-j\beta r_1}$$

$$= E_0\left[\frac{e^{-j\beta w/2}}{\sqrt{w/2}} D^{i,r}\left(\frac{w}{2}, \theta + \frac{\pi}{2}, n = 2\right)\right]\frac{e^{-j\beta r_1}}{r_1}$$

$$E_{\theta 1}^d(\theta) = E_0 V_B^{i,r}\left(\frac{w}{2}, \theta + \frac{\pi}{2}, n = 2\right)\frac{e^{-j\beta r_1}}{r_1}$$

Using a similar procedure, the field diffracted from wedge 2 can be written, by referring to the geometry of Figure 13-32b, as

$$E_{\theta 2}^d(\theta) = -E_0\left[\frac{e^{-j\beta w/2}}{\sqrt{w/2}} D^{i,r}\left(\frac{w}{2}, \xi_2, n = 2\right)\right]\frac{e^{-j\beta r_2}}{r_2}$$

$$E_{\theta 2}^d(\theta) = -E_0 V_B^{i,r}\left(\frac{w}{2}, \xi_2, n = 2\right)\frac{e^{-j\beta r_2}}{r_2}$$

where

$$\xi_2 = \psi_2 = \begin{cases} \dfrac{\pi}{2} - \theta, & 0 \le \theta \le \dfrac{\pi}{2} \\[2mm] \dfrac{5\pi}{2} - \theta, & \dfrac{\pi}{2} < \theta < \pi \end{cases}$$

For far-field observations

$$\left.\begin{array}{l} r_1 \simeq r - \dfrac{w}{2}\cos\left(\dfrac{\pi}{2} - \theta\right) = r - \dfrac{w}{2}\sin\theta \\[3mm] r_2 \simeq r + \dfrac{w}{2}\cos\left(\dfrac{\pi}{2} - \theta\right) = r + \dfrac{w}{2}\sin\theta \end{array}\right\} \quad \text{for phase terms}$$

$$r_1 \simeq r_2 \simeq r \qquad\qquad\qquad \text{for amplitude terms}$$

Therefore, the diffracted fields from wedges 1 and 2 reduce to

$$E_{\theta 1}^d(\theta) = +E_0 V_B^{i,r}\left(\frac{w}{2}, \theta + \frac{\pi}{2}, n = 2\right)e^{j(\beta w/2)\sin\theta}\frac{e^{-j\beta r}}{r}$$

$$E_{\theta 2}^d(\theta) = -E_0 V_B^{i,r}\left(\frac{w}{2}, \xi_2, n = 2\right)e^{-j(\beta w/2)\sin\theta}\frac{e^{-j\beta r}}{r}$$

It should be noted that there are oblique incidence diffractions from the other two edges of the ground plane that are parallel to the principal plane of observation. However, the diffracted field from these edges is primarily cross-polarized (E_ϕ component) to the incident E_θ field and to the E_θ field produced in the principal plane. The cross-polarized E_ϕ components produced by diffractions from these two sides cancel each other out so that in the principal plane there is primarily an E_θ component.

Using the total geometrical optics field and the field diffracted from wedges 1 and 2, a normalized amplitude pattern was computed for a $\lambda/4$ monopole mounted on a square ground plane of width $w = 4$ ft $= 1.22$ m at a frequency of $f = 1$ GHz. This pattern is shown in Figure 13-33 where it is compared with the computed GO (assuming an infinite ground plane) and measured patterns. A very good agreement is seen between the GO + GTD and measured patterns, which are quite different from that of the GO pattern.

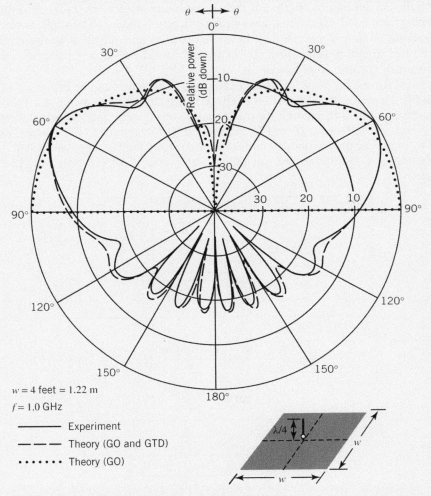

Figure 13-33 Measured and computed principal elevation plane amplitude patterns of a $\lambda/4$ monopole above infinite and finite square ground planes. (Source: C. A. Balanis, *Antenna Theory: Analysis and Design*, Third Edition, copyright © 2005, John Wiley & Sons, Inc. Reprinted by permission of John Wiley & Sons, Inc.)

13.3.4 Curved Edge Diffraction: Oblique Incidence

The edges of many practical antenna or scattering structures are not straight, as demonstrated in Figure 13-34 by the edges of a circular ground plane, a paraboloidal reflector, and a conical horn. In order to account for the diffraction phenomenon from the edges of these structures, even in their principal planes, curved edge diffraction must be utilized.

Curved edge diffraction can be derived by assuming an oblique wave incidence (at an angle β_0') on a curved edge, as shown in Figure 13-35, where the surfaces (sides) forming the curved edge in general may be convex, concave, or plane. Since diffraction is a local phenomenon, the curved edge geometry can be approximated at the point of diffraction Q_D by a wedge whose straight edge is tangent to the curved edge at that point and whose plane surfaces are tangent to the curved surfaces forming the curved edge. This allows wedge diffraction theory to be applied directly to curved edge diffraction by simply representing the curved edge by an equivalent wedge. Analytically, this is accomplished simply by generalizing the expressions for the distance parameter L that appear in the arguments of the transition functions.

The general form of oblique incidence curved edge diffraction can be expressed in matrix form as in (13-88). However, the diffraction coefficients, distance parameters, and spatial spreading factor must be modified to account for the curvature of the edge and its curved surfaces (sides).

The diffraction coefficients D_s and D_h are those of (13-89a) and (13-89b), where D^i and D^r can be found by imposing the continuity conditions on the total field across the incident and reflection shadow boundaries. Doing this, we can show that D^i and D^r of (13-90a) and (13-90b) take the form of [10]

$$
\begin{aligned}
D^i &(L^i, \phi - \phi', n) \\
&= -\frac{e^{-j\pi/4}}{2n\sqrt{2\pi\beta}\,\sin\beta_0'} \left\{ \cot\left[\frac{\pi + (\phi - \phi')}{2n}\right] F[\beta L^i g^+(\phi - \phi')] \right. \\
&\left. + \cot\left[\frac{\pi - (\phi - \phi')}{2n}\right] F[\beta L^i g^-(\phi - \phi')] \right\}
\end{aligned}
\tag{13-96a}
$$

$$
\begin{aligned}
D^r &(L^r, \phi + \phi', n) \\
&= -\frac{e^{-j\pi/4}}{2n\sqrt{2\pi\beta}\,\sin\beta_0'} \left\{ \cot\left[\frac{\pi + (\phi + \phi')}{2n}\right] F[\beta L^{rn} g^+(\phi + \phi')] \right. \\
&\left. + \cot\left[\frac{\pi - (\phi + \phi')}{2n}\right] F[\beta L^{r0} g^-(\phi + \phi')] \right\}
\end{aligned}
\tag{13-96b}
$$

where

$$
L^i = \frac{s(\rho_e^i + s)\rho_1^i \rho_2^i \sin^2\beta_0'}{\rho_e^i(\rho_1^i + s)(\rho_2^i + s)}
\tag{13-97a}
$$

$$
L^{r0,rn} = \frac{s(\rho_e^r + s)\rho_1^r \rho_2^r \sin^2\beta_0'}{\rho_e^r(\rho_1^r + s)(\rho_2^r + s)}
\tag{13-97b}
$$

ρ_1^i, ρ_2^i = radii of curvature of the incident wave front at Q_D

ρ_e^i = radius of curvature of the incident wave front in the edge fixed plane of incidence

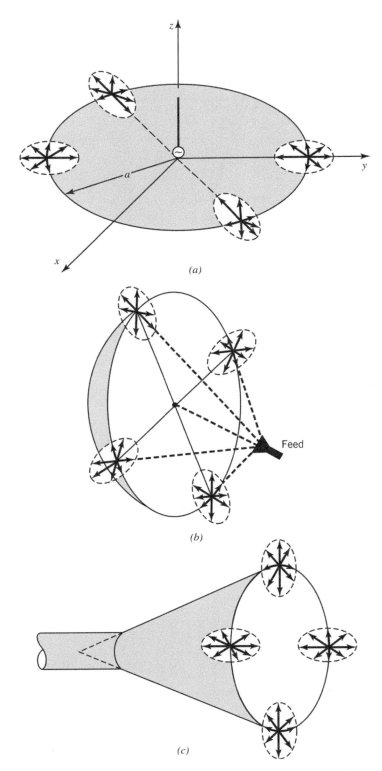

Figure 13-34 Diffraction by curved-edge structures. (*a*) Circular ground plane. (*b*) Paraboloidal reflector. (*c*) Conical horn.

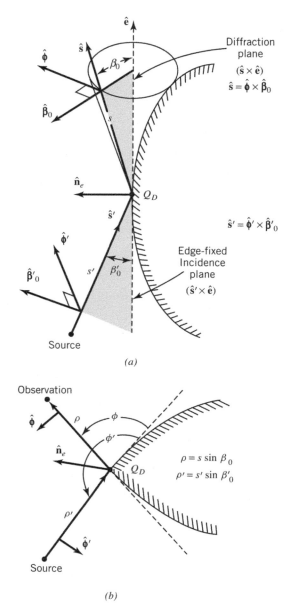

Figure 13-35   Oblique incidence diffraction by a curved edge. (*a*) Oblique incidence. (*b*) Top view.

ρ_1^r, ρ_2^r = principal radii of curvature of the reflected wave front at Q_D [found using (13-21a) and (13-21b)]

ρ_e^r = radius of curvature of the reflected wave front in the plane containing the diffracted ray and edge

The superscripts *ro* and *rn* of L in (13-96b) and (13-97a) denote that the radii of curvature ρ_1^r, ρ_2^r, and ρ_e^r must be calculated for *ro* at the reflection boundary $\pi - \phi'$ of Figure 13-24*a* and for rn at the reflection boundary $(2n-1)\pi - \phi'$ of Figure 13-24b. For far-field observation, where $s \gg \rho_e^i$, ρ_1^i, ρ_2^i, ρ_e^r, ρ_1^r, ρ_2^r (13-97a), (13-97b) simplify to

$$L^i = \frac{\rho_1^i \rho_2^i}{\rho_e^i} \sin^2 \beta_0' \tag{13-98a}$$

$$L^{r0,rn} = \frac{\rho_1^r \rho_2^r}{\rho_e^r} \sin^2 \beta_0' \tag{13-98b}$$

If the intersecting curved surfaces forming the curved edge in Figure 13-35 are plane surfaces that form an ordinary wedge, then the distance parameters in (13-97a) and (13-97b) or (13-98a) and (13-98b) are equal, that is,

$$L^{r0} = L^{rn} = L^i \tag{13-99}$$

Using the geometries of Figure 13-36, it can be shown that the spatial spreading factor $A(\rho_c, s)$ of (13-35) for the curved edge diffraction takes the form

$$A(\rho_c, s) = \sqrt{\frac{\rho_c}{s(\rho_c + s)}} \overset{s \gg \rho_c}{\simeq} \frac{1}{s}\sqrt{\rho_c} \tag{13-100}$$

$$\frac{1}{\rho_c} = \frac{1}{\rho_e} - \frac{\hat{\mathbf{n}}_e \cdot (\hat{\mathbf{s}}' - \hat{\mathbf{s}})}{\rho_g \sin^2 \beta_0'} \tag{13-100a}$$

where ρ_c = distance between caustic at edge and second caustic of diffracted ray
ρ_e = radius of curvature of incidence wave front in the edge-fixed plane of incidence which contains unit vectors $\hat{\mathbf{s}}'$ and $\hat{\mathbf{e}}$ (infinity for plane, cylindrical, and conical waves; $\rho_e = s'$ for spherical waves)
ρ_g = radius of curvature of the edge at the diffraction point
$\hat{\mathbf{n}}_e$ = unit vector normal to the edge at Q_D and directed away from the center of curvature
$\hat{\mathbf{s}}'$ = unit vector in the direction of incidence
$\hat{\mathbf{s}}$ = unit vector in the direction of diffraction
β_0' = angle between $\hat{\mathbf{s}}'$ and tangent to the edge at the point of diffraction
$\hat{\mathbf{e}}$ = unit vector tangent to the edge at the point of diffraction

For normal incidence, $\beta_0' = \pi/2$.

The spatial attenuation factor (13-100) creates additional caustics, other than the ones that occur at the points of diffraction. Each caustic occurs at a distance ρ_c from the one at the diffraction point. Diffracted fields in the regions of the caustics must be corrected to remove the discontinuities and inaccuracies from them.

To demonstrate the principles of curved edge diffraction, let us consider Example 13-7.

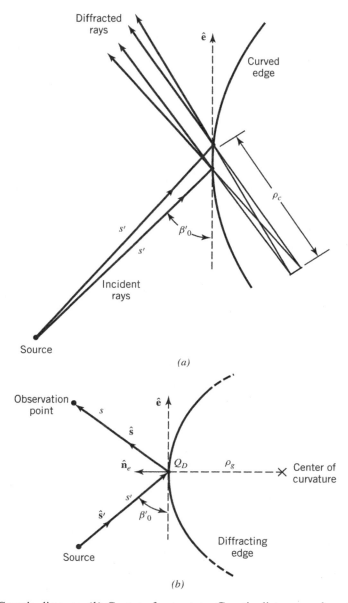

Figure 13-36   (*a*) Caustic distance. (*b*) Center of curvature. Caustic distance and center of curvature for curved-edge diffraction. (Source: C. A. Balanis, *Antenna Theory: Analysis and Design*, copyright © 1982, John Wiley & Sons, Inc. Reprinted by permission of John Wiley & Sons, Inc.)

Example 13-7

Determine the far-zone elevation plane pattern of a $\lambda/4$ monopole mounted on a circular electrically conducting ground plane of radius a, as shown in Figure 13-37*a*.

   Solution: Because of the symmetry of the structure, the diffraction mechanism in any of the elevation planes is the same. Therefore, the principal yz plane is chosen here. For observations made away from the symmetry axis of the ground plane ($\theta \neq 0°$ and $180°$), it can be shown [33] that most of the diffraction radiation from the rim of the ground plane comes from the two diametrically opposite points of the rim that coincide with the observation plane. Therefore, for points removed from the symmetry axis ($\theta \neq 0°$ and $180°$) the overall formulation of this problem, and that of Example 13-6, is identical other than the amplitude spreading factor, which now must be computed using (13-100) and (13-100a) instead of (13-95).

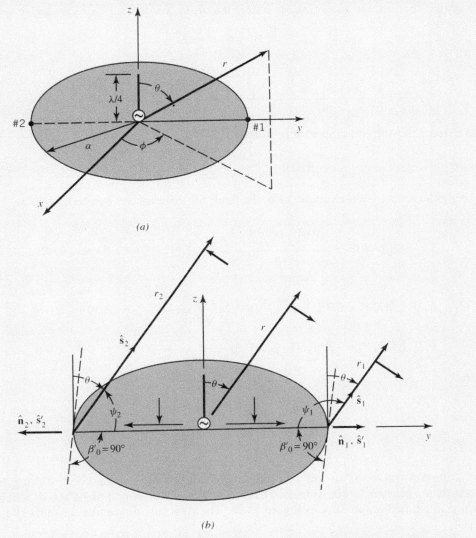

(a)

(b)

Figure 13-37 Quarter-wavelength monopole on a circular ground plane and diffraction mechanism. (a) $\lambda/4$ monopole. (b) Diffraction mechanism.

Referring to the geometry of Figure 13-37b, and using (13-100) and (13-100a), the amplitude spreading factor for wedge 1 can be written as

$$A_1(r_1, a) = \frac{1}{r_1}\sqrt{\rho_{c1}}$$

where

$$\frac{1}{\rho_{c1}} = \frac{1}{a} - \frac{\hat{\mathbf{n}}_1 \cdot (\hat{\mathbf{s}}_1' - \hat{\mathbf{s}}_1)}{a} = \frac{1 - \left[1 - \cos\left(\frac{\pi}{2} - \theta\right)\right]}{a} = \frac{\sin\theta}{a} \ \Rightarrow \ \rho_{c1} = \frac{a}{\sin\theta}$$

Therefore,

$$A_1(r_1, a) = \frac{1}{r_1}\sqrt{\frac{a}{\sin\theta}} \simeq \frac{1}{r}\sqrt{\frac{a}{\sin\theta}}$$

In a similar manner, the amplitude spreading factor for wedge 2 can be expressed as

$$A_2(r_2, a) = \frac{1}{r_2}\sqrt{\rho_{c2}}$$

where

$$\frac{1}{\rho_{c2}} = \frac{1}{a} - \frac{\hat{\mathbf{n}}_2 \cdot (\hat{\mathbf{s}}_2' - \hat{\mathbf{s}}_2)}{a} = \frac{1 - \left[1 - \cos\left(\frac{\pi}{2} + \theta\right)\right]}{a} = -\frac{\sin\theta}{a} \quad \Rightarrow \quad \rho_{c2} = -\frac{a}{\sin\theta}$$

This reduces the amplitude spreading factor to

$$A_2(r_2, a) = \frac{1}{r_2}\sqrt{-\frac{a}{\sin\theta}} \simeq \frac{1}{r}\sqrt{-\frac{a}{\sin\theta}}$$

Using the results from Example 13-6, the fields for this problem can be written as

$$E_{\theta G}(r, \theta) = E_0 \left[\frac{\cos\left(\frac{\pi}{2}\cos\theta\right)}{\sin\theta}\right]\frac{e^{-j\beta r}}{r}, \qquad 0 \le \theta \le \pi/2$$

$$E_{\theta 1}^d(r, \theta) = E_0 V_B^{i,r}\left(a, \theta + \frac{\pi}{2}, n=2\right)\frac{e^{j\beta a \sin\theta}}{\sqrt{\sin\theta}}\frac{e^{-j\beta r}}{r}, \qquad \theta_0 \le \theta \le \pi - \theta_0$$

$$E_{\theta 2}^d(r, \theta) = -E_0 V_B^{i,r}(a, \xi_2, n=2)\frac{e^{-j\beta a \sin\theta}}{\sqrt{-\sin\theta}}\frac{e^{-j\beta r}}{r}, \qquad \theta_0 \le \theta \le \pi - \theta_0$$

where

$$\xi_2 = \psi_2 = \begin{cases} \dfrac{\pi}{2} - \theta, & \theta_0 \le \theta \le \dfrac{\pi}{2} \\[2mm] \dfrac{5\pi}{2} - \theta, & \dfrac{\pi}{2} < \theta \le \pi - \theta_0 \end{cases}$$

It is noted that at $\theta = 0°$ or $180°$, the diffracted fields become singular because along these directions there are caustics for the diffracted fields. The rim of the ground plane acts as a *ring radiator*, which is illustrated graphically in Figure 13-38. The ring radiator can be formulated analytically

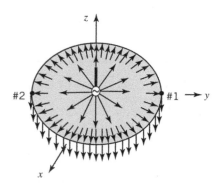

Figure 13-38   Uniform ring radiator representing diffractions around the rim of the circular ground plane.

(see Problem (13.49) as a continuous symmetrical and constant source of diffraction around the rim of the circular ground plane. This can be cast as an integral with uniform excitation around the rim of the circular ground plane, and it can be treated similarly as a circular loop of uniform current [38]. Toward and near $\theta = 0°$ and $180°$, the integral reduces to a Bessel function of the first kind of order one. However, it can be shown, using the method of steepest descent (saddle point method), that the integral representing the continuous ring radiator reduces to a two-point diffraction for angles away from $\theta = 0°$ and $180°$ [33]. Due to the ring radiator characteristics, the radiation of the monopole toward and near $\theta = 0°$ and $180°$ is much more intense for this geometry compared to when the monopole is mounted on a rectangular/square ground plane, as shown in Figure 13-33. Similarly, the *scattering* from circular ground planes and apertures is much more intense toward and near $\theta = 0°$ and $180°$ than that from rectangular/square ground planes and apertures. Therefore, to reduce the radar signature/visibility of the engine inlets (apertures) of the F-117, they may have been chosen to be rectangular/square, as shown in Figure 13-1. Toward $\theta = 0°$ and $180°$ the infinite number of diffracted rays from the rim are identical in amplitude and phase and lead to the caustics. Therefore, the diffracted fields from the aforementioned two points of the rim are invalid within a cone of half included angle θ_0, which is primarily a function of the radius of curvature of the rim. For most moderate size ground planes, θ_0 is in the range of $10° < \theta_0 < 30°$.

To make corrections for the diffracted field singularity and inaccuracy at and near the symmetry axis ($\theta = 0°$ and $180°$), due to axial caustics, the rim of the ground plane must be modeled as a ring radiator [32, 33]. This can be accomplished by using "equivalent" current concepts in diffraction, which will be discussed in the next section.

A pattern based on the formulations of the preceding two-point diffraction was computed for a ground plane of 4.064λ diameter. This pattern is shown in Figure 13-39 where it is compared with measurements. It should be noted that this pattern was computed using the two-point diffraction for $10° \lesssim \theta \lesssim 170°(\theta_0 \simeq 10°)$; the remaining parts were computed using equivalent current concepts that will be discussed next.

13.3.5 Equivalent Currents in Diffraction

In contrast to diffraction by straight edges, diffraction by curved edges creates caustics. If observations are not made at or near caustics, ordinary diffraction techniques can be applied; otherwise, corrections must be made.

One technique that can be used to correct for caustic discontinuities and inaccuracies is the concept of the *equivalent currents* [32–36, 54–61]. To apply this principle, the two-dimensional wedge of Figure 13-13 is replaced by one of the following two forms:

1. An equivalent two-dimensional electric line source of equivalent electric current I^e, for soft polarization diffraction.
2. An equivalent two-dimensional magnetic line source of equivalent magnetic current I^m, for hard polarization diffraction.

This is illustrated in Figure 13-40. The equivalent currents I^e and I^m are adjusted so that the field radiated by each of the line sources is equal to the diffracted field of the corresponding polarization.

The electric field radiated by a two-dimensional electric line source placed along the z axis with a constant current I_z^e is given by (11-10a), or

$$E_z = -\frac{\beta^2 I_z^e}{4\omega\varepsilon} H_0^{(2)}(\beta\rho) \overset{\beta\rho\to\infty}{\simeq} -I_z^e \frac{\eta\beta}{2}\sqrt{\frac{j}{2\pi\beta}}\frac{e^{-j\beta\rho}}{\sqrt{\rho}} \tag{13-101a}$$

where $H_0^{(2)}(\beta\rho)$ is the Hankel function of the second kind of order zero. The approximate form of (13-101a) is valid for large distances of observation (far field), and it is obtained by replacing

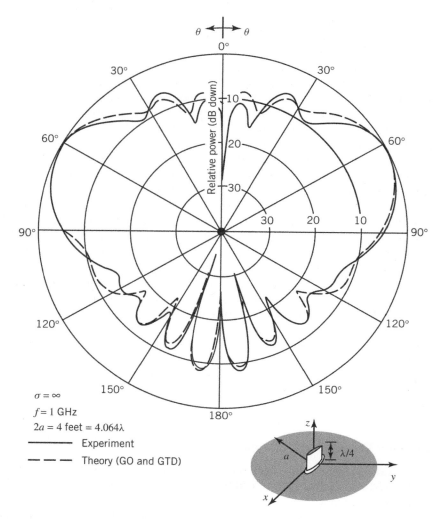

Figure 13-39   Measured and computed principal elevation plane amplitude patterns of a $\lambda/4$ monopole (blade) above a circular ground plane. (Source: C. A. Balanis, *Antenna Theory: Analysis and Design*, Third Edition, copyright © 2005, John Wiley & Sons, Inc. Reprinted by permission of John Wiley & Sons, Inc.)

the Hankel function by its asymptotic formula for large argument (see Appendix IV, Equation IV-17).

The magnetic field radiated by a two-dimensional magnetic line source placed along the z axis with a constant current I_z^m can be obtained using the duality theorem (Section 7.2, Table 7-2) and (13-101a). Thus,

$$H_z = -\frac{\beta^2 I_z^m}{4\omega\mu} H_0^{(2)}(\beta\rho) \overset{\beta\rho\to\infty}{\simeq} -I_z^m \frac{\beta}{2\eta}\sqrt{\frac{j}{2\pi\beta}}\frac{e^{-j\beta\rho}}{\sqrt{\rho}} \tag{13-101b}$$

To determine the equivalent electric current I_z^e, (13-101a) is equated to the field diffracted by a wedge when the incident field is of soft polarization. A similar procedure is used for the equivalent magnetic I_z^m of (13-101b). Using (13-34), (13-34a), (13-95), (13-101a), and (13-101b), and assuming normal incidence, we can write that

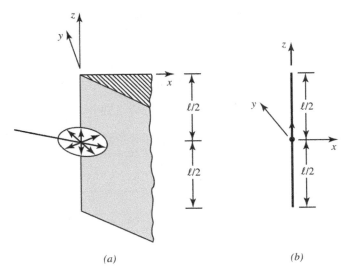

Figure 13-40   Wedge diffraction at normal incidence and its equivalent. (*a*) Actual wedge. (*b*) Equivalent.

$$E_z^i(Q_d)D_s(\xi^-,\xi^+,n)\frac{e^{-j\beta\rho}}{\sqrt{\rho}} = -I_z^e\frac{\eta\beta}{2}\sqrt{\frac{j}{2\pi\beta}}\frac{e^{-j\beta\rho}}{\sqrt{\rho}} \qquad (13\text{-}102a)$$

$$H_z^i(Q_d)D_h(\xi^-,\xi^+,n)\frac{e^{-j\beta\rho}}{\sqrt{\rho}} = -I_z^m\frac{\beta}{2\eta}\sqrt{\frac{j}{2\pi\beta}}\frac{e^{-j\beta\rho}}{\sqrt{\rho}} \qquad (13\text{-}102b)$$

where $E_z^i(Q)$ = incident electric field at the diffraction point Q_d
 $H_z^i(Q)$ = incident magnetic field at the diffraction point Q_d
 D_s = diffraction coefficient for soft polarization [(13-71a) or (13-71e)]
 D_h = diffraction coefficient for hard polarization [(13-71b) or (13-72f)]

Solving (13-102a) and (13-102b) for I_z^e and I_z^m respectively, leads to

$$I_z^e = -\frac{\sqrt{8\pi\beta}}{\eta\beta}e^{-j\pi/4}E_z^i(Q)D_s(\xi^-,\xi^+,n) \qquad (13\text{-}103a)$$

$$I_z^m = -\frac{\eta\sqrt{8\pi\beta}}{\beta}e^{-j\pi/4}H_z^i(Q)D_h(\xi^-,\xi^+,n) \qquad (13\text{-}103b)$$

If the wedge of Figure 13-40 is of finite length ℓ, its equivalent current will also be of finite length. The far-zone field radiated by each can be obtained by using techniques similar to those of Chapter 4 of [38]. Assuming the edge is along the z axis, the far-zone electric field radiated by an electric line source of length ℓ can be written using (4-58a) of [38] as

$$E_\theta^e = j\eta\frac{\beta e^{-j\beta r}}{4\pi r}\sin\theta\int_{-\ell/2}^{\ell/2}I_z^e(z')e^{j\beta z'\cos\theta}\,dz' \qquad (13\text{-}104a)$$

Using duality, the magnetic field of a magnetic line source can be written as

$$H_\theta^m = j\frac{\beta e^{-j\beta r}}{4\pi\eta r}\sin\theta \int_{-\ell/2}^{\ell/2} I_z^m(z')e^{j\beta z'\cos\theta}\,dz' \tag{13-104b}$$

For a constant equivalent current, the integrals in (13-104a) and (13-104b) reduce to a $\sin(\zeta)/\zeta$ form.

If the equivalent current is distributed along a circular loop of radius a and it is parallel to the xy plane, the field radiated by each of the equivalent currents can be obtained using the techniques of Chapter 5, Section 5.3, of [38]. Thus,

$$E_\phi^e = \frac{-j\omega\mu a e^{-j\beta r}}{4\pi r}\int_0^{2\pi} I_\phi^e(\phi')\cos(\phi-\phi')e^{j\beta a\sin\theta\cos(\phi-\phi')}\,d\phi' \tag{13-105a}$$

$$H_\phi^m = \frac{-j\omega\varepsilon a e^{-j\beta r}}{4\pi r}\int_0^{2\pi} I_\phi^m(\phi')\cos(\phi-\phi')e^{j\beta a\sin\theta\cos(\phi-\phi')}\,d\phi' \tag{13-105b}$$

If the equivalent currents are constant, the field is not a function of the azimuthal observation angle ϕ, and (13-105a) and (13-105b) reduce to

$$E_\phi^e = \frac{a\omega\mu e^{-j\beta r}}{2r}I_\phi^e J_1(\beta a\sin\theta) \tag{13-106a}$$

$$H_\phi^m = \frac{a\omega\varepsilon e^{-j\beta r}}{2r}I_\phi^m J_1(\beta a\sin\theta) \tag{13-106b}$$

where $J_1(x)$ is the Bessel function of the first kind of order 1.

For diffraction by an edge of finite length, the equivalent current concept for diffraction assumes that each incremental segment of the edge radiates as would a corresponding segment of a two-dimensional edge of infinite length. Similar assumptions are used for diffraction from finite-length curved edges. The concepts, although approximate, have been shown to yield very good results.

For oblique plane wave incidence diffraction by a wedge of finite length ℓ, as shown in Figure 13-41, the equivalent currents of (13-103a) and (13-103b) take the form

$$\begin{aligned} I_z^e &= -\frac{\sqrt{8\pi\beta}}{\eta\beta}e^{-j\pi/4}E_z^i(Q_D)D_s(\xi^-,\xi^+,n;\beta_0') \\ &\overset{s'\gg z'}{\simeq} -\frac{\sqrt{8\pi\beta}}{\eta\beta}e^{-j\pi/4}E_z^i(0)D_s(\xi^-,\xi^+,n;\beta_0')e^{-j\beta z'\cos\beta_0'} \end{aligned} \tag{13-107a}$$

$$\begin{aligned} I_z^m &= -\frac{\eta\sqrt{8\pi\beta}}{\beta}e^{-j\pi/4}H_z^i(Q_D)D_h(\xi^-,\xi^+,n;\beta_0') \\ &\overset{s'\gg z'}{\simeq} -\frac{\eta\sqrt{8\pi\beta}}{\beta}e^{-j\pi/4}H_z^i(0)D_h(\xi^-,\xi^+,n;\beta_0')e^{-j\beta z'\cos\beta_0'} \end{aligned} \tag{13-107b}$$

where $-\ell/2 \le z' \le \ell/2$ (ℓ = length of wedge) and D_s and D_h are formed by (13-96a) and (13-96b). The far-zone fields associated with the equivalent currents of (13-107a) and (13-107b) can be found using, respectively, (13-104a) and (13-104b).

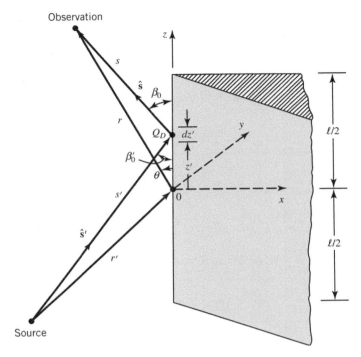

Figure 13-41   Oblique incidence diffraction by a finite-length wedge.

To demonstrate the technique of curved edge diffraction and the equivalent current concept, the radiation of a $\lambda/4$ monopole (blade) mounted on a circular ground plane was modeled. The analytical formulation is assigned as a problem at the end of the chapter. The computed pattern is shown in Figure 13-39 where it is compared with measurements.

To make corrections for the diffracted field discontinuity and inaccuracy at and near the symmetry axis ($\theta = 0°$ and $180°$), due to axial caustics, the rim of the ground plane was modeled as a ring radiator [32, 33]. Equivalent currents were used to compute the pattern in the region given by $0° \leq \theta \leq \theta_0$ and $180° - \theta_0 \leq \theta \leq 180°$. In the other space, a two-point diffraction was used. The two points were taken diametrically opposite to each other, and they were contained in the plane of observation. The value of θ_0 depends upon the curvature of the ground plane. For most ground planes of moderate size, θ_0 is in the range $10° < \theta_0 < 30°$.

A very good agreement between theory and experiment is exhibited in Figure 13-39. For a ground plane of this size, the blending of the two-point diffraction pattern and the pattern from the ring source radiator was performed at $\theta_0 \simeq 10°$. It should be noted that the minor lobes near the symmetry axis ($\theta \simeq 0°$ and $\theta \simeq 180°$) for the circular ground plane are more intense than the corresponding ones for the square plane of Figure 13-33. In addition, the back lobe nearest $\theta = 180°$ is of greater magnitude than the one next to it. These effects are due to the ring source radiation by the rim [33] of the circular ground plane toward the symmetry axis.

13.3.6 Slope Diffraction

Until now the field diffracted by an edge has been found based on (13-34a), (13-85), or (13-88) where $\mathbf{E}^i(Q_D)$ represents the incident field at the point of diffraction. This type of formulation indicates that if the incident field $\mathbf{E}^i(Q_D)$ at the point of diffraction Q_D is zero, then the diffracted field will be zero. In addition to this type of diffraction, there is an additional diffraction term that is based not on the magnitude of the incident field at the point of diffraction but rather on the slope (rate of change, or directional derivative) of the incident field at the point of diffraction. This is a higher-order diffraction, and it becomes more significant when the incident field at the point of diffraction vanishes. It is referred to as *slope diffraction*, and it creates currents on the wedge surface that result in a diffracted field [52].

By referring to the geometry of Figure 13-42, the slope diffracted field can be computed using

Soft Polarization

$$E^d = \frac{1}{j\beta} \left[\frac{\partial E^i(Q_D)}{\partial n} \right] \left(\frac{\partial D_s}{\partial \phi'} \right) \sqrt{\frac{\rho_c}{s(\rho_c + s)}} \, e^{-j\beta s} \tag{13-108}$$

$$\frac{\partial E^i(Q_D)}{\partial n} = \frac{1}{s'} \frac{\partial E^i}{\partial \phi'} \bigg|_{Q_D} = \text{slope of the incident field} \tag{13-108a}$$

$$\frac{\partial D_s}{\partial \phi'} = \text{slope diffraction coefficient} \tag{13-108b}$$

Hard Polarization

$$H^d = \frac{1}{j\beta} \left[\frac{\partial H^i(Q_D)}{\partial n} \right] \left(\frac{\partial D_h}{\partial \phi'} \right) \sqrt{\frac{\rho_c}{s(\rho_c + s)}} \, e^{-j\beta s} \tag{13-109}$$

$$\frac{\partial H^i(Q_D)}{\partial n} = \frac{1}{s'} \frac{\partial H^i}{\partial \phi'} \bigg|_{Q_D} = \text{slope of the incident field} \tag{13-109a}$$

$$\frac{\partial D_h}{\partial \phi'} = \text{slope diffraction coefficient} \tag{13-109b}$$

Therefore, in general, the total diffracted field can be found using

$$U^d = \left[U^i(Q_D) D_{s,h} + \frac{1}{j\beta} \frac{\partial U^i(Q_D)}{\partial n} \frac{\partial D_{s,h}}{\partial \phi'} \right] \sqrt{\frac{\rho_c}{s(\rho_c + s)}} \, e^{-j\beta s} \tag{13-110}$$

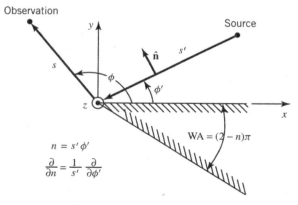

Figure 13-42   Wedge geometry for slope diffraction.

where the first term represents the contribution to the total diffracted field due to the magnitude of the incident field and the second accounts for the contribution due to the slope (rate of change) of the incident field. In (13-110), U represents the electric field for soft polarization and the magnetic field for hard polarization. Similarly $D_{s,h}$ is used to represent D_s for soft polarization and D_h for hard polarization.

The slope diffraction coefficients for soft and hard polarizations can be written, respectively, as [52]

$$
\frac{\partial D_s(\phi,\phi',n;\beta_0')}{\partial \phi'}
$$

$$
= -\frac{e^{-j\pi/4}}{4n^2\sqrt{2\pi\beta}\sin\beta_0'} \Bigg(\left\{ \csc^2\left[\frac{\pi+(\phi-\phi')}{2n}\right] F_s[\beta Lg^+(\phi-\phi')] \right.
$$

$$
\left. -\csc^2\left[\frac{\pi-(\phi-\phi')}{2n}\right] F_s[\beta Lg^-(\phi-\phi')] \right\}
$$

$$
-\left\{ \csc^2\left[\frac{\pi+(\phi+\phi')}{2n}\right] F_s[\beta Lg^+(\phi+\phi')] \right.
$$

$$
\left.\left. -\csc^2\left[\frac{\pi-(\phi+\phi')}{2n}\right] F_s[\beta Lg^-(\phi+\phi')] \right\} \right) \qquad \text{(13-111a)}
$$

$$
\frac{\partial D_h(\phi,\phi',n;\beta_0')}{\partial \phi'}
$$

$$
= -\frac{e^{-j\pi/4}}{4n^2\sqrt{2\pi\beta}\sin\beta_0'} \Bigg(\left\{ \csc^2\left[\frac{\pi+(\phi-\phi')}{2n}\right] F_s[\beta Lg^+(\phi-\phi')] \right.
$$

$$
\left. -\csc^2\left[\frac{\pi-(\phi-\phi')}{2n}\right] F_s[\beta Lg^-(\phi-\phi')] \right\}
$$

$$
-\left\{ \csc^2\left[\frac{\pi+(\phi+\phi')}{2n}\right] F_s[\beta Lg^+(\phi+\phi')] \right.
$$

$$
\left.\left. -\csc^2\left[\frac{\pi-(\phi+\phi')}{2n}\right] F_s[\beta Lg^-(\phi+\phi')] \right\} \right) \qquad \text{(13-111b)}
$$

where

$$
F_s(X) = 2jX\left[1 - j2\sqrt{X}e^{jX}\int_{\sqrt{X}}^{\infty} e^{-j\tau^2}\,d\tau\right] = 2jX[1-F(X)] \qquad \text{(13-111c)}
$$

A Fortran and MATLAB computer subroutine designated as **SWDC**, for Slope Wedge Diffraction Coefficients, computes the normalized (with respect to $\sqrt{\lambda}$) slope diffraction coefficients based on (13-111a) through (13-111c). It was developed and reported in [53]. This program uses the complex function FTF (Fresnel transition function) to complete its computations.

To use the subroutine, the user must specify $R = L$ (in wavelengths), PHID $= \phi$ (in degrees), PHIPD $= \phi'$ (in degrees), BTD $= \beta_0'$ (in degrees), and FN $= n$ (dimensionless) and the subroutine computes the normalized (with respect to $\sqrt{\lambda}$) slope diffraction coefficients CSDCS $= \partial D_s/\partial\phi'$ and CSDCH $= \partial D_h/\partial\phi'$.

13.3.7 Multiple Diffractions

Until now we have considered single-order diffractions from each of the edges of a structure. If the structure is composed of multiple edges (as is the case for infinitely thin strips, rectangular and circular ground planes, etc.), then coupling between the edges will take place. For finite thickness ground planes coupling is evident not only between diametrically opposite edges but also between edges on the same side of the ground plane. Coupling plays a bigger role when the separation between the edges is small, and it should then be taken into account.

A. Higher-Order Diffractions   For structures with multiple edges, coupling is introduced in the form of higher-order diffractions. To illustrate this point, let us refer to Figure 13-43a, where a plane wave of hard polarization, represented by a magnetic field parallel to the edge of the wedges, is incident upon a two-dimensional PEC structure composed of three wedges.

The diffraction mechanism of this system can be outlined as follows: The plane wave incident on wedge 1, represented by wedge angle WA$_1$, will be diffracted as shown in Figure 13-43a. This is referred to as *first-order diffraction*. The field diffracted by wedge 1 in the direction of wedge 2 (WA$_2$) will be diffracted again, as shown in Figures 13-43a and 13-43b. This is referred to as *second-order diffraction*, because it is the result of diffraction from diffraction. In turn, the field diffracted from wedge 2 toward wedges 1 and 3 will be diffracted again. The same procedure can be followed for second-order diffractions from wedge 3 due to first-order diffractions from wedge 1. Second- and higher-order diffractions are all referred to as *higher-order diffractions*, and they account for coupling between the edges and are more important for bistatic than monostatic scattering.

Following the procedures that have been outlined for diffractions from two-dimensional PEC wedges, the first-order diffractions from wedge 1, first-order diffractions from wedge 1 toward

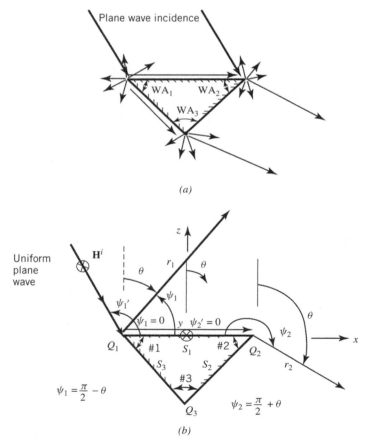

(a)

(b)

Figure 13-43   Higher-order diffractions from a two-dimensional wedged geometry. (a) Plane wave incidence and diffraction by wedges. (b) Second-order diffraction by wedge 2 due to diffractions from wedge 1.

wedge 2, and second-order diffractions from wedge 2 due to first-order diffractions from wedge 1, can be written, using the geometries of Figures 13-43a and 13-43b, as:

First-order diffractions from wedge 1

$$\mathbf{H}_{y1}^{d1} = \mathbf{H}_1^i(Q_1) \cdot \left[\hat{\mathbf{a}}_y \hat{\mathbf{a}}_y D_1^h(r_1, \psi_1, \psi_1', n_1) \right] \frac{1}{\sqrt{r_1}} e^{-j\beta r_1}$$

$$\mathbf{H}_{y1}^{d1} = \hat{\mathbf{a}}_y H_1^i(Q_1) \cdot \left[\hat{\mathbf{a}}_y \hat{\mathbf{a}}_y \begin{Bmatrix} D_1^i(r_1, \psi_1 - \psi_1', n_1) \\ + D_1^r(r_1, \psi_1 + \psi_1', n_1) \end{Bmatrix} \right] \frac{1}{\sqrt{r_1}} e^{-j\beta r_1}$$

$$\mathbf{H}_{y1}^{d1} = +\hat{\mathbf{a}}_y H_1^i(Q_1) \begin{Bmatrix} D_1^i(r_1, \psi_1 - \psi_1', n_1) \\ + D_1^r(r_1, \psi_1 + \psi_1', n_1) \end{Bmatrix} \frac{e^{-j\beta r_1}}{\sqrt{r_1}} \qquad (13\text{-}112a)$$

First-order diffractions from wedge 1 toward wedge 2

$$\mathbf{H}_{y1}^{d1}(r_1 = s_1, \psi_1 = 0, n_1)\Big|_{r_1=s_1, \psi_1=0} = +\hat{\mathbf{a}}_y H_1^i(Q_1) \begin{Bmatrix} D_1^i(s_1, -\psi_1', n_1) \\ + D_1^r(s_1, \psi_1', n_1) \end{Bmatrix} \frac{e^{-j\beta s_1}}{\sqrt{s_1}}$$

$$\mathbf{H}_{y1}^{d1}(r_1 = s_1, \psi_1 = 0, n_1)\Big|_{r_1=s_1, \psi_1=0} = +\hat{\mathbf{a}}_y H_1^i(Q_1) \begin{Bmatrix} V_1^i(s_1, -\psi_1', n_1) \\ + V_1^r(s_1, \psi_1', n_1) \end{Bmatrix} \qquad (13\text{-}112b)$$

Equation 13-112b represents the total diffracted field; half of it is the incident diffracted field and the other half is the reflected diffracted field.

Second-order diffractions from wedge 2 due to first-order diffractions from wedge 1

$$\mathbf{H}_{y2}^{d2} = \mathbf{H}_{21}^i(Q_2) \cdot \hat{\mathbf{a}}_y \hat{\mathbf{a}}_y D_2^h(s_2, \psi_2, \psi_2', n_2) \frac{1}{\sqrt{r_2}} e^{-j\beta r_2}\Big|_{\psi_2'=0,\ \psi_2=\frac{\pi}{2}+\theta}$$

$$\mathbf{H}_{y2}^{d2} = \frac{\mathbf{H}_{y1}^{d1}(r_1 = s_1, \psi_1 = 0, n_1)}{2} \cdot \left[\hat{\mathbf{a}}_y \hat{\mathbf{a}}_y D_2^h(s_2, \psi_2, \psi_2', n_2) \right] \frac{e^{-j\beta r_2}}{\sqrt{r_2}}$$

$$\mathbf{H}_{y2}^{d2} = +\hat{\mathbf{a}}_y \frac{H_1^i(Q_1)}{2} \left\{ V_{B1}^i(s_1, -\psi_1', n_1) + V_B^r(s_1, \psi_1', n_1) \right\}$$

$$\cdot \hat{\mathbf{a}}_y \hat{\mathbf{a}}_y \left[D_2^i(s_2, \psi_2, n_2) + D_2^r(s_2, \psi_2, n_2) \right] \frac{e^{-j\beta r_2}}{\sqrt{r_2}}$$

$$\mathbf{H}_{y2}^{d2} = +\hat{\mathbf{a}}_y \frac{H_1^i(Q_1)}{2} [V_{B1}^i(s_1, -\psi_1', n_1) + V_B^r(s_1, \psi_1', n_1)]$$

$$\cdot [D_2^i(s_2, \psi_2, n_2) + D_2^r(s_2, \psi_2, n_2)] \frac{e^{-j\beta r_2}}{\sqrt{r_2}}$$

$$\mathbf{H}_{y2}^{d2} = +\hat{\mathbf{a}}_y \frac{1}{2} H_1^i(Q_1) [V_{B1}^i(s_1, -\psi_1', n_1) + V_B^r(s_1, \psi_1', n_1)]$$

$$\cdot \left[2 D_2^i(s_2, \psi_2, n_2) \right] \frac{e^{-j\beta r_2}}{\sqrt{r_2}}$$

$$\mathbf{H}_{y2}^{d2} = +\hat{\mathbf{a}}_y H_1^i(Q_1) \left[V_B^i(s_1, -\psi_1', n_1) + V_B^r(s_1, \psi_1', n_1) \right]$$

$$\cdot D_2^i(s_2, \psi_2, n_2) \frac{e^{-j\beta r_2}}{\sqrt{r_2}} \qquad (13\text{-}112c)$$

The $\frac{1}{2}$ factor in the development of (13-112c) is used to represent the incident diffracted field of (13-112b) from wedge 1 toward wedge 2.

The procedure needs to the repeated for first- and second-order diffractions due to direct wave incidence to wedge 2. The method was developed for hard polarization as there are no higher-order diffractions for soft polarization, based on *regular* diffraction, since the diffracted field from any of the wedges toward the others will be zero due to the vanishing of the tangential electric field along the PEC surface of the structure.

B. Self-Consistent Method   It becomes apparent that the procedure for accounting for higher-order diffractions, especially for third and higher orders, can be very tedious, although straightforward. It is recommended that when third- and even higher-order diffractions are of interest, a procedure be adopted that accounts for all (infinite) orders of diffraction. This procedure is known as the *self-consistent method* [62], which is used in scattering theory [63]. It can be shown that the interactions between the edges can also be expressed in terms of a geometrical progression, which in scattering theory is known as the *successive scattering procedure* [63].

Let us now illustrate the self-consistent method as applied to the diffractions of Figure 13-43a. According to Figures 13-43a and 13-43b diffractions by wedge 1 that are due to radiation from the source and that are due to all orders of diffraction from wedge 2 can be written as

$$U_1^{s,h}(r_1,\phi)=U_0^{s,h}(Q_1)D_{10}^{s,h}\left[L_{10},\psi_{10}=\frac{\pi}{2}+\phi,\psi_{10}'=\delta,n_1\right]A_{10}(r_1)e^{-j\beta r_1}$$

$$+\frac{1}{2}[U_2^{s,h}(r_2=d,\phi=0)]D_{12}^{s,h}(L_{12},\psi_{12}=\pi-\phi,\psi_{12}'=0,n_1)$$

$$\times A_{12}(r_1)e^{-j\beta r_1} \tag{13-113a}$$

where $U^{s,h}$ is used to represent here the electric field for soft polarization and the magnetic field for hard polarization. In (13-113a),

$U_1^{s,h}(r_1,\phi)$ = total diffracted field by wedge 1

$U_0^{s,h}(Q_1)$ = field from source at wedge 1

$U_2^{s,h}(r_2=d,\phi=0)$ = total diffracted field (including all orders of diffraction) by wedge 2 toward wedge 1

$D_{10}^{s,h}$ = diffraction coefficient (for soft or hard polarization) of wedge 1 that is due to radiation from the source

$D_{12}^{s,h}$ = diffraction coefficient (for soft or hard polarization) of wedge 1 that is due to radiation from wedge 2

A_{10} = amplitude spreading factor of wedge 1 that is due to radiation from the source

A_{12} = amplitude spreading factor of wedge 1 that is due to radiation from wedge 2

The unknown part in (13-113a) is $U_2^{s,h}(r_2=d,\phi=0)$, and the self-consistent method will be used to determine it.

Using a similar procedure and referring to Figure 13-44c, the total diffracted field by wedge 2 that is due to all orders of diffraction from wedge 1 can be written as

$$U_2^{s,h}(r_2,\phi)=\frac{1}{2}[U_1^{s,h}(r_1=d,\phi=\pi)]$$

$$\times D_{21}^{s,h}(L_{21},\psi_{21}=\phi,\psi_{21}'=0,n_2)A_{21}(r_2)e^{-j\beta r_2} \tag{13-113b}$$

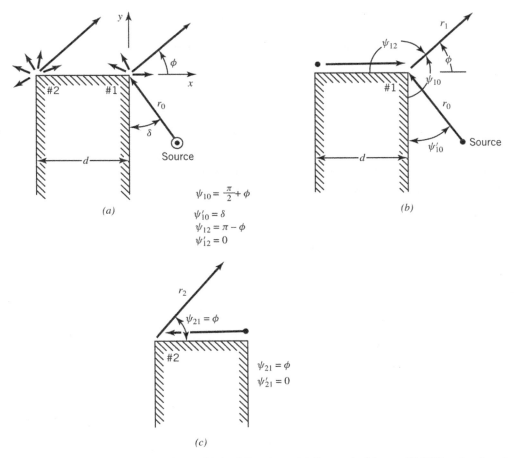

$$\psi_{10} = \frac{\pi}{2} + \phi$$

$$\psi'_{10} = \delta$$
$$\psi_{12} = \pi - \phi$$
$$\psi'_{12} = 0$$

$$\psi_{21} = \phi$$
$$\psi'_{21} = 0$$

Figure 13-44 Finite thickness edge for multiple diffractions. (*a*) Source incidence. (*b*) Diffraction by edge 1. (*c*) Diffraction by edge 2.

where $U_2^{s,h}(r_2,\phi)$ = total diffracted field by wedge 2

$U_1^{s,h}(r_1=d,\phi=\pi)$ = total diffracted field (including all orders of diffraction) by wedge 1 toward wedge 2

$D_{21}^{s,h}$ = diffraction coefficient (for soft or hard polarization) of wedge 2 due to radiation from wedge 1

A_{21} = amplitude spreading factor of wedge 2 due to radiation from wedge 1

In (13-113b), the unknown part is $U_1^{s,h}(r_1=d,\phi=\pi)$, and it will be determined using the self-consistent method.

Equations 13-113a and 13-113b form a consistent pair where there are two unknowns, that is, $U_2^{s,h}(r_2=d,\phi=0)$ in (13-113a) and $U_1^{s,h}(r_1=d,\phi=\pi)$ in (13-113b). If these two unknowns can be found, then (13-113a) and (13-113b) can be used to predict the diffracted fields from each of the wedges taking into account all (infinite) orders of diffraction. These two unknowns can be found by doing the following. At the position of wedge 2 ($r_1 = d, \phi = \pi$) the total diffracted field by wedge 1, as given by (13-113a), can be reduced to

$$[U_1^{s,h}(r_1 = d, \phi = \pi)]$$

$$= U_0^{s,h}(Q_1)\left\{ D_{10}^{s,h}\left[L_{10}, \psi_{10} = \frac{3\pi}{2}, \psi_{10}' = \delta, n_1 \right] A_{10}(r_1 = d)e^{-j\beta d} \right\}$$

$$+ [U_2^{s,h}(r_2 = d, \phi = 0)]\left\{ \frac{1}{2}D_{12}^{s,h}(L_{12}, \psi_{12} = 0, \psi_{12}' = 0, n_1)A_{12}(r_1 = d)e^{-j\beta d} \right\} \quad (13\text{-}114a)$$

In a similar manner, at the position of wedge 1 ($r_2 = d, \phi = 0$) the total diffracted field by wedge 2, as given by (13-113b), can be reduced to

$$[U_2^{s,h}(r_2 = d, \phi = 0)] = U_1^{s,h}(r_1 = d, \phi = \pi)$$

$$\times \left\{ \frac{1}{2}D_{21}^{s,h}(L_{21}, \psi_{21} = 0, \psi_{21}' = 0, n_2)A_{21}(r_2 = d)e^{-\beta d} \right\} \quad (13\text{-}114b)$$

Equations 13-114a and (13-114b) can be rewritten, respectively, in simplified form as

$$\boxed{[U_1^{s,h}(r_1 = d, \phi = \pi)] = U_0^{s,h}(Q_1)T_{10}^{s,h} + [U_2^{s,h}(r_2 = d, \phi = 0)]R_{12}^{s,h}} \quad (13\text{-}115a)$$

$$\boxed{[U_2^{s,h}(r_2 = d, \phi = 0)] = [U_1^{s,h}(r_1 = d, \phi = \pi)]R_{12}^{s,h}} \quad (13\text{-}115b)$$

where

$$T_{10}^{s,h} = D_{10}^{s,h}\left(L_{10}, \psi_{10} = \frac{3\pi}{2}, \psi_{10}' = \delta, n_1 \right) A_{10}(r_1 = d)e^{-j\beta d}$$
$$= \text{transmission coefficient from wedge 1 toward} \quad (13\text{-}115c)$$
$$\text{wedge 2 due to radiation from main source}$$

$$R_{12}^{s,h} = \frac{1}{2}D_{12}^{s,h}(L_{12}, \psi_{12} = 0, \psi_{12}' = 0, n_1)A_{12}(r_1 = d)e^{-j\beta d}$$
$$= \text{reflection coefficient from wedge 1 toward} \quad (13\text{-}115d)$$
$$\text{wedge 2 due to diffractions from wedge 2}$$

$$R_{21}^{s,h} = \frac{1}{2}D_{21}^{s,h}(L_{21}, \psi_{21} = 0, \psi_{21}' = 0, n_2)A_{21}(r_2 = d)e^{-j\beta d}$$
$$= \text{reflection coefficient from wedge 2 toward} \quad (13\text{-}115e)$$
$$\text{wedge 1 due to diffractions from wedge 1}$$

The self-consistent pair of (13-115a) and (13-115b) contains the two unknowns that are needed to predict the total diffracted field as given by (13-113a) and (13-113b). Solving (13-115a) and (13-115b) for $U_1^{s,h}(r_1 = d, \phi = \pi)$ and $U_2^{s,h}(r_2 = d, \phi = 0)$, we can show that

$$\boxed{U_1^{s,h}(r_1 = d, \phi = \pi) = U_0^{s,h}(Q_1)\frac{T_{10}^{s,h}}{1 - R_{21}^{s,h}R_{12}^{s,h}}} \quad (13\text{-}116a)$$

$$\boxed{U_2^{s,h}(r_2 = d, \phi = 0) = U_0^{s,h}(Q_1)\frac{T_{10}^{s,h}R_{21}^{s,h}}{1 - R_{21}^{s,h}R_{12}^{s,h}}} \quad (13\text{-}116b)$$

When expanded, it can be shown that (13-116a) and (13-116b) can be written as a geometric series of the form

$$U_1^{s,h}(r_1 = d, \phi = \pi) = U_0^{s,h}(Q_1)T_{10}^{s,h}[1 + x_0 + x_0^2 + \cdots] \tag{13-117a}$$

$$U_2^{s,h}(r_2 = d, \phi = 0) = U_0^{s,h}(Q_1)T_{10}^{s,h}R_{21}^{s,h}[1 + x_0 + x_0^2 + \cdots] \tag{13-117b}$$

where

$$x_0 = R_{21}^{s,h}R_{12}^{s,h} \tag{13-117c}$$

Each term of the geometric series can be related to an order of diffraction by the corresponding wedge.

In matrix form, the self-consistent set of equations as given by (13-115a) and (13-115b) can be written as

$$\begin{bmatrix} 1 & -R_{12}^{s,h} \\ -R_{21}^{s,h} & 1 \end{bmatrix}\begin{bmatrix} U_1^{s,h} \\ U_2^{s,h} \end{bmatrix} = \begin{bmatrix} U_0^{s,h}T_{10}^{s,h} \\ 0 \end{bmatrix} \tag{13-118}$$

which can be solved using standard matrix inversion methods.

The outlined self-consistent method can be extended and applied to the interactions between a larger number of edges. However, the order of the system of equations to be solved will also increase and will be equal to the number of interactions between the various edge combinations.

C. Overlap Transition Diffraction Region   The UTD diffraction coefficients fail to predict accurately the field diffracted near grazing angles. This is best illustrated in Figure 13-45, where a uniform plane wave is incident on a two-dimensional PEC strip.

The field diffracted by wedge 1 toward wedge 2 creates a Transition Region (TR), shown cross-hatched, over which the diffracted field is non-ray optical and the second-order and successive diffractions are not accurately predicted using the traditional GTD/UTD procedure outlined previously. The same is true for diffractions from other wedges with similar angles of incidence. However, as the angle of incidence moves away from grazing, the GTD/UTD diffractions become more valid. This is illustrated with some examples that follow.

To overcome the issue of the non-ray optical nature of the first-order diffractions and the inaccurate predictions by standard GTD/UTD of the higher-order successive diffractions by wedges, the following two methods can be used.

1. Extended Spectral Theory of Diffraction (ESTD) [64]
2. Extended Physical Theory of Diffraction (EPTD) [65]

While the GTD and UTD are considered to be somewhat heuristic, they are more general and less cumbersome in their application to multiple diffractions. The ESTD and EPTD are more

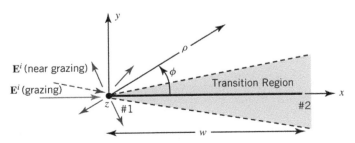

Figure 13-45   Diffraction by strip at and near grazing.

rigorous and more accurate but less general and more complex. The ESTD is an extension of the STD method [66]. Using the ESTD, the current density induced on the scatter of interest is transformed in the spectral domain. The radiation integral is then asymptotically evaluated in the spectral domain, after the induced current density is multiplied by a spectral diffraction coefficient. The original STD was limited to a half-plane and aperture scattering for plane wave incidence. The ESTD extends the STD to general double-wedge configuration, and it can be used for plane, cylindrical, and spherical wave incidence for both normal and oblique incidences.

The EPTD is an alternative transition-region method based upon a different evaluation of the surface radiation integral. The induced current density is approximated using the PTD fringe currents [67], whereas the ESTD uses the UTD diffraction coefficients. The resulting radiation integral is evaluated asymptotically to obtain the second-order field diffracted by the double wedge structure. The EPTD formulation is limited to plane-wave incidence, far-field observation in the plane of the structure. As with the ESTD, the EPTD doubly-diffracted field expression can be greatly simplified for certain geometries, such as the strip.

To demonstrate the concepts of near grazing-angle-incidence diffraction, a numbers of examples are considered for monostatic and bistatic scattering for both soft (TM^z) and hard (TE^z) polarizations [68, 69]. For monostatic scattering, the patterns are illustrated in Figures 13-46a and 13-46b for a strip of width $w = 2\lambda$. A width of 2λ is chosen for all cases so that the GTD/UTD diffraction coefficients are valid. For the soft polarization, only first-order UTD diffractions are

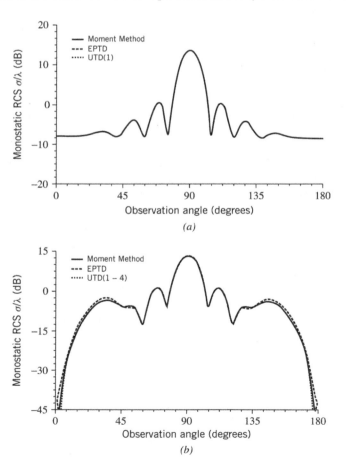

Figure 13-46   Monostatic RCS by a two-dimensional strip of width $w = 2\lambda$ [68, 69]. (a) Soft (TM^z) polarization. (b) Hard (TE^z) polarization.

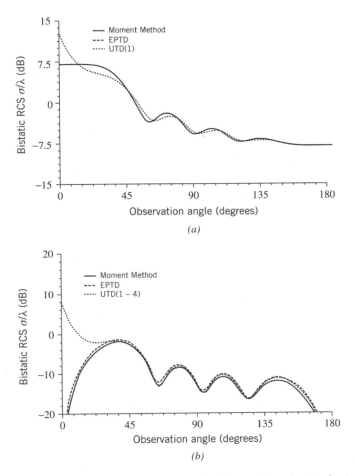

Figure 13-47 Bistatic RCS by a two-dimensional strip of width $w = 2\lambda$ at $\phi_i = 170°$ [68, 69]. (a) Soft (TM^z) polarization. (b) Hard (TE^z) polarization.

considered since higher-orders are not applicable; however, for the hard polarization, up to fourth-order UTD diffractions are included. It is clear that the results of all three methods (MM, EPTD, UTD) for both monostatic cases are in very good agreement.

The bistatic scattering results for incidence angles of near grazing ($\phi_i = 170°$) and away from grazing ($\phi_i = 135°$) are shown in Figures 13-47a, 13-47b, 13-48a, and 13-48b, respectively. As expected, because of the UTD diffracted fields near grazing angle ($\phi_i = 170°$) are non-ray optical, the patterns of the UTD results are not in very good agreement with those of the MM and EPTD as indicated in Figures 13-47a and 13-47b for both polarizations. However, the comparison of not near-grazing incidence ($\phi_i = 135°$) of all three methods is very good for both polarizations, as indicated in Figures 13-48a and 13-48b.

13.4 COMPUTER CODES

Using geometrical optics and wedge diffraction techniques, a number of computer codes have been developed over the years to compute the radiation and scattering characteristics of simple and complex antenna and scattering systems. Some are in the form of subroutines that are used primarily to compute wedge diffraction coefficients and associated functions. Others are very

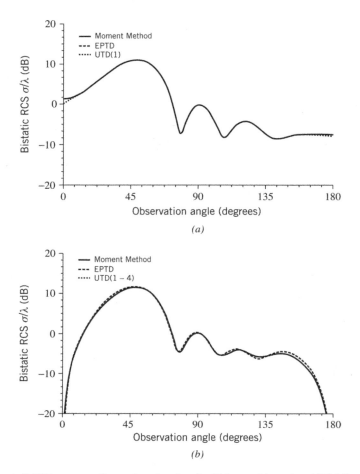

Figure 13-48   Bistatic RCS by a two-dimensional strip of width $w = 2\lambda$ at $\phi_i = 135°$ [68, 69]. (*a*) Soft (TM^z) polarization. (*b*) Hard (TE^z) polarization.

sophisticated codes that can be used to analyze very complex radiation and scattering problems. We will describe here two wedge diffraction subroutines and a subfunction, which the readers can utilize to either develop their own codes or to solve problems of interest. These subroutines were initially written in Fortran but were translated to MATLAB for this edition.

13.4.1   Wedge Diffraction Coefficients

The **Wedge Diffraction Coefficients** (**WDC**) subroutine computes the soft and hard polarization wedge diffraction coefficients based on (13-89a) through (13-90b). To complete the computations, the program uses the complex function FTF (Fresnel transition function). This program was developed at the ElectroScience Laboratory at the Ohio State University, and it was reported in [53].

To use this subroutine, the user must specify

$$R = L \quad \text{(in wavelengths)} = \text{distance parameter}$$
$$\text{PHID} = \phi \quad \text{(in degrees)} = \text{observation angle}$$
$$\text{PHIPD} = \phi' \quad \text{(in degrees)} = \text{incident angle}$$
$$\text{BTD} = \beta_0' \quad \text{(in degrees)} = \text{oblique angle}$$
$$\text{FN} = n \quad \text{(dimensionless)} = \text{wedge angle factor}$$

and the program computes the complex diffraction coefficients

$$\text{CDCS} = D_s = \text{complex diffraction coefficient (soft polarization)}$$
$$\text{CDCH} = D_h = \text{complex diffraction coefficient (hard polarization)}$$

13.4.2 Fresnel Transition Function

The Fresnel Transition Function (**FTF**) computes the Fresnel transition functions $F(X)$ of (13-68e), (13-68f), and (13-69e), (13-69f). The program is based on the asymptotic expression of (13-74a) for small arguments ($X < 0.3$) and on (13-74b) for large arguments ($X > 5.5$). For intermediate values ($0.3 \leq X \leq 5.5$), linear interpolation is used. The program was developed at the ElectroScience Laboratory of the Ohio State University, and it was reported in [53].

13.4.3 Slope Wedge Diffraction Coefficients

The Slope Wedge Diffraction Coefficients (**SWDC**) subroutine computes the soft and hard polarization wedge diffraction coefficients based on (13-111a) through (13-111c). To complete the computations, the program uses the complex function FTF (Fresnel transition function) of Section 13.4.2. This program was developed at the ElectroScience Laboratory at the Ohio State University, and it was reported in [53].

To use this subroutine, the user must specify

$$R = L \text{ (in wavelengths)} = \text{distance parameter}$$
$$\text{PHID} = \phi \text{ (in degrees)} = \text{observation angle}$$
$$\text{PHIPD} = \phi' \text{ (in degrees)} = \text{incident angle}$$
$$\text{BTD} = \beta_0' \text{ (in degrees)} = \text{oblique angle}$$
$$\text{FN} = n \text{ (dimensionless)} = \text{wedge angle factor}$$

and the program computes the complex slope diffraction coefficients

$$\text{CSDCS} = \frac{\partial D_s}{\partial \phi'} = \text{complex slope diffraction coefficient (soft polarization)}$$

$$\text{CSDCH} = \frac{\partial D_h}{\partial \phi'} = \text{complex slope diffraction coefficient (hard polarization)}$$

13.5 MULTIMEDIA

On the website that accompanies this book, the following multimedia resources are included for the review, understanding, and presentation of the material of this chapter.

- **MATLAB** computer programs (the first two also in Fortran):
  a. **WDC:** (Both MATLAB and Fortran). Computes the first-order wedge diffraction coefficient based on (13-89a) through (13-90b). The initial Fortran algorithm was developed and reported in [53].
  b. **SWDC:** (Both MATLAB and Fortran). Computes the first slope wedge diffraction coefficient based on (13-111a) through (13-111c). The initial Fortran algorithm was developed and reported in [53].

c. **PEC_Wedge:** Computes, based on the exact solution of (13-40a), the normalized amplitude pattern of a uniform plane wave incident upon a two-dimensional PEC wedge, as shown in Figure 13-13.

d. **PEC_Strip_Line_UTD:** Computes, using UTD, the normalized amplitude radiation pattern of a line source based on the UTD of Example 13-4. It is compared with that based of the Integral Equation (IE) of Sections 12.2.2 through 12.2.8 and Physical Optics (PO) of Section 11.2.3.

e. **PEC_Strip_SW_UTD:** Computes, using UTD, the TM^z and TE^z 2-D scattering width (SW), monostatic and bistatic, of a PEC strip of finite width, based on the UTD of Example 13-5. It is compared with that of the Integral Equation (IE) of Section 12.3.1 and Physical Optics of Section 11.3.1, and Figures 12-13 and 11-4.

f. **Monopole_GP_UTD:** Computes, using UTD, the normalized amplitude radiation pattern of a $\lambda/4$ monopole on a rectangular or circular ground plane based on UTD and Figures 13-32 and 13-37.

g. **Aperture_GP_UTD:** Computes, using UTD, the normalized amplitude radiation pattern of a rectangular or circular aperture, with either a uniform or dominant mode aperture field distribution, on a rectangular ground plane based on UTD and Figure 13-32, where the monopole is replaced by an aperture, as shown in Figure P13-41.

h. **PEC_Rect_RCS_UTD:** Computes, using UTD, the TE^x and TM^x bistatic and monostatic RCS of a PEC rectangular plate using UTD. It is compared with the Physical Optics (PO) of Section 11.2.3.

i. **PEC_Circ_RCS_UTD:** Computes the TE^x and TM^x monostatic RCS of a PEC circular plate using UTD. It is compared with the Physical Optics (PO) of Chapter 11 and Problem 11.24.

j. **PEC_Square_Circ_RCS_UTD.** Computes, using UTD, the TE^x and TM^x monostatic RCS of PEC square and circular plates, which have the same area and equal maximum monostatic RCS at normal incidence. The UTD patterns of the two plates, square and circular, are compared with the Physical Optics (PO) of Chapter 11.

• **PowerPoint (PPT)** viewgraphs, in multicolor.

REFERENCES

1. R. F. Harrington, "Matrix methods for field problems," *Proc. IEEE*, vol. 55, no. 2, pp. 136–149, Feb. 1967.

2. R. F. Harrington, *Field Computation by Moment Methods*, Macmillan, New York, 1968.

3. J. H. Richmond, "Digital computer solutions of the rigorous equations for scattering problems," *Proc. IEEE*, vol. 53, pp. 796–804, Aug. 1965.

4. J. Moore and R. Pizer, *Moment Methods in Electromagnetics*, Wiley, New York, 1984.

5. J. B. Keller, "Diffraction by an aperture," *J. Appl. Phys.*, vol. 28, no. 4, pp. 426–444, Apr. 1957.

6. J. B. Keller, "Geometrical theory of diffraction," *J. Opt. Soc. Amer.*, vol. 52, no. 2, pp. 116–130, Feb. 1962.

7. R. G. Kouyoumjian, "Asymptotic high-frequency methods," *Proc. IEEE*, vol. 53, pp. 864–876, Aug. 1965.

8. P. H. Pathak and R. G. Kouyoumjian, "The dyadic diffraction coefficient for a perfectly conducting wedge," Technical Report 2183-4 (AFCRL-69-0546), Ohio State University ElectroScience Lab., Jun. 5, 1970.

9. P. H. Pathak and R. G. Kouyoumjian, "An analysis of the radiation from apertures on curved surfaces by the geometrical theory of diffraction," *Proc. IEEE*, vol. 62, no. 11, pp. 1438–1447, Nov. 1974.

10. R. G. Kouyoumjian and P. H. Pathak, "A uniform geometrical theory of diffraction for an edge in a perfectly conducting surface," *Proc. IEEE*, vol. 62, no. 11, pp. 1448–1461, Nov. 1974.

11. G. L. James, *Geometrical Theory of Diffraction for Electromagnetic Waves*, Third Edition Revised, Peregrinus, London, 1986.

12. P. Y. Ufimtsev, "Method of edge waves in the physical theory of diffraction," translated by U.S. Air Force Foreign Technology Division, Wright-Patterson AFB, OH, Sept. 1971.

13. P. Y. Ufimtsev, "Approximate computation of the diffraction of plane electromagnetic waves at certain metal bodies," *Sov. Phys.—Tech. Phys.*, pp. 1708–1718, 1957.

14. P. Y. Ufimtsev, "Secondary diffraction of electromagnetic waves by a disk," *Sov. Phys.—Tech. Phys.*, vol. 3, pp. 549–556, 1958.

15. K. M. Mitzner, "Incremental length diffraction coefficients," Technical Report AFAL-TR-73-296, Northrop Corp., Aircraft Division, Apr. 1974.

16. E. F. Knott and T. B. A. Senior, "Comparison of three high-frequency diffraction techniques," *Proc. IEEE*, vol. 62, no. 11, pp. 1468–1474, Nov. 1974.

17. E. F. Knott, "A progression of high-frequency RCS prediction techniques," *Proc. IEEE*, vol. 73, no. 2, pp. 252–264, Feb. 1985.

18. T. Griesser and C. A. Balanis, "Backscatter analysis of dihedral corner reflectors using physical optics and physical theory of diffraction," *IEEE Trans. Antennas Propagat.*, vol. AP-35, no. 10, pp. 1137–1147, Oct. 1987.

19. P. M. Russo, R. C. Rudduck, and L. Peters, Jr., "A method for computing E-plane patterns of horn antennas," *IEEE Trans. Antennas Propagat.*, vol. AP-13, no. 2, pp. 219–224, 1965.

20. R. C. Rudduck and L. L. Tsai, "Aperture reflection coefficient of TEM and TE_{01} mode parallel-plate waveguide," *IEEE Trans. Antennas Propagat.*, vol. AP-16, no. 1, pp. 83–89, Jan. 1968.

21. C. A. Balanis and L. Peters, Jr., "Analysis of aperture radiation from an axially slotted circular conducting cylinder using geometrical theory of diffraction," *IEEE Trans. Antennas Propagat.*, vol. AP-17, no. 1, pp. 93–97, Jan. 1969.

22. C. A. Balanis and L. Peters, Jr., "Equatorial plane pattern of an axial-TEM slot on a finite size ground plane," *IEEE Trans. Antennas Propagat.*, vol. AP-17, no. 3, pp. 351–353, May 1969.

23. C. A. Balanis, "Radiation characteristics of current elements near a finite length cylinder," *IEEE Trans. Antennas Propagat.*, vol. AP-18, no. 3, pp. 352–359, May 1970.

24. C. A. Balanis, "Analysis of an array of line sources above a finite ground plane," *IEEE Trans. Antennas Propagat.*, vol. AP-19, no. 2, pp. 181–185, Mar. 1971.

25. C. L. Yu, W. D. Burnside, and M. C. Gilreath, "Volumetric pattern analysis of airborne antennas," *IEEE Trans. Antennas Propagat.*, vol. AP-26, no. 5, pp. 636–641, Sept. 1978.

26. G. A. Thiele and T. H. Newhouse, "A hybrid technique for combining moment methods with the geometrical theory of diffraction," *IEEE Trans. Antennas Propagat.*, vol. AP-23, no. 1, pp. 62–69, 1975.

27. W. D. Burnside, C. L. Yu, and R. J. Marhefka, "A technique to combine the geometrical theory of diffraction and the moment method," *IEEE Trans. Antennas Propagat.*, vol. AP-23, no. 4, pp. 551–558, Jul. 1975.

28. J. N. Sahalos and G. A. Thiele, "On the application of the GTD-MM technique and its limitations," *IEEE Trans. Antennas Propagat.*, vol. AP-29, no. 5, pp. 780–786, Sept. 1981.

29. R. K. Luneberg, *"Mathematical theory of optics,"* Brown University Notes, Providence, RI, 1944.

30. M. Kline, "An asymptotic solution of Maxwell's equations," in *The Theory of Electromagnetic Waves*, Interscience, New York, 1951.

31. M. Kline and I. Kay, *Electromagnetic Theory and Geometrical Optics*, Interscience, New York, 1965.

32. C. E. Ryan, Jr., and L. Peters, Jr., "Evaluation of edge-diffracted fields including equivalent currents for the caustic regions," *IEEE Trans. Antennas Propagat.*, vol. AP-17, pp. 292–299, May 1969; erratum, vol. AP-18, p. 275, Mar. 1970.

33. C. A. Balanis, "Radiation from conical surfaces used for high-speed spacecraft," *Radio Science*, vol. 7, pp. 339–343, Feb. 1972.

34. E. F. Knott, T. B. A. Senior, and P. L. E. Uslenghi, "High-frequency backscattering from a metallic disc," *Proc. IEEE*, vol. 118, no. 12, pp. 1736–1742, Dec. 1971.

35. W. D. Burnside and L. Peters, Jr., "Edge diffracted caustic fields," *IEEE Trans. Antennas Propagat.*, vol. AP-22, no. 4, pp. 620–623, Jul. 1974.

36. D. P. Marsland, C. A. Balanis, and S. Brumley, "Higher order diffractions from a circular disk," *IEEE Trans. Antennas Propagat.*, vol. AP-35, no. 12, pp. 1436–1444, Dec. 1987.

37. R. G. Kouyoumjian, L. Peters, Jr., and D. T. Thomas, "A modified geometrical optics method for scattering by dielectric bodies," *IRE Trans. Antennas Propagat.*, vol. AP-11, no. 6, pp. 690–703, Nov. 1963.

38. C. A. Balanis, *Antenna Theory: Analysis and Design*, Third Edition, Wiley, New York, 2005.

39. T. Griesser and C. A. Balanis, "Dihedral corner reflector backscatter using higher-order reflections and diffractions," *IEEE Trans. Antennas Propagat.*, vol. AP-35, no. 11, pp. 1235–1247, Nov. 1987.

40. W. Pauli, "On asymptotic series for functions in the theory of diffraction of light," *Physical Review*, vol. 34, pp. 924–931, Dec. 1938.

41. F. Oberhettinger, "On asymptotic series occurring in the theory of diffraction of waves by a wedge," *J. Math. Phys.*, vol. 34, pp. 245–255, 1956.

42. D. L. Hutchins, "Asymptotic series describing the diffraction of a plane wave by a two-dimensional wedge of arbitrary angle," Ph.D. dissertation, Dept. of EE, Ohio State University, 1967.

43. W. Franz and K. Deppermann, "Theorie der beugung am zylinder unter berücksichti gung der kriechwelle," *Ann. Phys.*, 6 Folge, Bd. 10, Heft 6–7, pp. 361–373, 1952.

44. B. R. Levy and J. B. Keller, "Diffraction by a smooth object," *Commun. Pure Appl. Math.*, vol. XII, no. 1, pp. 159–209, Feb. 1959.

45. J. B. Keller and B. R. Levy, "Decay exponents and diffraction coefficients for surface waves of nonconstant curvature," *IRE Trans. Antennas Propagat.*, vol. AP-7 (special suppl.), pp. S52–S61, Dec. 1959.

46. D. R. Voltmer, "Diffraction by doubly curved convex surfaces," Ph.D. dissertation, Dept. of EE, Ohio State University, 1970.

47. L. B. Felsen and N. Marcuvitz, *Radiation and Scattering of Waves*, Prentice-Hall, Englewood Cliffs, NJ, 1973.

48. F. B. Hildebrand, *Advanced Calculus for Applications*, Prentice-Hall, Englewood Cliffs, NJ, 1962.

49. J. Boersma, "Computation of Fresnel integrals," *J. Math. Comp.*, vol. 14, p. 380, 1960.

50. G. L. James, "An approximation to the Fresnel integral," *Proc. IEEE*, vol. 67, no. 4, pp. 677–678, Apr. 1979.

51. P. C. Clemmow, *The Plane Wave Spectrum Representation of Electromagnetic Fields*, Pergamon, Elmsford, NY, 1966.

52. R. G. Kouyoumjian, "The geometrical theory of diffraction and its application," in *Numerical and Asymptotic Techniques in Electromagnetics*, R. Mittra (Ed.), Springer, New York, 1975, Chapter 6.

53. "The modern geometrical theory of diffraction," vol. 1, *Short Course Notes*, ElectroScience Lab., Ohio State University.

54. R. F. Millar, "An approximate theory of the diffraction of an electromagnetic wave by an aperture in a plane screen," *Proc. IEE*, Monograph No. 152R, vol. 103 (pt. C), pp. 117–185, Sept. 1955.

55. R. F. Millar, "The diffraction of an electromagnetic wave by a circular aperture," *Proc. IEE*, Monograph No. 196R, vol. 104 (pt. C), pp. 87–95, Sept. 1956.

56. R. F. Millar, "The diffraction of an electromagnetic wave by a large aperture," *Proc. IEE*, Monograph No. 213R (pt. C), pp. 240–250, Dec. 1956.

57. W. D. Burnside and L. Peters, Jr., "Axial-radar cross section of finite cones by the equivalent-current concept with higher-order diffraction," *Radio Science*, vol. 7, no. 10, pp. 943–948, Sept. 1982.

58. E. F. Knott, "The relationship between Mitzner's ILDC and Michaeli's equivalent currents," *IEEE Trans. Antennas Propagat.*, vol. AP-33, no. 1, pp. 112–114, Jan. 1985.

59. A. Michaeli, "Equivalent edge currents for arbitrary aspects of observation," *IEEE Trans. Antennas Propagat.*, vol. AP-32, no. 3, pp. 252–258, Mar. 1984; erratum, vol. AP-33, no. 2, p. 227, Feb. 1985.

60. A. Michaeli, "Elimination of infinities in equivalent edge currents, Part I: Fringe current components," *IEEE Trans. Antennas Propagat.*, vol. AP-34, no. 7, pp. 912–918, Jul. 1986.

61. A. Michaeli, "Elimination of infinities in equivalent edge currents, Part II: Physical optics components," *IEEE Trans. Antennas Propagat.*, vol. AP-34, no. 8, pp. 1034–1037, Aug. 1986.

62. R. C. Rudduck and J. S. Yu, "Higher-order diffraction concept applied to parallel-plate waveguide patterns," Report No. 1691–16, The Antenna Lab. (now ElectroScience Lab.), Ohio State University, Oct. 15, 1965.

63. V. Twersky, "Multiple scattering of waves and optical phenomena," *J. Opt. Soc. Amer.*, vol. 52, no. 2, pp. 145–171, Feb. 1962.

64. R. Tiberio and R. G. Kouyoumjian, "A uniform GTD solution for the diffraction by strips illuminated at grazing incidence," *Radio Science*, vol. 14, pp. 933–941, Nov.-Dec. 1979.

65. A. Michaeli, "A closed form physical theory of diffraction solution for electromagnetic scattering by strips and 90° dihedrals," *Radio Science*, vol. 19, pp. 609–616, Mar.-Apr. 1984.

66. Y. Rahmat-Samii and R. Mittra, "Spectral domain interpretation of high frequency diffraction phenomena," *IEEE Trans. Antennas Propagat.*, vol. AP-25, pp. 676–687, 1977.

67. P. I. Ufimtsev, "Secondary diffraction of electromagnetic waves by a strip," *Sov. Phys.—Tech. Phys.*, vol. 28, pp. 535–548, 1958.

68. Lesley A. Polka, "Radar cross section prediction of strips and flat plates for grazing and nonprincipal-plane incidence," MS Thesis, Arizona State University, Aug. 1989.

69. L. A. Polka, C. A. Balanis, and A. C. Polycarpou, "High-frequency methods for multiple diffraction modeling: application and comparison," *Journal of EM Waves and Applications*, vol. 8, no. 9/10, pp. 1223–1246, 1994.

PROBLEMS

13.1. Using the geometry of Figure 13-5, derive (13-9).

13.2. An arbitrary surface of revolution can be represented by

$$z = g(u) \quad \text{where} \quad u = \frac{x^2 + y^2}{2}$$

Define

$$K^2 = 1 + 2u\left[\frac{dg(u)}{du}\right]^2$$

Then the unit vectors $\hat{\mathbf{u}}_1$ and $\hat{\mathbf{u}}_2$ of Figure 13-7 in the directions of the principal radii of curvature R_1 and R_2, respectively, can be determined using

$$\hat{\mathbf{u}}_1 = \frac{\hat{\mathbf{a}}_x y - \hat{\mathbf{a}}_y x}{\sqrt{x^2 + y^2}}$$

$$\hat{\mathbf{u}}_2 = \frac{\hat{\mathbf{a}}_x x + \hat{\mathbf{a}}_y y + \hat{\mathbf{a}}_z \left[(x^2 + y^2)\dfrac{dg(u)}{du}\right]}{K\sqrt{x^2 + y^2}}$$

and the R_1 and R_2 can be found using

$$\frac{1}{R_1} = \frac{1}{K}\frac{dg(u)}{du}$$

$$\frac{1}{R_2} = \frac{1}{K^3}\left[\frac{dg(u)}{du} + 2u\frac{d^2g(u)}{du^2}\right]$$

For a paraboloidal reflector (parabola of revolution), widely used as a microwave reflector antenna, whose surface can be represented by

$$z = f - \frac{x^2 + y^2}{4f}$$

where f is the focal distance, show that the principal radii of curvature are given by

$$\frac{1}{R_1} = -\frac{1}{2f}\frac{1}{\left[1 + \dfrac{x^2 + y^2}{4f^2}\right]^{1/2}}$$

$$\frac{1}{R_2} = -\frac{1}{2f} \frac{1}{\left[1 + \dfrac{x^2 + y^2}{4f^2}\right]^{3/2}}$$

13.3. The intensity radiated by an electromagnetic source is contained within a cone of total included angle $\alpha = 60°$. Assuming that the radiation density at a radial distance of 5 meters from the vertex of the cone is uniformly distributed over the spherical cap of the cone and it is $10\,\text{milliwatts/cm}^2$, determine, using classical geometrical optics and conservation of energy within a tube of rays, the:
(a) Total power radiated by the source.
(b) Radiation density (in milliwatts/cm^2) over the spherical cap of the cone at a radial distance of 50 meters from the vertex of the cone.

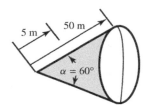

Figure P13-3

13.4. Derive (13-30) using the geometry of Figure 13-8.

13.5. An electric line source is placed in front of a 30° convex segment of an infinite length conducting circular arc, as shown in Figure P13-5. Assume the line source is placed symmetrically about the arc, its position coincides with the origin of the coordinate system, and it is parallel to the length of the arc. Using geometrical optics determine the following:

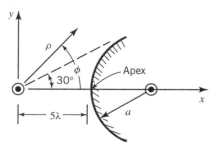

Figure P13-5

(a) The location of the caustic for fields reflected by the surface of the cap.
(b) The far-zone backscattered electric field at a distance of 50λ from the center of the apex of the arc.
   Assume that the radius of the arc is 5λ and the incident electric field at the apex of the arc is

$$\mathbf{E}^i(\rho = 5\lambda, \phi = 0°) = \hat{\mathbf{a}}_z 10^{-3}\ \text{V/m}$$

13.6. Repeat Problem 13.5 for a 30° concave segment of an infinite length conducting circular arc, as shown in Figure P13-6.

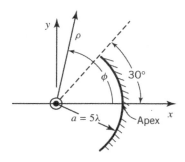

Figure P13-6

13.7. A uniform plane wave traveling in free space with an incident electric field given by

$$\mathbf{E}^i = \hat{\mathbf{a}}_x\, e^{-j\beta z}$$

is incident upon a smooth curved conducting surface as shown in Figure P13-7.

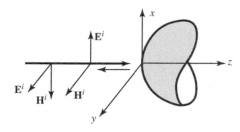

Figure P13-7

Assuming the radii of curvature in the two principal planes are: xz plane $= 10\lambda$ and yz plane $= 19.6\lambda$, utilize the simplest and sufficiently accurate method to:
(a) Write an expression, in vector form, of the backscattered electric field.

(b) Compute the normalized monostatic RCS (σ/λ^2) in dB.

(c) Repeat parts (a) and (b) when the incident electric field is given by

$$\mathbf{E}^i = \hat{a}_y \, e^{-j\beta z}$$

13.8. An infinite-length electric line source is placed a distance s' from a two-dimensional flat PEC strip of finite width in the x direction and infinite length in the z direction, as shown in Figure P13-8. To reduce the backscattered field, the flat strip is replaced with a two-dimensional semi-circular PEC arc of radius a, as shown dashed in the figure. Determine, using GO, the radius (in wavelengths) of the PEC arc so that the backscattered electric field from the arc is 20 dB down (or −20 dB) relative to the electric field backscattered from the flat PEC strip. Assume that $s' = 10\lambda$ and that $s \gg s' \gg a$.

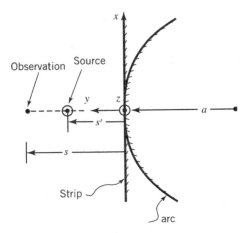

Figure P13-8

13.9. Repeat the calculations of Example 13-2 for $h = 0.25$ and $w = 2\lambda$.

13.10. Show that by combining (13-60a) and (13-60b) you get (13-61).

13.11. Show that evaluation of (13-48b) along C_T leads to (13-62a) and along SDP$_{\pm\pi}$ leads to (13-62b).

13.12. When the line source is in the vicinity of the edge of the wedge and the observations are made at large distances ($\rho \gg \rho'$) in Figure 13-38, show the following:

(a) Green's function of (13-38) can be approximated by

$$G \simeq \sqrt{\frac{2}{\pi\beta\rho}} \, e^{-j(\beta\rho - \pi/4)} F(\beta\rho')$$

where

$$F(\beta\rho') = \frac{1}{n} \sum_{m=0}^{\infty} \varepsilon_m J_{m/n}(\beta\rho') \, e^{+j(m/n)(\pi/2)}$$
$$\times \left[\cos\frac{m}{n}(\phi - \phi') \pm \cos\frac{m}{n}(\phi + \phi')\right]$$

(b) Geometrical optics fields of (13-65) can be written as

$$F_G(\beta\rho') = \begin{cases} e^{j\beta\rho'\cos(\phi-\phi')} \pm e^{j\beta\rho'\cos(\phi+\phi')} \\ \qquad\qquad \text{for } 0 < \phi < \pi - \phi' \\ e^{j\beta\rho'\cos(\phi-\phi')} \\ \qquad\qquad \text{for } \pi - \phi' < \phi < \pi + \phi' \\ 0 \qquad \text{for } \pi + \phi' < \phi < n\pi \end{cases}$$

(c) Diffracted fields of (13-67) can be written as

$$V_D^{i,r}(\rho', \phi \mp \phi', n) = V_D^i(\rho', \phi - \phi', n)$$
$$\pm V_D^r(\rho', \phi + \phi', n)$$

$$V_D^{i,r}(\rho', \phi \mp \phi', n) = \frac{e^{-j\pi/4}}{\sqrt{2\pi\beta}} \frac{1}{n} \sin\left(\frac{\pi}{n}\right)$$

$$\times \left[\frac{1}{\cos\left(\dfrac{\pi}{n}\right) - \cos\left(\dfrac{\phi-\phi'}{n}\right)} \pm \frac{1}{\cos\left(\dfrac{\pi}{n}\right) - \cos\left(\dfrac{\phi+\phi'}{n}\right)} \right]$$
$$\times \frac{e^{-j\beta\rho'}}{\sqrt{\rho'}}$$

13.13. A unity amplitude uniform plane wave of soft polarization is incident normally on a half-plane at an angle of 45°, as shown in Figure P13-13. At an observation point P of coordinates $\rho = 5\lambda$, $\phi = 180°$ from the edge of the half-plane, determine the following:

(a) Incident geometrical optics field.

(b) Reflected geometrical optics field.

(c) Total geometrical optics field.

(d) Incident diffracted field.

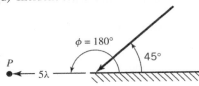

Figure P13-13

(e) Reflected diffracted field.
(f) Total diffracted field.
(g) Total field (geometrical optics plus diffracted).

13.14. Repeat Problem 13.13 for a hard polarization uniform plane wave.

13.15. An electric line source, whose normalized electric field at the origin is unity, is placed at a distance of $\rho' = 5\lambda$, $\phi' = 180°$ from the edge of a half-plane, as shown in Figure P13-15. Using the reciprocity principle of Figure 13-17 and the results of Problem 13.12, determine at large distances the following:
(a) Incident geometrical optics electric field.
(b) Reflected geometrical optics electric field.
(c) Total geometrical optics electric field.
(d) Incident diffracted electric field.
(e) Reflected diffracted electric field.
(f) Total diffracted electric field.
(g) Total electric field (geometrical optics plus diffracted).

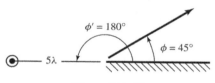

Figure P13-15

13.16. Repeat Problem 13.15 for a magnetic line source whose normalized magnetic field at the origin is unity. At each observation determine the magnetic field.

13.17. A uniform plane wave of soft polarization $(\mathbf{E} = \hat{\mathbf{a}}_z E^i)$, whose electric field normalized amplitude at the diffraction edge is 2, is incident upon a half-plane at an incidence angle of 60°. When $\rho = 5.5\lambda$, determine (in vector form) approximate values for the:
(a) Incident GO electric field at $\phi = 120°^-$ and $\phi = 120°^+$.
(b) Reflected GO electric field at $\phi = 120°^-$ and $\phi = 120°^+$.
(c) Incident diffracted electric field at $\phi = 120°^-$ and $\phi = 120°^+$.
(d) Reflected diffracted electric field at $\phi = 120°^-$ and $\phi = 120°^+$.
(e) Total electric field at $\phi = 120°^-$ and $\phi = 120°^+$.

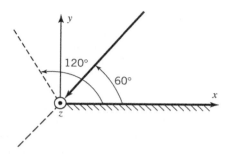

Figure P13-17

13.18. A uniform plane wave, of unity amplitude, is incident upon a PEC 90° wedge at an angle of $\phi' = 60°$. At an observation distance of $\rho = 81\lambda$ from the edge of the wedge and observation angle of 180°, compute the:
(a) Incident diffracted field using GTD.
(b) Approximate incident diffracted field using UTD.
(c) Reflected diffracted field using GTD.
(d) Approximate reflected diffracted field using UTD.

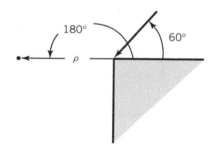

Figure P13-18

If you make any approximations, state as to why you think they are valid. Compare the corresponding answers of parts (a) through (d). Should they be approximately the same or different, and why?

13.19. A hard-polarized uniform plane wave, with the magnetic field directed in the $+z$ direction and traveling in the $+x$ direction, is incident upon a half-plane (knife edge), as shown in Figure P13-19. Assume an observation distance s from the edge of the wedge. Also assume that the magnitude of the incident magnetic field is unity.

Using exclusively the coordinate system shown in the figure, write vector expressions for the following fields:

(a) Incident magnetic field.
(b) Incident electric field.
(c) Backscattered diffracted magnetic field.
(d) Backscattered diffracted electric field.
(e) Magnitude (in dB) of the ratio of the backscattered magnetic field to the incident magnetic field at a distance of $s = 100\lambda$.

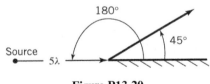

Figure P13-19

13.20. An electric line source, whose normalized electric field at the origin (leading edge of half-plane) is unity, is placed a distance 5λ from the edge of the half-plane, as shown in Figure P13-20. Determine, in the far field, the following:

(a) Incident GO electric field.
(b) Reflected GO electric field.
(c) Total GO electric field.
(d) Incident diffracted electric field.
(e) Reflected diffracted electric field.
(f) Total diffracted electric field.
(g) Total electric field (GO plus diffracted).

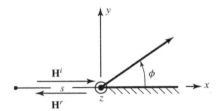

Figure P13-20

13.21. A uniform plane wave, with the electric/magnetic field directed in the $+z$ direction and traveling in the $+x$ direction, is incident upon a half-plane (knife edge), as shown in Figure 13-21. Assume an observation distance s from the edge of the wedge, and that the magnitude of the incident electric/magnetic field is unity at the leading edge.

Using exclusively the coordinate system shown in the figure, write vector expressions for the following fields. Assume $\phi = 90°$, $s = 9\lambda$.

(a) *Soft Polarization*: GTD diffracted electric field.
(b) *Hard Polarization*: GTD diffracted magnetic field.

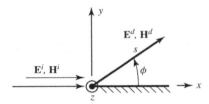

Figure P13-21

13.22. A unity amplitude uniform plane wave is incident normally on a 30° conducting wedge. Assume that the incident electric field is polarized in the z direction and the incident angle is 45°. Then determine at $\rho = 5.5\lambda$ the following:

(a) Incident GO electric field at $\phi = 225°^-$.
(b) Incident GO electric field at $\phi = 225°^+$.
(c) Approximate incident diffracted field at $\phi = 225°^-$.
(d) Approximate incident diffracted field at $\phi = 225°^+$.
(e) Approximate total electric field at $\phi = 225°^-$.
(f) Approximate total electric field at $\phi = 225°^+$.

Plus and minus refer to angles slightly greater or smaller than the designated values.

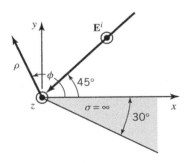

Figure P13-22

13.23. Repeat Problem 13.22 when the incident magnetic field of the unity amplitude uniform plane wave is polarized in the z direction. At each point determine the corresponding GO, diffracted, or total magnetic field.

13.24. A unity amplitude uniform plane wave of soft polarization is incident upon a half-plane, as shown in Figure P13-24. At a plane parallel and behind the half-plane perform the following tasks.
  (a) Formulate expressions for the incident and reflected geometrical optics fields, incident and reflected diffracted fields, and total field.
  (b) Plot along the observation plane the total field (geometrical optics plus diffracted fields) when $y_0 = 5\lambda$, $-5\lambda \le x_0 \le 5\lambda$.
  (c) Determine the total field at $y_0 = 5\lambda$ and $x_0 = 0$. Explain the result.

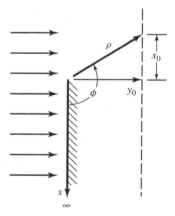

Figure P13-24

13.25. Repeat Problem 13.24 for a hard polarized uniform plane wave.

13.26. A unity amplitude uniform plane wave is incident normally on a two-dimensional conducting wedge, as shown in Figure P13-26.
  (a) Formulate expressions that can be used to determine the incident, reflected, and total diffracted fields away from the incident and reflected shadow boundaries.
  (b) Plot the soft and hard polarization normalized diffraction coefficients $(D_{s,h}/\sqrt{\lambda})$ as a function of ϕ for $n = 1.5$ and $n = 2$.
  (c) Simplify the expressions of part (a) when $n = 2$ (half-plane).

  (d) Formulate expressions for the two-dimensional scattering width of the half-plane $(n = 2)$ for soft and hard polarizations.
  (e) Simplify the expressions of part (d) for backscattering observations $(\phi = \phi')$. How can the results of this part be used to design low-observable radar targets?

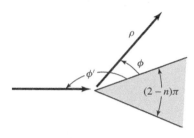

Figure P13-26

13.27. For the geometry of Problem 13.26 plot at $\phi = \phi'$ the normalized diffraction coefficient $(D_{s,h}/\sqrt{\lambda})$ as a function of the wedge angle $(1 \le n \le 2)$ when the polarization of the wave is (a) soft and (b) hard.

13.28. Repeat Problem 13.27 when the observations are made along the surface of the wedge $(\phi = 0°)$.

13.29. By approximating the integrand of the Fresnel integral of (13-63c) with a truncated Taylor series

$$e^{-j\tau^2} \simeq \sum_{n=0}^{M} \frac{(-j\tau^2)^n}{n!}$$

derive the small argument approximation of (13-74a) for the transition function.

13.30. By repeatedly integrating by parts the Fresnel integral of (13-63c), derive the large argument approximation of (13-74b) for the transition function.

13.31. Show that there is a finite discontinuity of unity amplitude with the proper polarity, similar to (13-83), along the following boundaries.
  (a) RSB of Figure 13-24a.
  (b) ISB of Figure 13-24b.
  (c) RSB of Figure 13-24b.

13.32. Using the Uniform Theory of Diffraction (UTD), the diffraction coefficient, assuming the incident wave illuminates the upper face of the PEC wedge, as shown in Figure P13-32, is given by (13-71a) and (13-71b). Identify which of the above cotangent functions and corresponding Fresnel transition functions F are used to eliminate the discontinuity introduced by the GO field along the:

(a) Incident Shadow Boundary (ISB).
(b) Reflection Shadow Boundary (RSB).

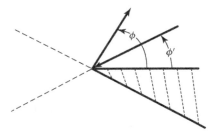

Figure P13-32

Identify the ISB and RSB in terms of angles, and indicate what happens individually, and as a product, to those two functions along the corresponding boundaries.

13.33. Repeat Problem 13.32 when the incident wave illuminates the lower face of the PEC wedge, as shown in Figure P13-33. The equations for the diffraction coefficients are the same, (13-71a) and (13-71b), as those given in Problem 13.32.

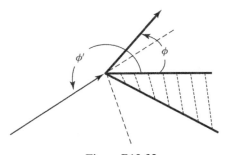

Figure P13-33

13.34. Compute the corresponding phases for the fields, GO and diffracted, for Figures 13-25a and 13-25b.

13.35. Repeat the calculations of Example 13-4 for a strip of width $w = 2\lambda$ and with the line source at a height of $h = 0.25\lambda$.

13.36. Repeat Example 13-5 for a hard-polarized uniform plane wave.

13.37. A uniform plane wave is incident upon a two-dimensional PEC strip, as shown in Figure 13-28. Formulate the problem, using GTD, for:

(a) Hard polarization (assume incident magnetic field of H_0 amplitude):
  • Backscattered/monostatic ($\phi = \phi'$) magnetic field and its backscattering/monostatic scattering width SW.
  • Bistatic scattered magnetic field and its bistatic scattering width SW.

(b) Soft polarization (assume incident electric field of E_0 amplitude).
  • Bistatic scattered electric field and its bistatic scattering width SW.

(c) Plot the bistatic and backscattering/monostatic patterns for $w = 2\lambda$, $f = 10$ GHz, and $\phi' = 120°$ for both soft- and hard-polarized uniform plane waves. For each polarization (soft and hard), the backscattering/monostatic patterns should be in two separate figures; one figure for the soft and the other for the hard. Similarly, the bistatic cases should be in two separate figures; one figure for the soft and the other for the hard. In each of the four figures, plot and compare, for the respective polarizations, the patterns based on EFIE/MoM SW/λ (dB) and SW(dBm).

13.38. A z-polarized electric-field uniform plane wave of unity amplitude is traveling in the negative y direction, as shown in Figure P13-38. In order to introduce a blockage to the wave, a half-plane (knife edge) is placed, as shown in the figure.
Assuming an observation point of $\phi = 330°$ and $s = 9\lambda$, determine, at that point, the:

(a) Vector electric field in the presence of the knife edge.

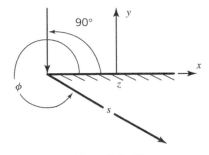

Figure P13-38

(b) Blockage (in dB) introduced to the wave by the knife-edge.

13.39. A unity amplitude uniform plane wave is incident normally on a two-dimensional strip of width w as shown in Figure P13-39.

(a) Formulate expressions for the fields when the observations are made below the strip along its axis of symmetry.

(b) Compute the field when $w = 3\lambda$ and $d = 5\lambda$ for soft and hard polarizations.

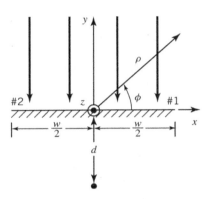

Figure P13-39

13.40. A transmitter and a receiver are placed on either side of a mountain that can be modeled as a perfectly conducting half-plane, as shown in Figure P13-40. Assume that the transmitting source is isotropic.

(a) Derive an expression for the field at the receiver that is diffracted from the top of the mountain. Assume the field is soft or hard polarized.

(b) Compute the power loss (in decibels) at the receiver that is due to the presence of the mountain when $h = 5\lambda$ and $d = 5\lambda$. Do this when the transmitter and receiver are both, for each case, either soft or hard polarized.

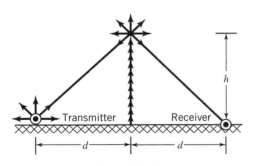

Figure P13-40

13.41. A rectangular waveguide of dimensions a and b, operating in the dominant TE_{10} mode, is mounted on a square ground plane with dimensions w on each of its sides, as shown in Figure P13-41. Using the geometry of Example 13-6 and assuming that the total geometrical optics field above the ground plane in the principal yz plane is given by

$$E_{\theta G}(\theta) = E_0 \left| \frac{\sin\left(\frac{\beta b}{2}\sin\theta\right)}{\frac{\beta b}{2}\sin\theta} \right| \frac{e^{-j\beta r}}{r}$$

$$0 \le \theta \le \frac{\pi}{2}$$

(a) Show that the fields diffracted from edges 1 and 2 in the principal yz plane are given by

$$E_{\theta 1}^d(\theta) = E_0 \frac{\sin\left(\frac{\beta b}{2}\right)}{\frac{\beta b}{2}}$$

$$\times V_B^i\left(\frac{w}{2}, \psi_1, n_1 = 2\right) e^{+j(\beta w/2)\sin\theta} \frac{e^{-j\beta r}}{r}$$

$$\psi_1 = \frac{\pi}{2} + \theta, \ \ 0 \le \theta \le \pi$$

$$E_{\theta 2}^d(\theta) = E_0 \frac{\sin\left(\frac{\beta b}{2}\right)}{\frac{\beta b}{2}}$$

$$\times V_B^i\left(\frac{w}{2}, \psi_2, n_2 = 2\right) e^{-j(\beta w/2)\sin\theta} \frac{e^{-j\beta r}}{r}$$

$$\psi_2 = \begin{cases} \frac{\pi}{2} - \theta, & 0 \le \theta \le \frac{\pi}{2} \\ \frac{5\pi}{2} - \theta, & \frac{\pi}{2} \le \theta \le \pi \end{cases}$$

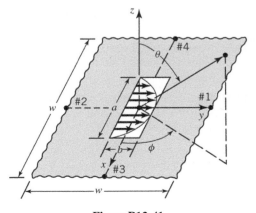

Figure P13-41

(b) Plot the normalized amplitude pattern (in decibels) for $0° \leq \theta \leq 180°$ when $w/2 = 14.825\lambda$ and $b = 0.42\lambda$.

13.42. A uniform plane wave is incident upon a square ground plane, of dimensions a and b in the x and y directions, respectively. The plane of incidence is the yz plane and the plane of observation is also the yz plane. For a hard-polarized (TE^x) shown in Figure 11-8b and Figure P13-42:

- Model/formulate the first-order diffractions from edges #1 and #2 for general bistatic scattering, and eventually 3-D RCS.
- Model/formulate the second-order bistatic diffractions from edges #1 and #2 due to first-order diffractions from edges #2 and #1, respectively, and eventually 3-D RCS.
- Plot, in one figure and three patterns, the 3-D RCS scattering bistatic patterns ($\sigma_{3\text{-}D}/\lambda^2$) (in dB) one due to first-order diffractions, and one due to first- plus second-order diffractions ($\sigma_{3\text{-}D}/\lambda^2$) (in dB) for an incidence angle of and $\theta_i = 30°$ and $a = b = 5\lambda$. On the same figure plot the 3-D RCS pattern based on Physical Optics ($\sigma_{3\text{-}D}/\lambda^2$) (in dB) shown in Figure 11-9a.
- Plot, in one figure and three patterns, the 3-D RCS scattering monostatic patterns ($\sigma_{3\text{-}D}/\lambda^2$) (in dB), in the upper $180°$ region, one due to first-order diffractions, and one due to first- plus second-order diffractions ($\sigma_{3\text{-}D}/\lambda^2$) (in dB) for $a = b = 5\lambda$. On the

same figure, plot the 3-D RCS pattern based on Physical Optics ($\sigma_{3\text{-}D}/\lambda^2$) (in dB) shown in Figure 11-9b.

Use (11-22e) to convert the 2-D RCS to 3-D RCS.

13.43. For Problem 13.42 and a soft-polarized (TM^x) shown in Figure 11-8(c), including the figures in Problem 13.42:

- Model/formulate the first-order diffractions from edges #1 and #2 for general bistatic scattering, and eventually 3-D RCS. There are no second- or higher-order regular diffractions for this polarization.
- Plot, in one figure and two patterns, the 3-D RCS scattering bistatic patterns ($\sigma_{3\text{-}D}/\lambda^2$) (in dB) due to first-order diffractions for an incidence angle of $\theta_i = 30°$ and $a = b = 5\lambda$. On the same figure plot the 3-D RCS based on PO ($\sigma_{3\text{-}D}/\lambda^2$) (in dB) shown in Fig. 11-9a.
- Plot, in one figure and two patterns, the 3-D RCS scattering monostatic patterns ($\sigma_{3\text{-}D}/\lambda^2$) (in dB), in the upper $180°$ region, due to first-order diffractions for $a = b = 5\lambda$. On the same figure, plot the 3-D RCS pattern based on Physical Optics ($\sigma_{3\text{-}D}/\lambda^2$) (in dB) shown in Fig. 11-9b.

Use (11-22e) to convert the 2-D RCS to 3-D RCS.

13.44. A soft-polarized spherical wave is incident, at a normal incidence angle, upon a half-plane with a straight edge. The wave emanates from a source a distance s' from the edge of the half-plane. In order to reduce the diffracted field, the straight edge is replaced with a curved edge of radius a.

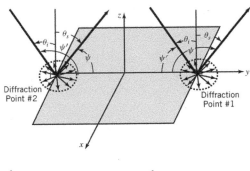

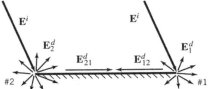

Figure P13-42

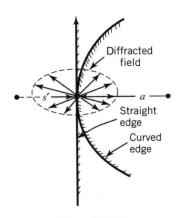

Figure P13-44

(a) Derive a simplified expression for the ratio of amplitude spreading factor reduction for the backscattered field, by a curved edge relative to that by the straight edge, at a large distance s from the edge of the half-plane.

(b) Determine, in dB, the relative amplitude reduction of the backscattered diffracted field by a curved edge of radius $a = 25\lambda$ and a source distance of $s' = 100\lambda$.

13.45. A uniform plane wave with an electric field given by

$$\mathbf{E}^i = \hat{\mathbf{a}}_z \, e^{-j\beta x}$$

is incident upon a flat conducting square plate, as shown in Figure P13-45. To reduce the backscattered diffractions from the leading edge, the square plate is replaced by a circular plate. Determine the radius of curvature a of the square plate so that the backscattered diffractions at a distance of 50λ from the leading edge diffraction point of the curved edge are -20 dB than those of the leading straight edge.

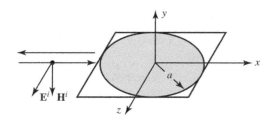

Figure P13-45

13.46. A unity amplitude uniform plane wave is incident at a grazing angle on a circular ground plane of radius a, as shown in Figure P13-46.

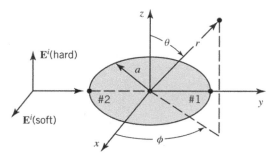

Figure P13-46

(a) Formulate expressions for the field diffracted from the leading edge (#2) of the ground plane along the principal yz plane for both soft and hard polarizations.

(b) Locate the position of the caustic for the leading edge.

(c) Assuming the incident electric field amplitude at the leading edge of the plate is 10^{-3} V/m and its phase is zero, compute for each polarization at $r = 50\lambda$ when $a = 2\lambda$ the backscattered electric field that is due to the leading edge of the plate.

13.47. Show that the fields radiated by a circular loop with nonuniform equivalent electric and magnetic current I_ϕ^e and I_ϕ^m are given, respectively, by (13-105a) and (13-105b). If the equivalent currents I_ϕ^e and I_ϕ^m are uniform, show that (13-105a) and (13-105b) reduce, respectively, to (13-106a) and (13-106b).

13.48. An infinitesimal dipole is placed on the tip of a finite cone, as shown in Figure P13-48. The total-geometrical optics magnetic field radiated by the source in the presence of the cone, referred to the center of the base of the cone, is given by

$$H_{\phi G} = R(\theta) e^{j\beta s \cos\theta} \cos(\alpha/2)$$

where $R(\theta)$ is the field distribution when the cone is infinite in length ($s = \infty$). For the finite-length cone there is also a diffracted field forming a ring source at the base of the cone.

(a) Using (13-105b), show that the diffracted magnetic field is given by

$$H_\phi^d = b\,R\!\left(\theta = \pi - \frac{\alpha}{2}\right) V_B^i(s, \psi, n)$$

$$\times \int_0^{2\pi} \cos\phi\, e^{j(\beta b)\sin\theta \cos\phi}\,d\phi, \quad \psi = \frac{\alpha}{2} + \theta$$

where $V_B^i(s, \psi, n)$ is the incident diffraction function and ϕ is the azimuthal observation angle.

(b) Assuming that the base of the cone is large ($b \gg \lambda$) and the observations are made away from the symmetry axis so that $\sin\theta > 0$, show that by using the method of steepest descent of Appendix VI the integral formed by the ring source diffracted field from the rim of the cone

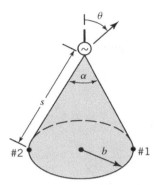

Figure P13-48

reduces to the field diffracted from two diametrically opposite points on the rim (i.e., points #1 and #2 of Figure P13-48).

13.49. For the $\lambda / 4$ monopole on the circular ground plane of Example 13-7, model the rim of the ground plane as a ring source radiator using equivalent current concepts of Section 13.3.5 to correct for the caustic formed when the observations are made near the axis ($\theta \simeq 0°$ and 180°) of the ground plane.

(a) Derive expressions for the diffracted field from the rim using the ring radiator of Figure 13-38.

(b) Show, using the method of steepest descent (saddle point method), that the ring radiator radiation of part a reduces to a two-point diffraction away from the symmetry axis.

(c) Compute and plot the pattern near the axis ($\theta \simeq 0°$ and 180°) of the ground plane when the diameter of the ground plane is $d = 4.064\lambda$. Use this part of the pattern to complement that computed using the two-point diffraction of Example 13-7.

13.50. We all have seen the general geometrical shapes of the stealth bomber and fighter (see Figure 13-1); most of the fuselage consists of wedge-type shapes. There must be a reason for that. To answer some of the questions, formulate the problem, and show why a wedge-type of geometry is desired, assume that a uniform plane wave, of either soft or hard polarization, impinges either upon a 2-D PEC wedge of included angle WA or a 2-D PEC cylinder of circular cross section and radius a, as shown in Figure P13-50:

(a) In one figure with two curves, plot the normalized 2-D monostatic scattering width SW$[\sigma_{2\text{-D}}/\lambda(\text{dB})]$ of the wedge vs. wedge angle WA (in degrees) $[0 \leq \text{WA} \leq 60°]$ for both hard and soft polarization.

(b) In one figure with two curves, plot the normalized 2-D monostatic scattering width SW$[\sigma_{2\text{-D}}/\lambda(\text{dB})]$ of the cylinder vs. cylinder radius a wavelengths) $[0 \leq a \leq 10\lambda]$ for both hard and soft polarization.

(c) In one figure with two curves, plot the normalized 2-D monostatic scattering width SW $[\sigma_{2\text{-D}}/\lambda(\text{dB})]$ of the wedge WA (in degrees) $[0 \leq \text{WA} \leq 60°]$ and that of cylinder with radius a $[0 \leq a \leq 10\lambda]$ for hard polarization.

(d) In one figure with two curves, plot the normalized 2-D monostatic scattering width SW $[\sigma_{2\text{-D}}/\lambda(\text{dB})]$ of the wedge WA (in degrees) $[0 \leq \text{WA} \leq 60°]$ and that of the cylinder with radius a $[0 \leq a \leq 10\lambda]$ for soft polarization.

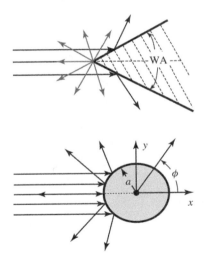

Figure P13-50

13.51. A uniform plane wave is incident upon a circular ground plane of radius a, as shown in the figure that follows. The plane of incidence is the yz plane and the plane of observation is also the yz plane. Referring to [36], Figure P13-51, and for hard-polarized (TE^x) wave:

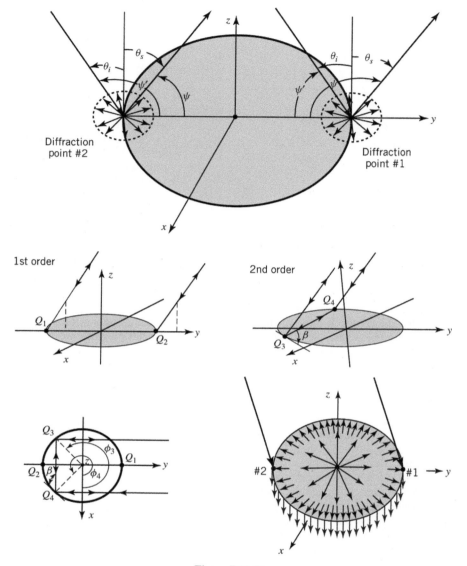

Figure P13-51

- Model the first-order 3-D diffractions from edges #1 and #2 for monostatic scattering. Use the middle of the plate as the reference point.
- Model the second-order monostatic 3-D diffractions from edges #1 and #2 due to first-order diffractions from edges #2 and #1, respectively.
- Model the second-order 3-D diffractions from the migrating points #3 and #4.
- Model the 'ring-radiator' contributions. In [36] they are referred to as

Axial Caustic Correction formulated using *Equivalent Currents*.
- Plot, in one figure and two patterns, the 3-D RCS scattering monostatic patterns $(\sigma_{3\text{-}D}/\lambda^2)$ (in dB), in the upper $180°$ region, due to first-order (#1 and #2) diffractions plus ring radiator for $a = 3.516\lambda$. On the same figure, plot the 3-D RCS pattern based on Physical Optics $(\sigma_{3\text{-}D}/\lambda^2)$ (in dB).
- Plot, in one figure and three patterns, the 3-D RCS scattering monostatic patterns $(\sigma_{3\text{-}D}/\lambda^2)$ (in dB), in the

upper 180° region, one due to first-order (#1 and #2) diffractions, and one due to first- (#1 and #2) plus all second-order (#1, #2, #3, and #4 points) diffractions plus ring radiator $(\sigma_{3\text{-D}}/\lambda^2)$ (in dB) for $a = 3.516\lambda$. On the same figure, plot the 3-D RCS pattern based on Physical Optics $(\sigma_{3\text{-D}}/\lambda^2)$ (in dB).

13.52. For Problem 13.51 and a soft-polarized (TM^x) wave, along with the figures of Problem 13.51:

- Model/formulate the first-order 3-D diffractions from edges #1 and #2 for monostatic scattering. For this polarization, there are no second-order regular diffractions from edges #1 and #2. Use the middle of the plate as the reference point.
- Model/formulate the second-order 3-D diffractions from the migrating points #3 and #4.
- Model/formulate the 'ring-radiator' contributions. In [36] they are referred to as *Axial Caustic Correction* formulated using *Equivalent Currents*.
- Plot, in one figure and two patterns, the 3-D RCS scattering monostatic patterns $(\sigma_{3\text{-D}}/\lambda^2)$ (in dB), in the upper 180° region, due to first-order diffractions (#1 and #2) plus the 'ring-radiator' for $a = 3.516\lambda$. On the same figure, plot the 3-D RCS pattern based on Physical Optics $(\sigma_{3\text{-D}}/\lambda^2)$ (in dB).
- Plot, in one figure and three patterns, the 3-D RCS scattering monostatic patterns $(\sigma_{3\text{-D}}/\lambda^2)$ (in dB), in the upper 180° region, one due to first-order diffractions (#1 and #2), and one due to first- (#1 and #2) plus second-order (#3 and #4) diffractions plus the ring radiator for $a = 3.516\lambda$. On the same figure, plot the 3-D RCS pattern based on Physical Optics $(\sigma_{3\text{-D}}/\lambda^2)$ (in dB).

13.53. According to PO, the maximum monostatic scattering RCS of flat plates, irrespective of polarization and plate configuration (rectangular, square, circular, elliptical, or any other shape), occurs at normal incidence and it is proportional to the square of the plate. Assuming a square ground plane and specifying the total length of one of its sides, w (in λ), you can determine the radius a (in λ) of

the circular ground plane so that both have the same area, as shown in Figure P13-53. For the circular ground plane the blending angle between the two-point diffraction and the 'ring radiator' is referred to as *thetao* (typically 10–30 degrees). The MATLAB program **Rect_Circ_Scat**, for both TE^x and TM^x polarizations, performs this task. Using the MATLAB **Rect_Circ_Scat**.

(a) Compute and plot the monostatic 3-D RCS $(\sigma_{3\text{-D}}/\lambda^2)$ (in dB), in the upper 180° region, for hard polarization and $w = 5\lambda$ that will include four curves on the same figure.

- One curve of a square ground plane using diffraction modeling that includes both first- and second-order diffractions from points #1 and #2.
- One curve of a square ground plane using PO.
- One curve of a circular ground plane using diffraction modeling that includes both first- and second-order diffractions from points #1 and #2, second-order diffractions from points #3 and #4, and ring radiator.
- One curve of a circular ground plane using PO.

(b) Compute and plot the monostatic 3-D RCS $(\sigma_{3\text{-D}}/\lambda^2)$ (in dB), in the upper 180° region, for soft polarization and $w = 5\lambda$ that will include four curves on the same figure.

- One curve of a square ground plane using diffraction modeling that includes first-order diffractions from points #1 and #2.

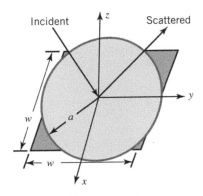

Figure P13-53

- One curve of a square ground plane using PO.
- One curve of a circular ground plane using diffraction modeling that includes first-order diffractions from points #1 and #2, second-order diffractions from points #3 and #4, and ring radiator.
- One curve of a circular ground plane using PO.

13.54. The normalized total geometrical optics field radiated in the principal xz plane (H plane; $\phi = 0°, 180°$) by the rectangular waveguide of Problem 13.41 (Figure P13-41) is given by

$$E_{\phi G} = E_0 \cos\theta \frac{\cos\left(\frac{\beta a}{2}\sin\theta\right)}{\left(\frac{\beta a}{2}\sin\theta\right)^2 - \left(\frac{\pi}{2}\right)^2} \frac{e^{-j\beta r}}{r}$$

$$0° \leq \theta \leq 90°$$

Use the slope diffraction concepts of Section 13.3.6.

(a) Formulate the field diffracted along the xz plane using two-point diffraction (points #3 and #4 of Figure P13-41).

(b) Plot (in decibels) the normalized amplitude pattern when $a = \lambda/2, w = 4\lambda$.

13.55. An infinite magnetic line source is placed on a two-dimensional square conducting cylinder at the center of its top side. Use successive single-order diffractions on each of the edges of the cylinder.

(a) Formulate expressions for the magnetic field that would be observed at the center of the bottom side of the cylinder.

(b) Compute the power loss (in decibels) at the observation point that is due to the presence of the cylinder when $w = 5\lambda$. Assume far-field observation approximations.

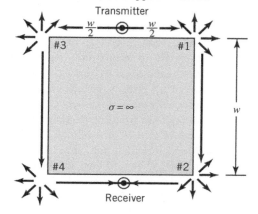

Figure P13-55

CHAPTER 14

𐂃𐂃𐂃𐂃𐂃𐂃𐂃𐂃𐂃𐂃𐂃𐂃

Diffraction by a Wedge with Impedance Surfaces

14.1 INTRODUCTION

In Chapter 11 we introduced the exact solution of wave scattering by a perfectly conducting wedge while in Chapter 13 we examined its asymptotic high-frequency solution and reduced it to geometrical optics (GO) fields, incident and reflected, and diffracted fields (GTD and UTD) (incident and reflected diffracted fields). The solution of electromagnetic scattering from a penetrable or lossy wedge has not been solved using exact boundary conditions based on Maxwell's equations. It has always been necessary to use some approximate boundary conditions from which the exact solution can be obtained. Hence, it is important to recognize that when an exact solution is discussed, it is for a boundary condition that in reality can only be approximated by physical materials or structures.

In recent years great interest has been generated, because of the design of low-profile (stealth) radar targets and the modeling of wave propagation for wireless communication in the presence of penetrable structures (such as buildings, ridges, hills, and various other structures), to extend the theories of perfectly conducting materials to include materials with penetrable characteristics. Many aircraft are now built with composite materials which, with appropriate shaping and integration of materials, can reduce the radar echo, RCS, and thus minimize the probability of detection. See Figure 13-1 of the F-117 Nighthawk, whose surface is composed primarily of faceted flat plates and wedges judiously oriented so that the maximum scattered field is toward specular direction and away from source of detection. In wireless communication, the wave propagation, direct and indirect (multipath), plays a pivotal role in the performance of the wireless mobile communication system using, for example, the Bit Error Ratio (BER) as a metric.

The most common of these boundary conditions is the *impedance boundary condition,* which relates the tangential components of the electric and magnetic fields by a surface impedance. The impedance surface boundary condition was first used in electromagnetic scattering by a number of Russian authors in the early 1940s [1], and it is generally attributed to Leontovich [2, 3], and commonly referred to as the *Leontovich boundary condition.* It was initially used to describe wave propagation over the earth's surface by specifying an impedance boundary for the ground surface. It was later used for other electromagnetics problems and was an effort to simplify very complex

Balanis' Advanced Engineering Electromagnetics, Third Edition. Constantine A. Balanis.
© 2024 John Wiley & Sons, Inc. Published 2024 by John Wiley & Sons, Inc.
Companion Website: www.wiley.com/go/balanis/advancedengineeringelectromagnetics3e

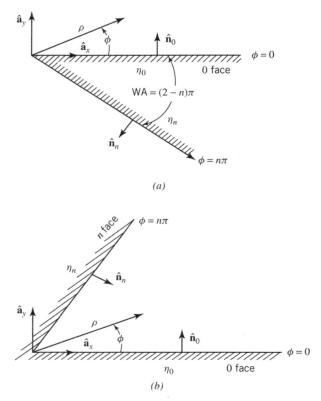

(a)

(b)

Figure 14-1   Exterior and interior wedges with impedance surfaces [28]. (*a*) Exterior wedge. (*b*) Interior wedge.

electromagnetics boundary-value problems, which were difficult to solve exactly. It is especially useful for surfaces that support surface waves.

The solution of a half-plane with impedance boundary conditions has been examined and reported in [4–8]. The more general problem of plane wave incidence on a wedge of arbitrary interior angle was solved by Maliuzhinets [9, 10]. The wedge was permitted to have different impedances on each face but only normal incidence was considered. No solution for oblique incidence on a general wedge has been obtained, except for a very few special cases. Maliuzhinets' exact solution of the wedge was accompanied by a steepest-descent analysis, very similar to Keller's theory for the PEC wedge, which decomposed the exact solution into *incident, reflected, diffracted,* and *surface wave* components. These Keller-type diffraction terms are not valid near the shadow boundaries and do not eliminate the discontinuities in the GO and surface wave terms; they had to be extended to include UTD type of formulations. There have been numerous publications [11–61] addressing various special cases of the diffraction by an impedance wedge, including that of the half plane. This chapter follows the presentation of [28, 30].

The impedance wedge, shown in Figure 14-1, is a canonical geometry whose faces can be represented by surface impedances. The exact solution for this geometry for the fields both interior and exterior to the wedge, similar to that of the PEC wedge of Chapter 11, does not exist because appropriate eigenfunctions and eigenvalues that satisfy Maxwell's equations and the boundary conditions cannot be found. However, an exact solution for the fields exterior to the wedge, for normal plane wave incidence and assuming uniform, but not identical surface impedances to the faces of the wedge, does exist and it was first presented by Maliuzhinets [9, 10]. The wedge, with wedge angle WA $= (2-n)\pi$ and $0 < n \le 2$, has two faces located at $\phi = 0$ and $\phi = n\pi$ with normalized uniform surface impedances of η_0 and η_n, respectively. An exterior wedge (Figure 14-1a) has values of n in the range of $1 \leq n \leq 2$ while for an interior wedge (Figure 14-1b) $0 < n < 1$. Maliuzhinets' solution can be used to extract geometrical optics, diffracted fields, and surface waves, as was done for the PEC wedge in Chapter 13.

14.2 IMPEDANCE SURFACE BOUNDARY CONDITIONS

The impedance surface boundary condition is simply a statement that the tangential electric and magnetic field vectors are related by a constant impedance that is related only to the properties and configurations of the material and is independent of the source illumination. Mathematically, it is essentially a boundary condition of the third kind, as it relates a function and its derivative at the surface through some constant. As such, it is only an approximation to the actual boundary conditions, developed through Maxwell's equations, which exist on complex structures. The accuracy of the approximation depends on the composition, geometry, and illumination of the actual surface of interest. The impedance boundary condition has been applied in the past to homogeneous and inhomogeneous materials, layered structures, lossy materials, and randomly rough surfaces.

Physically, it is easy to understand where a concept, like the impedance boundary condition, originates. For plane, cylindrical, and spherical waves in inhomogeneous media and sufficiently far from the source, the electric and magnetic field vectors are mutually orthogonal and are related to each other by the intrinsic impedance of the homogeneous media through which they propagate. For free space, this impedance is $Z_0 = 377$ ohms. In vector form, this statement is expressed as

$$\mathbf{E} = -Z_0 \hat{\mathbf{a}}_r \times \mathbf{H} \tag{14-1}$$

where $\mathbf{E}$ and $\mathbf{H}$ are the electric and magnetic field vectors, Z_0 is the intrinsic impedance of free space (377 ohms), and $\hat{\mathbf{a}}_r$ is a unit vector in the direction of wave propagation.

The impedance boundary condition simply extends this wave property to surface interfaces with which the wave interacts. The boundary condition requires that the tangential electric and magnetic field vectors at the interface be mutually orthogonal and can be related by the impedance Z of the surface. The normalized impedance η, relative to the free-space value Z_0, is commonly specified, and is given by $\eta = Z/Z_0$. This notation can cause confusion because many electromagnetics texts, as this one in previous chapters, use both η and Z as the total intrinsic impedance of the medium. Nonetheless, this notation has been utilized for surface impedances as early as 1959 [6] and has been used consistently since. Hence, in this chapter, η refers to a normalized impedance and Z refers to the total impedance, $Z = \eta Z_0$. For passive materials, the real part of η must be nonnegative.

In a vector equation form, the impedance surface boundary condition, referred to also as the *Leontovich Boundary Condition*, can be written as [1, 6]

$$\mathbf{E} - (\hat{\mathbf{n}} \cdot \mathbf{E})\hat{\mathbf{n}} = Z\,\hat{\mathbf{n}} \times \mathbf{H} = \eta Z_0 (\hat{\mathbf{n}} \times \mathbf{H}) \tag{14-2}$$

where

$\mathbf{E}$ = the electric field vector
$\mathbf{H}$ = the magnetic field vector
$\hat{\mathbf{n}}$ = the outward unit normal to the surface

It is noted that $(\hat{\mathbf{n}} \cdot \mathbf{E})\hat{\mathbf{n}}$ is simply the normal component of $\mathbf{E}$; hence, the left side of (14-2) represents the tangential electric field. The term $\hat{\mathbf{n}} \times \mathbf{H}$ is a vector whose magnitude equals the tangential magnetic field; however, its direction is perpendicular to both the tangential magnetic field and the unit normal. Hence, $\hat{\mathbf{n}} \times \mathbf{H}$ is also tangential to the surface. The tangential electric and magnetic fields are then related by the constant of proportionality Z, the surface impedance $Z = \eta Z_0$. In words, the tangential electric and magnetic fields are related by the surface impedance of the material.

14.3 IMPEDANCE SURFACE REFLECTION COEFFICIENTS

To determine the reflection coefficients appropriate for an impedance surface, the reflection of a plane wave at a planar interface is examined. At high frequencies, these reflection coefficients are also appropriate for curved boundaries for which the radius of curvature is relatively large compared to the wavelength. In Figure 14-2, a plane wave in free space and incident on an impedance boundary is illustrated. For soft (perpendicular) polarization, Figure 14-2a, **E** is perpendicular to the plane of incidence, and for hard (parallel) polarization **E** is parallel to it.

For *soft polarization*, Figure 14-2a, the impedance boundary condition of (14-2) (using the incident electric and magnetic fields as reference and $\phi_r = \phi_i$), reduces to

$$E^i + E^r = \eta Z_0 (H^i - H^r) \sin \phi_i \tag{14-3a}$$

while for the *hard polarization*, Figure 14-2b, the impedance boundary condition of (14-2) (using $\phi_r = \phi_i$), simplifies to

$$(E^i + E^r) \sin \phi_i = \eta Z_0 (H^i - H^r) \tag{14-3b}$$

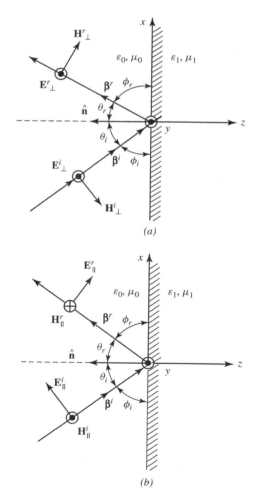

Figure 14-2 Oblique incident and reflected fields from a planar interface with impedance boundary for soft and hard polarizations and direction of their components. (*a*) Soft (perpendicular) polarization. (*b*) Hard (parallel) polarization.

where $\eta = Z_1/Z_0 = \eta_1/Z_0$ $(Z_1 = \eta_1)$ is the normalized surface impedance of the medium. The reader should pay attention to the directions of the incident and reflected fields, as indicated in Figures 14-2a and 14-2b. Using $E^i = Z_0 H^i$ and $E^r = Z_0 H^r$ in (14-3a) and (14-3b), the reflection coefficients for soft and hard polarization become, respectively

$$\Gamma_s = \frac{E^r}{E^i} = \frac{\eta \sin\phi_i - 1}{\eta \sin\phi_i + 1} = \frac{\sin\phi_i - \dfrac{1}{\eta}}{\sin\phi_i + \dfrac{1}{\eta}} \qquad (14\text{-}4a)$$

$$\Gamma_h = \frac{E^r}{E^i} = \frac{\eta - \sin\phi_i}{\eta + \sin\phi_i} = \frac{1 - \dfrac{\sin\phi_i}{\eta}}{1 + \dfrac{\sin\phi_i}{\eta}} \qquad (14\text{-}4b)$$

For a perfect electric conductor (PEC), $\eta = 0$, the reflection coefficients of (14-4a) and (14-4b) reduce, respectively, to $\Gamma_s = -1$ and $\Gamma_h = -1$. This implies, as it should, that the tangential electric field components vanish on the PEC surface. Similarly, for a nonphysical perfect magnetic conductor (PMC), $\eta = \infty$, the reflection coefficients of (14-4a) and (14-4b) reduce, respectively, to $\Gamma_s = +1$ and $\Gamma_h = +1$. This implies, as it should, that the tangential magnetic field components vanish on the PMC surface. At grazing incidence, $\sin\phi_i = 0$, $\Gamma_s = -1$ and $\Gamma_h = +1$ for the lossy impedance surface. These are based on the directions of the incident and reflected fields of Figures 14-2a and 14-2b.

The reflection coefficients of (14-4a) and (14-4b) are, respectively, the same as those of (5-17a) and (5-24c), which are derived based on Snell's laws, provided that $\sin\phi_t = \sin(\pi/2 - \theta_t) = 1$. This occurs when the constitutive parameters of the two media forming the interface are such that $\sin\phi_t = \sqrt{1 - \frac{\mu_0 \varepsilon_0}{\mu_1 \varepsilon_1}\cos^2\phi_i} \approx 1$, which is satisfied provided $\mu_1 \varepsilon_1 \gg \mu_0 \varepsilon_0$. Therefore, (14-4a) and (14-4b) are valid provided this relationship is satisfied.

For both polarizations, (14-4a) and (14-4b) show that there may be a particular Brewster angle ϕ_B [see definition in (5-33a), (5-33b)] for which the reflected field is zero. For the soft polarization, the Brewster angle $\phi_B = \sin^{-1}(1/\eta)$, while for the hard polarization $\phi_B = \sin^{-1}(\eta)$. If ϕ_B is complex, no physical Brewster angle exists. If η is complex, ϕ_B will be complex (and nonphysical) for both polarizations. If η is real and $0 \leq \eta \leq 1$, ϕ_B is real for the hard polarization and complex for the soft polarization. If η is real and $\eta \geq 1$, then ϕ_B is real for the soft polarization and complex for the hard polarization. Hence, a real Brewster angle cannot exist for both polarizations except when $\eta = 1$, and for this normalized impedance $(\eta = 1)$, $\phi_B = \pi/2$ for both polarizations. Obviously, this is just the case of normal incidence on a matched surface for which there is no reflection for either polarization. Since the real part of η must be nonnegative, and since the inverse sine function maps the right half-plane into the strip $0 \leq \text{Re}[\sin^{-1}\eta] < \pi/2$, then it must always be true that $0 \leq \text{Re}[\phi_B] < \pi/2$.

In practice, these reflection coefficients are quite accurate for many scattering geometries. However, for one special case, some intrinsic difficulties may arise; this special case is the hard polarization near grazing incidence for an imperfect electric conductor. Similar problems arise for the imperfect magnetic conductor. It is noted that for perfect conductors, $\eta = 0$ and $\Gamma_h = -1$. However, at grazing incidence $\sin\phi_i = 0$ and $\Gamma_h = +1$. For imperfect conductors (η complex) near grazing incidence ($\sin\phi_i \simeq 0$), the hard reflection coefficient may change very rapidly from $+1$ to -1 for very small changes in the grazing angle or the conductivity. In general, it is difficult to develop methods for the imperfectly conducting case that revert uniformly to the perfectly conducting case as the surface impedance approaches zero for the hard polarization at grazing incidence.

In passing, it is worth mentioning that the grazing incidence case is especially important to the radar community and hence deserves attention. It is well known that the lossy earth or sea surface near grazing incidence ($\sin \phi_i \simeq 0$) has reflection coefficients of $\Gamma_s = -1$ [i.e., the incident and reflected electric fields are in opposite directions using the direction designation of Figure 14-2a and definition of (14-4a)], and $\Gamma_h = +1$ [i.e., the incident and reflected electric fields are also in opposite directions using the direction designation of Figure 14-2b and definition of (14-4b)]. This implies that radar targets flying very near the ground or sea surface ($\sin \phi_i \simeq 0$) are especially difficult to detect because, at grazing incidence, the incident reflected electric fields are nearly equal in magnitude and opposite in phase. Hence, these two components effectively cancel each other, giving no radar echo/return, which makes low-flying radar targets difficult to track [62]. This physical phenomenon is exploited in many sea skimming missiles that fly within a few meters of the ocean surface.

14.4   THE MALIUZHINETS IMPEDANCE WEDGE SOLUTION

An exact solution for the fields exterior to the wedge, for normal plane wave incidence and assuming uniform, but not identical, surface impedances on the faces of the wedge, does exist and it was first presented by Maliuzhinets [9, 10]. The wedge has two faces located at $\phi = 0$ and $\phi = n\pi$, with normalized uniform surface impedances of η_0 and η_n, respectively. An interior wedge has values of n in the range of $0 < n < 1$ while for an exterior wedge $1 \leq n \leq 2$. Maliuzhinets' solution can be used to extract geometrical optics, diffracted fields, as was done for the PEC wedge, and surface waves.

Let us assume that a plane wave is incident from ϕ' and the observation point P is at a distance ρ from the edge of the wedge at an angle ϕ, where ϕ' and ϕ are both measured from the 0 face, as shown in Figure 14-3. Assuming a field of amplitude U_0, the exact solution for the total field U_t, including the incident and reflected fields is [9]

$$U_t(\rho,\phi) = jU_0 \frac{1}{2n\pi} \int_\gamma \frac{\Psi\left(z + \frac{n\pi}{2} - \phi\right)}{\Psi\left(\frac{n\pi}{2} - \phi'\right)} \frac{\sin\left(\frac{\phi'}{n}\right)}{\cos\left(\frac{z - \phi}{n}\right) - \cos\left(\frac{\phi'}{n}\right)} e^{j\beta\rho\cos z}\, dz \qquad (14\text{-}5)$$

where γ is the contour shown in Figure 14-4. *It should be remembered that this is an exact solution for an approximate boundary condition (i.e., the impedance boundary condition), and it is not an exact solution for the exact boundary conditions derived from Maxwell's equations.* Also in order to match the notation of Chapter 13 for the PEC wedge, the complex plane is represented by z, instead of α used in the corresponding published literature.

In (14-5), $\Psi(z)$ is the *auxiliary* Maliuzhinets function [9], and it depends explicitly on the integration variable z and implicitly on the parameters n, θ_0 and θ_n. Perhaps it would be more evident by writing $\Psi(z)$ as $\Psi(z; n, \theta_0, \theta_n)$; however, in the existing literature, it is exclusively represented by $\Psi(z)$. The function $\Psi(z)$ and its properties are discussed in [28]. Both the E_y (soft) and

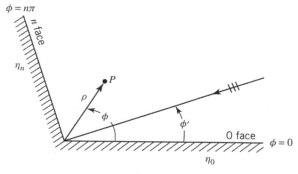

Figure 14-3   Interior wedge with impedance surfaces.

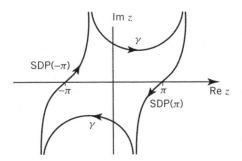

Figure 14-4 Integration contours for exact Maliuzhinets solution of wedge with impedance surfaces [28, 30]. (Source: T. Griesser and C. A. Balanis, "Reflections, diffractions, and surface waves for an interior impedance wedge of arbitrary angle," *IEEE Trans. Antennas Propagat.*, © 1989, IEEE.)

H_y (hard) are included in this representation [58] by simply using the normalized surface imped-ance, or its inverse, in determining the Brewster angles θ_0 and θ_n. The Brewster angles for each polarization are determined as follows.

For the *soft* (TM^z, E_z) *polarization*, set (14-4a) equal to zero and solve for ϕ_i; it leads to:

$$U_t(\rho,\phi) = E_z(\rho,\phi) \tag{14-6}$$

$$\theta_0 = \phi_{i0} = \sin^{-1}\left(\frac{1}{\eta_0}\right) \tag{14-6a}$$

$$\theta_n = \phi_{in} = \sin^{-1}\left(\frac{1}{\eta_n}\right) \tag{14-6b}$$

while for the *hard* (TE^z, H_z) *polarization*, set (14-4b) equal to zero and solve for ϕ_i; it leads to:

$$U_t(\rho,\phi) = H_z(\rho,\phi) \tag{14-7}$$

$$\theta_0 = \phi_{i0} = \sin^{-1}(\eta_0) \tag{14-7a}$$

$$\theta_n = \phi_{in} = \sin^{-1}(\eta_n) \tag{14-7b}$$

Since η_0 and η_n must be nonnegative real for passive materials, θ_0 and θ_n will always lie in the strip $(0 < \text{Re}[\theta_0] < \pi/2)$ and $(0 < \text{Re}[\theta_n] < \pi/2)$ due to the principal branch cuts of the inverse sine. For the soft polarization, θ will be real if η is real and $\eta \geq 1$. Similarly, for hard polarization, θ will be real if η is real and $0 \leq \eta \leq 1$. Otherwise the inverse sine will be complex and the corresponding Brewster angle θ_B will also be complex.

To calculate the inverse sine of a complex number or a real number with magnitude greater than unity, the following identities can be used since many computer software programs do not provide a complex inverse sine function.

$$\sin^{-1}(z) = \frac{\pi}{2} - \cos^{-1}(z) \tag{14-8}$$

$$\cos^{-1}(z) = \text{Arg}[G(z)] - j \ln|G(z)| \tag{14-8a}$$

$$G(z) = \begin{cases} z - \sqrt{z^2 - 1} & (\text{Re}[z])(\text{Im}[z]) < 0 \\ z + \sqrt{z^2 - 1} & (\text{Re}[z])(\text{Im}[z]) \geq 0 \end{cases} \tag{14-8b} \\ \tag{14-8c}$$

$$-\pi < \text{Arg}\, G(z) \leq \pi \tag{14-8d}$$

$$-\frac{\pi}{2} < \text{Arg}\left[\sqrt{z^2 - 1}\right] \leq \frac{\pi}{2} \tag{14-8e}$$

The exact wedge solution in its integral form of (14-5) cannot be easily evaluated. The integrand contains a ratio of *auxiliary Maliuzhinets functions*, each of which can be written as a product of four *Maliuzhinets functions*. Each Maliuzhinets function can only be written in a complex integral form or as an infinite product, and hence the Maliuzhinets function is not easily evaluated for arbitrary argument and wedge angle. Finally, the very complicated integrand must be integrated along the complex contour γ of Figure 14-4. Overall, the integral of (14-5) is difficult to evaluate efficiently.

To construct a more useful high-frequency asymptotic expansion of the integral, the *Method of Steepest Descent* (also known as the *Saddle Point Method*) of Chapter 13 will be used in this chapter. The steepest descent paths must first be located, and then the exact solution must be transformed to an integral along the steepest descent paths, as was done in Chapter 13 for the PEC wedge. Once the steepest descent paths are identified, the contour can then be closed to evaluate the exact solution. The sum of the integrals along the exact solution contour γ, plus along the steepest descent paths, must equal the sum of the residues enclosed. That is,

$$\int_\gamma (\text{integrand})\,dz + \int_{SDPs} (\text{integrand})\,dz = \int_{closed\ path} (\text{integrand})\,dz = \sum_{p=1}^{N} \text{Residues} \quad (14\text{-}9a)$$

where $SDPs = SDP(+\pi) + SDP(-\pi)$. Alternatively, the integral along the exact solution contour γ of (14-9a) can be written as the sum of the residues *minus* the contributions along the steepest descent paths, or

$$\int_\gamma (\text{integrand})\,dz = \sum_{p=1}^{N} \text{Residues} - \int_{SDPs} (\text{integrand})\,dz \quad (14\text{-}9b)$$

Therefore, the exact solution of (14-5) can now be written as

$$U_t(\rho,\phi) = jU_0 \frac{1}{2n\pi} \int_\gamma \frac{\Psi\left(z + \dfrac{n\pi}{2} - \phi\right)}{\Psi\left(\dfrac{n\pi}{2} - \phi'\right)} \frac{\sin\left(\dfrac{\phi'}{n}\right)}{\cos\left(\dfrac{z-\phi}{n}\right) - \cos\left(\dfrac{\phi'}{n}\right)} e^{j\beta\rho\cos z}\,dz$$

$$= \frac{U_0}{n} \sum_{p=1}^{N} \text{Res}\left[\frac{\Psi\left(z + \dfrac{n\pi}{2} - \phi\right)}{\Psi\left(\dfrac{n\pi}{2} - \phi'\right)} \frac{\sin\left(\dfrac{\phi'}{n}\right)}{\cos\left(\dfrac{z-\phi}{n}\right) - \cos\left(\dfrac{\phi'}{n}\right)} e^{j\beta\rho\cos z}, z_p \right]$$

$$- jU_0 \frac{1}{2n\pi} \int_{SDPs} \frac{\Psi\left(z + \dfrac{n\pi}{2} - \phi\right)}{\Psi\left(\dfrac{n\pi}{2} - \phi'\right)} \frac{\sin\left(\dfrac{\phi'}{n}\right)}{\cos\left(\dfrac{z-\phi}{n}\right) - \cos\left(\dfrac{\phi'}{n}\right)} e^{j\beta\rho\cos z}\,dz \quad (14\text{-}10)$$

where z_p are the poles of the integrand enclosed by the steepest descents paths while the notation $\text{Res}[f(z), z_p]$ represents the residue of $f(z)$ at the pole z_p.

14.5 GEOMETRICAL OPTICS

Using the canonical geometry of Figure 14-5, the solution of (14-10) can be decomposed into geometrical optics (incident and reflected), diffracted (incident and reflected), as was done in Chapter 13 for the PEC wedge, and surface wave fields. Because the wedge has impedance surfaces, surface waves must also be included. The region of interest is outside the wedge ($0 \le \phi \le n\pi$), which has been subdivided into three regions (I, II, III), as was done in Figure 13-13a of Chapter 13.

Maliuzhinets gives the geometrical optics terms as ratios of auxiliary Maliuzhinets functions. However, a more efficient and accurate method is to use a reflection coefficient at each reflection. In this section, based on the work of Griesser, et al. [28, 30], it is shown that the pole residue of the exact solution gives identically the same geometrical optics field as the simple ray tracing model for any number of interior reflections. Both methods give identical results in magnitude, phase, and also in angular range over which a particular reflection mechanism exists. Previously this was only performed for the two singly-reflected fields of the half plane [23, 24] and the general wedge [26].

For a geometrical analysis of the multiple reflected fields, the incident field is multiplied by the appropriate reflection coefficients. The surface impedance reflection coefficients for infinite planar boundaries and plane wave incidence are

$$\Gamma_{0,n}(\phi) = \frac{\sin\phi - \sin\theta_{0,n}}{\sin\phi + \sin\theta_{0,n}} \quad \text{soft polarization} \tag{14-11a}$$

$$\Gamma_{0,n}(\phi) = \frac{\sin\theta_{0,n} - \sin\phi}{\sin\theta_{0,n} + \sin\phi} \quad \text{hard polarization} \tag{14-11b}$$

where ϕ is the angle measured from the planar surface to the incident ray, as indicated in Figure 14-3.

Let us now consider a typical reflection mechanism. In particular, consider the third-order reflection, which is initially incident on face 0, as shown in Figure 14-6 for an interior wedge. The plane wave is incident and reflects at an angle ϕ' from the 0 face. Next it reflects at an angle $\pi - n\pi - \phi'$ from the n face, and lastly at an angle $\pi - 2n\pi - \phi'$ from the 0 face. Hence, the reflected field includes the product Γ_{Π} of these three reflection coefficients

$$\Gamma_{\Pi} = \Gamma_0(\phi')\Gamma_n[\pi - (n\pi + \phi')]\Gamma_0[\pi - (2n\pi + \phi')] \tag{14-12a}$$

or equivalently

$$\Gamma_{\Pi} = \Gamma_0(\phi')\Gamma_n(n\pi + \phi')\Gamma_0(2n\pi + \phi') \tag{14-12b}$$

The distance the third-order reflected ray has traveled can be determined by tracing the image ray through the 0 face. Taking the phase reference at the vertex of the wedge, the extra distance

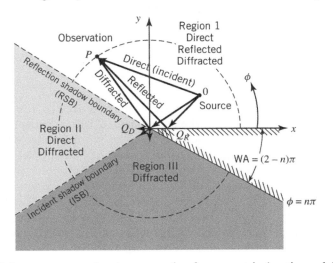

Figure 14-5 Wedge geometry and region separation for geometrical optics and diffracted fields.

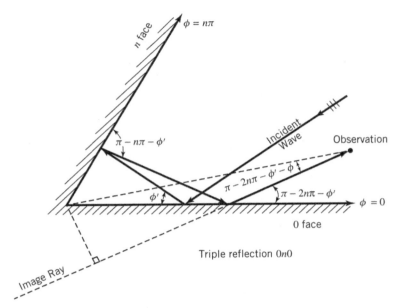

Figure 14-6 Typical third-order reflection from an interior wedge with impedance surfaces [28].

traveled is found by generating a perpendicular line to the image ray from the vertex, as shown in Figure 14-6, and it is equal to $\rho \cos(\pi - 2n\pi - \phi' - \phi)$.

Therefore, an additional phase factor of

$$e^{-j\beta\rho \cos(\pi - 2n\pi - \phi' - \phi)} = e^{-j\beta\rho \cos[\pi - (2n\pi + \phi' + \phi)]} = e^{j\beta\rho \cos(\phi + \phi' + 2n\pi)} \qquad (14\text{-}13)$$

must be included for the GO field. Finally, the total field for this third-order reflected component can be written as

$$U_{\text{GO}} = U_0 \Gamma_0(\phi') \Gamma_n(n\pi + \phi') \Gamma_0(2n\pi + \phi') e^{j\beta\rho \cos(\phi + \phi' + 2n\pi)} \qquad (14\text{-}14)$$

This third-order component has reflection shadow boundaries and does not exist for all ϕ. In fact, from the geometry of Figure 14-6, it exists only if $\phi < \pi - 2n\pi - \phi'$, or equivalently, if $\phi + \phi' < \pi - 2n\pi$.

In general, all the reflected components can be analyzed using the same procedure: ray tracing. The geometrical-optics reflected field for any multiple-reflected component C can be written as

$$U_{\text{GO}}^C = U_0 \Gamma_\Pi(\phi', \theta_0, \theta_n) e^{+j\beta\rho \cos(z_p)} \qquad (14\text{-}15)$$

where $\Gamma_\Pi(\phi', \theta_0, \theta_n)$ is a product of reflection coefficients and $\rho \cos(z_p)$ is a distance factor that yields the appropriate phase delay. Later, the component C will be shown to correspond to an angle $\xi^\pm$ and an integer m, and will be written as $U_{\text{GO}}^{m,\xi}$. The multiple-reflected field is identified by a sequence of 0's and n's indicating the order of reflection. As an example, component $0n0n$ is a quadruple-reflected field incident on face 0 which, in sequence, reflects from face 0 to n to 0 to n.

In Table 14-1, the term $\Gamma_\Pi(\phi', \theta_0, \theta_n)$ and z_p are listed for all reflection mechanisms up to fourth-order. In addition, the range over which these terms exist is listed, with the implied conditions that $0 \le \phi \le n\pi$ and $0 \le \phi' \le n\pi$. The number of terms presented is sufficient to identify the pattern by which the table can be expanded. The ordering of this table has been selected to correspond to the positions of the poles of the exact solution.

TABLE 14-1   Geometrical optics reflection coefficients and associated pole residues [28, 30]

Comp	Pole	$\Gamma_\pi(\phi', \theta_0, \theta_n)$	z_p	Existence
$0n0n$	$\xi^-, m=-2$	$\Gamma_0(\phi')\Gamma_n(n\pi+\phi')\Gamma_0(2n\pi+\phi')\Gamma_n(3n\pi+\phi')$	$\phi-\phi'-4n\pi$	$\phi-\phi'>4n\pi-\pi$
$n0n$	$\xi^+, m=-2$	$\Gamma_n(n\pi-\phi')\Gamma_0(2n\pi-\phi')\Gamma_n(3n\pi-\phi')$	$\phi+\phi'-4n\pi$	$\phi+\phi'>4n\pi-\pi$
$0n$	$\xi^-, m=-1$	$\Gamma_0(\phi')\Gamma_n(n\pi+\phi')$	$\phi-\phi'-2n\pi$	$\phi-\phi'>2n\pi-\pi$
n	$\xi^+, m=-1$	$\Gamma_n(n\pi-\phi')$	$\phi+\phi'-2n\pi$	$\phi+\phi'>2n\pi-\pi$
Inc	$\xi^-, m=0$	1	$\phi-\phi'$	$-\pi<\phi-\phi'<\pi$
0	$\xi^+, m=0$	$\Gamma_0(\phi')$	$\phi+\phi'$	$\phi+\phi'<\pi$
$n0$	$\xi^-, m=1$	$\Gamma_n(n\pi-\phi')\Gamma_0(2n\pi-\phi')$	$\phi-\phi'+2n\pi$	$\phi-\phi'<\pi-2n\pi$
$0n0$	$\xi^+, m=1$	$\Gamma_0(\phi')\Gamma_n(n\pi+\phi')\Gamma_0(2n\pi+\phi')$	$\phi+\phi'+2n\pi$	$\phi+\phi'<\pi-2n\pi$
$n0n0$	$\xi^-, m=2$	$\Gamma_n(n\pi-\phi')\Gamma_0(2n\pi-\phi')\Gamma_n(3n\pi-\phi')\Gamma_0(4n\pi-\phi')$	$\phi-\phi'+4n\pi$	$\phi-\phi'<\pi-4n\pi$

(Source: T. Griesser and C. A. Balanis, "Reflections, diffractions, and surface waves for an interior impedance wedge of arbitrary angle," *IEEE Trans. Antennas Propagat.*, © 1989, IEEE.)

When considering the exact solution of (14-5) for the impedance wedge, the geometrical optics poles must be identified. The GO poles of (14-5) are those for which

$$\cos\left(\frac{z_p - \phi}{n}\right) = \cos\left(\frac{\phi'}{n}\right) \tag{14-16}$$

This equation can be inverted to solve for z_p by considering every value of the multi-valued inverse cosine. The GO poles are represented by

$$z_p = \phi \pm \phi' + 2mn\pi = \xi^\pm + 2mn\pi, \qquad \xi^\pm = \phi \pm \phi' \tag{14-17}$$

where m is an integer, and n is real and depends on the wedge angle. The notation $\xi^\pm = \phi \pm \phi'$ is also conveniently used in the PEC wedge of Chapter 13. The choice of the notation z_p in this equation and in the GO phase factor will become evident when the pole residues are evaluated.

The geometrical optics poles of (14-17) appear in two sets of equally spaced poles corresponding to the upper and lower sign. For each set, the spacing between poles is $2n\pi$, or twice the exterior wedge angle $n\pi$. If ϕ' is considered to be fixed and ϕ varies from 0 to $n\pi$, then the GO poles move from $\pm\phi' + 2mn\pi$ to $\pm\phi' + 2mn\pi + n\pi$. These pole loci are located along the real axis of the complex z plane and are plotted in Figure 14-7 for $\phi' = 30°$ and an 85° interior wedge. The movement of the poles with increasing ϕ is indicated by the arrows, and it is noted that a given pole can only move half the distance to the next pole of the same set. In addition, the surface wave poles for $\eta_0 = \eta_n = 0.2 + j0.8$, although they have not yet been discussed, are indicated in Figure 14-7.

A particular pole contributes to the exact integral solution of (14-10) only if it lies between the steepest descent paths. Hence, it is possible to identify each pole with a specific reflection mechanism by the angular ranges of existence listed in Table 14-1. The example of the $0n0$ term considered previously will be examined here again. This third-order term exists for $\phi+\phi'<\pi-2n\pi$. Consider the pole for which $\xi = \xi^+ = \phi+\phi'$ and $m=1$. This pole contributes to the exact solution if it lies between the steepest descent paths that cross the real axis at $z=\pm\pi$. Since the steepest descent paths cross the real axis at $z_s = \pm\pi$, the pole contributes if

$$-\pi < z = \phi+\phi'+2n\pi < \pi \tag{14-18}$$

or equivalently if

$$-\pi - 2n\pi < \phi+\phi' < \pi - 2n\pi \tag{14-19}$$

Since $0 \le \phi \le n\pi$ and $0 \le \phi' \le n\pi$, then it is always true that $0 \le \phi+\phi' \le 2n\pi$. Therefore, the lower limit of (14-19) adds no new information. Hence, the pole contributes for $\phi+\phi' < \pi - 2n\pi$, which

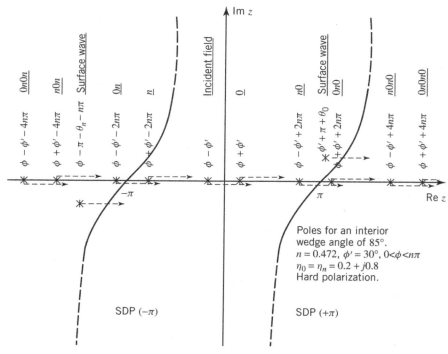

Figure 14-7   Geometrical optics and surface wave poles in complex plane [28, 30]. (Source: T. Griesser and C. A. Balanis, "Reflections, diffractions, and surface waves for an interior impedance wedge of arbitrary angle," *IEEE Trans. Antennas Propagat.*, © 1989, IEEE.)

is exactly the GO result determined after (14-14). The reflection component corresponding to each pole is labeled in Figure 14-7. For each GO reflection, the appropriate values of ξ and m are listed in Table 14-1.

From the exact solution, the contribution of the GO pole is $-j2\pi$ times the residue of the pole, where the minus $(-)$ sign is attributed to the clockwise contour encircling the pole in Figure 14-4. For the pole $z_p = \phi \pm \phi' + 2mn\pi$, the residue is

$$U_{GO}^C = \frac{U_0}{n} \operatorname{Res}\left[\frac{\Psi\left(z + \dfrac{n\pi}{2} - \phi\right)}{\Psi\left(\dfrac{n\pi}{2} - \phi'\right)} \frac{\sin\left(\dfrac{\phi'}{n}\right)}{\cos\left(\dfrac{z-\phi}{n}\right) - \cos\left(\dfrac{\phi'}{n}\right)} e^{j\beta\rho\cos z}, z_p \right] \tag{14-20}$$

Assuming that no poles of the Maliuzhinets function coincide with the GO poles, and since the exponential has no finite poles, the residue can be determined by

$$U_{GO}^C = \frac{U_0}{n} \lim_{z \to z_p}\left\{ (z - z_p)\left[\frac{\Psi\left(z + \dfrac{n\pi}{2} - \phi\right)}{\Psi\left(\dfrac{n\pi}{2} - \phi'\right)} \frac{\sin\left(\dfrac{\phi'}{n}\right)}{\cos\left(\dfrac{z-\phi}{n}\right) - \cos\left(\dfrac{\phi'}{n}\right)} e^{j\beta\rho\cos z} \right] \right\}$$

$$= \frac{U_0}{n} \frac{\Psi\left(z + \dfrac{n\pi}{2} - \phi\right)}{\Psi\left(\dfrac{n\pi}{2} - \phi'\right)} \sin\left(\dfrac{\phi'}{n}\right) e^{j\beta\rho\cos z} \lim_{z \to z_p} \operatorname{Res}\left[\frac{(z - z_p)}{\cos\left(\dfrac{z-\phi}{n}\right) - \cos\left(\dfrac{\phi'}{n}\right)} \right] \tag{14-21}$$

From (14-16) and (14-17)

$$\lim_{z \to z_p} \mathrm{Res} \left| \frac{(z - z_p)}{\cos\left(\dfrac{z - \phi}{n}\right) - \cos\left(\dfrac{\phi'}{n}\right)} \right| = \lim_{z \to \xi^\pm + 2mn\pi} \mathrm{Res} \left| \frac{[z - (\xi^\pm + 2mn\pi)]}{\cos\left(\dfrac{z - \phi}{n}\right) - \cos\left(\dfrac{\phi'}{n}\right)} \right| \qquad (14\text{-}22)$$

This is an indeterminate form that can be evaluated by L' Hopital's Rule. Then

$$\lim_{z \to z_p} \left| \frac{(z - z_p)}{\cos\left(\dfrac{z - \phi}{n}\right) - \cos\left(\dfrac{\phi'}{n}\right)} \right| = \lim_{z \to \xi^\pm + 2mn\pi} \left| \frac{1}{-\dfrac{1}{n}\sin\left(\dfrac{z - \phi}{n}\right)} \right|$$

$$= \frac{-n}{\sin\left(\dfrac{\pm\phi'}{n}\right)} = \frac{\mp n}{\sin\left(\dfrac{\phi'}{n}\right)} \qquad (14\text{-}23)$$

where the upper sign represents ξ^+ poles and the lower sign represents ξ^- poles.
Finally, (14-21) becomes

$$U_{\mathrm{GO}}^C = \mp U_0 \frac{\Psi\left(z_p + \dfrac{n\pi}{2} - \phi\right)}{\Psi\left(\dfrac{n\pi}{2} - \phi'\right)} e^{j\beta\rho \cos z}$$

$$= \mp U_0 \frac{\Psi\left(\pm\phi' + 2mn\pi + \dfrac{n\pi}{2}\right)}{\Psi\left(\dfrac{n\pi}{2} - \phi'\right)} e^{j\beta\rho \cos(\phi \pm \phi' + 2mn\pi)} \qquad (14\text{-}24)$$

corresponding to the poles $z_p = (\phi \pm \phi') + 2mn\pi$. By comparison with Table 14-1, it is clear that the phase factor matches the geometrical-optics phase term. However, it remains to be shown that the ratio of auxiliary Maliuzhinets functions in (14-24) is a product of reflection coefficients, as in the geometrical analysis of (14-15). For the incident GO field, which corresponds to $\xi = \xi^- = \phi - \phi'$ and $m = 0$, the ratio is

$$\frac{\Psi\left(\pm\phi' + 2mn\pi + \dfrac{n\pi}{2}\right)}{\Psi\left(\dfrac{n\pi}{2} - \phi'\right)} = \frac{\Psi\left(-\phi' + \dfrac{n\pi}{2}\right)}{\Psi\left(\dfrac{n\pi}{2} - \phi'\right)} = 1 \qquad (14\text{-}25)$$

So the incident field from the exact solution matches the GO incident field as expected, and no further consideration is necessary for $m = 0$.

To consider the many possible multiple reflections, the ratio of auxiliary Maliuzhinets functions (and leading sign) of (14-24) is denoted by $\Gamma_{\mathrm{II}}(\phi', \theta_0, \theta_n)$, since it is shown next that the ratio indeed reduces to the products of reflection coefficients of Table 14-1:

$$\Gamma_{\mathrm{II}}(\phi', \theta_0, \theta_n) = \mp \frac{\Psi\left(z_p + \dfrac{n\pi}{2} - \phi\right)}{\Psi\left(\dfrac{n\pi}{2} - \phi'\right)} \qquad (14\text{-}26)$$

where $z_p = (\phi \pm \phi') + 2mn\pi$. The numerator and denominator are expanded using [9, 26]

$$\Psi(z) = \Psi_n\left(z + \frac{n\pi}{2} + \frac{\pi}{2} - \theta_0\right)\Psi_n\left(z + \frac{n\pi}{2} - \frac{\pi}{2} + \theta_0\right)$$

$$\cdot \Psi_n\left(z - \frac{n\pi}{2} + \frac{\pi}{2} - \theta_n\right)\Psi_n\left(z - \frac{n\pi}{2} - \frac{\pi}{2} + \theta_n\right) \quad (14\text{-}27)$$

and (14-26) takes two forms depending on whether $\phi + \phi'$ or $\phi - \phi'$ is selected.

For $\xi = \xi^- = \phi - \phi'$:

$$\Gamma_{\Pi}(\phi', \theta_0, \theta_n) = \frac{\Psi_n\left(-\phi' + n\pi + \frac{\pi}{2} - \theta_0 + 2mn\pi\right)}{\Psi_n\left(-\phi' + n\pi + \frac{\pi}{2} - \theta_0\right)} \frac{\Psi_n\left(-\phi' + n\pi - \frac{\pi}{2} + \theta_0 + 2mn\pi\right)}{\Psi_n\left(-\phi' + n\pi - \frac{\pi}{2} + \theta_0\right)}$$

$$\cdot \frac{\Psi_n\left(-\phi' + \frac{\pi}{2} - \theta_n + 2mn\pi\right)}{\Psi_n\left(-\phi' + \frac{\pi}{2} - \theta_n\right)} \frac{\Psi_n\left(-\phi' - \frac{\pi}{2} + \theta_n + 2mn\pi\right)}{\Psi_n\left(-\phi' - \frac{\pi}{2} + \theta_n\right)} \quad (14\text{-}28a)$$

For $\xi = \xi^+ = \phi + \phi'$:

$$\Gamma_{\Pi}(\phi', \theta_0, \theta_n) = -\frac{\Psi_n\left(+\phi' + n\pi + \frac{\pi}{2} - \theta_0 + 2mn\pi\right)}{\Psi_n\left(\phi' - n\pi + \frac{\pi}{2} - \theta_0\right)} \frac{\Psi_n\left(\phi' + n\pi - \frac{\pi}{2} + \theta_0 + 2mn\pi\right)}{\Psi_n\left(\phi' - n\pi - \frac{\pi}{2} + \theta_0\right)}$$

$$\cdot \frac{\Psi_n\left(\phi' + \frac{\pi}{2} - \theta_n + 2mn\pi\right)}{\Psi_n\left(\phi' + \frac{\pi}{2} - \theta_n\right)} \frac{\Psi_n\left(\phi' - \frac{\pi}{2} + \theta_n + 2mn\pi\right)}{\Psi_n\left(\phi' - \frac{\pi}{2} + \theta_n\right)} \quad (14\text{-}28b)$$

The fact that $\Psi_n(z)$ is an even function of z has also been utilized here. Next, (14-28a) and (14-28b) are shown to reduce to products of reflection coefficients for multiple reflected rays.

The two single reflections are considered first, since both are exceptional cases. Each can be reduced to a single reflection coefficient using the trigonometric identity

$$\tan\left[\frac{1}{2}(x - y)\right]\cot\left[\frac{1}{2}(x + y)\right] = \frac{\sin x - \sin y}{\sin x + \sin y} \quad (14\text{-}29)$$

and the Maliuzhinets identity [15]

$$\frac{\Psi_n(z + n\pi)}{\Psi_n(z - n\pi)} = \cot\left[\frac{1}{2}\left(z + \frac{\pi}{2}\right)\right] \quad (14\text{-}30)$$

For the reflection from face 0 ($\xi = \xi^+$, $m = 0$), the last two terms in (14-28b) reduce to unity since $m = 0$. Then, using (14-30) in (14-28b), the ratio of Maliuzhinets functions can be reduced to the single reflection coefficient $\Gamma_{\Pi}(\phi', \theta_0, \theta_n) = \Gamma_0(\phi')$ from the 0 face by applying (14-29) and (14-11a), in that order. Similarly, for the single-reflected field from face n ($\xi = \xi^+$, $m = -1$), the first two terms in (14-28b) reduce to unity since $m = -1$. Again, using (14-30) in (14-28b), the ratio of Maliuzhinets functions can be reduced to the single reflection coefficient $\Gamma_{\Pi}(\phi', \theta_0, \theta_n) = \Gamma_n(n\pi - \phi')$ from the n face by applying (14-29) and (14-11b), in that order.

By comparing with Table 14-1, it is evident that the ratio of auxiliary Maliuzhinets functions reduces to the two single reflection coefficients for the $(\xi = \xi^+, m = 0)$ and $(\xi = \xi^+, m = -1)$ poles. This equality of the ratios of Maliuzhinets functions with the single reflection coefficients has been demonstrated in [23, 24, 26], but it is more difficult and less accurate for calculation purposes. The higher-order reflection coefficients were not considered elsewhere, and it was demonstrated in [28, 30] that they can be reduced to products of reflection coefficients for any order of reflections. The products appear in four forms, depending on whether the number of reflections is even or odd and whether the 0 and n face is initially illuminated for the multiple-reflected fields.

To obtain the appropriate products of reflection coefficients in (14-28a) and (14-28b) for the higher-order reflections, the following identities are utilized:

$$\frac{\Psi_n(x + 2mn\pi)}{\Psi_n(x)} = \prod_{p=1}^{m} \frac{\Psi_n[x + (2p-1)n\pi + n\pi]}{\Psi_n[x + (2p-1)n\pi - n\pi]}, \quad m > 0 \tag{14-31a}$$

$$\frac{\Psi_n(x + 2mn\pi)}{\Psi_n(x)} = \prod_{p=1}^{|m|} \frac{\Psi_n[x + (1-2p)n\pi - n\pi]}{\Psi_n[x + (1-2p)n\pi + n\pi]}, \quad m < 0 \tag{14-31b}$$

These are algebraic identities, valid for any function, which can be verified by expanding the product. Their value lies in the fact that they correctly separate the Maliuzhinets function into appropriate forms so that the recursion relations can be applied to reduce the Maliuzhinets ratios to products of reflection coefficients.

Using (14-29) through (14-31b) in (14-28a) and (14-28b), the GO residues are reduced to products of reflection coefficients. Four different forms are possible, depending on the choices of ξ and m. Consider first $\xi = \xi^- = \phi - \phi'$ and $m > 0$. Then (14-28a) can be written, using (14-31a), as

$$
\begin{aligned}
\Gamma_{\Pi}(\phi', \theta_0, \theta_n) = \; &\prod_{p=1}^{m} \frac{\Psi_n\left[-\phi' + n\pi + \dfrac{\pi}{2} - \theta_0 + (2p-1)n\pi + n\pi\right]}{\Psi_n\left[-\phi' + n\pi + \dfrac{\pi}{2} - \theta_0 + (2p-1)n\pi - n\pi\right]} \\[2mm]
\cdot \; &\prod_{p=1}^{m} \frac{\Psi_n\left[-\phi' + n\pi - \dfrac{\pi}{2} + \theta_0 + (2p-1)n\pi + n\pi\right]}{\Psi_n\left[-\phi' + n\pi - \dfrac{\pi}{2} + \theta_0 + (2p-1)n\pi - n\pi\right]} \\[2mm]
\cdot \; &\prod_{p=1}^{m} \frac{\Psi_n\left[-\phi' + \dfrac{\pi}{2} - \theta_n + (2p-1)n\pi + n\pi\right]}{\Psi_n\left[-\phi' + \dfrac{\pi}{2} - \theta_n + (2p-1)n\pi - n\pi\right]} \\[2mm]
\cdot \; &\prod_{p=1}^{m} \frac{\Psi_n\left[-\phi' - \dfrac{\pi}{2} + \theta_n + (2p-1)n\pi + n\pi\right]}{\Psi_n\left[-\phi' - \dfrac{\pi}{2} + \theta_n + (2p-1)n\pi - n\pi\right]}
\end{aligned}
\tag{14-32}
$$

Now using (14-30), (14-32) can be expressed as

$$\Gamma_{\Pi}(\phi', \theta_0, \theta_n) = \prod_{p=1}^{m} \cot\left\{\frac{1}{2}[-\phi' + n\pi + \pi - \theta_0 + (2p-1)n\pi]\right\}$$

$$\cdot \prod_{p=1}^{m} \cot\left\{\frac{1}{2}[-\phi' + n\pi + \theta_0 + (2p-1)n\pi]\right\}$$

$$\cdot \prod_{p=1}^{m} \cot\left\{\frac{1}{2}[-\phi' + \pi - \theta_n + (2p-1)n\pi]\right\}$$

$$\cdot \prod_{p=1}^{m} \cot\left\{\frac{1}{2}[-\phi' + \theta_n + (2p-1)n\pi]\right\} \quad (14\text{-}33)$$

Next, some of the cotangents are changed to tangents by trigonometric identities, and (14-33) can be written as

$$\Gamma_{\Pi}(\phi', \theta_0, \theta_n) = \prod_{p=1}^{m} -\tan\left\{\frac{1}{2}[-\phi' + n\pi - \theta_0 + (2p-1)n\pi]\right\}$$

$$\cdot \prod_{p=1}^{m} \cot\left\{\frac{1}{2}[-\phi' + n\pi + \theta_0 + (2p-1)n\pi]\right\}$$

$$\cdot \prod_{p=1}^{m} -\tan\left\{\frac{1}{2}[-\phi' - \theta_n + (2p-1)n\pi]\right\}$$

$$\cdot \prod_{p=1}^{m} \cot\left\{\frac{1}{2}[-\phi' + \theta_n + (2p-1)n\pi]\right\} \quad (14\text{-}34)$$

Using (14-29), all the terms in (14-34) can be combined, and (14-34) can then be expressed as

$$\Gamma_{\Pi}(\phi', \theta_0, \theta_n) = (-1)^{2m} \prod_{p=1}^{m} \frac{\sin(2pn\pi - \phi') - \sin\theta_0}{\sin(2pn\pi - \phi') + \sin\theta_0} \prod_{p=1}^{m} \frac{\sin[(2p-1)n\pi - \phi'] - \sin\theta_n}{\sin[(2p-1)n\pi - \phi'] + \sin\theta_n} \quad (14\text{-}35)$$

Finally, using (14-11a) and (14-11b), the two ratios of (14-35) reduce to products of reflection coefficients for the 0 and n faces, or

$$\Gamma_{\Pi}(\phi', \theta_0, \theta_n) = \prod_{p=1}^{m} \Gamma_0(2pn\pi - \phi') \prod_{p=1}^{m} \Gamma_n[(2p-1)n\pi - \phi'] \quad (14\text{-}36)$$

Again, (14-36) is only valid for $\xi = \xi^- = \phi - \phi'$ and $m > 0$. There are still three more cases to consider. However, all three cases reduce to similar products of reflection coefficients, and the final forms obtained from the GO pole residues can be written as:

For $\xi = \xi^- = \phi - \phi'$ and $m > 0$:

$$\Gamma_{\Pi}(\phi', \theta_0, \theta_n) = \prod_{p=1}^{m} \Gamma_0(2pn\pi - \phi') \prod_{p=1}^{m} \Gamma_n[(2p-1)n\pi - \phi'] \quad (14\text{-}37)$$

For $\xi = \xi^- = \phi - \phi'$ and $m < 0$:

$$\Gamma_{\Pi}(\phi', \theta_0, \theta_n) = \prod_{p=1}^{|m|} \Gamma_0[\phi' + 2(p-1)n\pi] \prod_{p=1}^{|m|} \Gamma_n[\phi' + (2p-1)n\pi] \quad (14\text{-}38)$$

For $\xi = \xi^+ = \phi + \phi'$ and $m > 0$:

$$\Gamma_{\Pi}(\phi', \theta_0, \theta_n) = \prod_{p=1}^{m+1} \Gamma_0[\phi' + 2(p-1)n\pi] \prod_{p=1}^{m} \Gamma_n[\phi' + (2p-1)n\pi] \quad (14\text{-}39)$$

For $\xi = \xi^+ = \phi + \phi'$ and $m < -1$:

$$\Gamma_{\Pi}(\phi', \theta_0, \theta_n) = \prod_{p=1}^{|m+1|} \Gamma_0(2pn\pi - \phi') \prod_{p=1}^{|m|} \Gamma_n[2(p-1)n\pi - \phi'] \qquad (14\text{-}40)$$

The cases $(\xi^-, m = 0)$, $(\xi^+, m = 0)$, and $(\xi^+, m = -1)$ are the incident and single reflected terms, which have already been considered. By comparison with Table 14-1, it is evident that (14-37) through (14-40) are identical to the geometrical ray-tracing analysis.

14.6 SURFACE WAVE TERMS

The surface wave is a wave that propagates along one face of the wedge and typically decays exponentially in a vertical direction away from the face. It is confined to a particular angular range from the wedge face, whenever it exists. Since the wave may decay slowly along the face, its contribution may be more dominant than other scattering mechanisms near the wedge surface. Hence, it is often important to include this contribution for reactive surfaces.

The surface wave is determined by considering the contributions of the residues of enclosed *complex* poles of the Maliuzhinets function between the steepest descent paths. The surface wave poles were identified by Maliuzhinets, and they are located at [9, 27]

$$z_0 = \phi + \pi + \theta_0 \qquad (14\text{-}41\text{a})$$

$$z_n = \phi - n\pi - \pi - \theta_n \qquad (14\text{-}41\text{b})$$

for the 0 and n faces, respectively. If ϕ is allowed to vary from 0 to $n\pi$, the z_0 pole moves from $\pi + \theta_0$ to $n\pi + \pi + \theta_0$ and the z_n pole moves from $-n\pi - \pi - \theta_n$ to $-\pi - \theta_n$. The loci are shown in Figure 14-7 for $\phi' = 30°$ for a wedge with an $85°$ interior angle and normalized surface impedances of $\eta_0 = \eta_n = 0.2 + j0.8$. The z_0 pole can only lie within the steepest descent paths for ϕ less than some maximum value. Similarly, the z_n pole can only lie within the steepest descent paths for ϕ greater than some minimum value. Hence, the surface wave term is bounded to a finite angular range near the corresponding face.

In general, the surface wave component corresponding to the pole z exists if it lies within the steepest descent paths of (14-18). Then

$$-\pi < z_r - \cos^{-1}\left(\frac{1}{\cosh z_i}\right) \operatorname{sgn}(z_i) < \pi \qquad (14\text{-}42)$$

where $z = z_r + jz_i$. To determine the pole residue, the following Maliuzhinets function identity is utilized [15, 18]:

$$\Psi_n\left[z \pm \left(n\pi + \frac{3\pi}{2}\right)\right] = \pm \sin\left(\frac{\pi \pm z}{2n}\right) \csc\left(\frac{z}{2n}\right) \Psi_n\left(n\pi - \frac{\pi}{2} \pm z\right) \qquad (14\text{-}43)$$

This expression isolates the singular part of the pole in the cosecant function, and hence the residue is readily calculated. The cosecant is singular at $z = 0$ and therefore $\Psi\left[\pm\left(n\pi + \frac{3\pi}{2}\right)\right]$ is also singular. The example of the z_0 pole is presented next, and the calculations for the z_n pole are similar.

Example 14-1

Evaluate the surface wave residue of U_{SW}^0.

Solution: The surface wave residue for the z_0 pole is given by

$$U_{SW}^0 = \frac{U_0}{n} \operatorname{Res} \left[\frac{\Psi\left(z + \frac{n\pi}{2} - \phi\right)}{\Psi\left(\frac{n\pi}{2} - \phi'\right)} \frac{\sin\left(\frac{\phi'}{n}\right)}{\cos\left(\frac{z-\phi}{n}\right) - \cos\left(\frac{\phi'}{n}\right)} e^{j\beta\rho\cos z}, z_0 \right]$$

Assuming that no GO poles coincide with the surface wave pole, and since the exponential has no finite poles, the residue can be determined as

$$U_{SW}^0 = \frac{U_0}{n} \lim_{z \to z_0} \left[(z - z_0) \frac{\Psi\left(z + \frac{n\pi}{2} - \phi\right)}{\Psi\left(\frac{n\pi}{2} - \phi'\right)} \frac{\sin\left(\frac{\phi'}{n}\right)}{\cos\left(\frac{z-\phi}{n}\right) - \cos\left(\frac{\phi'}{n}\right)} e^{j\beta\rho\cos z} \right]$$

In the limit, $(z - z_0)$ has a zero and $\Psi\left(z + \frac{n\pi}{2} - \phi\right)$ contains the complex pole. All the other terms can be brought outside the limit. Thus,

$$U_{SW}^0 = \frac{U_0}{n} \left[\frac{e^{j\beta\rho\cos(\phi+\pi+\theta_0)}}{\Psi\left(\frac{n\pi}{2} - \phi'\right)} \frac{\sin\left(\frac{\phi'}{n}\right)}{\cos\left(\frac{\pi+\theta_0}{n}\right) - \cos\left(\frac{\phi'}{n}\right)} \right] \lim_{z \to z_0} \left[(z - z_0)\Psi\left(z + \frac{n\pi}{2} - \phi\right) \right]$$

Next, the auxiliary Maliuzhinets function within the limit is decomposed into four Maliuzhinets functions using (14-27). That is,

$$\Psi\left(z + \frac{n\pi}{2} - \phi\right) = \Psi_n\left(z + n\pi - \phi + \frac{\pi}{2} - \theta_0\right)\Psi_n\left(z + n\pi - \phi - \frac{\pi}{2} + \theta_0\right)$$
$$\cdot \Psi_n\left(z - \phi + \frac{\pi}{2} - \theta_n\right)\Psi_n\left(z - \phi - \frac{\pi}{2} + \theta_n\right)$$

As $z \to z_0 = \phi + \pi + \theta_0$, it is evident that the first Maliuzhinets function becomes $\Psi_n\left(n\pi + \frac{3\pi}{2}\right)$, which was shown to be singular in (14-43). Since only the first Maliuzhinets function contains the complex pole, the other three can be moved outside the limit. Thus

$$U_{SW}^0 = \frac{U_0}{n} \left[\frac{e^{-j\beta\rho\cos(\phi+\theta_0)}}{\Psi\left(\frac{n\pi}{2} - \phi'\right)} \frac{\sin\left(\frac{\phi'}{n}\right)}{\cos\left(\frac{\pi+\theta_0}{n}\right) - \cos\left(\frac{\phi'}{n}\right)} \right] \Psi_n\left(n\pi + \frac{\pi}{2} + 2\theta_0\right) \cdot \Psi_n\left(\frac{3\pi}{2} + \theta_0 - \theta_n\right)$$
$$\cdot \Psi_n\left(\frac{\pi}{2} + \theta_0 + \theta_n\right) \lim_{z \to z_0} \left[(z - z_0)\Psi_n\left(z + n\pi - \phi + \frac{\pi}{2} - \theta_0\right) \right]$$

The limit of the above equation can be determined using (14-43). Thus, replacing $Z - Z_0 = \alpha$

$$\lim_{z \to z_0} \left[(z - z_0)\Psi_n\left(z + n\pi - \phi + \frac{\pi}{2} - \theta_0\right) \right] = \lim_{z \to z_0} \left[(z - z_0)\Psi_n\left(z - z_0 + n\pi + \frac{3\pi}{2}\right) \right]$$
$$= \lim_{\alpha \to 0} \left[\alpha\Psi_n\left(\alpha + n\pi + \frac{3\pi}{2}\right) \right] = \sin\left(\frac{\pi}{2n}\right)\Psi_n\left(n\pi - \frac{\pi}{2}\right) \lim_{\alpha \to 0} \left[\alpha \csc\left(\frac{\alpha}{2n}\right) \right]$$
$$= 2n\sin\left(\frac{\pi}{2n}\right)\Psi_n\left(n\pi - \frac{\pi}{2}\right)$$

which, when substituted into the previous equation, leads to the surface wave contribution. The final form of the surface waves from face 0 is

$$
U_{SW}^0 = U_0 \left[\frac{2\sin\left(\dfrac{\pi}{2n}\right) \sin\left(\dfrac{\phi'}{n}\right) e^{-j\beta\rho\cos(\phi+\theta_0)}}{\Psi\left(\dfrac{n\pi}{2}-\phi'\right)\left[\cos\left(\dfrac{\pi+\theta_0}{n}\right)-\cos\left(\dfrac{\phi'}{n}\right)\right]} \right] \Psi_n\left(n\pi-\frac{\pi}{2}\right)\Psi_n\left(n\pi+\frac{\pi}{2}+2\theta_0\right)
$$

$$
\cdot \Psi_n\left(\frac{3\pi}{2}+\theta_0-\theta_n\right)\Psi_n\left(\frac{\pi}{2}+\theta_0+\theta_n\right)
$$

In a similar manner, based on the procedure used in Example 14-1, the surface wave contribution U_{SW}^n from face n can be written as

$$
U_{SW}^n = U_0 \left[\frac{-2\sin\left(\dfrac{\pi}{2n}\right) \sin\left(\dfrac{\phi'}{n}\right) e^{-j\beta\rho\cos(\phi-n\pi-\theta_n)}}{\Psi\left(\dfrac{n\pi}{2}-\phi'\right)\left[\cos\left(\dfrac{n\pi+\pi+\theta_n}{n}\right)-\cos\left(\dfrac{\phi'}{n}\right)\right]} \right] \Psi_n\left(n\pi-\frac{\pi}{2}\right)\Psi_n\left(\frac{\pi}{2}+n\pi+2\theta_n\right)
$$

$$
\cdot \Psi_n\left(\frac{3\pi}{2}+\theta_n-\theta_0\right)\Psi_n\left(\frac{\pi}{2}+\theta_n+\theta_0\right) \tag{14-44}
$$

By the symmetry of the wedge geometry, it is noted that (14-44) can be obtained from the U_{SW}^0 of Example 14-1 by replacing ϕ by $n\pi-\phi$, ϕ' by $n\pi-\phi'$, θ_0 by θ_n, and θ_n by θ_0. The angular ranges over which U_{SW}^0 and U_{SW}^n exist are given by (14-41a) through (14-42).

The GO component of (14-21) and the surface wave U_{SW}^0 of Example 14-1 were obtained assuming the GO poles and surface wave poles did not coincide. This is a valid assumption because the GO poles are always real, and if a surface wave pole coincides with a GO pole, it must necessarily be real. However, from (14-41a) and (14-41b), it is evident that real surface wave poles always lie outside the steepest descent paths since $0 < \phi < n\pi$. Therefore, if a surface wave pole and a GO pole coincide, they must lie outside the steepest descent paths, and their residues do not contribute to the exact solution.

14.7 DIFFRACTED FIELDS

In the previous sections, Maliuzhinets' exact solution was examined for both interior and exterior impedance wedges. It was shown that the exact solution can be written as a sum of residues from the GO and surface wave poles plus contribution along the steepest descent paths. In this section, the steepest descent paths contribution is examined thoroughly using high-frequency asymptotic expansions. The high-frequency asymptotic expansion gives the diffracted field and the surface wave transition field. The diffracted field is similar in form to the diffraction coefficients for the PEC theory, with a suitable multiplying factor which includes ratios of auxiliary Maliuzhinets functions.

14.7.1 Diffraction Terms

Maliuzhinets' exact integral solution of (14-5) has been manipulated in (14-10) to follow the steepest descent paths through the saddle points at $z = \pm\pi$. The exact solution from (14-10) can be written as

$$U_t(\rho,\phi)=-jU_0\frac{1}{2n\pi}\int_\gamma \frac{\Psi\!\left(z+\dfrac{n\pi}{2}-\phi\right)}{\Psi\!\left(\dfrac{n\pi}{2}-\phi'\right)}\frac{\sin\!\left(\dfrac{\phi'}{n}\right)}{\cos\!\left(\dfrac{z-\phi}{n}\right)-\cos\!\left(\dfrac{\phi'}{n}\right)}e^{j\beta\rho\cos z}\,dz$$

$$=\frac{U_0}{n}\sum_{p=1}^{N}\mathrm{Res}\left[\frac{\Psi\!\left(z+\dfrac{n\pi}{2}-\phi\right)}{\Psi\!\left(\dfrac{n\pi}{2}-\phi'\right)}\frac{\sin\!\left(\dfrac{\phi'}{n}\right)}{\cos\!\left(\dfrac{z-\phi}{n}\right)-\cos\!\left(\dfrac{\phi'}{n}\right)}e^{j\beta\rho\cos z},z_p\right]$$

$$-jU_0\frac{1}{2n\pi}\int_{SDPs} \frac{\Psi\!\left(z+\dfrac{n\pi}{2}-\phi\right)}{\Psi\!\left(\dfrac{n\pi}{2}-\phi'\right)}\frac{\sin\!\left(\dfrac{\phi'}{n}\right)}{\cos\!\left(\dfrac{z-\phi}{n}\right)-\cos\!\left(\dfrac{\phi'}{n}\right)}e^{j\beta\rho\cos z}\,dz \qquad (14\text{-}45)$$

$$U_t(\rho,\phi)=\sum\left[U_{GO}^{m,\xi}+U_{SW}^{0}+U_{SW}^{n}\right]$$

$$-jU_0\frac{1}{2n\pi}\int_{SDPs} \frac{\Psi\!\left(z+\dfrac{n\pi}{2}-\phi\right)}{\Psi\!\left(\dfrac{n\pi}{2}-\phi'\right)}\frac{\sin\!\left(\dfrac{\phi'}{n}\right)}{\cos\!\left(\dfrac{z-\phi}{n}\right)-\cos\!\left(\dfrac{\phi'}{n}\right)}e^{j\beta\rho\cos z}\,dz \qquad (14\text{-}45a)$$

$$U_t(\rho,\phi)=\sum\left[U_{GO}^{m,\xi}+U_{SW}^{0}+U_{SW}^{n}+U_{SDP}\right]$$

$$=\sum\left[U_{GO}^{m,\xi}+U_{SW}^{0}+U_{SW}^{n}+U_{D}+U_{SWTR}^{0}+U_{SWTR}^{n}\right] \qquad (14\text{-}45b)$$

where

$\quad U_{GO}^{m,\xi}$ = GO incident or reflected field corresponding to integer m and $\xi=\xi^{+}$ or $\xi=\xi^{-}$ as given by (14-15) and in Table 14-1

U_{SW}^{0},U_{SW}^{n} = surface wave contributions, as given by Example 14-1 and (14-44)

$\quad\quad U_{D}$ = diffracted field, examined asymptotically in Section 14.7.2

$U_{SWTR}^{0},U_{SWTR}^{n}$ = surface wave transition field terms

The canonical wedge geometry of Figures 14-1a and 14-5 illustrate the shadow boundaries of the GO field for the exterior wedge. Similar shadow boundaries exist for multiple reflected fields for the interior wedge of Figures 14-1b and 14-6. All the GO shadow boundaries are compensated by the diffracted field. Similarly, the surface wave transition field compensates for discontinuities at the surface wave boundary.

14.7.2 Asymptotic Expansions

To determine an asymptotic expansion of a steepest descent integral, such as that of (14-45a), or

$$U_{SDP}(\rho,\phi)=-jU_0\frac{1}{2n\pi}\int_{SDPs} \frac{\Psi\!\left(z+\dfrac{n\pi}{2}-\phi\right)}{\Psi\!\left(\dfrac{n\pi}{2}-\phi'\right)}\frac{\sin\!\left(\dfrac{\phi'}{n}\right)}{\cos\!\left(\dfrac{z-\phi}{n}\right)-\cos\!\left(\dfrac{\phi'}{n}\right)}e^{j\beta\rho\cos z}\,dz$$

$$=-jU_0\frac{1}{2n\pi}\frac{\sin\!\left(\dfrac{\phi'}{n}\right)}{\Psi\!\left(\dfrac{n\pi}{2}-\phi'\right)}\int_{SDPs}\frac{\Psi\!\left(z+\dfrac{n\pi}{2}-\phi\right)}{\cos\!\left(\dfrac{z-\phi}{n}\right)-\cos\!\left(\dfrac{\phi'}{n}\right)}e^{j\beta\rho\cos z}\,dz \qquad (14\text{-}46)$$

requires careful consideration of poles that may lie close to the steepest descent paths. The contribution of the steepest descent integral should be discontinuous as a pole crosses the path to compensate for the addition or loss of the pole residue that corresponds to a GO or surface wave component. In this way, the total solution, which is the sum of the residues plus the integral contribution, is always continuous.

Two asymptotic expansions are used: the *Modified Pauli-Clemmow* method [63–67] for the GO poles and the *Felsen-Marcuvitz* method [68, 69] for the surface wave poles. The Modified Pauli-Clemmow method yields a diffracted field which is analogous to the PEC case for the real GO poles. A diffraction coefficient is formulated, which is similar to the UTD diffraction coefficient with multiplying factors that include suitable ratios of auxiliary Maliuzhinets functions. The method of Felsen-Marcuvitz is used for the complex surface wave poles to construct a surface wave transition field that yields the proper continuity. The method of Felsen-Marcuvitz [68] or that of [70] could be used for all the poles. However, there would be two disadvantages. First, the method would not be analogous to the PEC UTD, which is a powerful tool in modern diffraction. Second, the formulation would require that the contributions of all the poles for all the multiple reflections be added together. This is the approach used in [27], and the equations become very complicated after adding only the poles of the incident field, the two single-reflected fields, and the two surface waves. For an interior wedge with many multiple reflections, the Felsen-Marcuvitz method is more cumbersome than the Modified Pauli-Clemmow method for the GO poles.

The two methods of asymptotic approximation have been shown to be the same [70, 71], provided that the asymptotic expansions are complete; that is, they contain an infinite number of higher-order terms. In practice, this is usually not the case because only the first and second terms are retained and all others are omitted. Consequently, one or the other may be more appropriate for a given problem. However, either will give the correct discontinuity as the pole crosses the saddle point, and both reduce to the same expression when no pole is near the steepest descent path.

The Modified Pauli-Clemmow method of steepest descent was introduced in Chapter 13 to evaluate an integral of the (13-41) form, or

$$P(\beta\rho) = \int_C H(z)e^{\beta\rho h(z)}\, dz \overset{\beta\rho \to \text{large}}{\approx} e^{\beta\rho h(z_s)} H(z_s) e^{j\phi_s} \left| \sqrt{\frac{-2\pi}{\beta\rho h''(z_s)}} \right| F[\beta\rho g(\xi)] \qquad (14\text{-}47)$$

$$F[\beta\rho g(\xi)] \equiv j[h(z_s) - h(z_p)] = 2j\sqrt{\beta\rho g(\xi)} \int_{\sqrt{\beta\rho g(\xi)}}^{\infty} e^{-j\tau^2}\, d\tau$$

$$\equiv \text{a measure of separation between saddle points and poles} \qquad (14\text{-}47a)$$

and to reduce (13-60a)–(13-60b) to (13-63a)–(13-63b), and eventually to the UTD diffraction coefficients.

The Felsen-Marcuvitz method is used to evaluate an integral of the (13-41) or (14-47) form, and it is expressed as

$$P(\beta\rho) = \int_C H(z)e^{\beta\rho h(z)}\, dz \overset{\beta\rho \to \text{large}}{\approx} e^{\beta\rho h(z_s)} H(z_s) e^{j\phi_s} \left| \sqrt{\frac{-2\pi}{\beta\rho h''(z_s)}} \right|$$

$$+ \frac{h_1 \sqrt{\pi} e^{\beta\rho h(z_s)}}{\sqrt{\beta\rho[h(z_s) - h(z_p)]}} \left\{ F[\beta\rho g(\xi)] - 1 \right\} \qquad (14\text{-}48)$$

where h_1 is the residue of $H(z)$ at $z = z_p$; all other quantities are the same as for the Modified Pauli-Clemmow method.

14.7.3 Diffracted Field

Using a procedure similar to that of Section 13.3.2, it can be shown, through some trigonometric identities, that the integral of (14-46) can be written in a more convenient form as

$$
U_{SDP}(\rho,\phi) = -jU_0 \frac{1}{2n\pi} \int_{SDPs} \frac{\Psi\left(z + \frac{n\pi}{2} - \phi\right)}{\Psi\left(\frac{n\pi}{2} - \phi'\right)} \frac{\sin\left(\frac{\phi'}{n}\right)}{\cos\left(\frac{z-\phi}{n}\right) - \cos\left(\frac{\phi'}{n}\right)} e^{j\beta\rho\,\cos z}\, dz
$$

$$
= -jU_0 \frac{1}{4n\pi} \int_{SDPs} \frac{\Psi\left(z + \frac{n\pi}{2} - \phi\right)}{\Psi\left(\frac{n\pi}{2} - \phi'\right)} \left\{ \cot\left[\frac{(\phi+\phi')-z}{2n}\right] - \cot\left[\frac{(\phi-\phi')-z}{2n}\right] \right\} e^{j\beta\rho\,\cos z}\, dz
$$

$$
\tag{14-49}
$$

where the following substitution, through trigonometric identities, has been used.

$$
\frac{\sin\left(\frac{\phi'}{n}\right)}{\cos\left(\frac{z-\phi}{n}\right) - \cos\left(\frac{\phi'}{n}\right)} = \frac{1}{2}\left\{ \cot\left[\frac{(\phi+\phi')-z}{2n}\right] - \cot\left[\frac{(\phi-\phi')-z}{2n}\right] \right\}
\tag{14-50}
$$

By the convention of the PEC wedge formulation, $\xi^+ = (\phi+\phi')$ and $\xi^- = (\phi-\phi')$. The poles of the first cotangent occur at

$$
z = (\phi+\phi') - 2\pi Nn = \xi^+ - 2\pi Nn
\tag{14-51a}
$$

and correspond to the GO components that include an odd number of reflections (i.e., single reflections, triple reflections, etc.). The poles of the second cotangent occur at

$$
z = (\phi-\phi') - 2\pi Nn = \xi^- - 2\pi Nn
\tag{14-51b}
$$

and correspond to the GO terms that include an even number of reflections (i.e., incident, double reflected, etc.). In (14-51a) and (14-51b), N can be any integer, and hence there are an infinite number of poles of the integrand for both ξ^+ and ξ^-.

The PEC UTD diffraction considers only the four dominant poles of the integrand of (14-49). These four are the poles corresponding to ξ^+ that are the closest to the saddle points at $\pm\pi$, and the poles corresponding to ξ^- that are closest to the same saddle points. Hence, four values of N are chosen, where the N's are the integers that most closely satisfy

$$
2\pi n N_-^+ - \xi^- = +\pi
\tag{14-52a}
$$

$$
2\pi n N_-^- - \xi^- = -\pi
\tag{14-52b}
$$

$$
2\pi n N_+^+ - \xi^- = +\pi
\tag{14-52c}
$$

$$
2\pi n N_+^- - \xi^- = -\pi
\tag{14-52d}
$$

where the subscript in N indicates the choice of ξ. The superscript of N corresponds to the sign of $\pm\pi$ on the right side of (14-52). Therefore, based on (14-52a)–(14-52b), (14-49) can be decomposed into four integrals corresponding to the four selected poles; that is,

$$U_{SDP}(\rho,\phi)=+jU_0\frac{1}{4n\pi}\int_{SDP(-\pi)}\frac{\Psi\left(z+\frac{n\pi}{2}-\phi\right)}{\Psi\left(\frac{n\pi}{2}-\phi'\right)}\left\{\cot\left[\frac{\xi^- -z}{2n}\right]\right\}e^{j\beta\rho\cos z}\,dz$$

$$+jU_0\frac{1}{4n\pi}\int_{SDP(+\pi)}\frac{\Psi\left(z+\frac{n\pi}{2}-\phi\right)}{\Psi\left(\frac{n\pi}{2}-\phi'\right)}\left\{\cot\left[\frac{\xi^- -z}{2n}\right]\right\}e^{j\beta\rho\cos z}\,dz$$

$$-jU_0\frac{1}{4n\pi}\int_{SDP(-\pi)}\frac{\Psi\left(z+\frac{n\pi}{2}-\phi\right)}{\Psi\left(\frac{n\pi}{2}-\phi'\right)}\left\{\cot\left[\frac{\xi^+ -z}{2n}\right]\right\}e^{j\beta\rho\cos z}\,dz$$

$$-jU_0\frac{1}{4n\pi}\int_{SDP(+\pi)}\frac{\Psi\left(z+\frac{n\pi}{2}-\phi\right)}{\Psi\left(\frac{n\pi}{2}-\phi'\right)}\left\{\cot\left[\frac{\xi^+ -z}{2n}\right]\right\}e^{j\beta\rho\cos z}\,dz \qquad (14\text{-}53)$$

where the order of the terms of (14-53) corresponds to the order of the N's in (14-52). Each term is individually evaluated asymptotically using (14-47).

Example 14-2

To demonstrate the asymptotic procedure, evaluate the first term of (14-53).

Solution:

$$H_1(z)=\frac{\Psi\left(z+\frac{n\pi}{2}-\phi\right)}{\Psi\left(\frac{n\pi}{2}-\phi'\right)}\cot\left(\frac{\xi^- -z}{2n}\right)$$

$$h_1(z)=j\cos z$$
$$z_s=-\pi$$
$$z_p=\xi^- -2\pi N_-^+ n$$
$$g=j[-j-j\cos(\xi^- -2\pi N_-^+ n)]=[1+\cos(\xi^- -2\pi N_-^+ n)]$$
$$h''(z)=-j\cos z$$
$$\left|\sqrt{\frac{-2\pi}{\beta\rho h''(z_s)}}\right|=\left|\sqrt{\frac{-2\pi}{j\beta\rho}}\right|=\sqrt{\frac{2\pi}{\beta\rho}}$$

where the correct branch of the radical is determined by the angle of the integration path in the direction of integration. The asymptotic expansion for the first term of (14-53) can now be constructed using (14-47)–(14-47a) with $\phi_s=\pi/4$ and can be written, using the above, as

$$U_{SDP}^1(\rho,\phi)=+jU_0\frac{1}{4n\pi}e^{-j\beta\rho}\frac{\Psi\left(-\pi+\frac{n\pi}{2}-\phi\right)}{\Psi\left(\frac{n\pi}{2}-\phi'\right)}\cot\left(\frac{\xi^- +\pi}{2n}\right)\sqrt{\frac{2\pi}{\beta\rho}}e^{j\frac{\pi}{4}}$$

$$\cdot F\{\beta\rho[1+\rho\cos(\xi^- -2\pi N_-^+ n)]\}$$

The other three terms of (14-53) are evaluated in the same manner, remembering that for the steepest descent path through $+\pi$, the exponential $e^{j\phi_s}$ is written as $e^{j\phi_s} = e^{-j3\pi/4}$ because of the direction of the integration contour of Figure 14-4 (use $\phi_s = \pi/4$ for SDP through $-\pi$). The final total diffracted field can now be written as

$$U_{\text{SDP}}^{s,h}(\rho,\phi) = U_0 \frac{e^{-j\beta\rho}}{\sqrt{\rho}} \left[-\frac{e^{-j\frac{\pi}{4}}}{2n\sqrt{2\pi\beta}} \right]$$

$$\cdot \left\{ \left[\frac{\Psi\left(-\pi + \frac{n\pi}{2} - \phi\right)}{\Psi\left(\frac{n\pi}{2} - \phi'\right)} \cot\left(\frac{\pi + \xi^-}{2n}\right) F\{\beta\rho[1 + \cos(\xi^- - 2\pi N_-^+ n)]\} \right. \right.$$

$$+ \left. \frac{\Psi\left(\pi + \frac{n\pi}{2} - \phi\right)}{\Psi\left(\frac{n\pi}{2} - \phi'\right)} \cot\left(\frac{\pi - \xi^-}{2n}\right) F\{\beta\rho[1 + \cos(\xi^- - 2\pi N_-^- n)]\} \right]$$

$$- \left[\frac{\Psi\left(-\pi + \frac{n\pi}{2} - \phi\right)}{\Psi\left(\frac{n\pi}{2} - \phi'\right)} \cot\left(\frac{\pi + \xi^+}{2n}\right) F\{\beta\rho[1 + \cos(\xi^+ - 2\pi N_+^+ n)]\} \right.$$

$$+ \left. \left. \frac{\Psi\left(\pi + \frac{n\pi}{2} - \phi\right)}{\Psi\left(\frac{n\pi}{2} - \phi'\right)} \cot\left(\frac{\pi - \xi^+}{2n}\right) F\{\beta\rho[1 + \cos(\xi^+ - 2\pi N_+^- n)]\} \right] \right\} \quad (14\text{-}54)$$

where $F(z)$ is the Fresnel transition function. By comparing (14-54) with the PEC case of Chapter 13, it is evident that (14-54) is the same but with the introduction of suitable ratios of auxiliary Maliuzhinets functions as multiplying factors.

At the reflection and incident shadow boundaries, $\phi = +\pi \pm \phi' + 2n\pi N$ or $\phi = -\pi \pm \phi' + 2n\pi N$, the ratios of the auxiliary Maliuzhinets become

$$\left. \frac{\Psi\left(\pm\pi + \frac{n\pi}{2} - \phi\right)}{\Psi\left(\frac{n\pi}{2} - \phi'\right)} \right|_{\phi = \pm\pi \pm \phi' + 2\pi Nn} = \frac{\Psi\left(\mp\phi' - 2\pi Nn + \frac{n\pi}{2}\right)}{\Psi\left(\frac{n\pi}{2} - \phi'\right)} \quad (14\text{-}55)$$

In Section 14.5 it was verified that the ratio (14-55) is simply a product of reflection coefficients. The interesting conclusion drawn is that the diffraction coefficient is modified by the appropriate ratios that reduce the diffraction discontinuities to account for reflection discontinuities for the lossy wedge.

The diffracted field of (14-54) is a valid asymptotic expansion, but is not the best possible asymptotic expansion because it does not reduce to the PEC diffraction for the hard polarization. It has been verified, by numerical integration [28], that these expressions work well for the soft polarization or the imperfectly conducting hard polarization. However, for the hard polarization, the expressions fail as the conductivity increases since they do not revert to the PEC case.

To evaluate the auxiliary Maliuzhinets function ratio in (14-54), the following identity is utilized for the hard polarized PEC case $\theta_0 = \theta_n = 0$. That is,

$$
\frac{\Psi\left(\pm\pi + \dfrac{n\pi}{2} - \phi\right)}{\Psi\left(\dfrac{n\pi}{2} - \phi'\right)} = \frac{\cos\left(\dfrac{\pm\pi + n\pi - \phi}{2n}\right)\cos\left(\dfrac{\pm\pi - \phi}{2n}\right)}{\cos\left(\dfrac{n\pi - \phi'}{2n}\right)\cos\left(\dfrac{\phi'}{2n}\right)} = \frac{\cos\left(\dfrac{\pm\pi - \phi}{n} + \dfrac{\pi}{2}\right)}{\cos\left(\dfrac{\phi'}{n} - \dfrac{\pi}{2}\right)}
$$

$$
= \frac{-\sin\left(\dfrac{\pm\pi - \phi}{n}\right)}{\sin\left(\dfrac{\phi'}{n}\right)} \tag{14-56}
$$

Hence, as $\theta_0 = \theta_n \to 0$, this formulation differs from the PEC UTD by

$$
\frac{-\sin\left(\dfrac{\pm\pi - \phi}{n}\right)}{\sin\left(\dfrac{\phi'}{n}\right)} \tag{14-57}
$$

For the soft polarization, there is no problem because the ratio of the left side of (14-56) reduces, for $\theta_0 = \theta_n \to \infty$, to

$$
\lim_{\theta_0 = \theta_n \to \infty} \frac{\Psi\left(\pm\pi + \dfrac{n\pi}{2} - \phi\right)}{\Psi\left(\dfrac{n\pi}{2} - \phi'\right)} = 1 \tag{14-58}
$$

Example 14-3

Since (14-54) does not reduce to the PEC diffraction for the hard polarization, reformulate the asymptotic expansion in a different manner so that when (14-54) is written in a slightly different form, it reduces to the PEC case for both soft and hard polarizations.

Solution: To achieve diffraction coefficients that reduce to the PEC case for both polarizations, soft and hard, all that is needed is a different subdivision of the singular portions of the integrand of (14-49). Instead of substituting (14-50)

$$
\frac{\sin\left(\dfrac{\phi'}{n}\right)}{\cos\left(\dfrac{z - \phi}{n}\right) - \cos\left(\dfrac{\phi'}{n}\right)} = \frac{1}{2}\left\{\cot\left[\dfrac{(\phi + \phi') - z}{2n}\right] - \cot\left[\dfrac{(\phi - \phi') - z}{2n}\right]\right\}
$$

into (14-46), it was recommended in [26] that the following trigonometric identity be used. That is,

$$
\frac{\sin\left(\dfrac{\phi'}{n}\right)}{\cos\left(\dfrac{z - \phi}{n}\right) - \cos\left(\dfrac{\phi'}{n}\right)} = \frac{1}{2}\left[\frac{\sin\left(\dfrac{\phi'}{n}\right) + \sin\left(\dfrac{\theta_0}{n}\right)}{\sin\left(\dfrac{\phi - z}{n}\right) + \sin\left(\dfrac{\theta_0}{n}\right)} \cot\left[\dfrac{z - (\phi - \phi')}{2n}\right] \right. \\
\left. + \frac{\sin\left(\dfrac{\phi'}{n}\right) - \sin\left(\dfrac{\theta_0}{n}\right)}{\sin\left(\dfrac{\phi - z}{n}\right) + \sin\left(\dfrac{\theta_0}{n}\right)} \cot\left[\dfrac{z - (\phi + \phi')}{2n}\right] \right]
$$

where the saddle points are at $z_s = \pm \pi$. Also for $\theta_0 = \theta_n = 0$, the sine ratios become

$$\left. \frac{\sin\left(\dfrac{\phi'}{n}\right) \pm \sin\left(\dfrac{\theta_0}{n}\right)}{\sin\left(\dfrac{\phi - z}{n}\right) + \sin\left(\dfrac{\theta_0}{n}\right)} \right|_{\substack{z=\pm\pi \\ \theta_0 = 0}} = \frac{\sin\left(\dfrac{\phi'}{n}\right)}{-\sin\left(\dfrac{\pm\pi - \phi}{n}\right)}$$

which is exactly the proper term to remove the problematic multiplying factor (14-57). Hence, the resultant expression for the diffracted field will reduce to the PEC forms for the hard polarization as $\theta_0 = \theta_n \to 0$. For the soft polarization, where $\theta_0 = \theta_n \to \infty$ with $0 < \mathrm{Re}\,[\theta_0] < \pi/2$ and $0 < \mathrm{Re}\,[\theta_n] < \pi/2$, the sine ratios on the left side of the above equation reduce to ± 1 and hence, are identical to the second equation in this example. Therefore, the correct expressions, which have already been established for the soft polarization, are retained.

It is interesting that the asymptotic expansion is not unique but depends upon the subdivision of the poles into individual terms. While the diffracted field from the subdivision of (14-50) does not reduce to the PEC case for one polarization (the hard one), the diffracted field based on the second equation in Example 14-3 reduces to the PEC forms for both principal polarizations, and it is the proper identity to subdivide the singular parts of the Maliuzhinets function.

Substituting the correct expression (the second equation of Example 14-3) into (14-49), it is evident that the final expression for the diffracted field to replace (14-54) is

$$U_{\text{SDP}}^{s,h}(\rho,\phi) = U_0 \frac{e^{-j\beta\rho}}{\sqrt{\rho}} \left[-\frac{e^{-j\frac{\pi}{4}}}{2n\sqrt{2\pi\beta}} \right.$$

$$\cdot \left\{ \left[\frac{\Psi\left(-\pi + \dfrac{n\pi}{2} - \phi\right)}{\Psi\left(\dfrac{n\pi}{2} - \phi'\right)} \frac{\sin\left(\dfrac{\phi'}{n}\right) + \sin\left(\dfrac{\theta_0}{n}\right)}{\sin\left(\dfrac{\phi + \pi}{n}\right) + \sin\left(\dfrac{\theta_0}{n}\right)} \cot\left(\frac{\pi + \xi^-}{2n}\right) \right. \right.$$

$$\left. \cdot F\{\beta\rho[1 + \cos(\xi^- - 2\pi N_-^+ n)]\} \right.$$

$$\left. + \frac{\Psi\left(\pi + \dfrac{n\pi}{2} - \phi\right)}{\Psi\left(\dfrac{n\pi}{2} - \phi'\right)} \frac{\sin\left(\dfrac{\phi'}{n}\right) + \sin\left(\dfrac{\theta_0}{n}\right)}{\sin\left(\dfrac{\phi + \pi}{n}\right) + \sin\left(\dfrac{\theta_0}{n}\right)} \cot\left(\frac{\pi - \xi^-}{2n}\right) \right.$$

$$\left. \cdot F\{\beta\rho[1 + \cos(\xi^- - 2\pi N_-^- n)]\} \right]$$

$$+ \left\{ \frac{\Psi\left(-\pi + \dfrac{n\pi}{2} - \phi\right)}{\Psi\left(\dfrac{n\pi}{2} - \phi'\right)} \frac{\sin\left(\dfrac{\phi'}{n}\right) - \sin\left(\dfrac{\theta_0}{n}\right)}{\sin\left(\dfrac{\phi + \pi}{n}\right) + \sin\left(\dfrac{\theta_0}{n}\right)} \cot\left(\frac{\pi + \xi^+}{2n}\right) \right.$$

$$\cdot F\{\beta\rho[1 + \cos(\xi^+ - 2\pi N_+^+ n)]\}$$

$$\left. \left. + \frac{\Psi\left(\pi + \dfrac{n\pi}{2} - \phi\right)}{\Psi\left(\dfrac{n\pi}{2} - \phi'\right)} \frac{\sin\left(\dfrac{\phi'}{n}\right) - \sin\left(\dfrac{\theta_0}{n}\right)}{\sin\left(\dfrac{\phi + \pi}{n}\right) + \sin\left(\dfrac{\theta_0}{n}\right)} \cot\left(\frac{\pi - \xi^+}{2n}\right) \right. \right.$$

$$\left. \left. \cdot F\{\beta\rho[1 + \cos(\xi^+ - 2\pi N_+^- n)]\} \right] \right]$$

$$\tag{14-59}$$

Equation (14-59), although cast in a different form, is identical to those in [26]. However, in [26], a great deal of effort was focused on manipulating these expressions to a form that was symmetric with respect to θ_0 and θ_n, and to reduce the number of times the Maliuzhinets function needed to be calculated. The formulation of (14-59) does not appear explicitly symmetric with respect to θ_0 and θ_n, but it is symmetric because the $\Psi(z)$ auxiliary Maliuzhinets function includes θ_0 and θ_n implicitly. There are twelve Maliuzhinets functions that need to be calculated. Four are necessary for each auxiliary Maliuzhinets function, and there are three auxiliary functions:

$$\Psi\left(\frac{n\pi}{2}-\phi'\right), \ \Psi\left(\pi+\frac{n\pi}{2}-\phi'\right), \ \text{and} \ \Psi\left(-\pi+\frac{n\pi}{2}-\phi'\right)$$

In [26], it was shown that the Maliuzhinets functions reduce to a double-nested integration; however, this double-nested integral is exceedingly difficult to evaluate, even numerically. Hence, [26] only demonstrated results for three special wedge angles, corresponding to the half-plane (WA $= 0°$), the exterior right angle (WA $= 90°$), and the planar interface discontinuity (WA $= 180°$). It is shown in [28] that this double-nested integral can be written as eight Maliuzhinets functions, all of which are readily available for an arbitrary wedge angle by methods in Appendix B of [28].

For the integral of [26], which is denoted $M_n(\phi,\phi';\theta_0,\theta_n)$, the simpler expression is

$$M_n(\phi,\phi';\theta_0,\theta_n) = \frac{\Psi_n\left(n\pi-\phi'+\frac{\pi}{2}+\theta_n\right)\Psi_n\left(\phi'+\frac{\pi}{2}+\theta_n\right)}{\Psi_n\left(n\pi-\phi'+\frac{\pi}{2}-\theta_0\right)\Psi_n\left(\phi'+\frac{\pi}{2}-\theta_n\right)}$$

$$\cdot \frac{\Psi_n\left(n\pi-\phi-\frac{\pi}{2}-\theta_0\right)\Psi_n\left(\phi-\frac{\pi}{2}-\theta_n\right)}{\Psi_n\left(n\pi-\phi-\frac{\pi}{2}+\theta_0\right)\Psi_n\left(\phi'-\frac{\pi}{2}+\theta_n\right)} \qquad (14\text{-}60)$$

In this manner, the necessary evaluations of Maliuzhinets functions are reduced from twelve to eight, and the results are obtainable for arbitrary interior and exterior wedge angles rather than for only the three specific wedge angles [half-plane (WA $= 0°$), exterior right angle (WA $= 90°$), and planar interface discontinuity (WA $= 180°$)]. This allows for the modeling of more general structures.

The Keller-type diffraction coefficients, which are the nonuniform versions, are obtained from (14-59) by setting every Fresnel transition function to unity.

14.8 SURFACE WAVE TRANSITION FIELD

The surface waves of the exact impedance wedge solution exist only for a limited angular range near the associated wedge face. It was shown in Section 14.6 that this angular range was determined by the position of the surface wave pole relative to the steepest descent path. As the pole moves outside the region between the steepest descent paths, the surface wave term vanishes. This is completely analogous to the GO shadow boundaries that arise when a pole crosses a steepest descent path. The major difference is that the surface wave poles are generally complex, whereas the GO poles are always real.

The surface wave transition field provides continuity across the surface wave boundary by uniformly accounting for the complex pole in the vicinity of the steepest descent path. It performs the same function as the diffracted field in providing continuity for the GO field. This component of the exact solution has also been referred to as the *surface ray field* [27].

To determine the contribution of the complex surface wave pole in the steepest descent integral, the method of Felsen-Marcuvitz [68] is utilized. In (14-48), the integral was shown to be asymptotically approximated by the sum of two terms; the first corresponding to a first-order saddle point evaluation with no pole, and the second corresponding to the pole contribution. In (14-47), the modified Pauli-Clemmow method was shown to reduce precisely to the first term of (14-48) when the GO pole was not near the steepest descent path. Hence, a method to include the surface wave transition field is to use the second term of (14-48) added to the diffraction contribution of Section 14.7. This is valid provided the poles approach the steepest descent path individually.

The contribution of the surface wave transition field is then given by [28]

$$
U^0_{\text{SWTR}}(\rho,\phi) = U_0 \frac{e^{-j\beta\rho}}{\sqrt{\rho}} \left[\frac{-\sqrt{\dfrac{j}{\pi}}\sin\left(\dfrac{\pi}{2n}\right)}{\Psi\left(\dfrac{n\pi}{2}-\phi'\right)} \frac{\sin\left(\dfrac{\phi'}{n}\right)}{\cos\left(\dfrac{\pi+\theta_0}{n}\right)-\cos\left(\dfrac{\phi'}{n}\right)} \right]
$$
$$
\cdot \Psi_n\left(n\pi-\frac{\pi}{2}\right)\Psi_n\left(n\pi+\frac{\pi}{2}+2\theta_0\right)\Psi_n\left(\frac{3\pi}{2}+\theta_0-\theta_n\right)\Psi_n\left(\frac{\pi}{2}+\theta_0+\theta_n\right)
$$
$$
\cdot \frac{F\{\beta\rho[1-\cos(\phi+\theta_0)]\}-1}{\sqrt{\beta[\cos(\phi+\theta_0)-1]}} \tag{14-61a}
$$

$$
U^n_{\text{SWTR}}(\rho,\phi) = U_0 \frac{e^{-j\beta\rho}}{\sqrt{\rho}} \left[\frac{\sqrt{\dfrac{j}{\pi}}\sin\left(\dfrac{\pi}{2n}\right)}{\Psi\left(\dfrac{n\pi}{2}-\phi'\right)} \frac{\sin\left(\dfrac{\phi'}{n}\right)}{\cos\left(\dfrac{n\pi+\pi+\theta_n}{n}\right)-\cos\left(\dfrac{\phi'}{n}\right)} \right]
$$
$$
\cdot \Psi_n\left(n\pi-\frac{\pi}{2}\right)\Psi_n\left(n\pi+\frac{\pi}{2}+2\theta_n\right)\Psi_n\left(\frac{3\pi}{2}+\theta_n-\theta_0\right)\Psi_n\left(\frac{\pi}{2}+\theta_0+\theta_n\right)
$$
$$
\cdot \frac{F\{\beta\rho[1-\cos(\phi-n\pi-\theta_n)]\}-1}{\sqrt{\beta[\cos(\phi-n\pi-\theta_n)-1]}} \tag{14-61b}
$$

The pole residue has been determined as in (14-43) and Example 14-1. By the symmetry of the wedge geometry, it is noted that (14-61b) can be determined from (14-61a) by replacing ϕ by $n\pi-\phi$, ϕ' by $n\pi-\phi'$, θ_0 by θ_n, and θ_n by θ_0. This Fresnel transition function is as defined in [66] but extended for complex arguments. The $\sqrt{z}$ in the definition of $F(z)$ should have a branch cut along the positive imaginary axis so that $-3\pi/4 < [\text{Arg}\sqrt{z}] < \pi/4$. It is the discontinuity in $F(z)$ at the branch cut that provides the discontinuity in the integral contribution. When the pole is far from the steepest descent path, the Fresnel transition function is approximately unity, and hence, the surface wave transition field contribution is zero.

A troublesome case occurs when the surface wave pole coincides with a GO pole. Since the GO poles are always real, this can only occur for real surface impedances. Real surface impedances, however, cannot support surface waves, as the surface wave pole will never lie within the steepest descent paths. However, if the surface wave transition field is calculated by blindly applying (14-61a) and (14-61b), erroneous results will occur whenever the surface wave pole is far from the steepest descent path, yet near a GO pole. Indeed, the surface wave transition field should be zero when the surface wave pole is far from the steepest descent paths. *Hence, (14-61a) and (14-61b) should only be used when the surface wave pole is closer to a steepest descent path than to a GO pole. When the surface wave pole is near both a GO pole and the steepest descent path, then (16) of [26] should be used. When the surface wave pole is far from the steepest descent path, the surface wave transition field is taken as zero.*

A complete asymptotic expansion, which uniformly accounts for coalescing GO and surface wave poles, can be found in [70]. However, this type of formulation would not retain the form of the PEC UTD diffraction coefficient [26, 66], in which the four dominant poles are accounted for in four cotangent-Fresnel products. Retaining only the four dominant poles is sufficient for all shadow boundaries of all the multiple reflected fields.

14.9 COMPUTATIONS

Based on the analytical formulations developed in the previous sections, a number of computations were performed and a sample of them are presented here. Some of the others will be assigned as end-of-chapter problems.

The amplitude radiation patterns of an infinitesimal dipole placed at a height of 6λ above a square ground plane (10.6λ on each side) with impedance surface are shown in Figure 14-8. We see that the pattern variations are not severely affected by the changes in the normalized surface impedance, ranging from 0.001 to 0.5 (unnormalized impedances of $0.377-188.5$ ohms). Most of the variations occur above the ground plane within $\theta = \pm 60°$, where the field intensity is weaker. The amplitude radiation patterns, predicted [28] and measured [49], of a $\lambda/4$ monopole above a square graphite ground plane [normalized surface impedance of $0.001668(1+j)$] are exhibited in Figure 14-9. It is evident that there is a very good agreement between simulations and measurements.

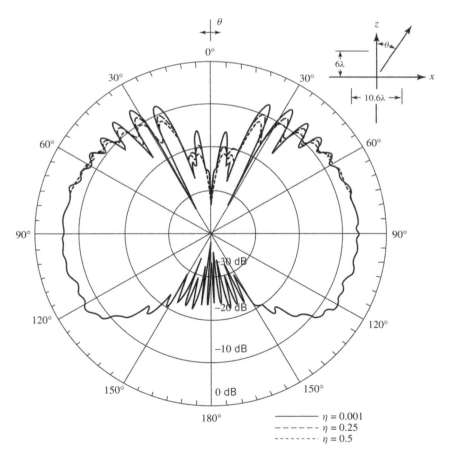

Figure 14-8 Amplitude radiation patterns of infinitesimal vertical dipole above square ground plane with normalized impedance surface [28].

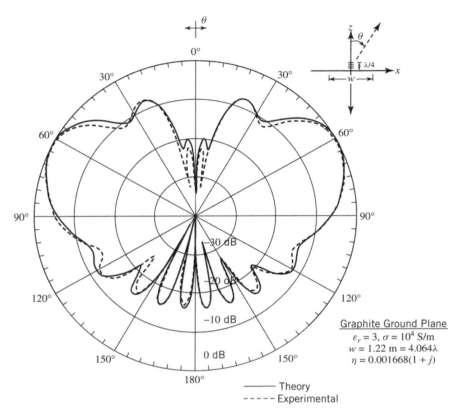

Figure 14-9   Amplitude radiation patterns of $\lambda/4$ vertical monopole above square graphite ground plane [28, 49]. (Source: C. A. Balanis and D. DeCarlo, "Monopole antenna patterns on finite size composite ground planes," *IEEE Trans. Antennas Propagat.*, © 1982, IEEE.)

A dihedral corner reflector is often used as a reference for RCS measurements. The geometry of a dihedral corner reflector, and its dimensions, are shown in Figure 14-10. A very thorough examination of a PEC dihedral corner reflector can be found in [72–74]. To reduce the RCS, the interior faces of the dihedral corner reflector can be coated with Radio Absorbing Material (RAM) with

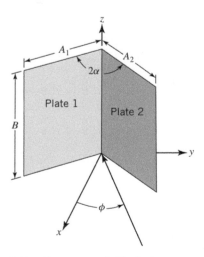

Figure 14-10   Geometry of dihedral corner reflector.

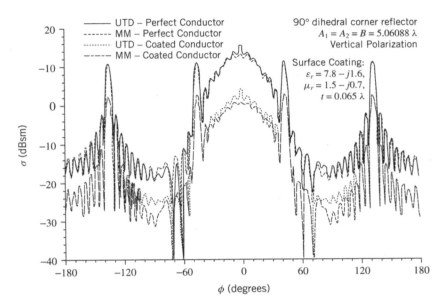

Figure 14-11   RCS of PEC and coated $90°$ corner reflector [28, 29]. (Source: T. Griesser, C. A. Balanis, and K. Liu, "RCS analysis and reduction for lossy dihedral corner reflectors," *Proceedings of the IEEE,* © 1989, IEEE.)

complex permittivity and permeability. This is a procedure used in practice to reduce the RCS of radar targets and make them more stealthy. The RCS patterns of a $90°$ corner reflector, PEC and coated, are shown in Figure 14-11 and are based on computations using the Uniform Theory of Diffraction (UTD) and Method of Moments (MoM). A $90°$ dihedral corner reflector possesses a very high RCS near its axis primarily because of the second reflection, which is directed toward the incident direction when the interior angle of the dihedral corner reflector is $90°$. This is not the case for interior angles other than $90°$. This can be verified by simply using Snell's law of reflection. It is also apparent from the patterns in Figure 14-11 that there is a very good agreement between the two methods, and both predict reductions of nearly 10 dB at the flare spots (directions where the RCS is the most intense; along the axis of the $90°$ corner reflector and in directions perpendicular to its faces; both interior and exterior). The RCS of $77°$ and $98°$ dihedral corner reflectors were also performed, and they are left as end-of-chapter exercises. We should expect that for these two dihedral corner reflectors ($77°$ and $98°$) the RCS at and near the axis is not very intense because the second reflection is not directed toward the direction of incidence.

14.10 MULTIMEDIA

On the website that accompanies this book, the following multimedia resources are included for the review, understanding, and presentation of the material of this chapter.

- **MATLAB** computer programs:
  a. **Monopole:** Computes the normalized amplitude pattern of a monopole on a rectangular ground plane with an impedance surface. The geometry is the same as that of Figure 13-32, where the PEC ground plane is replaced by one with an equivalent surface impedance. The pattern is computed using the UTD formulation of Chapter 14.

  b. **Dipole_Vertical:** Computes the normalized amplitude pattern of a vertical dipole placed at a height h above a rectangular ground plane with an equivalent surface impedance. The pattern is computed using the UTD formulation of Chapter 14.

  c. **Dipole_Horizontal_H_Plane:** Computes the normalized H-plane amplitude pattern of a horizontal dipole placed at a height h above a rectangular ground plane with an equivalent surface impedance. The pattern is computed using the UTD formulation of Chapter 14.

- **PowerPoint (PPT)** viewgraphs, in multicolor.

REFERENCES

1. T. B. A. Senior, "Impedance boundary conditions for imperfectly conducting surfaces," *Appl. Sci. Res.*, sec. B, vol. 8, nos. 5–6, pp. 418–436, 1960.

2. M. A. Leontovich, *Investigations of Propagation of Radio Waves*, Par II, Moscow, 1948.

3. M. A. Lentovich, *Diffraction, Refraction and Reflection of Radio Waves* (edited by V. A. Fock, N. Logan, and P. Blacksmith), U.S. GPO, Washington, DC, 1957, Appendix.

4. T. B. A. Senior, "Diffractions by a semi-infinite metallic sheet," *Proc. Roy. Soc.* (London), vol. 213, pp. 436–458, Jul. 22, 1952.

5. B. Noble, *Methods Based on the Wiener-Hopf Technique*, New York, Pergamon, 1958.

6. T. B. A. Senior, "Diffraction by an imperfectly conducting half-plane at oblique incidence," *Appl. Sci. Res.*, sec. B, vol. 8, no. 1, pp. 35–61, 1959.

7. W. E. Williams, "Diffractions of an electromagnetic plane wave by a metallic sheet," *Proc. Roy. Soc.* (London), vol. 257, pp. 413–419, Sept. 20, 1960.

8. P. C. Clemmow, "A method for the exact solution of a class of two-dimensional diffraction problems," *Proc. Roy. Soc.* (London), vol. 205, pp. 286–308, Feb. 7, 1951.

9. G. D. Maliuzhinets, "Excitation, reflection and emission of surface waves from a wedge with a given face impedances," *Soviet Physics, Doklady*, vol. 3, pp. 752–755, 1958.

10. G. D. Malyughinets (Maliuzhinets), "Das Sommerfeldsche integral und die losung von Beugungsaufgaben in Winkelgebieten," *Annalen der Physik*, Folge 7, Band 6, Heft 1–2, pp. 107–112, 1960.

11. T. B. A. Senior, "Diffraction by an imperfectly conducting wedge," *Comm. Pure Appl. Math.*, vol. 12, no. 2, pp. 337–372, May 1959.

12. W. E. Williams, *Proc. Roy. Soc.* (London), ser. A., vol. 252, p. 376, 1959.

13. R. A. Hurd and S. Przezdziecki, "Diffraction by a half-plane with different face impedances—a re-examination," *Canadian J. Physics*, vol. 59, pp. 1337–1347, 1981.

14. E. Lüneburg and R. A. Hurd, "On the diffraction problem of a half plane with different face impedances," *Canadian J. Physics*, vol. 62, pp. 853–860, 1984.

15. J. Shmoys, "Diffraction by a half-plane with a special impedance variation," *IRE Trans. Antennas Propagat.*, pp. S88–S90, Dec. 1959.

16. T. R. Faulkner, "Diffraction of an electromagnetic plane-wave by a metallic strip," *J. Inst. Maths. Applics.*, vol. 1, no. 2, pp. 149–163, Jun. 1965.

17. T. B. A. Senior, "Skew incidence on a right-angled impedance wedge," *Radio Science*, vol. 13, no. 4, pp. 639–647, July-Aug. 1978.

18. T. B. A. Senior, "Solution of a class of imperfect wedge problems for skew incidence," *Radio Science*, vol. 21, no. 2, pp. 185–191, Mar.-Apr. 1986.

19. T. B. A Senior and J. L. Volakis, "Scattering by an imperfect right-angled wedge," *IEEE Trans. Antennas Propagat.*, vol. AP-34, no. 5, pp. 681–689, May 1986.

20. R. G. Rojas, "Wiener-Hopf analysis of the EM diffraction by an impedance discontinuity in a planar surface and by an impedance half plane," *IEEE Trans. Antennas Propagat.*, vol. AP-36, no. 1, pp. 71–83, Jan. 1988.

21. R. G. Rojas, "Electromagnetic diffraction of an obliquely incident plane wave field by a wedge with impedance faces," *IEEE Trans. Antenna Propagat.*, vol. 36, pp. 956–970, Jul. 1988.

22. V. G. Vaccaro, "The generalized reflection method in electromagnetism," *Arch. Elektron and Ueber-taragungstech* (Germany), vol. 34, no. 12, pp. 493–500, 1980.

23. J. J. Bowman, "High-frequency backscattering from an absorbing infinite strip with arbitrary face impedances," *Canadian J. Phys.*, vol. 45, pp. 2409–2430, 1967.

24. O. M. Bucci and G. Franceschetti, "Electromagnetic scattering by a half plane with two face impedances," *Radio Sci.*, vol. 11, no. 1, pp. 49–59, Jan. 1976.

25. S. Sanyal and A. K. Bhattacharyya, "Diffraction by a half-plane with two face impedances, uniform asymptotic expansion for plane wave and arbitrary line source incidence," *IEEE Trans. Antennas Propagat.*, vol. AP-34, no. 5, pp. 718–723, May 1986. Corrections vol. AP-35, no. 12, p. 1499, Dec. 1987.

26. R. Tiberio, G. Pelosi, and G. Manara, "A uniform GTD formulation for the diffraction by a wedge with impedance faces," *IEEE Trans. Antennas Propagat.*, vol. AP-33, no. 8, pp. 867–873, Aug. 1985.

27. M. I. Herman and J. L. Volakis, "High frequency scattering from canonical impedance structures," University of Michigan Radiation Lab Technical Report 389271-T, Ann Arbor, MI, May 1987.

28. Timothy Griesser, "High-frequency electromagnetic scattering from imperfectly conducting surfaces," PhD dissertation, Arizona State University, Aug. 1988.

29. T. Griesser, C. A. Balanis, and K. Liu, "RCS analysis and reduction for lossy dihedral corner reflectors," *Proc. IEEE*, vol. 77, no. 5, pp. 806–814, May 1989.

30. T. Griesser and C. A. Balanis, "Reflections, diffractions, and surface waves for an interior impedance wedge of arbitrary angle," *IEEE Trans. Antennas Propagat.*, vol. 37, no. 7, pp. 927–935, Jul. 1989.

31. J. L. Volakis, "A uniform geometrical theory of diffraction for an imperfectly conducting half-plane," *IEEE Trans. Antennas Propagat.*, vol. AP-34, no. 2, pp. 172–180, Feb. 1986. Corrections vol. AP-35, no. 6, pp. 742–744, Jun. 1987.

32. T. B. A. Senior, "Half plane edge diffraction," *Radio Science*, vol. 10, no. 6, pp. 645–650, Jun. 1975.

33. T. B. A. Senior, "Diffraction tensors for imperfectly conducting edges," *Radio Science*, vol. 10, no. 10, pp. 911–919, Oct. 1975.

34. T. B. A. Senior, "Some problems involving imperfect half planes," in *Electromagnetic Scattering*, P. L. E. Uslenghi (Ed.), New York, Academic, pp. 185–219, 1978.

35. T. B. A. Senior, "The current induced in a resistive half plane," *Radio Sci.*, vol. 16, no. 6, pp. 1249–1254, Nov.-Dec. 1981.

36. T. B. A. Senior, "Combined resistive and conducting sheets," *IEEE Trans. Antennas Propagat.*, vol. AP-33, no. 5, pp. 57;7–579, May 1985.

37. J. L. Volakis and T. B. A. Senior, "Diffraction by a thin dielectric half-plane," *IEEE Trans. Antennas Propagat.*, vol. AP-35, no. 12, pp. 1483–1487, Dec. 1987.

38. R. Tiberio and R. G. Kouyoumjian, "A uniform GTD solution for the diffraction by strips illuminated at grazing incidence," *Radio Sci.*, vol. 14, no. 6, pp. 933–941, Nov. 1979.

39. R. Tiberio, F. Bessi, G. Manara, and G. Pelosi, "Scattering by a strip with two face impedances at edge-on incidence," *Radio Science*, vol. 17, no. 5, pp. 1199–1210, Sept.-Oct. 1982.

40. R. Tiberio, "A spectral extended ray method for edge diffraction," in *Hybrid Formulation of Wave Propagation and Scattering*, L. B. Felsen (Ed.), NATO ASI Series, Aug. 1983, pp. 109–130.

41. R. Tiberio and G. Pelosi, "High-frequency scattering from the edges of impedance discontinuities on a flat plane," *IEEE Trans. Antennas Propagat.*, vol. AP-31, no. 4, pp. 590–596, Jul. 1983.

42. M. I. Herman and J. L. Volakis, "High frequency scattering by a double impedance wedge," *IEEE Trans. Antennas Propagat.*, vol. 36, no. 5, pp. 664–678, May 1988.

43. M. I. Herman and J. L. Volakis, "High frequency scattering from polygonal impedance cylinders and strips," *IEEE Trans. Antennas Propagat.*, vol. AP-36, no. 5, pp. 679–689, May 1988.

44. T. B. A. Senior, "Scattering by resistive strips," *Radio Sci.*, vol. 14, no. 5, pp. 911–924, Sept.-Oct. 1979.

45. T. B. A. Senior, "Backscattering from resistive strips," *IEEE Trans. Antennas Propagat.*, vol. AP-27, no. 6, pp. 808–813, Nov. 1979.

46. M. I. Herman and J. L. Volakis, "High-frequency scattering by a resistive strip and extensions to conductive and impedance strips," *Radio Sci.*, vol. 22, no. 3, pp. 335–349, May-Jun. 1987.

47. T. B. A. Senior and V. V. Liepa, "Backscattering from tapered resistive strips," *IEEE Trans. Antennas Propagat.*, vol. AP-32, no. 7, pp. 747–751, Jul. 1984.

48. A. K. Bhattacharyya and S. K. Tandon, "Radar cross section of a finite planar structure coated with a lossy dielectric," *IEEE Trans. Antennas Propagat.*, vol. AP-32, no. 9, pp. 1003–1007, Sept. 1984.

49. C. A. Balanis and D. DeCarlo, "Monopole antenna patterns on finite size composite ground planes," *IEEE Trans. Antennas Propagat.*, vol. AP-30, no. 4, pp. 764–768, Jul. 1982.

50. J. L. Volakis, "Simple expressions for a function occurring in diffraction theory," *IEEE Trans. Antennas Propagat.*, vol. AP-33, no. 6, pp. 678–680, Jun. 1985.

51. P. Corona, G. Ferrara, and C. Gennarelli, "Backscattering by loaded and unloaded dihedral corners," *IEEE Trans. Antennas Propagat.*, vol. AP-35, no. 10, pp. 1148–1153, Oct. 1987.

52. V. Y. Zavadskii and M. P. Sakharora, "Application of the special function $\Psi_\Phi(z)$ in problems of wave diffraction in wedge shaped regions," *Soviet Phys. Acoustics*, vol. 13, no. 1, pp. 48–54, Jul.-Sept. 1967.

53. O. M. Bucci, "On a function occurring in the theory of scattering from an impedance half-plane," Rep. 75-1, Instituto Universitario Navale, via Acton 38, Napole, Italy, 1974.

54. J. L. Volakis and T. B. A. Senior, "Simple expressions for a function occurring in diffraction theory," *IEEE Trans. Antennas Propagat.*, vol. AP-33, no. 6, pp. 678–680, Jun. 1985.

55. K. Hongo and E. Najajima, "Polynomial approximation of Maliuzhinets' function," *IEEE Trans. Antennas Propagat.*, AP-34, no. 7, pp. 942–947, Jul. 1986.

56. M. I. Herman, J. L. Volakis, and T. B. A. Senior, "Analytic expressions for a function occurring in diffraction theory," *IEEE Trans. Antennas Propagat.*, vol. AP-35, no. 9, pp. 1083–1086, Sept. 1987.

57. T. B. A. Senior and J. L. Volakis, *Approximate Boundary Conditions in Electromagnetics*, IEE Press, New York and London, 1995.

58. T. B. A. Senior, "A note on impedance boundary conditions," *Canadian J. Phys.*, vol. 40, no. 5, pp. 663–665, May 1962.

59. N. G. Alexopoulos and G. A. Tadler, "Accuracy of the Lentovich boundary condition for continuous and discontinuous surface impedances," *J. Appl. Phys.*, vol. 46, no. 8, pp. 3326–3332, Aug. 1975.

60. D. S. Wang, "Limits and validity of the impedance boundary condition on penetrable surfaces," *IEEE Trans. Antennas Propagat.*, vol. AP-35, no. 4, pp. 453–457, Apr. 1987.

61. S. W. Lee and W. Gee, "How good is the impedance boundary condition?" *IEEE Trans. Antennas Propagat.*, vol. AP-35, no. 11, pp. 1313–1315, Nov. 1987.

62. T. Griesser and C. A. Balanis, "Oceanic low-angle monopulse radar tracking errors," *IEEE J. Appl. Phys. Oceanic Eng.*, vol. OE-12, pp. 289–295, Jan. 1987.

63. W. Pauli, "On asymptotic series for functions in the theory of diffraction of light," *Phys. Rev.*, vol. 54, pp. 924–931, Dec. 1938.

64. P. C. Clemmow, "Some extensions to the method of integration by steepest descents," *Quart. J. Mech. Appl. Math.*, vol. 3, pp. 241–256, 1950.

65. P. C. Clemmow, *The Plane Wave Spectrum Representation of Electromagnetic Fields*, Oxford, Pergamon Press, 1966, pp. 43–58.

66. R. G. Kouyoumjian and P. H. Pathak, "A uniform geometrical theory of diffraction for an edge in a perfectly conducting surface," *Proc. IEEE*, vol. 62, no. 11, pp. 1448–1461, Nov. 1974.

67. R. G. Kouyoumjian, "Asymptotic high-frequency methods," *Proc. IEEE*, vol. 53, no. 8, pp. 864–876, Aug. 1965.

68. L. B. Felsen and N. Marcuvitz, *Radiation and Scattering of Waves*, New Jersey, Prentice-Hall, 1973, p. 399.

69. L. B. Felsen, "Asymptotic methods in high-frequency propagation and scattering," in *Electromagnetic Scattering*, P. L. E. Uslenghi (Ed.), New York, Academic Press, 1978, pp. 29–65.

70. R. G. Rojas, "Comparison between two asymptotic methods," *IEEE Trans. Antennas Propagat.*, vol. AP-35, no. 12, pp. 1489–1492, Dec. 1987.

71. E. L. Yip and R. J. Chiavetta, "Comparison of uniform asymptotic expansions of diffraction integrals," *IEEE Trans. Antennas Propagat.*, vol. AP-35, no. 10, pp. 1179–1180, Jul. 1986.

72. T. Griesser and C. A. Balanis, "Backscatter analysis of dihedral corner reflectors using physical optics and the physical theory of diffraction," *IEEE Trans. Antennas Propagat.*, vol. AP-35, no. 10, pp. 1137–1147, Oct. 1987.

73. T. Griesser and C. A. Balanis, "Dihedral corner reflector backscatter using higher order reflections and diffractions," *IEEE Trans. Antennas Propagat.*, vol. AP-35, no. 11, pp. 1235–1247, Nov. 1987.

74. Timothy Griesser, "Backscatter cross sections of a dihedral corner reflector using GTD and PTD," MS Thesis, Arizona State University, Dec. 1985.

PROBLEMS

14.1. Using duality, derive the dual of (14-2) for the other polarization. Show that that the impedance boundary condition has the same duality as Maxwell's equations.

14.2. The impedance boundary condition can be written in a different form than the vector form of (14-2) to conveniently solve the wedge scattering problem. For a wedge, in which the wedge vertex is along the z axis in a cylindrical coordinate system, the wedge faces are located at $\phi = 0$ and $\phi = n\pi$. This wedge geometry is illustrated in Figure 14-1, for both the exterior and interior wedges. For the soft polarization (TM^z), the impedance boundary conditions are

$$E_z = -\eta_0 Z_0 H_\rho \quad \text{for} \quad \phi = 0$$
$$E_z = +\eta_n Z_0 H_\rho \quad \text{for} \quad \phi = n\pi$$

on the faces 0 and n, respectively. Rewrite/reduce each of these equations in terms of the E_z and its partial derivative on the respective faces.

14.3. Repeat Problem 14.2 for the hard polarization Polarization (TE^z) for which, using Figure 14-1, the impedance boundary conditions are

$$E_\rho = +\eta_0 Z_0 H_z \quad \text{for} \quad \phi = 0$$
$$E_\rho = -\eta_n Z_0 H_z \quad \text{for} \quad \phi = n\pi$$

on the faces 0 and n, respectively. Rewrite/reduce each of these equations in terms of the H_z and its partial derivative on the respective faces, usually referred to as a boundary condition of the third kind.

14.4. For the soft (perpendicular) polarization of Figure 14-2a, derive an expression for the reflection coefficient for oblique incidence assuming the reflecting planar surface is a PEC ground plane covered with a lossless dielectric slab of thickness t and constitutive parameters ε_1 and μ_1. Use transmission line theory to represent the normalized equivalent surface impedance at the leading interface formed by free space and the dielectric slab.

14.5. Repeat Problem 14.4 for the hard (parallel) polarization of Figure 14-2b.

14.6. For the method of steepest descent (saddle point method), the saddle point z_s (assuming only one saddle point) is the point in the complex plane at which $|e^{\beta \rho h(z)}|$ is a maximum of $|e^{\beta \rho h(z)}|$ along one direction and a minimum of $|e^{\beta \rho h(z)}|$ along a perpendicular direction in the complex plane. In three dimensions, a plot of $|e^{\beta \rho h(z_s)}|$ would look like a saddle with z_s at the center of the saddle. Assuming $H(z)$ in (14-71) is smoothly varying near the saddle point, the integrand achieves its maximum at the saddle point, and it is basically negligible elsewhere. The steepest descent and ascent paths can be found if the relationship $\text{Im}[h(z)] = \text{Im}[h(z_s)]$ is a constant. Derive equations for the steepest descent and ascent paths in the complex z plane.

14.7. In contrast to the steepest descent and ascent paths of Problem 14.6, the stationary phase paths are those for which $\text{Re}[h(z)] = \text{Re}[h(z_s)]$ is a constant. Derive equations for the stationary paths in the complex z plane.

14.8. A uniform plane wave of unity amplitude is incident upon a 90° dielectric wedge at an angle of $\phi' = 60°$ (Figure P14-8). At an observation distance of $\rho = 81\lambda$ from the edge of the wedge and an observation angle of 180°, compute for *hard* polarization the approximate:
(a) Incident diffracted field.
(b) Reflected diffracted field.
The electrical properties of the dielectric are: $\varepsilon_r = 4$, $\mu_r = 1$. Assume that the surface impedance of the faces of the wedge are each

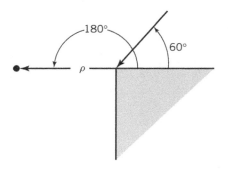

Figure P14-8

equal to the intrinsic impedance of the dielectric material. Clearly indicate what equations you are using. If you make any approximations, state as to why you think they are valid.

14.9. Repeat Problem 14.8 for *soft* polarization.

14.10. Derive (14-44) for U_{SW}^n using a similar procedure as in Example 14-1 for U_{SW}^0.

14.11. Using the procedure of Example 14-2, derive the second term of (14-54) from the second term of (14-53).

14.12. Using the procedure of Example 14-2, derive the third term of (14-54) from the third term of (14-53).

14.13. Using the procedure of Example 14-2, derive the fourth term of (14-54) from the fourth term of (14-53).

CHAPTER 15

Green's Functions

15.1 INTRODUCTION

In the area of electromagnetics, solutions to many problems are obtained using a second-order, uncoupled partial differential equation, derived from Maxwell's equations, and the appropriate boundary conditions. The form of most solutions of this type is an infinite series, provided the partial differential equation and the boundary conditions representing the problems are separable in the chosen coordinate system. The difficulty in using these types of solutions to obtain an insight into the behavior of the function is that they are usually slowly convergent, especially at regions where rapid changes occur. It would then seem appropriate, at least for some problems and associated regions, that closed-form solutions would be desirable. Even solutions in the form of integrals would be acceptable. The technique known as the *Green's function* accomplishes this goal.

Before proceeding with the presentation of the Green's function solution, let us briefly describe what the Green's functions represent and how they are used to obtain the overall solution to the problem.

With the Green's function technique, a solution to the partial differential equation is obtained using a unit source (impulse, Dirac delta) as the driving function. *This is known as the Green's function.* The solution to the actual driving function is written as a superposition of the impulse response solutions (Green's function) with the Dirac delta source at different locations, which in the limit reduces to an integral. The contributions to the overall solution from the general source may be greater or smaller than that of the impulse response depending on the strength of the source at that given location. In engineering terminology, then, the Green's function is nothing more but the *impulse response* of a system; in system theory, this is better known as the *transfer function*.

For a given problem, the Green's function can take various forms. One form of its solution can be expressed in terms of finite explicit functions, and it is obtained based on a procedure that will be outlined later. This procedure for developing the Green's function can be used only if the solution to the homogeneous differential equation is known. Another form of the Green's function is to construct its solution by an infinite series of suitably chosen orthonormal functions. The boundary conditions determine the eigenvalues of the eigenfunctions, and the strength of the sources influences the coefficients of these eigenfunctions. Integral forms can also be used to represent the

Balanis' Advanced Engineering Electromagnetics, Third Edition. Constantine A. Balanis.
© 2024 John Wiley & Sons, Inc. Published 2024 by John Wiley & Sons, Inc.
Companion Website: www.wiley.com/go/balanis/advancedengineeringelectromagnetics3e

Green's function, especially when the eigenvalue spectrum is continuous. All solutions, although different in form, give the same results. The form of the Green's function that is most appropriate will depend on the problem in question. *The representation of the actual source plays a significant role as to which form of the Green's function may be most convenient for a given problem.*

Usually, there is as much work involved in finding the Green's function as there is in obtaining the infinite-series solution. However, the major advantages of the Green's function technique become evident when the same problem is to be solved for a variety of driving sources and when the sources are in the presence of boundaries [1–8].

In this chapter we shall initially study the one-dimensional differential equation

$$[L + \lambda r(x)]y(x) = f(x) \tag{15-1}$$

where L is the Sturm-Liouville operator and λ is a constant. It is hoped that an understanding of the Green's function method for this equation will lead to a better understanding of the equations that occur in electromagnetic field theory applied to homogeneous media such as

$$\nabla^2 \phi(\mathbf{R}) + \beta^2 \phi(\mathbf{R}) = p(\mathbf{R}) \qquad \text{(scalar wave equation)} \tag{15-2a}$$

$$\nabla \times \nabla \times \psi(\mathbf{R}) + \beta^2 \psi(\mathbf{R}) = \mathbf{F}(\mathbf{R}) \qquad \text{(vector wave equation)} \tag{15-2b}$$

particularly since (15-2a) and (15-2b) can often be reduced to several equations similar to (15-1) by the separation of variables technique. Before proceeding to the actual solution of (15-1) by the Green's function method, we shall first consider several examples of Green's functions in other areas of electrical and general engineering. These will be followed by some topics associated with the Sturm-Liouville operator L before we embark on the solution of (15-2a).

15.2 GREEN'S FUNCTIONS IN ENGINEERING

The Green's function approach to solution of differential equations has been used in many areas of engineering, physics, and elsewhere [9–18]. Before we embark on constructing Green's function solutions to electromagnetic boundary-value problems, let us consider two other problems, one dealing with electric circuit theory and the other with mechanics. This will give the reader a better appreciation of the Green's function concept.

15.2.1 Circuit Theory

Analysis of lumped electric element circuits is a fundamental method of electrical engineering. Therefore, we will relate the Green's function to the solution of a very simple lumped-element circuit problem.

Let us assume that a voltage source $v(t)$ is connected to a resistor R and inductor L, as shown in Figure 15-1a. The equation that governs the solution to that circuit can be written as

$$L\frac{di}{dt} + Ri = v(t) \tag{15-3}$$

where $v(t)$ is the excitation voltage source that is turned on at $t = t'$. Initially, $(t < t')$ the circuit is at rest and at $t = t'$ the voltage is suddenly turned on by an impulse V_0 of a very short duration $\Delta t'$. For $t > t' + \Delta t'$, when $v(t) = 0$, the circuit performance is governed by the homogeneous equation

$$L\frac{di}{dt} + Ri(t) = 0 \quad \text{for} \quad t > t' + \Delta t' \tag{15-4}$$

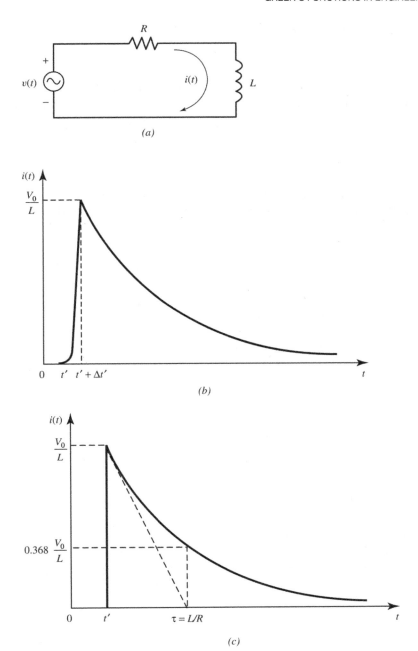

Figure 15-1   (*a*) *RL* series circuit. (*b*) Current response. (*c*) Time constant.

whose solution for $i(t)$ can be written as

$$i(t) = I_0 e^{-(R/L)t} \quad \text{for} \quad t > t' + \Delta t' \tag{15-4a}$$

where I_0 is a constant and L/R is referred to as the *time constant* τ of the circuit.

Since the voltage excitation $v(t)$ during $\Delta t'$ was an impulse V_0, then

$$\int_{t'}^{t'+\Delta t'} v(t)\, dt = V_0 \tag{15-5}$$

where V_0 is the voltage for duration $\Delta t'$ (volts$\cdot \Delta t =$ volts-sec). Therefore, between $t' \leq t \leq t'+\Delta t'$, (15-3) can be written using (15-5) as

$$L\int_{t'}^{t'+\Delta t'} di + R\int_{t'}^{t'+\Delta t'} i(t)\, dt = \int_{t'}^{t'+\Delta t'} v(t)\, dt$$

$$L[i(t'+\Delta t') - i(t')] + R\int_{t'}^{t'+\Delta t'} i(t)\, dt = V_0 \tag{15-6}$$

Because $\Delta t'$ is very small, we assume that $i(t)$ during the excitation $\Delta t'$ of the voltage source is not exceedingly large, and it behaves as shown in Figure 15-1b. Therefore, during $\Delta t'$

$$\lim_{\Delta t' \to 0} R\int_{t'}^{t'+\Delta t'} i(t)\, dt \simeq 0 \tag{15-7}$$

so that the terms on the left side of (15-6) reduce using (15-4a) to

$$i(t') = 0 \tag{15-8a}$$

$$i(t'+\Delta t') = I_0 e^{-(R/L)(t'+\Delta t')} \overset{\Delta t' \to 0}{\simeq} I_0 e^{-(R/L)t'} \tag{15-8b}$$

Using (15-7) through (15-8b), we can express (15-6) as

$$L I_0 e^{-(R/L)t'} = V_0 \tag{15-9}$$

or

$$I_0 = \frac{V_0}{L} e^{+(R/L)t'} \tag{15-9a}$$

Therefore, (15-4a) can be written, using (15-9a), as

$$i(t) = \begin{cases} 0 & t < t' & (15\text{-}10a) \\[2mm] \dfrac{V_0}{L} e^{-(R/L)(t-t')} & t \geq t' & (15\text{-}10b) \end{cases}$$

which is shown plotted in Figure 15-1c.

If the circuit is subjected to N voltage impulses each of duration Δt and amplitude V_i occurring at $t = t'_i$, $(i = 0,\ldots,N)$, then the current response can be written as

$$i(t) = \begin{cases} 0 & t < t'_0 \\[2mm] \dfrac{V_0}{L} e^{-(R/L)(t-t'_0)} & t'_0 < t < t'_1 \\[2mm] \dfrac{V_0}{L} e^{-(R/L)(t-t'_0)} + \dfrac{V_1}{L} e^{-(R/L)(t-t'_1)} & t'_1 < t < t'_2 \\[1mm] \vdots & \vdots \\[1mm] \displaystyle\sum_{i=0}^{N} \dfrac{V_i}{L} e^{-(R/L)(t-t'_i)} & t'_N < t < t'_{N+1} \end{cases} \tag{15-11}$$

If the circuit is subjected to a continuous voltage source $v(t)$ starting at t'_0 such that at an instant of time $t = t'$ and short interval $\Delta t'$ would produce an impulse of

$$dV = v(t')\, dt' \tag{15-12}$$

then the response of the system for $t \geq t'$ can be expressed, provided that $i(t) = v(t) = 0$ for $t < t'$, as

$$i(t) = \int_{t'}^{t} \left[\frac{v(t')\,dt'}{L} \right] e^{-(R/L)(t-t')} = \int_{t'}^{t} v(t') \frac{e^{-(R/L)(t-t')}}{L}\,dt'$$

$$\boxed{i(t) = \int_{t'}^{t} v(t')G(t,t')\,dt'} \tag{15-13}$$

where

$$\boxed{G(t,t') = \frac{e^{-(R/L)(t-t')}}{L} \quad \text{for} \quad t > t'} \tag{15-13a}$$

In (15-13), $G(t,t')$ of (15-13a), is referred to as the *Green's function,* and *it represents the response of the system for $t > t'$ when an excitation voltage $v(t)$ at $t = t'$ is an impulse (Dirac delta) function.* Knowing the response of the system to an impulse function, represented by the Green's function of (15-13a), the response $i(t)$ to any voltage source $v(t)$ can then be obtained by convolving the voltage source excitation with the Green's function according to (15-13).

15.2.2 Mechanics

Another problem that the reader may be familiar with is that of a string of length ℓ that is connected at the two ends and is subjected to external force per unit length (load) of $F(x)$. The objective is to find the displacement $u(x)$ of the string. If the load $F(x)$ is assumed to be acting down (negative direction), the displacement $u(x)$ of the string is governed by the differential equation

$$T\frac{d^2u}{dx^2} = F(x) \tag{15-14}$$

or

$$\frac{d^2u}{dx^2} = \frac{1}{T}\,F(x) = f(x) \tag{15-14a}$$

where T is the uniform tensile force of the string. If the string is stationary at the two ends, then the displacement function $u(x)$ satisfies the boundary conditions

$$u(x = 0) = u(x = \ell) = 0 \tag{15-15}$$

Initially, instead of solving the displacement $u(x)$ of the string subject to the load $F(x)$, let us assume that the load to which the string is subjected is a concentrated load (impulse) of $F(x = x') = \delta(x - x')$ at a point $x = x'$, as shown in Figure 15-2. For the impulse load, the differential equation 15-14a can be written as

$$\frac{d^2G(x,x')}{dx^2} = \frac{1}{T}\delta(x - x') \tag{15-16}$$

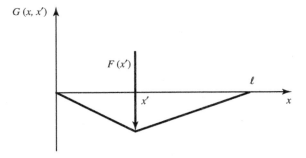

Figure 15-2 Attached string subjected to a load force.

subject to the boundary conditions

$$G(x=0,x') = G(0,x') = 0 \qquad (15\text{-}16a)$$

$$G(x=\ell,x') = G(\ell,x') = 0 \qquad (15\text{-}16b)$$

In (15-16), $G(x,x')$ represents the displacement of the string when it is subject to an impulse load of $1/T$ at $x=x'$, and it is referred to as the *Green's function* for the string. Once this is found, the displacement $u(x)$ of the string subjected to the load $F(x)$ can be determined by convolving the load $F(x)$ with the Green's function $G(x,x')$, as was done for the circuit problem by (15-13).

The solution to (15-16) is accomplished by following the procedure outlined here. Away from the load at $x=x'$, the differential equation 15-16 reduces to the homogeneous form

$$\frac{d^2 G(x,x')}{dx^2} = 0 \qquad (15\text{-}17)$$

which has solutions of the form

$$G(x,x') = \begin{cases} A_1 x + B_1 & 0 \le x \le x' & (15\text{-}17a) \\ A_2 x + B_2 & x' \le x \le \ell & (15\text{-}17b) \end{cases}$$

Applying the boundary conditions (15-16a) and (15-16b) leads to

$$G(x=0,x') = A_1(0) + B_1 = 0 \;\Rightarrow\; B_1 = 0 \qquad (15\text{-}18a)$$

$$G(x=\ell,x') = A_2\ell + B_2 = 0 \;\Rightarrow\; B_2 = -A_2\ell \qquad (15\text{-}18b)$$

Thus, (15-17a) and (15-17b) reduce to

$$G(x,x') = \begin{cases} A_1 x & 0 \le x \le x' & (15\text{-}19a) \\ A_2(x-\ell) & x' \le x \le \ell & (15\text{-}19b) \end{cases}$$

where A_1 and A_2 have not been determined yet.

At $x=x'$ the displacement $u(x)$ of the string must be continuous. Therefore, the Green's function of (15-19a) and (15-19b) must also be continuous at $x=x'$. Thus,

$$A_1 x' = A_2(x'-\ell) \;\Rightarrow\; A_2 = A_1 \frac{x'}{x'-\ell} \qquad (15\text{-}20)$$

According to (15-16), the second derivative of $G(x,x')$ is equal to an impulse function. Therefore, the first derivative of $G(x,x')$, obtained by integrating (15-16), must be discontinuous by an amount equal to $1/T$. Thus,

$$\lim_{\varepsilon \to 0} \left[\frac{dG(x'+\varepsilon,x')}{dx} - \frac{dG(x'-\varepsilon,x')}{dx} \right] = \frac{1}{T} \qquad (15\text{-}21)$$

or

$$\frac{dG(x'_+,x')}{dx} - \frac{dG(x'_-,x')}{dx} = \frac{1}{T} \qquad (15\text{-}21a)$$

Using (15-19a), (15-19b), and (15-20), we can write that

$$\frac{dG(x'_-,x')}{dx} = A_1 \qquad (15\text{-}22a)$$

$$\frac{dG(x'_+,x')}{dx} = A_2 = A_1 \frac{x'}{x'-\ell} \qquad (15\text{-}22b)$$

Thus, using (15-22a) and (15-22b), (15-21a) leads to

$$A_1 \frac{x'}{x' - \ell} - A_1 = \frac{1}{T} \quad \Rightarrow \quad A_1 \frac{\ell}{x' - \ell} = \frac{1}{T} \quad \Rightarrow \quad A_1 = \frac{1}{T} \frac{x' - \ell}{\ell} \tag{15-23}$$

Therefore, the Green's function of (15-19a) and (15-19b) can be written, using (15-20) and (15-23), as

$$G(x, x') = \begin{cases} \dfrac{1}{T}\left(\dfrac{x' - \ell}{\ell}\right) x & 0 \le x \le x' \tag{15-24a} \\[4mm] \dfrac{1}{T}\left(\dfrac{x - \ell}{\ell}\right) x' & x' \le x \le \ell \tag{15-24b} \end{cases}$$

The displacement $u(x)$ subject to the load $F(x)$, governed by (15-14a), can now be written as

$$\begin{aligned} u(x) &= \int_0^\ell F(x')\, G(x, x')\, dx' \\ &= \frac{1}{T} \int_0^x F(x')\left(\frac{x' - \ell}{\ell}\right) x\, dx' \\ &\quad + \frac{1}{T} \int_x^\ell F(x')\left(\frac{x - \ell}{\ell}\right) x'\, dx' \end{aligned} \tag{15-25}$$

15.3 STURM-LIOUVILLE PROBLEMS

Now that we illustrated the Green's function development for two specific problems, one for an *RL* electrical circuit and the other for a stretched string, let us consider the construction of Green's functions for more general differential equations subject to appropriate boundary conditions. Specifically in this section we want to consider Green's functions for the one-dimensional differential equation of the *Sturm-Liouville* form [1, 7, 13].

A one-dimensional differential equation of the form

$$\frac{d}{dx}\left[p(x)\frac{dy}{dx}\right] - q(x)y = f(x) \tag{15-26}$$

subject to homogeneous boundary conditions, is a *Sturm-Liouville* problem. This equation can also be written as

$$Ly = f(x) \tag{15-27}$$

where L is the Sturm-Liouville operator

$$L \equiv \left\{\frac{d}{dx}\left[p(x)\frac{d}{dx}\right] - q(x)\right\} \tag{15-27a}$$

Every general one-dimensional, source-excited, second-order differential equation of the form

$$A(x)\frac{d^2 y}{dx^2} + B(x)\frac{dy}{dx} + C(x)y = S(x) \tag{15-28}$$

or

$$Dy = S(x) \qquad (15\text{-}28a)$$

where

$$D \equiv \left[A(x)\frac{d^2}{dx^2} + B(x)\frac{d}{dx} + C(x) \right] \qquad (15\text{-}28b)$$

can be converted to a Sturm-Liouville form. This can be accomplished by following the procedure outlined here.

First, expand (15-26) and write it as

$$p(x)\frac{d^2 y}{dx^2} + \frac{dp}{dx}\frac{dy}{dx} - q(x)y = f(x) \qquad (15\text{-}29)$$

Dividing (15-28) by $A(x)$ and (15-29) by $p(x)$, we have that

$$\frac{d^2 y}{dx^2} + \frac{B(x)}{A(x)}\frac{dy}{dx} + \frac{C(x)}{A(x)}y = \frac{S(x)}{A(x)} \qquad (15\text{-}30a)$$

$$\frac{d^2 y}{dx^2} + \frac{1}{p(x)}\frac{dp}{dx}\frac{dy}{dx} - \frac{q(x)}{p(x)}y = \frac{f(x)}{p(x)} \qquad (15\text{-}30b)$$

Comparing (15-30a) and (15-30b), we see that

$$\frac{B(x)}{A(x)} = \frac{1}{p(x)}\frac{dp(x)}{dx} \qquad (15\text{-}31a)$$

$$\frac{C(x)}{A(x)} = -\frac{q(x)}{p(x)} \qquad (15\text{-}31b)$$

$$\frac{S(x)}{A(x)} = \frac{f(x)}{p(x)} \qquad (15\text{-}31c)$$

From (15-31a) we have that

$$\frac{dp(x)}{dx} = p(x)\frac{B(x)}{A(x)} \qquad (15\text{-}32)$$

which is a linear first-order differential equation, a particular solution of which is

$$\boxed{p(x) = \exp\left[\int^x \frac{B(t)}{A(t)}\,dt \right]} \qquad (15\text{-}32a)$$

From (15-31b)

$$\boxed{q(x) = -p(x)\frac{C(x)}{A(x)} = -\frac{C(x)}{A(x)}\exp\left[\int^x \frac{B(x)}{A(t)}\,dt \right]} \qquad (15\text{-}32b)$$

and from (15-31c)

$$\boxed{f(x) = p(x)\frac{S(x)}{A(x)} = \frac{S(x)}{A(x)}\exp\left[\int^x \frac{B(t)}{A(t)}\,dt \right]} \qquad (15\text{-}32c)$$

In summary then, *a one-dimensional, source-excited, second-order differential equation of the form* (15-28) *is converted to a Sturm-Liouville form* (15-26) *by letting* $p(x)$ *be that of* (15-32a), $q(x)$ *that of* (15-32b), *and* $f(x)$ *that of* (15-32c).

To demonstrate, let us consider an example.

Example 15-1

Convert the Bessel differential equation

$$x^2 \frac{d^2 y}{dx^2} + x \frac{dy}{dx} + (x^2 - \lambda^2) y = 0$$

to a Sturm-Liouville form.

Solution: Since

$$A(x) = x^2$$
$$B(x) = x$$
$$C(x) = x^2 - \lambda^2$$
$$S(x) = 0$$

and according to (15-32a), (15-32b), and (15-32c)

$$p(x) = \exp\left[\int^x \frac{B(t)}{A(t)} dt\right] = \exp\left[\int^x \frac{t}{t^2} dt\right] = \exp\left[\int^x \frac{dt}{t}\right] = e^{\ln(x)} = x$$

$$q(x) = -p(x) \frac{C(x)}{A(x)} = -x \frac{(x^2 - \lambda^2)}{x^2} = -\left[\frac{x^2 - \lambda^2}{x}\right]$$

$$f(x) = p(x) \frac{S(x)}{A(x)} = 0$$

Thus, using (15-26), Bessel's differential equation takes the Sturm-Liouville form

$$\frac{d}{dx}\left(x \frac{dy}{dx}\right) + \left[\frac{x^2 - \lambda^2}{x}\right] y = 0$$

As a check, when the preceding equation is expanded and is multiplied by x, it reduces to the usual form of Bessel's differential equation.

15.3.1 Green's Function in Closed Form

Now that we have shown that each general second-order, source-excited differential equation can be converted to a Sturm-Liouville form, let us develop a procedure to construct the Green's function of a Sturm-Liouville differential equation represented by (15-26) or more generally by

$$\left\{\frac{d}{dx}\left[p(x)\frac{dy}{dx}\right] - q(x)y\right\} + \lambda r(x) y = f(x) \tag{15-33}$$

which can be written as

$$\left[\left\{\frac{d}{dx}\left[p(x)\frac{d}{dx}\right] - q(x)\right\} + \lambda r(x)\right] y = f(x) \tag{15-33a}$$

or simply

$$\left[L + \lambda r(x)\right] y = f(x) \tag{15-33b}$$

where L is the Sturm-Liouville operator of (15-27a). In (15-33), $r(x)$ and $f(x)$ are assumed to be piecewise continuous in the region of interest ($a \le x \le b$) and λ is a parameter to be determined by the nature and boundary of the region of interest. *It should be noted that throughout this chapter λ is used to represent eigenvalues, and it should not be confused with wavelength. The use of λ to represent eigenvalues is a very common practice in Green's function theory.* The differential equations 15-33 through 15-33b possess a Green's function for all values of λ except those that are *eigenvalues* of the homogeneous equation

$$[L+\lambda r(x)]y=0 \tag{15-34}$$

For values of λ for which (15-34) has nontrivial solutions, a Green's function will not exist. This is analogous to a system of linear equations represented by

$$Dy=f \tag{15-35}$$

which has a solution of

$$y=D^{-1}f \tag{15-35a}$$

provided D^{-1} exists (i.e., D nonsingular). If D is singular (so D^{-1} does not exist), then (15-35) does not possess a solution. This occurs when $Dy=0$, which has a nontrivial solution when the determinant of D is zero [so D is written $\det(D)=0$].

According to (15-25), the solution of (15-33b) can be written as

$$y(x)=\int_a^b f(x')G(x,x')dx' \tag{15-36}$$

where $G(x,x')$ is the Green's function of (15-33) or (15-33b). Since (15-35a) is a solution to (15-35), and it exists only if D is nonsingular, then (15-36) is a solution to (15-33b) if the inverse of the operator $[L+\lambda r(x)]$ exists. Then (15-33b) can be written as

$$y(x)=[L+\lambda r(x)]^{-1}f \tag{15-36a}$$

By comparing (15-36) to (15-36a), then $G(x,x')$ is analogous to the inverse of the operator $[L+\lambda r(x)]$.

Whenever λ is equal to an eigenvalue of the operator $[L+\lambda r(x)]$, obtained by setting the determinant of (15-34) equal to zero, i.e.,

$$\det[L+\lambda r(x)]=0 \tag{15-37}$$

then the inverse of $[L+\lambda r(x)]$ does not exist, and (15-36) and (15-36a) are not valid. Thus, for values of λ equal to the eigenvalues of the operator $[L+\lambda r(x)]$, the Green's function does not exist.

For a unit impulse driving function, the Sturm-Liouville equation 15-33 can be written as

$$\frac{d}{dx}\left[p(x)\frac{dG}{dx}\right]-q(x)G+\lambda r(x)G=\delta(x-x') \tag{15-38}$$

where G is the Green's function. At points removed from the impulse driving function, (15-38) reduces to

$$\left\{\frac{d}{dx}\left[p(x)\frac{dG}{dx}\right]-q(x)G\right\}+\lambda r(x)G=0 \tag{15-38a}$$

As can be verified by the Green's function (15-24a) and (15-24b) of the mechanics problems in Section 15.2.2, the Green's functions of (15-38a), in general, exhibit the following properties:

Properties of Green's Functions

1. $G(x,x')$ satisfies the *homogeneous* differential equation *except* at $x = x'$.
2. $G(x,x')$ is symmetrical with respect to x and x'.
3. $G(x,x')$ satisfies certain *homogeneous* boundary conditions.
4. $G(x,x')$ is continuous at $x = x'$.
5. $[dG(x,x')]/dx$ has a discontinuity of $1/[p(x')]$ at $x = x'$.

The discontinuity of the derivative of $G(x,x')$ at $x = x'$ ($[dG(x',x')]/dx = 1/[p(x')]$) can be derived by first integrating the differential equation 15-38 between $x = x'-\varepsilon$ and $x = x'+\varepsilon$. Doing this leads to

$$\lim_{\varepsilon \to 0}\left\{\int_{x'-\varepsilon}^{x'+\varepsilon}\frac{d}{dx}\left[p(x)\frac{dG(x,x')}{dx}\right]dx + \int_{x'-\varepsilon}^{x'+\varepsilon}[-q(x)+\lambda r(x)]G(x,x')dx\right\}$$

$$= \int_{x'-\varepsilon}^{x'+\varepsilon}\delta(x-x')dx$$

$$\lim_{\varepsilon \to 0}\left\{p(x)\frac{dG(x,x')}{dx}\bigg|_{x'-\varepsilon}^{x'+\varepsilon} + \int_{x'-\varepsilon}^{x'+\varepsilon}[-q(x)+\lambda r(x)]G(x,x')dx\right\} = 1 \qquad (15\text{-}39)$$

Since $q(x)$, $r(x)$, and $G(x,x')$ are continuous at $x = x'$, then

$$\lim_{\varepsilon \to 0}\int_{x'-\varepsilon}^{x'+\varepsilon}[-q(x)+\lambda r(x)]G(x,x')dx = 0 \qquad (15\text{-}40)$$

Using (15-40) reduces (15-39) to

$$\lim_{\varepsilon \to 0}\left\{p(x)\left[\frac{dG(x'+\varepsilon,x')}{dx} - \frac{dG(x'-\varepsilon,x')}{dx}\right]\right\} = 1$$

$$p(x)\left[\frac{dG(x'_+,x')}{dx} - \frac{dG(x'_-,x')}{dx}\right] = 1 \qquad (15\text{-}41)$$

or

$$\frac{dG(x'_+,x')}{dx} - \frac{dG(x'_-,x')}{dx} = \frac{1}{p(x)} \qquad (15\text{-}41\text{a})$$

which proves the discontinuity of the derivative of $G(x,x')$ at $x = x'$.

The Green's function must satisfy the differential equation 15-38a, the five general properties listed previously, and the appropriate boundary conditions. We propose to construct the Green's function solution in two parts: one that is valid for $a \le x \le x'$ and the other for $x' \le x \le b$ where a and b are the limits of the region of interest. For the homogeneous equation 15-33, valid at all points except $x = x'$:

1. Let $y_1(x)$ represent a nontrivial solution of the homogeneous differential equation 15-33 in the interval $a \le x < x'$ satisfying the boundary conditions at $x = a$. Since both $y_1(x)$ and $G(x,x')$ satisfy the same differential equation in the interval $a \le x < x'$, they are related to each other by a constant, that is,

$$G(x,x') = A_1 y_1(x), \quad a \le x < x' \qquad (15\text{-}42\text{a})$$

2. Let $y_2(x)$ represent a nontrivial solution of the homogeneous differential equation of 15-33 in the interval $x' < x \leq b$ satisfying the boundary conditions at $x = b$. Since both $y_2(x)$ and $G(x, x')$ satisfy the same differential equation in the interval $x' < x \leq b$, they are related to each other by a constant, that is

$$G(x, x') = A_2 y_2(x), \quad x' < x \leq b \tag{15-42b}$$

Since one of the general properties of the Green's function is that it must be continuous at $x = x'$, then using (15-42a) and (15-42b)

$$A_1 y_1(x') = A_2 y_2(x') \Rightarrow -A_1 y_1(x') + A_2 y_2(x') = 0 \tag{15-43a}$$

Also one of the properties of the derivative of the Green's function is that it must be discontinuous at $x = x'$ by an amount of $1/p(x')$. Thus applying (15-42a) and (15-42b) into (15-41a) leads to

$$-A_1 y_1'(x') + A_2 y_2'(x') = \frac{1}{p(x')} \tag{15-43b}$$

Solving (15-43a) and (15-43b) simultaneously leads to

$$A_1 = \frac{y_2(x')}{p(x')W(x')} \tag{15-44a}$$

$$A_2 = \frac{y_1(x')}{p(x')W(x')} \tag{15-44b}$$

where $W(x')$ is the Wronskian of y_1 and y_2 at $x = x'$, defined as

$$\boxed{W(x') \equiv y_1(x')\, y_2'(x') - y_2(x')\, y_1'(x')} \tag{15-44c}$$

Using (15-44a) through (15-44c), the closed form Green's function of (15-42a) and (15-42b) for the differential equation 15-33 or (15-38) can be written as

$$G(x, x') = \begin{cases} \dfrac{y_2(x')}{p(x')W(x')} y_1(x), & a \leq x \leq x' \tag{15-45a} \\[3mm] \dfrac{y_1(x')}{p(x')W(x')} y_2(x), & x' \leq x \leq b \tag{15-45b} \end{cases}$$

where $y_1(x)$ and $y_2(x)$ are two independent solutions of the homogeneous form of the differential equation 15-33, each satisfying, respectively, the boundary conditions at $x = a$ and $x = b$.

The preceding recipe can be used to construct in closed form the Green's function of a differential equation of the form (15-33). It is convenient to use this procedure when the following conditions are satisfied.

1. The solution to the homogeneous differential equation is known.
2. The Green's function is desired in closed form, instead of an infinite series of orthonormal functions, which will be shown in the next section.

If this procedure is used for the mechanics problem of (15-14a), Section 15.2.2, the same answer [as given by (15-24a) and (15-24b)] will be obtained. In the next section we want to present an alternate procedure for constructing the Green's function. By this other method, the Green's function will be represented by an infinite series of orthonormal functions. Whether one form of the Green's function is more suitable than the other will depend on the problem in question. Remember, however, that the closed-form procedure just derived can only be used provided the solution to the homogeneous differential equation is known.

Before we proceed, let us illustrate that the Sturm-Liouville operator L exhibits *Hermitian* (or *symmetrical*) properties [1]. These are very important, and they establish the relationships that are used in the construction of the Green's function.

Example 15-2

Show that the Sturm-Liouville operator L of (15-27a) or (15-33) through (15-33b) exhibits Hermitian (symmetrical) properties. Assume that in the interval $a \le x \le b$ any solution $y_i(x)$ to (15-33b) satisfies the boundary conditions (where $i = 1, 2, \ldots, n$)

$$\alpha_1 y_i(x=a) + \alpha_2 \frac{dy_i(x=a)}{dx} = \alpha_1 y_i(a) + \alpha_2 y_i'(a) = 0$$

$$\beta_1 y_i(x=b) + \beta_2 \frac{dy_i(x=b)}{dx} = \beta_1 y_i(b) + \beta_2 y_i'(b) = 0$$

Solution: Let us assume that $y_1(x)$ and $y_2(x)$ are two solutions to (15-33) through (15-33b) each satisfying the boundary conditions. Then, according to (15-27a)

$$Ly_1(x) = \frac{d}{dx}\left[p(x)\frac{dy_1(x)}{dx}\right] - q(x)y_1(x)$$

$$Ly_2(x) = \frac{d}{dx}\left[p(x)\frac{dy_2(x)}{dx}\right] - q(x)y_2(x)$$

Multiplying the first by $y_2(x)$ and the second by $y_1(x)$, we can write each using a shorthand notation as

$$y_2 Ly_1 = y_2(py_1')' - y_2 q y_1$$
$$y_1 Ly_2 = y_1(py_2')' - y_1 q y_2$$

where $'$ indicates d/dx. Subtracting the two and integrating between a and b leads to

$$\int_a^b (y_2 Ly_1 - y_1 Ly_2)\, dx = \int_a^b [y_2(py_1')' - y_1(py_2')]\, dx$$

Since

$$(y_2 py_1')' = y_2(py_1')' + y_2' py_1'$$

and

$$(y_1 py_2')' = y_1(py_2')' + y_1' py_2'$$

then by subtracting the two,

$$(y_2 py_1') - (y_1 py_2')' = y_2(py_1')' - y_1(py_2')'$$

Thus, the integral reduces to

$$\int_a^b (y_2 Ly_1 - y_1 Ly_2)\, dx = \int_a^b [(y_2 py_1')' - (y_1 py_2')']\, dx = [p(y_2 y_1' - y_1 y_2')]_a^b$$

Each of the solutions, $y_1(x)$ and $y_2(x)$, satisfies the same boundary conditions that can be written as

$$\alpha_1 y_1(a) + \alpha_2 y_1'(a) = 0 \Rightarrow \alpha_1 y_1(a) = -\alpha_2 y_1'(a)$$
$$\beta_1 y_1(b) + \beta_2 y_1'(b) = 0 \Rightarrow \beta_1 y_1(b) = -\beta_2 y_1'(b)$$
$$\alpha_1 y_2(a) + \alpha_2 y_2'(a) = 0 \Rightarrow \alpha_1 y_2(a) = -\alpha_2 y_2'(a)$$
$$\beta_1 y_2(b) + \beta_2 y_2'(b) = 0 \Rightarrow \beta_1 y_2(b) = -\beta_2 y_2'(b)$$

Dividing the first by the third and the second by the fourth, we can write that

$$\frac{y_1(a)}{y_2(a)} = \frac{y_1'(a)}{y_2'(a)} \Rightarrow y_1(a)y_2'(a) = y_2(a)y_1'(a)$$

$$\frac{y_1(b)}{y_2(b)} = \frac{y_1'(b)}{y_2'(b)} \Rightarrow y_1(b)y_2'(b) = y_2(b)y_1'(b)$$

Using these relations, it is apparent that the right side of the previous integral equation vanishes, so we can write it as

$$\int_a^b (y_2 L y_1 - y_1 L y_2)\, dx = [p(y_2 y_1' - y_1 y_2')]_a^b = 0$$

or

$$\boxed{\int_a^b (y_2 L y_1)\, dx = \int_a^b (y_1 L y_2)\, dx}$$

This illustrates that the operator L exhibits Hermitian (symmetrical) properties with respect to the solutions $y_1(x)$ and $y_2(x)$.

15.3.2   Green's Function in Series

The procedure outlined in the previous section can only be used to derive in closed form the Green's function for differential equations whose homogeneous form solution is known. Otherwise, other techniques must be used. Even for equations whose homogeneous form solution is known, the closed-form representation of the Green's function may not be the most convenient one. Therefore, an alternate representation may be attractive even for those cases.

An alternate form of the Green's function is to represent it as a series of orthonormal functions. The most appropriate orthonormal functions would be those that satisfy the boundary conditions. To demonstrate the procedure, let us initially rederive the Green's function of the mechanics problem of Section 15.2.2 but this time represented as a series of orthonormal functions. We will then generalize the method to (15-33) through (15-33b).

A. Vibrating String   For the differential equation 15-14a, subject to the boundary conditions (15-15), its Green's function must satisfy (15-16) subject to the boundary conditions (15-16a) and (15-16b). Since the Green's function $G(x,x')$ must vanish at $x = 0$ and ℓ, it is most convenient to represent $G(x,x')$ as an infinite series of $\sin(n\pi x / \ell)$ orthonormal functions, that is

$$G(x,x') = \sum_{n=1}^{\infty} a_n(x') \sin\left(\frac{n\pi}{\ell}x\right) \tag{15-46}$$

where $a_n(x')$ represents the amplitude expansion coefficients that will be a function of the position x' of the excitation source.

Substituting (15-46) into (15-16), multiplying both sides by $\sin(m\pi x / \ell)$, and then integrating in x from 0 to ℓ leads to

$$-\sum \left(\frac{n\pi}{\ell}\right)^2 a_n(x') \int_0^\ell \sin\left(\frac{n\pi}{\ell}x\right)\sin\left(\frac{m\pi}{\ell}x\right) dx = \frac{1}{T}\int_0^\ell \delta(x-x')\sin\left(\frac{m\pi}{\ell}x\right) dx \tag{15-47}$$

Because the orthogonality conditions of sine functions state that

$$\int_0^\ell \sin\left(\frac{n\pi}{\ell}x\right)\sin\left(\frac{m\pi}{\ell}x\right)dx = \begin{cases} \dfrac{\ell}{2} & m=n \\ 0 & m\neq n \end{cases}$$

(15-48a)

(15-48b)

then (15-47) reduces to

$$-\left(\frac{n\pi}{\ell}\right)^2\frac{\ell}{2}a_n(x') = \frac{1}{T}\sin\left(\frac{n\pi}{\ell}x'\right)$$

(15-49)

that is,

$$a_n(x') = -\frac{2\ell}{\pi^2 T}\frac{1}{n^2}\sin\left(\frac{n\pi}{\ell}x'\right)$$

(15-49a)

Thus, the Green's function (15-46) can be expressed, using (15-49a), as

$$G(x,x') = -\frac{2\ell}{\pi^2 T}\sum_{n=1}^{\infty}\frac{1}{n^2}\sin\left(\frac{n\pi}{\ell}x'\right)\sin\left(\frac{n\pi}{\ell}x\right)$$

(15-50)

This is an alternate form to (15-24a) and (15-24b), but one that leads to the same results, even though its form looks quite different. The displacement $u(x)$ subject to the load $F(x)$, governed by (15-14a), can now be written as

$$u(x) = \int_0^\ell F(x')\,G(x,x')\,dx' = -\frac{2\ell}{\pi^2 T}\sum_{n=1}^{\infty}\frac{1}{n^2}\sin\left(\frac{n\pi}{\ell}x\right)\int_0^\ell F(x')\sin\left(\frac{n\pi}{\ell}x'\right)dx'$$

(15-51)

B. Sturm-Liouville Operator   Let us now generalize the Green's function series expansion method of the vibrating string, as given by (15-50) and (15-51), to (15-33b) where L is a Sturm-Liouville operator and λ is an arbitrary parameter to be determined by the nature and boundary of the region of interest. We seek a solution to solve the differential equation 15-33b

$$[L+\lambda r(x)]y(x) = f(x)$$

(15-52)

in the interval $a \leq x \leq b$ subject to the general boundary conditions

$$\alpha_1 y(x)|_{x=a} + \alpha_2\frac{dy(x)}{dx}\bigg|_{x=a} = \alpha_1 y(a) + \alpha_2\frac{dy(a)}{dx} = 0$$

(15-52a)

$$\beta_1 y(x)|_{x=b} + \beta_2\frac{dy(x)}{dx}\bigg|_{x=b} = \beta_1 y(b) + \beta_2\frac{dy(b)}{dx} = 0$$

(15-52b)

which are usually referred to as the *mixed* boundary conditions. In (15-52a) at least one of the constants α_1 or α_2, if not both of them, is nonzero. The same is true for (15-52b). The Green's function $G(x,x')$, if it exists, will satisfy the differential equation

$$[L+\lambda r(x)]\,G(x,x') = \delta(x-x')$$

(15-53)

subject to the boundary conditions

$$\alpha_1 G(a,x') + \alpha_2 \frac{dG(a,x')}{dx} = 0 \tag{15-53a}$$

$$\beta_1 G(b,x') + \beta_2 \frac{dG(b,x')}{dx} = 0 \tag{15-53b}$$

If $\{\psi_n(x)\}$ represents a complete set of orthonormal eigenfunctions for the Sturm-Liouville operator L, then it must satisfy the differential equation

$$\boxed{[L + \lambda_n r(x)]\psi_n(x) = 0} \tag{15-54}$$

subject to the same boundary conditions of (15-52a) or (15-52b) for the Green's function (15-53), that is,

$$\alpha_1 \psi_n(a) + \alpha_2 \frac{d\psi_n(a)}{dx} = 0 \tag{15-54a}$$

$$\beta_1 \psi_n(b) + \beta_2 \frac{d\psi_n(b)}{dx} = 0 \tag{15-54b}$$

The boundary conditions (15-54a) and (15-54b) of $\psi_n(x)$ are also used to determine the eigenvalues λ_n. In the finite interval $a \le x \le b$ the complete set of orthonormal eigenfunctions $\{\psi_n(x)\}$, and their amplitude coefficients, must satisfy the orthogonality condition of

$$\boxed{\int_a^b \psi_m(x)\psi_n(x)\, r(x)\, dx = \delta_{mn} = \begin{cases} 1 & m = n \\ 0 & m \ne n \end{cases}} \tag{15-55}$$

where δ_{mn} is the Kronecker delta function.

If the Green's function exists, it can be represented in series form in terms of the orthonormal eigenfunctions $\{\psi_n(x)\}$ as

$$G(x,x') = \sum_n a_n(x')\psi_n(x) \tag{15-56}$$

where $a_n(x')$ are the amplitude coefficients. These can be obtained by multiplying both sides of (15-56) by $\psi_m(x)r(x)$, integrating from a to b, and then using (15-55). It can be shown that

$$a_n(x') = \int_a^b G(x,x')\psi_n(x)r(x)\, dx \tag{15-56a}$$

Since $G(x,x')$ satisfies (15-53) and $\psi_n(x)$ satisfies (15-54), then we can rewrite each as

$$LG(x,x') = -\lambda r(x)G(x,x') + \delta(x - x') \tag{15-57a}$$

$$L\psi_n(x) = -\lambda_n r(x)\psi_n(x,x') \tag{15-57b}$$

Multiplying (15-57a) by $\psi_n(x)$, (15-57b) by $G(x,x')$, and then subtracting the two equations leads to

$$\psi_n(x)LG(x,x') - G(x,x')L\psi_n(x) = -(\lambda - \lambda_n)\,G(x,x')\psi_n(x)r(x) + \delta(x - x')\psi_n(x) \tag{15-58}$$

Integrating (15-58) between a and b, we can write that

$$\int_a^b [\psi_n(x)LG\,(x,x') - G(x,x')L\psi_n(x)]\,dx$$

$$= -(\lambda - \lambda_n)\int_a^b G(x,x')\psi_n(x)r(x)\,dx + \int_a^b \delta(x-x')\psi_n(x)\,dx \qquad (15\text{-}59)$$

which by using (15-56a) reduces to

$$\int_a^b [\psi_n(x)LG\,(x,x') - G(x,x')L\psi_n(x)]\,dx = -(\lambda - \lambda_n)a_n(x') + \psi_n(x') \qquad (15\text{-}59a)$$

By the symmetrical (Hermitian) property of the operator L, as derived in Example 15-2, with respect to the functions $y_1 = G(x,x')$ and $y_2 = \psi_n(x)$, the left side of (15-59) or (15-59a) vanishes. Therefore, (15-59)and (15-59a) reduce to

$$-(\lambda - \lambda_n)a_n(x') + \psi_n(x') = 0 \qquad (15\text{-}60)$$

and therefore,

$$a_n(x') = \frac{\psi_n(x')}{\lambda - \lambda_n}, \quad \text{where } \lambda \neq \lambda_n \qquad (15\text{-}60a)$$

Thus, the series form of the Green's function (15-56) can ultimately be expressed as

$$\boxed{G(x,x') = \sum_n \frac{\psi_n(x')\psi_n(x)}{(\lambda - \lambda_n)}} \qquad (15\text{-}61)$$

where $\{\psi_n(z)\}$ represents a complete set of orthonormal eigenfunctions for the Sturm-Liouville operator L, which satisfies the differential equation 15-54 subject to the boundary conditions (15-54a) and (15-54b). This is also referred to as the bilinear formula, and it represents, aside from (15-45a) and (15-45b), the second form that can be used to derive the Green's function for the differential equation 15-33 or 15-33b as a series solution in the finite interval $a \leq x \leq b$. It should be noted that at $\lambda = \lambda_n$ the Green's function of (15-61) possesses singularities. Usually, these singularities are simple poles although in some cases the λ_n's are branch points whose branch cuts represent a continuous spectrum of eigenvalues. In those cases the Green's function may involve a summation, for the discrete spectrum of eigenvalues, and an integral, for the continuous spectrum of eigenvalues.

Example 15-3

A very common differential equation in solutions of transmission-line and antenna problems (such as metallic waveguides, microstrip antennas, etc.) that exhibit rectangular configurations is

$$\frac{d^2\varphi(x)}{dx^2} + \beta^2\varphi(x) = f(x)$$

subject to the boundary conditions

$$\varphi(0) = \varphi(\ell) = 0$$

where $\beta^2 = \omega^2\mu\varepsilon$. Derive in closed and series forms the Green's functions for the given equation.

Solution: For the given equation, the Green's function must satisfy the differential equation

$$\frac{d^2G(x,x')}{dx^2} + \beta^2 G(x,x') = \delta(x-x')$$

subject to the boundary conditions

$$G(0) = G(\ell) = 0$$

The differential equation is of the Sturm-Liouville form (15-33) with

$$\left. \begin{array}{l} p(x) = 1 \\ q(x) = 0 \\ r(x) = 1 \\ \lambda = \beta^2 \\ y(x) = \varphi(x) \\ L = \dfrac{d^2}{dx^2} \end{array} \right\} \Rightarrow \frac{d^2\varphi}{dx^2} + \beta^2\varphi = f(x)$$

A. Closed-Form Solution: This form of the solution will be obtained using the recipe of (15-45a) and (15-45b) along with (15-44c). The homogeneous differential equation for $\varphi(x)$ reduces to

$$\frac{d^2\varphi}{dx^2} + \beta^2\varphi = 0$$

Two independent solutions, one $\phi_1(x)$ valid in the interval $0 \le x \le x'$ and that vanishes at $x=0$, and the other $\phi_2(x)$ valid in the interval $x' \le x \le \ell$ and that vanishes at $x=\ell$, take the form

$$\phi_1(x) = \sin(\beta x)$$
$$\phi_2(x) = \sin[\beta(\ell - x)]$$

According to (15-44c), the Wronskian can be written as

$$W(x') = -\beta \left\{ \sin(\beta x') \cos[\beta(\ell - x')] + \sin[\beta(\ell - x')] \cos(\beta x') \right\}$$
$$W(x') = -\beta \sin(\beta x' + \beta\ell - \beta x') = -\beta \sin(\beta\ell)$$

Thus, the Green's function in closed form can be expressed, using (15-45a) and (15-45b), as

$$G(x,x') = \begin{cases} -\dfrac{\sin[\beta(\ell - x')]}{\beta \sin(\beta\ell)} \sin(\beta x), & 0 \le x \le x' \\[3mm] -\dfrac{\sin(\beta x')}{\beta \sin(\beta\ell)} \sin[\beta(\ell - x)], & x' \le x \le \ell \end{cases}$$

This form of the Green's function indicates that it possesses singularities (poles) when

$$\beta\ell = \beta_r\ell = n\pi \Rightarrow \beta_r = \omega_r\sqrt{\mu\varepsilon} = 2\pi f_r\sqrt{\mu\varepsilon} = \frac{n\pi}{\ell}$$

that is,

$$f_r = \frac{n}{2\ell\sqrt{\mu\varepsilon}} \quad n = 1, 2, 3, \dots$$

B. Series-Form Solution: This form of the solution will be obtained using (15-61). The complete set of eigenfunctions $\{\psi_n(x)\}$ must satisfy the differential equation 15-54, or

$$\frac{d^2\psi_n(x)}{dx^2} + \beta_n^2\psi_n(x) = 0$$

subject to the boundary conditions

$$\psi_n(0) = \psi_n(\ell) = 0$$

The most appropriate solution of $\psi_n(x)$ is to represent it in terms of standing wave eigenfunctions which in a rectangular coordinate system are sine and cosine functions, as discussed in Chapters 3 and 8, Sections 3.4.1 and 8.2.1.

Thus, according to (3-28b) and (8-4a), we can write that

$$\psi_n(x) = A\cos(\beta_n x) + B\sin(\beta_n x)$$

The allowable eigenvalues of β_n are found by applying the boundary conditions. Since $\psi_n(0) = 0$, then

$$\psi_n(0) = A + B(0) = 0 \;\Rightarrow\; A = 0$$

Also since $\psi_n(\ell) = 0$, then

$$\psi_n(\ell) = B\sin(\beta_n\ell) = 0 \;\Rightarrow\; \beta_n\ell = \sin^{-1}(0) = n\pi$$

that is,

$$\beta_n = \frac{n\pi}{\ell}, \quad n = 1, 2, 3, \dots \quad \text{(for nontrivial solutions)}$$

Thus,

$$\psi_n(x) = B\sin(\beta_n x) = B\sin\left(\frac{n\pi}{\ell}x\right)$$

The amplitude constant B is such that (15-55) is satisfied. Therefore,

$$B^2\int_0^\ell \sin^2\left(\frac{n\pi}{\ell}x\right)dx = 1$$

$$\frac{B^2}{2}\int_0^\ell\left[1 - \cos\left(\frac{2n\pi}{\ell}x\right)\right]dx = B^2\left(\frac{\ell}{2}\right) = 1 \;\Rightarrow\; B = \sqrt{\frac{2}{\ell}}$$

Thus, the complete set of the orthonormal eigenfunctions of $\{\psi_n(x)\}$ is represented by

$$\psi_n(x) = \sqrt{\frac{2}{\ell}}\sin\left(\frac{n\pi}{\ell}x\right), \quad n = 1, 2, 3, \dots$$

with

$$\lambda = \beta^2 = \omega^2\mu\varepsilon$$

$$\lambda_n = \beta_n^2 = \left(\frac{n\pi}{\ell}\right)^2, \quad n = 1, 2, 3, \dots$$

In series-form, the Green's function (15-61) can then be written as

$$G(x,x') = \frac{2}{\ell}\sum_{n=1}^{\infty}\frac{\sin\left(\frac{n\pi}{\ell}x'\right)\sin\left(\frac{n\pi}{\ell}x\right)}{\beta^2 - \left(\frac{n\pi}{\ell}\right)^2}$$

which yields the same results as the closed-form solution of part A, even though it looks quite different analytically. It is apparent that the Green's function is symmetrical. Also it possesses a singularity, and it fails to exist when

$$\beta = \beta_r = \omega_r\sqrt{\mu\varepsilon} = 2\pi f_r\sqrt{\mu\varepsilon} = \frac{n\pi}{\ell}$$

that is,

$$f_r = \frac{n\pi/\ell}{2\pi\sqrt{\mu\varepsilon}} = \frac{n}{2\ell\sqrt{\mu\varepsilon}}$$

which is identical to the condition obtained by the closed-form solution in part A.

This is in accordance with (15-34), which states that the Green's function (15-33b) exists for all values of λ, in this case $\lambda = \beta^2 = \omega^2\mu\varepsilon$, except those that are eigenvalues of the homogeneous equation 15-34. For our case (15-34) reduces to

$$\frac{d^2\phi(x)}{dx^2} + \beta^2\phi(x) = 0$$

whose nontrivial solution takes the form

$$\phi(x) = C\sin\left(\frac{n\pi}{\ell}x\right)$$

with eigenvalues $\beta = n\pi/\ell, n = 1, 2, 3, \ldots$.

It should be noted that when

$$\beta = \beta_r = \omega_r\sqrt{\mu\varepsilon} = 2\pi f_r\sqrt{\mu\varepsilon} = \frac{n\pi}{\ell} \Rightarrow f_r = \frac{n}{2\ell\sqrt{\mu\varepsilon}}$$

the Green's function singularity consists of simple poles. At those frequencies the external frequencies of the source match the natural (characteristic) frequencies of the system, in this instance the transmission line. This is referred to as *resonance*. When this occurs, the field of the mode whose natural frequency matches the source excitation frequency (resonance condition) will continuously increase without any bounds, in the limit reaching values of infinity. For those situations, no steady-state solutions can exist. One way to contain the field amplitude is to introduce damping.

15.3.3 Green's Function in Integral Form

In the previous two sections we outlined procedures that can be used to derive the Green's function in closed and series forms. The bilinear formula (15-61) of Section 15.3.2 is used to derive the Green's function (15-33) through (15-33b) when the eigenvalue spectrum, represented by the λ_n's in (15-61), is discrete. However, often the eigenvalue spectrum is continuous, and it can be represented in (15-61) by an integral. In the limit, the infinite summation of the bilinear formula reduces to an integral. This form is usually desirable when at least one of the boundary conditions is at infinity. This would be true when a source placed at the origin is radiating in an unbounded medium.

To demonstrate the derivation, let us construct the Green's function of the one-dimensional scalar Helmholtz equation

$$\frac{d^2\varphi}{dx^2} + \beta_0^2\varphi = f(x) \tag{15-62}$$

subject to the boundary (radiation) conditions of

$$\varphi(+\infty) = \varphi(-\infty) = 0 \tag{15-62a}$$

The Sommerfield radiation condition [19] could also be used instead of (15-62a).

The Green's function $G(x,x')$ will satisfy the differential equation

$$\frac{d^2 G(x,x')}{dx^2} + \beta_0^2 G(x,x') = \delta(x - x') \tag{15-63}$$

subject to the boundary conditions

$$G(+\infty) = G(-\infty) = 0 \tag{15-63a}$$

The complete set of orthonormal eigenfunctions, represented here by $\{\psi(x)\}$, must satisfy the differential equation

$$\frac{d^2\psi}{dx^2} = -\lambda\psi = -\beta^2\psi \tag{15-64}$$

where $\lambda = +\beta^2$, subject to the boundary conditions of

$$\psi(+\infty) = \psi(-\infty) = 0 \tag{15-64a}$$

Since the source is radiating in an unbounded medium, represented here by the boundary (radiation) conditions, the most appropriate eigenfunctions are those representing traveling waves, instead of standing waves. Thus, a solution for (15-64), subject to (15-64a), is

$$\psi(x,x') = C(x')e^{\mp j\beta x} \qquad \begin{array}{l} - \text{ for } x > x' \\ + \text{ for } x < x' \end{array} \tag{15-65}$$

where for an $e^{j\omega t}$ time convention the upper sign (minus) represents waves traveling in the $+x$ direction, satisfying the boundary condition at $x = +\infty$, and the lower sign (plus) represents waves traveling in the $-x$ direction, satisfying the boundary condition at $x = -\infty$. Let us assume that the waves of interest here are those traveling in the $+x$ direction, represented in (15-65) by the upper sign, that is,

$$\psi(x,x') = C(x')e^{-j\beta x} \tag{15-65a}$$

which represents a plane wave of amplitude $C(x')$.

The Green's function can be represented by a continuous spectrum of plane waves or by a Fourier integral

$$G(x,x') = \frac{1}{\sqrt{2\pi}} \int_{-\infty}^{+\infty} g(\beta,x')e^{-j\beta x} d\beta \tag{15-66}$$

whose Fourier transform pair is

$$g(\beta,x') = \frac{1}{\sqrt{2\pi}} \int_{-\infty}^{+\infty} G(x,x')e^{+j\beta x} dx \tag{15-66a}$$

In (15-66) $G(x,x')$ is represented by a continuous spectrum of plane waves each of the form of (15-65a) and each with an amplitude coefficient of $g(\beta,x')$. Using (15-66a), we can write the transform $\hat{\delta}(\beta,x')$ of the Dirac delta function $\delta(x - x')$ as

$$\hat{\delta}(\beta,x') = \frac{1}{\sqrt{2\pi}} \int_{-\infty}^{+\infty} \delta(x,x')e^{+j\beta x} dx = \frac{1}{\sqrt{2\pi}} e^{+j\beta x'} \tag{15-67}$$

Thus, according to (15-66), $\delta(x,x')$ can then be written using (15-67) as

$$\delta(x,x') = \frac{1}{\sqrt{2\pi}} \int_{-\infty}^{+\infty} \tilde{\delta}(\beta,x') e^{-j\beta x} d\beta = \frac{1}{\sqrt{2\pi}} \int_{-\infty}^{+\infty} \left(\frac{1}{\sqrt{2\pi}} e^{+j\beta x'} \right) e^{-j\beta x} d\beta \qquad (15\text{-}67a)$$

The amplitude coefficients $g(\beta,x')$ in (15-66) can be determined by substituting (15-66) and (15-67a) into (15-63). Then it can be shown that

$$\frac{1}{\sqrt{2\pi}} \int_{-\infty}^{+\infty} \left(-\beta^2 + \beta_0^2 \right) g(\beta,x') e^{-j\beta x} d\beta = \frac{1}{\sqrt{2\pi}} \int_{-\infty}^{+\infty} \left[\frac{1}{\sqrt{2\pi}} e^{+j\beta x'} \right] e^{-j\beta x} d\beta$$

$$\frac{1}{\sqrt{2\pi}} \int_{-\infty}^{+\infty} \left\{ \left(\beta_0^2 - \beta^2 \right) g(\beta,x') - \frac{1}{\sqrt{2\pi}} e^{+j\beta x'} \right\} e^{-j\beta x} d\beta = 0 \qquad (15\text{-}68)$$

which is satisfied provided

$$g(\beta,x') = \frac{1}{\sqrt{2\pi}} \frac{e^{j\beta x'}}{\beta_0^2 - \beta^2} \qquad (15\text{-}68a)$$

Thus, the Green's function of (15-66) reduces to

$$\boxed{ G(x,x') = \frac{1}{2\pi} \int_{-\infty}^{+\infty} \frac{e^{-j\beta(x-x')}}{\beta_0^2 - \beta^2} d\beta } \qquad (15\text{-}69)$$

which is a generalization of the bilinear formula (15-61).

The integrand in (15-69) has poles at $\beta = \pm\beta_0$ and can be evaluated using residue calculus [5]. In the evaluation of (15-69), the contour along a circular arc C_R of radius $R \to \infty$ with center at the origin should close in the lower half plane for $x > x'$, as shown in Figure 15-3a, and should close in the upper half plane for $x < x'$, as shown in Figure 15-3b. This is necessary so that the contribution of the integral along the circular arc C_R of radius $R \to \infty$ is equal to zero. In general then, by residue calculus, the integral of (15-69) can be evaluated using the geometry of Figure 15-3, and it can be written as

$$G(x,x') = \frac{1}{2\pi} \int_{-\infty}^{+\infty} \frac{e^{-j\beta(x-x')}}{(\beta_0^2 - \beta^2)} d\beta$$

$$= \mp 2\pi j \left[\text{residue} \, (\beta = \pm\beta_0) \right] - \frac{1}{2\pi} \int_{C_R} \frac{e^{-j\beta(x-x')}}{\beta_0^2 - \beta^2} d\beta$$

$$G(x,x') = \mp 2\pi j \left[\text{residue} \, (\beta = \pm\beta_0) \right] \quad \begin{array}{l} \text{upper signs for} \;\; x > x' \\ \text{lower signs for} \;\; x < x' \end{array} \qquad (15\text{-}70)$$

since the contribution along C_R is zero.

It is apparent that the Green's function of (15-69) possesses pole singularities at $\beta = +\beta_0$ and $\beta = -\beta_0$. If these were allowed to contribute, then the exponentials in the integral of (15-69) for an $e^{j\omega t}$ time convention would be represented by either

$$e^{-j\beta_0(x-x')} e^{j\omega t} = e^{+j\beta_0 x'} e^{j(-\beta_0 x + \omega t)} \quad \text{for} \quad \beta = +\beta_0 \qquad (15\text{-}70a)$$

or

$$e^{+j\beta_0(x-x')} e^{j\omega t} = e^{-j\beta_0 x'} e^{j(\beta_0 x + \omega t)} \quad \text{for} \quad \beta = -\beta_0 \qquad (15\text{-}70b)$$

Thus, (15-70a) represents waves traveling in the $+x$ direction and (15-70b) represents waves traveling in the $-x$ direction. In the contour evaluation of (15-69), the contour should do the following:

1. For $x > x'$, pass the pole at $\beta = -\beta_0$ from below, the one at $\beta = +\beta_0$ from above, and then close down so that only the latter contributes, as shown in Figure 15-3a.
2. For $x < x'$, pass the pole at $\beta = -\beta_0$ from below, the one at $\beta = +\beta_0$ from above, and then close up so that only the former contributes, as shown in Figure 15-3b.

Sometimes an integral Green's function can be used to represent both discrete and continuous spectra whereby part of the integral would represent the discrete spectrum and the remainder would represent the continuous spectrum. Typically, the discrete spectrum would represent a finite number of propagating modes and an infinite number of evanescent modes in the closed regions, and the continuous spectrum would represent radiation in open regions.

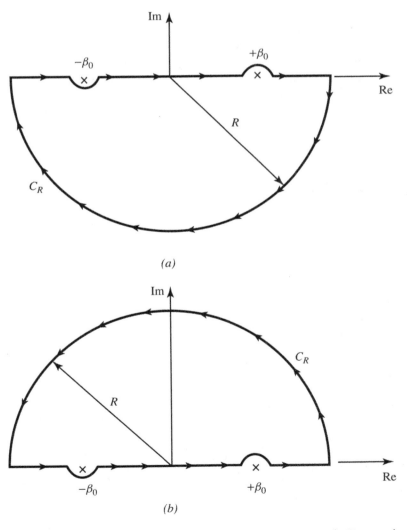

Figure 15-3 Residue calculus for contour integration (a) $x > x'$. (b) $x < x'$.

15.4   TWO-DIMENSIONAL GREEN'S FUNCTION IN RECTANGULAR COORDINATES

Until now we have considered the construction of Green's functions for problems involving a single space variable. Let us now consider problems involving two space variables, both for static and for time-varying fields.

15.4.1   Static Fields

A two-dimensional partial differential equation often encountered in static electromagnetics is Poisson's equation

$$\frac{\partial^2 V}{\partial x^2} + \frac{\partial^2 V}{\partial y^2} = f(x, y) = -\frac{1}{\varepsilon_0} q(x, y) \tag{15-71}$$

subject to the boundary conditions

$$V(x = 0, 0 \le y \le b) = V(x = a, 0 \le y \le b) = 0 \tag{15-71a}$$

$$V(0 \le x \le a, y = 0) = V(0 \le x \le a, y = b) = 0 \tag{15-71b}$$

In (15-71), V can represent the electric potential distribution on a rectangular structure of dimensions a along the x direction and b along the y direction and $f(x, y) = q(x, y)$ can represent the electric charge distribution along the structure. The objective here is to obtain the Green's function and ultimately the potential distribution. The Green's function $G(x, y; x', y')$ will satisfy the partial differential equation

$$\frac{\partial^2 G}{\partial x^2} + \frac{\partial^2 G}{\partial y^2} = \delta(x - x')\,\delta(y - y') \tag{15-72}$$

subject to the boundary conditions

$$G(x = 0, 0 \le y \le b) = G(x = a, 0 \le y \le b) = 0 \tag{15-72a}$$

$$G(0 \le x \le a, y = 0) = G(0 \le x \le a, y = b) = 0 \tag{15-72b}$$

and the potential distribution $V(x, y)$ will be represented by

$$V(x, y) = -\frac{1}{\varepsilon_0} \int_0^b \int_0^a q(x', y') G(x, y; x', y')\, dx'dy' \tag{15-73}$$

We will derive the Green's function here in two forms.

1. *Closed form*, similar to that of Section 15.3.1 but utilizing (15-44c) and (15-45a) through (15-45b) for a two space variable problem.
2. *Series form*, similar to that of Section 15.3.2 but using basically two-dimensional expressions for (15-54), (15-55), and (15-61).

A. Closed Form   The Green's function of (15-72) for the closed-form solution can be formulated by choosing functions that initially satisfy the boundary conditions either along the x direction, at $x = 0$ and $x = a$, or along the y direction, at $y = 0$ and $y = b$. Let us begin here the development of the Green's function of (15-72) by choosing functions that initially satisfy the boundary conditions along the x direction. This is accomplished by initially representing the Green's function by a normalized single function Fourier series of sine functions that satisfy the boundary conditions at $x = 0$ and $x = a$, that is,

$$G(x, y; x', y') = \sum_{m=1}^{\infty} g_m(y; x', y') \sin\left(\frac{m\pi}{a} x\right) \tag{15-74}$$

The coefficients $g_m(y; x', y')$ of the Fourier series will be determined by first substituting (15-74) into (15-72). This leads to

$$\sum_{m=1,2,\ldots} \left[-\left(\frac{m\pi}{a}\right)^2 g_m(y; x', y') \sin\left(\frac{m\pi}{a} x\right) + \sin\left(\frac{m\pi}{a} x\right) \frac{d^2 g_m(y; x', y')}{dy^2} \right]$$

$$= \delta(x - x') \delta(y - y') \tag{15-75}$$

Multiplying both sides of (15-75) by $\sin(n\pi x / a)$, integrating with respect to x from 0 to a, and using (15-48a) and (15-48b), we can write that

$$\frac{d^2 g_m(y; x', y')}{dy^2} - \left(\frac{m\pi}{a}\right)^2 g_m(y; x', y') = \frac{2}{a} \sin\left(\frac{m\pi}{a} x'\right) \delta(y - y') \tag{15-76}$$

Equation 15-76 is recognized as a one-dimensional differential equation for $g_m(y; x', y')$, which can be solved using the recipe of Section 15.3.1 as provided by (15-44c) and (15-45a) through (15-45b).

For the homogeneous form of (15-76), or

$$\frac{d^2 g_m(y; x', y')}{dy^2} - \left(\frac{m\pi}{a}\right)^2 g_m(y; x', y') = 0 \tag{15-77}$$

two solutions that satisfy, respectively, the boundary conditions at $y = 0$ and $y = b$ are

$$g_m^{(1)}(y; x', y') = A_m(x', y') \sinh\left(\frac{m\pi}{a} y\right) \quad \text{for } y \leq y' \tag{15-78a}$$

$$g_m^{(2)}(y; x', y') = B_m(x', y') \sinh\left[\frac{m\pi}{a}(b - y)\right] \quad \text{for } y \geq y' \tag{15-78b}$$

The hyperbolic functions were chosen as solutions to (15-77), instead of real exponentials, so that (15-78a) satisfies the boundary condition of (15-71b) at $y = 0$ and (15-78b) satisfies the boundary condition of (15-71b) at $y = b$.

Using (15-44c) where $y_1 = g_m^{(1)}$ and $y_2 = g_m^{(2)}$, we can write the Wronskian as

$$W(y; x', y') = -\left(\frac{m\pi}{a}\right) A_m B_m \left\{ \sinh\left(\frac{m\pi}{a} y'\right) \cosh\left[\frac{m\pi}{a}(b - y')\right] \right.$$

$$\left. + \cosh\left(\frac{m\pi}{a} y'\right) \sinh\left[\frac{m\pi}{a}(b - y')\right] \right\}$$

$$W(y; x', y') = -\left(\frac{m\pi}{a}\right) A_m B_m \sinh\left(\frac{m\pi b}{a}\right) \tag{15-79}$$

By comparing (15-76) with the form of (15-33), it is apparent that

$$p(y) = 1$$
$$q(y) = 0$$
$$r(y) = 1$$
$$\lambda = -\left(\frac{m\pi}{a}\right)^2 \tag{15-80}$$

Using (15-78a) through (15-80), the solution for $g_m(y; x', y')$ of (15-76) can be written, by referring to (15-45a) and (15-45b), as

$$g_m(y; x', y') = \begin{cases} -\dfrac{2}{m\pi} \sin\left(\dfrac{m\pi}{a} x'\right) \dfrac{\sinh\left[\dfrac{m\pi}{a}(b - y')\right]}{\sinh\left(\dfrac{m\pi b}{a}\right)} \sinh\left(\dfrac{m\pi}{a} y\right), & \tag{15-81a} \\[2em] \qquad\qquad 0 \le y \le y' & \\[2em] -\dfrac{2}{m\pi} \sin\left(\dfrac{m\pi}{a} x'\right) \dfrac{\sinh\left(\dfrac{m\pi}{a} y'\right)}{\sinh\left(\dfrac{m\pi b}{a}\right)} \sinh\left[\dfrac{m\pi}{a}(b - y)\right], & \tag{15-81b} \\[2em] \qquad\qquad y' \le y \le b & \end{cases}$$

Thus, the Green's function of (15-74) can be written as

$$G(x, y; x', y') = \begin{cases} -\dfrac{2}{\pi} \displaystyle\sum_{m=1}^{\infty} \dfrac{\sin\left(\dfrac{m\pi}{a} x'\right) \sinh\left[\dfrac{m\pi}{a}(b - y')\right]}{m \sinh\left(\dfrac{m\pi b}{a}\right)} \\[1.5em] \qquad \times \sin\left(\dfrac{m\pi}{a} x\right) \sinh\left(\dfrac{m\pi}{a} y\right) & \tag{15-82a} \\[1em] \qquad \text{for } 0 \le x \le a, \quad 0 \le y \le y' \\[1.5em] -\dfrac{2}{\pi} \displaystyle\sum_{m=1}^{\infty} \dfrac{\sin\left(\dfrac{m\pi}{a} x'\right) \sinh\left(\dfrac{m\pi}{a} y'\right)}{m \sinh\left(\dfrac{m\pi b}{a}\right)} \\[1.5em] \qquad \times \sin\left(\dfrac{m\pi}{a} x\right) \sinh\left[\dfrac{m\pi}{a}(b - y)\right] & \tag{15-82b} \\[1em] \qquad \text{for } 0 \le x \le a, \quad y' \le y \le b \end{cases}$$

which is a series of sine functions in x' and x, and hyperbolic sine functions in y' and y.

If the Green's function solution were developed by selecting and writing initially (15-74) by functions that satisfy the boundary conditions at $y = 0$ and $y = b$, then it can be shown that the Green's function can be written as a series of hyperbolic sine functions in x' and x, and ordinary sine functions in y' and y, or

$$G(x,y;x',y') = \begin{cases} -\dfrac{2}{\pi} \displaystyle\sum_{n=1,2,\dots}^{\infty} \dfrac{\sinh\left[\dfrac{n\pi}{b}(a-x')\right]\sin\left(\dfrac{n\pi}{b}y'\right)}{n\sinh\left(\dfrac{n\pi a}{b}\right)} \\ \quad \times \sinh\left(\dfrac{n\pi}{b}x\right)\sin\left(\dfrac{n\pi}{b}y\right) \\ \quad \text{for } 0 \le x \le x', \ 0 \le y \le b \\[6pt] -\dfrac{2}{\pi} \displaystyle\sum_{n=1,2,\dots}^{\infty} \dfrac{\sinh\left(\dfrac{n\pi}{b}x'\right)\sin\left(\dfrac{n\pi}{b}y'\right)}{n\sinh\left(\dfrac{n\pi a}{b}\right)} \\ \quad \times \sinh\left[\dfrac{n\pi}{b}(a-x)\right]\sin\left(\dfrac{n\pi}{b}y\right) \\ \quad \text{for } x' \le x \le a, \ 0 \le y \le b \end{cases}$$

(15-83a)

(15-83b)

The derivation of this is left to the reader as an end-of-chapter exercise.

Example 15-4

Electric charge is uniformly distributed along an infinitely long conducting wire positioned at $\rho = \rho'$, $\phi = \phi'$ and circumscribed by a grounded ($V = 0$) electric conducting circular cylinder of radius a and infinite length, as shown in Figure 15-4. Find series-form expressions for the Green's function and potential distribution. Assume free space within the cylinder.

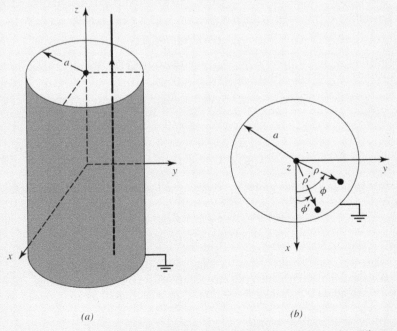

(a)   (b)

Figure 15-4 Long wire within a grounded circular conducting cylinder. (a) Wire and grounded cylinder. (b) Top view.

Solution: The potential distribution $V(p, \phi, z)$ must satisfy Poisson's equation

$$\nabla^2 V(\rho, \phi, z) = -\frac{1}{\varepsilon_0} q(\rho, \phi, z)$$

subject to the boundary condition

$$V(\rho = a, 0 \le \phi \le 2\pi, z) = 0$$

Since the wire is infinitely long, the solutions for the potential will not be functions of z. Thus, an expanded form of Poisson's equation reduces to

$$\frac{1}{\rho} \frac{\partial}{\partial \rho}\left(\rho \frac{\partial V}{\partial \rho}\right) + \frac{1}{\rho^2} \frac{\partial^2 V}{\partial \phi^2} = -\frac{1}{\varepsilon_0} q(\rho, \phi)$$

The Green's function $G(\rho, \phi; \rho', \phi')$ must satisfy the partial differential equation

$$\nabla^2 G(\rho, \phi; \rho', \phi') = \delta(\rho - \rho')$$

which in expanded form reduces for this problem to

$$\frac{1}{\rho} \frac{\partial}{\partial \rho}\left(\rho \frac{\partial G}{\partial \rho}\right) + \frac{1}{\rho^2} \frac{\partial^2 G}{\partial \phi^2} = \delta(\rho - \rho')$$

For the series solution of the Green's function, the complete set of orthonormal eigenfunctions $\{\psi_{mn}(\rho, \phi)\}$ can be obtained by considering the homogeneous form of Poisson's equation, or

$$\frac{1}{\rho} \frac{\partial}{\partial \rho}\left(\rho \frac{\partial \psi_{mn}}{\partial \rho}\right) + \frac{1}{\rho^2} \frac{\partial^2 \psi_{mn}}{\partial \phi^2} = -\lambda_{mn} \psi_{mn}$$

subject to the boundary condition

$$\psi_{mn}(\rho = a, 0 \le \phi \le 2\pi, z) = 0$$

Using the separation-of-variables method of Section 3.4.2, we can express $\psi_{mn}(\rho, \phi)$ by

$$\psi_{mn}(\rho, \phi) = f(\rho) g(\phi)$$

Following the method outlined in Section 3.4.2, the functions $f(\rho)$ and $g(\phi)$ satisfy, respectively, the differential equations (3-66a) and (3-66b), that is,

$$\rho^2 \frac{d^2 f}{d\rho^2} + p \frac{df}{d\rho} + (\lambda_{mn}\rho^2 - m^2)f = 0$$

and

$$\frac{d^2 g}{d\phi^2} = -m^2 g$$

whose appropriate solutions for this problem, according to (3-67a) and (3-68b), are, respectively,

$$f = A J_m\left(\sqrt{\lambda_{mn}}\,\rho\right) + B Y_m\left(\sqrt{\lambda_{mn}}\,\rho\right)$$
$$g = C\cos(m\phi) + D\sin(m\phi)$$

Since ψ_{mn} must be periodic in ϕ, then m must take integer values, $m = 0, 1, 2, \dots$, and both the $\cos(m\phi)$ and $\sin(m\phi)$ variations (modes) exist simultaneously; see Chapter 9. Also since ψ_{mn} must be finite everywhere, including $\rho = 0$, then $B = 0$. Thus, the eigenfunctions are reduced to either of two forms, that is,

$$\psi_{mn}^{(1)} = A_{mn} J_m\left(\sqrt{\lambda_{mn}}\,\rho\right)\cos(m\phi)$$

or

$$\psi_{mn}^{(2)} = A_{mn} J_m \left(\sqrt{\lambda_{mn}} \rho \right) \sin(m\phi)$$

The eigenvalues λ_{mn} are found by applying the boundary condition at $\rho = a$, that is,

$$\psi_{mn}(\rho = a, 0 \le \phi \le 2\pi) = A_{mn} J_m \left(\sqrt{\lambda_{mn}} a \right) = 0$$

or

$$\sqrt{\lambda_{mn}} a = \chi_{mn} \Rightarrow \lambda_{mn} = \left(\frac{\chi_{mn}}{a} \right)^2$$

where χ_{mn} represents the n zeroes of the Bessel function J_m of the first kind of order m. These are listed in Table 9-2.

The complete set of orthonormal eigenfunctions must be normalized so that

$$\int_0^{2\pi} \int_0^a \psi_{mn}^{(1)}(\rho,\phi) \psi_{mp}^{(1)}(\rho,\phi) \rho \; d\rho \; d\phi = \int_0^{2\pi} \int_0^a \psi_{mn}^{(2)}(\rho,\phi) \psi_{mp}^{(2)}(\rho,\phi) \rho \; d\rho \, d\phi = 1$$

Thus,

$$A_{mn}^2 \int_0^{2\pi} \int_0^a \rho J_m \left(\sqrt{\lambda_{mn}} \rho \right) J_m \left(\sqrt{\lambda_{mp}} \rho \right) \cos^2(m\phi) \; d\rho \, d\phi = 1$$

or

$$A_{mn}^2 \int_0^{2\pi} \int_0^a \rho J_m \left(\sqrt{\lambda_{mn}} \rho \right) J_m \left(\sqrt{\lambda_{mp}} \rho \right) \sin^2(m\phi) \; d\rho \, d\phi = 1$$

Since

$$\int_0^{2\pi} \cos^2(m\phi) \; d\phi = \begin{cases} 2\pi & m = 0 \\ \pi & m \ne 0 \end{cases}$$

$$\int_0^{2\pi} \sin^2(m\phi) \; d\phi = \begin{cases} 0 & m = 0 \\ \pi & m \ne 0 \end{cases}$$

and

$$\int_0^a \rho J_m \left(\sqrt{\lambda_{mn}} \rho \right) J_m \left(\sqrt{\lambda_{mp}} \rho \right) \; d\rho = \begin{cases} \dfrac{a^2}{2} \left[J_m' \left(\sqrt{\lambda_{mn}} a \right) \right]^2 & p = n \\ 0 & p \ne n \end{cases}$$

then

$$A_{mn}^2 \varepsilon_m \frac{\pi a^2}{2} \left[J_m' \left(\sqrt{\lambda_{mp}} a \right) \right]^2 = 1$$

hence,

$$A_{mn} = \sqrt{\frac{2}{\varepsilon_m \pi}} \frac{1}{a J_m' \left(\sqrt{\lambda_{mn}} a \right)}$$

where

$$\varepsilon_m = \begin{cases} 2 & m = 0 \\ 1 & m \ne 0 \end{cases}$$

Thus, the complete set of orthonormal eigenfunctions can be written as

$$\psi_{mn}^{(1)} = \sqrt{\frac{2\varepsilon_m}{\pi}} \frac{1}{a J_m' \left(\sqrt{\lambda_{mn}} a \right)} J_m \left(\sqrt{\lambda_{mp}} \rho \right) \cos(m\phi)$$

or

$$\psi_{mn}^{(2)} = \sqrt{\frac{2\varepsilon_m}{\pi}} \frac{1}{aJ_m'\left(\sqrt{\lambda_{mn}}a\right)} J_m\left(\sqrt{\lambda_{mp}}\rho\right) \sin(m\phi)$$

The Green's function can now be written using the bilinear formula (15-94) with $\lambda = 0$ as

$$G(\rho,\phi;\rho',\phi') = -\frac{2}{\pi}\varepsilon_m \frac{1}{a^2\left[J_m'\left(\sqrt{\lambda_{mn}}a\right)\right]^2} \sum_{m=0}^{\infty}\sum_{n=1}^{\infty} \frac{J_m\left(\sqrt{\lambda_{mn}}\rho'\right)J_m\left(\sqrt{\lambda_{mn}}\rho\right)}{\lambda_{mn}}$$

$$\times \left[\cos(m\phi)\cos(m\phi') + \sin(m\phi)\sin(m\phi')\right]$$

$$G(\rho,\phi;\rho',\phi') = -\frac{2}{\pi}\varepsilon_m \frac{1}{a^2\left[J_m'\left(\sqrt{\lambda_{mn}}a\right)\right]^2} \sum_{m=0}^{\infty}\sum_{n=1}^{\infty} \frac{J_m\left(\sqrt{\lambda_{mn}}\rho'\right)J_m\left(\sqrt{\lambda_{mn}}\rho\right)}{\lambda_{mn}}$$

$$\times \cos\left[m(\phi - \phi')\right]$$

where

$$\lambda_{mn} = \left(\frac{\chi_{mn}}{a}\right)^2$$

Both the $\cos(m\phi)$ and $\sin(m\phi)$ field variations were included in the final expression for the Green's function.

Finally, the potential distribution $V(\rho,z)$ can be written as

$$V(\rho,\phi) = -\frac{1}{\varepsilon_0} \int_0^{2\pi}\int_0^a q(\rho',\phi')\,G(\rho,\phi;\rho',\phi')\rho'd\rho'd\phi'$$

where $G(\rho,\phi;\rho',\phi')$ is the Green's function and $q(\rho',\phi')$ is the linear charge distribution.

The Green's function of Example 15-4 can also be developed in closed form. This is done in Section 15.6.2 for a time-harmonic electric line source inside a circular cylinder. The statics solution is obtained by letting $\beta_0 = 0$.

B. Series Form   For the series solution of the Green's function of (15-72), the complete set of orthonormal eigenfunctions $\{\psi_{mn}(x,y)\}$ can be obtained by considering the homogeneous form of (15-71), i.e.,

$$\frac{\partial^2 \psi_{mn}}{\partial x^2} + \frac{\partial^2 \psi_{mn}}{\partial y^2} = -\lambda_{mn}\psi_{mn} \tag{15-84}$$

subject to the boundary conditions

$$\psi_{mn}(x=0, 0 \leq y \leq b) = \psi_{mn}(x=a, 0 \leq y \leq b) = 0 \tag{15-84a}$$

$$\psi_{mn}(0 \leq x \leq a, y=0) = \psi_{mn}(0 \leq x \leq a, y=b) = 0 \tag{15-84b}$$

Using the method of separation of variables of Section 3.4.1, we can represent $\psi_{mn}(x,y)$ by

$$\psi_{mn}(x,y) = f(x)g(y) \tag{15-85}$$

Substituting (15-85) into (15-84) reduces to

$$\frac{1}{f}\frac{d^2 f}{dx^2} = -p^2 \;\Rightarrow\; \frac{d^2 f}{dx^2} = -p^2 f \;\Rightarrow\; f(x) = A\cos(px) + B\sin(px) \tag{15-85a}$$

$$\frac{1}{g}\frac{d^2 g}{dy^2} = -q^2 \;\Rightarrow\; \frac{d^2 g}{dy^2} = -q^2 g \;\Rightarrow\; g(y) = C\cos(qy) + D\sin(qy) \tag{15-85b}$$

where the system eigenvalues are those of

$$\lambda_{mn} = p^2 + q^2 \tag{15-85c}$$

Thus, (15-85) can be represented by

$$\psi_{mn}(x,y) = [A\cos(px) + B\sin(px)][C\cos(qy) + D\sin(qy)] \tag{15-86}$$

Applying the boundary conditions of (15-84a) on (15-86) leads to

$$\psi_{mn}(x=0, 0 \le y \le b) = [A(1) + B(0)][C\cos(qy) + D\sin(qy)] = 0 \;\Rightarrow\; A = 0 \tag{15-87a}$$

$$\psi_{mn}(x=a, 0 \le y \le b) = B\sin(pa)[C\cos(qy) + D\sin(qy)] = 0 \;\Rightarrow\; \sin(pa) = 0$$

$$pa = \sin^{-1}(0) = m\pi$$

$$p = \frac{m\pi}{a}, \quad m = 1, 2, 3, \dots \tag{15-87b}$$

Similarly, applying the boundary conditions of (15-84b) into (15-86), using (15-87a) and (15-87b), leads to

$$\psi_{mn}(0 \le x \le a, y=0) = B\sin\left(\frac{m\pi}{a}x\right)[C(1) + D(0)] = 0 \;\Rightarrow\; C = 0 \tag{15-88a}$$

$$\psi_{mn}(0 \le x \le a, y=b) = BD\sin\left(\frac{m\pi}{a}x\right)\sin(qb) = 0 \;\Rightarrow\; \sin(qb) = 0$$

$$qb = \sin^{-1}(0) = n\pi$$

$$q = \frac{n\pi}{b}, \quad n = 1, 2, 3, \dots \tag{15-88b}$$

Thus, the eigenfunctions of (15-85) reduce to

$$\psi_{mn}(x,y) = BD\sin\left(\frac{m\pi}{a}x\right)\sin\left(\frac{n\pi}{b}y\right) = B_{mn}\sin\left(\frac{m\pi}{a}x\right)\sin\left(\frac{n\pi}{b}y\right) \tag{15-89}$$

where the eigenvalues of (15-85c) are equal to

$$\lambda_{mn} = (p^2 + q^2) = \left(\frac{m\pi}{a}\right)^2 + \left(\frac{n\pi}{b}\right)^2, \quad \begin{array}{l} m = 1, 2, 3, \dots \\ n = 1, 2, 3, \dots \end{array} \tag{15-89a}$$

To form the Green's function, the eigenfunctions of (15-89) must be normalized so that

$$\int_0^b \int_0^a \psi_{mn}(x,y)\psi_{rs}(x,y)\,dx\,dy = \begin{cases} 1 & m=r, \quad n=s \\ 0 & m \ne r, \quad n \ne s \end{cases} \tag{15-90}$$

Equation 15-90 is similar, and is an expanded form in two space variables, of (15-55). Substituting (15-89) into (15-90), we can write that

$$B_{mn}^2 \int_0^b \int_0^a \sin\left(\frac{m\pi}{a}x\right)\sin\left(\frac{n\pi}{b}y\right)\sin\left(\frac{r\pi}{a}x\right)\sin\left(\frac{s\pi}{b}y\right)\,dx\,dy = 1 \tag{15-91}$$

Using (15-48a) and (15-48b) reduces (15-91) to

$$B_{mn}^2\left(\frac{ab}{4}\right) = 1 \tag{15-92}$$

hence,

$$B_{mn} = \frac{2}{\sqrt{ab}} \tag{15-92a}$$

Thus, (15-89) can be written as

$$\psi_{mn}(x,y) = \frac{2}{\sqrt{ab}}\sin\left(\frac{m\pi}{a}x\right)\sin\left(\frac{n\pi}{b}y\right) \tag{15-93}$$

The Green's function can be expressed as a double summation of two-dimensional eigenfunctions,

$$G(x,y;x',y') = \sum_m \sum_n \frac{\psi_{mn}(x',y')\psi_{mn}(x,y)}{\lambda - \lambda_{mn}} \tag{15-94}$$

Equation 15-94 is an expanded version in two space variables of the bilinear equation of (15-61). Thus, we can write the Green's function of (15-94), using (15-89a), (15-93), and $\lambda = 0$, as

$$G(x,y;x',y') = -\frac{4}{ab}\sum_{m=1}^{\infty}\sum_{n=1}^{\infty}\frac{\sin\left(\frac{m\pi}{a}x'\right)\sin\left(\frac{n\pi}{b}y'\right)}{\left(\frac{m\pi}{a}\right)^2 + \left(\frac{n\pi}{b}\right)^2}\sin\left(\frac{m\pi}{a}x\right)\sin\left(\frac{n\pi}{b}y\right) \tag{15-95}$$

The electric potential of (15-73), due to the static electric charge density of $q(x',y')$, can be expressed as

$$V(x,y) = -\frac{1}{\varepsilon_0}\frac{4}{ab}\sum_{m=1}^{\infty}\sum_{n=1}^{\infty}\frac{\sin\left(\frac{m\pi}{a}x\right)\sin\left(\frac{n\pi}{b}y\right)}{\left(\frac{m\pi}{a}\right)^2 + \left(\frac{n\pi}{b}\right)^2}$$
$$\times \int_0^b \int_0^a q(x',y')\sin\left(\frac{m\pi}{a}x'\right)\sin\left(\frac{n\pi}{b}y'\right)\,dx'\,dy' \tag{15-96}$$

15.4.2 Time-Harmonic Fields

For time-harmonic fields, a popular partial differential equation is

$$\frac{\partial^2 E_z}{\partial x^2} + \frac{\partial^2 E_z}{\partial y^2} + \beta^2 E_z = f(x,y) = j\omega\mu J_z(x,y) \tag{15-97}$$

subject to the boundary conditions

$$E_z(x=0, 0 \le y \le b) = E_z(x=a, 0 \le y \le b) = 0 \tag{15-97a}$$

$$E_z(0 \le x \le a, y=0) = E_z(0 \le x \le a, y=b) = 0 \tag{15-97b}$$

In (15-97), E_z can represent the electric field component of a TM^z field configuration (mode) with no z variations inside a rectangular metallic cavity of dimensions a, b, c in the x, y, z directions, respectively, as shown in Figure 15-5. The function $f(x,y) = j\omega\mu J_z$ can represent the normalized electric current density component of the feed probe that is used to excite the fields within the metallic cavity. The objective here is to obtain the Green's function of the problem and ultimately the electric field component represented in (15-97) by $E_z(x,y)$.

The Green's function $G(x, y; x', y')$ will satisfy the partial differential equation

$$\frac{\partial^2 G}{\partial x^2} + \frac{\partial^2 G}{\partial y^2} + \beta^2 G = \delta(x-x')\delta(y-y') \tag{15-98}$$

subject to the boundary conditions

$$G(x=0, 0 \le y \le b) = G(x=a, 0 \le y \le b) = 0 \tag{15-98a}$$

$$G(0 \le x \le a, y=0) = G(0 \le x \le a, y=b) = 0 \tag{15-98b}$$

and the electric field distribution $E_z(x,y)$ will be represented by

$$E_z(x,y) = j\omega\mu \int_0^b \int_0^a J_z(x',y') G(x,y;x',y') \, dx'dy' \tag{15-99}$$

The Green's function of (15-98) can be derived either in closed form, as was done in Section 15.4.1A for the statics problem, or in series form, as was done in Section 15.4.1B for the statics

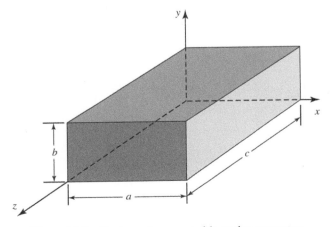

Figure 15-5 Rectangular waveguide cavity geometry.

problem. We will derive the Green's function here by a series form. The closed form is left as an end-of-chapter exercise for the reader.

For the solution of the Green's function, the complete set of eigenfunctions $\{\psi_{mn}(x,y)\}$ can be obtained by considering the homogeneous form of (15-97),

$$\frac{\partial^2 \psi_{mn}}{\partial x^2} + \frac{\partial^2 \psi_{mn}}{\partial y^2} + \beta_{mn}^2 \psi_{mn} = 0 \qquad (15\text{-}100)$$

subject to the boundary conditions

$$\psi_{mn}(x=0, 0 \leq y \leq b) = \psi_{mn}(x=a, 0 \leq y \leq b) = 0 \qquad (15\text{-}100a)$$

$$\psi_{mn}(0 \leq x \leq a, y=0) = \psi_{mn}(0 \leq x \leq a, y=b) = 0 \qquad (15\text{-}100b)$$

Using the separation of variables of Section 3.4.1, we can represent $\psi_{mn}(x,y)$ by

$$\psi_{mn}(x,y) = f(x)g(y) \qquad (15\text{-}101)$$

Substituting (15-101) into (15-100) and applying the boundary conditions of (15-100a) and (15-100b), it can be shown that $\psi_{mn}(x,y)$ reduces to

$$\psi_{mn}(x,y) = B_{mn} \sin(\beta_x x) \sin(\beta_y y) = B_{mn} \sin\left(\frac{m\pi}{a}x\right) \sin\left(\frac{n\pi}{b}y\right) \qquad (15\text{-}102)$$

where

$$\beta_x = \frac{m\pi}{a}, \quad m = 1, 2, 3, \ldots \qquad (15\text{-}102a)$$

$$\beta_y = \frac{n\pi}{b}, \quad n = 1, 2, 3, \ldots \qquad (15\text{-}102b)$$

The eigenvalues of the system are equal to

$$\lambda_{mn} = \beta_{mn}^2 = (\beta_x)^2 + (\beta_y)^2 = \left(\frac{m\pi}{a}\right)^2 + \left(\frac{n\pi}{b}\right)^2 \qquad (15\text{-}103a)$$

and

$$\lambda = \beta^2 \qquad (15\text{-}103b)$$

The eigenfunctions of (15-102) must satisfy an equation similar to (15-90) but over a volume integral. That is,

$$\int_0^c \int_0^b \int_0^a \psi_{mn}(x,y,z)\, \psi_{rs}(x,y,z)\, dx\, dy\, dz = \begin{cases} 1 & m=r, \quad n=s \\ 0 & m \neq r, \quad n \neq s \end{cases} \qquad (15\text{-}104)$$

which leads to

$$B_{mn} = \frac{2}{\sqrt{abc}} \qquad (15\text{-}104a)$$

Thus,

$$\psi_{mn}(x,y) = \frac{2}{\sqrt{abc}} \sin\left(\frac{m\pi}{a}x\right) \sin\left(\frac{n\pi}{b}y\right) \qquad (15\text{-}105)$$

The Green's function is obtained using (15-94), (15-103a), (15-103b), and (15-105). Doing this, we can write that

$$G(x,y;x',y') = \frac{4}{abc}\sum_{m=1}^{\infty}\sum_{n=1}^{\infty}\frac{\sin\left(\frac{m\pi}{a}x'\right)\sin\left(\frac{n\pi}{b}y'\right)}{\beta^2-\left[\left(\frac{m\pi}{a}\right)^2+\left(\frac{n\pi}{b}\right)^2\right]}$$
$$\times\sin\left(\frac{m\pi}{a}x\right)\sin\left(\frac{n\pi}{b}y\right) \tag{15-106}$$

and the electric field component of (15-99), due to the normalized electric current density represented by $J_n(x',y')$, can be written as

$$E_z(x,y) = \frac{4}{abc}\sum_{m=1}^{\infty}\sum_{n=1}^{\infty}\frac{\sin\left(\frac{m\pi}{a}x\right)\sin\left(\frac{n\pi}{b}y\right)}{\beta^2-\left[\left(\frac{m\pi}{a}\right)^2+\left(\frac{n\pi}{b}\right)^2\right]}$$
$$\times\int_0^b\int_0^a J_z(x',y')\sin\left(\frac{m\pi}{a}x'\right)\sin\left(\frac{n\pi}{b}y'\right)dx'\,dy' \tag{15-107}$$

It is apparent that the Green's function possesses a singularity, and it fails when

$$\beta = \beta_r = \omega_r\sqrt{\mu\varepsilon} = 2\pi f_r\sqrt{\mu\varepsilon} = \sqrt{\left(\frac{m\pi}{a}\right)^2+\left(\frac{n\pi}{b}\right)^2} \tag{15-108}$$

that is, when

$$f_r = \frac{1}{2\pi\sqrt{\mu\varepsilon}}\sqrt{\left(\frac{m\pi}{a}\right)^2+\left(\frac{n\pi}{b}\right)^2} \tag{15-108a}$$

This is in accordance with (15-34), which states that the Green's function of (15-33b) exists for all values of λ (here $\lambda = \beta^2 = \omega^2\mu\varepsilon$) except those that are eigenvalues λ_{mn} of the homogeneous equation 15-34 [here λ_{mn} given by (15-103a)]. At those eigenvalues for which (15-106) and (15-107) possess singularities (simple poles here), the frequencies of the excitation source match the natural (characteristic) frequencies of the system. As explained in Section 1.3.2, this is referred to as *resonance*, and the field will continuously increase without any bounds (in the limit reaching infinity). For these cases, no steady-state solutions exist. One way to contain the field is to introduce damping. In practice, for metallic cavities, damping is introduced by the losses due to nonperfectly conducting walls.

15.5 GREEN'S IDENTITIES AND METHODS

Now that we have derived Green's functions, both for single and two space variables in rectangular coordinates for the general Sturm-Liouville self-adjoint operator L, let us generalize the procedure for the development of the Green's function for the three-dimensional scalar Helmholtz partial differential equation

$$\nabla^2\phi(\mathbf{r})+\beta^2\phi(\mathbf{r}) = f(\mathbf{r}) \tag{15-109}$$

subject to the generalized homogeneous boundary conditions

$$\alpha_1\phi(\mathbf{r}_s)+\alpha_2\frac{\partial\phi(\mathbf{r}_s)}{\partial n_s}=0,\quad s=1,2,\ldots,N \tag{15-109a}$$

where $\mathbf{r}_s$ is on S and $\hat{\mathbf{n}}$ is an outward directed unit vector. In electromagnetics these are referred to not only as the *mixed* boundary conditions [7] but also as the *impedance* boundary conditions.
    The Green's function $G(\mathbf{r},\mathbf{r}')$ of (15-109) must satisfy the partial differential equation

$$\nabla^2 G(\mathbf{r},\mathbf{r}')+\beta^2 G(\mathbf{r},\mathbf{r}')=\delta(\mathbf{r}-\mathbf{r}') \tag{15-110}$$

subject to the generalized homogeneous boundary conditions

$$\alpha_1 G(\mathbf{r}_s,\mathbf{r}')+\alpha_2\frac{\partial G(\mathbf{r}_s,\mathbf{r}')}{\partial n_s}=0,\quad s=1,2,\ldots,N \tag{15-110a}$$

    To accomplish this, we will need two identities from vector calculus that are usually referred to as *Green's first* and *second identities*. We will state them first before proceeding with the development of the generalized Green's function.

15.5.1   Green's First and Second Identities

Within a volume V, conducting bodies with surfaces $S_1, S_2, S_3,\ldots,S_n$ are contained, as shown in Figure 15-6. By introducing appropriate cuts, the volume V is bounded by a regular surface S that consists of surfaces $S_1 - S_n$, the surfaces along the cuts, and the surface S_a of an infinite radius sphere that encloses all the conducting bodies. A unit vector $\hat{\mathbf{n}}$ normal to S is directed inward to the volume V, as shown in Figure 15-6.

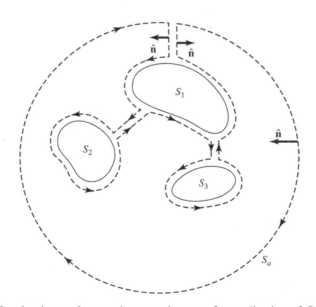

Figure 15-6   Conducting surfaces and appropriate cuts for application of Green's theorem.

Let us introduce within V two scalar functions ϕ and ψ, which, along with their first and second derivatives, are continuous within V and on the surface S. To the vector $\phi\nabla\psi$, we apply the divergence theorem (1-8),

$$\oiint_S (\phi\nabla\psi)\cdot d\mathbf{s} = \oiint_S (\phi\nabla\psi)\cdot\hat{\mathbf{n}}\, da = \iiint_V \nabla\cdot(\phi\nabla\psi)\, dv \qquad (15\text{-}111)$$

When expanded, the integrand of the volume integral can be written as

$$\nabla\cdot(\phi\nabla\psi) = \phi\nabla\cdot(\nabla\psi)+\nabla\phi\cdot\nabla\psi = \phi\nabla^2\psi+\nabla\phi\cdot\nabla\psi \qquad (15\text{-}112)$$

Thus (15-111) can be expressed as

$$\boxed{\oiint_S (\phi\nabla\psi\cdot d\mathbf{s}) = \iiint_V (\phi\nabla^2\psi)\, dv + \iiint_V (\nabla\phi\cdot\nabla\psi)\, dv} \qquad (15\text{-}113)$$

which is referred to as *Green's first identity*. Since

$$(\nabla\psi)\cdot\hat{\mathbf{n}} = \frac{\partial\psi}{\partial n} \qquad (15\text{-}114)$$

where the derivative $\partial\psi/\partial n$ is taken in the direction of positive normal, (15-113) can also be written as

$$\boxed{\oiint_S \left(\phi\frac{\partial\psi}{\partial n}\right) ds = \iiint_V (\phi\nabla^2\psi)\, dv + \iiint_V (\nabla\phi\cdot\nabla\psi)\, dv} \qquad (15\text{-}115)$$

which is an *alternate form of Green's first identity*.

If we repeat the procedure but apply the divergence theorem (15-111) to the vector $\psi\nabla\phi$, then we can write, respectively, Green's first identity (15-113) and its alternate form (15-115) as

$$\oiint_S (\psi\nabla\phi\cdot d\mathbf{s}) = \iiint_V (\psi\nabla^2\phi)\, dv + \iiint_V (\nabla\psi\cdot\nabla\phi)\, dv \qquad (15\text{-}116)$$

and

$$\oiint_S \left(\psi\frac{\partial\phi}{\partial n}\right) ds = \iiint_V (\psi\nabla^2\phi)\, dv + \iiint_V (\nabla\psi\cdot\nabla\phi)\, dv \qquad (15\text{-}117)$$

Subtracting (15-116) from (15-113), we can write that

$$\boxed{\oiint_S (\phi\nabla\psi - \psi\nabla\phi)\cdot d\mathbf{s} = \iiint_V (\phi\nabla^2\psi - \psi\nabla^2\phi)\, dv} \qquad (15\text{-}118)$$

which is referred to as *Green's second identity*. Its alternate form

$$\boxed{\oiint_S \left(\phi\frac{\partial\psi}{\partial n} - \psi\frac{\partial\phi}{\partial n}\right) ds = \iiint_V (\phi\nabla^2\psi - \psi\nabla^2\phi)\, dv} \qquad (15\text{-}119)$$

is obtained by subtracting (15-117) from (15-115).

Green's first and second identities expressed, respectively, either as (15-113) and (15-118) or (15-115) and (15-119), will be used to develop the formulation for the more general Green's function.

15.5.2 Generalized Green's Function Method

With the introduction of Green's first and second identities in the previous section, we are now ready to develop the formulation of the generalized Green's function method of φ for the partial differential equation 15-109 whose Green's function $G(\mathbf{r},\mathbf{r}')$ satisfies (15-110).

Let us multiply (15-109) by $G(\mathbf{r},\mathbf{r}')$ and (15-110) by $\varphi(\mathbf{r})$. Doing this leads to

$$G\nabla^2\varphi + \beta^2\varphi G = f\,G \tag{15-120a}$$

$$\varphi\nabla^2 G + \beta^2\varphi G = \varphi\delta(\mathbf{r}-\mathbf{r}') \tag{15-120b}$$

Subtracting (15-120a) from (15-120b) and integrating over the volume V, we can write that

$$\iiint_V \varphi\delta(\mathbf{r}-\mathbf{r}')\,dv - \iiint_V fG\,dv = \iiint_V (\varphi\nabla^2 G - G\nabla^2\varphi)\,dv \tag{15-121}$$

or

$$\varphi(\mathbf{r}=\mathbf{r}') = \varphi(\mathbf{r}') = \iiint_V f(\mathbf{r})G(\mathbf{r},\mathbf{r}')\,dv$$
$$+ \iiint_V [\varphi(\mathbf{r})\nabla^2 G(\mathbf{r},\mathbf{r}') - G(\mathbf{r},\mathbf{r}')\nabla^2\varphi(\mathbf{r})]\,dv \tag{15-121a}$$

Applying Green's second identity (15-118) reduces (15-121a) to

$$\varphi(\mathbf{r}') = \iiint_V f(\mathbf{r})G(\mathbf{r},\mathbf{r}')\,dv + \oiint_S [\varphi(\mathbf{r})\nabla G(\mathbf{r},\mathbf{r}') - G(\mathbf{r},\mathbf{r}')\nabla\varphi(\mathbf{r})]\cdot ds \tag{15-122}$$

Since $\mathbf{r}'$ is an arbitrary point within V and $\mathbf{r}$ is a dummy variable, we can also write (15-122) as

$$\boxed{\varphi(\mathbf{r}) = \iiint_V f(\mathbf{r}')G(\mathbf{r},\mathbf{r}')\,dv' + \oiint_S [\varphi(\mathbf{r}')\nabla' G(\mathbf{r},\mathbf{r}') - G(\mathbf{r},\mathbf{r}')\nabla'\varphi(\mathbf{r}')]\cdot d\mathbf{s}'} \tag{15-123}$$

where ∇' indicates differentiation with respect to the prime coordinates.

Equation 15-123 is a generalized formula for the development of the Green's function for a three-dimensional scalar Helmholtz equation. It can be simplified depending on the boundary conditions of φ and G, and their derivatives on S. The objective then will be to judiciously choose the boundary conditions on the development of G, once the boundary conditions on φ are stated, so as to simplify, if not completely eliminate, the surface integral contribution in (15-123). We will demonstrate here some combinations of boundary conditions on φ and G, and the simplifications of (15-123), based on those boundary conditions.

A. Nonhomogeneous Partial Differential Equation with Homogeneous Dirichlet Boundary Conditions

If the nonhomogeneous form of the partial differential equation of 15-109 satisfies the homogeneous Dirichlet boundary condition

$$\varphi(\mathbf{r}_s) = 0, \quad \text{where } \mathbf{r}_s \text{ is on } S \tag{15-124a}$$

then it is reasonable to construct a Green's function with the same boundary condition

$$G(\mathbf{r}_s, \mathbf{r}') = 0, \quad \text{where } \mathbf{r}_s \text{ is on } S \tag{15-124b}$$

so as to simplify the surface integral contributions in (15-123).

For these boundary conditions on φ and G, both terms in the surface integral of (15-123) vanish, so that (15-123) reduces to

$$\varphi(\mathbf{r}) = \iiint_V f(\mathbf{r}') G(\mathbf{r}, \mathbf{r}') dv' \tag{15-125}$$

The Green's function $G(\mathbf{r}, \mathbf{r}')$ needed in (15-125) can be obtained using any of the previous methods developed in Sections 15.3.1 through 15.3.2. In many cases, the bilinear form (15-61) or (15-94) or its equivalent, in the desired coordinate system and number of space variables, is appropriate for forming the Green's function. Its existence will depend upon the eigenvalues of the homogeneous partial differential equation, as discussed in Section 15.3.1.

B. Nonhomogeneous Partial Differential Equation with Nonhomogeneous Dirichlet Boundary Conditions   If the nonhomogeneous partial differential equation 15-109 satisfies the nonhomogeneous Dirichlet boundary condition

$$\varphi(\mathbf{r}_s) = g(\mathbf{r}_s), \quad \text{where } \mathbf{r}_s \text{ is on } S \tag{15-126a}$$

then we can still construct a Green's function that satisfies the boundary condition

$$G(\mathbf{r}_s, \mathbf{r}') = 0, \quad \text{where } \mathbf{r}_s \text{ is on } S \tag{15-126b}$$

For these boundary conditions on φ and G, the second term in the surface integral of (15-123) vanishes, so that (15-123) reduces to

$$\varphi(\mathbf{r}) = \iiint_V f(\mathbf{r}') \, G(\mathbf{r}, \mathbf{r}') \, dv' + \oiint_S \varphi(\mathbf{r}') \nabla' G(\mathbf{r}_s, \mathbf{r}') \cdot d\mathbf{s}' \tag{15-127}$$

The Green's function $G(\mathbf{r}, \mathbf{r}')$ needed in (15-127) can be determined using any of the previous methods developed in Sections 15.3.1 through 15.3.2.

C. Nonhomogeneous Partial Differential Equation with Homogeneous Neumann Boundary Conditions   When Neumann boundary conditions are involved, the solutions become more complicated primarily because the normal gradients of $\varphi(\mathbf{r})$ are not independent of the partial differential equation. If the nonhomogeneous form of the partial differential equation 15-109 satisfies the homogeneous Neumann boundary condition

$$[\nabla' \varphi(\mathbf{r}_s)] \cdot \hat{\mathbf{n}} = \frac{\partial \varphi(\mathbf{r}_s)}{\partial n} = 0 \quad \text{where } \mathbf{r}_s \text{ is on } S \tag{15-128}$$

then we *cannot*, in general, construct a Green's function with a boundary condition $[\nabla' G\,(\mathbf{r}_s, \mathbf{r}')]$ $\cdot \hat{\mathbf{n}} = [\partial G(\mathbf{r}_s, \mathbf{r}') / \partial n] = 0$. This is evident from what follows.

If we apply the divergence theorem (1-8) to the vector $\nabla G(\mathbf{r}, \mathbf{r}')$, we can write that

$$\oiint_S \nabla G(\mathbf{r}, \mathbf{r}') \cdot d\mathbf{s} = \iiint_V \nabla \cdot \nabla G(\mathbf{r}, \mathbf{r}') dv = \iiint_V \nabla^2 G(\mathbf{r}, \mathbf{r}') dv \tag{15-129}$$

Taking the volume integral of (15-110), we can express it as

$$\iiint_V \nabla^2 G(\mathbf{r},\mathbf{r}')\,dv + \beta^2 \iiint_V G(\mathbf{r},\mathbf{r}')\,dv = \iiint_V \delta(\mathbf{r}-\mathbf{r}')\,dv \qquad (15\text{-}130)$$

Using (15-129) reduces (15-130) to

$$\oiint_S \nabla G(\mathbf{r},\mathbf{r}')\cdot d\mathbf{s} + \beta^2 \iiint_V G(\mathbf{r},\mathbf{r}')\,dv = 1 \qquad (15\text{-}131)$$

If we choose

$$\left[\nabla G(\mathbf{r},\mathbf{r}')|_{\mathbf{r}=\mathbf{r}_s}\right]\cdot\hat{\mathbf{n}} = \left[\nabla G(\mathbf{r}_s,\mathbf{r}')\right]\cdot\hat{\mathbf{n}} = \frac{\partial G(\mathbf{r}_s,\mathbf{r}')}{\partial n} = 0 \qquad (15\text{-}132)$$

as a boundary condition for $G(\mathbf{r},\mathbf{r}')$, then (15-131) reduces to

$$\beta^2 \iiint_V G(\mathbf{r},\mathbf{r}')\,dv = 1 \qquad (15\text{-}133)$$

which cannot be satisfied if $\beta = 0$. When $\beta = 0$, (15-131) reduces to

$$\oiint_S \nabla G(\mathbf{r},\mathbf{r}')\cdot d\mathbf{s} = 1 \qquad (15\text{-}134)$$

or

$$\left|\nabla G(\mathbf{r}_s,\mathbf{r}')\right| S_0 = 1 \qquad (15\text{-}134a)$$

where S_0 is the area of the surface. This implies that a consistent boundary condition for the normal gradient of $G(\mathbf{r},\mathbf{r}')$ on S to satisfy (15-134) or (15-134a) would be

$$\nabla' G(\mathbf{r},\mathbf{r}')|_{\mathbf{r}=\mathbf{r}_s} = \frac{1}{S_0} = \nabla' G(\mathbf{r}_s,\mathbf{r}') \qquad (15\text{-}135)$$

Substituting (15-128) and (15-135) into (15-123) leads to

$$\boxed{\varphi(\mathbf{r}) = \iiint_V f(\mathbf{r}')G(\mathbf{r},\mathbf{r}')\,dv' + \frac{1}{S_0}\oiint_S \varphi(\mathbf{r}')\,ds'} \qquad (15\text{-}136)$$

The second term on the right side of (15-136) is a constant, and it can be dropped since $\varphi(\mathbf{r})$ is undetermined by the boundary conditions up to an additive constant.

D. Nonhomogeneous Partial Differential Equation with Mixed Boundary Conditions

When the boundary conditions on $\varphi(\mathbf{r})$ are such that $\varphi(\mathbf{r}_s)$ is specified in part of the surface S and $[\nabla'\varphi(\mathbf{r}_s)]\cdot\hat{\mathbf{n}} = \partial\varphi(\mathbf{r}_s)/\partial n$ is specified over the remaining part of S, it is referred to as having *mixed* boundary conditions. Then it is desirable to construct a Green's function so that it vanishes on that part of S over which $\varphi(\mathbf{r}_s)$ is specified, and its normal derivative $[\partial G(\mathbf{r}_s,\mathbf{r}')/\partial n]$ vanishes over the remaining part of S over which $\partial\varphi(\mathbf{r}_s)/\partial n$ is specified. Although this is a more complex procedure, it does provide a method to derive the Green's function even under those mixed boundary conditions.

15.6   GREEN'S FUNCTIONS OF THE SCALAR HELMHOLTZ EQUATION

Now that we have derived the development of the generalized Green's function, let us apply the formulation to the scalar Helmholtz equation in three-dimensional problems of rectangular, cylindrical, and spherical coordinates.

15.6.1   Rectangular Coordinates

The development of Green's functions in rectangular coordinates has already been applied for one and two space variables in almost all of the previous sections. In this section we want to derive it for a three-dimensional problem. Specifically, let us derive the Green's function for the electric field component E_y that satisfies the partial differential equation

$$\nabla^2 E_y + \beta_0^2 E_y = \left(\frac{\partial^2}{\partial x^2} + \frac{\partial^2}{\partial y^2} + \frac{\partial^2}{\partial z^2}\right) E_y + \beta_0^2 E_y = j\omega\mu J_y(x,y,z) \qquad (15\text{-}137)$$

subject to the boundary conditions

$$E_y(x=0, 0 \le y \le b, -\infty \le z \le +\infty) = E_y(x=a, 0 \le y \le b, -\infty \le z \le +\infty) = 0 \qquad (15\text{-}137\text{a})$$

In (15-137), E_y represents the electric field component of a TM^z field configuration (subject to $\nabla \cdot \mathbf{J} = 0$) inside a metallic waveguide of dimensions a,b in the x,y directions, respectively, as shown in Figure 15-7. Also $J_y(x,y,z)$ represents the electric current density of the feed probe that is used to excite the fields within the metallic waveguide. It is assumed that the wave is traveling in the z direction. The time-harmonic variations are of $e^{+j\omega t}$, and they are suppressed.

The Green's function must satisfy the partial differential equation

$$\nabla^2 G(x,y,z;x',y',z') + \beta_0^2 G(x,y,z;x',y',z') = \delta(x-x')\delta(y-y')\delta(z-z') \qquad (15\text{-}138)$$

subject to the boundary conditions

$$\begin{aligned} G(x=0, \ 0 \le y \le b, \ -\infty \le z \le +\infty) \\ = G(x=a, \ 0 \le y \le b, \ -\infty \le z \le +\infty) = 0 \end{aligned} \qquad (15\text{-}138\text{a})$$

Since the electric field and Green's function satisfy, respectively, the Dirichlet boundary conditions (15-124a) and (15-124b), then, according to (15-123) or (15-125), the electric field is obtained using

$$E_y(x,y,z) = j\omega\mu \iiint_V J_y(x',y',z') G(x,y,z;x',y',z')\,dx'dy'dz' \qquad (15\text{-}139)$$

The Green's function can be derived either in closed, series, or integral form. We will choose here the series form. We begin the development of the Green's function by assuming its solution can be represented by a two-function Fourier series of sine function in x and cosine function in y which satisfy, respectively, the boundary conditions at $x = 0, a$. Thus we can express $G(x,y,z;x',y',z')$ as

$$G(x,y,z;x',y',z') = \sum_{m=1}^{\infty} \sum_{n=1}^{\infty} g_{mn}(z;x',y',z') \sin\left(\frac{m\pi}{a}x\right) \cos\left(\frac{n\pi}{b}y\right) \qquad (15\text{-}140)$$

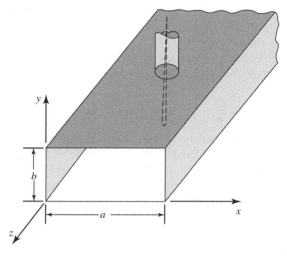

Figure 15-7   Rectangular waveguide excited by linear electric probe.

Substituting (15-140) into (15-138) leads to

$$\sum_{m=1}^{\infty}\sum_{n=1}^{\infty}\left[-\left(\frac{m\pi}{a}\right)^2 - \left(\frac{n\pi}{b}\right)^2 + \beta_0^2 + \frac{\partial^2}{\partial z^2}\right]$$

$$\times g_{mn}(z;x',y',z')\sin\left(\frac{m\pi}{a}x\right)\cos\left(\frac{n\pi}{b}y\right)$$

$$= \delta(x-x')\delta(y-y')\delta(z-z') \tag{15-141}$$

Multiplying both sides of (15-141) by $\sin(p\pi x/a)\cos(q\pi y/b)$, integrating from 0 to a in x and 0 to b in y, and using (8-56a), (8-56b), (15-48a), and (15-48b), we can reduce (15-141) to

$$\frac{ab}{4}\left(\frac{\partial^2}{\partial z^2} + \beta_z^2\right)g_{mn}(z;x',y',z') = \sin\left(\frac{m\pi}{a}x'\right)\cos\left(\frac{n\pi}{b}y'\right)\delta(z-z') \tag{15-142}$$

or

$$\boxed{\left(\frac{d^2}{dz^2} + \beta_z^2\right)g_{mn}(z;x',y',z') = \frac{4}{ab}\sin\left(\frac{m\pi}{a}x'\right)\cos\left(\frac{n\pi}{b}y'\right)\delta(z-z')} \tag{15-142a}$$

where

$$\beta_z^2 = \beta_0^2 - \left[\left(\frac{m\pi}{a}\right)^2 + \left(\frac{n\pi}{b}\right)^2\right] = \beta_0^2 - (\beta_x^2 + \beta_y^2) \tag{15-142b}$$

$$\beta_x = \frac{m\pi}{a}, \qquad m=1,2,3,\ldots \tag{15-142c}$$

$$\beta_y = \frac{n\pi}{b}, \qquad n=1,2,3,\ldots \tag{15-142d}$$

The function $g_{mn}(z;x',y',z')$ satisfies the single variable differential equation 15-142a, and it can be found by using the recipe of Section 15.3.1 as represented by (15-44c), (15-45a), and (15-45b). Two solutions of the homogeneous differential equation

$$\left(\frac{d^2}{dz^2}+\beta_z^2\right)g_{mn}(z;x',y',z')=0 \tag{15-143}$$

of (15-142a) are

$$g_{mn}^{(1)}=A_{mn}e^{+j\beta_z z} \qquad \text{for} \quad z<z' \tag{15-143a}$$

$$g_{mn}^{(2)}=B_{mn}e^{-j\beta_z z} \qquad \text{for} \quad z>z' \tag{15-143b}$$

Using (15-44c) where $y_1=g_{mn}^{(1)}$ and $y_2=g_{mn}^{(2)}$, we can write the Wronskian as

$$W(z')=A_{mn}B_{mn}(-j\beta_z)e^{j\beta_z z'}e^{-j\beta_z z'}-A_{mn}B_{mn}(j\beta_z)e^{-j\beta_z z'}e^{+j\beta_z z'}$$
$$W(z')=-j2\beta_z A_{mn}B_{mn} \tag{15-144}$$

By comparing (15-142a) to (15-33), it is apparent that

$$\begin{aligned} p(z)&=1\\ q(z)&=0\\ r(z)&=1\\ \lambda&=\beta_z^2 \end{aligned} \tag{15-145}$$

Using (15-143a) through (15-145), the solution for $g_{mn}(z;x',y',z')$ of (15-142a) can be written, by referring to (15-45a) and (15-45b), as

$$g_{mn}(z;x',y',z')=\begin{cases} j\dfrac{2}{ab}\dfrac{\sin\left(\dfrac{m\pi}{a}x'\right)\cos\left(\dfrac{n\pi}{b}y'\right)}{\beta_z}e^{-j\beta_z(z-z')} & \text{for} \quad z>z' \quad (15\text{-}146a)\\[6mm] j\dfrac{2}{ab}\dfrac{\sin\left(\dfrac{m\pi}{a}x'\right)\cos\left(\dfrac{n\pi}{b}y'\right)}{\beta_z}e^{-j\beta_z(z-z')} & \text{for} \quad z>z' \quad (15\text{-}146b) \end{cases}$$

or

$$g_{mn}(z;x',y',z')=j\frac{2}{ab}\frac{\sin\left(\dfrac{m\pi}{a}x'\right)\cos\left(\dfrac{n\pi}{b}y'\right)}{\beta_z}e^{-j\beta_z|z-z'|} \quad \text{for} \quad z<z', \; z>z' \tag{15-146c}$$

Thus, the Green's function of (15-140) can now be expressed as

$$\boxed{\begin{aligned} G(x,y,z;x',y',z')&=j\frac{2}{ab}\sum_{m=1,2,\ldots}^{\infty}\sum_{n=1,2,\ldots}^{\infty}\frac{\sin\left(\dfrac{m\pi}{a}x'\right)\cos\left(\dfrac{n\pi}{b}y'\right)}{\beta_z}\\ &\quad \times\sin\left(\frac{m\pi}{a}x\right)\cos\left(\frac{n\pi}{b}y\right)e^{-j\beta_z|z-z'|} \quad \text{for} \quad z<z', \; z>z' \end{aligned}} \tag{15-147}$$

where

$$\beta_z = \begin{cases} \sqrt{\beta_0^2 - (\beta_x^2 + \beta_y^2)} & \text{for} \quad \beta_0^2 > (\beta_x^2 + \beta_y^2) & \text{(15-147a)} \\ -j\sqrt{(\beta_x^2 + \beta_y^2) - \beta_0^2} & \text{for} \quad \beta_0^2 < (\beta_x^2 + \beta_y^2) & \text{(15-147b)} \end{cases}$$

It is evident from (15-147) through (15-147b) that when $\beta_0^2 > (\beta_x^2 + \beta_y^2)$ the modes are propagating and when $\beta_0^2 < (\beta_x^2 + \beta_y^2)$ the modes are not propagating (evanescent). The nonpropagating modes converge very rapidly when $|z - z'|$ is very large.

Once the Green's function is formulated as in (15-147), the electric field can be found using (15-139).

15.6.2 Cylindrical Coordinates

Until now we have concentrated on developing primarily Green's functions of problems dealing with rectangular coordinates. This was done to maintain simplicity in the mathematics so that the analytical formulations would not obscure the fundamental concepts. Now we are ready to deal with problems expressed by other coordinate systems, such as cylindrical and spherical.

Let us assume that an infinite electric line source of constant current I_z is placed at $\rho = \rho', \phi = \phi'$ inside a circular waveguide of radius a, as shown in Figure 15-8. The electric field component E_z satisfies the partial differential equation

$$\boxed{\nabla^2 E_z + \beta_0^2 E_z = f(\rho, \phi) = j\omega\mu I_z}$$
(15-148)

subject to the boundary condition

$$E_z(\rho = a, 0 \leq \phi \leq 2\pi, z) = 0$$
(15-148a)

The Green's function of this problem will satisfy the partial differential equation

$$\boxed{\nabla^2 G + \beta_0^2 G = \delta(\boldsymbol{\rho} - \boldsymbol{\rho}')}$$
(15-149)

The boundary condition for the Green's function can be chosen so that

$$G(\rho = a, 0 \leq \phi \leq 2\pi, z) = 0$$
(15-149a)

Since the boundary conditions (15-148a) and (15-149a) on E_z and G, respectively, are of the Dirichlet type, then according to (15-123) or (15-125)

$$E_z(\rho, \phi) = \iint_S f(\rho', \phi')G(\rho, \phi; \rho', \phi') \, ds' = j\omega\mu \iint_S I_z(\rho', \phi')G(\rho, \phi; \rho', \phi') \, ds'$$
(15-150)

Since both the current source and the circular waveguide are of infinite length, the problem reduces to a two-dimensional one. Thus, we can express initially the Green's function by an infinite Fourier series whose eigenvalues in ϕ satisfy the periodicity requirements. That is,

$$\boxed{G(\rho, \phi; \rho', \phi') = \sum_{m=-\infty}^{+\infty} g_m(\rho; \rho', \phi')e^{jm\phi}}$$
(15-151)

In cylindrical coordinates, the delta function $\delta(\boldsymbol{\rho} - \boldsymbol{\rho}')$ in (15-149) can be expressed, in general, as [10, 13]

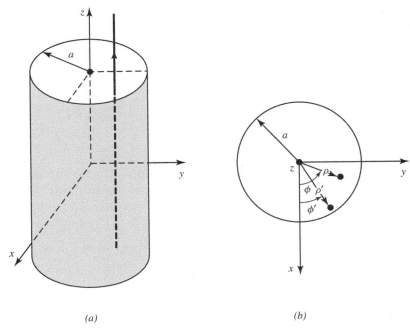

Figure 15-8 Electric line source within a circular conducting cylinder. (*a*) Line source and conducting cylinder. (*b*) Top view.

$$\delta(\rho - \rho') = \begin{cases} \dfrac{1}{\rho}\delta(\rho - \rho')\delta(\phi - \phi')\delta(z - z') & & \text{(15-152a)} \\[3mm] \dfrac{1}{2\pi\rho}\delta(\rho - \rho')\delta(z - z') & \text{for no } \phi \text{ dependence} & \text{(15-152b)} \\[3mm] \dfrac{1}{2\pi\rho}\delta(\rho - \rho') & \text{for neither } \phi \text{ nor } z \text{ dependence} & \text{(15-152c)} \\[3mm] \dfrac{1}{\rho}\delta(\rho - \rho')\delta(\phi - \phi') & \text{for no } z \text{ dependence} & \text{(15-152d)} \end{cases}$$

In expanded form, the Green's function of (15-149) can now be written, using (15-152d) and assuming no z variations, as

$$\frac{\partial^2 G}{\partial \rho^2} + \frac{1}{\rho}\frac{\partial G}{\partial \rho} + \frac{1}{\rho^2}\frac{\partial^2 G}{\partial \phi^2} + \beta_0^2 G = \frac{1}{\rho}\delta(\rho - \rho')\delta(\phi - \phi') \tag{15-153}$$

Substituting (15-151) into (15-153) leads to

$$\sum_{m=-\infty}^{+\infty}\left[\frac{\partial^2}{\partial \rho^2} + \frac{1}{\rho}\frac{\partial}{\partial \rho} - \frac{m^2}{\rho^2} + \beta_0^2\right]g_m(\rho; \rho', \phi')e^{jm\phi} = \frac{1}{\rho}\delta(\rho - \rho')\delta(\phi - \phi') \tag{15-154}$$

Multiplying both sides of (15-154) by $e^{-jn\phi}$, integrating both sides from 0 to 2π in ϕ, and using the orthogonality condition

$$\int_0^{2\pi} e^{j(m-n)\phi}\,d\phi = \begin{cases} 2\pi & m = n & \text{(15-155a)} \\ 0 & m \neq n & \text{(15-155b)} \end{cases}$$

reduces (15-154) to

$$2\pi\left[\frac{\partial^2 g_m}{\partial\rho^2}+\frac{1}{\rho}\frac{\partial g_m}{\partial\rho}+\left(\beta_0^2-\frac{m^2}{\rho^2}\right)g_m\right]=\frac{1}{\rho}e^{-jm\phi'}\delta(\rho-\rho')$$

or

$$\boxed{\rho\frac{d^2 g_m}{d\rho^2}+\frac{dg_m}{d\rho}+\left(\rho\beta_0^2-\frac{m^2}{\rho}\right)g_m=\frac{e^{-jm\phi'}}{2\pi}\delta(\rho-\rho')}$$ (15-156)

where the partial derivatives have been replaced by ordinary derivatives.

The function $g_m(\rho;\rho',\phi')$ satisfies the differential equation 15-156, and its solution can be obtained using the closed-form recipe of Section 15.3.1 represented by (15-44c) and (15-45a) through (15-45b). The homogeneous equation 15-156 can be written, by multiplying through by ρ, as

$$\rho^2\frac{d^2 g_m}{d\rho^2}+\rho\frac{dg_m}{d\rho}+\left(\beta_0^2\rho^2-m^2\right)g_m=0$$ (15-157)

or

$$\rho\frac{d^2 g_m}{d\rho^2}+\frac{dg_m}{d\rho}+\left(\beta_0^2\rho-\frac{m^2}{\rho}\right)g_m=0$$ (15-157a)

which is of the one-dimensional Sturm-Liouville form of (15-26) or (15-33) (see Example 15-1) where

$$p(\rho)=\rho$$
$$q(\rho)=\frac{m^2}{\rho}$$ (15-158)
$$r(\rho)=\rho$$
$$\lambda=\beta_0^2$$

Equation 15-157 is recognized as being Bessel's differential equation 3-64 whose two solutions can be written according to (3-67a) as

$$g_m^{(1)}=A_m J_m(\beta_0\rho)+B_m Y_m(\beta_0\rho)\quad\text{for }\rho<\rho'$$ (15-159a)
$$g_m^{(2)}=C_m J_m(\beta_0\rho)+D_m Y_m(\beta_0\rho)\quad\text{for }\rho>\rho'$$ (15-159b)

These two solutions were chosen because the fields within the waveguide form standing waves instead of traveling waves.

Since the Green's function of (15-151) must represent, according to (15-150), the field everywhere, including the origin, then $B_m=0$ in (15-159a) since $Y_m(\beta_0\rho)$ possesses a singularity at $\rho=0$. Also since the Green's function must satisfy the boundary condition (15-149a), then the solution of $g_m^{(2)}$ of (15-159b) must also satisfy (15-149a). Thus,

$$g_m^{(2)}(\rho=a)=C_m J_m(\beta_0 a)+D_m Y_m(\beta_0 a)=0$$

hence,

$$D_m=-C_m\frac{J_m(\beta_0 a)}{Y_m(\beta_0 a)}$$ (15-160)

Thus, (15-159a) and (15-159b) can be reduced to

$$g_m^{(1)}=A_m J_m(\beta_0\rho)\quad\text{for }\rho<\rho'$$ (15-161a)

$$g_m^{(2)} = C_m \left[J_m(\beta_0 \rho) - \frac{J_m(\beta_0 a)}{Y_m(\beta_0 a)} Y_m(\beta_0 \rho) \right] \quad \text{for } \rho > \rho' \tag{15-161b}$$

Using (15-44c) where $y_1 = g_m^{(1)}$ and $y_2 = g_m^{(2)}$, we can write the Wronskian as

$$W(\rho') = \beta_0 A_m C_m \frac{J_m(\beta_0 a)}{Y_m(\beta_0 a)} [J_m'(\beta_0 \rho') Y_m(\beta_0 \rho') - J_m(\beta_0 \rho') Y_m'(\beta_0 \rho')] \tag{15-162}$$

where the prime indicates partial with respect to the entire argument $[' \equiv \partial / \partial(\beta_0 \rho')]$. Using the Wronskian for the Bessel functions of (11-95), we can reduce (15-162) to

$$W(\rho') = -\frac{2}{\pi} A_m C_m \frac{J_m(\beta_0 a)}{Y_m(\beta_0 a)} \frac{1}{\rho'} \tag{15-162a}$$

Finally, $g_m(\rho; \rho', \phi')$ of (15-156) can be written using (15-158), (15-161a) through (15-161b), and (15-162a), by referring to (15-45a) and (15-45b), as

$$g_m(\rho; \rho', \phi') = \begin{cases} -\frac{1}{4}[J_m(\beta_0 \rho')Y_m(\beta_0 a) - J_m(\beta_0 a)Y_m(\beta_0 \rho')] \dfrac{J_m(\beta_0 \rho)}{J_m(\beta_0 a)} e^{-jm\phi'} \\ \qquad\qquad \text{for } \rho < \rho' \hfill (15\text{-}163a) \\[4pt] -\frac{1}{4}[J_m(\beta_0 \rho)Y_m(\beta_0 a) - J_m(\beta_0 a)Y_m(\beta_0 \rho)] \dfrac{J_m(\beta_0 \rho')}{J_m(\beta_0 a)} e^{-jm\phi'} \\ \qquad\qquad \text{for } \rho > \rho' \hfill (15\text{-}163b) \end{cases}$$

Thus, the Green's function (15-151) can then be written as

$$G(\rho, \phi; \rho', \phi') = -\frac{1}{4} \begin{cases} \displaystyle\sum_{m=-\infty}^{+\infty} [J_m(\beta_0 \rho')Y_m(\beta_0 a) - J_m(\beta_0 a)Y_m(\beta_0 \rho')] \\ \qquad \times \dfrac{J_m(\beta_0 \rho)}{J_m(\beta_0 a)} e^{jm(\phi-\phi')} \quad \text{for } \rho < \rho' \hfill (15\text{-}164a) \\[6pt] \displaystyle\sum_{m=-\infty}^{+\infty} [J_m(\beta_0 \rho)Y_m(\beta_0 a) - J_m(\beta_0 a)Y_m(\beta_0 \rho)] \\ \qquad \times \dfrac{J_m(\beta_0 \rho')}{J_m(\beta_0 a)} e^{jm(\phi-\phi')} \quad \text{for } \rho > \rho' \hfill (15\text{-}164b) \end{cases}$$

or in cosine terms as

$$G(\rho, \phi; \rho', \phi') = -\frac{1}{2} \begin{cases} \displaystyle\sum_{m=-\infty}^{+\infty} \dfrac{[J_m(\beta_0 \rho')Y_m(\beta_0 a) - J_m(\beta_0 a)Y_m(\beta_0 \rho')]}{\varepsilon_m} \\ \qquad \times \dfrac{J_m(\beta_0 \rho)}{J_m(\beta_0 a)} \cos[m(\phi-\phi')] \quad \text{for } \rho < \rho' \hfill (15\text{-}165a) \\[6pt] \displaystyle\sum_{m=-\infty}^{+\infty} \dfrac{[J_m(\beta_0 \rho)Y_m(\beta_0 a) - J_m(\beta_0 a)Y_m(\beta_0 \rho)]}{\varepsilon_m} \\ \qquad \times \dfrac{J_m(\beta_0 \rho')}{J_m(\beta_0 a)} \cos[m(\phi-\phi')] \quad \text{for } \rho > \rho' \hfill (15\text{-}165b) \end{cases}$$

where

$$\varepsilon_m = \begin{cases} 2 & m = 0 \\ 0 & m \neq 0 \end{cases} \qquad\qquad \text{(15-165c)}$$
$$\text{(15-165d)}$$

The Green's function for this problem can also be derived using the two space variable series expansion method whereby it is represented by orthonormal expansion functions. This is left to the reader as an end-of-chapter exercise.

Example 15-5

An infinite electric line source of constant current I_z is located at $\rho = \rho'$, $\phi = \phi'$, as shown in Figure 15-9, and it is radiating in an unbounded free-space medium. Derive its Green's function in closed form.

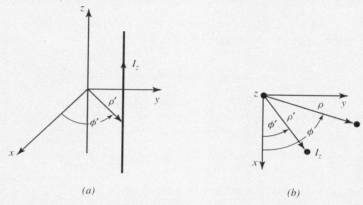

(a)   (b)

Figure 15-9   Electric line source displaced from the origin. (a) Perspective view. (b) Top view.

Solution: Since the line source is removed from the origin, its Green's function will be a function of ϕ and ϕ'. Thus, it takes the form of (15-151), and it satisfies the differential equations 15-153 through 15-157. However, since the Green's function must satisfy the radiation conditions at infinity ($G \to 0$ as $\rho \to \infty$), the two solutions to the homogeneous differential equation 15-157 can be written as

$$g_m^{(1)} = A_m J_m(\beta_0 \rho) + B_m Y_m(\beta_0 \rho) \qquad \text{for } \rho < \rho'$$
$$g_m^{(2)} = C_m H_m^{(1)}(\beta_0 \rho) + D_m H_m^{(2)}(\beta_0 \rho) \quad \text{for } \rho > \rho'$$

Because the fields must be finite everywhere, including $\rho = 0$, $g_m^{(1)}$ reduces to

$$g_m^{(1)} = A_m J_m(\beta_0 \rho) \quad \text{for } \rho < \rho'$$

In addition, for $\rho > \rho'$ the wave functions must represent outwardly traveling waves. Thus, for $e^{j\omega t}$ time variations, $g_m^{(2)}$ reduces to

$$g_m^{(2)} = D_m H_m^{(2)}(\beta_0 \rho) \quad \text{for } \rho > \rho'$$

Using (15-44c) where $y_1 = g_m^{(1)}$ and $y_2 = g_m^{(2)}$, we can write the Wronskian as

$$W(\rho') = \beta_0 A_m D_m \left[J_m(\beta_0 \rho') H_m^{(2)'}(\beta_0 \rho') - H_m^{(2)}(\beta_0 \rho') J_m'(\beta_0 \rho') \right]$$

$$= -j\beta_0 A_m D_m \left[J_m(\beta_0 \rho') Y_m'(\beta_0 \rho') - J_m'(\beta_0 \rho') Y_m(\beta_0 \rho') \right]$$

which by using the Wronskian of (11-95) for Bessel functions can be expressed as

$$W(\rho') = -j\frac{2}{\pi\rho'}A_m D_m$$

Thus, $g_m(\rho;\rho',\phi')$ of (15-156) can be written, using (15-45a) through (15-45b) and (15-158) along with the preceding expressions for $g_m^{(1)}$, $g_m^{(2)}$, and $W(\rho')$, as

$$g_m(\rho;\rho',\phi') = -\frac{1}{4j}\begin{cases} J_m(\beta_0\rho)H_m^{(2)}(\beta_0\rho')e^{-jm\phi'} & \text{for } \rho < \rho' \\ J_m(\beta_0\rho')H_m^{(2)}(\beta_0\rho)e^{-jm\phi'} & \text{for } \rho > \rho' \end{cases}$$

Thus, the Green's function of (15-151) can be written as

$$G(\rho,\phi;\rho',\phi') = -\frac{1}{4j}\begin{cases} \displaystyle\sum_{m=-\infty}^{+\infty} J_m(\beta_0\rho)H_m^{(2)}(\beta_0\rho')e^{jm(\phi-\phi')} & \text{for } \rho < \rho' \\ \displaystyle\sum_{m=-\infty}^{+\infty} J_m(\beta_0\rho')H_m^{(2)}(\beta_0\rho)e^{jm(\phi-\phi')} & \text{for } \rho > \rho' \end{cases}$$

which, by the addition theorem for Hankel functions of (11-69a) through (11-69b) or (11-82a) through (11-82b), can be expressed in succinct form as

$$G(\rho,\phi;\rho',\phi') = -\frac{1}{4j}H_0^{(2)}(\beta_0|\boldsymbol{\rho}-\boldsymbol{\rho}'|)$$

This is the well-known two-dimensional Green's function for cylindrical waves.

15.6.3 Spherical Coordinates

The development of the Green's function for problems represented by spherical coordinates is more complex, and it must be expressed, in general, in terms of spherical Bessel and Hankel functions, Legendre functions, and complex exponentials or cosinusoids (see Chapter 10). In order to minimize the mathematical complexities here, we will develop the Green's function of a source positioned at r',θ',ϕ', inside a sphere of radius a and with free-space, as shown in Figure 15-10.

The Green's function must satisfy the partial differential equation

$$\boxed{\nabla^2 G + \beta_0^2 G = \delta(\mathbf{r}-\mathbf{r}')} \tag{15-166}$$

subject to the boundary condition

$$G(r=a, 0\leq\theta\leq\pi, 0\leq\phi\leq 2\pi) = 0 \tag{15-166a}$$

In spherical coordinates the delta function $\delta(\mathbf{r}-\mathbf{r}')$ in (15-166) can be expressed, in general, as [10, 13]

$$\delta(\mathbf{r}-\mathbf{r}') = \begin{cases} \dfrac{1}{r^2\sin\theta}\delta(r-r')\delta(\theta-\theta')\delta(\phi-\phi') & & \text{(15-167a)} \\[2mm] \dfrac{1}{2r^2}\delta(r-r')\delta(\phi-\phi') & \text{for no } \theta \text{ dependence} & \text{(15-167b)} \\[2mm] \dfrac{1}{2\pi r^2\sin\theta}\delta(r-r')\delta(\theta-\theta') & \text{for no } \phi \text{ dependence} & \text{(15-167c)} \\[2mm] \dfrac{1}{4\pi r^2}\delta(r-r') & \text{for neither } \theta \text{ nor } \phi \text{ dependence} & \text{(15-167d)} \end{cases}$$

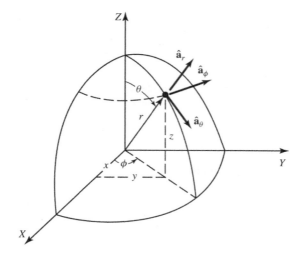

Figure 15-10 Spherical coordinate system.

In expanded form, the Green's function of (15-166) can now be written using (15-167a) as

$$\frac{1}{r^2}\frac{\partial}{\partial r}\left(r^2\frac{\partial G}{\partial r}\right)+\frac{1}{r^2\sin\theta}\frac{\partial}{\partial \theta}\left(\sin\theta\frac{\partial G}{\partial \theta}\right)+\frac{1}{r^2\sin^2\theta}\frac{\partial^2 G}{\partial \phi^2}+\beta_0^2 G$$

$$=\frac{1}{r^2\sin\theta}\delta(r-r')\delta(\theta-\theta')\delta(\phi-\phi') \tag{15-168}$$

Since the spherical harmonics form a complete set for functions of the angles θ and ϕ, the Green's function can be represented by a double summation of an infinite series

$$\boxed{\begin{aligned} G(r,\theta,\phi;r',\theta',\phi')&=\sum_{n=0}^{\infty}\sum_{m=-n}^{n}g_{mn}(r;r',\theta',\phi')P_n^m(\cos\theta)e^{jm\phi}\\ &=\sum_{n=0}^{\infty}\sum_{m=-n}^{n}g_{mn}(r;r',\theta',\phi')T_{mn}(\theta,\phi)\end{aligned}} \tag{15-169}$$

where $T_{mn}(\theta,\phi)$ represents the tesseral harmonics of (11-214a) and (11-214b), or

$$T_{mn}(\theta,\phi)=C_{mn}P_n^m(\cos\theta)e^{jm\phi} \tag{15-169a}$$

where

$$C_{mn}=\sqrt{\frac{(2n+1)(n-m)!}{4\pi(n+m)!}} \tag{15-169b}$$

Multiplying (15-168) by r^2 and then substituting (15-169) into (15-168), we can write that

$$\sum_{n=0}^{\infty}\sum_{m=-n}^{n}\left[T_{mn}\frac{\partial}{\partial r}\left(r^2\frac{\partial g_{mn}}{\partial r}\right)+\frac{g_{mn}}{\sin\theta}\frac{\partial}{\partial \theta}\left(\sin\theta\frac{\partial T_{mn}}{\partial \theta}\right)-\frac{m^2}{\sin^2\theta}g_{mn}T_{mn}+(\beta_0 r)^2 g_{mn}T_{mn}\right]$$

$$=\frac{1}{\sin\theta}\delta(r-r')\delta(\theta-\theta')\delta(\phi-\phi') \tag{15-170}$$

Dividing both sides of (15-170) by $g_{mn}T_{mn}$, we can write that

$$\sum_{n=0}^{\infty}\sum_{m=-n}^{\infty}\left[\frac{1}{g_{mn}}\frac{\partial}{\partial r}\left(r^2\frac{\partial g_{mn}}{\partial r}\right)+\frac{1}{T_{mn}\sin\theta}\frac{\partial}{\partial\theta}\left(\sin\theta\frac{\partial T_{mn}}{\partial\theta}\right)-\frac{m^2}{\sin^2\theta}+(\beta_0 r)^2\right]$$

$$=\frac{1}{g_{mn}T_{mn}\sin\theta}\delta(r-r')\delta(\theta-\theta')\delta(\phi-\phi')\tag{15-170a}$$

Using (3-86b), we can write that

$$\frac{1}{T_{mn}\sin\theta}\frac{\partial}{\partial\theta}\left(\sin\theta\frac{\partial T_{mn}}{\partial\theta}\right)-\frac{m^2}{\sin^2\theta}=-n(n+1)\tag{15-171}$$

hence,

$$\frac{1}{\sin\theta}\frac{\partial}{\partial\theta}\left(\sin\theta\frac{\partial T_{mn}}{\partial\theta}\right)+\left[n(n+1)-\left(\frac{m}{\sin\theta}\right)^2\right]T_{mn}=0\tag{15-171a}$$

Thus, (15-170a) reduces, by substituting (15-171) into it and then multiplying through by $g_{mn}T_{mn}$, to

$$\sum_{n=0}^{\infty}\sum_{m=-n}^{n}\left\{\frac{\partial}{\partial r}\left(r^2\frac{\partial g_{mn}}{\partial r}\right)+\left[(\beta_0 r)^2+n(n+1)\right]g_{mn}\right\}T_{mn}$$

$$=\frac{1}{\sin\theta}\delta(r-r')\delta(\theta-\theta')\delta(\phi-\phi')\tag{15-172}$$

From (11-214a) through (11-216d) and the definitions of the tesseral harmonics and Legendre functions, it can be shown that the orthogonality conditions of the tesseral harmonics are [20]

$$\int_0^{2\pi}\int_0^{\pi}T_{mn}(\theta,\phi)T_{pq}^*(\theta,\phi)\sin\theta\,d\theta\,d\phi=\delta_{mp}\delta_{nq}\tag{15-173}$$

where

$$T_{pq}^*(\theta,\phi)=(-1)^p T_{(-p)q}(\theta,\phi)\tag{15-173a}$$

$$\delta_{rs}=\begin{cases}1 & r=s\\0 & r\neq s\end{cases}\tag{15-173b}$$

Let

$$g_{mn}(r;r',\theta',\phi')=h_{mn}(r,r')T^*(\theta',\phi')\tag{15-173c}$$

Multiplying both sides of (15-172) by $T_{pq}^*(\theta,\phi)\sin\theta$, integrating from 0 to π in θ and 0 to 2π in ϕ, and using the orthogonality condition of (15-173), it can be shown that (15-172) reduces to

$$\boxed{\frac{d}{dr}\left(r^2\frac{dh_{mn}}{dr}\right)+\left[(\beta_0 r)^2-n(n+1)\right]h_{mn}=\delta(r-r')}\tag{15-174}$$

where the partial derivative $\partial/\partial r$ has been replaced by ordinary derivatives.

The function $h_{mn}(r, r')$ satisfies the differential equation 15-174, and its solution can be obtained using the closed-form recipe of Section 15.3.1 represented by (15-44c) and (15-45a) through (15-45b). The homogeneous equation 15-174 can be written as

$$\frac{d}{dr}\left(r^2 \frac{dh_{mn}}{dr}\right) + \left[(\beta_0 r)^2 - n(n+1)\right]h_{mn} = 0 \tag{15-175}$$

which is of the one-dimensional Sturm-Liouville form of (15-26) or (15-33) where

$$
\begin{aligned}
p(r) &= r^2 \\
q(r) &= -n(n+1) \\
r(x) &= r^2 \\
\lambda &= \beta^2
\end{aligned} \tag{15-176}
$$

Equation 15-175 is recognized as being (3-83) or (3-86a) whose solution can be represented by either (3-87a) or (3-87b). We choose here the form of (3-87a) since we need to represent the Green's function within the sphere by standing wave functions. Thus, the two solutions of (15-175) can be written as

$$h_{mn}^{(1)} = A_m j_n(\beta_0 r) + B_m y_n(\beta_0 r) \qquad \text{for } r < r' \tag{15-177a}$$

$$h_{mn}^{(2)} = C_m j_n(\beta_0 r) + D_m y_n(\beta_0 r) \qquad \text{for } r > r' \tag{15-177b}$$

where $j_n(\beta_0 r)$ and $y_n(\beta_0 r)$ are, respectively, spherical Bessel functions of the first and second kind.

Since the Green's function of (15-169) must be finite everywhere, including the origin, then $B_m = 0$ since $y_n(\beta_0 r)$ possesses a singularity at $r = 0$. Also the Green's function must satisfy the boundary condition (15-166a). Therefore, $h_{mn}^{(2)}$ of (15-177b) at $r = a$ reduces to

$$h_{mn}^{(2)}(r = a) = C_m j_n(\beta_0 a) + D_m y_n(\beta_0 a) = 0 \tag{15-178}$$

hence,

$$D_m = -C_m \frac{j_n(\beta_0 a)}{y_n(\beta_0 a)} \tag{15-178a}$$

Thus, (15-177a) and (15-177b) are reduced to

$$h_{mn}^{(1)} = A_m j_n(\beta_0 r) \qquad \text{for } r < r' \tag{15-179a}$$

$$h_{mn}^{(2)} = C_m \left[j_n(\beta_0 r) - \frac{j_n(\beta_0 a)}{y_n(\beta_0 a)} y_n(\beta_0 r) \right] \qquad \text{for } r > r' \tag{15-179b}$$

Using (15-44c) where $y_1 = h_{mn}^{(1)}$ and $y_2 = h_{mn}^{(2)}$, we can write the Wronskian as

$$W(r') = \beta_0 A_m C_m \frac{j_n(\beta_0 a)}{y_n(\beta_0 a)} [j_n'(\beta_0 r') y_n(\beta_0 r') - j_n(\beta_0 r') y_n'(\beta_0 r')] \tag{15-180}$$

Using the Wronskian for spherical Bessel functions of

$$j_n(\beta_0 r') y_n'(\beta_0 r') - j_n'(\beta_0 r') y_n(\beta_0 r') = \frac{1}{(\beta_0 r')^2} \tag{15-180a}$$

reduces (15-180) to

$$W(r') = -\frac{1}{\beta_0} A_m C_m \frac{j_n(\beta_0 a)}{y_n(\beta_0 a)} \frac{1}{(r')^2} \qquad (15\text{-}180b)$$

Finally, $h_{mn}(r, r')$ of (15-174) can be written using (15-176), (15-179a) through (15-179b), and (15-180b), by referring to (15-45a) and (15-45b), as

$$h_{mn}(r, r') = \begin{cases} -\beta_0 C_{mn}^2 [j_n(\beta_0 r') y_n(\beta_0 a) - j_n(\beta_0 a) y_n(\beta_0 r')] \dfrac{j_n(\beta_0 r)}{j_n(\beta_0 a)} \\ \qquad\qquad\qquad\qquad \text{for } r < r' \qquad (15\text{-}181a) \\[2em] -\beta_0 C_{mn}^2 [j_n(\beta_0 r) y_n(\beta_0 a) - j_n(\beta_0 a) y_n(\beta_0 r)] \dfrac{j_n(\beta_0 r')}{j_n(\beta_0 a)} \\ \qquad\qquad\qquad\qquad \text{for } r > r' \qquad (15\text{-}181b) \end{cases}$$

Thus, the Green's function of (15-169) can be written as

$$G(r, \theta, \phi; r', \theta', \phi') = -\beta_0 \begin{cases} \displaystyle\sum_{n=0}^{\infty} \sum_{m=-n}^{n} (-1)^m C_{mn}^2 [j_n(\beta_0 r') y_n(\beta_0 a) - j_n(\beta_0 a) y_n(\beta_0 r')] \\ \quad \times \dfrac{j_n(\beta_0 r)}{j_n(\beta_0 a)} P_n^m(\cos\theta) P_n^{-m}(\cos\theta') e^{jm(\phi-\phi')} \\ \qquad\qquad\qquad\qquad \text{for } r < r' \qquad (15\text{-}182a) \\[2em] \displaystyle\sum_{n=0}^{\infty} \sum_{m=-n}^{n} (-1)^m C_{mn}^2 [j_n(\beta_0 r) y_n(\beta_0 a) - j_n(\beta_0 a) y_n(\beta_0 r)] \\ \quad \times \dfrac{j_n(\beta_0 r')}{j_n(\beta_0 a)} P_n^m(\cos\theta) P_n^{-m}(\cos\theta') e^{jm(\phi-\phi')} \\ \qquad\qquad\qquad\qquad \text{for } r > r' \qquad (15\text{-}182b) \end{cases}$$

15.7 DYADIC GREEN'S FUNCTIONS

The Green's function development of the previous sections can be used for the solution of electromagnetic problems that satisfy the scalar wave equation. The most general Green's function development and electromagnetic field solution, for problems that satisfy the vector wave equation, will be to use *vectors and dyadics* [14–18]. Before we briefly discuss such a procedure, let us first introduce and define dyadics.

15.7.1 Dyadics

Vectors and dyadics are used, in general, to describe linear transformations *within a given orthogonal coordinate system,* and they simplify the manipulations of mathematical relations, compared to using tensors. For electromagnetic problems, where linear transformations between sources and fields *within a given orthogonal coordinate system* are often necessary, vectors and dyadics are very convenient to use.

A *dyad* is defined by the juxtaposition **AB** of the vectors **A** and **B**, with no dot or cross product between them. In general, a dyad has nine terms and in matrix form can be represented by

$$(\mathbf{AB}) = \begin{pmatrix} A_1 B_1 & A_1 B_2 & A_1 B_3 \\ A_2 B_1 & A_2 B_2 & A_2 B_3 \\ A_3 B_1 & A_3 B_2 & A_3 B_3 \end{pmatrix} \qquad (15\text{-}183)$$

A *dyadic* $\bar{\mathbf{D}}$ can be defined by the sum of N dyads. That is,

$$\bar{\mathbf{D}} = \sum_{n=1}^{N} \mathbf{A}^n \mathbf{B}^n \qquad (15\text{-}184)$$

In general, no more than three dyads are required to represent a dyadic, that is, $N_{max} = 3$.

Let us now define the vectors $\mathbf{A}$, $\mathbf{C}$, $\mathbf{D}_1$, $\mathbf{D}_2$, and $\mathbf{D}_3$ in a general coordinate system with unit vectors $\hat{\mathbf{a}}_1, \hat{\mathbf{a}}_2$, and $\hat{\mathbf{a}}_3$. That is,

$$\mathbf{A} = \hat{\mathbf{a}}_1 A_1 + \hat{\mathbf{a}}_2 A_2 + \hat{\mathbf{a}}_3 A_3 \qquad (15\text{-}185\text{a})$$

$$\mathbf{C} = \hat{\mathbf{a}}_1 C_1 + \hat{\mathbf{a}}_2 C_2 + \hat{\mathbf{a}}_3 C_3 \qquad (15\text{-}185\text{b})$$

$$\mathbf{D}_1 = \hat{\mathbf{a}}_1 D_{11} + \hat{\mathbf{a}}_2 D_{12} + \hat{\mathbf{a}}_3 D_{13} \qquad (15\text{-}185\text{c})$$

$$\mathbf{D}_2 = \hat{\mathbf{a}}_1 D_{21} + \hat{\mathbf{a}}_2 D_{22} + \hat{\mathbf{a}}_3 D_{23} \qquad (15\text{-}185\text{d})$$

$$\mathbf{D}_3 = \hat{\mathbf{a}}_1 D_{31} + \hat{\mathbf{a}}_2 D_{32} + \hat{\mathbf{a}}_3 D_{33} \qquad (15\text{-}185\text{e})$$

Let us now write that

$$\mathbf{C} = (\mathbf{A} \cdot \hat{\mathbf{a}}_1)\mathbf{D}_1 + (\mathbf{A} \cdot \hat{\mathbf{a}}_2)\mathbf{D}_2 + (\mathbf{A} \cdot \hat{\mathbf{a}}_3)\mathbf{D}_3$$
$$= \mathbf{A} \cdot (\hat{\mathbf{a}}_1 \mathbf{D}_1) + \mathbf{A} \cdot (\hat{\mathbf{a}}_2 \mathbf{D}_2) + \mathbf{A} \cdot (\hat{\mathbf{a}}_3 \mathbf{D}_3)$$
$$\mathbf{C} = \mathbf{A} \cdot (\hat{\mathbf{a}}_1 \mathbf{D}_1 + \hat{\mathbf{a}}_2 \mathbf{D}_2 + \hat{\mathbf{a}}_3 \mathbf{D}_3) \qquad (15\text{-}186)$$

or

$$\mathbf{C} = \mathbf{A} \cdot \bar{\mathbf{D}} \qquad (15\text{-}186\text{a})$$

where

$$\bar{\mathbf{D}} = \hat{\mathbf{a}}_1 \mathbf{D}_1 + \hat{\mathbf{a}}_2 \mathbf{D}_2 + \hat{\mathbf{a}}_3 \mathbf{D}_3 \qquad (15\text{-}186\text{b})$$

In (15-186) through (15-186b) $\bar{\mathbf{D}}$ is a dyadic, and it is defined by the sum of the three dyads $\hat{\mathbf{a}}_n \mathbf{D}_n, n = 1, 2, 3$. In matrix form (15-186) or (15-186a) can be written as

$$(C_1 \quad C_2 \quad C_3) = (A_1 \quad A_2 \quad A_3) \begin{pmatrix} D_{11} & D_{12} & D_{13} \\ D_{21} & D_{22} & D_{23} \\ D_{31} & D_{32} & D_{33} \end{pmatrix} \qquad (15\text{-}187)$$

where the dyadic $\bar{\mathbf{D}}$ has nine elements.

Just like vectors, dyadics satisfy a number of identities involving dot and cross products, differentiations, and integrations. The uninformed reader should refer to the literature [10–13] for such relations.

15.7.2 Green's Functions

In electromagnetics it is often desirable to solve, using the Green's functions approach, the linear vector problem of

$$\boxed{\mathcal{L}\mathbf{h} = \mathbf{f}} \qquad (15\text{-}188)$$

where $\mathcal{L}$ is a differential operator. Equation 15-188 is a more general and vector representation of (15-27). It should be noted here that the solution of (15-188) *cannot,* in general, be represented by

$$\mathbf{h}(\mathbf{r}) \neq \iiint_V \mathbf{f}(\mathbf{r}')G(\mathbf{r},\mathbf{r}')dv' \tag{15-189}$$

where $G(\mathbf{r},\mathbf{r}')$ is a single scalar Green's function. The relation (15-189) would imply that a component of the source $\mathbf{f}$ parallel to a given axis produces a response (field) $\mathbf{h}$ parallel to the same axis. This, in general, is not true.

A more appropriate representation of the solution of (15-188), in a rectangular coordinate system, will be

$$h_x(\mathbf{r}) = \iiint_V [f_x(\mathbf{r}')G_{xx}(\mathbf{r},\mathbf{r}') + f_y(\mathbf{r}')G_{xy}(\mathbf{r},\mathbf{r}') + f_z(\mathbf{r}')G_{xz}(\mathbf{r},\mathbf{r}')]dv' \tag{15-190a}$$

$$h_y(\mathbf{r}) = \iiint_V [f_x(\mathbf{r}')G_{yx}(\mathbf{r},\mathbf{r}') + f_y(\mathbf{r}')G_{yy}(\mathbf{r},\mathbf{r}') + f_z(\mathbf{r}')G_{yz}(\mathbf{r},\mathbf{r}')]dv' \tag{15-190b}$$

$$h_z(\mathbf{r}) = \iiint_V [f_x(\mathbf{r}')G_{zx}(\mathbf{r},\mathbf{r}') + f_y(\mathbf{r}')G_{zy}(\mathbf{r},\mathbf{r}') + f_z(\mathbf{r}')G_{zz}(\mathbf{r},\mathbf{r}')]dv' \tag{15-190c}$$

which, in a more compact form, can be written as

$$\boxed{\mathbf{h}(\mathbf{r}) = \iiint_V [f_x(\mathbf{r}')\mathbf{G}_x(\mathbf{r},\mathbf{r}') + f_y(\mathbf{r}')\mathbf{G}_y(\mathbf{r},\mathbf{r}') + f_z(\mathbf{r}')\mathbf{G}_z(\mathbf{r},\mathbf{r}')]dv'} \tag{15-191}$$

where, as in (15-185c) through (15-185e)

$$\mathbf{G}_x(\mathbf{r},\mathbf{r}') = \hat{a}_xG_{xx}(\mathbf{r},\mathbf{r}') + \hat{a}_yG_{yx}(\mathbf{r},\mathbf{r}') + \hat{a}_zG_{zx}(\mathbf{r},\mathbf{r}') \tag{15-191a}$$

$$\mathbf{G}_y(\mathbf{r},\mathbf{r}') = \hat{a}_xG_{xy}(\mathbf{r},\mathbf{r}') + \hat{a}_yG_{yy}(\mathbf{r},\mathbf{r}') + \hat{a}_zG_{zy}(\mathbf{r},\mathbf{r}') \tag{15-191b}$$

$$\mathbf{G}_z(\mathbf{r},\mathbf{r}') = \hat{a}_xG_{xz}(\mathbf{r},\mathbf{r}') + \hat{a}_yG_{yz}(\mathbf{r},\mathbf{r}') + \hat{a}_zG_{zz}(\mathbf{r},\mathbf{r}') \tag{15-191c}$$

In (15-190a) through (15-191c), the $G_{ij}(\mathbf{r},\mathbf{r}')$'s are the elements of the dyadic $\bar{\mathbf{G}}(\mathbf{r},\mathbf{r}')$, which is referred to here as the *dyadic Green's function.* In (15-191) through (15-191c), the $\mathbf{G}_i(\mathbf{r},\mathbf{r}')$'s are the column vectors of the dyadic Green's function $\bar{\mathbf{G}}(\mathbf{r},\mathbf{r}')$.

Using the notation of (15-186) through (15-186b), the solution of (15-191) can also be written as

$$\mathbf{h}(\mathbf{r}) = \iiint_V \{[\mathbf{f}(\mathbf{r}')\cdot\hat{a}_x]\mathbf{G}_x(\mathbf{r},\mathbf{r}') + [\mathbf{f}(\mathbf{r}')\cdot\hat{a}_y]\mathbf{G}_y(\mathbf{r},\mathbf{r}') + [\mathbf{f}(\mathbf{r}')\cdot\hat{a}_z]\mathbf{G}_z(\mathbf{r},\mathbf{r}')\}dv'$$

$$= \iiint_V \mathbf{f}(\mathbf{r}')\cdot[\hat{a}_x\mathbf{G}_x(\mathbf{r},\mathbf{r}') + \hat{a}_y\mathbf{G}_y(\mathbf{r},\mathbf{r}') + \hat{a}_z\mathbf{G}_z(\mathbf{r},\mathbf{r}')]dv'$$

$$\boxed{\mathbf{h}(\mathbf{r}) = \iiint_V \mathbf{f}(\mathbf{r}')\cdot\bar{\mathbf{G}}(\mathbf{r},\mathbf{r}')dv'} \tag{15-192}$$

where $\bar{\mathbf{G}}(\mathbf{r},\mathbf{r}')$ is the dyadic Green's function

$$\boxed{\bar{\mathbf{G}}(\mathbf{r},\mathbf{r}') = \hat{a}_x\mathbf{G}_x(\mathbf{r},\mathbf{r}') + \hat{a}_y\mathbf{G}_y(\mathbf{r},\mathbf{r}') + \hat{a}_z\mathbf{G}_z(\mathbf{r},\mathbf{r}')} \tag{15-192a}$$

The dyadic Green's function $\bar{\mathbf{G}}(\mathbf{r},\mathbf{r}')$ can be found by first finding the vectors $\mathbf{G}_x(\mathbf{r},\mathbf{r}')$, $\mathbf{G}_y(\mathbf{r},\mathbf{r}')$, and $\mathbf{G}_z(\mathbf{r},\mathbf{r}')$ each satisfying the homogeneous form of the partial differential equation 15-188, that is,

$$\mathscr{L}\mathbf{G}_x(\mathbf{r},\mathbf{r}') = \hat{\mathbf{a}}_x\,\delta(\mathbf{r}-\mathbf{r}') \tag{15-193a}$$

$$\mathscr{L}\mathbf{G}_y(\mathbf{r},\mathbf{r}') = \hat{\mathbf{a}}_y\,\delta(\mathbf{r}-\mathbf{r}') \tag{15-193b}$$

$$\mathscr{L}\mathbf{G}_z(\mathbf{r},\mathbf{r}') = \hat{\mathbf{a}}_z\,\delta(\mathbf{r}-\mathbf{r}') \tag{15-193c}$$

and the appropriate boundary conditions, and then using (15-192a) to form the dyadic Green's function. Any of the methods of the previous sections can be used to find the Green's function of (15-193a) through (15-193c).

An example for the potential use of the dyadic Green's function is the solution for the electric and magnetic fields due to a source represented by the electric current density $\mathbf{J}$. According to (6-32a) and (6-32b), the electric and magnetic fields can be written as

$$\mathbf{H}(\mathbf{r}) = \frac{1}{\mu}\boldsymbol{\nabla}\times\mathbf{A} \tag{15-194a}$$

$$\mathbf{E}(\mathbf{r}) = -j\omega\mathbf{A} - j\frac{1}{\omega\mu\varepsilon}\boldsymbol{\nabla}(\boldsymbol{\nabla}\cdot\mathbf{A}) \tag{15-194b}$$

where the vector potential $\mathbf{A}$ satisfies the partial differential equation 6-30 or

$$\nabla^2\mathbf{A} + \beta^2\mathbf{A} = -\mu\mathbf{J} \tag{15-195}$$

Using the dyadic Green's function approach, the vector potential $\mathbf{A}$ can be found using

$$\mathbf{A} = -\mu\iiint_V \mathbf{J}(\mathbf{r}')\cdot\bar{\mathbf{G}}(\mathbf{r},\mathbf{r}')\,dv' \tag{15-196}$$

where the dyadic Green's function must satisfy the partial differential equation

$$\nabla^2\bar{\mathbf{G}} + \beta^2\bar{\mathbf{G}} = \bar{\delta}(\mathbf{r}-\mathbf{r}') \tag{15-197}$$

and the appropriate boundary conditions.

Because of the complexity for the development of the dyadic Green's function, it will not be pursued any further here. The interested reader is referred to the literature [10–18] for more details.

15.8 MULTIMEDIA

On the website that accompanies this book, the following multimedia resources are included for the review, understanding, and presentation of the material of this chapter.

- **PowerPoint (PPT)** viewgraphs, in multicolor.

REFERENCES

1. R. Courant and D. Hilbert, *Methods of Mathematical Physics*, vol. I, Wiley, New York, 1937.
2. A. Wester, *Partial Differential Equations of Mathematical Physics*, S. Plimpton (Ed.), Second Edition, Hafner, New York, 1947.
3. P. M. Morse and H. Feshbach, *Methods of Theoretical Physics*, vols. I and II, McGraw-Hill, New York, 1953.
4. B. Friedman, *Principles and Techniques of Applied Mathematics*, Wiley, New York, 1956.

5. J. Dettman, *Mathematical Methods in Physics and Engineering*, McGraw-Hill, New York, 1962.

6. J. D. Jackson, *Classical Electrodynamics*, Third Edition, Wiley, New York, 1999.

7. H. W. Wyld, *Mathematical Methods for Physics*, Benjamin/Cummings, Menlo Park, CA, 1976.

8. I. Stakgold, *Green's Functions and Boundary Value Problems*, Wiley, New York, 1979.

9. R. E. Collin, *Field Theory of Guided Waves*, McGraw-Hill, New York, 1960.

10. J. Van Bladel, *Electromagnetic Fields*, Second Edition, IEEE Press, Wiley Interscience, Hoboken, NJ, 2007.

11. C.-T. Tai, *Dyadic Green's Functions in Electromagnetic Theory*, Intext Educational Publishers, Scranton, PA, 1971.

12. L. B. Felsen and N. Marcuvitz, *Radiation and Scatering of Waves*, Prentice-Hall, Englewood Cliffs, NJ, 1973.

13. D. C. Stinson, *Intermediate Mathematics of Electromagnetics*, Prentice-Hall, Englewood Cliffs, NJ, 1976.

14. C.-T. Tai, "On the eigenfunction expansion of dyadic Green's functions," *Proc. IEEE*, vol. 61, pp. 480–481, Apr. 1973.

15. C.-T. Tai and P. Rozenfeld, "Different representations of dyadic Green's functions for a rectangular cavity," *IEEE Trans. Microwave Theory Tech.*, vol. MTT-24, pp. 597–601, Sept. 1976.

16. A. Q. Howard, Jr., "On the longitudinal component of the Green's function dyadic," *Proc. IEEE*, vol. 62, pp. 1704–1705, Dec. 1974.

17. A. D. Yaghjian, "Electric dyadic Green's functions in the source region," *Proc. IEEE*, vol. 68, no. 2, pp. 248–263, Feb. 1980.

18. P. H. Pathak, "On the eigenfunction expansion of electromagnetic dyadic Green's functions," *IEEE Trans. Antennas Propagat.*, vol. AP-31, pp. 837–846, Nov. 1983.

19. I. Orlanski, "A simple boundary condition for unbounded hyperbolic flows," *J. Computational Physics*, 21, pp. 251–269, 1976.

20. R. F. Harrington, *Time-Harmonic Electromagnetic Fields*, McGraw-Hill, New York, 1961.

PROBLEMS

15.1. Using the procedure of Section 15.3.1, as represented by (15-44c) and (15-45a) through (15-45b), derive the Green's function of the mechanics problem of Section 15.2.2 as given by (15-24a) and (15-24b).

15.2. The displacement $u(x)$ of a string of length ℓ subjected to a cosinusoidal force

$$f(x,t) = f(x)\,e^{+j\omega t}$$

is determined by

$$u(x,t) = u(x)\,e^{+j\omega t}$$

where $u(x)$ satisfies the differential equation

$$\frac{d^2 u(x)}{dx^2} + \beta^2 u(x) = f(x), \qquad \beta^2 = \omega/c$$

Assuming the ends of the string are fixed

$$u(x=0) = u(x=\ell) = 0$$

determine, in closed, form the Green's function of the system.

15.3. Two PEC semi-infinite plates, which both are grounded ($V = 0$), are separated by a distance w in the x direction, as shown in Figure P15-3. A d. c. infinite line source of constant charge density $q(x')$ is positioned at x' between the two plates and extends to infinity in the y direction. Assume the plates are infinite in the y direction and free space exists between the two plates. For this problem:

(a) Specify the appropriate boundary conditions for the potential and the associated Green's function.

(b) Derive the Green's function in the x-y plane in series form, including an expression for the eigenvalues with the appropriate indices.

(c) Based on this Green's function and stated charge distribution, write an expression for the potential $V(x)$ between the plates. Show all the steps.

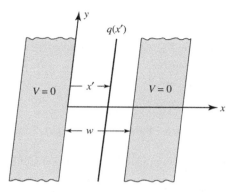

Figure P15-3

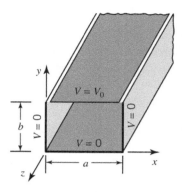

Figure P15-7

15.4. Derive in series form the Green's function of Problem 15.3.

15.5. Two infinite radial plates with an interior angle of α, as shown in Figure P15-5, are both maintained at a potential of $V = 0$. Determine in closed form the Green's function for the electric potential distribution between the plates.

15.8. The three sides of an infinite length and infinite height trough are maintained at a grounded potential of zero as shown in Figure P15-8. The width of the trough is a. Determine in closed form the Green's function for the electric field distribution within the trough.

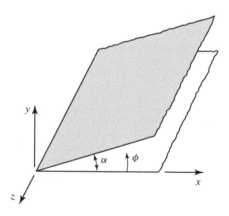

Figure P15-5

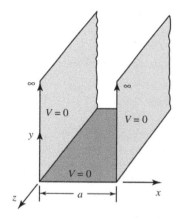

Figure P15-8

15.6. Repeat Problem 15.5 when the two plates are isolated from each other with the plate of $\phi = 0$ grounded while that at $\phi = \alpha$ is maintained at a constant potential V_0.

15.7. The top side of a rectangular cross-section, infinite length pipe is insulated from the other three, and it is maintained at a constant potential V_0. The other three are held at a grounded potential of zero as shown in Figure P15-7. Determine in closed form the Green's function for the electric potential distribution within the pipe.

15.9. Derive the Green's function of (15-83a) through (15-83b) by initially choosing a solution for (15-74) that satisfies the boundary condition at $y = 0$ and $y = b$.

15.10. An infinitely long conducting wire positioned at $\rho = \rho'$, $\phi = \phi'$ is circumscribed by a grounded $(V = 0)$ electric conducting circular cylinder of radius a and infinite length, as shown in Figure 15-4. Derive in closed form the Green's function for the potential distribution within the cylinder. Assume free space within the cylinder.

15.11. Derive in closed form the Green's function of the time-harmonic problem represented by (15-97) subject to the boundary conditions of (15-97a) and (15-97b). This would be an alternate representation of (15-106).

15.12. For Figure 15-7, derive the Green's function in closed form subject to the appropriate boundary conditions.

15.13. Repeat Example 15-5 for an infinite magnetic line source of constant current I_m located at $\rho = \rho'$, $\phi = \phi'$, as shown in Figure 15-9.

15.14. An annular microstrip antenna fed by a coaxial line is composed of an annular conducting circular strip, with inner and outer radii of a and b, placed on the top surface of a lossless substrate of height h and electrical parameters ε_s, μ_s as shown in Figure P15-14. The substrate is supported by a ground plane. Assuming the microstrip antenna can be modeled as a cavity with ideal open circuits of vanishing tangential magnetic fields at the inner ($\rho = a$) and outer ($\rho = b$) edges,

$$H_\phi(\rho = a, 0 \leq \phi \leq 2\pi, 0 \leq z \leq h)$$
$$= H_\phi(\rho = b, 0 \leq \phi \leq 2\pi, 0 \leq z \leq h) = 0$$

and vanishing tangential electric fields on its top and bottom sides, determine the Green's function for the TM^z modes (subject to $\nabla \cdot \mathbf{J} = 0$) with independent z variations within the cavity. For such modes the electric field must have only a z component of

$$\mathbf{E} = \hat{\mathbf{a}}_z E_z(\rho, \phi)$$

which must satisfy the partial differential equation

$$\nabla^2 E_z + \beta^2 E_z = j\omega\mu J_z(\rho_f, \phi_f), \quad \beta^2 = \omega^2\mu\varepsilon$$

15.15. Repeat Problem 15.14 for an annular sector microstrip antenna whose geometry is shown in Figure P15-15.

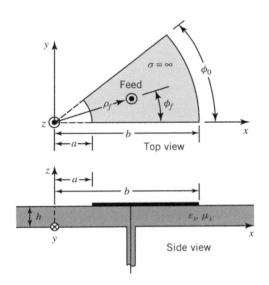

Figure P15-15

15.16. Repeat Problem 15.14 for a circular sector microstrip antenna whose geometry is shown in Figure P15-16.

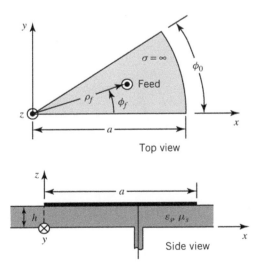

Figure P15-16

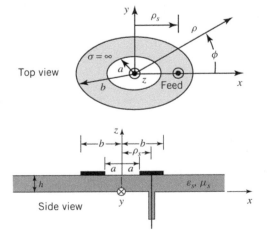

Figure P15-14

15.17. Derive the Green's function represented by (15-165a) and (15-165b) in terms of the two space variable series expansion method using orthonormal expansion functions.

15.18. An infinite length electric line source of constant current I_e is placed near a conducting circular cylinder of infinite length, as shown in Figure P15-18. Derive, in closed form, the Green's function for the fields in the space surrounding the cylinder.

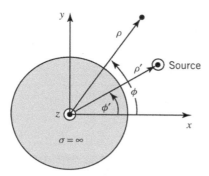

Figure P15-18

15.19. Repeat Problem 15.18 for a magnetic line source of constant current I_m.

15.20. An infinite length electric line source of constant electric current I_e is placed near a two-dimensional conducting wedge of interior angle 2α as shown in Figure P15-20. Derive, in closed form, the Green's function for the fields in the space surrounding the wedge. Compare with the expressions of (11-182a) and (11-182b).

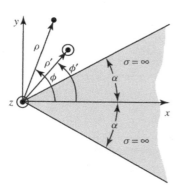

Figure P15-20

15.21. Repeat Problem 15.20 for an infinite length magnetic line source of constant magnetic current I_m. Compare the answers with the expressions of (11-192a) through (11-192b) or (11-193a) through (11-193b).

15.22. A point source placed at x', y', z' is radiating in free space as shown in Figure P15-22.
(a) Derive its Green's function of

$$G = -\frac{1}{4\pi} \frac{e^{-j\beta R}}{R}$$

where R is the radial distance from the point source to the observation point.
(b) By using the integral of (11-28a) or

$$\int_{-\infty}^{+\infty} \frac{e^{-j\beta R}}{R} dz = -j\pi H_0^{(2)}(\beta R)$$

show that the three-dimensional Green's function of the point source reduces to the two-dimensional Green's function of the line source derived in Example 15-5.

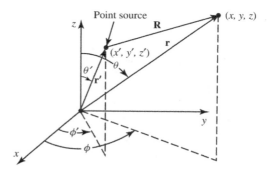

Figure P15-22

CHAPTER 16

Artificial Impedance Surfaces

16.1 INTRODUCTION

Metasurfaces, also referred to as artificial impedance surfaces and/or engineered electromagnetic surfaces, have been developed over the past few decades to alter the impedance boundary conditions of the surface of a structure and, thus, control the radiation characteristics, such as radiation efficiency and pattern, of antenna elements placed at or near them, or the scattering of impinging electromagnetic waves [1–6]. When electromagnetic waves interact with surfaces that exhibit geometrical periodicity, they result in some interesting and exciting characteristics, which typically have numerous applications that have captured the attention and imagination of engineers and scientists. Using a "broad brush" designation, these surfaces can also be referred to as metamaterials, which were discussed in Section 5.7. Since metamaterials may be used to designate double negative (DNG) type materials, there have been other designations of artificial impedance or engineered electromagnetic surfaces. These designations began initially as photonic bandgap (PBG) structures [7, 8], which targeted primarily optics type of structures and frequencies. PBG structures are 1-D, 2-D, and 3-D periodic configurations, both dielectric and conducting, which have the ability to manipulate the electromagnetic radiation so as to not allow it to propagate within certain frequency ranges or bandgaps. Most of the focus of PBG has been devoted to dielectric structures, although the applications have expanded to metallic structures, such as waveguides, resonators, filters, and antennas, which usually exhibit metallic losses that are more dominant at higher frequencies, compared to dielectric losses. The PBG structures are analogous to semiconductor materials that manipulate the electrons to exhibit electronic bandgaps. The PBG designation was expanded to include other type of structures and frequencies, such as electromagnetic bandgap (EBG) structures, frequency-selective surfaces (FSSs), high-impedance surfaces (HISs), artificial magnetic conductors (AMCs), perfect magnetic conductors (PMCs), etc. A comprehensive list of various EBG designations and references, organized by topics, can be found in the appendix of [6]. In [3], the EBG designation was introduced as a broader classification to encompass the others. Artificial impedance surfaces can be used for, but are not limited to:

- Changing the surface impedance
- Controlling the phase of the reflection coefficient

Balanis' Advanced Engineering Electromagnetics, Third Edition. Constantine A. Balanis.
© 2024 John Wiley & Sons, Inc. Published 2024 by John Wiley & Sons, Inc.
Companion Website: www.wiley.com/go/balanis/advancedengineeringelectromagnetics3e

- Manipulating the propagation of surface waves
- Controlling the frequency band (stopband, passband, bandgaps)
- Controlling the edge diffractions, especially of horns and reflectors
- Designing new boundary conditions to control the radiation pattern of small antennas
- Providing detailed control over the scattering properties
- Designing tunable impedance surfaces to be used as:
  a. steerable reflectors
  b. steerable leaky-wave antennas (LWAs)

This is accomplished by altering the surface of a structure, by modifying its geometry and/or adding other layers, so that the surface waves and/or the phase of the reflection coefficient of the modified surface can be controlled. Although the magnitude of the reflection coefficient will also be affected, it is the phase that primarily has the most significant impact. While an ideal PMC surface introduces, through its image, a zero-phase shift in the reflected field, in contrast to a perfect electric conductor (PEC), which presents a 180° phase shift, the reflection phase of an EBG surface can, in general, vary from $-180°$ to $+180°$, which makes the EBG more versatile and unique [9, 10]. This will be demonstrated in Section 16.3.

While, in general, PEC, PMC, and EBG surfaces possess individually attractive characteristics, they also exhibit shortcomings when electromagnetic radiating elements are mounted on such structures, especially when the designs are judged using aerodynamic, stealth, and conformal criteria. For example, when an electric element is mounted vertically on a PEC surface, its image reinforces its radiation and system efficiency; however, its geometry is not low-profile, an undesirable characteristic for aerodynamic, stealth, and conformal designs. However, when the same electric radiating element is placed horizontally on a PEC surface, its radiation efficiency suffers because its image possesses a 180° phase shift and its radiation cancels that of the actual element; however, the design exhibits low-profile characteristics usually desirable for aerodynamic, stealth, and conformal applications. In contrast, when the same electric radiating element is placed horizontally on a PMC surface, its image possesses a 0° phase and reinforces the direct radiation of the actual element, in addition to having low-profile characteristics. The characteristics of vertical and horizontal electric elements placed vertically and horizontally on PEC and PMC surfaces are based on the image theory of Figure 7-2, and they are visually contrasted in Figure 16-1 [11].

While EBG surfaces exhibit similar characteristics as PMCs when radiating elements are mounted on them, they also have the ability to suppress surface waves of low-profile antenna designs, such as microstrip arrays. Surface waves are introduced in microstrip arrays, which primarily travel within the substrate and are instrumental in developing coupling between the array elements. This can limit the beam scanning capabilities of the microstrip arrays; ultimately, surface waves and coupling may even lead to *scan blindness*, discussed in Section 16.10.5.

An EBG surface emulates a nearly PMC surface and suppresses surface waves only over a frequency range; thus, it is usually referred to as a bandgap structure. In general, the frequency range (bandgap) over which an EBG structure operates more efficiently depends upon the application. For example, the frequency range over which a radiating element in the presence of an EBG possesses a good impedance match, which in [5] is referred to as the input-match frequency band, may not be the same frequency range over which a microstrip array suppresses the surface waves, which in [5] is referred to as the surface-wave frequency bandgap. The operational band was introduced to define "the frequency region within which a low-profile wire antenna radiates efficiently, namely, having a good return loss and radiation patterns. The operational band is the overlap of the input-match frequency band and the surface-wave frequency bandgap" [5]. Generally, for EBG structures, the input-match frequency band and the surface-wave frequency bandgap are not necessarily the same. However, for a mushroom type of EBG surface, which will be introduced later, the input-match and surface-wave bands are nearly the same, which results in an overall

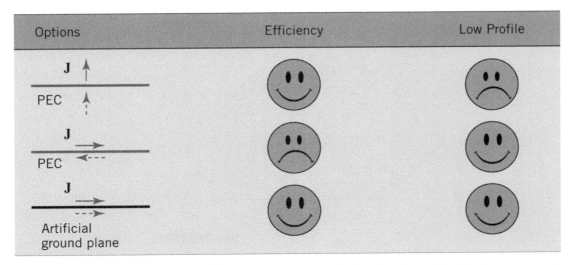

Options	Efficiency	Low Profile

Figure 16-1 Efficiency and conformal characteristics of a vertical and horizontal electric **J** current source at and near PEC and PMC surfaces. (Source: C. A. Balanis, *Antenna Theory: Analysis and Design*, Fourth Edition, John Wiley & Sons, New York, 2016, Reproduced with permission from John Wiley & Sons, Inc.)

operational band, which is a near overlap of the two other bands. It should be stated that for a mushroom type of EBG surface, the frequency band of its surface-wave suppression capability is determined by simulating and/or measuring its insertion loss amplitude, as shown later in Figure 16-7a, or by simulating the dispersion diagram of Figure 16-12. It should also be pointed out that the surface-wave suppression bandwidth of an EBG surface is not necessarily the same as the bandwidth over which the EBG surface behaves as a PMC type of surface. When a plane wave is normally incident upon a surface with a surface impedance Z_s, the $+90°$ to $-90°$ phase variation is also evident when the magnitude of the surface impedance exceeds the free-space intrinsic impedance, η [2]. An EBG surface that does not include the vias does not suppress the surface waves, even though its reflection phase changes between $+180°$ and $-180°$. Better representatives of the surface-wave suppression ability of an EBG surface is the dispersion diagram, which for Example 16-1 is displayed in Figure 16-12.

Now that we have introduced some of the basic definitions of artificial impedance surfaces, we examine some basic structures – corrugations in Section 16.2 and mushroom EBG surfaces in Section 16.3 – that exhibit such characteristics. Application of mushroom EBG surfaces to antenna technology is discussed in Section 16.10. A semiempirical procedure for the design of mushroom EBG surfaces is outlined in Section 16.4. The design leads, in some cases, to rather excellent results when compared to simulations based on a full-wave electromagnetic solver. The limitations of the design are summarized in Section 16.6.

16.2 CORRUGATIONS

There are a number of surfaces that have been developed over the years whose surface impedance has been altered by introducing changes on the surface of the structure. One such surface alteration, which has been in use for many decades, is the introduction of grooves, usually referred to as *corrugations*, with a depth at or near quarter of a wavelength, as shown in Figure 16-2. Since the width of the corrugations, w, is usually equal to or less than about $\lambda/10$ ($w \leq \lambda/10$ and the thickness t is also about 1/10 of the width, or $t \leq w/10 \leq \lambda/100$), each corrugation can

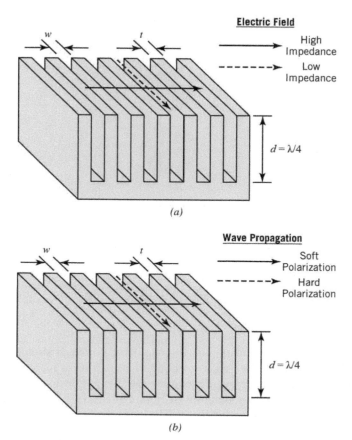

Figure 16-2   Corrugated surface and electric field direction for high- and low-impedance surface, and direction of wave propagation for soft and hard polarization impedance surfaces. (*a*) Electric field directions for high- and low-impedance surfaces. (*b*) Wave velocity for soft and hard polarization impedance surfaces. (Source: D. Sievenpiper, "Artificial impedance surfaces," Chapter 15 in *Modern Antenna Handbook*, C. A. Balanis (Ed.), John Wiley & Sons, pp. 737–777, 2008. Reproduced with permission from John Wiley & Sons, Inc.)

be treated as a shorted transmission line. Since the input impedance of a shorted transmission line of length l is

$$Z_{\text{in}} = jZ_c \tan(\beta l) \qquad (16\text{-}1)$$

then, for a corrugation of depth $d = l = \lambda/4$, the input impedance at the surface is ideally infinity ($Z_{\text{in}} = \infty$). Such a structure is classified as an anisotropic impedance surface since its impedance is high when the polarization of the electric field is perpendicular to the grooves, and it is a low-impedance surface when the electric field is parallel to the grooves, as shown in Figure 16-2*a*.

No matter what the polarization of the wave is, the surface is referred to as "hard" when the wave propagates parallel to the grooves and "soft" when the wave propagates perpendicular to the grooves, as shown in Figure 16-2*b* [2, 12, 13]. This type of designation of the surface is used to match the corresponding boundary conditions from acoustics, and it has been used extensively to design corrugated horns (see Figure 16-3) whose radiation characteristics are controlled by the design of the corrugations [9].

For a pyramidal horn, for example, the corrugations on the upper and lower walls of the horn are introduced to create a surface that nearly nulls the vertical electric field components of the

Figure 16-3   Conical corrugated horn antenna. (Source: C. A. Balanis, *Antenna Theory: Analysis and Design*, Fourth Edition, John Wiley & Sons, New York, 2016. Reproduced with permission from John Wiley & Sons, Inc.)

wave [9]. Therefore, the impedance of the upper and lower walls of the pyramidal horn nearly match those of the side walls, and they can be used to create nearly identical E- and H-plane patterns and nearly rotational by symmetric patterns in all the planes, especially in conical horns such as the one in Figure 16-3. Such an antenna element is widely used in many practical applications, especially as a feed for reflector antennas (dishes). A number of designs of corrugations can be found in [9, 10, 12–20], and they have been used to control the radiation characteristics of the horns, especially to lower the minor lobes, provide better impedance match, minimize the diffractions from the edges at the aperture of the horn, and attempt to synthesize a nearly symmetrical amplitude pattern by equalizing those of the E- and H-planes.

Just like corrugations/grooves have been introduced on PEC surfaces, conducting strips have also been placed circumferentially (T-strips) and axially (L-strips) on the surface of dielectric circular cylinders to create anisotropic boundary conditions and polarization-selective surfaces to control the scattering from cylinders [21]. Electric fields that are perpendicular to the length of the strips are ideally transmitted, while those parallel to the length of the strips are ideally reflected. This procedure creates an impedance surface that can perform as dichroic polarization-sensitive filter for electromagnetic waves.

16.3   ARTIFICIAL MAGNETIC CONDUCTORS, ELECTROMAGNETIC BANDGAP, AND PHOTONIC BANDGAP SURFACES

While perfect electric conductors (PECs) exist in nature, perfect magnetic conductors (PMCs) do not. However, it will be of benefit to fabricate PMCs, even artificially. From the electromagnetic boundary conditions, PEC surfaces are those over which the tangential components of electric fields vanish. Therefore, this precludes placement at, or even near, their surfaces radiating elements such as horizontal electric dipoles, spirals, etc., because their radiated fields over them will be shorted out, or nearly so. This of course is obvious from image theory where the actual source and its image are next to each other but are oriented in opposite directions; this is illustrated in Figure 7-2. Such arrangements even exhibit low radiation efficiency for low heights because of the 180° phase reversal of its image. In fact, when locating a horizontal electric element, such as a horizontal dipole, next to a PEC, it must be placed at a height $h = \lambda/4$ above it in order for the radiation, in a direction normal to the surface, to be maximum. Such an arrangement is usually not desired, especially when the elements are placed on spaceborne platforms, because of aerodynamic considerations. Also, for stealth type of targets, such configurations are quite 'visible' to radar and create a large radar cross-section (RCS) signature. Therefore, it is very beneficial if PMC surfaces can be created, even artificially.

Within recent years, PMC surfaces have been synthesized and fabricated artificially and exhibit PMC-type properties within a frequency range; therefore, these surfaces often are referred to as bandgap or band-limited surfaces. There have been many such surfaces – too numerous to mention here. The reader is referred to the literature, especially [1–8, 11–13, 22–24]. One of the first and most widely utilized PMC surfaces is that shown in Figure 16-4. This surface consists of an array of periodic patches of different shapes, in this case hexagons, placed above a very thin dielectric (which could be air) and connected to the ground plane by posts through vias. The height of the substrate is usually less than a tenth of a wavelength ($h < \lambda/10$). The vias are necessary to suppress surface waves within the substrate.

This structure is also referred to as EBG and PBG. It is a practical form of engineered textured surfaces or metamaterials, discussed in Section 5.7. Because of the directional characteristics of EBG/PBG structures, the integration of antenna elements with such structures can have some unique characteristics [7, 8, 25]. A semiempirical model of the mushroom EBG surface of Figure 16-4 was developed in [1, 2]. The presentation here follows that of [2].

Of the mushroom AMC/EBG/PBG structure shown in Figure 16-4, a unit cell of its structure is displayed in Figure 16-5a. When a wave impinges upon an array of such unit cells, electric fields are created across the gap of the unit cells that can be represented by an effective capacitance C.

Also, such impinging fields create currents that circulate between adjacent unit cells. The effects of these current paths through the neighboring walls or vias can be represented by an equivalent inductance L. Therefore, the equivalent circuit of the unit cell of Figure 16-5a is shown in Figure 16-5b, which consists of a capacitance C in parallel with an inductance L [2].

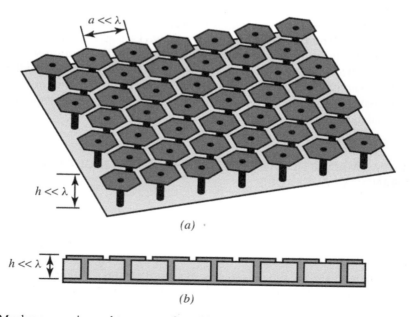

(a)

(b)

Figure 16-4   Mushroom engineered texture surface. (a) Perspective view. (b) Side view. (Source: D. Sievenpiper, "Artificial impedance surfaces," Chapter 15 in *Modern Antenna Handbook*, C. A. Balanis (Ed.), John Wiley & Sons, pp. 737–777, 2008. Reproduced with permission from John Wiley & Sons.)

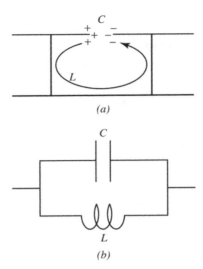

(a)

(b)

Figure 16-5 Unit cell and equivalent circuit of mushroom textured surface. (Source: D. Sievenpiper, "Artificial impedance surfaces," Chapter 15 in *Modern Antenna Handbook*, C. A. Balanis (Ed.), John Wiley & Sons, pp. 737–777, 2008. Reproduced with permission from John Wiley & Sons.)

The surface impedance of the individual unit cell of Figure 16-5 is given by

$$Z_s = j\frac{\omega L}{1 - \omega^2 LC} \tag{16-2}$$

while its resonant frequency is represented by

$$\omega_0 = \frac{1}{\sqrt{LC}} \tag{16-3}$$

However, for design purposes, as it will be shown in Section 16.4, sheet inductance L_s and sheet capacitance C_s must be used to define the resonant frequency. The sheet inductance and capacitance take into account not only the geometry of the individual unit cells but also the geometrical arrangement of the unit cells [2]. It is apparent from (16-2) that the surface of the unit cell is inductive below the resonant frequency, capacitive above the resonant frequency, infinity at resonance, and very high near resonance. Based on (8-158c) of Section 8.7.2, inductive surfaces support transverse magnetic (TM) types of surface waves, while based on (8-169i) of Section 8.7.3, capacitive surfaces support transverse electric (TE) type of surface waves.

The support of either TE or TM surface-wave modes, or both, was verified by measuring the transmission between a pair of coaxial probes placed near the surface of a fabricated 12-cm high-impedance surface reported in [2]. The artificially fabricated PMC surface consisted of a triangular lattice of metallic hexagons placed on the surface of a grounded substrate with a dielectric constant of 2.2, as shown in Figure 16-6. The excitation of the surface waves, TE or TM modes, is controlled by the orientation of the probes. The amplitude of the transmission between the probes is shown in Figure 16-7a while the phase, based on a plane wave normal incidence, is displayed in Figure 16-7b [2]. It is apparent from Figure 16-7a that the fabricated and tested surface exhibits high impedance between approximately 11 and 16 GHz (bandgap), while it supports TM surface waves (inductive surface) below the bandgap and TE surface waves (capacitive surface) above the bandgap. By examining the plane wave incidence phase response of the mushroom surface in Figure 16-7b, it is apparent that the edges of the bandgap occur where the phase varies nearly from $+90°$ to $-90°$, and it is basically zero at resonance. While the transmission amplitude

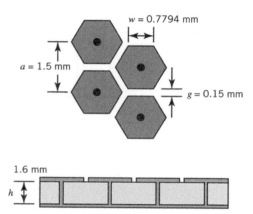

Figure 16-6 Triangular lattice of hexagons built on a grounded substrate with a relative dielectric constant of 2.2. (Source: D. Sievenpiper, "Artificial impedance surfaces," Chapter 15 in *Modern Antenna Handbook*, C. A. Balanis (Ed.), John Wiley & Sons, pp. 737–777, 2008. Reproduced with permission from John Wiley & Sons.)

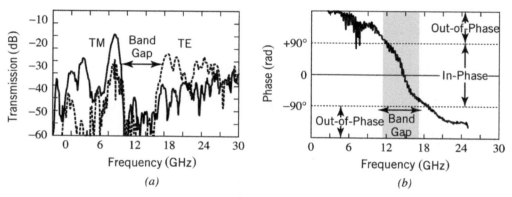

Figure 16-7 Transmission characteristics, amplitude and phase, of mushroom textured surface. (*a*) Amplitude: TM modes (solid), TE (dotted). (*b*) Phase. (Source: D. Sievenpiper, "Artificial impedance surfaces," Chapter 15 in *Modern Antenna Handbook*, C. A. Balanis (Ed.), John Wiley & Sons, pp. 737–777, 2008. Reproduced with permission from John Wiley & Sons.)

and phase characteristics shown in Figures 16-7a and 16-7b are those of a mushroom type of high-impedance surface, they may be different for other type of surfaces, particularly those that may not include vertical vias [2]. A semiempirical design procedure for high-impedance surfaces can be found in [2], and it will be presented here in Section 16.4 that follows.

16.4 DESIGN OF MUSHROOM AMC

Now that the metasurface and its general properties and applications have been introduced, an objective is how to design such a surface based on desired specifications. Given the procedure reported in [1, 2], the design is outlined here. Other design methods can be found in [6].

The design center-radian frequency, ω_0 of (16-3), where the phase of the reflection coefficient is zero and where the AMC surface exhibits PMC characteristics, can be written as

$$\omega_0 = \frac{1}{\sqrt{L_s C_s}}$$

(16-4)

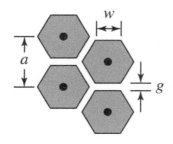

Figure 16-8 Geometry of hexagons lattice built on a substrate of height h. (Source: D. Sievenpiper, "Artificial impedance surfaces," Chapter 15 in *Modern Antenna Handbook*, C. A. Balanis (Ed.), John Wiley & Sons, pp. 737–777, 2008. Reproduced with permission from John Wiley & Sons.)

where L_s and C_s are, respectively, the sheet inductance and capacitance, which take into account not only the geometry of the individual unit cells of Figure 16-5 but also the geometrical arrangement of the unit cells, as shown in Figure 16-8.

The fractional bandwidth BW, over which the phase of the reflection coefficient is between $+90°$ and $-90°$, can be expressed as

$$BW = \frac{\Delta\omega}{\omega_0} = \frac{\sqrt{L_s/C_s}}{\sqrt{\mu_2/\varepsilon_2}} \qquad (16\text{-}5)$$

where ε_2 and μ_2 are, respectively, the permittivity and permeability of the superstrate (upper layer). Specifying the design radian frequency of (16-4) and the fractional bandwidth of (16-5), the sheet inductance L_s and sheet capacitance C_s can be determined.

Equations 16-4 and 16-5 can be solved for L_s and C_s in terms of ω_0 and $\Delta\omega/\omega_0$, that is,

$$L_s = f(\omega_0, \Delta\omega/\omega_0) = \frac{BW}{\omega_0}\sqrt{\mu_2/\varepsilon_2} \qquad (16\text{-}6a)$$

$$C_s = g(\omega_0, \Delta\omega/\omega_0) = \frac{1}{\omega_0 BW \sqrt{\mu_2/\varepsilon_2}} \qquad (16\text{-}6b)$$

The sheet inductance L_s can also be written as

$$L_s = \mu_1 h \qquad (16\text{-}7)$$

Equating (16-6a) to (16-7), it can be shown that the fractional bandwidth can be written in terms of the substrate height h, or

$$BW = \frac{\Delta\omega}{\omega_0} = \left(\frac{\mu_1}{\mu_2}\right)\beta_2 h = \left(\frac{\mu_1}{\mu_2}\right)\omega_0\sqrt{\mu_2\varepsilon_2} \;\Rightarrow\; h = \left(\frac{\mu_2}{\mu_1}\right)\frac{BW}{\omega_0\sqrt{\mu_2\varepsilon_2}} \overset{\mu_1=\mu_2}{=} \frac{BW}{\omega_0\sqrt{\mu_2\varepsilon_2}} \qquad (16\text{-}8)$$

Substituting (16-8) for the height in (16-7), the sheet inductance can be written as

$$L_s = \mu_1 h = \mu_1\left(\frac{\mu_2}{\mu_1}\right)\frac{BW}{\omega_0\sqrt{\mu_2\varepsilon_2}} = \frac{\mu_2}{\sqrt{\mu_2\varepsilon_2}}\left(\frac{BW}{\omega_0}\right) \overset{\mu_1=\mu_2}{=} \eta_2\left(\frac{BW}{\omega_0}\right) \qquad (16\text{-}9)$$

In turn, substituting (16-9) into (16-6b), the sheet capacitance can be expressed as

$$C_s = \frac{1}{(\omega_0)^2 L_s} = \frac{1}{\omega_0 \eta_2 \cdot BW} \tag{16-10}$$

To design each unit cell of Figure 16-8 (i.e., find its dimensions), the capacitance C of each unit cell is related to the sheet capacitance C_s by

$$C_s = C \times F \tag{16-11}$$

where F is a geometrical correction factor given by [1, 2].

Geometry	Geometrical correction factor (F)
Square	1
Triangle	$\sqrt{3}$
Hexagon	$1/\sqrt{3}$

Once the individual capacitance C of each unit cell is determined using (16-10) and the geometrical correction factor of the desired unit geometry, then the dimensions of the unit cell of Figure 16-8, for a two-layer design (superstrate with permittivity ε_2 and substrate with permittivity ε_1), can be determined using [1, 2]

$$C = \frac{w(\varepsilon_1 + \varepsilon_2)}{\pi} \cosh^{-1}\left(\frac{a}{g}\right) \tag{16-12}$$

If the upper layer (superstrate) of the two-layer structure is free space ($\varepsilon_2 = \varepsilon_o$), then (12-12) reduces to

$$C = \frac{w\varepsilon_0(\varepsilon_r + 1)}{\pi} \cosh^{-1}\left(\frac{a}{g}\right) \tag{16-13}$$

where ε_r is the dielectric constant of the substrate. Using Figure 16-8, the designer chooses one of the three dimensions (a, g, and w) and determines the other two using either (16-12) or (16-13) and one of the two following relationships:

$$\text{For square patch:} \qquad a = w + g \tag{16-14a}$$

$$\text{For hexagonal patch:} \quad a = \sqrt{3}w + g \tag{16-14b}$$

According to [1, 2], the thickness of the substrate is usually chosen to be much smaller than the operating wavelength, which reduces the bandwidth since the inductance is related by (16-9) to the thickness and inversely proportional to the capacitance for a constant center frequency. Because the thickness of a two-layer structure is very small, large inductance values cannot be achieved. Thus, low frequencies are achieved by loading the structure with large capacitances. On the other hand, large capacitances usually cannot be achieved by two-layer structures. Therefore, the designer is encouraged to use three-layer structures, with overlapping plates/patches, for low-frequency applications [1, 2]. The trade-off in designs is between thickness and bandwidth. By selecting different geometries and materials with dielectric constants in the range of 2–10, it is possible to obtain capacitances on the order of 0.01–1 pF. With conventional printed circuit fabrication facilities and techniques, minimum gap separations between metallic regions should be around 100–200 μm (microns). The design of a two-layer mushroom textured surface based on a desired resonant frequency is demonstrated with a MATLAB program titled **AMC_Designs** that is included with this chapter.

Example 16-1

Design a two-layer mushroom textured surface with square patches and air as the upper layer ($\varepsilon_2 = \varepsilon_0$, $\mu_2 = \mu_0$), as shown in Figure 16-9, to exhibit PMC characteristics between 10 and 14 GHz with a center frequency of $f_0 = 12$ GHz. Use a Rogers RT/Duroid 5880 with a dielectric constant of $\varepsilon_r = 2.2$. The square patches are supported by metallic circular posts that connect the patches to the bottom ground plane through vias in the substrate.

Solution: Based on the specifications with air as the upper layer and a fractional bandwidth of

$$BW = \frac{\Delta f}{f_0} = \frac{(14-10)10^9}{12 \times 10^9} = \frac{1}{3}, \quad \varepsilon_r = 2.2, \text{ and using (16-8), the height of the substrate is}$$

$$h = \frac{BW}{\omega_0 \sqrt{\mu_2 \varepsilon_2}} = \frac{BW}{\omega_0 \sqrt{\mu_0 \varepsilon_0}} = \frac{3 \times 10^8}{3(2\pi \times 12 \times 10^9)} = 1.3263 \times 10^{-3} = 1.3263 \text{ mm}$$

The sheet inductance and sheet capacitance based, respectively, on (16-9) and (16-10) are

$$L_s = \mu_1 h = \mu_0 h = 4\pi \times 10^{-7}(1.3263 \times 10^{-3}) = 16.6668 \times 10^{-10} = 1.66668 \text{ nH}$$

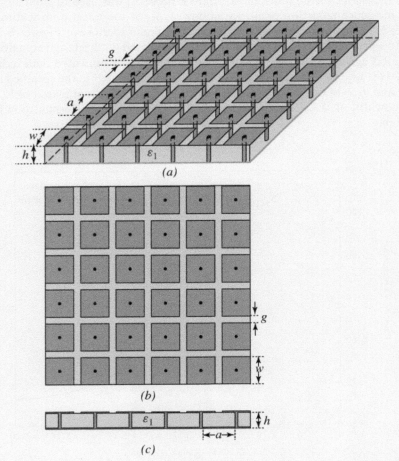

Figure 16-9 Geometry of PMC textured surface of square patches. (*a*) Perspective view. (*b*) Top view. (*c*) Side view.

$$C_s = \frac{1}{\omega_0 \eta_2 \cdot BW} = \frac{1}{\omega_0 \eta_0 \cdot BW} = \frac{3}{2\pi \times 12 \times 10^9 (377)}$$
$$= 0.10554 \times 10^{-2} = 0.10554 \text{ pF}$$

Since we are using square patches, the geometrical factor F is unity and, according to (16-11), the capacitance of each unit cell C is equal to the sheet capacitance C_s, that is, $C = C_s$.

A suitable value for the gap spacing g for integrated circuit technology between patches is $100\,\mu m$ or $g = 100\ \mu m$. Using the previously obtained values and the geometrical relation of (16-14a), $a = w + g$ for a square patch, the nonlinear design of (16-13)

$$C = \frac{w\varepsilon_0(\varepsilon_r + 1)}{\pi} \cosh^{-1}\left(\frac{a}{g}\right) = \frac{w\varepsilon_0(2.2 + 1)}{\pi} \cosh^{-1}\left(\frac{w + 0.1 \times 10^{-3}}{0.1 \times 10^{-3}}\right)$$

can be solved, using a nonlinear solver, for w of the PMC surface. Doing this leads to

$$w = 2.85 \text{ mm, and then } a = w + g = 2.95 \text{ mm}$$

To verify the performance of the mushroom PMC surface based on the specified and obtained geometrical dimensions, the commercial software HFSS [26] was used to simulate it. The plane wave normal incidence reflection phase variations of S_{11} of the mushroom textured surface of square patches of Figure 16-9 between $+90°$ and $-90°$, similar to those of Figure 16-7b, are shown in Figure 16-10 where they are compared with the results based on the design equations of Section 16.4; a very good agreement is indicated between the two. The simulated data indicate a bandwidth of 3.9 GHz ($f_0 = 10.35$ GHz and $f_h = 14.25$ GHz), compared to the specified one of 4 GHz ($f_l = 10$ GHz and $f_h = 14$ GHz), a center frequency of 12.15 GHz (compared to 12 GHz), and a fractional bandwidth of 0.321 (compared to 0.333). Overall, the performance indicates a very favorable design.

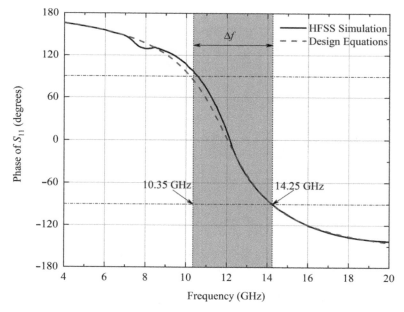

Figure 16-10   Phase of reflection coefficient S_{11} of PMC textured surface with square patches simulated using HFSS and design equations.

16.5 SURFACE-WAVE DISPERSION CHARACTERISTICS

The plane wave normal incidence reflection phase characteristics of the mushroom surface of Figure 16-9, as displayed in Figure 16-10, can be simulated by using the geometrical arrangement illustrated in Figure 16-11a. Since the structure is periodic, the problem can be solved by considering only a unit cell and by assigning the proper boundary conditions. To emulate periodicity based on the polarization of the plane wave incidence indicated in Figure 16-11a, with the electric field **E** parallel to the *yz* plane and the magnetic field **H** parallel to the *xy* plane, PEC boundary conditions are assigned on the front and rear walls of the unit cell, and PMC boundary conditions are assigned on the left and right walls of the unit cell. The assignment of the boundary conditions should be based on the polarization of the incident electric (**E**) and magnetic (**H**) fields. The images of **E** and **H** fields should be in the same direction as the actual field.

In addition to their unique plane wave normal incidence reflection phase characteristics, mushroom surfaces generate forbidden bandgaps within which the propagation of surface waves is suppressed. These forbidden bands can be observed via dispersion diagrams or directly by the amplitude variation of surface waves. A dispersion diagram is a plot that displays the relation between the frequency of the propagating modes and the wave vector. The dispersion diagram of a mushroom surface can be obtained by computing the allowable frequencies of different modes for certain values of the amplitude of the wave vector. This can be achieved by computing the eigenmodes of the periodic structure.

To understand the numerical approach, let us consider the vector wave equation for the electric field $\nabla^2 \mathbf{E} = -\omega^2 \mu\varepsilon \mathbf{E}$ as a starting point, where $-\omega^2\mu\varepsilon$ are the eigenvalues and the solutions to the wave equation are the corresponding eigenvectors. For the case of an infinite medium, the eigenvalues, as well as the ω's, have a continuous spectrum on the complex plane. On the other hand, in the case of a periodic structure, particularly a mushroom surface and because of the translational symmetry of the geometry, the solutions to the wave equation can be written in the form $\mathbf{E} = \mathbf{E}_n e^{j\phi}$, where $\mathbf{E}_n$ is a periodic function, which is referred to as *Bloch mode* [27]. The periodic

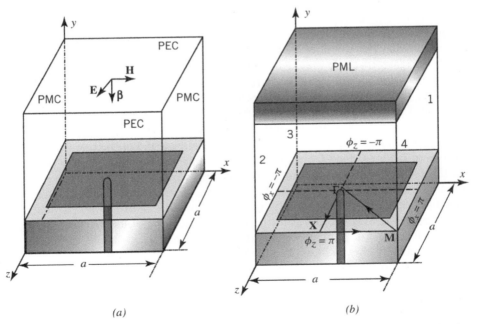

(a)   (b)

Figure 16-11   Unit cells for reflection phase and dispersion diagram simulations. (*a*) Reflection phase. (*b*) Dispersion diagram.

function $\mathbf{E}_n$ is a solution to another problem referred to as the *reduced Hermitian eigenproblem* [28]. Again, due to the periodicity of the structure, the wave vectors in the x and z directions can be written, respectively, as $\beta_x^n = \beta_x + m\left(\dfrac{2\pi}{a}\right)$ and $\beta_z^m = \beta_z + n\left(\dfrac{2\pi}{a}\right)$. The wave vectors can also be obtained by the spatial harmonic expansion of $\mathbf{E}$ [29]. Since $\mathbf{E}_n$ is periodic, we can solve the eigenproblem only over the finite domain referred to as the *Brillouin zone*, in which β_x and β_z take values between $-\pi/a$ and $+\pi/a$. As a consequence of the finite nature of this problem, the eigenvalues would generate discrete frequency spectrum bands. Furthermore, due to the other types of symmetries including rotation, mirror reflection, and inversion, the domain of the problem can be reduced into a smaller one that is called the *irreducible Brillouin zone* [27]. Figure 16-11*b* shows the Brillouin and irreducible Brillouin zones of a unit cell of a mushroom EBG structure. Surface 1 represents the rear wall, surface 2 the front wall, surface 3 the left wall, and surface 4 the right wall of the unit cell. The unit cell is truncated in the y direction by a PML design. To be able to plot the band structure in a regular two-dimensional (2-D) diagram, it is sufficient to consider only the extrema of the frequency bands. This can be obtained by solving the problem on the boundary of the irreducible Brillouin zone.

For numerical computation of the eigenvalues, surfaces 1 (rear) and 2 (front) of Figure 16-11*b* should be connected to each other through the relation $\mathbf{E}_2 = \mathbf{E}_1 e^{j\phi_z}$. Similarly, surfaces 3 (left) and 4 (right) can be related by $\mathbf{E}_4 = \mathbf{E}_3 e^{j\phi_x}$. This is accomplished by assigning Bloch boundary conditions [sometimes referred to as linked boundary conditions (LBCs)] to each pair. In contrast to the PEC or PMC boundary conditions, LBC supports both tangential and perpendicular components of the electric ($\mathbf{E}$) and magnetic ($\mathbf{H}$) fields, over the surface where they are defined. For Figure 16-11*b*, ϕ_x and ϕ_z should take values between $-\pi$ and π, which is a direct consequence of the periodicity of the structure, within the Brillouin zone. Since the wave vector is also related to the phase of the fields, the inputs of the numerical simulation can be these phase terms. In the path from $\boldsymbol{\Gamma}$ to $\mathbf{X}$, ϕ_z should be varied from 0 to π, while ϕ_x is maintained constant at 0. Similarly, in the path from $\mathbf{X}$ to $\mathbf{M}$, ϕ_x should be varied from 0 to π, while ϕ_z is kept constant at π. Finally, from $\mathbf{M}$ back to $\boldsymbol{\Gamma}$ both ϕ_x and ϕ_z should be varied from 0 to π, simultaneously. Using the HFSS software for simulations, surface 1 is referred to as the *master* and surface 2 as the *slave* for the z variations. Similarly, for the x variations, surface 3 is referred to as the *master* and surface 4 as the *slave*.

The dispersion diagram, obtained by HFSS [26] simulation for the geometry of Example 16-1, is illustrated in Figure 16-12. It is evident that the bandgap for the surface-wave suppression is approximately between 9.5 and 13.5 GHz, which nearly matches that based on the phase diagram of Figure 16-10. That is, the bandwidth over which the mushroom EBG structure of Figure 16-9 and Example 16-1 behaves nearly as a PMC, which is based on the phase diagram of a plane wave normal incidence (see Figure 16-10), is nearly the same as the surface-wave suppression bandwidth indicated in the dispersion diagram of Figure 16-12. The lower end of the bandgap of the surface-wave suppression is determined from the dispersion diagram of Figure 16-12, when the group velocity of the TM surface waves becomes zero. The upper bound of the bandgap is the frequency at which the TE mode crosses the light line because the surface waves can propagate when their group velocity is smaller than the velocity of light (slow waves). However, particularly for this example, the upper bound is selected as the point where the group velocity of the TE mode significantly deviates from that of light. Since the attenuation constant of (8-179) will be very small up to this point, practically, there will not be any surface-wave propagation.

To create dispersion diagrams, such as the one of Figure 16-12 and similar ones, which shed insight into the dispersion characteristics of a metasurface, two programs have been developed and are included in the **Multimedia** of Section 16.13. The first one, entitled **Dispersion_Diagram_Matlab**, is a MATLAB program that generates a dispersion diagram based on the analytical expressions of (16-2), (16-9), and (16-13), and on the formulation reported in [1]. The second dispersion diagram, entitled **Dispersion_Diagram_HFSS**, is based on Figure 16-11 using the Floquet

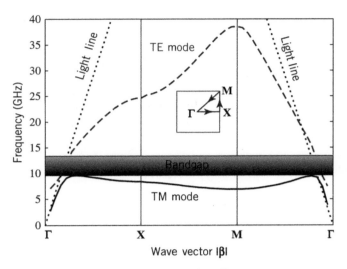

Figure 16-12 Dispersion diagram.

port simulation in Ansys HFSS. As part of the package of this program, two files, *Unit Cell Design* and *Graphical User Interface (GUI)*, are included to aid in the generation of the dispersion diagram.

Another evidence of the EBG structure bandgap characteristics is to model and simulate the amplitude transmission of the TM and TE modes, as was accomplished in Figure 16-7 through measurements.

It is worth pointing out that the dispersion diagram is not limited to metasurface structures, but it can be utilized to graphically describe the propagation constant for any bounded or unbounded medium. Figures 8-16 and 8-17 illustrate the dispersion behavior in a partially filled rectangular waveguide. Furthermore, the dispersion diagram is plotted in Figure 8-23 for a dielectric slab bounded above and below by air. Similarly, the dispersion diagram for a dielectric-covered ground plane can be calculated utilizing the characteristic equations 8-178a and 8-178*b* for TM and TE modes, respectively.

The characteristic/transcendental equation can be formulated based on the solution of the wave equation or by applying the transverse resonance method (TRM). Since the objective here is to find the propagation constants, the formulation based on the TRM is much simpler to apply. Section 8.6 outlined the procedure to calculate the transcendental equation for a partially filled waveguide utilizing the TRM. Furthermore, an approximate model using the TRM was developed in [30] to calculate the dispersion diagram for a metasurface in absence of vias. To illustrate the process, Example 16-2 follows for a dielectric-covered PEC [31].

Example 16-2

Using the TRM, determine for a PEC ground plane covered with a dielectric slab:

1. Derive the characteristic equation, for the TM mode, to calculate the propagation constants when the dielectric slab thickness is h. To simplify the analysis, assume an infinite flat PEC on the xy plane, no field variations in the x direction ($\frac{\partial}{\partial x} = 0$), and the waves traveling in the y direction.

2. Plot the dispersion diagram for a dielectric slab of thickness $h = 5.08$ mm and dielectric constant $\varepsilon_r = 2.2$.

‌

Solution:

1. The dielectric-covered PEC can be represented as a shorted transmission line displayed in Figure 16-13a. The input impedance looking downward (Z_{in}^-) can be approximated as a TM wave impedance ($Z_{wd}^{TM^z}$) of a semi-infinite shorted dielectric, while the input impedance looking upward (Z_{in}^+) is the TM free-space wave impedance ($Z_{w0}^{TM^z}$) as shown in Figure 16-13b. Hence, the characteristic equation for the grounded dielectric can be written as

$$Z_{in}^+ = -Z_{in}^- \tag{16-15}$$

$$\frac{\beta_{z0}}{\omega\varepsilon_0} = -j\left(\frac{\beta_{zd}}{\omega\varepsilon_d}\right)\tan(h\beta_{zd}) \tag{16-16}$$

$$\frac{\sqrt{\beta_0^2 - \beta_y^2}}{\omega\varepsilon_0} = -j\frac{\sqrt{\beta_d^2 - \beta_y^2}}{\omega\varepsilon_d}\tan\left(h\sqrt{\beta_d^2 - \beta_y^2}\right). \tag{16-17}$$

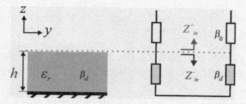

Figure 16-13 Geometry of a grounded dielectric, with its equivalent transmission line model. (Source: M. Alharbi, "High-directive printed antennas for low-profile applications," Ph.D. dissertation, School of Electrical, Computer, and Energy Engineering, Arizona State University, 2020.)

2. The solution of (16-17) is numerically evaluated and the data are compared to those obtained by HFSS, as presented in Figure 16-14. An excellent agreement is indicated between the two sets of data.

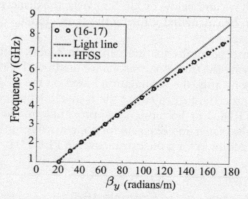

Figure 16-14 Dispersion diagram of a dielectric covered PEC. (Source: M. Alharbi, "High-directive printed antennas for low-profile applications," Ph.D. dissertation, School of Electrical, Computer, and Energy Engineering, Arizona State University, 2020.)

Two MATLAB programs, for the calculation and development of a dispersion diagram, are provided in this chapter. The first one, entitled **TRM_Metasurface_Dispersion**, is based on the model developed in [30] to calculate the dispersion of a metasurface, and the second one, entitled **TRM_Grounded_Dielectric_Dispersion**, is based on (16-17) and plots the dispersion diagram of a dielectric-covered PEC. For details, see their description in the **Multimedia** of Section 16.13.

16.6 LIMITATIONS OF THE DESIGN

The design procedure, which has been outlined and demonstrated by Example 16-1, is limited by the bandwidth that is independent of all of the physical dimensions of the structure except the thickness of the substrate. Therefore, special care must be exercised to design the physical geometry of the mushroom type surface for a desired center frequency and bandwidth. Indeed, for a fixed substrate thickness, we do not have simultaneous control of both the center frequency and bandwidth. The design is based primarily on three parameters: *substrate thickness, center frequency*, and *bandwidth*. Once two of these three parameters are specified, we have no control of the third one. In addition to this, if the thickness of the substrate is fixed, the entire frequency range cannot be covered by using reasonable patch dimensions. Furthermore, the design equations, outlined in the design procedure, are valid only if the wavelength within the substrate is much larger than the dimensions of the unit cell. Hence, if very large patches are used for the design to cover a large frequency range, the design equations will lead to less accurate designs.

To overcome some of the limitations mentioned above, there are other methods that can be used but are more complex. One method is the so-called *dynamic model* [32]. In this method, a different expression is used for the surface capacitance while the expression for the inductance is the same as in [1, 2]. In this technique, the capacitance is expressed in terms of an infinite summation without assuming that the dimensions of the unit cell are much smaller than the wavelength. This model can also take into account the influence of the higher-order modes generated within the high-impedance surface. If the dimensions of the unit cell are much smaller than the wavelength, the dynamic model reduces to a simpler expression for the surface capacitance, which leads to similar results as [1, 2]. However, because of its complexity and space limitations, it will not be presented here. The reader is directed to [32] for the details.

16.7 APPLICATIONS OF AMCs

Given that perfect magnetic conductors do not exist in nature, the significance of these AMCs becomes even more pivotal in some applications. Image theory indicates that a perfect magnetic conductor (PMC) surface would be an efficient candidate in the applications where the radiating element is very close to the ground plane, unlike a perfect electric conductor (PEC) surface whose radiation efficiency is very poor [5, 33]. Thus, in recent years, AMCs have been considered a major breakthrough in antenna and electromagnetics engineering. This chapter reviews some of the applications of AMCs, particularly:

- *RCS reduction using checkerboard artificial impedance surfaces*: Conventional methods of radar cross section (RCS) reduction are primarily based on two mechanisms:
  - Absorption of the incoming waves using bulky radar-absorbing material (RAM).
  - Reduction of the scattered waves by altering the physical geometry of the original structure; such geometries and analytical techniques are discussed in Chapter 14.

The functionality of the former mechanism is limited to a narrow bandwidth, and it requires bulky lossy material; the latter mechanism is not aerodynamically efficient. However, the introduction of AMCs not only eliminates some of the limitations of the conventional RCS reduction techniques but also improves their RCS reduction performance. For example, in [34], AMC-based thin RAMs are presented, which eliminate their bulky nature. Later, in [35], the scattered waves are redirected using AMC-based tailored metasurfaces without altering the physical geometry of the original structure. However, such metasurfaces reduce the RCS only in a narrow frequency band. Recent developments [36–41] lead to ultrabroadband RCS reduction.

- *Curvilinear AMCs for low-profile superdirective loop and spiral antennas*: AMCs are judicious ground planes to be used in low-profile applications. These structures are also utilized for miniaturization and bandwidth enhancement [42–46]. Although most of the literature focuses on rectangular AMCs, curved and circularly symmetric ground planes are addressed in [47–53]; however, they mostly concentrate on the electromagnetic bandgap properties rather than on reflection characteristics, which take place at different frequency bands. Loop and spiral elements are placed above curvilinear-designed AMCs to illustrate the superior bandwidth and gain of such ground planes, compared to their rectangular counterparts [54–56]. It should also be noted that when AMCs are used as antenna ground planes, they are usually referred to as high-impedance surfaces (HISs).
- *High-gain leaky-wave antennas using holography*: Holographic artificial impedance surfaces (HAISs) are leaky-wave antennas that are designed by applying the holographic principle to AMC-based surfaces [57–61]. Along with achieving high-gain pencil beam formation, the unidirectional beam scanning ability of the leaky waves grants HAISs the advantage over conventional arrays. The surface-wave propagation control using electrically small patches can be further exploited and extended to form two-dimensional surface waveguides [62].

16.8 RCS REDUCTION USING CHECKERBOARD METASURFACES

16.8.1 Introduction

Since the radar cross section (RCS) is a critical parameter in characterizing a target's scattering properties, there has been an increasing interest in the development of RCS reduction techniques utilizing metasurfaces. Conventionally, the RCS of radar targets can be reduced primarily by two main techniques: coating the target with *radar-absorbing materials* (RAMs) and shaping the target geometry. RAMs reduce the RCS by absorbing the incident power by converting the electromagnetic energy into thermal energy (heat) that is dissipated by the lossy material. A classic example of a RAM is the Salisbury screen [63], which is realized by covering the target with a lossy resistive sheet placed a quarter-wavelength above the dielectric ground plane; ideally, the total absorption of the incident wave is achieved by impedance matching. However, Salisbury screens have a narrow bandwidth, and they add an undesired thickness to the target. The thickness of a Salisbury screen can be reduced by placing it above a magnetic surface [34, 64]. However, such designs are sensitive to the angle of incidence and operate over narrow frequency bandwidth. The second technique, namely target geometrical shaping, reduces the RCS by redirecting the scattered fields away from the radar receiver. However, this often conflicts with the aerodynamic requirements of the structure/scatterer.

An alternative technique that reduces the RCS of a target by redirecting the waves away from the radar receiver, without geometrical reshaping of the target, is by placing an array of patches of perfect magnetic conductors (PMCs) and perfect electric conductors (PECs) on the same plane to achieve scattering cancellation between the fields scattered from each of the two surfaces.

16.8.2   Plane Wave Scattering by PEC-PMC Hybrid Surfaces

To demonstrate how an array of PEC and PMC patches on the same ground plane can reduce the RCS of a radar target, let us initially consider scattering by two different rectangular flat plates, which are pure PEC (see Figure 16-15a) and pure PMC (see Figure 16-15b) [65]. To illustrate the process and reduce the details, we assume that the incident wave is a TE^x polarized uniform plane wave on the yz plane, as shown in Figure 16–15a. The associated incident electric and magnetic fields can be written as

$$\mathbf{E}^i = \eta H_0(\hat{\mathbf{a}}_y \cos\theta_i + \hat{\mathbf{a}}_z \sin\theta_i) e^{-j\beta(y\sin\theta_i - z\sin\theta_i)} \tag{16-18a}$$

$$\mathbf{H}^i = \hat{\mathbf{a}}_x H_0 \, e^{-j\beta(y\sin\theta_i - z\sin\theta_i)} \tag{16-18b}$$

where H_0 is a constant which represents the amplitude of the incident magnetic field. The overall finite dimensions of the plates are the same for both cases. The scattered fields, and thus the associated RCS, are obtained using Physical Optics (PO), as detailed in Section 11.3.2 of Chapter 11.

The first scattering structure is a PEC rectangular flat surface of dimensions a and b, as shown in Figure 16-15a. The closed-form bistatic RCS, based on PO, is given by (11-44a) and repeated here as

$$\sigma_{\text{3-D}} = 4\pi \left(\frac{ab}{\lambda}\right)^2 \cos^2\theta_s \left|\frac{\sin\left[\frac{\beta b}{2}(\sin\theta_s \mp \sin\theta_i)\right]}{\frac{\beta b}{2}(\sin\theta_s \mp \sin\theta_i)}\right|^2 \quad \begin{array}{l} - \text{ for } \phi_s = \dfrac{\pi}{2}, \quad 0 \leqslant \theta_s \leqslant \pi/2 \\[2mm] + \text{ for } \phi_s = \dfrac{3\pi}{2}, \quad 0 \leqslant \theta_s \leqslant \pi/2 \end{array} \tag{16-19}$$

The details are outlined in Section 11.3.2.

The associated plots for $a = b = 5\lambda$ and $\phi_i = 270°$ ($\phi_s = 90°, 270°$ with $0 \leqslant \theta_s \leqslant 90°$) are shown by the solid line ($-$) curves in Figure 16-15c for $\theta_i = 0°$ and Figure 16-15d for $\theta_i = 15°$. It is clear that the bistatic RCS pattern of the PEC plate follows *Snell's law of reflection* (see Section 5.3) where its maximum, for electrically large large plates, is near the specular direction (i.e., $\theta_i = \theta_s = 0°$ for Figure 16-15c and $\theta_i = \theta_s = 15°$ for Figure 16-15d).

Now, let us repeat the procedure but for a rectangular and flat pure PMC ground plane ($\Gamma = 1$), as shown in Figure 16-15b. The closed-form principal plane bistatic RCS can be obtained by following the same steps outlined in Section 11.3.2, and the radar cross section can then be written as

$$\sigma_{\text{3-D}} = 4\pi \left(\frac{ab}{\lambda}\right)^2 \cos^2\theta_i \left|\frac{\sin\left[\frac{\beta b}{2}(\sin\theta_s \mp \sin\theta_i)\right]}{\frac{\beta b}{2}(\sin\theta_s \mp \sin\theta_i)}\right|^2 \quad \begin{array}{l} - \text{ for } \phi_s = \dfrac{\pi}{2}, \quad 0 \leqslant \theta_s \leqslant \pi/2 \\[2mm] + \text{ for } \phi_s = \dfrac{3\pi}{2}, \quad 0 \leqslant \theta_s \leqslant \pi/2 \end{array} \tag{16-20}$$

The associated plots for $a = b = 5\lambda$ and $\phi_i = 270°$ ($\phi_s = 90°, 270°$ with $0 \leqslant \theta_s \leqslant 90°$) are shown by the dashed line (- - -) curves in Figure 16-15c for $\theta_i = 0°$ and Figure 16-15d for $\theta_i = 15°$. The maximum RCS of the PMC ground plane is exactly along the specular direction following Snell's law of reflection (i.e., $\theta_i = \theta_s = 0°$ for Figure 16-15c and $\theta_i = \theta_s = 15°$ for Figure 16-15d).

From (16-19), it is apparent that the maximum RCS occurs in the principal plane (yz plane) approximately when ($\theta_s = \theta_i$) for large values of $b(b \gg \lambda)$. When b is large, the values of the $\sin(z)/z$ function of (16-19) vary more rapidly compared to the $\cos\theta_s$ function, such that $\cos\theta_s$

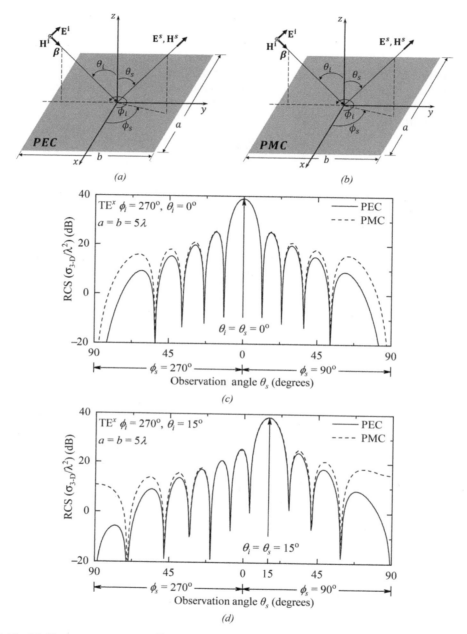

Figure 16-15 Uniform plane wave (TE^x) incident on a rectangular conducting plate (a) PEC and (b) PMC. Principal plane (yz plane) bistatic RCS ($\phi_i = 270°$; $\phi_s = 90°, 270°$) for (c) $\theta_i = 0°$ and (d) $\theta_i = 15°$.

is essentially constant at and near the maximum of $\sin(z)/z$. However, for the smaller values of b, the exact maximum of (16-19) can be found iteratively. Accordingly, the maximum bistatic RCS of a flat PEC of TE^x polarization does not occur exactly at the specular direction, but it approaches that value as the dimensions of the plate become large compared to the wavelength. However, from (16-20), since $\cos^2 \theta_i$ is a function of the incident angle θ_i and not the scattering angle θ_s, the maximum value of (16-20) occurs when $\theta_s = \theta_i$ no matter what the size of the scattering plate.

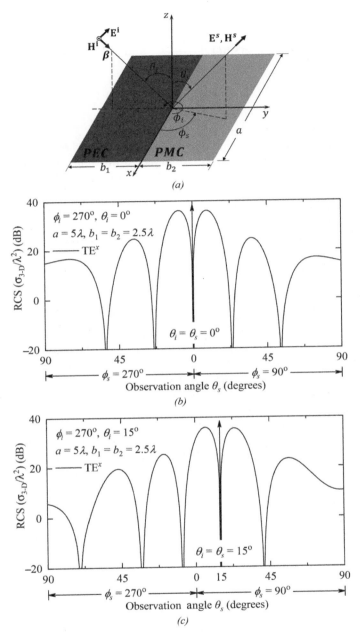

Figure 16-16 (*a*) Uniform plane wave incident on a rectangular hybrid conducting plate (PEC/PMC). And its principal plane (*yz* plane) bistatic RCS ($\phi_i = 270°, \phi_s = 90°, 270°$) for (*b*) $\theta_i = 0°$ and (*c*) $\theta_i = 15°$.

Now let us consider a hybrid scattering plate, combining both surfaces (i.e., PEC and PMC) on the same scattering plate, as shown in Figure 16-16a, and by following the steps detailed in Section 11.3, the closed-form bistatic RCS of the hybrid ground plane combination in the principal plane (*yz* plane) can be written as [65]

$$\sigma_{3\text{-D}} = 4\pi \left(\frac{a}{\lambda}\right)^2 |b_2 \cos\theta_i [M_2 - jN_2] - b_1 \cos\theta_s [M_1 + jN_1]|^2 \qquad (16\text{-}21a)$$

where

$$M_1 = \frac{\sin(2Y_1)}{2Y_1}, \quad N_1 = \sin(Y_1)\frac{\sin(Y_1)}{Y_1} \quad \text{and} \quad Y_1 = \frac{\beta b_1}{2}(\sin\theta_s \mp \sin\theta_i)$$

$$M_2 = \frac{\sin(2Y_2)}{2Y_2}, \quad N_2 = \sin(Y_2)\frac{\sin(Y_2)}{Y_2} \quad \text{and} \quad Y_2 = \frac{\beta b_2}{2}(\sin\theta_s \mp \sin\theta_i)$$

(16-21b)

with $-$ for $\phi_s = 90°, 0 \leqslant \theta_s \leqslant 90°$, and $+$ for $\phi_s = 270°, 0 \leqslant \theta_s \leqslant 90°$.

Plots of (16-21a) for a hybrid scattering plate with of $a = 5\lambda$ and $b_1 = b_2 = 2.5\lambda$, having overall dimensions equal to the previous two cases, are depicted in Figure 16-16b for $\theta_i = 0°, \phi_i = 270°$ and Figure 16-16c for $\theta_i = 15°, \phi_i = 270°$ ($\phi_s = 90°, 270°$ with $0 \leqslant \theta_s \leqslant 90°$). Due to the 180° phase difference between the reflection coefficients of the PEC ($\Gamma = -1$) and the PMC ($\Gamma = 1$) parts of the hybrid scattering plate, a destructive interference occurs along the specular direction; $\theta_s = 0°$ for Figure 16-16b and $\theta_s = 15°$ for Figure 16-16c. This destructive interference has led to the concept of RCS reduction using hybrid scattering surfaces where two different AMCs are arranged in a checkerboard architecture to create destructive interference between the fields scattered by the hybrid scattering plate.

16.8.3 Fundamentals of Conventional Checkerboard Metasurfaces

The concept of hybrid PEC/PMC scattering plates can be extended in two orthogonal directions and introduces the architecture of conventional checkerboard metasurfaces illustrated in Figure 16-17a; PMC scattering was synthesized by AMC-type metasurface unit cells. This leads to metasurfaces that consist of PEC and AMC patches configured in checkerboard pattern, as shown in Figure 16-17b. The primary appealing feature of this technique is its ability to provide RCS reduction over a broadband frequency range [36–38].

When a plane wave is incident, at normal incidence, on a PEC surface, as demonstrated in Figure 16-15a, the major lobe of the scattered fields is along the normal direction, as illustrated by the three-dimensional RCS pattern of Figure 16-18a. However, when the checkerboard surface of Figure 16-17b is illuminated by a broadside incident plane wave, the fields scattered from each of the AMC supercells are 180° out of phase and result in destructive interferences along two orthogonal planes. This forces the scattered waves to be redirected, for $d_x = d_y = 2.5\lambda$, along four quadrants with four major lobes along $\theta_0 = 12°$; $\phi_0 = 45°, 135°, 225°$, and 315°, as exhibited in Figure 16-18b. Figure 16-19b shows the two-dimensional RCS patterns along the principal plane (yz plane; $\phi_s = 90°, 270°$) and also along the two planes where the major lobes are directed (i.e., $\phi_s = 45°, 225°$

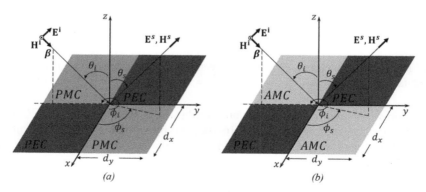

(a)   (b)

Figure 16-17 Comparison between (a) generic design of checkerboard surface of PEC/PMC and (b) conventional checkerboard surface of PEC/AMC.

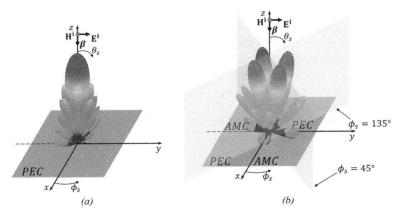

Figure 16-18 Simulated 3-D bistatic RCS patterns for (*a*) PEC and (*b*) conventional checkerboard surface with the same physical dimensions [66].

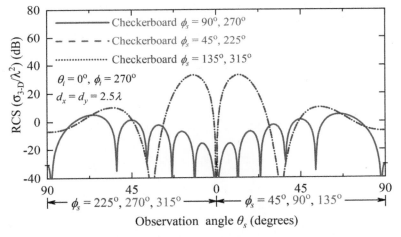

Figure 16-19 Bistatic RCS patterns of a plane wave incident on conventional checkerboard surface (see Figure 16-18*b*) along the principal plane (*yz* plane) $\phi_s = 90°, 270°$, $\phi_s = 45°, 225°$ and $\phi_s = 135°, 315°$ [66].

and $\phi_s = 135°, 315°$). The direction of all the maxima of the major lobes (θ_0, ϕ_0) can be determined approximately using array theory, particularly using [9]

$$\tan \phi_0 = \frac{\sin \theta_i \sin \phi_i \pm \dfrac{\delta_y}{\beta d_y}}{\sin \theta_i \cos \phi_i \pm \dfrac{\delta_x}{\beta d_x}} \tag{16-22}$$

$$\sin^2 \theta_0 = \left[\sin \theta_i \sin \phi_i \pm \frac{\delta_y}{\beta d_y} \right]^2 + \left[\sin \theta_i \cos \phi_i \pm \frac{\delta_x}{\beta d_x} \right]^2 \tag{16-23}$$

where (θ_0, ϕ_0) defines the angular direction of the major lobes, while (θ_i, ϕ_i) represents the angle of incidence; δ_x and δ_y represent the phase difference of the reflection coefficients of the adjacent supercells in the x and y directions, respectively, while d_x and d_y are the center-to-center separation distances between the adjacent supercells in the x and y directions, respectively. A MATLAB program titled **Angular_Direction_Checkerboard,** utilizing (16-22) and (16-23), determines the angular directions of the scattered fields and is included at the end of this chapter.

The four major lobes formed by the scattering from the checkerboard surface of Figure 16-18b are attributed to the fields scattered by each patch having 180° phase difference between them; $\Gamma = -1$ (PEC) and $\Gamma = +1$ (PMC). This can also be verified by examining the electric current density induced on the surface of the checkerboard structure. Based on boundary conditions (see Section 1.7.2) and the incident fields of (16-18), the tangential components of the magnetic field vanish on the PMC surface and double on the PEC, such that the electric current density induced on the PEC patches of the plate is represented by

$$\mathbf{J}_s|_{\text{PMC}(z=0)} \approx 0 \tag{16-24a}$$

$$\mathbf{J}_s|_{\text{PEC}(z=0)} \approx 2\hat{\mathbf{n}} \times \mathbf{H}^{\text{total}} = 2\hat{\mathbf{a}}_z \times \hat{\mathbf{a}}_x H_x = \hat{\mathbf{a}}_y \, 2H_0 e^{-j\beta y' \sin\theta_i} \tag{16-24b}$$

To verify the performance of the conventional checkerboard surfaces (see Figure 16-17b) with $d_x = d_y = 2.5\lambda$ and $\delta_x = \delta_y = 180°$, the induced electric current density $\mathbf{J}_s$ was simulated by the commercial software HFSS [26], and it is exhibited in Figure 16-20. It can be seen that the electric current density basically exists only on the PEC patches while that on the PMC patches is of very low intensity. This induced current density creates four major lobes in the RCS pattern where the angular directions of their maxima can be determined approximately using array theory [i.e., (16-22) and (16-23)]. Thus, the angular directions of the major lobes are $\theta_0 = 16.43°$ and $\phi_0 = 45°, 135°, 225°$, and 315°. However, by using the finite-element method solution of HFSS [26], the angular locations of the four maxima are directed along $\theta_0 = 12°$ and $\phi_0 = 45°, 135°, 225°$, and 315°.

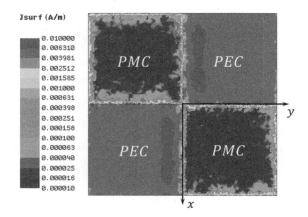

Figure 16-20   Induced surface current density on a conventional checkerboard surface (see Figure 16-18b) with $d_x = d_y = 2.5\lambda$.

Example 16-3

Determine the angular elevation direction (θ_0) of the four major lobes when a plane wave incident normally ($\theta_i = \phi_i = 0°$) on a checkerboard surface (Figure 16-17b, $\delta_x = \delta_y = 180°$) when the spacing/size of both PEC and PMC surfaces is uniform ($d_x = d_y$) and equal to

   a. $d_x = d_y = 2\lambda$

   b. $d_x = d_y = 3\lambda$

   c. $d_x = d_y = 5\lambda$

Explain the dependency of the angular directions of the major lobes on the size of the scattering surfaces.

Solution: The approximate elevation angle (θ_0) of the maxima of the four formed scattering lobes can be found based on the array theory of (16-23) and is tabulated below for each case:

$d_x = d_y$	θ_0 based on (16-23)
2λ	$20.70°$
3λ	$13.63°$
5λ	$8.13°$

It is observed that as the size of the PEC and PMC cells of the hybrid scattering ground plane increases, the major lobes' angular direction approaches broadside ($\theta_0 = 0°$). This is expected because as $d_x = d_y \Rightarrow$ large and $\theta_i = \phi_i = 0°$, (16-23) reduces to $\sin^2 \theta_0 \approx 0 \Rightarrow \sin \theta_0 \approx 0 \Rightarrow \theta_0 \approx 0°$.

16.8.4 Broadband RCS Reduction Metasurfaces

A. Conventional Broadband Checkerboard Metasurfaces   As illustrated in the previous section, when PEC and PMC patches are placed in a checkerboard pattern on a rectangular flat plate, the scattered fields are redirected and form four main beams, and the direction of each of the main beams can be predicted by antenna array theory [9]. One can also employ PO for more accurate predictions as recently reported in [67, 68]. The other key parameter is the RCS reduction bandwidth of these metasurfaces. Since most of the radar systems operate over a large frequency range, it will be desirable to design and realize RCS reduction metasurfaces with large frequency bandwidth. As reported in [36, 37], for checkerboard metasurfaces, this can be achieved by configuring two AMC supercells with each other in checkerboard-type architectures, as illustrated in Figure 16-17b. The primary appealing feature of this technique is its ability to provide RCS reduction over a broadband frequency range [36–38].

For such RCS reduction checkerboard metasurfaces, an analytical expression that approximates the 10-dB RCS reduction in its simplest form is given by [36, 37]

$$\text{RCS}_{\text{red}} = \left| \frac{A_1 e^{j\Phi_1} + A_2 e^{j\Phi_2}}{2} \right|^2 \tag{16-25}$$

where Φ_1 and Φ_2 are the phases of the fields reflected by AMC-1 and AMC-2, respectively, while $A1$ and $A2$ are their respective reflection coefficient amplitudes [36]. Therefore, to reduce the RCS by at least 10 dB, a phase difference of about $(180 \pm 37)°$ must be obeyed between the AMC supercells [36]. The RCS reduction versus the phase difference between the two AMCs patches is illustrated in Figure 16-21. Designing a conventional checkerboard surface for RCS reduction utilizing AMC surfaces of Section 16.8.4 and (16-25) is demonstrated with a MATLAB program entitled **Conventional_Checkerboard_RCS_Reduction** included at the end of this chapter.

As a result, the selection of the AMCs is a critical step in realizing a broadband RCS reduction metasurface. In [36], two single-band AMCs were considered for designing a wideband checkerboard surface (see Figure 16-22), for which a nearly 60% bandwidth of 10-dB RCS reduction was reported (see Figure 16-23). The induced electric surface current density is illustrated in Figure 16-24 at 4.65 GHz when the phase difference between the selected AMCs is 180°. Such a checkerboard architecture is classified as *conventional* design [37]. Similarly, in [38, 39], two dual-band AMCs were utilized to reduce the RCS in two separate wide frequency bands, for which about 60% and 26% 10-dB RCS reduction bandwidths were achieved.

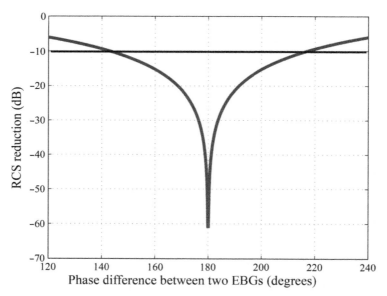

Figure 16-21 RCS reduction using (16-25) versus the phase difference between two AMC supercells. (Source: W. Chen, C. A. Balanis, and C. R. Birtcher, "Checkerboard EBG surfaces for wideband radar cross section reduction," *IEEE Trans. Antennas Propagat.*, vol. 63, no. 3, pp. 2636–2645, 2015. Reproduced with permission from IEEE.)

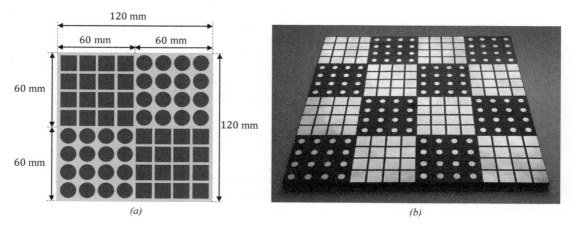

Figure 16-22 (*a*) Design and (*b*) fabricated prototypes of conventional checkerboard surface for wideband RCS reduction. (Source: W. Chen, C. A. Balanis, and C. R. Birtcher, "Checkerboard EBG surfaces for wideband radar cross section reduction," *IEEE Trans. Antennas Propagat.*, vol. 63, no. 3, pp. 2636–2645, 2015. Reproduced with permission from IEEE.)

Once the size of the AMC unit cells is selected, the number of unit cells in each individual AMC supercell defines the distances (d_x, d_y) between adjacent supercells. According to (16-22) and (16-23), since the distances between adjacent supercells are not critical for the RCS reduction under normal incidence, the number of unit cells in an individual AMC supercell is not as pivotal. However, according to (16-25), these distances still determine the angular direction of the lobe maxima (θ_0, ϕ_0) of the scattered fields [37].

B. Judicious Selection of AMC Unit Cells A judicious selection of AMCs was investigated [37, 40], and a guideline for the selection of AMCs was presented to synthesize ultrabroadband

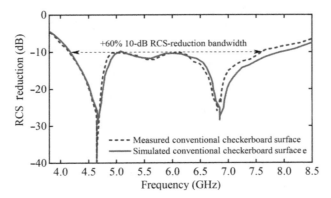

Figure 16-23 Simulated and measured RCS reduction using conventional checkerboard surface. (Source: W. Chen, C. A. Balanis, and C. R. Birtcher, "Checkerboard EBG surfaces for wideband radar cross section reduction," *IEEE Trans. Antennas Propagat.*, vol. 63, no. 3, pp. 2636–2645, 2015. Reproduced with permission from IEEE.)

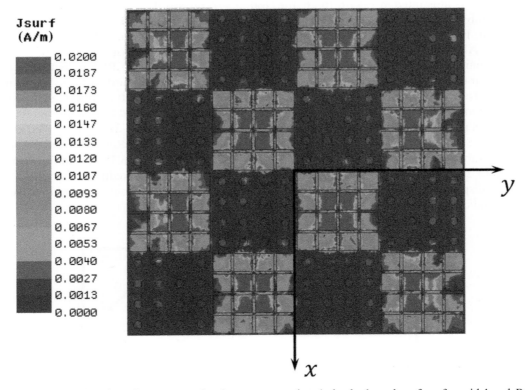

Figure 16-24 Induced surface current density on conventional checkerboard surface for wideband RCS reduction of Figure 16-22*b* at f = 4.65 GHz.

RCS reduction checkerboard surfaces. It was concluded that a combination of single- and dual-band AMCs, instead of two single-band AMCs, provides broader RCS reduction bandwidth. Such checkerboard surfaces with a combination of single- and dual-band AMCs are called *blended* designs [37]. For validation, the conventional checkerboard surface reported in [36] was transformed to a blended design, with a combination of single- and dual-band AMCs [37, 40].

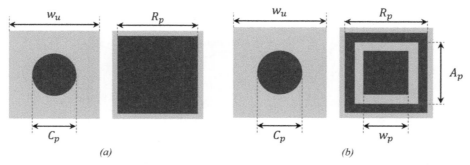

Figure 16-25   Selection of AMC unit cell designs to convert a conventional checkerboard surface into a blended checkerboard surface: (*a*) AMC-1 and (*b*) AMC-2. (Source: Adapted from A. Y. Modi, C. A. Balanis, C. R. Birtcher, and H. Shaman, "Novel design of ultrabroadband radar cross section reduction surfaces using artificial magnetic conductors," *IEEE Trans. Antennas Propagat.*, vol. 65, no. 10, pp. 5406–5417, 2017.)

The single-band AMC (AMC-1) design consisting of single circular and square patches [36], and shown in Figure 16-25*a*, was converted to a dual-band AMC (AMC-2), as shown in Figure 16-25*b*, by introducing an outer ring to the single-band square AMC patch; the circular patch is not altered. As illustrated in Figure 16-26, the $(180 \pm 37)°$ phase difference of the single band is satisfied only from 3.6 to 6.5 GHz, which results in about 60% fractional 10-dB RCS reduction bandwidth. However, after the transformation of AMC-1 from single-band AMC (AMC-1) to dual-band AMC (AMC-2), as illustrated in Figure 16-25, the required phase difference of $(180 \pm 37)°$ was maintained from 3.6 to 9.4 GHz, an increase of +23% in 10-dB RCS reduction bandwidth (from 60% to 83%). This suggests that the conventional checkerboard surface should first be designed and optimized with two single-band AMCs. After its optimization, to extend the RCS reduction bandwidth, one of the single-band AMCs should be transformed to a dual-band AMC.

The RCS reduction of the blended checkerboard surface (see Figure 16-27*a*) was verified using a prototype displayed in Figure 16-27*b*, which was fabricated on a standard Rogers RT/duroid-5880 dielectric substrate backed by a PEC ground plane (with thickness of 6.35 mm and dielectric constant of 2.2) [37]. The simulated and measured RCS data, with a 10-dB reduction bandwidth of 83%, are displayed in Figure 16-27*c*; an excellent agreement between simulations and measurements is indicated [37].

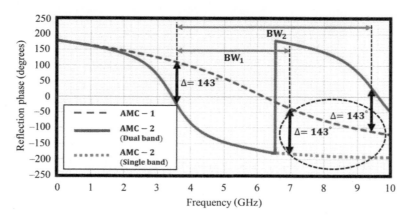

Figure 16-26   Typical reflection phase for AMC-1 and AMC-2 as a function of frequency. (Source: A. Y. Modi, C. A. Balanis, C. R. Birtcher, and H. Shaman, "Novel design of ultrabroadband radar cross section reduction surfaces using artificial magnetic conductors," *IEEE Trans. Antennas Propagat.*, vol. 65, no. 10, pp. 5406–5417, 2017. Reproduced with permission from IEEE.)

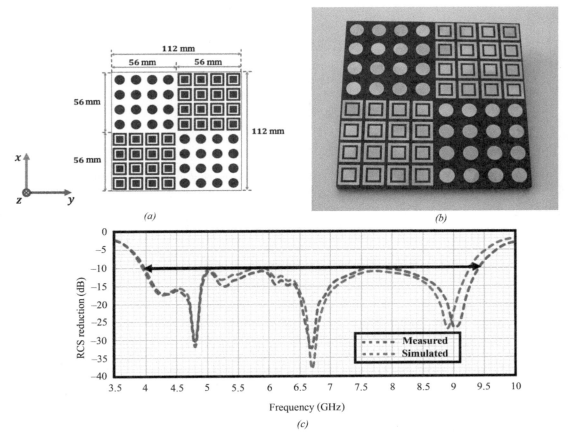

Figure 16-27 Blended checkerboard surface: (*a*) Design and (*b*) fabricated prototypes. (*c*) Measured and simulated RCS reduction results. (Source: A. Y. Modi, C. A. Balanis, C. R. Birtcher, and H. Shaman, "Novel design of ultrabroadband radar cross section reduction surfaces using artificial magnetic conductors," *IEEE Trans. Antennas Propagat.*, vol. 65, no. 10, pp. 5406–5417, 2017. Reproduced with permission from IEEE.)

However, in both the conventional and blended designs, the checkerboard surfaces have the same design architecture. In each of them, only the selection of the building supercells is different to achieve the required destructive interference (i.e., two different single-band AMCs in conventional; a combination of single- and dual-band AMCs in the blended). Therefore, in both of these checkerboard metasurfaces, to obtain greater than 10-dB RCS reduction, the $(180 \pm 37)°$ phase difference still must be obeyed between the building supercells. This is a primary limitation of such checkerboard designs.

Example 16-4

Design a checkerboard metasurface by synthesizing an AMC unit cell such that it exhibits zero reflection phase at 8 GHz. Use a Rogers RT/duroid 5880 with a dielectric constant of $\varepsilon_r = 2.2$ and substrate thickness of $h = 2$ mm. Plot the 10-dB RCS reduction bandwidth of the synthesized checkerboard metasurface comprised of PEC (metallic surface) and AMC surface. Repeat the same exercise for substrates with thicknesses of $h = 2$ mm and $h = 4$ mm. Explain the relationship between the thickness of the substrate and the 10-dB RCS reduction bandwidth. *Use the MATLAB computer program* ***Conventional_Checkerboard_RCS_Reduction*** *to verify the analytical design.*

Solution: Specifying a resonant frequency of $f_0 = 8$ GHz and a substrate thickness $h = 2$ mm, the sheet inductance and sheet capacitance based, respectively, on (16-9) and (16-10) are

$$L_s = \mu_1 h = \mu_0 h = 4\pi \times 10^{-7}(2.0 \times 10^{-3})$$

$$= 2.5133 \times 10^{-9} = 2.5133 \text{ nH}$$

$$C_s = \frac{1}{(\omega_0)^2 L_s} = \frac{1}{(2\pi \times 8 \times 10^9)^2 \times 2.5133 \times 10^{-9}}$$

$$= 1.5748 \times 10^{-13} = 0.15748 \text{ pF}.$$

Then, the dimensions of the of AMC surface (i.e., gap spacing g, patch width w, and unit cell size a) can be determined by choosing an initial gap spacing g of 1mm and determining the patch width w from (16-13), or

$$C = \frac{w\varepsilon_0(\varepsilon_r + 1)}{\pi} \cosh^{-1}\left(\frac{a}{g}\right) = \frac{w\varepsilon_0(2.2 + 1)}{\pi} \cosh^{-1}\left(\frac{w + 1 \times 10^{-3}}{1 \times 10^{-3}}\right)$$

Using a nonlinear solver to the above equation leads to a patch width of $w = 6.5$mm and $a = w + g = 7.5$ mm.

For the same design requirements and with a substrate thickness of $h = 4$ mm and $h = 6$ mm, the dimensions of the of AMC surface can be determined following the same procedure, and they are:

Substrate thickness h	Gap spacing g	Patch width w	Unit cell size a
(mm)	(mm)	(mm)	(mm)
2.0	1.0	6.5	7.5
4.0	1.5	4.3	5.8
6.0	2.0	3.47	5.47

Figure 16-28 illustrates the phase of the reflection coefficient S_{11} of the designed metasurfaces for $h = 2, 4$, and 6 mm.

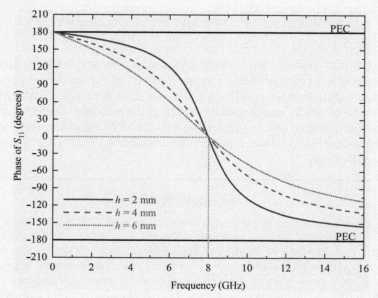

Figure 16-28 Phase versus frequency of reflection coefficient S_{11} of AMC surface with square patches of the three different substrate thicknesses.

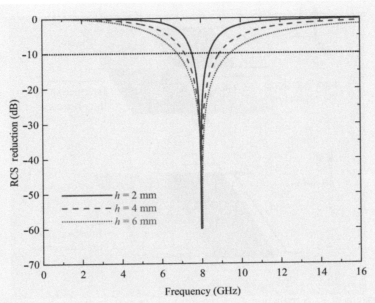

Figure 16-29   RCS reduction versus frequency of the designed AMCs using (16-25).

The corresponding 10-dB RCS reductions of the designed AMCs, when implemented with PEC ground plane in a checkerboard patterned design, are illustrated in Figure 16-29. It is clear that as the substrate thickness increases, the 10-dB RCS reduction bandwidth increases. This can be explained by examining the bandwidth on the designed AMCs. As the substrate thickness increases, the sheet inductance L_s of (16-9) increases and, thus, expands the bandwidth of the AMC surface as represented by $S_{11} = \pm 90°$. Therefore, slower variance in the reflection phase over the frequency range translates into broader 10-dB RCS reduction bandwidth.

C. Modified Checkerboard Surfaces   To overcome the limitation of maintaining the required phase difference of $(180 \pm 37)°$, the design architecture of these surfaces can be modified [41]. As exhibited in Figure 16-30a, applicable for both *conventional* and *blended* checkerboard surfaces, the destructive interference is not necessary in both the orthogonal planes. To provide the same RCS reduction under normal incidence, only one of these two destructive interferences, along the x and y directions, is required [37]. Consequently, the surface of Figure 16-30b, where destructive interference is developed only along the x direction, is equivalent to the original conventional/blended checkerboard surface of Figure 16-30a for reducing the RCS under normal incidence [37]. As depicted in Figure 16-31a, this then allows the introduction of another type of destructive interference. To further increase the RCS reduction bandwidth, the second combination of AMCs (AMC-3 and AMC-4) should be selected such that the destructive interference produced by them reduces the RCS by more than 10 dB outside the 10-dB RCS reduction bandwidth of the first combination (AMC-1 and AMC-2) [41]. In addition, the frequency ranges of the two combinations should overlap each other over the majority of their respective frequency ranges. The RCS reduction bandwidth can be enhanced even further by altering the respective portions of the total area covered by the individual combination of AMCs, as suggested in Figure 16-31b.

In [37], the same technique was implemented on a blended checkerboard surface with a combination of single- and dual-band AMCs, as illustrated in Figure 16-32a, to convert it from a *blended* design of equal size supercells into a *modified* checkerboard surface of unequal size

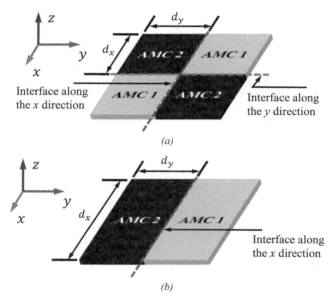

Figure 16-30   Designs of RCS reduction. (*a*) Conventional/blended checkerboard surface. (*b*) Surface with only one destructive interference. (Source: A. Y. Modi, "Metasurface-based techniques for broadband radar cross-section reduction of complex structures," Ph.D. dissertation, School of Electrical, Computer, and Energy Engineering, Arizona State University, 2020.)

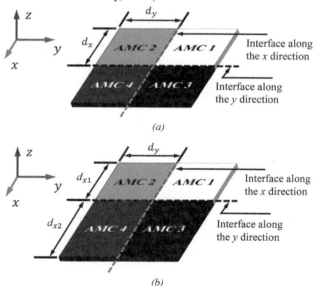

Figure 16-31   Designs of RCS reduction. (*a*) Modified surface with equal sized supercells. (*b*) Modified surface with unequal sized supercells. (Source: A. Y. Modi, "Metasurface-based techniques for broadband radar cross-section reduction of complex structures," Ph.D. dissertation, School of Electrical, Computer, and Energy Engineering, Arizona State University, 2020.)

supercells shown in Figure 16-32*d*. To understand the rationale and evolution process, the design is broken down into four (4) steps leading from Figure 16-32*a* to Figure 16-32*d*, and the design steps are illustrated in Figure 16-32. To increase the bandwidth, the second combination of AMCs (AMC-3 and AMC-4) with orthogonal destructive interference, shown in Figure 16-32*c*, was added to the first combination of AMCs (AMC-1 and AMC-2) of Figure 16-32*b*. When implemented as an independent blended checkerboard surface, the second combination of AMCs

(AMC-3 and AMC-4) attains an RCS reduction of more than 27 dB near 9.7 GHz. Therefore, in the final design of Figure 16-32d, the area covered by the second combination was increased. As a result of the two AMC combinations, the RCS reduction bandwidth was extended beyond 9.3 GHz, which was the limit of the first AMC combination (AMC-1 and AMC-2). Due the (180 ± 37)° phase difference limitation, the 10-dB RCS reduction bandwidth of the first combination of AMS (AMC-1 and AMC-2), of Figure 16-32a, is 83%. However, after employing the proposed design technique using both AMC combinations, (AMC-1 + AMC-2) and (AMC-3 + AMC-4) with two different orthogonal destructive interferences and unequal areas between them, of Figure 16-32d, the −10-dB RCS reduction fractional bandwidth was increased from 83% to 91%, an 8% enhancement from the design of Figure 16-32a to that of Figure 16-32d. A fabricated prototype is shown in Figure 16-33. To validate the design process, a prototype was fabricated, simulated, and measured; the data are illustrated and compared in Figure 16-34 [69] for both TM and TE polarizations; a very good agreement is indicated.

D. Generalized Approach to Synthesize Ultrabroadband RCS Reduction Checkerboard Surfaces

In summary, the steps of a generalized approach to synthesize an ultrabroadband RCS reduction checkerboard surface are outlined in what follows:

1. Initially design and optimize two single-band AMC surfaces to achieve maximum 10-dB RCS reduction bandwidth, with a combination of two single-band AMC structures, as shown in Figure 16-25a.
2. Alter one of these single-band AMCs to a dual-band AMC (see Figure 16-25b) to convert a *conventional* checkerboard surface into *blended* design (see Figure 16-32a) to improve the RCS reduction bandwidth. Then, convert the design of Figure 16-32a to Figure 16-32b, since both give identical 10-dB RCS reduction.

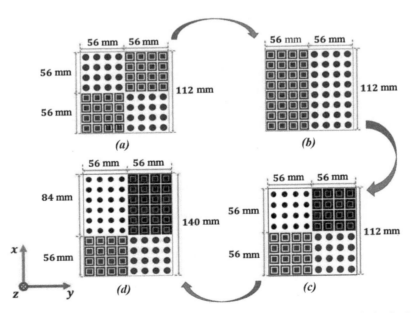

Figure 16-32 Evolution of modified checkerboard surfaces. Designs of (a) blended checkerboard surface, (b) surface with only one destructive interference, (c) modified surface with equal sized supercells, and (d) modified surface with unequal-sized supercells. (Source: A. Y. Modi, C. A. Balanis, and C. R. Birtcher, "Novel technique for enhancing RCS reduction bandwidth of checkerboard surfaces," in *Proc. IEEE Int. Symp. Antennas Propagat.*, San Diego, CA, pp. 1911–1912, 2017. Reproduced with permission from IEEE.)

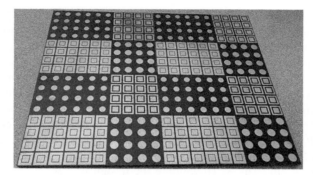

Figure 16-33   Fabricated prototype of modified checkerboard surface. (Source: A. Y. Modi, C. A. Balanis, C. R. Birtcher, and H. Shaman, "Novel design of ultrabroadband radar cross section reduction surfaces using artificial magnetic conductors," *IEEE Trans. Antennas Propagat.*, vol. 65, no. 10, pp. 5406–5417, 2017. Reproduced with permission from IEEE.)

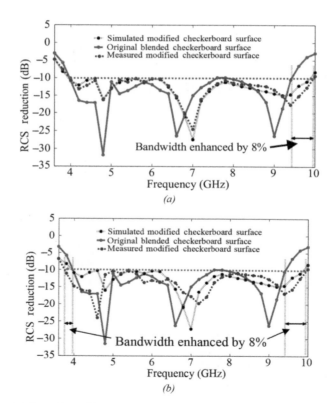

Figure 16-34   RCS reduction of modified checkerboard surface under normal wave incidence with (*a*) TM and (*b*) TE polarizations. (Source: C. A. Balanis, M. A. Amiri, A. Y. Modi, S. Pandi, and C. R. Birtcher, "Applications of AMC-based impedance surfaces," *EPJ Applied Metamaterials*, vol. 5, no. 3, pp. 1–15, 2018.)

3. Design and introduce another *blended* checkerboard surface with the second combination of AMCs (i.e., AMC-3 and AMC-4) having orthogonal destructive interference, as shown in Figure 16-32*c*, where the selection criterion for the second combination was established in the previous subsection.

4. Finally, to further extend the RCS reduction bandwidth, alter the respective areas covered by the two individual AMC combinations of these *blended* checkerboard surfaces to develop a *modified* checkerboard surface (see Figure 16-32d).

16.8.5   Broadband RCS Reduction of Complex Targets

As we have discussed in Section 16.8.4, the simplicity of the checkerboard concept makes it easy to modify the designs to achieve ultrabroadband RCS reduction bandwidth. For the same reason, it also makes it easier to implement it on complex structures to reduce their RCS. For example, the concept of checkerboard metasurfaces can be generalized and utilized to reduce the radar signature of antennas [70]. Further, such metasurfaces have also been implemented and investigated on curved structures, like cylindrical geometries as well as corner reflectors [68, 71]. In the subsections that follow, each of these complex structures is reviewed.

A. Antenna and Antenna Array   Antennas are primary contributors to high RCS signatures of radar targets. Consequently, the synthesis of low-RCS antennas has been of paramount interest. The checkerboard metasurface concept was generalized to synthesize low-RCS antennas [69, 72, 73]. However, it is important that modifications made to reduce the RCS of an antenna should not impact its radiation performance. In Figure 16-35, the method is implemented to reduce the RCS of a high-gain microstrip-patch antenna array of 10 elements in a (5 × 2) arrangement. As illustrated in Figure 16-35, the appealing feature of the proposed method is that it enables both in-band and out-of-band RCS reductions, without disturbing the radiation performance of the original antenna/antenna array [70, 73].

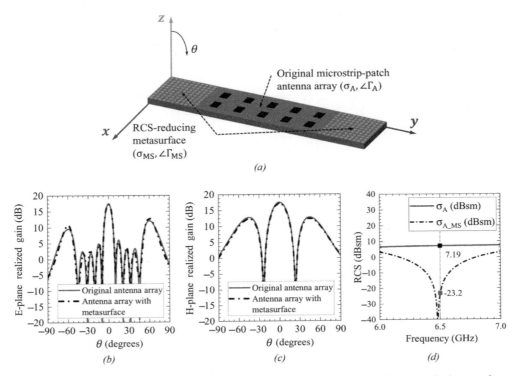

Figure 16-35   (*a*) Geometry of the final antenna array with metasurface, (*b*) its radiation performance comparison in E-plane, (*c*) its radiation performance comparison in H-plane, and (*d*) its RCS reduction performance. (Source: A. Y. Modi, C. A. Balanis, C. R. Birtcher, and H. N. Shaman, "Robust method for synthesizing low-RCS high-gain antennas using metasurfaces," in *Proc. of IEEE Int. Symp. Antennas Propagat.*, Atlanta, GA, pp. 627–628, 2019. Reproduced with permission from IEEE.)

B. Cylindrical Structures   Most of our previous discussions focused on planar metasurfaces. However, for practical applications, these metasurfaces have to perform on conformal structures. Therefore, checkerboard metasurfaces were investigated on flexible curvilinear structures [71]. To examine such structures, a prototype checkerboard metasurface was built on a flexible substrate. The prototype was then bent and mounted on the structure with two different radii of curvature. The metasurface was evaluated under normal incidence for both polarizations (i.e., HH and VV).

In [71], a flat checkerboard surface (see Figure 16-36a) was designed to obtain an RCS reduction of more than 5 dB over a bandwidth of 25% (as illustrated in Figure 16-36b). As summarized in Table 16-1, the same 5-dB RCS reduction bandwidth of 25% was approximately maintained (for both HH and VV polarizations) for the cylindrically curved checkerboard metasurface of Figure 16-36b with two orthogonal radii of curvature. It was observed, based on the basic laws of GO for curved surfaces, that the separation angle between the two adjacent innermost lobes on either side of the xy plane is inversely proportional to the radius of curvature.

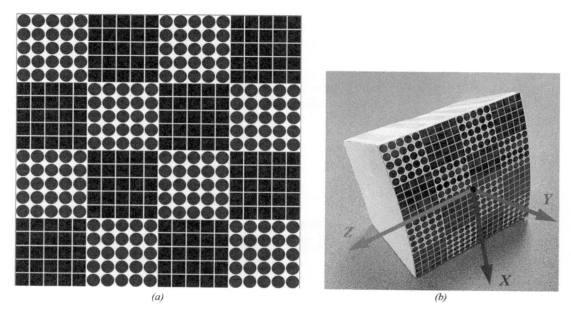

(a)   *(b)*

Figure 16-36   (a) Design of flat checkerboard surface. (b) Fabricated prototype of curved checkerboard architecture mounted on cylindrical structure. (Source: W. Chen, C. A. Balanis, C. R. Birtcher, and A. Y. Modi, "Cylindrically curved checkerboard surfaces for radar cross-section reduction," *IEEE Antennas Wireless Propagat. Lett.*, vol. 17, no. 2, pp. 343–346, Feb. 2018. Reproduced with permission from IEEE.)

TABLE 16-1   Summary of RCS reduction bandwidth and separation angle between the center two lobes

Radius of curvature		10λ	5λ
HH polarization	5-dB RCS reduction BW	22.2%	25%
	Separation angle	40°	52°
VV polarization	5-dB RCS reduction BW	23.6%	22.2%
	Separation angle	40°	52°

(Source: W. Chen, C. A. Balanis, C. R. Birtcher, and A. Y. Modi, "Cylindrically curved checkerboard surfaces for radar cross-section reduction," *IEEE Antennas Wireless Propagat. Lett.*, vol. 17, no. 2, pp. 343–346, Feb. 2018.)

C. Corner Reflectors   Since corners are building blocks of complex targets, their scattering behavior was investigated [68, 74]. Further, it is extremely important to reduce the RCS of corner reflectors since they possess large RCS [68]. While RCS reduction methods, like checkerboard metasurfaces, can successfully reduce the RCS over a broad frequency range when applied to planar or convex-curved cylindrical surfaces, they fail to reduce the RCS of complex targets like corner reflectors. This is demonstrated for the 90° corner reflector of Figure 16-37a with simulated results illustrated in Figure 16-38a using the shooting and bouncing rays (SBR) method [26]. For complex structures like corner reflectors, the incident fields undergo multiple bounces [68], and these methods require additional treatment to address unique challenges presented by multiple bounces [68]. It is demonstrated that these additional challenges introduced by the multiple-bounce mechanism can successfully be addressed using orthogonality principles and array theory, as shown in Figure 16-37b [68] and illustrated in Figure 16-38b using SBR-based simulated RCS data.

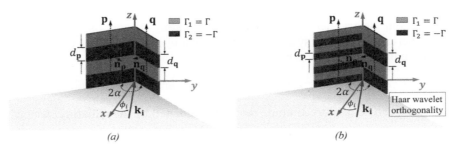

(a)   (b)

Figure 16-37   (a) Implementation of the one-dimensional checkerboard arrangement with the metasurfaces oriented perpendicularly to the plane of incidence (Configuration-2). (b) Checkerboard covered dihedral corner reflector with one-dimensional Haar wavelet orthogonality, in addition to the metasurfaces oriented perpendicularly to the plane of incidence (Configuration-4). (Source: A. Y. Modi, M. A. Alyahya, C. A. Balanis, and C. R. Birtcher, "Metasurface-based method for broadband RCS reduction of dihedral corner reflectors with multiple bounces," *IEEE Trans. Antennas Propagat.*, vol. 68, no. 3, pp. 1436–1447, Mar. 2020. Reproduced with permission from IEEE.)

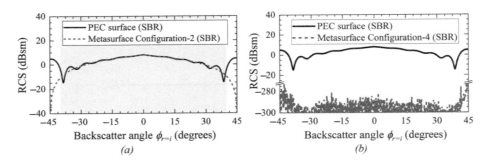

(a)   (b)

Figure 16-38   Comparison between the SBR-predicted RCSs of the target covered with PEC (solid line), and with the synthesized metasurfaces (dashed line). (a) Configuration-2. (b) Configuration-4. (Source: A. Y. Modi, M. A. Alyahya, C. A. Balanis, and C. R. Birtcher, "Metasurface-based method for broadband RCS reduction of dihedral corner reflectors with multiple bounces," *IEEE Trans. Antennas Propagat.*, vol. 68, no. 3, pp. 1436–1447, Mar. 2020. Reproduced with permission from IEEE.)

16.9 ANTENNA FUNDAMENTAL PARAMETERS AND FIGURES-OF-MERIT

Antenna performance is described in terms of fundamental parameters and figures-of-merit. These are many and are described in detail in Chapter 2 of [9]. Four of these fundamental parameters (figures-of-merit), which will be used in this section to judge the performance of metasurface-based radiating elements, are: *radiation resistance, maximum directivity, maximum gain,* and *maximum realized gain.* To illustrate some of them, we display in Figure 16-39 an equivalent circuit of a radiating element (antenna) connected as a load to a generator through a transmission line. The antenna (load) is represented by impedance Z_A

$$Z_A = (R_r + R_L) + X_A \tag{16-26}$$

where R_r = radiation resistance through which the antenna radiates

R_L = antenna loss resistance, representing the losses of the antenna

X_A = antenna reactance

Once the far-zone electric field components of a radiating element (antenna) are determined, which in the spherical coordinate system are represented by E_θ and E_ϕ, we can compute the normalized radiation intensity $U_n(\theta,\phi)$ using [9]

$$U_n(\theta,\phi) = [E_\theta(\theta,\phi)]^2 + [E_\phi(\theta,\phi)]^2 \quad \text{watts/steradian (unit solid angle)} \tag{16-27}$$

In turn, (16-27) can be used to compute the power radiated by the radiating element (antenna) using [9]

$$P_{\text{rad}} = \int_0^{2\pi}\int_0^{2\pi} U_n(\theta,\phi)\sin\theta\, d\theta\, d\phi \quad \text{watts} \tag{16-28}$$

where $(0 \le \theta \le \pi)$ and $(0 \le \phi \le 2\pi)$ are, respectively, the elevation and azimuthal angles of a spherical coordinate system.

With $U_n(\theta,\phi)$ and P_{rad} defined by (16-27) and (16-28), respectively, we can represent the radiation resistance R_r and the maximum directivity $D_0 = D_{\text{max}}$ by [9]

$$R_r = \frac{2P_{\text{rad}}}{|I_o|^2} \tag{16-29}$$

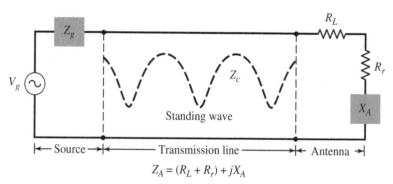

$$Z_A = (R_L + R_r) + jX_A$$

Figure 16-39 Equivalent circuit of an antenna connected to a generator. (Source: C. A. Balanis, *Antenna Theory: Analysis and Design*, Fourth Edition, John Wiley & Sons, New York, 2016. Reproduced with permission from John Wiley & Sons.)

$$D_{\text{max}} = D_0 = \frac{U_{n\text{max}}(\theta,\phi)}{U_0} = \frac{4\pi U_{n\text{max}}(\theta,\phi)}{P_{\text{rad}}} \quad \text{dimensionless} \tag{16-30a}$$

$$U_0 = \frac{P_{\text{rad}}}{4\pi} \quad \text{watts/steradian} \tag{16-30b}$$

where I_0 is the maximum amplitude of the current, and U_0 is the radiation intensity of an iso-tropic source which radiates equally well in all directions. The maximum directivity D_0 is the parameter that describes the *directional* characteristics (collimating capabilities) of the radiating element (antenna).

Once the directivity has been established, another important parameter is the maximum gain G_0, which is related to the maximum directivity D_0 by [9]

$$G_0 = e_{cd} D_0 = \frac{R_r}{R_r + R_L} D_0 \quad \text{dimensionless} \tag{16-31}$$

$$e_{cd} = \frac{R_r}{R_r + R_L} \tag{16-32}$$

where e_{cd} = *radiation efficiency* of the radiating element (antenna). For a lossless (no losses) radi-ating element, $R_L = 0$, the maximum gain G_0 is identical to the maximum directivity D_0 ($G_0 = D_0$).

The maximum directivity D_0 is referenced at the output terminals of the antenna, while the maximum gain G_0 is referenced at the input terminals of the antenna (power that has crossed over the intersection from the transmission line to the input side of the antenna), as illustrated in Figure 16-40a. The losses of the antenna, represented in Figure 16-39 by R_L, are highlighted in Figure 16-40b and can be either conducting losses (represented by i_c) or dielectric losses (represented by i_d).

The maximum gain G_0, as defined by (16-31), does not account for the matching losses at the intersection between the transmission line and the radiating element (antenna). To account for the mismatch losses between the transmission line (represented by the characteristic impedance Z_c)

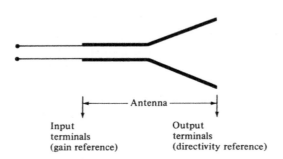

Input
terminals
(gain reference)

Output
terminals
(directivity reference)

(a) Antenna reference terminals

(b) Reflection, conduction, and dielectric losses

Figure 16-40 Reference terminals and losses of an antenna. (Source: C. A. Balanis, *Antenna Theory: Analysis and Design*, Fourth Edition, John Wiley & Sons, New York, 2016. Reproduced with permission from John Wiley & Sons.)

and the input impedance of the antenna represented by $Z_{in} = (R_r + R_L) + jX_A$, as illustrated in Figure 16-39, we introduce the reflection (mismatch) efficiency e_r defined as [9]

$$e_r = 1 - |\Gamma|^2 = \text{reflection efficiency (dimensionless)} \qquad (16\text{-}33a)$$

$$\Gamma = \frac{Z_{in} - Z_c}{Z_{in} + Z_c} = \text{reflection coefficient (dimensionless)} \qquad (16\text{-}33b)$$

$$Z_{in} = (R_r + R_L) + jX_A = \text{antenna input impedance} \qquad (16\text{-}33c)$$

Once the reflection (mismatch) efficiency is introduced by (16-33a), the maximum realized gain G_{re0} is defined as [9]

$$G_{re0} = \left[1 - |\Gamma|^2\right] G_0 = \left[1 - |\Gamma|^2\right] e_{cd} D_0 = e_r e_{cd} D_0 \qquad (16\text{-}34)$$

The reflection (mismatch) efficiency e_r between the antenna and the transmission line is represented in Figure 16-40b by the reflection coefficient Γ, defined by (16-33b).

It is apparent from (16-34) that for a lossless ($e_{cd} = 0$) and a matched to the transmission line ($e_r = 1, \Gamma = 0$) radiating element (antenna), the maximum directivity D_0, maximum gain G_0, and maximum realized gain G_{re0} are all equal ($D_0 = G_0 = G_{re0}$).

16.10 ANTENNA APPLICATIONS

As indicated previously, artificially fabricated surfaces have many applications, especially related to suppression or enhancement of surface waves and/or controlling the phase characteristics of the reflection coefficient. In this section, we will consider some basic radiating elements and their integration and interaction with metasurfaces.

16.10.1 Monopole

A basic illustrative example is to examine the radiation characteristics of a monopole mounted on a finite-size ground plane, PEC, and high-impedance surface (HIS). The monopole and its measured radiation patterns are shown in Figures 16-41a–16-41d. The basic geometry, shown in Figure 16-41a, consists of a 3-mm monopole mounted on a 5-cm ground plane; its amplitude patterns were measured at 35 GHz. When mounted on a PEC ground plane, the amplitude pattern displayed in Figure 16-41b is basically that which is expected and shown in Figure 13-33. It exhibits radiation not only in the upper hemisphere but also in the lower, due to diffractions from the edges of the finite ground plane. However, when the monopole is mounted on a high-impedance surface designed at the bandgap frequency of 35 GHz, the radiation in the lower hemisphere, as displayed in Figure 16-41c, is diminished as the diffractions from the edges of the ground plane have been reduced because the high-impedance surface suppresses the surface waves as they travel from the center of the ground plane toward its edges. Even in the upper hemisphere, the pattern is very smooth compared with that of Figure 16-41b, because the diffractions from the edges of the ground plane are basically insignificant compared to the direct radiation (geometrical optics) from the radiating element and its image. However, when the monopole is operated below the bandgap ($f = 26$ GHz), as shown in Figure 16-41d, the pattern is scalloped both in the upper and lower hemispheres, with considerable radiation in the lower hemisphere. The measured patterns in Figure 16-41 exhibit slight asymmetries, probably due to system errors introduced by cables and mounting structures.

Based on orientation and placement on ground planes, vertical monopole antennas of Figure 16-41a are not suitable for low-profile applications; however, horizontally oriented dipoles, or other radiators, may provide an alternative. To assess alternate geometries, let us first examine the

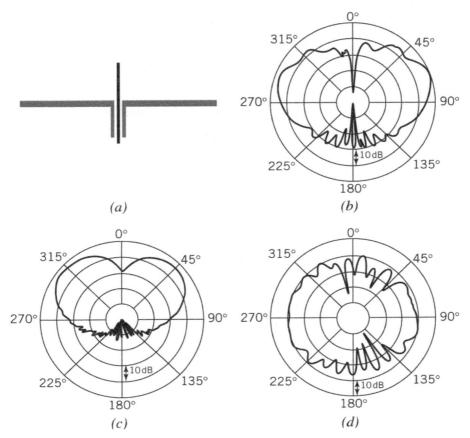

(a)

(b)

(c)

(d)

Figure 16-41 Monopole and its patterns when mounted on different ground planes; PEC and PMC. (*a*) Monopole geometry (3-mm monopole, 5-cm ground plane). (*b*) PEC ground plane ($f = 35$ GHz). (*c*) EBG ground plane ($f = 35$ GHz). (*d*) EBG ground plane ($f = 26$ GHz). (Source: D. Sievenpiper, "Artificial impedance surfaces," Chapter 15 in *Modern Antenna Handbook*, C. A. Balanis (Ed.), John Wiley & Sons, pp. 737–777, 2008. Reproduced with permission from John Wiley & Sons.)

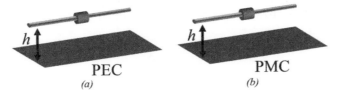

(a) (b)

Figure 16-42 Horizontal electric dipole above PEC and PMC ground planes.

performance of a horizontal electric dipole placed above PEC and PMC ground planes, as shown in Figure 16-42; their performance will be examined based on the radiation resistance R_r and maximum directivity D_0.

16.10.2 Horizontal Dipole

A dipole antenna is a classical radiating element, and it has been throughly investigated when it is isolated in free space [9]. The radiation performance and the 50-Ω matching are positively/negatively disturbed when another object/surface is placed in proximity to the radiating element. The influence of the surface on the dipole performance depends on different parameters such as size of the surface, the height of the dipole above the surface, and the composition of the surface material.

To demonstrate how the spacing and the ground plane influence the dipole performance, based on radiation resistance R_r and maximum directivity D_0, a horizontal electric dipole in the vicinity of infinite PEC and PMC is depicted in Figure 16-42.

For the PEC ground plane, the radiation resistance of a horizontal dipole is represented by [9]

$$R_r = \eta\pi\left(\frac{l}{\lambda}\right)^2\left[\frac{2}{3} - \frac{\cos(2\beta h)}{(2\beta h)^2} - \frac{\sin(2\beta h)}{2\beta h} + \frac{\sin(2\beta h)}{(2\beta h)^3}\right] \tag{16-35}$$

and the maximum directivity, for $h \leq \lambda/4$, by [9]

$$D_o = \frac{4\sin^2(\beta h)}{\left[\frac{2}{3} - \frac{\cos(2\beta h)}{(2\beta h)^2} - \frac{\sin(2\beta h)}{2\beta h} + \frac{\sin(2\beta h)}{(2\beta h)^3}\right]}, \quad h \leq \frac{\lambda}{4} \tag{16-36}$$

However, when the horizontal dipole is positioned in close proximity to an infinite PMC ground, the radiation resistance and the maximum directivity, for $h \leq \lambda/4$, are [9]

$$R_r = \eta\pi\left(\frac{l}{\lambda}\right)^2\left[\frac{2}{3} + \frac{\cos(2\beta h)}{(2\beta h)^2} + \frac{\sin(2\beta h)}{2\beta h} - \frac{\sin(2\beta h)}{(2\beta h)^3}\right] \tag{16-37}$$

$$D_0 = \frac{4}{\left[\frac{2}{3} + \frac{\cos(2\beta h)}{(2\beta h)^2} + \frac{\sin(2\beta h)}{2\beta h} - \frac{\sin(2\beta h)}{(2\beta h)^3}\right]}, \quad h \leq \frac{\lambda}{4} \tag{16-38}$$

The radiation resistance and maximum directivity [see (16-35)–(16-38)] for a $\lambda/10$ horizontal dipole above PEC and PMC ground planes are plotted in Figure 16-43. As illustrated, the PEC ground plane severely impacts the radiation resistance when the dipole is positioned very close to it ($h < 0.1\lambda$); however, the peak directivity is at maximum for smaller spacings. Unlike the PEC surface, the dipole in the presence of a PMC ground plane possesses a greater radiation resistance with small maximum directivity when h is less than 0.1λ. For low-profile applications, a preferable ground plane is one that leads to a radiation resistance comparable to the PMC surface, and at the

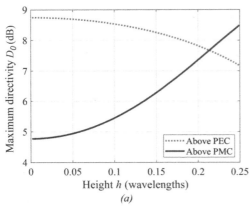

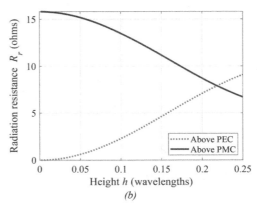

Figure 16-43 Maximum directivity and radiation resistance of horizontal electric dipole above PEC and PMC ground planes.

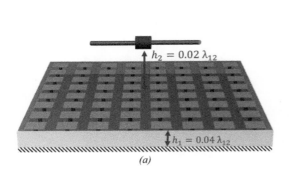

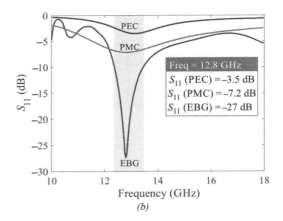

Figure 16-44 Geometry and S_{11} of horizontal electric dipole above PEC, PMC, and EBG surfaces. (*a*) Geometry. (*b*) S_{11}. (Source: Y. Rahmat-Samii and F. Yang, "Development of complex artificial ground planes in antenna engineering," Chapter 12 in *Metamaterials: Physics and Engineering Explorations*, N. Engheta and R. W. Ziolkowski (Eds.), IEEE Press, Wiley Inter-Science, New York, 2006. Reproduced with permission from John Wiley & Sons, Inc.)

same time, it results in high directivity, as the one attained with a PEC surface. This can be realized by the introduction of artificial magnetic conductors (AMCs) as ground planes.

Whether a PEC, PMC, or EBG surface outperforms the others as a ground plane depends upon the application. This is best illustrated by a basic example; a $l = 0.4\lambda_{12}$ dipole (λ_{12} is the free-space wavelength at $f = 12$ GHz) placed horizontally above PEC, PMC, and EBG surfaces, as shown in Figure 16-44; the EBG surface has a height of $0.04\lambda_{12}$. The dipole is placed at a height h of $0.06\lambda_{12}$ ($h_2 = 0.06\lambda_{12}$) above $\lambda_{12} \times \lambda_{12}$ PEC and PMC square surfaces.

The S_{11} of this system was simulated, using the FDTD method, over a frequency range of 10–18 GHz [5, 11]. Based on a 50-Ω line impedance, the results are shown in Figure 16-44*b*, where it is clear that the EBG surface (which has a reflection coefficient phase variation from $+180°$ to $-180°$) exhibits a best return loss of -27 dB, while the PMC (which has a reflection phase of $0°$) has a best return loss of -7.2 dB and the PEC (which has a reflection phase of $180°$) has a best return loss of only -3.5 dB. For the PMC surface, the return loss is influenced by the mutual coupling, due to the close proximity between the main element and its in-phase image, whereas for the PEC the return loss is influenced by the $180°$ phase reversal, which severely impacts the radiation efficiency. In this example, the EBG surface, because of its $+180°$ to $-180°$ phase variation over the frequency bandgap of the EBG design, outperforms the PEC and PMC and serves as a good ground plane. The other two, the PEC and PMC, possess constant out-of-phase and in-phase phase characteristics, respectively, over the entire frequency range.

Now that we have examined a horizontal oriented infinitesimal electric dipole above PEC and PMC ground planes based on their radiation resistance R_r and maximum directivity D_o, as well as an $l = 0.4\lambda$ dipole above PEC, PMC, and EBG surfaces, the results are summarized on a graphical illustration in Figure 16-45 based on radiation efficiency, maximum directivity, overall height, and low profile. It is clear that a horizontal electric radiating element above a PMC/EBG surface is the most attractive low-profile configuration. This configuration will now be pursued in detail with other radiating elements, such as an electric circular loop, placed above square and circular AMC surfaces.

16.10.3 Circular Loop

Another popular and classical radiating element is the circular loop. Depending on the electrical size of their circumference, loops can be classified as electrically small or large. Electrically small loops ($C < \lambda / 10$) suffer from low radiation resistance; thus, they are classified as poor radiators [9].

Structure	Source and image current	For low-profile configuration ($h \leq 0.1\,\lambda$)			
		Efficiency	Directivity	Overall height	Low-profile
Vertical dipole above infinite PEC	↑	✓	Moderate	Large	✗
Horizontal dipole above infinite PEC	→	✗	High	Low	✓
Horizontal dipole above infinite PMC	→	✓	Moderate	Low	✓

Figure 16-45 Performance summary of vertical and horizontal electric dipoles above PEC and PMC ground plane.

As the circumference approaches λ, the radiation resistance increases, and the loop becomes a better match to a 50-ohm feed line.

An alternate approach to alter the radiation resistance, besides increasing the circumference, is to introduce an AMC metasurface ground plane [54, 75]. The metasurface ground plane can favorably be utilized to enhance the radiating element performance over a certain operational bandwidth. One way to examine the aforementioned bandwidth is to place the loop near the metasurface and observe the input reflection coefficient, as the circumference of the loop is varied. Following this approach, the loop reflection coefficient has been examined when it is positioned at 1 mm above two different metasurface ground planes: circular and square [54], as depicted in Figures 16-46a and 16-47a.

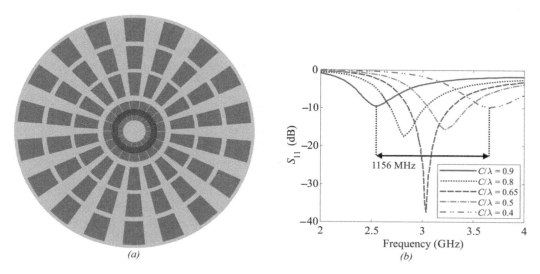

(a)   *(b)*

Figure 16-46 (a) Wire circular loop above a circular HIS. (b) Associated S_{11}. (Source: Adapted from M. A. Amiri, C. A. Balanis, and C. R. Birtcher, "Analysis, design and measurements of circularly symmetric high impedance surfaces for loop antenna applications," *IEEE Trans. Antennas Propagat.*, vol. 64, pp. 618–629, 2016.)

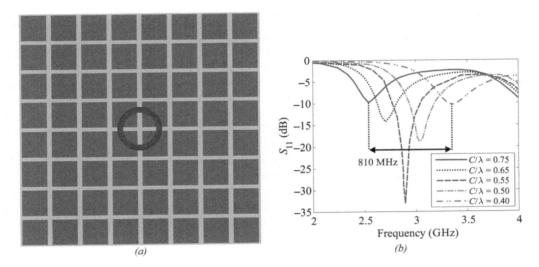

(a)   (b)

Figure 16-47   (a) Wire circular loop above a square HIS with square patches. (b) Associated S_{11}. (Source: M. A. Amiri, C. A. Balanis, and C. R. Birtcher, "Analysis, design and measurements of circularly symmetric high impedance surfaces for loop antenna applications," *IEEE Trans. Antennas Propagat.*, vol. 64, pp. 618–629, 2016. Reproduced with permission from IEEE.)

As illustrated in Figures 16-46*b* and 16-47*b*, the operational bandwidth of the circular loop above circular high-impedance surfaces (CHIS) is greater than their square counterparts. For further comparison, the realized gain pattern at 3 GHz for a circular loop with a radius of 1 cm, placed above circular and square HISs, was simulated, and it is shown in Figure 16-48. As observed from Figure 16-48, the gain patterns of the circular loop are more symmetric on different elevation planes when the loop is placed above the circular HIS. This is expected as the fields generated by the circular loop interact more efficiently with the circular HIS [54] and result in more symmetric patterns.

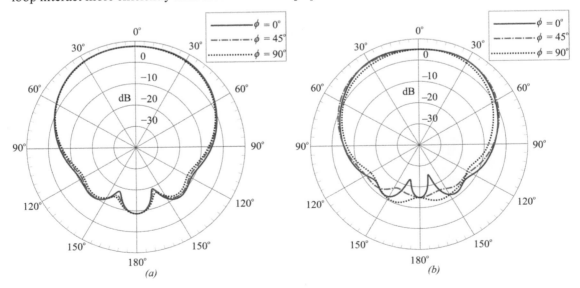

(a)   (b)

Figure 16-48   Realized gain of a circular loop along different ϕ planes, when it is located above circular and square HISs. (Source: M. A. Amiri, C. A. Balanis, and C. R. Birtcher, "Analysis, design and measurements of circularly symmetric high impedance surfaces for loop antenna applications," *IEEE Trans. Antennas Propagat.*, vol. 64, pp. 618–629, 2016. Reproduced with permission from IEEE.)

The broadside realized gain for the loop element is investigated for different configurations: in free space, and above PMC, *square* high impedance (SHIS), *circular* high-impedance (CHIS) ground planes; the simulated data is plotted in Figure 16-49. As expected, at 3 GHz, the realized gain is around 2 dB when the loop is isolated in free space [9]. However, as the different ground planes are introduced adjacent to the loop, the magnitude and the phase of the reflected fields change accordingly and lead to different realized gains. When the PMC ground plane is placed below the loop, the broadside realized gain increases by 1.3 dB. However, when the square HIS and circular HIS are utilized, the gain increases by 3 and 5.5 dB, respectively. From these results, it is clear that the circular HIS is a better ground plane for circular loops.

To validate the simulated results, a prototype of the circular loop above a circular metasurface was fabricated and measured, as depicted in Figure 16-50. The S_{11} of two different loops, with radii of 3 and 2.7 cm, was simulated and measured (see Figure 16-51a). Furthermore, the input impedance and the angular gain pattern for the 2.7-cm loop radius were also plotted in Figures 16-51b and 16-52, respectively. As observed, an excellent agreement is indicated between simulations and measurements.

To further enhance the radiation performance, a hybrid circular ground plane was proposed as an alternate ground plane for loop radiators [75]. The loop element was positioned on the top

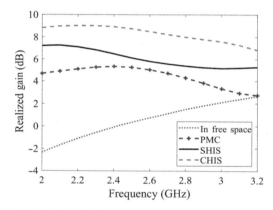

Figure 16-49 Broadside realized gain of a circular loop in free space and above different ground planes.

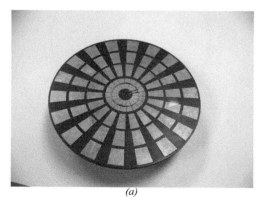 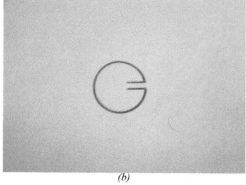

(a) (b)

Figure 16-50 Fabricated circular HIS with the loop antenna at a height of 0.01λ. (Source: M. A. Amiri, C. A. Balanis, and C. R. Birtcher, "Analysis, design and measurements of circularly symmetric high impedance surfaces for loop antenna applications," *IEEE Trans. Antennas Propagat.*, vol. 64, pp. 618–629, 2016. Reproduced with permission from IEEE.)

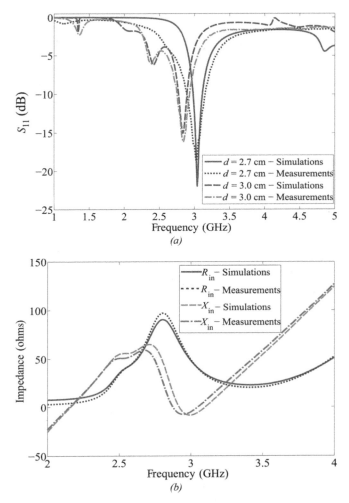

Figure 16-51 Simulations and measurements of the loop element above a CHIS. (Source: M. A. Amiri, C. A. Balanis, and C. R. Birtcher, "Analysis, design and measurements of circularly symmetric high impedance surfaces for loop antenna applications," *IEEE Trans. Antennas Propagat.*, vol. 64, pp. 618–629, 2016. Reproduced with permission from IEEE.)

surface of a hybrid ground with no spacing between them, as demonstrated in Figure 16-53. The loop was designed with a circumference of λ_0, since broadside radiation is desired at 2.9 GHz. At one-wavelength circumference, the loop exhibits capacitive reactance [9]; thus, an inductive load is required to resonate the loop. The hybrid structure incorporates two different metallic rings, as shown in Figure 16-53, printed on a grounded dielectric: circular HIS and annular circular ring. The radial length of the HIS ring (r_{HIS}) is 22 mm, while it is 24 mm for the annular ring ($r_{annular}$). The circular HIS ring was designed to provide an inductive impedance at 2.9 GHz; therefore, by placing the HIS ring close to the loop radiating element, resonance can be achieved. Once the mismatch losses are minimized, the next step was to enhance the broadside directivity. This can be realized by increasing the effective area of the structure. Therefore, the gaps were omitted in the annular ring, and it was designed with a slightly larger radial length. The combination of the two rings surrounding the loop element enhances both the matching and the broadside realized gain.

To further demonstrate the contribution of each ring of the hybrid ground plane, the antenna performance was investigated for the three designs shown in Figure 16-54. The S_{11} and broadside directivity for the three designs are presented in Figure 16-55. As observed, the presence of the HIS

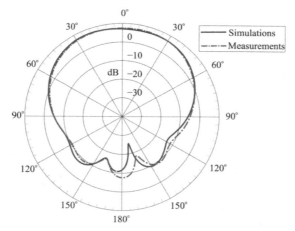

Figure 16-52   Simulations and measurements of the loop element above CHIS: Gain pattern at 3 GHz. (Source: M. A. Amiri, C. A. Balanis, and C. R. Birtcher, "Analysis, design and measurements of circularly symmetric high impedance surfaces for loop antenna applications," *IEEE Trans. Antennas Propagat.*, vol. 64, pp. 618–629, 2016. Reproduced with permission from IEEE.)

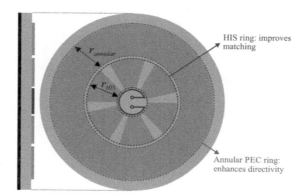

Figure 16-53   Geometry of a hybrid ground plane with printed loop element.

ring alleviates the mismatch losses and improves the reflection coefficient (S_{11}). However, at resonance, it only contributes to slight improvement in the broadside directivity. When the two rings are utilized to form the hybrid ground plane, in addition to possess better matching bandwidth, a 3-dB directivity increase is observed at resonance, primarily due to the presence of the outside PEC ring.

The hybrid ground plane was fabricated along with a printed loop element, as illustrated in Figure 16-56. The comparison between simulated and measured results for the reflection coefficient and broadside realized gain are plotted in Figure 16-57; an excellent agreement is indicated.

The superior performance of the circular metasurface ground plane inspired interest to investigate circular HISs when utilized with different curvilinear elements, such as a spiral and circular patch [55, 76].

16.10.4   Aperture Antenna

Another example of diffraction control that can be provided by a textured high-impedance surface is to examine the radiation of an aperture mounted on a high-impedance surface ground

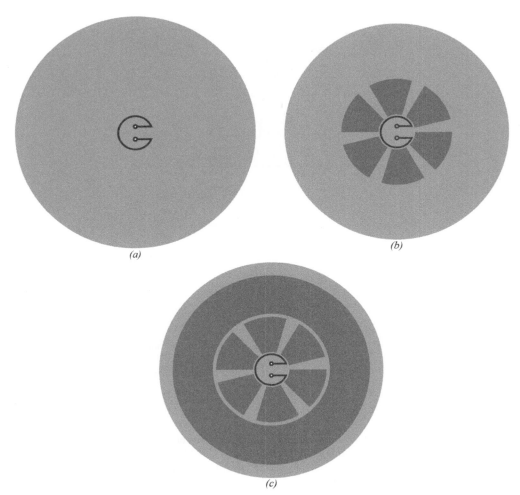

Figure 16-54 Printed loop antenna: (*a*) above grounded dielectric, (*b*) surrounded by a HIS section on grounded dielectric, and (*c*) surrounded by hybrid rings on grounded dielectric. (Source: M. Alharbi, C. A. Balanis, C. R. Birtcher, and H. N. Shaman, "Hybrid circular ground planes for high-realized-gain low-profile loop antennas," *IEEE Antennas Wireless Propagat. Lett.*, vol. 17, no. 8, pp. 1426–1429, Aug. 2018. Reproduced with permission from IEEE.)

plane. The geometry of the radiator is shown in Figure 16-58, and it consists of a rectangular aperture mounted on a 12.7-cm ground plane, both PEC and high impedance [2]. Each unit cell measures 3.7 mm and the textured high-impedance surface has been designed for a bandgap in the range of 12–18 GHz. The measured patterns, both for the PEC and high-impedance surface, are shown in Figure 16-59, and they were measured at 13 GHz, which is within the designed bandgap. In general, the shape of the radiation pattern is influenced not only by the shape and size of the aperture but also by the ground plane and its texture.

The patterns in Figure 16-59*a*, E- and H-planes, are those of the aperture on a PEC ground plane, and they are representative, as shown in Figures 16-59*a* and 16-59*b* [9]. The E-plane is usually broader than the H-plane, because for the E-plane its vertical polarized fields are not shorted out by the PEC ground plane, while for the H-plane its horizontally polarized fields are, ideally, nulled. When the aperture is mounted on a textured high-impedance surface and operated within its bandgap (specifically 13 GHz), its patterns in both the E- and H-planes, shown in Figure 16-16*b*,

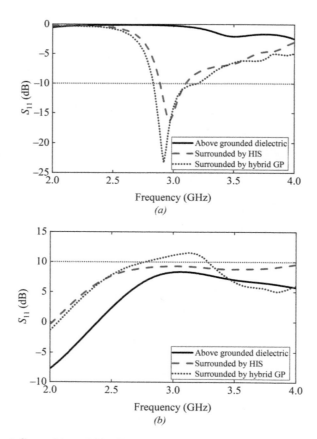

Figure 16-55   Simulated S_{11} and broadside directivity of loop element with three ground plane configurations. (Source: M. Alharbi, C. A. Balanis, C. R. Birtcher, and H. N. Shaman, "Hybrid circular ground planes for high-realized-gain low-profile loop antennas," *IEEE Antennas Wireless Propagat. Lett.*, vol. 17, no. 8, pp. 1426–1429, Aug. 2018. Reproduced with permission from IEEE.)

Figure 16-56   A photograph of the fabricated antenna. (Source: M. Alharbi, C. A. Balanis, C. R. Birtcher, and H. N. Shaman, "Hybrid circular ground planes for high-realized-gain low-profile loop antennas," *IEEE Antennas Wireless Propagat. Lett.*, vol. 17, no. 8, pp. 1426–1429, Aug. 2018. Reproduced with permission from IEEE.)

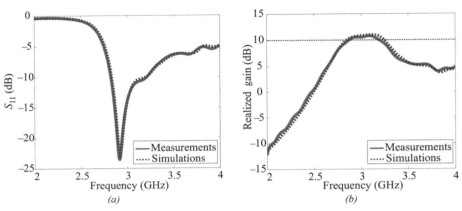

Figure 16-57 Comparisons between the simulated and measured results for the loop element above the hybrid ground plane. (Source: M. Alharbi, C. A. Balanis, C. R. Birtcher, and H. N. Shaman, "Hybrid circular ground planes for high-realized-gain low-profile loop antennas," *IEEE Antennas Wireless Propagat. Lett.*, vol. 17, no. 8, pp. 1426–1429, Aug. 2018. Reproduced with permission from IEEE.)

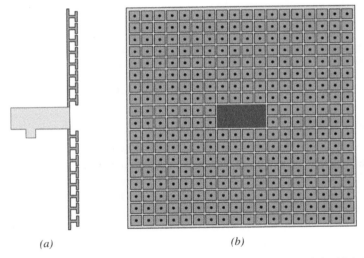

Figure 16-58 Aperture antenna in a high-impedance surface. The unit cells of the high-impedance surface measure 3.7 mm, and the size of the ground plane (not shown to scale) is 12.7 cm. The aperture is fed by a coax to Ku-band rectangular waveguide transition. (*a*) Side view. (*b*) Front view. (Source: D. Sievenpiper, "Artificial impedance surfaces," Chapter 15 in *Modern Antenna Handbook*, C. A. Balanis (Ed.), John Wiley & Sons, pp. 737–777, 2008. Reproduced with permission from John Wiley & Sons.)

are nearly the same and symmetrical because the textured surface suppresses both the TM and TE surface waves near the resonant frequency. However, when the aperture is mounted on the same textured surface but operated at the leading edge of the TE band where TM waves are suppressed, the H-plane pattern is broader than the E-plane (see Figure 16-59*c*), which is the opposite of that observed for the patterns in Figure 16-59*a* for the PEC ground plane. Since the behavior of the E- and H-plane patterns in Figure 16-59*c* is opposite of that of the patterns in Figure 16-59*a*, the textured ground plane acts as a PMC (ideally shorts out the tangential magnetic fields), which is the opposite of that for the PEC.

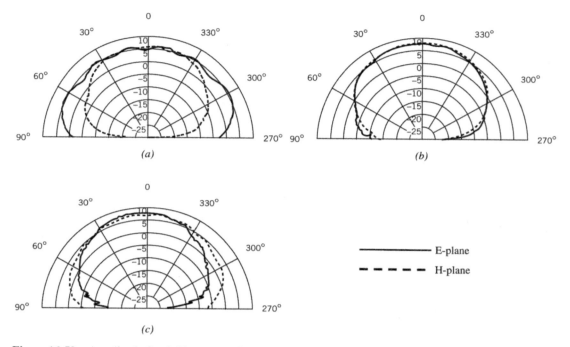

Figure 16-59 Amplitude far-field pattern of an aperture on a ground plane; PEC and AMC. (*a*) PEC. (*b*) AMC (*f* = 13 GHz) (high-impedance). (*c*) AMC (high-impedance near edge of TE band). (Source: D. Sievenpiper, "Artificial impedance surfaces," Chapter 15 in *Modern Antenna Handbook*, C. A. Balanis (Ed.), John Wiley & Sons, pp. 737–777, 2008. Reproduced with permission from John Wiley & Sons.)

16.10.5 Microstrip Array

In microstrip arrays, a detrimental phenomenon that leads to an increase in the input reflection coefficient of the array, as a function of the scan angle, is the surface waves created, sustained, and traveling within the substrate. Space waves also contribute but they are not as dominant as the surface waves, which can also lead to *scan blindness* [77]. One way to eliminate the surface waves and reduce the input reflection coefficient, and even eliminate scan blindness, is to use cavities to surround each of the patches [78]. This is a rather expensive design but it works.

Another way to minimize the surface waves, without the use of cavities, is to mount the patches on EBG-textured surfaces, as shown in Figure 16-60a for a 3 × 3 array. A 2 × 2 unit cell of the EBG surface, with a dipole in its middle, is displayed in Figure 16-60b [79]. Such surfaces have the ability to control and minimize surface waves in substrates. A microstrip dipole element, placed within a textured high-impedance surface, was designed, simulated, fabricated, and measured [79]. It consists of a dipole patch of length 9.766 mm placed in the middle of 4 × 4 EBG unit cells; each cell had dimensions of $w = l = 1.22$ mm and a separation gap between them of $g = 1.66$ mm. The substrate had a dielectric constant of $\varepsilon_r = 2.2$ and a height of $h = 4.771$ mm. The scanned magnitude of the simulated reflection coefficient at 13 GHz of such a design is shown in Figure 16-61; the bandgap frequency range was 9.7–15.1 GHz. The E-plane curves are presented by the *E* curves, the H-plane by the *H* curves, and the diagonal (45°) plane by the *D* curves. It is clear that using conventional substrates, the reflection coefficient varies as a function of the scan angle and, in fact, creates scan blindness around 50°. However, when the same elements were placed on a textured high-impedance surface, the reflection coefficient was reduced, especially in the E-plane, and in fact, the scan blindness was eliminated.

There are many other applications where textured high-impedance surfaces can be used to control the radiation characteristics of electromagnetic problems. Such applications include, but are

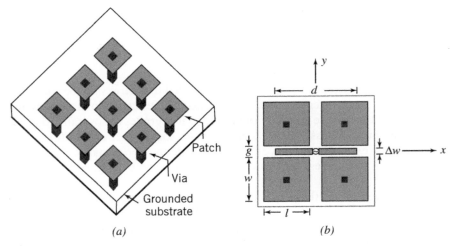

Figure 16-60 Dipole phased array on a PBG surface. (*a*) 3 × 3 array. (*b*) Dipole between 2 × 2 unit cell. (Source: L. Zhang, J. A. Castaneda, and N. G. Alexopoulos, "Scan blindness free phased array design using PBG materials," *IEEE Trans. Antennas Propagat.*, vol. 52, no. 8, pp. 2000–2007, Aug. 2004. Reproduced with permission from IEEE.)

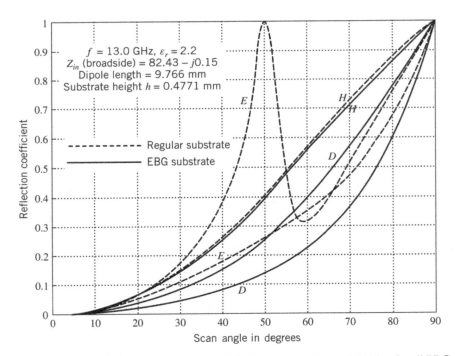

Figure 16-61 Reflection coefficient of phase array of dipoles on a regular and 4 × 4 unit cell PBG substrate. (Source: L. Zhang, J. A. Castaneda, and N. G. Alexopoulos, "Scan blindness free phased array design using PBG materials," *IEEE Trans. Antennas Propagat.*, vol. 52, no. 8, pp. 2000–2007, Aug. 2004. Reproduced with permission from IEEE.)

not limited to, reflective beam steering, microwave holography, low-profile antennas (see Section 16.10.3), and leaky-wave beam steering (see Section 16.12 that follows) . These are accomplished by using textured high-impedance surfaces to control, and even suppress, the surface waves within the bandgap and/or the phase of the reflection coefficient, which can even be made zero (PMC surface) at resonance. Other examples of interest utilizing EBG surfaces to suppress surface waves can be found in [6].

16.10.6 Surface-Wave Antennas

When a radiating element is placed adjacent to a grounded dielectric slab, besides radiating into free space, some of the power excites surface waves into the dielectric. For some antenna applications, the excitation of the surface waves is not desired as it degrades the overall radiation behavior, as the case for monopole discussed earlier. However, in some other applications, these surface waves are intentionally excited to advance the overall antenna performance [3, 80–83]. As metasurfaces can be utilized to suppress surface waves, they can also be designed to support and guide surface waves.

With proper arrangement of the periodic patches, the guided waves radiate as they propagate along the ground plane and dictate the radiation pattern. To achieve an omnidirectional, or monopole-like radiation pattern, a 0.26λ horizontal dipole antenna is placed 0.02λ above a metasurface ground plane to launch surface waves [81]. Since most of the radiation is contributed by the radiation of surface waves, the dipole length can be smaller than a half wavelength as it is only used as a surface-wave exciter. Nevertheless, the small radiation contributed by the dipole introduces a higher cross-polarization level compared to the monopole. To overcome this limitation, the dipole is replaced with a center-fed circular patch, and the square geometry of the grounded slab is replaced with a circular-shaped ground plane [82]. The overall radiation possesses a symmetrical omnidirectional pattern in the principal and diagonal planes. The attained patterns are very similar to those of a monopole with only a 0.05λ overall height profile.

Surface-wave antennas can also be designed to achieve a broadside radiation. This was accomplished by exciting surface waves utilizing different radiators such as dipole [84], truncated patch [85], diamond-slotted patch [85], and square ring elements [80]. As the field generated by different radiators varies, the launch and the excitation of surface waves will differ accordingly leading to unique radiation patterns. Moreover, the surface-wave resonances change according to the size of the metasurfaces and the dimensions of the periodic patches. As the size of the metasurfaces increases, more surface-wave resonances begin to contribute. For example, a dipole was positioned above square periodic patches (see Figure 16-62a) of different sizes (5×5, 7×7, and 9×9), and it was observed that the surface-wave resonances shift toward lower frequencies and become more closely spaced as the size of the metasurface increases. In addition to the metasurface size, the dimensions of the periodic patches control the surface-wave resonances, as was demonstrated in [79]. A square ring element was placed within the metasurface, as depicted in Figure 16-62b. The metasurface is formed with rectangular patches to possess different surface-wave resonances along two orthogonal planes. The resonance that is attributed to the square ring element, along with the surface-wave resonances, is designed to be closely spaced to improve the matching bandwidth by 26%. Furthermore, the proper excitation of surface waves in the design resulted in stable broadside radiation within the designed frequency range [79].

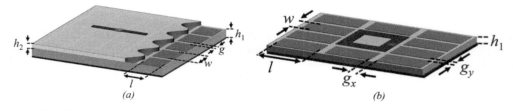

Figure 16-62 Geometry of surface-wave antennas for broadside radiation. (*a*) Dipole above metasurface; (*b*) Square ring within metasurface. (Source: M. Alharbi, C. A. Balanis, and C. R. Birtcher, "Performance enhancement of square-ring antennas exploiting surface-wave metasurfaces," *IEEE Antennas Wireless Propagat. Lett.*, vol. 18, no. 10, pp. 1991–1995, Oct. 2019, Reproduced with permission from IEEE.)

16.11 HIGH-GAIN PRINTED LEAKY-WAVE ANTENNAS USING METASURFACES

Leaky-wave antennas (LWAs) are traveling-wave antennas [9] that can be considered as waveguide structures. Radiation is achieved in the form of leakage from the propagating wave as it travels through the guiding structure. The propagation constant of leaky waves is complex ($\gamma = \alpha + j\beta$), where α is the attenuation constant that corresponds to the leakage (or radiation), and β is the phase constant.

LWAs can be classified as *one-dimensional* (1-D) and *two-dimensional* (2-D) [86]. In 1-D LWAs, the leaky waves are guided along a single direction, while in 2-D LWAs, the waves are guided in two-dimensional guiding structures. One-dimensional LWAs can be either *uniform* or *periodic* [86, 87]. Uniform leaky-wave antennas are uniform throughout their length, with no periodic modulation on their geometries. An example of a uniform leaky-wave antenna is the meandering long slot leaky-wave antenna shown in Figure 16-63.

The antenna is realized by the continuous long slot along the broad wall of a rectangular waveguide. The slot should be offset from the center of the rectangular waveguide to achieve radiation [88]. While straight-line slots have higher sidelobe levels, the offset of the meandering long slot, from the waveguide centerline toward the side wall, controls the coupling and, thus, the aperture distribution over the slot that is responsible for the shape of the amplitude radiation pattern, especially the minor lobe structure.

The fundamental mode of a uniform leaky-wave antenna is a *fast wave*, where the longitudinal phase constant (β) is smaller than the free-space phase constant (β_0), and the antenna radiates from the discontinuities. Contrary to uniform LWAs, periodic leaky-wave antennas have some form of a periodic modulation throughout their structure. Examples of such periodic LWAs are undulating conductors [89], periodic holes or slits on waveguides [90], periodically asymmetric trough waveguides [91], and strip gratings [58]. A classical waveguide LWA with discrete slots is shown in Figure 16-64 [9].

Figure 16-63 A continuous slot leaky-wave rectangular waveguide antenna.

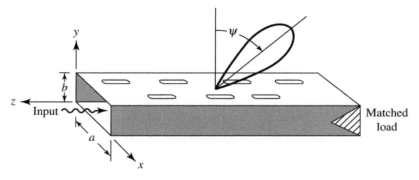

Figure 16-64 A discrete-slot leaky-wave rectangular waveguide antenna. (Source: C. A. Balanis, *Antenna Theory: Analysis and Design*, Fourth Edition, John Wiley & Sons, New York, 2016. Reproduced with permission from John Wiley & Sons.)

The antenna is based on a rectangular waveguide of lateral dimensions a and b, similar to the waveguide shown in Figure 8-3. The LWA forms a beam along an angle ψ from broadside, which can be scanned as the frequency is varied.

16.11.1 Floquet-Bloch Modes

In periodic structures, the periodic modulation at a given frequency causes the modal electromagnetic fields of the guided structure at any two points separated by the modulation period to be related by the same constant. In other words, the periodicity in the structure enables the electromagnetic fields to be represented as the product of an exponential factor corresponding to the traveling-wave nature of the fields and a periodic function with the same periodicity as that of the structure. For instance, a one-dimensionally modulated structure along the y axis with a periodicity of p is exhibited in Figure 16-65.

The electric field at any point on the surface can be represented by

$$\mathbf{E}(x,y,z) = \mathbf{E_p}(x,y,z)\, e^{-\gamma y} \tag{16-39}$$

where $\mathbf{E_p}(x,y,z)$ is periodic with a period p, which can be expressed as

$$\mathbf{E_p}(x,y+p,z) = \mathbf{E_p}(x,y,z) \tag{16-40}$$

$\mathbf{E_p}(x,y,z)$ can be expanded using Fourier series and represented as

$$\mathbf{E_p}(x,y,z) = \sum_{n=-\infty}^{+\infty} \mathbf{E_{pn}}(x,z)\, e^{-j\frac{2\pi}{p}ny} \tag{16-41}$$

Substituting (16-41) into (16-39) yields

$$\mathbf{E}(x,y,z) = \sum_{n=-\infty}^{+\infty} \mathbf{E_{pn}}(x,z)\, e^{-\left[\alpha+j\left(\beta+\frac{2\pi}{p}n\right)\right]y} \tag{16-42}$$

From (16-42), the electromagnetic fields can be characterized as an infinite number of space harmonics ($n = -\infty, \ldots, -1, 0, 1, \ldots, \infty$) referred to as *Floquet-Bloch modes*: this representation is referred to as the *Floquet-Bloch theorem*. The phase constant of the nth Floquet-Bloch mode can be expressed, using (16-42), as

$$\beta_n = \beta + \frac{2\pi}{p}n \tag{16-43}$$

Note the following notation for leaky-wave propagation and radiation: β_n for Floquet modes, β for the ($n = 0$) Floquet mode, and β_0 for the free-space constant.

In a periodic leaky-wave antenna, the fundamental Floquet-Bloch mode ($n = 0$) is usually a slow wave, where the longitudinal phase constant (β) is greater than the free-space phase constant (β_0), and it does not radiate continuously along the structure of the antenna. However, one or many of the space harmonics (for instance, $n = -1$) introduced by the periodic modulation are designed to be

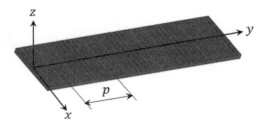

Figure 16-65   One-dimensional periodic structure of periodicity p.

fast waves, which radiate. The attenuation constant of the radiating harmonic controls the beam-width, while the phase constant determines the angular direction of the formed beam [92]. Because the phase constant β is a function of frequency, the angular direction of the formed beam varies with the frequency; hence, leaky-wave antennas are capable of *frequency scanning*.

Uniform leaky-wave antennas can beam-form only in the forward quadrant (near broadside to near forward endfire), while periodic LWAs can frequency scan in both the forward and backward quadrants. However, conventional 1-D leaky-wave antennas fed from one end cannot radiate along the broadside direction. This is due to the *open stopband phenomenon*, where, as the beam approaches the broadside direction, the attenuation constant varies rapidly and eventually reduces to zero, due to mode coupling. The impedance of the Floquet-Bloch mode is reactive and causes a stopband; this phenomenon is similar to scan blindness due to mutual coupling in arrays [77]. This issue has been addressed by adding impedance transformers between radiating elements to keep the impedance real, and by employing metamaterials [93, 94]. Broadside radiation can be achieved by other configurations such as feeding the antenna at the center or both ends, thereby forming two beams close to the broadside, which combine to form a single broadside beam [95]. Endfire radiation by conventional leaky-wave antennas is usually limited by the element pattern of the radiating elements.

16.12 METASURFACE LEAKY-WAVE ANTENNAS

Printed leaky-wave antennas can be implemented using high-impedance surfaces (HISs) with the surface impedance modulated with a periodic function; these are referred to as *metasurface leaky-wave antennas*.

16.12.1 Holographic Principle on Antennas

The surface-impedance modulation function required to form the desired beam in the desired direction can also be obtained from the *holographic principle*. This technique was inspired by the concept of optical holography [96] in which the interference pattern of the reflection from the object when illuminated by a laser source (object wave) and the illumination from the source (reference wave) is captured on a photographic plate, called a *hologram*, as illustrated in Figure 16-66.

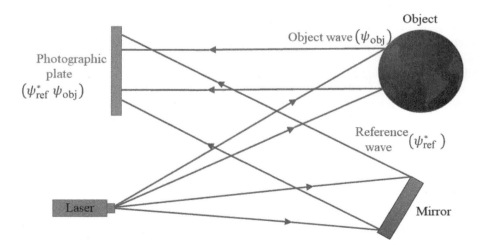

Figure 16-66 Optical holography: making a hologram. (Source: S. Ramalingam, "Impedance modulated metasurface antennas," Ph.D. dissertation, School of Electrical, Computer, and Energy Engineering, Arizona State University, 2020.)

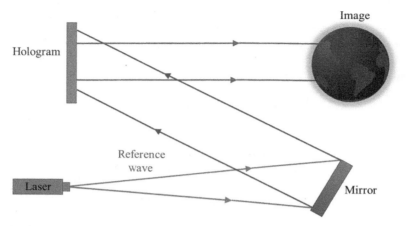

Figure 16-67   Optical holography: illuminating the hologram. (Source: S. Ramalingam, "Impedance modulated metasurface antennas," Ph.D. dissertation, School of Electrical, Computer, and Energy Engineering, Arizona State University, 2020.)

When the hologram is illuminated by the laser source an image of the object is produced, as illustrated in Figure 16-67.

Employing this technique on antennas, the surface impedance of the metasurface can be modulated as an interference pattern of the phase of the reference wave (Ψ_{ref}) and that of the object wave (Ψ_{obj}). The reference wave corresponds to the surface waves excited on the surface of the LWA, and the object wave corresponds to the aperture fields needed to form a fan beam or a pencil beam in the desired direction. The surface impedance is modulated as the interference pattern, which contains a term proportional to $\Psi_{\text{obj}}\Psi_{\text{ref}}^{*}$. When this interference pattern (impedance modulated surface) is illuminated by the reference wave (surface wave) Ψ_{ref}, the interference pattern produces a copy of the object wave as [57]

$$(\Psi_{\text{obj}}\Psi_{\text{ref}}^{*})\Psi_{\text{ref}} = |\Psi_{\text{ref}}|^2\,\Psi_{\text{obj}} \tag{16-44}$$

16.12.2   One-Dimensional Periodic Metasurface LWAs

One-dimensionally periodic metasurface LWAs can be realized as arrays of metallic patches or strips on a dielectric-covered ground plane with varying gap sizes between them. The surface-impedance profile of the metasurface is modulated with a periodic function like a sinusoidal wave or a square wave, and it is realized using HISs. A periodic modulation function can be represented as a sum of sinusoids using Fourier series. Each fast-wave mode of every sinusoid contributes to a fan beam.

The application of the holographic principle for a 1-D planar LWA fed at one end to form a fan beam at the desired angular direction results in a sinusoidal impedance modulation function, as illustrated in Figure 16-68.

The surface-impedance modulation function $Z_s(x, y)$ is expressed as

$$Z_s(x,y) = \pm jX_a\left[1 + M\cos\left(\frac{2\pi}{p}y\right)\right] \tag{16-45}$$

where X_a is the average surface reactance, M is the modulation index, and p is the modulation period. The surface reactance needs to be inductive to obtain radiation from TM-mode leaky waves and capacitive for TE-mode leaky waves. The average surface reactance and the modulation index control the gain of the fan beam, while the period of sinusoidal modulation, along with the

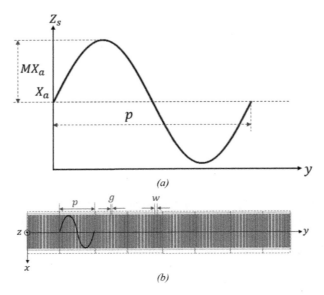

Figure 16-68 (*a*) The parameters in a sinusoidal impedance modulation. (*b*) The top view of a sinusoidally modulated planar 1-D LWA.

average surface reactance, controls the phase constant of the Floquet-Bloch mode and, therefore, controls the angular direction of the formed fan beam. The phase constant β_n of the nth Floquet-Bloch mode is related approximately to the modulation parameters by

$$\beta_n = \beta_0 \sqrt{1+\left(\frac{X_a}{\eta_0}\right)^2} + \frac{2\pi n}{p} \quad \text{(TM mode)} \tag{16-46}$$

$$\beta_n = \beta_0 \sqrt{1+\left(\frac{\eta_0}{X_a}\right)^2} + \frac{2\pi n}{p} \quad \text{(TE mode)} \tag{16-47}$$

A fan beam formed by a sinusoidally modulated 1-D metasurface LWA along the desired angular direction θ_d from broadside is demonstrated in Figure 16-69; θ_d is related to the phase constant of the Floquet mode by

$$\theta_d = \sin^{-1}\left(\frac{\beta_n}{\beta_0}\right) \tag{16-48}$$

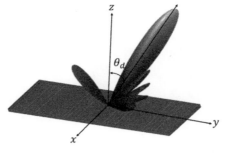

Figure 16-69 A fan beam formed by a 1-D metasurface LWA. (Source: S. Ramalingam, "Impedance modulated metasurface antennas," Ph.D. dissertation, School of Electrical, Computer, and Energy Engineering, Arizona State University, 2020.)

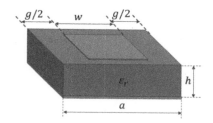

Figure 16-70   A HIS unit cell. The width of the patch is $w = a - g$. (Source: S. Ramalingam, "Impedance modulated metasurface antennas," Ph.D. dissertation, School of Electrical, Computer, and Energy Engineering, Arizona State University, 2020.)

The design of one-dimensional periodic metasurface LWAs to form a fan beam in the desired angular direction is also demonstrated using the MATLAB program entitled **One_Dim_LWA** included at the end of this chapter.

Each period of the modulation function is sampled at multiple points. The required surface impedance at each point is realized using unit cells, which consist of metallic strips of width w, separated by a gap g on a dielectric-covered ground plane of thickness h, as illustrated in Figure 16-70.

The gap size g can be varied to achieve the desired surface impedance at each unit cell. For a unit cell of length a and gap size g, the surface reactance X_s, at a given frequency of operation f and a phase constant along the surface β, can be computed using the transverse resonance method (TRM) [59] of Section 8.6 and illustrated in Figure 16-71.

According to the transverse resonance method the impedance looking upward (Z_{up}) and the impedance looking downward (Z_{down}) are equal in magnitude but of opposite signs. The impedance looking downward Z_{down} is the surface impedance of the PEC terminated high-impedance surface given by [30, 59]

$$Z_{down} = jX_s(f, \beta) = \frac{j\eta_0 \beta_{z1} \tan(\beta_{z1}h)}{\beta_0 \varepsilon_r - \beta_0(\varepsilon_r + 1)\delta\beta_{z1}\tan(\beta_{z1}h)} \qquad (16\text{-}49)$$

where η_0 is the free-space intrinsic impedance and h is the thickness of the substrate, as shown in Figures 16-70 and 16-71, and

$$\beta_{z1} = \sqrt{\beta_0^2 \varepsilon_r - \beta^2} \qquad (16\text{-}50)$$

$$\delta = \left(\frac{a}{\pi}\right) \ln\left\{\frac{1}{\sin[\pi(a-g)/2a]}\right\} \qquad (16\text{-}51)$$

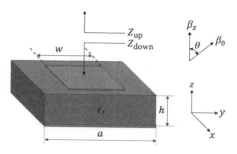

Figure 16-71   Transverse resonance method to obtain the surface impedance of a unit cell of 1-D LWA. (Source: S. Ramalingam, "Impedance modulated metasurface antennas," Ph.D. dissertation, School of Electrical, Computer, and Energy Engineering, Arizona State University, 2020.)

Equation (16-49) is applicable to both TE and TM modes. Z_{up} is the wave-impedance for both TE and TM modes, as discussed in Section 4.2.2B, and it is expressed as

$$Z_{up}^{TE} = \frac{\eta_0}{\cos\theta} = \frac{\eta_0\beta_0}{\beta_z} \tag{16-52}$$

$$Z_{up}^{TM} = \eta_0\cos\theta = \frac{\eta_0\beta_z}{\beta_0} \tag{16-53}$$

where

$$\beta_z = \sqrt{\beta_0^2 - \beta^2} \tag{16-54}$$

A relation between the surface reactance X_s and the gap sizes g can be derived by plotting the magnitude of the impedances looking upward Z_{up} and downward Z_{down} against the longitudinal phase constant β. The intersecting point of the two curves is the surface reactance of the unit cell with a gap size g. Using this technique on a unit cell of size $a = 3$ mm, with a substrate of dielectric constant $\varepsilon_r = 3$ and thickness $h = 1.524$ mm, the variation of the TM-mode surface reactance X_s as a function of the gap size g is displayed in Figure 16-72 [97]. A similar plot can be obtained for TE-mode surface waves, where the surface reactance is capacitive.

By using the relationship between X_s and g, the surface reactance of the LWA can be realized as an array of unit cells to build the 1-D metasurface LWA. The application of the transverse resonance method to determine the surface reactance of a high-impedance surface unit cell is also demonstrated with a MATLAB program entitled **TRM_HIS** that is included at the end of this chapter.

Conventionally, one-dimensional periodic metasurface LWAs have been impedance-modulated with a sinusoidal function. However, a square-wave modulated metasurface, shown in Figure 16-73, is a proposed design for low-observable applications [98].

These metasurface leaky-wave antennas can form high-gain fan beams while inherently having a low monostatic radar cross section (RCS) for parallel polarization. The modulation parameters such as the average surface reactance, the modulation index, and the modulation period control both the radiation and RCS characteristics. In this way, these low-observable LWAs are codesigned for both leaky-wave behavior and RCS reduction. High-resolution artwork used to fabricate prototypes of such antennas, with square-wave modulation, is displayed

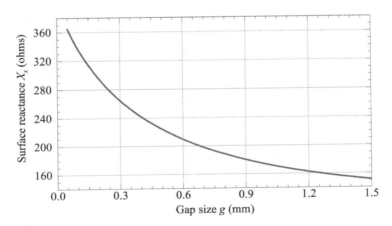

Figure 16-72   Plot of TM-mode surface reactance X_s versus gap size g for a unit cell of size $a = 3$ mm (the width of the patch $w = a - g$) at 15 GHz, with a substrate of dielectric constant $\varepsilon_r = 3$ and thickness $h = 1.524$ mm. (Source: S. Ramalingam, C. A. Balanis, C. R. Birtcher, and S. Pandi, "Analysis and design of checkerboard leaky-wave antennas with low radar cross section," *IEEE Open J. Antennas Propagat.*, vol. 1, pp. 26–40, 2020.)

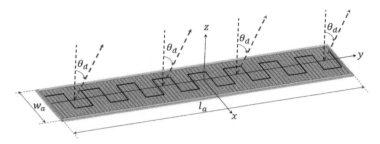

Figure 16-73   A square-wave modulated one-dimensionally periodic metasurface LWA. (Source: S. Ramalingam, "Impedance modulated metasurface antennas," Ph.D. dissertation, School of Electrical, Computer, and Energy Engineering, Arizona State University, 2020.)

Figure 16-74   Artwork used to fabricate a prototype of the square-wave modulated LWA. (Source: S. Ramalingam, C. A. Balanis, C. R. Birtcher, and S. Pandi, "Analysis and design of checkerboard leaky-wave antennas with low radar cross section," *IEEE Open J. Antennas Propagat.*, vol. 1, pp. 26–40, 2020.)

in Figure 16-74 [97]. The measured and simulated E-plane radiation patterns and the monostatic RCS patterns are shown in Figure 16-75 [97].

While sinusoidally modulated metasurfaces usually form fan beams with linear polarization, circularly polarized fan beams can be formed by multilayered *cascaded metasurfaces*, which consist of patterned anisotropic metallic layers with tensor impedances separated by dielectrics on a dielectric-covered ground plane [99].

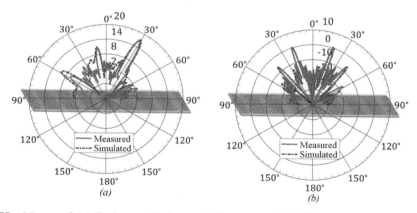

Figure 16-75   Measured (*a*) E-plane radiation and (*b*) monostatic RCS patterns compared to simulations. (Source: S. Ramalingam, C. A. Balanis, C. R. Birtcher, and S. Pandi, "Analysis and design of checkerboard leaky-wave antennas with low radar cross section," *IEEE Open J. Antennas Propagat.*, vol. 1, pp. 26–40, 2020.)

Example 16-5

Design a TM-mode planar metasurface leaky-wave antenna to form a fan beam along 45° from the broadside direction at 15 GHz. The variation of the surface impedance with the gap size is shown in Figure 16-72. Determine the modulation period required to form a fan beam using the $n = -1$ Floquet-Bloch mode.

Solution: From Figure 16-72, a value of the surface reactance in the middle of the plot is 230 ohms. Assume a modulation index $M = 0.25$.

Based on these and a frequency of 15 GHz,

$$\beta_0 = \omega\sqrt{\mu_0\varepsilon_0} = 2\pi f\sqrt{\mu_0\varepsilon_0} = 100\pi = 314.16 \text{ rad/m}$$

Based on (16-48)

$$\beta_{-1} = \beta_0 \sin 45° = 222.14 \text{ rad/m}$$

Using (16-46)

$$222.11 = 314.16\sqrt{1 + \left(\frac{230}{377}\right)^2} - \frac{2\pi}{p}$$

leads to a modulation period of $p = 43.1$ mm.

Metasurface LWAs can be conformed on curved surfaces, which makes them suitable for vehicular applications. One of the features of such an application, referred to as *automotive radar technology*, is illustrated in Figure 16-76.

Automotive radar technology is employed for driver assistance and safety applications. Some important features of this technology include collision avoidance, lane change assistance, parking aid, adaptive cruise control, and blind spot detection.

When the metasurface LWA is conformed on a curved aperture, where the direction of surface-wave propagation is nonplanar, the curvature causes defocusing. The phase constant can be varied locally to compensate for the defocusing introduced by the curvature [100, 101]. Alternatively, the impact of curvature can be compensated by using the holographic principle to design the conformal LWAs [102, 103]. On a cylindrically curved surface, the surface impedance can be modulated along the axis or the circumference of the cylinder, as shown in Figures 16-77a and 16-77b.

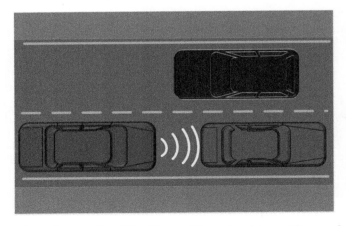

Figure 16-76 Automotive radars employed for driver safety and assistance. (Source: S. Ramalingam, C. A. Balanis, C. R. Birtcher, and S. Pandi, "Analysis and design of checkerboard leaky-wave antennas with low radar cross section," *IEEE Open J. Antennas Propagat.*, vol. 1, pp. 26–40, 2020.)

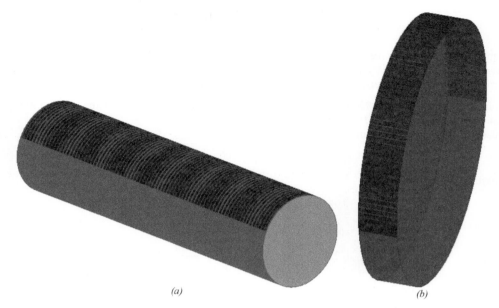

(a)   *(b)*

Figure 16-77   (*a*) Axially and (*b*) circumferentially modulated LWA. (Source: S. Ramalingam, "Impedance modulated metasurface antennas," Ph.D. dissertation, School of Electrical, Computer, and Energy Engineering, Arizona State University, 2020.)

In a circumferentially modulated LWA, the surface impedance is modulated along the *circumference* of the cylindrical surface. The electrically long flat metallic strips are placed transverse to the direction of surface-wave propagation. In an axially modulated LWA, the metasurface is impedance modulated along the *axis* of the cylindrical surface. In this case, the electrically long metallic strips are curved, and the antenna is flat along the direction of propagation of the surface waves. A fabricated axially modulated metasurface and a comparison between its measured and simulated E-plane radiation patterns are displayed in Figure 16-78 [103].

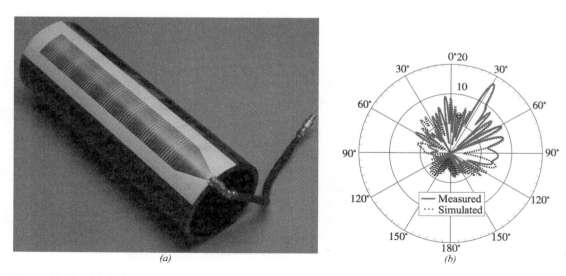

(a)   *(b)*

Figure 16-78   (*a*) A fabricated prototype of an axially modulated LWA and (*b*) its measured E-plane radiation pattern. (Source: S. Ramalingam, C. A. Balanis, C. R. Birtcher, S. Pandi, and H. N. Shaman, "Axially modulated cylindrical metasurface leaky-wave antennas," *IEEE Antennas Wireless Propagat. Lett.*, vol. 17, no. 1, pp. 130–133, Jan. 2018. Reproduced with permission from IEEE.)

16.12.3 Two-Dimensional Holographic Metasurface LWAs

In two-dimensional (2-D) LWAs, the leaky waves are guided in a 2-D guiding structure. The LWAs are usually excited at the center using a monopole or a patch, and the leaky waves propagate radially, as shown in Figure 16-79.

For 2-D metasurface LWAs, the period of radial surface-impedance modulation p needed to form a pencil beam along an angle θ_d, in the $\phi = \phi_d$ plane, is represented by [104, 105]

$$p(\phi) = \frac{2\pi}{\beta_n - \beta_0 \sin\theta_d \cos(\phi - \phi_d)} \tag{16-55}$$

The surface-impedance modulation function Z_s needed to form a pencil beam along the desired angular direction (θ_d, ϕ_d) can be derived using the holographic principle [57]. The reference wave Ψ_{ref} is the radially propagating leaky-wave mode expressed as

$$\Psi_{ref} = e^{-j\beta\sqrt{(x-x_c)^2 + (y-y_c)^2}} \tag{16-56}$$

while the objective wave is a plane wave propagating along the desired direction represented by

$$\Psi_{obj} = e^{-j\beta_0(x\sin\theta_d\cos\phi_d + y\sin\theta_d\sin\phi_d)} \tag{16-57}$$

The surface-impedance modulation function $Z_s(x,y)$ can be obtained as an interference pattern of the reference and objective waves using [57]

$$Z_s(x,y) = \pm jX_s(x,y) = \pm jX_a\left[1 + M\,\mathrm{Re}\left(\Psi_{ref}^*\Psi_{obj}\right)\right] \tag{16-58}$$

where X_a and M are the average surface reactance and modulation index, as defined in (16-45) and Figure 16-68, and (x_c, y_c) represents the location of the source. The holographic pattern can be viewed as concentric ellipses, whose major axis and the normal to the surface form the elevation plane of the formed pencil beam; this can also be observed in Figure 16-79. The surface-reactance profile derived using the holographic principle can be realized using high-impedance surface unit cells, of the form shown in Figure 16-70 as detailed in Section 16.12.2.

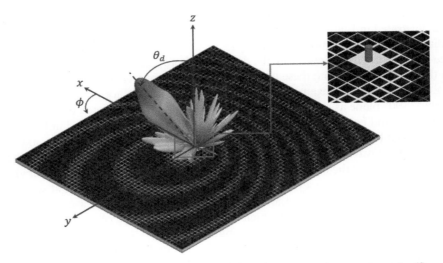

Figure 16-79 A 2-D holographic metasurface LWA fed at the center using a monopole. (Source: Adapted from S. Ramalingam, "Impedance modulated metasurface antennas," Ph.D. dissertation, School of Electrical, Computer, and Energy Engineering, Arizona State University, 2020.)

Example 16-6

Derive the surface-reactance modulation function for a TM-mode two-dimensional holographic meta-surface to form a pencil beam along ($\theta_d = 45°$, $\phi_d = 0°$). The substrate used is the same as that in Example 16-5, $\varepsilon_r = 3$ and $h = 1.524$ mm, whose surface-reactance plot is shown in Figure 16-72. The metasurface aperture is centered at the origin, and it is excited using a monopole, as shown in Figure 16-79. Assume a frequency of 15 GHz.

Solution: From Example 16-5, X_a = 230 ohms and M = 0.25. At a frequency of 15 GHz,

$$\beta_0 = 100\pi = 314.16 \text{ rad/m}$$

The phase constant β of the fundamental Floquet mode can be determined from (16-46) as

$$\beta = 314.16\sqrt{1 + \left(\frac{230}{377}\right)^2} = 368 \text{ rad/m}$$

From (16-56)

$$\Psi_{\text{ref}} = e^{-j368r}, r = \sqrt{x^2 + y^2}$$

From (16-57)

$$\Psi_{\text{obj}} = e^{-j314.16\sin 45°x} = e^{-j222.14x}$$

For the TM mode, the surface reactance is inductive. Therefore, the surface-reactance modulation function can be expressed as

$$X_s(x,y) = 230 + 57.5\cos(368r - 222.14x)$$
$$= 230 + 57.5\cos\left(368\sqrt{x^2 + y^2} - 222.14x\right) \text{ ohms}$$

While mapping the impedance profile $Z_s(x,y)$ of the metasurface, the position of any point (x,y) must be expressed in meters measured from the center/origin of the metasurface shown in Figure 16-79.

16.12.4   Radiation Mechanism of 2-D Holographic Metasurfaces

When a holographic metasurface is excited at the center to form a pencil beam in the desired direction (θ_d, ϕ_d), as illustrated in Figure 16-79, the pencil beam is formed from the constructive interference of the radiation from the radially propagating leaky waves. This necessitates the leaky waves in one half of the metasurface to radiate along the direction of propagation, and those in the other half to radiate in the opposite direction; these types of leaky waves are referred to as *forward* and *backward leaky waves*, respectively. For example, if the holographic metasurface is designed to form a pencil beam in the $\phi = 90°$ plane as shown in Figure 16-79, radiation is facilitated by forward leaky waves in the $y > 0$ plane and backward leaky waves in the $y < 0$ plane, as illustrated in Figure 16-80.

It can be shown from (16-48) that, with an increase in frequency, the beam maximum formed by the forward leaky-waves shifts toward the endfire direction, while that formed by the backward leaky waves shifts toward the broadside direction. This phenomenon forces the holographic meta-surface to operate at the frequency, referred to as *phase-crossover frequency*, at which the respective maxima of the radiation from the forward and backward leaky waves are along the same angular direction [105]. Frequencies at which the respective maxima of the forward and backward leaky-wave radiation are not along the same angular direction are referred to as *nonphase-cross-over frequencies*. When a holographic metasurface is operated at a nonphase-crossover frequency, the pencil beams formed by forward and backward leaky waves shift in opposite directions, causing a null, instead of a maximum, along the desired angular direction θ_d [60, 105].

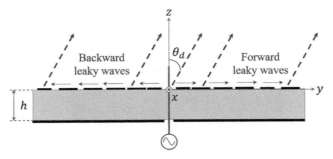

Figure 16-80 Forward and backward leaky waves in a holographic metasurface forming a pencil beam along θ_d from broadside.

16.12.5 Polarization-Diverse Holographic Metasurfaces

The polarization of the pencil beam formed by the antenna depends on the propagating leaky-wave mode, which in turn is a function of the frequency of operation and the high-impedance surface geometry. The first two modes of high-impedance surfaces are TM and TE, as detailed in this section. TM-polarized leaky waves form pencil beams with vertical polarization, while TE-polarized leaky waves form pencil beams with horizontal polarization [57]. Scalar impedance surfaces cannot independently control both TM- and TE-mode leaky waves. At the dominant TM mode, the horizontal polarized (E_ϕ) fields are canceled due to the symmetry of the metasurface along the major axis of the ellipses. However, as detailed in the previous section, the operation of conventional holographic metasurfaces is limited to the phase-crossover frequency. The concepts to control the polarization of the pencil beam formed by TM-polarized leaky waves in scalar impedance surfaces will be discussed.

A. Vertical Polarization Holographic metasurfaces form a pencil beam with vertical (E_θ) polarization while operating at a non-phase crossover frequency when the surface-reactance profile on one part of the surface (split by the line passing through the position of the feed and along the minor axis of the ellipses) is shifted by 180° with respect to the other. As discussed in the previous section, one part of the metasurface radiates using forward leaky waves, while the other uses backward leaky waves; these are 180° out of phase. Hence, the additional 180° phase difference reinforces radiation from forward and backward leaky waves.

The objective wave in (16-57) to form a pencil beam with vertical polarization can be rewritten as either (16-59) or (16-60) [60, 104]

$$\Psi_{obj} = \begin{cases} e^{-j\beta_0 (x\sin\theta_d \cos\phi_d + y\sin\theta_d \sin\phi_d)} & \forall y \geq y_c \\ -e^{-j\beta_0 (x\sin\theta_d \cos\phi_d + y\sin\theta_d \sin\phi_d)} & \forall y < y_c \end{cases} \tag{16-59}$$

or

$$\Psi_{obj} = \begin{cases} -e^{-j\beta_0 (x\sin\theta_d \cos\phi_d + y\sin\theta_d \sin\phi_d)} & \forall y \geq y_c \\ e^{-j\beta_0 (x\sin\theta_d \cos\phi_d + y\sin\theta_d \sin\phi_d)} & \forall y < y_c \end{cases} \tag{16-60}$$

The holographic surface-reactance pattern that includes the 180° phase difference to form a pencil beam along the desired direction θ_d in the $\phi = 90°$ plane is shown in Figure 16-81. It can be seen from Figure 16-81 that the symmetric halves (split along the major axis of the ellipses) are modulated with the same surface-reactance modulation. Hence, the E_ϕ components cancel in the far-field region resulting in a vertically (E_θ) polarized pencil beam along the desired angular direction.

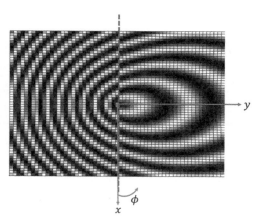

Figure 16-81   The holographic surface-reactance pattern to form a vertically polarized pencil beam along the $\phi = 90°$ plane. (Source: S. Pandi, C. A. Balanis, and C. R. Birtcher, "Design of scalar impedance holographic meta-surfaces for antenna beam formation with desired polarization," *IEEE Trans. Antennas Propagat.*, vol. 63, pp. 3016–3024, 2015. Reproduced with permission from IEEE.)

B. Horizontal Polarization   Horizontally (E_ϕ) polarized pencil beams are obtained by disturbing the symmetry, along the major axis of the ellipses, to avoid the cancellation of the E_ϕ components in the far-field region. The surface-reactance profile in one half (split along the major axis of the ellipses) is shifted by $180°$, with respect to the other to reinforce the E_ϕ components.

The corresponding objective wave can be represented by either (16-61) or (16-62) [104].

$$\Psi_{obj} = \begin{cases} e^{-j\beta_0(x\sin\theta_d\cos\phi_d + y\sin\theta_d\sin\phi_d)} & \forall x \geq x_m \\ -e^{-j\beta_0(x\sin\theta_d\cos\phi_d + y\sin\theta_d\sin\phi_d)} & \forall x < x_m \end{cases} \tag{16-61}$$

or

$$\Psi_{obj} = \begin{cases} -e^{-j\beta_0(x\sin\theta_d\cos\phi_d + y\sin\theta_d\sin\phi_d)} & \forall x \geq x_m \\ e^{-j\beta_0(x\sin\theta_d\cos\phi_d + y\sin\theta_d\sin\phi_d)} & \forall x < x_m \end{cases} \tag{16-62}$$

where $x_m = (y - y_c)\cot\phi_d + x_c$ is the major axis of the ellipses. The holographic surface-reactance pattern to obtain a horizontally polarized pencil beam, at a desired angular direction θ_d in the $\phi = 90°$ plane, is shown in Figure 16-82.

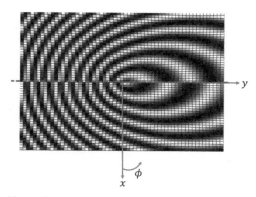

Figure 16-82   The holographic surface-reactance pattern to form a horizontally polarized pencil beam along the $\phi = 90°$ plane. (Source: Adapted from S. Pandi, C. A. Balanis, and C. R. Birtcher, "Design of scalar impedance holographic meta-surfaces for antenna beam formation with desired polarization," *IEEE Trans. Antennas Propagat.*, vol. 63, pp. 3016–3024, 2015. Reproduced with permission from IEEE.)

As the metasurface is operated at a nonphase-crossover frequency, the radiation from the forward and backward leaky waves are out of phase. Hence, E_θ fields cancel in the far-field region, resulting in a horizontally (E_ϕ) polarized pencil beam.

C. Circular Polarization   As discussed in Section 4.4.2, circular polarization can be achieved if there are two orthogonal linearly polarized field components, equal in magnitude and with a 90° phase difference. Hence, the conventional metasurface is geometrically modified to reinforce both E_θ and E_ϕ components in the far-field region, and a 90° phase difference included in their objective functions. Hence, the resulting objective function can be expressed by [60]

$$\Psi_{obj} = 0.5(\Psi_{obj1} \pm j\Psi_{obj2}) \tag{16-63}$$

$$\Psi_{obj1} = \begin{cases} e^{-j\beta_0(x\sin\theta_d\cos\phi_d + y\sin\theta_d\sin\phi_d)} & \forall y \geq y_c \\ -e^{-j\beta_0(x\sin\theta_d\cos\phi_d + y\sin\theta_d\sin\phi_d)} & \forall y < y_c \end{cases} \tag{16-64}$$

$$\Psi_{obj2} = \begin{cases} e^{-j\beta_0(x\sin\theta_d\cos\phi_d + y\sin\theta_d\sin\phi_d)} & \forall x \geq x_m \\ -e^{-j\beta_0(x\sin\theta_d\cos\phi_d + y\sin\theta_d\sin\phi_d)} & \forall x < x_m \end{cases} \tag{16-65}$$

A $\pm j$ is included in (16-63) to introduce a phase difference of 90° between the E_θ and E_ϕ components; the $\pm$ sign also determines the sense of rotation. The holographic pattern needed to form a pencil beam along ($\theta = \theta_d, \phi = 90°$) with circular polarization is shown in Figure 16-83.

D. Multiple Polarizations   The concepts discussed so far can be used to design a single holographic metasurface that can form, one at a time, a pencil beam with a vertical, horizontal, or circular polarization. The antenna is fed by two or more sources instead of one, to achieve polarization diversity. Each source or a combination of sources is designed to enable the metasurface to form a pencil beam with the desired polarization. When a holographic metasurface is fed by multiple sources the total surface-reactance modulation function $X_s(x, y)$ can be obtained as [106]

$$X_s(x, y) = X_a + M_1\text{Re}\left(\Psi^*_{ref\_1}\Psi_{obj\_1}\right) + M_2\text{Re}\left(\Psi^*_{ref\_2}\Psi_{obj\_2}\right) \tag{16-66}$$

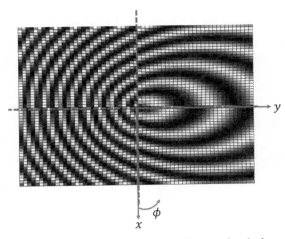

Figure 16-83   The holographic surface-reactance pattern to form a circularly polarized pencil beam along the $\phi = 0$ plane. (Source: S. Pandi, C. A. Balanis, and C. R. Birtcher, "Design of scalar impedance holographic metasurfaces for antenna beam formation with desired polarization," *IEEE Trans. Antennas Propagat.*, vol. 63, pp. 3016–3024, 2015. Reproduced with permission from IEEE.)

where subscripts 1 and 2 denote the parameters pertaining to the first and second sources, respectively. Hence, a polarization-diverse metasurface can be designed by feeding the metasurface with two sources and modulating the surface reactance with the function [104]

$$X_s(x,y) = X_a + M_1 \text{Re}\left(\Psi_{\text{ref\_1}}^* \Psi_{\text{obj\_1}}\right) \pm jM_2 \text{Re}\left(\Psi_{\text{ref\_2}}^* \Psi_{\text{obj\_2}}\right) \qquad (16\text{-}67)$$

where $\Psi_{\text{ref\_1}}$ and $\Psi_{\text{ref\_2}}$ are defined similar to (16-56), and $\Psi_{\text{obj\_1}}$ and $\Psi_{\text{obj\_2}}$ are expressed similar to (16-59)–(16-65). The metasurface can be designed to form a pencil beam with vertical polarization when excited by source 1 and horizontal polarization when excited by source 2, as illustrated in Figure 16-84.

The $\pm j$ term is included to enable the metasurface to form a circularly polarized pencil beam when excited by both sources and the $\pm$ sign also determines the sense of rotation; clockwise or counterclockwise. Holographic metasurfaces can be designed to include circular polarization with both senses of rotation, besides vertical and horizontal, by feeding them with three sources. A fabricated prototype of a polarization-diverse holographic metasurface and the measured circular polarized patterns, compared to the simulations, are shown in Figure 16-85. The design of two-dimensional periodic metasurface LWAs to form a pencil beam in the desired angular direction with the desired polarization is demonstrated using the MATLAB program entitled **Two_Dim_LWA** that is included at the end of this chapter.

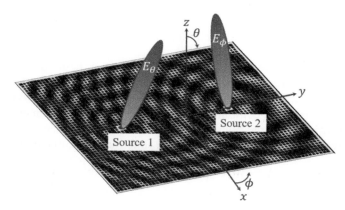

Figure 16-84 A polarization-diverse holographic metasurface. (Source: S. Ramalingam, "Impedance modulated metasurface antennas,"Ph.D. dissertation, School of Electrical, Computer, and Energy Engineering, ASU, 2020.)

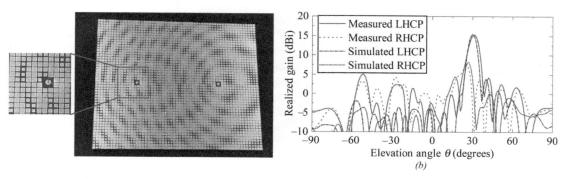

Figure 16-85 (*a*) A fabricated prototype of a polarization-diverse metasurface LWA. (*b*) Comparison of measured and simulated circularly polarized radiation patterns. (Source: S. Ramalingam, C. A. Balanis, C. R. Birtcher, and H. N. Shaman, "Polarization-diverse holographic metasurfaces," *IEEE Antennas Wireless Propagat. Lett.*, vol. 18, no. 2, pp. 264–268, Feb. 2019. Reproduced with permission from IEEE.)

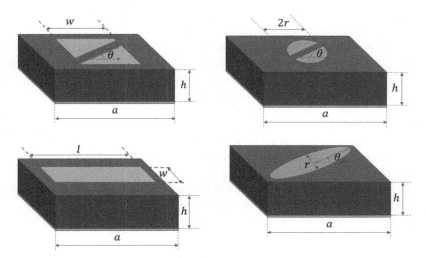

Figure 16-86 Anisotropic unit cells, proposed in the literature, that are used to realize a tensor surface reactance. (Source: S. Ramalingam, "Impedance modulated metasurface antennas," Ph.D. dissertation, School of Electrical, Computer, and Energy Engineering, Arizona State University, 2020.)

16.12.6 Tensor Impedance Surfaces

In tensor impedance surfaces, the electric and magnetic fields are related by an impedance tensor expressed as

$$\overline{\mathbf{Z}} = \begin{pmatrix} Z_{xx} & Z_{xy} \\ Z_{yx} & Z_{yy} \end{pmatrix} \tag{16-68}$$

The impedance tensor can be determined using the holographic principle and realized using anisotropic impedance surfaces. Some anisotropic unit cells that have been employed in metasurface antennas [57, 107] are exhibited in Figure 16-86.

Tensor impedance surfaces support both TE and TM modes as degenerate modes and can control both modes independently [108]. Hence, the aperture efficiency of anisotropic metasurfaces is much higher than that of scalar impedance metasurfaces.

16.13 MULTIMEDIA

On the website that accompanies this book, the following multimedia resources are included for the review, understanding, and presentation of the material of this chapter.

- **MATLAB** computer programs:
  a. **HIS_Mush** designs and computes the characterstics of a high-impedance surface (HIS) of Figures (16-4) and (16-9) based on the procedure and designs closed-form equations of Section 16.3.
  b. **TRM_Metasurface_Dispersion** plots the dispersion diagram for metasurface using the approximate TRM model developed in [30].
  c. **TRM_Grounded_Dielectric_Dispersion** develops and plots the dispersion diagram for dielectric-covered PEC based on the TRM as formulated and represented by (16-17).
  d. **AMC_Designs** designs a two-layer mushroom textured surface based on a desired resonant frequency and following Section 16.4. The program also plots the reflection phase versus frequency of the designed AMC surface.

e. **Dispersion_Diagram_Matlab** generates a dispersion diagram based on (16-2), (16-9), and (16-13), and on the formulation reported in [1].

f. **Dispersion_Diagram_HFSS**: As part of the package of this program, two files, *Unit Cell Design* and *Graphical User Interface (GUI)*, are included to aid in the generation of the dispersion diagram in HFSS. It is based on a simulation setup shown in Figure 16-11 using the Floquet ports and periodic boundary conditions in Ansys HFSS.

g. **Angular_Direction_Checkerboard** determines the angular directions of the scattered fields based on array theory of (16-22) and (16-23), and similar to Example 16-3.

h. **Conventional_Checkerboard_RCS_Reduction** determines RCS reduction of a conventional checkerboard surface based on an AMC design that is also entered by the user. This program is similar to Example 16-4.

i. **TRM_HIS** computes the surface reactance of a high-impedance surface unit cell based on the transverse-resonance method detailed in Section 16.12.2.

j. **One_Dim_LWA** determines the modulation period of a 1-D LWA to form a fan beam in the desired angular direction and plots the surface-reactance profile for both TM and TE modes.

k. **Two_Dim_LWA** computes and plots the surface-reactance profile to form a pencil beam in the desired angular direction with the desired polarization, with polarization diversity, for both TM and TE modes.

- **PowerPoint (PPT)** viewgraphs, in multicolor.

REFERENCES

1. D. Sievenpiper, "High-impedance electromagnetic surfaces," Ph.D. dissertation, Department of Electrical Engineering, UCLA, 1999.

2. D. Sievenpiper, "Artificial impedance surfaces," Chapter 15 in *Modern Antenna Handbook*, C. A. Balanis (Ed.), John Wiley & Sons, pp. 737–777, 2008.

3. Y. Rahmat-Samii and H. Mosallaei, "Electromagnetic band-gap structures: Classification, characterization, and applications," in *11th Int. Conf. Antennas Propagat. (ICAP 2001)*, Manchester, UK, Apr. 17-20, 2001.

4. H. Mosallaei and Y. Rahmat-Samii, "Periodic bandgap and effective materials in electromagnetics: Characterization and applications in nanocavities and waveguides," *IEEE Trans. Antennas Propagat.*, vol. 51, no. 3, pp. 549–563, Mar. 2003.

5. F. Yang and Y. Rahmat-Samii, "Reflection phase characterization of the EBG ground plane for low profile wire antenna applications," *IEEE Trans. Antennas Propagat.*, vol. 51, no. 10, pp. 2691–2703, Oct. 2003.

6. F. Yang and Y. Rahmat-Samii, *Electromagnetic Band Gap Structures in Antenna Engineering*, Cambridge University Press, UK, 2008.

7. E. Yablonovitch, "Photonic band-gap structures," *J. Opt. Soc. Amer. B*, vol. 10, no. 2, pp. 283–294, Feb. 1993.

8. E. R. Brown, D. D. Parker, and E. Yablonovitch, "Radiation properties of a planar antenna on a photonic-crystal substrate," *J. Opt. Soc. Amer. B*, vol. 10, no. 2, pp. 404–407, Feb. 1993.

9. C. A. Balanis, *Antenna Theory: Analysis and Design*, Fourth Edition, John Wiley & Sons, New York, 2016.

10. J. K. M. Jansen and M. E. J. Jeuken, "Surface waves in corrugated conical horn," *Electron. Lett.*, vol. 8, pp. 342–344, 1972.

11. Y. Rahmat-Samii and F. Yang, "Development of complex artificial ground planes in antenna engineering," Chapter 12 in *Metamaterials: Physics and Engineering Explorations*, N. Engheta and R. W. Ziolkowski (Eds.), IEEE Press, Wiley Inter-Science, New York, 2006.

12. P.-S. Kildal, "Artificially soft and hard surfaces in electromagnetics," *IEEE Trans. Antennas Propagat.*, vol. 38, no. 10, pp. 1537–1544, Oct. 1990.

13. "Special issue on "Artificial magnetic conductors, soft/hard surfaces, and other complex surfaces," *IEEE Trans. Antennas Propagat.*, vol. 53, no. 1, Jan. 2005.

14. C. A. Mentzer and J. L. Peters, "Properties of cutoff corrugated surfaces for corrugated horn design," *IEEE Trans. Antennas Propagat.*, vol. 22, no. 2, pp. 191–196, Mar. 1974.

15. C. A. Mentzer and J. L. Peters, "Pattern analysis of corrugated horn antennas," *IEEE Trans. Antennas Propagat.*, vol. 22, no. 3, pp. 304–309, May 1976.

16. B. M. Thomas, "Design of corrugated conical horns," *IEEE Trans. Antennas Propagat.*, vol. 26, no. 2, pp. 367–372, Mar. 1978.

17. W. D. Burnside and C. W. Chuang, "An aperture-matched horn design," *IEEE Trans. Antennas Propagat.*, vol. 30, no. 4, pp. 790–796, July 1982.

18. G. L. James, "TE_{11}-to-HE_{11} mode converters for small-angle corrugated horns," *IEEE Trans. Antennas Propagat.*, vol. 30, no. 6, pp. 1057–1062, Nov. 1982.

19. B. M. Thomas and K. J. Greene, "A curved-aperture corrugated horn having a very low cross-polar performance," *IEEE Trans. Antennas Propagat.*, vol. 30, no. 6, pp. 1068–1072, Nov. 1982.

20. B. M. Thomas, G. L. James, and K. J. Greene, "Design of wide-band corrugated conical horns for Cassegrain antennas," *IEEE Trans. Antennas Propagat.*, vol. 34, no. 6, pp. 750–757, Jun. 1986.

21. A. A. Kishk and P.-S. Kildal, "Asymptotic boundary conditions for strip-loaded scatterers applied to circular dielectric cylinders under oblique incidence," *IEEE Trans. Antennas Propagat.*, vol. 45, no. 1, pp. 51–56, Jan. 1997.

22. A. Monorchio, G. Manara, and L. Lanuaza, "Synthesis of artificial magnetic conductors by using multilayered frequency selective surfaces," *IEEE Antennas Wireless Propagat. Lett.*, vol. 1, no. 11, pp. 196–199, Nov. 2002.

23. D. J. Kern, D. H. Werner, A. Monorchio, L. Lanuzza, and M. J. Wilhelm, "The design synthesis of multiband artificial magnetic conductors using high impedance frequency selective surfaces," *IEEE Trans. Antennas Propagat.*, vol. 1, no. 53, pp. 8–17, Jan. 2005.

24. D. Sievenpiper, "Forward and backward leaky wave radiation with large effective aperture from an electronically tunable textured surface," *IEEE Trans. Antennas Propagat.*, vol. 1, no. 53, pp. 236–247, Jan. 2005.

25. T. Suzuki and P. L. Yu, "Experimental and theoretical study of dipole emission in the two-dimensional photonic band structures of the square lattice with dielectric cylinders," *J. Appl. Phys.*, vol. 49, no. 2, pp. 582–594, Jan. 1996.

26. ANSYS. HFSS 2019. Canonsburg, PA, USA, 2019. [Online]. Available: http://www.ansys.com.

27. J. D. Joannopoulos, R. D. Meade, and J. N. Winn, *Photonic Crystals Molding the Flow of Light*, Princeton University Press, NJ, 1995.

28. S. G. Johnson and J. D. Joannopoulos, *Photonic Crystals: The Road from Theory to Practice*, Kluwer Academic Publishers, Boston, 2002.

29. R. E. Collin, *Foundations for Microwave Engineering*, McGraw-Hill, New York, 1992.

30. O. Luukkonen, C. Simovski, G. Granet, G. Goussetis, D. Lioubtchenko, A. Risnen, and S. A. Tretyakov, "Simple and accurate analytical model of planar grids and high-impedance surfaces comprising metal strips or patches," *IEEE Trans. Antennas Propagat.*, vol. 56, no. 6, pp. 1624–1632, 2008.

31. M. Alharbi, "High-directive printed antennas for low-profile applications," Ph.D. dissertation, School of Electrical, Computer, and Energy Engineering, Arizona State University, 2020.

32. S. A. Tretyakov and C. R. Simovski, "Dynamic model of artificial reactive impedance surfaces," *J. of EM Waves and Applications*, vol. 17, no. 1, pp. 131–145, 2003.

33. A. Vallecchi, J. R. D. Luis, F. Capolino, and F. D. Flaviis, "Low profile fully planar folded dipole antenna on a high impedance surface," *IEEE Trans. Antennas Propagat.*, vol. 60, no. 1, pp. 51–62, 2012.

34. N. Engheta, "Thin absorbing screens using metamaterial surfaces," in *Proc. IEEE Antennas Propagat. Soc. Int. Symp.*, pp. 392–395, 2002.

35. M. Paquay, J. C. Iriarte, I. Ederra, R. Gonzalo, and P. de Maagt, "Thin AMC structure for radar cross-section reduction," *IEEE Trans. Antennas Propagat.*, vol. 55, no. 12, pp. 3630–3638, 2007.

36. W. Chen, C. A. Balanis, and C. R. Birtcher, "Checkerboard EBG surfaces for wideband radar cross section reduction," *IEEE Trans. Antennas Propagat.*, vol. 63, no. 3, pp. 2636–2645, 2015.

37. A. Y. Modi, C. A. Balanis, C. R. Birtcher, and H. Shaman, "Novel design of ultrabroadband radar cross section reduction surfaces using artificial magnetic conductors," *IEEE Trans. Antennas Propagat.*, vol. 65, no. 10, pp. 5406–5417, 2017.

38. W. Chen, C. A. Balanis, and C. R. Birtcher, "Dual wide-band checkerboard surfaces for radar cross section reduction," *IEEE Trans. Antennas Propagat.*, vol. 64, no. 9, pp. 4133–4138, 2016.

39. W. Chen, C. A. Balanis, and C. R. Birtcher, "Dual frequency band RCS reduction using checkerboard surfaces," in *Proc. of IEEE Int. Symp. Antennas Propagat.*, San Diego, CA, pp. 1913–1914, 2017.

40. A. Y. Modi, C. A. Balanis, and C. R. Birtcher, "AMC cells for broadband RCS reduction checkerboard surfaces," in *Proc. of IEEE Int. Symp. Antennas Propagat.*, San Diego, CA, pp. 1915–1916, 2017.

41. A. Y. Modi, C. A. Balanis, and C. R. Birtcher, "Novel technique for enhancing RCS reduction bandwidth of checkerboard surfaces," in *Proc. of IEEE Int. Symp. Antennas Propagat.*, San Diego, CA, pp. 1911–1912, 2017.

42. M. Z. Azad and M. Ali, "Novel wideband directional dipole antenna on a mushroom like EBG structure," *IEEE Trans. Antennas Propagat.*, vol. 56, pp. 1242–1250, 2008.

43. L. Akhoondzadeh-Asl, D. J. Kern, P. S. Hall, and D. H. Werner, "Wideband dipoles on electromagnetic bandgap ground planes," *IEEE Trans. Antennas Propagat.*, vol. 55, pp. 2426–2434, 2007.

44. H. Mosallaei and K. Sarabandi, "Antenna miniaturization and bandwidth enhancement using a reactive impedance substrate," *IEEE Trans. Antennas Propagat.*, vol. 52, pp. 2403–2414, 2004.

45. D. J. Kern, D. H. Werner, A. Monorchio, L. Lanuzza, and M. J. Wilhelm, "The design synthesis of multiband artificial magnetic conductors using high impedance frequency selective surfaces," *IEEE Trans. Antennas Propagat.*, vol. 53, pp. 8–17, 2005.

46. D. J. Kern, D. H. Werner, A. Monorchio, L. Lanuzza, and M. J. Wilhelm, "Reconfigurable ultra-thin EBG absorbers using conducting polymers," in *Antennas Propag. Soc. Int. Symp.*, pp. 204–217, 2005.

47. A. C. Durgun, C. A. Balanis, and C. R. Birtcher, "Reflection phase characterization of curved high impedance surfaces," *IEEE Trans. Antennas Propagat.*, vol. 61, pp. 6030–6038, 2013.

48. J. Sarrazin, A. C. Lepage, and X. Begaud, "Circular high-impedance surfaces characterization," *IEEE Antennas Wireless Propagat. Lett.*, vol. 11, pp. 260–263, 2012.

49. N. Llombart, A. Neto, and G. Gerini, "Planar circularly symmetric EBG structures for reducing surface waves in printed antennas," *IEEE Trans. Antennas Propagat.*, vol. 53, pp. 3210–3218, 2005.

50. A. Neto, N. Llombart, G. Gerini, and P. de Maagt, "On the optimal radiation bandwidth of printed slot antennas surrounded by EBGs," *IEEE Trans. Antennas Propagat.*, vol. 54, pp. 1074–1083, 2006.

51. M. Ettore, S. Bruni, G. Gerini, A. Neto, N. Llombart, and S. Maci, "Sector PCS-EBG antenna for low-cost high-directivity applications," *IEEE Antennas Wireless Propagat. Lett.*, vol. 6, pp. 537–539, 2007.

52. M. S. Rahimi, J. Rashed-Mohassel, and M. Edalatipour, "Radiation properties enhancement of a GSM/WLAN microstrip antenna using a dual band circularly symmetric EBG substrate," *IEEE Trans. Antennas Propagat.*, vol. 60, pp. 5491–5494, 2012.

53. T. A. Dendini, Y. Coulibaly, and H. Boutayeb, "Hybrid dielectric resonator antenna with circular mushroom-like structure for gain improvement," *IEEE Trans. Antennas Propagat.*, vol. 57, pp. 1043–1049, 2009.

54. M. A. Amiri, C. A. Balanis, and C. R. Birtcher, "Analysis, design and measurements of circularly symmetric high impedance surfaces for loop antenna applications," *IEEE Trans. Antennas Propagat.*, vol. 64, pp. 618–629, 2016.

55. M. A. Amiri, C. A. Balanis, and C. R. Birtcher, "Gain and bandwidth enhancement of spiral antenna using circularly symmetric HIS," *IEEE Antennas Wireless Propagat. Lett.*, vol. 16, pp. 1080–1083, 2017.

56. M. A. Amiri, C. A. Balanis, and C. R. Birtcher, "Notable gain enhancement of curvilinear elements using a circular HIS ground plane," in *Proc. of IEEE Int. Symp. Antennas Propagat.*, San Diego, CA, pp. 1671–1672, 2017.

57. B. H. Fong, J. S. Colburn, J. J. Ottusch, J. L. Visher, and D. F. Sievenpiper, "Scalar and tensor holographic artificial impedance surfaces," *IEEE Trans. Antennas Propagat.*, vol. 58, pp. 3212–3221, 2010.

58. A. M. Patel and A. Grbic, "A printed leaky-wave antenna based on a sinusoidally-modulated reactance surface," *IEEE Trans. Antennas Propagat.*, vol. 59, pp. 2087–2096, 2011.

59. S. Maci, G. Minatti, M. Casaletti, and M. Bosiljevac, "Metasurfing: Addressing waves on impenetrable metasurfaces," *IEEE Antennas Wireless Propagat. Lett.*, vol. 10, pp. 1499–1502, 2011.

60. S. Pandi, C. A. Balanis, and C. R. Birtcher, "Design of scalar impedance holographic metasurfaces for antenna beam formation with desired polarization," *IEEE Trans. Antennas Propagat.*, vol. 63, pp. 3016–3024, 2015.

61. G. Minatti, S. Maci, P. D. Vita, A. Freni, and M. Sabbadini, "A circularly-polarized isoflux antenna based on anisotropic metasurface," *IEEE Trans. Antennas Propagat.*, vol. 60, pp. 4998–5009, 2012.

62. R. Quarfoth and D. Sievenpiper, "Artificial tensor impedance waveguides," *IEEE Trans. Antennas Propagat.*, vol. 61, pp. 3597–3606, 2013.

63. W. W. Salisbury, "Absorbent body for electromagnetic waves," U.S. Patent 2 599 944, June 10, 1952.

64. R. L. Fante and M. T. McCormack, "Reflection properties of the salisbury screen," *IEEE Trans. Antennas Propagat.*, vol. 36, no. 10, pp. 1443–1454, Oct. 1988.

65. M. A. Alyahya, "Physical Optics modeling of AMC checkerboard surfaces for RCS reduction and low backscattering retrodirective array," Ph.D. dissertation, School of Electrical, Computer, and Energy Engineering, Arizona State University, 2020.

66. M. A. Alyahya, Personal communication, 2022.

67. M. Alyahya, C. A. Balanis, C. R. Birtcher, H. N. Shaman, and W. A. Alomar, "Physical Optics modeling of scattering by checkerboard structure for RCS reduction," in *Proc. IEEE Int. Symp. Antennas Propagat.*, Atlanta, GA, pp. 1693–1694, 2019.

68. A. Y. Modi, M. A. Alyahya, C. A. Balanis, and C. R. Birtcher, "Metasurface-based method for broadband RCS reduction of dihedral corner reflectors with multiple bounces," *IEEE Trans. Antennas Propagat.*, vol. 68, no. 3, pp. 1436–1447, Mar. 2020.

69. C. A. Balanis, M. A. Amiri, A. Y. Modi, S. Pandi, and C. R. Birtcher, "Applications of AMC-based impedance surfaces," *EPJ Appl. Metamaterials*, vol. 5, no. 3, pp. 1–15, 2018.

70. A. Y. Modi, C. A. Balanis, C. R. Birtcher, and H. Shaman, "New class of RCS-reduction metasurfaces based on scattering cancellation using array theory," *IEEE Trans. Antennas Propagat.*, vol. 67, no. 1, pp. 298–308, 2018.

71. W. Chen, C. A. Balanis, C. R. Birtcher, and A. Y. Modi, "Cylindrically curved checkerboard surfaces for radar cross-section reduction," *IEEE Antennas Wireless Propagat. Lett.*, vol. 17, no. 2, pp. 343–346, Feb. 2018.

72. A. Y. Modi, "Metasurface-based techniques for broadband radar cross-section reduction of complex structures," Ph.D. dissertation, School of Electrical, Computer, and Energy Engineering, Arizona State University, 2020.

73. A. Y. Modi, C. A. Balanis, C. R. Birtcher, and H. N. Shaman "Robust method for synthesizing low-RCS high-gain antennas using metasurfaces," in *Proc. IEEE Int. Symp. Antennas Propagat.*, Atlanta, GA, pp. 627–628, 2019.

74. A. Y. Modi, M. A. Alyahya, C. A. Balanis, and C. R. Birtcher "Modification to checkerboard metasurfaces for reducing the RCS of complex structures," in *IEEE Indian Conf. Antennas Propagat. (InCAP)*, Ahmedabad, India, pp. 1–2, 2019.

75. M. Alharbi, C. A. Balanis, C. R. Birtcher, and H. N. Shaman, "Hybrid circular ground planes for high-realized-gain low-profile loop antennas," *IEEE Antennas Wireless Propagat. Lett.*, vol. 17, no. 8, pp. 1426–1429, Aug. 2018.

76. D. Wen, Y. Hao, M. O. Munoz, H. Wang, and H. Zhou, "A compact and low-profile MIMO antenna using a miniature circular high-impedance surface for wearable applications," *IEEE Trans. Antennas Propagat.*, vol. 66, no. 1, pp. 96–104, Jan. 2018.

77. D. M. Pozar and D. H. Schaubert, "Scan blindness in infinite phased arrays of printed dipoles," *IEEE Trans. Antennas Propagat.*, vol. 32, no. 6, pp. 602–610, Jun. 1984.

78. F. Zavosh and J. T. Aberle, "Infinite phased arrays of cavity-backed patches," *IEEE Trans. Antennas Propagat.*, vol. 42, no. 3, pp. 390–398, Mar. 1984.

79. L. Zhang, J. A. Castaneda, and N. G. Alexopoulos, "Scan blindness free phased array design using PBG materials," *IEEE Trans. Antennas Propagat.*, vol. 52, no. 8, pp. 2000–2007, Aug. 2004.

80. M. Alharbi, C. A. Balanis, and C. R. Birtcher, "Performance enhancement of square-ring antennas exploiting surface-wave metasurfaces," *IEEE Antennas Wireless Propagat. Lett.*, vol. 18, no. 10, pp. 1991–1995, Oct. 2019.

81. F. Yang, A. Aminian, and Y. Rahmat-Samii, "A novel surface-wave antenna design using a thin periodically loaded ground plane," *Microw. Opt. Technol. Lett.*, vol. 47, no. 3, pp. 240–245, Sept. 2005.

82. F. Yang, Y. Rahmat-Samii, and A. Kishk, "Low-profile patch-fed surface wave antenna with a monopole-like radiation pattern," *IET Microw. Antennas Propagat.*, vol. 1, no. 1, pp. 261–266, Feb. 2007.

83. W. F. Croswell, T. Durham, M. Jones, D. Schaubert, P. Friederich, and J. G. Maloney, "Wideband arrays," Chapter 12 in *Modern Antenna Handbook*, C. A. Balanis (Ed.), John Wiley & Sons, New York, pp. 581–629, 2008.

84. F. Costa, O. Luukkonen, C. R. Simovski, A. Monorchio, S. A. Tretyakov, and P. M. de Maagt, "TE surface wave resonances on high-impedance surface based antennas: Analysis and modeling," *IEEE Antennas Wireless Propagat. Letters*, vol. 59, no. 10, pp. 3588–3596, Oct. 2011.

85. S. X. Ta and I. Park, "Low-profile broadband circularly polarized patch antenna using metasurface," *IEEE Antennas Wireless Propagat. Letters*, vol. 63, no. 12, pp. 5929–5934, Dec. 2015.

86. D. R. Jackson and A. A. Oliner, "Leaky-Wave Antennas," Chapter 7 in *Modern Antenna Handbook*, C. A. Balanis (Ed.), John Wiley & Sons, pp. 325–368, 2008.

87. A. A. Oliner and D. R. Jackson, "Leaky-wave antennas," Chapter 11 in *Antenna Engineering Handbook*, J. L. Volakis (Ed.), McGraw-Hill, New York, 2007.

88. F. L. Whetten and C. A. Balanis, "Meandering long slot leaky-wave waveguide-antennas," *IEEE Trans. Antennas Propag.*, vol. 39, no. 11, pp. 1553–1560, Nov. 1991.

89. W. Rotman and N. Karas, "The sandwich wire antenna: A new type of microwave line source radiator," *1958 IRE Int. Conv. Rec.*, New York, NY, pp. 166–172, 1957.

90. L. Goldstone and A. Oliner, "Leaky-wave antennas I: Rectangular waveguides," *IRE Trans. Antennas Propagat.*, vol. 7, no. 4, pp. 307–319, Oct. 1959.

91. W. Rotman and A. A. Oliner, "Periodic structures in trough waveguide," *IRE Trans. Microw. Theory Tech.*, vol. 7, no. 1, pp. 134–142, Jan. 1959.

92. A. A. Oliner and A. Hessel, "Guided waves on sinusoidally-modulated reactance surfaces," *IRE Trans. Antennas Propagat.*, vol. 7, pp. 201–208, 1959.

93. S. Paulotto, P. Baccarelli, F. Frezza, and D. R. Jackson, "Novel technique for open-stopband suppression in 1-D periodic printed leaky-wave antenna," *IEEE Trans. Antennas Propagat.*, vol. 57, no. 7, pp. 1894–1906, July 2009.

94. C. Caloz, T. Itoh, and A. Rennings, "CRLH metamaterial leaky-wave and resonant antennas," *IEEE Antennas Propagat. Mag.*, vol. 50, no. 5, pp. 25–39, Oct. 2008.

95. D. R. Jackson, C. Caloz, and T. Itoh, "Leaky-wave antennas," *Proceedings of the IEEE*, vol. 100, no. 7, pp. 2194–2206, Jul. 2012.

96. P. Hariharan, *Optical Holography: Principles Techniques and Applications*, Cambridge Univ. Press, Cambridge, UK, 1996.

97. S. Ramalingam, C. A. Balanis, C. R. Birtcher, and S. Pandi, "Analysis and design of checkerboard leaky-wave antennas with low radar cross section," *IEEE Open J. Antennas Propagat.*, vol. 1, pp. 26–40, 2020.

98. S. Ramalingam, "Impedance modulated metasurface antennas," Ph.D. dissertation, School of Electrical, Computer, and Energy Engineering, Arizona State University, 2020.

99. B. B. Tierney and A. Grbic, "Controlling leaky waves with 1-D cascaded metasurfaces," *IEEE Trans. Antennas Propagat.*, vol. 66, no. 4, pp. 2143–2146, Apr. 2018.

100. J. L. Gómez-Tornero, "Analysis and design of conformal tapered leaky-wave antennas," *IEEE Antennas Wireless Propagat. Lett.*, vol. 10, pp. 1068–1071, 2011.

101. A. Foroozesh, R. Paknys, D. R. Jackson, and J.-J. Laurin, "Beam focusing using backward-radiating waves on conformal leaky-wave antennas based on a metal strip grating," *IEEE Trans. Antennas Propagat.*, vol. 63, no. 11, pp. 4667–4677, Nov. 2015.

102. S. Pandi, C. A. Balanis, and C. R. Birtcher, "Curvature modeling in design of circumferentially modulated cylindrical metasurface LWA," *IEEE Antennas Wireless Propagat. Lett.*, vol. 16, pp. 1024–1027, 2017.

103. S. Ramalingam, C. A. Balanis, C. R. Birtcher, S. Pandi, and H. N. Shaman, "Axially modulated cylindrical metasurface leaky-wave antennas," *IEEE Antennas Wireless Propagat. Lett.*, vol. 17, no. 1, pp. 130–133, Jan. 2018.

104. S. Ramalingam, C. A. Balanis, C. R. Birtcher, and H. N. Shaman, "Polarization-diverse holographic metasurfaces," *IEEE Antennas Wireless Propagat. Lett.*, vol. 18, no. 2, pp. 264–268, Feb. 2019.

105. M. Nannetti, F. Caminita, and S. Maci, "Leaky-wave based interpretation of the radiation from holographic surfaces," in *IEEE Antennas Propagat. Int. Symp.*, Honolulu, HI, pp. 5813–5816, June 2007.

106. D. González-Ovejero, G. Minatti, G. Chattopadhyay, and S. Maci, "Multibeam by metasurface antennas," *IEEE Trans. Antennas Propagat.*, vol. 65, no. 6, pp. 2923–2930, Jun. 2017.

107. G. Minatti, M. Faenzi, E. Martini, F. Caminita, P. De Vita, D. González-Ovejero, M. Sabbadini, and S. Maci, "Modulated metasurface antennas for space: synthesis, analysis and realizations," *IEEE Trans. Antennas Propagat.*, vol. 63, no. 4, pp. 1288–1300, Apr. 2015.

108. R. Quarfoth, "Anisotropic artificial impedance surfaces," Ph.D. dissertation, Department of Electrical Engineering, University of California at San Diego, 2014.

PROBLEMS

16.1. Design a two-layer mushroom textured surface with hexagonal patches and air as the upper layer ($\varepsilon_2 = \varepsilon_0$, $\mu_2 = \mu_0$), as shown in Figures 16-4 and 16-8 and outlined in Section 16.4 to exhibit PMC characteristics between:
(a) 14 and 16 GHz
(b) 12 and 18 GHz

with a center frequency of $f_0 = 15$ GHz. Use a Rogers RT/duroid 3006 with a dielectric constant of $\varepsilon_r = 6.15$. The hexagonal patches are supported by metallic circular posts that connect the patches to the bottom ground plane through vias in the substrate. *Use the* **MATLAB** *computer program* **AMC_Designs** *to verify the solution design.*

16.2. Derive, showing all the details, the RCS expression in closed form for the PMC ground plane of Figure 16-15b as given by (16-18) for TE^x polarization.

16.3. Derive, showing all the details, the RCS expression in closed form for the PMC ground plane of Figure 16.15b for the perpendicular polarization (TM^x), as shown in Figure 11-8, where incident electric and magnetic fields can now be written as

$$\mathbf{E}^i = \hat{\mathbf{a}}_x E_0\, e^{-j\beta(y\sin\theta_i - z\sin\theta_i)}$$

$$\mathbf{H}^i = -\frac{E_0}{\eta}\left(\hat{\mathbf{a}}_y \cos\theta_i + \hat{\mathbf{a}}_z \sin\theta_i\right)$$
$$\times e^{-j\beta(y\sin\theta_i - z\sin\theta_i)}$$

16.4. Derive, showing all the details, the RCS expression in closed form for the PEC/PMC ground plane of Figure 16-16a as given by (16-21a) for TE^x polarization.

16.5. The electric field radiated by an electrical horizontal dipole in free space is given by:

$$E_\psi^d = j\eta \frac{kI_0 le^{-jkr_1}}{4\pi r}\sqrt{1-(\sin\theta\,\sin\phi)^2}$$

Determine the:
(a) Reflected electric field when the dipole is placed a height h above an infinite PEC plate.
(b) Total electric field.
(c) Power radiated (P_{rad}) by the total radiated field.

Hint: $P_{\text{rad}} = \dfrac{1}{2\eta} \displaystyle\int_0^{2\pi} \int_0^{\pi/2} |E_\psi|^2$

$$\times r^2 \sin\theta \, d\theta \, d\phi$$

16.6. Repeat Problem 16.5 when the horizontal electric dipole is placed a height h above an infinite PMC.

16.7. The radiation resistance of a radiating element is defined as

$$R_r = \frac{2P_{\text{rad}}}{(I_0)^2}$$

Find the radiation resistance of a horizontal electric dipole when it is placed a height h above:
(a) Infinite PEC ground plane.
(b) Infinite PMC ground plane.

16.8. For Rogers RT/duroid 5880 slab ($\varepsilon_r = 2.2$) covered PEC, perform the following:
(a) Plot the dispersion diagram when the slab thickness h is 1 mm.
(b) Highlight the tightly and loosely bound regions.

16.9. For the design in Example 16-6, examine if the second Floquet mode ($n = -2$) radiates. If so, determine (in degrees) the angular direction (from the normal to the strip) of the formed fan beam. (*Note*: If the second Floquet mode radiates, the formed fan beam is a minor lobe while the radiation from the first Floquet mode is the major lobe.)

16.10. Derive (16-45) using the holographic principle for a planar one-dimensional (1-D) periodic metasurface leaky-wave antenna.

16.11. Prove that for an axially modulated metasurface LWA, whose geometry is shown in Figure P16-11, the surface-impedance modulation function derived using the holographic principle is same as that of a planar LWA given by

$$Z_s(y,z) = jX_a \left[1 + M \cos\left(\frac{2\pi}{p} z \right) \right]$$

(*Note*: The effect of curvature can be modeled using the projected-aperture method. For further details, refer to [103].)

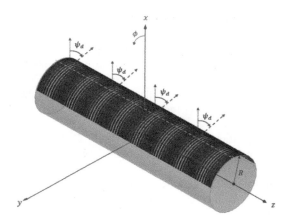

Figure P16-11 (Source: S. Ramalingam, C. A. Balanis, C. R. Birtcher, S. Pandi, and H. N. Shaman, "Axially modulated cylindrical metasurface leaky-wave antennas," *IEEE Antennas Wireless Propagat. Lett.*, vol. 18, no. 2, pp. 264–268, Feb. 2019. Reproduced with permission from IEEE.)

16.12 Show that the surface-impedance modulation function for a circumferentially modulated LWA, whose geometry is shown in Figure P16-12, to form a fan beam in the desired angular direction $\phi = \phi_d$ (ϕ is

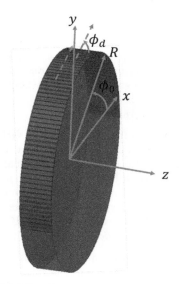

Figure P16-12 (Source: S. Pandi, C. A. Balanis, and C. R. Birtcher, "Curvature modeling in design of circumferentially modulated cylindrical metasurface LWA," *IEEE Antennas Wireless Propagat. Lett.*, vol. 16, pp. 1024–1027, 2017. Reproduced with permission from IEEE.)

measured from the x axis along the xy plane) is represented by

$$X_s(\phi) = X_a\{1 + M\cos[\beta R(\phi_0 - \phi)$$
$$-\beta_0 R\cos(\phi_d - \phi)]\}$$

where β is the phase constant along the direction of leaky-wave propagation in the LWA.

16.13. Repeat Example 16-6, if the source exciting the metasurface is off-centered by y_c along the y axis [location of the source is $(0, y_c)$].

16.14. Repeat Example 16-6 for a holographic metasurface to form a pencil beam with:
(a) Vertical polarization.
(b) Horizontal polarization.
(c) Circular polarization.

16.15. For the polarization-diverse metasurface shown in Figure 16-84, derive the surface-impedance modulation function to form a pencil beam on the yz plane at 45° from broadside ($\phi = 90°$). Assume that the position of sources 1 and 2 to be at $(0, y_{c1})$ and $(0, y_{c2})$, respectively.

16.16. The metasurface of Example 16-5 is impedance modulated with a square wave of

$$Z_s(y) = jX_a\left\{1 + M\mathrm{sgn}\left[\cos\left(\frac{2\pi y}{p}\right)\right]\right\}$$

Assuming that X_a, M, and p are the same values as in Example 16-5, determine the angular directions (in degrees) of the fan beams formed by the first and the third Fourier harmonics. Assume that only the first Floquet-Bloch mode of each harmonic radiates. *Hint:* Using Fourier series, the square-wave surface impedance modulation can be rewritten as

$$Z_s(y) = jX_a\left\{1 + M\left(\frac{4}{\pi}\right)\right.$$
$$\left.\times \sum_{q=1,3,5,\ldots}^{\infty} \frac{1}{q}\sin\left(\frac{2q\pi y}{p}\right)\right\}$$

(For further details, refer to [97]).

16.17. Derive the surface-impedance modulation function of a polarization-diverse holographic metasurface that forms a pencil beam with a left-hand circular polarization when excited by source 1, and a right-hand circular polarization for source 2, as illustrated in Figure 16-84. Assume that the position of sources 1 and 2 are at $(0, -y_c)$ and $(0, y_c)$, respectively. What is the polarization of the pencil beam when the metasurface is excited by both the sources? Assume that the two sources are orthogonal to each other ($\Psi_{ref2}^*\Psi_{ref1} = \Psi_{ref1}^*\Psi_{ref2} = 0$).

16.18. Derive the surface-impedance modulation function of a polarization-diverse holographic metasurface excited by three sources as shown in Figure P16-18. The metasurface should form a pencil beam with vertical polarization when excited by source 1, and horizontal polarization for the other two sources. It should also form a left-hand circularly polarized pencil beam when excited by both sources 1 and 2, and a right-hand circular polarization for excitation by sources 1 and 3. Assume that the position of sources 1, 2, and 3 are to be at $(0,0)$, $(-x_c, 0)$, and $(-x_c, 0)$, respectively.

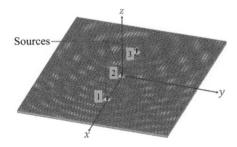

Figure P16-18 (Source: S. Ramalingam, C. A. Balanis, C. R. Birtcher, and H. N. Shaman, "Polarization-diverse holographic metasurfaces," *IEEE Antennas Wireless Propagat. Lett.*, vol. 18, no. 2, pp. 264–268, Feb. 2019. Reproduced with permission from IEEE.)

APPENDIX I

Identities

I.1 TRIGONOMETRIC

1. Sum or difference:
   a. $\sin(x+y) = \sin x \cos y + \cos x \sin y$
   b. $\sin(x-y) = \sin x \cos y - \cos x \sin y$
   c. $\cos(x+y) = \cos x \cos y - \sin x \sin y$
   d. $\cos(x-y) = \cos x \cos y + \sin x \sin y$
   e. $\tan(x+y) = \dfrac{\tan x + \tan y}{1 - \tan x \tan y}$
   f. $\tan(x-y) = \dfrac{\tan x - \tan y}{1 + \tan x \tan y}$
   g. $\sin^2 x + \cos^2 x = 1$
   h. $\tan^2 x - \sec^2 x = -1$
   i. $\cot^2 x - \csc^2 x = -1$

2. Sum or difference into products:
   a. $\sin x + \sin y = 2 \sin\frac{1}{2}(x+y) \cos\frac{1}{2}(x-y)$
   b. $\sin x - \sin y = 2 \cos\frac{1}{2}(x+y) \sin\frac{1}{2}(x-y)$
   c. $\cos x + \cos y = 2 \cos\frac{1}{2}(x+y) \cos\frac{1}{2}(x-y)$
   d. $\cos x - \cos y = -2 \sin\frac{1}{2}(x+y) \sin\frac{1}{2}(x-y)$

3. Products into sum or difference:
   a. $2 \sin x \cos y = \sin(x+y) + \sin(x-y)$
   b. $2 \cos x \sin y = \sin(x+y) - \sin(x-y)$
   c. $2 \cos x \cos y = \cos(x+y) + \cos(x-y)$
   d. $2 \sin x \sin y = -\cos(x+y) + \cos(x-y)$

Balanis' Advanced Engineering Electromagnetics, Third Edition. Constantine A. Balanis.
© 2024 John Wiley & Sons, Inc. Published 2024 by John Wiley & Sons, Inc.
Companion Website: www.wiley.com/go/balanis/advancedengineeringelectromagnetics3e

4. Double and half-angles:

    a. $\sin 2x = 2\sin x \cos x$

    b. $\cos 2x = \cos^2 x - \sin^2 x = 2\cos^2 x - 1 = 1 - 2\sin^2 x$

    c. $\tan 2x = \dfrac{2\tan x}{1 - \tan^2 x}$

    d. $\sin\dfrac{1}{2}x = \pm\sqrt{\dfrac{1 - \cos x}{2}}$ or $2\sin^2\theta = 1 - \cos 2\theta$

    e. $\cos\dfrac{1}{2}x = \pm\sqrt{\dfrac{1 + \cos x}{2}}$ or $2\cos^2\theta = 1 + \cos 2\theta$

    f. $\tan\dfrac{1}{2}x = \pm\sqrt{\dfrac{1 - \cos x}{1 + \cos x}} = \dfrac{\sin x}{1 + \cos x} = \dfrac{1 - \cos x}{\sin x}$

5. Series:

    a. $\sin x = \dfrac{e^{jx} - e^{-jx}}{2j} = x - \dfrac{x^3}{3!} + \dfrac{x^5}{5!} - \dfrac{x^7}{7!} + \cdots$

    b. $\cos x = \dfrac{e^{jx} + e^{-jx}}{2} = 1 - \dfrac{x^2}{2!} + \dfrac{x^4}{4!} - \dfrac{x^6}{6!} + \cdots$

    c. $\tan x = \dfrac{e^{jx} - e^{-jx}}{j(e^{jx} + e^{-jx})} = x + \dfrac{x^3}{3} + \dfrac{2x^5}{15} + \dfrac{17x^7}{315} + \cdots$

I.2   HYPERBOLIC

1. Definitions:

    a. Hyperbolic sine: $\sinh x = \frac{1}{2}(e^x - e^{-x})$

    b. Hyperbolic cosine: $\cosh x = \frac{1}{2}(e^x + e^{-x})$

    c. Hyperbolic tangent: $\tanh x = \dfrac{\sinh x}{\cosh x}$

    d. Hyperbolic cotangent: $\coth x = \dfrac{1}{\tanh x} = \dfrac{\cosh x}{\sinh x}$

    e. Hyperbolic secant: $\operatorname{sech} x = \dfrac{1}{\cosh x}$

    f. Hyperbolic cosecant: $\operatorname{csch} x = \dfrac{1}{\sinh x}$

2. Sum or difference:

    a. $\cosh(x + y) = \cosh x \cosh y + \sinh x \sinh y$

    b. $\sinh(x - y) = \sinh x \cosh y - \cosh x \sinh y$

    c. $\cosh(x - y) = \cosh x \cosh y - \sinh x \sinh y$

    d. $\tanh(x + y) = \dfrac{\tanh x + \tanh y}{1 + \tanh x \tanh y}$

    e. $\tanh(x - y) = \dfrac{\tanh x - \tanh y}{1 - \tanh x \tanh y}$

    f. $\cosh^2 x - \sinh^2 x = 1$

    g. $\tanh^2 x + \operatorname{sech}^2 x = 1$

    h. $\coth^2 x - \operatorname{csch}^2 x = 1$

i. $\cosh(x \pm jy) = \cosh x \cos y \pm j \sinh x \sin y$

j. $\sinh(x \pm jy) = \sinh x \cos y \pm j \cosh x \sin y$

3. Series:

a. $\sinh x = \dfrac{e^x - e^{-x}}{2} = x + \dfrac{x^3}{3!} + \dfrac{x^5}{5!} + \dfrac{x^7}{7!} + \cdots$

b. $\cosh x = \dfrac{e^x + e^{-x}}{2} = 1 + \dfrac{x^2}{2!} + \dfrac{x^4}{4!} + \dfrac{x^6}{6!} + \cdots$

c. $e^x = 1 + x + \dfrac{x^2}{2!} + \dfrac{x^3}{3!} + \dfrac{x^4}{4!} + \cdots$

I.3 LOGARITHMIC

1. $\log_b(MN) = \log_b M + \log_b N$

2. $\log_b(M/N) = \log_b M - \log_b N$

3. $\log_b(1/N) = -\log_b N$

4. $\log_b(M^n) = n \log_b M$

5. $\log_b(M^{1/n}) = \dfrac{1}{n} \log_b M$

6. $\log_a N = \log_b N \cdot \log_a b = \log_b N / \log_b a$

7. $\log_e N = \log_{10} N \cdot \log_e 10 = 2.302585 \log_{10} N$

8. $\log_{10} N = \log_e N \cdot \log_{10} e = 0.434294 \log_e N$

APPENDIX **II**

Vector Analysis

II.1 VECTOR TRANSFORMATIONS

In this appendix we will indicate the vector transformations from rectangular to cylindrical (and vice versa), from cylindrical to spherical (and vice versa), and from rectangular to spherical (and vice versa). The three coordinate systems are shown in Figure II-1.

II.1.1 Rectangular to Cylindrical (and Vice Versa)

The coordinate transformation from rectangular (x, y, z) to cylindrical (ρ, ϕ, z) is given, referring to Figure II-1(b), by:

$$
\begin{aligned}
x &= \rho \cos \phi \\
y &= \rho \sin \phi \\
z &= z
\end{aligned}
\tag{II-1}
$$

In the rectangular coordinate system, we express a vector **A** as

$$
\mathbf{A} = \hat{\mathbf{a}}_x A_x + \hat{\mathbf{a}}_y A_y + \hat{\mathbf{a}}_z A_z
\tag{II-2}
$$

where $\hat{\mathbf{a}}_x$, $\hat{\mathbf{a}}_y$, $\hat{\mathbf{a}}_z$ are the unit vectors and A_x, A_y, A_z are the components of the vector **A** in the rectangular coordinate system. We wish to write **A** as

$$
\mathbf{A} = \hat{\mathbf{a}}_\rho A_\rho + \hat{\mathbf{a}}_\phi A_\phi + \hat{\mathbf{a}}_z A_z
\tag{II-3}
$$

where $\hat{\mathbf{a}}_\rho$, $\hat{\mathbf{a}}_\phi$, $\hat{\mathbf{a}}_z$ are the unit vectors and A_ρ, A_ϕ, A_z are the vector components in the cylindrical coordinate system. The z axis is common to both of them.

Referring to Figure II-2, we can write

$$
\begin{aligned}
\hat{\mathbf{a}}_x &= \hat{\mathbf{a}}_\rho \cos \phi - \hat{\mathbf{a}}_\phi \sin \phi \\
\hat{\mathbf{a}}_y &= \hat{\mathbf{a}}_\rho \sin \phi + \hat{\mathbf{a}}_\phi \cos \phi \\
\hat{\mathbf{a}}_z &= \hat{\mathbf{a}}_z
\end{aligned}
\tag{II-4}
$$

Balanis' Advanced Engineering Electromagnetics, Third Edition. Constantine A. Balanis.
© 2024 John Wiley & Sons, Inc. Published 2024 by John Wiley & Sons, Inc.
Companion Website: www.wiley.com/go/balanis/advancedengineeringelectromagnetics3e

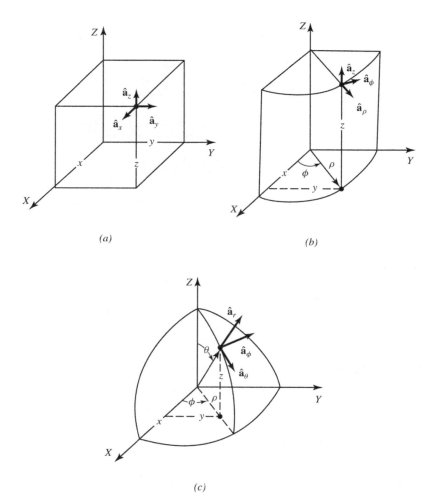

Figure II-1   Coordinate systems. (*a*) Rectangular. (*b*) Cylindrical. (*c*) Spherical. (Source: C. A. Balanis, *Antenna Theory: Analysis and Design*; Third Edition, copyright © 2005, John Wiley & Sons, Inc.; reprinted by permission of John Wiley & Sons, Inc.)

Using (II-4) reduces (II-2) to

$$\mathbf{A} = (\hat{\mathbf{a}}_\rho \cos\phi - \hat{\mathbf{a}}_\phi \sin\phi)A_x + (\hat{\mathbf{a}}_\rho \sin\phi + \hat{\mathbf{a}}_\phi \cos\phi)A_y + \hat{\mathbf{a}}_z A_z$$
$$\mathbf{A} = \hat{\mathbf{a}}_\rho (A_x \cos\phi + A_y \sin\phi) + \hat{\mathbf{a}}_\phi (-A_x \sin\phi + A_y \cos\phi) + \hat{\mathbf{a}}_z A_z \tag{II-5}$$

which when compared with (II-3) leads to

$$A_\rho = A_x \cos\phi + A_y \sin\phi$$
$$A_\phi = -A_x \sin\phi + A_y \cos\phi \tag{II-6}$$
$$A_z = A_z$$

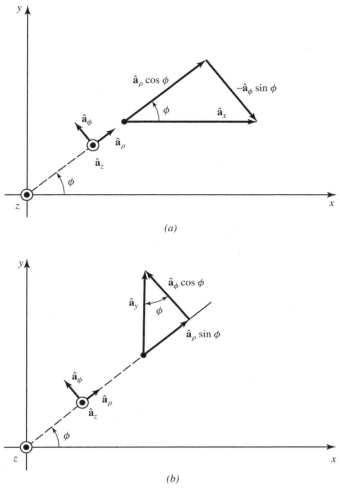

Figure II-2 Geometrical representation of transformation between unit vectors of rectangular and cylindrical coordinate systems. (*a*) Geometry for unit vector $\hat{\mathbf{a}}_x$. (*b*) Geometry for unit vector $\hat{\mathbf{a}}_y$. (Source: C. A. Balanis, *Antenna Theory: Analysis and Design*; Third Edition, copyright © 2005, John Wiley & Sons, Inc.; reprinted by permission of John Wiley & Sons, Inc.)

In matrix form, (II-6) can be written as

$$
\begin{pmatrix} A_\rho \\ A_\phi \\ A_z \end{pmatrix} = \begin{pmatrix} \cos\phi & \sin\phi & 0 \\ -\sin\phi & \cos\phi & 0 \\ 0 & 0 & 1 \end{pmatrix} \begin{pmatrix} A_x \\ A_y \\ A_z \end{pmatrix}
\tag{II-6a}
$$

where

$$
[A]_{rc} = \begin{pmatrix} \cos\phi & \sin\phi & 0 \\ -\sin\phi & \cos\phi & 0 \\ 0 & 0 & 1 \end{pmatrix}
\tag{II-6b}
$$

is the transformation matrix for rectangular-to-cylindrical components.

Since $[A]_{rc}$ is an orthonormal matrix (its inverse is equal to its transpose), we can write the transformation matrix for cylindrical-to-rectangular components as

$$[A]_{cr} = [A]_{rc}^{-1} = [A]_{rc}^{t} = \begin{bmatrix} \cos\phi & -\sin\phi & 0 \\ \sin\phi & \cos\phi & 0 \\ 0 & 0 & 1 \end{bmatrix} \tag{II-7}$$

or

$$\begin{pmatrix} A_x \\ A_y \\ A_z \end{pmatrix} = \begin{pmatrix} \cos\phi & -\sin\phi & 0 \\ \sin\phi & \cos\phi & 0 \\ 0 & 0 & 1 \end{pmatrix} = \begin{pmatrix} A_\rho \\ A_\phi \\ A_z \end{pmatrix} \tag{II-7a}$$

or

$$\begin{aligned} A_x &= A_\rho \cos\phi - A_\phi \sin\phi \\ A_y &= A_\rho \sin\phi + A_\phi \cos\phi \\ A_z &= A_z \end{aligned} \tag{II-7b}$$

II.1.2   Cylindrical to Spherical (and Vice Versa)

Referring to Figure II-1c, we can write that the cylindrical and spherical coordinates are related by

$$\begin{aligned} \rho &= r\sin\theta \\ z &= r\cos\theta \end{aligned} \tag{II-8}$$

In a geometrical approach, similar to the one employed in the previous section, we can show that the cylindrical-to-spherical transformation of vector components is given by

$$\begin{aligned} A_r &= A_\rho \sin\theta + A_z \cos\theta \\ A_\theta &= A_\rho \cos\theta - A_z \sin\theta \\ A_\phi &= A_\phi \end{aligned} \tag{II-9}$$

or in matrix form by

$$\begin{pmatrix} A_r \\ A_\theta \\ A_\phi \end{pmatrix} = \begin{pmatrix} \sin\theta & 0 & \cos\theta \\ \cos\theta & 0 & -\sin\theta \\ 0 & 1 & 0 \end{pmatrix} \begin{pmatrix} A_\rho \\ A_\phi \\ A_z \end{pmatrix} \tag{II-9a}$$

Thus, the cylindrical-to-spherical transformation matrix can be written as

$$[A]_{cs} = \begin{bmatrix} \sin\theta & 0 & \cos\theta \\ \cos\theta & 0 & -\sin\theta \\ 0 & 1 & 0 \end{bmatrix} \tag{II-9b}$$

The $[A]_{cs}$ matrix is also orthonormal so that its inverse is given by

$$[A]_{sc} = [A]_{cs}^{-1} = [A]_{cs}^{t} = \begin{bmatrix} \sin\theta & \cos\theta & 0 \\ 0 & 0 & 1 \\ \cos\theta & -\sin\theta & 0 \end{bmatrix} \tag{II-10}$$

and the spherical-to-cylindrical transformation is accomplished by

$$
\begin{pmatrix} A_\rho \\ A_\phi \\ A_z \end{pmatrix} = \begin{pmatrix} \sin\theta & \cos\theta & 0 \\ 0 & 0 & 1 \\ \cos\theta & -\sin\theta & 0 \end{pmatrix} \begin{pmatrix} A_r \\ A_\theta \\ A_\phi \end{pmatrix}
\tag{II-10a}
$$

or

$$
\begin{aligned}
A_\rho &= A_r \sin\theta + A_\theta \cos\theta \\
A_\phi &= A_\phi \\
A_z &= A_r \cos\theta - A_\theta \sin\theta
\end{aligned}
\tag{II-10b}
$$

This time the component A_ϕ and coordinate ϕ are the same in both systems.

II.1.3   Rectangular to Spherical (and Vice Versa)

Many times it may be required that a transformation be performed directly from rectangular to spherical components. By referring to Figure II-1c, we can write that the rectangular and spherical coordinates are related by

$$
\begin{aligned}
x &= r\sin\theta\cos\phi \\
y &= r\sin\theta\sin\phi \\
z &= r\cos\theta
\end{aligned}
\tag{II-11}
$$

and the rectangular and spherical components by

$$
\begin{aligned}
A_r &= A_x \sin\theta\cos\phi + A_y \sin\theta\sin\phi + A_z \cos\theta \\
A_\theta &= A_x \cos\theta\cos\phi + A_y \cos\theta\sin\phi - A_z \sin\theta \\
A_\phi &= -A_x \sin\phi + A_y \cos\phi
\end{aligned}
\tag{II-12}
$$

which can also be obtained by substituting (II-6) into (II-9). In matrix form (II-12) can be written as

$$
\begin{pmatrix} A_r \\ A_\theta \\ A_\phi \end{pmatrix} = \begin{pmatrix} \sin\theta\cos\phi & \sin\theta\sin\phi & \cos\theta \\ \cos\theta\cos\phi & \cos\theta\sin\phi & -\sin\theta \\ -\sin\phi & \cos\phi & 0 \end{pmatrix} \begin{pmatrix} A_x \\ A_y \\ A_z \end{pmatrix}
\tag{II-12a}
$$

with the rectangular-to-spherical transformation matrix being

$$
[A]_{rs} = \begin{bmatrix} \sin\theta\cos\phi & \sin\theta\sin\phi & \cos\theta \\ \cos\theta\cos\phi & \cos\theta\sin\phi & -\sin\theta \\ -\sin\phi & \cos\phi & 0 \end{bmatrix}
\tag{II-12b}
$$

The transformation matrix of (II-12b) is also orthonormal so that its inverse can be written as

$$
[A]_{sr} = [A]_{rs}^{-1} = [A]_{rs}^{t} = \begin{bmatrix} \sin\theta\cos\phi & \cos\theta\cos\phi & -\sin\phi \\ \sin\theta\sin\phi & \cos\theta\sin\phi & \cos\phi \\ \cos\theta & -\sin\theta & 0 \end{bmatrix}
\tag{II-13}
$$

and the spherical to rectangular transformation is accomplished by

$$
\begin{pmatrix} A_x \\ A_y \\ A_z \end{pmatrix} = \begin{pmatrix} \sin\theta\cos\phi & \cos\theta\cos\phi & -\sin\phi \\ \sin\theta\sin\phi & \cos\theta\sin\phi & \cos\phi \\ \cos\theta & -\sin\theta & 0 \end{pmatrix} \begin{pmatrix} A_r \\ A_\theta \\ A_\phi \end{pmatrix}
\tag{II-13a}
$$

or

$$A_x = A_r \sin\theta \cos\phi + A_\theta \cos\theta \cos\phi - A_\phi \sin\phi$$
$$A_y = A_r \sin\theta \sin\phi + A_\theta \cos\theta \sin\phi + A_\phi \cos\phi \qquad \text{(II-13b)}$$
$$A_z = A_r \cos\theta - A_\theta \sin\theta$$

II.2   VECTOR DIFFERENTIAL OPERATORS

The differential operators of gradient of a scalar $(\nabla\psi)$, divergence of a vector $(\nabla \cdot \mathbf{A})$, curl of a vector $(\nabla \times \mathbf{A})$, Laplacian of a scalar $(\nabla^2\psi)$, and Laplacian of a vector $(\nabla^2\mathbf{A})$, frequently encountered in electromagnetic field analysis, will be listed in the rectangular, cylindrical, and spherical coordinate systems.

II.2.1   Rectangular Coordinates

$$\nabla\psi = \hat{\mathbf{a}}_x \frac{\partial\psi}{\partial x} + \hat{\mathbf{a}}_y \frac{\partial\psi}{\partial y} + \hat{\mathbf{a}}_z \frac{\partial\psi}{\partial z} \qquad \text{(II-14)}$$

$$\nabla \cdot \mathbf{A} = \frac{\partial A_x}{\partial x} + \frac{\partial A_y}{\partial y} + \frac{\partial A_z}{\partial z} \qquad \text{(II-15)}$$

$$\nabla \times \mathbf{A} = \hat{\mathbf{a}}_x \left(\frac{\partial A_z}{\partial y} - \frac{\partial A_y}{\partial z} \right) + \hat{\mathbf{a}}_y \left(\frac{\partial A_x}{\partial z} - \frac{\partial A_z}{\partial x} \right) + \hat{\mathbf{a}}_z \left(\frac{\partial A_y}{\partial x} - \frac{\partial A_x}{\partial y} \right) \qquad \text{(II-16)}$$

$$\nabla \cdot \nabla\psi = \nabla^2\psi = \frac{\partial^2\psi}{\partial x^2} + \frac{\partial^2\psi}{\partial y^2} + \frac{\partial^2\psi}{\partial z^2} \qquad \text{(II-17)}$$

$$\nabla^2\mathbf{A} = \hat{\mathbf{a}}_x \nabla^2 A_x + \hat{\mathbf{a}}_y \nabla^2 A_y + \hat{\mathbf{a}}_z \nabla^2 A_z \qquad \text{(II-18)}$$

II.2.2   Cylindrical Coordinates

$$\nabla\psi = \hat{\mathbf{a}}_\rho \frac{\partial\psi}{\partial\rho} + \hat{\mathbf{a}}_\phi \frac{1}{\rho}\frac{\partial\psi}{\partial\phi} + \hat{\mathbf{a}}_z \frac{\partial\psi}{\partial z} \qquad \text{(II-19)}$$

$$\nabla \cdot \mathbf{A} = \frac{1}{\rho}\frac{\partial}{\partial\rho}(\rho A_\rho) + \frac{1}{\rho}\frac{\partial A_\phi}{\partial\phi} + \frac{\partial A_z}{\partial z} \qquad \text{(II-20)}$$

$$\nabla \times \mathbf{A} = \hat{\mathbf{a}}_\rho \left(\frac{1}{\rho}\frac{\partial A_z}{\partial\phi} - \frac{\partial A_\phi}{\partial z} \right) + \hat{\mathbf{a}}_\phi \left(\frac{\partial A_\rho}{\partial z} - \frac{\partial A_z}{\partial\rho} \right)$$
$$+ \hat{\mathbf{a}}_z \left(\frac{1}{\rho}\frac{\partial(\rho A_\phi)}{\partial\rho} - \frac{1}{\rho}\frac{\partial A_\rho}{\partial\phi} \right) \qquad \text{(II-21)}$$

$$\nabla^2\psi = \frac{1}{\rho}\frac{\partial}{\partial\rho}\left(\rho\frac{\partial\psi}{\partial\rho} \right) + \frac{1}{\rho^2}\frac{\partial^2\psi}{\partial\phi^2} + \frac{\partial^2\psi}{\partial z^2} \qquad \text{(II-22)}$$

$$\nabla^2\mathbf{A} = \nabla(\nabla \cdot \mathbf{A}) - \nabla \times \nabla \times \mathbf{A} \qquad \text{(II-23)}$$

or in an expanded form

$$\nabla^2 \mathbf{A} = \hat{\mathbf{a}}_\rho \left(\frac{\partial^2 A_\rho}{\partial \rho^2} + \frac{1}{\rho} \frac{\partial A_\rho}{\partial \rho} - \frac{A_\rho}{\rho^2} + \frac{1}{\rho^2} \frac{\partial^2 A_\rho}{\partial \phi^2} - \frac{2}{\rho^2} \frac{\partial A_\phi}{\partial \phi} + \frac{\partial^2 A_\rho}{\partial z^2} \right)$$

$$+ \hat{\mathbf{a}}_\phi \left(\frac{\partial^2 A_\phi}{\partial \rho^2} + \frac{1}{\rho} \frac{\partial A_\phi}{\partial \rho} - \frac{A_\phi}{\rho^2} + \frac{1}{\rho^2} \frac{\partial^2 A_\phi}{\partial \phi^2} + \frac{2}{\rho^2} \frac{\partial A_\rho}{\partial \phi} + \frac{\partial^2 A_\phi}{\partial z^2} \right)$$

$$+ \hat{\mathbf{a}}_z \left(\frac{\partial^2 A_z}{\partial \rho^2} + \frac{1}{\rho} \frac{\partial A_z}{\partial \rho} + \frac{1}{\rho^2} \frac{\partial^2 A_z}{\partial \phi^2} + \frac{\partial^2 A_z}{\partial z^2} \right) \tag{II-23a}$$

In the cylindrical coordinate system $\nabla^2 \mathbf{A} \neq \hat{\mathbf{a}}_\rho \nabla^2 A_\rho + \hat{\mathbf{a}}_\phi \nabla^2 A_\phi + \hat{\mathbf{a}}_z \nabla^2 A_z$ because the orientation of the unit vectors $\hat{\mathbf{a}}_\rho$ and $\hat{\mathbf{a}}_\phi$ varies with the ρ and ϕ coordinates.

II.2.3 Spherical Coordinates

$$\nabla \psi = \hat{\mathbf{a}}_r \frac{\partial \psi}{\partial r} + \hat{\mathbf{a}}_\theta \frac{1}{r} \frac{\partial \psi}{\partial \theta} + \hat{\mathbf{a}}_\phi \frac{1}{r \sin \theta} \frac{\partial \psi}{\partial \phi} \tag{II-24}$$

$$\nabla \cdot \mathbf{A} = \frac{1}{r^2} \frac{\partial}{\partial r} (r^2 A_r) + \frac{1}{r \sin \theta} \frac{\partial}{\partial \theta} (\sin \theta A_\theta) + \frac{1}{r \sin \theta} \frac{\partial A_\phi}{\partial \phi} \tag{II-25}$$

$$\nabla \times \mathbf{A} = \frac{\hat{\mathbf{a}}_r}{r \sin \theta} \left[\frac{\partial}{\partial \theta} (A_\phi \sin \theta) - \frac{\partial A_\theta}{\partial \phi} \right] + \frac{\hat{\mathbf{a}}_\theta}{r} \left[\frac{1}{\sin \theta} \frac{\partial A_r}{\partial \phi} - \frac{\partial}{\partial r} (r A_\phi) \right]$$

$$+ \frac{\hat{\mathbf{a}}_\phi}{r} \left[\frac{\partial}{\partial r} (r A_\theta) - \frac{\partial A_r}{\partial \theta} \right] \tag{II-26}$$

$$\nabla^2 \psi = \frac{1}{r^2} \frac{\partial}{\partial r} \left(r^2 \frac{\partial \psi}{\partial r} \right) + \frac{1}{r^2 \sin \theta} \frac{\partial}{\partial \theta} \left(\sin \theta \frac{\partial \psi}{\partial \theta} \right) + \frac{1}{r^2 \sin^2 \theta} \frac{\partial^2 \psi}{\partial \phi^2} \tag{II-27}$$

$$\nabla^2 \mathbf{A} = \nabla (\nabla \cdot \mathbf{A}) - \nabla \times \nabla \times \mathbf{A} \tag{II-28}$$

or in an expanded form

$$\nabla^2 \mathbf{A} = \hat{\mathbf{a}}_r \left(\frac{\partial^2 A_r}{\partial r^2} + \frac{2}{r} \frac{\partial A_r}{\partial r} - \frac{2}{r^2} A_r + \frac{1}{r^2} \frac{\partial^2 A_r}{\partial \theta^2} + \frac{\cot \theta}{r^2} \frac{\partial A_r}{\partial \theta} + \frac{1}{r^2 \sin^2 \theta} \frac{\partial^2 A_r}{\partial \phi^2} \right.$$

$$\left. - \frac{2}{r^2} \frac{\partial A_\theta}{\partial \theta} - \frac{2 \cot \theta}{r^2} A_\theta - \frac{2}{r^2 \sin \theta} \frac{\partial A_\phi}{\partial \phi} \right)$$

$$+ \hat{\mathbf{a}}_\theta \left(\frac{\partial^2 A_\theta}{\partial r^2} + \frac{2}{r} \frac{\partial A_\theta}{\partial r} - \frac{A_\theta}{r^2 \sin^2 \theta} + \frac{1}{r^2} \frac{\partial^2 A_\theta}{\partial \theta^2} + \frac{\cot \theta}{r^2} \frac{\partial A_\theta}{\partial \theta} \right.$$

$$\left. + \frac{1}{r^2 \sin^2 \theta} \frac{\partial^2 A_\theta}{\partial \phi^2} + \frac{2}{r^2} \frac{\partial A_r}{\partial \theta} - \frac{2 \cot \theta}{r^2 \sin \theta} \frac{\partial A_\phi}{\partial \phi} \right)$$

$$+ \hat{\mathbf{a}}_\phi \left(\frac{\partial^2 A_\phi}{\partial r^2} + \frac{2}{r} \frac{\partial A_\phi}{\partial r} - \frac{1}{r^2 \sin^2 \theta} A_\phi + \frac{1}{r^2} \frac{\partial^2 A_\phi}{\partial \theta^2} \right.$$

$$\left. + \frac{\cot \theta}{r^2} \frac{\partial A_\phi}{\partial \theta} + \frac{1}{r^2 \sin^2 \theta} \frac{\partial^2 A_\phi}{\partial \phi^2} + \frac{2}{r^2 \sin \theta} \frac{\partial A_r}{\partial \phi} + \frac{2 \cot \theta}{r^2 \sin \theta} \frac{\partial A_\theta}{\partial \phi} \right) \tag{II-28a}$$

Again note that $\nabla^2 \mathbf{A} \neq \hat{\mathbf{a}}_r \nabla^2 A_r + \hat{\mathbf{a}}_\theta \nabla^2 A_\theta + \hat{\mathbf{a}}_\phi \nabla^2 A_\phi$ since the orientation of the unit vectors $\hat{\mathbf{a}}_r, \hat{\mathbf{a}}_\theta,$ and $\hat{\mathbf{a}}_\phi$ varies with the r, θ, and ϕ coordinates.

II.3 VECTOR IDENTITIES

II.3.1 Addition and Multiplication

$$\mathbf{A} \cdot \mathbf{A} = |\mathbf{A}|^2 \tag{II-29}$$

$$\mathbf{A} \cdot \mathbf{A}^* = |\mathbf{A}|^2 \tag{II-30}$$

$$\mathbf{A} + \mathbf{B} = \mathbf{B} + \mathbf{A} \tag{II-31}$$

$$\mathbf{A} \cdot \mathbf{B} = \mathbf{B} \cdot \mathbf{A} \tag{II-32}$$

$$\mathbf{A} \times \mathbf{B} = -\mathbf{B} \times \mathbf{A} \tag{II-33}$$

$$(\mathbf{A} + \mathbf{B}) \cdot \mathbf{C} = \mathbf{A} \cdot \mathbf{C} + \mathbf{B} \cdot \mathbf{C} \tag{II-34}$$

$$(\mathbf{A} + \mathbf{B}) \times \mathbf{C} = \mathbf{A} \times \mathbf{C} + \mathbf{B} \times \mathbf{C} \tag{II-35}$$

$$\mathbf{A} \cdot \mathbf{B} \times \mathbf{C} = \mathbf{B} \cdot \mathbf{C} \times \mathbf{A} = \mathbf{C} \cdot \mathbf{A} \times \mathbf{B} \tag{II-36}$$

$$\mathbf{A} \times (\mathbf{B} \times \mathbf{C}) = (\mathbf{A} \cdot \mathbf{C})\mathbf{B} - (\mathbf{A} \cdot \mathbf{B})\mathbf{C} \tag{II-37}$$

$$\begin{aligned}(\mathbf{A} \times \mathbf{B}) \cdot (\mathbf{C} \times \mathbf{D}) &= \mathbf{A} \cdot \mathbf{B} \times (\mathbf{C} \times \mathbf{D}) \\ &= \mathbf{A} \cdot (\mathbf{B} \cdot \mathbf{D}\mathbf{C} - \mathbf{B} \cdot \mathbf{C}\mathbf{D}) \\ &= (\mathbf{A} \cdot \mathbf{C})(\mathbf{B} \cdot \mathbf{D}) - (\mathbf{A} \cdot \mathbf{D})(\mathbf{B} \cdot \mathbf{C}) \end{aligned} \tag{II-38}$$

$$(\mathbf{A} \times \mathbf{B}) \times (\mathbf{C} \times \mathbf{D}) = (\mathbf{A} \times \mathbf{B} \cdot \mathbf{D})\mathbf{C} - (\mathbf{A} \times \mathbf{B} \cdot \mathbf{C})\mathbf{D} \tag{II-39}$$

II.3.2 Differentiation

$$\nabla \cdot (\nabla \times \mathbf{A}) = 0 \tag{II-40}$$

$$\nabla \times \nabla \psi = 0 \tag{II-41}$$

$$\nabla(\phi + \psi) = \nabla\phi + \nabla\psi \tag{II-42}$$

$$\nabla(\phi\psi) = \phi\nabla\psi + \psi\nabla\phi \tag{II-43}$$

$$\nabla \cdot (\mathbf{A} + \mathbf{B}) = \nabla \cdot \mathbf{A} + \nabla \cdot \mathbf{B} \tag{II-44}$$

$$\nabla \times (\mathbf{A} + \mathbf{B}) = \nabla \times \mathbf{A} + \nabla \times \mathbf{B} \tag{II-45}$$

$$\nabla \cdot (\psi\mathbf{A}) = \mathbf{A} \cdot \nabla\psi + \psi\nabla \cdot \mathbf{A} \tag{II-46}$$

$$\nabla \times (\psi\mathbf{A}) = \nabla\psi \times \mathbf{A} + \psi\nabla \times \mathbf{A} \tag{II-47}$$

$$\nabla(\mathbf{A} \cdot \mathbf{B}) = (\mathbf{A} \cdot \nabla)\mathbf{B} + (\mathbf{B} \cdot \nabla)\mathbf{A} + \mathbf{A} \times (\nabla \times \mathbf{B}) + \mathbf{B} \times (\nabla \times \mathbf{A}) \tag{II-48}$$

$$\nabla \cdot (\mathbf{A} \times \mathbf{B}) = \mathbf{B} \cdot \nabla \times \mathbf{A} - \mathbf{A} \cdot \nabla \times \mathbf{B} \tag{II-49}$$

$$\nabla \times (\mathbf{A} \times \mathbf{B}) = \mathbf{A}\nabla \cdot \mathbf{B} - \mathbf{B}\nabla \cdot \mathbf{A} + (\mathbf{B} \cdot \nabla)\mathbf{A} - (\mathbf{A} \cdot \nabla)\mathbf{B} \tag{II-50}$$

$$\nabla \times \nabla \times \mathbf{A} = \nabla(\nabla \cdot \mathbf{A}) - \nabla^2\mathbf{A} \tag{II-51}$$

II.3.3 Integration

$$\oint_C \mathbf{A} \cdot dl = \iint_S (\boldsymbol{\nabla} \times \mathbf{A}) \cdot d\mathbf{s} \quad \text{Stokes' theorem} \tag{II-52}$$

$$\oiint_S \mathbf{A} \cdot d\mathbf{s} = \iiint_V (\boldsymbol{\nabla} \cdot \mathbf{A}) \, dv \quad \text{divergence theorem} \tag{II-53}$$

$$\oiint_S (\hat{\mathbf{n}} \times \mathbf{A}) \, ds = \iiint_V (\boldsymbol{\nabla} \times \mathbf{A}) \, dv \tag{II-54}$$

$$\oiint_S \psi \, ds = \iiint_V \boldsymbol{\nabla} \psi \, dv \tag{II-55}$$

$$\oint_C \psi \, dl = \iint_S \hat{\mathbf{n}} \times \boldsymbol{\nabla} \psi \, ds \tag{II-56}$$

APPENDIX **III**

Fresnel Integrals

$$C_0(x) = \int_0^x \frac{\cos(\tau)}{\sqrt{2\pi\tau}}\, d\tau \tag{III-1}$$

$$S_0(x) = \int_0^x \frac{\sin(\tau)}{\sqrt{2\pi\tau}}\, d\tau \tag{III-2}$$

$$C(x) = \int_0^x \cos\left(\frac{\pi}{2}\tau^2\right) d\tau \tag{III-3}$$

$$S(x) = \int_0^x \sin\left(\frac{\pi}{2}\tau^2\right) d\tau \tag{III-4}$$

$$C_1(x) = \int_x^\infty \cos(\tau^2)\, d\tau \tag{III-5}$$

$$S_1(x) = \int_x^\infty \sin(\tau^2)\, d\tau \tag{III-6}$$

$$C(x) - jS(x) = \int_0^x e^{-j(\pi/2)\tau^2}\, d\tau = \int_0^{(\pi/2)x^2} \frac{e^{-j\tau}}{\sqrt{2\pi\tau}}\, d\tau$$

$$C(x) - jS(x) = C_0\left(\frac{\pi}{2}x^2\right) - jS_0\left(\frac{\pi}{2}x^2\right) \tag{III-7}$$

$$C_1(x) - jS_1(x) = \int_x^\infty e^{-j\tau^2}\, d\tau = \sqrt{\frac{\pi}{2}} \int_{x^2}^\infty \frac{e^{-j\tau}}{\sqrt{2\pi\tau}}\, d\tau$$

$$C_1(x) - jS_1(x) = \sqrt{\frac{\pi}{2}}\left\{ \int_0^\infty \frac{e^{-j\tau}}{\sqrt{2\pi\tau}}\, d\tau - \int_0^{x^2} \frac{e^{-j\tau}}{\sqrt{2\pi\tau}}\, d\tau \right\}$$

$$C_1(x) - jS_1(x) = \sqrt{\frac{\pi}{2}}\left\{ \left[\frac{1}{2} - j\frac{1}{2}\right] - [C_0(x^2) - jS_0(x^2)] \right\}$$

$$C_1(x) - jS_1(x) = \sqrt{\frac{\pi}{2}}\left\{ \left[\frac{1}{2} - C_0(x^2)\right] - j\left[\frac{1}{2} - S_0(x^2)\right] \right\} \tag{III-8}$$

Balanis' Advanced Engineering Electromagnetics, Third Edition. Constantine A. Balanis.
© 2024 John Wiley & Sons, Inc. Published 2024 by John Wiley & Sons, Inc.
Companion Website: www.wiley.com/go/balanis/advancedengineeringelectromagnetics3e

x	$C_1(x)$	$S_1(x)$	$C(x)$	$S(x)$
0.0	0.62666	0.62666	0.0	0.0
0.1	0.52666	0.62632	0.10000	0.00052
0.2	0.42669	0.62399	0.19992	0.00419
0.3	0.32690	0.61766	0.29940	0.01412
0.4	0.22768	0.60536	0.39748	0.03336
0.5	0.12977	0.58518	0.49234	0.06473
0.6	0.03439	0.55532	0.58110	0.11054
0.7	−0.05672	0.51427	0.65965	0.17214
0.8	−0.14119	0.46092	0.72284	0.24934
0.9	−0.21606	0.39481	0.76482	0.33978
1.0	−0.27787	0.31639	0.77989	0.43826
1.1	−0.32285	0.22728	0.76381	0.53650
1.2	−0.34729	0.13054	0.71544	0.62340
1.3	−0.34803	0.03081	0.63855	0.68633
1.4	−0.32312	−0.06573	0.54310	0.71353
1.5	−0.27253	−0.15158	0.44526	0.69751
1.6	−0.19886	−0.21861	0.36546	0.63889
1.7	−0.10790	−0.25905	0.32383	0.54920
1.8	−0.00871	−0.26682	0.33363	0.45094
1.9	0.08680	−0.23918	0.39447	0.37335
2.0	0.16520	−0.17812	0.48825	0.34342
2.1	0.21359	−0.09141	0.58156	0.37427
2.2	0.22242	0.00743	0.63629	0.45570
2.3	0.18833	0.10054	0.62656	0.55315
2.4	0.11650	0.16879	0.55496	0.61969
2.5	0.02135	0.19614	0.45742	0.61918
2.6	−0.07518	0.17454	0.38894	0.54999
2.7	−0.14816	0.10789	0.39249	0.45292
2.8	−0.17646	0.01329	0.46749	0.39153
2.9	−0.15021	−0.08181	0.56237	0.41014
3.0	−0.07621	−0.14690	0.60572	0.49631
3.1	0.02152	−0.15883	0.56160	0.58181
3.2	0.10791	−0.11181	0.46632	0.59335
3.3	0.14907	−0.02260	0.40570	0.51929
3.4	0.12691	0.07301	0.43849	0.42965
3.5	0.04965	0.13335	0.53257	0.41525
3.6	−0.04819	0.12973	0.58795	0.49231
3.7	−0.11929	0.06258	0.54195	0.57498
3.8	−0.12649	−0.03483	0.44810	0.56562
3.9	−0.06469	−0.11030	0.42233	0.47521
4.0	0.03219	−0.12048	0.49842	0.42052
4.1	0.10690	−0.05815	0.57369	0.47580
4.2	0.11228	0.03885	0.54172	0.56320
4.3	0.04374	0.10751	0.44944	0.55400
4.4	−0.05287	0.10038	0.43833	0.46227
4.5	−0.10884	0.02149	0.52602	0.43427
4.6	−0.08188	−0.07126	0.56724	0.51619
4.7	0.00810	−0.10594	0.49143	0.56715
4.8	0.08905	−0.05381	0.43380	0.49675
4.9	0.09277	0.04224	0.50016	0.43507
5.0	0.01519	0.09874	0.56363	0.49919
5.1	−0.07411	0.06405	0.49979	0.56239

x	$C_1(x)$	$S_1(x)$	$C(x)$	$S(x)$
5.2	−0.09125	−0.03004	0.43889	0.49688
5.3	−0.01892	−0.09235	0.50778	0.44047
5.4	0.07063	−0.05976	0.55723	0.51403
5.5	0.08408	0.03440	0.47843	0.55369
5.6	0.00641	0.08900	0.45171	0.47004
5.7	−0.07642	0.04296	0.53846	0.45953
5.8	−0.06919	−0.05135	0.52984	0.54604
5.9	0.01998	−0.08231	0.44859	0.51633
6.0	0.08245	−0.01181	0.49953	0.44696
6.1	0.03946	0.07180	0.54950	0.51647
6.2	−0.05363	0.06018	0.46761	0.53982
6.3	−0.07284	−0.03144	0.47600	0.45555
6.4	0.00835	−0.07765	0.54960	0.49649
6.5	0.07574	−0.01326	0.48161	0.54538
6.6	0.03183	0.06872	0.46899	0.46307
6.7	−0.05828	0.04658	0.54674	0.49150
6.8	−0.05734	−0.04600	0.48307	0.54364
6.9	0.03317	−0.06440	0.47322	0.46244
7.0	0.06832	0.02077	0.54547	0.49970
7.1	−0.00944	0.06977	0.47332	0.53602
7.2	−0.06943	0.00041	0.48874	0.45725
7.3	−0.00864	−0.06793	0.53927	0.51894
7.4	0.06582	−0.01521	0.46010	0.51607
7.5	0.02018	0.06353	0.51601	0.46070
7.6	−0.06137	0.02367	0.51564	0.53885
7.7	−0.02580	−0.05958	0.46278	0.48202
7.8	0.05828	−0.02668	0.53947	0.48964
7.9	0.02638	0.05752	0.47598	0.53235
8.0	−0.05730	0.02494	0.49980	0.46021
8.1	−0.02238	−0.05752	0.52275	0.53204
8.2	0.05803	−0.01870	0.46384	0.48589
8.3	0.01387	0.05861	0.53775	0.49323
8.4	−0.05899	0.00789	0.47092	0.52429
8.5	−0.00080	−0.05881	0.51417	0.46534
8.6	0.05767	0.00729	0.50249	0.53693
8.7	−0.01616	0.05515	0.48274	0.46774
8.8	−0.05079	−0.02545	0.52797	0.52294
8.9	0.03461	−0.04425	0.46612	0.48856
9.0	0.03526	0.04293	0.53537	0.49985
9.1	−0.04951	0.02381	0.46661	0.51042
9.2	−0.01021	−0.05338	0.52914	0.48135
9.3	0.05354	0.00485	0.47628	0.52467
9.4	−0.02020	0.04920	0.51803	0.47134
9.5	−0.03995	−0.03426	0.48729	0.53100
9.6	0.04513	−0.02599	0.50813	0.46786
9.7	0.00837	0.05086	0.49549	0.53250
9.8	−0.04983	−0.01094	0.50192	0.46758
9.9	0.02916	−0.04124	0.49961	0.53215
10.0	0.02554	0.04298	0.49989	0.46817
10.1	−0.04927	0.00478	0.49961	0.53151
10.2	0.01738	−0.04583	0.50186	0.46885
10.3	0.03233	0.03621	0.49575	0.53061
10.4	−0.04681	0.01094	0.50751	0.47033

x	$C_1(x)$	$S_1(x)$	$C(x)$	$S(x)$
10.5	0.01360	−0.04563	0.48849	0.52804
10.6	0.03187	0.03477	0.51601	0.47460
10.7	−0.04595	0.00848	0.47936	0.52143
10.8	0.01789	−0.04270	0.52484	0.48413
10.9	0.02494	0.03850	0.47211	0.50867
11.0	−0.04541	−0.00202	0.52894	0.49991
11.1	0.02845	−0.03492	0.47284	0.49079
11.2	0.01008	0.04349	0.52195	0.51805
11.3	−0.03981	−0.01930	0.48675	0.47514
11.4	0.04005	−0.01789	0.50183	0.52786
11.5	−0.01282	0.04155	0.51052	0.47440
11.6	−0.02188	−0.03714	0.47890	0.51755
11.7	0.04164	0.00962	0.52679	0.49525
11.8	−0.03580	0.02267	0.47489	0.49013
11.9	0.00977	−0.04086	0.51544	0.52184
12.0	0.02059	0.03622	0.49993	0.47347
12.1	−0.03919	−0.01309	0.48426	0.52108
12.2	0.03792	−0.01555	0.52525	0.49345
12.3	−0.01914	0.03586	0.47673	0.48867
12.4	−0.00728	−0.03966	0.50951	0.52384
12.5	0.02960	0.02691	0.50969	0.47645
12.6	−0.03946	−0.00421	0.47653	0.50936
12.7	0.03445	−0.01906	0.52253	0.51097
12.8	−0.01783	0.03475	0.49376	0.47593
12.9	−0.00377	−0.03857	0.48523	0.51977
13.0	0.02325	0.03064	0.52449	0.49994
13.1	−0.03530	−0.01452	0.48598	0.48015
13.2	0.03760	−0.00459	0.49117	0.52244
13.3	−0.03075	0.02163	0.52357	0.49583
13.4	0.01744	−0.03299	0.48482	0.48173
13.5	−0.00129	0.03701	0.49103	0.52180
13.6	−0.01421	−0.03391	0.52336	0.49848
13.7	0.02639	0.02521	0.48908	0.47949
13.8	−0.03377	−0.01313	0.48534	0.51781
13.9	0.03597	−0.00002	0.52168	0.50737
14.0	−0.03352	0.01232	0.49996	0.47726
14.1	0.02749	−0.02240	0.47844	0.50668
14.2	−0.01916	0.02954	0.51205	0.51890
14.3	0.00979	−0.03357	0.51546	0.48398
14.4	−0.00043	0.03472	0.48131	0.48819
14.5	−0.00817	−0.03350	0.49164	0.52030
14.6	0.01553	0.03052	0.52113	0.50538
14.7	−0.02145	−0.02640	0.50301	0.47856
14.8	0.02591	0.02168	0.47853	0.49869
14.9	−0.02903	−0.01683	0.49971	0.52136
15.0	0.03103	0.01217	0.52122	0.49926

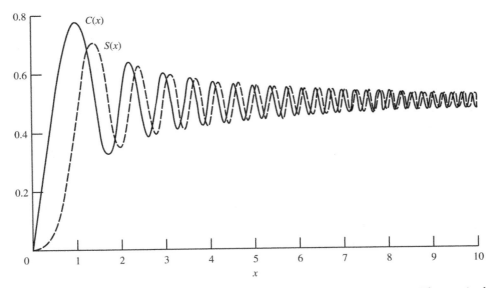

Figure III-1   Plots of $C(x)$ and $S(x)$ Fresnel integrals. (Source: C. A. Balanis, *Antenna Theory: Analysis and Design*, Third Edition, copyright © 2005, John Wiley & Sons, Inc. Reprinted by permission of John Wiley & Sons, Inc.)

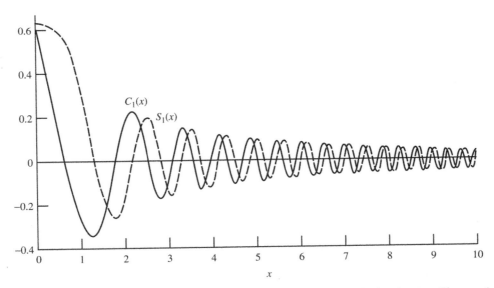

Figure III-2   Plots of $C_1(x)$ and $S_1(x)$ Fresnel integrals. (Source: C. A. Balanis, *Antenna Theory: Analysis and Design*, Third Edition, copyright © 2005, John Wiley & Sons, Inc. Reprinted by permission of John Wiley & Sons, Inc.)

APPENDIX **IV**

Bessel Functions

IV.1 BESSEL AND HANKEL FUNCTIONS

Bessel's equation can be written as

$$x^2 \frac{d^2 y}{dx^2} + x \frac{dy}{dx} + (x^2 - p^2)y = 0 \tag{IV-1}$$

Using the method of Frobenius, we can write its solutions as

$$y(x) = A_1 J_p(x) + B_1 J_{-p}(x) \quad p \text{ not an integer} \tag{IV-2}$$

or

$$y(x) = A_2 J_n(x) + B_2 Y_n(x) \quad n \text{ an integer} \tag{IV-3}$$

where

$$J_p(x) = \sum_{m=0}^{\infty} \frac{(-1)^m (x/2)^{2m+p}}{m!(m+p)!} \tag{IV-4}$$

$$J_{-p}(x) = \sum_{m=0}^{\infty} \frac{(-1)^m (x/2)^{2m-p}}{m!(m-p)!} \tag{IV-5}$$

$$Y_p(x) = \frac{J_p(x)\cos(p\pi) - J_{-p}(x)}{\sin(p\pi)} \tag{IV-6}$$

$$m! = \Gamma(m+1) \tag{IV-7}$$

$J_p(x)$ is referred to as the Bessel function of the first kind of order p, $Y_p(x)$ as the Bessel function of the second kind of order p, and $\Gamma(x)$ as the gamma function.

When $p = n =$ integer, using (IV-5) and (IV-7), it can be shown that

$$J_{-n}(x) = (-1)^n J_n(x) \tag{IV-8}$$

Balanis' Advanced Engineering Electromagnetics, Third Edition. Constantine A. Balanis.
© 2024 John Wiley & Sons, Inc. Published 2024 by John Wiley & Sons, Inc.
Companion Website: www.wiley.com/go/balanis/advancedengineeringelectromagnetics3e

and no longer are the two Bessel functions independent of each other. Therefore, a second solution is required, and it is given by (IV-3). It can also be shown that

$$Y_n(x) = \lim_{p \to n} Y_p(x) = \lim_{p \to n} \frac{J_p(x)\cos(p\pi) - J_{-p}(x)}{\sin(p\pi)} \tag{IV-9}$$

When the argument of the Bessel function is negative and $p = n$, using (IV-4) leads to

$$J_n(-x) = (-1)^n J_n(x) \tag{IV-10}$$

In many applications, Bessel functions of small and large arguments are required. Using asymptotic methods, it can be shown that

$$\left.\begin{array}{l} J_0(x) \simeq 1 \\[2mm] Y_0(x) \simeq \dfrac{2}{\pi}\ln\left(\dfrac{\gamma x}{2}\right) \\[2mm] \gamma = 1.781 \end{array}\right\} \quad x \to 0 \tag{IV-11}$$

$$\left.\begin{array}{l} J_p(x) \simeq \dfrac{1}{p!}\left(\dfrac{x}{2}\right)^p \\[3mm] Y_p(x) \simeq -\dfrac{(p-1)!}{\pi}\left(\dfrac{2}{x}\right)^p \end{array}\right\} \begin{array}{l} x \to 0 \\ p > 0 \end{array} \tag{IV-12}$$

and

$$\left.\begin{array}{l} J_p(x) \simeq \sqrt{\dfrac{2}{\pi x}}\cos\left(x - \dfrac{\pi}{4} - \dfrac{p\pi}{2}\right) \\[4mm] Y_p(x) \simeq \sqrt{\dfrac{2}{\pi x}}\sin\left(x - \dfrac{\pi}{4} - \dfrac{p\pi}{2}\right) \end{array}\right\} \quad x \to \infty \tag{IV-13}$$

For wave propagation, it is often convenient to introduce Hankel functions defined as

$$H_p^{(1)}(x) = J_p(x) + jY_p(x) \tag{IV-14}$$

$$H_p^{(2)}(x) = J_p(x) - jY_p(x) \tag{IV-15}$$

where $H_p^{(1)}(x)$ is the Hankel function of the first kind of order p and $H_p^{(2)}(x)$ is the Hankel function of the second kind of order p. For large arguments

$$H_p^{(1)}(x) \simeq \sqrt{\dfrac{2}{\pi x}}\, e^{j[x - p(\pi/2) - \pi/4]} \quad x \to \infty \tag{IV-16}$$

$$H_p^{(2)}(x) \simeq \sqrt{\dfrac{2}{\pi x}}\, e^{-j[x - p(\pi/2) - \pi/4]} \quad x \to \infty \tag{IV-17}$$

A derivative can be taken using either

$$\frac{d}{dx}[Z_p(\alpha x)] = \alpha Z_{p-1}(\alpha x) - \frac{p}{x} Z_p(\alpha x) \tag{IV-18}$$

or

$$\frac{d}{dx}[Z_p(\alpha x)] = -\alpha Z_{p+1}(\alpha x) + \frac{p}{x} Z_p(\alpha x) \tag{IV-19}$$

where Z_p can be a Bessel function (J_p, Y_p) or a Hankel function $(H_p^{(1)}$ or $H_p^{(2)})$. A useful identity, relating Bessel functions and their derivatives, is given by

$$J_p(x)Y_p'(x) - Y_p(x)J_p'(x) = \frac{2}{\pi x} \tag{IV-20}$$

and it is referred to as the Wronskian. The prime (′) indicates a derivative. Also

$$J_p(x)J_{-p}'(x) - J_{-p}(x)J_p'(x) = -\frac{2}{\pi x}\sin(p\pi) \tag{IV-21}$$

Some useful integrals of Bessel functions are

$$\int x^{p+1}J_p(\alpha x)\,dx = \frac{1}{\alpha}x^{p+1}J_{p+1}(\alpha x) + C \tag{IV-22}$$

$$\int x^{1-p}J_p(\alpha x)\,dx = -\frac{1}{\alpha}x^{1-p}J_{p-1}(\alpha x) + C \tag{IV-23}$$

$$\int x^3 J_0(x)\,dx = x^3 J_1(x) - 2x^2 J_2(x) + C \tag{IV-24}$$

$$\int x^6 J_1(x)\,dx = x^6 J_2(x) - 4x^5 J_3(x) + 8x^4 J_4(x) + C \tag{IV-25}$$

$$\int J_3(x)\,dx = -J_2(x) - \frac{2}{x}J_1(x) + C \tag{IV-26}$$

$$\int x J_1(x)\,dx = -x J_0(x) + \int J_0(x)\,dx + C \tag{IV-27}$$

$$\int x^{-1}J_1(x)\,dx = -J_1(x) + \int J_0(x)\,dx + C \tag{IV-28}$$

$$\int J_2(x)\,dx = -2J_1(x) + \int J_0(x)\,dx + C \tag{IV-29}$$

$$\int x^m J_n(x)\,dx = x^m J_{n+1}(x) - (m-n-1)\int x^{m-1}J_{n+1}(x)\,dx \tag{IV-30}$$

$$\int x^m J_n(x)\,dx = -x^m J_{n-1}(x) + (m+n-1)\int x^{m-1}J_{n-1}(x)\,dx \tag{IV-31}$$

$$J_1(x) = \frac{2}{\pi} \int_0^{\pi/2} \sin(x\sin\theta)\sin\theta \, d\theta \tag{IV-32}$$

$$\frac{1}{x}J_1(x) = \frac{2}{\pi} \int_0^{\pi/2} \cos(x\sin\theta)\cos^2\theta \, d\theta \tag{IV-33}$$

$$J_2(x) = \frac{2}{\pi} \int_0^{\pi/2} \cos(x\sin\theta)\cos 2\theta \, d\theta \tag{IV-34}$$

$$J_n(x) = \frac{j^{-n}}{2\pi} \int_0^{2\pi} e^{jx\cos\phi} e^{jn\phi} \, d\phi \tag{IV-35}$$

$$J_n(x) = \frac{j^{-n}}{\pi} \int_0^{\pi} \cos(n\phi) e^{jx\cos\phi} \, d\phi \tag{IV-36}$$

$$J_n(x) = \frac{1}{\pi} \int_0^{\pi} \cos(x\sin\phi - n\phi) \, d\phi \tag{IV-37}$$

$$J_{2n}(x) = \frac{2}{\pi} \int_0^{\pi/2} \cos(x\sin\phi)\cos(2n\phi) \, d\phi \tag{IV-38}$$

$$J_{2n}(x) = (-1)^n \frac{2}{\pi} \int_0^{\pi/2} \cos(x\cos\phi)\cos(2n\phi) \, d\phi \tag{IV-39}$$

The integrals

$$\int_0^x J_0(\tau) \, d\tau \quad \text{and} \quad \int_0^x Y_0(\tau) \, d\tau \tag{IV-40}$$

often appear in solutions of problems but cannot be integrated in closed form. Graphs and tables for each, obtained using numerical techniques, are included.

IV.2   MODIFIED BESSEL FUNCTIONS

In addition to the regular cylindrical Bessel functions of the first and second kind, there exists another set of cylindrical Bessel functions that are referred to as the *modified* Bessel functions of the first and second kind, denoted respectively by $I_p(x)$ and $K_p(x)$. These modified cylindrical Bessel functions exhibit ascending and descending variations for increasing argument as shown, respectively, in Figures IV-5 and IV-6. For real values of the argument, the modified Bessel functions exhibit real values.

The modified Bessel functions are related to the regular Bessel and Hankel functions by

$$I_p(x) = j^{-p}J_p(jx) = j^p J_{-p}(jx) = j^p J_p(-jx) \tag{IV-41}$$

$$K_p(x) = \frac{\pi}{2} j^{p+1} H_p^{(1)}(jx) = \frac{\pi}{2}(-j)^{p+1} H_p^{(2)}(-jx) \tag{IV-42}$$

Some of the identities involving modified Bessel functions are

$$I_{-p}(x) = j^p J_{-p}(jx) \tag{IV-43}$$

$$I_{-n}(x) = I_n(x) \quad n = 0, 1, 2, 3, \ldots \tag{IV-44}$$

$$K_{-n}(x) = K_n(x) \quad n = 0, 1, 2, 3, \ldots \tag{IV-45}$$

For large arguments, the modified Bessel functions can be computed using the asymptotic formulas

$$I_p(x) \simeq \frac{e^x}{\sqrt{2\pi x}}$$ (IV-46a)

$$\left. \right\} \quad x \to \infty$$

$$K_p(x) \simeq \sqrt{\frac{\pi}{2x}} e^{-x}$$ (IV-46b)

Derivatives of both modified Bessel functions can be found using the same expressions, (IV-18) and (IV-19), as for the regular Bessel functions.

IV.3 SPHERICAL BESSEL AND HANKEL FUNCTIONS

There is another set of Bessel and Hankel functions, which are usually referred to as the *spherical* Bessel and Hankel functions. These spherical Bessel and Hankel functions of order n are related, respectively, to the regular cylindrical Bessel and Hankel of order $n+1/2$ by

$$j_n(x) = \sqrt{\frac{\pi}{2x}} J_{n+1/2}(x)$$ (IV-47a)

$$y_n(x) = \sqrt{\frac{\pi}{2x}} Y_{n+1/2}(x)$$ (IV-47b)

$$h_n^{(1)}(x) = \sqrt{\frac{\pi}{2x}} H_{n+1/2}^{(1)}(x)$$ (IV-47c)

$$h_n^{(2)}(x) = \sqrt{\frac{\pi}{2x}} H_{n+1/2}^{(2)}(x)$$ (IV-47d)

where j_n, y_n, $h_n^{(1)}$, and $h_n^{(2)}$ are the spherical Bessel and Hankel functions. These spherical Bessel and Hankel functions are used as solutions to electromagnetic problems solved using spherical coordinates.

For small arguments

$$j_n(x) \simeq \frac{x^n}{1\cdot3\cdot5\cdots(2n+1)}$$ (IV-48a)

$$\left. \right\} \quad \begin{matrix} n=0,1,2,\ldots \\ x \to 0 \end{matrix}$$

$$y_n(x) \simeq -1\cdot3\cdot5\cdots(2n-1)x^{-(n+1)}$$ (IV-48b)

Another set of spherical Bessel and Hankel functions, which appear in solutions of electromagnetic problems, is that denoted by $\hat{B}_n(x)$ where $\hat{B}_n$ can be used to represent $\hat{J}_n$, $\hat{Y}_n$, $\hat{H}_n^{(1)}$, or $\hat{H}_n^{(2)}$. These are related to the preceding spherical Bessel and Hankel functions [denoted by b_n to represent j_n, y_n, $h_n^{(1)}$, or $h_n^{(2)}$] and to the regular cylindrical Bessel and Hankel functions [denoted by $B_{n+1/2}$ to represent $J_{n+1/2}, Y_{n+1/2}, H_{n+1/2}^{(1)}$, or $H_{n+1/2}^{(2)}$] by

$$\hat{B}_n(x) = xb_n(x) = \sqrt{\frac{\pi x}{2}} B_{n+1/2}(x)$$ (IV-49)

x	$J_0(x)$	$J_1(x)$	$Y_0(x)$	$Y_1(x)$
0.0	1.00000	0.00000	$-\infty$	$-\infty$
0.1	0.99750	0.04994	−1.53424	−6.45895
0.2	0.99003	0.09950	−1.08110	−3.32382
0.3	0.97763	0.14832	−0.80727	−2.29310
0.4	0.96040	0.19603	−0.60602	−1.78087
0.5	0.93847	0.24227	−0.44452	−1.47147
0.6	0.91201	0.28670	−0.30851	−1.26039
0.7	0.88120	0.32900	−0.19066	−1.10325
0.8	0.84629	0.36884	−0.08680	−0.97814
0.9	0.80752	0.40595	0.00563	−0.87313
1.0	0.76520	0.44005	0.08826	−0.78121
1.1	0.71962	0.47090	0.16216	−0.69812
1.2	0.67113	0.49829	0.22808	−0.62114
1.3	0.62009	0.52202	0.28654	−0.54852
1.4	0.56686	0.54195	0.33789	−0.47915
1.5	0.51183	0.55794	0.38245	−0.41231
1.6	0.45540	0.56990	0.42043	−0.34758
1.7	0.39799	0.57777	0.45203	−0.28473
1.8	0.33999	0.58152	0.47743	−0.22366
1.9	0.28182	0.58116	0.49682	−0.16441
2.0	0.22389	0.57673	0.51038	−0.10703
2.1	0.16661	0.56829	0.51829	−0.05168
2.2	0.11036	0.55596	0.52078	0.00149
2.3	0.05554	0.53987	0.51807	0.05228
2.4	0.00251	0.52019	0.51041	0.10049
2.5	−0.04838	0.49710	0.49807	0.14592
2.6	−0.09681	0.47082	0.48133	0.18836
2.7	−0.14245	0.44161	0.46050	0.22763
2.8	−0.18504	0.40972	0.43592	0.26354
2.9	−0.22432	0.37544	0.40791	0.29594
3.0	−0.26005	0.33906	0.37686	0.32467
3.1	−0.29206	0.30092	0.34310	0.34963
3.2	−0.32019	0.26134	0.30705	0.37071
3.3	−0.34430	0.22066	0.26909	0.38785
3.4	−0.36430	0.17923	0.22962	0.40101
3.5	−0.38013	0.13738	0.18902	0.41019
3.6	−0.39177	0.09547	0.14771	0.41539
3.7	−0.39923	0.05383	0.10607	0.41667
3.8	−0.40256	0.01282	0.06450	0.41411
3.9	−0.40183	−0.02724	0.02338	0.40782
4.0	−0.39715	−0.06604	−0.01694	0.39793
4.1	−0.38868	−0.10328	−0.05609	0.38459
4.2	−0.37657	−0.13865	−0.09375	0.36801
4.3	−0.36102	−0.17190	−0.12960	0.34839
4.4	−0.34226	−0.20278	−0.16334	0.32597
4.5	−0.32054	−0.23106	−0.19471	0.30100
4.6	−0.29614	−0.25655	−0.22346	0.27375
4.7	−0.26933	−0.27908	−0.24939	0.24450
4.8	−0.24043	−0.29850	−0.27230	0.21356
4.9	−0.20974	−0.31470	−0.29205	0.18125
5.0	−0.17760	−0.32758	−0.30852	0.14786
5.1	−0.14434	−0.33710	−0.32160	0.11374
5.2	−0.11029	−0.34322	−0.33125	0.07919

x	$J_0(x)$	$J_1(x)$	$Y_0(x)$	$Y_1(x)$
5.3	−0.07580	−0.34596	−0.33744	0.04455
5.4	−0.04121	−0.34534	−0.34017	0.01013
5.5	−0.00684	−0.34144	−0.33948	−0.02376
5.6	0.02697	−0.33433	−0.33544	−0.05681
5.7	0.05992	−0.32415	−0.32816	−0.08872
5.8	0.09170	−0.31103	−0.31775	−0.11923
5.9	0.12203	−0.29514	−0.30437	−0.14808
6.0	0.15065	−0.27668	−0.28819	−0.17501
6.1	0.17729	−0.25587	−0.26943	−0.19981
6.2	0.20175	−0.23292	−0.24831	−0.22228
6.3	0.22381	−0.20809	−0.22506	−0.24225
6.4	0.24331	−0.18164	−0.19995	−0.25956
6.5	0.26009	−0.15384	−0.17324	−0.27409
6.6	0.27404	−0.12498	−0.14523	−0.28575
6.7	0.28506	−0.09534	−0.11619	−0.29446
6.8	0.29310	−0.06522	−0.08643	−0.30019
6.9	0.29810	−0.03490	−0.05625	−0.30292
7.0	0.30008	−0.00468	−0.02595	−0.30267
7.1	0.29905	0.02515	0.00418	−0.29948
7.2	0.29507	0.05433	0.03385	−0.29342
7.3	0.28822	0.08257	0.06277	−0.28459
7.4	0.27860	0.10962	0.09068	−0.27311
7.5	0.26634	0.13525	0.11731	−0.25913
7.6	0.25160	0.15921	0.14243	−0.24280
7.7	0.23456	0.18131	0.16580	−0.22432
7.8	0.21541	0.20136	0.18723	−0.20388
7.9	0.19436	0.21918	0.20652	−0.18172
8.0	0.17165	0.23464	0.22352	−0.15806
8.1	0.14752	0.24761	0.23809	−0.13315
8.2	0.12222	0.25800	0.25012	−0.10724
8.3	0.09601	0.26574	0.25951	−0.08060
8.4	0.06916	0.27079	0.26622	−0.05348
8.5	0.04194	0.27312	0.27021	−0.02617
8.6	0.01462	0.27276	0.27146	0.00108
8.7	−0.01252	0.26972	0.27000	0.02801
8.8	−0.03923	0.26407	0.26587	0.05436
8.9	−0.06525	0.25590	0.25916	0.07987
9.0	−0.09033	0.24531	0.24994	0.10431
9.1	−0.11424	0.23243	0.23834	0.12747
9.2	−0.13675	0.21741	0.22449	0.14911
9.3	−0.15765	0.20041	0.20857	0.16906
9.4	−0.17677	0.18163	0.19074	0.18714
9.5	−0.19393	0.16126	0.17121	0.20318
9.6	−0.20898	0.13952	0.15018	0.21706
9.7	−0.22180	0.11664	0.12787	0.22866
9.8	−0.23228	0.09284	0.10453	0.23789
9.9	−0.24034	0.06837	0.08038	0.24469
10.0	−0.24594	0.04347	0.05567	0.24902
10.1	−0.24903	0.01840	0.03066	0.25084
10.2	−0.24962	−0.00662	0.00558	0.25019
10.3	−0.24772	−0.03132	−0.01930	0.24707
10.4	−0.24337	−0.05547	−0.04375	0.24155
10.5	−0.23665	−0.07885	−0.06753	0.23370

x	$J_0(x)$	$J_1(x)$	$Y_0(x)$	$Y_1(x)$
10.6	-0.22764	-0.10123	-0.09042	0.22363
10.7	-0.21644	-0.12240	-0.11219	0.21144
10.8	-0.20320	-0.14217	-0.13264	0.19729
10.9	-0.18806	-0.16035	-0.15158	0.18132
11.0	-0.17119	-0.17679	-0.16885	0.16371
11.1	-0.15277	-0.19133	-0.18428	0.14464
11.2	-0.13299	-0.20385	-0.19773	0.12431
11.3	-0.11207	-0.21426	-0.20910	0.10294
11.4	-0.09021	-0.22245	-0.21829	0.08074
11.5	-0.06765	-0.22838	-0.22523	0.05794
11.6	-0.04462	-0.23200	-0.22987	0.03477
11.7	-0.02133	-0.23330	-0.23218	0.01145
11.8	0.00197	-0.23229	-0.23216	-0.01179
11.9	0.02505	-0.22898	-0.22983	-0.03471
12.0	0.04769	-0.22345	-0.22524	-0.05710
12.1	0.06967	-0.21575	-0.21844	-0.07874
12.2	0.09077	-0.20598	-0.20952	-0.09942
12.3	0.11080	-0.19426	-0.19859	-0.11895
12.4	0.12956	-0.18071	-0.18578	-0.13714
12.5	0.14689	-0.16549	-0.17121	-0.15384
12.6	0.16261	-0.14874	-0.15506	-0.16888
12.7	0.17659	-0.13066	-0.13750	-0.18213
12.8	0.18870	-0.11143	-0.11870	-0.19347
12.9	0.19885	-0.09125	-0.09887	-0.20282
13.0	0.20693	-0.07032	-0.07821	-0.21008
13.1	0.21289	-0.04885	-0.05692	-0.21521
13.2	0.21669	-0.02707	-0.03524	-0.21817
13.3	0.21830	-0.00518	-0.01336	-0.21895
13.4	0.21773	0.01660	0.00848	-0.21756
13.5	0.21499	0.03805	0.03008	-0.21402
13.6	0.21013	0.05896	0.05122	-0.20839
13.7	0.20322	0.07914	0.07169	-0.20074
13.8	0.19434	0.09839	0.09130	-0.19116
13.9	0.18358	0.11653	0.10986	-0.17975
14.0	0.17108	0.13338	0.12719	-0.16664
14.1	0.15695	0.14879	0.14314	-0.15198
14.2	0.14137	0.16261	0.15754	-0.13592
14.3	0.12449	0.17473	0.17028	-0.11862
14.4	0.10649	0.18503	0.18123	-0.10026
14.5	0.08755	0.19343	0.19030	-0.08104
14.6	0.06787	0.19986	0.19742	-0.06115
14.7	0.04764	0.20426	0.20252	-0.04079
14.8	0.02708	0.20660	0.20557	-0.02016
14.9	0.00639	0.20688	0.20655	0.00053
15.0	-0.01422	0.20511	0.20546	0.02107

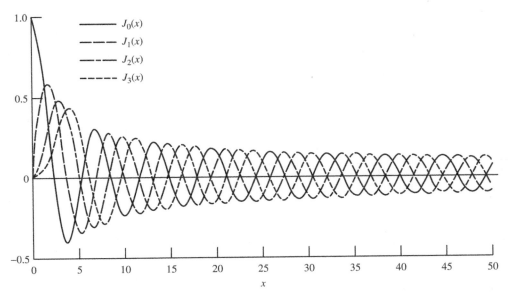

Figure IV-1 Bessel functions of the first kind [$J_0(x)$, $J_1(x)$, $J_2(x)$, and $J_3(x)$]. (Source: C. A. Balanis, *Antenna Theory: Analysis and Design*, Third Edition, copyright © 2005, John Wiley & Sons, Inc. Reprinted by permission of John Wiley & Sons, Inc.)

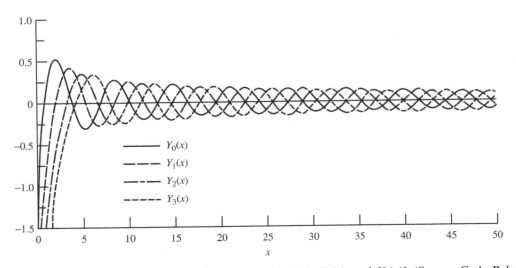

Figure IV-2 Bessel functions of the second kind [$Y_0(x)$, $Y_1(x)$, $Y_2(x)$, and $Y_3(x)$]. (Source: C. A. Balanis, *Antenna Theory: Analysis and Design*, Third Edition, copyright © 2005, John Wiley & Sons, Inc. Reprinted by permission of John Wiley & Sons, Inc.)

$J_1(x)/x$ function

x	$J_1(x)/x$	x	$J_1(x)/x$	x	$J_1(x)/x$
0.0	0.50000	5.0	−0.06552	10.0	0.00435
0.1	0.49938	5.1	−0.06610	10.1	0.00182
0.2	0.49750	5.2	−0.06600	10.2	−0.00065
0.3	0.49440	5.3	−0.06528	10.3	−0.00304
0.4	0.49007	5.4	−0.06395	10.4	−0.00533
0.5	0.48454	5.5	−0.06208	10.5	−0.00751
0.6	0.47783	5.6	−0.05970	10.6	−0.00955
0.7	0.46999	5.7	−0.05687	10.7	−0.01144
0.8	0.46105	5.8	−0.05363	10.8	−0.01316
0.9	0.45105	5.9	−0.05002	10.9	−0.01471
1.0	0.44005	6.0	−0.04611	11.0	−0.01607
1.1	0.42809	6.1	−0.04194	11.1	−0.01724
1.2	0.41524	6.2	−0.03757	11.2	−0.01820
1.3	0.40156	6.3	−0.03303	11.3	−0.01896
1.4	0.38710	6.4	−0.02838	11.4	−0.01951
1.5	0.37196	6.5	−0.02367	11.5	−0.01986
1.6	0.35618	6.6	−0.01894	11.6	−0.02000
1.7	0.33986	6.7	−0.01423	11.7	−0.01994
1.8	0.32306	6.8	−0.00959	11.8	−0.01969
1.9	0.30587	6.9	−0.00506	11.9	−0.01924
2.0	0.28836	7.0	−0.00067	12.0	−0.01862
2.1	0.27061	7.1	0.00354	12.1	−0.01783
2.2	0.25271	7.2	0.00755	12.2	−0.01688
2.3	0.23473	7.3	0.01131	12.3	−0.01579
2.4	0.21674	7.4	0.01481	12.4	−0.01457
2.5	0.19884	7.5	0.01803	12.5	−0.01324
2.6	0.18108	7.6	0.02095	12.6	−0.01180
2.7	0.16356	7.7	0.02355	12.7	−0.01029
2.8	0.14633	7.8	0.02582	12.8	−0.00871
2.9	0.12946	7.9	0.02774	12.9	−0.00707
3.0	0.11302	8.0	0.02933	13.0	−0.00541
3.1	0.09707	8.1	0.03057	13.1	−0.00373
3.2	0.08167	8.2	0.03146	13.2	−0.00205
3.3	0.06687	8.3	0.03202	13.3	−0.00039
3.4	0.05271	8.4	0.03224	13.4	0.00124
3.5	0.03925	8.5	0.03213	13.5	0.00282
3.6	0.02652	8.6	0.03172	13.6	0.00434
3.7	0.01455	8.7	0.03100	13.7	0.00578
3.8	0.00337	8.8	0.03001	13.8	0.00713
3.9	−0.00699	8.9	0.02875	13.9	0.00838
4.0	−0.01651	9.0	0.02726	14.0	0.00953
4.1	−0.02519	9.1	0.02554	14.1	0.01055
4.2	−0.03301	9.2	0.02363	14.2	0.01145
4.3	−0.03998	9.3	0.02155	14.3	0.01222
4.4	−0.04609	9.4	0.01932	14.4	0.01285
4.5	−0.05135	9.5	0.01697	14.5	0.01334
4.6	−0.05578	9.6	0.01453	14.6	0.01369
4.7	−0.05938	9.7	0.01202	14.7	0.01389
4.8	−0.06219	9.8	0.00947	14.8	0.01396
4.9	−0.06423	9.9	0.00691	14.9	0.01388
				15.0	0.01367

$\int_0^x J_0(\tau)\,d\tau$ and $\int_0^x Y_0(\tau)\,d\tau$ functions

x	$\int_0^x J_0(\tau)\,d\tau$	$\int_0^x Y_0(\tau)\,d\tau$	x	$\int_0^x J_0(\tau)\,d\tau$	$\int_0^x Y_0(\tau)\,d\tau$
0.0	0.00000	0.00000	5.0	0.71531	0.19971
0.1	0.09991	−0.21743	5.1	0.69920	0.16818
0.2	0.19933	−0.34570	5.2	0.68647	0.13551
0.3	0.29775	−0.43928	5.3	0.67716	0.10205
0.4	0.39469	−0.50952	5.4	0.67131	0.06814
0.5	0.48968	−0.56179	5.5	0.66891	0.03413
0.6	0.58224	−0.59927	5.6	0.66992	0.00035
0.7	0.67193	−0.62409	5.7	0.67427	−0.03284
0.8	0.75834	−0.63786	5.8	0.68187	−0.06517
0.9	0.84106	−0.64184	5.9	0.69257	−0.09630
1.0	0.91973	−0.63706	6.0	0.70622	−0.12595
1.1	0.99399	−0.62447	6.1	0.72263	−0.15385
1.2	1.06355	−0.60490	6.2	0.74160	−0.17975
1.3	1.12813	−0.57911	6.3	0.76290	−0.20344
1.4	1.18750	−0.54783	6.4	0.78628	−0.22470
1.5	1.24144	−0.51175	6.5	0.81147	−0.24338
1.6	1.28982	−0.47156	6.6	0.83820	−0.25931
1.7	1.33249	−0.42788	6.7	0.86618	−0.27239
1.8	1.36939	−0.38136	6.8	0.89512	−0.28252
1.9	1.40048	−0.33260	6.9	0.92470	−0.28966
2.0	1.42577	−0.28219	7.0	0.95464	−0.29377
2.1	1.44528	−0.23071	7.1	0.98462	−0.29486
2.2	1.45912	−0.17871	7.2	1.01435	−0.29295
2.3	1.46740	−0.12672	7.3	1.04354	−0.28811
2.4	1.47029	−0.07526	7.4	1.07190	−0.28043
2.5	1.46798	−0.02480	7.5	1.09917	−0.27002
2.6	1.46069	0.02420	7.6	1.12508	−0.25702
2.7	1.44871	0.07132	7.7	1.14941	−0.24159
2.8	1.43231	0.11617	7.8	1.17192	−0.22392
2.9	1.41181	0.15839	7.9	1.19243	−0.20421
3.0	1.38756	0.19765	8.0	1.21074	−0.18269
3.1	1.35992	0.23367	8.1	1.22671	−0.15959
3.2	1.32928	0.26620	8.2	1.24021	−0.13516
3.3	1.29602	0.29502	8.3	1.25112	−0.10966
3.4	1.26056	0.31996	8.4	1.25939	−0.08335
3.5	1.22330	0.34090	8.5	1.26494	−0.05650
3.6	1.18467	0.35775	8.6	1.26777	−0.02940
3.7	1.14509	0.37044	8.7	1.26787	−0.00230
3.8	1.10496	0.37896	8.8	1.26528	0.02451
3.9	1.06471	0.38335	8.9	1.26005	0.05078
4.0	1.02473	0.38366	9.0	1.25226	0.07625
4.1	0.98541	0.38000	9.1	1.24202	0.10069
4.2	0.94712	0.37250	9.2	1.22946	0.12385
4.3	0.91021	0.36131	9.3	1.21473	0.14552
4.4	0.87502	0.34665	9.4	1.19799	0.16550
4.5	0.84186	0.32872	9.5	1.17944	0.18361
4.6	0.81100	0.30779	9.6	1.15927	0.19969
4.7	0.78271	0.28413	9.7	1.13772	0.21360
4.8	0.75721	0.25802	9.8	1.11499	0.22523
4.9	0.73468	0.22977	9.9	1.09134	0.23448
			10.0	1.06701	0.24129

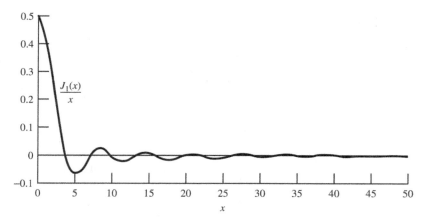

Figure IV-3   Plot of $J_1(x)/x$ function. (Source: C. A. Balanis, *Antenna Theory: Analysis and Design*, Third Edition, copyright © 2005, John Wiley & Sons, Inc. Reprinted by permission of John Wiley & Sons, Inc.)

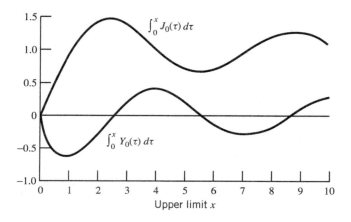

Figure IV-4   Plots of $\int_0^x J_0(\tau)\,d\tau$ and $\int_0^x Y_0(\tau)\,d\tau$ functions. (Source: C. A. Balanis, *Antenna Theory: Analysis and Design*, Third Edition, copyright © 2005, John Wiley & Sons, Inc. Reprinted by permission of John Wiley & Sons, Inc.)

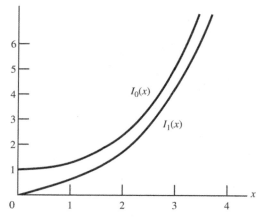

Figure IV-5   Modified Bessel functions of the first kind [$I_0(x)$ and $I_1(x)$].

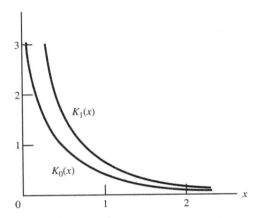

Figure IV-6   Modified Bessel functions of the second kind [$K_0(x)$ and $K_1(x)$].

APPENDIX **V**

▨▨▨▨▨▨▨▨▨▨▨▨▨

Legendre Polynomials and Functions

V.1 LEGENDRE POLYNOMIALS AND FUNCTIONS

The *ordinary* Legendre differential equation can be written as

$$(1-x^2)\frac{d^2y}{dx^2} - 2x\frac{dy}{dx} + p(p+1)y = 0 \tag{V-1}$$

Its solution can be written as

$$y(x) = A_1 P_p(x) + B_1 P_p(-x) \qquad p \text{ not an integer} \tag{V-2}$$

where $P_p(x)$ is referred to as the *Legendre function of the first kind*. If p is an integer ($p=n$), then $P_n(x)$ and $P_n(-x)$ are not two independent solutions because

$$P_n(-x) = (-1)^n P_n(x) \tag{V-3}$$

Therefore, two independent solutions to (V-1) for $p=n$, an integer, are

$$y(x) = A_2 P_n(x) + B_2 Q_n(x) \tag{V-4}$$

where $Q_n(x)$ is referred to as the *Legendre function of the second kind*.

When $p=n$, an integer, $P_n(x)$ are also referred to as the *Legendre polynomials* of order n, and are defined by

$$P_n(x) = \sum_{m=0}^{M} \frac{(-1)^m (2n-2m)!(x)^{n-2m}}{2^n m!(n-m)!(n-2m)!} \tag{V-5}$$

where $M = n/2$ or $(n-1)/2$, whichever is an integer.

The Legendre functions $Q_n(x)$ of the second kind are defined by

$$Q_n(x) = \lim_{p \to n} Q_p(x) = \lim_{p \to n} \frac{\pi}{2} \frac{P_p(x)\cos(p\pi) - P_p(-x)}{\sin(p\pi)} \tag{V-6}$$

The Legendre polynomials (or Legendre functions of the first kind) $P_n(x)$ can also be obtained more conveniently using *Rodrigues' formula*

$$P_n(x) = \frac{1}{2^n n!} \frac{d^n}{dx^n} (x^2 - 1)^n \tag{V-7}$$

which when expanded leads (for $n = 0, 1, 2, \ldots, 7$) to

$$\begin{aligned}
P_0(x) &= 1 \\
P_1(x) &= x \\
P_2(x) &= \tfrac{1}{2}(3x^2 - 1) \\
P_3(x) &= \tfrac{1}{2}(5x^3 - 3x) \\
P_4(x) &= \tfrac{1}{8}(35x^4 - 30x^2 + 3) \\
P_5(x) &= \tfrac{1}{8}(63x^5 - 70x^3 + 15x) \\
P_6(x) &= \tfrac{1}{16}(231x^6 - 315x^4 + 105x^2 - 5) \\
P_7(x) &= \tfrac{1}{16}(429x^7 - 693x^5 + 315x^3 - 35x)
\end{aligned} \tag{V-8}$$

If $x = \cos\theta$, the Legendre polynomials (or Legendre functions of the first kind) of (V-8) can be written as

$$\begin{aligned}
P_0(\cos\theta) &= 1 \\
P_1(\cos\theta) &= \cos\theta \\
P_2(\cos\theta) &= \tfrac{1}{4}(3\cos 2\theta + 1) \\
P_3(\cos\theta) &= \tfrac{1}{8}(5\cos 3\theta + 3\cos\theta) \\
P_4(\cos\theta) &= \tfrac{1}{64}(35\cos 4\theta + 20\cos 2\theta + 9) \\
P_5(\cos\theta) &= \tfrac{1}{128}(63\cos 5\theta + 35\cos 3\theta + 30\cos\theta) \\
P_6(\cos\theta) &= \tfrac{1}{512}(231\cos 6\theta + 126\cos 4\theta + 105\cos 2\theta + 50) \\
P_7(\cos\theta) &= \tfrac{1}{1024}(429\cos 7\theta + 231\cos 5\theta + 189\cos 3\theta + 175\cos\theta)
\end{aligned} \tag{V-9}$$

The Legendre functions $Q_n(x)$ of the second kind exhibit singularities at $x = \pm 1$ or $\theta = 0, \pi$ and can be obtained from the Legendre functions $P_n(x)$ of the first kind using the formula

$$Q_n(x) = P_n(x)\left\{\frac{1}{2}\ln\left(\frac{1+x}{1-x}\right) - \psi(n)\right\} + \sum_{m=1}^{n} \frac{(-1)^m(n+m)!}{(m!)^2(n-m)!}\psi(m)\left(\frac{1-x}{2}\right)^m \tag{V-10}$$

where

$$\psi(n) = 1 + \frac{1}{2} + \frac{1}{3} + \cdots + \frac{1}{n} \tag{V-10a}$$

When (V-10) is expanded, it leads (for $n = 0, 1, 2, 3$) to

$$Q_0(x) = \frac{1}{2}\ln\left(\frac{1+x}{1-x}\right)$$

$$Q_1(x) = \frac{x}{2}\ln\left(\frac{1+x}{1-x}\right) - 1$$

$$Q_2(x) = \frac{3x^2-1}{4}\ln\left(\frac{1+x}{1-x}\right) - \frac{3x}{2} \tag{V-11}$$

$$Q_3(x) = \frac{5x^3-3x}{4}\ln\left(\frac{1+x}{1-x}\right) - \frac{5x^2}{2} + \frac{2}{3}$$

or for $x = \cos\theta$ to

$$Q_0(\cos\theta) = \ln\left(\cot\frac{\theta}{2}\right)$$

$$Q_1(\cos\theta) = \cos\theta\ln\left(\cot\frac{\theta}{2}\right) - 1$$

$$Q_2(\cos\theta) = \frac{1}{4}(1+3\cos2\theta)\ln\left(\cot\frac{\theta}{2}\right) - \frac{3}{2}\cos\theta \tag{V-12}$$

$$Q_3(\cos\theta) = \frac{1}{8}(3\cos\theta+5\cos3\theta)\ln\left(\cot\frac{\theta}{2}\right) - \frac{5}{4}\cos2\theta - \frac{7}{12}$$

The Legendre functions of the first $P_n(x)$ and second $Q_n(x)$ kind obey the following recurrence relations:

$$(n+1)R_{n+1}(x) - (2n+1)xR_n(x) + nR_{n-1}(x) = 0 \tag{V-13a}$$

$$\frac{dR_{n+1}(x)}{dx} - x\frac{dR_n(x)}{dx} = (n+1)R_n(x) \tag{V-13b}$$

$$x\frac{dR_n(x)}{dx} - \frac{dR_{n-1}(x)}{dx} = nR_n(x) \tag{V-13c}$$

$$\frac{dR_{n+1}(x)}{dx} - \frac{dR_{n-1}(x)}{dx} = (2n+1)R_n(x) \tag{V-13d}$$

$$(x^2-1)\frac{dR_n(x)}{dx} = nxR_n(x) - nR_{n-1}(x) = -(n+1)(xR_n - R_{n+1}) \tag{V-13e}$$

where $R_n(x)$ can be either $P_n(x)$ or $Q_n(x)$.

Some other useful formulas involving Legendre polynomials $P_n(x)$ and $Q_n(x)$ are

$$\int_{-1}^{1} P_m(x)P_n(x)\,dx = 0 \qquad m \neq n \tag{V-14a}$$

$$\int_{-1}^{1} [P_n(x)]^2\,dx = \frac{2}{2n+1} \tag{V-14b}$$

$$\int_{0}^{1} [Q_n(x)]^2\,dx = \frac{1}{2n+1}\left[\frac{\pi^2}{4} - \frac{1}{(n+1)^2} - \frac{1}{(n+2)^2} - \cdots\right] \tag{V-14c}$$

which indicate that the Legendre polynomials are orthogonal in the range $-1 \le x \le 1$. Also

$$P_n(-x) = (-1)^n P_n(x) \tag{V-15a}$$

$$P_n(x) = P_{-n-1}(x) \tag{V-15b}$$

$$Q_n(-x) = (-1)^{n+1} Q_n(x) \tag{V-15c}$$

$$P_n(0) = \begin{cases} 0 & n = \text{odd} \\ (-1)^{n/2} \dfrac{1 \cdot 3 \cdot 5 \cdots (n-1)}{2 \cdot 4 \cdot 6 \cdots n} & n = \text{even} \end{cases} \tag{V-15d}$$

$$P_n(1) = 1 \tag{V-15e}$$

$$P_n(-1) = \begin{cases} 1 & n = \text{even} \\ -1 & n = \text{odd} \end{cases} \tag{V-15f}$$

$$Q_n(1) = +\infty \tag{V-15g}$$

$$Q_n(-1) = \begin{cases} -\infty & n = \text{even} \\ +\infty & n = \text{odd} \end{cases} \tag{V-15h}$$

$$P_n(x) = \frac{1}{\pi} \int_0^\pi (x + \sqrt{x^2 - 1} \cos \psi)^n \, d\psi \tag{V-15i}$$

$$\int P_n(x)\, dx = \frac{P_{n+1}(x) - P_{n-1}(x)}{(2n+1)} \tag{V-15j}$$

Plots of $P_n(x)$ and $Q_n(x)$ for $n = 0$, 1, 2, 3 in the range $-1 \le x \le 1$ are shown in Figures V-1 and V-2.

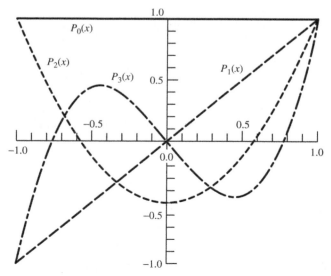

Figure V-1 Legendre functions of the first kind [$P_0(x)$, $P_1(x)$, $P_2(x)$, and $P_3(x)$].

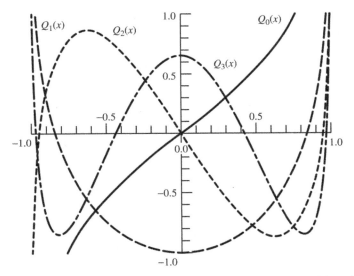

Figure V-2   Legendre functions of the second kind [$Q_0(x)$, $Q_1(x)$, $Q_2(x)$, and $Q_3(x)$].

V.2   ASSOCIATED LEGENDRE FUNCTIONS

In addition to the *ordinary* Legendre differential equation V-1, there also exists the *associated* Legendre differential equation

$$(1-x^2)\frac{d^2 y}{dx^2} - 2x\frac{dy}{dx} + \left[n(n+1) - \frac{m^2}{1-x^2}\right]y = 0 \tag{V-16}$$

whose solution, for nonnegative integer values of n and m, takes the form

$$y(x) = A_1 P_n^m(x) + B_1 Q_n^m(x) \tag{V-17}$$

where $P_n^m(x)$ and $Q_n^m(x)$ are referred to, respectively, as the *associated Legendre functions of the first and second kind.*

The associated Legendre functions $P_n^m(x)$ and $Q_n^m(x)$ of the first and second kind are related, respectively, to the Legendre functions $P_n(x)$ and $Q_n(x)$ of the first and second kind by

$$P_n^m(x) = (-1)^m (1-x^2)^{m/2} \frac{d^m P_n(x)}{dx^m}$$

$$= (-1)^m \frac{(1-x^2)^{m/2}}{2^n n!} \frac{d^{m+n}(x^2-1)^n}{dx^{m+n}} \tag{V-18a}$$

$$Q_n^m(x) = (-1)^m (1-x^2)^{m/2} \frac{d^m Q_n(x)}{dx^m} \tag{V-18b}$$

The associated Legendre functions $Q_n^m(x)$ of the second kind are singular at $x = \pm 1$, as are the Legendre functions $Q_n(x)$.

When (V-18a) and (V-18b) are expanded, we can write, using (V-8) and (V-11), the first few orders of $P_n^m(x)$ and $Q_n^m(x)$ as

$$P_0^0(x) = P_0(x) = 1 \qquad\qquad P_0^2(x) = 0$$

$$P_1^0(x) = P_1(x) = x \qquad\qquad P_1^2(x) = 0$$

$$P_2^0(x) = P_2(x) = \tfrac{1}{2}(3x^2 - 1) \qquad P_2^2(x) = 3(1 - x^2)$$

$$P_3^0(x) = P_3(x) = \tfrac{1}{2}(5x^3 - 3x) \qquad P_3^2(x) = 15x(1 - x^2)$$

$$\vdots \qquad\qquad\qquad \vdots$$

$$P_0^1(x) = 0 \qquad\qquad P_0^3(x) = 0 \qquad\qquad \text{(V-19a)}$$

$$P_1^1(x) = -(1 - x^2)^{1/2} \qquad\qquad P_1^3(x) = 0$$

$$P_2^1(x) = -3x(1 - x^2)^{1/2} \qquad\qquad P_2^3(x) = 0$$

$$P_3^1(x) = -\tfrac{3}{2}(5x^2 - 1)(1 - x^2)^{1/2} \quad P_3^3(x) = -15(1 - x^2)^{3/2}$$

$$\vdots \qquad\qquad\qquad \vdots$$

$$Q_0^0(x) = Q_0(x) = \frac{1}{2}\ln\left(\frac{1+x}{1-x}\right)$$

$$Q_1^0(x) = Q_1(x) = \frac{x}{2}\ln\left(\frac{1+x}{1-x}\right) - 1$$

$$Q_2^0(x) = Q_2(x) = \frac{3x^2 - 1}{4}\ln\left(\frac{1+x}{1-x}\right) - \frac{3x}{2}$$

$$\vdots$$

$$Q_0^1(x) = 0$$

$$Q_1^1(x) = -(1 - x^2)^{1/2}\left[\frac{1}{2}\ln\left(\frac{1+x}{1-x}\right) + \frac{x}{1-x^2}\right] \qquad \text{(V-19b)}$$

$$Q_2^1(x) = -(1 - x^2)^{1/2}\left[\frac{3x}{2}\ln\left(\frac{1+x}{1-x}\right) + \frac{3x^2 - 2}{1-x^2}\right]$$

$$\vdots$$

$$Q_0^2(x) = 0$$

$$Q_1^2(x) = 0$$

$$Q_2^2(x) = (1 - x^2)^{1/2}\left[\frac{3}{2}\ln\left(\frac{1+x}{1-x}\right) + \frac{5x - 3x^2}{(1-x^2)^2}\right]$$

$$\vdots$$

It should be noted that

$$P_n^0(x) = P_n(x) \qquad\qquad\qquad\qquad \text{(V-20a)}$$

$$Q_n^0(x) = Q_n(x) \qquad\qquad\qquad\qquad \text{(V-20b)}$$

$$P_n^m(x) = 0 \qquad m > n \qquad\qquad\qquad \text{(V-20c)}$$

$$Q_n^m(x) = 0 \qquad m > n \tag{V-20d}$$

$$P_n^m(-x) = (-1)^{n-m} P_n^m(x) \tag{V-20e}$$

$$P_n^m(x) = P_{-n-1}^m(x) \tag{V-20f}$$

$$Q_n^m(-x) = (-1)^{n+m+1} Q_n(x) \tag{V-20g}$$

$$P_n^m(1) = \begin{cases} 1 & m = 0 \\ 0 & m > 0 \end{cases} \tag{V-20h}$$

$$P_n^m(0) = \begin{cases} (-1)^{(n+m)/2} \dfrac{1 \cdot 3 \cdot 5 \cdots (n+m-1)}{2 \cdot 4 \cdot 6 \cdots (n-m)} & n+m = \text{even} \\[4mm] 0 & n+m = \text{odd} \end{cases} \tag{V-20i}$$

$$Q_n^m(0) = \begin{cases} 0 & n+m = \text{even} \\[2mm] (-1)^{(n+m+1)/2} \dfrac{2 \cdot 4 \cdot 6 \cdots (n+m-1)}{1 \cdot 3 \cdot 5 \cdots (n-m)} & n+m = \text{odd} \end{cases} \tag{V-20j}$$

$$\left. \frac{d^q P_n^m(x)}{dx^q} \right|_{x=0} = (-1)^q P_n^{m+q}(0) \tag{V-20k}$$

$$\left. \frac{d^q Q_n^m(x)}{dx^q} \right|_{x=0} = (-1)^q Q_n^{m+q}(0) \tag{V-20l}$$

Orthogonality relations of $P_n^m(x)$ in the range of $-1 \le x \le 1$ are

$$\int_{-1}^{1} P_n^m(x) P_l^m(x)\, dx = 0 \qquad n \ne l \tag{V-21a}$$

$$\int_{-1}^{1} [P_n^m(x)]^2\, dx = \frac{2}{2n+1} \frac{(n+m)!}{(n-m)!} \tag{V-21b}$$

$$\int_{-1}^{1} \left[\frac{dP_n(x)}{dx} \right]^2 dx = n(n+1) \tag{V-21c}$$

and useful recurrence formulas are

$$(n+1-m) R_{n+1}^m(x) - (2n+1)x R_n^m(x) + (n+m) R_{n-1}^m(x) = 0 \tag{V-22a}$$

$$R_n^{m+2}(x) + \frac{2(m+1)x}{(1-x^2)^{1/2}} R_n^{m+1}(x) + (n-m)(n+m+1) R_n^m(x) = 0 \tag{V-22b}$$

where $R_n^m(x)$ can be either $P_n^m(x)$ or $Q_n^m(x)$.

When m is not an integer in the associated Legendre differential equation V-16, the solutions become more complex and can be expressed in terms of *hypergeometric functions*. These solutions are beyond the scope of this book, and the reader is referred to the literature.

x	$P_0(x)$	$P_1(x)$	$P_2(x)$	$P_3(x)$
−1.00	1.00000	−1.00000	1.00000	−1.00000
−0.99	1.00000	−0.99000	0.97015	−0.94075
−0.98	1.00000	−0.98000	0.94060	−0.88298
−0.97	1.00000	−0.97000	0.91135	−0.82668
−0.96	1.00000	−0.96000	0.88240	−0.77184
−0.95	1.00000	−0.95000	0.85375	−0.71844
−0.94	1.00000	−0.94000	0.82540	−0.66646
−0.93	1.00000	−0.93000	0.79735	−0.61589
−0.92	1.00000	−0.92000	0.76960	−0.56672
−0.91	1.00000	−0.91000	0.74215	−0.51893
−0.90	1.00000	−0.90000	0.71500	−0.47250
−0.89	1.00000	−0.89000	0.68815	−0.42742
−0.88	1.00000	−0.88000	0.66160	−0.38368
−0.87	1.00000	−0.87000	0.63535	−0.34126
−0.86	1.00000	−0.86000	0.60940	−0.30014
−0.85	1.00000	−0.85000	0.58375	−0.26031
−0.84	1.00000	−0.84000	0.55840	−0.22176
−0.83	1.00000	−0.83000	0.53335	−0.18447
−0.82	1.00000	−0.82000	0.50860	−0.14842
−0.81	1.00000	−0.81000	0.48415	−0.11360
−0.80	1.00000	−0.80000	0.46000	−0.08000
−0.79	1.00000	−0.79000	0.43615	−0.04760
−0.78	1.00000	−0.78000	0.41260	−0.01638
−0.77	1.00000	−0.77000	0.38935	0.01367
−0.76	1.00000	−0.76000	0.36640	0.04256
−0.75	1.00000	−0.75000	0.34375	0.07031
−0.74	1.00000	−0.74000	0.32140	0.09694
−0.73	1.00000	−0.73000	0.29935	0.12246
−0.72	1.00000	−0.72000	0.27760	0.14688
−0.71	1.00000	−0.71000	0.25615	0.17022
−0.70	1.00000	−0.70000	0.23500	0.19250
−0.69	1.00000	−0.69000	0.21415	0.21373
−0.68	1.00000	−0.68000	0.19360	0.23392
−0.67	1.00000	−0.67000	0.17335	0.25309
−0.66	1.00000	−0.66000	0.15340	0.27126
−0.65	1.00000	−0.65000	0.13375	0.28844
−0.64	1.00000	−0.64000	0.11440	0.30464
−0.63	1.00000	−0.63000	0.09535	0.31988
−0.62	1.00000	−0.62000	0.07660	0.33418
−0.61	1.00000	−0.61000	0.05815	0.34755
−0.60	1.00000	−0.60000	0.04000	0.36000
−0.59	1.00000	−0.59000	0.02215	0.37155
−0.58	1.00000	−0.58000	0.00460	0.38222
−0.57	1.00000	−0.57000	−0.01265	0.39202
−0.56	1.00000	−0.56000	−0.02960	0.40096
−0.55	1.00000	−0.55000	−0.04625	0.40906
−0.54	1.00000	−0.54000	−0.06260	0.41634
−0.53	1.00000	−0.53000	−0.07865	0.42281
−0.52	1.00000	−0.52000	−0.09440	0.42848
−0.51	1.00000	−0.51000	−0.10985	0.43337

x	$P_0(x)$	$P_1(x)$	$P_2(x)$	$P_3(x)$
−0.50	1.00000	−0.50000	−0.12500	0.43750
−0.49	1.00000	−0.49000	−0.13985	0.44088
−0.48	1.00000	−0.48000	−0.15440	0.44352
−0.47	1.00000	−0.47000	−0.16865	0.44544
−0.46	1.00000	−0.46000	−0.18260	0.44666
−0.45	1.00000	−0.45000	−0.19625	0.44719
−0.44	1.00000	−0.44000	−0.20960	0.44704
−0.43	1.00000	−0.43000	−0.22265	0.44623
−0.42	1.00000	−0.42000	−0.23540	0.44478
−0.41	1.00000	−0.41000	−0.24785	0.44270
−0.40	1.00000	−0.40000	−0.26000	0.44000
−0.39	1.00000	−0.39000	−0.27185	0.43670
−0.38	1.00000	−0.38000	−0.28340	0.43282
−0.37	1.00000	−0.37000	−0.29465	0.42837
−0.36	1.00000	−0.36000	−0.30560	0.42336
−0.35	1.00000	−0.35000	−0.31625	0.41781
−0.34	1.00000	−0.34000	−0.32660	0.41174
−0.33	1.00000	−0.33000	−0.33665	0.40516
−0.32	1.00000	−0.32000	−0.34640	0.39808
−0.31	1.00000	−0.31000	−0.35585	0.39052
−0.30	1.00000	−0.30000	−0.36500	0.38250
−0.29	1.00000	−0.29000	−0.37385	0.37403
−0.28	1.00000	−0.28000	−0.38240	0.36512
−0.27	1.00000	−0.27000	−0.39065	0.35579
−0.26	1.00000	−0.26000	−0.39860	0.34606
−0.25	1.00000	−0.25000	−0.40625	0.33594
−0.24	1.00000	−0.24000	−0.41360	0.32544
−0.23	1.00000	−0.23000	−0.42065	0.31458
−0.22	1.00000	−0.22000	−0.42740	0.30338
−0.21	1.00000	−0.21000	−0.43385	0.29185
−0.20	1.00000	−0.20000	−0.44000	0.28000
−0.19	1.00000	−0.19000	−0.44585	0.26785
−0.18	1.00000	−0.18000	−0.45140	0.25542
−0.17	1.00000	−0.17000	−0.45665	0.24272
−0.16	1.00000	−0.16000	−0.46160	0.22976
−0.15	1.00000	−0.15000	−0.46625	0.21656
−0.14	1.00000	−0.14000	−0.47060	0.20314
−0.13	1.00000	−0.13000	−0.47465	0.18951
−0.12	1.00000	−0.12000	−0.47840	0.17568
−0.11	1.00000	−0.11000	−0.48185	0.16167
−0.10	1.00000	−0.10000	−0.48500	0.14750
−0.09	1.00000	−0.09000	−0.48785	0.13318
−0.08	1.00000	−0.08000	−0.49040	0.11872
−0.07	1.00000	−0.07000	−0.49265	0.10414
−0.06	1.00000	−0.06000	−0.49460	0.08946
−0.05	1.00000	−0.05000	−0.49625	0.07469
−0.04	1.00000	−0.04000	−0.49760	0.05984
−0.03	1.00000	−0.03000	−0.49865	0.04493
−0.02	1.00000	−0.02000	−0.49940	0.02998
−0.01	1.00000	−0.01000	−0.49985	0.01500
0.00	1.00000	0.00000	−0.50000	0.00000

x	$P_0(x)$	$P_1(x)$	$P_2(x)$	$P_3(x)$
0.01	1.00000	0.01000	−0.49985	−0.01500
0.02	1.00000	0.02000	−0.49940	−0.02998
0.03	1.00000	0.03000	−0.49865	−0.04493
0.04	1.00000	0.04000	−0.49760	−0.05984
0.05	1.00000	0.05000	−0.49625	−0.07469
0.06	1.00000	0.06000	−0.49460	−0.08946
0.07	1.00000	0.07000	−0.49265	−0.10414
0.08	1.00000	0.08000	−0.49040	−0.11872
0.09	1.00000	0.09000	−0.48785	−0.13318
0.10	1.00000	0.10000	−0.48500	−0.14750
0.11	1.00000	0.11000	−0.48185	−0.16167
0.12	1.00000	0.12000	−0.47840	−0.17568
0.13	1.00000	0.13000	−0.47465	−0.18951
0.14	1.00000	0.14000	−0.47060	−0.20314
0.15	1.00000	0.15000	−0.46625	−0.21656
0.16	1.00000	0.16000	−0.46160	−0.22976
0.17	1.00000	0.17000	−0.45665	−0.24272
0.18	1.00000	0.18000	−0.45140	−0.25542
0.19	1.00000	0.19000	−0.44585	−0.26785
0.20	1.00000	0.20000	−0.44000	−0.28000
0.21	1.00000	0.21000	−0.43385	−0.29185
0.22	1.00000	0.22000	−0.42740	−0.30338
0.23	1.00000	0.23000	−0.42065	−0.31458
0.24	1.00000	0.24000	−0.41360	−0.32544
0.25	1.00000	0.25000	−0.40625	−0.33594
0.26	1.00000	0.26000	−0.39860	−0.34606
0.27	1.00000	0.27000	−0.39065	−0.35579
0.28	1.00000	0.28000	−0.38240	−0.36512
0.29	1.00000	0.29000	−0.37385	−0.37403
0.30	1.00000	0.30000	−0.36500	−0.38250
0.31	1.00000	0.31000	−0.35585	−0.39052
0.32	1.00000	0.32000	−0.34640	−0.39808
0.33	1.00000	0.33000	−0.33665	−0.40516
0.34	1.00000	0.34000	−0.32660	−0.41174
0.35	1.00000	0.35000	−0.31625	−0.41781
0.36	1.00000	0.36000	−0.30560	−0.42336
0.37	1.00000	0.37000	−0.29465	−0.42837
0.38	1.00000	0.38000	−0.28340	−0.43282
0.39	1.00000	0.39000	−0.27185	−0.43670
0.40	1.00000	0.40000	−0.26000	−0.44000
0.41	1.00000	0.41000	−0.24785	−0.44270
0.42	1.00000	0.42000	−0.23540	−0.44478
0.43	1.00000	0.43000	−0.22265	−0.44623
0.44	1.00000	0.44000	−0.20960	−0.44704
0.45	1.00000	0.45000	−0.19625	−0.44719
0.46	1.00000	0.46000	−0.18260	−0.44666
0.47	1.00000	0.47000	−0.16865	−0.44544
0.48	1.00000	0.48000	−0.15440	−0.44352
0.49	1.00000	0.49000	−0.13985	−0.44088
0.50	1.00000	0.50000	−0.12500	−0.43750
0.51	1.00000	0.51000	−0.10985	−0.43337

x	$P_0(x)$	$P_1(x)$	$P_2(x)$	$P_3(x)$
0.52	1.00000	0.52000	−0.09440	−0.42848
0.53	1.00000	0.53000	−0.07865	−0.42281
0.54	1.00000	0.54000	−0.06260	−0.41634
0.55	1.00000	0.55000	−0.04625	−0.40906
0.56	1.00000	0.56000	−0.02960	−0.40096
0.57	1.00000	0.57000	−0.01265	−0.39202
0.58	1.00000	0.58000	0.00460	−0.38222
0.59	1.00000	0.59000	0.02215	−0.37155
0.60	1.00000	0.60000	0.04000	−0.36000
0.61	1.00000	0.61000	0.05815	−0.34755
0.62	1.00000	0.62000	0.07660	−0.33418
0.63	1.00000	0.63000	0.09535	−0.31988
0.64	1.00000	0.64000	0.11440	−0.30464
0.65	1.00000	0.65000	0.13375	−0.28844
0.66	1.00000	0.66000	0.15340	−0.27126
0.67	1.00000	0.67000	0.17335	−0.25309
0.68	1.00000	0.68000	0.19360	−0.23392
0.69	1.00000	0.69000	0.21415	−0.21373
0.70	1.00000	0.70000	0.23500	−0.19250
0.71	1.00000	0.71000	0.25615	−0.17022
0.72	1.00000	0.72000	0.27760	−0.14688
0.73	1.00000	0.73000	0.29935	−0.12246
0.74	1.00000	0.74000	0.32140	−0.09694
0.75	1.00000	0.75000	0.34375	−0.07031
0.76	1.00000	0.76000	0.36640	−0.04256
0.77	1.00000	0.77000	0.38935	−0.01367
0.78	1.00000	0.78000	0.41260	0.01638
0.79	1.00000	0.79000	0.43615	0.04760
0.80	1.00000	0.80000	0.46000	0.08000
0.81	1.00000	0.81000	0.48415	0.11360
0.82	1.00000	0.82000	0.50860	0.14842
0.83	1.00000	0.83000	0.53335	0.18447
0.84	1.00000	0.84000	0.55840	0.22176
0.85	1.00000	0.85000	0.58375	0.26031
0.86	1.00000	0.86000	0.60940	0.30014
0.87	1.00000	0.87000	0.63535	0.34126
0.88	1.00000	0.88000	0.66160	0.38368
0.89	1.00000	0.89000	0.68815	0.42742
0.90	1.00000	0.90000	0.71500	0.47250
0.91	1.00000	0.91000	0.74215	0.51893
0.92	1.00000	0.92000	0.76960	0.56672
0.93	1.00000	0.93000	0.79735	0.61589
0.94	1.00000	0.94000	0.82540	0.66646
0.95	1.00000	0.95000	0.85375	0.71844
0.96	1.00000	0.96000	0.88240	0.77184
0.97	1.00000	0.97000	0.91135	0.82668
0.98	1.00000	0.98000	0.94060	0.88298
0.99	1.00000	0.99000	0.97015	0.94075
1.00	1.00000	1.00000	1.00000	1.00000

x	$Q_0(x)$	$Q_1(x)$	$Q_2(x)$	$Q_3(x)$
−1.00	−∞	+∞	−∞	+∞
−0.99	−2.64665	1.62019	−1.08265	0.70625
−0.98	−2.29756	1.25161	−0.69109	0.29437
−0.97	−2.09230	1.02953	−0.45181	0.04408
−0.96	−1.94591	0.86807	−0.27707	−0.13540
−0.95	−1.83178	0.74019	−0.13888	−0.27356
−0.94	−1.73805	0.63377	−0.02459	−0.38399
−0.93	−1.65839	0.54230	0.07268	−0.47419
−0.92	−1.58903	0.46190	0.15708	−0.54880
−0.91	−1.52752	0.39005	0.23135	−0.61091
−0.90	−1.47222	0.32500	0.29736	−0.66271
−0.89	−1.42193	0.26551	0.35650	−0.70582
−0.88	−1.37577	0.21068	0.40979	−0.74148
−0.87	−1.33308	0.15978	0.45803	−0.77066
−0.86	−1.29334	0.11228	0.50184	−0.79415
−0.85	−1.25615	0.06773	0.54172	−0.81259
−0.84	−1.22117	0.02579	0.57810	−0.82653
−0.83	−1.18814	−0.01385	0.61131	−0.83641
−0.82	−1.15682	−0.05141	0.64164	−0.84264
−0.81	−1.12703	−0.08711	0.66935	−0.84555
−0.80	−1.09861	−0.12111	0.69464	−0.84544
−0.79	−1.07143	−0.15357	0.71770	−0.84259
−0.78	−1.04537	−0.18461	0.73868	−0.83721
−0.77	−1.02033	−0.21435	0.75774	−0.82953
−0.76	−0.99622	−0.24288	0.77499	−0.81973
−0.75	−0.97296	−0.27028	0.79055	−0.80799
−0.74	−0.95048	−0.29665	0.80452	−0.79447
−0.73	−0.92873	−0.32203	0.81699	−0.77931
−0.72	−0.90764	−0.34650	0.82804	−0.76265
−0.71	−0.88718	−0.37010	0.83775	−0.74460
−0.70	−0.86730	−0.39289	0.84618	−0.72529
−0.69	−0.84796	−0.41491	0.85341	−0.70481
−0.68	−0.82911	−0.43620	0.85948	−0.68328
−0.67	−0.81074	−0.45680	0.86446	−0.66078
−0.66	−0.79281	−0.47674	0.86838	−0.63739
−0.65	−0.77530	−0.49606	0.87130	−0.61321
−0.64	−0.75817	−0.51477	0.87326	−0.58830
−0.63	−0.74142	−0.53291	0.87431	−0.56275
−0.62	−0.72500	−0.55050	0.87446	−0.53662
−0.61	−0.70892	−0.56756	0.87378	−0.50997
−0.60	−0.69315	−0.58411	0.87227	−0.48287
−0.59	−0.67767	−0.60018	0.86999	−0.45537
−0.58	−0.66246	−0.61577	0.86695	−0.42754
−0.57	−0.64752	−0.63091	0.86319	−0.39942
−0.56	−0.63283	−0.64561	0.85873	−0.37107
−0.55	−0.61838	−0.65989	0.85360	−0.34254
−0.54	−0.60416	−0.67376	0.84782	−0.31387
−0.53	−0.59015	−0.68722	0.84141	−0.28510
−0.52	−0.57634	−0.70030	0.83441	−0.25628
−0.51	−0.56273	−0.71301	0.82682	−0.22745

x	$Q_0(x)$	$Q_1(x)$	$Q_2(x)$	$Q_3(x)$
−0.50	−0.54931	−0.72535	0.81866	−0.19865
−0.49	−0.53606	−0.73733	0.80997	−0.16992
−0.48	−0.52298	−0.74897	0.80075	−0.14129
−0.47	−0.51007	−0.76027	0.79102	−0.11279
−0.46	−0.49731	−0.77124	0.78081	−0.08446
−0.45	−0.48470	−0.78188	0.77012	−0.05634
−0.44	−0.47223	−0.79222	0.75898	−0.02844
−0.43	−0.45990	−0.80224	0.74740	−0.00080
−0.42	−0.44769	−0.81197	0.73539	0.02654
−0.41	−0.43561	−0.82140	0.72297	0.05357
−0.40	−0.42365	−0.83054	0.71015	0.08026
−0.39	−0.41180	−0.83940	0.69695	0.10658
−0.38	−0.40006	−0.84798	0.68338	0.13251
−0.37	−0.38842	−0.85628	0.66945	0.15803
−0.36	−0.37689	−0.86432	0.65518	0.18311
−0.35	−0.36544	−0.87209	0.64057	0.20773
−0.34	−0.35409	−0.87961	0.62565	0.23187
−0.33	−0.34283	−0.88687	0.61041	0.25552
−0.32	−0.33165	−0.89387	0.59488	0.27864
−0.31	−0.32055	−0.90063	0.57907	0.30124
−0.30	−0.30952	−0.90714	0.56297	0.32328
−0.29	−0.29857	−0.91342	0.54662	0.34474
−0.28	−0.28768	−0.91945	0.53001	0.36563
−0.27	−0.27686	−0.92525	0.51316	0.38591
−0.26	−0.26611	−0.93081	0.49607	0.40558
−0.25	−0.25541	−0.93615	0.47876	0.42461
−0.24	−0.24477	−0.94125	0.46124	0.44301
−0.23	−0.23419	−0.94614	0.44351	0.46074
−0.22	−0.22366	−0.95080	0.42559	0.47781
−0.21	−0.21317	−0.95523	0.40748	0.49420
−0.20	−0.20273	−0.95945	0.38920	0.50990
−0.19	−0.19234	−0.96346	0.37075	0.52490
−0.18	−0.18198	−0.96724	0.35215	0.53918
−0.17	−0.17167	−0.97082	0.33339	0.55275
−0.16	−0.16139	−0.97418	0.31450	0.56559
−0.15	−0.15114	−0.97733	0.29547	0.57769
−0.14	−0.14093	−0.98027	0.27632	0.58904
−0.13	−0.13074	−0.98300	0.25706	0.59964
−0.12	−0.12058	−0.98553	0.23769	0.60948
−0.11	−0.11045	−0.98785	0.21822	0.61856
−0.10	−0.10034	−0.98997	0.19866	0.62687
−0.09	−0.09024	−0.99188	0.17903	0.63440
−0.08	−0.08017	−0.99359	0.15932	0.64115
−0.07	−0.07011	−0.99509	0.13954	0.64711
−0.06	−0.06007	−0.99640	0.11971	0.65229
−0.05	−0.05004	−0.99750	0.09983	0.65668
−0.04	−0.04002	−0.99840	0.07991	0.66027
−0.03	−0.03001	−0.99910	0.05996	0.66307
−0.02	−0.02000	−0.99960	0.03999	0.66507
−0.01	−0.01000	−0.99990	0.02000	0.66627

x	$Q_0(x)$	$Q_1(x)$	$Q_2(x)$	$Q_3(x)$
0.00	0.00000	−1.00000	0.00000	0.66667
0.01	0.01000	−0.99990	−0.02000	0.66627
0.02	0.02000	−0.99960	−0.03999	0.66507
0.03	0.03001	−0.99910	−0.05996	0.66307
0.04	0.04002	−0.99840	−0.07991	0.66027
0.05	0.05004	−0.99750	−0.09983	0.65668
0.06	0.06007	−0.99640	−0.11971	0.65229
0.07	0.07011	−0.99509	−0.13954	0.64711
0.08	0.08017	−0.99359	−0.15932	0.64115
0.09	0.09024	−0.99188	−0.17903	0.63440
0.10	0.10033	−0.98997	−0.19866	0.62687
0.11	0.11045	−0.98785	−0.21822	0.61856
0.12	0.12058	−0.98553	−0.23769	0.60948
0.13	0.13074	−0.98300	−0.25706	0.59964
0.14	0.14092	−0.98027	−0.27632	0.58904
0.15	0.15114	−0.97733	−0.29547	0.57769
0.16	0.16139	−0.97418	−0.31450	0.56559
0.17	0.17167	−0.97082	−0.33339	0.55275
0.18	0.18198	−0.96724	−0.35215	0.53919
0.19	0.19234	−0.96346	−0.37075	0.52490
0.20	0.20273	−0.95945	−0.38920	0.50990
0.21	0.21317	−0.95523	−0.40748	0.49420
0.22	0.22366	−0.95080	−0.42559	0.47782
0.23	0.23419	−0.94614	−0.44351	0.46075
0.24	0.24477	−0.94125	−0.46124	0.44301
0.25	0.25541	−0.93615	−0.47876	0.42461
0.26	0.26611	−0.93081	−0.49607	0.40558
0.27	0.27686	−0.92525	−0.51316	0.38591
0.28	0.28768	−0.91945	−0.53001	0.36563
0.29	0.29857	−0.91342	−0.54662	0.34474
0.30	0.30952	−0.90714	−0.56297	0.32328
0.31	0.32054	−0.90063	−0.57907	0.30124
0.32	0.33165	−0.89387	−0.59488	0.27865
0.33	0.34283	−0.88687	−0.61041	0.25552
0.34	0.35409	−0.87961	−0.62565	0.23187
0.35	0.36544	−0.87210	−0.64057	0.20773
0.36	0.37689	−0.86432	−0.65518	0.18311
0.37	0.38842	−0.85628	−0.66945	0.15803
0.38	0.40006	−0.84798	−0.68338	0.13251
0.39	0.41180	−0.83940	−0.69695	0.10658
0.40	0.42365	−0.83054	−0.71015	0.08026
0.41	0.43561	−0.82140	−0.72297	0.05357
0.42	0.44769	−0.81197	−0.73539	0.02654
0.43	0.45990	−0.80225	−0.74740	−0.00080
0.44	0.47223	−0.79222	−0.75898	−0.02844
0.45	0.48470	−0.78189	−0.77012	−0.05633
0.46	0.49731	−0.77124	−0.78081	−0.08446
0.47	0.51007	−0.76027	−0.79102	−0.11279
0.48	0.52298	−0.74897	−0.80075	−0.14129
0.49	0.53606	−0.73733	−0.80997	−0.16992
0.50	0.54931	−0.72535	−0.81866	−0.19865
0.51	0.56273	−0.71301	−0.82682	−0.22746

x	$Q_0(x)$	$Q_1(x)$	$Q_2(x)$	$Q_3(x)$
0.52	0.57634	−0.70030	−0.83441	−0.25628
0.53	0.59014	−0.68722	−0.84141	−0.28510
0.54	0.60416	−0.67376	−0.84782	−0.31387
0.55	0.61838	−0.65989	−0.85360	−0.34254
0.56	0.63283	−0.64561	−0.85873	−0.37107
0.57	0.64752	−0.63091	−0.86319	−0.39942
0.58	0.66246	−0.61577	−0.86695	−0.42754
0.59	0.67767	−0.60018	−0.86999	−0.45537
0.60	0.69315	−0.58411	−0.87227	−0.48286
0.61	0.70892	−0.56756	−0.87378	−0.50997
0.62	0.72500	−0.55050	−0.87446	−0.53662
0.63	0.74142	−0.53291	−0.87431	−0.56275
0.64	0.75817	−0.51477	−0.87327	−0.58830
0.65	0.77530	−0.49606	−0.87130	−0.61321
0.66	0.79281	−0.47674	−0.86838	−0.63739
0.67	0.81074	−0.45680	−0.86446	−0.66078
0.68	0.82911	−0.43620	−0.85948	−0.68328
0.69	0.84795	−0.41491	−0.85341	−0.70481
0.70	0.86730	−0.39289	−0.84618	−0.72529
0.71	0.88718	−0.37010	−0.83775	−0.74460
0.72	0.90765	−0.34649	−0.82804	−0.76265
0.73	0.92873	−0.32203	−0.81699	−0.77931
0.74	0.95048	−0.29665	−0.80452	−0.79447
0.75	0.97296	−0.27028	−0.79055	−0.80799
0.76	0.99622	−0.24288	−0.77499	−0.81973
0.77	1.02033	−0.21435	−0.75774	−0.82953
0.78	1.04537	−0.18461	−0.73868	−0.83721
0.79	1.07143	−0.15357	−0.71770	−0.84259
0.80	1.09861	−0.12111	−0.69464	−0.84544
0.81	1.12703	−0.08711	−0.66935	−0.84555
0.82	1.15682	−0.05141	−0.64164	−0.84264
0.83	1.18814	−0.01385	−0.61131	−0.83641
0.84	1.22117	0.02579	−0.57810	−0.82653
0.85	1.25615	0.06773	−0.54172	−0.81259
0.86	1.29334	0.11227	−0.50184	−0.79415
0.87	1.33308	0.15978	−0.45803	−0.77066
0.88	1.37577	0.21068	−0.40979	−0.74148
0.89	1.42192	0.26551	−0.35651	−0.70582
0.90	1.47222	0.32500	−0.29737	−0.66271
0.91	1.52752	0.39005	−0.23135	−0.61091
0.92	1.58903	0.46191	−0.15708	−0.54880
0.93	1.65839	0.54231	−0.07268	−0.47419
0.94	1.73805	0.63376	0.02458	−0.38400
0.95	1.83178	0.74019	0.13888	−0.27357
0.96	1.94591	0.86807	0.27707	−0.13540
0.97	2.09230	1.02953	0.45182	0.04409
0.98	2.29755	1.25160	0.69107	0.29435
0.99	2.64664	1.62017	1.08264	0.70624
1.00	$+\infty$	$+\infty$	$+\infty$	$+\infty$

APPENDIX VI

The Method of Steepest Descent (Saddle-Point Method)

The method of steepest descent (saddle-point method) is used to evaluate for large values of β, in an approximate sense, integrals of the form

$$I(\beta) = \int_C F(z)e^{\beta f(z)} \, dz \tag{VI-1}$$

where $f(z)$ is an analytic function and C is the path of integration in the complex z plane, as shown in Figure VI-1. The philosophy of the method is that, within certain limits, the path of integration can be altered continuously without affecting the value of the integral provided that, during the deformation, the path does not pass through singularities of the integrand. The new path can also be chosen in such a way that most of the contributions to the integral are attributed only to small segments of the new path. The integrand can then be approximated by simpler functions over the important parts of the path and its behavior can be neglected over all other segments. If during the deformation from the old to the new paths, singularities for the function $F(z)$ are encountered, we must add (a) the residue when crossing a pole and (b) the integral, when encountering a branch point, over the edges of an appropriate cut where the function is single-valued.

In general, we can write (VI-1) as

$$I(\beta) = \int_C F(z)e^{\beta f(z)} \, dz = I_{\text{SI}} + I_{\text{SDP}} \tag{VI-2}$$

where I_{SI} takes into account the contributions from the singularities and I_{SDP} from the steepest-descent path. In this Appendix our concern will be the I_{SDP} contribution of (VI-2) or

$$I_{\text{SDP}} = \int_{\text{SDP}} F(z)e^{\beta f(z)} \, dz \tag{VI-3}$$

where now $F(z)$ is assumed to be a well-behaved function and $f(z)$ to be analytic in the complex z plane ($z = x + jy$).

Balanis' Advanced Engineering Electromagnetics, Third Edition. Constantine A. Balanis.
© 2024 John Wiley & Sons, Inc. Published 2024 by John Wiley & Sons, Inc.
Companion Website: www.wiley.com/go/balanis/advancedengineeringelectromagnetics3e

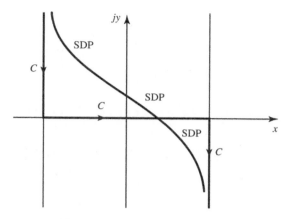

Figure VI-1 C and SDP paths.

Assuming that β is real and positive, we can write

$$f(z) = U(z) + jV(z) = U(x,y) + jV(x,y) \tag{VI-4}$$

where U and V are real functions, so that the integrand of (VI-3) can be written as

$$F(z)e^{\beta f(z)} = F(x,y)e^{\beta U(x,y)}e^{j\beta V(x,y)} \tag{VI-5}$$

If $f(z)$ is an analytic function, the Cauchy-Riemann conditions state that

$$\frac{df}{dz} = \frac{\partial U}{\partial x} + j\frac{\partial V}{\partial x} = -j\frac{\partial U}{\partial y} + \frac{\partial V}{\partial y} \tag{VI-6}$$

or

$$\frac{\partial U}{\partial x} = \frac{\partial V}{\partial y} \tag{VI-6a}$$

$$\frac{\partial U}{\partial y} = -\frac{\partial V}{\partial x} \tag{VI-6b}$$

If there exists a point $z_s = x_s + jy_s$ where

$$\left.\frac{df}{dz}\right|_{z=z_s} \equiv f'(z=z_s) = f'(z_s) = 0 \tag{VI-7}$$

then

$$\frac{\partial U}{\partial x} = \frac{\partial V}{\partial y} = \frac{\partial U}{\partial y} = \frac{\partial V}{\partial x} = 0 \quad \text{at} \quad x = x_s, y = y_s \tag{VI-8}$$

The surfaces $U(x,y) = $ constant and $V(x,y) = $ constant satisfy (VI-8) but do not have an absolute maximum or minimum at (x_s, y_s). The Cauchy-Riemann conditions (VI-6a) and (VI-6b) also tell us that, for a first-order saddle point $[f''(z_s) \neq 0]$

$$\frac{\partial^2 U}{\partial x^2} = \frac{\partial^2 V}{\partial x\,\partial y} = \frac{\partial}{\partial y}\left(\frac{\partial V}{\partial x}\right) = \frac{\partial}{\partial y}\left(-\frac{\partial U}{\partial y}\right) = -\frac{\partial^2 U}{\partial y^2}$$

$$\frac{\partial^2 U}{\partial x^2} = -\frac{\partial^2 U}{\partial y^2} \tag{VI-9a}$$

$$\frac{\partial^2 V}{\partial y^2} = \frac{\partial^2 U}{\partial y \partial x} = \frac{\partial}{\partial x}\left(\frac{\partial U}{\partial y}\right) = \frac{\partial}{\partial x}\left(-\frac{\partial V}{\partial x}\right) = -\frac{\partial^2 V}{\partial x^2}$$

$$\frac{\partial^2 V}{\partial y^2} = -\frac{\partial^2 V}{\partial x^2} \tag{VI-9b}$$

Because of (VI-8), (VI-9a), and (VI-9b) neither $U(x,y)$ nor $V(x,y)$ has a maximum or a minimum at such a point z_s, but a *minimax* or saddle point. If $U(x,y)$ has an extremum at z_s, then ΔU is positive for some changes in x and y and negative for others (a positive slope in one direction and negative at right angles to it), whereas ΔV remains constant. The same holds if $V(x,y)$ has an extremum. Thus, the lines of most rapid increase or decrease of one part of the complex function $f(z) = U(x,y) + jV(x,y)$ are constant lines of the other.

The magnitude of the exponential factor $e^{\beta U(x,y)}$ of (VI-5) may increase, decrease, or remain constant depending on the choice of the path through the saddle point z_s. To avoid $U(x,y)$ contributing in the exponential of (VI-5) over a large part of the path, we must pass the saddle point in the fastest possible manner. This is accomplished by taking the path of integration through the saddle point and leaving it along the line of the most rapid decrease (steepest descent) of the function $U(x,y)$.

Referring to Figure VI-2, let us choose a path P through the saddle point z_s with differential length ds. Then

$$\frac{dU}{ds} = \frac{\partial U}{\partial x}\frac{\partial x}{\partial s} + \frac{\partial U}{\partial y}\frac{\partial y}{\partial s} = \frac{\partial U}{\partial x}\cos\gamma + \frac{\partial U}{\partial y}\sin\gamma \tag{VI-10}$$

where γ is the angle between ds and the x axis. The function dU/ds is a maximum for values of γ defined by

$$\frac{\partial}{\partial\gamma}\left(\frac{\partial U}{\partial s}\right) = \frac{\partial^2 U}{\partial s^2} = \frac{\partial}{\partial\gamma}\left[\frac{\partial U}{\partial x}\cos\gamma + \frac{\partial U}{\partial y}\sin\gamma\right] = 0 \tag{VI-11}$$

or

$$\frac{\partial}{\partial\gamma}\left(\frac{\partial U}{\partial s}\right) = -\sin\gamma\left(\frac{\partial U}{\partial x}\right) + \cos\gamma\left(\frac{\partial U}{\partial y}\right) = 0 \tag{VI-11a}$$

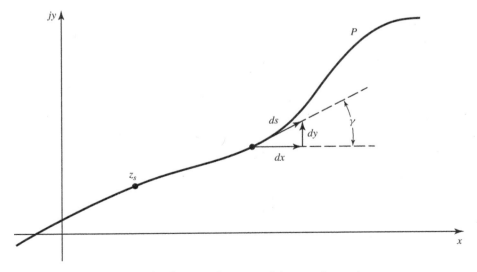

Figure VI-2   Steepest descent path in complex z plane.

Using the Cauchy-Riemann conditions (VI-6a) and (VI-6b), we can write (VI-11a) as

$$\frac{\partial}{\partial\gamma}\left(\frac{\partial U}{\partial s}\right) = -\sin\gamma\left(\frac{\partial V}{\partial y}\right) + \cos\gamma\left(-\frac{\partial V}{\partial x}\right) = -\left[\frac{\partial V}{\partial y}\sin\gamma + \frac{\partial V}{\partial x}\cos\gamma\right] = 0$$

$$\frac{\partial}{\partial\gamma}\left(\frac{\partial U}{\partial s}\right) = -\left(\frac{\partial V}{\partial y}\frac{\partial y}{\partial s} + \frac{\partial V}{\partial x}\frac{\partial x}{\partial s}\right) = -\left(\frac{dV}{ds}\right) = 0 \tag{VI-12}$$

Thus $V = $ constant for paths along which $U(x,y)$ changes most rapidly (and vice versa), so the *steepest amplitude path* is a *constant phase path*. These are known as *steepest ascent* or *descent* paths. We choose the steepest descent path, thus the name *method of steepest descent*. Since β is real and positive, the exponential $\exp[\beta U(x,y)]$ of (VI-5) will decrease rapidly with distance from the saddle point and only a small portion of the integration path, including the saddle point, will make any significant contributions to the value of the entire integral.

To find the path of steepest descent, we form a function

$$f(z) = f(z_s) - s^2 \tag{VI-13}$$

where z_s is the saddle point and s is real $(-\infty \leq s \leq +\infty)$. The saddle point corresponds to $s = 0$. Using (VI-4), we can write (VI-13) as

$$U(z) = U(z_s) - s^2 \tag{VI-13a}$$

$$V(z) = V(z_s) \quad \text{steepest descent path} \tag{VI-13b}$$

Since the imaginary part remains constant, while the real part attains maximum at $s = 0$ and decreases for other values, the path of steepest descent is described by (VI-13b).

To evaluate the integral of (VI-3), we first find the saddle point z_s by (VI-7). Next, we express $f(z)$, around the saddle point z_s, by a truncated Taylor series

$$f(z) \simeq f(z_s) + \frac{1}{2}(z - z_s)^2 f''(z_s) \tag{VI-14}$$

since $f'(z_s) = 0$. The double prime indicates a second derivative with respect to z. Substitution of (VI-14) into (VI-3) leads to

$$I(\beta) = \int_{\text{SDP}} F(z)e^{\beta f(z)}\,dz \simeq e^{\beta f(z_s)}\int_{\text{SDP}} F(z)e^{(\beta/2)(z-z_s)^2 f''(z_s)}\,dz \tag{VI-15}$$

Letting

$$-\beta(z - z_s)^2 f''(z_s) = \xi^2 \tag{VI-15a}$$

$$dz = \frac{d\xi}{\sqrt{-\beta f''(z_s)}} \tag{VI-15b}$$

we can write (VI-15), by extending the limits to infinity, as

$$I(\beta) \simeq \frac{e^{\beta f(z_s)}}{\sqrt{-\beta f''(z_s)}}\int_{-\infty}^{+\infty} F(z)e^{-\xi^2/2}\,d\xi \tag{VI-16}$$

Assuming that $F(z)$ is a slow-varying function in the neighborhood of the saddle point, we can write (VI-16), by replacing $F(z)$ by $F(z_s)$, as

$$I(\beta) \simeq \frac{e^{\beta f(z_s)}}{\sqrt{-\beta f''(z_s)}} 2F(z_s) \int_0^\infty e^{-\xi^2/2} \, d\xi$$

$$I(\beta) \simeq \frac{e^{\beta f(z_s)}}{\sqrt{-\beta f''(z_s)}} F(z_s) 2\sqrt{\frac{\pi}{2}} = \sqrt{\frac{2\pi}{-\beta f''(z_s)}} F(z_s) e^{\beta f(z_s)} \qquad \text{(VI-17)}$$

If more than one saddle point exists, then (VI-16) can be written as

$$I(\beta) \simeq \sqrt{\frac{2\pi}{\beta}} \sum_{s=1}^N \frac{F(z_s)}{\sqrt{-f''(z_s)}} e^{\beta f(z_s)} \qquad \text{(VI-18)}$$

where N is equal to the number of saddle points. The summation assumes, through the principle of superposition, that the contribution of each saddle point is not affected by the presence of the others.

Equation (VI-18) accounts for the contribution to the integral (VI-3) from first-order saddle points $[f'(z_s) = 0$ but $f''(z_s) \neq 0]$. For second-order saddle points $[f'(z_s) = 0$ and $f''(z_s) = 0]$, the expression is different. For general forms of $f(z)$, the determination of all the steepest descent paths may be too complicated.

If a constant level path is chosen such that $|\exp[\beta U(x, y)]|$ remains constant everywhere and $\exp[j\beta V(x, y)]$ varies most rapidly away from the saddle points, the evaluation of the integral can be carried out from contributions near the saddle points. Since the phase factor $\exp(j\beta V)$ is stationary at and near the saddle points, and oscillates very rapidly in the remaining parts of the path, it makes the net contributions from the other parts, excluding the saddle points, very negligible. This is known as the method of *stationary phase*, and it may not yield the same result as the method of steepest descent because their corresponding paths are different. The two will lead to identical results if the constant level path can be continuously deformed to the steepest descent path. This is accomplished if the two paths have identical terminations and there are no singularities of $f(z)$ in the region between the two paths. The Method of Stationary Phase is described in detail in Appendix VIII of *Antenna Theory: Analysis and Design* by C.A. Balanis, Fourth Edition, copyright © 2016, John Wiley & Sons, Inc.

Glossary of Symbols, Units, and Names (continued at back)

Symbol	Unit	Name
A	Wb/m	Magnetic vector potential
$A(s,s')$	——	Amplitude spreading factor
AR	——	Axial Ratio $\left[\dfrac{\text{major axis of ellipse}}{\text{minor axis of ellipse}}; +\text{ for CCW}, -\text{ for CW}\right]$
B	siemens	Susceptance
$\mathcal{B}$, **B**	Wb/m^2	Magnetic flux density
$b_n(\beta r)$	——	Spherical Bessel $[j_n(\beta r)]$ and Hankel $[h_n^{(1,2)}(\beta r)]$ functions
$\hat{B}_n(\beta r)$	——	Schelkunoff spherical Bessel $[\hat{J}_n(\beta r)]$ and Hankel $[\hat{H}_n^{(1,2)}(\beta r)]$ functions
C	F	Capacitance
$C_0(x), S_0(x)$		
$C(x), S(x)$	——	Fresnel integrals
$C_1(x), S_1(x)$		
cm	centimeter	Distance
CP	——	Circular Polarization
CW	——	Clock Wise
CCW	——	Counter Clock Wise
$\mathcal{D}$, **D**	C/m^2	Electric flux density
$\overline{\mathbf{D}}, \widetilde{\mathbf{D}}$	——	Dyadic diffraction coefficient (tensor)
$\mathcal{E}$, **E**	V/m	Electric field intensity
$\mathcal{F}$, **F**	Q/m	Electric vector potential
$F(X)$	——	Fresnel transition function
f	Hz	Frequency
f_c	Hz	Cutoff frequency
f_r	Hz	Resonant frequency
G	siemens	Conductance
$G(x,x')$	——	Green's function
$\mathcal{H}$, **H**	A/m	Magnetic field intensity
$H_n^{(1)}(x)$	——	Hankel function of first kind of order n
$H_n^{(2)}(x)$	——	Hankel function of second kind of order n
I_e	A	Electric current
I_m	V	Magnetic current
$I_n(x)$	——	Modified Bessel function of first kind of order n
$\mathcal{J}$, **J**	A/m^2	Volume electric current density
$\mathcal{J}_c$, **J**$_c$	A/m^2	Conduction electric current density
$\mathcal{J}_d$, **J**$_d$	A/m^2	Displacement electric current density
$\mathcal{J}_s$, **J**$_s$	A/m^2	Surface electric current density
$J_n(x)$	——	Bessel function of first kind of order n
$K_n(x)$	——	Modified Bessel function of second kind of order n
L	H	Inductance

Balanis' Advanced Engineering Electromagnetics, Third Edition. Constantine A. Balanis.
© 2024 John Wiley & Sons, Inc. Published 2024 by John Wiley & Sons, Inc.
Companion Website: www.wiley.com/go/balanis/advancedengineeringelectromagnetics3e

Symbol	Unit	Name
LH	——	Left Hand
LSE	——	Longitudinal Section Electric
LSM	——	Longitudinal Section Magnetic
$\mathcal{M}$	A/m	Magnetic polarization vector
$\mathcal{M}$, $\mathbf{M}$	V/m^2	Volume magnetic current density
$\mathcal{M}_i$, $\mathbf{M}_i$	V/m^2	Impressed magnetic current density
$\mathcal{M}_d$, $\mathbf{M}_d$	V/m^2	Displacement magnetic current density
m	meters	Distance
$\mathbf{m}$	A-m^2	Magnetic vector dipole moment
N_e	Electrons/m^3	Free electron density
N_h	Holes/m^3	Bound hole density
$\mathbf{p}$	C-m	Electric vector dipole moment
$\mathcal{P}$, $\mathbf{P}$	C/m^2	Electric polarization vector
$P_n^m(\cos\theta)$	——	Associated Legendre function of first kind
$P_n(\cos\theta)$	——	Legendre polynomial/function of first kind
$\wp_d, p_d$	W/m^3	Dissipated power density $\left(\sigma\,\mathcal{E}^2, \frac{1}{2}\sigma\,|\mathbf{E}|^2\right)$
$\mathcal{P}_d, P_d$	W	Dissipated power $\left(\iiint_V \sigma\,\mathcal{E}^2 dv, \frac{1}{2}\iiint_V \sigma\,|\mathbf{E}|^2\,dv\right)$
$\mathcal{P}_e, P_e$	W	Exiting power
$\wp_s, p_s$	W/m^3	Supplied power density $\left[-(\mathcal{H}\cdot\mathcal{M}_i + \mathcal{E}\cdot\mathcal{J}_i), -\frac{1}{2}(\mathbf{H}^*\cdot\mathbf{M}_i + \mathbf{E}\cdot\mathbf{J}_i)\right]$
$\mathcal{P}_s, P_s$	W	Supplied power $\left[-\left(\iiint_V(\mathcal{H}\cdot\mathcal{M}_i + \mathcal{E}\cdot\mathcal{J}_i)dv\right), -\left(\frac{1}{2}\iiint_V(\mathbf{H}^*\cdot\mathbf{M}_i + \mathbf{E}\cdot\mathbf{J}_i)dv\right)\right]$
$\mathcal{Q}, Q$	C	Electric charge
Q	——	Quality factor $\left(\omega\,\dfrac{\text{stored energy}}{\text{dissipated power}}\right)$
$Q_n^m(\cos\theta)$	——	Associated Legendre function of second kind
$Q_n(\cos\theta)$	——	Legendre polynomial/function of second kind
q_{ev}, q_{ev}	C/m^3	Electric volume charge density
q_{es}, q_{es}	C/m^2	Electric surface charge density
q_{mv}, q_{mv}	W/m^3	Magnetic volume charge density
q_{ms}, q_{ms}	W/m^2	Magnetic surface charge density
R	ohms	Resistance
RH	——	Right Hand
sec	seconds	Time
$\mathcal{S}$	W/m^2	Poynting vector $(\mathcal{S} = \mathcal{E}\times\mathcal{H})$
$\mathbf{S}$, $\mathbf{S}_{ave}$	W/m^2	Time-average Poynting vector $\left[\mathbf{S}_{av} = \mathbf{S} = \frac{1}{2}\text{Re}\left(\mathbf{E}\times\mathbf{H}^*\right)\right]$
SW	m	Scattering Width (two-dimensional RCS)
T_c	K	Critical temperature
$T_n(z)$	——	Tschebyscheff polynomial of order n
t_r	sec	Relaxation time
$\tan\delta$	——	Total electric loss tangent $\left(\dfrac{\sigma_e}{\omega\varepsilon'} = \dfrac{\sigma_s + \sigma_a}{\omega\varepsilon'} = \dfrac{\sigma_s}{\omega\varepsilon'} + \dfrac{\sigma_a}{\omega\varepsilon'}\right)$
$\tan\delta_a$	——	Alternating electric loss tangent $\left(\dfrac{\sigma_a}{\omega\varepsilon'}\right)$
$\tan\delta_s$	——	Static electric loss tangent $\left(\dfrac{\sigma_s}{\omega\varepsilon'}\right)$
T	——	Transmission coefficient
TE	——	Transverse Electric
TM	——	Transverse Magnetic
TEM	——	Transverse ElectroMagnetic
TRM	——	Transverse Resonance Method

Symbol	Unit	Name
SWR	——	Standing Wave Ratio $\left(\dfrac{\|E\|_{\max}}{\|E\|_{\min}} = \dfrac{1+\|\Gamma\|}{1-\|\Gamma\|}\right)$
$v,\ \mathbf{v}$	m/sec	Speed/velocity of wave
$w_e,\ w_e$	joules/m^3	Electric energy density $\left(\frac{1}{2}\varepsilon\left\|\boldsymbol{\mathscr{E}}^2\right\|,\ \frac{1}{4}\varepsilon\left\|\mathbf{E}\right\|^2\right)$
$w_m,\ w_m$	joules/m^3	Magnetic energy density $\left(\frac{1}{2}\mu\left\|\boldsymbol{\mathscr{H}}^2\right\|,\ \frac{1}{4}\mu\left\|\mathbf{H}\right\|^2\right)$
$\mathscr{W}_e,\ \overline{W}_e$	joules	Electric energy $\left[\iiint_V \frac{1}{2}\varepsilon\left\|\boldsymbol{\mathscr{E}}^2\right\|dv;\ \iiint_V \frac{1}{4}\varepsilon\|E\|^2\,dv\right]$
$\mathscr{W}_m,\ \overline{W}_m$	joules	Magnetic energy $\left[\iiint_V \frac{1}{2}\mu\left\|\boldsymbol{\mathscr{H}}^2\right\|dv;\ \iiint_V \frac{1}{4}\mu\|H\|^2\,dv\right]$
X	ohms	Inductance
Y	siemens	Admittance
$Y_n(x)$	——	Bessel function of second kind of order n
Z	ohms	Impedance
$Z_c,\ Z_0$	ohms	Characteristic impedance

Glossary of Symbols, Units, and Names (continued from front)

Symbol	Unit	Name
α	Np/m	Attenuation constant
β	Rad/m	Phase constant($\beta = \omega\sqrt{\mu\varepsilon} = 2\pi/\lambda$)
γ	——	Propagation constant($\gamma = \alpha + j\beta$)
γ	——	Gyromagnetic ratio $\left(\dfrac{m}{P} = \dfrac{\text{magnetic dipole moment}}{\text{angular momentum}}\right)$
Γ	——	Reflection coefficient$\left(\dfrac{Z_{in} - Z_c}{Z_{in} + Z_c}; \dfrac{\eta_2 - \eta_1}{\eta_2 + \eta_1}\right)$
δ	m	Skin depth
ε	F/m	Permittivity
$\dot{\varepsilon}$	F/m	Complex permittivity ($\dot{\varepsilon} = \varepsilon' - j\varepsilon''$)
ε_r	——	Relative permittivity (dielectric constant) ($\varepsilon/\varepsilon_o$)
η	ohms	Intrinsic impedance $\left[\eta = \sqrt{\dfrac{j\omega\mu}{\sigma + j\omega\varepsilon}} \overset{\sigma=0}{=} \sqrt{\dfrac{\mu}{\varepsilon}}\right]$
θ_c	rads	Critical angle
θ_B	rads	Brewster angle
θ_i	rads	Angle of incidence
θ_r	rads	Angle of reflection
θ_t	rads	Angle of refraction/transmission
λ	m	Wavelength
μ	H/m	Permeability
$\dot{\mu}$	H/m	Complex permeability ($\dot{\mu} = \mu' - j\mu''$)
μ_e	m^2/(V-s)	Mobility of electron
μ_h	m^2/(V-s)	Mobility of hole
μ_r	——	Relative permeability (μ/μ_o)
σ	S/m	Electric conductivity
σ	m^2	Radar Cross Section (RCS)
χ_e	——	Electric susceptibility
χ_m	——	Magnetic susceptibility
ω	rad/s	Angular (radian) frequency
ω_c	rad/s	Angular (radian) cutoff frequency
ω_r	rad/s	Angular (radian) resonant frequency
(x, y, z)	m	Rectangular coordinates
(ρ, θ, z)	m	Cylindrical coordinates
(r, θ, ϕ)	m	Spherical coordinates
$\hat{\mathbf{a}}_x, \hat{\mathbf{a}}_y, \hat{\mathbf{a}}_z$	——	Rectangular unit vectors
$\hat{\mathbf{a}}_\rho, \hat{\mathbf{a}}_\phi, \hat{\mathbf{a}}_z$	——	Cylindrical unit vectors
$\hat{\mathbf{a}}_r, \hat{\mathbf{a}}_\theta, \hat{\mathbf{a}}_\phi$	——	Spherical unit vectors
A_x, A_y, A_z	——	Rectangular vector components
A_ρ, A_ϕ, A_z	——	Cylindrical vector components
A_r, A_θ, A_ϕ	——	Spherical vector components

Constants

Symbol	Unit	Name
ε_0	F/m	Permittivity of free space ($\approx 8.854 \times 10^{-12}$)
η_0, Z_0	ohms	Intrinsic impedance of free space (≈ 377)
μ_0	H/m	Permeability of free space ($4\pi \times 10^{-7}$)
π	——	pi (3.1415927)
e	C	Electron charge ($1.60217646 \times 10^{-19}$)
G	10^9	Giga
k	10^3	Kilo
M	10^6	Mega
m_e	kg	Mass of electron at rest ($9.10938188 \times 10^{-31}$)
p	10^{-12}	Pico
rad	degrees	Radian ($180/\pi° = 57.296°$)
sr	(degrees)^2	Square radian $[(180/\pi)^2 = (57.296)^2 = 3,282.806]$
υ_0, c	m/sec	Velocity of light in free space ($2.9979 \times 10^8 \approx 3 \times 10^8$)

Index

A

Acceptors, 61

A.C. variations in materials, 70–91
  complex permeability, 82–83
  complex permittivity, 70–82
  ferrites, 83–91

Addition theorems
  Bessel functions, 602–604
  Hankel functions, 599–602

Air-substrate modes, 437

Aluminum
  atom, 42
  charge density, 64
  conductivity, 60
  mobility, 61

AMC_Designs program, 952, 1013

AMCs *see* Artificial magnetic conductors

Ampere's law, 3, 5, 147

Amplitude relation, 743–746

Angle of incidence
  Brewster angle, 179, 192–194
  critical angle, 179, 186, 194–204
  lossless media, 183–204
  lossy media, 208–212
  normal, 179–183

Angular_Direction_Checkerboard program,
    966, 1014

Angular frequency, 72, 74, 156, 158
  natural, 74
  resonant, 74

Anisotropic dielectrics, 69

Anisotropic impedance surfaces,
    945–947
  corrugations, 946

Anomalous (abnormal) dispersion, 80

Antennas
  aperture, 990–994
  circular loop, 985–990
  circular polarization, 1011
  figures of merit, 980–982
  holographic principle, 999–1000
  horizontal dipole, 983–985
  horizontal polarization, 1010–1011
  leaky-wave, 997–1012
  maximum directivity, 981
  maximum gain, 981–982
  maximum realized gain, 982
  metasurfaces, 960–1013
  microstrip arrays, 994–996
  monopole, 982–983
  multiple polarization, 1011–1012
  radar-cross section reduction, 977
  radiation efficiency, 981
  radiation resistance, 980–981
  scan blindness, 944, 994
  surface-wave, 996
  tensor impedance surfaces, 1013
  vertical polarization, 1009–1010

Antiferromagnetic materials, 55–56

Aperture antennas

Balanis' Advanced Engineering Electromagnetics, Third Edition. Constantine A. Balanis.
© 2024 John Wiley & Sons, Inc. Published 2024 by John Wiley & Sons, Inc.
Companion Website: www.wiley.com/go/balanis/advancedengineeringelectromagnetics3e

analysis
  far-field, 295–296, 298–317, 299
  near-field, 296–298, 299
  metamaterials, 990–994
Aperture_GP_UTD program, 830
AR *see* Axial ratio
Array factors, 332, 334
Array multiplication rule, 183, 332
Arrays, radar-cross section reduction,
    977
Artificial impedance surfaces, 943–1021
  anisotropic, 946
  antennas, 980–1013
    aperture, 990–994
    circular loop, 985–990
    holographic principle, 999–1000
    horizontal dipole, 983–985
    leaky-wave, 997–1013
    microstrip arrays, 994–996
    monopole, 982–983
    polarization, 1009–1012
    radar cross-section reduction, 977
    surface-wave, 996
    tensor impedance surfaces, 1013
  applications, 959–1013
  bandgap structure, 944–945
  basic principles, 943–945
  cascaded, 1004
  checkerboard, 959, 960–979
  corner reflectors, 979
  corrugations, 945–947
  cylindrical structures, 978
  hard and soft polarization, 946–947
  holographic principle, 999–1000
  mushroom AMCs, 947–960
    design, 950–954
    limitations, 959–960
    resonant frequency, 949
    surface impedance, 949
    surface-wave dispersion, 955–959
  operational band, 944–945
  plane wave scattering, 961–964
  polarization-diverse holographic,
    1009–1012
  pyramidal and conical horns, 946–947
  radar-asborbing materials, 959–960
  scan blindness, 944, 994
  tensor impedance surfaces, 1013
  unit cell selection, 968–973
Artificial magnetic conductors
    (AMCs), 943

antennas, 980–1013
  aperture, 990–994
  circular loop, 985–990
  holographic principle, 999–1000
  horizontal dipole, 983–985
  leaky-wave, 997–1013
  microstrip arrays, 994–996
  monopole, 982–983
  polarization, 1009–1012
  surface-wave, 996
  tensor impedance surfaces, 1013
applications, 959–1013
circular loop antennas, 985–990
horizontal dipole antennas, 983–985
mushroom, 947–960
  design, 950–954
  limitations, 959–960
  resonant frequency, 949
  surface impedance, 949
  surface-wave dispersion, 955–959
plane wave scattering, 961–964
resonant frequency, 949
unit cell selection, 968–973
Associated Legendre functions, 124
Astigmatic rays, 745–746, 750–751,
    760–761
Asymptotic expansions, 867–868
Atomic models
  equivalent circular electric loop, 50
  equivalent square electric loop, 50
  orbiting electrons, 50
Atomic number, 41
Atoms, 41–42
  aluminium, 42
  Bohr model, 42
  electron shells, 42–43
  germanium, 42
  hydrogen, 42
  silicon, 42
Attenuation
  circular waveguides, 491–496, 528
  conduction losses, 388–392
  coupling, 395–396
  cylindrical coordinate systems, 119
  dielectric losses, 392–395
  Fiber optics cables, 528
  lossless media, 131–132, 139–140
  lossy media, 144, 147–150
  rectangular coordinate systems, 110
  rectangular waveguides, 388–396, 464
  ridged waveguides, 464

standing waves, 110, 119
traveling waves, 109, 119
Automotive radar technology, 1005–1006
Auxiliary vector potentials, 271–317
    construction of solutions, 277–294
    cylindrical coordinate systems, 281–284,
        287–288, 290, 311–317
    far-field radiation, 295–296, 298–317
    inhomogeneous wave equation, 291–294
    near-field radiation, 296–298
    rectangular coordinate systems, 277–281,
        285–287, 288–290, 302–311
    vector potential **A**, 272–274, 275–277
    vector potential **F**, 274–277
    vector potentials **A** and **F**, 275–277
Axial ratio (AR), 165

B

Backscattering, 351–353, 741
    planar surfaces, 582, 594
Backward wave (BW) media
    transmission lines, 251–252
    see also Double negative materials
Bandgap structure, 944–945
Basis functions, 680, 689–693
    entire domain, 691–693
    physical equivalent method, 350
    subdomain, 689–691
Beam-forming, 999, 1001–1006
Beam translator, 248
Bednorz, J. G., 67
Bent wire, 682–684
BER *see* Bit error ratio
Bessel functions
    addition theorem, 602–604
    circular/cylindrical, 117–120, 533
    differential equation, 117, 932
    first kind, 118–119
    modified, 514
    second kind, 118–119
    spherical, 123–124, 547, 554–555, 559
    zeroes, 487
Biconical transmission lines, 555–559
    characteristic impedance, 558–559
    transverse electric modes, 555–557
    transverse electromagnetic modes, 557–559
    transverse magnetic modes, 557
Bilinear formula, 899
Binomial (maximally flat) transformers, 219,
    222–224

Bistatic/Bistatic scattering width, 582, 585
    circular cylinder, 605–637
    rectangular plate, 589–597
    spheres, 650–663
    strip, 582–595
Bit error ratio (BER), 847
Blended designs, checkerboard metasurfaces,
    975–977
Bloch mode, 955–956
Boundary conditions, 2, 12–18
    finite conductivity media, 12–14
    geometry, 13
    infinite conductivity media, 15–17
    sources along boundaries, 17–18
    time-harmonic electromagnetic field, 22–25
    vector wave equations, 104–105
Bound electrons, 61
Brewster angle, 179, 190, 192–194
Brillouin zone, mushroom EBGs, 956
BW *see* Backward wave

C

Canonical problem, 740
Capacitance, 8–11, 28
    mushroom AMCs, 952–954
    striplines, 451–454
Cascaded metasurfaces, 1004
Caustic rays, 328–329, 580, 745
Cavities
    circular, 496–504
        see also Circular cavities
    spherical, 559–566
        see also Spherical cavities
Center-radian frequency, mushroom AMCs,
    950–951
Charge density
    electric, 2, 11, 15, 292
    magnetic, 2–3, 28, 84–87
Charge distribution, 678–685
    bent wire, 682–684
    electrostatic, 678–684
    straight wire, 678–682
Checkerboard metasurfaces, 959, 960–979
    antennas and arrays, 977
    broadband, 967–979
    complex targets, 977–979
    conventional, 967–968
    corner reflectors, 979
    cylindrical structures, 978
    fundamentals, 964–967

introduction, 960
modified, 973–975
plane wave scattering, 961–964
ultrabroadband, 975–977
unit cell selection, 968–973
CHISs *see* Circular high-impedance surfaces
Chu, P. C. W., 67–68
CircDielGuide program, 537
Circuit equations, 7
Circuit-field relations, 7–11
    element laws, 10–11
    Kirchhoff's current Law, 8–10
    Kirchhoff's node current law, 9
    Kirchhoff's voltage Law, 7–8
Circuit theory
    field theory and relations between, 11
    Green's functions, 884–887
Circular..., *see also* Cylindrical...
Circular cavities, 496–504
    coupling, 496
    dissipated power, 503
    dominant mode, 501
    modes, 496–501
    quality factor, 501–504
    resonant frequency, 501–502
    stored energy, 503
    Transverse electric modes, 499–500
    Transverse magnetic modes, 500–501
    see also Circular dielectric resonators;
        Circular waveguides
Circular cylinders, scattering by, 605–637
    far-zone field, 608–610, 613–615, 619–621, 641
    line sources, 626–637
    normal incidence plane waves, 605–615
    oblique incidence plane waves, 615–624
    small radius approximation, 608, 612–613
Circular dielectric resonators, 522–530
    optical fiber cable, 528–530
    $TE_{01\delta}$ mode, 526–527
    transverse electric modes, 524–527
    transverse magnetic modes, 525–526
Circular dielectric waveguides, 512–522
Circular high-impedance surfaces (CHISs),
        985–990
Circular loop antennas, 985–990
Circular polarization, 151, 155–160, 166–167
    leaky-wave antennas, 1011
    left-handed, 151, 158–160, 166–167
    necessary and sufficient conditions, 160
    reflection, 228–235
    right-handed, 151, 155–158, 166–167

Circular waveguides, 479–545
    attenuation, 491–496, 528
    cavities, 496–504
    cutoff frequency, 481, 506–507, 509, 513
    dielectric, 512–530
    dielectric-covered, 530–537
    fiber optics cables, 494, 528–530
    hybrid modes, 512–522
    parallel plates, 505–509
    quality factor, 501–504
    radial, 505–512
    resonant frequencies, 500–501, 525–527
    transverse electric modes, 479–484, 492,
        499–500, 505–508, 510–521,
        524–527, 536–537
    transverse magnetic modes, 485–491, 492,
        500–501, 508–509, 510–521,
        525–526, 530–536
    wave impedance, 483
    wedged plates, 509–512
    see also Circular cavities; Circular cylinders;
        Circular dielectric resonators
Circumferentially modulated leaky-wave
        antennas, 1006
Classification
    leaky-wave antennas, 997
    metamaterials, 236–237
Clausius–Mosotti equation, 81
Closed form, Green's functions, 891–896,
        906–912
Coaxial transmission lines, rectangular
        waveguide coupling, 395–396
Coercive electric field, 45
Collocation method *see* Point matching
        method
Complex angles, 198, 226
Complex fields, uniform plane waves,
        129–130
Complex permeability, 82–83
Computer codes, 730–733, 827
    electromagnetic surface patch, 733
    geometrical theory of diffraction, 827–830
    mini-numerical electromagnetics code, 732
    Numerical Electromagnetic Code, 732–734
    Pocklington's wire radiation and scattering,
        730, 732
    two-dimensional radiation and scattering,
        730–732
Conducting circular cylinders
    line-source scattering by, 624–637
    normal incidence scattering, 605–615

transverse electric modes, 610–615
transverse magnetic modes, 605–610
oblique incidence scattering, 615–626
transverse electric polarization, 621–626
transverse magnetic polarization, 615–621
Conducting strips, dielectric materials, 947
Conducting wedges
diffraction, 847–877
boundary conditions, 849
computations, 875–877
diffracted fields, 866–873
geometrical optics, 855–863
Maliuzhinets solution, 852–854
reflection coefficients, 850–852
surface wave terms, 863–865
surface wave transition field, 873–875
scattering, 637–648
electric line-sources, 637–642, 646–648
magnetic line-sources, 642–648
plane waves, 642
Conduction band, 62–63
Conduction (ohmic) losses, rectangular waveguides, 388–392
Conductivity, 59–70
constitutive parameters, 69–70
dielectrics, 64
finite conductivity media, 12–14
infinite conductivity media, 15–17
metals, 64
metamaterials, 68–69
plasmas, 64
semiconductors, 61–66
superconductors, 66–68
Conductor–conductor interfaces
normal incidence, 205–207
oblique angles, 212
Conductors, 58–59
conductivity conditions, 64
energy levels, 63
good, 81, 146, 147–148
Conservation of energy, 19, 26
Conservation of power law, 21
Constant phase planes, 140–141
Constitutive parameters, 6–7
Constitutive relations, 5–6
Constraint (dispersion) equation, 108, 113, 117
Construction of solutions, 277–294
Inhomogeneous vector potential wave equation, 291–294

scalar Helmholtz wave equation, 554–555
spherical geometries, 547–555
transverse electric modes, 277, 288–290, 551–552
transverse electromagnetic modes, 277–284
transverse magnetic modes, 277, 285–288, 553
vector potential **A**, 550
vector potential **F**, 548–550
Continuity equation, 3
differential form, 6, 22–23
integral form, 6, 22–25
Continuous derivatives, 2
Convection current density, 58
Conventional broadband checkerboard metasurfaces, 967–968
Conventional_Checkerboard_RCS_Reduction program, 967, 1014
Conventional method of steepest descent, 768, 773
Coplanar waveguides (CPW), 252–253
Copper
charge density, 64
mobility, 64
Corner reflectors, 979
Corrugations, metasurfaces, 945–947
anisostropic, 946
Coupled equation, 115
Coupling
coaxial transmission lines, 395–396
electric field, 395
magnetic field, 395
rectangular waveguides, 395–396
CPW *see* Coplanar waveguides
Critical angle, 179, 186, 194–204
parallel polarization, 204
perpendicular polarization, 195–204
wave propagation along an interface, 204
Critically damped solution, 73
Critical temperature (Tc), 66–68
Curie temperature, 45
Curie–Weiss law, 45
Current density, 4, 58
conduction electric, 2, 75
conduction magnetic, 82
convection, 58
displacement electric, 2, 74–75
displacement magnetic, 2–3, 82
impressed electric, 2–3, 75–76
impressed magnetic, 2–5, 12, 82
surface equivalence theorem, 340–345

Curved edge diffraction, 806–813
  oblique incidence, 806–813
Curved surfaces, leaky-wave antennas,
      1005–1006
Cutoff frequency
  circular waveguides, 481, 506–507, 509, 513
  dominant mode, 373–374
  partially filled waveguides, 411–413, 415–418
  radial waveguides, 506–507, 509
  rectangular waveguides, 370–386, 429
Cutoff wave number, 370
Cylinder_RCS program, 663
Cylindrical, *see also* Circular...
Cylindrical coordinate systems, 926–931
  far-field radiation, 311–317
  Hankel functions, 118–120, 599–602, 605–630
  scattering, 597–604
  transverse electric modes, 290
  transverse electromagnetic modes, 281–284
  transverse magnetic modes, 287–288
  wave function solutions, 119
Cylindrical structures
  radar-absorption, 978
  see also Circular cavities; Circular dielectric
      resonators; Circular waveguides
Cylindrical waves
  addition theorem, 599–602
  attenuation, 119
  electric line sources, 574–578
  Hankel functions, 118–120, 599–602,
      605–630
  magnetic line sources, 578
  plane waves in terms of, 597–599
  transformations and theorems, 597–604
  wave function solutions, 114–119
Cyl_Resonator program, 537
Cyl_Waveguide program, 537

D

Damped
  critically, 73
  over, 73, 80
  under, 73, 80
Debye equation, 81
Degenerate modes
  circular waveguides, 487
  rectangular waveguides, 370
  spherical cavities, 561–564
Delta-gap sources, 721, 727
Design
  aperture antennas, 990–994
  circular loop antennas, 985–990
  holographic principle, 999–1000
  horizontal dipole antennas, 983–985
  leaky-wave antennas, 999–1012
  microstrip arrays, 994–996
  monopole antennas, 982–983
  mushroom AMCs, 950–954
Design center-radian frequencies, 950–951
Diagonalization, 70
Diamagnetic materials, 55–56
Dielectric–conductor interfaces, oblique
      incidence reflection, 208–211
Dielectric constants, 48–49, 77
Dielectric-covered conducting rods, 530–537
  geometry, 531
  transverse electric modes, 536–537
  transverse magnetic modes, 530–536
Dielectric covered ground plane, 447–450
Dielectric hysteresis, 75–76
Dielectric losses, rectangular waveguides,
      392–395
Dielectric materials, 7, 43–50, 61
  anisotropic, 69
  conducting strips, 947
  conductivity, 64
  electric susceptibility, 78–82
  good, 77–78, 144, 147
  hysteresis, 75–76
  metamaterials, 41, 68–69, 233–251
  metasurfaces, 943–1021
  mushroom AMCs, 947–960
  oblique incidence reflection, 208–211
  spherical scattering, 661–663
Dielectric resonators
  circular, 522–530
  transverse electric mides, 524–527
  transverse magnetic modes, 525–526
Dielectric waveguides
  circular, 512–530
  covered ground plane, 447–450
  eigenvalue equation, 515–517
  geometry, 423
  oblique incidence reflection, 208–211
  ray-tracing method, 437–447
  rectangular waveguides, 422–450
  slab, 422–424
  transverse electric modes, 433–437, 512–521
  transverse magnetic modes, 424–433, 512–521
Differential form of Maxwell's equations,
      2–3, 102, 103–115

coupled, 103–115, 120
uncoupled, 104–115, 552
Diffracted fields, 776, 865–873
asymptotic expansions, 866–867
diffraction terms, 865–866
incident, 774, 779, 788
modeling, 867–873
reflected, 771, 779
Diffraction, 739–830
curved edges, 806–813
edge, 759–817
amplitude, phase and polarization
relations, 759–763
curved, 806–813
equivalent currents, 813–817
normal incidence, 763–798
oblique incidence, 798–813
straight, 763–778
geometrical optics, 740–759
geometrical theory of, 637, 648, 677, 699,
739–830
Hankel functions, 766
normal incidence, 763–798
oblique angles, 798–813
slope, 797, 817–827
higher-order, 820–822
overlap transition regions, 825–827
self-consistent method, 822–825
straight edges, 763–798
wedge with impedance surfaces, 847–882
computations, 875–877
diffracted fields, 866–873
geometrical optics, 855–863
impedance surface boundary conditions,
849
impedance surface reflection coefficients,
850–852
Maliuzhinets impedance wedge solution,
852–854
surface wave terms, 863–866
surface wave transition field, 873–875
Diffraction coefficients, 778–798
hard polarization, 781–782
incident, 774
Keller's, 783
reflection, 780–781
soft polarization, 781–782
Diffraction plane, 799
Dipole_Horizontal_H_Plane program, 878
Dipoles
electric, 43–44, 46, 57, 70, 72

horizontal, 333–335, 983–985
image theory, 327–335
magnetic, 50–52, 57, 82–85
moment, 43–45
torque, 83–84
vertical, 329–333, 648–650
Dipole_Vertical program, 877
Dirac delta weighting functions, 697–698
Disperive materials, microstrips, 458–461
Dispersion (constraint) equation,
108, 113, 117
Dispersion_Diagram_HFSS program,
956–957, 1014
Dispersion_Diagram_Matlab program, *956*,
1014
Dispersion equation, 74
Dispersive materials, 7
Dissipation
circular cavities, 503
spherical cavities, 564
Distance parameter (L), 797, 801–802,
806, 828
Divergence theorem, 3–5, 919, 922
DNG *see* Double negative materials
Domains, 56
Dominant modes, 374–377, 449, 513,
557, 564
Donors, 61
Doping, 61
Double negative (DNG) materials,
41, 236–253, 943
classification, 236–237
history of, 238
Poynting vectors, 238, 242, 246, 248–249
propagation characteristics, 239–241
refraction, 242–249
transmission lines, 249–253
Double positive (DPS) materials, 236–237
history of, 238
transmission lines, 249–252
DPS *see* Double positive materials
Duality theorem, 323–324, 578
Dyadic Green's functions, 936–939
dyadics, 936–937
Dyadic reflection coefficient, 749, 761

E

EBG *see* Electromagnetic bandgap surfaces
Echo area, 581
see also Radar cross-section

Edge diffraction, 759–827
amplitude, phase and polarization relations, 759–763
curved edges, 806–813
equivalent currents, 813–817
normal incidence, 763–798
oblique incidence, 798–813
straight edges, 763–805
Effective dielectric constant, microstrip transmission line, 455–457
Effective radius, 536
EFIE *see* Electric field integral equation
Eigenfunction, 848, 898–903, 912–916
Eigenfunctions, Rectangular waveguides, 369
Eigenvalue equation, 515–517
Eigenvalues, 892
circular waveguides, 517–519
Rectangular waveguides, 370
spherical waveguides, 556–561
Eikonal equation, 742
Eikonal surfaces, 742–747, 759
cylindrical, 743
plane, 743
spherical, 743
Electrets, 45
Electrical properties of matter, 41–91
Electric current density, 3–4
Electric displacement current density, 3
Electric field coupling, 395
Electric field integral equation (EFIE), 350, 686, 701–711
perfectly electric conducting surfaces, 702
Transverse electric modes, 707–711
Transverse magnetic modes, 703–707
two-dimensional, 703–711
Electric fields
circular dielectric waveguides, 512–522
circular polarization, 155–159
corrugations, 946–947
elliptical polarization, 160–165
integral equation, 702–711
leaky-wave antennas, 998
linear polarization, 152–155
lossless media
oblique angles, 136–139
principal axis, 128–130
lossy media
oblique angle, 148–149
principal axis, 143–144
Electric line sources
above infinite plane electric conductor, 578–581
circular scattering, 626–630
conducting wedge scattering, 637–642, 646–648
cylindrical radiation, 574–578
wave impedance, 576
Electric polarization vector $\mathbf{P}$, 44
Electric potential, 678, 906, 914
Electric susceptibility, 48, 78–82
Electromagnetic bandgap (EBG) surfaces
applications, 944, 959–979
bandgap structure, 944–945
checkerboard metasurfaces, 959–979
microstrip arrays, 994–996
mushroom AMCs, 947–960
operational band, 944–945
principles, 943–945, 947–950
scan blindness, 944, 994
shortcomings, 944
surface waves, 944–945
see also Metasurfaces
Electromagnetic Surface Patch (ESP) code, 733
Electromagnetic theorems and principles, 323–363
backscattering, 351–353
duality theorem, 323–324
horizontal electromagnetic dipoles, 333–335
Huygens's principle, 340–345
image theory, 327–335, 352
induction theorem, 345–348, 351–356
Love's equivalence principle, 341–342
physical equivalent method, 349–356
physical optics equivalent method, 349–351
reaction theorem, 337–338
reciprocity theorem, 335–337
surface equivalence theorem, 340–345
uniqueness theorem, 325–327
vertical electric dipole, 329–333
volume equivalence theorem, 338–340
Electronic polarization, 45
Electrons
bound, 61–62
free, 58, 61–62
metamaterials, 68–69
mobility, 61
orbits, 42–43
semiconductors, 61–65
spin, 54–57, 83–91
superconductors, 66–68
Electrostatic charge distribution, 678–684
Element laws, 10–11
Elliptical polarization, 151, 160–165, 167
left-handed, 161
reflection, 228–235

right-handed, 161
Energy, 18–21, 25–29
  conservation, 3, 19, 21, 25–27, 26, 29,
      740–743, 746, 762
  density, 20–21
  electric, 21, 134
  inductor, 11
  magnetic, 21, 134
Energy density
  lossless media, 133, 141–142
  lossy media, 144, 149–150
Energy velocity, 133, 141–142, 149–151
ENG see Epsilon negative materials
Entire-domain functions, 691–693
E polarization see Perpendicular
      polarization
Epsilon negative (ENG) materials, 236–237
EPTD see Extended physical theory of diffraction
Equal amplitudes, 158, 160, 163
Equiphase plane, 127
Equiphase point, 111
Equivalence
  duality theorem, 323–324
  edge diffraction, 813–817
  induction theorem, 345–348, 351–356
  Love's principle, 341–342
  physical, 349–356
  reaction theorem, 337–338
  reciprocity theorem, 335–337
  surface theorem, 340–345
  volume theorem, 338–340
Equivalent currents, edge diffraction, 813–817
Equivalents, 3
ESP see Electromagnetic Surface Patch
ESTD see Extended spectral theory of diffraction
Evanescent waves
  cylindrical coordinate systems, 119
  parallel plates, 508–509
  rectangular coordinate systems, 110
  rectangular waveguides, 371
Expansion functions, 679, 688, 693, 699, 718
Extended physical theory of diffraction
      (EPTD), 825–828
Extended spectral theory of diffraction
      (ESTD), 825–826
Exterior wedges, 646, 777, 848

F

Faraday's law, 5
Far-field approximation, 580
Far-field radiation, 295–296, 298–317

conducting wedges, 641
cylindrical coordinate systems, 311–317
cylindrical scattering, 608–610, 613–615,
      619–621, 641
rectangular coordinate systems, 302–311
Felsen–Marcuvitz method, 867–868, 873
Fermat's principle, 740–741
Ferrimagnetic materials, 55–57, 81–91
Ferrites, 56, 83–91
Ferroelectric Curie temperature, 45
Ferroelectric materials, 45
Ferromagnetic materials, 55–56
Fiber optics cable, 423, 494, 528–530
  attenuation, 528
  discrete modes, 528
  graded-index, 528–530
  multimode, 528–530
  normalized diameter, 522
  single-mode, 528
  step-index, 528, 530
Field equations, 7
Field intensity
  electric, 2, 6, 14, 17–18, 24, 47–48, 271
  magnetic, 2, 6, 14, 17, 25, 28, 271
Field theory, circuit theory and relations
      between, 11
Figures of merit, 980–982
Finite conductivity media, 12–14
Finite diameter wires, 721–730
  delta-gap source modeling, 727
  Hallén's integral equation, 725–726
  magnetic frill generators, 727–730
  Pocklington's integral equation, 722–725
  source modeling, 721–730
Finite straight wire, 678–682
Floquet–Bloch modes, 998–999
Floquet Theorem, 234
Forbidden band, 62–63
Fourier–Legendre series, 650
Fractional bandwidth, mushroom AMCs, 951
Free electrons, 58, 61–62
Frequency-selective surfaces (FSSs), 943
Fresnel integrals, 774, 776, 782–783
Fresnel reflection coefficients, 186, 190
Fresnel transition functions (FTF), 782–783,
      787, 801, 819, 829
  large argument, 829
  small argument, 829
Fresnel transmission coefficient, 186, 190
Fringe wave, 677
FSSs see Frequency-selective surfaces
FTF see Fresnel transition functions

G

Galerkin's method, 695–697
Gallium arsenide, 64–65, 65
Gauss's law, 5
Generalized Green's function method, 920–923
    homogeneous Dirichlet boundary
        conditions, 921
    homogeneous Neumann boundary
        conditions, 922–923
    mixed boundary conditions, 923
    nonhomogeneous Dirichlet boundary
        conditions, 921
Geometrical optics (GO), 699, 740–759, 775
    amplitude relation, 743–746
    astigmatic rays, 745, 746, 750, 760
    caustic, 745, 746, 755
    conservation of energy flux, 741, 762
    divergence factor, 761, 762
    dyadic reflection coefficient, 749
    eikonal surface, 742–746
    Luneberg–Kline series, 747–748, 758–759
    normal section, 752–753
    phase, 759–763
    primary wave front, 741–742
    principal radii of curvature, 750–753
    ray optics, 740
    reflection from surfaces, 749–759
    region, 763–776
    secondary wave front, 741–742
    spatial attenuation, 748, 750
    spreading factor, 748, 750
    variational differential, 741
    wedge diffraction with impedance surfaces,
        854–863
Geometrical theory of diffraction (GTD),
        637, 648, 677, 699, 739–830
    computation, 827–830
    curved edges, 806–813
    edge diffraction, 759–817
    geometrical optics, 740–759
    slope diffraction, 817–827
    straight edges, 763–798
Germanium, 61, 64–65, 65
GO *see* Geometrical optics
Good conductors, 24, 81, 146, 147–148
Good dielectrics, 77–78, 144, 147
Graded index, 423, 528–530
Green's functions, 648, 883–942
    circuit theory, 884–887
    closed form, 891–896, 906–912

    in engineering, 884–889
    Green's identities and methods, 917–923
    in integral form, 902–905
    mechanics, 867–869
    rectangular coordinates, 906–917
    of scalar Helmholtz equation, 923–935
    series form, 896–902, 912–914
    static fields, 906–914
    Sturm–Liouville problems, 884, 889–905
    Time-harmonic fields, 915–917
Ground_TE_TM_Graph program, 466
Ground_TE_TM_Ray program, 466
Group velocity, 129, 133–134, 141–142, 384
GTD *see* Geometrical theory of diffraction
Guide wavelength (λ_g), 372, 384, 482
Gunn Diode oscillator, 365
Gyromagnetic ratio, 84

H

Hallén's integral equation, 721, 725–727
    source modeling, 727
Hankel functions, 508
    addition theorem, 599–602
    cylindrical, 118–120, 599–602, 605–630
    derivative, 606
    diffraction, 766
    electric line sources, 575–576
    integral equation, 684–688, 694
    spherical, 123–124, 547, 554–555, 558, 649,
        656, 658, 661
Hard polarization, 648, 766–767, 815,
        818–823, 850–853, 946
Helmholtz equation, 106, 547–548, 552–555,
        902
    cylindrical, 926–931
    rectangular, 923–926
    spherical, 931–935
Hermite–Gaussian functions, 530
Hermitian properties, 70, 895
Higher-order diffraction, slopes, 820–822
High-frequency asymptotic solution, 766–770
High-gain printed leaky-wave antennas, 944,
        997–1012
High-impedance surfaces (HISs), 943
    antennas
        aperture, 990–994
        circular loop, 985–990
        holographic principle, 999–1000
        horizontal dipole, 983–985
        leaky-wave, 997–1012

microstrip arrays, 994–996
monopole, 982–983
polarization, 1009–1012
surface-wave, 996
tensor impedance surfaces, 1013
aperture antennas, 990–994
circular loop antennas, 985–990
leaky-wave antennas, 997–1012
monopole antennas, 982–983
see also Artificial impedance surfaces
High-pass filters, 251–252
High temperature superconductivity (HTS), 68
HIS_Mush program, 1013
HISs *see* High-impedance surfaces
Holes, 62
Holographic metasurfaces
circular polarization, 1011
horizontal polarization, 1010–1011
multiple polarizations, 1011–1012
one-dimensional, 997, 1000–1006
tensor impedance surfaces, 1013
two-dimensional, 997, 1007–1009
vertical polarization, 1009–1010
Holographic principle, leaky-wave antennas,
  999–1000
Homogeneous materials, 7, 69–70
Horizontal dipoles
antennas, 983–985
electric, 333–335
Horizontal polarization, leaky-wave antennas,
  1010–1011
H polarization *see* Parallel polarization
HTS *see* High temperature
  superconductivity
Huygen's principle, *see also* Surface
  equivalence theorem
Huygens's principle, 340–345
Hybrid modes
circular waveguides, 512–522
rectangular waveguides, 404–407
Hydrogen atom, 42
Hysteresis, dielectric, 75–76
Hysteresis loop, 45, 56

I

IE *see* Integral equation
Image theory, 327–335, 352
horizontal electric dipoles, 333–335
vertical electric dipoles, 329–333
IMPATT diode, 365

Impedance
Biconical transmission lines, 558–559
directional, 140, 149
intrinsic, 132, 138, 140, 144, 148–149,
  205, 212, 220–222, 295, 301, 452
mushroom AMCs, 949
striplines, 454–455
surface boundary conditions, 849
surface reflection coefficients, 850–852
see also Artificial impedance surfaces;
  Wave impedance
Impedance boundary condition, 847
Impedance surfaces
wedge diffraction, 847–877
boundary conditions, 849
computations, 875–877
diffracted fields, 866–873
geometrical optics, 855–863
Maliuzhinets solution, 852–854
reflection coefficients, 850–852
surface wave terms, 863–865
surface wave transition field, 873–875
Incident diffracted field, 774, 778
Incident diffraction coefficient, 782
Incident geometrical optics, 773
Index of refraction, 48, 75, 235–236,
  238–240, 243, 423, 528–530, 741
Inductance, mushroom AMCs, 951, 953–954
Induction equivalents *see* Induction theorem
Induction theorem, 345–348
approximations, 351–356
field geometry, 346
Inductor, 8–11, 28, 483, 884
Infinite conductivity media, 15–17
Infinite line-source cylindrical wave radiation,
  573–581
above infinite plane conductor, 578–581
electric line sources, 574–578
magnetic line sources, 578
Infinitesimal dipole, 294, 649
Infinities
cylindrical coordinates, 119
rectangular coordinates, 110
spherical coordinates, 123
Inhomogeneous vector potential wave
  equation, 291–294
Instantaneous fields, 129–130
Insulators, 62
conductivity, 64
energy levels, 63
Integral equation (IE), 677–701

basis functions, 689–693
computation, 730–733
electric fields, 702–711
electrostatic charge distribution, 678–684
finite-diameter wires, 721–730
Hallén's, 725–726
Hankel functions, 684–688, 694
integral equation, 684–686
magnetic fields, 711–721
moment method, 695–701
Pocklington's, 722–725
point-matching method, 687–695
radiation patterns, 686–687
source modeling, 727–730
weighting functions, 695
Integral forms
    Green's functions, 902–905
    Maxwell's equations, 3–5
Interior wedges, 646, 848, 852–853
Intrinsic impedance, 132, 138, 140, 144,
    148–149, 205, 205–206, 212,
    220–222, 295, 849
Intrinsic (pure) semiconductor, 61
Intrinsic reflection coefficients, 214–215
Ionic or molecular polarization, 45
Irreducible Brillouin zone, 956
Isolated poles, 768, 773–774
Isotropic materials, 7, 69–70

K

Keller's diffraction coefficients/functions,
    776, 783
Kirchhoff's current Law, 8–10
Kirchhoff's voltage Law, 7–8
Kramers–Kronig relations, 81
Kronecker delta function, 719, 898

L

Larmor precession frequency, 83–89
Law of diffraction, 763
Leaky-wave antennas (LWAs), 944, 997–1012
automotive radar technology, 1005–1006
beamforming, 999
circular polarization, 1011
circumferentially modulated, 1006
classification, 997
fast wave mode, 997
Floquet–Bloch modes, 998–999
holographic principle, 999–1000

horizontal polarization, 1010–1011
multiple polarization, 1011–1012
(non-)phase crossover frequencies, 1008
one-dimensional, 997, 1000–1006
periodic, 998–999
polarization-diverse, 1009–1012
sinusoidal modulation, 1000–1005
surface-impedance modulation function,
    1000
tensor impedance surfaces, 1013
transverse resonance method, 1002
two-dimensional, 997, 1007–1009
uniform, 997–999
vertical polarization, 1009–1010
Left-handed (counterclockwise) polarization,
    151, 158–160, 161
Left-handed materials (LHM), 237, 238
    see also Double negative materials
Legendre differential equation, 122
Legendre polynomials, 650, 931, 933
associated functions, 124, 547, 554, 650, 657–658
differential equation, 121
function of first kind, 554
functions, 650
Leontovich boundary condition, 847, 849
Linear integral operator, 688
Linear materials, 7, 69–70
Linear polarization, 151, 152–155, 166–167
    reflection, 228
Line sources
circular scattering, 626–637
electric, 574–581, 626–630, 637–641
finite diameter wires, 721–730
infinite, 573–581
magnetic, 578, 630–637, 642–645
strip, 731
Longitudinal section electric (LSE) modes,
    404–413, 421
concepts, 404–407
filled rectangular waveguide, 412
partially filled waveguide, 407–413
transverse resonance method, 421
Longitudinal section magnetic (LSM) modes
concepts, 404, 407
filled rectangular waveguide, 411
partially filled waveguide, 414–419
rectangular waveguides, 404, 407, 414–419,
    422, 424–430
transverse resonance method, 422
Lorentz reciprocity theorem, 336–337
Lorenz conditions, 274, 548

Lossless media
    perpendicular reflection, 184–188, 192–193,
        195–204
    rectangular wave equations, 107–111
    reflection, 179–203
        Brewster angle, 179, 192–194
        critical angle, 194–204
        normal incidence, 204–207
        oblique incidence, 183–184
        parallel polarization, 188–192, 193–194, 204
    transverse electromagnetic modes, 128–142
        oblique angle, 136–142
        principle axis, 128–136
        standing waves, 134–136
Loss tangent
    electric, 76–77
    magnetic, 82
Lossy media
    conductor-conductor interfaces, 205–207, 212
    dielectric–conductor interfaces, 208–211
    dielectric spheres, 661–663
    rectangular wave equations, 112–114
    reflection, 204–211
        normal incidence, 205–207
        oblique incidence, 207–211
    transverse electromagnetic modes, 142–151
        oblique angle, 148–151
        principal axis, 143–148
        standing waves, 144–146
Love's equivalence principle, 341–342
Lowest-order modes, 277, 531, 536, 563, 574
Low-pass filters, 251–252
Low-radar cross-section antennas, 977
LS_TE_TM_X program, 466
LS_TE_TM_Y program, 466
L-strips, 947
Lumped-circuit theory, 389
Luneberg–Kline high-frequency expansion,
        747–748, 758–759
LWAs *see* Leaky-wave antennas

M

Macroscopic scale models of materials, 46
Magnetic current density, 3–5, 7, 11, 340,
        342, 348–349, 727
Magnetic dipoles, 50–52
Magnetic field coupling, 395
Magnetic field integral equation (MFIE), 350,
        711–721

transverse electric modes, 715–721
    transverse magnetic modes, 713–715
    two-dimensional, 713–721
Magnetic fields
    circular dielectric waveguides, 512–522
    circular polarization, 155–159
    elliptical polarization, 160–165
    integral equation, 711–721
    linear polarization, 152–155
    lossless media
        oblique angle, 136–139
        principal axis, 128–130
    lossy media
        oblique angle, 148–149
        principal axis, 143–144
Magnetic flux density, 6, 14, 28, 52
Magnetic frill generators, 727–730
Magnetic line sources
    circular scattering, 630–637
    conducting wedge scattering,
        642–648
    cylindrical wave radiation, 578
Magnetic materials, 3–28, 50–70, 83
    anisotropic, 7, 69
    antiferromagnetic, 55–56
    atomic model, 83
    diamagnetic, 55, 55–56
    dispersive, 7, 69
    ferrimagnetic, 55–57, 81–91
    ferromagnetic, 45, 55–56, 69, 81–82
    homogeneous, 7, 10, 69–70
    inhomogeneous, 7, 69
    isotropic, 7, 69–70
    linear, 7, 15, 45, 50, 69–70
    nondispersive, 7–8, 73
    nonhomogeneous, 7, 69
    nonisotropic, 7, 69
    nonlinear, 7, 69–70
    paramagnetic, 55–56, 81
    phenomenological model, 83–91
    precession, 83–89
    resonance, 91
    torque, 52, 83–85
Magnetic susceptibility, 53, 56, 90
Magnetization, 7, 50–57
    current, 50–52
    current density, 51–52, 54–55
    magnetization vector, 52
Maliuzhinets impedance wedge solution,
        852–854

Master surfaces, mushroom AMCs, 956
Maximally flat transformers, 219, 222–224
Maximum directivity, 981
Maximum gain, 981–982
Maximum realized gain, 982
Maxwell–Ampere equation, 54, 75
Maxwell's equations, 2–5
    time-harmonic, 22
    time-varying, 2–5
Meissner effect, 66
Metals, conductivity, 64
Metamaterials, 41, 68–69, 233–251
    classification, 236–237
    double negative, 237, 239–253
    double positive, 236–237, 249–252
    epsilon negative, 237
    historical perspective, 238
    mu negative, 237
    Poynting vectors, 238, 242, 246, 248–249
    refraction, 241–249
    transmission lines, 249–253
    see also Metasurfaces
Metasurface leaky-wave antennas, 997–1013
    see also Leaky-wave antennas
Metasurfaces, 943–1021
    antennas, 980–1013
        aperture, 990–994
        circular loop, 985–990
        holographic principle, 999–1000
        horizontal dipole, 983–985
        leaky-wave antennas, 997–1013
        microstrip arrays, 994–996
        monopole, 982–983
        polarization, 1009–1012
        radar cross-section reduction, 977
        surface-wave antennas, 996
        tensor impedance surfaces, 1013
    applications, 959–1013
    bandgap structure, 944–945
    cascaded, 1004
    checkerboard, 959, 960–979
    corner reflectors, 979
    corrugations, 945–947
    cylindrical structures, 978
    hard and soft polarization,
        946–947
    holographic principle, 999–1000
    mushroom AMCs, 947–960
        design, 950–954
        limitations, 959–960
        resonant frequency, 949

        surface impedance, 949
        surface-wave dispersion, 955–959
    operational band, 944–945
    plane wave scattering, 961–964
    polarization-diverse, 1009–1012
    principles, 943–945
    pyramidal horns, 946–947
    scan blindness, 944, 994
    stealth applications, 944
    tensor impedance surfaces, 1013
    unit cell selection, 968–973
Method of steepest descent, 637, 765,
        768–775, 854
    conventional, 768, 773–774
    Pauli–Clemmow, 774–775, 867–868
    steepest descent path, 769, 771, 773, 775
Method of weighted residuals, 695, 699
MFIE *see* Magnetic field integral equation
Microstrip transmission lines, 455–461
    arrays
        design, 994–996
        scan blindness, 944, 994
    boundary-value problem, 461
    effective dielectric constant, 456–457
    evolution, 452
    shielded configuration, 461
Microwave cooking, 76
Mie region, 660, 755
MININEC *see* Mini-Numerical
        Electromagnetic Code
Mini-Numerical Electromagnetic Code
        (MININEC), 732
MM *see* Moment method
MMICs *see* Monolithic microwave integrated
        circuits
MNG *see* Mu negative material
Mobility
    electron, 61
    holes, 62–63
Modal solutions, 765–766
Modes
    definition, 127
    polarization, 151–170
    transverse electric, 136
        construction of solutions, 277, 288–290
    transverse electromagnetic, 127–177
        circular polarization, 151, 155–160,
            166–167
        construction of solutions, 277–284
        elliptical polarization, 151, 160–165,
            166–167

linear polarization, 153–155, 166–167
lossless media
  oblique angle, 136–142
  principal axis, 129–136
lossy media
  oblique angle, 148–151
  principal axis, 143–148
plane waves, 128
Poincaré sphere, 165–170
transverse magnetic, 136
  construction of solutions, 277, 285–288
Modified checkerboard metasurfaces, 973–975
Modified Pauli–Clemmow method, 867–868
Molecules, 41
Moment method (MM), 695–733, 739
  basis functions, 680, 689–693
  collocation, 681, 687–689, 693–695, 705,
    709, 731
  delta-gap, 721, 727
  diagonal terms, 694
  expansion functions, 679, 693, 699, 731
  Galerkin's method, 695
  Hallén's integral equation, 721, 725–727
  linear integral operator, 688
  magnetic frill generator, 722, 727–730
  nondiagonal terms, 695
  self terms, 694
  testing functions, 695
  weighted residual, 695, 699
  weighting functions, 695
Monolithic microwave integrated circuits
    (MMICs), 523
Monopole antennas, 982–983
Monopole_GP_UTD program, 830
Monopole program, 877
Monostatic scattering width, 588, 731
  circular cylinder, 731–732
  rectangular cylinder, 731–732
  strip, 731
Mueller, K. A., 67
Multimode fibers, 528
Multiple diffractions, 819–827
  first-order diffraction, 820
  higher-order diffractions, 820
  overlap transition diffraction region,
    825–827
  second-order diffraction, 820
  self-consistent method, 822–825
Multiple interfaces, 212–228

binomial designs, 219, 222–224
Intrinsic reflection coefficients, 214–215
multiple layers, 220–228
oblique incidence, 226–228
quarter-wavelength transformer, 220–221
single slab layer, 212–220
Tschebyscheff design, 219, 224–226
Multiple layers
  reflection coefficients, 220–228
    binomial design, 219, 222–224
    oblique angles, 226–228
    quarter-wavelength transformer,
      221–222
    Tschebyscheff design, 219, 224–226
Mu negative (MNG) material, 236–237
Mushroom artificial magnetic conductors
    (mushroom AMCs), 947–960
  bandgap structure, 945
  checkerboard metasurfaces, 959, 960–979
    antennas and arrays, 977
    broadband, 967–979
    complex targets, 977–979
    conventional, 967–968
    corner reflectors, 979
    cylindrical structures, 978
    fundamentals, 964–967
    modified, 973–975
    plane wave scattering, 961–964
    principles, 960
    ultrabroadband, 975–977
    unit cell selection, 968–973
  design, 950–954
  design center-radian frequency, 950–951
  fractional bandwidth, 951
  irreducible Brillouin zone, 956
  limitations, 959–960
  master/slave surfaces, 956
  operational band, 945
  principles, 947–950
  resonant frequency, 949
  sheet capacitance, 952–954
  sheet inductance, 951, 953–954
  surface impedance, 949
  surface waves, 945, 955–959
  transverse electric modes, 949, 956–957
  transverse magnetic modes, 949, 956–957
  transverse resonance method, 957
  unit cell capacitance, 952
  unit cell selection, 967–968

N

Natural angular frequency, 74
Near-field radiation equation, 296–298
NEC *see* Numerical Electromagnetic Code
Negative (refractive) index materials (NIM/ NRI), 236–253
  historical perspective, 238
  Poynting vectors, 238, 242, 246, 248–249
  propagation characteristics, 239–241
  refraction, 241–249
  transmission lines, 249–253
Negative-refractive-index-transmission lines (NRI-TL), 249–253
Neutrons, 41
NIM *see* Negative (refractive) index materials
Nondispersive media, 69–70
Nonhomogeneous materials, 7, 69
Nonisotropic materials, 7, 69
Nonlinear materials, 7, 69
Non-phase crossover frequency, 1008
Nonpolar materials, 45–46
Normal dispersion, 80
Normal incidence
  circular scattering, 605–615
  conductor–conductor interfaces, 205–207, 212
  definition, 180
  dielectric–conductor interfaces, 208–211
  lossless media, 179–183
  lossy media, 205–207
  straight edge diffraction, 763–798
  transmission in, 180–181
Normal section, 752–753
NRI *see* Negative (refractive) index materials
N-type semiconductors, 61, 64
Nuclear spin, 54
Null field approach, 713
Numerical Electromagnetic Code (NEC), 732–733

O

Object waves, 999–1000
Oblique angles
  circular scattering, 615–626
  conductor–conductor interfaces, 212
  curved edge diffraction, 806–813
  dielectric–conductor interfaces, 208–211
  edge diffraction, 798–813
  lossless media, 183–204

multiple layers, 226–228
  parallel polarization, 188–192, 193, 204
  perpendicular polarization, 184–188, 192–193, 195–204
  phase velocities, 199–204
  reflection, 183–204, 212, 226–228
  straight edge diffraction, 798–805
  uniform plane waves
    lossless media, 136–142
    lossy media, 148–151
Ohmic losses *see* Conduction losses
Ohm's law, 10–11, 28
One-dimensional leaky-wave antennas, 997, 1000–1006
One_Dim_LWA program, 1002, 1014
Onnes, H. K., 66
Open stopband phenomenon, 999
Operational band, 944–945
Optical holography, 999–1000
Optics, geometrical *see* Geometrical optics
Orientational polarization, 43–45
Orthogonality relationships, 648, 650–651
Orthonormal eigenfunctions, 898–899, 901, 903, 910–912
Overdamped solution, 73
Overlap regions, slope diffraction, 825–827

P

Parallel plates, radial waveguides, 505–509
Parallel polarization, ray-tracing method, 442–445
Parallel (vertical/H) polarization, 184, 188–192, 193–194, 204
Paramagnetic materials, 55–56
Partially filled waveguides, 407–419
  cutoff frequency, 411–413, 415–418
  electric modes, 407–413
  magnetic modes, 414–419
Pattern multiplication rule, 332
Pauli–Clemmow modified method of steepest descent, 774
PBG *see* Photonic band-gap surfaces
PEC_Circ_Plate_RCS program, 663
PEC_Circ_RCS_UTD program, 830
PEC_Cyl_Normal_Fields program, 663
PEC_Cyl_Normal_SW program, 663
PEC_Cyl_Oblique_Fields program, 663
PEC_Cyl_Oblique_RCS program, 663
PEC_Cyl_Oblique_SW program, 663

PEC_DIEL_Sphere_Fields program, 663
PEC_Rect_Plate_RCS program, 663
PEC_Rect_RCS_UTD program, 830
PECs *see* Perfect electric conductors
PEC_Square_Circ_RCS_UTD program, 830
PEC_Strip_Line_MoM program, 733
PEC_Strip_SW_MoM program, 733
PEC_Strip_SW program, 663
PEC_Strip_SW_UTD program, 830
PEC_Wedge program, 830
Penetration depth, 66
Perfect electric conductors (PECs), 653–661
    horizontal dipole antennas, 983–985
    induction theorem, 347–348
    monopole antennas, 982–983
    mushroom AMCs, 955–959
    physical equivalent method, 349–356
    principles, 944
    radar absorbing materials, 959, 960–979
    radar cross-section, 948
    shortcomings, 944
    spherical scattering, 653–661
    stealth applications, 944
    surface equivalence theorem, 341
Perfect magnetic conductors (PMCs), 15, 18,
    523, 851
    design center-radian frequency, 950–951
    horizontal dipole antennas, 983–985
    mushroom AMCs, 947–960
    plane wave scattering, 961–964
    principles, 943–944
    radar absorbing materials, 959, 960–979
    radar-absorbing materials, 959–979
    Radar cross-section, 948
    shortcomings, 944
    stealth applications, 944
    surface equivalence theorem, 342–344
Periodic leaky-wave antennas, 998–999,
    1000–1006
Permeability, 50–57
    complex, 82–83
    effective, 88
    mushroom AMCs, 951
    relative, 53–56
    static, 53, 70
    tensor, 91
Permittivity, 43–50
    complex, 70–82
    mushroom AMCs, 951
    principal, 70

    relative, 48–49, 79–82
    static, 48
    tensor, 69–70
Perpendicular (horizontal/E) polarization,
    184–188, 192–193, 195–204,
    445–447
Phase constants, 106, 112, 133, 146, 148
    rectangular waveguides, 394
    reflection and transmission, 205, 211,
    237, 240
Phase crossover frequency, 1008
Phase velocity, 111, 210, 251, 384
    definition, 133
    oblique angle, 140–141, 149
    oblique angles, 199–204
    principal axis, 133–134, 146
    transmission lines, 251–252
Phenomenological model, 83–91
Photonic band-gap (PBG) surfaces, 235, 237
    mushroom AMCs, 947–960
    principles, 943–944, 948–950
    radar-absorbing materials, 959–979
    see also Metasurfaces
Physical equivalent method, 349–356
Physical optics (PO) technique, 349–351, 582,
    592, 594, 699, 739
Physical theory of diffraction (PTD), 573, 739
Planar surfaces
    plane wave scattering, 581–604
      transverse electric from rectangular plate,
      589–597
      transverse magnetic from a strip, 582–589
Plane of incidence, 183–184, 732, 749
    Brewster angle, 179, 192–194
    critical angle, 179, 194–204
    lossless media, 183–204
    lossy media, 208–212
    see also Normal incidence; Oblique angles
Plane waves, 743
    cylindrical, 597–599
    definition, 127–128
    lossless media
      oblique angle, 136–139
      principal axis, 128–136
    lossy media
      oblique angle, 148–151
      principle axis, 143–148
    planar surface scattering, 561–597
    Poincaré sphere, 165–170
    polarization, 151–170

circular, 151, 155–160, 166–167
elliptical, 151, 160–165, 166–167
linear, 151, 152–155, 166–167
Poincaré sphere, 165–170
scattering, 581–637, 642
by circular cylinders, 605–637
by conducting wedges, 642
by cylindrical structures, 597–637
by planar surfaces, 581–597
rectangular plates, 589–597
strips, 582–589
radar-absorbing materials, 961–964
standing waves, 134–136, 144–146
wave impedance, 131–132, 139–140,
144, 150
Plasma, 238
Plasmas, 63–64
PMCs *see* Perfect magnetic conductors
PO *see* Physical optics
Pocklington's integro-differential equation,
721, 722–730, 732–733
Pocklington's wire radiation and scattering
(PWRS) program, 732–733
Poincaré sphere
polarization states, 165–170
reflection, 228–235
Point matching (collocation) method,
687–689, 693–695
Polarization, 43–50, 151–170
circular, 151, 155–160, 166–167, 228–235
dipole, 44
electric polarization vector **P**, 44
electronic polarization, 45
elliptical, 151, 160–165, 166–167, 228–235
ionic or molecular, 45
left-handed, 151, 158–160, 161, 166–167
linear, 151, 152–155, 166–167, 228
metasurfaces, 946–947, 1009–1012
orientational, 43–44
parallel, 184, 188–192, 193–194, 204
perpendicular, 184–188, 192–193, 195–204
Poincaré spheres, 165–170, 228–235
reflection, 184–204, 228–235
right-handed, 152, 155–158, 161, 166–167
Polarization_Propag program, 167, 171
Polarization_Refl_Trans program, 251
Polar materials, 44–46
Power, 18–21, 25–29, 386–388
conservation, 21
density, 20–21, 386–388

dissipated, 20–21, 27, 389–390, 564
exiting, 21, 27
supplied, 21
Power density
lossless media, 133, 141–142
lossy media, 144, 149–150
Poynting vector, 20, 25–26, 133
Poynting vectors, metamaterials, 238,
242, 246, 248–249
Precession frequency, 83–89
Primary wave front surface ($[phi]_0$), 741
Principal axis
reflection
lossless media, 179–183
lossy media, 205–207
uniform plane waves
lossless media, 128–136
lossy media, 143–148
see also Normal incidence
Principal coordinates, 70
Principal permittivities, 70
Principal planes, 752–753, 756, 798,
802–805
Principal radii of curvature, 750–753, 808
incident wave front, 751, 753, 801, 806
reflected wave front, 750–751, 753–754, 808
Printed leaky-wave antennas, 944, 997–1012
Propagation, 127–177
circular polarization, 151, 155–160,
166–167
constant, 147–149, 412, 419
constants, 106, 112
elliptical polarization, 151, 160–165,
166–167
linear polarization, 151, 152–155, 166–167
lossless media, 128–143, 183–204
lossy media, 142–151, 204–211
metamaterials, 239–249
oblique angle, 136–143, 148–150, 183–204,
226–228
polarization, 151–170
principal axis, 128–143, 142–151
rectangular waveguides, 371
standing waves, 134–136, 144–146
Protons, 41–42
PTD *see* Physical theory of diffraction
P-type semiconductors, 61, 64
PWRS program *see* Pocklington's wire
radiation and scattering program
Pyramidal horns, corrugations, 946–947

Q

Quality factor (Q), 396, 401
  circular cavities, 501–504, 566
  rectangular cavities, 366, 504
  spherical cavities, 564–566
Quanta, 42–43
Quarter-wavelength transformer, 221–222
QuarterWave_Match program, 251

R

Radar-absorbing materials, 876
Radar-absorbing materials (RAMs), 959–960
Radar cross-section (RCS), 581–582, 589,
      594–597, 660, 663, 740, 826–829,
      876
  conversion to two-dimensional, 582, 621
  definition, 581
  perfect electrical conductors, 948
  perfect magnetic conductors, 948
  see also Radar-absorbing materials
Radial waveguides, 505–512
  cutoff frequency, 506–507, 509
  parallel plates, 505–509
  transverse electric modes, 506–508,
      510–511
  transverse magnetic modes, 508–509, 511–512
  wedged plates, 509–512
Radial waves
  cylindrical wave equations, 114–119
  rectangular wave equations, 107–114
  spherical wave equations, 120–125
Radiation efficiency, 981
Radiation equation, 296–317
  cylindrical coordinate systems, 311–317
  far field, 298–317
  near field, 296–298
  rectangular coordinate systems, 302–311
Radiation fields, coordinate systems, 291
Radiation pattern, 581, 686–687
Radiation resistance, 980–981
Radio absorbing material *see* Radar-absorbing
      materials
RAM *see* Radar-absorbing materials
Rate of change (slope), 798
Rayleigh region scattering, 660–661
Ray-tracing method, 437–447
  reflecting plane wave representation, 441
  transverse electric modes, 442, 445–447
  transverse magnetic modes, 442–445

wave beam properties, 437
RCS *see* Radar cross-section
Reaction theorem, 337–338
Reciprocity theorem, 335–337
Rectangular coordinate systems, 277–281,
      285–287, 923
  far-field radiation, 302–311
  infinities, 109
  transverse electric modes, 288–290
  transverse electromagnetic modes, 277–281
  transverse magnetic modes, 285–287
  two-dimensional Green's functions, 906–917
    static fields, 906–914
    time-harmonic fields, 915–917
  wave equations, 106–113
    attenuation, 110
    lossless media, 106–111
    lossy media, 111–113
  wave functions, 109
  zeroes, 109
Rectangular plates
  scattering, 589–597
    backscattered, 594
    bistatic, 590, 594–597
    monostatic, 595–597
Rectangular resonant cavities, 396–404
  geometry, 399
  transverse electric modes, 399–402
  transverse magnetic modes, 403–404
Rectangular waveguides, 365–478
  attenuation, 388–396, 464
  Conduction losses, 388–392
  coupling, 395–396
  cutoff frequency, 370–386
  cutoff wave number, 370
  degenerate modes, 370
  dielectric, 422–450
  dielectric loaded, 405
  dielectric losses, 392–395
  eigenfunctions, 369
  eigenvalues, 370
  evanescent waves, 371
  guide wavelength, 373, 376
  hybrid modes, 404–407
  microstrip, 452, 455–461
  normalized wavelength, 373
  partially filled, 407–419
  phase constant, 394
  power density, 386–388
  ray-tracing method, 437–447

reference table, 397–398
resonant cavities, 396–404
ridged, 461–464
stripline, 451–455
TE$_{10}$ mode, 373–374, 378–386, 395–396
transverse electric modes, 367–375, 379–386, 399–402, 404–406, 433–437
transverse magnetic modes, 375–378, 403–404, 407, 414–419, 424–433
transverse resonance method, 419–422
wave impedance, 372–373
Rect_Resonator program, 464
Rect_Waveguide program, 464
Reflection, 179–269
binomial transformers, 219, 222–224
Brewster angle, 179, 192–194
coefficients, 180–194, 199, 204
conductor–conductor interfaces, 205–207, 212
critical angle, 179, 186, 194–204
Dielectric–conductor interfaces, 208–211
diffraction coefficient, 780
double negative materials, 237–253
intrinsic coefficients, 214–215
lossless media, 179–204
lossy media, 204–211
metamaterials, 235–253
multiple interfaces, 212–228
normal incidence, 179–183, 205–207
oblique angles, 183–204, 212, 226–228
parallel polarization, 184, 188–192, 193–194, 204
perpendicular polarization, 184–188, 192–193, 195–204
phase velocity, 199–204, 210, 251–252
Poincaré spheres, 228–235
polarization, 184–204, 228–235
quarter-wavelength transformers, 221–222
tightly bound slow surface waves, 202–203
total, 194–204
Tschebyscheff transformers, 219, 224–226
Refl_Trans_Multilayer program, 254
Refraction, double negative materials, 241–249
Relative permittivity, 48–49, 53, 55, 79
Relaxation time constant, 58, 81
Remnant polarization, 45
Residue calculus, 771–772, 904–905
Resistivity, 61
Resistor, 3, 10–11, 28
Resonance, 91, 564, 621, 660, 755, 902, 917

circular cavities, 501–502
circular waveguides, 501–502, 525–527
mushroom AMCs, 949
ridged waveguides, 462–464
Resonant angular frequency, 74
Resonators, circular dielectric, 522–527
RHM *see* Right-handed materials
Ridged waveguides, 461–464
bandwidth, 461
cross sections, 463
geometry, 461, 463
normalized attenuation, 464
resonance, 462–464
transverse resonance method, 462
Right-handed (clockwise) polarization, 151, 155–158, 161
Right-handed materials (RHM), 238
see also Double positive materials
Ring radiator, 812–813, 817

S

Saddle points, 768, 770–771, 773–775, 866–869
Scalar Helmholtz equation, Green's functions, 923–935
Scalar Helmholtz wave equation, 106, 554–555
Scalar wave equation, 884
Scan blindness, 944, 994
Scattered field, 571
Scattering, 573–676
backscattering, 351–353, 582, 594, 731, 741, 826–827
bistatic, 582, 586, 589, 594, 731–732
circular cylinders, 607–637
normal incidence, 605–615
oblique incidence, 615–624
conducting wedges, 637–648
transverse electric modes, 642–648
transverse magnetic modes, 637–642, 646–648
cylindrical structures, 597–637
far-field, 298–317
cylindrical coordinates, 311–317
rectangular coordinates, 302–311
field, 345, 573, 606, 685, 694, 702–705
infinite line-source cylindrical wave, 573–581
monostatic, 351, 353, 582, 594, 731, 741, 826–827
near-field, 296–298

orthogonality relationships, 650–651
plane waves, 581–604
radar cross section, 581–582
rectangular plates, 589–597
    bistatic, 590, 594–597
    monostatic, 595–597
specular, 582
by spheres, 648–663
strips, 581–589
vertical dipole spherical waves, 648–650
width, 581–582
Schelkunoff, S. A., 124, 340, 554
Secondary wave front surfaces, 741
Self-consistent method, slope diffraction,
    822–825
Semiconductors, 61–66
  acceptors, 61
  bound electrons, 61
  conductivity, 64
  donors, 61
  doping, 61
  energy levels, 63
  forbidden bands, 62
  free electrons, 61
  germanium, 61, 64–65
  holes, 62
  intrinsic, 61
  n-type, 61, 64
  p-type, 61, 64
  silicon, 61
Separation of variables method, 106–125,
    285, 287, 367
  cylindrical coordinates, 114–120
  rectangular coordinates, 107–114
  spherical coordinates, 120–125
Series form, Green's functions, 896–902,
    912–914
Shadow boundaries
  incident, 764, 774, 776, 785–788
  reflected, 764, 776
Sheet capacitance, mushroom AMCs, 952–954
Sheet inductance, mushroom AMCs, 951,
    953–954
Shells, electrons, 42–43
Shielded configurations, microstrip trans-
    mission lines, 461
Silicon, 42, 61
  atoms, 42
  charge density, 64
  conductivity, 62–64
  mobility, 63

Silver
  charge density, 64
  conductivity, 60
  mobility, 63
Single-mode step-index, 528
Single slab layers, reflection coefficients, 212–219
Single_Slab program, 251
Skin depth, 66, 145–148, 207, 450
Slab layers, reflection coefficients, 212–219
Slab_TE_TM_Graph program, 466
Slab_TE_TM_Ray program, 466
Slab waveguides, dielectric, 422–424
Slave surfaces, mushroom AMCs, 956
Slope diffraction, 797, 817–819
  coefficients, 798, 818–819
  hard, 818
  higher-order, 820–822
  multiple, 819–827
  overlap transition region, 825–827
  self-consistent method, 822–825
  soft, 818
Slope wedge diffraction coefficient (SWDC),
    819, 829, 830
Small radius approximation, 608, 612–613
Snell's law of reflection, 185, 189
Snell's law of refraction, 185, 189, 443, 446
Soft polarization, 815, 850, 853, 946
Solid-state microwave sources, 365
Source-free media
  rectangular vector wave equations, 107–114
    lossless media, 107–111
    lossy media, 111–114
  spherical transmission lines, 548–553
  transverse electric modes, 288–290,
    551–552, 555–557
  transverse electromagnetic modes, 277–284,
    557–559
  transverse magnetic modes, 285–288, 548,
    557, 561–562
Sources
  along boundaries, 17–18
  delta-gap, 721, 727
  finite-diameter wires, 721–730
  infinite, cylindrical, 573–581
  see also Electric line sources; Magnetic line
    sources
Sphere_RCS program, 664
Spheres, scattering by, 648–663
  lossy dielectric, 661–663
  Mie region, 660
  monostatic, 659–663

orthogonality relationships, 650–651
perfect electric conductors, 653–661
plane wave incidence, 654
radar cross section, 658–660
Rayleigh region, 660–661
resonance region, 660
transformations and theorems, 651–653
vertical dipoles, 648–650
Spherical Bessel functions, 123, 547, 561,
    563, 934–935
Spherical cavities, 559–566
    degenerate modes, 561–564
    dissipated power, 564
    dominant mode, 564
    quality factor, 564–566
    resonant frequency, 560–564
    stored energy, 564
    transverse electric modes, 560–562
    transverse magnetic modes, 562–564
    wave equation solutions, 120–125
Spherical coordinate systems, 931–935
    wave equation solutions, 119–125
Spherical Hankel functions, 123–124, 649,
    656, 658, 661
Spherical transmission lines, 547–559
    biconical, 555–559
    construction of solutions, 547–555
    Scalar Helmholtz wave equation, 554–555
    transverse electric modes, 551–552, 555–557
    transverse electromagnetic modes, 557–559
    transverse magnetic modes, 548, 557, 561–562
    vector potential $\mathbf{A}$, 550
    vector potential $\mathbf{F}$, 548–550
Spherical wave orthogonalities
    orthogonality relationships, 650–651
    transformations, and theorems, 648–653
    vertical dipoles, 648–650
    wave transformations and theorems, 651–653
Spin, 54–57, 83–91
Standing wave ratio (SWR), 135–136, 183
Standing waves, 547
    attenuation, 110, 119
    cylindrical wave equations, 119
    lossless media, 134–136
    lossy media, 144–146
    rectangular coordinate systems, 110
    rectangular waveguides, 371
    reflection, 183
    spherical wave equations, 123
Static (d.c.) conductivity, 61
Static fields, 58, 66, 85, 906–914

closed form, 906–912
precession, 85–91
series form, 912–914
Static permeability, 53
Static permittivity, 48
Stealth applications, 944, 959–979
Steepest descent path, 769, 771, 854–858,
    866–870, 873–874
Step index, 423
    multimode, 528
    single-mode, 528–529
Stokes' theorem, 3–4
Straight edge diffraction, 763–778
    normal incidence, 763–798
    oblique incidence, 798–805
Stray capacitance, 9
Stray inductance, 8
Stripline transmission lines, 451–455
    capacitance model, 453
    evolution, 452
    geometry, 451
    impedance, 454–455
Strip scattering, 582–589, 731
Sturm-Liouville problems/operator, 884,
    889–905
    equation, 864, 889–891
    Hermitian properties, 895
    symmetrical properties, 895, 896
Subdomain functions, 689–693
Substrate modes, 437, 439
Successive scattering procedure, 822
Superconductors, 66–68
Surface equivalence theorem, 340–345
    electric current density, 340
    equivalence principle models, 342
    Love's equivalence principle, 341–342
    magnetic current density, 340, 342
Surface impedance, mushroom AMCs, 949
Surface-impedance modulation function, 1000
Surface ray field, 873
Surface-wave antennas, 996
Surface waves, 427, 433, 449–452, 847
    critical angle, 196–198
    electromagnetic bandgap surfaces, 944–945
    frequency bandgaps, 452
    modes, 422
    mushroom AMCs, 945, 955–959
    terms, 863–886
    tightly bound slow, 202–203
    transition fields, 873–875
Susceptibility

electric, 48, 78
ionic, 78
magnetic, 53, 90
tensor, magnetic, 90
SWDC *see* Slope wedge diffraction coefficient
SWDC program, 830
SWR *see* Standing wave ratio
SWR_Animation_Γ_SWR_Impedance
    program, 253

T

Tc *see* Critical temperature
TDRS *see* Two-dimensional radiation and
    scattering
TE *see* Transverse electric modes
TEM *see* Transverse electromagnetic modes
$TE_{01\delta}$ mode, 526–527
TE_{10} mode, 378–386
    coupling, 395–396
    cutoff frequency, 373–374
    electric current density patterns, 383
    electric field patterns, 382
    group velocity, 385
    magnetic field patterns, 383
    phase velocity, 385
    uniform plane wave representation, 384
Tensor impedance surfaces, 1013
Tesseral harmonics, 650, 933
Testing (weighting) functions, 695
Tightly bound slow surface waves, 202–203
Time constant t, 884–885
Time-harmonic electromagnetic fields, 21–30
    boundary conditions, 22–25
    energy, 25–29
    Green's functions, 915–917
    Maxwell's equations, 22
    power, 25–29
    wave equation, 105–106
Time-harmonic fields, linear polarization, 155
Time-varying electromagnetic fields, 1–21
    boundary conditions, 12–18
    circuit-field relations, 7–11
    constitutive parameters and relations, 5–7
    finite conductivity media, 12–14
    infinite conductivity media, 15–17
    Maxwell's equations, 2–5
    power and energy, 18–21
    wave equation, 103–105
TM *see* Transverse resonance method;
        Transverse magnetic modes

TM_HIS program, 1003, 1014
Total transmission, 192–194
    parallel polarization, 193–194
    perpendicular polarization, 192–193
TR *see* Transition region
Transfer function, 883
Transition function, 774–775
Transition region (TR), 776, 825–827
Transmission, 179–269
    binomial transformers, 219, 222–224
    coefficients, 179–182, 186–191, 204, 213,
        219, 225, 226, 229, 245
    double negative materials, 237–253
    lossless media, 179–204
    lossy media, 204–211
    metamaterials, 235–253
    multiple interfaces, 212–228
    normal incidence, 180–183
    oblique angles, 226–228
    parallel polarization, 184, 188–192,
        193–194, 204
    perpendicular polarization, 184–188,
        192–193, 195–204
    phase velocity, 199–204, 210, 251–252
    quarter-wavelength transformer, 221–222
    total, 192–194
    Tschebyscheff impedance transformer, 219,
        224–226
Transmission lines, 249–253
    capacitance, 451–454
    coupling, 395–396
    impedance, 454–455
    microstrip, 452, 455–461
    spherical, 547–559
    stripline, 451–455
Transverse direction wave number, 420
Transverse electric (TE) modes, 136
    biconical transmission lines, 555–557
    circular cavities, 499–500
    circular waveguides, 479–484, 492, 499–500,
        505–508, 510–511, 512–521, 536–537
    construction of solutions, 277, 288–290
    corrugations, 946–947
    cylindrical coordinate systems, 290
    dielectric-covered conducting rod, 536–537
    dielectric resonators, 524–527
    dielectric waveguides, 433–437, 512–521
    electric field integral equation, 707–711
    filled rectangular waveguides, 412
    hybrid, 404–406, 512–522
    leaky-wave antennas, 1001–1003

magnetic field integral equation, 715–721
mushroom AMCs, 949, 956–957
parallel plates, 505–508
partially filled waveguides, 407–413
radial waveguides, 506–508, 510–511
rectangular coordinate systems, 288–290
rectangular waveguides, 367–375, 379–386, 399–402, 404–406, 433–437
scattering, 589–597, 610–615, 621–626, 630–637
spherical cavities, 560–562
spherical transmission lines, 551–552, 555–557
TE_{10}, 373–374, 378–386, 395–396
$TE_{01\delta}$, 526–527
transverse resonance metod, 421
wedged plates, 510–511, 642–648
Transverse electric to y (TE^y), 136–139
Transverse electromagnetic (TEM) modes, 127–177
    biconical transmission lines, 557–559
    circular polarization, 151, 155–160, 166–167
    construction of solutions, 277–284
    cylindrical coordinate systems, 281–284
    elliptical polarization, 151, 160–165, 166–167
    equiphase plane, 128
    linear polarization, 152–155, 166–167
    lossless media, 129–142
    lossy media, 142–151
    modes, 128
    oblique angle, 136–143, 148–151
    plane waves, 128
    Poincaré sphere, 165–170
    polarization, 151–170
    principal axis, 128–136, 143–148
    rectangular coordinate systems, 277–281
    spherical transmission lines, 557–559
    standing waves, 134–136, 144–146
    uniform plane waves, 128–134
    wave impedance, 131–132, 139–140, 144, 149
Transverse magnetic (TM) modes, 136
    biconical transmission lines, 557
    circular cavities, 500–501
    circular waveguides, 485–491, 492, 499–500, 508–509, 511–521, 530–536
    construction of solutions, 277, 285–288
    cylindrical coordinate systems, 287–288
    dielectric-covered conducting rod, 530–536

dielectric resonators, 525–526
dielectric waveguides, 424–433, 512–521
electric field integral equation, 703–707
filled rectangular waveguide, 411
hybrid, 407, 414–419, 512–522
leaky-wave antennas, 1001–1003
magnetic field integral equation, 715–721
mushroom AMCs, 949, 956–957
parallel plates, 508–509
partially filled waveguide, 415–419
radial waveguides, 508–509, 511–512
Rectangular coordinate systems, 285–287
rectangular waveguides, 375–378, 403–404, 407, 414–419, 424–433
scattering, 582–589, 605–610, 615–621, 626–630
spherical cavities, 562–564
spherical transmission lines, 548, 557, 561–562
transverse resonance method, 422
wedged plates, 511–512, 637–642, 646–648
Transverse magnetic to y (TM^y), 139
Transverse resonance method (TRM), 419–422
    equation, 420
    leaky-wave antennas, 1002
    modes, 421–422
    mushroom AMCs, 957
    wave number, 419–420
Transverse wave equation, 420
Traveling waves, 108, 119, 123, 131
    attenuation, 110, 119
    cylindrical wave equations, 119
    rectangular wave equations, 110
    reflection, 183
    spherical wave equations, 123
Trigonometric identity, 861–862, 868
TRM_Grounded_Dielectric_Dispersion program, 1013
TRM_Metasurface_Dispersion program, 1013
Tschebyscheff (equal-ripple) transformers, 219, 224–226
T-strips, 947
Two-dimensional electric field integral equations
    transverse electrical modes, 704–711
    transverse magnetic modes, 704–707
Two-dimensional leaky-wave antennas, 997, 1007–1009
Two-dimensional magnetic field integral equations
    transverse electrical modes, 715–717
    transverse magnetic modes, 713–715

Two-dimensional radiation and scattering (TDRS), 730–732
  circular, elliptical, or rectangular cylinder, 731–732
  strip, 731
Two-Dimensional Radiation and Scattering (TDRS) program, 733
Two_Dim_LWA program, 1014

U

Ultrabroadband checkerboard metasurfaces, 975–977
Uncoupled wave equations, 104–105, 116, 552
Underdamped solution, 73–74, 80
Uniform leaky-wave antennas, 997–999
Uniform plane waves, 127–151
  complex fields, 129–130
  definition, 127–128
  instantaneous fields, 130
  lossless media
    oblique angles, 136–142
    principal axis, 129–136
  lossless normal reflection, 182–183
  lossy media
    oblique angle, 148–151
    principal axis, 143–148
  standing waves, 134–136, 144–146
  wave impedance, 131–132, 139–140, 144, 149
Uniform theory of diffraction (UTD), 776, 782, 877
Uniqueness theorem, 325–327
Unit cell capacitance, mushroom AMCs, 952
Unit cell selection, checkerboard metasurfaces, 967–968
Unknown current density, 351, 687
UTD *see* Uniform theory of diffraction

V

Valence
  bands, 42
  electrons, 42, 62–63
  shells, 42
Variational differential, 741
Vector potential **A**, 272–274, 275–277, 550
Vector potential **F**, 274–277, 548–549
Vector potentials **A** and **F**, 275–277
Vector wave equation, 884

Vector wave equations, 104–105
  see also Wave equations
Vertical dipoles
  electric, 329–333
  spherical waves, 648–650
Vertical polarization, leaky-wave antennas, 1009–1010
Veselago planar lens, 239–249
Vibrating string, 896–897
Virtual source, 327–329, 578
Virtual sources, 580
Voltage, 7–8, 11
Voltage standing wave ratio (VSWR), 135–136
Volume equivalence theorem, 338–340
VSWR *see* Voltage standing wave ratio

W

Watson transformation, 637
Wave equations
  cylindrical coordinate systems, 114–120
  lossless media, 106, 114
  lossy media, 111–113
  rectangular coordinate systems, 107–114
  solutions, 106–125
  spherical coordinate systems, 120–125
  time-harmonic fields, 105–106
  time-varying fields, 103–105
  transverse wave equation, 420
  vector wave equations, 104
Wave functions
  cylindrical coordinates, 119
  rectangular coordinates, 110
  spherical coordinates, 123
Waveguides
  biconical transmission lines, 555–559
  cavities, 496–504, 559–566
  circular, 479–545
    see also Circular waveguides
  coplanar, 252–253
  dielectric, 422–450, 512–530
    see also Dielectric waveguides
  dielectric covered, 530–537
  fiber optics cables, 494, 528–530
  modes, 437
  parallel plates, 505–509
  radial, 505–512
    see also Radial waveguides

rectangular, 365–478
see also Rectangular waveguides
ridged, 461–464
spherical, 547–572
see also Spherical cavities; Spherical
transmission lines
transmission lines, 249–253, 395–396,
451–461, 547–559
wedged plates, 509–512
Wave impedance
circular waveguides, 483
electric line sources, 576
lossless media
oblique angle, 139–140
principal axis, 131–132
lossy media
oblique angle, 150
principle axis, 144
rectangular waveguides, 372–373
Wave propagation, 127–177
circular polarization, 151, 155–160,
166–167
definition, 127
elliptical polarization, 151, 160–165,
166–167
linear polarization, 153–155, 166–167
lossless media
oblique angle, 136–142
principal axis, 129–136
lossy media
oblique angle, 148–151
principal axis, 143–148
plane waves, 128
Poincaré sphere, 165–170
polarization, 151–170
Wave transformations, 597–598, 651–653
cylindrical, 597–604
Wave velocity
lossless media
oblique angle, 140–141

principal axis, 132–133
WDC program, 803–804, 829
Wedge
exterior, 646, 740
interior, 646
Wedged plates, 509–512
diffraction, 739–829, 847–877, 847–882
boundary conditions, 849
computations, 875–877
diffracted fields, 866–873
geometrical optics, 855–863
Maliuzhinets solution, 852–854
reflection coefficients, 850–852
surface wave terms, 863–865
surface wave transition field, 873–875
diffraction coefficients, 829
scattering, 637–648
transverse electric modes, 642–648
transverse magnetic modes, 637–642,
646–648
transverse electric modes, 510–511, 642–648
transverse magnetic modes, 511–512,
637–642, 646–648
Weighting (testing) functions, 695
Wire_Charge program, 733
Wronskian of Bessel functions, 607

X

X-band microwave sources
Gunn diode wafer, 366
X-13 klystron, 366

Z

Zeroes
cylindrical coordinates, 119
rectangular coordinates, 110
spherical coordinates, 123
Zonal harmonics, 650